THE NEUROSCIENCES

THIRD STUDY PROGRAM

THE

THIRD STUDY

The MIT Press

CAMBRIDGE, MASSACHUSETTS, AND

LONDON, ENGLAND

NEUROSCIENCES

PROGRAM Francis O. Schmitt and Frederic G. Worden *Editors-in-Chief*

Associate Editor: George Adelman

Contributing Editors: F. E. Bloom

G. M. Edelman

E. V. Evarts

B. Milner

D. W. Pfaff

C. S. Pittendrigh

K. H. Pribram

H. M. Shein

S. H. Snyder

L. Sokoloff

G. Werner

C. A. G. Wiersma

Second printing, 1979

Library of Congress Cataloging in Publication Data
Main entry under title:

The Neurosciences: third study program.

   Based on the NRP conference held in Boulder, Colo., summer 1972.
   1. Neurobiology—Congresses. 2. Nervous system—Congresses. 3. Psychobiology—Congresses. I. Schmitt, Francis Otto, 1903– ed. II. Worden,
Frederic G., ed. III. Neurosciences Research Program. [DNLM: 1. Neurology—Congresses. WL100 N496 1972]
QP361.N484   591.1'88   73–11116
ISBN 0–262–19112–1

INTRODUCTION

THIS BOOK reports the proceedings of the Third Intensive Study Program (ISP) organized by the Neurosciences Research Program (NRP) to promote progress in neuroscience. Approximately one hundred and fifty scientists from around the world participated in a three-week program at Boulder, Colorado, from July 24 through August 11, 1972.

In several ways the 1972 ISP differed from the first one, held in 1966, and the second one, held in 1969. Some of these differences are reflected in this book, and comment about them may be useful.

This volume contains the papers given under the 12 topics that were included in the formally scheduled part of the 1972 ISP. The rationale for the selection of these topics grew out of a searching evaluation by NRP Staff, Associates, and consultants. The 1966 ISP had attempted to identify a coherent field of research activities relevant to a scientific understanding of the nervous system, especially the brain and mind of man, the field collectively called *neuroscience*. In 1969, the Second ISP augmented the impact of the First and contributed further to the rapid acceleration of neuroscience, but, with its spectacular growth and reception, it became necessary to ask what purpose would be served by having another ISP. This was a critical question because the staging of an ISP requires a substantial investment not only of money but also of the time and energies of many scientists. Moreover, the stringent cutbacks in funds available for the support of research projects necessitated evaluation of whether non-laboratory activities such as an ISP are a justifiable use of grant funds. In resolving these questions, the NRP Staff solicited opinions from participants in the first two ISPs, from NRP Associates and consultants, and from other neuroscientists. In addition, sales of the first two volumes, as well as feedback from the readers, suggested that these books not only contribute significantly to the research and teaching of working scientists but also amplify the direct impact of the ISP on its participants.

Emerging from these considerations was the concept of the ISP as a periodic collaborative effort by carefully selected experts to recognize and evaluate critically important advances in neuroscience research, as well as —perhaps even more importantly—to identify opportunities for further research efforts. The need for such a mechanism becomes more urgent as the acceleration of scientific publications overtaxes the ability of individual scientists to keep up with the flood of new reports. The NRP maintains an active information network among neuroscientists, one purpose of which is to promote the evaluation and interpretation of research in order to facilitate conceptual and theoretical progress. These ongoing activities provide a unique background resource upon which to base the planning of periodic ISPs.

The development of the conceptual model of the ISP as a recurrently continuing exercise resulted in a slight shift of emphasis away from the effort to have the program cover neuroscience as broadly as possible and toward the effort to select topics judged to represent points of particularly significant and promising research progress. Insofar as it was compatible with this primary criterion, an attempt was made to avoid overlap with the contents of the previous programs and also to include topics spanning the different levels of organization of the nervous system (molecular, cellular, neural, and behavioral).

Another new feature of this third volume is that, along with the authoritative review-type articles, there are shorter papers in the form of research reports. These reflect an experimental change in the Fellowship Program for the 1972 ISP. From a large field of well-qualified nominees, 48 were selected for 1972 ISP Fellowships. For 36 of those chosen, a requirement was that their research activities be relevant to one of the 12 topics in the program, so that these Fellows could be assigned as speakers in the scheduled presentations. The other 12 Fellowships specifically did not include this criterion, so that the traditional purpose of the ISP Fellowships—to recruit scientists from other fields into the neurosciences—could be continued. Each topic in the program was presented by a team of eight scientists, four of whom were widely recognized experts and four of whom were younger scientists participating as ISP Fellows. The hope was that front-line reports of research by the Fellows could illustrate and enliven the presentation of the topic as treated in the scholarly overview used in the first two volumes.

Like its predecessors, the third volume represents compromises between the goal of an exposition at the most advanced and challenging level suitable for experts at the frontier of a given research and the goal of a treatment that will be comprehensible and tutorial for scientists in other research areas or at slightly less advanced stages of training and experience. The editors hope that the Third ISP book will prove as useful to working scientists and their students as have the first two volumes and that it will prove another milestone in the fruitful exploration of the opportunities to catalyze progress in neuroscience through refinement of the ISP idiom.

The editors acknowledge with thanks the contributions of all those who helped to plan and organize the 1972 ISP and to all participants for the dedication, talent, and hard work with which they accepted the challenge and responsibility of establishing and maintaining the highest possible level of scientific excellence of the program.

Professor Walter A. Rosenblith, Provost of the Massachusetts Institute of Technology, a neuroscientist himself, opened the Third ISP and attested to the deep interest of M.I.T. in neuroscience and in the ISP idiom. Dr. Howard W. Johnson, Chairman of the Corporation of M.I.T., also brought greetings. Their presence and their words of encouragement were deeply appreciated.

As in previous ISPs, the Fellows were selected from a large list of highly qualified candidates by a committee consisting of Drs. Dana L. Farnsworth, Seymour S. Kety, and William H. Sweet of Harvard University, and Dean Irwin W. Sizer and Dr. John B. Goodenough of M.I.T. We thank these gentlemen for their dedication and tireless efforts.

To the 12 Topic Chairmen and the like number of NRP Associates who served as Topic Advisers, we extend our thanks not only for organizing and chairing the activities of their topics, but also for editing the manuscripts that represented the scholarly grist of the lectures.

For this Third ISP, as for the First and the Second, the smooth operation of the rather complex program depended upon the experienced, energetic, and cooperative work of the administration and staff of the University of Colorado at Boulder. We particularly thank Mr. William E. Wright, Bureau of Conferences and Institutes, and his colleagues in other university services for their effective provision for all our needs during the three-week program.

Without the generous support of the Federal and the philanthropic agencies, acknowledged on page viii, the Third ISP could not have been held nor the book written and published; the Associates and the Staff of NRP acknowledge their deep indebtedness for this financial help.

Throughout the many months of planning and exchanges of ideas, the Associates and the Staff of NRP labored to maximize the scientific originality and high professional quality of the ISP. To them the editors extend their special thanks and appreciation. We are also indebted to Paul Andriesse for his excellent art work.

FRANCIS O. SCHMITT AND FREDERIC G. WORDEN

Brookline, Massachusetts
1 February 1973

Acknowledgment of Sponsorship and Support

The Neurosciences Research Program is sponsored by the Massachusetts Institute of Technology and supported in part by the U.S. Public Health Service, National Institutes of Health, Grant Nos. GM 10211, MH 23132, and NS 09937; the National Science Foundation, Grant No. GB 32782X; The Grant Foundation; Alfred P. Sloan Foundation; The Rogosin Foundation; and Neurosciences Research Foundation, Inc. The 1972 Intensive Study Program in the Neurosciences was supported by National Institutes of Health, Grant No. GM 12568; North Atlantic Treaty Organization; and Neurosciences Research Foundation, Inc.

CONTENTS

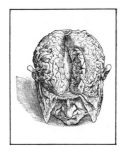

HEMISPHERIC SPECIALIZATION AND INTERACTION

Chapter

FEATURE EXTRACTION BY NEURONS AND BEHAVIOR

CENTRAL PROCESSING OF SENSORY INPUT

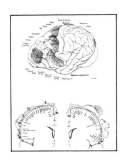

CENTRAL PROCESSING OF SENSORY INPUT LEADING TO MOTOR OUTPUT

INVERTEBRATE NEURONS AND BEHAVIOR

CIRCADIAN OSCILLATIONS AND ORGANIZATION IN NERVOUS SYSTEMS

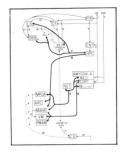

HORMONAL FACTORS IN BRAIN FUNCTION

BIOCHEMISTRY AND BEHAVIOR

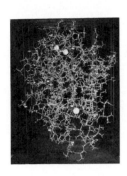

MOLECULAR MACHINERY OF THE MEMBRANE

REGULATORY BIOCHEMISTRY IN NEURAL TISSUES

DYNAMICS OF SYNAPTIC MODULATION

INTERACTION OF BRAIN CELLS AND VIRUSES

Photograph by Mark Solomon

AHARON (KATZIR) KATCHALSKY
15 September 1914—30 May 1972

Molecules, Information, and Memory:
From Molecular to Neural Networks[*]

MANFRED EIGEN

THIS LECTURE is dedicated to the memory of a friend,

Aharon (Katzir) Katchalsky.

He was supposed to speak in this place tonight, but his voice, a great voice for science and humanity, has been stilled.

I never heard him say a word of hatred; so I will not profane his memory by bringing back to your mind the terrible events of the 30th of May 1972—results of hatred and contempt for the dignity of man.

Aharon spent the last week of his life at Göttingen, where he lectured, discussed science, and hiked with us in the woods around Berlepsch Castle during the "Israeli-German Workshop on Membranes," 24–27 May 1972. Tonight I should like to communicate to you some of his last ideas, reflected in my own thinking. I wish I could also communicate to you only a fraction of the excitement he brought us during that unforgettable week.

As you know, Aharon liked mathematics. He was never really satisfied unless he could put an idea into the clear and unequivocal form of an equation. I would draw an incomplete picture, if I would spare you such a puristic pleasure, although I can't hope that I could persuade you, as well as he could, to believe that you, too, have understood everything. Therefore, let me distill all the mathematics into a sequence of tables, where you will find it as a highly concentrated essence that you may use to season the text according to your taste.

In his lecture on "Network Thermodynamics," Aharon Katchalsky (Oster, Perelson, and Katchalsky, 1971) stressed the analogy between a condenser discharge and a chemical equilibration process (see Table I where some of his handwritten notes are reproduced). He was intrigued by the idea of an abstract description of networks, because he hoped to utilize the huge amount of

experience gained with electrical networks in order to treat complex biochemical reaction systems as well as neural nets. In the discussion of this lecture, which Aharon gave as an opening address to our Membrane Workshop, I asked him whether he could quote any chemical property that would resemble an electric inductance. His spontaneous answer was: "Just go back to the definition of the inductance in an electric network and apply it to the fluxes and forces in a chemical system." Such analogies are compiled in more detail in Table II, and Table III demonstrates the significance of some "chemical" circuit parameters with examples of elementary reactions. For the capacitance and the resistance, the analogy is indeed more than a formal one. The decay of affinity or of the (perturbed) extent of reaction in a chemical relaxation

TABLE I

Original lecture notes of Aharon Katchalsky on "Network Thermodynamics" (Göttingen, 24 May 1972)

$$\text{Capacitors} \quad C\frac{dX}{dt} = J \qquad J = \frac{dn}{dt} = \frac{dn}{d\mu}\cdot\frac{d\mu}{dt} = C\frac{d\mu}{dt}$$

$$\text{Resistors} \quad R = \frac{dX}{dJ} \qquad R = \frac{d(\Delta\mu)}{dJ}$$

$$\text{Relaxation time} \quad \int - \quad \frac{d\xi}{dt} = -\frac{\xi - \bar{\xi}}{\tau} = J_R = \frac{A}{R_R}$$

$$A = \bar{A} + \left(\frac{\partial \bar{A}}{\partial \xi}\right)(\xi - \bar{\xi}) = \left(\frac{\partial \bar{A}}{\partial \xi}\right)(\xi - \bar{\xi})$$

$$-\frac{\xi - \bar{\xi}}{\tau} = \frac{1}{R_R}\left(\frac{\partial \bar{A}}{\partial \xi}\right)(\xi - \bar{\xi}) \qquad -\frac{1}{\tau} = \frac{1}{R_R}\left(\frac{\partial \bar{A}}{\partial \xi}\right)$$

$$\text{since} \quad A = -\sum \nu_i \mu_i \quad \left(\frac{\partial \bar{A}}{\partial \xi}\right) = -\sum \nu_i \frac{\partial \bar{\mu}}{\partial \xi} = -\sum \nu_i \left(\frac{\partial \bar{\mu}_i}{\partial \xi_i}\right)$$

$$\left(\frac{\partial \bar{A}}{\partial \xi}\right) = -\sum \frac{\nu_i^2}{c_i} = -\frac{1}{C_R}$$

$$\boxed{\tau = R_R \cdot C_R}$$

$$\text{which for} \quad A \underset{k_r}{\overset{k_f}{\rightleftharpoons}} B \qquad \tau = \frac{1}{k_f + k_r}$$

[*] Extended abstract of the "Aharon Katchalsky Memorial Lecture" given at Rehovot, Israel, 3 July 1972, and at Boulder, Colorado, 2 August, 1972.

MANFRED EIGEN Max-Planck-Institut für Biophysikalische Chemie, Göttingen, Germany

process has physically much in common with the decay of voltage or current in a condenser discharge, whereas the "formally" possible analogy with an LR-circuit does not make much sense.

TABLE II

Analogies between electrical and chemical terms

Flux (J)

Electric current: I Chemical turnover rate: $\dfrac{d\xi}{dt}$

(nonvectorial flux)

Force (X)
Voltage: U Affinity (reaction "force"): A

Linear Relations

$\delta I = \dfrac{\delta U}{R_E}$ (Ohm's law) $\dfrac{d\xi}{dt} = \dfrac{1}{R_{Ch}}\left(\dfrac{\partial A}{\partial \xi}\right)_{T,P} \delta\xi$

(e.g., at const T, P)

Resistance (R)

$R_E = \dfrac{dU}{dI}$ $R_{Ch} = \dfrac{[RT]}{[\bar{v}]}$

Capacitance (C)

$C_E = \dfrac{\int_0^t I\, dt}{U}$ $C_{Ch} = -\dfrac{1}{(\partial A/\partial \xi)_{T,P}}$

$= \left[[RT] \sum_i \dfrac{v_i^2}{C_i} \right]^{-1}$

Inductance (L)

$L_E = \dfrac{U}{dI/dt}$ Meaningless for systems close to equilibrium

Time Constants (τ)

$\tau_E = R_E C_E$ $\tau_{Ch} = -\left[\dfrac{[\bar{v}]}{RT}\left(\dfrac{\partial A}{\partial \xi}\right)_{T,P} \right]^{-1}$

$\tau'_E = \dfrac{L_E}{R_E}$ $= -\left[[\bar{v}] \sum_i \dfrac{v_i^2}{C_i} \right]^{-1}$

$[RT]$ = thermal energy (i.e., R = gas constant, T = absolute temperature), t = time. The extent of reaction ξ is defined as concentration change δC_i divided by the stoichiometric number v_i : $\xi = \delta C_i / v_i$. The stoichiometric numbers are positive for reaction products and negative for reactants.

The affinity is defined as $A = -(\partial G/\partial \xi)_{T,P}$ where G is the Gibbs free energy. It can be expressed in terms of chemical potentials $\mu_i = (\partial G/\partial C_i)_{T,P}$ as $-A = \sum_i v_i \mu_i$ (note the signs of v_i). The equilibrium exchange rate $[\bar{v}]$ is defined as the turnover (dC/dt) of reactants that equals the reverse turnover of products at equilibrium.

The term $[\sum_i v_i^2/C_i]^{-1}$ is the "relaxational amplitude" factor. The meaning of the different terms is exemplified in Table III.

The next morning, Aharon and I, in company with Frank Schmitt and Harden McConnell, drove from Göttingen to Berlepsch Castle, where the sessions were held. We immediately began a vivid discussion. Both Aharon and I had not slept well, for we both realized the difficulties of such a formal assignment. We all agreed that the best way to solve our problem was to analyze an

TABLE III

Significance of chemical circuit parameters as shown with examples of elementary reactions

$$A \underset{k_{BA}}{\overset{k_{AB}}{\rightleftharpoons}} B \qquad (1)$$

$$\tau_{Ch} = [k_{AB} + k_{BA}]^{-1} = R_{Ch}C_{Ch}$$

$$R_{Ch} = [RT]/[\bar{v}] = \frac{[RT]}{k_{AB}\bar{C}_A} = \frac{[RT]}{k_{BA}\bar{C}_B}$$

$$C_{Ch} = \frac{1}{[RT]} \frac{\bar{C}_A \bar{C}_B}{\bar{C}_A + \bar{C}_B} \quad (\text{maximum at } \bar{C}_A = \bar{C}_B)$$

$$A + B \underset{k_d}{\overset{k_f}{\rightleftharpoons}} AB \qquad (2)$$

$$\tau_{Ch} = [k_f(\bar{C}_A + \bar{C}_B) + k_d]^{-1} = R_{Ch}C_{Ch}$$

$$R_{Ch} = \frac{[RT]}{k_f \bar{C}_A \bar{C}_B} = \frac{[RT]}{k_d \bar{C}_{AB}}$$

$$C_{Ch} = \frac{1}{[RT]} \frac{\bar{C}_A \bar{C}_B}{\bar{C}_A + \bar{C}_B + K_{AB}^{-1}}$$

$$K_{AB} = \frac{\bar{C}_{AB}}{\bar{C}_A \bar{C}_B}$$

oscillatory chemical reaction system, if possible a linear one that ought to exhibit a harmonic oscillation. The electric analog of such a system is shown in Table IV. The solutions are characterized by eigenvalues $\lambda_{1,2}$ that may become complex (i.e., involving an imaginary part) if the critical resistance R becomes smaller than $2\sqrt{L/C}$. Suppose the oscillatory circuit of Table IV (with closed switch S) were in a "black box" and it were possible to measure (e.g. via some antenna) only the resonance frequency, the damping constant (or, in a different experiment, the phase shift of the circuit's response to a forced excitation). Then all we could determine would be the combinations: LC and RC or $L/R = LC/RC$. This is most easily seen from the rewritten expressions in the lower part of Table IV. To determine the single quantities L, C, and R would require a "probe" that could measure phases at defined positions. (If, for instance the switch in the circuit diagram in Table IV were open and connections with the circuit could be made, R may be easily detected with the help of an impedance bridge using the resonance frequency of the circuit.) I mention this fact, which for an electrical engineer would certainly be trivial, because we shall encounter a similar "black box" situation for those chemical systems that show harmonic oscillations and hence may be represented by the analog of such a circuit. Are there any chemical systems that do show harmonic oscillations? Coupled chemical reactions of higher than first-order are required. Let us look at coupled nonlinear reaction systems, which can be linearized by expansion around a reference state of zero net turnover. The resulting properties, of course, are not simply "material" qualities but refer to a particular state of coupling (equilibrium or steady state). We have already considered such a one-step system (see the second example in Table III) that upon linearization allows the assignment of a "reaction-resistance" and "capacitance," in analogy to the elements of an electric RC-circuit.

TABLE IV
Damped electric oscillatory circuit

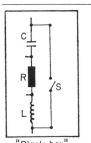

"Black box"

Differential equation:

$$L\frac{dI}{dt} + RI + \frac{1}{C}\int_0^t I\,dt = \text{const}$$

$$L\frac{d^2I}{dt^2} + R\frac{dI}{dt} + \frac{1}{C}I = 0$$

Solution:

$$I = A_1 e^{\lambda_1 t} + A_2 e^{\lambda_2 t}$$

Eigenvalues:

$$\lambda_{1,2} = -\frac{R}{2L} \pm \sqrt{\frac{R^2}{4L^2} - \frac{1}{LC}}$$

Oscillation, if $R < 2\sqrt{L/C}$

For periodically forced oscillation ($\omega = 2\pi\nu$, $\nu = $ frequency) phase angle:

$$tg\,\varphi = \frac{L\omega - 1/\omega C}{R}$$

The expressions of λ and $tg\,\varphi$ contain only the combinations of LC and RC or L/R:

$$\lambda_{1,2} = -\frac{RC}{2LC}\left[1 \pm \sqrt{1 - \frac{4L/R}{RC}}\right]$$

$$tg\,\varphi = \omega\left(\frac{L}{R}\right) - \frac{1}{\omega(RC)}$$

For any system close to equilibrium, regardless of its reaction order and complexity, an exact solution can be presented, as shown in Table V (Meixner, 1953; Eigen and De Maeyer, 1963). We conclude that for any reaction system near equilibrium the solutions contain only negative and real eigenvalues; therefore, oscillation (characterized by complex eigenvalues) or growth (represented by positive real eigenvalues) is excluded. Equilibrium represents a stable state, the perturbation of which will always result in a relaxational response, i.e., an exponential decay of the "perturbation." The negative reciprocal eigenvalues represent a spectrum of relaxation times. The fact that only real eigenvalues occur is a consequence of Onsager's relations, which are based on the "principle of detailed balance" (or microscopic reversibility) applying to any system near equilibrium. The negative sign of the eigenvalues follows from the sign of the individual rate terms, which in the characteristic equation always yield coefficients of uniform sign.

With respect to our original question, we arrive at the following conclusions:

1. For any reaction system close to equilibrium there is no meaningful analog of an "inductance."

2. If harmonic oscillations do exist in a chemical system, they should be present only in nonequilibrium

TABLE V
General reaction network near equilibrium

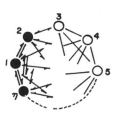

In this reaction system, any compound might react with any other compound and the reaction may involve any molecularity, i.e., any number of partners.

We have a system of n rate equations. At equilibrium, dC_i/dt reaches zero. Since the sum of all concentrations is constant (conservation of matter), one may reduce to $(n-1)$ or less independent variables. After expansion, the variables represent the deviations of concentrations from the reference values: $x_i = C_i - C_{i0}$ yielding a system of $n - 1$ (or less, depending on conservation conditions) linear differential equations for dx_i/dt. The rate coefficient matrix of this system is

		Contribution from component			
		1	2	$\cdots$	n
Temporal	1	α_{11}	α_{12}	$\cdots$	α_{1n}
change of	2	α_{21}	α_{22}		
compound	$\vdots$	$\vdots$		$\ddots$	$\vdots$
	n	α_{n1}		$\cdots$	α_{un}

It relates the temporal change (row) of a certain component to the contributor of the corresponding reaction step (column). The trace of the matrix $\sum_i \alpha_{ii}$ is the sum of all contributions, i.e., all rate terms. The α_{ik} contain rate constants as well as equilibrium concentrations (resulting from the expansion). Near equilibrium one can always write the system in a form that $\alpha_{ik} = \alpha_{ki}$ (Onsager's relations).

The determinant equation $|A - \lambda E| = 0$ yields the real and negative eigenvalues λ_i. A is the matrix (α_{ik}) and E the unity matrix. The negative reciprocal eigenvalues represent the spectrum of relaxation times. If we call the diagonal matrix

$$\begin{pmatrix} \lambda_1 & 0 & \cdots & 0 \\ 0 & \lambda_2 & \ddots & \vdots \\ \vdots & \ddots & \ddots & 0 \\ 0 & \cdots & 0 & \lambda_n \end{pmatrix} = B,$$

then $B = MAM^{-1}$ where M is a transformation matrix (and M^{-1} its inverse) that transforms the concentration variables x_i into the "normal modes" y_i:

$$\vec{y} = M\vec{x}.$$

The symmetry of α_{ik} due to Onsager's relations ensures real eigenvalues. The characteristic equation $|A - \lambda E| = 0$ has only terms with uniform sign, causing negative eigenvalues.

systems where a linearization procedure can be applied (i.e., around a steady state) and where the reaction properties are such as to yield complex eigenvalues.

3. These eigenvalues, like the "real" eigenvalue spectrum of the equilibrium process, will be related to certain "normal modes." Any assignment of a "resistance," "capacitance," and (possibly) "inductance" would refer to a combination of rate terms, involving the contributions of different species and applying only to the neighborhood of the reference state.

Before we proceed to look for such a system, let us try to understand the physical meaning of a "normal mode."

The transformation that leads to a normal mode actually effects a decoupling of variables. The normal mode then is such a (linear) combination of original concentration shifts δC_i, which is uniquely assigned only to one eigenvalue and independent of all the others. In Table VI an example is shown. The reaction system—the famous model of substrate binding to an allosteric enzyme, according to Monod, Wyman, and Changeux (1965)—involves nine normal modes (Kirschner, Eigen, Bittman et al., 1966; Kirschner, 1968), which, under the peculiar conditions (e.g. "fast" binding and "slow" conformational change, as often encountered in nature), have a relatively simple physical interpretation. Far from equilibrium such a reaction network can be used for all types of basic control processes as is illustrated in Figure 1 (Eigen, 1967).

TABLE VI

Allosteric enzyme reaction network

$R_0 \rightleftharpoons T_0$		R and T characterize 2 different conformations
$\updownarrow \quad \updownarrow$		of the enzyme that has 4 identical subunits. The con-
$R_1 \rightleftharpoons T_1$		formational change is an all-or-none change, i.e.,
$\updownarrow \quad \updownarrow$		it involves all 4 subunits simultaneously. The num-
$R_2 \rightleftharpoons T_2$		ber index indicates the number of substrate (or
$\updownarrow \quad \updownarrow$		effector) molecules bound. The intrinsic binding
$R_3 \rightleftharpoons T_3$		constants are degenerate, i.e., the same for each
$\updownarrow \quad \updownarrow$		subunit, independent of the binding to other sub-
$R_4 \rightleftharpoons T_4$		units but different for the R (relaxed) and T (tight)

configuration. The particular assumption made here is that of a fast binding in R_i and T_i (but for R_i fast as compared to T_i) and of a comparatively slow conformational change $R_i \rightleftharpoons T_i$.

The system has 10 states and therefore 9 normal modes (the sum of the concentrations of all states is constant). These normal modes describe the following processes:

	1.	Free substrate + sum of free R sites $\rightleftharpoons$ sum of occupied R sites.
Fastest	2.⎫ 3.⎬ 4.⎭	Redistribution of substrate among 4 R sites at constant free substrate and $\sum R_i$ concentration.
	5.	Free substrate + sum of free T sites $\rightleftharpoons$ sum of occupied T sites with reequilibration of substrate binding in R (due to change of substrate concentration).
Intermediate	6.⎫ 7.⎬ 8.⎭	Redistribution of a bound substrate among all T and R sites at constant free substrate concentration.
Slow	9.	Conformational change: Sum of all $R \rightleftharpoons$ sum of all T states, while binding equilibria are readjusted.

Because of the equivalence of binding sites, the steps 2, 3, 4 and 6, 7, 8 are not detectable. Therefore the relaxation spectrum involves only 3 detectable time constants of which 2 relate to binding normal modes and one (the slowest) to the normal mode of conformational changes.

① Transducer element (Amplifier):

② Logic element (Switch):

③ Feedback and control elements:

Negative Feedback: Activator = Product, favoring T-State
Positive Feedback: Activator = Product, favoring R-State
Threshold Function: Activator = Substrate, favoring R-State
Negative Resistance: Activator = Substrate, favoring T-State

FIGURE 1   Allosteric control functions.

Now let us look at reaction networks far from equilibrium. Here we can encounter all sorts of oscillatory systems. The simplest was described by Lotka (1910), 62 years ago. It is treated in Table VII.

Any oscillatory system will require some feedback in the form of an autocatalytic or cyclic-catalytic process. This condition will always imply a basically nonlinear reaction system. Hence these systems as such cannot resemble the linear electric oscillatory network of Table IV. One usually finds that the behavior depends on the amplitude of perturbation; some systems may resemble "limit cycle" behavior.* Recall the linearization procedure we applied to systems near equilibrium. If we can reach a stable steady state with a constant distribution of concentrations, we may as well carry out such an expansion and linearize.

In Table VII this procedure is applied to the Lotka system. The solution yields two complex eigenvalues in perfect analogy to the electric circuit treated in Table IV. Upon any perturbation of the steady state, the system will return in the form of a damped harmonic oscillation, within the "linear range" independent of the amount of perturbation, i.e., with a constant characteristic fre-

* The limit cycle represents the relation between the variables x_1 and x_2 of a two-state system that, if plotted against each other, yields a closed loop, which the system approaches asymptotically, independent of the initial conditions (Glansdorff and Prigogine, 1971).

TABLE VII

The Lotka reaction scheme

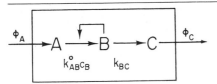

Reactions: $A + B \rightarrow 2B$

Influx of A: ϕ_A

$B \rightarrow C$

Outflux of C: ϕ_C

(B catalyzes its own production)

$$\frac{dC_A}{dt} = \phi_A - (k^0_{AB}C_B)C_A$$

$$\frac{dC_B}{dt} = (k^0_{AB}C_B)C_A - k_{BC}C_B$$

Steady state:

$$\phi_A = k^0_{AB}\bar{C}_B\bar{C}_A$$

$$k^0_{AB}\bar{C}_A = k_{BC}$$

Linearization: (around steady state)

$$\frac{dx_A}{dt} = \alpha_{11}x_A + \alpha_{12}x_B$$

$$\frac{dx_B}{dt} = \alpha_{21}x_A + \alpha_{22}x_B$$

matrix: $\begin{pmatrix} -k^0_{AB}\bar{C}_B & -k^0_{AB}\bar{C}_A \\ +k^0_{AB}\bar{C}_B & 0 \end{pmatrix}$

Reciprocal time constants:

$$\tau^{-1}_{1,2} = -\frac{k^0_{AB}\bar{C}_B}{2}\left[1 \pm \sqrt{1 - \frac{4\bar{C}_A}{\bar{C}_B}}\right]$$

$$\bar{C}_A/\bar{C}_B = \frac{k^2_{BC}}{k^0_{AB}\bar{C}_B} \equiv \kappa$$

(damped oscillation) if

$$\kappa > 1/4$$

Assignments:

$$\frac{1}{LC} = \omega^2_0 = k^0_{AB}\bar{C}_B k_{BC}$$

$$RC = k^{-1}_{BC}$$

$$\frac{L}{R} = (k^0_{AB}\bar{C}_B)^{-1}$$

quency and damping constant. However, we can only assign values of LC, RC, or L/C, in analogy to the electric circuit enclosed in a "black box." We see clearly the limitations of such a formal description. In this case, the ambiguity of assignment is genuine. Only the combined terms have a real physical meaning: The larger the turnover ($k^0_{AB}\bar{C}_B$ and k_{BC}), the smaller is LC, i.e., the higher the frequency. The faster the disappearance of B relative to that of A, the smaller is R relative to $\sqrt{L/R}$, i.e., the smaller is the damping constant.

Frequently in science, one starts out with a definite question but eventually arrives at an answer that, although having no relation to the original question, nevertheless turns out to be of more general interest. After discussing chemical oscillation for several days, both Aharon and I had the feeling that such systems involve more exciting problems than the formalism of analog representation. Indeed, many other formalistic assign-

ments, such as negative resistances, could be used to describe particular chemical networks; some examples are given in Figure 1. No chemical oscillatory system as simple as the Lotka scheme has been found in nature so far. On the other hand, some years ago we encountered a class of autocatalytic reaction mechanisms, involving nucleic acids as well as proteins, which is closely related to the Lotka scheme. It shows the same type of oscillations, but, in addition, it is distinguished by properties that are quite typical for a living system. Reaction networks of this kind, for instance, can evolve to any degree of complexity and hence may be considered the prototypes of Darwinian evolutionary systems. They demonstrate how biological information may have originated.

Table VIII shows how the Lotka scheme has to be modified in order to apply to a polymerization mechanism. The symbol A may represent a set of energy-rich

TABLE VIII

Modification of the Lotka scheme

A polymerizes to B, which decomposes.

$\mathfrak{F}$ = formation rate constant,

$f(C_A)$ = stoichiometric function of C_A,

D = decomposition rate constant.

The production of the polymer is autocatalytic, i.e., the polymer acts as template.

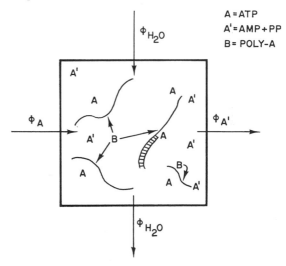

e.g.,

$$A = \text{ATP}; \quad A^1 = \text{AMP}; \quad B = \text{poly-A} + \text{PP}$$

$$\frac{dC_B}{dt} = \mathfrak{F} \cdot f(C_A)C_B - DC_B$$

$$\frac{dx_B}{dt} = \mathfrak{F}\frac{\partial f(C_A)}{\partial C_A}\bar{C}_B x_A + \mathfrak{F}f(\bar{C}_A)x_B - Dx_B$$

This equation is formally identical with the linearized equation of the Lotka scheme and can be solved accordingly.

monomers (such as ATP) that spontaneously aggregate to form a polymer B (e.g. polyadenylic acid [poly-A]) using the reaction product as a template. Of course, now we are no longer dealing with a simple process of defined molecularity, unless we know in detail the mechanism of polymerization (e.g. the stoichiometry of the rate limiting step); but if we again expand around a steady state, we need not know the explicit form of the stoichiometry function (see $f(C_A)$ in Table VIII). The calculations show that the linearized rate equations have the same form as those of the simple Lotka scheme (see Table VII).

As long as A represents a class of uniform monomers and B, correspondingly, a uniform polymer, the system does not offer any new clues. If, however, we substitute for A by a set of at least 2 different "digits," or better 4 (such as ATP, UTP, GTP, CTP), or even 20 (such as the natural amino acids), we arrive at a system that now is characterized by competitive selection behavior. Given a

sufficient degree of complexity to start with (see the "hypercycle" described in Table X), these systems may undergo an almost unlimited evolution; an example, resembling S. Spiegelman's famous "test-tube evolution" experiments using Qβ-phage-RNA and -replicase, is represented in Table IX. The detailed theoretical treatment of such self-organizing systems has recently been worked out (Eigen, 1971).

The conclusions are that Darwinian evolutionary behavior starts at the level of macromolecular synthesis, given systems with inherently self-instructive behavior. The selection principle appears to be derivable from more fundamental physical principles, such as the "principle of minimum entropy production," which applies to steady states, and the resulting stability criteria as derived by P. Glansdorff and I. Prigogine (1971). In his lecture at Göttingen, Aharon put special emphasis on this principle that he derived from his general network theory (Oster, Perelson, and Katchalsky, 1973). On the basis of this principle a "value" parameter can be introduced into the dynamic theory of matter (Eigen, 1971). This value parameter defines a preferred direction

TABLE IX

Generalization of the modified Lotka scheme, resembling competitive template instructed polymerization

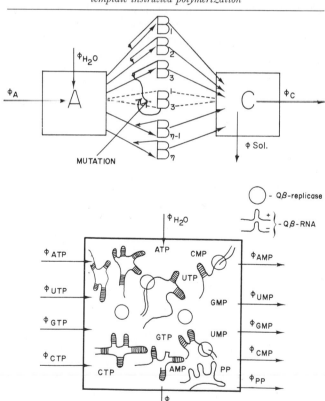

Each B_i catalyzes or instructs its own reproduction.
Example Qβ-evolution experiment, according to S. Spiegelman.

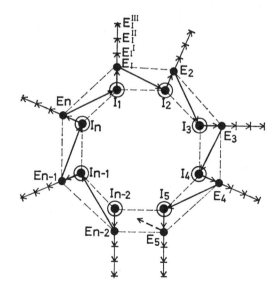

FIGURE 2 The self-instructive catalytic hypercycle. I_i represents information carriers, i.e. complementary single strands of RNA. Each small cycle indicates the self-instructive property of the I_i collective involving the two complementary strands. E_i (encoded by I_i) represents catalytic function. Each E_i branch may include several functions (e.g., polymerization, translation, control), one of which has to provide a coupling to the information carrier I_{i+1} (e.g., enhancement of the formation of I_{i+1} by specific recognition). The trace representing all couplings must close up, i.e., there must be an E_n that enhances the formation of I_1. The hypercycle is described by a system of nonlinear differential equations.

TABLE X
Properties of hypercycle

(a) Growth and Competitive Selection Properties
1. Autocatalytic growth of cycle as a whole.
2. Competition among different cycles.
3. Sharp selection all-or-none due to nonlinearity (involving singularities).
4a. Utilization of small selective advantages; fast evolution.
4b. Universal code and chirality.
5. Large information capacity (optimized) with small individual sequences.
6. Evolution (utilization of genotypic mutations in cycle).
7. Selection against parasitic branches; toleration of one branch.
8. Gene and operon structure.

(b) Normal Mode Analysis of Oscillatory Behavior (M. Eigen and P. Schuster, in preparation)

The analysis involves a linearization by expansion around the reference state. In absence of any constraint, such as constant flows or constant forces, the hypercycle always includes a "growth" mode, characterized by a positive eigenvalue which can be normalized to $+1$. This mode, of course, will disappear under suitable constraints effecting growth limitation. For a two-membered cycle (involving I_1 and I_2) a second negative eigenvalue (-1) will appear describing a relaxation into a fixed ratio of the population densities I_1 and I_2. Oscillation is introduced with the third member (eigenvalues $+1$, $-\frac{1}{2}(1 \pm i\sqrt{3})$). It is, however, a damped oscillation, decaying again to a stable distribution. The 4-member system is unique, because it is the only system showing normal modes of undamped harmonic oscillation (eigenvalues $+1$; -1; $+i$; $-i$). It can be shown that the origin of translation requires the population of a minimum of four states ($I_1 \to E_1$; $I_2 \to E_2$; $I_3 \to E_3$; $I_4 \to E_4$). (A detailed description can be found in M. Eigen and P. Schuster, in preparation.) In any hypercycle with more than 4 members, even under the constraints suppressing the growth mode (eigenvalue $+1$), the rest of the eigenvalues involves further complex growth modes, i.e., instabilities. The (linear) system will always depart from the reference state but will eventually be levelled off and reversed via the nonlinear mechanism. The resulting oscillations show quite complex patterns, as we see them also in the discharge patterns of a neural net (examples of oscillatory patterns in Figures 3 and 4).

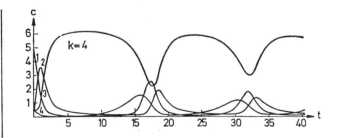

FIGURE 3 Stationary oscillation in an equilibrated 4-membered hypercycle with unsymmetrical rate distribution. The formation rate constant of the 4th member is 10 times smaller than that of all other members. (From Eigen, 1971.)

of evolutionary processes and hence renders them "inevitable" under the given physical and chemical conditions. On the other hand, the source of this information production is a "noise generator" represented by the randomly occurring mutations (i.e., "error copies"), the temporal sequence of which cannot be predicted, because only a minute fraction of all possible states can be populated within finite limits of time and space. How relevant self-organizing chemical networks are with respect to the problem of the origin of life may be demonstrated by a last example. The reaction hypercycle shown in Figure 2 exhibits essential features of the reproduction mechanism of a "living" cell. In particular the hypercycle explains how the gene and operon structure of the long DNA chain, comprising the total information content of a cell, may have come about via protein-linked cooperation of shorter nucleotide sequences that encode certain functional features. The origin of such a hypercycle is linked to a nucleation of some primitive code and translation machinery. For an alphabet of not more than eight codons or amino acids, translation might even originate from a random assignment involving the functional coincidence of previously undetermined sequences of polypeptides and polynucleotides (M. Eigen and P. Schuster, in preparation). The properties relevant for the evolutionary behavior of the hypercycle are summarized in Table X. As an appendix to our previous discussion of chemical oscillations, a normal mode analysis of this reaction network is discussed also in Table X and the resulting oscillatory behavior is shown in Figures 3 and 4.

If we consider biological evolution as a grandiose process of self-organization of a genetic memory of enormous functional sophistication, we may as well ask whether analogous principles also are responsible for the self-organization of memory in our central nervous system. I certainly do not pretend that this process would take place at the same level of molecular organization. The neural networks certainly include, as their elements, cells such as neurons and glia, as well as the manifold interconnections of these cells, the various types of synapses. Aharon's interest in an abstract description of networks certainly was aimed to include this central issue of our science, the central issue of this meeting. He was well aware how little is explained by formalisms. Even if we know the principles according to which "memory" could organize itself, any clue as to which of the possible alternatives was chosen by nature would have to come from the detailed analysis of the elementary processes, relevant to this particular network level, just as the clues in molecular biology came only from the detailed analysis of the double helix, of the hemoglobin structure, of the lac operon, etc.

MANFRED EIGEN   XXV

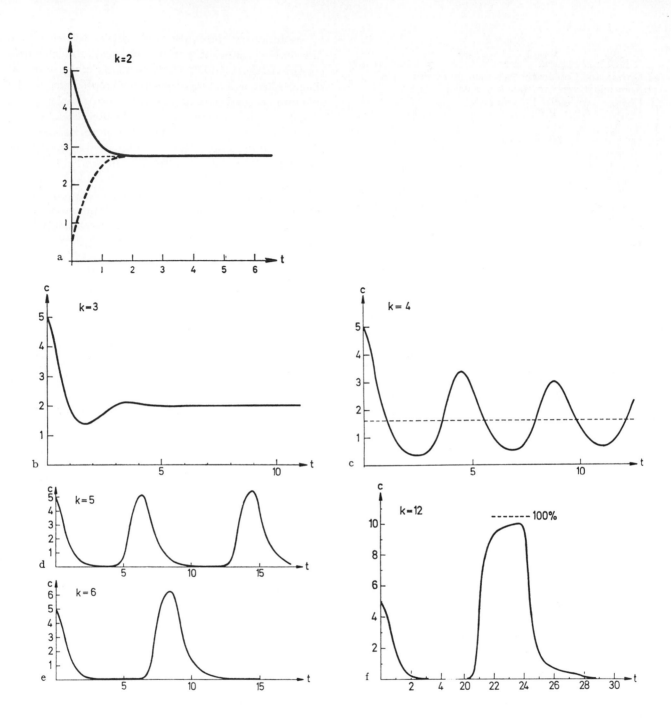

FIGURE 4 Solutions describing the selection of k-member hypercycles under the constraint of constant overall organization ($k = 2, 3, 4, 5, 6, 12$). The reaction system is described by a simple second-order formation term—identical for all members —as well as by a first-order "removal" term to maintain the condition $\sum_k x_k = $ const. The solutions are shown for one member only. The equilibrium value is constant for $k \leq 3$. For $k = 3$ the approach to selection equilibrium is represented by a damped oscillation, whereas for $k > 3$ stationary oscillations occur. This can be shown by starting from a constant distribution and introducing a small perturbation at $t = 0$. The oscillation then builds up (calculations by P. Schuster [Eigen and Schuster, in preparation]). (From Eigen, 1971.)

TABLE XI

*What is the sequence of events of information storage in
neural networks?*

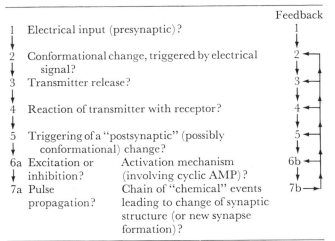

It was the clear realization of this fact that long ago led Aharon to start experimental work on membranes, i.e., on models, as well as on true biological membranes. During his stay at Göttingen, he participated in discussions that finally resulted in the drafting of a programmatic paper (Neumann, Nachmansohn, and Katchalsky, 1973) on the interpretation of nerve excitability, a challenge for a more precise study of the sequence of physical events in nerve membrane excitation. Tables XI and XII refer to these

TABLE XII

Basic questions about information processing by neural networks

1. How do cells interact with each other to build up specific connections in the network?

2. After connections are formed, how do cells communicate with each other in the network?

3. How does this communication provide feedback to alter or reinforce the connections?

4. What principle is behind the self-organization in such a network: (a) internal-redundancy, feedback, and competition, (b) superimposed-"value"?

ultimate ideas of Aharon, in which we all share a common interest. These tables contain only questions. It is our tribute to the memory of this great universal scientist to try to bring these problems closer to a solution.

REFERENCES

EIGEN, M., 1967. Kinetics of reaction control and information transfer in enzymes and nucleic acids. In *Fast Reactions and Primary Processes in Chemical Kinetics.* Nobel Symposium 5, S. Claesson, ed. Stockholm: Almquist and Wiksell, pp. 333–369 (see especially p. 354).

EIGEN, M., 1971. Selforganization of matter and the evolution of biological macromolecules. *Naturwissenschaften* 58:465–523.

EIGEN, M., and L. DE MAEYER, 1963. Relaxation methods. In *Technique of Organic Chemistry,* Vol. VIII, 2nd Ed., A. Weissberger, ed. New York: Wiley-Interscience, pp. 895–1054.

GLANSDORFF, P., and I. PRIGOGINE, 1971. *Thermodynamic Theory of Structure, Stability and Fluctuations.* New York: John Wiley & Sons.

KIRSCHNER, K., 1968. Allosteric regulation of enzyme activity: An introduction to the molecular basis of and the experimental approaches to the problem. *Curr. Top. Microbiol. Immunol.* 44:123–146.

KIRSCHNER, K., M. EIGEN, R. BITTMAN, and B. VOIGT, 1966. The binding of nicotinamide-adenine dinucleotide to yeast *d*-glyceraldehyde-3-phosphate dehydrogenase: Temperature-jump relaxation studies on the mechanism of an allosteric enzyme. *Proc. Natl. Acad. Sci. USA* 56:1661–1667.

LOTKA, A. J., 1910. Contribution to the theory of periodic reactions. *J. Phys. Chem.* 14:271–274.

MEIXNER, J., 1953. Die thermodynamische Theorie der Relaxationserscheinungen .... *Kolloid Z.* 134:3–20.

MONOD, J., J. WYMAN, and J.-P. CHANGEUX, 1965. On the nature of allosteric transitions. A plausible model. *J. Molec. Biol.* 12:88–118.

NEUMANN, E., D. NACHMANSOHN, and A. KATCHALSKY, 1973. An attempt at an integral interpretation of nerve excitability. *Proc. Natl. Acad. Sci. USA* 70:727–731.

OSTER, G., A. PERELSON, and A. KATCHALSKY, 1971. Network thermodynamics. *Nature* 234:393–399.

OSTER, G., A. PERELSON, and A. KATCHALSKY, 1973. Network thermodynamics: Analysis of biophysical systems. *Quart. Rev. Biophys.* 6:1–138.

SPIEGELMAN, S., 1970. Extracellular evolution of replicating molecules. In *The Neurosciences: Second Study Program,* F. O. Schmitt et al., eds. New York: The Rockefeller University Press, pp. 927–945.

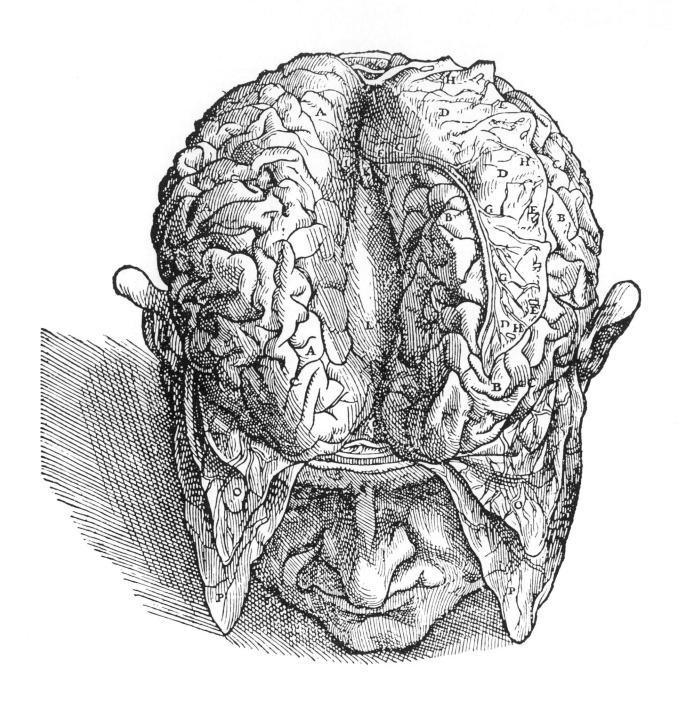

HEMISPHERIC SPECIALIZATION AND INTERACTION

The two hemispheres of the human brain, as depicted in the seventh book of the anatomy of Vesalius (De Humani Corporis Fabrica, "On the Workings of the Human Body"), *completed at Padua and Venice in 1542 and printed at Basel in that and the following year. It is only quite recently that serious attempts have been made to define the complementary roles of man's cerebral hemispheres in complex behavior, and to understand the functions of the major pathways connecting the two sides of the brain. (From the reproduction on p. 93 in Charles Singer, 1952,* Vesalius on the Human Brain. *London: Oxford University Press.)*

HEMISPHERIC SPECIALIZATION
AND INTERACTION

Introduction

BRENDA MILNER

IT HAS BEEN realized now for many years that the left and right cerebral hemispheres of adult man are not functionally equivalent despite a near identity of structure. Already by the early 1860's, observations such as those of Dax and of Broca on patients with unilateral brain disease pointed to a major role for the left hemisphere in the speech processes of right-handed subjects. By a curious extrapolation this idea of the dominance of the left side of the brain for speech and skilled movements soon led to the more general view that the left hemisphere was prepotent for most cognitive processes. In spite of Hughlings Jackson's warning, in 1874, that the right side may play a special role in what he termed "visual ideation," the possibility of a complementary specialization was largely overlooked. It was not until the time of World War II that a series of careful studies on patients with well-lateralized brain lesions began to reveal the full importance of the posterior part of the right hemisphere for spatial functions. This evidence was later strengthened by work in Montreal on the temporal lobes (which I shall describe), delineating specific memory deficits after unilateral temporal lobectomy that vary with the side of the removal.

New evidence has come in recent years from normal subjects, but the most direct and striking demonstration of dual functional asymmetry is that provided by the work of Roger Sperry, with a small group of patients in whom the main interhemispheric commissures have been severed as a treatment for epilepsy. By appropriate techniques, he and his colleagues have been able to confine sensory input and motor output to one cerebral hemisphere and thus bring out the contrasting specializations of the two sides of the adult human brain.

In addition to considering neurosurgical patients, one must take the normal nervous system into account, especially because it is normal behavior that one is ultimately trying to understand. Fortunately, it is now possible by special techniques (embodying bilateral stimulation of the two ears or the two hemiretinae) to put the left and right sides of the normal brain in competition and thus reveal the underlying asymmetries. For audition, some of the most fruitful results have come from the dichotic-listening technique devised by Donald Broadbent. In this procedure, different inputs are channeled simultaneously to the two ears, and the subject simply reports, or otherwise indicates, what he has heard. Christopher Darwin discusses the significance of such experiments, in which a right-ear advantage is found for most speech sounds and a left-ear advantage for music. Similarly, Giovanni Berlucchi describes asymmetries in normal vision, which he has demonstrated by very brief exposure of competing stimuli in left and right visual fields. He finds greater accuracy of recognition and greater speed of response to nonverbal patterns, such as faces, when these are exposed on the left but to letters when exposed on the right. These findings again reflect the contrasting specializations of the two halves of the human brain.

In the intact organism, the two sides work together and, if they are asked to do conflicting jobs, there may be interference between them. Broadbent has marshalled a considerable amount of data from normal subjects to show that two tasks can rarely be carried out simultaneously without loss of efficiency. He argues that if the two hemispheres were really so different in their modes of action, then it should be fairly easy to find tasks that can be done in parallel, one by the left side and one by the right. The fact that interference is the rule, instead, suggests to him that, at some critical stage of decision-making, man behaves as a single-channel organism. It follows that one should not be content with delineating the contrasting specializations of the two sides but must also be prepared to tackle the more difficult problem of how they interact in normal behavior.

At this point we have to admit to almost complete ignorance about the second half of our topic, that of hemispheric interaction, in so far as it pertains to man. Although we can make a few inferences from reaction-time studies, such purely behavioral measures cannot tell us the form in which information is transmitted from one hemisphere to the other, nor any details as to which pathways may be essential for the performance of specific tasks. If we are to begin to answer these questions, we must first look at experiments in lower species (not only mammals but also birds and fish), where direct physiological interventions with environmental controls may provide a clue to elementary mechanisms that have some parallel in man. This approach is exemplified by Michel Cuénod who takes the pigeon as his experimental animal and, by selectively blocking different interhemispheric commissures, determines the particular route by which the interocular transfer of visual information is achieved.

These animal experiments have highlighted the importance of commissural pathways in the integration of complex behavior. They cannot, however, be expected to disclose functional asymmetries of the scope and importance of those seen in man. Among these, the most conspicuous (and the one that most clearly distinguishes man from subhuman primates) is the emergence of speech and its lateralization to the left hemisphere in the typical right-handed subject. Alvin Liberman addresses himself to the question of what listening to speech and producing speech demand of the nervous system, and more specifically of the left cerebral hemisphere. In a discussion based on his own recent work and that of his colleagues at the Haskins Laboratories, he argues that speech perception differs radically from other perceptual processes (though others might disagree). Speech, in his view, requires much higher degrees of encodedness: quite dissimilar sounds produced by a speaker have to be interpreted by the listener as representing one and the same phoneme.

As Hans-Lukas Teuber points out in his interim summary, similar working hypotheses are needed for dealing with the specializations of the right side of the brain, and he asks whether the perception of faces, in which the right hemisphere excels, may not also require a special form of encoding. Finally, the question is raised, in Teuber's chapter and in mine, as to how one might approach the developmental history of this dual functional asymmetry of the human brain. Here important clues may be derived from atypical forms of representation, particularly those associated with early brain lesions of the right or left hemisphere.

1 Lateral Specialization in the Surgically Separated Hemispheres

R. W. SPERRY

ABSTRACT New evidence concerning cerebral dominance, left-right specialization of verbal vs. perceptual functions, and the localization of consciousness in the human brain comes from follow-up studies on people who have undergone cerebral commissurotomy for severe intractable epilepsy. In commissurotomy patients, it is possible to measure and compare the positive performance of each hemisphere of the same individual functioning independently on the same test task, and the two hemispheres can be pitted against each other for response dominance. Fine differences can be measured under these conditions and qualitative distinctions discerned that are much more difficult or even impossible in comparisons involving different persons. Results indicate two distinct modes of central processing in the disconnected left and right hemispheres.

Introduction

UNLIKE THE brains of most mammalian species, the human brain already at birth possesses an intrinsic left-right differentiation that appears to be predetermined to a considerable extent by inheritance, and for which various genetic models have been suggested. The latest of these proposed by Levy and Nagylaki (1972) involves two genes with four alleles. One gene determines which hemisphere will be language dominant, with the allele for the right hemisphere being recessive, and a second gene determines whether hand control will be contralateral or ipsilateral to the language hemisphere, with ipsilateral control being recessive. This gives a total of nine possible different allelic combinations or genotypes for handedness in man, with the weaker mixed combinations presumed to be more easily reversible by training.

Handedness in animals, or paw preference, has been shown not to be genetically determined in rats (Peterson, 1934) and mice (Collins, 1969), and such determination in the higher subhuman primates is still uncertain. Recently Collins (1970) in a widely cited article, poetically entitled, "The sound of one paw clapping," has concluded from an analysis of published human data that in man, also, the evidence does not support a genetic basis for handedness. This conclusion is contested by Nagylaki and

Levy (1973), who claim that "the sound of one paw clapping" is not sound and is based on a misinterpretation of the sibling-sibling relationship. They reaffirm their two gene-four allele model as the best fit for the evidence now available.

Most of our information regarding the nature and functional role of hemispheric specialization comes historically from the kinds of functional impairments produced by asymmetric brain damage (Zangwill, 1964; Mountcastle, 1962; Milner, 1971). In the last eleven years, this evidence has been strengthened and extended by observations on patients with midline cerebral commissurotomy. This is an operation that eliminates direct cross-communication between the hemispheres but leaves both hemispheres otherwise intact and functioning independently. In persons having had this operation, each hemisphere can be tested separately for its positive as well as its negative competence, and direct comparisons can be made for the independent performance of the left and right hemisphere in the same individual.

The following comes mainly from studies by a long line of colleagues and me on a group of such patients operated upon for treatment of severe, intractable epilepsy. These are all patients of P. J. Vogel, Chief of Neurosurgery at the White Memorial Medical Center, and J. E. Bogen, Ross-Loos Medical Group, in Los Angeles. The patient group is small, because this is a kind of surgery that is undertaken only as a last resort measure in an effort to control advancing life-threatening epilepsy where it cannot be contained by medication nor by simpler unilateral ablations.

In a total of only 16 persons operated upon to date, the majority have complications including preoperative asymmetric brain damage that makes them unsuitable for many studies of hemispheric specialization, depending on the specific problem. Furthermore, a new form of the operation that involves only a partial commissurotomy, i.e., section of the anterior two-thirds of the corpus callosum along with the anterior commissure (Gordon et al., 1971), promises to render the complete disconnection unnecessary in most future cases.

The surgery with which we are here concerned includes

R. W. SPERRY Division of Biology, California Institute of Technology, Pasadena, California

complete section of the large corpus callosum in its entirety plus, also, the smaller anterior commissure (Bogen et al., 1965). The thin hippocampal commissure subjacent to the callosum is not separately visualized but is presumed to be sectioned along with the corpus callosum. The massa intermedia, which is variable in man, is also sectioned when it is found. Only the subordinate right hemisphere is exposed and slightly retracted. All the cross-connections are sectioned completely in a single-stage operation, and it is this mainly that distinguishes the operation as reintroduced by Vogel in 1961 from that used earlier in the late 1930s by Van Wagenen (Akelaitis, 1943), in which the deconnections were either only partial or performed in stages, or both.

As the brain is basically bisymmetric in its structural plan, each disconnected hemisphere retains a full set of cerebral centers and interconnections for all the different kinds of cerebral function, excepting, of course, those that involve cross-connections between the hemispheres (Figure 1). Some of the more obvious problems created by the elimination of cross connections are indicated in Figure 2. Note that the optic image of the outside world on its way into the brain is divided down the middle into right and left halves, which then are projected separately into the disconnected left and right hemispheres, respectively. The same applies to the cerebral representation for the right and left hands and legs. This includes both the sensory and the primary motor control centers for the limbs. Note especially that the surgery separates the entire right hemisphere, and all that goes on in that hemisphere, from the speech and main language centers located in the left hemisphere.

Considering the enormous size of the sectioned neocortical systems, estimated to contain over 200 million fibers cross-connecting nearly all regions of the cerebral cortex, with some decussations to subcortical centers, it is generally agreed that one of the more remarkable effects of this kind of operation is the seeming lack of effect insofar as ordinary daily behavior goes. A person two years recovered from the operation and otherwise without complications might easily go through a routine medical check-up without revealing that anything was particularly wrong to someone not acquainted with his surgical history. Speech, verbal intelligence, calculation, established motor coordination, verbal reasoning and recall, personality, and temperament are all preserved to a surprising degree in the absence of hemispheric interconnection.

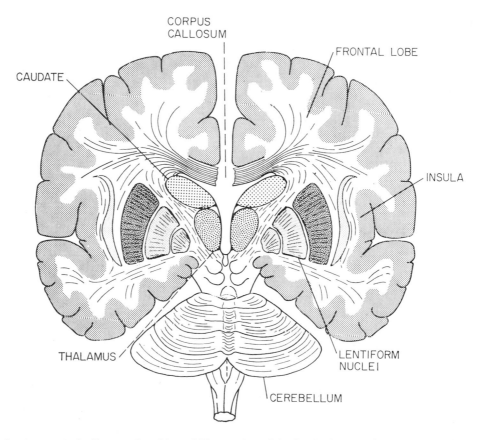

FIGURE 1 Anatomical effect produced by midline section of the forebrain commissures shown schematically.

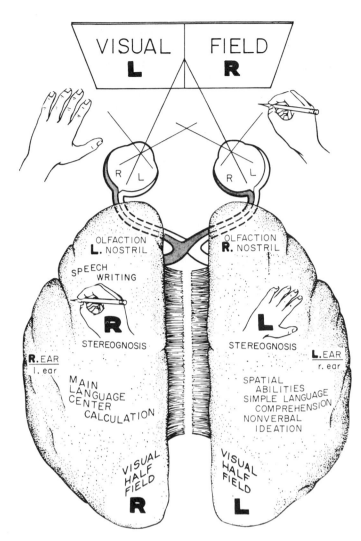

FIGURE 2 Functions separated by the surgery: A simplified summary combined from known neuroanatomy, cortical lesion data, and postoperative testing.

The amazing paucity of symptoms with complete surgical section and also with congenital absence of the great cerebral commissure had become one of the major challenges for brain research back in the 1940s and 1950s and was being cited to support various radical views of how brains operate at their upper levels without depending on specific fiber connections. We know better today, of course, and now recognize a wide and still growing array of interhemispheric functions mediated by the corpus callosum and disrupted by its transection.

Brain bisection and conscious awareness

Despite the apparent normality of these people under ordinary conditions and in contradiction to the prevailing neurological doctrine of the 1950s we can now demon-

strate with appropriate tests a whole array of distinct impairments (Sperry et al., 1969) that are most simply summarized by saying that the left and right hemispheres, following their disconnection, function independently in most conscious mental activities. Each hemisphere, that is, has its own private sensations, perceptions, thoughts, and ideas all of which are cut off from the corresponding experiences in the opposite hemisphere. Each left and right hemisphere has its own private chain of memories and learning experiences that are inaccessible to recall by the other hemisphere. In many respects each disconnected hemisphere appears to have a separate "mind of its own."

This presence of two rather separate streams of conscious awareness is manifested in many ways in different kinds of testing situations. For example, following surgery, these people are unable to recognize or remember a visual stimulus item that they have just looked at if it is presented across the vertical midline in the opposite half visual field; that is, the normal perceptual transfer that one expects to find between the left and right halves of the field of vision is lacking. Similarly, objects identified by touch with one hand cannot be found or recognized with the other hand. Odors identified through one nostril are not recognized through the other nostril (Gordon and Sperry, 1969). Following surgery, these people are unable to name or to describe verbally objects seen in the left half field of vision, objects felt with left hand or foot, odors smelled through the right nostril, or sounds heard by the right hemisphere.

Most of these deficits are easily compensated or concealed under ordinary conditions by exploratory movements of the eyes, shifting of the hands, and through auditory and other cues that bilateralize the sensory information. Their demonstration thus requires controlled lateralized testing procedures. Mainly, these involve the restriction of sensory input to one or the other hemisphere, with or without lateralized motor readout, combined often with processes like speech or writing, the lateralization of which has already been established (see Figure 3). These testing procedures and the various manifestations of the general syndrome of hemisphere deconnection have already been described elsewhere in some detail (Sperry, 1968a; Sperry et al., 1969; Gordon and Sperry, 1969; Levy et al., 1972). They are illustrated in Figures 4 and 5.

The conflict and disruption of behavior that might otherwise be expected from having two separate cerebral systems competing for control of the one body is counteracted by a variety of unifying mechanisms. For example, the two retinal half-fields of the eyeball move as one, and eye movements are conjugate, so that when one hemisphere directs the gaze to a given target the other hemisphere is automatically locked in at all times on the

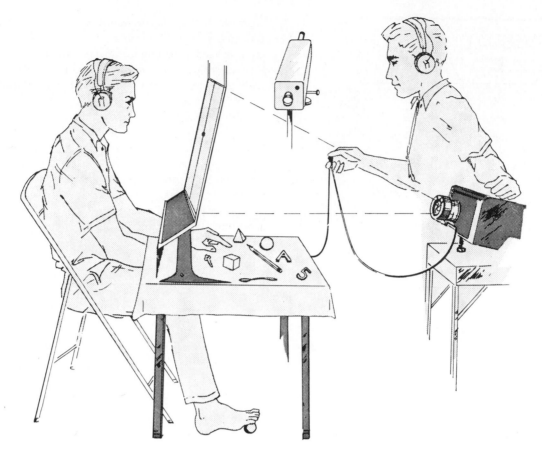

FIGURE 3   Diagrammatic sketch of set-up for lateralized testing after hemisphere deconnection.

same target. Also, the uncrossed fiber systems of the brain allow a certain degree of bilateral sensory representation and motor control within each hemisphere that helps greatly to keep behavior unified.

Emotional effects triggered through one hemisphere as by lateral presentation of an offensive odor, embarrassing photo, etc., tend to spread easily into the opposite hemisphere, presumably through intact brainstem routes and through feedback from peripheral changes. Vascular reactions like blushing and alterations in blood pressure, facial expressions, giggling, frowning, exclamations, and the like all help to bilateralize an emotion and to counteract attempts to establish concurrent conflicting emotional sets on left and right sides. Whether a central emotional state could survive in one hemisphere in the presence of conflicting emotional expression imposed peripherally from a different emotional state in the other hemisphere has yet to be determined. The possibility that mild emotional experiences may be confined temporarily to one hemisphere is by no means ruled out, but this is difficult to demonstrate (Gordon and Sperry, 1969; Gazzaniga, 1970) and would seem to be more the exception than the rule. Emotional processes thus appear

to exert a general unifying influence, and this applies also when affect is involved as reinforcement in learning. Reinforcement along with other feedback effects must be laterally controlled and strong emotional responses avoided.

Mechanisms of orientation and attention likewise tend to bring hemispheric activity into a unified focus. Active task performance on the part of one disconnected hemisphere will ordinarily interfere in varying degree with simultaneous performance on a different task by the opposite hemisphere. Such interference is, of course greatly reduced over that obtained with the commissures intact, and can be largely nullified with practice under the right conditions (Gazzaniga, 1970). Interference between the disconnected hemispheres is reduced when a common motor posture is utilized for both left and right tasks, or when both tasks involve the same general mental set. Separate parallel performance on different tasks, though possible under special facilitating conditions, seems not to be the general rule. The fact that attention in many tests seems to become focused in one separate hemisphere and simultaneously become repressed in the other would appear to reflect lateral differentiation in the

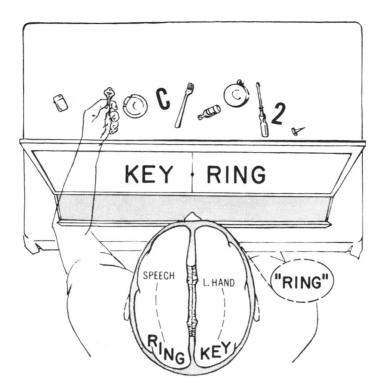

FIGURE 4 Sample split-brain responses: Subject reports (through speaking hemisphere) having seen only the visual stimulus flashed to right half of screen and denies having seen left-field stimulus or recognizing objects presented to left hand. At the same time, left hand correctly retrieves objects named in left field for which subject verbally denies having any knowledge. When asked to name object selected by left hand, speaking hemisphere refers it to stimulus shown in right field.

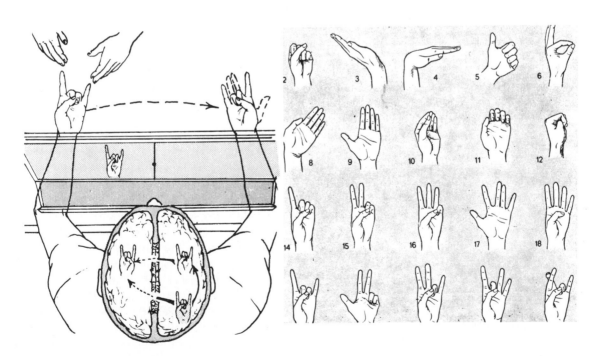

FIGURE 5 Additional tests used by author to detect lack of interhemispheric integration following commissurotomy. Subject attempts to replicate complex hand and finger postures flashed to left and right visual half-fields or imposed directly on one hand by examiner. Intrahemispheric combinations are performed successfully but crossed combinations fail.

thalamo-brainstem attentional mechanisms. The sleep-wake cycle and associated arousal mechanisms also retain strong bilateral unity after commissurotomy, presumably centered in midbrain structures. They too involve metabolic, blood-pressure, and blood-chemistry cycles that together help reinforce a unified bisymmetric state of basal arousal level in both hemispheres.

It is important that a distinction be made in the behavioral unity that survives commissurotomy between that which is attributable to bilateral representation of the given function in each hemisphere, like facial sensibility, and the unity that results from representation at lower undivided levels, like that of the brainstem orienting mechanisms and the cerebellar controls for motor coordination. Functional unity of the former type involving hemispheric reduplication is not a contradiction of the general conclusion that each of the disconnected hemispheres has its own separate domain of conscious awareness. With most of our research interest naturally concentrated on the divided aspects of brain function, it is easy to underemphasize the many components of behavior that remain unified.

Compensatory readjustment and agenesis

Postoperative use of the above unifying mechanisms and associated cross-cuing strategies can be significantly strengthened by practice. We find evidence of this particularly in the youngest patient operated upon at age 13. At 7 years after surgery this patient (L.B.) shows improvements in cross-integration to such an extent that many of the basic phenomena of hemisphere deconnection are hardly evident (Zaidel and Sperry, 1972a). For example, he can report verbally 1 to 3 numbers or letters flashed to the left visual field or presented to the left hand. Given a series of 8 objects for blind tactual identification, he can cross-retrieve with the left hand a given item initially identified with the right hand. He is able to report verbally the sum, difference, or product of 2 numbers flashed one to the left and one to the right visual half-field. Although his facility at such tasks is still well below that seen with congenital absence of the callosum, described below, the general picture at 7 years is very different from that displayed during the first 1 to 2 years after surgery, when this same patient was totally unable to report numbers or letters flashed to the left half-field.

Another patient (N.G.), also with a minimum of known preoperative brain damage but operated upon much later in life (at age 30), shows much less improvement after 8 years. She occasionally is able to report single large numbers and geometric shapes from the left half-field of vision and to align bars across the visual midline, which she could not do previously, but such performance

is rare and seems to depend on transient shifts in attentional set, and related factors not consistently under control.

The extreme to which this sort of functional compensation for loss of the commissures can go is illustrated in cases of congenital absence or agenesis of the corpus callosum resulting from accidents of development occurring usually at about the fourth month of gestation (Unterharnscheidt et al., 1968). A person with congenital total absence of the callosum was studied by Saul and Sperry, 1968; Sperry, 1968c; 1970c), and compared on similar tests with the commissurotomy patients. This was an "asymptomatic" case of callosal agenesis, one of about 18 such cases on record in medical history (Unterharnscheidt et al., 1968). This 19-year-old college sophomore, with an average scholastic record and previously presumed to be normal until hospitalized and X-rayed for a series of headaches, easily went through the entire battery of cross-integration tests that we had devised for the surgical patients, performing essentially like a normal control subject. Even words projected half to one and half to the other hemisphere, near the threshold of visual acuity, were read off promptly with no difficulty. This was true even where the pronunciation depended on cross-integration between the two halves of the word.

The anterior commissure appeared to be present in the X-rays and somewhat enlarged, and presumably the functions of the uncrossed or bilateral fiber systems of the brain had been enriched by use. Further, lateral anesthetization of one hemisphere at a time by intracarotid Amytal injection revealed that speech had been developed in both hemispheres. All of this emphasizes the greater plastic compensatory capacities of the still-growing, developing brain as compared to the adult brain.

However, when we went further and compared this patient's performance on more complex tasks with that of normal subjects (Sperry, 1970c), she was found to be selectively subnormal on a variety of perceptuomotor, spatial, nonverbal reasoning tasks that together added up to what appeared to be a mild "minor hemisphere syndrome." These deficits were in contrast to her above-normal score of 111-112 on the verbal part of the Wechsler Adult Intelligence Scale (WAIS). With the language and the nonlanguage functions both necessarily crowded together within the same hemisphere on both sides, instead of the normal right-left division of labor, the verbal faculties had apparently become dominant and had developed at the expense of the nonverbal. As will be evident below, this tendency for language to develop at the expense of the competing nonverbal functions is more the rule than the exception. Even though the spatial-perceptual-performance functions appear to have primacy in terms of evolution and also seem to get a head start over language during development, there seems to be something about the growth and maturational processes in the human brain that tends to favor the elaboration of language. All of which brings us now more directly to the question of the nature of the functional differences between the left and right hemispheres.

Hemispheric dominance and differentiation

The commissurotomy patient, as mentioned, offers special advantages for the study of hemispheric specialization in that it is possible to measure and to compare the positive performance of each hemisphere functioning independently. Direct comparisons can be made for performance on the same task in the same individual where the life history and other background factors are all equated, and the two hemispheres can be pitted against each other for response dominance. Fine differences can be measured under these conditions and qualitative distinctions discerned that are much more difficult or even impossible in comparisons involving different persons. Earlier caution and conservative uncertainties regarding the extent and the importance of hemispheric specialization still being voiced in the early 1960s (Mountcastle, 1962) tend to dissolve in the face of direct cross-comparisons of this kind. It is most compelling to see the same individual performing the same test task in very different ways, and with different strategies, depending on whether he is using his left or his right hemisphere.

Repeated examination during the past 10 years has consistently confirmed the strong lateralization and dominance for speech, writing, and calculation in the disconnected left hemisphere in these right-handed patients. The minor, right hemisphere, by contrast, is unable to respond in speech or writing in the great majority of test situations. Nor can it typically perform calculations except for simple additions up to sums less than 20. The language-dominant hemisphere is also the more aggressive, executive, leading hemisphere in the control of the motor system. After the surgery, these patients seem to run primarily on the left hemisphere. This is the hemisphere that we mainly see in action and the one with which we regularly communicate. It is the highly developed and dominant capacities of the left hemisphere apparently that are largely responsible for the earlier impressions that cerebral functions persevere with little impairment in the absence of the corpus callosum.

The mute, minor hemisphere, by contrast, seems to be carried along much as a passive, silent passenger who leaves the driving of behavior mainly to the left hemisphere. Accordingly, the nature and quality of the inner mental world of the silent right hemisphere remains relatively inaccessible to investigation, requiring special testing measures with nonverbal forms of expression. Although some authorities have been reluctant to credit the disconnected minor hemisphere even with being conscious, it is our own interpretation based on a large number and variety of nonverbal tests, that the minor hemisphere is indeed a conscious system in its own right, perceiving, thinking, remembering, reasoning, willing, and emoting, all at a characteristically human level, and that both the left and the right hemisphere may be conscious simultaneously in different, even in mutually conflicting, mental experiences that run along in parallel.

Though predominantly mute and generally inferior in all performances involving language or linguistic or mathematical reasoning, the minor hemisphere is nevertheless clearly the superior cerebral member for certain types of tasks. If we remember that in the great majority of tests it is the disconnected left hemisphere that is superior and dominant, we can review quickly now some of the kinds of exceptional activities in which it is the minor hemisphere that excels. First, of course, as one would predict, these are all nonlinguistic, nonmathematical functions. Largely they involve the apprehension and processing of spatial patterns, relations, and transformations. They seem to be holistic and unitary rather than analytic and fragmentary, and orientational more than focal, and to involve concrete perceptual insight rather than abstract, symbolic, sequential reasoning. However, it yet remains for someone to translate in a meaningful way the essential right-left characteristics in terms of the brain process, and accordingly we still do well to refer to the actual test activities.

The superior minor hemisphere

It was shown very early in the study of these patients by Bogen and Gazzaniga (1965; Bogen, 1969; Gazzaniga, 1965) that the right hemisphere is superior to the left in the construction of block designs, and also in copying and drawing test figures like a Necker cube, a house, a Greek cross, etc. This is consonant with prior evidence for visuospatial constructional apraxia following right hemisphere lesions (Arrigoni and DeRenzi, 1964; Hécaen, 1962; Piercy et al., 1960; Zangwill, 1964). These differential hemispheric capacities were shown by Levy (1969a) to involve more than just praxic ability. She developed a test for these patients involving intermodal spatial transformations in which praxic or motor requirements were kept extremely simple, namely, manual pointing, but which required complex perceptual and cognitive central processes. Three-dimensional blocks of various shapes were placed in the subject's right or left hand to be perceived blindly by touch and then matched to two-dimensional patterns, perceived in free vision, of what the blocks would look like if made of cardboard and unfolded.

In addition to a striking quantitative superiority of the minor hemisphere, Levy also found qualitative differences, the performance with the left hand being rapid, silent, and direct, whereas that with the right hand was more hesitant and accompanied by a running verbal commentary that was difficult to inhibit. The final response of the right hand was apparently dependent on a sequence of verbal reasoning. It was concluded that the disconnected left hemisphere was applying a verbal analytic mode of thinking in contrast to the right hemisphere that had reasoned by direct perceptual, synthetic, or Gestalt, processing. With further analysis of which items were most easy and which most difficult for each hemisphere, she inferred that the left and right modes of central processing would mutually interfere within the same hemisphere and that this would give a rationale for the evolution of cerebral dominance.

The left-hand performance of patient L.B., who scored highest on this cross-modal test, was found to be better than 31% of college sophomores when the test was later standardized at the University of Southern California. However, on a similar visual test for spatial relations presented in free vision with unrestricted hand use, this same patient failed completely, rating lower than 99% of the population of his age and education. This striking contradiction seems best explained as another example of the suppression and interference caused by left hemisphere activity upon the expression of abilities centered in the right hemisphere.

Similar qualitative differences in the problem-solving strategies of the separate hemispheres were later observed on a modification of the Raven's Progressive Matrices Test administered separately to the left and right hemispheres by Zaidel and myself (1973). The choice array of figures containing the correct answer was presented in the form of raised patterns for tactual discrimination by left and right hands separately. It was necessary to reduce the standard six-item choice array to three items under these conditions as in the above-mentioned test used by Levy. The commissurotomy patients became too confused when obliged to keep track of more than three items in blind touch. Apparently, they have a reduced mental grasp for holding in mind a series of newly identified tactual stimuli. Although the colored stage of the Raven's test in particular appears to require mainly spatial insight, the left hemisphere (right-hand performance) was nevertheless found to be able to attain a score only slightly below that of the right hemisphere by using a slower process of verbal reasoning. Thus both hemispheres could find the correct answers but by employing different strategies. We concluded that the scores for the Progressive Matrices Test do not in themselves eminently distinguish left and right hemispheric capabilities.

The special spatial aptitude of the disconnected minor hemisphere is not confined to the visual modality. This was shown in tests by Milner and Taylor (this volume) that were based entirely on tactual discrimination and memory for nonsense shapes made of bent wire. In tests involving the perception of part-whole relations worked up by Nebes (1971), a strong superiority of the right hemisphere was also evident in a task presented entirely through touch. The task involved the matching of a given sample circle-segment or arc to the correct one of an array of three whole circles of different sizes. Both sample and array were presented by touch with vision excluded. The test was also given cross-modally, going from vision to touch, or from touch to vision, with similar results in all three modes of presentation. Scores for normal control subjects using both hemispheres and without medication or a history of epilepsy (and with considerably higher IQs than these patients) averaged only 10 to 15% better than those for the disconnected minor hemisphere.

In another test used by Nebes (1971), the subjects examined a fractured or exploded figure in free vision and then searched blindly by touch behind a screen for the correct, intact pattern to match, from among an array of three raised figures, using one or the other hand. Again the left-hand right-hemisphere system came out far ahead. The scores for the left hemisphere in this and the preceding task were hardly above chance; apparently this kind of task was almost too difficult for the disconnected left hemisphere.

In a modification of the Kasanin-Haufmann Concept Formation Test used by Kumar (1971), the subjects had to sort 16 blocks into four categories using blind tactual identification. It was required that the subject discover by trial and error which of several possible criteria, like shape or size or weight or height, etc., was the correct one for the classification. The minor hemisphere scored almost twice as well as the major in terms of time and correct choices, whereas in a similar test involving 16 familiar objects easily named, the score standings were reversed.

Duplicate and competitive right-left processing

Figure 6 illustrates how normal scanning movements of the eyes from right to left edge of an object being examined, would give two sensory representations or percepts of the object, one in each hemisphere (Sperry, 1970a). The constancy of the visual field in the presence of eye movements must be taken into account, and inversions of the visual pathways assumed. We have long

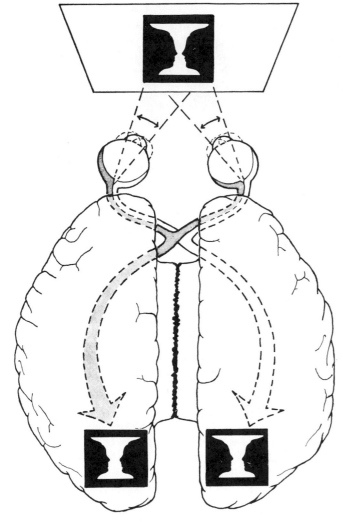

FIGURE 6 Diagram showing how exploratory eye movements give two full percepts of an examined object, one in each hemisphere. Position constancy in the presence of eye movement must be taken into account and image inversion in the optic pathways assumed.

wondered about the functional utility of this kind of right-left redundancy in the normal brain, but if the two hemispheres process sensory information in distinctly different ways as we now think, then perhaps this left-right doubling makes sense.

In recent studies (Levy et al., 1972), the two sensory images perceived in right and left hemispheres are arranged to be different and conflicting. A series of visual stimuli are split down the middle and then recombined and joined at the midline to make composite right-left chimeras (see Figure 7). These chimeric figures are then flashed to the subject while the gaze is centered, as indicated in the figure. The missing half of each image tends to be filled in by each hemisphere according to a

general rule for perceptual completion (Trevarthen and Kinsbourne, 1973). Since each hemisphere is cut off from the conscious experience of the other, the split-brain subjects remain blandly unaware of even gross discordance between the left and right halves of these composite stimuli.

The result is that the two hemispheres are induced to see two different things occupying the same position in space at the same time, something that the normal brain of course rejects. With rival, competing perceptual processes thus set up on left and right sides, it becomes possible to determine which hemisphere is the more proficient in dominating the response. The response may be verbal or manual, like pointing to a matching item among a choice of alternative possibilities. This same principle has been used with nondescript and geometric figures, faces, objects, perception of movement, words, clock faces, serial patterns, colors, and combinations of these, under different testing conditions, that is, with different mental and motor sets, and with verbal as well as manual readout.

In general, the results with composite input conform with earlier findings in that if linguistic processing is involved, the subject's response is dominated by the disconnected left hemisphere selecting in favor of the right half of the composite stimulus. On the other hand, with nonlinguistic, manual readout either side may dominate. With the perception of faces and of complex or nondescript patterns, and with any direct visual-visual matching of shape or pattern, the right hemisphere is dominant. This applies even for words, provided that no interpretation of word meaning is involved. When a conceptual translation or verbal reply is required, the dominance promptly shifts to the opposite left hemisphere. Nondescript shapes proved to be extremely difficult for the left hemisphere, even without concurrent right hemispheric stimulation.

An unexpected minor-hemisphere dominance for letter and word percepts has been found with bilateral competitive input in the tactual mode (Preilowski and Sperry, 1972). The task involved tactual discrimination and retention of competing pairs of three and four letter nouns and verbs presented simultaneously to the left and right hands one letter at a time. The response consisted of pointing to a choice array of words presented in free vision. A change to purely verbal response brought a prompt shift of dominance to the left hemisphere.

The chimeric results also demonstrated that the disconnected minor hemisphere is capable of capturing and controlling the motor system under conditions where the two hemispheres are in equal and free competition, that is, where the sensory input is equated and the subject is quite free to use either left or right hand. Although this had been seen sporadically in occasional tests before

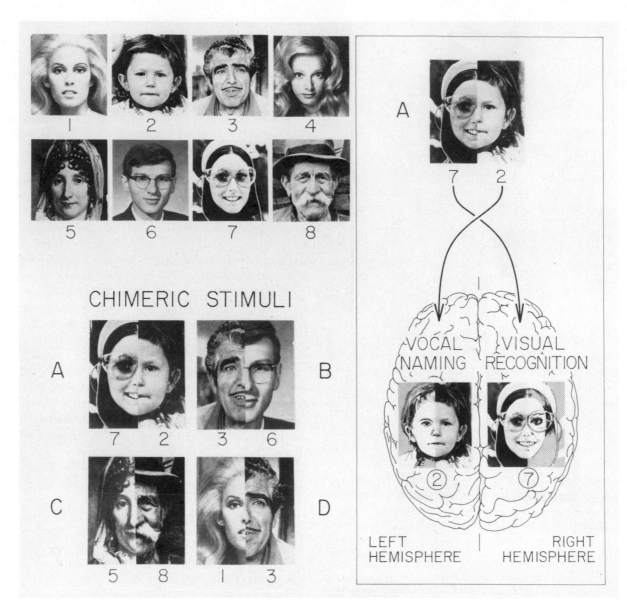

FIGURE 7 Composite face stimuli for testing hemispheric specialization for facial recognition. From Levy et al. (1972). See explanation in text.

under special conditions, we had not previously obtained the phenomenon so consistently or convincingly. The same was found in the above-mentioned study with competitive bilateral tactual stimulation. These observations favor the view that in the normal brain volitional movements may be directly controlled from either hemisphere depending on which is superior and dominant for a given activity. The chimeric findings also reaffirm the impression that the disconnected left and right hemispheres apprehend and process things in different ways. In dealing with faces, for example, the right hemisphere seems to respond to the whole face directly as a perceptual unit,

whereas the left seems to focus on salient features and details to which verbal labels are easily attached and then used for discrimination and recall.

Verbal vs. nonverbal in dominance relations

If a special mutual antagonism exists between the mode of cognitive processing used for language and that used for spatial perceptual function, as postulated by Levy, this ought to show up statistically, she predicted, as a perceptual deficit among left-handed persons, because sinistrals are known to be more bilateralized for language competence than dextrals as shown in the way such persons recover from cerebral injury. Supporting evidence has been found (reviewed in Levy, 1973) not only among sinistrals in their IQ subtest profiles but also in perceptual tests with persons with right-hemisphere speech resulting from early injuries to the left hemisphere, and in the case of callosal agenesis described above. The inverted handwriting seen in many left-handers and very rarely in dextrals is tentatively attributed by Levy to the development in left-handers of ipsilateral motor control from a left language-dominant hemisphere. Normal writing in left-handers is accordingly thought to reflect contralateral control from a language-dominant right hemisphere.

A comprehensive literature on human sex differences (Garai and Scheinfeld, 1968) supports the generalization that adult women do better statistically, on the average, in tasks involving verbal facility, whereas adult males do better in spatial and mathematical tasks. Appraising the evidence in the perspective of hemispheric specialization, Levy (1973) finds the lower spatial ability in females to be similar to that in sinistrals and describes it as probably a sex-linked genetically determined factor that possibly results from hemispheres less well laterally specialized than those of males. On the other hand, women are more consistently right-handed than men (Annett, 1970).

Though the foregoing story may be roughly correct, as far as it goes, we can expect the total picture regarding cerebral dominance, sinistrality, and sexuality to be less simple than here outlined when all the facts are in, and particularly so in regard to implications for any individual brain. It should be remembered that all of this is highly statistical. The brains of individual left-handers, for example, exhibit many different forms and degrees of right-left asymmetry; a complete mirror switch in cerebral dominance should leave no effect on the pattern of abilities. With variations of this kind innately predetermined to a considerable extent, it may be seen that differential balancing of these left and right hemispheric abilities could provide quite a spectrum of individual variation in the structure of human intellect (Sperry, 1971).

Regardless of the indications for an intrahemispheric antagonism between the verbal and spatial modes of thinking, the two modes apparently integrate quite harmoniously under normal conditions, i.e., in the presence of the corpus callosum. This harmony could be achieved conceivably through a distinct right-left, back and forth, on and off switching between the two hemispheres, but there are reasons to favor the view (Sperry, 1973) that a smooth integration of the two into a unified process is the more general rule. If so, this would suggest that the interference obtained intrahemispherically may be a result of competition for the same cellular systems and

circuit mechanisms as much as, or more than, incompatibility in organizational dynamics per se. In this same connection, the possible operational advantages of having a single unilateral control center for speech and language (Jones, 1966) should not be overlooked as part of the rationale for hemispheric specialization.

Effects on memory, mental grasp, and well-being

One symptom that has shown up consistently in ordinary behavior following commissurotomy is a pronounced memory impairment (Sperry, 1968a). Especially during the first year and a half after surgery these patients have been notably unable to remember appointments or telephone messages, where they have put things, which cards have been played, how to get back to a parked car, etc. This memory deficit is also combined with a tendency to fill in memory gaps by confabulation. Improvement in the first few years after operation suggested that the memory defect might largely disappear with time, but this has not been the case. Administration of the Wechsler Memory Scale and Benton's Visual Retention Test in patients 4 to 9 years after surgery, and comparisons of these scores with those of the same patients on the WAIS scale and with others published for groups of epileptics indicate a selective memory deficit (Zaidel and Sperry, 1972b). A specific role of the forebrain commissures in mnemonic functions is suggested.

Logically, any memory laid down solely in the right hemisphere would be inaccessible after commissurotomy to verbal recall in patients with typical language dominance. Similarly, any storage, encoding, or retrieval process dependent normally on integration between symbolic functions in the left hemisphere and spatio-perceptual mechanisms in the right would also be disrupted by commissurotomy. Neither the patients nor their families on being questioned have noticed or complained about any general loss after the surgery of long-term memory. In some instances, however, subnormal retention for events during the several years preceding the operation is suspected. The absence of any conspicuous loss in old memory could mean that firmly established, frequently used memories become in time sufficiently verbalized in the left hemisphere to make the loss of the right hemispheric component pass unnoticed.

Under testing conditions favoring performance with the right hemisphere, it appears that the minor hemisphere can readily learn and remember such things as spatial relationships and related sorting and assembly tasks. Similarly simple verbal memory seems to be well preserved in the left hemisphere. Thus, processing and laying down of memory within each hemisphere may be little affected as such, implying the general impairment to be a reflection primarily of transactional aspects of interhemispheric integration. The memory impairment as displayed in ordinary unrestricted conditions is tied in part to the problem of the extent to which memory lodged in the minor hemisphere is unable to gain expression in ordinary behavior because of the prevailing left-hemisphere dominance, something about which our information remains scanty. It is interesting that the patients with only partial commissurotomy (as described in the next section) score almost as low on specific memory tests as do those with complete callosal section and during the first 3 to 4 years at least show comparable mnemonic impairments in general behavior.

Studies of bimanual motor coordination (Preilowski, 1972) show the commissurotomy patients to be severely impaired in performing new movements that require interdependent regulation of speed and timing between right and left hands. This is in contrast to habitual long-established bimanual coordinations, like tying neckties or shoelaces, which have been found to survive the surgery with little or no detectable deficit. The contrast between new and old motor coordinations suggests that prolonged use brings a change in the higher control mechanisms such as a shift to lower undivided centers like the cerebellum, or a shift to a cerebral mechanism that governs both contralateral and ipsilateral hands from the same hemisphere.

A postoperative loss in general unrestricted performance of what might be called "mental grasp" is indicated in the patients' consistently low scores after operation on the digit-symbol and arithmetic subtests of the Wechsler Adult Intelligence Scale and on the Verbal and Abstract Reasoning parts of the Differential Aptitude Test. It was evident also in various lateralized tests as for example in the subjects' inability to keep track of more than three patterns perceived sequentially by touch in the Progressive Matrices Test and in Nebes' Circle-Arc Test, and may have been a factor in low performance on the Milner-Taylor (1972) Bent Wire Test. In regard to subnormal scores on any minor hemisphere task, however, one must also rule out interference effects from the dominant hemisphere. This latter and the consequent inability of the minor hemisphere to gain unhampered motor expression under ordinary conditions seem to be responsible in large part for the minor hemisphere syndrome that is typical of the commissurotomy patient.

In addition to measurable data from specific tests, one gets also a general impression from working with these patients over long periods that their overall mental potential is affected by the commissurotomy. Perseverance in tasks that are mentally taxing remains low in most of the patients, as does also the ability to grasp broad, long-

term, or distant implications of a situation. With the exception of the youngest patient, their conversation tends to be restricted mainly to what is immediate and simple. Undue repetition of the same information or anecdote is common. There are indications of a tendency to fantasize, and a mild logorrhea seems to be present in some cases. Most of these symptoms, however, have not yet been subjected to specific study, and also most of the patients have substantial extracommissural damage contributing to such effects.

Although the bisected brain may perform certain simple double tasks better than the normal intact brain, under very selected conditions (Gazzaniga, 1970), implications that brain bisection produces an advantageous increase over the normal channel capacity of the brain have not been regarded seriously by most investigators who have worked with these subjects. The great bulk of the evidence continues to support our general impression (Sperry, 1970a) that two hemispheres united are much to be preferred over two hemispheres divided.

It has been noted from the start that the commissurotomy patients, following recovery from the surgery, are inclined to experience an improved sense of well-being and generally do not display particular anxiety about their symptoms even when the testing conditions evoke and force attention to conflicting experiences in left and right hemispheres. Part of this can be attributed to the reduction in medication and in the incidence of seizures, and also to the apparent obliviousness on the part of each hemisphere to the presence and experiences of the other (Sperry, 1968a, b). The following factor may also be involved: Local electrical lesions of the genu of the corpus callosum that interrupt the transcallosal cingulostriate pathways have been used to treat chronic psychiatric symptoms of tension and anxiety and are reported to result in a feeling of well-being and relaxation (Laitinen, 1972). Such lesions would of course be included as part of the fiber transection in the surgery in all of these patients.

Commissurotomy with partial sparing of callosum

In an effort to avoid the more pronounced symptoms now known to result from hemisphere deconnection while at the same time hoping to preserve the therapeutic effect on seizure control, Vogel has sectioned in the last three patients operated only the anterior commissure and front two-thirds of the corpus callosum, sparing the posterior callosum. Study of two of these latter partial commissurotomy patients (Gordon et al., 1971) has shown them to be remarkably free of the basic disconnection symptoms described above that involve cross-integration between the hands, visual half-fields, and the language centers.

One is impressed that a little callosum, at least at its posterior end, goes a long way functionally.

Presumably many functions yet to be delineated are mediated through the sectioned anterior half of the callosum for which proper tests have not yet been found. The partial commissurotomy subjects score almost as poorly as those with complete deconnection on standard memory tests including the Wechsler Memory Scale and the Benton Revised Visual Retention Tests (Zaidel and Sperry, 1972b). Also, in comparison with normal subjects they show definite impairment in the learning of bimanual motor coordinations (Figure 8) that require mutually dependent timing of movements of the two hands (Preilowski, 1972). The initial bimanual motor performance is much better than in the patients with complete callosal section, and early stages of motor learning were found to not be markedly retarded so long as the movement was relatively slow and subject to visual guidance. However, more advanced stages of motor

FIGURE 8 Bimanual motor task of Preilowski (1972) impaired in subjects with partial commissurotomy. Cranking speed for left and right handles is mutually interdependent.

learning that appeared with practice in the normal controls failed to occur in the partial commissurotomy patients. These involved further smoothing and speeding of the hand coordinations and seemed to depend on an advancement from visual to kinesthetic control and a closer integration between the hemispheric control centers for right and left hands.

Replication and interaction at conscious levels

The general observation that hemisphere deconnection creates two distinct left and right realms of conscious experience leaves unexplained many details regarding the extent and nature of this separation. Bilateral sensory representation like that for the face and for other axial structures, along with crude sensibility even in the extremities, makes for a large common denominator of identical experience in right and left hemispheres (Sperry, 1968a). There also is strong bilateral projection in the auditory pathways and in the proprioceptive system for movement and position sense (McKloskey, 1973). Even in the visual system bilateral projections have recently been described that involve pathways through the superior colliculus, pulvinar, and inferior temporal area (see Graybiel, this volume). The latter would in effect bring the ipsilateral half-field of vision into each hemisphere along with the main contralateral representation transmitted through the geniculostriate system. The more robust geniculate half-field is favored by attentional and other factors that seem to suppress actively the second system most of the time. Under the right conditions, however, it seems the subordinate half-field may be perceived and integrated with the main contralateral activity (Trevarthen, 1970).

Attentional reversals are seen on exceptional occasions that bring the subordinate half-field of vision or the subordinate hand into focus, while at the same time suppressing the usually dominant contralateral system to the point where the subject can talk about objects in the left visual half-field of vision but not in the right (Trevarthen and Sperry, 1973) or can perform linguistic operations like spelling with the left hand but not with the right (Preilowski and Sperry, 1972). To what extent these may involve shifts of attention between hemispheres as well as between subsystems within the same hemispheres is not clear. With reference to the latter, it would appear that the ipsilateral and contralateral systems within the same hemisphere tend to be used quite separately in these kinds of situations with active attention to one switching off the other.

The large common overlap of identical experience within the hemispheres that results from bilateral sensory projection and other factors described above represents a complicating and misleading element in descriptions of conscious mental or functional unity in the bisected brain. Though the experience within each hemisphere is presumed to be cut off from its identical counterpart in the other, the demonstration of this duplication must rest always on extrapolation from experiences that are not identical but in some way differ in the two hemispheres. Further, with respect to those facets of conscious experience that remain unified, it is not now possible to determine how much of such conscious unity is a result of the foregoing bilateral representation within each hemisphere and how much alternatively may come from representation at lower undivided neural levels like the brain stem and cerebellum.

The fact that surgical section of the forebrain commissures produces such a profound left-right separation in conscious awareness would seem to indicate that conscious experience is not centered in the mesencephalon, cerebellum, or other lower structures. It tells us further that the mediating cerebral mechanisms are in principle restricted and localizable and may in time be identified. The fact that the two separated hemispheres may each mediate conscious experience independently and concurrently does not mean that this happens under normal conditions. The intact callosum might a priori act merely to harmonize and complete duplicate conscious processes, one in each hemisphere. It seems more probable, however, especially with each hemisphere processing its input in fundamentally different ways, that callosal excitation serves to span and unite a single unified process with parts in each hemisphere. Out of these and related concerns has come a modified concept of the mind-brain relation in which the properties of subjective experience are conceived to be an integral part of the brain process and to play a causal control role in cerebral function (Sperry, 1969, 1970b). More than any other cerebral system, the interhemispheric commissures and their cortical associations continue to offer promise in the search for an eventual direct correlation between the phenomena of complex subjective experience and known variables in specified neural structures.

ACKNOWLEDGMENTS Work by the writer and his co-workers cited here has been supported by Grant No. 03372 from the National Institute of Mental Health, U.S. Public Health Service and by the Hixon Fund of the California Institute of Technology.

REFERENCES

AKELAITIS, A. J., 1943. Studies on the corpus callosum. VII: Study of language functions (tactile and visual lexia and graphia) unilaterally following section of corpus callosum. *J. Neuropathol. Exp. Neurol.* 2: 226–262.

ANNETT, M., 1970. Handedness, cerebral dominance and the growth of intelligence. In *Specific Reading Disability*, D. J. Bakker and P. Satz, eds. Rotterdam: Rotterdam University Press.

ARRIGONI, G., and E. DeRENZI, 1964. Constructional apraxia and hemispheric locus of lesion. *Cortex* 1:170–197.

BOGEN, J. E., 1969. The other side of the brain. I: Dysgraphia and dyscopia following cerebral commissurotomy. *Bull. L.A. Neurol. Soc.* 34:73–105.

BOGEN, J. E., and M. S. GAZZANIGA, 1965. Cerebral commissurotomy in man. Minor hemisphere dominance for certain visuospatial functions. *J. Neurosurg.* 23:394–399.

BOGEN, J. E., E. D. FISHER, and P. J. VOGEL, 1965. Cerebral commissurotomy: A second case report. *J. Amer. Med. Assoc.* 194:1328–1329.

COLLINS, R. L., 1969. On the inheritance of handedness. II: Selection for sinistrality in mice, *J. Hered.* 60:117–119.

COLLINS, R. L., 1970. The sound of one paw clapping: An inquiry into the origin of left-handedness. In *Contributions to Behavior Genetic Analysis: The Mouse as a Prototype*, G. Lindzey and D. D. Thiessen, eds. New York: Appleton-Century-Crofts, pp. 115–136.

GARAI, J. E., and A. SCHEINFELD, 1968. Sex differences in mental and behavioral traits. *Gen. Psych. Mon.* 77:189–299.

GAZZANIGA, M. S., 1965. Some effects of cerebral commissurotomy on monkey and man. Ph.D. Thesis, California Institute of Technology, Pasadena, California.

GAZZANIGA, M. S., 1970. *The Bisected Brain*. New York: Appleton-Century-Crofts.

GORDON, H. W., and R. W. SPERRY, 1969. Lateralization of olfactory perception in the surgically separated hemispheres in man. *Neuropsychologia* 7:111–120.

GORDON, H. W., J. E. BOGEN, and R. W. SPERRY, 1971. Absence of deconnexion syndrome in two patients with partial section of the neocommissures. *Brain* 94:327–336.

GOTT, P., 1973. Language following dominant hemispherectomy. *J. Neurol. Neurosurg. Psychiat,* (in press).

HÉCAEN, H., 1962. Clinical symptomatology in right and left hemispheric lesions. In *Interhemispheric Relations and Cerebral Dominance*, V. B. Mountcastle, ed. Baltimore: The Johns Hopkins Press, pp. 215–243.

JONES, R. K., 1966. Observations on stammering after localized cerebral injury. *J. Neurol. Neurosurg. Psychiat.* 29:192–195.

KUMAR, S., 1971. Lateralization of concept formation in human cerebral hemispheres. *Biol. Ann. Rep.* (Calif. Inst. Techn.), p. 118.

LAITINEN, L. V., 1972. Stereotactic lesions in the knee of the corpus callosum in the treatment of emotional disorders. *Lancet*, Feb. 26.

LEVY, JERRE, 1969a. Information processing and higher psychological functions in the disconnected hemispheres of human commissurotomy patients. Ph.D. Thesis, California Institute of Technology, Pasadena, California.

LEVY, JERRE, 1969b. Possible basis for the evolution of lateral specialization of the human brain. *Nature (Lond.)* 224:614–615.

LEVY, JERRE, 1973. Lateral specialization of the human brain: Behavioral manifestations and possible evolutionary basis. In *The Biology of Behavior*, J. Kiger, ed. Corvallis: Oregon State University Press, (in press).

LEVY, J., and T. NAGYLAKI, 1972. A model for the genetics of handedness. *Genetics* 72:117–128.

LEVY, J., C. TREVARTHEN, and R. W. SPERRY, 1972. Perception

of bilateral chimeric figures following hemisphere deconnection. *Brain* 95:61–78.

McCloskey, D. I., 1973. Position sense following surgical disconnexion of the cerebral hemispheres in man. (in press).

Milner, B., 1971. Interhemispheric differences in the localization of psychological processes in man. *Brit. Med. Bull.* 27: 272–277.

Milner, B., and L. Taylor, 1972. Right hemisphere superiority in tactile pattern-recognition after cerebral commissurotomy: Evidence for nonverbal memory. *Neuropsychologia* 10:1–15.

Mountcastle, V. B., 1962. *Interhemispheric Relations and Cerebral Dominance.* Baltimore: The Johns Hopkins Press.

Nagylaki, T., and J. Levy, 1973. "The sound of one paw clapping" isn't sound. *Behav. Genet.* (in press).

Nebes, R. D., 1971. Investigations on lateralization of function in the disconnected hemispheres of man. Ph.D. Thesis, California Institute of Technology, Pasadena, California.

Nebes, R. D., and R. W. Sperry, 1971. Hemispheric deconnection syndrome with cerebral birth injury in the dominant arm area. *Neuropsychologia* 9:247–259.

Peterson, G. M., 1934. Mechanisms of handedness in the rat. *Comp. Psychol. Monogr.*, No. 46.

Piercy, M., H. Hécaen, and J. Ajuriaguerra, 1960. Constructional apraxia associated with unilateral cerebral lesions—left and right cases compared. *Brain* 83:225–242.

Piercy, M., and V. O. G. Smyth, 1962. Right hemisphere dominance for certain non-verbal intellectual skills. *Brain* 85:775–790.

Preilowski, B., 1972. Interference between limbs during independent bilateral movements. *1972 Proc. Amer. Psych. Assoc.*, Washington, D.C.

Preilowski, U., and R. W. Sperry, 1972. Minor hemisphere dominance in a bilateral competitive tactual word recognition task. *Biol. Ann. Rep.* (Calif. Inst. Tech.), 83–84.

Saul, R., and R. W. Sperry, 1968. Absence of commissurotomy symptoms with agenesis of the corpus callosum. *Neurology* 18:307.

Sperry, R. W., 1968a. Mental unity following surgical disconnection of the cerebral hemispheres. *The Harvey Lectures, Series 62.* New York: Academic Press, pp. 293–323.

Sperry, R. W., 1968b. Hemisphere deconnection and unity in conscious awareness. *Amer. Psychol.* 23:723–733.

Sperry, R. W., 1968c. Plasticity of neural maturation. *Develop. Biol.*, Supplement 2, 27th Symposium. New York: Academic Press, pp. 306–327.

Sperry, R. W., M. S. Gazzaniga, and J. E. Bogen, 1969. Interhemispheric relationships: The neocortical commissures; syndromes of hemisphere disconnection. In *Handbook of Clinical Neurology*, P. J. Vinken and G. W. Bruyn, eds. Vol. 4, Ch. 14, pp. 273–290.

Sperry, R. W., 1969. A modified concept of consciousness. *Psych. Rev.* 76:532–536.

Sperry, R. W., 1970a. Perception in the absence of the neocortical commissures. *Percept. Disord.* (A.R.N.M.D.) 48:123–138.

Sperry, R. W., 1970b. An objective approach to subjective experience: Further explanation of a hypothesis. *Psych. Rev.* 77:585–590.

Sperry, R. W., 1970c. Cerebral dominance in perception. In *Early Experience in Visual Information Processing in Perceptual and Reading Disorders*, F. A. Young and D. B. Lindsley, eds. Washington, D.C.: Nat. Acad. Sci., pp. 167–178.

Sperry, R. W., 1971. How a developing brain gets itself properly wired for adaptive function. In *The Biopsychology of Development*, E. Tobach, E. Shaw, and L. R. Aronson, eds. New York: Academic Press, pp. 27–44.

Sperry, R. W., 1972. Hemispheric specialization of mental faculties in the brain of man. 36th Yearbook Claremont Reading Conference, M. P. Douglas, ed. Pub. by Claremont Graduate School, Claremont, California.

Sperry, R. W., 1973. Lateral specialization of cerebral function in the surgically separated hemispheres of man. In *Psychophysiology of Thinking*, J. McGuigan, ed., (in press).

Trevarthen, C. B., 1970. Experimental evidence for a brainstem contribution to visual perception in man. *Brain, Behav. Evol.* 3:338–352.

Trevarthen, C. B., and R. W. Sperry, 1973. Perceptual unity of the ambient visual field in human commissurotomy patients. *Brain.*, (in press).

Trevarthen, C. B., and M. Kinsbourne, 1973. Perceptual completion of words and figures by commissurotomy patients. (Personal commun.)

Unterharnscheidt, F., D. Jachnik, and H. Gött, 1968. *Der Balkenmangel*, Monog. Neurol. Psychiat. 128. New York: Springer-Verlag.

Zaidel, D., and R. W. Sperry, 1973. Performance on the Raven's Colored Progressive Matrices Following Hemisphere Deconnection. *Cortex* 9 (in press).

Zaidel, D., and R. W. Sperry, 1972a. Functional reorganization following commissurotomy in man. *Biol. Ann. Rep.* (Calif. Inst. Tech.), p. 80.

Zaidel, D., and R. W. Sperry, 1972b. Memory following commissurotomy. *Biol. Ann. Rep.* (Calif. Inst. Tech.), p. 79.

Zangwill, O. L., 1964. The current status of cerebral dominance. In *The Brain and Disorders of Communication, Res. Publ., Assoc. Nerv. Ment. Dis.* 42:103–118.

2 Commissural Pathways in Interhemispheric Transfer of Visual Information in the Pigeon

MICHEL CUÉNOD

ABSTRACT Integration of visual input from the two eyes of the pigeon occurs in the wulst. The supraoptic decussation (DSO) is critical for mediating interhemispheric transfer of electrophysiological and behavioral responses. DSO section produces a split-brain preparation in which unilateral wulst lesions cause deficits for learning with the corresponding eye. Other commissures cannot compensate for the loss of transmission. These anatomic, electrophysiologic, and behavioral studies are being supplemented by biochemical experiments that delineate the contribution of cell bodies to synaptic structure and function along the pigeon's visual pathways.

Introduction

EVERY VERTEBRATE has two relatively symmetric "half" brains. Sperry (1961) and his collaborators have demonstrated by means of the so-called "split-brain" preparation that visual learning can be restricted to one hemisphere, in the cat and in the monkey, if one cuts the optic chiasm, the corpus callosum, and the anterior commissure along the midline. Such a disconnection leaves the two halves of the brain relatively independent of each other in their ability to receive, analyze, and store sensory information and to control motor performance. Interhemispheric cross-communication ceases; interhemispheric transfer of conditioned habits acquired unilaterally is impaired or abolished, and interhemispheric integration of material presented simultaneously and bilaterally is perturbed or impossible.

The "split-brain" preparation has not only shed much light on the function of the commissures but it permits the study of perception, learning, and memory specifically restricted to one hemisphere, leaving the other as a control. Additionally, it may prove very useful for the investigation of the biochemical correlates of learning behavior (Cuénod, 1972).

MICHEL CUÉNOD Brain Research Institute, Zürich University, Zürich, Switzerland

Birds provide a particularly suitable preparation for the application of the split-brain technique; they possess a highly developed visual system and present various degrees of interhemispheric transfer. Furthermore, one advantage they have over mammals is that their optic pathways are completely crossed in the optic chiasm; thus, a split-brain preparation could be obtained without cutting the optic chiasm, thereby keeping the primary input intact.

The anatomy of the visual pathways has been extensively studied by Karten (1969) during the last several years. Two main projections to the mesodiencephalon and thence to the telencephalon have been described (Figure 1). One pathway, after leaving the retina, crosses the midline in the optic chiasm and projects to the dorsolateral thalamus. From there, fibers reach the wulst region (hyperstriatum accessorium, intercalatum superius and dorsale) of the telencephalon. This phylogenetically younger pathway can be called the *retino-thalamo-hyperstriatal system*. The second and older pathway also crosses in the optic chiasm. The next relay is in the nucleus rotundus thalami, and from there fibers reach the ectostriatum in the telencephalon. This pathway can be called the *retino-tecto-rotundo-ectostriatal system*. Two major pathways crossing the midline seem to be related to each of these visual connections, the supraoptic decussation on the one hand and the tectal and posterior commissures on the other. Some aspects of visual organization, particularly the interhemispheric transfer of information, have also been investigated using anatomical, electrophysiological, and behavioral approaches.

The thalamo-hyperstriatal system and the supraoptic decussation

ANATOMOPHYSIOLOGY The general distribution in the pigeon telencephalon of potentials evoked by electric stimulation of the optic nerves shows that large areas of the telencephalon, mainly in the wulst and the

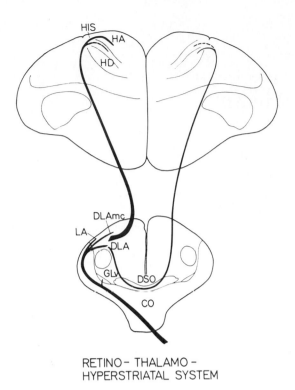

RETINO– THALAMO –
HYPERSTRIATAL SYSTEM

RETINO– TECTO – ROTUNDO –
ECTOSTRIATAL SYSTEM

FIGURE 1 Diagram of the visual pathways in the pigeon (modified from Karten, 1969). *Abbreviations:* CO, chiasma opticum; CP, commissura posterior; CT, commissura tectalis; DLA, nucleus dorsolateralis anterior; DLAmc, nucleus dorsolateralis anterior pars magnocellularis; DSO, decussatio supraoptica; E, ectostriatum; GLv, nucleus geniculatum lateralis pars ventralis; HA, hyperstriatum accessorium, wulst; HD, hyperstriatum dorsale, wulst; HIS, hyperstriatum intercalatum superius, wulst; LA, nucleus lateralis anterior; N, neostriatum; Rt, nucleus rotundus; TeO, tectum opticum; TrO, tractus opticus; W, wulst.

ectostriatum, are activated by both contralateral and ipsilateral optic nerve stimulation.

Stimulation of either eye of the pigeon evoked bilateral slow potentials and single unit discharges in the wulst regions of the pigeon (Figures 2 and 3). After electrical stimulation of the optic nerve papilla, the mean latency to the first component of the evoked response was 10.1 msec for the contralateral response and 14.3 msec for the ipsilateral one. Out of 125 wulst units activated by electrical stimulation of the optic nerve, 52% responded only to contralateral stimulation with a first spike latency of 7 to 39 msec; these were located dorsally in the wulst. 27% responded exclusively to ipsilateral stimulation with first spike latencies of 10 to 60 msec and were located in the ventral part of the wulst. A population of neurons found in the intermediate zone (21%) was activated by both contralateral and ipsilateral stimulation, the discharge latencies being always longer to the ipsilateral than to the contralateral eye stimulation (Figure 2). These data show that each optic nerve, despite its total crossing in the chiasm, has bilateral oligosynaptic projections to the

wulst; they give evidence for binocular convergence in this region of the brain. They imply the existence of a pathway responsible for the recrossing of the visual input to the wulst ipsilateral to the stimulated eye. Indeed, electrolytic lesions or reversible cooling with the cryogenic probe of Bénita (1970) in the rostral and dorsal part of the supraoptic decussation (DSO) induced a significant decrease in the amplitude of ipsilaterally evoked potentials; these procedures had no effect on contralaterally evoked responses (Figure 3) (Perišić, Mihailović, and Cuénod, 1971).

The relay involved in the visual projection to the wulst region was then investigated. Anatomical results, particularly the degeneration experiments of Karten and Nauta (1968) in the owl, point to the dorsolateral thalamic region as the main relay in this pathway (see also Nauta and Karten, 1970). Using the autoradiographic detection of proteins transported by rapid axoplasmic flow in the retinal fibers, Meier and Cuénod (1973) confirmed the existence of exclusively contralateral projections to the following diencephalic areas: N. lateralis anterior, N.

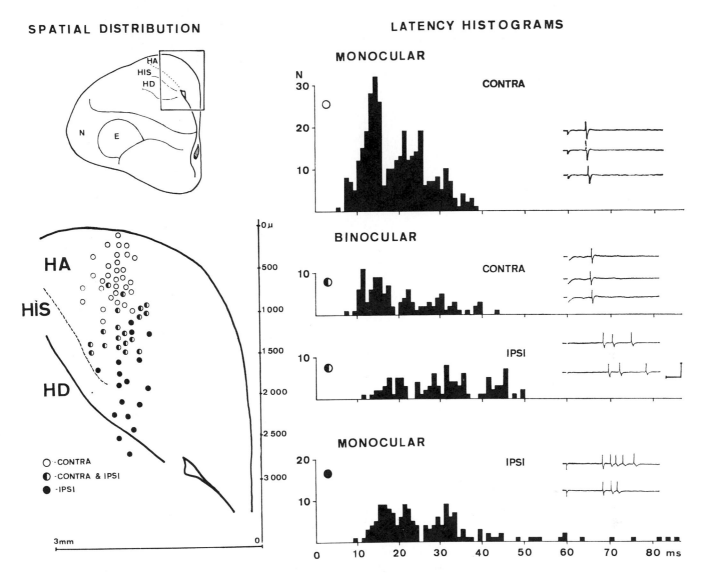

FIGURE 2 Topographic distribution and latency histograms of single units in the wulst (HA, HIS, HD) activated by electrical stimulation of either the contralateral, or ipsilateral or both optic nerves. Calibration: 10 msec and 1 mV, positivity upward. (From Perišić et al., 1971.)

dorsolateralis anterior pars magnocellularis, N. dorsolateralis anterior pars lateralis, and several subnuclei of N. dorsolateralis anterior, dorsally and partly rostrally of N. rotundus. Similarly, electrical stimulation of the optic nerve evoked potentials in the contralateral area anterior and anterodorsal to the N. rotundus. No response could be recorded on the side ipsilateral to the stimulated optic nerve. The latency of the contralateral response ranged from 2 to 3 msec and was the same in most thalamic areas investigated. The shapes of the evoked potentials were fairly similar in N. dorsolateralis anterior pars lateralis, N. lateralis anterior, N. intercalatum, and N. triangularis where they differ from the one recorded in the N. rotundus. In all dorsal thalamic areas, the responses followed stimulation up to 100 Hz, whereas in the N. rotundus they did not follow high frequencies. These findings confirm the results of Revzin and Karten (1966/1967).

In the owl, Karten and Nauta (1968) showed the existence of an ascending pathway from the dorsolateral thalamus to the wulst. Ablation of this telencephalic structure in the pigeon, made according to the

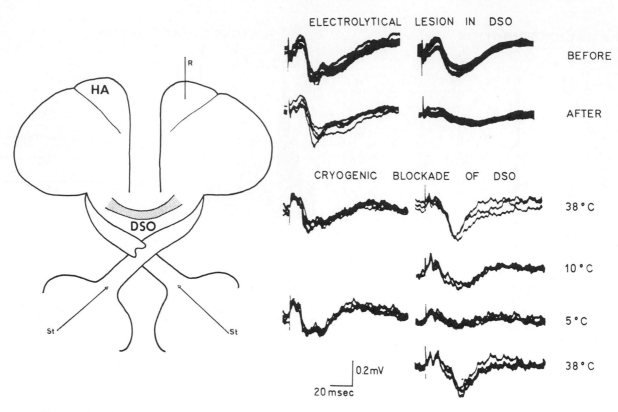

CONTRA   IPSI

ELECTROLYTICAL LESION IN DSO

BEFORE

AFTER

CRYOGENIC BLOCKADE OF DSO

38°C

10°C

5°C

38°C

0.2mV
20 msec

FIGURE 3   Effect of electrolytic lesion or reversible cooling of the region of the supraoptic decussation (DSO) on wulst responses evoked by electrical stimulation of the optic nerve. (From Perišić et al., 1971.)

electrophysiological maps, leads to severe retrograde degeneration in the ipsilateral dorsal thalamus, in the N. dorsolateralis anterior, overlapping the zone of projection of the optic nerve fibers and slightly medial to it. No differences were observed whether the ablation was superficial or deep in the wulst area (Meier et al., 1972b). Electrical stimulation of the thalamic region evoked large complex potentials in the wulst on the same side and a smaller potential on the opposite side. The latencies were, respectively, 2.4 to 4.5 and 3 to 6 msec. Only the ipsilateral response followed stimulation frequencies up to 100 Hz.

When the left thalamic region was reversibly cooled down to 5° C, the responses evoked by electrical stimulation of the right optic nerve in both the left and right wulst were strongly depressed. In contrast, the response evoked in the right wulst by stimulation of the left optic nerve was enhanced during the cooling of the left thalamic area, while the left wulst response was unchanged. Cooling of the N. rotundus had no effect on the wulst responses to optic nerve stimulation (Mihailović, Perišić, Bergonzi, in preparation). Thus, the dorsolateral thalamic region appears clearly as a relay station in the visual projections

to the wulst in the pigeon. This is in full agreement with the conclusions of Karten et al. (1973).

Cooling of the DSO induced the following effects: The thalamic response to contralateral optic nerve stimulation was unaffected. The left wulst response to left thalamic stimulation was delayed by 1 msec, and the amplitude of the late potential was very much increased. The right wulst response to left thalamic stimulation completely disappeared (Figure 4).

The DSO thus plays an important role in mediating a crossed visual input from one-retino-thalamo-hyperstriatal pathway to the opposite one. This interpretation has recently found support in the anatomical observations of Hunt and Webster (1972) and of Karten et al. (1973).

BEHAVIOR   Intact animals were trained monocularly to discriminate two simultaneously presented visual stimuli in an instrumental conditioning situation (Skinner), using a continuous reinforcement schedule. Monocular learning took about the same number of trials as binocular learning. For pattern discrimination, (V/Λ or ∪/∩), the criterion was reached with the first eye in 500 to 600 trials

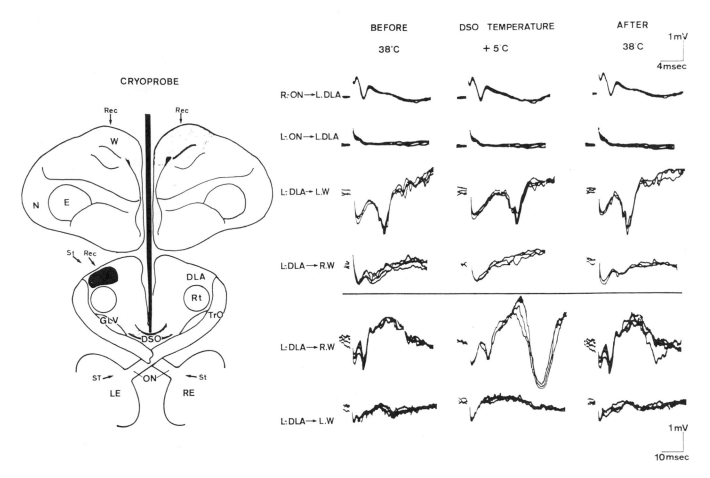

FIGURE 4 Effect of reversible cooling of region of the supra-optic decussation (DSO) on thalamic responses evoked by optic nerve stimulation and on wulst responses evoked by dorso-lateral thalamus stimulation. *Abbreviations:* L, left; R, right; E, eye; others, see Figure 1. Note change in calibration between first four and last two lines. (From Mihailović et al., in preparation.)

or in 6 to 8 days. Comparison of the learning curves obtained monocularly with the two eyes trained successively showed that sometimes, with the second eye, the animal performed immediately at criterion without any further learning. More often, a short period of training was necessary with the second eye but, in all cases, the criterion was reached in fewer trials than necessary with the first eye (Figure 5). These results show that the training performed with one eye is somehow transferred to the other (Meier, 1971). Similar interhemispheric transfer of color, pattern, and movement discrimination has been observed in birds by many authors (Kohler, 1917; Diebschlag, 1940; Levine, 1945; Siegel, 1953; Menkhaus, 1957; Moltz and Stettner, 1962; Mello et al., 1963; Mello, 1966; Catania, 1965; Konermann, 1966; Ogawa and Ohinata, 1966). Lesions were next placed in the DSO prior to training to see if they would induce a deficit in this interhemispheric transfer of visual information. Indeed, a section involving at least 75% of the DSO fibers severely

impaired the transfer of both color and pattern discriminations. The animals needed as many trials to reach criterion with the second eye free as with the first (Figure 5). There was no statistically significant difference between the number of trials to criterion with each eye, indicating a complete lack of transfer (Meier, 1971).

When the two patterns to be discriminated were left-right mirror images, the transfer in normal animals was frequently paradoxical, that is, with the second eye, the animal preferred the pattern unrewarded with the first. This paradoxical transfer of left-right mirror images is also impaired by section of the DSO (Meier, 1971).

The effect of lesions along the retino-thalamo-hyper-striatal pathway on monocular learning was tested. Unilateral lesions were placed in the wulst with or without section of the DSO. The discrimination was presented in alternate sessions to each eye individually. The postoperative retention of discriminations learned preoperatively was the same with both eyes in all animals. In

CONTROL

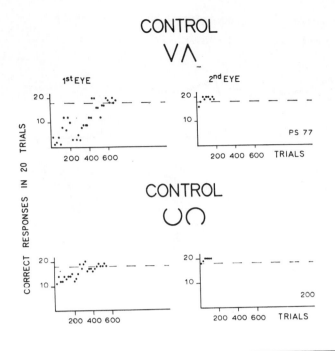

CT + CP

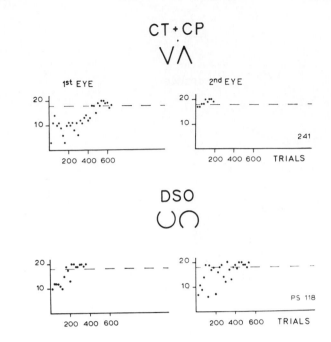

COLOR

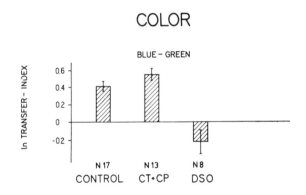

PATTERN

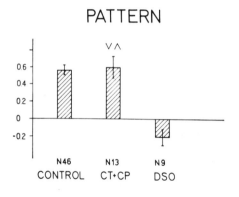

FIGURE 5 *Above:* Individual learning curves in commissurotomized pigeons. Notice that after section of the tectal and posterior commissures (CT + CP), interhemispheric transfer remains as in control animals. After section of the supraoptic decussation (DSO), the learning period for the second eye is almost as long as that for the first. *Below:* Interhemispheric transfer plotted as ratio (mean ± standard error) of the number of trials to criterion for presentation to the first eye and the second eye. N, number of animals. (From Meier, 1971.)

contrast, the animals having a wulst lesion associated with a DSO section learned slightly more slowly a new discrimination with the eye contralateral to the lesion than with the ipsilateral. Superficial as well as deep wulst lesions had the same consequences. The pigeons subjected to a wulst lesion or to a DSO section alone or to a sham operation presented no significant deficit (Figure 6). This suggests that when the DSO is intact, the opposite hemisphere can compensate for the missing wulst region; when however, this decussation is interrupted, a slight delay in

learning is induced for the eye contralateral to the wulst lesion (Meier et al., 1972a).

When unilateral lesions were placed in the dorsolateral thalamus, then important deficits appeared in the monocular learning of difficult patterns with the eye contralateral to the lesion. More than twice as many errors were made with this eye, compared to the ipsilateral eye; some animals never reached criterion, although they did not give signs of blindness with this eye and learned normally an intensity discrimination (Maier and Tanaka, 1972).

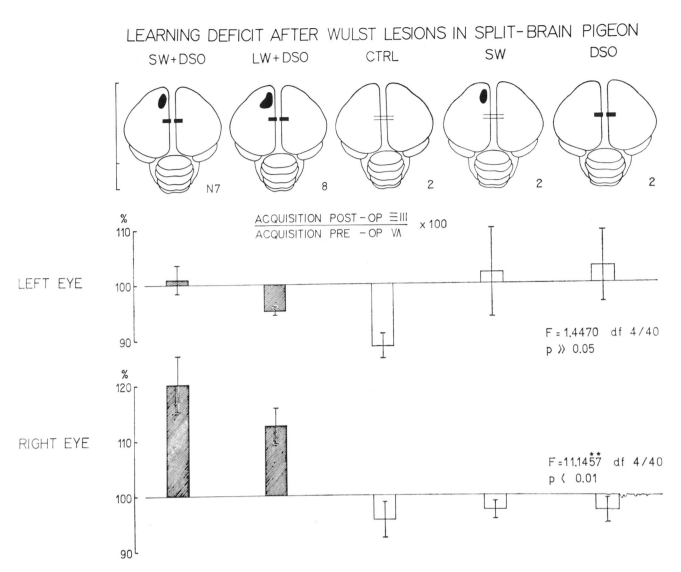

FIGURE 6   Effects of small (SW) and large (LW) wulst lesions with and without section of the supraoptic decussation (DSO) on monocular learning of pattern discrimination. *Above:* Schematic representation of each group with number of animals, including control group (CTRL). *Below:* Ratio of the number of trials to criterion after and before the lesion for each eye. (From Meier et al., 1972a.)

Thus, both anatomophysiological and behavioral techniques converge to attribute to the DSO an important role in mediating interhemispheric transfer of visual information, particularly for stationary pattern and color discriminations. Section of the DSO produces a "split-brain" preparation in this respect. The dorsolateral thalamic region seems to be the main relay involved in this system.

Tectal commissure

ELECTROPHYSIOLOGY   The tectal commissure connects the two optic lobes together. Electrical stimulation of one optic tectum evoked a short latency response in the depth of the contralateral tectum, which was abolished by section of the tectal and posterior commissures. With both photic stimulation and electrical stimulation of the optic nerve, a response could be recorded not only in the contralateral optic tectum but also in the ipsilateral one. This ipsilateral response was maximum in the middle and deep layers of the tectum and was delayed by 10 to 16 msec in comparison to the contralateral one. It was strongly diminished or abolished by section of the tectal and posterior commissures (Robert and Cuénod, 1969a).

The slow potential evoked in the optic tectum by stimulation of the contralateral eye was strongly diminished by the application of a conditioning train of pulses to the opposite tectum. The negative (N) wave evoked in the middle layers of the tectum by electrical stimulation of the optic nerve was reduced by 80% under these conditions. This inhibiting effect lasted from 10 msec to 80 msec and was maximal around 25 msec. It was abolished by section of the tectal and posterior commissures.

At the unit level, the same phenomenon could be observed: Out of 222 tectal units activated by electrical stimulation of the contralateral optic nerve, 49 units or 22% had their discharge suppressed, diminished, or delayed by a conditioning train of pulses applied 20 to 30 msec before to the opposite tectum. The units that could be inhibited were found mainly between 300 and 1000 μ from the surface. As for the N-wave, unit inhibition was effective with conditioning-test intervals of 5 to 70 msec; the maximal inhibition was at 20 to 30 msec (Robert and Cuénod, 1969b).

BEHAVIOR Complete section of the tectal and posterior commissures resulted in no significant deficit in the transfer of stable color or pattern discriminations. The animals always learned more quickly with the second eye free than with the first; thus they did not differ from the control animals (Meier, 1971). It is noteworthy that most of the fastest-learning pigeons belonged to the group that had been subjected to section of the tectal and posterior commissures. This might be due to the loss of some inhibitory action from one tectum to the other.

The pigeon visual system as a biological model

Independently of the commissurotomized preparation, the synaptic proteins of the retinotectal pathway have been studied, particularly with respect to their origin in the cell body and their transport along the axon (Barondes, 1969; Grafstein, 1969; Weiss, 1969). After intraocular injection of radioactive amino acids, the existence of a rapid and a slow migration of proteins has been confirmed: The first is largely destined to the nerve ending, while the second renews material in both the axon and the ending (Cuénod and Schonbach, 1971; Schonbach and Cuénod, 1971; Schonbach et al., 1971, 1973). The labeling of fast flowing material appears predominantly in the high-molecular-weight protein subunits, while the slowly flowing material labels predominantly the lower-molecular-weight subunits (Marko et al., 1971). The rapidly flowing material is mainly found in the membranous subfraction of synaptosomes (Marko and Cuénod, in preparation). Reversible block of the fast flow by colchicine (Boesch et al., 1972) induces reversible ultrastructural changes in the nerve endings, consisting of enlarged vesicles (Cuénod et al., 1972); this alteration correlates with a reversible deficit in synaptic transmission

(Perišić and Cuénod, 1972). Destruction of the retina induces in the optic nerve terminals a similar early vesicular enlargement, but it is irreversible (Cuénod et al., 1970). These degenerative alterations are characterized by the disappearance of some protein bands in a membranous fraction, presumably specific for the retinotectal endings (Cuénod et al., 1973).

This rapid survey invites consideration of the possibilities provided by the pigeon visual system and of the correlative investigations to which the split-brain pigeon can be applied. Complex behavioral experiments are always difficult to control because of the large number of unknown variables. In the commissurotomized animal, one side of the brain can undergo the processes of learning while the other does not. Any diffuse, nonspecific consequences of the experimental situation are presumably equally distributed throughout both hemispheres. This advantage has two limitations: Only some, and not all, variables can be controlled in this preparation, and the control value of the experiment rests on the functional symmetry of the two halves of the brain. This symmetry has been assumed for most subhuman vertebrate species and seems to obtain for the pigeon.

ACKNOWLEDGMENTS The essential participation of my collaborators, Drs. R. E. Meier, J. Mihailović, M. Perišić and Miss V. Maier is gratefully acknowledged. I thank Dr. K. Akert for his constant support and Dr. R. B. Livingston and Dr. H. Karten for reading the manuscript. This work was supported by the Swiss National Foundation for Scientific Research (Grant Nos. 3.329.70 and 3.133.69) and by the Slack-Gyr Foundation.

REFERENCES

BARONDES, S., 1969. Axoplasmic transport. In *Handbook of Neurochemistry*, A. Lajtha, ed. New York: Plenum Press, pp. 435–446.

BÉNITA, M., 1970. Démonstration d'un appareil et de sondes cryogéniques, isolées thermiquement par gaine de vide, et fonctionnant en circuit fermé. *J. Physiol. (Paris)* 62:333–334.

BOESCH, J., P. MARKO, and M. CUÉNOD, 1972. Effects of colchicine on axonal transport of proteins in the pigeon visual pathways. *Neurobiology* 2:123–132.

CATANIA, A. C., 1965. Interocular transfer of discriminations in the pigeon. *J. Exp. Anal. Behav.* 8:147–155.

CUÉNOD, M., C. SANDRI, and K. AKERT, 1970. Enlarged synaptic vesicles as an early sign of secondary degeneration in the optic nerve terminals of the pigeon. *J. Cell. Sci.* 6:605–613.

CUÉNOD, M., and J. SCHONBACH, 1971. Synaptic proteins and axonal flow in the pigeon visual pathway. *J. Neurochem.* 18:809–816.

CUÉNOD, M., 1972. Split-brain studies. Functional interaction between bilateral central nervous structures. In *The Structure and Foundation of Nervous Tissue*, Vol. 5., G. H. Bourne, ed. New York: Academic Press, pp. 455–506.

CUÉNOD, M., C. SANDRI, and K. AKERT, 1972. Enlarged synaptic vesicles in optic nerve terminals induced by intraocular injection of colchicine. *Brain Res.* 39:285–296.

CUÉNOD, M., P. MARKO, and E. NIEDERER, 1973. Disappearance of particulate tectal protein during optic nerve degeneration in the pigeon. *Brain Res.* 49:422–426.

DIEBSCHLAG, E., 1940. Ueber den Lernvorgang bei der Haustaube. *Z. Vergl. Physiol.* 28:67–104.

GRAFSTEIN, B., 1969. Axonal transport: Communication between soma and synapse. In *Advances in Biochemical Psychopharmacology*,Vol. 1, E. Costa and P. Greengard, eds. New York: Raven Press, 11–25.

HUNT, S. P., and K. E. WEBSTER, 1972. Thalamo-hyperstriate interrelations in the pigeon. *Brain Res.* 44:647–651.

KARTEN, H. J., and W. J. H. NAUTA, 1968. Organization of retino-thalamic projections in the pigeon and owl. *Anat. Rec.* 160:373.

KARTEN, H. J., 1969. The organization of the avian telencephalon and some speculations on the phylogeny of the amniote telencephalon. *Ann. N.Y. Acad. Sci.* 167:164–179.

KARTEN, H. J., W. HODOS, W. J. H. NAUTA, and A. M. REVZIN, 1973. Telencephalic projections of the retino-diencephalic pathway in the pigeon (*Columba livia*) and the burrowing owl (*Speotyto conicularia*). *J. Comp. Neurol.* (in press).

KOHLER, W., 1917. Die Farbe der Sehdinge beim Schimpansen und beim Haushuhn. *Z. Psychol.* 77:248–255.

KONERMANN, G., 1966. Monokulare Dressur von Hausgänsen, z.T. mit entgegengesetzter Merkmalsbedeutung für beide Augen. *Z. Tierpsychol.* 23:555–580.

LEVINE, J., 1945. Studies on the interrelations of central nervous structures in binocular vision. II. The conditions under which interocular transfer of discriminative habit takes place in the pigeon. *J. Genet. Psychol.* 67:131–142.

MAIER, V., and M. TANAKA, 1972. Monocular pattern discrimination deficits in pigeons after unilateral lesions of the dorsolateral region of the thalamus. *Brain Res.* (in press).

MARKO, P., J.-P. SUSZ, and M. CUÉNOD, 1971. Synaptosomal proteins and axoplasmic flow: Fractionation by SDS polyacrylamide gel electrophoresis. *FEBS Letters* 17:261–264.

MEIER, R. E., 1971. Interhemisphärischer Transfer visueller Zweifachwahlen bei kommissurotomierten Tauben. *Psychol. Forsch.* 34:220–245.

MEIER, R. E., and M. CUÉNOD, 1973. Autoradiographic demonstration of retinal projections in the pigeon (*Columba livia*). *Acta Anat. (Basel)*, (in press).

MEIER, R. E., V. MAIER, and M. CUÉNOD, 1972a. Visual learning following unilateral telencephalic lesions in the split-brain pigeon. *Brain Res.* 37:356.

MEIER, R. E., J. MIHAILOVIĆ, M. PERIŠIĆ, and M. CUÉNOD, 1972b. The dorsal thalamus as a relay in the visual pathways of the pigeon. *Experientia* 28:730.

MELLO, N. K., F. R. ERWIN, and S. COBB, 1963. Intertectal integration of visual information in pigeon: Electrophysiological and behavioral observations. *Bol. Estud. Méd. Biol. (Méx.)* 21:519–533.

MELLO, N. K., 1966. Concerning the inter-hemispheric transfer of mirror-image patterns in pigeon. *Physiol. Behav.* 1:293–300.

MENKHAUS, I., 1957. Versuche über einäugiges Lernen und Transponieren beim Haushuhn. *Z. Tierpsychol.* 14:210–230.

MOLTZ, H., and L. J. STETTNER, 1962. Interocular mediation of the following-response after patterned-light deprivation. *J. Comp. Physiol. Psychol.* 55:626–632.

NAUTA, W. J. H., and H. J. KARTEN, 1970. A general profile of the vertebrate brain, with sidelights on the ancestry of cerebral cortex. In *The Neurosciences: Second Study Program*, F. O. Schmitt, ed. New York: The Rockefeller University Press, pp. 7–26.

OGAWA, T., and S. OHINATA, 1966. Interocular transfer of color discriminations in a pigeon. *Ann. Anim. Psychol.* 16:1–9.

PERIŠIĆ, M., and M. CUÉNOD, 1972. Synaptic transmission depressed by colchicine blockade of axoplasmic flow. *Science* 175:1140–1142.

PERIŠIĆ, M., J. MIHAILOVIĆ, and M. CUÉNOD, 1971. Electrophysiology of contralateral and ipsilateral visual projections to the wulst in pigeon (*Columba livia*). *Intern. J. Neurosci.* 2:7–14.

REVZIN, A. M., and H. KARTEN, 1966/1967. Rostral projections of the optic tectum and the nucleus rotundus in the pigeon. *Brain Res.* 3:264–276.

ROBERT, F., and M. CUÉNOD, 1969a. Electrophysiology of the intertectal commissures in the pigeon. I. Analysis of the pathways. *Exp. Brain Res.* 9:116–122.

ROBERT, F., and M. CUÉNOD, 1969b. Electrophysiology of the intertectal commissures in the pigeon. II. Inhibitory interaction. *Exp. Brain Res.* 9:123–136.

SCHONBACH, J., and M. CUÉNOD, 1971. Axoplasmic migration of proteins: A light microscopic autoradiographic study in the avian retino-tectal pathway. *Exp. Brain Res.* 12:275–282.

SCHONBACH, J., C. SCHONBACH, and M. CUÉNOD, 1971. Rapid phase of axoplasmic flow and synaptic proteins: An electron microscopical autoradiographic study. *J. Comp. Neurol.* 141:485–498.

SCHONBACH, J., C. SCHONBACH, and M. CUÉNOD, 1973. Distribution of transported proteins in the slow phase of axoplasmic flow. An electron microscopical autoradiographic study. *J. Comp. Neurol.* (in press).

SIEGEL, A. I., 1953. Deprivation of visual form definition in the ring dove. II. Perceptual-motor transfer. *J. Comp. Physiol. Psychol.* 46:249–252.

SPERRY, R. W., 1961. Cerebral organization and behavior. *Science* 133:1749–1757.

WEISS, P., 1969. Neuronal dynamics and neuroplasmic ("axonal") flow. In *Cellular Dynamics of the Neuron*, S. H. Barondes, ed. New York: Academic Press, pp. 3–34.

3 Division of Function and Integration of Behavior

DONALD E. BROADBENT

ABSTRACT Differences between the cerebral hemispheres in man are shown by the results of callosal section, by the results of unilateral damage, and by asymmetries of performance in normal people. However, the performance of normal people shows a surprising amount of interference between stimuli and responses on one side of the body and those on the other. Furthermore, the superior function of the right ear for speech stimuli applies only for certain kinds of phoneme and for certain acoustic cues. The two hemispheres must be seen therefore as performing different parts of an integrated performance, rather than completely separate and parallel functions.

The current position

EVIDENCE ABOUT the differences between the hemispheres comes from three lines of inquiry; one of the clearest sources of evidence is that of patients who have their corpus callosum sectioned, so that the hemispheres cannot communicate. This evidence comes mostly from the work of Sperry and his collaborators, so that it would be presumptuous to add anything to the chapter already presented. The second line of evidence comes from the effects of damage to one side of the brain, with relatively little involvement to the other side. Here one of the leading authorities is Milner, so that it would be a pity to anticipate her chapter. The third line of evidence concerns normal people and the way in which they can perform tasks better with one side of their body than they can with the other. This, therefore, is the area from which some points will be selected for the present chapter.

The majority of recent work on these asymmetries has fallen into three classes. First, there are studies showing that reaction time is faster with one hand than with the other, and that this difference in speed depends upon the particular sense organ stimulated. That is, a sound reaching the right ear produces a faster reaction with the right hand than with the left, and vice versa for stimuli to the other ear (Simon and Rudell, 1967). Similarly, if the conditions are appropriate, a stimulus to the right side of the visual fixation point will give a faster reaction with the right hand, and a stimulus to the left side will give a faster reaction with the left hand. This relationship may be complicated by the nature of the stimulus, as we shall see shortly, but at least in some cases it suggests that the hand controlled from one hemisphere reacts faster to stimuli that are projected to the same hemisphere rather than to the opposite side.

A second source of evidence concerns the accuracy with which a stimulus can be perceived if it is delivered to one ear while other information is going simultaneously to the other ear. If for example, three spoken digits are heard one after another in the right ear, and three different digits simultaneously in the other ear, it is possible for the experimental subject to reproduce most or all of the items. If errors do take place, they are with right-handed people more likely to occur on the left ear than the right. Right-handed people (and of course a substantial number of left-handed people) will normally show greater suppression of speech by Sodium Amytal injected into the carotid on the left side rather than on the right; but one can find the occasional individual, probably left handed, who will show greater sensitivity to Sodium Amytal on the right side. If such people are given the task of listening to simultaneous digits, they make more errors on the right side rather than the left. Thus although each ear seems to work equally well when there is no conflicting information from the other ear, yet when the two ears are in conflict, the one that does best is the one on the opposite side of the body to the hemisphere that seems to have prime responsibility for speech. This is true, however, primarily when the listening task uses speech sounds such as spoken digits. If one does the same kind of experiments with passages of music, it has been reported that the left ear rather than the right does better in the majority of cases (Kimura, 1961, 1964; Broadbent, 1971, pp. 201–203).

The third line of evidence concerns the accuracy of perception of visual stimuli delivered to the right or left of the fixation point. Here the results are a shade more complicated, because in most countries where the experiments are done, people are very much used to reading from left to right. Correspondingly, if you give them a

DONALD E. BROADBENT Medical Research Council Applied Psychology Unit, Cambridge, England brief exposure containing several letters or similar stimuli, they will reproduce these letters in an order from left to right; and the first items to be written down or otherwise reported will stand a better chance of being correct than later ones will. On the other hand, this effect is reversed among people who read Hebrew, and it seems therefore simply to be a matter of habitual strategy rather than biological asymmetry. If one presents simply one letter, or other familiar nameable stimulus, by itself, then it is more likely to be correctly reported if it comes to the right of the fixation point rather than the left. Furthermore, this effect is more marked in right-handed people than in left-handed ones. It seems unlikely therefore to have anything to do with reading habits, since those are presumably in the same direction whatever the handedness of the particular person (Kimura, 1966; Broadbent, 1971, pp. 203–205). The best conclusion about these visual experiments seems therefore to be that there is *both* a scanning pattern across the visual field from left to right in those who are used to reading in that direction *and* greater efficiency at naming objects or alphabetical material in the part of the visual field that projects to the major hemisphere. These two tendencies conflict and interact in rather complicated ways, depending upon the exact conditions of the experiment.

These three lines of evidence about normal subjects, from reaction time tasks, from dichotic listening, and from tachistoscopic visual presentation, all agree in broad outline with the data from callosal section and from lesions. It is tempting to produce a sweeping and simplified picture in which we say that for the majority of people the left hemisphere contains the mechanisms responsible for speech and language, as well as those used in the more complex motor skills. The right hemisphere on the other hand contains musical functions and those involving space perception, such as drawing maps. One might argue that a stimulus has a faster and more determinate effect upon centers close to its projection area, and therefore performance is more efficient if the centers controlling response are near that area. If the task is one of using the hands, and the stimulus is projected to one hemisphere only, the more efficient hand is the one for which the cortical representation is in the same hemisphere as the projection area of the stimulus. In perceptual experiments involving speech, the center controlling response is in a different hemisphere from that concerned if the task involves music. (In experiments that use auditory stimuli, we can only detect this asymmetry by swamping the ipsilateral projection with competing stimuli from the other ear and thus confining the sensory projection to the contralateral hemisphere.) Again keeping to this view, more complex results can be explained by simple exten-

sions of this basic set of concepts. For example Klatzky (1970) has used a task in which a subject learns a small set of letters and then is given one of two kinds of stimuli. In one case, he is given a letter and has to decide whether it is one of those in the set, according to a well-known technique of Sternberg (1966). In the other condition, the subject is shown a picture of an object, and his task is to decide whether the name of the object starts with one of the letters in the set or not. The time taken to reach a decision depends, as is usual with the Sternberg technique, upon the number of letters in the set. It is as if the subject were running through the list of possibilities to see whether the present stimulus were in the list or not. If the stimulus was a letter, then the time taken was greater if the stimulus was on the right of the fixation point than on the left, by a short time that was the same regardless of the size of the set of letters being considered. If on the other hand the stimulus was a picture, then for very small sets the reaction time was actually faster when the stimulus was on the right, but the amount of increase was greater as the size of the set increased. In terms of the broad view already stated, one could interpret these results as Klatzky did, by supposing that letters are matched on a purely visual basis and because this is a visuospatial function it is predominantly in the right hemisphere, so that any stimulus presented on the left of the fixation point takes less time to reach the appropriate brain area. Once having done so, however, the process proceeds in the same way as if the stimulus had been elsewhere in the visual field. For pictures, on the other hand, the picture must first be named, and this process is faster if the stimulus is delivered to the hemisphere containing the speech centers; although subsequent comparisons with the set of letters are then slower.

In this kind of way, one can understand a number of such studies, but there remain certain difficulties and discrepancies. While it is not clear how these difficulties are to be resolved, they do at least suggest that the simple version already given needs further expansion and complication, and it is with these difficulties therefore that the rest of this chapter is concerned. Broadly speaking, the difficulties involve the very large amount of interaction and integration of function found in normal people, despite the evidence for asymmetry; and the fact that various details of the experiments suggest that the activities localized in one hemisphere cannot be adequately described as "speech" or "spatial perception." They must rather be more abstract aspects of information processing, of which some may indeed be particularly important in speech but also present in other tasks. Others perhaps are more important in spatial or musical performance but also to some extent present in speech.

Difficulties: The integration of performance

INTERFERENCES OF WIDELY DIFFERENT TASKS If it were indeed the case that given tasks could be allocated clearly and distinctly to the different hemispheres, then each function should be capable of carrying on unimpaired by the simultaneous performance of the other. Perhaps the most obvious model for cortical function may once have been something modeled on the reflex arc, in which activity arising from one particular stimulus sparked over into the correct motor pathway, regardless of what might be going on elsewhere in the nervous system. For many years now, most investigators have discarded such a model, because of the demonstrable way in which reaction time and other measures of performance on one combination of stimulus and response vary with other factors, such as the number of other stimuli and responses that might have been presented, their nature, and so on. As we all know, the delay with which a key can be pressed to a light stimulus is increased if there were other stimuli that might have occurred, even though on this occasion they did not. One cannot therefore regard the stimulus as having been connected to the manual response by its own private wire independent of the processes that connect other lights to other responses.

One might still argue however that the control of hand movements in response to light might be carried out within a single region of the brain, even though within that region the consequences of any one stimulus were interdependent with those of others. The interaction between the different stimuli and responses of one task might well exist because of their use of the same general portion of the nervous system, much as the fact that the eyes are fixated on one point excludes their being fixated simultaneously upon another. The stimuli and actions of one task need not however necessarily interfere with those of another, just as the eyes may fixate a point while simultaneously the legs of the same person are carrying him over rough ground. In a very naive form, this theoretical position appears in the occasional attempts of engineers to improve the display of information to pilots, automobile drivers, and similar highly "loaded" individuals by presenting information to the periphery of the eye or to the ears while other information is being presented to central vision. Of course the difficulty is that the eye and ear are quite likely to involve common central processes when a man is engaged in the single task of flying an aircraft or driving a car, and consequently it is not surprising that this kind of technique usually shows some interference with central vision when the additional information is being presented. But if we consider regions of the central nervous system that are as anatomically distinct, and as apparently

different in function, as the two hemispheres, then we might reasonably suppose that we could find tasks that could be performed simultaneously by the man without mutual interference, perhaps one in one hemisphere and the other in the other. Thus far, nobody has to my knowledge been successful in demonstrating such simultaneous performance, and it is interesting to consider some of the failures to do so.

EXPERIMENTS COMBINING VERY DIFFERENT TASKS *Tactile reaction time and acoustic memory for letters.* An example of interference between two very different tasks is provided by Broadbent and Gregory (1965). They were studying reaction time to stimulation of the finger tips, and in the case of greatest interest the reaction was to press downwards with the finger stimulated. Such a task is highly "compatible," shows very little increase in response time as the number of possible stimuli is increased, and from some points of view is little more complicated than a reflex. The subjects in these particular experiments were performing one such reaction every 5 sec and also were receiving every 5 sec a spoken letter of the alphabet. This letter was normally different at each 5-sec interval, but within each minute some one letter occurred twice. Every minute the subject had to report verbally which letter was the most frequent. He was thus performing simultaneously the manual reaction to touch, and a spoken reaction based on memory of speech stimuli. Yet these two tasks interfered with each other: Perhaps particularly striking was the fact that memory for the speech was impaired by variation in the number of fingers being stimulated in the reaction time task. Furthermore, the speed of manual reaction was slower when the memory task was in progress, and this was true for the left hand considered alone. Even functions as apparently different as these, therefore, interfere with each other in some way.

Music and speech. While speech is associated with the left hemisphere in most people, there is no particular reason to associate tactile reaction time with the right hemisphere, even though the nature of the task is logically apparently very different from that of speech. In this respect, the task of Broadbent and Gregory does not differ from most of those that have been used to show interference between tasks; therefore, it is particularly interesting to find a case in which the shadowing of speech (that is, listening to it and repeating it aloud as fast as it comes in) has been combined with the sight reading of music. Allport et al. (1972) practised experimental subjects at shadowing until they reached a criterion of two successive trials without making any omissions. The accuracy of shadowing was later measured when the subjects were simultaneously playing the piano, sight-reading pieces that they had not previously seen. The subjects then averaged at least two errors per trial, depending upon the difficulty of the simultaneous sight reading and of the prose they were shadowing. The accuracy of the piano playing, as assessed by a scorer who did not know which condition the subjects were under, was relatively little affected. Nevertheless, for the most difficult grade of music, there was a significant increase in the errors of timing when simultaneously shadowing, in a session in which the dual task was performed after the shadowing task; in another session in which the two tasks were performed in the opposite order, the difference was in the same direction but not significant. These results are particularly interesting, because the authors were avowedly anxious to show that the two tasks could be carried out independently and had given every opportunity to the subjects to do as well as possible. They had, for instance, used a shadowing task on which perfect performance was possible in isolation so that an unknown margin was left before any interference could show itself; they had used a musical task that allowed the subject to process the input information at times that were in detail under his own control, and they had adequately practised the subjects. It is therefore particularly impressive that there are still residual problems for people who try to cope with speech and music simultaneously. It is, of course, clearly the case that the interference is much less than would be expected from two simultaneous speech messages; the point of interest with regard to differentiation of the two hemispheres is that there is still a difficulty in combining the two tasks.

Memory and the difficulty of auditory detection. A possible clue to the sources of interference when two such different tasks are combined is to be found in considering yet another situation that shows interference between two functions. This is to be found in a task that involves presentation of a series of digits to one ear, and a detection of a pure tone in noise in the other ear. It was shown by Broadbent and Gregory (1963) that the need to report the digits from memory interferes in this case with the detection of the signal. Once again, we have a speech task delivered to one ear apparently interfering with a non-speech task delivered to the other ear. In this case however the tone was quite difficult to hear against the noise background. A similar experiment was performed by Lindsay and Norman (1969), using heavily practised subjects and a number of different intensities of tone. When the tone was relatively intense, discrimination was almost perfect, and there was no mutual interference between the two tasks. But at the lowest intensity of the tone, however, errors in its detection began to appear, and correspondingly there was an interference between the two tasks. It does indeed seem possible therefore for two functions to go on in parallel, provided that in one task there is a

simple and determinate link between each of the possible stimuli and its appropriate response.

This kind of simple and determinate linkage, in which there is no uncertainty involved, is of course one that will save time in any processing mechanism. If we act as statisticians and are given some number that has been drawn either from one distribution or another, then if the two distributions are widely separated in mean value and have very little variance, we can be confident of the correct answer on only one sample of evidence. If however, the two distributions overlap very considerably, then to get the right answer will demand a continuing accumulation of a large number of samples of evidence. Thus it seems to be this process of accumulating evidence about the correct reaction to make that cannot be carried out simultaneously on two tasks. Rather similar conclusions can be drawn from the work of Trumbo and Noble (1970), who have analyzed the nature of tasks that people can combine and those that they cannot.

The foregoing experiments have taken us some way from the problem of the hemispheres; but one might summarize them as showing that people have great difficulty in carrying out two functions in true independence, even when they appear quite different, and in particular that one cannot separate speech tasks from others such as musical ones. The common process that seems to produce the interference is one of building up the evidence toward a decision on the next course of action. At this stage however, it is appropriate to get closer again to the asymmetries of the two sides of the body.

TRANSMISSION OF INFORMATION BETWEEN RIGHT AND LEFT
The use of speech and music as tasks to be performed simultaneously begs the question of the extent to which each is associated with a particular cerebral hemisphere. A closer link is provided by certain reaction-time tasks in which one hand responds to signals coming from one side of the visual field, while the other hand responds to signals from the other side. Interference between these two tasks has frequently been demonstrated in experiments under the heading of the "psychological refractory period." By this is meant experiments in which one stimulus is presented and another occurs closely subsequent to it. It is usually found that the second signal produces a reaction only after a considerable delay, unless the interval between the two signals is rather more than a single reaction time.

Some of the experiments have been contaminated by expectancy effects, because in any one session the subject might be regarded as anticipating the second signal at about the average of the preceding intervals and thus being unprepared for it if it occurs sooner. Yet if one uses a simple reaction time, it is necessary to vary the interval between the signals, as otherwise the subject will simply initiate the second reaction at a fixed time interval after the first reaction and thus may quite well show zero reaction times to the second stimulus! This difficulty was overcome by Broadbent and Gregory (1967), who employed choice reactions. The first stimulus was one of two lights on one side of the visual field, and the correct response was to move either the first or the second finger of the corresponding hand. The second stimulus was on the other side of the visual field and, similarly, was one of two lights each requiring response by an appropriate finger on the second hand. The interval between the two stimuli could then be kept completely fixed, as the subject would never know which light was coming in the second stimulus, even though he knew perfectly when it would arrive. Nevertheless, there was a substantial interference between the two tasks if the second stimulus arrived before the first response had been made. Thus efficient reaction to one side of the visual field using the corresponding hand (which one might expect to be conducted all within one hemisphere) interfered with efficient reaction on the other side of the visual field using the other hand (Figures 1, 2).

So far as it goes, this result might merely be regarded as due to a general tension produced by the first reaction, or something of that sort, perhaps comparable to the knitting of the brows by an inexperienced typist or learning car driver. However, closer examination of the individual responses shows that this is not the whole story. In particular, the amount of delay on the second reaction depends upon which reaction it was, and which was the first reaction. If the first reaction used the index finger of one hand, then the delay to the second reaction was greater if that reaction was to be with the middle finger than if it was to be with the index finger. If the first reaction used the middle finger, then the reverse was the case. In other words, a stimulus on one side of the visual field requiring a movement of one particular finger on that side seems to give a bias during the next third of a second or so for reactions with the symmetrical finger on the other hand. Therefore, far from the two sides of the body acting independently, the stimulus on one side seems to some extent to produce a reaction with the appropriate finger, even on the inappropriate side of the body. Even under these conditions, the two hemispheres do not seem to work separately. Notice that the nature of the interference between the two sides of the body is what one might expect if the two tasks had in common some process similar to that mentioned earlier as a likely source of interference. That is, the allocation of the correct response to the particular stimulus is the source of the problem.

In a task of this kind, one might conceivably argue that it is misleading to look only at the average response time, which mixes together inextricably the different individual responses. By the average score, some interference would be recorded even if there were only a few instances where reaction on one side interfered with reaction on the other. One might well argue that the use of the two hemispheres independently might result only from long practice, so that it would appear only in some rather than all cases; and on this view, the key question is whether any instances occur in which the two reactions, one on each side, are made with the same efficiency as either would show if it were performed alone. Such analyses have been carried out in the past, for example, by Elithorn and his associates, and in those papers it was argued that occasionally there were pairs of reactions in which both hands responded with normal speed (see Halliday et al., 1960; Broadbent, 1971, pp. 314–316).

However, the technique used by Elithorn and his associates is a rather special one, in which there were only two possible stimuli, one on each side. The correct response in each case was to press with the corresponding hand, but the subject did not know which stimulus would come first. Once one stimulus had arrived, it then became completely certain that the second stimulus would be on the other side. Thus a fast reaction to the second stimulus might be due either to the two reactions occurring independently, as Elithorn and his colleagues suggest, or, alternatively, to the information about occurrence of the first stimulus being transmitted across to the other side. The situation used by Broadbent and Gregory (1967) allows some check on these possibilities, because the subject always knew perfectly which hand would have to react first. Consequently, we can analyze out individual pairs of reactions, on the lines suggested by Elithorn, by comparing each individual reaction time with the average of times obtained when only one hand is being used. Thus we can look at all the cases in which the first response time took less than the average for isolated reactions and see if there are any cases in which the second reaction was also of normal duration. In fact, I have reanalyzed the data of Broadbent and Gregory (1967), and it is clear that there are no such cases. Thus it seems to me plausible that there is no genuine independence of action of the two hands.

These results, like those given earlier, show that the intact person does have quite marked interference between tasks involving the two sides of the body. We cannot therefore take the simple-minded view that reactions of the right hand to the right ear or right visual field involve only the left hemisphere, and vice versa, just as we saw earlier that the understanding and utterance of speech does not leave unaffected the reading and playing of music, the performance of tactual reactions, and so on. Yet we know that there is some differentiation of function between the hemispheres. Why therefore should there be interference of tasks that are concentrated each to one side?

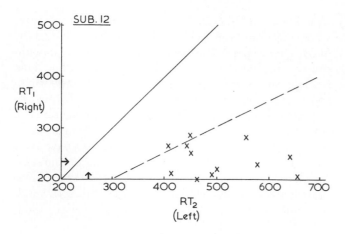

FIGURE 1   Individual pairs of response times for a single subject receiving two stimuli, one on each side of the body, at a known interval of 100 msec. Each reaction was a choice between two alternatives; the arrows indicate the mean time for control stimuli given in isolation. The solid line represents the performance expected if the two reactions were exactly equal, i.e., parallel function on the two sides. The dotted line represents completely serial function, i.e., the second reaction time being started only when the first is complete. Note that not one pair of reactions shows parallel processing.

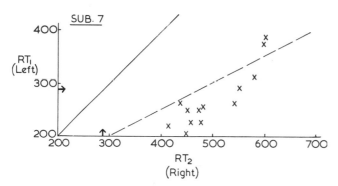

FIGURE 2   A second subject analyzed as in Figure 1 but reacting with the hands in the opposite order. These subjects are typical of those tested by Broadbent and Gregory (1967), that is, not one reaction by any of the 12 subjects showed parallel processing.

One rather trivial possibility might be that, when a region of the nervous system is active in carrying out some task, there is a general depression or inhibition of all other regions, not because functions carried out there have anything to do with the functions involved in the main task but simply to try and confine the person to one action at a time, to avoid any general or random disturbance of the main task. This would be an extension of the principle by which one holds one's breath when squeezing a trigger or listening for a burglar, not because of physiological in-

compatibility between breathing and finger movement or listening but because the noise and vibration of breathing might physically interfere with the desired activities. It would seem rather wasteful to have such a mechanism universally, however, and it does not explain the fact shown by Trumbo and Noble that one can indeed carry out a predictable series of actions with one limb while performing some other task with another; it is only choices or decisions about actions in one task that interfere with choices or decisions in another.

Another possibility is that there are particular peripheral components in each task that overlap, even though the main centers are quite separate. For example, playing the piano does involve the right hand, which is presumably controlled from regions in the left hemisphere not very distant from those involved in understanding speech, and it may be this particular component of reading music that is involved in the interference between the two tasks. We shall see shortly reasons for thinking that this is too simple a view of the control of hand movements; once again it raises the difficulty that tasks using the same limbs and stimuli may show markedly different amounts of interference with each other.

A personal choice of the most likely explanation, therefore, is that the different tasks involve a central process that is common even to tasks which appear to involve the two sides of the body, or to functions as apparently distinct as speech and music. This suggests, in fact, that the single term "speech" includes a variety of information-processing operations, some of which are indeed occurring in the understanding of music as well, and which may not all be localized in the same hemisphere. The nature of such processes can perhaps be illuminated by looking at some of the more curious details of the experiments that show asymmetries in performance.

Difficulties: Some experimental detail

THE RIGHT OF THE BODY OR THE RIGHT OF THE WORLD? We saw earlier that reaction to a sound in the right ear may be faster with the right hand than the left, and vice versa. This seems easy to explain if we suppose that it is quicker for some signals to travel from a projection area to a motor area in the same hemisphere than in the opposite hemisphere. However, a sound reaching the right ear to a man facing north appears to him to come from the east, and in the real world it is unusual for a sound to stimulate only one ear. Under natural conditions, a sound from the east would of course stimulate the left ear as well, although rather more faintly and with a slight time delay. In the laboratory, we can arrange that the same sound stimulates both ears at the same intensity, but with a slight difference in time of arrival, and the listener will then report a sound coming from one side of him, even though it has in fact stimulated both ears. It has been shown by Simon et al. (1970) that a sound which appears to come from a certain direction in space in this way will give a faster reaction with the hand nearer that direction in space, even though the stimulus reaches both ears. What matters, in fact, is not that the sense organ should project to the hemisphere that controls the responding hand but that the percept should be on the same side as that hand.

One might explain this by supposing that the two stimuli that differ in time of arrival will give a larger effect in the hemisphere opposite to the earlier ear stimulated, a suggestion for which there is some tentative evidence; but unfortunately there are other experiments both in the visual case and in dichotic listening that suggest this explanation is inadequate. The dichotic experiment has been conducted by myself and Mrs. Gregory especially for this meeting and has not previously been published. We prepared lists of digits of the type mentioned previously, in which four digits could be presented to one ear and four to the other. The tape recordings we employed gave the usual advantage for the right ear when presented through headphones. We then presented them through loudspeakers, each loudspeaker being placed on one side of the experimental subject. Thus the speech sounds were now heard by both ears, but with natural time and intensity differences. As when using headphones, we turned the subject round halfway through the experiment so that each loudspeaker and tape recording was used on each ear an equal number of times. Yet the advantage for the digits on the right completely disappeared (Table I).

TABLE I

Results for simultaneous presentation of digits from different apparent locations

	Items Correct 4 Digits on Each Ear	
	L (%)	R (%)
Earphones	63.8	74.5
Loudspeakers	57.7	58.9
Earphones with improved response compatibility	59.1	68.2

Each condition was measured on a separate group of subjects; in each case, instructions were given for half the lists to be written down with the right-hand message first, and for half in the opposite order. For the loudspeaker case and for one of the earphone groups, responses were written from left to right, whichever message was being reported first. For the other earphone group, the right-hand message was always written on the right of the paper, whether it was written first or last. Whatever the response instructions, earphones give a significant advantage for the right ear, but loudspeakers do not.

If therefore one explains the result of Simon as due to a greater activity in the hemisphere near the hand that is reacting, why does not this greater activity produce greater efficiency in the hemisphere that reacts to speech?

The third experiment to be mentioned in this section makes matters even worse. Wallace (1971) showed subjects visual shapes, which appeared either to the right or left of a fixation point. They had to press the right hand for one shape, and the left hand for another. The right hand was faster when the shape was to the right of the fixation point, the left when it was to the left. Wallace then had the idea of getting the subject to cross his hands, so that the key on the right was being pressed by the left hand, while the key on the left hand was being pressed by the right hand. If the man was moving faster with the hand controlled from the same hemisphere as the one where the stimulus had arrived, then we should now find that the key on the left was pressed more rapidly when the stimulus was on the right. But this was not the case. It was always the key on the same side as the stimulus that was pressed faster, even though the hand pressing it was coming from the other side of the body.

Wallace's result suggests a rather different interpretation of the original finding. Contemporary psychologists do not think so much of a man as reacting to individual isolated stimuli but rather as establishing somewhere within his nervous system a representation of the whole of his environment distributed in space and time, and utilizing the evidence arriving from his senses to change this representation or model of the outside world. Thus a visual stimulus of a particular objective size will be seen as having different subjective sizes and being located at different distances from the eye, depending upon perspective, texture, and other features of other parts of the visual field. Thus the various features of the outside world are grouped into objects, represented at different places and in varying relations to each other; a change at one part of this representation of the world most naturally produces a reaction at the same part, whichever limb is used to produce the effect. It is not at all clear therefore that the original experiments, showing faster reaction with the hand on the same side as the stimulus, are due to shortness of pathway between the projection area and the motor area; we really have no grounds from these experiments for thinking of the hemispheres as operating independently.

The experiments on listening to speech, however, might still be explained by proximity of the projection areas of the right ear and the response centers for speech. There remains therefore the question why the right ear is better in dichotic listening, provided that one uses headphones rather than loudspeakers.

CONSONANTS OR VOWELS? The advantage of the right ear is not to be found for all aspects of speech. One exception need cause us little surprise; it has been found by Darwin (1969) that under suitable conditions the intonation pattern or melodic line of the pitch of a spoken voice is better detected on the left ear. This however fits in with the findings already mentioned about music and could simply be interpreted as meaning that the response mechanisms for music are in the opposite hemisphere from those that identify the intellectual content or meaning of speech. A more surprising difference is that found by Shankweiler and Studdert-Kennedy (1967), who found that consonants were identified better on the right ear, but vowels were not. In general, this finding has been supported by other work, although there are some interesting exceptions that are considered later. When the finding was first reported, furthermore, there were some relatively simple explanations of it which could not be excluded; for example, the fact that two vowel sounds, one on each ear, may tend to fuse in the middle of the head if they have the same fundamental pitch and thus be extremely difficult to separate. Later experiments have however ruled out this kind of explanation; for example, Darwin (1971a) has shown that the vowels showed an advantage on the right ear even in an experiment in which the same pitch of vowel was presented on each ear. It seems clear therefore that we cannot suppose that "speech" as a whole is located in one hemisphere but rather some particular functions concerned with speech that appear more readily in the case of consonants than they do in the case of vowels.

Another and rather desperate explanation might be that the response mechanisms concerned with identifying consonants are in the left hemisphere while those for vowels are not. This however is ruled out through experiments by Haggard (1971) and by Darwin (1971a), each of which showed a right ear advantage for vowels, under certain particular conditions. In the original experiments on vowels, the same physical stimulus was used for each vowel, that is, the parts of the spectrum that showed the greatest energy were always the same if the same response was required. This corresponds in ordinary life to a vowel being said always by the same person; but we can of course understand the same vowel as spoken by different people, when the vocal tracts are different and the resonances are therefore systematically displaced. If vowels apparently spoken by different voices in this way are used as the stimuli, then an advantage appears for the right ear.

Thus although one often gets equality of performance with the two ears when trying to identify vowels, one does get the advantage for the right ear if during the same experiment different physical stimuli must be identified

as the same vowel response. Similarly, the right-ear advantage for consonants can be changed in size by changing the conditions of the experiment. Darwin (1971a) used fricative consonants that differed both in place of articulation (as in the distinction between s and f) and also in the presence or absence of voicing (as in the distinction between f and v). In natural speech, fricatives contain a burst of noise whose peak is at a different part of the spectrum for different fricatives and is relatively unaffected by the nature of the vowel that follows. There is also in natural speech however a sliding transition of the spectral peaks from the noise burst into the succeeding vowel, and these transitions naturally depend very much upon the nature of the subsequent vowel. Thus a response of the same nature may be appropriate to quite different acoustical transitions, depending upon the nature of the subsequent vowel. If one uses artificial speech sounds, it is possible to produce fricatives that possess both the spectral peak and the transition, or else the peak alone without the transition, or the transition without the spectral peak. When Darwin did this, he found that the identification of the fricative showed an advantage for the right ear if transitions were present but not if the identification was based solely upon the burst of noise. Thus once again as in the case of vowels, it is not the nature of the speech response that matters but the fact that the same response should be elicited by any one of a set of different stimuli.

The usual finding, therefore, that consonants show a bigger advantage for the right ear than vowels do, is probably because in most experiments the consonants are the more likely to show this variety of acoustic cues relevant to the particular task that the subject is performing. It is tempting at this point to draw attention to another recent difference between consonants and vowels, which has been pointed out by Crowder (1971). This difference concerns the short-lasting form of memory that persists like an echo for a short time after some sounds have been heard and that can be washed out by later sounds. Thus for example, if you say to a telephone operator "Get me Cambridge 55294, please," you are more likely to get a mistake than if you say "Please, could you get me Cambridge 55294?" If such an error does occur, it will be particularly on the last digit of the telephone number. The word following the last digit has apparently destroyed a stored representation of it. The digits differ from each other both in consonants and vowels, but Crowder points out one can conduct such experiments using items that differ only in consonants or, alternatively, only in vowels. If this is done, one finds that the memory for the last acoustic event, which is destroyed by subsequent events, appears to apply only to vowels and not to consonants.

To summarize, we can see that it is misleading to talk of

"speech" as if it were a single unanalyzable function. When one is listening to speech, the same response may be appropriate to any one of a whole set of stimuli, depending upon the other speech sounds that come before or after. The occurrence of the correct percept depends therefore upon a number of stages: First, there is a stored representation of the recent past. Second, there is an encoding that allocates the correct response to the particular stimulus, using information about the past context. Only last is there a response process as such. Each of these stages can be broken down yet further; the important point for our purposes is that the first and last stages do not seem to be specially related to the left hemisphere, and it is rather the middle stage of encoding or allocating of responses to stimuli that shows such a relationship. It will be recalled that this is the same stage that we mentioned earlier as possibly being the source of difficulty in combining tasks, which on the surface should be capable of being performed independently.

SCANNING, PERCEPTUAL SELECTIVITY, AND ERASURE   The experiments on dichotic listening place great emphasis on the heterogeneity of the stimuli that may be met in a particular experiment. If a resonance at a particular frequency is always appropriate to the response of naming a particular vowel, then there may be no advantage to the right ear; it is only when the same vowel may be represented by one resonance in one context, and a different resonance in another context, that the advantage for the right ear appears. In the visual field also there are similar effects of the range of stimuli employed. Thus for example Egeth (1971) has looked at the speed with which two visual forms could be identified as the same or different. When a block of similar trials were presented in which, say, both forms were on one side of the fixation point, then the speed of decision was unaffected by the location in which the forms were seen. If however the same materials were presented in a mixed and heterogeneous series, so that the subject never knew where the shapes were to appear, then the more usual result of faster reaction to spatial stimuli on the left side of the visual field was obtained. It seems reasonable to interpret these results as showing that, if the experimental subject knows which part of the world is to deliver the stimulus, he can be selectively prepared for it; but that when there are different possible positions, he has to scan those positions selectively in order. This sequential analysis of the different parts of the sensory field recalls the habitual left-to-right order of scanning in reading, which we mentioned earlier, and there are other visual experiments that show a similar scanning pattern. For example, if a row of different stimuli is presented visually and then very rapidly followed by a complex masking pattern, the subject can report items along the row of stimuli for a distance that is connected with the time interval before the masking pattern arrives (Weisstein, 1966). This kind of backward masking of one stimulus by another arriving subsequently has of course been a field of considerable experimentation in recent years; it is rather complicated by the fact that there can be peripheral masking of one visual or auditory stimulus by another, which has little to do with any supposed serial scanning of features from different parts of the sensory array. Peripheral masking is greatest when the mask arrives at the time of the stimulus and falls off evenly as the mask is delivered either shortly before or shortly after the target stimulus. If however one presents the target stimulus to one retina and the mask to the other, one can demonstrate some central interference that is greatest when the mask follows the target (Kietzman et al., 1971). Thus it seems that the information from the visual scene reaches some central mechanism where it can be disrupted subsequently by information coming from another sense organ.

If now we consider again the case of dichotic listening, Darwin (1971b) presented consonant-vowel combinations in one ear and tried to mask them with various sounds in the other ear. The two most notable results were, first, that the consonant was heard less accurately if the mask *followed* in the opposite ear by about 60 msec than if the mask preceded the target. Second, this was particularly true if the mask was another irrelevant consonant-vowel sound rather than any other of the sounds used. Thus once again it seems that the central process that identifies the consonant can be erased or blotted out by a subsequent stimulus of a similar kind coming from a different sense organ.

These results raise several possible explanations for the general advantage of the right ear in dichotic listening experiments. For example, if we take it that the later information reaching some central stage washes out the earlier, then we might argue perhaps that for synchronized stimulation of the two ears, there is a habitual scanning pattern that first extracts information from the left ear, and then subsequently takes that from the right. The latter would then produce a more effective response because it was the later information to reach the central mechanisms. This however is rather unlikely, because, as already mentioned, the advantage of the right ear disappears in some left-handed individuals, and it is not at all clear why those individuals should have a different scanning pattern from right-handed people.

A second possibility is that the advantage of the right ear consists in some preferential setting of a selective mechanism, designed to prevent erasure from occurring. On this view, two sources of information could not both be admitted to the central mechanism because of the erasure effect, and for some reason a bias would normally be applied to admit the right ear information with higher priority. A view of this sort might be consistent with results found by Oxbury et al. (1967). It is now rather unlikely however, because various investigators have found that the right ear advantage appears equally with subjects instructed to attend only to one ear (see Darwin, 1971a).

A third possibility is more plausible, although very speculative. We have mentioned the hypothesis that the right ear is better in listening to speech when both ears are receiving different signals, because the ipsilateral projections are then each swamped by the contralateral ones, and the right ear gives the major projection close to the centers involved with speech. This view is at first sight attractive, because close proximity of the incoming and outgoing pathways would surely suggest that information would travel more effectively or faster between one and the other. We have however seen a number of difficulties in the general concept that proximity of sensory and projection areas gives greater efficiency. In the light of the experiments on erasure that we have most recently considered, it is tempting however to try an exact reversal of the apparently obvious interpretation and to suppose that each ear projects by a shorter pathway to the ipsilateral hemisphere than it does to the contralateral one. If the central mechanism, which has something to do with speech and which shows erasure effects, is in the left hemisphere, then it might be that information from the right ear reaches it *later* than information from the left ear, and that this is the source of the right-ear advantage.

Conclusions

The foregoing results are scattered and divergent, and their main impact is to make us sceptical of simple interpretations of the role of the two hemispheres. The main reasons for scepticism can be summarized as follows:

1. If it were really the case that certain functions were performed solely in the major hemisphere and others in the minor one, then it ought to be possible to find combinations of tasks that can be performed without mutual interference. This has proved extremely difficult, and therefore it is likely that functions such as speech perception, no matter how unitary they appear to ordinary inspection, actually involve processes in both hemispheres.

2. We cannot adopt the simple approach of supposing that a fast reaction to a stimulus given on one side supposes that the projection area for that stimulus is adjacent to the motor area for that type of response, because the important factor appears in some cases to be the part of the environment affected rather than the limb that is producing the effect.

3. When tasks are performed simultaneously, and peripheral interference at the sense organs or limbs is removed as far as possible, there appears to be some central interference in a mechanism that categorizes uncertain situations in terms of the response that they ought to produce.

4. In so far as there is evidence for better handling of speech delivered to the left hemisphere, it is in those parts of speech perception that require categorical identification of speech sounds from stimuli that may take many different forms.

5. Some of the asymmetries in performance are certainly due to preferred habits of scanning incoming information and to erasure. Because these habits may be very influenced by past experience and by the momentary situation, they confuse understanding of permanent biological mechanism.

These conclusions are therefore somewhat destructive. If however one tries to formulate some general account consistent with these findings, it might be of the following form: Integrated action by a human being requires a representation of the world, in which each feature depends upon many of the stimuli reaching individual sense organs. For this reason, each stimulus cannot affect each part of the representation independently and produce responses that have no relation to each other. Rather the current state of the environment has to be reassessed whenever any change occurs and a single decision reached about the appropriate action now necessary. At stages before this categorization of events, different processes can go on independently, and this is also the case once decisions have been reached. Different substages and activities are therefore involved in any function such as the perception of speech, even though from the point of view of an outside observer it may appear a single unity. Correspondingly, such processes are also involved in tasks of music or of space perception.

Some tasks, however, may be critically dependent upon the stored representation of the world that retains its features in parallel for quite long periods of time. Other tasks may be more dependent upon the successive analysis of a series of changes that take place from one instant to another. Obvious candidates for the first category are music and space perception, and for the second, speech. My speculation would therefore be that the processes that are differentiated between the so-called major and minor hemispheres are those of categorizing changes in the environment, on the one hand, and sustaining the continuing representation of the environment, on the other. This differentiation might then appear in crude terms as a differentiation between speech on the one hand and music or space perception on the other, but those tasks each in fact involve functions of both kinds even though to differing extents.

This speculation is perhaps consistent with some of the

evidence from callosal section and from lesions, which show certain simple speech functions in the minor hemisphere. If it is to be extended to a more adequate and testable level, it is clear that we need more information about the sort of details of speech perception that have been briefly touched upon above. As much of the most distinguished work in this area comes from the Haskins Laboratories, one can therefore look forward to Dr. Liberman's subsequent chapter for signposts of this kind.

REFERENCES

ALLPORT, D. A., B. ANTONIS, and P. REYNOLDS, 1972. On the division of attention: A disproof of the single channel hypothesis. *Q. J. Exp. Psychol.* 24:225–235.

BROADBENT, D. E., 1971. *Decision and Stress.* New York: Academic Press.

BROADBENT, D. E., and M. GREGORY, 1963. Division of attention and the decision theory of signal detection. *Proc. R. Soc. Lond. (Biol.)* 158:222–231.

BROADBENT, D. E., and M. GREGORY, 1965. On the interaction of S-R compatibility with other variables affecting reaction time. *Br. J. Psychol.* 56:61–67.

BROADBENT, D. E., and M. GREGORY, 1967. Psychological refractory period and the length of time required to make a decision. *Proc. R. Soc. Lond. (Biol.)* 168:181–193.

CROWDER, R. G., 1971. The sound of vowels and consonants in immediate memory. *J. Verbal Learn. Verbal Behav.* 10:587–596.

DARWIN, C. J., 1969. The relationship between auditory perception and cerebral dominance. Ph.D. Thesis, University of Cambridge, Cambridge, England.

DARWIN, C. J., 1971a. Ear differences in the recall of fricatives and vowels. *Q. J. Exp. Psychol.* 23:46–62.

DARWIN, C. J., 1971b. Dichotic backward masking of complex sounds. *Q. J. Exp. Psychol.* 23:386–392.

EGETH, H., 1971. Laterality effects in perceptual matching. *Percept. Psychophys.* 9 (No. 4):375–376.

HAGGARD, M. P., 1971. Encoding and the REA for speech signals. *Q. J. Exp. Psychol.* 23:34–45.

HALLIDAY, A. M., M. KERR, and A. ELITHORN, 1960. Grouping of stimuli and apparent exceptions to the psychological refractory period. *Q. J. Exp. Psychol.* 12:72–89.

KIETZMAN, M. L., R. C. BOYLE, and D. B. LINDSLEY, 1971. Perceptual masking: Peripheral vs. central factors. *Percept. Psychophys.* 9 (No. 4):350–352.

KIMURA, D., 1961. Cerebral dominance and the perception of verbal stimuli. *Can. J. Psychol.* 15:166–171.

KIMURA, D., 1964. Left-right differences in the perception of melodies. *Q. J. Exp. Psychol.* 16:355–358.

KIMURA, D., 1966. Dual functional asymmetry of the brain in visual perception. *Neuropsychology* 4:275–285.

KLATZKY, R. L., 1970. Interhemispheric transfer of test stimulus representation in memory scanning. *Psychonomic Sci.* 21:201–203.

LINDSAY, P. H., and D. A. NORMAN, 1969. Short-term retention during a simultaneous detection task. *Percept. Psychophys.* 5:201–205.

OXBURY, S., J. OXBURY, and J. GARDINER, 1967. Laterality effects in dichotic listening. *Nature (Lond.)* 214:742–743.

SHANKWEILER, D. P., and M. STUDDERT-KENNEDY, 1967. Identification of consonants and vowels presented to left and right ears. *Q. J. Exp. Psychol.* 19:59–63.

SIMON, J. R., and A. P. RUDELL, 1967. Auditory S-R compatibility: The effect of an irrelevant cue on information processing. *J. Appl. Psychol.* 51:300–304.

SIMON, J. R., A. M. SMALL, R. A. ZIGLAR, and J. L. CRAFT, 1970. Response interference in an information processing task: Sensory versus perceptual factors. *J. Exp. Psychol.* 85:311–314.

STERNBERG, S., 1966. High-speed scanning in human memory. *Science* 153:652–654.

TRUMBO, D., and M. NOBLE, 1970. Secondary task effects on serial verbal learning. *J. Exp. Psychol.* 85:418–424.

WALLACE, R. J., 1971. S-R compatibility and the idea of a response code. *J. Exp. Psychol.* 88:354–360.

WEISSTEIN, N., 1966. Backward masking and models of perceptual processing. *J. Exp. Psychol.* 72:232–240.

4 The Specialization of the Language Hemisphere

A. M. LIBERMAN

ABSTRACT The language hemisphere may be specialized to deal with grammatical recodings, which differ in important ways from other perceptual and cognitive processes. Their special function is to make linguistic information differentially appropriate for otherwise mismatched mechanisms of storage and transmission. At the level of speech we see the special nature of a grammatical code, the special model that rationalizes it, and the special mode in which it is perceived.

THE FACT THAT language is primarily on one side of the brain implies the question I will ask in this paper: How does language differ from the processes on the other side? I will suggest, as a working hypothesis, that the difference is grammatical recoding, a conversion in which information is restructured, often radically, as it moves between the sounds of speech and the messages they convey. To develop that hypothesis, I will divide it into four more specific ones: Grammatical codes have a special function; they restructure information in a special way; they are unlocked by a special key; and they are associated with a special mode of perception.

Language is the only cerebrally lateralized process I will be concerned with. I will not try to deal with its relation to other processes that may be in the same hemisphere, such as those underlying handedness or perception of fine temporal discriminations (Efron, 1965), though I surely agree that we understand cerebral specialization better when we see all the activities of a hemisphere as reflections of the same underlying design. (See, for example, Semmes, 1968.)

In talking about the function of grammatical codes, the subject of the first of my more specific hypotheses, I will be concerned with language in general. Otherwise, I will limit my attention to speech and, even more narrowly, to speech perception. I do this partly because I know more about speech perception than about other aspects of language. But I am motivated, too, by the fact that more is known about speech perception that bears on the purposes of this seminar. This becomes apparent

when, in interpreting research on hemispheric specialization, we must separate processes that are truly linguistic from those that may only appear so. It becomes even more apparent when we try to frame experimental questions that might help us to discover, quite exactly, what the language hemisphere is specialized for. In any case, not so much is lost by this restriction of attention as might be supposed, since, if recent arguments are accepted, speech perception is an integral and representative part of language, both functionally and formally (Liberman, 1970; Mattingly and Liberman, 1969).

The special function of grammatical codes: Making linguistic information differentially appropriate for transmission and storage

Perhaps the simplest way to appreciate the function of grammar is to consider what happens when we remember linguistic information. Should you try tomorrow to recall this lecture, we might expect, if what I say is sensible, that you would manage very well. But we can hardly conceive that you would reproduce exactly the strings of consonants and vowels, words, or sentences you will have heard. Nor can we suppose that your performance would be evaluated by any reasonable person in terms of the percentage of such elements you correctly recalled, or by the number of times your failure to recall lay merely in the substitution of a synonym for the originally uttered word. A judge of your recall would be concerned only with the extent to which you had captured the meaning of the lecture; he would expect a paraphrase, and that is what he would get.

Paraphrase is not a kind of forgetting but a normal part of remembering. It reflects the conversions that must occur if that which is communicated to us by language is to be well retained (and understood) or if that which we retain (and understand) is to be efficiently communicated. In the course of those conversions, linguistic information has at least three different shapes: An acoustic (or auditory) vehicle for transmission; a phonetic representation, consisting of consonants and vowels, appropriate for processing and storage in a short-term memory; and a semantic representation (or its less linguistically

A. M. LIBERMAN Haskins Laboratories; and Yale University, New Haven, Conn.; University of Connecticut, Storrs, Connecticut structured base) that fits a nonlinguistic intellect and long-term memory. Of course, the conversions among these shapes would be of no special interest if they meant no more than the substitution of one unit for another—for example, a neural unit for an acoustic one—give or take the sharpening, distortions, and losses that must occur. But the facts of paraphrase imply far more than that kind of alphabetic encipherment. Since an accurate paraphrase need not, and usually does not, bear any physical resemblance to the originally presented acoustic (or auditory) signal, we must suppose that the information has been thoroughly restructured. It is as if the listener had stored a semantic representation that he synthesized or constructed out of the speech sounds, and then, on the occasion of recall, used the semantic representation as a base for synthesizing still another set of sounds. Plainly, these syntheses are not chaotic or arbitrary; they are, rather, constrained by rules of a kind that linguists call grammar. There is, therefore, a way to see the correspondence between the original and recalled information, or, indeed, between the transmitted and stored forms. But this can be done only by reference to the grammar, not by comparison of the physical properties of the two sets of acoustic events or of transforms performed directly on them. An observer who does not command the grammar cannot possibly judge the accuracy of the paraphrase.

Since my aim is to raise questions about the distinctiveness of language, I should pause here to ask whether paraphrase is unique. In visual memory, for example, is paraphrase even conceivable? Of course, the remembered scene one calls up in his mind's eye will usually differ from the original. But cannot the accuracy of recall always be judged by reference to the physical properties of the remembered scene, allowing, of course, for reversible transformations performed directly on the physical stimuli themselves? Except in the case of the most abstract art, about which there is notorious lack of agreement, can we ever say of two visual patterns that they correspond only in meaning, and, accordingly, that the correspondence between them can be judged only by reference to rules like those of grammar?

But I should return now to the function of grammatical recoding, which is the question before us. Why must the linguistic information be so thoroughly restructured if it is to be transmittable in the one case and storable in the other? The simple and possibly obvious answer is that the components for transmission and storage are grossly mismatched; consequently, they cannot deal with information in anything like the same form. I should suppose that the reason for the mismatch is that the several components developed separately in evolution and in connection with different biological activities. At the one end of the system

is long-term memory, as well as the nonlinguistic aspects of meaning and thought. Surely, these must have existed before the development of language, much as they exist now in nonspeaking animals and, I dare say, in the non-language hemisphere of man. At the other end of the system, the components most directly concerned with transmission—the ear and the vocal tract—had also reached a high state of development before they were incorporated as terminals in linguistic communication. [Important adaptations of the vocal tract did presumably occur in the evolution of speech, as has been shown (Lieberman, 1968, 1969; Lieberman et al., 1969; Lieberman and Crelin, 1971; Lieberman et al., 1972); however, these did not wholly correct the mismatch we are considering.] We might assume, then, following Mattingly (1972), that grammar developed as a special interface, joining into a single system the several components of transmission and intellect that were once quite separate. What is conceivably unique to language, to man, and to his language hemisphere is only grammatical codes. These are used to reshape semantic representations so as to make them appropriate, via a phonetic stage, for efficient transmission in acoustic form.

We should recognize, of course, that the consequences of being able to make those grammatical conversions might be immense, not merely because man can then more efficiently communicate his semantic representations to others but also because he can, perhaps more easily than otherwise, move them around in his own head. If so, there may be thought processes that can be carried out only on information that has gone into the grammatical system, at least part way. We should also see that the nonlinguistic intellectual mechanisms might themselves have been altered in the course of evolutionary adaptations associated with the development of grammar. Indeed, exactly analogous adaptations did apparently take place at the other end of the system where, as has already been remarked, the vocal tract underwent structural changes that narrowed the gap between its repertory of shapes (and sounds) and that which was required by the nature of the phonetic representation at the next higher level. But such considerations do not alter my point, however much they may complicate it. We may reasonably suppose that the basic function of grammatical codes is to join previously independent components by making the best of what would otherwise be a bad fit.

At this point I should turn again to our question about the distinctiveness of language and ask whether the function of grammatical codes, as I described it here, is unique. Are there other biological systems in which different structures, having evolved independently, are married by a process that restructures the information passing between them? If not, then grammatical codes solve a biologically novel problem, and we should wonder whether it was in connection with such a solution that a new functional organization evolved in the left hemisphere.

But if we are to view grammar as an interface, we ought to see more clearly how bad is the fit that it corrects. For that purpose I will deal separately with two stages of the linguistic process: The interconversion between phonetic message and sound, which I will refer to throughout this paper as the *speech code*, and then briefly with the part of language that lies between phonetic message and meaning.

THE PHONETIC REPRESENTATION VS. THE EAR AND THE VOCAL TRACT   At the phonetic level, language is conveyed by a small number of meaningless segments— roughly three dozen in English—called *phones* by linguists and well known to all as consonants and vowels. These phonetic segments are characteristic of all natural human languages and of no nonlinguistic communication systems, human or otherwise. Their role in language is an important one. When properly ordered, these few dozen segments convey the vastly greater number of semantic units; thus, they take a large step toward matching the demands of the semantic inventory to the possibilities of the vocal tract and the ear. They are important, too, because they appear to be peculiarly appropriate for storage and processing in short-term memory (Liberman et al., 1972). In the perception of speech the phonetic segments are retained in short-term memory and somehow organized into the larger units of words and phrases; these undergo treatment by syntactic and semantic processes, yielding, if all goes well, something like the meaning the speaker intended. But if the larger organizations are to be achieved, the phonetic units must be collected at a reasonably high rate. (To see how important rate is, try to understand a sensible communication that is spelled to you slowly, letter by painful letter.) In fact, speaking speeds produce phonetic segments at rates of 8 to 20 segments per second, and research with artificially speeded speech (Orr et al., 1965) suggests that it is possible to perceive phonetic information at rates as high as 30 segments (that is, about seven words) per second.

Now if speech had developed from the beginning as a unitary system, we might suppose that the components would have been reasonably well matched. In that case there would have been no need for a radical restructuring of information—that is, no need for grammar—but only the fairly straightforward substitution of an acoustic segment for each phonetic one. Indeed, just that kind of substitution cipher has commonly been assumed to be an important characteristic of speech. But such a simple conversion would not work, in fact, because the requirements of phonetic communication are not directly met either by the ear or by the vocal tract.

Consider first the ear. If each phonetic unit were represented, as in an alphabet or cipher, by a unit of sound, the listener would have to identify from 8 to 30 segments per second. But such rates would surely strain, and probably overreach, the temporal resolving power of the ear. Consider next the requirement that the order of the segments be preserved. Of course, the listener could hardly be expected to order the segments if, at high rates, he could not even resolve them. We should note, however, that even at slower rates, and in cases where the identity of the sound segments is known, there is some evidence that the ear does not identify order well. Though this question has not been intensively investigated, data from the research of Warren et al. (1969) suggest that the requirements for ordering in phonetic communication would exceed the psychoacoustically determined ability of the ear by a factor of 5 or more.

Apparently, then, the system would not work well if the conversion from phonetic unit to sound were a simple one. We should suppose that this would be so for the reasons I just outlined. But the case need not rest on that supposition. In fact, there is a great deal of confirming evidence in the experience gained over many years through the attempts to develop and use acoustic (nonspeech) alphabets. That experience has been in telegraphy—witness Morse code, which is a cipher or alphabet as I have been using the terms here—and much more comprehensively in connection with the early attempts to build reading machines for the blind. Even after considerable practice, users do poorly with those sound alphabets, attaining rates no better than one-tenth those that are achieved in speech (Freiberger and Murphy, 1961; Coffey, 1963; Studdert-Kennedy and Cooper, 1966; Nye, 1968).

Nor does the vocal tract appear to be better suited to the requirements of phonetic communication. If the sounds of speech are to be produced by movements of the articulatory organs, we should wonder where in the vocal tract we are going to find equipment for three dozen distinctive gestures. Moreover, we should wonder, since the order of the segments must be preserved, how a succession of these gestures can be produced at rates as high as one gesture every 50 msec.

THE PHONETIC VS. THE SEMANTIC REPRESENTATIONS
Though appropriate for storage over the short term, the

phonetic representation apparently does not fit the requirements of the long-term store or of the essentially nonlinguistic processes that may be associated with it. Those requirements are presumably better met by the semantic representation into which the phonetic segments are converted. Because of its inaccessibility, we do not know the shape of the information at the semantic level, which is a reason we do well, for our purposes, to concentrate our attention on the acoustic and phonetic levels where we can more readily experiment. Still, some characteristics of the semantic representation can be guessed at. Thus, given the innumerable aspects of our experience and knowledge, we should suppose that the inventory of semantic units is very large, many thousands of times larger than the two or three dozen phonetic segments that transmit it. We should suppose, further, that, however the semantic units may be organized, it is hardly conceivable that they are, like the phonetic segments, set down in ordered strings. At all events, the phonetic and semantic representations must be radically different, reflecting, presumably, the differences between the requirements of the processes associated with short- and long-term memory.

The special restructuring produced by the speech code: Simultaneous transmission of information on the same cue

We can usefully think of grammatical coding as the restructuring of information that must occur if the mismatched components I have talked about are to work together as a single system. In developing that notion, I have so far spoken of three levels of linguistic information —semantic, phonetic, and acoustic—connected, as it were, by grammars that describe the relation between one level and the next. The phonetic and acoustic levels are linked by a grammar my colleagues and I have called the speech code. That is the grammar I shall be especially concerned with. But we should first place that grammar in the larger scheme of things and establish some basis for demonstrating its resemblance to grammars of a more conventional kind. That has been done in some detail in recent reviews already referred to (Liberman, 1970; Mattingly and Liberman, 1969). Here I will offer the briefest possible account.

Exactly what we say about the more conventional grammars depends, of course, on which linguistic theory we choose. Fortunately, the choice is, for us, not crucial. Our purposes are well served by a very crude approximation to the transformational or generative grammar that is owing to Chomsky (1965). On his view, the conversion

from semantic to phonetic levels is accomplished through two intermediate levels called deep structure and surface structure. At each level—including also the phonetic, to which I have already referred—there are strings of segments (phones, words) organized into larger units (syllables, phrases). From one level to the next the organized information is restructured according to the rules of the appropriate grammar: Syntax for the conversion from deep to surface, phonology for the conversion from surface to phonetic. It is not feasible to attempt an account of these grammars, even in broad terms. But I would point to one of the most general and important characteristics of the conversions they rationalize: Between one level and the next, there is no direct or easily determined correspondence in the number or order of the segments. Taking a simple example, we suppose that in the deep structure, the level closest to meaning, there are strings of abstract, wordlike units that, when translated into the nearest kind of plain English, might say: *The man is young. The man is tall. The man climbs the ladders. The ladders are shaky.* According to the rules of syntax, and by taking advantage of referential identities, we should delete and rearrange the segments of the four deep sentences, emerging in due course at the surface with the single sentence: *The tall young man climbs the shaky ladders.* It is as if the first, second, and fourth of the deep sentences had been folded into the third, with the result that information about all four sentences is, at the surface, transmitted simultaneously and on the same words.

The information at the level of surface structure is in turn converted, often by an equally complex encoding, to the phonetic level. But I will only offer an example of one of the simplest aspects of that conversion that nevertheless shows that the information does change shape in its further descent toward the sounds of speech and also illustrates a kind of context-conditioned variation that grammatical conversions often entail. Consider in the word *ladders* the fate of the segment, spelled *s*, that means "more than one." Its realization at the phonetic level depends on the segmental context: In our example, *ladders*, it becomes [z]; in a word like *cats*, it would be [s]; and in *houses* it would be [$əz$].

The more obvious parts of grammar, and of the paraphrase which so strikingly reflects it, occur in the conversion between phonetic and semantic representations. But, as I have already suggested, there is another grammar, quite similar in function and in form, to be found in the speech code that connects the phonetic representation to sound. The characteristics of this code have been dealt with at some length in several recent papers (Liberman et al., 1967, 1972). I will only briefly describe some of those characteristics now to show how they might mark speech perception and, by analogy, the rest of language as different from other processes.

How the Phonetic Message is Articulated: Matching the Requirements of Phonetic Communication to the Vocal Tract Consider, again, that there are several times more segments than there are articulatory muscle systems capable of significantly affecting the vocal output. A solution is to divide each segment into features, so that a smaller number of features produces a larger number of segments, and then to assign each feature to a significant articulatory gesture. Thus, the phonetic segment [b] is uniquely characterized by four articulatory features: Stop manner, i.e., rapid movement to or from complete closure of the buccal part of the vocal tract, which [b] shares with [d, g, p, t, k, m, n, ŋ] but not with other consonants; orality, i.e., closure of the velar passage to the nose, which [b] shares with [d, g, p, t, k], but not with [m, n, ŋ]; bilabial place of production, i.e., closure at the lips, which [b] shares with [p, m] but not with [d, g, t, k, n, ŋ]; and voiced condition of voicing, i.e., vocal fold vibration beginning simultaneously with buccal opening, which [b] shares with [d, g], but not with [p, t, k].

It remains, then, to produce these segments at high rates. For that purpose the segments are first organized into larger units of approximately syllabic size, with the restriction that gestures appropriate to features in successive segments be largely independent and therefore capable of being made at the same time or with a great deal of overlap. In producing the syllable, the speaker takes advantage of the possibilities for simultaneous or overlapping articulation, perhaps to the greatest extent possible. Thus, for a syllable like [bæg], for example, the speaker does not complete the lip movement appropriate for [b] before shaping the tongue for the vowel [æ] and then, only when that has been accomplished, moves to a position appropriate for [g]. Rather, he overlaps the gestures, sometimes to such an extent that successive segments, or their component features, are produced simultaneously. In this way, coarticulation produces segments at rates faster than individual muscle systems must change their states and is thus well designed, as Cooper (1966) has put it, to get fast action from relatively slow-moving machinery.

How the Coarticulation of the Phonetic Message Produces the Peculiar Characteristics of the Speech Code The grouping of the segments into syllables and the coarticulation of features represents an organization of the phonetic message, but not yet a very

drastic encoding, since it is still possible to correlate isolable gestures with particular features. It is in the further conversions, from gestures to vocal-tract shapes to sounds, that the greater complications of the speech code are produced. For it is there that we find a very complex relation of gesture to vocal-tract shape and then, in the conversion from vocal-tract shape to sound, a reduction in the number of dimensions. The result is that the effects of several overlapped gestures are impressed on exactly the same parameter of the acoustic signal, thus producing the most important and complex characteristic of the speech code. That characteristic is illustrated in Figure 1, which is intended to demonstrate how several segments of the phonetic message are encoded into the same part of the sound. For that purpose, we begin with a simple syllable comprising the phonetic string [b] [æ] [g] and then, having shown its realization at the level of sound, we determine how the sound changes as we change the phonetic message, one segment at a time. The schematic spectrogram in the left-most position of the row at the top would, if converted to sound, produce an approximation to [bæg], which is our example. In that spectrogram the two most important formants—a formant is a concentration of acoustic energy representing a resonance of the vocal tract—are plotted as a function of time. Looking at only the second (i.e., higher) formant, so as to simplify our task, we try to locate the information about the vowel [æ]. One way to do that is to change the message from [bæg] to [bɔg] and compare the acoustic representations. The spectrogram for the new syllable [bɔg] is shown in the next position to the right, where, in order to make the comparison easier, the second formant of [bæg] is reproduced in dashed lines. Having in mind that [bæg] and [bɔg] differ only in their middle segments—that is, only in the vowels—we note that the difference between the acoustic signals is not limited, correspondingly, to their middle sections but rather extends from the beginning of the acoustic signal to the end. We conclude, therefore, that the vowel information is everywhere in the second formant of the sound. To find the temporal extent of the [b] segment of our original syllable [bæg], we should ask, similarly, what the acoustic pattern would be if only the first segment of the phonetic message were now changed, as it would be, for example, in [gæg]. Looking, in the next position to the right, at that new syllable [gæg], we see that the change has produced a second-formant that differs from the original through approximately the first two-thirds of the temporal extent of the sound. A similar test for [g], the final consonant of our example, is developed at the right-hand end of the row; information about that segment exists in the sound over all of approximately the last two-thirds of its time course.

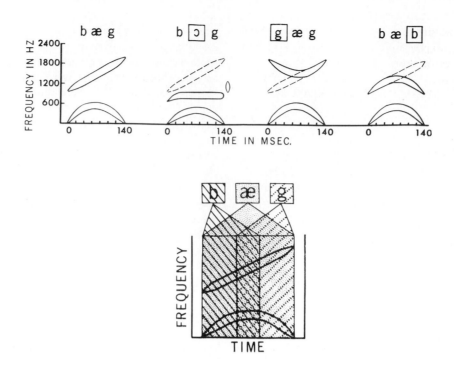

FIGURE 1   Schematic spectrograms showing how the segments of the phonetic message are conveyed simultaneously on the same parameter of the sound.

The general effect is illustrated in the single pattern in the lower half of the figure, which shows over what parts of the sound each of the three message segments extends. We see that there is no part of the sound that contains information about only one phonetic segment: At every point, the sound is carrying information simultaneously about at least two successive segments of the message, and there is a section in the middle where information is simultaneously available about all three. It is as if the initial and final consonants [b] and [g] had been folded into the vowel, much as the flanking deep-structure sentences of the earlier syntactic example were folded into that middle sentence that served, like the vowel, as a core or carrier.

Given that information about successive segments of the message is often carried simultaneously on the same parameter of the signal, the acoustic shape of a cue for a particular segment (or feature) will necessarily be different in different contexts. To see that this is so we should look again at the figure, but instead of noting, as we did before, that changing only the middle segment caused a change in the entire acoustic signal, we should see now that, though we retained two of the three original message segments, we nevertheless left no part of the acoustic signal intact. That is to say that the acoustic cues for [b] and [g] are very different in the contexts of the different

vowels into which they are encoded. Such context-conditioned variation, similar perhaps to that we noted in the phonology, is often very great, not only for a consonant segment with different vowels, as in the example offered here, but also for different positions in the syllable, different kinds of syllable boundaries, different conditions of stress, and so on. (See, for example, Liberman et al., 1967.)

Thus, as in the conversions between other levels of the language structure, the connection between phonetic message and sound is that of a very complex code, not an alphabet or substitution cipher. Information about successive segments of the message is often encoded into a single acoustic event with the result that there is no direct or easily calculated correspondence in segmentation, and the resulting variation in the shape of the acoustic cue can be extreme. At the levels of phonetic and acoustic representations, those characteristics define what I mean by a grammatical code.

But is the speech code—and, by extension, the other grammatical codes—unique? In visual and auditory perception, the relations between stimulus pattern and perceived response may be just as complex as those of speech, but they appear, as a class, to be different. I find it difficult to characterize the difference in general terms beyond saying that, apart from speech perception, we do not find the kind of simultaneous transmission that requires the perceiver to process a unitary physical event so as to recover the two or three discrete perceptual events that are encoded in it.

How the Speech Code Matches the Requirements of Phonetic Communication to the Properties of the Ear   I remarked earlier that we can and do hear speech at rates that would appear to overreach the resolving power of the ear if each phone were transmitted by a unit sound. But we have seen that the phones are not transmitted in that direct way; they are, rather, converted so as to encode several phones into the same acoustic unit. Though this produces a great complication in the relation between signal and message, and one that will have to be dealt with by a correspondingly complex decoder, it serves the important purpose of reducing significantly the number of discrete acoustic events that must be heard, and thus makes it possible to perceive phonetic information at reasonable rates. Given that the segments are encoded into units of approximately syllabic size, we should suppose that the limit on perception is set, not by the number of phonetic segments per unit time, but more nearly by the number of syllables.

I also remarked earlier on another way in which the ear appears to be ill suited to the requirements of phonetic communication: A listener must identify the order of phonetic segments, yet in ordinary auditory perception he cannot do that well. The solution to this problem that is offered by the speech code is that order is often marked, not only by time of occurrence but also by context-conditioned variations in the shape of the cue. Thus, because of the kind of encoding that occurs, a primary acoustic cue for the two *b*s in [*bæb*] will be mirror images of each other. In words like [*tæks*] and [*tæsk*] the acoustic cues for [*k*] will have very different shapes, again because of coarticulation. Hence, the speech code offers the listener the possibility of constructing (or, more exactly, reconstructing) the order of the segments out of information which is not simply, or even primarily, temporal.

More and Less Encoded Aspects of Speech   An important characteristic of the speech code, especially in relation to questions about hemispheric specialization, is that not all parts of the speech signal bear a highly encoded relation to the phonetic message. In slow to moderate articulation, vowels and fricatives, for example, are sometimes represented by a simple acoustic alphabet or cipher: There are isolable segments in which information about only one phonetic segment is carried, and there may be little variation in the shape of the acoustic cues with changes in context. Segments belonging to the classes liquids and semivowels can be said to be gram-matically encoded to an intermediate degree. Though these segments cannot be isolated in the speech signal (except for *r*-colored vowels), they do have brief steady-state portions, even in rapid articulation.

Nongrammatical Complications in the Relation between Acoustic and Phonetic Levels   There are several characteristics of speech apart from its encodedness that might require special treatment in perception. One is that the speech signal seems very poorly designed, at least from an engineering point of view. The acoustic energy is not concentrated in the information-bearing parts of the sound but is, rather, spread quite broadly across the spectrum. Moreover, the essential acoustic cues are, from a physical point of view, among the most indeterminate. Thus, the formant transitions, so important in the perception of most consonants, are rapid changes in the frequency position of a resonance that, by their nature, scatter energy.

Another kind of difficulty arises from the gross variations in vocal tract dimensions among men, women, and children. A consequence is that the absolute values of the formant cues will be different depending on the sex and size of the speaker. Obviously, some kind of calibration is necessary if the listener is to perceive speech properly.

What, Then, Is the Language Hemisphere Specialized for?   I have suggested, as a working hypothesis, that the distinctive characteristic of language is not meaning, thought, communication, or vocalization, but, more specifically, a grammatical recoding that reshapes linguistic information so as to fit it to the several originally nonlinguistic components of the system. That hypothesis may be useful in research on hemispheric specialization for language, because it tells how we might make the necessary distinction between that which is linguistic and that which is not. Our aim, then, is to discover whether it is, in fact, the processes of grammatical recoding that the language hemisphere is specialized for. That will be hard to do at the level closest to the semantic representation because we cannot, at that end of the language system, so easily define the boundary between grammatical coding and the presumably nonlinguistic processes it serves. But in speech, and especially in speech perception, we can be quite explicit. As a result, we can ask pointed questions and, because appropriate techniques are available, get useful answers. I will offer a few examples of such questions and answers. Because the experiments I will talk about represent a large and rapidly growing class, I should emphasize that, for the special purposes of this chapter, I will describe only a few.

Speech vs. nonspeech   After investigations of people with cortical lesions, including especially the studies by

Brenda Milner (1954, 1958), had indicated that perception of speech and nonspeech might be primarily on opposite sides of the head, Doreen Kimura (1961a; 1961b) pioneered the development of an experimental technique that permits us to probe this possibility with normal people. Adapting for her purposes a method that had been used earlier by Broadbent (1956), Kimura presented spoken digits dichotically, one to one ear and a different one to the other. She discovered that most listeners heard better the digits presented to the right ear. It was subsequently found, by her and others, that the same effect is obtained with nonsense syllables, including those that differ in only one phonetic segment or feature (Kimura, 1967; Shankweiler and Studdert-Kennedy, 1967). When the stimuli are musical melodies or complex nonspeech sounds, the opposite effect, a left-ear advantage, is obtained (Kimura, 1964). On the assumption that the contralateral auditory representation is stronger than the ipsilateral, especially under conditions of dichotic competition, Kimura interpreted these findings to reflect left-hemisphere processing of the speech signals and right-hemisphere processing of the others. In any case, many studies now support the conclusion that the ear advantages are reliable reflections of the functional asymmetry of the cerebral hemispheres. (For summaries, see Kimura, 1967; Shankweiler, 1971; Studdert-Kennedy and Shankweiler, 1970.)

Auditory vs. phonetic processing If, as seems reasonable, the right-ear advantage for speech is interpreted to reflect the work of some special device in the left hemisphere, we should ask whether that device is specialized for grammatical decoding or for something else. Consider, then, a case such as the stop consonants. As I pointed out earlier, these phonetic segments are encoded grammatically in the exact sense that there is no part of the acoustic signal that carries information only about the consonant; the formant transitions, which contain all the information about the consonant, are simultaneously providing information about the following vowel. Any device that would perceive the segments correctly must deal with that grammatical code. Conceivably, that is what the device in the language hemisphere is specialized for. But there are other, nongrammatical jobs to be done and, accordingly, other possibilities. Among these are the tasks I referred to earlier when I spoke of the need to clean up the badly smeared speech signal, to track the very rapid frequency modulations (formant transitions) that are such important cues, and to calibrate for differences in vocal-tract size. Though not grammatical according to our definition, these tasks confront the listener only in connection with speech. They might, therefore, be more closely associated with the language hemisphere than those other auditory processes that must underlie

the perception of all sounds, speech and nonspeech alike. But that is precisely the kind of issue that can be settled experimentally.

Several investigators (for example Darwin, 1971; Haggard, 1971; Shankweiler and Studdert-Kennedy, 1967; Studdert-Kennedy and Shankweiler, 1970; Studdert-Kennedy et al., 1972) have suggested and considerably refined questions like those I posed in the preceding paragraph and, in a number of ingenious experiments, found answers. I cannot here describe, or even summarize, these generally complex studies except to say that they provide some support for the notion that in speech perception the language hemisphere extracts phonetic features, which is to say in the terminology of this paper that it does grammatical decoding. There is, however, an experiment by Darwin (1971) which suggests that the language hemisphere may also be responsible for normalizing the acoustic signal to take account of the complications produced by the differences among speakers in length of the vocal tract. That finding indicates that our hypothesis is, at best, incomplete. Of course, we can hope to discover a mechanism general enough to include both vocal-tract normalization and grammatical decoding, the more so since these processes are so intimately associated with each other and with nothing else. Meanwhile, we can proceed to find out by experiment whether the language hemisphere is responsible for the other nongrammatical tasks, however closely or remotely they may be associated with speech. Perhaps an example of such an experiment will clarify the question and also our hypothesis.

Imagine a set of stop-vowel syllables [ba, da, ga] synthesized in such a way that the only distinguishing acoustic cue is the direction and extent of the first 50 msec of the second-formant transition, the rapid frequency modulation referred to earlier. Suppose, now, that we present these dichotically—that is, [ba], for example, to one ear, [da] to the other—in randomly arranged pairs and, as usual, get the right-ear advantage that is presumed to reflect left-hemisphere processing. On the hypothesis proposed here, we should say that these signals were being processed in the language hemisphere because they required grammatical decoding. In that case, we should have in the language hemisphere a device that is quite properly part of a linguistic system. There is, however, an alternative, as I have already implied, which is that the language hemisphere is specialized not for grammatical decoding but for responding to a particular class of auditory events, specifically the rapid frequency modulations of the second-formant transitions that are, in the stimuli of the experiment, the only acoustic cues. In that case, the left hemisphere would be said, at least in this respect, to be specialized for an auditory task, not a linguistic one. An experiment that helps to decide between these possibilities would go as follows: First, we remove from the synthetic syllables the second-formant transition cues and present them in isolation. When we do that we hear, not speech, but more or less distinguishable pitch glides or bird-like chirps. Now, given that there is a right-ear (left hemisphere) advantage when these formant-transition cues are in a speech pattern, we determine the ear advantage when they are presented alone and not heard as speech. Donald Shankweiler, Ann Syrdal, and I (personal communication) have been doing that experiment. The results so far obtained are not wholly convincing, because, owing largely to the difficulty our listeners have in identifying the transition cues alone, the data are quite noisy. So far as the results can be interpreted, however, they suggest that the second-formant transitions in isolation produce a left-ear advantage, in contrast to the right-ear advantage obtained when those same transitions cued the perceived distinctions along [ba, da, ga]. If that result proves reliable, we should infer that the language hemisphere is specialized for a linguistic task of grammatical decoding, not for the auditory task of tracking formant transitions.

A clearer answer to essentially the same question, arrived at by a very different technique, is to be found in a recent doctoral dissertation by Wood (1973). He first replicated an earlier study (Wood et al., 1971) in which it had been found that evoked potentials were exactly the same in the right hemisphere whether the listener was distinguishing two syllables that differed only in a linguistically irrelevant dimension (in this case [ba] on a low pitch vs. [ba] on a high pitch) or in their phonetic identity ([ba] vs. [da] on the same pitch), but the evoked potentials in the left hemisphere were different in the two cases. From that result it had been inferred that the processing of speech required a stage beyond the processing of the nonlinguistic pitch parameter and, more important, that the stage of speech processing occurred in the left hemisphere. Now, in his dissertation, Wood has added several other conditions. Of particular interest here is one in which he measured the evoked potentials for the isolated acoustic cues, which were, as in the experiment described above, the second-formant transitions. The finding was that the isolated cue behaved just like the linguistically irrelevant pitch, not like speech. This suggests, as does the result we have so far obtained in the analogous dichotic experiment, that the processor in the language hemisphere is specialized, not for a particular class of auditory events, but for the grammatical task of decoding the auditory information so as to discover the phonetic features.

More vs. less encoded elements As we saw earlier, only some phonetic segments—for example, [b, d, g]—are always grammatically encoded in the sense that information about them is merged at the acoustic level with information about adjacent segments. Others, such as the fricatives and the vowels, can be, and sometimes are, represented in the sound as if in a substitution cipher; that is, pieces of sounds can be isolated which carry information only about those segments. Still others, the liquids and semivowels, appear to have an intermediate degree of encodedness. We might suppose that only the grammatically encoded segments need to be processed by the special phonetic decoder in the left hemisphere; the others might be dealt with adequately by the auditory system. It is of special interest, then, to note the evidence from several studies that the occurrence or magnitude of the right-ear advantage does depend on encodedness (Darwin, 1971; Haggard, 1971; Shankweiler and Studdert-Kennedy, 1967). Perhaps the most telling of these experiments is a very recent one by Cutting (1972). He presented stop-liquid-vowel syllables dichotically—e.g. [kre] to one ear, [glæ] to the other—and found for the stops that almost all of his subjects had a right-ear advantage, while for the vowels the ear advantage was almost equally divided, half to the right ear and half to the left; the results with the liquids were intermediate between those extremes.

We might conclude, again tentatively, that the highly encoded aspects of speech—those aspects most in need of grammatical decoding—are always (or almost always) processed in the language hemisphere. The unencoded or less highly encoded segments may or may not be processed there. We might suppose, moreover, that some people tend to process all elements of language linguistically while others use nonlinguistic strategies wherever possible. If that is so, could it account for at least some of the individual differences in "degree" of ear advantage that turn up in almost all investigations?

Primary vs. secondary speech codes; cross-codes People ordinarily deal with the complications of the speech code without conscious awareness. But awareness of some aspects of speech, such as its phonetic structure, is sometimes achieved. When that happens secondary codes can be created, an important example being language in its alphabetically written form. This written, secondary code is not so natural as the primary code of speech but neither is it wholly unnatural, since it presumably makes contact with a linguistic physiology that is readily accessible when reading and writing are acquired. Research suggests that the contact is often (if not always) made at the phonetic level (Conrad, 1972); that is, that which is read is recoded into a (central) phonetic representation. If so, then we might expect to see the consequences in studies of hemispheric specialization for the perception of written language, as indeed we do (Milner, 1967; Umiltà et al., 1972).

In addition to the complications of secondary codes, there are special problems arising out of the tendency, under some conditions, to cross-code nonlinguistic experience into linguistic form. As found in a recent experiment by Conrad (1972), for example, confusions in short-term memory for pictures of objects were primarily phonetic, not visual (or optical). Such results do not reveal the balance of nonlinguistic and linguistic processes, but they make it nonetheless evident that in the perception of pictures and, perhaps, of other kinds of nameable patterns, too, some aspects of the processing might be linguistic and therefore found in the language hemisphere.

A special key to the code: The grammar of speech

If the speech code were arbitrary—that is, if there were no way to make sense of the relation between signal and message—then perception could only be done by matching against stored templates. In that case there could be no very fundamental difference between speech and nonspeech, only different sets of templates. Of course, the number of templates for the perception of phonetic segments would have to be very large. It would, at the least, be larger than the number of phones because of the gross variations in acoustic shape produced by the encoding of phonetic segments in the sound; but it would also be larger than the number of syllables, because the effects of the encoding often extend across syllable boundaries and because the acoustic shape of the syllable varies with such conditions as rate of speaking and linguistic stress.

But grammatical codes are not arbitrary. There are rules—linguists call them grammars—that rationalize them. Thus, in terms of the Chomsky-like scheme I sketched earlier, the grammar of syntax tells us how we can, by rule, reshape the string of segments at the level of deep structure so as to arrive at the often very different string at the surface. In the case of the speech code, we have already seen the general outlines of the grammatical key: A model of the articulatory processes by which the peculiar but entirely lawful complications of the speech code come about. The chief characteristic and greatest complication of the speech code, it will be recalled, is that information about successive segments of the message is carried simultaneously on the same acoustic parameter. To rationalize that characteristic, we must understand how it is produced by the coarticulation I described earlier. Though crude and oversimple, that account of coarticulation may nevertheless have shown that a proper model of the process would explain how the phonetic message is encoded in the sound. Such a proper model would be a grammar, the grammar of speech in this case. It would differ from other grammars—for example, those of syntax and phonology—in that the grammar of speech

would be a grammar done in flesh and blood, not, as in the case of syntax, a kind of algebra with no describable physiological correlates. Because the grammar of speech would correspond to an actual process, it is tempting to suppose that the understanding of the speech code it provides is important, not just to the inquiring scientist but also to the ordinary listener who might somehow use it to decode the complex speech sounds he hears. To yield to that temptation is to adopt what has been called a *motor theory of speech perception*, and then to wonder if the language hemisphere is specialized to provide a model of the articulatory processes in terms of which the decoding calculations can be carried out.

One finds more nearly direct evidence for a motor theory when he asks which aspect of speech, articulatory movement or sound, is more closely related to its perception. That question is more sensible than might at first appear because the relation between articulation and sound can be complex in the extreme. Thus, as I have already indicated, the section of sound that carries information about a consonant is often grossly altered in different vowel contexts, though the consonant part of the articulatory gesture is not changed in any essential way. Following articulation rather than sound, the perception in all these cases is also unchanged (Liberman et al., 1952; Liberman, 1957; Lisker et al., 1962). Though such findings support a motor theory, I should note that only a weak form of the theory may be necessary to account for them. That is, they may only suggest that the perception of phonetic features converges, in the end, on the same neural units that normally command their articulation; in that case, the rest of the processes underlying speech perception and production could be quite separate.

Evidence of a different kind can be seen in the results of a recent unpublished study by L. Taylor, Brenda Milner, and C. Darwin. Testing patients with excisions of the face area in the sensori-motor cortex of the left hemisphere, these investigators found severe impairments in the patients' ability to identify stop consonants (by pointing to the appropriate letter printed on a card) in nonsense-syllable contexts, though the pure-tone audiograms and performance on many other verbal tasks were normal. Patients with corresponding damage in the right hemisphere, and those with temporal or frontal damage in either hemisphere, were found not to differ from normal control subjects. It is at least interesting from the standpoint of a motor theory that lesions in the central face area did produce an inability to identify encoded stop consonants, though, as the investigators have pointed out, the exact nature of the impairment, whether of perception or of short-term memory, will be known only after further research.

The idea that the left hemisphere may be organized appropriately for motor control of articulation is in the theory of hemispheric specialization proposed by Josephine Semmes (1968). It is, perhaps, not inconsistent with her theory to suppose, as I have, that the organization of the language hemisphere makes a motor model more available to perceptual processes. But one might, on her view, more simply assume that lateralization for language arose primarily for reasons of motor control. This would fit with the suggestion by Levy (1969) that, to avoid conflict, it would be well not to have bilaterally issued commands for unilateral articulations. In that respect, speech may be unique, as Evarts (personal communication) has pointed out, because other systems of coordinated movements ordinarily require different commands to corresponding muscles on the two sides. Conceivably, then, motor control of speech arose in one hemisphere in connection with special requirements like those just considered, and then everything else having to do with language followed. This assumption has the virtue of simplicity, at least in explaining how language got into one hemisphere in the first place. Moreover, it is in keeping with a conclusion that seems to emerge from the research on patients with "split" brains, which is that, of all language functions, motor control of speech is perhaps, most thoroughly lateralized (Sperry and Gazzaniga, 1967).

At all events, though the grammar of speech makes sense of the complexly encoded relation between phonetic message and sound, it does not tell us how the decoding might be carried out. Like the other grammars of phonology and syntax, the grammar of speech works in one direction only, downward; the rules that take us from phonetic message to sound do not work in reverse. Indeed, we now know the downward-going rules well enough, at least in acoustic form, to be able to use them (via a computer) to generate intelligible speech automatically from an input of (typed) phonetic segments (Mattingly, 1968, 1971). But we do not know how to go automatically in the reverse direction, from speech sounds to phonetic message, except perhaps via the roundabout route of analysis by synthesis—that is, by guessing at the message, generating (by rule) the appropriate sound, and then testing for match (Stevens, 1960; Stevens and Halle, 1967).

Still, I should think that we decode speech with the aid of a model that is, in some important sense, articulatory. If so, we might suppose that the functional organization of the left hemisphere is peculiarly appropriate for the conjoining of sensory and motor processes that such a model implies.

Having said that the speech code is rationalized by a production model, I should ask whether in this respect it differs from the relations between stimulus and perception in other perceptual modalities. I think perhaps it does. In visual and auditory perception of nonverbal material, the complex relations between stimulus and perception are also "rule-bound" rather than arbitrary, but the rules are different from those of the speech code if only because the complications between stimulus and perception in the nonspeech case do not come about as a result of the way the human perceivers produce the stimulus: The very great complications of shape constancy, for example, are rationalized, not in terms of how a perceiver makes those shapes but by the rules of projective geometry. This is not to say that motor considerations are unimportant in nonspeech perception. Obviously, we must, in visual perception, take account of head and eye movements, else the world would appear to move when it should stand still (Teuber, 1960, pp. 1647–1648). But in those cases the motor components must be entered only as additional data to be used in arriving at the perception; the perceptual calculations themselves would be done in other terms.

I wonder, too, if the fact that the speech rules work in only one direction makes them different from those that govern other kinds of perception. In the case of shape constancy, for example, we know that one can, by the rules of geometry, calculate the image shape on the retina if he knows the shape of the stimulus object and its orientation. That would be analogous to using the grammar of speech to determine the nature of the sound, given the phonetic message. But in shape constancy, it would appear that the calculations could be made in reverse— that is, in the direction of perception. Knowing the image shape on the retina and the cues for orientation, we ought to be able to calculate directly the shape of the object. If so, then there would be no need in shape constancy, and conceivably in other kinds of nonspeech perception, for a resort to analysis by synthesis if the perceptual operations are to be done by calculation; rather, the calculations could be performed directly.

Special characteristics of perception in the speech mode

A commonplace observation about language is that it is abstract and categorical. That means, among other things, that language does not fit in any straightforward or isomorphic way onto the world it talks about. We do not use longer words for longer objects, or, less appropriately, louder words for bluer objects. If we change only one phonetic segment out of four in a word, we do not thereby create a word less different in meaning than if we had changed all four. Apart from onomatopoeia and phonetic symbolism, which are among the smallest and least typical parts of language, we do not use continuous linguistic variations to represent the continuous variations of the outside world.

It is of interest, then, to note that in the case of the encoded phonetic segments, speech perception, too, is abstract. In listening to the syllable [*ba*], for example, one hears the stop consonant as an abstract linguistic event, quite removed from the acoustic and auditory variations that underlie it. He cannot tell that the difference between [*ba*] and [*ga*] in simplified synthetic patterns is only a rising frequency sweep in the second formant of [*b*] compared with a falling frequency sweep in the second formant of [*g*]. But if those frequency sweeps are removed from the syllable context and sounded alone, they are heard as rising and falling pitches, or as differently pitched "chirps," just as our knowledge of auditory psychophysics would lead us to expect. Perception in that auditory mode follows the stimulus in a fairly direct way; in that sense, and in contrast to the perception of speech, it is not abstract.

Perception of the encoded segments of speech is, as a corollary of its abstractness, also categorical. Thus, if we vary a sufficient acoustic cue for [*b, d, g*] in equal steps along a physical continuum, the listener does not hear step-wise changes but more nearly quantal jumps from one perceived category to another. This categorical perception has been measured by a variety of techniques and has been given several different but not wholly unrelated interpretations (Conway and Haggard, 1971; Fry et al., 1962; Fujisaki and Kawashima, 1969; Liberman et al., 1957; Stevens et al., 1969; Vinegrad, 1972; Pisoni, 1971). It characterizes the grammatically encoded segments (e.g. stop consonants), as I have indicated, but not the segments (e.g. the vowels in slow articulation) that are, as I noted earlier, represented in the acoustic signal as if by an alphabet or substitution cipher. Moreover, categorical perception cannot be said to be characteristic of a class of acoustic (and corresponding auditory) events, because the acoustic cues are perceived categorically only when they cue the distinctions among speech sounds; when presented in isolation and heard as nonspeech, their perception is more nearly continuous (Mattingly et al., 1971).

At all events, the grammatically encoded aspects of speech do appear to be perceived in a special mode. That mode is, like the rest of language, abstract, categorical, and, perhaps more generally, nonrepresentational. Does this not present a considerable contrast to the perception of nonverbal material? For all the abstracting that special detector mechanisms may do in vision or hearing, perception in those modes seems nevertheless to be more nearly isomorphic with the physical reality that occasions it. If that is truly a difference between grammatical and nongrammatical perception, it may be yet another reflection of the different organizations of the cerebral hemispheres.

Summary

The aim of this paper is to suggest that the language hemisphere may be specialized to deal with grammatical coding, a conversion of information that distinguishes language from other perceptual and cognitive processes. Grammatical coding is unique, first, in terms of its function, which is to restructure information so as to make it appropriate for long-term storage and (non-linguistic) cognitive processing at the one end of the system and for transmission via the vocal tract and the ear at the other.

To see further how grammatical restructurings are unique, we should look, more narrowly, at the speech code, the connection between phonetic message and sound. There we see a grammatical conversion that produces a special relation between acoustic stimulus and perception: Information about successive segments of the perceived phonetic message is transmitted simultaneously on the same parameter of the sound. On that basis we can tentatively distinguish that which is grammatical or linguistic from that which is not. Then, by taking advantage of recently developed experimental techniques, we can discover to what extent our hypothesis about hemispheric specialization is correct and how it needs to be modified.

The speech code is unique in still other ways that may be correlates of the special processes of the language hemisphere. Thus, the speech code requires a special key. To understand the relation between acoustic stimulus and perceived phonetic message, we must take account of the manner in which the sound was produced. Conceivably, the language hemisphere is specialized to provide that "understanding" by making available to the listener the appropriate articulatory model.

The speech code is unique, too, in that it is associated with a special mode of perception. In that mode, perception is categorical, digital, and most generally, nonrepresentational. Perhaps these perceptual properties reflect the specialized processes of the language hemisphere.

ACKNOWLEDGMENT The preparation of this paper, as well as much of the research on which it is based, was aided by grants from the National Institute of Child Health and Human Development and the Office of Naval Research. I am indebted to my colleagues at Haskins Laboratories, especially Franklin S. Cooper, Ignatius G. Mattingly, Donald Shankweiler, and Michael Studdert-Kennedy, for ideas, suggestions, and criticisms. Hans-Lukas Teuber, Brenda Milner, and Charles Liberman have also been very helpful. None of these people necessarily agrees with the views I express here.

REFERENCES

BROADBENT, D. E., 1956. Successive responses to simultaneous stimuli. *Q. J. Exp. Psychol.* 8:145–162.

CHOMSKY, N., 1965. *Aspects of the Theory of Syntax.* Cambridge, Mass.: MIT Press.

COFFEY, J. L., 1963. The development and evaluation of the Batelle aural reading device. *Proceedings of the International Congress on Technology and Blindness I.* New York: American Foundation for the Blind, pp. 343–360.

CONRAD, R., 1972. Speech and reading. In *Language by Ear and by Eye,* J. F. Kavanagh and I. G. Mattingly, eds. Cambridge, Mass.: MIT Press, pp. 205–240.

CONWAY, D. A., and M. P. HAGGARD, 1971. New demonstrations of categorical perception. In *Speech Synthesis and Perception Progress Report No. 5.* Cambridge, England: Cambridge University Psychological Laboratory, pp. 51–73.

COOPER, F. S., 1966. Describing the speech process in motor command terms. *J. Acoust. Soc. Amer.* 39:1121-A. (Text in *Status Report on Speech Research* SR-5/6. New Haven: Haskins Laboratories, pp. 2.1–2.27.)

CUTTING, J. E., 1972. A parallel between encodedness and the magnitude of the right ear effect. In *Status Report on Speech Research* SR-29/30. New Haven, Conn.: Haskins Laboratories, pp. 61–68.

DARWIN, C. J., 1971. Ear differences in recall of fricatives and vowels. *Q. J. Exp. Psychol.* 23:46–62.

EFRON, R., 1965. The effect of handedness on the perception of simultaneity and temporal order. *Brain* 86:261–284.

FREIBERGER, J., and E. G. MURPHY, 1961. Reading machines for the blind. *IRE Professional Group on Human Factors in Electronics* HFE-2:8–19.

FRY, D. B., A. S. ABRAMSON, P. D. EIMAS, and A. M. LIBERMAN, 1962. The identification and discrimination of synthetic vowels. *Language and Speech* 5:171–189.

FUJISAKI, H., and T. KAWASHIMA, 1969. On the modes and mechanisms of speech perception. In *Annual Report No. 1.* Tokyo: University of Tokyo, Division of Electrical Engineering, Engineering Research Institute, pp. 67–73.

HAGGARD, M. P., 1971. Encoding and the REA for speech signals. *Q. J. Exp. Psychol.* 23:34–45.

KIMURA, D., 1961a. Some effects of temporal-lobe damage on auditory perception. *Can. J. Psychol.* 15:156–165.

KIMURA, D., 1961b. Cerebral dominance and perception of verbal stimuli. *Can. J. Psychol.* 15:166–171.

KIMURA, D., 1964. Left-right differences in the perception of melodies. *Q. J. Exp. Psychol.* 16:355–358.

KIMURA, D., 1967. Functional asymmetry of the brain in dichotic listening. *Cortex* 3:163–178.

LEVY, J., 1969. Possible basis for the evolution of lateral specialization of the human brain. *Nature (Lond.)* 224:614–615.

LIBERMAN, A. M., 1957. Some results of research on speech perception. *J. Acoust. Soc. Amer.* 29:117–123.

LIBERMAN, A. M., 1970. The grammars of speech and language. *Cognitive Psychology* 1:301–323.

LIBERMAN, A. M., F. S. COOPER, D. P. SHANKWEILER, and M. STUDDERT-KENNEDY, 1967. Perception of the speech code. *Psychol. Rev.* 74:431–461.

LIBERMAN, A. M., P. C. DELATTRE, and F. S. COOPER, 1952. The role of selected stimulus variables in the perception of the unvoiced stop consonants. *Amer. J. Psychol.* 65:497–516.

LIBERMAN, A. M., K. S. HARRIS, H. S. HOFFMAN, and B. C. GRIFFITH, 1957. The discrimination of speech sounds within and across phoneme boundaries. *J. Exp. Psychol.* 54:358–368.

LIBERMAN, A. M., I. G. MATTINGLY, and M. T. TURVEY, 1972. Language codes and memory codes. In *Coding Processes in Human Memory,* A. W. Melton and E. Martin, eds. Washington, D.C.: V. H. Winston, pp. 307–334.

LIEBERMAN, P., 1968. Primate vocalizations and human linguistic ability. *J. Acoust. Soc. Amer.* 44:1574–1584.

LIEBERMAN, P., 1969. On the acoustic analysis of primate vocalizations. *Behav. Res. Meth. Instrument.* 1:169–174.

LIEBERMAN, P., and E. S. CRELIN, 1971. On the speech of Neanderthal man. *Linguistic Inquiry* 2:203–222.

LIEBERMAN, P., E. S. CRELIN, and D. H. KLATT, 1972. Phonetic ability and related anatomy of the newborn and adult human, Neanderthal man and chimpanzee. *Amer. Anthropol.* 74:287–307.

LIEBERMAN, P., D. H. KLATT, and W. A. WILSON, 1969. Vocal tract limitations of the vocal repertoires of Rhesus monkey and other nonhuman primates. *Science* 164:1185–1187.

LISKER, L., F. S. COOPER, and A. M. LIBERMAN, 1962. The uses of experiment in language description. *Word* 18:83–106.

MATTINGLY, I. G., 1968. Synthesis by rule of General American English. In *Supplement to Status Report on Speech Research.* New Haven: Haskins Laboratories, pp. 1–223.

MATTINGLY, I. G., 1971. Synthesis by rule as a tool for phonological research. *Language and Speech* 14:47–56.

MATTINGLY, I. G., 1972. Speech cues and sign stimuli. *Amer. Sci.* 60:327–337.

MATTINGLY, I. G., and A. M. LIBERMAN, 1969. The speech code and the physiology of language. In *Information Processing in the Nervous System,* K. N. Leibovic, ed. New York: Springer-Verlag, pp. 97–117.

MATTINGLY, I. G., A. M. LIBERMAN, A. K. SYRDAL, and T. HALWES, 1971. Discrimination in speech and nonspeech modes. *Cognit. Psychol.* 2:131–157.

MILNER, B., 1954. Intellectual functions of the temporal lobe. *Psychol. Bull.* 51:42–62.

MILNER, B., 1958. Psychological defects produced by temporal lobe excision. *Proc. Assoc. Res. Nerv. Ment. Disord.* 36:244–257.

MILNER, B., 1967. Brain mechanisms suggested by studies of temporal lobes. In *Brain Mechanisms Underlying Speech and Language,* F. L. Darley, ed. New York: Grune and Stratton, pp. 122–132.

NYE, P., 1968. Research on reading aids for the blind—a dilemma. *Med. Biol. Eng.* 6:43–51.

ORR, D. B., H. L. FRIEDMAN, and J. C. C. WILLIAMS, 1965. Trainability of listening comprehension of speeded discourse. *J. Educ. Psychol.* 56:148–156.

PISONI, D., 1971. On the nature of categorical perception of speech sounds. Doctoral dissertation, University of Michigan; in *Supplement to Status Report on Speech Research.* New Haven: Haskins Laboratories, pp. 1–101.

SEMMES, J., 1968. Hemispheric specialization: A possible clue to mechanism. *Neuropsychologia* 6:11–27.

SHANKWEILER, D., 1971. An analysis of laterality effects in speech perception. In *The Perception of Language,* D. L. Horton and J. J. Jenkins, eds. Columbus, Ohio: Chas. E. Merrill, pp. 185–200.

SHANKWEILER, D., and M. STUDDERT-KENNEDY, 1967. Identification of consonants and vowels presented to left and right ears. *Q. J. Exp. Psychol.* 19:59–63.

SPERRY, R. W., and M. S. GAZZANIGA, 1967. Language following surgical disconnection of the hemispheres. In *Brain Mechanisms Underlying Speech and Language,* C. H. Millikan and F. L. Darley, eds. New York: Grune and Stratton, pp. 108–121.

STEVENS, K. N., 1960. Toward a model for speech recognition. *J. Acoust. Soc. Amer.* 32:47–55.

STEVENS, K. N., and M. HALLE, 1967. Remarks on analysis by synthesis and distinctive features. In *Models for the Perception of Speech and Visual Form*, W. Wathen-Dunn, ed. Cambridge, Mass.: MIT Press, pp. 88–102.

STEVENS, K. N., A. M. LIBERMAN, S. E. G. OHMAN, and M. STUDDERT-KENNEDY, 1969. Cross-language study of vowel perception. *Language and Speech* 12:1–23.

STUDDERT-KENNEDY, M., and F. S. COOPER, 1966. High-performance reading machines for the blind; psychological problems, technological problems, and status. *Proceedings of Saint Dunstan's Conference on Sensory Devices for the Blind.* London, pp. 317–342.

STUDDERT-KENNEDY, M., and D. SHANKWEILER, 1970. Hemispheric specialization for speech perception. *J. Acoust. Soc. Amer.* 48:579–594.

STUDDERT-KENNEDY, M., D. SHANKWEILER, and D. PISONI, 1972. Auditory and phonetic processes in speech perception:

Evidence from a dichotic study. *Cognit. Psychol.* 3:455–466.

TEUBER, H.-L., 1960. Perception. In *Handbook of Physiology Section 1: Neurophysiology*, Vol. 3, J. Field, ed. Washington, D.C.: American Physiological Society, pp. 1595–1669.

UMILTA, C., N. FROST, and R. HYMAN, 1972. Interhemispheric effects on choice reaction times to one-, two-, and three-letter displays. *J. Exp. Psychol.* 93:198–204.

VINEGRAD, M., 1972. A direct magnitude scaling method to investigate categorical versus continuous modes of speech perception. *Language and Speech* 15: 114–121.

WARREN, R. M., C. J. OBUSEK, R. M. FARMER, and R. T. WARREN, 1969. Auditory sequence: Confusion of patterns other than speech or music. *Science* 164:586–587.

WOOD, C. C., 1973. Levels of processing in speech perception: Neurophysiological and cognitive analyses. Unpublished Ph.D. Dissertation, Yale University.

WOOD, C. C., W. R. GOFF, and R. S. DAY, 1971. Auditory evoked potentials during speech perception. *Science* 173: 1248–1251.

5 Ear Differences and Hemispheric Specialization

C. J. DARWIN

ABSTRACT Recent experiments on ear-difference effects in dichotic listening are critically reviewed, and some problems in their interpretation are pointed out. In particular, the notion of *functional decussation* and the possibly related psychological concept of echoic memory are introduced to account for the failure of particular types of sound to yield reliable ear differences.

FOR ABOUT 10 years now, there has been evidence that in normal right-handed adults verbal stimuli presented to the right ear or to the right half of visual space tend to be reported more accurately than if they had been presented to the left ear or to the left half of visual space. We also know that the opposite is true (left better than right) for some nonverbal stimuli. We have good reasons, which I will describe later, for believing that these asymmetries are related to a complementary specialization of the two hemispheres, the left for verbal tasks and the right for nonverbal, the clinical evidence for such specialization being now overwhelming (Sperry, this volume; Milner, this volume). It is obviously very convenient to be able to study these cerebral asymmetries in normal intact subjects, and such experiments give valuable corroboration of the conclusions drawn from the clinical material. Such corroboration is particularly important, because the most reliable clinical evidence is based on patients who have a history of early brain damage and who consequently could have an abnormal distribution of function between the two hemispheres. However, the convenience of studying normal subjects is not bought cheaply. The price we pay is that we must make many more assumptions in interpreting these data than is necessary for interpreting at least some of the clinical data. The present chapter will try to clarify some of the problems we face in interpreting the data on asymmetries in auditory perception derived from work with normal subjects.

To those familiar with the essentially bilateral projection of the auditory system, it may come as some surprise that it is possible to reveal asymmetries in cortical function by differences in performance for the two ears.

C. J. DARWIN Laboratory of Experimental Psychology, University of Sussex, Brighton, England

Unlike the visual system, where each hemisphere receives direct sensory input from only the contralateral half of space, the afferent fibers of the auditory system project from either cochlea to both hemispheres. Nevertheless, it is clear from electrophysiological work in the cat with both macroelectrodes (Rosenzweig, 1951) and microelectrodes (Hall and Goldstein, 1968) that rather more cells fire to the contralateral ear than to the ipsilateral. When a sound is played into one ear, there will thus be a slight tendency for the hemisphere contralateral to that ear to respond more than the ipsilateral. This tendency is probably not sufficient, though, for differences in hemisphere function to be detected reliably from differences in the way sounds entering the two ears are treated; there have, however, been isolated reports of ear differences being obtained when only one ear is stimulated at a time. In order to obtain reliable differences between the ears, it is usually necessary to play two different messages simultaneously, one into each ear. This technique was originally used in a different context by Broadbent (1954), but the discovery that it could be used to detect differences between the ears was made by Kimura (1961a, 1961b). Kimura showed that the right ear was more efficient for the recall of digits, whereas the left ear became superior when nonverbal material such as music was used (Kimura, 1964). Her claim that this ear difference was due to functional differences between the two hemispheres was supported by evidence that subjects known (from the Wada Sodium-Amytal test) to have reversed dominance for speech also showed reversed ear-advantages.

One of the questions that these experiments raised was the nature of the distinction between stimuli that gave an advantage for the right ear and those that gave an advantage for the left. An experiment that I performed showed that the left ear can be superior even for a "verbal" stimulus, provided that the task the subject has to perform is not linguistically significant. In this experiment (Darwin, 1969) the subjects had to report which two of four possible pitch contours had been played to the two ears. These four contours were rising, falling, rising-falling and falling-rising and were quite short in duration (225 msec). The pitch contours were carried on a number of different timbres, one of which was the word *tea*. This condition gave just as large an advantage for the left ear as other conditions in which the pitch contours were carried on a nonverbal sound.

Another question raised by Kimura's experiments is why dichotic presentation is so much more effective at revealing hemispheric differences than is monotic presentation. Rosenzweig (1951) had suggested on the basis of his evoked-potential studies on the cat that the ipsilateral population might be contained within the contralateral, so that very few cells respond only to the ipsilateral stimulus. Kimura proposed that if this were the case, the contralateral population might occlude the ipsilateral when they were in competition for the same cells. These two suggestions have been supported by later work. Hall and Goldstein (1968) showed directly, using microelectrodes, that Rosenzweig's speculation was substantially correct, and a study by Milner, Taylor, and Sperry (1968) on commissurotomized patients showed very clearly that the contralateral pathway dominates the ipsilateral very much more under dichotic presentation than under monaural (see Milner, this volume). Commissurotomized patients were able to report monaurally presented digits perfectly well from either ear but, with dichotic presentation, could report few or none from the left ear, although continuing to report those from the right ear without difficulty. This interesting finding was confirmed by a paper on one additional commissurotomized patient by Sparks and Geschwind (1968). Their paper adds a further point: Their subject was able to report progressively more of the speech signal on the left ear as the speech signal on the right became more and more distorted.

This latter finding suggests that the extent to which the ipsilateral pathways are occluded by the contralateral may depend on the types of sound used at the two ears. A related experiment by Baru amplifies this point. Baru (1966) found in electrophysiological experiments on the cat that the preponderance of the contralateral over the ipsilateral pathways was much greater for brief sounds than for long ones. These two experiments lead us to suppose that there are at least two variables that can influence the extent to which the ipsilateral input is occluded. The first, from Baru's experiment in the cat, is the nature of the sound itself, so that the longer the sound, the less is the ipsilateral occlusion; the second, from the Sparks and Geschwind paper, is the relationship between the sounds to the two ears, the ipsilateral pathway being occluded more when similar sounds are played dichotically than when very different ones are. Let us now introduce the term *functional decussation* to refer to the extent to which the ipsilateral input is occluded by the contralateral. Although the Sparks and Geschwind study

used a commissurotmized patient, we do not intend to restrict this term to the direct auditory pathway from the ipsilateral ear but wish also to include the indirect input from the contralateral hemisphere by way of the corpus callosum.

In any experiment that yields a difference in performance between the two ears, the magnitude of ear difference obtained will in general depend on two things. First, and most interestingly, it will depend on the relative contribution that the two hemispheres make to the task the subject is performing; but the ear difference will also depend on the amount of functional decussation of the auditory pathways. If there were no functional decussation, it would be impossible to obtain any ear difference no matter what hemispheric differences were present.

I will now turn to discussing how this variable of functional decussation could be responsible for some of the reported changes in the ear-difference effect. In a germinal paper published in 1967, Shankweiler and Studdert-Kennedy reported the finding that under dichotic presentation, the right ear was superior to the left for recall of consonant-vowel-consonant nonsense syllables differing only in the initial stop consonant. However, these authors were unable to detect any significant difference between the ears for the recall of similar syllables differing only in the vowel. Rather than involving differences in the cortical treatment of these two classes of sound, it is possible to suppose that the reason the vowels failed to give any clear indication of superior recall from the right ear is because vowels are considerably longer in duration than stop consonants and so might be expected to show less functional decussation than the brief stops.

A similar interpretative difficulty arises in an experiment of my own (Darwin, 1971). This experiment attempted to dissociate stimulus and response factors as possible determiners of the ear difference. It asked the question whether the ear-difference effect for consonants depended only on the response given to a particular kind of stimulus, or whether the particular acoustic cues that give rise to the response are also important. The question is an interesting one, because the split-brain studies by Sperry and his associates (Gazzaniga and Sperry, 1967) have stressed the virtually complete inability of the right hemisphere to speak, despite the fact that this hemisphere seems to be able to understand speech to a considerable extent. This means that the ear-difference effects for verbal material could be attributable entirely to response factors, providing we assume that the right hemisphere has some facility for emitting vowels but is unable to articulate consonants. I tackled this question by comparing the ear differences for a fixed set of consonants but with varying acoustic cues, hoping to find some change in the ear difference while the response of naming the consonants was held constant. Such a change, if obtained, would leave open the possibility that there are differences between the hemispheres' abilities to perceive speech, as well as the acknowledged differences in outputting the resultant percept.

The stimuli I used in this experiment were fricative consonants, in which there are two main types of acoustic cue. The primary cue that distinguishes fricatives from other consonants is the friction itself. The spectral qualities of this friction also distinguish between fricatives, so that a high-pass filtered noise will change from being appropriate for [ʃ] with the filter set at 1000 Hz to being appropriate for [ʃ] and then [s] as the cut-off is raised. Although the formant transitions into the adjacent vowel provide a secondary cue that can make some contribution to intelligibility (Harris, 1958), these transitions are not

as important a cue as the friction itself. Thus, when the friction appropriate for [ʃ] is paired with the transition appropriate for [s], the impression will generally be of [ʃ].

Figure 1 gives stylized spectrograms of the fricative [ʃ] in the various forms used in my experiment. In the upper part (Condition 1) both the friction and formant transitions are present and the fricative is followed by a final syllable, to give the nonsense word *fep*. In Condition 2, the formant transitions have been removed and the vowel extended into the time slots previously occupied by the transitions. This sound is still an intelligible version of *fep* though slightly less intelligible than the version in Condition 1. In Condition 3 of Figure 1 the vowel has been removed altogether, leaving just the friction. If ear differences were determined simply by response or output mechanisms, then we should not expect to find any

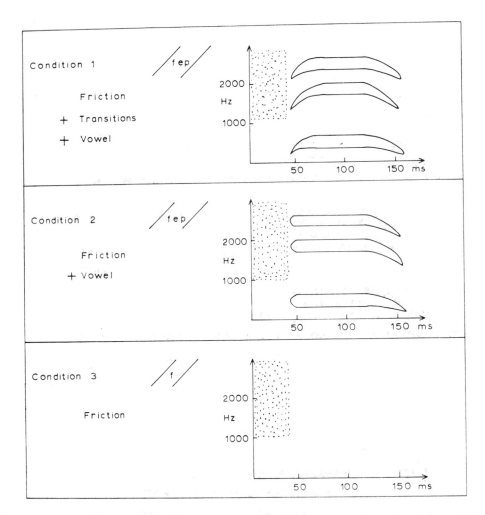

FIGURE 1 Stylized spectrograms of the stimuli used in fricatives experiment. Conditions 1 and 2 are identical except that Condition 1 has appropriate formant transitions from the friction into the succeeding vowel, whereas Condition 2 has no transitions but an abrupt change into the vowel. Both these conditions are an acceptable version of [*fep*], although Condition 1 is slightly more intelligible. Condition 3 has the vowel and final stop removed and is an acceptable version of [ʃ].

difference in the ear advantage for these three stimulus conditions, because in each case the fricative is being identified by the same response. In fact, I found that there was an advantage for the right ear only in the first stimulus condition, the one with the formant transitions. This result is shown in Figure 2. Performance on the left ear is essentially the same across all three stimulus conditions, but performance on the right ear is significantly better in Condition 1 than it is in either Condition 2 or Condition 3. Clearly then, the stimulus conditions are important in determining whether there will be an advantage for the right ear or not, so that the ear difference effect cannot be entirely attributed to response mechanisms.

But we are still faced with a dilemma. If we want to extend our conclusion from an ear difference to a *hemisphere* difference, we must be sure that the auditory decussation was adequate for Conditions 2 and 3, which did not show any ear differences. The information that the left hemisphere obtains from the left ear, either along the direct ipsilateral pathway or via the callosum, may be sufficient to enable it to perceive steady state friction but not the rapidly changing formant transitions.

Here it might be appropriate to mention some results from a very different sort of experiment, which may be related to this problem. The amount of information that a particular hemisphere can obtain from either ear may depend not only on the amount of functional decussation but also on the amount of time it has in which to obtain that information. It is a common experience that if you do not perceive a sound when it arrives because your attention was elsewhere (reading a novel, when called to the table), provided you switch attention soon enough you can still get the message even though you switch after the sound has ended. For this, and many other more substantial reasons (see for example Guttman and Julesz, 1963; Crowder and Morton, 1969; Darwin, Turvey, and Crowder, 1972; Glucksberg and Cowen, 1970), experimental psychologists find useful the concept of an *echoic memory* (Neisser, 1967) in which auditory stimuli are held in a relatively crude uncategorized form for perhaps 1 or 2 sec after their arrival. From what we know of this type of memory, it seems to be rather like a tape recording that gets progressively more and more distorted as time passes (Baddeley and Darwin, forthcoming). Some recent work by Crowder (1971) has shown that stop consonants are rather less well represented in echoic memory than are vowels, as one would expect on the distorted tape-recording analogy. Crowder (in press) also showed that fricative consonants, identical to the ones that I used in

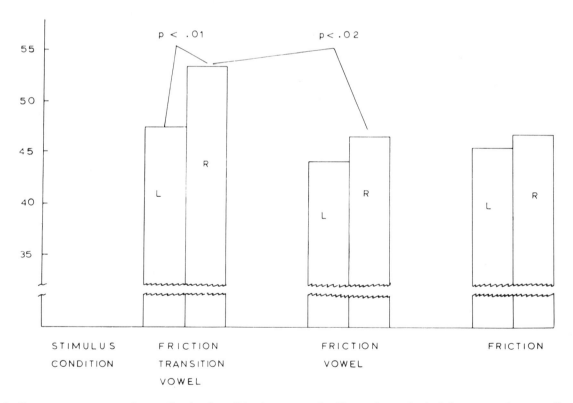

FIGURE 2 Percent correct scores by ear for the three fricative conditions in Figure 1 presented dichotically. There is a significant right-ear advantage only for Condition 1. There is no significant change in the left-ear score between Conditions 1 and 2, but there is a significant change in the right-ear score.

my dichotic experiment, are rather better preserved in echoic memory than are stop consonants, irrespective of whether they contain formant transitions or not, although they are not as well preserved as vowels. The frictional component of the fricatives may thus be adequately preserved in echoic memory while the rapid formant transitions quickly become valueless with the progressing degradation.

If we imagine that this echoic memory is something to which both hemispheres have access (although they may have better access to that part representing the contralateral half of auditory space), then evidently each hemisphere will have much more time to extract the necessary information from the steady-state sounds of vowels and friction than from rapid formant transitions. This added time could compensate the less able hemisphere and make it more difficult for us to detect its relative incompetence.

Some independent support for this kind of notion comes from an experiment by Weiss and House (1970), who find that the right ear becomes significantly better than the left for dichotically presented vowels when noise is added to them. This already distorted signal may have a shorter life in echoic memory and thus tax the right hemisphere more. It may also be significant in this context to point out that I found it much harder to obtain a significant advantage for the *left* ear for slowly moving pitch contours than for relatively rapid ones (Darwin, 1969).

An experiment that goes some of the way towards answering the question whether changes of functional decussation and life in echoic memory are responsible for the failure of some sounds to give ear differences is one I recently performed on vowels (Darwin, 1971). I attempted to show with this experiment that the very same stimuli in a dichotic pair may or may not give an ear difference depending on the complexity of the perceptual discrimination which the subject has to perform on them. Any factors that are simply a matter of the stimulus can be ruled out as contributing to the ear difference here, because exactly the same stimuli are compared. This experiment took advantage of the finding by Haggard (1971) that when a subject does not know which of two talkers will produce a particular vowel, there is an advantage for the right ear. This suggests that the process of vocal-tract normalization (Ladefoged and Broadbent, 1957) is carried out better in the left hemisphere than in the right. My experiments used two different groups of subjects. The first group listened dichotically to a set of vowels that came from only one (synthetic) talker. These subjects showed no advantage for the right ear. The second group of subjects heard another tape on which were the original trials that the first group heard, mixed

in randomly with more trials that used vowels from another synthetic talker, so that the subject could not tell which talker would appear on which ear on any trial. These subjects did show an advantage for the right ear overall, and they also showed a right-ear advantage on those trials which were the same as the ones that the first group took (Figure 3). Exactly the same sounds are here giving a different ear advantage, depending on the other sounds present in the experiment. This evidence strengthens the interpretation that there are true perceptual differences between the hemispheres but is still not quite immune from the echoic memory objection that I raised earlier.

It is unlikely of course that the nature of the perceptual discrimination can influence the rate at which a signal is degraded in echoic memory, but it may be that in order to perform well in the condition with two vocal tracts, more detailed information about the acoustic properties of the vowel is needed than when there is only one vocal tract. If this is the case, then a given amount of degradation may have affected the performance of the two-vocal-tract group more than that of the one-vocal-tract group. This possibility could be tested by comparing the ear difference for subjects listening to a set of, say, eight vowels from one vocal tract with that of a group listening to four vowels from two vocal tracts. This would distinguish the discriminability of the stimulus set from the particular perceptual operation of compensating for vocal-tract size.

In this final section I turn to a rather different area of dichotic-listening experiments. There is now some interesting evidence that linguistically more complex processes than those of phonetic perception may be able to mediate ear differences. Zurif and Sait (1970) played strings of nonsense words dichotically to their subjects. In one condition, the nonsense words had bound morphemes attached in such a way that the string sounded syntactically reasonable, although semantically quite anomalous (e.g. "The wak jud shendily"). In another condition, they used the same nonsense words but without the syntactic structuring. Their subjects recalled more nonsense words from the right ear than the left in the syntactically structured condition but not in the unstructured condition, suggesting that the left hemisphere is better than the right at using syntactic structure to organize a memorandum.

There are also some experiments (Jarvella, Herman, and Pisoni, 1970) that claim to obtain significant advantages for the right ear for the recall of real sentences presented *monaurally* either in noise or without noise but with a period of distraction before recall was allowed. Perhaps perceiving and remembering a real sentence uses much more the specialities of the left hemisphere

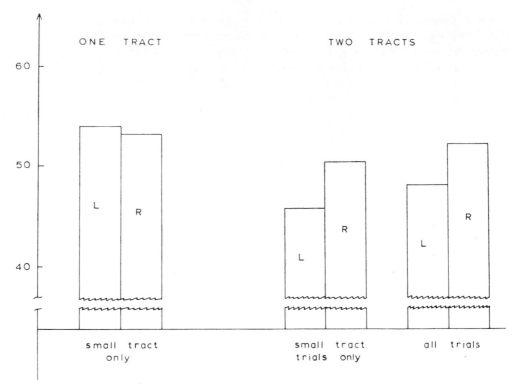

FIGURE 3 Percent correct scores by ear for dichotically presented vowels. The "one-tract" scores were obtained when there was only one speaker in the experiment. The "two tracts" scores when there were two. There is a significant right-ear advantage for the latter, even when only the stimuli used in the "one-tract" experiment are considered. There is no right-ear advantage for these stimuli when they are the only ones present in the experiment.

than does the more restricted task of simply perceiving phonemes, and so ear differences can be obtained even without recourse to dichotic stimulation. The possibility of being able to find correlates of syntactic and semantic variables in the differential specialization of the two hemispheres is an intriguing one that would repay careful study.

REFERENCES

BARU, A. V., 1966. On the role of the temporal parts of the cerebral cortex in the discrimination of acoustic signals of different duration. *Zh. Vyssh. Nerv. Deiat.* 16:655–665, abstracted in *Dsh. Abstracts* 7:219 (1967).

BROADBENT, D. E., 1954. The role of auditory localization in attention and memory span. *J. Exp. Psychol.* 47:191–196.

CROWDER, R. G., 1973. The representation of speech sounds in precategorical acoustic storage (PAS). *J. Exp. Psychol.* (in press).

CROWDER, R. G., and J. MORTON, 1969. Pre-categorical acoustic storage (PAS). *Percept. Psychophys.* 5:365–373.

CROWDER, R. G., 1971. The sound of vowels and consonants in immediate memory. *J. Verb. Learn. Verb. Behav.* 10:587–596.

DARWIN, C. J., 1969. Auditory perception and cerebral dominance. Unpublished Ph.D. Thesis, University of Cambridge, Cambridge, England.

DARWIN, C. J., 1971. Ear differences in the recall of fricatives and vowels. *Q. J. Exp. Psychol.* 23:46–62.

DARWIN, C. J., TURVEY, M. T., CROWDER, R. G., 1972. An auditory analogue of the Sperling partial report procedure: Evidence for brief auditory storage. *Cognit. Psychol.* 3:255–267.

GAZZANIGA, M. S., and R. W. SPERRY, 1967. Language after section of the cerebral commissures. *Brain* 90:131–148.

GLUCKSBERG, S., and G. N. COWEN, 1970. Memory for non-attended auditory material. *Cognit. Psychol.* 1:149–156.

GUTTMAN, N., and B. JULESZ, 1963. Lower limits of periodicity analysis. *J. Acoust. Soc. Amer.* 35:610.

HAGGARD, M. P., 1971. Encoding and the REA for speech signals. *Q. J. Exp. Psychol.* 23:34–45.

HALL, J. L., and M. H. GOLDSTEIN, 1968. Representations of binaural stimuli by single units in primary auditory cortex of unanesthetized cats. *J. Acoust. Soc. Amer.* 43:456–461.

HARRIS, K. S., 1958. Cues for the discrimination of American English fricatives in spoken syllables. *Language and Speech* 1:1–17.

JARVELLA, R. J., S. J. HERMAN, and D. B. PISONI, 1970. Laterality factors in the recall of sentences varying in semantic content. *J. Acoust. Soc. Amer.* 48:84(A).

KIMURA, D., 1961a. Some effects of temporal lobe damage on auditory perception. *C. J. Psychol.* 15:156–165.

KIMURA, D., 1961b. Cerebral dominance and the perception of verbal stimuli. *Can. J. Psychol.* 15:166–171.

KIMURA, D., 1964. Left-right differences in the perception of melodies. *Q. J. Exp. Psychol.* 14:355–358.

LADEFOGED, P., and D. E. BROADBENT, 1957. Information conveyed by vowels. *J. Acoust. Soc. Amer.* 29:98–104.

MILNER, B., L. TAYLOR, and R. W. SPERRY, 1968. Lateralized suppression of dichotically presented digits after commissural section in man. *Science* 161:184–186.

NEISSER, U., 1967. *Cognitive Psychology*. New York: Appleton-Century-Croft.

ROSENZWEIG, M. R., 1951. Representations of the two ears at the auditory cortex. *Amer. J. Physiol.* 167:147–158.

SHANKWEILER, D. P., and M. STUDDERT-KENNEDY, 1967. Identification of consonants and vowels presented to left and right ears. *Q. J. Exp. Psychol.* 19:59–63.

SPARKS, R., and N. GESCHWIND, 1968. Dichotic listening in man after section of neocortical commissures. *Cortex* 4:3–16.

WEISS, M. J., and A. S. HOUSE, 1970. Perception of dichotically presented vowels. *J. Acoust. Soc. Amer.* 49:96(A).

ZURIF, E. B., and P. E. SAIT, 1970. The role of syntax in dichotic listening. *Neuropsychologia* 8:239–243.

6 Cerebral Dominance and Interhemispheric Communication in Normal Man

G. BERLUCCHI

ABSTRACT Most of the evidence on hemispheric interaction and specialization in man comes from studies of patients with unilateral brain damage or with split cerebral commissures. This paper presents some results of an alternative approach to the problem: The analysis of hemispheric differences and processes of interhemispheric communication in normal man by measuring reaction time to lateralized visual stimuli. Data are presented to support the hypothesis that certain laterality effects in this task are at least in part related to hemispheric asymmetries and to commissural functions.

BEHAVIORAL studies in animals (Sperry, 1961; Myers, 1961) and man (Sperry, 1968) have clearly indicated that sectioning the forebrain commissures prevents or greatly interferes with the normal bihemispheric utilization for visual perception and learning of visual information channeled into one single hemisphere. These remarkable behavioral findings have prompted a series of anatomical and physiological investigations dealing with the neural bases of the exchange of visual information between the cerebral hemispheres of several animal species (see reviews in Berlucchi, 1972; Cuénod, 1972). A similar analysis of the visual functions of the forebrain commissures in man would be of considerable interest, particularly in view of the fact that the functional asymmetry between the right and left cerebral hemispheres in several visual tasks—a peculiar, if not exclusive feature of the human brain—should at least in principle be associated with highly specialized and perhaps unique mechanisms of interhemispheric communication of visual information. Since obvious ethical considerations preclude the use of the most basic neurophysiological methods in the normal subject, the approach to the study of visual interhemispheric differentiation and interaction in man has necessarily been rather indirect. In particular, attention has been largely centered on the so-called laterality effects in visual perception, which have been known to experi-

mental psychologists for several years (see review in White, 1969). In the present chapter, I describe a special class of these laterality effects, i.e., certain right-left differences in discriminative reaction time to lateralized visual stimuli and discuss their possible relation to the factors of hemispheric specialization and commissural functions in vision.

Opposite right-left visual field superiorities for discriminative reaction time to alphabetical and physiognomical material

According to the laws of physiological optics and the pattern of retinocerebral connections, stimuli lying in the left part of the visual field (that is, to the left of the fixation point) are projected to the right halves of the retinae and to the right hemisphere, whereas stimuli lying in the right part of the visual field (that is to the right of the fixation point) are projected to the left halves of the retinae and to the left hemisphere. If a visual stimulus is tachistoscopically presented on one side of the fixation point, the exposure time being too brief to allow foveation of the stimulus, visual information is restricted to the hemisphere contralateral to the hemifield of stimulation. By using this technique, Sperry and his associates (see Sperry, 1968) have demonstrated that, in the split-brain subject, stimuli presented in one visual hemifield are perceived and responded to correctly only by the contralateral hemisphere, the ipsilateral hemisphere being practically unaware of the occurrence and nature of the stimulus.

Since this gross dissociation is obviously not the case in the normal subject, it means that visual information selectively channeled into one hemisphere of the intact brain is in some way made available to the opposite hemisphere by the forebrain commissures. Neurophysiologists, in general, believe that all exchanges of information within the brain must involve the conduction

G. BERLUCCHI Istituto di Fisiologia dell'Università di Pisa e Laboratorio di Neurofisiologia del CNR, Pisa, Italy

of nervous impulses along neuronal axons and their transmission across synapses. Since these processes take time, it follows that a measurable amount of time must intervene between the arrival of visual information to the directly stimulated hemisphere and the reception by the other hemisphere of the corresponding information, transmitted via the commissures.

Think now of a task in which the subject is asked to produce, as fast as possible, a discriminative response to visual patterns presented in one hemifield or the other. Suppose also that the recognition of these patterns is subserved by neural activities taking place exclusively or predominantly in one hemisphere. A logical hypothesis would predict that the stimuli presented in the visual field projecting to the specialized hemisphere should be responded to faster than stimuli presented in the other visual field. Indeed, in the first case, the visual input is directly received and processed by the specialized hemisphere, while in the latter case it has to cross from the nonspecialized hemisphere to the other side of the brain for analysis.

Rizzolatti et al. (1971) submitted this hypothesis to an experimental control in a study in which one group of subjects was tested for choice reaction time to lateralized presentations of single capital letters, and another group of subjects was tested for choice reaction time to lateralized presentations of photographs of faces of unknown persons. Each one of the two groups consisted of 12 right-handed young male subjects; the task was to press a key as fast as possible following the appearance of positive stimuli and not to press the key following the appearance of negative stimuli. There were four discriminative stimuli (two positive and two negative) for both groups; each stimulus was presented for 100 msec, 5 degrees to the right or the left, and on a level with the fixation point. The stimuli are shown at 2/3 of their actual size in Figure 1; they were viewed monocularly by the subjects from a distance of 50 cm. In both groups, half of the subjects

used the right eye and half the left eye throughout the testing. Each subject was tested during four sessions and each session consisted of four blocks of 35 (letters) and 41 (faces) trials. The four blocks corresponded to all possible hand/visual field combinations: Right-field–right-hand; right-field–left-hand; left-field–left-hand; left-field–right-hand. In each block positive and negative stimuli were presented in a quasi-random sequence, and the order of the four experimental conditions was varied over sessions according to a Latin square design; further, the sequence of conditions was different for each subject in any one group, but the pattern for a subject in one group was in every case matched by that of a subject in the other.

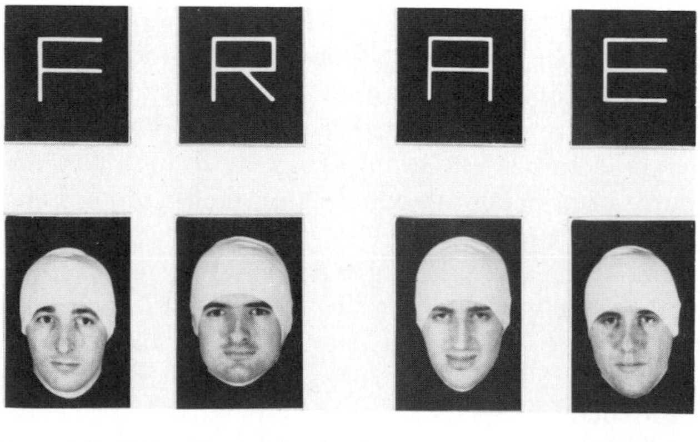

The clinical evidence indicates that left-hemisphere lesions interfere with verbal abilities, leading in general to various forms of aphasia, and right-hemisphere lesions interfere with the apprehension of complex configurational properties of visual stimuli, leading sometime to the inability to recognize faces (see Milner, 1971). Accordingly, one would expect that in our experimental conditions normal subjects would show a reaction time superiority of the right visual field, subserving the left hemisphere, over the left visual field, when the stimulus is a letter, and a reaction-time superiority of the left visual field, subserving the right hemisphere, over the right visual field, when the stimulus is a face.

These opposite visual field superiorities are clearly borne out by the results presented in Table I. In the letter group, average reaction time was 18.5 msec faster for the right field, whereas in the face group average reaction time 14.5 msec faster for the left field. These differences, tested by a combined analysis of variance, were highly significant ($p < 0.005$ in both cases). In neither group were there significant differences in reaction time between the hands or between the various hand-field situations, nor could the distribution of the few errors of omission or commission be correlated with any one of the experimental variables, including the side of stimulation.

Are the right-left visual field asymmetries in reaction time related to hemispheric differences and interhemispheric communication?

The results are in complete accord with the hypothesis, but before regarding these differences in discriminative reaction time between the visual fields as the direct expression of an interhemispheric transfer of visual information, some considerations are in order.

First of all, it would be totally erroneous to think that the slower speed of response to stimuli in the subordinate field can be entirely accounted for by the time lost for impulse conduction along commissural fibers. Since the

FIGURE 1 Stimuli used in the letter and face discrimination tasks. (From Rizzolatti et al., 1971.)

TABLE I

Average reaction time for the letter discrimination task and the face discrimination task, as a function of the side of stimulation, the responding hand, and the four interactions between these two factors. Only the side of stimulation is a significant source of variance

	Side of Stimulus*		Responding Hand*		Interaction Side of Stimulus/ Responding Hand*			
					Right field Right hand	Right field Left hand	Left field Left hand	Left field Right hand
	Right	Left	Right	Left				
Letter discrimination	431.5	450	441.5	440	431	432	448	452
Face discrimination	610	594.5	602	602.5	606	614	591	598

*Reaction time in msec.

information-carrying capacity is likely to be less for the commissural input than for the direct visual input, it follows that the information arriving at the specialized hemisphere from the contralateral visual field is not only faster but also more precise and detailed than the one issuing from the ipsilateral visual field. Hence both factors—loss of time and loss of information—may at least in principle contribute to the differences in reaction time between the visual fields. [Throughout this chapter I have used the word *information* in the lay sense rather than the technical sense. Thus, when I say that transmission of a visual message from one hemisphere to the other across the commissures implies a loss of information, I simply mean that the original message is deteriorated during the interhemispheric transfer.]

But other explanations for these differences are also at hand. Kinsbourne (1970) believes that laterality effects in perception can be attributed to an attentional bias toward the field subserving the specialized hemisphere, rather than to a more efficient transmission of information by the shorter pathway to this hemisphere. According to his model, a subject performing a visual task that calls for the special ability of one hemisphere would undergo a higher degree of activation in that same hemisphere, and this would in turn result in a facilitation of the appropriate input from the contralateral visual field. Further, even when the visual task would not produce an asymmetrical hemispheric activation by itself, and thus no laterality effects in perception should ensue, such effects may be brought about by asking the subject to perform simultaneously the task and some other activity that selectively depends on one particular hemisphere. In a later paper, Kinsbourne (1972) has corroborated his proposition by showing that mental operations presumably requiring a selective right- or left-hemisphere activation produce an ocular deviation toward the visual field contralateral to the activated hemisphere.

Accordingly, our results could be explained by the assumption of an orientational bias, based on both a

process of selective attention and, possibly, an uncontrolled shift in gaze, toward the right visual field in the subjects performing the letter discrimination, and by a similar bias in the reverse direction in the subjects performing the face discrimination. The nature of the tasks is in fact such that the left hemisphere (so the argument runs) would be preactivated in the first group and the right hemisphere in the second group of subjects.

We cannot be sure that our subjects did not move their eyes toward one particular visual field during the presentation of the stimulus, but we have preliminary evidence indicating that the explanation proposed by Kinsbourne cannot account for our results. The findings are straightforward: The superiority in choice reaction time of the right field for letters and of the left field for faces could still be observed when these two types of lateralized stimuli were presented, intermixed in a random sequence, to the same subject in the same experiment. A third type of stimulus, a square patch of light, which was also randomly intermixed with the other two types of stimuli and had to be responded to in all cases, produced no field differences. Since on no trial did the subjects know what kind of stimulus was going to appear on the screen, the differences in reaction time between the visual fields cannot be the result of a hemispheric preactivation set up by prior knowledge of the nature of the stimulus. Instead, this difference must have been at least in part caused by a substantial inequality in stimulus analysis between the hemispheres and in information transmission between the direct and the commissural inputs.

Right-left visual field differences in choice reaction time to line orientation

White (1971) has questioned the hypothesis that the right-field superiority for letter recognition is due to the lateralization of the speech centers to the left hemisphere. He has tested the ability to report the orientation of a thin line flashed for 20 to 25 msec, 3 degrees to the left or the right of the fixation point. Four angles of orientation relative to the vertical were employed: 0, 45, 90, and 135 degrees. Then 47.4% of stimulus orientations were correctly identified in the left visual field and 84.2% in the right visual field. This right-field superiority was significantly correlated with the right-field superiority for single letter recognition observed in the same subjects. From these results White concludes that both the right-field superiority for recognition of letters and the right-field superiority for recognition of line orientations "are explicable in terms of a selective contour-tuning apparatus which favors stimuli (at a peripheral retinal or central level) shown in the right visual hemifield rather than in terms of the superior functioning of the left-dominant language areas."

The finding that the left hemisphere is dominant for the perception of line orientation is a bit perplexing, in view of the clinical evidence that it is the patients with *right* hemisphere damage that are selectively impaired on a rod orientation test (see De Renzi et al., 1971). Yet, Umiltà et al. (1973) have obtained results quite comparable to those of White, by measuring choice reaction time to single rectangles presented in the right and left fields with the four orientation also used by White. The experiment was carried out with the same technique as described above for the letter discrimination and face discrimination experiment. The size of the rectangle was 5.5 × 1 cm. Reaction time was significantly faster with the stimulus in the right visual field (see Figure 2A). But in two other similar experiments, in which the orientation of the line stimuli was changed to, respectively, 30, 45, 120, and 135 degrees from the vertical in one experiment, and 15, 45, or 60 degrees from the vertical in the other experiment, there was a clear shift toward a right-hemisphere superiority, which reached a large significance in the second experiment (see Figure 2B and 2C).

We believe that these results indicate that the vertical, horizontal, and intermediate orientations are perceived and responded to faster by the left hemisphere exactly because these orientations are easily analyzed and categorized in verbal terms (e.g. vertical, horizontal, tilted to the right, tilted to the left). On the contrary, with the other orientation discriminations, it is difficult to encode the orientation of each particular stimulus by itself, and probably it is necessary to proceed by an internal comparison between the orientation of the present stimulus and those of the other discriminative stimuli. Under these conditions, the superior ability of the right hemisphere in analyzing spatial relationships would emerge and reverse the visual field asymmetry.

ACKNOWLEDGMENT The experiments reported here were carried out with Drs. R. Camarda, C. Franzini, C. A. Marzi, G. Rizzolatti, C. Umiltà, and G. Zamboni, to all of whom I express my gratitude for their advice and criticism. The preparation of this manuscript has been aided by the Consiglio Nazionale delle Ricerche Contratto N.70.01687/18.

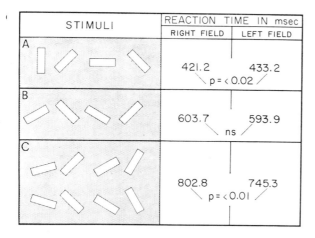

FIGURE 2 Discriminative reaction time to line orientation. Twelve subjects performed the discrimination task shown in A, 18 the discrimination task shown in B, and 12 the discrimination task shown in C. With each task, the two stimuli on the left were positive for half of the subjects, and the two stimuli on the right were positive for the other subjects. In C, for symmetry reasons the four stimuli at the top were presented in the left visual field, and the four stimuli at the bottom were presented in the right visual field. Average time of correct responses as a function of side of stimulation is shown at right for each task.

REFERENCES

BERLUCCHI, G., 1972. Anatomical and physiological aspects of visual functions of corpus callosum. *Brain Res.* 37:371–392.

CUÉNOD, M., 1972. Split-brain studies. Functional interaction between bilateral central nervous structures. In *The Structure and Function of Nervous Tissue*, Vol. 5, G. H. Bourne, ed. New York: Academic Press.

DE RENZI, E., P. FAGLIONI, and G. SCOTTI, 1971. Judgement of spatial orientation in patients with focal brain damage. *J. Neurol. Neurosurg. Psychiat.* 34:489–495.

KINSBOURNE, M., 1970. The cerebral basis of lateral asymmetrics in attention. *Acta Psychol.* 33:193–201.

KINSBOURNE, M., 1972. Eye and head turning indicates cerebral lateralization. *Science* 176:539–541.

MILNER, B., 1971. Interhemispheric differences in the localization of psychological processes in man. *Brit. Med. Bull.* 27:272–277.

MYERS, R. E., 1961. Corpus callosum and visual gnosis. In *Brain Mechanisms and Learning*, J. T. Delasfresnaye, ed. Oxford: Blackwell, pp. 481–505.

RIZZOLATTI, G., C. UMILTÀ, and G. BERLUCCHI, 1971. Opposite superiorities of the right and left cerebral hemispheres in discriminative reaction time to physiognomical and alphabetical material. *Brain* 94:431–442.

SPERRY, R. W., 1961. Cerebral organization and behavior. *Science* 133:1749–1757.

SPERRY, R. W., 1968. Mental unity following surgical disconnection of the cerebral hemispheres. *Harvey Lect.* 62: 273–323.

UMILTÀ, C., G. RIZZOLATTI, C. A. MARZI, G. ZAMBONI, C. FRANZINI, R. CAMARDA, and G. BERLUCCHI, 1973. Hemispheric differences in normal human subjects: Further evidence from study of reaction time to lateralized visual stimuli. *Brain Res.* 49:499–500.

WHITE, M. J., 1969. Laterality differences in perception: A review. *Psychol. Bull.* 72:387–405.

WHITE, M. J., 1971. Visual hemifield differences in the perception of letter and contour orientation. *Can. J. Psychol.* 25:207–212.

7 Why Two Brains?

HANS-LUKAS TEUBER

ABSTRACT   Current studies agree on the facts of complementary specialization of man's right and left cerebral hemispheres but leave open the questions of what it is, precisely, that characterizes the special functions of either hemisphere, how the interhemispheric connections act, and whence the specialization arises in phylogeny and ontogeny.

ONLY A FEW years ago, the topic of hemisphere specialization and interaction, in human and animal brains, might have seemed an ill-fitting intrusion into a series of chapters on basic neural science. Why then does it seem so timely to deal with this topic now and to risk juxtaposing considerations of membranes and immunochemistry with the question of how the two halves of our own brain divide their roles between them?

The problem of dual functional asymmetry of man's cerebral hemispheres involves some of the most complex aspects of neuronal organization. Yet we believe that these complex achievements of our brain need to be kept in view if one wants to give aim to the inquiries into its elementary constituents. Furthermore, the study of hemisphere specialization has now reached a stage where there is surprising agreement on principles, even though there is still perplexity about mechanism.

As nearly every contributor to the topic has stressed, the concept of unilateral dominance of left over right hemisphere in man has been abandoned and replaced by one of complementary specialization. This congruence of views is all the more remarkable because it has been reached, as Professor Milner, the topic organizer, has pointed out, by three quite different routes: First, by continuing analysis of unilateral cortical lesions and their effects on human performance (see, e.g., Hécaen, 1962; Milner, 1962, 1967); and second, by intensive study of those behavioral effects that follow surgical division of major connections between the paired hemispheres, as in the work of Professor Sperry. These studies on split-brain man have brought a magnificent vindication of the distinctions reached earlier from analyses of unilateral lesions. Third, and last, there is the additional evidence derived from all those phenomena that show how the two halves of the brain can get in each other's way.

HANS-LUKAS TEUBER   Department of Psychology, Massachusetts Institute of Technology, Cambridge, Massachusetts

I have often stressed the clues that one might derive from abnormal hemispheric interaction, as in the processes of "extinction" of sensations upon double simultaneous stimulation of the two sides of the body (Teuber, 1962), but a particularly powerful tool has been developed by Professor Broadbent (1954; Broadbent and Gregory, 1964), with his dichotic listening technique, where the two ears get simultaneous but different messages. The extension of this approach to patients with unilateral cerebral lesions (by Doreen Kimura in Professor Milner's laboratory, Kimura, 1967, 1973) has launched a veritable torrent of studies involving dichotic listening tasks, simultaneous stimulation of the opposite halves of a visual field or of tactile surfaces on the body, all in normal subjects and in patients. The results are, again, essentially congruent with those derived from studies of convexity lesions or division of major interhemispheric commissures.

With so much current agreement on principles, it deserves to be stressed how little we know about the fundamental questions of what, how, and whence. Yet we need to know (1) *what* it is that characterizes the specific function of the right and left hemispheres in the normal adult; (2) *how* the commissures act in providing information transfer, between the hemispheres, and in constraining, or modulating, the activities in the parallel halves of the brain, in such a way that a functional asymmetry arises and is maintained; (3) last, one should reopen the question of *whence* the asymmetries arise in phylogeny and ontogeny. For man, one should ask whether it is not after all a genetic predisposition that produces somewhat dissimilar hemispheres in the course of embryonic development, and, if so, is this bias initially limited to one hemisphere, determining the fate of the other only secondarily? Or do we start with a double but opposite bias of the two sides? It is remarkable that these latter questions, those pertaining to the ontogeny of hemispheric specialization, have hardly been touched upon at all in the past and current work on our two hemispheres.

To begin then with the first issue, it is evident that the question of *what* is not fully settled. Much time and effort has already been spent on characterizing the crucial aspects of functions that distinguish right and left hemispheres in the majority of cases in adult man. Yet, somehow the appropriate description remains elusive.

Professor Liberman has asked his favorite question: "Is speech special?" He has put it in terms of what it is about speech that would make it fitting for the left hemisphere to have it rather than the right. He proposed the high degree of encodedness of speech signals as the crucial factor that a left hemisphere can treat better than a right (see, also, Liberman et al., 1967). Some of the most recent extensions of Professor Broadbent's techniques as modified by Doreen Kimura seemed to point in similar directions. But there are still some considerable difficulties.

Christopher Darwin stressed how certain aspects of speech sounds, such as rapid transitions, may be crucial, thus refining the earlier distinctions made by Shankweiler and others at the Haskins Laboratories between consonantal phonemes (lateralized to the left hemisphere) and steady-state vowels (not lateralized; see also, Darwin, 1971; Studdert-Kennedy and Shankweiler, 1970). Professor Broadbent preferred to describe the left-sided dominance for speech in altogether different terms, as a form of habitual distribution of attention.

Professor Sperry characterized the distinction boldly as analytic (left) versus global or configurational (right), and in some of the subsequent discussions, left was called "categorical," right, "representational" or "isomorphic" in its mapping of the world—a distinction made already a decade ago by Young (1962) in an earlier meeting on hemisphere specialization (Mountcastle, 1962). There even were overtones of ancient Chinese polarities: Left hemisphere (thus right side of body) was deemed masculine (hence good), and right hemisphere, feminine (hence not so good), and that clearly will not do.

With all that, it is probably true that the *what* of complementary specialization remains incompletely characterized, and the contrast between hemispheres remains in danger of being overstated. Yet, since it is undesirable to have diagnosis without remedy, we should try and see how the present state of affairs might be improved. For one, I would suggest that we look with more care at the kinds of things at which the right hemisphere excels, such as recognition of faces. As could be seen from Professor Berlucchi's paper, reaction times for letters are better in the right half of the visual field and for faces in the left half; hence the converse inside the head: letters, left, and faces, right.

But it is of course precisely here that we must ask Professor Liberman's question, though with a difference: Not only, is speech special, but are faces special, and if so, in what way? What makes faces special, so that it should be appropriate for face recognition to be more dependent on the right hemisphere? It is difficult to believe that the process of face recognition is as global, as different from

the grammars of linguistic material, as some of the contemporary workers seem to assume. Yet, obviously, speech is better characterized than anything that pertains to the right hemisphere—which need not mean that the right hemisphere does not categorize its inputs, only that we have found out less, so far, about the ways in which these categorizations proceed.

One of our recent doctoral students in the department at M.I.T., Robert Yin, asked the question very directly by presenting a series of snapshots of unfamiliar faces to his observers (young normal adults), holding the snapshots either in normal positions or upside down. Compared with other mono-oriented patterns (such as pictures of houses), the inversion produced a severe handicap for recognition, and this handicap was greatest for those people who had done best on the task of recognizing faces right-side up (Yin, 1969).

From this, Yin predicted that inversion would be even less of a handicap for patients with penetrating brain wounds in the right posterior sector of their hemispheres and this was exactly what happened: Those patients with right-sided lesions did very poorly with faces presented in upright orientation, worse than patients with lesions elsewhere in the brain, and much worse than the normal control subjects. But, by the same token, these men with penetrating right-sided lesions were *less* handicapped than any other group when the snapshots were shown to them upside down (Yin, 1970). We interpret this finding as suggesting a very special form of processing of faces by normal observers—a schematizing that is destroyed by inversion, so that those brain-injured persons who lack the schematism would look at faces as if they dealt with any ordinary array of shadings or contours. Yet, we obviously do not know enough about face perception to come to grips with the essential process.

Even for the left side, the ultimate answer may be simpler and less directly related to language than it seems so far. One of the least expected outcomes of a recent study of nearly 100 men with various penetrating battle injuries of the brain in our laboratory has been the (unpublished) finding, by James Lackner and myself, of a permanent difficulty in two-click resolution after left posterior (temporoparietal) lesions (Lackner and Teuber, 1973). It seems as if fine time discriminations, even in the absence of any special verbal components, were particularly difficult for such patients (see also, Efron, 1963; Swisher and Hirsh, 1972; Tallal and Piercy, 1973). Could it suffice for the left hemisphere to be genetically predisposed, in most of us, for the achievement of fine temporal resolutions to attract speech into that rather than the other hemisphere?

The second unresolved issue relates to the role of the commissures between the hemispheres. Are they necessary for the existence of those complementary patterns that have been found? How do the interhemispheric commissures act when they do? Little of this came into the discussions except perhaps for Dr. Cuénod's paper. It is widely believed that the functions of the callosal fiber system, in particular, are almost entirely inhibitory; that may be true, but there appear to be much more intricate aspects. Professor Berlucchi refrained from touching upon his important data on the kinds of visuosensory information that are transmitted by the posterior sector of the callosum in cats where definite visual receptor fields with specialization for particular orientations seem to exist (Berlucchi et al., 1967). Dr. Cuénod pointed out how much might be learned by experimenting on the critical timing of reversible commissure sections (through cold probes), permitting one to interrupt the information flow between hemispheres at critical moments during and after learning. He also stressed that only in animals might we have a chance, so far, of coming closer to the physiologic mechanism of hemispheric interaction, as in his studies of radio-labeled amino acid flow through the connecting pathways. Nevertheless, at this stage, the *how* of interaction remains as one of the major gaps in the coverage of our topic.

Still the biggest gap by far, and the most important issue, is the question of *whence* the dual asymmetry of our brain arises, both in phylogeny and ontogeny. An answer to this question of origins would go far toward a general solution of the central issues of purpose and mechanisms. What selection pressures, in early evolution, created, first, bisymmetrical organisms and nervous systems, and what selective advantage is conferred by decussations? Cajal (1917) and Coghill (1929) struggled with this issue, and no clear answer is at hand.

Beyond that, what evolutionary pressures created a division of roles between paired halves of a nervous system? Undoubtedly, there are evolutionary prodromes for an arrangement where bilateral symmetry of the two halves of a nervous system gives way to a partial asymmetry, as in the habenular complex of frogs, but the only clear instance of unilateral "dominance" known to me at present is the amazing story of Nottebohm's song birds, whose song patterns are badly mutilated if one destroys the descending nerve to the syrinx, the organ of song, on the left, but much less affected if the destruction is inflicted on the right (Nottebohm, 1970).

The parallel with the "dominance" of man's left hemisphere for speech is even more striking because of the developmental aspects of this story: If the transection is done before song has crystallized, then the effects of left-sided lesions are minimal, almost foreshadowing the relative sparing of speech with injuries to the left hemisphere in young children (Lenneberg, 1967). But, so far,

no corresponding relative dominance of the right side, for functions other than song, has been demonstrated for the avian brain, and there are of course very serious doubts about the extent to which bird song—even in those birds where most of the pattern is learned (Marler, 1970; Thorpe, 1960)—should be considered as anything but a very distant analog of human speech (Chomsky, 1967).

There is little question, for our own species, that language tends to be relatively spared after early damage to the left hemisphere. Yet it is less clear how complete such an escape or rapid recovery of language can be, after very early lesions, or whether other aspects of performance suffer when language seems to escape. Nor do we know whether the functions of the right hemisphere would show similar sparing after early damage to the right, as an assumption of greater equipotentiality of the infantile hemispheres would have to suggest.

To look into these issues, Bryan Woods, Sue Carey-Block, and I have recently begun to examine some 50 children with congenital or early postnatal hemipareses (1973). Those with left hemisphere involvement in early infancy show some language retardation and marked deficits on visuospatial tasks when tested in their teens and early twenties. Those with very early right-hemisphere lesions show massive losses on visuoconstructive tasks and only slight difficulties with linguistic development.

All in all, these findings suggest a definite hemisphere specialization at birth, with a curiously greater vulnerability to early lesions for those capacities that depend, in the adult, on the right hemisphere—as if speech were relatively more resilient or simply earlier in getting established. Yet this resiliency is purchased at the expense of nonspeech functions as if one had to admit a factor of competition in the developing brain for terminal space, with consequent crowding when one hemisphere tries to do more than it had originally been meant to do.

A possible animal model for these processes suggests itself: Gerald Schneider has worked for nearly 7 years in our departmental laboratories on the postnatal development of the hamster's visual system (1969, 1973). He has recently shown that destruction of one superior colliculus soon after birth will be followed by an entire array of redirections of terminals within the visual pathways. Among these are abnormally growing fibers from the eye contralateral to the destroyed colliculus across the midline to the medial border of the colliculus that remains. These abnormal connections are functionally competent, since these hamsters turn wrongly in the corresponding sector of their visual field. If Schneider now varies this experiment by removing not only one superior colliculus, say, the right, at birth, but also the right eye, then the fibers from the remaining eye can be seen to cover, later on, almost the entire surface of the remaining colliculus. This invasion of vacant terminal space, and the precedence effect implied in these experiments, may contain clues as to how we might picture the ontogeny of hemisphere interaction and specialization, both in the normal and the injured child.

In summary, the *why* of hemisphere specialization still eludes us, but we shall come closer to an answer if we perfect our understanding of the *what*, the *how* and the *whence*. Perhaps it is a personal bias if I express the belief that much more can still be learned in the last of these areas, the evolutionary and ontogenetic aspects; wherever we see differentiated living structures, we understand them much better if we can show by what stages their differentiation has come about.

So put, the question of why we have two brains takes on the awesome simplicity of Kepler's question about crystals. In his lovely New Year's gift of 1611, "On Hexagonal Snow," he asks insistently, "Why six?" Whence the six points of the snow flake? And he goes on to say, of course, if atomism were correct, then an explanation might be at hand; but, Kepler adds, everyone knows that atomism is wrong, and so there is no explanation. Are we getting as close (and as far) as Kepler got to the problem of crystals, in our attempts at explaining the dual asymmetry of the brain?

REFERENCES

BERLUCCHI, G., M. S. GAZZANIGA, and G. RIZZOLATTI, 1967. Microelectrode analysis of transfer of visual information by the corpus callosum. *Arch. Ital. Biol.* 105:583–596.

BROADBENT, D. E., 1954. The role of auditory localization in attention and memory span. *J. Exp. Psychol.* 47:191–196.

BROADBENT, D. E., and MARGARET GREGORY, 1964. Accuracy of recognition for speech presented to the right and left ears. *Q. J. Exp. Psychol.* 16:359–360.

CAJAL, S. R., 1917. Contribución al conocimiento de la retina y centros ópticos de los Cefalópodos. *Trab. Lab. Invest. Biol. Univ. Madrid* 15:1–82.

CHOMSKY, N., 1967. The general properties of language. In *Brain Mechanisms Underlying Speech and Language*, F. L. Darley, ed. New York: Grune and Stratton, pp. 73–88.

COGHILL, G. E., 1929. *Anatomy and the Problem of Behavior.* Cambridge: Cambridge University Press.

DARWIN, C., 1971. Ear differences in the recall of fricatives and vowels. *Q. J. Exp. Psychol.* 23:46–62.

EFRON, R., 1963. Temporal perception, aphasia and *déjà vu.* *Brain* 86:403–424.

HÉCAEN, H., 1962. Clinical symptomatology in right and left hemispheric lesions. In *Interhemispheric Relations and Cerebral Dominance.* V. B. Mountcastle, ed. Baltimore: Johns Hopkins Press, pp. 215–243.

KEPLER, J., 1966. *Strena seu de nive sexangula.* (A New Year's Gift or The Six-Cornered Snowflake, 1611.) Edited and translated by Colin Hardie, Oxford: Clarendon Press.

KIMURA, DOREEN, 1967. Functional asymmetry of the brain in dichotic listening. *Cortex* 3:163–178.

KIMURA, DOREEN, 1973. The asymmetry of the human brain. *Sci. Am.* 228:70–78.

LACKNER, J., and H.-L. TEUBER, 1973. Alterations in auditory fusion thresholds after cerebral injury. *Neurospsychologia,* submitted.

LENNEBERG, E. H., 1967. *Biological Foundations of Language.* New York: John Wiley & Son.

LIBERMAN, A. M., F. S. COOPER, D. P. SHANKWEILER, and M. STUDDERT-KENNEDY, 1967. Perception of the speech code. *Psychol. Rev.* 74:431–461.

MARLER, P., 1970. A comparative approach to vocal learning: Song development in white-crowned sparrows. *J. Comp. Physiol. Psychol., Monogr.,* (May) 71 (No. 2, pt. 2):1–25.

MILNER, B. 1962. Laterality effects in audition. In *Interhemispheric Relations and Cerebral Dominance.* V. B. Mountcastle, ed. Baltimore: Johns Hopkins Press, pp. 177–195.

MILNER, B., 1967. Brain mechanisms suggested by studies of temporal lobes. In *Brain Mechanisms Underlying Speech and Language.* F. L. Darley, ed. New York: Grune and Stratton.

MOUNTCASTLE, V. B., ed., 1962. *Interhemispheric Relations and Cerebral Dominance.* Baltimore: Johns Hopkins Press.

NOTTEBOHM, F., 1970. Ontogeny of bird song: Different strategies in vocal development are reflected in learning stages, critical periods, and neural lateralization. *Science* 167:950–956.

SCHNEIDER, G. E., 1969. Two visual systems: Brain mechanisms for localization and discrimination are dissociated by tectal and cortical lesions. *Science* 163:895–902.

SCHNEIDER, G. E., 1973. Early lesions of superior colliculus: Factors affecting the formation of abnormal retinal projections. *Brain, Behav. Evol.* (in press).

STUDDERT-KENNEDY, M., and D. SHANKWEILER, 1970. Hemispheric specialization for speech perception. *J. Acoust. Soc. Amer.* 48:579–594.

SWISHER, LINDA, and I. J. HIRSH, 1972. Brain damage and the ordering of two temporally successive stimuli. *Neurospsychologia* 10:137–152.

TALLAL, P., and M. PIERCY, 1973. Effects of non-verbal auditory perception in children with developmental aphasia. *Nature* (16 February) 241:468–469.

TEUBER, H.-L., 1962. Effects of brain wounds implicating right or left hemisphere in man. In *Interhemispheric Relations and Cerebral Dominance.* V. B. Mountcastle, ed. Baltimore: Johns Hopkins Press, pp. 131–157.

THORPE, W. H., 1960. *Bird Song: The Biology of Vocal Communication and Expression in Birds.* Cambridge: Cambridge University Press.

WOODS, B. T., and H.-L. TEUBER, 1973. Early onset of complementary specialization of cerebral hemispheres in man. In *Program of the 98th Meeting of the American Neurological Association.*

YIN, R. K., 1969. Looking at upside-down faces. *J. Exp. Psychol.* 81:141–145.

YIN, R. K., 1970. Face recognition by brain-injured patients: A dissociable disability? *Neurospsychologia* 8:395–402.

YOUNG, J. Z., 1962. Why do we have two brains? In *Interhemispheric Relations and Cerebral Dominance.* V. B. Mountcastle, ed. Johns Hopkins Press, pp. 7–24.

8 Hemispheric Specialization: Scope and Limits

BRENDA MILNER

ABSTRACT Earlier views on the relations between man's cerebral hemispheres were limited to concepts of dominance; new evidence abundantly shows that there is complementary specialization. This is manifested in normal subjects by the way the hemispheres interact when competing information simultaneously enters the two ears or the two visual half-fields, and it becomes still more evident when the interhemispheric connections have been divided. Similarly, after unilateral focal lesions of frontal or temporal cortex, different patterns of functional representation emerge, but here the alterations in behavior reflect not only the side but also the site of the lesion. Clues to the ontogeny of such functional specialization come from examining the effects of differently located early lesions of the left hemisphere upon cerebral organization at maturity.

AMONG THE three ways of approaching the complementary specialization of the left and right halves of the human brain, the oldest is that of analyzing the effects of circumscribed unilateral lesions. This method, on which my chapter will be focussed, still has certain advantages, because it allows one to examine the question of hemispheric differences in greater detail by looking at the form these asymmetries take in the case of particular cortical areas. Nevertheless, the appeal of the notion of a dual asymmetry derives from the convergence of findings from all three sources: The continuing study of unilateral lesions, the observations on functional asymmetry as disclosed by appropriate stimulus-presentation to normal subjects, and the striking effects of commissure section. The two more recent methods lead one to emphasize the overall contrast between right and left halves of the brain; comparison of the effects of unilateral lesions can, however, help one to balance such evidence of contrast against the evidence of a parallel organization of function on the two sides. It is this parallelism that is at times in danger of being overlooked.

Accordingly, I shall first describe briefly some of my own experience within the two more recently opened domains by recounting findings in the field of normal adult behavior and observations made on a small group

BRENDA MILNER Montreal Neurological Institute, McGill University, Montreal, Canada

of patients with cerebral commissurotomy whom Sperry kindly allowed me to study on two occasions. Then I shall turn to my main source, a long series of patients with stable, unilateral cortical lesions seen at the Montreal Neurological Institute for surgical treatment of focal epilepsy. Included in this population, there will be a special series of early lesions of the left hemisphere, which were treated by more extensive cortical removals. These cases permit one to make certain conjectures about the ontogeny of hemispheric specialization for language and about some of the preconditions for anomalous speech-representation.

Dichotic listening in normal subjects I shall begin by reviewing the early dichotic-listening experiments of Doreen Kimura, because they stimulated so much subsequent work (see Darwin, this volume) as well as providing the first clear demonstration of dual functional asymmetry in the normal brain (Kimura, 1964). In her original study, Kimura used the technique devised by Broadbent (1954), in which different strings of digits are presented simultaneously to the two ears by means of a dual-channel tape recorder and stereophonic earphones. In what proved to be the most discriminating condition, two digits, say 4 and 7, would be presented together, 4 to the left ear and 7 to the right, followed by another pair a half-second later and then by a third pair; after this the subject reported all the numbers he had heard, in any order. With this competing verbal input, most normal right-handed subjects, and most right-handed patients with epileptogenic lesions of frontal or temporal lobes, report more digits correctly for the right ear than for the left (Kimura, 1961a).

To account for this asymmetry, Kimura supposed that the contralateral pathway to the speaking left hemisphere (from right ear to left temporal lobe) was more effective than the ipsilateral one. This interpretation in terms of privileged access to the dominant hemisphere for speech was strengthened by the finding of left-ear superiority on the same task in patients who were known to be right-hemisphere dominant for speech, as determined by the Wada technique of intracarotid injection of sodium Amytal (see below).

The next step was to look for an opposite asymmetry if a musical input was used instead of a verbal one. This prediction arose from the fact that right anterior temporal lobectomy had been shown to impair the discrimination of tonal patterns and tone quality whereas left anterior temporal lobectomy has no effect on the performance of these tasks (Milner, 1962). These findings for the right temporal lobe of man recall the auditory deficits seen after bilateral removal of the corresponding cortical areas in cat and monkey; in neither case is simple pitch discrimination affected, but there are losses in the discrimination of complex sounds (Dewson, 1964) and of tonal sequences (Neff, 1961; Evans, this volume).

TABLE I

Dichotic listening: Results for 20 right-handed normal control subjects. (From Kimura, 1964.)

Dichotic Task	Mean Number Correct Responses		Number of Subjects		
	Left Ear	Right Ear	Better on Left	No Difference	Better on Right
Digits	86.7 (90%)	90.3 (94%)	3	2	15
Melodies	13.6 (75%)	11.3 (63%)	16	2	2

Taking the right-temporal auditory deficits as her starting-point, Kimura constructed a dichotic-melodies task, in which two different snatches of music (excerpts from concerti grossi) are presented simultaneously, one to the left ear and the other to the right, followed by successive binaural presentation of four such snatches, among which the subject has to select the two that he has just heard. With this nonverbal material, she obtained a left-ear superiority in the same normal subjects who showed a right-ear superiority for digits (Table I). This finding cannot be explained away by the fact that a recognition procedure was used for melodies and a recall procedure for digits, because Broadbent and Gregory (1964) have demonstrated a right-ear superiority for digits under either condition of testing, recognition, or recall. It seems that as the acoustic task changes from a verbal to a musical one, the ear superiority also shifts, reflecting the contrasting specialization of the two temporal lobes.

Dichotic listening after cerebral commissurotomy We have seen that, for both digits and music, the direction of the ear advantage in normal subjects implies some dominance of the contralateral over the ipsilateral auditory pathway when the two are in competition. These ear differences, however, though statistically reliable, are quite small. In contrast, a striking laterality effect emerges when digits are presented dichotically to patients with complete section of the midline commissures (Milner et al., 1968; Sparks and Geschwind, 1968). Despite the fact that both ears are represented in both hemispheres, five of the seven patients tested by us complained that they could hear no numbers at all in the left ear and registered almost no correct responses for that ear; the other two patients reported only a third as many digits correctly for the left ear as for the right (Figure 1). Yet none of these patients had any difficulty reporting digits from the left ear under monaural conditions, thereby showing that the ipsilateral pathway could be utilized.

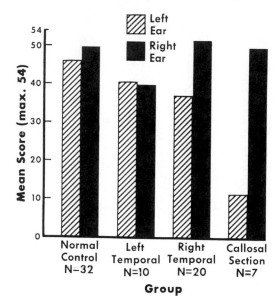

FIGURE 1 Mean number of digits correctly reported for each ear, when different digits are presented simultaneously to the two ears. Results for normal control subjects and for patients tested after either left or right temporal lobectomy (including Heschl's gyrus) or after section of the corpus callosum and other interhemispheric commissures. (From Milner et al., 1968; copyright 1968 by the American Association for the Advancement of Science.)

In Figure 1 the performance of the commissurotomized patients on the dichotic task can be compared with that seen after removal of one temporal lobe including the primary auditory cortex of Heschl's gyrus. Whereas the normal control group shows a slight but significant right-ear superiority, the effect of a temporal lobectomy is to decrease the efficiency of report for the ear contralateral to the lesion (Kimura, 1961b). On this verbal task, patients with left temporal lobectomy in the dominant hemisphere for speech show a mild impairment on the test as a whole, but those with right temporal lobectomy merely have an accentuation of the right-ear superiority found in normal subjects. What is perhaps unexpected in these results is that the effect of commissure section should be so much greater than that of a right temporal lobectomy; yet the finding is consistent with the fact that, at least in the monkey, evoked responses to acoustic stimuli can be elicited from a wide area of the posterior cortex, including parts of the parietal lobe (Pribram et al., 1954). It is evident that when deprived of input from the entire right hemisphere (and not merely from the right temporal lobe), the speaking left hemisphere exhibits to a remarkable degree the suppression of the ipsilateral by the contralateral projection system. That this dominance should be so much less apparent when the commissures are intact suggests that normally a significant fraction of the input from the left ear to the left hemisphere comes via the callosum.

We had hoped to be able to demonstrate the converse effect in the commissurotomized patients if simple melodies of three or four notes were presented to them dichotically with the instruction to sing or hum the melodies they heard. Instead, it proved impossible to obtain a consistent vocal read-out from the right hemisphere except in the case of one patient (C.C.), who hummed only the melodies that had been presented to the left ear, although showing a complete suppression of that ear on the verbal dichotic task. For the other five patients tested, the results were quite disorderly, as if the right hemisphere could not compete effectively for control of the vocal musculature, even though speech was not involved. My example of right-hemisphere specialization after commissurotomy will therefore be taken from the tactile mode, where it is much easier than in audition to test one hemisphere without interference from the other. The task chosen assesses pattern recognition by touch, the stimulus objects being presented to one hand at a time, behind a screen, and retrieved later with the same hand from a group of several similar ones. The purpose of the experiment was to explore the memory capacities of the verbally silent hemisphere and hence discover how well a complex pattern can be retained without the use of words.

Right-hemisphere superiority in tactile pattern recognition after cerebral commissurotomy For this investigation test material was chosen that did not lend itself readily to verbal coding; in this way we sought to put the left hemisphere at a disadvantage while bringing out the specialization of the right. The tactual stimuli used are shown in Figure 2. They consisted of four irregular wire figures, each made by bending a 10-inch length of heavy wire into the desired flat shape. To test memory after commissurotomy, a matching-to-sample procedure was followed in which the patient first felt one of the patterns

carefully with the left hand and then after a variable delay tried to find it again by touch from the set of four, still using the same hand. The initial training involved matching at zero delay until the patient could recognize all the patterns reliably. Delayed matching was then introduced with gradually increasing intratrial intervals up to an arbitrary limit of 120 sec, or until the patient failed to meet criterion. The whole procedure was then repeated with the right hand.

The main finding is depicted in Figure 3, which shows the longest delay interval successfully bridged with each

Examiner

Subject

FIGURE 2 Four wire figures used to test tactile pattern matching, in the orientation in which they are presented to the subject (Milner, 1971; Milner and Taylor, 1972). In the delayed-matching procedure, a sample figure from the set is given to the subject to explore with one hand for a few seconds. It is then replaced among the other figures to form a horizontal array, as shown here. After a variable delay, the subject uses the same hand to find the sample again. The order of the figures in the array is changed at random from trial to trial.

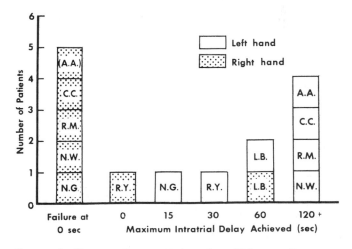

FIGURE 3 Bar graphs comparing the efficiency of pattern matching with right and left hands, respectively, for each of seven commissurotomized patients (indicated by their initials). All subjects, except L.B., did better with the left hand than with the right. (From Milner, 1971; Milner and Taylor, 1972.)

hand, for the seven commissurotomized patients who took part in the dichotic-digits task. With the left hand (and right hemisphere), four patients succeeded in bridging the 2 min interval and the other three had partial success. With the right hand, five patients failed to reach criterion at zero delay even with extensive training and one other patient (R.Y.) failed completely as soon as a delay was introduced. Only one patient, L.B. (the intelligent young boy whom Sperry has so extensively studied), was as successful with the right hand as with the left, matching correctly at 60 sec but failing at 120 sec. This patient tries to solve most problems verbally and in his case we had the impression that the left hemisphere was at times interfering with the performance of the right.

These results confirm what we set out to prove: namely, that complex perceptual material can be remembered accurately without the use of words. The sharp contrast between left-hand and right-hand performance also illustrates beyond dispute the major role of the right cerebral hemisphere in the appreciation of spatial patterns. The degree of disability shown by the commissurotomized patients when working with the right hand (and left cerebral hemisphere) surprised us, because only one of these patients (A.A.) had any sensory defects on the right hand that could account for the difficulty in pattern recognition; yet the left hemisphere seemed unable to obtain a clear enough representation of the form palpated to be able to distinguish it reliably from similar ones, even with practice. These failures thus confirm Semmes's (1965) findings, based on the study of men with penetrating brain wounds, that tactile form perception can be disturbed in the absence of primary sensory defects.

Had our study stopped at this point, we might have concluded that tactile pattern recognition is solely a function of the right cerebral hemisphere, but this would have been premature. On testing a series of control patients with various unilateral cortical excisions for epilepsy but with intact commissures and no sensory defects, we observed practically errorless performance from the beginning for all patients with either hand, whereas the commissurotomy patients required considerable training even with the left hand. Thus the performance of the control subjects far surpassed that of the most proficient of the patients with commissure section. This means either that the separation of the hemispheres reduces their functional efficiency or that the left hemisphere normally participates in tasks of this kind, perhaps by adding some useful verbal tag, once the pattern has been accurately perceived by the right hemisphere (De Renzi et al., 1971; Milner, 1971; Milner and Taylor, 1972).

Hemispheric specialization as revealed by unilateral brain lesions

The abnormalities shown by the commissurotomized patients on dichotic listening, their low performance on tactual pattern recognition, and some of the imperfections of memory alluded to by Sperry (this volume) support Broadbent's claim that the efficient solution of most tasks requires the concerted action of the two sides of the brain. Even on strictly verbal learning tasks, Zaidel and Sperry (personal communication) find impairment after cerebral commissurotomy, although one would not necessarily have predicted this to be the case. This being said, one can still maintain that different kinds of processing take place on the two sides. Yet only studies of the effects of unilateral lesions can tell us whether these differences outweigh the many instances of parallelism in the functional organization of the two hemispheres. Such parallelism is most conspicuous for the primary sensory projection areas and for the motor cortex, the two sides of the body and the two halves of the external world being represented in mirror-image fashion in corresponding areas, right and left, in the brain.

Analyses of the deficits resulting from unilateral lesions of other cortical areas suggest at least a family resemblance between the functions of the two sides. In the neurological literature, the importance of both left and right parietal lobes for spatial skills has been particularly stressed (Critchley, 1953; Semmes et al., 1955), though it is now generally conceded that the right plays a more important role and perhaps one that is qualitatively different (Hécaen, 1962; Warrington, 1969). Both similarities and differences in the functions of corresponding areas can also be shown for the anterior temporal region and for the frontal lobes. For each, it has been possible to demonstrate that the basic disturbance, after a lesion in either hemisphere, involves similar dimensions of performance (e.g., particular aspects of memory after a lesion of the temporal lobes), even though, as I shall point out, these memory disorders take somewhat different forms depending on whether the lesion is on the left or on the right.

Such a situation is in obvious contrast to that in lower primates, where a bilateral lesion is usually required to bring out such deficits as the impairment in visual discrimination learning that follows destruction of inferotemporal cortex in the monkey (see Gross, this volume; Weiskrantz, this volume). It may be true that very special experimental tasks can be constructed that are sensitive to unilateral lesions even in monkeys (for an example involving difficult auditory discriminations, see Dewson et al., 1970), but no one has so far shown convincingly that such tasks would uncover instances of complementary specialization akin to those seen in man.

The homologous cortical regions in monkeys are thus much more nearly equivalent in their function, so that one side can largely substitute for the other. In man, this basic pattern has become modified by further functional differentiation; yet in spite of the dual asymmetry that results, the basic similarity of function of corresponding regions must not be underrated either. I shall try to document these claims by giving illustrations from work on the temporal lobes and then on the frontal lobes.

TEMPORAL LOBES A comparison of the effects of left and right anterior temporal lobectomy in epileptic patients has revealed certain specific memory defects that vary with the side of the lesion. Unlike the global amnesia that follows bilateral damage in the hippocampal zone, these effects of unilateral lesions are restricted to a particular kind of stimulus material, though not necessarily to a particular sensory mode. Thus, left temporal lobectomy, in the dominant hemisphere for speech, selectively impairs the learning and retention of verbal material (Meyer and Yates, 1955; Milner, 1958), regardless of whether the material is heard or read (Blakemore and Falconer, 1967; Milner, 1967) and regardless of whether a recall or a recognition procedure is used (Milner and Teuber, 1968). Yet left-temporal lesions do not affect memory for perceptual material, such as places, faces, melodies or nonsense patterns. Conversely, removal of the right, non-dominant temporal lobe leaves verbal memory intact but impairs the recognition and recall of visual and auditory patterns that do not lend themselves easily to verbal coding (Kimura, 1963; Milner, 1962; 1967; 1968a; Shankweiler, 1966; Warrington and James, 1967). Right temporal lobectomy also retards the learning of stylus mazes, whether visually or proprioceptively guided (Corkin 1965; Milner, 1965), whereas left temporal lobectomy does not. Thus, within the sphere of memory there is a double dissociation (Teuber, 1955) between the effects of these two lesions.

The typical unilateral operation for temporal-lobe epilepsy includes not only the lateral neocortex but also the amygdala and parts of the hippocampus and hippocampal gyrus on the medial aspect of the hemisphere. Because profound and generalized learning deficits result from bilateral medial temporal-lobe lesions (Milner, 1970), the question naturally arises as to whether the severity of the material-specific memory changes seen after unilateral temporal lobectomy may not also depend on the medial extent of the removal. We have known for some time that the deficit in maze learning after right temporal lobectomy is related to

excision of the hippocampus and adjacent medial temporal cortex (Corkin, 1965; Milner, 1965), and the same may be true for the deficit in the recognition of unfamiliar photographed faces (Milner, 1968a), although not for the recognition or recall of nonsense figures or complex geometric designs. More direct evidence comes from recent experiments by Corsi (1972), who has used formally similar verbal and nonverbal tasks to bring out the importance of the hippocampal lesion in the material-specific memory disorders and to show the differential effects of left and right hippocampal removals.

Impaired recall of alphabetical material after left hippocampal excision The verbal memory task on which Corsi has amassed the most data is a simplified version of the Peterson and Peterson (1959) technique in which the subject has to recall a group of three consonants (for example, *XBJ*) after a short interval occupied with counting backwards from a given number (for example, 357). The purpose of the interpolated activity is simply to prevent rehearsal of the consonants during the delay. Peterson and Peterson observed that under such conditions the decline in correct recall of the trigrams becomes quite marked as the interval gets longer. As can be seen from Figure 4, Corsi's patients with left temporal-lobe lesions show abnormally rapid forgetting (consonant with their poor performance on other verbal memory tasks), whereas the performance of patients with right temporal-lobe lesions is indistinguishable from that of normal control subjects.

In order to explore the contribution of the hippocampal lesion to the verbal memory impairment seen after left temporal lobectomy, Corsi subdivided the patients with left temporal removals into four groups,

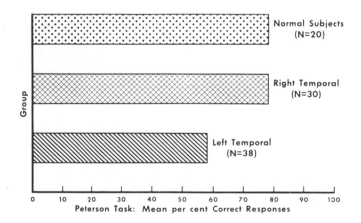

FIGURE 4 Verbal memory defect after left temporal lobectomy. Corsi's (1972) results for the recall of three consonants after a short distracting activity. Data averaged across different retention intervals. (From Milner, 1972.)

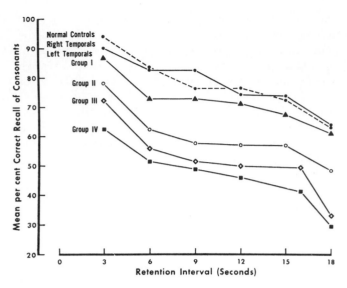

FIGURE 5 Recall of consonant trigrams as a function of intratrial interval. Results for normal control subjects and for patients tested after right or left temporal lobectomy. The subjects with left temporal lobectomy have been further subdivided into four groups according to the extent of hippocampal removal (Group I, hippocampus spared; Group II, pes hippocampi removed; Group III, pes plus 1 cm body removed; Group IV, radical hippocampal excision). (From Corsi, 1972.)

based on the surgeon's estimates of the amount of hippocampus destroyed in each case. From Figure 5, which shows recall as a function of intratrial interval, it can be seen that performance deteriorates progressively from Group I (with the hippocampus completely spared) to Group IV, the group with the most radical excisions of the left hippocampus.

Impaired recall of visual location after right hippocampal excision Corsi next set out to determine whether with appropriate material a similar effect could be demonstrated for the right temporal lobe. The task chosen derives from Posner (1966) and involves memory for visual location. It is illustrated in Figure 6, where what has to be recalled is the exact position of a $\frac{1}{4}$-inch-diameter circle situated at a variable distance along an 8-inch line. The line is exposed for 5 sec, permitting the subject to mark the centre of the circle (Figure 6A) and thus indicate that he has attended to the display. The inspection window is then covered and after a short delay the recall window is opened to expose an 8-inch response line without a circle on it. The subject marks this line (Figure 6B), so as to reproduce as well as he can the position of the circle that had been displayed a few seconds earlier. On half the trials (work condition), the subject is prevented from thinking about the position

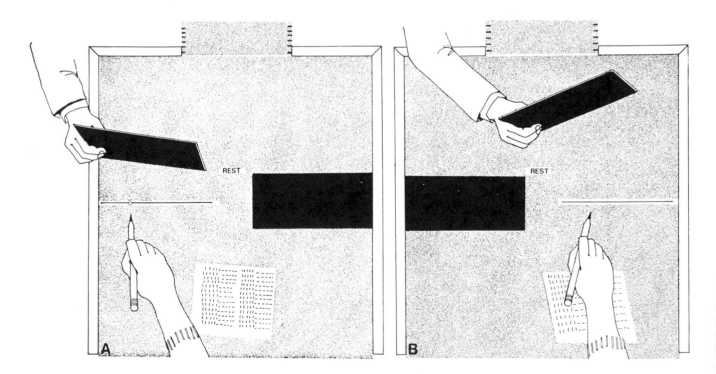

FIGURE 6 A and B, sketches illustrating the procedure used by Corsi (1972) to test memory for visual location. (A) The patient marks the circle indicated on the exposed 8-inch line. (B) After a short delay, he tries to reproduce this position from

memory as accurately as possible, on a similar 8-inch line. The sign "REST" means that the patient does nothing during the interval, in contrast to "WORK" trials in which a distracting activity is interpolated. (Task adapted from Posner, 1966.)

of the circle during the retention interval by having to arrange random strings of digits in ascending order; on the other trials (rest condition), he sits quietly, fixating a point midway between the two windows.

Because as many as 24 different positions of the circle are sampled in each condition, it is impossible to do well on this task by making a rough verbal estimate, such as "about one third of the way along," and using this as a mnemonic. Instead, one has to achieve a precise visual impression of the diplay and retain it. As can be seen from Figure 7, Corsi finds the predicted deficit after right

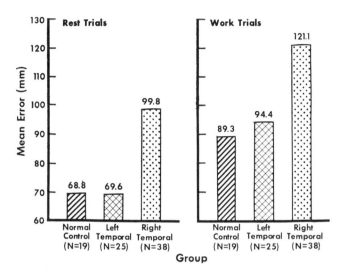

FIGURE 7 Corsi's results for the spatial memory task, showing impairment after right temporal lobectomy but not after left. The score is the total error (in mm) for four trials, without regard to sign, averaged across three retention intervals. (From Milner, 1972.)

temporal lobectomy in both the rest and work conditions, whereas the mean performance of his left temporal group does not differ significantly from that of normal control subjects (Milner, 1972). Corsi (1972) notes that on this nonverbal task there is no relationship between the extent of left hippocampal removal and test performance.

In Figure 8, the patients tested after right temporal lobectomy have also been subdivided in terms of the amount of hippocampus excised and their mean error scores plotted as a function of the length of the intratrial interval. These data are for the work condition, which is closely analogous to the conditions of the verbal-recall task previously described. One observes that the rate of forgetting for this simple visual material is steeper for Group 4 (with the most extensive right hippocampal ablations) than for any other group.

Taken together, these two experiments, one verbal and one nonverbal, implicate the hippocampus and

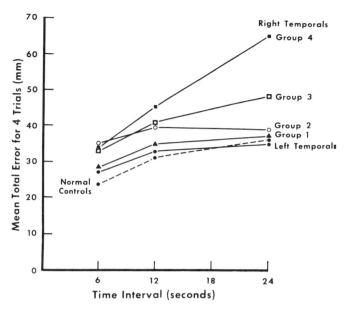

FIGURE 8 Recall of visual location as a function of retention interval filled with distracting activity (work condition). Mean scores for normal control subjects and for patients tested after right or left temporal lobectomy. In this case the right temporal-lobe subjects have been subdivided according to the estimated amount of hippocampus excised, varying from complete sparing (Group 1) to radical hippocampal excision (Group 4). Group 4 shows the most rapid forgetting. (From Corsi, 1972.)

neighboring medial-temporal regions in the holding of simple items of information in the face of distraction. Whether the ablation from the right or the left side is critical depends on the nature of the memoranda. Surgical removal limited to amygdala, hippocampus, and parahippocampal gyrus bilaterally produces a more severe memory disturbance on both these tasks than that seen in the most impaired patients with unilateral lesions.

FRONTAL LOBES The study of patients undergoing unilateral frontal lobectomy for the relief of epilepsy has by now yielded evidence for functional differences between the two frontal lobes of man (Corkin, 1965; Milner, 1964; 1971), although the results are perhaps not as clearcut as for the temporal lobes. In these left-right comparisons, the frontal excisions in the left hemisphere spare Broca's area but otherwise can be as extensive as those seen on the right; the removal typically goes back on the lateral surface to within one gyrus of the motor cortex and on the medial surface extends to include the anterior cingulate and sub-callosal gyri.

It has long been recognized that frontal-lobe lesions

can cause peculiar alterations of behavior that appear against a background of normal performance on many cognitive tasks (Milner, 1964; Teuber, 1964). The fact that some of these changes can be seen after either a left or a right frontal-lobe lesion, and that they are still more noticeable in cases of bilateral damage, has until recently tended to obscure any differential effects related to side of lesion. Yet careful scrutiny of the data reveals numerous laterality effects (for recent reviews, see Milner, 1971; Teuber, 1972), only one of which will be described here. It is again taken from the field of memory disorder.

Patients with frontal-lobe lesions have been shown to perform normally on a wide variety of memory tasks involving both verbal and nonverbal memoranda (Ghent et al., 1962; Milner, 1967; 1968); the present experiments were stimulated by the clinical observation that these same patients do extremely poorly on those tasks in which a few different stimuli recur in different pairings throughout the test, so that the subject must be able to suppress the memory of previous trials and compare the present stimulus only with the one that immediately preceded it (Prisko, in Milner, 1964; and Milner and Teuber, 1968). It looked to me as though the frontal lobectomy had interfered with the patients' ability to keep the different trials apart (Pribram and

Tubbs, 1967), and hence that they could not distinguish between the most recently presented stimulus and one shown some trials before. Yntema and Trask (1963) have argued that items in memory normally carry time-tags that permit the discrimination of the more from the less recent; if so, then this time-marking process could be disturbed after frontal lobectomy so that temporal distinctions become blurred. Corsi has now tested this hypothesis by measuring the discrimination of recency in patients with different unilateral cortical excisions, using a verbal and a nonverbal task (Milner, 1971).

In the verbal form of the recency task, the subject is given a deck of 184 cards on each of which two spondaic words are inscribed (e.g., *cowboy, railroad*). He reads the words aloud and then turns to the next card. From time to time a test card appears bearing two words with a question mark between them, and the subject must then indicate which of these words he read more recently. Usually both words on the test card will have appeared before (say, 4 cards ago as compared with 8) but in the limiting condition one of the words will be new. The task then reduces to a very simple test of recognition memory in which the subject merely chooses the word that he has seen before. Under these simple conditions, patients with left temporal-lobe lesions show a slight deficit (consistent with their impairment on other verbal

FIGURE 9 Sample item from the nonverbal recency-discrimination task (original in color). The question mark means that the subject has to indicate which of the two paintings he has seen more recently. (Material taken by Corsi from Gasser, M., 1966, *Miro*, Blandford Press Ltd., London, and from Muller, J. E., 1957, *Klee: Magic Squares*, Tudor Publishing Co., New York.)

memory tasks), but patients with frontal-lobe lesions do not. For the recency discriminations, which are more difficult, the position is reversed, with the left temporal group unimpaired and both frontal-lobe groups showing a deficit. Here the point to emphasize is that the impairment is significantly greater after left frontal lobectomy than after right, despite the fact that the removals on the left are smaller than those on the right.

Figure 9 shows a test card from the nonverbal form of the recency task. The stimulus items are reproductions of abstract paintings and the procedure is similar to that used in the verbal task. The subject is allowed to inspect each card for 6 sec before turning to the next one and, whenever a question mark appears (as in Figure 9), he must point to the picture that he saw more recently. Again, the presence of some cards in which only one picture of the pair has appeared before provides a built-in control for recognition memory. On this task, patients with right temporal-lobe lesions show the impairment in recognition that would be expected on the basis of their faulty memory for visual patterns but they do not seem to have any difficulty with temporal order as such. After right frontal lobectomy, in contrast, recency judgments are at, or near, the chance level, although recognition is essentially normal. On this nonverbal task, patients with left frontal-lobe lesions are not significantly impaired, although their mean score is slightly lower than that of either the left temporal or the normal control group. Thus, both these experiments support the notion that the temporal ordering of events is disturbed after frontal lobectomy, but they also indicate a complementary specialization of the left and right frontal lobes, in the form of material-specific effects related to the side of the lesion.

Some determinants of typical and atypical speech representation

The results presented thus far have been derived primarily from the study of adult right-handed subjects in whom the left cerebral hemisphere was believed to be dominant for speech. In this situation we see in its clearest form the differential specialization of the two halves of the brain, with the left dominant for language and the right playing a major role in many still rather ill-defined nonverbal processes, including some aspects of space perception and of music. Though typical, this pattern of cerebral organization admits of considerable individual variation in normal subjects and still more in patients with early brain lesions. Understandably, because speech is such a salient feature of human behavior, the evidence for atypical representation is better documented for language than for nonverbal capacities;

therefore, in seeking to uncover possible determinants of hemispheric specialization and its anomalies, I shall focus on the cerebral representation of speech. The discussion will include the question of speech lateralization in left-handers as well as some effects of early left-hemisphere lesions on the pattern of cerebral organization at maturity.

CEREBRAL DOMINANCE FOR SPEECH AS RELATED TO HAND PREFERENCE Although there is by now overwhelming clinical proof that the left hemisphere is dominant for speech in most right-handed persons of right-handed ancestry, this dominance does not exclude the possibility of a limited participation by the right in verbal processes. As Sperry has pointed out, the isolated right hemisphere shows some rudimentary verbal comprehension, even though it appears to be largely incapable of speech or writing (Gazzaniga and Sperry, 1967; Gazzaniga and Hillyard, 1971). Furthermore there are well-attested instances of right-handers in whom the right hemisphere, not the left, was dominant for speech, thus reversing the normal pattern; these are, however, believed to be extremely rare (Zangwill, 1967; but see Milner et al., 1964).

When one looks further and considers evidence from left-handed and ambidextrous subjects, one finds that the pattern of cerebral organization is harder to predict, but that it is unlikely to be a simple mirror image of that seen in most right-handers (Hécaen and Sauguet, 1971). Contrary to traditional doctrine (Broca, 1865), left-handers are more likely to have speech on the left than on the right, left-handedness being far more common than right-hemisphere dominance for speech (Goodglass and Quadfasel, 1954; Penfield and Roberts, 1959; Zangwill, 1960; Branch et al., 1964). Nevertheless, speech is certainly represented in the right hemisphere more frequently in left-handed and ambidextrous subjects than in right-handers (see below). What is perhaps more interesting is the suggestion that speech may be less strongly lateralized to one hemisphere in left-handers, with the possibility, in some subjects at least, of a more nearly equal participation of the two sides in language processes (Chesher, 1936; Conrad, 1949; Subirana, 1958; Zangwill, 1960).

The early evidence for such bilateral speech representation was all indirect; it came from recording, for large series of unselected cases, the incidence and time course of aphasia following unilateral brain lesions in adult life. These studies suggested that in left-handers, unlike right-handers, speech disturbance is rarely severe or lasting but that the incidence is fairly high after lesions of either hemisphere (Chesher, 1936; Conrad, 1949). More direct knowledge about speech lateralization has been obtained in recent years by the use in Montreal of the Wada (1949) technique of intracarotid injection of sodium Amytal, the right and left sides being injected on different days. By thus temporarily interfering with the functioning of each hemisphere in turn, it has been possible to compare the speech functions of the left and right sides of the brain in the same individual (Branch et al., 1964; Milner et al., 1964; 1966). The main results, which are summarized below, demonstrate some bilateral speech representation in roughly 15 percent of the non-right-handers studied but at the same time reaffirm quite strikingly the importance of the left cerebral hemisphere for speech processes of both sinistrals and dextrals.

Patterns of speech representation as revealed by intracarotid injection of sodium Amytal   For the past 12 years, we have been using the Amytal method routinely to determine the side of speech representation in all our left-handed and ambidextrous patients who were being considered for brain surgery, as well as in those few right-handers for whom there was reason to question the lateralization of speech. Latterly, we have built up a larger and more representative group of right-handed subjects, because we now use the Amytal technique to assess memory, and not only speech, employing it as a screening device in the preoperative evaluation of some of our patients with temporal-lobe epilepsy, most of whom are right-handed (Milner, 1972).

In our procedure, 200 mg of 10 percent sodium Amytal (amobarbital) are injected rapidly into the common carotid artery of one side, while the patient counts aloud slowly, with legs flexed, arms raised, and fingers moving. The immediate effect of the drug is to inactivate most of one hemisphere, producing a contralateral hemiplegia, hemianopia, and partial hemianaesthesia. If the injection is made on the side of the dominant hemisphere for speech, the patient is usually mute for about two minutes and then makes many dysphasic errors; these include both misnaming of common objects and mistakes in repeating well-known series, such as the days of the week, forward, and backward. In contrast, when the nondominant side is injected, speech is rarely interrupted for more than a few seconds and no mistakes of naming or serial order are seen. In those patients whom we classed as having some bilateral speech representation, dysphasia was observed after both the left and the right-sided injections but the speech defects were mild from both sides, although the mean duration of contralateral hemiparesis was the same as for the cases classed as unilateral.

In analysing our results, we took into account the fact that many neurological patients are mandatory left-handers, because severe injury to the left hemisphere has deprived the right hand of its normal strength and dexterity. In such cases, if the lesions have occurred early enough in life, one might well expect speech to be represented in the right hemisphere more often than in a normal population of left-handers. For this reason, we have subdivided our left-handed and ambidextrous patients into those with clinical and radiological evidence of damage to the left hemisphere sustained before the age of 6 and those for whom there is no such indication. The early-lesion group is not restricted to cases of infantile right hemiplegia, but includes, for example, patients with right visual-field defects or with right-sided sensory loss but normal power. In Table II, which gives the findings for 212 consecutively studied patients (95 of whom were right-handed and 117 left-handed or ambidextrous), the results for these two sub-groups of left-handers are listed separately.

TABLE II

Handedness and carotid-Amytal speech lateralization: Results for 212 consecutively studied patients. (Incorporating data from Milner, Branch, and Rasmussen, 1966.)

Handedness	No. of Cases	Speech Representation		
		Left	Bilateral	Right
Right	95	87 (92%)	1 (1%)	7 (7%)
Left or ambidextrous WITHOUT early left-hemisphere damage	74	51 (69%)	10 (13%)	13 (18%)
WITH early left-hemisphere damage	43	13 (30%)	7 (16%)	23 (54%)

Several findings invite comment here. First, one notes the surprisingly high incidence of right-hemisphere dominance for speech among the right-handers. Although the figure of 7 percent is almost certainly an artefact of selection, there may well be more such cases among the general population than is usually assumed. In support of this, I would point out that our short series includes one instance of right hemisphere trauma in adult life, one right frontal meningioma, and one patient with an early lesion of the *right* posterior cortex (Milner et al., 1964).

Second, these Amytal results confirm that cerebral dominance for speech is indeed more variable among left-handers than among right-handers, with a substantial number of them showing evidence of some bilateral speech representation (Table II). In these cases, not only were the speech defects mild, from both the right and the left-sided injections (as was noted above), but in 9 of the 17 patients a totally unexpected dissociation between defects of naming and defects of serial order was observed: Seven patients made mistakes in saying the days of the week and in counting forward and backward after right-sided injection, whereas after left-sided injection they made mistakes in naming but no mistakes in series; the other two patients showed the converse pattern. This qualitative distinction, which has been validated in some of our postoperative findings, contrasts sharply with what is seen after injection into the dominant (left) hemisphere of right-handed patients, where naming and serial order are typically disturbed together and recover together. Whether this asymmetric distribution of speech functions between the hemispheres in certain left-handers is a normal phenomenon or whether it is a secondary consequence of early brain lesions is still unknown.

When all this has been said, the main outcome of these Amytal studies has still to be discussed: namely, the impressive tendency for speech functions to become organized in the left cerebral hemisphere. Admittedly, this is less true of left-handers than of right-handers, suggesting the play of genetic factors; yet, if we consider only the group of left-handers without evidence of early left brain injury (and these are presumably more representative of the normal population), we note from Table II that over two-thirds of them have speech on the left. In the other group, most of the subjects had sustained gross injuries to the left hemisphere at or near the time of birth; yet we observe that in 30 percent of the cases this damaged left hemisphere is still responsible for speech. In the light of such findings, it seems important to try to find some clues as to how this remarkable left-hemisphere specialization comes about and under what conditions it is aborted.

THE PROBLEM OF ONTOGENY   It is generally conceded that in childhood, during the early acquisition of language, injury to the right hemisphere affects speech much more frequently than it does in adults but that there is extremely rapid recovery (Basser, 1962). This kind of observation suggests that at this early stage both hemispheres are participating in the development of language, though probably not to an equal extent. Such bilaterality is consonant with the fact that severe injury to the left side of the brain before the age of 6, though it may cause a temporary loss of speech, is usually followed by an orderly development of language processes in the right hemisphere, so that no permanent dysphasia results; this is not the case with similar injuries later in life. It looks then as though the early left-hemisphere lesion leaves the right hemisphere free to develop its innate language capacities, whereas, in the intact brain, these capacities are actively suppressed by the left hemisphere at some critical stage of early development.

Role of intrinsic morphologic asymmetries   If one asks what gives the left hemisphere its developmental advantage, one may find part of the answer in actual morphological differences between the two sides of the brain. Thus, Geschwind and Levitzky in 1968 examined the brains of 100 adult patients who had died of non-neurologic causes and found that the supratemporal plane (the area just behind the auditory cortex in the temporal lobe) was about 1 cm longer in the left hemisphere than in the right for 65 percent of the cases; in 11 percent it was longer on the right, and in 24 percent it was equal on the two sides. Since then, these findings have been replicated by Wada (personal communication) for another 100 adult brains and also demonstrated for a series of 100 normal infant brains. In Wada's two series, the asymmetry was if anything more striking than in the earlier study, with 90 percent of both the adult and the infant brains showing a substantially larger planum temporale on the left than on the right. He also reports that this asymmetry is visible at the gestational age of 20 weeks. Such an anatomical difference in the temporal speech zone could well provide an advantage for the left hemisphere in language acquisition from birth onwards.

At this point one naturally wonders if there is some corresponding asymmetry which will favor the right hemisphere and perhaps provide a clue to the critical zones in the development of the nonverbal skills in which that hemisphere excels. It is therefore of interest to note that Wada finds that the area of the frontal operculum is systematically larger on the right than on the left, both in the infant and the adult brains. McRae et al. (1968) have also reported that the occipital horn of the lateral ventricle tends to be longer on the left than on the right, which could mean that the actual mass of brain tissue in that posterior part of the hemisphere is greater on the right than on the left. Whether either of these measurements has any relevance for behavior is still undetermined, whereas it is difficult to dismiss the asymmetries in the temporal region as irrelevant to the asymmetric representation of language.

Site of early lesions as a critical factor   At this point it becomes important to find out which early left-hemisphere injuries cause speech to develop on the other side (or perhaps bilaterally) and which are compatible with the continued specialization of the left hemisphere in verbal processes (see Table II). Age at injury and severity of injury are known to be major determinants (Lansdell, 1969), but it is perhaps less generally realized that the locus of the injury is also a crucial factor (Milner, 1968b). This will be illustrated below with reference to the speech zones of the left cerebral cortex (Penfield and Roberts, 1959).

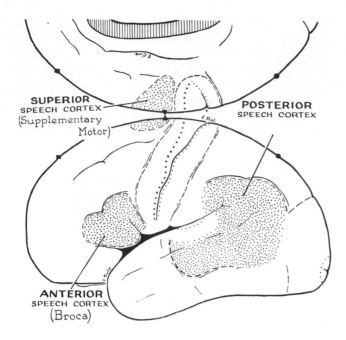

FIGURE 10 Brain chart showing areas in the left cerebral cortex where electrical stimulation during surgery has caused a speaking patient to become momentarily dysphasic. Lateral view below; medial view above. (Adapted from Penfield and Roberts, 1959, p. 201. Reproduced by permission of Princeton University Press.)

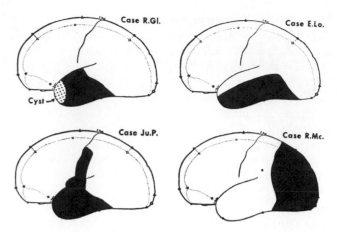

FIGURE 11 Cortical excisions for epilepsy in patients with speech representation on the left despite early injury to the left cerebral hemisphere.

Figure 10 shows the traditional speech areas in the inferior frontal and posterior temporoparietal regions of the left hemisphere, as mapped out by electrical stimulation of the exposed cortex in conscious adults. Penfield and Roberts found that stimulation in these areas of the dominant hemisphere interferes with speech and thus they are always spared in elective surgery for fear of causing permanent speech disability. The superior speech cortex in the supplementary motor area can safely be excised, although stimulation here also arrests speech. The results of our Amytal studies for cases of early left-hemisphere injury indicate that lesions outside these two critical zones are without effect upon the lateralization of speech, but that comparable destruction within these zones is likely to result in right-hemisphere dominance for speech or, more rarely, in a bilateral representation, in which the damaged area can be removed without risk to language.

The cortical excisions shown diagrammatically in Figures 11 and 12 were carried out for the treatment of epilepsy, in patients who had sustained early lesions of the left hemisphere and in whom preoperative intra-carotid Amytal tests had been performed to determine the side of speech representation. The patients whose removals are illustrated in Figure 11 were all found to have speech still represented in the left hemisphere. They include a case of early destruction of the occipital lobe associated with a right homonymous hemianopia (R.Mc.), as well as one in which the face area of the pre- and postcentral gyri (between the two speech zones) was damaged, together with the anterior temporal cortex (Ju.P.).

Figure 12 shows, instead, removals that have been carried out in patients who had been found to have speech represented on the right. Some are extensive lesions destroying the anterior or posterior two-thirds of the hemisphere, so that it is hardly surprising that speech developed on the other side. Others, however, are limited to the posterior parietotemporal region (J.Pe. and G.Ra.) but clearly invade the posterior speech zone as depicted in Figure 10. In Y.P., the posterior cortex is spared but Broca's area was grossly damaged together with the rest of the frontal lobe and the motor cortex. A comparison of these brain maps with those shown in Figure 11, suggests to me that, in cases of early lesion, patterns of speech representation at maturity may depend quite critically on whether or not the lesion involves the primary speech areas.

The intellectual consequences of these two types of early lesions are quite different. Damage outside the primary speech areas is associated with highly specific defects which reflect the specialization of the hemisphere and the site of the lesion. Thus, regardless of whether the lesion occurs early or late in life, left anterior temporal damage is likely to cause a specific impairment of verbal memory but no general intellectual loss. With an early injury so located that speech has to develop on the other

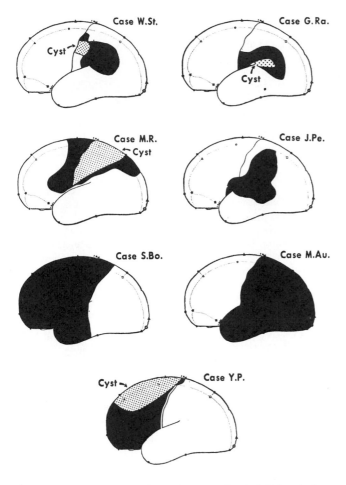

FIGURE 12. Cortical excisions in cases of early left-hemisphere injury with speech lateralized to the right. Note that anomalous representation of language is associated with early lesions encroaching significantly on the language areas as depicted in Figure 10.

side, the situation is radically different, because there is always an intellectual price to pay for such plasticity. As both Sperry and Teuber have indicated, people with most of their cognitive abilities "crowded" into one hemisphere are likely to be lower in general intelligence than the normal population (Hebb, 1942; Basser, 1962; Lansdell, 1969). Again, both Sperry and Teuber emphasize that verbal skills tend to develop at the expense of nonverbal ones in this kind of intrahemispheric competition; but the fact remains that both are low.

As study of the ontogeny of hemispheric specialization continues, much will be gained if one can find some index of right-hemisphere specialization that could be as readily identified as speech. Given such a behavioral marker, one could then refine one's questions as to the extent and limits, not only of the division of roles between the hemispheres, but also of possible re-

organization within each hemisphere following early or late lesions of either side.

ACKNOWLEDGMENTS   I wish to thank Dr. R. W. Sperry for making it possible for me to study the commissurotomized patients at the California Institute of Technology, and Dr. H.-L. Teuber for his help in the organization of the program, as well as for valuable critical comment during the writing of the manuscript. The work was supported by the Medical Research Council of Canada through Operating Grant MT 2624 and through the award of a Medical Research Associate-ship to the author.

REFERENCES

BASSER, L. S., 1962. Hemiplegia of early onset and the faculty of speech with special reference to the effects of hemispherectomy. *Brain* 85:427–460.

BLAKEMORE, C. B., and M. A. FALCONER, 1967. Long-term effects of anterior temporal lobectomy on certain cognitive functions. *J. Neurol. Neurosurg. Psychiat.* 30:364–367.

BRANCH, C., B. MILNER, and T. RASMUSSEN, 1964. Intracarotid Amytal for the lateralization of cerebral speech dominance. *J. Neurosurg.* 21:399–405.

BROADBENT, D. E., 1954. The role of auditory localization in attention and memory. *J. Exp. Psychol.* 47:191–196.

BROADBENT, D. E., and M. GREGORY, 1964. Accuracy of recognition for speech presented to the right and left ears. *Quart. J. Exp. Psychol.* 16:359–360.

BROCA, P., 1865. Sur la faculté du langage articulé. *Bull. Soc. d'Anthropol.* (*Paris*) 6:337–393.

CHESHER, E. D., 1936. Some observations concerning the relation of handedness to the language mechanism. *Bull. Neurol. Inst., N.Y.* 4:556–562.

CONRAD, K., 1949. Uber aphasische Sprachstörungen bei hirnverletzten Linkshändern. *Nervenarzt* 20:148–154.

CORKIN, S., 1965. Tactually-guided maze learning in man: Effects of unilateral cortical excisions and bilateral hippocampal lesions. *Neuropsychologia* 3:339–351.

CORSI, P. M., 1972. *Human Memory and the Medial Temporal Region of the Brain.* Unpublished Ph.D. thesis, McGill University.

CRITCHLEY, M., 1953. *The Parietal Lobes.* London: Edward Arnold (Reprinted 1966, New York: Hafner Publishing Company).

DE RENZI, E., P. FAGLIONI, and G. SCOTTI, 1971. Judgment of spatial orientation in patients with focal brain lesions. *J. Neurol. Neurosurg. Psychiat.* 34:489–495.

DEWSON, III, J. H., 1964. Speech sound discrimination by cats. *Science* 144:555–556.

DEWSON, III, J. H., A. COWEY, and L. WEISKRANTZ, 1970. Disruptions of auditory sequence discrimination by unilateral and bilateral cortical ablations of superior temporal gyrus in the monkey. *Exp. Neurol.* 28:529–548.

GAZZANIGA, M. S., and S. A. HILLYARD, 1971. Language and speech capacity of the right hemisphere. *Neuropsychologia* 87:415–422.

GAZZANIGA, M. S., and R. W. SPERRY, 1967. Language after section of the cerebral commissures. *Brain* 90:131–148.

GESCHWIND, N., and W. LEVITZKY, 1968. Human brain: Left-right asymmetries in temporal speech region. *Science* 161: 186–189.

GHENT, L., M. MISHKIN, and H.-L. TEUBER, 1962. Short-term memory after frontal-lobe injury in man. *J. Comp. Physiol. Psychol.* 55:705–709.

GOODGLASS, H., and F. A. QUADFASEL, 1954. Language laterality in left-handed aphasics. *Brain* 77:521–548.

HEBB, D. O., 1942. The effect of early and late brain injury upon test scores, and the nature of normal adult intelligence. *Proc. Am. Philos. Soc.* 85:275–292.

HÉCAEN, H., 1962. Clinical symptomatology in right and left hemispheric lesions. In: *Interhemispheric Relations and Cerebral Dominance*, V. B. Mountcastle, ed. Baltimore: The Johns Hopkins Press, pp. 215–243.

HÉCAEN, H., and J. SAUGUET, 1971. Cerebral dominance in left-handed subjects. *Cortex* 7:19–48.

KIMURA, D., 1961a. Cerebral dominance and the perception of verbal stimuli. *Canad. J. Psychol.* 15:166–171.

KIMURA, D., 1961b. Some effects of temporal-lobe damage on auditory perception. *Canad. J. Psychol.* 15:156–165.

KIMURA, D., 1963. Right temporal-lobe damage. *Arch. Neurol.*, 8:264–271.

KIMURA, D., 1964. Left-right differences in the perception of melodies. *Quart. J. Exp. Psychol.* 16:355–358.

LANSDELL, H., 1969. Verbal and nonverbal factors in right-hemisphere speech. *J. Comp. Physiol. Psychol.* 69:734–738.

McRAE, D. L., C. L. BRANCH, and B. MILNER, 1968. The occipital horns and cerebral dominance. *Neurology* 18:95–98.

MEYER, V., and A. J. YATES, 1955. Intellectual changes following temporal lobectomy for psychomotor epilepsy. *J. Neurol. Neurosurg. Psychiat.* 18:44–52.

MILNER, B., 1958. Psychological defects produced by temporal-lobe excision. *Res. Publ. Assoc. Res. Nerv. Ment. Dis.* 36:244–257.

MILNER, B., 1962. Laterality effects in audition. In: *Interhemispheric Relations and Cerebral Dominance*, V. B. Mountcastle, ed. Baltimore: Johns Hopkins Press, pp. 177–195.

MILNER, B., 1964. Some effects of frontal lobectomy in man. In: *The Frontal Granular Cortex and Behavior*, J. M. Warren and K. Akert, eds. New York: McGraw-Hill, pp. 313–334.

MILNER, B., 1965. Visually-guided maze learning in man: Effects of bilateral hippocampal, bilateral frontal, and unilateral cerebral lesions. *Neuropsychologia* 3:317–338.

MILNER, B., 1967. Brain mechanisms suggested by studies of the temporal lobes. In: *Brain Mechanisms Underlying Speech and Language*, F. L. Darley, ed. New York: Grune and Stratton, pp. 122–145.

MILNER, B., 1968a. Visual recognition and recall after right temporal-lobe excision in man. *Neuropsychologia* 6:191–209.

MILNER, B., 1968b. Paper read at XIVth International Symposium of Neuropsychology. Abstracted by E. De Renzi and M. Piercy, 1969. *Neuropsychologia* 7:383–386.

MILNER, B., 1970. Memory and the medial temporal regions of the brain. In: *Biological Bases of Memory*, K. H. Pribram and D. E. Broadbent, eds. New York: Academic Press, pp. 29–50.

MILNER, B., 1971. Interhemispheric differences in the localization of psychological processes in man. *Brit. Med. Bull.* 27:272–277.

MILNER, B., 1972. Disorders of learning and memory after temporal-lobe lesions in man. *Clin. Neurosurg.* 19:421–446.

MILNER, B., C. BRANCH, and T. RASMUSSEN, 1964. Observations on cerebral dominance. In: *Disorders of Language (Ciba Foundation Symposium)*, A. V. S. De Reuck and Maeve

O'Connor, eds. London: J. & A. Churchill Ltd., pp. 200–214.

MILNER, B., C. BRANCH, and T. RASMUSSEN, 1966. Evidence for bilateral speech representation in some non-right-handers. *Trans. Am. Neurol. Assoc.* 91:306–308.

MILNER, B., L. TAYLOR, and R. W. SPERRY, 1968. Lateralized suppression of dichotically presented digits after commissural section in man. *Science* 161:184–186.

MILNER, B., and L. TAYLOR, 1972. Right-hemisphere superiority in tactile pattern-recognition after cerebral commissurotomy: Evidence for nonverbal memory. *Neuropsychologia* 10:1–15.

MILNER, B., and H.-L. TEUBER, 1968. Alteration of perception and memory in man: Reflections on methods. In: *Analysis of Behavioral Change*, L. Weiskrantz, ed. New York: Harper and Row, pp. 268–375.

NEFF, W. D., 1961. Neural mechanisms of auditory discrimination. In: *Sensory Communication*, W. A. Rosenblith, ed. New York: John Wiley & Sons, pp. 259–278.

PENFIELD, W., and L. ROBERTS, 1959. *Speech and Brain Mechanisms*. Princeton: Princeton University Press.

PETERSON, L. R., and M. S. PETERSON, 1959. Short-term retention of individual verbal items. *J. Exp. Psychol.* 58:193–198.

POSNER, M. I., 1966. Components of skilled performance. *Science* 152:1712–1718.

PRIBRAM, K. H., B. S. ROSNER, and W. A. ROSENBLITH, 1954. Electrical responses to acoustic clicks in monkeys: Extent of neocortex activated. *J. Neurophysiol.* 17:336–344.

PRIBRAM, K. H., and W. E. TUBBS, 1967. Short-term memory, parsing, and the primate frontal cortex. *Science* 156:1765–1767.

SEMMES, J., 1965. A non-tactual factor in astereognosis. *Neuropsychologia* 3:295–315.

SEMMES, J., S. WEINSTEIN, L. GHENT, and H.-L. TEUBER, 1955. Spatial orientation in man after cerebral injury—I. Analyses by locus of lesion. *J. Psychol.* 39:227–244.

SHANKWEILER, D. P., 1966. Effects of temporal-lobe damage on perception of dichotically presented melodies. *J. Comp. Physiol. Psychol.* 62:115–119.

SPARKS, R., and N. GESCHWIND, 1968. Dichotic listening in man after section of neocortical commissures. *Cortex* 4:3–16.

SUBIRANA, A., 1958. The prognosis in aphasia in relation to cerebral dominance and handedness. *Brain* 81:415–425.

TEUBER, H.-L., 1955. Physiological psychology. *Ann. Rev. Psychol.* 6:267–296.

TEUBER, H.-L., 1964. The riddle of frontal-lobe function in man. In: *The Frontal Granular Cortex and Behavior*, J. M. Warren and K. Akert, eds. New York: McGraw-Hill, pp. 410–444.

TEUBER, H.-L., 1972. Unity and diversity of frontal lobe functions. *Acta Neurobiol. Exp.* 32:615–656.

WADA, J., 1949. [A new method for the determination of the side of cerebral speech dominance. A preliminary report on the intracarotid injection of Sodium Amytal in man.] *Igaku to Seibutsugaku* [*Medicine and Biology*] 14:221–222 (Japanese).

WARRINGTON, E. K., 1969. Constructional apraxia. In: *Handbook of Clinical Neurology Vol. 4: Disorders of Speech, Perception and Symbolic Behavior*, P. J. Vinken and G. W. Brown, eds. Amsterdam: North-Holland Publishing, pp. 67–83.

WARRINGTON, E. K., and M. JAMES, 1967. An experimental

investigation of facial recognition in patients with unilateral cerebral lesions. *Cortex* 3:317–326.

YNTEMA, D. B., and F. P. TRASK, 1963. Recall as a search process. *J. Verb. Learn. Verb. Behav.* 2:65–74.

ZANGWILL, O. L., 1960. *Cerebral Dominance and its Relations to Psychological Function.* Edinburgh: Oliver and Boyd.

ZANGWILL, O. L., 1967. Speech and the minor hemisphere. *Acta Neurol. Psychiat. Belg.* 67:1013–1020.

FEATURE EXTRACTION BY NEURONS AND BEHAVIOR

The processing of images by computers as a means of determining the informational requirements for visual object recognition. A transparency of a conventional portrait is flying-spot scanned, and the resultant signals are converted to digital form on magnetic tape. A high-speed digital computer stores the dissected image in the form of 1024 × 1024 points, to each of which one of 1024 possible grey levels is assigned. In the illustrated example of Lincoln's portrait, the computer then divided the entire picture frame into 20 × 20 squares and computed for each square an average brightness value. Next, the brightness of each square is requantified, typically to 16 different values lying between all black and all white. Finally, the computer reads the information contained in this coarse-grained processed picture onto a digital tape from which it can be displayed very much like a conventional television image. Viewed closeup, the pictures produced in this fashion look like cubistic paintings. Viewed remotely (e.g. 30 to 40 picture diameters), faces are perceived and recognized. (Leon D. Harmon, Professor and Chairman, Department of Biomedical Engineering, Case Western Reserve University, kindly made this illustration available. For a detailed account from which part of the above legend is quoted verbatim, see: L. D. Harmon, Some aspects of recognition of human faces. In Pattern Recognition in Biological and Technical Systems, *O. J. Grüsser and R. Klinke, eds. Berlin: Springer-Verlag, 1971.)*

FEATURE EXTRACTION BY NEURONS
AND BEHAVIOR

Introduction

GERHARD WERNER

As an alternative to the exhaustive specification of sensory input by neural activity in afferent pathways, coding schemes that reduce redundancy in sensory stimuli have attracted attention since the mid-1950s (Attneave, 1954; Barlow, 1959). The underlying notion was that evolutionary adaptation of the organisms to certain types of redundancies, which are always present in the environment, would have occurred. The guiding principle in these considerations was Shannon's concept of "optimal codes," which match the statistical structure of regularities in the available repertoire of messages.

As far as it was known at that time, there were only two mechanisms available to the nervous system to reduce redundancy of information in the stimulus. These were, in the temporal domain, mechanisms specifically sensitive to onset and cessation of a stimulus; and in the spatial domain, lateral inhibition. The potential significance of such redundancy reducing codes consists of economy in signal transmission, because these codes exploit lawful regularities in the stimulus source. For instance, the duration of a stationary stimulus is uniquely defined by the moments of its beginning and end, and thus there is no need for generating neuronal impulses in the interval between these points in time.

The concept of the "stimulus feature" can be considered a generalization of this principle. This concept becomes applicable whenever a neuronal discharge signals with relatively high selectivity the occurrence of an input state that contains in its specifications the concomitant occurrence of certain regularities in the stimulus space, in addition to being specific for a certain place on the receptor sheet and stimulus modality. For instance, because matter is cohesive, objects can be fully characterized in terms of their boundaries; hence, boundaries (that is, edges, corners, and angles) become the "features" in terms of which the spatial layout of a stimulus object can be unambiguously represented.

Such features may be purely of a spatial nature, consisting of stationary contours or patterns, or they may combine spatial with temporal information in the form of a stimulus motion. The economy consists, then, of limiting the characterization of a shape to the signaling of its boundaries, or of emphasizing change over stationarity.

Carried to its logical consequence, this principle implies that the central nervous system takes the information available in the proximal stimulus on receptor sheets out of its original context and imposes a classification into disjunctive entities. The general principle appears to be that the number of neurons available to signal stimulus information increases with progression along the afferent pathway but that any given neuron is less frequently activated as the constraints of its stimulus feature become more severe (Barlow, 1969).

The neurophysiological reality of feature detection is by now amply substantiated; however, it also raises many problems. This is most aptly captured in the concluding sentence of a paper by Hubel and Wiesel (1968):

Specialized as the cells of area 17 are, compared with rods and cones, they must, nevertheless, still represent a very elementary stage in handling complex forms, occupied as they are with a relatively simple region-by-region analysis of retinal contours. How this information is used at later stages in the visual path is far from clear and represents one of the most tantalizing problems for the future.

These problems range from clarifying in neurophysiological experiments the synaptic mechanisms enabling feature detection to establishing the perceptual relevance of feature-sensitive neurons by psychological means.

This latter issue comes into sharper focus if we ask the question which place and what role feature-signaling neurons in the nervous system could play in a mechanistic account of behavioral and perceptual functions and in a psychological theory of information processing. To illustrate contrasting possibilities of their involvement, we may consider features as a hierarchy of elements in terms of which entire scenes of sensory input are analyzed and described (Barlow, Narasimhan, and Rosenfeld, 1972); alternatively, we may consider that feature elements stand at the interface between perception and action, in the sense that they trigger or "release" behavioral acts in an automatalike fashion.

Irrespective of our theoretical biases and inclinations, these are fundamental problems in trying to bring behavioral performance into register with current neurophysiological concepts. To some extent, at least, these problems are a consequence of the manner in which neurophysiological experiments characterize and identify stimulus feature-sensitive neurons, namely, by observing electrical activity in individual neurons, one at a time. The correlated perceptual activity, on the other hand, involves global activity in complex neural systems.

These comments are intended to underscore some of the principal issues to which the following papers in this part are addressed.

REFERENCES

ATTNEAVE, F., 1954. Informational aspects of visual perception. *Psychol. Rev.* 61:183–193.

BARLOW, H. B., 1959. Sensory mechanisms, the reduction of redundancy and intelligence. Symposium on the mechanization of thought processes at the National Physical Laboratory, London. H.M. Stationery Office, Symp. No. 10, 535–539.

BARLOW, H. B., 1969. Trigger features, adaptation and economy of impulses. In *Information Processing in the Nervous System*, K. N. Leibovic, ed. Berlin: Springer-Verlag.

BARLOW, H. B., R. NARASIMHAN, and A. ROSENFELD, 1972. Visual pattern analysis in machines and animals. *Science* 177: 567–575.

HUBEL, D. H., and T. N. WIESEL, 1968. Receptive fields and functional architecture of monkey striate cortex. *J. Physiol. (London)* 195:215–243.

9 The Transmission of Spatial Information through the Visual System

FERGUS W. CAMPBELL

ABSTRACT The contrast threshold of grating patterns with a variety of wave forms is examined. The thresholds, over a wide range of spatial frequency, can be accounted for by the simple application of Fourier theory. It is suggested that there may be in the visual system channels tuned to spatial frequency because it has been shown neurophysiologically, in the cat and monkey, that the neurons discovered in the cortex by Hubel and Wiesel are indeed sharply tuned to spatial frequency as well as orientation. The analogy of this mechanism with the one present in the auditory system for pitch transmission has led to many psychophysical experiments to test how useful this paradigm is in accounting for established visual phenomena and for predicting the visibility of other test patterns.

THE THEOREM OF Fourier (1768–1830) enabled Helmholtz (1821–1894) to make his contribution to our understanding of the physical nature of music and to the first scientific attempt to unravel the physiology of hearing. We know from his biographer, Koenigsberger (1965), that Helmholtz "had for years gone to bed and got up again with Fourier's series in his mind," and it is therefore surprising that he did not apply with equal success Fourier's technique to optics and to the physiology of the eye.

Helmholtz came very near to considering it, for Koenigsberger (1965) reports that Helmholtz thought that "the eye has no harmony in the same sense as the ear; it has no music." In making this comparison between the eye and the ear, Helmholtz was contrasting color vision with that of harmony; for the ear can analyze a compound sound into its frequency components while the eye cannot resolve a compound color into its component spectral wavelengths. Half a century had to pass before the first step was taken by Duffieux (1946) to apply Fourier theory to physical optics. It could well have been done by Helmholtz.

I hope to show you that harmony is certainly present in the visual sense, but it is in the domain of space and not of color.

We know from experience that as an object recedes from us it becomes increasingly difficult to perceive its details until, at a sufficiently great distance, the object itself disappears from view. Alice, in *Through the Looking-Glass*, once remarked "I see nobody on the road," and the White King replied in a fretful tone "I only wish I had such eyes, to be able to see nobody and at that distance too."

Many factors have been put forward to account for this everyday experience. For example, the limits of visual acuity may be restricted by the optical properties of the eye, or by the dimensions of the foveal mosaic of cones, or by the rate at which light quanta are captured by individual photoreceptors (Rose, 1970) or even by limitations within the visual nervous system itself. In order to solve this problem in an analytical manner, it is necessary to measure the transmission of spatial signals through each part of the system. To do this it is essential to choose an input signal that produces a fairly simple output signal. We have chosen to use the mathematically simplest spatial signal of all—a grating whose luminance varies sinusoidally along one axis (see Campbell, 1968, and Figure 1 for illustration). This is the homologue of a pure tone in the auditory system.

We are going to consider the following elements to be in cascade and investigate the transmission properties of each section in turn; object → image → ganglion cell → geniculate fiber → visual cortex neuron → psychophysics → ?

This approach for investigating the transmission of information through a system is really borrowed from electrical engineering where the attenuation and amplification of temporal signals is important. Hopkins (1962), O'Neill (1963), and Linfoot (1964) have already successfully developed Fourier theory to a point where many optical systems can be precisely and analytically described in this manner.

Arnulf and Dupuy (1960) and Westheimer (1960) were the first to realize the power of this approach in visual optics and they used a technique invented earlier by Le Grand (1937). They passed coherent beams of light into the eye and set up a sinusoidal interference grating on the retina using the principle of Thomas Young (Young, 1800). This ingenious technique bypasses the defects that might arise from aberrations in the optical components of

FERGUS W. CAMPBELL The Physiological Laboratory, Cambridge, England

FIGURE 1 The grating has a sinusoidal wave form of constant spatial frequency. Its contrast is changing logarithmically from about 0.4 to about 0.004. Due to nonlinearities in the photographic process, the wave form is not sinusoidal at high contrast levels. The reader should observe the grating from different distances and note the change in the threshold contrast with spatial frequency.

the eye, and it permitted them to establish the resolving power of the retina, coupled to the brain, in isolation from the dioptrics. Unfortunately, the complete success of this experiment depends upon having a monochromatic source with a high degree of coherence and also high luminance.

Such a source became available when the neon-helium laser was invented. Using it, Campbell and Green (1965) were able to show that the fundal image formed by a well-focused eye, with a normal-sized pupil, was surprisingly good and that most of the loss of contrast in the perception of fine gratings was due to the properties of the retina and/or brain. The transmission properties of the dioptrics have also been measured directly using an objective method (Campbell and Gubisch, 1966) and the results obtained by these two fundamentally different methods agree well.

The results of change of pupil size as well as those of focus have been investigated (Green and Campbell, 1965). Green (1967) has demonstrated the effects of off-axis aberrations and Campbell and Gubisch (1967) have

shown the effect of the chromatic aberration present in the eye on the contrast sensitivity function.

All of these experiments indicate that the quality of the retinal image in an emmetrope is very good. Thus at high light levels, the resolution is limited mainly by the properties of the retina and/or brain. At low light levels, the main limit is the number of photons being captured per cone per integration time (about 1/10 sec).

These approaches have so far been useful in finding out the nature of the fundal image (Gubisch, 1967); but can these be used to further our understanding of how spatial signals are transmitted and transformed by the nervous system itself?

The first convenient level at which spatial signals can be detected is at the ganglion cells that transmit the signals from the retina to the geniculate body. Enroth-Cugell and Robson (1966), using sinusoidal gratings generated on the face of an oscilloscope, have studied the response of these cells in the cat. In this animal, the direction of movement of a grating is unimportant as the cells respond equally to movement in all directions. They found one class of cells that responded in a linear manner to their stimuli. A finding, which may be of great significance, was that each cell responded only over a limited range of spatial frequency. The properties of these linear ganglion cells have been further studied by Cleland, Dubin, and Levick (1971).

The fibers from the geniculate body terminate in the striate visual cortex, and their activity can readily be monitored by microelectrodes (Hubel, 1957). Cooper and Robson (1968) and Campbell, Cooper, and Enroth-Cugell (1969) find that at this level the geniculate units again respond only to limited bands of spatial frequency. Like the ganglion cells, the geniculate units respond to movement in all directions.

However, the cells in the striate visual cortex behave quite differently, as has been most elegantly shown by Hubel and Wiesel (1959, 1962, and 1965), in the cat and also in the monkey (Hubel and Wiesel, 1968). Here the cells are very sensitive to the orientation of an edge or bar when it is moved across the receptive field. Using grating patterns Campbell, Cleland, Cooper, and Enroth-Cugell (1968) have measured quantitatively this selectivity to orientation in the cat. These orientationally selective cells will, of course, also respond to a grating providing it is moved at the optimum orientation. These cortical cells also respond to only limited bands of spatial frequencies, each cell responding to a different range in the spectrum of spatial frequencies (Cooper and Robson, 1968; Campbell, Cooper, and Enroth-Cugell, 1969). There is some preliminary evidence (Campbell, Cooper, Robson, and Sachs, 1969) that neurons in the cortex of the monkey are similarly organized.

These neurophysiological findings in the cat and monkey suggest that two important properties of an image have been coded. First, the information about orientation of an edge, bar, or grating is extracted. Second, the spatial frequency content of the image is also extracted.

This organization is strikingly similar to that found in the auditory system where units are found that respond to pure-tone bursts over a limited range of pitch (Kiang, 1966), rather like the neurons of the visual cortex that respond over a limited range of spatial frequency. Again, in the auditory cortex neurons are found that only respond to either a rise or a fall of pitch (Whitfield and Evans, 1965); indeed, because of their resemblance to the visual cortical cells, they have been called "directional cells" (Whitfield, 1967).

Is there any evidence that the visual system of man is similarly organized? If such organization can be demonstrated, it will greatly strengthen the argument that these neurophysiological findings are directly relevant properties of the mechanism by which we perceive, and possibly recognize, objects.

Campbell and Kulikowski (1966) attempted to show that man has channels sensitive to the orientation of a grating by measuring psychophysically the threshold of a test grating in the presence of a high-contrast masking grating. They changed the orientation of the masking grating relative to the test grating and found that maximum masking occurred when the two gratings had the same orientation and that no masking occurred when the masking grating was at right angles to the test grating. For intermediate angles, the masking effect decreased exponentially; the masking effect was reduced by half when the angle between the gratings was only 12° to 15°. Man seems to have a slightly higher orientational selectivity than the cat (Campbell, Cleland, Cooper, and Enroth-Cugell, 1968) and may resemble more closely the monkey (Hubel and Wiesel, 1968) where the angular selectivity was found to be higher than that of the cat. Thus, as far as orientation performance is concerned, there is a striking agreement between the neurophysiology and the psychophysics.

Can we show that man also has channels selectively sensitive to spatial frequency? In a preliminary note, Campbell and Robson (1964) suggested that Fourier theory might be applied to the psychophysics of spatial vision. In their main paper (Campbell and Robson, 1968), they measured the contrast threshold for a number of gratings each with a different wave form. They found that the threshold is determined by the amplitude of the fundamental Fourier component in the grating and that the higher harmonics do not contribute to the threshold, providing they are below their own threshold. In like manner, a square-wave grating can be distinguished from a sinusoidal-wave grating when the contrast is sufficiently high for the third harmonic to be detected in the presence of the fundamental. Their findings led them to suggest that there must be a number of channels in the human visual system each tuned to different spatial frequencies. They thought that the effective bandwidth of each channel is probably not greater than about one octave of spatial frequency.

In 1966, Campbell and Kulikowski used the technique of masking a low-contrast test grating with a high-contrast grating of the same orientation and spatial frequency. In undertaking these experiments they found that it was necessary to have the two gratings of the same spatial frequency or the masking effect was much less. Gilinsky (1967, 1968) developed this masking method further by adapting for some time to a grating of a given frequency and then subsequently inspecting the test grating that was at a different orientation. Using her adaptation method, Pantle and Sekuler (1968) studied the influence of gratings with the same orientation but with different spatial frequencies on each other. They summarize their findings as: "These conclusions are similar to those reached by Campbell and Robson (1968) from a Fourier analysis of the visibility of gratings of different spatial frequencies and wave forms."

Unfortunately, Pantle and Sekuler (1968) used square-wave gratings to produce their adaptation effects, and this complicates the interpretation of their data, for the higher Fourier components in their gratings also influenced the results. By using sinusoidal gratings covering a wide range of spatial frequency, Blakemore and Campbell (1969) were able to measure accurately the bandwidth of each channel and also to demonstrate that there were many channels. In these experiments the subject adapted to a given frequency, and the loss in contrast sensitivity was measured at, and on either side of, that frequency, (open circles, and right-hand and upper scale of Figure 2). The "bandwidth" of the adaptation effect is about ±1 octave. Also, the response of a single neuron from the cortex of the cat is plotted in this figure (closed circles, and left-hand and lower scale). The adaptation technique of Blakemore and Campbell shows that there are channels tuned to spatial frequencies ranging from 3 c/deg up to the upper limit of resolution at about 48 c/deg.

Blakemore and Campbell (1969) noticed that, if a subject adapted to a square-wave grating, the threshold for detecting both the fundamental and the third harmonic was elevated. Tolhurst (1972b) has examined this phenomenon more quantitatively by adapting to both square waves and also the fundamental with its third harmonic. He finds that, for high levels of adapting contrast, the presence of more than one grating reduces the

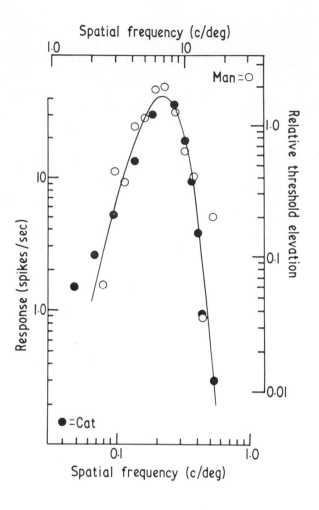

FIGURE 2 The closed circles (●) are the responses (spikes/sec) from an orientational cortical neuron from a cat. The sinusoidal grating stimulus had a contrast of 0.5 and was drifting in the preferred direction so that one bar of the grating passed the receptive field each second (Results supplied by J. G. Robson and G. F. Cooper, the Physiological Laboratory, Cambridge).

The open circles (○) are the relative threshold elevations in man produced by adapting to a sinusoidal grating of contrast 0.7 and spatial frequency 7 c/deg (Blakemore and Campbell, 1969).

To compare the response of this particular cat neuron with the psychophysical adaptation, the data has been superimposed to obtain the best fit by eye and the appropriate scales attached. It must be emphasized that there is a range of cortical neurons each responding maximally to a different spatial frequency (Campbell, Cooper, and Enroth-Cugell, 1969; Campbell, Cooper, Robson, and Sachs, 1969). Likewise, in the human, adaptation to different spatial frequencies reveals a comparable range of tuned channels (Blakemore and Campbell, 1969).

The curve through the results is

$$s = (e^{-f^2} - e^{-(2f)^2})^3$$

where s = contrast sensitivity and f = spatial frequency.

adaptation effect. He postulates that there is mutual inhibition between the fundamental and its third harmonic at these suprathreshold contrast levels; at or near threshold, there is no such reciprocal inhibition.

Sharpe (1972) has studied the fading of retinal capillaries and arterioles when they are stabilized on the retina. He finds that the fading is influenced not only by the orientation of a given capillary but also by its spatial frequency or size. His results can be explained in terms of adaptation (Blakemore and Campbell, 1969) of spatial frequency and orientation channels.

The basic idea that there might be, in vision, channels tuned to spatial frequency has been taken up in audition, and adaptation to frequency modulated tones has been discovered (Kay and Matthews, 1971, 1972).

Campbell and Maffei (1970) and Maffei and Campbell (1970) have developed a technique for using the evoked potential recorded from the visual area of the scalp to obtain an objective measure for the existence of these channels. In this manner, the response of the observer is not required, for thresholds can be determined objectively. They have effectively removed the psycho from psychophysics. They confirm, at the level where the evoked potential arises, that there is a mechanism generating an electrical signal that is selectively sensitive to spatial frequency and orientation, just as has been shown in the previous single neuron and psychophysical studies.

Campbell and Kulikowski (1972) have shown that this evoked potential technique predicts objectively the 50% probability of seeing in the contrast domain.

Blakemore, Nachmias, and Sutton (1970) have shown that if one adapts to a grating of a slightly lower or higher frequency, the lower frequency appears of even lower frequency; conversely the higher frequency appears to be of even higher frequency (Figure 3). This supports the original suggestion by Blakemore and Campbell (1969) that these channels, selective to spatial frequency, may be used for coding the sizes of objects. Sachs, Nachmias, and Robson (1971) have shown psychophysically that each channel is independent from its neighbor by testing the effect on the threshold of mixing two different spatial frequencies. Even when the frequencies are quite close in period, they only add according to the laws of probability summation; in other words the channels are functionally separate. An implication from their measurements is that the bandwidth of the channels being measured in their way is very much narrower than ± 1 octave.

Graham and Nachmias (1971) measured contrast thresholds for gratings containing two superimposed sinusoidal components; the frequency of one component was always three times that of the other, but the phase between the components could be varied so that the sum

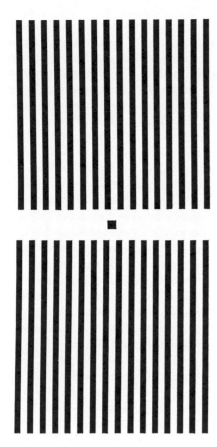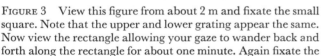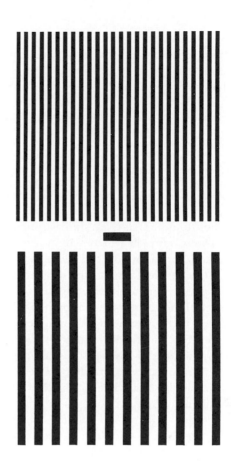

FIGURE 3   View this figure from about 2 m and fixate the small square. Note that the upper and lower grating appear the same. Now view the rectangle allowing your gaze to wander back and forth along the rectangle for about one minute. Again fixate the square and note the change in the appearance of the gratings which were previously identical. (From Blakemore and Sutton, 1969.)

of their contrasts took on several values. Two models of pattern vision were then tested: (1) a single-channel model in which pattern vision is a function of a single neural network and (2) a multiple-channel model in which the stimulus information is processed by many channels, each selectively sensitive to a narrow range of spatial frequencies. Their clear-cut results support the multiple-channel and convincingly reject the single-channel model.

Graham (1972) went on to study the sensitivity of these channels for patterns at low mean luminance or high drift rate. She found that the channels did not behave in the way expected from retinal ganglion cell neurophysiology. Instead, the channels remained selectively sensitive to narrow ranges of spatial frequency even when the luminance was low and the drift rate was high—conditions that tend to affect the inhibitory surround of ganglion cells.

A long-lasting color aftereffect, specific to edge orien-

tation was first demonstrated by McCollough (1965). She found that the effect could be built up by viewing a vertical grating of black stripes upon one color that alternated with a horizontal grating on another color. Complementary colored aftereffects were seen on test gratings of black and white stripes, with the provision that the retinal orientation was similar to the adapting gratings. Stromeyer (1972) has found that the McCollough effect is spatial frequency specific; that is, the effect gets weaker the greater the difference in spatial frequency between the adapting and test frequency. He also found that the same orientation could be used and the color effects be generated by adapting at two different spatial frequencies. The aftereffects could then be varied by the apparent frequency shift (Blakemore and Sutton, 1969); however, the interocular transfer of the frequency shift had no effect on the color aftereffect.

Using this wide range of experimental approaches in the cat, monkey, and man, it has been possible to show that the original paradigm of Campbell and Robson (1964, 1968) is productive in the sense that it has suggested a number of novel experiments, each of which confirmed their original suggestion:

Thus a picture emerges of functionally separate mechanisms in the visual nervous system each responding maximally at some particular spatial frequency and hardly at all at spatial frequencies differing by a factor of two. The frequency selectivity of these mechanisms must be determined by integrative processes in the nervous system and they appear to a first approximation at least, to operate linearly (Campbell and Robson, 1968).

It is important to note that these studies confine themselves to high spatial frequencies from about 1 to 50 c/deg in the human and from about 0.1 to 5 c/deg in the cat. At lower spatial frequencies another picture emerges. It seems unlikely that we will understand spatial vision until the performance of the visual system at both high and low spatial frequencies is fully understood and integrated.

If we observe a bipartite field whose luminance changes gradually in the middle of the field from one level to a slightly different one (a ramp profile), we see an extra narrow dark, or light, band at either end of the ramp where the luminance gradient changes abruptly. These spurious bands are called after the discoverer, Ernst Mach (Ratliff, 1965). A related illusion was described by Craik (1940) and was published posthumously (Craik, 1966). Here the two half-fields have the same luminance, but a single sawtoothed notch is introduced between the half-fields. This spatial transient makes the half-fields appear unequally bright. Cornsweet (1970) enhanced the illusion by using adjacent positive and negative sawtoothed luminance notches (Brindley, 1970, Fig. 6.3). The remarkable finding about the Craik-Cornsweet illusion is that having disturbed the uniform luminance by this localized spatial transient the visual system continues to be misled about the luminance of the field for a considerable distance away from the transient.

Campbell, Howell, and Robson (1971) have developed another way of showing this effect. A grating pattern with uniform light and dark bars is generated on the face of an oscilloscope. The spatial frequency of the grating must be less than 1 c/deg and the contrast must not be too high. As expected, the grating looks as if it had a square-wave distribution of luminance. Now we subtract from the square wave its fundamental Fourier component, leaving intact all the higher components—the third, fifth, and higher odd harmonics. Surprisingly, the pattern still looks as though it had a square wave luminance distribution. It seems that at these low spatial frequencies it is the higher harmonics alone that generate the square wave appearance of the grating. The missing luminance due to the absent fundamental has been "filled in."

Now consider what would be seen if we were to remove the higher harmonics of the square wave grating leaving only the fundamental spatial frequency. When this is done we observe that the screen is of uniform luminance; the low frequency fundamental is not perceived at all. This failure to see the fundamental is not unexpected in view of the well-known increase in the contrast threshold that occurs for spatial frequencies less than 3 c/deg (Robson, 1966; Campbell and Robson, 1968). When this illusion is considered in terms of harmonic analysis, it is no longer surprising that removing the fundamental from a square wave does not appreciably change its appearance, for the fundamental on its own is not detectable.

If we repeat the observations at higher spatial frequencies, we do notice the absence of the fundamental, presumably because the fundamental is not being attenuated relative to the higher harmonics.

It could be argued that there is another type of neuron with odd symmetry in its line-weighting function. These neurons may signal the presence of a transient of contrast, that is, an edge (Tolhurst, 1972a).

The availability of lasers and high-capacity digital computers has made it feasible to apply spatial filtering techniques to the problem of picture analyses. A well-illustrated introduction to these techniques has been prepared by Andrews (1972), and Chapter 10 of Lipson (1972) is very lucid. The practical problems that arise in making an object recognition device are of direct relevance to solving the problem of how we recognize familiar objects in complex scenes. Conversely, pattern recognition engineers are very interested in the visual system, for here is a device that actually works surprisingly well. Ginsburg (1971) has based his approach on the Kabrisky model of the visual system. He has shown that we only require quite a narrow range of low spatial frequency information to recognize many objects. Campbell, Carpenter, and Switkes (1971) are developing methods for filtering scenes using the transmission characteristics of single, and combinations of, neurons.

Using gratings with a sinusoidal luminance profile Campbell and Howell (1972) have been able to demonstrate monocular rivalry. If two gratings with sinusoidal luminance profiles are projected upon a white screen and if they are at right angles to each other, the appearance of the gratings changes continuously. If the gratings are of different color, the effect is seen even more dramatically. Say the gratings are red and green and their intensities are matched so that yellow is perceived where they cross, one then observes that sometimes the red grating is seen on its own and at other times only the green grating is observed. There are periods when both gratings are seen together, but there are never periods when both disappear. In other words, the gratings are seen to alternate.

If the two gratings have the same orientation, and the red bars of one are superimposed on the green bars of the other, then a yellow and black grating is observed, just as one would expect from the rules of color mixing. If the gratings are interdigitated by putting them out-of-phase by 180°, a red-green grating is perceived; no yellow is observed between the red and green sections. This is surprising, for in this region of the sinusoidal distribution the amounts of red and green are such that they previously produced yellow.

With the interdigitated grating, there is no alternation; the appearance is quite stable and unified. Thus, the alternation observed when the gratings were crossed cannot be due to factors such as chromatic aberration of fluctuations of accommodation. Likewise, the alternations cannot be caused by chromatic stereopsis or changes in convergence, for the alternations are observed when one eye is covered. It is not necessary to fixate on any part of the grating, and the pattern of eye movement does not significantly affect the alternation.

The interesting question arises, what is the angle required between the two gratings to produce alternation? If the phenomenon had some simple physical explanation, we might expect the rate of the alternations to increase gradually from 0° to 90°, say as a sine function. It is easy to demonstrate that this is not so; the gratings do not begin to alternate until they are tilted relative to each other by about 15–20°. This orientational selectivity is similar to that found psychophysically by Campbell and Kulikowski (1966) and shown in the cortical neurons of the cat (Campbell, Cleland, Cooper, and Enroth-Cugell, 1968).

The alternations that occur with crossed sinusoidal gratings do not depend upon the colors used, and the phenomenon can be demonstrated with either widely separated wavelengths, such as blue and red, or closely spaced colors such as yellow and yellow-green. While the effect is very dramatic when the gratings differ in color, color difference is not essential to obtain the alternation. It will work if the two gratings are the same color or are white.

If gratings with a square-wave luminance profile are used, the rate of alternation is slower. In the case of sinusoidal gratings, one grating can disappear completely, but with square-wave gratings, complete disappearance does not occur; only a weak attenuation of apparent contrast is observed.

Similar alternations occur with two crossed bars, with the provision that the bars are sufficiently blurred to remove the higher spatial frequencies in the image of the cross. This makes a useful and simple lecture-room demonstration of the orientational selectivity of the visual system discovered by Hubel and Wiesel in 1959.

One could explain this monocular rivalry on the assumption that orientations and spatial frequencies are processed in independent channels that are highly selective ($\pm 15°$ and ± 1 octave). For some unknown reason, first one channel dominates and then the other. It would be interesting to see if this alternation occurs in individual neurons in the cat or monkey.

Why has the experimental application of elementary Fourier theory so far worked so well? Its strict application demands that the system being studied be linear; that is, that the principle of superposition holds. We have used the theory only because it is the queen of all description and also because it is easy to explain the results and their implications to a wide public. It does not follow that it is the best one, or even ultimately the correct one. Since the visual system has neurons tuned to each orientation, it seems empirically sensible to use a one-dimensional function for studying these later stages of signal transmission.

It is true that the fact that a threshold exists means that there is indeed a nonlinearity. However, for threshold measurements this nonlinearity is easy to handle mathematically. The visual system is grossly nonlinear when the dynamics of dark and light adaptation are involved. We have always eschewed this complication by restricting our studies to the performance of the eye at one mean light level. Likewise, the visual system is nonlinear if it is overloaded with very high contrast levels. We have confined ourselves to levels of contrast less than 0.7, which covers almost all of the range used in normal sight, providing one does not look directly at light sources or subject the eye to glare. The contrast of the print that you are now reading is 0.7.

It would be rather disappointing if this internal consistency for the prediction of thresholds for different types of objects was restricted to repetitive patterns, such as gratings, and not to other more interesting and realistic objects such as surround us in daily life. Campbell, Carpenter, and Levinson (1969) have attempted to predict the thresholds of thin lines and bars from the contrast sensitivity function. They find that it is possible to do this, again using a linear theory (Fourier).

Sullivan, Georgeson, and Oatley (1972) have studied the effect of adapting to gratings on the subsequent threshold for bars of different width. They find that bar width and spatial frequency are not equivalent and that there is no evidence for width selective channels. Bars seem to be detected when their frequency components most easily detected by the visual system (near 5 c/deg) rise above their independent thresholds.

Carter and Henning (1971) have designed some experiments to test whether the behavior of the eye can be considered as similar to the behavior of Helmholtz's model of the ear. Their experiments are analogous to some auditory experiments of Wightman and Leshowitz (1970) and Leshowitz and Wightman (1971). The detectibility of sinusoidal gratings comprised of either one or many cycles was measured in veiling luminances, the spatial frequencies of which were either narrow or broad-band. In narrow-band noise, the single-cycle grating was detected with approximately 0.6 log units less contrast than the many-cycle grating. On the other hand, when both broad-band and narrow-band noise were present, there was no measureable difference in the detectability of the two types of grating. Carter and Henning (1971) consider critically a number of detection models to account for their results and conclude

In any case, it is clear that the concept of the energy density spectrum of the object grating, together with the Campbell and Robson (1968) hypothesis that the visual system analyses spatial frequencies in separate bands, leads to qualitatively predictable results.

I may not have convinced you that the visual system really does have harmony, but you may now agree that it will be in the domain of spatial frequency and contrast, and not color, where we will appreciate the harmony of which Helmholtz might well have dreamed. Today it would be difficult to imagine our concepts of sound, music, and audition if Fourier had not gifted Helmholtz with this approach. In audition it has brought some order out of chaos (Helmholtz, 1877), although much remains to be understood. The history of the application of the Fourier series in vision lies in the future. It may be short, it may be long; at least it should be interesting.

NOTES ADDED IN PROOF Maffei and Fiorentini (1972) showed that the information about the amplitude and phase of two sinusoidal stimuli, presented separately to the two eyes, can be synthesized by the visual system. Results indicate that this process of synthesis can be described in terms of Fourier theory, at least for relatively low contrast, and nervous operators exist that are able to rebuild the image from its sinusoidal components. (L. Maffei, and A. Fiorentini, 1972, *Nature* 240:479–481.) Bodis-Wollner (1972) measured the contrast sensitivity function on two patients with neurological disorders in the visual cortex and finds that, although visual acuity is only slightly decreased, they have markedly decreased contrast sensitivity at all spatial frequencies particularly at intermediate and high spatial frequencies. He coins the term visuogram for this clinical method. (I. Bodis-Wollner, 1972, *Science* 178:769–771.)

REFERENCES

ANDREWS, H. C., 1972. Digital computers and image processing. *Endeavour* 31 (113): 88–94.

ARNULF, A., and O. DUPUY, 1960. La transmission des contrastes par le système optique de l'oeil et les seuils des contrastes rétiniens. *Compt. Rend. Acad. Sci.* 250: 2727–2759.

BLAKEMORE, C., and F. W. CAMPBELL, 1969. On the existence in the human visual system of neurons selectively sensitive to the orientation and size of retinal images. *J. Physiol. (London)* 203:237–260.

BLAKEMORE, C., J. NACHMIAS, and P. SUTTON, 1970. The perceived spatial frequency shift: Evidence for frequency selective neurons in the human brain. *J. Physiol. (London)* 210: 727–750.

BLAKEMORE, C., and P. SUTTON, 1969. Size adaption: The new aftereffect. *Science* 166:245–247.

BRINDLEY, G. S., 1970. *Physiology of the Retina and Visual Pathway*, 2nd ed. London: Edward Arnold (Publishers) Ltd.

CAMPBELL, F. W., 1968. The human eye as an optical filter. *Proc. IEEE* 56:1009.

CAMPBELL, F. W., R. H. S. CARPENTER, and J. Z. LEVINSON, 1969. Visibility of aperiodic patterns compared with that of sinusoidal gratings. *J. Physiol. (London)* 204:283–298.

CAMPBELL, F. W., R. H. S. CARPENTER, and E. SWITKES, 1971. Simple scanning devices for computer modelling of visual processes. *J. Physiol. (London)* 217:18–19.

CAMPBELL, F. W., B. C. CLELAND, G. F. COOPER, and CHRISTINA ENROTH-CUGELL, 1968. The angular selectivity of visual cortical cells to moving gratings. *J. Physiol. (London)* 198: 237–250.

CAMPBELL, F. W., G. F. COOPER, and CHRISTINA ENROTH-CUGELL, 1969. The spatial selectivity of the visual cells of the cat. *J. Physiol. (London)* 203:223–235.

CAMPBELL, F. W., G. F. COOPER, J. G. ROBSON, and M. B. SACHS, 1969. The spatial selectivity of visual cells of the cat and the squirrel monkey. *J. Physiol. (London)* 204: 120–121.

CAMPBELL, F. W., and D. G. GREEN, 1965. Optical and retinal factors affecting visual resolution. *J. Physiol. (London)* 181: 576–593.

CAMPBELL, F. W., and R. W. GUBISCH, 1966. Optical quality of the human eye. *J. Physiol. (London)* 186:558–578.

CAMPBELL, F. W., and R. W. GUBISCH, 1967. The effect of chromatic aberration on visual acuity. *J. Physiol. (London)* 192:345–359.

CAMPBELL, F. W., and E. R. HOWELL, 1972. Monocular alternation: A method for the investigation of pattern vision. *J. Physiol. (London)* 222:19–21.

CAMPBELL, F. W., E. R. HOWELL, and J. G. ROBSON, 1971. The appearance of gratings with and without the fundamental Fourier component. *J. Physiol. (London)* 217:17–18.

CAMPBELL, F. W., and J. J. KULIKOWSKI, 1966. Orientational selectivity of the human visual system. *J. Physiol. (London)* 187:437–445.

CAMPBELL, F. W., and J. J. KULIKOWSKI, 1972. The visual evoked potential as a function of contrast of a grating pattern. *J. Physiol. (London)* 222:345–356.

CAMPBELL, F. W., and L. MAFFEI, 1970. Electrophysiological evidence for the existence of orientation and size detectors in the human visual system. *J. Physiol. (London)* 207:635–652.

CAMPBELL, F. W., J. NACHMIAS, and J. JUKES, 1970. Spatial-frequency discrimination in human vision. *J. Opt. Soc. Am.* 60:555–559.

CAMPBELL, F. W., and J. G. ROBSON, 1964. Application of Fourier analysis to the modulation response of the eye. *J. Opt. Soc. Am.* 54:581.

CAMPBELL, F. W., and J. G. ROBSON, 1968. Application of Fourier analysis to the visibility of gratings. *J. Physiol. (London)* 197:551–566.

CARTER, B. E., and G. B. HENNING, 1971. The detection of gratings in narrow-band visual noise. *J. Physiol. (London)* 219: 355–365.

CLELAND, B. G., M. W. DUBIN, and W. R. LEVICK, 1971. Sustained and transient neurons in the cat's retina and lateral geniculate nucleus. *J. Physiol. (London)* 217:473–496.

COOPER, G. F., and J. G. ROBSON, 1968. Successive transformations of spatial information in the visual system. Conference on pattern recognition, N.P.L. Inst. of Elec. Eng., London.

CORNSWEET, T. N., 1970. *Visual Perception*. New York and London: Academic Press.

CRAIK, K. J. W., 1940. Visual Adaptation. Ph.D. dissertation, University of Cambridge.

CRAIK, K. J. W., 1966. *The Nature of Psychology*, Sherwood, S. L., ed. Cambridge University Press.

DUFFIEUX, P. M., 1946. L'integrale de Fourier et ses applications a l'optique. Privately printed (Besancon).

ENROTH-CUGELL, CHRISTINA, and J. G. ROBSON, 1966. The contrast sensitivity of retinal ganglion cells of the cat. *J. Physiol. (London)* 187:517–552.

GILINSKY, ALBERTA S., 1967. Masking of contour-detectors in the human visual system. *Psychon. Sci.* 8:395–396.

GILINSKY, ALBERTA S., 1968. Orientation-specific effects of patterns of adapting light on visual acuity. *J. Opt. Soc. Am.* 58:13–18.

GINSBURG, A. P., 1971. Psychological Correlates of a Model of the Human Visual System. Masters thesis, Air Force Institute of Technology, Wright-Patterson AFB, Ohio 45433.

GRAHAM, NORMA, 1972. Spatial frequency channels in the human visual system: Effects of luminance and pattern drift rate. *Vision Res.* 12:53–68.

GRAHAM, N., and J. NACHMIAS, 1971. Detection of grating patterns containing two spatial frequencies: A comparison of single-channel and multiple-channel models. *Vision Res.* 11:251–259.

GREEN, D. G., 1967. Visual resolution when light enters the eye through different parts of the pupil. *J. Physiol. (London)* 190:583–593.

GREEN, D. G., and F. W. CAMPBELL, 1965. Effect of focus on the visual response to a sinusoidally modulated spatial stimulus. *J. Opt. Soc. Am.* 55:1154.

GUBISCH, R. W., 1967. Optical performance of the human eye. *J. Opt. Soc. Am.* 57:407–415.

HELMHOLTZ, H., 1877. *On the Sensations of Tone.* (Reprinted 1954.) New York: Dover Publications.

HOPKINS, H. H., 1962. 21st Thomas Young Oration. The application of frequency response techniques in optics. *Proc. Phys. Soc. (London)* 79:889–919.

HUBEL, D. H., 1957. Tungsten microelectrode for recording from single units. *Science* 125:549–550.

HUBEL, D. H., and T. N. WIESEL, 1959. Receptive fields of single neurons in the cat's striate cortex. *J. Physiol. (London)* 148:574–591.

HUBEL, D. H., and T. N. WIESEL, 1962. Receptive fields, binocular interaction and functional architecture in the cat's visual cortex. *J. Physiol. (London)* 160:106–154.

HUBEL, D. H., and T. N. WIESEL, 1965. Receptive fields and functional architecture in two nonstriate visual areas (18 and 19) of the cat. *J. Neurophysiol.* 28:229–289.

HUBEL, D. H., and T. N. WIESEL, 1968. Receptive fields and functional architecture of monkey striate cortex. *J. Physiol. (London)* 195:215–243.

KAY, R. H., and D. R. MATTHEWS, 1971. Temporal specificity in human auditory conditioning by frequency-modulated tones. *J. Physiol. (London)* 218:104–106.

KAY, R. H., and D. R. MATTHEWS, 1972. On the existence in human auditory pathways of channels selectively tuned to the modulation present in frequency-modulated tones. *J. Physiol. (London)* 225:657–677.

KIANG, N. Y.-S., 1966. *Discharge Patterns of Single Fibers in the Cat's Auditory Nerve.* Cambridge, Mass., MIT Press.

KOENIGSBERGER, L., 1965. *Hermann von Helmholtz.* New York: Dover Publications.

LE GRAND, Y., 1937. La formation des images retiniennes. Sur un mode de vision éliminant les défauts optiques de l'oeil. Paris: 2e Reunion de l'Institute d'Optique.

LESHOWITZ, B., and F. L. WIGHTMAN, 1971. On-frequency tonal masking. *J. Acoust. Soc. Am.* 49:1180–1190.

LINFOOT, E. H., 1964. *Fourier Methods for Optical Image Evaluation.* London and New York: The Focal Press.

LIPSON, H., 1972. *Optical Transforms.* New York and London: Academic Press.

McCOLLOUGH, C., 1965. Color adaptation of edge-detectors in the human visual system. *Science* 149:1115–1116.

MAFFEI, L., and F. W. CAMPBELL, 1970. Neurophysiological localization of the vertical and horizontal visual coordinates in man. *Science* 167:386–387.

MITCHELL, D. E., R. D. FREEMAN, and G. WESTHEIMER, 1967. Effect of orientation on the modulation sensitivity for interference fringes on the retina. *J. Opt. Soc. Am.* 57:246–249.

O'NEILL, E. L., 1963. *Introduction to Statistical Optics.* Reading, Mass.: Addison-Wesley Publishing Company.

PANTLE, A., and R. SEKULER, 1968. Size-detecting mechanisms in human vision. *Science* 162:1146–1148.

RATLIFF, F., 1965. Mach Bands: Quantitative studies on neural networks in the retina, 1st ed. San Francisco: Holden-Day, 37–169.

ROBSON, J. G., 1966. Spatial and temporal contrast-sensitivity functions of the visual system. *J. Soc. Opt. Am.* 56:1141–1142.

ROSE, A., 1970. Quantum limitations to vision at low light levels. *Image Tech.* 12:13–31.

SACHS, M. B., J. NACHMIAS, and J. G. ROBSON, 1971. Spatial-frequency channels in human vision. *J. Opt. Soc. Am.* 61:1176–1186.

SHARPE, C. R., 1972. The visibility and fading of thin lines visualized by their controlled movement across the retina. *J. Physiol. (London)* 222:113–134.

STROMEYER, C. F., 1972. Edge-contingent color after effects: spatial frequency specificity. *Vision Res.* 12:717–733.

SULLIVAN, G. D., M. A. GEORGESON, and K. OATLEY, 1972. Channels for spatial frequency selection and the detection of single bars by the human visual system. *Vision Res.* 12:383–394.

TOLHURST, D. J., 1972a. On the possible existence of edge detector neurons in the human visual system. *Vision Res.* 12:797–804.

TOLHURST, D. J., 1972b. Adaptation to square-wave gratings: inhibition between spatial frequency channels in the human visual system. *J. Physiol. (London)* 226:231–248.

WESTHEIMER, G., 1960. Modulation thresholds for sinusoidal light distributions on the retina. *J. Physiol. (London)* 152:67–74.

WHITFIELD, I. C., 1967. *The Auditory Pathway.* London: Edward Arnold Ltd.

WHITFIELD, I. C., and E. F. EVANS, 1965. Responses of auditory cortical neurons to stimuli of changing frequency. *J. Neurophysiol.* 28:655–672.

WIGHTMAN, F. L., and B. LESHOWITZ, 1970. Off-frequency tonal masking. *J. Acoust. Soc. Am.* 47A:107.

YOUNG, T., 1800. Outlines of experiments and enquiries respecting sound and light. *Phil. Trans.* 106–150.

10 Developmental Factors in the Formation of Feature Extracting Neurons

COLIN BLAKEMORE

ABSTRACT   The final organization of pattern detecting neurons in the visual cortex of a cat is fundamentally determined by the kitten's early visual experience. Exposure, for as little as 1 hour or less, to vertical or horizontal stripes causes virtually every cortical neuron to adopt the experienced orientation as its preferred stimulus. This drastic modification by the environment has a distinct critical period from about 3 weeks to about 14 weeks. Prolonged exposure in an older kitten or an adult has no such effect. This mechanism may be adaptive, the feature-detecting apparatus of the visual system being optimally matched to the animal's visual environment.

Features: Universal or specific?

WHAT COULD BE more compelling than the idea of sensory systems matched precisely to their different sensory worlds? Of course, Johannes Müller (1842) realized that the receptors themselves are specialized to transduce particular forms of energy; but what of the patterns or combinations of stimuli within each modality, which we call *features*? Does a sensory system take an elementary, reductionist approach, treating its input entirely in the simplest possible terms (point-by-point analysis of light on the retina or tone-by-tone analysis of sound in the cochlea, for instance)? Or does it actively search out these complicated patterns of stimuli and disregard rare or meaningless inputs?

We now think that there is, at least within the visual system, a limited repertoire of specific, feature-extracting neurons that the animal uses to deal with its visual world. These neurons work on the signals from the receptors and detect specific spatiotemporal patterns that represent particularly important forms of stimuli. The frog's retina has its bug detectors, edge detectors, and dimming detectors (Lettvin, Maturana, McCulloch, and Pitts, 1959); the rabbit has direction-selective, orientation-selective, and uniformity-detecting ganglion cells in its retina (Barlow and Hill, 1963; Levick, 1967); cats and monkeys, in the visual cortex, have orientation-detecting neurons, which are also sensitive to the stereoscopic dis-

tance of the stimulus (Hubel and Wiesel, 1962, 1970a; Barlow, Blakemore, and Pettigrew, 1967). In fact every species whose visual pathway has been probed with microelectrodes has been found to possess a limited number of classes of neuron, each sensitive to a particular combination of elementary sensory events, which one could call a feature.

An examination of the results of these experiments shows that there seem to be two kinds of detector neurons:

1. *Universal feature detectors.* These are classes of cell that are found in almost all visual systems, and it seems reasonable to assume that the features they detect are fundamental. These include *temporal transients* (most detector neurons only respond at the beginning or the end of a sudden stimulus); *spatial transients* or *edges* (the process of lateral inhibition in the retina ensures that visual cells respond mainly to luminance contrasts at the edges of stimuli and not to uniform illumination of the retina, Barlow, 1961); *orientation of edges* and *direction of movement.*

2. *Species-specific feature detectors.* Some classes of visual neuron are restricted to certain species and, in a teleological sense, might be concerned more with those particular elements of the visual environment that are especially relevant to the animal in question. The frog's bug detectors are a good example; so are the stereoscopic disparity detecting neurons in cat and monkey cortex; color sensitive cells in the goldfish (Wagner et al., 1960), the ground squirrel (Michael, 1968), and the monkey (Wiesel and Hubel, 1966); detectors of rotational movement in the fly (Mimura, 1970); these are all species-specific feature detectors, not found in every visual system.

The formation of feature-detecting neurons

This remarkable selectivity of visual neurons, whether for universal or for species-specific features, puts enormous demands on the anatomical construction of the animal's visual pathway. The properties of each cell depend on the precise nature of all the connections from the retinal rods and cones through to the neuron itself. It is difficult to believe that the staggering complexity of this circuitry could all be preprogrammed exactly by the genetic code.

COLIN BLAKEMORE   The Physiological Laboratory, University of Cambridge, Cambridge, England

Indeed it would be eminently reasonable (particularly for species-specific features) for the actual visual environment in which a developing visual system finds itself to play some part in sustaining, validating, or even specifying the connections that the visual system finally adopts.

In this chapter I shall present some evidence that early visual experience is crucially important for determining the feature-detecting properties of neurons in the cat's visual cortex. Early in a kitten's life, its vision is helping it to build a visual system appropriate to the world in which it lives.

The visual cortex of normal, adult cats

The primary visual cortex, in the occipital lobe, is just four synapses from the photoreceptors themselves. Hubel and Wiesel (1962) showed that, although the neurons in the cat's visual cortex are of several different kinds in their detailed properties, they are nearly all orientation selective and binocularly driven (Figure 1). Each cell responds

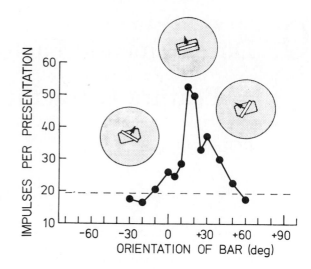

FIGURE 2   The orientational *tuning curve* for a *complex* cortical cell. The stimulus was a bright bar, generated on a display oscilloscope, as shown in the inset diagrams where the large rectangle is the receptive field. Each point is the mean number of impulses produced during six successive sweeps at that orientation. The dashed line is the mean spontaneous discharge in the same period of time, in the absence of any stimulus.

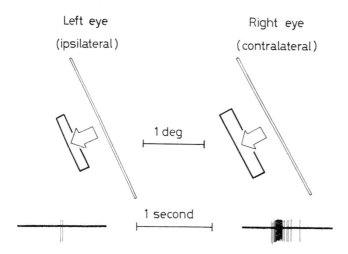

FIGURE 1   This binocular cortical neuron, recorded in the primary visual cortex of an adult cat, falls into ocular dominance group 2. The responses, in the records at the bottom photographed from an oscilloscope, show that a thin white bar moved across the receptive field (the thickly bordered rectangle) in the right eye is much more effective than the same stimulus shown to the left eye. (Reproduced, by permission, from Blakemore and Pettigrew, 1970.)

to a moving edge, or a light or dark bar, of a particular orientation, moving across the receptive field in either eye. The response is rapidly attenuated if the angle of the stimulus is altered (Figure 2). Different neurons prefer different orientations and, in the normal animal, every orientation is equally represented. Figure 3 is a polar diagram of the distribution of preferred orientations for a sample of neurons from the primary visual cortex of one

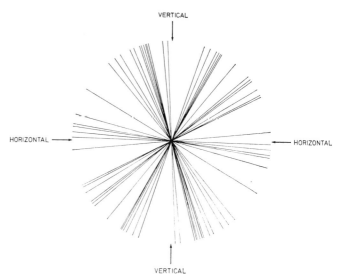

FIGURE 3   In this polar diagram, each line is the best orientation for one cortical neuron. This is a sample of 34 neurons from a normal adult cat.

cat. Each line is the best orientation for one cell; therefore, the cat has an armory of orientation detectors enabling it to deal with and analyze the shape of any object that it sees.

Although nearly all these cells are binocular, and they are almost certainly providing a mechanism for the recognition of stereoscopic distance (Barlow et al., 1967; Blakemore, 1970; Bishop, 1970), they are not all equally influenced by the two eyes. Hubel and Wiesel (1962) designed a simple qualitative scheme for categorizing the cells according to their ocular dominance (Figure 4),

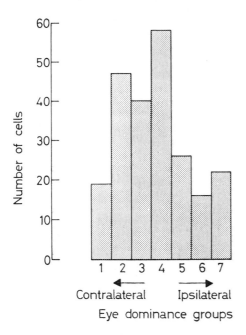

FIGURE 4 This histogram shows the distribution of ocular dominance among 228 cells from normal adult cats. The dominance groups describe the relative influence of the two eyes with group 4 cells being equally excited by the two eyes, group 1 cells driven only by the contralateral eye and group 7 cells only by the ipsilateral eye. (Reproduced, with permission, from Blakemore and Pettigrew, 1970.)

classifying them into seven dominance groups ranging from 1 (monocular cells only driven by the contralateral eye) through 4 (equally driven by both eyes) to 7 (monocular cells only driven by the ipsilateral eye).

The visual cortex of very young kittens

One obvious way of answering the whole question of the relative contributions of genetic and environmental factors would be to record from neurons in the visual cortex of very young kittens that have never had any visual experience. Both Hubel and Wiesel (1963) and Barlow and Pettigrew (1971) have studied young kittens and they agree about several things:

1. The neurons are inherently binocular with a normal ocular dominance distribution.

2. It is generally difficult to influence young cortical cells with visual stimuli; they respond very weakly and there is often rapid habituation during successive presentations of any stimulus.

3. A large number of cells certainly have *no* preference for any particular orientation and will respond equally for moving spots or edges of any angle.

There is strong disagreement about whether *any* cells are orientation detectors, like neurons in the adult. Hubel and Wiesel (1963) say there are many such cells; Barlow and Pettigrew (1971) say that there are none in the visually inexperienced kitten, although some are direction selective in the sense that they will respond to any target moving through the receptive field in a particular direction.

Therefore, some properties of cortical cells, such as their binocularity and their responses to movement, are certainly built in. Others seem to develop, or at least become sharpened up, during the first few weeks of vision. Now we must ask whether this process is passive maturation or whether it is a forceful influence of the environment.

Binocular deprivation

If both eyelids of a kitten are sutured shut at about 10 days of age (the time of natural eye opening) and its visual cortex is studied after several weeks of binocular deprivation, many of the neurons are still very infantile in their properties (Figure 5), many are not orientation selective, and a large number are completely visually unresponsive

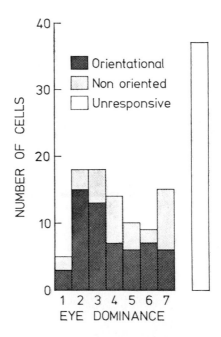

FIGURE 5 This ocular dominance histogram summarizes Wiesel and Hubel's experiments on binocular deprivation. There are 126 cells from 4 kittens deprived of vision from about 8 days until $2\frac{1}{2}$ to $4\frac{1}{2}$ months. Apparently normal orientation selective cells, cells with no orientational preference, and visually unresponsive neurons are shown separately. (From Blakemore, 1973.)

(Wiesel and Hubel, 1965). In fact, Barlow and Pettigrew (1971) maintain that a binocularly deprived cortex again has no orientation-selective cells at all. This certainly suggests that visual experience is essential for the establishment of the normal feature-detecting properties of neurons in the visual cortex and that they do not merely mature. On the other hand, the binocularity generally persists without any binocular visual experience.

Monocular deprivation

How different the result is if only one eyelid is closed during development. After such a period of monocular deprivation, the cortical cells that respond to visual stimuli can usually be influenced only through the experienced eye (Wiesel and Hubel, 1965). Apart from this monocular dominance, the cells are adult in their pattern-detecting properties.

These dramatic effects of monocular deprivation can occur only during a distinct *sensitive* or *critical period* (Figure 6). Covering one eye before the beginning of the

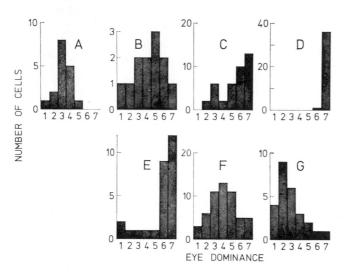

FIGURE 6   These histograms, redrawn from Hubel and Wiesel (1963, 1970b), illustrate the duration of the sensitive period for the results of monocular deprivation. (A) Results for two normal, very young, visually inexperienced kittens. In B to G the recordings were taken from the left hemisphere after the following periods of monocular deprivation in the right (contralateral) eye: (B) 9 to 19 days, (C) 10 to 31 days, (D) 23 to 29 days, (E) 2 to 3 months, (F) 4 to 7 months, (G) A previously normal adult monocularly deprived for 3 months. (From Blakemore, 1973.)

third week (Figure 6B) or after the fourth month (Figures 6F and 6G) has no influence on the dominance distribution; but as little as a few days of monocular deprivation during the fourth week (Figure 6D) leads to total dominance by the other eye (Hubel and Wiesel, 1970b).

During this remarkably short period, there must be intense competition for synaptic sites on cortical cells between the thalamocortical fibers from the two eyes, and it seems that simultaneous activity (or inactivity) in the two inputs is the requirement for the maintenance of binocular connections.

Environmental modification of orientation selectivity

If the degree of binocularity of cortical neurons can be influenced by visual experience, then possibly orientation selectivity itself can be modified by the visual environment. Hirsch and Spinelli (1971) combined the strategy of monocular deprivation with a procedure that dramatically limited the type of pattern that the cortex could experience. They reared kittens wearing an apparatus that allowed one eye to see three vertical lines, the other three horizontal ones. Thus the two eyes never saw similar orientations, and one might expect the binocularity of the cortex to suffer as a consequence. Indeed it did. All the cells with oriented receptive fields were monocular and all but one had a preferred orientation that closely matched the visual experience of the eye to which they were connected. There were no neurons, binocular or monocular, with diagonally-oriented receptive fields. This latter observation, and this alone, is incompatible with the simple notion that orientation selectivity is genetically specified and that there is merely competition between the inputs from the two eyes. If this were the case, all the diagonal detectors in Hirsch and Spinelli's cats should have been binocularly deprived, and therefore they should have survived. Because they did not, we must suspect that orientation selectivity itself can be *changed* by visual experience.

Blakemore and Cooper (1970) tested this idea with a much less sophisticated, but somewhat more naturalistic, apparatus. We reared two kittens in the dark except for a few hours each day when they were put into special chambers that essentially restricted their normal, binocular visual experience to edges of one orientation, vertical for one kitten, horizontal for the other (Figure 7). They stood on a glass plate suspended in a tall cylinder that was painted with black and white stripes and illuminated from above. A ruff around their necks prevented them from seeing their own bodies, so wherever they looked they simply saw stripes of one orientation.

The first two kittens had almost 300 hours of experience in this apparatus from 2 weeks to $5\frac{1}{2}$ months of age. After another 2 months, during which they were occasionally taken from the dark into a normal room to study their visual behavior, we recorded from the visual cortex of

FIGURE 7 The apparatus for exposure to vertical edges. In this photograph the cat is almost full-grown.

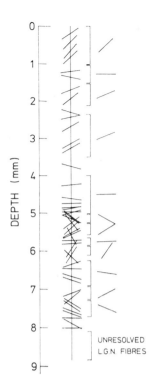

UNRESOLVED
L.G.N. FIBRES

DEPTH (mm)

FIGURE 8 A reconstruction of the preferred orientations of neurons encountered during a long penetration through the visual cortex of a kitten reared in horizontal edges from 2 weeks to $5\frac{1}{2}$ months. The columnar aggregates of cells with similar orientational preferences are shown by the bracketed regions on the right.

each animal. Despite their very strange early visual diet, we found virtually no visually unresponsive neurons or nonoriented cells and no regions of silent cortex. The neurons were quite normal and adult in their properties. Only one thing was really unusual: We could find no neurons that had a preferred orientation within 20 deg of the angle orthogonal to the stripes that the kitten had been reared in. Nearly all the cells responded best to orientations within 45 deg of that which the kitten had seen early in life.

Figure 8 is a reconstruction of a long penetration through the primary visual cortex of the first cat reared in horizontal stripes. The position of each neuron recorded in the penetration is shown by a line at the best orientation for that cell. Just as in the normal adult's cortex, neighboring neurons tended to prefer very similar orientations. These *columns* of cells are shown in Figure 8 as bracketed regions with the average preferred orientation for each column on the right.

There is, then, little doubt that early visual experience (albeit rather odd) can modify the orientation selectivity as well as the binocularity of cortical cells. By analogy with Wiesel and Hubel's hypothesis of competition between the inputs from the two eyes, one could imagine the newborn, naked cortical neuron besieged by groups of afferent fibers, each group representing a different orientation of edge on the retina. The set of input fibers most often used might win the battle for synaptic space.

The sensitive period for environmental modification

I have recently completed a series of similar experiments in which kittens were exposed to stripes for a limited period of time at different ages, in an attempt to define whether there is a crucial sensitive period for environmental modification of orientation selectivity, as there is for the effects of monocular deprivation. The results, illustrated in Figure 9, showed that there is such a sensitive period and that it coincides exactly with that for the modification of binocularity.

All but three of the kittens who produced the results of Figure 9 were kept in the dark until a particular age and then were exposed in a striped environment for 2 to 4 hours each day for a period of time from a few days to a few weeks. After the period of visual experience, they were kept continuously in the dark until I recorded from their cortical neurons, some days or weeks later. The abscissa of Figure 9 is the animal's age and each polar diagram is positioned at the kitten's age in the middle of the period of exposure. The upper half of Figure 9 shows kittens exposed to vertical stripes, the lower half those

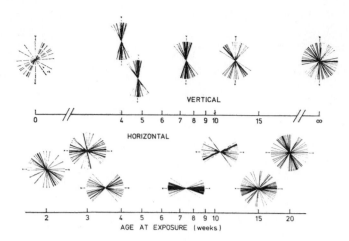

FIGURE 9 The critical period for environmental modification of the visual cortex. These kittens, with the exceptions described in the text, were kept in the dark and exposed to vertical (upper half) or horizontal (lower half) for a few hours each day, at various ages. On the upper abscissa the results that appear above zero age refer to a binocularly deprived kitten where the orientational preferences (dashed lines) were very vague, and those at ∞ on the abscissa are for a normal adult animal kept in vertical stripes over a 4 month period. The logarithmic abscissa carries no theoretical implications: It is merely for graphical convenience.

exposed to horizontal. The three exceptions to this basic paradigm are as follows:

1. The animal whose diagram lies above zero age on the upper abscissa had no visual experience. Its eyelids were sutured before natural eye opening and only re-opened just before the recording (Blakemore and Mitchell, 1972). This animal had many visually unresponsive and completely nonoriented neurons and even the cells illustrated in Figure 9 had only the merest bias toward the orientations shown. (That is why the lines are drawn interrupted rather than solid.) They all responded to some extent to every orientation of edge and almost as well for a moving spot. This kind of behavior is totally unlike that of adult neurons or indeed all the other cells shown in these polar diagrams. Therefore, on the basis of this very preliminary study, we must agree more with Barlow and Pettigrew (1971) than with Wiesel and Hubel (1965).

2. The kitten whose diagram appears at 15 weeks on the lower abscissa was kept in a normal lighted room until it was 6 weeks old, after which it was treated like all the others. This was done to preclude the possibility of simple degeneration due to disuse in the visual pathway during the long waiting period in the dark.

3. The results shown at ∞ on the upper abscissa refer to a normal adult cat that was kept in the dark for 4

months, during which it spent about 135 hours in the vertically striped environment.

From Figure 9 it is clear that the sensitivity to environmental modification increases suddenly at about 3 weeks of age, remains particularly high from 4 to 7 weeks, and then declines gradually, until after about 14 weeks of age no amount of experience of one orientation will influence the visual cortex.

However, kittens kept in the dark and exposed to stripes only before or after this critical period have many real orientation-selective neurons (quite unlike the binocularly deprived animal), even though there is no over representation of the experienced orientation. Two possible explanations spring to mind. Either there is another sensitive period, starting earlier and finishing later, during which *any* visual input will strengthen any innate predispositions in the neurons without modifying their preferences; or the few seconds of dim light that these kittens experienced each day in the dark room, during feeding and cleaning, was enough to set up many normal cortical cells, for all orientations.

The minimum necessary exposure

Even in the experiments described in Figure 9 the kittens were exposed for rather long periods of time, usually 30 to 50 hours, so one could argue that such a prolonged and unusual visual experience is so unlike a kitten's normal early vision that perhaps environmental modification is simply an artifact of a bizarre experimental manipulation. Blakemore and Mitchell (1973) very recently set out to define the minimum length of time in a striped environment that is necessary to influence the organization of the visual cortex.

We knew that the fourth week is a period of extraordinary sensitivity, so we kept six kittens in the usual manner and exposed them to vertical for various lengths of time (1, 3, 6, 18, 27, and 33 hours) on or around the 28th day and performed the recording at 6 weeks of age. The results, shown in Figure 10, surprised us and taught us that we had planned our experiment rather badly! The results for the binocularly deprived animal are reproduced (Figure 10A) for comparison, and it is patently clear that even 1 hour of vision totally changes the properties of the visual cortex. Not only were almost all the cells adult in their properties but the orientational preferences were mainly for angles close to vertical (Figure 10B). The few cells biased toward other orientations were generally more infantile in their properties, and we found them, together with a few nonoriented and visually unresponsive neurons, in small regions between the totally normal columns of cells that responded best to vertical.

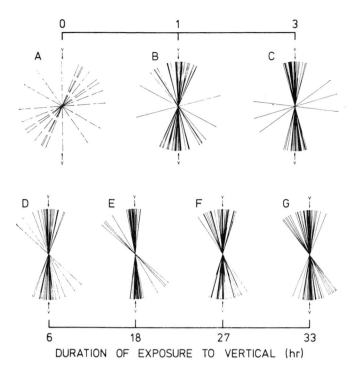

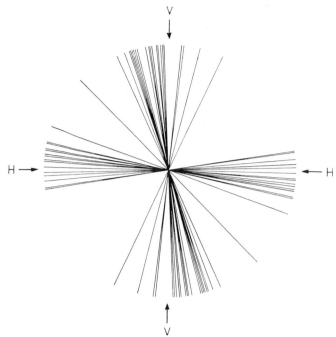

DURATION OF EXPOSURE TO VERTICAL (hr)

FIGURE 10 Orientational preferences for the cortical cells of six kittens (B to G) exposed to vertical for 1, 3, 6, 18, 27, and 33 hours, on or around the 28th day of age. The diagram with dashed lines (A) shows the vague orientational preferences of cells from a binocularly deprived kitten. (Reproduced, by permission, from Blakemore and Mitchell, 1973.)

There was a further slight refinement in the distribution of preferred orientations with longer exposure (Figures 10C to G), but the bulk of the process needs only 1 hour of experience.

An alternating visual environment

If 1 hour of vision is enough to modify cortical neurons, then what would be the effect of alternating exposure to the two orientations? Blakemore and Mitchell (in preparation) have allowed kittens to see horizontal and vertical stripes alternately, for various cycle lengths. Figure 11 shows what happened to the cortical neurons of a kitten reared for 2 hours in horizontal, 2 hours in vertical, 2 hours in horizontal, and so on, until it had accumulated more than 50 hours in each. It had two distinct populations of neurons, with their orientational preferences clustered around the two axes.

In general the columnar structure in this kitten, and others with a similar history, was as might be expected— columns of cells preferring vertical alternated with columns of neurons preferring horizontal orientation. However, there were also large regions of mixed orienta-

FIGURE 11 The orientational preferences for 37 cells from a kitten exposed alternatively for 2 hours to horizontal, 2 hours to vertical, until it had experienced more than 50 hours of each.

tion, with horizontal and vertical detecting neurons all jumbled up. In these strange columns we found some very peculiar cells: Some that were inhibited by one orientation and excited by the perpendicular, others that could not be excited and were only inhibited by one orientation, and still others that were excited by both axes (but not by diagonal edges) and, hence, responded best to a cross shape moving through the receptive field. These dual-axis cells are reminiscent of, but not identical to, Hubel and Wiesel's (1965) "higher-order hypercomplex cells" found only in area 19 of the cortex, not in area 17. Therefore, here is an indication that it is possible to synthesize feature-detecting neurons that are not normally found.

If, in such an alternating experiment, the *first* period of exposure comes in the fourth week and it lasts for 10 hours or more, it is difficult to reverse the arrangement by exposure to the opposite orientation, even for rather a long period. In such an animal, most cortical cells prefer the first orientation they experienced, and rather few are captured by the opposite.

Conclusions

Many questions about the genesis of the visual cortex remain unanswered. Is there, for instance, any contribution from the inbuilt predispositions of cortical cells?

Binocularity is innate, but it can certainly be modified. The position of the receptive field on the retina is also prewired, but even that is capable of being changed on the basis of visual experience (Shlaer, 1971).

Environmental modification becomes a simple phenomenon in the terribly reduced conditions of our experiments, but it is difficult to understand how it might operate in a normal kitten whose eye movements would expose each cortical cell to all orientations in quick succession. Perhaps it is merely the *probability* of experience that determines the final preference of a cell. Perhaps each neuron selects as its preferred stimulus the feature that it has seen most often. In that case environmental modification might have real adaptive value, because it would ensure that the animal builds for itself a visual system optimally matched to its particular visual world.

Humans have rather higher resolution acuity for vertical and horizontal patterns than for diagonal (Campbell, Kulikowski, and Levinson, 1966); perhaps this is due to the predominance of these orientations in the rectilinear environment of Western cities. More relevant, humans who grow up with astigmatism so severe that it greatly weakens the contrast of patterns of one orientation are left with "meridional amblyopia"—a reduced acuity for the orientation that was originally out of focus (Mitchell, Freeman, Millodot, and Haegerstrom, 1973). This meridional amblyopia, like normal reduced acuity for diagonals, cannot be rectified by perfect spectacle correction of the eye's optics.

Finally, it now seems valid to speculate that the actual physical changes that take place in the cortex during environmental modification are analogous to the fundamental process underlying learning and memory. We have no idea what is the physical correlate of, say, a visual memory of a complex object, but because total environmental modification of the visual cortex can occur after an hour or less of vision, it is worth considering the idea that memory itself depends on similar anatomical rearrangements to synthesize gnostic neurons in the brain.

ACKNOWLEDGMENTS   G. F. Cooper and D. E. Mitchell collaborated in some of the experiments on the development of visual cortex. This work was generously supported by research grants (Nos. G970/807/B and G972/463/B) from the Medical Research Council, London. It would all have been impossible without the excellent technical assistance of R. D. Loewenbein and J. S. Dormer.

REFERENCES

BARLOW, H. B., 1961. Possible principles underlying the transformations of sensory messages. In *Sensory Communication*, W. A. Rosenblith, ed. Cambridge, Mass.: MIT Press, pp. 217–234.

BARLOW, H. B., and R. M. HILL, 1963. Selective sensitivity to direction of movement in ganglion cells of the rabbit retina. *Science* 139:412–414.

BARLOW, H. B., and J. D. PETTIGREW, 1971. Lack of specificity of neurones in the visual cortex of young kittens. *J. Physiol.* (*London*) 218:98–100P.

BARLOW, H. B., C. BLAKEMORE, and J. D. PETTIGREW, 1967. The neural mechanism of binocular depth discrimination. *J. Physiol.* (*London*) 193:327–342.

BISHOP, P. O., 1970. Beginning of form perception and binocular depth discrimination in cortex. In *The Neurosciences: Second Study Program*, F. O. Schmitt, ed. New York: Rockefeller University Press, pp. 471–485.

BLAKEMORE, C., 1970. The representation of three-dimensional visual space in the cat's striate cortex. *J. Physiol.* (*London*) 209: 155–178.

BLAKEMORE, C., 1973. Environmental constraints on development in the visual system. In *Constraints on Learning: Limitations and Predispositions*, R. A. Hinde and J. S. Hinde, eds. London: Academic (in press).

BLAKEMORE, C., and G. COOPER, 1970. Development of the brain depends on the visual environment. *Nature* 228:477–478.

BLAKEMORE, C., and D. E. MITCHELL, 1973. Environmental modification of the visual cortex and the neural basis of learning and memory. *Nature* (in press).

BLAKEMORE, C., and J. D. PETTIGREW, 1970. Eye dominance in the visual cortex. *Nature* 225:426–429.

CAMPBELL, F. W., J. J. KULIKOWSKI, and J. LEVINSON, 1966. The effect of orientation on the visual resolution of gratings. *J. Physiol.* (*London*) 187:427–436.

HIRSCH, H. V. B., and D. N. SPINELLI, 1971. Modification of the distribution of receptive field orientation in cats by selective visual exposure during development. *Exp. Brain Res.* 12: 509–527.

HUBEL, D. H., and T. N. WIESEL, 1962. Receptive fields, binocular interaction and functional architecture in the cat's visual cortex. *J. Physiol.* (*London*) 160:106–154.

HUBEL, D. H., and T. N. WIESEL, 1963. Receptive fields of cells in striate cortex of very young, visually inexperienced kittens. *J. Neurophysiol.* 26:994–1002.

HUBEL, D. H., and T. N. WIESEL, 1965. Receptive fields and functional architecture in two non-striate visual areas (18 and 19) of the cat. *J. Neurophysiol.* 28:229–289.

HUBEL, D. H., and T. N. WIESEL, 1970a. Cells sensitive to binocular depth in area 18 of the macaque monkey cortex. *Nature* 225:41–42.

HUBEL, D. H., and T. N. WIESEL, 1970b. The period of susceptibility to the physiological effects of unilateral eye closure in kittens. *J. Physiol.* (*London*) 206:419–436.

LETTVIN, J. Y., H. R. MATURANA, W. S. McCULLOCH, and W. H. PITTS, 1959. What the frog's eye tells the frog's brain. *Proc. Inst. Radio Engr.* 47:1940–1951.

LEVICK, W. R., 1967. Receptive fields and trigger features of ganglion cells in the visual streak of the rabbit's retina. *J. Physiol.* (*London*) 188:285–307.

MICHAEL, C. R., 1968. Receptive fields of single optic nerve fibers in a mammal with an all-cone retina. III: Opponent color units. *J. Neurophysiol.* 31:268–282.

MIMURA, K., 1970. Integration and analysis of movement information by the visual system of flies. *Nature* 226:964–966.

MITCHELL, D. E., R. D. FREEMAN, M. MILLODOT, and G. HAEGERSTROM, 1973. Meridional amblyopia: evidence for modification of the human visual system by early visual experience. *Vision Res.* (In press.)

MÜLLER, J., 1842. *Elements of Physiology*, Book V, Vol. II, translated by W. Baly. London: Taylor and Walton, pp. 1059–1087. (Reproduced in *Visual Perception: the Nineteenth Century*, W. Dember, ed. 1964. New York: John Wiley & Sons, pp. 35–69.)

SHLAER, R., 1971. Shift in binocular disparity causes compensatory change in the cortical structure of kittens. *Science* 173: 638–641.

WAGNER, H. G., E. F. MacNICHOL, and M. L. WOLBARSHT, 1960. The response properties of single ganglion cells in the goldfish retina. *J. Gen. Physiol.* 43 (Suppl. 2):45–62.

WIESEL, T. N., and D. H. HUBEL, 1965. Comparison of the effects of unilateral and bilateral eye closure on cortical unit responses in kittens. *J. Neurophysiol.* 28:1029–1040.

WIESEL, T. N., and D. H. HUBEL, 1966. Spatial and chromatic interactions in the lateral geniculate body of the rhesus monkey. *J. Neurophysiol.* 29:1115–1156.

11 Processing of Spatial and Temporal Information in the Visual System

MICHAEL J. WRIGHT and HISAKO IKEDA

ABSTRACT *On* center and *off* center receptive fields of cat retinal ganglion cells can be divided into two categories: *sustained* or *X-cells* and *transient* or *Y-cells*. The receptive field organization of the two categories of cells was investigated by several independent methods and found to differ. These results suggested that *X*- and *Y*-cells might begin the separate processing of spatial and temporal features of visual stimuli. Evidence is presented that the information from *X*- and *Y*-cells is transmitted over separate central pathways, and that the thresholds for detection of spatial and temporal contrast are set by different populations of neurons.

Introduction

SEVERAL SOURCES of evidence suggest that the visual pathways contain neurons that are specifically tuned to different stimuli of submodalities, for example, orientation, size, and retinal disparity.

On the other hand, examination of the receptive field organization of retinal ganglion cells in the cat reveals two basic types, *sustained* and *transient* cells. This paper considers evidence that these two classes of retinal ganglion cells are the origin for the *separate* processing of spatial and temporal features of visual stimuli, respectively, and that the submodalities of vision may be grouped around this spatial/temporal dichotomy.

Retinal ganglion cell receptive fields in the cat

The classical description of the receptive field of the cat's retinal ganglion cell was given by Kuffler (1953). Within each receptive field, he described a central area with a low threshold to a small, flashing spot of light. The discharge pattern of this center region is opposite to that found in the surround or periphery. The center may give predominantly *off*, the surround *on* discharges, or vice

MICHAEL J. WRIGHT and HISAKO IKEDA Research Department of Ophthalmology, Royal College of Surgeons of England, London, England

versa. The familiar concentric arrangement of receptive field regions revealed by plotting the field with a small spot of light is shown in Figure 1(a).

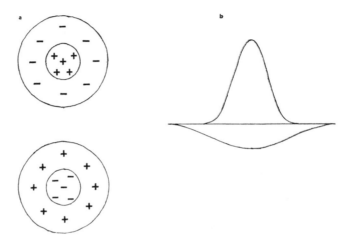

FIGURE 1 (a) Receptive field of an on-center (top) and an off-center (bottom) retinal ganglion cell as revealed by plotting with a small spot of light. The concentric, center-surround organization described by Kuffler (1953) is shown. (b) Model of retinal ganglion cell receptive field proposed by Rodieck (1965). The *X*-axis shows distance along a diameter of the receptive field, and the *Y*-axis shows sensitivity. Center and surround mechanisms have concentric, overlapping, Gaussian sensitivity distributions, the peak sensitivity of the surround being smaller than that of the center and its variance greater. The center and surround mechanisms sum their inputs separately and have opposite, antagonistic effects on the ganglion cell discharge. The model applies both to on-center and to off-center cells.

A receptive field plot of this kind is the result of a particular experimental procedure: It is obtained with a spot of particular diameter and intensity, rather than being a constant property of the cell. The receptive field plot alone does not permit a full prediction of the response of the cell to other visual stimuli. Indications of this fact were evident in the original experiments of Kuffler (1953), who found that the spatial extent of the receptive field increased when the intensity of the plotting spot was increased. Similarly, Ikeda and Wright (1972a) showed that the field expands if the area of the spot is increased while its intensity is kept constant. The obvious interpretation of these findings is that spots of greater intensity or area stimulate additional regions of lower sensitivity, so that a complete description of the receptive field must include the *spatial distribution of sensitivity* of the center and surround.

A model of this type, developed from Kuffler's findings, has been proposed by Rodieck (1965) (see Figure 1(b)). It assumes that the distributions of the center and surround sensitivities are Gaussian, concentric, and overlapping, with the surround distribution having a smaller peak and larger variance. Signals from photoreceptors in the center region and signals from photoreceptors in the surround region are summed separately, and the resulting signals have opposite, antagonistic effects on the ganglion cell response. In other words, there is linear spatial summation in both center and surround, and a subtractive interaction. This model applies both to on-center and to off-center fields.

In accord with this model, a luminance increment within the field will produce a response, either excitatory or inhibitory: But if the luminance changes in symmetrical halves of the field are equal and opposite, the signals reaching the cell from each half of the center and surround mechanism will be equal and opposite, giving no net response. This test of Rodieck's model was carried out by Enroth-Cugell and Robson (1966). One class of retinal ganglion cells (*X*-cells) behaved according to the model, but for another group (*Y*-cells) there was no null position of the stimulus border within the receptive field. This was not due to radial asymmetry of *Y*-cell fields, because they, too, have the concentric organization described by Kuffler (1953) but resulted from some form of nonlinearity of spatial summation. (A model of the *Y*-cell receptive field has recently been proposed by Ikeda and Wright, 1972b).

The *X/Y* classification, therefore, reflects a fundamental difference in receptive field organization; it is not dependent on a particular method of plotting the field. Cleland, Dubin, and Levick (1971) showed that *X*- and *Y*-cells also differed in their response to a small steady spot of appropriate contrast to excite the receptive field center.

R.F. RESPONSE PROFILE: ON-CENTRE SUSTAINED CELL

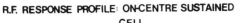

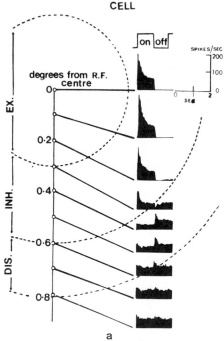

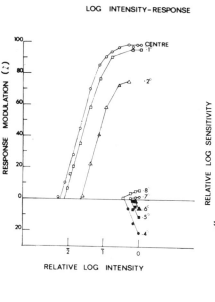

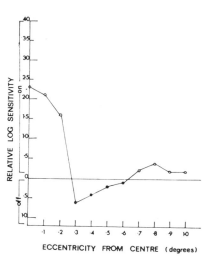

a   b   c

FIGURE 2 (a) Poststimulus histograms obtained from an on-center sustained cell at different distances from the receptive field center (0°–0.8°). The locations of the excitatory center and the inhibitory and outer excitatory surrounds are shown diagramatically. Spot size 25′, spot intensity 257 cd/m², background intensity 17.5 cd/m². (b) Response as a function of log intensity of the stimulus at different distances from the receptive field center. Unattenuated spot intensity, size, and background intensity as in Figure 2a. Note that the *off* response also shows a linear response to log intensity. (c) Sensitivity gradient of the cell derived from Figure 2b by reading off the intensity level required to produce a threshold response for each location. (From Ikeda and Wright, 1972c,d.)

The responses of Y-cells are phasic; there is a transient excitation when the spot is introduced which decays, in 5 sec or less, to a steady maintained level. In the X-cell response, however, there is a tonic component while the spot is present. On the basis of their response to a steady contrast, X- and Y-cells are also known by the descriptive names, *sustained* and *transient* cells (Cleland, Dubin, and Levick, 1971). In the following, we use the terms sustained and transient to refer to cells showing behavior characteristic of X- and Y-cells.

Receptive field organization of sustained and transient cells

In order to obtain a more complete characterization of ganglion cell behavior, we plotted the receptive field of each cell repeatedly, using spots of different diameters and a wide range of intensities extending to threshold. This method enabled us to determine both sensitivity gradients and spatial summation properties across the receptive fields of sustained and transient cells.

Figure 2 shows the results of a basic experiment of this type, for a single on-center sustained cell; Figure 2(a) shows the change in the poststimulus histogram as the spot position is moved from the excitatory center to the inhibitory surround. Beyond the inhibitory surround, we found in nearly all cells by averaging of the responses a small peak of excitation that constituted an outer excitatory (disinhibitory) surround (Ikeda and Wright, 1972c). Figure 2(b) shows the effect of varying the intensity of the spot. When we plotted the logarithm of the spot intensity against the percentage response modulation, we obtained a linear response at low intensities and a saturation at high intensities. The inhibitory surround and outer excitatory surround also have approximately a linear response to log spot intensity. Figure 2(c) shows the sensitivity gradient plotted from this data.

In Figure 3, the same experiment is carried out on an on-center transient cell. The poststimulus histograms obtained at different positions across the receptive field (with approximately the same stimulating conditions used in Figure 2) are shown in Figure 3(a). The histograms

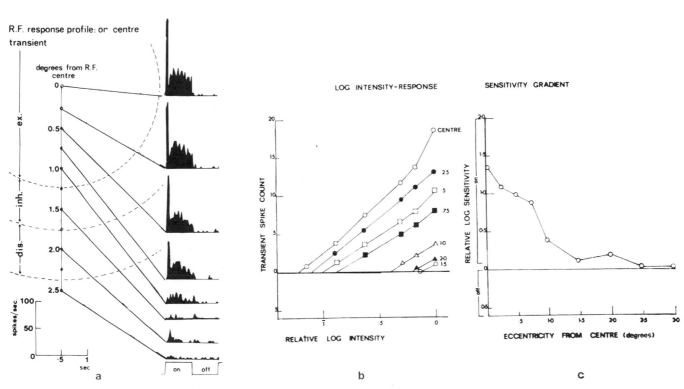

FIGURE 3 (a) Poststimulus histograms obtained from an on-center transient cell at different distances (0°–2.5°) from the receptive field center. The approximate locations of the excitatory center, inhibitory, and outer excitatory surrounds are shown diagramatically, but the divisions are in fact ill-defined. Spot size 28′, spot intensity 230 cd/m², background intensity 10 cd/m². No distinctive response is obtained from the inhibitory surround with this stimulus. (b) Response of this cell as a function of log spot intensity at different distances from the

receptive field center. Unattenuated spot intensity, size, and background intensity as in Figure 3a. The transition from the inhibitory to the outer excitatory zone appears as a reverse order of response-intensity curves at 1.5° and 2.0°. (c) Sensitivity gradient of the cell derived from Figure 3b, by reading off the intensity level required to produce threshold response at each location. (From Ikeda and Wright, 1972c, and unpublished data.)

have a characteristically different shape, with a very brief initial excitation. There is only a very small off response from the surround, and this co-exists with a much larger center-type response. The on response increases again beyond the inhibitory surround in the outer excitatory region. In Figure 3(b), the response/log intensity function is shown for different receptive field locations, and Figure 3(c) shows the sensitivity gradient reconstructed from this data. Unlike sustained cells, transient cells rarely showed a clear division of the sensitivity gradient into excitatory, inhibitory, and outer excitatory zones, because the surround did not give an appreciable response, its threshold to spot stimulation being 1–2 log units higher than that of sustained cells. If a larger spot was used, a clear surround response appeared. This is shown in Figure 4.

In transient cells, however, if the stimulus spot were enlarged so as to elicit a surround response, the center response was increased also. It was rarely possible to elicit a pure surround response in transient cells, because the receptive field center expands with increased stimulus flux (Ikeda and Wright, 1972a), and transient, on-off responses are then obtained from points in the field periphery (Figure 4). In sustained cells, this did not occur. There was always a diminution of the response if the spot were enlarged beyond a certain size, due to center-surround antagonism, and a definite limit to the expansion of the receptive field center with increasing spot size (Ikeda and Wright, 1972a).

The results of this series of experiments may be summarized under four headings.

First, we confirmed the observation of Cleland, Dubin, and Levick (1971) that transient cells gave strongly phasic responses while sustained cells gave tonic responses with an initial phasic component to a small spot in the receptive field center. Furthermore, the phasic response of transient cells was maintained throughout the receptive field, whereas the tonic response of sustained cells tended to decrease with increasing eccentricity (Ikeda and Wright, 1972d).

Second, we found that the antagonism of center and surround was much stronger in sustained than in transient cells. In sustained cells the regions from which mixed on-off discharges could be obtained were limited to the boundaries between excitatory and inhibitory regions, whereas in transient cells, a transient center-type response often persisted throughout the entire surround.

Third, we have shown that sustained cells had an optimum spot size beyond which the cell response decreased, whereas there was no marked decrease in response for larger-than-optimum stimuli in transient cells. The expansion of the receptive field with increasing spot size in transient cells (Figure 4) was associated with the presence of a periphery effect (McIlwain, 1964), which

R.F. response profile; on-centre transient

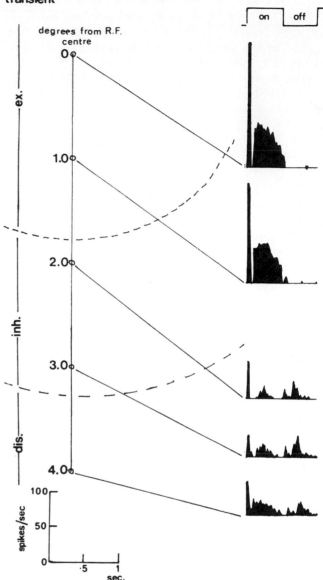

FIGURE 4 Poststimulus histograms obtained at different locations in the receptive field of an on-center transient cell. This is the same cell as shown in Figure 3, but the stimulus spot is larger (51′). Other stimulus conditions as in Figure 3. The excitatory, inhibitory, and outer excitatory zones all show an apparent increase in diameter with the larger spot, and a definite off response is obtained from the inhibitory surround, but there is still considerable overlap between receptive field zones, and mixed (on-off) responses are obtained from most of the receptive field periphery (Ikeda and Wright, unpublished observations.)

we have argued is an extension of receptive field properties into the far periphery. (Ikeda and Wright, 1972b).

Fourth, the spatial extent of the excitatory, inhibitory, and disinhibitory phases of the sensitivity gradient was greater in transient than in sustained cells. If the diameter of the sensitivity gradient were taken at $1/e$ of the sensitivity of the central point, the mean value was 0.7° for sustained cells at 3–8° from the area centralis, while it was 2.1° for transient cells from the same retinal region. The surrounds of transient cells were also larger than those of sustained cells by a factor of 3. A similar result was obtained by Cleland and Levick (1972).

Given these receptive field properties, we can make some generalizations about the types of visual information transmitted by sustained and transient cells. With their narrow, sharply peaked sensitivity gradients, sustained cells are able to resolve finer visual detail than transient cells. It is known that sustained cells preserve information about the phase of the finest resolvable grating patterns drifting across the receptive field whereas transient cells respond only to the onset and offset of movement of such gratings (Enroth-Cugell and Robson, 1966). The smaller surround diameter of sustained cells gives them sharper tuning to stimulus size. Although the response of transient cells is less critically dependent on the spatial distribution of illumination in the receptive field, the phasic nature of the response gives greater temporal definition than the tonic responses of sustained cells. Sustained cells would therefore seem to be concerned predominantly with the spatial aspects of visual stimuli, and transient cells with the temporal aspects.

Anatomical and physiological independence of sustained and transient pathways

The sustained/transient classification of visual neurons corresponds to classifications based on other physiological or anatomical criteria. The fastest-conducting fibers in the cat's optic nerve are the axons of Y-cells, and the axons of X-cells are slower conducting (Cleland, Dubin, and Levick, 1971; Fukada, 1971). In addition, there is a group of very slow-conducting fibers, W-cells (Hoffmann and Stone, 1972), which have suppressed-by-contrast (Stone and Fabian, 1966), or excited-by-contrast receptive fields.

The destinations of these fiber groups within the central nervous system are different. The lateral geniculate nucleus (LGN) receives X- and Y-cell afferents (Cleland, Dubin, and Levick, 1971) whereas the superior colliculus receives W- and Y-cell fibers (Hoffmann and Stone, 1972). Moreover, as Cleland, Dubin, and Levick (1971) showed, cells within the LGN may be classified as sustained or transient on the same criteria as retinal ganglion cells; transient LGN cells receive their excitatory input *only* from transient optic nerve axons, and sustained LGN cells only from sustained optic nerve axons (Hoffmann, Stone, and Sherman, 1972). There is thus little mixing of

sustained and transient information arriving at the visual cortex.

Hoffmann and Stone (1971) present preliminary evidence that sustained and transient optic radiation axons innervate different populations of cortical neurons. However the classificatory scheme for cortical receptive fields that they used (the simple, complex, and hypercomplex cells of Hubel and Wiesel, 1965) must be extended to include the *temporal* as well as the *spatial* organization of receptive fields if the fate of sustained and transient inputs to the cortex is to be understood. Noda, Freeman, Gies, and Creutzfeldt (1971), who recorded from visual cortex neurons in awake, unparalyzed cats, found a large group of neurons giving tonic responses to stationary gratings in a preferred orientation, and found a second group that responded only when the grating was moved at moderate speed in a preferred direction; it would certainly be consistent with the known properties of neurons in the visual pathway to suppose that the motion-sensitive units received inputs from transient LGN cells and that units responding to stationary stimuli received sustained inputs.

The lateral posterior nucleus (LP) of the cat thalamus contains two visual centers, one of which receives fibers from the superior colliculus, the other from the visual cortex (see Ann Graybiel, Part 3 of this volume). Neurons in both regions are characterized by phasic responses, and many respond to discontinuous rather than smooth movements: They appear to be higher-order, movement-sensitive neurons of various kinds (Wright, 1971). The responses of a discontinuous movement unit from the lateral (cortical) part of LP are shown in Figure 5. The response characteristics of these neurons and the pattern of their afferent connections suggest that these cells are driven by transient neurons in visual cortex or superior colliculus.

This anatomical and physiological evidence points to the conclusion that there are separate sustained and transient pathways within the visual projections. Apparently the sustained pathway is limited to the geniculostriate system, but the transient pathway has both cortical and subcortical branches and may subserve a corresponding multiplicity of functions.

There are differences in the distribution of sustained and transient ganglion cells across the retina. According to Cleland and Levick (1972), 90% of the cells in the area centralis were sustained and 10% transient, but for the remainder of the retinal area within 30° of the cat's area centralis, 62% were sustained and 38% transient. Ikeda and Wright (1972a) found that the proportion of transient cells increased steadily toward the periphery. The differing functions of the fovea and retinal periphery might be served by the filtering properties of sustained and transient

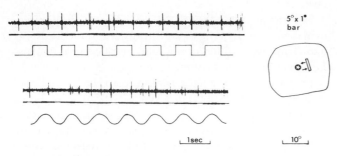

Square and sinusoidal movement

FIGURE 5 Response of a neuron in the lateral part of the lateral posterior nucleus (LP) of the cat to different modes of stimulus movement. In the upper trace, a 5° × 1° light rectangle is subjected to abrupt displacements of position within the receptive field (the rectangle is reflected off a mirror driven by a square wave of 1 Hz from a function generator). The neuron responds to each displacement of the rectangle with a short burst of spikes. In the lower trace, the rectangle is given a sinusoidal motion of the same frequency, and the cell no longer responds. Many units in LP showed this pattern of response, responding to discontinuous but not to smooth movements. (Wright, unpublished observations.)

cells. Transient cells are exquisitely sensitive to changes in the pattern of illumination over their receptive fields, such as may be produced by a moving object in the peripheral visual field. Attention-getting stimuli, which invariably involve such changes, initiate fixation reflexes resulting in the object of interest being centered on the fovea, where its spatial detail may be analyzed by the sustained pathway. Transient cells, particularly peripheral ones, are tolerant to defocusing of the stimulus by refractive errors (in a three-dimensional world, only objects lying along the visual axes are consistently kept in focus by accommodation), but central, sustained cells require a sharply focused image in order to respond (Ikeda and Wright, 1972a).

Role of sustained and transient cells in visual perception

In deciding the relevance of these arguments to general problems of visual perception, there are two questions to be answered. First, to what extent are findings in the cat characteristic of other species? Second, do the functional interpretations we have placed on the behavior of X- and Y-cells represent truly distinct modes of visual processing that can be identified in psychophysical experiments?

There is good evidence from comparative anatomy and physiology that the sustained/transient classification is applicable to the visual pathways of a wide range of vertebrates. Werblin and Dowling (1969) recorded sustained and transient retinal ganglion cells in Necturus and

identified amacrine cells as the element in the input network of ganglion cells responsible for the generation of transient responses. Dubin (1970) examined the inner plexiform layer of vertebrate retinas by electron microscopy and found greater number and complexity of amacrine synaptic connections in those species that have ganglion cells with refined temporal discrimination. Movement-sensitive ganglion cells may often be classified as transient on the basis of their response to a stationary spot, and a large majority of ganglion cells in the retina of the pigeon (L. Holden, unpublished results) and rabbit (H. Ikeda, unpublished results) are transient on the usual criteria; these species have a larger number of complex amacrine synapses than animals that rely on stereoscopic vision, such as the cat and monkey (Dubin, 1970).

Sustained and transient neurons also exist in primate visual systems. Gouras (1969) found that in the monkey retina, transient cells, as in the cat, have faster-conducting axons than sustained cells and are relatively more common in peripheral retina. In addition, transient cells mix signals from different cone mechanisms in both center and surround, but sustained cells receive excitatory signals from one cone mechanism in the center, and inhibitory signals from another cone mechanism in the periphery of the receptive field (Gouras, 1968).

The use of stimuli of spatially periodic patterns (gratings) with variable contrast and constant mean luminance, and the application of Fourier analysis to the results, have considerably advanced the psychophysics of form perception (Campbell, earlier in this part). The same methods may be extended to include temporal periodicities in stimuli, and some investigations of this type do provide evidence that spatial and temporal features of visual stimuli are to some extent separately encoded in the nervous system.

Kulikowski (1971) used a sinusoidal grating, of variable spatial frequency, that was alternated 180° in phase at temporal frequencies of 0.5–30 Hz, and viewed monocularly, with central fixation. At intermediate spatial and temporal frequencies, such a grating is in apparent movement appearing to drift in either direction at constant speed. However, if the spatial frequency were above 20 c/deg, Kulikowski's subjects reported that the grating appeared to be stationary. When the contrast of the grating was varied, it was found that the threshold for detecting the grating was lower than the threshold for detecting the apparent movement, at all spatial frequencies. As this experiment proves, it is possible to detect a grating without being aware of its movement, whereas as soon as a grating is above threshold contrast, the subject is aware of its orientation and spatial frequency.

A plausible explanation for this phenomenon is that even for patterns stimulating the central retina, the detection of purely spatial (orientation and spatial frequency) features and the detection of spatiotemporal (real and apparent movement) features of stimuli are accomplished by different mechanisms, which might be elements in the sustained and transient pathways, respectively. The low sensitivity of the transient pathway to patterns of high spatial frequency and low contrast would prevent its stimulation by such patterns, resulting in elimination of the sensation of movement, while the sustained pathway continued to detect the grating itself.

Experiments by Pantle and Sekuler (1969) support the conclusion that the detection of stationary patterns and of movement is accomplished by separate mechanisms. They found that prolonged viewing of a slowly moving grating of high contrast led to a reduction in contrast sensitivity for specific test targets. By varying the contrast of the adapting grating, they separated two components of the response. The first was specific to the orientation of the grating and was independent of its movement, and the second was specific to the direction of movement. The orientation- and direction-specific mechanisms revealed in these experiments showed different changes in contrast sensitivity with an adapting stimulus of equal contrast and duration.

There is evidence that many purely spatial features of visual stimuli are encoded by a *single* population of cells: Detectors that are tuned to spatial frequency are also orientation specific (Blakemore and Campbell, 1969) and even specific for retinal disparity (Blakemore and Hague, 1972). Blakemore (1973) suggests that each cortical cell encodes a variety of (spatial) submodalities and that a single cortical cell is thus *multichannel* in nature, so that the same population of cells would encode *all* the orientations, *all* the spatial frequencies, and *all* the disparities.

On the other hand, the psychophysical experiments reviewed here would suggest that *different* populations of cells are concerned with the detection of a pattern and the detection of its movement. It would appear that for central vision, there are at least two populations of cells in the brain, one encoding the spatial features of stimuli, the other encoding movement. The characteristics of this second population of cells, which presumably receives input from transient neurons, and the role of transient cells in peripheral vision are in need of further investigation.

ACKNOWLEDGMENTS The work in this laboratory was supported by the Medical Research Council and the R.N.I.B. We thank Janet Nuza, Sheena Dyer and John Dench for excellent technical assistance. The experiments on the lateral posterior nucleus by M. J. Wright were done in the Psychological Laboratory, Cambridge, and supported by the Medical Research Council. Thanks are due to Dr. G. Horn for help at all stages of this work.

REFERENCES

BLAKEMORE, C., 1973. Central visual processing. In *Foundations of Psychobiology*, M. S. Gazzaniga and C. Blakemore, eds. London and New York: Academic Press.

BLAKEMORE, C., and F. W. CAMPBELL, 1969. On the existence in the human visual system of neurons selectively sensitive to the orientation and size of retinal images. *J. Physiol. (London)* 203:237-260.

BLAKEMORE, C., and B. HAGUE, 1972. Evidence for disparity-detecting neurons in the human visual system. *J. Physiol. (London)* 225:437-456.

CLELAND, B. G., M. W. DUBIN, and W. R. LEVICK, 1971. Sustained and transient neurons in the cat retina and lateral geniculate nucleus. *J. Physiol. (London)* 217:473-497.

CLELAND, B. G., and W. R. LEVICK, 1972. Physiology of cat retinal ganglion cells. *Invest. Ophthal.* 11:285-290.

DUBIN, M. W., 1970. The inner plexiform layer of the retina: a quantitative and comparative electron microscopic analysis. *J. Comp. Neurol.* 140:479-506.

ENROTH-CUGELL, C., and J. G. ROBSON, 1966. The contrast sensitivity of retinal ganglion cells of the cat. *J. Physiol. (London)* 187:517-552.

FUKADA, Y., 1971. Receptive field organization of cat optic nerve fibers with special reference to conduction velocity. *Vision Res.* 11:209-226.

GOURAS, P., 1968. Identification of cone mechanisms in monkey ganglion cells. *J. Physiol. (London)* 199:533-547.

GOURAS, P., 1969. Antidromic responses of orthodromically identified ganglion cells in monkey retina. *J. Physiol. (London)* 204:407-419.

HOFFMANN, K.-P., and J. STONE, 1971. Conduction velocity of afferents to cat visual cortex: a correlation with cortical receptive field properties. *Brain Res.* 32:460-466.

HOFFMAN, K.-P., J. STONE, and M. SHERMAN, 1972. Relay of receptive field properties in dorsal lateral geniculate nucleus of the cat. *J. Neurophysiol.* 35:518-531.

HUBEL, D. H., and T. N. WIESEL, 1962. Receptive fields, binocular interaction and functional architecture in the cat's visual cortex. *J. Physiol. (London)* 160:106-154.

HUBEL, D. H., and T. N. WIESEL, 1965. Receptive fields, binocular interaction and functional architecture in two non-striate visual areas (18 and 19) of the cat. *J. Neurophysiol.* 28:229-289.

IKEDA, H., and M. J. WRIGHT, 1972a. Differential effects of refractive errors and receptive field organization of central and peripheral ganglion cells. *Vision Res.* 12:1465-1476.

IKEDA, H., and M. J. WRIGHT, 1972b. Functional organization of the periphery effect in retinal ganglion cells. *Vision Res.* 12:1857-1879.

IKEDA, H., and M. J. WRIGHT, 1972c. The outer disinhibitory surround of the retinal ganglion cell receptive field. *J. Physiol. (London)* 226:511-544.

IKEDA, H., and M. J. WRIGHT, 1972d. Receptive field organization of "sustained" and "transient" retinal ganglion cells which subserve different functional roles. *J. Physiol. (London)* 227:769-800.

KUFFLER, S. W., 1953. Discharge patterns and functional organization of the mammalian retina. *J. Neurophysiol.* 16: 37-68.

KULIKOWSKI, J. J., 1971. Effect of eye movements on the contrast sensitivity of spatio-temporal patterns. *Vision Res.* 11:261-273.

MCILWAIN, J. T., 1964. Receptive fields of optic tract axons and lateral geniculate neurons: peripheral extent and barbiturate sensitivity. *J. Neurophysiol.* 27:1154-1173.

NODA, H., R. B. FREEMAN, JR., B. GIES, and O. D. CREUTZFELDT, 1971. Neuronal responses in the visual cortex of awake cats to stationary and moving targets. *Exp. Brain Res.* 12:389–405.

PANTLE, A., and R. SEKULER, 1969. Contrast response of human visual mechanisms sensitive to orientation and direction of motion. *Vision Res.* 9:397–406.

RODIECK, R. W., 1965. Quantitative analysis of cat retinal ganglion cell response to visual stimuli. *Vision Res.* 5:583–601.

STONE, J., and M. FABIAN, 1966. Specialized receptive fields of the cat's retina. *Science* 152:1277–1279.

STONE, J., and K.-P. HOFFMANN, 1972. Very slow-conducting ganglion cells in the cat's retina: a major, new functional type? *Brain Res.* 43:610–616.

WERBLIN, F. S., and J. E. DOWLING, 1969. Organization of the retina of the mudpuppy, *Necturus maculosus*. II. Intracellular recording. *J. Neurophysiol.* 32:339–355.

WRIGHT, M. J., 1971. Responsiveness to visual stimuli of single neurons in the pulvinar and lateral posterior nuclei of the cat's thalamus. *J. Physiol. (London)* 219:32–33P.

12 The Psychophysics of Visually Induced Perception of Self-Motion and Tilt

JOHANNES DICHGANS and TH. BRANDT

ABSTRACT Visual stimuli that move in a horizontal plane may lead to two different perceptual interpretations: The observer may perceive himself either as being stationary in space while the visual stimulus appears to move or he may experience an illusion of self-motion while the moving surround appears at rest. Furthermore, when an observer views a wide-angled display rotating around his line of sight, he feels his body tilted and sees a vertical straight edge tilted in the direction opposite to the moving stimulus. Psychophysical and neurophysiological observations suggest that the sensation of optokinetically induced self-motion and tilt are attributable to interactions of visual and vestibular inputs within the vestibular system.

Introduction

A SUBJECT exposed to a large horizontally moving pattern may experience an apparent self-motion in the direction opposite to that of the moving visual stimulus (Mach, 1885; Helmholtz, 1896; Fischer and Kornmüller, 1930; Gurnee, 1931). At the same time, the moving pattern may seem to be stationary in space. This compelling illusion will invariably occur if the entire visual surroundings are moving and if enough time is allowed for stimulation. With stimulation in a horizontal plane, the perceived self-motion, which in the following will be called circular vection (CV), cannot be distinguished subjectively from true passive body motion (Brandt et al., 1971). This holds even with respect to *Coriolis effects* which, during real body rotation, are caused by bending the head toward the shoulder. Vestibular Coriolis effects are generally assumed to be mainly due to the effects of cross-coupling of angular acceleration applied to different semi-circular canals (Groen, 1961). Therefore, it is quite puzzling that during the visually induced illusion of self-rotation *pseudo-Coriolis effects* arise from similar head movements, which eventually lead to motion sickness. Their symptoms (apparent tilt, dizziness, drowsiness, and nausea) have been shown to be qualitatively the same as in Coriolis effects (Dichgans and

JOHANNES DICHGANS and TH. BRANDT Department of Neurology, University of Freiburg, West Germany; Department of Psychology, Massachusetts Institute of Technology, Cambridge, Massachusetts.

Brandt, 1972).

A second phenomenon also underscores in our opinion the importance of visual motion information for orientation with respect to gravity: If exposed to a visual pattern that rotates in a vertical plane around the observer's line of sight, the subject will not only experience body motion but also tilt of both the visual and postural vertical (Dichgans et al., 1972). Similar observations have been made by Wood (1895) and Helmholtz (1896).

In this paper, both phenomena will be described in more detail. In each case, the role of stimulus time, stimulus velocity, and stimulus area and its location within the visual field will be examined. Special emphasis will be given to clarification of the stimulus features that determine whether or not a visual stimulus is perceived as moving in the outer world (egocentric motion perception) and that determine the perceived orientation of the body in space (exocentric motion perception).

Some of the results described in this paper led us to the assumption of visual-vestibular interaction within the vestibular system. Among these are the phenomenal equality of self-motion as perceived either by exclusive visual or by vestibular stimulation, the existence of pseudo-Coriolis effects, and the direction-specific modulation of vestibular thresholds (Young et al., 1972) for the perception of body acceleration during CV. The hypothesis will be discussed in reference to supporting evidence from a few neurophysiological experiments.

We feel justified in putting so much emphasis on these *illusions*, as they must be called under our experimental conditions, because the underlying physiological mechanisms participate in spatial orientation under real-life conditions. This will be shown in the last paragraph of the chapter.

Visually induced perception of self-motion

The *experimental apparatus* consisted of a rotating chair located in the center of a closed cylindrical drum 1.5 m in diameter whose inner walls were painted with alternating vertical black-and-white stripes subtending 7° of visual angle. Both the chair and the drum could be rotated separately or simultaneously in either the same or opposite directions and at the same or different speeds. Optokinetic stimulation by the moving drum could be restricted to any desirable spatial extent and could be presented at any location within the visual field. This was achieved by using black masks that were mounted immediately adjacent to the inner wall of the drum. The masks were fixed on poles connected to the back of the chair. To stabilize the direction of the visual axis, subjects were asked to focus on a 1° luminous spot mounted on the chair and presented in a position straight ahead of the subject. Eye movements were recorded by electronystagmography. In the different experiments described, a total of 68 students, who were previously unfamiliar with the investigated phenomena, took part in the study.

The *prolonged time course* of CV after onset and termination of the visual stimulus was studied with constant speeds of drum-rotation ranging between 10° and 180°/sec. Subjects initially sitting in the dark were suddenly exposed to the moving pattern (in the light) that stimulated the entire visual field. Invariably, the initial experience was that of surround motion; however, within an average of 3 to 4 sec, an apparent body acceleration opposite to the direction of drum-rotation began, during which the surroundings seemed to move progressively more slowly. Within an average of 8 to 12 sec after stimulus onset, an exclusive self-rotation (CV) was perceived, and the surroundings seemed to be stationary. After switching off the illumination inside the drum, CV never stopped immediately but continued in the same direction, outlasting the visual stimulus by an average of 8 to 11 sec. The time course of CV shows considerable interindividual variability but is relatively constant for each subject over trials. In our experiments, latencies are barely influenced by stimulus velocities. This observation, and the fact that even with drum accelerations up to 15°/sec² subjects still experienced self-rotation, show that one cannot infer from the lack of actual vestibular stimulation that only the visual surround is moving.

The *velocity range* in which the sensation of exclusive self-motion can be elicited by a moving visual stimulus has its upper limit at approximately 90°/sec (Figure 1). Within this range, the perceived velocity of CV is linearly related to stimulus speed, and the apparent velocity is independent of whether the subject fixates on a stationary target or tracks the moving pattern. This is in contrast to egocentric motion perception in which the perceived velocity of the stimulus is greater by a factor of 1.6 when the eyes are kept stationary than when optokinetic nystagmus occurs (Dichgans et al., 1969). One may conclude

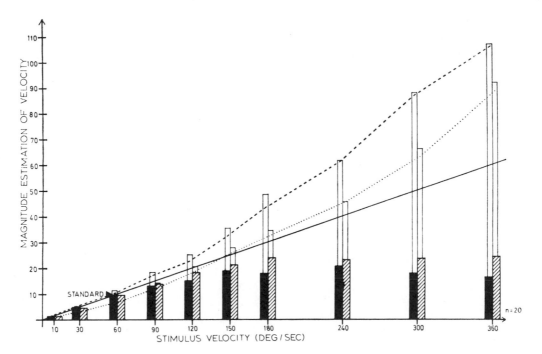

FIGURE 1 Magnitude estimations of velocity in egocentric (interrupted lines) and exocentric motion perception (dark columns). In each of these, velocities were scaled with fixation of a small stationary target (dashed line, black column) or with optokinetic nystagmus (dotted line, shaded column). The white columns symbolize estimates of apparent surround motion that with higher stimulus velocities occurs in addition to CV. The standard stimulus (modulus) is indicated by a black triangle. For further explanations see text.

that the visual motion information that leads to self-rotation sensation is abstracted before the differentiation between the two modes of egocentric velocity evaluation occurs.

With stimulus speeds exceeding 120°/sec, a mixed sensation occurs whereby the drum appears to rotate in the direction opposite to the perceived CV. The sum of the velocities of these apparent motions seems to correspond roughly to the phenomenal velocity of a stimulus that appears to move in reference to the stationary observer. The apparent velocity of CV moderately increases with stimulus speeds from 90° to 180°/sec and remains approximately constant with further increase of stimulus velocity up to the highest speeds tested (360°/sec).

Stimulus area and location within the visual field are the major factors in determining whether sensation of self-motion occurs or whether the stimulus is correctly perceived as moving relative to the observer. Experiments reported in detail by Brandt et al. (1973) showed that the peripheral retina dominates dynamic spatial orientation whereas the more central parts of the retina subserve egocentric motion perception, visual grasping, and eye tracking. As demonstrated in Figure 2, stimulation of the central parts of the visual field up to 30° in diameter scarcely ever leads to CV and even up to 60° yields only moderate CV. Moreover, central masking up to 120° in diameter has only a slightly diminishing effect on CV. The data in Figure 2 represent magnitude estimations of the subjective velocity of CV, but since both "intensity" and velocity strongly vary with each other, they are also representative for the rather ill-defined "compellingness" of the phenomenon. Obviously, an increase in the area stimulated by a moving pattern augments its effect on spatial orientation. But this is at least to a considerable extent due to the concurrent increase in the amount of peripheral retina included in stimulation. The predominating importance of the peripheral retina was further illustrated in an experiment where stimuli of equal area (30° in diameter) were exposed to the center and the periphery. It was evident that the peripheral stimuli yield stronger effects (Brandt et al., 1973).

In the experiment depicted in Figure 3, contradictory stimuli are applied that move in opposite directions across the center (30°) and the periphery of the visual field. In this case, exocentric and egocentric motion perception occur simultaneously. Again, the dynamic spatial orientation depends on the peripheral stimulus while the central stimulus is perceived as moving in reference to the observer. Optokinetic nystagmus subserving pattern perception is guided by the central stimulus despite its small area. The results not only illustrate the different functions of the center and periphery of the visual field but provide conclusive evidence that dynamic spatial orientation does not depend on the direction of eye movements.

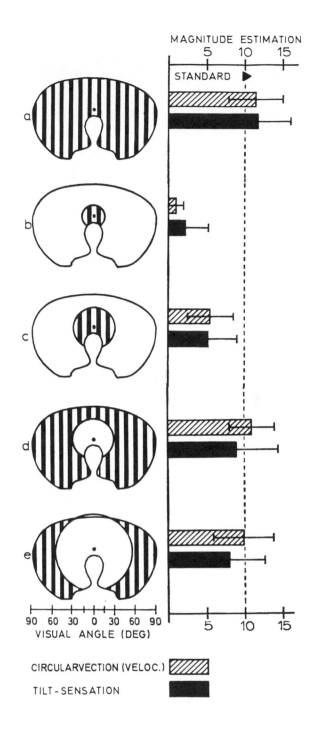

FIGURE 2 Magnitude estimates of subjective velocity (shaded columns) and tilt perceived during pseudo-Coriolis effects (black columns) in relation to variations of stimulus area and location within the visual field. Optokinetic stimuli (60°/sec) are schematically depicted on the left. Masks are symbolized by white areas. Particular importance of stimulation of the peripheral retina is shown in d and e in which central masks up to 120° in diameter (e) scarcely diminish CV and tilt. Stimulation of the more central retina yields rather small effects (b and c). Variations in perceived velocity and tilt sensation are strongly related to each other.

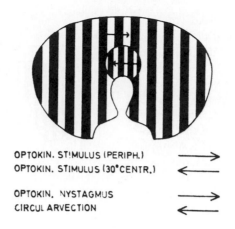

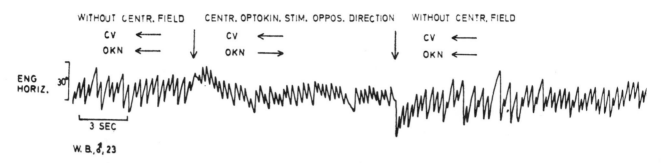

OPTOKIN. STIMULUS (PERIPH.)	→
OPTOKIN. STIMULUS (30°CENTR.)	←
OPTOKIN. NYSTAGMUS	→
CIRCULARVECTION	←

FIGURE 3   Opposite optokinetic stimulation of center and periphery. In this situation, CV is determined by the peripheral stimulus while optokinetic nystagmus (OKN) (lower graph) is guided by the central stimulus. As soon as the central stimulus appears, the direction of OKN reverses. The direction of CV is unaltered.

Moving visual scenes shift the apparent direction of gravity

Using visual displays without cues of visual orientation that could possibly conflict with cues of gravity, it has been demonstrated that seen motion as such may induce tilt of the apparent upright of visual contours and body posture (Dichgans et al., 1972). If a large visual display (130° of visual angle) is rotated around the observer's line of sight, a stationary central test edge, initially set to the vertical, seems to be tilted in the direction opposite to the rotation of the field. Subjects were asked to readjust continuously the orientation of the central test edge to vertical. The corrections made to compensate for perceived tilt were continuously recorded. The tilt illusion starts shortly after initiation of the movement, increases rapidly at first, then progressively more slowly, and reaches its steady state within an average of 18 sec. The amount of apparent tilt increases with angular velocity of the visual stimulus up to 30°/sec (Figure 4A). The estimated tilt averaged 15° for 7 subjects and exceeded 40° in one subject. In a second series of experiments on which we will extensively

report elsewhere (Held et al., in preparation), we found that tilt increases with stimulus area and that motion stimulation of the more peripheral parts of the retina exerts a disproportionately strong influence on the perceived upright (Figure 4B). This result corresponds with the observations on CV in which stimulation of the periphery dominates the induction of an apparent body motion. This latter stimulus condition also elicits the sensation of self-motion but, as described by Dichgans et al. (1972), the paradoxical sensation of continuous motion with limited tilt of the body occurs. The magnitude of tilt militates against the notion that small rotatory eye movements can account for the effect.

In order to demonstrate that the tilt effect is not an exclusively visual phenomenon but is due to a shift of the internal representation of the vector of gravity, we designed a second experiment in which tilt of the perceived postural upright was induced by motion stimulation. Subjects were seated in a moving-base airplane trainer (Link GAT 1). Using a control stick, they were able to adjust the position of the trainer to the subjective upright. A visual pattern was projected onto the side

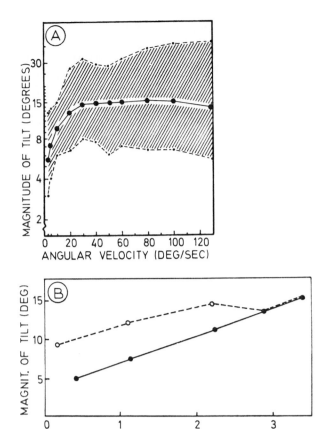

FIGURE 4   (A) Magnitude of apparent tilt of the visual vertical (geometric means on the solid line) in relation to velocity of the visual stimulus. The shaded region represents the range of single values of the 7 subjects. Interindividual data are highly replicable but individual variations are large. (B) Magnitude of apparent tilt in relation to stimulus area and location within the visual field (one subject). The size of the field was increased by either decreasing the outer radius or by increasing the inner radius of a white ring mask. Equal areas of stimulation (abscissa) yield greater tilt effects if exposed to the more peripheral parts of the visual field (open circles) than with stimulation of the more central retina (black dots).

windows of the trainer so as to move upward on one side and downward on the other. Shortly after the pattern was set into motion, the subjects started to roll the trainer's position off its initial vertical position. This was done in an attempt to compensate for the perceived tilt in postural orientation that, as in the case of the visual vertical, was opposite in direction to the moving stimulus. The resulting tilt of the cabin was therefore in the direction of the moving stimulus. It reached its steady stage after 17 sec and, with pattern velocities of 14 to 26°/sec, averaged 8.5° in 4 sub-

jects. The somewhat smaller effect on the perceived postural vertical may be due to the rather small area of motion stimulation and to graviceptive information originating from somatosensory pressure receptors that signal the induced change in body posture.

We have argued earlier that the tilt of both the visual and the postural vertical can be attributed to the visual motion stimulus alone (Dichgans et al., 1972). We have also stated that the perceptual effects are equivalent to the results of shifting the direction of gravity. Indeed, real displacement of the vector of gravity results in corresponding perceptions (Mach, 1875). This has been demonstrated in very thorough experiments by Clark and Graybiel (1951) and Graybiel (1952), who accomplished this effect in a human centrifuge. They found a "fairly good correspondence" between the subject's estimate of tilt in postural orientation and perceived visual vertical, on the one hand, and the angle between the direction of gravity and the vector of force resulting from gravity and centrifugal forces, on the other. We hypothesize that the orientation of the gravitational vector is computed by the central nervous system from graviceptive input originating in the labarynthine otoliths and somato sensory pressure receptors as well as from seen motion.

The possible neurophysiological mechanisms

Circularvection and the change in the apparent direction of gravity seem to be related phenomena: Both may be explained by the convergence of visual and vestibular information. So far, this possibility has only been examined in a few neurophysiological experiments in animals. Indeed, optokinetic stimulation can induce a direction-specific modulation of resting discharge in the vestibular nerve of the goldfish (Klinke and Schmidt, 1970) and the medial and lateral vestibular nuclei of the rabbit (Dichgans and Brandt, 1972). In the rabbit, some of the neurons in the vestibular nuclei respond not only to angular acceleration but also to exclusive visual motion stimulation without any head movement. The directional specificity of these neurons is opposite for visual and vestibular stimuli. This corresponds to the natural condition in which a rotation of the animal to the left is accompanied by relative motion of the environment to the right and vice versa. The prolonged summation and decay of frequency modulation after stimulus onset and termination roughly correspond to the time course of CV.

The anatomical pathways along which visual information is carried into the vestibular system are still to be investigated. Direct connections to the vestibular nuclei from either the lateral geniculate body or the superior colliculus are unknown. The long latencies for the induction of CV and apparent tilt of the vertical would suggest a multisynaptic pathway, possibly through the network of the reticular formation. The hypothesis that the superior colliculus would serve as an important relay within the circuit of convergence of visual information into the vertibular system for the purpose of dynamic spatial orientation seems to be attractive, for this structure reportedly represents a major center for visual and auditory orientation within a static environment. This notion, however, still lacks experimental evidence.

The preference of neurons in the superior colliculus for moving as opposed to stationary visual stimuli is evident in all mammalian species so far investigated. Sprague et al. (1973) pointed out that "it is only in the superior colliculus that movement is a specific trigger for the great majority of neurons," and he suggested "that behavioral performances involving detection of motion or rates of motion, are at least in part integrated by neuronal circuits taking an obligatory route through the superior colliculus."

At this point one should remember that the superior colliculus receives a clear-cut projection from the very peripheral retina (Apter, 1945; Forrester, 1967) whereas the projection to the lateral geniculate body originates mainly from the fovea and the more central parts of the retina (Bishop et al., 1962; recent review by Freund, 1973). It has been claimed that the superior colliculus of the monkey, in contrast to that of the cat, does not receive any direct projections from the fovea (Brouwer and Zeeman, 1926; Wilson and Toyne, 1970). Although still controversial, the latter finding would support the hypothesis of the superior colliculus being a major center for *ambient vision* and orientation (Trevarthen, 1968).

It has been indirectly shown by Klinke and Schmidt (1970) and Uemura and Cohen (1972) that the efferent vestibular pathway carries optokinetic information to the vestibular receptor. Nevertheless, numerous additional sites of visual-vestibular convergence are to be expected within the central nervous system.

The *functional importance* of visual motion information for the perception of self-motion may now be considered. With rotation or linear motion in a horizontal plane, the vestibular receptors signal only acceleration. Once constant velocity is reached, cupulae and otoliths slowly return to their resting position and the vestibular input dies out. Yet during constant velocity, even with passive movement, the perception of self-motion is maintained by visual motion information. This can easily be demonstrated. If, at constant velocity, one closes the eyes or if, in a rotating room, one is confronted with a seemingly stationary visual environment, motion is not perceived. With self-motion at constant velocity or at levels of acceleration below the vestibular threshold, visual input is obviously a most important source of information to

which kinesthesis may also contribute. The different functions of the peripheral retina and of its more central parts allow for adequate self-motion perception and for eye tracking and the perception of smaller objects moving in relation to the observer and his environment.

The functional importance of the second phenomenon in which the apparent direction of gravity is shifted by a visual stimulus moving in a vertical plane is not as well understood at the present time. It is generally accepted that "a coordinate system based on the direction of gravity forms the basis of a reference system by which man orients himself to the earth and to objects in space" (Graybiel, 1952). Since the internal estimate of the vector of gravity also depends on visual motion information, it is conceivable that the effect of relative motion of the visual surroundings caused by any body displacement from real upright would corroborate the actual information from gravireceptors for postural adjustment. Thus, the stability of visual and postural orientation with respect to gravity might depend in part upon motion within the observer's visual field. Visual stabilization of posture is evident in patients with a bilateral labyrynthine disease.

REFERENCES

APTER, J. T., 1945. The projection of the retina on superior colliculus of cats. *J. Neurophysiol.* 8:123–134.

BISHOP, P. O., W. KOZAK, W. R. LEVIK, and G. VAKKUR, 1962. The determination of the projection of the visual field onto the lateral geniculate nucleus of cat. *J. Physiol.* (*London*) 163:503–539.

BRANDT, TH., E. WIST, and J. DICHGANS, 1971. Optisch induzierte Pseudo-Coriolis-Effekte und Circularvektion: Ein Beitrag zur optisch-vestibulären Interaktion. *Arch. Psychiat. Nervenkr.* 214:365–389.

BRANDT, TH., J. DICHGANS, and E. KOENIG, 1973. Differential effects of central versus peripheral vision on egocentric and exocentric motion perception. *Exp. Brain Res.* (in press).

BROUWER, B., and W. P. C. ZEEMAN, 1926. The projection of the retina in the primary optic neurons in monkeys. *Brain* 49:1–35.

CLARK, B., and A. GRAYBIEL, 1951. Visual perception of the horizontal following exposure to radial acceleration on a centrifuge. *J. Comp. Physiol. Psychol.* 44:525–534.

DICHGANS, J., F. KÖRNER, and K. VOIGT, 1969. Vergleichende Skalierung des afferenten und efferenten Bewegungssehens beim Menschen: Lineare Funktionen mit verschiedener Anstiegssteilheit. *Psychol. Forsch.* 32:277–295.

DICHGANS, J., and TH. BRANDT, 1972. Visual-vestibular interaction and motion perception. In *Cerebral control of eye movements and motion perception*, J. Dichgans and E. Bizzi, eds. Basel and New York: S. Karger.

DICHGANS, J. R., R. HELD, L. R. YOUNG, and TH. BRANDT, 1972. Moving visual scenes influence the apparent direction of gravity. *Science* 178:1217–1219.

FISCHER, M. H., and A. E. KORNMÜLLER, 1930. Optokinetisch ausgelöste Bewegungswahrnehmungen und optokinetischer Nystagmus. *J. Psychol. Neurol.* (Lpz.) 41:273–308.

FORRESTER, J. M., and S. K. LAL, 1967. The projection of the rats visual field upon the superior colliculus. *J. Physiol. (London)* 189:25–26.

FREUND, H.-J., 1973. Neuronal mechanisms of the lateral geniculate body. In *Handbook of Sensory Physiology*, Vol. VII/3B, R. Jung, ed. Berlin, Heidelberg, New York: Springer-Verlag.

GRAYBIEL, A., 1952. Oculogravic illusion. *A.M.A. Arch. Ophthalmol.* 48:605–615.

GROEN, J. J., 1961. The problems of the spinning top applied to the semicircular canals. *Confin. Neurol. (Basel)* 21:454–455.

GURNEE, H., 1931. The effect of a visual stimulus upon the perception of bodily motion. *Amer. J. Psychol.* 43:26–48.

HELMHOLTZ, H. von, 1896. *Handbuch der physiologischen Optik.* Hamburg and Leipzig: Voss.

KLINKE, R., and C. L. SCHMIDT, 1970. Efferent influence on the vestibular organ during active movement of the body. *Pflügers Arch. ges. Physiol.* 318:325–332.

MACH, E., 1875. *Grundlinien der Lehre von den Bewegungsempfindungen.* Leipzig: Engelmann.

MACH, E., 1885. *Die Analyse der Empfindungen.* Jena: Fischer.

SPRAGUE, J. M., G. BERLUCCHI, and G. RIZZOLATTI, 1973. The role of the superior colliculus and pretectum in vision and visually guided behavior. In *Handbook of Sensory Physiology*, Vol. VII/3B, R. Jung, ed., Berlin, Heidelberg, New York: Springer.

TREVARTHEN, C. B., 1968. Two mechanisms of vision in primate. *Psychol. Forsch.* 31:229–337.

UEMURA, T., and B. COHEN, 1972. Vestibular-ocular reflexes: Effect of vestibular nuclei lesions. In *Progress in Brain Research, 37. Basic Aspects of Central Vestibular Mechanisms.* A. Brodal and O. Pompeiano, eds., Amsterdam: Elsevier.

WILSON, M. E., and M. J. TOYNE, 1970. Retino-tectal and corticotectal projections in Macaca Mulatta. *Brain Res.* 24:395–406.

WOOD, R. W., 1895. The "haunted swing" illusion. *Psychol. Rev.* 2:277–278.

YOUNG, L. R., J. DICHGANS, R. MURPHY, and TH. BRANDT, 1972. Influence of optokinetically induced self-rotation on perception of horizontal body acceleration. *Pflügers Arch.* 334 (Suppl. R): 78.

13 Neural Processes for the Detection of Acoustic Patterns and for Sound Localization

E. F. EVANS

ABSTRACT The processing of auditory stimulus information in the periphery of the auditory system rests largely on spectral analysis and the preservation of a certain amount of temporal information including that related to interaural differences. The auditory system appears to divide into two subsystems that can be differentiated anatomically and functionally, at the brainstem level at least, and that may be related to the processing of localization and pattern information, respectively. Central processing leads to a considerable complexity and diversity of neural response patterns and stimulus specificity. The latter manifests itself, particularly at the cortical level, as a preferential or specific selectivity for important temporal and spatial features of stimuli (including biologically significant sounds). There is some psychophysical, and considerable behavioral, evidence that such feature-sensitive neurons are required to abstract important features from a complex auditory environment.

Introduction

IN THIS BRIEF review, the emphasis and selection of material will be directed toward evidence for feature-sensitive mechanisms in the auditory nervous system. More comprehensive accounts of the neurophysiological data will be found elsewhere (e.g., Evans, 1971; Erulkar, 1972).

It is probably true that studies of the auditory system directed at this problem have not produced such coherent evidence for elements capable of feature detection as in the visual system. There are a number of identifiable reasons for this. First, there has been a preoccupation with the problem of pitch perception and its resolution in terms of rival *place* and *time* theories. There is now evidence that pitch perception itself may involve a type of central pattern recognition process (Houtsma and Goldstein, 1972; see Wilson, the following chapter). Then there is the difficulty that faces investigators wishing to use acoustic stimuli more complex than pure tones, of manip-

ulating and defining their parameters: For example, it is not possible to manipulate the temporal and frequency parameters of an acoustic stimulus completely independently. Figure 1 illustrates some of the stimuli that are used. Finally, there are important differences between the anatomical organization of the auditory and visual systems. Compared with the primary visual pathway—the retino-geniculo-striate pathway—impulses reach the auditory cortex by way of an indefinite number of synaptic interruptions (with four as a minimum, see Figure 3), and cross-connections. Hence the auditory cortex is more deleteriously affected by anaesthesia, particularly as far as neurons with feature-specific sensitivities are concerned.

Features of auditory stimuli

What constitutes a significant feature of a complex auditory stimulus is to a certain extent a matter of guess-work and intuition. A feature for one species may not be a feature for another. Thus, the relatively narrow band of supersonic energy in the bat vocalization is an extremely powerful stimulus feature for the noctuid moth (Roeder, 1971). Certain vocalizations of the squirrel monkey act as releasers of appropriate behavior patterns in that species alone (Winter et al., 1966).

Alternatively, a consideration of a stimulus itself may indicate distinctive features in the spectrum or wave form that we might expect to be significant (Figure 2). Thus, in many vocalizations, including human speech, several features are prominent, (Figure 2A), and indeed many have been shown to be important for speech perception (e.g., Liberman et al., 1959). Noise, of more or less limited bandwidth is the dominant component of certain consonants, transient bursts, or clicks. Frequency changes (Figures 1D and 1E) are characteristic of certain consonants and vowels, (such as the vowel "i" in Figure 2A). The relationship between the frequency components (formants) is another important feature of vowel sounds. Changes in amplitude (Figure 1B), or the envelope (as in

E. F. EVANS Medical Research Council Group in Neurophysiology, Department of Communication, University of Keele, Keele, Staffordshire, England

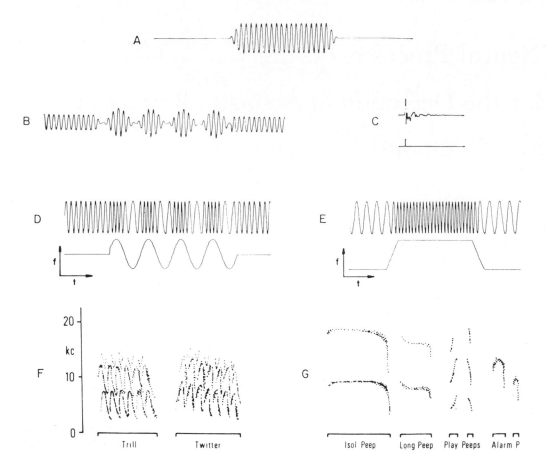

FIGURE 1 Simple and complex acoustic stimuli. A: Pure tone burst (onset and termination "shaped" to avoid click artifacts). B: Sinusoidally amplitude-modulated tone. C: Acoustic wave form of a "click" (upper trace) generated by a loudspeaker fed by an electrical pulse of short duration (lower trace). D: Sinusoidally frequency-modulated tone. Lower trace indicates excursions of frequency as a function of time. E: Linearly ("ramp") frequency-modulated tone. Lower trace as in D. F and G: Naturally occurring biologically significant stimuli; spectrograms of species-specific vocalizations of squirrel monkey. (From Winter et al., 1966.) F shows example of calls with approximately sinusoidal frequency modulation. G shows examples of calls with linear frequency modulation components.

Figure 2B) or "attack" of a sound can be distinctive features of sound complexes. The temporal patterning of the components of a stimulus can also be critical, as in the case of the distinctions between songs of different behavioral significance in the same species of cricket (Figure 2C). To this list of features should be added the dominant features for the localization of a sound source in space, namely, interaural differences in the time of arrival and the intensity of stimuli.

Alternatively, a number of acoustic features of complex sounds can be identified in that they are highly effective stimuli for neurons at the upper levels of the auditory system, as will be shown in a later section.

Two auditory systems for localization and form?

The complex anatomy of the auditory pathway may be made more comprehensible by recent evidence support-ing a division of the lower levels by the older anatomists (Poljak, 1926) into two separable subsystems (Figure 3). On anatomical grounds, Poljak suggested that the ventral pathway originating in the ventral cochlear nucleus and including the trapezoid body and superior olivary nuclei might subserve localization and reflex functions, whereas the dorsal pathway might subserve discriminatory function.

These two pathways first separate in the cochlear nucleus, in the ventral and dorsal divisions, respectively. Early experiments recording single unit activity in the cochlear nucleus (e.g., Rose et al., 1959) did not find significant differences between the morphologically very different ventral and dorsal nuclei. However, data obtained in the unanaesthetized cat (Evans and Nelson, 1973a) and in other preparations (reviewed in Evans and Nelson, 1973a) indicate that they are substantially different in their functional properties.

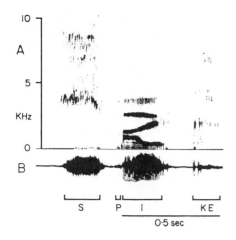

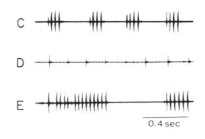

FIGURE 2 Features of naturally occurring acoustic stimuli. A and B: Human speech, illustrated by the utterance "SPIKE." A: Spectrogram. Note broad-band but differing distributions of energy in the "S," "P," and "KE" regions; multiple frequency components (formants) of the vowel "I" and the frequency transitions of the lowest two formants; the click-like transient at onset of K. B: Wave form. C, D, and E: Three types of songs in male crickets. C: Calling. D: Courtship. E: Rivalry song. (From Huber, 1972.)

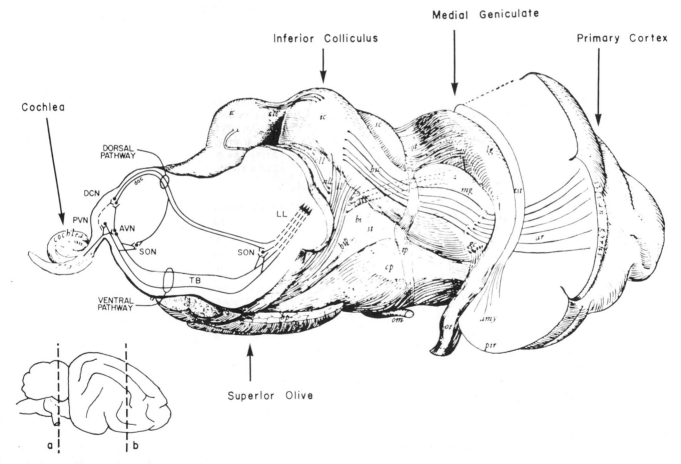

FIGURE 3 Simplified diagram of auditory pathways in cat. The brain has been sectioned across the brainstem (section a of inset), and the cerebellum and cerebral cortex removed up to incision b. The pathway consists, as a minimum, of the following steps: cochlear nerve; dorsal pathway via dorsal cochlear nucleus (DCN) to lateral lemniscus (LL); ventral pathway via anterior (AVN) and posterior (PVN) ventral cochlear nucleus subdivisions, trapezoid body (TB) and superior olivary nuclei (SON); lateral lemniscus (LL); inferior colliculus (ic); medial geniculate nucleus (mg); primary projection cortex; secondary areas (not shown). (After Papez, 1967, and Evans and Nelson, 1973b.)

To a first approximation, the responses of the *ventral pathway* reflect the properties of primary auditory neurons. An increase in firing rate (excitation) of short latency and with relatively little adaptation during sustained stimulation (Figure 4A) is characteristic of the responses of most neurons in the ventral cochlear nucleus (e.g., Rose et al., 1959; Evans and Nelson, 1973a) and superior olive (e.g., Tsuchitani and Boudreau, 1966). The responses occur over a restricted range of frequencies comparable to those of fibers in the cochlear nerve. Primary auditory neurons have a high degree of frequency

selectivity (Kiang et al., 1965, 1967; Evans, 1972), and this is retained for signals of greater complexity than pure tones (de Boer, 1969; Evans and Wilson, 1971; Wilson and Evans, 1971). This selectivity is approximately equivalent to that of the filters responsible for the spectrograms of Figures 1F, 1G, and 2A. The neurons of the ventral pathway are furthermore tightly organized on the basis of their optimum frequency sensitivity (*characteristic* or *best* frequency), that is, the pathway is tonotopically organized (Rose et al., 1959; Evans and Nelson, 1973a), and this is preserved in the superior olive (Tsuchitani and Boudreau, 1966). Hence, the distribution of neural activity across the ventral pathway in response to a complex signal will at least, to a first approximation, resemble a typical spectrographic analysis of that signal.

The ventral pathway also preserves a certain amount of temporal information. The spike discharges of many (but not all, see Rupert and Moushegian, 1970) neurons are time locked to a particular phase of the stimulus, for frequencies below about 4 to 5 kHz, and at least at the cochlear nerve level can more or less faithfully reproduce a rectified version of the wave form of complex low-frequency stimuli (reviewed by Hind, 1972). It is not by any means certain that the auditory system can make use of this temporal information for the analysis of simple and complex acoustic patterns (see Houtsma and Goldstein, 1972; Goldstein, 1972), but preservation of time information is necessary for the analysis by the superior olive of the differences in times of arrival of signals at the two ears, as is shown in the next section.

The *dorsal pathway* arises in the morphologically complex dorsal division of the cochlear nucleus, receiving its input at least in part by way of intranuclear connections from the ventral division (Evans and Nelson, 1968, 1973b). A wide variety of responses to tones can be obtained in the dorsal cochlear nucleus, particularly in the unanaesthetized preparation, analogous to those of retinal ganglion cells (especially the *Y*-cells: See Wright earlier in this part). Some neurons give excitatory responses to tones over a restricted range of frequencies and are inhibited by contiguous frequency bands of considerable extent (Figure 4B). Others exhibit only inhibitory responses to tones. The time course of the response can be a very complex sequence of excitation and inhibition (including "after discharges") dependent upon the stimulus frequency, and intensity (Figure 4B), and repetition rate.

The ventral and dorsal pathways converge in the lateral lemnisci. Whether they retain their separate identities at the higher levels of the auditory pathway remains to be shown; however, the responses of neurons in the inferior colliculi (e.g., Rose et al., 1963), medial geniculate nuclei (e.g., Adrian et al., 1966) and auditory

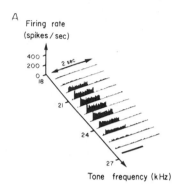

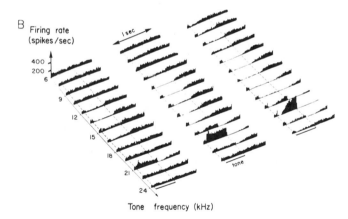

FIGURE 4 Ventral and dorsal pathway responses. A: Response characteristic of neurons in ventral pathway. Array of histograms of activity before, during and after 1 sec tone (indicated by bar), taken at the tonal frequencies indicated, at 15 dB above threshold. Each histogram represents the average of 7 tone presentations. Note that the only response is sustained excitation across the response band. B: Response characteristic of many neurons in dorsal pathway. Array of time histograms as in A, at three intensity levels: 10, 20, and 30 dB above threshold, respectively, from left to right. Each histogram represents the average of 10 tone presentations. Note extensive frequency band of inhibition sustained throughout and following the tone (indicated by bar and interrupted lines), particularly at higher intensities; small "island" of excitation arising from "sea" of inhibition, with complex time course, indicating delayed inhibition. Dorsal cochlear nucleus under chloralose. (After Evans and Nelson, 1973a.)

cortex (e.g., Bogdanski and Galambos, 1960; Katsuki et al., 1962; Evans and Whitfield, 1964; Goldstein et al., 1968) show great variety and complexity compared with those of the lower levels, particularly of the ventral pathway.

This division of the ascending auditory system into two anatomically and functionally separable subsystems may turn out to be analogous to the two major subsystems proposed for the visual system (e.g., by Held et al., 1967) subserving, respectively, the processing of place and form information.

Neural analysis of cues for sound location

Information in the ventral pathways from the two ears first converges in the superior olive, particularly in the accessory nucleus. The cells of the latter have two transverse dendrites that, respectively, receive input from each ear (Moushegian et al., 1964; Hall, 1965).

Hall (1965) described a number of types of response in the accessory nucleus of the cat, which were determined by excitatory input from the contralateral, and inhibitory input from the ipsilateral, ear. Of most interest, were responses from what he called "time-intensity trading" cells (Figure 5). The probability of discharge is related to both the interaural time delay and the interaural intensity differences of the signal in such a way that the effects of one can be traded against the other to maintain a constant level of response. Thus, to maintain the probability of discharge in the neuron of Figure 5 at a level corresponding to zero interaural delay at equal signal levels to both ears (i.e., ca. 0.1) when the level at the left ear is reduced by 10 dB (middle curve), requires the signal to the left ear to be advanced relative to the right ear by about

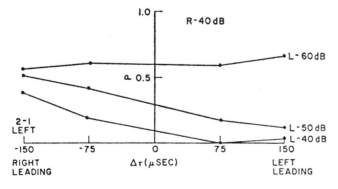

FIGURE 5 "Time-intensity trading" in a neuron in the accessory nucleus of the cat superior olive (see text). *Ordinate*: Probability of discharge of neuron in response to a click stimulus presented to each ear with the interaural delay indicated by the abscissa. Intensity at right ear constant at 40 dB below a reference level; to the left ear 40, 50, and 60 dB below the reference level, as indicated to the right of the respective plots. (From Hall, 1965.)

150 μsec (i.e., right-most point of middle curve). Hall has shown that the extent of this "time-intensity trading" is of the same order of magnitude as that observed psychophysically in man.

At the inferior colliculus level (again in the cat), Rose et al. (1966) have demonstrated that for some neurons of low (less than 3 kHz) characteristic frequency, their discharge is dependent upon the interaural time delay but is independent within wide limits of frequency and absolute intensity (Figure 6). The probability of discharge reaches a maximum at a so-called "critical delay" (140 μsec in the example of Figure 6) virtually irrespective of the stimulus frequency and of the stimulus intensity (over 60 dB in one unit). The "critical delay neuron" of Figure 6 could serve to signal sound source locations about 35° from the midline sagittal plane of the cat's head. The importance of this exciting observation has been called into question by Moushegian, Stillman, and Rupert (1972) on two grounds. First, they found "critical delays" in the inferior colliculus of the kangaroo cat restricted to interaural delays of 100 to 300 μsec. Compared with the maximum naturally occurring interaural stimulus delay in this species of 105 μsec, these "critical delays" are unlikely to be of behavioral significance for localization of the sound source. Second, they pointed out that the "critical delays" are not critical, that is, they represent a broad peak or plateau of discharge rate over a range of the order of 100 μsec, which is large compared with the maximum interaural delay, even in the cat (ca. 250 μsec).

Also in the inferior colliculus, Rose et al. (1966) found a few neurons of low and high characteristic frequencies that were exquisitely sensitive to very small binaural intensity differences (of a few dB). This sensitivity persisted over a wide range of absolute levels of the binaural stimuli. Such neurons would be important for the localization of continuous sounds of frequency in excess of that at which timing mechanisms would cease to operate.

Geisler et al. (1969) have shown that the above observations also obtain for wide-band noise stimuli.

A further level of processing appears to obtain in the cat inferior colliculus. Altman (1968) discovered neurons that were sensitive to the virtual *direction of movement* of a train of click stimuli. For these neurons to respond, virtual sound sources on one side of the head or in a small segment of space had to move in a certain direction, i.e., toward or away from the midline.

Comparing responses at both the collicular and superior olivary levels, Altman (1971) has concluded that collicular neurons demonstrate a higher degree of specificity to particular ranges of interaural time or intensity differences approaching an order of magnitude, in terms of change in impulse discharge rate, for equal changes in stimulus parameters.

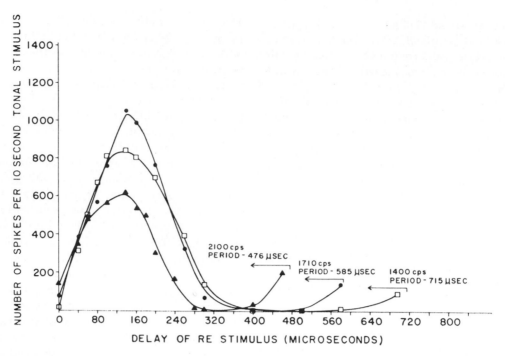

FIGURE 6 "Critical delay" neuron in inferior colliculus of cat. *Ordinate*: Number of spikes evoked by a 10 sec tone of the frequencies indicated as the parameters, presented to the two ears, the right ear (RE) signal being delayed relative to the left by the time indicated on the abscissa. (From Rose et al., 1966.)

Similar sensitivities to stimulus parameters related to the location of the sound source are shown by cortical neurons (Evans, 1968; Hall and Goldstein, 1968; Brugge et al., 1969). In a study in the unanaesthetized unrestrained cat, Evans (1968) found that 56% of the neurons in the primary cortices showed some preference for a certain location, and 31% *required* a certain location for any response. The majority required or preferred a contralateral location. A smaller number (9% and 4%, respectively) preferred or required an ipsilateral location for response. A few units responded only when the sound source was situated in the median sagittal plane anterior to the cat's head. Locations more than a few degrees on either side of this plane were ineffective. About 10% of the units were responsive to stimuli originating within the visual fields, and most of these responded only when the animal was both observing and listening to the sound source. Thus, the neurons that Hubel et al. (1969) described as "attention units" may merely represent neurons requiring simultaneous visual and auditory stimulation and having superimposable visual and auditory spatial receptive fields. It is apparent from the "case histories" of Hubel's unit that the criterion for "attention" was that the animal turned to look at the stimulus. The bimodally sensitive units appear to be analogous to those described by Spinelli et al. (1968) and Morrell (1972) in the visual cortex of the cat. Wickelgren (1971) has described similar neurons in the cat superior colliculus that had the ad-

ditional property of sensitivity to movement of the sound source.

Neural analysis of acoustic features

Probably the first evidence suggesting that the auditory system contained neural mechanisms sensitive to particular features of complex auditory stimuli came from an analysis of neurons in the primary auditory cortex of the unanaesthetized unrestrained cat by Bogdanski and Galambos (1960). Evans and Whitfield (1964; Whitfield and Evans, 1965; summarized in Evans, 1968), using the same preparation, Katsuki et al. (1962) in the monkey, Suga (e.g., 1965, 1972) in the anaesthetized bat, and Goldstein et al. (1968) using the unanaesthetized paralyzed cat preparation, have obtained similar but more comprehensive data.

Cortical neurons recorded under these conditions show striking differences from neurons at the lower levels of the auditory system discussed above. Thus, auditory cortical neurons exhibit a wide variety of stimulus sensitivities and response forms (Table I allows a comparison of results from a number of studies). In the first place, not all auditory cortical neurons respond to tonal stimuli. In the complete study of Evans and Whitfield, (Evans, 1965, 1968), less than three-quarters of the neurons responded to tones and many of these not consistently, unless certain manipulations of the spatial and temporal parameters of

Table I

Percentages of neurons with the characteristics indicated found in primary auditory cortex by various authors

Neurons in A1	Authors*				
	A (%)	B (%)	C (%)	D (%)	E (%)
1. Responding to sound	82	85–95	98		
2. Responding to tonal stimuli	70	57–77	82		
3. Not responding to tones but to some sound complex	12	21–18	16		
4. Responding to white noise	66		65		
5. Responding only to white noise			3.5		
6. Responding to click stimuli	26		52		
7. Responding only to click stimuli			1.5		
8. Not responding to steady tones but to frequency modulation		6–5		13	14

*A: Bogdanski and Galambos (1960), unanaesthetized un-restrained cat. B: Evans and Whitfield (Evans, 1965), unanaesthetized unrestrained cat. The first and second figures in the ranges given indicate respectively: the mean for all experiments and the figures obtained in the later experiments of the series where recording technique had been improved and where the location of the sound source in space was adjusted to give optimal responses. C: Goldstein, Hall, and Butterfield (1968), unanaesthetized paralysed cat. Each neuron was tested with stimuli presented binaurally and separately to each ear. D: Vardapetyan (1967), cat anaesthetized with chloralose. E: Suga (1965), bat anaesthetized with pentobarbitone. Includes unspecified number of neurons responding to steady tones but having pure tone thresholds more than 10 dB above those to frequency-modulated tones.

the stimuli were made. About 3% of the neurons responded only to a visual stimulus, generally movement of an edge. A further group of neurons (see the previous section) responded to both visual and auditory stimuli. About one-fifth of the population could not be driven by tonal stimuli but by complex sounds such as clicks, noise, and "kissing" sounds—that is, the "back-door" noises used for human communication with pet felines! Those neurons that did respond to tones did so in a variety of ways depending upon the neuron, and often upon the frequency and intensity of the tonal stimuli. These responses comprise sustained excitation, inhibition, and transient responses at the onset, termination, and onset and termination of the pure tones. About 5% of the total population (about 10% of the neurons responding to tones) responded only if the frequency was changing. The frequency selectivity of many cortical neurons is wide or indeterminate (Katsuki et al., 1959; Evans and

Whitfield, 1964), compared with the well-defined and restricted frequency selectivity at the level of the cochlear nerve (Kiang et al., 1965; Evans, 1972).

Many cortical neurons, then, at least in unanaesthetized preparations, appear to be less "interested" in the classically studied parameters of acoustic stimuli, namely, the frequency and intensity of continuous tones, than in certain less easily definable features of complex sounds. A considerable body of data has been collected, notably by Suga (in the bat: summarized, 1972), on the responses of neurons at the upper levels of the auditory system to noise, frequency- and amplitude-modulated stimuli, and to stimuli patterned in frequency and in time.

NOISE The above-mentioned studies by Bogdanski and Galambos (1960), Evans and Whitfield (1964), and Goldstein et al. (1968) indicated that broadband stimuli (such as rattling keys, kissing noises, etc.) were extremely effective for many cortical neurons in the cat, in the sense that these stimuli gave responses more consistently or at lower thresholds than did pure tones. Up to 20% of cortical neurons could be driven *only* by noise complexes. (Smaller percentages are recorded when unstructured white noise stimuli are used; Table I.) Such "noise-specialized" neurons have been described in the inferior colliculus and cortex of the anaesthetized bat (Suga, 1972). At the periphery, neurons respond to noise stimuli as if the latter were filtered in accordance with their frequency-threshold "tuning" curves, and consequently the noise threshold is always 10 to 20 dB *higher* than that of pure tones (Evans and Wilson, 1971; Moller, 1970). That the reverse is the case in these cortical neurons must presumably arise from central convergence of spectral information in the manner of a logical AND operation: The need for activation of a number of spectrally separate inputs would increase the tone threshold relative to that for noise.

FREQUENCY MODULATION Bogdanski and Galambos (1960) were the first to report the presence, in the auditory cortex of unanaesthetized cats, of neurons specifically sensitive to changes in the frequency of tones but which were unstimulable by steady tones. Independently, Evans and Whitfield (1964), in the cat cortex, and Suga (1964, 1965), in the inferior colliculus and cortex of the bat, observed similar neurons and subjected them to more detailed study (Whitfield and Evans, 1965). Many of the findings have been confirmed by Vardapetyan (1967) and by Goldstein et al. (1968) in the cat auditory cortex.

While many cortical neurons gave more consistent responses to frequency-modulated tones than to steady tones, in the study of Evans and Whitfield (1964) 10% of the units responding to tones did so *only* if the frequency was changing. Goldstein et al. (1968) found such neurons to be "in the minority." Vardapetyan (1967) and Suga

(1965) encountered such "FM specialized" properties in nearly one-eighth of the neurons in their studies of the anaesthetized cat and bat, respectively.

For these neurons preferentially or specifically sensitive to frequency change, the direction and the rate of change are important parameters. Thus some will respond only to frequency changes in an upward (low- to high-frequency) direction, others to downward frequency changes, and others to both directions of change. With sinusoidal modulation of frequency, these neurons respond to more or less restricted regions of modulation rates within the range 0.5 to 15 cycle/sec (Whitfield and Evans, 1965; see also Figure 7A; Evans and Jolley, unpublished data). This selectivity for modulation rate may relate to the recent psychophysical evidence for frequency modulation sensitive "channels" in the human auditory system of Kay and Matthews (1971, 1972; see also Figure 7B). These channels can be adapted by pre-exposure to a tone frequency modulated at a given rate so that the detection threshold for stimuli modulated at that rate is raised by as much as a factor of 3, whereas the threshold to neighboring rates is less affected. As far as the neurons are concerned, the upper limit of the range of effective sinusoidal modulation rates appears to be determined more by the rate of *repetition* of the frequency change rather than the rate of change *per se*. The use of intermittent linear ramps of

frequency change indicates that cortical units can respond to almost instantaneous changes in frequency (Whitfield and Evans, 1965).

Neurons sensitive to the direction of frequency change have been found in smaller numbers at lower levels of the auditory pathway: in the inferior colliculus (Suga, 1964, 1965; Nelson et al., 1966; Vartanian, 1971), in the superior olivary nucleus (Watenabe and Ohgushi, 1968), and to a much lesser extent in the cochlear nucleus (Evans and Nelson, 1966a, 1966b; Erulkar et al., 1968). However, neurons responding specifically (that is, *only*) to frequency-modulated tones have not been reported at levels lower than the inferior colliculus. Suga (1968, 1972) has made a detailed study of these neurons at this level in the bat and describes an additional selectivity, namely for the *form* of the frequency sweep (e.g., linear or exponential).

AMPLITUDE MODULATION   Nelson et al. (1966) found a number of neurons in the inferior colliculus of the anaesthetized cat that exhibited some selectivity to tonal stimuli sinusoidally modulated in amplitude. Some neurons would respond *only* to amplitude-modulated stimuli. Many were sensitive to the direction of modulation. In these cases, the nature and magnitude of the responses were critically dependent upon the rate and depth of modulation.

A

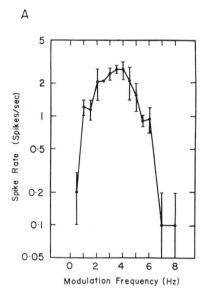

B

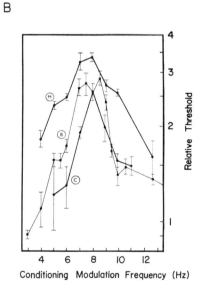

FIGURE 7   Selectivity of cat auditory cortical neurons and human auditory system for rate of frequency modulation. A: Mean spike rate (and range) of neuron in cat primary auditory cortex in response to tone at characteristic frequency modulated sinusoidally at rate indicated on abscissa. (Evans and Jolley, unpublished observations.) B: Psychophysical selectivity to sinusoidal frequency modulation. *Ordinate*: Elevation of detection threshold for frequency modulation of 250 Hz tone mod-

ulated at 8 Hz, after exposure to same tone modulated at the rates indicated on the abscissa. (Conditioning and test tones at 40 to 45 dB above pure tone threshold). M, B, and C: monaural, binaural and contra-aural testing, respectively, of three subjects; bars indicate 95% confidence limits. (After Kay and Matthews, 1971.) While the relationship between the ordinate scales of A and B is arbitrary, it is nevertheless apparent that the modulation frequencies and selectivities are comparable.

Vartanyan (1970) has confirmed these findings in the inferior colliculus of the anaesthetized rat. Selectivity was limited to a few neurons of the class that exhibited transient onset responses to tones. These neurons responded either by reproducing the modulation rhythm or by an increase in their mean discharge rate to stimuli modulated within a relatively restricted range of rates occurring above 10 cycle/sec. Thus, they illustrate a neuron that responded to modulation rates only within the range 20 to 30 cycle/sec. By contrast, neurons that gave sustained responses to tonal stimuli reproduced the modulation rhythm over a wide range of modulation rates, with little evidence of selectivity within ranges of 5 to 100 cycle/sec, in some cases.

Vartanyan (1971) has studied the thresholds of collicular neurons also as a function of the rise time of tonal stimuli. Whereas those for neurons responding in a sustained manner to tones were invariant for rise times up to as long as 70 msec, the thresholds for neurons responding transiently to tones were substantially greater for rise times above 5 msec.

In the inferior colliculus of the bat, Suga (1971) found responses dependent upon the rise time of noise and tone stimuli. In particular, some neurons had an "upper threshold" above which sounds failed to excite, determined by the rate and extent of the rise of the amplitude envelope. Below certain rates of rise, the upper threshold was absent. In other neurons, variations in rise time produced systematic changes in the sizes of the frequency threshold response areas: For some, increasing the rise time increased the extent; for other neurons, the response area was reduced. Suga considered these neurons to be specialized for responding to particular rise times. On the other hand, 26% of his sample had latency and response patterns that were not affected over a wide range of stimulus amplitudes and rise times.

More recently, Swarbrick and Whitfield (1972) described a few neurons in the primary auditory cortex of the unanaesthetized cat that appeared to be sensitive to the degree of symmetry of the amplitude envelope of noise bursts. Some neurons responded maximally to symmetrical envelopes, others to asymmetrical envelopes (e.g., where the rise versus fall time was less than 1:5), both types being independent of the duration of the stimulus over a tenfold range. These neurons appear to be selectively responsive to the "shape" of acoustic stimuli.

MULTIPLE FREQUENCY PATTERNS   There are relatively few reports of systematic studies of the responses of neurons to stimuli composed of more than one frequency component. Multicomponent stimuli (i.e., comb-filtered noise) have been utilized for examining the frequency resolving power of primary auditory neurons (Evans et al.,

1971; Wilson and Evans, 1971). At this level a straightforward spectral analysis appears to take place: The neurons respond as would be predicted on the basis that their pure tone frequency threshold characteristics represent linear (one-half to one-tenth octave) filtering.

Oonishi and Katsuki (1965) found neurons in the auditory cortex of the lightly anaesthetized cat that exhibited peaks of sensitivity at more than one band of frequencies. The most common frequency ratio between the peaks was 2:1 or 2.5:1. Abeles and Goldstein (1972) and Evans and Jolley (unpublished results) have observed similar multipeak frequency response areas in cortical neurons of unanaesthetized and lightly anaesthetized preparations, respectively.

Oonishi and Katsuki (1965) and Abeles and Goldstein (1972) subjected the cortical neurons under their study to two-component tonal stimuli. They observed various patterns of facilitation and inhibition, depending upon the frequency separation of the tones and the nature of the neurons' response area. Whereas the former authors encountered unexpected sensitivities to two simultaneous tones, Abeles and Goldstein were unable to find evidence of neurons with response magnitudes to two tone stimuli that could not be attained by single tones.

Feher and Whitfield (1966) attempted to drive, by combinations of stimuli, the significant proportion of neurons that in the unanaesthetized cat could not be stimulated by single tonal stimuli. They succeeded in producing responses in a few such neurons with a combination of a pure and a frequency-modulated tone. The effective combination was fairly specific: Similar differences of frequency in adjacent portions of the spectrum were ineffective.

TEMPORAL PATTERNING   It has already been mentioned that many of the neurons recorded from the auditory cortex of the unanaesthetized cat give labile responses to tonal stimuli unless manipulations are made of certain stimulus parameters. One of these parameters is the temporal patterning of the tones independent of any changes in frequency (Figure 8; Evans, 1968). The responses to tones of most cortical neurons in this preparation habituate rapidly (Figure 8A). By contrast, a few neurons require repeated presentations of the same stimulus for response (Figure 8B). Most cortical neurons exhibiting responses to the termination of tones ("off" responses) responded more vigorously to tones of longer duration than to short tones (Figures 8C, 8D); in a few others only short duration tones evoked consistent responses (Figure 8E).

The temporal patterning of component "chirps" is an essential feature distinguishing various calls of the cricket (Figures 3C to 3E). Huber and his colleagues (see Huber,

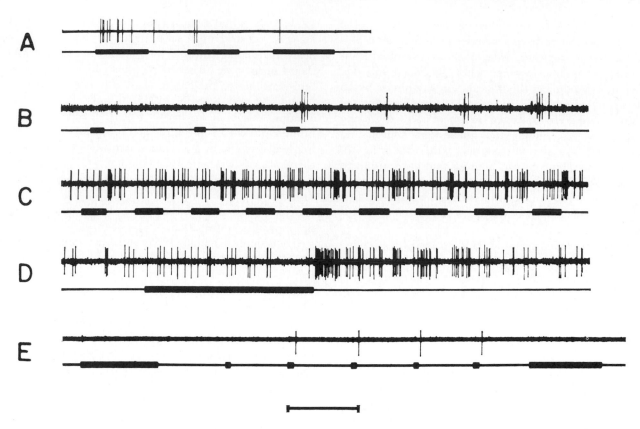

FIGURE 8 Sensitivity of auditory cortical neurons for temporal pattern. Neurons in AI of unanaesthetized cat. A: Rapid habituation of response to the repetitive presentation of a tone (at characteristic frequency of neuron). B: "Off" response re-quiring frequent repetitions of the same tone. C and D: Typical "off" response of a neuron; more vigorous response to tones of longer duration. E: "Off" response of a neuron to tones of short duration only. Time bar: 1 sec. (From Evans, 1968.)

1972) have identified two classes of interneurons in female crickets (*Gryllus*), one selective to the rate of occurrence and the duration of each chirp, and the other to the sequence of pulse components of individual chirps. With these interneurons, it is suggested that the female is able to distinguish the species-specific calls.

Neural analysis of biologically significant sound patterns

One of the first attempts to investigate the neural mechanisms underlying an organisms' sensitivity to meaningful or significant sounds, was that of Frishkopf et al. (1966, 1968). They observed that the mating call of mature male North American bullfrogs (*Rana catesbeiana*) consisted of two main spectral components with peak energies at 0.3 and 1.5 kHz. These frequencies corresponded to the characteristic frequencies of two groups of neurons found in the bullfrog's auditory nerve, arising from the amphibian and basilar papillae, respectively. Synthetic calls comprising two peaks of energy and even two tone stimuli at these frequencies evoked behavioural responses, i.e., responsive calling in other males. Since calls containing only one peak of energy (i.e., at a single frequency) were behaviorally ineffective, Frishkopf and his colleagues searched for neurons in the bullfrog's medulla and midbrain that were sensitive to the simultaneous presence of energy in the two frequency bands—a "mating call detector." Their search was unsuccessful. Related studies on the cricket frog (*Acris crepitans*; see Capranica and Frishkopf, 1972 for, review) however, indicated that the characteristic frequencies of auditory neurons in the medulla corresponded to the frequency spectrum of the call of male cricket frogs from that geographical region.

Konishi (1970) has assembled much evidence from studies of various birds to indicate that the spectral sensitivities of neurons at the medullary levels of the auditory system "match" to a greater or lesser extent the spectral energy distribution of the bird's own song. In particular, there is excellent correspondence between the upper limits of the neural and vocal frequency distributions.

Undoubtedly the most exciting work in this direction has been that on the squirrel monkey (*Saimiri sciureus*) by the late P. Winter in collaboration with Funkenstein and Nelson (1970, 1971, 1972) and their successors in Nelson's laboratory, Wollberg and Newman (1972). This monkey is found in the dense arboreal regions of South America and has a highly developed repertoire of vocalizations of social significance. Some 20 to 30 of these vocalizations have been spectrographically and behaviorally characterized in Ploog's laboratory in Munich (Winter et al., 1966). At least five main groups of calls have been established: "Distancing" calls emitted for example during feeding (see Figure 1F), acoustical contact calls (see Figure 1G), and calls related to directed aggression, to general aggression, and to high excitement. These calls consist of patterns of formants with characteristic frequencies and transitions (frequency modulation). Winter and his colleagues (Funkenstein et al., 1970; Winter and Funkenstein, 1971; Winter, 1972) demonstrated that many of the neurons in the primary auditory cortex of unanaesthetized squirrel monkeys were sensitive to certain of the vocalizations. Of 283 neurons tested, 41% responded to vocalizations and tonal as well as noise stimuli. Ten percent responded differentially between the natural and laboratory stimuli. Of these, 3.5% were selective for the vocalizations; that is, they did not respond to the laboratory stimuli tried. The selectivity was limited to one specific call type or calls with very similar acoustic properties (Figure 9A). It is noteworthy that in the example of Figure 9A, the discharges of the three neurons illustrated do not commence until as much as 200 msec after the onset of the nearly sinusoidal frequency modulation characteristic of the "trill" call. Wollberg and Newman (1972) have confirmed these important findings. Figure 9B illustrates a neuron from their study that did not respond to pure tones (note absence of response in the frequency array of dot displays in Figure 9F) but that was excited by calls from the "isolation peep" (acoustic contact) (plot B) and not by others (plots C to E; G to J).

Feature sensitive neurons and feature extraction

Sufficient evidence has been adduced above to indicate that a significant proportion of neurons in the upper levels of the auditory system are specifically or preferentially sensitive to certain features of complex auditory stimuli. The question arises whether these neurons subserve the function of feature detectors or extractors in the nervous system.

In the case of the acoustic neurons responsible for the detection of the presence of insectivorous bats, in the noctuid moth (reviewed in Roeder, 1971), the feature detection properties are obvious. It is debatable, however, whether many of the neurons whose properties have been

outlined above can justifiably be termed feature detectors. For the present it seems wise to reserve judgment and to use the operational term "feature sensitive."

Against an optimistic interpretation of the role of these neurons, it must be recognized that investigations involving recordings from single neurons at upper levels of the auditory system in unanaesthetized animals cannot entirely exclude or control other factors besides the "specific" stimulus features under study. Stimuli (particularly natural vocalizations) that evoke greater or lesser modifications of the state of "arousal" or "attention" of the animal may produce responses that, though appearing to be specific, are in fact not. Here, however, we appear to be faced with a paradox. Given that certain stimuli do evoke behavioral (e.g., "attention") responses, how can the investigator on the one hand completely exclude nonspecific effects on the cells under study, or on the other hand exclude that cell from the neural apparatus responsible in the first place for the modification in the animal's attention? Yet another difficulty arises from the recent report that the inconsistent responses given to steady tonal stimuli by cortical neurons in the unanaesthetized animal may become consistent if the animal is trained to respond to tonal stimuli, irrespective of whether the animal is actually responding at the time (Miller et al., 1972).

Against such objections, it is clear that many of the feature sensitivities reported above can be found (albeit in smaller numbers) in anaesthetized animals irrespective of the state of arousal. Furthermore, in unanaesthetized animals, in the case of the neurons specifically selective for one stimulus (e.g., one out of a number of species-specific vocalizations), stimuli that could be expected to evoke similar behavioral responses are ineffective, or produce inhibitory response (Figure 9B, plots B, C, E, and H).

Certainly, the neurons described in this chapter are endowed with properties that could be of analytical value to an organism. Specifically, cortical neurons are capable of providing "answers" to the following questions: Is the stimulus a noise, a click, a tone, or a species-specific call? Is the stimulus on? Has it just commenced? Has it just been terminated? What is its duration and its repetition rate? Is the frequency changing? If so, in which direction, at what rate? How rapidly is the amplitude rising? Where is the stimulus located in space? Is it moving? And so on. On the other hand, stimulus parameters such as frequency and intensity *per se* are not well represented in the responses of many cortical cells. There is a large body of data that suggests that it is the potentially abstractive functions that are preferentially lost in cases of cortical damage.

Ablation of the auditory cortex has been shown to interfere severely with the ability of animals (cat and monkey) to make discriminations based on the duration,

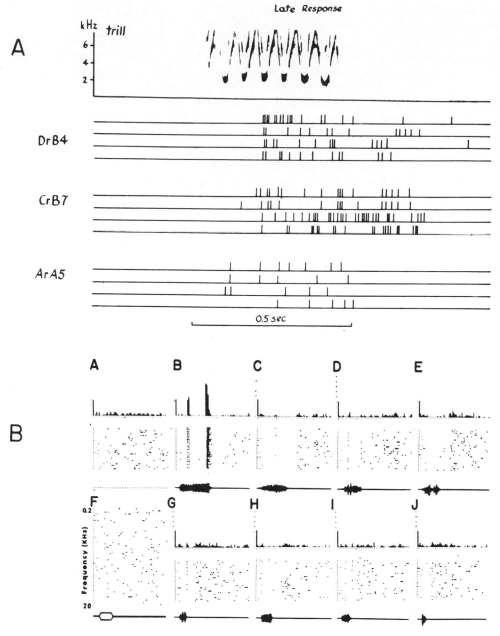

FIGURE 9 Responses of neurons in auditory cortex of squirrel monkey to the species-specific vocalizations. A: 3 neurons responding only to "trill" calls. *Above*: spectrogram of call. *Below*: discharge patterns of each neuron to 4 repetitions of the recorded vocalizations. (From Winter, 1972.) B: Neuron responding selectively to one out of 8 types of vocalization and not to tones. *Lower trace* in each case: envelope of stimulus. *Middle plot*: "dot" display of time of neuronal spike discharge during each of the 25 presentations of the same stimulus making up the average time histogram of the activity over a 2.4 sec sweep, shown in the upper plot. Plot A: spontaneous activity. Plot B: strong excitation by isolation peep call. Plots C to E, Plots G to J: no excitation by err, twitter, err-chuck, rough cackle, shriek, peep and vit calls, respectively. Plot F: "dot" display of results of tone stimulation frequencies between 0.2 and 20 kHz (no discernable response). (From Wollberg and Newman, 1972; copyright 1972 by the American Association for the Advancement of Science.)

temporal patterning, and spectral/temporal complexity and location of auditory stimuli, whereas the ability to discriminate differences of frequency and intensity *per se* are affected little or not at all (reviewed by Neff and Gold-berg, 1960; Neff, 1961). In the cat, bilateral ablation of the primary auditory cortex (AI) alone produces severe deficits in the performance of two-tone pattern discriminations (i.e., high-low-high versus low-high-low) and small but significant degradation of performance in spatial localization tasks (Neff, 1968). Enlargement of the area of ablation to include the primary and secondary auditory areas on both sides (i.e., AI, AII, Ep; for locations, see Woolsey, 1960), eliminates completely tone pattern discrimination and produces severe impairment of the localization of sounds in space. The inclusion of the insulotemporal area (IT) and suprasylvian gyrus in the bilateral ablations rendered the latter deficit complete. Bilateral loss of all auditory areas (AI, AII, Ep, IT, and the auditory area in the second somatic cortex, SII) produces complete loss of the ability to discriminate between tones of differing durations (Sharlock et al., 1965), as well as to localize sounds in space and to detect differences in the temporal patterning of tones. In addition, Kelly and Whitfield (1971) found that lesions of this extent produced a substantial deficit in the ability of cats to relearn to discriminate between the direction of frequency modulation of tones, i.e., the frequency of which were swept in an upward versus a downward direction. Cranford et al. (1971) have recently demonstrated a defect produced by *unilateral* ablation of auditory cortex, namely, the inability of cats to lateralize dichotically presented pulse pairs in a contralateral-ipsilateral (relative to the lesion) sequence to the contralateral ear. On the other hand, bilateral ablation of the primary and secondary auditory areas does not affect the ability of cats to make fine discriminations of differences in tone frequency and intensity. Meyer and Woolsey (1952) found that extending the area of ablation to SII eliminated their previously obtained differential response for frequency, but this failure turned out to be related to the training paradigms used (Thompson, 1960). Thus, later workers are agreed that normal discrimination limina for tone frequency and intensity can be obtained in cats in spite of the bilateral absence of AI, AII, Ep, IT, and SII (see Neff and Goldberg, 1960; Diamond, 1967; Oesterreich et al., 1971). Furthermore, measurements of pitch generalization in the cat before and after bilateral ablation of auditory cortex revealed no significant differences between preoperative and postoperative performance (Diamond, 1967).

Many of these results have been duplicated in the monkey (see Neff and Goldberg, 1960). In addition, in the Rhesus monkey, Dewson et al. (1969) were able to demonstrate that bilateral ablations of the auditory cortex eliminated the ability to relearn and retain discrimination between two vowel sounds, /i/ versus /u/, equated for intensity, whereas the animals could still discriminate tones from broad-band noise. Lesions to the temporal lobes in man also produce sensory deficits analogous to those described above. These produce impairment of localization of sounds in the contralateral acoustic field (Sanchez-Longo and Forster, 1958). Furthermore, lesions of the left temporal lobe in right-handed individuals impair the perception of speech, and lesions of the right lobe can produce loss of the discrimination of differences in timbre, and tonal sequence and pattern. Discrimination limina for frequency and intensity on the other hand are unaffected (summarized in Kimura, 1961).

There is thus a sufficient body of data from the study of single neuron behavior and of the results of cortical lesions to justify the continuation of the search for putative feature abstracting elements in the auditory system, and the mechanisms giving rise to them. We may not be able to substantiate the optimism of some who would see feature-sensitive elements providing the building blocks of such auditory processes as the recognition of music (Deutsch, 1969), but we may be able to uncover something of the "prewired" organization of the auditory system.

Finally, it should be emphasized that, in contrast to the single neuron studies of the visual system, investigations of neuronal properties in the auditory cortex have been almost entirely restricted to the primary projection area. What evidence has already emerged from the relatively restricted number of systematic studies at this level should encourage auditory physiologists to look with more sophisticated stimuli for elements and neuronal assemblies capable of feature extraction, particularly in the areas of auditory cortex as yet unexplored.

ACKNOWLEDGMENT   I am grateful to Dr. J. P. Wilson for helpful criticism of the manuscript.

REFERENCES

ABELES, M., and M. H. GOLDSTEIN, JR., 1972. Responses of single units in the primary auditory cortex of the cat to tones and to tone pairs. *Brain Res.* 42:337–352.

ADRIAN, H. O., W. M. LIFSHITZ, R. J. TAVITAS, and F. P. GALLI, 1966. Activity of neural units in medial geniculate body of cat and rabbit. *J. Neurophysiol.* 29:1046–1060.

ALTMAN, J. A., 1968. Are there neurones detecting direction of sound source motion? *Exp. Neurol.* 22:13–25.

ALTMAN, J. A., 1971. Neurophysiological mechanisms of sound-source localization. In *Sensory Processes at the Neuronal and Behavioural Levels*, G. V. Gersuni, ed. New York: Academic Press, pp. 221–244.

BOGDANSKI, D. F., and R. GALAMBOS, 1960. Studies of the auditory system with implanted electrodes. In *Neural Mechanisms of Auditory and Vestibular Systems*, G. L. Rasmussen and W. F. Windle, eds. Springfield, Ill.: Charles C Thomas, pp. 143–148.

BRUGGE, J. F., N. A. DUBROWSKY, L. M. AITKIN, and D. J. ANDERSON, 1969. Sensitivity of single neurones in auditory cortex of cat to binaural tonal stimulation; effects of varying interaural time and intensity. *J. Neurophysiol.* 32:1005–1024.

CAPRANICA, R. R., and L. S. FRISHKOPF, 1972. Cited in: *Auditory processing of biologically significant sounds. Neurosciences Res. Prog. Bull.* 10 (1):65.

CRANFORD, J., R. RAVIZZA, I. T. DIAMOND, and I. C. WHIT-

FIELD, 1971. Unilateral ablation of the auditory cortex in cat impairs complex sound localization. *Science* 172:286–288.

DE BOER, E., 1969. Reverse correlation II. Initiation of nerve impulses in the inner ear. *Proc. Kon. Nederl. Akad. Wet. [Biol. Med.]* 72:129–151.

DEUTSCH, D., 1969. Music recognition. *Psychol. Rev.* 76:300–307.

DEWSON, J. H., III, K. H. PRIBRAM, and J. C. LYNCH, 1969. Effects of ablation of temporal cortex upon speech sound discrimination in the monkey. *Exp. Neurol.* 24:579–591.

DIAMOND, I. T., 1967. The sensory neocortex. In *Contributions to Sensory Physiology*, W. D. Neff, ed. New York: Academic Press, pp. 51–100.

ERULKAR, S. D., 1972. Comparative aspects of spatial localization of sounds. *Physiol. Rev.* 52:237–359.

ERULKAR, S. D., R. A. BUTLER, and G. L. GERSTEIN, 1968. Excitation and inhibition in cochlear nucleus. II. Frequency-modulated tones. *J. Neurophysiol.* 31:537–548.

EVANS, E. F., 1965. Behaviour of neurones in the auditory cortex. Ph.D. thesis. University of Birmingham, England.

EVANS, E. V., 1968. Cortical representation. In *Hearing Mechanisms in Vertebrates*, A. V. S. de Reuck and J. Knight, eds. London: J. & A. Churchill, pp. 272–287.

EVANS, E. F., 1971. Central mechanisms relevant to the neural analysis of simple and complex sounds. In *Pattern Recognition in Biological and Technical Systems*, O.-J. Grusser and R. Klinke, eds. Heidelberg: Springer-Verlag.

EVANS, E. F., 1972. The frequency response and other properties of single fibers in the guinea-pig cochlear nerve. *J. Physiol.* 226:263–287.

EVANS, E. F., and P. G. NELSON, 1966a. Behaviour of neurones in cochlear nucleus under steady and modulated tonal stimulation. *Fed. Proc.* 25:463.

EVANS, E. F., and P. G. NELSON, 1966b. Responses of neurones in cat cochlear nucleus to modulated tonal stimuli. *J. Acoust. Soc. Amer.* 40:1275–1276.

EVANS, E. F., and P. G. NELSON, 1968. An intranuclear pathway to the dorsal division of the cochlear nucleus of the cat. *J. Physiol.* 196:76–78P.

EVANS, E. F., and P. G. NELSON, 1973a. The responses of single neurones in the cochlear nucleus of the cat as a function of their location and the anaesthetic state. *Exp. Brain Res.*, in press.

EVANS, E. F., and P. G. NELSON, 1973b. On the relationship between the dorsal and ventral cochlear nucleus. *Exp. Brain Res.*, in press.

EVANS, E. F., J. ROSENBERG, and J. P. WILSON, 1971. The frequency resolving power of the cochlea. *J. Physiol.* 216:58–59P.

EVANS, E. F., and I. C. WHITFIELD, 1964. Classification of unit responses in the auditory cortex of the unanaesthetized and unrestrained cat. *J. Physiol.* 171:476–493.

EVANS, E. F., and J. P. WILSON, 1971. Frequency sharpening of the cochlea: The effective bandwidth of cochlear nerve fibers. *Proc. 7th Internat. Cong. on Acoustics, Vol. 3.* Budapest: Akademiai Kiado, pp. 453–456.

FEHER, O., and I. C. WHITFIELD, 1966. Auditory cortical units which respond to complex tonal stimuli. *J. Physiol.* 182:39P.

FRISHKOPF, L. S., and R. R. CAPRANICA, 1966. Auditory responses in the medulla of the bullfrog: Comparison with eighth-nerve responses. *J. Acoust. Soc. Amer.* 40:1262.

FRISHKOPF, L. S., R. R. CAPRANICA, and M. H. GOLDSTEIN, JR., 1968. Neural coding in the bullfrog's auditory system: A teleological approach. *Proc. IEEE.* 56:969–980.

FUNKENSTEIN, H., P. WINTER, and P. G. NELSON, 1970. Unit responses to acoustic stimuli in the cortex of awake squirrel monkeys. *Fed. Proc.* 29:394.

GEISLER, C. D., W. S. RHODE, and D. W. HAZELTON, 1969. Responses of inferior colliculus neurones in the cat to binaural acoustic stimuli having wide-band spectra. *J. Neurophysiol.* 32:960–974.

GOLDSTEIN, J. L., 1972. Evidence from aural combination tones and musical notes against classical temporal periodicity theory. In *Hearing Theory*. Eindhoven: IPO, pp. 186–208.

GOLDSTEIN, M. H., JR., J. L. HALL, II, and B. O. BUTTERFIELD, 1968. Single unit activity in primary auditory cortex of unanaesthetized cats. *J. Acoust. Soc. Amer.* 43:444–455.

HALL, J. L., II, 1965. Binaural interaction in the accessory superior olivary nucleus of the cat. *J. Acoust. Soc. Amer.* 37:814–823.

HALL, J. L., II, and M. H. GOLDSTEIN, JR., 1968. Representation of binaural stimuli by single units in primary auditory cortex of unanaesthetized cats. *J. Acoust. Soc. Amer.* 43:456–461.

HELD, R., D. INGLE, G. E. SCHNEIDER, and C. B. TREVARTHEN, 1967–1968. Locating and identifying: Two modes of visual processing. *Psychol. Forsch.* 31:44–62; 299–348.

HIND, J. E., 1972. Physiological correlates of auditory stimulus periodicity. *Audiol.* 11:42–57.

HOUTSMA, A. J. M., and J. L. GOLDSTEIN, 1972. The central origin of the pitch of complex tones: Evidence from musical interval recognition. *J. Acoust. Soc. Amer.* 51:520–529.

HUBEL, D. H., C. O. HENSON, A. RUPERT, and R. GALAMBOS, 1959. "Attention" units in the auditory cortex. *Science* 129:1279–1280.

HUBER, F., 1972. Cited in: *Auditory processing of biologically significant sounds. Neurosciences Res. Prog. Bull.* 10. (1):67.

KATSUKI, Y., N. SUGA, and Y. KANNO, 1962. Neural mechanisms of the peripheral and central auditory systems in monkeys. *J. Acoust. Soc. Amer.* 34:1396–1410.

KATSUKI, Y., T. WATENABE, and N. MARUYAMA, 1959. Activity of and neurones in upper levels of brain of cat. *J. Neurophysiol.* 22:343–359.

KAY, R. H., and D. R. MATTHEWS, 1971. Temporal specificity in human auditory conditioning by frequency-modulated tones. *J. Physiol.* 218:104–106P.

KAY, R. H., and D. R. MATTHEWS, 1972. On the existence in human auditory pathways of channels selectively tuned to the modulation present in frequency-modulated tones. *J. Physiol.* 225:657–678.

KELLY, J. B., and I. C. WHITFIELD, 1971. Effects of auditory cortical lesions on discriminations of rising and falling frequency-modulated tones. *J. Neurophysiol.* 34:802–816.

KIANG, N. Y-S., M. B. SACHS, and W. T. PEAKE, 1967. Shapes of tuning curves for single auditory nerve fibers. *J. Acoust. Soc. Amer.* 42:1341–1342.

KIANG, N. Y-S., T. WATENABE, E. C. THOMAS, and L. F. CLARK, 1965. *Discharge Patterns of Single Fibers in the Cat's Auditory Nerve.* Cambridge, Mass.: The MIT Press.

KIMURA, D., 1961. Some effects of temporal lobe damage on auditory perception. *Canad. J. Psychol.* 25:156–165.

KONISHI, M., 1970. Comparative neurophysiological studies of hearing and vocalizations in songbirds. *Z. vergl. Physiologie.* 66:257–272.

LIBERMAN, A. M., F. INGEMANN, L. LISKER, P. DELATTRE, and F. S. COOPER, 1959. Minimal rules for synthesizing speech. *J. Acoust. Soc. Amer.* 31:1490–1499.

MEYER, D. R., and C. N. WOOLSEY, 1952. Effects of localized cortical destruction upon auditory discriminative conditioning in the cat. *J. Neurophysiol.* 15:149–162.

MILLER, J. M., D. SUTTON, B. PFINGST, A. RYAN, R. BEATON, and G. GOUREVITCH, 1972. Single cell activity in the auditory cortex of rhesus monkeys: Behavioural dependency. *Science* 177:449–451.

MOLLER, A., 1970. Unit responses in the cochlear nucleus of the rat to noise and tones. *Acta Physiol. Scand.* 78:289–298.

MORRELL, F., 1972. Visual system's view of acoustic space. *Nature (London)* 238:44–46.

MOUSHEGIAN, G., A. RUPERT, and M. A. WHITCOMB, 1964. Brain-stem neuronal response patterns to monaural and binaural tones. *J. Neurophysiol.* 27:1174–1191.

MOUSHEGIAN, G., R. D. STILLMAN, and A. L. RUPERT, 1972. Characteristic delays in superior olive and inferior colliculus. In *Physiology of the Auditory System.* M. B. Sachs, ed. Baltimore: National Educational Consultants, pp. 245–254.

NEFF, W. D., 1961. Neural mechanisms of auditory discrimination. In *Sensory Communication*, W. A. Rosenblith, ed. New York: John Wiley & Sons, Chap. 15.

NEFF, W. D., 1968. Behavioural studies of auditory discrimination: Localization of the sound source in space. In *Hearing Mechanisms in Vertebrates*, A. V. S. de Reuck and J. Knight, eds. London: J. & A. Churchill, pp. 207–231.

NEFF, W. D., and J. M. GOLDBERG, 1960. Higher functions of the central nervous system. *Ann. Rev. Physiol.* 22:499–524.

NELSON, P. G., S. D. ERULKAR, and J. S. BRYAN, 1966. Responses of units of the inferior colliculus to time-varying acoustic stimuli. *J. Neurophysiol.* 29:834–860.

OESTERREICH, R. E., N. L. STROMINGER, and W. D. NEFF, 1971. Neural structures mediating differential sound intensity discrimination in the cat. *Brain Res.* 27:251–270.

OONISHI, S., and Y. KATSUKI, 1965. Functional organization and integrative mechanism on the auditory cortex of the cat. *Jap. J. Physiol.* 15:342–365.

PAPEZ, J. W., 1967. *Comparative Neurology*, 1st Ed. 1929. New York: Hafner Publishing Co. Also, The connection of the acoustic nerve. *J. Anat.* 60:465–469.

POLJAK, S., 1926. The connections of the acoustic nerve. *J. Anat.* 60:465–469.

ROEDER, K. D., 1971. Acoustic alerting mechanisms in insects. *Ann. N.Y. Acad. Sci.* 188:63–79.

ROSE, J. E., R. GALAMBOS, and J. R. HUGHES, 1959. Microelectrode studies in the cochlear nuclei of the cat. *Bull. Johns Hopkins Hospital* 104:211–251.

ROSE, J. E., D. D. GREENWOOD, J. M. GOLDBERG, and J. E. HIND, 1963. Some discharge characteristics of single neurones in inferior colliculus of the cat. I. Tonotopic organization, relation of spike counts to tone intensity and firing patterns of single elements. *J. Neurophysiol.* 26:294–320.

ROSE, J. E., N. B. GROSS, C. D. GEISLER, and J. E. HIND, 1966. Some neural mechanisms in the inferior colliculus of the cat which may be relevant to localization of a sound source. *J. Neurophysiol.* 29:288–314.

RUPERT, A. L., and G. MOUSHEGIAN, 1970. Neuronal responses of kangaroo rat ventral cochlear nucleus to low frequency tones. *Expl. Neurol.* 26:84–102.

SANCHEZ-LONGO, L. P., and F. M. FORSTER, 1958. Clinical significance of impairment of sound localization. *Neurology* 8:119–125.

SHARLOCK, D. P., W. D. NEFF, and N. L. STROMINGER, 1965. Discrimination of tone duration after bilateral ablation of cortical auditory area. *J. Neurophysiol.* 28:673–681.

SPINELLI, D. N., A. STARR, and T. W. BARRETT, 1968. Auditory specificity in unit recordings from cat's visual cortex. *Exp. Neurol.* 22:75–84.

SUGA, N., 1964. Recovery cycles and responses to frequency-modulated tone pulses in auditory neurones of echo-locating bats. *J. Physiol.* 175:50–80.

SUGA, N., 1965. Functional properties of auditory neurones in the cortex of echo-locating bats. *J. Physiol.* 181:671–700.

SUGA, N., 1968. Analysis of frequency-modulated and complex sounds by single auditory neurones of bats. *J. Physiol.* 198:51–80.

SUGA, N., 1971. Responses of inferior collicular neurones of bats to tone bursts with different rise times. *J. Physiol.* 217:159–177.

SUGA, N., 1972. Analysis of information bearing elements in complex sounds by auditory neurones of bats. *Audiol.* 11:58–72.

SWARBRICK, L., and I. C. WHITFIELD, 1972. Auditory cortical units selectively responsive to stimulus "shape." *J. Physiol.* 224:68–69P.

THOMPSON, R. F., 1960. Function of auditory cortex of cat in frequency discrimination. *J. Neurophysiol.* 23:321–334.

TSUCHITANI, C., and J. C. BOUDREAU, 1966. Single unit analysis of cat superior olive S segment with tonal stimuli. *J. Neurophysiol.* 29:684–699.

VARDAPETYAN, G. A., 1967. Classification of single unit responses in the auditory cortex of cats. *Neurosci. Trans.* 1:1–11.

VARTANYAN, I. A., 1969. Unit activity in inferior colliculus to amplitude-modulated stimuli. *Neurosci. Trans.* 10:17–26.

VARTANYAN, I. A., 1971. Temporal characteristics of auditory neuron responses in rat to time varying acoustic stimuli. *Proc. 7th Internat. Cong. of Acoustics*, Vol. 3. Budapest: Akademiai Kiado, pp. 401–404.

WATENABE, T., and K. OHGUSHI, 1968. FM sensitive auditory neuron. *Proc. Jap. Acad.* 44:968–973.

WHITFIELD, I. C., and E. F. EVANS, 1965. Responses of auditory cortical neurons to stimuli of changing frequency. *J. Neurophysiol.* 28:655–672.

WICKELGREN, B. G., 1971. Superior colliculus: Some receptive field properties of bimodally responsive cells. *Science* 173:69–72.

WILSON, J. P., and E. F. EVANS, 1971. Grating acuity of the ear: Psychophysical and neurophysiological measures of frequency resolving power. *Proc. 7th Internat. Cong. on Acoustics*, Vol. 3. Budapest: Akademiai Kiado, pp. 397–400.

WINTER, P., 1972. Cited in "Auditory processing of biologically significant sounds." *Neurosciences Res. Prog. Bull.* 10(1):72–74.

WINTER, P., and H. FUNKENSTEIN, 1971. The auditory cortex of the squirrel monkey: Neuronal discharge patterns to auditory stimuli. Proc. 3rd Cong. Primat. Zurich, 1970. Basel: Kruger, Vol. 2, pp. 24–28.

WINTER, P., D. PLOOG, and J. LATTA, 1966. Vocal repertoire of the squirrel monkey (*Saimiri sciureus*), its analysis and significance. *Expl. Brain Res.* 1:359–384.

WOLLBERG, Z., and J. D. NEWMAN, 1972. Auditory cortex of squirrel monkey: Response patterns of single cells to species-specific vocalizations. *Science* 175:212–214.

WOOLSEY, C. N., 1960. Organization of cortical auditory system: A review and a synthesis. In *Neural Mechanisms of the Auditory and Vestibular Systems*, G. L. Rasmussen and W. F. Windle, eds. Springfield, Ill.: Charles C. Thomas. pp. 165–180.

14 Psychoacoustical and Neurophysiological Aspects of Auditory Pattern Recognition

J. P. WILSON

ABSTRACT Pitch perception can be considered as a specific example of spectral pattern recognition. Signals with several frequency components frequently give rise to a sensation of pitch that is anomalous on classical theories of pitch perception. Some experiments giving clues to a possible mechanism are described, together with a model of a neural network for the extraction of pitch information. A system of this kind requires inhibitory interaction in order to reject unstructured signals such as white noise and other patterns with components at inappropriate frequencies.

Pitch perception: Emergence of a pattern recognition theory

As PATTERN recognition models for pitch perception have only recently gained support, a brief review of previous theories is necessary, particularly as it is not clear that these can be rejected completely.

At least three kinds of pitch perception have been claimed. (a) Pitch *height* in mels is based on magnitude estimates and fractionation procedures (Stevens and Volkmann, 1940). (b) *Musical* pitch is based on a logarithmic scale of frequency, so that equal pitch intervals are given by equal ratios of frequencies. (It is considered by some to be a cyclical or helical scale, because frequencies an octave apart are more alike than smaller intervals). (c) *Absolute* pitch is usually considered to be a specific kind of musical pitch perception in which subjects can recognize and name pitch without any standard of reference, though frequently for only one note or on a specific musical instrument (Bachem, 1937). Theories have generally been directed toward the second, i.e., musical pitch perception.

Two kinds of theory have persisted under the general headings of *place* and *volley* theory. Place theories have been based on the principle that the separate frequency components of a signal can be analyzed and "mapped" out along some spatial dimension; for example, along the basilar membrane of the cochlea, so that the pitch perceived depends on the place of maximal excitation along this dimension. In theory such a tonotopic relationship can be preserved at higher levels of the auditory system, and it has been observed by Kiang et al. (1965) in the cochlear nerve and by Rose, Galambos, and Hughes (1959) in the cochlear nucleus of cat.

Against this simple view there is a considerable body of evidence that complex signals with components of high frequency can lead to the perception of low pitches. Schouten in 1940 named this phenomenon the "residue" (see Schouten et al., 1962). This led to the concept of "wave form periodicity detection" where pitch depended on the period between repetitions of certain features of the wave form. For this to be possible, there are two requirements: First, that two or more frequency components should interact at the same place to provide the necessary periodicity, contrary to "place" mechanism requirements that would filter the components into separate channels. Second, that temporal features of this wave form should be preserved up to the level of the auditory system at which analysis takes place. The latter does not necessarily mean that each nerve fiber must respond at the wave form repetition rate, but that over the population the wave form should be preserved. This is the basis of the *volley* theory (Wever, 1949) and requires that nerve impulses, although infrequent, should be at least partially phase locked to the stimulus. There is general agreement, based on both psychophysical and single unit studies, that the limit at which phase locking breaks down must be in the region of 5 kHz. Temporal periodicity mechanisms therefore cannot be responsible for what we hear at higher frequencies. This is also the approximate upper limit of *musical* pitch perception. Unfortunately for the auditory theorist, most of the signals giving a strong pitch sensation can be described either in temporal or spectral terms. This equivalence is of course the basis of Fourier analysis. Much ingenuity has been exercised in devising stimulus situations in which some temporal feature exists

J. P. WILSON Department of Communication, University of Keele, Staffordshire, England without a spectral correlate (e.g., the pitch of pulsed noise, Miller and Taylor, 1947; Harris, 1963; and dichotic pitch phenomena, Cramer and Huggins, 1958; Fourcin, 1970; and the sweep tone effect in mistuned consonances, Plomp, 1967). These special stimulus conditions, however, lead to very weak sensations of pitch and may be mediated by a different mechanism from that utilized for signals with strong pitch.

Some of the first evidence against periodicity pitch came from the ranks of its proponents: Ritsma (1962) found that a three-component tonal residue had lowest threshold (for side bands) when the components were resolvable, and later Ritsma (1967) that the dominant components for the pitch of a complex signal came not from the region of maximal interaction of components but from a region where they were well resolved. However, the significance of these results was not pointed out until later (Terhardt, 1970; Houtsma and Goldstein, 1972).

In attempting to account for the pitch of nonharmonic residues, de Boer (1956) suggested that for widely spaced components it corresponded to the frequency whose integral multiples resemble the given tones as much as possible (i.e., best fitting harmonic series); he also developed, for closely spaced components, an equivalent temporal model based on fine structure of the waveform. Thurlow proposed a *multicue mediation* theory in 1963, but it did not receive wide support at the time. In this theory he proposed that subjects perceived a low pitch by humming, or imagining that they hummed, a low note whose overtone structure matched the components of the stimulus. Such *active matching* or *analysis by synthesis* has, of course, been invoked in other perceptual tasks. Walliser (1969) proposed an empirical rule that perceived pitch corresponds with the subjective subharmonic of the lowest component of a complex that lies closest to the beat frequency between adjacent components (i.e., closest to the common frequency difference). Terhardt (1970) extended Walliser's observations and incorporated them into a *secondary sensation* model in which a low pitch is mediated by the perception of high-frequency components. This appears to be a passive version of Thurlow's model. Whitfield (1970) pointed out that we normally perceive a complex sound as a unitary experience with a specific pitch. This information is carried by the pattern of active and inactive nerve fibers arising from the frequency spectrum of the stimulus. The absence or distortion of a small part of this pattern should not entirely destroy the normal sensation. Wilson (1970), considering the pitch of broadband *comb-filtered* noise, (i.e., noise with multiple spectral peaks, see Figures 1 and 2), proposed a model based on *matching* the spectral peaks in the stimulus to a *best fitting* harmonic series over the *dominant* region of the spectrum. Although matching might be considered to

imply an active process such as Thurlow suggested, this was not specifically intended. The present account will attempt to highlight the important features of this class of matching model and provide evidence in its support. For simplicity a passive model will be formulated here, as there does not appear to be any evidence favoring an active system.

Pitch perception as pattern recognition: An experimental example

A stimulus consisting of noise plus delayed noise has a power spectrum with a series of evenly spaced peaks and troughs (Figure 1a) with the first peak position and common spacing frequency equal to the reciprocal of the time delay Δt. The pitch elicited by this stimulus corresponds with the frequency of the first peak and, therefore, also with the common difference frequency and the reciprocal of the delay time. If now the spectrum is *inverted* (Figure 1b) by phase inverting either the direct or delayed noise, the common difference frequency and time delay will remain the same but the first peak will be an octave lower. Remarkably, however, the perceived pitch neither remains the same nor falls by an octave, but drops by a small amount. According to Fourcin (1965), this pitch change is in the ratio 7:8, but Wilson (1967) found this ratio to depend upon Δt.

A consideration of the initial frequency analysis of this signal by the auditory system indicates that an intermediate region of the spectrum should be dominant in

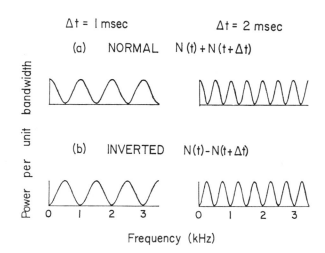

FIGURE 1 Frequency spectra of noise added to delayed noise for delay times (Δt) of 1 and 2 msec. The peaks in the spectral envelopes (a) are harmonically related so that the frequency of the first peak and the common spacing frequency each equal the reciprocal of the delay time ($1/\Delta t = 1$ kHz and 500 Hz). By inverting the phase of one of the noise components the spectral envelope is inverted (b) so that the peaks and valleys of the spectrum are interchanged.

perception. Figure 2 shows spectra similar to those of Figure 1 but plotted on a logarithmic frequency scale to represent better the way in which they are analyzed by the ear. The lower peaks are relatively broad ($\Delta f/f$ large) and cannot be expected to produce a well-defined pitch, whereas the very high peaks, although relatively very sharp, become progressively closer spaced so that eventually the individual peaks cannot be resolved from their neighbors by a system with finite bandwidth filtering as is found in the cochlear nerve (Wilson and Evans, 1971).

If for the moment it is assumed that this dominant region extends from $3/\Delta t$ to $5/\Delta t$ (Figure 2), then the expected pitch ratios can be calculated. On the basis of fitting a harmonic series to the components of the signal in this region, there are two possible series that fit quite closely. One of these has a lower fundamental frequency ($\Delta t' = 9/8\ \Delta t$) and the other higher ($\Delta t'' = 7/8\ \Delta t$) giving pitches in the ratios 9:8 and 7:8, respectively, compared with that of the normal spectrum. In order to be able to provide quantitative predictions for pitch, it is necessary to obtain measurements on the extent and location of the region of spectral dominance.

Ritsma (1967) has provided data on dominance for filtered pulse trains. But as all components in Ritsma's stimulus would be sharp and well defined, it is desirable to perform comparable experiments using comb-filtered

noise as the stimulus. Detection threshold was measured rather than the pitch criterion used by Ritsma in view of the possibility that the fourth peak had been found to be dominant because it was a double octave. Subjects were required to determine the detection threshold value of spectral modulation depth by adjustment of the ratio of direct and delayed noise for the whole range of spectral positions and for a variety of spectral peak spacings. The signal was made up from two sources combined through filters. In the first half of the experiment the high-frequency part of the spectrum comprised comb-filtered noise of adjustable peak to valley ratio and the lower part of the spectrum was "filled-in" with noise at the same mean spectral density (see Figure 3, left inset). The

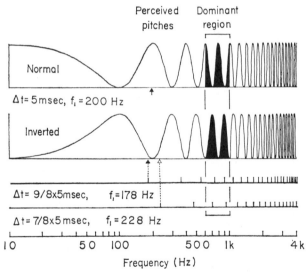

FIGURE 2 Hypothetical scheme to explain the pitch of a *comb-filtered* noise signal. It is assumed to correspond to the fundamental component of a harmonic complex that *fits* the stimulus spectrum over the *dominant* region. These spectra appear different from those of Figure 1, because they are now plotted with a logarithmic frequency abscissa to represent the way they would be *mapped* along the cochlear partition. Below the inverted spectrum are shown two harmonic series that best fit the shaded peaks of the inverted spectrum. Thus the inverted spectrum has a pitch of either 178 Hz or, less frequently, of 228 Hz compared with 200 Hz for the normal spectrum.

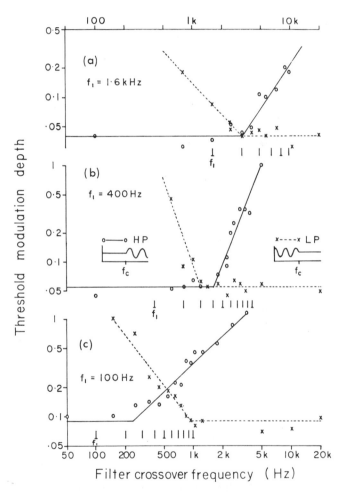

FIGURE 3 Experiments to determine the *dominant* spectral region for three peak spacings (first peak, $f_1 =$ (a) 1.6 kHz, (b) 400 Hz and (c) 100 Hz). The threshold modulation depth (peak to valley ratio of the spectral envelope) measured in the experiment is plotted as a function of the frequency of crossover between a flat spectrum and the comb-filtered noise (see insets that apply to all graphs). The dominant region is that common region where the thresholds obtained under the HP and LP filtered conditions rise (see text). (Reproduced by permission of A. W. Sijthoff.)

threshold peak-to-valley ratio was determined, using a two alternative forced choice (2AFC) technique. With a low value of crossover frequency, f_c, a low threshold was found. Increasing the crossover frequency, however, did not lead to an increased threshold. This means that frequencies between the first and second crossover points were not contributing to the low threshold observed that was determined at a higher frequency region of the signal. Further increments of crossover frequency were made, until eventually a point was found beyond which the threshold increased rapidly. From this it would appear that frequencies above this turnover point were necessary to obtain a low threshold. In the second half of the experiment, the upper and lower parts of the spectrum were interchanged (Figure 3, right inset) and analogous determinations of threshold were made. A turnover point was again found below which the threshold rose. The dominant region is that delimited by the two turnover points. This was repeated for a wide range of peak spacings in addition to those of (a), (b), and (c) in Figure 3.

The mean or optimum dominant frequency is plotted for two subjects as a function of peak spacing in Figure 4. The same data are also expressed as dominant harmonic numbers by reference to the open symbols and right-hand

axis. For comparison, Ritsma's mean values are also included. It will be observed that the dominant peak number measured in this way decreases with an increase in the spacing of the spectral peaks. Thus we must expect the pitch difference between the *inverted* and *normal* spectra to become greater as the peak spacing is increased. Measurements of the perceived pitch difference between normal and inverted spectra at two widely separated values of peak spacing, namely at 40 Hz ($\Delta t = 25$ msec) and 1 kHz ($\Delta t = 1$ msec), gave ratios of 1.06 and 1.2, respectively (Wilson, 1967). The corresponding values of dominant peak numbers for these two spacings were 10 and $2\frac{1}{2}$, respectively (Figure 4). This gives predicted pitch ratios of $10:10\frac{1}{2}$ and $2\frac{1}{2}:3$, i.e., 1.05 and 1.2, respectively, in good agreement with measurement.

On the other hand a similar experiment by Fourcin (1965) found the pitch ratio to be approximately 7:8 (1.14) over the frequency range from 110 Hz to 1.2 kHz. This result would be consistent with a *constant* dominant peak number equal to the mean value over this range given by Figure 4. It is possible that Fourcin's subject was listening specifically for this interval. If attention were directed to a particular region of the spectrum, the pitch difference could be expected to correspond to the change

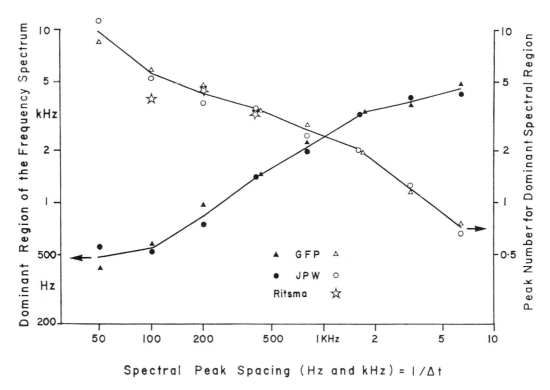

FIGURE 4 Dominant region as a function of peak spacing for two observers. Each point represents a determination such as (a), (b), and (c) of Figure 3. The filled symbols express the dominant regions in frequency terms as spectral position and are referred to the left ordinate. The same data are expressed in terms of the dominant harmonic (peak) number by the open symbols referred to the right ordinate. The mean values obtained by Ritsma (1967) for filtered pulse trains are shown as stars. (Reproduced by permission of A. W. Sijthoff.)

in peak position at that point. This is what Bilsen and Ritsma (1970) found when the range was restricted using filtered signals. They also investigated intermediate shifts in peak position, which gave correspondingly smaller pitch ratios. Although they explained their results in terms of periodicity theory, identical results would be predicted on a pattern recognition model. In the case of high-pass filtering the results would also reduce to Walliser's (1969) rule.

It was seen from Figure 2 that two harmonic series give a good fit to the dominant region of the inverted spectrum. These were obtained by a minimal contraction or expansion of the harmonic series based on 200 Hz and would give rise to pitches corresponding to 178 and 228 Hz, respectively. In fact, both can be observed although the lower one is usually more prominent.

Neural models

In order to account for some of the known properties of pitch perception, a simple neural *pitch unit* can be developed, based on the principle, outlined above, of spectral pattern recognition. Thus if the appropriate harmonically related combination of spectral components is present, a response should occur. At first sight it might appear that inputs to the model could be limited to the dominant region. When the dominant part of the spectrum is removed, however, say by high-pass filtering, pitch can still be perceived. The upper limit may be bounded by what Ritsma (1962) termed the *existence* region of the

tonal residue. The model pitch unit must also be capable of responding to a single pure tone at the fundamental frequency.

Figure 5a shows such a unit in which the frequency spectrum has been mapped into neural space. For simplicity the space dimension has been pictured as linear corresponding to Figure 1 rather than the more realistic logarithmic scale of Figure 2. Such a unit represents one out of a population covering the whole range of possible pitches. The excitatory regions indicated by shading would be determined by the effective bandwidths of the appropriate cochlear nerve fiber inputs (Evans and Wilson, 1971). This simple model would respond to the appropriate signals, but it would also respond to signals that have no pitch such as white noise. To eliminate this it is necessary for the response to be inhibited by the presence of power in the intermediate zones (Figure 5b). Unfortunately, even this system would not be satisfactory, because a single tone at any of the higher harmonies would lead to perception of the fundamental frequency. With the exception of the fundamental input, then, it is necessary to insert a multiplicative process or AND gates so that a response can be obtained only if two adjacent inputs are stimulated simultaneously (Figure 5c). Dominance has not been specifically included in the model shown, but it would be possible to apply either simple weightings, or inhibitory interactions, so that power in the dominant region inhibited responses from surrounding regions.

This model has been derived on the assumption that

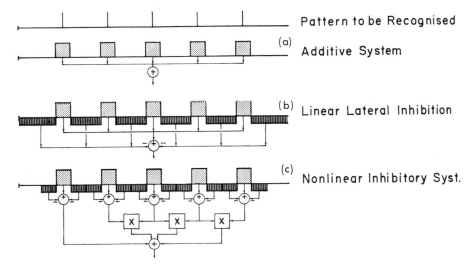

FIGURE 5 Three models for a *pitch unit* for spectral pattern recognition. It is assumed that the spectral pattern to be recognized (a harmonic series in this case) is transformed into neural space according to cochlear analysis and tonotopic organization (upper line). The dotted areas represent excitatory regions corresponding to the effective bandwidths of different regions of the cochlea and the vertically shaded areas represent inhibition. Addition or subtraction of inputs is indicated by the symbols + and −. Here X represents a multiplicative interaction or the operation of a logical AND gate. For the relative merits of each system see text. Each *pitch unit* represents one out of a population covering the whole range of perceivable pitches.

harmonically related signals are the norm and that non-harmonically related signals are detected by the most closely fitting *pitch unit*. It is, of course, possible to postulate other relationships between the receptors to recognize other spectral patterns of particular biological significance. An obvious example would be for the patterns of formants in speech.

At the moment there is little neurophysiological evidence to support in detail a recognizer of this type (see Evans, in the preceding chapter). There is some psychophysical evidence that the pattern recognizer must be central in location, because Houtsma and Goldstein (1972) found that the pitch sensation from a two-tone complex could still be elicited if the tones were presented separately to the two ears.

Clearly a neural model of this type must become quite complex to accommodate all known properties of pitch perception. It would not have been any easier to develop an active matching model in detail. The general concept of pattern recognition, however, is attractive and allows a given percept to be triggered in a number of different ways. Perception is frequently richer than the stimulus array giving rise to it and depends on learning and prior expectation.

Smoorenburg (1970) found that subjects could be divided into two groups in the way that they perceived a two-component stimulus. If the lower component were reduced in frequency, one group consistently heard the pitch fall whereas the other group heard the pitch rise, presumably corresponding to the consequent increase in fundamental frequency. Individual differences of this kind may also be the basis of the controversy concerning the effect of relative phase of the components on the pitch of a residue. Mathes and Miller (1947) and Ritsma and Engel (1964) reported phase effects in accordance with the requirements of periodicity theory whereas Wightman (1972) reported that he and Patterson had failed to do so in these and other circumstances. Wightman in fact incorporated the latter findings in a pattern recognition model in which the pattern of output of the cochlea undergoes a Fourier analysis. The output of this second stage is then equivalent to an autocorrelation function and a further stage picks the highest peak, which represents the pitch. Such a model, like the present one, is quite insensitive to phase.

In view of the experiments mentioned earlier in which pitch was elicited purely by temporal information (e.g. the pitch of gated noise and dichotic pitch), it would appear desirable to incorporate an additional "box" into the pattern recognition model to allow this. Similar or further boxes may be required for phase sensitivity and the *roughness* detection of Terhardt (1970), and indeed any other attribute that has presumably become associa-

ted with pitch through learning. Nevertheless it is only when the ear is denied a pattern of frequency components over the dominant region that it may need to fall back on other such mechanisms.

ACKNOWLEDGMENT   I am grateful to Dr. E. F. Evans for helpful criticism of the manuscript.

REFERENCES

BACHEM, A., 1937. Various types of absolute pitch. *J. Acoust. Soc. Amer.* 9:146–151.

BILSEN, F. A., and R. J. RITSMA, 1968. Repetition pitch mediated by temporal fine structure at dominant spectral regions. *Acustica* 19:114–115.

BILSEN, F. A., and R. J. RITSMA, 1970. Some parameters influencing the perceptibility of pitch. *J. Acoust. Soc. Amer.* 47:469–475.

CRAMER, E. M., and W. H. HUGGINS, 1958. Creation of pitch through binaural interaction. *J. Acoust. Soc. Amer.* 30:413–417.

DE BOER, E., 1956. On the *residue* in hearing. Ph.D. thesis. University of Amsterdam.

EVANS, E. F., and J. P. WILSON, 1971. Frequency sharpening of the cochlea: the effective bandwidth of cochlear nerve fibres. *Proc. 7th Internat. Cong. Acoustics.* Budapest: Akademiai Kiado, 453–456.

FOURCIN, A. J., 1965. The pitch of noise with periodic spectral peaks. *Rapports 5e Congrès International d'Acoustique*, Liège, Ia, B.62.

FOURCIN, A. J., 1970. Central pitch and auditory lateralisation. In *Frequency Analysis and Periodicity Detection in Hearing.* R. Plomp and G. F. Smoorenburg, eds., Leiden: Sijthoff, 319–328.

HARRIS, G. G., 1963. Periodicity perception by using gated noise. *J. Acoust. Soc. Amer.* 35:1229–1233.

HOUTSMA, A. J. M., and J. L. GOLDSTEIN, 1972. The central origin of the pitch of complex tones: evidence from musical interval recognition. *J. Acoust. Soc. Amer.* 51:520–529.

KIANG, N. Y.-s, T. WATANABE, E. C. THOMAS, and L. F. CLARK, 1965. *Discharge Patterns of Single Fibres in the Cat's Auditory Nerve.* Research Monogram No. 35. Cambridge, Mass.: MIT Press.

MATHES, R. C., and R. L. MILLER, 1947. Phase effects in monaural perception. *J. Acoust. Soc. Amer.* 19:780–797.

MILLER, G. A., and W. G. TAYLOR, 1947. The perception of repeated bursts of noise. *J. Acoust. Soc. Amer.* 20:171–181.

PLOMP, R., 1967. Beats of mistuned consonances. *J. Acoust. Soc. Amer.* 42:462–474.

RITSMA, R. J., 1962. Existence region of the tonal residue. I. *J. Acoust. Soc. Amer.* 34:1224–1229.

RITSMA, R. J., and F. L. ENGEL, 1964. Pitch of frequency-modulated signals. *J. Acoust. Soc. Amer.* 36:1637–1644.

RITSMA, R. J., 1967. Frequencies dominant in the perception of pitch of complex sounds. *J. Acoust. Soc. Amer.* 42:191–199.

ROSE, J. E., R. GALAMBOS, and J. R. HUGHES, 1959. Microelectrode studies of the cochlear nuclei of the cat. *Bull. Johns Hopkins Hospital* 104:211–251.

SCHOUTEN, J. F., R. J. RITSMA, and B. L. CARDOZO, 1962. Pitch of the residue. *J. Acoust. Soc. Amer.* 34:1418–1424.

SMOORENBURG, G. F., 1970. Pitch perception of two-frequency stimuli. *J. Acoust. Soc. Amer.* 48:924–942.

STEVENS, S. S., and J. VOLKMANN, 1940. The relation of pitch to frequency. *Amer. J. Psychol.* 53:329–353.

TERHARDT, E., 1970. Frequency analysis and periodicity detection in the sensations of roughness and periodicity pitch. In *Frequency Analysis and Periodicity Detection in Hearing*. R. Plomp, and G. F. Smoorenburg, eds. Leiden: Sijthoff, 278–290.

THURLOW, W. R., 1963. Perception of low auditory pitch: a multicue, mediation theory. *Psychol. Rev.* 70:461–470.

WALLISER, K., 1969. Uber ein Funktions schema für die Bildung der Periodentonhöhe aus dem Schallreiz. *Kybernetik* 6:65–72.

WEVER, E. G., 1949. *Theory of Hearing*. New York: John Wiley & Sons.

WHITFIELD, I. C., 1970. Central nervous processing in relation to spatio-temporal discrimination of auditory patterns. In *Frequency Analysis and Periodicity Detection in Hearing*, R. Plomp, and G. F. Smoorenburg, eds. Leiden: Sijthoff, 136–152.

WIGHTMAN, F. L., 1972. Pitch as auditory pattern recognition. *Hearing Theory, 1972.* Eindhoven: IPO, 161–171.

WILSON, J. P., 1967. Psychoacoustics of obstacle detection using ambient or self-generated noise. In *Animal Sonar Systems*, R. G. Busnel, ed. Gap, Hautes-Alpes, France: Louis-Jean, pp. 89–114.

WILSON, J. P., 1970. An auditory after-image. In *Frequency Analysis and Periodicity Detection in Hearing*, R. Plomp, and G. F. Smoorenburg, eds. Leiden: Sijthoff, 303–318.

WILSON, J. P., and E. F. EVANS, 1971. Grating acuity of the ear: psychophysical and neurophysiological measures of frequency resolving power. *Proc. 7th Internat. Cong. Acoustics* Budapest: Akademai Kiado, 397–400.

15 Parallel "Population" Neural Coding in Feature Extraction

ROBERT P. ERICKSON

ABSTRACT   The development of an understanding of the neural bases of complex neural functions, such as feature detection, is facilitated if a systematic and general understanding of the more primitive processes from which these derive is first established. This chapter is an attempt to examine the extent to which the known "neural" facts of vision, kinesthesis, taste, and so forth may be included under one set of rules, including as guiding principles not only the previously given introspective principles but the known characteristics of neural activity. An initial assumption is that the neural possibilities afforded to one sense by the structure and physiology of its neurons and their connections are also available to the other senses, and thus the encoding parameters available to one are also, to some degree, available to another. This attempt to force the neural processes of all sensory systems into the same small group of mechanisms, by its successes and failures, gives us a view of a general systematic. These principles are then shown to be relevant also to more complex functions, such as feature extraction.

OUR APPROACH to the problem of the neural representation of stimuli in general, including stimulus features, is influenced by at least two very strong and pervasive factors. These influences are radical enough to determine the basic form of our experimental design, as well as the form of our conclusions; thus it would seem reasonable to evaluate these influences from time to time. The first factor concerns our interpretive vocabulary, and the second, our techniques.

Our vocabulary is based on formulations developed by introspective psychologists of the late nineteenth century, mainly Wundt. This formulation was that sensations or perceptions are made up of two kinds of *attributes*, sensory quality and intensity. To these attributes a few others have been added, notable among these being sensory location and duration. The basic requirements for sensory attributes were that variations in each could be accomplished continuously without variations in another (i.e., that they were independent) and that each was necessary for a perception, (i.e., if any one was reduced to zero the sensation would not exist).

It appears that modern neurophysiologists have in large degree accepted this formulation in the sense that the neural underpinnings for one attribute of stimulation (location, for example) are expected to be separate and distinct from those of another (quality, for example). However, some stimulus parameters, such as kinesthesis, do not have a clear definition in this system. Also the structure of the system has been loose enough to allow incorporation of other attributes, with their formal relation to the system remaining unclear; stimulus movement is an important example. The acceptance of this system of stimulus attributes seems to be only informal and implicit, but it appears to be our operating system none the less.

I believe that Wundt's effort to provide a framework that could accommodate all sensory systems may continue to be of use at this time. However, I believe that due to Wundt's lack of understanding of neural processes and later tendencies either to leave a good idea alone or to work without a framework, the present concept of attributes may now hinder rather than facilitate work in the neurophysiology of sensation.

We may usefully reevaluate the concept of attributes in the context of a framework for sensory neural functioning in general; the first part of this chapter, which finds its origin in an earlier paper on the neural representation of simple stimuli (Erickson, 1968), will be devoted to such a systematic approach. In the second half, this system is applied to problems of feature extraction.

To reevaluate the concept of attributes, let us cast aside for the moment all the facts known to Wundt and the psychophysicists after him who were concerned primarily with the description and classification of stimuli. We shall turn to the sensory neural data now available and ask if there is not a simple system of rules that might be used to formalize these events. The basic heuristic will be to include together all similar neural activities into classes and, as consistently as possible, to give similar labels and derive similar encoding principles for all members in a class. Stimulus attributes will be defined neurally, and these definitions will account well for stimuli as diverse in complexity as points, features, and objects.

There are two basic assumptions that will be used in this formulation, assumptions that I believe may be made clearly, but beyond which theoretical travel is perilous.

ROBERT P. ERICKSON   Departments of Psychology and Physiology, Duke University, Durham, North Carolina

First, any process of discrimination, including feature extraction, requires that any stimulus may be defined only in terms of a different or changed stimulus: For example, red would not exist without some other color for comparison. Thus the repeated emphasis in this paper will be on changes in or differences between neural states. Second, the minimal neural requirement for representing discriminable stimulus change is that the neural response must in *some* manner be different for the two discriminably different stimuli: Assumptions are often made about the nature of these changes, for example, that they are accomplished by individual *specific* neurons, such as *feature-detector* neurons. However, the nature of these neural changes is unknown, especially those involved in complex processes such as feature extraction, and form the issue of concern here.

In developing this framework, we must keep perspective on a second factor that has guided our research and thinking, the bias introduced by our techniques. The primary guiding methodology in sensory neurophysiology is the microelectrode, and the basic data consist of the responsiveness of individual neurons. Since our data are largely based on the activity of individual neurons, we are encouraged to conclude that the looked-for functions reside in the activity of these structures. That the individual neuron is the appropriate unit of study for some levels of neural function, such as conduction and synaptic transmission, is beyond question. That it is the appropriate unit of analysis for all neural functions, including more complex functions such as feature extraction, is not clear and should be held open to question. To avoid this bias, consideration will be given to the response of populations of neurons as well as those of individual neurons.

Neural definitions of sensory attributes

In this section an attempt will be made to formulate classes of change in neural activity; these classes will be useful in the neural and psychophysical description of stimulus attributes.

With any given stimulus situation, there are many neurons in activity. To describe or classify this activity in a systematic way would seem at the outset a terribly difficult and complex task. However, we may easily note at least two simple and general kinds of changes in the neural activity, following stimulus changes, that would provide information about the basic parameters involved. For one, the general appearance of the activity might remain the same but simply change in amount, growing greater or smaller. The second general type of change would be that the appearance of the activity could change with the total amount remaining the same; the change in

appearance might be either a change occurring within the same population of neurons, or it might involve a partial or total shift to a new population of neurons.

These types of changes certainly do not accommodate all the candidate parameters of relevant change in neural activity, one notable omission being that of the temporal characteristics of the response and the sequence of activity between neurons; these are treated by Werner (later in this part). Nevertheless, this simplified account of neural changes will suffice to show how such a systematization might facilitate our understanding of complex neural processes, such as feature extraction. It will be used to distill all sensory attributes into two general classes, intensity and quality; the latter class will include such diverse categories of sensory function as color vision, temperature, kinesthetic sensitivity, and location, as well as a range of complexities of sensory function including punctate stimuli and stimulus features.

Intensity

The first type of neural change, that of the total amount of evoked activity, is closely related to the attribute of intensity. Intensive attributes are based on stimulus dimensions producing roughly parallel and monotonic changes in the rate of neural response in all involved neurons (as seen below, nonparallel and nonmonotonic changes in the activity of the involved neurons denote qualitative changes).

In most examples, intensive changes relate in a very straightforward way to existing sensory neural data. Several interesting problems arise with the Wundtian view of intensity, a view based on the psychophysical "more or less"-ness of the stimulus.

Temperature and kinesthesis as intensive attributes

In the temperature sense, when stimulus manipulations are accomplished that have usually been understood as increases in intensity, we know that the steady-state neural activity in at least the $A\delta$ fibers is decreasing. An example of this is shown in Figure 1 (derived from Poulos, 1970). The 15 curves show the responses of individual first-order temperature-sensitive neurons in the trigeminal system of the monkey. I refer to such tuning of curves from all sensory systems as *neural response functions* (NRFs) to call attention to the hypothesized basic similarity in their actions and also use the term *just noticeable differences* (JNDs).

As is clear, the more extreme temperatures evoke the least activity, whereas the more central temperatures evoke the most activity. It would appear that to define neural mechanisms, or attributes, in terms of neural activity, we should label the colder and warmer stimuli as *less* intense than the intermediate stimuli. I would like to return to a discussion of this problem in a moment.

A similar problem in this area arises when we consider the kinesthetic sense. Werner (1968) has successfully used kinesthesis as a model of an intensity attribute. In line with this view, a number of investigators (Andrew and Dobt, 1953; Boyd and Roberts, 1953; Cohen, 1955; Mountcastle and Powell, 1957; Mountcastle et al., 1963; Burgess and Clark, 1969) have found that as the extremes of extension and flexion are approached, the neural activity increases for some neurons and decreases for others such that some have their maximum rate of response at extension and others at flexion. In Figure 2 are schematically represented the NRFs of individual kinesthetic neurons. For the sake of clarity, only those peaking in activity at full flexion are shown; the mirror-image family of curves, those that peak at extension, has been omitted.

Intuitively, it seems strange to classify kinesthesis as an intensive parameter; common sense would suggest another term such as *position*. However, since these neurons do change their rate of firing monotonically as joint position is varied, the term *intensive* seems a good choice.

Mountcastle et al. (1963) have also shown that the rate of neural response across this parameter follows Steven's power law; this law is very useful in describing intensive functions. However, considering both those neurons that give their greatest response at flexion and extension, it is not clear that the *total* neural input increases as either of these extremes is approached. Therefore, to classify kinesthesis as an intensive continuum at least raises some questions; I would like to discuss this further when I return to considerations of the temperature sense.

Quality

The second major case of neural changes that we observe are those in which the general "appearance" of the neural activity changes; these will be referred to here as changes in the stimulus attributes of quality. The appearance changes because the neurons are tuned in a nonmonotonic and nonparallel manner along their parameters.

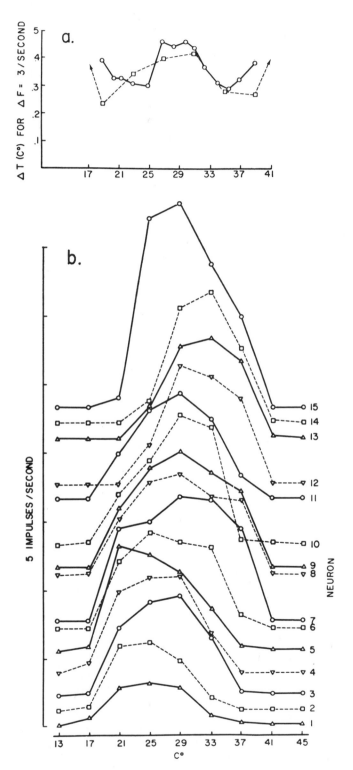

FIGURE 1 (a) Solid line: amount of temperature change necessary to produce a JND. After 8 min adaptation, two subjects' difference limens between the hands were obtained by the method of constant stimuli. Peltier refrigerators were used as temperature stimuli for each hand. (Erickson, unpublished.) Dashed line (derived from lower part of figure): amount of temperature change necessary to produce a constant change (here, three impulses/second; arbitrary) in the population of neural response functions given below. (b) Responses to steady temperatures of 15 first-order temperature-sensitive neurons from the face of the monkey (redrawn from Poulos, 1970); extrapolated to 13°C and 45°C. The response function for each neuron is displaced along the ordinate for clarity of illustration; the end points for each curve represent zero impulses/second.

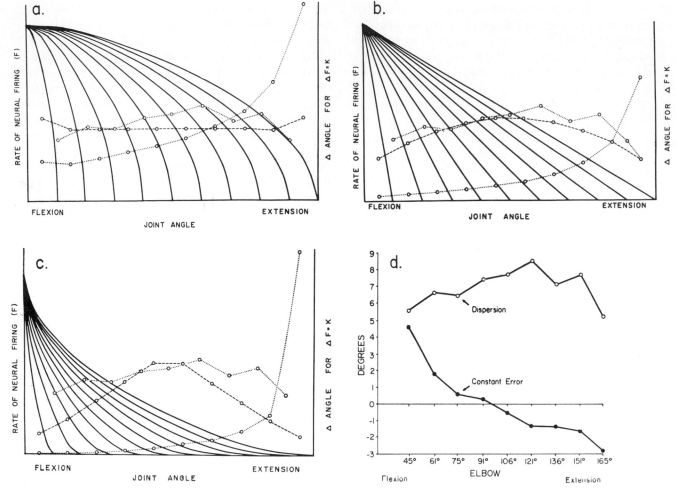

FIGURE 2 (a, b, c) Neural coding of joint positions. *Solid lines:* three probable forms of neural response function (NRFs) of individual kinesthetic neurons (see text for reference). Only those peaking in activity at flexion shown: for clarity, *extensor* neurons not shown. *Dotted line with open circles:* JND curves based on equal changes in the NRFs as in Figures 1 and 5. Curve peaking at *extension* derived from *flexion* NRFs shown. *Dashed line with open circles:* curves derived from the population of both *extensor* and *flexor* NRFs. In b and c, better discrimination is predicted towards the end of the continuum. *Dotted line with* *closed circles:* Identical in a, b, c; also appears in d. Experimentally determined reliability of localization of elbow joint in humans. Note correspondence with derived curves based on both flexors and extensors. (Erickson, unpublished.) (d) *Accuracy of positioning of human elbow.* The six subjects were asked to duplicate the position of one arm with the other arm. The *average* position taken is given as the *constant error* from the correct position. The reliability (standard deviation) is given as the *dispersion;* this curve also appears in a, b, c. Reliability is greatest towards the ends of the dimension. (Erickson, unpublished.)

Erickson has described two general overlapping classes of stimulus quality coding, topographic and nontopographic (1968). In the nontopograpic modalities, broad sensitivity of individual neurons along the relevant dimension (as color coding in visual neurons) is necessary to accomplish the representation of the complete dimensions with a few neurons. In the topographic modalities (as "locations" in vision or in somesthesis) the large quantities of available neurons permit narrow tuning of each. Their inclusion together here as stimulus qualities derives from the similarities in their encoding processes detailed below.

Quality coding: nontopographic

Hue as a Nontopographic Quality A classic example of nontopographic quality coding would be that occasioned by changes in the wavelength of the stimulating light. If the position and intensity of the visual stimulus are held constant with only the wavelength varying, then it is approximately true that the same population of neurons will respond, but the general appearance or profile of the activity across these neurons will be modified. That this *across-fiber pattern* (Erickson, 1968) will adequately encode the wavelength of the stimulus follows the logic of Helmholtz (and Hering, for that matter).

In neural terms, this general plan is that hue is represented by certain *relative amounts of activity within a small population* of broadly tuned color-coded neurons. This is illustrated in Figure 3a in which the three NRFs characterize the activity elicited in three types of color-coded neurons; whether these curves are of this particular form or some other broadly tuned form (e.g., opponent process

cells) is not of importance here. In Figure 3b is shown the fact that the wavelength of the stimulus is confounded within any one neural element (and may be confused with intensity changes). In Figure 3c it is shown that the wavelength is clearly encoded when the relative activity among the neurons of the population, the across-fiber pattern, is considered; a unique population response obtains for each wavelength-intensity combination. The

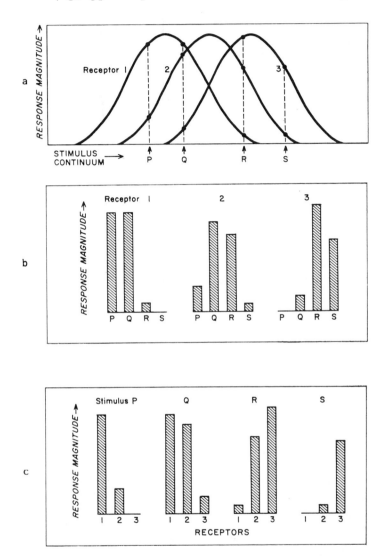

FIGURE 3 Afferent fiber types and patterns of neural activity. (a) Afferent fiber types (or receptor types). NRF curves 1, 2, and 3 represent the responsiveness of three hypothetical afferent fiber types (or receptor types) along a hypothetical stimulus dimension. *P*, *Q*, *R*, and *S* represent four stimuli along this stimulus dimension. The responsiveness of a fiber type to one of these stimuli is indicated by the intersection of the response curve and the ordinate erected as the stimulus. (b) Responsiveness of the three fiber (or receptor) types to the four stimuli in (a). In each of the bar graphs is shown the responsiveness of one of the three types to each stimulus in (a). One neuron cannot adequately represent a stimulus, that is, it cannot differentiate

between stimuli, because equivalent responses can be given to different stimuli (response of receptor 1 to stimuli *P* and *Q*), and the variations in response magnitude could also be affected by variations in stimulus intensity. (c) Across-fiber patterns. In these bar graphs are shown the patterns of activity across the three fiber types produced by the four stimuli in (a). Each stimulus produces a characteristic pattern across the three fiber types. There would be as many across-fiber patterns as stimuli. With changes in stimulus intensity, the height of the pattern changes but not its form. The quality of stimulus is given in the form of this pattern across a population of neurons. (From Erickson, 1963.)

differences between the four across-fiber pattern responses shown in Figure 3c indicate what is meant in the present context by the encoding of qualitative changes within the same population of neurons.

This view shares similarities with Rock's (1970) speculations about the act of perception in which the meaning of the activity of any neuron is clear only after a "taking account" of the activity of other neurons. This is another way of stating Helmholtz's view that in the sensation of hue, the meaning of the activity of the *red* neurons is only clear after an evaluation (or "taking account") of the activity of the *green* and *blue* neurons. Many others have expressed similar views (for example, Adrian et al., 1931; Hahn, 1971; Hartline, 1940; Mountcastle, 1966, 1968; Mesarovic and Macko, 1969; Poggio et al., 1969; Spinelli et al., 1970; Tower, 1940, 1942; Vastola, 1968; Wright, 1947; Bishop, 1970; O'Connell and Mozell, 1969).

Following Wundt's definition closely, stimulus wavelength changes with the accompanying changes in perceived hue have classically been considered to be an example of the stimulus attribute of quality. He specified that stimulus quality changes should be capable of occurring without changes in any other attributes, such as intensity or duration, and that they should be capable of being accomplished without any discontinuities. Thus, in color, we may move from red through orange and green to blue smoothly and without interruption, and without changes in the other attributes.

TEMPERATURE AND KINESTHESIS AS NONTOPOGRAPHIC QUALITIES   Now I would like to return to a consideration of the problem of temperature and kinesthesis earlier left unresolved. I would first like to point out the general similarities between the NRFs along the temperature dimension (Figure 1) and those of the color-coded neurons along the wavelength dimension (Figure 3). In both cases, each NRF is broadly tuned across the dimension, and its peak activity occurs toward the center of its range. Perhaps, then, the neural coding processes in these systems share some similarities. If true, then the rubric most useful for psychophysical and neural analysis for the temperature sense might be the same as the rubric used for color coding, i.e., sensory quality.

The analysis suggested here for color coding, following the leads of Helmholtz and Hering, will thus be applied to temperature coding; the difference between two temperatures will be presented as neurally and psychophysically analogous to the differences between two colors. What is to be derived is the neural situation obtaining for different temperatures, analogous to the neural situations for differing color codes in Figure 3. This analysis can be obtained from the NRFs of Figure 1 as is

shown in Figure 4 for several different temperatures. In this figure is shown the amount of activity in each of the 15 neurons of Figure 1 at a given temperature; each temperature produces a distinctive pattern of neural activity across this population of neurons.

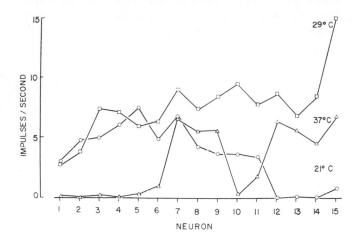

FIGURE 4   Neural representation of temperature. The data points and neuron numbers are from Figure 1. Whereas in Figure 1, the data are presented by neurons, here they are presented by temperatures. This population response, or *across-fiber pattern*, should signal the quality of the stimuli, i.e., the temperature.

That the stimulus changes in hue and temperature are similarly given by the change in the across-fiber pattern, suggesting a communality of the encoding process, may be shown in relation to Figures 1 and 5. Hypothesizing that a JND along these dimensions requires a certain constant amount of change in the population response (Erickson, 1968), the size of the step taken along these dimensions to give a constant change in response is plotted below the curve for the receptor processes in Figure 5; the same derivation may be obtained from data published on opponent-process color-coded neurons. The familiar *w*-shaped function depends on the fact that the size of the step for the constant change in neural activity need not be as large where the slopes of the receptor function are great. For temperature, the same type of derivation is shown in Figure 1a where the two minima of the *w*-shaped function correspond with the more steeply sloping parts of the neural functions. Included also with this is a JND curve obtained psychophysically from the skin of the hand of human subjects. The correspondence between the derived and experimentally determined functions suggests that a similar across-fiber pattern analysis of the quality code is occurring in both cases.

To return to kinesthesis, as illustrated in Figure 2, these NRFs seem to be something like a reversal of the temperature or color-coded NRFs; that is, they show the

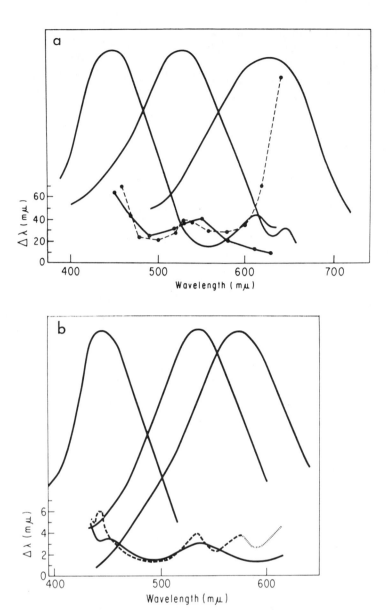

FIGURE 5 Comparison of color JND curves derived from changes in receptor across-fiber patterns with behaviorally determined JND curves. (a) Goldfish. Trichromatic receptor NRFs, from Marks (1965). Dashed curve: derived from receptor NRFs, giving changes in wavelength ($\Delta\lambda$) necessary to produce a constant total change across all three receptor types. Solid curve: behaviorally determined JNDs for goldfish (Yarczower and Bitterman, 1965). Note correspondence between derived and behaviorally determined JND curves (except at long wavelengths). (b) Humans; as upper portion of figure. Receptor NRFs from Marks et al. (1964). Behavioral JND curve from Wright (1947). Note correspondence between general form of JND curves (except at long wavelengths).

same general tuning characteristics except that their peaks are at the extremes of the continuum rather than at the center. There are striking similarities in the types of neural changes occurring with joint movement and with changes in color or temperature stimuli. As the stimuli are varied, the *total* amount of neural activity across all involved neurons varies slightly, but not monotonically as would be expected with intensive changes; the most striking change is in the relative amounts of activity within the population of responding neurons. In this particular case the major changes would be in the pattern of activity between the two general classes of neurons, the extensors and the flexors; these would be analogous in action to two classes of color-coded neurons, or two classes of temperature neurons. This would classify kinesthesis with color and temperature as a qualitative sense.

We would have our choice here either of considering the neural activity from the joint as two intensive parameters (flexion and extension, and perhaps rotation), or as one qualitative parameter. The qualitative parameter is perhaps more appropriate, because the analysis appears to be that of *comparison* of the activity among all participating neurons, rather than an analysis of one population of these neurons *or* the other. In Figure 2 is shown an analysis of the JND function that would occur considering flexors or extensors alone, and an analysis of the population response analogous to those done of the population response in Figures 1 and 5 for temperature and color. In each part of Figure 2 are shown two derived JND functions, one depending simply on one population of neurons, and one that would derive considering both extensor and flexor neurons simultaneously. Considering only one group of neurons, in each of the three types of curves given (Figure 2a, b, c) the discrimination would become extremely poor at one end of this dimension. Considering both families of neurons, it can be seen that although the JND function would depend on the exact form of the kinesthetic NRFs, it would be much flatter across the continuum than if it were based on only one family of these NRFs. The kinesthesis JNDs that we have derived psychophysically (human elbow) are also shown in Figure 2a–d. The similarity of these "flat" JND functions to those derived from a *population* response of flexors and extensors simultaneously suggests that both families of these neurons are used in the analysis. The most likely form of the NRFs is also suggested, i.e., that shown in Figure 2b or c where discrimination would be better at the extremes of the continuum rather than Figure 2a, where discrimination would be poorer at the extremes.

This suggests that the analysis occurs in a manner similar to that for color and temperature, and that for a simple systematic approach to neural function, kinesthesis should be grouped with them as a qualitative modality.

TASTE AS NONTOPOGRAPHIC QUALITY A number of facts point to differences between tastes as representing sensory quality in a nontopographic manner (see Erickson, 1968, for review). In mammals in general, each stimulus activates nearly all receptors with which it comes in contact, and each neuron is sensitive to nearly all taste stimuli. Small populations of neurons, those serving one taste papilla for example, are capable of carrying the message for a variety of tastes (Harper, Jay, and Erickson, 1966); this would be expected with each neuron being broadly tuned, and the aggregate of neurons functioning in an across-fiber pattern manner (Figure 6) as in the color system. Also, although diffuse regional differences in sensitivities to various tastes occur across the tongue, and

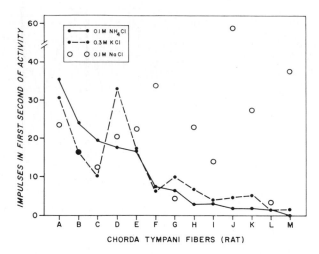

FIGURE 6 Coding of taste quality in a population of neurons. These records from 13 chorda tympani neurons in the pentobarbital anesthetized rat are typical of several hundred obtained. Each neuron responds to all three stimuli, and thus responses in any individual neuron are equivocal with respect to taste quality. The across-fiber pattern of response to each stimulus (at various intensities) uniquely expresses that stimulus. (From Erickson, 1963; see Erickson 1968, Ganchrow and Erickson 1970, and Doetsch and Erickson 1970, for reviews.)

perhaps across areas of neural tissue, the point of stimulation is not a primary cue to the identification of taste quality as would be expected in a topographic system.

SUMMARY It appears that in order to treat systematically the neural activity resulting from stimulus changes, certain classes of changes may be conveniently grouped together on the basis of their similarities. These include color, temperature, taste, and kinesthetic stimulus changes; other possible sensory mechanisms in these categories not discussed here may include auditory pitch (see below), olfaction (O'Connell and Mozell, 1969), and vestibular sensitivity. The main requirements for this type of stimulus coding are that they may occur within the same set of neurons, that they may occur continuously along their dimension, that they need not be accompanied by intensive changes, and that the main change encoding the stimulus occurs in the pattern of activity evoked across the population of responding neurons (*across-fiber pattern*). The term used here for such change is sensory *quality*. Since these changes do not necessarily require changes in the sets of neurons involved, they are considered nontopographic quality changes.

Quality coding: topographic

The second major class of changes of neural activity involving a change in the "appearance" of the neural activity is that in which there is a partial or total change in the population of neurons involved. The proportion of neurons changed in the two neural situations may vary continuously from no change (which would be the nontopographic class of events just discussed), through slight amounts of change in the population, to a complete change to another set of neurons. *That these types of changes are not entirely disjoint will be emphasized here.* Due to the similarity in the encoding processes, both of these types of neural change will be included under the general classification of "qualitative" changes. With the population of neurons involved changing, we may note a movement of the locus of activity across the neural sheet, such as occurs with movement of the visual stimulus across a visual field or movement of a mechanical stimulus across the skin; these stimulus changes may thus be termed topographic. Since the previously discussed group of neural changes did not involve such a movement across the neural sheet (the change may be accomplished within the same set of neurons), they were termed nontopographic.

TOPOGRAPHIC QUALITY IN AUDITION   Auditory coding frequency was previously suggested as a *topographic* qualitative system (Erickson, 1968). At the cortical level, frequency coding may be more appropriately placed in the *nontopographic* group. Several considerations prompt this suggestion. The NRFs for frequency in audition seem to be very broad, at least hazy, at the cortical level. Also, the tonotopic organization of the neural tissue, at the cortical level, appears to be very rough. Perhaps this tissue is given in the topographic way to auditory localization, the rough tonotopic organization seen there being a reflection of an appropriate high-frequency sensitivity for the stimuli in the area of the neuron's sensitivity—on the close side. Narrow tuning for location rather than frequency could be expected of these neurons.

LOCATION AS TOPOGRAPHIC QUALITY   It would undoubtedly be objected that there are not many similarities between neural coding in color vision on the one hand and somesthetic location or visual space on the other and that therefore it is not helpful to include them together under the same general coding classification. For example, it seems more likely that a finely tuned individual neuron could encode topographic location in somesthesis or vision more easily than an individual broadly tuned neuron could encode the nontopographic parameters of color and, thus, that a different mechanism, *specificity*, might be expected for location. However, even in these topographic systems in which the neurons may be narrowly tuned across the stimulus dimensions, the specificity of tuning does not account for the fineness of discrimina-

tion any better than it does for hue; discriminations can be easily made over much smaller ranges than the width of the NRFs. Perhaps in both cases discriminability depends upon certain constant amounts of change in the across-fiber patterns encoding the stimuli, independent of the width of the NRF (discussed in Erickson, 1968).

An example of what is meant by coding of this nature in somesthesis is given in Figure 7. These records were obtained from first-order neurons from the forepaw of the rat. Here it is seen that stimuli at two different locations may be encoded by the relative amounts of activity of the population of neurons responding. Stimulus location and intensity are confounded in the activity of any given neuron but are unequivocally given in the population response. Many previous investigators, for over three decades (for example, Adrian et al., 1931; Tower, 1940, 1942), have advocated this kind of analysis in somesthesis. In this, the coding is identical to that in the nontopographic systems.

It is relevant to point out here that Hartline et al. (1961), in an analysis similar to that given for somesthesis above, were able to disclose aspects of neural activity related to spatial edge effects through inferences about relative amounts of activity across a population of neurons.

SUMMARY   The main point to be made here is that the coding process in the topographic systems does not appear to be entirely distinct and different from that in the nontopographic systems. In our search for a general simple vocabulary to account for observable neural changes that accompany stimulus changes, we have the choice either of specifying these two types of coding as different, i.e., quality and location, or of taking note of the similarities of the coding process as seen from the point of view of the nervous system, and including them together as two subtypes of the same coding process, that of stimulus quality.

It is interesting to note here that *stimulus location* follows closely the definition set forth by Wundt for stimulus quality; that is, changes may be made along this stimulus dimension in a continuous fashion and independent of other stimulus attributes. It may have been our concern with the physical description of the stimulus, rather than a clear difference in neural process, which has led us to deal with location and quality as involving separate sensory mechanisms.

Coding for other neural functions

Before bringing this rather simple formulation to bear on the problem of feature extraction, I would like to point out one other result of this systematization of neural

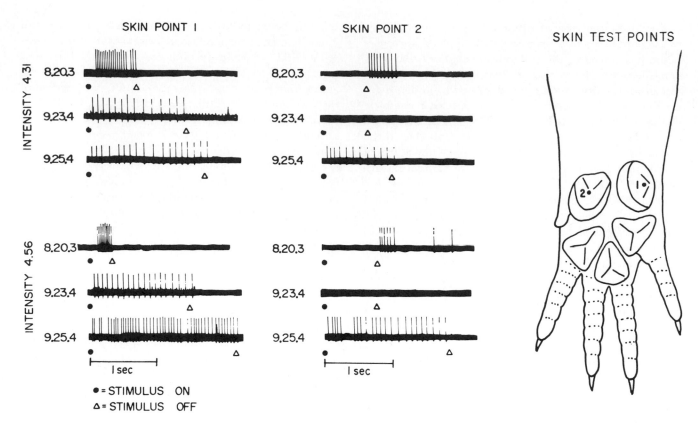

SKIN POINT 1   SKIN POINT 2   SKIN TEST POINTS

INTENSITY 4.31

8.20.3
9.23.4
9.25.4

INTENSITY 4.56

8.20.3
9.23.4
9.25.4

I sec

● = STIMULUS ON
△ = STIMULUS OFF

FIGURE 7 Parallel neural coding of stimulus location and intensity in somesthesis. Responses of three (representative of recordings from 50) individual neurons in the median nerve of the phenobarbital anesthetized rat to graded stimulation by von Frey hairs (Semmes and Weinstein, et al., 1960) at two skin locations. The rate of response of any individual neuron may be similarly influenced by changes in location, intensity, and in some cases (neuron 8.20.3) in timing—stimulus *on* vs. *off*. However, considering all three neurons together, there is an unequivocal representation of position, intensity, and timing. (Cassel and Erickson, unpublished.)

activity. The neural changes in a sheet of neural tissue involved in motor response (see Figure 8, part 3) and those in a sensory sheet (see Figure 8, part 1) are thus seen to involve many similarities. To classify movements neurally, many parallel neurons would be in activity for each movement. That is, no movement, at any neural level, is ever represented by activity in one neuron or one group of neurons; at the very least, a complex pattern of activity in flexors and extensors is involved. As the organism changes from one movement to another, the general across-fiber pattern of activity of many efferents would be seen to shift in amplitude and shape.

I make these suggestions for several reasons; first, on the intuitive level it would be rather surprising if the nervous system evolved different principles of encoding mechanisms for sensory and motor functions—it would be uneconomical in light of the similarities of the machinery used (neurons and their interconnections). Other functions such as memory storage or other biasing mechanisms might also share some of these principles. (See Figure 8, part 2.)

Second, these various functions, i.e., sensation, movement, and memory, face many of the same problems. The basic problem is one of economics; that is, many different events must be expressed with a limited number of neurons. The color system, in which a very few types of color-coded neurons can represent the total wavelength dimension, expresses the solution to this problem admirably. In the motor system, many different movements must be expressed with a limited number of motor neurons. Even in a simple nervous system with *command* neurons, this principle applies (see Davis and Kennedy, 1972). In memory, the problem seems to be particularly staggering. The memories of a lifetime must be represented somehow with a rather limited number of neurons.

Also, in each case, the function is given in a distributed manner to the cooperative responsibility of many neurons, and thus each function has the neural safety and weight of being expressed by a large portion of the brain. It is to be noted that although this distributed manner of coding gives each function the mass action and equipotentiality of a hologram, the similarities do not necessarily include

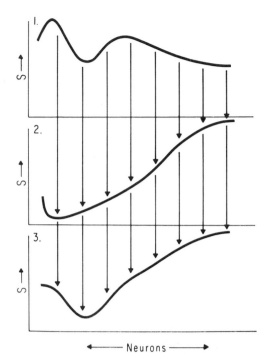

FIGURE 8 General conception of the form and interrelationships of the representations of neural events at several levels. Ordinates give level of activity (e.g., graded potentials, biochemical state) of the neurons represented along abscissa. Coding at all levels takes the same general parallel "population" form, and influences from one level to the next are also parallel. The intermediate level 2 represents mechanisms such as memory storage and motivational states, modifying the response 3 to the input level 1. The level 3 is any efferent or resultant function, such as perception or movement. The presence of convergence, divergence, feedback, etc. does not qualitatively alter the general form of the coding process and is not illustrated.

the mechanisms producing these effects (see Gabor, 1972).

Finally, if the same general principles are utilized by these several interlocking processes, then there need be no change in the general *form* of the message from one level to another. For example, the nervous system need not change the form of its information for the purpose of intermediary *gnostic* or *pontifical* neurons.

These considerations are not entirely novel; some have very long histories. They are briefly reiterated here to indicate the generality and feasibility of the proposed mechanism of neural coding.

A final note might be added about the feasibility of studying population responses and the improbability that all functions utilize the same mechanisms. Perkel and Bullock (1969) have suggested that the encoding of sensory events probably depends on a population response but were doubtful about the possibility of the researcher accomplishing such an analysis. It was shown above that such an analysis is only slightly more complex than analysis of the activity of individual neurons and that the latter

may be misleading in the investigation of sensory neural processes.

Concerning the generality of sensory mechanisms, some factors are probably unique to each sensory system. For example, the problem of stereopsis may be unique to vision, the solution of phase-locked volleying and the use of interaural Δt unique to audition, and a descending afferent anatomy unique to olfaction. Still, the nervous system uses much the same neural machinery in all the sense systems and thus probably did not evolve entirely different coding processes for each of them. Their similarities should prove useful, and at least encourage the illumination of one area of sensory neurophysiology by discoveries in other areas.

Application to feature extraction

The general systematic for neural sensory representation presented in the first half of this paper is of use in two problems related to feature extraction. It bears on the question of (1) how stimulus features are represented neurally, and on the issue of (2) the level of feature "specificity" which we may expect from individual neurons participating in feature extraction.

Neural representation of stimulus features

In considering feature extraction, we generally turn to the area of vision. Here, there are certainly more extractable features than in some of the simpler afferent systems. Of the two problem areas of color and form, probably those of color, including adaptation and contrast effects, compose the simplest categories; these are probably encoded simply by the across-fiber pattern of activity within the responding population as suggested by Helmholtz and Hering. The more complex problems of feature extraction probably relate to visual form.

The basic requirements for the encoding of a complex array of visual form is that each form elicits neural responses that differ from others in some respect; this is met by a camera-like nervous system of simple point detectors. Why should there have evolved in the visual system any process of feature extraction more complicated than this?

To approach this problem, let us inspect the utility of some fairly well accepted feature-extractor neurons, the line-sensitive neurons of Hubel and Wiesel. In brief, these central visual neurons respond best to straight lines, and the response of each of these neurons drops off *gradually* as the line is rotated from its best orientation. This situation is schematized in Figure 9 with the representation of the NRFs of three neurons responsive to the orientation of straight lines. Neither the fact that the stimulus was a line nor the orientation of that line is portrayed in the activity of any individual neuron, just as in color vision the wavelength of the stimulus is not unequivally given in the responsiveness of any color-coded visual neuron. Within each neuron, equivalent responses may be obtained at either of two line orientations, and the amount of this

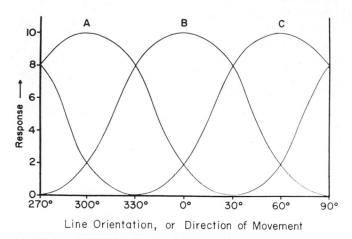

FIGURE 9 Parallel neural coding for all line orientation by three broadly tuned, line sensitive neurons. The mechanism is presumed to be the same as described for the encoding of all visual wavelengths by a few broadly tuned color-coded neurons (see Figure 3). This figure also is used to describe, in a similar manner, the encoding of direction of movement of stimuli in somesthesis and vision. Table 1 is derived from this figure.

response is also influenced by the intensity and the wavelength of the light; thus all stimulus parameters are confounded in any one neuron's response. The neuron will also respond, although to a lesser degree, to points, curved lines, lines of various width and so forth, and complex forms; these latter problems will be returned to after consideration of the coding of the stimulus as a straight line.

The three curves given in Figure 9 were chosen to emphasize the similarities between line-orientation coding and coding in color vision (Figures 3 and 5); the continuity of curves in line coding, rather than being confined to three groups, does not qualitatively influence the argument concerning their similarities, nor does the fact that the breadth of tuning may be somewhat narrower than for color. If this analogy is pursued, the advantages of this type of neural responsiveness are immediately clear: All line orientations and their brightnesses may be encoded by considering the relative amounts of activity across a few neurons. Thus, coding for line orientation could to great advantage follow the same principle as seen for color vision, temperature, and other members of the nontopographic quality series.

An example from Figure 9 of how simultaneous encoding of these stimulus features could be supported in a small population of nine neurons is given in Table I. Here unequivocal and simultaneous encoding of stimulus location (given by the topographic identity of the group), intensity,

TABLE I

Each cell gives the responses of the same nine color-coded neurons with line orientation properties as illustrated in Figure 9. The curves in Figure 9 give the response at the most stimulating wavelength (at intensity). As an example, a red line at 0° orientation (center cell, upper row) stimulates maximally (10) a *red* neuron tuned optimally to 0°. The *green* neuron tuned maximally to 0° would give a lesser response (e.g. 5), and a *blue* neuron still less (e.g. 2.5). (The latter 2 responses not illustrated in Figure 9.) This ratio across R-G-B neurons (10-5-2.5) indicates *red* under all conditions of line orientation and intensity. Changing wavelength to green at the same line orientation (moving from center to right cell) changes pattern across R-G-B neurons to 5-10-5 (G neurons now give maximal response).

Line orientation is given across A-B-C neurons (2-10-2 ratio in upper center cell). This pattern is unchanged in changing wavelength (moving from center to right cell). However, changing orientation but not wavelength (moving from center to left cell) changes only A-B-C pattern, (from 2-10-2 to 8-0-8) leaving color pattern across R-G-B neurons unchanged.

Changes in intensity (moving from upper to lower row of cells) changes level of responding, leaving patterns unchanged.

				Straight-Line Stimulus							
			90° Red			0° Red			0° Green		
	Neurons		R	G	B	R	G	B	R	G	B
	"120°"	A	8	4	2	2	1	0.5	1	2	1
Intensity 1	"0°"	B	0	0	0	10	5	2.5	5	10	5
	"60°"	C	8	4	2	2	1	0.5	1	2	1
	"120°"	A	16	8	4	4	2	1	2	4	2
Intensity 2	"0°"	B	0	0	0	20	10	5	10	20	10
	"60°"	C	16	8	4	4	2	1	2	4	2

hue, and line orientation are condensed into the activities of nine neurons.

To the extent that this example is correct in principle, the breadth of tuning along several stimulus parameters indicates that stimulus features may not be extracted in the activities of individual neurons. Rather, this breadth of tuning makes possible the process of feature extraction by a small population of neurons. Thus, although the stimulus feature of a straight line could be adequately represented by point detectors, it is also adequately represented by a much smaller population of neurons broadly tuned to straight lines.

A major advantage of representing this stimulus feature with a small number of neurons is that it may be represented redundantly over many small populations of neurons and in a distributed manner across neural tissue. Redundancy becomes a major issue in sensory representation. In a camera-like system composed of point detectors, the activity of each neuron is crucial in the representation of the stimulus, and loss of one small point in the sheet of neural "film" would cause distortion of the perceived figure. Due to redundancy of figure representation in the broadly tuned line detector system, poor reliability (in the rate of neural firing), or even neuron loss, becomes much less critical.

The development of multiple neural sensitivities

In an attempt to determine what further or different processes of feature extraction may be expected, let us state the rules that characterize what has happened in the above very simple and concrete example of "feature extraction." At the simpler level we have a stimulus encoding process in which intensity is given by the total amount of activity across the responding population. In addition, one stimulus quality, i.e., location, is encoded topographically with narrowly tuned NRFs and another stimulus quality, i.e., wavelength, is encoded nontopographically with broadly tuned NRFs by the same neurons. It was possible and fruitful to add another sensitivity, that is, the tilt of the straight lines at that point. This process could be characterized simply as the adding of another nontopographic parameter of sensitivity to the NRFs of these neurons.

Without belaboring the encoding of these aspects of stimulus in detail, it may be pointed out that as was the case with line orientation, nontopographic responsiveness along other dimensions such as direction and rate of movement may be added to these neurons. Figure 9, with relabeled axes, may be used to demonstrate coding of direction and rate of movement by the conjoint responses of a few neurons (the abscissa now scales direction of movement; different curve heights would result from different rates of movement). The similarity of neural coding between stimulus movement and the other sensory qualities would indicate that movement should be classified with them.

In this model, the *simultaneous* activity of a small group of movement-sensitive neurons is used to signal the rate and direction of movement. Considered over time, the postulate that any stimulus aspect could be encoded by *any* differences in neural activity indicates that an adequate code for stimulus movement also resides in the *sequential* activity of point detectors that are not individually sensitive to movement. Werner (later in this part) has also indicated the possible role of movement detectors in the sequential buildup of perceptions from elementary feature detectors.

As long as the additional sensitivities are broadly tuned, then each stimulus (in this example a line of any color, orientation, width, length, curvature, movement, etc.) will evoke a unique response in the *population* of responding neurons, and thus will be adequately encoded. If the neurons become narrowly tuned along any of these dimensions, such as stimulus location, then the size of the population of neurons responding will be limited. If the NRFs along more than one parameter become narrowly tuned, the population response would be severely limited. For example, if neurons were tuned narrowly to both stimulus location and line orientation, or any other pair of dimensions, then the response to any stimulus would necessarily be restricted to a very small number of neurons: The activity in such a system would be close to zero, and we know that this is not the case.

Presumably if narrow tuning along some "new" parameter is introduced at higher CNS levels, as Gross has shown for visual neurons in inferotemporal cortex (Part 3, this volume), then the narrow tuning along other parameters—as *location specificity*—must be abandoned. On the other hand, whenever narrow tuning is abandoned, as Evans has shown for cortical auditory neurons along the frequency dimension (later in this part), then we may expect narrow tuning, possibly with topographic organization, along some other "new" parameter.

To summarize, each neuron may be sensitive along many broadly tuned (nontopographic) dimensions but sensitive to only one narrowly tuned (topographic) dimension.

The development of complex sensitivities

Stimulus features could, in the manner described above, be encoded by any of a large range of complexities of sensitivities. What degree of complexity would be the most effective? Would sensitivity of a neuron to a particular human face be feasible? Probably few neurons would develop such a sensitivity to this stimulus, because this stimulus would not be important enough to a species or occur frequently enough in the experience of an individual to command the sensitivity of a neuron. The occurrence of "face" neurons would be uneconomical, because such neurons would seldom, if ever, be brought into play. They could be considered "stupid" neurons in that their usefulness would be very limited; on the one hand we might equate narrowness of responsiveness with inflexibility (e.g., the fixed and limited usefulness of some narrowly tuned invertebrate neurons), and on the other we might relate breadth of responsiveness with versatility.

At the other extreme of complex sensitivity, a *point-sensitive* neuron would often be in use, responding to any stimulus falling on its receptive field. However, a neuron sensitive to any other common, or frequent, stimulus feature would also be assured an important role in the encoding of visual stimuli. These sensitivities could be of *any* form, but they would have to match common stimuli in the experience of the animal to keep the mass of the neural response high. The mass of neural response to any stimulus must be kept fairly high to make redundance possible. A great variety of sensitivities would be appropriate for this requirement; however, neurons sensitive to points and approximate line shapes would be, by far, the most numerous, because they would be most heavily activated.

Presumably, these complexities of sensitivities of the visual neurons would develop only because of the complexities common in the stimuli. In kinesthesis and temperature, no further, higher-level processes of feature detection are developed, because there simply are no further useful complexities to be encoded. The feature-extraction process which evolves, then, depends on the useful complexities in the stimulus situation as well as on their frequency of occurrence.

This statement must be modified somewhat, because each species, in the face of a similar world of stimuli, develops different neural sensitivities. Each species evolves its own unique general neural machinery that is maximally useful in its niche. The departure from a point-detecting neural sheet of "film" toward a system selective to important aspects of the stimulus is different for each species. The amount of complexity of sensitivity of the individual neurons would depend both on the number of neurons available and the behavioral requirements of the organism. The frog certainly does not have as much neural tissue to devote to vision as does the cat. If the frog expends these few neurons in the development of complex sensitivities such that heavy inputs occur only for a few specialized stimuli, it does so at the expense of versatility in its output; it has only a few responses related to a correspondingly limited number of classes of stimuli, and this is appropriate for a frog.

Also, since the sensitivities of these neurons may depend to a certain degree upon the experience of the organism

(Spinelli et al., 1972; Blakemore, earlier in this part), the responsiveness developed probably results from whatever neural input has most often been involved (has been "common") in evoking their activity. This would make it probable that each stimulus produces activity in a large population of neurons. The particular sensitivities to aspects of the stimulus may consist of decrements in sensitivity from some innate baseline due to nonoccurrence of a particular feature, or acquisition of sensitivities beyond this baseline due to frequent occurrence of a particular feature, or both. In any case, the final sensitivity suggested here is one that would find the neuron frequently responding to the environment.

Thus, the precise and subtle sensitivities, as influenced by the experience of the organism, are probably impressed upon each neuron's rough innate sensitivities. The thrust of both these processes, evolutionary and experiential, would be to develop neural sensitivities as rich or complex as would be consistent with heavy neural responding to each stimulus of importance to the organism.

From this I would like to suggest that there probably exists a rich array of neural sensitivities, ranging from approximate straight-line sensitivity through increasing levels of complexity, to perhaps a "few" face neurons. The frequency of occurrence of these cells would be greatest for the more simply formed sensitivities, dropping rapidly with greater levels of complexity; face cells would be in a very extreme minority. This progression toward cells of great complexity of sensitivity, however, would not be the process of feature extraction; the feature extraction process would consist of the population response. At no level of specific complexity of response could a neuron by itself be a feature extractor, because the mass of the response and the redundancy would be so critically limited.

Summary

In order to gain perspective on the neural processes of feature extraction, a systematic model for the neural encoding of stimuli was developed; this model was based on the known characteristics of the responsiveness of afferent systems to simple stimuli. The model was incomplete in detail and was used merely to suggest how a systematic approach to problems of feature extraction might be useful.

Within the context of this model, it was concluded that feature extraction is not necessarily as complex a neural process as the complexity of the percept would suggest, i.e., that there may not be an isomorphism of complexity between percepts and neural processes. It may be incorrect to treat as qualitatively distinct such diverse processes as sensory location and classical quality, feature detection, and the perception of objects. All these processes probably depend in a qualitatively similar manner upon the relative amounts of activity of a population of neurons, each member of which would show rather simple stimulus sensitivity, rather than the activity of individual highly complex extractor neurons. It was suggested that in general the sensitivities of the neurons involved in feature extraction would be to rather general and simple stimulus features. The frequency of occurrence of various levels of complexity of sensitivity should be inversely related to the extent of these complexities. Demonstration of the feasibility and usefulness of the analysis of the activity of population of neurons were shown for a variety of sensory systems.

ACKNOWLEDGMENT Supported in part by National Science Foundation research grant GB-33464X.

REFERENCES

ADRIAN, E. D., McK. CATTELL, and H. HOAGLAND, 1931. Sensory discharges in single cutaneous nerve fibers. *J. Physiol. (London)* 72:377–391.

ANDREW, B. L., and E. DOBT, 1953. The deployment of sensory nerve endings at the knee of the cat. *Acta Physiol. Scand.* 28:287–296.

BISHOP, P. O., 1970. Beginning of form vision and binocular depth discrimination in cortex. In *The Neurosciences, Second Study Program*, G. C. Quarton, T. Melnechuk and G. Adelman, eds. New York: Rockefeller University Press, pp. 471–485.

BOYD, I. A., and T. D. M. ROBERTS, 1953. Proprioceptive discharges from stretch receptors in the knee-joint of the cat. *J. Physiol.* 122:38–58.

BURGESS, R. P., and F. L. CLARK, 1969. Characteristics of knee-joint receptors in the cat. *J. Physiol.* 203:317–335.

COHEN, L. A., 1955. Activity of knee-joint proprioceptors recorded from posterior articular nerve. *Yale J. Biol. Med.* 28:225.

DAVIS, W. V., and D. KENNEDY, 1972. Command interneurons controlling swimmeret movements in lobster. II. Interaction of effect on motoneurons. *J. Neurophysiol.* 35:13–19.

DOETSCH, G., and R. P. ERICKSON, 1970. Synaptic processing of taste quality information in the nucleus tractus of the rat. *J. Neurophysiol.* 33:490–507.

ERICKSON, R. P., 1963. Sensory neural patterns and gustation. In *Olfaction and Taste*. Y. Zotterman, ed. New York: Pergamon Press, pp. 205–213.

ERICKSON, R. P., 1968. Stimulus coding in topographic and non-topographic afferent modalities: On the significance of the activity of individual sensory neurons. *Psychol. Rev.* 75: 447–465.

GABOR, DENNIS, 1972. Holography. *Science* 177(4046):299–313.

GANCHROW, J. R., and R. P. ERICKSON, 1970. Neural correlates of gustatory intensity and quality. *J. Neurophysiol.* 33:768–783.

HAHN, J. F., 1971. Stimulus-response relationships in first-order sensory fibers from cat vibrissae. *J. Physiol.* 213:215–226.

HARPER, H. W., J. R. JAY, and R. P. ERICKSON, 1966. Chemically evoked sensations from single human taste papillae. *Physiol. and Behav.* 1:319–325.

HARTLINE, H. K., 1940. The receptive field of the optic nerve fibers. *Am. J. Physiol.* 130:690–699.

HARTLINE, H. K., F. RATLIFF, and W. H. MILLER, 1961. Inhibitory interaction in the retina and its significance in vision. In *Nervous Inhibition*, E. Florey, ed. New York: Pergamon Press, pp. 241–284.

MARKS, W. B., 1965. Visual pigments of single goldfish cones. *J. Physiol.* 178:14–32.

MARKS, W. B., W. H. DOBELLE, and E. F. MACNICHOL, 1964. Visual pigments of single primate cones. *Science* 143:1181–1183.

MESAROVIC, M. D., and D. MACKO, 1969. Foundations for a scientific theory of hierarchical systems. In *Hierarchical Structures*, L. L. Whyte, A. G. Wilson, and D. Wilson, eds. New York: Elsevier Publishing Company, pp. 29–50.

MOUNTCASTLE, V. B., and T. P. S. POWELL, 1957. Central nervous mechanisms subserving position sense and kinesthesis. *Bull. Johns Hopkins Hosp.* 105:173–200.

MOUNTCASTLE, V. B., G. F. POGGIO, and G. WERNER, 1963. The relation of thalamic cell response to peripheral stimuli varied over an intensity continuum. *J. Neurophysiol.* 26:807.

MOUNTCASTLE, V. B., 1966. The neural replication of sensory events in the somatic afferent system. In *Brain and Conscious Experience*, J. C. Eccles, ed. Berlin: Springer-Verlag, p. 85.

MOUNTCASTLE, V. B., 1968. Physiology of sensory receptors: Introduction to sensory processes. In *Medical Physiology*, vol. II. St. Louis: The C. V. Mosby Co., 12th ed. pp. 1345–1371.

O'CONNELL, R. J., and M. M. MOZELL, 1969. Quantitative stimulation of frog olfactory receptors. *J. Neurophysiol.* 32:51–63.

PERKEL, D. H., and T. H. BULLOCK, 1969. Neural coding. In *Neurosciences Research Symposium Summaries*, vol. 3, F. O. Schmitt, T. Melnechuk, G. C. Quarton, and G. Adelman, eds. Cambridge, Mass.: MIT Press.

POGGIO, G. R., Y. LAMARRE, F. BAKER, and E. R. SANSEVERINO, 1969. Afferent inhibition at input to visual cortex of the cat. *J. Neurophysiol.* 32:892–915.

POULOS, D. A., and R. A. LENDE, 1970. Response of trigeminal ganglion neuron to thermal stimulation of oral-facial regions: 1. Steady-state response. *J. Neurophysiol.* 33:508–517.

ROCK, I., 1970. Perception from the standpoint of psychology. In *Perception and its Disorders*, Res. Publ. Assoc. Res. Nervous and Mental Disease, Vol. 48. Baltimore: Williams & Wilkins.

SEMMES, J., S. WEINSTEIN, L. GHENT, and H. L. TEUBER, 1960. *Somatosensory Changes after Penetrating Brain Wounds in Man.* Cambridge, Mass.: Harvard University Press.

SPINELLI, D. N., H. V. B. HIRSCH, R. W. PHELPS, and J. METZLER, 1972. Visual experiences as a determinant of the response characteristics of cortical receptive fields in cats. *Exp. Brain Res.* 15:289–304.

SPINELLI, D. N., K. H. PRIBRAM, and B. BRIDGEMAN, 1970. Visual receptive field organization of single units of the visual cortex of monkey. *Intern. J. Neurosc.* 1:67–74.

TOWER, S. S., 1940. Unit for sensory reception in cornea. *J. Neurophysiol.* 3:486–500.

TOWER, S. S., 1942. Pain: Definition and properties of the unit for sensory reception. Research Publications. *Assoc. for Research in Nervous and Mental Disease*, 23:16–43.

VASTOLA, E. F., 1968. Localization of visual functions in the mammalian brain: A review. *Brain, Behav., and Evol.* 1:420–471.

WERNER, G., 1968. The study of sensation in physiology: Psychophysical and neurophysiologic correlations. In *Medical Physiology*, vol. II. St. Louis: The C. V. Mosby Co., 12th ed. pp. 1643–1671.

WRIGHT, W. D., 1947. *Researches on Normal and Defective Color Vision.* St. Louis: The C. V. Mosby Co., pp. 167–172.

YARCZOWER, M., and M. E. BITTERMAN, 1965. Stimulus generalizations in the goldfish. In *Stimulus Generalizations*. D. Mostofsky, ed. Stanford University Press, pp. 179–192.

16 Neural Information Processing with Stimulus Feature Extractors

GERHARD WERNER

ABSTRACT The complex stimulus properties to which certain classes of neurons are sensitive fall into two broad categories: One, consisting of movement or temporal modulation of stimuli; the other, representing stationary spatial patterns. The comparison between stimulus feature-sensitive neural mechanisms in the visual and the somesthetic system suggests that the stimulus features reflecting movement and change of stimuli play a role in directing the receptor sheet to positions relative to the stimulus object, which enable other feature detectors to respond to their appropriate stationary patterns. This conception emphasizes the serial aspect of information acquisition in perception: Each sample of stationary sense impressions would in this view be the result of neural activity in many sensory channels in parallel, and these samples would be acquired by movement of the receptor sheet into successive positions under guidance by neurons that signal stimulus motion and change. There is some evidence that this form of conjoint activity of motion and pattern-sensitive feature detectors would generate successions of sense impressions in a form suited for the encoding of spatial representations of stimulus objects.

Introduction

THE STUDY OF perception, like that of any other field of inquiry, starts from some observational facts and seeks explanations for them. The facts in perception are the way things appear to the naive observer. Hence, they are subjective and personal. In this, the subject matter of perception is unique among the sciences, for subjective experience is not publicly observable as the facts of science are generally required to be (Rock, 1970b).

In spite of this special status of the subject matter, the research methodology in perception is entirely comparable to that of other sciences. Once the conditions for a perceptual experience (for instance, the occurrence of a particular illusion) are established on which several observers agree, planned perturbations of the stimulus pattern can be employed as experimental tools. Stimuli that are randomly patterned in space (e.g., the familiar Rorschach test) or in time are among the most prolific procedures for this purpose. Such stimuli fulfill in the study of perception the role of the communication engineer's test signals that

GERHARD WERNER Department of Pharmacology, University of Pittsburgh, School of Medicine, Pittsburgh, Pennsylvania

reveal any bias in the information handling proclivities of a network (MacKay, 1965a). Furthermore, such stimuli may to some extent also permit the characterization of the information flow in the nervous system that must underlie the perceptual experience, and such stimuli enable the distinction between processes at the receptor level and in the central nervous system. The ingenious random dot stereograms of Julesz (1971) are a notable example of this approach.

An alternative approach starts from the experimentally established data of neurophysiology and attempts to examine the role that demonstrated neural events may play in perception. In this, we will encounter the three levels of discourse that MacKay (1965b) termed the *mind-talk*, the *brain-talk*, and the *computer-talk*: The language of self-observation is in this view irreducibly mentalistic, whereas the language of mechanistic description is appropriate only to the observer of another brain. The language and the ideas of the theory of information and control provide in this situation the right kind of hybrid status to serve as a conceptual bridge "enabling data gleaned in physiological terms to bear upon hypotheses in psychological terms, and *vice versa*" (MacKay, 1965b).

Beyond merely providing an *interlingua* for expressing relations between psychological and physiological phenomena (MacKay, 1967), the theory of information processing also promoted in recent years the emergence of a new paradigm in psychology (Neisser, 1972); theoretical terms such as *storage*, *retrieval*, *recoding*, and others as well, are now frequently applied conceptual aids. In this framework, perceptual activity is understood to encompass all processes beginning with the translation of stimulus energy on receptors, and leading to reports of experiences or responses to that stimulation, as well as memory persisting beyond the termination of the stimulation. Hence, sensation, perception, memory, and cognition are viewed as a continuum in the processing of sensory information, as the latter acts upon and is transformed by the nervous system (Haber, 1969). Moreover, from the moment the concept of information was introduced, it became impossible to consider in the study of perceptual processes isolated stimuli apart from others that, although physically absent, belonged together with the one present in the same class (Broadbent, 1971).

These general comments are intended to set the stage and delineate a framework for examining the role of feature-extracting neurons in perception. The central question is whether the functional characteristics of feature-extracting neurons, as a neural subsystem, correspond with those required by the antecedent psychological theory of perceptual information processing; and what aspects of perceptual functions would correspond to the neural system of feature extractors (see Fodor, 1968).

Perceptual phenomena suggestive of feature extraction

Starting with an accidental observation in a BBC studio lined with slotted wall paper, MacKay (1957, 1961) was led to discover (and, in some instances, rediscover) a wide range of effects with the perception of regular, spatially repetitive patterns. As explanation, MacKay suggested that the direction of a contour is signaled in the nervous system in a manner such that the presence of many contours with the same direction results, as it were, in a *directional satiation*; concomitantly, the perception of contours in the complementary direction would prevail.

Subsequently, many other perceptual phenomena were described that seemed amenable to similar interpretation. For instance, Blakemore and Campbell (1969) and Campbell and Maffei (1970) discovered that prolonged inspection of a grid elevates the detection threshold for grids with the same and similar orientation and spatial frequency. This rise of perceptual threshold was also accompanied by a reduction of the stimulus evoked cortical potentials (see Campbell, earlier in this part). Similarly, specific adaptation for the direction in which a grating of low spatial frequency moves across the visual field can also occur (Pantle and Sekuler, 1969).

These observations, as well as the evidence for differential color adaptation of orientation specific edge detectors (McCollough, 1965; Held and Shattuck, 1971) in human visual perception, lend considerable plausibility to the idea that the proximal stimulus on the retina is, at some stage in the visual information processing, decomposed into certain parts (or features); that these features engage independent information transmitting elements; and that these elements can be equated with the stimulus feature-sensitive neurons which had been demonstrated in the visual cortex of the cat (Hubel and Wiesel, 1962, 1965; Nikara et al., 1968; Campbell et al., 1968), monkey (Hubel and Wiesel, 1968) and man (Marg et al., 1968).

An intriguing observation by Richards (1971) adds an element of drastic realism to this conjecture: The visual displays that arise during ophthalmic migraines can have the appearance of serrated arcs illustrated in Figure 1,

whereby each line segment in the perceived display would be the perceptual counterpart of the neural discharges in a cluster of cortical cells for which the same line orientation is the trigger stimulus.

Lest we commit the "fallacy of early success," some

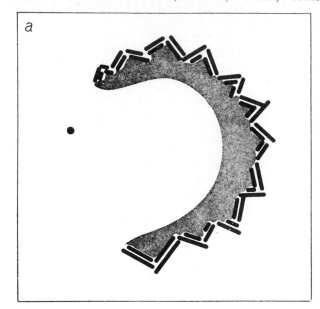

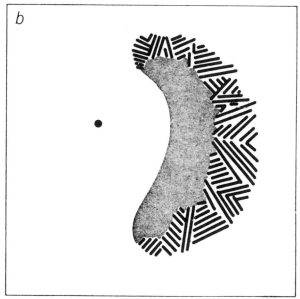

FIGURE 1   Fortification illusions of migraines, showing in (a) and (b) two appearances of their expanding boundaries. In the simplest type (a), the arc encircling a transient region of blindness (gray) is serrated and consists of two lines about as bright as an overhead fluorescent lamp; the lines oscillate in brightness at about 5 cycle/sec, with all the inside lines *on* when all the outside lines are *off*, and vice versa. In (b), a more complex appearance of a fortification illusion is shown. In each case, the gray area represents a transient region of blindness that moves outward from the point of fixation (black dot in figures), roughly parallel to the expanding arc. (From W. Richards, 1971.)

complicating issues need to be raised. Among these is the startling finding that colored aftereffects produced by parallel stripes or spirals may last for as long as six weeks (Stromeyer and Mansfield, 1970): It strains the neurophysiologist's imagination to assume that the activity of cortical feature-detecting neurons would, likewise, be affected throughout this period of time. Also, why does the adaptation to a certain orientation in a stimulus pattern produce as aftereffect a pattern with orthogonal orientation? Why did the adaptation not lead to a diffuse image, sparing only the adapted pattern, as is actually the case in audition? After exposure to auditory white noise from which a narrow frequency band is filtered out, the subjects report hearing a faint pitch at the center frequency of that filter (Zwicker, 1964). Thus, to account for the different adaptation response in the visual system, some special assumptions about interactions between feature-detecting neurons need to be introduced.

These unresolved issues barely detract from the elegance and conceptual simplicity of Sutherland's (1959, 1968) outline of a theory of visual pattern recognition. This theory anticipated and now rests, partly, on the well-known properties of feature-detecting neurons in the visual cortex (Hubel and Wiesel, 1962, 1965, 1968) that signal stimulus properties of different degrees of complexity. This complexity increases further with the projection from striate and prestriate cortex to the inferotemporal lobe (Gross et al., 1972). Two chief organizational principles seem to prevail: The first is the increase of receptive field size, leading with the inferotemporal neurons to the inclusion of the fovea into each neuron's receptive field; as a result, neurons signal the occurrence of their adequate stimulus with increasing independence from the actual site of stimulus impingement on the retina, as one progresses along the visual pathway. The second principle consists of increasing specificity of the adequate stimulus and, hence, progressive gain of information conveyed by a neural response (see Whitsel et al., 1972): The discharge of a neuron that responds to an edge in a particular orientation provided that edge ends at a certain point (hypercomplex cells of Hubel and Wiesel) is, informationally, much richer than is the discharge of a neuron that merely signals the presence of a light bar with a particular orientation (simple cells of Hubel and Wiesel). In somatic sensibility, a similar increase of complexity of stimulus features occurs: Some neurons in somatic sensory area 1 signal direction of stimulus motion over circumscript cutaneous receptive fields (Whitsel et al., 1972), whereas neurons in cytoarchitectural area 5 respond to more complex stimulus patterns, involving multiple joints and skin areas (Duffy and Burchfield, 1971).

Sutherland's outline of a theory also introduces the concept of an abstract description of the relation between the features signaled by these processing neural elements, which would take the form of an abstract symbolism, not unlike that of a formal language for processing and identifying pictorial structures by computers (Clowes, 1967, 1971); the relation of *join*, *next-to*, *inside*, *larger than*, etc., are examples of the presumed descriptive vocabulary.

According to this theory, when a shape is shone on the retina, it is analyzed into component parts by the feature-signaling neurons that function as the processor elements; memorizing or recognition of that shape requires then a recoding of the processor output according to some rules for description of pictures. This description is thought to be of sufficient generality that size, brightness, and position invariance of the perceived stimuli can be attained. The detailed logic of this process is expected to be so complex that a refinement of the theory, notably with the help of computer simulation, would have to precede any attempt to study its implementation in the nervous system.

The two components of Sutherland's proposal place an essential difference between available neurophysiological data and perceptual performance into clear focus: The former consist at present of descriptions of *local* properties of neural elements, or small groups of such elements, but the latter deals with the *global* aspects of perception. This difference prompts three questions that we will address in sequence: First, is the signaling of stimulus features by neurons as invariant and as independent of global stimulus properties as is often tacitly assumed? Second, what can be learned about the contribution of feature-detecting neurons to the global mechanisms of perception from the rules governing stimulus equivalence in transposition experiments? And, third, what place could feature-extracting neurons occupy in the acquisition of stimulus information and in perceptual learning?

Short-term changes of neural stimulus features

The observations of Horn and Hill (1969), Spinelli (1970), and Denney and Adarjani (1972) put the notion of immutability of visual stimulus feature detectors in doubt: In some neurons of the cat's visual cortex, the axis of an elongated receptive field appears to tilt concomitantly with head tilt, suggesting that the receptive field can maintain within limits its orientation in space, or at least undergo some less specific alteration with changes in excitation of vestibular statoreceptors (Schwartzkroin, 1972).

The observation by Hubel et al. (1959) that a class of neurons in the auditory cortex responds to acoustical stimuli only if the subject "pays attention" introduced the notion that the behavioral context in which a certain stimulus is applied determines its adequacy to elicit a neuronal response. The apparent significance of this modulation of neural responses by behavioral context is underscored by the finding of Miller et al. (1972) that the intensity of a neural response in macaques trained in an auditory discrimination task depends on whether or not the subjects are actually required to perform the learned response to the sensory cue. There is now also some accumulating evidence that the state of attention and vigilance can affect the sensitivity of neurons to stimulus features: Roppolo et al. (1972) have shown that neurons, which signal in the alert state the direction in which a cutaneous stimulus moves across the receptive field, may lose this directional specificity during periods of light sleep or drowsiness. However, it is not yet established whether more subtle behavioral contingencies, like the importance of a certain stimulus feature as cue in a behavioral task, can affect the specificity of the neural response to that same stimulus feature.

Morrell (1967) achieved a different type of modification of neural responses in studies that involved the repeated pairing of a trigger stimulus with a previously ineffective stimulus; eventually the latter stimulus becomes also effective in eliciting a neural response. Moreover, clusters of neurons were identified in parastriate cortex that responded to visual *and* auditory stimuli, provided the sound source was located in the neuron's receptive field for visual stimuli (Morrell, 1972). One might argue that such neurons are specific for events occurring at a certain locus in space and disregard the modality of the stimulus.

There is, finally, one more ambiguity in the signaling capability of feature-sensitive neurons: In the experimental studies, such neurons are conventionally classified according to their *best stimulus*, which is that stimulus configuration which elicits the maximal response; a less than maximal·response merely indicates departure from the optimal stimulus but not the direction of departure, nor does it indicate which component of the stimulus configuration was altered. Moreover, at least in one case, namely the class of neurons in primate somatic sensory area I, which signal directions of stimulus motion on the skin, there appears an entire spectrum of *tuning*. These neurons differ widely in the range of orientations along which their responses differ with stimulus movement in opposing directions (Figure 2).

At least to some extent, these considerations tend to obscure the neatness of operational segregation into a class of stable processing elements on the one hand, and the hypothetical generator of an abstract description of the stimulus content; rather, it appears that the processing elements themselves would already be capable of performing some of the tasks required for insuring a stable perceptual world and for taking into account the context in which the stimulus occurs. In that sense, activity in *individual* stimulus feature detecting neurons hardly meets the requirements for a neural code of stimulus properties, although a *population* of such neurons may do so, albeit in a more complex manner that escapes experimental analysis with current methods (Perkel and Bullock, 1968).

Lessons from stimulus transposition

An interesting conceptualization of certain forms of perceptual constancies was proposed by Hoffman (1970); it rests on the generalization that neurons comprising certain cortical projection areas are not only arranged to topological maps of their respective receptor sheets but, in addition, are also sensitive and, therefore, represent in some sense particular local orientations and movement directions in the stimulus field. Hence, these cortical projection patterns may be viewed as vector fields, each stimulus orientation and direction singling out a set of neurons for which the particular orientation or direction is a "feature." Accordingly, stimulus contours and textures would be represented as appropriate alignments of orientation indicating field elements. Hoffman (1970) then proposed that certain basic invariances of perception, such as size or shape constancy, can be interpreted in terms of the operation of continuous transformation groups on the cortical representation of the field of view.

FIGURE 2   The contribution of a selected region of the receptive field to the cortical neuronal response as a function of the direction of stimulus motion. To generate each of the polar plots of this figure, the moving stimuli were applied at a variety of intersecting orientations within the receptive field, as is indicated in the accompanying figurine. The response profiles obtained of each stimulus orientation were arranged to permit identification of those *bins* that correspond to the point of chord intersection in the receptive field. The mean discharge rate per stimulus (expressed in impulse/sec) in these bins reflects the contribution of that portion of the receptive field to the neuronal response as the direction of brush advance is varied. Each direction of stimulus motion (identified by the direction of the arrow heads on the figurines) is assigned a number that corresponds to one of the heavy points on the circumference of the polar plots. The distance of these points from the origin of the coordinate system measures the neuronal response (in impulse/sec) generated by traversing the receptive field center in the direction identified by the appropriate number. Each stimulus was replicated 25 times. The circle around the origin of the coordinate system represents the level of spontaneous activity. The length of the interrupted radial line in each plot represents a calibration value of 50 impulse/sec/stimulus. (A) A "symmetrical" S-I neuron for the magnitude of neuronal response is independent of stimulus direction. (B, C, and D) "Asymmetric" S-I neurons that displayed direction selectivity at certain chord orientations. The stimulus velocities employed were 63 mm/sec (neuron A); 56 mm/sec (neuron B); 39 mm/sec (neuron C); and 236 mm/sec (neuron D). (From Whitsel et al., 1972, with the permission of the American Physiological Society.)

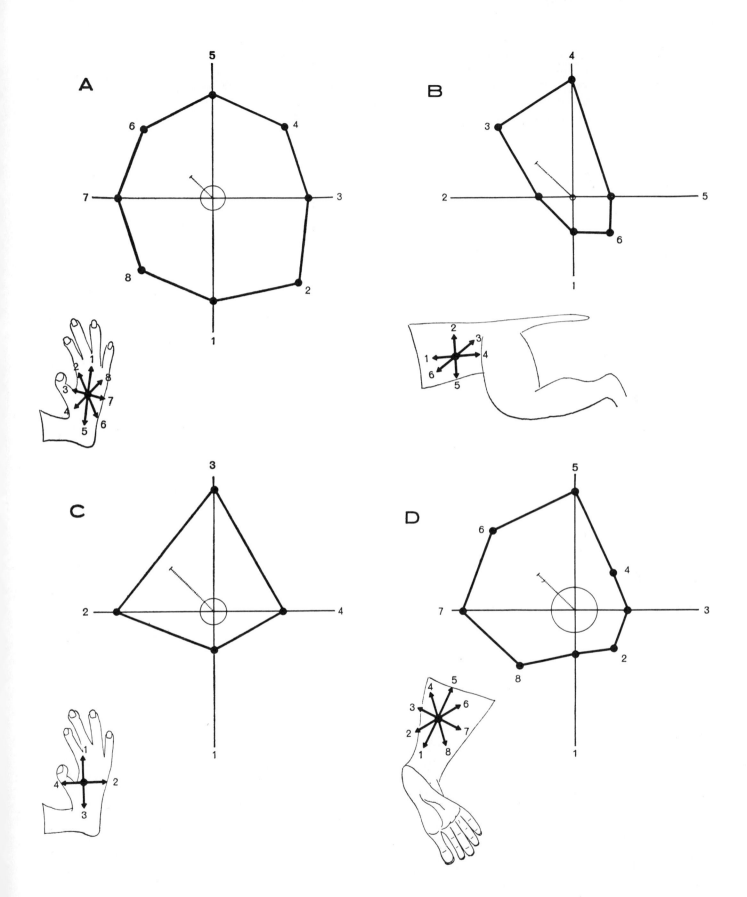

There are, however, limitations to the general applicability of this concept or, at least, situations in which additional factors come to play a role, for some stimulus transformations do not yield perceptual constancies, in spite of the fact that these stimulus transformations themselves possess group structure by virtue of the fact that each stimulus transformation has an inverse and that consecutive execution of transformations is again a transformation. One example is of stimulus discrimination when patterns are rotated. Recent work by Carlson (1972) with macaques, shown in Figure 3 illustrates that orientation plays a part in the definition of what we mean by the form or shape of an object (Hake, 1966).

This along with similar evidence from studies with subhuman primates suggests that the comparison of discriminanda transposed by rotation or reflection (mirror-image symmetry) is primarily based on some localized part of the figure; the site of the shape most significantly involved in the judgment of similarity or difference of the transposed shapes is, most often, the top or the bottom of the figures (Riopelle et al., 1969). This might also be related to the numerous observations with children, some of which date back to Ernst Mach (1886), that some kinds of rotations and reversals have more effect on transposition than others (for review, see Reese, 1968).

In short, patterns *can* be recognized despite rotation; but this accomplishment depends on rather complex mechanisms. The perceiver seems to construct within the figure an axis that defines some part as the top, and another as the bottom. Only then is he able to perform the comparison with another pattern previously presented. This circumstance precludes any simple account of the role of feature-sensitive neurons in pattern recognition and identification (see Neisser, 1966; Rock, 1970a). Moreover, Blakemore and Campbell (1969) have emphasized the difficulties that a feature-analyzing mechanism for spatial frequency—while accomplishing generalization for size most effectively—would encounter if it also had to generalize orientation.

Active processes in perception

There are two broad senses in which an organism has been said to be an *active perceiver* and perception to be an active process (Gyr et al., 1966). In the first sense, the perceiver's contribution consists of exploratory search of stimulation and variation in sensory input; activity in this sense involves movements of parts of the body (including the eyes) that lead to variations in the position of receptors relative to the stimuli. Thus, interaction between sensory and motor events becomes part of the information acquisition in perception. In the second sense, active perception stresses transactions between the observer and

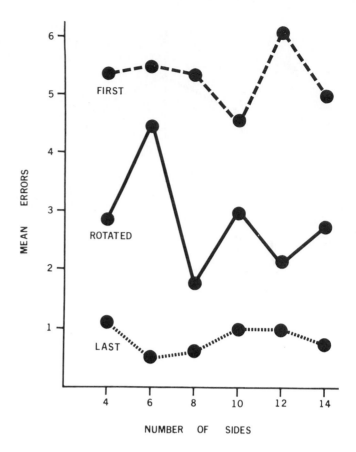

FIGURE 3 Monkeys were taught successive visual discriminations between pairs of random figures, both figures in a pair having the same number of sides. After criterion performance was achieved on each problem, the figures were rotated 180° and an additional 20 trials were given. The figure shows the mean errors as a function of the number of sides, on the first 20 trials of a problem (FIRST), the last 20 trials (LAST), and the subsequent 20 trials with the rotated figures (ROTATED). Errors after rotation were intermediate between and significantly different from errors on both the first and last 20 trials. This suggests that rotation only partially disrupts discrimination performance; subjects neither treated the rotated figures as comprising an entirely new problem, nor were they unaffected by the change in orientation. (From Carlson, 1972, with the permission of the Psychonomic Society.)

the environment to which the former may contribute a judgment-like activity (Helmholtz, 1867; Gregory, 1969), or memories from past experiences (Postman, 1955), or some form of "trial-and-check" process that leads to the structuring of percepts (Solley and Murphy, 1960; Miller et al., 1960).

Clearly, these two aspects of active perception are in no way mutually exclusive, for its appears plausible that past experiences can provide the strategies for employing sensory-motor mechanisms in the service of acquisition of stimulus information. To sharpen the issue, we should contrast this with sensory inputs simply impinging as a result of accidental orientation of sense organs, or with inputs provided by some outside source, such as the experimenter.

What role can we attribute to stimulus feature signaling neurons in this complex information transaction between an observer, who has become familiar with certain redundancies and regularities in his environment, and changing patterns of stimulus energy?

Vision and movement

Broadly speaking, the past decade has led to the emergence of the principle of *separability of pattern and place* (Horridge, 1968) in the ways in which visual information is processed in the nervous system. One line of evidence for this comes from complete lesions of the striate cortex in monkeys. Such lesions produce deficits in object recognition and pattern discrimination but leave detection and localization of stationary and moving visual stimuli, as well as visually guided movements in relation to those stimuli, intact (Humphrey, 1970). This residual capacity implies at least some form of figure-ground differentiation within the sensory field; yet, this field remains perceptually unstructured, because forms are not recognized. Hence, Humphrey's (1970) imaginative proposal that spatial information comes in this case indirectly: Eye movements would be guided by the sensory field, and there would be a tendency automatically to fixate visually differentiated figures; the cortically deprived monkey, thus, knows of the location of an object by monitoring, internally, where her eyes are looking. In effect, the monkey has learned to "use her hands as an extension of her eyes; in a real sense she sees *with*, not through, the eye" (Humphrey, 1970).

The second source of evidence is that each hemisphere can function independently of the other in pattern identification and discrimination when the commissural connections between the two cortices are cut; thus, a stimulus identification learned through one eye can no longer be performed when the same stimulus is presented to the other eye, as is the case with intact commissural connection (Trevarthen, 1967).

The conclusions from these and similar studies point to an anatomical differentiation between two neural systems with different competence for processing visual information: One, involved in locating and orienting toward targets in the visual field, which is not affected by severance of hemispheric connections and, thus, composed of retinomesencephalic projections; the other involving retinotelencephalic projections capable of processing visual information of patterns and shapes (Schneider, 1967, 1969).

Does stimulus feature detection in the retinostriate and in the mesencephalic system differ in a manner that could account for their different role in visual perception?

The cellular stratification of the superior colliculi is suggestive of a hierarchy of neural functions within them: In the macaque, the large majority of neurons in the upper layers respond to small stationary spots of light though their receptive fields are large, and they are strongly excited by movement of a light spot in any direction within the receptive field; but they are not affected by shape, size, or orientation of a stimulus. In addition, there is emphasis on novel stimuli, because responses decline with repeated stimulus application (Humphrey 1968). Thus, local change is the *one* aspect about a stimulus that these neurons signal. A more specific operation is carried out by a minority of neurons in this layer because they respond only to certain directions of stimulus motion and not to others (Wurtz and Goldberg, 1972).

Finally, there are neurons that discharge with increased vigor whenever a visual stimulus becomes the target of an eye movement: Some are related to saccadic eye movements, and others, notably in the intermediate gray layers, fire just prior to an eye movement made to fixate a particular area of the visual field. Each of these latter neurons has a certain direction of eye movement associated with its most intense discharge, which is independent of the absolute position of the eye, either prior to or at the end point of the movement. Some of these neurons also have visual receptive fields that, before movement, correspond roughly to the spatial position to which the eye is directed by the neuron's discharge (Wurtz and Goldberg, 1972).

In its entirety, this neuronal assembly consistently filters change and novelty in the visual field as "parcels of information" (Lettvin and Maturana; see Ploog, 1971), which lead to the issuing of specific commands. Whether these commands are related to the mechanism of foveation (i.e., target acquisition by saccades and target maintenance by smooth pursuit) (Schiller and Koerner, 1971), or whether they signal to other regions of the brain that an eye movement is impending (i.e., a corollary discharge; Sperry, 1950; Teuber, 1960) is still uncertain. In any case, however, there is a clear difference between the nature of the events in the visual field that affect the neurons of the superior colliculi and the stimuli for which a substantial contingent of neurons in the striate cortex is *tuned*. The emphasis of these latter neurons is on the spatial layout of contours, present in *stationary* patterns in the visual field (Wurtz, 1969; Noda et al., 1971).

Although so clearly distinct in neuronal response characteristics, there is strong support for the concept that cortex and midbrain (i.e., superior colliculi and pretectum) jointly contribute to the learning of visual pattern discrimination tasks. Anatomical evidence establishes the existence of ascending tectal pathways, supplied mostly from the peripheral retina (Wilson and Toyne, 1970), to certain parts of visual cortex via posterior thalamic nuclei (Diamond and Hall, 1969; Graybiel, Part 3 of this volume); and the behavioral studies of Berlucchi et al. (1972) demonstrated the essential role of superior colliculi and pretectum in the acquisition of visual discrimination performance. It appears that visual cortical areas (notably circumstriate cortex) have come into possession of the best of two worlds: The stimulus motion detection in the peripheral retina as source of eye movement commands, and the stationary contour coding in the foveal component of the geniculostriate system. Does this juxtaposition of extrafoveal, eye-movement related and foveal, pattern sensitive cortex in the circumstriate belt contain a clue for its role in pattern vision?

Familiar visual scenes can be recognized if presented in a brief flash, too short to enable scanning by eye movements. However, when eye movements are permitted to occur, and when they are required for the perception of a novel stimulus pattern, they externalize, as it were, the information processing in the visual system (Yarbus, 1967; Jeannerod et al., 1968). An extreme position in this line of thought is taken in the *scan path theory* of Noton (1970; see also, Noton and Stark, 1971), which proposes that the central representation of a visual pattern consists of a sequence of sensory and motor memory traces, recording individual features of the pattern and the eye movements required to pass from feature to feature across the pattern. More generally, and in keeping with the ability for tachistoscopic perception of familiar visual scenes, is the alternative suggestion that entire samples of visual scenes would be acquired at each eye position. Each individual sample would in this view be the result of information processing in many visual channels in parallel, and these samples would be acquired by successive eye fixations.

The question is, could the visual cortex (notably the circumstriate area) implement the acquisition of sequencies of "snapshots" of the visual scene by bringing one image sample after the other into foveal vision? It could accomplish this in some sense automatically, at least on first presentation of a novel pattern when no prior strategies for inspection are available from past experience. As an eye movement brings a new visual scene into foveal "focus," other components of the visual field would activate neurons in the peripheral retina; their activity would generate in the colliculo-pulvinar-extrastriate system the "program" for the next eye movement, and so on. Between eye movements, the geniculostriate projection system could apply its "static" feature analyzing operation *seriatim* to one sample from the visual scene after the other.

In some sense, we have made a full circle: While "separability of pattern and place" was suggested by the differences in the central routing of feature signals for structural (i.e., purely spatial) and for time-space dependent (i.e., motion) aspects of the stimulus, we have come to see ways in which the neural responses to these two classes of visual stimuli can jointly constitute a *description* of a stimulus pattern. This description has some formal analogy to the requirements that were found to suffice for mimicking certain forms of perceptual and cognitive functions in artificial intelligence studies (Simon and Newell, 1971). The analogy consists of (a) serial information processing and (b) the individual data items being "linked" to lists (in an abstract sense) in which the sequential eye movement commands are the "pointers" marking the order in which the list is structured.

The tactile apprehension of object quality and shape

Identification of objects by palpation is based on the concurrent sensory influx from cutaneous touch and articular position and motion. Palpation encompasses at least two distinct activities: One, consisting of the shaping of the hand for grasping the object, or parts of it; the other consisting of displacing the hand by successive movements along the object's contours and surfaces.

As one would expect, the manipulation plays a major role as the observer seems to be trying to obtain mechanical events at the skin in various combinations and at various places. Gibson (1962) quantitated the superiority of "active" over "passive" touch: Shapes were to be identified when they were either pressed into the palm of the hand or accessible to exploration by the finger tips. The result was highly impressive, for in passive touch, correct matches were only half as often attained as in active touch.

Neurophysiological and neuroanatomical investigation characterized in recent years the neural system that evolution specialized for active touch. The evidence comes from the currently ongoing reassessment of the role of the dorsal column, medial lemniscal pathway and its cortical projection in somesthesis. It has been known for some time that lesions of somatic sensory area I (S-I) in the monkey impair shape discrimination by palpatory exploration of the stimulus object (Orbach and Chow, 1959); yet, tactile thresholds are not altered (Schwartzman and Semmes, 1971). Furthermore, lesions of area 5 in the parietal lobe that receives a heavy projection from S-I (Jones and Powell, 1969) also affect active tactual discrimination selectively (Ettlinger et al., 1966). In contrast, discrimination of passively received cutaneous stimuli involves primarily the second somatic sensory area (Glassman, 1970; Schwartzman and Semmes, 1971).

As regards the dorsal columns themselves, evidence has accumulated that they are indispensable for the execution of tasks in which accurate timing and sequencing of limb projections into extracorporeal space are important aspects (Dubrovsky et al., 1971; Melzack and Bridges, 1971). Earlier, Gilman and Denny-Brown (1966) concluded that the dorsal columns constitute an essential part of the neural mechanisms for projected limb movement into ambient space and, in particular, for the fine contactual orienting reactions of forelimbs. In contrast, there are numerous studies of tactile discrimination tasks with passively applied cutaneous stimuli that do not attribute more than, at best, a secondary role to the dorsal columns (for review, see Semmes, 1972).

Therefore, Semmes (1969) and Wall (1970) can muster strong support for their contention that sensory information in the dorsal column pathway and its principal cortical receiving area in S-I is available in a form that is primarily suited for guiding object manipulation and exploration by the limbs. In this conception, the dorsal columns function essentially as an alerting system, setting into motion mechanisms for analyzing sensory information arriving through parallel channels and directing motor search for additional data. The cutaneous afferents from the *distal* limbs receive in this a preferential treatment, for they reach their destination in S-I almost exclusively *via* the dorsal columns, while afferents from more proximal portions of the limbs and from the trunk ascend also in the dorsolateral tract of the spinal cord: Thus, the cortical projection fields of the distal portions of hand and feet appear, essentially, as islands of dorsal column projection within S-I (Dreyer et al., 1973).

Is this an analogy to the juxtaposition of foveal and extrafoveal representations in the visual cortex, separating also in the somesthetic system movement detection from pattern analysis? The answer is incomplete, but as far as it is available, it supports at least the expectation of movement signaling neurons in somesthetic cortex: It has recently been possible to identify in S-I a class of neurons that single out certain orientations of chords in their cutaneous receptive fields for which responses to stimuli moving in opposite directions differ substantially. Direction and orientation of stimulus motion on the skin assume, thereby, the character of a stimulus feature (Whitsel et al., 1972). Figure 4 illustrates that the orientational and directional specificity of these neurons increases, as a rule, at higher stimulus velocities and

reaches an optimum in the range of 100–200 mm/sec. This is also the velocity range for the best texture identification of objects moving over the skin of humans (Katz, 1925).

Adherence to the prototype of the visual system would suggest that these directionally selective detectors for stimulus movement on the skin surface play a role for triggering manual grasping automatisms. In both cases a moving stimulus would direct a receptor sheet to "capture" the stimulus. However, much of somesthesis, being a phylogenetic latecomer, introduces an intriguing difference. As far as is known, there are no mesencephalic centers for dynamic stimulus features in somesthesis, which would be comparable to tectum and superior colliculi in vision, or inferior colliculi in audition. Instead, movement related stimulus features appear to any appreciable extent for the first time as far rostral in the projection path as S-I, exclusively as a result of intracortical information processing (Whitsel et al., 1972). It may be safe to predict that movement related features occur in greater diversity and specificity in cortical areas beyond S-I.

Of the various forms of grasping automatisms (Seyffarth and Denny-Brown, 1948), the *instinctive grasp reaction* appears as a promising candidate upon which voluntary prehension and palpatory exploration of the objects of touch could build (see Twitchell, 1965): The moving contact stimulus to any part of the hand elicits a highly integrated response that places the hand in a position of readiness for a series of light palpating and groping movements and is eventually followed by a final grasping of the object. The conjecture is that such automatisms are the building blocks of haptic sensibility.

The unity of the phenomenal object in perception

The preceding sections assembled some of the arguments in favor of the general concept that different classes of neurons "attend" to different aspects of a stimulus. Some of the stimulus features reflect movement and change of the stimulus and appear to take part in directing the receptor sheet to positions relative to the stimulus, which enable other feature detectors to register glimpses of stationary patterns. This conception emphasizes the *serial* aspect of information acquisition in perception.

In an imaginative experiment that consisted of moving objects or patterns behind a hole whose diameter was such that at most one corner could be seen at the time, Hochberg (1968) proved at least for vision that the serial presentation of sequential views of a shape was *sufficient* for its identification. However, this experiment also underscores the complexity of the further processing of

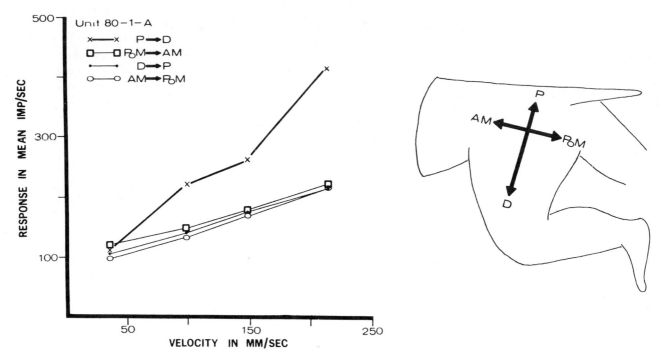

FIGURE 4 Contribution of a selected region of the receptive field to cortical neuronal response as a function of velocity and direction of stimulus motion. To obtain these plots, the response profiles resulting from 25 replications of each of the four directions of stimulus motion were arranged to permit identification of those bins that correspond to the point of chord intersection in the receptive field. The mean discharge rate per stimulus (expressed in impulses per second) in these bins is plotted against velocity. The velocity dependence of the response at the point of chord intersection varies dramatically with the direction of stimulus motion; i.e., a preferred direction of the stimulus motion became increasingly predominant (namely, the P → D direction) at the higher stimulus velocities. AM, anterior midline; PoM, posterior midline; P, proximal; D, distal. (From Whitsel et al., 1972, with the permission of the American Physiological Society.)

the serially obtained views that must be assumed. There must be perceptual memory for retaining the sequential views in some short-term memory and some recoding to larger "perceptual units" (see Neisser, 1966). These later stages of information processing are accessible to identification and characterization by perceptual masking and figural aftereffects (Haber, 1969).

Upon entering these stages of information processing, differences between sense modalities play a rapidly diminishing role (Massaro, 1972), except for a clear distinction in the handling of spatial as opposed to verbal information (Brooks, 1968). This enables the trading of spatial information acquired by different sense modalities

in, at least, two forms: First, sense information acquired in one modality can become organized in the spatial schema of another modality; an example of this is the preference for representing spatial location and shapes in visual terms, even when based on tactile input (Scholtz, 1957; Attneave and Benson, 1969). Second, man and some subhuman primates (Davenport and Rogers, 1970) become capable of crossmodal comparisons of shapes. But, although freed from constraints by sense modalities, this supramodal mechanism is still sensitive to the manner of stimulus presentation: It only operates effectively if the patterns are presented in spatiotemporal succession, for instance by describing them with a stylus as paths in space and time (Krauthamer, 1968). Hence, this perceptual coding mechanism requires for its optimal effectiveness that stimulus patterns be presented as successions of sense impressions.

This circumstance puts the function of the motion and pattern sensitive neural elements and pathways into a new light: For, according to the conception presented earlier, their conjoint activity would generate sense information in just the form that matches the information handling proclivity of the coding system for spatial representation.

REFERENCES

ATTNEAVE, F., and B. BENSON, 1969. Spatial coding of tactual stimulation. *J. Expl. Psychol.* 81:216–222.

BERLUCCHI, G., J. M. SPRAGUE, J. LEVY, and A. C. DiBERARDINO, 1972. Protection and superior colliculus in visually guided behavior and in flux and form discrimination in the cat. *J. Comp. Physiol. Psychol.*, Monogr. 75, No. 1, p. 123–172.

BLAKEMORE, C., and F. W. CAMPBELL, 1969. On the existence of neurons in the human visual system selectively sensitive to the orientation and size of retinal images. *J. Physiol.* (*London*) 203:237–260.

BROADBENT, D. E., 1971. *Decision and Stress.* New York: Academic Press.

BROOKS, L. R., 1968. Spatial and verbal components of the act of recall. *Canad. J. Psychol./Rev. Canad. Psychol.* 22:349–368.

CAMPBELL, F. W., B. G. CLELAND, G. F. COOPER, and C. ENROTH-CUGELL, 1968. The angular selectivity of visual cortical cells to moving gratings. *J. Physiol.* (*London*) 198:237–250.

CAMPBELL, F. W., and L. MAFFEI, 1970. Electrophysiological evidence for the existence of orientation and size detectors in the human visual system. *J. Physiol.* (*London*) 207:635–652.

CARLSON, K. R., 1973. Visual discrimination of random figures by rhesus monkeys. *Anim. Learn. Behav.* 1:23–35.

CLOWES, M. B., 1967. Perception, picture processing and computers. In *Machine Intelligence*, 1:181–197. N. L. Collins and D. Michie, eds. New York: Elsevier Publishing Company.

CLOWES, M. B., 1971. On seeing things. *Artif. Intell.* 2:79–116.

DAVENPORT, R. K., and C. M. ROGERS, 1970. Intermodal equivalence of stimuli in apes. *Science* 168:279–280.

DENNEY, D., and C. ADARJANI, 1972. Orientation specificity of visual cortical neurons after head tilt. *Exp. Brain Res.* 14:312–317.

DIAMOND, I. T., and W. C. HALL, 1969. Evolution of neocortex. *Science* 164:251–262.

DREYER, D. A., R. SCHNEIDER, C. METZ, and B. L. WHITSEL, 1973. Differential contributions of spinal pathways to the body representation in the postcentral gyrus. *J. Neurophysiol.* (in press).

DUBROVSKY, B., E. DAVELAAR, and E. GARCIA-RILL, 1971. The role of dorsal columns in serial order acts. *Expl. Neurol.* 33:93–102.

DUFFY, F. H., and J. L. BURCHFIELD, 1971. Somatosensory system-organizational hierarchy from single units in monkey area 5. *Science* 172:273–275.

ETTLINGER, G., H. B. MORTON, and A. MOFFETT, 1966. Tactile discrimination performance in the monkey: The effect of bilateral posterior parietal and lateral frontal ablations, and of callosal section. *Cortex* 2:6–29.

FODOR, J. A., 1968. *Psychological Explanation: An introduction to the Philosophy of Psychology.* New York: Random House.

GIBSON, J. J., 1962. Observations on active touch. *Psychol. Rev.* 69:477–491.

GILMAN, S., and D. DENNY-BROWN, 1966. Disorders of movement and behavior following dorsal column lesions. *Brain* 89:397–418.

GLASSMAN, R., 1970. Cutaneous discrimination and motor control following somatosensory cortical ablations. *Physiol. Behav.* 5:1009–1019.

GREGORY, R. L., 1969. On how so little information controls so much behavior. In *Towards a Theoretical Biology*, (an IBUS Symposium) Vol. 2, C. H. Waddington, ed. Chicago: Aldine Publishing Company, pp. 236–247.

GROSS, C. G., C. E. ROCHA-MIRANDA, and D. B. BENDER, 1972. Visual properties of neurons in inferotemporal cortex of macaque. *J. Neurophysiol.* 35:96–111.

GYR, J. W., J. S. BROWN, R. WILLEY, and A. ZIVIAN, 1966. Computer simulation and psychological theories of perception. *Psychol. Bull.* 65:174–192.

HABER, R. N., 1969. Introduction. In *Information-Processing Approaches to Visual Perception*. New York: Holt, Rinehart and Winston.

HAKE, H. W., 1966. Form discrimination and the invariance of form. In *Pattern Recognition*, L. Uhr, ed. New York: John Wiley & Sons, pp. 142–173.

HELD, R., and S. R. SHATTUCK, 1971. Color- and edge-sensitive channels in the human visual system: Tuning for orientation. *Science* 174:317–316.

HELMHOLTZ, H., 1867. *Handbuch der Physiologischen Optik.* Leipzig: Voss.

HIRSCH, H. V. B., 1972. Visual perception in cats after environmental surgery. *Exp. Brain Res.* 15:405–423.

HOCHBERG, JULIAN, 1968. In the mind's eye. In *Contemporary Theory and Research in Visual Perception*, R. N. Haber, ed. New York: Holt, Rinehart and Winston, pp. 309–331.

HOFFMAN, W. C., 1970. Higher visual perception as prolongation of the basic lie transformation group. *Math. Biosci.* 6:437–471.

HORN, J., and R. M. HILL, 1969. Modification of receptive fields of cells in the visual cortex occurring spontaneously and associated with bodily tilt. *Nature* (*London*) 221:186–188.

HORRIDGE, G. A., 1968. *Interneurons.* San Francisco: W. H. Freeman and Company.

HUBEL, D. H., C. O. HENSON, A. RUPPERT, and R. GALAMBOS, 1959. Attention units in the auditory cortex. *Science* 129:1279–1280.

HUBEL, D. H., and T. N. WIESEL, 1962. Receptive fields, binocular interaction and functional architecture in the cat's visual cortex. *J. Physiol.* (*London*) 160:106–154.

HUBEL, H., and T. N. WIESEL, 1965. Receptive fields and functional architecture in two non-striate visual areas (18 and 19) of the cat. *J. Neurophysiol.* 28:229–289.

HUBEL, D. H., and T. N. WIESEL, 1968. Receptive fields and functional architecture of monkey striate cortex. *J. Physiol.* (*London*) 195:215–243.

HUMPHREY, N. K., 1968. Responses to visual stimuli of units in the superior colliculus of rats and monkeys. *Exp. Neurol.* 20:312–340.

HUMPHREY, N. K., 1970. What the frog's eye tells the monkey's brain. *Brain, Behav., Evolut.* 3:324–337.

JEANNEROD, M., P. GERIN, and J. PERNIER, 1968. Deplacements et fixation du regard dans l'exploration libre d'une scène visuelle. *Vision Res.* 8:81–97.

JONES, E. G., and T. P. S. POWELL, 1969. Connexions of the somatic sensory cortex of the rhesus monkey. I. Ipsilateral cortical connexions. *Brain* 92:477–502.

JULESZ, B., 1971. *Foundations of Cyclopean Perception.* University of Chicago Press.

KATZ, D., 1925. Der Anfbau der Tast welt. Leipzig: Barth.

KRAUTHAMER, G., 1968. Form perception across sensory modalities. *Neuropsychologia* 6:105–113.

McCOLLOUGH, C., 1965. Color adaptation of edge-detectors in the human visual system. *Science* 149:1115–1116.

MACH, E., 1907. *The Analysis of Sensations and the Relation of the Psychical.* (Translated by C. M. Williams from the 1st ed. of *Die Analyse der Empfindungen und das Verhaltnis des Psychischen zum Physischen,* 1914). La Salle, Ill.: Open Court Publishing Company.

MACKAY, D. M., 1957. Moving visual images produced by regular spatial patterns. *Nature (London)* 180:849–850.

MACKAY, D. M., 1961. Interactive processes in visual perception. In *Sensory Communication,* W. A. Rosenblith, ed. Cambridge, Mass.: The MIT Press and John Wiley & Sons, pp. 339–355.

MACKAY, D. M., 1965a. Visual noise as a tool of research. *J. Gen. Psychol.* 72:181–197.

MACKAY, D. M., 1965b. A mind's eye view of the brain. In *Cybernetics of the Nervous System,* Progress in Brain Research, Vol. 17. N. Wiener and J. P. Schade, eds. New York: Elsevier Publishing Company, pp. 321–332.

MACKAY, D. M., 1967. Report on the symposium on neural communication. *IBRO Bull.* 6:5–18.

MARG, E., J. E. ADAMS, and B. RUTKIN, 1968. Receptive fields of cells in the human visual cortex. *Experientia* 24:348–350.

MASSARO, D. W., 1972. Preperceptual images, processing time, and perceptual units in auditory perception. *Psychol. Rev.* 79:124–145.

MELZACK, R., and J. A. BRIDGES, 1971. Dorsal column contribution to motor behavior. *Exp. Neurol.* 33:53–68.

MILLER, G. A., E. GALANTER, and K. H. PRIBRAM, 1960. *Plans and the Structure of Behavior.* New York: Holt, Rinehart and Winston.

MILLER, J. M., D. SUTTON, B. PFINGST, A. RYAN, and R. BEATON, 1972. Single cell activity in the auditory cortex of Rhesus monkeys: Behavioral dependency. *Science* 177:449–451.

MORRELL, F., 1967. Electrical signs of sensory coding. In *The Neurosciences,* a Study Program. G. C. Quarton, T. Melnechuk, and F. O. Schmitt, eds. New York: The Rockefeller University Press.

MORRELL, F., 1972. Integrative properties of parastriate neurons. In *Brain and Human Behavior,* A. G. Karczmar, and J. C. Eccles, eds. New York: Springer-Verlag, pp. 259–289.

NEISSER, U., 1966. *Cognitive Psychology.* New York: Appleton-Century-Crofts.

NEISSER, U., 1972. A paradigm shift in psychology (Book review). *Science* 176:628–630.

NIKARA, T., P. O. BISHOP, and J. D. PETTIGREW, 1968. Analysis of retinal correspondence by studying receptive fields of binocular single units in cat striate cortex. *Exp. Brain Res.* 6:353–372.

NODA, H., R. B. FREEMAN, JR., B. GIES, and O. D. CREUTZFELDT, 1971. Neuronal responses in the visual cortex of awake cats to stationary and moving targets. *Exp. Brain Res.* 12:389–405.

NOTON, D., 1970. A theory of visual pattern perception. *IEEE Trans. Syst. Sci. Cybern.* 6:349–357.

NOTON, D., and L. STARK, 1971. Scanpaths in saccadic eye movements while viewing and recognizing patterns. *Vision Res.* 11:929–942.

ORBACH, J., and K. CHOW, 1959. Differential effects of resections of somatic areas I and II in monkeys. *J. Neurophysiol.* 22:195–203.

PANTLE, A., and R. SEKULER, 1969. Contrast response of human visual mechanisms sensitive to orientation and direction of motion. *Vision Res.* 9:397–406.

PERKEL, D. H., and T. H. BULLOCK, 1968. Neural coding. *Neurosciences Research Program Bulletin* 6(3).

PLOOG, D., 1971. The relevance of natural stimulus patterns for sensory information processes, *Brain Res.* 31:353–359.

POSTMAN, L., 1955. Association theory and perceptual learning. *Psychol. Rev.* 62:438–446.

REESE, H., 1968. *The Perception of Stimulus Relations.* New York: Academic Press.

RICHARDS, W., 1971. The fortification illusions of migraines. *Sci. Amer.* 224(5):88–96.

RIOPELLE, A. J., U. RAHM, N. ITOIGAWA, and W. A. DRAPER, 1964. Discrimination of mirror-image patterns by rhesus monkeys. *Percept. Motor Skills* 19:383–389.

ROCK, I., 1970a. Perception from the standpoint of psychology. *Res. Publ. Ass. Res. Nerv. Ment. Dis.* 48:139–149.

ROCK, I., 1970b. Toward a cognitive theory of perceptual constancy. In *Contemporary Scientific Psychology,* A. R. Gilgen, ed. New York: Academic Press.

ROPPOLO, J. R., D. DREYER, B. WHITSEL, and G. WERNER. Phencyclidine action on neural mechanism of somesthesis. *Neuropharmacology* (in press).

SCHILLER, P. H., and F. KOERNER, 1971. Discharge characteristics of single units in superior colliculus of the alert rhesus monkey. *J. Neurophysiol.* 34:920–936.

SCHNEIDER, G., 1967. Contrasting visuomotor functions of tectum and cortex in the golden hamster. *Psychol. Forsch.* 31:52–62.

SCHNEIDER, G. E., 1969. Two visual systems. *Science* 163:895–902.

SCHOLTZ, D. A., 1957. Die Grundsatze der Gestaltwahruehmung in der Haptik. *Acta Psychologica (Amst.)* 13:299–333.

SCHWARTZKROIN, P. A., 1972. The effect of body tilt on the directionality of units in cat visual cortex. *Exp. Neurol.* 36:498–506.

SCHWARTZMAN, R. J., and J. SEMMES, 1971. The sensory cortex and tactile sensitivity. *Expl. Neurol.* 33:147–158.

SEMMES, J., 1969. Protopathic and epicritic sensation: A reappraisal. In *Contribution to Clinical Neuropsychology.* A. L. Benton, ed. Chicago: Aldine Publishing Company, pp. 142–169.

SEMMES, J., 1972. Somesthetic effects of damage to the central nervous system. In *Handbook of Sensory Physiology,* A. Iggo, ed. Berlin: Springer-Verlag.

SEYFFARTH, H., and D. DENNY-BROWN, 1948. The grasp reflex and the instinctive grasp reaction. *Brain* 71(2):9–183.

SIMON, H. A., and A. NEWELL, 1971. Human problem solving: The state of the theory in 1970. *Amer. Psychol.* 26:145–159.

SOLLEY, C. M., and G. M. MURPHY, 1960. *Development of the Perceptual World.* New York: Basic Books.

SPERRY, R. W., 1950. Neural basis of the spontaneous optokinetic response produced by visual inversion. *J. Comp. Physiol. Psychol.* 43:482–489.

SPINELLI, D. N., 1970. Recognition of visual patterns. *Res. Publ. Ass. Res. Nerv. Ment. Dis.* 48:139–149.

STROMEYER, III, C. F., and R. J. W. MANSFIELD, 1970. Colored aftereffects produced with moving edges. *Perception and Psychophysics* 7:108–114.

SUTHERLAND, N. S., 1959. Stimulus analyzing mechanisms. *Proc. Symp. on Mechanization of Thought Processes*, Vol. 2, p. 575–609. London: H.M. Stationery Office.

SUTHERLAND, N. S., 1968. Outlines of a theory of visual pattern recognition in animals and man. *Proc. Roy. Soc. (Biol.)* 171:297–317.

TEUBER, H., 1960. Perception. In *Handbook of Physiology*, Vol. 3, Section 1, J. Field, ed. Washington, D.C.: American Physiological Society, pp. 1595–1668.

TREVARTHEN, C. B., 1967. Two mechanisms of vision in primates. *Psychol. Forsch.* 31:299–337.

TREVARTHEN, C. B., 1970. Experimental evidence for a brain stem contribution to visual perception in man. *Brain, Behav., Evol.* 3:338–352.

TWITCHELL, T. E., 1965. The automatic grasping responses of infants. *Neuropsychologia* 3:247–259.

WALL, P. D., 1970. The sensory and motor role of impulses traveling in the dorsal columns towards cerebral cortex. *Brain* 93:505–524.

WERNER, G., B. L. WHITSEL, and L. M. PETRUCELLI, 1972. Data structure and algorithms in the primate somatosensory cortex. In *Brain and Human Behavior*, A. G. Karczmar, and J. C. Eccles, eds. Berlin: Springer-Verlag, pp. 164–186.

WHITSEL, B. L., J. R. ROPPOLO, and G. WERNER, 1972. Cortical information processing of stimulus motion on primate skin. *J. Neurophysiol.* 35:691–717.

WILSON, M. E., and M. J. TOYNE, 1970. Retino-tectal and corticotectal projection in *Macaca Mulatta. Brain Res.* 24:395–406.

WURTZ, R. H., 1969. Visual receptive fields of striate cortex neurons in awake monkeys. *J. Neurophysiol.* 37:727–741.

WURTZ, H., and M. E. GOLDBERG, 1972. The role of the superior colliculus in visually evoked eye movements. In *Cerebral Control of Eye Movements and Motion Perception*, J. Dichgans and E. Bizzi, eds. Basel: Karger pp. 149–158.

YARBUS, A. L., 1967. *Eye Movements and Vision.* New York: Plenum Press.

ZWICKER, E., 1964. Negative afterimage in hearing. *J. Acoust. Soc. Amer.* 46:805–811.

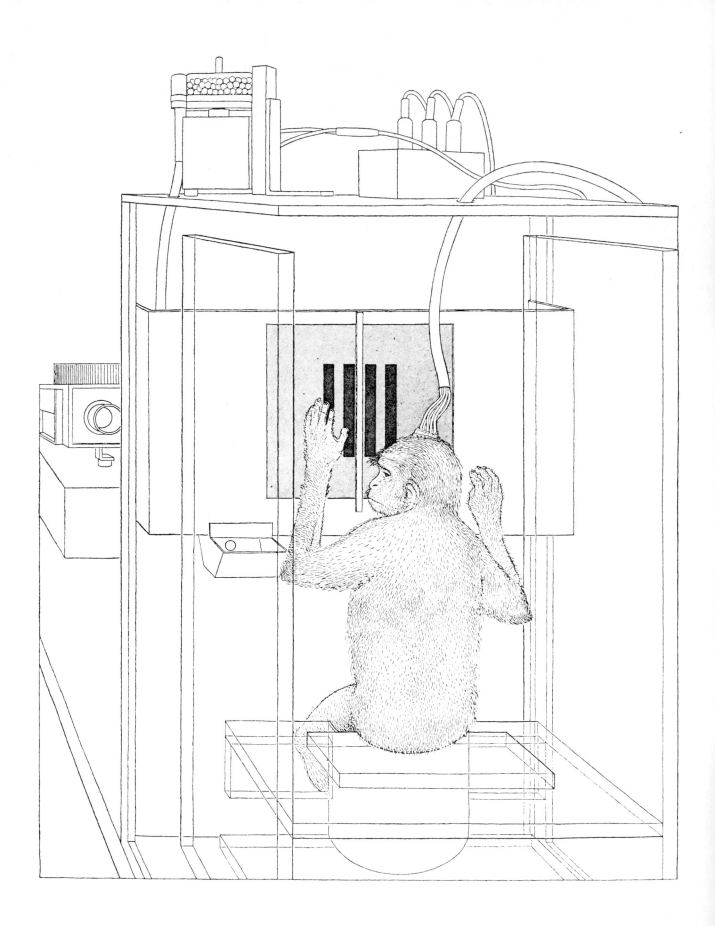

CENTRAL

PROCESSING

OF SENSORY INPUT

Set-up of an experiment demonstrating the functions of the visual cortex. A monkey initiates a flashed stimulus display and responds by pressing either the right or left half of the display panel to receive a reward while electrical brain recordings are made on line with a small general purpose computer (PDP-8). On the translucent panel in front of him the monkey sees either a circle or a series of vertical stripes, which have been projected for 0.1 msec from the rear. He is rewarded with a peanut, which drops into the receptacle at his left elbow, if he presses the right-half of the panel when he sees the circle or the left-half when he sees the stripes. Electrodes record the wave forms that appear in the monkey's visual cortex as he develops skill at this task. Early in the experiments, the stimulus-locked wave forms show whether the monkey sees the circle or stripes. Eventually they reveal in advance which half of the panel the monkey will press. Each trace sums 300 trials of 500 msec of electrical activity following the stimulus flash.

CENTRAL PROCESSING OF SENSORY INPUT

Introduction

KARL H. PRIBRAM

MOST OF US HERE are aware that neuroscience is in a period of unprecedented activity. Contributions of historic proportions are continually revising cherished views. In another part of this volume, we are privileged to read of two such important advances in knowledge that have a direct bearing on the topic we are about to explore—the problem of how decisions are achieved by the brain.

The first of these is contained in the paper by Donald Broadbent, in the first part of this volume. Broadbent presented us with evidence that a good deal of the output machinery of the organism, as well as its input mechanism, is organized in parallel. Yet as Sherrington so beautifully stated: "The singleness of action from moment to moment is the keystone to the construction of the individual." Sherrington placed the locus of this keystone in the final common path, the spinal motorneuron. Broadbent's evidence makes it necessary to revise our views upward: Singleness of action is the consequence of central not peripheral processing. We must search the brain not the spinal cord for the locus of the decisional machinery. Three of this volume's parts—Feature Extraction, Detection, and Behavior; this part, Central Processing of Sensory Input; and the next part, Central Processing of Sensory Input leading to Motor Output—are devoted to this search.

The second contribution that has bearing on the problems discussed in this section is Fergus Campbell's, in the second part of this volume. Campbell detailed for us his evidence on the nature of parallel processing in the input systems. He showed us that the primary visual projection system (as well as the auditory) displays the characteristics of a spatial frequency analyzer. His data make it necessary to re-examine the usual interpretation placed on the discovery by Hubel and Wiesel of line orientation sensitive neurons in the striate cortex of the brain. On Campbell's evidence, these neurons are probably not elements from which cartoon-like caricatures are composed at a subsequent level by convergence. Rather, they are proving to be elements of a parallel processing spatial frequency mechanism of true image reconstruction in which the full richness of input information is maintained with minimal distortion and degradation (at least for frequencies above one cycle per degree—below this frequency another mechanism may be operative).

Our charge is defined by these two contributions: Our topic is to serve as the focus for a workshop that concentrates on the issue of *how it is that sensing so much, we can do so little*. We limit ourselves to the visual mechanism, because the data on brain function in vision are complicated enough to challenge our full efforts. As you will see from the reports of Graybiel and Jones, at least three separate visual systems have been identified anatomically. Our task therefore is (1) to attempt to specify the precise function of each system—here the contribution by Weiskrantz is critical; (2) then to ask just how these functions are related to one another, the subject to which Gross primarily addresses himself; and finally, (3) to inquire into the possible neural mechanisms that make the relationship possible—which is the aim of Pellen's presentation and mine.

17 The Interaction between Occipital and Temporal Cortex in Vision: An Overview

L. WEISKRANTZ

ABSTRACT Current information regarding anatomical connections, electrophysiological single-unit tuning functions, and lesions is reviewed for the visual, prestriate, and temporal lobe cortices of the monkey in the form of "tables of ignorance" showing the gaps that still remain to be studied. Similar tables are also presented for functional interpretations of various components of the system. Empirical studies suggest that, in the absence of striate cortex, animals can still master pattern discriminations but only within the limits of significantly reduced acuity. In the absence of striate cortex, visual capacity appears to depend on the integrity of prestriate cortex. Prestriate cortex appears to be involved in the elaboration of spatial frameworks, and deficits associated with discrete lesions may be understood in terms of alterations in specific regions of the visual fields. Analysis of inferotemporal cortex suggests that it is important for categorization of visual events, which could account for why impaired visual memory and visual attention deficits are sometimes associated with lesions in this area.

The problem

THE SPECIFICATION of an overview with respect to altitude is not precise and depends also on the density of cloud. At the extremes, an overview is in danger of becoming either an oversight or a low pass. The question of level was resolved for me when I tried to list what is known about occipital and temporal lobe interactions in vision, because what emerged was a set of "ignorance tables," and these may provide a basis for deciding where the major problems lie. There is a difficulty about demonstrating ignorance: Like the universal negative, it cannot be proved. I trust that there will not be too many cases where my claims of ignorance are based on lack of knowledge rather than knowledge of a lack.

The problem will be considered in terms of five main aspects—connections, tuning characteristics, lesions, functions, and the requirements that a black-box analysis might impose on the real neural gray box. An attempt will be made to decide where it is agreed that we think we know

something, where there is real and honest disagreement, or where we are ignorant or at least rather uncertain. Of course, this is not only arbitrary but it is not possible to attach equal degrees of confidence to all the data; in some cases we have only a single investigation on which to rely, and thus no disagreement, and perhaps even no inclination to disagree.

CONNECTIONS Let us consider connections first of all, and let us start with the monkey retinal outputs. (This paper will be largely restricted to the monkey.) There are at least six known outputs, the best known of which is the output to the dorsal lateral geniculate nucleus of the thalamus, as shown in Table I. Some of the next steps in the relay are unknown and appear in our table of ignorance, but the major output from the lateral geniculate nucleus is to the occipital lobe of the cerebral hemisphere to a cytoarchitecturally distinct cortex called striate because of its readily visible stripes.

Outputs from the lateral striate cortex (V_1) can be traced to two parallel, long ribbons of cortex buried in the lunate sulcus and the inferior occipital sulcus (see review by Jones, later in this part). These ribbons widen and emerge on to the lateral surface of the brain in the region just anterior to the foveal region of the striate cortex. The first of these ribbons, area 18 of Brodmann (V_2), connects directly with the striate cortex. There is some difficulty in identifying the second ribbon area 19 (V_3), because, according to Brodmann, it extends well forward of the lunate sulcus. Thus, I prefer the more neutral terminology of V_2 and V_3, which is based on silver staining technique and describes the relay of neurons beginning in the striate cortex. The connections between V_1 and both V_2 and V_3 maintain a good measure of topological regularity (Zeki, 1969; Figure 1). In addition, striate cortex projects to the posterior portion of the inferior bank of the superior temporal sulcus. Another projection from V_1 has been reported by Kuypers et al. (1965) to the depths of a part of

L. WEISKRANTZ Department of Experimental Psychology, University of Oxford, England

TABLE I
Connections

	Agreement	Disagreement	Ignorance or Uncertainty
1st Order	Retina to: Dorsal lateral geniculate nucleus (LGN) Ventral LGN Superior colliculus (SC) Pretectum Pulvinar Accessory optic tract nuclei (AOTN)		
2nd Order	D.LGN to area 17 (V_1) Pulvinar to posterior association cortex Superior colliculus to pulvinar		V.LGN to ? AOTN to ?
3rd Order	Area 17 (V_1) to V_2 and V_3 V_1 to inferior caudal bank of superior temporal sulcus (ST) V_1 to SC	V_1 to depths of intra- parietal sulcus (IP)	
4th Order	V_2 and V_3 to Prelunate gyrus (PLG) V_2 and V_3 (especially vertical meridian projections) to corpus callosum PLG, ST, V_2 and/or V_3 to frontal eye fields		ST to ?
5th Order	PLG widely to inferotemporal cortex (IT)		

the intraparietal sulcus, but this has not been confirmed by Zeki or Cowey (personal communication). All of these projections from the striate cortex V_1 form what Kuypers et al. called the "circumstriate belt."

These workers considered that there is a projection

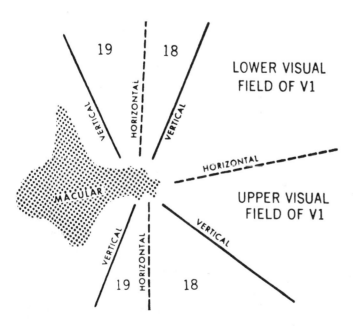

directly from "circumstriate belt" to inferotemporal cortex. According to both Zeki and Cowey, another synapse intervenes first: The V_2 and V_3 project to the prelunate gyrus, but it is too soon to know how faithfully topological aspects of the fields are maintained. Lesions in the prelunate gyrus, at least that part that is anterior to the foveal projections to V_2 and V_3, cause silver stainable degeneration that is distributed widely to the inferotemporal cortex. This region also receives a thalamic input from the pulvinar. The superior colliculus receives an input from V_1 and also is known to send fibers to the inferior pulvinar (Mathers, 1971) in the squirrel monkey. There is also a claim that the pulvinar also receives a

FIGURE 1 Schematic remapping of striate cortex onto areas 18 (V_2) and 19 (V_3) in rhesus monkey. The vertical meridian of V_1 lies just caudal to the lunate sulcus (lower visual field) and just dorsal to the inferior occipital sulcus (upper visual field). V_2 and V_3 lie buried within these sulci, except for the hatched macular region that is exposed on the lateral surface of the brain. Note that points lying increasingly peripheral along the horizontal meridian of V_1 project to increasingly more dorsal or more ventral points in V_2 and V_3, depending upon whether they are in the lower or upper field representation, and that the lower and upper fields of representation are separate in V_2 and V_3. From Zeki (1969) reprinted by permission of Macmillan Journals, Ltd.

projection, although not a heavy one, directly from the optic tract. Therefore the inferotemporal cortex receives inputs over two different routes, but these two routes are by no means independent.

TUNING CHARACTERISTICS   An approach to function can be made by considering the tuning characteristics of cells at these various by-stations (Table II). We are not concerned only, or primarily, with topological distributions, i.e., maps, or with receptive field size as such, but with what features of stimulation most excite the cell. The properties of units in the dorsal lateral geniculate nucleus (LGN) and area 17 are too well known to require elaboration here, although there is by no means agreement on all details in other species, or even in the framework within which the descriptions should be placed. For the next stage, some of the tuning properties have been isolated. Hubel and Wiesel (1970) report that more than 40% of the units in area V_2 respond to disparity and about 50% are binocular complex or hypercomplex cells. Units in the superior colliculus have been shown by Schiller and

TABLE II
"Tuning" characteristics

Agreement	Disagreement	Ignorance or Uncertainty
D.LGN		Pulvinar
V_1		AOTN
SC		IP
ST		V.LGN
IT		V_3
V_2		PLG

Koerner (1971) to respond to provocative stimuli such as moving and flashing lights but are very poorly tuned—indeed, are untuned—to their shape or direction of movement. Some of the units discharge in advance of a rapid eye movement of a characteristic size and direction for each unit, and significantly, the receptive fields of these units tend to lie in just the target area to which the fovea is directed by the eye movement. More recently inferior temporal cortex units have been isolated by Gross and his colleagues (this volume) and are reported characteristically to be large, to include the fovea, and often to be bilateral in field representation. Even the mysterious patch in the superior temporal sulcus has been shown by Dubner and Zeki (1971) to contain cells most of which react to movement. Recently, Zeki (personal communication) has found evidence for a high preponderance of color-tuned cells in a region of the prelunate gyrus (PLG) that appear to be columnarly organized. This is interesting in view of the relative paucity of color units in those other areas of cortex, including V_1, that have been studied so

far. But our ignorance is starting to be enlarged. The tuning properties of the pulvinar, the accessory optic tract nuclei (AOTN), intraparietal sulcus (IP), and the ventral LGN are not well studied. (See Table I for definitions of abbreviations.)

LESIONS   Considering the complexity of our knowledge on the anatomical and electrophysiological fronts, it is surprising how sparse our knowledge is about lesions (Table III). The geniculostriate system has been well

TABLE III
Lesions

Agreement	Disagreement	Uncertainty or Ignorance
V_1	V_1	D.LGN
IT—anterior vs. posterior	V_2	V.LGN
	V_3	AOTN
	SC and pretectum	ST
		IP
		Pulvinar
		PLG

studied for years, in fact at least since Schäfer and Ferrier in the last century. There is agreement that very significant changes in behavior occur after area 17 lesions, but as yet no agreement on the most parsimonious description of the monkey without striate cortex. This is why V_1 is listed in both the left-hand and middle columns of Table III. Evidence is still accumulating on this point, some of which we will discuss soon. It is now universally agreed that inferotemporal cortex lesions give rise to difficulties in visual discrimination learning. More recently it has been found that there are two dissociable foci, one in the middle and anterior portions of the inferotemporal region, the other in the posterior portion and perhaps including the foveal portions of V_2 and V_3. Lesions in either region give rise to visual discrimination deficits but of different kinds as we shall see soon. But notice how our areas of disagreement and uncertainty have increased. Some of this is due to technical difficulties; it is extremely difficult, for example, to lesion the ventral LGN without damaging other pathways. Some is due simply to lack of trying or to the recency of the implication of certain structures in vision. Some of it, however, is just plain disagreement.

The most serious disagreement concerns the effects of lesioning V_2 and V_3 and the prelunate gyrus, thereby severing most of the cortical-cortical outputs from V_1. According to Mishkin, the effects on visual discrimination learning are devastating (Mishkin, 1972). According to Pribram et al. (1969), they are more or less negligible. Not surprisingly such a fundamental disagreement over facts makes it more than a little difficult to reach agreement on interpretations of the function of these regions of the brain and also of the inferotemporal cortex, to which they eventually project. We will return to this conflict later.

The disagreement regarding superior colliculus and pretectum is less blatant, because of differences in species and in extent of lesions in this not easily accessible region of the brain, and the matter is perhaps better characterized as lack of agreement rather than disagreement. But because of the recent speculation in "two visual systems" the facts are of more than a little interest.

FUNCTIONS   Indeed, when we turn to the putative *functions* of various regions, we have now reached the point where the agreement column is entirely barren (Table IV). You may find this a bit sobering. You may even

TABLE IV
Functions

Agreement	Disagreement	Ignorance or Uncertainty
	V_1	ST
	V_2	IP
	V_3	Pulvinar
	PLG	AOTN
	IT	
	SC and pretectum	

disagree with the conclusion that there is such ubiquitous disagreement or ignorance. But for every claim of a final assignment of function to a region or combination of regions or a system, it is possible to find a credible counterclaim. The disagreement or ignorance stems from different sources in the different cases, and it would be rash to try to generalize. In part it is based, as we have seen, on an outright disagreement over the facts. Even where there is agreement on the facts, there are many possible interpretations as to function, and this reflects the fundamental difficulties that are inherent in attempting to proceed from observation to interpretation in the nervous system, or to do so in a way that limits the degrees of freedom for alternative interpretations. And these difficulties, in turn, have at least two sources: First is that with none of our physiological or behavioral techniques is it possible, or at least is it more than very rarely possible, to study pure and isolated variables. Thus, no one has yet studied any of the six known outputs from the retina to the rest of the brain in complete *isolation* by lesioning the remaining five. And no one has invented, nor I believe ever will invent, a pure behavioral task that measures only a single behavioral capacity. Second, there is the difficulty of drawing inferences from the facts of abnormality to the mechanisms of normality.

ANALYSIS   One might be led by this last consideration to attempt to view the problem from the other end—to isolate aspects of normal functioning as inferred from common sense, psychophysics, experimental psychology or black-box theory, and see how well we have isolated such functions in the nervous system either in terms of location or mode of operation (Table V). Here we must rescale our

TABLE V
Black boxes to grey boxes

End of Tunnel May Be in Sight	Beginning of Tunnel Has Disappeared	Where is the Tunnel?
	Noticing	
	Locating	
Examining—feature extracting		
		Synthesizing
		Stabilizing (constancies)
		Recognizing
	Classifying	
	Selectively attending	
	Associating and remembering	

categories, because we are really talking solely about various degrees of ignorance. The functions, which are deliberately put in the active form rather than terms of static entities, will not be universally accepted as reasonable, but anyone is welcome to try another set to see if it will work better for present purposes. Only in the field of feature extraction is there, I believe, any reason to feel that the end of the tunnel may be in sight. In the middle column of Table V are those functions about which much work is in progress, but it is too soon to say when we will emerge or even if we are in the right tunnel. For the all-important functions of synthesizing, i.e., concocting perceptual entities such as objects and people out of detection of features, and for the imparting of perceptual constancies, I believe there has been no greying of the black box—that is, finding an assignment in the gray box of the real nervous system as opposed to speculating entirely in the realm of abstract black boxes.

The striate cortex

Having forced our current state of knowledge into categories, and necessarily having emphasized its more negative character, let us look in great detail at a few of the research areas that appear to be pushing back the frontiers; in the course of the discussion some of the methodological issues that arise in this area of brain and behavior research may become exposed. Let us start with V_1 of monkey. That it is involved in feature extraction would seem to be a reasonable supposition, given the information we now have on its tuning characteristics. There is, however, a considerable conflict of opinion as to the most economical description of the code, which need not concern us here as it will undoubtedly receive intensive treatment in another session. Lesions of V_1 typically produce a severe visual disablement. But is it fair to infer that V_1 is *necessary* for feature extraction, or that feature extraction is its only function? There are good reasons for thinking that the answer is "no" to both questions.

EFFECTS OF EXCISION   Taking the first question first, it is interesting to note how long it has remained unanswered. Ferrier argued trenchantly 86 years ago that monkeys were more or less unaffected by occipital lobectomy, whereas Schäfer and Munk argued that occipital lobectomy was sufficient to produce grave visual disorders. From the extensive work of Klüver much later, (1936, 1941, 1942), it seemed that total removal of V_1 destroyed pattern vision and perceptual constancies, the animal being left with only total luminous flux discrimination. But within the past few years good grounds have emerged for challenging this conclusion. Firstly, if all or part of V_1 is removed we know that brief light flashes can still be detected even in the defective region of space. We know, with subtotal lesions, that the field defect has a shallow gradient rather than a sharp edge, that it gradually shrinks, that such shrinkage depends on experience, that the animal's residual acuity is better than would be predicted by the function relating acuity to predicted eccentricity of fixation (Weiskrantz and Cowey, 1970). But I do not want to dwell on the subtotal lesion story here, except to point to the resilience of the nervous system and the importance of postoperative training for the display of this resilience.

Some years ago it was found that a young monkey with just a few remaining LGN cells could discriminate patterns (Figure 2) that differed in total contour but were equal in flux (Weiskrantz, 1963). Other workers found similar results in the rat (Mize et al., 1971; Cowey and Weiskrantz, 1971) and the cat (Dalby et al., 1970; Wetzel, 1969). From

various subsidiary experiments involving movement, contour and flux, it seemed reasonable to extend Klüver's conclusion from total flux to total retinal ganglionic activity, such that the animal could discriminate any two stimuli differing in total ganglionic activity. It was possible to predict, for example, that there should be a trade-off between flux and contour such that at some point two stimuli should become indistinguishable to the animal even if they differed from each other both in total contour and in total flux (Figure 3). But this was not inconsistent with saying that feature detection was impossible for the destriated monkey; rather, the claim was that the animal was simply reacting to the integral of all ganglionic activity but could not react differentially to the spatial distribution of that activity. Even this claim, however, has been shown to be too conservative. Humphrey has shown, with simple but intensive training methods, that monkeys with intended total V_1 removal can still localize certain visual events in space with impressive accuracy (Humphrey and Weiskrantz, 1967; Humphrey, 1970). It seems unlikely that the monkeys are cheating in reacting differentially to the spatial distribution of light by sweeping their eyes across the array, which is a method by which even a photocell might be made to react differentially to "patterns," because in current experiments now in progress by Cowey, C. Passingham, and myself we find that monkeys with intended total V_1 removal can reach accurately for randomly positioned stimuli even when they are of very brief duration (50 msec).

TWO VISUAL SYSTEMS?   Until this point it still seemed parsimonious to argue that the destriated monkey was not extracting features, despite its accurate localizing of small stimuli in space. Various authors, indeed, argued that the results supported a "two visual systems" theory wherein removal of V_1 led to the loss of identification of patterns but left unimpaired the ability to localize stimuli. Schneider (1969), for example, has argued that the V_1 in the hamster is necessary for determining "what" a stimulus is, whereas the midbrain system is necessary for saying "where" it is. Humphrey (1970) has argued that the destriated monkey is localizing stimuli on the basis of their attention-getting properties, or their saliency, but when these are matched the destriated monkey cannot learn to discriminate two stimuli even though he can localize both of them accurately. The monkey, so to speak, knows that something interesting has occurred and where it has occurred, but not what it is.

But more recently the Pasiks and colleagues (Pasik and Pasik, 1971; Schilder et al., 1971) have reported that destriated monkeys with confirmed total degeneration of the dorsal LGN can learn pattern, brightness and color discrimination (Figure 4). It seems highly unlikely that

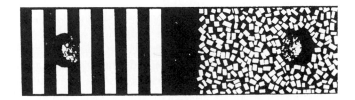

FIGURE 2   Stimulus board used to test destriated monkeys. The total flux of the left and right halves of the board are equal. The animal reaches through the holes to uncover the food wells. From Weiskrantz (1963), reprinted by permission of Pergamon Press.

BRIGHTNESS X CONTOUR TRADE-OFF

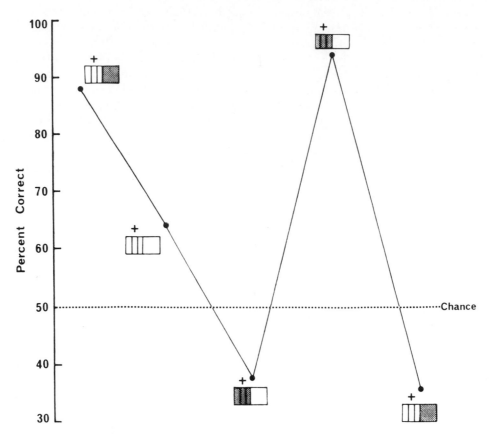

FIGURE 3 Performance of monkey "J-12" on discrimination between stripes and homogeneous area under different conditions of relative flux. In the first condition, total flux of stripes and patch are equal, and discrimination is presumably based on the greater ganglionic activity produced by larger total contour of the positive stimulus. In the second condition, flux has been added to the patch, and performance falls even though total flux of the two stimuli is unequal. In the third condition, flux is subtracted from the stripes by placing them behind a neutral density filter, and performance falls below chance. It may be inferred that the total ganglionic activity produced by the dim stripes is now less than that produced by the relatively bright patch. In the fourth condition, the animal was retrained on the stimuli of the third condition until criterion performance was achieved, so that the animal was rewarded for choosing the stimulus producing relatively less ganglionic activity. When, in the fifth condition, the animal was given the original discrimination, the performance fell below chance as predicted. From Weiskrantz (1972), reprinted by permission of the Royal Society, London.

their results could be accounted for on the basis of differences of stimulus "saliency," although that has not been tested for specifically. Some results emerging now from an experiment in progress with destriated rats would also make it difficult to argue that the animals had lost their ability to identify patterns. This experiment is an extension of a finding published by Cowey and myself (1971) that destriated rats could discriminate vertical stripes from a homogeneous grey matched in total luminous flux. In the current experiment I am interested in discovering what stimuli the destriated animal will treat as equivalent to either the stripes or the grey, and hence each of these stimuli was paired randomly with three others after the animal relearned the original discrimination. The results indicate that destriated rats are as good as controls in all the transfer tests except those where visual acuity may be inadequate (Figure 5). It seems that these animals *can* see "what" a stimulus is. Conversely, we judge "where" stimuli are in part by perspective and other visual frameworks and also in part by stereoscopy. It seems difficult to believe that the cortex is not critical for these judgments. For example, no noncortical retinal disparity units or orientational units have been found in the monkey. It is interesting that Humphrey's destriated animals, while

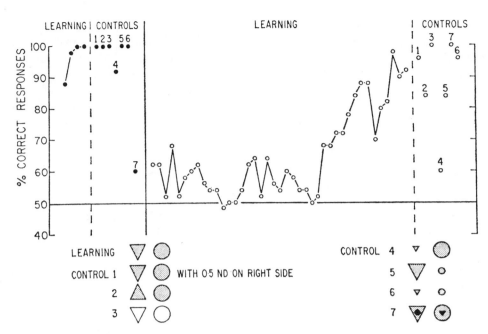

FIGURE 4 Results of experiment on totally destriated monkey by Pasik and Pasik (1971). Preoperative learning and control sessions are shown on the two left-hand panels. Postoperative relearning was retarded, but final performance level was high. Control 1, in which a 0.5 neutral density filter was added to whichever stimulus randomly was in the right-hand position, ruled out possible minimal flux differences as the basis of the performance. High performance was maintained on this and most other control conditions. Reprinted by permission of Pergamon Press.

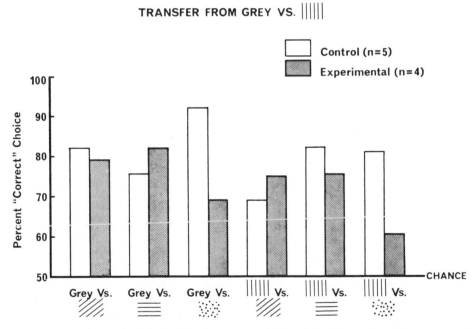

TRANSFER FROM GREY VS. ||||||

FIGURE 5 Performance on transfer trials by destriated and unoperated control rats. Animals were first trained on a discrimination between vertical stripes and a grey homogeneous patch, equated for total luminous flux. For approximately half the animals the stripes were positive, for the rest the grey patch was positive. Transfer trials for stimuli shown on the abscissa were then given, randomly interspersed between additional training trials. All transfer trials were equated for total luminous flux. Each transfer trial contained either the original positive or negative training stimulus in combination with a new stimulus. On transfer trials both stimuli were rewarded. A "correct" choice is one that was appropriate to the sign of the original training stimuli. From Weiskrantz (1972), reprinted by permission of the Royal Society, London.

able to localize the direction of stimuli in relation to their own bodies, were apparently unable to determine the distance of the stimuli.

The description that, for the moment at least, seems most parsimonious is that the destriated animal is like a normal animal with poor acuity, i.e., he may be merely amblyopic (Weiskrantz, 1972). Such a hypothesis would also be suggested by the findings of Ward and Masterton (1970) on destriated tree shrews. It would also accommodate the observations by Gross and myself (1959) that in far peripheral vision, where acuity is poor, stimuli are matched by human subjects on the basis of their total luminous flux and not on their brightness, at least for stimuli up to about 3° in diameter. Localization is still reasonably good under conditions of poor acuity, as is the detection of small spots unless they are masked by surrounding stimuli. Other features of our earlier monkey results also fit with such a description. Whether the behavior of the animal will reflect an "integration" of retinal stimuli or "discrimination" of them will depend on the properties of those stimuli in relation to relatively large receptive fields and not on the invocation of any novel type of mechanism in the nervous system or a multiplication of visual systems.

If the animal without striate cortex is like one with only ambient vision, to use Trevarthen's (1968) distinction between "ambient" and "focal" vision, then it follows that ambient vision is, within certain limits at least, capable of extracting stimulus features that will serve as the basis for pattern discrimination. Indeed, no one has suggested that foveal retinal lesions seriously impair feature extraction if the features do not exceed the limits of acuity. Therefore the striate cortex is not necessary for feature extraction, although in the normal intact animal it may be actively involved in the fine-grained examination of stimuli. On the other hand, if it is argued that the distinction between "ambient" and "focal" is useful for other purposes of interpretation of normal visual function, then the striate cortex is clearly involved in "ambient" as well as "focal" vision to the extent that the peripheral fields of vision are "ambient." As William James put it "The main function of the peripheral part of the retina is that of sentinels which, when beams of light move over them, cry: 'Who goes there?' and call the fovea to the spot." (1892, p. 73). In fact, Cowey (personal communication) has pointed out that in both the midbrain system and the geniculostriate system there are anatomical distinctions between foveal and peripheral inputs and so if there are two visual systems they may not be uniquely associated with either the cortical or the midbrain pathways. In my view the distinction between the two types of vision would tend to be relative rather than absolute—as the receptive field increases one sacrifices acuity for

sensitivity. One can trade off sensitivity for acuity, just as the astronomer may learn to use an eccentricity of fixation that produces an optimal compromise of the two. His sensitivity increases with increasing eccentricity of fixation, but if the fixation is too extreme he will no longer be able to identify the constellation that he can more readily detect.

The circumstriate belt

How do we relate these findings on V_1 to other regions of cortex? First, we can ask if *other* regions of cortex are necessary for sustaining the capacity of the animal from whom V_1 has been removed. Here the recent evidence by the Pasiks suggests that if an extrastriate lesion is added to a striate lesion then the monkey is indeed reduced to the level of a flux detector. For such a preparation they also claim that if the AOTN arc lesioned as well total blindness results. The importance of extrastriate tissue has also been pressed in their work with tree shrews by Ward and Masterton, whose results are in rough accord with the Pasiks' (1971). Klüver's results could be accounted for if his lesions included extrastriate as well as striate tissue. Just what the route is over which the cortex receives information in the absence of V_1, and what the critical extrastriate regions are, remain unknown. But it is interesting that recent work has brought us back to the position originally advanced by Ferrier almost 90 years ago.

It is when we turn to the consideration of lesions *restricted* to the circumstriate belt, the prestriate cortex alone (i.e., those cortical regions receiving a direct input from V_1), that the mystery is at its height. In fact, about some cortical regions receiving a direct input from V_1, we can say nothing whatever; e.g. the small region buried in the superior temporal sulcus, to which the whole ipsilateral visual half-field projects, has never been the specific object of a lesion study. Restricting ourselves to V_2 and V_3, the discovery of retinal disparity units and of the anatomical interhemispheric links, especially of the vertical meridia projections, lead to direct predictions about lesion effects, but these have not deliberately been tested.

Is there any evidence that the prestriate cortex is necessary, and if so for what? There are three views. The first is that it is quite dispensible; this may be Pribram's position later in this part. The second is that it is important as a relay to the inferotemporal cortex, a region that everyone admits *is* important. The third is that it is important in its own right, whether or not it serves as a relay. It is the last view that we will examine first. Aside from the physiological evidence suggesting retinal disparity detection and interhemispheric integration, some clue as to its role might be gained from the fact that topological relations of V_1 are also preserved in V_2 and V_3. The large foveal representation in V_1 is reflected in a large foveal prestriate region, and more peripheral representations in V_1 appear to have corresponding projections to V_2 and V_3.

THE VENTRAL PORTION   Do restricted lesions of parts of V_2 and V_3 give rise to perceptual impairments in restricted parts of the visual fields? We unfortunately have very little information, surprising as it may seem. Lesions in the foveal prestriate region have been shown (Iwai and Mishkin, 1968; Cowey and Gross, 1970; Gross et al., 1971; Mishkin, 1972) to cause severe effects on object and pattern discrimination, and to this we can add brightness discrimination as well if the stimuli are small and located on the response key (unpublished data). But there is an unsettled point about the foveal prestriate lesions of Cowey and Gross, because the lesions appear to have extended anteriorly beyond the actual limits of foveal V_2 and V_3 into the posterior portion of inferotemporal cortex, [Mishkin and Iwai's posterior strips 0, I, II (see Figures 6 and 7)]. It is still not possible to say whether the severity of the deficits reported by them is due to the posterior inferotemporal or the prestriate segments of their lesion, or to both combined.

THE DORSAL PORTION   The only visual impairment that has been reported for any other subpart of the prestriate region includes the dorsal portions of the prelunate gyrus and V_2 and V_3. Monkeys with bilateral lesions in this area are defective in a so-called "landmark" discrimination, where the stimuli are placed a little way away from the response plaque. Mishkin, who reported this defect (1972), suggests that the lesion may upset the "perception of spatial relations among objects scattered in the visual field. This visual spatial ability, and the extrafoveal mechanisms on which it probably depends, is . . . perhaps a primitive visual ability." One can suggest perhaps a more conservative view, that it is upsetting processing of stimuli falling mainly in the lower parafoveal region of the visual field and that the foveal prestriate lesion is upsetting processing of stimuli falling mainly in the fovea, but the character of that processing may be similar within each region. Lesions placed elsewhere ought to affect other regions of the field in a predictable manner. We are currently testing some of these ideas with results that are encouraging, but the study is not yet completed. On this notion it would not be surprising that typically there is no defect in pattern vision in animals with dorsal prestriate lesions as patterns are preferentially processed foveally, and this is especially true for the mon-

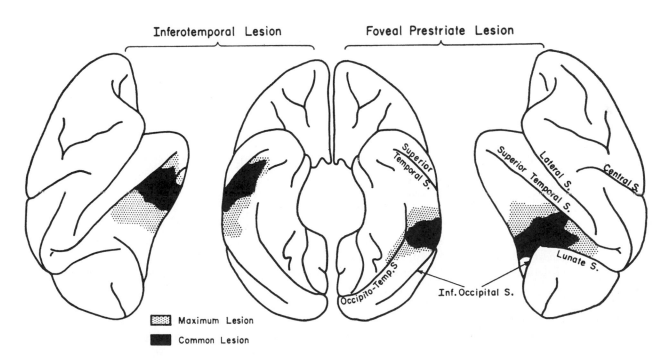

FIGURE 6   Lateral and ventral views of cerebral hemispheres showing placement of foveal prestriate (right) and inferotemporal (left) lesions in study by Cowey and Gross (1970) and Gross et al. (1971). Labeling indicates the largest lesion of each type and the lesion common to all animals in each group. From Gross et al. (1971), reprinted by permission of the American Psychological Association.

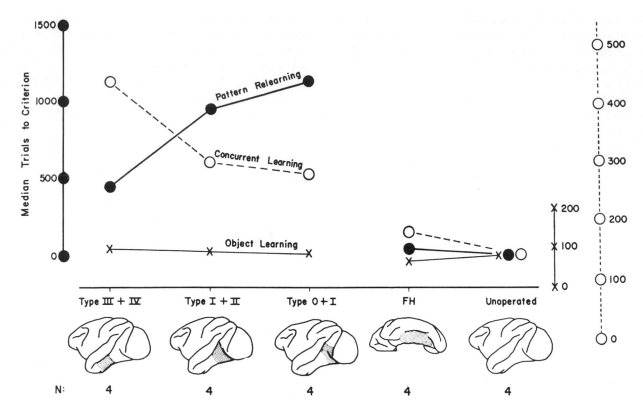

FIGURE 7   Effects of various temporal lobe lesions on three types of learning tasks. Roman numerals refer to arbitrarily designated strips of cortex at various anterior-posterior loci as defined by Iwai and Mishkin (1968). Note that posterior lesion produces a deficit on pattern discrimination learning, and anterior lesion a deficit on concurrent learning (which refers to task in which the animal is given several different discrimination tasks in each testing session, with the trials for all tasks arranged in a random sequence. In pattern discrimination, only a single task is given in a session.) FH: fusiform-hippocampal gyrus control lesion. From Mishkin (1972).

key. At any rate, we at least now have two deficits with which to go to work, namely the foveal prestriate discrimination learning impairment and the dorsal prestriate "landmark" deficit.

THE PERCEPTION OF SPACE?   More speculatively, one can ask whether there is any way of uniting these two deficits beyond saying that they might reflect impaired processing in different parts of the visual fields. It is interesting to note two kinds of "schizophrenia" in the visual nervous system. The first arises in the geniculostriate system and involves a sharp division between the left and right half-fields, which are kept anatomically segregated from each other even in V_1. The bifurcation of the fields has generally been linked with forward positioned eyes and the possibility that this offers for stereopsis. In V_2 and V_3 it seems that a second split is introduced, between upper and lower portions of the half-fields. One of the consequences of a double remapping of V_1 on to V_2 and V_3 is that as distance from the fovea increases along the horizontal meridian in V_1, pairs of points placed equally and symmetrically below and above the horizontal meridian become more and more separated from each other in V_2 and V_3 (see Figure 1). In other words, the projections from the upper and lower fields are segregated from each other and are separated by the foveal projection. Such an arrangement might be important, especially for a tree-climbing animal, for maintaining a separation of perspective cues from above and below the perceptual horizon. Prestriate cortex might provide a region, therefore, where both spatial perspective framework and stereopsis-generated information can be exploited, and the finding in the cat that prestriate units show spatial congruence for both the visual and auditory modes (Morrell, 1972) suggests that such a region might be important for directing the animal's behavior in a multimodal spatial framework. Even more speculatively, one can argue that certain of the perceptual constancies, particularly size and shape constancy, require a stable spatial framework. Such constancies normally apply to patterns in all parts of the visual fields within the limits imposed by acuity, and moreover we know that in man preoccipital damage gives rise to instabilities in visual perception. It would be instructive to know whether impaired constancy could be demonstrated that is restricted to one portion of the visual field with normal constancy in the remainder of the field.

In considering the possible *relay* functions of prestriate cortex, we are back to the conflict of results reported by Pribram (Pribram, 1971; Pribram et al., 1969) and Mishkin (1972); there may be slight but important differences between the two lesions in the foveal prestriate region, which appears to have been more completely damaged in Mishkin's work. More importantly, the banks and depths of inferior occipital sulcus, which include portions of V_2 and V_3, are apparently spared in Pribram's animals. On one point both investigators agree with each other, and indeed with the earlier literature: Apparently monkeys with extensive prestriate lesions are not permanently pattern-blind; they can display visually guided behavior. There is, however, perhaps a point worth mentioning here, because the monkey of Pribram's with the most complete lesion *was* apparently blind for some weeks after surgery and only gradually recovered. Pribram attributes this to the massive field defects caused by cutting the radiations and not to the prestriate lesion directly. An alternative is, in view of our earlier discussion, that the prestriate cortex was by no means irrelevant to the severity of the field defect, but that the prestriate lesion significantly accentuated the effects of inadvertent optic radiation and striate cortex damage. This would give prestriate cortex a supportive and not a critical role and, hence, is not important as a main issue in the present context. But one point that needs raising is whether Mishkin's animals had recovered as fully as Pribram's before formal testing started. All we know is that Mishkin states that the animal did "show adequate visually-directed behavior during testing." Far more important, perhaps, in Pribram's animals parallel pathways allowed recovery of a capacity normally dependent upon the corticocortical pathways.

Inferotemporal cortex

But even if we were to assume for the moment that prestriate cortex is not important for maintenance of pattern vision as such, it does not follow that it does not serve a necessary and important relay function between striate cortex and inferotemporal cortex for *some* aspect of discrimination learning. That there is a dependence of the inferotemporal region upon striate cortex, and that the route by which the dependence is exerted is a cortical one seems inescapable from Mishkin's combined unilateral striate-corpus callosum-unilateral inferotemporal lesion study (Mishkin, 1966). The inferotemporal units studied by Gross (this volume) are directly influenced by cortical connections via corpus callosum. Thus, the

anatomical, electrophysiological and lesion material all allow one to claim a dependence of inferotemporal cortex upon the integrity of V_1 and its transcortical outputs, although admittedly several synapses intervene.

But if so, what aspect of vision might be upset, and what can we speculate is the role of the inferotemporal region? We now know from the work of Iwai and Mishkin (1968) and of Cowey and Gross (1970) that the classical inferotemporal region contains two foci, one anteriorly and the other posteriorly placed (Figures 6 and 7). (We still do not know whether the V_2 and V_3 components of the posterior lesion are important.) The posterior inferotemporal plus foveal prestriate lesion (which henceforth will be called *foveal prestriate*) and the more anterior inferotemporal lesion (which will be called *inferotemporal*) produce doubly dissociable behavioral deficits (Figure 7). The foveal prestriate lesion causes an impairment on object and pattern discrimination learning and relearning, the inferotemporal lesion on concurrent object discrimination learning, color discrimination, and also on memory for learned discriminations when tested under certain conditions of interference.

THE POSTERIOR SUBDIVISION One hypothesis of the posterior deficit that is very compelling is that it reflects faulty selection of relevant visual cues and maintenance of attention to them, and this would allow a reconciliation with the provocative ideas and electrophysiological results of Pribram (1971), who attributes a downstream or gating function to the whole of the inferotemporal region including its most posterior portion. He reports, for example, changes in receptive fields of units in V_1 produced by stimulation of inferotemporal cortex. (It is only fair to point out, however, that one specific form of the efferent control model—that inferotemporal cortex modulates recovery cycles in the striate cortex—could not be confirmed in a stringent test of it by Schwartzkroin, Cowey, and Gross, 1969). The experiment by Manning, Gross, and Cowey (1971) comparing partial reinforcement with continuous reinforcement is a sophisticated test of an attentional defect and strongly suggests that the foveal prestriate animals have a more widely dispersed spread of attention and hence are not disturbed by partial reinforcement, unlike normal controls and inferotemporals. As Manning et al. put it, the most plausible hypothesis is "that foveal prestriate lesions and partial reinforcement impair discrimination learning in a similar fashion, by preventing the animal from consistently attending to a single dimension or aspect of the visual stimuli." An alternative hypothesis is one that has already been mentioned—that the foveal prestriates are impaired in processing foveal information, and that this will, under certain circumstances, deflect attention to nonfoveal inputs. The attention defect would then be secondary to some other defect. No attention defect should appear in these animals, according to this hypothesis, for nonfoveal stimulation. Either hypothesis, the diffusion of attention or changes in processing in local regions of the visual fields, would be relevant to the conflict between Pribram and Mishkin over the effect of cutting the cross-cortical prestriate connections. In Pribram's study the stimuli could appear in any of 16 positions in a large array of stimulus panels, whereas in Mishkin's study they were always in the same positions. It seems likely that in Pribram's animals the foveal and dorsal prestriate area was more completely invaded than the ventral prestriate area, and hence the animals should have been most impaired in the bottom half of the field: this evidently was the case.

THE ANTERIOR SUBDIVISION For the inferotemporal region it seems wise to seek some function that is no longer dependent upon which region of the retina happens to be stimulated by objects in space. That topology is no longer preserved seems indicated by Cowey's demonstration that a small prelunate lesion near the foveal V_2 and V_3 region causes degeneration widely throughout the inferotemporal region, which incidentally provides a basis for understanding why all of Gross's inferotemporal units include the fovea. From the work of Jones and Powell (1970), we might be permitted to speculate that the inferotemporal region is one in which there is multisynaptic convergence of visual pathways from all regions of V_1 to each subregion of inferotemporal cortex. In fact, Cowey and I (1967) have failed to find any evidence of restricted field defects in monkeys with inferotemporal lesions. Roughly speaking there seem to be three hypotheses about function that might fit at least some of the facts. The first is that there is a defect of visual memory. For a long time we seriously entertained this view as a result of a series of studies by Iversen and myself, but with considerable reluctance abandoned the view that it had at least to do with more rapid forgetting of visual information or failure at consolidation, akin to what used to be believed of human medial temporal deficits (Weiskrantz, 1970). It could be, of course, an impairment getting visual information *into* long-term store, with normal storage of what happens to crash the barrier. But it is difficult to understand, if this is so, why inferotemporals should have such peculiar difficulty with color discriminations, as has been found both by Wilson et al. (1972) and by Dean (personal communication). And the interference results of Gross et al. (1971) run counter to normal expectations because the greatest interference in their experiment was produced by dissimilar rather than similar intervening stimuli.

In fact, their results would fit more neatly with the second hypothesis, that the inferotemporal defect is also an attentional one but of too restricted rather than too diffuse a sampling of cues, as has been argued for the foveal prestriate animals (Butter, 1968; Manning, 1971). But again, it is difficult to account for the color discrimination difficulty, where the stimuli allow for only a minimal range of sampling of stimulus attributes. The third hypothesis is that there is a disturbance of perceptual classification. Before visual information can achieve significance through its associations particularly with reward and punishment, it is presumably abstracted and refined for convenient handling by those systems that allow access to the hypothalamus, in this case the more anteromedial portions of the temporal lobe. We may be able to discriminate thousands of different shades of color or shape in paired comparisons, but we do not remember that we were bitten by a dog that was one JND larger than our own pet and so many JNDs removed from it in hue, etc. What we remember is that we were bitten by a big, brown dog. In other words, discriminations tend to be relative and visual memories absolute and to be classified in terms of templates concocted from experience. The number of categories used in classification of judgments by humans has been shown to be rather severely restricted (Miller, 1956). Some evidence that would support the view that categorization is important was provided by Howlett, a student working in our laboratory, who suffered a tragic death before his results could be published. He compared generalization gradients for orientation of lines of inferotemporal operates and foveal prestriate operates with lateral striate and normal controls. After training the animals to respond to a vertical line for reward, he presented lines at various orientations to determine the extent to which the animals would treat them as equivalent to the original stimulus. It will be seen that inferotemporal operates show broader generalization gradients, whereas the foveal prestriate operates are unimpaired (Figure 8). These results confirm and extend the earlier and similar findings by Butter et al. (1965), produced before the two foci were known. A special point in the methodology of Howlett's work is that he took special pains to match his different groups of animals for level of performance on the original discrimination, so that we can be confident that the generalization results are not an artifact of slower learning, or poor memory. There are other possible approaches to the same question, some of which are currently being developed and pursued by Paul Dean at Oxford. The hypothesis may not be entirely without difficulty in accounting for all the results, e.g., the interference data of Gross et al. (1971), but one would require more detailed analysis of their findings for this objection to be critical. It would, on the other hand, fit the interference results of Iversen (1970) without difficulty.

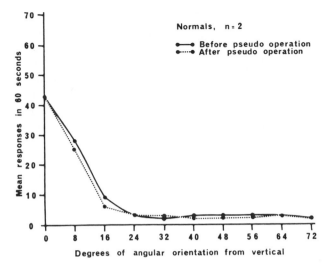

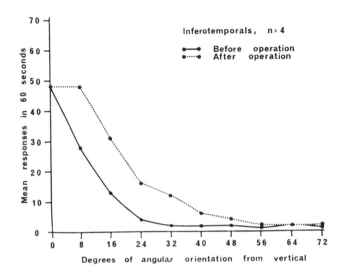

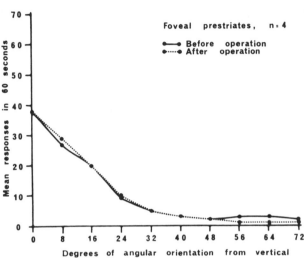

FIGURE 8   Effects of foveal prestriate and inferotemporal and lateral striate lesions on generalization gradients. Each animal served as its own control. Animals were first trained to respond to vertical line for reward, and the response rate then deter-mined for lines of various orientations. Only the inferotemporal group shows a significant shift. (From unpublished research by C. Howlett.)

Conclusion

It is too soon to attempt to arbitrate among these alternatives. The object is really to try to summarize the current issues and the current ignorance and to indicate how research may develop. A number of general points are perhaps worth making. First, very few of the observations from any of the disciplines reduce the degrees of freedom sufficiently to induce fear of sudden redundancy of research workers. Even in the field of anatomy where the facts seem most secure, there is still not insignificant ignorance. For example, we are woefully ignorant of the exact disposition of the fibers leaving V_1 on the medial surface and where the corresponding medial V_2

and V_3 are located. Indeed, the likely anatomical ignorance is even greater, because anatomists by their very nature tend to offer us additional new connections rather than to force the rejection of old alleged connections, so that a cynic may say that soon there will be sufficient anatomy to sustain any conceivable hypothesis. Second, the time has come for our lesion studies to be guided much more finely by our newly acquired anatomical knowledge. Separate analyses of V_2, V_3 (both foveal and nonfoveal), superior temporal sulcus, and prelunate gyrus are now called for, and the time has come when we can begin to consider cutting all but one at a time of the known outputs from V_1. Until this can be accomplished, we will be bedeviled in a fundamental way by parallel pathways sidestepping the disturbances caused by lesions to just one of the outputs.

Third, the hypotheses about function are often not mutually incompatible. The inferotemporal region may be both processing throughputs and affecting inputs by downstream control. Fourth, the behavioral phenomena themselves make it difficult but crucially important to tease apart primary from secondary phenomena. Thus, an animal with altered categorization will almost certainly as a result display altered memory under certain conditions, and vice versa. An animal with unstable foveal space will almost certainly select stimulus cues differently from a control, and so forth. Nonpsychologists may consider that this is the fault of our psychological concepts and categories, but the tearing apart of primary from secondary phenomena would remain just as serious a fundamental problem regardless of what psychological system we happen to favor. The same logical problem exists just as patently in other endeavors, such as pharmacology and neurochemistry. There is no bypass around these issues; nor is there a bypass around the necessity for behavioral validation itself. Finally, assignment of function, like acceptance of any hypothesis, is ultimately not a matter of proof but of parsimony, cohesion, and convergence. We are still in the main trying to assign guilt by circumstantial evidence. No one has discovered how to extract a confession.

Let us try to put an overall schema on to our cortical system. Activity proceeds in a generally rostral direction from the striate cortex, interacting at all stages with activity in the midbrain-posterior thalamic system. As it does so, the retinotopic aspect becomes less and less important until, when we reach the temporal lobe, it has disappeared altogether. It is difficult to credit the nervous system with throwing away information on such a colossal scale unless it is gaining some advantage in doing so. And, in any case, the retinotopic information is not thrown away if we consider relays to the frontal eye fields and the midbrain. What is gained is an increasing abstraction of the nonretinotopic features. What might be the utility of such an abstraction? Consider the following experiment, which is a real one that Richard Platt and I have just completed. Monkeys are trained to discriminate between two simple 3D objects that are novel to them. But, we insist that the monkey always view the objects from the same angle by requiring him to look at them through a peephole. The animal is rewarded if he reaches under one object, the positive stimulus, but not the other. He learns to do this consistently in a few trials. Now we transform the stimuli in a variety of ways—we greatly alter their size, their orientation, color, or even present a 2D photograph. With the exception sometimes of the photograph the monkey unhesitatingly recognizes the correct object,

even though we can be quite certain that his retina has *never* been stimulated by the precise configuration in his entire history.

I see only one way to describe this. The animal has stored a model or a template of the object, if you will, to which an extremely large set of possible transformations can be related even though they have never been seen before. Moreover, it would be economical if it were the case that the reward value has attached itself to the model and not merely to an additive combination of feature extractors. This is, after all, the world we, and presumably our primate relatives, live in—a world of good, bad, and indifferent objects and creatures that intrude into our memory and can be juggled in our imagination, free from the tyranny of precise retinotopic localization or a unique retinal image. As John Locke (1690) put it, in the language of the 17th century, "... If every particular idea we take in should have a distinct name, names must be endless. To prevent this, the mind makes the particular ideas received from particular objects to become general." There is good reason to believe that even the young infant may be an efficient model builder. There are two processes you will note. First the model must be built up, and second its value established. Perhaps as we proceed forward in the temporal lobe processing appropriate to these two functions is carried out, the posterior doing the constructing, and the anterior allowing it to be associated with reward and punishment systems of the hypothalamus through the connections that the anterior inferotemporal cortex has with the amygdala and pyriform cortex. Some credibility for these ideas comes from the finding of Humphrey and myself (1969) showing that inferotemporal lesions extending well into the posterior region produce a definite and persistent defect in size constancy in the monkey, even though the monkey can discriminate size as such when distance is equal. Some support for the speculation about anterior inferotemporal cortex and amygdala comes from work of Jones and Mishkin (1972) relating the effects of anterior inferotemporal lesions and amygdala lesions.

The IT region, we suggest, then, is a network that permits the establishment of cohesive categories of visual events in the form of abstractions that can be exploited with economy by other parts of the nervous system. Not only that, but allows such categories to be manipulated and recombined, which is a prerequisite for thinking and imagination. The temporal lobe according to this approach is the organ of categorization par excellence. We have already heard of the acoustic categorization account in earlier sessions, and we know that in man the critical region is part of the temporal lobe. I am suggesting that the visual system has its own categorization requirements and that another region of the temporal lobe may be necessary for these. This is, of course, speculation and although it is, I believe, testable speculation, we all know full well the dangers and the seductiveness of the contemplative life. What is certain is that the problem of visual model building is a real one for which, in higher mammals at least, the nervous system must have exquisite mechanisms that have so far eluded us.

REFERENCES

BUTTER, C. M., 1968. The effect of discrimination training on pattern equivalence in monkeys with inferotemporal and lateral striate lesions. *Neuropsychologia* 6:27–40.

BUTTER, C. M., M. MISHKIN, and H. E. ROSVOLD, 1965. Stimulus generalization in monkeys with inferotemporal lesions and lateral occipital lesions. In *Stimulus Generalization*, D. Mostofsky, ed. Stanford, California: Stanford University Press, pp. 119–133.

COWEY, A., and L. WEISKRANTZ, 1967. A comparison of the effects of inferotemporal and striate cortex lesions on the visual behavior of rhesus monkeys. *Quart. J. Exp. Psychol.* 19:246–253.

COWEY, A., and C. G. GROSS, 1970. Effects of foveal prestriate and inferotemporal lesions on visual discrimination by rhesus monkeys. *Exp. Brain Res.* 11:128–144.

COWEY, A., and L. WEISKRANTZ, 1971. Contour discrimination in rats after frontal and striate cortical ablations. *Brain Res.* 30:241–252.

DALBY, D. A., D. R. MEYER, and P. M. MEYER, 1970. Effects of occipital neocortical lesions upon visual discrimination in the cat. *Physiol. Behav.* 5:727–734.

DUBNER, R., and S. M. ZEKI, 1971. Response properties and receptive fields of cells in an anatomically defined region of the superior temporal sulcus in the monkey. *Brain Res.* 35:528–532.

GROSS, C. G., and L. WEISKRANTZ, 1959. A note on the perception of total luminous flux in monkeys and man. *Quart. J. Exp. Psychol.* 11:49–53.

GROSS, C. G., A. COWEY, and F. J. MANNING, 1971. Further analysis of visual discrimination deficits following foveal prestriate and inferotemporal lesions in rhesus monkeys. *J. Comp. Physiol. Psychol.* 76:1–7.

HUBEL, D. H., and T. N. WIESEL, 1970. Cells sensitive to binocular depth in area 18 of the macaque monkey cortex. *Nature, (London)* 225:41–42.

HUMPHREY, N. K., 1970. What the frog's eye tells the monkey's brain. *Brain, Behav. and Evol.* 3:324–337.

HUMPHREY, N. K., and L. WEISKRANTZ, 1967. Vision in monkeys after removal of the striate cortex. *Nature (London)* 215:595–597.

HUMPHREY, N. K., and L. WEISKRANTZ, 1969. Size constancy in monkeys with inferotemporal lesions. *Quart. J. Exp. Psychol.* 21:225–238.

IVERSEN, S. D., 1970. Interference and inferotemporal memory deficits. *Brain Res.* 19:277–289.

IWAI, E., and M. MISHKIN, 1968. Two visual foci in the temporal lobe of monkeys. In *Neurophysiological Basis of Learning and Behavior*, N. Yoshii and N. A. Buchwald, eds. Japan: Osaka University Press.

JAMES, W., 1892. *Text-book of Psychology*. London: Macmillan & Co., p. 73.

JONES, B., and M. MISHKIN, 1972. Limbic lesions and the problem of stimulus-reinforcement associations. *Exp. Neurol.* 36: 362–377.

JONES, E. G., and T. P. S. POWELL, 1970. An anatomical study of converging sensory pathways within the cerebral cortex of the monkey. *Brain* 93:793–820.

KLÜVER, H., 1936. An analysis of the effects of removal of the occipital lobes in monkeys. *J. Psychol.* 2:49–61.

KLÜVER, H., 1941. Visual functions after removal of the occipital lobes. *J. Psychol.* 11:23–45.

KLÜVER, H., 1942. Functional significance of the geniculostriate system. *Biological Symposia* 7:253–299.

KUYPERS, H. G. J. M., M. K. SZWARCBART, M. MISHKIN, and H. E. ROSVOLD, 1965. Occipitotemporal corticocortical connections in the rhesus monkey. *Exp. Neurol.* 11:245–262.

LOCKE, J., 1690. *Essay Concerning Human Understanding*. Book 2, Chap. 11, paragraph 9 of edited version by A. C. Fraser (1894). Oxford: Clarendon Press.

MANNING, F. J., 1971. The selective attention "deficit" of monkeys with ablations of foveal prestriate cortex. *Psychon. Sci.* 25(5):291–292.

MANNING, F. J., C. G. GROSS, and A. COWEY, 1971. Partial reinforcement: effects on visual learning after foveal prestriate and inferotemporal lesions. *Physiol. Behav.* 6:61–64.

MATHERS, L. A., 1971. Tectal projection to the posterior thalamus of the squirrel monkey. *Brain Res.* 35:295–298.

MILLER, G. A., 1956. The magical number seven, plus or minus two: some limits on our capacity for processing information. *Psychol. Rev.* 63:81–97.

MISHKIN, M., 1966. Visual mechanisms beyond the striate cortex. In *Frontiers in Physiological Psychology*, R. Russell, ed. New York: Academic Press, pp. 93–119.

MISHKIN, M., 1972. Cortical visual areas and their interaction. In *The Brain and Human Behavior*, A. G. Karczmar, and J. C. Eccles, eds. Berlin: Springer-Verlag, pp. 187–208.

MIZE, R. R., A. B. WETZEL, and V. E. THOMPSON, 1971. Contour discrimination in the rat following removal of posterior neocortex. *Physiol. Behav.* 6:241:246.

MORRELL, F., 1972. Visual system's view of acoustic space. *Nature (London)* 238:44–46.

PASIK, T., and P. PASIK, 1971. The visual world of monkeys deprived of striate cortex: effective stimulus parameters and the importance of the accessory optic system. *Vision Res.* Supplement No. 3, pp. 419–435.

PRIBRAM, K. H., 1971. *Languages of the Brain*. Englewood Cliffs, New Jersey: Prentice-Hall.

PRIBRAM, K. H., D. N. SPINELLI, and S. L. REITZ, 1969. The effects of radical disconnexion of occipital and temporal cortex on visual behavior of monkeys. *Brain* 92:301–312.

SCHILDER, P., T. PASIK, and P. PASIK, 1971. Extrageniculostriate vision in the monkey. II. Demonstration of brightness discrimination. *Brain Res.* 32:383–398.

SCHILLER, P. H., and F. KOERNER, 1971. Discharge characteristics of single units in superior colliculus of the alert rhesus monkey. *J. Neurophysiol.* 34:920:936.

SCHNEIDER, G. E., 1969. Two visual systems. *Science* 163:895–902.

SCHWARTZKROIN, P. A., A. COWEY, and C. G. GROSS, 1969. A test of an "efferent model" of the function of inferotemporal cortex in visual discrimination. *Electroenceph. Clin. Neurophysiol.* 27:594–600.

TREVARTHEN, C. B., 1968. Two mechanisms of vision in primates. *Psychol. Forsch.* 31:299–337.

WARD, J. P., and B. MASTERTON, 1970. Encephalization and visual cortex in the tree shrew (*Tupaia glis*). *Brain, Behav. Evol.* 3:421–469.

WEISKRANTZ, L., 1963. Contour discrimination in a young monkey with striate cortex ablation. *Neuropsychologia* 1:145–164.

WEISKRANTZ, L., 1970. Visual memory and the temporal lobe of the monkey. In *The Neural Control of Behavior*, R. E. Whalen, ed. New York: Academic Press, pp. 239–256.

WEISKRANTZ, L., 1972. Behavioural analysis of the monkey's visual nervous system. *Proc. R. Soc. Lond. B.* 182:427–455.

WEISKRANTZ, L., and A. COWEY, 1970. Filling in the scotoma: A study of residual vision after striate cortex lesions in monkeys. In *Progress in Physiological Psychology, Vol. 3*, E. Stellar, and J. M. Sprague, eds. New York: Academic Press, pp. 237–260.

WETZEL, A. B., 1969. Visual cortical lesions in the cat. *J. Comp. Physiol. Psychol.* 68:580–588.

WILSON, M., H. M. KAUFMAN, R. E. ZIELER, and J. P. LIEB, 1972. Visual identification and memory in monkeys with circumscribed inferotemporal lesions. *J. Comp. Physiol. Psychol.* 78:173–183.

ZEKI, S. M., 1969. Representation of central visual fields in prestriate cortex of monkey. *Brain Res.* 14:271–291.

18 Studies on
the Anatomical Organization of
Posterior Association Cortex

ANN M. GRAYBIEL

ABSTRACT   The parieto-temporo-occipital association cortex, in addition to its long-recognized participation in information transfer from the sensory to motor areas, is related to major afferent systems emanating from the thalamus and from the frontal lobe. Together with the efferent systems of this cortical expanse, these connections suggest a division of the posterior association cortex into proximal, distal, and prelimbic categories that are differentially related (1) to various subdivisions of the lateral thalamic group and (2) to premotor, prefrontal, and limbic areas of the cerebral hemisphere.

THE PIONEERING work of Adrian (1940, 1941), Woolsey (1942, 1947, 1952, 1958), Talbot and Marshall (1941), Ades (1943) and their colleagues established by means of evoked potential methods the occurrence of orderly representations of the receptor surfaces on the cerebral cortex of the cat as well as many other mammalian forms. For each modality, two adjoining recipient regions were identified, set off sharply from the surrounding and as yet uncharted association cortex. Subsequent studies, while confirming the presence of these "primary" and "secondary" maps, quickly uncovered an almost bewildering multitude of additional representational maps within the posterior cortex, as well as an apparent gradation of modal specificity within these areas that made it very difficult to draw a line between "sensory" and "association" cortex. In the case of the extensively studied neocortex of the cat, for example, third and fourth visual areas, an additional somatic sensory area, and no less than six representations of the auditory sensorium have been found. As a result, the secondary sensory areas, together with a variety of other areas having representational maps, have come to be grouped in a general "perisensory belt" that cuts deeply into the territory once simply identified as association cortex.

These findings, extending the realm of representation of sensory space far beyond the major sensory fields,

ANN M. GRAYBIEL   Department of Psychology, Massachusetts Institute of Technology, Cambridge, Mass.

would not in themselves seem in conflict with the classic notion of posterior association cortex as foremost a mechanism forwarding sensory information toward the motor cortex. For, with only minor exceptions, the areas of "belt cortex" defined electrophysiologically have been shown in neuroanatomical studies to receive direct projections from the sensory fields and, in turn, to project into a series of pathways leading to the frontal lobe. Nonetheless, these findings indirectly led to a major revision in our thinking about the organization of the posterior association cortex. In particular, they prompted studies that revealed efferent connections of these regions far more widespread than would be consistent with the notion of a "simple" information relay to the motor cortex. Moreover, particular findings made in the course of such studies appeared to attach new meaning to the long-familiar thalamic projection received by this region. Last, these studies supplied evidence of a second massive system of transcortical afferents to the posterior association cortex, namely, from the frontal lobe.

Thalamic afferents to the parieto-temporo-occipital association cortex

A thalamic input to the posterior association fields has been known for many years, but until very recently the thalamic areas in question—the lateral thalamic group and its swollen caudal part, the pulvinar-posterior system —were thought to represent no more than a dependent auxiliary association system developing in parallel with the great posterior association fields of the neocortex. These thalamic regions, as also the frontal-lobe-dependent mediodorsal nucleus, were considered to be *intrinsic* thalamic nuclei, that is to say, thalamic nuclei receiving no independent ascending afferents but rather deriving their inputs only from the neocortex and, perhaps, other thalamic nuclei. The major surge of interest in this thalamic territory and, consequently, in its influence upon posterior association cortex, has come in the wake of

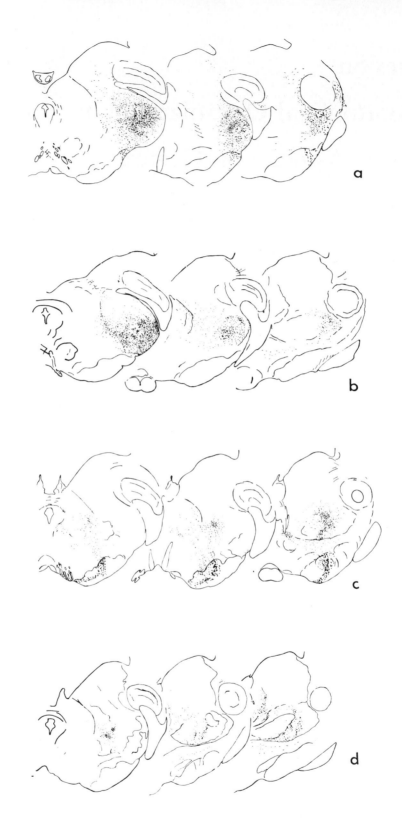

FIGURE 1 Semidiagrammatic chartings of selected transverse sections through the pulvinar-posterior system illustrating fiber degeneration elicited by lesions of (a) auditory cortex, (b) inferior colliculus, (c) somatic sensory cortex, and (d) antero-lateral column of spinal cord. Degenerating axons are shown by dotted lines, preterminal degeneration by smaller, irregularly placed dots. (Based on Figures 9 and 12 from Graybiel, 1973b.)

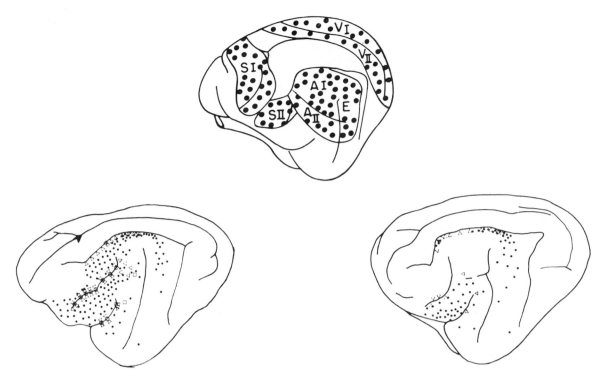

FIGURE 2 *Above:* Schematic diagram of the cat's cerebral hemisphere illustrating the major auditory (AI, AII, Ep) somatic sensory (SI and SII) and visual (VI and VII) areas of the neocortex based upon Woolsey's (1952) evoked potential studies. *Below:* Fields of neocortical fiber degeneration resulting from two lesions of the posterior nuclear group. Degeneration within sulcal cortex is denoted by outline symbols, squares in the case of sulcal fundus, triangles in the case of sulcal banks. Solid dots denote degeneration affecting cortex forming the crown of a gyrus. Differential density of degeneration has been indicated schematically by the spacing of the symbols.

recent studies showing that a variety of ascending channels do in fact reach subdivisions of this region and thus suggesting that the posterior association cortex receives information independent of that reaching the classic sensory nuclei of the thalamus.

ASCENDING AFFERENT PATHWAYS RELATED TO THE AUDITORY, SOMATIC SENSORY AND VISUAL SYSTEMS   Few of these thalamic-afferent pathways have as yet been traced out in detail in the primate, but a number of reports, beginning with the studies of Mehler (1957) on the somatic sensory system, Altman and Carpenter (1961) on the visual system and Moore and Goldberg (1963) on the auditory system, have shown that in the cat, the pulvinar-posterior system stands in a nodal position with respect to a variety of nonlemniscal pathways ascending toward the posterior association cortex. Combined with more recent information about the thalamocortical projections of the pulvinar-posterior system in the cat (Graybiel, 1970, 1972a, 1972b, 1973a; Heath and Jones, 1971a), these findings provide some clues as to the nature of the system's major input-output relationships.

The ventral part of the pulvinar-posterior system—the so-called posterior nuclear group—stands in closest relation to the auditory and somatic sensory systems. Not only does this region receive descending influx from auditory and somatic sensory cortices but also from ascending pathways that largely appear to parallel the lemniscal channels directed towards the corresponding classic sensory nuclei (Figure 1). This characteristic disposition of afferent channels, in which both the ventrobasal complex and the ventral lobe of the medial geniculate body are flanked by thalamic cell groups that also receive somatic sensory and auditory afflux, is reflected by the efferent projections of the posterior nuclear group (Figure 2). The fiber pathways traced from lesions of this region are distributed almost exclusively to rim areas adjoining the major auditory and somatic sensory areas. These rim areas comprise virtually the whole of the regions of auditory and somatic sensory "belt cortex" defined by electrophysiological studies but exclude the primary and secondary sensory fields known to receive thalamocortical projections from the ventrobasal complex and ventral lobe of the medial geniculate body (Figure 3).

A similar plan of organization may hold for the ancillary visual pathways in the cat, for it has become clear from recent studies (Graybiel, 1972b) that the whole of the

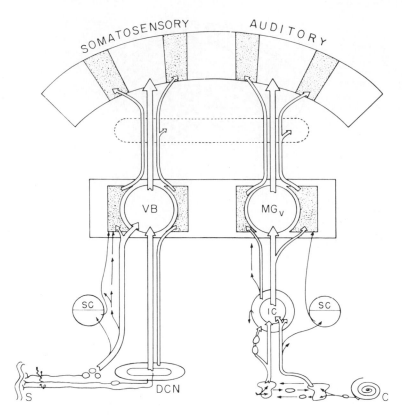

FIGURE 3 Highly schematic diagram illustrating the ascending somatic sensory and auditory channels discussed in the text. At the bottom are the peripheral elements of the somatosensory (S = skin) and auditory (C = cochlea) systems. The thalamus is shown centrally and the neocortex at the top of the diagram. Arising from the secondary sensory nuclei are: (1) main sensory channels or *lemniscal line systems* ascending toward the ventrobasal complex (VB) and medial geniculate body (MG_v), from which the thalamocortical limbs of these channels issue toward the major somatosensory and auditory areas of the neocortex. (2) Ancillary sensory channels or *lemniscal adjunct systems* (Gray-

biel, 1973a, b) ascending to the regions of the pulvinar-posterior system neighboring the classic sensory nuclei (stippled) and thence to proximal association areas of the neocortex (stippled). Offshoots of the thalamocortical limbs of these channels are shown entering the basal ganglia (dotted oblong below cortex). *Not shown* are the descending corticofugal projections to thalamic and prethalamic cell stations along these pathways. *Other abbreviations:* SC = superior colliculus, IC = inferior colliculus, DCN = dorsal column nuclei. (Adapted from Figure 13 from Graybiel, 1973b.)

more dorsal part of the pulvinar-posterior system, lying nearby the lateral geniculate body, is dominated by afferents related to the visual modality. In addition to the well-known tectothalamic pathway directed toward the medial and ventral parts of the nucleus lateralis posterior (LP), a second major fiber system, arising in the pretectal region, reaches the more laterally situated pulvinar. The areas defined by the projections of these midbrain visual systems are shown in the charts of Figure 4. Intercalated between and probably interdigitating with the two thalamic regions in question is a cell zone that receives descending projections from the visual cortex. Because of the problems of method that inevitably beset lesion studies, it has proven difficult to determine whether a significant degree of overlap occurs along the borders of these distribution areas and, in particular, whether the pretecto-recipient zone may extend beyond the pulvinar to include

part of the intermediate subdivision of the lateral nucleus, LI, or even a part of LP itself. It may be noted, however, that Wright (see the preceding part of this volume) has recently gathered electrophysiological evidence suggesting that the three zones illustrated in Figure 4 may each have distinctive modes of visual input processing.

Broad areas of the posterior association cortex receive the thalamocortical projections of these vision-related dorsal zones of the pulvinar-posterior system. Like those of the more ventral auditory and somatic sensory regions, these thalamocortical projections are restricted to rim areas neighboring the major sensory fields. A marked distinction is apparent, however, in the efferent connections of the zones related to the pretectum and superior colliculus. As shown in the diagrams of Figure 5, the pretecto-recipient zone of the pulvinar appears to project mainly to a restricted region of perivisual cortex lying in

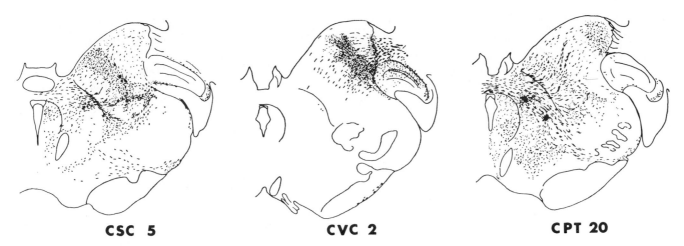

CSC 5 **CVC 2** **CPT 20**

FIGURE 4   Charts at approximately match levels through the pulvinar-posterior system illustrating fiber degeneration elicited by lesions of the superior colliculus (left), visual cortex (center), and pretectal region (right). Lesions of the pretectal region inevitably transect fibers of passage ascending from the superior colliculus; accordingly, the pattern of fiber degeneration seen after tectal lesions is shown in diminished intensity in the case of lesion to the pretectum illustrated to the left (see Graybiel, 1972b). Symbols are the same as those used in Figure 1. (From Graybiel, 1972b.)

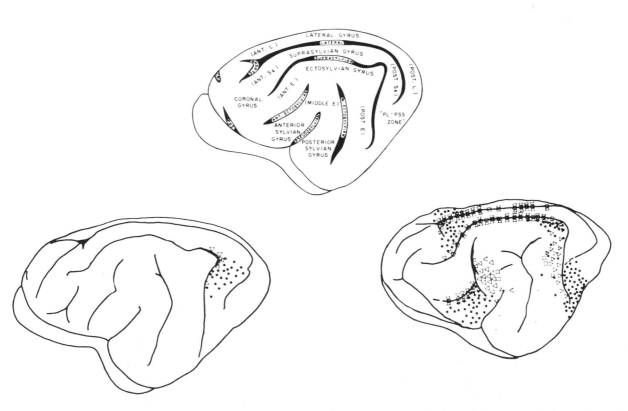

FIGURE 5   *Above:* Schematic diagram of the cat's cerebral hemisphere illustrating the major sulci and gyri. *Below:* Fields of neocortical fiber degeneration resulting from a lesion of the pulvinar (left) and of the nucleus lateralis posterior (right). (Symbols are the same as those used in Figure 1.)

the posterior suprasylvian cortex nearby area 19. By contrast, the tecto-recipient zone, probably together with its flanking fields, sends fiber projections not only to the visually responsive Clare-Bishop area and area 19 but also to cortical strips adjoining auditory and somatic sensory cortex. The contribution of the intercalated visuocortex-receptive zone to this widespread projection is as yet unclear, but it would nevertheless appear from these results that the dorsal pulvinar-posterior system includes synaptic way-stations for at least two ascending conduction routes that, in addition to the geniculostriate pathway, carry visual information toward the neocortex (Figure 6).

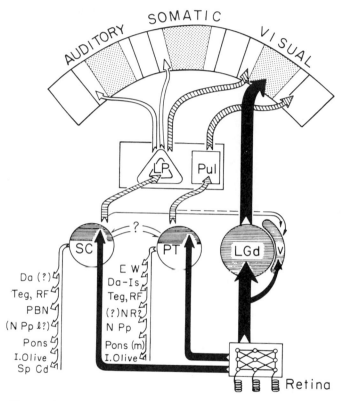

FIGURE 6 Highly schematic diagram summarizing vision-related pathways discussed in the text. *Right:* classic retino-geniculo-striate pathway. *Left:* tectal and pretectal channels. *Abbreviations:* SC = superior colliculus; PT = pretectal region: LGd, v = dorsal and ventral nuclei of lateral geniculate body; LP = nucleus lateralis posterior; Pul = pulvinar; EW = nucleus of Edinger-Westphal; Da = nucleus of Darkschewitsch; Is = nucleus interstitialis of Cajal; Teg-RF = tegmentum, reticular formation; NR = nucleus ruber, perirubral fields; NPp = nucleus papilioformis; Sp Cd = spinal cord; I. olive = inferior olive. (Modified from Graybiel, 1972b.)

OTHER ASCENDING AFFERENT SYSTEMS Taken together, the results reviewed above suggest a plan of organization of the somatic sensory, auditory, and visual systems whereby the classic lemniscal conduction routes are paralleled by ancillary sensory pathways synapsing in regions ad-

jacent to the corresponding major sensory nuclei of the thalamus, and in areas adjoining the major sensory fields of the neocortex. The regions of posterior association cortex singled out by these thalamic pathways receive, in addition to these ancillary sensory inputs, direct projections from the sensory fields that they adjoin.

Beyond these regions of "proximal association cortex," however, lie still other posterior association fields that appear at once to be more remote from the major sensory fields, and to lack a massive input from the auditory, somatic sensory, and visual subdivisions of the pulvinar-posterior system. In the cat, this "distal association cortex" (Graybiel, 1972a; 1973b) comprises much of the crown of the middle suprasylvian gyrus considered from electrophysiological evidence to represent processing mechanisms more general or "multimodal" than those of the perisensory belt appear to be. This region receives projections from several areas of proximal association cortex (see Heath and Jones, 1971b) and, on the basis of the corticocortical connections described by Pandya and Kuypers (1969) and Jones and Powell (1970), would appear to correspond most closely to the inferior parietal region of the cortex in the monkey, including the depths of the superior temporal sulcus. Although information about the thalamic affiliations of distal association cortex is still far from complete, a number of clues suggest that in the cat, this cortical district may be related to those parts of the lateral nucleus that rostrally adjoin the pulvinar-posterior system.

Figure 7 illustrates two cases of lesion to this rostral zone affecting the nucleus lateralis intermedius (LI) and the ventral part of the nucleus lateralis dorsalis (LD). Dense fiber degeneration appears on the crown of the middle suprasylvian gyrus and, hidden on the medial aspect of the hemisphere, in the cingulate gyrus. Because of the mutual proximity of these thalamic zones, it is difficult, in practice, to determine their relative contribution to the thalamoparietal system observed in these experiments. The LD has been reported to project principally to a dorsocaudal region of the cingulate gyrus (Locke et al., 1964) and to receive descending projections from this cortical area (Larson, 1962; Domesick, 1969). Several other studies suggest, however, that not only LI (Graybiel, 1972a) but also LD may project to the crown of the suprasylvian gyrus in the cat (Nauta and Whitlock, 1954; Niimi et al., 1971) and to the inferior parietal cortex in the monkey (Walker, 1938), thus reciprocating the parietothalamic projections traced to this nucleus in both species (Heath and Jones, 1971b; Petras, 1971). The significance of a thalamoparietal projection from LI must remain in doubt so long as the afferent connections of this thalamic district are unknown; but findings of at least two studies suggest that the neighboring LD lies within the

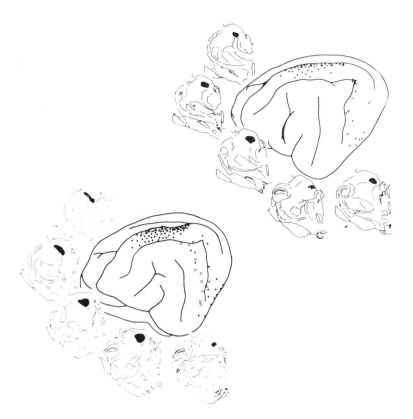

FIGURE 7 Two cases of lesion to the nuclei lateralis intermedius and lateralis dorsalis. For each case, the lesion is indicated semi-schematically in solid black on outline drawings of transverse sections through the pulvinar-posterior system. Resulting neo-cortical fiber degeneration is shown in accompanying diagrams of the cerebral hemispheres. (Symbols the same as those used in Figure 2.)

efferent domain of limbic forebrain mechanisms by virtue of inputs from the fornix (Valenstein and Nauta, 1959) and cingulate gyrus (Domesick, 1969). The possibility cannot be ignored, therefore, that by means of afferent connections from rostral parts of the lateral thalamic nucleus, the posterior association cortex, in particular the regions here distinguished as distal association cortex, could receive ascending impulse afflux from the limbic system.

Connections from the frontal lobe

No less remarkable than the highly differentiated thalamic input systems of the posterior association cortex are its recently demonstrated afferent connections from the frontal lobe (Pandya and Kuypers, 1969; Jones and Powell, 1970). The details of these conduction routes are considered in Jones's contribution to this volume (see the following chapter), and it is only necessary here to review the general organization of these newly described pathways and the reciprocating forward projections arising from the posterior association cortex.

It has already been mentioned above that from the major sensory areas of the cat's cerebral cortex there emanate pathways leading to proximal association fields and that the next step in this progression involves distal association cortex. The observations of Pandya and Kuypers (1969) and Jones and Powell (1970) have emphasized that in the monkey, these conduction routes—more highly differentiated in the monkey than in the cat—are accompanied by direct projections toward the visceral and somatic effector mechanisms of the frontal lobe (see Nauta, 1971) and toward limbic forebrain areas such as the cingulate cortex and parahippocampal gyrus. It is noteworthy that this sequential order independently supports the distinction between proximal and distal association areas that was proposed on the basis of the contrasting thalamocortical affiliations of these areas (Graybiel, 1972a; 1973b). These studies have shown that the areas of posterior association cortex most closely linked to the major sensory fields project principally to various regions of premotor cortex (areas 6, 8 and PrCO). By contrast, the posterior association areas more remote from the sensory fields, and receiving connections from proximal association cortex, distribute parietofrontal projections preferentially to prefrontal limbic areas and—with the notable exception of a projection from area 20 to the amygdala (Whitlock and Nauta, 1956; Jones and Powell, 1970)—provide the first massive inputs to the cingulate and parahippocampal gyri and other limbic forebrain regions.

These forward projections of the posterior association cortex provide an important setting for the elegant analyses of the frontofugal fiber systems carried out by Pandya and Kuypers (1969) and by Jones and Powell (1970). Each of these studies can be interpreted as showing that in the monkey, the posterior association cortex receives a massive afferent system from the frontal lobe that is divisible into a *premotor* component directed preferentially toward areas most closely linked to the major sensory fields and a *prefrontal* component distributed to the parietotemporal regions that represent distal association cortex.

(Recent experimental findings of Pandya, personal communication, suggests that in the monkey, one particular part of distal association cortex may form an exception to this categorization. Pandya's results indicate that the zone of sensory convergence in the depths of the superior temporal sulcus may entertain reciprocal connections

with the frontal eye fields rather than with regions of prefrontal cortex. This parietotemporal zone nevertheless appears to project to the cingulate gyrus, as do other regions of distal association cortex and, in addition, to the parahippocampal gyrus.)

Conclusion

The foregoing account has reviewed evidence that the parieto-temporo-occipital association cortex, in addition to its long-recognized participation in information transfer from the sensory to motor areas, is related to major afferent systems emanating from the thalamus and from the frontal lobe. In good accord with the efferent connections of this vast cortical expanse these results suggest a division of posterior association cortex into proximal and distal association areas that form consecutive stepping stones in the forward progression of pathways leading out from the sensory fields, and that seem to possess distinct afferent and efferent connections. The areas of posterior association cortex lying nearest the major sensory fields appear to be reciprocally related, on the one hand, to

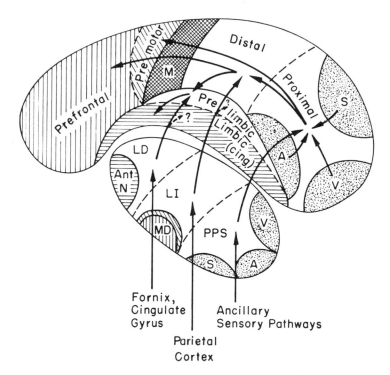

FIGURE 8 Highly simplified and schematic summary diagram illustrating some of the neuronal pathways discussed in the text. Thalamocortical affiliations of proximal, distal and prelimbic categories of the posterior association cortex are largely based upon experiments in the cat (Graybiel, 1972a, b) while transcortical pathways are translations of data provided by Pandya and Kuypers (1969) and Jones and Powell (1970) for the monkey and by Heath and Jones (1971b) for the cat. Conduction routes known to reciprocate many of these connections have been omitted (see text). *Abbreviations:* neocortex: S, V, A = somatosensory, visual, and auditory areas; M = motor cortex. Thalamus: S, V, A = somatosensory, visual, and auditory nuclei; LD = nucleus lateralis dorsalis; LI = nucleus lateralis intermedius; PPS = pulvinar posterior system; Ant N = anterior nuclear group; MD = mediodorsal nucleus.

multiple visual, auditory, and somatic sensory pathways that funnel through the pulvinar-posterior system of the caudal thalamus and, on the other hand, to the supplementary motor and premotor areas of the frontal lobe. The second general category, by contrast, appears to be preferentially related to more rostral parts of the lateral nucleus and to more rostral "prefrontal" limbic areas of the frontal lobe (Figure 8).

Together, these two subdivisions of parieto-temporo-occipital association cortex comprise most of the areas that lie outside the projection range of the specific thalamic nuclei related to the posterior neocortex: the classic sensory nuclei and the anterior nuclear group. There remain in the posterior neocortex, however, two further areas that are not encompassed by this classification but that may nevertheless, by virtue of their afferent affiliations, bear close systemic relation to the parieto-temporo-occipital association fields discussed above. Both of the districts in question, parts of the posterior agranular cingulate cortex and the perirhinal temporal cortex, e.g., area TH of von Bonin and Bailey (1947), receive connections from regions of distal association cortex and, at least in the case of the cingulate area, a thalamic projection from the lateral nuclear group. The possibility arises, therefore, that these regions may comprise a third, "prelimbic" category of posterior association cortex that is at once more remote from the fields of distal association cortex from which they receive transcortical afferents, and more intimately related to the areas of limbic forebrain that they border (Figure 8).

In the face of uncertainties about the details of these pathways, it is unclear whether the notion of such a dual— or even triple—pattern of neuronal connections could eventually lead toward a functional distinction between areas of posterior association cortex differentially involved in the sensory-motor and affective sides of perceptual experience. It could be suspected, however, that such widespread neuronal affiliations may at least provide an anatomical corollary to the complex behavioral syndromes associated with dysfunction of the posterior association cortex.

ACKNOWLEDGMENTS It is a pleasure to thank Dr. Walle J. H. Nauta and Dr. Hans-Lukas Teuber for their guidance and encouragement and Dr. Depak N. Pandya for his generous permission to cite unpublished findings from his laboratory.

REFERENCES

ADES, H. W., 1943. A second acoustic area in the cerebral cortex of the cat. *J. Neurophysiol.* 6:59–63.

ADRIAN, E. D., 1941. Afferent discharges to the cerebral cortex from peripheral sense organs. *J. Physiol.* 100:159–191.

ADRIAN, E. D., 1940. Double representation of the feet in the sensory cortex of the cat. *J. Physiol.* 98:16P–18P.

ALTMAN, J., and M. B. CARPENTER, 1961. Fiber projections of the superior colliculus in the cat. *J. Comp. Neurol.* 116:157–178.

DOMESICK, V. B., 1969. Projections from the cingulate cortex in the rat. *Brain Res.* 12:296–320.

GRAYBIEL, A. M., 1973a. Efferent connections of the so-called posterior nuclear group of the thalamus in the cat. *Brain Res.* 49:229–244.

GRAYBIEL, A. M., 1973b. Some fiber pathways related to the posterior thalamic region in the cat. In *Thalamic Structure and Function*, W. Riss, ed. *Brain, Behav., Evol.* (in press).

GRAYBIEL, A. M., 1972a. Some ascending connections of the pulvinar and nucleus lateralis posterior of the thalamus in the cat. *Brain Res.* 44:99–125.

GRAYBIEL, A. M., 1972b. Some extrageniculate visual pathways in the cat. In *US-Australian Symposium on Vision*, P. Gouras, and P. O. Bishop, eds. *Invest. Ophthal.* 11:322–332.

GRAYBIEL, A. M., 1970. Some thalmocortical projections of the pulvinar-posterior system of the thalamus in the cat. *Brain Res.* 22:131–136.

HEATH, C. J., and E. G. JONES, 1971a. An experimental study of ascending connections from the posterior group of thalamic nuclei in the cat. *J. Comp. Neur.* 141:397–426.

HEATH, C. J., and E. G. JONES, 1971b. The anatomical organization of the suprasylvian gyrus of the cat. *Ergebn. Anat. Entwicklungsgesch.* 45:1–64.

JONES, E. G., and T. P. S. POWELL, 1970. An anatomical study of converging sensory pathways within the cerebral cortex of the monkey. *Brain* 93:793–820.

LARSON, S. J., 1962. Efferent connections of the cingulate gyrus in macaque. *Anat. Rec.* 142:251.

LOCKE, S., J. B. ANGEVINE, JR., and P. I. YAKOVLEV, 1964. Limbic nuclei of thalamus and connections of limbic cortex. VI. Thalamocortical projection of lateral dorsal nucleus in the cat and monkey. *Arch. Neurol. (Chicago)* 11:1–12.

MEHLER, W. R., 1957. The mammalian "pain tract" in phylogeny. *Anat. Rec.* 127:332.

MOORE, R. Y., and J. M. GOLDBERG, 1963. Ascending projections of the inferior colliculus in the cat. *J. Comp. Neurol.* 121:109–136.

NAUTA, W. J. H., 1971. The problem of the frontal lobe: a reinterpretation. *J. Psychiat. Res.* 8:167–187.

NAUTA, W. J. H., and D. G. WHITLOCK, 1954. An anatomical analysis of the non-specific thalamic projection system. In *Brain Mechanisms and Consciousness*. Springfield, Ill.: Charles C. Thomas, pp. 81–116.

NIIMI, K., and H. INOSHITA, 1971. Cortical projections of the lateral thalamic nuclei in the cat. *Proc. Jap. Acad.* 47: 664–669.

PANDYA, D. N., and H. G. J. M. KUYPERS, 1969. Corticocortical connections in the rhesus monkey. *Brain Res.* 13:13–36.

PETRAS, J. M., 1971. Connections of the parietal lobe. *J. Psychiat. Res.* 8:189–201.

TALBOT, S. A., and W. H. MARSHALL, 1941. Physiological studies on neural mechanisms of visual localization and discrimination. *Amer. J. Ophthal.* 24:1255–1263.

VALENSTEIN, E. S., and W. J. H. NAUTA, 1959. A comparison of the distribution of the fornix system in the rat, guinea pig, cat and monkey. *J. Comp. Neur.* 113:337–363.

VON BONIN, G., and P. BAILEY, 1947. *The Neocortex of Macaca Mulatta*. Urbana, Ill.: The University of Illinois Press.

WALKER, A. E., 1938. *The Primate Thalamus*. Chicago: University of Chicago Press, p. 185.

WHITLOCK, D. G., and W. J. H. NAUTA, 1956. Subcortical

projections from the temporal neocortex in *Macaca mulatta*. *J. Comp. Neur.* 106:183–212.

Woolsey, C. N., 1958. Organization of somatic sensory and motor areas of the cerebral cortex. In *Biological and Biochemical Bases of Behavior*. H. F. Harlow and C. N. Woolsey, eds. Madison: University of Wisconsin Press, pp. 63–81.

Woolsey, C. N., 1952. Patterns of localization in sensory and motor areas of the cerebral cortex. In *The Biology of Mental Health and Disease*. New York: Hoeber-Harper, pp. 193–206.

Woolsey, C. N., 1947. Patterns of sensory representation in the cerebral cortex. *Fed. Proc.* 6:437–441.

Woolsey, C. N., and E. M. Walzl, 1942. Topical projection of nerve fibers from local regions of the cochlea to the cerebral cortex of the cat. *Bull. Johns Hopkins Hosp.* 71:315–344.

19 The Anatomy of Extrageniculostriate Visual Mechanisms

E. G. JONES

ABSTRACT An anatomical basis for visual mechanisms outside the striate cortex exists in the peristriate and inferotemporal cortex, which receive afferents from the lateroposterior-pulvinar complex of the thalamus. This complex is the main thalamic recipient of efferent fibers from the superior colliculus. However, in the normal brain, this system cannot be considered as separate from the geniculostriate system, since there is a progressive, stepwise convergence upon the inferotemporal cortex of corticocortical connections emanating from the striate cortex. Moreover, all areas, including the striate, project upon the lateroposterior-pulvinar complex and superior colliculus.

IN 1669 AT A meeting in Paris, the anatomist Stensen expressed the wish that the brain were as well understood as the majority of philosophers and anatomists imagined it to be. Though not all would agree with this sort of scientific pessimism, it is well known that workers in different disciplines view the visual system, as it were, through different eyes.

The psychophysicist is concerned, naturally, with man; many experimental psychologists would be probably more inclined to think first of the inferotemporal cortex of the monkey, for it is here that a great deal of behavioral work has been done. Those who would consider the possible electrophysiological correlates of cortical processing of visual information would probably think in the first instance of the cat, since the information available in this species lends itself to an interpretation of the visual cortex in terms of an hierarchically organized neuronal system (Hubel and Wiesel, 1965). In the following sections, an attempt will be made to bring together that anatomical data which is available on the connections of the striate and extrastriate visual areas of the cat and monkey and which may be relevant to these different fields of interest. There are some major gaps in our knowledge of the cortical and thalamic visual apparatus in both the cat and the monkey and, unfortunately, the emphasis in experimental studies of the two has frequently been different so that a cross-species correlation is often difficult.

E. G. JONES Department of Anatomy, Washington University School of Medicine, St. Louis, Missouri

Inevitably, therefore, the following synthesis, in drawing information from work in the two species, may need to be modified as further data become available. Most of the anatomical work summarized here has been carried out with the technique of anterograde axonal degeneration. All the experimental studies considered are recent ones, and all have made use of the Nauta technique or one of its several variants. Unless otherwise specified, the information presented has been drawn from the papers of Kuypers, Szwarcbart, Mishkin, and Rosvold (1965), Cragg (1969), Zeki (1969a, 1971a, b), Pandya and Kuypers (1969), Jones (1969), Jones and Powell (1970), Graybiel (1970, 1972), and Heath and Jones (1970, 1971)—all of which are based on the use of silver staining techniques.

Visual areas of the cat and monkey

Though widely used in experimental studies, the cat has generally been regarded as somewhat aberrant in that it was said to have, on the basis of cytoarchitectural differentiation (Otzuka and Hassler, 1962), four primary visual cortical areas—areas 17, 18, and 19 and the lateral suprasylvian (or Clare-Bishop) area all of which apparently received fibers from the dorsal lateral geniculate nucleus—but virtually no association cortex (Figures 1A and 2A). In the monkey, on the other hand, only area 17 receives fibers from the dorsal lateral geniculate nucleus, and the other areas are usually regarded as forming part of an extensive association cortex (Figures 1B and 2B). This also includes areas in the inferotemporal region that were labeled 20, 21, and 38 by Brodmann (1905, 1909), Mauss (1908), and others, but which are cytoarchitectonically very similar. The difficulty of defining the boundaries of these areas as well as those of area 19 (Figure 2B) (see also von Bonin and Bailey, 1947) has led to some confusion in recent experimental studies of the region. The experimental observations reported in the succeeding sections may help to clarify some of these problems.

Figure 3 is representative of experiments in the cat in which small parts of the dorsal lateral geniculate nucleus were destroyed by electrolytic lesions made by a

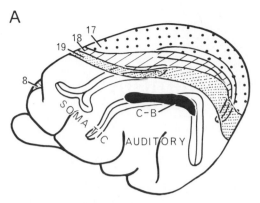

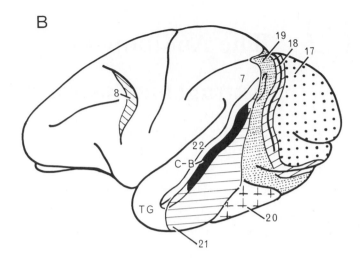

FIGURE 1 The visual cortical and adjoining fields of the cat (A) and rhesus monkey (B) as commonly understood, drawn after several authors (see text).

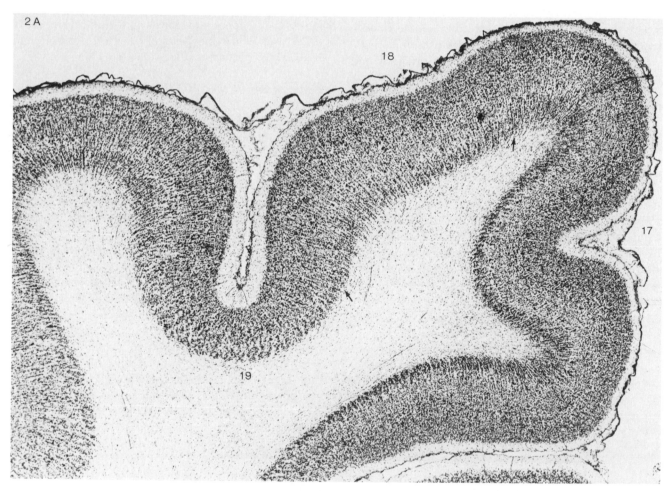

FIGURE 2 (A) Photomicrograph of a thionin stained, frontal section through the lateral gyrus of the cat showing the architectonic fields, areas 17, 18 and 19 (X21).

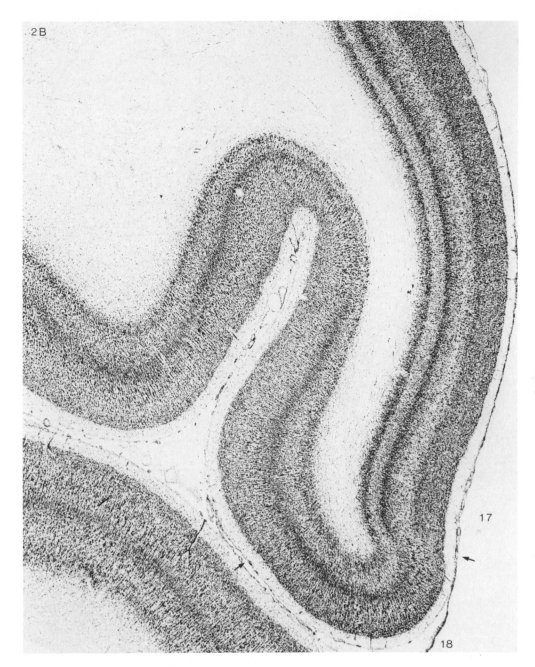

FIGURE 2 (B) Photomicrograph of a thionin stained, oblique frontal section through the inferior occipital sulcus of the rhesus monkey. Note that areas 17 and 18 are quite distinct cyto-architectonically but that it is difficult on structural grounds alone to subdivide the remaining cortex into separate fields (X12).

stereotaxic approach that avoided interference with the overlying cerebral cortex or with adjacent thalamic nuclei (Heath and Jones, 1971). The distribution in the cortex of Nauta-stained degeneration in these experiments shows that the cortical projection of the lateral geniculate nucleus is restricted to areas 17 and 18. (In the experiment illustrated, only parts of areas 17 and 18 contain degeneration, and this is due to the fact that only a small part of the retinal representation in the lateral geniculate nucleus was destroyed). It is now apparent that previous reports of additional projections to area 19 and the Clare-Bishop area (Glickstein, King, Miller, and Berkley, 1967; Wilson and Cragg, 1967) were based upon interruption of fibers derived from more medially situated parts of the thalamus and traversing the vicinity of the lateral geniculate nucleus en route to the cortex. The experiment

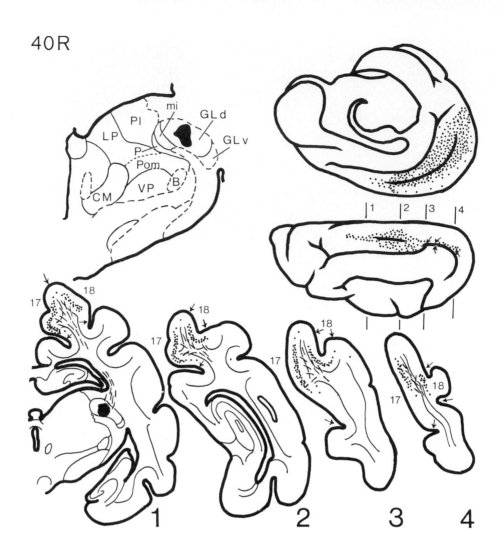

FIGURE 3 An experiment showing the distribution of Nauta-stained axonal degeneration (stipple) in the cortex of a cat following a small electrolytic lesion (black) of the dorsal lateral geniculate nucleus. In this case, where there has been no inter-ference with the overlying cortex or with other thalamic nuclei, the degeneration is confined to areas 17 and 18. (Reproduced from Heath and Jones (1971) with the permission of the Springer-Verlag.)

illustrated in Figure 4 is representative of those which show that area 19 and the Clare-Bishop area receive their thalamic input from the lateral nuclear complex and/or the adjacent posterior nucleus of Rioch.

Despite these observations, the thalamocortical visual systems of the cat and monkey are obviously still not identical. However, the difference is less than has pre-viously been supposed, and it may be that a clue to under-standing the remaining difference (the projection of the dorsal lateral geniculate nucleus of the cat to area 18) lies in recognizing in the monkey the homologue of the feline medial interlaminar nucleus which Burrows and Hayhow (1971) regard as projecting preferentially to area 18.

A progression of function in the temporal lobe of the monkey?

Figure 5 is a schematic summary of experiments carried out with the Nauta technique and designed to demonstrate the sequence of corticocortical connections passing out-wards from the striate cortex of the rhesus monkey. The initial steps in the sequence are shown separately in the two experiments illustrated in Figure 6.

If area 17 is damaged, the Nauta-stained degeneration outlines a number of other cortical fields. There is, first, a peristriate band essentially in the banks of the lunate and inferior occipital sulci, in which two retino-topically organized fields are embedded. One of these corresponds

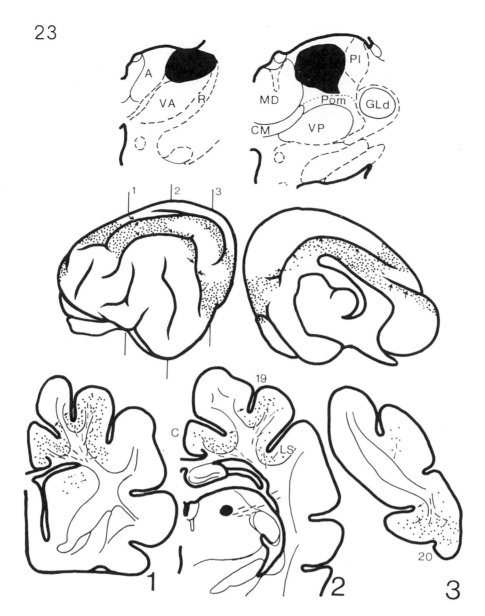

FIGURE 4 An experiment showing the distribution of Nauta-stained degeneration in the cortex following a lesion of the lateral nuclear complex of the cat's thalamus. Note that area 19 and the Clare-Bishop (LS) area receive their thalamic input from the lateral nuclear complex and not from the lateral geniculate nucleus. (Reproduced from Heath and Jones (1971) with the permission of the Springer-Verlag.)

to area 18 (V_2) and the other may be referred to as area 19 (V_3), although this latter is narrower than the area 19 shown by Brodmann (1909) on his schematic surface drawing of the monkey brain. The bare area shown in the prelunate gyrus in Figure 6, for example, would fall within area 19 of Brodmann's surface map. Another cortical area not readily distinguished on architectonic criteria but clearly outlined by degeneration, is divorced from the peristriate band and lies in the posterior bank of the superior temporal sulcus. This area has receptive field properties and connections similar to those of the Clare-Bishop area of the cat and has been tentatively identified as its equivalent.

If now the extrastriate areas are damaged (Figure 6B), in addition to reciprocal connections to area 19, another area contains degeneration. This is in the posteromedial aspect of the temporal lobe: On the lateral surface it has a small part lying below the inferior occipital sulcus, but it expands medially along the rhinal fissure, and most of its extent is on the inferior surface of the temporal lobe; it approximately corresponds to what Brodmann called area 20. (According to Zeki (1971b), there may be another step

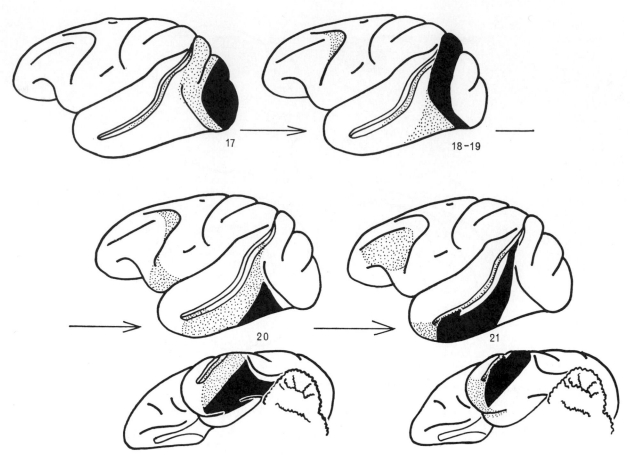

FIGURE 5 Schematic summary of the outward progression of corticocortical connections emanating from the striate cortex. Note that the progression in the temporal lobe is paralleled by a similar one in the frontal lobe. The final step at the tip of the temporal lobe is connected with the orbito-frontal cortex and area 20 projects to the amygdala (not shown). According to Zeki (1971b), there may be an additional step intercalated between areas 19 and 20.

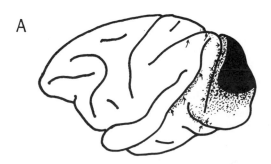
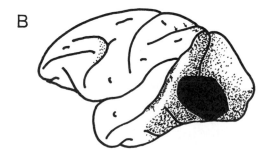

FIGURE 6 Summary reconstructions of two Nauta experiments showing the distribution of degenerating corticocortical fibers following lesions of area 17 (A) and of area 17 together with the peristriate regions to which it projects (B). Degeneration in the banks of sulci is indicated by small arrows pointing to the appropriate bank. (See text for further details.)

intercalated between the extrastriate areas and area 20, but it is not clear from his published data whether or not the area which he calls Visual 4 is not merely a dorsal extension of the area called 20 here.) All of these connections are reciprocal. In the experiments with lesions of the peristriate regions, there is additional degeneration in the frontal eye field (Figure 6).

If area 20 is next damaged (Figure 5), a further local area is outlined by degeneration more anteriorly. This corresponds approximately to Brodmann's area 21, but it may be noted that it extends dorsally in the prelunate gyrus so that it includes part of what on his surface map he labeled area 19. There is an additional frontal projection to the precentral agranular field of Roberts and Akert (1963), which is a part of the premotor cortex. Area 20 is also heavily connected with the basolateral amygdala (Whitlock and Nauta, 1956; Jones and Powell, 1970).

If area 21 is damaged (Figure 5), there is a further local projection to the tip of the temporal lobe, to area TG of von Bonin and Bailey (1947) who followed the cytoarchitectural nomenclature of von Economo; this is an area which in man Brodmann labeled area 38. The area of degeneration also extends upwards along the fundus of the superior temporal sulcus and medially into the perirhinal cortex. There is also a further frontal projection, now to a prefrontal area in the inferior bank of and below the principal sulcus. The temporal tip region has not been damaged selectively, but its main connections are with the orbitofrontal cortex and with other areas on the orbital surface of the frontal lobe. These areas on the orbital surface of the frontal lobe are connected with the hypothalamus (Nauta, 1964).

Finally, it can be shown that some of the frontal connections of the inferotemporal cortex are reciprocal, as indicated by experiments with lesions of the premotor and prefrontal cortex.

There is, thus, an outward progression of corticocortical steps from the striate cortex (the number of steps being for the moment perhaps debatable), with each new local step being paralleled by a similar stepwise sequence in the frontal lobe and with interconnection by reciprocal projections of many of the local and frontal steps. Most of the steps can be correlated to some extent with previously identified architectonic divisions of the cortex, but, perhaps more importantly, they are also related to areas which seem to be of significance in behavioral experiments.

Behavioral correlations

It is known that lesions of the posterior part of the temporal lobe— approximately area 20—give rise to defects of visual discrimination, whereas those of the more anterior part—approximately area 21—give rise to defects in an animal's ability to perform several discriminative tasks interspersed among one another (Iwai and Mishkin, 1967). These, according to Wilson, Kaufman, Zieler, and Lieb (1972) are, respectively, defects of identification and of retention; though, as pointed out by Weiskrantz (this volume), the anterior temporal defect may be more one of visual categorization.

The final step, in the floor of the superior temporal sulcus and particularly its upper part, is a region of convergence between the corticocortical pathway emanating from the visual areas and a similar one proceeding outward from the somatic sensory cortex (Jones and Powell, 1970) and might be significant as a region of visuospatial interaction.

The direct connection of area 20 with the amygdala furnishes a route into the hypothalamus for visual cortical influences. With the exception of the olfactory regions, no other cortical sensory area has such a direct input to the hypothalamus. It is difficult to correlate this with known deficits in man or animals, although Geschwind (1965) attributes the tameness to visual stimuli of monkeys with a Klüver-Bucy syndrome to disconnection of visually related areas from the amygdala.

In the frontal lobe, bilateral lesions in and about the principal sulcus cause defects of delayed alternation in monkeys (Rosvold and Mishkin, 1961; Gross and Weiskrantz, 1964; Mishkin, 1966; Stamm and Weber-Levine, 1971). Such studies have usually concentrated on both banks of the sulcus, although, from the point of view of connections, it appears to be only the lower bank and regions below the sulcus that are related to the visual system. The areas above and in front of the sulcus are, respectively, related to corticocortical pathways arising ultimately from the somatic sensory and auditory cortices (Jones, 1969; Jones and Powell, 1970). This may be of relevance to the fact that large lesions of this vicinity cause alternation defects referable to all three modalities (Gross and Weiskrantz, 1964).

A similar corticocortical sequence in the cat

A virtually identical sequence of corticocortical connections is found arising from area 17 of the cat (Heath and Jones, 1970, 1971). The cortical fields delineated on the basis of connections are summarized in Figure 7. Areas 17, 18, and 19, and the Clare-Bishop area are reciprocally connected by corticocortical fibers that follow the representation pattern found in these areas (Hubel and Wiesel, 1962, 1965, 1969; Wright, 1969). This reciprocity of connections makes it impossible to regard the geniculocortical and extrageniculocortical visual pathways as two separate visual systems.

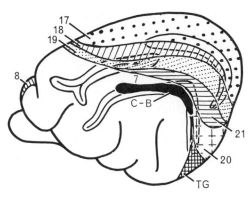

FIGURE 7 A diagram summarizing work of Heath and Jones (1971) and showing the extent and relative dispositions of cortical fields in the cat delineated on the basis of cortical and thalamic connections. Between these fields there is a progression of corticocortical connections virtually identical to that seen in the monkey.

Area 19 also projects outside the visual areas, to what has been called area 20 in the inferior part of the posterior suprasylvian gyrus (Heath and Jones, 1970); this area, in turn, projects to area 21, which lies along the lateral boundary of area 19 at the junction of the middle and posterior suprasylvian gyri. Area 21 sends axons into the fundus of the posterior suprasylvian sulcus, to an area which expands below the sulcus and into the perirhinal cortex. This last area, therefore, may be the equivalent of area 38 or area TG of Bailey and von Bonin (TG). All areas are connected, in some cases reciprocally, with the frontal cortex, and, as in the monkey, area 20 projects to the amygdala.

Thus, a similar sequence of corticocortical connections is present in the cat as in the monkey, though the later steps in the sequence are not necessarily in what has sometimes been regarded in the cat as an incipient temporal lobe. The main difference is that, in the cat, the frontal cortex seems to be less elaborately arranged; all projections to it seem to converge (mainly in the ventral bank of the cruciate sulcus) and not on the gyrus proreus as would be expected from the data on the monkey. Further, the several steps seen in the frontal cortex of the monkey are not so readily discernible.

A question of retinotopy

The topographic organization inherent in the visual and indeed in all sensory systems except possibly the olfactory is well known. This is present up to and including at least the first cortical level, and the anatomical connections follow the representation pattern (see Figure 3). Although there is not a great deal of work available on this, it would appear that beyond area 19 topography starts to become less important in the sense that different parts of

the representation in some of the earlier areas send fibers to the same part of the Clare-Bishop area and of area 20, and small parts of area 20 project diffusely throughout area 21. This convergence of topographic representations probably has a bearing on the fact that the receptive fields of neurons in the Clare-Bishop and inferotemporal regions are very large (Hubel and Wiesel, 1969; Wright, 1969; Gross, this volume; Gross et al., 1972; Dubner and Zeki, 1972). This also appears as though it should be the next step in the much quoted hierarchical cortical processing system of Hubel and Wiesel, and it would be interesting, therefore, to see their work extended to areas 20 and 21 of the cat.

From their work on the response properties of single neurons in areas 17, 18, and 19 of the cat, Hubel and Wiesel (1965) suggested that each could be dealing with progressively more complex aspects of the analysis of visual stimuli. Although it is clear from subsequent work, at least in the monkey (Hubel and Wiesel, 1968) that cells with response features once thought to be characteristic of area 18 or 19 can also be found in area 17, the suggestion of serial processing may still be valid from other points of view. For example, the convergence of topographic representations in certain parts of the extrastriate visual cortex, as shown by both anatomical and physiological methods, could possibly be thought of as indicating a progressive convergence of *place information* necessary for building up a total picture of an object or a concept of external space. The convergence of topographic representations might, therefore, be thought of as a *generalizing sequence*. Nonetheless, the response properties of neurons, so far as they have been studied in the extrastriate areas, seem to be remarkably *specific*, especially in regard to the number of stimulus parameters that are required to make the cell fire. Thus, a mechanism would also have to exist for retaining, within any generalizing sequence, many of the specific features of the original stimulus.

This process has much in common with the logical sequence postulated by many philosophers since the time of John Locke (see for example, Cassirer, 1923) for the formulation of generic concepts. The formulation of any concept is said to involve a progressive generalization from particular cases but in such a way that the final generalization embodies, as in an all-embracing mathematical formula, all the information derived from the individual cases and the logical steps which have been used in its formulation. It might also be relevant to the observations of Diamond and Hall (1969) that, in the tree shrew, the visual cortex is necessary for enabling the animal to perform a task involving the abstraction of one simple pattern unit from another but that the parieto-temporal cortex is necessary for more complex total pattern discriminations.

The view put forward here might be reinforced by drawing attention to the analogous situation in the somatic sensory system. Figure 8 illustrates two degeneration experiments (Jones and Powell, 1969) in which the commencing overlap in area 5 of projections from the trunk and hand areas of the somatic sensory area lying in the postcentral gyrus is demonstrated. The connections of the different parts of the body representation have up to this level remained quite distinct. This commencing convergence of different parts of the body representation is also

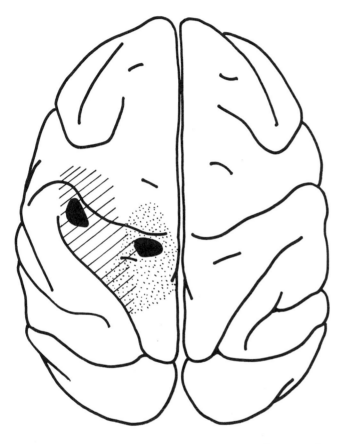

FIGURE 8 Two Nauta experiments reconstructed on a tracing of the same monkey brain and demonstrating the commencing overlap in area 5 of corticocortical connections emanating from the trunk (medial) and arm (lateral) representation in the first somatic sensory cortex. (After Jones and Powell, 1969.)

recognizable at the single unit level. For example, some single neurons in the somatic cortex are activated only from single, specific joints of the body, whereas joint neurons in area 5 can be activated from several joints, even from joints in different limbs (Duffy and Burchfiel, 1971). In other words, the sign of "joint" is retained, but the position of the joint appears to have become less relevant.

A sequence of thalamocortical connections

From the point of view at least of topographic representations, the corticocortical sequence in the temporal lobe can be thought of as a generalizing one. In addition, the steps in the sequence are interrelated with similar ones in the frontal lobe and may finally channel information into the limbic system. It is possible, however, that there is also an anatomical basis for conserving within the generalizing sequence, some of the specifics of the original stimulus as would be necessary in terms of the type of logical analysis postulated above.

Each new step in the corticocortical sequence passing through the temporal lobe cannot be considered as though it were becoming more and more divorced from the primary visual cortex and, therefore, from the original stimulus. This is because every step in the corticocortical sequence also receives a thalamic input passing through the lateroposterior and pulvinar nuclear complex of the thalamus and the old belief that large areas of the temporal cortex are athalamic is no longer tenable. Most of the available information on this aspect using silver staining techniques (and discussion here is limited to such studies) is derived from the cat (Graybiel, 1970, 1972a, b; Heath and Jones, 1970, 1971) (Figures 9 and 10), but complementary data are available in certain other species (Diamond, personal communication).

In the cat, the anteromedial part of the lateroposterior nucleus consists of a homogeneous population of small cells (Figure 10A). It receives fibers from the deep layers of the superior colliculus and projects to area 5 of the suprasylvian cortex (Figure 9). Area 5 is related to the somatic sensory system and is out of the immediate visual sequence. So too is the adjacent area 7 that (with the cingulate cortex) receives its thalamic input from the laterodorsal nucleus (Figures 9 and 10A).

The middle and lateral part of the lateroposterior nucleus *is relevant*. It consists of a mixed population of large and small cells, and it has a specific subnucleus related to it both topographically and from the point of view of connections: the posterior nucleus of Rioch (Figures 9, 10B, and 10C). Each of these receive afferents from all visual areas of the cortex—areas 17 and 18 as well as area 19 and the Clare-Bishop area—but they project only to area 19 and the Clare-Bishop area. Whether these parts of the lateral nuclear complex have any ascending connections is uncertain, but the superficial layers of the superior colliculus have been suggested.

The posterodorsal part of the lateroposterior nucleus consists mainly of large cells (Figure 10C). Its ascending afferents appear to be derived from the superior colliculus, and it projects to area 20 (Figure 9). The so-called *pulvinar* of the cat is a single homogeneous entity (Figure 10C)

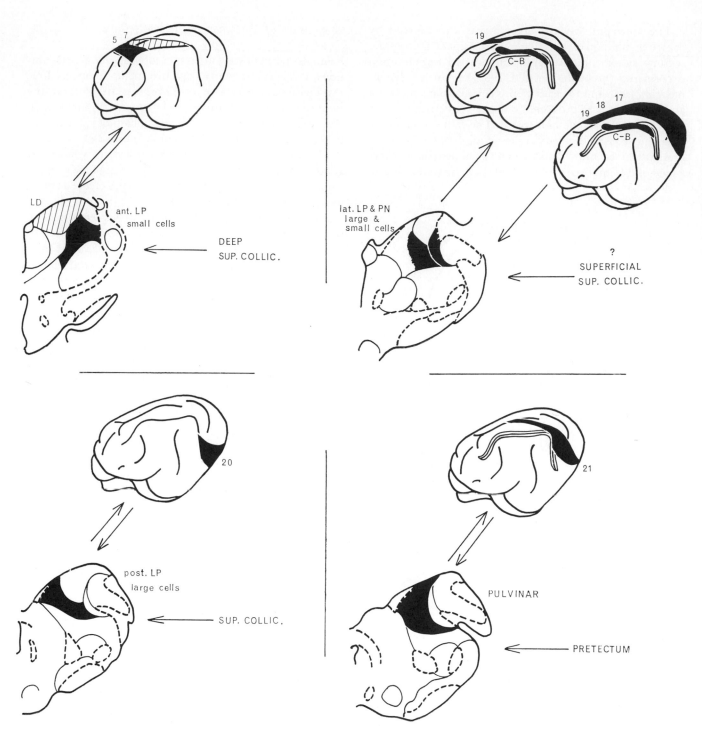

FIGURE 9   A diagram summarizing work of Graybiel (1970; 1972, this volume), Heath and Jones (1970, 1971) and Jones (unpublished observations) and showing the cortical relationships and afferent connections (so far as they are known) of the lateral nuclear complex and the pulvinar of the cat.

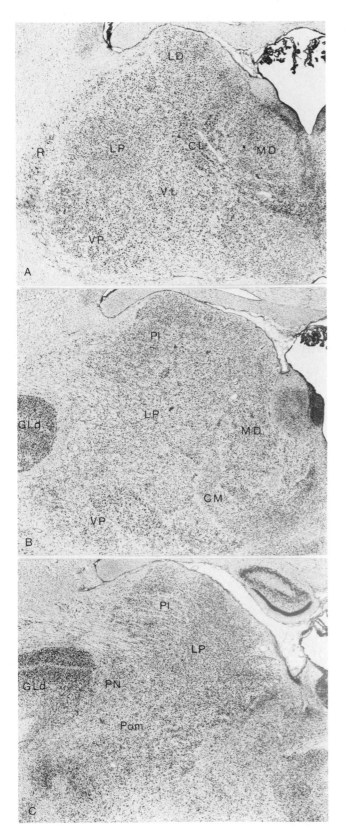

that receives afferents from the pretectum and projects to area 21 (Figure 9). All the thalamocortical cortical connections mentioned are reciprocal (Figure 9).

It has been customary in the past to regard the lateroposterior nucleus of the cat as an entity, but it is clear from the points of view of both cytoarchitecture and connections and possibly also of physiological properties (see Wright, this volume) that it consists of at least three parts. Leaving aside the anteromedial—somatic—part but including the posterior nucleus and the pulvinar, there are, thus, four (visual) parts of the thalamus related to the middle and posterior suprasylvian gyri. This compares quite favorably with the several subdivisions of the pulvinar related to the parieto-temporo-occipital region of the primate, and though the connections of the pulvinar have not been fully explored with silver techniques in the primate, it may be assumed, on the basis of retrograde degeneration studies (Chow, 1950), that the system has similarities with the cat. In which case, it would receive visual and possibly other sensory information from the superior colliculus and pretectum, as well as from all areas of the visual and inferotemporal cortex (Figure 11). It then projects upwards to the various steps in what was referred to as a generalizing sequence of corticocortical connections. It may be suggested, therefore, that there is the possibility at all steps in the generalizing sequence for the thalamus to add data, which in not having passed through the cortex, embody as it were many specific features of the stimulus. The question as to whether these new data are related predominantly to eye movements has been considered by Werner (this volume).

One visual system or several?

Although from the operational point of view it may be useful to consider the visual system in terms of a geniculostriate pathway, perhaps concerned primarily with spatial aspects of visual behavior, and an extra-geniculostriate pathway, perhaps concerned primarily with orientation towards visual stimuli (Schneider, 1969), from the anatomical point of view the duality is less apparent. The various areas are so intimately interconnected that it is impossible to conceive of activity in one being uninfluenced by activity in the others. The main points of interaction are illustrated in the summarizing flow diagram (Figure 11). These are: first, the strong interconnections between areas 17, 18, and 19 and the

FIGURE 10 Photomicrographs of thionin stained sections through the anterior (A), middle (B), and posterior (C) parts of the lateral nuclear complex and pulvinar of the cat's thalamus and showing the different subnuclear regions indicated in Figure 9 (X10).

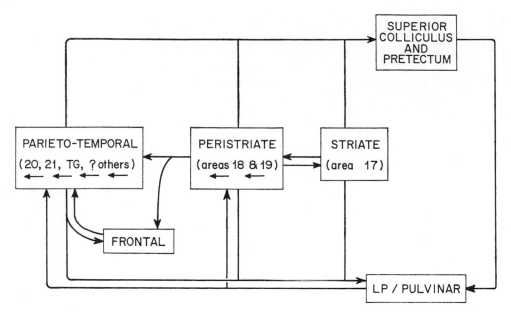

FIGURE 11 Schematic flow diagram indicating the major connectional relationships of the thalamic and cortical visual apparatus. Note the strong interconnections and non-reciprocal feedbacks which make it difficult to separate two or more visual systems.

Clare-Bishop area; second, the projections of areas 17 and 18 to the source of afferents to the "second visual system," the lateral part of the lateroposterior nucleus. (These corticofugal projections make it impossible in any experimental procedure to disconnect the striate area from the extrastriate areas merely by severing their corticocortical connections.) Third, all visually related areas—striate, extrastriate, and inferotemporal (or its homologue)—project to the superficial layers of the superior colliculus (Garey, Jones, and Powell, 1966; Wilson and Toyne, 1970) and some of them to the pretectum; these are, in turn, the main sources of afferents to the extra-geniculate, visual parts of the thalamus.

There are currently many gaps in our understanding of the anatomy of the thalamic and cortical visual apparatus. The present account, has attempted to provide some sort of rational basis for experimental approaches to the visual system in two of the commonest laboratory animals. However, there are many remaining uncertainties and new studies, particularly those of the primate pulvinar, are sorely needed in order to resolve these.

REFERENCES

BRODMANN, K., 1905. Beiträge zur histologishen Lokalisation der Grosshirnrinde. *J. Psychol. Neurol. Lpz.* 4:176–226.

BRODMANN, K., 1909. *Vergleichende Lokalisationslehre der Grosshirnrinde in ihren Prinzipien dargestellt auf Grund des Zellenbaues.* Leipzig: Barth.

BURROWS, G. R., and W. R. HAYHOW, 1971. The organization of the thalamo-cortical visual pathways in the cat. An experimental degeneration study. *Brain, Behav., Evol.* 4:220–259.

CASSIRER, R., 1923. *Substance and Function and Einstein's Theory of Relativity.* Chicago and London: Open Court Publishing Co., pp. 3–26.

CHOW, K. L., 1950. A retrograde cell degeneration study of the cortical projection field of the pulvinar in the monkey. *J. Comp. Neurol.* 93:313–339.

CRAGG, B. G., 1969. The topography of the afferent projections in the circumstriate visual cortex of the monkey studied by the Nauta method. *Vision Res.* 9:733–748.

DIAMOND, I. T., and W. C. HALL, 1969. Evolution of neocortex. *Science* 164:251–262.

DUBNER, R., and S. M. ZEKI, 1972. Response properties and receptive fields of cells in an anatomically defined region of the superior temporal sulcus of the monkey. *Brain Res.* 35: 528–532.

DUFFY, F. H., and J. L. BURCHFIEL, 1971. Somatosensory system: organizational hierarchy from single units in monkey area 5. *Science* 172:273–275.

GAREY, L. J., E. G. JONES, and T. P. S. POWELL, 1968. Interrelationships of striate and extrastriate cortex with the primary relay sites of the visual pathway. *J. Neurol. Neurosurg. Psychiat.* 31:135–157.

GESCHWIND, N., 1965. Disconnexion syndromes in animals and man. *Brain* 88:237–294; 585–644.

GLICKSTEIN, M., R. A. KING, J. MILLER, and M. BERKLEY, 1967. Cortical projections from the dorsal lateral geniculate nucleus of cats. *J. Comp. Neurol.* 130:55–76.

GRAYBIEL, A. M., 1970. Some thalamocortical projections of the pulvinar-posterior system of the thalamus in the cat. *Brain Res.* 22:131–136.

GRAYBIEL, A. M., 1972. Some extrageniculate visual pathways in the cat. *Invest. Ophthal.* 11:322–332.

GROSS, C. G., C. E. ROCHA-MIRANDA, and D. B. BENDER, 1972. Visual properties of neurons in inferotemporal cortex of the macaque. *J. Neurophysiol.* 35:96–111.

GROSS, C. G., and WEISKRANTZ, L., 1964. Some changes in behavior produced by lateral frontal lesions in the macaque. In *The Frontal Granular Cortex and Behavior*. J. M. Warren, and K. Akert, eds. New York: McGraw-Hill, pp. 74–101.

HEATH, C. J., and E. G. JONES, 1970. Connexions of area 19 and the lateral suprasylvian area of the visual cortex of the cat. *Brain Res.* 19:302–305.

HEATH, C. J., and E. G. JONES, 1971. The anatomical organization of the suprasylvian gyrus of the cat. *Ergebn. Anat. Entwicklungsgesch.* 45:1–64.

HUBEL, D. H., and T. N. WIESEL, 1962. Receptive fields, binocular interaction and functional architecture in the cat's visual cortex. *J. Physiol.* 160:106–154.

HUBEL, D. H., and T. N. WIESEL, 1965. Receptive fields and functional architecture in two non-striate visual areas (18 and 19) of the cat. *J. Neurophysiol.* 28:229–289.

HUBEL, D. H., and T. N. WIESEL, 1969. Visual area of the lateral suprasylvian gyrus (Clare-Bishop area) of the cat. *J. Physiol.* 202:251–260.

IWAI, E., and M. MISHKIN, 1967. Quoted by Wilson et al., 1972.

JONES, E. G., 1969. Interrelationships of parieto-temporal and frontal cortex in the rhesus monkey. *Brain Res.* 13:412–415.

JONES, E. G., and T. P. S. POWELL, 1969. Connexions of the somatic sensory cortex of the rhesus monkey. I. Ipsilateral cortical connexions. *Brain* 92:477–502.

JONES, E. G., and T. P. S. POWELL, 1970. An anatomical study of converging sensory pathways in the cerebral cortex of the monkey. *Brain* 93:793–820.

KUYPERS, H. G. J. M., M. K. SZWARCBART, M. MISHKIN, and H. E. ROSVOLD, 1965. Occipitotemporal corticocortical connections in the rhesus monkey. *Exp. Neurol.* 11:245–262.

MAUSS, T., 1908. Die faserarchitektonische Gliederung der Grosshirnrinde bei den niederen Affen. *J. Psychol. Neurol. Lpz.* 13:263–325.

MISHKIN, M., 1966. Visual mechanisms beyond the striate cortex. In *Frontiers in Physiological Psychology*, R. W. Russell, ed. New York: Academic Press, pp. 93–119.

NAUTA, W. J. H., 1964. Some efferent connections of the prefrontal cortex in the monkey. In *The Prefrontal Granular Cortex and Behavior*, J. M. Warren, and K. Akert, eds. New York: McGraw-Hill, pp. 397–407.

OTZUKA, R., and R. HASSLER, 1962. Über Aufbau und Gliederung der corticalen Sehsphäre bei der Katze. *Arch. Psychiat.*

Nervenkr. 203:212–247.

PANDYA, D. N., and H. G. J. M. KUYPERS, 1969. Corticocortical connections in the rhesus monkey. *Brain Res.* 13:13–26.

ROBERTS, T. S., and K. AKERT, 1963. Insular and opercular cortex and its thalamic projection in *Macaca mulatta*. *Schweiz. Arch. Neurol. Neurochir. Psychiat.* 92:1–43.

ROSVOLD, H. E., and M. MISHKIN, 1961. Non-sensory effects of frontal lesions on discrimination learning and performance. In *Brain Mechanisms and Learning*, J. L. Delafresnaye, ed. Oxford: Blackwell, pp. 555–567.

SCHNEIDER, G. E., 1969. Two visual systems. *Science* 163:891–895.

STAMM, J. S., and M. L. WEBER-LEVINE, 1971. Delayed alternation impairments following selective prefrontal cortical ablations in monkeys. *Exp. Neurol.* 33:263–278.

VON BONIN, G., and P. BAILEY, 1947. *The Neocortex of Macaca Mulatta*. Urbana, Ill.: University of Illinois Press.

WHITLOCK, D. G., and W. J. H. NAUTA, 1956. Subcortical projections from the temporal neocortex in *Macaca mulatta*. *J. Comp. Neurol.* 106:183–212.

WILSON, M., H. M. KAUFMAN, R. E. ZIELER, and J. P. LIEB, 1972. Visual identifications and memory in monkeys with circumscribed inferotemporal lesions. *J. Comp. Physiol. Psychol.* 78:173–183.

WILSON, M. E., and B. G. CRAGG, 1967. Projections from the lateral geniculate nucleus in the cat and monkey. *J. Anat. (London)* 101:677–692.

WILSON, M. E., and M. J. TOYNE, 1970. Retino-tectal and corticotectal projections in *Macaca mulatta*. *Brain Res.* 24:395–406.

WRIGHT, M. J., 1969. Visual receptive fields of cells in a cortical area remote from the striate cortex in the cat. *Nature (London)* 223:973–975.

ZEKI, S. M., 1969a. Representation of central visual fields in prestriate cortex of monkey. *Brain Res.* 14:271–291.

ZEKI, S. M., 1969b. The secondary visual areas of the monkey. *Brain Res.* 13:197–226.

ZEKI, S. M., 1971a. Convergent input from the striate cortex (area 17) to the cortex of the superior temporal sulcus in the rhesus monkey. *Brain Res.* 28:338–340.

ZEKI, S. M., 1971b. Cortical projections from two prestriate areas in the monkey. *Brain Res.* 34:19–36.

20 Inferotemporal Cortex: A Single-Unit Analysis

CHARLES G. GROSS, DAVID B. BENDER, and CARLOS E. ROCHA-MIRANDA

ABSTRACT Inferotemporal cortex, the cortex on the inferior surface of the primate temporal lobe, is involved in complex visual functions. Neurons in inferotemporal cortex respond only to visual stimuli and have large receptive fields that always include the center of gaze and often extend into both visual half-fields. Most neurons were sensitive to the size, shape, orientation, or direction of movement of the stimulus. Their visual responsiveness was dependent on input from striate cortex but not on input from the pulvinar. Inferotemporal cortex appears to extend the feature-detecting mechanisms found in the geniculo-striate system and may have additional functions.

LITTLE IS KNOWN of the neuronal mechanisms underlying perception, cognition, and memory: There is still an enormous gap between the phenomena of human consciousness and our knowledge of neuron physiology. The study of inferotemporal cortex may provide an opportunity for reducing this gap. Inferotemporal cortex is "association cortex" on the inferior convexity of the primate temporal lobe (Figure 1A); its removal produces a specifically visual cognitive disorder. Since inferotemporal cortex receives afferents from several visual areas and sends efferents both to other association cortices and to various limbic structures, an understanding of its functions should provide a bridge between sensory physiology and the physiology of cognition.

Inspired in part by the success of single neuron recording in analyzing other cortical areas, we set out to explore inferotemporal cortex at the neuronal level. The first part of this essay outlines the problem by summarizing the behavioral effects of inferotemporal lesions and the anatomic relations of inferotemporal cortex with other brain structures. In the second part, we describe our experiments on the response properties of neurons in inferotemporal cortex (Gross et al., 1967, 1969, 1972). The third part is devoted to our experiments on the anatomical pathways that contribute to these properties (Bender et al., 1972).

CHARLES G. GROSS, DAVID B. BENDER, and CARLOS E. ROCHA-MIRANDA Department of Psychology, Princeton University; Instituto de Biofísica, Universidade Federal do Rio de Janeiro, Brazil

The problem

Both the behavioral effects of inferotemporal lesions and the anatomical connections of inferotemporal cortex provide some clues to the functions of this tissue.

In monkeys bilateral inferotemporal lesions produce a severe deficit in visual discrimination performance. After such lesions, monkeys are impaired in the retention and acquisition of visual discrimination tasks involving pattern, brightness, or color cues. (See reviews by Pribram, 1967; Gross, 1972a, b; Mishkin, 1966). This deficit is specific to vision; learning in other modalities remains normal. However, inferotemporal lesions do not affect basic visual functions such as visual acuity, visual perimetry, backward masking, and several psychophysical thresholds (Cowey, and Weiskrantz, 1967; Gross, 1972b). A similar loss of visual recognition and visual memory follows lesions of the temporal cortex of the nondominant hemisphere in man (Milner, 1968). Thus, in man and monkey, inferotemporal cortex appears to be involved in higher order perceptual and cognitive processes underlying visual recognition.

Inferotemporal cortex receives afferents from two areas that process visual information. The first is the circumstriate belt, which in turn receives projections from striate cortex (Figures 1A and 2). Hubel and Wiesel have proposed a hierarchy of visual feature detectors extending through the geniculostriate system into the circumstriate belt (Hubel and Wiesel, 1962, 1965, 1968, 1970). The projection of the circumstriate belt onto inferotemporal cortex suggests that inferotemporal cortex may be a further stage in this hierarchy. A second source of visual information for inferotemporal cortex is the projection it receives from the pulvinar. The pulvinar, in turn, receives a projection from the superior colliculus (Figure 2). The superior colliculus and perhaps the pulvinar are involved in visual orientating and localizing functions (Denny-Brown and Chambers, 1958; Schneider, 1969; Schiller and Koerner, 1971; Goldberg and Wurtz, 1972; Humphrey, 1970). Thus, inferotemporal cortex may integrate the orienting mechanisms of a tecto-pulvinar system with the feature detection mechanisms of the geniculo-striate-circumstriate system.

Inferotemporal cortex sends outputs to limbic structures such as the amygdala, entorhinal cortex, and hippocampus, which are thought to be involved in motivational and mnemonic aspects of behavior (Figure 2). These connections suggest that inferotemporal cortex

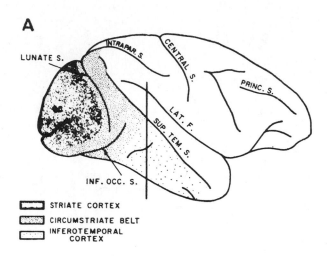

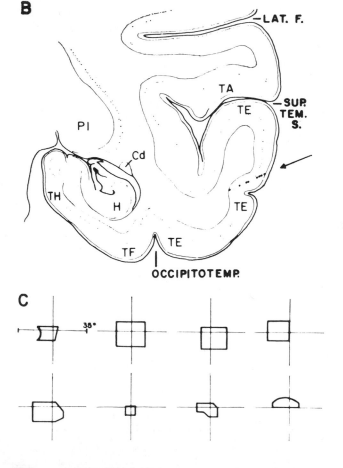

might also play a role in the storage and retrieval of visual information and in giving it gnostic significance (see Weiskrantz, 1970). This possibility is supported by the behavioral analysis of the inferotemporal deficit. The deficit depends not only on the visual discriminanda but also on several nonsensory factors such as the animal's prior experience, the training procedure used, and the type of reinforcement (Gross, 1972a, b).

In summary, the effects of inferotemporal lesions and the connections of inferotemporal cortex indicate that this cortex plays a unique and integrative role in visual pattern recognition.

Visual properties of inferotemporal neurons

Our methods have been described in detail elsewhere (Gross et al., 1972). Monkeys were anesthetized with a mixture of nitrous oxide and oxygen, paralyzed to eliminate eye movements, and artificially respired. Their eyes were focused on a 64° × 64° tangent screen onto which visual stimuli were projected. Discharges of isolated neurons were recorded with platinum-iridium electrodes and displayed on an oscilloscope, played over a loudspeaker, recorded on magnetic tape, and analyzed with a PDP-12 computer. The computer was also used to control the presentation of the visual stimuli on the screen.

MODAL SPECIFICITY AND LATENCY   All the neurons we encountered were spontaneously active. We were able to drive about three-quarters of them with visual stimuli, whereas none were responsive to a variety of auditory, somesthetic, and olfactory stimuli. This demonstration

FIGURE 1   (A) Diagram of lateral view of cerebral hemisphere of *Macaca mulatta*. The circumstriate belt is shown after Kuypers et al. (1965). In this chapter "circumstriate belt" and "prestriate cortex" are used synonymously. Inferotemporal cortex is defined as corresponding to von Bonin and Bailey's (1947) area TE. The vertical line indicates the level of the coronal section shown in B. (B) Coronal section through the temporal lobe. The arrow shows the site of entry of a typical pass and the dots indicate the approximate location of representative cells recorded on that pass. (C) Receptive fields of the inferotemporal neurons whose locations are shown in B. The receptive fields recorded at increasing depth are shown from left to right starting at the top left. The axes represent the horizontal and vertical meridia of the visual field, and the half-field contralateral to the recording electrode is on the left. The scale is in degrees of visual angle. Note that all receptive fields include the fovea and that some extend well into both visual half-fields. INTRAPAR. S. = intraparietal sulcus; PRINC. S. = principal sulcus; LAT. F. = lateral fissure; SUP. TEM. S. = superior temporal sulcus; INF. OCC. S. = inferior occipital sulcus; Cd = caudate nucleus; H = hippocampus; Pl = pulvinar; TE (inferotemporal cortex), TA, TF and TH refer to cytoarchitectonic areas of von Bonin and Bailey.

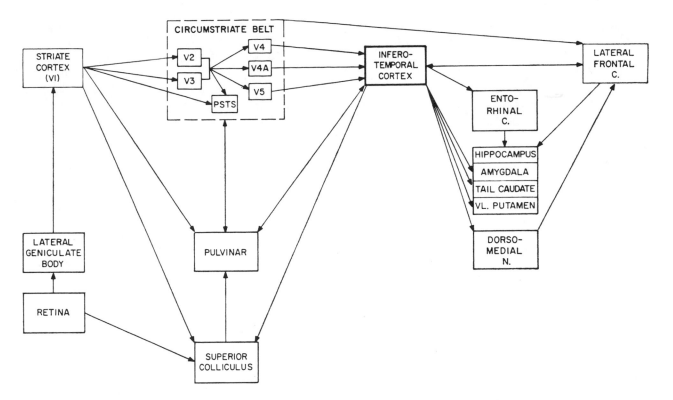

FIGURE 2   Major ipsilateral afferent connections of infero-temporal cortex in *Macaca mulatta*. Each arrowhead is not intended to indicate a single synapse. For example, axons from the superior colliculus may not terminate in the pulvinar on the same neurons that project from the pulvinar to inferotemporal

cortex. V2, V3, V4, V4A, V5, and the posterior bank of the superior temporal sulcus (PSTS) are subdivisions of the cir-cumstriate belt based on the anterograde degeneration studies of Zeki (1971b) and others. (See also Figure 1.)

that inferotemporal neurons respond only to visual stimuli parallels the previous findings that inferotemporal lesions impair only visual learning.

The response latency of inferotemporal neurons to visual stimuli was surprisingly long. No neurons respon-ded before 70 msec, and the mean latency of the earliest part of the response was about 120 msec (Figure 3). In many units, time-locked activity continued for 400 msec or longer after the stimulus offset.

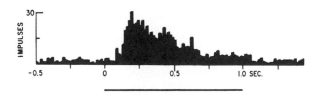

FIGURE 3   Poststimulus time histogram for a unit in infero-temporal cortex. The stimulus was a vertical $1° \times 70°$ black slit centered on the fovea. It was presented twenty times for a duration of one second with an interstimulus interval of 10 sec. The horizontal line indicates the stimulus duration. The vertical scale indicates total impulses per bin. The bin width was 15.6 msec. The long latency of the response was typical of infero-temporal neurons.

RECEPTIVE FIELDS   The activity of almost all neurons was altered only by stimulation of a restricted portion of the visual field; i.e., the neurons had "receptive fields." Every receptive field included the center of gaze, the fovea. Over half of the receptive fields extended well into both visual half-fields, about one-third were in the half-field contralateral to the electrode, and the rest were in the ipsilateral half-field (Figures 1C, 4, and 5A). This is in striking contrast to the geniculostriate system in which the receptive fields are confined, within a few degrees, to the contralateral half-field and do not invariably include the center of gaze (e.g., Hubel and Wiesel, 1960, 1965, 1968). Another unusual feature of the inferotemporal receptive fields was their large size. They were usually more than $10° \times 10°$ with a median area of 418 square degrees and an interquartile range of 150 to 1410 square degrees, (Figures 1C and 4).

EFFECTS OF STIMULUS PARAMETERS   The strength of a response or even its existence was usually dependent on several parameters of the stimulus. Among these param-eters were contrast, wavelength, size, shape, orien-tation, and direction of movement. Some neurons were

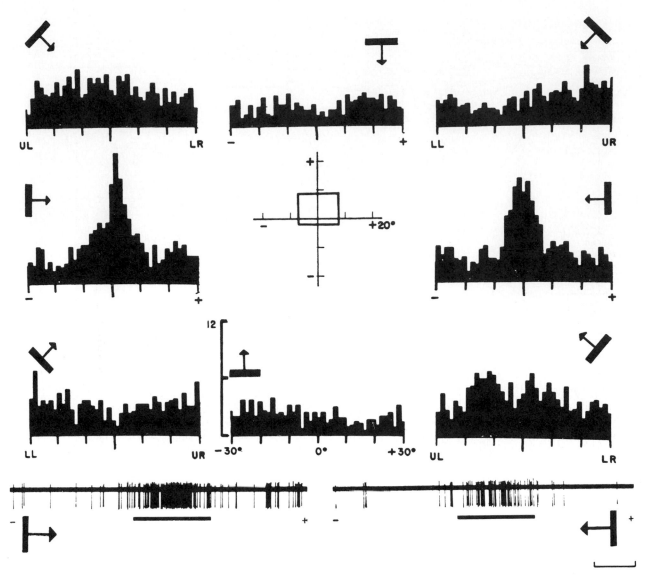

FIGURE 4 Receptive field and responses of an inferotemporal neuron that was sensitive to the orientation of a white slit but not to the direction of its movement. The histograms indicate unit activity as a function of the retinal locus of a 1° × 70° white slit moving at 5° per second in the direction indicated above each histogram. Each histogram was generated by seven sweeps of the stimulus. For the eight histograms the vertical scale indicates total number of impulses per bin. The bin width was 293 msec. The horizontal scale indicates degrees of visual angle. The middle of each horizontal scale (0°) represents the center of gaze. The receptive field of this unit is shown in the center of the array of histograms. Plus (+) in all parts of the figure indicates upper or right of the visual field; minus (−) indicates lower or left; UL, upper left; LR, lower right; LL, lower left; UR, upper right. The lower part of the figure shows responses of the unit to a single sweep of the stimulus in the direction indicated. The lines below each trace correspond to the horizontal extent of the receptive field. The marker indicates 8° or 1.6 sec. The histograms and trace in which the arrow is shown on the left were generated from left to right, whereas the converse was true where the arrow is shown on the right. See also legend to Figure 1 (After Gross et al., 1972).

sensitive to all of these parameters, whereas other neurons were sensitive to only a few. Most neurons responded more vigorously to a moving stimulus than a stationary one. About three-quarters of the responsive neurons could be driven by dark stimuli on a white background, about three-quarters by light stimuli on a dark background, and about half of them were responsive to both contrast conditions. For the great majority of neurons, a white, black, or colored slit about 1° in width was a more potent stimulus than other rectangular or circular stimuli. About half of the units were sensitive to the orientation of the slit. Of these, some responded best or exclusively to one direction of movement of the slit in its optimal orientation while other neurons responded equally well to both directions of movement, e.g. Figure 4. We found a number of neurons that were more responsive to a colored stimulus than to a white stimulus presented over a great range of intensities. Red sensitive neurons were far more common than either blue or green ones. The optimal stimulus or *trigger feature* for a cell was optimal throughout the entire receptive field, although stimulation of the receptive field at the fovea usually produced the most vigorous response.

A minority of the cells we studied had extremely specific and complex trigger features, most of which were discovered accidentally. Among such neurons were ones whose best stimulus appeared to be a monkey hand, the shadow of a hemostat, a bottle brush, or a specific curvature. Some of these neurons responded more vigorously to a particular three-dimensional object than to any two-dimensional representation of it. It is possible, of course, that some of the apparently nonspecific neurons in our sample actually had highly specific trigger features that we never found. That is, a neuron that responded best to a 1° × 5° red slit oriented at 45° within its receptive field may not have been coding this size, shape, color, and orientation. Rather, its trigger feature might have been a far more specific, complex, and perhaps meaningful stimulus that we never used and that happened to share some of the stimulus parameters of the stimulus we did use.

In summary, we found that neurons in inferotemporal cortex were exclusively visual and had large receptive fields that always included the fovea and often extended into both visual half-fields. They coded (i.e., differentially responded to) many aspects of the stimulus and some of them had highly specific trigger features.

Functional anatomy

AFFERENT CONNECTIONS OF INFEROTEMPORAL CORTEX How do neurons in inferotemporal cortex receive the visual information that is the basis of their visual properties? Inferotemporal cortex receives projections from both the circumstriate belt (Kuypers et al., 1965; Jones, and Powell, 1970) and from the pulvinar (Chow, 1950). Both these structures are known to process visual information, and both, in turn, receive projections from visual areas that have been well studied.

Striate cortex projects to at least three areas of the circumstriate belt, which Zeki has named V2, V3, and the "posterior superior temporal sulcus field" (Zeki 1969, 1971a, b; Cragg and Ainsworth, 1969; Cowey, personal communication). These projections appear to be retinotopically organized. Both V2 and V3 project to the same posterior superior temporal area to which striate cortex projects and also to three regions in the anterior portion of the circumstriate belt, two of these Zeki has designated

V4 and V4A, and the third may be called V5 (Zeki, 1971b; Cowey, personal communication). Retinotopic organization, at least anatomically, is no longer discernable in V4, V4A, and V5. Finally, areas V4, V4A, and V5 project in a diffuse fashion to inferotemporal cortex (Cowey, personal communication). There are extensive interhemispheric connections between the circumstriate belts carried by the splenium of the corpus callosum (Zeki, 1970; Pandya et al., 1971). These interhemispheric connections could bring representations of the ipsilateral visual field to each hemisphere. Thus, each inferotemporal cortex could receive visual information about the entire visual field from striate cortex by way of a minimum of two synapses in the circumstriate belt. However, the long latency of the visual responses of inferotemporal neurons suggests that there is more processing in the circumstriate belt than this minimum of two synapses implies. Perhaps prior to its arrival in inferotemporal cortex, there is considerable recycling of visual information (see Barlow et al., 1972).

In addition to this corticocortical input, inferotemporal cortex also receives a projection from the pulvinar (Chow, 1950). Neurons in the inferior pulvinar are responsive to visual stimuli, and a retinotopic organization of this area has been described (Allman et al., 1972). The inferior pulvinar, in turn, receives projections both from striate cortex (Myers, 1962) and the superior colliculus (Mishkin, 1972; Mathers, 1971). Thus, inferotemporal cortex could also receive visual information by way of synapses in the pulvinar from either striate cortex or the superior colliculus. The afferent and efferent connections of inferotemporal cortex are summarized in Figure 2.

FUNCTIONS OF THE AFFERENTS TO INFEROTEMPORAL COR-TEX What are the functional roles of the cortical and subcortical projections to inferotemporal cortex? Normal visual pattern discrimination learning is dependent on the input to inferotemporal cortex from striate cortex by way of synapses in the circumstriate belt. Direct or indirect interruption of this corticocortical pathway or of its foveal component produces a deficit in visual learning at least as severe as that after inferotemporal lesions (Mishkin, 1966, 1972; Cowey, and Gross, 1970). By contrast, destruction of the pulvinar does not impair visual pattern learning (Chow, 1954; Mishkin, 1972).

In three experiments we directly studied the contribution of the afferent pathways to inferotemporal cortex (Bender et al., 1972). In the first experiment, we totally removed the striate cortex of one hemisphere in several monkeys. If the visual properties of inferotemporal cortex were dependent on striate cortex, then after unilateral ablation of striate cortex inferotemporal units should have been responsive only to visual stimulation in the half-field contralateral to the remaining striate cortex (Figure 5).

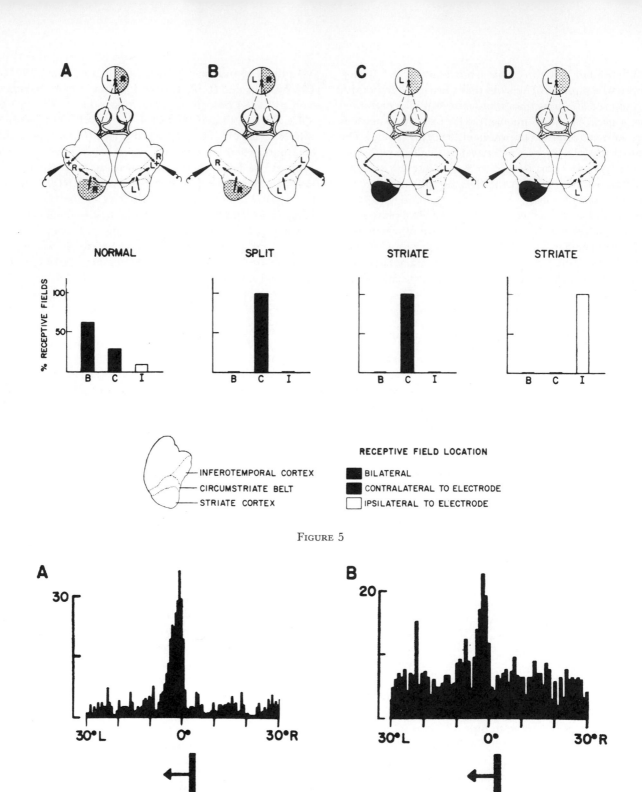

NORMAL   SPLIT   STRIATE   STRIATE

RECEPTIVE FIELD LOCATION

INFEROTEMPORAL CORTEX
CIRCUMSTRIATE BELT
STRIATE CORTEX

■ BILATERAL
■ CONTRALATERAL TO ELECTRODE
□ IPSILATERAL TO ELECTRODE

FIGURE 5

FIGURE 6

This hypothesis was confirmed: In the animals with unilateral striate lesions, all the receptive fields in both inferotemporal cortices were unilateral and confined to the visual half-field contralateral to the intact striate cortex, whereas in normal animals the majority of receptive fields had extended well into both visual half-fields (see Figures 5A, 5C, 5D, and 6B).

In the second experiment, the corpus callosum and anterior commissure were sectioned. If these interhemispheric connections were the basis of the responses to stimulation in the ipsilateral half-field that characterized two-thirds of the inferotemporal neurons, then sections of these connections should eliminate such responses. This hypothesis was also confirmed as shown in Figure 5B and 6A.

Although these experiments demonstrated that inferotemporal cortex receives visual information from striate cortex, they did not establish the specific route, except that it must include the corpus callosum or anterior commissure. As shown in Figure 2, inferotemporal cortex could receive input from striate cortex either by way of synapses in the circumstriate belt or by way of synapses in the pulvinar (the pulvinar route could also involve the superior colliculus, the circumstriate belt, or both areas).

In the third experiment, we studied the effects of bilateral pulvinar lesions to decide between a corticocortical route and a route involving the pulvinar. After pulvinar lesions, we found the incidence of responsive neurons in inferotemporal cortex and their trigger features to be the same as those in intact monkeys. There was, however, a dramatic and unexpected effect of the pulvinar lesions: Virtually none of the inferotemporal neurons had discrete receptive fields; rather, they responded to visual stimulation throughout the $64° \times 64°$ area of the tangent screen on which the visual stimuli were projected (see Figure 7). Thus the receptive field boundaries of inferotemporal neurons appear dependent on the input from the pulvinar, whereas their trigger features are dependent on the corticocortical input.

The devastating effects of interrupting the corticocortical pathway from striate cortex to inferotemporal cortex, both on the visual properties of inferotemporal neurons and on the performance of visual discrimination tasks, indicate the crucial nature of this pathway for the functioning of inferotemporal cortex. The topography of this multisynaptic pathway suggests explanations for two unusual properties of inferotemporal neurons. The invariable inclusion of the fovea in the receptive fields is presumably due to the heavy projections inferotemporal cortex receives from the regions of V4, V4A, and V5 onto which the foveal representation in V2 and V3 projects (Cowey, personal communication). The responsiveness to stimuli in the ipsilateral visual field presumably derives from the interhemispheric connections of V4, V4A, and V5 through the corpus callosum or possibly the interhemispheric connections of inferotemporal cortex itself.

The unexpected loss of discrete receptive field organization after the pulvinar lesions has interesting implications. It suggests that the receptive field boundaries of neurons in inferotemporal cortex may be determined, at least in part, by the pulvinar. This hypothesis is consistent with both the apparent absence of retinotopic organization in the areas of the circumstriate belt that project to inferotemporal cortex and with the presence of retinotopic organization in the pulvinar. The apparent dependence of the inferotemporal receptive field boundaries on the pulvinar suggests that in the behaving animal the pulvinar may modulate receptive field size as a function of attention or some other state of the animal (see Goldberg, and Wurtz, 1972).

In summary, the visual responsiveness and stimulus specificities of inferotemporal neurons are dependent on

---

FIGURE 5 Location of receptive fields of inferotemporal neurons: (A) in normal monkeys; (B) after section of the interhemispheric commissures; (C) and (D), after ablation of striate cortex. For each of the four conditions, the bar graphs show the percent of the receptive fields that extended into both visual half-fields (Bilateral), the percent that were confined to the half-field contralateral to the recording site (Contralateral to electrode), and the percent that were confined to the half-field ipsilateral to the recording site (Ipsilateral to electrode). The data shown in A and B were obtained from recordings in both hemispheres, the data in C from the inferotemporal cortex contralateral to the striate lesion, and the data in D from the inferotemporal cortex ipsilateral to the striate lesion. The brain diagrams (upper) show how information from the right (R) and the left (L) visual half-fields could reach inferotemporal cortex along a corticocortical route and how each lesion interferes with this pathway. The brain diagrams are after Mishkin (1966).

FIGURE 6 Typical responses of neurons in the right inferotemporal cortex in an animal with previous section of the corpus callosum and anterior commissure (A) and in an animal with previous ablation of the left striate cortex (B). See Figures 5B and 5D. The histograms (upper) were generated by ten sweeps of the stimulus, which was a $0.5° \times 70°$ white slit (A) or a $1° \times 70°$ white slit (B) moving in the indicated direction at 5° per second. The vertical scale indicates the total number of impulses per bin. The bin width was 123 msec (A) or 176 msec (B). The horizontal scale indicates degrees of visual angle. The oscilloscope trace (lower) shows the response to a single presentation of the stimulus. The scale for the trace indicates 10° or 2 sec. In both the upper and lower figures 0° represent the center of gaze. Both the histogram and trace were generated from right to left. Note that after the commissure section (A) the neuron responds only in the half-field contralateral to the electrode and that after the striate lesion (B) the neuron responds only in the half-field contralateral to the intact striate cortex.

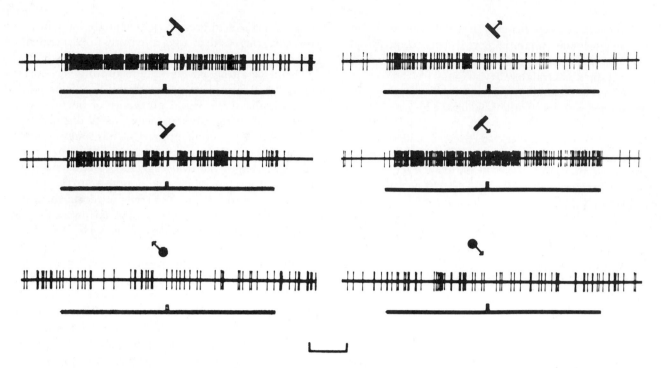

FIGURE 7 Typical responses of a neuron in inferotemporal cortex in an animal that had previously received bilateral pulvinar lesions. The upper four traces show the responses to a 1° × 64° white slit moving at 7° per second in the directions indicated. The lower two traces show responses to a white disc of 7.5° diameter also moving at 7° per second. The horizontal line be- low each trace indicates when the stimulus was traversing the 64° tangent screen and visible to the animal. The marker indicates 10° or 1.4 sec. Note that the unit responds to the slits of light everywhere on the tangent screen independent of the direction of movement but does not respond to the disc.

the input they receive from striate cortex by way of synapses in the circumstriate belt. The receptive field boundaries of inferotemporal neurons appear dependent on the projections they receive from the pulvinar.

Concluding comments

The behavioral effects of inferotemporal lesions indicate that inferotemporal cortex is involved in visual pattern recognition. Broadly viewed, pattern recognition includes (a) detection of stimulus features, (b) their synthesis across time and space and integration with eye and body movements, and finally, (c) the storage and retrieval of visual information. In the introductory portion of this chapter we speculated that the anatomical connections of inferotemporal cortex indicated that it might be involved in all three aspects of pattern recognition. How do our experimental results bear on these speculations?

The projections inferotemporal cortex receives from the circumstriate belt indicate that inferotemporal cortex might be a further stage in the hierarchy of feature detection mechanisms described by Hubel and Wiesel in striate and circumstriate cortex of the cat and monkey. They have proposed that the geniculostriate system con- sists of a series of converging and diverging connections such that at each successive tier of processing mechanisms, single neurons respond to increasingly specific visual stimuli falling on an increasingly wider area of the retina (Hubel and Wiesel, 1962, 1968). This hierarchy continues into the circumstriate belt (or prestriate cortex) where cells with still more specific trigger features and larger receptive fields are found (Hubel and Wiesel, 1965, 1970). Our finding that the visual responsiveness of inferotemporal neurons depends on striate cortex certainly indicates that inferotemporal cortex further processes visual information received from the geniculostriate system. Some of the neurons in inferotemporal cortex had properties that continue the trends seen in ascending the geniculo-striate-circumstriate system: They had larger receptive fields and more specific trigger features than neurons in striate and circumstriate cortex. Thus, the properties of at least some inferotemporal neurons support the idea that inferotemporal cortex is a further stage in the hierarchy of stimulus feature detection that begins in the geniculostriate system.

However, the majority of inferotemporal neurons, although they had large receptive fields, appeared less sensitive to such stimulus parameters as length, width, and orientation than the cells in striate and circumstriate cortex described by Hubel and Wiesel (see, however, Poggio, 1972). This apparent lack of specificity may have been because these cells had complex and specific trigger features that we never found. The existence of cells in our sample with very complex trigger features is consistent with this possibility. On the other hand, these cells could have other functions than to continue the hierarchy of feature detectors begun in the geniculostriate system. They might be involved in some fundamentally different type of stimulus analysis mechanism, such as one involving Fourier-like analysis (see Pollen et al., 1971) or one involving ensembles of cells. Or, they might be concerned with the two other facets of pattern recognition, integration with head and eye movement, and storage and retrieval.

In the introduction, we proposed that the tecto-pulvinar input to inferotemporal cortex implied that inferotemporal cortex might integrate the orienting or "where it is" functions of the tectum with the feature detection or "what it is" functions of the geniculostriate system. The finding that receptive field boundaries disappear after pulvinar lesions lends some support to this possibility. We also speculated that inferotemporal cortex might be involved in gnostic and mnemonic functions. The extreme specificity of a few inferotemporal units, perhaps the result of experience, is consistent with this possibility. However, the present experiments are not really adequate tests of these hypotheses. The speculation that the activity of inferotemporal neurons depends on more than the retinal stimulus—that it also depends on the orientation of the animal toward the stimulus and on the meaning of the stimulus for the animal must be directly tested in behaving and thinking monkeys not in paralyzed anesthetized ones.

Summary

In primates inferotemporal cortex is a "higher" visual area. Its removal produces a severe deficit in visual learning while leaving basic visual functions and learning in other modalities unaffected.

We have studied the visual properties of single neurons in inferotemporal cortex. They had receptive fields that always included the fovea and were relatively large. Many fields extended into both visual half-fields; some were confined to the contralateral or ipsilateral half-field. Most inferotemporal neurons were sensitive to several parameters of the stimulus such as contrast, wavelength, size, shape, orientation, and direction of movement. Some had highly complex stimulus requirements.

Inferotemporal cortex receives a corticocortical input from striate cortex by way of synapses in the circumstriate belt and a subcortical input from the pulvinar. We have shown that the visual responsiveness and trigger features of inferotemporal neurons depend on the corticocortical input from striate cortex whereas the receptive field boundaries appear to depend on the subcortical input.

The results suggest that at least some inferotemporal neurons form a further stage in the hierarchy of pattern recognition mechanisms that begin in the geniculostriate system. Other possible functions for inferotemporal cortex were proposed.

ACKNOWLEDGMENTS Mortimer Mishkin, Laura Frishman, and Susan Volman participated in some of the experiments described, and the manuscript was typed by Susan F. Ring. Preparation of the paper and research of the laboratory were supported by grants from the National Institute of Mental Health and the National Science Foundation.

REFERENCES

ALLMAN, J. M., J. H. KAAS, R. H. LANE, and F. M. MIEZIN, 1972. A representation of the visual field in the inferior nucleus of the pulvinar in the owl monkey (*Aotus trivirgatus*). *Brain Res.* 40:291–302.

BARLOW, H. B., R. NARASIMHAN, and A. ROSENFELD, 1972. Visual pattern analysis in machines and animals. *Science* 177: 567–574.

BENDER, D. B., C. E. ROCHA-MIRANDA, C. G. GROSS, S. VOLMAN, and M. MISHKIN, 1972. Effects of striate lesions and commissure section on the visual response of neurons in inferotemporal cortex. *Physiologist* 15:84.

CHOW, K. L., 1950. A retrograde cell degeneration study of the cortical projection field of the pulvinar in the monkey. *J. Comp. Neurol.* 93:313–340.

CHOW, K. L., 1954. Lack of behavioral effects following destruction of some thalamic association nuclei in monkey. *A.M.A. Arch. Neurol. Psychiat.* 71:762:771.

COWEY, A., and C. G. GROSS, 1970. Effects of foveal prestriate and inferotemporal lesions on visual discrimination by rhesus monkeys. *Exp. Brain Res.* 11:128–144.

COWEY, A., and L. WEISKRANTZ, 1967. A comparison of the effects of inferotemporal and striate cortex lesions on the visual behavior of rhesus monkeys. *Quart. J. Exp. Psychol.* 19:246–253.

CRAGG, B. G., and A. AINSWORTH, 1969. The topography of the afferent projections in the circumstriate visual cortex of the monkey studied by the Nauta method. *Vision Res.* 9:733–747.

DENNY-BROWN, D., and R. A. CHAMBERS, 1958. Visual orientation in the macaque monkey. *Trans. Amer. Neurol. Assoc.* 20:37–40.

GOLDBERG, M. E., and R. H. WURTZ, 1972. Activity of superior colliculus in behaving monkey. II. Effect of attention on neuronal responses. *J. Neurophysiol.* 35:560–574.

GROSS, C. G., D. B. BENDER, and C. E. ROCHA-MIRANDA, 1969. Visual receptive fields of neurons in inferotemporal cortex of the monkey. *Science* 166:1303–1306.

GROSS, C. G., P. H. SCHILLER, C. WELLS, and G. L. GERSTEIN, 1967. Single unit activity in temporal association cortex of the monkey. *J. Neurophysiol.* 30:833–843.

GROSS, C. G., C. E. ROCHA-MIRANDA, and D. B. BENDER, 1972.

Visual properties of neurons in inferotemporal cortex of the macaque. *J. Neurophysiol.* 35:96–111.

GROSS, C. G., 1972a. Visual functions of inferotemporal cortex. In *Handbook of Sensory Physiology*, R. Jung, ed. Berlin: Springer-Verlag, Vol. 7: part 3B (in press).

GROSS, C. G., 1972b. Inferotemporal cortex and vision. In *Progress in Physiological Psychology*, E. Stellar and J. M. Sprague, eds. New York: Academic Press, Vol. 5 (in press).

HUBEL, D. H., and T. N. WIESEL, 1960. Receptive fields of optic nerve fibers in the spider monkey. *J. Physiol. (London)* 154: 572–580.

HUBEL, D. H., and T. N. WIESEL, 1962. Receptive fields, binocular interaction and functional architecture in the cat's visual cortex. *J. Physiol. (London)* 160:106–154.

HUBEL, D. H., and T. N. WIESEL, 1965. Receptive fields and functional architecture in two nonstriate visual areas (18 and 19) of the cat. *J. Neurophysiol.* 28:229–289.

HUBEL, D. H., and T. N. WIESEL, 1968. Receptive fields and functional architecture of monkey striate cortex. *J. Physiol. (London)* 195:215–243.

HUBEL, D. H., and T. N. WIESEL, 1970. Stereoscopic vision in macaque monkey. *Nature (London)* 225:41–42.

HUMPHREY, N. K., 1970. What the frog's eye tells the monkey's brain. *Brain, Behav., Evol.* 3:324–337.

JONES, E. G., and T. P. S. POWELL, 1970. An anatomical study of converging sensory pathways in the cerebral cortex of the monkey. *Brain* 93:793–820.

KUYPERS, H. G. J. M., M. K. SZWARCBART, M. MISHKIN, and H. E. ROSVOLD, 1965. Occipito-temporal cortico-cortical connections in the rhesus monkey. *Exp. Neurol.* 11:245–262.

MATHERS, L. H., 1971. Tectal projection to the posterior thalamus of the squirrel monkey. *Brain Res.* 35:295–298.

MILNER, B., 1968. Visual recognition and recall after right temporal-lobe excision in man. *Neuropsychol.* 6:191–209.

MISHKIN, M., 1966. Visual mechanisms beyond the striate cortex. In *Frontiers in Physiological Psychology*, R. Russell, ed. New York: Academic Press, pp. 93–119.

MISHKIN, M., 1972. Cortical visual areas and their interaction. In *The Brain and Human Behavior*, A. G. Karczmar and J. C. Eccles, eds. Berlin: Springer-Verlag, pp. 187–208.

MYERS, R. E., 1962. Striate cortex connections in the monkey. *Fed. Proc.* 21:352.

PANDYA, D. N., E. A. KAROL, and D. HEILBRONN, 1971. The topographical distribution of interhemispheric projections in the corpus callosum of the rhesus monkey. *Brain Res.* 32: 31–43.

POGGIO, G. F., 1972. Spatial properties of neurons in striate cortex of unanesthetized macaque monkey. *Invest. Ophthalmol.* 11:368–376.

POLLEN, D. A., J. R. LEE, and J. H. TAYLOR, 1971. How does the striate cortex begin the reconstruction of the visual world? *Science* 173:74–77.

PRIBRAM, K. H., 1967. Neurophysiology and learning: I. Memory and the organization of attention. In *Brain Function, Vol. 4: Brain Function and Learning*, D. B. Lindsley and A. A. Lumsdaine, eds. Berkeley: University of California Press.

SCHILLER, P. H., and F. KOERNER, 1971. Discharge characteristics of single units in the superior colliculus of the alert rhesus monkey. *J. Neurophysiol.* 34:920–936.

SCHNEIDER, G. E., 1969. Two visual systems. *Science* 163:895–902.

VON BONIN, G., and P. BAILEY, 1947. *The Neocortex of Macaca Mulatta*. Urbana, Ill.: University of Illinois Press.

WEISKRANTZ, L., 1970. Visual memory and the temporal lobe of the monkey. In *The Neural Control of Behavior*, R. E. Whalen, R. F. Thompson, M. Verzeano, and N. M. Weinberger, eds. New York: Academic Press, pp. 239–256.

WURTZ, R. H., and M. E. GOLDBERG, 1972. Activity of superior colliculus in behaving monkey. IV. Effects of lesions on eye movements. *J. Neurophysiol.* 35:587–596.

ZEKI, S. M., 1969. Representation of central visual fields in prestriate cortex of monkey. *Brain Res.* 14:271–291.

ZEKI, S. M., 1970. Interhemispheric connections of prestriate cortex in monkey. *Brain Res.* 19:63–75.

ZEKI, S. M., 1971a. Convergent input from the striate cortex (area 17) to the cortex of the superior temporal sulcus in the rhesus monkey. *Brain Res.* 28:338–340.

ZEKI, S. M., 1971b. Cortical projections from two prestriate areas in the monkey. *Brain Res.* 34:19–35.

21 The Striate Cortex and the Spatial Analysis of Visual Space

DANIEL A. POLLEN and JOSEPH H. TAYLOR

ABSTRACT Our experimental and theoretical findings, together with the psychophysical and electrophysiological evidence of many workers, support (but are not yet sufficient to prove) a theory that a neural population sufficient to specify two-dimensional spatial frequency decompositions of subdomains of visual space exists at the complex cell output stage in the striate cortex. An hypothesis that the brain may construct global Fourier transforms of visual space beyond the striate cortex is considered. Some advantages of a Fourier transform representation for pattern recognition, cross-correlation, and information storage are considered. The properties of a neural substratum that could subserve such transforms are defined in terms of electrophysiologically measurable variables. Some critical experimental tests that could either confirm or invalidate these ideas are discussed.

ELSEWHERE IN this volume Fergus Campbell has reviewed the remarkable work that he, Robson, and other coworkers have carried out over the past decade. These studies along with the classical work of Hubel and Wiesel (1959, 1962, 1965, 1968) in cat and monkey comprise a large part of the foundation for further probes of the problem of visual pattern recognition. In the present chapter we would like first to consider what the main "boundary-value" conditions are, and we would go back to Lashley (1942) who considered that the fundamental common problem of sensory physiology and neuropsychology was to explain how an organism can identify an object as the same object, despite changes of its position in visual space, its apparent size, its brightness, and certain other qualities that apparently make little difference in the ability to recognize the object. A brief report of our approach to this problem has been published (Pollen et al., 1971).

For the present we shall ignore the complexities of binocular vision and color sensitivity, and we shall consider only the case of responses to stationary visual stimuli. In this sort of study the usual method of the sensory physiologist is to record the activity of single neurons while varying a wide range of stimulus parameters. The region of visual space over which a cell responds to stimulation (either by an increase or decrease in firing rate) is called its receptive field. Using this method, the physiologist sees just a tiny part of the picture of how the entire population of neurons is responding.

Receptive field properties

Let us first briefly review the receptive field properties of neurons from retina to striate cortex. In the retina, ganglion cells have approximately concentric *on* and *off* regions; a spot of light anywhere within the circular center produces either an increase or decrease in cell firing rate (Kuffler et al., 1953). If the spot extends into an antagonistic surrounding area, the response is diminished. These two types of concentric cells tend to provide a point-to-point or topographic representation of visual space. In the lateral geniculate body, the topographic spatial representation still holds, but the penalty for encroaching upon the surround increases (Hubel and Wiesel, 1959). At the input stage of the striate cortex, the middle cortical layers, we come to one of the most notable discoveries of Hubel and Wiesel (1959, 1962). Here there are *simple* cells where spots of light produce excitatory responses anywhere within a long and narrow rectangular area which is flanked by an inhibitory surround. Both *on* and *off* simple cells are found, as well as cells which respond preferentially to light-dark borders. Cell responses are sharply selective to line orientation and to translational displacements of the stimulus.

There is another type of cell, which seems to be the next cell in a serial processing scheme. These *complex* cells have receptive fields 1° to 3° across, even in the central region. In these cells, an appropriately oriented slit stimulus gives a response of more or less the same amplitude regardless

DANIEL A. POLLEN and JOSEPH H. TAYLOR Massachusetts General Hospital, Boston, Massachusetts; Department of Physics and Astronomy, University of Massachusetts, Amherst, Massachusetts

of its position in the field. Complex cells, which are found in both the upper and lower cortical layers, are believed to receive a good part (if not most) of their information from the simple cells in the middle layers (Hubel and Wiesel, 1962). Although there may be monosynaptic geniculate inputs to some complex cells (Hoffman and Stone, 1971), there is as yet no evidence that such inputs impart any of the spatial and orientation selectivity of complex cell responses.

There are several types of complex cells (Hubel and Wiesel, 1962); one category prefers dark-light edges rather than rectangular slits, and another one responds selectively to moving slits. The complex cells that we have studied tend to be in or close to the area centralis. They are usually responsive to stationary stimuli and in some cases fire better to stationary than to moving stimuli.

The function of simple cells

The main point of departure in receptive field structure is the orientation selectivity of both simple and complex cell responses. Until the simple cell stage, there had been a point-to-point representation of visual space. Now we would like to discuss what information can be carried by the simple cell stage, and then look farther to see what information about the retinal representation of visual space can be held in the complex cell stage.

Let's ask some fundamental questions about what the simple and complex cells in the striate cortex of cats (lightly anesthetized with pentothal) are doing when we vary such parameters as the luminance and size of the stimulus (Pollen et al., 1971, 1973). When a spot of light is flashed onto the receptive field center of a simple cell, there is a primary neural response in about 30 msec, which lasts another 20 msec. Following this response, we find a series of diminishing secondary peaks that recur at alpha frequency (Figure 1A). Complex cells show similarly spaced peaks while the stimulus is on but also show pronounced *off* responses just after the stimulus ceases (Figure 1B).

If one plots the number of responses in the primary peak versus the logarithm of the light intensity, one finds a linear increase in the number of responses per second per stimulus (Figure 2). At high light intensities the response saturates, but the described relationship holds over a 1.0 to 1.5 log unit range. If we increase the size of a spot or extend it to a rectangular slit within the receptive field center of a simple cell, or if in a complex cell we increase a slit along the preferred direction, we find an increase in the number of responses. Thus the output of simple and complex cells depends upon both the extent of an area along the preferred direction stimulated as well as upon stimulus luminance. A single simple cell cannot convey

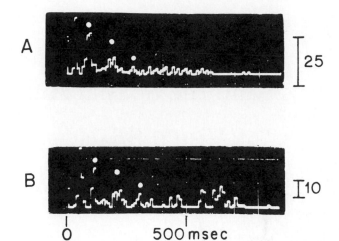

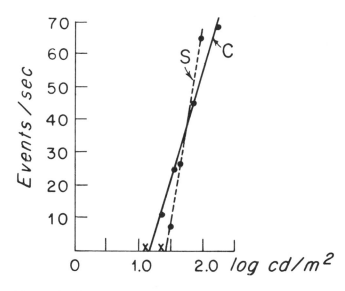

FIGURE 1 Poststimulus histogram responses (bin width, 10 msec) of a simple cell (A) and a complex cell (B) to discrete light stimuli 500 msec long. Dots are placed over the first three secondary peaks occurring at alpha frequency. (Pollen et al., 1971.)

FIGURE 2 Average unit response per second for the first 100 msec of response (averaged for 10 responses) for a simple cell (S) and a complex cell (C). A square 20° × 20 min of arc at the intensities indicated on the abscissa was used to stimulate the simple cell within its vertically oriented receptive field center, 2° × 20 min of arc. Stimulus area for the complex cell was 30° × 30 min of arc, and the receptive field covered a 3° × 3° area. Background illumination for each cell indicated by X. (Pollen et al., 1971.)

whether it is responding to one bright spot, two somewhat dimmer spots, a still dimmer but more elongated slit, or an indefinite number of combinations of stimulus luminance and area within its receptive field. The information available for further stages of information processing appears to be ambiguous.

However, when one considers the family of *all* simple cells that may be scanning the stimulus, it becomes clear that all of the information may be preserved after all. The information present in the simple cells (Figure 3) is analogous to that sometimes obtained experimentally in X-ray crystallography, radio astronomy, and other fields, and the problem has received extensive theoretical treatment (Bracewell, 1956; Bracewell and Riddle, 1967; Taylor, 1967). Simple-cell receptive fields, indicated by v_1 to v_6 in Figure 3, may overlap or be closely adjacent, scanning any given region of visual space. Other sets of cells scan visual space in a horizontal direction and still others at oblique angles. Figure 3 represents an over-

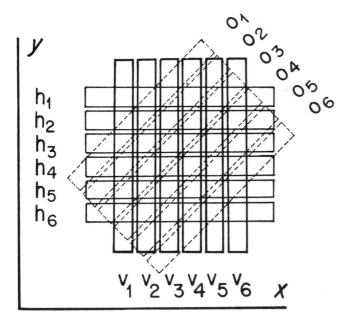

FIGURE 3 Adjacent simple cells selectively respond to stimulation within such receptive field centers as $v_1, v_2 \ldots v_6$. The same region of space is also covered by simple cells with receptive field centers favoring horizontal (h_1 to h_6) and oblique (o_1 to o_6) orientations. In the cat, space is scanned by sets of obliques every 10 to 15° in rotation apart. (Pollen et al., 1971.)

simplified picture in that there are not sharp rectangular cutoffs; there may be overlapping receptive fields, and there may be different widths of simple cells present. However, the data set is one in which there is a certain typical aperture width for the elements that scan a certain region of space. For simple cells, we found that the response increased both with the logarithm of the intensity and also with an increase in area; for low intensities the function relating cell response to increase in area may be approximately linear. In effect, then, the simple cells are performing a strip integration of the luminance distribution. They do not resolve where the luminance is distrib-

uted within their excitatory receptive field; they simply report a summation.

One method of analyzing a luminance distribution is to use a long and narrow aperture to scan across it at a number of orientations sequentially. Such a procedure yields a set of data with amplitude values for each position at each orientation—exactly the same set of data as that present in the simple cortical cells. A mathematical solution for the luminance distribution may be obtained by taking the amplitude values for v_1 to v_n and computing the Fourier transform of each such one-dimensional function. (The Fourier transform of a wave form is simply its decomposition into sine wave components, which may then be linearly superimposed to reconstruct the original wave form. A one-dimensional wave form may be decomposed into a set of sine waves of different frequencies, amplitudes, and phases. A two-dimensional brightness distribution may similarly be decomposed into sets of sine wave gratings.)

By assembling the amplitude and phase data for each orientation into a composite array, one obtains an approximation to the two-dimensional transform of the original luminance distribution. The original itself may be obtained by a Fourier inversion, which shows (among other things) that no information was lost during the process of strip integration. The only requirement for this to be true is that a sufficient number of orientations must be present in the set of one-dimensional data.

Bracewell (1956) has quantitatively discussed the limits of accuracy obtainable in a luminance distribution reconstructed in the manner just described, taking into account the fact that one does not have infinitely thin strips doing the integrating or an infinite number of orientations. His results show that the angular resolution of the reconstructed picture is equal to the narrowest dimension of the slit (which Hubel and Wiesel had also realized) as long as scans at a sufficient number of angles were present. The number of angles required is essentially equal to the overall angular diameter of the region of interest, divided by twice the angular resolution. Thus, for a 1 min angular resolution over the main part of the foveal region, which is about 1 degree, 30 angles would be required to avoid any loss of resolution. This is just about the number of angles that Hubel and Wiesel (1968) found in the monkey.

Thus one can answer one of the questions frequently asked: How does an orientation-selective system resolve objects that are not orientationally selective? Given a limit of angular resolution, it doesn't matter whether objects are circular, elliptical, or of any arbitrary shape; there has not been any loss of information in the aggregate of orientation selective neurons. And so we come to a general principle: Any *single* simple cell provides only ambiguous information but the *assembly* of such cells contains all the information required to define a two dimensional luminance distribution.

The function of complex cells

The radio astronomer may wish to carry out Fourier transforms on his strip-integrated data set, but are there any reasons to suppose that the visual system might do likewise? Certainly there is no evidence to suggest that a Fourier analysis is occurring within the retina, the lateral geniculate body, or the simple cell stage, because in the precortical stages, cell responses fail to show orientation selectivity, nor is there translational invariance of cell response in any of these stages. But could Fourier transformation occur after the simple cell stage, which would seem the logical place to start looking in view of the just completed process of strip integration? There are a number of reasons why we had to consider the Fourier possibility further. The main reason is the long-known advantage of the Fourier transform for pattern recognition (which we shall discuss in more detail shortly). Following publication of our first report (Pollen et al., 1971) we learned that Kabrisky (1965) had made what to our knowledge is the earliest suggestion that the brain (specifically area 18) might carry out a two-dimensional Fourier transform of visual space. These ideas have been further developed by Kabrisky and his colleagues (Kabrisky et al., 1970). We also had in mind the fact that Fourier transform pairs describe physically real relationships between many quantities that exist in nature, such as the relationship between a wave form and its frequency spectrum. More than a century ago, Ohm (1843) suggested quite correctly that the ear hears the same components as those predicted by Fourier analysis.

We know too that whatever computational methods the brain might subsequently employ, the striate cortex is working toward some transform relationship starting with a process of integration along lines at numerous orientations. Moreover, it is entirely feasible for a neural network to derive a two-dimensional Fourier transform. The transforms of the strip-integrated data at any given orientation angle could be handled one-dimensionally and, as can be shown mathematically (Bracewell, 1956; Taylor, 1967), it is only necessary to have these one-

dimensional transforms at a finite number of orientations to specify sufficient information for a two-dimensional transform. Finally, the translational invariance of the amplitude of complex cell responses (Figure 4) suggests that an appropriate neuronal substratum for carrying out such a transform might be present.

Thus, starting from this set of boundary conditions, we have reached a point of view similar to that of Campbell and Robson (1968) and Blakemore and Campbell (1969) based on their psychophysical studies. Their work and that of their coworkers has been reviewed in the preceding part of this volume. Three of their discoveries are of overwhelming significance in the present context. First, their psychophysical studies established that the human central visual system contains a large number of channels, each of which responds maximally to a selectively oriented sine wave grating but hardly at all to gratings with spatial frequencies differing by a factor of two (Blakemore and Campbell, 1969). The independence of these channels has recently been demonstrated in detection experiments (Sachs et al., 1971.) Second, electrophysiological studies with moving grating patterns have demonstrated that there exists a class of neurons in the striate cortex of cats (Campbell et al., 1969a) and monkeys (Campbell et al., 1969b) each member of which is maximally sensitive to a given spatial frequency and orientation. There were some orientation-selective neurons in which the response fell to about half amplitude an octave away from the center frequency. Third, Blakemore and Campbell (1969) showed that when a subject looks at square wave (rather than sine wave) gratings he adapts most strongly to the very channels that would be predicted by application of Fourier theory, namely the first and third harmonics. Blakemore et al. (1970) went on to show convincingly the independence of the third harmonic in studies of the perceived spatial frequency shift.

Let us therefore focus our attention on the complex cell as perhaps the earliest orientation selective stage in the visual system that might show both spatial frequency selectivity and translational invariance of response amplitude. At this point, we are assuming that complex cell activity is indicative of the amplitude alone and not the phase of a Fourier component. Rybicki suggested to us that this latter assumption (as well as the degree of frequency selectivity) could be tested by an experiment

---

FIGURE 4 Each line in A and C shows the summed poststimulus histogram response for the primary peak to ten successive stimuli for slits whose positions within the receptive field are indicated in the insert in B. Here B and D show 10 responses to 2 slits of light simultaneously flashed according to the positions noted to the left of each line and in accordance with the positions indicated in the receptive field insert. In E, line 3 shows the summation of 19 responses and the other 4 lines show the summation of 20 responses to single slits with relative positions shown in the insert in F. In F, 20 double slit responses in the positions indicated are shown. The lowest line, labeled 1, shows responses to fixed slit position 1 again after double slit study had been completed. (Rybicki et al., 1972.)

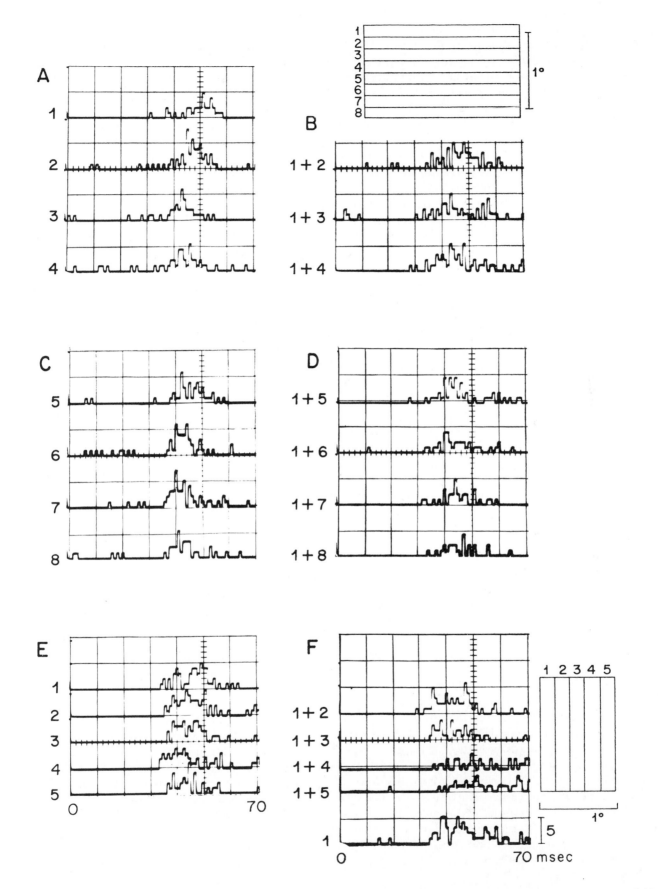

somewhat analogous to Thomas Young's classic experiment in the interference of light. Specifically, Rybicki suggested that we determine whether two stimuli, each producing a uniform response in itself, can produce interference when presented simultaneously. The use of paired stimuli with varying spacing would seem to be the simplest method of investigating interference effects and has the advantage that it does not depend on any preconceived ideas (such as a Fourier model, for which a sine wave grating would be the most natural *optimal* stimulus). Thus we proceeded to carry out detailed interslit spacing studies on nine complex cells in the deep layers of the striate cortex (Rybicki et al., 1972.) Complex cell responses were found to depend systematically upon the spacing between paired slits (Figure 4). Although the data were consistent with a Fourier model, they were not sufficient to establish a particular model; this was primarily because no cells were found that exhibited more than two maxima in the curve of response versus spacing.

More recently we have improved our experimental tactics in an attempt to probe the logical framework of spatial interactions within the complex cell receptive field. As before, we have attempted to do this in a neutral manner, independent of any a priori hypotheses. We have begun the following set of studies that should help to resolve the nature of the complex cell's spatial transform, whatever it may be, provided that there are no overwhelming spatial nonlinearities. We concentrate on those neurons that respond well to stationary stimuli rather than to neurons that might be tuned primarily to stimulus velocity and direction.

The first objective is to determine whether a particular visual neuron is selective to orientation. Neurons that are sensitive to one stimulus orientation are then tested as a function of varying widths of an appropriately oriented slit. In theory if complex cells were *slit-width detectors* one might expect a peak response at some given width with a progressive response fall-off for slits successively narrower or broader than an optimal width. If, on the other hand, complex cells are Fourier analyzers, the response to a slit of any given width will depend upon the spatial frequency to which the cell is tuned. In such a case, relative maxima would be expected for slit widths at odd multiples of the narrowest "optimal" width and relative minima at even multiples. If some other form of spatial interaction were operating in these cells, other experiments such as multiple-slit interaction studies would be needed to determine the correct transform function.

Spatial frequency selectivity can be tested by measuring the number of cell responses for a wide range of sinusoidally modulated test gratings. If neurons were indeed frequency selective, then the sharpness of tuning should be greater in those cases where a larger number of grating

cycles are contained within the receptive field. Cell response as a function of the parallel displacement of a single slit can also be tested. In a Fourier-like system, the amplitudes of spatial frequency coefficients should be essentially invariant to displacements of this type, at least within the broad central part of each receptive field. In two-slit interaction studies, cell responses for paired narrow test slits can be plotted as a function of interslit spacing. Such studies could demonstrate either a spatially periodic interaction as in a Fourier-like transform or define some other general transform function. Except for the limitations of time and a low signal-to-noise ratio with very thin slits, this is perhaps the simplest method of investigating interference that does not presume any a priori model. Finally, if in the previous studies spatial frequency content seems to be the selective parameter, then we plan another experiment that will maximize the signal-to-noise ratio. One or two cycles of the optimal grating are *removed* from the center and cell response is tested as one hemigrating is moved in and out of phase with the other. If the neuron is selectively responsive to spatial frequency, the response should reach a minimum or null point when the gratings are out-of-phase and a maxima when the gratings are in-phase.

Recently Pollen and Ronner have started carrying out this program on complex cells in the striate cortex and hope to extend the work to the wide receptive field complex cells found in higher visual areas. Thus far the first four experimental steps have been done on some 25 neurons. Though these studies are still in preliminary stages, it may be significant that a number of cells have shown spatial frequency selectivity. Cells with either one or two full grating cycles within the receptive field have tuning curves such that the response falls to about one-half amplitude an octave away from the center frequency. The tuning curves for these cells resemble those for the two most sharply tuned cells of Campbell et al. (1969a) in Figure 6B. Recently several much more sharply tuned cells have been found. In these cells four full cycles of the preferred grating fell within the receptive field, and the response fell to half amplitude or less only one half octave away from the center frequency. Moreover, the slit-width studies for these cells showed an optimal response when the single bar width was equal to just half a cycle of the optimal grating. Though the slit-width and spatial frequency studies are self-consistent in terms of a Fourier model, similar results will have to be obtained in many more cells and the cells subjected to the further experimental steps outlined above before one can know whether the actual transform is Fourier-like or some other transform.

Nevertheless, the evidence reviewed by Campbell (this volume) and in our work (Pollen et al., 1971,

1973; Rybicki et al., 1972) is sufficiently consistent with a Fourier-like model to warrant some discussion of the advantages of the Fourier transform domain for pattern recognition. Of course, whatever transform is *begun* in the striate cortex, we assume that it must be extended over larger regions of visual space and extend across the vertical meridian somewhere in higher visual areas such as the inferotemporal cortex (Gross et al., 1972; Gross, this volume), for only such a *global* transform could account for our ability to recognize objects independently of their precise position on the central retina. Furthermore, any linear or nonlinear transform, itself, will not always be sufficient to uniquely and correctly identify patterns, especially if the context is ambiguous or the background is confusing. In these situations some mechanisms of selective attention must be involved (Julesz, 1971). Perhaps the visual system may be able to selectively attend or set a window to one small part of a pattern at a time as in reading a page.

No matter how an object is encoded, it can be identified or recognized only by some sort of match or cross-correlation with a previous representation of a similar object to which some meaning has been conjoined by a related experience. The main advantage of a global Fourier transform is the translational invariance of spatial frequency information. Cross-correlation in the Fourier domain is probably the simplest type of cross-correlation, because information at any given spatial frequency and orientation need only be multiplied by or *filtered through* the corresponding term in the memory. In theory, the phase might be encoded within a group of orientation and frequency selective neurons with overlapping receptive fields, the value of the phase depending upon which particular neurons were firing. Alternatively, phase might be encoded in the time domain, or it might be independently encoded in another group of neurons. Second, simple objects can be recognized by the lower harmonic content alone (Kabrisky et al., 1970; Ginsburg, 1971). Third, the "distributedness" of information in the Fourier domain protects against losses due to visual scotomata and brain lesions beyond the striate cortex and might explain the degree of memory retained after brain tissue has been damaged or removed (Pribram, 1969.) A fourth advantage is that for identifying symmetrical objects only the Fourier components in a one-dimensional scan are necessary to specify an object (Bracewell, 1956).

There are two other advantages of a Fourier representation that deserve special mention. A considerable degree of nonlinear compression can be tolerated in a system making use of frequency analysis. A nonlinear operation that severely distorts the amplitude relations of a signal may have a remarkably small effect on the signal's spectrum (Licklider and Pollack, 1948; Oppenheim et

al., 1968). Moreover, Ginsburg (1971) and Ginsburg et al. (1973) have shown that the Gestalt principles of closure, similarity, and proximity can be nicely explained assuming that the brain can make its earliest judgments in a *decision space* that first analyzes or utilizes the lower frequency Fourier components. For example, they have shown in explaining closure that a sharply defined letter *G*, a blurred letter *G*, and a *G* arranged out of a series of dots have much the same lower spatial frequencies. The *Gestalt* comes out of the common low spatial frequency components.

Our ability to recognize objects independently of their apparent size evidently requires some sort of scaling mechanism. A change in size of an object produces an inversely proportional change in the set of spatial frequencies represented in the transform domain; therefore, cross-correlation of an object's representation at one image size with a template of the object stored at some other size cannot be done directly. However, in certain types of temporal lobe epilepsy, the visual scene may seem to undergo an increase or a succession of increases in apparent size (macropsia), or a decrease (micropsia) (Penfield and Jasper, 1954). Thus, the normal temporal lobe may contain a mechanism that scans or zooms over the representation of visual space at a finite number of sizes, so that a number of different object sizes may be cross-correlated with the memory. Whether such zooming occurs within the temporal lobe or via temporal efferent influences on other levels of the visual system that have been shown to exist by Spinelli and Pribram (1966, 1967) is unknown. Because object size and apparent distance are so closely related, we might also suggest that two dimensional cuts of three-dimensional space at a number of focal planes would need to be separately processed, a possibility that is consistent with experimental work on binocular depth perception (Barlow et al., 1967; Pettigrew et al., 1968; Blakemore, 1970).

Conclusion

We shall now very briefly consider the problem of information storage, because should it turn out that the brain is developing a spatial frequency code for pattern recognition, it would seem logical that the storage form might involve coding in terms of spatial frequency data or information easily accessible to a readout in this form. This carries us to consideration of the holographic analogy to human memory made independently by several workers including van Heerden, 1963, 1968; Julesz and Pennington, 1965; Pribram, 1966, 1969, 1971, this volume; Westlake, 1970. A hologram is a total recording of both amplitude and phase information (Gabor, 1951). In the optical case, the amplitude and phase information is stored in a suitable medium, usually a photographic emulsion. Should the neural hypothesis for a global Fourier transform of visual space be confirmed, then a Fourier holographic hypothesis may become amenable to experimental testing.

To us the challenge of the spatial pattern-recognition problem is not simply the problem of coding or cue utilization, although that is important enough. It is not just the problems posed by Lashley, although those too are important. The broader challenge, though very long range, is that the form of the code itself will lead to the problem of storage of that code and hence to the sites and mechanisms of memory. This may lead on to the challenge of trying to understand the cross-correlation process, beyond which are those mechanisms by which we derive the higher thought processes upon which our lives as sentient and intelligent beings depend.

ACKNOWLEDGMENT   This work was supported by the United States Public Health Service Grants 5-K3-NS-14, 353 and EY-00597.

REFERENCES

BARLOW, H. B., C. BLAKEMORE, and J. D. PETTIGREW, 1967. The neural mechanisms of binocular depth discrimination. *J. Physiol.* 193:327–342.

BLAKEMORE, C., 1970. The representation of three-dimensional visual space in the cat's striate cortex. *J. Physiol.* 209:155–178.

BLAKEMORE, C., and F. W. CAMPBELL, 1969. On the existence of neurons in the human visual system selectively sensitive to the orientation and size of retinal images. *J. Physiol.* 203:237–260.

BLAKEMORE, C., J. NACHMIAS, and P. SUTTON, 1970. The perceived spatial frequency shift: Evidence for frequency selective neurons in the human brain. *J. Physiol.* 210:727–750.

BRACEWELL, R. N., 1956. Strip integration on radio astronomy. *Australian J. Physics* 9:198–216.

BRACEWELL, N. R., and A. C. RIDDLE, 1967. Inversion of fan-beam scans in radio astronomy. *Astrophys. J.* 150:427–434.

CAMPBELL, F. W., G. F. COOPER, and C. ENROTH-CUGELL, 1969a. The spatial selectivity of the visual cells of the cat. *J. Physiol.* 203:223–235.

CAMPBELL, F. W., G. F. COOPER, J. G. ROBSON, and M. B. SACHS, 1969b. The spatial selectivity of visual cells of the cat and the squirrel monkey. *J. Physiol.* 204:120P.

CAMPBELL, F. W., and J. G. ROBSON, 1968. Application of Fourier analysis to the visibility of gratings. *J. Physiol.* 197:551–566.

GABOR, D., 1951. Microscopy by reconstructed wave-fronts. *Proc. Roy. Soc. (London) A* 197:454–487.

GINSBURG, A. P., 1971. *Psychological Correlates of a Model of the Human Visual System.* M.S. Thesis GE/EE/715-2 Wright-Patterson AFB, Ohio: Air Force Institute of Technology, DDC AD 731197, May 1971.

GINSBURG, A. P., J. W. CARL, M. KABRISKY, C. F. HALL, and R. A. GILL, 1973. Psychological aspects of a model for the classification of visual images. *Proc. 1972 Int. Cong. Cyb.* (in press).

GROSS, C. G., C. E. ROCHA-MIRANDA, and D. B. BENDER, 1972.

Visual properties of neurons in inferotemporal cortex of the macaque. *J. Neurophysiol.* 35:96–111.

HOFFMAN, K. P., and J. STONE, 1971. Conduction velocity of afferents to cat visual cortex: A correlation with cortical receptive field properties. *Brain Res.* 32:460–466.

HUBEL, D. H., 1959. Single unit activity in striate cortex of unrestrained cats. *J. Physiol.* 147:226–238.

HUBEL, D. H., and T. N. WIESEL, 1959. Receptive fields of single neurons in the cat's striate cortex. *J. Physiol.* 148:574–591.

HUBEL, D. H., and T. N. WIESEL, 1962. Receptive fields, binocular interaction and functional architecture in the cat's visual cortex. *J. Physiol.* 160:106–154.

HUBEL, D. H., and T. N. WIESEL, 1965. Receptive fields and functional architecture in two non-striate visual areas (18 and 19) of the cat. *J. Neurophysiol.* 28:229–289.

HUBEL, D. H., and T. N. WIESEL, 1968. Receptive fields and functional architecture of monkey striate cortex. *J. Physiol.* 195:215–243.

JULESZ, B., 1971. *Foundations of Cyclopean Perception.* Chicago and London: University of Chicago Press.

JULESZ, G., and K. S. PENNINGTON, 1965. Equidistributed information mapping: an analogy to holograms and memory. *J. Opt. Soc. Am.* 55:604.

KABRISKY, M., 1965. A proposed model for visual information processing in the brain. In *Models for the Perception of Speech and Visual Form*, W. Wathen-Dunn, ed. Cambridge, Mass.: MIT Press, pp. 354–361.

KABRISKY, M., O. TALLMAN, C. M. DAY, and C. M. RADOY, 1970. A theory of pattern recognition based on human physiology. In *Contemporary Problems in Perception*, A. T. Welford and L. H. Houssiadas, eds. London: Taylor and Francis, Ltd., pp. 129–147.

KUFFLER, S. W., 1953. Discharge patterns and functional organization of mammalian retina. *J. Neurophysiol.* 16:37–68.

LASHLEY, K. S., 1942. The problem of cerebral organization in vision. *Biol. Symp.* 7:301–322.

LICKLIDER, J. C. R., and I. POLLACK, 1948. Effects of differentiation, integration, and infinite peak clipping upon the intelligibility of speech. *J. Acoust. Soc. Am.* 20:42–51.

OHM, G. S., 1843. Über die Definition des Tones, nebst daran geknüpfter Theorie der Sirene und ähnlicher tonbildender Vorrichtungen. *Ann. Physik Chem.* 59:513–565.

OPPENHEIM, A. V., R. W. SHAFER, and T. G. STOCKHAM, JR., 1968. Non-linear filtering of multiplied and convoluted signals. *Proc. IEEE* 56:1264–1291.

PENFIELD, W., and H. JASPER, 1954. *Epilepsy and the Functional Anatomy of the Human Brain.* Boston: Little, Brown.

PETTIGREW, J. C., T. NIKARA, and P. O. BISHOP, 1968. Binocular interaction on single units in cat striate cortex: simultaneous stimulation by single moving slit with receptive fields in correspondence. *Exp. Brain Res.* 3:391–410.

POLLEN, D. A., J. R. LEE, and J. H. TAYLOR, 1971. How does the striate cortex begin the reconstruction of the visual world? *Science* 173:74–77.

PRIBRAM, K. H., 1966. Some dimensions of remembering: steps toward a neuropsychological model of memory. In *Macromolecules and Behavior*, J. Gaito, ed. New York: Academic Press, pp. 165–187.

PRIBRAM, K. H., 1969. The neurophysiology of remembering. *Sci. Amer.* 220(1):73–86.

PRIBRAM, K. H., 1971. *Languages of the Brain.* Englewood Cliffs: Prentice-Hall.

RYBICKI, G., D. R. TRACY, and D. A. POLLEN, 1972. Complex

cell response depends upon interslit spacing. *Nature New Biol.* 240:77–78.

SACHS, M. B., J. NACHMIAS, and J. G. ROBSON, 1971. Spatial-frequency channels in human vision. *J. Opt. Soc. Am.* 61:1176–1186.

SPINELLI, D. N., and K. H. PRIBRAM, 1966. Changes in visual recovery functions produced by temporal lobe stimulation in monkeys. *Electroenceph. Clin. Neurophysiol.* 20:44–49.

SPINELLI, D. N., and K. H. PRIBRAM, 1967. Changes in visual recovery function and unit activity produced by frontal cortex stimulation. *Electroenceph. Clin. Neurophysiol.* 22:143–149.

TAYLOR, J. H., 1967. Two-dimensional brightness distributions of radio sources from lunar occultation observations. *Astrophys. J.* 150:421–426.

VAN HEERDEN, P. J., 1963. Theory of optical information storage in solids. *Appl. Opt.* 2:393–400.

VAN HEERDEN, P. J., 1968. The basic principles of artificial intelligence. In *The Foundation of Empirical Knowledge*, P. J. van Heerden, ed. Wassenaar, The Netherlands: N. V. Uitgeveriz Wistik (Royal Van Gorcum, Ltd.) p. 143.

WESTLAKE, P. R., 1970. The possibilities of neural holographic processes within the brain. *Kybernetik* 7:129–153.

22 How Is It That Sensing So Much We Can Do So Little?

KARL H. PRIBRAM

ABSTRACT   I summarize in this chapter the reports made by the others and interpret their results in terms of my own experimental findings. These are (a) information is distributed in the striate cortex; (b) nonvisual information becomes encoded in the visual cortex; (c) resections of the inferior temporal cortex produce devastating deficits in visual discriminations involving choices but do not interfere with ordinary visual processing; (d) radical resections of the circumstriate cortex do *not* interfere with the performance of behavior involving such visual choices. On the basis of these data, a proposal is entertained that the functions of the inferior temporal cortex are carried out by addressing in parallel (attending) the information relevant to the decision, information that is encoded in a distributed (holographic) fashion in the primary visual system. Experimental evidence to support this proposal is adduced.

Introduction

MY STARTING point in delineating the neural mechanisms that relate the separate visual functions of the anatomically separable visual systems is a set of experimental data that do not fit what I was taught (see, e.g., Pribram, 1960). Kornhuber elsewhere in this volume reviews with you the classical view of the functions of the cerebrum as composed of transcortical reflex arcs: Beginning in the sensory projection areas, converging on association cortex and leaving the brain via motor cortex, these processes were initially couched in terms of the association of ideas but today are still put forward in the language of information processing. Nor are they completely false: Other contributors to this 1972 Third Study Program have reviewed the evidence that in fact the discrete organization that characterizes the primary sensory and motor projection systems gives way to a broader organization in the perisensory areas and that, often beyond this perimeter, cellular electrical response depends on input from more than a single sensory modality. Yet, as Kornhuber details for us, the transcortical reflex arc has been found wanting as a model for explaining not only the results of experiments on the motor systems but also of clinical observations on the agnosias and apraxias.

KARL H. PRIBRAM   Professor of Psychiatry and Psychology, Department of Psychology, Stanford University, California

It is just these clinical entities we set out to study a quarter of a century ago by making animal models, producing brain damage hopefully comparable to that found in man (Pribram, 1954). The results of these experiments are the ones that have raised the need for a more useful model.

The model that has gradually emerged from the data, some of which I will present now, might be called a *recognition program-tape model*. At the brain level of organization, the model resembles in many respects that presented for the cellular level by Edelman elsewhere in this volume. As he pointed out, the antigen-antibody problem is that of Maxwell's recognition demon: Is specificity due to selection or associative instruction, or perhaps both? As I understood his presentation, there is overwhelming evidence for an alphabet of preformed antibodies within the cell that becomes assembled into a specific "receptor" at the cell surface by a series of steps involving feedback at each step between antigen and the cell's response to that antigen.

I want now to present evidence that a similar stepwise process characterizes the construction of Maxwell's psychological recognition demon by the brain. The apparatus consists of an alphabet of image elements distributed in the primary sensory systems. This alphabet becomes assembled by steps involving feedback at each step between a particular input and certain elements of the alphabet into a specific program tape, the analogue of Edelman's cell membrane receptor. The program tape then preprocesses, that is, is specifically sensitive to, subsequent occurrences of that particular input.

There are five separate classes of empirical questions that are generated by the model: (1) those that characterize the alphabet; (2) those that testify to its distributed nature; (3) those that inquire into the mechanism of distribution; (4) those that specifically concern the assembly of the program tape; and (5) those that delineate the functions of such a program. There is considerable independence between the data sets that constitute the domains of each of these questions—the degree to which one data set is supportive or destructive to hypotheses within a domain should therefore, at this stage of model building, not influence credibility too greatly in another domain. Thus, I am able to spell out fairly precisely both the data that gave rise to the overall model and the limitations exposed in each domain when specific hypotheses were tested.

Characterizing the alphabet

The alphabet of image elements is characterized by the receptive field properties of units in the primary visual system. I want to discuss especially that part of the alphabet that shows the most striking properties, the orientation sensitive neurons discovered by Hubel and Wiesel. Pollen showed us in his preceding chapter that the output from any one of these neurons is actually ambiguous with regard to orientation: Changes in number, width, and contrast of lines influence output as much as does orientation. Only a population of neurons can code orientation. Pollen distinguishes, as do Hubel and Wiesel, between the properties of *simple* and *complex* cells; but overall, both his work and that of Fergus Campbell, in the preceding part, show that the job of this population of orientation sensitive elements is to respond selectively to spatial frequency that specifies not only orientation but also number, width, and luminance contrast of the input lines (gratings) used as stimuli.

Specification of spatial frequency can produce reconstructions of images in detail far beyond any that can be obstructed using orientation-sensitive mechanisms only.

In terms of the analogy to a verbal alphabet, we might think of the spatial frequency elements as the *vowel* part of our receptive field alphabet. Vowels, of course, are the carriers of speech onto which the consonants are grafted. Let me now show you what this vowel part of the alphabet of image elements looks like by visualizing for you the receptive fields of cortical neurons in cat and monkey (Figure 1). These receptive field maps were made by a technique devised by Spinelli (1966).

A small white spot is held against a black background by a magnet, which is attached to an *X-Y* plotter. The *X-Y* plotter is controlled by a small computer that therefore *knows* where the spot is. The spot is moved about the background, and a record is made with tungsten microelectrodes of the number of impulses evoked in a neuron in the visual system for every location of the spot. The computer then displays this record either in a three-dimensional contour diagram, a two-dimensional cross-section of that contour diagram usually taken two standard deviations above background activity, or a series of histograms. The most useful display for us has been the cross-sectional display. Here are some orientation sensitive receptive fields portrayed by this method. Note, as Colin Blakemore remarked in one of the discussions, that each

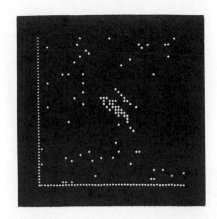

FIGURE 1 Bar shaped receptive field. Direction of scan: Vertical. Firing levels: 1 or greater. Note adjacent inhibitory bar and secondary excitatory field.

is characterized not only by an excitatory bar but by an inhibitory region to one side of that bar and often by another somewhat less distinct excitatory bar. It is this configuration that suggests spatial frequency sensitivity, and Pollen has just recently demonstrated with a different technique that complex cells are maximally sensitive to four such bars.

But let us remind ourselves at this point (Figure 2) that these *vowel* parts of the *alphabet* are not the only receptive fields demonstrable. The orientation sensitive units—simple and complex cells—are only part of population, about 10% at the foveal representation of the rhesus monkey cortex (Jung, 1961; Crcutzfeldt, 1961; Spinelli, Pribram, and Bridgeman, 1970). Ernst Pöppel made this point in one of the discussions by registering his surprise on finding so few orientation sensitive units when he first mapped visual cortex. We tend to neglect the many other sensitivities of cells (for example, those to color, De Valois, 1960; or those to auditory stimuli, Spinelli, Starr, and Barrett, 1968) in the primary visual cortex. It remains an open and important question, however, whether the pattern recognition mechanism is dependent solely on the spatial frequency analyzing neurons in the visual system.

Evidence for the distributed nature of the alphabet

Just as an alphabet of antibodies appears to be distributed relatively randomly within a cell, the alphabet of image elements appears to be randomly distributed within the striate cortex. This is not to deny the columnar organization of related orientation sensitive elements (Powell and Mountcastle, 1959; Hubel and Wiesel, 1968; Werner, 1970) but only to point out that the information encoded within any block of columns is repeatedly replicated in other blocks considerably removed anatomically from one

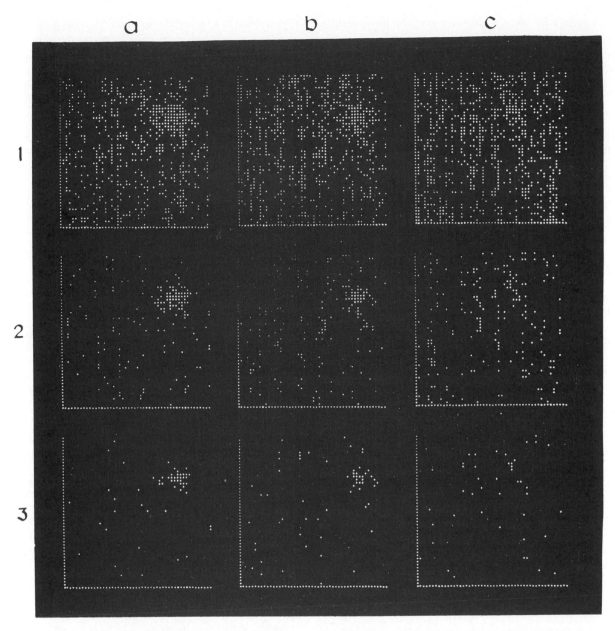

FIGURE 2   This figure shows a disk-shaped receptive field. In column a the unit was mapped with both eyes open; in columns b and c with the left and the right eye respectively. Rows 1, 2 and 3 represent regions where the unit fired 1, 2, 3 times or more, respectively.

another. Indirect evidence for such wide distribution of encoded information comes from observations and experiments on patients and animals who have suffered damage to their visual brain. When a patient suffers a stroke that wipes out half or more of his visual system, he does not go home to recognize only half of his family. With whatever visual field he has left he is able to recognize all that he ever recognized. Weiskrantz earlier in this part delineated the refinement of the laboratory model of this clinical fact; he reviewed the earlier work of Lashley (1929) and of Klüver (1941) on this topic. Perhaps less well known are the recent experiments of Galambos et al. (1967) and of Chow (1970). Galambos cut as much as 98% of the optic tracts of cats bilaterally—the 98% was verified anatomically—and the cats showed remarkable retention of the ability to discriminate figures such as the letter *F* from an upside down *F*, even when changes in size or a reversal of the figure-ground relationship (black-on-white to white-on-black) were made. In order to control for the possibility that scanning with a small tunnel of remaining visual system would account for these results, Chow took these experiments one step further by combining such lesions of the optic tract with extensive removals of the cat's striate cortex and again demonstrated remarkable retention of the animal's ability to make visual discriminations.

But there is also direct evidence for anatomical distribution of information in the primary visual system (Figure 3). In a series of experiments Spinelli and I showed that

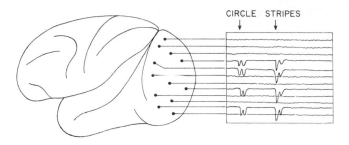

FIGURE 3 A diagrammatic representation of the finding that the differences in the potentials evoked by circles and stripes are distributed over the striate cortex. Note that not every lead shows the differences.

we could record, with small macroelectrodes implanted in the striate cortex of awake monkeys, different configurations of electrical responses evoked when the animal was exposed to brief (10 μsec) flashes of circles and of stripes. Of interest, here is the fact that these differential configurations were not recorded from every electrode, rather they were recorded from apparently random locations over the extent of striate cortex. However, each electrode location that showed the differential evoked response did so reliably for weeks and months.

The mechanism of distribution

The question arises as to how information becomes distributed in input systems. One simple way would be if the brain were like a randomly connected net, but the exquisite anatomical sensory-topic organization of the visual, auditory, and somatosensory systems rules this out. The random-net explanation has been utilized extensively and almost to the exclusion of any other by the computer simulation community, and even there has not fulfilled the earlier promise (see Minsky and Papert, 1969). For years, therefore, neuroscientists were baffled by this problem, if they faced the problem at all. For me the issue of the mechanism of distribution became an experimental one with the advent of a realizable alternative to the random-net proposal. This realizable alternative is provided by the holographic process.

What is the holographic hypothesis of brain function in

perception? The weak form of the hypothesis simply states that percepts (images of objects) are reconstructed by activation from input of a distributed information store. A writeout from the core memory of a computer would satisfy this definition. The strong form of the hypothesis is much more interesting, however, because it specifies process and therefore experiments to test its validity. The strong form of the holographic hypothesis states that images are reconstructed from a distributed information store by a transform of the input, a transform that on a prior occasion was responsible for the distribution of the information. Fourier holograms and their equivalents in the spatial frequency domain are the models for this strong form of the hypothesis (Pribram, 1966, 1972).

A hologram arises in any system, whether optical, computer, or neural, when neighborhood interactions among elements (e.g., spatial frequency) become encoded in the process of transformation. This chapter is not the one that permits my detailing for you the mechanism of neighborhood interaction, but this is extensively covered in the first half or so of my book *Languages of the Brain* (Pribram, 1971). Essentially, I make the case that lateral inhibitory networks are involved in organizing a microstructure of slow potentials occurring at synapses and in dendrites and show that the resulting interactions can be described in spatial frequency terms. Fundamental to the proposal are the findings (1) that prior to the ganglion cell layer in the retina, practically no nerve impulses are generated; thus, receptive field organization at the optic nerve level is structured by interactions among slow potentials. (2) At the cortex, both intracellular (Benevento, Creutzfeldt, and Kuhnt, 1973) and extracellular (Phelps, 1972) recordings have demonstrated inputs to be excitatory (depolarizing) while horizontal interactions appear to be exclusively inhibitory (hyperpolarizing).

What, then, would be the advantages to holographic encoding? They are (1) Equivalence of functional parts (the distribution of information) and therefore resistance to damage. From very small parts of the hologram, the entire image can be reconstructed. (2) Large memory storage capacity: In physical holograms 100 million bits of information have been stored in 1 mm^3. (3) Associative recall: When only part of the input that originally constituted the hologram recurs, the remainder of the scene is reconstructed as a *ghost* image. (4) Translational invariance: Recognition and recall can take place irrespective of the position or size of the input. This provides a mechanism for a zoom effect which when pathological becomes macropsia or micropsia. (5) Instantaneous cross-correlation between stored and input patterns and among input patterns. (6) Reversibility (invertibility): The transform restores the original in all its textural detail.

Thus holographic processes can serve as catalysts to other brain mechanisms. A corollary to this is that there are other brain mechanisms; even the strongest form of the holographic hypothesis does not suggest that these are the only transformations that occur in the input systems or elsewhere in the brain. May I again resort to my analogy of vowels in our verbal alphabet. They are the essential binding elements that make speech possible; they do not, however, completely specify the entire range of the alphabet or its combinatorial powers.

And where do we stand with regards to neurophysiological and neurobehavioral data that relate this model to brain function? What are the virtues and the limitations encountered when tests of the model are made? You have already been exposed to the evidence presented in support of the existence of a series of spatial frequency sensitive mechanisms operative in the visual (and auditory) systems (Campbell, earlier in this volume; Pollen, in the preceding chapter). But these mechanisms appear to be relatively broadly tuned. Any invertible process such as the Fourier transformation demands independent, narrowly tuned channels to be effective. Pollen has proposed that simple cells function as strip integrators that provide some independence. Whitman and Spitzberg (1972) have suggested, on the basis of their evidence, that the spatial frequency domain functions much as does the color domain: that three fairly broadly tuned retinal processes become neurally analyzed into a spectrum of narrowly tuned spatial frequencies at the cortical level. To subject these suggestions to neurological test is feasible; e.g., can opponent processes be demonstrated to operate for visual cells in the spatial frequency domain as DeValois (1960) has shown them to operate in the color domain? Henry and Bishop (1971) have devised an interesting technique using binocular stimulation to demonstrate opponent properties of simple cells.

Pollen, in his contribution to this section, has made several additional suggestions. In our laboratory as well as his, experiments are completed or under way to investigate the sensitivity of cortical units to bars spaced at different distances, the effect of presenting several spots or lines in various orientations simultaneously, etc. The issue is: How closely do the quantitative descriptions of these interactions come to expressions of invertible transforms? During our discussions, MacKay suggested that a modification of a Fourier process called a *logon* may be expected to fit better the interactive receptive field characteristics of visual neurons than any simple invertible transform. Gabor (1969) has already published alternative mathematics that could accomplish a holographic process as does the Fourier transform.

But perhaps the most critical limitation on the strong form of the holographic hypothesis to date is the evidence presented to us by Pollen in the preceding chapter. This limitation comes from the small size of the visual receptive fields of striate cortical neurons—especially in the foveal representation. What is necessary to make the holographic hypothesis swing is a mechanism that simultaneously covers a large number of receptive field elements.

Assembly of recognition program

The anatomical and lesion evidence presented by Weiskrantz and by Jones earlier in this part suggests that the circumstriate belt (peristriate or prestriate cortex, Brodmann's areas 18 and 19, Zeki's areas V_2 and V_3) is the locus of neural elements that could provide this mechanism by assembling the input from a number of striate cortex neurons. Recall that striate plus circumstriate cortical resections lead to a monkey sensitive primarily to luminous flux and that the organization of the circumstriate belt is such that as one moves forward within the belt, larger and larger visual receptive fields become organized. It remains to be shown what are the spatial frequency sensitivities of neurons in this belt.

Sounds simple, doesn't it? All we should now have to do is relate the functions of the temporal lobe cortex to this mechanism and then have a holiday. This is the truth table we have been presented so far, which is presented by Jones earlier in this part. Note the unidirectional arrows. Professor Jones assures me that truth table gremlins are responsible for some omitted arrows: In fact, corticocortical connections are for the most part reciprocal over short distances. Alas, the truth table conflicts with, to use Weiskrantz's challenging homily, *a table of ignorance* that has been generated by lesion experiments performed on the circumstriate belt. Removal of the belt without damage to the striate cortex does not irretrievably destroy visual pattern recognition: In fact, in this monkey (Figure 4) formal testing showed a complete sparing of the ability to recognize.

But I must pause here a moment to analyze a discrepancy in results that has plagued those of us working on this problem. Mishkin (1966), you will recall from Weiskrantz's presentation, has shown that partial lesions restricted to the upper or lower half of the circumstriate belt produce no effect on discrimination performance, while total lesions of the entire belt do. This evidence apparently conflicts with that of Gross, Cowey, and Manning (1971) who report that lesions of the *foveal* (i.e., ventral) portion of the circumstriate belt *do* result in a deficit. The results of both of these investigations are seemingly at odds with my observations that I have just shown you. A closer look at the data and techniques goes a long way toward resolving these discrepancies. First,

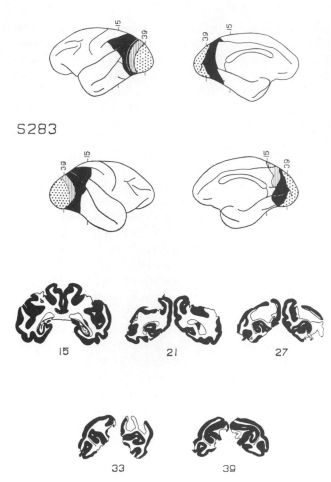

S283

FIGURE 4 Reconstruction of bilateral prestriate lesions after which monkey could still perform a visual discrimination (the numerals 3 versus 8) at 90% criterion.

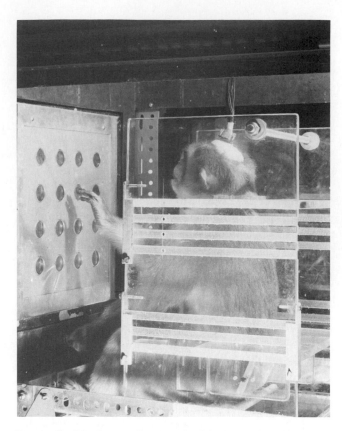

FIGURE 5 Monkey performing in Discrimination Apparatus for Discrete Trial Analysis (DADTA). A general purpose computer (PDP-8) programs stimulus presentation, records behavioral and electrophysiological results on magnetic tape, and provides typed or oscilloscope display readouts. Simple collations of data are performed on-line. More complex analyses are performed on taped data store.

in a current series of monkeys I have replicated Gross and Cowey's results exactly; these results are that the monkeys do have difficulty in relearning the discrimination after surgery (recall that even the best monkey whose lesion I demonstrated was essentially blind for weeks and had to be slowly retrained to respond visually). However, all monkeys do relearn and therefore the foveal circumstriate cortex cannot be, by itself, essential to the recognition process. With respect to the total lesions, Weiskrantz has suggested during our sessions here, that my technique (Figure 5) of randomizing the position of cues over 16 vertically presented locations aids recovery; Mishkin used a horizontal two-choice situation (Figure 6) for testing. This technique maximizes the disturbance because a ventral hemianopia is almost invariably produced in resecting the circumstriate belt by interruption of the dorsal part of the geniculostriate radiations that lie close to the surface just under the circumstriate cortex. Such interruption of geniculostriate radiation does not by itself produce any visual discrimination deficit (Wilson,

1957), but the combination of striate and peristriate removals might. Weiskrantz is now testing this interpretation in his laboratory at Oxford.

Another possibility remains: That the functions of the circumstriate belt are more widely distributed and that the functions of the cortex on the inferior convolutions of the temporal lobe overlap those of the circumstriate system. Our discovery of the visual functions of this temporal lobe cortex, in fact, included the entire extent of circumstriate preoccipital-temporal cortex (Blum, Chow, and Pribram, 1950). Only later did we find that the circumstriate portion of the lesion was dispensable in producing the effect on visual discrimination.

With regard to the temporal lobe cortex, another major discrepancy needs to be resolved. Charles Gross (1969) has elegantly demonstrated that the visual receptive field characteristics of cells in the inferior temporal gyrus are dependent both on the presence of the ipsilateral striate cortex and on an input from the thalamus (pulvinar). This contrasts sharply with the facts

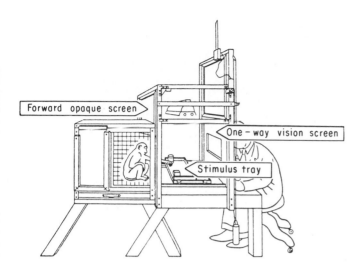

FIGURE 6 Wisconsin General Testing Apparatus (WGTA). Used to test monkeys in a variety of discrimination and learning problems.

I have just presented that radical resections of the cortico-cortical connections between striate and temporal cortex only temporarily impair visual recognition and Mishkin's recent experiments (35 monkeys) that visual recognition remains intact after massive posterior thalamic lesions that destroy the entire pulvinar (unpublished, Mishkin).

I would like to suggest that this discrepancy is also amenable to possible resolution by experiment. Some years ago Rosenblith, Rosner, and I (Pribram et al., 1954) showed that the responses evoked in Auditory Area III were dependent on the presence of the medial geniculate nucleus and not on connections from Auditory Areas I and II. In acute experiments, removal of Auditory Areas I and II did not alter the responses evoked in Auditory III, but when, after surgical removals in chronic preparations, time for degeneration of the medial geniculate nucleus was allowed, the responses disappeared. We inferred that a collateral projection from the medial geniculate nucleus to Auditory III could account for the results. A parallel experiment could be tried to see whether the receptive field of Gross's cells in the temporal cortex would be affected in acute experiments (his data so far are based on chronic preparations in which the lateral geniculate nucleus has degenerated). Perhaps, as for Auditory III, the receptive field characteristics of cells in the inferior temporal cortex are dependent on a booster from direct or indirect collaterals from the geniculate nucleus.

Where then does all this evidence leave us with regard to the problem of the mechanism of assembly of a recognition mechanism? My view is that these data unequivocally tell us what cannot be, but they leave us in open ignorance as to what actually does constitute the assembly mechanism. The mechanism, and I must emphasize again that this is my opinion based on my data, cannot be solely a transcortical hierarchical process. My response to the experimental results has been to emphasize an alternative to a simple transcortical reflex model: Over the years, I have suggested that a corticofugal efferent control system emanates from the temporal cortex downward to subcortical structures, there to influence by a parallel processing mechanism the visual input (Pribram, 1958, 1960, 1969, 1971). (Arbib has facetiously made the point that just because the brain looks like a bowl of porridge does not mean that it is a serial [cereal] computer.)

Functions of the program

Let us therefore look briefly at some of the electrophysiological data that provide evidence for the existence of such a parallel processing efferent system and how it functions before attempting a synthesis. In one of these experiments, we stimulated the inferior temporal cortex of cats to determine the effect on visual receptive field organization. We found changes occurring as far peripheral as the optic nerve, but the most cleancut and systematic effects were shown at the lateral geniculate level (Figure 7). Note the marked shrinking of the excitatory center and expansion of the inhibitory surround produced in this unit. (Note also that frontal lobe stimulation has an opposite effect

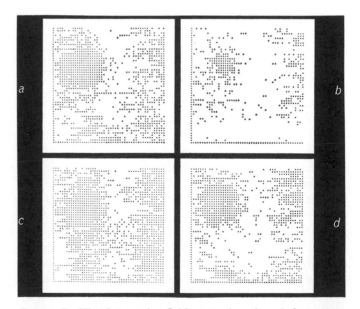

FIGURE 7 Visual receptive field maps show how information flowing through the primary visual pathway is altered by stimulation elsewhere in the brain. Map a is the normal response of a cell in the geniculate nucleus when a light source is moved through a raster-like pattern. Map b shows how the field is contracted by stimulation of the inferior temporal cortex. Map c shows the expansion produced by stimulation of the frontal cortex. Map d is a final control taken 55 min after recording a.

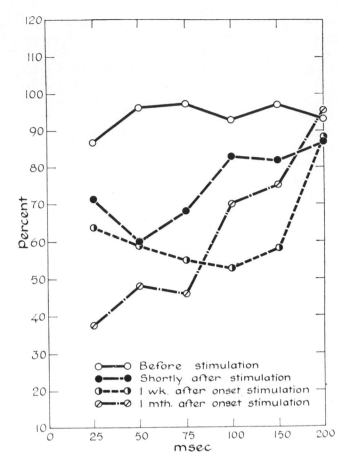

FIGURE 8 A plot of the recovery functions obtained in one monkey before and during chronic stimulation of the infero-temporal (IT) cortex.

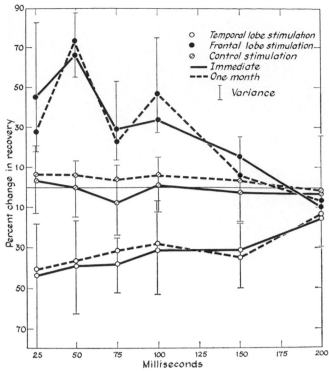

FIGURE 9 This figure plots the percent change in recovery for all subjects in the various experiments. It is thus a summary statement of the findings.

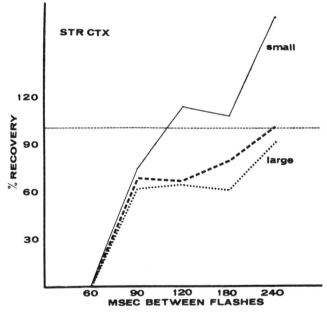

FIGURE 10 Comparison of flash recovery functions obtained when the probe stimulation of the lateral geniculate nucleus results in small (solid line) or large (dotted line) striate cortex response. Control without probe stimulation is indicated by dashed line. Note that when the probe stimulation produces a small response (solid line), i.e., when the monkey is attending, recovery is speeded. This same effect is obtained with temporal lobe (IT cortex) stimulation.

and that parietal and precentral stimulation has no effect on the receptive field organization of these cells.)

In another experiment we demonstrated changes in recovery cycles recorded from the striate cortex of fully awake monkeys sitting in restraining chairs (Figures 8 and 9). Electrical stimulation of inferior temporal lobe cortex shortens the recovery cycle of the response evoked in striate cortex by brief flashes of light (whereas frontal cortex stimulation lengthens it and no effect is produced by parietal and precentral stimulation). We interpreted this effect on recovery as indicating reduction (or, in the case of frontal stimulation, on enhancement) in redundancy of the visual channel. It is interesting to note that Waterman, collaborating with Wiersma (1966), found a similar redundancy control mechanism in invertebrates.

We quickly found that this effect could be demonstrated only when the monkeys were not attending to some other aspect of their environment. We therefore designed experiments to test specifically the relationship between the functions of the inferotemporal cortex and attention and showed in fact that attentional factors were critical (Figure 10; Gerbrandt, Spinelli, and Pribram, 1970). Gross (1972) also has shown that monkeys must be attending his experiment if he is to obtain unit responses in this cortex. But probably the most clear-cut demonstration of the process involved comes from one such experiment in which monkeys were trained to discriminate between flashed (10 μsec) cues that varied in two dimensions: color and form (Rothblat and Pribram, 1972). The monkeys were first trained to respond to red

by differentially reinforcing a response to the color dimension. They were then subjected to a discrimination reversal procedure, green was now the rewarded cue. Next, responses to the stripes were differentially reinforced and finally circles became the rewarded cue. In each stage the electrical activity from inferior temporal, circumstriate, and striate cortex was recorded for three days of criterion performance, 90% and 100 consecutive trials (Figures 11 and 12). Note the pattern of electrical responses evoked in the inferior temporal cortex when the monkey selectively responds to color and the different pattern when he responds to form. These evoked potentials are discernible only when the records are correlated to the time of response; in this respect (and several others)

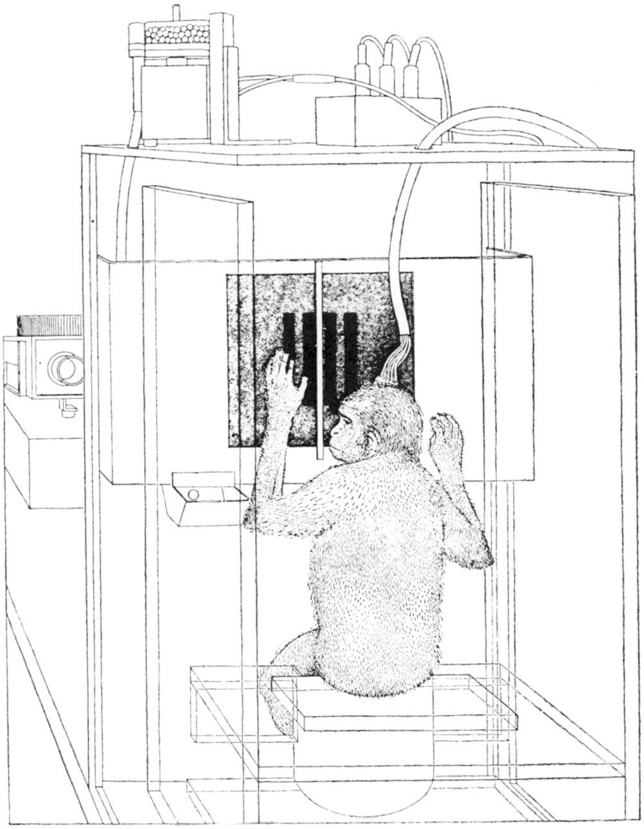

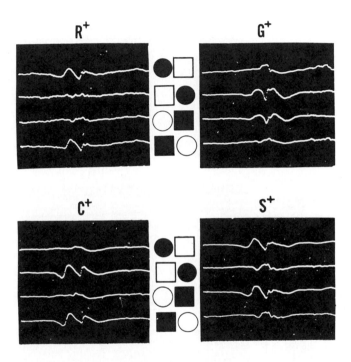

FIGURE 11 Set-up of an experiment demonstrating the functions of the visual areas. A monkey initiates a flashed stimulus display and responds by pressing either the right or left half of the display panel to receive a reward while electrical brain recordings are made on line with a small general purpose computer (PDP-8). On the translucent panel in front of him the monkey sees either a circle or a series of vertical stripes, which have been projected for 0.1 msec from the rear. He is rewarded with a peanut, which drops into the receptacle at his left elbow, if he presses the right-half of the panel when he sees the circle or the left-half when he sees the stripes. Electrodes record the wave forms that appear in the monkey's visual cortex as he develops skill at this task. Early in the experiments, the stimulus-locked wave forms show whether the monkey sees the circle or stripes. Eventually they reveal in advance which half of the panel the monkey will press. Each trace sums 300 trials of 500 msec of electrical activity following the stimulus flash.

FIGURE 12 Results of an experiment demonstrating the functions of the inferotemporal cortex using a set-up similar to that shown in Figure 11. Comparison of *response-locked* activity evoked in temporal cortex (IT) when monkeys are performing (90% correct) color (top panels) and pattern (bottom panels) discrimination. Each tracing sums, over 300 consecutive trials, the activity recorded when the stimulus configuration presented to the monkey appeared as in the diagrams between the panels. Each tracing includes 500 msec of electrical activity—250 prior to and 250 just after each response. Note that during the color discriminations the 1st and 4th (and the 2nd and 3rd) traces are similar, while during the pattern discriminations the 1st and 3rd (and 2nd and 4th) traces are alike. These similarities reflect the position of the color cues in the color task and the position of the patterns in the pattern task. Position per se, however, is not encoded in these traces. Note that this difference occurs despite the fact that the retinal image formed by the flashed stimulus is identical in the pattern and color problems.

they are different from the stimulus-locked evoked potentials in the striate cortex I showed earlier.

Perhaps the most interesting finding, however, came when we traced the emergence of the evoked response differences as a function of changing the reward from one cue dimension to the other (Figure 13). Note in the lower panel of tracings that the left-most and right-most patterns are almost identical to those I showed on the last slide. These are from a different electrode in a different monkey, however. Note now what happens while the animal shifts from selectively responding to (attending) the color dimension to responding to (attending) the form dimension. While the monkey is performing at chance, the evoked electrical activity shows no regularity. When he performs at about 75% his temporal lobe activity begins to take on the pattern related to the form discrimination. This pattern becomes enhanced at criterion performance and now, for the first time, appears also in the record made from the striate cortex. With overtraining, the striate cortex record becomes almost as striking as that obtained from the temporal cortex.

Conclusion

This experimental result, more than any other single finding, has led to my conviction that, for pattern per-

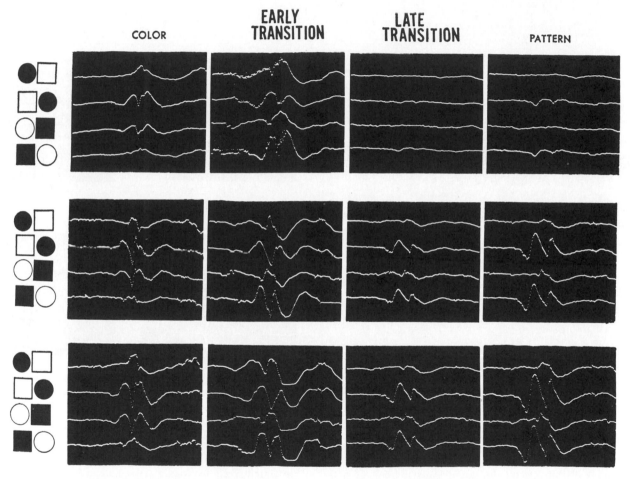

FIGURE 13 Experiment shows the development of the response-locked activity. In this experiment the flashed stimulus again consisted of colored (red and green) stripes and circles, exactly as in Figure 12. Reinforcing contingencies determined whether the monkeys were to attend and respond to the color (red versus green) or pattern (circle versus stripes) dimension of the stimulus. As in the earlier experiment, shown in Figure 11, stimulus, response, and reinforcement variables were found to be encoded in the primary visual cortex. In addition, this experiment showed that the association between stimulus dimension (pattern or color) and response shown in Figure 12 occurs first in the infero-temporal cortex. This is shown in the lower panels where the electrophysiological data averaged (summed) from the time of response (forward for 250 msec and backward 250 msec from center of record) again show clear differences in wave form depending on whether pattern or color is being reinforced. Note that in these tracings the response-locked difference in recorded activity can already be seen in the temporal lobe recording when the monkey is performing at 75% correct but does not appear in the striate cortex recording until criterion performance is attained. Overtraining enhances this difference in the striate cortex recording.

ception to occur, a program tape must become assembled to address the input when called for. It really looks as if the activity of the inferior temporal cortex, having become organized by the reinforcing contingencies of the situation, throws a programmed filter or program tape into the visual system that thereupon addresses and organizes (categorizes) the electrical responses, and presumably the image elements of the striate cortex. The assembling of such a program tape involves a great amount of processing; no wonder that sensing so much, we can do so little.

It is, of course, imperative to know the pathways by which such an assembling of a program tape can occur, and I remind you of Graybiel's Figure 6 showing the three visual systems and their corticocortical interactions. Recall that the three systems are distinguished on the basis of their afferent connections. However, the question remains whether any corticocortical organization supports the trichotomy: The effects of lesions of the various visual systems are completely dissociable from those of the premotor, frontal, and limbic systems—as are the effects of each of these systems from those of any of the others.

No such dissociation occurs when the effects of lesions of these areas are compared with those produced in subcortical structures to which efferents project. For instance, the behavioral deficit that follows frontal cortex resection can be produced as well from lesions of the head of the caudate nucleus. And lesions deep to the inferotemporal cortex in the region of the tail of the caudate nucleus and ventral putamen (Rosvold and Szwarcbart, 1964) produce deficits in visual discrimination. Thus, we should have been alerted (but were still surprised by the size and extent) when we obtained strong evidence of efferent connections to the entire ventral putamen by mapping the electrical responses evoked by stimulation of the inferotemporal cortex. (See Figures 14 and 15.) Here were powerful connections from association cortex to a nucleus usually identified with the motor system (Reitz and Pribram, 1969).

However, as stated by Kornhuber and Ito and others elsewhere in this volume, the motor functions of the non-pyramidal motor systems function in large part as organizers of programs controlling more reflex levels of function. Some of this control is exerted via the γ system and is thus receptor, i.e., input, control. A fascinating task ahead is to determine experimentally whether the basal ganglia influence the input from the special senses as well as those of the motor systems—and if so, just where and how.

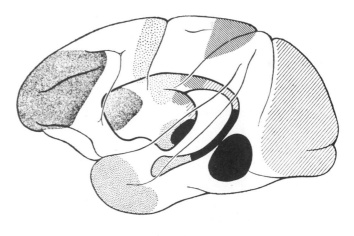

Projections of Cerebral Cortex Onto Basal Ganglia

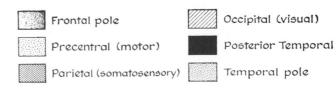

FIGURE 14 (a) Side view of the brain showing stimulation sites in experiment that traced the subcortical connections of the inferotemporal cortex. (b) Selected cross-section showing sites (‡) where response was evoked by inferotemporal cortex stimulation. Note especially the responses in putamen and superior colliculus.

FIGURE 15 Corticofugal connections to the basal ganglia. (Drawn from study by Kemp and Powell, 1970.)

The model developed in these pages reads, therefore: Input becomes distributed in sensory systems through the action of lateral inhibitory networks into an *alphabet* of spatial frequency sensitive elements at the striate cortex by a more or less invertable transformation. This alphabet becomes temporarily assembled for the purposes of any specific recognition, not by some hierarchical process leading to a pontifical "grandfather" neuron in the circumstriate cortex. Rather, a parallel processing mechanism initiated in the inferotemporal cortex addresses (categorizes) the elements of the alphabet via motor structures (e.g., the putamen) much as a program tape organizes a program by addressing elements in the memory of a computer. The model thus constitutes a progressively differentiating self-organizing system. We have seen that, as a heuristic, the model has had considerable merit. A great amount of otherwise conflicting data are subsumed and perhaps more important, five *areas of ignorance* can now be detailed sufficiently to generate specific experiments. What more can we ask when just a few years ago there was only enigma?

REFERENCES

BENEVENTO, L. A., O. D. CREUTZFELDT, and U. KUHNT, 1973. Significance of intracortical inhibition in the visual cortex. *Nature (London)*, (in press).

BLUM, J. S., K. L. CHOW, and K. H. PRIBRAM, 1950. A behavioral analysis of the organization of the parieto-temporo-preoccipital cortex. *J. Comp. Neurol.* 93:53–100.

CHOW, K. L., 1970. Integrative functions of the thalamocortical visual system of cat. In *Biology of Memory*, K. H. Pribram and D. Broadbent, eds. New York: Academic Press, pp. 273–292.

CREUTZFELDT, O. D., 1961. General physiology of cortical neurons and neuronal information in the visual system. In *Brain and Behavior*, M. A. Brazier, ed. Washington, D.C.: American Institute of Biological Sciences, pp. 299–358.

DEVALOIS, R. L., 1960. Color vision mechanisms in monkey. *J. Gen. Physiol.* 43:115–128.

GABOR, D., 1969. Information with coherent light. *Opt. Acta (London)* 16:519–533.

GALAMBOS, R., T. T. NORTON, and C. P. FROMMER, 1967. Optic tract lesions sparing pattern vision in cats. *Exp. Neurol.* 18: 8–25.

GERBRANDT, L. K., D. N. SPINELLI, and K. H. PRIBRAM. 1970. The interaction of visual attention and temporal cortex stimulation on electrical activity evoked in the striate cortex. *Electroenceph. Clin. Neurophysiol.* 29:146–155.

GROSS, C. G., D. B. BENDER, and C. E. ROCHA-MIRANDA, 1969. Visual receptive fields of neurons in inferotemporal cortex of the monkey. *Science* 166:1303–1305.

GROSS, C. G., A. COWEY, and F. J. MANNING, 1971. Further analysis of visual discrimination deficits following foveal prestriate and inferotemporal lesions in rhesus monkeys. *J. Comp. Physio. Psychol.* 76:1–7.

GROSS, C. G., C. E. ROCHA-MIRANDA, and D. B. BENDER, 1972. Visual properties of neurons in inferotemporal cortex of the macaque. *J. Neurophysiol.* 35:96–111.

HENRY, G. H., and P. O. BISHOP, 1971. Simple cells of the striate cortex. In *Contributions to Sensory Physiology*, W. D. Neff, ed. New York: Academic Press.

HUBEL, D. H., and T. N. WIESEL, 1968. Receptive fields and functional architecture of monkey striate cortex. *J. Physiol.* 195:215–243.

JUNG, R., 1961. Neuronal integration in the visual cortex and its significance for visual information. In *Sensory Communication*, W. A. Rosenblith, ed. New York: John Wiley & Sons, pp. 627–674.

KLÜVER, H., 1941. Visual functions after removal of the occipital lobes. *J. Psychol.* 11:23–45.

LASHLEY, K. S., 1929. *Brain Mechanisms and Intelligence*. Chicago: University of Chicago Press.

MINSKY, M. L., and S. PAPERT, 1969. *Perceptrons*. Cambridge: MIT Press.

MISHKIN, M., 1966. Visual mechanisms beyond the striate cortex. In *Frontiers of Physiological Psychology*, R. Russell, ed. New York: Academic Press, pp. 93–119.

PHELPS, R. W., 1972. Inhibitory interactions in the visual cortex of the cat. Ph.D. Thesis, Stanford University, California.

POWELL, T. P. S., and V. B. MOUNTCASTLE, 1959. Some aspects of the functional organization of the cortex of the postcentral gyrus of the monkey: A correlation of findings obtained in a single unit analysis with cytoarchitecture. *Bull. Johns Hopkins Hosp.* 105:133.

PRIBRAM, K. H., 1954. Toward a science of neuropsychology. In *Current Trends in Psychology and the Behavioral Sciences*, R. A. Patton, ed. Pittsburgh: University of Pittsburgh Press, pp. 115–142.

PRIBRAM, K. H., 1958. Neocortical function in behavior. In *Biological and Biochemical Bases of Behavior*, H. F. Harlow and C. N. Woolsey, eds. Madison: University of Wisconsin Press, pp. 151–172.

PRIBRAM, K. H., 1960. The intrinsic systems of the forebrain. In *Handbook of Physiology, Neurophysiology II*, J. Field, H. W. Magoun, and V. E. Hall, eds. Washington: American Physiological Society, pp. 1323–1344.

PRIBRAM, K. H., 1966. Some dimensions of remembering: steps toward a neuropsychological model of memory. In *Macromolecules and Behavior*, J. Gaito, ed. New York: Academic Press, pp. 165–187.

PRIBRAM, K. H., 1969. The amnestic syndromes: Disturbances in coding? In *Pathology of Memory*, G. A. Talland and N. C. Waugh, eds. New York: Academic Press, pp. 127–157.

PRIBRAM, K. H., 1971. *Languages of the Brain*. Englewood Cliffs: Prentice-Hall.

PRIBRAM, K. H., 1972. Some dimensions of remembering: Steps toward a neuropsychological model of memory. In *Macromolecules and Behavior*, 2nd Ed., J. Gaito, ed. New York: Appleton-Century-Crofts, pp. 367–393.

PRIBRAM, K. H., M. NUWER, and R. BARRON, 1973. The holographic hypothesis of memory structure in brain function and perception. In *Contemporary Developments in Mathematical Psychology* (in press).

PRIBRAM, K. H., B. S. ROSNER, and W. A. ROSENBLITH, 1954. Electrical responses to acoustic clicks in monkey: Extent of neocortex activated. *J. Neurophysiol.* 17:336–344.

PRIBRAM, K. H., D. N. SPINELLI, and M. C. KAMBACK, 1967. Electrocortical correlates of stimulus response and reinforcement. *Science* 157:3784.

REITZ, S. L., and K. H. PRIBRAM, 1969. Some subcortical connections of the inferotemporal gyrus of monkey. *Exp. Neurol.* 25:632–645.

ROSVOLD, H., and M. K. SZWARCBART, 1964. Neural structures involved in delayed-response performance. In *The Frontal Granular Cortex and Behavior*, J. M. Warren and K. Akert, eds. New York: McGraw-Hill, pp. 1–15.

ROTHBLAT, L., and K. H. PRIBRAM, 1972. Selective attention: Input filter or response selection? *Brain Res.* 39:427–436.

SPINELLI, D. N., 1966. Visual receptive fields in the cat's retina: Complications. *Science* 152:1768–1769.

SPINELLI, D. N., K. H. PRIBRAM, and B. BRIDGEMAN, 1970. Visual receptive field organization of single units in the visual cortex of monkey. *Int. J. Neurosci.* 1:67–74.

SPINELLI, D. N., A. STARR, and T. W. BARRETT, 1968. Auditory specificity in unit recordings from cat's visual cortex. *Exp. Neurol.* 22:75–84.

WATERMAN, T. H., 1966. Systems analysis and the visual orientation of animals. *Amer. Sci.* 54:15–45.

WERNER, G., 1970. The topology of the body representation in the somatic afferent pathway. In *The Neurosciences, Second Study Program*, F. O. Schmitt, Editor-in-Chief. New York: Rockefeller University Press, Vol. 2, pp. 605–616.

WHITMAN, R., and R. SPITZBERG, 1972. Spatial-frequency channels: Many or few? Paper presented at the Annual Meeting of the Optical Society of America, 1972.

WILSON, M., 1957. Effects of circumscribed cortical lesions upon somesthetic discrimination in the monkey. *J. Comp. Physiol. Psychol.* 50:630–635.

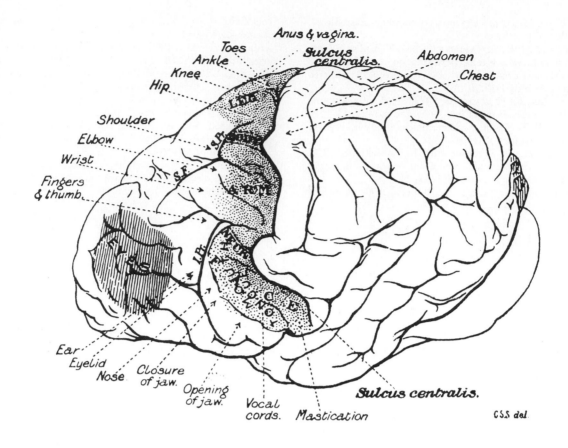

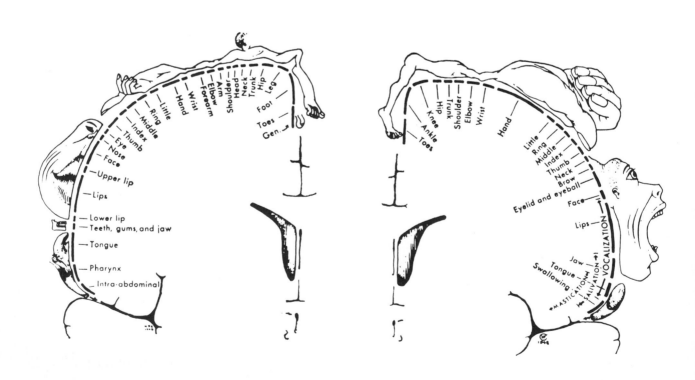

CENTRAL PROCESSING OF SENSORY INPUT LEADING TO MOTOR OUTPUT

Top: *The motor area of the chimpanzee brain. (From
A. S. F. Grünbaum and C. S. Sherrington, 1902. Observations on the physiology of the cerebral cortex of some higher
apes. Preliminary communication.* Proc. Roy. Soc.
69: 206–209.) Bottom: *The pattern of localization in
the sensory cortex (left) and the motor cortex (right) of the
human brain. (From W. Penfield and T. Rasmussen, 1950.*
The Cerebral Cortex of Man: A Clinical Study of
Localization of Function. *London: Macmillan &
Company.)*

CENTRAL PROCESSING OF SENSORY INPUT LEADING TO MOTOR OUTPUT

Introduction

EDWARD V. EVARTS

Two YEARS AGO a work session of the Neurosciences Research Program was held to consider central control of movement, and at this work session one of the questions considered was: How and to what extent is motor output controlled by sensory input? As Weiss (1941) has written, "Nobody in his senses would think of questioning the importance of sensory control of movement, but just what is the precise scope of that control? Is the sensory influx a constructive agent, instrumental in building up the motor patterns, or is it a regulative agent, merely controlling the expression of autonomous patterns without contributing to their differentiation?" For movements under sensory guidance, what are the mechanisms whereby the sensory input is transformed in such a way as to set up appropriate patterns of motoneuron output? These issues aroused sufficient interest to give rise to this part on Central Processing of Sensory Input Leading to Motor Output. The order of the chapters presented in this part corresponds in part to the historical order in which ideas on motor control have developed. The initial discoveries on cerebral control of movement were made by clinical neurologists, and the first chapter in this series is Kornhuber's overview based in large measure on data from clinical neurology. Henneman deals with that division of the sensorimotor system concerning which Sherrington made his initial contributions: The spinal cord and reflex control of the motoneuron. Henneman discusses the size principle, presenting evidence that the motoneurons innervating a single muscle can be viewed as a set of elements that become active only in one particular order, this order being determined by motoneuronal size. Ito's chapter concerns the cerebellum, a component of the motor control system that receives powerful sensory inputs, but which sends its output to other parts of the brain rather than to motoneurons. Ito hypothesizes that the cerebellum has a special role in relation to feedforward control. Additional consideration is given to the cerebellum by Aschoff, and to the cerebral cortex by Hyvärinen. DeLong considers an additional major component of the motor system, the corpus striatum, and Evarts deals with activity of sensorimotor cortex in primates carrying out learned movements. It is apparent that this review of the way in which sensory input may be processed so as to give rise to motor output involves consideration not only of the sensorimotor cortex, but of the spinal cord, basal ganglia, and cerebellum as well. Indeed, the central goal of current studies of motor control is to discover the way in which these several parts of the brain are functionally interconnected in the control of movement.

REFERENCE

WEISS, P., 1941. Does sensory control play a constructive role in the development of motor coordination? *Schweiz. Med. Wschr.* 71:406–407.

23 Cerebral Cortex, Cerebellum, and Basal Ganglia: An Introduction to Their Motor Functions

H. H. KORNHUBER

ABSTRACT The motor commands of the cerebral cortex that result from information processing in sensory, association, and motivation areas must be converted by subcortical function generators into spatiotemporal motor patterns. Rapid movements are preprogrammed by the cerebellum with regard to timing and duration of activity. The basal ganglia serve as a ramp generator for slow voluntary smooth movements of different speed. For those movements that need sophisticated analysis of tactile objects, the output patterns of the cerebellum and basal ganglia are further processed in the motor cortex.

MOVEMENTS ARE parts of actions, and actions have to satisfy the needs of the organism and secure the survival of the species. Therefore, they must be guided by messages from the internal milieu as well as from the environment. Motor mechanisms without regulation via sensory information are nonsense. In the case of a voluntary aimed action, this regulation must, of course, be more complex than that for a simple segmental reflex at the level of the spinal cord.

Strategy and tactics of action

If we are hungry, the strategy of our action is to search for food. This strategy is determined by the glucose concentration in the blood and by other factors of the internal milieu, which are centrally sensed in the hypothalamus. The tactics, however, of our search depend on the environment. This is centrally represented in the sensory, primary, and association areas of the occipital, temporal, and parietal lobes, in which the analysis and synthesis of the raw sensory data are performed.

Let us first consider the tactical aspect of action. It has long been known that lesions of the sensory association cortex in the parietal, temporal, and occipital lobes—

H. H. KORNHUBER Abteilung Neurologie, Universität Ulm, Ulm, West Germany

especially bilateral lesions—result first in a disorder called agnosia, which is a disturbance in recognition of tactile, visual, or auditory stimuli, and second in apraxia, an inability to perform skilled, aimed movements although there is no paralysis of the limb or eye movements.

Evidence from cerebral potentials preceding voluntary movements in man

These results from lesions in sensory association cortex of man agree well with evidence from cerebral potentials preceding voluntary movements in man obtained by the method of reverse analysis (Kornhuber and Deecke, 1965; Deecke et al., 1969; Becker et al., 1972). In such experiments, the first event preceding a movement of the right index finger is not a potential over the left motor cortex but rather a bilateral potential, the Bereitschaftspotential (readiness potential), which is widespread over precentral and parietal regions. The second potential, the premotion positivity (PMP), which starts 90–80 msec prior to movement, is also bilateral and similarly widespread. Only the third potential, the motor potential, starting about 50 msec before the movement, is limited to the hand area of the contralateral motor cortex.

A corticocortical reflex as the basis of voluntary movement: An old mistake

When apraxia and aphasia were discovered, it was assumed that in analogy to the spinal reflex arcs, the disturbance might be caused by a disruption of an intracortical reflex arc. The corticocortical pathways were thought in the case of apraxia to lead from the occipital and parietal lobe to the motor cortex and, in the case of aphasia, from the temporal lobe to the frontal speech area.

The cerebral cortex is not a function generator

When the neurologists of the nineteenth century jumped to the conclusion of a corticocortical reflex arc as the basis of voluntary movement, they did not consider that, from the technical point of view, voluntary movement is radically different from reflex regulation, because it requires active production of a spatiotemporal pattern. In order to guide a movement, e.g., of a light spot on an oscilloscope screen, a function generator is needed not just a feedback regulation. As I hope to show, there are good reasons to believe that the function generators for

voluntary movements in the vertebrate brain are the basal ganglia and the cerebellum. The cerebral cortex is an analyzer for feature detection and for data reduction. In addition it is a synthesizer of patterns and a memory core, but it is not a function generator for spatiotemporal patterns of movement. There is no support from anatomy, physiology, or clinical data for the hypothesis of a corticocortical reflex as the basis of voluntary movement. There are no direct corticocortical connections from the visual association areas to the motor cortex (Pandya and Kuypers, 1969). Neurons of the motor cortex do not respond to visual stimuli (Kornhuber and Aschoff, 1964). Voluntary movement is virtually impossible without the basal ganglia and cerebellum—structures that I shall now refer to as *function generators.*

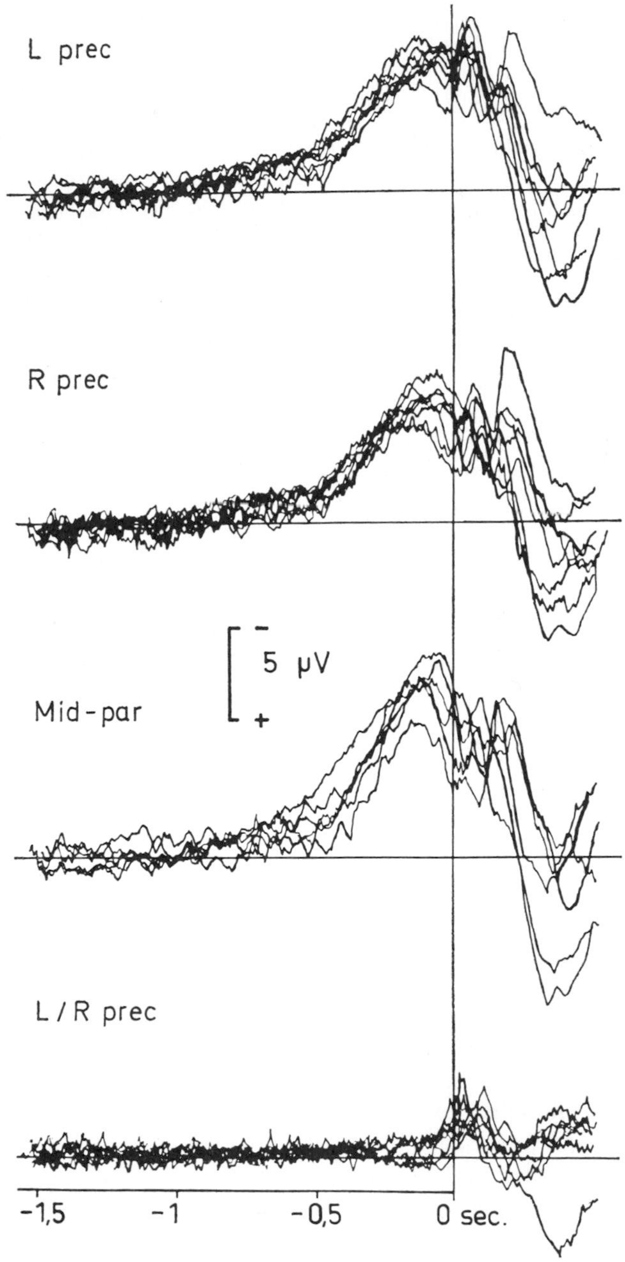

Older views about cerebellar and basal ganglia functions

As far as I can see, there exists no real theory about the function of the basal ganglia. On the cerebellum, there is the recent theory of Brindley (1964), elaborated by Marr (1969) and by Ito (1970), that the cerebellum is an instrument for learning movements. Although it might be that the neuronal circuitry of the cerebellum is capable of learning, this is certainly not a function unique to the cerebellum. On the contrary, learning capacity belongs to the outstanding characteristics of the cerebral cortex, and from the quantitative point of view, most motor learning in man is certainly done in the forebrain. Furthermore, the findings after lesions of the cerebellum cannot be explained as loss of motor learning capacity but rather they show a stereotyped pattern of disability to perform certain simple movements or to hold postures. However, there was one element of theory common to all considerations of cerebellar and basal ganglia function:

FIGURE 1   Cerebral potentials, recorded from the human scalp, preceding voluntary rapid flexion movements of the right index finger. The potentials are obtained by the method of reverse analysis (Kornhuber and Deecke). Eight experiments on different days with the same subject; about 1000 movements per experiment. Upper three rows: monopolar recording, with both ears as reference; the lowermost trace is a bipolar record left versus right precentral hand area. The Bereitschaftspotential (readiness potential) starts about 0.8 sec prior to onset of movement; it is bilateral and widespread over precentral and parietal areas. The premotion positivity, bilateral and widespread too, starts about 90 msec before onset of movement. The motor potential appears only in the bipolar record; it is unilateral over the left precentral hand area, starting 50 msec prior to onset of movement in the electromyogram. (Experiment of W. Becker, L. Deecke, B. Grözinger, and H. H. Kornhuber, presented at the German Physiological Society Meeting 1969, *Pflügers Arch. Physiol.* 312:R108.)

Namely, the belief that the cerebellum and the basal ganglia come into play after triggering of the movement by the motor cortex. Furthermore it was believed that the cerebellum and the basal ganglia contribute to motor function in making movements smooth and in avoiding oscillations. Cerebellar *intention tremor* and the resting tremor present in some cases of Parkinson's disease were taken as evidence.

Fiber connections between cerebral cortex, cerebellum, and basal ganglia

Anatomically, the cerebellum (Nyby and Jansen, 1951) and the basal ganglia (the strio-nigro-pallidum) (Kemp and Powell, 1970) are connected in parallel between nearly the whole cerebral cortex as the source of input and the motor cortex as the goal of output. (I will come later to the additional output of the cerebellum and basal ganglia to the red nucleus and reticular formation and to the output of the cerebellum to the vestibular nuclei). Although there is some input from the motor cortex to the cerebellum and basal ganglia, most afferents to these structures come from the association cortex so that for the most part the motor cortex follows the basal ganglia and cerebellum in the direction of information flow. On the other hand, it is not too surprising that the motor cortex also projects to the cerebellum and basal ganglia, for the motor cortex is from the sensory point of view a somatosensory and vestibular association area (Kornhuber and Aschoff, 1964) (see Figure 9 later in this chapter).

Timing of rapid movements by the short time clock of the cerebellar cortex

My belief, that the cerebellum does not merely regulate movements, which have been programmed and initiated by the cortex, but that the cerebellum is necessary for the design of rapid movements prior to their initiation—movements that are preprogrammed because they are too fast to be regulated continuously by feedback—this belief originated in the observation of dysmetria of saccadic eye movements due to atrophy of the cerebellar cortex (Kornhuber, 1968). Fast saccadic, ballistic eye movements are the only active eye movements we have. The smooth pursuit movements depend on a moved stimulus. To understand the cerebellar dysmetria of saccadic eye movement, we must know that saccadic eye movements are too fast to be regulated continuously by visual feedback. The course of a saccadic eye movement in time follows automatically out of its preset angle, and cannot be changed at will. The velocity of the saccade is a function of its amplitude. The angle to be moved is measured by the visual system prior to onset of movement. Within the visual association cortex of the forebrain, the angle to be moved is obviously represented as spatial data; it is probably coded in the connections of a nerve cell column, because the information transmission for length depends only on the endpoints and not a line connecting them (Bechinger et al., 1972). For the eye movement, however, these spatial data must be translated into time—namely, into the duration of the burst of high-frequency action potentials in the

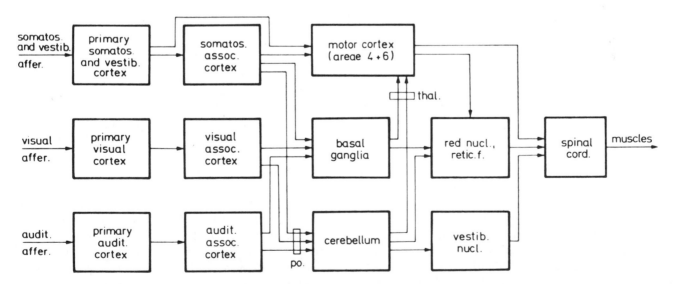

FIGURE 2   Information flow diagram for the *tactics* of voluntary movements. The sensory messages, processed in the primary and association areas of the cerebral cortex, give rise to motor commands that are transformed into spatiotemporal patterns by means of the function generators in the cerebellum and basal ganglia. The output of these goes for the simpler movements directly via the old supranuclear motor centers of the brain stem (the vestibular nuclei and the red nucleus-reticular formation) to the spinal cord but for the movements that need sophisticated tactile analysis of objects via the motor cortex.

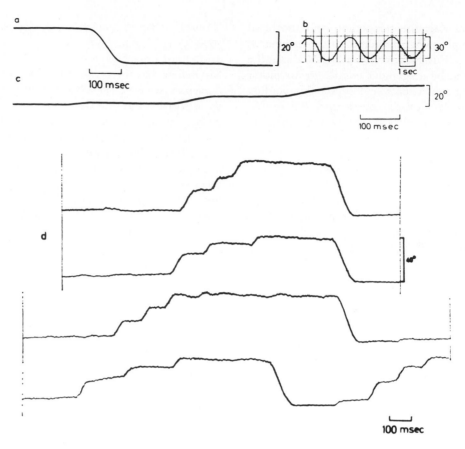

FIGURE 3 Dysmetria (hypometria) of human horizontal saccadic eye movements due to lesions of the cerebellar cortex. (a) Normal saccadic eye movement to the left of a normal subject, with a small correction saccade. (b) Normal smooth pursuit movement (to a sinusoidal stimulus) of the patient with the dysmetria shown in c. (c) Hypometria and slowing down of saccadic eye movements in a case of cerebellar atrophy. The patient was asked to perform a one-step eye movement to a goal 20° to the right. (d) Unilateral hypometria of saccadic eye movements to the right in a case of a cerebellar tumor; contrary to c, the speed of the hypometric eye movements is normal in this case. (a to c) from Kornhuber, (1969), *Arch. Ohrenheilk.* 194: 111–148; (d) from Kornhuber, (1971b).

oculomotor nuclei since the size of the saccadic eye movement depends entirely on that duration and not on impulse frequency; extraocular motor neurons always fire at highest possible rate during the saccades except during very small saccades (Schiller, 1970; Fuchs and Luschei, 1970). (For a discussion of the conflicting evidence, given by Robinson, 1970, see Kornhuber, 1971a.) The duration of saccadic eye movements (Robinson, 1964) as well as of oculomotor unit burst (Schiller, 1970; Fuchs and Luschei, 1970) is linearly related to the amplitude of the saccades. Inability to determine the duration of the saccadic bursts must, therefore, result in a disturbance of the size of the saccade; and that is what I found in cases of cerebellar cortical atrophy: Instead of a single large movement to the goal (or a large movement with one small correction saccade), there are several small or medium size saccades. It looks as if the brain no longer knows how to program the full angle. It initiates eye movements of the appro-

priate direction, but cannot find the precise angle other than by trial and error. In unilateral lesions, dysmetria is ipsilateral. In pure cases of cerebellar cortical atrophy with dysmetria of saccadic eye movement, there is no gaze paresis and no gaze nystagmus. The smooth pursuit movements of the eyes, which are continuously regulated by vision, and the vestibular eye movements are normal. The pure picture of hypometria of saccadic eye movements is best seen in cases of rapidly advancing cerebellar cortical atrophy as in the Louis-Bar syndrome in children or the subacute cerebellocortical atrophy in elderly people, which is sometimes due to vitamin B_1 deficiency. In the late stages of slowly progressing cerebellar atrophy, there is (probably due to transneuronal atrophy) a slowing down of the remaining saccadic eye movements. In cases of rapidly advancing cerebellar cortical atrophy, however, the velocity of the hypometric saccade is normal and corresponds to their amplitude in the normal way. In the final stage of cerebellocortical atrophies, all saccadic eye movements (including large and small ones) disappear.

The inability of the brain to generate saccadic eye movements of normal size in cases of cerebellar cortical atrophy shows that the cerebellar cortex participates in preprogramming the course of the saccade by determining its duration. How does the cerebellum do this? Probably it does it by means of the arrangement of the parallel fibers and the Purkinje cells—a structure unique in the nervous system. I believe we should imagine that the axons of a row of Purkinje cells converge onto a common neuron in the cerebellar nuclei, as shown in Figure 4. As long as the action potentials travel in the parallel fibers, a neuron in the cerebellar nuclei will be inhibited or (by means of an inhibitory interneuron) disinhibited. The duration of the travel of an action potential depends on diameter and length of the fiber. With a fiber length up to 5 mm and conduction velocities down to 0.05 m/sec movements up to 100 msec are possible. The neurons in the cerebellar nuclei involved in timing saccadic eye movements then project directly or via an interneuron to the oculomotor nuclei in the brainstem that innervate the extraocular muscles, while those responsible for timing fast hand movements are connected via the thalamus to the motor cortex. The parallel fibers of the cerebellar cortex as a delay line with the ability to transform spatial into temporal patterns was the conclusion reached independently by Braitenberg (Braitenberg and Atwood, 1958; Braitenberg, 1961), based on purely anatomical grounds. Unfortunately, Braitenberg's ingenious papers, which have often been quoted for the anatomical facts, have been largely disregarded in the subsequent discussions with regard to functional interpretation, perhaps because of the lack of two additional concepts: First, discontinuous (saccadic) timing, and second, preprogramming of fast movements by the cerebellum prior to their onset.

The action of the parallel fibers as a delay line explains why we usually see hypometria and not hypermetria in cases of cerebellar cortical atrophy. Any metabolic deficiency affects first the thinnest and ·longest parallel fibers, that is, those which are responsible for the large amplitude saccades.

Eye movements are represented within the cerebellar cortex in the middle part of the vermis, mainly lobuli VI and VII A, but also V b and c. The afferent projection from the eye muscles terminates in these lobuli (Fuchs and Kornhuber, 1969). Electrical stimulation of these lobuli yields saccadic eye movements (Hampson et al., 1952), and to this region project the afferents from the visual areas of the forebrain (Snider and Eldred, 1952). Lesions of this part of the cerebellar cortex in the monkey result in dysmetria of saccadic eye movement (Aschoff and Cohen, 1971).

In the pathway from the paravisual areas of the

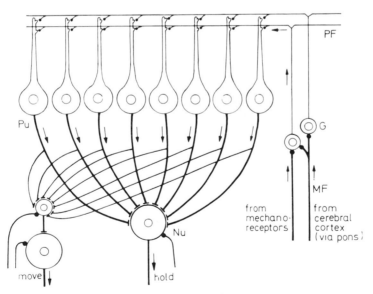

FIGURE 4 The parallel fiber (PF) delay line in the cerebellar cortex with the convergence of a row of Purkinje cells (Pu) on a neuron of the cerebellar nuclei (Nu). This arrangement acts as a clock, timing the duration of fast preprogrammed movements that are too quick to be regulated continuously (saccadic or ballistic movements). Also, the interval between different fast movements can be timed by this clock. At the granular cells (G)

that give rise to the parallel fibers, the afferents converge from the cerebral cortex (via the pons) and from the spinal cord and mechanoreceptors. The inhibitory action of the Purkinje cell output stops the holding function of a hold neuron and disinhibits (by means of an inhibitory interneuron) a move neuron. (From Kornhuber, 1971b.)

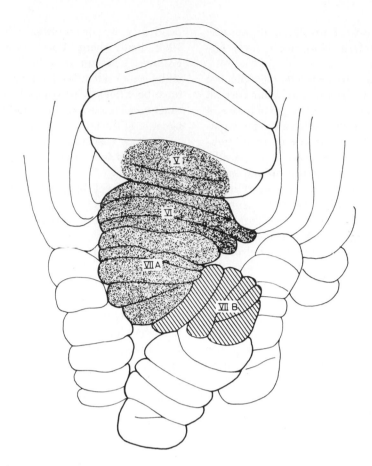

FIGURE 5   The projection of eye muscle afferents to the cerebellar cortex of the cat is located in lobuli V B + C, VI, and
VII A. There is a weaker projection to VII B. Based on potentials
evoked by small step stretch of extraocular muscles in barbiturate anesthesia. (From Fuchs and Kornhuber, 1969.)

forebrain (where the measurement of the angle to be moved
takes place and where probably the command for the
saccadic eye movements is given) via the pons to the
cerebellum, there is, at least in a part of the fibers, a
synapse in the optic tectum. Here are cells that are active
only preceding saccadic eye movement of a certain direction and of a certain magnitude not before smaller or
larger saccades and not before smooth pursuit movements
of the eyes (Wurtz and Goldberg, 1971). The duration of
the saccades is not, however, coded in the discharge of
these neurons, because many of them fire 2 to 3 times
longer than the duration of the corresponding saccade.
The time of onset of the discharge is only irregularly related to the start of movement. The interval between the
neuronal response and the eye movement ranges up to
150 msec. This surprisingly long latency corresponds well
to the latency of the premotion positivity of the human
cerebral cortex, which precedes saccadic eye movements
by about 150 msec (Becker et al., 1968, 1972), while
finger movements by 90 to 80 msec (Deecke et al., 1969;
Becker et al., 1969). (The details of the pathway for sac-

cadic eye movement from the cerebral cortex through the
cerebellum to the oculomotor nuclei are presented later
in this part by Jürgen Aschoff.)

The discontinuous, saccadic nature of cerebellocortical
action throws some light on the function of the elaborate
inhibitory network of the cerebellar cortex found by
Eccles and co-workers (1967). Discontinuous action necessitates a clearing command after each operation, which
ensures an exact stop of the movement and erases the
traces of previous action.

Another fact agrees well with the idea that the cerebellar cortex contains a short time clock. This is the high
discharge frequency of cerebellar Purkinje and nuclear
cells during movement (Thach, 1968). For fine gradation
of time a high-frequency clock is needed.

Although I came to my interpretation of cerebellocortical function when observing eye movements, it is
obvious that the saccadic clock theory holds true for hand
movements as well. In the first place, dysmetria of aimed
hand movements, and ultimately loss of the rapid aimed
movements, are stages in the course of cerebellocortical

atrophy. Another symptom is adiadochokinesis, that is, inability to perform rapid repetition of fast hand movements.

The functions of the peripheral afferents to the cerebellar cortex and the temporal coordination of fast movements

This brings us to the question of the function of the peripheral afferents to the cerebellum, coming from receptors in skin, joints, muscles, and perhaps from spinal motor centers monitoring their activity.

If the cerebellar cortex is a clock, preprogramming the duration of fast movements, why does this clock or step generator need these afferents from peripheral receptors and spinal events? Obviously, the function of these afferents cannot be to provide negative feedback in a continuous servoregulation. If we consider the more important cerebellocortical mechanism (the granular cells with their parallel fibers connected to a row of Purkinje cells), the influence of the spinal input could be: First, the timing of the start of the parallel fiber volley, and second, the selection of the correct set of granular cells.

Timing the start of the next movement by reafferent signals from the preceding movement, in other words the temporal coordination of successive rapid movements, is certainly important for actions like writing or speaking or playing the piano or (in birds) flying. The activity of many small birds consists mainly of rapid movements that are too fast to be regulated continuously by sensory feedback. When a bird winds through the branches of trees, survival depends on temporal coordination of rapid movements. It is perhaps because of this that birds have

such a large cerebellum. Adiadochokinesis, the slowing down and dysmetria of hand writing and a hesitant, slow type of speech, are signs of cerebellocortical disturbance in man. It is, of course, easier to use internal feedback from preceding motor actions or reafferent sensory signals for the timing of successive movements than to construct an exact temporal superprogram that takes care of the start of each single movement component independent of the occurrence of the preceding and following movements. We might therefore imagine that, while the sequential program of movements is provided by the cerebral cortex, the exact timing depends on reafferent volleys, converging with the messages from the forebrain at the granular cells of the cerebellar cortex.

Another function of the afferents to the cerebellar cortex from mechanoreceptors might be to participate in the selection of the correct granular cells, which is a function of choice or decision making. The energies needed (and this means, in the case of maximal force, the durations needed) for movements of different starting positions are different. For instance, saccadic eye movements from the midposition to a lateral position have a longer duration than saccades from the lateral to the midposition. The selection of granular cells is, of course, mainly a task for the afferents from the cerebral cortex (via the pontine nuclei). In addition, however, afferents from mechanoreceptors should have an influence on the selection of granular cells in order to take into account the starting position of the limb to be moved.

By contrast, the afferents to the cerebellar nuclei from the mechanoreceptors provide negative feedback for the continuous hold regulation.

In retrospect we can see why it was difficult to arrive at an adequate functional interpretation of the cerebellar

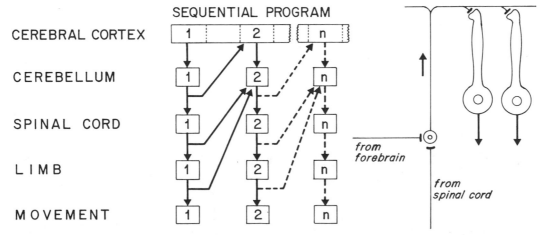

FIGURE 6   Timing of a series of rapid movements by reafferent volleys. Instead of constructing an exact temporal superprogram that could start each single component of movement independent of the occurrence and timing of the preceding movement, the

cerebral cortex probably provides only a sequential program with specification of direction, extent, etc., and the cerebellum takes care of the timing, using reafferent signals from lower centers of receptors.

cortex. The problems originated from two prejudices: That the function should be continuous, and that cerebellar activity should follow (not precede) the activity of the motor cortex.

The hold regulation in the cerebellar nuclei

The concept of relay nuclei has turned out to be oversimplified in all cases where the function of a group of nerve cells has been understood.

The cerebellar nuclei participate in the timing function of the cerebellar cortex by integrating the output of a row of Purkinje cells. But in addition, the cerebellar nuclei have a function of their own, which is radically different from that of the cerebellar cortex. This function is to hold a position between the rapid movements preprogrammed by the cerebellar cortex. In contrast to the discontinuous function of the cerebellar cortex, the holding function of the cerebellar nuclei is a continuous one with the standard principle of negative feedback. The evidence is as follows:

1. Contrary to lesions of the cerebellar cortex, lesions of the cerebellar nuclei or their efferent connections to the brainstem and thalamus result in oscillations (tremor), a fact that points to a disturbance of a continuous regulation.

2. These oscillations occur during holding only, not while at rest with relaxed muscles: Cerebellar tremor is a holding tremor, and it is maximal when holding a position after a fast movement (see Figure 7). I would like to stress that cerebellonuclear function is the same for eye and limb movements, because, when later considering the basal ganglia, I have to point out fundamental differences between limb and eye movements. The eye tremor due to cerebellonuclear lesions is called pendular nystagmus; it is a holding tremor that disappears when the eyes are closed (Aschoff et al., 1970, see Figure 7). Precise holding of a position after a fast movement is an important

function; without a holding regulation, the eye would tend to drift because of viscous forces, and clear vision would be impossible.

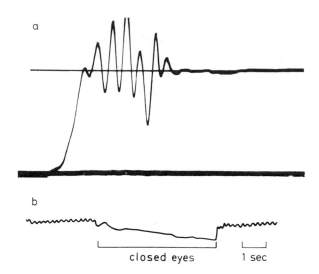

FIGURE 7 Holding tremor of the arm (a) or eye (b) in man due to lesions of the cerebellar nuclei or their efferent connections with the thalamus (a) or brainstem (b). In a, the arm was rested on a lever to allow movements in the horizontal plane only. The oscillation appears when holding a position after a rapid step movement. In b, the pendular eye nystagmus (which is an acquired symptom in a case of multiple sclerosis) is present during fixation and disappears when the eyes are closed. (From H. H. Kornhuber, 1971b) (a is a record done by Dr. B. Conrad; b from Aschoff et al., 1970.)

3. The intracerebellar nuclei are analogous (in function and in their connections with the cerebellar cortex) to the vestibular nuclei. Phylogenetically, the vestibular nuclei are the model for the cerebellar nuclei, because the oldest part of the cerebellar cortex was derived from the vestibular nuclei. The function of the vestibular nuclei is

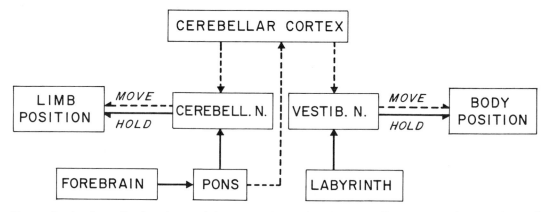

FIGURE 8 Analogy (in function and in connections with the cerebellar cortex) of vestibular and cerebellar nuclei. Phylogenetically, the vestibular nuclei have been the model for the cerebellar nuclei.

definitely to hold a position. The difference is, that the vestibular nuclei hold the position of the whole body as determined by the labyrinths, while the cerebellar nuclei hold positions of parts of the body as determined by the cerebral cortex.

You may demonstrate to yourself the difference between the *saccadic* (rapid move) and *hold* functions by a simple experiment, the finger-to-nose test. If it is performed as rapidly as possible, three components of the movement are clearly distinguishable: (1) a rapid preprogrammed movement bringing the finger near to the nose, (2) a hold with oscillation, and (3) a slow ramp movement to the nose. It is the second component that shows maximal oscillation in cases of *intentional* tremor due to lesions of the cerebellar nuclei—a better term is *holding* tremor. The first component shows much less oscillation, but it is dysmetric. The spatiotemporal pattern for the third, the slow component, is not generated by the cerebellum, it is produced by the basal ganglia.

In phylogenetic perspective, the story of the cerebellum is as follows: After evolution of the vestibular system with its holding function, it was necessary to give up holding from time to time in order to make a movement, especially a fast aimed or an escape movement. In order to do so, the archicerebellar cortex was developed with some visual afferents. It enabled the animal to inhibit holding and to perform rapid preprogrammed movements. After proving that this model worked well, the rest of the cerebellum was designed according to it, with a set of new holding nuclei: The intracerebellar nuclei.

From this we can understand why the cerebellar Purkinje cells are inhibitory in action, as found by Ito and co-workers (1970). Certainly, inhibition of holding is not identical to motion. In the vestibular nuclei as well as in the cerebellar nuclei, however, there are small interneurons in addition to large neurons. The small neurons are, perhaps, inhibitory in function. In this case, by means of inhibition of these inhibitory interneurons, the main neuron could be released for action (see Figure 4).

On the basis of this theory of an independent (hold) function of the cerebellar nuclei, we might predict that the cerebellar nuclei have separate afferents at least from the pontine nuclei (not only collaterals from fibers to the cortex).

Archicerebellum and motion sickness

There is little reason to assume that the function of the archicerebellar cortex (the flocculonodular lobe) is principally different from the rest of the cerebellar cortex. This implies that it is also essentially saccadic (discontinuous) in nature (that is timing short inhibitions) despite the fact that the main symptom of lesions of the floc-

culonodular lobe is a tonic one, namely, positional nystagmus. Like the visual or the somatosensory system, the vestibular system has the problem of distinguishing between messages due to the environment and stimuli due to active movements. Apparently it solved this problem in a manner similar to that of other sensory systems by inhibiting certain sensory effects of active movement. This inhibition should be as exactly timed as the movement. While, in the cat, the vestibular nuclei are inhibited by the cerebellum, in the frog, the vestibular labyrinthine receptors are under direct efferent inhibitory control of the cerebellum (Llinás and Precht, 1969). Regarding a similar hair cell mechanoreceptor, the lateral line stream receptor, there is a figure in the literature (Schmidt, 1965) that shows a short burst of efferent inhibitory impulses to the receptor preceding each active gill movement. Extirpation of the lower vermis including the nodulus relieves the susceptibility to motion sickness (Bard et al., 1947). Perhaps this fact may be explained as follows: Motion sickness is mainly due to head movements during rotation (Johnson and Mayne, 1953). In this situation, the Coriolis effect causes labyrinthine stimulation in a plane different from that of head movement. Therefore the short vestibular inhibitions associated with head movements have an inappropriate direction, and this causes motion sickness.

The basal ganglia as a ramp generator

If the cerebellum is a generator for step movements, we may ask where the continuous, slow, smooth movements of our limbs are produced. The spatiotemporal pattern for these movements are, in my opinion, generated by the basal ganglia. In other words, the basal ganglia may be regarded as a generator for smooth movements of voluntary speed or, in technical terms, a ramp generator. For this belief there are three pieces of evidence: A negative one and two positive.

The negative evidence is that the basal ganglia contribute importantly to the production of hand, foot, trunk, and head movements but not for eye movements. An additional ability of hand movement as compared to eye movement should be a contribution of the basal ganglia input through the motor cortex, because cerebellar function is the same in both hand and eye movements. Such an additional ability of hand movement exists: The ramp function or the voluntary speed smooth movement. We can move our hand, arm, leg, head, trunk, etc., at any speed (within limits) at will, but we cannot do the same with the eyes. Voluntary eye movements are saccadic; a smooth pursuit movement of the eyes is possible only with a moving target. Voluntary speed ramp eye movements independent of a moved visual stimulus are not only unnecessary but would disturb visual perception. Therefore, in the interval between the microsaccades, which are used to overcome retinal adaptation, the eye is kept relatively steady. In phylogenetic development, a very long refractory period (about 80 times longer than the refractory period of a neuron) after each saccadic eye movement was introduced in order to secure undisturbed visual perception and information processing.

The statement that the basal ganglia are needed for other voluntary movements but not for saccadic eye movements is supported by four facts concerning the structure to which the basal ganglia project, that is, the motor cortex. First, the motor potential of the human motor cortex, which follows the readiness potential and the premotion positivity and precedes the onset of hand movement by about 50 msec, does not appear before eye movements. Eye movements are preceded only by the readiness potential and the premotion positivity (Becker et al., 1968, 1972). Second, contrary to neurons of the motor cortex preceding hand movement, the neurons of the frontal eye field do not fire previous to the onset of eye movement (Bizzi, 1968). Third, while the precentral motor cortex receives its thalamic afferents from the VL (or in nomenclature for man V.o.), which receives afferents from the basal ganglia and cerebellum, the frontal eye field receives thalamic afferents from the dorsomedial nucleus, which is not known to receive basal ganglia afferents. Fourth, contrary to the hand, there is no long lasting paresis of gaze after lesions of the frontal eye field. All these facts show that the neural mechanisms underlying voluntary eye movements are fundamentally different from the other voluntary movements.

Some positive evidence for a ramp generator in the basal ganglia comes from human pathology. Most of the symptoms of lesions in the basal ganglia are signs of release: Rigor, resting tremor, chorea, athetosis, torsion dystonia, and ballism. The only deficiency symptom is akinesia. Therefore, akinesia is the only symptom of the basal ganglia that cannot be relieved by stereotaxic surgery in the pallidum or thalamus. On the contrary, akinesia of the arm, leg, or speech is usually more severe after bilateral lesions in the VL or pallidum. Athetosis and torsion dystonia consist clearly of released ramp movements due to gross lesions in the basal ganglia. Akinesia in the Parkinson syndrome, however, consists of a defect in ramp generation. It affects predominantly the following actions: Rising from sitting or lying positions, turning when lying, starting to walk, stopping walking, and turning during walking. In all these actions, the initiation or the change of ramp movements is essential. As may be seen from my motion pictures, in Parkinson akinesia some ramp movements are performed very slowly; the large ramp movements of the legs during

walking, however, (which would endanger equilibrium if performed very slowly) are replaced by small rapid (saccadic) movements.

While for the cerebellum, the theory presented here still awaits more direct proof by single unit recording during fast and slow movements, this type of evidence has very recently been obtained for the basal ganglia by Mahlon DeLong (later in this part). He found that neurons in the putamen, which are inactive during fast movements, show strong activity preceding and during ramp movements of the monkey arm.

The motor cortex: A device for adjustment of motor patterns generated by the cerebellum and basal ganglia to tactile objects

It has often been argued that the motor cortex is specialized in producing voluntary movements. While it is true that the motor cortex participates in generating certain types of voluntary movement, it is by no means the only structure necessary for such movement. The sensory association cortex, the basal ganglia and the cerebellum, the frontal cortex, and even the hypothalamus are equally important for such movements, a statement that is not meant to deny that there are automatic movements like breathing or the oculocephalic reflex that depend only on brainstem mechanisms, not to speak of spinal reflexes. The motor cortex does not have the functions of decision making about movement, of timing the course of rapid movements, or of generating ramps.

The specialty of the motor cortex is not voluntary movement, but the sophisticated somatosensory regulation of those movements that need such a regulation. The following facts support this hypothesis: (1) Somatic afferents are the most common sensory afferents at neu-

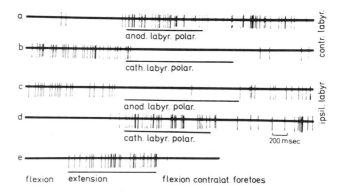

FIGURE 9 Typical neuron of the cat's motor cortex showing convergence of deep somatic and vestibular afferents. In the awake cat, out of 101 neurons of the areas 4 and 6, 65 responded to somatic, 53 to vestibular, 6 to auditory, but none to visual stimuli. (From Kornhuber and Aschoff, 1964.)

rons of the motor cortex; the next common sensory afferents are the vestibular ones (Kornhuber and Aschoff, 1964). (2) The somatosensory and the vestibular cortical areas are located in juxtaposition to the motor cortex (Frederickson et al., 1966). (3) The somatosensory and vestibular cortical areas are the only sensory projection

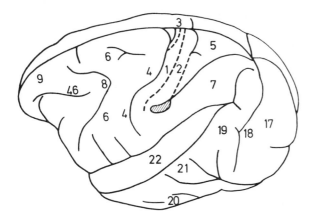

FIGURE 10 The cortical vestibular area in the rhesus monkey is located in the postcentral gyrus in juxtaposition to the face projection and the area 2 hand projection. In area 2, somatic position sense and kinesthesia are represented. Experiment of J. M. Fredrickson and H. H. Kornhuber, after Kornhuber: *Physiologie und Klinik des zentralvestibulären Systems* (Blick- und Stützmotorik). In *Hals-Nasen-Ohrenheilkunde, ein kurzgefaßtes Handbuch*, Berendes et al., eds. Vol. III, pt. 3, pp. 2150–2351. Stuttgart: G. Thieme Verlag, 1966.

areas of the cortex which have direct fiber connections to the motor cortex (Pandya and Kuypers, 1969; Jones and Powell, 1970). (4) There is a quantitative correspondence between representations of the different parts of the body in the somatic and motor cortical areas. (5) The only reactions induced by sensory stimuli that are lost after extirpation of the motor cortex are the somatic ones: Tactile placing and hopping (Bard, 1938), and tactile grasping (Denny-Brown, 1960). (The influence of the vestibular afferents on the motor cortex is, in agreement with the small vestibular representation in the cortex, much smaller than the somatic influence). (6) Movements that are not regulated by somatic messages (eye movements) do not depend on the precentral motor cortex. (7) Movements that need most exteroceptive somatosensory control (finger movements) are most corticalized.

From the phylogenetic point of view, only those movements that could take advantage of the enormously elaborated analysis of tactile messages provided by the development of the somatosensory cortex ascended in evolution to the motor cortex, while those movements that worked as well without sophisticated tactile regulation maintained their centers at the older, lower levels of the central nervous system: In the vestibular nuclei and

in the red nucleus and reticular formation, all of these getting afferents from the cerebellum and the latter two also from the basal ganglia (see Figure 2).

Neural mechanisms in the strategy of action

It is known since the experiments of W. R. Hess (1949) that some of the nervous substrates underlying the various drives to act are located in the diencephalon. Hess was able to elicit exploratory behavior, feeding, flight, aggression, vomiting, relaxation, sleep, micturition, and other activities by means of hypothalamic stimulation. Unilateral hypothalamic lesions produce inattention to visual, tactile, and proprioceptive stimuli on the opposite side. Orienting, feeding, or attack behavior to contralateral stimuli is lost for several weeks after lateral hypothalamic lesions. In the clinical literature, the loss of the drives to act is entitled *akinetic mutism*. The syndrome is characterized by absolute mutism and complete immobility except for the eyes, which are usually kept open and moved in all directions. The lesions that lead to akinetic mutism are bilateral and located in the central gray matter, the hypothalamus, the medial thalamus, or the cingulate gyrus and nearby structures. Lesions of the frontal lobes, to which the hypothalamus via the mediodorsal thalamic nucleus projects, do not result in akinetic mutism, but bilateral lesions of the frontal convexity may cause apathy and reduction of all activities while lesions of the orbital cortex often result in disinhibition of primitive drives. Obviously the needs for homeostasis of the internal milieu and survival of the species are represented in the mesencephalon, the hypothalamus and the limbic system, and these needs are mediated into action, at least in part, by way of the frontal cortex and the cingulate gyrus.

The frontal cortex has its own access to the cerebellum (via the pontine nuclei), to the basal ganglia, and directly to area 6 of the motor cortex, but we have to imagine that a good deal of influence on behavior occurs via the frontotemporal, frontoparietal, and fronto-occipital connections, as well as via fibers from the cingulate gyrus to the lateroposterior thalamus that in turn projects to the parietal lobe. In other words, the mechanisms of strategy of action, represented at cortical level in the frontal and limbic cortex, and the mechanisms of tactics, represented in the temporal and parietal lobes, communicate with each other.

Strategy and tactics in the neural mechanisms of speech: Broca's and Wernicke's aphasia

The findings in aphasia agree with this communication. The frontal aphasia due to lesions of Broca's field is characterized by loss of the drive to speak and of the strategy of

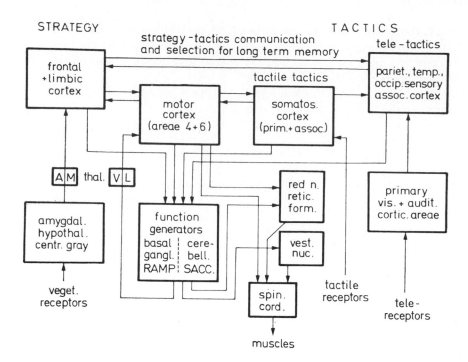

STRATEGY   TACTICS

FIGURE 11 Information flow diagram of strategic and tactical mechanisms in voluntary actions. The mechanisms of strategy (represented at cortical level in the frontal lobe and cingulate gyrus) as well as the mechanisms of tactics (in the posterior association cortex) have their own access to the motor function generators in cerebellum and basal ganglia. A anterior, M medial, VL ventrolateral thalamus. The communication lines between frontal and sensory association cortex are also important for data selection from short- to long-term memory (see Kornhuber, in Zippel, ed.: *Memory and Transfer of Information.* Plenum Publ. Comp. New York 1973.) The neural mechanisms underlying the strategy of actions (e.g. hypothalamus, amygdala, and orbitofrontal cortex) receive, of course, exteroceptive in addition to vegetative information; the former, however, in a preprocessed format from the sensory association areas; the hypothalamus gets this information indirectly via the cingulate gyrus-hippocampus loop.

speaking: The language of Broca's aphasia is agrammatic and lacks any prosody or speech melody. By contrast, the temporal lobe type of aphasia, due to lesions of Wernicke's field, is characterized by destruction of the lexicalic details, the nouns, in other words, the tactics of speaking. In Broca's aphasia, the nouns are preserved but cannot be activated and organized in sentences; this demonstrates that normally Broca's field acts on Wernicke's. In Wernicke's aphasia the strategy of speaking is preserved and often disinhibited, which results in a lot of meaningless speech; this shows that normally Wernicke's field acts on Broca's as well.

The old view of a corticocortical reflex arc underlying speech from Wernicke's to Broca's field and then via the motor cortex down to the motoneurons is not tenable, because lesions in the cerebellum, basal ganglia, and thalamus are able to produce characteristic speech disorders. Obviously Broca's and Wernicke's fields have direct access to the basal ganglia and (via the pons) to the cerebellum, and the corticocortical interaction between them is probably at least as much in the direction from Broca to Wernicke as back.

Shift of the guiding focus during action

If the hand reaches for a pencil, the movement is first guided by visual mechanisms; but if it writes, the movement depends largely on tactile inputs. However, the proportion of the words, the keeping on a straight line, and the finding of the beginning of the next line depend again on visual guidance. In everyday life, the leading focus for the hand often shifts between all the four lobes of the forebrain.

What is in charge of this shift is still partly a matter of speculation. But one point seems clear: It is not a single superhomunculus in the brain, nor is it a single superdrive like Freud's libido. As ethology—the science of animal and human behavior—reveals, man has many drives. And as Sperry's investigations in split-brain patients show, the unity of the human personality depends on many fiber connections.

Biochemical differentiation of the motor subsystems

The motor subsystems differ not only in function but also in their synaptic transmitters. Therefore diseases of some subsystems can now be differently treated by drugs. In particular, Parkinson's disease (which is the most common disease of the basal ganglia) can effectively be treated by L-dopa, a precursor of dopamine, and spasticity, a motor release phenomenon, can be treated by a derivative of GABA.

REFERENCES

ASCHOFF, J. C., B. CONRAD, and H. H. KORNHUBER, 1970. Acquired pendular nystagmus in multiple sclerosis. Amsterdam: *Proc. Bárány Soc.* (1. Extraordinary Meeting), pp. 127–132.

ASCHOFF, J. C., B. COHEN, 1971. Changes in saccadic eye movements produced by cerebellar cortical lesions. *Exp. Neurol.* 32:123–133.

BARD, P., 1938. Studies in the cortical representation of somatic sensibility. *Harvey Lec.* 33:143–169.

BARD, P., C. N. WOOLSEY, R. S. SNIDER, V. B. MOUNTCASTLE, and R. B. BROMILEY, 1947. Delimitation of central nervous mechanisms involved in motion sickness. *Fed. Proc.* 6:72.

BECHINGER, D., G. KONGEHL, und H. H. KORNHUBER, 1972. Eine Hypothese für die physiologische Grundlage des Größensehens: Quantitative Untersuchungen der Informationsübertragung für Längen und Richtungen mit Punkten und Linien. *Arch. Psychiat. Nervenkr.* 215:181–189.

BECKER, W., L. DEECKE, O. HOEHNE, K. IWASE, H. H. KORNHUBER, und P. SCHEID, 1968. Bereitschaftspotential, Motorpotential und prämotorische Positivierung der menschlichen Hirnrinde vor Willkürbewegungen. *Naturwissenschaften* 55:550.

BECKER, W., L. DEECKE, B. GRÖZINGER, and H. H. KORNHUBER, 1969. Further investigations on cerebral potentials preceding voluntary movements in man. *Pflügers Arch.* 312:R108.

BECKER, W., O. HOEHNE, K. IWASE, und H. H. KORNHUBER, 1972. Bereitschaftspotential, prämotorische Positivierung und andere Hirnpotentiale bei sakkadischen Augenbewegungen. *Vision Res.* 12:421–436.

BIZZI, E., 1968. Discharge of frontal eye field neurons during saccadic and following eye movements in unanesthetized monkeys. *Exp. Brain Res.* 6:69–80.

BRAITENBERG, V., 1961. Funktionelle Deutung von Strukturen in der grauen Substanz des Nervensystems. *Naturwissenschaften* 48:489–496.

BRAITENBERG, V., and R. P. ATWOOD, 1958. Morphological observations on the cerebellar cortex. *J. Comp. Neurol.* 109:1–34.

BRINDLEY, G. S., 1964. The use made by the cerebellum of the information that it receives from sense organs. *Int. Brain Res. Org. Bull.* 3:80.

DEECKE, L., P. SCHEID, and H. H. KORNHUBER, 1969. Distribution of readiness potential, pre-motion positivity and motor potential of the human cerebral cortex preceding voluntary finger movements. *Exp. Brain Res.* 7:158–168.

DENNY-BROWN, D., 1960. The general principles of motor integration. In *Handbook of Physiology*, Section I: Neurophysiology, J. Field, H. W. Magoun, and V. E. Hall, eds. Washington: Amer. Physiological Society.

ECCLES, J. C., M. ITO, and J. SZENTAGOTHAI, 1967. *The Cerebellum as a Neuronal Machine*. Berlin-Heidelberg-New York: Springer.

FREDRICKSON, J. M., U. FIGGE, P. SCHEID, and H. H. KORN-HUBER, 1966. Vestibular nerve projection to the cerebral cortex of the Rhesus monkey. *Exp. Brain Res.* 2:318.

FUCHS, A. F., and H. H. KORNHUBER, 1969. Extraocular muscle afferents ιo the cerebellum of the cat. *J. Physiol.* (*London*) 200:713–722.

FUCHS, A. F., and E. S. LUSCHEI, 1970. Firing patterns of abducens neurons of alert monkeys in relationship to horizontal eye movement. *J. Neurophysiol.* 33:382–392.

HAMPSON, J., C. HARRISON, and C. WOOLSEY, 1952. Cerebro-cerebellar projections and the somatotopic localization of motor function in the cerebellum. *Res. Publ. Ass. Res. Nerv. Ment. Dis.* 30:299–316.

HESS, W. R., 1949. *Das Zwischenhirn. Syndrome, Lokalisationen, Funktionen.* Basel: B. Schwabe.

ITO, M., 1970. Neurophysiological aspects of the cerebellar motor control system. *Int. J. Neurol.* 7:162–176.

ITO, M., M. YOSHIDA, and K. OBATA, 1964. Monosynaptic inhibition of the intracerebellar nuclei induced from the cerebellar cortex. *Experientia* (Basel) 20:575–576.

JOHNSON, W. H., and J. W. MAYNE, 1953. Stimulus required to produce motion sickness. Restriction of head movement as a preventive of air sickness. *J. Aviat. Med.* 24:400.

JONES, E. G., and T. P. S. POWELL, 1970. An anatomical study of converging sensory pathways within the cerebral cortex of the monkey. *Brain* 93:793–820.

KEMP, J. M., and T. P. S. POWELL, 1970. The cortico-striate projection in the monkey. *Brain* 93:525–546.

KORNHUBER, H. H., 1968. Neurologie des Kleinhirns. *Zbl. Ges. Neurol. Psychiat.* 191:13.

KORNHUBER, H. H., 1971a. Motor functions of the cerebellum and basal ganglia: the cerebello-cortical saccadic (ballistic) clock, the cerebello-nuclear hold regulator, and the basal ganglia ramp (voluntary speed smooth movement) generator. *Kybernetik* 8:157–162.

KORNHUBER, H. H., 1971b. Das vestibuläre System, mit Exkursen über die motorischen Funktionen der Formatio reticularis, des Kleinhirns, der Stammganglien und des motorischen Cortex sowie über die Raumkonstanz der Sehdinge. In *Vorträge der Erlanger Physiologentagung 1970*, W. D. Keidel and K. -H. Plattig, eds. Berlin, Heidelberg, New York: Springer-Verlag, pp. 173–204.

KORNHUBER, H. H., und J. C. ASCHOFF, 1964. Somatisch-vestibuläre Integration an Neuronen des motorischen Cortex. *Naturwissenschaften* 51:62–63.

KORNHUBER, H. H., und L. DEECKE, 1965. Hirnpotentialänderungen bei Wilkürbewegungen und passiven Bewegungen des Menschen: Bereitschaftspotential und reafferente Potentiale. *Pflügers Arch. Ges. Physiol.* 284:1–17.

LLINAS, R., and W. PRECHT, 1969. The inhibitory vestibular efferent system and its relation to the cerebellum in the frog. *Exp. Brain Res.* 9:16–29.

MARR, D., 1969. Theory of cerebellar cortex. *J. Physiol.* (*London*) 202:437–470.

NYBY, O., and J. JANSEN, 1951. An experimental investigation of the cortico-pontine projection in *Macaca mulatta*. *Norske Vid. Akad. SKR I. Math. Naturw. Kl.* Nr. 3, 47 S.

PANDYA, D. N., and H. G. J. M. KUYPERS, 1969. Cortico-cortical connections in the rhesus monkey. *Brain Res.* 13:13–36.

ROBINSON, D. A., 1964. The mechanics of human saccadic eye movement. *J. Physiol.* (*London*) 174:245.

ROBINSON, D. A., 1970. Oculomotor unit behavior in the monkey. *J. Neurophysiol.* 33:393.

SCHILLER, P. H., 1970. The discharge characteristics of single

units in the oculomotor and abducens nuclei of the unanesthetized monkey. *Exp. Brain Res.* 10:347–362.

SCHMIDT, R. S., 1965. Amphibian acoustico-lateralis efferents.
J. Cell. Comp. Physiol. 65:155–162.

SNIDER, R. S., and E. ELDRED, 1952. Cerebro-cerebellar relationship in the monkey. *J. Neurophysiol.* 15:27–40.

SPERRY, R. W., M. S. GAZZANIGA, and J. E. BOGEN, 1969.
Interhemispheric relationships: the neocortical commissures:
syndromes of hemisphere disconnection. In *Handbook of*
Clinical Neurology, P. J. Vinken and G. W. Bruyn, eds.
Amsterdam: North Holland Publishing, Vol. IV, pp. 273–
291.

THACH, W. T., 1968. Discharge of Purkinje and cerebellar
nuclear neurons during rapidly alternating arm movements
in the monkey. *J. Neurophysiol.* 31:785–797.

WURTZ, R. H., and M. E. GOLDBERG, 1971. Superior colliculus
cell responses related to eye movements in awake monkeys.
Science 171:82.

24 Principles Governing Distribution of Sensory Input to Motor Neurons

ELWOOD HENNEMAN

ABSTRACT Input to motor neurons has been studied by recording intracellularly the responses evoked by impulses in single Ia fibers. A single Ia fiber from the medial gastrocnemius muscle supplies input to all or nearly all of its 300 alpha motor neurons and conversely, a single motor neuron of that muscle receives terminals from all or nearly all of its 60 Ia fibers. The principles governing this distribution of input are discussed and reasons are given for suggesting that terminals of Ia fibers are distributed nonpreferentially to all the cells in a homonymous pool, not in equal numbers but in proportion to their surface areas, so that each motor neuron has an equal density of Ia endings. Evidence indicating that input from internuncial cells is organized in a similar fashion is discussed.

THE PERIPHERAL and central nervous systems provide motor neurons with a variety of inputs that combine to control the discharge of these cells. Little is known about the manner in which the inputs are organized to achieve control. For many years, thinking about input to pools of motor neurons was dominated by the concepts of Sherrington. From his studies of the effects of combining different inputs to a pool, Sherrington concluded that each afferent fiber terminates on a localized group of cells in a different portion of the pool, as illustrated in Figure 1A. Depending upon the level of *background* excitation reaching the pool, he believed that each input could result in a separate *discharge zone* with a surrounding *subliminal fringe* of cells that were excited but not discharged (Figure 1B). The overlapping of adjacent subliminal fringes could result in the firing of their cells due to summation of subliminal effects in a zone of *convergence*. Overlapping of two discharge zones, as shown in Figure 1A, could result in *occlusion*. Sherrington's experiments were unimpeachable, and his ingenious inferences found their way into almost all textbooks where they have held sway for many years. Unfortunately, the organization of input that Sherrington inferred is not consistent with a number of recent observations on recruitment of motor neurons (Henneman et al., 1965a and 1965b; Somjen et al., 1965), and it is contradicted by direct evidence of how afferent

ELWOOD HENNEMAN Harvard Medical School, Boston, Massachusetts

terminals of group Ia fibers are distributed (Mendell and Henneman, 1971).

Input to motor neurons must satisfy certain requirements which Sherrington was not aware of. It has become apparent that a pool is not just a uniform collection of all the cells that supply a given muscle. It is a population of cells statistically distributed with respect to the sizes of the cells, their excitabilities and inhibitabilities and other properties that depend upon size. An important corollary of variability in size is variability in total surface area. According to Gelfan (1963) the mean surface area of all neurons in the seventh lumbar segment of a dog's spinal cord is 4160 μ^2. A large motor neuron with a cell body 65 μ in diameter has a total surface area of approximately 54,000 μ^2. The consensus of opinion from several independent anatomical studies (Gelfan and Rapisarda, 1964; Wyckoff and Young, 1956; David et al., 1959) is that the density of endings is approximately equal for cells of widely differing surface areas. Large motor neurons, therefore, offer proportionately more sites for synapses than small ones, and this must be taken into account in any formulation of input.

Control over each pool is exerted by a variety of inputs from muscles, joints, and skin in the periphery and from all levels of the central nervous system. We tend to think of these inputs as exerting different types of synaptic effects, some excitatory, some inhibitory, some weak, some strong. Potency of input may be determined by a number of factors: (1) location of endings on the different parts of the receptive surface, (2) size of endings, (3) density of endings, and (4) distribution of endings throughout a pool. In this discussion, experimental evidence regarding input to motor neuron pools will be examined in order to identify general principles governing this input.

The most readily accessible of all the inputs to motor neurons is that coming from its own muscle. Group Ia fibers innervating the primary endings in muscle spindles are the only ones from muscle that establish direct, monosynaptic connections with motor neurons, hence they offer great advantages for electrophysiological investigations of input. Limitations in existing anatomical techniques make it difficult to follow all the axonal ramifications of a

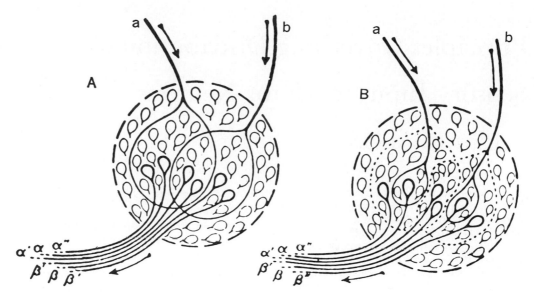

FIGURE 1 Sherrington's representation of a motor neuron *pool*, showing (B) *summation* of subliminal effects due to overlap of two subliminal fields and (A) *occlusion* of reflex discharge, due to overlap of two discharge zones. In B, fiber a and fiber b each discharge one unit separately; together they discharge four units. In A, fiber a and fiber b each discharge four units separately; together they discharge not eight, but six units. (From Creed et al., 1932.)

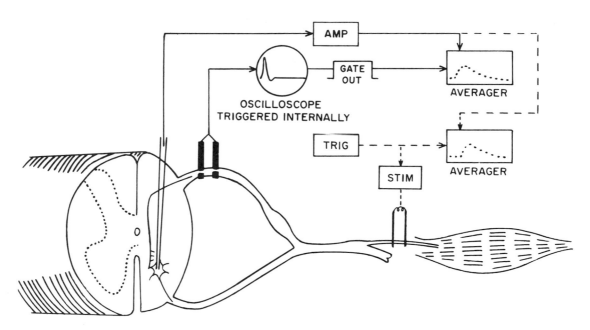

FIGURE 2 Schematic diagram of the two methods of triggering the averaging computer. The averager is represented twice: *below*, it is triggered at the same time as the electrical stimulus to the muscle nerve (electrical stimulation of afferent fibers); *above*, it is triggered by stretch-evoked impulses in a dorsal root filament (natural stimulation of afferent fibers). The signal from the microelectrode is amplified and then led to the averager. (From Mendell and Henneman, 1971.)

single neuron in the central nervous system, especially if the branches are widely distributed. Physiological techniques offer a number of ways of studying this problem. Several approaches were explored before adopting the one to be described. In brief, recordings were made of the excitatory postsynaptic potentials (EPSPs) produced in motor neurons of the gastrocnemius muscle by impulses in a single afferent fiber (group Ia) from that muscle. Since these *individual EPSPs* were usually too small to distinguish among the other spontaneously occurring responses of the motor neuron, an averaging computer was used to summate a few hundred of them. The presence or absence of a monosynaptic EPSP in a motor neuron was used to determine whether or not the afferent fiber under study sent terminals to that particular cell. The amplitude and time course of the EPSPs permitted inferences regarding the location of the active synapses. This technique is fully described in a recent paper by Mendell and Henneman (1971). The following account and the illustrations accompanying it are drawn largely from that paper.

Figure 2 illustrates schematically the two techniques that were used to summate the tiny electrical potentials representing the *individual EPSPs*. Figure 3 illustrates a summated EPSP, too small to detect in single sweeps, that was *extracted* from the background activity by repeating it many times. Natural (i.e., stretch-evoked) stimulation was used. Traces A, B, and C represent summated EPSPs developed after 100, 500, and 1000 repetitions, respectively. They demonstrate that the amplitude of the response grows linearly with increasing numbers of repetitions. Comparison of the summated EPSPs with the summated square waves (200 μV) in the same tracings reveals

that the calibrated amplitude of the EPSPs was about 190 μV in each of the three traces. The absence of any response in trace D, which was recorded after the microelectrode had been withdrawn to a point just outside the cell, indicates that the technique used in these experiments did not detect responses of nearby motor neurons and interneurons or activity in branches of the afferent fiber itself.

In general, it was found that each Ia fiber from the medial gastrocnemius muscle projected to almost all (93%) of the motor neurons of that muscle. Intracellular recordings were obtained from 122 homonymous motor neurons. The number of cells studied in each experiment and the result obtained are given in Table I. Individual EPSPs were recorded from 114 of these 122 cells. Of the 22 afferent fibers studied, 14 were found to project to every motor neuron impaled. Each of the other 8 afferent fibers projected to all but 1 of the motor neurons from which intracellular recordings were obtained.

The shape of individual EPSPs varied over a wide

TABLE I

Summary of experimental results

Preparation	Trigger Mode	Ia Conduction Velocity (msec)	Number of Motoneurons Recorded	Number of Positive Responses	Longitudinal Limits (mm)
Anesthetized	electrical	76	4	4	2.0
Anesthetized	electrical	103	10	10	2.5
Anesthetized	electrical	88	10	9	1.9
Anesthetized	stretch	—	2	2	—
Anesthetized	stretch	72	3	3	0.1
Anesthetized	stretch	87	8	8	2.2
Anesthetized	stretch	92	13	12	2.4
Anesthetized	stretch	82	6	5	3.6
Anesthetized	stretch	95	3	3	1.9
Anesthetized	stretch	85	5	4	1.1
Anesthetized	stretch	—	9	8	3.8
Anesthetized	stretch	100	4	4	2.7
Anesthetized	stretch	95	7	7	3.9
Anesthetized	stretch	—	1	1	—
Anesthetized	stretch	103	2	2	0.1
Anesthetized	stretch	103	8	7	3.8
Anesthetized	stretch	84	3	3	0.9
Anesthetized	stretch	83	8	7	2.4
Spinal	stretch	—	5	4	1.5
Spinal	stretch	99	3	3	—
Spinal	stretch	—	2	2	1.2
Spinal	stretch	87	6	6	—

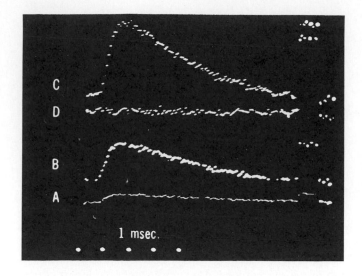

FIGURE 3 Summated EPSPs produced by repetition of many individual EPSPs in a gastrocnemius motoneuron. Traces A, B, and C illustrate the results of summing 100, 500, and 1000 sweeps, respectively; trace D is the sum of 1000 sweeps recorded after the microelectrode had been withdrawn to a point just outside the motoneuron, as indicated by the sudden disappearance of the resting membrane potential. A calibrating pulse of 200 μV was injected at the end of each sweep and summated in the same manner as the signal. (From Mendell and Henneman, 1971.)

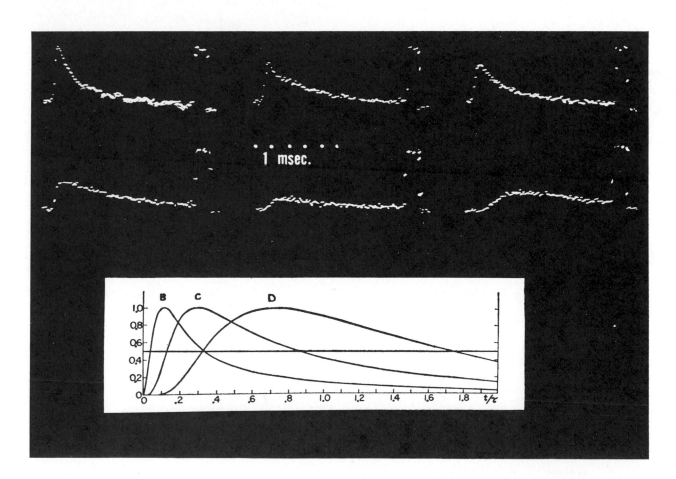

FIGURE 4 *Above:* Summated EPSPs evoked by impulses in the same Ia fiber terminating on six different motoneurons. Calibration pulse is 200 μV in each case. (From Mendell and Henneman, 1971.) *Below:* Computed EPSPs for input distributed to three different parts of the motoneuron. The three curves represent the EPSPs produced by inputs limited to (B) the soma, (C) proximal dendrites, and (D) distal dendrites. They are reproduced from Figure 2 of Rall, 1967.

range. Examples of this variation are reproduced in Figure 4. The six EPSPs shown there were evoked by impulses in the same afferent fiber terminating on six different motor neurons. There was no apparent tendency in this or in other experiments for the EPSPs elicited by activity in a particular afferent fiber to have similar time courses. As illustrated in Figure 4, EPSPs with fast-rising phases had fast-falling phases as well. Slow-rising phases were invariably accompanied by slow-falling phases.

In their shape and time course these individual EPSPs were surprisingly similar to the theoretical EPSPs obtained by Rall (1967) when inputs were applied to different parts of a model motor neuron. The model developed by Rall was based on anatomical and physiological data (Rall, 1959, 1960, 1962, 1964). The soma and branching dendrites were represented as a chain of 10 compartments with equal surface areas, corresponding to equal increments of electrotonic length, as illustrated in Figure 5.

With a digital computer, simulated inputs were applied to each of these compartments. The computed EPSPs for inputs to compartments 2 and 3 are reproduced in part A of Figure 5. Parts B, C, and D illustrate the computed EPSPs for inputs to compartments 4 and 5, 6 and 7, and 8 and 9. There was no anatomical evidence indicating that the endings of a Ia fiber are concentrated in single, limited areas as Rall's theoretical inputs were. Hence the close resemblance of the experimental EPSPs to the computed EPSPs was quite unexpected. It strongly suggests that the individual EPSPs were caused by Ia inputs concentrated within limited areas of motoneuronal surface. A more detailed comparison of Rall's computed EPSPs and the individual EPSPs in our experiments, as illustrated in Figure 6, again suggests that a particular Ia fiber terminates on different parts of the various motor neurons to which it projects and that the *boutons terminaux* it gives to each cell are, in general, not widely separated but densely congregated on a small portion of the receptive surface.

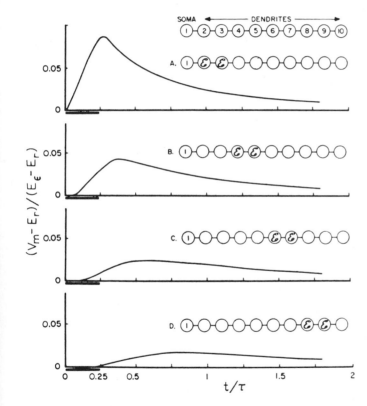

FIGURE 5 Effect of location of an excitatory postsynaptic potential upon transient soma-membrane depolarization. The soma-dendrite receptive surface is divided into 10 equal compartments (upper right) from soma (1) to dendritic tip (10). As the stimulus (the E pulse) is applied at loci progressively more distal to the somal trigger area, the depolarization waves, as computed to appear at the soma, show increased latency of onset, with progressively lower and later peaks and slower rise times. The curves were drawn through computed values for time steps of 0.05 τ. (For further details see Rall, 1964.) (From Rall, 1964.)

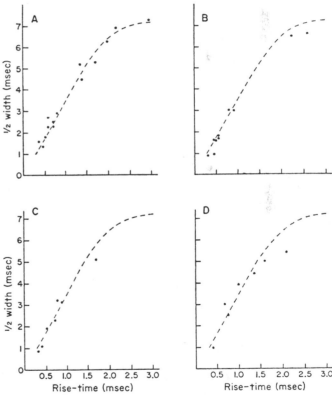

FIGURE 6 Time course of individual EPSPs in four experiments. The rise time of each EPSP (from foot to peak) is plotted against its half-width (duration at half of peak amplitude). The dashed curve reproduced in each graph shows how the time course of theoretical EPSPs in a model motoneuron (Rall et al., 1967) varied with the location of the input. (From Mendell and Henneman, 1971.)

Scheibel and Scheibel (1969) have recently published a Golgi study of the primary afferent collaterals in the cat's spinal cord that provides an anatomical basis for some of these findings. It is now agreed (Scheibel and Scheibel, 1969; Sterling and Kuypers, 1967a, 1967b) that, contrary to earlier views, the soma and dendrites of motor neurons are disposed in a predominantly longitudinal column running for a considerable distance in the rostrocaudal axis of the spinal cord. This orientation is clearly evident in the motor neurons labeled f and g in Figure 7. Scheibel and Scheibel (1969) show that each primary afferent fiber in the dorsal columns gives off a series of primary afferent collaterals at intervals of 100 to 200 μ. As illustrated in Figure 7, these collaterals leave the dorsal columns *at right angles and drop ventrally with almost plumb-line precision.* A variable number of collaterals run together, forming a *microbundle.* As shown in Figures 7 and 8, these microbundles pass perpendicularly through the dendrites of the motor neurons. The terminals of these fibers ramify almost exclusively in the mediolateral plane

of the spinal cord. As a result, the terminal field of each collateral is virtually two dimensional and makes contact with a very limited extent of the longitudinally disposed motor neurons. These terminal fields are 10 to 50 μ in thickness. Comparison of the anatomical and electrical observations reveals a surprising fact about the connections that are established, which is not apparent from the anatomical studies alone. Although each dendrite must traverse a series of about 10 terminal fields from the same Ia fiber, judging from the time courses of the individual EPSPs, it receives endings from only one of its collaterals. The shape of the EPSPs, that is, indicates that they are generated by inputs concentrated within one *compartment* (Rall, 1967) of the motor neuron. The possibility of inputs from two or more primary afferent collaterals of the same Ia fiber going to different compartments of the same motor neuron is essentially excluded by the findings, although, as illustrated in Figure 8, a primary afferent collateral might make synaptic contacts with two or more dendrites from the same cell. From the electrophysio-

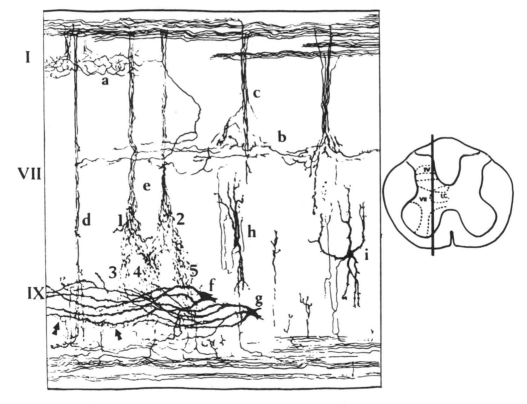

FIGURE 7 Sagittal section through lumbosacral cord of a 40-day-old cat showing selected details of presynaptic and postsynaptic neuropil. (a) Longitudinal neuropil of dorsal horn; (b) elements of longitudinal neuropil of intermediate nucleus generated by short primary afferent collaterals such as those in microbundle c; (e) microbundles made up of long primary afferent collaterals generating *primary* mix (1, 2) and secondary mix (3, 4, 5) neuropil fields; (f and g) motoneuron somata in staggered position with overlapping systems of sagittally running dendrites; (h) interneuron of lamina VII with typical single dendritic system; (i) interneuron of lamina VII with multiple dendrite systems. Arrows point to terminating axon collaterals from propriospinal fibers of ventral column that run parallel to motoneuron dendrites and apparently establish multiple synaptic contacts. Rapid Golgi modification. (X200) (From Scheibel and Scheibel, 1969.)

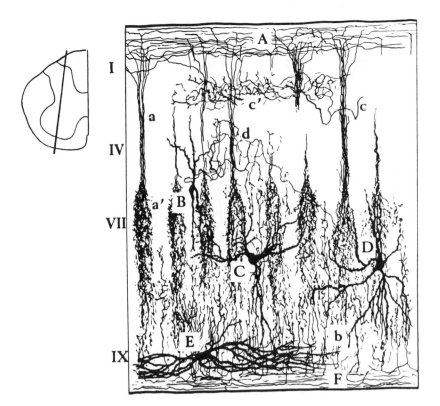

FIGURE 8 Sagittal section through lumbosacral cord of 30-day-old cat showing organization of afferent neuropil and a few postsynaptic elements. (A) Fibers of dorsal column; (B) interneuron of lamina VI; (C and D) dorsally situated motoneurons with radiating dendrite patterns; (E) motoneuron with sagittally running dendrites; (F) fibers of ventral column; (a) microbundle generating terminal neuropil field a′; (b) area of intermingling of preterminals from primary afferent collaterals with those from ventral column; (c) presumed coarse cutaneous afferent generating longitudinal gelatinosal plexus c′; (d) local axonal collateral elaborations generated by interneuron B. Rapid Golgi modification. (X200) (From Scheibel and Scheibel, 1969.)

logical observations, it appears that after one set of endings from a particular Ia fiber is established on a motor neuron, no others from different primary collaterals of that fiber can develop, even on other dendrites of the motor neuron.

The functional significance of this arrangement is not entirely clear. It appears to be designed to prevent the synchronization of input that would result if several primary afferent collaterals from the same fiber made contact with the same motor neuron. Dispersing the terminals of each Ia fiber to 300 cells probably has the effect of instantaneously averaging the input throughout the pool, for it ensures that the timing of excitatory events (EPSPs) and thus, the degree of temporal convergence is the same for all members of the pool. This is an important consideration if the size principle is not to break down due to transitory inequalities in input resulting from different degrees of temporal convergence.

One other feature in the fine structure of this system deserves comment. As illustrated in Figure 9, the longitudinally running dendrites of interneurons and motor neurons are covered with spines, whereas the vertically oriented dendrites are almost devoid of spines. The significance of this difference is not known.

The afferent fibers isolated in different experiments varied widely in conduction velocity. This provided an opportunity to determine whether an impulse in a large Ia fiber exerts a greater synaptic effect on a motor neuron than one in a smaller fiber. Relevant data for individual, stretch-evoked EPSPs in homonymous motor neurons are plotted in Figure 10. Since EPSPs are influenced by several neural and experimental factors that cause large variations in their amplitude, all the EPSPs obtained with fibers of the same size were averaged to minimize these influences. The average amplitudes are plotted with small filled circles. In order to utilize larger numbers of responses for averaging, the EPSPs in each 5 msec interval of conduction velocity were also grouped. The average for each pentad is plotted with open circles. Presented in this way, the data in Figure 10 reveal a clear relation between the conduction velocity of the afferent fiber and the size of the EPSPs elicited by its impulses.

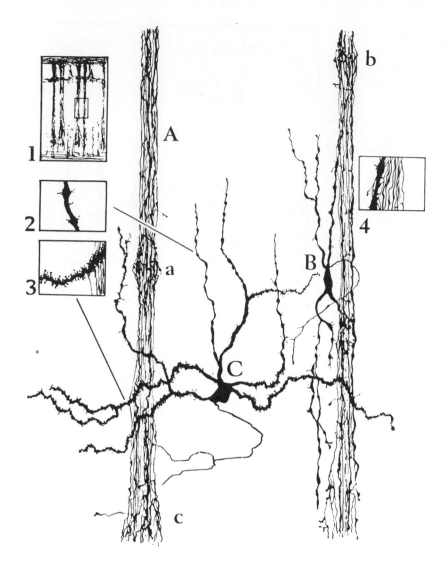

FIGURE 9 Some details of synaptic organization as seen in sagittal sections of lumbosacral spinal cord in 30-day-old kitten. (A) Microbundle traversing the central portion of the spinal gray (laminae VI–VII); (B) interneuron with single, vertically oriented dendrite system; (C) large interneuron with several vertically oriented dendrite systems and an unusually extensive dendritic spread in the rostrocaudal (horizontal) dimension; (a and b) area of massed intrafascicular specializations resulting in elbow or plumber's joint appearance; (c) initial bifurcations of microbundle fibers which will produce the terminal neuropil field. *Inset:* Diagram 1 shows approximate position of these structures; 2 shows nodular, almost spine-free appearance of vertically oriented dendrite shafts; 3 shows spine-rich appearance of horizontally oriented dendrites and their *cross-over* relation with vertical axonal elements; 4 shows the nature of the relationship between parallel-lying dendritic and axonal elements. A very small number of immediately adjacent presynaptic elements may effect repeated axodendritic contacts. Rapid Golgi modifications. (X600) (From Scheibel and Scheibel, 1969.)

Density of Ia endings on motor neurons

Of greater functional significance than the absolute numbers of Ia endings are their densities and locations on individual motor neurons. By histological examination the total density of endings of all types is approximately constant from cell to cell. Using several methods of counting synaptic knobs, Gelfan and Rapisarda (1964) found that there are about 20 knobs per 100 μ^2 of cell surface and

that there is little difference in their density on soma and dendrites. As Figure 2 of their paper illustrates, there is no correlation between density of endings and cell size.

Nothing in the study described above or in previous studies of Lloyd (1943) or Lloyd and McIntyre (1955) indicates that Ia input is distributed preferentially to certain cells. In fact, the evidence is quite the contrary. Regardless of the number of Ia receptors actively discharging in a muscle, all of the motor neurons of that

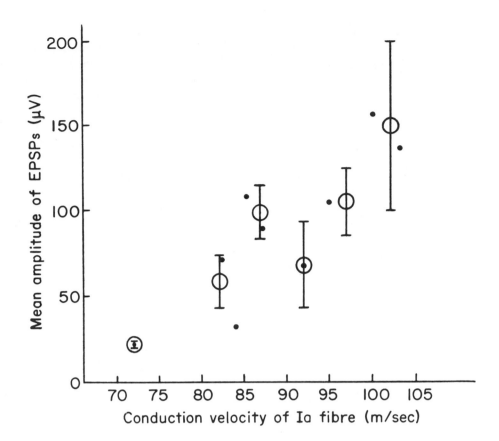

FIGURE 10 Relation between conduction velocity of Ia fibers and the average amplitude of the individual EPSPs elicited by their impulses. Small filled circles represent average amplitudes of all EPSPs obtained with fibers of the same conduction velocity. Open circles represent average amplitudes of all EPSPs in each 5 msec interval of axonal conduction velocity. Vertical bars indicate mean deviations of these values. (From Mendell and Henneman, 1971.)

muscle apparently receive signals from essentially all endings. The small percentage of negative results did not occur predominantly in cells of a particular size. Moreover, there is no evidence that Ia terminals tend to be located more proximally on small cells than on large ones. In order to examine this question, the sample of motor neurons was divided into two groups, one with axonal conduction velocities of 62 to 89 msec, the other with velocities of 90 to 117 msec. The average rise time of the EPSPs in the group of 36 smaller cells was 1.1 msec; that of the group of 33 larger cells was 1.0 msec. If the endings on the smaller cells had been located more proximally, this tendency should have been reflected in shorter rise times.

Burke (1968) proposed that the greater susceptibility of small cells to reflex discharge, reported by Henneman et al. (1965a, 1965b) is due to preferential distribution of Ia input to small cells. It must be emphasized, however, that small cells are the most excitable regardless of the source of the input. Any hypothesis advanced to explain orderly recruitment must, therefore, apply not only to Ia fibers but to all types of input. A small motor neuron con-

ceivably might have a greater density of endings from some one type of afferent fiber. Due to the limitation of surface area, however, this greater density could occur only if there were correspondingly fewer endings from some other type of fiber. Since the total density of endings is the same on motor neurons of all sizes, a small cell could not accommodate a greater density of excitatory endings from all of its various afferent fibers. Similarly, all the different types of afferents could not be distributed on the surface of a small cell so that each of them had a maximally effective arrangement of its endings.

If preferential distribution is ruled out, what is the alternative? It would seem to be some kind of *equalization* of input for all cells. On the basis of the clues in previous work and the findings described above, it is suggested that the terminals of Ia fibers are distributed to all the cells in a homonymous pool, not in equal numbers but to each cell in proportion to its surface area. Each motor neuron would then have an equal density of Ia endings. This appears to be the only way of arranging Ia terminals without distributing input preferentially. The *equal density hypothesis* appears to offer the only rational explanation for the available electrophysiological data. Until objective evidence for equal density can be obtained, however, it must be regarded as a logical possibility rather than an established fact.

Some functional implications of this arrangement of input

Control of skeletal muscle depends on graded differences in the excitability of the motor neurons themselves. These differences in excitability do not arise from preferential distribution of afferent terminals, at least in the case of the Ia fibers, but they must be preserved in the face of great variations in excitatory and inhibitory input if control is to remain precise under all circumstances. The *equal density hypothesis* helps to explain why the normal order of firing in motor neurons is not altered when an inhibitory input is added to an excitatory input. The fact that each motor neuron receives terminals from all or nearly all Ia fibers indicates that as Ia input grows in intensity due to recruitment of additional stretch receptors and more rapid firing of active endings, it increases *pari passu* for all the homonymous motor neurons of the pool, including those already firing. By the time that the input is sufficiently intense to discharge the large motor neurons, there must be a great excess of excitation for the small cells beyond that required to reach their firing level. If an inhibitory input is now added to the excitatory influx, it silences the largest of the active cells first and the smallest ones last. The order in which motor neurons are silenced probably depends on the relative amounts of excess excitation they are receiving. The largest cells with the smallest excess or safety margin are the least resistant to inhibition, whereas the smallest cells with the largest excess are the most resistant. This hypothesis explains why an inhibitory input of increasing intensity produces a net effect that is similar to a reduction in excitatory input.

Organization of input in other systems

There is much less information available concerning input arriving at motor neurons over routes other than the Ia system. Some fibers in descending spinal tracts establish direct connections with motor neurons, but most input reaches them by way of internuncial cells in the spinal cord. The distribution of this internuncial input within motor neuron pools has not been adequately studied with anatomical techniques. Recent physiological studies (Gillies et al., 1971; Clamann et al., 1971), however, indicate that the organization of these inputs may be similar in principle to that of the Ia system.

In studies not yet reported except in abstract form, a technique was devised to measure the percentage of cells in a pool that was discharged monosynaptically by an afferent volley. With the aid of posttetanic potentiation (PTP), this percentage could be varied from 0 to 100. The responses of individual motor neurons belonging to the same pool were recorded simultaneously in ventral root filaments. The firing *thresholds* of these units were obtained by determining what percentage of the entire pool had to be discharged monosynaptically in order to fire the particular motor neuron under study. The threshold obtained in this way was a measure of the relative excitability or rank order of the cell in question as compared with all other cells in the pool.

The thresholds determined with this technique were extremely constant, varying on repeated trials by less than 3%, often by no more than the accuracy of the equipment itself. The results, therefore, suggested that there is a stable rank order of the motor neurons in a pool when their excitability is tested by means of monosynaptic reflexes. If this rank order were the same for other inputs, it would be reasonable to infer that they were organized and distributed in a manner at least analogous to that of the Ia system. Previous experiments (Henneman et al., 1965a, 1965b; Somjen et al., 1965) on order of recruitment with different types of inputs had indicated that weak excitatory inputs discharge the smallest motor neurons first and that stronger inputs recruit progressively larger motor neurons. The techniques of filament recording used in those experiments, however, did not permit quantitative measurement of thresholds or establishment of rank order. In order to retain the advantage of measuring threshold precisely, it was decided to examine the effects of various other inputs upon threshold measured with monosynaptic reflexes.

In order to avoid certain technical difficulties, inhibitory inputs were used in the great majority of the experiments. By observing the gross effects of these inputs upon test monosynaptic reflexes at all levels from 0 to 100%, it was easy to demonstrate that the endings responsible for the inhibition were distributed to all portions of the motor neuron pool. This could be shown with weak as well as strong inhibitory inputs. The effect of such inputs on the threshold of motor neurons was examined, using inhibition elicited by stimulation of the ipsilateral common peroneal nerve or the brainstem reticular formation. The effects of these inputs upon the thresholds of the individual neurons were used to determine how the input was distributed among the members of the pool. If the effect of an inhibitory input, for example, was to lower the threshold of a motor neuron, it would be inferred that the inhibitory endings were distributed preferentially to cells that were smaller and more excitable than the unit under study. If the effect was to raise the threshold, it would be inferred that the inhibitory endings were distributed preferentially to the unit being investigated. If no change in threshold was observed, it would be inferred that the inhibitory input was distributed preferentially to cells larger than the one under study or to all cells in proportion to their surface area. By studying a sufficient number of individual motor neurons in this way, the distribution of input could be worked out quite satisfactorily.

In general, the inhibitory inputs were found to have little or no effect on the threshold of gastrocnemius motor neurons. In most instances the threshold range during inhibition overlapped that of the previously determined *control* range. The maximum separation between the mean thresholds with and without inhibition was 3.5%, and the maximum separation between their ranges was 3.5%.

The findings in these preliminary experiments indicate that inhibitory endings are widely distributed to all parts of the motor neuron pool and not exclusively to restricted portions of it. It is not clear how this wide distribution is achieved. A recent suggestion by Jankowska and Roberts (1972) is worth noting. In a study of internuncial activity activated by collaterals of Ia fibers from the quadriceps muscle, it was estimated that each internuncial projected to about 20% of the motor neurons of the closest antagonists, posterior biceps, and semitendinosus. If the Ia fibers of quadriceps projected to all of the internuncial cells that send axons to PBST, the overall system might satisfy the requirements for wide distribution. Jankowska and Roberts point out that

If each Ia afferent fiber from Q terminates on several of these interneurons, as they do on motoneurons (Mendell and Henneman, 1971), which in turn project randomly to about 20% of the PBST motoneurons, then each Ia fiber from Q might influence most of the population of PBST motoneurons. The emerging picture of the connexions between the Ia afferents and the motoneurons of the antagonist muscles would thus closely resemble that of the direct connexions of the Ia afferents on the homonymous motoneurons described by Mendell and Henneman.

Whatever the organization of the various inhibitory systems may prove to be, it is clear from the foregoing experiments that these inputs are distributed with equal density to all members of the motor neuron pool. It is a reasonable conjecture, therefore, that the equal density hypothesis may apply to all inputs to motor neuron pools.

REFERENCES

BURKE, R. E., 1968. Group Ia synaptic input to fast and slow twitch motor units of cat triceps surae. *J. Physiol. (London)* 196:605–630.

CLAMANN, H. P., J. D. GILLIES, R. D. SKINNER, and E. HENNEMAN, 1971. Two methods of measuring the percentage discharge of a motoneuron pool in a monosynaptic reflex. *Fed. Proc.* 30(2):E55.

CREED, R. S., D. DENNY-BROWN, J. C. ECCLES, E. G. T. LIDDELL, and C. S. SHERRINGTON, 1932. *Reflex Activity of the Spinal Cord.* London: Oxford University Press.

DAVID, G. B., A. W. BROWN, and K. B. MALLION, 1959. On the distribution of synaptic end-feet (*boutons terminaux*) in the central nervous system of the cat. *J. Physiol.* 147:55–56P.

GELFAN, S., 1963. Neuron and synapse population in the spinal cord: indication of role in total integration. *Nature (London)* 198:162–163.

GELFAN, S., and A. F. RAPISARDA, 1964. Synaptic density on spinal neurons of normal dogs and dogs with experimental hind-limb rigidity. *J. Comp. Neurol.* 123:73–95.

GILLIES, J. D., H. P. CLAMANN, R. D. SKINNER, and E. HENNEMAN, 1971. Thresholds of individual motoneurons in monosynaptic reflexes defined in terms of the percent discharge of the entire pool. *Fed. Proc.* 30(2):132D.

HENNEMAN, E., G. SOMJEN, and D. O. CARPENTER, 1965a. Functional significance of cell size in spinal motoneurons. *J. Neurophysiol.* 28:560–580.

HENNEMAN, E., G. SOMJEN, and D. O. CARPENTER, 1965b. Excitability and inhibitibility of motoneurons of different sizes. *J. Neurophysiol.* 28:599–620.

JANKOWSKA, E., and W. J. ROBERTS, 1972. Synaptic actions of single interneurons mediating reciprocal Ia inhibition of motoneurons. *J. Physiol.* 222:623–642.

LLOYD, D. P. C., 1943. Neuron patterns controlling transmission of ipsilateral hind-limb reflexes in cat. *J. Neurophysiol.* 6:293–315.

LLOYD, D. P. C., and A. K. McINTYRE, 1955. Transmitter potentiality of homonymous and heteronymous monosynaptic reflex connections of individual motoneurons. *J. Gen. Physiol.* 38:789–799.

MENDELL, L. M., and E. HENNEMAN, 1971. Terminals of single Ia fibers: location, density, and distribution within a pool of 300 homonymous motoneurons. *J. Neurophysiol.* 34:171–187.

RALL, W., 1959. Branching dendritic trees and motoneuron membrane resistivity. *Exp. Neurol.* 1:491–527.

RALL, W., 1960. Membrane potential transients and membrane time constant of motoneurons. *Exp. Neurol.* 2:503–532.

RALL, W., 1962. Theory of physiological properties of dendrites. *Ann. N.Y. Acad. Sci.* 96:1071–1092.

RALL, W., 1964. Theoretical significance of dendritic trees for neuronal input-output relations. In *Neural Theory and Modeling*, R. F. Reisa, ed. Stanford, Calif.: Stanford University Press, pp. 73–79.

RALL, W., 1967. Distinguishing theoretical synaptic potentials computed for different soma-dendritic distributions of synaptic input. *J. Neurophysiol.* 30:1138–1168.

SCHEIBEL, M. E., and A. B. SCHEIBEL, 1969. Terminal patterns in cat spinal cord. III. Primary afferent collaterals. *Brain Res.* 13:417–443.

SOMJEN, G., D. O. CARPENTER, and E. HENNEMAN, 1965. Responses of motoneurons of different sizes to graded stimulation of supraspinal centers of the brain. *J. Neurophysiol.* 28:958–965.

STERLING, P., and H. G. J. M. KUYPERS, 1967a. Anatomical organization of the brachial spinal cord of the cat. I. The distribution of dorsal root fibers. *Brain Res.* 4:1–15.

STERLING, P., and H. G. J. M. KUYPERS, 1967b. Anatomical organization of the brachial spinal cord of the cat. II. The motoneuron plexus. *Brain Res.* 4:16–32.

WYCKOFF, W. G., and J. Z. YOUNG, 1956. The motoneuron surface. *Proc. Roy. Soc. B. London* 144:440–450.

25 The Control Mechanisms of Cerebellar Motor Systems

MASAO ITO

ABSTRACT In order to reveal the significance of the cerebellar motor control, the neuronal construction of vestibular reflex arcs in connection with the cerebellum has been analyzed. Comparison between the vestibulo-ocular, vestibulospinal, and spinovestibulospinal reflex arcs thus suggests that an essential feature of cerebellar function is the elaboration of an open-loop feedforward motor control.

THE CEREBELLUM has been regarded as an organ responsible for coordination of movements. This view was first presented by Flourens (1842) and substantiated by later workers, particularly by Chambers and Sprague (1955a, 1955b). The cerebellum may be divided into several portions, each of which acts as a coordinator for a certain type of motor action. The vermis (medial part of the cerebellum) is thus concerned mainly with the body equilibrium, the paravermis (intermediate part) with the locomotion, and the hemisphere (lateral part) with the volitional movements. The vestibulocerebellum is the phylogenetically oldest part of the cerebellum that has been developed in connection with the vestibular organ (see Herrick, 1924).

In the past decade, a wealth of knowledge has been obtained concerning the neuronal compositions and activities in the cerebellum and related structures (Eccles, Ito, and Szentágothai, 1967; Evarts and Thach, 1969; Llinás, 1969). Thus there have been opened the three following new approaches to the mechanisms of the cerebellar coordination. The first approach is to analyze the neuronal connectivity in the cerebellar cortex and to suppose what sort of computation may occur there. The regular spacing of Purkinje cells led Braitenberg and Onesto (1962) to conceive the cerebellum as a timing device, as the impulse conduction along parallel fiber beams would excite a group of Purkinje cells in a sequential manner such as the ticking of a clock. This delay-line hypothesis was reexamined and supported by Freeman (1969), but the significance of the relatively short delay (up to 10 msec in frog, Freeman, 1969) in the actual

MASAO ITO Department of Physiology, Faculty of Medicine, University of Tokyo, Japan

cerebellar function has not been clear. By contrast, Marr's (1969) theory emphasizes spatial patterns of cerebellar cortical nerve net and its modifiability and assumes the cerebellar cortex to be a sort of learning machine. Albus's (1971) theory is similar to Marr's (1969) but is closer to the idea of "perceptrons" (Rosenblatt, 1962). In spite of their charm, these theories have not yet been accepted, because their basic assumptions are not supported experimentally (see below). Apart from these theories, computer simulation techniques have been utilized to visualize the possible signal traffics in the cerebellar cortex (Pellionisz, 1970). The second approach is to observe impulse activities in cerebellar cortical neurons and to speculate what sort of information processing occurs there. Discharges of Purkinje cells have been recorded in awake monkeys as they maintain postures and perform movements (Thach, 1972). The third approach is to analyze how the cerebellum is linked with subcerebellar motor centers and to grasp the significance of the cerebellar control of various motor functions (Eccles, 1969). Moruzzi (1950) pointed out long ago that the cerebellar anterior lobe is a circuit parallel with postural reflex arcs. To advance this line of investigation, the author has tried to combine the classic concepts of reflexes with the modern concepts of control theory. To provide the experimental basis, the vestibulo-cerebellum and vermis were chosen as the material in the hope that these phylogenetically older parts of the cerebellum may contain a prototype of the cerebellar control system in a relatively simple fashion (Ito, 1970, 1972).

Reflex and control

The understanding of neural motor control mechanisms has been advanced through studies of reflexes. The reflex is a unit reaction in the central nervous system having a stereotyped and repeatable input-output relationship. Sherrington (1906) took the view that the neural control mechanisms of coordinated movements are explicable in terms of the compounding reflexes. Modern meaning may be added to this classical physiological view of neural motor control when it is compared with recent theoretical and practical developments in engineering control systems. A reflex arc executing a reflex is composed of a receptor, afferent path, reflex center, efferent path, and effector. This corresponds to an engineering control system composed of an input, controller, process to be controlled, and output (Figure 1A). Reflex arcs are often equipped with a feedback pathway, corresponding to the classic feedback, or closed-loop, control system. Application to biological systems of these concepts of feedback and closed-loop control has been quite fruitful. Reflex arcs, however, often lack such feedback and form a feedforward control system, or more generally, an open-loop control system. The usefulness of the feedforward control in biological systems has been discussed by McKay (1966).

Many control systems may be combined to form a large-scale control system, each control system serving as one building block of the whole system, just as does a reflex arc in the central nervous system. In fact in a large-scale plant, such as a hydraulic electric generator station, many control systems with different designs are combined in a pyramid-like organization. At the bottom of this pyramid, there may be many systems performing the classical feedback control. To higher hierarchical levels belong those control systems that perform the feedforward control, cascade control, adaptive control, etc. At the

peak of the pyramid there will be a final control that holds the conditions of the whole system optimum.

In the neural motor control system as well, a similar hierarchical structure should be considered. Integrative action of the central nervous system (Sherrington, 1906) is not due to a simple amalgamation of reflexes, but it is due to an organization of reflexes with certain specific principles. Classic concepts of reflexes thus could be expanded to the whole neural motor control.

Feedback control

Feedback control is the most basic form of engineering control wherein the output is returned to the input. It has been well known that the stretch reflex is carried out through a neuronal chain containing two neurons in series, as shown in Figure 1B, and has the features of a feedback control system. It maintains the muscle length constant by the follow-up length servomechanism. The reference input for this control is provided by γ motoneurons (Figure 1B). Recently, the stretch reflex arc in cat was analyzed to determine its control system parameters (Roberts, Rosenthal, and Terzuolo, 1971).

Another typical example of feedback control may be

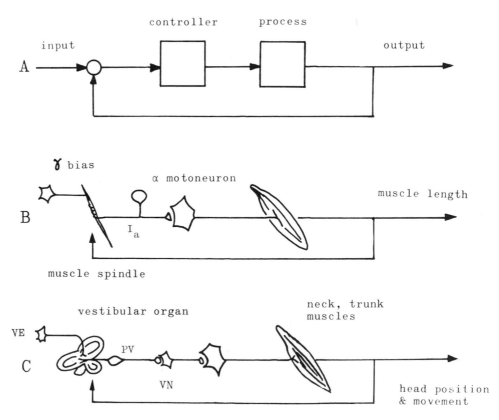

FIGURE 1   Reflex arc as a feedback control system. (A) Block diagram for a feedback control system. (B) Neuronal connection for the stretch reflex. I_a, group I_a muscle afferent neuron.

(C) That for the vestibulospinal reflex. PV, primary vestibular neuron. VN, vestibular nuclei cell. VE, vestibular efferent cell.

performed by the vestibulospinal reflex arc to maintain the head position constant. The major portion of this reflex arc is formed of three neurons in series, as shown in Figure 1C. Any sign of changes in the head position will be detected by the vestibular organ, signals from which will in turn evoke contraction in neck and trunk muscles to prevent the head position from being altered. The vestibular organ is equipped with efferent innervation (VE in Figure 1C), but it is not certain whether this is to function as the set point for the control. A beautiful demonstration of the vestibulospinal reflex action has recently been given in owl (Money and Correia, 1972). The control system performance of the spinovestibular reflex system in fish has been studied by Howland (1971).

Feedforward control

The feedforward control is a kind of open-loop control and is based on the principle of load compensation. As shown in Figure 2A, the controller works out the effect of an external input to the process and generates another input to cancel it. A typical example is the vestibulo-ocular reflex arc. As shown in Figure 2B, when a positional change or movement of the head is signalled by the vestibular organ, this reflex arc will evoke a compensatory positional change or movement of the eyes to maintain constant retinal input. The final output of the vestibulo-ocular reflex, i.e., position and movement of eyes relative to the external environment, will be detected by vision, but there is no simple way to return the visual information to the vestibular organ. The vestibulo-ocular reflex is *open-loop* in this sense and *feedforward* in its mode of operation. The vestibulo-ocular reflex is developed particularly well in those animals with one-sided eyes, i.e., fish, birds, rabbits, etc.

The well-known *tonic neck reflex* would also have an aspect of the feedforward control. In this reflex, signals arising from the neck region are forwarded to influence limb muscles, and this plays an important role in the maintenance of posture. Another typical example of the

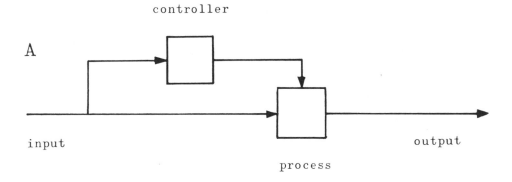

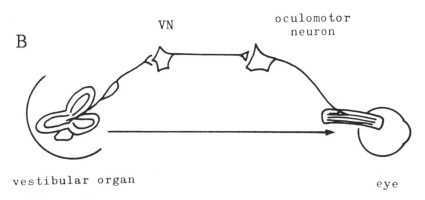

FIGURE 2 Reflex arc as a feedforward control system. (A) Block diagram for a feedforward control system. (B) Vestibulo-ocular reflex arc. VN, vestibular nuclei cell.

feedforward control may be given by the blink response to a glabella tap; mechanical stimuli to the glabella evoke quick blinking so as to avoid a possible damage to eyes (Woody and Brozek, 1969).

Importance of the feedforward control in motor systems

The feedback control system is characterized by its insensitiveness to external disturbances and changes in control parameters. However, an obvious limitation of the feedback-control system is the requirement of an uncomplicated feedback pathway from the output to the input. In the actual motor control, such a pathway as this is not always available, because of the following reasons: First, as in the cases of the vestibulo-ocular and glabella tap reflexes, the input and output signals are often represented with different modalities, i.e., vestibular vs. visual, or mechanical vs. visual. Second, even when the modality is similar, spatial separation of the input and output would make the feedback difficult. This is the case in the tonic neck reflex. Third, even if there is a pathway connecting the output to the input, its performance as a feedback pathway could be incomplete if the control system operates at a high speed as in quick and complex motions; the so-called *loop time* will be the limiting factor.

Because of these reasons, the feedforward control appears to be involved at many instances in the motor control. Compared with the feedback control, the feedforward control is much more sensitive to external disturbances and changes in parameters. If an elaborate, precise performance is required for a feedforward control system, this drawback should be removed in some way. A computer is used in engineering control systems. An essential point of the present study is to point out that a part of the cerebellum is likewise utilized to assist a reflex arc that has the feedforward control performance.

The vestibulo-ocular reflex (VOR) arc and the vestibulocerebellum

The major pathway for the VOR is the histologically well-defined three-neuron arc: Primary vestibular neurons, secondary vestibular neurons, oculomotor neurons (Figure 1C). Signals arising from the semicircular canal and otolith organs thus are relayed by secondary vestibular neurons located in four anatomical structures and forwarded toward six groups of oculomotor neurons, as shown in Figure 3. The secondary vestibular neurons have either inhibitory or excitatory action. The inhibitory action for most oculomotor neuron groups is relayed by the superior vestibular nucleus (Highstein, Ito, and

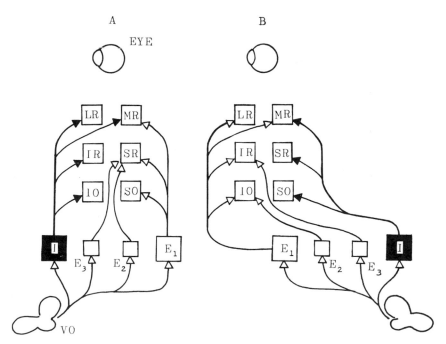

FIGURE 3 Connection in the vestibulo-ocular reflex arc in rabbits. (A) From the left vestibular organ to the left eye. (B) From the right vestibular organ to the left eye. VO, vestibular organ. LR, motoneurons for lateral rectus muscle. MR, medial rectus. IR. inferior rectus. SR, superior rectus. IO, inferior oblique. SO, superior oblique. I, inhibitory vestibulo-ocular relay cells. E_1, excitatory relay cells in the medial nucleus. E_2, those in the y group of the vestibular nuclear complex. E_3, those in the cerebellar lateral nucleus. This diagram is based on the reports by Highstein et al. (1971) and Highstein (1971, 1973a, 1973b).

Tsuchiya, 1971; Highstein, 1971; Precht and Baker, 1972). However, the possibility is raised that the inhibition for the abducens motoneuron group is relayed by the rostral extreme of the medial vestibular nucleus (Baker, Mano, and Shimazu, 1969; Highstein, 1973b). On the other hand, the excitation is mediated by the rostral half of the medial vestibular nucleus, the *y* group of the vestibular nuclear complex, and the cerebellar lateral nucleus (Highstein et al., 1971; Highstein, 1973a). The spatial pattern of projection from the four groups of secondary vestibular neurons to the six groups of oculomotor neurons has thus been determined on both anatomical and physiological basis.

The relationship between this VOR arc with the cerebellum has been subjected to anatomical and physiological investigations. It turned out that one part of the vestibulocerebellum, i.e., the flocculus, is linked with the VOR arc in the following two ways as illustrated in Figure 4. First, the flocculus receives the same primary vestibular afferent signals that are fed to the secondary vestibular neurons. In the cerebellar cortex of the floc-

culus, the primary vestibular fibers form the mossy fiber terminals (Brodal and Hoivik, 1964). Second, the flocculus sends Purkinje cell axons to the lateral cerebellar nucleus as well as the vestibular nuclei (Angaut and Brodal, 1967). These Purkinje cell axons have specific inhibitory action upon the secondary vestibular neurons as tested in rabbits (Fukuda, Highstein, and Ito, 1972).

Thus, the flocculus forms a side path to the VOR arc. It is conceivable that the flocculus elaborates the feedforward control by the VOR arc on the basis of processing the primary vestibular information. Lesion of the flocculus indeed impairs the VOR (Di Giorgio and Giulio, 1949; Manni, 1950; Carpenter, 1972).

The vestibulospinal reflex (VSR) arc and the vestibulocerebellum

The view that the vestibulocerebellum is related to the VOR in order to improve its feedforward control performance is strongly supported by the analysis of the VSR that performs the closed-loop control.

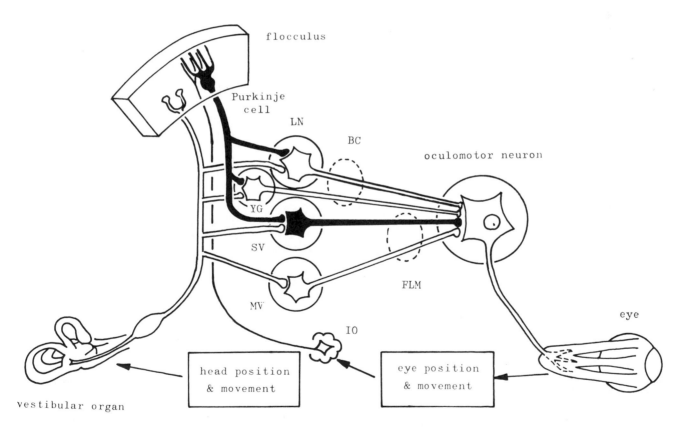

FIGURE 4 Neuronal connection between the vestibulo-ocular reflex arc and the flocculus. LN, cerebellar lateral nucleus. YG, the *y* group of the vestibular nuclear complex. SV, superior vestibular nucleus. MV, medial vestibular nucleus. BC, brachium conjunctivum. FLM, fasciculus longitudinalis medialis. IO, inferior olive. In this as well as in Figure 5, inhibitory neurons are filled in black, while excitatory neurons are indicated by hollow structure. (Modified from Figure 8 of Fukuda et al., 1972.)

The major aspect of the VSR arc is the three-neuron chain composed of the primary vestibular afferents, secondary vestibular neurons, and spinal motoneurons (Figure 1C). The axons of the secondary vestibular neurons for the VSR pass through two separate fiber bundles, the lateral (LVST) and medial (MVST) vestibulospinal tracts.

Three types have been distinguished from the secondary vestibular neurons for the VSR, as shown in Figure 5: (1) those with LVST axons conducting with a relatively fast velocity and exerting an excitatory action upon spinal neurons (Lund and Pompeiano, 1968; Grillner, Hongo, and Lund, 1970); (2) those with excitatory and fast conducting MVST axons (Akaike, Fanardjian, Ito, Kumada, and Nakajima, 1973a; Akaike, Fanardjian, Ito, and Ohno, 1973b); (3) those with inhibitory and slowly conducting MVST axons (Wilson and Yoshida, 1969; Wilson, Yoshida, and Shor, 1970; Akaike et al., 1973a, b). In any of these three types of cells, there is no evidence indicating that Purkinje cell inhibition is derived from the vestibulocerebellum (Akaike, Fanardjian, Ito, and Nakajima, 1973c).

Thus, in spite of their similar trineuronal construction, the VSR arc is free of the inhibitory influences from the vestibulocerebellum as contrasted to the VOR arc that is closely connected to the flocculus. It is corollary that those reflex arcs performing the feedback control do not require assistance by the cerebellar cortex.

The spinovestibulospinal reflex (SVSR) arc and the cerebellum

Those cells located in the dorsal portion of cat's Deiters' nucleus do not receive the primary vestibular impulses, but they receive a synaptic input from the spinal cord (Pompeiano and Brodal, 1957). The spinal input to Deiters's nucleus has been shown to be excitatory (Ito, Hongo, Yoshida, Okada, and Obata, 1964). In rabbits, it has been demonstrated that this excitatory action is mediated by those fibers ascending the lateral funiculus of the cervical cord and passes into the cerebellar vermis (Akaike et al., 1973c). The spinovestibulospinal reflex (SVSR) arc will thus be formed through a population of LVST and MVST cells as shown in Figure 6. The follow-

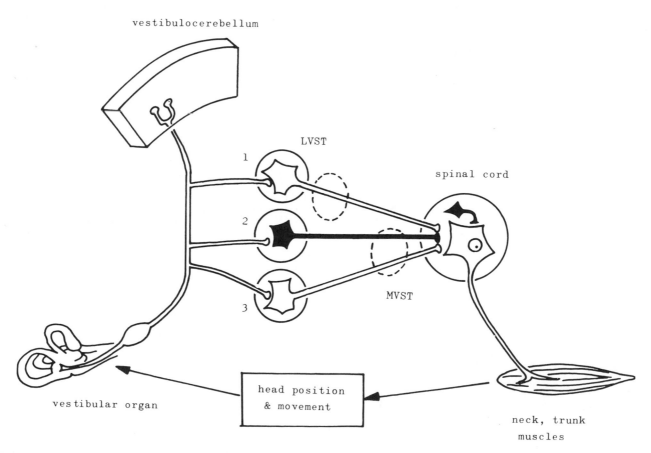

FIGURE 5   Neuronal construction of the vestibulospinal reflex arc. LVST, lateral vestibulospinal tract. MVST, medial vestibulospinal tract, 1, fast excitatory LVST cell. 2, slow inhibitory MVST cell. 3, fast excitatory MVST cell. (This diagram is based on the reports by Akaike et al., 1973a, 1973b, 1973c.)

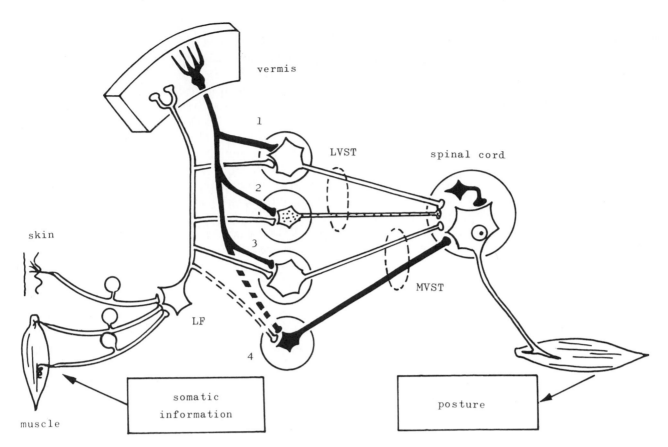

FIGURE 6 Relationship between the spinovestibulospinal reflex arc and the cerebellar vermis. LF, spinocerebellar afferents ascending through the lateral funiculus. 1, fast excitatory LVST cell. 2, slow LVST cell with unknown synaptic action. 3. fast excitatory MVST cell. 4, slow inhibitory MVST cell. Interrupted lines indicate uncertainty for the connections to group 4 neuron. (Based on the reports by Akaike et al., 1973a, 1973b, 1973c.)

ing four types of cells appear to be involved in rabbit's SVSR arc: (1) those with fast conducting and excitatory axons; (2) those with slowly conducting LVST axons, the action of which is unknown; (3) those with fast conducting and excitatory MVST axons; (4) those with slowly conducting and inhibitory MVST axons.

On the other hand, it has been demonstrated that Purkinje cells in the vermal cortex send axons into Deiters' nucleus, exerting inhibitory action (Ito and Yoshida, 1966; Ito, Obata, and Ochi, 1966; Ito, Kawai, and Udo, 1968). This Purkinje cell inhibition occurs in good correlation with the spinal excitation (Akaike et al., 1973c). Thus, the same two-way relationship as that seen between the VOR arc and the flocculus can be obtained between the SVSR arc and the cerebellar vermis.

The SVSR arc may have an aspect of the feedforward control system, as its input and output would separate spatially, i.e., between the neck and limbs, between forelimbs and hindlimbs, or between the left and right sides. There is evidence indicating that the LVST impulses not only activate ipsilateral extensor motoneurons but also

influence the contralateral side via interneuronal relays (Ten Bruggencate, Burke, Lundberg, and Udo, 1969; Hongo, Kudo, and Tanaka, 1971; Aoyama, Hongo, Kudo, and Tanaka, 1971). It is also known that individual cells of cat's Deiters' nucleus are excited from a relatively broad receptive field over many peripheral nerves (Allen, Sabah, and Toyama, 1972). The close linkage of the SVSR arc with the cerebellar vermis thus seems to be related to the presumed feedforward-control performance of the SVSR arc.

Coupling of the VSR and SVSR arcs

The contribution of the vestibulospinal tract neurons to postural maintenance can now be represented by combination of the two reflex arcs with different control performance, i.e., VSR and SVSR arcs. These arcs have their major relaying sites at the ventral and dorsal portions of Deiters' nucleus, respectively, and would perform the vestibular operated feedback control and the spinal operated feedforward control separately (VST-1 and -3 in Figure 7). However, there is an overlap between the two groups of cells involved in these reflex arcs (Walberg and Mugnaini, 1969). The overlap is more prominent in rabbits; almost all cells in the VSR arc are involved also in the SVSR arc (Akaike et al., 1973c). The two types of the control systems are coupled as illustrated in Figure 7 (VST-2). Such a coupling of different control systems would be an important aspect of the integration in the neural motor control.

Operation of the VOR arc with a visual feedback

The effect of visual stimulation upon the vestibulo-cerebellum has been studied by Maekawa and Simpson (1972). It emerged that there is a pathway that mediates the visual information to the flocculus, via the accessory optic tract, the central tegmental tract, and the inferior olive, as indicated in Figure 4. As contrasted to the mossy fiber terminals formed by the primary vestibular afferents, the axons of the inferior olive cells form climbing fiber terminals which exert a powerful excitatory action upon Purkinje cells.

The operation of the VOR arc can be demonstrated by stimulating the vestibular nerve electrically and recording the impulses evoked in the secondary vestibular neurons. In Figure 8, two vestibular nerve stimuli were paired, with various time intervals. At intervals of less than 3 msec, the second response was reduced, apparently due to refractoriness in both the primary and secondary vestibular neurons. With increasing the time interval, the second response recovers, but at about 3 msec it is depressed again to exhibit an inhibition curve having a duration of 15 msec or more. This depression appears to be produced by Purkinje cell impulses evoked by the primary vestibular fibers that excite Purkinje cells via the mossy fiber granule cell pathway (Eccles et al., 1967). Direct stimulation of the inferior olive likewise evokes a prominent inhibition of the secondary vestibular volley at time intervals of more than 3 msec (Figure 9). This inhibition is apparently due to Purkinje cell discharges evoked via the inferior olive-climbing fiber pathway. In accordance with Maekawa and Simpson's (1972) observation in the flocculus, electric stimulation of the optic disk of the retina evoked an inhibition with a time interval of 12 msec (Figure 9 open circle, hollow arrow), which is approximately the time for the climbing fiber activation of Purkinje cells. In Figure 9, some depression is seen at shorter intervals of 3 to 7 msec after the optic disk stimulation, the reason for which has not yet been given. Thus, the VOR arc appears to be equipped with a visual feedback pathway in order to improve its control performance. In an engineering control system, the feedforward control is often combined with the feedback control and improves the control performance tremendously.

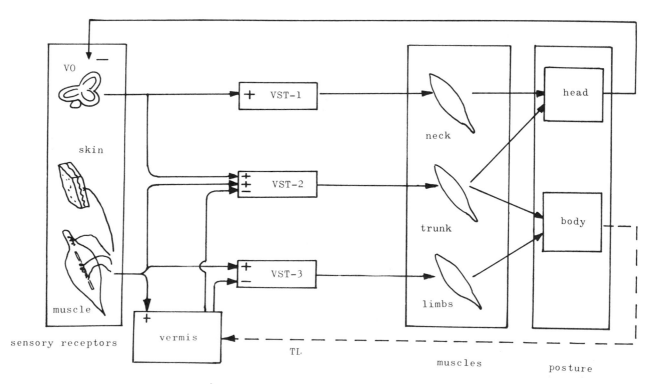

FIGURE 7   Coupling of the vestibulospinal and spinovestibulo-spinal reflex arcs. VO, vestibular organ. TL, line informing the results of the posture control in the body to the cerebellar vermis. Explanation is in the text.

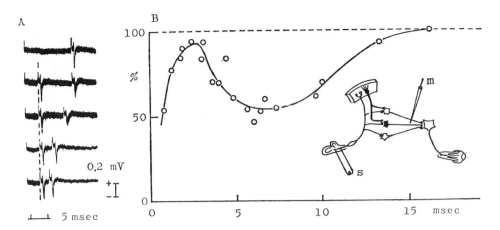

FIGURE 8 Vestibular-evoked cerebellar inhibitory control on VOR as demonstrated by double shock to the vestibular nerve. Inset figure indicates the recording (m) and stimulating (s) electrodes on the diagram modified from Figure 4. (A) Secondary vestibular volley recorded from the fasciculus longitudinalis medialis just ventral to IIIrd nucleus. Vertical interrupted line marks the moment of conditioning stimulation. (B) Plots the amplitude of the testing secondary vestibular volley relative to the control (ordinate) as a function of testing-conditioning time interval (abscissa). (Unpublished observation by Ito.)

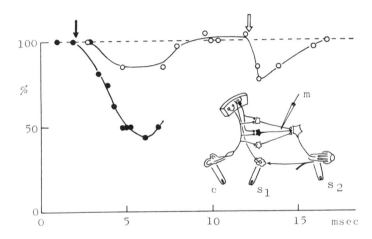

FIGURE 9 Cerebellar inhibition in VOR evoked through the visual system. Illustrated similarly to Figure 8B, but conditioning stimuli were given either to the inferior olive (s_1, filled circles) or the optic disk (s_2, open circles). (c) Stimulating electrode on the vestibular nerve. Filled-in arrow indicates the moment of the inferior olive evoked inhibition. Hollow arrow marks that of optic disk evoked inhibition that onset in phase with the climbing fiber activation of flocculus Purkinje cells. (Unpublished observation by Ito.)

Possible learning process in the cerebellovestibular system

In Figure 4 when the activity in the visual feedback pathway to the flocculus lasts sufficiently long, signaling inadequate performance of the vestibular operated feedforward control, will there be a change in parameters involved in activation of flocculus Purkinje cells by primary vestibular impulses so that the vestibular operated feedforward control will be improved? Lorente de Nó (1931) previously described that even though newborn rabbits exhibit the VOR before their eyes are open, the performance of the reflex is abnormal. Its final develop-

ment is achieved about 40 days after their birth, and apparently visual influence plays a role in this development. Very recently, Gonshor (1971) has shown that in human subjects wearing prism glasses through which the right-left relationship in the visual field is reversed, the VOR is modified in the following way: The extent of the reflex in the horizontal direction is reduced gradually and becomes null at about 5 days after the beginning of the experiment, thereafter there being reversal in the polarity of the reflex movement. When the glasses are removed, the normal VOR is restored relatively slowly; about 4 weeks are needed for complete recovery.

One of the attractive points in the above idea of the

possible cerebellar learning is that it accords with the hypothesis proposed by Marr (1969) and Albus (1971). Both of these authors assumed that synapses supplied by the cerebellar granule cells to Purkinje cells and other intracortical cells are modifiable and the modification occurs when the activation via the granule cells and the activation via the climbing fibers coincide in one and the same Purkinje cell and other intracortical cells. Unfortunately, there is as yet no experimental evidence supporting this postulate. Possibly, the cerebellar learning is such a slow process that it cannot be detected in usual acute experiments. It may be worth mentioning that in training experiments with daily repeated rotary, centrifugal, see-saw and pendulum-like motions, it takes many days for the animals to learn the vestibular operated postural adjustment (Fukuda, Tokita, Hinoki, and Kitahara, 1963).

Summary

The specific relationship between certain elementary reflex arcs and the phylogenetically old parts of the cerebellum appears to be determined by the feedback and feedforward control performance of these reflex arcs. With the aid of the cerebellum, feedforward control systems may manage complex input–output relationships that cannot be managed by feedback-control systems. This view is consistent with the neurophysiological concept that the cerebellum acts as a coordinator of complex movements. It is desirable to extend similar considerations to other reflex systems linked with phylogenetically newer parts of the cerebellum.

REFERENCES

AKAIKE, T., V. V. FANARDJIAN, M. ITO, M. KUMADA, and H. NAKAJIMA, 1973a. Electrophysiological analysis of the vestibulospinal reflex pathway of rabbit. I. Classification of tract cells. *Exp. Brain Res.* (in press).

AKAIKE, T., V. V. FANARDJIAN, M. ITO, and T. OHNO, 1973b. Electrophysiological analysis of the vestibulospinal reflex pathway of rabbit. II. Synaptic actions upon spinal neurons. *Exp. Brain Res.* (in press).

AKAIKE, T., V. V. FANARDJIAN, M. ITO, and H. NAKAJIMA, 1973c. Cerebellar control of the vestibulospinal tract cells of rabbit. *Exp. Brain Res.* (in press).

ALBUS, J. S., 1971. A theory of cerebellar function. *Math. Biosci.* 10:25–61.

ALLEN, G. I., N. H. SABAH, and K. TOYAMA, 1972. Synaptic actions of peripheral nerve impulses upon Deiters neurones via the mossy fibre afferents. *J. Physiol.* 226:335–351.

ANGAUT, P., and A. BRODAL, 1967. The projection of the vestibulo cerebellum onto the vestibular nuclei in the cats. *Arch. Ital. Biol.* 105:441–479.

AOYAMA, M., T. HONGO, N. KUDO, and R. TANAKA, 1971. Convergent effects from bilateral vestibulospinal tracts on spinal interneurons. *Brain Res.* 35:250–253.

BAKER, R. G., N. MANO, and H. SHIMAZU, 1969. Postsynaptic potentials in abducens motoneurons induced by vestibular stimulation. *Brain Res.* 15:577–580.

BRAITENBERG, V., and N. ONESTO, 1962. The cerebellar cortex as a timing organ. Discussion of a hypothesis. Proc. 1st Congr. Int. Med. Cibernetica. Giannini, Naples.

BRODAL, A., and B. HOIVIK, 1964. Site and mode of termination of primary vestibulo-cerebellar fibers in the cat. *Arch. Ital. Biol.* 102:1–21.

CARPENTER, R. H. S., 1972. Cerebellectomy and the transfer function of the vestibulo-ocular reflex in the decerebrate cat. *Proc. Roy. Soc. Lond.* B 181:353–374.

CHAMBERS, W. W., and J. M. SPRAGUE, 1955a. Functional localization in the cerebellum. I. Organization in longitudinal cortico-nuclear zones and their contribution to the control of posture, both extra pyramidal and pyramidal. *J. Comp. Neurol.* 103:105–129.

CHAMBERS, W. W., and J. M. SPRAGUE, 1955b. Functional localization in the cerebellum. II. Somatotopic organization in cortex and muscle. *Arch. Neurol. Psychiat.* 74:653–680.

DI GIORGIO, A. M., and L. GIULIO, 1949. Reflessi oculari di origire otolitica ed influenza del cervalletto. *Boll. Soc. Ital. Biol. Sper.* 25:145–146.

ECCLES, J. C., 1969. The dynamic loop hypothesis of movement control. In *Information Processing in the Nervous System*, K. N. Leibovic, ed. Berlin, Heidelberg, New York: Springer-Verlag. pp. 245–269.

ECCLES, J. C., M. ITO, J. SZENTÁGOTHAI, 1967. *The Cerebellum as a Neuronal Machine.* New York: Springer-Verlag.

EVARTS, E. V., and W. T. THACH, 1969. Motor mechanisms of the CNS: cerebrocerebellar interrelations. *Ann. Rev. Physiol.* 32:451–498.

FLOURENS, P., 1842. *Récherches expérimentales sur les Propriétés et les Fonctions du Systeme Nerveau dans les Animaux Vertébres.* Edition 2. Paris: Bailliere.

FREEMAN, J. A., 1969. The cerebellum as a timing device: An experimental study in the frog. In *Neurobiology of Cerebellar Evolution and Development*, R. Llinás, ed. Chicago: American Medical Association, pp. 397–420.

FUKUDA, J., S. M. HIGHSTEIN, and M. ITO, 1972. Cerebellar inhibitory control of the vestibulo-ocular reflex investigated in rabbit IIIrd nucleus. *Exp. Brain Res.* 14:511–526.

FUKUDA, T., T. TOKITA, M. HINOKI, and M. KITAHARA, 1963. The physiology of training. The functional development of the labyrinthine function through the daily repetition of rotary, centrifugal, see-saw and pendulum-like motions. *Acta Otolaryng.* 56:239–250.

GONSHOR, A., 1971. Habituation of natural head movement in man. *Proc. Internat. Physiol. Sci.* 9:211.

GRILLNER, S., T. HONGO, and S. LUND, 1970. The vestibulospinal tract. Effects on alpha-motoneurons in the lumbosacral spinal cord in the cat. *Exp. Brain Res.* 10:94–120.

HERRICK, C. J., 1924. Origin and evolution of the cerebellum. *Arch. Neurol. Psychiat.* 11:621–625.

HIGHSTEIN, S. M., 1971. Organization of the vestibulo-ocular pathways to rabbits IIIrd and IVth nuclei. *Brain Res.* 32:218–224.

HIGHSTEIN, S. M., 1973a. The organization of the vestibulo-ocular and trochlear reflex pathways in the rabbit. *Exp. Brain Res.* (in press).

HIGHSTEIN, S. M., 1973b. Synaptic linkage in the vestibulo-ocular and cerebello-vestibular pathways to the VIth nucleus in the rabbit. *Exp. Brain Res.* (in press).

HIGHSTEIN, S. M., M. ITO, and T. TSUCHIYA, 1971. Synaptic linkage in the vestibulo-ocular reflex pathway of rabbit. *Exp. Brain Res.* 13:306–326.

HONGO, T., N. KUDO, and R. TANAKA, 1971. Effects from the vestibulospinal tract on the contralateral hindlimb motoneurons in the cat. *Brain Res.* 31:220–223.

HOWLAND, H. C., 1971. The role of the semicircular canals in the angular orientation of fish. *Ann. N.Y. Acad. Sci.* 188:202–216.

ITO, M., 1970. Neurophysiological aspects of the cerebellar motor control system. *Internat. J. Neurol.* 7:162–176.

ITO, M., 1972. Neural design of the cerebellar motor control system. *Brain Res.* 40:81–84.

ITO, M., T. HONGO, M. YOSHIDA, Y. OKADA, and K. OBATA, 1964. Antidromic and transsynaptic activation of Deiters' neurons induced from the spinal cord. *Jap. J. Physiol.* 14:638–658.

ITO, M., N. KAWAI, and M. UDO, 1968. The origin of cerebellar-induced inhibition of Deiters neurone. III. Localization of the inhibitory zone. *Exp. Brain. Res.* 4:310–320.

ITO, M., K. OBATA, and R. OCHI, 1966. The origin of cerebellar-induced inhibition of Deiters neurones. II. Temporal correlation between the trans-synaptic activation of Purkinje cells and the inhibition of Deiters' neurones. *Exp. Brain Res.* 2:350–364.

ITO, M., and M. YOSHIDA, 1966. The origin of cerebellar-induced inhibition of Deiters neurones. I. Monosynaptic initiation of the inhibitory postsynaptic potentials. *Exp. Brain Res.* 2:330–349.

LLINÁS, R., 1969. Neuronal operations in cerebellar trans-actions. In *The Neurosciences: Second Study Program*, F. O. Schmitt, editor-in-chief. New York: Rockefeller University Press, pp. 409–426.

LORENTE DE NÓ, R., 1931. Ausgewälte Kupitel aus der vergleichenden physiologie des Labyrinthes. Die Augenmuskel-reflexe beim Kaninchen und ihre Grundlagen. *Ergebn. Physiol.* 32:73–242.

LUND, S., and O. POMPEIANO, 1968. Monosynaptic excitation of alpha motoneurons from supraspinal structures in the cat. *Acta Physiol. Scand.* 73:1–21.

MAEKAWA, K., and J. I. SIMPSON, 1972. Climbing fiber activation of Purkinje cells in the flocculus by impulses transferred through the visual pathway. *Brain Res.* 39:245–251.

MANNI, D. E., 1950. Localizzazioni cerebellari corticali della cavia nota 29; Effetti di lesioni delle "parti vestibolari" del cervelletto. *Arch. Fisiol.* 50:110–123.

MARR, D., 1969. A theory of cerebellar cortex. *J. Physiol.* (*London*) 202:437–470.

MCKAY, D. M., 1966. Cerebral organization and the conscious control of action. In *Brain and Conscious Experience*, J. C. Eccles, ed. Berlin, Heidelberg, New York: Springer-Verlag, pp. 422–440.

MONEY, K. E., and M. J. CORREIA, 1972. The vestibular system of the owl. *Comp. Biochem. Physiol.* 42(A):353–358.

MORUZZI, G., 1950. *Problems in Cerebellar Physiology*. Springfield, Ill.: Charles C Thomas.

PELLIONISZ, A., 1970. Computer simulation of the pattern transfer of large cerebellar neuronal fields. *Acta Biochim. Biophys. Acad. Sci. Hung.* 5:71–79.

POMPEIANO, O., and A. BRODAL, 1957. The origin of vestibulo-spinal fibers in the cat. An experimental-anatomical study, with comments on the descending medial longitudinal fasciculus. *Arch. Ital. Biol.* 95:166–195.

PRECHT, W., and R. BAKER, 1972. Synaptic organization of the vestibulo-trochlear pathway. *Exp. Brain Res.* 14:158–184.

ROBERTS, W. J., N. P. ROSENTHAL, and A. TERZUOLO, 1971. A control model of stretch reflex. *J. Neurophysiol.* 34:620–634.

ROSENBLATT, F., 1962. *Principles of Neurodynamics: Perceptrons and the Theory of Brain Mechanisms*. Washington, D.C.: Spartan Books.

SHERRINGTON, C. S., 1906. *Integrative Action of the Nervous System*. New Haven: Yale University Press.

TEN BRUGGENCATE, G., R. BURKE, A. LUNDBERG, and M. UDO, 1969. Interaction between the vestibulospinal tract, contra-lateral flexor reflex afferents and Ia afferents. *Brain Res.* 14:529–532.

THACH, W. T., 1972. Cerebellar output: Properties, synthesis and uses. *Brain Res.* 40:89–97.

WALBERG, F., and E. MUGNAINI, 1969. Distinction of degenerating fibers and boutons of cerebellar and peripheral origin in the Deiters' nucleus of the same animal. *Brain Res.* 14:67–75.

WILSON, V. J. and M. YOSHIDA, 1969. Monosynaptic inhibition of neck motoneurons by the medial vestibular nucleus. *Exp. Brain Res.* 9:365–380.

WILSON, V. J., M. YOSHIDA, and R. H. SHOR, 1970. Supraspinal monosynaptic excitation and inhibition of thoracic back motoneurons. *Exp. Brain Res.* 11:282–295.

WOODY, C. D., and G. BROZEK, 1969. Gross potential from facial nucleus of cat as an index of neural activity in response to glabella tap. *J. Neurophysiol.* 32:704–716.

26 Reconsideration of the Oculomotor Pathway

JÜRGEN C. ASCHOFF

ABSTRACT   There are two basically different sets of associated eye movements: (a) smooth pursuit movements that are continuously regulated and that depend on the presence of a moving stimulus; and (b) rapid search movements called saccades that are too fast to be continuously controlled by visual feedback. The supranuclear pathways for these two kinds of eye movements are described whereby special interest has been paid to the saccadic pathway. This pathway was thought in the classical concept to originate in the frontal eye field (receiving information from visual cortices), to cross in the mesencephalic reticular formation, and to project finally onto the immediate supranuclear gaze center in the paramedian pontine reticular formation. Evidence is presented that saccadic eye movements utilize a different pathway: They originate in the visual and paravisual association cortices, from where the projection is homolaterally onto the lateral pontine nuclei (pes pontis). The pathway terminates in the supranuclear gaze centers for horizontal and vertical gaze only after passing through the cerebellum (cortex and nuclei). The so-called oculomotor decussation, therefore, does not take place in the mesencephalic reticular formation but within the cerebellum.

THE SUPRANUCLEAR pathway for oculomotor control has been intensively investigated by gross potential and single-unit studies and by ablation and stimulation experiments, both in animals as well as in man. Despite the large body of literature and our increasing knowledge, the exact pathways for supranuclear control of eye movements have yet to be defined.

Let us first stress that we deal with two basically different sets of associated eye movements, which were described for the first time some 70 years ago (Dodge, 1903): First, there are the so-called saccades, whereby our eyes jump between two points of interest in our surroundings. These are the most rapid eye movements one can produce with maximal velocities of up to 600° per second. Second, the oculomotor system is capable of smooth pursuit movements. Its task consists in matching the angular velocity of the eyes with the velocity of a moving object, whereby the maximal velocity does not exceed 90°/sec.

A short introduction into the normal characteristics of saccadic eye movements, their similarities and dissimilarities to smooth pursuit movements as well as to limb movements might be useful. In general we are able to move our limbs voluntarily at any desired speed from rest through a continuous range of velocities up to the most rapid ballistic movements such as encounted in boxing or piano playing. For limb movements, the velocity is independent of the amplitude or of any external stimulus and can be changed and regulated at will. But not so our eye movements. The velocity and amplitudes of smooth pursuit movements depend upon a moving visual target without which they cannot be executed at all. On the other hand, saccadic eye movements can be performed at will and at any desired amplitude, but their velocity cannot be changed at will. It is entirely dependent on the magnitude of the eye movement and is influenced only involuntarily by the state of alertness.

These two sets of associated eye movements are more or less independent of each other, and they utilize different parts of the brain in their supranuclear pathways. The classical concept for supranuclear control of saccadic eye movements may be summarized as follows: Peripheral visual stimuli reach area 17 of the occipital lobe from where association fibers communicate with areas 18 and 19, where an error signal will probably be formed. These will then reach the frontal eye field, that is Brodman's cortical area 8, where a saccadic motor command signal is thought to be generated. This motor command signal would then be sent via the corticobulbar tract ipsilaterally to the mesencephalic reticular formation, where a decussation of the oculomotor pathway was thought to occur at the level of the fourth nerve nucleus. The signal then would reach the contralateral pontine reticular formation that acts as the immediate supranuclear center for horizontal gaze. Smooth pursuit movements are generated in the occipital lobe and reach the same supranuclear pontine gaze center by way of pretecto- or tecto-mesencephalic-pontine fibers. Finally, the two signals would be transmitted to the appropriate oculomotor nuclei.

Even though the participation of the cerebellum in eye movements has long been known, the cerebellum was not included in this concept of oculomotor pathways. From a thorough search through the literature as well as from

---

JÜRGEN C. ASCHOFF   Department of Neurology and Neurophysiology, University of Ulm, Germany

experimental data, the pathways for supranuclear control of eye movements are reconsidered. The difference between the classical concept and the proposed concept is as follows: The most important point is the fact that the cerebellum is involved in all saccadic eye movements and will be activated prior to the initiation of a saccade rather than during or after its initiation. The oculomotor signal, therefore, will not be directly transmitted from the mesencephalic to the pontine reticular formation but will travel from the cerebral cortex to the lateral pontine nuclei, from here to the cerebellum, and only after passing through the cerebellum will it reach the pontine reticular formation as the immediate supranuclear gaze center. The oculomotor decussation, which has been placed in the mesencephalic reticular formation at the level of the fourth nerve nucleus but which has never been demonstrated anatomically, therefore, does not exist at this level but occurs in the brachium pontis. For the production of smooth pursuit movements, on the other hand, the cerebellum is not needed, and an ipsilateral corticopontine pathway is proposed for these movements.

Eye movements at the level of the cerebral cortex

Like other motor systems saccadic eye movements are supposed to originate from somewhere in the cerebral cortex. The still unsolved question is from where. Contraversive eye movements can be elicited by stimulating the frontal, parietal, temporal (Ferrier, 1876), and occipital lobe (Crosby and Henderson, 1948). Apart from upward and oblique directions, all of these eye movements are contraversive. Increased interest has been paid to the frontal eye field, which also has been called a center for "command eye movements." Robinson and Fuchs (1969), stimulating the frontal eye fields in monkeys, could elicit saccades ranging from 1 to 70°. The eye movements started roughly 25 msec after the stimulus and revealed all the characteristic saccadic properties. On the other hand, single unit studies by Bizzi (1968) within the frontal eye field of the monkey failed to detect a single neuron with activity *prior* to saccadic eye movements. His units always fired *after* the initiation of a saccade. In neither of these two studies could differences be found between saccades in the light and in the dark. It seems worth mentioning that removal of the frontal cortex results in a deviation of gaze toward the side of the injury but that this is always a temporary phenomenon. Ablation of the frontal cortex in addition does not prevent eye movements on stimulation of occipital and parietal eye fields and vice versa. Bilateral frontal lesions in man produce a more enduring loss of saccades and particularly enhance fixation, a syndrome that Holmes (1936) called

"spasm of fixation." This has also been seen in monkeys by Henderson and Crosby (1952). Holmes has explained this "spasm of fixation" as a disinhibition of occipital lobe fixation function, which normally depends on a proper input from the frontal lobe.

If the frontal eye field is not the center of command for saccadic eye movements, the next best cortical structure would be the occipital lobe. Stimulation of area 17, and especially of areas 18 and 19, will elicit eye movements that are always contraversive with possible vertical components. Single unit studies by Wurtz (1968) in the monkey were restricted to the striate cortex and, like Bizzi's work in the frontal eye field, failed to detect units active *prior* to saccadic eye movements. The more interesting study on units areas 18 and 19 has yet to be done. Cerebrovascular disease in man often results in unilateral lesions of the occipital lobe or underlying white matter. As in unilateral frontal lesions, damage to the occipital lobe will produce loss of contralateral saccades for some time but will also markedly affect the smooth pursuit system toward the ipsilateral side. In the occipitomesencephalic pathway, eye movements are therefore arranged according to directions in such a way that unilateral lesions will affect contralateral saccades and ipsilateral smooth pursuit movements. The conclusion is, that each occipital hemisphere mediates movements of pursuit toward the ipsilateral side as well as saccadic rapid eye movements in the opposite direction.

Stimulation as well as ablation experiments have so far provided only incomplete knowledge about where on the cerebral cortex saccadic eye movements originate. Another technique has therefore been used by Becker et al. (1968, 1972) and by Barlow and Ciganek (1969). Time reversed and averaged EEG data demonstrate potentials, which precede voluntary movements (Deecke et al., 1969). Three different potentials precede a voluntary limb movement: The readiness potential, a slow negative change of cortical potential spread bilaterally over a large area and starting about 1 sec prior to the onset of movement; second, the premotor positivity, a rapid positivation occurring independently of the readiness potential bilaterally over large areas; and third, the motor potential, which is a rapid unilateral negativity over the motor cortex contralateral to the moved limb. The premotor positivity is thought to represent the initiation of the movement in association areas, while the motor potential reflects only the final activity of the contralateral motor cortex. In experiments with human saccadic eye movements using this technique, neither Barlow and Ciganek (1969) nor Becker et al. (1968, 1972) could detect a motor potential similar to that over the motor cortex prior to limb movements. But a bilateral premotor positivity about

150 msec prior to the saccade has been recorded bilaterally over large areas. Most usually it occurred over the vertex and, with an equal probability, over the precentral and the occipitoparietal lobes.

The relatively short recovery period for saccadic eye movements after unilateral cortical lesions as well as the bilateral premotor positivity prior to a saccadic eye movement points to a bilateral representation of saccadic movements, whereby each hemisphere seems to be able to produce contralateral as well as ipsilateral eye movements. But the particular location on the cerebral cortex where voluntary saccadic eye movements have their origin has still not been established. Not one single unit within the cerebral cortex has been found as yet with activity preceding a saccadic movements. But taking together all available data the most probable of all locations where saccadic eye movements are initiated seem to be the association areas in the parietal-occipital lobes where visual and other inputs are integrated.

Corticopontine pathway

Even though the command center for voluntary eye movements has not yet been definitely established, we do know that unilateral lesions in different parts of the cerebral cortex and further down as far as the level of the mesencephalic reticular formation produce a *gaze paresis* but no gaze palsy to the *contralateral* side. Below the level of the fourth nerve nucleus, that is, below the mesencephalic in the pontine reticular formation, a unilateral lesion will produce *ipsilateral* conjugate *gaze palsy*. Above this "decussation" of the oculomotor pathway contraversive eye movements can be induced by electrical stimulation with latencies of about 20 msec (Komatsuzaki et al., 1972), but just below the "decussation" only 3 msec delays have been observed between stimulation and ipsiversive eye movements (Cohen and Komatsuzaki, 1972). This sudden change in latencies from contraversive to ipsiversive eye movements just above and below this assumed oculomotor decussation, as well as a sudden change from gaze paresis to a complete gaze palsy, lead to the conclusion that the major pathway for voluntary saccades reaches the pontine gaze center only by way of another loop, and this loop consists of the lateral pontine nuclei (pes pontis) and the cerebellum.

Corticopontine connections (pontine nuclei) are well known from the work of Mettler (1935), Sunderland (1940), and from Nyby and Jansen (1951). All cortical areas from which eye movements can be elicited project ipsilaterally onto the pontine nuclei, in particular the dorsolateral nuclei in the pes pontis. On the other hand neither the frontal eye field (area 8) nor the striate cortex

(area 17) project to these pontine nuclei, which also receive a strong ipsilateral input from the tectum. If this ipsilateral projection is important for voluntary eye movements, contraversive eye movements should be elicited by stimulating these lateral pontine structures in the same way as they are by stimulating the cerebral cortex. This stimulation in lateral pontine structures has been done, and contraversive eye movements were seen by Bender and Shanzer (1964). Glickstein (1972) in single-cell recordings from these pontine nuclei found the pontine cells to respond best to moving visual stimuli, whereby cortical projection to these cells originates mainly in area 18. The authors suggest that the pons relays information about direction and velocity of moving objects to the cerebellum, but the experiments can also be interpreted as demonstrating that these cells are responsible for saccadic eye movements.

Cerebellar cortex

The dorsolateral pontine nuclei in turn project through the brachium pontis to the contralateral vermis of the cerebellum, but within the vermis some bilateral projection has been established (Nyby and Jansen, 1951). To prove the point that the cerebellar vermis is involved in saccadic eye movements, and especially prior to their onset, in addition to the anatomical connections, the following points seem important: Gross potentials recorded from the vermian cerebellar cortex in the cat precede saccadic eye movements by 28 to 30 msec (Wolfe, 1971). A similar result in the goldfish has been published by Hermann (1971). In addition, electrical stimulation of very special parts of the vermis elicit ipsilateral eye movements (Cohen et al., 1965). Cerebellar lesions in monkeys (Aschoff and Cohen, 1971) produce changes in ipsiversive saccadic eye movements in such a way that dysmetria of saccades was produced toward the ipsilateral side, but only lesions, which were centered over a relatively restricted area that included Larsell's lobuli V, VI, and VII, interfered with ipsilateral saccadic eye movements. These regions of cerebellar vermis are unique in a number of ways: Their predominant afferent input comes from the cerebral cortex by way of the lateral pontine nuclei, especially from those cerebral areas, from which eye movements can be elicited by electrical stimulation. In contrast, adjacent parts of the cerebellar cortex are more heavily supplied by spinocerebellar pathways and receive no corticocerebellar projections (Jansen and Brodal, 1954). In addition, potential changes that arise from activity in eye muscle proprioceptors are found over these areas (Fuchs and Kornhuber, 1969). Slow phases of optokinetic nystagmus and smooth pursuit movements remained normal after vermian lesions as did the amplitude velocity relationship of saccades in all the monkeys tested (Aschoff and Cohen, 1971). The cerebellar cortex, therefore, is not involved in the production of smooth pursuit movements. This is confirmed by Wadia and Swami (1971), who observed a new form of heredofamiliar spinocerebellar degeneration where all saccadic eye movements were absent, but smooth pursuit movements appeared to be normal. All available evidence leads to the conclusion that the cerebellar cortex acts as a structure preprogramming *saccadic* eye movements as well as limb movements (Kornhuber, 1971a, b; this volume).

Pathways to the pontine gaze center

As known from clinical data and stimulation and ablation experiments, the cerebellar vermis mediates ipsilateral saccadic eye movements. The final destination of the saccadic oculomotor pathway must be the supranuclear gaze center. For horizontal eye movements, this center has been definitely established as part of the nucleus gigantocellularis in the paramedian zone of the pontine reticular formation (Goebel et al., 1971). This paramedian zone of the pontine reticular formation lies behind the 3rd and 4th nerve nuclei and rostral to the 6th nerve nucleus. It occupies medial portions of the nucleus gigantocellularis between 0.5 and 2 mm from the midline of either side of the pons. Lesions within this pontine reticular formation produce permanent loss of saccadic eye movements toward the side of the lesion and, as mentioned earlier, stimulation in this region produces horizontal eye movements to the ipsilateral side that, depending upon stimulus characteristics, resemble either saccadic or smooth pursuit movements (Cohen and Komatsuzaki, 1972).

As both cerebellar vermian and pontine reticular formation lesions lead to disturbances in eye movements toward the side of the lesions, an ipsilateral uncrossed or twofold crossed efferent connection is needed from the vermis to this pontine gaze center by way of the deep cerebellar nuclei. There is no question that the vermis projects onto the fastigial nucleus, while the paramedian zone and more lateral parts of the hemispheres project onto the interpositius and dentate nuclei. Therefore, the cerebellar nucleus in question for horizontal eye movements is the fastigial nucleus that when stimulated leads to ipsilateral, and sometimes contralateral, eye movements. This is explained by the fact that each fastigial nucleus sends fibers through the contralateral fastigial nucleus. Depending upon whether or not nerve cells of the nucleus or passing fibers from the contralateral nucleus are stimulated, ipsilateral or contralateral eye movements are produced. Fibers that pass through the

contralateral fastigial nucleus leave by way of the brachium conjunctivum (Jansen and Jansen, 1955), cross back to the ipsilateral side (the same side from which they originate), join the descending part of the brachium conjunctivum, and project onto nuclei in the pontine reticular formation, which in turn project onto the pontine reticular formation that acts as the immediate supranuclear center for horizontal gaze. This last synapse is not yet established anatomically but has to be postulated from all known stimulation experiments (Cohen et al., 1965; Cohen et al., 1966; Cohen and Komatsuzaki, 1972).

The cerebellar nuclei

In between the cerebellar cortex and the brachium conjunctivum, all signals have to be relayed through the cerebellar nuclei. Their function (see Kornhuber, this volume) has been deduced from the following observation (Aschoff, Conrad, Kornhuber, 1970): In a large population of patients suffering from multiple sclerosis, 4% exhibited an acquired pendular nystagmus. This pendular nystagmus is present only when the patient fixates a target and disappears when the eyes are closed. It therefore has something to do with a lesion within the fixation system, that is, the hold system for the eyes. By comparing 30 patients with pendular nystagmus with some 600 patients without pendular nystagmus, the predominant difference occurs with respect to the accompanying cerebellar symptoms especially to head tremor and truncal ataxia. These cerebellar symptoms are highly significant when correlated with patients with pendular nystagmus. As it is well known that lesions of the cerebellar cortex do not lead to pendular nystagmus (Dow and Manni, 1964; Wadia and Swami, 1971; Aschoff and Cohen, 1971), the cerebellar nuclei or their efferent pathways, as the only other important structures, must be made responsible for this pendular nystagmus occurring in the course of multiple sclerosis. Autopsies in patients with acquired pendular nystagmus of other origin than multiple sclerosis (Alajouanine et al., 1935; Guillain, 1938; Nathanson, 1956) have always shown lesions in the cerebellar nuclei and/or the brachium conjunctivum. In addition, Nashold, Slaughter, and Gills (1969) could produce oscillating eye movements by stimulating the fastigial nucleus in man. Our clinical data, with pendular nystagmus occurring only while the subject fixates but disappearing in the dark or when the eyes are closed, as well as the stimulation experiments and the autopsies lead to the conclusion that the cerebellar nuclei act as structures responsible for the hold function of the eyes. During rapid eye movement, this hold function of the cerebellar nuclei will be inhibited by the known inhibitory effect of the Purkinje cells on the cerebellar nuclei (Ito et al., 1965).

Vertical eye movements

Vertical or rotatory eye movements can be elicited by the stimulation of more lateral parts of the cerebellum (Cohen et al., 1965), which project to the interpositus and dentate nuclei that in turn send fibers to the nucleus of Darkschewitsch, nucleus of Cajal, and other pretectal nuclei that are known to act as supranuclear centers for vertical and rotatory eye movements (Carpenter, 1971). These pathways have been established by Mehler, Vernier, and Nauta (1958). In addition, direct connections between these more lateral deep cerebellar nuclei and the oculomotor nuclei responsible for vertical eye movements have been described by Carpenter and Strominger (1964).

Summary and conclusion

The oculomotor pathway for smooth pursuit movements and for rapid saccadic eye movements utilizes different parts of the brain. The classical concept with a mesencephalic-pontine pathway and an oculomotor decussation (never seen anatomically) at the level of the 4th nerve nucleus acts as an ipsilateral (no decussation) cortico-pretecto-mesencephalic-pontine pathway for smooth pursuit movements originating from the striate cortex. The cerebellum, at least at the cortical level, is not involved in smooth pursuit movements. For saccadic eye movements, a new pathway has been proposed. Saccades are most probably initiated in visual association areas, and evidence is presented that the oculomotor pathway for saccadic eye movements reaches the supranuclear gaze center for horizontal eye movements (pontine reticular formation) or vertical eye movements (nucleus of Darkschewitsch, nucleus of Cajal) only by way of the lateral pontine nuclei, the cerebellar cortex, and deep cerebellar nuclei. While the cerebellar cortex is needed for preprogramming proper amplitudes, the cerebellar nuclei provide the hold mechanism of the eyes during fixation. The oculomotor decussation for saccadic eye movements, therefore, does not occur in the mesencephalic reticular formation but within the cerebellum. The pathways are summarized in Figure 1.

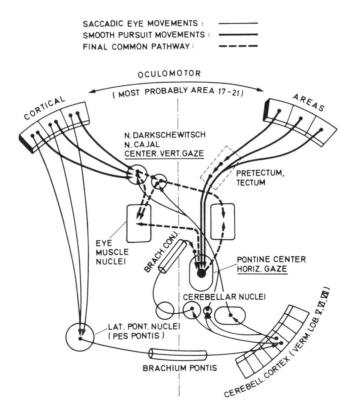

FIGURE 1 Pathway for supranuclear control of saccadic and smooth pursuit eye movements in the horizontal and vertical plane.

REFERENCES

ALAJOUANINE, TH., R. THUREL, and TH. HORNET, 1935. Un cas anatomo-clinique de myoclonies vélo-pharyngées et oculaires. *Rev. Neurol. (Paris)* 64:853.

ASCHOFF, J. C., B. CONRAD, and H. H. KORNHUBER, 1970. Acquired pendular nystagmus in multiple sclerosis. Amsterdam: *Proc. Bárány Soc.* pp. 127–135.

ASCHOFF, J. C., and B. COHEN, 1971. Changes in saccadic eye-movements produced by cerebellar cortical lesions. *Exp. Neurol.* 32:123.

BARLOW, J. S., and L. CIGANEK, 1969. Lambda responses in relation to visual evoked responses in man. *Electroenceph. Clin. Neurophysiol.* 26:182.

BECKER, W., O. HOEHNE, K. IWASE, H. H. KORNHUBER, and R. TÄUMER, 1968. Bereitschaftspotential und evozierte Potentiale der menschlichen Großhirnrinde bei willkürlichen Blickeinstellbewegungen. *Pflüger Arch.* 300:R105.

BENDER, M., S. SHANZER, 1964. Oculomotor pathways defined by electrical stimulation and lesions in the brain stem of monkey. In *The Oculomotor System*, M. Bender, ed. New York: Harper and Row.

BIZZI, E., 1968. Discharge of frontal eye field neurons during saccadic and following eye movements in unanesthetized monkeys. *Exp. Brain Res.* 6:69–80.

CARPENTER, M. B., 1971. Central oculomotor pathways. In *The Control of Eye Movements*, P. Bach-y-Rita, C. Collins, and J. Hyde, eds. New York: Academic Press, pp. 67–104.

CARPENTER, M. B., and N. STROMINGER, 1964. Cerebello-oculomotor fibers in the rhesus monkey. *J. Comp. Neurol.* 123:211.

COHEN, B., K. GOTO, S. SHANZER, and A. WEISS, 1965. Eye movements induced by electrical stimulation of the cerebellum in the alert cat. *Exp. Neurol.* 13:145.

COHEN, B., A. KOMATSUZAKI, and S. SHANZER, 1966. Quantitative oculomotor effects of lesions of the pontine reticular formation. *Trans. Amer. Neurol. Ass.* 211.

COHEN, B., and A. KOMATSUZAKI, 1972. Eye movements induced by stimulation of the pontine reticular formation. *Exp. Neurol.* (in press).

CROSBY, E. C., and J. W. HENDERSON, 1948. The mammalian midbrain and isthmus regions. Part II. Fiber connections of the superior colliculus. *J. Comp. Neurol.* 88:53–91.

DEECKE, L., P. SCHEID, and H. H. KORNHUBER, 1969. Distribution of readiness potential, pre-motor positivity and motor potential of the human cerebral cortex preceding voluntary finger movements. *Exp. Brain Res.* 7:158–168.

DODGE, R., 1903. Five types of eye movements in the horizontal meridian plane of the field of regard. *Am. J. Physiol.* 8:307–329.

DOW, R., and E. MANNI, 1964. The relationship of the cerebellum to extraocular movement. In *The Oculomotor System*, Bender, ed. New York: Harper & Row, pp. 280–292.

FERRIER, D., 1876. *The Functions of the Brain.* New York: G. P. Putnam.

FUCHS, A. F., and H. H. KORNHUBER, 1969. Extraocular muscle afferents to the cerebellum of the cat. *J. Physiol.* 200:713.

GLICKSTEIN, M., J. STEIN, and R. KING, 1972. Visual input to the pontine nuclei. *Science* 78:1110.

GOEBEL, H., A. KOMATSUZAKI, M. BENDER, and B. COHEN, 1971.

Lesions of the pontine tegmentum and conjugate gaze paralysis. *Arch. Neurol.* 24:431.

GUILLAIN, G., 1938. The syndrome of synchronous and rhythmic palato-pharyngo-laryngo-diaphragmatic myoclonus. *Proc. Roy. Soc. Med.* 31:41.

HENDERSON, J., and E. CROSBY, 1952. An experimental study of optokinetic responses. *Arch. Ophthal. (Chicago)* 47:43–54.

HERMANN, H. T., 1971. Saccade correlated potentials in optic tectum and cerebellum of *Carassius auratus*. *Brain Res.* 26:273.

HOLMES, G., 1936. Looking and seeing. *Irish J. Med. Sci.* 129:565.

ITO, M., N. KAWA, and M. UDO, 1965. The origin of cerebellar-induced inhibition and facilitation in the neurons of Deiters and intracerebellar nuclei. *23rd Inter. Congr. Physiol. Sci. (Abstr.)*, Tokyo, 1965, p. 997.

JANSEN, J., and A. BRODAL, ed. (1954): *Aspects of Cerebellar Anatomy.* Oslo: Johan Grundt Tanum Verlag, 1954.

JANSEN, J., and J. JANSEN, 1955. On the efferent fibers of the cerebellar nuclei in the cat. *J. Comp. Neurol.* 102:607.

KOMATSUZAKI, A., J. ALPERT, H. HARRIS, and B. COHEN, 1972. Effects of mesencephalic reticular formation lesions on optokinetic nystagmus. *Exp. Neurol.* 34:522.

KORNHUBER, H. H., 1971a. Motor functions of cerebellum and basal ganglia: the cerebellocortical saccadic (ballistic) clock, the cerebellonuclear hold regulator, and the basal ganglia ramp (voluntary speed smooth movement) generator. *Kybernetik* 8:157.

KORNHUBER, H. H., 1971b. Das vestibulare System, mit Exkursen über die motorischen Funktionen der Formatio reticularis, des Kleinhirns, der Stammganglien und des motorischen Cortex sowie über die Raumkonstanz der Sehdinge. In *Vorträge der Erlanger Physiologentagung* 1970, W. D. Keidel and K.-H. Plattig, eds. Berlin, Heidelberg, New York: Springer-Verlag.

MEHLER, W. R., V. VERNIER, and W. NAUTA, 1958. Efferent projections from dentate and interpositus nuclei in primates. *Anat. Rec.* 130:430.

METTLER, F. A., 1935. Corticifugal fiber connections of the cortex of *Macaca mulatta*. The frontal region. *J. Comp. Neurol.* 61:509.

NASHOLD, B., D. SLAUGHTER, and J. GILLS, 1969. Ocular reactions in man from deep cerebellar stimulation and lesions. *Arch. Ophthal. (Chicago)* 81:538.

NATHANSON, M., 1956. Palatal myoclonus. *Arch. Neurol. Psychiat.* 75:285.

NYBY, O., and J. JANSEN, 1951. An experimental investigation of the cortico-pontine projection in *Macaca mulatta*. *Norske Vid. Akad. Skr.*, I. Math. Naturwiss. K1.3.

ROBINSON, D., and A. FUCHS, 1969. Eye movements evoked by stimulation of frontal eye fields. *J. Neurophysiol.* 32:637.

SUNDERLAND, S., 1940. The projection of the cerebral cortex on the pons and cerebellum in the macaque monkey. *J. Anat.* 74:201.

WADIA, N. H., and R. K. SWAMI, 1971. A new form of heredofamilial spinocerebellar degeneration with slow eye movements (nine families). *Brain* 94:359–374.

WOLFE, J. W., 1971. Relationship of cerebellar potentials to saccadic eye movements. *Brain Res.* 30:204.

WURTZ, R., 1968. Visual cortex neurons: response to stimuli during rapid eye movements. *Science* 162:1148.

27 Central Sensory Activities between Sensory Input and Motor Output

JUHANI HYVÄRINEN, ANTTI PORANEN, and YRJÖ JOKINEN

ABSTRACT Using cutaneous vibration as a sensory stimulus to which a conscious monkey responds for reward, one can map with microelectrodes the structures activated in this task. On the sensory side, vibration is coded in the time structure of unit responses; on the motor side, in mean rate of impulse firing only. Convergence of afferent information is seen in the associative areas and the cellular responses are often very specific to stimulus quality. Prior to response, attention toward the stimulus augments sensory unit responses in some areas.

EXPERIMENTALLY the question posed in the broad title suggested for this chapter could hardly be asked a few years ago. Because of recent microphysiological advances, we are now able to study the processing in the brain between the input and the output in behaving animals. Our group has studied the somesthetic processing with microelectrode recordings from trained monkeys. As a cutaneous signal, we used vibration because the neural mechanisms involved in transmission of cutaneous vibration have been extensively worked out by Talbot et al. (1968) and Mountcastle et al. (1969, 1972). The anatomical locations chosen for study included the primary somesthetic cortex (SI) comprised of Brodmann's areas 3, 1, and 2, the motor cortex (area 4), the parietal association area (area 7), the thalamic ventrobasal complex and the secondary somesthetic cortex (SII).

Methods

Stumptail monkeys (*Macaca speciosa*) are trained in a specially designed chair (Figure 1). For head immobilization during recordings, the monkeys wear a halo fixed with pins to their skulls in the same way that human patients chronically carry such halos for immobilization of cervical spine after surgery or fractures (Nickel et al., 1968). The base cylinder of an Evarts-type hydraulic manipulator (Evarts, 1966) is fixed over a hole made in the skull above the target area for moving metal microelectrodes in the brain.

The monkeys are taught the following task (Figure 2):

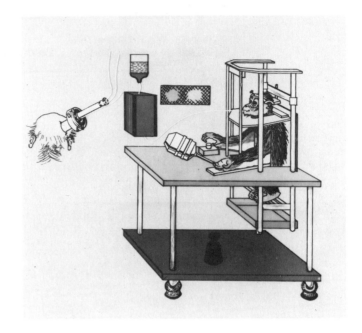

FIGURE 1 A stumptail monkey sitting in a specially designed training and recording apparatus. The two signal lights, yellow (*relevant*) and red (*irrelevant*) are shown in the background. The monkey is feeling the vibrator tip with his left hand and pressing the response button with his right hand. The head is fixed with a halo in the chair framework and a hydraulic microdrive positioned above the target area (inset).

A 4 sec period of vibration is given on the skin of the hand and within 0.6 sec from the end of vibration the monkey should press a key with his opposite hand to be rewarded with fruit juice. However, he is taught that he will be rewarded only when a yellow light (the *relevant* light) is on and that he is never rewarded when a red light (the *irrelevant* light) is on. Thus during the yellow light, the monkey has to be attentive toward the stimulus at the end of vibration during the yellow light in order to succeed in pressing the key quickly enough. On the other hand, he quickly learns never to respond when the red signal light is on.

The cellular responses are analyzed by making two kinds of histograms. In a per-stimulus-time (pst) histogram

JUHANI HYVÄRINEN, ANTTI PORANEN, and YRJÖ JOKINEN, Institute of Physiology, University of Helsinki

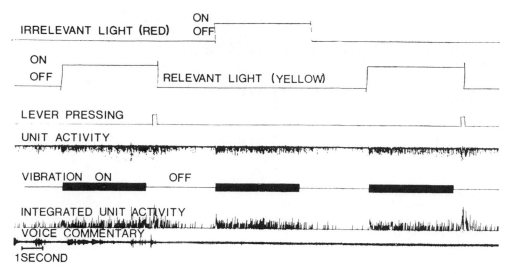

FIGURE 2 A polygraphic record of the monkey's task. At the end of vibration (channel 5) the monkey presses the lever (channel 3) provided that the yellow light is on (channel 2) to get a juice reward. When the vibration is presented during the red light (channel 1), no reward is given and the monkey learns not to respond. At the same time neuronal unit activity is recorded (channels 4 and 6), and a voice commentary of the animal's behavior is dictated on tape (channel 7). (From Hyvärinen et al., 1973 with the permission of the Georg Thieme Verlag.)

(Figure 3), a number of 4 sec responses are summed. The time base of a cycle histogram equals the length of the sine wave cycle, and the numbers of impulses that occur at different times during the cycle are plotted in the histogram. Thus the cycle histogram reveals the degree of phase locking to the sinusoidal stimulus (Talbot et al., 1968; Mountcastle et al., 1969). A μ-Linc computer is used on-line or off-line for these analyses.

Results

In the ventrobasal thalamus we have found regions with good rapidly adapting responses to cutaneous vibration and other regions with pacinian-type high-frequency responses. Mostly, these regions are invariant in their responses to the vibratory stimuli, but occasionally a cell group that gives clearly stronger responses to relevant than irrelevant stimuli may be found.

Also in the secondary somesthetic cortex (SII), both kinds of cell groups are found. Figure 4 shows relevant and irrelevant cycle histograms and pst histograms of a cell in SII that gives a stronger response in the attentive state. The difference is clearly seen toward the end of the vibration period at two different vibration amplitudes.

In the hand area of the primary somesthetic cortex (SI) the reactions of populations of cells on the average appear different in the three cytoarchitectural areas. Area 3b located anteriorly in the posterior bank of central sulcus appears to contain almost exclusively cell groups related either to rapidly or slowly adapting cutaneous receptors of the skin. In this region, cells respond faithfully to all cutaneous stimuli on the receptive fields and always seem to do so no matter what the functional state of the animal is. Typical responses of a cell in area 3 are shown in

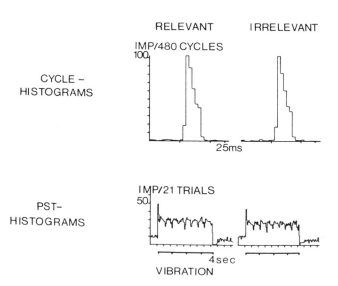

FIGURE 3 Responses to vibration of a neuron in Brodmann's area 3. The receptive field was on the glabrous skin of the distal phalanx of the forefinger. The monkey's performance was 100% correct with 40 Hz vibration of 370 μ peak-to-peak amplitude. There is no difference between the relevant and the irrelevant responses. (The oscillation on top of pst histograms is a beat phenomenon between the bin size and the cycle length.)

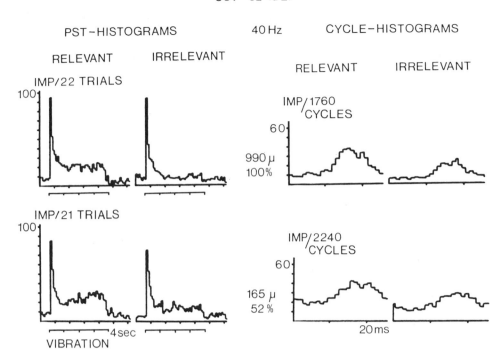

FIGURE 4   An example of responses recorded in the secondary somesthetic cortex (SII). The left side shows per-stimulus-time histograms constructed separately of relevant and irrelevant responses and the right side shows the corresponding cycle histograms. This cell shows an increase of the response rate toward the end prior to the behavioral response.

Figure 3. This figure shows that responses are identical for relevant and irrelevant stimuli.

In the next areas posteriorly, areas 1 and 2, the cellular responses to cutaneous vibration are often less clear than in area 3. Even here the responses of a larger proportion of cells appears as not mutually different for the relevant and irrelevant stimuli. However, a smaller proportion of cells resemble the one illustrated for SII in Figure 4. In such cells, more impulses are discharged for the relevant than the irrelevant stimuli, but in most such cells there is no clear phase locking to the stimulus, although there is in some.

Posteriorly on the gyrus in areas 1 and 2 also cells that require more specific stimuli are seen. Figure 5 shows an example of a directionally selective cell. Only stimuli moving from proximal to distal over the receptive field evoke clear responses, whereas movement over the same receptive field in the opposite direction evoke no response, and responses to punctate stimuli are equivocal. There are also some cells that are even more difficult to activate. Some of them seem to require active manipulation by the monkey coupled with cutaneous excitation over the receptive field.

Neurons that are difficult to drive are also encountered

anterior to the central sulcus in the area 4 or motor cortex. There is a clear sensory driving in the motor cortex, but the neurons are much less sensitive than the ones in SI and they do not replicate the varying time structure of the stimuli. Also after a change in stimulus location, a motor cortex cell is likely to respond only a few times before it adapts and ceases to respond to the stimulus. After a change in stimulus location, responses can be evoked again a few times from the new locus. Such responsiveness and adaptation seems to follow closely the monkey's tendency to make small finger movements when a stimulus changes position on the skin. Typical for the discharges of the motor cortex cells illustrated in Figure 6 is a rapid total adaptation to the vibratory stimulus, although a strong discharge is elicited by the beginning of the stimulus. There is a marked difference between the relevant and the irrelevant responses, but unlike the sensory cortex the difference is greatest in the initial part of the response several seconds before the motor reaction that is performed with the hand ipsilateral to the recording.

We have made a few observations of convergence of visual and somesthetic input to the cells of the parietal associative area 7 in single neuron recordings. The results

KA2-14

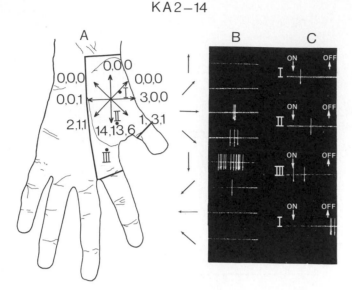

hand in the optimal direction over the cutaneous receptive field trying to keep the speed of movement and pressure equal every time. As a visual stimulus, a moving white card (15 × 21 cm) was introduced from the right at a distance of 30 cm to the center of the monkey's visual field. The arrows below the histograms show the timing of the stimuli. At the left end of the arrow in the visual histogram, the card was moved rapidly into the visual field, then kept

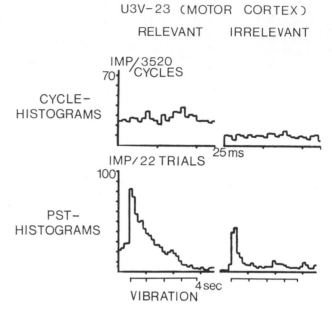

FIGURE 5   An example of a directionally selective neuron from area 1 related to cutaneous receptors. (A) shows the receptive field. In (C) punctate stimuli were applied to points I, II, III, and I again. (B) shows responses to a manually held stimulating probe moving in various directions with the same speed and pressure. The duration of each record is 2 sec. In (A) the numbers of impulses are plotted for various directions of movement on three consecutive rounds of stimuli. (From Hyvärinen et al., 1973 with permission of Georg Thieme Verlag.)

FIGURE 6   An example of the responses of neurons in the motor cortex contralateral to the sensory stimulus. The response is greatly augmented during the relevant task, although there is a rapid adaptation before the motor response performed with the hand ipsilateral to the recording.

for two neurons recorded in area 7 are shown in Figures 7 and 8. The neuron illustrated in Figure 7 was discovered to be visually activated when it responded to movement of a shadow over the monkey's hand. As is illustrated in Figure 7, this unit responded also to cutaneous vibration although the response was weak. There is a clearly better response to the relevant stimulus than to the irrelevant. However, no cyclic entrainment of the neuron by vibratory stimuli is visible. In general, it seems that behavioral attention is an important factor in the activation of units in the associative area.

Figure 8 illustrates another unit with somesthetic and visual convergence. This unit had a cutaneous receptive field on the right shoulder and upper arm, and it was activated preferentially by a cutaneous movement over this receptive field toward the periphery (arrow). Also visual stimuli in the direction of the right arm appeared to produce some responses and gave rise to a series of visual tests. In Figure 8A, 20 trials were made each consisting of a cutaneous and a visual stimulus. The cutaneous stimulation was performed by passing the experimenter's

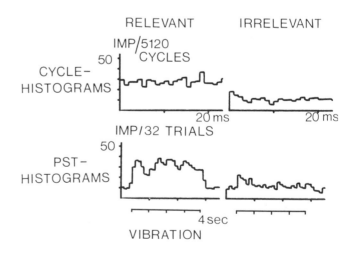

FIGURE 7   Recordings from a neuron in the parietal association area, Brodmann's area 7. There is no cyclic following of vibration but a clear difference between the relevant and the irrelevant responses in the pst histogram.

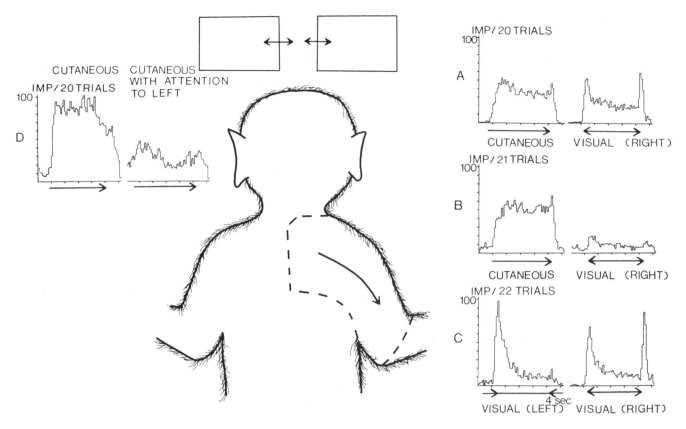

FIGURE 8 Responses of a cell in area 7 to cutaneous and visual stimuli. The cutaneous receptive field on the right shoulder and upper arm was directionally selective for movement toward the periphery. Visual responses were recorded to movements of white cards in and out of the visual field as described in the text.

there and at the end of a 4 sec period (indicated by the arrowhead to the right), the card was rapidly removed from the visual field to the right. We can see that both the cutaneous and the visual stimuli produced good responses. Moreover, in the visual response, there is a discharge both at the introduction of the card to the visual field (movement toward the left) and at the withdrawal of the card (movement toward the right).

To show that this response was really visual, we performed the same experiment again, now using blinders in front of the monkey's eyes. The blinders left only a small part of the peripheral visual field uncovered. As shown in Figure 8B after the monkey's sight was blocked with blinders the cutaneous response remains, but the visual response is greatly reduced although a barely detectable response is still visible because of the uncovered peripheral field.

There seemed to be some differences between the responses to card movements to the right and to the left. Therefore, we did a study of this neuron by introducing the card to the visual field alternately from the left and from the right. As illustrated in Figure 8C a good response was elicited in both cases, but there is that difference that a

card withdrawal to the left did not produce a response. Thus from both sides a movement to the right produced a response, but movement to the left produced a response only when that movement occurred in the beginning of the stimulus.

In Figure 8D this unit was studied by presenting regularly the optimum cutaneous stimulus, a movement to the right over the receptive field on the shoulder and upper arm. Every other time, this stimulus was accompanied by a spoonful of juice presented simultaneously from the left. The monkey had to look at the juice in order to lick it up from the spoon. Figure 8D shows the responses to identical cutaneous stimuli with no distraction (left histogram) and with the distractive stimulus on the left side (right histogram). Looking in the opposite direction reduced the response elicited from movement to the right over the shoulder and upper arm.

Discussion

This study demonstrates once again the functional differences between the cytoarchitectural regions of the cortex. This principle has previously been demonstrated for SI by Powell and Mountcastle (1959) and Phillips et al. (1971), for audition by Rose and Woolsey (1949), and for vision by Hubel and Wiesel (1962, 1965). According to Jones and Powell (1970a), the anatomical projection from the ventrobasal complex of the thalamus to Brodmann's area 3 appears particularly heavy when compared with the projections to areas 1 and 2. Thus, there is an anatomical correlation to the finding of high sensitivity to cutaneous stimuli of cells in area 3. The thalamic projection to areas 1 and 2 is less heavy, and peripheral driving is also more difficult. Moreover, there are practically no cells in area 3 that would require specific stimuli other than just touching the skin to be activated. Such cells can be encountered in areas 1 and 2. According to Jones and Powell (1969), these areas project to each other. Such projections could form more complex connections in posterior SI. Whitsel et al. (1972) have made a thorough study of the directionally sensitive cells in SI and demonstrated that cells with differential sensitivity to movement in opposite directions seem to be absent in the peripheral nerves.

According to Jones and Powell (1970b), the further connections from SI are to areas 4 and 5, SII, and the supplementary motor region in area 6. Area 5 projects to areas 6 and 7. Area 7 projects to the limbic cortex and to the temporal and frontal lobes. In view of these anatomical connections, some recent physiological findings are of interest. Directionally selective cutaneous units appear to be more common toward the posterior part of SI, because most have been described in areas 1 and 2 (Mountcastle et al., 1969; Schwarz and Fredrickson, 1971; Whitsel et al., 1972; Sakata, 1973; Hyvärinen et al., 1973). If such units are the result of cortical convergence, one would expect to find more of them posteriorly. Recently Duffy and Burchfield (1971) described multiple joint convergence in area 5 of paralyzed monkeys. Sakata (1973) demonstrated that such multiple joint convergence occurs in a specific manner so that certain positions and directions of movement in several joints have to be specifically combined for optimum activation of neurons in area 5. Moreover, in this area Sakata also demonstrated convergence of joint and skin input in a similar specific manner. Thus the anatomical projection from SI to area 5 would appear to contribute to a specific convergence of joint and skin information to units in area 5.

So far the question remains open of the route along which the visual information arrives to area 7. In human patients with lesions in the posterior parietal region of the minor hemisphere, a characteristic syndrome called apractognosia results (Denny-Brown and Banker, 1954; Hécaen et al., 1956). In this syndrome, the patients have a characteristic difficulty to direct movements of the opposite hand according to visual guidance. Our pre-

liminary findings on the visual input to area 7 would appear to complement the results of observations on such lesions in patients. Our results suggest that the visual input to this region is not a simple projection, because only such visual stimuli appear effective that occur in the space close to the monkey. The anatomical work on the pulvinar-posterior system is of interest in this regard (see Graybiel and Jones, this volume).

In conclusion it appears that the anatomical projection in the somesthetic system from one cytoarchitectural area to the next is accompanied by differences in the functional properties of cells in these areas. Toward the posterior part of the parietal lobe, the anatomical connections become more complex and so do the physiological responses.

ACKNOWLEDGMENTS This work has been supported by the Foundations' Fund for Research in Psychiatry, the United States Public Health Service (Grant 1RO3 MH 18277–01 MSM) and the Finnish National Research Council for Medical Sciences. We wish to thank Mr. Ilkka Linnankoski and Miss Katriina Yläkotola for able technical assistance.

REFERENCES

DENNY-BROWN, D., and B. BANKER, 1954. Amorphosynthesis from left parietal lesion. Arch. Neurol. Psychiat. 71:302–313.

DUFFY, F. H., and J. L. BURCHFIELD, 1971. Somatosensory system: Organizational hierarchy from single units in monkey area 5. Science 172:273–275.

EVARTS, E. V., 1966. Methods for recording activity of individual neurons in moving animals. In Methods in Medical Research, R. F. Rushmer, ed. Chicago, Ill.: Year Book Medical Publishers, pp. 241–250.

HÉCAEN, H., W. PENFIELD, C. BERTRAND, and R. MALMO, 1956. The syndrome of apractognosia due to lesions of the minor cerebral hemisphere. Arch. Neurol. Psychiat. 75:400–434.

HUBEL, D. H., and T. N. WIESEL, 1962. Receptive fields, binocular interaction and functional architecture in the cat's visual cortex. J. Physiol. (London) 160:106–154.

HUBEL, D. H., and T. N. WIESEL, 1965. Receptive fields and functional architecture in two non-striate visual areas (18 and 19) of the cat. J. Neurophysiol. 28:229–289.

HYVÄRINEN, J., A. PORANEN, Y. JOKINEN, R. NÄÄTÄNEN, I. LINNANKOSKI, 1973. Observations on unit activity in the primary somesthetic cortex of behaving monkeys. In The Somatosensory System, H. H. Kornhuber, ed. Stuttgart: Georg Thieme Verlag, (in press).

JONES, E. G., and T. P. S. POWELL, 1970a. Connections of the somatic sensory cortex of the rhesus monkey. I. Ipsilateral cortical connections. Brain 92:477–502.

JONES, E. G., and T. P. S. POWELL, 1970a. Connections of the somatic sensory cortex of the rhesus monkeys. III. Thalamic connections. Brain 93:37–53.

JONES, E. G., and T. P. S. POWELL, 1970b. An anatomical study of converging sensory pathways within the cerebral cortex of the monkey. Brain 93:793–820.

MOUNTCASTLE, V. B., R. H. LaMOTTE, and G. CARLI, 1972. Detection thresholds for stimuli in humans and monkeys: Comparison with threshold events in mechanoreceptive afferent nerve fibers innervating the monkey hand. *J. Neurophysiol.* 35:122–136.

MOUNTCASTLE, V. B., W. H. TALBOT, H. SAKATA, and J. HYVÄRINEN, 1969. Cortical neuronal mechanisms in flutter-vibration studied in unanesthetized monkeys. Neuronal periodicity and frequency discrimination. *J. Neurophysiol.* 32:452–484.

NICKEL, V. L., J. PERRY, A. GARRETT, and M. HEPPENSTALL, 1968. The halo. A spinal skeletal traction fixation device. *J. Bone Joint Surg. (Amer.)* 50A:1400–1409.

PHILLIPS, C. G., T. P. S. POWELL, M. WIESENDANGER, 1971. Projection from low-threshold muscle afferents of hand and forearm to area 3a of baboon's cortex. *J. Physiol. (London)* 217:419–446.

POWELL, T. P. S., and V. B. MOUNTCASTLE, 1959. Some aspects of the functional organization of the cortex of the postcentral gyrus of the monkey: A correlation of findings obtained in a single unit analysis with cytoarchitecture. *Bull. Johns Hopkins Hosp.* 105:133–162.

ROSE, J. E., and C. N. WOOLSEY, 1949. The relations of thalamic connections, cellular structure and evocable electrical activity in the auditory region of the cat. *J. Comp. Neurol.* 91:441–466.

SAKATA, H., 1973. Somatosensory responses of neurons in the parietal association area (area 5) in monkeys. In *The Somatosensory System*, H. H. Kornhuber, ed. Stuttgart: Georg Thieme Verlag, (in press).

SCHWARZ, D. W. F., and J. M. FREDRICKSON, 1971. Tactile direction sensitivity of area 2 oral neurons in the rhesus monkey cortex. *Brain Res.* 27:397–401.

TALBOT, W. H., J. DARIAN-SMITH, H. H. KORNHUBER, and V. B. MOUNTCASTLE, 1968. The sense of flutter-vibration: Comparison of the human capacity with response patterns of mechano-receptive afferents from the monkey hand. *J. Neurophysiol.* 31:301–334.

WHITSEL, B. L., J. R. ROPPOLO, and G. WERNER, 1972. Cortical information processing of stimulus motion on primate skin. *J. Neurophysiol.* 35: 691-717.

28 Motor Functions of the Basal Ganglia: Single-Unit Activity during Movement

MAHLON R. DeLONG

ABSTRACT   Studies on the activity of single units in the basal ganglia and cerebellum of the monkey during rapid arm movements have failed to reveal any significant differences in the role of these structures in movement. Recent studies, however, comparing the activity of putamen neurons during slow and rapid arm movements indicate a preferential involvement of this structure in the control of slow movements. This finding supports the view of Kornhuber that the major motor function of the basal ganglia is to generate slow (ramp) rather than rapid (ballistic) movements.

THE BASAL ganglia provide a major subcortical link between sensory and motor areas of the cerebral cortex and, like the cerebellum, are a site of convergence and integration of diverse inputs. The importance of the basal ganglia in sensorimotor integration is evident to anyone who has observed the extreme disturbances of movement, posture, and muscle tone that result when these nuclei are damaged through injury or disease (e.g., Parkinsonism). Clinicopathologic studies in man strongly implicate these structures in the control of movement and posture (see Jung and Hassler, 1960; Denny-Brown, 1962; Martin, 1967). In order to determine how these structures normally function, the activity of single basal ganglia neurons has been studied in the intact monkey during conditioned limb movements (DeLong, 1971; 1972). This essay will discuss the results of these and related studies on the motor system within the framework of relevant anatomical, physiological, and clinical data.

Anatomical considerations

The term *basal ganglia* will be used to include the striatum (i.e., the caudate and putamen) and the pallidum (globus pallidus), as well as the anatomically closely related substantia nigra and subthalamic nucleus. Recent neuroanatomic studies have helped to clarify the major features of the afferent and efferent pathways as well as the intrinsic relations of those nuclei. These anatomical re-

MAHLON R. DeLONG   Laboratory of Neurophysiology, National Institute of Mental Health, U.S. Department of Health, Education and Welfare, Bethesda, Maryland

lationships have been the subject of several recent comprehensive reviews (Nauta and Mehler, 1966; Nauta and Mehler, 1969; Kemp and Powell, 1971). The major anatomical pathways as presently understood are summarized in Figure 1.

The striatum receives the majority of extrinsic afferents to the basal ganglia; these come from three main sources: (1) the cerebral cortex (Webster, 1961, 1965; Carman et al., 1963; Kemp and Powell, 1970), (2) the intralaminar nuclei of the thalamus (Nauta and Whitlock, 1954; Mehler, 1966), and (3) the substantia nigra (Andén et al., 1964, 1965). The corticostriate projection in the monkey has been shown (Kemp and Powell, 1970) to arise from all areas of the cerebral cortex and to be topographically organized. Of particular interest are dense, overlapping projections from the sensory (S_I) and motor (M_I) cortices as well as a bilateral projection upon the striatum from the supplementary motor area (M_{II}) and area 5. The projection from the sensorimotor cortex is directed largely upon the dorsolateral parts of the putamen and to a slight extent the body of the caudate. The center median, the largest of the intralaminar nuclei, also projects largely to the putamen (see Mehler, 1966). The putamen is thus of considerable interest from the standpoint of sensorimotor integration. The projection from the substantia nigra to the striatum was first identified by the fluorescent-histochemical staining technique (Andén et al., 1964, 1965). This now well-known dopamine-containing pathway appears to be responsible for the high concentration of dopamine within the striatum. The anatomical relations between the nigra and striatum are reciprocal, because the striatum also projects to the nigra.

The bulk of efferents from the striatum converge upon the pallidum, which gives rise to the major output from the basal ganglia. The pallidum is composed of an external and an internal division. The external division projects upon the subthalamic nucleus, which in turn appears to send its output largely back to the internal division of the pallidum (Carpenter and Strominger, 1967). It is thus from the internal division of the pallidum that the major efferent projections from the basal ganglia take origin. Pallidal efferents terminate largely in portions

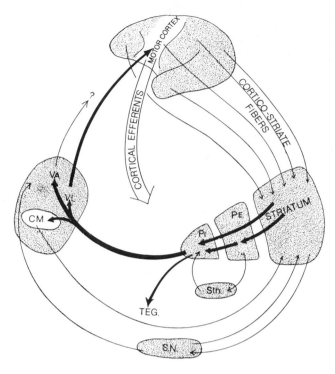

FIGURE 1   A summary of the major afferent, efferent, and intrinsic connections of the basal ganglia. The striatum (caudate-putamen) receives the majority of afferents to the basal ganglia. These come from three major sources: (1) the entire cerebral cortex, (2) the intralaminar nuclei of the thalamus, i.e., the center-median (CM), central lateral and parafasicular nuclei, and (3) the substantia nigra (SN). The striatum sends its output to the pallidum and the substantia nigra. The nigra gives rise to the well known dopaminergic pathway to the striatum and also projects to the medial portions of the ventrolateral (VL) and ventroanterior (VA) nuclei of the thalamus. The internal division of the pallidum (P$_I$) gives rise to the major efferents from the basal ganglia, which terminate in the lateral portions of VL, VA, and CM and in the midbrain tegmentum (TEG). The projections from VA and VL to the prefrontal cortex provide a route whereby the basal ganglia can influence the motor and prefrontal cortex. Anatomically the basal ganglia are in a manner of speaking, wired between the entire cortex on the one hand and the motor cortex on the other; i.e., they are "afferent" to the motor cortex. Direct descending influences from these nuclei do not extend below the midbrain.

of the thalamus, i.e., ventralis anterior (VA) and ventralis lateralis (VL), and center median (CM), and in the midbrain tegmentum (n. Tegmenti pontinus) (Nauta and Mehler, 1966). While the sites of termination of the projection from VA to the frontal cortex remain unclear (Carmel, 1971), the VL projects directly to the motor cortex (Walker, 1938; Strick, 1971, 1973). These thalamic nuclei constitute the thalamic gateway to the precentral cortex, because they receive not only the output from the basal ganglia but also the projection from the cerebellum (the brachium conjunctivum) to the cortex.

Overall similarities in the input–output connections of the basal ganglia and the cerebellum have been pointed out by Nauta and Mehler (1966) and more recently have been expanded upon by Kemp and Powell (1971). For our purposes, suffice it to say that both the basal ganglia and the cerebellum receive input from all areas of the cerebral cortex, and both send a major portion of their output to the motor cortex via the ventral thalamus (VA and VL). In a sense, both the basal ganglia and the cerebellum are wired in parallel between the entire cortex on the one side and the motor cortex on the other, i.e., these structures are *afferent* to the motor cortex. Anatomically, it would seem that activity in the basal ganglia and cerebellum may influence or even be largely responsible for the patterns of activity in the motor cortex.

While the similarities in the overall anatomical features of the basal ganglia and cerebellum are intriguing, it is well known that damage to these structures in man and experimental animals produces clearly distinguishable clinical syndromes. Generally speaking, basal ganglia disorders are characterized by combinations of akinesia, tremor at rest, rigidity or involuntary movements, whereas cerebellar disorders are characterized by tremor during movement, hypotonia, and dysmetria. Surely the functional roles of these two different regions of the brain in the control of movement and posture must be fundamentally different. The elucidation of these differences is a major goal of experimental studies on the motor system.

Single-unit studies: rapid movements

Studies on the activity of both cerebellar and basal ganglia neurons in the intact animal have been largely directed at determining the specific contributions of these two subcortical systems to the control of limb movement. For the cerebellum, Thach (1968) first showed that both cerebellar Purkinje and nuclear cells of the monkey changed their rate of discharge phasically in relation to self-paced, rapidly alternating limb movements. In an experiment similar to that of Thach, it was shown that units in the pallidum also changed their discharge phasically in relation to rapidly alternating limb movements (DeLong, 1971). A unit from the external pallidal segment is shown in Figure 2 during rest, and during rapidly alternating movements of the contralateral and ipsilateral limbs. In this study pallidal units were found to discharge at high rates even during rest and to modulate their discharge above and below resting levels in relation to the phasic arm movements. Unit activity was best correlated with movements of the contralateral limb. Units related to the arm movements were found largely in the lateral portions of both pallidal segments but were

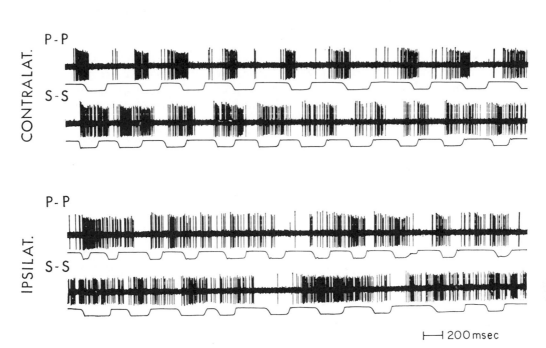

┣━┫ 200 msec

FIGURE 2 Activity of a unit from the pallidum (external segment) during rest and during rapidly alternating push-pull (P-P) and side-to-side (S-S) movements of the contralateral and ipsilateral arms. The line below each unit record represents the position of the lever moved by the monkey and thus provides a record of arm movement. The discharge of this unit is consistently correlated with movements of the contralateral but not the ipsilateral arm. The pattern of discharge is different for push-pull and side-to-side movements. The high rate of discharge during rest is characteristic of pallidal neurons. (Reprinted from DeLong, 1971.)

intermingled to some extent with units related to leg movement. No clear evidence for somatotopic organization was obtained. Thach, likewise, in the dentate found no evidence for somatotopic organization of movement-related units. Comparison of the two studies on unit activity in the cerebellum and pallidum during rapidly alternating limb movements thus failed to reveal any clue as to the functional differences between these structures.

In later experiments the issue of *timing* of unit activity in relation to movement was investigated in order to determine where and when the earliest changes in neural activity occurred in relation to the onset of a rapid, conditioned arm movement. Evarts (1966) had shown that neurons in the arm area of the monkey motor cortex changed their discharge as early as 100 msec prior to the onset of a rapid arm movement (i.e., visuomotor-reaction time task). Thach (1970) in a similar experiment subsequently showed that cells in the dentate nucleus of the cerebellum also changed their discharge prior to the onset of movement, with a distribution of onset times similar to

those obtained for PT cells in the arm area of the motor cortex. For many units the changes in activity occurred prior to the earliest changes in muscle activity and therefore could not have resulted from sensory feedback from the movement. This study showed clearly that activity in the cerebellum could influence the motor cortex prior to the onset of movement but provided no answer to the important question of where (i.e., cerebellum or motor cortex) the first change in activity occurred.

In view of these observations, it was of considerable interest to determine whether the onset of unit activity in the basal ganglia also preceded the onset of limb movement and whether differences in temporal onset of activity between these structures could be demonstrated. Recent studies (DeLong, 1972; in preparation) have shown that, as for the cerebellum, neurons in the pallidum also change their discharge prior to the onset of rapid conditioned arm movements. The distribution of onset times for pallidal neurons was found to be similar to that of neurons in the arm area of the motor cortex. Assuming that differences

MAHLON R. DeLONG   321

in the timing of activity exist between motor cortex, cerebellum, and basal ganglia prior to a motor response of this type, it would appear that these techniques are unable to resolve such differences.

One of the major difficulties in comparing unit activity in motor cortex with that in cerebellum and basal ganglia is the lack of clearly defined somatotopic organization in these subcortical structures. Whereas one can be reasonably confident that an arm movement-related unit in the arm area of the motor cortex is involved in some aspect of contralateral arm movement, one has no such confidence about an arm movement-related unit in the cerebellum or basal ganglia. Thus, while a correlation between unit discharge and arm movement in these regions may indeed result from the unit's involvement in the initiation or control of the arm movement, it may also result from the unit's involvement in an associated movement of some other body part or the unit's role in some other central process (e.g., corollary discharge) occurring synchronously with the limb movement. An obvious example of this problem arises when a conditioned arm movement is accompanied by a "consumatory response" directed at obtaining a reinforcement. Many units in the basal ganglia and cerebellum are clearly involved in the mediation of this consumatory response, which may occur more or less synchronously with the conditioned limb movement. Whereas in the motor cortex one can select either population of units (i.e., those involved in the arm or the consumatory response) for study by simply placing the recording chamber over the proper area of somatotopically structured motor cortex, this is not possible in the basal ganglia or cerebellar nuclei where the degree of somatotopic organization is much less clear. In essence, the issues of causality and sampling biases become much more of a concern where one leaves the somatotopically organized motor cortex and attempts to correlate unit activity and behavior in regions of the brain located several synapses from the output and lacking clearly defined somatotopic organization.

Slow versus rapid movements

The studies discussed thus far have all involved rapid, "ballistic" limb movements (i.e., either self-paced, rapidly alternating movements, or visuomotor reaction time tasks). Kornhuber, (1971; this volume), however, has recently put forth the view, based on clinical and experimental evidence, that the basal ganglia function primarily in the generation of slow, "ramp" movements, whereas the cerebellum preprograms and initiates rapid, "ballistic" movements. These two types of movement are generally considered to differ in that, whereas slow movements may be regulated by sensory feedback, rapid movements must be entirely preprogrammed by the central nervous system, because there is inadequate time for feedback modification (Stetson, 1935). From the hypothesis of Kornhuber, one would predict that neurons in the basal ganglia would discharge preferentially during slow movements and that neurons in the cerebellum would be preferentially related to ballistic movements. To test this hypothesis for the basal ganglia, the activity of neurons in the striatum was recorded in a monkey trained to make *both* slow and rapid arm movements. Although the results are preliminary, this experiment will be discussed, because the findings lend experimental support to the view of Kornhuber.

For this study a monkey was trained to grasp a lever and make both slow and fast movements between two 1 cm zones spaced 5 cm apart. The animal had to hold first in one of the two zones for 2 to 6 sec. Then, when a green lamp came on, the animal had to move the lever slowly to the opposite zone (movement time between the zones had to be greater than 0.7 sec). After holding again for several seconds, a red lamp came on; this was the stimulus for a rapid movement back to the starting zone. This fixed sequence of slow (ramp) and fast (ballistic) movements was maintained throughout the experiment. The starting position (zone) could be changed after several trials in order to obtain ramp and ballistic movements in both directions (i.e., pushing and pulling). The monkey was rewarded with a drop of juice following each successful rapid movement (i.e., movement time < 140 msec). A potentiometer was coupled to the axis of the lever in order to monitor the position of the lever during the task. Single-unit activity was recorded from the contralateral putamen using techniques previously described (DeLong, 1971). The monkey performed the task while the electrode was advanced in search of movement-related units.

Of 95 movement-related putamen units studied, more than half discharged either solely or preferentially in relation to the slow movements, while less than 10% of units discharged preferentially in relation to the rapid movements. The remaining units discharged in relation to both movements. The discharge pattern of a representative ramp-related unit is shown in Figure 3. This unit was active during slow movements in both directions but discharged only occasionally during the ballistic movements. While some units discharged for both directions of the slow movement, others discharged for only one direction, as illustrated by the unit in Figure 4, which discharged for pulling but not pushing ramp movements. This unit also frequently discharged with one or more spikes during the pulling ballistic movement (B). A weak discharge of this type was often seen in ramp-related units during the ballistic movement in the same direction as the related ramp movement. Many of the ramp-related units became active prior to any change in the position of the arm, as can be seen in the units in Figures 3 and 4.

The finding that the majority of movement-related units in the putamen are preferentially related to slow movements is consistent with the view of Kornhuber that the basal ganglia function as "ramp generators" and that

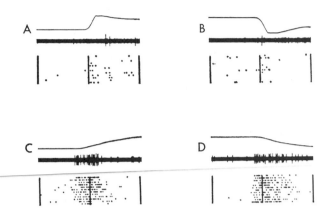

FIGURE 3 Activity of a unit from the putamen during pushing and pulling ballistic (A and B) and ramp (C and D) movements. For each case the upper trace represents the position of the arm during a single trial; the middle trace the unit discharge for the same trial; and the lower portion the activity of the unit during 12 successive trials shown in raster form. Each trial is aligned on the time of leaving the zone (center bar). The interval from the center bar to the margin of the raster is 1 sec. This unit discharged prior to and during the first portion of both pushing and pulling ramp movements (C and D) but not during pushing or pulling ballistic movements (A and B). This unit, like most putamen neurons, was inactive during rest.

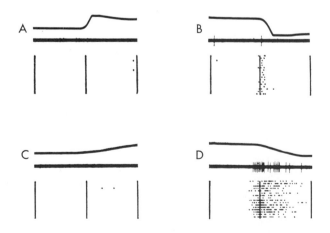

FIGURE 4 Activity of a unidirectional ramp-related putamen unit during performance of the task. The conventions are the same as for Figure 3. This unit discharged during the pulling ramp (D) movements but not during pushing ramp movements. The unit discharged weakly during the pulling ballistic movement as well, usually with only one or two spikes.

separate neural mechanisms may control these two types of movement. Kornhuber's view that the basal ganglia function as ramp generators predicts not only that units there should be preferentially related to ramp movements but also that the changes in unit activity should precede the onset of movement, i.e., the earliest changes in muscle activity. This requirement is obviously necessary if the basal ganglia are actually initiating these slow movements, i.e., generating the control signals. Whether the observed changes in activity do precede the earliest changes in the muscle remains uncertain at present, because although many ramp-related units showed changes in activity prior to any change in the position of the lever, precise determinations of the earliest change in muscle activity associated with the ramp movements have not yet been made. Conceivably, the activity observed in these units might be a result of sensory feedback (via muscle, joint, or cutaneous afferents) from the moving limb. Driving of striate units by somatosensory stimulation, particularly from skin and muscle, has been shown previously (Segundo and Machne, 1956; Albe-Fessard et al., 1960; Sedgwick and Williams, 1967). In the present experiment as well, some units in the putamen with clearly demonstrable responses to somatic stimulation were observed, but these units typically showed changes in activity during both ramp and ballistic movements as well as in response to passive joint movement or mechanical stimulation of the skin and deep structures. Against the possibility that the changes observed in ramp-related putamen units might result from sensory feedback is the fact that, in the units tested, it was not possible to drive them by passive limb movement or other stimulation.

The present experiment has revealed specific requirements for activation of striatal neurons during movement and has shown the importance of studying the activity of basal ganglia neurons during different types of movement. In earlier experiments in which the activity of putamen units was studied solely in relation to ballistic movements, only a small number of units were found to discharge during the movements. Presumably, the units that were selectively related to ramp movements were inactive during the movements. One can only speculate about the specific conditions necessary for the activation of the many silent units passed by in these experiments. The fact that striatal neurons normally are extremely inactive is well known. Intracellular recordings (Hull et al., 1969) have shown them to be hyperpolarized to a high degree and extremely difficult to fire even when excitatory postsynaptic potentials (EPSPs) of considerable amplitude are evoked by stimulation of their inputs. When activated, the response of striatal units to electrical stimulation is characteristically brief. The selective and sustained activation of these cells observed in relation to slow movements must therefore depend on simultaneous and maintained activity in one or more of the cell's inputs (i.e., corticostriatal, nigrostriatal, and thalamostriatal afferents).

Clinical relevance

These studies on the normal functioning of the basal ganglia in the monkey during movement may contribute to our understanding of the pathophysiology of certain "extrapyramidal" movement disorders in man. Each study has clearly indicated that these structures play a role in voluntary movement. The recent study comparing unit activity during slow and fast movement, moreover, indicates that at least a large portion of the striatum (the putamen) is preferentially involved in the control of slow rather than rapid movements. While the exact nature of this control is not yet clear, one might expect that damage to this structure would result in a preferential disturbance of slow movements. Certain findings in man and animals support this notion. In Parkinsonism, for example, there occurs a peculiar inability to initiate movement, referred to as *akinesia*. Kornhuber (1971; this volume) argues from clinical observations of patients that the akinesia results largely from a selective inability to initiate slow (ramp) movements, while rapid (ballistic) movements are less disturbed. Consistent with this is the fact that simple reaction times in Parkinson patients are nearly normal (Butz, 1970).

The single-unit studies are consistent with considerable evidence indicating that akinesia results from loss of the functional integrity of the striatum. Thus, it is now widely believed that the akinetic syndrome of Parkinsonism results from degeneration of the dopamine-containing cells of the nigra that terminate in the striatum, with resultant decrease in striatal dopamine (Hornykiewicz, 1966). Thus, treatment of Parkinsonism with L-dopa, the precursor of dopamine, is effective in relieving akinesia as well as tremor and rigidity, while treatment with stereotactic lesions in the pallidum or thalamus (which interrupt the output from the striopallidum) is not and, indeed, may actually exacerbate the akinesia (Schwab et al., 1959). It thus appears that L-dopa acts to restore the functional integrity of the striatum, and thus to relieve akinesia, whereas stereotactic lesions, which further reduce the output from the striopallidum, may actually worsen akinesia while relieving tremor and rigidity. In monkeys, lesions of the nigrostriatal dopamine neurons result in hypokinesia of the limbs contralateral to the lesion (Poirier and Sourkes, 1965). Recent studies by Ungerstadt (1971; this volume) give further evidence that the etiology of akinesia results from loss of striatal function,

because selective destruction of the nigrostriate pathways in the rat by 6-OH-dopamine results in a profound hypokinesia that, like the akinesia of Parkinsonism, can be reversed by L-dopa. Thus, from the studies on striatal activity in the monkey during movement as well as from clincopathologic and pharmacologic evidence, it appears that akinesia results from a loss of function at the level of the striatum, due to a loss of input from the nigrostriate dopamine system.

Loss of function may occur when a neural mechanism is damaged or blocked, whereas "release" of function may result where normal controlling influences are removed. Kornhuber has suggested that whereas one might view akinesia as a loss of the ramp-generating function of the basal ganglia, certain basal ganglia movement disorders characterized by spontaneously occurring involuntary movements (e.g. chorea, athetosis, dystonia), as well as the rigidity of Parkinsonism, may be due to various forms of release of the ramp-generating mechanisms from their normally restraining influences. Again, while the pathology is not clear in all these disorders, chorea and athetosis are generally thought to result from damage to cells in the striatum.

Further studies on the activity of single neurons in the basal ganglia of the intact animal during specific behavioral tasks and in conjunction with physiological and pharmacologic manipulation may help to further clarify the normal physiology of the basal ganglia as well as the pathophysiology of the various basal ganglia disorders.

REFERENCES

ALBE-FESSARD, D., C. ROCHA-MIRANDA, and E. OSWALDO CRUZ, 1960. Activités évoquées dans le noyau caudé du chat en response à des types divers d'afférences. II. Étude microphysiologique. *Electroenceph. Clin. Neurophysiol.* 12:649–661.

ANDÉN, N.-E., A. CARLSSON, A. DAHLSTRÖM, K. FUXE, N.-A. HILLARP, and K. LARSSON, 1964. Demonstration and mapping out of nigro-neostriatal dopamine neurons. *Life Sci.* 3:523–530.

ANDÉN, N.-E., A. DAHLSTRÖM, K. FUXE, and K. LARSSON, 1965. Further evidence for the presence of nigro-neostriatal dopamine neurons in the rat. *Am. J. Anat.* 116:329–333.

BUTZ, P., W. KAUFMANN, and M. WIESENDANGER, 1970. Analyse einer raschen Willkürbewegung bei Parkinsonpatienten vor und nach stereotaktischem Eingriff am Thalamus. *Z. Neurol.* 198:105–119.

CARMAN, J. B., W. M. COWAN, and T. P. S. POWELL, 1963. The organization of cortico-striate connexions in the rabbit. *Brain* 86:525–562.

CARMEL, P. W., 1971. Efferent projections of the ventral anterior nucleus of the thalamus in the monkey. *Amer. J. Anat.* 128:159–184.

CARPENTER, M. B., and N. L. STROMINGER, 1967. Efferent fibers of the subthalamic nucleus in the monkey. A comparison of the efferent projections of the subthalamic nucleus, substantia nigra and globus pallidus. *Am. J. Anat.* 121:471–472.

DeLong, M. R., 1971. Activity of pallidal neurons during movement. *J. Neurophysiol.* 34:414–427.

DeLong, M. R., 1972. Activity of basal ganglia neurons during movement. *Brain Res.* 40:127–135.

Denny-Brown, D., 1962. *The Basal Ganglia.* London: Oxford University Press.

Evarts, E. V., 1966. Pyramidal tract activity associated with a conditioned hand movement in the monkey. *J. Neurophysiol.* 29:1011–1027.

Hornykiewicz, O., 1966. Metabolism of brain dopamine in human Parkinsonism: Neurochemical and clinical aspects. In *Biochemistry and Pharmacology of Basal Ganglia*, E. Costa, L. J. Coté, and M. D. Yahr, eds. New York: Raven-Hewlett, pp. 171–181.

Hull, C. D., N. A. Buchwald, and L. M. Vernon, 1969. Intracellular responses in caudate and cortical neurons. In *Psychotropic Drugs and Dysfunctions of the Basal Ganglia*, G. E. Crane and R. Gardner, Jr., eds. PHS Publ. No. 1938, Washington, D.C.: U.S. Govt. Printing Office, pp. 92–97.

Jung, R., and R. Hassler, 1960. The extrapyramidal motor system. In *Handbook of Physiology. Neurophysiology*. Washington, D.C.: Amer. Physiol. Soc., sect. 1, vol. II, chap. 35, pp. 863–927.

Kemp, J. M., and T. P. S. Powell, 1970. The cortico-striate projection in the monkey. *Brain* 93:525–546.

Kemp, J. M., and T. P. S. Powell, 1971. The connexions of the striatum and globus pallidus: synthesis and speculation. *Phil. Trans. Roy. Soc. (London)* (B) 262:441–457.

Kornhuber, H. H., 1971. Motor functions of cerebellum and basal ganglia: the cerebellocortical saccadic (ballistic) clock, the cerebellonuclear hold regulator, and the basal ganglia ramp (voluntary speed smooth movement) generator. *Kybernetik* 8:157–162.

Martin, J. P., 1967. *The Basal Ganglia and Posture.* London: Pitman Medical.

Mehler, W. R., 1966. Further notes on the center median nucleus of Luys. In *The Thalamus*, D. P. Purpura and M. D. Yahr, eds. New York: Columbia University Press, pp. 109–127.

Nauta, W. J. H. and W. R. Mehler, 1966. Projections of the lentiform nucleus in the monkey. *Brain Res.* 1:3–42.

Nauta, W. J. H., and W. R. Mehler, 1969. Fiber connections of the basal ganglia. In *Psychotropic Drugs and Dysfunctions of the Basal Ganglia*, G. E. Crane and R. Gardner, Jr., eds. PHS Publ. No. 1938, Washington, D.C.: U.S. Govt. Printing Office, pp. 68–74.

Nauta, W. J. H., and D. G. Whitlock, 1954. An anatomical analysis of nonspecific thalamic projection system. In *Brain Mechanisms and Consciousness*, J. F. Delafresnaye, ed. Oxford: Blackwell, pp. 81–104.

Poirier, L. J., and T. L. Sourkes, 1965. Influence of the substantia nigra on the catecholamine content of the striatum. *Brain* 88:181–192.

Poirier, L. J., T. L. Sourkes, G. Bouvier, T. Boucher, and S. Carabin, 1966. Striatal amines, experimental tremor and the effect of harmaline in the monkey. *Brain* 89:37–52.

Schwab, R. S., A. C. England, and E. Peterson, 1959. Akinesia in Parkinson's disease, *Neurology* 9:65–72.

Sedgwick, E. M., and T. D. Williams, 1967. The response of single units in the caudate nucleus to peripheral stimulation. *J. Physiol.* 189:281–298.

Segundo, J. P., and H. Machne, 1956. Unitary responses to afferent volleys in lenticular nucleus and claustrum. *J. Neurophysiol.* 19:325–339.

Stetson, R. H., and II. D. Bouman, 1935. The coordination of simple skilled movements. *Arch. Neerl. Physiol.* 20:179–254.

Strick, P. L., 1971. Functional zones in the cat ventrolateral thalamus. *Proc. First Annual Meeting Soc. Neuroscience*, p. 122.

Strick, P. L., 1973. Light microscopic analysis of the cortical projection of the thalamic ventrolateral nucleus in the cat. *Brain Res.* (in press).

Thach, W. T., 1968. Discharge of Purkinje and cerebellar nuclear neurons during rapidly alternating arm movements in the monkey. *J. Neurophysiol.* 31:785–797.

Thach, W. T., 1970. Discharge of cerebellar neurons related to two maintained postures and two prompt movements. I. Nuclear cell output. *J. Neurophysiol.* 33:527–536.

Ungerstedt, U., 1971. Adipsia and aphagia after 6-hydroxy-dopamine induced generation of the nigro-striatal dopamine system. *Acta Physiol. Scand., Suppl.* 367:95–122.

Walker, A. E., 1938. *The Primate Thalamus.* Chicago, Ill.: University of Chicago Press.

Webster, K. E., 1961. Cortico-striate interrelations in the albino rat. *J. Anat.* 95:532–544.

Webster, K. E., 1965. The cortico-striate projection in the cat. *J. Anat.* 99:329–337.

29 Sensorimotor Cortex Activity Associated with Movements Triggered by Visual as Compared to Somesthetic Inputs

EDWARD V. EVARTS

ABSTRACT   This report contrasts input-output relations for two different inputs leading to a similar output. In one set of experiments, monkeys were trained to push or pull a handle in response to a visual stimulus. In the second set of experiments, monkeys pushed or pulled in response to a perturbation of the handle. In both cases, neuronal activity in hand area of sensorimotor cortex was studied in relation to the learned movement. For the visual modality, the interval from stimulus to pyramidal tract neuron (PTN) response of precentral motor cortex was about 100 msec, but when the stimulus was delivered via the responding hand the interval from stimulus to PTN response was as little as 25 msec. It thus appears that there are fundamental differences in the processing depending on the modality of input.

Introduction

THIS CHAPTER will describe differences in the activity of the cerebral sensorimotor cortex neurons for movement carried out in response to visual as compared to somesthetic inputs. It is well known that somesthetic inputs to the motor cortex are prominent and of relatively short latency (Albe-Fessard and Liebeskind, 1966; Brooks and Stoney, 1971; Oscarsson and Rosen, 1963, 1966; Rosen and Asanuma, 1971; Towe, 1968). In contrast, inputs from the visual system are less massive and reach the motor cortex by more circuitous routes. Are these differences in the projection of visual and somesthetic inputs to motor cortex reflected in differences of motor cortex output in response to these sensory stimuli in association with learned movement? This was the problem that the experiments described in this chapter were designed to answer.

EDWARD V. EVARTS   Laboratory of Neurophysiology, National Institute of Mental Health, Bethesda, Maryland

Methods

The experiments were begun by training animals to carry out movements using operant conditioning techniques. Figure 1 illustrates the arrangement used for initial training. As indicated in this figure, the monkey received a juice reward for correct manipulation of a handle mounted on an apparatus attached to his cage. For visually triggered movements, the monkey was required to hold the handle against one of the two stops shown in Figure 1, and then quickly to move the handle into contact with the opposite stop in response to the appearance of a visual stimulus (a lamp). The monkey was unable to predict the time of occurrence of the visual stimulus. The task was similar to visual reaction time in man. For the movement initiated by a somatosensory stimulus, the monkey was required to hold the handle in a fixed position until (at a time that was unpredictable) the handle was deflected by an external force; the monkey's task was to return the handle to its original position as quickly as possible. Thus, the animal made rapid ballistic arm movements in response to either visual or somatosensory stimuli, and it was possible to observe the activity of central neurons during the performance of movements triggered by these two sorts of stimuli. Different monkeys were used in studies of movements initiated via the two different modalities.

When animals had completed learning, a chamber was attached to the skull. A microelectrode advancer could be attached to this chamber, and a microelectrode could then be introduced into sensorimotor cortex with the animal seated in a primate chair. The microelectrode was lowered into either the precentral motor area or the postcentral sensory area, and activity of single nerve cells was picked up during movements of the hand

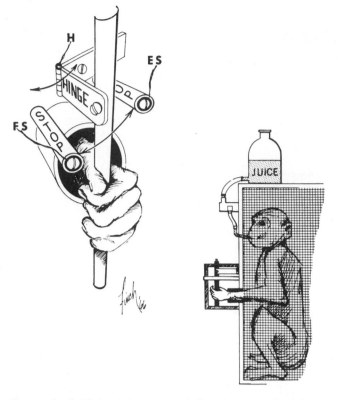

FIGURE 1 Initial training was carried out in the monkey's home cage. Here the monkey's left hand is seen to be protruding from a tube in a lucite panel attached to the front of the cage. In order to receive a fruit-juice reward, the monkey was required to grasp the vertical rod attached to a hinge and to move it into contact with one stop or the other. The stops are labeled FS (flexor stop) and ES (extensor stop). The monkey was required to maintain contact with the flexor stop until a lamp was turned on, at which time he was rewarded for a quick movement into contact with extensor stop, etc. A narrow slit, just large enough to accommodate the monkey's wrist, prevented side-to-side arm movement and required flexion-extension at the wrist. (Reprinted from Evarts, 1968.)

contralateral to the site of single-cell recording. At the end of each recording session, the monkey was returned to his cage. Each day for a number of weeks, a microelectrode was introduced into a slightly different location in the sensorimotor cortex, electrode penetrations being made at distances of 1 mm from adjacent electrode penetrations. Stimulating electrodes were permanently implanted in the medullary pyramid, and cells showing antidromic responses to medullary stimulation could be identified as pyramidal tract neurons (PTNs).

DATA ANALYSIS Figure 2 illustrates activity picked up from a PTN in the precentral motor cortex in association with performance of a visually triggered movement. This figure shows data for 12 trials. It may be seen that there is considerable variation in the interval between the visual

stimulus (the traces of the figure begin with this stimulus) and the occurrence of the single-unit discharge. Variability such as this requires that these data be subjected to statistical analysis and that many trials be taken for each cell that is investigated. Analysis and display of data such as these may be facilitated by means of computer techniques, as illustrated in Figure 3, where each of the rows corresponds to one trial that an animal has made, and each dot in the row corresponds to one action potential of a single nerve cell. The vertical line in each row corresponds to the motor response. With the discharge of the single neuron represented as a dot, it is possible to display much data in a small space. Thus, Figure 3 displays results for 25 trials. This display, photographed from the oscilloscope associated with the computer, has the additional advantages of presenting the discharge of the neuron prior to, as well as after, the stimulus.

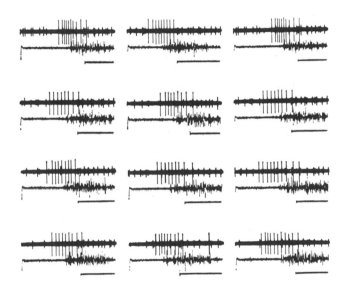

FIGURE 2 The PTN and extensor EMG response to a visual stimulus. This figure illustrates a series of 12 trials for a PTN that was silent during flexion and that consistently discharged prior to visually triggered extension of the contralateral wrist. All traces start at the onset of the light. The minimum response latency for this PTN was about 120 msec. This latency of PTN response was associated with an EMG latency of 170 msec. In general, the shortest latency PTN responses were associated with the shortest latency EMG responses. Sweep duration is 500 msec. (Reprinted from Evarts, 1966.)

A further advantage of the computer analysis is shown in Figure 4, where it can be seen that the computer can display histograms corresponding to the dot displays of Figure 3. An additional feature shown in Figure 4 is the realignment of rows so that neuronal activity can be shown in relation to the occurrence of the motor response as well as in relation to occurrence of the stimulus. In the

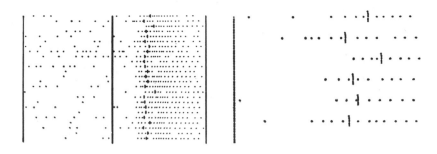

30

FIGURE 3   At the left is shown a dot display with 25 rows of dots corresponding to 25 trials; in the center of this 25-row display is a vertical line indicating the time of occurrence of the stimulus; the dots to the left of this central line represent neuronal activity occurring 500 msec prior to the stimulus (S). Each heavy dot following (i.e., to the right of) S represents the time at which R was detected. To the right of the 25-row display is shown an enlarged view of the poststimulus activity for the first six trials of the display at the left.

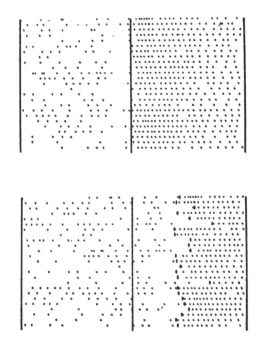

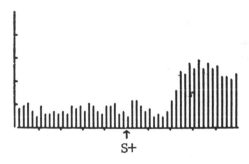

R+

S+

FIGURE 4   Dot displays may be aligned either with respect to the stimulus (S) or response (R). In the lower dot display the central vertical line represents S, and the single heavy dot in each row to the right of S represents R. The histogram corresponding to this dot display (lower right) is centered on S, whose time of occurrence is indicated by the arrow. In the upper dot display each row of dots has been shifted to the left until R reaches the center of the display. The corresponding histogram (upper right) is thus centered about R, whose time of occurrence is indicated by the arrow.

bottom dot display of Figure 4, the rows of dots are aligned with respect to the stimulus; the time of occurrence of the stimulus is shown by the solid line at the center of the dot display in the lower left-hand section of the figure. In the right half of this figure the heavy dots (one in each row) illustrate occurrence of the motor response. After these heavy dots (i.e., after each motor response), there is an increase of discharge in the neuron. If the rows of dots in the lower section of this figure are realigned so that each row moves to the left until the response reaches the center, one derives the display of neuronal activity aligned with respect to response (R) instead of stimulus

EDWARD V. EVARTS   329

(S). This is shown in the upper left-hand section of Figure 4. Of course, the upper section of Figure 4 shows an increase of neuronal activity following R, but in addition the realigned rows of dots reveal that prior to this increase of neuronal discharge with the response, there is a decrease prior to the response. This decrease can be seen in the histogram labeled R +. When activity of this neuron is aligned with respect to S, the decrease fails to appear (histogram labeled S +). This failure is due to the variable latency between the S and R. It is clear that for neuronal activity, which is related in time to R rather than S, it is useful to be able to align discharges of the neuron in relation to R.

Another advantage of the computer display is that it allows separation of different sorts of motor responses that may actually have occurred intermixed. The result of such a separation of different sorts of responses is illustrated in Figure 5, where there are two sets of dot displays, one corresponding to 25 flexions of the wrist and the other corresponding to 25 extensions. As was pointed out in the introduction to this Methods section, monkeys were trained to maintain wrist flexion and abruptly extend and then to maintain extension and abruptly flex in response to a light stimulus. Thus for each neuron, two sorts of motor responses were alternately obtained,

and for each of these motor responses neuronal activity was observed. In quantitating neuronal discharge, it was of course necessary to carry out the analysis separately for the two sorts of movements, and the computer provides for separation of the trials. For the neuron shown in Figure 5, flexion of the wrist (referred to as R +) was associated with a decrease of activity prior to the motor response. This decrease is shown in the left-hand section of Figure 5, where the dot display and the histogram above it reveal a reduction of neuronal discharge in association with the movement. For the dot display at the right-hand section of the figure, there is an increase of activity for the same neuron in association with wrist extension (referred to as R −). Thus, for this neuron, there is an opposite change in discharge for the two oppositely directed movements of the wrist.

A different, but equally important, feature of data analysis involves making a decision as to the time of occurrence of the movement to which activity of neurons is related. Figure 6 illustrates recordings from the biceps and triceps muscles for an arm movement in which the animal pushed its arm forward in response to the appearance of a light. During the period prior to the light stimulus, the animal was pulling toward itself, and, in the upper record of electromyographic activity, tonic activity

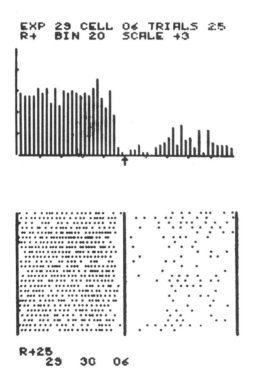
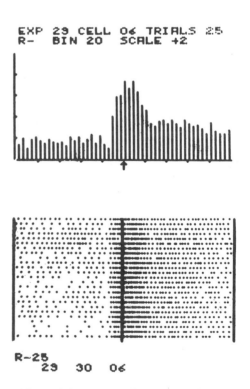

FIGURE 5 Computer analysis allows separation of two sorts of movements that may have occurred alternately. In the display shown here, R + means flexion and R − means extension. During the actual experiment these two movements occurred

alternately, and the neuron whose activity is represented in this figure showed a decrease of activity prior to R + and an increase of activity prior to R −.

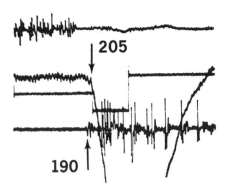

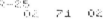
205

190

FIGURE 6 Biceps EMG (top trace) and triceps EMG (bottom trace) are shown in relation to changes in force associated with a ballistic pushing movement. The change in force (indicated by arrow at 205 msec) was detected 15 msec following the increase of triceps EMG activity (at 190 msec) associated with the push. A decrease of biceps EMG occurred earlier, about 35 msec prior to detection of the force change. The negative-going pulse in the trace just above triceps EMG provides a measure of the time of occurrence of R, and it is this time which is used in computer analysis.

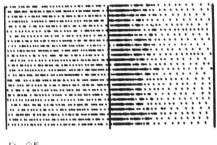

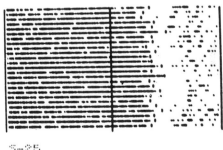

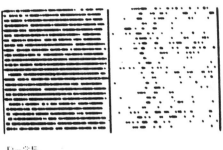

EXTENSORS

FLEXORS

FIGURE 7 The EMG activity of flexor and extensor muscles associated with wrist extension has been rectified and integrated by an integrator that, when a certain value has been reached, resets to zero and emits a pulse. Each dot in the dot display corresponds to the emitted pulse and thus represents one resetting of the integrator. The displays at the left show an increase of extensor activity occurring simultaneously with R. The displays at the right show that for exactly the same 25 trials, flexor EMG decreases prior to R. In the lower displays, activity is aligned with respect to S. For the upper displays, activity is aligned with respect to R.

is seen in the biceps muscle; during this same period the triceps muscle is inactive. About 170 msec following the light stimulus, there was a decrease in the activity of the biceps muscle, and 20 msec later (190 msec following the light stimulus) the triceps activity appeared. At 205 msec following stimulus the record of force exerted on the handle changed sufficiently to cross a threshold set on a level detector. It is this *crossing of the threshold on a level dectector* by the output of the force transducer that we have used as a measure of response. It should thus be noted that this measure of motor response necessarily lags the first change in muscle activity. The change in force is reliably related to changes in a number of different muscles, however, and it can be correlated with muscle activity recorded during performance of the motor act. By means of the computer, it is possible to display the activity of a muscle in much the same format that is used to display the activity of single nerve cells. This is shown in Figures 7 and 8, which illustrate dot-displays for activity of flexor and extensor muscles in association with flexion and extension of the wrist. The dots displaying muscle activity are not single motor unit discharges but correspond to integrated electromyographic (EMG) activity; a dot occurs each time the EMG integrator resets. Figure 7 shows activity of extensor muscles and flexor muscles in association with extension of the wrist. On the left are dot-displays for extensor

muscle EMG aligned with respect to the stimulus (below) and the response (above). It may be seen that for this particular recording, extensor muscle increase was detected approximately simultaneously with the detection of the change in force that defines the occurrence of the response (R). However, for this same movement (extension of the wrist), it may be seen that there is a decrease of flexor activity before R has been detected. This prior decrease has been reported previously (Hufschmidt and Hufschmidt, 1954). Figure 8 shows that muscle activity can also be presented as a histogram in much the same way as single nerve cell activity. Figure 8 shows activity of flexor muscle in relation to wrist flexion rather than wrist extension. It may be seen that the increase of activity of the flexor muscles occurs approximately simultaneously with the detection of the response (R +) of flexion on the basis of change in force. Finally, Figure 9 illustrates activity of a precentral cortex neuron in relation to the changes in force which serve as a basis for response (R) detection. It may be seen in this figure that the change in activity of the precentral cortex neuron precedes changes in force quite considerably and that for the two directions of movement shown (wrist extension and wrist flexion) the neuron shows an increase of activity for one (wrist flexion) and a reciprocal decrease for the other. In the analyses that follow, we have aligned nerve cell activity both with respect to stimulus (S) and response (R). We

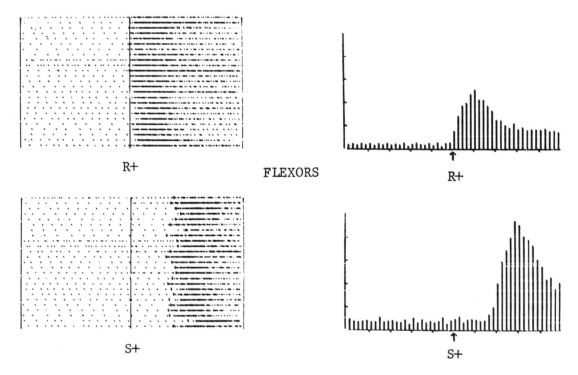

FIGURE 8 The dot displays of EMG activity may also be represented as histograms, aligned either with respect to S or to R. In these displays, flexor muscle EMG is shown for a flexor movement of the wrist.

have made a similar alignment of the activity of muscles with respect to stimulus and response. Knowing the temporal relationship between activity of nerve cells and S or R, and knowing the temporal relationship between changes of muscle activity and S or R, it is possible to infer the temporal relations between changes of neuronal activity in the motor cortex and changes of muscle activity.

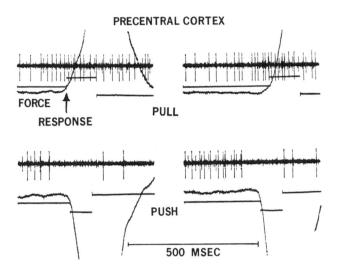

PRECENTRAL CORTEX

FORCE
RESPONSE
PULL
PUSH
500 MSEC

FIGURE 9   This shows a precentral PTN which was oppositely related to pulling and pushing movements. For further explanation see text.

Results

MOTOR CORTEX ACTIVITY ASSOCIATED WITH VISUALLY TRIGGERED MOVEMENT   For these studies, the input was a light stimulus to which the monkey's response was an arm movement. In some cases the movement was wrist flexion or extension, in other cases a push or pull. Figure 10 summarizes results for several hundred precentral neurons recorded from a monkey performing a wrist movement. The figure shows that for some neurons, activity began approximately 100 msec prior to response detection. The first changes of muscle activity also preceded R but by a considerably smaller amount than was the case for precentral neurons, and therefore one can infer that precentral neurons are active prior to the muscle activity. Figure 10 shows that the distribution of onset times for precentral PTNs and non-PTNs was the same. Thus, the discharge of motor cortex neurons prior to movement is seen both for pyramidal and nonpyramidal tract neurons.

Granted that there are changes in activity of precentral neurons prior to movement, the question remains as to whether these changes are restricted to the pre-

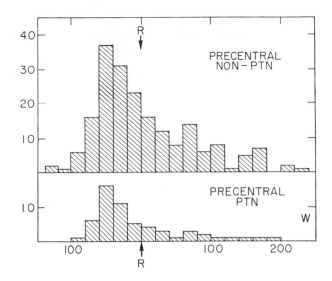

FIGURE 10   This figure shows onset times of motor cortex neurons in relation to visually triggered movement. An onset time for each neuron related to the visually triggered arm movement was computed on the basis of dot displays and histograms. The distribution shown above summarizes the onset times for several hundred motor cortex neurons. There is no difference in onset times for PTNs as compared to non-PTNs.

central area or may also occur in the postcentral region of the sensorimotor cortex. Results on this question are illustrated in Figure 11, which contrasts neural response onset times of pre- and postcentral neurons. It is seen that the distribution of onset times for precentral neurons leads the distribution of onset times for postcentral

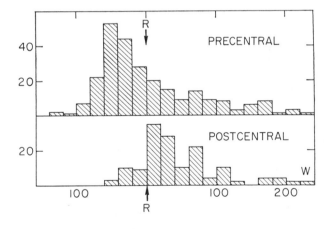

FIGURE 11   Precentral and postcentral neuron onset times. Periresponse latencies were determined for several hundred precentral and postcentral neurons which showed a change (either increase or decrease) of activity in association with visually triggered movement. This figure shows two distributions of periresponse latencies, one for precentral neurons and the other postcentral neurons. The abscissa gives time in msec before and after R. Ordinate is number of units. For further details see text. (Reprinted from Evarts, 1972.)

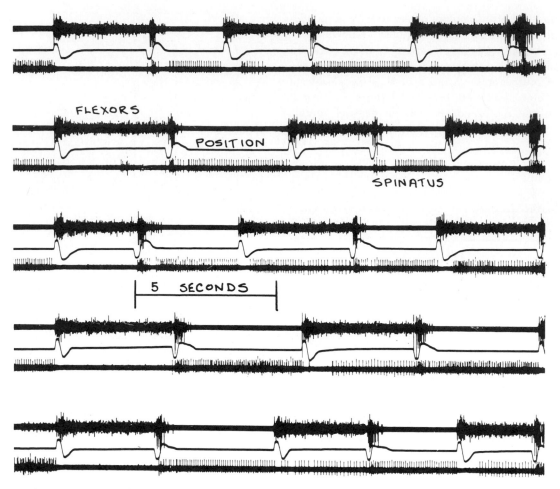

FLEXORS

POSITION

SPINATUS

5 SECONDS

FIGURE 12   Muscle activity and handle position are shown for a series of stimuli and responses for the response triggered by input to the hand. The uppermost trace in each set of three traces is activity of forearm flexor muscles, the middle trace represents handle position, and the lowermost trace is activity of supraspinatus muscle. During periods of tonic supraspinatus activity, an external force maintained on the handle requires that the monkey push against the handle to maintain it in the correct position. At a time which is unpredictable to the monkey, the force acting on the handle is reversed, and as a result the monkey's arm moves forward. There is then a prompt shift in the balance of muscle activity as the monkey must now pull the handle rather than push. Tonic supraspinatus discharge disappears and is replaced by tonic flexor discharge. After another unpredictable delay, the external forces acting on the handle are again reversed, and the handle moves toward the monkey. This stimulus results in reappearance of tonic supraspinatus activity, as the monkey must now push against the handle to maintain it in the correct position.

neurons by approximately 60 msec. Thus, there is a clear functional differentiation between precentral and postcentral neurons in relation to when they fire in association with movement. Though precentral activity occurs earlier than postcentral, there are nevertheless a few neurons in the postcentral area that show activity prior to R. Of course, it must be recalled that some muscle activity is already occurring in advance of R, and it is thus possible that most of the activity in postcentral may occur in response to sensory feedback. But it is also possible that some postcentral activity may occur prior to feedback from the periphery consequent upon the movement.

MOTOR CORTEX ACTIVITY ASSOCIATED WITH HAND MOVEMENT TRIGGERED BY SOMESTHETIC INPUT   For experiments on movements initiated by an input via the hand, the monkey held the handle in the correct position (a light indicated when the handle was in the correct position) for a sufficient length of time (2 to 5 sec), and the handle was then moved abruptly forward or backward by means of an external power source. During the holding period prior to the stimulus, a steady-state force was exerted by the external power source, and therefore the monkey was either pushing or pulling as he waited for the stimulus. The stimulus consisted of a reversal of the external forces applied to the handle. Thus, if the animal was pulling during the waiting period, the stimulus involved a disappearance of the force against which he was pulling with replacement of this force by an opposite force. As a result, the animal's hand would be displaced, and in order to receive a reward he had to change the pattern of muscle activity so as to restore the handle to the correct position. When he had restored the handle to the correct position, a new holding period began; the steady-state force was reversed in alternate holding periods and each was terminated by a shift in the position of the handle at a time that was unpredictable to the animal. The changes in handle position and associated changes in electromyographic activity associated with these two sorts of stimuli are illustrated in Figure 12. Figure 13 illustrates activity of sensorimotor cortex neurons in association with hand movements initiated by the input to the hand. Whereas sensorimotor cortex activity in response to a visual stimulus occurred approximately 100 msec after the visual stimulus, discharges for the somatosensory stimulus occurred in motor cortex PTNs in as short a time as 25 msec following stimulus. Activity

in the postcentral somatic receiving area occurred even sooner, the interval from stimulus to neuronal discharge being as short as 10 msec. Thus, when the input initiating a movement enters via the hand, discharge in the postcentral area occurs *prior* to discharge in the precentral area, whereas the reverse order occurs for movement initiated by a visual stimulus. Finally, it should be pointed out that there are latency differences between pyramidal and nonpyramidal tract neurons in the motor area. For the visually elicited movement, no difference was found between onset times of PTNs and non-PTNs in the precentral motor cortex. However, for the movement elicited by the input to the hand, differences were found in motor cortex for PTNs and non-PTNs. The differences were slight but clearly significant; motor cortex non-PTNs were found to discharge at latencies of less than 20 msec from the stimulus to the hand. The difference between pyramidal tract neurons and nonpyramidal tract neurons is slight (about 5 msec), and such difference might easily have been missed due to variability in the case of the visually triggered movements where the onset times of

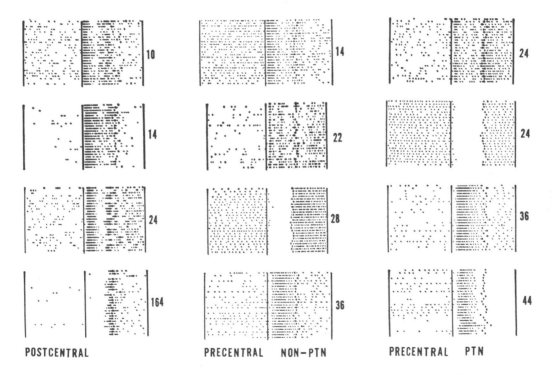

POSTCENTRAL **PRECENTRAL   NON−PTN** **PRECENTRAL   PTN**

FIGURE 13   Each of the 12 rasters shows neuronal activity for 500 msec before and 500 msec after the abrupt handle movement whose time of occurrence is indicated by the vertical line at the center of the raster. In each raster there are 25 rows of dots, corresponding to 25 successive trials. The individual dots in each row correspond to individual neuronal impulses. Neuronal response latency from handle perturbation to first change in neuronal discharge frequency (either increase or decrease) was computed for each neuron on the basis of 25 trials, and this latency is shown at the right of each raster. In the right half of

each raster a single heavy dot indicates the time at which the handle was returned to the correct zone by the monkey's motor response. This heavy dot may be seen most clearly in the PTN raster at the lower right. Rasters in columns at left, center, and right correspond to postcentral neurons, precentral non-PTNs and precentral PTNs, respectively. The latency values for one non-PTN (28) and one PTN (24) refer to *decreases* of activity; remaining latencies refer to increases of neuronal activity. (Reprinted from Evarts, 1973.)

neuronal discharge were less tightly locked to the stimulus. The results for the movement initiated by the somesthetic stimulus thus reveal three classes of neurons discharging at successively greater delays from the stimulus. The earliest discharge occurs in neurons of the postcentral gyrus; presumably these are classical sensory responses to the inputs that are delivered to the tactile, pressure, joint, and muscle receptors of the hand and arm when the handle is abruptly deflected by the external power source. The second set of neurons to discharge are nonpyramidal tract neurons in the precentral motor cortex. Finally, at an interval of about 25 msec following the stimulus, pyramidal tract neurons of motor cortex begin to discharge.

Figure 14 is a schematic representation of the relationship between motor cortex PTN activity and the muscle response occurring as a result of the movement of the handle versus a visual input. When the handle grasped by the monkey is suddenly moved (and the muscle is stretched), there is a muscle discharge occurring at a latency of 15 msec, which is less than the latency of the first PTN discharge of the motor cortex. This first response of the muscle actually occurs quickly enough to suggest that it may be mediated by monosynaptic inputs to motoneurons from spindle afferents. A second discharge of muscle occurs with a latency of approximately 35 msec, though this figure varies considerably depending upon the muscle from which recording is made and depending upon the "set" and expectancy of the monkey.

Recordings of arm muscle EMG during electrical stimulation of the medullary pyramid of monkeys in these experiments showed that action potentials occur in arm muscle at latencies as short as 6 msec following a single shock to the medullary pyramid. The large PTNs of motor cortex discharge antidromically in response to this same stimulus at a latency of less than 1 msec. Thus, PTN discharge might have a role in arm muscle EMG discharge occurring at a latency of as little as 7 msec following PTN activity. It may be concluded that though motor cortex activity cannot play a role in the very earliest muscle response triggered by hand displacement, monosynaptic inputs of PTNs to motoneurons probably do play a role in the later phases of the monkey's corrective response.

Discussion

The findings presented in the Results section showed that the timing of motor cortex output depends upon the modality of the input. Thus, a visual stimulus gives rise to pyramidal tract neuron activity beginning about 100 msec following the stimulus and in this situation the activity of postcentral neurons begins about 60 msec later. This timing of activity in motor cortex neurons of monkey corresponds to the timing of evoked potentials recorded from the scalp of human subjects in association with reaction time performance (Vaughan et al., 1965; Vaughan et al., 1968; Vaughan and Costa, 1968).

Reaction times in human subjects for visual inputs are not as fast as reaction times to kinesthetic inputs. Fast reaction times for kinesthetic inputs were demonstrated in man by Hammond (1956), who recorded motor responses in subjects instructed either to resist or not to resist displacement of the arm. Hammond found that the EMG reponses of the subjects showed a difference depending on instructions at a latency as short as 50 msec from arm displacement. These very short latencies suggested that the instruction given to the subject resulted in a presetting of spinal cord mechanisms such that the reflex activity evoked by this displacement of the arm in

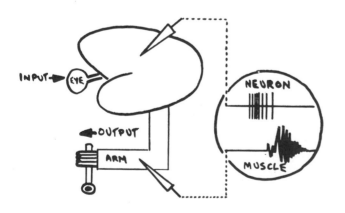

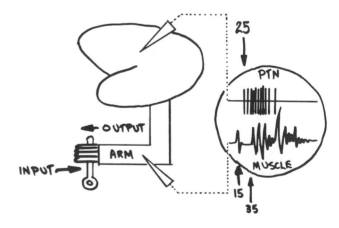

FIGURE 14   When input is visual (left), PTN activity occurs at a latency of about 100 msec, and muscle response occurs even later. For inputs via the responding hand, the muscle and PTN responses occur much earlier, at times indicated in the right-hand section of the figure.

human subjects could vary depending on the set of the subject. It was with these results of Hammond in mind that the observations on monkey responses to displacements of the hand were observed.

The finding that in monkeys EMG responses at a latency of 35 msec are preceded by activity in sensorimotor cortex pyramidal tract neurons with a latency of 25 msec indicates that the cerebral cortex may be involved in certain short latency responses to kinesthetic inputs. Of course, these findings in no way argue against a pre-setting of the spinal cord reflex mechanisms in addition. But the very short latency with which pyramidal tract output occurs in response to input via the hand makes it clear that one need not exclude cortical mediation of short latency reflex processes merely on the assumption that the input-output time requirements for the cortex are too great.

The present experiments do not cast any light on the pathway traversed from input to cortical pyramidal tract neurons. However, it might be pointed out that the observation of pyramidal tract activity at short latency in response to kinesthetic inputs to the hand points to the possibility that these short latency sensory inputs to the cerebral motor area may be important not only in certain reflex movements (see Brooks, 1972) but in voluntarily controlled postural activity and learned movement as well. It remains for a number of further experiments to work out the details of the way in which these short latency inputs to precentral cortex are involved in movement.

REFERENCES

ALBE-FESSARD, D., and J. LIEBESKIND, 1966. Origine des messages somato-sensitifs activant les cellules du cortex moteur chèz le singe *Exp. Brain Res.* 1:127–146.

BROOKS, V. B., 1972. Tight input-output coupling. In *Neuro-sciences Research Symposium Summaries*, Vol. 6, F. O. Schmitt et al., eds. Cambridge, Mass.: MIT Press, pp. 51–56.

BROOKS, V. B., and S. R. STONEY, JR., 1971. Motor mechanisms: the role of the pyramidal system in motor control. *Ann. Rev. Physiol.* 33:337–392.

EVARTS, E. V., 1966. Pyramidal tract activity associated with a conditioned hand movement in the monkey. *J. Neurophysiol.* 29:1011–1027.

EVARTS, E. V., 1968. Relation of pyramidal tract activity to force exerted during voluntary movement. *J. Neurophysiol.* 31:14–27.

EVARTS, E. V., 1972. Contrasts between activity of precentral and postcentral neurons of cerebral cortex during movement in the monkey. *Brain Res.* 40:25–31.

EVARTS, E. V., 1973. Motor cortex reflexes associated with learned movement. *Science* 179:501–503.

HAMMOND, P. H., 1956. The influence of prior instruction to the subject on an apparently involuntary neuro-muscular response. *J. Physiol. (London)* 132:17–18P.

HUFSCHMIDT, H. I., and T. HUFSCHMIDT, 1954. Antagonist inhibition as the earliest sign of a sensory motor reaction. *Nature (London)* 174:607.

OSCARSSON, O., and I. ROSEN, 1963. Projection to cerebral cortex of large muscle spindle afferents in forelimb nerves of the cat. *J. Physiol. (London)* 169:924–945.

OSCARSSON, O., and I. ROSEN, 1966. Short latency projections to the cat's cerebral cortex from skin and muscle afferents in the contralateral forelimb. *J. Physiol. (London)* 182:164–184.

ROSEN, I., and H. ASANUMA, 1971. Peripheral afferent inputs to the forelimb area of the monkey motor cortex: Input-output relations. *Exp. Brain Res.* 14:257–273.

TOWE, A. L., 1968. Neuronal population behavior in the somato-sensory systems. In *The Skin Senses*, D. R. Kenshalo, ed., Springfield, Ill.: Charles C Thomas, pp. 552–574.

VAUGHAN, JR., H. G., L. D. COSTA, L. GILDEN, and H. SCHIMMEL, 1965. Identification of sensory and motor components of cerebral activity in simple reaction time tasks. *Proc. 73rd Conv. Am. Psychol. Assoc.* 1:179–180.

VAUGHAN, JR., H. G., L. D. COSTA, and W. RITTER, 1968. Topography of the human motor potential. *Electroenceph. Clin. Neurophysiol.* 25:1–10.

VAUGHAN, JR., H. G., and L. D. COSTA, 1968. Analysis of electroencephalographic correlates of human sensorimotor processes. *Electroenceph. Clin. Neurophysiol.* 24:288.

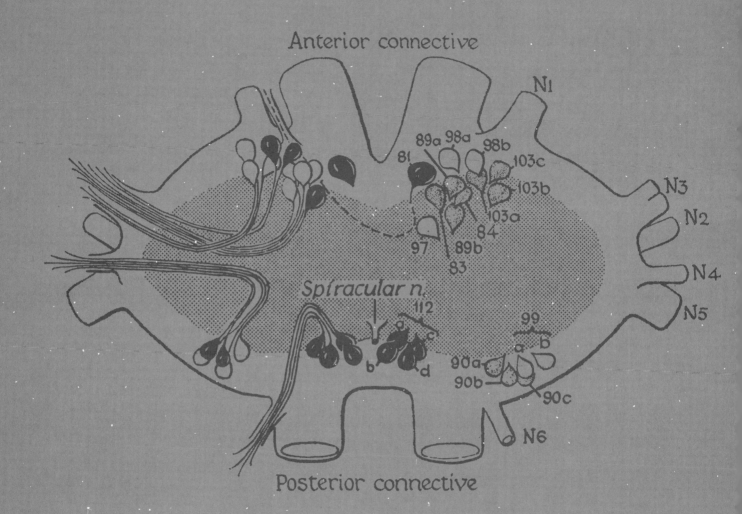

INVERTEBRATE NEURONS AND BEHAVIOR

Locations of identified neurons of locust (Schistocerca gregaria) *ganglia. Mesothoracic ganglion indicating flight motorneurons. Numbers refer to muscles innervated. From D. R. Bentley, A topological map of the locust flight system motor neurons.* J. Insect. Physiol. *16: 905–918.*

INVERTEBRATE NEURONS AND BEHAVIOR

Introduction

C. A. G. WIERSMA

THE FORBIDDING numbers of individual nerve cells that form the central nervous system of all vertebrates and that are usually still formidable, though considerably smaller, in those invertebrates that can be handled by present techniques, have led to the view that single-cell reactions are of little or no value for the analysis of brain function. But, as amply demonstrated in this volume, there are now many examples in both invertebrates and vertebrates of aggregations of cells, in which each member appears to have a circumscribed functional connectivity and mode of reaction. Especially in the invertebrates, a number of such cells can be unambiguously recognized and named, being present in all normal individuals of a given species. Identifiable interneurons on the sensory side are interesting, because their responses indicate which features in the environment are selected as important for the animal. Interneurons on the motor side often illustrate by their patterned output the existence of preprogrammed patterns of muscle contractions. In both cases modulation by other pathways than their main input channel can be studied, considerably enhancing the understanding of how central and local factors are integrated. Such influences are present in the most simple of all behavioral reactions, the monosynaptic reflex, and the study of cases in which input and output are provided by one or more identified neurons will be very rewarding, as indicated by several of the following chapters. Interneurons that are high-level links in complex behavioral reactions are good monitors of the mechanism by which such reactions are obtained, and their study may be expected to provide important insights for correlating neural and behavioral events. Concerning the applicability of these concepts to even more complex nervous systems, it looks at present that although they abide by very similar rules (see especially Bullock in this part), such systems do not have the same extreme degree of predestination. Even when no neuron is quite like any other in a large population of interneurons subserving a given overall function, the differences among neighboring cells might be so small that loss of a number of them would not result in noticeable change in output. In such cases a statistical approach might be the most fruitful methodology. But it should be noted that the underlying principle could be the same, enhancing the value of studying the simpler invertebrate systems.

The different chapters in this part are divided according to the main divisions of the higher invertebrates. There are sound reasons for such a treatment, because even though nervous systems are among the least variable organs in evolution and show pronounced overall similarities, differences, especially important to methods of investigation, are also present. For instance, in molluscs, the presence of peripheral connections that allow reflex activity in the absence of the central ganglia represent difficulties for evoked behavioral responses, especially because it is not yet very clear how these systems interact. The cephalopod molluscs are even more complex in that the number of their central nervous elements is very large (Young, 1971), so that their investigation presents the same type of difficulties as vertebrates and may thus form the same type of hindrance for investigations at the cellular level. On the other hand, they share with vertebrates a very readily established conditioned behavior for various stimuli and are furthermore distinguished by the presence of two memories, one for visual and one for tactile stimuli. Annelids appear to have integration of sensory input at a very early stage (Nicholls and Baylor, 1968), so that sensory fibers entering their central nervous systems become comparable to those of arthropod interneurons, representing large surface areas only. Insects, though resembling crustaceans in many respects, are usually capable of more complex acts and also demand, if for no other reason than their number and that of their investigators, special treatment. The reader should therefore be aware that certain findings in one group of animals are not necessarily transferable to others.

REFERENCES

NICHOLLS, J. G., and D. A. BAYLOR, 1968. Specific modalities and receptive fields of sensory neurons in CNS of the leech. *J. Neurophysiol.* 31:740–756.

YOUNG, J. Z., 1971. *The Anatomy of the Nervous System of Octopus vulgaris.* London: Oxford University Press.

30 Comparisons between Vertebrates and Invertebrates in Nervous Organization

THEODORE HOLMES BULLOCK

ABSTRACT Some differences between vertebrate and higher invertebrate nervous systems are itemized. These concern myelin, dendrites, somata and concentrations of them, and neuropile and electrical activity. The control of muscle and several other physiological and anatomical items are alluded to.

The phenomenon of identifiable neurons is particularly discussed. It embraces both contrast and commonality between vertebrates and invertebrates. The claim is advanced that identifiable neurons are not peculiar to invertebrates or lower vertebrates. Three propositions are made: (1) In both vertebrates and invertebrates equivalence of neurons comes in various degrees, down to a few and even single, unique cells. (2) Among masses of neurons, formerly thought to be so nearly equivalent that large population statistics best characterize their actions, nonequivalent subsets and units will be discerned and more and more hierarchies will emerge. (3) There is no known prohibition that limits the integrative level to which the preceding propositions can operate. Some corollaries are stated.

WHY A SYMPOSIUM on invertebrate nervous systems in 1972? It is long since they needed to be sold or to have attention called to their advantageous preparations.

For my part, the reason for proposing and defending this topic is the conviction that besides important similarities there are significant contrasts between invertebrates and vertebrates in the organization of the substrates of behavior. How these contrasts are reflected in behavior, even to identify the contrasts satisfactorily let alone explain their consequences or adaptive significance, is still an exercise in groping, as it was in our treatise some years ago (Bullock and Horridge, 1965, pp. 10–24). But, on the eclectic principle that we cannot wait for one level of inquiry to be cleared up before proceeding at higher levels, it was my conviction that the special features of nervous organization in the higher protostomes— annelids, arthropods, and molluscs—justified a considera-

tion of exemplary current work. Mutual exposure, I continue to hope, will build bridges across the unfortunate chasm often separating students of vertebrate and invertebrate nervous systems.

There are two hazards to this position, one on each side. Some colleagues, who are experts on vertebrates, expected the exercise to provide the elusive, ideal, "simple system." This is not altogether wrong, for an important spin-off, to be sure, is the advertisement of useful preparations. Other colleagues, on the other hand, including some of my fellow invertebrate zoologists, accuse me of infidelity for yielding to the siren song of vertebrates and even mammals! Such a one can no longer be trusted!

Rather than appealing to the record to make the case for following certain neural problems without regard to taxa or spines, this seems to be the place to plead guilty of the sin of comparing. I would like to lift up for emphasis just a few of the themes underlying our symposium and to place them in the perspective of general issues subsumed under the question: What can we say from vertebrate-invertebrate comparisons? Cohen (1970) has given a valuable essay on nearly the same theme in the last "Boulder volume."

I will not here catalog the extensive and profound *similarities* especially in signaling and integrative functions but will simply refer to Bullock and Horridge (1965). Time and again a feature discovered in one group and thought to be peculiar to it has then been found in the other—in both directions.

A. First, let us look at some *differences* between vertebrate and higher (more complex) invertebrate nervous organization.

1. Since some apparent exceptions have been resolved, it now seems correct to say that only vertebrates have true myelin—tightly and spirally wrapped, impeding current flow, forcing saltation and providing for conduction velocities five or ten times higher than the usual invertebrate maxima. (Shrimps achieve even higher velocity but

THEODORE HOLMES BULLOCK Department of Neurosciences, School of Medicine, and Neurobiology Unit, Scripps Institution of Oceanography, University of California, San Diego, La Jolla, California

by a different means and without true myelin as defined. Amphioxus may have some myelin.)

2. With respect to afferent and integrative branches of neurons it seems likely—and, possibly, profoundly significant—that invertebrates essentially lack dendrites on the structural criteria for those processes in vertebrates for which the term was introduced. Invertebrate afferent processes are typically not made of cytoplasm so closely resembling that of the soma as is true in the vertebrates. Though some have short side branches, they are not readily equated with the spines of vertebrate dendrites. Invertebrate afferent processes do not arise from the soma but are borne on the axon or its collaterals. They are less profusely branched than many vertebrate dendrites and probably receive far fewer presynaptic contacts from far fewer types of impinging neurons, *as a rule* (i.e., overlapping frequency histograms but distinct modes).

I recommend a more rigorous usage of the term dendrite, returning to purely cytological defining characters as in the historical origin of the concept. This is a change from our former definition (Bullock and Horridge, 1965; see also Bodian, 1962), which mixed classical structural criteria and receptive and nonimpulse-conducting physiological criteria. Bodian and I now agree that advances in knowledge make the physiological criteria untenable for purposes of definition, although they are probably very common properties of many dendrites. True (cytological) dendrites can be both presynaptic and postsynaptic; receiving is commonplace in both somata and axons as well as dendrites; impulses can arise in or invade some good dendrites. The opening statement of the paragraph is based on this neoclassical definition that excludes receiving input as a defining criterion for dendrite. Invertebrate afferent processes or proximal or dendriform ramifications should be so termed, according to whether the context refers to functional or structural features, reserving dendrite for the case where one intends to imply that specified cytological criteria have been met. The first sentence of this paragraph is not meant to imply that we know enough to be at all confident of it. In fact it is striking how little we know of the cytology of proximal ramifications, or the localization of synaptic contacts, or the origin and limitation of impulses in invertebrates.

3. A significant difference in role of the perikaryon or cell body is closely related to the differences between the vertebrates and invertebrates in the cytology and the source of the afferent processes. In the invertebrates unipolar neurons are dominant and axosomatic synapses rare; the cell body is not in the path of excitation, is not needed for neural functions, and primarily serves to maintain the processes. Correspondingly, we see conspicuous histologic differences in central tissue, charac-

teristic of these major divisions of the animal kingdom. Invertebrate ganglia have a central core of pure neuropil and a rind of cell bodies; vertebrates have gray matter in which the neuropil is usually broken up by cell bodies with axosomatic synapses. Here and there extensive neuropile, sparse in cell bodies (for example, molecular layer of cortex of cerebrum and cerebellum), is locally developed in vertebrates.

4. The ongoing electrical activity typical of higher centers in the two groups is possibly related to these general differences. Vertebrates have smooth, slow brain waves, as seen by macroelectrodes, almost without energy above 50 Hz, whether large mammal or tiny fish, whether with or without cortex, in highly ordered centers with geometrically regular arrays of cells and processes (cortex, hippocampus, etc.), and in centers not notable for such patterned structure (basal ganglia, hypothalamus, etc.). As seen by the same type of electrodes, invertebrates of many kinds in all three phyla sampled (annelids, arthropods, molluscs), with large or small brains, are typified by fast, spiky activity, very small slow waves, and plenty of energy above 100 Hz. (Cephalopods, for which we have only preliminary information, form a possible exception.) I suspect this is not due to size or number of cells but is a superficial manifestation of some important underlying difference inherent in items 2 and 3 above.

5. There are additional contrasts. The efferent control of muscle and integration in the periphery is a congeries of them. Degeneration and regeneration of motor axons is vastly different at least in some well-studied species, perhaps generally. Types of neurons, organelles and inclusions, types of glial cells, and vascular relations are generally different. The form and position of most sensory neurons is also, and the use of nonnervous sense cells in certain vertebrate receptors (acoustico-lateralis and taste) contrasts with the rule in invertebrates. Cephalization in invertebrates generally means movement of ganglia forward, whereas vertebrates add to existing rostral ganglia and shift functions forward—to oversimplify for the sake of brevity.

Note that the items of our list so far do not obviously predict, explain, or make sense in terms of general vertebrate-invertebrate differences in behavior or nervous achievement. The anatomical and physiological differences we know of to date do not themselves prove the existence of significant differences in integrative behavior, but they do strongly suggest that the mechanisms employed in integration are different in detail.

B. Let me now turn to a motif that played a prominent role in the chapters of this part and is often thought to be a feature of contrast between vertebrates and invertebrates, namely, the phenomenon of *identifiable neurons*. Identifiable neurons are unique cells that can be defined by distinguishing criteria, anatomical, chemical, and physiological and recognized in each individual of the species. Position of cell body is sometimes a good criterion, sometimes not. Processes and connections, cytoplasmic texture and pharmacologic type, synaptic responses, and spontaneous pattern are better criteria. Referring to the first of these, Pitman, Tweedle, and Cohen (1972) write "We are impressed by the precise reproduction of the branching pattern from one animal to the next for any given cell thus far examined." The number of identifiable neurons known is increasing every year and involves arthropods, molluscs, annelids, nematodes, and at least fish among the vertebrates.

The phenomenon of identifiable neurons embraces both contrast and commonality between vertebrates and invertebrates. In what degree or respect there is contrast is not yet known. The usual view today emphasizes the contrast, holding that vertebrates have large numbers of essentially equivalent and quite unreliable neurons operating probabilistically in masses (see Bullock, 1970). In this chapter I want to propose an important degree of *resemblance* between vertebrates and invertebrates, without denying at all that there may be important differences in respect to identifiable neurons.

Essentially, the claim I would like to advance is that (a) identifiable neurons are *not* peculiar to invertebrates or lower vertebrates but will also be found, with much greater difficulty of course, because of sheer numbers, in higher (more complex) forms and higher centers and that (b) other cells, though not individually identifiable, will be found to be essentially equivalent in larger and smaller groups down to very small numbers. For example, there are probably not many, if any, lateral geniculate or visual cortical cells with the same receptive field and the same size, threshold, contrast sensitivity, input-output function, time constants, and destination of axon. All of these would have to be the same for complete redundancy. A similar statement could be made for many other regions of the brain. Obviously, even if there were unique cells, this does not preclude overlapping properties; both can be graded from just barely to extensively.

Let me state the argument and its implications in the form of three propositions and their corollaries:

PROPOSITION 1: In both vertebrates and invertebrates equivalence of neurons comes in various degrees, from at least dozens of equivalent or virtually equivalent cells down to a few and even to zero equivalence (single, unique cells).

Corollaries:

(a) Nature does not require duplication of neurons.

(b) We may not assume as axiomatic that reserves against the contingency of injury or disease are an inevitable or ubiquitous good (in terms of survival value), stronger than opposing values, for example, for economy, adaptability or complexity.

(c) Large numbers of neurons do not necessarily mean high degrees of essential equivalence of neurons.

(d) Degrees of equivalence come in both dimensions: (1) gradation in numbers of fully equivalent cells and (2) gradation in equivalence (overlap of afferent fields and of efferent fields).

(e) Uniqueness of a unit may be due either to incomplete overlap of afferent field or of efferent field (or, of course, its transactional properties).

(f) All these statements apply to the highest and to the lowest nervous systems.

(g) The existence of potentially identifiable neurons is not antithetical to the highest vertebrate brains.

(h) Nothing is said or implied about the principles that govern or the facts of relative incidence of different degrees of equivalence in invertebrates vis à vis vertebrates. We cannot assume that even the direction of a preponderance, let alone the principles, are generally agreed to.

(i) The proposition predicts a one-way development of our knowledge, i.e., increasing numbers of instances of unique neurons and increasing discovery of differentiation among supposedly equivalent neurons even among vertebrates, and of course among invertebrates.

PROPOSITION 2: As masses of neurons are better understood and we sort them into nonequivalent subsets and units, not all of these will be hierarchically on a level but more and more hierarchies will emerge, with and without feedback. In other words, the prediction is that research on nucleus X will not only disprove the view that the population of cells is a mass of identical units with unspecified connections but will also reveal that some are higher than others in the same sense as Area 4 pyramidal cells are higher than ventral horn cells or Purkinje cells are higher than granule cells. The limitations of the concept of hierarchy are real, but the point here is that cells are likely to have nonparallel input and influence.

PROPOSITION 3: Working in from the periphery (*forward* from receptors, *backward* from effectors) to higher hierarchical levels, there is no known prohibition that limits the level to which the foregoing propositions can operate.

Corollaries:

(a) Nothing is said or implied about the principles or the facts of relative incidence of unique cells at higher and lower levels beyond the denial that such cells are a priori excluded.

(b) The proposition predicts a one-way development

of our knowledge, i.e., increasing numbers of instances of neurons with decreasing degrees of equivalence will be found *even at higher*, and of course at lower levels of all nervous systems.

These proposals represent severe extrapolations from the present state of knowledge. Nevertheless they are heuristic. The proposals not only suggest interesting lines of inquiry, but by their explicit statement, they also allow disproof or support of very general postulates by fragmentary evidence from special cases, as it comes in.

REFERENCES

BODIAN, D., 1962. The generalized vertebrate neuron. *Science* 137:323–326.

BULLOCK, T. H., 1970. Reliability in neurons. *J. Gen. Physiol.* 55:565–584.

BULLOCK, T. H., and G. A. HORRIDGE, 1965. *Structure and Function in the Nervous Systems of Invertebrates.* (2 vols.) San Francisco: W. H. Freeman & Company.

COHEN, M. J., 1970. A comparison of invertebrate and vertebrate central neurons. In *The Neurosciences: Second Study Program*, F. O. Schmitt, G. C. Quarton, T. Melnechuk, and G. Adelman, eds. New York: Rockefeller University Press, pp. 798–812.

PITMAN, R. M., C. D. TWEEDLE, and M. J. COHEN, 1972. The form of nerve cells: Determination by cobalt impregnation. *Symposium on Staining Techniques in Neurobiology*, Iowa City, Iowa, 1972. New York: Springer-Verlag, (in press).

31 An Invertebrate System for the Cellular Analysis of Simple Behaviors and Their Modifications

ERIC R. KANDEL

ABSTRACT   The abdominal ganglion of *Aplysia* controls several simple behaviors. One of these, the gill-withdrawal reflex, undergoes both short- and long-term behavioral modifications. Studies of the cellular bases of these behaviors and of their mechanisms of modification are reviewed.

Introduction

The study of the cellular mechanisms of behavior and its modification has recently received considerable impetus from two types of advances. One is the continuing improvement in techniques for analyzing the morphology, function, and biochemistry of individual nerve cells. The other is the utilization of the nervous systems of certain invertebrate animals (leeches, crayfish, lobsters, insects, and snails) that permit one to relate more directly the function of individual cells to behavior and to its modification. The central nervous systems of these invertebrates contain only 10^4 to 10^5 neurons, compared to the 10^{12} neurons of most vertebrates. Indeed some invertebrate ganglia contain only 1000 to 2000 neurons yet are capable of mediating several different behaviors so that the number of cells committed to a single behavior may be only 100 or less. Because of the small number of central neurons, the task of analyzing the functional architecture of behavior is significantly reduced. An additional advantage of invertebrates is that their nervous systems contain invariant cells that can be identified from preparation to preparation. The same cell can, therefore, be examined with the animal in different functional states.

As a result of these advantages, it is technically feasible in invertebrates to combine cellular neurophysiological and behavioral approaches and to attempt to clarify a number of interrelated neurophysiological and behavioral problems. For example, one can examine questions related to neuronal specialization. It is becoming evident that neurons may differ considerably in their physiological and biochemical properties. These differences are implied in the existence of uniquely identifiable cells. For what are these differences? Why are nervous systems not made up of identical units? By relating individual nerve cells to neural circuits controlling particular behaviors, one can examine how the distinctive properties of different types of neurons express themselves in mediating behavior. Using a combined approach, one can also examine behavioral questions. In particular one can analyze the relationships between behavioral modifications such as habituation and dishabituation, dishabituation and sensitization, and short-term and long-term memory. In the past, these processes have been compared only phenomenologically; with cellular techniques it may be possible to determine their mechanistic interrelationships.

By examining some of these questions I will here try to illustrate how studies of nerve cells and behavior can complement each other. I will first consider briefly some studies on neuronal specialization that have shown that different nerve cells and synapses have different plastic capabilities (for detailed reviews see Kandel and Spencer, 1968; von Baumgarten, 1970; Weight, this volume). Then I will consider recent experiments in a particular invertebrate, the marine mollusc *Aplysia*, which have delineated the neural circuits of two simple behaviors. One of the behaviors can be modified in several ways and provides an opportunity for examining the relationship of neuronal plasticity to behavioral modifications. Finally I will consider the mechanistic interrelationships between these behavioral modifications.

The plastic capabilities of neurons

Behavioral modifications require a prolonged, and sometimes a permanent change in neuronal function. How does this come about? As far as one can tell many connections between nerve cells are at least relatively specific.

ERIC R. KANDEL   Department of Neurobiology and Behavior, The Public Health Research Institute of the City of New York; The Departments of Physiology and Psychiatry, New York University School of Medicine

They appear to be determined largely by genetic and developmental processes, and are present soon after birth (for alternative views on connection specificity see Sperry, 1965; Hubel and Wiesel, 1965; Gaze and Keating, 1972; and Blakemore, this volume). Furthermore, learning can occur quite rapidly often in a matter of minutes or seconds. These findings suggest that at least some forms of learning do not require the development of new nerve cells or the sprouting of new connections. It seems more likely that these forms of learning involve a change in the functional properties of previously existing cells or in the effectiveness of previously existing connections (Konorski, 1948; Hebb, 1949; Eccles, 1953; Kandel and Spencer, 1968). If this view is correct, we would expect that nerve cells or their synapses should be capable of changing their functional effectiveness and should show a certain amount of plasticity following repeated or patterned stimulation. Stated another way, if learning involves a change in the functional properties of neurons, then the very least we must demand is that some neurons show plastic capabilities.

This prediction has now been well supported: Work in a number of laboratories during the last three decades has shown that chemical (but not electrical) synapses can undergo a variety of plastic changes lasting minutes and even hours in response to repeated presynaptic stimulation (Larrabee and Bronk, 1947; del Castillo and Katz, 1954; Hughes, Evarts, and Marshall, 1956; Martin and Pilar, 1964; Kandel and Tauc, 1965b; Bruner and Tauc, 1966b; Spencer and April, 1970; Sherman and Atwood, 1971). Some synapses show facilitation, some show depression, some show facilitation at one frequency of stimulation and depression at other frequencies. Moreover, changes in synaptic plasticity can be homosynaptic or heterosynaptic. In homosynaptic plastic changes (posttetanic facilitation and depression; low-frequency facilitation and depression) the plastic change in functional effectiveness occurs in a given pathway as a result of a change in the frequency of activity in that pathway. In heterosynaptic plastic change (heterosynaptic facilitation and depression) the plastic change occurs in a pathway as a result of activity in another pathway.

The plastic capacities of neurons are not limited to their synaptic region. In certain instances the endogenous spike rhythms of pacemaker neurons can be modulated for relatively long periods of time and in other instances the thresholds of nonpacemaker neurons can change (Strumwasser, 1965; Frazier, Waziri and Kandel, 1965; Kandel, 1967; Stephens, 1972).

How can we relate these well-established instances of synaptic plasticity and other existing mechanisms of neuronal plasticity to behavior and behavioral modifica-

tions? To bridge this gap it is first necessary to develop suitable preparations in which behavior can be studied at the cellular level so that one can specify the neural circuit, or wiring diagram, that controls the behavior. This involves knowing which nerve cells control the different components of a behavioral sequence, and how these nerve cells interact with each other and with their respective sensory and motor neurons. Ultimately, it will also be important to know the transmitter biochemistry and synaptic morphology of all the cells that participate in the behavior. These stringent requirements cannot, at the moment, be satisfactorily met in the mammalian nervous system, although a good beginning is being made in several non-mammalian vertebrates (see for example Rovainen, 1967; Diamond, 1971; Martin and Wickelgren, 1971). However, these requirements are increasingly being met in certain arthropods, annelids, and molluscs (Kennedy, Selverston, and Remler, 1969; Nicholls and Purves, 1970; Kupfermann and Kandel, 1969).

To serve as a suitable preparation for relating behavioral modifications to cellular and synaptic processes, an invertebrate should be capable of generating interesting behaviors, comparable to those of vertebrates, and have a nervous system accessible to cellular analysis. For these reasons my colleagues Irving Kupfermann, Harold Pinsker, Vincent Castellucci, Thomas Carew, and I have worked on the role of one ganglionic complex, the abdominal ganglion, in controlling the behaviors of the marine mollusc *Aplysia*.

Aplysia is a large opisthobranch mollusc that generally lives only one year but grows to more than a foot in length and to several pounds in weight during its brief lifetime. This large marine snail has the four morphotypic features (Figure 1A4) that characterize the molluscan phylum: (1) a *head* carrying a mouth, tentacles, and eyes; (2) a *foot*; (3) a *mantle cavity*—a respiratory space containing the gill, and the discharge site for several excretory glands and for the renal and digestive systems; and (4) the *visceral* hump, which lies below the mantle cavity, and contains the heart, stomach, digestive gland and sexual apparatus. The mantle cavity and gill of *Aplysia* are covered by a protective layer, the *mantle shelf*, which contains a small, vestigial shell (Figure 1A1). The posterior margins of the mantle shelf are flared into a fleshy spout, the siphon, that serves as an exhalent funnel for sea water and debris circulating through the mantle cavity.

The *Aplysia* central nervous system contains about 15,000 to 20,000 neurons that are grouped into 9 major (and several minor) clusters called ganglia (Figure 1C1). Each ganglion contains only 1000 to 2000 neurons, yet individual ganglia are capable of generating specific behaviors. Eight of the nine major ganglia are symmetrical

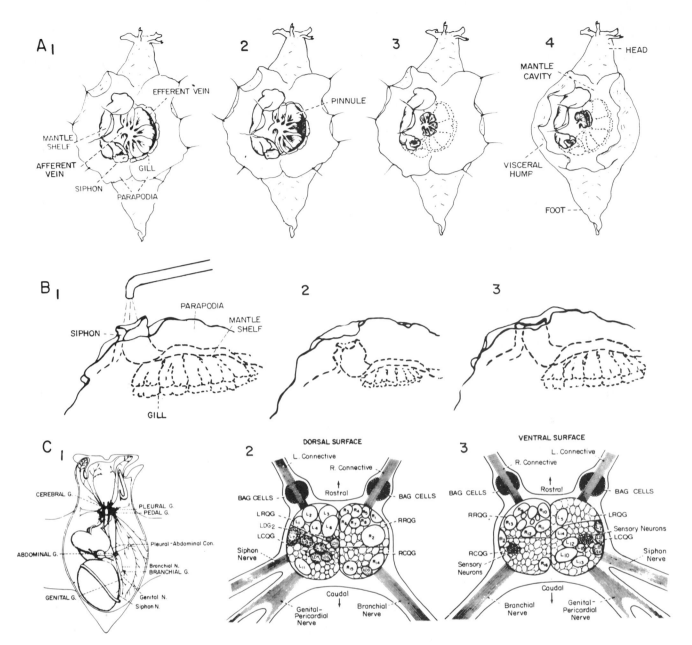

FIGURE 1 Different behaviors controlled by the abdominal ganglion of *Aplysia*. (A) Dorsal view of an intact *Aplysia* indicating the four features of the molluscan morphotype: Head, foot, mantle, and visceral hump (A4). To illustrate the three behaviors of the gill and siphon system, the parapodia and mantle shelf have been retracted so as to reveal the gill better. (A1) Position of organs in the unstimulated condition. (A2) The pinnule response to direct tactile stimulation of the gill. (A3) The defensive-withdrawal reflex to tactile stimulation of the siphon or mantle shelf. (A4) The centrally commanded gill, siphon and parapodial movements. (Modified from Kupfermann and Kandel, 1969; and Kupfermann, Pinsker, Castellucci, and Kandel, 1971). (B) Side view of an intact *Aplysia*, illustrating withdrawal of the siphon between the parapodia and the concomitant withdrawal of the gill into the mantle cavity during the defensive withdrawal reflex. The gill is seen through the para-

podia. (C) The central nervous system of *Aplysia*. (C1) Four paired symmetrical (circumesophageal) ganglionic masses form a ring around the esophagus (the paired buccal, cerebral, pleural, and pedal ganglia); one ganglionic complex, the abdominal ganglion, lies below the esophagus. The circumesophageal ganglia innervate the head and foot. The abdominal ganglion innervates the mantle and its organs (gill, siphon, osphradium, purple and opaline glands) and the visceral hump (intestine, stomach, heart, sexual apparatus). The paired buccal ganglia are not indicated in the drawing; they lie on the ventral surface of the buccal mass, approximately in the position of the cerebral ganglia (from Eales, 1921). (C2) Dorsal and (C3) ventral views of the abdominal ganglion, illustrating identified cells (modified from Frazier, Waziri, and Coggeshall, 1967). The motor neurons for siphon and gill and the two sensory clusters have been stippled.

(paired cerebrals, buccals, pleurals and pedals) and form a *circumesophageal* ring of ganglia that control the organs of the head and foot. A single, ninth, unpaired *infraesophageal* or abdominal ganglion controls the organs of the mantle cavity and visceral hump. In addition to the central ganglia there also exists a peripheral nervous system that controls local responses (Bethe, 1903; Bullock and Horridge, 1965; Kandel and Spencer, 1968; Peretz, 1970; Kupfermann, Pinsker, Castellucci, and Kandel, 1971).

The cells of the *Aplysia* central nervous system are exceptionally large, particularly in the abdominal ganglion, where the smallest cells (50 to 70μ) are comparable to the larger cells in the mammalian brain and the largest cells reach almost 1 mm in diameter (Figures 1C2, 1C3). These giant cells of opisthobranch nervous systems are the largest known nerve cell bodies. Because of their size, they are easy to impale with microelectrodes, and one can readily record from, or stimulate, several cells simultaneously. They can also be individually dissected for biochemical studies (Giller and Schwartz, 1968, 1971a, 1971b; McCaman and Dewhurst, 1970; Wilson, 1970; Lasek and Dower, 1971). Some cells are so characteristic in their position, appearance, and electrophysiological, morphological and biochemical properties (Figures 1C1, 1C2) that they can be reliably identified from preparation to preparation (Arvanitaki and Tchou, 1942; Arvanitaki and Chalazonitis, 1958; Frazier et al., 1967). These advantages have made *Aplysia* useful for studying the biophysical properties of nerve cells (Tauc, 1962, 1966; Strumwasser, 1967, this volume), the pharmacology and biochemistry of synaptic transmission (Tauc, 1967; Tauc and Gerschenfeld, 1961; Kandel, 1968; Kehoe, 1972), and the plastic properties of neurons and synapses (Kandel and Tauc, 1965a; Bruner and Tauc, 1966b; Strumwasser, 1967; Bruner and Kehoe, 1970). Recently it has also proved possible to map the interconnections between neurons in *Aplysia* ganglia on a cell-to-cell basis (Kandel, Frazier, Waziri, and Coggeshall, 1967; Kehoe, 1969; Gardner, 1971; Gardner and Kandel, 1972). This permits one to examine how cells are connected to each other and to the sensory and motor periphery, and to evaluate, at times quantitatively, what a given cell contributes to a particular behavior. I will consider below one well-studied example of a cellular analysis of behavior.

The gill and siphon of Aplysia: *Multiple controls of a single effector system*

The simple behaviors of vertebrates and invertebrates can be divided into two classes: (1) reflexive behavior; and (2) centrally initiated (or commanded) behavior (see Thorpe, 1963; Hinde, 1970). A reflex behavior is elicited by a specific sensory stimulus, and its response amplitude is a function of the intensity of the sensory stimulus. A centrally commanded behavior can occur spontaneously or can be triggered by a sensory stimulus. When triggered, the response amplitude is independent of stimulus strength. Because the amplitude and sequence of these responses are preprogrammed, centrally commanded behaviors are often called "fixed action patterns" (Tinbergen, 1951).

Some effector systems can be controlled both reflexly and by central commands. The siphon and gill system of *Aplysia* is of this type and provides a useful preparation for examining the interrelationships of reflexive and centrally commanded behaviors. This system also turns out to be useful for comparing behaviors mediated by the central nervous system and peripherally mediated local responses, characteristic of molluscs.

As described earlier, the gill of *Aplysia* is housed in the mantle cavity that is covered by a fold of skin, the mantle shelf (Figure 1A1). Posteriorly the mantle shelf forms a fleshy spout, the siphon, that normally protrudes from the mantle cavity between the parapodia, the wing-like extensions of the foot (Figures 1A1, 1B1). The gill is a large external respiratory organ composed of about 16 individual pinnules supplied by an afferent and an efferent vein. When the siphon or the mantle shelf is touched, both the siphon and the gill withdraw reflexly into the mantle cavity (Figures 1A3, 1B2). As the gill withdraws its volume is reduced, its halves tend to close and the pinnules separate slightly (Figure 1A3). This response has the features of a simple defensive withdrawal reflex; it has a short latency and its amplitude is a function of the intensity of the tactile stimulus. This reflex permits the animal to respond to a potentially threatening stimulus by withdrawing gill and siphon into the mantle cavity where they are protected by the mantle shelf and the parapodia. The contraction of the gill and siphon occur concomitantly so that observing one component of the system provides an index of the activity of the other component. This is convenient because gill movement can only be measured in restrained animals in which the parapodia and mantle shelf are retracted to expose the gill (Figure 2B). Siphon movements can be measured, however, in the completely unrestrained animal (Figure 1B).

In response to a central command from the abdominal ganglion, a superficially similar but actually quite different behavioral response occurs (Figure 1A4). It consists of a contraction with reduction of gill volume, accompanied by rostral rotation of the gill, bunching together of the pinnules and separation of the two gill halves (flaring) exposing the efferent vein. The parapodia

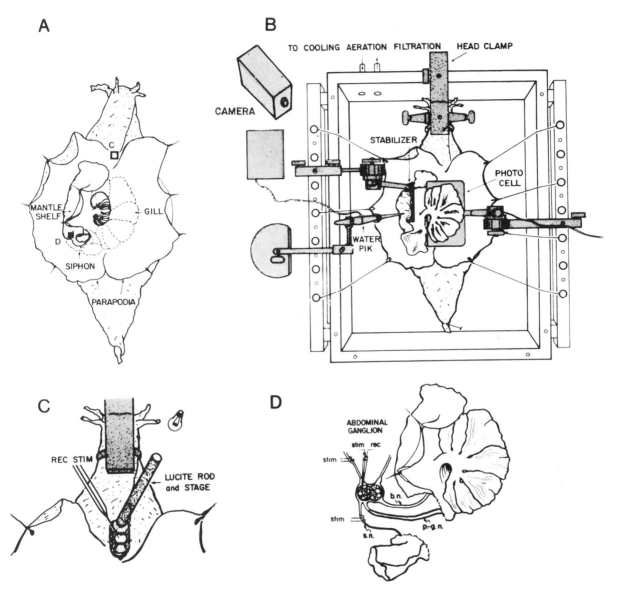

FIGURE 2   Different preparations used to study gill and siphon withdrawal reflexes. (A) Dorsal view of the intact animal with parapodia and mantle shelf retracted. (B) Experimental set-up for behavioral studies of gill withdrawal in the intact animal. *Aplysia* immobilized in a small aquarium containing cooled, filtered and aerated circulating sea water. The edge of the mantle shelf is pinned to a substage and a constant and quantifiable tactile stimulus is delivered by brief jets of sea water from a commercially available Water Pik. The behavioral responses were monitored with a photocell placed under the gill or by means of 16 mm movies (modified from Pinsker et al., 1970). (C) Preparation for simultaneous study of cellular and behavioral responses. Intact animal immobilized as in (B). A small slit is made at the most posterior end of the spermatic groove (see C in A), and the abdominal ganglion with its peripheral nerves and connectives intact is externalized and pinned to a lucite stage. Identified motor cells impaled with double-barrelled micro-electrodes for intracellular stimulation and recording. Gill movements recorded with photocell and camera (from Kupfermann et al., 1970). (D) Test system of gill-withdrawal reflex in isolated abdominal ganglion. In some cases, the abdominal ganglion remained connected to the gill via the branchial (b.n.) and pericardial-genital (p-g.n.) nerves and to the siphon via the siphon nerve (s.n.). Intracellular recordings are obtained from motor neurons (usually L7); tactile stimuli are applied to the siphon skin or electrical stimuli to the siphon nerve or connective. (Modified from Castellucci et al., 1970.)

also often contract, forcing sea water through the mantle cavity. These spontaneous contractions sometimes recur at intervals of several minutes and serve to pump sea water through the mantle cavity thereby aerating the gill and cleansing the cavity.

In addition to these two behavioral sequences, individual pinnules of the gill contract when a tactile stimulus is applied to them directly (Figure 1A2). The receptive field of the pinnule response is centered on the pinnule itself. The strongest contraction occurs immediately under the site of stimulation (Peretz, 1970; Kupfermann, Pinsker, Castellucci, and Kandel, 1971).

To relate these behaviors to cellular neuronal mechanisms four different preparations have been used. The *first* is a completely intact unrestrained animal in which siphon responses to natural stimuli can be measured (Figures 1B, 2A). The *second* preparation is an intact animal restrained in a small aquarium, containing cooled circulating sea water, in which gill responses to natural stimuli can be measured by means of a photocell. For a more detailed and qualitative analysis, motion pictures can be taken of gill movements (Figure 2B). Quantifiable and constant tactile stimuli are provided by jets of sea water from a Water Pik. The *third* preparation is an intact animal prepared as above, but with a slit in its neck (Figure 2C) through which the abdominal ganglion is externalized and gill responses and intracellular responses from single nerve cells are recorded or photographed simultaneously. The *fourth* preparation consists of the isolated abdominal ganglion with or without the siphon skin and gill and can be used as a test system of the behavior for more detailed cellular studies (Figure 2D).

GILL MOTOR NEURONS   *The movements produced by different motor neurons*   The motor components of the reflex response and of the centrally commanded movement (Figure 2C) were investigated in the intact preparation with a slit in the animal's neck (Kupfermann and Kandel, 1968, 1969; Kupfermann, Carew, and Kandel, 1973). One or more nerve cells were impaled with microelectrodes and stimulated with intracellular current pulses while the movements of the different external organs of the mantle cavity were monitored with a photocell and, in some cases, with a 16mm movie camera. Six cells were found—five in the left caudal quarter-ganglion ($L7$, LD_{G1}, LD_{G2}, $L9_{G1}$, and $L9_{G2}$) and one in the right caudal quarter-ganglion (RD_G)—that produced movements of the gill (Figures 1C2, 1C3, 3A; see also Peretz, 1969). Stimulation of three cells ($L7$, LD_{G1}, and LD_{G2}) produced particularly large and brisk gill movements. The other three cells ($L9_{G1}$, $L9_{G2}$, and RD_G) produced smaller contractions.

Stimulation of L7 typically elicits a gill contraction that results in a great reduction in volume and concomitant closure of the two gill halves with separation of the pinnules at their base (Figure 1A3). Cells $L9_{G1}$ and $L9_{G2}$ produce similar but weaker movements. Like L7, stimulation of LD_{G1} also causes a large contraction with a great reduction of volume, but it also causes rostral rotation of the gill and bunching together of the pinnules. Stimulation of LD_{G2} produces flaring of the two gill halves (Figure 1A4); RD_G produces similar but weaker movements. Thus the two different gill behaviors are based on the characteristic movements produced by the three major motor cells L7, LD_{G1}, and LD_{G2}. The reduction in gill volume and separation of the pinnules with no flaring or even antiflaring of the gill halves, characteristic of the defensive reflex, is based upon the combined actions of all three cells with the L7 actions slightly predominant. The reduction in gill volume with rostral rotation, bunching of the pinnules, and flaring of the two gill halves characteristic of the central command is based upon the unopposed actions of LD_{G1} and LD_{G2} with the actions of LD_{G2} usually predominant.

Four neurons in the left caudal quarter-ganglion (LB_{S1}, LB_{S2}, LD_{S1}, and LD_{S2}) were found to be specific for siphon contraction. In addition, the gill motor neuron L7 also causes contraction of the siphon as well as of the mantle shelf. The motor component of the total defensive withdrawal reflex thus consists of at least ten cells, organized in a hierarchical pattern. Five of the ten motor cells produce movements limited to the gill (LD_{G1}, LD_{G2}, RD_G, $L9_{G1}$, and $L9_{G2}$). Four cells produce movements largely limited to the siphon (LB_{S1}, LB_{S2}, LD_{S1}, and LD_{S2}) and one cell (L7) produces movements of the gill, siphon, and mantle shelf. Thus, the motor components of this reflex consist of individual elements with both a restricted and an overlapping distribution.

Neuromuscular connections of gill motor neurons   All six motor cells to the gill send their axons out of the ganglion via peripheral nerves to innervate the gill. To determine whether the major motor neurons produced their effect directly on gill muscle or via peripheral interneurons in the gill itself, Carew, Pinsker, Rubinson, Schwartz, and Kandel (1973) studied the neuromuscular connections of the major motor cells, L7, LD_{G1}, and LD_{G2}. They impaled motor neurons in the ganglion, and individual muscle fibers in the walls of the efferent or afferent blood vessel of the gill, and found that action potentials in each of the motor neurons produced excitatory junction potentials (EJPs) in the muscle fibers (Figures 3B1, 3B2). These EJPs followed the motor neuron spikes one for one up to relatively high frequencies (15/sec). The latencies of the EJPs were constant even in high Ca^{++} solutions that would raise the threshold of interneurons.

These results indicated that these major motor neurons innervate gill muscle directly. The junctions are chemical and are reversibly blocked by sea water solutions containing high Mg^{++}. The neuromuscular connections of the three minor motor neurons have not yet been examined.

The chemical transmitters produced by LD_{G1} and L7 As a beginning in the study of the transmitter biochemistry of the motor cells for gill movement, Carew, Schwartz, and Kandel (1972) examined the biochemical and pharmacological properties of LD_{G1} and L7. They found that the gill contraction caused by LD_{G1} is reversibly inhibited by hexamethonium, which blocks the excitatory actions of acetylcholine (ACh) in *Aplysia* (Tauc and Gerschenfeld, 1961). By contrast hexamethonium did not affect the contractions produced by L7. These results suggested that LD_{G1} is cholinergic and that L7 is not. Moreover, when the cells were dissected from the ganglion and assayed for the presence of choline acetyl-transferase, the enzyme required for the synthesis of ACh, cell LD_{G1} was found to contain the transferase whereas L7 did not. Next, radioactive choline was injected into the cell body of L7 or LD_{G1} to determine the amount of radioactive choline that each cell converted to ACh. In L7 the ^3H-choline was converted to substances other than ACh, predominantly phosphoryl choline, but in LD_{G1} 85% of the choline was converted to acetylcholine. Taken together, the pharmacological and biochemical data support the idea that LD_{G1} is cholinergic and L7 is not.

Some muscle fibers were found to receive EJPs from each of these two cells (Figure 3B3). This finding suggests that two excitatory motor neurons, using different chemical transmitters, can make direct connections with the same muscle fiber. However, the possibility that L7 and LD_{G1} innervate different fibers, which are electrically coupled, is not ruled out.

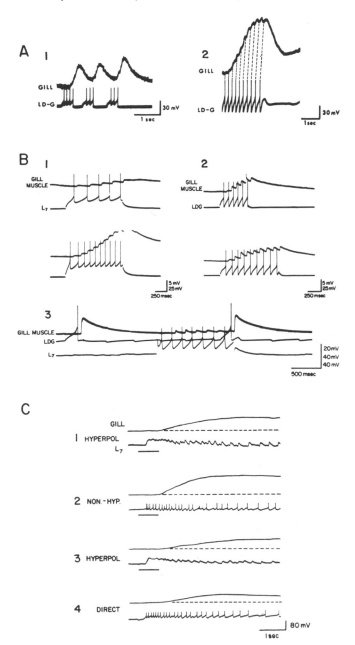

FIGURE 3 Responses of gill and individual gill muscle fibers to stimulation of individual motor cells and sensory input. (A) Response of the gill. Upper traces represent the output of a photocell placed under the gill, lower traces intracellular recordings from a motor neuron causing gill contraction. (A1) Smooth contraction produced by small number of spikes in cell LD_{G1}. (A2) Individual twitches produced by LD_{G1} after it was first stimulated at high frequency to potentiate its effects. (From Kupfermann and Kandel, 1969.) (B) Intracellular recordings of excitatory junction potentials (EJPs) in gill muscle fibers (top trace) due to intracellular stimulation of L7 or LD_{G1} (bottom traces). (B_1) Spikes in L7 were elicited by depolarizing pulses of different intensities. The EJPs in the gill followed L7 spikes one for one with a fixed latency at both low frequencies (5/sec) and higher frequencies (10/sec). At the higher frequency the EJPs potentiate. (B2) Response of gill muscle fiber to intracellular stimulation of LD_{G1}. (B3) Intracellular recording from LD_{G1}, L7, and a single muscle fiber showing two types of EJPs, one from LD_{G1} and the other from L7. (From Carew, Pinsker, Schwartz, Rubinson and Kandel, 1973.) (C) The sensory input to the gill motor neuron. The gill-withdrawal reflex was elicited every 5 min by a jet of sea water (indicated by solid line, under L7 record) applied to the siphon. Hyperpolarization of L7 on alternate trials (C1 and C3) unmasked the large excitatory synaptic input that normally underlies the repetitive discharge (C2). Comparisons of hyperpolarized to nonhyperpolarized trials showed that the gill contraction was reduced by about 40%. This reduction was approximately equal to the size of the gill contraction produced by L7 when it was fired intracellularly by a long depolarizing pulse (C4) that caused L7 to fire in a pattern comparable to that produced by the normal excitatory input. (From Kupfermann, Pinsker, Castellucci, and Kandel, 1971.)

THE SENSORY INPUT TO GILL MOTOR NEURONS CAUSING DEFENSIVE WITHDRAWAL REFLEX   To study the sensory input to motor neurons Kupfermann and Kandel (1969) and Kupfermann, Pinsker, Castellucci, and Kandel (1971) stimulated the body surface of the animal with tactile stimuli while recording from gill motor neurons, one at a time. These tactile stimuli produced excitatory postsynaptic potentials (EPSPs) in each of the motor neurons that caused the cell to fire repetitively (Figure 3C). The receptive fields of the motor neurons were identical with that for the gill-withdrawal reflex.

To examine the EPSPs the motor neuron was hyperpolarized to prevent spike generation. In this way, the gill contraction to a constant siphon stimulus could be measured under two conditions: (1) with the motor neuron at its resting potential and firing normally (Figure 3C2), and (2) with the motor neuron selectively removed from the reflex pathway by hyperpolarizing it to prevent spike generation (Figures 3C1, 3C3). By seeing how much the gill contraction was reduced by the temporary removal of one cell from the reflex pathway, it was possible to determine the relative contributions of the individual motor neurons to the total gill-withdrawal reflex. This could be checked by showing that when the motor neuron was fired by intracellular depolarization, in a manner that simulated its synaptic activation during the reflex (Figure 3C), the gill movement produced was comparable to that component of the total reflex response that was removed when the motor neuron was hyperpolarized. By this means, the average contribution of L7 was estimated to be 35 to 40%, and that of LD_{G1} about 30 to 35%. The contractions produced by LD_{G1} and L7 were additive. Thus, these two cells account for a major portion of the total reflex. Each of the other motor neurons (LD_{G2}, RD_G, $L9_{G1}$, and $L9_{G2}$) contributes less to the reflex contractions, but their quantitative contributions have not yet been determined.

The reflexly produced EPSPs unmasked in these motor neurons by hyperpolarization (Figures 3C1, 3C3) are due to the activity of both monosynaptic and polysynaptic pathways. The monosynaptic contribution from the siphon and mantle shelf is mediated by two distinctive clusters of sensory neurons. One of these is located in the left caudal quarter-ganglion near the motor neurons (Figure 1B2) and conveys tactile information from the siphon skin via the siphon nerve. The other is located in the right caudal quarter-ganglion and conveys tactile input from the mantle shelf via the branchial nerve (Castellucci, Pinsker, Kupfermann, and Kandel, 1970; Byrne, Castellucci, and Kandel, 1973). Only the cluster containing the siphon sensory cells has been studied in detail. Each of these sensory neurons has about the same threshold to mechanical stimulation and a rather discrete receptive field. At least some of the fields are invariant. The receptive fields fall into three size classes: Some fields are small, covering about 20% of the siphon skin; some are medium sized, covering about 40% of the siphon skin; and two are large, covering 80% of the siphon. Thus, like the motor neurons, the sensory neurons are arranged in a field size hierarchy.

The sensory neurons from the siphon skin connect to the motor neurons directly (Figure 4A). In addition, they connect to three types of interneurons (Figure 4B). Two of these are excitatory (Figure 4B1), and produce EPSPs in motor neuron L7. One interneuron, cell L16, is inhibitory (Figure 4B2). It inhibits the sensory neurons as well as other interneurons (Castellucci and Kandel, in preparation).

SYNAPTIC INPUT TO GILL MOTOR NEURONS DURING CENTRALLY COMMANDED GILL MOVEMENTS   During defensive withdrawal all the gill motor cells fire synchronously (Figure 4C1) but during the centrally commanded movement, they fire in a specific sequence (Figure 4C2). The central command manifests itself at the level of the motor neurons as a characteristic PSP burst (Kupfermann and Kandel, 1968, 1969; Peretz, 1969). This is attributed to the activity of several closely coupled interneurons, one of which is Interneuron II (Kandel et al., 1967). The PSP burst does not require sensory feedback from the gill for its generation and can occur in the completely isolated abdominal ganglion (Waziri and Kandel, 1969). In LD_{G1}, LD_{G2}, and RD_G, the PSP burst consists of EPSPs that produce excitation. In LD_{G1}, the EPSP burst is preceded by a brief period of inhibition (Figure 4C2). In the three other motor cells (L7, $L9_{G1}$, $L9_{G2}$), the PSP burst consists of IPSPs that produce inhibition followed by rebound. The rebound excitation adds to the last stages of the contraction. Synaptic inhibition therefore provides a means for sequencing the firing of the motor neurons so that the discharge of LD_{G2} and RD_G precedes that of LD_{G1}, which in turn precedes that of L7, $L9_{G1}$, and $L9_{G2}$. As a consequence, a pumping action can be produced that seems to circulate blood through the gill while the concomitant parapodial contractions move fresh water through the mantle cavity.

The siphon motor neurons are also excited in synchrony by the defensive reflex and differentially excited (LD_{S1} and LD_{S2}) and inhibited (L7, LB_{S1}, and LB_{S2}) during the central command.

NEURAL CIRCUITS OF THE DEFENSIVE-WITHDRAWAL REFLEX AND THE CENTRALLY COMMANDED MOVEMENTS   The above results allow a wiring diagram for both reflex and centrally commanded responses to be drawn

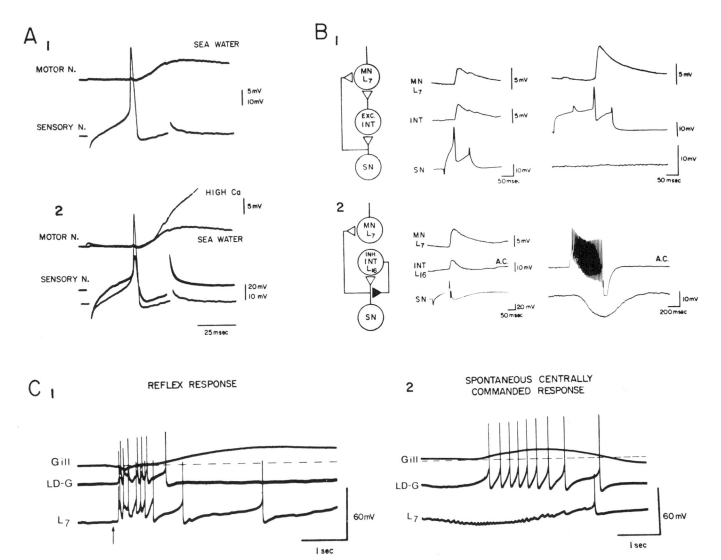

FIGURE 4   The synaptic input for the (A–C1) gill-withdrawal reflex and for the (C2) centrally commanded gill movements. (A) The EPSP in gill motor neuron L7 produced by intracellular stimulation of a sensory neuron. (A1) Fast sweep recording to illustrate the short latency of the EPSP. (A2) Superimposed traces of EPSP elicited in artificial sea water with normal Ca^{++} content (9.1 mM) and in a solution of high Ca^{++} content (91 mM). Because of the increased threshold in high Ca^{++}, the intracellular stimulation had to be increased considerably, resulting in a large current artifact that could not be completely balanced with the Wheatstone bridge; the gain was therefore reduced (see calibration). The latency of the EPSP remained the same. (From Carew, Castellucci, and Kandel, 1971.) (B) Central connections of sensory neurons (SN) with interneurons.

(B1) Intracellular stimulation of sensory neuron produces an EPSP in an excitatory interneuron (middle). An action potential in the excitatory interneuron, in turn, causes an excitatory potential in the motor neuron L7 (right). These connections are summarized in the drawing on the left. (B2) Intracellular stimulation of a sensory neuron produces an EPSP in inhibitory interneuron L16 (middle column). Repetitive discharge of the interneuron inhibits this and other sensory neurons as shown by the hyperpolarization (right). These connections are summarized in the drawing on the left. (From Castellucci and Kandel, in preparation.) (C) Comparison of synaptic input to L7 and LD_{G1} during gill-withdrawal reflex (C1) and centrally commanded contraction (C2). (From Kupfermann and Kandel, 1969.)

(Figures 5A, 5B1). The mechanoreceptor sensory neurons involved in the defensive withdrawal reflex make excitatory connections with both gill and siphon motor neurons directly and indirectly via excitatory interneurons (Figure 5B1). In addition, they connect with an inhibitory interneuron which feeds back on the sensory neurons and forward to the excitatory interneurons. In the spontaneous movements (Figure 5A) the central command elements excite some gill and siphon motor neurons and inhibit others. The central command interneurons can fire spontaneously but also receive either a direct or indirect excitatory input from the siphon and the mantle shelf. Thus, as a result of multiple synaptic inputs, the different members of the same population of motor cells can be switched between either a simple reflex or a centrally commanded behavior.

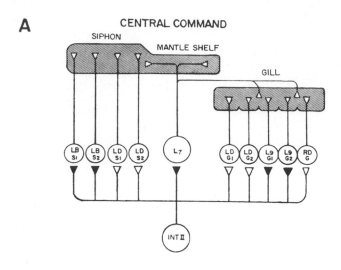

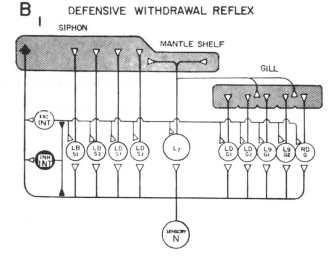

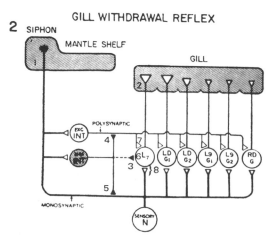

Two main findings are illustrated by these circuit diagrams. (1) The various motor neurons of the mantle organs are independent, parallel elements that are not interconnected with each other. (2) Both motor neurons and sensory neurons are arranged in a hierarchical manner. Some sensory neurons receive from small areas of siphon skin, some from medium sized areas, and one cell receives from the whole siphon. Similarly, some motor neurons move only the gill, some move only the siphon, and one moves the siphon together with the gill and mantle shelf. As a result of these two factors, reflex actions or central commands can play on different combinations of a common motor pool to generate different behavioral combinations and different temporal sequences of the same combination. This permits an effector organ, or a whole effector system, to be controlled in a number of different ways. In addition, the whole mantle system (gill, siphon, and mantle shelf) can also be controlled in different combinations with other effector systems, such as heart and kidney (Mayeri, Kupfermann, Koester, and Kandel, 1971).

Whereas Figure 5B1 illustrates the total defensive withdrawal reflex, Figure 5B2 illustrates the neural circuit of the gill component on which is based most of the work considered below.

THE RELATIONSHIP OF THE PINNULE RESPONSE TO THE GILL-WITHDRAWAL REFLEX Since both pinnule response and gill-withdrawal reflex involve movements of the gill, it became of interest to compare them to see whether they involve similar or different processes (Peretz, 1970). The gill-withdrawal reflex is elicited by stimulating not the gill itself but a remote receptive field

FIGURE 5   Neural system controlling centrally commanded gill movements and defensive withdrawal reflex. Dark triangles indicate inhibition; light triangles, excitation. (A) Centrally commanded gill movements. Interneuron II probably represents several closely coupled interneurons which excite some gill and siphon motor neurons (RD_G, LD_{G1}, LD_{G2}, LD_{S1}, LD_{S2}) and inhibit others ($L9_{G1}$, $L9_{G2}$, $L7$, LB_{S1}, LB_{S2}). (B) Defensive withdrawal reflex. (B1) The total reflex, illustrating both siphon and gill components. (B2) The gill-withdrawal component of the reflex. The sensory input from the siphon is mediated directly and via interneurons. (Exc. Int. and Sensory N represent classes of cells.) Motor neurons and inhibitory interneurons are identified cells. Small numbers indicate possible sites of plasticity. The large triangles represent the output of major motor cells that contribute importantly to the gill withdrawal, small triangles that of cells producing weaker effects. In this, and subsequent figures, synapses are schematically indicated as being on the cell body. Actually they are on the initial segment of the axon and its branches. (Modified from Kupfermann and Kandel, 1969; Kupfermann, Carew, and Kandel, 1973.)

consisting of the siphon and mantle shelf (Figures 1A3, 1B1). This reflex is centrally mediated. When individual central motor neurons are hyperpolarized and prevented from spiking so that their contributions to the reflex are eliminated, the total reflex is significantly reduced (Figure 3C1). When chemical synaptic transmission in the abdominal ganglion is reversibly blocked by bathing it in a solution of isotonic magnesium chloride, the reflex response to moderate tactile stimuli is fully but reversibly blocked (Kupfermann et al., 1971). Surgical removal of the ganglion also eliminates the gill-withdrawal reflex. However, when the siphon is surgically isolated from the rest of the animal and left attached to the abdominal ganglion by only the siphon nerve (Figure 2D), so that the central pathways are preserved while parallel peripheral pathways via the skin are totally eliminated, tactile stimulation of the siphon still produces a brisk gill-withdrawal reflex.

The pinnule response differs from the gill-withdrawal reflex in several ways. First, the most effective site for eliciting the pinnule response is not a remote site but the gill pinnule itself (Figure 1A2). Second, although the pinnule response can be elicited in the intact animal, it can also be elicited when all central ganglia are removed. This indicates that central pathways are not necessary for the pinnule response. The response is mediated by peripheral pathways either by the neurons in the gill, by collaterals of central neurons, or by direct stimulation of the muscle itself (Peretz, 1970; Kupfermann, Pinsker, Castellucci, and Kandel, 1971).

Neural plasticity and behavioral modifications of the gill-withdrawal reflex: Habituation and dishabituation

Since gill (and siphon) movements occur spontaneously as well as reflexly, they are, in principle, amenable to both operant and classical conditioning training procedures. To begin the study of behavioral modifications of this system, Pinsker, Kupfermann, Castellucci, and Kandel (1970) and Carew, Pinsker, and Kandel (1972) focused on the reflex response and utilized an even simpler behavioral paradigm, habituation and dishabituation.

Habituation, sometimes considered the most elementary form of learning, is a decrease in a behavioral response that occurs when an initially novel stimulus is presented repeatedly (Figure 6A). When a novel stimulus, such as a sudden noise, is presented for the first time, attention is immediately drawn to it and one's heart rate and respiratory rate may increase. However, if the same noise is repeated, one's attention, as well as

one's bodily responses, gradually diminish. As a result of habituation, one can become accustomed to initially distracting sounds and work effectively even in a noisy environment. One also becomes habituated to the clothes he wears and to his own bodily sensations. These enter awareness rarely, only under special circumstances. In this sense habituation is learning to ignore recurrent external or internal stimuli that have lost novelty or meaning. Habituation develops not only in reflex response systems but in instinctive response systems as well, with the result that an animal can respond more effectively to other stimuli, such as those necessary for his survival and that of his species (Peeke, Hertz, and Gallagher, 1971). For example, a fish will defend his territory against a nonspecific intruder. Upon repeated exposure to an intruder, to whom he has no access, the fish will gradually suppress his aggressive behavior, permitting him to respond more frequently to sexually prepared females.

Besides being important in its own right, habituation is also frequently involved in more complex learning, which consists not only in acquiring new responses but also in learning to reduce errors by eliminating responses to inappropriate stimuli.

Once a response is habituated, two processes can lead to its restoration: (1) *spontaneous recovery*, which occurs as a result of withholding the habituating stimulus (Figure 6A); and (2) *dishabituation*, which occurs as a result of changing the stimulus pattern, for example, by presenting another, stronger stimulus to another pathway (Figure 6B).

With repeated stimulation of the siphon at half-minute to three-minute intervals, the gill-withdrawal reflex undergoes progressive habituation to roughly 30% of control. Typically the major decrement occurs as a result of the first five stimuli (trials). After a training session of ten trials, there is a rapid phase of recovery from habituation lasting 10 to 20 min that accounts for about 75 to 85% of reflex restoration. This is followed by a slow and highly variable return of the reflex response to its original level, often lasting many hours. But almost immediate restoration of reflex responsiveness can be obtained by presenting a single strong dishabituatory stimulus to another part of the receptive field (Figure 6B).

RELATION OF HABITUATION OF THE WITHDRAWAL REFLEX TO HABITUATION OF THE LOCAL PINNULE RESPONSE AND TO CENTRALLY COMMANDED GILL CONTRACTION Peretz (1970) has recently shown that the pinnule response also undergoes habituation. Kupfermann et al. (1971) therefore examined whether habituation of the gill-withdrawal response and of pinnule

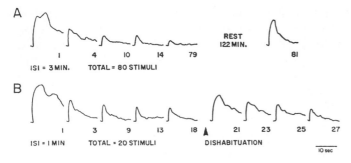

FIGURE 6 Short-term habituation and dishabituation of the gill-withdrawal reflex. Behavioral data from the intact animal (as in Figure 2B) illustrating habituation, recovery, and dishabituation. A contraction of the gill results in an upward deflection of the polygraph record (A1). With repeated stimulation at 3 min (*top*) or 1 min (*bottom*) intervals, gill response habituated to roughly 30% of control as is shown in the two separate habituation runs on the left. When recovery from habituation was tested following complete rest, without interposed stimuli, almost complete recovery occurred following 2 hr of rest (right half of A). Following rehabituation of the response (left half of B), a single strong stimulus presented to another part of the receptive field produced dishabituation (right half of B). ISI = interstimulus intervals. (From Pinsker et al., 1970.)

withdrawal are related. They found that habituation of the gill-withdrawal reflex did not affect the pinnule response and that habituation of the pinnule response did not affect the magnitude of the gill-withdrawal reflex elicited by siphon stimulation. Although one cannot rule out a small contribution from peripheral pathways to the habituation of the reflex response, the data indicate that the two behavioral modifications are independent and that habituation of the gill-withdrawal reflex is largely if not totally dependent on central pathways. Indeed the failure of generalization of habituation from one pathway to another provides further support for the independence of the central and peripheral pathways in the gill.

Similarly, Pinsker et al. (1970) examined the interaction of the reflex response and the centrally commanded contractions. They found that the centrally commanded gill contractions were of identical size before reflex habituation, during maximal response decrement, and following recovery. These experiments provide indirect evidence that habituation of the gill-withdrawal reflex is not due to muscle fatigue, as directly shown below.

LOCUS AND NATURE OF THE FUNCTIONAL CHANGES IN THE NEURAL CIRCUIT OF THE GILL-WITHDRAWAL REFLEX DURING HABITUATION AND DISHABITUATION   The schematic drawing of the circuit of the gill-withdrawal reflex

(Figure 5B2) shows that even within a simple neuronal circuit such as this, there are several possible sites (Figure 5B2: 1 to 8) for plasticity that could account for behavioral modifications.

The first question to answer was whether or not habituation involved peripheral changes, external to the CNS, either in the muscle (Figure 5B2: 2), in the sensory receptors, or in the skin (Figure 5B2: 1). The role of sensory adaptation was investigated by recording extracellularly from single afferent fibers in peripheral nerves and intracellularly from sensory neurons (Kupfermann et al., 1970; Castellucci et al., 1970; Byrne, Castellucci, and Kandel, 1973). When the skin was stimulated at intervals that produce habituation, no significant changes were found in the response of the sensory neurons. Additional evidence that changes in receptor responsiveness could not explain habituation was provided by experiments in which stimulation of the receptors was circumvented by applying a constant electrical stimulus to the afferent nerve. Under these circumstances rapid reflex habituation was still obtained (see below, Figure 7B).

The next possibility was that habituation occurred at either the neuromuscular junction or the muscle itself (Figure 5B2: 2). However, direct stimulation of the motor neurons or efferent nerves at intervals that produced habituation also did not produce significant decrement of gill contraction. Moreover the size of a directly evoked gill response produced by intracellular stimulation of a gill motor neuron was the same before and after habituation (Kupfermann et al., 1970).

Habituation must therefore result from some change within the central nervous system (Figure 5B2: 3 to 7). To determine its nature, Kupfermann et al. (1970) recorded from the major motor neurons in the intact animal (Figure 2C) during the course of habituation. They found that in the rested state, tactile stimulation of the siphon produced large and effective excitatory synaptic potentials (EPSPs) that caused repetitive discharges in the identified motor neurons that innervate the gill. When the tactile stimulus was repeated, however, the number and frequency of evoked spikes decreased (Figure 7A1). Recovery of reflex responsiveness produced either by rest (Figure 7A2) or a dishabituatory stimulus (Figure 7A3) was associated with an increase in the number and frequency of spikes in the gill motor neurons. Since the gill contraction is determined by the output of several motor neurons, the magnitude of the gill contraction was not perfectly correlated with spiking in any one motor neuron. The correlation was reasonably close, however, and a decrease of one or two spikes in the initial discharge following tactile stimulation of L7 or LD_{G1} was often associated with a measurable decrease of the gill contraction. This is consistent with the results obtained with intracellular stimulation of these cells in the same experiments where the magnitude of the gill contraction was also found to be quite sensitive to changes in spike output of the motor cells.

To determine the cellular changes underlying the alterations in firing frequency of the motor neuron, the motor cells were hyperpolarized and the changes in PSP amplitude were examined. These experiments showed that during habituation there was a progressive decrease in the amplitude of the PSPs evoked in the motor neuron by siphon stimulation that accounted for the changes in firing rate (Figure 7B1). With rest (Figure 7B2) and

dishabituation (Figure 7B3) there was a restoration of the PSP amplitude (Castellucci, Pinsker, Kupfermann, and Kandel, 1970).

The decrease in the PSP during habituation was found to be due to a decrease in excitatory synaptic transmission (Figure 5B2: 7, 8) and not to a progressive increase of an underlying inhibitory postsynaptic potential (Figure 5B2: 3). Similar PSP decrement occurred when the membrane was hyperpolarized well beyond the equilibrium potential of the spontaneous IPSPs in these motor neurons. Additional evidence that IPSP potentiation does not produce the PSP decrement was provided by experiments on monosynaptic inputs discussed below. The same experiments also made it unlikely that the PSP decrement is due to presynaptic or remote inhibition.

Analysis of habituation The decrease in EPSP amplitude that occurs during habituation could result from a change either in the input resistance of the motor neuron (Figure 5B2: 6) or in the synaptic input to the motor neuron (Figure 5B2: 4, 5, 7, 8). To distinguish between these possibilities, Castellucci et al. (1970) developed a test system in which the neural circuit of the reflex could be examined in greater detail. They isolated the ganglion, with or without the siphon and the gill, and worked with it in an experimental chamber perfused with sea water (Figure 2D). To monitor the reflex output they recorded intracellularly from L7, one of the major motor neurons. To elicit the reflex response, the siphon nerve, which carries the afferent input from the siphon skin, was electrically stimulated. The complex EPSPs elicited by

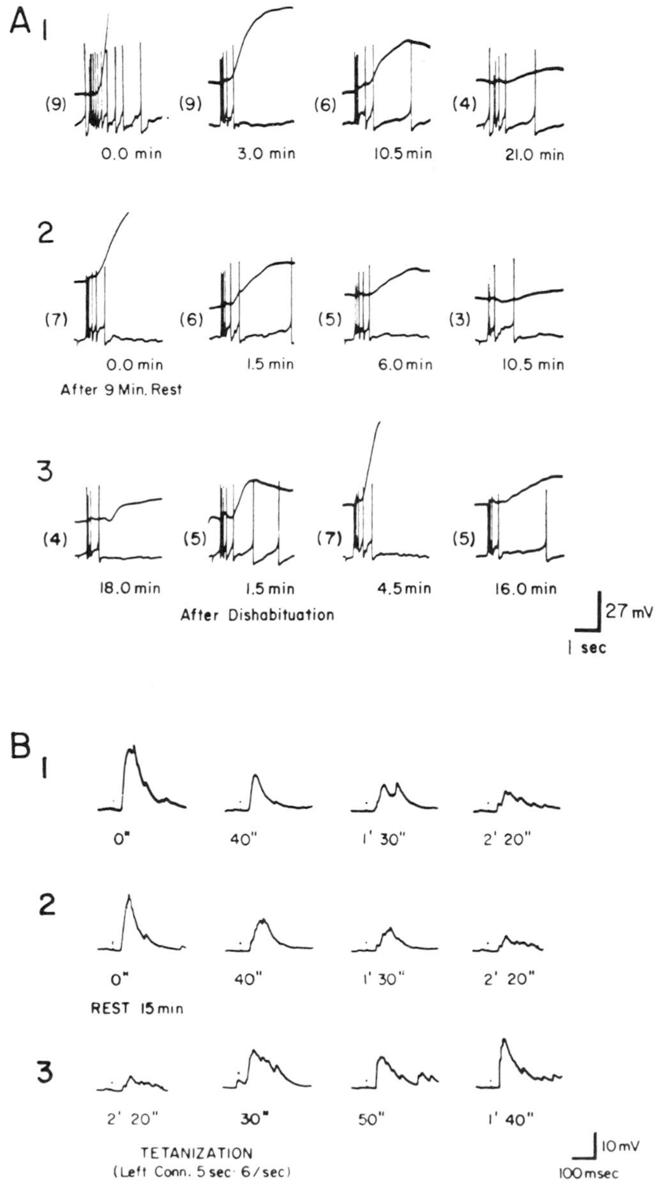

FIGURE 7 Central synaptic changes accompanying habituation, recovery and dishabituation. (A) Correlation between gill contraction and motor neuron response in a relatively intact preparation as in Figure 2C. Gill contractions (*top trace*) and simultaneous intracellular recordings from an identified motor neuron (L7) (*bottom trace*). (A1) Habituation run: Stimuli were presented every 90 sec over a period of 21 min. (A2) Partial recovery after a 9 min rest and subsequent rehabituation of the reflex. (A3) Following the last habituation trial shown in the first record of the row, a dishabituatory stimulus was introduced. Numbers in parentheses indicate numbers of spikes during first second after beginning of response. (From Kupfermann et al., 1970.) (B) Test system of the gill-withdrawal reflex in isolated ganglion as in Figure 2D. Change in complex EPSP produced in motor neuron L7 following electrical stimulation of siphon nerve at a rate of 1 per 10 sec. The membrane potential of L7 was hyperpolarized to the reversal level of the IPSP to prevent spike generation. (B1) With repeated stimulation of the siphon nerve the EPSP decreased in amplitude. (B2) It recovered with 15 min rest and then decreased again when stimulation was resumed. (B3) Tetanization (6/sec for 5 sec) of the left connective produced an immediate increase in EPSP amplitude paralleling the behavioral dishabituation. (From Schwartz et al., 1971.)

stimulating the siphon nerve resemble those produced in L7 by natural stimuli to the siphon (Figure 2B1). When the siphon nerve is stimulated repeatedly at intervals (10 sec to 2 min) that produce habituation in the whole animal, the complex PSPs decrease (Figure 7B1) and they recover their amplitude after a period of rest (Figure 7B2).

During the decrement of the EPSP no changes occurred in the input resistance of the motor neuron. These measurements are not completely conclusive because they do not rule out resistance changes at a site remote from the microelectrode in the cell body. They do, however, rule out a gross change in input resistance near the cell body and suggest that PSP alterations are more likely due to changes in the synaptic input.

A decrease in the synaptic input to the motor neuron could, in turn, be caused by an increase in central inhibition that alters the number or the firing frequency of the active afferent elements contributing to the complex EPSP (Figure 5B2: 3 or 5), or by a decrease in synaptic efficacy of individual afferent excitatory elements (Figure 5B2: 7 or 8). A demonstration of PSP decrement when a single afferent fiber is stimulated would provide an answer to this question. Castellucci et al. (1970) therefore further simplified the afferent limb of the reflex pathway by intracellular stimulation of a single mechanoreceptor neuron. Each spike in such a cell produced an elementary, presumably monosynaptic, EPSP of constant and short latency in the motor neuron (Figure 4A). When stimulated repeatedly, the changes in the magnitude of these EPSPs paralleled the habituation of the reflex; they showed decrement with repeated stimulation and recovery of their amplitude with rest (Figure 8A).

These experiments provide evidence that response decrement can occur in an excitatory monosynaptic system. Because a single sensory neuron was stimulated, the likelihood is reduced that the EPSP decrement is due to presynaptic inhibition resulting from the activation of parallel elements that act upon and inhibit the presynaptic terminals of the primary sensory neurons (Figure 5B2: 4 and 5). Rather, the findings suggest that decrement of the EPSP is the result of homosynaptic (low-frequency) depression of the excitatory synapse due either to an alteration in the mechanism of transmitter release or to a change in sensitivity of the postsynaptic receptor (Figure 5B2: 8). More recently Castellucci and Kandel (in preparation) have found that similar changes occur in the PSPs produced by the sensory neurons on the excitatory interneurons, on the inhibitory interneurons (L16), and on some motor neurons to the siphon. When two branches of the same sensory neuron, innervating different motor or interneurons, were ex-

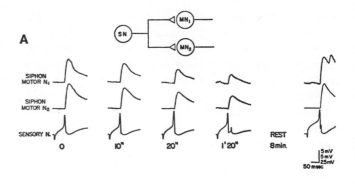

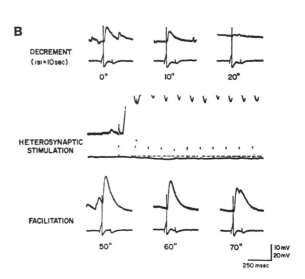

FIGURE 8   Changes in the efficacy of an elementary presumably monosynaptic connection between sensory and motor neurons due to repetitive stimulation and heterosynaptic facilitation. (A) Simultaneous recording of the EPSPs produced by repeated stimulation of one sensory neuron (interstimulus interval 10 sec) in two siphon motor neurons (MN). Note the decrement and the recovery of PSP amplitude after a rest. (From Castellucci and Kandel, in preparation.) (B) Decrement and facilitation of EPSP in the gill motor neuron L7. Intracellular stimulation of a sensory cell (lower trace of each pair) was repeated every 10 sec and produced decrement of the EPSP on L7 (top set of traces). In the middle set of traces heterosynaptic stimulation was applied to the left connective (7/sec for 5.5 sec). The heterosynaptic stimulus did not fire the sensory neuron (note the stimulus artifact in SN trace) but caused a sustained hyperpolarization in it. After the heterosynaptic stimulation (bottom set of traces) the EPSP was facilitated for several minutes. All indicated times are referred to the first stimulus (0″). The decrement illustrated is the third training session of this experiment and is therefore very rapid and profound. (From Carew, Castellucci, and Kandel, 1971.)

amined simultaneously, both showed comparable, although not always identical, decrements (Figure 8A). These results indicate that the major loci of habituation of the gill-withdrawal reflex are at the synapses made by branches of the sensory neurons on the interneurons and motor neurons responsible for the gill-withdrawal reflex. A similar locus for habituation of defensive reflexes has now been found in crayfish (Zucker, 1972) and in insects (Callec et al., 1971).

Analysis of dishabituation Castellucci et al. (1970) also examined dishabituation in the test system of the isolated ganglion and found that a strong stimulus to the pleuro-abdominal connective, equivalent to one leading to dis-habituation, produced facilitation of the decremented complex EPSP (Figure 7B3). This facilitation was not accompanied by a measurable change in the input re-sistance of the motor neuron, nor could it be ascribed to an action on the presynaptic terminals of the sensory neurons or interneurons by spike generation in the motor neuron. Spike generation triggered by intra-cellular stimulation of the motor neuron produced no facilitation of the EPSP. The facilitation occurred in an elementary EPSP produced by direct stimulation of the cell body of the mechanoreceptor neuron (Figure 8B). The facilitation did not involve a change in the frequency of firing of the sensory neuron, thereby excluding post-tetanic potentiation as a mechanism for facilitation. These data suggest that the restoration of the EPSP by a dishabituatory stimulus is due to heterosynaptic facilitation (Kandel and Tauc, 1965a, 1965b).

The simplified circuit diagram of Figure 9 illustrates the locus and the mechanism for these plastic changes as suggested by Castellucci et al. (1970). Only the motor neuron (L7) on which most of the work with the ele-mentary EPSP was done is shown. Habituation occurs as a result of repetitive stimulation of tactile receptors that leads to a homosynaptic plastic change at the indicated synapse between the afferent fiber and the motor neuron. The exact mechanism of the synaptic change is uncertain. A change in postsynaptic receptor sensitivity could not be excluded, although this mech-anism seems somewhat less likely in view of the ease with which heterosynaptic facilitation occurs. By analogy with brief low-frequency depression that has now been analyzed in detail at vertebrate and crayfish neuro-muscular junction (del Castillo and Katz, 1954; Dudel and Kuffler, 1961), Castellucci et al. 1970 suggested that the decrement of the elementary EPSP represents a de-crease in the release of excitatory transmitter from the pre-synaptic terminal (see also Bruner and Tauc, 1966a and 1966b; Bruner and Kehoe, 1970). However, unlike the brief low-frequency depression evident at many synapses, those in the habituation pathway show a remarkably large and prolonged decrement. This basic difference from the usual forms of low-frequency depression re-quires that additional features be operative in synapses of the habituating pathway. Such factors might be a highly

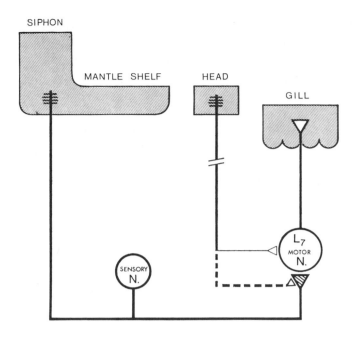

FIGURE 9 Simplified schematic diagram to indicate the postulated plastic changes in the wiring diagram of the gill reflex underlying habituation and dishabituation. See text for details. (From Castellucci et al., 1970.)

limited reservoir of immediately available transmitter substance and a slow mobilization of transmitter into this reservoir from the storage pool. Alternatively, there could be a progressive decrease in the probability of transmitter release or a decrease in the number of active presynaptic release sites.

Dishabituation occurs as a result of heterosynaptic facilitation. An attractive model for heterosynaptic facilitation is that of presynaptic facilitation at the same synapse resulting in an enhanced release of transmitter substance. Such a mechanism might operate by mobi-lizing transmitter substance from the storage pool to the immediately available transmitter reservoir, by increas-ing the probability of releasing transmitter, or by increasing the number of active release sites. Since the elementary PSPs in the presumed monosynaptic pathways are quite large, changes in their efficacy may account fully for at least the early component of habituation and dishabituation of the gill response in *Aplysia*.

This analysis demonstrates that habituation and dis-habituation of this reflex involves plastic changes in the excitatory transmission of previously existing connec-tions. No fundamental changes occurred in the wiring diagram; no new connection appeared and no existing connections disappeared. Instead, the functional effec-tiveness of certain previously existing connections was changed. The synapses between the sensory neuron and inter- and motor-neuron are endowed with remarkable plastic properties so that transmission across the synapse is greatly decreased during habituation and recovers only slowly after rest; it is restored immediately, however, by a dishabituating stimulus. These results are therefore consistent with the idea that genetic and developmental processes determine the properties of the individual cells and the anatomical interconnections between them. These processes leave unspecified the degree of effectiveness of certain of these connections. Environmental factors can play upon these plastic capabilities such as short-term learning to produce behavioral events.

SHORT-TERM HABITUATION AND DISHABITUATION IN THE ABSENCE OF PROTEIN SYNTHESIS Once a behavioral modification has been specified in cellular terms, it should be possible to bring biochemical techniques to bear on the analysis of molecular mechanisms of learning. As a start, Schwartz, Castellucci and Kandel (1971) examined the action on the neuronal correlates of habituation and dishabituation of three antibiotics (anisomycin, sparsomycin and pactamycin) that inhibit protein synthesis in the abdominal ganglion by more than 95%. The effect of these antibiotics was studied in the isolated ganglion using the test system of the gill-withdrawal reflex (Figure 2D). Repeated stimulation was applied to the siphon nerve, and the neuronal correlates of habituation and dishabituation were monitored in the absence or in the presence of anisomycin. Neither the change in synaptic efficacy associated with habituation or that associated with dishabituation were found to be affected by prior incubation in anisomycin for up to 6 hr. Longer periods of incubation were not examined.

Although long-term habituation may depend on new protein synthesis, these results indicate that short-term habituation and dishabituation do not, nor do they depend on the synthesis of proteins that have remarkably fast turn-over rates. This result is consistent with experiments in vertebrates, which indicates that new protein synthesis is not required for short-term learning (see for example, Agranoff, 1967; Barondes, 1970). These findings also draw attention to a need for biochemical models other than alterations in macromolecular synthesis to explain the mechanisms underlying short-term neuronal plasticity. As I have mentioned above, the synaptic changes associated with habituation and dishabituation might be due to prolonged alterations in transmitter release. Alternatively, but perhaps less likely, one or both of these processes could reflect a change in receptor sensitivity. Alterations in molecular conformation might not be expected to persist for long periods of time, although molecular changes lasting for several minutes have been observed (see Frieden, 1970). More likely, the

biochemical mechanisms underlying these short-term plastic changes are composed of a series of sequential reactions that result in a new distribution of transmitter substance or its availability. Mechanisms involving cyclic 3', 5' AMP might serve as one example of a series of reactions that result in transient enhancement in the activity of a critical enzyme system (Rall and Gilman, 1970; Rasmussen, 1970). A pathway of this kind might trigger the mobilization of transmitter from one compartment (a long-term store) to another (an immediately releasable store) or might control the number and availability of release sites. Several recent reports of altered synaptic activity might reasonably be explained by this mechanism (Dudel, 1965; Goldberg and Singer, 1969; Kuba, 1970). Thus, sequential reactions that include the formation of degradable small molecules but that do not involve the production of new structural, receptor, or enzymatic components, could account at least for short-term plastic changes.

Whether or not new protein synthesis is essential for the transition to long-term memory could in principle also be examined in simple systems.

Interrelationships between different behavioral modifications

Cellular studies can also help clarify questions about the relationship between different behavioral modifications. Two examples are considered here: (1) the relationship of habituation to dishabituation, and (2) the relationship of short- to long-term memory.

RELATIONSHIP OF HABITUATION TO DISHABITUATION There is disagreement among students of behavior as to whether dishabituation is a reversal of habituation (Pavlov, 1927; Humphrey, 1930, 1933; Konorski, 1967; Sokolov, 1963; Wickelgren, 1967; Wall, 1970) or a different, and in some ways independent, facilitatory process superimposed upon habituation (Sharpless and Jasper, 1956; Spencer, Thompson, and Neilson, 1966; Groves and Thompson, 1970). This question was examined by Carew, Castellucci, and Kandel (1971), at both the behavioral and neurophysiological levels, using another property of the gill-withdrawal reflex, the lack of generalization of habituation from one pathway to another (Figure 10A).

The receptive field of the gill-withdrawal reflex consists of the siphon and the edge of the mantle shelf that contains the purple gland (Figure 10A1). The anterior portion of the mantle shelf and the purple gland are innervated by a branch of the branchial nerve. The siphon and the posterior third of the purple gland and mantle shelf are innervated by the siphon nerve (Figures

10A1, 10A2). The siphon and the anterior third of the mantle shelf (purple gland) therefore provide two limited distinct afferent pathways for a common motor response. In intact animals habituation of the reflex response via one pathway does not alter responsiveness via the other pathway. Similarly, decrement of the PSPs produced by repeated electrical stimulation of one pathway does not affect PSP responsiveness from the other pathway (Figures 10A2, 10A3).

To examine the relationship of dishabituation to habituation, Carew, Castellucci, and Kandel (1971)

compared the effects of a common "dishabituatory" stimulus on both habituated and nonhabituated responses using these two different reflex pathways. If the dishabituation is due to a reversal of habituation, the "dishabituatory" stimulus should facilitate only the habituated pathway and bring it up to its original value. If dishabituation is a superimposed facilitation (sensitization), then a "dishabituatory" stimulus might facilitate a nonhabituated response. In addition, the habituated responses might be facilitated *beyond* their control value.

In behavioral experiments a single test stimulus was presented to either the siphon or the anterior third of the purple gland of an intact restrained animal. The reflex was then habituated by repeated stimulation of the other pathway. Following a second test stimulus to the nonhabituated pathway, a strong tactile stimulus was presented to the head or neck region of the animal. Dishabituation of the habituated pathway was observed in every preparation and was occasionally *larger* than the initial control value. Sensitization of the nonhabituated pathway was only found in one-third of the animals but in 5 of 6 cases where the neck stimulus produced sensitization, it also produced dishabituation. In the sixth case dishabituation could not be examined. As we shall

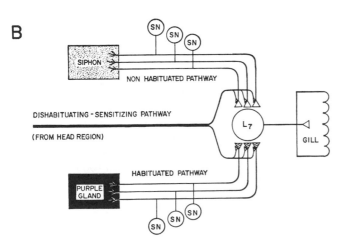

FIGURE 10 Relation of habituation of the gill-withdrawal reflex to dishabituation and sensitization. (A) Behavioral and cellular neurophysiological preparations. (A1) Dorsal view of an intact *Aplysia*. The parapodia and mantle shelf have been retracted to reveal the gill. The anterior part of the mantle shelf and purple gland are indicated by solid black lines. The siphon and the posterior part of the mantle shelf are stippled. The two sites have a separate innervation and were used as independent reflex inputs for gill withdrawal. (A2) Diagram illustrating test system in isolated ganglion. In the test system the branchial nerve, which innervates the anterior part of the mantle shelf and purple gland, or the siphon nerve, which innervates the posterior part of the purple gland, and siphon were electrically stimulated. Stimulation of the right or left connective provided stimuli for heterosynaptic facilitation. (A3) Summary graph of 9 runs from 4 different preparations showing absence of generalization of EPSP decrement between decremented and test inputs to the motor neuron (L7). Following a single train of stimuli (6 per sec for 6 sec) to the connective (connective stimulation ↑), facilitated both decremented and nondecremented responses. (B) Schematic diagram illustrating the difference in spatial extent of the pathways mediating habituation (shaded synapses) and sensitization (or dishabituation). Diagram explains the lack of generalization of habituation as well as the evident generalization of sensitization (dishabituation). The model postulates that the sensitizing pathway (which is not necessarily monosynaptic) mediates presynaptic facilitation on the synapse between the primary sensory neurons (SN) and the motor neuron L7 (here used as a model for the entire reflex). (From Carew, Castellucci, and Kandel, 1971.)

see below (Figure 13), unrestrained animals showed profound sensitization (Pinsker, Carew and Kandel, 1972), and its absence in restrained animals may have resulted from sensitization caused by the restraint.

In cellular neurophysiological experiments, Carew et al. (1971) examined how the neuronal correlates of dishabituation relate to those of sensitization, using the test system of the isolated ganglion. They stimulated either the branchial or siphon nerve repeatedly and produced decrement of the EPSP. They then examined the effect of a heterosynaptic stimulus, applied to a connective, on the amplitude of the decremented EPSP produced by the previously stimulated pathway and on the amplitude of the nondecremented EPSP produced by the pathway not previously stimulated. They found that the decremented EPSP was invariably facilitated and in 70% of the cases the facilitated EPSP was larger than the initial (nondecremented) control. The same stimulus to the connective also facilitated the nondecremented EPSP in 80% of the cases (Figure 10A3). Both decremented and nondecremented PSPs showed similar increases (about 120%) compared to their *initial* controls. The decremented EPSPs obviously showed proportionately more facilitation (224%) when the last observed (decremented) response was used as control, instead of the first of the series.

Thus, in behavioral experiments, the same strong stimulus that produced dishabituation of a habituated pathway was also capable of producing sensitization of a nonhabituated pathway, at least some of the time. In the neurophysiological experiments the same strong stimulus that produced facilitation of a decremented EPSP usually also produced facilitation of a nondecremented EPSP. Taken together, these two sets of data support the notion that dishabituation of the gill-withdrawal reflex is not simply the removal of habituation but is an independent facilitation. Dishabituation is thus a special case of sensitization (Spencer, Thompson, and Neilson, 1966) that differs from it only in involving a previously habituated response. The neuronal correlates of habituation and dishabituation (sensitization) are thus not mirror image processes but reflect two separate regulatory mechanisms acting on a common set of synapses. Indeed, these two processes have different spatial distributions (Figure 10B). Whereas habituation is limited to the stimulated pathway, sensitization has a more widespread distribution involving both stimulated and unstimulated pathways. The difference in the extent of the pathways mediating habituation and sensitization (dishabituation) explains the lack of generalization of habituation as well as the evident generalization of sensitization to nonhabituated pathways.

RELATIONSHIP OF SHORT- TO LONG-TERM HABITUATION

A particularly important current problem in behavioral psychology is the relation of short- to long-term memory. Experiments that interfere with normal brain function, such as inhibition of protein synthesis or electroconvulsive shock, suggest that short- and long-term memory are distinct. Short-term memory is thought to be impaired by electroconvulsive shock whereas long-term memory is obliterated by inhibition of protein synthesis. (For review, see Agranoff, 1967; Barondes, 1970). These experiments support the dual-trace theory of memory advocated by Hebb (1949), which postulates that a short-term "holding" process somehow leads to long-term storage. However, the interpretation of the electroconvulsive shock and protein inhibition experiments is not clear cut (Weiskrantz, 1970). To determine whether long-term memory is a distinct process or an extension of the short-term process requires a system in which these questions can be examined on the cellular level.

Toward this end, Carew, Pinsker, and Kandel (1972) have developed a training procedure for studying long-term habituation that permits a mechanistic analysis of the transition from short- to long-term habituation. As we have seen, ten habituation trials in one session lead to habituation that is largely dissipated several hours later. To examine long-term habituation, one cannot work on the restrained animal, because animals do not survive for several days if restrained. Carew et al. (1972) thus studied unrestrained animals using the siphon component instead of the gill component of the defensive reflex as an index of reflex withdrawal (Figure 1B). When habituation training (10 trials/day) is repeated daily for 4 days, habituation of the reflex response builds up progressively across days so that habituation occurs more rapidly on each subsequent day. Thus on day 5 the median net duration of reflex response during ten trials is only 20% of its duration on day 1 (Figures 11A1, 11A2). The habituation is long lasting and persists unchanged for a week, recovering only partially after three weeks (Figures 11A1, 11A2). This makes it possible to use this system as a potential model for studying the acquisition of long-term memory.

Whereas the siphon component of the withdrawal reflex offers the advantage that it can be studied in the freely moving animal, it has potential disadvantages. Unlike the gill-withdrawal component, which is centrally mediated, the siphon withdrawal component is known to involve some contribution from the peripheral nerve net (Kupfermann and Kesselman, 1968; Lukowiak and Jacklet, 1972; Newby, 1972), because, to elicit siphon withdrawal, the stimulus is applied to the siphon itself, which optimizes the recruitment of peripherally mediated responses (Bethe, 1903). To determine the relative contribution of peripheral and central components of the siphon withdrawal in the intact animal, Pinsker, Carew, and Kandel (1973) examined siphon withdrawal both in normal animals and in animals from which the abdominal ganglion had been surgically removed. They found that whereas normal (unhabituated) animals routinely withdrew the siphon beyond the parapodia (Figure 1B2), deganglionated animals did not and gave only very slight responses.

To obtain an independent evaluation of the central contribution, Carew et al. (1972) undertook another series of experiments, this time on the gill component of the reflex, in order to ascertain whether the type of siphon stimulation that produced long-term habituation of the siphon component of the defensive reflex also

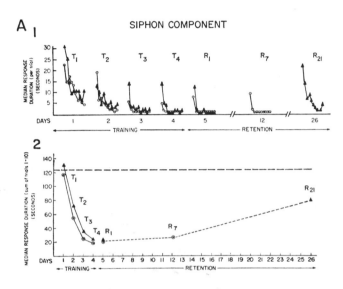

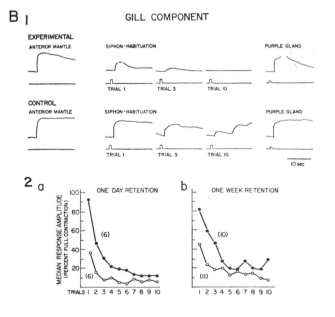

FIGURE 11 Long-term habituation of both siphon and gill components of the defensive withdrawal reflex. (A) Long-term habituation of siphon component. (A$_1$) Build-up of habituation during training for 4 days (T1 to T4) and retention after 24 hr (R1), 1 week (R7), and 3 weeks (R21). Data from three experiments ($n = 19$) are presented: Two independent, identical replications in which retention was tested at 24 hr and 1 week have been pooled (⊙——⊙); in the third experiment ($n = 14$), retention was tested at 24 hr and 3 weeks (△——△). Each data point is the median duration of siphon withdrawal for each of ten habituation trials. (Parallel control data ($n = 20$, $n = 14$) are presented in Carew et al., 1971. Figure 1.) (A2) Time course of habituation. Habituation within each daily session is expressed as a single score, the sum of ten trials (the net amount of time of siphon withdrawal during the entire habituation training session). The upper dashed line indicates control habituation (day 1 of training). Compare retention tested at 24 hr (day 5) and 1 week, day 12 (⊙——⊙) and retention at 24 hr (day 5) and 3 weeks, day 26 (△——△) with control (day 1) habituation. (B) Long-term habituation of gill component. (B1) Gill-withdrawal reflexes were recorded with a photocell. One week previously, experimental animals had shown significant 24-hr retention of habituation of siphon withdrawal as compared to controls. Retention of habituation of gill withdrawal (produced by the previous siphon habituation training) was measured 7 days later in restrained animals. A single test gill contraction was first produced by vigorous tactile stimulation of the anterior mantle region (first column). After a brief period of rest, habituation of gill withdrawal was produced by repetitive siphon stimulation with a jet of sea water (ten trials; interstimulus interval, 30 sec), and this was followed by another single test stimulus to the purple gland. Anterior mantle or purple gland stimulation produced comparable contractions in both experimental and control animals. However, siphon stimulation produced a significantly reduced initial gill withdrawal in experimental than in control animals (compare trials 1, 3, and 10 of both records). Time calibration, 10 sec. (B2) Comparison of experimental and control habituation of gill withdrawal at 1 day and 1 week. Median amplitude of gill responses expressed as a percentage of a full contraction. The number of animals in each group is indicated in parentheses. (B2a) Gill withdrawal 24 hr after siphon habituation training. Experimental animals (○——○) exhibited significantly greater habituation (P < 0.001) of gill withdrawal (sum of trials 1 to 10) than controls (○——○) and were significantly lower than controls on 9 out of 10 trials. (B2b) Gill withdrawal 7 days after siphon habituation training. Experimental animals again exhibited significantly greater habituation (P < 0.001) and were significantly lower than controls on 8 out of 10 trials. These data indicate that stimuli that produce long-term habituation of siphon withdrawal also produce long-term habituation of gill withdrawal. (From Carew, Pinsker, and Kandel, 1972.)

produced long-term habituation of the gill component. Either 1 or 7 days after the 4-day habituation training sessions in freely moving animals, experimental animals were mixed with controls, and the gill-withdrawal reflex of both groups was measured (by a blind procedure) with the animals restrained and a photocell placed under the gill (Figure 2B). In the testing procedure, as in the training procedure, habituation was produced by ten trials of siphon stimulation. Both 1-day and 7-day experimental animals exhibited significantly greater habituation of gill withdrawal than controls (Figure 11B), indicating that repeated siphon stimulation produces long-term habituation not only of the siphon component of this reflex but also of the gill component.

Whereas the reflex responsiveness to siphon stimulation was significantly reduced in experimental animals, there was no statistical difference between gill-withdrawal reflexes produced by purple gland or anterior mantle shelf stimulation in both experimental and control animals (Figure 11B). These findings illustrate that, like short-term habituation, long-term habituation does not generalize from the siphon to nonstimulated parts of the receptive field; it is restricted specifically to the stimulated afferent pathway.

Recently Carew and Kandel (1973) have simplified the training procedure for long-term habituation by presenting the four spaced training sessions within one day, separated from each other by only $1\frac{1}{2}$ hr. This shortened spaced training procedure still leads to long-term habituation and has the advantage of making it possible to study the cellular mechanisms of acquisition and even of retention of the long-term process. In experiments with the isolated ganglion test system, stimulation of an afferent nerve with the same temporal patterning as in the behavioral experiments produced a progressive decrease in the complex EPSP that persisted for at least 24 hr. These changes in the complex EPSP look similar to the changes that accompany the short-term habituation. However, changes in the amplitude of the complex EPSP can be caused by several different mechanisms. Whether or not the long-term process is indeed an extension of the short-term one will therefore require further analysis.

RELATION OF LONG-TERM HABITUATION TO MORE COMPLEX LEARNING   The kinetics and time course of long-term habituation are interesting because they resemble certain apsects of higher learning found in vertebrates and man. For example, when a human subject is asked to tap a Morse code key and hold it down for 0.7 sec, his performance improves with repeated trials (Figure 12A). Although a comparison of a few parametric features of human and invertebrate behavior is by itself not very meaningful, it is interesting to note that in *Aplysia*, as in human learning, there is a slight "forgetting" between the last performance of one day and the first performance the next day, despite the progressive improvement over days (Figure 12B).

Since, in these parametric properties, long-term habituation in *Aplysia* bears at least a superficial resemblance to behavioral modifications characteristic of vertebrates, Carew, Pinsker and Kandel (1972) looked for other features shared by long-term habituation and more complex learning. In most types of complex learning, temporally spaced training usually produces better learning than massed training (Woodworth and Schlosberg, 1954). Carew et al. therefore compared a spaced-training group (four training sessions of ten habituation trials each, separated by anywhere from 1.5 to 24 hr) with a massed-training group (one training session consisting of 40 consecutive habituation trials). Despite the fact that both groups had received the same number of training trials, animals receiving spaced training exhibited significantly greater habituation one day and one week later than did those receiving massed training (Figures 12B1 to 12B4). These data indicated that spacing habituation training by even as much as one hour is much more effective in producing long-term habituation than massing the habituation training.

Thus, despite its simplicity, habituation of gill and siphon withdrawal in *Aplysia* shows a number of interesting parallels to the complex learning of vertebrates. It not only provides an instructive paradigm for studying the neural mechanisms of short- and long-term behavioral modifications but it may also be useful for examining the mechanisms of pattern sensitivity in behavioral modifications. In particular it would be interesting to know whether the pattern of stimulation is the essential feature that is responsible for the transition from short- to long-term memory. Studies along these lines could specify how long-term memory is established and how it relates to the short-term process.

Summary and perspectives

The studies that I have outlined here and parallel ones that are now being carried out with arthropods and insects (Zucker, 1972; Callec et al., 1971) do not, as yet, increase understanding of the neural mechanisms of behavior and of learning. These studies involve relatively simple behavioral modifications in simple animals. Only the merest beginnings have been made.

Moreover, although we can now specify the locus of the neural change during short-term habituation, as in

Aplysia, and know that the critical change results from an alteration in synaptic transmission, we still have only a superficial understanding of the detailed mechanisms of the synaptic change. To obtain more meaningful insights into these mechanisms, we will need more extensive electrophysiological analysis of synaptic transmission at the critical synapses between the sensory neurons and the central neurons they innervate. In addition, morphological techniques for studying the fine structure of these synapses are needed.

If these techniques can be developed, however, it may become possible to analyze the mechanisms of more complex behaviors and more interesting instances of behavioral modifications. In addition, these preparations may ultimately be useful for studying the cellular mechanisms of behavioral abnormalities. For example, habituation occurs readily in a safe environment where an *Aplysia* learns quickly to ignore a novel but innocuous stimulus that is repeatedly presented. But animals cannot afford to habituate their withdrawal reflexes readily in a hostile environment. Pinsker, Carew, and Kandel

(1973) have found that animals do not readily habituate even to an innocuous siphon stimulus if they have previously received noxious stimulation to the head. Animals receiving 2 such stimuli a day for 10 days become highly sensitized and show reflex responsiveness that differs radically from that of a normal animal (Figure 13). Animals so treated live in a dangerous environment and show the reflex behavior of an overly responsive or aroused animal. Particularly interesting is the finding that the heightened responsiveness persists for many days *after* the noxious stimulation has stopped. Although the heightened responsiveness was appropriate in the dangerous environment, it is quite inappropriate for the neutral environment in which the animal now again lives. In so far as an animal's responsiveness over the long term is less appropriately related to the immediate stimulus that elicits it than it is to certain moments in its past history, the animal can be thought of as responding inappropriately, or abnormally, to the demands of its current environment. Thus, as we gain a better understanding of the learning process, we might also be able to develop animal model systems that may permit us insights into certain behavioral abnormalities that result from learning.

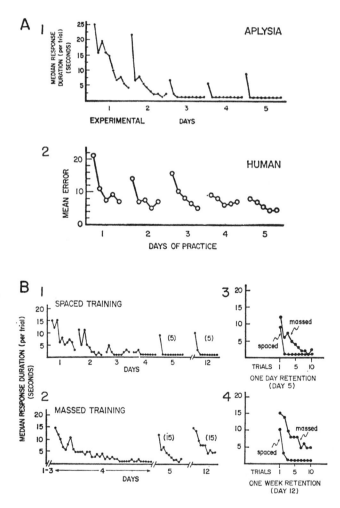

FIGURE 12   Comparison of *Aplysia* habituation to higher order behavioral modifications. (A) Comparison of habituation in *Aplysia* (A1) to human learning (A2). The human subject was asked to depress a Morse key for a duration of 0.7 sec. Subject's error from the correct duration was shown to him after each attempt providing immediate feedback. As in the *Aplysia* experiment 10 daily trials were given but at 12-sec intervals instead of 30-sec intervals. The mean error of 10 subjects for 2 successive trials are plotted. The data in *Aplysia* are taken from one of the experiments (⊙——⊙) illustrated in Figure 11A1. The human data are replotted from MacPherson, Dees, and Grindley, 1949. (B) Comparison of spaced (B1) with massed (B2) habituation training. Two groups of experimental animals were given either 4 days of habituation training (10 trials per day, spaced training) or 1 day of habituation training (40 trials, massed training). Both groups were tested for retention after 24 hr (day 5) and 1 week (day 12). The spaced training group (B1) was significantly lower than the massed-training group (B2) at both 24 hr (p < 0.01) and 1 week (p < 0.01), despite the fact that both groups had received the same number of training trials. The massed-training group was not significantly different from unstimulated control in either retention test. In the 1 week retention test (day 12) the massed training group was no longer significantly lower than their initial performance (day 4, trials 1 to 10) indicating that this group showed no retention of training after 1 week while the spaced-training group still exhibited significant retention of habituation. Number of animals in each group is indicated in parentheses. (B3 and B4.) Comparison of spaced to massed habituation on each retention day. (From Carew, Pinsker, and Kandel, 1972.)

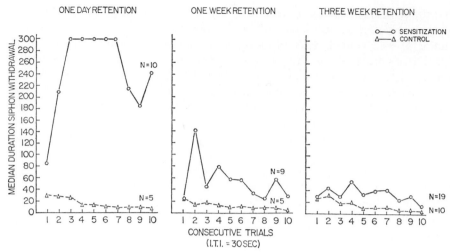

ONE DAY RETENTION   ONE WEEK RETENTION   THREE WEEK RETENTION

FIGURE 13 Long-term sensitization. Comparison of control animals (N=10) who received no prior training (△—△) to experimental (N=19) animals (O—O) who had received two noxious stimuli per day for 10 days (sensitization) training prior to the first day of retention. The experimental and control groups were each divided into two after training was completed. One experimental and control group was examined for one day retention, the other two groups for one week retention. The two experimental and control groups were again combined for the 3 week retention tests. On test days (1 day, 1 week, and 3 weeks after sensitization training), both control and sensitized animals received a series of ten weak tactile stimuli to the siphon by means of a jet of sea water applied by a Water Pik. The experimental animals showed an increased response, and their habituation curve had different kinetics than that of the controls showing a buildup (particularly on 1 day and 1 week retention tests) rather than the usual negatively accelerating decrement. The flat peak of the sensitization group on the one-day retention test results from the fact that more than half of the animals exceeded the 300 sec limit on those trials.

ACKNOWLEDGMENTS I thank Drs. Vincent Castellucci and Irving Kupfermann for their helpful comments on an earlier draft of this manuscript and Kathrin Hilten for her usual expert help with the figures. I am supported by a Career Scientist Award from NIMH #MH 18558-06 and the work of my laboratory is supported by grants NS 09261-03 and NIMII 19795-02.

REFERENCES

AGRANOFF, B. W., 1967. Agents that block memory. In *The Neurosciences*, G. C. Quarton, T. Melnechuk, and F. O. Schmitt, eds. New York: Rockefeller University Press, pp. 756–764.

ARVANITAKI, A., and S. H. TCHOU, 1942. Les lois de la croissance relative individuelle des cellules nerveuses chez l'*Aplysie*. *Bull. Histol. Appl. Tech. Microscop.* 19:244–256.

ARVANITAKI, A., and N. CHALAZONITIS, 1958. Configurations modales de l'activité propres à différents neurons d'un même centre. *J. Physiol.* (*Paris*) 50:122–125.

BARONDES, S. H., 1970. Multiple steps in the biology of memory. In *The Neurosciences: Second Study Program*, F. O. Schmitt, ed. New York: Rockefeller University Press, pp. 272–278.

BAUMGARTEN, R. J. VON, 1970. Plasticity in the nervous system at the unitary level. In *The Neurosciences: Second Study Program*, F. O. Schmitt, ed. New York: Rockefeller University Press, pp. 260–271.

BETHE, A., 1903. *Allgemeine Anatomie und Physiologie des Nervensystems.* Leipzig: Thieme.

BRUNER, J., and J. KEHOE, 1970. Long term decrements in the efficacy of synaptic transmission of molluscs and crustaceans. In *Short-Term Changes in Neural Activity and Behavior*, G. Horn and R. Hinde, eds. London: Cambridge University Press, pp. 323–359.

BRUNER, J., and L. TAUC, 1966a. Habituation at the synaptic level in *Aplysia. Nature (Lond.)* 210:37–39.

BRUNER, J., and L. TAUC, 1966b. Long-lasting phenomena in the molluscan nervous system. *Symp. Soc. Exp. Biol.* 20:457–475.

BULLOCK, T. H., and G. A. HORRIDGE, 1965. *Structure and Function in the Nervous Systems of Invertebrates.* San Francisco: W. H. Freeman.

BYRNE, J., V. CASTELLUCCI, and E. R. KANDEL, 1973. Receptive fields and response properties of mechanoreceptor neurons in the siphon skin of *Aplysia*. (In preparation.)

CALLEC, J. J., J. C. GUILLET, Y. PICHON, and J. BOISTEL, 1971. Further studies on synaptic transmission in insects. II. Relations between sensory information and its synaptic integration at the level of a single giant axon in the cockroach. *J. Exp. Biol.* 55:123–149.

CAREW, T. J., V. F. CASTELLUCCI, and E. R. KANDEL, 1971. An analysis of dishabituation and sensitization of the gill-withdrawal reflex in *Aplysia. Intern. J. Neuroscience.* 2:79–98.

CAREW, T. J., and E. R. KANDEL, 1973. Rapid acquisition of long-term habituation: Behavioral and cellular neurophysiological correlation. (In preparation.)

CAREW, T. J., H. M. PINSKER, and E. R. KANDEL, 1972. Long-term habituation of a defensive withdrawal reflex in *Aplysia. Science* 175:451–454.

CAREW, T. J., H. PINSKER, K. RUBINSON, J. H. SCHWARTZ, and E. R. KANDEL, 1973. Physiological and biochemical properties of neuromuscular transmission between identified motor neurons and gill muscle in *Aplysia*. (In preparation.)

CAREW, T. J., J. H. SCHWARTZ, and E. R. KANDEL, 1972. Innervation of *Aplysia* gill muscle fibers by two identified excitation motor neurons using different chemical transmitters. *Physiologist* 15(3):100.

Castellucci, V., H. Pinsker, I. Kupfermann, and E. R. Kandel, 1970. Neuronal mechanisms of habituation and dishabituation of the gill-withdrawal reflex in *Aplysia*. *Science* 167:1745–1748.

del Castillo, J., and B. Katz, 1954. The effect of magnesium on the activity of motor nerve endings. *J. Physiol.* (*Lond.*) 124: 553–559.

Diamond, J., 1971. The Mauthner Cell. In *Fish Physiology*, Vol. 5, W. S. Hoar and D. J. Randall, eds. New York: Academic Press, pp. 265–346.

Dudel, J., 1965. Facilitatory effects of 5-hydroxy-tryptamine on the crayfish neuromuscular junction. *Arch. Exptl. Pathol. Pharmakol.* 249:515–528.

Dudel, J., and S. W. Kuffler, 1961. Mechanism of facilitation at the crayfish neuromuscular junction. *J. Physiol.* (*Lond.*) 155:530–542.

Eales, N. B., 1921. *Aplysia. Proc. Trans. Liverpool Biol. Soc. Mem.* 35:183–266.

Eccles, J. C., 1953. *The Neurophysiological Basis of Mind*. Oxford: Clarendon Press.

Frazier, W. T., E. R. Kandel, I. Kupfermann, R. Waziri, and R. E. Coggeshall, 1967. Morphological and functional properties of identified neurons in the abdominal ganglion of *Aplysia californica*. *J. Neurophysiol.* 30:1288–1351.

Frazier, W. T., R. Waziri, and E. R. Kandel, 1965. Alterations in the frequency of spontaneous activity in *Aplysia* neurons with contingent and non-contingent nerve stimulation. *Fed. Proc.* 24(2171):522.

Frieden, C., 1970. Kinetic aspects of regulation of metabolic processes. The hysteretic enzyme concept. *J. Biol. Chem.* 245: 5788–5799.

Gardner, D., 1971. Bilateral symmetry and interneuronal organization in the buccal ganglia of *Aplysia*. *Science* 173: 550–553.

Gardner, D., and E. R. Kandel, 1972. Diphasic postsynaptic potential: A chemical synapse capable of mediating conjoint excitation and inhibition. *Science* 176:675–678.

Gaze, R. M., and M. J. Keating, 1972. The visual system and "neuronal specificity." *Nature* (*Lond.*) 237:375–378.

Giller, Jr., E., and J. H. Schwartz, 1968. Choline acetyltransferase: Regional distribution in the abdominal ganglion of *Aplysia*. *Science* 161:908–911.

Giller, Jr., E., and J. H. Schwartz, 1971a. Choline acetyltransferase in identified neurons of abdominal ganglion of *Aplysia californica*. *J. Neurophysiol.* 34:93–107.

Giller, Jr., E., and J. H. Schwartz, 1971b. Acetylcholinesterase in identified neurons of abdominal ganglion of *Aplysia californica*. *J. Neurophysiol.* 34:108–115.

Goldberg, A. L., and J. J. Singer, 1969. Evidence for a role of cyclic AMP in neuromuscular transmission. *Proc. Natl. Acad. Sci. USA* 64:134–141.

Groves, P. M., and R. F. Thompson, 1970. Habituation: A dual-process theory. *Psychol. Rev.* 77(5):419–450.

Hebb, D. O., 1949. *The Organization of Behavior*. New York: John Wiley & Sons.

Hinde, R. A., 1970. *Animal Behavior: A Synthesis of Ethology and Comparative Psychology*. (2nd edition), New York: McGraw-Hill.

Hubel, D. H., and T. N. Wiesel, 1965. Binocular interaction in striate cortex of kittens reared with artificial squint. *J. Neurophysiol.* 28:1041–1059.

Hughes, J. R., E. V. Evarts, and W. H. Marshall, 1956. Post-tetanic potentiation in the visual system of cats. *Am. J. Physiol.* 186:483–487.

Humphrey, G., 1930. Le Chatelier's rule and the problem of habituation and dishabituation in *Helix albolabris*. *Psychol. Forsch.* 13:113–127.

Humphrey, G., 1933. *The Nature of Learning in its Relation to Living Systems*. New York: Harcourt, Brace.

Kandel, E. R., 1967. Cellular studies of learning. In *The Neurosciences*, G. C. Quarton, T. Melnechuk, and F. O. Schmitt, eds. New York: Rockefeller University Press, pp. 666–689.

Kandel, E. R., 1968. Dale's principle and the functional specificity of neurons. In *Psychopharmacology: A Review of Progress 1957–1967*, D. E. Efron, ed. Public Health Service Publ. No. 1836. U.S. Govt. Printing Office, pp. 385–398.

Kandel, E. R., W. T. Frazier, R. Waziri, and R. E. Coggeshall, 1967. Direct and common connections among identified neurons in *Aplysia*. *J. Neurophysiol.* 30:1352–1376.

Kandel, E. R., and W. A. Spencer, 1968. Cellular neurophysiological approaches in the study of learning. *Physiol. Rev.* 48:65–134.

Kandel, E. R., and L. Tauc, 1965a. Heterosynaptic facilitation in the neurons of the abdominal ganglion of *Aplysia depilans*. *J. Physiol.* (*Lond.*) 181:1–27.

Kandel, E. R., and L. Tauc, 1965b. Mechanism of heterosynaptic facilitation in the giant cell of the abdominal ganglion of *Aplysia depilans*. *J. Physiol.* (*Lond.*) 181:28–47.

Kehoe, J. S., 1969. Single presynaptic neuron mediates a two component postsynaptic inhibition. *Nature* (*Lond.*) 221:866–868.

Kehoe, J. S., 1972. Three acetylcholine receptors in *Aplysia* neurons. *J. Physiol.* (*Lond.*) 225:115–146.

Kennedy, D., A. I. Selverston, and M. P. Remler, 1969. Analysis of restricted neural networks. *Science* 164:1488–1496.

Konorski, J., 1948. *Conditioned Reflexes and Neuron Organization*. London: Cambridge University Press.

Konorski, J., 1967. *Integrative Activity of the Brain; an Inter-Disciplinary Approach*. Chicago: University of Chicago Press.

Kuba, K., 1970. Effects of catecholamines on the neuromuscular junction in the rat diaphragm. *J. Physiol.* (*Lond.*) 211:551–570.

Kupfermann, I., T. J. Carew, and E. R. Kandel, 1973. Local, reflex and central commands controlling gill and siphon movements in *Aplysia*. (In preparation.)

Kupfermann, I., V. Castellucci, H. Pinsker, and E. Kandel, 1970. Neuronal correlates of habituation and dishabituation of the gill-withdrawal reflex in *Aplysia*. *Science* 167:1743–1745.

Kupfermann, I., and E. R. Kandel, 1968. Reflex function of some identifiable cells in *Aplysia*. *Physiologist* 27:348.

Kupfermann, I., and E. R. Kandel, 1969. Neuronal controls of a behavioral response mediated by the abdominal ganglion of *Aplysia*. *Science* 164:847–850.

Kupfermann, I., and M. Kesselman, 1968. Withdrawal reflexes in intact and "deganglionated" *Aplysia californica*. Cited in Kandel and Spencer, 1968.

Kupfermann, I., H. Pinsker, V. Castellucci, and E. R. Kandel, 1971. Central and peripheral control of gill movements in *Aplysia*. *Science* 174:1252–1256.

Larrabee, M. G., and D. W. Bronk, 1947. Prolonged facilitation of synaptic excitation in sympathetic ganglia. *J. Neurophysiol.* 10:139–154.

Lasek, R. J., and W. J. Dower, 1971. *Aplysia californica*: Analysis of nuclear DNA in individual nuclei of giant neurons. *Science* 172:278–280.

Lukowiak, K., and J. W. Jacklet, 1972. Habituation and

dishabituation: Interactions between peripheral and central nervous systems in *Aplysia. Science* 178:1306–1308.

MACPHERSON, S. J., V. DEES, and G. C. GRINDLEY, 1949. The effect of knowledge of results on learning and performance. III. The influence of the time interval between trials. *Quart. J. Exp. Psych.* 1:167–174.

MARTIN, A. R., and G. PILAR, 1964. Presynaptic events during post-tetanic potentiation and facilitation in the airan ciliary ganglion. *J. Physiol.* 175:740–746.

MARTIN, A. R., and W. O. WICKELGREN, 1971. Sensory cells in the spinal cord of the sea lamprey. *J. Physiol. (Lond.)* 212: 65–83.

MAYERI, E., I. KUPFERMANN, J. KOESTER, and E. R. KANDEL, 1971. Neural coordination of heart rate and gill contraction in *Aplysia. Am. Zool.* 11(244):667.

McCAMAN, R. E., and S. A. DEWHURST, 1970. Choline acetyltransferase in individual neurons of *Aplysia californica. J. Neurochem.* 17:1421–1426.

NEWBY, N. A., 1972. Habituation to light and spontaneous activity in the isolated siphon of *Aplysia*: The effects of synaptically active pharmacological agents. Ph.D. Thesis. Case Western Reserve University.

NICHOLLS, J. G., and D. PURVES, 1970. Monosynaptic chemical and electrical connexions between sensory and motor cells in the central nervous system of the leech. *J. Physiol. (Lond.)* 209:647–667.

PAVLOV, I. P., 1927. *Conditioned Reflexes.* (G. V. Anrep, trans.) London: Oxford University Press.

PEEKE, H. V. S., M. J. HERZ, and J. E. GALLAGHER, 1971. Changes in aggressive interaction in adjacently territorial convict cichlids (*Cichlasoma nigrofasciatum*): A study of habituation. *Behavior* 40:43–54.

PERETZ, B., 1969. Central neuron initiation of periodic gill movements. *Science* 166:1167–1172.

PERETZ, B., 1970. Habituation and dishabituation in the absence of a central nervous system. *Science* 169:379–381.

PINSKER, H., T. CAREW, and E. R. KANDEL, 1973. Long-term sensitization of a defensive withdrawal reflex in *Aplysia*. (In preparation.)

PINSKER, H., and E. R. KANDEL, 1967. Contingent modification of an endogenous bursting rhythm by monosynaptic inhibition. *Physiologist* 10:279.

PINSKER, H., I. KUPFERMANN, V. CASTELLUCCI, and E. KANDEL, 1970. Habituation and dishabituation of the gill-withdrawal reflex in *Aplysia. Science* 167:1740–1742.

RALL, T. W., and A. G. GILMAN, 1970. The role of cyclic AMP in the nervous system. *Neurosci. Res. Prog. Bull.* 8:221–323.

RASMUSSEN, H., 1970. Cell communication, calcium ion, and cyclic adenosine monophosphate. *Science* 170:404–412.

ROVAINEN, C. M., 1967. Physiological and anatomical studies on large neurons of central nervous system of the sea lamprey (*Petromyzon marinus*). I. Müller and Mauthner cells. *J. Neurophysiol.* 30:1000–1023.

SCHWARTZ, J. H., V. F. CASTELLUCCI, and E. R. KANDEL, 1971. Functioning of identified neurons and synapses in abdominal ganglion of *Aplysia* in absence of protein synthesis. *J. Neurophysiol.* 34:939–953.

SHARPLESS, S., and H. JASPER, 1956. Habituation of the arousal reaction. *Brain* 79:655–680.

SHERMAN, R. G., and H. L. ATWOOD, 1971. Synaptic facilitation: Long term neuromuscular facilitation in crustacea. *Science* 171:248–250.

SOKOLOV, E. N., 1963. *Perception and the Conditioned Reflex.* (S. W. Waydenfeld, trans.) Oxford: Pergamon Press.

SPENCER, W. A., and R. S. APRIL, 1970. Plastic properties of monosynaptic pathways in mammals. In *Short-Term Changes in Neural Activity and Behavior*, G. Horn and R. A. Hinde, eds. London: Cambridge University Press, pp. 433–474.

SPENCER, W. A., R. F. THOMPSON, and D. R. NEILSON, JR., 1966. Response decrement of the flexion reflex in the acute spinal cat and transient restoration by strong stimuli. *J. Neurophysiol.* 29:221–239.

SPERRY, R. W., 1965. Selective communication in nerve nets: Impulse specificity vs. connection specificity. *Neurosci. Res. Prog. Bull.* 3(5):37–43.

STEPHENS, C. L., 1972. Progressive decrements in the activity of *Aplysia* neurons following repeated intracellular stimulation. Ph.D. Thesis, University of California, Los Angeles.

STRUMWASSER, F., 1965. The demonstration and manipulation of a circadian rhythm in a single neuron. In *Circadian Clocks*, J. Aschoff, ed. Amsterdam: North-Holland Publishing, pp. 442–462.

STRUMWASSER, F., 1967. Types of information stored in single neurons. In *Invertebrate Nervous Systems: Their Significance for Mammalian Neurophysiology*, C. A. G. Wiersma, ed. Chicago: University of Chicago Press, pp. 291–319.

STRUMWASSER, F., 1973. This volume.

TAUC, L., 1962. Identification of active membrane areas in the giant neuron of *Aplysia. J. Gen. Physiol.* 45:1099–1115.

TAUC, L., 1965. Presynaptic inhibition in the abdominal ganglion of *Aplysia. J. Physiol. (Lond.)* 181:282–307.

TAUC, L., 1966. Physiology of the nervous system. In *Physiology of Mollusca*, Vol. 2, K. M. Wilbur and C. M. Yonge, eds. New York: Academic Press, pp. 387–454.

TAUC, L., 1967. Transmission in invertebrate and vertebrate ganglia. *Physiol. Rev.* 47:521–593.

TAUC, L., and H. M. GERSCHENFELD, 1961. Cholinergic transmission mechanisms for both excitation and inhibition in molluscan central synapses. *Nature (Lond.)* 192:366–367.

THORPE, W. H., 1963. *Learning and Instinct in Animals*, 2nd Ed. London: Methuen.

TINBERGEN, N., 1951. *The Study of Instinct.* Oxford: Clarendon Press.

WALL, P. D., 1970. Habituation and post-tetanic potentiation in the spinal cord. In *Short-Term Changes in Neural Activity and Behavior*, G. Horn and R. A. Hinde, eds. London: Cambridge University Press, pp. 181–210.

WAZIRI, R., and E. R. KANDEL, 1969. Organization of inhibition in abdominal ganglion of *Aplysia*. III. Interneurons mediating inhibition. *J. Neurophysiol.* 32:520–539.

WEISKRANTZ, L., 1970. A long-term view of short-term memory in psychology. In *Short-Term Changes in Neural Activity and Behavior*, G. Horn and R. A. Hinde, eds. London: Cambridge University Press, pp. 63–74.

WICKELGREN, B. G., 1967. Habituation of spinal interneurons. *J. Neurophysiol.* 30:1424–1438.

WILSON, D. L., 1970. Molecular weight distribution of proteins synthesized in single identified neurons of *Aplysia. J. Gen. Physiol.* 57:26–40.

WOODWORTH, R. S., and H. SCHLOSBERG, 1954. *Experimental Psychology*, 2nd Ed. New York: Holt, Rinehart and Winston.

ZUCKER, R. S., 1972. Crayfish escape behavior and central synapses. II. Physiological mechanisms underlying behavioral habituation. *J. Neurophysiol.* 35:621–637.

32 Characterization of Connectivity among Invertebrate Motor Neurons by Cross Correlation of Spike Trains

WILLIAM B. KRISTAN, Jr.

ABSTRACT The cross-correlation histograms generated from spike trains of neuron pairs can detect connections between cells and common input to them. They can help distinguish between these modes of connectivity, even when the spikes occur in bursts. Such histograms were used to detect and quantitate the connections among and to motor neurons in invertebrate preparations. One of these is the group of motor cells involved in swimming in the leech; another is the motor fibers for slow movements of the tail in the crayfish. Electrotonic connections among motor neurons are correlated with behavior in these two systems but not to the extent shown by similar connections in motor systems with a more limited, stereotyped behavioral repertoire.

MANY INVERTEBRATE motor neurons, unlike their vertebrate counterparts, are individually distinguishable. In the leech, for example, Ann Stuart (1970) showed that the five different groups of muscles controlling the behavior of the body wall are differently innervated by identifiable motor neurons. Each segment has two homologous sets of 14 excitatory motor and 3 inhibitory neurons, one set controlling each side. No two motor neurons on each side have exactly the same properties. For instance, of the six motor neurons supplying the layer of longitudinal muscles of each half-segment of the body wall, one called the *L cell* innervates longitudinal muscle fibers in the entire half-segment; another innervates fibers only in the dorsolateral quadrant and a third only those in the ventrolateral quadrant; three others supply only dorsal, lateral, or ventral longitudinal muscle fibers. Of the two inhibitory neurons, one innervates the dorsolateral and the other the ventrolateral quadrant.

Different behavioral acts must necessarily involve different sets of these motor neurons. For instance, Nicholls and Purves (1970) have shown that the L cell is probably responsible for the symmetrical shortening reflex seen in leeches when they are touched. Tactile sensory neurons strongly excite the L cells and cause shortening of all longitudinal muscles on both sides of the segment. The symmetric shortening of the two sides of a segment is assured by a very effective electrotonic junction between the two homologous L cells in each segment (Stuart, 1970).

Swimming in the leech is the result of repeated passages of a head-to-tail wave of alternate contractions of the dorsal and ventral longitudinal muscles in successive segments. This behavior must be caused by phase-locked alternating activity in each set of motor neurons that innervate only dorsal or ventral longitudinal muscles. A study has been recently begun on the neural basis of swimming in the leech (Kristan, Stent, and Ort, unpublished). Extracellular recordings in peripheral regions of segmental nerves in swimming animals repeatedly revealed neurons that generate bursts of spikes only while either the dorsal or the ventral longitudinal muscles are contracting. Intracellular recordings identified some as motor neurons that innervate only the dorsal or ventral longitudinal muscles. To ascertain whether connections among motor neurons are responsible for the pattern of spike activity during swimming, simultaneous intracellular recordings from two neurons at a time are helpful. But, since the spike amplitude recorded from cell bodies of many of the leech motor neurons is quite small, even quite effective synaptic potentials might not be detected, and hence existing connections might be missed. Cross-correlation histograms are generated from the extracellularly recorded spike trains to overcome this problem.

Generation and interpretation of cross-correlation histograms

Cross-correlation histograms measure the likelihood that spikes in two neurons occur at particular times relative to one another (Perkel et al., 1967; Gerstein and Perkel, 1972). For example, two neurons that have excitatory synapses on one another tend to generate spikes at the same time. The strength of this tendency depends on the

WILLIAM B. KRISTAN, JR. Department of Molecular Biology, University of California, Berkeley strengths of the excitatory connections, whereas the relative timing depends upon synaptic delays and the time course of the synaptic potentials; both can be obtained from cross-correlation histograms. Common excitatory input also causes two neurons to generate spikes at nearly the same time.

Figure 1A shows data from two cells which are mutually excitatory, whereas Figure 1B shows data from two cells receiving a common excitatory input. These patterns of connections had been determined by intracellular recordings which, in these pairs of neurons, clearly show synaptic potentials (Hagiwara and Morita, 1962; Kristan, 1971).

Mutually excitatory connections

The paired *colossal* or Retzius cell bodies in each ganglion of the medicinal leech are electrotonically coupled. Nearly every spike in one cell causes a spike in the other, unless the latter is in a refractory state. This strong connection is demonstrated by their nearly superimposable spike time records as shown in Figure 1A1.

In order to generate the cross-correlation histogram between two spike trains, the intervals between each spike in one cell and every spike in the other cell are scored only during some finite time period before and after each reference spike. Positive values are given to cross-intervals in which the second cell's spikes occur later than the reference spike and negative values when the second cell's spikes occur earlier.

The resolution to which cross-intervals are measured and the length of time sampled before and after each spike depend upon the data to be analyzed. For instance, inspection of long spike train records of two Retzius cells showed that the cross-intervals were all less than 13 msec or more than 100 msec. Hence, the part of the histogram that bears on the connectivity of these two cells was the range from -13 to $+13$ msec. In this case, the time resolution of 0.5 msec best showed the time course of the excitation of each cell on the other.

As can be seen in Figure 1A2, there is little tendency for the two cells to generate spikes within 2 msec of one another. There is, however, a strong tendency for cell B spikes to occur 2 to 6.5 msec after those in cell A, and for cell A spikes to occur 3 to 12 msec after cell B spikes. These tendencies are due to the depolarizing electrotonic potentials generated in each cell by spikes in the other (Figure 1A3). These intracellular recordings show that both cells

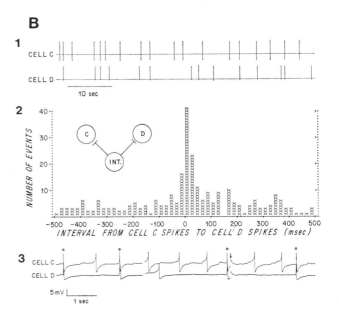

FIGURE 1 Recordings and cross-correlation histograms of the spike activity in pairs of neurons which show nearly synchronous spike times. (A1) Times of occurrence of spikes in the two Retzius cells in the same ganglion of a medicinal leech (simulation of extracellular record using data from intracellular recording). (A2) Cross-correlation histogram generated from 280 spike pairs from the same two cells. The inset shows the manner of connection between the cells, the contacts being excitatory electrical synapses. Bin width = 0.5 msec. (A3) Intracellular recordings of the same two cells, showing the activity caused in one cell by an action potential in the other. (B1) Spike occurrence times from recordings of two cells C and D, in the right pleural ganglion of *Aplysia*. (B2) Cross-correlation histogram using 252 spikes in cell C and 234 spikes in cell D. Inset shows the cause of the peak in the histogram, a common interneuron which provides strong excitatory input to both cells. Bin width = 20 msec. (B3) Intracellular recordings from cells C and D. The arrow marks input from the common interneuron, which causes spikes in both cells on the occurrences marked with dots.

had a progressive, spontaneous depolarization of their membrane potential. In the present data, this slow depolarization caused 148 spikes in cell A (exemplified by the first and third spikes in the upper record in Figure 1A3) and 216 spikes in cell B (exemplified by the second spike in the lower record in Figure 1A3). These spontaneously generated spikes caused concurrent depolarizations in the other cell, which were usually sufficient to cause it to fire. The spikes in the follower cell also generated depolarizations in the leader cell that never caused spikes because of refractoriness in the leader cell. In the 148 cases in which cell A fired first, the resulting depolarization always caused a spike in cell B, usually on the early part of the rising phase of the depolarization. When cell B fired first (216 times), the resulting depolarization was less effective, causing cell A spikes in only 132 cases, which usually occurred near the peak of the excitatory potential. This asymmetry in synaptic effectiveness is reflected in the cross-correlation histogram shown in Figure 1A2. There are 148 points in the positive-valued peak and 132 points in the negative-valued peak. In addition, the negative-valued peak has a longer delay and is more dispersed than the positive-valued peak.

These data were selected to show the appearance of excitatory potentials of different effectiveness in a cross-correlation histogram. The more usual finding is that the more posterior Retzius cell (B) always reaches threshold first and is, therefore, generally the leader cell (Lent, 1972). In such cases, there would be only a negative-valued peak in the cross-correlation histogram. The asymmetry in synaptic effectiveness may be due to the fact that the more anterior Retzius cell (A) in each ganglion has a longer proximal axon (Charles Lent, personal communication). Spike potentials generated by the posterior cell (B) might, therefore, be more attenuated because they are entering a larger cell volume than similar potentials passing in the other direction.

Common excitatory input

Many neurons in the pleural ganglia of the sea hare *Aplysia* show simultaneously occurring large synaptic potentials (EPSPs) believed to result from spikes in a single presynaptic interneuron (Kristan, 1971). EPSPs in two such cells are shown in Figure 1B3. Both cells usually generated spikes (dots), but not always (arrow), when this potential appeared. The postsynaptic cells are not interconnected nor do they receive any other major common input. These cells do generate spikes not caused by the common input, however, and thus show a lesser tendency to fire at the same time (Figure 1B1) than do electrotonically coupled cells (Figure 1A1).

The cross-correlation histogram (Figure 1B2) shows that the long duration of the common excitatory potential produces a broad peak. If the common excitatory potential arrived simultaneously at both cells and was equally effective, the peak would be centered at zero. The peak is centered at 17 msec to the right because the common input, whose arrival time at each cell of the pair varied by less than 1 msec in intracellular records, was more effective in cell C than in cell D.

A peak produced by common input is distinguishable from one produced by synaptic excitation between two cells because it includes very short cross-intervals, e.g. 0.5 msec, or less.

Not every point in the central peak resulted from the common input, because some spikes were due to unrelated excitation of the two cells. The cross-correlation histogram cannot be used to tell *which* of the near-synchronous spikes were chance occurrences, but it can be used to tell *how many* there were. Since all unrelated spikes in the two cells will occur at all cross-intervals with equal likelihood, the cross-correlation, when extended to times that are long relative to the known duration of the input, will provide this number per time unit.

The number due to common input is therefore the total number of points in the peak less the number due to chance. In Figure 1B2, the average number of points per column beyond 200 msec is 3.23. This means that of the 175 spikes generated in cell D within 200 msec of cell C spikes, 65 were due to unrelated excitation and 110 to the common input. The number of mutual spikes caused by the common input, based on counts from the intracellular records, was 104, thus checking well with that of the cross-correlation histogram. Common input which causes a spike in only one or neither cell is not detected by the cross-correlation histogram. Therefore, without intracellular recordings, the 110 common input spikes might be assigned to either 110 common excitatory potentials that were 100% effective or to many more individually less effective ones, or a mixture of the two. When two cells have all their input in common, but all EPSPs are small, the chances of any particular presynaptic impulse bringing both cells simultaneously to threshold would be small, and the cross-correlation histogram would show no peak at zero. Consequently, the size of peaks in the cross-correlation histogram indicates only the lower limit of the number of common excitatory potentials.

Synergistic leech motor neurons

During swimming, synergistic leech motor neurons generate bursts of spikes at one phase of the swim cycle and are silent between bursts. In different cells the bursts differ in duration but overlap in time. Such bursts probably result primarily from periodic bursts in presynaptic fibers common to all synergistic motor neurons (see the discussion section).

Figure 2a shows extracellular recordings from three peripheral nerves on the same side of one segment in a swimming leech. Spikes from only one neuron appear in the nerve A recording, while both the nerve B and C recordings show many classes, each from a single neuron, of spikes distinguishable by their size and shape. The spikes in nerve A are from a motor neuron, cell A, that fires during contraction of the dorsal longitudinal muscles (Lent, unpublished). The largest negative-going spikes in nerve B and the largest spikes in C are also from motor neurons (cells B and C) that generate bursts at the same time.

The shape of the broad peak in the cross-correlation histogram of two of these cells depends upon the pattern of spikes within the burst in each cell. If each neuron generated spikes at a nearly constant but different frequency, the peak would be binomial in appearance. The spike frequency in leech motor neuron bursts is nearly constant at the beginning of the burst but decreases sharply at the end. This gives long tails to the peaks, making them appear more normally distributed.

Burst duration varied from 50 to more than 500 msec in the three cells, and relative onset times of the bursts also varied somewhat. However, the means of the broad peaks in their cross-correlation histograms, shown in Figures 2b to 2d (data from different preparations), are each within 15 msec of zero. Therefore, the center of the bursts in all three cells is consistently at nearly the same time.

As is demonstrated in Figure 1, large common excitatory potentials or interconnections will cause one or more peaks at or near zero that are no wider than the duration of the individual synaptic potentials. The central area of the cross-correlation histogram of cells A and C (Figure 2b) does not show any significant narrow peaks. Therefore, these cells probably have neither an excitatory interconnection nor a common source of large excitatory potentials. However, in the cross-correlation histogram of B and C, a narrow peak occurs in the −5 to 10 msec cross-interval range. Since this peak crosses zero, cells B and C probably receive large excitatory potentials from a common source. The 84 points in this peak represent the minimum number of the 695 spikes in B and 1432 spikes in C that were caused by this common input.

The cross-correlation histogram of cells A and B (Figure 2d) has a narrow peak at +5 to +10 msec, implying an excitatory input from A to B. The 52 points mean that 6.5% of the activity in cell B resulted from this input by cell A.

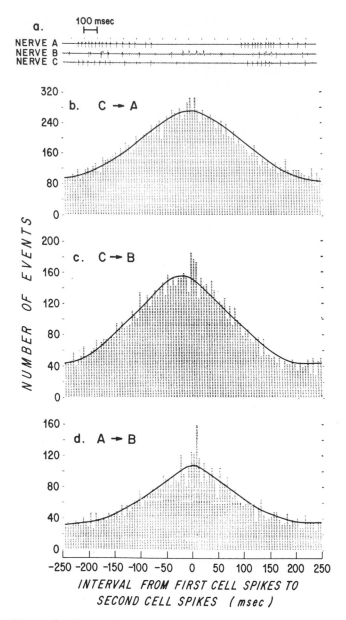

FIGURE 2 Connections among three synergistic leech motor neurons. (a) Extracellular recordings showing three motor neurons that generate bursts of spikes during dorsal longitudinal muscle contraction. Spikes in the nerve A recording are from a single cell called "A," the most negative-going spikes in the nerve B recording are from the cell called "B," and the largest spikes in the nerve C recording are from the cell called "C." (b) Cross-correlation histogram of 1990 A spikes and 2187 C spikes. For this histogram and that of Figures 2c, 2d, and 4c, the solid line enveloping the broad peak was determined by drawing a symmetric bell-shaped curve on the histogram. The actual value in each column was then subtracted from the value predicted by the curve. The whole curve was then shifted up or down by the amount necessary to make the sum of these differences zero. (c) Cross-correlation histogram of 695 B spikes and 1432 C spikes. (d) Cross-correlation histogram of 1125 A spikes and 803 B spikes. Each column in all histograms represents 5 msec.

Synergistic crayfish motor neurons

Figure 3 shows data of William Tatton which were obtained by extracellular recordings from motor neurons which control posture and slow movements of the crayfish tail. The anatomy of this motor system is similar in many respects to the leech longitudinal muscle system. Each of the first five segments in the crayfish tail has a dorsal and a ventral longitudinal set of slow muscles. Contraction in the dorsal muscles causes a lifting or extension of the tail, whereas ventral muscle contraction causes dropping or flexion of the tail. There are six neurons innervating each flexor and extensor muscle on one side of one segment, five being excitatory, with different areas of innervation, and one being inhibitory (Kennedy and Takeda, 1965).

Figure 3a is a recording in which spikes from one group of six flexor neurons can be seen. Typical spikes from the two largest excitatory neurons are marked. Cell 6 fires in bursts of 2 to 4 spikes. Cell 4 is more tonically active, but increases its frequency during cell 6 bursts.

Figure 3b shows a cross-correlation histogram of spike activity in cells 6 and 4 while random electrical shocks were delivered to a sensory nerve. The 0 to +20 msec cross-interval column constitutes a highly significant narrow peak containing 165 points. When a 1 msec time resolution was used, the peak was restricted to the +3 to +5 msec columns, suggesting an excitatory connection from cell 6 to cell 4. Cell 6 generated 368 spikes and 165 of them caused spikes in cell 4.

The solid horizontal line in the histogram marks the average number of events per column at long cross-interval times. There is a broad peak above this line, implying that there are cell 4 bursts that overlap with those of cell 6. The total width of this peak is about 1300 msec. Since cell 6 bursts last no longer than 100 msec, the cell 4 bursts must be considerably longer and are, therefore, only partially due to the excitation from cell 6.

Bilateral homologous leech motor neurons

When a few middle segments of a leech are opened and prevented by fixation from contracting, the free front and

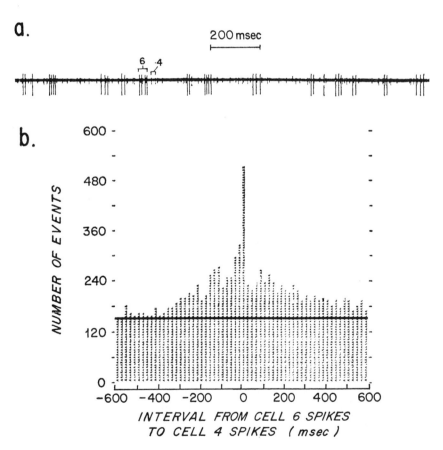

FIGURE 3 Connections between two crayfish tail motor neurons. (a) Extracellular recordings from a peripheral nerve to the flexor muscles showing spikes in all six motor neurons. Typical spikes from the two largest excitatory flexor motor neurons have been labeled "6" and "4," according to the convention of Kennedy and Takeda, 1965. (b) Cross-correlation histogram of 368 cell 6 spikes and 4810 cell 4 spikes. Each column represents 20 msec. (Courtesy of W. Tatton.)

back parts of the animal can still swim in unison. Under these conditions homologous motor neurons on the two sides of a single segment generate very similar bursts of spikes during swimming episodes, as shown in Figure 4a.

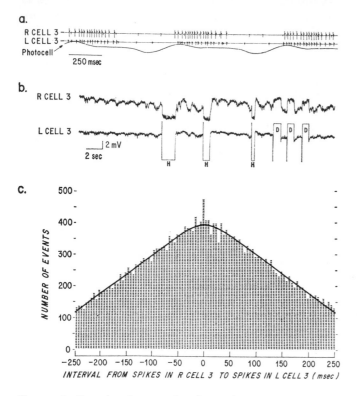

FIGURE 4 Functional strength of an electrotonic junction between bilaterally symmetric motor neurons. (a) Extracellular recordings from homologous peripheral nerves on opposite sides of the same segment in a swimming leech. The largest spikes in each recording are from bilaterally symmetric motor neurons called "cell 3." The bottom trace shows variations in the voltage from a photocell that indicate swimming movements in the anterior part of the leech. (b) Intracellular recordings from the cell bodies of the same two cells in another preparation. Hyperpolarizing (H) and depolarizing (D) currents were passed through the recording electrode into the left cell 3. Since both polarities of current pulses produce corresponding membrane potential changes in the right cell 3, the two cells are electrotonically coupled. (c) Cross-correlation of 2659 spikes in the right cell 3 and 2549 spikes in the left cell 3. Each column represents 5 msec.

The largest spikes in each extracellular record are from two bilaterally homologous motor neurons in a midbody ganglion, whose somata positions within the ganglion are known. Their names, "cell 3," derive from an arbitrary numbering of the motor cells (Kristan, Stent and Ort, unpublished). Intracellular recordings from the two cell bodies show that they are electrotonically coupled (Figure 4b). Both hyperpolarizing and depolarizing potentials induced by passing current into the left cell spread to the right one. Therefore, those two cells are connected in the same manner as the Retzius cells.

If this interconnection were as effective for mutual excitation as in the Retzius cells, the similarity of the bursts in the two cells would be explained. Intracellular recordings were of little use in determining the effectiveness, because in the cell bodies there is little or, as in Figure 4b, no evidence of the spikes recordable in the peripheral nerves.

A cross-correlation histogram was generated from extracellular recordings to measure the effectiveness of spikes in each cell on spike generation in the other. The data from 280 bursts were used to reduce variability in the height of columns in the histogram so that a narrow peak at zero time would be detectable. Such a narrow peak of 5 msec duration was found but contained only 65 points, whereas over 5200 spikes were generated in the two cells. Thus only 1.3% of them could be attributed to excitation passing across the electrotonic junction, showing that in this instance mutual excitation by an electrotonic junction is only of very minor functional importance and that common synaptic input must be the reason for the concurrent bursts.

Discussion

Neither intracellular recording nor cross-histogram analysis can, under all circumstances, unambiguously describe the relationships between individual cells. For instance on the basis of Figure 2, cell A is thought to excite cell B whereas B and C are thought to receive effective common excitation. These conclusions derive from comparing the timing of the narrow peaks in their cross-correlation histograms with those in Figure 1, which were obtained from cells of known connectivity. Unfortunately, alternate interpretations are easily imagined for either type of histogram. The single peak in positive time in Figure 2d might result from common excitation which takes longer to reach cell B. The wider peak in Figure 2c might result from a combination of excitation and an excitatory synapse from cell C to B, with the separate peaks merging into a single one which crosses zero cross-interval time.

Efforts have been made to distinguish between such possibilities on the basis of further analysis of the same spike trains (Gerstein, 1970; Moore et al., 1970). Another technique is possible if the neurons can be individually stimulated. If the frequency of spikes were increased solely in cell A and if it were presynaptic to cell B, the amplitude of the peak in their cross-correlation histogram would be increased. If the peak is due to common input, it would remain the same size. Likewise, activation of cell B or C by themselves should not change the amplitude of their cross-correlation peak if they receive only common input and are not interconnected.

The connections among motor neurons responsible for swimming in the leech are weaker than those found among other motor neuron systems in invertebrates (see Selverston later in this part). Strong synaptic connections among motor neurons severely limit their behavioral repertoire. For instance, the strong excitation of cell 6 on cell 4 in crayfish abdominal system makes it impossible to activate cell 6 without increasing activity in cell 4. In fact, both cells cause contraction in the same muscle, with cell 6 typically being active only during strong contractions (Kennedy and Takeda, 1965). The synaptic contact between them increases the amount of excitation that cell 6 spikes deliver to the muscle. Similarly the very effective electrotonic junction between L cell of the leech (Stuart, 1970) that insures symmetric shortening of a segment does so at the expense of being able to use these motor neurons in turning.

The same muscles the leech uses to swim are used in different combinations to walk, turn, and maintain an upright position in swimming. Therefore, these motor neurons must be less fettered by interconnections, and most of the excitation generating bursts in leech motor neurons, therefore, must come from common input. However, the amount of common input due to individually effective synaptic excitations is small, as shown in Figure 2. Presumably, the common excitation results from summed potentials that are individually weak. Alternately, the motor neurons could receive their excitation from a common presynaptic cell that generates only slow oscillations of its membrane potential but no spikes (Mendelson, 1971).

ACKNOWLEDGMENTS I would like to thank William Tatton and Charles Lent for the use of unpublished data and Donald Perkel for assistance in using his spike train analysis programs. I am also grateful to Donald Kennedy, C. A. G. Wiersma, Theodore Bullock, and especially to Gunther Stent for suggestions in the preparation of this manuscript. This work was done at both Stanford University and the University of California at Berkeley. It was supported by National Science Foundation Research Grant GB 31933X, and a National Science Foundation Postdoctoral Fellowship.

REFERENCES

GERSTEIN, G. L., 1970. Functional association of neurons: Detection and interpretation. In *The Neurosciences Second Study Program*, F. O. Schmitt, ed. New York: Rockefeller University Press, pp. 648–660.

GERSTEIN, G. L., and D. H. PERKEL, 1972. Mutual temporal relationships among neuronal spike trains. *Biophys. J.* 12: 453–473.

HAGIWARA, S., and H. MORITA, 1962. Electrotonic transmission between two nerve cells in leech ganglion. *J. Neurophysiol.* 25:721–731.

KENNEDY, D., and K. TAKEDA, 1965. Reflex control of abdominal flexor muscles in the crayfish. II. The tonic system. *J. Exp. Biol.* 43:229–246.

KRISTAN, W. B., JR., 1971. Plasticity of firing patterns in neurons of *Aplysia* pleural ganglion. *J. Neurophysiol.* 34:321–336.

LENT, C. M., 1972. Electrophysiology of Retzius cells of segmental ganglia in the horse leech, *Haemopus marmorata* (Say). *Comp. Biochem. Physiol.* 42A:857–862.

MENDELSON, M., 1971. Oscillator neuron in crustacean ganglia. *Science* 171:1170–1173.

MOORE, G. P., J. P. SEGUNDO, D. H. PERKEL, and H. LEVITAN, 1970. Statistical signs of synaptic interaction in neurons. *Biophys. J.* 9:876–900.

NICHOLLS, J. G., and D. PURVES, 1970. Monosynaptic chemical and electrical connections between sensory and motor cells in the central nervous system of the leech. *J. Physiol.* 209:647–667.

PERKEL, D. H., G. L. GERSTEIN, and G. P. MOORE, 1967. Neuronal spike trains and stochastic point processes. II. Simultaneous spike trains. *Biophys. J.* 7:419–440.

STUART, A. E., 1970. Physiological and morphological properties of motoneurons in the central nervous system of the leech. *J. Physiol.* 209:627–646.

33 Connections among Neurons of Different Types in Crustacean Nervous Systems

DONALD KENNEDY

ABSTRACT Crustacean nervous systems share many of the advantages found in other invertebrate preparations: Large cell size, restricted numbers of neurons, and consistent neurogeography. In addition, their motor systems are unusually well charted and accessible, and as a result control neurons have been identified not only in terms of their connection with the sensory periphery but also in terms of the coordinated motor outputs they release. These *command fibers* appear, like the sensory interneurons, to be unique entities, each with a set of connections appropriate to an element of behavior. The identification of elements has proceeded to a point at which we are able to describe identified connections at five levels in the nervous system that are responsible for a behavioral act. Sensory neurons differ, however, from the constant, connection-specific central elements: They are added constantly during life, degenerate relatively rapidly following axonotomy, and make less rigidly specified connections with other cells. The central neurons must adjust continually to the presence of new input from this source. The fact that a relatively greater proportion of axons in the central nervous system of the adult comes from this second cell population indicates that earlier estimates of the number of interneurons and motor neurons are too high.

ALMOST EXACTLY twenty years ago, the chairman of our study team delivered a paper, *Neurons of Arthropods*, at the Cold Spring Harbor Symposium (Wiersma, 1952). Both the symposium and Dr. Wiersma's contribution proved to be bench marks in early postwar neurophysiology. The new techniques of intracellular recording were for the first time being applied to the mammalian spinal cord, and it can properly be said that our inventory of the integrative capacity of central neurons was just beginning. Wiersma was able to outline a number of respects in which the study of arthropod nerve cells might help to advance that goal. First, he pointed out that arthropods have a markedly reduced number of neurons compared with that found in vertebrates and suggested that single cells in arthropods may therefore have to perform a more complex set of integrative functions. He reviewed evidence that the

DONALD KENNEDY   Department of Biological Sciences, Stanford University, Stanford, California

lateral giant fibers of decapod Crustacea, for example, frequently conduct in both directions as a consequence of their anatomical organization, and he pointed to the broad variety of synaptic contacts made by these identified cells. Finally, he showed that the diversity of output junctions made by a single neuron can make it the trigger for a complex, highly differentiated behavioral output.

Since we have come to expect a good deal of Wiersma's crystal ball, it is perhaps not surprising that his remarks in 1952 provide such a useful outline for a treatment of the arthropod central nervous system in 1972. The major conceptual advances that have been made in the intervening two decades can, I think, be summarized as follows: (1) The concept of *identified cells* has been extended from its application to a few, highly specialized large neurons—the giant fibers and a few others—to embrace, if not the entire central nervous system, at least a reasonably large proportion of the central interneurons. Based mainly on studies in decapod Crustacea, it seems clear that each central interneuron in the arthropod nervous system is a unique entity, connecting with some special set of input fibers and having an equally specific array of output connections. (2) It is now widely believed that this specificity of connections in the arthropod nervous system is established in part by stereotyped branching by the neurons during development. This conclusion has been reached through the application of new dye-injection techniques that allow the spread of a fluorescent or electron-dense label throughout the processes of the cell and permit a complete reconstruction. Other neurogeographic studies in arthropods, annelids, and molluscs have shown that the arrangement of identified cell somata with respect to one another is also highly constant, as shown in the communications by Kandel, Hoyle, and Selverston elsewhere in this part. (3) Synaptic organization in arthropod neuropil, though it differs in some morphological respects from that in the vertebrate central nervous system, builds upon the same synaptic mechanisms found in vertebrates. Presynaptic inhibition, electrotonic junctions, conventional chemical EPSPs and IPSPs all occur—although in different proportions and sometimes in unexpected functional roles. (4) Many of the "complex" features of patterned motor output—reciprocity, phase-constancy, and the like—are the products not of continuous reflex control from the periphery during the execution of the behavior, but, instead, of central connections among a relatively small population of neural elements.

It is in connection with this last issue that arthropod nervous systems have seen their hardest experimental service. My late colleague, Don Wilson, reviewed this problem at the last Boulder conference (Wilson, 1970), and I therefore intend to pay relatively little attention to the important subject of pattern-generating networks. It does seem appropriate, however, to summarize the evidence for the other generalities that have emerged during the past twenty years, and then to deal with some challenges to them. I intend to focus on crustacean nervous systems entirely, for these have been a kind of "proving ground" for new technologies and insights that could then be applied to insects. In the long run, one must concede that the insects are the most promising of the organisms with "reduced" nervous systems, both because they possess a wider range of behaviors and—most important—because they have accessible genetics. Crustaceans, however, although they don't fly, sing, or reproduce well in the laboratory, do one thing better than insects: They grow. Because of their size, one can do things with them that one would hardly dream of trying on an insect unless emboldened by success with a similar but easier preparation. Thus the major technical breakthroughs in understanding neuropil—intracellular recording and dye injection—have both been first accomplished in Crustacea and then successfully exported to insects.

Identified cells

Although it was known by 1950 that a variety of animals possessed fixed numbers of unusually large, highly specialized neurons that could be identified in one preparation after another, it was not established even for the smallest divisions of these nervous systems that they consisted of arrays of "unique" cells. In his 1952 Cold Spring Harbor paper, Wiersma alluded to the stimulation of several different central neurons that evoked distinctive behavioral effects, and over the next ten years he produced a series of papers demonstrating for a much larger number of cells that there were also unique input connections. First in the circumesophageal connectives (Wiersma, 1958) then in the abdominal nervous system (Wiersma

and Hughes, 1961; Wiersma and Bush, 1963), it was shown that nerve fibers with identical peripheral receptive fields could repeatedly be isolated from a particular part of the cross section of a central connective. Each such entity is unique: For example, of the set of interneurons receiving input from abdominal tactile hairs, one may connect with hairs on both sides of all six segments, another from a particular dorsal patch on only one segment, another from three, another from five, and so on; a particular locus in the periphery is multiply represented centrally, but each time it converges with a unique combination of other peripheral loci. Although the position of identified interneurons in the central connectives is not exactly constant, there is little difficulty in identifying the large ones, and the precision with which their receptive fields are repeated from animal to animal is a fact we can fully confirm. A total of about 120 interneurons (in addition to some sensory fibers) have been identified and mapped by this technique.

For the same purpose, central elements can be tested by stimulating them and seeing what happens. Wiersma first showed that stimulating fine filaments of nerve from central connectives could produce complex, stereotyped behavioral output; Wiersma and Ikeda (1964) described several such neurons that control the swimmeret rhythm and called them *command* fibers. Since then it has been shown (Kennedy et al., 1967) that impulse activity restricted to a single cell can produce coordinated motor output involving some 300 motor neurons.

In our laboratory we have attempted to establish the functional uniqueness of the premotor elements that produce the same general class of behavioral result. I will cite four systems of increasing complexity in which command interneurons can be differentiated by the details of their peripheral effect. (1) In slow abdominal flexor muscles of the crayfish, stimulation of a single central neuron produces excitation of the motor neuron, central inhibition of the motor neuron to the antagonistic muscle, and reciprocal effects on the peripheral inhibitory axons that also innervate these muscles. If one records independently from several segments, it is clear that each one produces a special segmental ratio of effect and thus encodes a particular geometry of movement (Evoy and Kennedy, 1967). (2) Extensor command fibers may excite motorneurons that go exclusively to "working" extensor muscles, or they may excite motorneurons that also innervate the receptor muscle of the muscle receptor organ. Like the activity of gamma motorneurons, this engages a myotatic reflex loop that provides for load compensation by supplying additional excitation to the extensor muscles (Fields, 1966; Fields, Evoy, and Kennedy, 1967). Thus two different command routes exist, one the equivalent of alpha activation, the other the equivalent of alpha-gamma coactivation; both are pre-programmed, and the differentiation between them is on the basis of output connections from the command interneurons. (3) In the motor systems that control positioning of an appendage in three dimensions, single command interneurons may affect output to several antagonistic sets of muscles. Command fibers produce different positions by evoking different ratios of activity in the several sets (Larimer and Kennedy, 1969). (4) Finally, interneurons commanding complex, cyclical locomotor programs in a series of appendages may produce special effects on such different parameters of the output as ratios of the alternating flexor and extensor discharges, specific portions of the dynamic range, and so on (Davis and Kennedy, 1972). For present purposes, the interesting outcome is that each central element is again unique; that is, it makes a set of connections not quite like any other interneuron and thus produces a special output pattern. Interneuronal organization on the motor side thus mirrors that on the sensory side: The cells are connected with great specificity to their output elements, and no one is just like any other.

Constancy of form

In the Cold Spring Harbor paper mentioned earlier, Wiersma described arthropod neuropil as "a complicated felt-work of nerve fibers, in which it is not possible to trace the connections." This situation was very difficult to deal with until the development of dyes that, after injection into single cells, could be expected to stay inside them and diffuse to the tips of even their finer processes. Stretton and Kravitz (1968) found that the fluorescent dye, Procion yellow, fulfilled both of these criteria and also survived subsequent fixation. They used it to reconstruct the branching pattern of a flexor inhibitor neuron in the lobster. We applied it to a series of flexor motor neurons and interneurons in the crayfish in order to test proposals about their uniqueness and the constancy of their branching and connection patterns (Remler, Selverston, and Kennedy, 1968; Kennedy, Selverston, and Remler, 1969; Selverston and Kennedy, 1969; Selverston and Remler, 1972). At least to the level of major branches, the patterns proved quite consistent for individual motor neurons, and it moreover proved possible to visualize direct connections between the fast flexor motor neurons and the central giant fibers: Connections could be physiologically verified when branches from the motor neurons appeared to contact the axis cylinders of the giant fibers and were found to be absent when such contacts were not apparent. The question of constancy in such cells is dealt with in more depth by Selverston elsewhere in this part.

Nature of the connections

The first intracellular recordings in arthropod neuropil (Preston and Kennedy, 1960; Kennedy and Preston, 1960) demonstrated all the graded forms of inhibitory and excitatory transmission that one had come to expect from central nervous structures elsewhere. Although some cell bodies of motor neurons show passive invasion by spikes (Takeda and Kennedy, 1964), most are electrically inexcitable. It is now clear that most of the interneuron somata, like the lateral giant cell illustrated in Figure 1, are so far from the sites of electrical activity in the neuropil that they are quite useless for monitoring; thus a method that has worked so well in molluscan ganglia—that of using the neurogeography of somata as the basis for a physiological identification of cells—has failed for most crustacean cells except the large motor neurons.

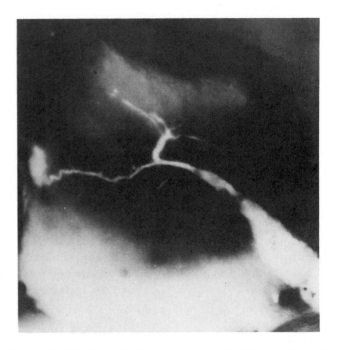

FIGURE 1 The lateral giant neuron in the third abdominal ganglion of the crayfish, injected with Procion yellow and photographed in a whole mount by fluorescence microscopy (Remler, Selverston, and Kennedy, 1968).

More recently, two more novel features of connections in neuropil have emerged. First, it is clear that dendritic branches of motor neurons and interneurons are capable of supporting all-or-none activity, and that such branch spikes summate with graded postsynaptic potentials to determine firing (Takeda and Kennedy, 1964, 1965; Sandeman, 1969). Second, a large proportion of the synaptic activity generated in sensory neuropil by afferent stimulation, or by activity in identified interneurons, is mediated by electrical connections (Zucker et al., 1971).

At this point, the correlations between structure and function in neuropil begin to break down. Although on physiological grounds many identified neurons in the crayfish central nervous system must receive electrical connections, it has not yet been possible to reveal appropriate morphological substrates for them by electron microscopy. Although "gap junctions" appear at such known sites of electrical transmission as the septum between lateral giant fibers and the junctions between giant and motor giant axons, they have not been found on the lateral giant axons in neuropil, even though electrical transmission is known to occur there (Krasne and Stirling, 1972; our own observations). Neither do they appear at other known sites of electrical transmission (Sandeman and Mendum, 1971). There are other difficulties: Even in some of the situations where electrical transmission is known to occur, synaptic vesicles are selectively concentrated at the presynaptic side of the junction. Methods of separating electrical from chemical synaptic action that are considered reliable in other situations fail to make the discrimination here: Lowered calcium and increased magnesium *do not* interfere with chemical synaptic transmission in crustacean neuropil, and conversely lowered calcium *does* affect electrotonic junctions (Asada and Bennett, 1971).

These features may perhaps best be described in connection with a specific piece of behavior; the account is taken from the work of Zucker et al. (1971). When the abdomen of a rested crayfish is tapped, the animal responds with a tail flip, incomplete in the most caudal segments, that propels it up and away from the stimulus. This behavior is known to result from a single impulse in the lateral giant fibers (Wine and Krasne, 1972), which make different connections upon motor neurons than do the medial giant fibers and produce a distinctive response (Larimer et al., 1971). It has been possible to analyze identified neurons at half a dozen hierarchial levels concerned with this behavior (Figure 2), and I treat each of them here briefly.

1. *Hair receptors* associated with the exoskeleton are deformed by the stimulus and carry impulses centrally; these have a central action at two levels. They provide an excitation to lower-order interneurons that is essentially chemical in nature, probably mediated by acetycholine (Barker et al., 1972); this response "follows the stimulus" only at relatively low frequencies. The same set of sensory neurons also makes electrical contact with higher-order interneurons, including identified cells each of which receive input in a number of adjacent body segments.

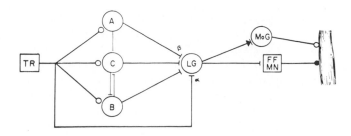

FIGURE 2 Circuit for excitation of the "tail flip" reflex by tactile stimuli. TR: tactile receptors; A: primary interneuron; B and C: second order interneurons; LG: lateral giant neuron; FFMN: fast flexor motor neuron; MoG: motor giant neuron. Circles signify chemical synapses, which show rapid depression in all cases where the circles are open; bars signify electrical synapses. (From Zucker, Kennedy, and Selverston, 1971.)

2. *Primary interneurons* receive exclusively chemical excitation from the afferents. They in turn synapse electrically on all interneurons of higher order.

3. *Multisegmental interneurons.* These receive electrical connections from primary afferents and from at least some primary interneurons and activate the lateral giant neuron electrically.

4. *Lateral giant interneuron.* This "motor" interneuron receives an exclusively electrical input from all the above elements.

5. *Motor neurons.* These are excited by the lateral giant fibers, and probably by other elements of lower order. The majority of their input is electrical. They supply chemical excitation to phasic flexor muscles of the abdomen. At the neuromuscular junctions, additional temporal labilities are introduced (Bruner and Kennedy, 1970).

This brief account emphasizes several features of the neural network involved in escape behavior. First, it is a kind of cascade, with short loops featuring electrical excitation and longer ones that are begun by a chemical stage. This last feature makes them subject to depression in response to repetition of the stimulus, and for that reason this part of the circuit shows substantial lability. The electrical component of the primary afferent volley is, by itself, unable to activate the final common neuron that acts as the behavioral trigger. Sensory neurons have a dual synaptic action, electrically exciting some neurons and chemically exciting others; both roles may be played by branches of the same sensory neuron. Finally, temporal lability is evident only at the first and last stages of the circuit.

Unfortunately, however, the only explicit statements we can make about cell identity and connectivity in this circuit involve the centrally located interneurons and motor neurons. What about the specificity of relations between this center and the sensory periphery?

Afferent axons in the arthropods develop from peripherally located cell bodies, which arise by the division and differentiation of epithelial cells. We have found that these cells differ from central neurons in their specificity of connection, in their response to axonotomy, and in their pattern of growth and development. While these new findings threaten one or two generalities about connectivity in these nervous systems, they save us from at least one potential embarrassment.

The first-stage interneuron described above, which responds to water-borne vibrations that activate exoskeletal hairs on the tail fan, has provided an opportunity to analyze the connections made by sensory fibers. A map of the peripheral region that supplies axons to the sixth ganglion via the fourth root is shown in Figure 3. Each hair sends one and sometimes two axons to the central nervous system, and these may either stop in the ganglion of entry or continue to higher levels. Hairs in the more rostral portion of the field (especially P_1, P_2, P_3, L_1 and S_1 in the figure) show higher thresholds and give more phasic responses. In experiments begun in collaboration with Allen Selverston, we have measured the conduction velocity of afferent fibers from single hairs to determine their relative size and recorded the unitary excitatory postsynaptic potentials they produced in the interneuron (Figure 4). The four or five largest proximal cells are very likely to make connections with the interneuron; they do so in at least 80 or 90% of the cases. More distally, predictions are less easy to make; cells in the central region may show a hit-or-miss pattern of connection, and in the most distal or lateral areas only an occasional hair will produce a unitary synaptic potential. There is *no* correlation between fiber diameter and synaptic efficacy, but the larger afferents are more likely to make connections.

For present purposes, the interesting feature is that the neurons belonging to this system do not show the rigorously specified connection program that seems to be characteristic of central neurons.

Sensory and central neurons differ in other ways. Several years ago, we first showed that crustacean motor axons have a remarkable response to axonotomy. For up to several months following section of the axon, the distal, enucleated fiber will still conduct impulses and release transmitter in normal amounts—even after what should be exhausting tetani containing 100,000 impulses (Hoy, Bittner, and Kennedy, 1967). Observations made in our laboratory, in collaboration with George Bittner, agree with those of Nordlander and Singer (1972) in showing that the distal axons survive but have expanded glial enwrappings.

The mechanisms used by the central stumps of such

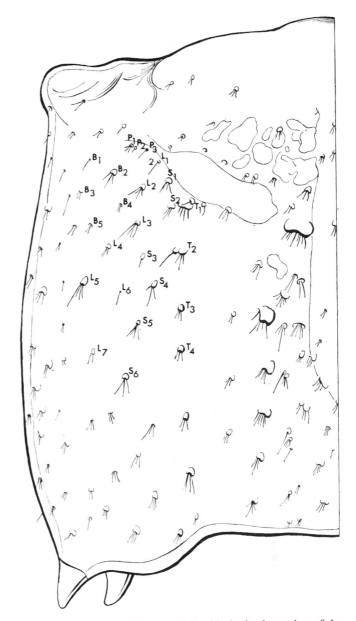

FIGURE 3 Drawing of the exoskeletal hairs in the region of the crayfish telson innervated by root 4 of the sixth ganglion. The hairs make connections with identified interneurons, as analyzed in Figure 4.

axons to reconnect with the muscle are not known. However, our observations, and subsequent ones by Hoy (1969) and by Bittner (unpublished), are consistent with the idea that the outgrowing central axon reconnects with the surviving distal segment. Nordlander and Singer (1972) disagree, primarily on the basis of having seen multiple axonal processes in a region distal to the point at which the cut was made. The matter is not yet resolved; what is clear is that the regeneration is correct, in the sense that outgrowing motor axons reconnect with their own muscles so that adaptive reflex responses are restored.

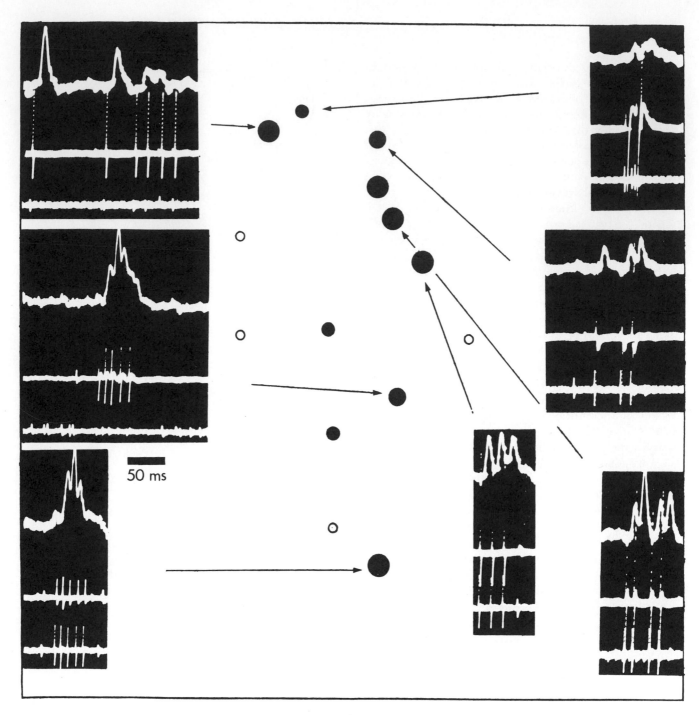

FIGURE 4 Receptive field or "connection" map for sensory hairs in the region shown in Figure 3. *Lower traces:* Proximal and distal *en passant* recordings from the fourth root, sixth ganglion. *Top traces:* Unitary EPSPs recorded from a first-stage interneuron. The size of the spot marking the hair's position reflects the relative amplitude of the first unitary EPSP in a train. For further details see Kennedy (1971).

Central interneurons respond in much the same fashion; they take a very long time to degenerate after sectioning, as has been observed by Wine (1973) for the medial giant axon, and they restore approximately appropriate central connections upon regeneration.

We have now shown, both by electrophysiological tests and by electron microscopy, that sensory neurons behave in an entirely different way. Clear signs of degeneration are apparent in axons separated from their peripherally located cell bodies for two weeks, and after four weeks they are nearly unrecognizable as neurons. Between these times, they lose their capacity to activate other neurons synaptically. Outgrowth of the axons still attached to their perikarya is vigorous, so that limited restoration of function in a crushed root takes place within three weeks.

There is now abundant evidence that these restored connections are appropriate *in a general way*. After severing the sensory nerves to the sixth abdominal ganglion, we find (Kennedy and Wine, work in progress) that tactile afferents make new connections with ascending interneurons even after taking tortuous, ectopic routes into the ganglion. These are roughly correct, in that hairs in the appropriate region of the tail reestablish their normal pattern of contacts upon particular interneurons. Edwards and Sahota (1967) and Edwards and Palka (1971) have shown that regenerating afferents from cricket cerci also reconnect with ascending "giant" interneurons. Here, too, the analysis is relatively coarse: It does not speak to the question of whether particular identified cells reconnect with other specific neurons. Thus we can only conclude that in arthropods the extent of connection specification for sensory fibers relates a peripheral *region* to particular central neurons; as far as we now know, the sensory axons display no more specificity than that during the reconnection process.

In addition to these differences in connection specificity and regeneration patterns, the sensory neurons show another unique feature: They are continually added during life. Crayfish hatch from the egg in nearly the adult form, but with a length of only 5 mm or so. They then undergo a series of molts, at first at intervals of a few days and eventually of weeks or months, that takes them to adult size in one to two years—an increase of about twentyfold in length. As far as we know, the very young animal has a full complement of central and motor neurons—at least all the large and obvious ones can be found. But the periphery is adding new sensory structures all the time: The number of ommatidia in the eye is increasing, more cuticular hairs are being produced, and the number of sensory fibers in peripheral roots is multiplying. Figure 5 gives some values measured on montages prepared from electron micrographs of the fourth root of ganglion 6 (a sensory root) and of the connective between abdominal ganglia 5 and 6. The number of sensory axons increases by nearly tenfold during development.

We now believe that these findings can be applied to the receptive field organization discussed above. Although some addition of new exoskeletal sensory structures takes place between already formed elements, much of it occurs at the lateral and distal margins of the segment. The axons innervating these marginal hairs are small, judging from the amplitudes of their extracellularly recorded impulses.

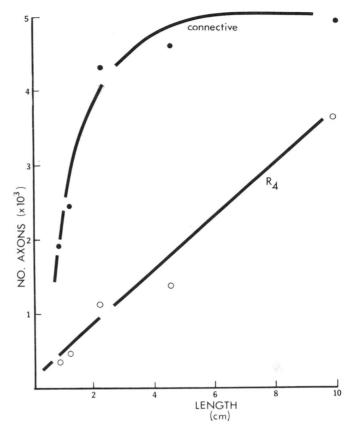

FIGURE 5　Number of axon profiles in the 5–6 connective and the fourth root of the sixth ganglion as a function of size: All points are counts from montages such as that shown in Figure 6.

We suppose that the largest axons belong to the "pioneer" hairs—those that developed earliest. These are also the elements that have the highest probability of connecting with the identified interneurons, though the occasional connections made by axons from the marginal hairs are just as effective as those formed by the larger ones. According to this hypothesis, age both increases cell size (the diameter enlargement is of the order of tenfold for the largest axons) and confers an improved probability that connection will have occurred.

The other curve in Figure 5 makes a separate point that may provide comfort for those of us who believe that arthropod nervous systems hold out some hope for complete circuit analysis. We have all been worried, as Wiersma was in 1952, about the number of fine fibers in the central nervous system. After Cold Spring Harbor, things got worse instead of better: The electron microscope turned up previously uncounted small axons in the central connectives, and the estimates were revised upward (see Figure 6). In young animals, we find that the number of neurons in a 5–6 abdominal connective is only about 1800, as compared with the adult count of 5000. The increase we presume to be due mainly to the addition of sensory axons, and the main difference in the numbers seems to come in several discrete pockets of very small axons (Figures 6 and 7). If these assumptions are correct, we can use the ratios of increase in the sensory roots and in the CNS to estimate the number of sensory axons present in the CNS at hatching. This number turns out to be about 1200 cells, leaving only 600 cells per connective as the complement of "central" neurons.

In several respects, then, the sensory cells have a different biology from the central neurons. The former are epithelial in origin, and thus arise from a continuous source of newly differentiated products. The sensory cells

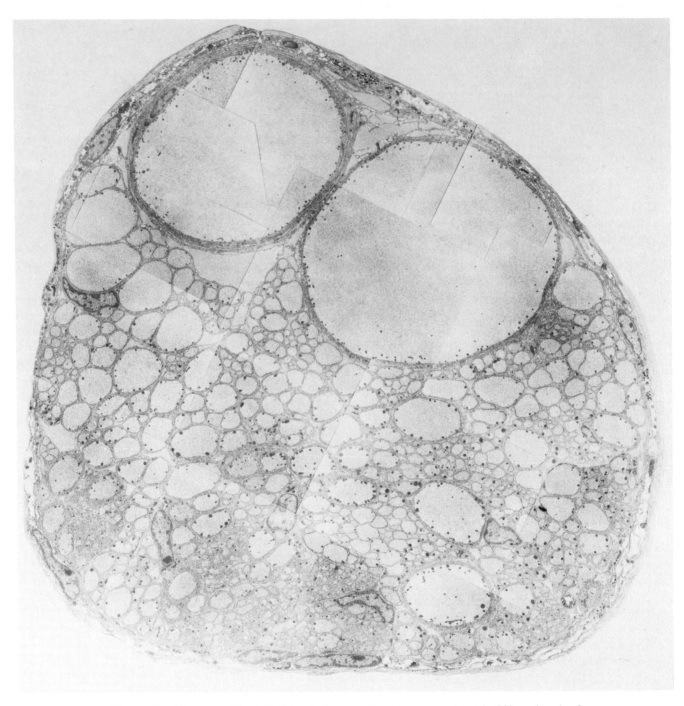

FIGURE 6 Montage of the 5–6 abdominal connective in an approximately fifth-molt animal. The largest axon profile (top right) is the lateral giant; its diameter is approximately 20 μ.

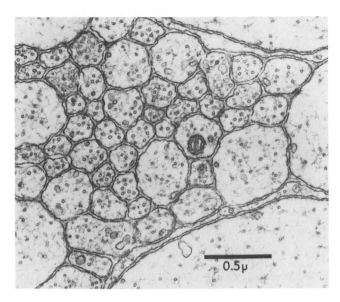

FIGURE 7 Detail of a region of small axons from the 5–6 connective illustrated in Figure 6.

exhibit an "open" developmental history that contrasts with that of the central cells that display fixed number and early specification. Sensory neurons show a much heavier dependence upon the presence of a nucleus and may well regenerate by an entirely different mechanism. Their connections are apparently less individually specified than those of the central elements. (Parenthetically it should be noted that some sensory cells in annelids, molluscs, and certain arthropods have been dealt with as unique, identified cells with individually specified connections. Virtually all of these cases involve neurons having central, rather than peripheral, somata. This important difference suggests that such cells may have originally been central neurons and developed their sensory function secondarily.)

The fixed central elements must adjust continuously to the growth of input provided by the late-developing sensory axons. The duality of the neuronal population raises important questions about the nature of this accommodation. Is late-arriving input concentrated in distal dendritic regions and, hence, less effective? Our experiments to date suggest not. Are the connection probabilities of late-arriving fibers improved by accidental denervation of inputs established earlier? Indeed, are there any alterations in connectivity as a result of derangements in the developmental pattern? We are now taking advantage of the unique identity of the central elements to ask these questions of a second set of nerve cells that, though less unique, may just on that account be more interesting.

ACKNOWLEDGMENT Recent work from the author's laboratory was supported by Grant NB 02944 from the U.S.P.H.S. I acknowledge the help of Joan Misch with the electron microscopy; Joanna Hanawalt provided unflagging technical assistance and criticized this manuscript. I also benefited from helpful discussions with colleagues who deserve mention quite apart from their contributions discussed in the text: Jeffrey Wine, Robert Zucker, Ronald Calabrese, Richard Roth, Jan Letourneau, and Paul Grobstein.

REFERENCES

ASADA, Y., and M. V. L. BENNETT, 1971. Experimental alteration of coupling resistance at an electrotonic synapse. *J. Cell Biol.* 49:159–172.

BARKER, D. L., E. HERBERT, J. G. HILDEBRAND, and E. A. KRAVITZ, 1972. Acetylcholine and lobster sensory neurons. *J. Physiol.* (*London*), (in press).

BRUNER, J., and D. KENNEDY, 1970. Habituation: Occurrence at a neuromuscular junction. *Science* 169:92–93.

DAVIS, W. J., and D. KENNEDY, 1972. Command interneurons controlling swimmeret movements in the lobster. I. Types of effects on motoneurons. *J. Neurophysiol.* 35(1):1–12.

EDWARDS, J. S., and J. PALKA, 1971. Neural regeneration: Delayed formation of central contacts in insect sensory cells. *Science* 172:591–594.

EDWARDS, J. S., and T. S. SAHOTA, 1967. Regeneration of a sensory system: The formation of central connections by normal and transplanted cerci of the house cricket *Acheta domesticus. J. Exp. Zool.* 166(3):387–396.

EVOY, W. H., and D. KENNEDY, 1967. The central nervous organization underlying control of antagonistic muscles in the crayfish. I. Types of command fibers. *J. Exp. Zool.* 165:223–238.

FIELDS, H. L., 1966. Proprioceptive control of posture in the crayfish abdomen. *J. Exp. Biol.* 44:455–468.

FIELDS, H. L., W. H. EVOY, and D. KENNEDY, 1967. Reflex role played by efferent control of an invertebrate stretch receptor. *J. Neurophysiol.* 30:859–874.

HOY, R. R., 1969. Degeneration and regeneration in abdominal flexor motor neurons in the crayfish. *J. Exp. Zool.* 172(2):219–232.

HOY, R. R., G. D. BITTNER, and D. KENNEDY, 1967. Regeneration in crustacean motoneurons: Evidence for axonal refusion. *Science* 156:251–252.

KENNEDY, D., 1971. Crayfish interneurons. *The Physiologist* 14:5–30.

KENNEDY, D., W. H. EVOY, B. DANE, and J. T. HANAWALT, 1967. The central nervous organization underlying control of antagonistic muscles in the crayfish. II. Coding of position by command fibers. *J. Exp. Zool.* 165:239–248.

KENNEDY, D., and J. B. PRESTON, 1960. Complex responses of central neurons in the crayfish to presynaptic and direct stimulation. *Anat. Rec.* 148:360.

KENNEDY, D., A. I. SELVERSTON, and M. P. REMLER, 1969. Analysis of restricted neuronal networks. *Science* 164:1488–1496.

KRASNE, F. B., and CH. A. STIRLING, 1972. Synapses of crayfish abdominal ganglia with special attention to afferent and efferent connections of the lateral giant fibers. *Z. Zellforsch.* 127:526–644.

LARIMER, J. L., A. C. EGGLESTON, L. M. MASUKAWA, and D. KENNEDY, 1971. The different connections and motor outputs of lateral and medial giant fibers in the crayfish. *J. Exp. Biol.* 54:391–402.

LARIMER, J. L., and D. KENNEDY, 1969. The central nervous control of complex movements in the uropods of crayfish. *J. Exp. Biol.* 51:135–150.

NORDLANDER, R. H., and M. SINGER, 1972. Electron microscopy of severed motor fibers in the crayfish. *Z. Zellforsch.* 126:157–181.

PRESTON, J. B., and D. KENNEDY, 1960. Integrative synaptic mechanisms in the caudal ganglion of the crayfish. *J. Gen. Physiol.* 43:671–681.

REMLER, M. P., A. I. SELVERSTON, and D. KENNEDY, 1968. Lateral giant fibers of crayfish: Location of somata by dye injection. *Science* 162:281–283.

SANDEMAN, D. C., 1969. The site of synaptic activity and impulse initiation in an identified motoneuron in the crab brain. *J. Exp. Biol.* 50:771–784.

SANDEMAN, D. C., and C. M. MENDUM, 1971. The fine structure of the central synaptic contacts on an identified crustacean motoneuron. *Z. Zellforsch.* 119(4):515–525.

SELVERSTON, A. I., and D. KENNEDY, 1969. Structure and function of identified nerve cells in the crayfish. *Endeavour* 27:105–108.

SELVERSTON, A. I., and M. P. REMLER, 1972. Neural geometry and activation of crayfish fast flexor motorneurons. *J. Neurophysiol.* 35:797–814.

STRETTON, A. O. W., and E. A. KRAVITZ, 1968. Neuronal geometry: Determination with a technique of intracellular dye injection. *Science* 162:132–134.

TAKEDA, K., and D. KENNEDY, 1964. Some potentials and modes of activation of crayfish motoneurons. *J. Cell Comp. Physiol.* 64:165–182.

TAKEDA, K., and D. KENNEDY, 1965. The mechanism of discharge pattern formation in crustacean interneurons. *J. Gen. Physiol.* 48:435–453.

WIERSMA, C. A. G., 1952. Neurons of arthropods. *Cold Spring Harb. Symp. Quant. Biol.* 17:155–163.

WIERSMA, C. A. G., 1958. On the functional connections of single units in the central nervous system of the crayfish, *Procambarus clarkii* (Girard). *J. Comp. Neurol.* 110:421–471.

WIERSMA, C. A. G., and B. M. H. BUSH, 1963. Functional neuronal connections between the thoracic and abdominal cords of the crayfish, *Procambarus clarkii* (Girard). *J. Comp. Neurol.* 121:207–235.

WIERSMA, C. A. G., and G. M. HUGHES, 1961. On the functional anatomy of neuronal units in the abdominal cord of the crayfish, *Procambarus clarkii* (Girard). *J. Comp. Neurol.* 116:209–228.

WIERSMA, C. A. G., and K. IKEDA, 1964. Interneurons commanding swimmeret movements in the crayfish, *Procambarus clarkii* (Girard). *Comp. Biochem. Physiol.* 12:509–525.

WILSON, D. M., 1970. Neural operations in arthropod ganglia. In *The Neurosciences: Second Study Program,* F. O. Schmitt, ed. New York: Rockefeller University Press, pp. 397–408.

WINE, J., 1973. Invertebrate central neurons: Orthograde degeneration and retrograde changes following axonotomy. *Experimental Neurol.* (in press).

WINE, J., and F. B. KRASNE, 1972. The organization of escape behavior in crayfish. *J. Exp. Biol.* 56:1–18.

ZUCKER, R. S., D. KENNEDY, and A. I. SELVERSTON, 1971. Neuronal circuit mediating escape responses in crayfish. *Science* 173:645–650.

34 Synaptic and Structural Analysis of a Small Neural System

ALLEN I. SELVERSTON and BRIAN MULLONEY

ABSTRACT The functional connectivity patterns in the crayfish fast flexor and the lobster stomatogastric system are constant from animal to animal. Studies utilizing new histological methods indicate that the topology of the individual neurons involved in these motor systems is less consistent. The way in which such anatomical variation can produce functionally equivalent circuits is discussed.

BECAUSE IT HAS been so difficult to determine the detailed circuitry of complex neuronal networks, many neurobiologists have instead sought to analyze their global properties by correlating changes in electrical fields with different behavioral states of the animal. This approach has not been very successful.

If we consider smaller neuronal systems, such as those found in the central nervous systems of some invertebrates, we see that it is neither necessary nor desirable to look for such global properties as EEGs. One can instead describe these systems in terms of the properties of their individual elements and the synaptic relationships between individual elements. Perhaps such a reductionist methodology may never be applicable to the nervous systems of higher animals, because invertebrate nerve cells might somehow differ in their integrative properties. Similarly, a detailed analysis of single invertebrate cells may fail to consider adequately any unique properties of the network as a whole. We take the view that understanding the mechanisms by which small neural networks produce patterned motor output will reveal principles that apply to more complex networks. If all constraints between the nerve cells in a system can be described quantitatively, they can be used effectively to model the system, that is, to describe and predict the output of the whole system. We think that such a component analysis provides an adequate description of the system.

Many nervous systems can generate complex patterns of activity in sets of motor neurons even when isolated completely from all sensory input. Such patterns resemble more or less the patterns associated with particular modes of behavior in the intact animal. We are now close to a description of the mechanisms that generate stereotyped behavior in a few small nervous systems in arthropods, relying heavily on the presence of identifiable neurons.

Identifiable neurons

An identifiable neuron is a neuron that can be found in, and that performs the same physiological function in, each individual of a species. Often these neurons exist in bilaterally symmetric pairs, each member being a mirror image of the other with regard to its anatomy. Vertebrate neurons that can be identified as to class—for example, Purkinje cells or motor neurons innervating the extensor digitorum longus—can be reliably assigned as belonging to a particular functional class, but each member cannot reliably be distinguished in different animals. They are not, therefore, identifiable neurons. Identifiable neurons, singly or in combination, permit the experimenter to control experiments properly. Identifiable cells are thus as important to the physiologist as pure chemicals to the molecular biologist.

The stomatogastric ganglion

The stomatogastric ganglion (St G) of the lobster *Panulirus interruptus* generates all the coordinated movements of the animal's stomach. Unlike vertebrates, the stomach of the lobster originates from ectodermal tissue, is lined with chitin, contains no smooth muscle, and has a set of internal teeth that chew food after it enters the stomach. Striated muscles move the stomach, and the motor neurons of the stomatogastric ganglion control their contractions. Early workers (Huxley, 1880; Parker, 1876), in a burst of over-enthusiasm for comparative anatomy, gave the various parts of the crustacean stomach names identical to those of the vertebrate stomach; all these homologies are false. Worse, the nervous system that controls the stomachs of Crustacea has been incorrectly homologized with the

ALLEN I. SELVERSTON and BRIAN MULLONEY Department of Biology, University of California, San Diego, La Jolla, California sympathetic nervous system of vertebrates (e.g., Orlov, 1927). These cumulative errors have produced a nomenclature for the system that obscures reality and misleads the reader but that is firmly established by precedent and has recently been standardized (Maynard and Dando, in preparation).

The St G of *Panulirus* contains about thirty neurons. Twenty-four of these are motor neurons, two are interneurons, and four are of unknown function but have distinctive physiological characteristics. There may also be some sensory neurons (Larimer and Kennedy, 1966); we have not investigated this point. All the neurons are monopolar, and their somata form a cortex around a central neuropil, where all synaptic interactions occur. Unlike many arthropod monopolar neurons, both attenuated spikes and subthreshold synaptic activity can be recorded from their somata. All can be penetrated with microelectrodes under visual control since their somata vary in diameter from about 15 μm to 60 μm.

The isolated St G produces two independent rhythmic motor patterns in discrete sets of motor neurons (Figure 1) (Maynard, 1972). The slower rhythm occurs in motor neurons innervating muscles in the anterior wall of the stomach that move the heavy teeth. These motor neurons form the gastric subsystem of the St G. The faster rhythm occurs in motor neurons that innervate muscles in the posterior pyloric region of the stomach. This region acts as a filter to sort food particles for entry into either the midgut or the hepatopancreas (Yonge, 1924). These neurons form the pyloric subsystem of the St G. The two rhythms can occur simultaneously, or either one can be present in the absence of the other, or both may be absent. The isolated ganglion can change irregularly from one of these operational states to another. In the intact lobster, all four states do occur (Cynthia Mayer, personal communication).

Both rhythms can be modulated by input from the central nervous system. Two unpaired fibers from the supraesophageal ganglion enter the St G via the inferior ventricular nerve and the stomatogastric nerve (SGN). These fibers selectively excite elements of the pyloric subsystem (Dando and Selverston, 1972) and interneuron 1 of the gastric subsystem (Mulloney and Selverston, in preparation). In addition to these two fibers, there are numerous other fibers in the SGN that excite the gastric subsystem (D. K. Hartline, personal communication).

What are the ways by which these motor patterns are generated in the St G? We have recorded the integrative activity of all but one of its cells with intracellular microelectrodes while monitoring extracellularly the activity of all the other neurons. The cells can be positively

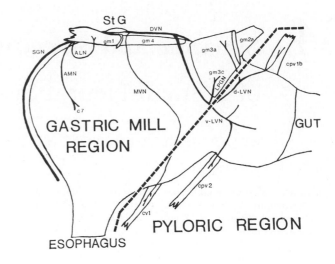

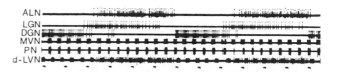

FIGURE 1 *Above:* Side view of lobster stomach. The two principal functional divisions, the gastric mill region and the pyloric region, are separated by a dashed line. Part of the stomatogastric nervous system is shown together with a few of the stomach muscles that they innervate. The stomatogastric ganglion (St G) can be seen on the dorsal surface of the stomach just above gastric mill muscle 1 (gm 1). *Below:* The two basic rhythms produced by the isolated ganglion can be seen in the extracellular recordings from nerves supplying the two different regions of the stomach. The top three nerves (ALN, LGN, DGN), not all shown here, supply muscles that operate the gastric mill. Note that the bursts of activity bear a particular phase relationship with each other and that the duration of each burst lasts several seconds. The three lower traces contain axons of motor neurons (MVN, PN, d-LVN) supplying pyloric muscles. The burst durations are much shorter, and the overall frequency is about seven times as fast as the gastric mill. Note also that the bursts of activity maintain a particular phase relationship. The d-LVN trace contains axons innervating both regions (see above), and the long bursts seen in this trace are from axons to muscle gm 3a. (See Figure 4 for abbreviations.)

identified by antidromic stimulation via the extracellular electrodes and by correlating spikes in the soma with spikes in peripheral nerves. An example of the data obtained is shown in Figures 2 and 3. In these figures the cells are firing spontaneously. When the cells are not firing, synaptic relationships can be ascertained by passing current through one microelectrode while recording simultaneously the intracellular responses of the other cells. The synthesis of such data has produced an almost complete synaptic network diagram for the entire system

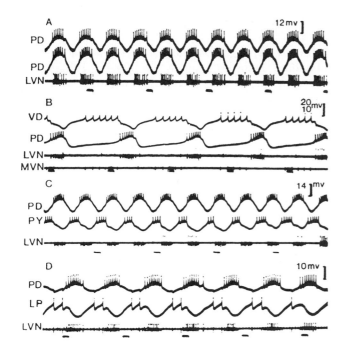

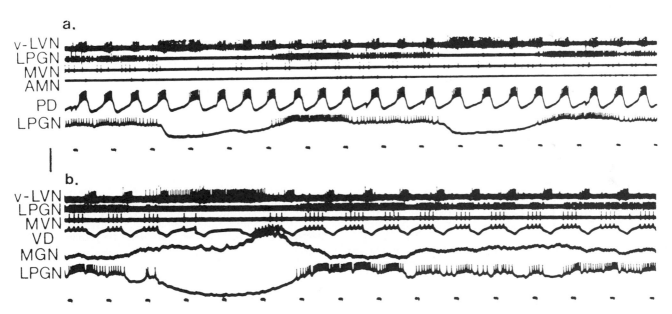

FIGURE 2 Pyloric cycle; intracellular recordings from the somata of pairs of pyloric motor neurons (upper traces) and extracellular recording from the nerves containing the axons (lower traces). (A) Shows the sequential firing of the two PD cells. Note that the spikes occur during the depolarizing phase of the oscillating DC potentials and that the spikes are not locked in 1:1. The occasional large spike in the extracellular trace is from the axon of another motor neuron. (B) Shows the alternation between a PD and a VD cell. (C) Shows alternation between a PD and a PY cell. Only the PD cell is also being recorded extracellularly. (D) Shows the relationship between a PD and a LP cell. Axonal spikes from both units can be seen in the LVN recording. Inhibitory postsynaptic potentials occur in each cell during activity in the other. Time bars are at 1/sec. See text for more details.

FIGURE 3 Gastric cycle; two examples of recordings from an isolated stomatogastric ganglion, obtained during one experiment. They show alternating bursts between neurons of the gastric system (LPGN, AMN, spikes of LGN, PD, VD, the very large LP spikes on the v-LVN). (a) Shows two intracellular recordings, one from a PD neuron and one from an LPGN. The spikes of each of these neurons were recorded simultaneously on the extracellular traces. The LPGN is alternating with units of the v-LVN trace. The PD is alternating much more frequently with units on the MVN and v-LVN traces. Calibration of intracellular traces: 20 mV. (b) Shows three intracellular traces, one from a VD neuron, whose spikes are also recorded on the MVN trace, one from an MGN, whose spikes are recorded as very small spikes on the v-LVN, and the same LPGN as in (a). During the long burst of spikes on the v-LVN, both the VD and LPGN are inhibited, while the MGN is depolarized and finally fires a short burst. The inhibition of the VD is probably due to its electrotonic connection with the LPGN. When the LPGN is inhibited, the hyperpolarizing current crosses the electrotonic junction and reduces the firing frequency of the VD. Calibration of intracellular traces: VD, 50 mV; MGN, 10 mV; LPGN, 20 mV. Time marks for both records: 1/sec.

(Figure 4). The network contains 123 inhibitory synapses, 6 chemical excitatory synapses, and 29 electrotonic junctions. The following classes of interactions occur:

1. Unilateral inhibition (DGN and GM)
2. Reciprocal inhibition (MGN and LPGN; VD and IC)
3. Unilateral inhibition with electrotonic coupling (LGN and GM; PD and VD)
4. Reciprocal inhibition with electrotonic coupling (LGN and MGN)
5. Unilateral excitation (Int 2 and Int 1)
6. Electrotonic coupling alone (LPGN and GM)

We have evidence for three additional interactions that we have not yet been able to confirm by simultaneous intracellular recording from the cells involved, and these are not included in Figure 4.

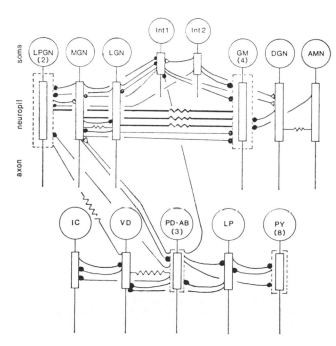

FIGURE 4 Neuronal connectivity diagram for the lobster stomatogastric ganglion. All the cells except Interneurons 1 and 2 are motor neurons. The top ten neurons operate the gastric mill cycle and the bottom fourteen cells operate the pyloric rhythm. The axonal pathways, as well as the muscles innervated by the cells, are known. Soma, neuropile, and axonal parts of the cells are indicated on the left. Dashed lines around some of the neuropile areas indicate that cells of that group are electrotonically connected and can be considered together. Round dots indicate chemical inhibitory synapses; triangles, chemical excitatory synapses; and resistors, electrotonic junctions. Abbreviation of cells indicated in the figure are: LPGN, lateral posterior gastric neuron; MGN, median gastric neuron; LGN, lateral gastric neuron; Int 1 and 2, Interneuron neuron 1 and 2; GM, gastric mill neuron; DGN, dorsal gastric neuron; AMN, anterior median neuron; IC, inferior cardiac; VD, ventricular dilator; PD, pyloric dilator; AB, anterior burster; LP, lateral pyloric; PY, pyloric.

Pyloric rhythm

The network in combination with the properties of the individual neurons explains the observed pattern of motor output in the pyloric subsystem. Like certain neurons in *Aplysia* (Strumwasser, 1965), the PD-AB neurons are endogenous bursters (Maynard, 1972) and determine the periodicity of the pyloric rhythm. The synaptic connections of the neurons in the pyloric subsystem determine the observed pattern. Figure 2 shows examples of·intracellular recordings from different cells of the pyloric group.

When the PD-AB group fires, the VD, the LP and the PY cells are all strongly inhibited. After the PD burst the VD and LP cells reach threshold at about the same time even though they are mutually inhibitory. The inhibition from the LP to the PY is stronger, however, so the PY neurons are held below threshold for most of the LP burst, finally firing just before the next PD-AB burst. Each PD-AB burst starts a new cycle.

Gastric rhythm

There is no evidence for the presence of endogenously bursting neurons in this subsystem. The network itself may explain both the periodicity and the pattern of the gastric motor output. The spontaneous output of the ganglion (Figure 3) reflects burst patterns consistent with the synaptic connections.

Preliminary attempts to simulate the gastric network, in collaboration with Donald H. Perkel (Stanford University), show that even without any long time-constant assumptions about the properties of any of the neurons, we are able to get irregular alternating output that resembles remarkably that of the isolated ganglion. The simulated neurons fired regularly, without any bursting, when all synaptic and electrotonic connections were cut. We are now investigating the effects of fatigue and accumulating refractoriness in individual neurons on the stability of the total system.

Pyloric-gastric interactions

At least part of the interactions of the gastric and pyloric rhythms is well explained by the network. When the LGN and MGN fire intense bursts (Figure 3), the VD neuron is strongly inhibited, and the PD-AB neurons are more weakly inhibited. The burst frequency of the PD-AB neurons is decreased temporarily. The electrotonic coupling of the VD and LPGN neurons and the direct inhibition of the MGN onto the PD-ABs explain the results. The LPGNs are strongly inhibited by the LGN and MGN; their somata hyperpolarize some 25 mV when the latter cells fire. This hyperpolarization crosses the electrotonic junction with the VD and inhibits its spontaneous firing. This inhibition can be simulated by passing hyperpolarizing current into the soma of an LPGN; the VD neuron stops firing when the LPGN is hyperpolarized. The PD-ABs are not affected to the same degree by hyperpolarization of the LPGN, but stimulation of the MGN does affect them to an extent comparable to the spontaneous interaction. Discrete synaptic potentials from the MGN are not visible, however, in the PD-AB neurons. The LP neuron is also influenced by the gastric rhythm, but we do not yet know the source of this interaction.

Neuronal structure

The cell bodies in many arthropodan and molluscan ganglia occur in an ordered array. One can often draw a map of the cell bodies and use it to find particular neurons in different animals (Frazier et al., 1967; Otsuka, et al., 1967; Cohen and Jacklet, 1967; Selverston and Remler, 1972). The St G is not organized with this degree of regularity. The soma of a given neuron may be located in any quadrant of the ganglion. We were, therefore, curious to see how much the neuropilar structure of these neurons varied.

Earlier studies of the fast flexor motor neurons in the third abdominal ganglion of *Procambarus clarkii* (Selverston and Remler, 1972), using intracellular recording and injection of Procion dye (Stretton and Kravitz, 1968), showed that both physiological properties and the neuropilar structures of these neurons were very similar in different animals (Figure 5).

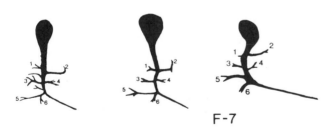

F-7

FIGURE 5   Examples of neurons with much more consistent structure can be seen in crayfish abdominal ganglia. Fast-flexor motor neurons have constant cell body locations and similar anatomy. Shown here are tracings from whole mounts of motor neuron F-7 from three different animals. Processes have been numbered and can be compared in each. (For other examples, see Selverston and Remler, 1972.)

The structures of the motor neurons of the St G are much less constant. We have begun to analyze the structural variability of the PD neurons to see if it has physio-

logical significance. The somata of these neurons have no constant location in the ganglion, but their major processes in the neuropil appear similar in each case (Figure 6). The major process leaving the soma bends and

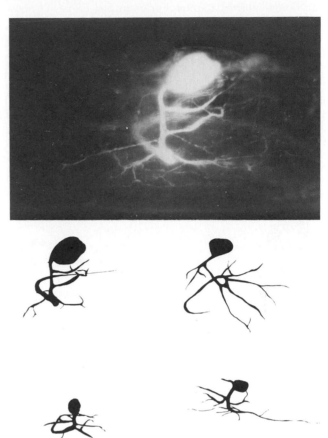

FIGURE 6   A fluorescent whole mount and drawings made from other whole mounts of different PD motoneurons in the stomatogastric ganglion. Although the drawings have been made from different angles of rotation, the basic structure of the cell, as described in the text, is apparent. There is a considerable amount of variation in the smaller processes.

descends into the neuropil, bends again and crosses to the other side of the neuropil and finally curves posteriorly and tapers to form the axon. This basic structure is common to all PD neurons (Selverston, 1973). The minor branches of these cells vary both in the point from which they leave the major process and in their orientation in the neuropil. Different PD neurons are not superimposable, even if one compensates for altered orientation of whole mounts. These differences may have no significant effect on the integrative properties of the cells, or some physiological compensation may occur to correct for structural deviation, because the physiological properties of PD neurons are highly consistent in different animals.

Paired injections

One major test of the network (Figure 4) consisted of injections of Procion dyes into pairs of cells with known physiological interactions. For example, Figure 7a shows

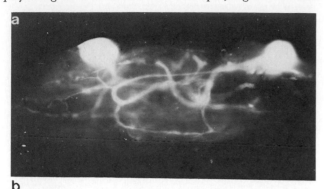

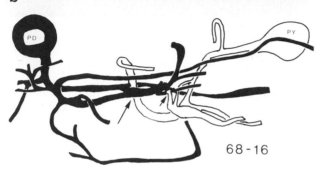

FIGURE 7 The *upper* fluorescent photomicrograph shows a paired Procion yellow injection of two pyloric cycle cells. The PD cell on the left has inhibitory connections to the PY cell on the right. Photographs of paired cells in whole mounts are usually taken from several angles of rotation. The *lower* drawing was made by examining the fluorescent profiles of the sectioned tissue. The arrows indicate points where the processes of the two cells appear to contact one another, and are likely sites of synaptic interaction. (Drawing courtesy of David King.)

a PD and a PY neuron injected with Procion yellow. Figure 7b is a reconstruction of these neurons made by David King (U.C., San Diego) that shows two points of contact between these neurons, as indicated by the arrows. Figure 4 shows only one synaptic interaction between each of the PD and PY neurons. Which of the two points of contact is the synapse whose physiology we know? What is happening at the other point? We do not have enough samples of paired injections of these two cells to answer this question, because it is not possible at the level of light microscopy to identify conclusively the presence and polarity of synapses. New dyes are being developed (see, for example, Pitman, et al., 1972) that will allow us to examine the ultrastructure of such contacts. As we build a library of injected sets of cells, we

may be able to assign with certainty the physiological function of an apparent structural contact.

Quantitative analysis of neuronal structure

The traditional methods of reconstructing neurons from serial sections are by the use of stacked sheets of plastic (Selverston and Kennedy, 1969; Mulloney, 1970) and by graphic reconstruction (Pusey, 1939; Iles, 1972). Neither method allows us to measure easily the structural parameters of a cell that are necessary for certain biophysical analyses. Therefore, in collaboration with Sheryl Glasser and N. H. Xuong (U. C., San Diego), we have developed a digital method of reconstructing a neuron from serial sections. The programs permit us to reconstruct and rotate a neuron in three dimensions and to measure accurately cell volume, surface area, the power relations and numbers of branch points, lengths of processes, and other parameters of this sort. This information combined with physiological data from the same cell may allow us to model and understand the integrative processes of these neurons with considerable accuracy.

Summary

We have described virtually the entire network of synaptic connections in the stomatogastric ganglion of *Panulirus interruptus*. This network explains in part the patterned firing of the motor neurons in the ganglion.

Structural evidence obtained with intracellular dyes corroborates the physiological results.

ACKNOWLEDGMENTS The authors' research is supported by grants number NS-09322 and NS-10629 from N.I.H. and by the Alfred P. Sloan Foundation. Brian Mulloney is a postdoctoral fellow of NINDS-NIH.

REFERENCES

COHEN, M. J., and J. W. JACKLET, 1967. The functional organization of motor neurons in an insect ganglion. *Phil. Trans. Roy. Soc. (London)* 252(B):501–527.

DANDO, M. R., and A. I. SELVERSTON, 1972. Command fibers from the supraoesophageal ganglion to the stomatogastric ganglion in *Panulirus argus. J. Comp. Physiol.* 78:138–175.

FRAZIER, W. T., E. R. KANDEL, I. KUPFERMANN, R. WAZIRI, and R. E. COGGESHALL, 1967. Morphological and functional properties of identified neurons in the abdominal ganglion of *Aplysia californica. J. Neurophysiol.* 30:1288–1351.

HUXLEY, T. H., 1880. *The Crayfish: An Introduction to the Study of Zoology.* London: Kegan Paul and Co.

ILES, J. F., 1972. Structure and synaptic activation of the fast coxal depressor motor neuron of the cockroach *Periplaneta americana. J. Exp. Biol.* 56:647–656.

LARIMER, J. L., and D. KENNEDY, 1966. Visceral afferent signals in the crayfish stomatogastric ganglion. *J. Exp. Biol.* 44:345–354.

MAYNARD, D. M., 1972. Simpler networks. *Ann. N.Y. Acad. Sci.* 193:59–72.

MULLONEY, B., 1970. The structure of the giant fibers of earthworms. *Science* 168:994–996.

ORLOV, J., 1927. Das Magenganglion des Fluszkrebses. *Z. Mikr. Anat. Forsch.* 8:73–96.

OTSUKA, M., E. A. KRAVITZ, and D. D. POTTER, 1967. Physiological and chemical architecture of a lobster ganglion with particular reference to GABA and glutamate. *J. Neurophysiol.* 30:725–752.

PARKER, T. J., 1876. On the stomach of the freshwater crayfish. *J. Anat. Physiol.* (quoted in Huxley, T. H., 1880).

PITMAN, R. M., C. D. TWEEDLE, and M. J. COHEN, 1972. Branching of central neurons: Intracellular cobalt injection for light and electron microscopy. *Science* 176:412–414.

PUSEY, H. K., 1939. Methods of reconstruction from microscope sections. *J. Roy. Microsc. Soc.* 59:232–244.

SELVERSTON, A. I., 1973. The use of intracellular dye injections in the study of small neural networks. In *Intracellular Staining Techniques in Neurobiology*, S. Kater, and C. Nicholson, eds. New York: Springer-Verlag (in press).

SELVERSTON, A. I., and D. KENNEDY, 1969. Structure and function of identified nerve cells in the crayfish. *Endeavour* 38:107–114.

SELVERSTON, A. I., and M. P. REMLER, 1972. Neural geometry and activation of crayfish fast flexor motorneurons. *J. Neurophysiol.* (in press).

STRETTON, A. O. W., and E. A. KRAVITZ, 1968. Neuronal geometry: Determination with a technique of intracellular dye injection. *Science* 162:132–135.

STRUMWASSER, F., 1965. The demonstration and manipulation of a circadian rhythm in a single neuron. In *Circadian Clocks*, J. Aschoff, ed. Amsterdam: North Holland, pp. 442–462.

YONGE, C. M., 1924. Studies on the comparative physiology of digestion. 2. The mechanism of feeding, digestion, and assimilation in *Nephrops norvegicus*. *J. Exp. Biol.* 1:343–389.

35 Neural Machinery Underlying Behavior in Insects

GRAHAM HOYLE

ABSTRACT Neuron somata of large insects, especially locusts, are being impaled with microelectrodes in preparations that permit movements of limbs. Motor neurons are identified and mapped after physiological testing and Procion dye injection. They are anatomically prescribed and can be relocated by reference to maps. Larger interneurons are also being studied, and by combining recordings from pairs of neuron somata with peripheral recording, the neural bases of integration and of control of locomotion are being examined. An identified tonic motor neuron is being entrained by computer to fire at significantly higher or lower rates than its normal average and the possible physiological basis of the change tested.

THE MAJOR AIM OF this chapter will be to try to show that after a late start insects are about to come into their own as major targets for neuroethological research. Only after the introduction of the fruit fly as a research subject and the concentration of effort on it by a large number of investigators was genetics revolutionized and put on a firm scientific foundation. It is all too obvious that a comparable target for attention is needed in neurobiology. Although the crayfish has been suggested as the standard (Wiersma, 1959), I feel that the subject eventually chosen will be an insect, for some of the same reasons that *Drosophila* is so useful in genetics. A major reason for the delay in using insects in research was the lack of success encountered in many early attempts to record intracellularly from insect ganglia. Within the last three years, however, several investigators have been able to break this barrier and, while it is regrettable that the nervous system of *Drosophila* itself is too small, electrophysiological analysis at levels down to, and including, individual neurons is now possible in larger insects. The great promise offered by insects is because of the likelihood that their behavior, both instinctive and learned, may be amenable to ultimate analysis in terms of underlying neuronal events.

This analysis requires a synthesis starting with ethology for accurate description of an act of behavior and breaking that behavior down into components that are ultimately sequences of muscular contractions and relaxations. Next there should be physiology, if such a distinction needs to be made in view of Wigglesworth's (1939) perceptive, though little known, dictum that "behavior is physiology." Closely correlated with the physiology should be anatomy —the elucidation of the neural circuitry, with attendant synaptology, and its development, especially those experimental approaches aimed at understanding neuronal specificity. Then there should be genetic analysis, to be closely coupled with the foregoing; finally there should be an evolutionary synthesis, relating behaviors in allied species and considering how they may have evolved and be evolving. The available methods will first be considered briefly and then the progress being made in each major category presented.

Methods for insect neuroethology

ELECTROMYOGRAMS Insect muscles are operated by a system completely different from that occurring in higher vertebrates, and while this may be a disadvantage from the point of view of transferring knowledge from one area to the other, the simplicity of the insect system offers a great experimental advantage (Figure 1). Many of their muscles are operated by a single motor neuron that innervates each muscle fiber at many points (multiterminal innervation). Tension is varied by altering the frequency of firing. Only a few muscles comprise more than one motor unit. When a second excitatory neuron sharing innervation of the same muscle fibers is present, it gives larger twitch contractions and is used only to boost the peak tension for extra-fast movements (Figure 1B; Hoyle, 1964; Pearson, 1972). The result is that the electromyogram is a digital sequence that is easily recorded and that can be studied quantitatively and from which the resulting movements, though not recorded, can be inferred (Hoyle, 1957). The situation is complicated, for some muscles, by an additional axon or axons of inhibitor type that cause peripheral polarization, but the functional role of these axons is obscure at the present time. They produce only very small, if any, myograms.

The myograms are obtained simply by attaching very

GRAHAM HOYLE   Department of Biology, University of Oregon, Eugene, Oregon

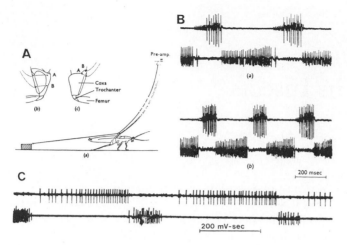

FIGURE 1 (A) Electromyographic method illustrated. (a) Experimental arrangement for recording activity in coxal levator and depressor muscles of a cockroach during walking. The insect's movements were restricted by a fine thread attached to a ring inserted through the dorsal cuticle and to an elevated support in the center of the area. Retraction of the hind legs was resisted by a sliding weight attached to the back. (b) Ventral view of the coxa showing the arrangement of electrodes for recording the depressor activity. (c) Dorsal view of the coxa showing positions for recording the levator activity. (B) Typical electromyograms from antagonists during walking at two different speeds; (a) slow, (b) fast. *Upper traces*: levator (182C) and *bottom traces*: depressor (177D). Discharge in depressor is due to single slow axon giving train of facilitating spikes. Electrical activity in levator includes single slow axon (small, facilitating responses) and single fast axon (large spikes). (C) Recording of spontaneously generated motor nerve spikes going to antagonistic muscles after total removal of leg receptors. Reciprocal output patterns are still present, although at rather low rates. *Upper trace*: levator; *lower trace*: depressor. From Pearson (1972).

fine insulated wires to the insect, with cut ends inserted through the cuticle into the relevant muscles. Larger insects can behave normally while trailing many such leads. The present maximum obtained by Elsner (1968) on a small grasshopper, is 15 pairs of wires. All neurobehavioral studies have used this method, specifically to examine insect flight (reviewed by Wilson, 1968), cricket singing (reviewed by Huber, 1971), grasshopper courtship (reviewed by Elsner, 1973), and cricket and locust walking (Bentley, 1969a; Hoyle, 1964). Wires can also be used to record nerve impulse activity. By correlating nerve and muscle activity in a freely walking locust, Runion and Usherwood (1966) were able to identify inhibitor nerve discharges. They made a check on the efficacy of timing of inhibitor impulses in causing increased relaxation rate by using tape-recorded discharges of the neurons obtained from free-walking preparations in order to excite the axons in slave preparations from which the tension was recorded (Usherwood and Runion, 1970).

MICROELECTRODE RECORDINGS In 1956 two papers reporting extremely promising results for insect neurophysiology were published. Hagiwara and Watanabe (1956) obtained 60 mV resting potentials with glass capillary electrodes placed in the motorneurons of the sound-evoking tymbal muscle of a cicada. Furthermore, as a response to a single afferent volley evoked in the auditory nerve, they could make the neurons fire bursts closely resembling the natural output causing sounds. Spikes were larger than 70 mV, overshooting, and preceded by excitatory postsynaptic potentials (EPSPs) that were sometimes below threshold in size. Maynard (1956), also using a capillary electrode and going straight to the highest place, a cockroach brain, obtained a sharp spike of 30 mV amplitude from the calyx of the corpora pedunculata following electrical stimulation of the antennal nerve. In spite of the size of the response, Maynard stated it was recorded not intracellularly but extracellularly. Several investigators had already tried microelectrodes in insect ganglia and obtained only resting potentials, with no synaptic or spike activity apparent. Stimulated by the above publications, they tried again but with indifferent results. Hagiwara and Watanabe were not sure from which part of the motorneurons they had recorded, nor were they able to extend their work to interneurons. No further progress was reported until 1960 when Rowe (1960) announced briefly that he had recorded inhibitory postsynaptic potentials (IPSPs) as well as EPSPs from a cockroach third thoracic ganglion. Later, after the kind of delay that has beset the subject, Rowe (1969) published pictures of spikes as well as synaptic activity; but since he did not attempt to identify those neurons, his results did not greatly promote the subject. In the preceding year Kerkut, Pitman, and Walker (1968) obtained EPSPs, IPSPs, and overshooting spikes from what was undoubtedly a cell body of the cockroach sixth abdominal ganglion. They used the preparation (Kerkut et al., 1969) to show that the neuron is excited by 1.3×10^{-13} mole acetylcholine and hyperpolarized by 1.05×10^{-13} mole gamma-aminobutyric acid.

Extensive explorations with intracellular electrodes in the mesothoracic ganglion of the cricket (*Gryllus campestris*) were made by Bentley (1969a) who obtained synaptic and spike activity of enormous variety. Some of this was evoked by stimulation of an evasion response; other activity was associated with ventilation and flight. But while showing that insect nervous systems could yield their secrets, the recordings were still not from identified neurons and therefore not sufficient to provide a firm foundation for detailed analysis of any of these activities. Bentley made remarkably fine use of a single preparation of a male cricket that, after localized brain lesions, sang continually for 12 hours. By combined intracellular recording from the ganglion, from unknown, chance sites, and nerve trunks, he obtained recordings of interesting, large prepotentials leading to motor output to specific wing muscles. Many of these were probably from motorneuron somata. He also obtained the first direct evidence of electrical coupling between wing motorneurons.

Meanwhile, the systematic study of both the anatomical and physiological characteristics of the same neuron was made possible by the introduction of the first useful intracellularly injectable dye, Procion Yellow (Stretton and Kravitz, 1968). Hoyle and Burrows (1970; 1973a, b;

Burrows and Hoyle, 1973) worked on the motorneurons of the locust third thoracic ganglion (Figure 2) and Bentley (1970) on the flight motorneurons of the locust second thoracic ganglion. Some large neurons on the dorsal surface of the locust and cockroach third thoracic ganglia were added to the list (Crossman et al., 1971) but were not functionally identified, and a few possible premotor interneurons of locusts were located and given preliminary identification by Hoyle and Burrows (1973a, b). Of these the smallest is just less than 20 μm in diameter. The ability to work with neurons of this diameter brings

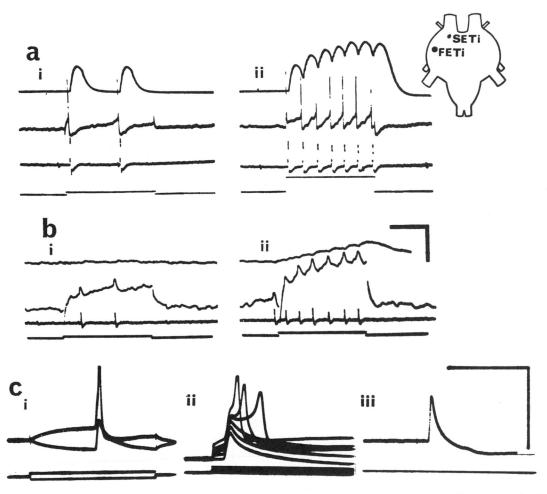

FIGURE 2   Identification of motorneurons and analysis of responses. An intracellular electrode is placed in a suspected motorneuron soma and a depolarizing current pulse passed across it through the microelectrode using a bridge circuit (current monitored on lowest traces) until the cell fires (second traces). Extracellular recordings are made from muscles until the correct one is found (third traces). Confirmation is made by attaching the muscle to a tension recorder (top traces). Locations of motorneurons indicated in insert. (a) i and ii: fast extensor tibiae at two current strengths. (b) i and ii: slow extensor tibiae at two current strengths. Calibration (a) and (b): vertical 10 mV, horizontal 20 msec. (c) Analysis of excitability

and response of fast neuron soma. Responses obtained by antidromic excitation. (i) and (ii) With slight hyperpolarization the response is slightly depressed but with slight depolarization it is enhanced. The enhancement is clearly due to the initiation of a graded electrogenic response, probably in the soma membrane. The initial antidromic response is purely electrotonic invasion from the principal neurite. (iii) Antidromic response at normal membrane potential. There is no, or very little, depolarizing electrogenesis, but a large undershoot indicates that strong repolarizing electrogenesis has been initiated. Calibration: 20 mV; (i) and (ii): 40 msec; (iii) 20 msec. From Hoyle and Burrows (1973a).

within range of experimental manipulation probably all of the motorneurons and about 25% of the interneurons of the locust. Preparations studied range from isolated ganglia, as used by Crossman et al. (1971), to nearly intact preparations. Hoyle and Burrows (1973a, b) used a preparation with the insect on its back, which permitted free movements of the legs but from an unnatural posture. In a preparation recently developed by Dagan and Hoyle, the ganglia are approached from the dorsal surface and the insect is able to assume a normal posture and to walk on a light turntable.

Each experiment is performed with a dye-filled electrode so that an anatomical study can be subsequently made. After fixation, ganglia are photographed before serial sections are made. Projections of these sections are made onto acetate sheets, that on stacking with appropriate spacing, make the reconstructed neuron, with its neurites and dendrites, visible in three dimensions. The fine branches of a neuron are much more clearly seen in whole mounts when filled with cobalt sulfide rather than Procion Yellow. But for three-dimensional reconstruction, the fluorescent dye is easier to use. However, the very finest branches of nerves are not visible. Each specific neuron occupies about the same site in different ganglia and has a similar dendritic topography. It has thus become possible to draw neuron soma location and neurite pathway maps (Figure 3) that are directly useful in further work. For instance, after preparing the map locations, simultaneous recordings from selected pairs of neurons became possible (Hoyle and Burrows, 1973b).

To provide preliminary classification of neuron types located, I shall refer to neurons within the ganglia as either motorneurons, peripheral inhibitor neurons, premotor neurons, sensory integrating interneurons, and all others as other interneurons. Only motorneurons causing excitatory junctional potentials in muscle fibers were thereby easy to define unequivocally. A neuron that makes direct synaptic contact with a motorneuron but has no axonal branch leaving the ganglion will be termed a premotor interneuron. All neurons that receive synaptic input directly from a sense organ and that are not motorneurons or peripheral inhibitors will be termed sensory integrating interneurons. Interneurons that receive input only from these interneurons, or ones of a still higher order that are not also premotor, will be simply termed interneurons.

Electrophysiology of insect neurons

The only insect nerve fibers that have been examined in detail electrophysiologically are the "giant" axons of the ventral nerve cord of the cockroach. These axons taper gradually from 30–60 μ to 20–30 μ as they pass from their

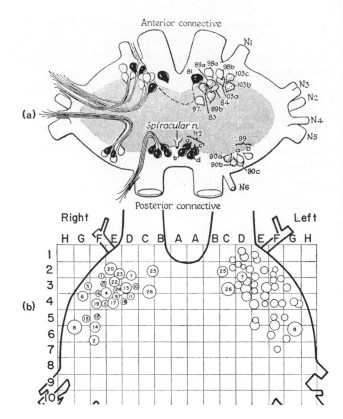

FIGURE 3 Locations of identified neurons of locust (*Schistocerca gregaria*) ganglia. *Upper:* Mesothoracic ganglion indicating flight motorneurons. Numbers refer to muscles innervated. From Bentley (1970). *Lower:* Metathoracic ganglion indicating leg motorneurons with diameters drawn to correct relative scale. Identifications as given on accompanying Table I. Identifications by Burrows and Hoyle (Burrows and Hoyle, 1973; Hoyle and Burrows, 1973a, b; unpublished).

origin in the last abdominal ganglion, and they also have an abrupt "waist" within each ganglion they pass through (Spira et al., 1969). They have conventional squid-axon type sodium-potassium overshooting spikes that are blocked by tetrodotoxin (Narahashi, 1965); this is probably true of insect axons in general.

The somata, by contrast, are not only unconventional electrically, they have been found to have a wide variety of responses. Even allowing for the possibility of damage during dissection, sheath removal (where involved), loss of tracheal supply, less than optimal bathing solutions, etc., some of the response types observed must be valid ones, because these same factors are common to the same neurons in all experiments. The membranes of neuron somata of insects range, within a single ganglion, from completely electrically inexcitable through weakly to strongly graded electrogenesis to all-or-none spike responsiveness. The latter groups will be dealt with separately.

TABLE I

Identified motorneurons of the metathoracic ganglion of Schistocerca gregaria. *Numbers of muscles are from Albrecht (1953) adapted from Snodgrass*

Muscle	Number	Type	Maximum Soma Diameter Microns (μm)	Identification No.	Grid Reference (all on right side of animal)
Retractor unguis	139a	slow	35	1	F2
		fast	65	2	F7
Levator tarsus	137	slow	40	3	F4
		fast	60	4	E3
Depressor tarsus	138	slow	45	5	G3
		fast	55	6	G4
Extensor tibiae	135	slow	55	7	D2
		fast	90	8	G6
Flexor tibiae	136	Anterior Group			
		slow	40	9	E4
		intermediate	35	10	D3
		fast	45	11	D4
		Posterior Group			
		slow	35	12	F5
		intermediate	45	13	G5
		fast	65	14	F6
Depressor trochanteris	133	fast	65	15	D3
Abductor coxa	125	slow	35	16	F4
		fast	55	17	E4
Adductor coxa	130	slow	35	18	D4
Anterior rotator coxa	121	slow	55	19	F4
		fast	75	20	E2
Posterior rotator coxa	122	slow	30	21	F3
		fast	55	22	E3
Tergosternal	113	fast	70	23	E2
Promotor tergum	118	fast	65	24	E3
Promotor extensor of hind wing (basalar)	128	fast	70	25	C2
Depressor extensor of hind wing (subalar)	29	fast	80	26	C3

ALL-OR-NONE SPIKES In the cockroach sixth abdominal ganglion (Kerkut et al., 1969), all-or-none spikes were recorded in unidentified somata, which may be sensory integrating interneurons of the phallic region. All-or-none spikes were also found in a group of seven or eight unpaired neurons located in the midline on the dorsal surface of locust and cockroach metathoracic ganglia (Crossman et al., 1971). These neurons branch equally to left and right sides and have axons that leave the ganglion in nerve 5, but again their function is unknown. Dagan and Hoyle (in preparation) determined in the locust that they are efferent neurons, although they do not innervate any leg muscles. They are either neurosecretory or innervate a peripheral sensory structure such as a chordotonal organ. Resting potentials in Hoyle's (1953) saline are 50 to 60 mV, overshoots are up to 20 mV, and undershooting after-potentials as much as 25 mV (Figure 4Ai). The height of the total excursion, from peak overshoot to peak undershoot, is constant and about 85 mV when the neuron fires after a rest, but the height wanes rapidly upon repetition, and after a few seconds of firing at a high rate the spike ceases. Both the progressive decay in response and

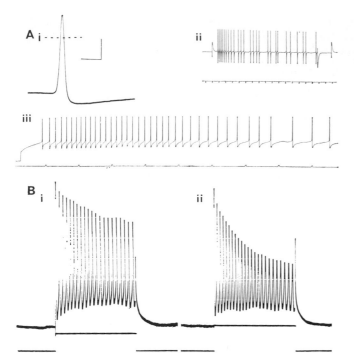

FIGURE 4 Examples of neurons giving overshooting spikes as recorded in the somas. (A) A large, dorsal, unpaired neuron of locust metathoracic ganglion. (i) Single impulse, 90 mV peak to trough, with a 25 mV overshoot from approximate 55 mV resting potential. Spike duration about 8 msec. Calibration: vertical, 20 mV; horizontal, 10 msec. (ii) Slow train from (i) with decreasing frequency initiated by directly applied depolarizing current through overbalanced bridge circuit. Time in seconds. (iii) Fastest train obtainable from (i), in response to break from large hyperpolarizing current pulse. Maximum firing rate 6 Hz. Time in seconds. (B) i and ii: Discharge trains in sensory integrating neuron of locust (auditory) to depolarizing pulses of different strengths for comparison with Figure 4A. Spike duration about 2 msec. Maximum firing rate 300 Hz. Figure 4A from Dagan and Hoyle (in preparation); Figure 4B from Hoyle and Burrows (unpublished).

the early block could be accounted for if these neurons have a tight sheath across which only limited ionic exchange is possible, but there is no evidence for this. The interneuronal space is very slight, with no more than a few millimicrons between axons, which should lead to a rapid local accumulation of potassium. A neuron excited by direct depolarization or by break from hyperpolarization of its soma still fires impulses, but these invade the soma only electrotonically. By additional depolarization soma spikes can still be elicited, but it is a property of these neurons that the electrically excitable response fatigues very rapidly. Nothing is yet known about the ionic permeability mechanisms underlying the spikes. Preliminary studies by Hoyle and Burrows (unpublished) indicate that sensory integrating interneurons, by contrast, fire all-or-none spikes at high rates, with little accommodation (Figure 4B).

GRADED RESPONSES In some large motorneurons of the locust, a spike of up to 30 mV magnitude appears in the soma after an impulse has occurred in the axon following either orthodromic or antidromic excitation. The response is probably antidromically initiated in both cases, because the threshold for firing an action potential in the impulse initiation zone, situated in neuropil some distance from the soma, is very low compared with that of the soma membrane. The response magnitude is fairly constant although it is of graded type, i.e., proportional to the stimulus strength. Its constancy reflects only the all-or-none aspect of the spike that initiates it by electrotonic invasion. If the latter fails to excite depolarizing electrogenesis because it is too weak, it nevertheless usually initiates some repolarizing electrogenesis, causing the electrotonic potential to undershoot. A small depolarizing pulse applied to the soma at the time of invasion leads to activation of such depolarizing electrogenesis (Figure 2c).

Developmental biology of insect nervous systems

In order to understand the nature of neural circuitry in general, it may be essential to follow its development. The use of degeneration and regeneration techniques will also continue to be of great importance as they have in mammals for over a hundred years. In those regards insects are more amenable to experiment than other forms. A wealth of information already exists on the developmental anatomy of insect nervous systems (Edwards, 1969). Now, as it becomes possible to work out the definitive circuitry in parts of adult insect nervous systems, it will be extremely fruitful to correlate developmental and adult regeneration studies with cellular studies.

Two pioneering studies deserve special mention, both on crickets. Edwards and Sahota (1967) found that the sensory nerve fibers of abdominal cerci regenerating after ablation establish normal functional connections with the giant fiber system in the terminal ganglion. This provides a superb model system for testing Sperry's (1943) hypothesis of neuronal specificity. In ganglia deprived of cercal afferents by ablation (Edwards and Palka, 1971), the functional connections are remade, and both the physiological and the behavioral responses can be studied directly. Both were identical to normal, even when the cricket had been deprived of the cercal input from both cerci throughout development.

These results seem to indicate an independent genetic control of the sensory system and the interneuron system (also, see chapter by Kennedy earlier in this part for similar indications in Crustacea), and Edwards pointed out that they appear to contrast strongly with the implications of the well-known findings of Wiesel and Hubel (1963, 1965) on kittens deprived of visual input that the development of the nervous system is dependent upon normal sensory inflow, especially in early stages. But it is compatible with the even more famous optic nerve regeneration experiments of Weiss (review, 1950) and Sperry (1943) on amphibia. There is one major difference technically in that in the Wiesel/Hubel experiments the nerves were not cut to obtain sensory deprivation. It may be that permanent inhibition results in this case, from connections which are made, but not used, compared with ones which are made late. The second set of studies are by Bentley and Hoy (1970) and Kutsch (1971) on the development of adult motor patterns of cricket flight and song. They used implanted leads to study the two motor neural patterns (as myograms). Flight was elicited by loss of tarsal contact while song was caused by making heat lesions in selected areas of the brains of males. Sixth instar nymphs were capable of generating flight patterns, but these were never maintained as they are in adults four instars later. There must therefore be some final anatomical or physiological addition to the system at the last instar. Song patterns of calling, courtship, and aggression of perfect adult form could be produced only in nymphs in the last instar. These nymphs never sing, but the experiments proved that the neural machinery has been developed and must await the onset of sexual maturity for adequate initiation or release from inhibition. The authors suggested that premature activation of sequentially developed circuits is suppressed by descending inhibition from the brain until maturity.

At the level of tissue culture, there have recently been two instances of marked progress. In cultures of 16-day-old cockroach embryos, intimate relations were established between nerve fibers and myocytes (Aloe and Levi-Montalcini, 1972). In comparable *Drosophila* cultures, junctions were proven to be functional (Seecof et al., 1972) by stimulating the axons and observing the newly formed myocytes contract. This is the first time for any in vitro system that complete differention of neurons, myocytes, and junctions has been obtained.

Genetics and insect neuroethology

There is a rich variety of behavior in many insect species, some of which may be subjected to genetic selection procedures. In *Drosophila*, pioneering work by Hirsch and Boudreau (1958) on the heritability of phototaxis and of Manning (1963) on mating behavior has led to a clear realization that genes influence behavior in highly discrete ways. Benzer (1971) recently listed 26 behavioral mutants of *Drosophila* in six categories, though each of these affects the manner in which the fly behaves rather than affecting Lorenzian "units" of behavioral sequence, which include uncoordinated, nonclimbing, arrhythmic,

nonoptomotor reactions. That behavioral units are subject to Mendelian laws has also been shown for certain strains of bees by Rothenbuhler (1964a, b). The bees, termed *hygienic*, normally remove dead larval honey bees from the comb. There are two subunits of the behavior: uncapping and removal of dead larvae. When crossed with a strain not showing this behavior, termed *unhygienic*, the progeny are all unhygienic. But a backcross of the hybrids to the recessive, hygienic strain leads to a simple segregation into four equal groups: (1) bees that uncap only and do not remove dead corpses; (2) bees that do not uncap but remove dead larvae if the cap is first removed by the investigator; (3) ones that do both, i.e., are hygienic; (4) ones that do neither, i.e., are unhygienic. These investigations clearly demonstrate genetic control not just of postural and taxic nervous mechanisms, but of meaningful bits of a complex behavior sequence having several components.

A few of the *Drosophila* mutants have been subjected to preliminary neurophysiological analysis, representing technical feats in view of the size of the fly. Excellent electroretinograms of eyes of the nonphototaxic forms were obtained and found to be markedly different from those of normal eyes (Hotta and Benzer, 1969). Hyperkinetic male mutants are distinguished by showing a specific rhythmic leg shaking during etherization and Ikeda and Kaplan (1970a, b) have shown that these movements are initiated by thoracic neurons. They recorded intracellularly from an unidentified neuron 50 mV action potentials associated with the rhythm.

Unfortunately, since *Drosophila* is so small, progress at the cellular level is not likely to be advanced very far. Of more promise as an experimental animal is the cricket, which is large enough for insertion of two or three electrodes at a time into identified cells. The generation time of some tropical species at 35°C is a mere six weeks! Bentley (1971) took advantage of different numbers of pulses in a trill within a call in the calling song of wild-type *Teleogryllus oceanicus* compared with those occurring in the closely related *T. commodus*. The two species can be mated, and although the F_1 females are sterile, the F_1 males are fertile and can be backcrossed to either parent type. Since male crickets lack a Y chromosome, genes located on the X chromosome received from the maternal parent are unduplicated, and phenotypic differences in the songs of the males can be attributed to X chromosomes. Bentley obtained a series of six types of males from wild-type *T. oceanicus* (A), which has the smallest number of pulses, to wild-type *T. commodus* (F), which has the most. A single motor nerve impulse can determine a pulse, and the number in a trill in the range studied was 2, 3, 4, 5, 6 or 2 × 7 (Figure 5). Hence he could make the important deduction that "genetically derived information

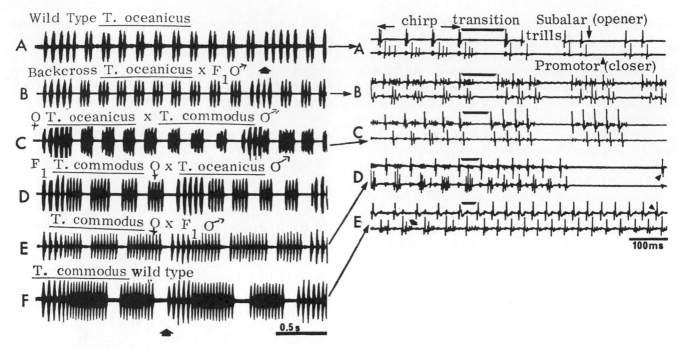

FIGURE 5 Genetics and physiology of sound production by male tropical crickets of the genus *Teleogryllus*. The calling song starts with a single chirp containing 4 to 6 pulses, followed after a transition by several trills. The species *T. oceanicus* has 2 pulses per trill whilst another species *T. commodus* has 14. F_1 males of the species cross are fertile and were backcrossed to parent types yielding a series with varying extents of parental characteristics. The series, presented above, resulted in trills containing 3 (B), 4 (C), 5 (D), and 6 (E) pulses. At left are shown sound patterns at a slow time base speed, while at right are shown electromyograms from a wing opener (*upper*: subalar) and a closer (*lower*: promotor) muscle unit on a faster time base. From Bentley (1971).

is adequate to specify the difference of a single impulse in the output of homologous neurons from different genotypes."

Central versus peripheral control

As investigators began to think about neural mechanisms underlying behavior, old controversies came to the fore-ground, the most compelling of which was whether or not the Pavlovian mechanistic view, "behavior is reflexes," is acceptable. In the thirties, the emerging ethologists, especially Lorenz (review, 1950) and Tinbergen (review, 1951), guided in part by von Holst, were finding that many acts of behavior appear long after a stimulus has ceased or, in many instances, in the complete absence of an external stimulus. Hess (1943) and von Holst and von St. Paul (1963) found centers in vertebrate brain stems that caused discrete, easily recognized complex behavioral acts to occur when electrically stimulated, whatever the environment. In invertebrates Wells (1950) found intrinsic rhythms related to feeding and defecation cycles in small pieces of isolated tissue of lugworms, and Huber (1962) evoked locomotor and song patterns in crickets by local brain lesions. With a lucky lesion, a male cricket will sing continually for more than 12 hr. Such experiences shattered forever simplistic behaviorist theories and led to searches to test whether given acts of behavior are endogenously generated or peripherally determined. Wilson (review, 1968) firmly established that the flight pattern of locusts does not require peripheral feedback, and Huber and his associates (e.g. Kutsch and Huber, 1970) have determined that cricket singing is likewise endogenous. The most subtle, complex behavior yet studied in this manner is the courtship by the males of the small grass-hopper *Gomphocerippus rufus*, in which elegant experiments have shown it to be largely determined by an endogenous program (Loher and Huber, 1966; Elsner, 1969; Huber and Elsner, 1969).

Now there is increasing agreement that in many behaviors there is a basic central nervous program that only needs to be "turned on" in order to ensure the emergence of a behavioral sequence. Correct performance of the relevant movements nevertheless requires proprio-ceptive feedback for control of the details of the motor impulse patterns.

WALKING The classical view was that walking consists of a series of resistance reflexes, with coordination between legs both within one half of the ganglion and across it. Following an amputation, insects switch very rapidly to a new appropriate gait, a fact that hardly seemed compatible with a central rhythm. Yet Wilson (1966) showed that the changes are not really highly complex and are compatible with control from a rather simple basic central program. When Pearson (1972) removed all sensory input from leg receptors in cockroach preparations, he still obtained reciprocal output in antagonistic leg motorneurons (Figure 1C). This observation does not prove the central program hypothesis but clearly shows that at least part of the necessary neuronal machinery is present entirely within the central nervous system.

Delcomyn (1971a, b) analyzed limb movements from high-speed motion picture film, both of normal and amputated cockroaches. He concluded from the observations that while there may be a central basic program, proprioceptive feedback is undoubtedly of great importance. Later, in a computer-aided electrophysiological study in which legs were forced to move sinusoidally (Delcomyn, 1971c), he found that the intraleg reflexes are such as to augment leg thrust. This is of considerable general interest, because several independent studies on crustaceans (e.g., MacMillan and Dando, 1973) have come to a similar conclusion. Thus it seems that there is a central inherent program that directs locomotory movements but that its output alone tends to evoke weak movements in some systems. In these systems, motor outputs are enhanced by positive feedback during movements. In the cockroach, the program's output is already approximately correctly phased for all legs, including alternating excitation of antagonists.

There are indications of a comparable situation in some mammalian behavior. For example, facial grooming in mice occurs in a basically normal manner following de-afferentation, even in the muscles of the stumps of amputated forelegs (Fentress, 1972).

It might be expected that sometimes in a generally excited preparation, a spontaneous turning on of one of the program generators, especially that for walking, would occur. If this happened at a time when an intracellular electrode was placed in a motorneuron or interneuron, one might get clues as to the mechanism involved. Most of the common movement patterns that insects make, such as respiration, walking, song, and other courtship and sexual movements, involve rhythmical repetitions of contractions. Accordingly, a major part of the endogenous neural machinery is often considered to consist of one or more "oscillators" (Wilson, 1967; Davis, 1969; Davis and Kennedy, 1972a, b). A specific search for a central flight oscillator in the locust failed to find one (Page, 1970). But interesting oscillatory membrane potential swings of large size occurring at the relevant natural rhythm have now been observed intracellularly in three situations: sound production, flight, and walking rhythm of crickets and locusts (Figure 6). Action potentials

in the motor axon of the relevant neurons were monitored by recording extracellularly from the muscle innervated. In all cases there has been a peripheral spike only if the depolarization recorded in the soma reached a critical value. If the soma potential was both large and steeply rising, a close pair, triplet, or quadruplet of impulses was produced. It is thus probable that the wave of depolarization is an electrotonically conducted correlate of the neuropilar dendrite/neurite potential that generates the

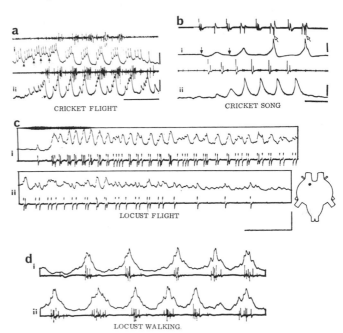

FIGURE 6 Rhythmic membrane potentials occurring in ganglion cells recorded intracellularly. (a) i and ii: Two sequences from cricket during flight patterns—possibly from wing elevator neurons, on lower traces. Extracellular electromyograms from subalar muscles on upper traces. Arrows indicate possible inhibitory activity at start of depressor bursts that drive the subalar muscles. Calibration: vertical, 10 mV; horizontal, 100 msec. From Bentley (1969a). (b) From cricket during singing. Records from unknown wing closers with an intracellular electrode (lower traces) and extracellular from motor nerves (upper traces). Filled arrows: onset of polarizing activity; open arrows: efferent impulses following ganglion cell spikes. Calibration: vertical, 10 mV; horizontal, (i) 25 msec; (ii) 50 msec. From Bentley (1969b). (c) i and ii: Activity in locust metathoracic flight motorneuron whose location is indicated in the insert, during flight pattern. Activity excited by an air puff applied to the head (monitored via microphone on upper beam). Lower beam: myogram from subalar muscles. Note exact correspondence with spikes on intracellular record. The entire sequence takes place during a plateau of depolarization that is maintained during the flight sequence but gradually subsides towards termination. Calibration: 10 mV; 250 msec. From Hoyle and Burrows (1973b). (d) i and ii: Continuous records from locust posterior fast flexor during spontaneously occurring flexions of the tibia. Electromyogram (upper trace) is from femur and shows that reciprocal extensor activity was very weak. Calibration: 10 mV; 500 msec. From Hoyle and Burrows (1973b).

spikes. The nature of the waves is at the present time uncertain. Either they are due to many small, summating EPSPs, to an electrotonically coupled oscillator neuron (Mendelson, 1971), or they are intrinsically generated in the neuropilar segment of the motorneuron itself.

Some spontaneously occurring potential swings of about 15 mV peak-to-peak causing output in flexor tibiae motorneurons recorded by Hoyle and Burrows (1973b) had a frequency of about 2 Hz, characteristic of locust stepping during walking. But in these cases the antagonistic slow extensor motorneuron (the fast extensor fires only in jumping) was silent and the leg moved hardly at all. In this case therefore, the action closely resembled what might be expected from a spontaneously firing "oscillator." The membrane potential swings in the flight motorneurons were more than 10 mV peak-to-peak in the soma and occurred at the flight rhythm of 19 Hz.

Neuronal circuitry

JUMP RESPONSE The simple grasshopper and locust escape behavior, jumping, is the one in which most progress is being made toward a reasonably complete analysis of underlying neuronal events. All the motorneurons involved, six flexors and the single fast extensor, have been located and examined intracellularly during elicitation of the reflex (Hoyle and Burrows, 1973a). Three interneurons associated with flexion have been located and recorded from, and the action of some others can be inferred.

In a jump response the left and right fast extensors fire simultaneously and at similar rates; this coordination is not achieved by direct excitatory connections between them but by independent simultaneous activation (Hoyle and Burrows, 1973a). This requires input from chordotonal organs located in the femora (Bassler, 1968) that are stimulated by the rapid strong flexion of the tibia, which is always the initial action in a jump response and is due to a fast flexor neuron. The next stage is a simultaneous inhibitory input to the flexor motorneurons and an excitatory one to the fast extensor motorneurons, of both sides. Once an action potential has been fired by the fast extensor it re-excites itself. As judged by the delay and number of summating, facilitating EPSPs involved, this is probably via collaterals and interneurons. The net result is that if the general excitatory level is high, as it usually is after stimuli of the kind that cause jumps, the two FETi fire bursts of impulses that fully tetanize the extensor muscles, causing take-off. The jump is completed by a flexion of the tibia. This is achieved in part by delayed excitation occurring automatically via an interneuron that synapses with the flexors and is excited by a fast extensor collateral.

Earlier stages in the initiation of the jump response are

obscure. It has been assumed for many years that the response, like the startle response of the cockroach, is due to action of giant fibers (Roeder, 1948; Cook, 1951). But recently, the giant fibers were shown to be unnecessary in the cockroach (Parnas and Dagan, 1971); their function may be simply to arrest other ongoing activity preparatory to execution of the escape reponse.

In the locust the only neurons so far discovered that have axons to both sides of the ganglion, which could therefore mediate a bilateral response, are the central dorsal giants (Crossman et al., 1971). These do always fire prior to the jump response (Dagan and Hoyle, unpublished), but their behavioral role is not yet known.

COMMAND FIBERS Hughes and Wiersma (1960) discovcovered interneurons in crayfish circumesophagel connectives that cause patterned, relatively complex movements to occur when stimulated by a simple train. The movements of some parts, for example the swimmerets, are rhythmical, continuing ones. Wiersma and Ikeda (1964) termed such fibers "command" fibers. Fibers that have some aspects in common with these have been found in only one insect thus far (Figure 7), a small grasshopper *Gastrimargus africanus* by Elsner (1969). The responses he obtained were rhythmical up and down leg movements.

The actual functional role of such fibers in the insect has not been established. It may be that these fibers are no more than general excitatory interneurons (Hoyle, 1964)

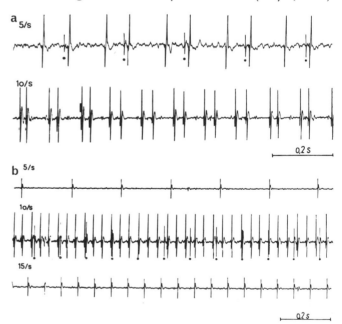

FIGURE 7 Rhythmical muscle activity not related 1:1 to stimulus elicited in a grasshopper (*Gastrimargus africanus*) by stimulation of axon bundles in cervical connectives. (a) Paired outputs for single inputs. (b) Sequences of bursts at fixed long intervals elicited only at 10 Hz not at 5 or 15 Hz. From Elsner (1969).

that cause autogenic circuits to come into operation. If a brain is shown to really use these fibers to "command" the lower motor centers to "do their thing" by issuing continuing trains, that will be a major step in our understanding of neuroethology.

EXCITATION OF ANTAGONISTS   The traditional, Sherringtonian view of the use of antagonists as simple reciprocators during movements has generally been accepted as the norm for operation in insect movements, although Hoyle (1964) found that in long-term recordings from freely walking locusts the electrical records suggested a high degree of cocontraction. The problem was easily resolvable with the aid of two microelectrodes placed in somas of antagonistic pairs of motorneurons because the activity in the somas cannot be misinterpreted. The data for several pairs of antagonists all showed clearly that the synaptic excitation causing movements would lead to cocontraction in at least 20% of the cases. About 50% of inputs are of simple reciprocal nature and in the remainder co-excitation and reciprocal excitation are mixed (Figure 8).

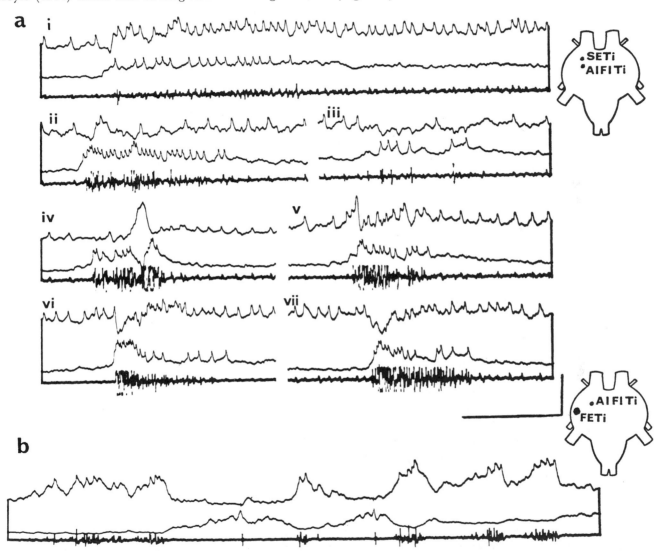

FIGURE 8   Examples of spontaneously occurring synaptic input to antagonists of the locust metathoracic ganglion during simultaneous intracellular recordings. (a) Slow extensor tibiae (upper traces) and anterior intermediate flexor (middle traces). Note that the slow extensor may be excited either: at the same time as the flexor leading to cocontraction [(a) i and (a) ii]; completely reciprocally [(a) iii and (a) vii]; or mixed coactivation and reciprocal activation [(a) ii and (a) vi]. (b) Fast extensor tibiae (upper traces) and anterior intermediate flexor (middle traces). The input is reciprocally related, but note the depolarizing response in the flexor following firing of the extensor. The locations of the neuron somas from which the intracellular recordings were made are shown in the two insert diagrams of the metathoracic ganglion from the ventral aspect. The lower traces denote extracellular electromyograms recorded with wires placed in the femur. Calibration: 10 mV; 250 msec. From Hoyle and Burrows (1973a).

ENTRAINMENT OF IDENTIFIED NEURONS Horridge (1962) found that it is easy to train headless cockroaches and locusts to hold up a leg in order to avoid receiving a shock. Hoyle (1965) made a successful attempt in the locust to train a tonically firing, postural motorneuron to alter its firing rate (Figure 9). The preparation is promising, because it offers the possibility of examining the cellular basis of a learning process. The relevant neuron, the anterior coxal adductor, was localized in the ganglion by Hoyle and Burrows (1973a). A computer program has now been developed that permits automatic entrainment. The discharge of the neuron is stored by a computer that sorts succeeding impulses into a series of bins of spike intervals. By preset instructions, the computer recognises the appearance in the discharge of any significant trend toward consecutive long intervals (equal to a low firing rate). In the whole insect, these would be followed by loss of muscle tone and lowering of the leg. The computer then triggers a stimulator that delivers a shock to a sensory nerve having indirect input to the motorneuron. The result of a few minutes of such training, in which only 5 to 10 shocks are received, is often a stable shift in the interval histogram toward shorter intervals. Examples of before- and after-training histograms photographed directly from the scope face of a Linc-8 computer are given in Figure 9.

Summary

Among all animals, insects with complex, varied behavior yet small total numbers of nerve cells and very short generation times offer the greatest opportunities for in-depth studies of neural circuits and the ways in which behavior is controlled. Simple electromyograms and nerve impulse recordings are easily obtained from freely moving intact animals, providing quantitative information about the motor output. Recently developed techniques have enabled intracellular recordings to be made from nerve cell bodies in the ganglia, in preparations in which there is at least limited behavior. The recordings are followed by dye injections that in turn permit both neuron localization and the following of dendritic and axonal pathways. Preliminary work on neural integration in locusts is described as well as experiments in which the frequency of firing of a tonic, postural motorneuron is modified by operant conditioning. Work under way in other laboratories on the genetics of insect behavior can be linked to possible underlying neural mechanisms. Developmental and tissue culture studies on cockroach ganglia seem likely to provide basic knowledge on neuronal specificity and the making of functional contacts between neurons.

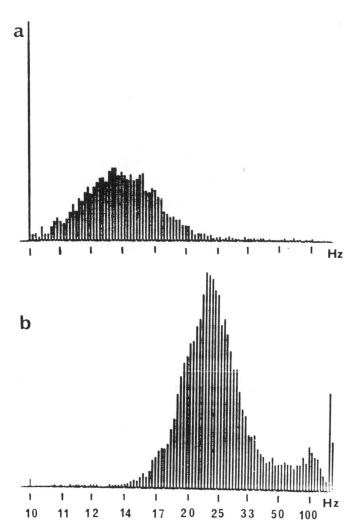

FIGURE 9 Entrainment of an identified neuron. Computer histogram (on oscilloscope) of discharge of single excitatory axon to anterior coxal adductor. (a) Fresh insect before training. (b) Same insect after applying training program. A weak electric shock was applied to the leg each time successive long intervals, which were accumulated in a single bin, left-hand side of record in (a), had occurred. The preparation responds by stable increased mean firing rate, with almost total lack of long intervals. Records made for equal times, hence more intervals stored in (b). From Tosney and Hoyle (in preparation).

ACKNOWLEDGMENTS The original research referred to in this article was supported by National Science Foundation Research Grant No. 32018.

I would like to acknowledge the technical assistance of Lee H. Vernon, Harrison Howard, Betty Moberley, and Julie Pohl, to praise the superb technical skills of my collaborators Dr. Malcolm Burrows and Dr. Dani Dagan, and to commend all fellow workers in the field, especially Dr. David Bentley, Dr. Franz Huber, Dr. Keir Pearson, and Dr. Norbert Elsner.

REFERENCES

ALBRECHT, F. O., 1953. *The Anatomy of the Migratory Locust.* London: Athlone Press.

ALOE, L., and R. LEVI-MONTALCINI, 1972. Interrelation and dynamic activity of visceral muscle and nerve cells from insect embryos in long-term culture. *J. Neurobiol.* 3:3–23.

BASSLER, U., 1968. Zur Steuerung des Springens bei der Wanderheuschrecke *Schistocerca gregaria. Kybernetik* 4:112.

BENTLEY, D. R., 1969a. Intracellular activity in cricket neurons during the generation of behaviour patterns. *J. Insect Physiol.* 15:677–699.

BENTLEY, D. R., 1969b. Intracellular activity in cricket neurons during generation of song patterns. *Z. Vergl. Physiol.* 62:267–283.

BENTLEY, D. R., 1970. A topological map of the locust flight system motor neurons. *J. Insect Physiol.* 16:905–918.

BENTLEY, D. R., 1971. Genetic control of an insect neuronal network. *Science* 174:1139–1141.

BENTLEY, D. R., and R. R. HOY, 1970. Postembryonic development of adult motor patterns in crickets: a neural analysis. *Science* 170:1409–1411.

BENZER, S., 1971. From gene to behavior. *J. Amer. Med. Ass.* 218:1015–1022.

BURROWS, M., and G. HOYLE, 1973. Neural mechanisms underlying behavior in the locust *Schistocerca gregaria.* 3. Topography of limb motorneurons in metathoracic ganglia. *J. Neurobiol.* 4:167–186.

COOK, P. M., 1951. Observations on giant fibers of the nervous system of *Locusta migratoria. Quart. J. Mic. Sci.* 92:297–305.

CROSSMAN, A. R., G. A. KERKUT, R. M. PITMAN, and R. J. WALKER, 1971. Electrically excitable nerve cell bodies in the central ganglia of two insect species *Periplaneta americana* and *Schistocerca gregaria.* Investigation of cell geometry and morphology by intracellular dye injection. *Comp. Biochem. Physiol.* 40(A):579–594.

DAVIS, W. J., 1969. The neural control of swimmeret beating in the lobster. *J. Exp. Biol.* 50:99–117.

DAVIS, W. J., and D. KENNEDY, 1972a. Command interneurons controlling swimmeret movements in the lobster. I. Types of effects on motorneurons. *J. Neurophysiol.* 35:1–12.

DAVIS, W. J., and D. KENNEDY, 1972b. Command interneurons controlling swimmeret movements in the lobster. II. Interaction of effects on motorneurons. *J. Neurophysiol.* 35:13–19.

DELCOMYN, F., 1971a. The locomotion of the cockroach *Periplaneta americana. J. Exp. Biol.* 54:443–452.

DELCOMYN, F., 1971b. The effect of limb amputation on locomotion in the cockroach *Periplaneta americana. J. Exp. Biol.* 54:453–469.

DELCOMYN, F., 1971c. Computer aided analysis of a locomotor leg reflex in the cockroach *Periplaneta americana. Z. Vergl. Physiol.* 74:427–445.

EDWARDS, J. S., 1969. Postembryonic development and regeneration of the insect nervous system. *Adv. Ins. Physiol.* 6:97–137.

EDWARDS, J. S., and J. PALKA, 1971. Neural regeneration: delayed formation of central contacts by insect sensory cells. *Science* 172:591–594.

EDWARDS, J. S., and T. S. SAHOTA, 1967. Regeneration of a sensory system: the formation of central connections by normal and transplanted cerci of the house cricket *Acheta domesticus. J. Exp. Zool.* 166:387–396.

ELSNER, N., 1968. Die neuromuskularen Grundlagen des Werbeverhaltens der roten Keulenheuschrecke *Gomphocerippurufus* (L.). *Z. Vergl. Physiol.* 60:308–350.

ELSNER, N., 1969. Kommandofasern im Zentralnervensystem der Heuschrecke *Gastrimargus africanus* (Oedipodinae). *Verh. Deutsch. Ges. Zool.* 33:465–471.

ELSNER, N., 1973. The central nervous control of courtship behavior in the grasshopper *Gomphocerippus rufus* (L.) (Orthoptera: Acrididae). 1971. *Sympos. Invert. Neurobiol.*

FENTRESS, J., 1972. Development of patterning of movement sequences in inbred mice. In *The Biology of Behavior,* J. Kiger, ed. Eugene, Oregon: Oregon State University Press.

HAGIWARA, S., and A. WATANABE, 1956. Discharges in motorneurons of cicada. *J. Cell. Comp. Physiol.* 47:415–428.

HESS, W. R., 1943. Das Zwischenhirn als Koordinationsorgan. *Helv. Physiol. Pharmacol. Acta* 1:549–565.

HIRSCH, J., and J. C. BOUDREAU, 1958. Studies in experimental behavior genetics. 1. The heritability of phototaxis in a population of *Drosophila melanogaster. J. Comp. Physiol. Psychol.* 51:647–651.

HORRIDGE, G. A., 1962. Learning leg position by the ventral nerve cord in headless insects. *Proc. Roy. Soc. [Biol.]* 157:33–52.

HOTTA, Y., and S. BENZER, 1969. Abnormal electroretinograms in visual mutants of *Drosophila. Nature (London)* 222:354–365.

HOYLE, G. 1953. Potassium ions and insect nerve muscle. *J. Exp. Biol.* 30:121–135.

HOYLE, G., 1957. Nervous control of insect muscles. In *Recent Advances in Invertebrate Physiology.* Eugene, Oregon: University of Oregon Publications, pp. 73–98.

HOYLE, G., 1964. Exploration of neuronal mechanisms underlying behavior of insects. In *Neural Theory and Modeling,* R. Reiss, ed. Stanford, Calif.: Stanford University Press, pp. 346–376.

HOYLE, G., 1965. Neurophysiological studies on "learning" in headless insects. In *The Physiology of the Insect Central Nervous System,* J. E. Treherne, and J. W. L. Beament, eds. London and New York: Academic Press, pp. 203–232.

HOYLE, G., and M. BURROWS, 1970. Intracellular studies on identified neurons of insects. *Fed. Proc.* 29:1920A.

HOYLE, G., and M. BURROWS, 1973a. Neural mechanisms underlying behavior in the locust *Schistocerca gregaria.* 1. Physiology of identified neurons in the metathoracic ganglion. *J. Neurobiol.* 4:3–41.

HOYLE, G., and M. BURROWS, 1973b. Neural mechanisms underlying behavior in the locust *Schistocerca gregaria.* 2. Integrative activity in metathoracic neurons. *J. Neurobiol.* 4:43–67.

HUBER, F. 1962. Central nervous control of sound production in crickets and some speculation on its evolution. *Evol.* 16:429–442.

HUBER, F., 1971. Nervose Grundlagen der akustiscken Kommunikation bei Insekten. *Acad. Wiss. Phein-West.* 205:41–84.

HUBER, F., and N. ELSNER, 1969. Die Organisation des Werbegesanges der Heuschrecke *Gomphocerippus rufus* L. in Abhangigkeit von zentralen und peripheren Bedingungen. *Z. Vergl. Physiol.* 65:389–443.

HUGHES, G. M., and C. A. G. WIERSMA, 1960. The co-ordination of swimmeret movements in the crayfish *Procambarus clarkii* (Girard). *J. Exp. Biol.* 37:657–670.

IKEDA, K., and W. D. KAPLAN, 1970a. Patterned neural activity of a mutant *Drosophila melanogaster. Proc. Nat. Acad. Sci. (USA)* 66:765–772.

IKEDA, K., and W. D. KAPLAN, 1970b. Unilaterally patterned

neural activity of gynandromorphs, mosaic for a neurological mutant of *Drosophila melanogaster*. *Proc. Nat. Acad. Sci. (USA)* 67:1480–1487.

KERKUT, G. A., R. M. PITMAN, and R. J. WALKER, 1968. Electrical activity in insect nerve cell bodies. *Life Sci.* 7:605–607.

KERKUT, G. A., R. M. PITMAN, and R. J. WALKER, 1969. Sensitivity of neurons of the insect central nervous system to iontophoretically applied acetylcholine or GABA. *Nature (London)* 222:1075–1076.

KUTSCH, W., 1971. The development of the flight pattern in the desert locust *Schistocerca gregaria*. *Z. Vergl. Physiol.* 74:156–168.

KUTSCH, W., and F. HUBER, 1970. Zentrale versus periphere Kontrolle des Gesanges von Grillen (*Gryllus campestris*). *Z. Vergl. Physiol.* 67:140–159.

LOHER, W., and F. HUBER, 1966. Nervous and endocrine control of behavior in a grasshopper (*Gomphocerippus rufus* L. Acrididae). *Sympos. Soc. Exp. Biol.* 20:381–400.

LORENZ, K. Z., 1950. The comparative method in studying innate behavior patterns. *Sympos. Soc. Exp. Biol.* 4:221–268.

MACMILLAN, D. L., and M. R. DANDO, 1973. Tension receptors associated with the apodemes of muscles in the walking legs of the crab *Cancer magister*. In *Marine Biology and Behavior* (in press).

MANNING, A., 1963. Selection for mating speed in *Drosophila melanogaster* based on the behavior of one sex. *Anim. Behav.* 11:116–120.

MAYNARD, D. M., 1956. Electrical activity in the cockroach cerebrum. *Nature (London)* 177:529–530.

MENDELSON, M., 1971. Oscillator neurons in crustacean ganglia. *Science* 171:1170–1173.

NARAHASHI, T., 1965. The physiology of insect axons. In *The Physiology of the Insect Central Nervous System*, J. E. Treherne, and J. W. L. Beament, eds. London and New York: Academic Press, pp. 1–20.

PAGE, C. H., 1970. Unit responses in the metathoracic ganglion of the flying locust. *Comp. Biochem. Physiol.* 37:565–571.

PARNAS, I., and D. DAGAN, 1971. Functional organizations of giant axons in the central nervous systems of insects. New aspects. *Adv. Insect Physiol.* 8:95–143.

PEARSON, K. G., 1972. Central programming and reflex control of walking in the cockroach. *J. Exp. Biol.* 56:173–193.

ROEDER, K. D., 1948. Organization of the ascending giant fiber system in the cockroach (*Periplaneta americana*). *J. Exp. Zool.* 108:243–262.

ROTHENBUHLER, W. C., 1964a. Behavior genetics of nest cleaning in honey-bees. 1. Responses of four inbred lines to disease-killed brood. *Anim. Behav.* 12:578–583.

ROTHENBUHLER, W. C., 1964b. Behavior genetics of nest cleaning in honey-bees. 4. Responses of F_1 and back-cross generations to disease-killed brood. *Amer. Zool.* 4:111–123.

ROWE, E. C., 1960. Activity of single nerve cells in an insect thoracic ganglion. *Anat. Rec.* 137:389 (Abstract).

ROWE, E. C., 1969. Microelectrode records from a cockroach thoracic ganglion: Synaptic potentials and temporal patterns of spike activity. *Comp. Biochem. Physiol.* 30:529–539.

RUNION, H. I., and P. N. R. USHERWOOD, 1966. A new approach to neuromuscular analysis in the intact free-walking insect preparation. *J. Insect Physiol.* 12:1255–1263.

SEECOF, R. L., R. L. TEPLITZ, I. GERSON, K. IKEDA, and J. J. DONADY, 1972. Differentiation of neuromuscular junctions in cultures of embryonic *Drosophila* cells. *Proc. Nat. Acad. Sci. (USA)* 69:566–570.

SPERRY, R. W., 1943. Effect of 180 degree rotation of the retinal field on visuomotor co-ordination. *J. Exp. Zool.* 92:263–279.

SPIRA, M. E., I. PARNAS, and F. BERGMANN, 1969. Histological and electrophysiological studies on the giant axons of the cockroach *Periplaneta americana*. *J. Exp. Biol.* 50:629–634.

STRETTON, A. O. W., and E. A. KRAVITZ, 1968. Neuronal geometry: Determination with a technique of intracellular dye injection. *Science* 162:132–134.

TINBERGEN, N., 1951. *The Study of Instinct*. London: Oxford University Press.

USHERWOOD, P. N. R., and H. I. RUNION, 1970. Analysis of the mechanical responses of metathoracic extensor tibiae muscles of free-walking locusts. *J. Exp. Biol.* 52:39–58.

VON HOLST, E., and U. VON ST. PAUL, 1963. On the functional organization of drives. *An. Behav.* 11:1–20.

WEISS, P., 1950. Experimental analysis of co-ordination by the disarrangement of central peripheral relations. *Sympos. Soc. Exp. Biol.* 4:92–111.

WELLS, G. P., 1950. Spontaneous activity cycles in polychaete worms. *Sympos. Soc. Exp. Biol.* 4:127–142.

WIERSMA, C. A. G., 1959. Coding and decoding in the nervous system. *Engineering and Science Monthly* (Calif. Inst. Technol.) 23:21–24.

WIERSMA, C. A. G., and K. IKEDA, 1964. Interneurons commanding swimmeret movements in the crayfish *Procambarus clarkii* (Girard). *Comp. Biochem. Physiol.* 12:509–525.

WIESEL, T. N., and D. H. HUBEL, 1963. Single-cell responses in striate cortex of kittens deprived of vision in one eye. *J. Neurophysiol.* 26:1003–1017.

WIESEL, T. N., and D. H. HUBEL, 1965. Comparison of the effects of unilateral and bilateral eye closure on cortical unit responses in kittens. *J. Neurophysiol.* 28:1029–1040.

WIGGLESWORTH, V. B., 1939. *The Principles of Insect Physiology*, 1st Ed., London: Methuen.

WILSON, D. M., 1966. Insect walking. *Ann. Rev. Entom.* 11:103–122.

WILSON, D. M., 1967. An approach to the problem of control of rhythmic behavior. In *Invertebrate Nervous Systems*, C. A. G. Wiersma, ed. Chicago: Chicago University Press, pp. 219–229.

WILSON, D. M., 1968. The nervous control of insect flight and related behavior. *Adv. Insect Physiol.* 5:289–338.

36 A Comparison of the Visual Behavior of a Predatory Arthropod with That of a Mammal

MICHAEL F. LAND

ABSTRACT Any predator has the tasks of locating prey, identifying it, and capturing it. These operations necessarily involve specializations: On the sensory side for high spatial acuity and on the motor side for speed of action. The chapter is concerned chiefly with the visual system of jumping spiders, which catch flies by stalking them. Each of the four pairs of simple eyes—the fixed side eyes and movable principal eyes—plays a different role in prey capture and courtship, and techniques are described by which the several kinds of control mechanism involved can be separated and examined. Specific examples are the control of turning by the side eyes during the localization of moving stimuli and the part played by scanning movements of the principal eyes in identifying seen stimuli as potential prey or potential mates.

Introduction

To DRAW USEFUL conclusions from a comparison between animals with totally different evolutionary histories, it is necessary to choose species whose life styles are similar. Then it is possible to begin to sort out how much the organization of an animal's behavior reflects the principles of construction of the nervous system peculiar to a phylum and how much is due to convergent evolution caused by similarities in the tasks performed. Predatory animals, especially those that hunt visually, always have interesting behavior that is usually comprehensible to an observer, because the goals of many of the different components are clear. They are thus attractive subjects on which to base a comparative study. There are relatively few terrestrial arthropods that catch active prey by stalking them, in the manner of a mammalian carnivore. There is, however, one group, the jumping spiders (*Salticidae*), that do so; the strategy they adopt in catching flies is sufficiently similar to that of a cat catching a bird—to creep toward the prey undetected until the chance of escape is small, and then spring—that it is worthwhile to

MICHAEL F. LAND School of Biological Sciences, University of Sussex, Falmer, Sussex, England

enquire further into the optical, neural, and behavioral specializations that make this possible. This chapter is concerned chiefly with examining parallels and contrasts in visually directed behavior; more detailed studies of salticid behavior are given by Drees (1952), Crane (1949), and Heil (1936); of visual anatomy by Land (1969a) and Eakin and Brandenburger (1971); of visual optics by Land (1969a, 1969b) and Homann (1928); and of motor control by Land (1971, 1972).

Optical organization

All spiders have either 8 or 6 simple eyes, each constructed with a lens that is part of the cuticle, a cellular but transparent *vitreous*, and a retina containing up to 10^4 primary receptors but with no ganglion cells or other neurons in the eye itself. In jumping spiders, 6 of the 8 eyes are particularly well developed and fall into two groups: The fixed lateral eyes (Figure 1) that between them cover a field of view extending the full 360° around the animal, including a 25° region of binocular overlap anteriorly, and the principal eyes with a much longer focal length but very restricted field of view—between 1 and 5° laterally by 20° vertically. However, the retinae of the principal eyes are movable, so that the range of vision of these eyes during movements is extended to approximately 35° on either side of the midline in the horizontal plane, and by a similar amount vertically (Land, 1969b).

The lateral eyes are movement detectors, and there is no behavioral evidence to indicate that they do anything else. When something moves in the visual environment—whether it is an insect, another spider, or even a car on the street—the spider turns to face it, and this turn is mediated by the lateral eyes. In the posterolateral eyes the receptor spacing is everywhere between 1 and 1.5°, and any object that subtends at least this angle and moves through a similar angle (i.e., from one receptor to the next) stands an equal chance of evoking a turn (Land, 1971). In the anterolateral eyes the situation is similar, except that, in

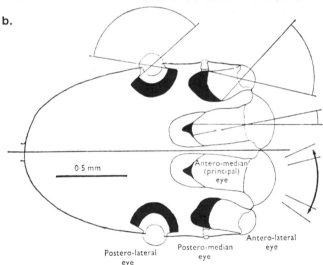

FIGURE 1 (a) Juvenile jumping spider (*Metaphidippus harfordi*: 2 mm long) showing the eyes on the prosoma. (b) Diagrammatic longitudinal section through the prosoma of a jumping spider, showing the fields of view of the eyes in the horizontal plane. For the right principal eye the field is shown as it is extended by lateral eye-movements. In some spiders (e.g., *Salticus*) the fields of the posterolateral eyes meet behind the animal.

the region of the retina close to the spider's axis, the receptor spacing decreases to 30 min, and in this region turns can be evoked by much smaller objects.

By contrast, the principal eyes have much higher resolution and, as Homann (1928) showed by blinding each set of eyes in turn, the principal eyes are concerned

with tracking objects that have already been located by the lateral eyes and with pattern recognition. These retinae have a complex structure with 4 layers of receptors, and it is argued elsewhere that this layering is used to compensate for longitudinal chromatic aberration (Land, 1969a) in a manner similar to the layered arrangement of cones in fish retinae (Eberle, 1967); there is now ample behavioral and physiological evidence that jumping spiders possess at least dichromatic color vision. However, of more interest to the present discussion is the finding that the retina is arranged in a *foveal* manner, with receptor separations in the central part of about 10 min increasing in the upper and lower extremities to nearly 45 min. During certain eye movements (see below) this central region of retina is directed to stimuli in the visual field, so that the use of the term fovea is functionally as well as anatomically correct.

In mammals, at least in carnivores and primates including man, the functions of the peripheral and central regions of the retina are quite different, the former being concerned largely with detecting objects to the side and directing the retinal center toward them by a saccadic eye movement, a head movement, or a combination of the two. Tasks requiring higher resolution—pattern recognition and tracking of small objects—are performed by the foveal region. In jumping spiders the same division of labor exists, but the different functions are performed by different eyes. It is interesting that the task of movement detection, which requires relatively low resolution and for which movement of the eye itself is actually a potential source of confusion, is performed here by fixed eyes of short focal length (in *Metaphidippus aeneolus* the focal length of the posterolateral eyes is 174 μm, nearly 3 times shorter than that of the principal eyes which is 512 μm). This optical division of labor results in an enormous economy of space, because a spherical eye built on the vertebrate plan would have to have a long focal length to provide the required resolution at the center and would occupy a volume nearly 3^3 times that of the posterolateral eyes, which provide wide-field coverage. As it is, this space, nearly the entire volume of the spider's prosoma, is saved, and the fovea of the system is confined to a long but narrow tubular eye. It is tempting to argue that the system evolved by jumping spiders is much the more efficient.

Last, because jumping spiders have simple eyes of wide aperture, their resolving power is both theoretically and practically greater than that of their prey with compound eyes. The Rayleigh criterion sets a limit of at best 30 min to the resolution achievable by a compound eye of the apposition type—3 times worse than that of the principal eyes of these spiders (see Barlow, 1952). Ophthalmoscopic examination of the retinae of jumping spiders, which

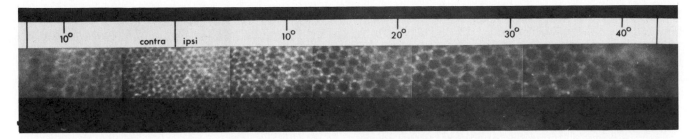

FIGURE 2 Photographic montage of a 5°-wide horizontal strip across the retina of the fixed anterolateral eye of *Salticus scenicus*, taken through the spider's own lens using an ophthalmoscope (Land, 1969b). The field of view of the eye extends from 43° to the ipsilateral side of the body axis to 13° across the axis. The lighter lattice is caused by light scattering from the melanin granules that partition the individual receptors from each other: These are contained in the dark hexagons. In the region of the body axis, the angular separation of individual receptors is 3 times less than at the peripheries: This is not the result of optical distortion.

makes use of, and is therefore limited by, the optical quality of the lens and ocular media, demonstrates convincingly that the eyes are optically excellent and that the limit to resolution is set by receptor density (see Land, 1969a, and Figure 2). The minimum receptor separation of 10 min in the principal retinae is just over one order of magnitude larger than in man but less than in any other invertebrate except the cephalopod molluscs.

Control of movement

Human eye movements have among other tasks the two of locating objects of interest and of tracking their movements. Locating usually involves a saccade, whose control system is *open* in nature in that the size of the movement is determined prior to its execution and is not modified by visual feedback during its progress. Smooth-pursuit eye movements, on the other hand, necessarily involve continuous feedback (see Bizzi and Evarts, 1971). In catching a fly, the spider has the same two tasks—location and tracking—and it is interesting that the two control systems concerned are similar to their mammalian analogues: Turns initiated by the lateral eyes do not require visual feedback for their accurate completion, whereas the small turns used in tracking, for which the principal eyes are responsible, are disturbed when the spider cannot see the visual consequences of its own motion.

This can be demonstrated with the simple apparatus shown in Figure 3, which makes use of the spider's ability to hold, walk on, and turn a small card ring of its own weight. The animal is suspended by a small piece of rigid card waxed to the prosoma; this suspension keeps the body stationary in space, although the legs can still move normally relative to the body. Stimuli, usually black dots on a white background, are then presented to the animal, and the kinds of turn that result under these visually open-loop conditions can be studied either by cinematography or by using a matching-pointed system to monitor the angular position of the ring (Land, 1971). It is an interesting comment on the inflexibility of spider behavior that the animals fail to react in any obvious way to the incongruity that results from interchanging a movable ring for the fixed world but will continue for many hours to make logically correct though totally inappropriate movements.

When a small spot is moved a short distance (1 to 5°) in the field of view of the lateral eyes, but outside that of the principal eyes, the spider executes a single rapid turn

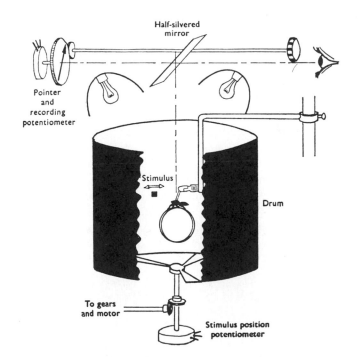

FIGURE 3 Apparatus used to evoke turns in a visually open-loop situation. The spider is fixed in the center of a 12 cm drum and turns its ring when the stimulus is moved a short distance. Stimulus position and angle turned are measured with 360° potentiometers.

whose magnitude is nearly equal to the angle between the stimulus and the body axis: i.e., the angle turned is that which would be required by a free spider to bring the stimulus into the field of vision of the principal eyes (Figure 4). However, the fact that this turn does not result in a change in the relative position of the stimulus demonstrates that, like a human saccade, the size of the turn is dictated in advance solely by the retinal location of the stimulus image. Interesting questions are thus raised about the mechanism of translation between the input of the system (retinal location) and output (stepping leg movements). During a turn the spider steps forward with the legs on one side and backward with those on the other, thereby spinning round within its own length. The accuracy of turning is independent of both the speed of turning (over a tenfold range) and the inertial load (over at least a three-hundredfold range), and so also are the lengths of the steps the spider makes. It seems likely, therefore, that the size of the turn to be made is specified by the retina in terms of the number of steps to be taken: In *Metaphidippus harfordi* each step rotates the animal by approximately 9°, which is a small enough quantum to account for the observed accuracy of turning (Land, 1972).

When a stimulus lies within the field of view of the principal eyes, the situation is quite different. Stationary stimuli are now capable of evoking turns, and under open-loop conditions the visual response to a spot or line 10 to 30° from the body axis is a series of up to 10 small turns, each of about 10°, separated by approximately 100 msec intervals (Figure 5). These turns are very similar to the *Tantalus-like* saccadic pursuit movements made by the human eye when a stabilized image is presented a few degrees away from the fovea (Robinson, 1965). By comparison with turns initiated by the lateral eyes, the control system now operating is formally a closed one (because opening the visual loop results in multiple turns). However, the essential difference between the two systems is simply that the principal eyes, being movable, can detect stationary objects whereas the lateral eyes cannot: A *continuously* moving stimulus presented to the lateral eyes evokes Tantalus-like turning also. Both systems can be said to be of the *sampled data* rather than *continuously monitored* type, but the restrictions on the way data are sampled are different in the two cases.

Eye movements

Perhaps the most extraordinary parallel between jumping spiders and mammals is the finding that the retinae of the principal eyes can be moved with the same three degrees of freedom as the human eye (Land, 1969b). When examined ophthalmoscopically (Figure 6), the retinae are

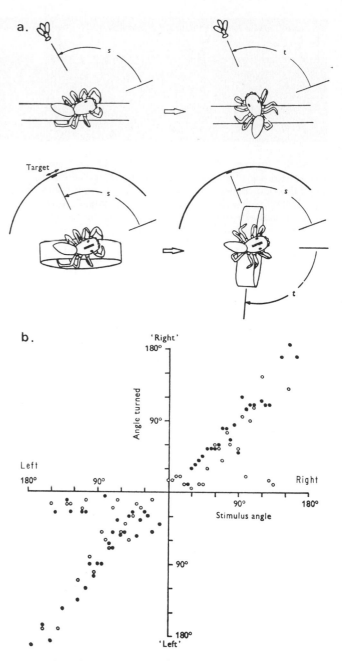

FIGURE 4 (a) *Upper row*: Spider standing on a twig, turning toward a fly 110° to its left. *Lower row*: Open-loop situation in which the spider sees a stimulus 110° to its left and turns the ring through the same angle in the opposite direction. The spider makes the same leg movements as in the upper figure. (b) Result of a single experiment using the apparatus shown in Figure 3. The stimulus was a 5° black spot moved through 5° at 72 different positions around the animal. Most of the points show that the spider makes a turn in which amplitude is nearly equal to the angle the stimulus makes with the animal's body axis. A small proportion of the turns are in the correct direction but are of small amplitude (10 to 20°), which is independent of stimulus position. Free spiders also make both kinds of turn. Open and filled circles indicate stimuli moving in opposite directions.

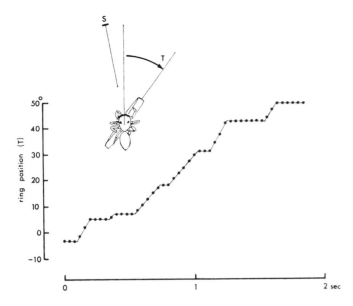

FIGURE 5 "Tantalus-like" turning toward a 2° wide stripe that is stationary, 10° to the left of the spider's axis. The points show the position of the ring on successive frames of film, with the spider in the apparatus shown in Figure 3. This is a fairly typical outcome in which the spider makes 6 turns, whose mean size is 9°, separated by variable intervals of around 100 msec. Multiple turning to stationary stimuli is only seen when the stimulus lies within the field of the movable principal eyes.

seen to move vertically and horizontally, or with a vector combination, and also to rotate about the visual axes by up to 25° in each direction. However, in contrast to the human eye, only the retina and eye tube move, the lens remains fixed so that the retina scans a stationary image. The muscles that move the eyes were first described by Scheuring in 1913; his description was confirmed by Land (1969b). Interestingly, the oculomotor nerve to each eye contains only 6 axons—1 per muscle—which contrasts with man where there are about 10^3 oculomotor axons.

The four kinds of eye movements that have been observed are shown diagrammatically in Figure 7. When the spider is looking at a 30° homogeneous Maxwellian field of view, the eyes make intermittent movements, usually to-and-fro in the horizontal plane but at very varying velocities. These movements can easily be observed in young transparent spiders without ophthalmoscopy, as Dzimirski showed in 1959. These are termed *spontaneous activity* for want of any precise information about their function. If a small dot (1°) is brought into the field of view of either the principal eyes or the anterolateral eyes, the principal retinae move rapidly toward the dot and fixate it with their central regions. The retinal fields of view do not overlap (as in mammals) but are nearly contiguous, as in Figure 7. This movement can be termed

a saccade, because it is exactly equivalent to its analogue in humans. If, following fixation, the dot is moved, the retinae will track it with a movement that looks like smooth pursuit, although it would be premature to say more about the control system of tracking at this stage. However, it is clear that when tracking, say a fly, a combination of eye movements (when the stimulus is within 10° of the axis) and small turns by the animal as discussed above are used cooperatively in maintaining fixation when the stimulus strays further from the axis.

The most interesting kind of eye movement is scanning, which occurs after a stimulus has been fixated and is no longer moving (Figure 8). It consists of a very regular oscillatory side-to-side motion of both retinae, from one edge of the stimulus to the other, with a period of 1 to 2 sec. Simultaneously, both retinae rotate about their visual axes by up to 25° in each direction. This movement is also repeated, but the period is much longer (5 to 15 sec). Scanning appears to be a very fixed and experimentally repeatable activity: It has been observed now in 5 species, and no discernable difference was found in its form from one to another. The duration of scanning varies greatly, from bouts lasting longer than a minute to ones consisting of only one or two side-to-side movements; a spider that has been presented with the same spot many times may simply make a saccade toward it and then allow the eyes to drift slowly back to the resting position without scanning at all. The only aspect of scanning that has been found to be affected consistently by the nature of the stimulus is the amplitude of the side-to-side movements: These increase from a minimum of about 1° to a maximum of 10° as the width of the stimulus increases. With stimuli larger than 10°, the retinae tend to drift across the image scanning each edge sequentially.

Although both saccades and tracking movements have obvious counterparts in mammalian vision, scanning appears to be unique. It is true that humans, for example, move their eyes a great deal while examining the visual world and learn to scan in quite specialized ways such as during reading. However, we have here a single very fixed kind of behavior, which appears to be a general purpose procedure for extracting certain important features from the stimulus being scrutinized. What these features may be is discussed in the next section.

Pattern recognition

Jumping spiders are repeatedly faced with the problem of deciding whether the small moving object to which they have just oriented is potential prey or another jumping spider: The importance of getting this decision right is obvious. The moment at which this decision is made is easily determined by an observer; because when con-

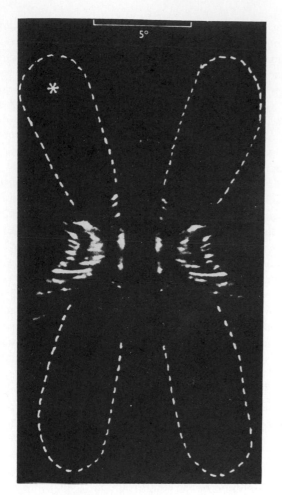

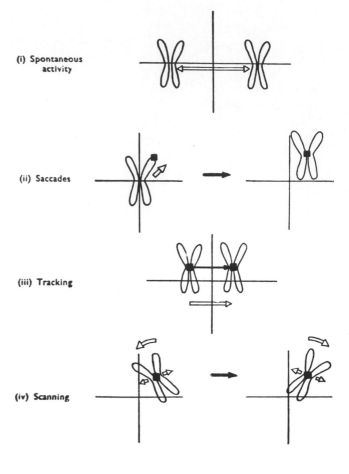

FIGURE 6   Ophthalmoscopic appearance of the retinae of the principal eyes at rest. The photograph was taken with the instrument described in Land (1969b); only the uppermost layer of each retina is visible, and the dotted outline of the rest of the retinae has been reconstructed from the histological structure. The ophthalmoscopic image has the geometrical configuration of the retinal fields of view, which are reversed and inverted with respect to the anatomical layout of the retinae in the body. Note that the fields of view are nearly contiguous but do not overlap. The asterisk (*) corresponds to the ventral region of the right retina.

FIGURE 7   Movements made by the retinae of the principal eyes, (i) without, and (ii - iv) with a stimulus object in the field of view. See text for further details.

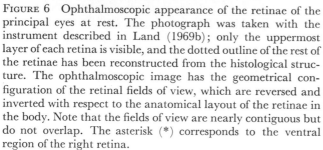

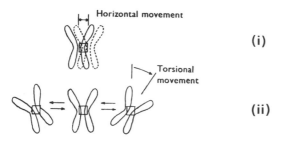

FIGURE 8   Recording of a long bout of "scanning" made using a recording technique in which a movable graticule line is used to track the position and orientation of the inner edge of one of the retinae (Figure 6), while the retina is scanning a 3° dot 5° to the left of the body axis. Scanning consists of a regular side-to-

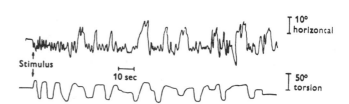

side motion of both retinae (upper trace), coupled to a much slower conjugate rotation (lower trace). After the first 20 sec, scanning becomes irregular and intermittent, merging into spontaneous activity after about a minute. This is an unusually long bout.

416   A COMPARISON OF VISUAL BEHAVIOR

fronted with a conspecific, the male spider's behavior changes from a slow approach toward the other animal into a sexual display. In adults this can be quite elaborate, usually involving raising the first pair of legs and proceeding in a series of sideways zigzag arcs that maximize visibility, rather than minimize it as in the slow approach toward prey (for descriptions of salticid sexual behavior see Crane, 1949, or Bristowe, 1958).

Since jumping spiders will display at each other when in separate airtight transparent containers, vision is the only sense involved at least until contact is made, and several workers (Heil, 1936; Crane, 1949; Drees, 1952) have attempted to define what aspects of the visual pattern that one spider presents to another are crucial for recognition. The most thorough analysis is that of Drees; one of his most convincing series of results is shown in Figure 9.

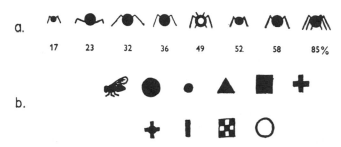

FIGURE 9 One of a series of experiments performed by Drees on *Epiblemum* (*Salticus*) to test which features of a two-dimensional stimulus a male jumping spider would respond to by sexual display, as opposed to preycatching behavior. (a) Figures responded to with sexual display, in order of effectiveness: An ideal "spider" has a number of lines, inclined to the vertical, on each side of a central body. The numbers are the percentages of presentations on which males responded to the figures with a sexual display. (b) All these figures evoked prey capture when moved. (Both figures are after Drees, 1952.)

He found that many different kinds of small object, provided that they had been moved and seen by the spider, elicited prey capture, but that sexual behavior required a rather specific configuration of the object with a central "body" and a pattern of lines—"legs"—on either side. Not surprisingly the optimal pattern obtained by Drees is that most reminiscent of a spider, but it is notable that this two-dimensional, very simplified pattern was almost as effective as a real spider.

From this evidence it is reasonably certain that what the spider is "looking for" during scanning is the presence of lines or edges with appropriate orientations in the target that has been fixated. Let us assume that receptors in the principal retinae are connected centrally in rows, forming line detectors with properties similar to the "simple" cells of the cat's visual cortex (Hubel and

Wiesel, 1959). We can then ask how such detectors should be moved in order to find out whether or not a stationary target contained appropriately oriented lines. The answer would be that the detectors must be moved to and fro to detect the presence of lines and rotated to determine their orientation. These movements correspond exactly with the horizontal and torsional components of scanning (Figure 7). It is interesting that scanning by the spider's principal retinae is the precise converse of the operation that a mammalian physiologist performs in trying to determine the orientation selectivity of a cell in the visual cortex: He moves a line stimulus to-and-fro, rotating it until the optimal orientation for that cell is found.

It is implied in the conclusions reached by Drees and by myself that pattern recognition is strictly limited to the single circumscribed tasks of distinguishing prey (which is small and moving but whose geometry is irrelevant) from conspecifics whose geometry is important. At least two other kinds of object are also discriminated: Predators, which are presumably large and whose image on the retinae increases rapidly in size and to which the spider responds by running away or jumping off its substrate, and stationary objects such as twigs, to which the animal may reach out and climb up, or jump onto. There is also some evidence from Drees and others that jumping spiders can discriminate and avoid certain sorts of prey that have proved distasteful. Nongeometrical cues, such as color, also play a part in the final stages of courtship, at least in some species (Crane, 1949). However, none of this modifies the basic conclusion that pattern recognition in these animals differs markedly from that of mammals in that it consists of going through a set procedure (scanning), which leads to a yes-no answer, as opposed to recognition via complex "gestalts," which must be learned and generalized, as is the case in man and presumably other mammals.

Conclusions

Because catching a fly by stalking is not different in principle from catching a bird or mouse, jumping spiders have evolved a range of visual mechanisms that are remarkably similar to those of predatory higher vertebrates. Chief among these are the following:

1. The use of a combination of open- and closed-loop visual control for locating and keeping the prey centrally fixated as the hunt proceeds. The use of different eyes, one set fixed and one with movable retinae, for the performance of these two functions can be regarded as an evolutionary trick that saves a great deal of space in a necessarily small animal, without compromising optical resolution.

2. The development of a versatile system of retinal movements that not only permits close tracking of targets

but also assists directly in making the vital distinction between prey and other spiders—a task that by extension from man's capacity for pattern recognition one simply assumes to present no problem for a mammalian carnivore.

The differences in behavior that result from the limitations of the arthropod nervous system only become apparent when the spider is taken out of the natural situation to which it is adapted and made to perform tricks in the laboratory. Unlike a cat or primate, the spider apparently cannot comprehend that there is something odd about turning a card ring rather than its own body nor that it is strange that stimuli seen to the side never appear in front as they should (Figure 3). The impression is of an animal very well equipped to do a few things well but unable to vary its repertoire. This conclusion is also confirmed by the ease with which jumping spiders can be made to catch dots on paper or display to very simplified representations of other spiders, as well as by the inflexible scanning procedure used to identify stimuli. However, the behavioral limitations are accompanied by an extraordinary degree of neural economy; as examples, each eye muscle is a single motor unit, and identification of another spider at 10 cm requires the participation of a maximum of about 30 receptors out of a total of less than 2000 in both principal eyes. Thanks to two decades of work on single units in crustacean and insect nervous systems, we are now accustomed to finding single cells in arthropods performing functions that require tens or hundreds in higher vertebrates. Nevertheless, it is interesting to find a similar degree of compression in an animal whose activities so closely resemble those of a mammal.

REFERENCES

BARLOW, H. B., 1952. The size of ommatidia in apposition eyes. *J. Exp. Biol.* 29:667–674.

BIZZI, E., and E. V. EVARTS, 1971. Central control of movement: Eye movements. *Neurosciences Res. Prog. Bull.* 9(1):36–40.

BRISTOWE, W. S., 1958. *The World of Spiders.* London: Collins.

CRANE, J., 1949. Comparative biology of salticid spiders at Rancho Grande, Venezuela. Part IV. An analysis of display. *Zoologica* 34:159–214.

DREES, O., 1952. Untersuchungen über die angeborenen Verhaltensweisen bei Springspinnen (*Salticidae*). *Z. Tierpsychol.* 9:169–207.

DZIMIRSKI, I., 1959. Untersuchungen über Bewegungssehen und Optomotorik bei Springspinnen (*Salticidae*). *Z. Tierpsychol.* 16:385–402.

EAKIN, R. M., and J. L. BRANDENBURGER, 1971. Fine structure of the eyes of jumping spiders. *J. Ultrastruct. Res.* 37:618–663.

EBERLE, H., 1967. Cone length and chromatic aberration in the eye of *Lebistes reticulatus*. *Z. Vergl. Physiol.* 57:172–173.

HEIL, K. H., 1936. Beiträge zur Physiologie und Psychologie der Springspinnen. *Z. Vergl. Physiol.* 23:1–25.

HOMANN, H., 1928. Beiträge zur Physiologie der Spinnenaugen. *Z. Vergl. Physiol.* 7:201–268.

HUBEL, D. H., and T. N. WIESEL, 1959. Receptive fields of single neurons in the cat's striate cortex. *J. Physiol.* (*London*) 148:574–591.

LAND, M. F., 1969a. Structure of the principal eyes of jumping spiders in relation to visual optics. *J. Exp. Biol.* 51:443–470.

LAND, M. F., 1969b. Movements of the retinae of jumping spiders in response to visual stimuli. *J. Exp. Biol.* 51:471–493.

LAND, M. F., 1971. Orientation by jumping spiders in the absence of visual feedback. *J. Exp. Biol.* 54:119–139.

LAND, M. F., 1972. Stepping movements made by jumping spiders during turns mediated by the lateral eyes. *J. Exp. Biol.* 57:15–40.

ROBINSON, D. A., 1965. The mechanics of human smooth pursuit eye movement. *J. Physiol.* (*London*) 180:569–591.

SCHEURING, L., 1913. Die Augen der Arachnoiden, II. *Zool. Jb.* (*Anat.*) 37:369–464.

37 Behavior of Neurons

C. A. G. WIERSMA

ABSTRACT At least for identifiable neurons, it is legitimate to talk about their "specific" behaviors. The viewpoint is here exposed and defended that the study of behavior of more complexly reacting interneurons can help to bridge the large gap between neurophysiological and behavioral reactions. The necessity of study at this specific level is pointed out. From examples mainly concerned with reactions to diverse visual stimuli, it is concluded that both *attention* and *memory* mechanisms of the animal as a whole are here not assignable to a single locus but have a diffuse representation, because these mechanisms are part of the behavior of individual interneurons, which have to combine their outputs for actual behavioral events to occur.

Introduction

A PREREQUISITE for the development of scientific principles that can lead to an understanding of natural phenomena is the establishment of valid hierarchical levels. It will be the purpose of this chapter to propose that what may be defined as neuronal behavior is such a level, which can bridge the gap between neurophysiology and psychobiology to an as yet unknown extent. Support for stressing the importance of neuronal behavior over connectivity of networks, synaptology, microanatomy, and the like is provided by studies on single cells. Such studies have shown that the reactions of a unit can be so complex as to defy description in terms of these factors. So many parameters would have to be known to predict the reactions, under all circumstances, of most identified interneurons that this is, as yet, well beyond our reach. However, my thesis is that by studying the behavior of such neurons, the behavior of the animal as a whole becomes more explicable. This view is directly opposed to the one in which neurophysiology is considered to be of little or no relevance to animal behavior. Though acknowledging that neurons can serve as telegraph wires, the latter view implies that neurons might be bypassed if a type of field communication were realized. Indeed, the latter concept is fundamental to the existence of extrasensory perception. It may be that this type of communication does exist and that mechanisms are present in the nervous system that cannot be explained on the basis of neuronal behavior. If so, they will be more readily delineated as failures of the proposed approach become evident.

Cell specificity

The assumption that neurons are individually recognizable and that any given one is present in all *normal* specimens of a given species, or at least in a pure line, is central to the view that neuronal behavior warrants study. This is contradictory to a good deal of evidence from higher mammals, e.g. to the proven existence of *plasticity* in function of nerve cells when their connections are changed by surgical means (see Yoon, 1972). I will try to show later why these objections, which on the surface appear to negate the whole thesis, may not be as fatal as it would seem.

Let me first state a rather obvious tenet: If the nervous system in a certain species has developed to a level that is *optimal* for its needs, the best way to ensure that the offspring will be as successful as the parent stock is to provide them with the same number and functional value of all constituent nervous elements. Of course, genetic variability will lead to differences between individuals; therefore, the question arises how large are these normally. Especially from invertebrate studies, it appears, as was shown in several of the preceding articles and will be reiterated here, that interneurons, even those with remarkably complex functions, can readily be individually recognized in all specimens.

Starting at the anatomical level, it has often been possible to identify a cell on the basis of the location of its cell body or its axon. However, minor deviations in this respect frequently occur; for instance, cross-sections of the abdominal cord of the crayfish, though showing overall similarity, are usually not strictly symmetrical, in that axons differ somewhat both in size and in location. The most extreme example that I have ever observed concerned the two medial giant fibers, which have their cell bodies in the brain and end blindly in the telson ganglion. In this specimen, both fibers ran from the brain in the same half of the ventral cord, one being smaller than the other. Notwithstanding this abnormal anatomical location, each made the correct connections in all

C. A. G. WIERSMA Division of Biology, California Institute of Technology, Pasadena, California ganglia, as far as could be ascertained electrophysiologically and anatomically; obviously the functional setup was only slightly distorted, if at all (see Selverston, this volume, for a more extensive discussion on this general subject).

How far is connectivity, by itself, sufficient to provide for the individuality of a neuron? That it is one important, but not exclusive, factor is readily obvious to any neurophysiologist, though not so much so to some modelers of neural networks. Neuromimes, though they have their uses, are dangerous in disregarding the fact that in any nervous system different types of neurons are present that, by their inherent properties, react quite specifically to their inputs, even though the latter may be identical. This is an important point, because it illustrates the danger of concluding what the functional significance of a presynaptic axon is from its firing pattern. Obviously the result depends both on the type of postsynaptic effect, excitatory or inhibitory, and on the synaptic efficiency. But connectivity in conjunction with transmitter sensitivity is not the whole story involved in neuronal behavior, as other known properties of neurons enter into the picture. The main one of these is the variation in tendency to fire *spontaneously*, an aspect that may be much more widespread than is generally assumed and appears to be present in various preset amounts. It leads to such features as a cell being driven to fire at a higher rate than its spontaneous one by an inhibitory input and the triggering of a number of discharges by a single presynaptic event. In addition, our ever-increasing knowledge of the complexities existing at the synaptic level itself, e.g. synaptic regulation of electrotonic coupling between cells (Bennett and Spira, 1972), presence of excitatory and inhibitory receptive sites sensitive to the same transmitter substance (Kandel, 1968) further illustrates the difficulties in describing input-output relationships in terms of synaptic events. Of course, this does not mean that it would not be worthwhile to determine for any given interneuron class its connectivities and types of presynaptic and postsynaptic influences, as well as its "amount" of spontaneity, and such additional factors as changes in firing level, whether or not hormonally induced. Since several of these variables may involve time factors of considerable duration, long-term changes in reactivity do not necessarily indicate "plastic" properties, in the sense that the cell acquires properties that it would not have had "normally." *Plasticity* becomes very much a question of definition, because different states merge into one another, without sharp boundaries. That neurons can show real plasticity cannot be denied. For instance, the experiments of Yoon (1971, 1972) on the goldfish optic tectum in which he was able to make retinal neurons form reversible connections with different parts of the tectum, by blocking and unblocking its posterior half, clearly demonstrate that interneurons can change their functional connection. This type of finding shows that predetermination as present in all cases of identifiable neurons is not absolute but can be changed by experimental and thus presumably developmental circumstances. It, however, is also indicated that the nervous system is still working on the principle of very deterministic types of interconnectivity, for when only about half of the normal complement of neurons are available, all appear to change, many giving up their normal inputs to make room for those that are "homeless," but in an orderly manner. Since functional recovery of optic reflexes from the retinal part that was disconnected is found, it seems indicated that these arrangements also involve the output channels of the optic tectum cells. These findings can therefore not be considered as in favor of a type of information transmission by whole populations as against cell specific activity.

The study of individual neuronal behavior has as one of its main goals the understanding of how the information flow between inputs and outputs is structured. On the input side one can ask how the firing rate is related to environmental changes. As we will see, this relationship may vary strongly due to factors other than the main input channel. On the output side one can obtain useful information about the significance of a given interneuron by stimulating it directly and observing the effect in terms of overall movements of body parts or of combinations of muscular contractions. Variations in this output due, for instance, to localized sensory stimulation of the body parts involved can further reveal the amount of modulation such programmed motor patterns undergo by local events. In preceding papers of this part, several examples and discussions of this topic have been provided. For interneurons on the sensory side, intracellular stimulation, which has so far been only occasionally used, appears to be much less revealing, because no readily observable output in the form of motor effects results. In such cases, it would therefore be necessary to discover the various interneurons toward the motor side for which such a fiber is presynaptic in order to evaluate its functional significance. Therefore, the study of the reactivity of such fibers to the presynaptic events has been so far the main approach.

Sensory representation in the CNS

An aspect present in invertebrates, which has so far no known equivalent in vertebrates, is the representation of large, mutually overlapping sensory fields, with an apparent absence of restricted fields immediately beyond the primary sensory level. It does appear as if the often quite good resolution present in the periphery due to the many sense cells is not centrally used, but whether this is actually the case is still problematic, because at least one case is known in which very accurate representation of primary sensory input is strongly indicated (see below, page 423). To illustrate the representation of large sensory fields by interneurons, that for the dorsal hairs on the crayfish abdomen may be briefly mentioned (Wiersma and Hughes, 1961; Wiersma and Bush, 1963). The minimum extent of the sensory areas to which the interneurons involved react consists of one-half of each dorsal abdominal segment. In addition there are interneurons responding to two, three, or more successive unilateral areas, some with also a contralateral representation, as well as interneurons with larger bilateral fields culminating in one that is responsive to touch of hairs anywhere on the whole body. As a result, touch of one dorsal abdominal area can activate as many as 20 interneurons. These findings illustrate a pronounced parallel computation of the information for which the functional significance is not immediately obvious. It should be noted that this system is a very poor one for retrieval of more exact stimulus localization, because in this case it could hardly be more precise than that provided by the interneurons with the smallest sensory fields. The likely explanation is that the information carried by the different neurons, which react to one local stimulation, feed into various output channels, causing more localized or widespread reactions, related to their sensory field size. Some evidence for this type of connectivity is indicated for the crayfish abdominal ganglia by the experimental results of Kennedy (1971), and also for visual interneurons as we will see later.

Feature detection by single units

To various degrees, reaction only to specific aspects of a given stimulus situation is provided by specialization of the primary sense cells. Many examples come readily to mind, and I will quote here one case, to illustrate the difficulties involved in deducing the consequences of this type of parallel input. Elastic strand organs in the joints of crustacean appendages have their origin usually on a tendon, and their insertion is either on a hard part of the more distal segment or on the joint membrane. The most peripheral joint of each thoracic appendage is spanned by such a strand, in which the sensory endings of numerous neurons are embedded (Whitear, 1962; Mill and Lowe, 1971). The sensory fibers are divided into four distinct classes; two indicate position in one or the other direction from the midpoint, and two are specifically triggered by movements, one type for each direction. Frequency changes occur in the movement fibers only within a small velocity range, because a "saturation" rate is reached when the velocity is only slightly above threshold; how-

ever, the saturation rates may vary with the load (Wiersma et al., 1970). Fibers with different thresholds are present in all classes (Wiersma and Boettiger, 1959; Wiersma, 1959). The total result is that the CNS receives very accurate information about the peripheral situation almost constantly, because both frequency and fiber number coding are available for position, and number coding is available for movement. Though the overall output has been shown to be of importance for the firing patterns of both motor and inhibitory fibers to the muscles moving the joint (Bush, 1965; Spirito et al., 1972), it is as yet not known in which way the different fiber types and the individual ones of each type are involved in the resulting myotatic reflexes. Rather unexpectedly it was found that this input is not of importance to the normal motions of either its own or other legs, as in walking (Barnes et al., 1972). For central fibers responsive to induced joint motions, of which many react to various joints of more than one leg (Wiersma, 1958), it is also unknown whether the position and movement signals are separately channeled or whether these two informational aspects are recombined. Whether different pathways are present for low- and high-threshold events in either type of input is another unsolved problem.

Especially in optic systems, it is certain that feature detection is mainly performed by the neuronal connections in the optic ganglia, though it is difficult to decide to what extent peripheral selectivity and interconnectivity of the primary sense cells are also involved. Photopic, scotopic, and color vision depend on such peripheral differentiation. Lateral inhibition between primary photoreceptors often appears to be present, but at least in some decapods there is, in addition, a similar type of inhibition that depends on neuronal connections. Here, light in an area far outside the excitatory field of an interneuron inhibits, as does light shining in the other eye, and both must depend on a neuronal network. Specific reactivity to movement involves spatial and temporal readout of neuronal collecting stations by neural networks, and it is remarkable that several types of interneurons responsive to movement are present, each optimally reacting to a different aspect of the movement and the physical features of the targets. For instance, in crustacean optic nerves one finds on the one hand the "jittery movement fibers" that react to a black object entering into the field from any direction, but which soon stop firing whenever large objects travel linearly through it. On the other hand, the unidirectional movement fibers usually fire more strongly the farther an object travels linearly in the preferred direction but are inhibited by opposite movements. This example shows that each fiber type has its specific inhibitory mechanism that must be provided by different networks regulating the same input channels. I quote this special instance to illustrate how difficult it may be to determine the connectivity involved in each type.

Table I lists the known types of interneurons reactive to visual stimuli in different species of decapods. Note that certain types of "feature extractors" appear to be present in all species, whereas others have been encountered in only one. Of course, the absence of a class in a species does not necessarily mean that it does not exist, for the methods used do preferentially show fibers of larger diameter. Nevertheless there does appear to be a relationship between the visual behavior of an animal and its visual interneuron equipment. A comparison of the crayfish and rock lobster, whose optic nerves have been studied most, shows that behaviorally the crayfish is visually less "astute," in accord with its fewer optic fiber classes. In a considerable number of insects, the indications are that the neurons in the optic system are very similar to those in the crustaceans, but perhaps because their numbers are much larger, identification of specific interneurons has not been made in most cases. Much current research concerns this problem in insects.

TABLE I

Table of known visual fibers in the different decapod types investigated

Class of Fiber	Crayfishes	Rock Lobsters	Crabs
Sustaining	+	+	+
Dimming	+	+	+
Jittery movement	+	+	+
Light movement		+	
Medium movement		+	+
Fast movement	+ ?	+	+
Seeing		+	?
Slow movement			+
Space constant	+	+	+
Unidirectional movement	+ ?	+	+
Multimodal	+	+	+

(From Wiersma and Yanagisawa, 1971.)

Centrifugal influences

Of the various classes of visually reactive interneurons in the crustacean optic nerve, the sustaining and dimming ones appear to be the "earliest" both with regard to their rather simple type of integration and to their apparently more peripherally located origin in the optic ganglion chain. Nevertheless, their discharges are strongly influenced by centrifugal factors, such as the inhibitory influence from contralateral eye neurons (T. Yamaguchi, personal communication) and the facilitatory effect of the general state of excitement of the animal (Wiersma and

Yamaguchi, 1966). An excited state may cause as much as a fivefold increase in the number of impulses elicited by a light flash (Figure 1), although the state by itself does not give rise to impulses during a dark period if no dark discharge is present and can at best raise it to a still low discharge rate (Aréchiga and Wiersma, 1969a). The effect can be pictured as due to a lowering of the threshold at the spiking locus by the excited state, but other explanations such as removal of inhibition are not excluded.

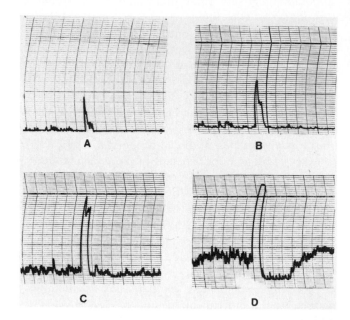

FIGURE 1 Changes in responsiveness of a crayfish sustaining fiber to a constant light stimulus with various states of activity of the animal. The electrode was implanted and the animal had been adapted for several hours to a dim light (0.01 candle/ft²). A light pulse of 4 candle/ft² and of 15 sec duration was applied successively during an experiment of several minutes duration. In (A) the animal was in a "sleep" attitude. In (B) it walked slowly in its container. In (C) the animal was more aroused by external mechanical stimuli, and in (D) it was in a "defense reflex" position, ready for attack. Note that the frequency of the background discharge to the low light level also increases with the state of excitement. (From Aréchiga and Wiersma, 1969a.)

It is interesting to note that light shining on the eye can bring about a higher state of excitement, but this may take up to a few seconds, whereas an increase of excited state causes an almost immediate increase in the visual response. A possible centrifugal pathway causing this increase is provided by the so-called activity fibers, whose discharge rate is proportional to the excited state level (Figure 2). Again, this mechanism is inferred and needs experimental confirmation of the predicted increase of sustaining fiber discharge on electrical stimulation of the activity fiber. Since it is possible to implant electrodes

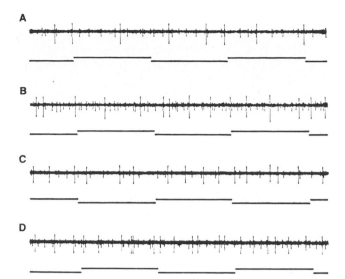

FIGURE 2 Illustration of the relationship in firing frequency between a centrifugal interneuron (small spikes) and a sustaining fiber (large spikes) during different levels of excitement and light intensity in a crayfish. (A) Low light level stimulus during a low state of excitement. The smaller centrifugal "activity fiber" fires at the same rate as it would in total darkness (12 impulses per sec). The sustaining fiber for the back upper rim (038) gives a sustained discharge of 4 impulses per sec. In (B) the light level was identical, but the state of excitement was increased by mechanical stimulation. The firing rate of the activity fiber increased to 28 impulses per sec, that of 038 to 8 per sec; both fibers fired therefore about two times more frequently. Here (C) and (D) are records of the same preparation at a later time, and with an increased light level. In (C) the 038 fires at a sustained rate of 7 per sec although the activity level is about the same as in (A), as indicated by the activity fiber that again fires at 12 per sec. In (D) this rate is 22 per sec (lower than in B) and 038 fires at 11 per sec. The percent increase in either is therefore less than from (A) to (B), however, the ratio is somewhat less similar. In (A) and (C) the tethered preparation showed no movements of the appendages, in (B) and (D) they were pronounced. (From Wiersma and Yamaguchi, 1967.)

that remain functional for extended periods of time (e.g. 1 month) (Aréchiga and Wiersma, 1969b), such experiments appear feasible.

While it is known that the excited state influences both sustaining and dimming fibers in the same way, its influence on the firing rates of other optic fiber classes appears to be much less, and not present in all. In the rock lobster, several of the fiber classes are, instead, specifically regulated by other factors. This is especially well illustrated by the seeing fibers (Wiersma and York, 1972). These interneurons are more complex in that their excitatory field boundaries, though in most instances not much different from those in the simpler classes, are vague and not necessarily the same for objects traveling in opposite directions. An additional feature is that the field consists

of two distinguishable but not distinct parts, one in which a moving target is perceived but no reaction to a stationary one occurs, and a more central area in which a stationary object also causes impulses for as long as 30 sec. These fibers seem to be clearly influenced by the "mood" of the animal in that they have a low reactivity when the animal is "drowsy," react well when the animal is in a waking state, and are actually inhibited if the animal is in a high state of excitement, as during a defense reaction. On the other hand, in this situation, but not during walking, the jittery movement fibers are highly sensitive and their normally high rate of habituation decreases greatly. It is not difficult to conjecture reasons for these differences when the behavior of the animal is considered. Seeing fibers are almost certainly the input pathways for the pointing reactions of the lobster's antennae that consist of rather accurately tracking a target, e.g. a fish (Lindberg, 1955), because the habituation of the pointing reflex and that of the seeing fibers to the same object often coincide; furthermore, both habituations can often be negated by moving the target through another trajectory. On the other hand, one specific jittery movement fiber most likely triggers and others sustain the defense reflex, as will be discussed next. Therefore the difference in behavior of the seeing and the jittery movement fibers under different circumstances relates well with their final output tasks.

The specific jittery movement fiber mentioned belongs to a special set of visual interneurons, the space-constant ones, of which one member occurs in the majority of the classes. The space-constant fibers have in common that in the normal eye position they respond to stimuli above the eye's horizontal axis and to a symmetrical field on the lower eye half when the animal is in the upside-down position. In many, but not all cases, the potential field encompasses the whole retina. Their maximum field size occurs when the middle of the potential field points straight upward (Figure 3). This control by gravity is mainly provided by the statocysts, and because their removal results in the sensory field becoming of maximal size even when the eye points downward, the influence is obviously inhibitory. We have here a good example of one input channel that is read out by a number of otherwise quite differently behaving interneurons. It, of course, also indicates that this space-constant mechanism is of great importance to the animal, presumably for spatial localization. Since the sizes of the visual fields change very accurately with statocyst position, the channel transmitting the information has to consist of the axons of the primary sense cells of that organ, and indications that these do, indeed, reach the optic ganglia have been obtained. Thus here is a case in which the accuracy of the primary sensory input appears to be used to its fullest extent.

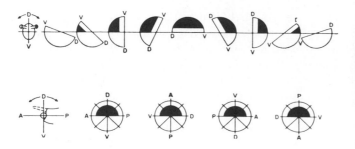

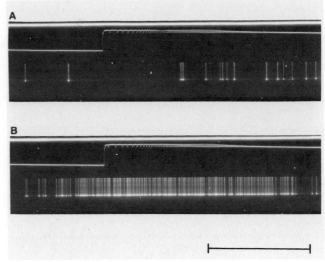

FIGURE 3 Illustration of how the field of a crayfish space-constant sustaining fiber travels over the surface of the eye as body position is changed. In (A) the animal was rotated along its longitudinal axis, and the sustaining fiber ceased to respond as soon as every part of its visual field looked at objects below the horizon, whereas the field was enlarged and circular when its whole potential field was exposed to light from above. In (B) the animal was rotated about its transverse axis, in which case the visual field did not change in extent, but rotated in the opposite direction over the eye surface. (From Wiersma and Yamaguchi, 1967.)

Attention

The seeing fibers show a number of features that can be described in psychological terms. This does not mean, for instance, that when a seeing fiber pays attention to a particular object that the *animal* itself does so. Under several circumstances the seeing fibers can completely neglect an object when introduced in their visual field. When an object enters and travels very slowly through the field it will not elicit any impulses from either the movement or the seeing area. When interesting visual stimuli occur in other eye regions during a "seeing" reaction, they often lead to a severe reduction in spike number, and the frequency may then increase again when the distracting stimulus stops. Another aspect of this type of inhibition is that while a moving target larger than the visual field may be ignored, reduction of its size can, on the next turn, give a vigorous discharge (Figure 4). At least some of the seeing fibers can be similarly "distracted" by purely mechanical stimulation, for instance, in the case of a seeing fiber looking in the forward direction, by stroking the animal's tail. However, an enhancement of this fiber's discharge can be obtained by appropriate movements of the antenna. One category of unidirectionally sensitive fibers was very reactive when an object moved over the "entrance" part of the field at a rather slow rate. Passage over the rest of the field would then give a strong response, but when the entrance part was quickly traversed, the response for the remainder of the field stayed low even for the speed that was optimal under the first condition. One may describe this as the presence of a trigger zone for attention.

FIGURE 4 Rock lobster seeing fiber looking at moving targets of different dimensions and reacting distinctly different to them. The seeing fiber used was LO 141, which has a rather complex binocular visual field. In (A) the visual stimulus was a vertical stripe made up of 5 black squares (15°) traveling at a speed of 8° per sec through the field. This led to an inhibition of its background discharge rate. In (B) only the middle of the 5 squares was left in place. With the same rotation speed, there was now a strong discharge as soon as the target entered the field, which remained until it left. Time base, 5 sec. The upper lines monitor both drum speed and position; the sudden jump occurs at one constant drum position. (Spike events were transcribed from tape to Grass polygraph by a spike height selective transducer.) (From Wiersma and York, 1972.)

Memory

The seeing fibers possess a short-term memory mechanism, which has been shown in two ways: One is to turn out the light while the fiber is giving a "seeing" discharge. It will then resume firing when the target is reilluminated, e.g. after 10 sec, but will be silent if the target has been removed in the dark. Conversely, when a target is placed in the seeing area during a dark period, the fiber will start to fire shortly after illumination, but only when the dark period is not longer than about 30 sec (Figure 5) (Wiersma and York, 1972).

A longer lasting and less distractable memory-like phenomenon is exhibited by the motor neurons that effect horizontal optokinetic movements of the outer eyecup. The discovery of such a memory mechanism was made in the crab, *Carcinus*, by Horridge (1965, 1966, 1968) who registered eyecup position. This was studied in considerable detail by Shepheard (1966), under stimulus conditions much like those we are now using. Our results agree fairly well for such factors as duration of memory. We have shown that the motor fibers of the rock lobster, which are driven by optokinetic input, are responsible for the phenomenon (York et al., 1972). The input for the optokinetic reaction is normally elicited by turns performed by the animal (though this is contrary to the view that only passive and not active turns elicit them) (Dijkgraaf, 1956) and is mimicked by turning a striped drum around the animal. In the crayfish the read-in for this reflex appears to be provided by two unidirectional movement fibers, responding antagonistically, that discharge only during motion. In the rock lobster on the other hand, two fibers with similar fields, also apparently located in the optic nerve, differ from those of the crayfish

in that they, like the optokinetic motor fibers, show after-discharges when movement stops. And besides the difference in localization, they are strictly responsive to ipsilateral input (York et al., 1972), whereas the motor fibers are usually reactive to both eyes. Unfortunately, it is therefore still uncertain which interneurons are involved in the optokinetic reflex.

All evidence, however, is against making the same input pathways responsible for both the optokinetic reflex and the memory read-in and read-out. We found that the memory, if any, was poor for a moving object; i.e., if stripes have been turning for some time when the light is turned off, their position at light "off" is not remembered. When stripes are turned during a dark period and the light is turned on up to 5 min after it went off, appropriate and clear positive or negative responses are often obtained (Figure 6). This is one reason why movement fibers as such cannot be regarded as the source for the input,

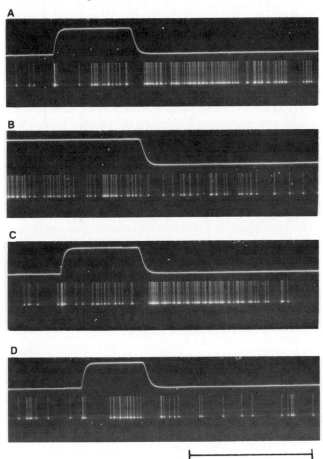

FIGURE 5 Memory response of a binocular seeing fiber (LO 121) of the rock lobster. A target (15° square) was placed or omitted in the seeing area of the fiber during periods of darkness, indicated by the upward deflection of the upper signal, monitoring light intensity. In (A) and (C), the target was introduced during a 3-sec dark period, and the fiber responded strongly to this situation when the light was turned on. In (B) the dark period was started 19 sec before the light was turned on and, notwithstanding the target's presence, no significant discharge resulted after relighting. In other instances the "memory-span" of a seeing fiber was found to be somewhat longer but did not exceed 30 sec in duration. In (D) no target was placed in the drum and no discharge resulted from relighting. (From Wiersma and York, 1972.)

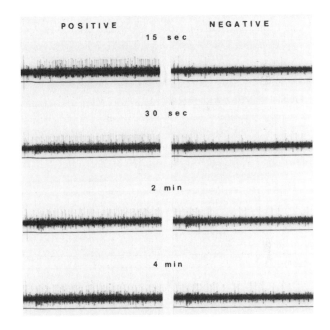

FIGURE 6 "Memory-span" of an horizontal optokinetic motor fiber of the shore crab, *Pachygrapsus crassipes*. The animal was in the center of a striped drum (stripe width was 12°), in which it was left at least 1 min before the light was turned out for 5 sec. During this dark time, the drum was rotated by 3°, either in the positive (that is, the direction that would increase the fiber's discharge when drum was turned in the light) or in the negative direction. The latter would silence the fiber except when a fast nystagmic eye motion occurred, when it would give a short burst. The left column shows a gradual decrease in the positive direction, as the period of darkness increased from 15 sec to 4 min; the right-hand column, a more or less corresponding increase in spike events. Significant differences between positive and negative results can be present for times somewhat exceeding 5 min. Lower lines, light intensity. Each record starts just before light is turned on. (Each record is 12½ sec long.) (Wiersma and Hirsh, unpublished.)

because they do not react to any extent under these conditions. Not only stripes but also randomly distributed targets can elicit both reflex and memory.

In an as yet unpublished series of experiments performed on optokinetic motor fibers with Dr. Hirsh and the late Dr. Reddy, it was found that the maximum angle of a dark turn giving rise to a positive or negative read-out depends on the stripe width. No read-out occurs when the drum is moved one stripe width, and from this point on the effect becomes negative for positive turns and the reverse (Figure 7). The minimum angle of rotation that gave rise to positive responses was somewhat different for the species used. In the lobster it is about 0.4°, in the crayfish 0.2°, and 0.1° in the shore crab (Table II). In all cases this resolution is far better than would be indicated by the angle neighboring ommatidia make with each other.

TABLE II
Visual acuity in Pachygrapsus

Angle of Turn	Number of Spikes		Difference Pos. − Neg.
	Positive	Negative	
1.6°	87	2	85
0.8°	59	15	44
0.4°	43	7	36
0.2°	34	4	30
0.1°	32	9	23
0.05°	16	—	—

Protocol: 12° stripes turned in indicated direction during 5 sec dark period; spikes counted during 5 sec period after light on.

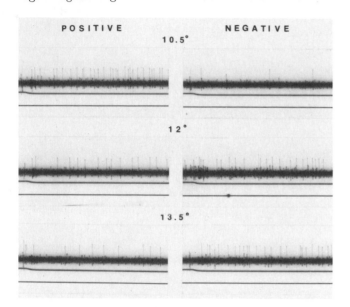

FIGURE 7 Reversal point for positive as against negative memory read-out for an optokinetic motor fiber, for 12° wide stripes, in *Pachygrapsus*. When turned in the dark period (5 sec) for 10.5° the read-out is positive for a positive turn, but at 13.5° it is negative. Negative turns show the reverse. At 12° the two turns give the same result. If the stripes were 6° wide, a turn of that value would give the "undecided" answer. Each record is 12½ sec long. (Wiersma and Hirsh, unpublished.)

The eyes can be prevented from changing position in the crab and crayfish by gluing them to the carapace. With regard to the duration of the memory and the minimum angle, the results were very similar to those obtained from the same fiber before this procedure. However, the duration of the change in discharge was invariably longer when the eyes were fixed. This phenomenon is difficult to explain, because the moving eye would counteract the discharge when it moved in the right direction, but back-

drift, as indicated by the gradual decrease in frequency, should enhance the discharge duration, because this would constitute a positive stimulus. Therefore, in the moving eye the frequency should be lower at first, which appears to be so, but should subsequently decrease only gradually, whereas in reality it stops considerably sooner than in the glued eye (Figure 8).

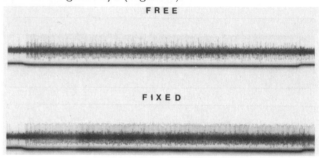

FIGURE 8 Comparison of read-out of the same optokinetic motor fiber in *Pachygrapsus*, before and after the eye was made immobile by gluing it to the carapace. A 3° turn was given during darkness in the positive direction for the fiber. The lower line shows the moment when light went on (and off after 20 sec). Note that read-out starts at the same time, is somewhat weaker in the beginning for the free eye, and remains at considerably higher frequency in the fixed one. Record durations, 25 sec. (Wiersma and Hirsh, unpublished.)

We did not obtain any strong effects of eye drift during the dark period. The three species differ with regard to the effect of darkness on the firing frequency. In the rock lobster the basic frequency is regularly lower in the dark than in light and, therefore changes, especially after dark turns in the positive direction. Both decreases and increases occur in the crab during darkness, whereas in the crayfish increases are usually obtained, so that negative turns are then often especially effective. We found that excited states and even eye withdrawal reflexes elicited during darkness are in general compatible with the correct amount of subsequent read-out, and we have practically never encountered read-outs that after these procedures were of the opposite sign from those expected. This is remarkable, especially for the rock lobster where the large changes in frequency for positive turns are present, notwithstanding that there must have been eye drift during the low-frequency dark discharge. Note that in this respect our results differ from those of Shepheard (1966) who frequently found reversed eye movements after longer dark periods. Dark discharge rates thus appear to be unrelated to the memory process, one reason to exclude afterimages as reference points for the read-out. Another reason for rejecting a peripheral location of the memory mechanism is that the intensity of the read-out is relatively independent of the read-in and read-out illumination levels. Furthermore, the strongest afterimage should be caused by the overhead light, which, because it was often as much as 20° from vertical, would change in position with regard to the eye only when the animal is turned within the drum. Experiments in which drum versus animal turning by equal amounts were compared usually showed no noticeable differences and therefore the afterimage of the light was not used as a reference point.

It should be realized that the memory input needs the stimulation of a number of collecting stations in order to be effective. For instance, a single black stripe will give at best a weak effect, whereas three black stripes are nearly as effective as a completely striped surround, even when widely separated. When such stripes were built from single square targets (15°) and subsequently distributed randomly within the same surface area, it was found that the responses were equivalent. If, afterwards, the number of targets was reduced to that of a single stripe, distributed over the sum total surface area, the reduced result was equivalent to that of a single stripe. This shows that the read-in stations in horizontal movements summate their reactions and that each contributes about equally whether distributed over the total surface area of the eye or arranged in stripes.

READ-IN TIMES By letting the animal look for preset times at a stationary background, it can be decided how long it takes before memory is established. This process is not all or none, and, as shown by Shepheard (1966), varies with the length of exposure. In *Carcinus*, maximal responses to small angles require a relatively long period of up to 70 sec. In our experience 60 sec read-in is only slightly better than 30 sec, but shorter times decrease the response. We have used another approach to determine this factor. After a new situation was exposed for a given number of seconds, it was noted whether a second turn in the opposite direction would give an appropriate dis-

charge to the shortly exposed position or would still signal this final position in relation to the first, well read-in situation. Two main variations have been used. In one, the drum was turned 3° in either direction and then turned back to the original position. Under these circumstances, the read-out would either be appropriate for the second turn or no effect would ensue if read-in had failed. The minimum read-in times thus found were usually quite short, especially in the crab, where appropriate responses appeared with exposures as short as one second. In the second method, the second turn was only half the size of the first, again in the opposite direction (Figure 9). Here the final result can be still of the same sign as the first, because with regard to the original situation the stripe position had changed, though less, in the same

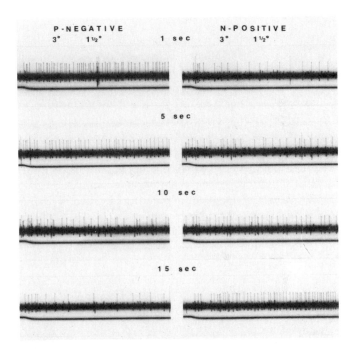

FIGURE 9 Read-in time of a 12° striped pattern after it was moved from a stationary position to which the eye had been exposed for 60 sec as determined by the read-out after a second turn in the opposite direction, following various exposure times. The second turn was only half that of the first, so that with insufficient read-in time the displacement is referred to the direction of the first turn (original memory) and is in the direction of the second turn, when the "new" memory is sufficiently read in. The record shows that the "old" and the "new" memories are mutually antagonistic. Where the second turn is negative (first column) the read-out after only 1 sec exposure is still positive and is not clearly negative until 15 sec of exposure. The reverse is true when the first turn was negative and the second positive (second column). At 10 sec there was no significant read-out. Of the two read-outs, only the one after the second turn is shown. Lower signal, light on after second dark turn. Each record is 15 sec in length. (Wiersma and Hirsh, unpublished.)

direction, or it can be reversed in sign, when the second position is sufficiently read-in. As expected, a longer time was needed before a reversal occurred, varying from $2\frac{1}{2}$ sec to as long as 15 sec. We often observed that these times increased during an experimental session, indicating a weakening of the memory mechanism. Such loss of memory can become total, when the mechanism deteriorates presumably due to oxygen lack, at a time when the optokinetic reflex discharges are still quite strong. In lobsters we have met with some individuals who seemed to lack optomotor fiber memory, though they were apparently in excellent condition; we have also seen lobsters that lost this memory after a few sessions but regained it after several weeks rest. I mention these facts because, although hardly solid data, they do indicate that memory and optokinetic mechanisms are different and also that afterimages are not likely to be the basic mechanism involved in memory.

Localization of the memory mechanism may be difficult. One fact is evident: Because there is a marked difference between the read-out of a driving eye against that of one being driven by the other's input, each seems to have its own mechanism with a weaker connection to the other eye. However, there may be a total of four memory locations. For, although the output is usually symmetrical, in that positive and negative responses are of similar magnitude for dark turns in opposite directions, in several animals either excitation or inhibition was clearly more pronounced, pointing to two memory locations, one for each direction.

It is clear that these are useful systems for further study, but how far the phenomenon can be resolved to its fundamental processes remains speculative. However, some worthwhile conclusions are evident from Shepheard's (1966) and our results. One is that there is competition between the old and the new memory, so read-in is not an all-or-none phenomenon. Another is that this memory system is present in most animals to an equivalent extent, and that it occurs under circumstances that are more comparable to those involved in imprinting than in most learning situations. What use the animal makes of the mechanism is uncertain. It might be one method of localization in space and especially useful when, e.g. an eye-withdrawal reflex upsets the relationship between retina and surround, because eyecup position is not monitored by feedback from joint position receptors (Horridge and Sandeman, 1964; Wiersma and Oberjat, 1968). This joint appears in this respect to be different from all other known arthropod joints that are provided with appropriate sense cells. Finally, it may be noted that the rabbit's optokinetic system, which is similar in reflex features, is without this type of memory (Collewijn and van der Mark, 1972).

Information flow in the CNS

Ideally, one wants to know which particular input channels are involved in specific behavioral reactions. In some of the preceding chapters, several examples were presented where it was possible to trace with more or less confidence the pathways involved. As stated, when an interneuron with known input can be stimulated electrically, the pathway can be traced with more certainty. This is possible in only a few preparations, but even now other methods can be used to indicate the relationships. Examples of such an approach may be quoted from the crustacean visual system. Obviously the main difficulty involved is that any natural input affects very diverse interneurons, and one has to go through an elimination process to determine the main input channel. Inactivation of the parallel pathways as used by Kandel for this purpose (this volume) is not necessary under favorable conditions. One can, by knowing which central elements are specifically stimulated by a given input situation leading to a characteristic output, obtain weaker or stronger indications of the information flow in the CNS, even though all elements of the input-output chain have not been found. One such instance has been presented above, with relation to the seeing fibers and the antennal pointing reaction. In this case the chain is very incomplete, as it does not include any elements on the motor output side.

The neural chain of the reflexes, triggered by the angle at which light falls on the eye to the head-up and head-down optomotor fibers, is more strongly indicated. Here we find that light on front and back rim areas of the eye gives rise to reactions in identified sustaining and dimming fibers and causes appropriate reflex discharges in the motor fibers, representing a push-pull type of connectivity. Possibly these visual interneurons are directly and appropriately connected to the motor fibers. Further indication that the sustaining and/or dimming fibers are involved in the reflex discharges of these optomotor fibers comes from the following considerations: First, the sensory areas causing the optomotor reflex are essentially the same as those of the "rim" interneurons. Second, a synchronization of the discharge rate of the motor fibers with shadowing at a frequency of 5 to 10 per sec often occurs, and of all the fiber classes present in the optic nerve only sustaining and dimming fibers show continued short bursts during such stimulation. Since the visual interneurons apparently involved are but a few of the members of the two classes, the question arises what the functions of the others are. Likely, all are involved in more than one functional input-output relationship, and one could well be participation in the read-in and read-out systems involved in optokinetic memory.

For two major reflexes that can be visually elicited, the

input-output pathways are indicated even more certainly, because the visual area that triggers them is space constant. The escape reflex, mediated by the medial giant fibers, is triggered optically by quickly approaching objects. Because of the high thresholds of the giant fibers, it is difficult to obtain such optically triggered inputs. The reflex cannot be obtained from below in a crayfish in a transparent vessel, either in the normal or eye-down position of the animal. But in the latter situation, it can be obtained from both the dorsal and ventral halves of the upward pointing eye. This is exactly the same change in visual field as shown by the fast space-constant movement fiber. In the defense reflex, the animal takes up a characteristic position in which the claws and the head are raised. This position can be triggered when an axon with known location in the circumesophageal commissure is repetitively stimulated (Wiersma, 1952). This fiber is known to respond with firing to jittery motion in the visual field, such as finger movements (Wiersma and Mill, 1965), but that this reaction is space constant has not yet been proved. For the defense reflex, space constancy was shown by the methods described for the escape response. This indicates that the space-constant jittery movement fibers are the main input channels. Whereas electrical stimulation of the defense reflex fiber causes completely symmetrical output, during natural execution the animal orients towards the location of the stimulus also, presumably due to activity of parallel pathways signaling other aspects of the visual surround. Incomplete as this evidence is, it nevertheless shows that in some cases it is possible to approach the goal of determining information flow along the total pathway for rather complex behavioral events.

Present and future prospects for the study of neuronal behavior

It has been shown in this and other papers in this part, that one may, indeed, speak of the behavior of a neuron. This behavior, like that of the whole organism, is influenced by both present and past events and is therefore not readily predictable. However in many cases, interneuron behavior depends on fewer factors than that of the animal as a whole and therefore will be more amenable to analysis. Each animal possesses several populations of neurons that come to specific levels of activity only under definite sets of circumstances and these therefore provide good indications of what specific changes in the environment are important to the organism. One main problem still remains, how the overall reaction is determined in situations in which more than one group is activated.

From these considerations it follows that study of interneuronal behavior will lead to a better understanding of how total behavior evolves. As an example let us discuss

the, as yet, incomplete facts concerning the attention mechanism of the lobster seeing fibers and its relation to the mechanism of the antennal pointing reaction. The pointing reaction may be regarded as behavior of the animal as a whole. Pointing does not arise from any specific center, whose inputs do not show an attention mechanism, but instead from a combination of inputs, each equipped with its own. Attention, then, is not a readily localized property but has a diffuse representation. Whether or not the additional integrative mechanism necessary for the pointing movement also has its own additional "attention" is hard to judge, although it would seem unnecessary. At any rate, the mechanism in single interneurons is more promising for detailed analysis of the underlying events than that of the whole animal. But before this analysis can be really fruitful, it will be necessary to obtain more information about the input connections of the seeing fibers in order to place accurately the location of the attention mechanism and of the events leading up to it. With regard to memory mechanisms, the same statement about diffuse locations can be made. Since the seeing fibers show short-term memories, the above argument also applies here. In addition, we have seen that for the optokinetic memory mechanisms more than one location is strongly indicated. The advantage of this system is that it is more consistent under experimental circumstances than that of the seeing fibers and that its output is easier to study. Again, the input channels are insufficiently known, and in both instances the prime question will be whether or not they can be resolved to the extent that their circuitry can be experimentally handled.

In general, two major difficulties are involved in studying the behavior of neurons. One is that their reactivity may be regulated by factors other than synaptic input, of which changes in chemical surround appear to be the most important. Such changes are likely to be gradual and should not be of major importance in the above systems when studied over short periods. Another difficulty is that communication between interneurons may depend partially or completely on input other than that caused by presynaptic spikes. A good example of such a system is the vertebrate retina, where all three neuronal elements involved in information transmission between the primary sense cells and the spike trains in the optic nerve fibers appear to communicate exclusively by electrotonic interactions. In this nerve net both feed-forward and feed-back aspects are present, with polarizing and depolarizing effects. Therefore, it will be difficult to decide, e.g. in the rabbit's retina, which ganglion cells are movement sensitive and which ones are "on"-center, "off"-surround from their anatomical connections with the presynaptic elements. When centrifugally provided information is present, as in the arthropod eye, this difficulty may be even greater, especially if such pathways end on more than one type of element. As far as information flow is concerned, such systems may have to remain black boxes until methods can be devised to deal with them and the underlying events thereby revealed. Nevertheless, studying their overall input-output relationships in terms of general input and spike discharge output will show what is being integrated. The study of such systems is therefore analogous to the study of the behavior of whole organisms. Note also that knowledge about which factors are important for the behavior of the diverse interneuronal classes may be of considerable aid in the design of behavioral experiments on the whole animal.

There has been a strong tendency to ignore the behavior of the neuron and emphasize the synaptic event level. I am convinced that this will not be successful in explaining behavior even though we may by these means be able to locate a certain aspect. For example, many cases of reflex habituation involve the synaptic transmission process between input fibers and a single postsynaptic cell (see Kandel for *Aplysia*, in this volume). But note here, that at other synapses such habituation appears to be completely absent, though the transmission processes seem very much alike. Certainly the reason for this difference is important from a neurophysiological point of view. However, it may not only take a long time to solve this problem but it is rather immaterial for behavior, as long as its consequences are understood. Aréchiga et al. (1973) found that habituation with the same characteristics as above is present in several crayfish interneurons. When the same habituating stimulus was presented to the common sensory area of interneurons with overlapping fields, different extents of habituated areas resulted. Realizing that such differences may depend on any of a number of factors, it is unrealistic to study all such subsystems in detail to provide for the final outcome. Also, as mentioned previously, synaptic connections differ so much from each other in function and properties that modelling at this level may be considered as a poor choice. Which relationships are present in any given case would not be known until each pathway was studied in great depth, and if one waited until this is done, no progress would be made in "neurological ethology." Certainly we should also be ill-advised to go the other way and neglect synaptology. However, in order to gain an understanding of what makes a central nervous system perform the functions it fulfills, and especially to gain insight into such problems as why certain cue stimuli are preferentially reacted to by certain species whereas they are replaced by different ones in closely related ones, the neuronal behavior approach is likely to provide answers more quickly.

One shortcoming in this general area that should be rather easy to remedy is lack of knowledge of the sum total of behavioral acts that a given species can perform. Such information is essential in order to be able to correlate the behavior of the neuron with that of the animal. This is true not only for invertebrates but also for most vertebrates. I hope I have shown that the neuronal behavioral approach is both interesting and intellectually challenging. I foresee an impressive growth for this field in the near future, as well as a long lifetime.

ACKNOWLEDGMENTS   I am very grateful to Drs. H. Aréchiga, R. Hirsh, T. Yamaguchi, K. Yanagisawa, and B. York and the late Dr. Reddy, all of whom contributed to the research reported. My very special thanks are due to J. Roach for editorial comments. The research reported here has been supported by grants from the National Science Foundation and the National Institutes of Health, Public Health Service.

REFERENCES

ARÉCHIGA, H., B. BARRERA-MERA, and B. FUENTES-PARDO, 1973. Habituation of mechanoreceptive neurons of the crayfish. *J. Neurobiol.* (in press).

ARÉCHIGA, H., and C. A. G. WIERSMA, 1969a. The effect of motor activity on the reactivity of single visual units in the crayfish. *J. Neurobiol.* 1:53–69.

ARÉCHIGA, H., and C. A. G. WIERSMA, 1969b. Circadian rhythm of responsiveness in crayfish visual units. *J. Neurobiol.* 1:71–85.

BARNES, W. J. P., C. P. SPIRITO, and W. H. EVOY, 1972. Nervous control of walking in the crab. *Cardisoma guanhumi.* II. Role of resistance reflexes in walking. *Z. Vergl. Physiol.* 76:16–31.

BENNETT, M. V. L., and M. E. SPIRA, 1972. Synaptic control of electrotonic coupling between neurons. *Brain Res.* 37:294–300.

BUSH, B. M. H., 1965. Leg reflexes from chordotonal organs in the crab, *Carcinus maenas. Comp. Biochem. Physiol.* 15:567–587.

COLLEWIJN, H., and F. VAN DER MARK, 1972. Ocular stability in variable visual feedback conditions in the rabbit. *Brain Res.* 36:47–57.

DIJKGRAAF, S., 1956. Kompensatorische Augenstieldrehungen und ihre Auslösung bei der Languste (*Palinurus vulgaris*). *Z. Vergl. Physiol.* 38:491–520.

HORRIDGE, G. A., 1965. A direct response of the crab *Carcinus* to the movement of the sun. *Nature (London)* 207:1413–1414.

HORRIDGE, G. A., 1966. Optokinetic memory in the crab, *Carcinus. J. Exp. Biol.* 44:233–245.

HORRIDGE, G. A., 1968. Five types of memory in crab eye response. In *Physiological and Biochemical Aspects of Nervous Integration,* F. D. Carlson, ed., Englewood Cliffs, N. J.: Prentice-Hall, pp. 245–265.

HORRIDGE, G. A., and D. C. SANDEMAN, 1964. Nervous control of optokinetic responses in the crab *Carcinus. Proc. Roy. Soc. (London)* B161:216–246.

KANDEL, E. R., 1968. Dale's principle and the functional specificity of neurons. In *Cellular Pharmacology,* E. Coella, ed. Springfield, Ill.: Charles C Thomas, pp. 385–398.

KENNEDY, D., 1971. Crayfish interneurons. *The Physiologist* 14:5–30.

LINDBERG, R. G., 1955. Growth, population dynamics, and

field behavior in the spiny lobster, *Panulirus interruptus* (Randall). *Calif. Univ. Publ. Zool.* 59:157–247.

MILL, P. J., and D. A. LOWE, 1971. Transduction processes of movement and position sensitive cells in a crustacean limb proprioceptor. *Nature (London)* 229:206–208.

SHEPHEARD, P. R. B., 1966. Optokinetic memory and the perception of movement by the crab, *Carcinus*. In *The Functional Organization of the Compound Eye*, C. G. Bernhard, ed. Oxford: Pergamon Press, pp. 543–557.

SPIRITO, C. P., W. H. EVOY, and W. J. P. BARNES, 1972. Nervous control of walking in the crab, *Cardisoma guanhumi*. I. Characteristics of resistance reflexes. *Z. Vergl. Physiol.* 76:1–15.

WHITEAR, M., 1962. The fine structure of crustacean proprioceptors. I. The chordotonal organs in the legs of the shore crab, *Carcinus maenas*. *Phil. Trans. Roy. Soc. (London)* B 245:291–325.

WIERSMA, C. A. G., 1952. Neurons of arthropods. *Cold Spring Harbor Symp. Quant. Biol.* 17:155–163.

WIERSMA, C. A. G., 1958. On the functional connections of single units in the central nervous system of the crayfish, *Procambarus clarkii* (Girard). *J. Comp. Neurol.* 110:421–471.

WIERSMA, C. A. G., 1959. Movement receptors in decapod crustacea. *J. Mar. Biol. Assoc., U.K.* 38:143–152.

WIERSMA, C. A. G., and E. G. BOETTIGER, 1959. Unidirectional movement fibres from a proprioceptive organ of the crab, *Carcinus maenas*. *J. Exp. Biol.* 36:102–112.

WIERSMA, C. A. G., and B. M. H. BUSH, 1963. Functional neuronal connections between the thoracic and abdominal cords of the crayfish, *Procambarus clarkii* (Girard). *J. Comp. Neurol.* 121:207–235.

WIERSMA, C. A. G., and G. M. HUGHES, 1961. On the functional anatomy of neuronal units in the abdominal cord of the crayfish, *Procambarus clarkii* (Girard). *J. Comp. Neurol.* 116:209–228.

WIERSMA, C. A. G., F. VAN DER MARK, and L. FIORE, 1970. On the firing patterns of the "movement" receptors of the elastic organs of the crab, *Carcinus*. *Comp. Biochem. Physiol.* 34:833–840.

WIERSMA, C. A. G., and P. J. MILL, 1965. "Descending" neuronal units in the commissure of the crayfish central nervous system; and their integration of visual, tactile and proprioceptive stimuli. *J. Comp. Neurol.* 125:67–94.

WIERSMA, C. A. G., and T. OBERJAT, 1968. The selective responsiveness of various crayfish oculomotor fibers to sensory stimuli. *Comp. Biochem. Physiol.* 26:1–16.

WIERSMA, C. A. G., and T. YAMAGUCHI, 1966. The neuronal components of the optic nerve of the crayfish as studied by single unit analysis. *J. Comp. Neurol.* 128:333–358.

WIERSMA, C. A. G., and T. YAMAGUCHI, 1967. Integration of visual stimuli by the crayfish central nervous system. *J. Exp. Biol.* 47:409–431.

WIERSMA, C. A. G., and K. YANAGISAWA, 1971. On types of interneurons responding to visual stimulation present in the optic nerve of the rock lobster, *Panulirus interruptus*. *J. Neurobiol.* 2:291–309.

WIERSMA, C. A. G., and B. YORK, 1972. Properties of the seeing fibers in the rock lobster: Field structure, habituation, attention, and distraction. *Vision Res.* 12:627–640.

YOON, M., 1971. Reorganization of retinotectal projection following surgical operations on the optic tectum in goldfish. *Exp. Neurol.* 33:395–409.

YOON, M., 1972. Reversibility of the reorganization of retinotectal projection in goldfish. *Exp. Neurol.* 35:565–577.

YORK, B., C. A. G. WIERSMA, and K. YANAGISAWA, 1972. Properties of the optokinetic motor fibres in the rock lobster: Build-up, flipback, afterdischarge and memory, shown by their firing patterns. *J. Exp. Biol.* 57:217–227.

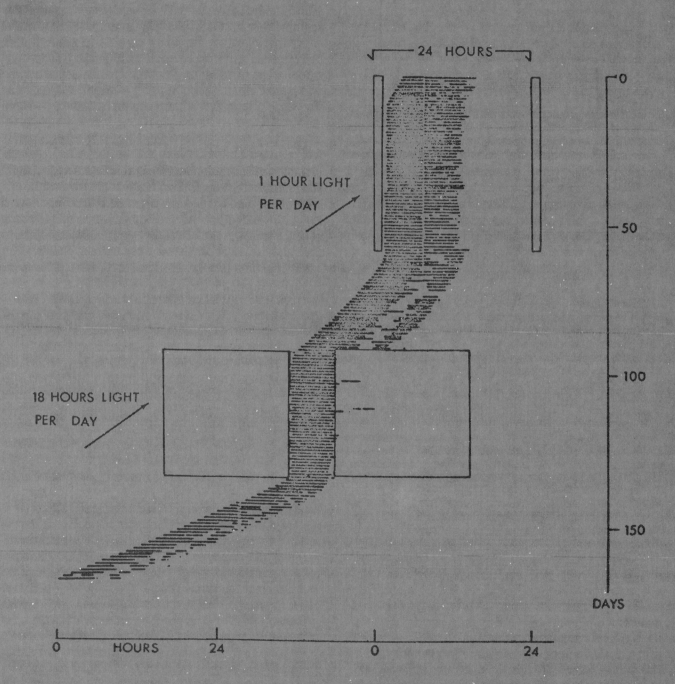

CIRCADIAN OSCILLATIONS AND ORGANIZATION IN NERVOUS SYSTEMS

Entrained and freerunning circadian rhythms of running-wheel activity in the deermouse Peromyscus leucopus. *The animal uses a running wheel during its nightly activity period which lasts about 10 hr and is recorded as a heavy black line. Successive days are plotted beneath each other. The observation lasts 6 months. On the 10th day of the experiment the animal had locked onto (was entrained by) the daily light pulse (1-hr duration). On the 60th day, the light was discontinued; the rhythm persisted with a circadian period (τ) of 23.6 hr. On the 90th day, it was again entrained by a 24-hr light cycle in which the light pulse lasted 18 hr. On the 140th day, the light was discontinued, and the rhythm again ran free with a circadian period of 23.0 hr. The value of τ shows an "after-effect" of the prior light pulse.*

CIRCADIAN OSCILLATIONS AND
ORGANIZATION IN NERVOUS SYSTEMS

Introduction

COLIN S. PITTENDRIGH

SCIENCE, IN Medawar's phrase, is the "Art of the soluble."
No matter how well founded, observations enter the
main stream of scientific concern only when they seem
tractable to analysis, or when, as unexplained phenomena,
their recognition facilitates the analysis of other problems.
It has been known since the early eighteenth century that
the familiar daily periodicity of biological activity may
persist in the absence of identifiable environmental cues.
The physiological challenge implicit in that fact did not
escape its original observer (de Mairan); but with few
outstanding exceptions (Wilhelm Pfeffer; Erwin
Bünning) it failed to attract serious attention until
approximately 20 years ago. Perhaps it seemed wholly
intractable to the current state of physiology, and its
implications for other analyses had still to be fully
appreciated.

In a series of independent, classic studies Gustav
Kramer and Karl von Frisch showed in 1950 that both
starlings and honeybees had excellent "clocks" permitting them to use the sun's azimuth as a compass,
compensating for its continuous displacement in the
course of the day. By the middle 1950s, it was clear that
de Mairan's phenomenon of persistent daily rhythmicity
reflected the presence in cells of self-sustaining oscillations
with a *circadian* (L. *circa*; *dies*) period close to—but
significantly different from—24 hours. It was shown, too,
that these same circadian oscillations, enjoying a remarkable homeostasis of their frequency in the face of temperature change, were in fact the "clocks" used by so many
animals in time-compensated sun orientation.

The outburst of interest in the 1950s cannot be
attributed to any new promise that de Mairan's observation was now more tractable to detailed explanation. In

its newly explicit form it seemed, if anything, more formidable: The time-constants involved, the precision and the homeostasis of frequency were—and still are—formidable challenges. Surely some of us were hopeful, even then, that the phenomenon was not too complex, but there were more tangible motives to pursue the matter.

First, there were clear dividends to reap in analyzing the formal properties of circadian clocks as self-sustaining cellular oscillators: An understanding of their entrainment by light and temperature cycles was essential in the analysis of animal navigation and the mechanisms of photoperiodic induction. Moreover, the purely formal problem was not only interesting in its own right, but its solution could well provide clues to the concrete nature of the oscillation.

Second, and perhaps more important, was the realization that, whether or not one knew its physical basis, circadian periodicity, innate to the cell's machinery, was a major, manipulable variable in all physiological analysis. Circadian oscillations have been found not only in single cells but cell fragments (a single case) and in enzyme systems. There was evidence that the reading of the genotype and protein synthesis were subject to circadian control. Thus, whether biochemist, cell physiologist, neurophysiologist, or student of behavior, the biologist could only ignore circadian oscillations at a needless price: They provide a framework for the temporal organization of biological activity; they create a "day-within" in the life of cells and organisms. To neglect the phase of the system's oscillation is to add—putting it mildly—wholly unnecessary variance in any biological analysis.

The neurophysiologist is entitled to no exemption from this warning. On the contrary, the nervous system as principal originator and coordinator of metabolic as well as behavioral control is surely where to look for the circadian pacemakers responsible for an animal's "internal day." The chapters in this part show that this search has already been fruitful. Circadian clocks have been anatomically localized in the nervous system of several animals, and several different circadian clocks have been localized in a single animal. In a truly elegant experiment by Truman, a (neural) circadian oscillation has been transplanted (retaining its phase) from one insect species to another where it reset the host's internal day. The state of sensory systems, including the rate of their adaptation, has been found to be a function of circadian phase. The neurochemistry of the pineal changes systematically with a circadian frequency. And the release of preprogrammed behavior seems everywhere to be timed by circadian oscillators to an "appropriate time" of day.

Especially intriguing and tractable questions concern the maintenance of temporal order within the nervous system as a population of circadian oscillators. Neurons and groups of coupled neurons may have slightly but significantly different intrinsic frequencies. How are they made synchronous? How important is such synchrony to normal function? Can the formalism of coupled circadian oscillators in the CNS provide a new approach to the obscure problems of sleep and its derangement?

The phenomenology of circadian rhythmicity, now well quantified, is in a state comparable to that of genetics before 1953. The concept of the gene and the formalism of its replication, transmission, and action played a major role in the clarification of biological problems even before we knew the structure of DNA. It would clearly be foolish to ignore circadian oscillations until their physical basis is fully elucidated.

38 Circadian Oscillations in Cells and the Circadian Organization of Multicellular Systems

COLIN S. PITTENDRIGH

ABSTRACT Properties of freerunning circadian oscillators are reviewed and the mechanism of their entrainment by exogenous *Zeitgebers* is analyzed. Multicellular systems, e.g., nervous systems, constitute a population of circadian oscillators, which are coupled to each other. If they are uncoupled, i.e., if the internal temporal order among the circadian periodicities is lost, physiological penalties may result.

Introduction

THE DIVERSITY of phenomena that currently concern students of circadian rhythmicity ranges from the control of gene transcription and replication to animal navigation. It is too broad for useful review in a single paper, and what follows suffers from necessary compromise. It seemed desirable in an introductory paper to indicate the origins and scope of a relatively new field, but I have paid most attention to the area where I think circadian phenomena are of greatest interest to the neurophysiologist. Clearly the neuron, as single cell, is a proper object for work intended to elucidate the concrete, subcellular mechanisms of circadian oscillations; but there are many other cells, more readily cultivated in vitro, where that goal can be pursued at least as easily. It is probably in the integrative physiology of the *nervous system*, as such, where circadian phenomena have, today, their greatest relevance to the neurophysiologist's special preoccupations: The couplings, mutual and hierarchical, of circadian oscillators in the nervous system must play a major role in its temporal organization. That is the theme to which I pay most attention. The section before the summary of the paper attempts to relate these principles of circadian (temporal) organization to the phenomena of "photoperiodism," which, in animals, are controlled by the nervous system.

COLIN S. PITTENDRIGH Department of Biological Sciences, Stanford University, Stanford, California

Circadian clocks: The development of a field

Figure 1, using contemporary data from a deermouse, illustrates a general biological phenomenon discovered in 1729 by the French geologist de Mairan, observing a plant. The familiar fact that organisms execute various functions at different times of day seems too trivial to merit analytic attention; yet when—as de Mairan did—one places an organism in an aperiodic environment (more specifically in constant temperature and illumination) its "daily" periodicity commonly persists indefinitely in the absence of known external driving cycles. This is a nontrivial fact, as de Mairan promptly perceived. The darkness of night is not the whole and adequate explanation, it seems, of the deceptively "obvious" fact that green plants photosynthesize only during the day: Cells like *Acetabularia* maintain a striking persistent daily rhythm of photosynthesis in constant temperature *and constant light*. All these systems—plant, deermouse, and single cell—when placed in an aperiodic environment behave as though they were autonomous, free-running, self-sustaining (or weakly damped) oscillators (Pittendrigh and Bruce, 1957). In fact they are: It is implicit in the de Mairan phenomenon (present at all levels of eukaryotic organization) that the controls of metabolism include self-sustaining oscillations with a free-running period (τ) that is close to 24 hr (hence their designation by Halberg, 1960, as *circadian*). They provide the framework for an environmentally oriented temporal organization of the cell; they organize "a day within," that is, an evolved match to the periodicity of the external world.

Neither de Mairan himself, nor any of his distinguished followers in the nineteenth century (e.g. Sachs, Darwin, Pfeffer), made what was implicit fully explicit. Through the middle of this century, the principal interest in de Mairan's phenomenon appears to have been the possibility that it was caused by some undetected external

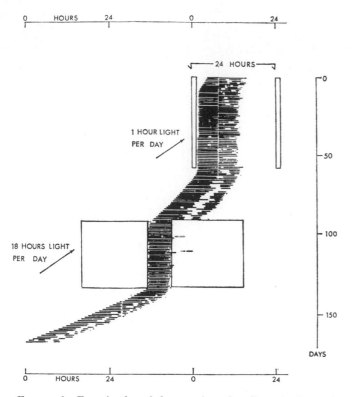

FIGURE 1 Entrained and freerunning circadian rhythms of running-wheel activity in the deermouse *Peromyscus leucopus*. The animal uses a running wheel during its nightly activity period which lasts about 10 hr and is recorded as a heavy black line. Successive days are plotted beneath each other. The observation lasts 6 months. On the 10th day of the experiment the animal had locked onto (was entrained by) the daily light pulse (1-hr duration). On the 60th day, the light was discontinued; the rhythm persisted with a circadian period (τ) of 23.6 hr. On the 90th day, it was again entrained by a 24-hr light cycle in which the light pulse lasted 18 hr. On the 140th day, the light was discontinued, and the rhythm again ran free with a circadian period of 23.0 hr. The value of τ shows an "after-effect" of the prior light pulse (see text).

periodicity and offered a hint, therefore, of unknown sensory modalities. The beginnings of the field as it now stands derive, in my opinion, from two sources.

The first was the utilization of automatic recording techniques and organisms adequate for the observation of very long (weeks, months) runs of unentrained daily rhythms. Only then was it rigorously established that their period was significantly different from 24 hr and hence not simply attributable to any external cycle deriving from the earth's precise 24-hr rotation. Their circadian period is by no means the only evidence of the endogenous nature of circadian oscillations; but it was, historically, the most compelling single factor persuading virtually all concerned that the possibility of external control was only a distraction from a major physiological problem.

The second root of the modern era was the demonstration made independently by Kramer (1950) and von Frisch (1950) that birds and bees, respectively, could use the sun's azimuth as a compass by virtue of possessing an *internal chronometer* to compensate for its continuous movement through the day. This discovery initiated search for the clock involved and caused several biologists in the 1950s to view de Mairan's phenomenon in a new light, not as a cue to undetected external cycles but as reflecting the presence of an internal oscillator functioning as a 24 hr clock (Pittendrigh, 1954).

The union of these two fields of animal behavior was consummated in a classic experiment by Klaus Hoffmann (1960). Starlings trained to orient to a given compass direction in search of food—using the sun's azimuth as compass—were kept in constant light in the laboratory for two weeks during which their activity/rest cycles were monitored to yield data comparable to those of Figure 1. Their activity cycle was found to have a short circadian period (~ 23.5 hr). After several days of such unentrained oscillation, they were again challenged to use the sun as compass, and their orientation reflected the "error" of their fast, freerunning chronometers. The clock utilized in sun orientation had the same circadian period as the oscillation responsible for their activity cycles.

While the Kramer-von Frisch discoveries undoubtedly played this seminal role in the early 1950s, it is true that they had historical antecedents that were less compelling in their day. Kalmus (1935) and Stein-Beling (1935) were both grasping for the relationship between persistent daily rhythms and biological time, but the most important earlier connection was made by the remarkable intuition of Erwin Bünning in 1936 that endogenous daily rhythmicity was causally involved in the recently discovered phenomena of photoperiodism. It was not until after 1950, however, that this intuition—now fully validated—was more explicitly formulated, as a time-measurement problem, in the new terminology of "circadian clocks," which had been evolved in response to the discoveries of time-compensated sun orientation.

Two significant developments in the period following Kramer and von Frisch were (1) the application of the formalism of self-sustaining oscillators to the discussion of de Mairan's phenomenon (Pittendrigh and Bruce, 1957) and (2) the demonstration that single cells could sustain such oscillators (Bruce and Pittendrigh, 1956; Ehret, 1959; Hastings and Sweeney, 1958).

The entrainability of self-sustaining oscillators by other periodicities to which they can couple is the central feature of the oscillator model in elucidating their functional role as cellular clocks for the recognition of local (sidereal) time. In its entrained steady state, a circadian oscillation assumes a unique phase relationship to the 24-hr light/dark cycle that entrains it: Each phase point in its cycle occurs at a specific time of day. By coupling any cellular event to a particular phase in the oscillation's motion, selection assures its occurrence at an "appropriate" time. The functional significance of this is many sided: It permits *initiation* of events in *anticipation* of the time at which their culmination most appropriately occurs, and it permits timing that cannot be entrusted to control by conditions for which the system has no adequate sensory modality. Both points are illustrated by the daily migration of *Plasmodium* and microfilariae into the peripheral blood stream of their vertebrate hosts: That migration is timed to coincide with the flight of their local mosquito vector about which no sensory equipment in the parasite could conceivably inform it. Circadian oscillations in the parasite evidently entrain to the marked circadian periodicity of its host's internal environment, which, in turn, is phased to local time by the entrainment of the host's circadian clocks to the light cycle. The initiation of migration is apparently coupled to the appropriate phase of the parasite's circadian oscillator, assuring its presence under the host's skin when the vector is flying. That time is different in different localities where the mosquito vectors differ (Hawking, 1962) in their flight time.

Circadian oscillations, phase locked to the external light cycle, provide a framework for the temporal organization of the entire system relative to local time: They are indeed "clocks." The coupling of different functions to different phase points of a common circadian pacemaker in a single cell is beautifully exemplified in the dinoflagellate *Gonyaulax polyedra*. Four functions—photosynthesis, mitosis, induced luminescence, and spontaneous glow— have been shown to occur at four different times of day in this cell when it is entrained to a 24-hr light/dark cycle. The rhythmicity of all four functions persists when the cell is transferred to constant dim light, and they maintain their characteristic phase relationships. More importantly, they still maintain them when the whole system is phase shifted by a single light pulse; all four functions are in some way coupled to different phase points of a common driving oscillation, which is what responds to the phasing action of the light pulse (McMurry and Hastings, 1972).

The general problems

At this date it is surely unnecessary to elaborate on the widespread presence of circadian phenomena in eukaryotic systems at all levels of organization. The bulk of the literature is still based on assays that exploit, for reasons of technical simplicity, behavioral phenomena too complex to interest the biochemist. One hopes, however, it is

necessary here only to note: (1) that circadian oscillations exist not only in single cells but in cell fragments: The isolated stem of *Acetabularia* (Schweiger et al., 1964) continues its circadian periodicity of photosynthesis in the absence of the nucleus; (2) that some individual enzyme systems, even in nongrowing cells, manifest a clear circadian periodicity of activity (e.g. Sulzmann and Edmunds, 1972); and (3) that the circadian periodicity of enzymes with very short half-lives (like tyrosine-aminotransferase in rat liver) suggests a circadian periodicity of gene transcription. The phenomenology we are concerned with pervades the entire spectrum of organizational levels, and while it raises a multitude of special questions of interest to students of each level, there are nevertheless some truly general problems, as follows: (1) What is the nature of the driving oscillation in the single cell; what is it, concretely? (2) How is it coupled to the broad array of cellular activities that it times differentially? (3) How is it coupled to (entrained by) the external environment? (4) What emergent properties characterize multicellular systems as populations of circadian oscillators?

Properties of freerunning oscillations

DEPENDENCE ON METABOLIC ENERGY SOURCES; CHARGE AND DISCHARGE FRACTIONS OF THE CYCLE Knowledge of the driving oscillation is still largely restricted to its formal properties. It is an endogenous, indeed innate, self-sustaining oscillation with a circadian period (τ) close enough to lock on to the period ($T = 24$ hr) of the external world. Its endogenous nature is attested by too many facts (in addition to its circadian period) to summarize here; but it is useful to add one that makes an additional point. The oscillation (Figure 2) that gates the emergence of adult *Drosophila* from their puparia can be stopped immediately if the system is placed in pure N_2 during its early "subjective night." Although the pupae survive many days of anoxia (mobilizing enough energy for maintenance metabolism by an anaerobic route), their gating oscillators stop; but they immediately resume their motion when O_2 is returned (Figure 3). The phase of the gating oscillator is displaced by (almost) precisely the duration of the anoxia. If, however, anoxia is initiated in the late subjective night, the oscillation proceeds (indeed accelerates) until it reaches the beginning of its oxygen-dependent phase, when again it stops until air is returned; a brief period (~ 6 hr) constitutes an energy-absorbing (O_2-dependent) "charge" phase, followed by an energy dissipating "discharge" phase. It is perhaps worth noting that the "charge" phase in the early subjective night is that part of the cycle where light pulses cause phase delays; during the "discharge" part of the cycle, light pulses cause phase advances (Figure 6). It is

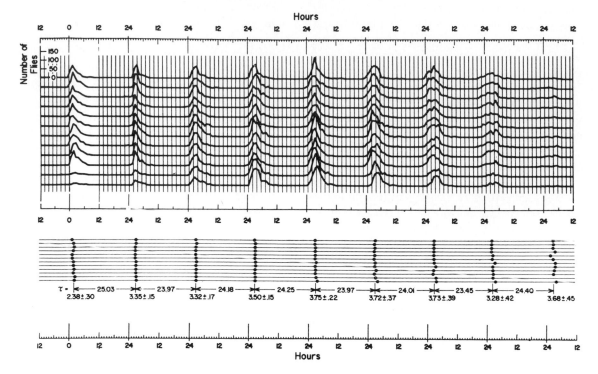

Drosophila pseudoobscura – Pupal Eclosion Rhythm. 20°C.

FIGURE 2 The circadian rhythm of adults emerging from a population of developmentally asynchronous *Drosophila pseudoobscura*. The emergence (eclosion) of each individual fly is limited to a narrow (~6 hr) fraction of a circadian "gating" oscillator in its brain. The gating oscillators in all individuals have been synchronized (entrained) by exposure to a 24-hr light/dark cycle (12 hr light:12 hr dark). The last 12 hr light period is shown. Then 12 replicate populations are allowed to freerun.

The circadian rhythm in the population reflects the synchrony of the gating oscillators in all individuals. Medians of each emergence peak (marking the phase of the gating system) are plotted below the raw data. The declining amplitude of the emergence peaks reflects the (trivial) fact that fewer flies remain in the population as time passes; it bears no relation to the amplitude of the gating oscillation.

as though whatever is synthesized during the charge phase could be destroyed by processes initiated photochemically. In *Aplysia*, aflatoxin Bl has the effect of annihilating the rhythm (Rothman and Strumwasser, 1973) presumably by blocking RNA and protein synthesis; its effectiveness is limited to the beginning of the "charge" phase as here defined.

GENETIC CONTROL The innateness of circadian oscillations is attested by a vast array of facts. The first direct demonstration of genetic control began with Bünning's discovery (1935) that the oscillations in two strains of *Phaseolus*, with different average frequencies, were polygenically controlled. Today the induction of point mutations affecting the oscillation has become a common tool for its analysis. Using the chemical mutagen EMS, Konopka and Benzer (1971) have produced mutants of a locus on the X-chromosome of *Drosophila melanogaster*. One of these changes τ from 24 hr to 29 hr; a second changes it to 19 hr; the third destroys the oscillation com-

pletely. These mutations affect the oscillations controlling the adult's activity cycle as well as its eclosion. Five mutations, also produced by EMS, have been found by Fox and Pittendrigh (unpublished experiments) in *Drosophila pseudoobscura*, again on the X-chromosome. All of these as hemi- or homozygotes are aperiodic in constant darkness although two of them show a clear, forced daily rhythm in a light/dark cycle. Some of the heterozygotes show incomplete complementation: They entrain like a normal system to a light/dark cycle but with highly atypical phase; and, while they show a subsequent free-running rhythm in constant darkness, its period is not only longer than that of wild type but much noisier (Figures 4 and 5). Feldman and Waser (1971) have produced a mutant *Neurospora* in which the oscillation is no longer temperature compensated. Bruce (1972) has produced several mutants lengthening τ in *Chlamydomonas*, and they are additive in their action. While genetic control, at least in some cases, is clearly polygenic, the dramatic effects of changing one locus in *Drosophila*

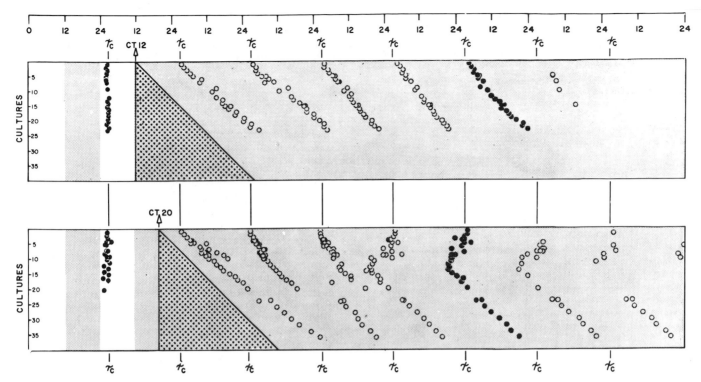

FIGURE 3 *Drosophila pseudoobscura*. The effect of anoxia on the circadian oscillations that gate eclosion. *Upper panel*: The gating oscillations in 24 populations (cultures) were synchronized by a light cycle before being released to freerun in darkness. The first population served as control; the remaining 23 were placed in pure N_2 at the beginning of the rhythm's freerun. This phase of the rhythms is ct-12 (the beginning of the subjective night) as defined in Figure 6. The 23 treated populations were exposed to successively longer duration of N_2 (heavy stipple) before being returned to air. The phase of the subsequent rhythm in each population is indicated by circles, which mark the medians of successive peaks of emergence activity. The fifth peak (solid circles) is taken to mark the phase of the rhythms after their exposure to anoxia. The phase shift relative to the untreated suggests that genetic dissection may prove fruitful here as it has in the analysis of other complex systems.

control (ψ_c) corresponds with the duration of N_2 exposure which clearly stopped the oscillation. *Lower panel*: A comparable experiment involving 36 cultures. In this case anoxia began at ct-20 in the late subjective night. Again N_2 affects the system which for several days is not in steady state. The 4th, 5th, 6th, and 7th peaks are in steady state. The 5th peak (solid circles) is again taken to mark the phase of the steady state. The driving oscillation (see text) not only remained in motion but accelerated under all N_2 exposures up to $\sim$13-hr duration when it had again reached an O_2-dependent phase and stopped until air was returned. (The transient behavior of the system preceding its steady state reflects the behavior of the driven *B*-oscillator (see text) regaining phase with the pacemaker whose response to the anoxia is only seen in the steady state: peak 4 onwards.

THE STABILITY AND LABILITY OF τ The frequency of circadian oscillators is generally remarkably stable; the standard error on τ is rarely more than a few minutes in individual rodents. But it is important to stress that, at least in multicellular systems, τ of an individual is open to some significant change even in a stable environment. Thus what is genetically prescribed is a narrow range of freerunning frequencies, not a single frequency. Many workers have reported abrupt "spontaneous" frequency change; it may also change systematically as a function of time. And, as I reported many years ago (1960), it is subject to "after effects," which merit attention here for several reasons: They are more widespread than the current literature suggests; they are not accounted for by any of the several mathematical models so far published; and they must be reckoned with in the mechanism of entrainment. In our laboratory, we have found after effects on the frequency of freerunning rhythms following: (1) phase shifts induced by light signals; (2) entrainment by cycles whose period is near the limit of entrainment (see Figure 12); (3) exposure to constant light; (4) change in photoperiod. In the *Peromyscus* of Figure 1 τ is shorter in the freerun following the long photoperiod than in the freerun following the short photoperiod.

THE HOMEOSTASIS OF τ Change in τ can be induced by changing the (aperiodic) level of those external variables capable of entraining it when they are experienced as cycles. This is, in itself, almost an analytic necessity; what is much less obvious—and predictable only when circadian

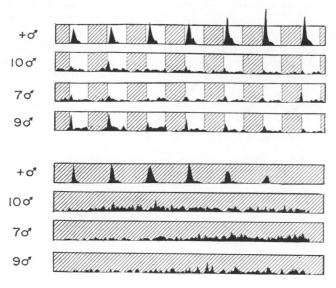

FIGURE 4 *Drosophila pseudoobscura*. EMS-induced mutations on the *X*-chromosome affecting the eclosion rhythm. Three mutants (cl^{10}, cl^7 and cl^9) and wild-type in males in a 24-hr light cycle are shown in the upper 4 panels and in constant darkness (lower 4 panels) following prior exposure to 24-hr light/dark cycle. All the mutants are aperiodic in darkness—they have no self-sustaining oscillation to gate eclosion. One of them (cl^9) is, however, susceptible to a "forced" rhythmicity in a light cycle.

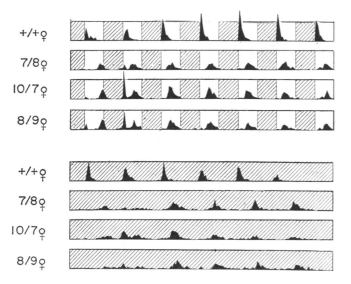

FIGURE 5 The EMS-induced *X*-chromosome mutants of *D. pseudoobscura*. Some of the mutants complement each other, partially, in heterozygotes (females). The 3 complements illustrated have a clear rhythm in a light cycle, but the phase of the gate is unstable initially and when it does stabilize its phase is atypical (5-hr later than wild-type). All the complements sustain a freerunning rhythm in darkness but the period is longer than that of wild-type, and the rhythm is noisier.

oscillations are perceived as clocks—is the fact that the range of τ-variation realizable by maximal changes of light and temperature is small; it is rarely more than 10%, and more often only 5%. It was the hypothesis that circadian rhythms were indeed the biological *clocks* of Kramer and von Frisch that led me in 1954 to look for and to find what I then, unfortunately, called the temperature "independence" of τ. But circadian oscillations are not, strictly speaking, independent of temperature: They can be phase shifted by temperature pulses or steps, and entrained by temperature cycles. They are, however, temperature *compensated*: Over a wide range of temperature, their freerunning frequency is homeostatically conserved within the narrow limits essential to an oscillator that must function as a clock.

Temperature compensation of τ has been stressed ever since 1954 as one of the most remarkable features of circadian oscillations, but in that emphasis we may have lost sight of a more important point—that their frequency (as that of a clock) is probably homeostatically protected in the face of *all* change likely to be encountered in the cell. Cyanide concentrations can be raised to the point of nearly killing the alga *Oedogonium* and reducing nearly to zero the amplitude of the function that the oscillator times, but without changing *its* period (Bühnemann, 1955). In *Drosophila* the eclosion gating system comprises two oscillations, one of which (the *A*-oscillator) is coupled to the external light cycle and has a temperature-compensated frequency. It drives, by a temperature-dependent coupling mechanism, a second (*B*-) oscillation that performs the actual gating. Because their coupling is temperature dependent, the phase relation of the *B*-oscillator's gate relative to both the *A*-oscillator and the external light cycle is also temperature dependent. When flies are raised on a medium containing 25% D$_2$O the phase of the temperature-dependent gate is displaced by many hours but the period of the driving *A*-oscillator is lengthened only trivially (Pittendrigh, Caldarola, and Cosbey, 1973). The circadian pacemaker is protected from change by D$_2$O as well as temperature. The mechanism of the oscillation's homeostasis is a major challenge and (until we understand it better) a major barrier to evaluating the *quantitative* effects of any agent used to perturb it. The general homeostasis of τ may well have contributed to the difficulty encountered in so many attempts to affect the oscillation chemically (Pittendrigh and Caldarola, 1973).

THE CONCRETE NATURE OF THE OSCILLATOR Interpretation of the few unequivocal changes produced by chemicals is a hazardous undertaking. Heavy water is the one agent that consistently produces a marked effect. In all published cases it lengthens τ. But this result is explicable by an unlimited array of hypotheses. Heavy water slows virtually every biological system exposed to it and for a diversity of reasons (Katz and Crespi, 1972). Its most consistent effect is to simulate a lower cellular temperature by stabilizing protein structure, but this gives no immediate clue to the nature of circadian oscillations.

Two general, and not mutually exclusive, intuitions seem to prompt most approaches to the concrete nature of the oscillation. The first, that it involves membranes, has led to finding effects due to potassium (Eskin, this volume), lithium (Engelmann, 1972), and valinomycin (Bünning and Moser, 1972). The bearing of D_2O effects on the "membrane hypothesis" (Enright, 1971) is unclear at best (Pittendrigh, Caldarola, and Cosbey, 1973). The second, that it involves protein synthesis, has led to studies with a variety of blocking agents. The clearest effects are those of Rothman and Strumwasser (1973), using aflatoxin Bl, and Feldman (1967), using cycloheximide. The cycloheximide case raises an important caveat. Feldman found that τ in actively growing *Euglena* was a clear linear function of the rate of protein synthesis that he manipulated with the drug. Brinkmann (1971) confirmed this result of Feldman's but added the complication that a circadian oscillation persists in nongrowing cells of *Euglena* and is totally insensitive to cycloheximide. Moreover, the formal properties of the temperature-compensating mechanism are different in growing and non-growing cells. One and the same cell can evidently develop circadian oscillations with significantly different properties and perhaps with different concrete mechanisms.

The available facts do not merit more extended discussion here. The plain fact is that we have little or no insight into the physical nature of circadian oscillations, and none of the concrete models proposed seem tractable to rigorous test. Circadian periodicity in cellular biochemistry is now commonplace. Control is probably exerted postranslationally in some cases, and in others at the transcriptional level. The prime problem is the lack of criteria that inform us when we are seeing chemical changes intrinsic to the oscillation's causal loop as distinct from changes driven by it. And until we know *something* about the concrete nature of the driver, we cannot even speculate on how it is coupled to the diversity of cellular events it times. The complexity of the problem, even in the single cell, is illustrated by the behavior of *Acetabularia*. Its cytoplasm can sustain an oscillation without a nucleus, but that cytoplasmic oscillation is itself a slave to another in the nucleus (Schweiger et al., 1964). The evidence from *Gonyaulax* that a single pacemaker controls four different rhythms is persuasive, but this picture may yet prove an oversimplification.

In this brief review, I have not attempted to distinguish between uni- and multicellular freerunning systems. That distinction may, however, be important. We do not know, for example, whether the important features of (1) after effects, and (2) spontaneous, abrupt frequency change are properties of the oscillation in a single cell: They may well be the properties of a population of coupled cells acting as a "single" pacemaker in a multicellular system.

Entrainment by environmental cycles

External cycles known to entrain circadian oscillations include light, temperature, and in a few cases social cues. The light cycle is by far the most general and effective entraining agent or *Zeitgeber*, and its importance surely derives from its relatively noise-free nature. Temperature cycles are effective only in poikilo- and heterotherms, and act, presumably, directly on the oscillation itself. Transduction through neural pathways must be involved in the case of social cues and is also involved in some instances of entrainment by light. Rodents, cockroaches, and crickets, in which eyes have been disconnected from the CNS, freerun in light/dark cycles. On the other hand, there are cases where light bypasses organized photoreceptors. This is the case in birds, several insects, and of course in all protistan and green plant systems. In some cases (*Drosophila*, *Pectinophora*, and *Neurospora*) where the eye is bypassed and action spectra have been obtained, they are interestingly similar: There is a broad maximum in the blue between 460 and 480 nm and a sharp drop off near 500 nm (Frank and Zimmerman, 1969; Bruce and Minis, 1969; Sargent and Briggs, 1967). Carotenoids, though compatible with this spectrum, have been eliminated in the case of *Drosophila*, and the pigment involved remains unknown (Zimmerman and Goldsmith, 1971).

All the phenomena of entrainment (Aschoff, 1960; Bruce, 1960) cannot be summarized here. Parametric entrainment is possible and probably occurs in nature; circadian oscillations entrain both in the laboratory and above the Arctic Circle to sine waves of light intensity and temperature (Swade and Pittendrigh, 1967). Parametric effects may also be involved to some extent when the action of light involves long-lasting photoperiods. On the other hand, the action of long light signals can be simulated remarkably by two short pulses that constitute what has been called a "skeleton photoperiod" (Pittendrigh and Minis, 1964) in which one signal simulates the onset of a long photoperiod and the second signal simulates its termination. Such nonparametric cases are the only ones convincingly explained so far and provide a convenient framework here for brief summary of the essential issues involved in entrainment more generally.

To be entrainable by any cycle of an external variable, an oscillation must be differentially responsive to it at different phases of its cycle. This analytic necessity has been most extensively explored experimentally in the case of light. Single brief light pulses applied to freerunning oscillations in darkness cause phase shifts ($\Delta\phi$), of which both the magnitude and direction (delay, advance) are a function of the phase (ϕ) pulsed; $\Delta\phi$ is a function of ϕ.

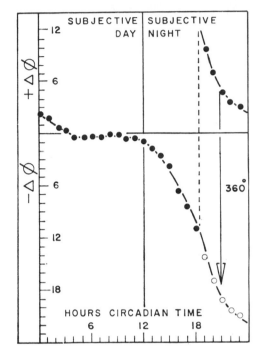

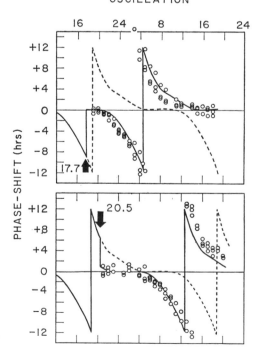

Figure 6 illustrates the phase-response curve for the *Drosophila* gating system obtained using 15-min pulses of white light. The phase scale is defined in terms of *circadian time*; the full cycle is divided into 24 circadian hours. Circadian time 0 (ct-0) is that phase of the oscillation where the beginning of a 12-hr light pulse falls in the entrained steady state. The first half-cycle between ct-0 and ct-12 is called the "subjective day" (SD); the second, between ct-12 and ct-24, is the "subjective night" (SN). The *Drosophila* curve exemplifies qualitatively the features of *all* circadian oscillators' phase-response curves for light. Thus, response is always minimal in the subjective daytime; the strong responses beginning in the late SD and continuing into the early SN are always phase delays; phase advances appear in the late SN and, with diminishing magnitude, continue into the beginning of the SD. The amplitude of the response curve is, however, not always as great as in *Drosophila pseudoobscura*. Winfree (1967) recognizes two types of phase-response curves and generalizes their relationships, but both have all the features noted here.

Phase-response curves plot the *steady-state* $\Delta\phi$ of the rhythm that may (especially in the case of phase advances) not be expressed until several transient cycles have passed since the rephasing pulse was seen. These transients have long been thought, especially in the case of *Drosophila*, to reflect the motion of a driven component in the system

FIGURE 6 The response to short (15 min) light pulses of the circadian oscillation that drives the eclosion gating system in *D. pseudoobscura. Upper panel*: The "Phase Response Curve" plots the steady state phase delays ($-\Delta\phi$) and phase advances ($+\Delta\phi$) caused by pulses applied at 24 different phases (or "circadian times," ct) of the oscillation. The phase-shift induced is dependent on the phase pulsed. (When $+\Delta\phi$ values are displaced 360°, one obtains a monotonic curve which is graphically useful, as in Figures 7 and 8). *Middle panel*: The driving oscillator is reset immediately by the light pulse and to the extent ultimately expressed in the steady state of the system. Rhythms were initiated synchronously in many populations by transferring them from constant light to constant darkness at the same time. The oscillations begin their motion at ct-12, the beginning of the subjective night. All populations receive an initial light pulse at ct-17.7, which is predicted to cause an *immediate* phase delay of 10 hr. Separate populations then receive a second pulse at successively later times. The $\Delta\phi$ caused by the second pulse is a measure of the oscillation's phase when it is given. The dashed line indicates the projected time course of the oscillation had it not received the first pulse; the solid line indicates its time course predicted on the assumption that the first pulse (heavy arrow) does indeed effect an immediate reset. The plotted points are experimental values of the $\Delta\phi$ caused by the second pulse. Clearly a 10 hr $-\Delta\phi$ was effected immediately by the initial pulse at ct-17.7. *Lower panel*: The same is true when a pulse (heavy arrow) is applied at ct-20.5: It causes an immediate phase advance of 5.9 hr.

that only gradually regains its steady-state phase relation to the circadian pacemaker that is thought to be fully reset, essentially instantaneously, by the light pulse (Pittendrigh and Bruce, 1959). The validity of these assumptions has been attested by the success of many predictions based on them (see Pittendrigh, 1966) and is now fully established by unpublished experiments exemplified in the lower panels of Figure 6.

This "simplicity" of action makes it easy to predict, in quantitative detail, many aspects of the oscillation's entrainability by short pulses. In the entrained steady state the oscillator assumes the period (T) of the driving cycle. The action of the light pulse in each cycle is to cause a phase-shift ($\Delta\phi$) equal to the difference between the period (τ) of the oscillator and that (T) of the light cycle: $\tau - T = \Delta\phi$. Thus in the steady state, the light pulse must fall at that phase point (ϕ) where the $\Delta\phi$ it causes equals $\tau - T$. It does so, as Figure 7 shows.

For nonparametric entrainment of this kind, one can, using phase-response curves, predict precisely the phase of the oscillation relative to any driving cycle whose frequency is known, and, given the fact that stable entrainment is possible only when the slope of the response curve is less than |2.0|, one can also predict successfully the limits of entrainment (Ottesen, 1965; Pittendrigh, 1966). The *Drosophila pseudoobscura* oscillator can be entrained down to about $T = 17$ hr and up to about $T = 30$ hr. When T is less than τ, the oscillation phase lags the light cycle; when T is greater than τ, it phase leads the light cycle. The complex characteristics of entrainment involving two or more pulses per cycle can be predicted with equal ease (Pittendrigh, 1966).

In a comparative study of many individual animals in three rodent species Pittendrigh and Daan (1973) have discovered a strong correlation between the period of a circadian oscillation and the shape of its phase-response curve for short light pulses. The ratio (D/A) of the areas

under the delay and advance sections of the curve is negatively correlated with τ. This rule even holds intraspecifically: D/A is greater in *individuals* with shorter

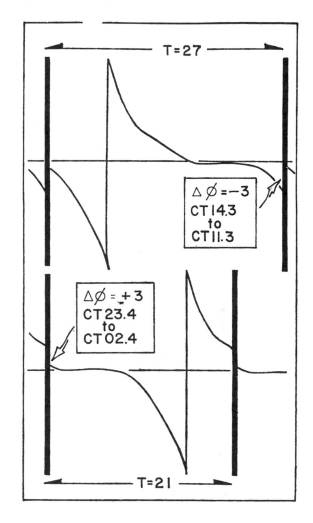

FIGURE 7 The entrainment of the *D. pseudoobscura* gating system by light cycles whose periods (T) are 27 and 21 hr (one 15' pulse per cycle). *Upper panel:* The successive light pulses in the 27-hr and 21-hr cycles are indicated by heavy black bars; the predicted steady-state phase-relation of the oscillation, relative to the entraining cycle, is indicated by the phase of its phase-response curve relative to the light. When $T = 27$ hr the light pulse must fall at ct-14.3 to cause the necessary $-\Delta\phi$ of 3 hr; when $T = 21$ hr it must fall at ct-23.4 to cause $+\Delta\phi$ of 3 hr. *Lower panel:* The predicted phase of the oscillation, relative to the light pulse is given for both $T = 21$ and $T = 27$ as solid lines; the curve is plotted in its monotonic form (see Figure 6). The plotted points indicate the phase shifts caused by 15 min pulses at successively later times after the system is released from its light cycle into a freerun. They confirm the predicted phase of the oscillation relative to its last seen entraining light.

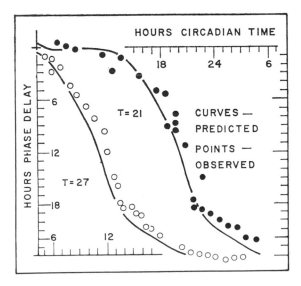

τ. The relationship is intuitively "sensible": The shorter τ, the greater is the net delaying action of the light needed to drive the oscillation at $T = 24$ hr. But its full implications must await better understanding of (1) the parametric effects of long photoperiods and (2) the evidence presented later in this chapter that the curves possibly—indeed probably—reflect the behavior of a system of two coupled oscillations.

There is no obvious a priori reason why circadian oscillators should not be true relaxation oscillators which can be rephased *only* by delays (or *only* by advances), irrespective of the phase perturbed. Entrainment could readily be effected by a system capable either of *only* delays were τ genetically fixed at less than 24 hr, or of *only* advances were τ set at > 24 hr. The fact that *all* circadian oscillations studies respond by *both* delays and advances at different phases of their cycle is thus a nontrivial fact to which no explicit attention has been given and whose meaning remains unclear. It could, however, be related to the propositions outlined below in the section on Separate Oscillators Coupled to Dawn and Sunset. If that is the case, we would have to consider the driving oscillation in *single* cells as itself a pair of coupled oscillators, because their phase response curves also include both delays and advances.

Circadian organization in multicellular systems

THE MULTIPLICITY OF OSCILLATORS IN METAZOA The fact that a single eukaryotic cell can develop a circadian oscillation raises questions with respect to multicellular systems: Do *all* their cells develop their own circadian oscillations, and does the circadian rhythmicity of function in virtually every mammalian tissue studied reflect the action of circadian oscillations autonomous to the cells of that tissue? Or, does the broad array of rhythmicity reflect the response of target tissues to central control exerted only by a few cells (even one) acting as circadian pacemaker for the entire metazoan system?

First, it is clear that *not all* cells in the metazoan act as autonomous oscillators—at least at every stage in their differentiation. For example, the oscillator gating egg-hatch in *Pectinophora* is not differentiated until midway through the approximately 12-day embryogenesis of that moth (Minis and Pittendrigh, 1968). The 10-day time course of imaginal disk differentiation in *Drosophila* is completely independent of the phase of any circadian oscillator either internal or external to the cells involved (Skopik and Pittendrigh, 1967). And Strumwasser (this volume) finds no circadian rhythmicity in impulse rate in most neurons in the same *Aplysia* ganglion that contains the parabolic burster which *is* a circadian oscillator. The potentiality of eukaryotic cells to differentiate a circadian

oscillation, presumably universal, is as open to developmental suppression or evocation as the rest of the genotype.

However, it is now equally clear that the metazoan is literally a population of coupled circadian oscillators as I suggested in 1960. In *Aplysia* there are at least two independent sets of circadian pacemakers, one in the eye (at the head of the optic nerve), the other in the parieto-visceral ganglion; Strumwasser (this vol.) concludes that separate cells in that ganglion (while in culture, isolated from hierarchial entrainment by other pacemakers and the external world) have slightly different circadian periods attesting to their autonomy as separate cellular oscillators. In *Drosophila persimilis* the oscillator gating the pharate adult's emergence act has a period of 24 hr, but the same insect's activity cycle is driven by an oscillator with a period of about 22 hr (Pittendrigh, 1973). There is now strong evidence that the suprachiasmatic nucleus in the hypothalamus contains the pacemaker for the circadian cycle of activity in rats (Stephan and Zucker, 1972; Moore, this volume); but the adrenals, known to be involved in the total physiology of activity, are just as clearly capable of autonomous oscillations in their steroid production (see Menaker, this volume).

Both Lobban (1960) and Aschoff (1969) report distinct frequencies in circadian rhythms freerunning concurrently in individual humans. Aschoff's cases (Figure 8) are especially clear involving the well-known rhythms of body temperature and activity/rest (sleep-wakefulness). The rhythm of body temperature is, surprisingly, not a simple consequence of changing metabolic level dependent on the activity/rest cycle: It reflects an independent

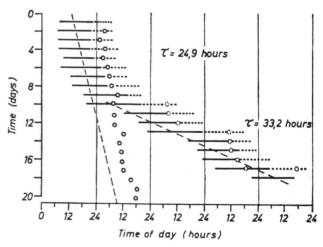

FIGURE 8 The dissociation of two circadian rhythms in a human subject enclosed in an underground bunker without time cues. Solid lines indicate activity time; dotted lines indicate rest. Circles indicate minima of body temperature, drawn twice from day 10 onward in order to indicate the true circadian period (closed circles) as well as the changes in phase relationship to the activity cycle (dotted circles). (From Aschoff, 1969.)

oscillation, presumably in the "set point" of the hypothalamic temperature regulation system. The period (τ) of the activity/rest cycle is evidently longer than that of the temperature cycle, and while they often freerun with the same frequency (due either to mutual coupling or submission to a common internal pacemaker driving them), they frequently uncouple with the result that activity periods eventually, transiently, coincide with the low point of the temperature cycle.

TEMPORAL ORGANIZATION: HIERARCHICAL AND MUTUAL ENTRAINMENT WITHIN THE SYSTEM The individual metazoan thus comprises many circadian oscillators, each of which is evidently subject to *some* frequency variation of its own. Different oscillators may, evidently, have different ranges of realizable frequency. This situation is markedly different from that in the single cell where diverse functions are apparently ordered by their coupling to a single, common pacemaker. How is temporal order maintained in a metazoan driven by multiple, frequency-labile pacemakers? It must depend, in large part, on entrainment phenomena—both hierarchical and mutual—among those constituent oscillators.

Hierarchical entrainment is unilateral: One oscillator drives another without feedback. In this respect, it is comparable to the entrainment of circadian oscillations by external periodicities. The gating of *Drosophila* eclosion has long been interpreted as involving a hierarchically coupled pair of oscillations (Pittendrigh and Bruce, 1959): One (the *A*-oscillator) is coupled to and unilaterally entrained by the light cycle; the second (*B*-oscillator), which does the actual gating, is coupled to and hierarchically entrained by the first. The phase-relation ($\psi_{B,A}$) between the oscillators, as in all hierarchical cases, depends on the ratio (τ_B/τ_A) of their free-running periods and the strength with which they are coupled which, as noted above, is temperature dependent. The effect of changing τ_B/τ_A is exemplified by this *Drosophila* system.

Selection has produced two strains in *Drosophila pseudoobscura* in which the gate of the entrained system relative to the gate of the unselected line (Stock) is *Early* and *Late*, respectively (Pittendrigh, 1967). This difference is not due to change in the light-sensitive *A*-oscillator, which maintains unchanged phase relative to the light cycle in all three strains (Figure 9). The change effected by selection is evidently on the period (τ_B) of the gating (*B*) oscillation itself: in *Early*, τ_B is shorter, and in *Late*, it is longer than τ_B of the unselected line: Change in τ_B/τ_A is responsible for the change in the gating oscillator's phase relative to the external world. Among other evidence leading to this interpretation is the confirmation of its predictions about the behavior of the system when

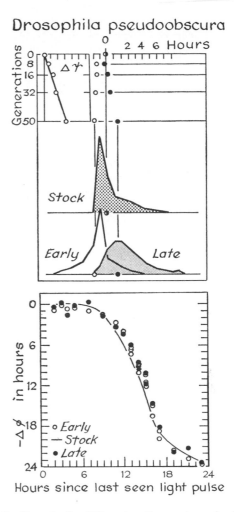

FIGURE 9 Genetically different gating systems in *D. pseudoobscura*, produced by selection through 50 generations. In *Early* the emergence peak precedes that of the unselected *Stock*; in *Late* it occurs some hours later. However the phase of the light sensitive *A*-oscillation (see text) relative to the light-cycle remains unchanged. Selection has acted on the actual gating (*B*) oscillation, probably by changing its period (τ_B). (See text and Figure 10.) (From Pittendrigh, 1967.)

the light-sensitive pacemaker is taken close to the limits of *its* entrainability by light pulses. The gating oscillator (*B*) follows its pacemaker (*A*) easily (better even than Stock) to the lowest limit because τ_B is short: On the other hand, it fails to follow the pacemaker to its upper limit—again because τ_B is short. Precisely the converse holds in *Late* (Figure 10) because τ_B has been lengthened.

This case exemplifies not only hierarchical control (unilateral entrainment) as a mechanism for the maintenance of internal timing but its potentialities for the adaptive adjustment of the phase of individual functions whose timing is entrusted to oscillators lower in the hierarchy. Comparable selection for early and late eclosion gating oscillations in *Pectinophora* left the phase of

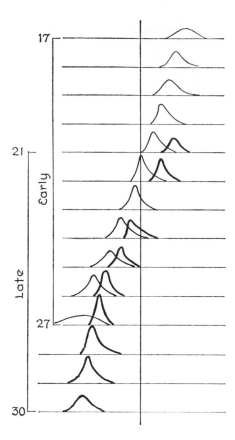

FIGURE 10 Differences in the limits of entrainment of the gating oscillator in the *Early* and *Late* strains of *D. pseudoobscura*. The vertical line marks the light pulse in cycles whose period (T) range from 17 to 30 hr. The actual peaks for *Early* (thin line), and *Late* (heavy line) are drawn. Both phase lag the light pulse when T is short and phase lead it when T is long, but *Late* always phase lags *Early*. The light-sensitive pacemaker can be driven down to $\sim T = 17$ and up to $\sim T = 30$. The gating oscillator (B) itself follows its pacemaker to the lower limit better in *Early* (because τ_B is shorter) than in *Late*; but *Late* follows it better (because τ_B is longer) to the pacemaker's upper limit. Where no peak is drawn, emergence activity is either nearly arhythmic or is driven directly by the light pulse: The gating oscillator is not entrained.

other oscillations responsible for the adult's oviposition rhythm totally unaffected (Pittendrigh and Minis, 1971).

In vertebrates, one envisages extensive hierarchical entrainment, involving both neural and humoral channels, of circadian oscillations in target organs by one or more central pacemakers in the brain. Stephan and Zucker (1972) and Moore (this volume) appear to have located one of these in the suprachiasmatic nucleus in the rat hypothalamus. Menaker (this volume) has probably identified another in the pineal of *Passer domesticus*. Pinealectomy leaves oscillators that entrain normally to a light cycle, but the birds rapidly become aperiodic in its absence. While other interpretations have still to be

excluded, it seems to me most likely that the periodicity of normal birds in constant conditions is dependent on hierarchical entrainment by the pineal of oscillations (damped or undamped) that uncouple and freerun with different frequencies in pinealectomized birds.

Hierarchical entrainment is evidently also involved in the complex of circadian oscillations in the mollusk *Aplysia*. When the eyes are removed, the animal becomes gradually aperiodic like Menaker's birds. It is also implicated by Eskin's finding (1971) that the rate at which the circadian pacemaker in *Aplysia* eyes can be reset by light is slower when all connection with the cerebral ganglion is cut.

Unilateral entrainment of rhythmicity in peripheral tissues by a central pacemaker is only part of the overall architecture of temporal order. Two or more oscillators can be coupled, with feedback, such that they mutually entrain each other. This leads to their sharing a *common* frequency, intermediate between their own freerunning frequencies. In large populations of mutually coupled oscillators, it also assures a greater *stability* of frequency (for the system) than any one of its components might itself maintain. Mutual entrainment is an important principle in the maintenance of temporal organization.

When two independent pacemakers for a circadian rhythm are bilaterally distributed in the CNS, the issue of mutual coupling is immediately raised. The lability of the period of cockroach activity cycles, even in fully controlled environments, is considerable. Were the two pacemakers located in the left and right optic lobes (Nishiitsutsuji-Uwo and Pittendrigh, 1968) to vary independently of each other, the system as a whole would rapidly lose overt periodicity. It does not, even in the course of many weeks. Frequency fluctuation is shared in common by the two optic lobes that must, therefore, be coupled and entrain each other. The same argument, à fortiori, must apply to the integrity of "single" pacemakers that are multicellular. Strumwasser and colleagues (this volume) show that the pacemaker at the base of the *Aplysia* eye probably consists of a few cells chemically synapsing with many electrotonically coupled output cells.

Mutual entrainment of cellular oscillations is the most likely explanation of the response of the hamster, illustrated in the right-hand panel of Figure 11, that was placed in constant light of high intensity (140 lux). It gradually became totally aperiodic with activity randomly scattered through the day. After weeks of such aperiodicity its activity began to "nucleate" into more concentrated packets, which in turn nucleated into a single clear rhythm. This behavior has all the marks of a population of (cellular) oscillators, initially uncoupled by their abrupt exposure to constant light, gradually again

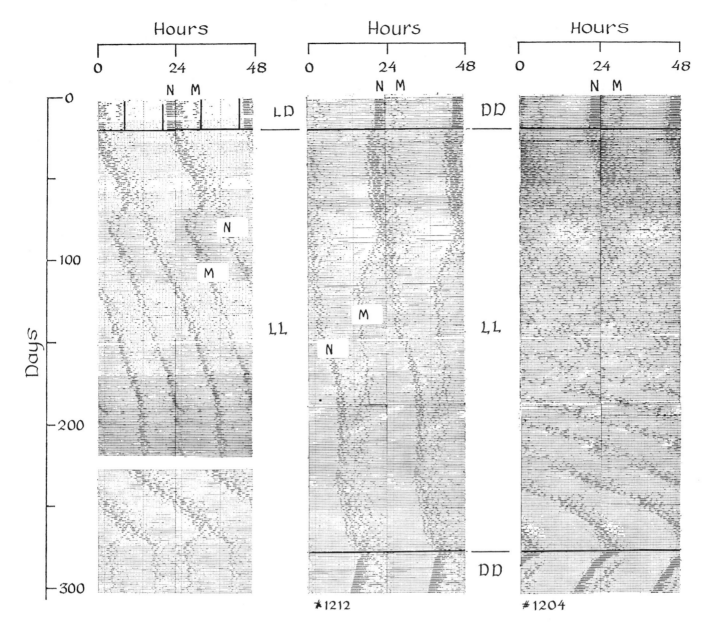

FIGURE 11 Constant light causes a dissociation of separable oscillatory components in the pacemaker responsible for the circadian activity cycle of hamsters (*Mesocricetus auratus*). Here N and M mark the two components of the total activity period (α) as discussed in the text. *Left*: after 18 days in a 24-hr light cycle (12-hr light/12-hr dark) the animal is placed in constant light (LL) of ~140 lux. The freerunning period is longer than 24 hr. On the 43rd day in LL, it is clear that the M component is freerunning at a higher frequency than N; it continues to free-run for 17 days when it assumes a new steady-state phase (actually 180° antiphase) relative to M. The two coupled components, in their 180° antiphase relation, freerun indefinitely thereafter. In the lower left panel, the same animal, having been re-entrained by a light cycle is again put into LL. Again the M component eventually splits and freeruns at a high frequency until it reaches the 180° antiphase with N when both components now share an intermediate frequency. *Middle*: In this case M continues its freerun, fails to lock onto N in the 180° antiphase relation, and ultimately resumes its normal phase relative to N. *Right*: A third hamster becomes totally aperiodic in LL after 30 days. Subsequently a very long period rhythm "nucleates" out of the noise (see text).

locking onto each other until the entire cellular population becomes synchronous. Winfree (1967) has made valuable computer simulations of such phenomena.

When humans are screened from external entraining agents, their circadian rhythms of body temperature and activity share a common τ (~ 25 hr) in all cases for many cycles. It is in only about 16% of cases (Aschoff, personal communication) that the two oscillators dissociate and reveal their intrinsically different frequencies. In other words, even without the aid of a common *external* driver, they show an identical frequency. This could be explained either by their mutual entrainment *or* by their submission to a common (hierarchical) pacemaker. The capacity of these two autonomous oscillators to couple is further demonstrated by the observations of both Aschoff (Aschoff et al., 1967) and Jouvet (Jouvet et al., this volume). The oscillator responsible for the activity cycle (whose τ is the larger) may—after an initial dissociation—again lock on to the body-temperature oscillator, but at half its frequency: Both authors report men in isolation showing ~ 50 hr activity cycles and a clear ongoing temperature cycle of ~ 25 hr. This is a case of frequency demultiplication, well known as a formal property of the entrainability of self-sustaining oscillations (Pittendrigh and Bruce, 1957).

SEPARATE OSCILLATORS COUPLED TO DAWN AND SUNSET, AND TO EACH OTHER   Many animals tend to be bimodal in their activity pattern. In hamsters, there is a major burst of activity (N in Figure 10) in the early night and a small burst (M) just before morning. In constant darkness, these bursts remain closely associated, constituting what Aschoff calls α—the total fraction of the cycle devoted to activity; the fraction devoted to rest is designated ρ; and $\tau = \alpha + \rho$. The period (τ) of hamsters in constant light lengthens slightly and the α/ρ ratio decreases (typical of nocturnal rodents). In some individuals, this pattern persists indefinitely. In the majority of cases, however, the activity period (α) splits into two distinct components (Figure 10) that *always*, in steady state, lie 180° apart. Sometimes the dissociation of these components is so gradual that one can trace a steady phase advance of the M component from its original close phase relation with N to a new phase relation 180° from N. It is a rule, without exception in over 15 hamsters analyzed, that the period (τ) of the "split" system is different from that of the unsplit system at any given light intensity. Sometimes (Figure 10) the uncoupled M component fails to lock on in its 180° antiphase and continues its transient freerun until it regains its original phase relative to M (Pittendrigh, 1967; and Figure 11, middle).

This splitting phenomenon is not unique to *Mesocricetus*. I noted it originally (Pittendrigh, 1960) in Swade's (1963) data for the diurnal ground squirrel *Spermophilus* where one component of α uncoupled and freeran until it again regained its usual phase with the rest. His data also show that an abrupt transfer from light to constant darkness induces a clear dissociation of two components (with stable 180° antiphase) in the squirrel *Eutamias*. Uncoupling of two components is also evident in Pohl's (1972) data for yet another squirrel (*Funambulus*), and we find it in deermice and cockroaches in our laboratory. But quite the clearest and most elegantly demonstrated confirmation of the hamster behavior comes from Hoffmann's (1971) study of *Tupaia*. Here the uncoupling of two components is induced by reducing light intensity and the new steady state always finds them in 180° antiphase. Indeed it was Hoffmann who first drew attention to this striking aspect of the split system. He, too, finds τ changes when the system splits.

The following propositions account for the "splitting" phenomena and, in addition, some important aspects of entrainment not previously reported. (1) The circadian pacemaker for the activity cycle comprises two separable oscillators, one responsible for the N component of α and the other for its M component. α (the total period of activity) is governed by the phase relation ($\psi_{M,N}$) between M and N. (2) τ_M (the freerunning period of M) is a positive function and τ_N is a negative function of light intensity. (3) When τ_M and τ_N are close enough (i.e., within a limited range of light intensities), they can *mutually* entrain each other and (as a coupled system) share a frequency between those that each oscillator would express were it freerunning at that intensity. Thus within this range, the system's frequency is intensity compensated, and consequently their mutual coupling constitutes a homeostatic device (Pittendrigh, 1958, p. 254). (4) Within this range, as τ of the coupled system changes slightly, the phase relation of the coupled oscillators ($\psi_{M,N}$) will necessarily change, because it is a function of τ_M/τ_N; thus α (which is equal to $\psi_{M,N}$) changes as τ changes. (5) Light intensity can, however, be raised to a level where *mutual* coupling fails and one of the oscillators (N in the hamster case) transiently freeruns at a very high frequency until it couples with M, showing the maximum phase lead (180°) possible in a coupled state. (6) The N-oscillator can only be phase-delayed by light pulses and is maximally responsive to them at or near the beginning of the (coupled) system's subjective night; M can only be phase-advanced by light falling at or near dawn. The observed phase-response curve for short light pulses (which shows both advances and delays) thus reflects the response of a system of two mutually coupled driving oscillators: Single pulses cause advances or delays depending on which oscillator is maximally affected at the time of the pulse.

This approach, stimulated by the splitting of the free-running system, is supported by peculiarities of its entrainment by the same short (15 min) pulses used to obtain its phase-response curve. Hamsters do entrain to cycles involving these short light pulses, and their general response is what theory demands: When $T > \tau$, the light pulse, in the steady state, falls at the beginning of the subjective night causing the required phase delays; when $T < \tau$, it falls at the end of the subjective night causing the necessary phase advances. Entrainment appears always to involve some *compression of* α—a reduction in the total duration of activity; and the present model accounts for, indeed predicts, this as a consequence of the repeated advance or retardation of one of the two constituent oscillators. The compression of α is most compellingly clear when the system is pushed to the limits of entrainability. Figure 12 shows animals exposed to cycles of $T = 23$ and $T = 25$ hr. In the majority of animals, entrainment fails at these T values, but only gradually over many cycles during which α gets progressively compressed as though ($T = 23$) the M component was fully entrained but progressively fails to drive N, which gets steadily later with α being eventually compressed almost to vanishing point. Eventually M itself is affected by the light, and its delay response abuptly carries the system beyond the point where light affects it and α promptly decompresses back to the long duration typical of the freerun in constant darkness (DD).

The compression of α (and hence change in α/ρ) is effected in cycles with $T = 24$ hr if two pulses are employed in each cycle, constituting skeleton photoperiods. When the interval between the two pulses is gradually reduced, α becomes steadily compressed by them: The model sees the N oscillator delayed by one pulse and the M oscillator advanced by the second. The activity time (α) has been reduced in this way down to 2 hr in some hamsters and deermice. This result on the one hand eludes explanation in terms of the level-threshold entrainment model of the Erling-Andechs school (e.g. Wever, 1962), because the total amount of light remains the same in each cycle, and on the other hand is just what the two-oscillator model for the driving system would predict.

Apart from their intrinsic interest, the phenomena summarized in this section have relevance to the problems of photoperiodic induction discussed below in the section on Photoperiodism.

Loss of Temporal Organization and the "Resonance" Phenomenon   The nexus of internal couplings is not always sufficient to maintain normal internal timing that is probably dependent, to greater or less extent, on control by external entraining cycles. This is especially clear when multicellular organisms are placed in constant light,

which always either damps the oscillation completely or changes τ. And it may well change it differentially in different oscillators as the previous section suggested. Significant physiological penalty follows exposure to constant light in the few cases where it has been deliberately looked for. It has been found in several plants (see Pittendrigh and Bruce, 1959, for review) and in *Drosophila melanogaster* (Pittendrigh and Minis, 1972). Its interpretation as the consequence of losing internal circadian organization is supported by Hillman's (1956) important observation that damage to plants can be avoided or repaired by superimposing (as an adequate *Zeitgeber*) a 24 hr temperature cycle on the constant illumination, which is otherwise so deleterious.

Went (1959) made the pioneering observation that their growth is optimal (in rate and amount) when plants are raised in light cycles whose period (T) is close to 24 hr and is significantly impaired when it is far from 24 hr. Pittendrigh's initial indication (1960) that comparable phenomena occurred in animal systems has since been extended by Pittendrigh and Minis (1972) and Aschoff (unpublished, personal communication) who show that the longevity of *Drosophila* and *Phormia*, respectively, is significantly impaired when the flies are maintained on 21-hr or 27-hr days; longevity is greatest at (or near) 24 hr (Figure 13). In other words, *the organism functions most effectively when, as an innately oscillatory system, it is driven close to its "natural frequency"—when the relation between organism and environment is close to resonance.* Saunders (1972) now reports comparable facts indicating that the growth rate of *Sarcophaga* larvae is a function of the relationship to T and τ.

Although these facts are still few and present knowledge is clearly inadequate for detailed explanation, a general qualitative conclusion seems inescapable: The impairment of growth and longevity must be attributed to disruption of the organism's "normal" internal temporal order. It is at least likely that this loss of normal organization is, more specifically, loss of normal phase relationships among constituents of the multioscillator system. Differing in their freerunning frequencies and the strength with which they are coupled—hierarchically and mutually, their phase relations *must* change when the system as a whole is driven at different frequencies. Their phase relations when $T = 24$ hr are those selection has produced as "normal" and they are lost when the system is driven, for example, at $T = 21$ or $T = 27$ (Pittendrigh and Minis, 1972): The system operates most effectively when it is close to resonance. Such disruption is, incidentally, predictable (and known to occur in man) following abrupt, major phase shifts of the system. Unless each constituent oscillator is coupled with identical strength to the entraining agent, they will regain their steady-state

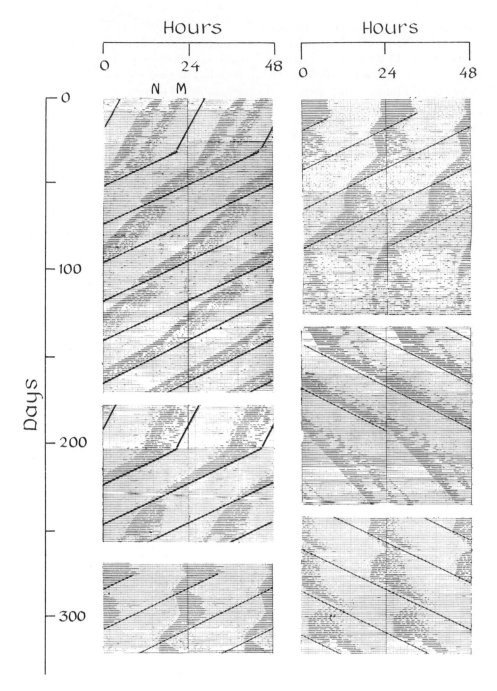

FIGURE 12 The compression of α (the activity period comprising *M* and *N* components) in hamsters driven to the limits of their entrainability. *Upper left*: The animal shows distinct *M* and *N* components and entrains successfully for 30 cycles to $T = 23.75$ hr. The light pulse in this and all other cycles used in the figure was of 15 min duration; its position on each successive day is indicated by the ink lines. On the 31st day, *T* is shortened to 23.000 hr. Then *M* continues to entrain to the light, but *N* has a longer period; α is consequently compressed almost to vanishing point; then the system "breaks free" and α promptly decompresses. The system again fails to entrain on the 110th day. Later it does achieve a steady state in which α is much shorter than when it was entrained to $T = 23.75$. *Middle left*: Initial entrainment at $T = 23.75$ followed by failure to entrain to $T = 23$. Again α compresses as entrainment fails, and decompresses when *M* is freed from the light. *Lower left*: Repeated failure to entrain to $T = 23$, with repeated compression and decompression of α. *Upper right*: The same. Note the marked decompression of α when the animal is released into darkness and that τ is much shorter in this freerun following release from $T = 23$ than in the freerun following release from $T = 25$ in the middle right panel. This is a case of "after-effects." (See text.) In the *middle panel* α is compressed while the animal was entrained to $T = 25$. *Lower right*: Repeated failure to entrain to $T = 25$ hr with an associated compression and decompression of α: In this case *N* follows the light well but fails to drive *M*.

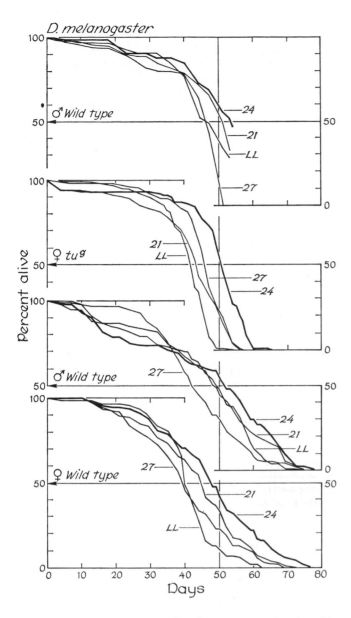

FIGURE 13 The "resonance" phenomenon. Survivorship curves for *D. melanogaster* maintained in continuous light (LL) or on 21, 24, and 27 hr days (light on for 50% of each cycle). The flies live longest on 24 hr days (Pittendrigh and Minis, 1972).

phase relative to it—*and hence to each other*—at different rates, thus generating a transient disruption of internal temporal organization. This is the probable explanation of the "stress" associated with air travel across many time zones. Longevity, a more easily quantified measure of abnormal function, again attests to the penalty paid for even these transient disruptions of circadian organization: Aschoff and colleagues (1971) have shown that oft-repeated phase shifts significantly reduce the longevity of *Phormia*.

Photoperiodism

CIRCADIAN ORGANIZATION AND PHOTOPERIODIC INDUCTION The control of seasonally appropriate change in metabolic pathways is commonly effected by the relative duration of light and darkness each day. The "photoperiod" (day length) is the environment's most noise-free marker of the time of year. In animals as diverse as insects, birds and mammals day length is the initial determinant in the causal chain leading to reproductive activity; it determines when birds moult and when they migrate; it is the photoperiod, again, that determines whether insect development proceeds to completion or stops in a "diapause" that is strategically timed in advance of unfavorable conditions; and in insects it is also known to exert seasonal change in temperature tolerance (Pittendrigh, 1961) and growth rate (Saunders, 1972). In a few insects (Beck, 1962; Lees, 1971, and earlier; Pittendrigh and Minis, 1971) the evidence indicates that the time measurement involved (how *long* is the photoperiod?) is effected by a "clock" that lacks periodicity—it has the formal properties of an hour glass that needs daily recharging by the light and actually measures the duration of darkness. But in the vast majority of cases analyzed, the day length (or night length) dictates seasonal change in metabolic state via its effect on the circadian oscillations internal to the organism concerned. How it does so, concretely, is not known: What we do know is almost entirely formal, and current discussions of the problem are necessarily abstract. However, the mechanisms in animals are neural and the neurophysiologist cannot ignore the challenge. We must in the meantime proceed with what Erich von Holst called "eine niveau adequate Terminologie" (a conceptual scheme adequate to the organizational level being analyzed) before the issues, sufficiently elucidated, can be reduced to the level of concreteness the neurophysiologist usually enjoys.

When (nearly always) circadian organization *is* involved in photoperiodic induction, it may play a diversity of roles. The only secure generalization we can make (Pittendrigh, 1972) is that photoperiodic induction (of metabolic change) is a function of the circadian system's entrained steady state: It is maximal in only a few entrained states, and these are all when the system is close to resonance. I have recently outlined three distinct possibilities encompassed by that generalization. They are not mutually exclusive, and some may be involved in one organism but not in another. My concern here is to relate them to the multioscillator nature of circadian systems.

THE RESONANCE EFFECT; CIRCADIAN SURFACES One of the few cases where a nonperiodic clock (what the literature calls an "hour glass") appears responsible for the time measurement we are concerned with is the photoperiodic induction of developmental arrest (diapause) in the moth *Ostrinia nubilalis*. Beck's (1962) data show that no matter what the duration of light in an exotic ($T \neq 24$) cycle, induction of diapause in *Ostrinia* is most successful when the dark duration is 12 hr: He concludes that the crucial measurement is that of darkness alone and is executed by an "hour glass." However, as noted elsewhere (Pittendrigh, 1966 and 1972), further analysis of his data indicates that circadian organization is involved, even if it is not responsible for the time measurement as such. When the amount of the induced effect (diapause) is plotted on coordinates indicating the duration of both light and dark, a striking result emerges: Any given level of induction can be effected by an infinite number of L/D cycles which form a closed loop—an *isoinduction contour*—on the L versus D coordinates. Moreover, the isoinduction contours for different levels of induction describe a "circadian surface" that peaks when the sum of $L + D$ ($= T$) is close to 24 hr and *the circadian system is therefore close to resonance*. Evidently the success with which the moth utilizes its information on night length is modulated—like growth rate and longevity in other systems—by its proximity to resonance (Figure 14).

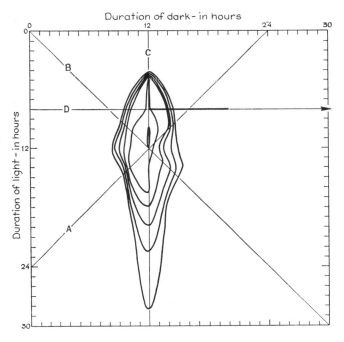

FIGURE 14 A circadian surface based on the induction of diapause in the moth *Ostrinia nubilalis*. Six iso-induction contours define the light cycles that induce 60, 70, 80, 90, 95, and 100% diapause. (From Pittendrigh, 1972, based on Beck's data, 1962.)

INTERNAL COINCIDENCE VS. EXTERNAL COINCIDENCE   In its two other possible involvements, the circadian system functions as the clock itself. They demand fundamentally

different models, for which the terms "external coincidence" and "internal coincidence" seem appropriate (Pittendrigh, 1972). *External coincidence* is what Bünning proposed in his now classic paper of 1936: Photoperiodic induction only occurs when a specific phase point in the circadian cycle of the organism is coincident in time with light—a specific phase in an external cycle. The resultant photochemical reaction is the initial, necessary step in the inductive process. This model has been discussed in detail by Pittendrigh and Minis (1964, 1971) and Pittendrigh (1966) in relation to the phenomena of entrainment by light cycles.

Essential features of the *internal coincidence* model were outlined in 1960 when I first developed the theme of multicellulars as populations of coupled oscillators. The proposal has been left undeveloped, however, until the recent work of Tyschenko and colleagues in Russia (see Danilevsky et al., 1970, for a summary in English). The proposition is that some oscillations in the system are coupled to light (or the light-off step) at sunset, and some to light (or the light-on step) at dawn. The phase relations of such oscillations will obviously change as the day length changes, and specific phase points in the two internal oscillations will coincide, in time, only when the system is entrained by certain photoperiods. Were short-lived reactants present only at those phase points, the system would constitute an internal-coincidence device, restricting occurrence of the initial step in photoperiodic induction to a limited array of phase relations between the two oscillations—that is, to a limited array of photoperiods. One reason why this model has received little attention has been lack of evidence that separate oscillators are in fact driven by dawn and sunset. The important studies of Takimoto and Hamner (1964) on the plant *Pharbitis nil* have yielded the only indication of their presence prior to the evidence presented for animals above in the section on Separate Oscillators Coupled to Dawn and Sunset.

THE UTILITY OF EXTENDED CIRCADIAN SURFACES   The circadian surface of Figure 14 implies that the utilization of an (apparently) hour glass time measurement is affected by the system's proximity to circadian resonance. This aspect of circadian organization cannot, then, be excluded in cases where the time measurement itself is effected by the circadian system. For example, it complicates the standard interpretation of the usual outcome of experiments employing the most widely used protocol (introduced by Nanda and Hamner in 1958) intended to test the external coincidence model. The technique is to assay the amount of some photoperiodically induced effect when, holding a given light duration fixed, the length of the cycle is extended by extension of the dark period. The now common result, found in plants, birds, mammals, and insects, is the occurrence of maximal induction by light cycles that are modulo-τ. This has always been interpreted in terms of "external coincidence" —as reflecting the recurrence (modulo-τ) through the long dark period of a photoinducible phase in the circadian cycle. However, the occurrence of inductive maxima under cycles that are modulo-τ is expected—because of the optimization of performance at circadian resonance— whether external *or* internal coincidence is responsible for the recognition of photoperiod.

The circadian surface developed for Beck's data suggested to me (Pittendrigh, 1972) that extended circadian topographies, in which the *D* coordinate is extended well beyond 24 hr, might help to unconfound the several possible roles that circadian organization can, in principle, play in photoperiodic induction. For example, one expects: (1) that for Beck's case—and other true "hour glasses" measuring dark duration—such a topography would show no extra peaks due to cycles whose periods (*T*) are modulo-τ; (2) that, if *external coincidence* is involved in the photoperiodic time measurement, second and third peaks should, however, appear on the surface but with a precise shape that is not immediately predictable; and (3) that, if *internal coincidence* effects the time measurement, second and third peaks should also be found on the surface with a predictable topography: One expects a triangular peak shape reflecting the changing phase angle between dawn and sunset.

Saunders (1972, 1973a, 1973b) has pursued this idea in our laboratory using the insects *Nasonia* and *Sarcophaga*. The *Nasonia* case, Figure 15, is indeed instructive: (1) The multiple peaks immediately implicate circadian organization in the time measurement as such; and (2) their shape is precisely what is expected if the phase

relationship of oscillators separately coupled to dawn and sunset is indeed responsible for the measurement of photoperiod. Here is the strongest evidence we have so far that photoperiodic effects on metabolism are due to seasonal change in the mutual phase relationship of constituent oscillations in the circadian system.

When, as in this *Nasonia* case, *internal coincidence* is implicated as the basis of sensing the duration between dawn and dusk, the only role of light is to entrain the circadian system; it lacks the dual role (Pittendrigh and Minis, 1964) that is so fundamental to the *external coincidence* model. It follows, therefore, as I have recently pointed out (Pittendrigh, 1972), that in *Nasonia* what we call "photoperiodic" induction should be possible *without the use of light* at all, because all known circadian oscillations are entrainable by temperature cycles: It should be possible to establish the critical phase relations between constituent oscillators within the system by varying "thermoperiod" as readily as by varying "photoperiod." This prediction, in fact, constitutes the most rigorous test open to us to discriminate between external- and internal-coincidence as the mechanism whereby circadian systems function as "clocks" to recognize "day length" and hence season.

PHOTOPERIODISM AS A NEUROPHYSIOLOGICAL PROBLEM
Photoperiodic induction is ripe for more attention from the neurophysiologist than it has so far received. There are two reasons. First, it seems likely (at least to this writer) that we are about to find many ways in which the *general* metabolic state of animals is affected by photoperiod: That the all-or-none effects currently known to be controlled by photoperiod have received attention largely because they are so easily assayed and quantified. The fact that temperature tolerance (Pittendrigh, 1961) and larval growth rates (Saunders, 1972) are photoperiodically modulated is a hint at the broader spectrum of phenomena probably involved. In short, photoperiodism is probably a much larger chapter in the general control of metabolism by the nervous system than we currently realize. Second, the time is rapidly passing when the analysis is necessarily restricted to the purely formal analysis exemplified by inferences from circadian topographies, useful as it has been. The mechanism of photoperiodic induction in animals is neural: Williams and Adkisson (1964) have shown convincingly that the photoperiodic clock in silkmoth larvae is in the brain; and if, as one cannot seriously doubt, that is also true of other animals, this means that the concrete basis of induction involves the phase relations of circadian pacemakers in the brain. These are now being identified in several animals, and the availability of long-term recording techniques makes their direct neurophysiological assay a realistic prospect in the immediate future.

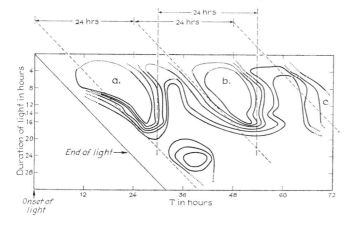

FIGURE 15 An extended circadian surface, on which contours define the light cycles that will induce 5, 10, 30, 50, 70, and 90% diapause in the wasp *Nasonia vitripennis*. (From Saunders, 1973a.)

Summary

1. The study of circadian oscillations is a relatively young field whose development, especially since 1950, is briefly outlined.

2. The general problems are formulated: (a) What is the concrete (physical) nature of circadian oscillations? (b) How are they coupled to the diversity of cellular events they time? (c) How are they coupled to (entrained by) environmental cycles? (d) What emergent properties characterize multicellular systems as populations of coupled oscillators?

3. The formal and concrete properties of freerunning circadian oscillations are reviewed. The homeostatic conservation of their frequency is one of their major features. It is also a major physiological challenge and, until understood, a major barrier to the evaluation of chemical and genetic attacks on the nature of the oscillator.

4. The mechanism of entrainment of circadian oscillations by light cycles is analyzed; other entrainment phenomena are briefly reviewed.

5. Multicellular systems (including nervous systems) constitute a population of circadian oscillations with slightly labile and slightly different frequencies. How they maintain synchrony and appropriate mutual phase relationships is discussed in general terms: It is an active field of inquiry.

6. Evidence is presented that loss of internal temporal order among circadian periodicites entails physiological penalty. Organisms as innately oscillatory systems function most effectively when they are close to "resonance," i.e., when they are driven by environmental cycles whose period (T) is close to (or modulo) τ, which is the organism's "natural" frequency.

7. The phenomena of photoperiodic induction in animals are shown to be an aspect of the circadian organization of the nervous system.

ACKNOWLEDGMENTS This paper is dedicated to my friend, Professor Jürgen Aschoff, on the occasion of his sixtieth birthday, more than twenty years since the beginning of his great and continuing contribution to present knowledge of circadian phenomena.

The preparation of this paper and the previously unpublished experiments it reports were supported with funds from several sources: The Whitehall Foundation, NASA, and NIMH.

REFERENCES

ASCHOFF, J., 1960. Exogenous and endogenous components in circadian rhythms. *Cold Spring Harbor Symp. Quant. Biol.* 25: 11–28.

ASCHOFF, J., 1969. Desynchronization and resynchronization of human circadian rhythms. *Aerospace Med.* 40:844–849.

ASCHOFF, J., U. GERECKE, and R. WEVER, 1967. Desynchroni-zation of human circadian rhythms. *Jap. J. Physiol.* 17:450–457.

ASCHOFF, J., E. v. SAINT-PAUL, and R. WEVER, 1971. Die Lebensdauer von Fliegen unter dem Einfluss von Zeit-Verschiebungen. *Naturwiss.* 58:574.

BECK, S. D., 1962. Photoperiodic induction of diapause in an insect. *Biol. Bull.* 122:1–12.

BRINKMANN, K., 1971. Metabolic control of temperature compensation in the circadian rhythm of *Euglena gracilis*. In *Biochronometry*, M. Menaker, ed. Washington, D.C.: National Academy of Sciences, pp. 567–593.

BRUCE, V. G., 1960. Environmental entrainment of circadian rhythms. *Cold Spring Harbor Symp. Quant. Biol.* 25:29–48.

BRUCE, V. G., 1972. Mutants of the biological clock in *Chlamydomonas reinhardi*. *Genetics* 70:537–548.

BRUCE, V. G., and D. H. MINIS, 1969. Circadian clock action spectrum in a photoperiodic moth. *Science* 163:583–585.

BRUCE, V. G., and C. S. PITTENDRIGH, 1956. Temperature independence in a unicellular "clock". *Proc. Natl. Acad. Sci. USA* 42:676–682.

BÜHNEMANN, F., 1955. Das endodiurnale system der Ocdogoniumzelle. *Bio. Zentralblatt.* 74:691–705.

BÜNNING, E., 1935. Zur Kenntnis der erblichen Tagesperiodizität bei den Primärblättern von *Phaseolus multiflorus*. *Jahrb. wiss. Botan.* 81:411–418.

BÜNNING, E., 1936. Die endogene Tagesrhythmik als Grundlage der photoperiodischen Reaktion. *Ber. Dtsch. Bot. Ges.* 54: 590–607.

BÜNNING, E., and I. MOSER, 1972. Influence of valinomycin on circadian leaf movements of *Phaseolus*. *Proc. Natl. Acad. Sci. USA* 69:2732–2733.

DANILEVSKY, A. S., N. I. GORYSHIN, and V. P. TYSCHENKO, 1970. Biological rhythms in terrestrial arthropods. In *Annual Review of Entomology*, R. F. Smith and T. E. Mittler, eds. Palo Alto, Calif.: Annual Reviews, pp. 201–244.

EHRET, C. F., 1959. Photobiology and biochemistry of circadian rhythms in non-photosynthesizing cells. *Fed. Proc.* 18:1232–1240.

ENGELMANN, W., 1972. Lithium slows down the *Kalanchoe* clock. *Z. Naturforsch.* 27b:477.

ENRIGHT, J. T., 1971. Heavy water slows biological timing processes. *Z. Vergl. Physiol.* 72:1–16.

ESKIN, A., 1971. Properties of the *Aplysia* visual system: In vitro entrainment of the circadian rhythm and centrifugal regulation of the eye. *Z. Vergl. Physiol.* 74:353–371.

FELDMAN, J. F., 1967. Lengthening the period of a biological clock in *Euglena* by cycloheximide, an inhibitor of protein synthesis. *Proc. Natl. Acad. Sci. USA* 57:1080–1087.

FELDMAN, J. F., and N. M. WASER, 1971. New mutations affecting circadian rhythmicity in *Neurospora*. In *Biochronometry*, M. Menaker, ed. Washington, D.C.: National Academy of Sciences, pp. 652–656.

FRANK, K. D., and W. F. ZIMMERMAN, 1969. Action spectra for phase shifts of a circadian rhythm in *Drosophila*. *Science* 163: 688–689.

FRISCH, K. VON., 1950. Die Sonne als Kompass im Leben der Bienen. *Experientia* 6:210–221.

HALBERG, F., 1960. Temporal coordination of physiologic function. *Cold Spring Harbor Symp. Quant. Biol.* 25:289–310.

HASTINGS, J. W., and B. M. SWEENEY, 1958. A persistent diurnal rhythm of luminescence in *Gonyaulax polyedra*. *Biol. Bull.* 115: 440–458.

HAWKING, F., 1962. Microfilaria infestation as an instance of periodic phenomena seen in host-parasite relationships. *Ann. N.Y. Acad. Sci.* 98:940–953.

HIGHKIN, H. R., and J. B. HANSON, 1954. Possible interaction between light-dark cycles and endogenous daily rhythms on the growth of tomato plants. *Plant Physiol.* 29:301–302.

HILLMAN, W. S., 1956. Injury of tomato plants by continuous light and unfavorable photoperiodic cycles. *Am. J. Bot.* 43:89–96.

HOFFMANN, K., 1960. Experimental manipulation of the orientational clock in birds. *Cold Spring Harbor Symp. Quant. Biol.* 25:379–384.

HOFFMANN, K., 1971. Splitting of the circadian rhythm as a function of light intensity. In *Biochronometry*, M. Menaker, ed. Washington, D.C.: National Academy of Science, pp. 134–150.

KALMUS, H., 1935. Periodizität und Autochronie als zeitregelnde Eigenschaften der Organismen. *Biol. Generalis*. 11:93–114.

KATZ, J. J., and H. L. CRESPI, 1972. Isotope effects in biological systems. In *Isotope Effects in Reaction Rates*, C. J. Collins and N. S. Bowman, eds. New York: Van Nostrand-Reinhold Book Co., Chap. 5, pp. 286–363.

KONOPKA, R., and S. BENZER, 1971. Clock mutants of *Drosophila melanogaster*. *Proc. Natl. Acad. Sci. USA* 68:2112–2116.

KRAMER, G., 1950. Orientierte Zugaktivität gekäfigter Singvogel. *Naturwissenschaften*, 37:188.

KRAMER, G., 1950. Weitere Analyse der Faktoren, welche die Zugaktivität des gekäfigten Vogels orientieren. *Naturwissenschaften*, 37:377–378.

LEES, A. D., 1971. The relevance of action spectra in the study of insect photoperiodism. In *Biochronometry*, M. Menaker, ed. Washington, D.C.: National Academy of Sciences, pp. 372–380.

LOBBAN, M. C., 1960. The entrainment of circadian rhythms in man. *Cold Spring. Harbor Symp. Quant. Biol.* 25:325–332.

McMURRY, L., and J. W. HASTINGS, 1972. No desynchronization among four circadian rhythms in the unicellular alga, *Gonyaulax polyedra*. *Science* 175:1137–1138.

MINIS, D. H., and C. S. PITTENDRIGH, 1968. Circadian oscillation controlling hatching: Its ontogeny during embryogensis of a moth. *Science* 159:534–536.

NANDA, K. K., and K. C. HAMNER, 1958. Studies on the nature of the endogenous rhythm affecting photoperiodic response of Biloxi soybean. *Bot. Gaz.* (*Chicago*) 120:14–25.

NISHIITSUTSUJI-UWO, J., and C. S. PITTENDRIGH, 1968. Central nervous system control of circadian rhythmicity in the cockroach. III. The optic lobes, locus of the driving oscillation? *Z. Vergl. Physiol.* 58:14–46.

OTTESEŃ E., 1965. Analytical studies on a model for the entrainment of circadian systems. Undergraduate thesis. New Jersey: Princeton University.

PITTENDRIGH, C. S., 1954. On temperature independence in the clock system controlling emergence time in *Drosophila*. *Proc. Natl. Acad. Sci. USA* 40:1018–1029.

PITTENDRIGH, C. S., 1958. Perspectives in the study of biological clocks. In *Symposium on Perspectives In Marine Biology*, A. A. Buzzati-Traverso, ed. Berkeley: University of California Press, pp. 239–268.

PITTENDRIGH, C. S., 1960. Circadian rhythms and the circadian organization of living systems. *Cold Spring Harbor Symp. Quant. Biol.* 25:159–184.

PITTENDRIGH, C. S., 1961. On temporal organization in living systems. *Harvey Lectures Series*. New York: Academic Press, pp. 93–125.

PITTENDRIGH, C. S., 1966. The circadian oscillation in *Drosophila pseudoobscura* pupae: A model for the photoperiodic clock. *Z. Pflanzenphysiol*. 54:275–307.

PITTENDRIGH, C. S., 1967. Circadian rhythms, space research, and manned space flight. In *Life Sciences and Space Research V*. Amsterdam: North Holland Publishing Co., pp. 122–134.

PITTENDRIGH, C. S., 1967. Circadian systems, I. The driving oscillation and its assay in *Drosophila pseudoobscura*. *Proc. Natl. Acad. Sci. USA* 58:1762–1767.

PITTENDRIGH, C. S., 1972. Circadian surfaces and the diversity of possible roles of circadian organization in photoperiodic induction. *Proc. Natl. Acad. Sci. USA* 69:2734–2737.

PITTENDRIGH, C. S., 1973. Circadian activity rhythms in adult *Drosophila* species (in preparation).

PITTENDRIGH, C. S., and V. G. BRUCE, 1957. An oscillator model for biological clocks. In *Rhythmic and Synthetic Processes in Growth*, D. Rudnick, ed. Princeton, New Jersey: Princeton University Press, pp. 75–109.

PITTENDRIGH, C. S., and V. G. BRUCE, 1959. Daily rhythms as coupled oscillator systems and their relation to thermoperiodism and photoperiodism. In *Photoperiodism and Related Phenomena in Plants and Animals*, R. B. Withrow, ed. Washington, D.C.: American Association for the Advancement of Science, pp. 475–505.

PITTENDRIGH, C. S., and P. C. CALDAROLA, 1973. On the general homeostasis of the frequency of circadian oscillations. *Proc. Natl. Acad. Sci. USA* (in press).

PITTENDRIGH, C. S., P. C. CALDAROLA, and E. S. COSBEY, 1973. A differential effect of heavy water on temperature-dependent and temperature-compensated aspects in the *Drosophila pseudoobscura* circadian system. *Proc. Natl. Acad. Sci. USA* (in press).

PITTENDRIGH, C. S., and S. DAAN, 1973. On a relationship between the frequency and the phase-response curves of circadian oscillations. In preparation.

PITTENDRIGH, C. S., and D. H. MINIS, 1964. The entrainment of circadian oscillations by light and their role as photoperiodic clocks. *American Nat.* 98:261–294.

PITTENDRIGH, C. S., and D. H. MINIS, 1971. The photoperiodic time measurement in *Pectinophora gossypiella* and its relation to the circadian system in that species. In *Biochronometry*, M. Menaker, ed. Washington, D.C.: National Academy of Sciences, pp. 212–250.

PITTENDRIGH, C. S., and D. H. MINIS, 1972. Circadian systems: Longevity as a function of circadian resonance in *Drosophila melanogaster*. *Proc. Natl. Acad. Sci. USA* 69:1537–1539.

POHL, H., 1972. Die Aktivitätsperiodik von zwei tagaktiven Nagern, *Funambulus palmarum* und *Eutamius sibiricus* unter Dauerlichtbedingungen. *J. Comp. Physiol.* 78:60–74.

ROTHMAN, B., and F. STRUMWASSER, 1973. Aflatoxin B1 blocks the circadian rhythm in the isolated eye of *Aplysia*, *Fed. Proc.* 32: (in press).

SARGENT, M. L., and W. R. BRIGGS, 1967. The effects of light on a circadian rhythm of conidiation in *Neurospora*. *Plant. Physiol.* 42:1504–1510.

SAUNDERS, D. S., 1972. Circadian control of larval growth rate in *Sarcophaga argyrostoma*. *Proc. Natl. Acad. Sci. USA* 69:2738–2740.

SAUNDERS, D. S., 1973a. Temporal organization in insects. II. Evidence for "dawn" and "dusk" oscillators in the *Nasonia* photoperiodic clock. In preparation.

SAUNDERS, D. S., 1973b. Temporal organization in insects. III. The photoperiodic clock in the flesh-fly, *Sarcophaga argyrostoma*. In preparation.

SCHWEIGER, E., H. WALLRAFF, and H. SCHWEIGER, 1964. Endogenous circadian rhythm in cytoplasm of *Acetabularia*: Influence of the nucleus. *Science* 146:658–659.

SKOPIK, S. D., and C. S. PITTENDRIGH, 1967. Circadian systems, II. The oscillation in the individual *Drosophila* pupa; its independence of developmental stage. *Proc. Natl. Acad. Sci. USA* 58:1862–1869.

STEIN-BELING, I., 1935. Über das Zeitgedächtnis bei Tieren. *Biol. Revs.* 10:18–41.

STEPHAN, F. K., and I. ZUCKER, 1972. Circadian rhythms in drinking behavior and locomotor activity of rats are eliminated by hypothalamic lesions. *Proc. Natl. Acad. Sci. USA* 69:1583–1586.

SULZMANN, F. M., and L. N. EDMUNDS, 1972. Persisting circadian oscillations in enzyme activity in non-dividing cultures of *Euglena. Biochem. Biophys. Res. Commun.* 47:1338–1344.

SWADE, R. H., 1963. Circadian Rhythms in the Arctic. Ph.D. thesis. New Jersey: Princeton University.

SWADE, R. H., and C. S. PITTENDRIGH, 1967. Circadian locomotor rhythms of rodents in the Arctic. *Amer. Nat.* 101:431–466.

TAKIMOTO, A., and K. C. HAMNER, 1964. Effects of temperature and preconditioning on photoperiodic response of *Pharbitis nil. Plant Physiol.* 39:1024–1030.

WENT, F. W., 1959. The periodic aspect of photoperiodism and thermoperiodicity. In *Photoperiodism and Related Phenomena in Plants and Animals*, R. B. Withrow, ed. Washington, D.C.: American Association for the Advancement of Science, pp. 551–564.

WEVER, R., 1962. Zum Mechanismus der biologischen 24-Stunden-Periodik. *Kybernetik* 1:139–154.

WILLIAMS, C. M., and P. L. ADKISSON, 1964. Physiology of insect diapause. XIV. An endocrine mechanism for the photoperiodic control of pupal diapause in the oak silkworm, *Antherea Pernyi. Bio. Bull.* 127:511–525.

WINFREE, A. T., 1967. Biological rhythms and the behavior of populations of coupled oscillators. *J. Theoret. Biol.* 16:15–42.

ZIMMERMAN, W. F., and T. H. GOLDSMITH, 1971. Photosensitivity of the circadian rhythm and of visual receptors in carotenoid-depleted *Drosophila. Science* 171:1167–1169.

39 Neuronal Principles Organizing Periodic Behaviors

FELIX STRUMWASSER

ABSTRACT   How do neurons produce and read out the free-running (nonreflexive) programs that control behavior? The answer to this question may involve certain specialized neurons that possess oscillators with long periods (minutes to hours). When these neurons proceed into the active phase of their cycle, they may turn on and turn off different populations of neurons controlling or triggering different behaviors. The property that behavioralists refer to as motivation may be associated with the activity of such sophisticated neurons.

All of the work that I will describe has been performed on the well-known marine mollusc, *Aplysia*. The behaviors that I study are periodic, e.g. sleep-waking and sexual cycles. The sleep-waking cycle is circadian in period, while sexual activity has a pronounced annual cycle in *Aplysia*. With the help of my colleagues, we are beginning to learn about the nature of the secretions that control one aspect of reproductive behavior, i.e., egg-laying (Toevs, 1970; Arch, 1972). Egg-laying behavior is entirely under the control of a 6000 dalton polypeptide synthesized by several hundred neurosecretory neurons, the bag cells, in the parietovisceral ganglion.

Sleep-waking behavior has been studied by time-lapse photography and, more recently, by an automatic tracking system involving a TV camera, a special video encoder that recognizes the animal, and a computer that stores the animal's position and computes its movements. The short- and long-term effects on the sleep-waking system of removing several parts of the nervous system (the eyes, and the parietovisceral ganglion) are described, which contain neurons possessing circadian oscillators. The results, in brief, suggest that the eyes mediate both *entrainment* and *synchronization* of the sleep-waking cycle and that the motor commands for movement emanate from the cerebral ganglion.

The frequency of optic nerve impulses in the isolated eye of *Aplysia* oscillates with a large-amplitude circadian period when kept in darkness (Jacklet, 1969a; Eskin, 1971). The eye is now a second example of a circadian neural oscillator, the other being the parabolic burster neuron (R15) in the parietovisceral ganglion (Strumwasser, 1965). The neural organization within the eye has been deduced, so far, from experiments involving blockers of chemical and electrical synapses (Audesirk, 1971, 1973) and from tissue reduction and microstimulation experiments (Sener, 1972). These findings will be reviewed, as they provide the best available (but indirect) evidence that special pacemakers with a circadian oscillator mechanism drive a larger population of electrically coupled output cells. Eskin, this

volume, demonstrates that membrane depolarization influences the circadian oscillation in the eye, which suggests that processes in the membrane are coupled to the internal oscillator. Other experiments have shown that in R15 a circadian cycle of membrane activity can be expressed when synaptic transmission and impulse production is blocked in the entire ganglion (Strumwasser, 1971). The circadian rhythm in this neuron is expressed as a modulation of faster pacemaker waves. The available evidence for the mechanism of pacemaker wave production will be reviewed at both the membrane electrophysiological and biochemical levels. The macromolecular constitution of some of these specialized pacemaker and circadian neurons has been studied by separating on gels the proteins synthesized in single identifiable neurons (Wilson, 1971).

Results from the organ-cultured parietovisceral ganglion (PVG) will be described. Multiunit discharges can be recorded from the nerve trunks of the PVG for periods of up to 6 weeks. A special digital template-sorter has been developed and is capable of reliably sorting the discharges of up to 8 single neurons recorded with a single electrode. Thus, the long-term activities of several neurons can be selectively and simultaneously tracked 24 hours a day for many weeks. Some of these single neurons show regular circadian periods in their discharge rates, allowing a study of the organization of a circadian system by direct observation of the unit elements.

THIS CHAPTER is concerned with an ongoing search for principles operating in the nervous system that organize behaviors, particularly those that are periodic. Most of the research work that I will describe has been carried out on the sea hare, *Aplysia californica* (Mollusca: Opisthobranchia). The fact that we study the neurophysiology, neurochemistry, and behavior of the sea hare raises the question as to whether the principles that organize periodic behaviors in this simpler organism are applicable to the higher vertebrates. I believe that the answer to this question is clearly yes. The arguments for this attitude are based on the available evidence which indicates that the nature of neuronal and glial mechanisms in nervous systems is amazingly conservative. The evidence presented (Strumwasser, 1973) that neuronal and glial mechanisms are basically conservative suggests that when new mechanisms are discovered, they should be considered to be of more general significance rather than specific to the species or phylum.

FELIX STRUMWASSER   Division of Biology, California Institute of Technology, Pasadena, California

Specialized neurons: Oscillators and neurosecretors

The remainder of this chapter is concerned with the presence of circadian and faster oscillators (pacemakers) within single neurons and groups of neurons in specialized parts of the sea hare nervous system. One of the circadian oscillators that will be described is known to play a role in the sleep-waking cycle of this organism. I would suggest that similar neurons exist in the vertebrates including mammals; these are of course more difficult to find and to study in such preparations. The molecular mechanisms of any circadian oscillator are really not known. However, some progress has been made in an analysis of those membrane and biochemical mechanisms that may be unique to pacemaker neurons. Since certain of these pacemaker neurons are also circadian in their discharge rates, the former mechanisms may be of eventual importance in understanding the nature of the circadian oscillator.

The second generalizable point of this paper is that specialized nerve cells can secrete polypeptides that control behavior. In the sea hare, a specific group of neurons secretes a polypeptide in response to copulation. This polypeptide has remarkable effects in organizing the subsequent temporal behavior of the sea hare, including the production of an egg string and its storage on the substrate. The fact that specialized nerve cells might secrete polypeptides to trigger certain behaviors may be general to other phyla including the vertebrates. There is some evidence that mammalian sleep may be mediated by a small molecule (<500 mol wt) released in the brain and that a brain peptide exists which causes hyper-activity (Fencl et al., 1971).

Sleep-waking behavior in the sea hare

As in most multicellular organisms, the sea hare, *Aplysia californica*, has a circadian rhythm of locomotor activity (Strumwasser, Lu, and Gilliam, 1966; Strumwasser, 1967b; Kupfermann, 1968; Jacklet, 1972). Time lapse photography has been used to study the behavior of the animal during its quiet and active period (Strumwasser, 1971). The sea hare is day active and becomes quiescent around the offset of light. During the night the sea hare makes postural adjustments but usually does not shift from a particular position in the home tank. Intermittent movements during the night are apparent in the head region of the animal, the rhinophores and the anterior tentacles. The initiation of locomotor activity occurs around the light onset. The anticipation of light by certain animals already suggests that there is built into the organism an endogenous oscillator. However, more convincing evidence for this point of view comes

from experiments in which the animal is suddenly exposed to a constant light or a constant dark schedule. Animals shifted from a light-dark cycle to constant light still show circadian sleep-waking cycles for at least several days (Strumwasser, 1971).

TRACKING LOCOMOTOR MOVEMENTS AND ANALYZING THEIR PERIODICITIES These earlier experiments raised the obvious question as to the location and nature of the endogenous oscillator in the organism. In order to study the effects of lesions of the nervous system on sleep-waking behavior in this or any organism, quantitative methods are needed for tracking the movements of the animal and analytic tools are needed to describe the nature of the periodicities that are present in the animal's rhythms. A noncontact method of locating the sea hare in its marine environment is utilized in my laboratory. A television camera transduces the visual scene into analog electrical information that is then analyzed by a special video encoder that essentially pattern recognizes the sea hare. The position coordinates X and Y, of the sea hare, are fed into a digital computer that writes the coordinate data for each day at midnight onto magnetic tape. This data on magnetic tape can then be analyzed in various ways. Typically, averages of several days of locomotor movements, periodograms and power spectra are used to analyze the data.

The average activity, the periodogram and the power spectra of two consecutive 9-day periods are shown in Figure 1 for an intact sea hare. The average activity, which is expressed in meters per hour on the ordinate, clearly shows that this animal is quiet in the dark and very active during the light period. During the first 9-day period (calendar days 152 to 160) the initiation of locomotor activity is very close to the onset of light, and the termination of activity starts prior to light offset. However, in the subsequent 9-day period (calendar days 161 to 169) the animal initiates its activity before light onset and although activity declines prior to the offset of light it is continued for a short while after the lights are off.

The periodogram in this illustration is expressed as a relative sigma and as a function of a trial period scan from approximately 2 hr to 50 hr with 72 min resolution. The periodogram is obtained by arranging each 9-day data array into a matrix whose width is the trial period. At each trial period the columns of data are averaged and the standard deviation of the column means are expressed relative to the standard deviation of the unordered matrix, giving rise to the relative sigma (Panofsky and Halberg, 1961; Enright, 1965; Strumwasser, Schlechte, and Streeter, 1967). The periodogram for this animal clearly shows two large relative sigma peaks for

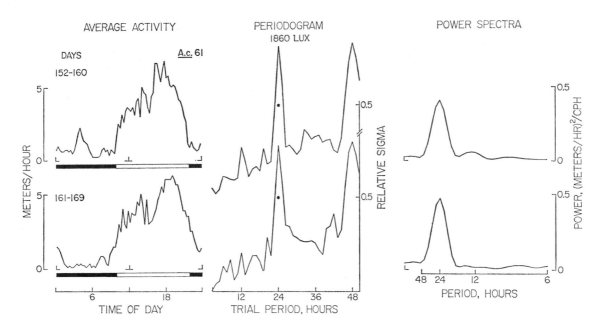

AVERAGE ACTIVITY   PERIODOGRAM   POWER SPECTRA
1860 LUX

FIGURE 1 Three methods of analysis of locomotor activity in *Aplysia californica*. The *left-most column* shows the average activity. The numbers under days are the calendar days over which the data frame was obtained. The average activity is expressed as meters per hour. The average activity curve has 18-min resolution and was generated by taking a stationary motion average from nine coordinate pairs spaced at 2-min intervals. The abscissa is time of day with 0 representing midnight of a calendar day. The photoperiod that the animal was exposed to was 12 hr of darkness followed by 12 hr of light, and the timing is shown immediately below the graph of average activity. The *middle column* is a periodogram analysis of the 9 inclusive days of move- ment. The ordinate of a periodogram is a relative sigma (stan- dard deviation)—see text. The trial period scans from 2 hr and 24 min (that is eight 18-min data points) to 50 hr and 24 min. A large dot has been positioned at 24 hr and 0.5 relative sigma. The *right-most column* is a power spectral analysis of the same data string. The ordinate is expressed as power in (meters per hour)^2 per cycle per hour. The abscissa scans from 216 hr to 6 hr. The data group size is 3 hr and the maximum lag is 2.25 days. The format of this figure is continued in Figures 2, 3, and 4 with the exception that the power spectral column has not been shown.

each of the 9-day periods. These peaks are located at 24 hr and 48 hr on the abscissa (trial period). The peaks for this animal at the 24-hr trial period are relatively high (0.82 and 0.78, respectively). This is clear evidence that the animal was well entrained to the light-dark cycle. Other than the harmonic peak at 48 hr there are no other undulations in the periodogram that suggest any other contributing (e.g. ultradian) oscillator to this sleep- waking cycle. The power spectrum which is a Fourier transform of the autocorrelation function (Blackman and Tukey, 1959) and has units of power expressed here as (meters/hr)^2/cycle/hr also shows the dominant frequency to be circadian. All of the results that I will mention in this paper have been checked by power spectra and periodograms, although for the sake of brevity in both the figures and the text I shall refer only to periodogram analysis.

THE EYES MEDIATE ENTRAINMENT AND SYNCHRONIZA- TION OF THE SLEEP-WAKING CYCLE The isolated eyes of *Aplysia* have been shown not only to respond to light but also to possess a circadian rhythm of optic nerve impulse discharge during sustained darkness (Jacklet, 1969a; Eskin, 1971). The fact that a circadian oscillator is al- ready built into the neural mechanisms of this eye sug- gested that the eye might play an important role in either driving or synchronizing the circadian rhythm of locomotion, besides being involved in entrainment of locomotion by light-dark cycles in the environment.

The stability of a sham operated sea hare is indicated in Figure 2. One eye, the right eye, was removed from the sea hare as well as a small piece of skin near the left eye. The animal was tracked in the locomotor system for some 100 days. The data shown in this figure spans 72 inclusive days. The topmost frame shows 21 days of data, while the second and third frames show the first 11 and the next 10 days of this 21 day period. The long-term average clearly shows anticipatory activity prior to light onset as previously mentioned. The periodograms clearly show that the animal is entrained since the peaks are all at 24 hr and are large. The last frame shows the last 10 days of this 72 day period. Anticipatory activity prior to light onset is not obvious but entrainment is still present and good. The loss of one eye clearly does not appear to

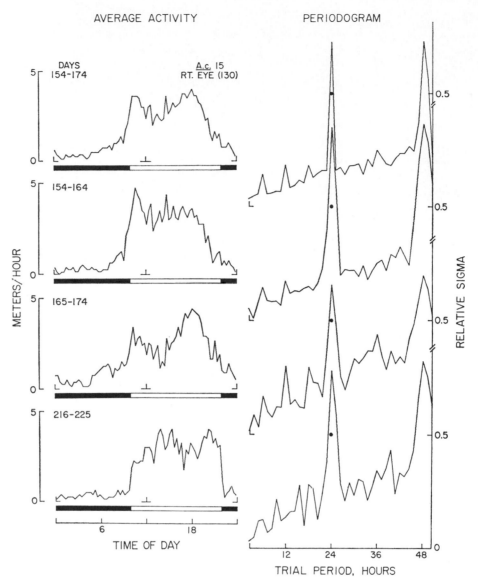

AVERAGE ACTIVITY

PERIODOGRAM

DAYS
154-174

A.c. 15
RT. EYE (130)

154-164

165-174

216-225

METERS/HOUR

TIME OF DAY

RELATIVE SIGMA

TRIAL PERIOD, HOURS

FIGURE 2 The average locomotor activity and periodogram analysis of a sea hare after removal of the right eye (on day 130). A small piece of skin in front of the left eye was also removed at the same time. The data in this figure spans 72 days. The data for days 175 to 215 are not shown. (The first frame starts 24 days after the operation. The light intensity was 390 lux.) (See Figure 1 for further details.)

influence entrainment and the ability of the animal to express a normal amplitude and day-active circadian rhythm.

EVENTUAL DESYNCHRONIZATION BY BLINDING The effects of removing the two eyes of the sea hare are quite different. Figure 3 shows four 6-day frames. The first two frames are control periods. The control period average activity clearly shows that the animal is normal, that is, day-active and quiet at night. The periodograms show that the animal is clearly entrained to the 24-hr photoperiod (relative sigmas are 0.69 and 0.76, respectively).

The last two frames show the behavior of the animal starting on the third day after eye removal. The first 6-day frame after eye removal shows quite clearly that the animal is no longer entrained to the 24-hr cycle, the period being 28 hr. The subsequent 6-day frame which starts at 13 days after eye removal shows no obvious circadian rhythm. All the peaks that are present are close to the noise level but there are small peaks at 12, 24, 27.5, 38.4, 42, and 48 hr. This pattern of periodogram is best described as *desynchronized*. More detailed case histories are to be found in a paper in preparation (Strumwasser and Schlechte). Typically, the results of

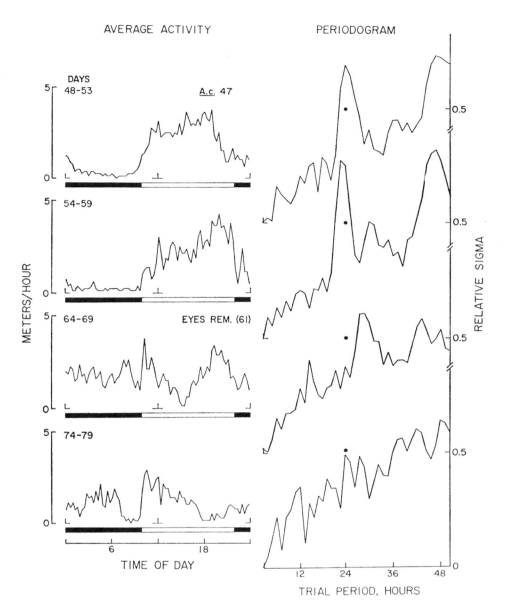

AVERAGE ACTIVITY

PERIODOGRAM

FIGURE 3 The average locomotor activity and periodogram analysis of a sea hare before and after the removal of both eyes. The first two rows of data were taken before removal of the eyes. Twelve inclusive days are contained in this period. This period terminates 2 days prior to eye removal. The experimental data frames began 3 days after eye removal. The light intensity was 350 lux. (See Figure 1 for further details.)

removing both eyes in moderate intensity light (approximately 400 lux) are that sea hares are no longer entrained and ultimately the circadian free running rhythm of sleep-waking becomes desynchronized.

There is some evidence for a secondary photoreceptor system in *Aplysia californica* that does not involve the eyes (that is, is extra-retinal; see Block, 1971). If a freshly blinded animal is exposed to bright light (approximately 1800 lux), the circadian rhythm can be entrained. However, animals that are blinded for more than 5 or 6 weeks do not show this response, suggesting that the secondary photoreceptor system is perhaps dependent for its integrity on the presence of the eyes. Perhaps the secondary photoreceptor system becomes nonfunctional or degenerates in the absence of the eyes and this process may take a few weeks.

THE PARIETOVISCERAL GANGLION IS NOT ESSENTIAL FOR THE CIRCADIAN LOCOMOTOR RHYTHM The parietovisceral ganglion (PVG) of the sea hare is known to contain a particular neuron with an endogenous circadian oscillator (Strumwasser, 1965, 1967a, 1967b; Lickey,

1969; Lickey et al. 1971). In addition, there are unidentified neurons in this ganglion that have been demonstrated to possess circadian oscillations in organ culture (Strumwasser, 1967b). It has recently been found quite possible to remove this ganglion in the sea hare and thus to determine the contribution of the ganglion toward the sleep-waking cycle (Strumwasser, Schlechte, and Bower, 1972). Figure 4 shows 42 days of locomotor activity in a sea hare in which the parietovisceral ganglion had been removed. The first frame starts 22 days after the ganglion had been removed and is a run in constant darkness. The 8 days of this frame clearly show in the periodogram that the animal is capable of producing a free running circadian rhythm with a period that is less than 24 hr. The second frame shows the behavior of this sea hare in a light-dark cycle. The animal is clearly entrained to the 24-hr photoperiod and is day active. The last frame shows the 25-day average and periodogram to indicate the stability of such a preparation. The parietovisceral ganglion is clearly not essential for either entrainment or the expression of the free-running circadian rhythm. Such experiments, however, do not exclude the possibility that the presence of the ganglion contributes something to the circadian locomotor rhythm.

Cellular organization of and the circadian rhythm in the eyes of Aplysia

The sea hare has two small (approximately 1 mm in diameter) eyes located just at the base and anterior to the two rhinophores, which are feelers located bilaterally on the head of the animal. The eyes were first studied by Jacklet (1969b) while in our laboratory. The eye is of the closed vesicle type containing a small lens. Histological sections reveal a layer of primary photoreceptor cells with microvillus processes which extend into the space surrounding the lens. Electron microscopic studies (Jacklet, Alvarez, and Bernstein, 1972) show that the receptor cells are surrounded by pigmented support cells. The primary photoreceptor is unusual in that it contains in its cell body a dense packing of clear vesicles approximately 400 Å in diameter (see Helix—Eakin and Brandenburger, 1967; Hermissenda—Eakin et al., 1967). Below the layer of photoreceptors and support cells is a

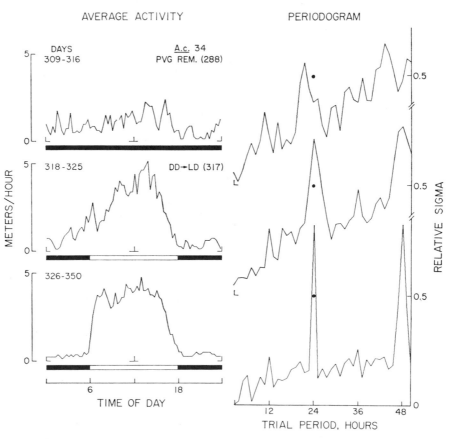

FIGURE 4 The average locomotor activity and periodogram analysis of a sea hare whose parietovisceral (or abdominal) ganglion had been removed. Note that the first data frame was a run in constant darkness (red fluorescent lights on continuously at about 60 lux when measured from a view of the entire tank). Entrainment to a photoperiod (390 lux) is shown on the next two frames. (See Figure 1 for further details.)

more diffusely organized layer of secondary neuronal cell bodies, and the processes of these, and receptor cells. The optic nerve, at the base of the eye, runs to the cerebral ganglion. Sener, Alvarez, and Strumwasser (in preparation) find that at the base of the eye there is a particularly thick neuropile.

Jacklet (1969b) showed that the normal mode of communication by the eye was a compound action potential which propagated toward the cerebral ganglion. Jacklet (1969a) found that the frequency of these compound action potentials (CAPs) in constant darkness fluctuated with a large amplitude circadian rhythm. The first cycle of this circadian rhythm of optic nerve impulses correlated well with the previous photoperiod to which the intact animal had been exposed. Activity in the eye anticipated the projected light onset and declined before the projected light offset. The eye was quiet during projected darkness. Jacklet (1969b) also found that the eye was responsive to light, exhibiting a transient fast discharge of the CAPs followed by a sustained tonic discharge that was essentially an acceleration of the dark discharge as long as the light was maintained on. Because the CAPs were graded during the transient light-on response and because the CAPs could be broken down into much smaller units by hypoosmotic sea water, Jacklet concluded that the compound action potential was possibly due to the simultaneous discharge of many neurons with emergent axons in the optic nerve.

More recent experiments in my laboratory by Audesirk (1971, 1973) utilizing agents that are known to block separately chemical and electrical synaptic transmission, allow the proposal of a simplified model of organization and connections in the eye of *Aplysia*. Figure 5 illustrates a diagram of this model. The photoreceptors are shown arranged radially around the lens. The second-order cells are shown immediately below the photoreceptor layer. Because high-magnesium, low-calcium solutions or a small amount of lanthanum salt added to the normal artificial sea water will block the dark discharge and the tonic response to maintained light, Audesirk concludes that the dark discharge is due to a specialized pacemaker cell that chemically synapses with the second order neuronal population. In high-magnesium, low-calcium solutions (or by adding lanthanum) chemical synapses between the pacemaker neuron and the second order cells are blocked, thus giving rise to the absence of discharge in sustained darkness and the absence of a tonic response in the presence of light.

Figure 6 shows the nature of the compound action potential discharge in constant darkness, the typical transient and tonic discharge due to sustained light and the effects of lanthanum chloride at 1.1 mM concentration. Shortly after the addition of lanthanum chloride

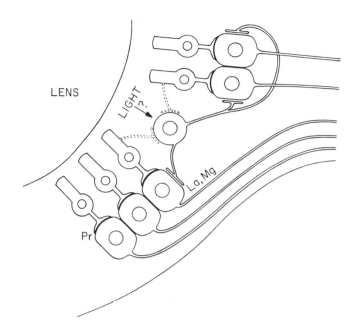

FIGURE 5 A model of the neuronal and synaptic organization within the eye of the sea hare. This model is based on electrophysiological and ion substitution experiments done by Audesirk (1973). The photoreceptors are the elements arranged radially closest to the lens. These cells make electrical synapses with the second-order neurons whose axons project into the optic nerve. The second-order neurons also make electrical junctions with one another. A single case is shown of a third type of neuron, a pacemaker cell, which makes a chemical synapse with second order neurons. The electrical synapses are blocked by propionate while the chemical synapses are blocked by high-magnesium, low-calcium or lanthanum ions. The tonic response to light (see text) is thought to be due either to the fact that there are chemical synapses between receptor cells and the pacemaker neuron or that the pacemaker neuron is itself responsive to light. (Modified from Audesirk, 1973.)

toward the end of line 2 the dark discharge is blocked. Also, as shown on line 3 the tonic but not the phasic component of the response to sustained light is blocked. About 4 hr after two rinses with artificial sea water the dark discharge eventually returns as on the last lines of this figure.

The model in Figure 5 also shows that photoreceptors are electrically coupled to the second-order neurons and that the second-order neurons are electrically coupled to each other. The evidence that the photoreceptors are electrically coupled to second-order neurons stems from the fact that propionate substituted for chloride will totally block all components of the light response as seen in the optic nerve. However, the ERG is unimpaired, strongly suggesting that transmission between the receptors and the second-order neurons have been blocked. Since high-magnesium, low-calcium solutions or the addition of lanthanum does not block this junction, while

1.1 mM LaCl₃ SW

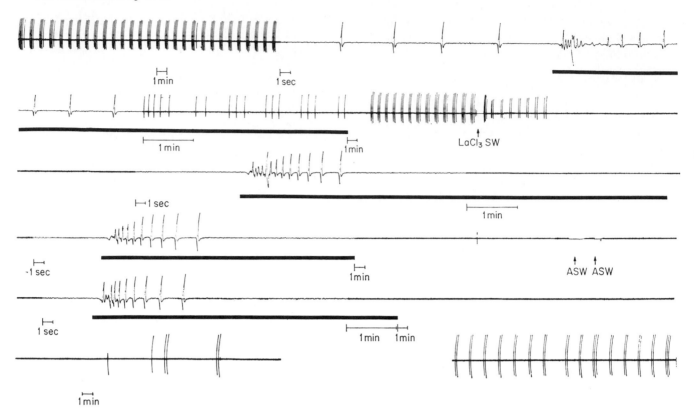

FIGURE 6 The influence of lanthanum chloride sea water on the dark discharge and the tonic component of the light evoked discharge in the eye of *Aplysia*. The eye is initially in ASW. The first four lines are consecutive recordings; line five follows with a delay of approximately 1 hr, line 6 follows with a delay of approximately 1.5 hr and the gap in line 6 represents approximately 1 hr. The total time to recovery of normal bursting patterns in the dark, after lanthanum chloride, is 3.5 hr in artificial sea water. Sustained illumination is signaled by a thick dark line immediately underneath the cell recording. Changes in paper speed are indicated by a time marker that starts with a long vertical bar and terminates with a shorter vertical bar. SW stands for sea water and ASW for artificial sea water. The spikes in this figure are the compound action potentials that are recorded from the optic nerve of the isolated eye. (From Audesirk, 1973.)

propionate does, Audesirk suggests that the coupling between photoreceptors and secondary cells is electrical. Asada, Bennett, and Pappas (1971) have found that propionate substituted for chloride blocks the electrotonic junctions in the lateral septate axons of the crayfish. The evidence that the second-order neurons are themselves coupled is suggested by the very fact that the discharges in the optic nerve in normal solutions or those containing lanthanum are compound action potentials.

Audesirk's model of the eye already suggests that the circadian rhythm of dark discharge might emanate from the hypothetical pacemaker cell or cells. It should be emphasized that this model can only account at present for the gross features of the responses of the eye to light and the dark discharge. Intracellular recordings by Jacklet (1969b) demonstrated greater complexity in the eye than this model is able to account for. However, at present it is not thought that the additional features that Jacklet observed, e.g. inhibition of certain neurons upon light presentation, are particularly relevant toward an understanding of the circadian discharge in this eye.

Manipulation of the circadian rhythm in the isolated eye

There are four methods that have been used to manipulate the circadian rhythm of the isolated eye—light, tissue reduction, high external potassium, and inhibitors of macromolecular metabolism. Eskin (1971) showed that isolated eyes could be entrained to light-dark cycles in vitro. Entrainment to 180° phase-shifted light-dark cycles in vitro took 3 to 5 days whereas in vivo entrainment could be obtained in 1 day.

Jacklet and Geronimo (1971) removed eye tissue and

claimed that if less than 0.2 of an eye had been left, after trimming, that the circadian rhythm could not be recorded. Sener (1972) has repeated these experiments trimming the eyes after a preliminary period of control recording and during projected darkness, whereas Jacklet and Geronimo had trimmed their eyes at the beginning of the experiment. Sener's results are nicely summarized in Figure 7. Thirteen pairs of eyes were used in these experiments, one for control recordings while the other eye was trimmed down to a very small piece. The average frequency of discharge of the 13 control eyes are compared with a similar average for the 13 trimmed eyes. Both curves show the typical circadian rhythm. The only difference between the two sets of eyes is that the intact eyes have a slightly higher minimum and maximum frequency of discharge; the times of onset and offset of the circadian rhythm are clearly not different between these two sets of eyes. The control eyes had a *peak* with respect to projected dawn at −1.4 ± 1.1 hr, whereas the experimental eyes possessed a peak at −1.5 ± 1.9 hr.

Sener estimates from light microscopy that under 50 receptor cells and perhaps less than this number of

second-order neurons are left in such fragments. Jacklet and Geronimo had used their results—the disturbance or absence of a circadian rhythm—to suggest that coupling between many cells each possessing ultradian rhythms were involved in the production of a circadian rhythm. Sener's results reopen the possibility that the circadian rhythm is produced by only a few cells and suggest that Jacklet and Geronimo's results might have been due to injury of some key neurons in the base of the eye.

A third way of manipulating the circadian rhythm in the isolated eye is to depolarize all of the cells of the eye with a high potassium pulse (100 mM) and to assay the shift of the rhythm with respect to a control eye. Eskin (1972) who performed these experiments in my laboratory found that the effects of a pulse of high potassium depended on the time in the circadian cycle at which the pulse was delivered. These findings are further discussed by Eskin in a chapter in this book. The mechanisms, however, by which membrane depolarization might couple to and influence the presumably intracellular circadian system are unknown.

A fourth method of manipulating the eye suggests that a certain pattern of RNA and protein synthesis may be necessary for maintaining the rhythm. Rothman has performed experiments in which a 3-hr pulse of aflatoxin Bl, a reversible inhibitor of RNA synthesis, has differential effects on the rhythm depending on the time at which it is administered (Rothman and Strumwasser, 1973). Figure 8 compares 5 control eyes with 5 experimental eyes. The

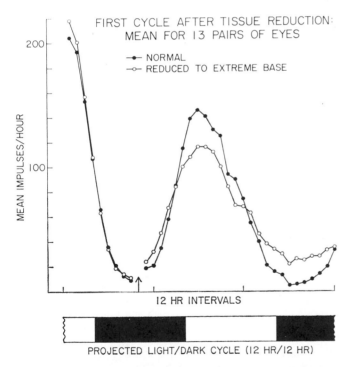

FIGURE 7   Comparison of the first cycle of the circadian rhythm in a group of reduced eyes and intact eyes. There were 13 eyes in each experiment. Points on the curves are hourly averages. The arrow indicates the time that both groups of eyes were exposed to illumination during which time the experimental eyes were reduced to the extreme base (remainder of eye estimated as less than 0.2 of its normal size). The projected light/dark cycle is the photoperiod that the intact *Aplysia* had been exposed to. (From Sener, unpublished; see Sener, 1972.)

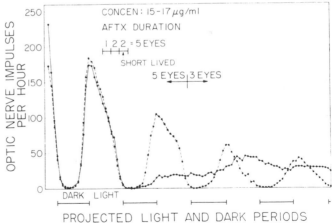

FIGURE 8   The effects of Aflatoxin Bl, administered for 3 hr, on the circadian rhythm of optic nerve impulses in the eye of *Aplysia*. The two curves were generated by taking hourly averages of the optic nerve impulses from a set of 5 control (dashed curve) and 5 experimental (solid curve) eyes. "Projected light and dark" refers to the photoperiod that the intact *Aplysia* had been exposed to. Aflatoxin (AFTX) 3-hr pulses ranged from about 6 hr prior to projected dusk to 3 hr after dusk. (From Rothman, unpublished; see Rothman and Strumwasser, 1973.)

control eyes have a clear circadian rhythm of optic nerve impulse discharge. The experimental eyes were allowed to cycle once and a pulse of aflatoxin B1 was delivered during the late projected afternoon and the early projected night. The experimental eyes turn on late for the next cycle, reach about one-fifth the amplitude of the control eyes for the second cycle, and are no longer rhythmic in that the optic nerve impulses do not turn off. The rhythm, if at all present, is clearly abnormal. Rothman finds that similar aflatoxin pulses delivered 180° out of phase (that is, in the late projected night and into the early projected day) have virtually no effects on the in progress cycle of the experimental eyes. The differential sensitivity of the eyes to aflatoxin pulses given at different times in the circadian cycle suggests that the control for the circadian cycle may involve particular species of RNA and protein that are synthesized at some particular temporal phase point.

Circadian rhythms in particular neurons of the parietovisceral ganglion

The eyes of the sea hare have a circadian rhythm of optic nerve impulses as has just been discussed. There are also neurons in the parietovisceral ganglion (PVG) of the sea hare that possess circadian oscillators. The eye has the disadvantage that up to now the circadian rhythm can only be recorded from the optic nerve over extended periods of time and not from individual cells within the eye. The PVG offers the advantage that at least one identifiable neuron in it is known to possess a circadian oscillator and can be recorded from intracellularly for several days. One of the factors contributing to the entrainment of the circadian rhythm in the parabolic burster or R15 neuron is photoperiod. A number of papers have described the circadian rhythm in this neuron in the isolated ganglion (Strumwasser, 1965, 1967a; Lickey, 1969). In this section, I will describe recent results in which nerve trunk recordings from the organ cultured PVG (Strumwasser and Bahr, 1966) have been obtained for periods up to 6 weeks (Strumwasser, 1967b, 1971). These recordings have allowed us to verify that there are other neurons in the ganglion with a similar circadian rhythm that can be expressed under organ culture conditions. The location, however, of the cell bodies of these neurons in the ganglion is not presently known.

The PVG can be organ cultured in Eagle's minimum essential medium made up in filtered sea water containing 20% *Aplysia* serum for periods up to 6 weeks (Strumwasser and Bahr, 1966; Strumwasser, 1971). During the first 4 to 5 of these 6 weeks, the resting and action potentials of the neurons appear normal as well as the spontaneous or evoked postsynaptic potentials. The ganglion can be

mounted in a special sterilizable chamber in which the nerve trunks are pulled through miniature tunnels containing a pair of recording platinum-iridium wires. Spontaneous unitary axon spikes can then be observed in the recordings (Strumwasser, 1967b, 1971). These spontaneous unitary spikes depend, for their maintained discharge, on connection with the ganglion. The presence of unitary spikes in a recording from a nerve trunk is usually associated with the presence of several large axons among the population of axons in the nerve trunk. It has been shown previously that stable unit recordings can be obtained from such preparations for up to 6 weeks (see Figures 13 and 14 in Strumwasser, 1971).

The major problem with such recordings has been to sort the units out by some automatic technique. In my laboratory special hardware has been constructed in association with a digital computer that allows the reliable sorting of up to eight different axonal discharges from a single nerve trunk. The sorting is done on-line and will not be described here. Figure 9 shows the wave forms of

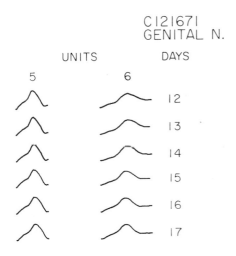

FIGURE 9   Wave forms of two axonal units in the genital nerve of the organcultured parietovisceral ganglion of *Aplysia*. The wave forms shown were sorted on-line by a digital pattern recognizing device operating in conjunction with a digital computer. Each wave form is an average of many wave forms sampled at 3-hr intervals throughout each day for a 10-min period.

two small units in the genital nerve between days 12 and 17 of an organ cultured PVG. The wave forms shown were classified by the on-line sorter in a highly reliable way. The wave forms are also found to be quite stable from day to day. Slow changes that normally occur in the wave forms are easily accommodated by an operator changing some of the parameters on which sorting is based.

Periodogram analysis of days 12 to 17 are shown for

thrcc units in the genital nerve of this same preparation in Figure 10. It can be seen that two of the three units (unit 0 and unit 5) have circadian rhythms whereas unit 6 does not appear to have any, at least over this period. Furthermore, it is interesting that unit 5 and unit 0 have peaks in the periodogram that are separated by about one hour. Inspection of the smoothed frequency of firing curves of these two units, as a function of time, clearly reveals that the wave forms of the circadian rhythms and the phases of the peaks are *not* synchronous between these two units and that there is relative motion between the two units, with respect to time. These facts indicate, most probably, that these two neurons are free running with slightly different periods. If these two units are coupled, the coupling must indeed be weak. The presence of circadian rhythms within individual neurons in the PVG in organ culture suggests that the eyes of *Aplysia* do not drive these oscillators but rather probably entrain them.

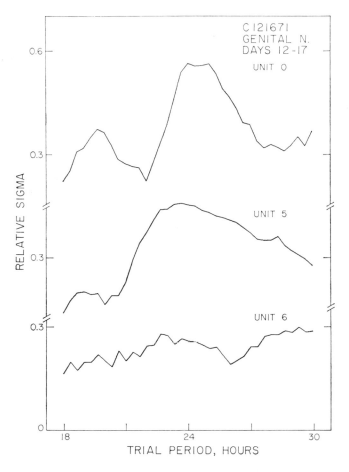

FIGURE 10 Periodogram analysis of 3 simultaneously recorded axonal units in the genital nerve of the organ-cultured parietovisceral ganglion. The wave forms of units 5 and 6 are shown in Figure 9. For further details on periodograms, see Figure 1 and associated text.

Pharmacological isolation of a neurosecretory neuron with a circadian rhythm

The parabolic burster of the PVG is a neurosecretory neuron with a circadian rhythm of impulse activity which is expressed in the isolated ganglion. Early studies were based on prolonged intracellular recordings (about 2 days) from this neuron in the isolated ganglion (Strumwasser, 1965, 1967a, 1967b; Lickey, 1969). These studies showed that the peak activity in the parabolic burster occurred usually around projected dawn of the light-dark cycle that had been administered to the intact organism, this phase angle being similar to that of the circadian rhythm in the isolated eye. Some parabolic bursters, however, showed major peaks at projected dusk. Strumwasser (1967a) and Lickey (1969; Lickey et al.,1971) have presented evidence that there is a seasonal influence on the phase angle of the parabolic burster with respect to the entraining photoperiod.

More recent studies have been aimed at investigating whether impulse activity is a requisite for the production of a circadian rhythm. Calcium-free artificial sea water and the addition of tetrodotoxin (TTX) abolish all the impulse activity in the ganglion as evidenced by extracellular recording from the nerve trunks and intracellular recording from several cell bodies in the ganglion. There are no recordable intracellular postsynaptic potentials under such conditions. The only electrical activity that can be obtained under such conditions is that of slow membrane oscillations in the pacemaker cells; the non-pacemaker cells of the ganglion, such as the giant cell R2, never show such membrane oscillations (Strumwasser, 1968, 1971; Strumwasser and Kim, 1969). The frequencies of these membrane oscillations agree well with the burst rate in the bursting pacemakers such as R15 just before pharmacological isolation. At least one full cycle of a circadian rhythm can be expressed in the parabolic burster in the absence of synaptic input and the inability to produce impulses.

Figure 11 illustrates the effects of pharmacologically isolating the parabolic burster about one hour after projected dusk. Each line in this record is 1-hr long. The endogenous oscillations are present for approximately 12 hr, terminating shortly after projected dawn; the subsequent 10 hr of recordings demonstrate that the parabolic burster is inactive. A convenient physiological test that can be performed during this period, to make sure that the cell is not dead, is to apply a small depolarizing pulse which at the termination of the current causes a long-lasting hyperpolarization (Strumwasser, 1968). Two examples are shown at "test" (lines 18 and 21). In addition, the pharmacological aspects of TTX and the calcium-free medium are completely reversible when the chamber is perfused with filtered sea water. Both spike generating mechanisms and postsynaptic potentials return and are elicitable from cells in the ganglion, including the parabolic burster.

However, TTX has a phase shifting effect on the circadian rhythm that is not presently understood. As can be noticed from Figure 11, the cell is oscillating during projected night and quiet during projected day. The 180° phase inverting effect of TTX can be induced at other times than around projected dusk, as can be seen from Figure 12. In this example, TTX was applied approximately 5 hr prior to projected dusk; the parabolic burster remained quiet during the last 5 hr of projected day and became active at projected dusk. In this experiment, activity terminated approximately 3.5 hr prior to projected dawn. These experiments indicate that impulse production and synaptic transmission are not required within the ganglion for at least the expression of one cycle of the circadian rhythm in the parabolic burster. More subtle methods of pharmacological isolation and/or recording perhaps are needed in order to allow the parabolic burster to express more than one cycle of its circadian rhythm when impulses and synaptic transmission are blocked in the ganglion. The phase shifting properties of

TTX may be due to the small but sustained depolarization that occurs in the presence of this agent. In the parabolic burster TTX causes a 5 to 10 mV depolarization along with its blocking effects on impulses. The speculation that this depolarization is the cause of the phase shift is reasonable in view of the demonstration by Eskin (1972) that potassium-induced depolarization in the eye of *Aplysia* causes predictable phase shifts in the circadian rhythm of that preparation.

Mechanisms of "slow" pacemaker oscillations

In the preceding section it was demonstrated that the circadian rhythm (CR) in the parabolic burster is expressed through oscillations of the membrane potential. These oscillations are here termed "slow" pacemaker oscillations to distinguish them from the pacemaker potentials that occur between spikes during repetitive firing. There are good arguments for believing that insight into these slow pacemaker oscillations will reveal useful information concerning the circadian mechanism (CRM). The CRM must couple to the neuronal membrane if it is to be expressed as a pattern of impulses and the slow pacemaker oscillations serve as the intermediary

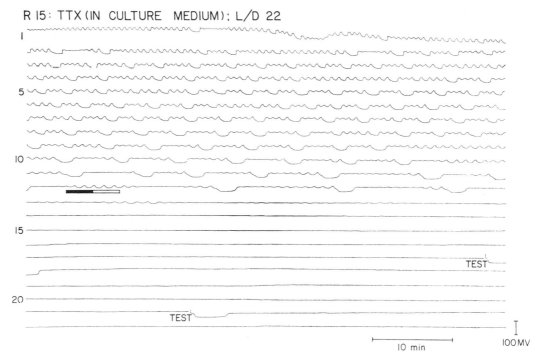

FIGURE 11 Long-term intracellular recording from the parabolic burster neuron (R15). The impulses in the PVG were blocked by tetrodotoxin (TTX, 25 μg/ml). The PVG was maintained in culture medium rather than filtered sea water (Strumwasser and Bahr, 1966). Each line of recording represents one hour. The PVG was obtained from an *Aplysia* entrained to 22 light/dark cycles, the projected dawn transition occurring on line 12. At "test" (lines 17 and 21), about 3×10^{-9} A of depolarizing current was passed across the cell membrane for about 15 sec resulting, *after* current flow, in a long-lasting hyperpolarizing response (referred to as a POBH, Strumwasser, 1968). Temperature was 14°C.

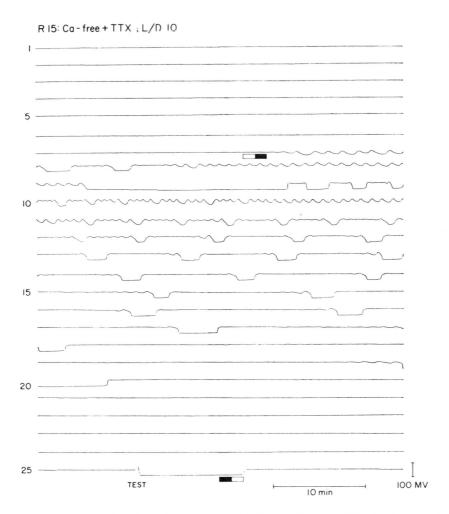

R 15: Ca-free + TTX ; L/D 10

TEST   10 min   100 MV

FIGURE 12 Long-term intracellular recording from the parabolic burster neuron (R15). The impulses in the PVG were blocked by tetrodotoxin (TTX, 25 μg/ml) and synaptic transmission by calcium-free artificial sea water. Each line of recording represents 40 min. The PVG was obtained from an *Aplysia* entrained to 10 light/dark cycles, the projected dusk transition occurring on line 7. A short depolarizing current was passed across the membrane at "test" (see Figure 11). Temperature was 14°C. (From Strumwasser, 1971.)

mechanism between the CRM and the pattern of impulses generated. This coupling is an output coupling of the CRM to the membrane. An input coupling of the membrane to the CRM also exists. This input coupling presumably mediates entrainment and the best evidence for this in *Aplysia* comes from the studies of Eskin (1972) on the effects of depolarization, induced by high potassium on the CR of the eyes of *Aplysia*.

Insights into the output coupling mechanism may be expected when there is more complete knowledge about the nature of the membrane mechanisms involved in slow pacemaker oscillations. The CRM presumably couples to one or more of the membrane mechanisms controlling slow pacemaker oscillations in order to express its output.

There are four important facts concerning the slow pacemaker oscillation in the parabolic burster. First, there is a special sodium channel utilized by the slow pacemaker mechanism, independent of the electrically excitable sodium channel used by the action potential mechanism. The evidence for this consists of the fact that slow pacemaker oscillations continue when the electrically excitable sodium channel is blocked by TTX (see Figures 10 and 11). However, these slow oscillations are sodium-sensitive for when either choline or Tris is substituted for sodium, the oscillations stop (Strumwasser, 1968; Strumwasser and Kim, 1969).

The second important fact about the pacemaker mechanism is that in the parabolic burster about one-half of the total membrane conductance is sodium dependent. Figure 13 compares the membrane resistance of a pacemaker neuron (the parabolic burster R15) and a non-pacemaker neuron (the giant cell R2) when in filtered sea

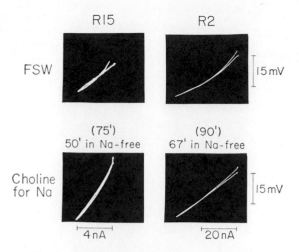

R15   R2

FSW   ]15 mV

(75')   (90')
50' in Na-free   67' in Na-free

Choline
for Na   ]15 mV

⊢4 nA⊣   ⊢20 nA⊣

FIGURE 13   Comparison of the membrane resistance changes of R15 and R2 in filtered sea water (FSW) and after substitution of choline for sodium. A slowly rising and falling "sawtooth" of hyperpolarizing current was passed across the cell membrane with one electrode while a second intracellular electrode measured the membrane potential changes. Voltage is on the ordinate and current is on the abscissa. The filled circle to the left of each frame is positioned to indicate the resting potential for the left hand frames; the resting potential for the right-hand frames are indicated by the tops of the voltage calibration bars. nA stands for 10^{-9} A. Electrode current was measured by an operational amplifier connected to the virtual ground in the bath.

water and after substitution of choline for sodium. There is a drastic change in membrane resistance in the parabolic burster (an increase of about 100%) while R2 shows little resistance change (less than 5%). This twofold membrane resistance increase in the parabolic burster, in choline or Tris substituted for sodium, implies that the resting membrane has a sodium leak that is one-half the total membrane conductance. In view of the fact that the nonpacemaker neuron, R2, does not possess this high leakage to sodium, this latter parameter may be uniquely built into the pacemaker membrane and a necessary requirement for sustained slow oscillations.

Thirdly, the pacemaker oscillations are very sensitive to interference with active transport of sodium across the membrane. A particularly convincing way to demonstrate this fact is to use the substitution of lithium for sodium. Lithium is presumably not recognized by the membrane ATPase involved in sodium transport (Flynn and Maizels, 1949; Ritchie and Straub, 1957; Keynes and Swan, 1959), but will generally substitute for sodium, at least in the electrically-excitable sodium channel supporting the action potential. Figure 14 shows the effect of lithium substituted for sodium on the parabolic burster slow pacemaker oscillations. Perfusion of lithium for sodium starts and is completed on line A. Within 3 min

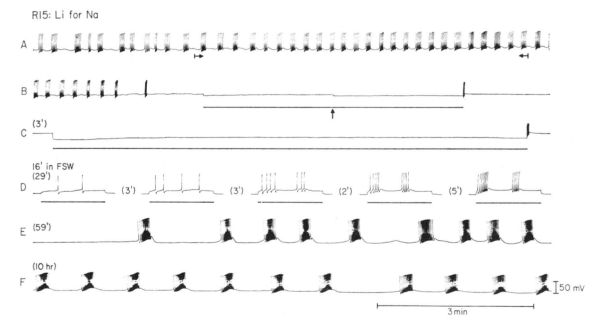

R15: Li for Na

FIGURE 14   Abolition of slow pacemaker oscillations by lithium substituted for sodium. In line A, 50 cc of lithium for sodium artificial sea water were perfused through an 8 ml chamber containing the PVG between the arrows. In line B after pacemaker oscillations had stopped, continuous hyperpolarization was applied as indicated by the heavier marker line; at the arrow the hyperpolarization was increased. Stronger hyperpolarization of the membrane is shown on line C. Spikes fired when hyperpolarization was released. Between lines C and D filtered sea water (FSW) was perfused through the chamber. All the spikes shown on line D were elicited by applying depolarizing currents for the duration shown by the heavier marker line. The first burst on line E is the first spontaneous burst that occurred after the pacemaker oscillations had been blocked by lithium for sodium in line B. Numbers in parentheses refer to the time between strips or traces.

after completion of the perfusion, slow pacemaker oscillations have stopped (line B). Immediate hyperpolarization of the neuron (middle of line B, solid marker line) is unable to restore slow pacemaker oscillations. Even stronger hyperpolarization a few minutes later (line C) does not restore slow pacemaker oscillations. The effects of lithium substituted for sodium are completely reversible as is shown in lines D, E, and F. It should be noted in this experiment, which is typical, that spontaneous burst production did not occur until 1.5 hr after returning to filtered sea water from Li for Na sea water. Once the slow pacemaker mechanism generating bursts of impulses recovered, however, it was essentially normal when continuously monitored over the next 12 hr (see line F).

A fourth important fact concerning the pacemaker mechanism in the parabolic burster is its dependence on a normal external chloride concentration. Other bursting

pacemakers in the same ganglion do not have this chloride sensitivity as demonstrated in Figure 15. This figure demonstrates a simultaneous intracellular recording from L3 (Frazier et al., 1967) and the parabolic burster (R15). When acetate is substituted for chloride (lines 2 and 3) the parabolic burster stops its oscillations within the first perfusion while the L3 neuron, although modified in the waveshape of the slow pacemaker oscillation, continues to oscillate even 10 hr later (see line 6). Quite typically the parabolic burster never fires during the substitution of acetate for chloride although, in a few experiments, slow repetitive firing has occurred for a short time after several hours of substitution.

A MODEL FOR SLOW PACEMAKER OSCILLATIONS The studies of Connor and Stevens (1971b, 1971c) show that *repetitive firing* in nudibranch neurons is dependent on a specific potassium channel that is activated by stepping

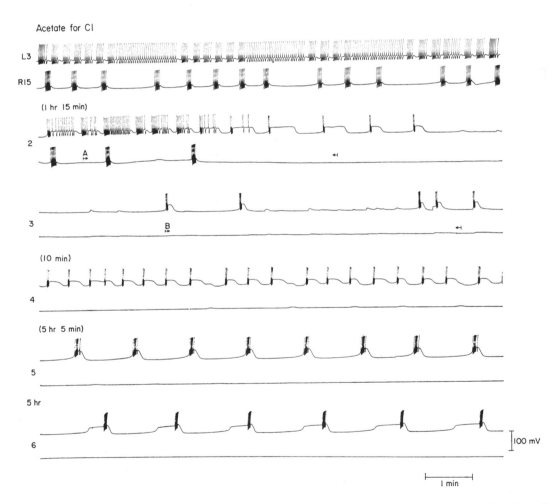

FIGURE 15 Abolition of pacemaker oscillations in R15 by acetate substituted for chloride. Simultaneous intracellular recording from an upper left quadrant burster (L3) and R15. In line 2 perfusion of acetate for chloride artificial sea water takes place between the arrows at A and again on line 3 between the arrows at B. The time between records on separate lines is shown in parentheses. Lines 2 and 3 are continuous.

from potentials negative to the resting potential toward the resting potential. This process is thought to be mediated by channels operationally distinct from the K^+ channels mediating "delayed rectification." Essentially similar findings have been recently described by Neher and Lux (1971) in *Helix* pacemaker neurons. Connor and Stevens (1971c) show, by computation, that the first two spikes generated by a sustained DC current are properly predicted by consideration of the two conventional channels (for Na^+ and K^+) and this new potassium channel. While this special potassium channel may have relevance for the precise timing of impulses in response to a slow pacemaker oscillation, my own view is that this channel does not contribute significantly to the actual slow pacemaker oscillations. The primary reason for this point of view is that lithium substituted for sodium abolished the slow pacemaker oscillation even in the presence of weak to strong hyperpolarization. Also, in the parabolic burster, substitution of acetate, propionate or butyrate (Strumwasser, 1968) for external chloride immediately stops slow pacemaker oscillations which again fail to occur even when the membrane is hyperpolarized or depolarized. The findings of Eaton (1972) that the decline of the outward (K^+) currents in the parabolic burster with long voltage clamp pulses is probably due to an accumulation of K^+ just outside the membrane, rather than a K^+ inactivation, as in nudibranch neurons (Connor and Stevens, 1971a), is probably of significance for the increased duration of the successive action potentials during a burst (Strumwasser, 1965, 1967a; Faber and Klee, 1972).

Strumwasser and Kim (1969; Strumwasser, 1971) have suggested and computed that membrane oscillations can occur in a membrane system with a high-sodium leak due to intermittent activation of a sodium electrogenic pump (in the case of the parabolic burster a sodium-chloride electrogenic pump). It should be noted that in this computation the Na^+ and K^+ channels were treated as purely passive (electrically inexcitable). Since the pacemaker oscillations in *Aplysia* are essentially slow, the rise and fall of a pacemaker depolarization taking about 15 to 30 sec at 14°C, this intermittent pump mechanism seems reasonable on purely kinetic grounds. Connor and Stevens (1971b) report time constants of decay of the special potassium channel between 220 to 600 msec at 5°C. Since they state that this process has a Q_{10} of approximately 3, the time constant at 14°C would be approximately 200 msec at maximum. This time constant is at least 2 to 3 orders of magnitude too fast to account for the very long interburst intervals that can be seen in the parabolic burster (see Strumwasser, 1967b, Figure 18; and Figures 10 and 11 of this chapter).

In conclusion, the mechanisms of slow pacemaker oscillations in the parabolic burster appear to depend on the following: (1) A special sodium channel that provides a relatively constant inward depolarizing current and that is not affected by tetrodotoxin. (2) A large contribution of this channel (about 50%) to the total "resting" conductance of the cell membrane. (3) A sodium (chloride-dependent) electrogenic pump that is probably activated by the concentration of sodium at the inner membrane surface (see Thomas, 1972).

Special biochemical characteristics of pacemaker cells

By investigating the nature of proteins unique to pacemaker cells and obtaining insights into the regulation of these molecules within the cell, one may be led eventually to the nature of the coupling between the CRM and the membrane. Autoradiographic experiments had shown that ^{3}H-leucine was incorporated into neuronal cell bodies of the parietovisceral ganglion during 4 hr incubation periods with minimal incorporation by the surrounding glia (Strumwasser and Bahr, 1966; Strumwasser, 1967a). Neurons adjacent to one another often showed markedly different incorporation that could be due to different leucine pools or different rates of protein synthesis within these cells. These experiments were extended, in our laboratory, by Wilson (1971) who studied the patterns of proteins synthesized within single neurons dissected from the ganglia at the end of an incubation period. A similar approach has been recently described by Gainer (1972) for the neurons of the terrestrial snail, *Otala*. Wilson used miniature sodium dodecyl sulfate (SDS) polyacrylamide gels to separate radioactive proteins from a single soma according to molecular weight. Pacemaker cells (such as R14, R15, and L11) in the PVG synthesize an excess of proteins around 12,000 mol wt, whereas nonpacemaker cells (such as R2 and the left pleural giant neuron) synthesize little protein in this region with peaks of synthesis around 50,000 to 60,000 mol wt (Wilson, 1971; Ram, 1972).

In order to determine the functional significance of certain of the large peaks of newly synthesized proteins, Wilson and I developed the following strategies for new experiments: We assumed that the synthesis of proteins might be regulated directly or indirectly in association with particular functions that they might subserve. For example, if one of the large peaks of newly synthesized protein were sensitive to some manipulation of pacemaker activity one might tentatively correlate this protein peak with pacemaker function. Two manipulations, one systematically giving a negative result and the other a clear positive result, will be discussed here.

Single experiments consisted of incubating one ganglion in ^{14}C-leucine and another in ^{3}H-leucine. One of these ganglia served as a control while the other was either pharmacologically treated or had certain ions in the artificial sea water substituted. One paradigm consisted of using TTX and calcium-free artificial sea water to abolish all impulses and synaptic transmission during the incubation period (see Figures 10 and 11 and associated text). Figure 16 compares a pair of R2 cells in the upper frame with a pair of R15 cells in the lower frame. No difference can be discerned between the pair of R2 cells,

one of which is a control. A similar result was obtained with the R15 pair of neurons. One should notice that the R15 patterns consist of strong peaks around 12,000 and 6,000 to 9,000 mol wt. This result in Ca-free, TTX medium was not particularly surprising in R2, because this neuron produces very few spontaneous impulses and hence the change in impulse rate would be quite minimal with this pharmacological manipulation. The result was

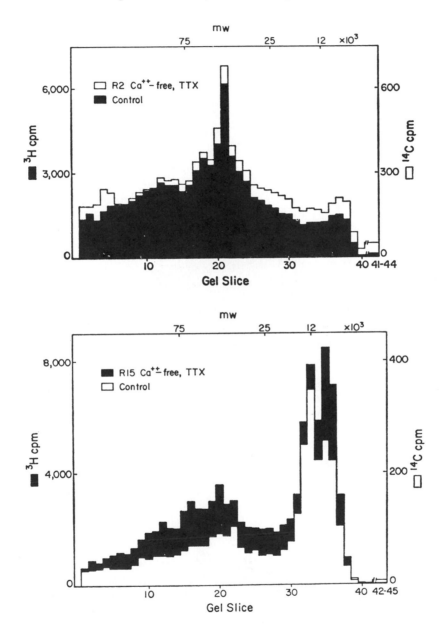

FIGURE 16   Comparison of the patterns of proteins synthesized in the R2 and R15 neurons in control conditions and during pharmacological blockade of impulses and synaptic activity. Ganglia were preincubated for 4 hr without radioactive label in either control (artificial sea water) or experimental (Ca-free, TTX) solutions. One ganglion was then shifted to ^{14}C-leucine (2.5 μC/ml) and the other to 4,5-^{3}H-leucine (100 μC/ml) media

during the 12 hr incubation period. At the end of the incubation period an R15 (or R2) neuron from each ganglion was dissected and the two R15 neurons were homogenized together. The experimental incubation medium was a Ca-free, TTX containing solution. (See text for further details.) (From Wilson, unpublished.)

somewhat surprising for R15 however, because a change in spike production is the most dramatic influence of this pharmacological manipulation; R15 produces spiking rates over 24 hr that range from 0 to 60 spikes/minute (see Strumwasser, 1965, Figures 2 to 6). It should be noted however that Ca-free, TTX medium does not block pacemaker oscillations (see Figures 10 and 11).

Wilson and I used either acetate or propionate substituted for chloride to block pacemaker oscillations in R15. Utilizing a pair of doubly labeled ganglia, we found that the 6,000 to 9,000 mol wt peak was considerably reduced in R15 in the acetate or propionate medium (Figure 17). Table I summarizes the results of 41 experi-

ments on R15. Three areas of the gels have been analyzed in detail (50,000; 12,000 and 6,000 to 9,000 mol wt regions). In control parabolic bursters approximately 10% of the counts appear in the 50,000 mol wt region while 21% and 15% appear in the 12,000 and 6,000 to 9,000 mol wt regions, respectively. It can be seen that either acetate or propionate substitutions decrease the counts in the 6,000 to 9,000 molecular weight region by 80% and 72%, respectively—a statistically very significant result. This result allows us to tentatively conclude that the 6,000 to 9,000 molecular weight protein may be correlated with a pacemaker function, perhaps related to a lower rate of the sodium-chloride dependent electrogenic

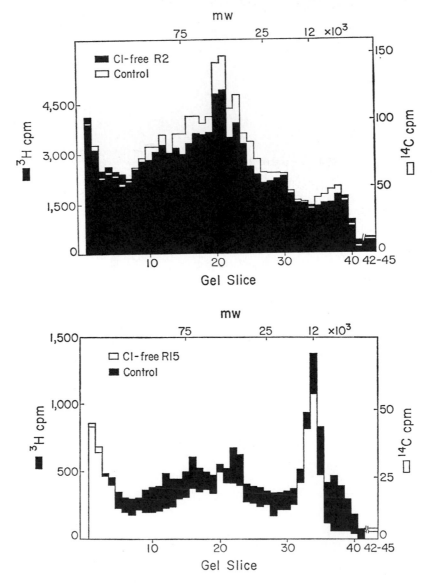

FIGURE 17   Comparison of the patterns of proteins synthesized in the R2 and R15 neurons in control conditions and during blockade of slow pacemaker oscillations by acetate substituted for chloride. Protocol as given in Figure 16 with experimental solution being an acetate for chloride substitution. (From Wilson, unpublished.)

TABLE I

The effects of ion replacements on the protein synthesis pattern of R15

	Average % cpm in gel at:*			
	50k mw	12k mw	6 to 9k mw	N
Control	9.8 ± 1.1	20.9 ± 3.7	14.7 ± 4.1	14
lo Ca^{++}, hi Mg^{++}	9.7	22.7	10.5	2
Ca^{++}-free, TTX	8.5 ± 1.1	24.2 ± 6.4	15.4 ± 1.0	3
Cl by acetate	10.0 ± 0.8	25.2 ± 4.9 ($.02 < p < .05$)	3.0 ± 0.5 ($p < .01$)	5
Cl by propionate	9.5 ± 0.7	25.2 ± 6.5	4.1 ± 0.4 ($p < .01$)	4
Na by Tris	9.0 ± 0.5	33.8 ± 5.9 ($p < .01$)	4.5 ± 0.6 ($p < .01$)	5
Na by choline	10.8	21.9	11.7	2
Na by arginine	9.4 ± 1.1	19.3 ± 3.9	19.6 ± 7.5	3
Na by lysine	10.1 ± 1.4	23.3 ± 2.0	13.7 ± 3.5	3

*Ratio of cpm in 3 gel slices at the region indicated (50k mw = 50,000 mol wt region, etc.) to total cpm in the gel times 100. Means ± standard deviations for N samples are shown.

pump under conditions of zero external chloride media. If the absence of a chloride influx in acetate or propionate artificial sea water reduces the rate of this pump and if the 6,000 to 9,000 mol wt proteins are either components or controllers of the pump, a lower rate of synthesis of these constituents might be expected and would be meaningful.

Conclusions

It appears that both slow and even circadian oscillators can occur as special mechanisms within single neurons. In the parabolic burster neuron (and the eye) of *Aplysia*, the slow 1/min and circadian oscillations already span three orders of magnitude, in terms of frequency; therefore it seems that single mechanisms governing both oscillators are unlikely. There are weakly electric fish (certain fresh water gymnotids) that generate electric pulses, used in navigation and communication, at rates around a few hundred Hz for all of their life (Lissmann, 1958). It seems unlikely that this high-frequency oscillation would have a mechanism similar to that of the parabolic burster slow pacemaker oscillator that is four orders of magnitude lower in frequency. It appears likely that different aspects of cellular organization (pumps and channels in membranes, transcriptional and translational controls in macromolecular metabolism) have evolved to cover these seven orders of magnitude of frequency.

The functional significance of oscillators can only be worked out when there is enough specific information in favorable cases. The eyes of *Aplysia*, from the evidence presented in this chapter, have a circadian rhythm of optic nerve impulses which appears to serve to synchronize other circadian oscillators. Macrobehaviors, such as

sleep, waking, and cycles of sexual activity, can be imagined to be controlled by such systems of circadian oscillators, but the details clearly need to be worked out before most of us will be convinced. Some circadian oscillators are neurosecretory. The nature of neurosecretory products and their physiological and behavioral effects is an area with sparse information but again in favorable cases, such as *Aplysia*, a single polypeptide neurosecretory product (Toevs and Brackenbury, 1969; Toevs, 1970; Arch, 1972) is known to organize behavioral egg-laying (Kupfermann, 1967; Strumwasser et al., 1969). The conservative nature of neuronal and glial mechanisms discussed in the introduction of this chapter is a constant reminder that the chances of encountering general principles from special cases, where the system is more accessible to analysis, are quite good indeed.

ACKNOWLEDGMENTS There are individuals who have helped in many phases of the original work that are not appropriately cited in the text. I am happy to use this occasion to express my gratitude to Miss Suzy Bower, Mr. Ben Ellert, Mr. J. J. Gilliam, Mr. Kent Gordon, Mrs. Sandra Smith, Miss Shelly Rempel, Mr. John Rupp and Mr. Floyd Schlechte. Original research reported in this paper has been supported by grants from the NIH (NS 07071), NASA (NGR 05-002-031) and the Sloan Foundation to the author. Computer facilities have been supported by a grant from the NSF to Professor G. Ingargiola (GJ 28424).

REFERENCES

ARCH, S., 1972. Polypeptide secretion from the isolated parietovisceral ganglion of *Aplysia californica*. *J. Gen. Physiol.* 59:47–59.

ASADA, Y., and M. V. L. BENNETT, 1971. Experimental alteration of coupling resistance at an electronic synapse. *J. Cell Biol.* 49:159–172.

AUDESIRK, G., 1971. Neuronal interactions in optic nerve impulse production in *Aplysia*. *Physiologist*. 14:105.

AUDESIRK, G., 1973. Spontaneous and light-induced compound action potentials in the isolated eye of *Aplysia*: Initiation and synchronization. *Brain Res.* (in press).

BLACKMAN, R. B., and J. W. TUKEY, 1959. *The Measurement of Power Spectra*. New York: Dover Publications.

BLOCK, G. D., 1971. Behavioral evidence for extraoptic entrainment in *Aplysia*. *Physiologist*. 14:112.

CONNOR, J. A., and C. F. STEVENS, 1971a. Inward and delayed outward membrane currents in isolated neural somata under voltage clamp. *J. Physiol. (Lond.)* 213:1–19.

CONNOR, J. A., and C. F. STEVENS, 1971b. Voltage clamp studies of a transient outward membrane current in gastropod neural somata. *J. Physiol. (Lond.)* 213:21–30.

CONNOR, J. A., and C. F. STEVENS, 1971c. Prediction of repetitive firing behavior from voltage clamp data on an isolated neuron soma. *J. Physiol. (Lond.)* 213:31–53.

EAKIN, R. M., and J. L. BRANDENBURGER, 1967. Differentiation in the eye of a pulmonate snail, *Helix aspersa*. *J. Ultrastruct. Res.* 18:391–421.

EAKIN, R. M., J. A. WESTFALL, and M. J. DENNIS, 1967. Fine structure of the eye of a nudibranch, mollusc, *Hermissenda crassicornis*. *J. Cell. Sci.* 2:349–358.

EATON, D. C., 1972. Potassium ion accumulation near a pace-making cell of *Aplysia*. *J. Physiol. (Lond.)* 224:421–440.

ENRIGHT, J. T., 1965. The search for rhythmicity in biological time-series. *J. Theoret. Biol.* 8:426–468.

ESKIN, A., 1971. Properties of the *Aplysia* visual system: In vitro entrainment of the circadian rhythm and centrifugal regulation of the eye. *Z. Vergl. Physiol.* 74:353–371.

ESKIN, A., 1972. Phase shifting a circadian rhythm in the eye of the *Aplysia* by high potassium pulses. *J. Comp. Physiol.* 80.

FABER, D. S., and M. R. KLEE, 1972. Membrane characteristics of bursting pacemaker neurons in *Aplysia*. *Nature (Lond.)* 240:29–31.

FENCL, V., G. KASKI, and J. R. PAPPENHEIMER, 1971. Factors in cerebrospinal fluid from goats that affect sleep and activity in rats. *J. Physiol. (Lond.)* 216:565–589.

FLYNN, F., and M. MAIZELS, 1949. Cation control in human erythrocytes. *J. Physiol. (Lond.)* 110:301–318.

FRAZIER, W. T., E. R. KANDEL, I. KUPFERMANN, R. WAZIRI, and R. E. COGGESHALL, 1967. Morphological and functional properties of identified cells in the abdominal ganglion of *Aplysia californica*. *J. Neurophysiol.* 30:1288–1351.

GAINER, H., 1972. Patterns of protein synthesis in individual, identified molluscan neurons. *Brain Res.* 39:369–385.

JACKLET, J., 1969a. A circadian rhythm of optic nerve impulses recorded in darkness from the isolated eye of *Aplysia*. *Science* 164:562–564.

JACKLET, J. W., 1969b. Electrophysiological organization of the eye of *Aplysia*. *J. Gen. Physiol.* 53:21–42.

JACKLET, J. W., 1972. Circadian locomotor activity in *Aplysia*. *J. Comp. Physiol.* 79:325–341.

JACKLET, J. W., R. ALVAREZ, and B. BERNSTEIN, 1972. Ultra-structure of the eye of *Aplysia*. *J. Ultrastruct. Res.* 38:246–261.

JACKLET, J. W., and J. GERONIMO, 1971. Circadian rhythm: Population of interacting neurons. *Science* 174:299–302.

KEYNES, R. D., and R. C. SWAN, 1959. The permeability of frog muscle fibers to lithium ions. *J. Physiol. (Lond.)* 147:626–638.

KUPFERMANN, I., 1967. Stimulation of egg laying: Possible neuroendocrine function of bag cells of abdominal ganglion of *Aplysia californica*. *Nature (Lond.)* 216:814–815.

KUPFERMANN, I., 1968. A circadian locomotor rhythm in *Aplysia californica*. *Physiol. Behav.* 3:179.

LICKEY, M. E., 1969. Seasonal modulation and non-twenty-four hour entrainment of a circadian rhythm in a single neuron. *J. Comp. Physiol. Psych.* 68:9–17.

LICKEY, M. E., S. ZACK, and P. BIRRELL, 1971. Some factors governing entrainment of a circadian rhythm in a single neuron. In *Biochronometry*, M. Menaker, ed. Washington, D.C.: National Academy of Sciences, pp. 549–564.

LISSMANN, H. W., 1958. On the function and evolution of electric organs in fish. *J. Exp. Biol.* 35:156–191.

NEHER, E., and H. D. LUX, 1971. Properties of somatic membrane patches of snail neurons under voltage clamp. *Pflügers Arch.* 322:35–38.

PANOFSKY, A., and F. HALBERG, 1961. II. Thermovariance spectra-simplified computational example and other method-ology. *Exp. Med. Surg.* 19:323–338.

RAM, J., 1972. Effects of high potassium media on radioactive leucine incorporation into *Aplysia* nervous tissue. *Physiologist.* 15:242.

RITCHIE, J. M., and R. W. STRAUB, 1957. The hyperpolarization which follows activity in mammalian non-medullated fibers. *J. Physiol. (Lond.)* 136:80–97.

ROTHMAN, B., and F. STRUMWASSER, 1973. Aflatoxin B1 blocks the circadian rhythm in the isolated eye of *Aplysia*. *Fed. Proc.* 32:365.

SENER, R., 1972. Site of circadian rhythm production in *Aplysia* eye. *Physiologist.* 15:262.

STRUMWASSER, F., 1965. The demonstration and manipulation of a circadian rhythm in a single neuron. In *Circadian Clocks*, J. Aschoff, ed. Amsterdam: North-Holland Publishing, pp. 442–462.

STRUMWASSER, F., 1967a. Types of information stored in single neurons: In *Invertebrate Nervous Systems*, C. A. G. Wiersma, ed. Chicago: University of Chicago Press, pp. 291–319.

STRUMWASSER, F., 1967b. Neurophysiological aspects of rhythms. In *The Neurosciences: First Study Program*, F. O. Schmitt, ed. New York: Rockefeller University Press, pp. 516–528.

STRUMWASSER, F., 1968. Membrane and intracellular mechanisms governing endogenous activity in neurons. In *Physiological and Biochemical Aspects of Nervous Integration*, F. D. Carlson, ed. New Jersey: Prentice-Hall, pp. 329–341.

STRUMWASSER, F., 1971. The cellular basis of behavior in *Aplysia*. *J. Psychiat. Res.* 8:237–257.

STRUMWASSER, F., 1973. Neural and humoral factors in the temporal organization of behavior (The Seventeenth Bowditch Lecture.) *Physiologist.* 16:9–42.

STRUMWASSER, F., and R. BAHR, 1966. Prolonged in vitro culture and autoradiographic studies of neurons in *Aplysia*. *Fed. Proc.* 25:512.

STRUMWASSER, F., J. W. JACKLET, and R. B. ALVAREZ, 1969. A seasonal rhythm in the neural extract induction of behavioral egg-laying in *Aplysia*. *Comp. Biochem. Physiol.* 29:197–206.

STRUMWASSER, F., and M. KIM, 1969. Experimental studies of a neuron with an endogenous oscillator and a quantitative model of its mechanism. *Physiologist.* 12:367.

STRUMWASSER, F., C. LU, and J. J. GILLIAM, 1966. Quantitative studies of the circadian locomotor system in *Aplysia*. *Calif. Inst. Tech. Biol. Ann. Report,* p. 153.

STRUMWASSER, F., F. R. SCHLECHTE, and S. BOWER, 1972. Distributed circadian oscillators in the nervous system of *Aplysia*. *Fed. Proc.* 31:405.

STRUMWASSER, F., F. R. SCHLECHTE, and J. STREETER, 1967. The internal rhythms of hibernators. In *Proceedings of the Third International Symposium on Mammalian Hibernation*, K. Fisher, ed. Edinburgh: Oliver and Boyd, pp. 110–139.

THOMAS, R. C., 1972. Electrogenic sodium pump in nerve and muscle cells. *Physiol. Rev.* 52:563–594.

TOEVS, L., 1970. Identification and characterization of the egg-laying hormone from the neurosecretory bag cells of *Aplysia*. Ph.D. Dissertation, California Institute of Technology, Pasadena, California.

TOEVS, L. A., and R. W. BRACKENBURY, 1969. Bag cell-specific proteins and the humoral control of egg laying in *Aplysia californica*. *Comp. Biochem. Physiol.* 29:207–216.

WILSON, D. L., 1971. Molecular weight distribution of protein synthesized in single, identified neurons of *Aplysia*. *J. Gen. Physiol.* 57:26–40.

40 Aspects of the Physiology of Circadian Rhythmicity in the Vertebrate Central Nervous System

MICHAEL MENAKER

ABSTRACT   The physiology of circadian organization in birds and mammals is compared with respect both to coupling of the organism with external light cycles and to the coupling involved in the maintenance of internal temporal order. Differences between the two groups are emphasized. In entrainment to light cycles birds utilize at least one, and possibly several, brain photoreceptors in addition to the eyes; mammals use the eyes alone. Internal temporal order in birds appears to depend heavily on the pineal organ that functions either as a master driving oscillator or as a central coupling device among various peripheral rhythmic processes; on the other hand, mammalian circadian organization depends on mutual coupling among many self-sustained oscillators and not on a single central driving or coupling mechanism. In spite of the existence of these apparently major differences between birds and mammals, their depth and significance cannot yet be fully assessed.

On the occasion of his sixtieth birthday, this paper is dedicated to Professor Jürgen Aschoff whose enthusiasm and insight have inspired all who know him.   The Author.

Introduction

THERE ARE AT present two major sets of problems in the study of circadian rhythms. The first concerns the nature of the subcellular oscillator that underlies the overt circadian rhythms that we can measure in many eukaryotic cells and that probably exists in them all. Within this set are questions about the physicochemical nature of the oscillator, its location within the cell, its genetic control, and its source of energy as well as the mechanism of its control by environmental cycles of light and temperature and therefore its "sensory physiology." The second set, with which this chapter attempts to deal, concerns the organization of the circadian *system* in complex multicellular organisms. Here we are primarily interested in questions of coupling; coupling within the organism among the multiple overt circadian rhythmicities that have been well documented at several

MICHAEL MENAKER   Department of Zoology, University of Texas, Austin

different levels of organization and coupling of the whole circadian system with the natural environment.

While we are entitled at least to hope that common mechanisms may exist among diverse organisms at the subcellular level, which will illuminate and unify our efforts to understand circadian oscillations within cells, such a hope applied to the circadian systems of multicellular organisms in groups as diverse as plants and mammals is almost certainly vain. It therefore seems expedient to take a comparative approach—studying in as much detail as possible the coupling relations among components of the circadian systems of organisms from different phylogenetic groups. In this way we can hope ultimately to define the range and variety of mechanisms employed in circadian organization and to make easier the analysis of any specific new system that becomes particularly interesting.

At the physiological level, the comparative study of circadian coupling relations has scarcely begun. There has been a good deal of descriptive work on circadian rhythms in the fields of sensory and neurophysiology and endocrinology, more than enough in fact to indicate that understanding of the circadian aspects of these disciplines will, when it comes, constitute a powerful analytic tool in the disciplines themselves and to verify the natural a priori assumption that within these three fields lie the keys to an understanding of the organization of circadian systems. On the other hand, work on any one group of organisms has not yet been intensive enough to lay bare more than a bone or two of the circadian skeleton. Thus in perhaps the best-known circadian subsystem, the mammalian hypothalamic-pituitary-adrenal axis, the hierarchy of circadian control mechanisms is far from understood, and the driving, coupling, and feedback relationships among the constituent rhythms can only be guessed at with the information at hand. Even the perception of environmental light cycles that entrain the circadian system has turned out to be much more complex than one would have naively expected. The recent demonstration of multiple interacting photoreceptors for entrainment in the reptiles and birds and of at least special anatomical routes conveying lighting information to the mammalian circadian system underlines our ignorance of exactly how light reaches or affects the clock of any vertebrate.

In spite of the difficulties, the field of circadian rhythms has by now arrived at the point at which explicit discussion of the physiology of coupling can and should be attempted for heuristic if for no other reasons. In this chapter, I have attempted such a discussion focused on some aspects of coupling relations in the circadian systems of birds and mammals—coupling of the circadian system to the external light cycles (external coupling) and coupling among the components within the system (internal coupling). Of necessity this is a speculative exercise, the primary goal of which is to identify the available experimental handles and to suggest lines for future thought and experiment.

External and internal coupling in the mammalian circadian system

EXTERNAL COUPLING   Although several other environmental parameters have been shown to be capable of entraining mammalian circadian rhythms, notably cyclic social cues in humans, the natural light cycle clearly constitutes the dominant environmental synchronizer for most species. Recent work with submammalian vertebrates (see below), demonstrating the existence of extraretinal photoreception in entrainment to light cycles, forces one to begin a discussion of the physiology of the coupling of organisms to light cycles by asking whether the photoreceptors involved have been positively identified.

Alone among the vertebrates, the mammals appear to employ exclusively ocular photoreception for entrainment of at least the majority of circadian rhythms. Studies from several different laboratories agree in the conclusion that the overt circadian rhythms of mammals blinded by bilateral enucleation fail to entrain to artificial light cycles (Browman, 1943; Bruss et al., 1958; Richter, 1965). In fact, although light has been shown to penetrate deep into the brain of a mammal with a skull as thick as that of a sheep (Ganong et al., 1963), there are only two known cases of extra-ocular photoreception by mammals. One concerns the effect of light on biochemical processes in the pineal of neonatal, but not of adult, rats (Zweig et al., 1966); the other reports an effect of light, piped directly to the hypothalamus of adult female rats with quartz rods, on their estrous cycles (Lisk and Kannwischer, 1964). Either or both of the end points used in these studies could involve circadian components; in neither case has this been directly demonstrated.

The identity of the photoreceptors within the mammalian eye that mediate entrainment is completely unknown. There is no information available that would help to assign this function either to the rods or cones or to particular restricted groups of these structures. Indeed recent work with laboratory rats deprived of their known retinal photoreceptors by prolonged exposure to continuous illumination suggests strongly that there may be ocular photoreceptors, at least in this species, the existence of which was previously unsuspected (Anderson and O'Steen, 1972; Bennett et al., 1972). Rats treated in this way retain a variety of responses to light, including undiminished resolving power (by behavioral measurement) and entrainability by light cycles (Dunn et al., 1972). In any case, and especially in view of the surprises to which we have been subjected in the past 4 or 5 years concerning photoreception for entrainment in both the invertebrates and the lower vertebrates, we are not entitled to any easy a priori assumptions about the identity of this functional component of the mammalian circadian system.

Like its exact identity, the physiology of the entrainment photoreceptor is largely unknown. Systematic studies of action spectra (Gordon and Brown, 1971) and thresholds are almost entirely lacking. The photopigments involved are unknown, and there is not even one careful study investigating the reciprocity, or lack of it, between the intensity of an entraining stimulus and its duration.

As a result of several recent studies, we do have some information on the anatomical routes by which ocular photoreceptors are coupled to regions of the brain that may exert control over the mammalian circadian system. Information about the external light cycles that affect biochemical rhythms in the pineal apparently arrives at this organ by way of the accessory optic tracts (Moore et al., 1967; Moore et al., 1968). Direct retino-hypothalamic fibers may be involved in the entrainment of eating, drinking, and locomotor rhythms (Moore and Eichler, 1972; Stephan and Zucker, 1972).

In summary, circadian photoreception in mammals appears to be "centralized" at least to the eye, whereas the routes by means of which the photoreceptors communicate with the brain are probably multiple, discrete, and different from those involved in pattern vision. Among the vertebrates this mode of organization is probably unique; it is, as we shall see, very different from the situation in birds.

INTERNAL COUPLING   Overt circadian rhythmicity can be observed simultaneously in many different functions of a single individual mammal (Halberg, 1960). Under most experimental conditions, these multiple rhythms have the same period length and bear fixed, presumably adaptive, phase relationships to each other and to environmental cycles. It is useful to consider the totality of circadian rhythms known and unknown in an individual as a set that we can then call the circadian system. Certain very abnormal environmental conditions, notably continuous light, may cause rhythms in particular functions (i.e., components of the system) to dissociate from each other—to free run with different periods and to take up abnormal or unstable phase relationships. The system can be artificially separated, at least partially, into its component parts.

Dissociation has been observed in mammalian circadian systems at several levels of organization. At the level of the whole organism, dissociation has been observed between the circadian rhythm of body temperature and the rhythm of locomotor activity in man (Aschoff, 1969). Different rhythmic functions of the same organ sometimes dissociate. The best example comes from studies of the multiple excretory rhythms of the kidney, which dissociate when human beings from temperate latitudes are exposed to the continuous light of the Arctic summer and may take months to regain their normal phase relationships when the subjects return home (Lobban, 1960). Components of a "single" circadian function are known to dissociate in some cases. Rhythms of both locomotor activity and body temperature sometimes split into multiple components that then behave quasi-independently of each other (Menaker, 1959; Hoffman, 1971; Pittendrigh, this volume).

These observations raise profound questions about the internal temporal organization of mammals, questions that can of course be asked of other groups of multicellular organisms and that by analogy can be extended to other noncircadian levels of rhythmic temporal order. Do all overt circadian rhythms passively follow the behavior of some master oscillator that forces rhythmicity on them? If so, how are we to understand dissociation? Is there, rather, a single endogenous driving oscillator that maintains the integrity of the circadian system by imposing phase and period on each of the component rhythms much as a light cycle imposes phase and period on the whole organism in the normally entrained situation? Are there separate driving oscillations for subsets of overt rhythms with related functions? Does each overt rhythm have its own autonomous, self-sustained driving oscillator? What role does feedback play in maintaining the temporal organization of the system?

In order to rephrase some of these questions in concrete form, I have attempted to apply them to a specific case, the mammalian hypothalamic-pituitary-adrenal axis, in which the rhythms are well described while the circadian coupling relations are still largely a matter of conjecture.

Control of the mammalian adrenal cortex is a complex subject with a literature far too large to review here. In order to keep the discussion within manageable bounds, I will confine it specifically to rhythmicity in three of the major components; corticotropin-releasing factor from the median eminence of the hypothalamus (CRF), adrenocorticotropic hormone from the anterior pituitary (ACTH), and the glucocorticoids of the adrenal cortex (for present purposes, corticosterone). The levels of each of these substances have been shown to vary rhythmically in light-dark (LD) cycles (Critchlow, 1963; Retiene, 1968; Hiroshige, 1971). [It is important to insert a caveat here. Because it has so far proved impossible to make frequent repeated measurements of hormone levels in single animals, published reports always refer to the rhythmicity of a population from which individuals are selected and sacrificed at intervals throughout the day. While this difficulty has little effect on data collected from animals held in LD cycles, it complicates enormously the interpretation of data from constant light or darkness and all but precludes the rigorous demonstration that the rhythms studied are in fact circadian.] All three rhythms are entrainable by light cycles and can be phase shifted by shifting the phase of the entraining cycle. They are differentially phased with respect to light and at least to a first approximation their phase relationships make sense with respect to their known physiological relationships: CRF peaks before the others, ACTH peaks either just before or coincident with the peak of corticosterone, corticosterone (in rats) peaks in the early part of the night. The simplest possible explanation of their rhythmic interrelationships would be that the rhythm in CRF levels forces rhythmicity on a passive pituitary which as a result secretes ACTH rhythmically in turn forcing rhythmic secretion of corticosterone from a passive adrenal cortex.

This is clearly not the case, for if the adrenal is isolated in organ culture, it continues rhythmic secretion of corticosterone for as long as 10 days in vitro (Andrews, 1971). The isolated adrenal clearly has a *circadian* rhythm, for a single gland is persistently rhythmic in culture. The phase of the in vitro rhythm can be shifted with pulses of ACTH applied at some times of day while identical pulses applied at other times are ineffective (Figure 1). This suggests that the adrenal per se has a phase response curve to ACTH. The cultured organ also has a circadian rhythm of O_2 consumption that does not shift phase when the gland is pulsed with ACTH and that has been shown to be temperature compensated (Andrews and Folk, 1964).

Neither can the rhythmicity of any component within this system be explained solely on the basis of feedback from the others, for the rhythm in ACTH persists after adrenalectomy (Cheifetz, 1968), and the rhythm in CRF persists following either hypophysectomy or

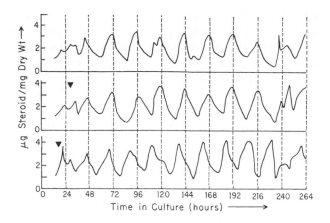

FIGURE 1 Rhythmic corticosteroid secretion by isolated, organ-cultured hamster adrenals for 10 consecutive days. The upper trace represents data from untreated controls; the middle and lower traces represent data from glands treated with single pulses of ACTH (1.0 i.u.) at the times indicated, solid triangle. Note that the phase of the rhythm is shifted by the pulse of ACTH in the lower but not in the middle trace. (Redrawn from Andrews, 1971.)

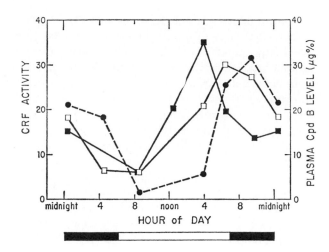

FIGURE 2 Effect of adrenalectomy on the phase of the rhythm of CRF in the median eminence. The CRF rhythm in normal animals is represented by the curve connecting the open squares; that in adrenalectomized animals by filled squares. The rhythm of plasma corticosterone level from normal animals, represented by filled circles, has been included for reference. (Redrawn from Hiroshige and Sakakura, 1971.)

adrenalectomy (Hiroshige and Sakakura, 1971; Takebe et al., 1971; Seiden and Brodish, 1972). On the other hand, feedback mechanisms almost certainly do influence phasing within this system, for in the presence of a light cycle, the peak of CRF is phase-advanced by adrenalectomy (Figure 2) and by hypophysectomy (Takebe et al., 1971). Further evidence that phase relationships within this system depend on the hormonal milieu comes from the observation of large differences between males and females in the phase of the ACTH peak with respect to the light cycle and probably also with respect to the corticosterone rhythm (Critchlow, 1963).

The observation of circadian rhythmicity in corticosterone secretion by the organ-cultured adrenal demonstrates beyond doubt that at least this individual organ contains all the machinery necessary for independent circadian function. Taken together with the experimental results outlined above, this suggests that the mammalian circadian system may be composed of independently oscillating functions that achieve adaptive significance through largely unexplored, hierarchically arranged, mutual coupling. As with photoreception, the avian circadian system may well be organized along different lines.

External and internal coupling in the avian circadian system

EXTERNAL COUPLING In birds as in mammals, light is the dominant environmental entraining agent, but its perception is accomplished in radically different ways in the two classes of vertebrates. The circadian rhythms of birds remain entrainable by light cycles following blinding by bilateral optic enucleation (Menaker, 1968a). Suitably controlled experiments have established that this is an effect of wavelengths in the visible region of the spectrum. Entrainment is mediated by as yet unidentified photoreceptors in the brain (Figure 3), which are surprisingly sensitive to light (Menaker, 1968b). The circadian rhythm of locomotor activity in about 50% of blind house sparrows tested, entrains to LD cycles consisting of 12 hr of light at 0.1 lux and 12 hr of complete darkness. The threshold for entrainment of blind sparrows thus approximates the intensity of bright moonlight. (In the field, this diurnal organism is of course exposed to light intensities many orders of magnitude higher.)

The understanding of circadian photoreception in birds is complicated by the fact that the eyes also contribute to the perception of entraining light cycles. The light intensity threshold for entrainment of normal sparrows has not been determined, but it is clearly lower than that for blind birds, as all normal sparrows will entrain to the same stimulus that entrains only 50% of the blind ones. Furthermore there is some evidence that suggests that the eyes and the brain photoreceptors play different roles in supplying information about lighting conditions to the circadian system. Arrhythmicity in continuous light, which is a common response of circadian rhythms, can be produced in normal sparrows at an intensity of about 50 lux. Despite the demonstrated sensitivity of the brain photoreceptor to entraining

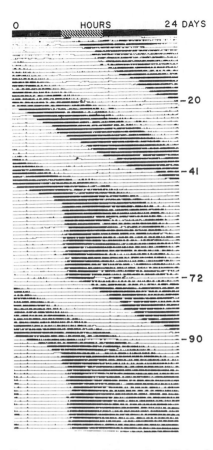

—20

—41

—72

—90

FIGURE 3 Entrainment to a light cycle utilizing brain photo-reception. The figure is a photograph of raw locomotor activity data collected from a blind sparrow continuously for a period of 118 days. The light cycle which is diagrammed at the top of the figure (L:D, 6:18—light of approximately 0.06 lux indicated by the shading; complete darkness indicated by the black bars) was present throughout the experiment. Initially (top of the figure) the bird's circadian rhythm of locomotor activity free runs in the presence of the light cycle—the light intensity used was below the threshold for entrainment of this blind bird. On day 20 (indicated by the number at the right), feathers were plucked from the bird's back without affecting the free running rhythm. On day 41 feathers were plucked from the head increasing the intensity of the light reaching the brain by an order of magnitude; the bird then entrained to the light cycle. On day 72, india ink was injected between the skull and skin of the head blocking light to the brain; the bird free-ran until day 90 when a small piece of skin and the ink below it was removed and the bird entrained once more. (From Keatts, 1968.)

stimuli, blind sparrows do not become arrhythmic at light intensities up to 2,000 lux (Menaker, 1968a).

Extraretinal photoreception has been shown to be involved in entrainment of circadian rhythms in fish (Reed, 1968), amphibians (Adler, 1969) and reptiles as well as in birds. In lizards (Underwood and Menaker, 1970) removal of the lateral eyes, the parietal eye, and the pineal organ fails to abolish the entrainment response to

light. The eyes also participate and appear to play a special role, although a somewhat different one from that which they play in the avian circadian system. The parietal eye and the pineal organ do not seem to be involved even though both organs have been clearly shown by neurophysiological criteria to function as photo-receptors. On the other hand, amphibians do appear to use the frontal organ (probably homologous to the parietal eye of lizards) for perception of entraining light cycles (Adler, 1971).

It has been known for some time that brain photo-receptors are involved in the perception of light in the day-length measuring system used by ducks to synchronize their reproductive cycles with season (photoperiodism) (Benoit, 1964). Recently this work has been extended to several other avian species, including the house sparrow (Menaker and Keatts, 1968; Homma and Sakakibara, 1971; Gwinner et al., 1971; Homma et al., 1972). In sparrows it has been possible to compare brain photoreception involved in photoperiodism with that involved in entrainment of circadian rhythms and two profound differences have emerged. First, the thresholds for the two responses to light are very different. While an intensity of 0.1 lux is sufficient for entrainment by brain photoreceptors, about 10 lux is needed to produce testis growth in response to long days in either normal or blind sparrows, a difference in threshold of two orders of magnitude. Second, while the eyes are clearly involved in some way in entrainment, they do not appear to be involved at all in photoperiodic photoreception (Menaker et al., 1970). No differences between blind and normal sparrows have been observed in the photoperiodic response to light. Moreover normal sparrows with fully functional eyes do not use them in photoperiodic photoreception (Figure 4). It may well be that, at least in sparrows, there are two or more brain photoreceptors mediating light perception for the two systems that are known to utilize extraretinally perceived cues.

Additional interest is lent to this situation by the demonstration of an intimate relationship in birds between circadian rhythmicity and day length measurement controlling the reproductive response. Experiments of the "resonance" type demonstrate that the circadian system influences the interpretation of photoperiodically active light signals (Hamner, 1963; Turek, 1972). A somewhat different experimental paradigm has been used in house sparrows to show that the circadian system performs the day length measurement (Menaker and Eskin, 1967). It thus appears that in birds two and possibly three distinct photoreceptors are coupled to different functional aspects of the circadian system.

The pineal organ is not the primary brain photo-ceptor involved either in entrainment or in photoperiodic

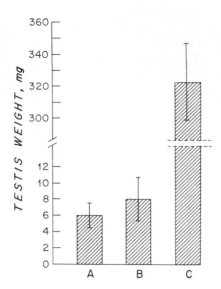

FIGURE 4 Eyes, although present and normal, do not contribute to photoperiodic phororeception in sparrows. Two groups of birds (B and C) with regressed testes were exposed for 39 days to a stimulatory light cycle (*L*:D*16*:8) in which the intensity of the light portion of the cycle (10 lux) was just above the threshold for the photoperiodic response. The eyes of both groups were intact but the birds in group B had received india ink injections between the skull and skin of the head, while those in group C had had feathers plucked from their heads. The testes of the birds in group C grew in response to the stimulatory photoperiod whereas those of the birds in group B did not grow significantly compared with controls taken at the beginning of the 39 day experimental period (group A). Thus, blocking light to the brain, even in the presence of normally functioning eyes, abolishes photoperiodic photoreception. (From Menaker et al., 1970.)

photoreception. Pinealectomy of house sparrows does not abolish the reproductive response to light; blind, pinealectomized sparrows still entrain to light cycles although their entrainment patterns are more variable than are those of blind or normal birds (Menaker, 1971). This variability could result from the removal of an ancillary photoreceptor but is also interpretable in other ways (see below). Although it is not always treated as such, the question of whether the avian pineal functions as a photoreceptor is completely separable from the above considerations.

Although there is no direct evidence in favor of assigning a photoreceptive function to the avian pineal, there are good evolutionary and morphological grounds and even some indirect physiological evidence for suggesting that this may be the case. The pineal of fish is almost certainly photoreceptive (Dodt, 1963), and there is solid electrophysiological evidence for photoreceptivity of the pineals of both amphibians and reptiles (Dodt and Heerd, 1962; Hamasaki and Dodt, 1969). In the pineals of birds (among which that of the house sparrow has been par-

ticularly well studied), there are found a number of structures comparable to those found in the pineal receptor cells of lower vertebrates (Oksche and Kirschstein, 1969; Ueck, 1970). Bulbous cilia of the 9-0 type protrude into the lumen of the organ and form concentric lamellar whorls (Figure 5). The ultrastructure of the inner segments of avian receptor-like pineal cells is very similar to that of the receptor cells in lower vertebrates. In the house sparrow, they contain synaptic ribbons that probably

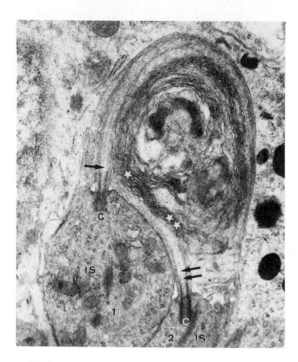

FIGURE 5 Structures that partly resemble outer segments of pineal photoreceptor cells of lower vertebrates can be observed in the pineal organ of *Passer domesticus*. The electron micrograph shows two pinealocytes (1, 2). These cells display an inner segment (IS) rich in mitochondria and granular endoplasmic reticulum. A cilium (arrow and double arrow) originates from a centriole (C) belonging to a pair of centrioles. Systems of lamellae (* and **) arise from both cilia and contribute to a circular, whorl-like body. If compared with the comblike (Anura) or dome-like (Teleostei) outer segments in pineal organs with proven photoreceptor function, the whorl body of *Passer domesticus* appears to be less regular. (From a manuscript in preparation by Oksche, Kirschstein, Binkley, Silver, and Menaker.)

synapse with a recently discovered abundant system of small acetylcholinesterase-positive neurons that run down the pineal stalk and into the brain (Ueck and Kobayashi, 1972).

Systematic exploration of the pineal of the pigeon with recording microelectrodes failed to reveal an effect of light applied, either directly to the pineal or to the lateral eyes, on electrical activity (Morita, 1966). Essentially the same results were obtained with the Japanese quail and the house sparrow (Ralph and Dawson, 1968). On the other hand, neither blinding nor superior cervical ganglionectomy (which presumably denervates the pineal) blocked the rise in HIOMT activity which normally occurs in the pineal of chicks in constant light (Lauber et al., 1968). Either there is an unknown route by means of which extraretinally perceived light is able to influence pineal HIOMT (hydroxyindole-O-methyltransferase) levels or the chick pineal is itself photoreceptive.

If the avian pineal is photoreceptive, it could, like the eyes, play some special role in circadian photoreception or be involved in a completely different system. It will be remembered that the photoreceptive pineal complex of lizards does not appear to participate in the entrainment of the circadian system and in fact has no known photoreceptive function.

In contrast to that of mammals, the avian circadian system receives information about environmental light from at least two anatomically distinct photoreceptors (eyes and brain photoreceptor) and possibly from as many as four such structures (eyes, brain photoreceptor for entrainment, brain photoreceptor for reproductive system, and pineal organ). Of course there may be others, the existence of which is not yet suspected. The poikilo-thermic vertebrates and many invertebrates also appear to have multiple photoreceptors with inputs to the circadian system.

INTERNAL COUPLING Although pinealectomy in sparrows does not abolish entrainment by light cycles, it does have other, profound effects on the circadian system. Pinealectomy completely abolishes the free-running circadian rhythm of locomotor activity that is invariably observed in normal birds held in constant darkness or constant dim light (Gaston and Menaker, 1968). In birds entrained to light-dark cycles, it also changes the phase relationship of the onset of locomotor activity to the entraining cycles (Gaston, 1971).

Without their pineals, sparrows are continuously active in constant darkness, and their locomotor records are arrhythmic both by visual inspection and by power spectrum analysis. However, the absolute amount of activity is not increased by pinealectomy; only its distribution in time is affected (Binkley et al., 1972). Arrhythmicity occurs within one or at the most two cycles in birds that are pinealectomized while free running in constant darkness (of course they must be briefly exposed to light for surgery), but if a pinealectomized bird is entrained to an LD cycle and then placed in constant darkness, arrhythmicity often develops slowly over as many as ten cycles by progressive daily increase in the duration of activity (Figure 6).

Other free running circadian rhythms are also abolished by removing the pineal organ of sparrows. No circadian fluctuations can be extracted from continuously telemetered body temperature records of pinealectomized birds held in constant darkness although normal sparrows under these conditions display unusually clear rhythms with amplitudes of 3 to 4°C (Figure 7). Recent results indicate that the excretion of uric acid in the urine of normal sparrows in constant darkness has a clear circadian rhythm and that this rhythm too is abolished by pinealectomy (Mackey and Menaker, unpublished).

Preliminary evidence indicates that severing the nerve tract leaving the pineal through its stalk (the only known neural output from the organ) has no effect on the circadian locomotor rhythm (Rouse and Menaker, unpublished). It has occasionally been possible to maintain rhythmicity in birds whose own pineals have been removed and replaced with the pineal of a donor sparrow implanted in the pineal site (Gaston, 1971). In the latter case, the high failure rate is probably due to technical difficulties which are, in the context of this discussion, trivial. These observations suggest the tentative hypothesis that the sparrow pineal exerts its effect on the circadian system hormonally. A detailed study of the anatomy of the sparrow pineal and its vasculature suggests further that the hormone(s) concerned are probably released into the general circulation rather than into the cerebrospinal fluid or into a specialized portal system (Silver, 1972).

The effects of pineal removal on the circadian system of sparrows clearly relates to the general questions about internal coupling raised earlier. In birds there is no evidence for internal desynchronization among components of the circadian system as there is for mammals (see however, Aschoff and Pohl, 1970; Pohl, 1971). Neither are there any reports in the literature of persistent circadian rhythmicity in isolated avian organs in culture. As convincing reports of both phenomena are rare even in mammals, with which a great deal more research has been done, the absence of such reports for the avian system must not be overinterpreted. At the moment, however, we have no grounds whatsoever for assuming that circadian organization in these two vertebrate classes is identical or even very similar.

The arrhythmicity induced by pinealectomy of sparrows cannot be due directly to removal of a putative photoreceptor. The major effect of the operation is to produce a change in the behavior of the bird in continuous darkness. In any case, the effect is in the wrong direction; arrhythmicity can be induced by exposure to bright constant light, but one would hardly expect to mimic this effect simply by removing a photoreceptor. It could be argued that pinealectomy is in some physiological sense equivalent to exposing sparrows to bright constant light. Three lines of reasoning argue, though not conclusively, against this hypothesis: There are differences between

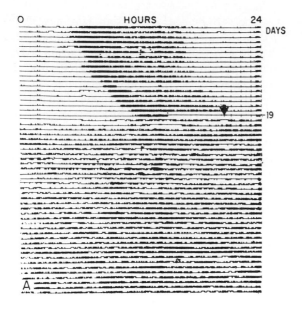

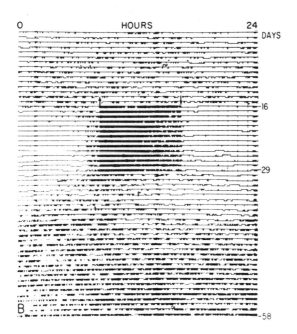

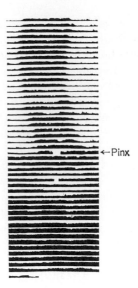

FIGURE 7   Effect of pinealectomy on the circadian rhythm of body temperature in sparrows. The figure shows 46 days of continuously telemetered body temperature data from a single sparrow held in constant darkness. On the day indicated (← Pinx) the bird was pinealectomized. As with locomotor activity (Figure 6A) rhythmicity disappears almost immediately. Several artifacts can be seen in the data during the 12 days following pinealectomy. (From Binkley et al., 1971.)

FIGURE 6   Effect of pinealectomy on the circadian rhythm of locomotor activity of sparrows. The raw data have been handled as in Figure 3. In A, a sparrow free running in constant darkness was pinealectomized on day 19 at the time indicated by the arrow. Note the almost immediate occurrence of arrhythmicity. In B, a pinealectomized bird, arrhythmic in constant darkness was exposed to a light cycle (↑ lights on; ↓ lights off) for 14 days beginning on day 16 and then returned to constant darkness on day 29. Note: (1) that the pinealectomized bird entrains to the light cycle; (2) that, during entrainment the onset of locomotor activity phase leads the onset of the light portion of the cycle; and (3) that the return to arrhythmicity following entrainment is gradual, requiring about 8 days. (From Gaston and Menaker, 1968.)

birds made arrhythmic by pinealectomy and those made arrhythmic by exposure to bright constant light both in the absolute amount of activity and in the relationship of the activity to body temperature (Binkley et al., 1971; Binkley et al., 1972); the gonads of pinealectomized birds are still maintained in the regressed condition by short days under the same conditions in which they would grow if exposed to constant light (Donham and Wilson, 1969); although pinealectomized birds are continuously active in constant darkness, they remain entrainable by light cycles and locomotor activity is suppressed during the major part of the dark portion of the cycle.

The effects of pinealectomy on circadian rhythmicity in sparrows are completely consistent with either of two formulations concerning the organ's role in maintaining the integrity of the circadian system. The pineal may be the seat of a master driving oscillation that normally entrains numerous other oscillators each of which is responsible for circadian rhythmicity in a particular function. Since all rhythms measured so far disappear following pinealectomy, we must at present assume that the oscillators driven by the pineal are not self-sustained but rather are damped and persist for only a few cycles when deprived of their driver. We must further assume that these damped oscillations can be driven directly by light cycles as well as by the pineal organ. On the basis of this interpretation, one can see the phase change with respect to LD cycles that is produced by pinealectomy as reflecting the differences in the entrained steady state between the system with and without its driver. The gradual loss of rhythmicity of pinealectomized birds transferred from LD to DD directly reflects the damping of the now driverless system; this occurs more slowly than it does when the pineal is removed from a bird free running in DD because the driving influence exerted directly by light cycles is in some way stronger than that exerted by the pineal in the absence of light.

Alternatively the pineal could be a coupling device between a master driving oscillator and other damped, light sensitive oscillators that in turn drive overt circadian rhythms.

Neither of these formulations fits with what is known of the organization of the mammalian circadian system. The facts that (1) an isolated organ maintains circadian rhythmicity in culture and (2) overt rhythms can be dissociated from one another suggest that in the mammalian system there *is no master driving oscillation* in the sense that we have suggested here for birds. The most that could be expected at the top of the mammalian circadian hierarchy would be an oscillator that entrains a population of self-sustained suboscillations that would dissociate rather than damp out following its removal. Note, in this connection, that the three circadian rhythms that have been shown to damp out following pinealectomy in sparrows—locomotor activity, body temperature, and an excretory rhythm of the kidney—are precisely those that have been shown to dissociate from one another under special circumstances in mammals.

The evidence at hand, incomplete though it is, forces us toward the conclusion that the avian circadian system is quite different in both its external and internal coupling relations from that of the mammals. On the other hand, it may well be that I have overstated the case in favor of the existence of fundamental differences between the circadian systems of these two vertebrate classes: Multiple photoreceptors with discrete functions may yet be discovered within the mammalian eye, and other circadian subsystems, not under pineal control, may be found in birds. In any case the approach to circadian organization outlined here has been at least heuristically valuable in our hands. I should like to illustrate this by describing an hypothesis concerning the function of the mammalian pineal to which this approach together with some recent data, has led us.

The mammalian pineal

There are almost certainly circadian aspects to the functioning of the mammalian pineal; several steps in the synthesis of biogenic amines have been shown to be rhythmic (Snyder et al., 1965; Klein, 1973). On the other hand, pinealectomy has only subtle if any effects on the circadian rhythms of locomotor activity and body temperature in mammals (Kincl et al., 1970; Quay, 1970, 1972).

The mammalian pineal has long been known to influence the reproductive system. It appears to exert this control by secreting an antigonadal substance(s), not as yet positively identified, in the absence of light sensed by the eyes and transmitted to the pineal by way of the superior cervical ganglion (Wurtman et al., 1968). Although this effect has been demonstrated in several mammalian species it is clearest in the golden hamster (Reiter and Sorrentino, 1970).

The reproductive system of hamsters is under photoperiodic control and recent experiments (Elliott et al., 1972) have shown that photoperiodic time is measured in hamsters, as in birds, with reference to a circadian rhythm of photoperiodic photosensitivity (Figure 8).

Considered within the framework that has been developed above, these facts lead naturally to the hypothesis

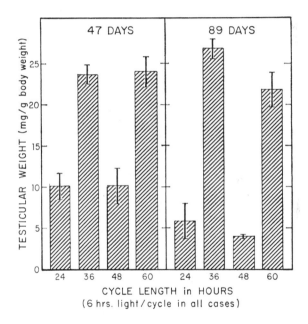

FIGURE 8 Resonance in the photoperiodic control of reproductive state in hamsters. Hamsters require long days to maintain the weight and spermatogenic capacity of the testis. In the experiment diagrammed here, 4 groups of hamsters with fully mature gonads were exposed to photoperiods in which 6 hr of light were coupled with 18, 30, 42, and 54 hr of darkness (*LD* 6:18, *LD* 6:30, *LD* 6:42 and *LD* 6:54). Animals were killed after 47 and 89 days of exposure to these cycles. Note that *LD* 6:30 and *LD* 6:54 maintain the testis for the entire duration of the experiment while in *LD* 6:18 and *LD* 6:42 there is a progressive decline in testis weight, indicating the involvement of the circadian system in photoperiodic time measurement. (For a full discussion of the meaning of resonance experiments see Pittendrigh, this volume.) (From Elliott et al., 1972).

that the mammalian pineal plays an important role in the circadian system but one that may be primarily limited to the circadian aspects of reproduction. The mammalian pineal may contain one of a population of normally coupled, self-sustained oscillators, specifically that one which performs the day length measurement in the control of reproductive state.

Virtually nothing is known about the physiology of the reptilian pineal. It has not been implicated in the control either of circadian rhythmicity or of reproduction. However, the fact that it is photoreceptive and has neural connections with the brain renders it almost certain that it functions as a mediator of some kind between environmental light conditions and the organism's internal state (Stebbins and Wilhoft, 1966). Each in its own way, the birds and the mammals appear to have taken evolutionary advantage of the constitutional opportunities afforded by the pineal organ as inherited from the reptiles.

ACKNOWLEDGMENTS   Much of the experimental work reported here was supported by NIH grant HD 03803 and NIH career development award HD 9327 to M.M. I thank N. Headrick and J. Rogers for technical assistance.

REFERENCES

ADLER, K., 1969. Extraoptic phase shifting of circadian locomotor rhythm in salamanders. *Science* 164:1290–1291.

ADLER, K., 1971. Pineal end organ: Role in extraoptic entrainment of circadian locomotor rhythm in frogs. In *Biochronometry*, M. Menaker, ed. Washington, D.C.: Nat. Acad. Sci., pp. 342–350.

ANDERSON, K. V., and W. K. O'STEEN, 1972. Black-white and pattern discrimination in rats without photoreceptors. *Exp. Neurol.* 34:446–454.

ANDREWS, R. V., and G. E. FOLK, JR., 1964. Circadian metabolic patterns in cultured hamster adrenal glands. *Comp. Biochem. Physiol.* 11:303–409.

ANDREWS, R. V., 1971. Circadian rhythms in adrenal organ cultures. *Gegenbaurs Morphol. Jahrb.* 117:89–98.

ASCHOFF, J., 1969. Desynchronization and resynchronization of human circadian rhythms. *Aerospace Med.* 40:844–849.

ASCHOFF, J., and H. POHL, 1970. Rhythmic variations in energy metabolism. *Fed. Proc.* 29:1541–1552.

BENNETT, M. H., R. F. DYER, and J. D. DUNN, 1972. Light induced retinal degeneration: Effect upon light-dark discrimination. *Exp. Neurol.* 34:434–445.

BENOIT, J., 1964. The role of the eye and of the hypothalamus in the photostimulation of gonads in the duck. *Ann. N.Y. Acad. Sci.* 117:204–215.

BINKLEY, S., E. KLUTH, and M. MENAKER, 1971. Pineal function in sparrows: Circadian rhythms and body temperature. *Science* 174:311–315.

BINKLEY, S., 1972. Pineal and locomotor activity: Levels and arrhythmia in sparrows. *J. Comp. Physiol.* 77:163–169.

BROWMAN, L. G., 1943. The effect of bilateral optic enucleation upon activity rhythms of the albino rat. *J. Comp. Psychol.* 36:33.

BRUSS, R. T., R. JACOBSON, F. HALBERG, H. A. ZANDER, and J. J. BITTNER, 1958. Effects of lighting regimen and blinding upon gross motor activity of mice. *Fed. Proc.* 17:21.

CHEIFETZ, P., N. GAFFUD, and J. F. DINGMAN, 1968. Effects of bilateral adrenalectomy and continuous light on the circadian rhythm of corticotropin in female rats. *Endo.* 82:1117–1124.

CRITCHLOW, V., R. A. LIEBELT, M. BAR-SELA, W. MOUNTCASTLE, and H. S. LIPSCOMB, 1963. Sex difference in resting pituitary-adrenal function in the rat. *Amer. J. Physiol.* 205:807–815.

DODT, E., and E. HEERD, 1962. Mode of action of pineal nerve fibers in frogs. *J. Neurophysiol.* 25:405–429.

DODT, E., 1963. Photosensitivity of the pineal organ in the teleost, *Salmo irideus* (Gibbons). *Experientia* 19:642–643.

DONHAM, R. S., and F. E. WILSON, 1969. Pinealectomy in Harris' Sparrow. *Auk* 86:553–555.

DUNN, J., R. DYER, and M. BENNETT, 1972. Diurnal variation in plasma corticosterone following long-term exposure to continuous illumination. *Endocrinology* 90:1660–1663.

ELLIOTT, J. A., M. H. STETSON, and M. MENAKER, 1972. Regulation of testis function in golden hamsters: A circadian clock measures photoperiodic time. *Science* 178:771–773.

GANONG, W. F., M. D. SHEPHARD, J. R. WALL, E. E. VAN BRUNT, and M. T. CLEGG, 1963. Penetration of light into the brain of mammals. *Endocrinology* 72:962–963.

GASTON, S., and M. MENAKER, 1968. Pineal function: The biological clock in the sparrow? *Science* 160:1125–1127.

GASTON, S., 1971. The influence of the pineal organ on the circadian activity rhythm in birds. In *Biochronometry*, M. Menaker, ed. Washington, D.C.: Nat. Acad. Sci., pp. 541–548.

GORDON, S. A., and G. A. BROWN, 1971. Observations on spectral sensitivities for the phasing of circadian temperature rhythms in *Perognathus penicillatus*. In *Biochronometry*, M. Menaker, ed. Washington, D.C.: Nat. Acad. Sci., pp. 363–371.

GWINNER, E. B., F. W. TUREK, and S. D. SMITH, 1971. Extraocular light perception in photoperiodic responses of the white-crowned sparrow (*Zonotrichia leucophrys*) and of the golden-crowned sparrow (*Z. atricapilla*). *Z. Vergl. Physiologie* 75:323–331.

HALBERG, F., 1960. Temporal coordination of physiologic function. *Cold Spring Harbor Symp. Quant. Biol.* 25:289–310.

HAMASAKI, D. I., and E. DODT, 1969. Light sensitivity of the lizard's *epiphysis cerebri*. *Pflügers Arch.* 313:19–29.

HAMNER, W. M., 1963. Diurnal rhythm and photoperiodism in testicular recrudescence of the house finch. *Science* 142:1294–1295.

HIROSHIGE, T., and M. SAKAKURA, 1971. Circadian rhythm of corticotropin-releasing activity in the hypothalamus of normal and adrenalectomized rats. *Neuroendocrin.* 7:25–36.

HOFFMAN, K., 1971. Splitting of the circadian rhythm as a function of light intensity. In *Biochronometry*, M. Menaker, ed. Washington, D.C.: Nat. Acad. Sci., pp. 134–151.

HOMMA, K., and Y. SAKAKIBARA, 1971. Encephalic photoreceptors and their significance in photoperiodic control of sexual activity in Japanese quail. In *Biochronometry*, M. Menaker, ed. Washington, D.C.: Nat. Acad. Sci., pp. 333–341.

HOMMA, K., W. O. WILSON, and T. D. SIOPES, 1972. Eyes have a role in photoperiodic control of sexual activity of coturnix. *Science* 178:421–423.

KEATTS, Henry C., 1968. Extra-retinal photoreception by the brain of the sparrow (*Passer domesticus*). Masters Thesis, University of Texas, Austin, pp. 48.

KINCL, F. A., C. C. CHANG, and V. ZBUZKOVA, 1970. Observations on the influence of changing photoperiod on spontaneous wheel-running activity of neonatally pinealectomized rats. *Endocrinology* 87:38–42.

LAUBER, J. K., J. E. BOYD, and J. AXELROD, 1968. Enzymatic synthesis of melatonin in avian pineal body: Extraretinal response to light. *Science* 161:489–490.

LISK, R. D., and L. R. KANNWISCHER, 1964. Light: Evidence for its direct effect on hypothalamic neurons. *Science* 146:272–273.

LOBBAN, M. C., 1960. The entrainment of circadian rhythms in man. *Cold Spring Harbor Symp. Quant. Biol.* 25:325–332.

MENAKER, M., 1959. Endogenous rhythms of body temperature in hibernating bats. *Nature (Lond.)* 4694:1251–1252.

MENAKER, M., and A. ESKIN, 1967. Circadian clock in photoperiodic time measurement: A test of the Bünning hypothesis. *Science* 157:1182–1185.

MENAKER, M., and H. KEATTS, 1968. Extraretinal light perception in the sparrow, II: Photoperiodic stimulation of testis growth. *Proc. Natl. Acad. Sci. USA* 60:146–151.

MENAKER, M., 1968a. Extraretinal light perception in the sparrow, I: Entrainment of the biological clock. *Proc. Natl. Acad. Sci. USA* 59:414–421.

MENAKER, M., 1968b. Light perception by extraretinal receptors in the brain of the sparrow. *Proc. 76th Ann. Convention, Amer. Psychol. Assoc.* 3:299–300.

MENAKER, M., R. ROBERTS, J. ELLIOTT, and H. UNDERWOOD, 1970. Extraretinal light perception in the sparrow, III: The eyes do not participate in photoperiodic photoreception. *Proc. Natl. Acad. Sci. USA* 67:320–325.

MENAKER, M., 1971. Rhythms, reproduction and photoreception. *Biol. Reprod.* 4:295–308.

MOORE, R. Y., A. HELLER, R. J. WURTMAN, and J. AXELROD, 1967. Visual pathway mediating pineal response to environmental light. *Science* 155:220–223.

MOORE, R. Y., A. HELLER, R. K. BHATNAGER, R. J. WURTMAN, and J. AXELROD, 1968. Central control of the pineal gland: Visual pathways. *Arch. Neurol.* 18:208–218.

MOORE, R. Y., and V. B. EICHLER, 1972. Loss of a circadian adrenal corticosterone rhythm following suprachiasmatic lesions in the rat. *Brain Res.* 42:201–206.

MORITA, Y., 1966. Absence of electrical activity of the pigeon's pineal organ in response to light. *Experientia* 22:402–404.

OKSCHE, A., and H. KIRSCHSTEIN, 1969. Elektronenmikroskopische untersuchunger am pinealorgan von *Passer domesticus*. *Z. Zellforsch.* 102:214–241.

POHL, v. H., 1971. Über beziehungen zwischen circadianen rhythmen bei vögeln. *J. Ornithol.* 112:266–278.

QUAY, W. B., 1970. Precocious entrainment and associated characteristics of activity patterns following pinealectomy and reversal of photoperiod. *Physiol. Behav.* 5:1281–1290.

QUAY, W. B., 1972. Pineal homeostatic regulation of shifts in the circadian activity rhythm during maturation and aging. *Trans. N.Y. Acad. Sci.* 34:239–253.

RALPH, C. L., and D. C. DAWSON, 1968. Failure of the pineal body of two species of birds (*Coturnix coturnix japonica* and *Passer domesticus*) to show electrical responses to illumination. *Experientia* 24:147–148.

REED, B. L., 1968. The control of circadian pigment changes in the pencil fish: A proposed role for melatonin. *Life Sci.* 7:961–973.

REITER, R. J., and S. SORRENTINO, JR., 1970. Reproductive effects of the mammalian pineal. *Amer. Zool.* 10:247–258.

RETIENE, K., E. ZIMMERMAN, W. J. SCHINDLER, J. NEUENSCHWANDER, and H. S. LIPSCOMB, 1968. A correlative study of endocrine rhythms in rats. *Acta Endocrinology* 57:615–622.

RICHTER, C. P., 1965. *Biological Clocks in Medicine and Psychiatry.* Springfield, Ill.: Charles C Thomas, pp. 109.

SEIDEN, G., and A. BRODISH, 1972. Persistence of a diurnal rhythm in hypothalamic corticotrophin-releasing factor (CRF) in the absence of hormone feedback. *Endocrinology* 90:1401–1403.

SILVER, J., 1972. The morphology of the pineal of the house sparrow, *Passer domesticus*. Masters Thesis, Univ. of Texas, Austin, pp. 160.

SNYDER, S. H., M. ZWEIG, J. AXELROD, and J. E. FISCHER, 1965. Control of the circadian rhythm in serotonin content of the rat pineal gland. *Proc. Nat. Acad. Sci. USA* 53:301–305.

STEBBINS, R. C., and D. C. WILHOFT, 1966. Influence of the parietal eye on activity in lizards. In *The Galápagos: Proceedings of the Symposia of the Galápagos International Scientific Project*, R. I. Bowman, ed. Becheley: University of California Press, pp. 258–268.

STEPHAN, F. K., and I. ZUCKER, 1972. Circadian rhythms in drinking behavior and locomotor activity of rats are eliminated by hypothalamic lesions. *Proc. Nat. Acad. Sci. USA* 69:1583–1586.

TAKEBE, K., M. SAKAKURA, Y. HORIUCHI, and K. MASHIMO, 1971. Persistence of diurnal periodicity of CRF activity in adrenalectomized and hypophysectomized rats. *Endocrinol. Jap.* 18:451–455.

TUREK, F. W., 1972. Circadian involvement in termination of the refractory period in two sparrows. *Science* 178:1112–1113.

UECK, M., 1970. Weitere untersuchungen zur feinstruktur und innervation des pinealorgans von *Passer domesticus*. *Z. Zellforsch.* 105:276–302.

UECK, M., and H. KOBAYASHI, 1972. Vergleichende untersuchungen über acetylcholinesterase-haltige neurone im pinealorgan der vögel. *Z. Zellforsch.* 129:140–160.

UNDERWOOD, H., and M. MENAKER, 1970. Extraretinal light perception: Entrainment of the biological clock controlling lizard locomotor activity. *Science* 170:190–193.

WURTMAN, R. J., J. AXELROD, and D. E. KELLY, 1968. *The Pineal.* New York: Academic Press, p. 199.

ZWEIG, M., S. H. SNYDER, and J. AXELROD, 1966. Evidence for a nonretinal pathway of light to the pineal gland of newborn rats. *Proc. Nat. Acad. Sci. USA* 56:515–520.

41 Toward a 48-Hour Day:

Experimental Bicircadian Rhythm in Man

MICHEL JOUVET, JACQUES MOURET,
GUY CHOUVET, and MICHEL SIFFRE

ABSTRACT In free-running experiments involving no time cues it is possible for some subjects to desynchronize and manifest a bicircadian rhythm (34-hr waking and 14-hr sleep). Such a rhythm cannot be maintained spontaneously for a long time. However, an external light Zeitgeber permits sustaining such a bicircadian rhythm for as long as 30 cycles (2 months) without any physiological or psychological discomfort. In such conditions, the polygraphic studies of sleep have revealed a striking fixity of the ultradian period of paradoxical sleep and a constant ratio between the duration of paradoxical sleep and the duration of the preceding period of waking.

THERE IS A large amount of data that indicate that desynchronization of a 24-hr cycle may occur in the so-called *free run* condition when subjects are isolated in a cave or in a bunker without any external time cue. However, the amount of desynchronization is rather limited, and the sleep-waking periodicity remains around 24 hr (circadian). Moreover most of these experiments have been done without monitoring the EEG, so that one does not know the possible alteration of the ultradian rhythm of paradoxical sleep (PS) that might occur in these conditions (Aschoff, 1965; Aschoff et al., 1969; Kleitman, 1949; Siffre et al., 1966; Mills, 1964).

In order to study the internal organization of sleep with continuous polygraphic recordings in free run conditions (and thus during desynchronization), a preliminary pilot experiment was first undertaken in 1966 on J. P. Mairetet, a 23-yr-old volunteer, under the technical direction of M. Siffre. This subject spent 6 months, without time cue, isolated in a cave and was thereafter credited with the world record. He could be recorded polygraphically continuously during 34 cycles. The results obtained were rather striking. At the beginning, the periods of the cycle of activity (and of the estimated cycle by the subject) were circadian (25 and 27 hr). Then, sharply, after 8 cycles and a few oscillations, the periods reached values

varying from 48 to 50 hr (bicircadian or circabidian rhythm). There were 15 bicircadian cycles followed by a spontaneous return to circadian rhythm (27.25 to 24.39 hr). The analysis of the polygraphic sleep recordings that were obtained during this experiment showed that there was a very significant shortening of the intervals between sleep onset and the first PS episode (as in narcoleptic patients, Rechtschaffen et al., 1963) when the subject could no longer maintain a bicircadian rhythm.

Thus, for the first time, it was shown that man could adapt in free run conditions to a 48-hr day (34 hr of activity, 14 hr of sleep) which was subjectively estimated to be a 24-hr day, and that the return to circadian rhythm was accompanied by sleep disturbances similar to narcolepsy.

Then, another experiment was undertaken in order to answer the following questions (see references in Siffre, 1972 for technical details):

1. Was J. P. Mairetet an exception? I.e., is it possible to obtain again, in another subject, a total desynchronization with a bicircadian rhythm?

2. If a subject spontaneously reaches a bicircadian rhythm in free-run conditions, is it possible to synchronize him under such a rhythm (i.e., 34 hr waking, 14 hr sleep) for a long period with an external Zeitgeber (34 hr light, 14 hr dark)?

3. What happens to the different parameters of sleep under such conditions, and particularly what happens to the ultradian rhythm of PS?

4. Does the same alteration of PS occur when the subject can no longer adapt to bicircadian rhythm?

The experiments which were performed in 1968–1969 were quite successful and answered positively these questions. They are summarized briefly here:

Two young male volunteers, Jacques Chabert (23) and Philippe Englender (28) were placed in isolation in two different caves (Aven Ollivier, Massif de l'Audibergue, 30 km NW Grasse, France). These caves were isolated from each other and from the surface. At this depth (−65 m, −85 m), the temperature (6°C) and relative humidity (100%) remained constant. Throughout this

MICHEL JOUVET, JACQUES MOURET, GUY CHOUVET, and MICHEL SIFFRE Department of Experimental Medicine, University Claude Bernard, Lyon, France

experiment (3080 hr), the subjects could communicate by telephone to the surface at any time. They were asked to call everytime they were changing their activity and after urine collection; at these moments some tests were performed (time estimation). The EEG, EKG, EMG, EOG, rectal temperature and respiration rate were monitored either by telemetry (J.C.) or through cables (P.E.). The high quality of the recordings allowed the perfect recognition of the sleep stages (57 recordings for P.E., 46,111 min; 58 for J.C., 42,869 min).

Except for the test instructions, they were completely free in their choice of activity (exploring the cave, reading, writing, listening to music). According to the periodicity and to the light schedules, the experiments were divided as follows (Figure 1).

a. The subject P.E. spontaneously reached a bicircadian rhythm on the 13th cycle, which he could maintain for 9 cycles. He was then given (from the 21st cycle to the 51st cycle, i.e., during 30 cycles, or 2 months) an artificial Zeitgeber of 34 hr of light (a 500 watt light

bulb) from the surface, at the end of each sleep period (whose duration was around 14 hr). The subject, without time cue, did not know the duration of the light and always estimated subjectively its bicircadian cycle to be a 24 hr day. After the 51st cycle, the subject was again free running during 25 cycles up to the end of the experiment.

b. Because the subject J.C. was still living a circadian cycle after the 33rd cycle and could not spontaneously reach a bicircadian rhythm, he was given a continuous light (500 w bulb) up to the end of the experiment (108th cycle).

c. Both subjects were recorded for 6 days immediately before and after the experiments in the Sleep Laboratory of the Neurological Hospital in Lyons.

Results

SLEEP-ACTIVITY CYCLE   Figure 1 illustrates the history of the two subjects.

During the free-run period, the duration of the cycle of

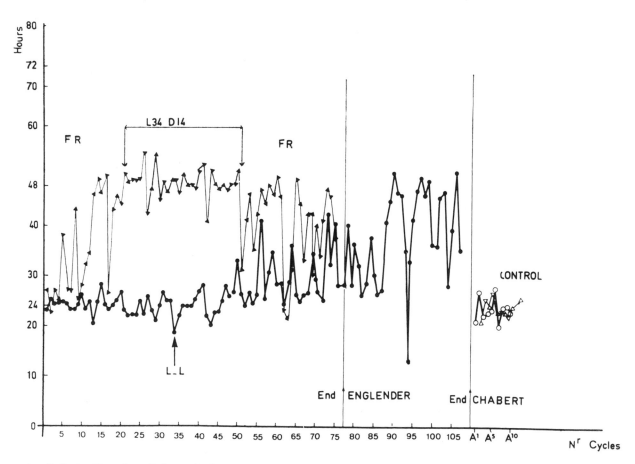

FIGURE 1   *Ordinate*: Duration of the cycle (in hours) for both subjects. *Heavy line*: Subject J.C. *Light line*: Subject P.E. *Abscissa*: Numbers of the cycle. FR: Free running. LL: Continuous light for J.C. L34, D14: Artificial Zeitgeber for P.E. Controls immediately after the end of experiment. Note the small variability of the bicircadian rhythm during the external Zeitgeber for P.E. as compared with the free-running condition (see also Table I).

492   BICIRCADIAN RHYTHM IN MAN

subject P.E. increases very sharply (as in the preliminary experiment performed on J.P.M.) and after 13 cycles reaches and maintains a bicircadian pattern. As shown in Table I the variability of this rhythm was more pronounced during the free run conditions than during the artificial synchronization. Thus light could really act as a true Zeitgeber. The phase shifting of sleep onset (Figure 2), which is observed during the artificial synchronization period, is due to the fact that the subject could stay awake even when the 500-w light was turned off from the surface (the subject had a dim light (40 w), which permitted some reading).

As shown in Figure 1, the subject J.C. stayed on a circadian rhythm for a very long time and the persistent light did not affect, at least immediately, his sleep-waking cycle. However this subject could secondarily reach a bicircadian rhythm, which persisted for about 10 cycles.

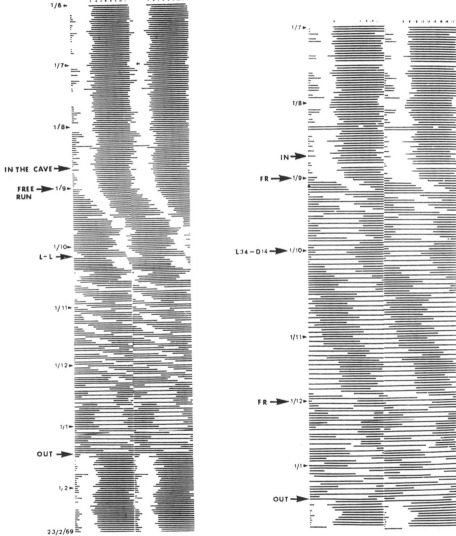

FIGURE 2 Desynchronization of the circadian rhythm of activity (black) and sleep (white) in the subjects J.C. and P.E. The beginning of the free-running experiment in the cave is marked by the second horizontal arrow. Other arrows signal the permanent illumination (LL) for J.C., and the end of the experiment (out). For P.E., the external Zeitgeber (L 34, D 14) is also indicated and is followed by a return to free running conditions (FR). Other details in the text. The dates (real time) are marked in the left margin.

Table I
Sleep-activity cycle

	Cycles	Number of Cycles	Mean Duration of Sleep-Activity Cycle (min)	Conditions
	(a)	15	1451 ± 28	Controls
	1–33	33	1476 ± 35	FR
J.C.	34–79	45	1722 ± 63	L – L
	80–91	11	2108 ± 116	L – L
	92–108	16	2322 ± 222	L – L
	(a)	16	1460 ± 17	Controls
	1–20	20	2270 ± 200	FR
P.E.	21–51	30	2918 ± 38	L 34 – D 14
	52–75	23	2482 ± 105	FR

(a) Pre + postexperiments.

Duration of the sleep activity cycle in both subjects and in different conditions (see legends in Figure 1). The results are expressed in mean ± SEM.

Table II
Paradoxical sleep cycle

	J.C.		P.E.	
	A	B	A	B
Cycle duration (min)	1451 ± 28	2661* ± 73	1460 ± 17	2804* + 105
Number of cycles chosen for analysis	15	15	16	26
Mean interval between PS phase onset (min)	85.7 ± 3.6	88.9 ± 2.2	102.4 ± 4.9	109.4 ± 2
Maximums of binary autoc. (min)	91 ± 2.9	92 ± 2.8	113.8 ± 2.8	114 ± 3.5
Mean number of PS phase	5.1 ± 0.6	8.4* ± 0.6	4 ± 0.31	6.1* ± 0.14

*Significantly different from control group ($p < 0.05$).

Temporal characteristics of paradoxical sleep during the circadian conditions (A) and the bicircadian conditions (B) for both subjects.

Table III
Stages of sleep

	J.C.		P.E.	
	A	B	A	B
PPW (min)	1004 ± 19.8	1968* ± 75.2	1000 ± 26.8	2175* ± 34
TST per cycle (min)	453 ± 30	744* ± 38	446 ± 28	651* ± 15.5
PS per cycle (min)	118 ± 2.9	225* ± 5.7	96 ± 8	173* ± 6.7
Latency (min) of PS onset	73 ± 10	30* ± 5	94 ± 11	77 ± 3
Mean PS (min) phase duration	26.5 ± 0.2	31.8* ± 0.4	26.8 ± 0.5	32* ± 0.2
Stages I + II (min)	262 ± 26.4	306.2* ± 27.3	275.5 ± 15.8	378* ± 14.2
Stage III (min)	55.3 ± 4.9	64 ± 4.6	29.70 ± 3.3	54* ± 3.5
Stage IV per cycle (min)	39.3 ± 1.1	124.5* ± 1.4	40 ± 1	49 ± 0.8

*Significantly different from control group ($p < 0.05$).

Parameters of sleep during circadian (A) and bicircadian (B) conditions in both subjects.

Note that PS duration increases almost twofold when the preceding period of waking (PPW) is increased twofold, while other stages of sleep do not increase accordingly. (TST: Total sleep time.)

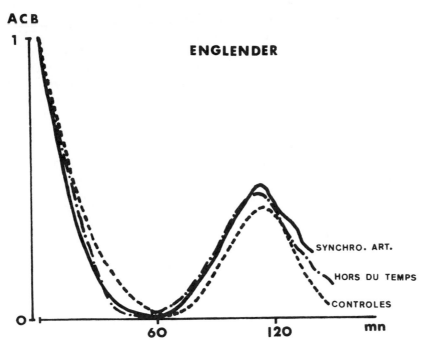

ACB

ENGLENDER

SYNCHRO. ART.

HORS DU TEMPS

CONTROLES

60   120   mn

FIGURE 3 Mean binary autocorrelation of the PS episodes. During control pre- and postexperiment: dotted line (n = 15). During free running in the cave (hors du temps): interrupted line. During the external Zeitgeber (synch. art.: L 34, D 14): solid line. Note that the maximum of the binary autocorrelation on the abscissa (lags in minutes) indicates that the duration of the ultradian PS cycle does not depend upon the external conditions and remains around 114 min.

PARADOXICAL SLEEP CYCLE The analysis of PS ultradian rhythm (T') was performed through two different techniques: Binary autocorrelation (Globus, 1970) or measurement of the intervals between the onset of each PS episode. As shown in Table II, the two methods give similar results and emphasize the striking constancy of this rhythm in both subjects, i.e., there is no significant alteration of T' when the subjects live in a 48 hr day (Figure 3).

RELATIONSHIP BETWEEN ACTIVITY AND SLEEP The total sleep duration is not proportional to the duration of the preceding period of wakefulness (PPW), i.e., the duration of sleep does not increase twofold when the waking duration is doubled. The results concerning the stages of sleep are summarized in Table III. They are different for each subject in regard to stage IV. However, the general trend consists of a relative decrease of the proportion of stage I, II, and III. In contrast to stages I, II, III, and IV, a striking fixed relation appears between PPW and PS duration so that the ratio between PPW and PS remains constant (about $\frac{1}{10}$) (such a relationship had been noted also in the preliminary experiment with J.P.M.).

Since PS duration increases relatively more than other stages of sleep, it is evident that the ratio between PS and the total sleep time increases (from 21.5% to 26.6% for P.E. and from 26% to 33% for J.C.). As previously shown, the increase of PS duration is not due to the decrease of the duration of the PS cycle but is established by the increase of duration of each PS episode.

The latency of the first PS episode remains relatively constant for subject P.E. who could well adapt to the bicircadian rhythm, whereas a striking decrease of the latency occurs in subject J.C. (Table III). In some cases, PS could occur less than 5 min after sleep onset (Figure 4).

Discussion

BICIRCADIAN RHYTHM Since the number of subjects is limited, it is difficult to ascertain what are the factors responsible for such a large desynchronization of the sleep-waking rhythm. Such a phenomenon was not observed in the numerous subjects run by Aschoff or Lewis, in bunker or in cave, who remained in circadian rhythm. The living conditions are certainly different in our experimental conditions: Cold, humidity. The duration of the experiment does not seem to be an important factor, because two subjects (J.P.M. and P.E.) reached a circadian rhythm after only 8 and 13 cycles (which is less than the duration of many experiments performed by others).

Whatever might be the cause of the appearance of a well-sustained bicircadian rhythm, these results favor some endogenous clock that will adapt to a much slower rhythm. It is interesting to note that in the first

MICHEL JOUVET, JACQUES MOURET, GUY CHOUVET, AND MICHEL SIFFRE   495

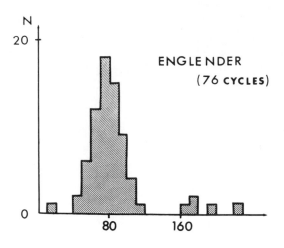

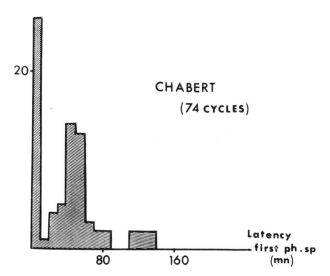

FIGURE 4 Histograms of the latencies between sleep onset and the first PS episode in subjects P.E. (up) and J.C. (down). Note the large number of short latencies (less than 5 min) in subject J.C. who could not sustain bicircadian rhythm. Abscissae: Class of 5 min.

pilot experiment, the rhythm of the rectal temperature remained circadian (24.03 and 24.42 hr between the maxima of rectal temperature, or 24.07 and 24.48 hr between the minima) (Colin et al., 1968) even when the sleep-waking rhythm was bicircadian. Therefore, it can be said that the rhythm of rectal temperature does not follow the sleep-activity cycle when this latter goes too far beyond some limit.

Our results are in accordance with those of Colin et al. (1968) and suggest that the rectal temperature is linked to some "internal influences" more strongly than to the sleep-activity cycle.

Another result should be emphasized: Our two subjects had a theoretical sleep debt that amounted to between 3 and 4 hours of sleep per cycle, if there is a

correlation between waking and sleep. Despite such a debt, which would amount to 150 to 200 hr of sleep for subject P.E. who stayed in a bicircadian rhythm of sleep-waking activity during 50 cycles, there was no evidence of any "rebound" of either slow wave sleep or PS immediately after the experiment and during the entire week when both subjects were recorded in Lyon. During this period their sleep patterns were found to be quantitatively and qualitatively normal.

Finally, the good psychological and physiological adaptation of subject P.E. during more than 36 bicircadian cycles (i.e., at least 72 days) is in sharp contrast with the study of Meddis (1968). He reported the "absence of any real sign of adapting to a new 48-hr routine in either a psychological or a physiological sense," which was observed in seven subjects who tried to adapt (as long as 56 days) to a 48-hr cycle on the surface.

PARADOXICAL SLEEP   On one hand, the striking constancy of T' (PS cycle duration) is interesting, because one should have expected that such a change in sleeping habits would have altered this rhythm. In fact, there is some indirect evidence in mice that some internal factors of sleep (cortical electrical pattern, circadian repetition of PS) might be genetically determined (Valatx et al., 1972). If this is the case in man, this could explain the fixity of T'. In such a case, epigenetic events could mostly affect the length of each episode but not its periodicity. On the other hand, the fixed ratio between the duration of PS and the preceding waking seems to indicate that there must be some relation between the waking and the succeeding PS so that roughly 1 min of PS corresponds to 10 min of preceding waking in adult men. Under these conditions, such a ratio might indicate that, during PS, some sequential events occur which could be related directly or indirectly with preceding waking events. It is tempting to speculate that these sequential events are represented by the phasic phenomena that occur centrally during PS. Some peripheral indexes could be represented by the rapid eye movements of PS.

The importance of T' and PS should also be taken into consideration when studying the adaptation to bicircadian rhythm. It is worth noting that J.P.M. and J.C., who could not sustain a 48-hr rhythm, enter PS almost directly after sleep onset. This phenomenon has also been described by Meddis in a subject who directly entered PS during an attempt to live a 48-hr day under surface conditions. The "need for PS" and possibly the duration of the PS cycle seem to be an important factor for the success or failure of any physiological or psychological adaptation to a 48-hr rhythm.

Finally, whatever might be the intimate mechanisms that do or do not permit the adaptation to a 48-hr day, our results strongly suggest that PS appears to be a very important variable that should be taken into consideration by continuous polygraphic recordings.

Conclusion

In free running experiments performed in caves, three young male subjects were able to reach and maintain a bicircadian rhythm (34-hr waking–14-hr sleep) (which is subjectively felt as a 24 hr day).

Such a bicircadian rhythm cannot be maintained spontaneously for a long time (6.12 cycles). However, an external light Zeitgeber (34 hr light, 14 hr dark) permits sustaining such a rhythm as long as 60 days (30 cycles) without any physiological or psychological discomfort.

Polygraphic sleep recordings were made continuously during most of these experiments. They demonstrate some pattern of PS during desynchronization of the sleep-waking rhythm. (i) The duration of PS cycle (T') remains strikingly constant during desynchronization. (ii) The ratio between PS and preceding waking is also relatively constant ($\frac{1}{10}$) while the ratio between other stages of sleep and waking decreases. (iii) Alterations of sleep onset (very short latency between sleep onset and the first PS episodes) are found whenever the subject can no longer sustain bicircadian cycle.

These results emphasize the need for recording all the parameters of sleep and PS whenever experiments are effectuated concerning the sleep-waking cycle in free running conditions.

ACKNOWLEDGMENTS These experiments have been supported by the Délégation Générale pour la Recherche Scientifique et Technique et la Direction des Recherches et Moyens d'Essais.

REFERENCES

ASCHOFF, J., 1965. Circadian rhythms in man. *Science* 148: 1427–1432.

ASCHOFF, J., E. PÖPPEL, and R. WEVER, 1969. Circadiane periodik des menschen unter dem Einfluss von Licht-Dunkel-Wechseln unter Schiedlicher periode. *Pflügers Arch.* 306:58–70.

COLIN, J., J. TIMBAL, C. BARTELIER, Y. HONDAS, and M. SIFFRE, 1968. Rhythm of the rectal temperature during a 6 month free-running experiment. *J. Appl. Physiol.* 25:170–176.

GLOBUS, G. G., 1970. Quantification of the REM sleep cycle as a rhythm. *Psychophysiol.* 7:248–253.

KLEITMAN, N., 1949. Biological rhythms and cycles. *Physiol. Rev.* 29:1–30.

LEWIS, P. R., and M. C. LOBBAN, 1957. Dissociation of diurnal rhythms in human subjects living on abnormal time routines. *Quart. J. Exp. Physiol.* 42:371–386.

MEDDIS, R., 1968. Human circadian rhythms and the 48 hour day. *Nature (Lond.)* 218:964–965.

MILLS, J. N., 1964. Circadian rhythms during and after three months in solitude underground. *J. Physiol. (Lond.)* 174: 217–231.

RECHTSCHAFFEN, A., E. A. WOLPERT, W. C. DEMENT, S. A. MITCHELL, and C. FISHER, 1963. Nocturnal sleep of narcoleptics. *Electroenceph. Clin. Neurophysiol.* 15:599–609.

SIFFRE, M., 1972. *Expériences Hors du Temps.* Paris: Fayard, p. 463.

SIFFRE, M., A. REINBERG, F. HALBERG, J. GHABA, G. PERDRIEL, and R. SLIND, 1966. L'isolement souterrain prolongé. Étude de deux sujets adultes sains, avant, pendant et après cet isolement. *Presse Méd.* 18:915–919.

VALATX, J. L., R. BUGAT, and M. JOUVET, 1972. Genetic study of sleep in mice. *Nature (Lond.)* 238:226–227.

42 Monoaminergic Regulation of the Sleep-Waking Cycle in the Cat

MICHEL JOUVET

ABSTRACT The hypothesis that the serotonin-containing neurons of the raphe system are involved in the induction of slow-wave sleep and in the priming of paradoxical sleep is strongly supported by the following facts: (a) The inhibition of the synthesis of serotonin with p-chlorophenylalanine leads to an insomnia that is immediately reversed to normal sleep by a subsequent injection of small doses of 5-hydroxytryptophan, the immediate precursor of serotonin, which by-passes the inhibition of the synthesis. (b) The destruction of the serotonin-containing perikarya located in the raphe system leads to insomnia, the intensity of which is correlated with the decrease of cerebral serotonin.

It is likely that there is some interaction between the serotonin-containing neurons and catecholaminergic neurons and most of the data obtained in the cat favor the hypothesis that the sleep-waking cycle is regulated by two antagonistic systems of neurons: The serotonin-containing neurons for sleep and the catecholaminergic (and possibly cholinergic neurons) for waking and paradoxical sleep.

THE STUDY OF the possible role of monoamines in the regulation of the sleep-waking cycle has been recently helped by three technical developments:

1. Long-term continuous polygraphic recordings are now routinely used in most laboratories. These recordings can provide us with the indispensable information about the quantity of waking, slow-wave sleep (SWS), and paradoxical sleep (PS). Thus, these quantitative data can be used as dependent variables after any central or pharmacological alteration of brain monoamines.

2. The biochemical steps that lead to the biosynthesis or to the catabolism of monoamines in the brain are now known with accuracy (although there are still some uncertainties). It is now also possible to develop drugs that act more-or-less specifically upon each step of the metabolism of brain monoamines. Thus a neuropharmacological dissection of the sleep-waking cycle is, at least theoretically, feasible.

3. Finally, the development of histochemistry and histofluorescent techniques have permitted us to narrow the gap between neuropharmacology and neurophysiology in demonstrating the topography of mono-

MICHEL JOUVET Department of Experimental Medicine, University Claude Bernard, Lyon, France

amine-containing perikarya and their axonal connections. New systems of serotonin or catecholamine containing neurons have been mapped out in the complicated circuitry of the reticular formation. Many of these systems are located in places where lesions had been shown, by classical neurophysiology, to impair waking or sleep. Thus, it is now possible to approach the mechanism of the sleep-waking cycle by a systematic study of these histochemical systems through lesion, stimulation, recording, neuropharmacological alteration, and biochemical analysis.

We summarize here some experimental evidence that favors the role of serotonin, 5-hydroxytryptamine (5-HT) in sleep mechanisms and the role of catecholamines (CA) in both waking and paradoxical sleep. Most of the data that are reviewed have been obtained in the neurophysiologist's most favored animal, the cat, in which unfortunately the circadian rhythm of sleep is not well developed (as compared with rat or man). The preceding chapter (Jouvet et al.) was devoted to some recent experiments on the sleep-waking cycle in isolated men. These experiments demonstrate that under special conditions, man can adapt himself for a long period of time to a 48-hr day.

Monoaminergic regulation of the sleep-waking cycle in the cat

SEROTONINERGIC REGULATION OF SLEEP

1. *Neuropharmacological alteration of 5-HT*. Since 5-HT does not cross the blood-brain barrier, since it is not certain that 5-hydroxytryptophan (5-HTP) is the physiological precursor of 5-HT in normal conditions, and since other drugs (reserpine, inhibitors of monoamine oxidase) may interfere with other monoamines, one of the best pharmacological tools that is now available to alter "selectively" 5-HT metabolism is p-chlorophenylalanine (PCPA) (Koe and Weissman, 1966). This drug inhibits tryptophan hydroxylase in the cat brain as shown in the following experiment (Pujol et al., 1971): Slices of cat cerebral cortex or brain stem were incubated in tritiated tryptophan (TRY) or ^3H 5-HTP and the amount of

³H 5-HTP synthetized was subsequently measured. There is a 80 to 90% decrease of the ³H 5-HT synthetized from ³H TRY in PCPA pretreated cats (as compared with the control), whereas there is no alteration or ³H 5-HT synthetized from ³H 5-HTP (Figure 1A). The action of PCPA upon the states of sleep of the cat has been the subject of many studies (Delorme et al., 1966; Mouret et al., 1967; Koella et al., 1968; Pujol, 1970; Pujol et al., 1971; Cohen et al., 1970; Hoyland et al., 1970), which can be summarized as follows:

After a single intraperitoneal injection of 400 mg/kg of PCPA, no apparent alteration of behavior nor of polygraphic recordings is observed during the first 18 to 24 hr. This fact demonstrates that the drug, in itself, has no direct pharmacological action upon the brain. Following this period, an abrupt decrease of both slow-wave sleep (SWS) and paradoxical sleep (PS) occurs, and after about 30 to 40 hr, an almost total insomnia appears (Figure 1B) as shown by a permanent and quiet waking behavior, mild mydriasis, and by an almost permanent low-voltage fast cortical activity. The increase of waking is accompanied by the appearance of almost permanent discharges of pontogeniculo occipital (PGO) activity (similar to the PGO activity that is observed during SWS immediately preceding PS or during PS in normal cats). The mechanism of the appearance of PGO activity during waking after PCPA is still argued. It may be related to the decrease of 5-HT at some serotoninergic terminals, because the same phenomenon is observed after lesion of the raphe system (see below). The recovery of sleep begins after the 40th hr and qualitatively and quantitatively normal patterns of sleep are resumed after about 200 hr.

Under the influence of PCPA, a significant correlation has been found to exist between the decrease of SWS and the decrease of cerebral 5-HT and 5-hydroxyindoleacetic acid (5-HIAA) (Koella et al., 1968; Pujol et al., 1971) whereas there was no alteration of catecholamine level.

Since PCPA inhibits only the first step of the synthesis of 5-HT at the level of tryptophan hydroxylase and since

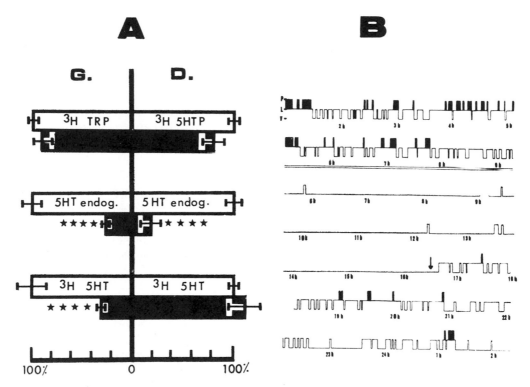

FIGURE 1 Effects of the administration of p-chlorophenylalanine (PCPA) upon the synthesis of cerebral 5-HT and the states of sleep of the cat. (A) The biosynthesis of cerebral 5-HT is effected in vitro 40 hr after the intraperitoneal administration of 400 mg/kg of PCPA. Slices of cerebral cortex are incubated respectively into ³H-tryptophan or DL ³H-5-HTP. After 30 min of incubation, the quantities of ³H-TRP, ³H-5-HTP, 5-HT, and ³H-5-HT are calculated. The results are expressed as percentage of the mean value (±SEM) obtained in control cats (white bars). The black bars represent the PCPA treated cats.

$P < 0.01$. (B) Effect of secondary injection of 5-HTP during insomnia produced by p-chlorophenylalanine in the cat. *Top two lines*: Normal pattern of sleep in a cat during control recordings. *Bottom five lines*: Total insomnia 96 hr after the last of the three daily injections of p-chlorophenylalanine (400 mg/kg), then, following the injection (see arrow) of 5-HTP (5 mg/kg), reappearance of the normal pattern of sleep. Insomnia returns 8 hr after the injection. (*Base line of each record*: Waking. *White rectangles*: Slow wave sleep. *Black rectangles*: Paradoxical sleep.) (Modified from Pujol, 1971, and Jouvet, 1969.)

the synthesis of 5-HT is still possible from 5-HTP, it is now possible to by-pass the blocking action of PCPA and thus to re-establish a higher level of 5-HT by injection of 5-HTP. With this procedure, it is possible to manipulate at will the state of sleep of the animal: Thus, a single injection of a very small dose of 5-HTP (2 to 5 mg/kg) given when the insomnia has reached its maximum (30 hr following the administration of 400 mg/kg of PCPA) is able to restore a quantitatively and qualitatively normal pattern of both states of sleep for 6 to 8 hr (Figure 1B) (Mouret et al., 1967; Koella et al., 1968; Jouvet, 1969; Pujol et al., 1971). If a larger dose of 5-HTP is injected (30 to 50 mg/kg), only cerebral synchronization accompanied by sedation reappears during the first hours and PS is delayed for 4 to 6 hr. This suggests that, after the inhibition of tryptophan hydroxylase, in the absence of endogenous substrate, the 5-HT-containing neurons are able to synthetize 5-HT rapidly from *small* amounts of exogenous 5-HTP, whereas it appears possible that with larger quantities, exogenous 5-HTP may interfere at some other site with the normal process of PS.

Other experiments have shown that a cat receiving balanced daily doses of 5-HTP with the dose of PCPA that would induce insomnia in a control cat may present normal or even hypersomnia during at least one week (Mouret et al., 1967). These experiments show that sleep mechanisms can be manipulated by interfering *only* with the synthesis of 5-HT.

2. *Neurophysiological alteration of 5-HT.* Thanks to the development of the histofluorescent technique, the serotonin containing perikarya have been mapped out in the rat (Dahlström and Fuxe, 1964) and cat brain (Pin et al., 1968). Most of them are concentrated in the raphe system. Thus a direct approach of the serotoninergic neurons has been made possible.

Destruction of 5-HT-containing perikarya of the raphe system. The destruction of the raphe system was performed stereotaxically in chronically implanted cats (Renault, 1967; Jouvet, 1969). Following the operation, the animals were continuously recorded for a period of 10 to 13 days (this being the critical duration for the voiding of the serotoninergic terminals). On the 13th day, the cats were sacrificed at the same hour in order to avoid possible circadian variation of brain monoamines. With this method, the following information was obtained: A valid quantification of the sleep states (obtained by the mean percentages of SWS and PS for 10 to 13 days recording), a measurement of the volume of the lesion by topographical analysis (represented by the percentage of the total raphe system destroyed), and an analysis of the monoamine level in the brain (expressed as the percentage of 5-HT, 5-HIAA, NA, and DA) found in normal cats sacrificed under the same conditions.

Following incomplete (80 to 90%) coagulation of the raphe system, a state of permanent behavioral and EEG arousal is observed during the first 3 to 4 days. In the period that follows (up to 3 weeks), the percentage of SWS does not exceed 10 to 15% of the nycthemeron. In these preparations, PS is never observed. However continuous discharges of PGO spikes, recorded from the lateral geniculate or occipital cortex appear at least 3 hr after the destruction of the raphe system at a rate of 30 to 40 min. The pattern of discharge is similar to the one that follows injection of reserpine (Delorme et al., 1965) or PCPA in the normal cat and has been called "reserpinic syndrome." The rate of discharge of the PGO spikes diminishes on the third day to 10 min. The injection of a low dose (5 mg/kg) of 5-HTP does not alter the permanent arousal that follows raphe lesion (Figure 2B). Larger doses (30 to 50 mg/kg) induce a state of cortical synchronization accompanied by a waking behavior. Partial lesions of the raphe system result in an insomnia which is less pronounced. As shown in Table I, there is some topographical organization of the raphe nuclei concerning their role in both SWS and PS. The rostral group is mainly responsible for the induction of SWS (and for the serotoninergic innervation of the telediencephalon) while the intermediary group (n. raphe pontis and magnus) seems to be involved with the "priming" mechanisms of PS. Since the volume of the rostral group is greater than that of other groups, there is a significant correlation between the amount of destruction of the raphe system,

TABLE I

Correlations between the lesion of each nucleus of the raphe system, the amount of 5-HT in the tele-diencephalon, slow-wave sleep (SWS), and paradoxical sleep (PS)

	5-HT ($n = 25$)	SWS ($n = 40$)	PS ($n = 40$)
N. raphe dorsalis	0.5847 XXX	0.4687 XXX	0.30 NS
N. centralis superior	0.4651 XX	0.4755 XXX	0.24 NS
N. raphe pontis	0.3911 X	0.3219 X	0.5251 XXX
N. raphe magnus	0.006 NS	0.3723 X	0.4845 XXX
N. raphe pallidus	0.10 NS	0.2958 NS	0.2935 NS

N. raphe obscurus was never destroyed.

(n: number of animals in each group. The Bravais-Pearson coefficient was used.) It is apparent that only the rostral group (n. raphe dorsalis and centralis superior) is involved in the 5-HT innervation of the telediencephalon and in SWS mechanism, whereas n. raphe pontis and magnus are involved in the priming of PS.

XXX: $P < 0.01$; XX: $P < 0.02$; X: $P < 0.05$; NS: Nonsignificant.

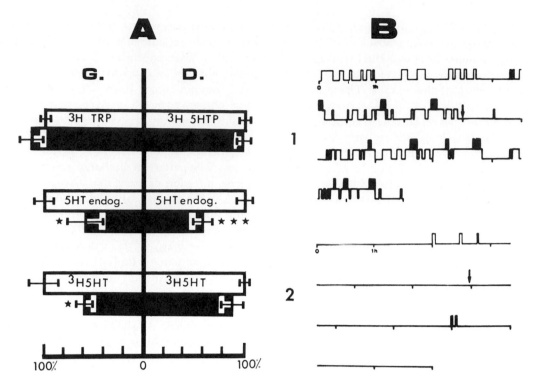

FIGURE 2 Effect of the destruction of the raphe system upon the biosynthesis of cerebral 5-HT and the states of sleep of the cat. (A) Biosynthesis of cerebral 5-HT effectuated in vitro 8 days after the destruction of the rostral raphe system (nucleus raphe dorsalis, centralis superior, and raphe pontis). The brain is treated and the results are expressed in the same way as in Figure 1. (X $P < 0.05$; XXX $P < 0.01$.) In the left cortex (G) there is a significant decrease of endogenous 5-HT which is similar to the decrease of ^{3}H-5-HT synthetized from ^{3}H-TRP. However, in the right cortices where the decrease of endogenous

5-HT is similar to the left side, there is no alteration of the biosynthesis of ^{3}H-5-HT from ^{3}H-5-HTP. This result indicates that only the synthesis of 5-HT from TRP is specific for serotoninergic terminals whereas 5-HTP may be decarboxylated in other cells. (B) Control recordings of sleep (1) in a normal cat and (2) 2 days after the destruction of the rostral raphe system. The permanent insomnia is *not* altered by the injection of 5 mg/kg of DL 5 HTP (arrow). Compared with Figure 1B. Time mark in hours. (Modified from Pujol et al., 1971.)

the intensity of the resulting insomnia, and the selective decrease of cerebral 5-HT.

Metabolism of 5-HT in raphe-destroyed cats. By incubating cortical or brain stem slices in labeled precursor of 5-HT, some additional information has been obtained concerning the metabolism of 5-HT at the level of the terminals at different intervals after the destruction of the 5-HT containing perikarya (Pujol et al., 1971, 1972).

18 hr after the lesion, there was a decrease of the synthesis of ^{3}H 5-HT from ^{3}H TRY and also a decrease of the catabolism of ^{3}H 5-HT, because ^{3}H 5-HIAA is diminished. This parallel decrease of both synthesis and catabolism explains why the concentration of ^{3}H 5-HT is not altered. In contrast, there is a significant (60%) decrease of the spontaneous release in vitro of ^{3}H 5-HT present in the incubation medium. This fact suggests that there might also be a very significant decrease of the release of 5-HT in vivo at the terminal level after the destruction of the 5-HT-containing perikarya at the time of maximum insomnia.

10 days after the lesion of 5-HT-containing perikarya there is a very significant decrease of endogenous 5-HT and 5-HIAA at the level of the terminals which is correlated with the decrease in labeled 5-HT synthesized in vitro from ^{3}H TRY. In contrast, there is no significant alteration in ^{3}H 5-HT synthesized in vitro from ^{3}H 5-HTP (Figure 2B). This can be explained by the fact that 5-HTP can be decarboxylated in other neurons, presumably catecholaminergic or in brain capillaries, where the "nonspecific" 5-HTP-DOPA decarboxylase is located.

This later finding is in agreement with other experiments that show that only tryptophan is the physiological precursor of 5-HT in the brain (see Moir and Eccleston, 1968). Apparently, exogenous 5-HTP at a very low dose is able to be decarboxylated preferentially in 5-HT-containing neurons only when endogenous 5-HTP is lacking (as after PCPA). It is thus possible that high doses of 5-HTP (50 mg/kg) (which are usually given) do not really induce a physiological increase in 5-HT but might induce a nonphysiological state due either to exogenous 5-HTP itself or to the presence of 5-HT in other monoamine-containing neurons. This result demonstrates also that 5-HT *must be released from 5-HT neurons in order to induce sleep*, since the injection of a small dose of 5-HTP has no effect upon the permanent arousal that follows the destruction of the raphe (whereas it restores sleep in PCPA pretreated cat).

The insomnia following the destruction of the raphe system has also been obtained by a midsagittal split of the brainstem that destroys most of the raphe neurons (Michel and Roffwarg, 1967). According to Mancia (1969), the insomnia could be explained not by the destruction of the 5-HT perikarya but by the interruption of synchronizing pathways, ascending from the medulla and crossing at the level of the pons. However, this hypothesis is unlikely, because even in a split brainstem preparation, homolateral cortical synchronization can still be elicited by stimulation of the vagoaortic nerves (Puizillout and Ternaux, 1971).

3. *Mechanisms of action of 5-HT during sleep.* Different hypotheses are possible when considering the presynaptic and postsynaptic mechanisms that are involved in the serotoninergic induction of sleep.

Presynaptic mechanisms: What triggers 5-HT neurons? The mechanisms by which 5-HT neurons are activated in the cat are still obscure. The waking system is almost indefatigable, because a subtotal insomnia of at least 2 weeks duration may be obtained after raphe lesion. Thus the onset of sleep has to be triggered by the activity of the 5-HT sleep system (and not just by a possible circadian dampening of the turnover of the CA waking neurons). It must be recognized that almost nothing is known concerning the mechanisms that activate the 5-HT system. It is possible that 5-HT-synapses may exist from n. paragiganto cellularis to the medial raphe whereas raphe cells (but not necessarily 5-HT perikarya) may apparently also respond to iontophoretic administration of NA and ACh (Couch, 1970). Thus the 5-HT perikarya could be either serotoniceptive, catecholaminoceptive, or cholinoceptive. In such a case, the "doors of SWS" might also be opened by other transmitters or neural systems belonging to "synchronizing structures of the lower brainstem." The fact that the stimulation of the vagoaortic nerves is still able to induce phasic cortical synchronization (and myosis) after destruction of the raphe system (Puizillout and Ternaux, 1971) is an indication that some neural mechanisms are able to trigger EEG signs of sleep even in the probable absence of 5-HT in the terminals. It remains to be shown, however, that true physiological sleep outlasting the duration of the stimulus can be obtained under these conditions. It is also possible that the cortical synchronization induced by stimulation of the vagoaortic

nerves is not related to 5-HT mechanisms per se but instead to the phasic inhibition of the waking system. In any case, some significant advances in the knowledge of the process of the onset of sleep will probably come from the study of the interactions between the raphe system and the nucleus of the solitary tract or its afferents. Finally, some relationships are also possible between the raphe system and the area postrema, which plays a role in the synchronizing mechanisms. According to Koella (1969), 5-HT could act via the area postrema as an agent which facilitates or increases "gain" in the "transfer function" located in the solitary tract nucleus. This would enhance the inhibitory feedback from this nucleus and depress the arousal level and possibly enhance the output from the "thalamic hypnogenic area." Such an hypothesis is difficult to reconcile with the fact that the destruction of the thalamus (Naquet et al., 1965; Angeleri et al., 1969) does not interfere significantly with either state of sleep.

Besides true neural mechanisms, it is also quite likely that the triggering of the 5-HT sleep system could be facilitated (at the perikarya or at the terminal level) by true humoral influences. Among them, the level of blood tryptophan might play a role since tryptophan hydroxylase is not a true rate-limiting enzyme, and it is possible that a study of the intimate mechanism regulating tryptophan uptake by the synaptosomes might elucidate some forms of pathological insomnia or hypersomnia.

Post-synaptic mechanisms: Where are the serotoninergic receptors involved in sleep located? It is likely that the onset of sleep could be obtained through the release of 5-HT upon the neurons of the mesencephalic reticular formation, upon the CA-containing neurons responsible for the prolongation of cortical arousal (see below), or both. This would explain the persistence of the ocular behavior of slow-wave sleep in the chronic decorticated or high mesencephalic cat (Villablanca, 1966). This would also explain the striking arousal that follows the mediopontine pretrigeminal transection (Batini et al., 1959). Indeed, in such a case, most activating structures are situated just in front of the transection while most raphe neurons are situated caudally to the transection. In fact, the selective destruction of the raphe system caudal to the plane of the mediopontine transection induces a similar insomnia (Renault, 1967).

It is also possible that the release of 5-HT in some forebrain structures might be involved in slow-wave sleep, especially in the preoptic region. In this case the release of 5-HT would facilitate (or modulate) the postsynaptic synchronizing mechanisms that are apparently triggered from this region (Sterman and Clemente, 1962). This would explain why the direct injection of 5-HT in the preoptic area may trigger SWS (Yamaguchi et al., 1963) and why the pretreatment of cats with PCPA suppresses the synchronizing action of the stimulation of the preoptic area (Wada and Terao, 1970). The secondary and mild insomnia (as compared with the immediate and total insomnia produced by the destruction of the raphe system) that follows the destruction of the preoptic region (McGinty and Sterman, 1968) could indicate the involvement of 5-HT mechanisms. That is, the destruction of the 5-HT terminals located in this area would secondarily affect the 5-HT perikarya of the raphe system and decrease 5-HT turnover. Finally, one puzzling question should be asked. Is 5-HT a true neurotransmitter or a neuromodulator? The fact that 6 to 8 hr of physiological sleep can be restored by such a small dose of 5-HTP (2-5 mg/kg) in a PCPA pretreated cat is difficult to explain by our classical view of neurotransmitter release and reuptake, whereas the possible binding of 5-HT to some receptors (and a neuromodulator role upon other transmitters) might be more compatible with such a long action.

Catecholaminergic regulation of waking and paradoxical sleep

Since its discovery, the paradoxical similarities between the polygraphic aspects of paradoxical sleep and those of the most excited waking have often been pointed out. Thus it is not a surprise for the neurobiologist to find out that most of the drugs that impair waking also impair PS while the same (or almost the same) group of CA-containing neurons seems to play a paramount importance in the regulation of waking and PS. However, it must be admitted that many systems are still obscure, that many delicate regulations are not yet understood, and that the interplay of cholinergic neurons in both waking and PS deserves more study (see the review of the role of cholinergic neurons in Jouvet, 1967).

CA-Containing Neurons and Waking   The intervention of catecholaminergic mechanisms in the control of tonic cortical arousal in the cat is suggested by converging experimental evidence that can be summarized as follows (see references in Jouvet, 1972):

1. The increase of cerebral catecholamine (after L-dopa) induces a long-lasting arousal in the cat.

2. The inhibition of the synthesis of catecholamine with alpha-methyl-P-tyrosine decreases waking in normal cats and also suppresses the behavioral and EEG arousal which normally follows the injection of DL-amphetamine.

Norepinephrine and not dopamine seems to be responsible for the increase of cortical arousal for the following reasons:

On the one hand, the destruction of the substantia nigra (and thus of dopamine containing neurons), which decreases telediencephalic dopamine levels by more than 90%, does not interfere significantly with cortical arousal but strongly alters behavioral waking (Jones et al., 1968).

On the other hand, the destruction of the dorsal norepinephrine (NE) pathway in the mesencephalon, or of groups A8 in the mesencephalic reticular formation, strongly reduces telediencephalic NE and increases cortical synchronization. There is also a significant correlation between the decrease of cortical arousal and the decrease of NE in the mesencephalon and telediencephalon (Jones et al., 1968).

CA-Containing Neurons and Paradoxical Sleep   The intervention of CA-containing neurons of the laterodorsal pontine tegmentum in the "executory mechanisms" of PS are suggested by these neuropharmacological and neurophysiological facts (see references in Jouvet, 1972):

1. Reserpine-pretreated cats receiving dopa (50 mg/kg) exhibit PS much earlier (2 to 4 hr) than animals not receiving dopa (24 hr). This suggests that the refilling of some pools in the CA terminals could be a condition for the reappearance of PS.

2. The administration of α-methyl-dopa (200 mg/kg), which results in the synthesis of the false transmitter α-methyl-NE, selectively suppresses PS in the cat for 12 to 16 hr and decreases waking.

3. The destruction of the caudal third of the nucleus locus coeruleus suppresses the motor inhibition which occurs during PS, while other ascending phenomena are still present (PGO, fast cortical activity). The total destruction of the locus coeruleus and of the nucleus subcoeruleus selectively suppresses all the central and peripheral components of PS (including PGO waves). This same selective suppression of PS is also obtained secondarily after microinjections of 6-hydroxydopamine in the latero dorsal pontine tegmentum (Buguet et al., 1970).

In summary, both the fast electrical activity of waking and most of the components of paradoxical sleep are under the influence of a group of CA-containing neurons located in the coeruleus complex. (N. locus coeruleus and subcoeruleus.) It is likely that, on one hand, the thin NE terminals that innervate the cerebral cortex are involved in the control of cortical arousal and depend upon the dorsal NE bundle that ascends in the mesencephalon. On the other hand, it appears possible that most of the central tonic and phasic events of PS depend mostly upon the thick CA terminals that are issued from the subcoeruleus group and the axons of which ascend in the intermediary pathway (Maeda and Pin, 1971; Maeda et al., 1973). Thus two possible complementary systems of CA terminals, innervating monosynaptically most of the cerebrum, are controlling the efficiency of waking (at the cortical level) and paradoxical sleep. The fact that such a small group of neurons containing CA (and also possibly acetylcholine) located in the dorsolateral part of the pontine tegmentum may be responsible for most, if not all, the events occurring during PS may seem unbelievable. In fact, the organization of this system (a few perikarya innervating monosynaptically hundreds of thousands of other cells and also intracerebral vesicles) is perfectly adapted to the regulation of the activity of all the cerebrum. The main problem that remains to be solved concerns the postsynaptic second-messenger effect mediated through the release of NE (and possibly ACh) during waking or during PS (at a time when 5-HT terminals might be acting differently). There is, in fact, some indirect evidence which suggests that 5-HT is antagonistic to the waking CA mechanism, while it may act agonistically with the CA mechanism involved in PS.

INTERACTION BETWEEN 5-HT AND CA MECHANISM

Possible inhibitory role of 5-HT neurons upon the catecholaminergic neurons involved in waking. The increased waking that follows the destruction of the rostral raphe is probably mediated by catecholaminergic mechanisms as shown by the following experiment (see Jouvet, 1971).

An injection of 200 mg/kg of α-methyl-P-tyrosine was given to cats with raphe destruction at a time when behavioral and EEG waking was almost permanent (2 to 6 days after the lesion). The running movements stopped, myosis appeared, and there was a significant increase of cortical synchronization that was maximum after 12 hr. This lasted for 24 hr after which time there was a rapid return to behavioral and EEG insomnia. The experiment provides some indirect neuropharmacological evidence that the almost permanent arousal which follows the destruction of the 5-HT-containing neurons of the raphe system might be related to the increased turnover of central catecholaminergic neurons.

Thus, it is possible that under normal conditions, 5-HT neurons might exert tonic inhibitory action directly or indirectly upon some CA-containing neurons at the onset of sleep.

Possible inhibitory role of catecholaminergic neurons upon 5-HT neurons. The destruction of the caudal part of the ascending dorsal NE pathway at the level of the isthmus induces a striking hypersomnia with increases in both SWS and PS (up to 300%) for 3 to 8 days. Biochemical analyses of the brain show a decrease of NE in the forebrain, and a significant increase of both tryptophan and 5-HIAA in the telediencephalon, mesencephalon, and also the spinal cord (Petitjean and Jouvet, 1970). It is likely that the destruction of some NE fibers, coming from the anterior part of the locus coeruleus, has suppressed some tonic inhibition that could be maintained upon the rostral raphe (n. raphe dorsalis) where many CA terminals are located (Loizou, 1970). This, in turn, could increase 5-HT turnover (as demonstrated by the increase of 5-HIAA). It is interesting to note that the possible increase of 5-HT turnover is accompanied by an increase of tryptophan. In accordance with other pharmacological or neurophysiological studies, the uptake of tryptophan seems to play an important role in the regulation of the biosynthesis of brain 5-HT. Finally, after the demonstration of a total insomnia following the inactivation of 5-HT neurons, it should be emphasized that the only true hypersomnia induced by a central lesion is accompanied by an increase of the turnover of 5-HT neurons. These two opposite results strongly favor the hypothesis that 5-HT neurons are involved in both states of sleep in the cat.

Interaction of 5-HT neurons with the catecholaminergic neurons involved in paradoxical sleep. The hypothesis that the 5-HT-containing neurons play a role in the "priming" of PS is based upon the following:

There must exist a critical minimum of slow wave sleep in order for PS to appear (after either subtotal lesion of the raphe system or inhibition of 5-HT synthesis with PCPA) (Figure 3). Thus, a minimum of 16% of daily SWS (which is roughly one third of the normal amount) is necessary to prime PS. Since there is some relationship between the amount of SWS and 5-HT turnover, this would suggest that any important decrease of 5-HT turnover (either induced by lesion or drug) would decrease or suppress PS. This hypothesis might also explain why inhibitors of monoamine oxidase (nialamide, phenylisopropylhydrazine), which decrease the 5-HT turnover, are very potent suppressors of PS. But other interpretations are possible and MAO inhibitor might suppress PS by inhibiting the formation of deaminated metabolites of 5-HT like 5-hydroxy-tryptophol or 5-hydroxy-acetaldehyde (see Jouvet, 1969). In fact, these two latter compounds may have a sleep-inducing effect (Sabelli et al., 1969).

It must be admitted that many of these mechanisms are still hypothetical. Direct or indirect determination of the turnover of both indoleamine and catecholamine in those discrete brain regions implicated in the regulation of sleep-waking are needed. However, most of the data obtained in the cat favor the hypothesis that the sleep-waking cycle is regulated by two antagonistic ascending systems of neurons: The 5-HT neurons for inducing sleep and priming PS and the CA neurons for waking and PS (Figure 4).

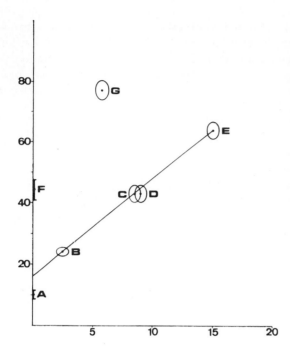

FIGURE 3 Relationships between paradoxical sleep and slow-wave sleep after different lesions of the monoamine systems. Results are expressed in mean ±SEM. (A) Total or subtotal lesions of the raphe system (16 cats). (B) Partial lesions of the raphe system (13 cats). (C) Control lesions of the brain stem outside the monoaminergic system in the ventral pons, caudally to the locus coeruleus or in the tectum (11 cats). (D) Sham-operated cats recorded under the same conditions as the experimental animals during 10 to 13 days (11 cats). (E) Hypersomniac cats with increase of both SWS and PS after lesion of the isthmus destroying the dorsal NE bundle (9 cats). (F) Subtotal lesion of the locus coeruleus complex selectively suppressing paradoxical sleep (15 cats). (G) Lesion of the mesencephalic tegmentum destroying group A 8 of CA-containing neurons (decrease of waking but not true hypersomnia since PS is slightly decreased). In the groups of cats, B, C, D, E, the correlation between PS and SWS may be obtained with the following formula: % PS = (%SWS/3.2) − 16. This means that under a minimum amount of SWS, which is equal to 16% of recording time, no PS can occur. *Ordinate*: Percentage of cortical synchronization during the 10 to 13 days of the postoperative survival. *Abscissa*: Percentage of paradoxical sleep during the same duration. (From Jouvet, 1972.)

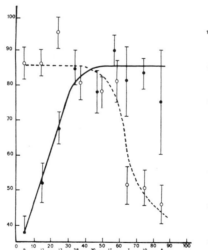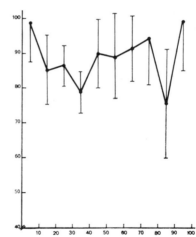

FIGURE 4 Monoaminergic (MA) control of the cortical activity in chronic cats. Two antagonistic systems are strongly implicated in the mechanisms of cortical synchronization and desynchronization. The figure summarizes all the biochemical data obtained from 133 cats (which were all operated, recorded, sacrificed, and biochemically analyzed by the same technique). They were subjected to lesions of the midline raphe, or of the bulbar, pontine, mesencephalic tegmentum, or of the substantia nigra. Control lesions were also made outside MA-containing neurons, and 12 sham-operated animals served as controls. (Groups of animals are classified according to the percentage of EEG synchronization during 10 to 13 days of the postoperative survival and the mean value of the percentage of 5-HT and norepinephrine in the telencephalon and diencephalon for each group is given.) Whereas there is no significant alteration of dopamine (in the telediencephalon) in the right, it is clear that 5-HT and norepinephrine fit two opposite curves: Cats with insomnia (less than 30% of cortical synchronization) have a decreased telediencephalic 5-HT and normal telencephalic norepinephrine content. Cats with normal or subnormal levels of synchronization (30 to 60%) have normal or subnormal values for both 5-HT and noradrenaline, whereas cats with increased synchronization have a decreased norepinephrine with normal or subnormal 5-HT level in the telediencephalon. *Ordinates*: Percentage of 5-HT (black circle) and NE (white circle) in the telediencephalon relative to control cat (mean and SEM). *Abscissae*: Percentage of cortical synchronization during the 10 to 13 days of survival after the lesion. The smaller numbers below refer to the number of cats in each group. The absolute values (100%) for 50 control cats are, respectively, 622 ng/g ± 22 for 5-HT, 399 ng/g = ± 10 for NE, and 947 ± 41 for DA. (Modified from Jouvet, 1972.)

Summary and conclusion

Although no direct evidence has yet been presented, there are many converging data that strongly favor the hypothesis that in the cat the 5-HT-containing neurons of the rostral part of the raphe system are involved in sleep mechanisms. The most convincing evidence may be briefly summarized as follows: (1) the inhibition of the synthesis of 5-HT (at the level of tryptophan hydroxylase) with p-chlorophenylalanine leads to an insomnia that is immediately reversed to normal sleep by a subsequent injection of small doses of 5-HTP, the immediate precursor of 5-HT, which by-passes the inhibition of the synthesis. (2) The destruction of the 5-HT-containing perikarya located in the rostral part of the raphe system leads to insomnia, the intensity of which is correlated with the decrease of cerebral 5-HT. (3) The hypersomnia (increase of both slow wave sleep and paradoxical sleep), which follows the destruction of the dorsal norepinephrine bundle at the level of the isthmus, is accompanied by an increase of 5-HT turnover.

Some possible interactions between 5-HT-containing neurons and other monoaminergic neurons are summarized. Most of the data obtained in the cat favor the hypothesis that the sleep-waking cycle is regulated by two antagonistic ascending systems of neurons: The 5-HT neurons for inducing sleep and priming paradoxical sleep, the catecholaminergic neurons of the ponto-mesencephalic reticular formation for waking and paradoxical sleep.

REFERENCES

ANGELERI, F., G. F. MARCHESI, and A. QUATTRINI, 1969. Effects of chronic thalamic lesions on the electrical activity of the neocortex and on sleep. *Arch. Ital. Biol.* 107:633–668.

BATINI, C., G. MORUZZI, M. PALESTINI, G. F. ROSSI, and A. ZANCHETTI, 1959. Effects of complete pontine transections on the sleep-wakefulness rhythm: The midpontine pretrigeminal preparation. *Arch. Ital. Biol.* 97:1–12.

BUGUET, A., F. PETITJEAN, and M. JOUVET, 1970. Suppression des pointes PGO du sommeil par lésion ou injection in situ de 6-hydroxydopamine au niveau du tegmentum pontique. *C. R. Soc. Biol. (Paris)* 164:2293–2298.

COHEN, H., J. FERGUSON, S. HENRIKSEN, J. M. STOLD, V. J. ZARCONE, J. BARCHAS, and W. DEMENT, 1970. Effects of chronic depletion of brain serotonin on sleep and behavior. *Proc. Amer. Psychol. Ass.* 78:831–832.

COUCH, J. R., 1970. Responses of neurons in the raphe nuclei to serotonin, norepinephrine and acetylcholine and their correlation with an excitatory synaptic input. *Brain Res.* 19:137–150.

DAHLSTROM, A., and K. FUXE, 1964. Evidence for the existence of monoamine neurons in the central nervous system. I. Demonstration of monoamines in the cell bodies of brain stem neurons. *Acta Physiol. Scand.* 62 (supp.):232.

DELORME, F., J. L. FROMENT, and M. JOUVET, 1966. Suppression du sommeil par la p-chlorometamphetamine et la p-chlorophenylalanine. *C. R. Soc. Biol. (Paris)* 160:2347–2351.

DELORME, F., M. JEANNEROD, and M. JOUVET, 1965. Effects remarquables de la réserpine sur l'activité EEG phasiques ponto-géniculo-occipitale. *C. R. Soc. Biol. (Paris)* 159:900–903.

HOYLAND, V. J., E. E. SHILLITO, and M. VOGT, 1970. The effect of parachlorophenylalanine on the behavior of cats. *Brit. J. Pharmacol.* 40:659–677.

JONES, B., P. BOBILLIER, and M. JOUVET, 1968. Effets de la destruction des neurons contenant des catécholamines du mésencéphale sur le cycle veille-sommeil du chat. *C. R. Soc. Biol. (Paris)* 163:176–180.

JOUVET, M., 1967. Neurophysiology of the states of sleep. *Physiol. Rev.* 47:117–177.

JOUVET, M., 1969. Biogenic amines and the states of sleep. *Science* 163:32–41.

JOUVET, M., 1971. Some monoaminergic mechanisms controlling sleep and waking. In *Brain and Human Behavior*, A. Karczmar and J. Eccles, eds. New York: Springer-Verlag.

JOUVET, M., 1972. The role of monoamines and acetylcholine containing neurons in the regulation of the sleep-waking cycle. *Ergeb. Physiol.* 64:166–307.

KOE, B. K., and A. WEISSMAN, 1966. P-chlorophenylalanine, a specific depletor of brain serotonin. *J. Pharmacol. Exp. Ther.* 154:499–516.

KOELLA, W. P., 1969. Neurohumoral aspects of sleep control. *Biol. Psychiat.* I:161–177.

KOELLA, W. P., A. FELDSTEIN, and J. S. CZICMAN, 1968. The effect of para-chlorophenylalanine on the sleep of cats. *Electroenceph. Clin. Neurophysiol.* 25:481–490.

LOIZOU, L. A., 1970. Uptake of monoamines into central neurons and the blood brain barrier in the infant rat. *Brit. J. Pharmacol.* 40:800–813.

MAEDA, T., and C. PIN, 1971. Organization et projections des systèmes catécholaminergiques du pont chez le chaton. *C. R. Soc. Biol. (Paris)* 165:2137–2141.

MAEDA, T., C. PIN, D. SALVERT, M. LIGIER, and M. JOUVET, Les neurons contenant des catécholamines du tegmentum pontique et leurs voies de projection chez le chat. *Brain Res.*, (in press).

MANCIA, M., 1969. EEG and behavioral changes owing to splitting of the brain stem in cats. *Electroenceph. Clin. Neurophysiol.* 27:487–503.

MCGINTY, D. J., and M. B. STERMAN, 1968. Sleep suppression after basal forebrain lesions in the cat. *Science* 160:1253–1255.

MICHEL, F., and H. P. ROFFWARG, 1967. Chronic split brain stem preparation: Effect on the sleep-waking cycle. *Experientia (Basel)* 23:126–128.

MOIR, A. T. B., and D. ECCLESTON, 1968. The effects of precursor loading in the cerebral metabolism of 5-hydroxyindoles. *J. Neurochem.* 15:1093–1109.

MOURET, J., J. L. FROMENT, P. BOBILLIER, and M. JOUVET, 1967. Etude neuropharmacologique et biochimique des insomnies provoquées par la p-chlorophenylalanine. *J. Physiol. (Paris)* 59:463–464.

NAQUET, R., M. DENAVIT, J. LANOIR, and D. ALBE-FESSARD, 1965. Altérations transitoires ou définitives de zones diencéphaliques chez le chat. Leurs effets sur l'activité électrique corticale et le sommeil, M. Jouvet, ed. Paris: Centre National de la Recherche Scientifique.

PETITJEAN, F., and M. JOUVET, 1970. Hypersomnie et augmentation de l'acide 5-hydroxy-indolacétique cérébral par lésion isthmique chez le chat. *C. R. Soc. Biol. (Paris)* 164: 2288–2293.

PIN, C., B. JONES, and M. JOUVET, 1968. Topographie des neurones monoaminergiques du tronc cérébral du chat: Étude par histofluorescence. *C. R. Soc. Biol. (Paris)* 162: 2136–2141.

PUIZILLOUT, J., and J. P. TERNAUX, 1971. Persistance d'un endormement vagoaortique après destruction chirurgicale et pharmacologique des noyaux du raphé. *J. Physiol. (Paris)* 63:272.

PUJOL, J. F., 1970. Contribution à l'étude des modifications de la régulation du métabolisme des monoamines centrales pendant le sommeil et la veille. Thèse de Doctorat ès-Sciences, Paris, p. 192.

PUJOL, J. F., A. BUGUET, J. L. FROMENT, B. JONES, and M. JOUVET, 1971. The central metabolism of serotonin in the cat during insomnia: A neurophysiological and biochemical study after p-chlorophenylalanine or destruction of the raphe system. *Brain Res.* 29:195–212.

PUJOL, J. F., F. SORDET, F. PETITJEAN, D. GERMAIN, and M. JOUVET, 1972. Insomnie et métabolisme cérébral de la séro-tonine chez le chat: Etude de la synthèse et de la libération de la sérotonine mesurées in vitro 18 heures après destruction du système du raphé. *Brain Res.* 39:137–149.

RENAULT, J., 1967. Monoamines et sommeil. Rôle du système du raphé et de la sérotonine cérébrale dans l'endormissement. Thèse de Médecine, Université de Lyon, Tixier, ed., p. 140.

SABELLI, H. C., S. G. ALIVISATOS, P. K. SETH, and F. UNGAR, 1969. Aldehydes of brain amines affect central nervous system. *Nature (Lond.)* 223:73–74.

STERMAN, M. B., and C. D. CLEMENTE, 1962. Forebrain inhibitory mechanisms: Sleep patterns induced by basal forebrain stimulation in the behaving cat. *Exp. Neurol.* 6:103–117.

VILLABLANCA, J., 1966. Behavioral and polygraphic study of "sleep" and wakefulness in chronic decerebrate cats. *Electroenceph. Clin. Neurophysiol.* 21:562–577.

WADA, J. A., and A. TERAO, 1970. Effect of parachlorophenyl-alanine on basal forebrain stimulation. *Exp. Neurol.* 28:501–506.

YAMAGUCHI, T., T. J. MARCZINSKI, and M. LINGI, 1963. The effects of electrical and chemical stimulation of the preoptic region and some non specific thalamic nuclei in unrestrained, waking animals. *Electroenceph. Clin. Neurophysiol.* 15:145–166.

43 Circadian Rhythms in Indole Metabolism in the Rat Pineal Gland

DAVID C. KLEIN

ABSTRACT There are circadian rhythms in the concentration of serotonin, N-acetylserotonin, and melatonin in the rat pineal gland. These rhythms are regulated by the rhythm in the activity of the enzyme that converts serotonin to N-acetylserotonin, N-acetyltransferase. The rhythmic changes in the activity of this enzyme are controlled transsynaptically by norepinephrine. The effects of norepinephrine are mediated intracellularly by a cyclic AMP mechanism requiring protein synthesis. The endogenous neural signals that drive this system appear to originate in the CNS, possibly in the suprachiasmatic nucleus. The generation or the transmission of these signals can be blocked by environmental lighting acting via the eye.

INDOLE METABOLISM in the rat pineal gland is remarkably dynamic during the course of a typical day. Under conditions that provide alternating periods of light and darkness of about 12 hr, daily rhythms occur in the activities of the enzymes and the concentration of the compounds in the pathway that converts serotonin to melatonin. Most of these rhythms have been found to be truly circadian: They persist when animals are in constant darkness or are blinded. These rhythms and the questions of how these rhythms are generated and integrated are discussed in this chapter.

Rhythms in the serotonin-melatonin pathway in the rat pineal gland

The conversion of serotonin (5-hydroxytryptamine) to melatonin (N-acetyl 5-methoxytryptamine) involves two enzymes and the intermediate, N-acetylserotonin (N-acetyl 5-hydroxytryptamine). The enzyme that converts serotonin to N-acetylserotonin is serotonin N-acetyltransferase (E.C.2.3.1.5); the acetyl group comes from acetyl coenzyme A (Weissbach et al., 1960; 1961). The enzyme that converts N-acetylserotonin to melatonin is hydroxyindole-O-methyltransferase (E.C.2.1.1.4); the methyl group donor is S-adenosyl methionine (Axelrod and Weissbach, 1961). Details of this pathway appear in Figure 1.

DAVID C. KLEIN Laboratory of Biomedical Sciences, National Institute of Child Health and Human Development, National Institutes of Health, Bethesda, Maryland

During the hours of daylight, the concentration of serotonin in the rat pineal gland is about 0.5 mM (Quay, 1963; Snyder et al., 1965; Illnerová, 1971). At night the concentration gradually falls to about one-half this value. Concurrently, the activity of serotonin N-acetyltransferase increases and reaches values that are 15- to 70-fold higher than the day values (Klein and Weller, 1970a; Ellison et al., 1972; Deguchi and Axelrod, 1972a). This increase in enzyme activity is accompanied by an increase in the concentration of N-acetylserotonin to values that are 10- to 30-fold greater than day values (Klein and Weller, 1973a). The concentration of melatonin also increases at night and reaches values that are about 7- to 10-fold greater than the day values (Quay, 1964; Lynch, 1971). The activity of hydroxyindole-O-methyltransferase has been reported to increase at night (Axelrod et al., 1965; Klein and Lines, 1969). This increase is much smaller than that of N-acetyltransferase, N-acetylserotonin, or melatonin and has been difficult to observe consistently (Quay, 1967; Lynch and Ralph, 1970; Reiter and Klein, 1971).

When the dark-light transition takes place there is a rapid decrease (halving time $\cong$ 3 min) in the activity of N-acetyltransferase (Klein and Weller, 1972; Deguchi and Axelrod, 1972a) and a rapid increase in the concentration of serotonin to daytime values within 14 min (Illnerová, 1971). A rapid decrease in the concentration of N-acetylserotonin also occurs (Klein and Weller, 1973a). The melatonin content of the pineal gland also decreases at this time (Lynch, 1971), but the rate of this change has not been reported.

The data describing these rhythms have been abstracted from the original reports referenced above or unpublished data from this laboratory as detailed below and are presented in Figure 1. The lighting cycles were not precisely the same in the original reports. For the sake of comparison, we have altered the length of the light and dark periods (no more than 2 hr) to normalize the lighting cycles used in the original studies. The serotonin graph is based on the reports of Snyder et al.(1965), and Illnerová (1971). The N-acetyltransferase graph is based on several

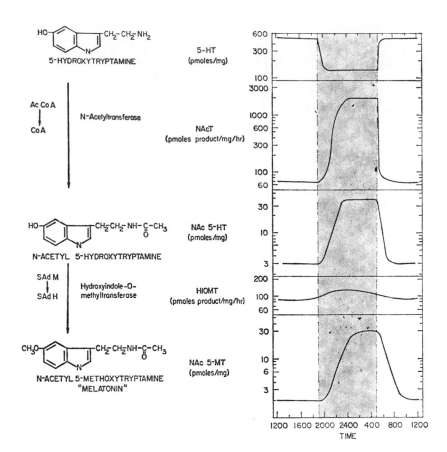

FIGURE 1 Rhythms in indole metabolism in the rat pineal gland. The metabolic pathway from 5-hydroxytryptamine to melatonin is on the left. The daily variations in the concentrations of metabolites and activities of enzymes are on the right. The shaded portion indicates the dark period of the lighting cycle. The data have been abstracted from reports in the liter- ature as detailed in the text. AcCoA, acetyl coenzyme A; CoA, coenzyme A; S AdM, S-adenosyl methionine; S AdH, S-adenosyl homocysteine; 5-HT, 5-hydroxytryptamine, serotonin; NAcT, N-acetyltransferase; HIOMT, hydroxyindole-O-methyltransferase; NAc 5-MT, N-acetyl 5-methyoxytryptamine, melatonin.

reports from this laboratory (Klein and Weller, 1970a; Klein et al., 1971b; Klein and Weller, 1972). The N-acetylserotonin graph is based on the report of Klein and Weller (1973a). The hydroxyindole-O-methyltransferase graph is based on the reports of Klein and Lines (1969) and Reiter and Klein (1971) and on unpublished reports from this laboratory. The melatonin graph is based on the report of Lynch (1971).

The following hypothesis has been proposed to explain the integrated regulation of the rhythms in indole metabolism (Klein et al., 1971a). The initial perturbation in the system is the increase in the activity of N-acetyltransferase. This causes a 20- to 70-fold increase in the rate of conversion of serotonin to N-acetylserotonin and results in the 50% decrease in the steady state levels of serotonin and the 10- to 30-fold increase in the concentration of N-acetylserotonin. The latter effect causes an increase in the production of melatonin by mass action: With more substrate available more product is produced, because the amount of N-acetylserotonin in the pineal gland, assuming uniform distribution, is never at concentrations that saturate hydroxyindole-O-methyltransferase (Klein and Weller, 1973a). The increased production of melatonin causes a 7- to 10-fold increase in the gland content of melatonin and the increased release of melatonin. It is probable that large tonic changes in hydroxyindole-O-methyltransferase activity produced by constant light or darkness (Axelrod et al., 1965) will modify the effects of large changes in N-acetyltransferase activity on melatonin production.

The rapid changes in indole metabolism also appear to be due primarily to the change in the activity of N-acetyltransferase. When it drops, the production of N-acetylserotonin is apparently cut off, which results in a decrease in the concentration of N-acetylserotonin and melatonin. The content of serotonin increases apparently because serotonin is not being used in the N-acetylation pathway.

THE CIRCADIAN VERSUS DAILY NATURE OF THE RHYTHMS IN INDOLE METABOLISM IN THE PINEAL GLAND The term *circadian* is used by the student of biological periodicities to describe a rhythm that has two characteristics: a complete cycle of about a day's length, and the persistence of the rhythm in the absence of lighting cues under conditions of either constant darkness or constant lighting. This sets the endogenously generated circadian rhythms apart from the exogenously generated daily rhythms that are totally dependent upon environmental cues of lighting transitions (light→dark; dark→ light) and disappear in the absence of lighting cues in both constant darkness and constant lighting. The rhythms in the concentration of serotonin, N-acetylserotonin, melatonin, and the activity of N-acetyltransferase are all circadian; they persist in sighted animals in the dark or in blinded animals in constant lighting (Snyder et al., 1965; Klein and Weller, 1970a; Ralph et al., 1971; Klein et al., 1971b; Reiter et al., 1971). In contrast, the rhythm in hydroxyindole-O-methyltransferase is a daily rhythm; it is absent in conditions of constant lighting and constant darkness (Axelrod et al., 1965).

THE EFFECT OF LIGHT ON THE CIRCADIAN RHYTHMS IN PINEAL INDOLE METABOLISM Although the circadian rhythms in indole metabolism in the pineal gland do not depend upon lighting for their generation, all are known to be influenced by light. The changes that are associated with the light→dark transition do not occur if lighting is maintained, and the circadian rhythms are never detected when animals are kept in constant light conditions (Snyder et al., 1965; Klein and Weller, 1970a; Reiter and Klein, 1971; Ralph et al., 1971; Klein and Weller, 1973a). In all cases light has been shown to act via the eye. In the next section the mechanism through which light appears to act on the generation of the circadian rhythms is discussed.

GENERATION OF THE CIRCADIAN RHYTHMS IN PINEAL INDOLE METABOLISM AND INTEGRATION OF LIGHTING SIGNALS The circadian rhythms in pineal serotonin, N-acetylserotonin, and N-acetyltransferase activity depend upon neural input (Fiske, 1964; Klein and Weller, 1973a; Klein et al., 1971a). All the neural input to the pineal gland comes from the superior cervical ganglia (Ariens-Kappers, 1960). Removal of these ganglia, which leads to the disappearance of pineal nerve fibers, or decentralization of the ganglia, which leaves the postganglionic connections to the pineal gland intact but cuts input to the ganglia, abolishes the rhythms in serotonin and N-acetyltransferase activity (Snyder et al., 1965; Klein et al., 1971a). The N-acetylserotonin rhythm is also abolished by removal of the ganglia (Klein and Weller,

1973a). Electrical stimulation of the sympathetic nerve chain leading to the ganglia increases the activity of N-acetyltransferase (Volkman and Heller, 1971). These observations indicate that the rhythms are not generated in the pineal gland or in the superior cervical ganglia but rather that they are driven by a generator located distal to the superior cervical ganglia.

The superior cervical ganglia receive inputs from cell bodies in the upper thoracic spinal segment, which are in turn controlled by nerve fibers originating in central structures. In studies on the central structures involved in the regulation of the N-acetyltransferase rhythms that are detailed by Moore elsewhere in this volume, it has been found that lesions in the medial forebrain bundle, which transmits impulses to the superior cervical ganglia, block the rhythm in pineal N-acetyltransferase. This indicates that the endogenous driving mechanism is probably in the brain. It has also been shown that lesions of the suprachiasmatic nuclei, a pair of cell groupings immediately above the optic chiasm in the anterior hypothalamus, also block the pineal N-acetyltransferase rhythm. When all the input to this region appears to be obliterated, the rhythm in N-acetyltransferase activity persists, indicating that the suprachiasmatic nuclei may be the originating site of the circadian rhythm generator driving the N-acetyltransferase rhythm.

As we have discussed above, constant light can block the rhythm in N-acetyltransferase activity. Moore et al., in this volume, have demonstrated the existence of a unique neural pathway from the retina to the suprachiasmatic nuclei. This retinohypothalamic projection consists of unmyelinated fibers that course with the primary optic nerves and branch off at the level of the optic chiasm. Since this projection is the only apparent direct input from the retina to the suprachiasmatic nucleus, it is possible that it is by these fibers that light can act to either block the generation of the circadian rhythm signals or their transmission. Present collaborative investigations by this laboratory and Moore's are directed at testing the hypothesis that the retinohypothalamic projection mediates the effects of light on the N-acetyltransferase rhythm and that the suprachiasmatic nucleus is the site of the endogenous circadian rhythm generator.

TRANSSYNAPTIC GENERATION OF INDOLE RHYTHMS As mentioned above, the rhythms in N-acetyltransferase activity, serotonin concentration, and N-acetylserotonin concentration are blocked when neural input to the pineal gland is blocked by either decentralization or removal of the superior cervical ganglia. The nerve fibers that leave the superior cervical ganglia and terminate in the pineal gland contain norepinephrine (Pellegrino de Iraldi and Zieher, 1966; Bondareff and Gordon, 1966; Wolfe et al.,

1962). The possibility that norepinephrine mediates the neural regulation of indole metabolism by being released from nerve fibers in the pineal gland and stimulating pinealocytes has been investigated with organ culture of the pineal gland. This technique makes it possible to maintain a pineal gland under defined conditions (Shein et al., 1967; Klein and Weller, 1970b). When cultured pineal glands are treated with L-norepinephrine there is an increase in the activity of N-acetyltransferase (Klein et al., 1970a; Klein and Weller, 1973b), an increase in the N-acetylserotonin in pineal gland (Klein and Weller, 1973a), and a decrease in the amount of serotonin in pineal glands (Klein et al., 1973). L-Norepinephrine treatment also causes an increase in the conversion of radiolabeled tryptophan to radiolabeled N-acetylserotonin (Klein and Berg, 1970) and melatonin (Axelrod et al., 1969; Klein and Berg, 1970) in the culture medium and an increase in the total amount of N-acetylserotonin in the culture medium (Klein and Weller, 1973a). These effects are mediated by a highly specific receptor that is not responsive to indoleamines and is in general less responsive to closely related catecholamines and to the D-isomeric form of norepinephrine. In addition, all these effects of norepinephrine are blocked by a beta-adrenergic receptor blocking agent, propranolol, indicating that this is a beta-receptor (Wurtman et al., 1971; Klein et al., 1973b; Klein and Weller, 1973a, b). Deguchi and Axelrod (1972a, b) have also presented in vivo evidence indicating that the activity of N-acetyltransferase is regulated by a beta-adrenergic receptor. From these in vivo and in vitro findings, it seems highly probable that norepinephrine is the neurochemical involved in the transsynaptic transmission of neural information regulating the activity of N-acetyltransferase, which in turn causes the changes in the concentration of serotonin and the production and concentration of N-acetylserotonin and melatonin to occur. The adrenergically regulated changes that occur in organ culture are similar to those that occur at night in the dark. This means that during the night in the dark the release of norepinephrine occurs. Conversely, light must block this release.

Recent studies on the regulation of the rapid decrease in N-acetyltransferase activity, which appears to be responsible for the rapid changes in serotonin and N-acetylserotonin in the pineal gland, indicate this decrease may be under adrenergic control. A single injection of a beta-adrenergic blocking agent to an animal in the dark results in a rapid decrease in enzyme activity (Deguchi and Axelrod, 1972a). Thus it would appear that simple displacement of the norepinephrine by the blocking agent is sufficient to cause a rapid reversal of the norepinephrine effects. However, in unpublished studies, we attempted similar experiments in organ culture using compounds known to compete with norepinephrine at postsynaptic sites. All were used at 10- or 100-fold higher concentrations than was norepinephrine. Treatment with these compounds did not reverse the effects of a 6 hr treatment with norepinephrine. This failure may only be a result of the artificial nature of organ culture. It is also possible that these findings are an indication that the displacement and removal of norepinephrine alone may not be sufficient to reverse the effects of norepinephrine and that another active mechanism is involved in the reversal of the stimulatory effects of norepinephrine.

THE MECHANISM OF ACTION OF NOREPINEPHRINE The transsynaptic regulation by norepinephrine of the dark induced changes in indole metabolism appears on an intracellular level to involve adenosine 3′,5′-monophosphate (cyclic AMP). Norepinephrine, acting via a beta-adrenergic receptor, stimulates the activity of adenyl cyclase in broken cell preparations (Weiss and Costa, 1967; 1968) and increases the concentration of cyclic AMP in cultured pineal glands (Strada et al., 1972). The effects of norepinephrine on indole metabolism are mimicked by ^{6}N, 2′O-dibutyryl cyclic AMP (dibutyryl cyclic AMP), an analog of cyclic AMP that can inhibit the breakdown of cyclic AMP by pineal phosphodiesterase, and by theophylline (Klein and Berg, 1970; Berg and Klein, 1971). The following effects of norepinephrine on indole metabolism are mimicked by dibutyryl cyclic AMP: stimulation of the conversion of radiolabeled tryptophan and serotonin to radiolabeled N-acetylserotonin and melatonin (Shein and Wurtman, 1969; Klein et al., 1970b; Berg and Klein, 1971; Klein and Weller, 1973b), increased amount of N-acetylserotonin in the glands and medium (Klein and Weller, 1973a), increased activity of N-acetyltransferase (Klein et al., 1970a; Klein and Weller, 1973b), and decreased amount of serotonin in the gland (Klein et al., 1973).

The mechanism of action of cyclic AMP is not clear. The effect of both norepinephrine and dibutyryl cyclic AMP are blocked by cycloheximide but not by actinomycin D (Klein and Berg, 1970), which indicates that new synthesis of RNA is not required but that new synthesis of protein is necessary. This may mean that new molecules of active N-acetyltransferase are made. However, it is also possible that inactive N-acetyltransferase molecules are made continually and that the effect of norepinephrine is to stabilize the newly formed enzyme. It is also possible that inactive N-acetyltransferase molecules are always available and that there is new synthesis of an activating enzyme stimulated by norepinephrine. The cyclic AMP activated protein kinase in the pineal gland (Fontana and Lovenberg, 1971) may be involved, but this has not been proven yet.

As indicated above, the regulation of the rapid decrease in N-acetyltransferase activity stimulated by light appears to involve the adrenergic receptor (Deguchi and Axelrod, 1972a). The intracellular mechanism involved is not known; it may involve the spontaneous degradation of N-acetyltransferase. We have found (Binkley et al., 1973) that when homogenates of pineal glands obtained at night are incubated at 37°C, there is a rapid disappearance of enzyme activity. This disappearance can be prevented by the cosubstrate of the enzyme, acetylcoenzyme A, and a small fragment of acetylcoenzyme A, cysteamine. Perhaps the rapid decrease in N-acetyltransferase is regulated via this stabilizing mechanism. Light may act by blocking the release of norepinephrine and allowing reuptake into nerve fibers to occur, and perhaps by stimulating a second mechanism that blocks the stabilization of the enzyme, by a cysteamine-like structure.

POSSIBLE FUNCTIONS OF THE RHYTHMS IN PINEAL INDOLE METABOLISM. The large changes in indole metabolism that occur in the pineal gland during the course of the day provide a means of turning environmental light signals into biochemical messages that are measurements of the *duration* of the dark and light periods. Long dark periods would allow longer periods of high melatonin and N-acetylserotonin production to occur, and these would presumably result in periods during which blood levels of these compounds would be elevated. In this role of translating environmental lighting information into chemical information, the pineal gland has been termed a neuroendocrine transducer (Wurtman and Anton-Tay, 1969). The rapid effects of light on N-acetyltransferase resulting in rapid effects on the production of N-acetylserotonin and the short blood half-life of melatonin (Kopin et al., 1961) could greatly enhance the precision of this system and allow small differences in the duration of the dark period to be transduced as significant differences in the duration of the high blood level of melatonin. The pineal gland has been implicated in the gonadal atrophy that occurs when hamsters are deprived of light (Reiter and Fraschini, 1969). When these animals are switched from a lighting cycle, which provides 14 hr of light and 10 hr of darkness, to one that provides 14 hr of darkness and 10 hr of light, gonadalatrophy is initiated. Large differences in gonadal weight are detected in about 6 to 10 weeks. It is possible that the effect of extended darkness on gonadal weight is mediated in part by changes in the production and release of N-acetylserotonin and melatonin.

General implications

The experiments discussed here indicate that, in the pineal gland, norepinephrine can regulate the metabolism of serotonin by a cyclic AMP mechanism requiring protein synthesis, and that this regulation involves the increased formation of N-acetylated derivatives, which is accompanied by a decrease in the steady-state levels of serotonin. This mechanism appears to be the basis of the circadian rhythms in indole metabolism in the pineal gland. Reis et al. (1969) have detected rhythms in the concentration of serotonin in many areas of the cat brain. These rhythms are not synchronized with each other. The N-acetyltransferase activity has been detected in several areas of the brain (Ellison et al., 1972), and this enzyme might be involved in regulating serotonin rhythms in some of these areas. The N-acetylation mechanism could function to modulate the amount of serotonin available for release as a neurotransmitter. Alternatively, the N-acetylated derivatives of serotonin could function as transmitters for local hormones.

Jouvet has implicated serotonin in the mechanism underlying sleep (see Jouvet, this volume). Disturbances in serotonin metabolism in the raphe nuclei, which contain all the cell bodies of the serotonergic neurons in the brain, result in disturbances in sleep. The raphe nuclei also receive adrenergic input. These areas are similar therefore to the pineal gland in two respects: both have serotonin containing cell bodies and both receive adrenergic input. However, it is not known if the cells in the raphe nuclei are similar to the pineal gland in the manner in which the concentration of serotonin is regulated by norepinephrine via the N-acetylation mechanism. Perhaps they are.

REFERENCES

ARIENS-KAPPERS, J., 1960. The development, topographical relations and innervation of the epiphysis cerebri in the albino rat. *Z. Zellforschung. Mickrosk. Anat.* 52:163–215.

AXELROD, J., H. M. SHEIN, and R. J. WURTMAN, 1969. Stimulation of C^{14}-melatonin synthesis from C^{14}-tryptophan by norepinephrine in rat pineal in organ culture. *Proc. Natl. Acad. Sci. USA* 62:544–549.

AXELROD, J., and H. WEISSBACH, 1961. Purification and properties of hydroxyindole-O-methyltransferase. *J. Biol. Chem.* 236:211–215.

AXELROD, J., R. J. WURTMAN, and S. H. SNYDER, 1965. Control of hydroxyindole-O-methyltransferase activity in the rat pineal gland by environmental lighting. *J. Biol. Chem.* 240:949–954.

BERG, G. R., and D. C. KLEIN, 1971. Pineal gland in organ culture II: The role of adenosine 3′,5′-monophosphate in the regulation of radiolabeled melatonin production. *Endocrinol.* 89:453–464.

BINKLEY, S. A., J. L. WELLER, and D. C. KLEIN, 1973. Protection of induced pineal N-acetyltransferase activity. *Fed. Proc.* 32: 251.

BONDAREFF, W., and B. GORDON, 1966. Submicroscopic localization of norepinephrine in sympathetic nerves of rat pineal. *J. Pharm. Exp. Ther.* 153:42–47.

DEGUCHI, T., and J. AXELROD, 1972a. Control of circadian change in serotonin N-acetyltransferase activity in the pineal organ by the β-adrenergic receptor. *Proc. Natl. Acad. Sci. USA* 69:2547–2550.

DEGUCHI, T., and J. AXELROD, 1972b. Induction and super-induction of serotonin N-acetyltransferase by adrenergic drugs and denervation in the rat pineal. *Proc. Natl. Acad. Sci. USA* 69:2208–2211.

ELLISON, N., J. WELLER, and D. C. KLEIN, 1972. Development of a circadian rhythm in the activity of pineal serotonin N-acetyltransferase. *J. Neurochem.* 19:1335–1341.

FISKE, V. M., 1964. Serotonin rhythm in the pineal organ: Control by the sympathetic nervous system. *Science* 146:253–254.

FONTANA, J. A., and W. LOVENBERG, 1971. A cyclic AMP-dependent protein kinase of the bovine pineal gland. *Proc. Natl. Acad. Sci. USA* 68:2787–2790.

ILLNEROVÁ, H., 1971. Effect of light on the serotonin content of the pineal gland. *Life Sci.* 10(I):955–961.

KLEIN, D. C., and G. R. BERG, 1970. Pineal gland: Stimulation of melatonin production by norepinephrine involves cyclic AMP mediated stimulation of N-acetyltransferase. *Adv. Biochem. Psychopharmacol.* 3:241–263.

KLEIN, D. C., G. R. BERG, and J. WELLER, 1970a. Melatonin synthesis: Adenosine 3′,5′-monophosphate and norepinephrine stimulate N-acetyltransferase. *Science* 168:979–980.

KLEIN, D. C., G. R. BERG, J. WELLER, and W. GLINSMANN, 1970b. Pineal gland: Dibutyryl cyclic adenosine monophosphate stimulation of labeled melatonin production. *Science* 167:1738–1740.

KLEIN, D. C., and S. V. LINES, 1969. Pineal hydroxyindole-O-methyltransferase activity in the growing rat. *Endocrinol.* 89:1523–1525.

KLEIN, D. C., R. J. REITER, and J. L. WELLER, 1971a. Pineal N-acetyltransferase activity in blinded and anosmic male rats. *Endocrinol* 89:1029–1023.

KLEIN, D. C., and J. L. WELLER, 1970a. Indole metabolism in the pineal gland: A circadian rhythm in N-acetyltransferase. *Science* 169:1093–1095.

KLEIN, D. C., and J. WELLER, 1970b. Input and output signals in a model neural system: The regulation of melatonin production in the pineal gland. *In Vitro* 6:197–204.

KLEIN, D. C., and J. L. WELLER, 1972. Rapid light-induced decrease in pineal N-acetyltransferase activity. *Science* 177: 532–533.

KLEIN, D. C., and J. L. WELLER, 1973a. Regulation of pineal N-acetylserotonin. *J. Neurochem.* (in press).

KLEIN, D. C., and J. L. WELLER, 1973b. Adrenergic-adenosine 3′,5′-monophosphate regulation of serotonin N-acetyltransferase activity and the temporal relationship of serotonin N-acetyltransferase activity to synthesis of ^{3}H-N-acetylserotonin and ^{3}H-melatonin in the cultured rat pineal gland, (in press).

KLEIN, D. C., J. L. WELLER, and R. Y. MOORE, 1971a. Melatonin metabolism: Neural regulation of pineal serotonin N-acetyltransferase activity. *Proc. Natl. Acad. Sci. USA* 68:3107–3110.

KLEIN, D. C., A. YUWILER, J. WELLER, and S. PLOTKIN, 1973. Postsynaptic adrenergic-cyclic AMP control of serotonin in the cultured rat pineal gland. *J. Neurochem.* (in press).

KOPIN, I. J., C. M. B. PARE, J. AXELROD, and H. WEISSBACH, 1961. The biological fate of melatonin. *J. Biol. Chem.* 236: 3072–3075.

LYNCH, H. J., 1971. Diurnal oscillations in pineal melatonin content. *Life Sci.* (I):791–795.

LYNCH, H. J., and C. L. RALPH, 1970. Diurnal variation in pineal melatonin and its nonrelationship to HIOMT activity. *Am. Zool.* 10:300.

PELLEGRINO DE IRALDI, A., and L. M. ZIEHER, 1966. Norepinephrine and dopamine content of normal, decentralized, and denervated pineal glands of the rat. *Life Sci.* 5(I):149–154.

QUAY, W. B., 1963. Circadian rhythm in rat pineal serotonin and its modification by estrous cycle and photoperiod. *Gen. Comp. Endocrinol.* 3:473–479.

QUAY, W. B., 1964. Circadian and estrous rhythms in pineal melatonin and 5-hydroxyindole-3-acetic acid. *Proc. Soc. Exp. Biol. Med.* 115:710–713.

QUAY, W. B., 1967. Lack of rhythm and effect of darkness in rat pineal content of N-acetylserotonin-O-methyltransferase. *Physiologist.* 10:286.

RALPH, C. L., D. MULL, H. J. LYNCH, and L. HEDLUND, 1971. A melatonin rhythm persists in rat pineals in darkness. *Endocrinol.* 89:1361–1366.

REIS, D. J., A. CORVELLI, and J. CONNERS, 1969. A circadian rhythm of serotonin regionally in cat brain. *J. Pharmacol. Exp. Ther.* 167:328–333.

REITER, R. J., and F. FRASCHINI, 1969. Endocrine aspects of the mammalian pineal gland. A review. *Neuroendocrinol.* 5:219–255.

REITER, R. J., and D. C. KLEIN, 1971. Observations on the pineal gland, the Harderian glands, the retina, and the reproductive organs of adult female rats exposed to continuous light. *J. Endocrinol.* 51:117–125.

REITER, R. J., S. SORRENTINO, JR., C. L. RALPH, H. J. LYNCH, D. MULL, and E. JARROW, 1971. Some endocrine effects of blinding and anosmia in adult male rats with observations on pineal melatonin. *Endocrinol.* 88:895–900.

SHEIN, H. M., and R. J. WURTMAN, 1969. Cyclic adenosine monophosphate: Stimulation of melatonin and serotonin synthesis in cultured rat pineals. *Science* 166:519–520.

SHEIN, H. M., R. J. WURTMAN, and J. AXELROD, 1967. Synthesis of serotonin by pineal glands of the rat in organ culture. *Nature (Lond.)* 213:730.

SNYDER, S. H., M. ZWEIG, J. AXELROD, and J. E. FISCHER, 1965. Control of the circadian rhythm in serotonin content of the rat pineal gland. *Proc. Natl. Acad. Sci. USA* 53:301–305.

STRADA, S. J., D. C. KLEIN, J. L. WELLER, and B. WEISS, 1972. Effect of norepinephrine on the concentration of adenosine 3′,5′-monophosphate of rat pineal gland in organ culture. *Endocrinol.* 90:1470–1476.

VOLKMAN, P. H., and A. HELLER, 1971. Pineal N-acetyltransferase activity: Effect of sympathetic stimulation. *Science* 173:839–840.

WEISS, B., and E. COSTA, 1967. Adenyl cyclase activity in rat pineal gland: Effects of chronic denervation and norepinephrine. *Science* 173:1750–1752.

WEISS, B., and E. COSTA, 1968. Selective stimulation of adenyl cyclase of rat pineal gland by pharmacologically active catecholamines. *J. Pharm. Exp. Ther.* 161:310–319.

WEISSBACH, H., B. G. REDFIELD, and J. AXELROD, 1960. Biosynthesis of melatonin: Enzymic conversion of serotonin to N-acetylserotonin. *Biochim. Biophys. Acta.* 43:352–353.

WEISSBACH, H., B. G. REDFIELD, and J. AXELROD, 1961. The enzymatic acetylation of serotonin and other naturally occurring amines. *Biochim. Biophys. Acta.* 54:190–192.

WOLFE, D., L. POTTER, K. RICHARDSON, and J. AXELROD, 1962. Localizing norepinephrine in sympathetic axon by electron microscopic autoradiography. *Science* 138:440–442.

WURTMAN, R. J., and F. ANTON-TAY, 1969. The mammalian pineal as a neuroendocrine transducer. *Rec. Prog. Horm. Res.* 25:493–522.

WURTMAN, R. J., H. M. SHEIN, and F. LARIN, 1971. Mediation by a β-adrenergic receptor of effects of norepinephrine on pineal synthesis of ^{14}C-serotonin and ^{14}C-melatonin. *J. Neurochem.* 18:1683–1687.

44 Circadian Rhythm of Sensory Input in the Crayfish

HUGO ARÉCHIGA

ABSTRACT   Long-term records of single sensory units of the crayfish show the presence of a circadian rhythm of responsiveness in various sensory modalities with a diurnal phase of low reactivity and a nocturnal phase of high reactivity. Such rhythm is detectable at all levels of sensory integration, from primary receptors to high-order interneurons and is driven through (a) efferent neural influences to sensory units, and (b) neurosecretory products acting upon sense organs and interneurons.

WITHIN THE LAST two decades, a considerable amount of information has been gathered about the modulation of sensory inflow. Efferent influences are channeled to sensory neurons by (a) efferent fibers reaching sensory receptors or interneurons in specific pathways, (b) motor control of nonneural structures in sense organs that modify the admittance of stimuli to receptive cells, and (c) humoral influences modulating the activity of various elements in sense organs and sensory pathways.

At the cellular level, though, the bulk of data pertains to the short-term range, given the hindrances of a technical nature to maintaining, for prolonged periods, the record of activity in single neurons. In invertebrates, techniques have recently been developed to record over long periods from single elements in isolated ganglia (Strumwasser, 1965), from small populations of cells in isolated eyes (Jacklet, 1968, also see Eskin, this volume), and from single neurons of healthy, unrestrained animals (Aréchiga and Wiersma, 1969a).

The present chapter is concerned with circadian variations of activity in sensory elements of the crayfish, *Procambarus*.

Circadian rhythm of visual input

MODULATION OF RETINAL ACTIVITY   The visual input of the crayfish is mediated by the compound eyes, although light sensitivity is also known to exist in neurons of the sixth abdominal ganglion (Prosser, 1934). The photosensitive element in the compound eye is the rhabdom, consisting of the fused photopigment-containing membranes of seven neighboring retinula cells (Parker, 1895). The light reaches the rhabdom through the transparent elements, the corneal cells and cristalline cones. Both transparent structures and rhabdom are surrounded by three sets of pigmentary elements: The proximal and distal pigments are dark-colored, and their position, as shown in Figure 1, is related to the light received by the eye. Under intense illumination, both pigments come close to each other by protoplasmic flow inside their respective cells (Kleinholz, 1961), thus shielding the rhabdom from scattered light. In darkness, the reverse occurs. The third set of pigmentary elements are the reflecting cells, whitish in color, whose position, in crayfish, is independent of illumination. The position of proximal and distal pigments can be determined in the living animal by the glow produced by light reflected on the whitish pigment (Day, 1911). The area of glow depends upon the separation between both pigments (Figure 1).

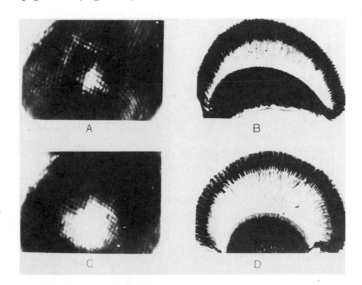

FIGURE 1   Correlation between the area of glow (clear zone) produced by light reflected on the retina of the crayfish *P. bouvieri* (Ortmann) (left) and the position of retinal pigments in sections (right). A and B correspond to light adapted eye; C and D correspond to dark-adapted eye.

HUGO ARÉCHIGA   Departamento de Fisiología, Facultad de Medicina, Universidad Nacional Autónoma de México

In crayfish kept in continuous darkness, it has been shown that both proximal (Bennitt, 1932) and distal retinal pigments (Welsh, 1939) display a diurnal rhythm of migration, adopting the light-adapted position during day-time, and the dark-adapted position during night-time. The area of glow changes accordingly (Jahn and Crescitelli, 1940). The response of retinal photoreceptors to light varies in a correlative manner, and Figure 2A shows records taken along a 24 hr cycle of the electro-retinogram (ERG), which in the crayfish is entirely generated by retinal cells (Naka and Kuwabara, 1959) without contribution of the optic ganglia or efferent elements. In animals kept in darkness except for the test light pulses applied at regular intervals, the responsiveness to light is at a low level during day-time and attains a high plateau during night-time. The duration of the circadian period of this oscillation of responsiveness bears a direct relationship to the amount of light received by the animal, as might be predicted by Aschoff's law (Aschoff 1960) for a nocturnal animal like the crayfish. The range of oscillation varies from 22 to 26 hr (Aréchiga, Fuentes, and Barrera, 1973) for programs between one flash (1 msec duration and 120 candle/ft^2 intensity) every 30 min to one every second. These two programs were found to

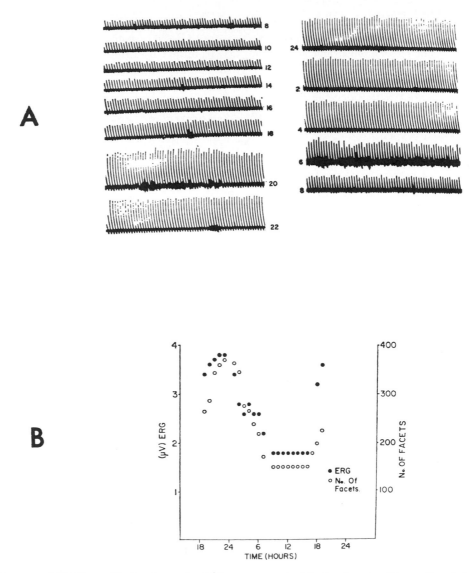

FIGURE 2   (A) Changes of ERG amplitude along the 24-hr cycle in an animal receiving one flash of 1 msec duration and 120 candle/ft^2 every 10 sec, against a dark background. Each trace is a sample of 10 min recording, taken at the hour of day labeled at one side of record. (B) Correlation between number of facets comprised in the glow (see Figure 1) and the amplitude of ERG simultaneously recorded along a 24-hr cycle, in a crayfish kept in continuous darkness, interrupted only by a flash of 120 candle/ft^2 of intensity and 1 msec duration, applied every hour. (From Aréchiga, Fuentes-Pardo, and Barrera-Mera, unpublished.)

cover the span of variation for the length of the circadian period. The rhythm persists for a large number of cycles under "free-running conditions" and gradually wanes but can be reinstated with minor changes in the program of stimulation.

The rhythm of amplitude is strictly parallel to that of retinal shielding pigments (Figure 2B) and any manipulation on short- or long-term basis, which affects the position of pigments, concurrently alters the ERG rhythm. Therefore, although it is not possible at present to rule out an intrinsic rhythm of sensitivity to light in the photoreceptors, the data available indicate that retinal shielding pigment position plays a key role in this modulation.

CIRCADIAN RHYTHM OF RESPONSIVENESS IN VISUAL INTERNEURONS   The highest level of visual integration in the crayfish is provided by the interneurons whose axons run in the optic peduncle (Wiersma, 1967), which is a tract connecting the visual ganglia with the supraoesophageal ganglion. There are about 18,000 axons in the peduncle (Nunnemacher, Camougis, and McAlear, 1962), but only 48 of them have been so far identified as visual interneurons, the rest being efferents to the visual ganglia (Wiersma and Yamaguchi, 1966). The units of one of the groups present in the peduncle, the *sustaining fibers*, are activated by changes in light intensity, and the F-log I curve for a given unit is parallel, over the dynamic range, to the V-log I curve for the photoreceptors in its field (Glantz, 1971), thus suggesting that such fibers are true channels for information about light intensity.

Long-term records of the activity in single sustaining units, show a circadian rhythmicity in responsiveness to light. As seen in Figure 3, the duration of the circadian period is different from 24 hr (about 22.5 in that experiment), and, after a number of days under a regular program of test light pulses, the amplitude of cycles diminishes in a similar fashion described for the receptor rhythm. Yet, even after several weeks, when the rhythm is no longer apparent, it can be reinstated by slight changes in the program of regular stimulation, as described by Aréchiga and Wiersma (1969b).

The rhythm of responsiveness in sustaining fibers is correlated with a change in the organization of the retinal receptive fields. Each unit receives excitatory input from a certain area of the retina and is inhibited by illumination of any other retinal region (Wiersma and Yamaguchi, 1967). After 30 to 60 min of dark-adaptation, an expansion of the excitatory receptive fields is observed, and regions of the eye, which in light-adapted conditions induce only inhibition, are able, in the dark-adapted eye, to evoke an initial excitation followed by inhibition (Aréchiga and Yanagisawa, 1973). In animals kept in

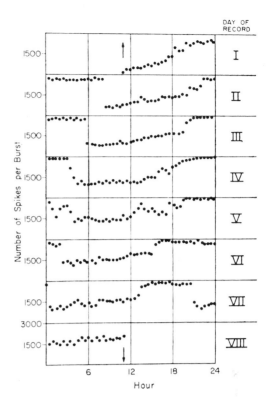

FIGURE 3   Circadian variations in the response to light of a sustaining fiber (0–38) during one week of continuous recording. The animal was in darkness except for a test light pulse applied every 30 min (from Aréchiga and Wiersma, 1969b).

darkness except for the test light pulses, a reduction of excitatory fields occurs in the day phase and an expansion in the night phase. This phenomenon is most likely another consequence of the migration of retinal shielding pigments, which at night-time, when separated, increase the light admittance for any group of photoreceptors.

Simultaneous recording of ERG and sustaining fiber activity manifest a close correlation along the 24-hr cycle, only the night-to-day transition being steeper in the sustaining fibers. This difference may be related to another input received by these fibers, of a facilitatory nature and dependent on motor activity. Presumably it is conveyed by one of the efferent axons in the optic peduncle, whose activity is proportional to the overall motor excitation of the animal (Aréchiga and Wiersma, 1969b). The activation of such neurons (0-71 in Wiersma and Yamaguchi, 1966, notation) precedes by 20 to 40 msec the facilitation of the sustainings, and the duration of both phenomena is the same as that of the bursts of motor excitement. Simultaneous records of the "activity fiber" and of sustaining units, show that the slope at night-to-day transition is even greater in 0-71. Actograms taken along several 24-hr cycles have confirmed early observations (Kalmus, 1938; Roberts, 1941; and

Schalleck, 1942) of the persistence of a diurnal rhythm of locomotor activity with a nocturnal peak, although a secondary peak is often observed at mid-day. The duration of the locomotor circadian period ranges from 22 hr under darkness, to 26 hr under constant illumination (Aréchiga, Fuentes, and Barrera, 1973).

From the aforementioned data, it is clear that the various inputs to sustaining fibers are affected by a circadian oscillation, but the excitability of the sustaining neuron itself may show the same oscillation. That this is the case is suggested by the observation that after the animal has been in darkness for several hours, a spontaneous activity of 1 to 5 impulse/sec appears in the sustaining fibers, and its rate varies in a circadian manner (Aréchiga and Wiersma, 1969b). Although this activity is under the facilitatory influence of motor action, when the movements of the animal are monitored it is seen that even during periods of quiescence the rate is higher during night than during day-time.

Because the sustaining fibers are the final channels for information about light intensity, the circadian variation of their activity should lead to a circadian rhythm of light-induced motor responses, and, indeed, when simultaneously recorded along the 24-hr cycle, close correlation is found in the motor response to light and the activity in sustaining fibers. Moreover, the sustaining units turn out to be fine monitors for the magnitude of motor excitation induced by light. Yet, some differences exist between the locomotor and the visual circadian rhythms: the dissimilar time courses at night-to-day transitions, the observation that phase-shifts are commonly seen between the ERG rhythm and that of

locomotion when simultaneously recorded, and the fact that the damping of both rhythms may proceed differently, suggest that more than one circadian pacemaker may be acting.

Circadian rhythm in nonvisual sensory interneurons

Most of the sensory axons so far characterized electrophysiologically in the various connective tracts of the crayfish receive mechanoreceptive input (Wiersma, 1958; Wiersma and Yamaguchi, 1966). In the optic peduncle, most of the interneurons are mechanoreceptive and 12,000 of the total 18,000 axons are primary sensory axons arising in the base of the hairs of the body surface. In tethered animals, probing at primary mechanoreceptors and on high-order interneurons, a circadian rhythm is apparent both in responsiveness to specific sensory stimuli and in the level of activity in those units that show spontaneous discharges. Figure 4 illustrates the differences in the habituation of a mechanoreceptive interneuron, repetitively activated on the same sensory spot of the antenna at different times of day. When the spontaneous activity in nonvisual interneurons is correlated with that in sustaining fibers, both show similar variations along the 24-hr cycle.

Humoral modulation of circadian rhythm in the nervous system

The occurrence of a circadian rhythm of activity in the various sensory elements, and even in nonsensory structures such as the retinal shielding pigments that are not

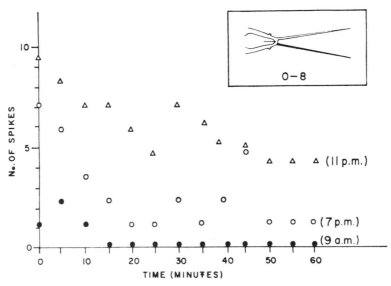

FIGURE 4   Differences in responsiveness of a mechanoreceptive interneuron (0–8), whose receptive field, as indicated on inset, covers the hairs of homolateral antenna, to a repetitive mech-

anical stimulation, applied at different times of day. (From Aréchiga, Barrera-Mera, and Fuentes-Pardo, in preparation.)

innervated, suggests that a common oscillator is driving the afferent neuronal activity through a hormonal channel. The central organ for hormonal release in the crayfish is the so-called *sinus gland*, which is located in the eyestalk, between the *medulla externa* and the *medulla interna*, and whose relation to endocrine activity has long been established (see review by Knowles and Carlisle, 1956). Among other functions, the sinus gland has been related to the light-induced migration of both proximal and distal retinal pigments (Welsh, 1939; 1941). The locomotor rhythm is also altered by sinus gland excision (Kalmus, 1938; Roberts, 1941; Schalleck, 1942). The sinus gland is a neurohemal organ, where bulbous nerve endings, laden with secretory granules are distributed around a blood sinus. Five different types of granules have been characterized on the basis of their staining properties, size, and shape (Bunt and Ashby, 1967) at the electron microscopic level. There are no secretory cell bodies in the gland and the origin of the endings composing it has been traced to neurosecretory cells in visual ganglia, particularly the *medulla terminalis*, in the supraoesophageal ganglion and even in the ganglion chain (Bliss, Durand, and Welsh, 1954; Durand, 1956; Fingerman and Aoto, 1959).

The injection of sinus gland extracts in dark-adapted crayfish induces a decrement both in the amplitude of ERG and the firing rate of sustaining fibers when the eye is illuminated. This effect is correlated to a migration of retinal shielding pigments toward a light adapted position (Aréchiga and Fuentes, 1970) and, hence, to a reduction in the light admittance to retinal photoreceptors, as can be seen in Figure 5. The expanded receptive fields of sustaining fibers shrink under the influence of sinus gland extracts. But in addition to those effects, all of which can be accounted for by the mere change in position of shielding pigments, the spontaneous activity of sustaining fibers, in darkness, is depressed as well, and so are the spontaneous activity and the induced responses in sensory nonvisual neurons after injection of the extract of sinus gland. In in vitro experiments, when the extract is topically applied to ganglia or single neurons such as the abdominal stretch receptors, then a similar inhibition is apparent (Aréchiga, Fuentes, Barrera, and Abreu, 1971).

As might be expected, effects similar to those of sinus gland extracts are obtained with extracts from all the ganglia where neurosecretory cells are present. These observations do support the contention that the neural-inhibiting substance, stored and released at the sinus gland, is produced elsewhere. The mechanisms whereby release undergoes a circadian rhythm are at present unknown. The chemical nature of the neurosecretory products released by the sinus gland, and which regulate the retinal pigments position in the crayfish, has not yet

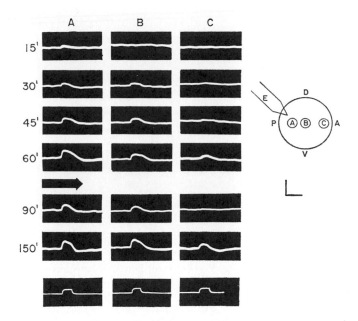

FIGURE 5   Depression of responsiveness to light in the eye of the crayfish, after injection (arrow) of sinus gland extract, in an animal dark-adapted for 60 min. The illuminated areas are shown in inset, which represents the surface of the eye (D, dorsal; V, ventral; P, posterior; A, anterior); E, indicates position of recording electrode. Duration of light pulses indicated by signals from photocells at bottom. Cal., 2 mV × 2 sec. (From Aréchiga, Fuentes-Pardo, and Barrera-Mera, unpublished.)

been clarified, although for some crustaceans, a considerable amount of information is available (Kleinholz, 1966). An attempt to isolate from crude extracts of sinus gland an active neural-inhibiting fraction has led to the identification, using disk gel electrophoresis after the technique of Nicoll et al. (1969), of a distinct protein band, which when injected is capable of reproducing the inhibitory effects of the whole extract and the actions on the retinal shielding pigments as well (Aréchiga and Mena, in preparation). The density of the band decreases with illumination and increases in darkness as shown in Figure 6. When animals are kept in continuous darkness and sacrificed at intervals along the 24-hr cycle, while monitoring the position of retinal shielding pigments, a change is apparent: the density is greater in the afternoon and night when the retinal pigments are in dark-adapted position and suddenly decreases at the night-to-day transition (Figure 6B). This suggests that during nighttime, the release of hormone from the sinus gland is depressed, and that it increases during day-time. The same fraction has been found in extracts from supraoesophageal ganglion, although in lesser amounts, and in even lesser amounts in thoracic and abdominal chain. Further work of purification of the various neurosecretory products of the sinus gland is necessary in order to decide

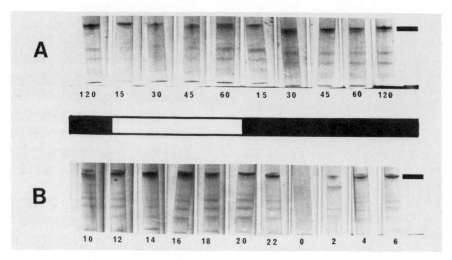

FIGURE 6 Separation by gel electrophoresis from sinus gland extracts of a band (labeled at right) which, when injected into crayfish, induces migration of retinal shielding pigments toward a light adapted position and inhibits neural activity. Changes in density of the band are shown at A, as a consequence of illumination (light bar at bottom) during 60 min after 120 min of darkness and of readapting to darkness during another 120 min (dark bar). Times of sampling, in minutes, shown at bottom. At B, after 72 hr in darkness, the animals were sacrificed in the sequence shown (hours of day). Notice the abrupt decrement in density at the predicted dawn (0 hours, see text). (From Aréchiga and Mena, unpublished.)

whether or not the neural-inhibiting substance is different from some other putative neurohormones it stores, and what correlation exists among physiologically active fractions and particular types of secretory granules. On the other hand, if an inhibitory hormone is responsible for the day phase of low reactivity in sensory units, it remains undecided whether the nocturnal phase is caused merely by the lack of release of the inhibitory hormone, as suggested by our experiments (Figure 6), or if there is also a neural-exciting substance.

The data so far presented point to the existence, in the crayfish, of two channels whereby modulatory actions are exerted upon sensory input. One is the neurally conveyed influence upon sensory interneurons, and the other, the humoral influences driving slow fluctuations (e.g. circadian) of responsiveness, acting on peripheral sense organs and interneurons as well.

ACKNOWLEDGMENT The collaboration of Drs. B. Fuentes-Pardo and B. Barrera-Mera in many of the experiments herewith reported is gratefully acknowledged.

REFERENCES

ARÉCHIGA, H., and B. FUENTES, 1970. Correlative changes between retinal shielding pigments position and electroretinogram in crayfish. *Physiologist* 13:137.

ARÉCHIGA, H., B. FUENTES, and B. BARRERA, 1973. Circadian rhythm of responsiveness in the visual system of the crayfish. In *Biological Rhythms*, J. Salanki, ed. Plenum Press (in press).

ARÉCHIGA, H., B. FUENTES, B. BARRERA, and L. F. ABREU, 1971. On the nature of the circadian rhythm of activity in crayfish nervous system. *Proc. Int. Union Physiol. Sci.* 9:23.

ARÉCHIGA, H., and C. A. G. WIERSMA, 1969a. The effect of motor activity on the reactivity of single visual units. *J. Neurobiol.* 1:53–69.

ARÉCHIGA, H., and C. A. G. WIERSMA, 1969b. Circadian rhythm of responsiveness in crayfish visual units. *J. Neurobiol.* 1:71–85.

ARÉCHIGA, H., and K. YANAGISAWA, 1973. Inhibition of visual units in the crayfish. *Vis. Res.* 13:731–744.

ASCHOFF, J., 1960. Exogenous and endogenous components in circadian rhythms. *Cold Spring Harbor Symp. Quant. Biol.* 25: 11–28.

BENNITT, R., 1932. Diurnal rhythm in the proximal cells of the crayfish retina. *Physiol. Zool.* 5:65–69.

BLISS, D. E., J. B. DURAND, and J. H. WELSH, 1954. Neurosecretory systems in decapod crustacea. *Z. Zellforsch. Mikrosk. Anat.* 39:520–536.

BUNT, A. H., and E. A. ASHBY, 1967. Ultrastructure of the sinus gland of the crayfish *Procambarus clarkii*. *Gen. Comp. Endocr.* 9:334–342.

DAY, E. C., 1911. The effect of colored light on pigment-migration in the eye of the crayfish. *Bull. Mus. Comp. Zool.* 53:305–312.

DURAND, I. B., 1956. Neurosecretory cell types and their secretory activity in the crayfish. *Biol. Bull.* 111:62–76.

FINGERMAN, M., and T. AOTO, 1959. The neurosecretory system of the dwarf crayfish, *Cambarellus shufeldti*, revealed by electron and light microscopy. *Trans. Am. Micro. Soc.* 78:305–317.

GLANTZ, R. M., 1971. Peripheral versus central adaptation in the crustacean visual system. *J. Neurophysiol.* 34:485–492.

JACKLET, J., 1968. Circadian rhythm in isolated eye of *Aplysia*. *Science* 164:562–563.

JAHN, T. L., and F. CRESCITELLI, 1940. Diurnal changes in the electrical response of the compound eye. *Biol. Bull.* 78:42–52.

KALMUS, H., 1938. Das aktogram des Flusskrebses und seine Beeinflussung durch Organextracte. *Z. Vergl. Physiol.* 25: 798–802.

KLEINHOLZ, L. H., 1961. Pigmentary Effectors. In *The Physiology of Crustacea*, T. H. Waterman, ed. New York: Academic Press, pp. 133–169.

KLEINHOLZ, L. H., 1966. Hormonal regulation of retinal pigment migration in crustaceans. In *The Functional Organization of the Compound Eye*, C. G. Bernhard, ed. New York: Pergamon Press, pp. 89–101.

KNOWLES, F. G. W., and D. B. CARLISLE, 1956. Endocrine control in crustacea. *Biol. Rev.* 31:396–473.

NAKA, K., and M. KUWABARA, 1959. Two components from the compound eye of the crayfish. *J. Exp. Biol.* 36:51–61.

NICOLL, C. S., J. A. PARSONS, R. P. FIORINDO, and C. W. NICHOLS, JR., 1969. Estimation of prolactin and growth hormone levels by polyacrylamide disc electrophoresis. *J. Endocrinol.* 45:183–196.

NUNNEMACHER, R. F., G. CAMOUGIS, and J. H. McALEAR, 1962. The fine structure of the crayfish nervous system. *5th. Int. Congr. Elect. Mic.* 2:N–11.

PARKER, G. H., 1895. The retina and optic ganglia in decapods, especially in *Astacus. Mitt. Zool. Stat. Naepel.* 12:1–73.

PROSSER, C. L., 1934. Action potentials in the nervous system of the crayfish. II. Responses to illumination of the eye and caudal ganglion. *J. Cell. Comp. Physiol.* 4:363–377.

ROBERTS, T. W., 1941. Evidences that hormonal inhibition of locomotion occurs for the crayfish *Cambarus virilis*, Hagen. *Anat. Rec.* (Suppl.) 81:46–47.

SCHALLECK, W., 1942. Some mechanisms controlling locomotor activity in the crayfish. *J. Exp. Zool.* 91:155–156.

STRUMWASSER, F., 1965. The demonstration and manipulation of a circadian rhythm in a single neuron. In *Circadian Clocks*, J. Aschoff, ed. Amsterdam: North-Holland Publishing, pp. 442–462.

WELSH, J. H., 1939. The action of eye-stalk extracts on retinal pigment migration in the crayfish, *Cambarus bartoni. Biol. Bull.* 77:119–125.

WELSH, J. H., 1941. The sinus gland and 24-hour cycles of retinal pigment migration in the crayfish. *J. Exp. Zool.* 86:35–49.

WIERSMA, C. A. G., 1958. On the functional connections of single units in the central nervous system of the crayfish *Procambarus clarkii*, Girad. *J. Comp. Neurol.* 110:421–471.

WIERSMA, C. A. G., 1967. Visual central processing in crustaceans. In *Invertebrate Nervous Systems*, C. A. G. Wiersma, ed. Chicago, Ill.: The University of Chicago Press, pp. 269–284.

WIERSMA, C. A. G., and T. YAMAGUCHI, 1966. The neuronal components of the optic nerve of the crayfish, as studied by single unit analysis. *J. Comp. Neurol.* 128:333–358.

WIERSMA, C. A. G., and T. YAMAGUCHI, 1967. Integration of visual stimuli by the crayfish central nervous system. *J. Exp. Biol.* 47:409–431.

45 Circadian Release of a Prepatterned Neural Program in Silkmoths

JAMES W. TRUMAN

ABSTRACT   In the giant silkmoths the control of adult emergence involves two separable neural components. The first consists of a brain-centered, circadian clock that directs the time of release of a neurosecretory hormone (the eclosion hormone) from the brain. The second component includes the neural information that generates the pre-eclosion behavior—a stereotyped sequence of abdominal movements that assist the moth in the escape from the pupal skin. Electrophysiological studies show that the pre-eclosion behavior is prepatterned into the abdominal ganglia. At the prescribed time of day, the eclosion hormone is released into the blood and acts on neural centers in the abdomen to trigger the read-off of this 1.25-hr motor program.

THE FACT THAT animals perform certain types of behavior during specific times of day has long been known to students of animal behavior. However, it has only been relatively recently that this daily organization of behavior has been shown to be due to internal clocks within the organism itself. The preceding chapters by Strumwasser and Eskin in this volume have described neurons or groups of neurons which can act as circadian pacemakers in the mollusc, *Aplysia*. This chapter examines mechanisms by which neuronal clocks in insects may be coupled to the final motor behavior.

Figure 1 gives two simplified examples of how circadian clocks are involved in the control of behavior patterns. In the simplest example, a hypothetical control center for the behavior is tightly coupled to the output of the clock. Many specialized pieces of behavior that have restricted environmental or developmental contexts apparently have this type of control. In the Pernyi silkmoth (*Antheraea pernyi*), the onset of cocoon spinning by the larva (L. P. Lounibos, unpublished) and the eclosion of the adult moth (Truman, 1971a) occur during a specific time of day and apparently cannot be prematurely triggered by any stimulus that does not work through the clock. Another such behavior is the daily assumption of the pheromone

JAMES W. TRUMAN   The Biological Laboratories, Harvard University, Cambridge, Massachusetts

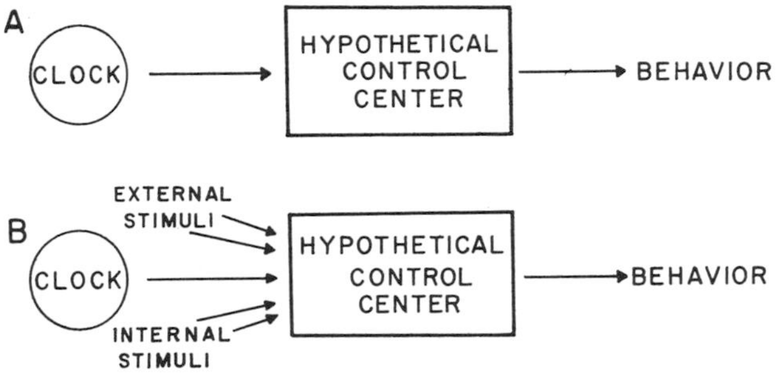

FIGURE 1   Two extreme representations of the influence of the driving clock and of other neural inputs on the control of (A) specialized types of behavior or (B) generalized types of behavior.

release posture by virgin, Pernyi female moths (Truman, Lounibos, and Riddiford, unpublished).

More generalized types of behavior that may serve various diverse functions appear to have a more complex control (Figure 1B). For example, a cockroach whose hiding place is disturbed in the middle of the day will not remain still even though its locomotor clock is in the middle of the rest period. The control center for locomotor activity has many inputs, only one of which is the phase of its circadian clock. This multiplicity of inputs greatly complicates an experimental determination of the pathway from the clock to the locomotor centers. Therefore, this chapter will consider what is known of the chain of events leading from the clock to the motor output of a specialized behavior—the eclosion behavior of silkmoths.

The eclosion response of silkmoths

The transformation of the caterpillar into the moth culminates with the eclosion (ecdyses, emergence) of the "pharate" moth from the old pupal exuviae. This event is accompanied by a sequence of stereotyped behaviors that are displayed only at this time. The first two behaviors in this sequence will be considered here. The pre-eclosion behavior is a species-specific program of abdominal movements that lasts approximately $1\frac{1}{4}$ hr and that immediately precedes eclosion. The latter is comprised of the thoracic and abdominal movements used in the actual shedding of the exuviae.

Eclosion occurs after adult development is completed but only during temporal "gates," which are dictated by the insect's circadian clock (Pittendrigh and Skopik, 1970). Insects that attain developmental competence after the gate has closed on a given day then wait for the opening of the next gate on the following day. Thus, the behavioral event of eclosion is somewhat independent of the developmental processes that formed the adult insect.

As stated above, the eclosion behavior of Pernyi moths is very tightly coupled to the eclosion clock. Under a 17L:7D regimen, where L = light and D = dark, this species emerges during the last 5.5 hr before lights-off (Figure 2). If approximately 10 hr before the opening of the gate, a range of stimuli (vigorous prodding; partial removal of the cuticle; exposure of the male to female sex pheromone) are given to pharate moths, none provokes precocious eclosion. Even the complete removal of the pupal cuticle only yields a helpless moth that shows few adult motor patterns. With the subsequent arrival of the gate, these peeled moths perform a pantomine eclosion and then assume the full adult repertoire of behavior (Truman, 1971b). Thus, eclosion cannot be prematurely triggered by nonclock-related stimuli and, moreover, at the time of the gate the behavior is performed even if environmental conditions make it pointless.

The silkmoth brain and the hormonal control of eclosion

The importance of the brain to the gating of eclosion was first indicated by the surgical removal of this ganglion from the silkmoth pupae (Truman and Riddiford, 1970). The resultant brainless moths showed aberrant pre-eclosion and eclosion behaviors (Truman, 1971b) and emerged without regard to time of day or night (Figure 2B). These effects of brain extirpation were completely repaired by implanting a brain into the abdomen of de-brained animals. These "loose-brain" moths showed the proper gating of eclosion (Figure 2C) and the normal performance of the associated behaviors. Thus the silkmoth brain has an important role in the circadian timing and performance of the complex behavioral acts that occur during eclosion.

The nature of the role of the brain in eclosion was demonstrated by interchanging brains between *Hyalophora cecropia* and Pernyi pupae (Truman and Riddiford, 1970). As seen in Figure 2D, the time of emergence of the resulting moths was characteristic of the species that donated the brain. But the pre-eclosion behavior displayed was always typical of the host species. Clearly, the brain is responsible for the gating of eclosion, but the detailed pattern of the behavior is coded elsewhere in the nervous system. Moreover, the linkage between these two systems must be hormonal.

As would be expected from the above results, eclosion-stimulating activity can be demonstrated in homogenates

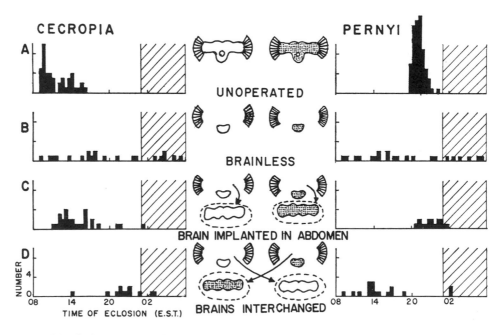

FIGURE 2   The eclosion of *Hyalophora cecropia* and *Antheraea pernyi* moths in a 17L:7D regimen showing the effects of brain removal, the transplantation of the brain to the abdomen, and the interchange of brains between the two species (from Truman, 1971c).

prepared from the brain and corpora cardiaca of pharate moths. When injected into pharate moths, these homogenates triggered the precocious display eclosion and the other associated behaviors (Truman and Riddiford, 1970; Truman, 1971b). Homogenates prepared from other portions of the nervous system were completely inactive when tested in this manner (Truman, 1973).

The brains and corpora cardiaca removed from various stages in the moth life history were assayed for the presence of the "eclosion hormone" (Truman, 1973). During the larval stages and in the pupa, only traces of hormonal activity could be found in these structures. As adult development proceeded, the hormone in the brain and corpora cardiaca gradually built up to a high titer. Then, on the last day of development, the onset of the gate signaled the rapid depletion of 70 to 80% of the stored hormonal material and its coincident appearance in the blood (Truman, 1973). Thus the eclosion hormone is apparently produced and released only at one time in the moth life history for the principal purpose of triggering one specific sequence of behavior.

The relationship between the circadian clock and the neuroendocrine system

The Cecropia brain contains two major clusters of neurosecretory cells. These medial and lateral cell groups are symmetrically paired in the cerebral lobes and contain 11 and 7 cells, respectively, in each hemisphere (Herman and Gilbert, 1965). The eclosion hormone appears to be produced by neurosecretory cells located in the medial cluster (Truman, 1973).

The pupal brain was subdivided and the various fragments assayed for their ability to gate eclosion by implantation into a debrained animal. A piece consisting only of the intact cerebral lobes adequately gated eclosion (Truman, 1972). When the cerebral lobes were then subdivided by cuts made lateral to each medial neurosecretory cell group, the resulting median piece did not gate eclosion (Truman, 1972), even though it contained the cells that produce the eclosion hormone. Although many interpretations can be attached to this last result, it does suggest the possibility of an interaction of the medial neurosecretory cells with a driving clock located in the lateral areas of the cerebral lobes.

The response of the moth nervous system to the eclosion hormone

The action of the eclosion hormone on the moth nervous system was studied with respect to the triggering of the pre-eclosion behavior. In Cecropia, this behavior is made up of three distinct parts: (1) an initial hyperactive period that lasts about 0.5 hr and that involves primarily rotational movements of the abdomen; (2) a quiescent period of similar length during which time the frequency of abdominal rotations is markedly reduced; and (3) a short period of hyperactivity consisting of peristaltic contractions that travel in a posterior to anterior direction along the abdomen. This last phase signals the onset of eclosion and assists the moth in shedding the pupal cuticle.

The pre-eclosion behavior can be displayed even after higher neural centers in the head and thorax have been removed. Manifestly, when abdomens are severed from pharate moths and then injected with the eclosion hormone, they perform the complete pre-eclosion behavior and may even succeed in completely shedding the surrounding pupal cuticle (Truman, 1971b). Therefore, the centers that respond to the hormone and direct the performance of the pre-eclosion behavior are located in the moth abdomen.

The action of the eclosion hormone was further examined using the semidissected preparation shown in Figure 3 (Truman and Sokolove, 1972). Isolated abdomens were opened ventrally, the abdominal nerve cord exposed, and all peripheral nerves severed. Suction

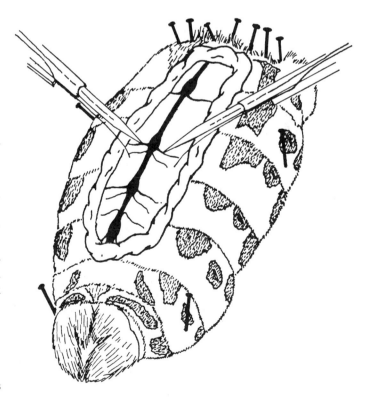

FIGURE 3   A drawing of a semidissected, isolated abdomen of a Cecropia moth showing the placement of suction electrodes on the dorsal roots of the second abdominal ganglion.

electrodes were then placed on the proximal stumps of the dorsal nerves—the nerves that supply the intersegmental muscles (Libby, 1961). The latter are dense, longitudinal bands of muscle that line the 4th through 6th abdominal segments of pharate moths and are the primary muscle groups responsible for the movements observed during eclosion.

In these deafferented preparations addition of the eclosion hormone typically elicited a response such as that seen in Figure 4. Twenty to 40 min after hormone addition, spontaneous efferent bursts appeared. Bursting

continued at a relatively high frequency for approximately 0.5 hr and then declined to a low level of relative or complete quiescence. After an additional 0.5 hr bursting abruptly resumed. Manifestly, in the temporal arrangement of efferent bursts the activity of the deafferented nerve cord closely mimicked the temporal sequence of movements seen in the intact insect.

The motor bursts seen during the first two periods showed a rotational patterning. These bursts were typically composed of many volleys of firing with a clear right-left alternation (Figure 5D to 5F) but with essen-

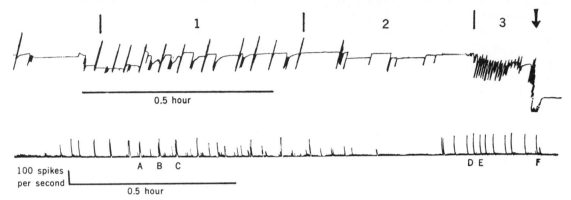

FIGURE 4   Behavioral and electrical activity showing the three phases of the Cecropia pre-eclosion behavior. *Top*: Record of abdominal movements obtained by attaching the tip of the abdomen to a lever writing on a revolving drum. Rotary movements produced complete excursions of the trace. The downward deflections were produced by both ventrally directed twitches and peristaltic movements of the abdomen. (1) The first hyperactive period, (2) the quiescent period, and (3) the second hyperactive period. Arrow identifies the moment of adult emergence. *Bottom*: Integrated motor activity from the right dorsal nerve of the second abdominal ganglion after addition of the eclosion hormone to the deafferented nerve cord. Letters refer to bursts represented in an expanded form in Figure 5 (from Truman and Sokolove, 1972). (Copyright 1972 by the American Association for the Advancement of Science.)

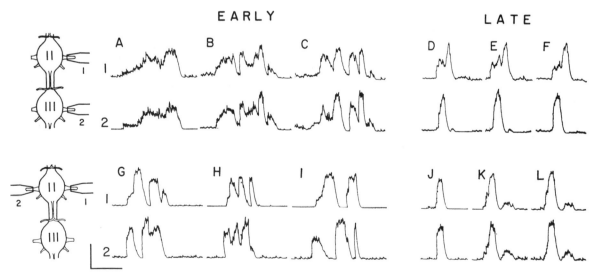

FIGURE 5   Examples of integrated spontaneous bursts recorded from the deafferented abdominal ganglia after addition of the eclosion hormone. *Top*: Electrodes placed sequentially on the right dorsal roots of the second and the third abdominal ganglia. *Bottom*: Electrodes placed bilaterally on right and left roots of second ganglion. Bursts A to C and G to I are of the rotational pattern typically observed during the first (early) hyperactive period. Bursts D to F and J to L are of the peristaltic pattern observed during the second (late) hyperactive period. Vertical line equals 100 spikes per second; horizontal line, 10 seconds (from Truman and Sokolove, 1972). (Copyright 1972 by the American Association for the Advancement of Science.)

tially synchronous firing from roots on the same side of sequential ganglia (Figure 5A to 5C). This pattern of efferent output would generate the rotational movements observed during the first hyperactive period and the quiet period.

With the onset of the third phase, one observed a new patterned burst. These "peristaltic bursts" were characterized by a right-left symmetry (Figure 5J to 5L) and by the temporal sequence of major bursts starting at the more posterior ganglion and moving anteriorly (Figure 5G to 5I). This pattern would generate the peristaltic contractions which are observed during the last phase of the pre-eclosion behavior.

Therefore, the entire pre-eclosion behavior is centrally programmed. The rotational and peristaltic movements are generated by prepatterned motor scores, as originally described for flight behavior in the locust (Wilson, 1961). Moreover, there is also a central timing mechanism that (1) represses the rotational burst after the program has progressed through the first 0.5 hr and (2) activates the peristaltic pattern after 1 hr has elapsed. The mechanism of this long-term timing is unknown.

Thus, eclosion is brought about through an interaction of the head with the abdomen. The synchronization of this behavior with environmental signals is accomplished through a totally brain-centered circadian mechanism. The pattern and timing of movements that comprise the pre-eclosion behavior are derived from a motor program that is built into the neural circuitry of the abdominal ganglia. At the proper time of day the brain then triggers the pre-eclosion program through the release of a neurosecretory hormone.

ACKNOWLEDGMENTS   I thank Prof. Lynn M. Riddiford for a critical reading of the manuscript.

REFERENCES

HERMAN, W. S., and L. I. GILBERT, 1965. Multiplicity of neurosecretory cell types and groups in the brain of the saturniid moth *Hyalophorea cecropia* (L.). *Nature (Lond.)* 205:926–927.

LIBBY, J. L., 1961. The nervous system of certain abdominal segments and the innervation of the male reproductive system and genitalia of *Hyalophora cecropia* (Lepidoptera: Saturniidae). *Ann. Ent. Soc. Amer.* 54:887–896.

PITTENDRIGH, C. S., and S. D. SKOPIK, 1970, Circadian systems. V. The driving oscillation and the temporal sequence of development. *Proc. Nat. Acad. Sci. USA* 65:500–507.

TRUMAN, J. W., 1971a. Hour-glass behavior of the circadian clock controlling eclosion of the silkmoth *Antheraea pernyi*. *Proc. Nat. Acad. Sci. USA* 68:595–599.

TRUMAN, J. W., 1971b. Physiology of insect ecdysis. I. The eclosion behavior of silkmoths and its hormonal control. *J. Exp. Biol.* 54:805–814.

TRUMAN, J. W., 1971c. Circadian rhythms and physiology with special reference to neuroendocrine processes in insects. In *Proceedings of the International Symposium on Circadian Rhythmicity*. Wageningen, Netherlands: Pudoc Press, pp. 111–135.

TRUMAN, J. W., 1972. Physiology of insect rhythms. II. The silkmoth brain as the location of the biological clock controlling eclosion. *J. Comp. Physiol.* 81:99–114.

TRUMAN, J. W., 1973. Physiology of insect ecdysis. II. The assay and occurrence of the eclosion hormone in the Chinese oak silkmoth, *Antheraea pernyi Biol. Bull.* 114:200–211.

TRUMAN, J. W., and L. M. RIDDIFORD, 1970. Neuroendocrine control of ecdysis in silkmoths. *Science* 167:1624–1626.

TRUMAN, J. W., and P. G. SOKOLOVE, 1972. Silkmoth eclosion: Hormonal triggering of a centrally programmed pattern of behavior. *Science* 175:1491–1493.

WILSON, D. M., 1961. The central nervous control of flight in a locust. *J. Exp. Biol.* 38:471–490.

46 Circadian Rhythmicity in the Isolated Eye of *Aplysia*: Mechanism of Entrainment

ARNOLD ESKIN

ABSTRACT The isolated eye of *Aplysia* exhibits a circadian rhythm of spontaneous optic nerve impulses. This preparation may be useful for investigating the mechanism underlying entrainment of rhythms by light-dark cycles. The eye can be maintained in vitro for long periods of time and its rhythm can be entrained in vitro by light-dark cycles. Treating the eye with a depolarizing stimulus (elevation of K_0^+) produces phase shifts in its circadian rhythm.

Introduction

CIRCADIAN RHYTHMS are normally entrained by environmental oscillations such as light-dark (LD) cycles. For entrainment to occur, information related to the environmental entraining cycle must be translated and conducted to the cellular level where it affects (entrains) the oscillating mechanism responsible for the rhythm. A systematic investigation of the processes by which LD cycles couple to and affect the oscillating mechanism on the cellular level should enlarge our understanding of central questions in the field of circadian rhythms. In tracing the pathway taken by light cycle information to the cellular oscillating mechanism, at least a portion of that mechanism should be elucidated.

The use of circadian systems for approaching central questions in the neurosciences such as learning and the relative importance of genetic and environmental factors has previously been discussed (Strumwasser, 1967; Lickey, 1967; Marler and Hamilton, 1966). Entrainment of circadian rhythms might serve as a useful model for investigations of the cellular mechanisms by which environmental events modify behavior in general.

The eye of Aplysia *and its rhythm*

Their simplicity and the ease with which experimental conditions can be specified make organ culture systems of particular use in experiments on rhythms. One such system that appears to have potential for entrainment studies is the isolated eye of a gastropod mollusk, *Aplysia californica*. This eye exhibits a circadian rhythm of spontaneous compound optic nerve impulses when isolated in filtered sea water under conditions of constant darkness and temperature (Jacklet, 1969a). The rhythm can be assayed by recording the impulses from the optic nerve with suction electrodes. In addition, the rhythms are reproducible and have a good degree of resolution (Figure 1A; Eskin, 1971).

The neurons of the eye of *Aplysia californica* have been somewhat characterized electrophysiologically (Jacklet, 1969b), and some characteristics of the morphology of the eye are known (Jacklet et al., 1972; Hughes, 1970). The eye is 600μ in diameter and is of the closed vesicle type with a spherical lens surrounded by a layer of receptor neurons and another peripheral layer of neurons. It is composed of thousands of small cells but there are only 4 or 5 cell types present (Jacklet, 1969b). On the basis of extracellular recordings from the optic nerve under a variety of ionic conditions, Audesirk (1971, 1973) proposed that the primary photoreceptors are electrically coupled to secondary neurons that are electrically coupled among themselves. The spontaneous compound optic nerve impulses, according to Audesirk, come from the secondary neurons that are driven in this case through chemical synapses by a group of cells that are spontaneously active.

The eye is attached to the cerebral ganglion by about 10 mm of nerve. The cerebral ganglion influences the eye through efferent optic nerve activity (Eskin, 1971). The pattern of afferent optic nerve impulses and the wave form of the impulse rhythm from the eye depend on its attachment to the cerebral ganglion.

The eye's circadian rhythm of spontaneous compound nerve impulses could be driven by a circadian oscillation in a single pacemaker neuron or neurosecretory cell or could emerge from the mutual entrainment of a population of coupled oscillators (neurons); although both of these possibilities have been examined, it is still not clear which is correct (Jacklet and Geronimo, 1971; Sener, 1972).

The specific questions to which the experiments

ARNOLD ESKIN Department of Biology, Rice University, Houston, Texas

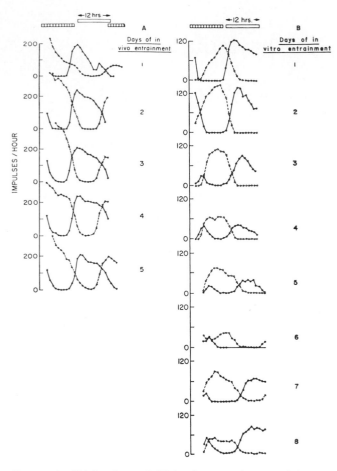

FIGURE 1   (A) In vivo and (B) in vitro entrainment of the eye. The open bars on the tops of the figures represent the projected light portion of the control LD cycle, and the dashed bar represents the projected light portion of the phase advanced LD cycle. The top curves were obtained after one day of experimental treatment, and the curves just below these were obtained after 2 days of experimental treatment, etc. The rhythms shown were obtained from isolated eyes maintained in constant darkness (Eskin, 1971).

reported here have been addressed are: Can the in vitro rhythm of the eye be maintained for long periods of time? Can the rhythm of the eye be entrained by LD cycles in vitro? If so, are the characteristics of in vitro entrainment of the eye the same as those of in vivo entrainment? Can a stimulus which depolarizes membrane potentials of neurons phase shift the rhythm?

In vitro maintenance and entrainment

The rhythm from the eye is entrainable in vitro by LD cycles, but eyes entrained in vitro require 3 to 4 more cycles for entrainment than do eyes entrained in vivo. In the in vivo entrainment experiment, rhythms were recorded in vitro from eyes of animals that had been

exposed to a control LD cycle and from eyes of other animals that had been exposed to a LD cycle the phase of which was advanced 13 hr relative to that of the control cycle. The phase shift of the rhythm of the experimental eyes was essentially complete after exposure to one phase-advanced light cycle (Figure 1A).

In the in vitro entrainment experiment, isolated eyes were cultured in a slightly modified Eagle's minimum essential medium (Eskin, 1971). One group of these eyes was exposed to a control LD cycle and another group was exposed to an LD cycle the phase of which was advanced 13 hr relative to that of the control cycle. Complete phase shifting of the isolated eye in vitro required exposure to 4 or 5 phase-advanced light cycles (Figure 1B). At present it is not possible to decide whether the difference in the rates of in vivo and in vitro entrainment are due to differences in the free-running periods of the in vivo and in vitro rhythms, to neural and/or humoral factors absent in vitro that normally contribute to the entrainment of the eye in vivo, or to a nutritional factor absent in the culture medium that is necessary for the normal maintenance of the eye.

The fact that the rhythm from the eye is entrainable in vitro indicates that the mechanism of entrainment can be studied in this isolated organ. The in vitro entrainment experiment also demonstrates that rhythmicity in the isolated eye can be maintained for at least 9 days (Figure 1B).

Phase shifting by high potassium

It was postulated that LD cycles appear as cycles of depolarization at the membrane of the cells in which the "clock" mechanism resides and that the clock is normally entrained through coupling between it and the transmembrane potential difference. If this is the case, exposure of the eye to a depolarizing stimulus should produce a phase shift in its rhythm. Since raising the concentration of potassium in the medium surrounding neurons depolarizes them, high-potassium treatments of various durations were tested for phase shifting effect. High potassium pulses produced either delay or advance phase shifts or arrhythmicity depending on the phase of the rhythm at which they were administered.

The concentration of K^+ in the high-potassium medium (hi-K) was about 10-fold greater than normal (106.7 mM compared with 9.7 mM), and the concentration of Na^+ was 382.4 mM (normally 492 mM) to maintain proper osmolality. Control experiments in which eyes were treated with a low Na^+ (382.4 mM) normal K^+ medium resulted in no phase shifts in the rhythm. A delay phase shift of 6 hr produced by a 4 hr hi-K pulse given during the early projected night is shown in Figure 2A

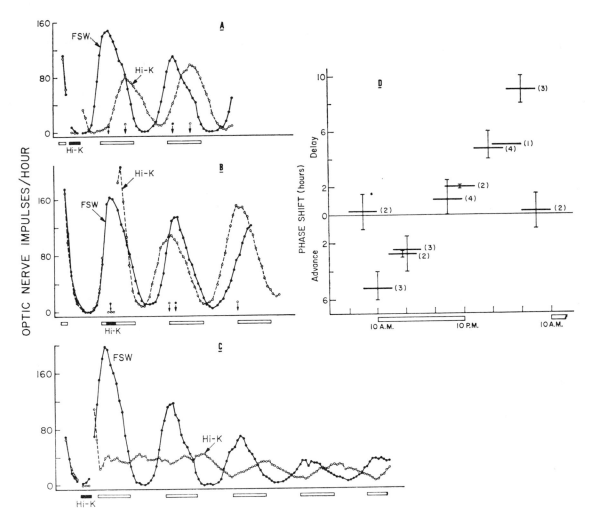

FIGURE 2 (A) Delay phase shift, (B) advance phase shift, and (C) arrhythmicity, produced by 4-hr hi-K pulses during the time shown by the solid bars at the bottom of the graphs. The open bars at the bottom of the graphs represent the projected light portion of the LD cycle. The solid curves (FSW) are from control eyes and the broken curves are from eyes which were exposed to hi-K pulses. The eyes remained in constant dark throughout the experiment. (D) Phase shifts of the impulse rhythm as a function of the time of day (on the abscissa) that the eyes were exposed to 4-hr hi-K pulses. The horizontal bars are mean responses (number of experiments shown in parentheses) and span the time of exposure to hi-K. The vertical bars indicate the range of phase shifts obtained (Eskin, 1972). (FSW: filtered sea water.)

and an advance phase shift of 2 hr produced by a similar pulse given during the early projected day is shown in Figure 2B. Hi-K pulses administered during the latter half of the projected night produced modifications in the wave form of the rhythm. Following pulses given at this time, the impulse rate became arrhythmic for 1 or 2 days in some cases (Figure 2C), and in others the amplitude of the rhythm was greatly reduced and a constant background level of impulses was present.

A summary of the data from the experiments in which eyes were treated with 4 hr hi-K pulses at different phases of their rhythms is shown as a response curve in Figure 2D. The magnitude of the phase shift varies in a linear manner (slope = 0.63) over most of the curve. The existence of such a response curve indicates that the stimulus used to produce it could serve as an entraining agent for the rhythm.

Mode of action of hi-K

The action of hi-K on the rhythm may be mediated in several ways: (1) by increasing K_i^+, which in turn could influence a number of intracellular processes (Kernan, 1965); (2) by affecting the rate of an ionic or metabolite pump, which would in turn affect the concentration of intracellular ions or metabolites; (3) by depolarizing the transmembrane potential difference, leading either to permeability changes in the membrane, release or activation of an intracellular substance, or release of a transmitter or neurohormone that might mediate the effect on other cells. Alternatively, the effect might be directly mediated by a voltage sensitive substance or mechanism within the membrane (e.g. the mechanisms underlying the excitable Na^+ and K^+ channels found in the membranes of nerve and muscle cells).

These possibilities have been discussed elsewhere (Eskin, 1972) and will be dealt with only briefly here. It is unlikely that the increase in K_0^+ concentrations used in these experiments resulted in a significant K_i^+ increase. The K_i^+ of nerve cells in the abdominal ganglion of *Aplysia* is 232 mM (Sato et al., 1968) and the K_i^+ of the nerve cells of the eye is likely to be similar. Having K_0^+ as low as 50 mM (a 40 mM increase over normal) has produced significant phase shifts (Eskin, unpublished). Furthermore, hi-K pulses in a medium in which 20% to 80% of the chloride had been substituted with acetate produced phase shifts of the same magnitude as did hi-K pulses with normal chloride. Assuming a Donnan equilibrium across the cell membrane, the change in K_i^+ expected for a hi-K pulse with the acetate substituted for Cl^- would be less than that expected for a hi-K pulse with normal Cl^-.

It is also unlikely that rates of ionic pumps were altered by the increase in K_0^+ during hi-K pulses, for such effects of K_0^+ thus far observed saturate slightly beyond $K_0^+ = 10$ mM (Baker, 1968; Sjodin and Beauge, 1968).

I believe that the effect of hi-K on the rhythm is probably mediated by membrane depolarization although preliminary investigation of possible mechanisms have thus far yielded only negative results. Hi-K in the presence of ionic conditions ($2.5 \times Mg^{++}$ or $0.1 \times Ca^{++}$), which should block the release of transmitters and neurohormones, produced phase shifts very similar to those produced by the normal hi-K medium (Figure 3). Cyclic AMP does not appear to be involved for 4-hr treatments of CAMP (10 mM) or its dibutyryl derivative (dibutyryl-3′,5′-cyclic adenosine monophosphate, 10^{-3} mM to 10 mM) did not produce phase shifts. Furthermore, CAMP (10 mM) with theophylline (1 mM) did not produce phase shifts nor did theophylline added to the hi-K medium increase its phase shifting effect. Finally, hi-K treatments in a medium with zero Ca^{++} (Mg^{++} raised to 125 mM) produced phase shifts similar to those produced by the normal hi-K medium, indicating that Ca^{++} is probably not involved.

Do hi-K pulses act like light pulses?

Exposure of isolated eyes to single 4-hr light pulses (150 to 630 ft candle) early in their subjective night produced at most a small advance phase shift, whereas 4-hr hi-K

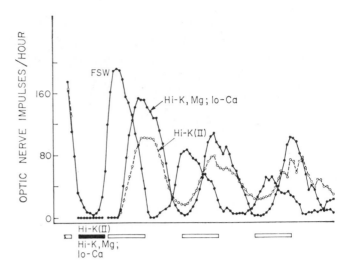

FIGURE 3   Comparison of phase shifting by hi-K and by hi-K in the presence of ions that should block neurotransmitter and neurohormone release. The eyes were exposed to 9-hr pulses of hi-K and hi-K, Mg; lo-Ca during the time indicated by the solid bar at the bottom of the graph. The sea water control eye (FSW) was from a different animal but was run at the same time as the other 2 eyes (Eskin, 1972).

pulses given at the same phase produced 5-hr delay phase shifts. This result seems to indicate that light pulses do not act on the mechanism underlying the rhythm in the same manner as hi-K pulses, but there are several other possible explanations for the observed differences: (1) Light and hi-K pulses may act similarly in the intact animal, but in in-vitro conditions the photoreceptor degenerates. (2) The photoreceptor in the eye may have a rhythmic sensitivity to light. If this were true, light would pass through a rhythmic filter before affecting the underlying oscillator, whereas hi-K might not pass through this filter but could act on the oscillator more directly. (3) As was suggested in the discussion of the difference in the rates of entrainment for in vivo and in vitro exposure to light, the normal pathway of light to the oscillating mechanism in the eye may involve neural or humoral agents not present in the isolated eye. Since LD cycles can entrain the eye in vitro, there must be some pathway, though perhaps an indirect one, by which light information reaches the oscillator. On the other hand hi-K may act directly on the neurons housing the oscillating mechanism and thereby produce a greater phase shift than light does.

Conclusion

Exposure of isolated eyes to hi-K pulses results in phase shifts in their circadian rhythms. The results obtained thus far are consistent with the idea that the mechanism underlying the rhythm is coupled in some way to the transmembrane potential difference. Understanding of the mechanism of the hi-K effect should help elucidate the physiology of the underlying oscillator.

The rhythm of the eye is entrainable in vitro by LD cycles but at a significantly lower rate than that of in vivo entrainment. There is also a major difference between the response of the rhythm of the eye to hi-K pulses and to light pulses. An understanding of the bases of these differences should help define the pathway by which light-dark information reaches the mechanism responsible for the rhythm.

Some evidence has accumulated indicating the involvement of membrane permeability and membrane potential changes in mechanisms underlying rhythms, though in no case has such involvement been unequivocally established (Enright, 1971; Englemann, 1972; Bünning and Moser, 1972; Jacklet, 1972, 1973; Eskin, 1972). Sweeney raised this possibility in 1969, concluding her book with the sentence: "Perhaps as membranes become better understood, the solution of the problem of the mechanism of the cellular clocks will become apparent." The isolated eye of *Aplysia californica* appears to be an excellent preparation in which to pursue this possibility.

ACKNOWLEDGMENTS I am very grateful to Dr. Felix Strumwasser for providing excellent facilities, a stimulating environment, and guidance during the course of this work. I thank Dr. G. F. Gwilliam for his encouragement and support. I also thank Mr. D. Knowles for his fine help in the construction of equipment and Mr. L. Harf for his able assistance in performing experiments. This work was supported in part by U.S.P.H.S. Postdoctoral Fellowships to A.E., NIH, NASA, and the Sloan Foundation grants to Dr. Strumwasser, and NSF and Sloan grants to Reed College.

REFERENCES

AUDESIRK, G., 1971. Neuronal interactions in optic nerve impulse production in *Aplysia*. *Physiologist* 14:105.

AUDESIRK, G., 1973. Spontaneous and light-induced compound action potentials in the isolated eye of *Aplysia*: Initiation and synchronization. *Brain Res.* (in press).

BAKER, P. F., 1968. Recent experiments on the properties of the Na efflux from squid axons. *J. Gen. Physiol.* 51 (Part 2):172s–179s.

BÜNNING, E., and I. MOSER, 1972. Influence of Valinomycin on circadian leaf movements of *Phaseolus*. *Proc. Nat. Acad. Sci. USA* 69:2732–2733.

ENGELMANN, W., 1972. Lithium slows down the *Kalanchoe* clock. *Z. Naturforsch.* 27b:477.

ENRIGHT, J. T., 1971. Heavy water slows biological timing processes. *Z. Vergl. Physiologie* 72:1–16.

ESKIN, A., 1971. Properties of the *Aplysia* visual system: In vitro entrainment of the circadian rhythm and centrifugal regulation of the eye. *Z. Vergl. Physiologie* 74:353–371.

ESKIN, A., 1972. Phase shifting a circadian rhythm in the eye of *Aplysia* by high potassium pulses. *J. Comp. Physiol.* 80:353–376.

HUGHES, H. P. I., 1970. A light and electron microscope study of some opisthobranch eyes. *Z. Zellforsch.* 106:79–98.

JACKLET, J. W., 1969a. Circadian rhythm of optic nerve impulses recorded in darkness from isolated eye of *Aplysia*. *Science* 164:562–563.

JACKLET, J. W., 1969b. Electrophysiological organization of the eye of *Aplysia*. *J. Gen. Physiol.* 53:21–42.

JACKLET, J. W., 1972. A circadian rhythm in neuronal activity depends on chloride ions. *Physiologist* 15:179.

JACKLET, J. W., 1973. Circadian neuronal activity: Abolition by propionate. *J. Gen. Physiol.* 61:270.

JACKLET, J. W., R. ALVAREZ, and B. BERNSTEIN, 1972. Ultrastructure of the eye of *Aplysia*. *J. Ultrastruct. Res.* 38:246–261.

JACKLET, J. W., and J. GERONIMO, 1971. Circadian rhythm: Population of interacting neurons. *Science* 174:299–302.

KERNAN, R. P., 1965. *Cell K*. Washington, D.C.: Butterworth.

LICKEY, M., 1967. Effect of various photoperiods on a circadian rhythm in a single neuron. In *Invertebrate Nervous Systems*, C. A. G. Wiersma, ed. Chicago: The University of Chicago Press, pp. 321–328.

MARLER, P., and W. J. HAMILTON, III, 1966. Circadian rhythms: Exogenous or endogenous control. In *Mechanisms of Animal Behavior*. New York: John Wiley & Sons, pp. 25–72.

SATO, J., G. AUSTIN, H. YAI, and J. MARUHASHI, 1968. The ionic permeability changes during acetylcholine-induced responses of *Aplysia* ganglion cells. *J. Gen. Physiol.* 51:321–345.

SENER, R., 1972. Site of circadian rhythm production in *Aplysia* eye. *Physiologist* 15:261.

SJODIN, R. A., and L. A. BEAUGE, 1968. Coupling and selectivity of sodium and potassium transport in squid giant axons. *J. Gen. Physiol.* 51 (Part 2):152s–161s.

STRUMWASSER, F., 1967. Neurophysiological aspects of rhythms. In *The Neurosciences: A Study Program*, G. C. Quarton, T. Melnechuk, F. O. Schmitt, eds. New York: The Rockefeller University Press, pp. 516–528.

SWEENEY, B. M., 1969. *Rhythmic Phenomena in Plants*. New York: Academic Press, p. 134.

47 Visual Pathways

and the Central Neural Control

of Diurnal Rhythms

ROBERT Y. MOORE

ABSTRACT   Three functionally separate components of optic nerve axons emerge from the optic chiasm. The largest of these, the primary optic tracts, innervate the lateral geniculate nuclei, pretectal area, and tectum and mediate visually guided behavior. The accessory optic tracts innervate three terminal nuclei of the midbrain tegmentum and appear to mediate a tonic neuroendocrine response to light in the pineal gland. The last component is a direct projection from the retina to the suprachiasmatic nuclei of the hypothalamus. Available evidence indicates that these fibers are essential to the maintenance of the phasic timing of circadian rhythms to the diurnal pattern of environmental light. In addition, recent observations suggest that the suprachiasmatic nuclei are a critical central neural mechanism in the regulation of circadian rhythms.

THE PARTICIPATION of central neural mechanisms in the regulation of diurnal rhythms has received little attention. These mechanisms would appear to require at least two components. Diurnal rhythms are controlled by the pattern of environmental lighting (Halberg, 1969), and the visual pathways mediating critical visual information would be the first component. The function of these pathways is to maintain the precise relationship of the rhythms to the diurnal pattern of environmental light. In the absence of diurnal visual stimuli, as in a blinded animal, most diurnal rhythms continue in a free-running fashion with a periodicity unsynchronized to the pattern of environmental light. The central functions that maintain diurnal rhythms in the absence of visual cues are the second neural regulatory mechanism. Although little has been done either to localize or analyze these mechanisms, data are now available suggesting that both the critical visual pathways and a central cell group participating in the generation of diurnal rhythms can be specified for one mammalian species, the albino rat. This work began with an attempt to designate central visual pathways participating in the control of the pineal melatonin-forming

ROBERT Y. MOORE   Departments of Pediatrics, Medicine and Anatomy and the Joseph P. Kennedy Jr. Mental Retardation Research Center, The University of Chicago, Chicago, Illinois

enzyme, hydroxyindole-O-methyltransferase (HIOMT). This enzyme, found only in the pineal in mammals (Axelrod et al., 1961), exhibits a diurnal variation in its content but, unlike most 24-hr rhythms, the pineal content of HIOMT is not maintained by central mechanisms in the absence of lighting cues (Axelrod et al., 1965). In blinded animals, or ones kept in continuous darkness, pineal HIOMT content remains high, whereas in animals kept in continuous lighting, it is continuously low (Axelrod et al., 1965). The diurnal rhythm in pineal HIOMT content, then, can be viewed as a response to light, and the visual pathways mediating this response can be designated experimentally. In order to do this it is necessary to know the morphology of the central retinal projections. These have been studied extensively in the rat by Hayhow and his collaborators (Hayhow et al., 1960, 1962) and are shown diagrammatically in Figure 1. All visual input from the eye enters the optic chiasm through the optic nerves. Two principal pathways emerge from the chiasm. The largest of these is the primary optic tract that contains both crossed and uncrossed retinal projections to the lateral geniculate, pretectal area, and superior colliculus. Running within the primary optic tract also is a small, crossed pathway, the superior accessory optic tract that terminates in the midbrain tegmentum. The other visual pathway leaving the optic chiasm is the inferior accessory optic tract. This tract is made up entirely of crossed retinal fibers that run caudally through the lateral hypothalamus among the fibers of the medial forebrain bundle to terminate in the medial terminal nucleus of the accessory optic system. Also in Figure 1, in dotted lines, is a retinohypothalamic tract. This had been described in nonmammalian vertebrates, but its existence in mammals was unsubstantiated (Nauta and Haymaker, 1969) and Hayhow et al. (1960) could find no evidence for such a projection in the rat.

On the basis of this information, the following experiment was designed to determine the visual pathways mediating the pineal HIOMT response to light: Five

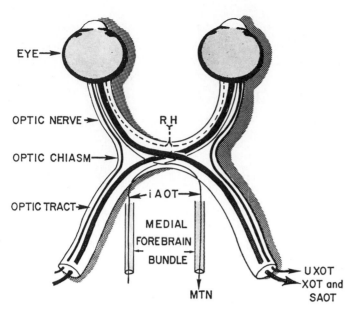

CENTRAL VISUAL PROJECTIONS IN THE RAT

Figure 1 The central retinal projections in the rat; XOT, crossed component of the primary optic tract; UXOT, uncrossed component of the primary optic tract; SAOT, superior accessory optic tract: IAOT, inferior accessory optic tract—completely crossed bundle of fibers leaving the primary optic tract to enter the lateral hypothalamus and run caudally within the medial forebrain bundle to end in the medial terminal nucleus (MTN) of the accessory optic system; RH, retinohypothalamic tract—dotted lines indicate that the existence of this projection was in doubt when the figure was made (from Moore et al., 1968).

groups of operated animals were prepared. The first was a sham-operated control group. In the second the animals were blinded and all visual pathways from the eye were destroyed. In the third group, large bilateral lesions were made to transect all of the primary optic tract fibers as they entered the diencephalon. These lesions ablated all visual input to the lateral geniculate nuclei, pretectal nuclei, and superior colliculus and severed the superior accessory optic tracts. The lesions in the fourth group transected both inferior accessory optic tracts. This was accomplished by removing one eye and, because the tracts completely cross in the optic chiasm, severing the medial forebrain bundle on the side ipsilateral to the removed eye. This left one primary optic tract intact in these animals. In the fifth group, both the primary and accessory optic tracts were transected. This would leave intact a retinohypothalamic projection, if such existed. The results of this study are shown in Table I (see Moore et al., 1967, 1968). The sham-operated animals exhibit the expected low pineal HIOMT levels in constant light and high levels in constant dark and, as in the study of

TABLE I

Visual pathway lesions: Effects on the pineal hydroxyindole-O-methyltransferase response to light

Operated Group	Pineal Hydroxyindole-O-Methyltransferase Content	
	Environmental Lighting Conditions	
	Constant Light	Constant Dark
Sham operated	low	high
Blinded	high	high
Primary optic tract section	low	high
Accessory optic tract section	high	high
Primary and accessory optic tract section	high	high

Adult female rats were maintained in continuous lighting conditions, constant light or dark, for 30 days following visual pathway lesions (original data in Moore et al., 1967).

Axelrod et al. (1965), blinding abolishes the response to light. Bilateral transection of the primary optic tracts does not affect the response to light even though such animals are behaviorally blind and have fixed, dilated pupils (Moore et al., 1967, 1968; Chase et al., 1969). In contrast to this, accessory optic tract lesions eliminated the pineal HIOMT response to light but the animals showed normal behavioral responses (Moore et al., 1967, 1968; Chase et al., 1969). Combined accessory and primary optic tract destruction eliminated both functions. From these observations, it was suggested that there is a separation of functions among the visual pathways in the rat with the primary optic tracts mediating behavioral responses to light and the inferior accessory optic tracts mediating neuroendocrine responses to light.

This hypothesis was tested by examining the effects of primary and accessory optic tract lesions on two other diurnal rhythms in the rat. These were the rhythm in adrenal corticosterone content (Critchlow, 1963) and in the pineal enzyme, N-acetyltransferase (Klein and Weller, 1970). In contrast to the HIOMT rhythm, each of these is a true circadian rhythm (Wurtman, 1967) and becomes free-running in the absence of visual input. The effects of visual pathway lesions on these rhythms are shown in Table II. Sham-operated animals exhibit normal rhythms in both adrenal corticosterone content and pineal N-acetyltransferase activity. Blinded animals show a shift in the timing of peak values in both rhythms, indicating that the rhythms had become free-running, but neither visual pathway lesion alters the timing of the rhythms. From this it appeared that neither the primary nor the accessory optic tracts can mediate the effects of light in entraining these circadian rhythms. Since no other central retinal projections were known with certainty in mammals, this led to a re-evaluation of the

TABLE II

Visual pathway lesions: Effects on diurnal rhythms in the adrenal and pineal glands

Operated Group	Adrenal Corticosterone Rhythm			Pineal N-Acetyltransferase Rhythm		
	Time of Peak Value	Time of Trough Value	Rhythm Status	Time of Peak Value	Time of Trough Value	Rhythm Status
Sham operation	1900	0700	entrained	2400	0700	entrained
Blinded	0700	1900	free-running	0700	1900	free-running
Primary optic tract section	1900	0700	entrained	2400	0700	entrained
Accessory optic tract section	1900	0700	entrained	2400	0700	entrained

Adult female rats were maintained in diurnal light (lights on 0700 to 1900) prior to operation and until sacrificed 3 weeks after operation. Samples were obtained at 0700, 1300, 1900, and 2400 hrs for each group. (Data from Moore and Eichler, 1972; Moore and Klein, 1972.)

possibility that a direct retinohypothalamic tract might exist. This pathway could not be shown with conventional neuroanatomic methods (Hayhow et al., 1960; Moore, 1969), but the development of a new method for tracing central pathways (Lasek et al., 1968; Cowan et al., 1972) provided another possibility. This method is based on the observations that nerve cells will take up labeled amino acids, incorporate them into proteins, and transport the labeled proteins along their axons to the axon terminals (Grafstein, 1969). Autoradiography then allows microscopic localization of the labeled axons and terminals. This technique has been employed successfully in tracing several central pathways (Cowan et al., 1972), and it was applied to the problem of retinohypothalamic projections in the rat (Moore and Lenn, 1972). Following the injection of tritiated leucine or proline into the posterior chamber of the eye, labeled protein is found autoradiographically in all of the known terminal nuclei of the primary and accessory optic systems. In addition, labeled protein is evident in the suprachiasmatic nuclei of the medial hypothalamus (Figure 2), both ipsilateral and contralateral to the injected eye (Moore and Lenn, 1972). No other hypothalamic nucleus is labeled, and it was concluded that there is a retinohypothalamic projection terminating bilaterally in the suprachiasmatic nuclei. This was confirmed by electron-microscopic observation of degenerating terminals in these nuclei following section of the optic nerve (Moore and Lenn, 1972), and an identical projection has been found in a series of mammals including a marsupial, insectivores, carnivores, and prosimian and anthropoid primates (Moore, 1973). From these observations a further study on the central control of diurnal rhythms was planned.

The purpose of this study was to examine the effects of retinohypothalamic tract section on the adrenal corticosterone and pineal N-acetyltransferase rhythms. Since the suprachiasmatic nuclei are embedded in the optic chiasm (Figure 2), it is not possible to section the retinohypothalamic tract without ablating these nuclei. In addition, lesions destroying the suprachiasmatic nuclei

may transect descending pathways running through the medial hypothalamus, and lesions behind the optic chiasm are known to affect the adrenal corticosterone rhythm (Halász et al., 1967). For that reason, three lesion groups were prepared to compare with sham operated animals. The location of these lesions is shown in Figure 3. The first two lesions were made using a modified Halász (Halász and Pupp, 1965) knife. These produce rostral deafferentation of the medial hypothalamus at two different levels, one rostral to the optic chiasm (prechiasmatic cut) and one caudal to the optic chiasm (postchiasmatic cut). The other lesion was produced electrolytically and destroys the suprachiasmatic nuclei. Some of these lesions spare the optic chiasm whereas others involve it to a varying extent (Figure 3; Moore and Eichler, 1972). Since the extent of chiasmal damage does not alter the lesion effect, all of these were placed in a single group. Partial deafferentation of the medial hypothalamus by a prechiasmatic cut does not affect either the adrenal corticosterone rhythm or the pineal N-acetyltransferase rhythm (Table III). Both rhythms are abolished by the postchiasmatic cut. This effect on the adrenal rhythm confirms that reported by Halász et al. (1967) for a similar lesion. Taken together these observations indicate that the rhythms are not generated by centers rostral to the chiasm, but they do depend upon the integrity of influences passing caudally from the region of the chiasm. Since the retinohypothalamic projection terminates in the suprachiasmatic nuclei lying just above the chiasm, and the visual pathways extending beyond the chiasm do not control the rhythms (Table II), the suprachiasmatic nuclei appear critical to the generation of these influences. Evidence for this is obtained from the effects of ablating these nuclei (Table III). Regardless of the extent of chiasmal involvement, destruction of the suprachiasmatic nuclei eliminates both the adrenal corticosterone and pineal N-acetyltransferase rhythms. This effect cannot be viewed as due to section of a critical visual pathway. If this were the case, the rhythms should become free-running as occurs with blinding. The

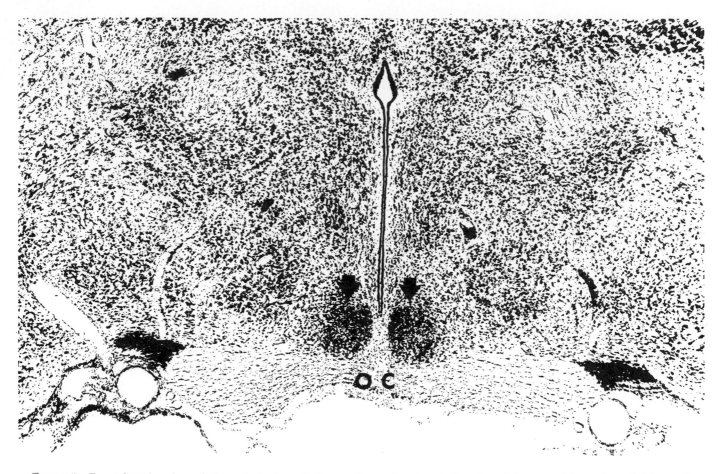

FIGURE 2 Frontal section through the anterior hypothalamus of a rat showing the location of the suprachiasmatic nuclei (arrows) with their characteristic closely compacted, small cells lying just above the optic chiasm (OC). Cresyl violet stain, × 64.

TABLE III

Hypothalamic lesions: Effects on diurnal rhythms in the adrenal and pineal glands

	Adrenal Corticosterone Rhythm			Pineal N-Acetyltransferase Rhythm		
Operated Group	Time of Peak Value	Time of Trough Value	Rhythm Status	Time of Peak Value	Time of Trough Value	Rhythm Status
Sham operation	1900	0700	entrained	2400	0700	entrained
Suprachiasmatic nucleus lesion	none	none	abolished	none	none	abolished
Prechiasmatic cut	1900	0700	entrained	2400	0700	entrained
Postchiasmatic cut	none	none	abolished	none	none	abolished

Adult female rats were maintained in diurnal light (lights on 0700 to 1900) prior to operation and until sacrificed 3 weeks after operation. Samples were obtained at 0700, 1300, 1900, and 2400 hrs for each group. (Data from Moore and Eichler, 1972; Moore and Klein, 1972.)

tentative interpretation made from these observations is that visual input from the retina to the suprachiasmatic nuclei is critical to the maintenance of an entrained rhythm *and* that the suprachiasmatic nuclei are necessary to endogenous generation of the rhythm. Stephan and Zucker (1972a, 1972b) have also found that section of the primary or accessory optic tracts in the rat does not affect

circadian rhythms, in their case in activity and drinking behavior, whereas suprachiasmatic lesions abolish the rhythms. Our findings and those of Stephan and Zucker (1972b) are the first to clearly localize a central area participating in the endogenous generation of diurnal rhythms.

On the basis of all of the data reviewed here, it would

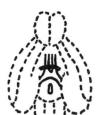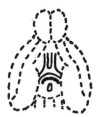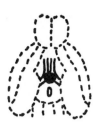

PRE-CHIASMATIC   POST-CHIASMATIC   SUPRACHIASMATIC
CUT   CUT   LESION

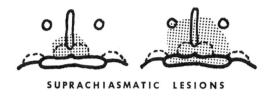

SUPRACHIASMATIC LESIONS

FIGURE 3 Lesions affecting diurnal adrenal and pineal rhythms in the rat. The top three figures illustrate the locations, respectively, of the prechiasmatic cut, the postchiasmatic cut, and the suprachiasmatic lesion. The prechiasmatic cut is placed rostral to the optic chiasm and suprachiasmatic nuclei. It does not involve either the optic nerves or chiasm. The postchiasmatic cut lies between the chiasm and the median eminence. The suprachiasmatic lesions are placed to be centered in these nuclei. The bottom two drawings, corresponding to Figure 2, show the extent of a small suprachiasmatic nucleus lesion on the left (dotted-in area) with a large lesion on the right.

appear possible to make the following generalizations about visual pathways and the central neural control of diurnal rhythms in the rat: There are three major components of the central retinal projection. The primary optic tract mediates visually guided behavior. The accessory optic tracts appear to mediate tonic responses to continuous visual stimulation. The retinohypothalamic projection and its site of termination, the suprachiasmatic nucleus, appear necessary for the mediation of information concerning the diurnal pattern of environmental light and utilization of this in the generation of rhythms. Influences from the suprachiasmatic nuclei pass caudally in the medial hypothalamus. However, there are many questions that remain for future investigation. These include such problems as the nature of the rhythm generating mechanism, how the coupling between visual input and the rhythm generating mechanism occurs, how information from the suprachiasmatic region regulates a diversity of rhythms, and whether these observations can be generalized to other species.

REFERENCES

AXELROD, J., P. D. MACLEAN, R. W. ALBERS, and H. W. WEISSBACH, 1961. Regional distribution of methyltransferase enzymes in the nervous system and glandular tissues. In *Regional Neurochemistry*, S. S. Kety and J. Elkes, eds. Oxford: Pergamon Press, pp. 307–311.

AXELROD, J., R. J. WURTMAN, and S. H. SNYDER, 1965. Control of hydroxyindole-O-methyltransferase activity in the rat pineal gland by environmental lighting. *J. Biol. Chem.* 240: 949–954.

CHASE, P. A., L. S. SEIDEN, and R. Y. MOORE, 1969. Behavioral and neuroendocrine responses to light mediated by separate visual pathways in the rat. *Physiol. Behav.* 4:949–952.

COWAN, W. M., D. I. GOTTLIEB, A. E. HENDRICKSON, J. L. PRICE, and T. A. WOOLSEY, 1972. The autoradiographic demonstration of axonal connections in the central nervous system. *Brain Res.* 37:21–51.

CRITCHLOW, V., 1963. The role of light in the neuroendocrine system. In *Advances in Neuroendocrinology*, A. Nalbandov, ed. Urbana, Ill.: University of Illinois Press, pp. 377–401.

GRAFSTEIN, B., 1969. Axonal transport: Communication between soma and synapse. In *Advances in Biochemical Psychopharmacology*, Vol. 1, E. Costa and P. Greengard, eds. New York: Raven Press, pp. 11–25.

HALÁSZ, B., and L. PUPP, 1965. Hormone secretion of the anterior pituitary gland after physical interruption of all nervous pathways to the hypophysiotrophic area. *Endocrinology* 77:553–562.

HALÁSZ, B., M. SLUSHER, and R. GORSKI, 1967. Adrenocorticotrophic hormone secretion in rats after partial or total deafferentation of the medial basal hypothalamus. *Neuroendocrinology* 2:43–55.

HALBERG, F., 1969. Chronobiology. *Ann. Rev. Physiol.* 31:675–725.

HAYHOW, W. R., C. WEBB, and A. JERVIE, 1960. The accessory optic system in the rat. *J. Comp. Neurol.* 113:281–314.

HAYHOW, W. R., A. SEFTON, and C. WEBB, 1962. Primary optic centers of the rat in relation to the terminal distribution of the crossed and uncrossed optic fibers. *J. Comp. Neurol.* 118: 295–322.

KLEIN, D. C., and J. L. WELLER, 1970. Indole metabolism in the pineal gland: A circadian rhythm in N-acetyltransferase. *Science* 169:1093–1095.

LASEK, R., B. J. JOSEPH, and D. G. WHITLOCK, 1968. Evaluation of an autoradiographic neuroanatomical tracing method. *Brain Res.* 8:319–336.

MOORE, R. Y., A. HELLER, R. J. WURTMAN, and J. AXELROD, 1967. Visual pathway mediating pineal response to environmental light. *Science* 155:220–223.

MOORE, R. Y., A. HELLER, R. K. BHATNAGAR, R. J. WURTMAN, and J. AXELROD, 1968. Central control of the pineal gland: Visual pathways. *Arch. Neurol. (Chicago)* 18:208–218.

MOORE, R. Y., 1969. Visual pathways controlling neuroendocrine function. In *Progress in Endocrinology*, C. Gual and F. J. G. Ehling, eds. Amsterdam: Excerpta Medica, pp. 490–494.

MOORE, R. Y., 1973. Retinohypothalamic projection in mammals: A comparative study. *Brain Res.* 49:403–409.

MOORE, R. Y., and V. B. EICHLER, 1972. Loss of a circadian adrenal corticosterone rhythm following suprachiasmatic lesions in the rat. *Brain Res.* 42:201–206.

MOORE, R. Y., and N. J. LENN, 1972. A retinohypothalamic projection in the rat. *J. Comp. Neurol.* 146:1–14.

MOORE, R. Y., and D. C. KLEIN, 1972. Central control of the pineal N-acetyltransferase rhythm. In preparation.

NAUTA, W. J. H., and W. HAYMAKER, 1969. Retino-hypothalamic connections. In *The Hypothalamus*, W. Haymaker,

E. Anderson, and W. J. H. Nauta, eds. Springfield, Ill.: Charles C Thomas, pp. 187–189.

STEPHAN, F. K., and I. ZUCKER, 1972a. Rat drinking rhythms: Central visual pathways and endocrine factors mediating responsiveness to environmental illumination. *Physiol. Behav.* 8:315–326.

STEPHAN, F. K., and I. ZUCKER, 1972b. Circadian rhythms in drinking behavior and locomotor activity of rats are eliminated by hypothalamic lesions. *Proc. Natl. Acad. Sci. USA* 69:1583–1586.

WURTMAN, R. J., 1967. Ambiguities in the use of the term circadian. *Science* 156:104.

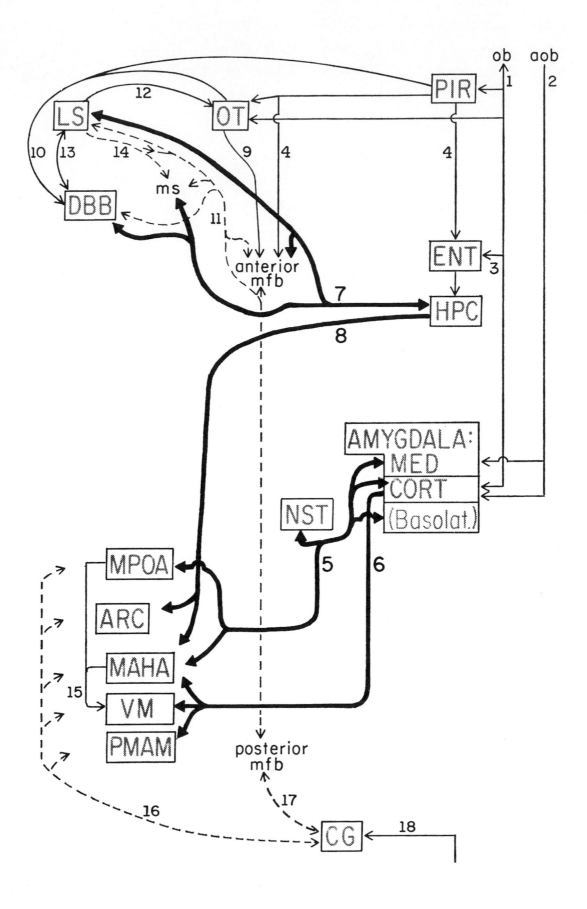

HORMONAL

FACTORS IN

BRAIN FUNCTION

Schematic diagram showing large number of established anatomical connections among brain structures that contain peak numbers of estrogen-concentrating neurons. These pathways could provide an anatomical substrate for coordinated physiological action by estrogen-sensitive cells. Structures with peak estrogen concentration are designated by capital letters in boxes. LS, lateral septum; OT, olfactory tubercle; DBB, cells in diagonal band of Broca; PIR, prepiriform cortex; ENT, entorhinal cortex; HPC, ventral hippocampus; NST, nucleus of stria terminalis; MPOA, medial preoptic area; ARC, arcuate nucleus of hypothalamus; MAHA, medial anterior hypothalamic area; VM, ventromedial hypothalamic nucleus; PMAM, ventral premammillary nucleus; CG, mesencephalic central grey. (Reproduced from Figure 9–5 in B. S. McEwen and D. W. Pfaff, 1973. Chemical and physiological approaches to neuroendocrine mechanisms: Attempts at integration. In Frontiers in Neuroendocrinology, W. F. Ganong and L. Martini, eds. New York: Oxford University Press, in press.)

HORMONAL FACTORS IN BRAIN FUNCTION

Introduction

DONALD PFAFF

HORMONES ACT on the brain, thereby affecting pituitary function and behavior. They complete endocrine feedback loops by influencing neural events that control secretions from the pituitary. For controlling behavioral responses, steroid hormones apparently operate on systems of limbic and hypothalamic neurons, affecting forebrain control over reflex loops located in the lower brainstem and spinal cord.

Some of these hormone effects occur in adulthood. Others occur early in development and influence neuroendocrine and behavioral controls throughout much of the life of the organism.

Chapters in this section present a variety of technical approaches to studying effects of hormones on the brain. Ganong and Scapagnini review some of the basic facts concerning neural control over the pituitary and report some recent experimental results which indicate a neuroendocrine role for aminergic neural transmission in the hypothalamus. Goy reviews factors that control the development of copulatory responses. Chapters by Feder and Goldman cover the use of certain endocrine techniques in determining exact temporal relationships among protein gonadotrophic hormones, gonadal steroid hormones, and mating behavior. While most neuroendocrine investigations have been carried out using mammals, Hutchison has obtained clear behavioral effects of brain-implanted sex hormones in birds. McEwen's chapter covers his investigations of presumptive steroid receptors in the brain and considers developmental effects of hormones from a biochemical viewpoint. Ultimately, we hope to understand how steroid hormones, acting through hormone-sensitive neurons, produce neurophysiological alterations in reflex loops such that behavioral changes are observed. The chapter by Pfaff et al. reviews recent studies on the locations and physiological properties of estrogen-concentrating neurons in the rat brain and describes a "reflexological" approach to the discovery of neural pathways subserving lordosis, an estrogen-sensitive mating response of the female rodent.

Thus, the following chapters deal with how hormone-sensitive pituitary functions and behavioral events are regulated—and synchronized with each other—by mechanisms in the mammalian forebrain.

48 Brain Mechanisms Regulating the Secretion of the Pituitary Gland

WILLIAM F. GANONG

ABSTRACT The nervous system and the endocrine system interact in multiple, complex ways. In this chapter, attention is focused on neural regulation of the secretion of the anterior lobe of the pituitary gland. A family of anterior pituitary-regulating peptides is secreted in the hypothalamus and passes to the anterior pituitary via the portal-hypophyseal vessels. Two of these releasing and inhibiting factors, a tripeptide and a decapeptide, have been characterized and synthesized, and it has been suggested that their synthesis is enzymatic rather than ribosomal. The secretion of the factors is controlled by feedback of hormones secreted by the regulated glands, and by converging neural inputs. Norepinephrine-containing neurons in these inputs increase the secretion of some hormones and decrease the secretion of others. Prolactin appears to be unique in that its secretion is inhibited by dopamine-containing neurons. The role of serotonin is less well defined, but circuits in which it is a mediator appear to be involved in the patterning of endocrine secretion.

Introduction

THE RELATION between the nervous system and the endocrine system is reciprocal: The brain, largely but not exclusively via the hypothalamus, regulates the secretion of a large portion of the endocrine system, and the hormones secreted by the endocrine glands act back in turn on the brain to modify its function in a variety of ways.

The effects of hormones on the brain are remarkably widespread and diverse, but they can be roughly grouped into five categories. There are positive and negative feedback effects on the secretion of glands such as the pituitary, an example being stimulation of luteinizing hormone (LH) secretion by estrogen (Knobil et al., 1973). Hormones also initiate or adjust certain specific patterns of instinctive behavior: Control of sexual behavior by gonadal hormones is probably the best-known example.

There are also effects on other types of behavior, such as the changes in conditioned avoidance reflexes produced by ACTH. Many hormones exert general metabolic effects on the brain: Examples include the action of thyroid hormones on the developing brain, and the well-known effect of hypoglycemia induced by insulin on cerebral function. Lastly, there are very interesting, long-lasting *inductive* effects of hormones on brain development. The effect of early exposure to male sex hormone in the female rat is possibly the most dramatic example. If a female rat is given a single small dose of androgen in the 5th day of life, she subsequently appears to mature normally, but in adult life, she shows the male rather than the female pattern of secretion of gonadotropins from the pituitary. She also behaves sexually as a male rather than a female. Conversely, male rats castrated at birth show the female pattern of gonadotropin secretion and appreciable amounts of female sexual behavior. These changes in the rat and other species have been extensively studied and have been the subject of a number of reviews (Harris, 1964; Goy, later in this part).

The best known example of the other aspect of brain-endocrine relations, neural regulation of endocrine function, is hypothalamic control of pituitary secretion. However, neural factors also influence the pineal body, and they affect the secretion of the adrenal medulla, many of the hormone-secreting cells in the gastrointestinal tract, the renin-secreting juxtaglomerular cells of the kidney, and the β and possibly the α cells of the pancreatic islets as well. This does not mean that the neural component is the only or even the major regulator of the various glands, but in each, it does appear to play a role. Since the anterior pituitary, via its tropic hormones, regulates the secretion of the adrenal cortex, thyroid, and gonads, only the parathyroids and thymus as endocrine organs are left free of known neural regulation.

This chapter is limited to consideration of the regulation of pituitary secretion. The 10 hormones found in this

WILLIAM F. GANONG Department of Physiology, University of California, San Francisco, California gland are listed in Table I. Intermediate lobe tissue contains two hormonally active substances, α and β-MSH, but at least in man, it appears that only β-MSH is

TABLE I

Hormones of the pituitary gland in mammals

Posterior lobe	oxytocin
	vasopressin
Intermediate lobe tissue	α-MSH
	β-MSH
Anterior lobe	adrenocorticotropin (ACTH)
	growth hormone
	prolactin
	thyrotropin (TSH)
	luteinizing hormone (LH)
	follicle-stimulating hormone (FSH)

secreted. The posterior lobe hormones are nonapeptides, if one counts each half-cystine as one amino acid residue. The intermediate lobe hormones are peptides containing 13 to 22 amino acid residues. The anterior lobe hormones are larger polypeptides (ACTH, growth hormone, prolactin) or glycoproteins (TSH, FSH, LH). The three glycoprotein hormones are each secreted by a distinct cell type, but each is made up to two subunits, and the structure of one of the subunits is remarkably similar in all three hormones. (Pierce, 1971.)

Secretions of the posterior and intermediate lobes of the pituitary

The posterior lobe hormones are, of course, known to be secreted at the endings of supraoptic and paraventricular neurons directly into the general circulation, with vasopressin acting primarily on the kidney and oxytocin acting primarily on the breast. The physiology of their secretion will not be considered in detail, but it is worth reviewing their formation and secretion as a model of hormone secretion by neurons. A good deal of evidence indicates that posterior lobe hormones are synthesized within the endoplasmic reticulum in the cell bodies of supraoptic and paraventricular neurons (see Bloom et al., 1970). Two binding polypeptides, neurophysin I and neurophysin II, are closely associated with oxytocin and vasopressin, respectively, but it is not yet settled whether oxytocin and vasopressin are synthesized combined to the neurophysins—i.e., whether the synthesized products are prohormones, like proinsulin—or whether the binding occurs after synthesis. The bound hormones are transported along the axons of the supraoptic and paraventricular neurons to their endings, where they are stored and

released in response to action potentials proceeding along these neurons. There has been considerable debate about the release process, but recent electron microscopic observations and related data indicate that the hormones are released by exocytosis. The neurophysins appear to be released in a quantal fashion with the posterior lobe hormones, an observation that adds support to the concept of a release by exocytosis, and neurophysins are found in the general circulation. Indeed, radioimmunoassays have been developed for their measurement, while it has proved difficult to develop radioimmunoassays for vasopressin and oxytocin themselves, and it has been suggested that circulating levels of neurophysins may be measured as an index of oxytocin and vasopressin secretion (Bloom et al., 1970; Cheng et al., 1972).

The intermediate lobe is an enigma for a number of reasons. Not the least of these is the fact that despite abundant speculation, the function of intermediate lobe hormones in mammals, if any, is unknown. Furthermore, the lobe is rudimentary or absent in a number of mammalian species, and in man, intermediate lobe tissue is found in the anterior lobe. In species with a well-developed intermediate lobe, there appears to be no special blood supply, but there is an adrenergic innervation. Current evidence indicates that the neurons innervating intermediate lobe cells in these species contain dopamine (Nobin et al., 1972) and that the innervation is inhibitory in function (see Ganong, 1970).

Hypothalamic control of anterior pituitary secretion

In most species, including mammals, the anterior lobe of the pituitary gland does not receive a major innervation from the hypothalamus, but it is supplied by blood that comes directly from the hypothalamus via the portal-hypophyseal vessels (Figure 1). These vessels begin in capillary loops that penetrate the median eminence of the hypothalamus and pass down the pituitary stalk to the anterior lobe where they end in the sinusoids supplying the anterior lobe cells. In many species including man, the portal vessels supply almost all the blood that reaches the anterior lobe (Daniel, 1966). The presence of these vessels, plus experiments involving lesions, stimulation, hypothalamic extracts, and pituitary stalk section, led to the hypothesis that the secretion of the anterior pituitary was controlled by chemical agents transported from the hypothalamus to the pituitary in the portal vessels. Much of the credit for the formulation of this neurovascular hypothesis goes to the late Geoffrey Harris (1955). Its essential correctness is now established, and fractions obtained from hypothalamic extracts have been shown to increase the secretion of all six anterior lobe hormones (Vale et al., 1973). There are also two factors that exert

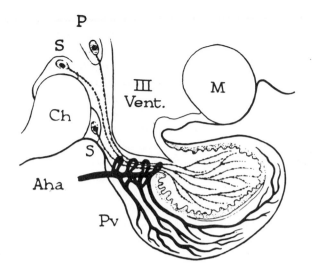

inhibitory effects. The known and probable releasing and inhibiting factors are listed in Table II. The structure of two of them, TRF and LRF, is known, and they have been synthesized. Some authors are now using the term

hormone instead of factor when referring to these substances; thus, TRF is called TRH and LRF is called LHRH. There are arguments for both terminologies. However, a semantic debate is beyond the scope of the present discussion, and the term *factor* has the merit of historical precedent; therefore it is used in this review.

It is worth noting that substances claimed to increase and decrease the secretion of MSHs from the intermediate lobe have also been extracted from hypothalamic tissue (see Vale et al., 1973). Indeed, the three C-terminal amino acids of oxytocin are said to inhibit MSH secretion and have been called MSH-inhibiting factor (MIF). It has been claimed that the ring portion of oxytocin also

has MIF activity (Hruby et al., 1972) and a pentapeptide with MIF activity has been isolated and synthesized (Kastin et al., 1973). However, in view of the fact that this area of research is still clouded in controversy and that at least in many mammalian species there is no portal blood supply to the intermediate lobe, it seems wise to reserve judgment on the question of whether physiologically important intermediate lobe-regulating peptides exist.

TRF and LRF

TRF was originally isolated from porcine and ovine hypothalami, but it appears that it has the same structure in a variety of different mammals including man (Vale et al., 1973). It is a tripeptide with a pyroglutamyl configuration at the N-terminal end and an amide group at the C-terminal end (Figure 2). The LRF in ovine and

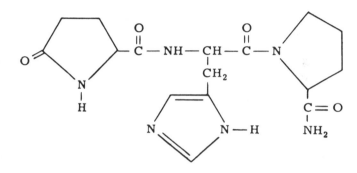

FIGURE 2 Structure of TRF, (pyro) Glu-His-Pro-NH$_2$.

porcine hypothalami is a decapeptide that also has a pyroglutamyl configuration at one end and an amide group at the other; its structure is (pyro) Glu-His-Trp-Ser-Tyr-Gly-Leu-Arg-Pro-Gly-NH$_2$. The structure of LRF in other species is not as yet known, but the synthetic decapeptide is active in all mammals tested, including man (see Vale et al., 1973). The pyroglutamyl amide configuration is interesting, because it is also found in the gastrointestinal hormone gastrin and in some other peptides of biological interest.

Elucidation of the structure of TRF and LRF led rapidly to the synthesis of various analogs in an effort to find antagonists and explore structure-function relationships. The potential for LRF antagonists in population control is obvious, because it is a short burst of LH secretion triggered by LRF that is responsible for ovulation.

The TRF appears not to affect the secretion of ACTH, FSH, LH, or growth hormone. In humans and a number of domestic animal species, it stimulates prolactin secretion, but the significance of this stimulation is uncertain. Evidence is accumulating that there is a PRF in the hypothalamus and that this PRF is separate from TRF (Frantz, 1973).

Also LRF has FRF activity, and it has been suggested that LRF and FRF are the same substance. However, FSH and LH secretion are not always parallel, and various authors have reported rapid diverging changes in the rate of secretion of the two hormones. This and other evidence suggest that there is a separate FRF as well (Vale et al., 1973).

Other releasing and inhibiting factors

The structure of the PRF, the FRF, and the other releasing and inhibiting factors is unknown, although they are probably polypeptides. A decapeptide GRF that is capable of lowering the content of a material, which gives a positive reaction in the bioassay for growth hormone, has been isolated and synthesized. However, it does not produce any increase in circulating growth hormone as measured by the highly specific and sensitive radioimmunoassay for this hormone. For this reason and because the peptide has a structure that resembles the first ten amino acids of the β chain of porcine hemoglobin, there is some uncertainty about its physiologic significance (see Vale et al., 1973).

It is interesting to note that ACTH, TSH, FSH, and LH cause the secretion from their target glands of hormones, which feed back to inhibit further secretion of their specific tropic hormones. Prolactin, on the other hand, elicits the secretion of no known target gland hormone, and it is regulated by both an excitatory (PRF) and an inhibitory (PIF) hypothalamic factor. Growth hormone also appears to be regulated by two hypothalamic factors, one excitatory and one inhibitory. However, there is evidence that growth hormone also causes the secretion of somatomedin, a polypeptide factor that is responsible for many of the effects of the hormone and that may feed back to regulate growth hormone secretion (Garland and Daughaday, 1972).

What hypothalamic structures secrete the releasing and inhibiting factors? They are found in highest concentration in the median eminence, a region that contains nerve endings, nerve fibers en route to the posterior pituitary, and ependymal cells, but very few nerve cell bodies. There are many reasons to believe that the factors are secreted from the nerve endings, but as yet there is no proof. Indeed, it has been suggested that they may be secreted, or at least transported by ependymal cells (see Kobayashi and Matsui, 1969). The large tanycytes, which are apparently ependymal cells that extend all the way through the median eminence, have processes that end on or near the portal capillary loops. Some neurons

have been reported to end synaptically on these tanycytes (Kobayashi and Matsui, 1969), and we have confirmed the occurrence of close associations between neurons and tanycytes, with subsynaptic thickenings at some of these "junctions" (Cuello et al., 1972a). However, it is a long step from morphological description to proof that these associations are physiologically significant or that the ependymal cells secrete releasing and inhibiting factors. Evidence in favor of the concept of neural secretion of the factors includes the biologic analogy of the secretion of posterior lobe peptides by neurons. In addition, the factors can be found in hypothalamic tissue some distance from the ventricle (Quijada et al., 1971). However, additional work, including histochemical and other histological analyses, is clearly in order finally to localize and identify the cells secreting the releasing and inhibiting factors.

Whatever the actual nature of the releasing and inhibiting factor-secreting elements may be, there is evidence that they are for the most part located entirely within the hypothalamus. This evidence, which is summarized in the review by Halász (1969), stems in part from experiments in which a knife is inserted stereotaxically and turned in such a way that it interrupts all the neural connections between the hypothalamus and the rest of the brain; this process has come to be called *deafferentation* of the hypothalamus, even though it obviously cuts efferent fibers as well.

Synthesis of releasing and inhibiting factors

The synthesis of releasing factors is beginning to receive attention, and in this regard the findings of Reichlin and his associates are of considerable interest, along with confirmatory evidence from Guillemin's laboratory (see Reichlin and Mitnick, 1973). These investigators argue that TRF synthesis is enzymatic, like that of glutathione. They report that TRF synthesis is not blocked by puromycin or cycloheximide, and that it occurs in ribosome-free hypothalamic extracts if the appropriate amino acids and cofactors are added. Recently, they have presented evidence suggesting that not only the tripeptide TRF but the decapeptide LRF may be synthesized enzymatically without dependence on the genetic mechanism regulating the synthesis of most polypeptides. If this hypothesis of enzymatic nonribosomal synthesis of releasing factors is confirmed, it will add a new dimension to our concept of neurochemical processes.

Secretion of releasing and inhibiting factors

What controls the secretion of the releasing and inhibiting factors? Feedback regulation by the hormones whose secretion is stimulated is one important factor. The types of known and postulated feedback involved in the regulation of releasing and inhibiting factor secretion are summarized in Figure 3. The most common form is negative feedback of the steroid or other hormone secreted by the target gland of a particular anterior pituitary tropic hormone. The site of feedback is often the anterior pituitary,

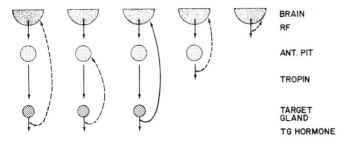

BRAIN
RF

ANT. PIT

TROPIN

TARGET
GLAND

TG HORMONE

FIGURE 3 Known and proposed types of feedback involved in the regulation of anterior pituitary secretion. Solid lines indicate stimulation; dashed lines inhibition. RF, releasing factor; TG, target gland; ant. pit., anterior pituitary.

but feedback at the hypothalamic level is also known to be present in some instances and has been postulated to occur in others. Positive feedback also occurs (Knobil et al., 1973). In the instance in which this is best established, the site of feedback appears to be the anterior hypothalamus. There is also considerable evidence that large amounts of pituitary hormones can act directly on the hypothalamus to inhibit their own secretion via appropriate effects on the releasing and inhibiting factors involved (Motta et al., 1969), although there is some uncertainty about the physiologic significance of such *short-loop feedback*. Inhibition by releasing factors of their own secretion (*minifeedback*) has also been postulated.

Another factor regulating releasing and inhibiting factor secretion is neural input. There are excitatory and inhibitory pathways converging on the median eminence and adjacent hypothalamus from several parts of the brain. Among these pathways, the dopamine-containing and norepinephrine-containing pathways are particularly prominent.

Aminergic innervation of the hypothalamus

The work of numerous investigators has demonstrated that there is an intrahypothalamic system of dopamine-containing neurons with their cell bodies in the arcuate nucleus and their endings in the external layer of the median eminence close to the capillary loops of the portal vessels (Ungerstedt, 1971). A dopaminergic innervation of the intermediate and posterior lobes of the pituitary by fibers with cell bodies in the arcuate and periventricular nuclei has also been described (Nobin et al., 1972). There are norepinephrine-containing nerve fibers in the hypothalamus, including the median eminence, but these fibers appear to be primarily the axons of neurons with their cell bodies in the brainstem. The main input to the ventral hypothalamus comes from the locus ceruleus and locus subceruleus (Hökfelt and Fuxe, 1972). Deafferentation with the Halász knife has no significant effect on the dopamine concentration in the mediobasal hypothalamus but reduces the norepinephrine concentration to zero (Weiner et al., 1972). If parallel cuts are made along the lateral borders of the hypothalamus (Cuello et al., 1972a), the norepinephrine concentration is reduced and degenerative changes are observed in nerve endings in the arcuate nucleus and median eminence, including its external layer (Figure 4). Systematically administered 6-hydroxydopamine also produces a significant change in hypothalamic norepinephrine concentration without a significant change in dopamine concentration and produces degenerative changes in nerve endings in the external layer of the median eminence (Cuello et al., 1972b). Taken together, these observations suggest that norepinephrine-containing as well as dopamine-containing neurons end in the external layer of the median eminence. Serotonin-containing neurons are also found in the hypothalamus, but they are generally located above the median eminence (Ungerstedt, 1971).

Excitation of CRF secretion

The excitatory paths responsible for increasing ACTH secretion in response to trauma have been mapped in some detail (Gibbs, 1969; Greer et al., 1970; C. F. Allen et al., 1972; J. P. Allen et al., 1972). The path mediating the ACTH response to a leg break in the rat crosses the midline within several segments of the site of entry of the afferent fibers into the spinal cord and ascends contra-laterally in the spinal cord. It is still unilateral in the hypothalamus, where it appears to run in the medial forebrain bundle. Interestingly, posterolateral cuts with the Halász knife do not interrupt the response to leg break, while anterolateral cuts do. Thus, it appears that the fibers responsible for increasing ACTH secretion leave the medial forebrain bundle and enter the median eminence from a lateral and even an anterior direction.

The nature of the synaptic mediators in this excitatory pathway is unknown. The catecholamines do affect ACTH secretion, but the evidence from my laboratory indicates that central norepinephrine-containing fibers inhibit rather than excite ACTH secretion (for references, see Ganong, 1972a, 1972b, 1972c).

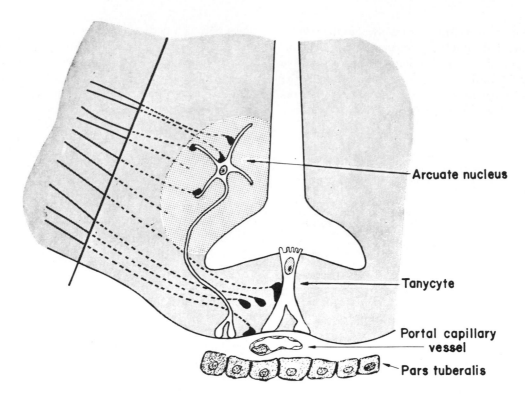

Arcuate nucleus

Tanycyte

Portal capillary vessel

Pars tuberalis

FIGURE 4 Distribution of degenerated endings after a knife cut along the lateral border of the mediobasal hypothalamus. Note that some of the neurons with cell bodies outside the mediobasal hypothalamus appear to end on tanycytes and others end in the external layer of the median eminence (based on data of Cuello et al., 1972a).

Inhibition of CRF secretion

We first became interested in the possibility that a central adrenergic system inhibited ACTH secretion through our study of α-ethyltryptamine. This compound had been shown by Tullner and Hertz (1964) to inhibit stress-induced ACTH secretion in dogs. We confirmed their observations and found that amphetamine, methamphetamine, L-dopa, and a number of other catecholamine-releasing compounds shared this property (Table III). The time course of the inhibition produced by α-ethyltryptamine is shown in Figure 5. In this and other experiments, the index of ACTH secretion used is the output of 17-hydroxycorticoids from the adrenal. In the case of the rat, the index is the peripheral plasma concentration of corticosterone. The decrease in corticoid output produced by α-ethyltryptamine and related drugs is not due to an action at the adrenal level, because exogenous ACTH overcomes it.

That the action of the ACTH-inhibiting compounds is central rather than peripheral is indicated by the fact that, upon systemic administration, neither norepinephrine nor dopamine inhibited ACTH secretion; unlike the other agents, these amines do not cross the blood-brain barrier. In addition, it was found that guanethidine, a drug that blocks catecholamine release and does not cross the blood-brain barrier, failed to increase ACTH secretion upon systemic administration in rats but did when administered directly into the third ventricle. Small doses of the catecholamine-depleting agent α-methyl-p-tyrosine had a similar effect upon intraventricular administration. Large doses administered systemically also increased plasma corticosterone, and this effect was largely overcome by simultaneous administration of L-dopa.

We attempted to determine the site of action of

TABLE III

Sympathomimetic agents that inhibit stress-induced ACTH secretion in pentobarbital-anesthetized dogs (from Ganong, 1972a)

α-Ethyltryptamine
α-Methyltryptamine
Amphetamine
Methamphetamine
2-Aminoheptane
Clopane
L-dopa

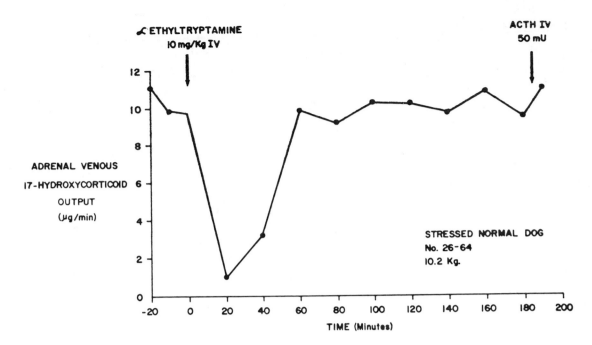

FIGURE 5   Time course of inhibiting effect of α-ethyltryptamine on ACTH secretion, as shown by changes in output of 17-hydroxycorticoids in the adrenal vein of a dog subjected to continuous surgical stress.

aminergic inhibition by stimulating various brain regions known to increase ACTH secretion and seeing if this increase was overcome by α-ethyltryptamine (Ganong et al., 1965). We found that stimulation along the dorsal longitudinal fasciculus and mammillary peduncle produced an increase that was inhibited, while stimulation of the hypothalamus produced an increase that was not (Figure 6). Subsequently, we found that the increase in ACTH secretion produced by stimulation of the amygdala was also blocked by α-ethyltryptamine. These results suggest that the site of inhibition is not the CRF-secreting cells themselves, but the pathways that converge on them

or the junctions between the afferent pathways and the CRF-secreting cells.

To investigate whether norepinephrine or dopamine mediated the inhibition of ACTH, we treated rats with the dopamine-β-hydroxylase inhibitor, FLA-63 [bis-(1-methyl-4-homopiperazinyl-thiocarbonyl)-disulphide]. Hypothalamic norepinephrine content was reduced, while dopamine content was unaffected and plasma corticosterone was increased. Conversely, treatment with α-methyl-p-tyrosine and dihydroxyphenylserine (DOPS) produced depletion of dopamine with relatively little depletion of norepinephrine, and corticosterone was not as

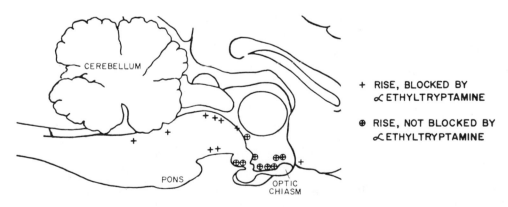

FIGURE 6   Location of points at which stimulation increased 17-hydroxycorticoid secretion, projected on a midsagittal section of the brain of the dog. Plus signs indicate a rise blocked by α-ethyltryptamine, circled plus signs a rise not blocked by α-ethyltryptamine (from Ganong et al., 1965).

TABLE IV

Treatment	Hypothalamic Concentration		Plasma Corticosterone Concentration
	Norepinephrine	Dopamine	
α-methyl-p-tyrosine	↓	↓	↑
α-methyl-p-tyrosine and L-dopa	N	N	N
FLA-63	↓	N	↑
α-methyl-p-tyrosine and DOPS	N	↓	N

↑ increased
↓ decreased
N normal (or more nearly normal than with α-methyl-p-tyrosine alone)

high as it was with α-methyl-p-tyrosine alone. These results, which are summarized in Table IV, indicate that the probable mediator is norepinephrine rather than dopamine.

The problem of the receptor for the inhibitory effect on ACTH secretion has been approached by injecting blocking agents directly into the third ventricle and administering L-dopa systemically. The β-adrenergic blocking agent propranolol had no effect in this situation (Table V), but relatively small amounts of α-adrenergic blocking agent, phenoxybenzamine, inhibited the inhibitory effect of L-dopa. Surprisingly, phentolamine had no effect, but there may be precedent for this; it has been reported that dibenamine, a close relative of phenoxybenzamine, blocks

the release of pituitary hormones that cause ovulation, while phentolamine does not (Sawyer, 1963). In addition, Scapagnini, in the following chapter, finds that phentolamine increases ACTH secretion in the rat. Phenoxybenzamine in high concentrations has anticholinergic activity so anticholinergic agents were tested. The muscarinic blocker atropine had no effect. Except in the case of propranolol, negative experiments were repeated, using larger doses of blocking agents. The inhibitory effect of L-dopa in dogs treated with the nicotinic blocking agent pentolinium may have been reduced, but the inhibition was still appreciable. Thus it appears that the receptor involved is an α-receptor.

TABLE V

Effect of adrenergic and cholinergic blocking drugs in the third ventricle on the adrenal response to surgical stress and L-dopa (50 mg/kg)

Blocker Third Ventricle	17-Hydroxycorticoid Output (μg/min)†			
	Stress	Stress + Blocker	Stress + Blocker + L-dopa	ACTH
Phenoxybenzamine 0.04,* 0.01‡	8.5 ± 1.5	8.0 ± 1.4	6.6 ± 0.7	8.4 ± 0.7
Phentolamine 0.04-0.4, 0.07-0.27	11.0 ± 1.2	6.4 ± 2.0	0.7 ± 0.5**	12.0 ± 1.7
Propranolol 0.05, 0.01	8.5 ± 1.2	9.9 ± 0.6	2.5 ± 1.1**	10.2 ± 0.9
Atropine 0.05, 0.025	10.6 ± 0.9	8.3 ± 1.1	2.7 ± 1.0**	10.3 ± 0.9
Pentolinium 0.3, 0.7-0.9	9.3 ± 0.8	9.9 ± 0.7	5.8 ± 1.5	10.7 ± 0.7

†The values are means ± standard errors of 17-hydroxycorticoid output in the right adrenal vein in pentobarbital-anesthetized dogs 5 min after laparotomy.
*Load dose, mg/kg.
‡Constant infusion dose, mg/kg/hr.
**Significantly lower than stress value, $P < 0.05$.

TABLE VI

Effect of 6-hydroxydopamine* in the third ventricle on resting plasma corticosterone levels
and hypothalamic norepinephrine and dopamine concentration in rats

| | Days after Last Injection of 6-Hydroxydopamine or Sodium Bromide Control | Hypothalamic Concentration | | Plasma |
		Norepinephrine (μg/g)	Dopamine (μg/g)	Corticosterone (μg/100 ml)
6-0HDA	1	0.22 ± 0.07	—	28.8 ± 6.3
control	1	1.99 ± 0.15	—	11.2 ± 2.1
6-0HDA	15	0.19 ± 0.03	0.42 ± 0.03	8.3 ± 1.5
control	15	1.13 ± 0.09	0.41 ± 0.03	5.5 ± 0.8

*2 Doses of 200 μg 24 hr apart.

Effects of 6-hydroxydopamine on ACTH secretion

An obvious next step was to test the effect of 6-hydroxydopamine on ACTH secretion. This compound, which destroys large numbers of catecholamine-containing neurons, was injected into the third ventricle in rats by Shoemaker and Cuello in my laboratory. Twenty-four hours after injection, there was an 89% decrease in hypothalamic norepinephrine content (Table VI) and a significant elevation in plasma corticosterone concentration. Fifteen days after injection, there was an 83% decrease in hypothalamic norepinephrine, but plasma corticosterone concentration had returned to normal. Similar results have been obtained by Scapagnini (see the following chapter). It is worth noting in this regard that although treatment with α-methyl-p-tyrosine acutely elevates plasma corticosterone, chronic treatment does not cause adrenal hypertrophy in rats (Weiner and Ganong, 1971).

We have studied the earliest changes produced by 6-hydroxydopamine upon constant infusion into the third ventricle of anesthetized dogs. The initial response to this infusion is inhibition of stress-induced ACTH secretion (Figure 7). This result is a reasonable one, because it is

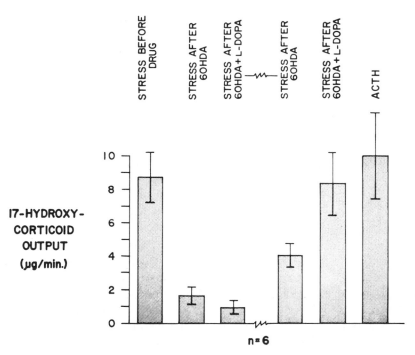

FIGURE 7   Effect of 6-hydroxydopamine (6-OHDA) on stress-induced 17-hydroxycorticoid secretion and on the response to L-dopa in the dog. The animals were tested 30 min (2nd and 3rd bars) and 130 min (4th and 5th bars) after the start of a constant infusion of 6-hydroxydopamine into the 3rd ventricle. The dogs weighed approximately 10 kg, and each received 500 μg of 6-hydroxydopamine followed by 25 μg/min.

known that the initial response to 6-hydroxydopamine is release of catecholamines contained in adrenergic neurons (Malmfors and Thoenen, 1971). These acute experiments have been repeated in rats, and the results are similar. However 2 and 4 hours after the start of the infusion in the dog, the response to stress is returning toward normal and the inhibitory effect of L-dopa is significantly reduced. Thus, the response to 6-hydroxydopamine appears to be triphasic: Initial inhibition of ACTH secretion, presumably due to release of norepinephrine from neurons; subsequent transient increased ACTH secretion and refractoriness to L-dopa, presumably due to diffuse disruption of the function of the adrenergic neurons; and finally a return of ACTH secretion to normal. The explanation of the return of normal functions may be that a small number of surviving adrenergic neurons are sufficient to maintain normal ACTH secretion. Another possibility is that there is a pool of catecholamines that with time becomes resistant to 6-hydroxydopamine and α-methyl-p-tyrosine. A third possibility is that the action of the adrenergic system on ACTH secretion is phasic, and that other mechanisms take over in the absence of adrenergic function.

Adrenergic effects on the secretion of other hormones

Central adrenergic systems also affect the secretion of the other anterior pituitary hormones. In man, L-dopa increases growth hormone secretion (Frantz, 1973), and we have confirmed this finding in the dog. There is considerable evidence in various species that secretion of this hormone is increased by a neural system in which norepinephrine is a mediator and an α-adrenergic receptor is involved. The system resembles that involved in the control of ACTH in that stimulation of the dorsal hippocampus and amygdala produce an increase in growth hormone secretion that is blocked by α-methyl-p-tyrosine and reserpine, whereas direct stimulation of the hypothalamus produces an increase that is not (Martin, 1972a, 1972b).

At least in some situations, regulation of FSH and LH secretion is also affected by an adrenergic system in which norepinephrine is the apparent mediator. For example, in experiments reported by Kalra and McCann (1972), stimulation of the medial preoptic area produced an increase in LH secretion that was blocked by α-methyl-p-tyrosine or a dopamine-β-hydroxylase inhibitor, but restored in α-methyl-p-tyrosine-treated rats given DOPS. However, stimulation of the median eminence-arcuate area produced an LH increase that was unaffected by α-methyl-p-tyrosine. Dopamine had previously been reported to increase LH and FSH secretion, but this response was reported to be blocked by phenoxyben-

zamine (Kamberi et al., 1970), and this agent is a better norepinephrine than dopamine blocker (Fuxe and Hökfelt, 1970).

To date, TSH has not been as extensively studied, but there is evidence suggesting that its secretion is increased by norepinephrine (see Ganong, 1972c).

Adrenergic discharge inhibits prolactin secretion, but the inhibition appears to be mediated by dopamine. In rats and humans, prolactin secretion is inhibited by L-dopa and markedly stimulated by chlorpromazine and related phenothiazine tranquilizers. However, dopamine-β-hydroxylase inhibitors do not affect prolactin secretion (Donoso et al., 1971). There is some evidence that the effect of dopamine on prolactin secretion is due to increased secretion of PIF. Prolactin is probably unique among anterior pituitary hormones in that, with disruption of the connections between the hypothalamus and the pituitary, its secretion rate rises. Prolactin secretion is decreased by addition of hypothalamic tissue to pituitaries incubated in vitro, and chlorpromazine prevents this decrease, while the drug has no effect on the pituitary alone (Danon et al., 1968).

It is interesting in terms of the developing nervous system that catecholamines appear to be involved in the regulation of FSH, LH, and prolactin secretion before as well as after puberty (Müller et al., 1972). Puberty is a complex neuroendocrine event that may occur because there is a decrease in the sensitivity of the hypothalamus to the negative feedback effects of gonadal steroid hormones on gonadotropin secretion (see Critchlow and Bar-Sella, 1967). However, aminergic mechanisms do not appear to play a decisive role in the onset of puberty. Reserpine has been reported to inhibit the onset of puberty, but it also decreases food intake, and Weiner and I (1971) found that puberty was equally delayed without a change in brain norepinephrine by simply reducing the food intake of untreated rats to that of reserpine-treated animals. Disulfarim and p-chlorophenylanine also delay puberty, but again their effect appears to be due to decreased food intake rather than a change in brain concentration of norepinephrine or serotonin (Scapagnini, Baker, Moberg, Kragt, and Ganong, unpublished observations). Furthermore, chronic treatment with α-methyl-p-tyrosine decreases brain norepinephrine concentration but has no effect on food intake or the onset of puberty (Weiner and Ganong, 1971).

The presently available data on amine regulation of anterior pituitary secretion are summarized in Table VII. I have also included some tentative conclusions about posterior pituitary and intermediate lobe secretion and the effects of amines on other endocrine glands. The data on the other glands and the posterior and intermediate lobes of the pituitary are discussed elsewhere (Ganong,

TABLE VII

Known and postulated catecholamine-mediated effects on endocrine glands. In the case of anterior pituitary hormones, the changes are secondary to changes in the secretion of hypothalamic releasing and inhibiting factors

Secretion Stimulated			Secretion Inhibited		
Hormone	Mediator	Receptor	Hormone	Mediator	Receptor
FSH and LH	n	α	Prolactin	da	?
GH	n	α	ACTH	n	α
TSH	n	?			
Oxytocin	?	?	Vasopressin	n	?
			MSHs	da	?
Pineal*	n,e	β			
Juxtaglomerular cells	n,e	β			
β Cells of pancreatic islets	n,e	β	β Cells of pancreatic islets	n,e	α

*Melatonin synthesis.

n, norepinephrine; e, epinephrine; da, dopamine.

1972c). The table serves to emphasize the interesting observation that most of the effects on pituitary secretion appear to be mediated via α-adrenergic receptors, whereas the excitatory effects on peripheral endocrine systems are mediated via β-receptors.

Mechanism of adrenergic effects

How do the catecholamines exert their effects on anterior pituitary secretion? In the case of norepinephrine inhibition of ACTH secretion, there are at least four possibilities (Figure 8). The dopamine-containing neurons, and possibly some of the norepinephrine-containing neurons, end on the capillary loops of the portal vessels in the median eminence. Norepinephrine could be released into these vessels and act directly on the hormone-secreting cells of the anterior pituitary. However, addition of norepinephrine directly to anterior pituitary tissue in vitro has no direct effect on ACTH secretion and has even been reported to facilitate the action of CRF (Saffran et al., 1955). Furthermore, systemically administered norepinephrine, which should penetrate the anterior pituitary gland with ease, has no effect on stress-induced ACTH secretion in dogs (Ganong, 1972a, 1972b). Thus, a direct action on the pituitary seems unlikely.

Another possibility is an action on portal blood flow. A mass constriction of the portal vessels with reduced transport of CRF to the pituitary seems unlikely, because norepinephrine increases secretion of other hormones while inhibiting ACTH secretion. Furthermore, we have found that very large doses of the vasoconstrictor peptide angiotensin II have no inhibitory effect on ACTH

secretion when administered into the third ventricle in dogs (Caren, Boryczka, and Ganong, unpublished observations). However, there is some evidence for anatomic localization within the median eminence of the zones where each of the releasing factors is secreted, and constriction or dilatation of the capillary loops in a particular area might increase or decrease selectively the input of the

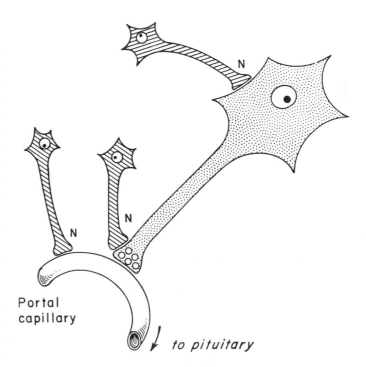

FIGURE 8 Possible sites at which norepinephrine could be acting to decrease ACTH secretion (redrawn from Ganong, 1972b).

releasing factor secreted in that area to the portal blood. Against this conclusion is the fact that hypothalamic stimulation overcomes the blockade of ACTH secretion produced by α-ethyltryptamine (see above); this inhibition should still be present if it was due to limited access of CRF to the portal blood.

A third possibility is modification of the release of CRF into the portal vessels from the ends of the secretory elements in the hypothalamus. If the CRF-secreting elements are neurons, this could be by way of a process akin to presynaptic inhibition. Definite axo-axonal synapses have not been seen in the median eminence despite a diligent search, but it is possible that the effects could be mediated without an anatomically distinct synaptic junction. Somewhat against this possibility is the previously mentioned fact that hypothalamic stimulation overcomes the inhibition; one would expect that inhibition at the secretory terminals would still block release in the stimulated animal. In addition, adrenergic agents that fail to cross the blood-brain barrier are ineffective in inhibiting ACTH secretion in the dog when administered systemically, and the ventral portion of the median eminence where such axo-axonal endings would be expected to be located is "outside the blood-brain barrier" (Wislocki and King, 1936).

The fourth and apparently most likely possibility is termination of inhibitory norepinephrine-containing neurons on the cell bodies of the CRF-secreting neurons, or at least on the nonterminal portions of the CRF-secreting cells. Such an explanation would fit with the stimulation experiments and with the failure of amines, which do not penetrate the blood-brain barrier to inhibit ACTH secretion. The precedent of the supraoptic neurons may be pertinent; these neurons have a relatively dense adrenergic innervation on their cell bodies, and it has been shown that iontophoretic application of norepinephrine onto single supraoptic neurons frequently produces inhibitory electrical changes (Barker et al., 1971; Cross, 1973).

Much the same reasoning applies to the secretion of the releasing factors affecting FSH, LH, growth hormone, and TSH, except that in these instances, the effects of norepinephrine appear to be excitatory. Excitatory as well as inhibitory effects of norepinephrine on hypothalamic neurons have been reported (Cross, 1973).

In the case of prolactin, dopamine appears to stimulate secretion of PIF. However, although systemically administered L-dopa increases prolactin secretion, systemically administered dopamine does not (Lu et al., 1970). This suggests that the action of dopamine, like that of norepinephrine, may be above the median eminence. If this hypothesis is correct, it will explain a number of endocrine phenomena but it leaves the dopamine-con-taining and the norepinephrine-containing neurons in the median eminence without an apparent function.

Role of serotonin

There remains the question of the role of serotonin in the regulation of anterior pituitary secretion. There is serotonin in the median eminence, but most of the serotonin-containing neurons are more dorsally located (see Ungerstedt, 1971). Scapagnini has been interested in the relation of serotonin to circadian fluctuations in ACTH secretion, and with several associates, he investigated this problem when he was in my laboratory. In the rat, a nocturnal animal, concentration is low in the morning and high in the evening (Figure 9). There is a similar but reversed pattern of cortisol secretion in humans with the peak in the early morning. There is evidence that the circadian fluctuation in ACTH secretion is controlled by a "biological clock" in the limbic structures of the brain. Scapagnini et al., (1971) demonstrated a circadian fluctuation in the serotonin concentration of the brain stem and limbic region in rat and showed that this fluctuation parallels the plasma corticosterone concentration. There was no such circadian fluctuation in hypothalamic serotonin. In rats treated with the serotonin-depleting agent, p-chlorophenylanine, brain serotonin was reduced and the plasma corticosterone fluctuations were abolished. Furthermore, the plasma corticosterone concentration was constant throughout the day at levels that were higher than normal in the morning but lower than normal in the evening (Figure 9). A similar flat curve is produced by deafferentation of the hypothalamus (Halász, 1969). Additional evidence that the mechanism responsible for the normal circadian fluctuation is in the limbic area is provided by the observation that section of the fornix also produces intermediate, constant corticosterone levels (Moberg et al., 1971; and Figure 10). Scapagnini has pursued this matter further and has found that lesions of the raphe nuclei also abolish the circadian fluctuation in corticosterone (Scapagnini, personal communication). The relation of changes in plasma corticosterone levels to changes in sleep-wakefulness cycles needs to be studied further. There is scattered evidence indicating that serotonin and related indole amines affect the secretion of FSH, LH, and prolactin, but no clear case has yet been made for a physiological role of serotonin in the regulation of gonadal function.

Conclusion

This review has focused primarily on the neuroendocrine control of pituitary secretion, and within that topic on the regulation of anterior pituitary secretion. Special atten-

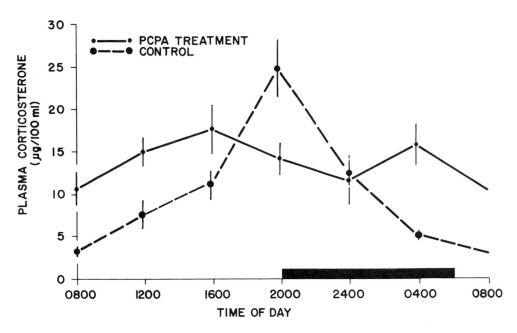

FIGURE 9   Effect of p-chlorophenylalanine (PCPA) treatment on plasma corticosterone concentration in rats. The black bar on the horizontal axis represents the period of darkness. Vertical bars are standard errors (from Scapagnini et al., 1971).

tion has been paid to the relation of neurons containing norepinephrine, dopamine, and serotonin to the regulatory processes. It is now established that a family of anterior pituitary-regulating peptides is secreted in the hypothalamus and passes to the pituitary via the portal-hypophyseal vessels. Two of these releasing and inhibiting factors, a tripeptide and decapeptide, have been characterized and synthesized, and it has been suggested that

their synthesis is enzymatic rather than ribosomal. The cells that secrete the factors are probably neurons, but they could be ependymal or other cells. The secretion of the factors is controlled by feedback of hormones secreted by the regulated glands, and by converging neural inputs. In these inputs, neurons containing norepinephrine and dopamine play a prominent role. Norepinephrine-containing neurons appear to inhibit the secretion of CRF

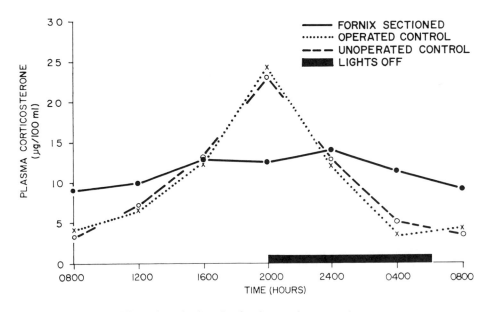

FIGURE 10   Effect of sectioning the fornix on plasma corticosterone in rats. See legend for Figure 9 (data from Moberg et al., 1971).

and to stimulate the secretion of LRF, FRF, GRF, and TRF. Prolactin secretion is apparently inhibited by dopamine-containing neurons, probably because they increase PIF secretion. The role of serotonin is less well defined, but circuits in which it is a mediator appear to be involved in the patterning of endocrine secretion.

ACKNOWLEDGMENT This paper includes reports of research, some previously unpublished, supported by USPHS Grants AM06704 and AM05613. The research was conducted with numerous collaborators, including particularly Drs. Umberto Scapagnini, Augusto Cuello, and William Shoemaker, Mr. Roy Shackelford, and Miss Angela Boryczka.

REFERENCES

ALLEN, C. F., J. P. ALLEN, and M. A. GREER, 1972. Diencephalic localization of the pathways through which a leg break stimulates ACTH secretion (Abstract). *Excerpta Medica International Congress Series* 256:202.

ALLEN, J. P., C. F. ALLEN, and M. A. GREER, 1972. Two discrete spinal cord pathways through which stress stimulates ACTH secretion (Abstract). *Excerpta Medica International Congress Series* 256:202.

BARKER, J. L., J. W. CRAYTON, and R. A. NICOLL, 1971. Supraoptic neurosecretory cells: Adrenergic and cholinergic sensitivity. *Science* 171:208–209.

BLOOM, F. E., L. L. IVERSEN, and F. O. SCHMITT, 1970. Macromolecules in synaptic function. *Neurosciences Research Program Bull.* 8:327–455.

CHENG, K. W., J. B. MARTIN, and H. G. FRIESEN, 1972. Studies of neurophysin release. *Endocrinology* 91:177–184.

CRITCHLOW, V., and M. E. BAR-SELLA, 1967. Control of the onset of puberty. In *Neuroendocrinology*, Vol. II, L. Martini, and W. F. Ganong, eds. New York: Academic Press, pp. 101–162.

CROSS, B. A., 1973. Unit responses in the hypothalamus. In *Frontiers in Neuroendocrinology, 1973,* W. F. Ganong, and L. Martini, eds. New York: Oxford University Press (in press).

CUELLO, A. C., R. I. WEINER, and W. F. GANONG, 1972a. Effect of lateral deafferentation on the norepinephrine concentration and ultrastructure of the mediobasal hypothalamus (Abstract). *Excerpta Medica International Congress Series* 256:22.

CUELLO, A. C., W. F. GANONG, and J. DE GROOT, 1972b. Effect of 6-hydroxydopamine on axons in the median eminence of the rat (Abstract). *Anat. Rec.* 172:297.

DANIEL, P. M., 1966. The anatomy of the hypothalamus and pituitary gland. In *Neuroendocrinology*, Vol. I, L. Martini, and W. F. Ganong, eds. New York: Academic Press, pp. 15–80.

DANON, A., S. D. DIKSTEIN, and F. G. SULMAN, 1968. Stimulation of prolactin secretion by perphenazine in pituitary-hypothalamus organ culture. *Proc. Soc. Exp. Biol. Med.* 114:366–368.

DONOSO, A. O., W. BISHOP, C. P. FAWCETT, L. KRULICH, and S. M. McCANN, 1971. Effects of drugs that modify brain monoamine concentrations on plasma gonadotropin and prolactin levels in the rat. *Endocrinology* 89:774–784.

FRANTZ, A., 1973. The regulation of prolactin secretion in humans. In *Frontiers in Neuroendocrinology, 1973,* W. F. Ganong, and L. Martini, eds. New York: Oxford University Press (in press).

FUXE, K., and T. HÖKFELT, 1970. Central monoaminergic systems and hypothalamic function. In *The Hypothalamus,* L. Martini, M. Motta, and F. Fraschini, eds. New York: Academic Press, pp. 123–138.

GANONG, W. F., 1970. Control of adrenocorticotropin and melanocyte-stimulating hormone secretion. In *The Hypothalamus,* L. Martini, M. Motta, and F. Fraschini, eds. New York: Academic Press, pp. 313–335.

GANONG, W. F., 1972a. Brain amines and ACTH secretion. *Excerpta Medica International Congress Series* 219:814–821.

GANONG, W. F., 1972b. Evidence for a central noradrenergic system that inhibits ACTH secretion. In *Brain-Endocrine Interaction,* K. M. Knigge, D. E. Scott, and A. Weindl, eds. Basel, Switzerland: S. Karger, pp. 254–266.

GANONG, W. F., 1972c. Pharmacological aspects of neuroendocrine integration. *Progr. Brain Res.* (in press).

GANONG, W. F., 1973. *Review of Medical Physiology,* 6th ed. Los Altos, California: Lange Medical Publications (in press).

GANONG, W. F., B. L. WISE, R. SHACKELFORD, A. T. BORYCZKA, and B. ZIPF, 1965. Site at which α-ethyltryptamine acts to inhibit the secretion of ACTH. *Endocrinology* 72:526–530.

GARLAND, J. T., and W. J. DAUGHADAY, 1972. Feedback inhibition of pituitary growth hormone in rats infected with *Spirometra mansonoides. Proc. Soc. Exp. Biol. Med.* 139:497–499.

GIBBS, F. P., 1969. Central nervous system lesions that block the release of ACTH by traumatic stress. *Amer. J. Physiol.* 217:78–83.

GREER, M. A., C. F. ALLEN, F. P. GIBBS, and K. GULLICKSON, 1970. Pathways at the hypothalamic level through which traumatic stress activates ACTH secretion. *Endocrinology* 86:1404–1409.

HALÁSZ, B., 1969. The endocrine effects of isolation of the hypothalamus from the rest of the brain. In *Frontiers in Neuroendocrinology, 1969,* W. F. Ganong, and L. Martini, eds. New York: Oxford University Press, pp. 307–342.

HARRIS, G. W., 1955. *Neural Control of the Pituitary Gland.* London: Edward Arnold.

HARRIS, G. W., 1964. Sex hormones, brain development, and brain function. *Endocrinology* 75:627–648.

HÖKFELT, T., and K. FUXE, 1972. On the morphology and the neuroendocrine role of the hypothalamic catecholamine neuron. In *Brain-Endocrine Interaction,* K. M. Knigge, D. E. Scott, and A. Weindl, eds. Basel, Switzerland: S. Karger, pp. 181–223.

HRUBY, V. J., C. W. SMITH, A. BOWER, and M. E. HADLEY, 1972. Melanophore-stimulating hormone: Release inhibition by ring structures of neurohypophysial hormone. *Science* 176:1331–1332.

KALRA, S. P., and S. M. McCANN, 1972. Modification of brain catecholamine levels and LH release by preoptic stimulation (Abstract). *Excerpta Medica International Congress Series* 256:202–203.

KAMBERI, I. A., R. S. MICAL, and J. C. PORTER, 1970. Possible role of α-adrenergic receptors in mediating the response of the hypothalamus to dopamine (Abstract). *Physiologist* 13:329.

KASTIN, A. J., N. P. PLOTNIKOFF, T. W. REDDING, R. M. G. NAIR, and M. S. ANDERSON, 1973. MSH-release inhibiting factor. In *Proceedings of Serono Foundation Conference on Hypothalamic Hypophysiotropic Hormones,* C. Gual, and E. Rosenberg, eds. (in press).

KNOBIL, E., D. J. DIERSCHKE, T. YAMAJI, F. J. KARSCH, J. HOTCHKISS, and R. F. WEICK, 1973. Role of estrogen in the positive and negative feedback control of luteinizing hormone secretion during the menstrual cycle of the rhesus monkey. In *International Symposium on Gonadotropins*, B. B. Saxena, ed. New York: John Wiley & Sons (in press).

KOBAYASHI, H., and T. MATSUI, 1969. Fine structure of the median eminence and its functional significance. In *Frontiers in Neuroendocrinology, 1969*, W. F. Ganong, and L. Martini, eds. New York: Oxford University Press, pp. 3–46.

LU, K.-H., Y. AMENOMORI, C.-L. CHEN, and J. MEITES, 1970. Effects of central acting drugs on serum and pituitary prolactin levels in rats. *Endocrinology* 87:667–672.

MALMFORS, T., and H. THOENEN, eds., 1971. *6-Hydroxydopamine and Catecholamine Neurons*. Amsterdam: North Holland Publishing Company.

MARTIN, J. B., 1972a. Mechanisms of GH release induced by electrical stimulation in the rat (Abstract). *Excerpta Medica International Congress Series* 256:204.

MARTIN, J. B., 1972b. Plasma growth hormone (GH) response to hypothalamic or extrahypothalamic electrical stimulation. *Endocrinology* 91:107–115.

MOBERG, G. P., U. SCAPAGNINI, J. DE GROOT, and W. F. GANONG, 1971. Effect of sectioning of the fornix on diurnal fluctuation and plasma corticosterone levels in the rat. *Neuroendocrinology* 7:11–15.

MOTTA, M., F. FRASCHINI, and L. MARTINI, 1969. "Short" feedback mechanisms in the control of anterior pituitary function. In *Frontiers in Neuroendocrinology, 1969*, W. F. Ganong, and L. Martini, eds. New York: Oxford University Press, pp. 211–254.

MÜLLER, E. E., D. COCCHI, A. VILLA, F. ZAMBOTTI, and F. FRASCHINI, 1972. Involvement of brain catecholamines in the gonadotropin-releasing mechanism(s) before puberty. *Endocrinology* 90:1267–1276.

NOBIN, A., A. BJÖRKLUND, and U. STENEVI, 1972. Origin of the hypophyseal catecholamine innervation (Abstract). *Excerpta Medica International Congress Series* 256:50.

PIERCE, J. G., 1971. The subunits of pituitary thyrotropin—their relationship to other glycoprotein hormones. *Endocrinology* 89:1331–1344.

QUIJADA, M., L. KRULICH, C. P. FAWCETT, D. K. SUNDBERG, and S. M. MCCANN, 1971. Localization of TSH-releasing factor (TRF), LH-RF and FSH-RF in rat hypothalamus (Abstract). *Fed. Proc.* 30:197.

REICHLIN, S., and M. MITNICK, 1973. Biosynthesis of hypothalamic hypophysiotropic factors. In *Frontiers in Neuroendocrinology, 1973*, W. F. Ganong, and L. Martini, eds. New York: Oxford University Press (in press).

SAFFRAN, M., A. V. SCHALLY, and B. G. BENFEY, 1955. Stimulation of the release of corticotropin from the adenohypophysis by an neurohypophysial factor. *Endocrinology* 57:439–444.

SAWYER, C. H., 1963. Discussion. In *Advances in Neuroendocrinology*, A. V. Nalbandov, ed. Urbana, Illinois: University of Illinois Press, pp. 445–457.

SCAPAGNINI, U., G. P. MOBERG, G. R. VAN LOON, J. DE GROOT, and W. F. GANONG, 1971. Relation of brain 5-hydroxytryptamine content to the diurnal variation in plasma corticosterone in the rat. *Neuroendocrinology* 7:90–96.

TULLNER, W. W., and R. HERTZ, 1964. Suppression of corticosteroid production in the dog by Monase. *Proc. Soc. Exp. Biol. Med.* 116:837–840.

UNGERSTEDT, U., 1971. Stereotaxic mapping of the monoamine pathways in the rat brain. *Acta Physiol. Scand.* 367:1–95.

VALE, W., G. GRANT, and R. GUILLEMIN, 1973. Chemistry of the hypothalamic releasing factors—studies on structure function relationships. In *Frontiers in Neuroendocrinology, 1973*, W. F. Ganong, and L. Martini, eds. New York: Oxford University Press (in press).

WEINER, R. I., and W. F. GANONG, 1971. The effect of the depletion of brain catecholamines on puberty and the estrous cycle in the rat. *Neuroendocrinology* 8:125–135.

WEINER, R. I., J. E. SHRYNE, R. A. GORSKI, and C. H. SAWYER, 1972. Changes in the catecholamine content in the rat hypothalamus following deafferentation. *Endocrinology* 90:867–873.

WISLOCKI, G. B., and L. S. KING, 1936. The permeability of the hypophysis and hypothalamus to vital dyes, with the study of the hypophyseal vascular supply. *Amer. J. Anat.* 58:421–472.

49 Pharmacological Studies of Brain Control over ACTH Secretion

UMBERTO SCAPAGNINI

ABSTRACT   The chapter gives evidence in favor of a central noradrenergic system that tonically inhibits ACTH secretion via α-receptor stimulation. In rats, drugs that decrease brain catecholamine levels provoke adrenocortical activation, even when injected directly into the third ventricle at systemically ineffective doses. Specific depletion of norepinephrine but not of dopamine causes an increase of plasma corticosterone. Blockade of α, but not of β receptors removes the tonic noradrenergic inhibition. Finally, experiments concerning long-term catecholamine depletion suggest the presence of a small, defined, functional pool of norepinephrine responsible for the tonic inhibition of ACTH secretion.

THE POSSIBILITY that a monoaminergic system may take part in the central mechanism that controls the secretion of anterior pituitary hormones has produced, in recent years, a great deal of interest. Although many results have been reported in favor of a catecholaminergic or of an indolaminergic role in the regulation or modulation of tonic or phasic secretion of hypophyseal trophins throughout hypothalamic releasing factors, the matter still appears to be unclear and in evolution.

In presenting the results dealing with the catecholaminergic involvement in the regulation of corticotropin releasing factor (CRF) and ACTH secretion, I want to stress the fact that each single pharmacologic experiment can be criticized and interpreted in many ways but that from the complex of many experiments one can gain an overall view that may elucidate a physiologic system.

Several groups of investigators have reported increased adrenal corticosteroid secretion following the injection of catecholamines directly into the brain (Naumenko, 1968; Krieger and Krieger, 1965, 1970). On the other hand, numerous studies have failed to support an excitatory role of brain catecholamines in the regulation of ACTH secretion (Smelik, 1967; Carr and Moore, 1968; De Schaepdryver et al., 1969).

UMBERTO SCAPAGNINI   Department of Pharmacology, University of Naples, Naples, Italy

It is difficult to interpret these discordant results, but it is worth remembering that most of them have been obtained by performing microinjections of transmitters or depletors and obtaining, therefore, a very high or very low concentration of amines in a very unphysiological way. A reappraisal of the literature suggests a different hypothesis as a basic one for an interaction between brain catecholamines and ACTH secretion.

Reserpine does not release active catecholamines (Kopin and Gordon, 1963). It produces depletion of brain catecholamines in association with increased secretion of ACTH (Maickel et al., 1961; Munson, 1963) and CRF (Bhattacharya and Marks, 1969a). Also, drugs interfering with the synthesis (Carr and Moore, 1968; Vernikos-Danellis, 1968; Bhattacharya and Marks, 1970) and with the sites of action (De Wied, 1967; Bhattacharya and Marks, 1969a) of brain catecholamines increase the activity of the hypothalamic-pituitary-adrenal axis.

On the contrary, drugs such as amphetamine, which appears to release active catecholamines from nerve endings (Glowinski and Axelrod, 1966), and MAO inhibitors, which decrease catabolism of catecholamines, decrease ACTH secretion (Lorenzen and Ganong, 1967; Hirsch and Moore, 1968; Bhattacharya and Marks, 1969b).

Stresses such as electric shock, hemorrhage, and hypoglycemia are associated with decreased brain norepinephrine content (Bliss et al., 1968; Thierry et al., 1968) and increased ACTH secretion (Lorenzen and Ganong, 1967).

Moreover, the intravenous injection of L-dopa, a catecholamine precursor that crosses the blood-brain barrier (Wurtman, 1966), is able to inhibit, in the dog, adrenocortical activation due to laparotomy stress. Norepinephrine and dopamine, unable to cross the blood-brain barrier, are able to produce the inhibition of the adrenocortical activation only when injected directly into the third ventricle of the dog brain but not when injected systemically. Pretreatment with drugs affecting the available store of endogenous catecholamines modifies the amount of systemic L-dopa or intraventricular norepinephrine necessary to produce the inhibition (Van Loon et al., 1971a, b).

In a first series of experiments (Scapagnini et al., 1970), we found that the intraperitoneal administration of α-methyl-p-tyrosine (α-MpT), an inhibitor of catecholamine synthesis, caused an increase of adrenocortical activity; when L-dopa was injected simultaneously with α-MpT, the mean increase of plasma corticosterone was significantly lower. Moreover, a negative linear correlation was found between plasma corticosterone and hypothalamic concentrations of both norepinephrine and dopamine after α-MpT plus L-dopa (Van Loon et al., 1971c).

The intraventricular administration of a systemically ineffective dose of α-MpT or of guanethidine (a catecholamine depletor that does not cross the blood-brain barrier) produced an elevation of plasma corticosterone and a depletion of hypothalamic norepinephrine and dopamine (Scapagnini et al., 1971). These data suggest that, in the rat, the adrenergic system inhibiting ACTH secretion is a central one.

Using two pharmacologic tools, we depleted selectively brain norepinephrine or dopamine in order to investigate which amine is the mediator of the inhibition.

Four hours after FLA-63, an inhibitor of dopamine-β-oxidase, a fall of brain norepinephrine was present accompanied by adrenocortical activation. Such an activation did not occur when only dopamine was reduced using α-MpT to deplete both catecholamines followed by dihydroxy-phenyl-serine to replete norepinephrine (Scapagnini et al., 1972). Although the specificity of DOPS as norepinephrine repletor is under debate, these findings once more emphasize that large amounts of hypothalamic norepinephrine are needed in order to control the activity of the hypothalamic-pituitary-adrenal axis.

The nature of the adrenergic receptors involved in the ACTH inhibiting system was also considered in our study (Scapagnini and Preziosi, 1972).

In rats the systemic (intraperitoneal) administration of phentolamine, an α-blocking agent, and not of propranolol, a β-blocking agent, is able to produce an increase of the plasma corticosterone level.

The same effect is provoked by the injection into the third ventricle of the brain of a systemically ineffective dose of phentolamine (200 μg/kg).

Finally, further evidence favoring adrenergic α-mediation of the ACTH inhibiting system is provided by the finding that, in laparotomized rats, phentolamine, but not propranolol, is able to remove in part the dexamethasone + iproniazid induced inhibition of ACTH secretion. This kind of inhibition has been demonstrated by Dallman and Yates (1968), who found that the laparatomy-induced adrenocortical activation can be prevented by pretreatment with dexamethasone plus iproniazid, while either drug given alone is ineffective. Our results with the blocking agents indicate that, at least in part, this kind of inhibition is due to the increased levels of catecholamines, because a block of specific adrenergic receptor can remove it.

Furthermore we can speculate that since the increase of catecholamines is potentiating the feedback inhibition due to exogenous glucocorticoids, such a mechanism could be suggested also for the feedback inhibition due to the endogenous release of corticosteroids.

In another group of experiments, we investigated the apparent discrepancy between the above hypothesis and the results obtained by several authors in long-term experiments. In fact, both after single (Maickel et al., 1961; Carr and Moore, 1968) and repeated (Montanari and Stockman, 1962) injections of reserpine, in spite of the prolonged depletion of catecholamines, the adrenocortical activation lasts only for a few hours. In our experiments, daily intraperitoneal (ip) administration in rats of reserpine (0.5 mg/kg for 9 days) strongly depletes hypothalamic norepinephrine while the corticosteroid level progressively decreases to reach the control values at the fifth day of treatment (Table I). Both in peripheral (Kopin, 1966) and central (Segal et al., 1971) studies, the presence of functionally active stores of norepinephrine

TABLE I

Effect of daily treatment with reserpine on plasma corticosterone levels 5 hr after the injection. Values are means ± standard error

Treatment	Plasma Corticosterone (μg/100 ml) at the Following Days								
	1	2	3	4	5	6	7	8	9
Saline, 0.5 ml/kg ip	10.0 ± 1.8	11.2 ± 2.2	9.3 ± 1.9	8.8 ± 2.2	10.3 ± 1.4	11.1 ± 2.0	12.8 ± 1.8	9.5 ± 2.0	12.4 ± 1.3
Reserpine, 0.5 ml/kg ip	23.5* ± 2.0	26.1* ± 3.1	15.9† ± 1.2	18.3† ± 1.8	13.2 ± 2.8	11.9 ± 1.8	11.1 ± 1.3	10.8 ± 1.8	11.3 ± 1.4

*Significantly different from saline injected animals at P < 0.01.
†Significantly different from saline injected animals at P < 0.05.

after reserpine treatment has been suggested; this should be consequent to an increase of tyrosine hydroxylase activity related to feedback stimulation.

In our experiments, long-term treatment with low doses of reserpine progressively increases tyrosine hydroxylase activity in the brain stem (Figure 1). This increase parallels the disappearance of adrenocortical activation.

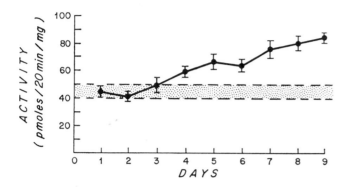

FIGURE 1   Effect of a daily treatment with reserpine (0.5 mg/kg ip for 9 days) on the brain stem tyrosine hydroxylase activity (pmole/20 min/mg).

Furthermore the injection of α-MpT (ip, 9 hr prior to sacrifice) in rats treated for 9 days with reserpine provokes adrenocortical activation and further decrease of hypothalamic norepinephrine; the animals, on the other hand, had a normal rise of plasma corticosterone after exog-enous ACTH (100 mU/kg 25 min prior to sacrifice) revealing an unimpaired adrenocortical reactivity after prolonged treatment with reserpine (Figure 2).

Therefore, in light of these results, we may suggest that the lack of adrenocortical activation in the presence of norepinephrine depletion following long-term treatment with reserpine is due to the presence of a small functional active pool of norepinephrine available for the tonic inhibition of ACTH secretion.

In a similar series of experiments, long-lasting reduction in the brain concentration of catecholamines was obtained by using 6-hydroxydopamine (6-OHDA), a compound that causes a selective degeneration of nerve terminals containing catecholamines (Bloom et al., 1969). This drug, injected into the third ventricle or the lateral ventricle of the rat brain, caused 24 hr later a pronounced reduction of hypothalamic norepinephrine, accompanied by adrenocortical activation. On the other hand, after 3 to 15 or 30 days, in spite of the significant decrease of hypothalamic norepinephrine, the corticosterone levels were unmodified (Table II). The injection of α-MpT (250 mg/kg ip prior to sacrifice) in rats pretreated with 6-OHDA provokes adrenocortical activation (Figure 3). Therefore, even after selective degeneration of catecholamine terminals, there should still be present in the hypothalamus a pool of norepinephrine responsible for the inhibition of the hypothalamus-pituitary-adrenal axis. The availability of a sufficient pool could be due to the fact that those neurons containing norepinephrine that survive the degenerative effects of 6-OHDA, show

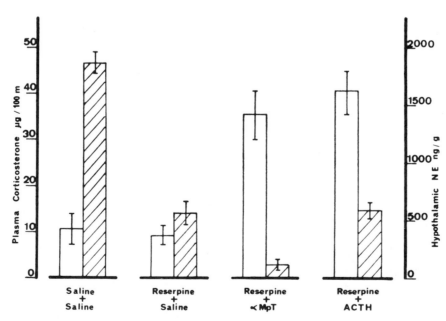

FIGURE 2   Effect of α-MpT (250 mg/kg ip 9 hr prior to sacrifice) and ACTH (100 mU/kg 25 min prior to sacrifice) on plasma corticosterone (open bars) and hypothalamic norepinephrine (dashed bars) in rats pretreated with reserpine (0.5 mg/kg ip for 9 days).

TABLE II

Correlation between plasma corticosterone and hypothalamic content of norepinephrine and dopamine after 6-hydroxydopamine (6-OHDA). Values are means ± standard error

Treatment	Place of Injection	Days after Injection	Plasma Corticosterone (μg/100 ml)	Hypothalamic Content (μg/g)	
				Norepinephrine	Dopamine
6-OHDA 300 μg/ animal + 200 μg/animal after 72 hr	L.V.	1	15.5 ± 1.8*	1.38 ± 0.09*	—
		3	12.8 ± 1.8	1.36 ± 0.08*	0.30 ± 0.03
		15	3.2 ± 2.2	1.01 ± 0.12*	0.28 ± 0.04*
		30	4.2 ± 3.0	0.82 ± 0.07*	—
Saline 30 μl/ animal + 20 μl/animal after 72 hr	L.V.	1	7.2 ± 0.5	1.90 ± 0.09	—
		3	8.8 ± 1.4	1.74 ± 0.11	0.39 ± 0.08
		15	7.9 ± 2.0	1.84 ± 0.09	0.41 ± 0.03
		30	9.2 ± 1.5	1.87 ± 0.10	0.41 ± 0.11
6-OHDA 50 μg/ animal	III V.	1	26.8 ± 2.2*	1.32 ± 0.11*	—
		3	14.8 ± 2.1	1.26 ± 0.10	—
		15	7.3 ± 2.0	1.06 ± 0.07	—
		30	6.6 ± 1.8	0.99 ± 0.06	—
Saline 5 μl/ animal	III V.	1	6.5 ± 1.8	1.86 ± 0.11	—
		3	9.0 ± 1.2	1.96 ± 0.08	—
		15	8.1 ± 1.3	1.88 ± 0.10	—
		30	6.5 ± 1.4	1.89 ± 0.09	—

*Significantly different from corresponding saline control at P < 0.01.
L.V.: lateral ventricle; III V.: third ventricle.

an increase in their physiologic state of activity to compensate for the loss of neuronal function after degeneration (Uretsky et al., 1971).

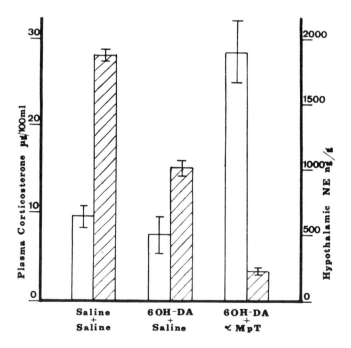

FIGURE 3 Effect of α-MpT (250 mg/kg ip 9 hr prior to sacrifice) on plasma corticosterone (open bars) and hypothalamic norepinephrine (dashed bars) in rats pretreated 30 days before with 6-OHDA (50 μg/animal into the third ventricle).

However we can not disregard the possibility of the appearance of receptor supersensitivity following the functional denervation due to 6-OHDA (see Ungerstedt in this volume).

In summary, pharmacologic evidence seems to suggest the existence of a central adrenergic system that tonically inhibits ACTH secretion. The physiologic significance of this system is not yet defined; however, our experiments indicate that norepinephrine rather than dopamine is the catecholamine involved and that its inhibitory action is mediated via adrenergic α-receptors. The long-term depletion experiments suggest the presence of a small, defined, functional pool of norepinephrine responsible for the tonic inhibition of ACTH secretion.

ACKNOWLEDGMENTS This paper includes the results of research supported in part from U.S.P.H.S. grant AM 0604 and from the Brooks Fund, and in part, from the Consiglio Nazionale delle Ricerche no. 71.00231.04.

The research was carried out primarily in the laboratory of W. F. Ganong by G. Moberg, U. Scapagnini, and G. R. Van Loon (Department of Physiology, San Francisco Medical Center) and in the laboratory of P. Preziosi by L. Annunziato and U. Scapagnini.

REFERENCES

BHATTACHARYA, A. N., and B. H. MARKS, 1969a. Reserpine and chlorpromazine-induced changes in hypothalamo-hypophyseal-adrenal system in rats in the presence and absence of hypothermia. J. Pharmacol. Exp. Ther. 165:108–116.

BHATTACHARYA, A. N., and B. H. MARKS, 1969b. Effects of pargyline and amphetamine upon acute stress response in rats. *Proc. Soc. Exp. Biol. Med.* 130:1194–1198.

BHATTACHARYA, A. N., and B. H. MARKS, 1970. Effects of alpha-methyltyrosine and *p*-chlorophenylalanine on the regulation of ACTH secretion. *Neuroendocrinology* 6:49–55.

BLISS, E. L., J. AILION, and J. ZWANZIGER, 1968. Metabolism of norepinephrine, serotonin and dopamine in rat brain with stress. *J. Pharmacol. Exp. Ther.* 164:122–134.

BLOOM, F. E., S. ALGERI, A. GROPPETTI, A. REVELTA, and E. COSTA, 1969. Lesion of central norepinephrine terminals with 6-hydroxydopamine: Biochemistry and fine structures. *Science* 166:1284–1286.

CARR, L. A., and K. E. MOORE, 1968. Effects of reserpine and α-methyltyrosine on brain catecholamines and the pituitary adrenal response to stress. *Neuroendocrinology* 3:285–302.

DALLMAN, M. F., and F. E. YATES, 1968. Anatomical and functional mapping of central neural input and feedback pathways of the adrenocortical system. In *Memoirs of the Society for Endocrinology (no. 17)*, V. H. T. James, and J. London, eds. Cambridge: Cambridge University Press, pp. 39–71.

DE SCHAEPDRYVER, A. F., P. PREZIOSI, and U. SCAPAGNINI, 1969. Brain monoamines and stimulation or inhibition of ACTH release. *Archs. Int. Pharmacodyn.* 180:11–18.

DE WIED, D., 1967. Chlorpromazine and endocrine function. *Pharmacol. Rev.* 18:251–288.

GLOWINSKI, J., and J. AXELROD, 1966. Effects of drugs on the disposition of H^3-norepinephrine in the rat brain. *Pharmacol. Rev.* 18:775–785.

HIRSCH, G. H., and K. E. MOORE, 1968. Brain catecholamines and the reserpine-induced stimulation of the pituitary-adrenal system. *Neuroendocrinology* 3:398–405.

KOPIN, I. J., and E. K. GORDON, 1963. Metabolism of administered and drug-released norepinephrine 7-H^3 in the rat. *J. Pharmacol. Exp. Ther.* 140:207–216.

KOPIN, I. J., 1966. Metabolism and disposition of catecholamines in the central and peripheral nervous system. In *Endocrines and Central Nervous System.* (Res. Publ. Ass. Nerv. Ment. Dis.) R. Levine, ed. Baltimore: Williams and Wilkins, pp. 343–353.

KRIEGER, D. T., and H. P. KRIEGER, 1965. The effect of intrahypothalamic injection of drugs on ACTH release in the cat. *Proc. Int. Congr. Endocrinol.*, Excerpta Med. Found., pp. 640–645.

KRIEGER, H. P., and D. T. KRIEGER, 1970. Chemical stimulation of the brain; effect of adrenal corticoid release. *Am. J. Physiol.* 218:1632–1641.

LORENZEN, L. C., and W. F. GANONG, 1967. Effect of drugs related to α-ethyltryptamine on stress-induced ACTH secretion in the dog. *Endocrinol.* 80:889–892.

MAICKEL, R. P., E. O. WESTERMANN, and B. B. BRODIE, 1961. Effects of reserpine and cold exposure on pituitary adrenocortical function in rats. *J. Pharmacol. Exp. Ther.* 134:167–175.

MONTANARI, R., and M. A. STOCKMAN, 1962. Effects of single and repeated doses of reserpine on the secretion of adrenocorticotrophic hormone. *Brit. J. Pharmacol.* 18:337–345.

MUNSON, P. L., 1963. Pharmacology of neuroendocrine blocking agents. In *Advances in Neuroendocrinology*, A. V. Nalbandov, ed.

Urbana: University of Illinois Press, pp. 427–444.

NAUMENKO, E. V., 1968. Hypothalamic chemoreactive structure and the regulation of pituitary-adrenal function. Effects of local injection of norepinephrine, carbachol and serotonin into the brain of guinea pigs with intact brains and after mesencephalic transection. *Brain Res.* 11:1–10.

SCAPAGNINI, U., G. R. VAN LOON, G. P. MOBERG, and W. F. GANONG, 1970. Effects of α-methyl-*p*-tyrosine on the circadian variation of plasma corticosterone in rats. *Eur. J. Pharmacol.* 11:266–270.

SCAPAGNINI, U., G. R. VAN LOON, G. P. MOBERG, P. PREZIOSI, and W. F. GANONG, 1971. Evidence for a central adrenergic inhibition of ACTH secretion in rat. *Archiv. Pharm.* 269:408.

SCAPAGNINI, U., G. R. VAN LOON, G. P. MOBERG, P. PREZIOSI, and W. F. GANONG, 1972. Evidence for a central norepinephrinergic inhibition of ACTH secretion in the rat. *Neuroendocrinol.* 10:155–160.

SCAPAGNINI, U., and P. PREZIOSI, 1972. Receptor involvement in the central noradrenergic inhibition of ACTH secretion in rat. *Neuropharmacol.* (in press).

SEGAL, D. S., J. L. SULLIVAN, R. T. KUCRENSKI, and A. J. MANDELL, 1971. Effects of long-term reserpine treatment on brain tyrosine-hydroxylase and behavioral activity. *Science* 173:847–849.

SMELIK, P. G., 1967. ACTH secretion after depletion of hypothalamic monoamines by reserpine inplants. *Neuroendocrinol.* 2:247–254.

THIERRY, A. M., F. JAVOY, J. GLOWINSKI, and S. S. KETY, 1968. Effects of stress on the metabolism of norepinephrine, dopamine and serotonin in the central nervous system on the rat. I. Modification of norepinephrine turnover. *J. Pharmacol. Exp. Ther.* 163:163–171.

URETSKY, N. J., M. A. SIMMONDS, and L. L. IVERSEN, 1971. Changes in the retention and metabolism of ^{3}H-l-norepinephrine in rat brain *in vivo* after 6-hydroxydopamine pretreatment. *J. Pharmacol. Exp. Ther.* 176:489–496.

VAN LOON, G. R., A. B. HILGER, A. T. KING, A. T. BORITZKA, and W. F. GANONG, 1971a. Inhibitory effect of L-dihydroxyphenylalanine on the adrenal venous 17-hydroxy-corticosteroid response to surgical stress in dogs. *Endocrinol.* 88:1404–1414.

VAN LOON, G. R., U. SCAPAGNINI, R. COHEN, and W. F. GANONG, 1971b. Intraventricular administration of adrenergic drugs on the adrenal venous 17-hydroxycorticosteroid response to surgical stress in the dog. *Neuroendocrinol.* 8:257–272.

VAN LOON, G. R., U. SCAPAGNINI, G. P. MOBERG, and W. F. GANONG, 1971c. Evidence for central adrenergic neural inhibition of ACTH secretion in the rat. *Endocrinol.* 89:1464–1469.

VERNIKOS-DANELLIS, J., 1968. The pharmacological approach to the study of the mechanism regulating ACTH secretion. In *Pharmacology of Hormonal Polypeptides and Proteins*, N. Back, L. Martini, and R. Paoletti, eds. New York: Plenum Press, pp. 175–189

WURTMAN, R. J., 1966. *Catecholamines.* Boston: Little, Brown.

50

Experiential and Hormonal Factors
Influencing Development
of Sexual Behavior
in the Male Rhesus Monkey

ROBERT W. GOY and DAVID A. GOLDFOOT

ABSTRACT   Rearing conditions have been shown to influence the development of sex-related behavior and sexual behavior in rhesus monkeys. When rhesus monkeys were taken from their mothers at 3 months of age and exposed to peers of both sexes for one-half hour per day, the development of mature patterns of mounting behavior occurred during the second and third years of life. Psychological preferences for mounting partners of the opposite sex did not appear until the animals were 3 to 4 years of age. Moreover, many male monkeys reared in this manner failed to display a full pattern of sexual behavior as adolescents and adults. When rearing conditions were altered so that infants were raised in social groups that allowed unrestricted access to both mothers and peers for the first 9 months of age, the development of sexual and sex-related patterns was facilitated. Males reared in this manner displayed the mature mounting pattern at 3 to 4 months of age, and by 6 months of age, they showed marked preference for mounting partners of the opposite sex. One pseudohermaphroditic female, reared in the condition of continuous exposure to mothers and peers, showed a corresponding facilitation of mounting behavior and partner preference. Findings suggest that prenatal hormonal conditions determine the basic psychosexual orientation of the individual. The experiential history of heterosexually inadequate male rhesus monkeys shows marked deficits in the development of mounting behavior, but no measurable deficits in other social behaviors normally associated with the juvenile male gender role.

Introduction

THE DEVELOPMENT of social interaction patterns in young nonhuman primates has been studied both in the laboratory and in natural environments using several approaches and theoretical frameworks. The great interest shown in the area of social development in the monkey is in part due to the rather strong possibility that both monkeys and humans are influenced by similar experiential and hor-

monal variables governing the development of social and sexual behavior. The opportunity to identify specific temporal, experiential, and hormonal factors that affect these responses of nonhuman primates is therefore of importance for clinical psychology, child psychology, and psychiatry, as well as for students of animal behavior.

While the specific social behaviors of monkey and man differ greatly, the similarity of conditions influencing the formation of social relations is remarkable. Both species are severely impaired in their respective societies if normal social skills are not acquired at fairly specific, early ages. Isolation or less severe social deprivation at early ages is a basic source of maladaptation in later years (Bowlby, 1952; Harlow, 1962). The influence of prenatal hormones on later social interactions is similarly of great importance to both species. Abnormally high concentrations of hormones with androgenic properties during fetal development, for example, can markedly influence sexual and social development of genetic female humans as well as genetic female rhesus monkeys. In a series of controlled experiments with rhesus monkeys, it was found that females exposed to testosterone in utero displayed essentially male patterns of socialization, including male-like play, aggression, and sexual behavior (Goy, 1970a, b). The situation for the human is less clear in this regard, but evidence has been presented from case histories that suggests that similar effects occur, at least in terms of the patterns of play shown by masculinized girls (Money and Ehrhardt, 1968). A complicating factor, more obvious in man but also valid for monkey, is that hormonal and experiential factors can interact to produce a wide array of individual social patterns. By using the monkey as an experimental animal, the influence of several experiential factors can be manipulated, and their interactions with hormonal influences can be assessed. The present chapter attempts to identify the effects of two specific rearing

ROBERT W. GOY and DAVID A. GOLDFOOT   Wisconsin Regional Primate Research Center, Madison, Wisconsin conditions on the psychosexual development of the rhesus monkey. Hormonal manipulations include exposure of fetal female monkeys to exogenous testosterone propionate, castration of infant males and females at birth, and a comparison of intact male and female infants with these experimentally altered animals. The major behavior studied in this chapter is the development of mounting by young males. Aspects of play and aggression are also reported, but not in detail, nor with exhaustive analysis of the data. We have concentrated on the male in this chapter, because several researchers have found the male to be more susceptible than the female to traumatic environmental deficits during infancy (Senko, 1966; Jensen, 1970; Harlow and Harlow, 1971; and Sackett, 1972).

Effects of social deprivation on sexual behavior

In those species dependent upon social experience for the development of sexual behavior, our current understanding of the factors related to reproductive success or failure is limited to a comparison of cases at the extremes of the continuum, extending from social isolation of laboratory subjects on the one hand to the complex social life of wild-born individuals on the other. The importance of the factor of experience was recognized early in the history of work with primates (Bingham, 1928). Experiments by Mason (1960) and Harlow (1962) with rhesus monkeys utilized techniques of social isolation during early development that prevented any direct social contact between animals until nearly two and one-half years of age.

Mason (1960) observed three males that had been separated from their mothers during the first month of life and were then housed in single cages for 28 to 29 months. No social experience except for seeing and hearing monkeys in adjacent cages was possible during this period. The sexual behavior of these males at 29 months was assessed in a series of 3-min pair tests with feral females that had adequate social histories, and with laboratory reared females deprived of social experience. Deficiencies were observed in a number of sexual parameters for the males regardless of the type of stimulus partner employed. Considering only the tests with feral females, the socially restricted males attempted to mount as frequently as feral males of the same age, but they used the adult double foot clasp mount less than 3% of the time. In contrast, the feral males used the adult mounting pattern exclusively. In tests with socially deprived females as partners, sexual behavior was even more severely deficient, reflecting the obvious effects of social deprivation for females as well as males.

Subsequent work extended the observations of Mason for socially deprived males and females and similarly demonstrated that severe social deprivation for a minimum of the first 6 months of life resulted in a nearly complete inability to copulate in adulthood (Harlow and Harlow, 1966). Senko (1966) demonstrated that the deleterious effects of social deprivation during the first year of life could not be reversed by later prolonged social experience (but see Suomi, Harlow, and McKinney, 1972, for possible exceptions). Senko found that only 2 of the 16 males deprived of social contact for the first 19 months of life copulated, even though they lived continuously with oppositely sexed peers for as long as 2 years before the sexual tests were conducted. With additional work, it appeared that absence of social experience with peers was even more critical than mother-infant relations for normal development of sexual behavior (Harlow, Joslyn, Senko, and Dopp, 1966). When infants were allowed to stay in a single cage with their mothers but were afforded no social contact with peers for 1 year, development of subsequent sexual behavior was deficient. As in Mason's original findings, deficiencies were not characterized by an absence of mount attempts, but rather by incorrect orientation to the partner, improper responses to partner solicitations, negative responses of the female partner to the male's ineptness, and the nearly complete absence of the double foot clasp mount.

In contrast, infants deprived of mothers but not of peers were characterized by Harlow as displaying essentially normal behavioral patterns of sexual development. Males that were in isolation for the first 80 days of life and then placed in peer groups for 10 mins a day showed "almost normal" sexual development at $10\frac{1}{2}$ months of age (Harlow and Harlow, 1962). Similarly, monkeys separated from mothers at birth but placed in large live-in pens with peers were described as displaying normal sexual development after an initial period of maladjustment during the first year. Unfortunately, animals reared in this manner (Harlow calls them "together-together monkeys") have not been studied for sexual behavior in adulthood.

The effects of moderate social deprivation were studied in the chimpanzee by Nissen. Riesen (1971) has summarized the findings of Nissen's extensive studies in a recent publication. Of eight males brought to the nursery within 2 weeks of birth and raised there, and for which direct social contact with other chimpanzees was prevented, only one displayed sexual behavior in adulthood without extensive "tutoring" by experienced female chimpanzees. Probably not irrelevant to the success of this particular male was the circumstance that he was given an opportunity to interact socially with slightly older peers at an earlier age than any other male studied. In the work with the chimpanzee, as in that dealing with the rhesus, reproductive deficiencies were not observed in wild-born primates captured and brought to the laboratory. Our own experience with feral captive rhesus monkeys is similar, and most, if not all, will copulate in captivity after adaptation to the laboratory environment.

The effects of rearing under conditions of limited access to peers

A less extreme form of social deprivation than those used in any of the studies described above has recently been shown in our own laboratory to be associated with deficiencies in adult male sexual behavior of the rhesus monkey. The conditions of rearing for these animals were markedly different from those imposed in the studies involving social isolation. All males were left with their mothers for the first 3 months of life. Following weaning they were housed in individual cages and placed daily with four to five peers for periods up to one-half hour each day for 100 successive weekdays. Thus, the opportunity for social interaction with the mother was unrestricted during the first 3 months of life, and opportunities for interactions with peers, though limited in duration, were provided frequently thereafter. In each subsequent year through the fourth year of life, these monkeys were allowed 50 successive weekdays of restricted peer interaction with periods of continuous interaction of 1 or 2 months interspersed between the 50-day runs. During the periods of restricted interaction, the social and sexual behaviors displayed by each animal were recorded.

Twenty-three males brought up in this peer-rearing system were tested six times each for sexual behavior with mature, receptive females. At the time of these tests the males varied in age from 48 to 80 months, and physical signs of adolescence were evident in all subjects. The youngest animal tested was 6 months older than the oldest age estimated for the onset of testosterone secretion in a developmental study of the rhesus by Resko (1967) that utilized males from the peer-rearing system. Only nine of these males displayed intromission during the tests, and only three ejaculated. Copulatory success, leniently defined as the display of intromission rather than ejaculation, was unrelated to age. Two of three males over 6 years of age met this criterion of copulatory success, and two of four males 48 months of age also met the criterion.

Characteristics of early behavior and their relation to copulatory success in males reared with limited access to peers

DEVELOPMENT OF MOUNTING PATTERNS We have recently analyzed the data for 13 males reared in the fashion described above, of which 7 showed intromission in

mating tests during adulthood and 6 did not. The most striking difference found between the histories of successful and unsuccessful copulators was in the developmental changes in the immature and mature mounting patterns. The term *immature mount* refers strictly to the pattern illustrated in Figure 1. The male stands with both feet

FIGURE 1 The immature mounting posture of the rhesus monkey. Both feet remain on the floor, but orientation is posterior. This type of mount is scored only when pelvic thrusting is displayed. See text for complete definition.

on the floor and with hands lightly touching the partner's back or rump. Pelvic thrusting is exhibited in all mounts classified as immature, and orientation to the rear of the partner is also required. The mature mount (Figure 2) may or may not include pelvic thrusting. Its essential criteria are posterior orientation and clasping of the partner's hocks with one or both feet. Special note should be taken of the fact that the term *mature mount* does not refer to or imply the occurrence of intromission. In fact, the mature mount appears developmentally 2 to 3 years before intromission is displayed by the male monkeys in the situation in which we have studied its development.

Among successful copulators, there were consistent increases over the first, second, and third years in the ratios of mature to immature mounts (Table I). A corresponding trend was not found for most of the males that failed to achieve intromission during the standardized tests. The difference between the groups emerged gradually, and during the first year of life copulators and noncopulators could not be predicted on the basis of the ratio of mature to immature mounts. During the second year, three males displayed high ratios of mature to immature mounts, and all proved to be successful copulators. The development of a high ratio beginning in the third year was associated with later copulatory success in 4 of 6

FIGURE 2 The mature mounting posture of the rhesus monkey. The male clasps the partner's hocks with his hind feet. The mount is classified as mature regardless of the display of pelvic thrusting.

males. None of the remaining males (4) displayed high ratios of mature mounts, and all of these males failed to achieve intromission during the tests at later ages.

The developmental deficiencies in mounting shown by unsuccessful copulators were not strictly limited to a failure to achieve a high ratio of mature to immature mounts. For two of the six unsuccessful males, mounting of either type was not displayed during the third year, although it had been displayed in earlier years. Thus this system of rearing was actually associated with two distinct developmental syndromes linked with copulatory failure: (1) a gradual reduction and/or loss of mounting behavior, and (2) fixation of an immature mounting pattern.

DEVELOPMENT OF SOCIAL PLAY, THREAT, AND PARTNER PREFERENCES A striking difference between the effects of moderate social deprivation of the peer-rearing system and that of the more severe social isolation is that male sexual behavior was, at least superficially, the principal

TABLE I
Relationship between development of mature mounting pattern and later copulatory success in rhesus males

	Subject	Ratio of Mature to Immature Mounts		
		First Year	Second Year	Third Year
Males displaying intromission at 4–6 years of age	1558	0.00	90.00	94.00
	1618	0.84	19.43	71.00
	1636	0.00	1.67	120.00
	1958	0.22	0.67	20.50
	1960	0.03	0.17	27.50
	1966	0.00	0.50	1.20
	4845	0.03	0.00	21.17
Mean		0.16	16.06	50.84
Males failing to display intromission at 4–6 years of age	1954	0.00	0.00	15.00
	2354	0.40	0.00	0.00
	2356	0.25	0.00	0.00
	2358	0.00	0.00	0.02
	2359	0.60	0.37	1.16
	4850	0.00	0.12	0.03
Mean		0.21	0.08	2.70

category of responses severely affected in the former group of animals. Monkeys in our peer-reared situation did not show the bizarre stereotypic motor patterns nor the hyperaggression and absence of play reported for the isolate (Mason, 1960; Harlow, 1962), nor did they show pronounced timidity and fear of either familiar or strange peers. The development of various forms of play has been described previously for males in the peer-reared situation (Goy, 1970b), and deficiencies comparable to those described by Harlow for social isolates were never observed.

Furthermore, peer-reared animals deficient in sexual behavior were not necessarily deficient in any other aspect of social interactions. Frequencies of social behavior displayed during the first year of life were entirely unrelated to copulatory success in young adulthood (Table II). Although this is not surprising, because mounting performance during the first year of life was also found to be unrelated to sexual performance later in life, the similarity of the social behavior scores for copulators and

TABLE II
Social behaviors during the first year of life for successful and unsuccessful copulators

Copulatory Ability	N	Gape	Play Initiation	Rough & Tumble Play	Pursuit Play	Aggressive Threat
			Median Frequency per 100 Days of Observation			
Successful	7	152	264	412	75	90
Unsuccessful	6	426	241	370	67	86

noncopulators persisted, even through the third year of life. The median frequencies for the 50 days during which the males were observed in the third year are as follows for copulators and noncopulators, respectively: gape, 116, 154; play initiation, 108, 107; rough and tumble play, 99, 78; pursuit play, 23, 5; aggressive threat, 17, 17. None of the differences between medians were statistically significant, and the overlap between the two groups was extensive for every measure of behavior except pursuit play.

As knowledge was gained of the high proportion of peer-reared males that were sexually inadequate in adulthood, the relationship between a preference for female partners and copulatory success was examined. In part our interest in the question was stimulated by the careful observations of Hanby (1972) showing that under conditions of troop-rearing, young male *Macaca fuscata* developed strong preferences to mount females. By the time the troop-reared males reached 18 months of age, 90% of their mounts were of the mature type, and female partners were distinctly preferred. A quite different pattern emerged for rhesus males in the peer-reared situation. At all times studied during the first, second, and third years of life, the majority of the 13 males mounted male partners slightly more frequently than they mounted female partners. In the third year, in fact, only 3 males displayed distinct preferences for mounting female partners. One male was successful at copulation, and the other 2 were not.

The failure of partner preferences to predict copulatory success or failure in these tests should not be accepted as a general principle of their irrelevance to adult sexual performance. The sex tests given in adulthood required neither choices between partners nor high levels of sexual performance. Both of these requirements exist in natural troops of monkeys, and the age at which males develop a strong and persistent preference for mounting female partners may be strongly related to copulatory success in that situation.

Prenatal hormonal influences on the development of mounting patterns under conditions of limited access to peers

The deleterious effects on mounting patterns of the peer-rearing system described are not dependent on the gonadal status of the individual. When male monkeys are castrated either on the day of birth or at the time of weaning prior to the initiation of peer experience, entirely comparable results are obtained. Six males castrated at these early ages were studied developmentally and were injected with testosterone at 5 to 7 years of age before they were tested for sexual behavior with females. Three of the males copulated successfully during these tests, and 3 did

not. During the second year of life the successful copulators showed ratios of mature to immature mounts of 7.0, 2.2, and 3.2. Ratios as high or higher were obtained during the third year. The 3 castrated males that failed to copulate successfully during replacement therapy showed ratios in the second year of 1.0, 0.0, and 0.1. Moreover, during the subsequent years of social testing, 2 of the noncopulating males failed to display any mounting behavior.

Pseudohermaphroditic females reared with the peer system showed corresponding deficits in adult male sexual behavior. Three pseudohermaphroditic subjects produced by identical treatments with testosterone prior to birth (Goy, 1970a) were studied. One of these females became a successful copulator with normal female partners during the reinstatement of testosterone treatment in adulthood. The other 2, given similar treatments as adults, did not. Ratios of mature to immature mounts during the first, second, and third years of life are presented for these subjects as well as for the 6 males castrated within 3 months following birth in Table III.

TABLE III

Relationship between development of mature mounting and later copulatory success in castrated male and pseudohermaphroditic female rhesus

Copulatory Ability	Subjects	Ratio of Mature to Immature Mounts		
		First Year*	Second Year	Third Year
Castrated males				
Successful	♂824	—	7.0	4.1
	♂825	—	2.2	2.3
	♂827	—	3.2	7.7
Unsuccessful	♂826	—	1.0	0.0
	♂837	—	0.0	0.0
	♂838	—	0.1	0.0
Pseudohermaphroditic females				
Successful	♀1640	0.0	10.2	31.0
Unsuccessful	♀836	0.3	0.0	0.2
	♀1616	0.0	0.0	0.0

*Data not available for castrated males during the first year of life.

Effects of rearing under conditions of unrestricted access to mothers and peers upon development of mounting and social patterns

Although monkeys reared with peers and provided with limited daily periods of social interaction are deficient in copulatory behavior compared to feral males, they perform better than do males raised in social isolation. The proportion of males copulating and the overall quantity of their positive social interactions with other monkeys are greater than the corresponding values for those males described by Mason (1960), Harlow (1962), and Senko (1966).

In recent work we have attempted to evaluate the effectiveness of a rearing system designed to permit a greater amount of time for social interactions during early development. Kim Wallen has studied the development of mounting behavior in monkeys provided with an opportunity for unrestricted social interaction throughout the first 10 months of life. Under these conditions (which we designate the mother-infant rearing system) groups of 5 mothers with infants were placed in a large pen and allowed to live together continuously. Observations of social behavior were carried out daily beginning at approximately 3 months of age. To date we have studied 8 males, 6 females, and 1 pseudohermaphroditic female reared under these conditions.

The results demonstrate that males reared in mother-infant groups develop a predominance of mature mounts within the first 100 days of observation. Data have been analyzed for this 100-day run by dividing the period into 5 blocks of 20 daily trials (Figure 3). Average frequencies

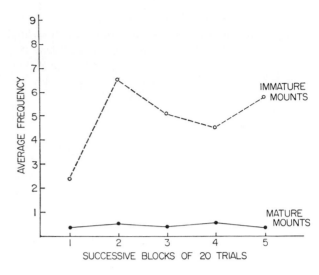

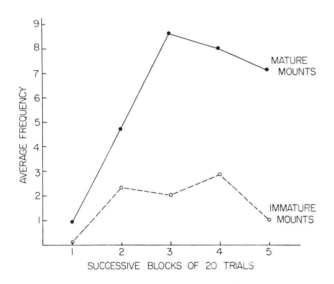

FIGURE 3 The development of mounting behavior in the first year of life for eight males reared with their mothers and peers.

of the mature mount are higher than average frequencies of the immature mount throughout the 100 days but become distinctly higher by the end of the second block. By the end of the third block, the mature pattern clearly predominates. Contrasting results for peer-reared males are illustrated in Figure 4. For all five blocks of trials the immature mount is the predominant pattern, and the mature mount is only infrequently displayed.

FIGURE 4 The development of mounting behavior in the first year of life for 16 males reared only with peers.

The development of mature mounting seen in the mother-infant rearing condition was a nearly uniform effect. In contrast to the results for the peer-rearing system in which only some males developed the mature mounting pattern by the second or third year of life, 7 of 8 males developed a predominance of mature mounting in the mother-infant rearing system. Moreover, the development of this predominance of the mature mount occurred at an earlier age in the mother-infant reared males compared to the peer-reared males. The magnitude of the difference between groups can be gauged by the observation that, during the first year of mother-infant rearing, males showed ratios of mature to immature mounts equal to or above those shown by peer-reared males 18 months older. We interpret the difference not as a precocity of the behavior in mother-infant males, which resemble in this respect males studied in natural troops (Lindburg, 1971), but as a retardation of the development of this behavior in peer-reared males. The influence of rearing condition upon the age at which the mature mounting pattern becomes predominant suggests that the developmental schedule is not maturationally, but rather experientially, determined.

In addition to the development of the mature mount within the first 100 days, a definite mounting partner preference developed for the majority of mother-reared males. Table IV illustrates this preference and contrasts it with the lack of preference for female partners in the peer-reared situation, which was referred to earlier. The development of the preference for females as mounting partners is of great interest, because, in both rearing conditions, the preference for male partners for rough forms of contact play is very marked. Mother-reared males

TABLE IV

Effect of rearing conditions on the display of mounting and partner preferences during the first year of life*

Rearing Condition	N	Not Mounting %	Mounting Only Male Partners %	Mounting Only Female Partners %	Mounting Both Sexes %	Frequency of Mounts per Male Partner†	Frequency of Mounts per Female Partner†	Showing Preference for Female Partner %
Peers	25	12	20	16	52	10.6	6.7	45 (10/22)
Mothers and peers	8	0	0	12.5	87.5	9.3	11.8	75 (6/8)

*Subjects were observed for 25 min per day for 100 consecutive days beginning at 3 months of age. Mounting includes both mature and immature patterns but not abortive mounts.

†These estimates are based on data obtained for males mounting both male and female partners.

choose males to play with and females to mount; peer-reared males choose males to play with and display no sex preference for a partner to mount.

Studies proving that the experiential influences upon mounting of male monkeys are not mediated by alterations in testicular function have not been performed. The possibility is rendered unlikely, however, by our findings that the early development of a predominance of mature mounting was evident in a pseudohermaphroditic female monkey similarly reared (Figure 5). While the pre-

It is manifested only in genetic males that produce testosterone endogenously during prenatal life (Goy and Resko, 1972) and in pseudohermaphroditic females exposed prenatally to an adequate amount of exogenous testosterone. Normal genetic females reared in the mother-infant system do not develop a preponderance of the mature mounting pattern at these early ages, and both mature and immature mounts are displayed less frequently by the females than by the males or the pseudohermaphroditic female (Figure 6). There is a

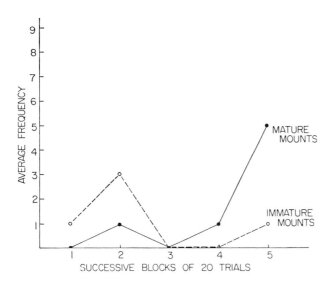

FIGURE 5 The development of mounting behavior in the first year of life for one pseudohermaphroditic female reared with mothers and peers.

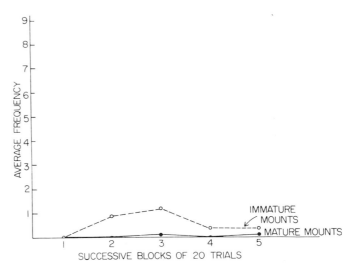

FIGURE 6 Mounting behavior of females in the first year of life reared with mothers and peers.

dominance of the mature mount was not consistent across all five blocks as it was on the average for males, the ratio of mature to immature mounts was 1.40 over all trials. The increased frequency of display of mature mounts in the mother-infant rearing condition is sexually specific.

suggestion from the data, however, that the frequencies of mature mounts displayed by females in the mother-reared condition are elevated relative to females in the peer-reared groups. Nevertheless, the effects of the mother-reared condition are much more profound and apparent on genetic males and pseudohermaphroditic females than on normal genetic females. When the patterns of mounting behavior shown by males and females reared in isolation, in the peer-condition, and in the mother-infant condition are compared, the possibility is suggested that these experiential influences affect primarily the frequency and form of the behavior and not the underlying psychosexual orientation of the individual or the predisposition to acquire sex-related responses. This possibility is also strongly supported by the observation that males and females reared in isolation continue to manifest mounting behaviors dimorphically even though the form is abnormal and the frequency of expression is atypical.

Our studies are incomplete in the sense that the expression of adult sexual behavior has not been evaluated in mother-infant reared males. A prediction of outcomes can be ventured, however, based on the information obtained from the study of peer-reared males. The results presented earlier for such males suggested a trend: The earlier in life that mature mounting became predominant, the greater the likelihood of copulatory success in adulthood. Assuming the validity of this trend for other rearing conditions, the prediction of successful copulation in adulthood for mother-infant reared males does not seem unreasonable.

Patterns of behavioral development in different rearing conditions and their relation to sexual outcome in adulthood

Although social deprivation has a pronounced inhibitory influence upon the development of sexual behavior in the male chimpanzee, this species apparently can recover from the effects of a restrictive environment to a greater degree than can the rhesus monkey. Following extensive practice in adulthood with experienced female partners, male chimpanzees deprived of social contact during early infancy have some chance of copulating (Davenport and Rogers, 1970).

The work with the chimpanzee seems consistent with Nissen's (1954) view of the situation, based on his early work with the problem. The organization of components into an effective pattern of mating behavior was regarded by him as a process of trial-and-error learning. More essential to success in the trial-error process than specific forms of play behavior, in Nissen's view, was activity level *per se*. The higher the individual animal's activity level, the more likely the occurrence of the pattern of responses critical for reproductive success. The factor of age was of secondary importance, and it influenced outcomes only in the sense that activity in general was lower in older individuals and therefore decreased the probability of occurrence of the correct sequence of responses. As a corollary of this view of the age factor, socially iso-

lated males should be capable of learning normal sexual behavior at any age if enough time is allowed. This corollary apparently is valid for the chimpanzee but is not supported by any of the data currently available for isolated male rhesus studied by Mason, by Harlow and associates, or by Missakian (1969).

Although Missakian has reported recently (1972) that isolated male rhesus monkeys show improvement in performance when, as adults, they are placed with younger animals for periods of 4 to 6 months, it should be emphasized that the improvement was minimal and limited to increases in the frequency of "mount attempts." We should like to point out that this behavior recorded by Missakian does not correspond to the behaviors we have defined as mature and immature mounts, and, in our experience, that its relevance to copulatory success is questionable.

On the basis of information presently available on the rhesus monkey, it is possible to gain the impression that the total amount of time allowed for social interaction is directly related to development of a pattern of sexual responses at an early age that is critical for later copulatory success. Isolates, which have no opportunity to develop this critical response in a social context, are the most uniformly and completely deficient in their copulatory behavior in adulthood. A higher proportion of males reared first with mothers and then with peers show copulatory behavior in adulthood, and the successful individuals develop the mature mounting pattern early in life, usually within the second year. Males reared with mothers and peers present continually during the first year of life develop the mature mounting pattern during this time, and in this respect, as well as in the development of partner preferences for mounting, their behavior resembles that of males in captive troops. While the copulatory success of mother-reared males cannot now be known because of their young age, a prediction of normal copulatory performance would be consistent with the notion that the amount of time permitted for social interaction and the early development of a critical response are the essential factors.

In terms of the details of the character of the behavior displayed by monkeys reared under these different conditions, the generalization of a quantitative continuum of experience is not readily acceptable. Among other difficulties with such a generalization is the underlying assumption that the situations employed to suggest this hypothesis have differed only in respect to the factor of time and that other variables influencing the development of reproductive adequacy have been held constant.

Differences exist in the social conditions investigated that extend beyond a simple formulation of quantity of social experience, and these differences are not as easy to specify as their effects on behavior are to describe. Monkeys reared in social isolation early in life show profound disturbances of affect that may be the basis for interference with many kinds of social behavior in addition to sexual behavior. For isolated monkeys as for isolated chimpanzees, fear of other monkeys and timidity in social situations are conspicuous characteristics (Riesen, 1971; Senko, 1966). Furthermore, several interfering habits become established that are very persistent through time. These include motor stereotypy as well as unpredictable aggressive attacks, despite the social timidity that generally characterizes these isolates. In addition, deficits in positive social relations with peers have been observed frequently. Such deficits interfere with play behavior at early ages, which in turn may lead to inadequacies in sexual behavior later in life. It is the conspicuous nature of these affectional deficiencies in isolates compared with peer-reared monkeys (Harlow and Harlow, 1966) that has permitted insight into one aspect of the dynamics of primate sexuality. The importance of play behavior in the establishment of affectional relations is paralleled by its inclusion of "basic response patterns which are not directly heterosexual (but which) may nevertheless insure that normal heterosexual posturing will be acquired as long as normal . . . affectional responses are given an adequate opportunity to develop." (Harlow, 1965, p. 243).

Generalized deficiencies in affectional relationships seem very likely related to the sexual inadequacies of socially isolated monkeys, but more subtle and specific affectional deficiencies may be relevant to the sexual inadequacies of males reared with peers. These deficiencies cannot be those that are associated with, or derived from, low levels or absence of play, because play behaviors were high in frequency among peer-reared males. The possible affectional disorders may be reflected in the early behavioral differences of peer-reared compared with mother-reared males. Two of these differences have been discussed at length above, but for ease of summarization they are repeated here. They are (1) delayed development of mature mounting in peer-reared compared with mother-reared males, and (2) the failure of uniform development of preference for mounting female partners in peer-reared males.

Aside from the early differences in mounting, peer and mother-infant conditions differ most in the quality of the dominance interactions between peers during the first year of life. In the mother-rearing condition infants rarely displayed submissive or aggressive reactions to one another, and the dominance contests defined by these stereotyped responses occurred almost exclusively between mothers. Peer-reared infants, in contrast, established dominance relations based on typical aggression-submission behavior.

The average frequency of aggressive behavior displayed toward other peers was 13.5 for 13 peer-reared males and only 0.2 for the 8 males reared with mothers and peers during the 100-day run of the first year. For submissive behaviors, indicated by the facial expression referred to as the *fear grimace*, the differences were just as marked. The average fear grimace frequencies were 51.6 for the peer-reared males, and 2.1 for mother-reared males. These data are shown separately in Table V for each group of peer-

TABLE V

Aggressive and submissive responses for males in two different rearing conditions during the first year

Rearing Condition	Group Number	N	Average Frequency of Behavior per Male	
			Aggression	Fear Grimace
With peers	30	3	18.3	50.7
	32	4	17.5	69.0
	36	4	11.8	48.5
	44	2	2.0	24.5
With mothers and peers	1	3	0.7	4.7
	2	3	0.0	0.3
	3	2	0.0	1.0

and mother-reared males. The submissive behaviors of peer-reared males and males interacting with peers in the presence of their mothers differed not only in frequency but also in a critical dimension presumably related to dominance. Of all the social behaviors recorded during observations of interactions of young animals, only the fear grimace showed a hierarchical organization based on partner discrimination rather than simple frequency of occurrence. Typically in peer-reared individuals, the hierarchical organization of fear grimacing occurs early and persists strongly through time. An example is shown in Table VI. The hierarchical organization is not perfect, but it is much more nearly so than the corresponding

TABLE VI

Frequency of submissive gestures performed by each subject distributed according to partner (Peer-reared monkeys, Group 44)

	Performer					
Partner	♂4845	♂4878	♂4850	♀4862	♀4890	♀4875
♂4845	—	15	11	55	42	38
♂4878	0	—	13	21	18	45
♂4850	0	0	—	84	39	89
♀4862	0	0	25	—	4	13
♀4890	0	0	0	0	—	1
♀4875	0	0	0	1	0	—

matrices obtained for mother-infant peer groups (Table VII). The question that these observations raise concerns the consequences to development brought about by displaying these stereotyped dominance behaviors precociously by peer-reared males. The fact of high frequencies of aggressive and submissive behaviors for peer-reared compared with mother-infant reared males implies qualitative differences in the emotional relationships among peers in these two conditions. The exact role of the

TABLE VII

Frequency of submissive gestures performed by each subject distributed according to partner
(Mother-infant-reared subjects, Group 3)

Partner	Performer				
	♂S37	♀S34	♀S32	♂S27	♀S21
♂S37	—	0	0	0	0
♀S34	0	—	0	0	0
♀S32	0	0	—	2	3
♂S27	0	0	0	—	1
♀S21	0	0	0	0	—

mother in reducing the frequency of submissive and aggressive encounters cannot be analyzed from our present data collection techniques, although perhaps by interfering with infant-infant interactions before overt aggression or intense threat developed, the mothers might have reduced conflicts that would have led to high levels of tension and fear. Second, the presence of the mother was surely a highly effective source of security to which the infant could flee in any situation that produced tension. With increased attention paid to mother-infant interactions, such as by employing the analyses used by Hinde and Spencer-Booth (1967), it is hoped that answers to these questions will be available in the near future.

A confounding factor in evaluating emotional differences between the two types of groups studied and their impact upon the development of mounting behavior is the circumstance that mounting behavior itself is, under special circumstances, an expression of dominance relationships. The diverse ways in which mounting can be used to express dominance has been most carefully examined by Maslow (1936a, b). We were unable in the situation studied to find a clear hierarchical organization of any class of mounting behavior that corresponded to the degree of hierarchical organization typically exhibited for submissive gestures. The possibility exists, nevertheless, that on particular occasions the mounting behaviors displayed are motivated by what Maslow referred to as "the dominance drive" and more frequently by this mechanism in peer-reared than in mother-infant reared males. If our inference is correct that the dominance drive

is less frequently aroused or less necessary to infant-infant interactions in the mother-reared group and strongly manifested in the peer-reared groups, then the mounting behavior displayed by males in these two conditions may not be related to similar underlying motivational conditions. Maslow has described differences in the form of mounting behavior exhibited in dominance-provoking situations and in reproductive contexts. It may not be a coincidence that the forms of mounting that he describes as motivated by dominance drive include the type that we have called the immature mount. Accordingly, the differences noted for peer- and mother-reared males in rates of development of the mature mounting pattern may, in fact, be due to the motivational differences underlying mounting behavior in these two rearing conditions. The possibility of the selection and fixation of a specific response pattern to a single motivational condition intrigues us as a mechanism for the establishment of sexual inadequacies of peer-reared males and for their developmental deficiencies of the mounting pattern.

ACKNOWLEDGMENTS We wish to acknowledge the financial support of several grants that have helped not only in the conduct of the work but also in the preparation of the manuscript itself: Grant MH 21312 from the National Institute of Mental Health; Grant RR 00167 to the Wisconsin Regional Primate Research Center; and Grant RR 00163 to the Oregon Regional Primate Research Center from the National Institutes of Health.

Publication number 12-010 of the Wisconsin Regional Primate Research Center.

REFERENCES

BINGHAM, H. C., 1928. Sex development in apes. *Comp. Psychol. Monogr.* 5:1–165.

BOWLBY, J., 1952. *Maternal care and mental health*, 2nd Ed. Monograph Series No. 2. Geneva: World Health Organization.

DAVENPORT, R. K., and C. M. ROGERS, 1970. Differential rearing of the chimpanzee: A project survey. *The Chimpanzee* 3:337–360.

GOY, R. W., 1970a. Experimental control of psychosexuality. *Phil. Trans. Roy. Soc.*, London B. 259:149–162.

GOY, R. W., 1970b. Early hormonal influences on the development of sexual and sex-related behavior. In *The Neurosciences: Second Study Program*, F. O. Schmitt, Editor-in-Chief. New York: The Rockefeller University Press, pp. 196–207.

GOY, R. W., and J. A. RESKO, 1972. Gonadal hormones and behavior of normal and pseudohermaphroditic nonhuman female primates. *Rec. Prog. Horm. Res.* 28:707–733.

HANBY, J. P., 1972. The sociosexual nature of mounting and related behaviors in a confined troop of Japanese Macaques. Doctoral dissertation, Department of Psychology, University of Oregon, Eugene, Oregon.

HARLOW, H. F., 1962. The heterosexual affectional system in monkeys. *Am. Psychol.* 17:1–19.

HARLOW, H. F., 1965. Sexual behavior in the rhesus monkey. In *Sex and Behavior*, F. A. Beach, ed. New York: John Wiley & Sons, pp. 234–265.

HARLOW, H. F., and M. K. HARLOW, 1962. The effect of rearing conditions on behavior. *Bull. Menninger Clinic* 26:213–224.

HARLOW, H. F., and M. K. HARLOW, 1966. Learning to love. *Amer. Sci.* 54:244–272.

HARLOW, H. F., and M. K. HARLOW, 1971. Psychopathology in monkeys. In *Experimental Psychopathology: Recent Research and Theory*, H. D. Kimmel, ed. New York: Academic Press, pp. 203–229.

HARLOW, H. F., W. D. JOSLYN, M. G. SENKO, and A. DOPP, 1966. Behavioral aspects of reproduction in primates. *J. Anim. Sci.* 25:49–67.

HINDE, R. A., and Y. SPENCER-BOOTH, 1967. The behavior of socially living rhesus monkeys in their first two and a half years. *Anim. Behav.* 15:169–196.

JENSEN, G. D., 1970. Environmental influences on sexual differentiation: Primate studies. In *Environmental Influences on Genetic Expression*, N. Kretchmer, and D. N. Walcher, eds. Bethesda: National Institutes of Health, pp. 129–139.

LINDBURGH, D. G., 1971. The rhesus monkey in north India: An ecological and behavioral study. In *Primate Behavior, Developments in Field and Laboratory Research*, Vol. 2, L. A. Rosenblum, ed. New York: Academic Press, pp. 1–106.

MASLOW, A. H., 1936a. The role of dominance in the social and sexual behavior of infra-human primates: I. Observations at Vilas Park Zoo. *J. Genet. Psychol.* 48:261–277.

MASLOW, A. H., 1936b. The role of dominance in the social and sexual behavior of infra-human primates: III. A theory of sexual behavior of infra-human primates. *J. Genet. Psychol.* 48:310–338.

MASON, W. A., 1960. The effects of social restriction on the behavior of rhesus monkeys: I. Free social behavior. *J. Comp. Physiol. Psychol.* 53:582–589.

MISSAKIAN, E. A., 1969. Reproductive behavior of socially deprived adult male rhesus monkeys (*Macaca mulatta*). *J. Comp. Physiol. Psychol.* 69:403–407.

MISSAKIAN, E. A., 1972. Effects of adult social experience on patterns of reproductive activity of socially deprived male rhesus monkeys (*Macaca mulatta*). *J. Personal. Soc. Psychol.* 21:131–134.

MONEY, J., and A. A. EHRHARDT, 1968. Prenatal hormonal exposure: Possible effects on behaviour in man. In *Endocrinology and Human Behaviour*, R. P. Michael, ed. London: Oxford University Press, pp. 32–48.

NISSEN, H. W., 1954. Development of sexual behavior in chimpanzees. Conference on Genetic, Psychological and Hormonal Factors in the Establishment of Sexual Behavior in Mammals. W. C. Young, Chairman. Lawrence, Kansas: University of Kansas Libraries.

RESKO, J. A., 1967. Plasma androgen levels of the rhesus monkey: Effects of age and season. *Endocrinology* 81:1203–1212.

RIESEN, A. H., 1971. Nissen's observations on the development of sexual behavior in captive-born, nursery-reared chimpanzees. *The Chimpanzee* 4:1–18.

SACKETT, G. P., 1972. Exploratory behavior of rhesus monkeys as a function of rearing experiences and sex. *Developmental Psychol.* 6:260–270.

SENKO, M. G., 1966. The effects of early, intermediate, and late experiences upon adult macaque sexual behavior. Unpublished Master's Thesis, Madison, Wisconsin: University of Wisconsin Libraries, pp. 1–102.

SUOMI, S. J., H. F. HARLOW, and W. I. MCKINNEY, JR., 1972. Monkey psychiatrists. *Amer. J. Psychiat.* 128:927–931.

51 Integrative Actions of Perinatal Hormones on Neural Tissues Mediating Adult Sexual Behavior

HARVEY H. FEDER and G. N. WADE

ABSTRACT Perinatal testicular secretions aid in integrating reproductive function in adult males by insuring that central and peripheral tissues mediating reproduction are able to respond simultaneously to the activating effects of androgens secreted by the adult testis. The sequential nature of the female reproductive cycle obviates the necessity for a perinatal organizational action of steroid hormones. Both estrogen and progesterone participate in regulating the timing of the sequential events in reproductive cycles of adult rats, guinea pigs, and hamsters. The properties of the estrogen and progesterone uptake systems in brain differ in several respects.

SECRETION, OR absence of secretion, of testicular hormones during perinatal life in mammals has comprehensive repercussions for the individual. In this paper we discuss some implications of the effects of perinatal secretions for unified functioning of behavioral and gonadotropic systems responsive to steroids in adult mammals.

Integrative influence of perinatal hormones in adult males

Under normal circumstances, genetic males of mammalian species produce androgenic hormones during fetal or neonatal life (Resko, Feder, and Goy, 1968). Among many effects of these perinatal secretions are the following: (a) They guide differentiation of Wolffian duct, urogenital sinus, and external genital primordia into masculine genital duct system, seminal vesicles, prostate, and phallus. In addition to their effects on differentiation, perinatal secretions increase sensitivity of these tissues to androgens administered in adulthood (Beach, 1971). (b) By their action on diencephalic tissues, they cause a tonic rather than a cyclic pattern of gonadotropin release in adulthood (Harris, 1964). (c) They decrease responsiveness to estradiol (Gerall and Kenney, 1970) and progesterone (Clemens, Shryne, and Gorski, 1970) of adult

HARVEY H. FEDER and G. N. WADE Rutgers University, Institute of Animal Behavior, Newark, New Jersey

neural tissues mediating sexual behavior. They may also increase responsiveness to androgen in adult spinal cord (Hart, 1968), but their effects on adult androgen-sensitive brain tissues have not been established.

Obviously, the effects of fetal hormones on the genitalia are important, because these tissues must be adequately formed for reproduction to occur. A more subtle aspect of perinatal hormone action on the male genitalia is illustrated by the following: Penile, seminal vesicular, and prostatic tissues of male rats castrated on the day of birth do not respond well to testosterone administered in adulthood, although these tissues are structurally complete in all their component parts (Beach, 1971). The quantity of testosterone administered in adulthood to such animals is sufficient to activate the display of some components (mounting and thrusting) of male sexual behavior. Thus, removal of perinatal hormones by castration favors a wide divergency between neural and genital tissues in adult sensitivity to androgen. Perinatal hormones apparently decrease the range of sensitivity of various target tissues to adult androgens. This range presumably must be narrow enough so that testicular androgens secreted by the adult will be sufficient to stimulate both neural and genital tissues.

Actions of perinatal hormones integrate not only the genital with the neurobehavioral tissues but also the gonadotropic and neurobehavioral tissues. Two features of male sexual behavior are (a) a fairly long latency (ca. 7 days) between onset of exposure to androgen and display of complete sexual responses, and (b) after their establishment, male sexual responses can be displayed day after day for prolonged periods (Young, 1961). These two behavioral characteristics are most compatible with a gonadotropic system that can mediate a relatively constant secretion of androgen over prolonged periods. Perinatal testicular secretions assure development of such a system by causing development of a tonic, acyclic pattern of gonadotropin release in the adult (Harris, 1964).

Naftolin and colleagues propose that the androgen thus tonically secreted must be aromatized to estrogen in the brains of adults before masculine behavior is displayed (Naftolin, Ryan, and Petro, 1971). This hypothesis appears untenable. First, perinatal hormones decrease responsiveness to estrogen in the adult. It would seem uneconomical for perinatal males to produce substances that would decrease responsiveness to a required hormone in adulthood. Second, castrated adult male rats given testosterone plus antiestrogen display masculine behavior (Whalen, Battie, and Luttge, 1972). Finally, castrated adult males given estrogen plus dihydrotestosterone did not show high levels of male behavior, even though the dihydrotestosterone insured adequate growth of peripheral tissues (Feder and Naftolin, unpublished).

Influence of absence of perinatal androgens on adult females

All tissues, central and peripheral, mediating reproduction in males are active simultaneously. In females, on the other hand, there is a sequential activation of the various tissues affecting reproduction. Growth of uterine tissues precedes a major burst of gonadotropic activity, which precedes display of sexual behavior, which precedes ovulation. The sequential, rather than simultaneous, nature of female reproductive function is compatible with divergent thresholds to estrogen activation among the various tissues mediating reproduction. This, in turn, is compatible with the absence of perinatal hormones serving to render such thresholds more uniform in adults. In fact, administration of hormonal steroids to perinatal females increases to unacceptable levels the divergences in tissue responsiveness to estrogens in adults (Whalen and Nadler, 1965).

Another consequence of absence of perinatal hormones in spontaneously ovulatory females is development of a cyclic gonadotropin release system (Harris, 1964). In nonprimate females, the cyclic release of gonadotropin results in cyclic ovarian secretion of estrogen and therefore cyclic display of sexual behavior (for example, rats show estrous behavior once every 4 or 5 days). A different relationship between sexual behavior and ovulation exists in primates, because some female primates are capable of showing sexual receptivity throughout a menstrual cycle (Zuckerman, 1932). Therefore, if receptive behavior in such primates has a hormonal basis, it would have to be one that is distinct from the fluctuations in ovarian estrogen secretion that regulate the occurrence of ovulation. Herbert and colleagues have shown that androgens of adrenal origin maintain sexual receptivity in adult female rhesus monkeys (Everitt, Herbert, and Hamer, 1972). Because the adrenal system develops whether or not androgens were present in fetal life, an adrenal behavior-regulating system would permit the male-like characteristic of persistent display of sexual behavior to occur in female primates. Another implication of the Herbert work is that androgens need not be converted to estrogens in order to stimulate sexual receptivity.

Female mammals are quite responsive to estrogen and progesterone in adulthood, and this steroid sensitivity is also dependent on the lack of secretion of perinatal hormones (Grady, Phoenix, and Young, 1965). The cellular basis of sensitivity to estrogen lies in the presence of estrogen-binding macromolecules in target tissues. It is possible that in the absence of estrogen these macromolecular receptors will not develop. However, recent data indicate that brain and uterine tissues of perinatal guinea pigs have estradiol receptors even though such animals presumably are not yet capable of secreting estrogen (Wade and Feder, unpublished). Furthermore, administration of exogenous estradiol to perinatal rats apparently decreases rather than increases the number of estrogen receptors in adult brain (Lisk and Ciaccio, 1971). It has also been implied that the inability of many adult female rats to display lordosis behavior after long-term ovariectomy is attributable to breakdown of estrogen-deficient brain receptors (McGuire and Lisk, 1968). The validity of this idea requires further testing in rats, but it seems not to have applicability to other species. Even after long-term ovariectomy (6 months), adult female guinea pigs display excellent sexual receptivity following a single exposure to estrogen (Peretz and Feder, unpublished). Ovariectomized ewes given repeated estrogen injections show less intense female behavior than animals given a single estrogen injection (Scaramuzzi, Lindsay, and Shelton, 1972). In neither the guinea pig nor the ewe is there evidence supporting the view that brain macromolecules mediating estrous behavior degenerate in the absence of prolonged estrogen stimulation (macromolecules in other target tissues of these species may or may not share this property).

Maintenance of a *progesterone receptor* mechanism in guinea pigs also appears to be independent of presence of progesterone. Thus, there are no differences among neonatal females, adult males, ovariectomized adult females, and adult females given exogenous progesterone with regard to uptake of radioactive progesterone (Wade and Feder, 1972).

Progesterone occupies a position distinct from estrogen in relation to neural tissues mediating female behavior. Estrogen alone can induce female behavior in many species when administered in adequate quantities, whereas progesterone alone cannot (Young, 1961). Instead, progesterone synergizes with estrogen in several nonprimate spontaneous ovulators to induce sexual receptivity. One significant aspect of this steroid interaction is that it synchronizes the short-lived cyclic display of sexual receptivity with the event of ovulation. For example, a proposed synchronization method in rats, hamsters, and guinea pigs may operate as follows (Joslyn, Feder, and Goy, 1972): When a certain level of estrogen is secreted in proestrus, a release of gonadotropin results that will lead to ovulation in a matter of about 10 hours. However, the quantity of estrogen secreted is insufficient, by itself, to induce sexual receptivity. Receptive behavior is synchronized with ovulation by means of a surge of ovarian progesterone released in response to the secretion of gonadotropin. Progesterone then synergizes with the estrogen produced earlier to induce sexual behavior at precisely the time when the ovulatory process has already been set irreversibly into motion.

This particular action of progesterone as a "fine tuner" of reproductive cyclicity requires that progesterone trigger display of sexual behavior rapidly. Injection experiments suggest that latency of progesterone action on behavior is not more than 4 hours and may be as little as 15 minutes (Lisk, 1960). What cellular mechanisms might mediate such a rapid behavioral response? Although a latency of 4 hours is adequate time for an effect of progesterone on uptake of amino acid into protein in brain (Wade and Feder, unpublished), the synergistic effect of progesterone with estrogen on sexual behavior does not seem to depend on protein synthesis. First, protein synthesis inhibitors fail to block the facilitatory effect of progesterone on estrous behavior in guinea pigs (Wallen, Goldfoot, Joslyn, and Paris, 1972). Second, although estrogen and progesterone have synergistic effects on incorporation of labeled amino acids into protein in guinea pig uterus, a similar synergy is not apparent in diencephalic, mesencephalic, or cerebral cortical tissues (Wade and Feder, unpublished).

Additional biochemical studies indicate that less than 1% of radioactivity is found in guinea pig brain cell nuclei after administration of tritiated progesterone. In contrast, more than 10% of radioactivity is found in hypothalamic cell nuclei after injection of tritiated estradiol (Wade and Feder, unpublished). Furthermore, the brain uptake mechanism for progesterone in guinea pigs is nonsaturable, indicating absence of a limited capacity receptor system for progesterone in brain (Wade and Feder, 1972). None of these data is incompatible with the notion that the facilitatory action of progesterone on female sexual behavior is mediated by a nonnuclear mechanism distinct from that mediating estrogen action. In this context, the finding that perinatal androgens decrease behavioral responsiveness of adult female rats to progesterone (Clemens, Shryne, and Gorski, 1970) takes on additional significance, because it implies that peri-

natal hormones can affect behavioral systems by means other than differentiation or destruction of saturable nuclear receptor molecules.

ACKNOWLEDGMENT Unpublished data reported here were from experiments supported by NIH-HD-04467 and Research Scientist Development Award KO 2-MH-29006 from NIMH. This essay is respectfully dedicated to the memory of Professor G. W. Harris and is Contribution No. 159 of the Institute of Animal Behavior.

REFERENCES

BEACH, F. A., 1971. Hormonal factors controlling the differentiation, development, and display of copulatory behavior in the ramstergig and related species. In *The Biopsychology of Development*, E. Tobach, L. R. Aronson, and E. Shaw, eds. New York: Academic Press, pp. 249–296.

CLEMENS, L. G., J. SHRYNE, and R. A. GORSKI, 1970. Androgen and development of progesterone responsiveness in male and female rats. *Physiol. Behav.* 5:673–678.

EVERITT, B. J., J. HERBERT, and J. D. HAMER, 1972. Sexual receptivity of bilaterally adrenalectomized female rhesus monkeys. *Physiol. Behav.* 8:409–415.

GERALL, A. A., and A. M. KENNEY, 1970. Neonatally androgenized females' responsiveness to estrogen and progesterone. *Endocrinology* 87:560–566.

GRADY, K. L., C. H. PHOENIX, and W. C. YOUNG, 1965. Role of the developing rat testis in differentiation of the neural tissues mediating mating behavior. *J. Comp. Physiol. Psychol.* 46:138–144.

HARRIS, G. W., 1964. Sex hormones, brain development and brain function. *Endocrinology* 75:627–648.

HART, B. L., 1968. Neonatal castration: Influence on neural organization of sexual reflexes in male rats. *Science* 160:1135–1136.

JOSLYN, W. D., H. H. FEDER, and R. W. GOY, 1972. Estrogen conditioning and progesterone facilitation of lordosis in guinea pigs. *Physiol. Behav.* 7:477–482.

LISK, R. D., 1960. A comparison of the effectiveness of intravenous as opposed to subcutaneous injection of progesterone for the induction of estrous behavior in the rat. *Can. J. Biochem.* 38:1381–1383.

LISK, R. D., and L. A. CIACCIO, 1971. The physiology of hormone receptors: Patterns of uptake and retention of estradiol and progesterone in relation to reproductive capability. In *Steroid Hormones and Brain Function*, C. H. Sawyer and R. A. Gorski, eds. Los Angeles: University of California Press, pp. 227–236.

McGUIRE, J. L., and R. D. LISK, 1968. Estrogen receptors in the intact rat. *Proc. Natl. Acad. Sci. (USA)* 61:497–503.

NAFTOLIN, F., K. J. RYAN, and Z. PETRO, 1971. Aromatization of androstenedione by the diencephalon. *J. Clin. Endocr.* 33:368–370.

RESKO, J. A., H. H. FEDER, and R. W. GOY, 1968. Androgen concentrations in plasma and testis of developing rats. *J. Endocr.* 40:485–491.

SCARAMUZZI, R. J., D. R. LINDSAY, and J. N. SHELTON, 1972. Effect of repeated oestrogen administration on oestrous behaviour in ovariectomized ewes. *J. Endocrinol.* 52:269–278.

WADE, G. N., and H. H. FEDER, 1972. 1,2-^{3}H-Progesterone uptake by guinea pig brain and uterus: Differential localization, time-course of uptake and metabolism, and effects of age, sex, estrogen-priming and competing steroids. *Brain Res.* 45:525–543.

WALLEN, K., D. A. GOLDFOOT, W. D. JOSLYN, and C. A. PARIS, 1972. Modification of behavioral estrus in the guinea pig following intracranial cycloheximide. *Physiol. Behav.* 8:221–223.

WHALEN, R. E., C. BATTIE, and W. G. LUTTGE, 1972. Anti-estrogen inhibition of androgen induced sexual receptivity in rats. *Behav. Biol.* 7:311–320.

WHALEN, R. E., and R. D. NADLER, 1965. Modification of spontaneous and hormone-induced sexual behavior by estrogen administered to neonatal female rats. *J. Comp. Physiol. Psychol.* 60:150–152.

YOUNG, W. C., 1961. The hormones and mating behavior. In *Sex and Internal Secretions*, 3rd ed., W. C. Young, ed. Baltimore: Williams & Wilkins, pp. 1173–1239.

ZUCKERMAN, S., 1932. *The Social Life of Monkeys and Apes.* London: Kegan Paul, Trench, Trubner.

52 The Hypothalamic-Pituitary-Gonadal Axis and the Regulation of Cyclicity and Sexual Behavior

BRUCE D. GOLDMAN

ABSTRACT The mammalian pituitary gland produces gonadotropic hormones (LH and FSH) that regulate a variety of physiological processes related to reproduction. These include (1) direct effects of pituitary hormones upon morphogenetic and hormone secretory processes in the gonads and (2) indirect effects which result from stimulation of gonadal steroidogenesis. Thus, sexual behavior, which is directly dependent upon gonadal steroids, can be inhibited not only by gonadectomy but also by blocking the required pituitary input to the gonads. The observations discussed in this paper indicate that the pituitary gonadotropic hormones are important both in adult life and during early development. Of particular interest is the observation that pituitary gonadotropins are involved in the regulation of early sexual differentiation of the brain.

THE HYPOTHALAMIC-PITUITARY axis is of central importance in the regulation of reproductive cycles in vertebrates and forms a part of a hormonal feedback loop along with the gonads. The role of the pituitary as a part of this feedback loop is the subject of this chapter. It is emphasized here that the pituitary is important not only during adulthood but also in early life.

The ovulatory cycle

In the laboratory rat and the golden hamster, ovulatory cycles are spontaneous (i.e., occurring in the absence of coital stimulation) and continuous unless interrupted by pregnancy. In these species, two temporal factors can be distinguished related to the release of the ovulatory surge of luteinizing hormone (LH) from the pituitary gland. These are, respectively, the particular *day of the cycle* during which the release occurs and the specific *time of day* of LH release.

DAY OF LH RELEASE The particular day on which the ovulatory surge of LH is to be released appears to be

determined as a result of the action of ovarian steroid hormone(s) on the hypothalamic-pituitary axis. Early studies implicated both estrogen and progesterone as *triggering* hormones for LH release, but the most recent evidence strongly favors estrogen. In the hamster, the blood estradiol concentration rises abruptly during the day prior to proestrus (Labhsetwar, 1972). If the ovaries are removed early on that day—i.e., prior to the increased secretion of estradiol—the sharp increase in serum LH concentration that normally occurs on the following day fails to appear. However, if ovariectomy is delayed until a few hours after the increase in plasma estradiol (i.e., 6:00 PM on the second day of diestrus) then a surge of LH is released on the following day (Goldman, Mahesh, and Porter, 1971). Since the blood progesterone concentration is relatively constant during the day preceding proestrus and rises sharply only late on the proestrous day (Leavitt and Blaha, 1970; Lukaszewska and Greenwald, 1970), the presence of an LH surge following ovariectomy at late diestrus suggests that estradiol may be the primary trigger for LH release in the hamster. Furthermore, it appears that by 6:00 PM on the second day of diestrus—i.e., approximately 20 hours prior to LH release—sufficient estrogen has been secreted to prime the hypothalamic-pituitary axis. Several additional studies in rats, hamsters, and sheep have pointed to estrogen as the primary trigger for ovulatory release of LH, and in each case the increase in serum LH follows the increase in estrogen by from several hours to one day, depending upon the species (Ferin et al., 1969; Goding et al., 1969; Labhsetwar, 1972; Scaramuzzi et al., 1971).

TIME OF DAY OF LH RELEASE Under normal conditions, the time of day during which LH release occurs appears to be regulated primarily by the diurnal regimen of illumination. Thus, rats and hamsters normally release LH during the afternoon, but the time of release can be altered merely by an appropriate phase shift in the times at which the lights in the animal room turn on and shut

BRUCE D. GOLDMAN Department of Biobehavioral Sciences, University of Connecticut, Storrs, Connecticut off. Ovulation occurs at night, approximately 10 to 12 hours after LH release. Under a lighting regimen with lights on from 5:00 AM to 7:00 PM the hamster shows peak levels of serum LH at 4:00 PM; the serum LH concentration at 4:00 PM is 50 to 100 times greater than at 12:30 PM or at 7:00 PM (Goldman, Mahesh, and Porter, 1971).

Hamsters that were ovariectomized at late diestrus or early proestrus still released LH on the afternoon of proestrus (see above), but the peak serum LH concentrations were only about 50% as high as the levels observed in intact animals (Goldman, Mahesh, and Porter, 1971). At present the failure of the acutely ovariectomized animals to show the full normal increase in serum LH remains unexplained. Nevertheless, it has been reported that in the rabbit LH stimulates ovarian progestin and that the progestin exerts a positive feedback to cause the release of additional LH (Hilliard et al., 1967). A similar mechanism may exist in the hamster and, if so, might account for the decrement in LH release following acute ovariectomy.

A second unexplained observation in the course of these experiments was as follows: Hamsters that were ovariectomized and serially bled on days other than the projected day of proestrus usually showed a slight increase in serum LH concentration between 12:30 PM and 4:00 PM, followed by a decrease between 4:00 PM and 7:00 PM. These fluctuations were much smaller than those observed in hamsters that were ovariectomized late during the second diestrous day or on early proestrus and bled on proestrus. Nevertheless, when data from several such groups were combined, the increase in serum LH at 4:00 PM was statistically significant. This observation raises the question of a possible diurnal rhythm in the LH release mechanism similar to the rhythm that is apparent on the proestrous day. However, the issue is confounded by the fact that such diurnal fluctuations were not apparent in intact hamsters on days other than proestrus. Perhaps the overt expression of the rhythm—i.e., LH release—is suppressed on these days by the presence of ovarian hormones. If the neural elements that control the cyclic release of LH can be localized and studied via a variety of approaches, we may be able to learn more about the nature of the rhythmicity of LH secretion. Recent reports suggest that in the rat an increase in neural firing in the arcuate region may be correlated with LH release (Teresawa and Sawyer, 1969); however, further work is required to elucidate possible causal relationships between neural activity and hormone release.

ROLE OF OVULATORY LH RELEASE   It has long been known that the surge of gonadotropin is required for the ovulation of mature follicles and the subsequent formation of corpora lutea. In most species that show spontaneous ovulatory cycles, the period of behavioral estrus coincides with the ovulatory period; that is, a period of sexual receptivity ensues shortly following the time of ovulatory LH release. This raises the question of whether the release of LH is involved in initiating the sexual behavior. Since many species are effectively brought into heat by sequential exposure to estrogen and progesterone and since LH stimulates the secretion of progesterone by the preovulatory follicle, it seemed likely that the LH spike could trigger a phase of progestin secretion, leading to the initiation of a period of sexual receptivity. To test this hypothesis, hamsters were injected with either gonadotropin antiserum (GTH A/S) or normal rabbit serum (NRS) shortly prior to the expected period of LH release on proestrus. (This GTH antiserum was prepared by immunizing rabbits to ovine LH; the antiserum neutralizes the biological activity of both LH and FSH of hamster origin.) These animals were then tested for sexual receptivity by exposing them to stud males. The NRS controls began to show lordotic responses at 7:00 to 8:00 PM of the proestrous day, but the females that received GTH A/S failed to show estrous behavior even when tested as late as 3:00 AM of the following morning (Table I). Thus, neutralization of endogenous LH by the

TABLE I

Blockade of sexual behavior and ovulation in the hamster by administration of gonadotropin antiserum (GTH A/S)

Treatment	Lordosis	Ovulation
NRS[+] at 1:30 PM, proestrus	4/4*	4/4†
GTH A/S at 1:30 PM, proestrus	0/7	0/7

[+] Normal rabbit serum.
*Number of hamsters showing lordosis/total.
†Number of hamsters ovulating/total.

antibody appeared to result in a complete block to estrous behavior. As expected, sexual receptivity could be restored in the antiserum-treated hamsters by injecting large doses of LH. More interestingly, behavior could also be restored by injecting 200 μg progesterone. Estradiol benzoate, even in extremely large doses (8 to 50 μg), failed to effectively restore sexual behavior. Neither steroid had any significant effect on ovulation, which was prevented by treatment with the GTH A/S. These results appear to support the LH-progesterone-sexual receptivity hypothesis.

Progesterone exerts a biphasic effect upon the expression of sexual behavior. That is, initially progesterone (in the estrogen-primed animal) induces a period of heat that lasts for several hours; this is followed by a period during which the animal is unreceptive and is refractory to further exposure to sex steroids. Previous evidence for this biphasic effect was derived from experiments with chronically ovariectomized, steroid-treated animals (Zucker, 1966). To determine whether endogenous progesterone exerts a biphasic effect during the normal cycle, the paradigm described above was employed with modification as follows: Female hamsters received either NRS or GTH A/S shortly prior to the onset of the proestrous LH surge and were injected with 4 μg estradiol benzoate later on the same day. On the following day, all the animals received 800 μg progesterone and were tested for lordotic responses 6 to 8 hours later. Hamsters which had been treated with GTH A/S were highly responsive, while those that had received NRS were either totally unresponsive or only weakly receptive (Table II). These results, when

TABLE II

Induction of sexual receptivity by sequential injection of estradiol benzoate (EB) and progesterone (P) following ovulation blockade by GTH A/S

Treatment			Results	
1:30 PM, Proestrus	4:00 PM, Proestrus	2:00 PM, "Estrus"	Lordosis	Ovulation
GTH A/S	EB	P	8/8	0/8
NRS	EB	P	3*/9	9/9

*Lordotic responses of short duration were observed.

considered in conjunction with those presented above (which indicate that GTH A/S blocks the normal period of estrous behavior by preventing LH-dependent progesterone secretion), suggest that ovarian progesterone exerts a biphasic effect upon sexual behavior in the cyclic hamster.

The pituitary and sexual differentiation

While the role of the pituitary in the regulation of gonadal steroidogenesis during adulthood is well known, its functions during early stages of development have received much less attention. However, the testis is hormonally active early in life and its secretions are of paramount importance for the process of sexual differentiation. This raises the question of possible involvement of gonadotropic hormones. Indeed, recent reports indicate that a hypothalamic-pituitary-gonadal feedback axis is operative early in life and that, at least in the rat, pituitary gonadotropin has a role in the control of the process of sexual differentiation.

BLOOD LEVELS OF GONADOTROPINS IN EARLY LIFE   Serum levels of both LH and FSH as measured by radioimmunoassays are low in neonatal male rats (1 to 12 days of age) as compared to adult males. This is not due to an inability of the pituitary to secrete these hormones, because levels of both FSH and LH in blood were greatly increased by 1 day following castration in neonates. Indeed, the response to castration was almost as dramatic in the neonatal male as in the adult (Goldman et al., 1971). Also, after hemicastration at 10 days of age, compensatory hypertrophy of the testis was evident within 2 days and plasma FSH levels had risen to 240% above the control level (Ojeda and Ramirez, 1972). These observations suggest that the hypothalamic-pituitary system of the neonate is under strong inhibition by testicular hormone(s). Further evidence for the early existence of such a feedback system was provided by the observation that both exogenous androgen and estrogen were capable of reducing the blood levels of gonadotropins in castrated neonatal male rats (Goldman and Gorski, 1971).

The female neonatal rat shows a very different pattern of gonadotropin secretion as compared to the male. Blood concentrations of both LH and FSH are high during the first 15 days of life (Goldman et al., 1971; Kragt and Dahlgren, 1972). Little change in hormone levels was observed one day following ovariectomy, although some suppression occurred after administration of testosterone propionate and, at least for FSH, with estradiol benzoate. Furthermore, compensatory hypertrophy was not observed until 20 to 25 days of age following hemiovariectomy on day 10, and no significant elevation of plasma LH or FSH levels was observed in the hemiovariectomized rats (Ojeda and Ramirez, 1972). Similar results have been obtained in studies of the sheep. In this species, serum LH levels are low in both sexes for 10 days after birth but then rise in the female only. Ovariectomized (at 5 days of age) and intact female lambs showed similar patterns of LH in the blood, and the LH levels were suppressed by estradiol in both groups (Foster, Cook, and Nalbandov, 1972; Liefer, Foster, and Dziuk, 1972). These observations suggest that the neonatal ovary may be relatively unresponsive to gonadotropins, at least with respect to steroidogenesis. Thus, the state of the hypothalamic-pituitary axis in the neonatal female probably approximates that in the gonadectomized animal, whereas in the male testicular steroid, feedback is evident at birth.

GONADOTROPINS AND SEXUAL DIFFERENTIATION   The action of endogenous gonadotropins was blocked in neonatal rats by administering antibodies. It was found that GTH A/S resulted in permanent sterility if administered to male rats with treatment beginning on the day of birth and extending to day 5 of life. The same amount of anti-serum failed to induce sterility if administered on days 7 to 11 of life. The *critical period* observed for these experiments (days 1 to 5 after birth) is the same period during which exogenous androgen is most effective in *masculinizing* the CNS in female rats and in neonatally castrated males. Also, administration of testosterone propionate at birth partially prevented the effect of the GTH A/S. These observations led to the hypothesis that GTH A/S acted by preventing gonadotropin-dependent testicular steroidogenesis during a *critical* developmental period (Goldman and Mahesh, 1970).

Further studies showed that the sterility of the GTH A/S-treated male rats resulted from an incomplete pattern of sexual behavior in adulthood. The experimental males frequently mounted receptive females, but they achieved few intromissions and rarely ejaculated. It was also observed that the size of the penis was reduced in the antiserum-sterilized males. Since sensory feedback from the phallus is believed to be important in the regulation of masculine sexual behavior, it seems that the incomplete pattern of behavior might have been a result of reduced phallus size. This hypothesis would appear to be in harmony with the postulated mechanisms for action of GTH A/S, since testicular secretions in early life have been shown to be important in *programming* phallus development.

Unlike normal male rats, males castrated at birth respond to estradiol-progesterone treatment in adulthood by displaying frequent lordotic responses when mounted by a stud male. This has been attributed to a failure of CNS masculinization in the neonatal castrates. Male rats treated with GTH A/S during days 1 to 5 of life also showed strong feminine sexual behavior following estrogen-progesterone treatment in adulthood. Thus, a failure of masculinization of behavioral control centers in the brain is indicated in this preparation as well (Goldman et al., 1972).

A male rat that is castrated in adulthood and given an ovarian transplant secretes gonadotropin only in a tonic fashion, so that the transplanted ovary shows follicular development but no evidence of ovulations. However, a male that is castrated at birth is capable of cyclic release of gonadotropins in adulthood, implicating testicular hormone in the masculinization of the GTH control system. Therefore, it was of interest to determine whether the required testicular hormones were dependent upon the neonatal pituitary, as proved to be the case for the system controlling sex behavior. Thus, male rats treated with GTH A/S on days 1 to 5 of life were castrated in adulthood and given ovarian grafts. Five weeks later the grafts showed follicular development but no evidence of ovulations, just as in normal males. However, these antiserum-treated males showed the same behavioral alterations as described above. Two explanations seemed possible: (1) The amount of GTH A/S may have been insufficient to counteract all endogenous gonadotropin and the remaining gonadotropin may have stimulated sufficient androgen for masculinization of the GTH release mechanism. (2) The neonatal testis may be capable of secreting some androgen even in the absence of GTH. Either explanation would require that the mechanism for control of the cyclic secretion of GTH be more sensitive to the developmental influence of androgen than are the mechanisms for control of sexual behavior. The literature lends some support to such a difference in relative sensitivities (Barraclough and Gorski, 1962). The results described here indicate the need to study the hypothalamic-pituitary-gonadal axis during early life in order to understand the development of the pertinent control systems.

ACKNOWLEDGMENT   The work described here was supported in part by a Research Grant from the National Institutes of Health, Division of Child Health and Human Development.

REFERENCES

BARRACLOUGH, C. A., and R. A. GORSKI, 1962. Studies on mating behavior in the androgen-sterilized female rat in relation to the hypothalamic regulation of sexual behavior. *J. Endrocrinol.* 25:175–182.

EDWARDS, D. A., 1970. Induction of estrus in female mice: Estrogen-progesterone interactions. *Hormone Behav.* 1:299–304.

FERIN, M., A. TEMPONE, P. ZIMMERING, and R. VANDE WIELE, 1969. Effect of antibodies to 17β estradiol and progesterone on the estrous-cycle of the rat. *Endocrinology* 85:1070–1078.

FOSTER, D. L., B. COOK, and A. V. NALBANDOV, 1972. Regulation of luteinizing hormone (LH) in the fetal and neonatal lamb: Effects of castration during the early postnatal period on levels of LH in sera and pituitaries of neonatal lambs. *Biol. Reprod.* 6:253–257.

GODING, J. R., K. J. CATT, J. M. BROWN, C. C. KALTENBACK, I. A. CUMMING, and B. J. MOLE, 1969. Radioimmunoassay for ovine luteinizing hormone. Secretion of luteinizing hormone during estrus and following estrogen administration in the sheep. *Endocrinology* 85:133–142.

GOLDMAN, B. D., and V. B. MAHESH, 1970. Induction of infertility in male rats by treatment with gonadotropin antiserum during neonatal life. *Biol. Reprod.* 2:441–451.

GOLDMAN, B. D., and R. A. GORSKI, 1971. Effects of gonadal steroids on the secretion of LH and FSH in neonatal rats. *Endocrinology* 89:112–115.

GOLDMAN, B. D., Y. R. GRAZIA, I. A. KAMBERI, and J. C. PORTER, 1971. Serum gonadotropin concentrations in intact and castrated neonatal rats. *Endocrinology* 88:771–776.

GOLDMAN, B. D., V. B. MAHESH, and J. C. PORTER, 1971. The role of the ovary in control of cyclic LH release in the hamster, *Mesocricetus auratus. Biol. Reprod.* 4:57–65.

GOLDMAN, B. D., D. M. QUADAGNO, J. SHRYNE, and R. A. GORSKI, 1972. Modification of phallus development and sexual behavior in rats treated with gonadotropin antiserum neonatally. *Endocrinology* 90:1025–1031.

HILLIARD, J., R. PENARDI, and C. H. SAWYER, 1967. A functional role for 20-α-hydroxypregn-4-en-3-one in the rabbit. *Endocrinology* 80:901–909.

KRAGT, C. L., and J. DAHLGREN, 1972. Development of neural regulation of follicle stimulating hormone (FSH) secretion. *Neuroendocrinology* 9:30–40.

LABHSETWAR, A. P., 1972. Role of estrogens in spontaneous ovulation: Evidence for positive feedback in hamsters. *Endocrinology* 90:941–946.

LEAVITT, W. W., and G. C. BLAHA, 1970. Circulating progesterone levels in the golden hamster during the estrous cycle, pregnancy and lactation. *Biol. Reprod.* 3:353–361.

LIEFER, R. W., D. L. FOSTER, and P. DZIUK, 1972. Levels of LH in the sera and pituitaries of female lambs following ovariectomy and administration of estrogen. *Endocrinology* 90: 981–985.

LUKASZEWSKA, J. H., and G. S. GREENWALD, 1970. Progesterone levels in the cyclic and pregnant hamster. *Endocrinology* 86: 1–9.

OJEDA, S. R., and V. D. RAMIREZ, 1972. Plasma level of LH and FSH in maturing rats: Response to hemigonadectomy. *Endocrinology* 90:466–472.

SCARAMUZZI, R. J., S. A. TILLSON, I. H. THORNEYCROFT, and B. V. CALDWELL, 1971. Action of exogenous progesterone and estrogen on behavioral estrus and luteinizing hormone levels in the ovariectomized ewe. *Endocrinology* 88:1184–1191.

TERESAWA, E., and C. H. SAWYER, 1969. Changes in electrical activity in the rat hypothalamus related to electrochemical stimulation of adenohypophyseal function. *Endocrinology* 85: 143–149.

ZUCKER, I., 1966. Facilitatory and inhibitory effects of progesterone on sexual responses of spayed guinea pigs. *J. Comp. Physiol. Psychol.* 62:376–381.

53 Differential Hypothalamic Sensitivity to Androgen in the Activation of Reproductive Behavior

JOHN B. HUTCHISON

ABSTRACT Androgens influence reproductive behavior by direct action on the anterior hypothalamus. Studies to determine the mode of action of androgen using intrahypothalamic implants of testosterone propionate in castrated male doves show that aggressive components of courtship behavior require higher hypothalamic concentrations of androgen for their activation than nest-orientated components, indicating that mechanisms within the androgen-sensitive areas of the hypothalamus associated with courtship behavior are differentially sensitive to androgen; mechanisms associated with aggressive behavior have a higher threshold of sensitivity to androgen than those associated with nest-orientated behavior. Since the behavioral effects of hypothalamic implants of testosterone propionate diminish as the period between castration and implantation is lengthened, these sensitivity thresholds change according to the endocrine state of the animal. It is concluded that the activating effects of androgen on mechanisms underlying reproductive behavior depend both on effective hormone concentration in the anterior hypothalamus and the threshold of the hypothalamic mechanism underlying the particular behavior pattern.

Androgenic activation of brain mechanisms mediating behavior

STEROID SEX hormones influence brain mechanisms underlying reproductive behavior in two major ways: in perinatal development, hormones "organize" the neural substrate that later integrates reproductive behavior (Goy, 1970); in adulthood, hormones "activate" these brain mechanisms in order that reproductive behavior may be elicited by the appropriate sensory stimulation. It is becoming clear that the activational process may involve the influence of hormones on peripheral sensory systems that modify afferent input to the brain. Thus the sensory field of the pudendal nerve in female rats, corresponding to external genital areas important in tactile elicitation of lordosis, is significantly larger, when determined using a multiunit recording technique (Komisaruk, Adler, and

JOHN B. HUTCHISON   Medical Research Council Unit on the Development and Integration of Behaviour, University Sub-Department, Madingley, Cambridge, England

Hutchison, 1972), in ovariectomized females treated with estrogen than in females receiving no hormone treatment. The most important influence of hormones in relation to behavior is on the brain itself with consequent direct modification of brain function. Numerous studies (reviewed by Hutchison, 1970a) have demonstrated, using intracerebral crystalline steroid implants with a limited diffusion range, that areas in the hypothalamus are hormone sensitive and closely linked with the mechanisms underlying reproductive behavior.

While there is some knowledge of the areas of the brain that mediate the effects of sex steroids on behavior, the activational process is little understood. Two aspects of the central activation of behavior in the adult will be considered: first, that the hypothalamus plays a part in determining which behavior pattern is displayed, and it does so by means of a system that is differentially sensitive to androgen concentration within the hypothalamus; second, that this system is unstable and organized so that the sensitivity of the hypothalamus to androgen in relation to behavior is modified according to the endocrine state of the animal.

Birds are particularly useful in the investigation of these problems in view of their elaborate visual courtship displays, which are stereotyped and easily quantified. The courtship behavior of the male Barbary dove consists initially of a rapid alternation of aggressive displays (termed chasing and bowing), which cause the female to retreat, and nest-orientated behavior (termed nest soliciting), where the male selects a potential site for the nest, which causes the female to approach.

All courtship displays of the male decline and disappear within 20 days of castration and are rapidly reinstated by daily 300 μg intramuscular injections of testosterone propionate, indicating that these displays depend on gonadal hormones and that testosterone may be the effective steroid (Hutchison, 1970b). Qualitatively normal courtship displays can also be reinstated by means of 20 μg, fused, crystalline testosterone propionate implants inserted stereotaxically into the preoptic or anterior hypothalamic areas but not into other areas of the brain (Figure 1), suggesting that a localized area sensitive to testosterone is closely associated with the coordination of courtship behavior (Hutchison, 1971). Although studies are in progress, the distribution of cells in the hypothalamus that take up testosterone has not been estab-

lished yet in the male dove. However, the distribution of preoptic and anterior hypothalamic cells, which take up radioactivity after injection of [³H]-testosterone, determined in the male chaffinch by Zigmond, Nottebohm, and Pfaff (1972), is similar to the area delimited by testosterone propionate implants in male doves (Figure 1).

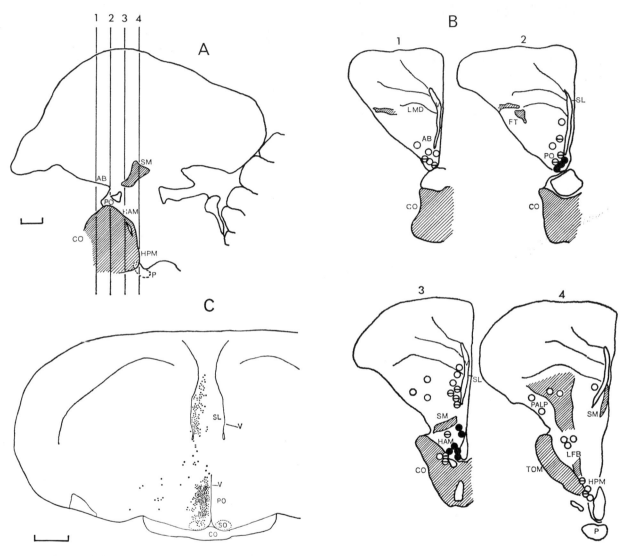

FIGURE 1 (A) (*upper left*). Sagittal view of the brain of a Barbary dove approximately 1 mm lateral to the midline showing position and plane of coronal sections 1–4 in Figure 1B. (B) (*upper* and *lower right*). Maps showing locations of intracerebral implants of testosterone propionate in castrated male Barbary doves. Each symbol denotes the position of the tip of the implant in the coronal section and the type of behavior elicited. (●) complete courtship with aggressive and nest-orientated behavior; (⊖) incomplete courtship lacking in either aggressive behavior or nest-orientated behavior; (○) no courtship—behavior similar to castrate. (C) (*lower left*). Map showing locations of labeled cells in the male chaffinch forebrain following systemic administration of [³H]-testosterone. Chaffinches were decapitated 1 hr following intramuscular injection of the labeled hormone. Auto-

radiograms were prepared by mounting unfixed, unembedded frozen brain sections (6μ thick) in the darkroom directly onto emulsion-coated slides. In the map, each dot shows the position of a labeled cell in a coronal section at the level of the preoptic area (PO) in one representative chaffinch. (From Zigmond, Nottebohm, and Pfaff, 1972).

AB, area basalis; CO, optic chiasma; FT, tractus fronto-thalamicus; HAM, nucleus hypothalamicus anterior medialis; HPM, n. hypothalamicus posterior medialis; LFB, lateral forebrain bundle; LMD, lamina medullaris dorsalis; V, lateral forebrain ventricle; PALP, paleostriatum primitivum; P, pituitary; PO, n. preopticus medialis; SL, lateral septum; SO, n. supraopticus; SM, tractus septomesencephalicus; TOM, tractus opticus marginalis. (Scales represent 1 mm.)

Hypothalamic androgen concentration and behavior

The significant feature about the action of testosterone on the male dove brain is that the type of courtship displayed appears to depend on the concentration of testosterone in the anterior hypothalamus. This is suggested by consistent differences between the rates of decline of the courtship patterns after castration (Hutchison, 1970b). Whereas bowing disappears immediately after castration, chasing and nest soliciting decline at approximately the same rate in terms of number of days on which these patterns are displayed after castration. However, quantitative analysis shows that chasing occupies progressively less of the declining courtship displays than does nest soliciting. Since it can be assumed that endogenous androgens are metabolized and disappear from the peripheral plasma within minutes after castration, aggressive behavior, which disappears first, may depend more on current plasma levels and relatively high concentrations of androgen in the hypothalamus than does nest-orientated behavior.

The view that the concentration of hormone in the hypothalamus is related to the type of behavior displayed can be tested by manipulating hypothalamic androgen level directly. First, testosterone can be elevated to different concentrations in the hypothalamus using solid implants of differing surface areas because the rate of release of hormone from a solid implant in the hypothalamus is proportional to the surface area (Michael, 1965). Second, the effects of testosterone on the hypothalamus can be reduced by means of a testosterone antagonist. Using the first of these experimental approaches, 3 types of implants with differing surface areas were compared: (a) spherical high-diffusion implants (55 to 85 μg); (b) spherical medium-diffusion implants (25 to 55 μg); (c) low diffusion implants (hormone contained in bore of 27 gauge stainless steel tubing). Both high- and medium-diffusion implants restored the display of behavior, but a larger proportion of males with high diffusion implants displayed aggressive behavior for consistently longer periods than males with medium diffusion implants. The low-diffusion implants resulted in behavior that was strikingly different; the aggressive components of courtship were absent, whereas nest-orientated behavior was restored to precastration levels (Hutchison, 1970a).

Progesterone acts as a testosterone antagonist in doves and will block the courtship-inducing effects of testosterone by acting at the level of the hypothalamus (Hutchison, in preparation). This progesterone-testosterone antagonism was used to reduce the effectiveness of testosterone acting on the hypothalamus and consequently to change the balance of courtship components. The prediction was made that as progesterone concentration in the hypothalamus increased, selectively blocking the effects of testosterone, the aggressive components of courtship would decline relative to the nest-orientated behavior. This was found to be the case in males treated with progesterone and testosterone propionate (300 μg of each hormone/day). The effects on courtship were very similar to those obtained by castration. Thus aggressive behavior declined rapidly and disappeared within 3 to 4 days of initial treatment. Nest-orientated behavior continued to be displayed by the majority of males until the tenth day after initial treatment.

Taken together, the results of the two methods of manipulation of hypothalamic testosterone level are consistent with the hypothesis that when effective concentrations of testosterone in the anterior hypothalamus are high both aggressive and nest-orientated behavior will be displayed, whereas when testosterone concentrations are lower, aggressive behavior will be absent. It can be suggested that the anterior hypothalamus differentiates between testosterone concentrations by means of a threshold system organized so that mechanisms in the anterior hypothalamus associated with aggressive behavior have a higher threshold of sensitivity to androgen than those associated with nest-orientated behavior. In fact, the hypothalamic mechanism associated with courtship behavior may not have an absolute "all-or-none" threshold of response to testosterone, which requires a finely tuned sensitivity to testosterone concentration in the anterior hypothalamus. Rather, the mechanism associated with aggressive behavior may be activated by testosterone concentrations falling within a range that differs from that required for the activation of nest-orientated behavior. Further work will, however, be required to determine the limits of these ranges.

At present, it is only possible to point to a cellular mechanism that would form the basis for the threshold system. Steroid uptake of hypothalamic cells may be differentially sensitive to androgen concentration. Thus certain cells may have a greater affinity for binding androgen than other androgen-sensitive cells. The linear relationship between brain tissue, including hypothalamus, and plasma uptake of [³H]-testosterone (McEwen, Pfaff, and Zigmond, 1970) and estradiol (McEwen and Pfaff, 1970; McEwen, 1972) in male and female rats indicates that a mechanism of this sort may operate physiologically.

Changes in hypothalamic sensitivity to androgen

The second point relating to the activational effects of hormones on the hypothalamus is that the thresholds of sensitivity to testosterone of brain mechanisms underlying behavior patterns may change according to the endocrine state of the animal. Two observations suggest that hypothalamic sensitivity to testosterone declines

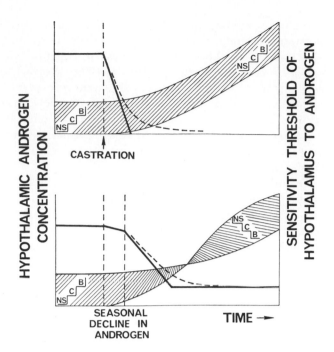

FIGURE 2 Model of changes in hypothalamic sensitivity to androgen underlying male courtship behavior in doves. Following elimination of testicular androgen after castration, hypothalamic thresholds of sensitivity to androgen are elevated and interrelationships between thresholds of individual behavior patterns are maintained. Following seasonal decline of endogenous androgen, thresholds of sensitivity are elevated, but interrelationships between thresholds of individual behavior patterns are inverted. Heavy line, hypothalamic androgen concentration; heavy dotted line, residual effects of androgen on hypothalamic cells; hatched area, threshold of sensitivity of the hypothalamus to androgen, stepped line, relationships between sensitivity thresholds of mechanisms underlying aggressive behavior (bowing, B; chasing, C); and nest-orientated behavior (nest soliciting, NS).

following castration: First, courtship is seldom restored to precastration levels by systemic testosterone therapy in the majority of castrates; second, 30% of all castrates do not respond to testosterone therapy unless the dosage normally required to elicit courtship behavior (300 µg/day) is doubled. These effects may, of course, be due to peripheral metabolic changes that might alter the rate of transport of hormone to the brain. Therefore, the hypothesis that a specifically hypothalamic change underlies these behavioral differences can be tested only by implanting testosterone propionate directly into the hypothalamus at different periods after castration. When this was carried out at 15, 30, and 90 days after castration, there were distinct qualitative differences between the behavioral responses of the groups of castrates. Whereas implants of testosterone propionate were highly effective in restoring courtship in 15 day castrates, their effective-

ness was lower in 30 day castrates, particularly with respect to bowing behavior. Implants were almost completely ineffective in initiating the display of courtship patterns in 90 day castrates (Hutchison, 1969). The behavioral effects of testosterone propionate implants in the anterior hypothalamus are therefore inversely related to the duration of the period between castration and implantation, indicating that the threshold of sensitivity of the anterior hypothalamus to testosterone rises after castration. That this is not an irreversible change can be demonstrated by injecting a group of 90 day castrates intramuscularly with testosterone propionate (300 µg/day). The display of courtship was re-established after a delay of 3 to 4 days. This latency to the initial display of courtship behavior was significantly longer in males treated systemically 90 days after castration than in males treated at 30 days where the delay was only 1 to 2 days. The longer delay in the 90 day males can be understood on the hypothesis that sensitization of the anterior hypothalamus, by the continuous influence of testosterone, is necessary before the activating effects of the hormone can be mediated by the hypothalamus.

A further question is whether the interrelationship between the sensitivity thresholds of mechanisms underlying aggressive and nest-orientated behavior to testosterone remains stable after castration. This would appear to be the case, because nest soliciting with the lowest threshold of sensitivity to testosterone was elicitable from a few 90 day castrates, whereas bowing was not evoked at all. However, when sexually inactive males that showed no courtship were tested with hypothalamic implants in winter, when their testes were atrophic and plasma testosterone levels were presumably at their lowest, it was found that some chasing and bowing was elicited, but no nest soliciting. This suggests that threshold relationships were reversed in these animals such that mechanisms associated with aggressive behavior had a lower threshold of sensitivity to testosterone than those associated with nest-orientated behavior (Figure 2). Supplementary evidence for this was provided by determination of plasma testosterone levels in a sample of intact males undergoing normal seasonal development that consistently showed the aggressive components of courtship unaccompanied by nest-orientated behavior. These males had significantly lower plasma testosterone levels (Katongole and Hutchison, in preparation) than males that displayed both aggressive and nest-orientated components. The sensitivity threshold relationships of the aggressive males would appear to be the reverse of the group that displayed complete courtship consisting of both aggressive and nest-orientated components.

In considering the cellular mechanisms involved in changes in threshold of hypothalamic sensitivity to testosterone, it is relevant to view the hypothalamus as an endocrine target organ. Cell fractionation studies in male doves indicate the hypothalamic cell nuclei accumulate 14 times as much [^{3}H]-testosterone as the cerebrum (Zigmond, Stern, and McEwen, 1972). Moreover, some of the testosterone is converted to 5α-dihydrotestosterone, the metabolite thought to play a part in mediating the effects of testosterone on target cells. Do testosterone receptor cells change their steroid uptake and retention characteristics after castration? Experiments to answer this question are in progress with regard to the dove brain. The latency to peak uptake of [^{3}H]-estradiol in female rat brain is proportional, however, to the period after castration (Lisk, 1971). In the absence of estrogen, it would appear that the molecules which bind steroids degenerate or become inactivated. Therefore, the activational effects of a hormone on behavioral mechanisms may depend on two variables, regeneration of steroid binding "receptors" in hypothalamic cells and retention of hormone by these cellular "receptors."

A final question concerns the development of individual differences in the expression of adult male behavior. Does the development of particular hypothalamic thresholds of sensitivity to circulating testosterone determine the type of adult behavior an individual will display? In doves (Hutchison, 1970a) as in other species, individual behavior varies quantitatively. Although individual behavioral differences are probably due in part to differences in experience and learning during development, it is possible that the ways in which increasing levels of gonadal steroids influence the hypothalamus during development are critical in determining the individual's later responsiveness to the activational effects of testosterone. Since perinatal steroid treatment of female rats has been shown to affect the development of hypothalamic steroid binding molecules in adulthood (Flerko, Mess, and Illei-Donhoffer, 1969; McGuire, and Lisk, 1969; McEwen, and Pfaff, 1970; see also the chapter in this part by McEwen), and because a maturational sequence can be traced in the development of hypothalamic oestradiol "receptors" (Kato, 1972), it can be suggested that sex hormone concentration may be important at critical perinatal and pubertal stages of development in determining the adult activation threshold of the system, by affecting the generation of steroid "receptors."

REFERENCES

FLERKO, B., B. MESS, and A. ILLEI-DONHOFFER, 1969. On the mechanism of androgen sterilization. *Neuroendocrinology* 4: 164–169.

GOY, R. W., 1970. Experimental control of psychosexuality. *Phil. Trans. Roy. Soc. London B.* 259:149–162.

HUTCHISON, J. B., 1969. Changes in hypothalamic responsiveness to testosterone in male Barbary doves (*Streptopelia risoria*). *Nature (London)* 222:176–177.

HUTCHISON, J. B., 1970a. Influence of gonadal hormones on the hypothalamic integration of courtship behaviour in the Barbary dove. *J. Reprod. Fert. Suppl.* 11:15–41.

HUTCHISON, J. B., 1970b. Differential effects of testosterone and oestradiol on male courtship in Barbary doves (*Streptopelia risoria*). *Anim. Behav.* 18:41–52.

HUTCHISON, J. B., 1971. Effects of hypothalamic implants of gonadal steroids on courtship behaviour in Barbary doves (*Streptopelia risoria*). *J. Endocrinol.* 50:97–113.

KATO, J., 1972. Maturation of estradiol receptors in female rat hypothalamus and the onset of puberty. Proceedings of the 4th International Congress of Endocrinology (abstract).

KOMISARUK, B. R., N. T. ADLER, and J. B. HUTCHISON, 1972. Genital sensory field: enlargement by oestrogen treatment in female rats. *Science* 178:1295–1298.

McEWEN, B. S., and D. W. PFAFF, 1970. Factors influencing sex hormone uptake by rat brain regions. 1. Effects of neonatal treatment, hypophysectomy, and competing steroids on estradiol uptake. *Brain Res.* 21:1–16.

McEWEN, B. S., D. W. PFAFF, and R. E. ZIGMOND, 1970. Factors influencing sex hormone uptake by rat brain regions. II. Effects of neonatal treatment and hypophysectomy on testosterone uptake. *Brain Res.* 21:17–28.

McEWEN, B. S., 1972. Sites of steroid binding and action in the brain. In *Structure and Function of Nervous Tissue*, G. Bourne, ed. Vol. V. New York: Academic Press.

McGUIRE, J. L., and R. D. LISK, 1969. Oestrogen receptors in the androgen or oestrogen sterilized female rat. *Nature (London)* 221:1068–1069.

MICHAEL, R. P., 1965. Oestrogens in the central nervous system. *Br. Med. Bull.* 21:87–90.

LISK, R. D., 1971. The physiology of hormone receptors. *Am. Zoologist* 11:755–767.

ZIGMOND, R. E., F. NOTTEBOHM, and D. W. PFAFF, 1972. Distribution of androgen-concentrating cells in the brain of the male chaffinch. Progress in Endocrinology: Proceedings of the IVth International Congress of Endocrinology. *Exerpta Medica*, International Congress Series No. 256 (abstract No. 340).

ZIGMOND, R. E., J. M. STERN, and B. S., McEWEN, 1972. Retention of radioactivity in cell nuclei in the hypothalamus of the ring dove after injection of ^{3}H-testosterone. *Gen. Comp. Endocrinol.* 18:450–458.

54 Chemical Studies of the Brain as a Steroid Hormone Target Tissue

BRUCE S. McEWEN, CARL J. DENEF, JOHN L. GERLACH, and LINDA PLAPINGER

ABSTRACT The brain is a "target" tissue for the action of steroid hormones on neural processes underlying neuroendocrine function and certain behaviors. Neurochemical aspects of steroid hormone action are discussed in this chapter. Where steroids regulate function of mature cells, those target cells contain stereospecific intracellular binding sites for the active hormone that are located in the soluble portion of the tissue and in the cell nuclei. Such sites have been found for estradiol and corticosterone in anatomically defined brain regions. In certain instances, the target tissue contains enzymes that convert the hormone to a more active metabolite, which interacts with intracellular receptor sites. Such metabolism is illustrated by the conversion of testosterone to 5α-dihydrotestosterone in different brain regions. Where steroids interact with differentiating tissues, certain characteristics of those tissues often become permanently altered. Such "organizing" influences of steroid hormones also occur in the brain and are considered from a neurochemical point of view and compared with regulatory effects of hormones on mature brain cells.

Introduction

STEROID HORMONES regulate certain cell functions in differentiated *target* cells and in special instances alter permanently the developmental course of differentiating cells. Both types of phenomena are manifested in the interaction of steroid hormones with the brain and have important consequences for neuroendocrine physiology and behavior. Since other chapters in this volume are devoted to the functional aspects of hormone action on the brain, the emphasis in this chapter will be on a cellular and molecular mechanism by which certain steroid hormones may act.

A particularly fruitful approach has been to study the fate of a radioactively labeled hormone in the target tissue. Such studies have demonstrated in differentiated target cells of brain and other tissues the existence of high affinity, stereospecific hormone binding proteins—presumptive *receptors*—located in the soluble portion of the tissue and in the cell nuclei. Evidence will be reviewed

BRUCE S. McEWEN, CARL J. DENEF, JOHN L. GERLACH, and LINDA PLAPINGER The Rockefeller University, New York

showing that certain regions of the rat brain that respond to implanted estradiol contain nuclear and soluble estradiol receptors. A different pattern of soluble and cell nuclear binding exists for corticosterone, principal adrenal glucocorticoid in the rat. Corticosterone binding sites occur brain-wide and are present in high levels in two limbic structures, hippocampus and amygdala. Evidence is reviewed pointing to glucocorticoid influence upon a number of behavioral, neuroendocrine, and neurophysiological events, at least two of which can be tentatively associated with hippocampal function.

While most steroid hormones bind to receptors without being metabolized, testosterone is transformed intracellularly to 5α-dihydrotestosterone (5α-DHT) by certain of its target tissues before it binds to intracellular receptors and influences cell function. In brain, 5α-DHT is effective in suppressing gonadotrophin secretion. In certain nonneural tissues, and apparently in rat brain as well, androgen is aromatized to estrogen, and aromatizable androgen, but not 5α-DHT, is capable of producing the permanent effects on neuroendocrine function and behavior that occur during the *critical period* of sexual differentiation. The mechanism of the developmental effects of testosterone and estrogen is briefly considered at the end of the chapter.

Interaction of hormones with differentiated cells

We recognize as hormones a wide variety of chemical substances, including amino acid derivatives, polypeptides, proteins, and steroids. In view of this chemical diversity, no single cellular mechanism of action would seem to account for the effect of all hormones on target cells. In fact, two mechanisms are now recognized by which hormones regulate cell function. These are illustrated in Figure 1. Certain amino acid derivatives, e.g. epinephrine, and polypeptides, e.g. glucagon and adrenocorticotrophic hormone (ACTH), are known to interact with receptors on the cell surface and set in motion the conversion of ATP to cyclic AMP within the cell (Figure 1a). The cyclic AMP is believed to combine with receptors and to stimulate the action of a phospho-protein phosphokinase which phosphorylates a number of proteins (Figure 1a). It is these phosphorylated proteins that are believed to complete the action of the hormone on cell metabolism (see Robison et al., 1971).

Steroid hormones are recognized to interact with target cells in quite a different way, combining with intracellular protein receptors located in the soluble portion of the tissue and moving into the cell nucleus, where the hormone is believed to modify expression of the genome (Figure 1b). Evidence for entry of the steroid into the target cell has been obtained indirectly in studies with

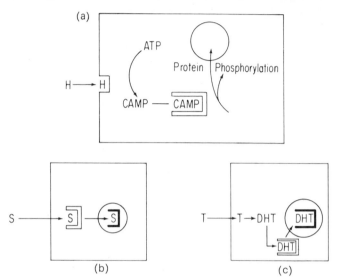

FIGURE 1 Representation of mode of interaction of hormones with target cells. (a) Certain amino-acid and polypeptide hormones (H) interact with "receptors" on the surface to stimulate intracellular production of cyclic AMP (CAMP) from ATP. CAMP binds to an intracellular "receptor" and stimulates phosphorylation of intracellular proteins. Steroid hormones (S) enter target cells and bind to intracellular "receptors" and enter the cell nucleus. (b) Binding of steroid without metabolism. (c) Conversion of testosterone (T) to 5α-dihydrotestosterone (DHT), which binds to intracellular "receptors."

sulfhydryl blocking agents (Levinson et al., 1972), and directly by autoradiography (see, for example, Stumpf, 1968; Cameron et al., 1969), and cell fractionation techniques (Jensen et al., 1969; Liao and Fang, 1969; Swaneck et al., 1969). A representative list of the hormones and target tissues is presented in Table I.

Although some evidence has been presented for cyclic AMP involvement in steroid hormone effects on the uterus (Hechter et al., 1967; Szego and Davis, 1967) and seminal vesicles (Singhal et al., 1970), it seems reasonably clear that cyclic AMP is not an obligatory intermediate in steroid hormone action (Granner et al., 1968; Wicks et al., 1969; Rosenfeld and O'Malley, 1970; deVellis et al., 1971).

TABLE I

Target tissues for steroid hormones

Target Tissue	Hormone	References
Uterus	estradiol-17β	Hamilton, 1968; Jensen and DeSombre, 1972
Oviduct	progesterone	O'Malley et al., 1972
Prostate Seminal vesicles Kidney	5α-dihydrotestos-terone	Liao and Fang, 1969; Williams-Ashman and Reddi, 1972
Thymus Liver Hepatoma	glucocorticoids	Tomkins et al., 1969 Litwack and Singer, 1972
Kidney Bladder	Mineralocorti-coids	Swaneck et al., 1969 Sharp, 1972
Intestinal epithelium	1, 25-dihydroxy-cholecalciferol	Boyle et al., 1972 DeLuca and Melancon, 1972

A major means of controlling intracellular hormone levels is intracellular metabolism of the hormone to more or less active forms, as has been demonstrated both for testosterone (see Table II) and for progesterone (Karavo-las and Herf, 1971; Snipes and Shore, 1972; Massa et al., 1972). Testosterone is known to be converted in the prostate to 5α-dihydrotestosterone (5α-DHT) before binding to intracellular receptors and entering the cell nucleus (Figure 1c). Since 5α-DHT is also considerably more active than testosterone in a number of bioassays (for review see Williams-Ashman and Reddi, 1972), the target tissue metabolism is thought to be a factor regulat-ing the intracellular level of "active" androgens.

Intracellular conversion of testosterone to 5α-DHT and other 5α-reduced metabolities, such as 3α-androstanediol, has also been observed in other androgen target tissues, in which tissues these metabolites are active androgens (see Table II). However, not all androgen target tissues re-spond to 5α-reduced metabolites of testosterone. During early differentiation of the Wolffian ducts in rats, little or no formation of 5α-DHT can be demonstrated, nor is 5α-DHT active in inducing differentiation of this tissue (Table II). Similarly, androgenization of the neonatal rat brain and stimulation of male sexual behavior in adult rats cannot be produced by 5α-DHT but are produced by testosterone (Table II). As will be discussed below, infant and adult rat brain does have the capacity to produce 5α-DHT. Production of this metabolite might be regarded as a means of inactivating the hormone. In this connection, an alternate route of metabolism, namely, aromatization of testosterone to estradiol, has been demonstrated in brain tissue (Naftolin et al., 1972) and has been suggested as a possible mechanism for androgen effects that are insensitive to 5α-DHT (McDonald et al., 1970;

Target Organ	Conversion to 5α-Reduced Metabolites	Reference	DHT	Androstanediol	Reference
			\<center\>Response to\</center\>		
\<center\>During Fetal/Neonatal Sexual Differentiation\</center\>					
Wolffian ducts (rat, rabbit)	Only after initial sexual differentiation, DHT	Wilson and Lasnitzki, 1971	yes (< T)†	yes (< T)	Schultz and Wilson, 1972
Urogenital sinus and tubercle (rat, rabbit)	DHT	Wilson and Lasnitzki, 1971	yes (> T)	yes (> T)	Goldman and Baker, 1971
Brain areas of gonadotrophin release (rat)	DHT, DIOL	see text	no		Whalen and Luttge, 1971b; Ulrich et al., 1972
Brain areas of sexual behavior (rat)	DHT, DIOL	see text	no		Whalen and Luttge, 1971b
Liver steroid metabolizing enzymes (rat)	DHT, DIOL	Denef and De Moor, 1968			
\<center\>In Mature Animals\</center\>					
Prostate, seminal vesicles, epididymis, penis, preputial gland, scrotum (several species)	DHT, DIOL	Bruchovsky and Wilson, 1968; Gloyna and Wilson, 1969; Tveter and Aakvaag, 1969; Bottiglioni et al., 1971; Richardson and Axelrod, 1971	yes (> T)	yes (< T)	Dorfman and Shipley, 1956; Liao and Fang, 1969; Baulieu et al., 1968
Testicular seminiferous tubules (rat)	DIOL	Rivarola and Podesta, 1972			
Submaxillary gland (mouse)	DHT, DIOL	Coffey, 1972			
Sebaceous gland (hamster)	DHT, DIOL	Takayasu and Adachi, 1972			
Exorbital lacrimal gland			yes (< T)	yes (> T)	Liao and Fang, 1969
Brain areas of gonadotrophin release (rat, dog)	DHT, DIOL	see text	yes (> T)	yes (> T)	Beyer et al., 1971; Davidson, 1973
Brain areas of sexual behavior (rat, dog)	DHT, DIOL	see text	no		Whalen and Luttge, 1971a; McDonald et al., 1970
Kidney (rat, mouse)	DHT, DIOL	Wilson and Glyona, 1970			

*DHT: 5α-dihydrotestosterone; DIOL: androstanediol: 3 α,17 β-dihydroxy-5α-androstane.

† > < T: response greater or smaller than that to testosterone.

Beyer et al., 1970), because 5α-DHT cannot be aromatized.

Another example of metabolism in the action of steroid hormones is the case of cholecalciferol, a steroid-like hormone classified as Vitamin D, which is produced naturally by the action of ultraviolet light on 7-dehydrocholesterol in the skin (DeLuca and Melancon, 1972). Cholecalciferol is next converted to 25-hydroxycholecalciferol in the liver (DeLuca and Melancon, 1972). This metabolite is converted to 1,25-dihydroxycholecalciferol in the kidney (Boyle et al., 1972). This latter metabolite then stimulates calcium transport in the intestine (Myrtle and Norman, 1971).

Evidence that steroid-hormone binding proteins are receptors

Intracellular proteins that bind steroid hormones are frequently referred to as *receptors*, thereby implicating them as mediators of the physiological response of the tissue to the hormone. It is interesting to consider some of the evidence that these proteins are indeed receptors, for it provides special insights into the biology of steroid hormone action. The first category of evidence is that binding proteins are present in tissues where physiological effects of the hormone are observed (i.e., so-called *target tissues*). Some of these target tissues are listed in Table I; for example, the uterus where estrogen stimulates uterine growth and functional maturation (Hamilton, 1968), the kidney and urinary bladder where aldosterone stimulates active sodium transport (Swaneck et al., 1969; Sharp, 1972), the thymus where corticosteroids suppress cell function and lead to cell death (Makman et al., 1967). We shall be considering below the occurrence of hormone binding in brain in those regions where demonstrated physiological and behavioral effects are observed and the absence of hormone binding in brain regions where hormone effects are absent.

A second type of evidence for receptor function of binding proteins is that in certain strains of animals, a target tissue does not respond to the appropriate hormone. Studies of the cellular basis of this nonresponsiveness indicates that in each case the binding protein is deficient and cell nuclear binding of the hormone is markedly reduced. In certain mutant strains of mice, which are insensitive to androgen, binding proteins for dihydrotestosterone are deficient and cell nuclear binding of the labeled hormone is reduced in the kidney (Bullock et al., 1971; Dofuku et al., 1971; Gehring et al., 1971). In lymphoma cells, which are insensitive to glucocorticoid killing action, fewer glucocorticoid binding sites are present than in sensitive lymphoma cell lines (Rosenau et al., 1972). Mouse mammary tumors that grow in the absence of estrogens contain some soluble estrogen binding proteins but, unlike the estrogen-dependent tumors, lack the ability to bind the hormone in the cell nucleus (Shyamala, 1972).

A third type of evidence that binding proteins are receptors is that, in general, physiologically active hormones bind to these proteins, and inactive steroids do not bind. This has been discussed in detail for estrogens and the uterus (Terenius, 1966; Stone and Baggett, 1965; Korenman, 1970), aldosterone and the kidney or urinary bladder (Swaneck et al., 1969; Alberti and Sharp, 1970), and corticosteroids and the hepatoma cell (Baxter and Tomkins, 1971). Certain steroids, which are by themselves weakly active, act as antagonists to the action of the more active steroids (Alberti and Sharp, 1970; Korenman, 1970; Baxter and Tomkins, 1971; Kenney and Lee, 1971). Other steroids, inactive by themselves, are also able to block the action of potent steroids (Makman et al., 1967; Alberti and Sharp, 1970; Baxter and Tomkins, 1971; Kenney and Lee, 1971). Both classes of antagonists compete with active steroids for binding to the soluble receptor proteins (Alberti and Sharp, 1970; Korenman, 1970; Baxter and Tomkins, 1971).

Cell nuclear uptake of hormone and action on the genome

As indicated in Figures 1b and 1c, the most generally accepted view of steroid hormone uptake by target cells is that the initial association with the receptors occurs in the cytoplasm after which the hormone moves into the cell nucleus (Jensen et al., 1968). Support for this idea comes from time-course experiments in which nuclear and cytosol fractions are prepared after incubation of the tissue with radioactive hormone (Jensen et al., 1968; Shyamala and Gorski, 1969), although probable artifacts in the overestimation of cytosol binding have recently been pointed out (Williams and Gorski, 1971). Of particular interest is the observation that after the radioactive hormone has had time to enter the nucleus, the levels of cytosol binding protein are depleted (Shyamala and Gorski, 1969). In dissociated uterine cells exposed to radioactive estradiol for more than 5 min at 37°C, more than 90% of the bound estradiol is in the nucleus (Williams and Gorski, 1971). Temperature reduction from 37 to 0–4°C results in a reduction in cell nuclear uptake that is greater than the reduction in cytosol binding (Jensen et al., 1968; Williams and Gorski, 1971), supporting the idea that there is a sequential transfer from cytosol to nucleus, which is temperature sensitive.

Another way of studying the interaction between soluble and cell nuclear hormone binding is to study the transfer in vitro in a cell-free system. This approach has led in the case of the chick oviduct system to the demonstration of tissue-specific sites of association between the cytosol binding protein and chromatin, prepared from isolated cell nuclei (Spelsburg et al., 1971). These *acceptor* sites appear to be acidic cell nuclear proteins (O'Malley et al., 1972), although there appears to be a role for the DNA (Musliner et al., 1971) and histones (King et al., 1971) in another system.

Several preliminary reports indicate that the cytosol receptor-hormone complex is able to stimulate RNA polymerase activity in isolated nuclei and on template activity of chromatin with added RNA polymerase (for summary, see Jensen and DeSombre, 1972; McEwen et al., 1972b). It is interesting to note that, while in most tissues these effects of the hormone are stimulatory, in the thymus gland, where glucocorticoids suppress cell function, the effect on RNA polymerase and template activity is to decrease it (Drews and Wagner, 1970; Nakagawa and White, 1970).

The mature brain as a hormone target tissue

Not only is the brain an essential part of the mechanisms governing various endocrine secretions, it is also a target tissue for the action of these hormones in regulating pituitary secretion and influencing certain behaviors (see chapters by Ganong, Goy, and Pfaff in this part and by de Wied in the following part). It is the purpose of the following sections to characterize the mature brain from a chemical point of view as other target tissues have been investigated. Specifically, we and those in other laboratories have studied the fate of radioactive steroid hormones entering the brain from the blood: Uptake, binding to soluble and cell nuclear receptors; metabolism to more or less active forms. This approach has the advantage of pointing to sites of action of these hormones within the brain cells and suggesting a possible mechanism of action of the hormone on the genome. It also has provided information as to probable anatomical sites of hormone action within the brain. This information is useful in designing biochemical and physiological experiments to study hormone action. It is also valuable in corroborating information from lesioning, stimulation, and recording experiments that point to a role for particular neural pathways or structures in hormone-mediated behaviors and in regulation of pituitary secretion (see chapter by Pfaff in this part).

Subcellular distribution of binding sites for corticosterone and estradiol in rat brain

Many steroid hormones have been shown to enter the brain from the blood. Progesterone, cortisol, and corticosterone have been detected in brain tissue by direct chemical analysis (Touchstone et al., 1966; Henkin et al., 1968; Raisinghani et al., 1968; Butte et al., 1972), while uptake from blood of radioactive steroids has been demonstrated for estradiol (Eisenfeld and Axelrod, 1965; Kato and Villee, 1967a; McEwen and Pfaff, 1970), testosterone (McEwen et al., 1970a, 1970b), progesterone (Seiki et al., 1969), aldosterone (Swaneck et al., 1969), and cortisol and corticosterone (Eik-Nes and Brizzee, 1965; McEwen et al., 1969; 1970a) by liquid scintillation counting. Very early in the studies of estradiol uptake it was recognized that a certain portion of the uptake could be prevented by administering an excess of unlabeled steroid together with the radioactive estradiol (Eisenfeld and Axelrod, 1966; Kato and Villee, 1967b). This was interpreted to mean that limited-capacity binding sites exist in the brain for estradiol. Similar observations were subsequently made for corticosterone (McEwen et al., 1969).

Considering the extensive work on soluble and cell nuclear binding proteins in steroid hormone target tissues (see above), it was reasonable to search for similar molecules in the brain as the explanation for limited-capacity uptake of radioactive estradiol and corticosterone. Soluble estradiol-binding proteins were found in brain

and pituitary (Kahwanago et al., 1969; Eisenfeld, 1970; Kato et al., 1970a; Notides, 1970; Vertes and King, 1972). Cell nuclear binding of estradiol was also discovered in brain (Chader and Villee, 1970; Zigmond and McEwen, 1970; Mowles et al., 1971; Vertes and King, 1972) and pituitary (Kato et al., 1970b). Soluble and cell nuclear corticosterone-binding proteins were found in brain (McEwen and Plapinger, 1970; McEwen et al., 1970c; Grosser et al., 1971; McEwen et al., 1972a). Both soluble and cell nuclear binding of radioactive corticosterone and estradiol could be eliminated by competition with unlabeled hormone, and competition studies with structurally related steroids revealed that the binding of estradiol and corticosterone is stereospecific for each hormone (Eisenfeld, 1970; McEwen et al., 1970c; Zigmond and McEwen, 1970). Reduction by competition of the cell nuclear binding of ^{3}H-corticosterone is illustrated by autoradiographs presented in Figure 2. Figures 2A and 2B show a number of neurons from a control brain demonstrating heavy accumulation of silver grains over the cell nucleus, while Figures 2C and 2D show comparable neurons from a competition experiment in which cell nuclear labeling has been virtually eliminated. From such evidence it appears likely that the brain contains receptor proteins for estradiol and corticosterone which perform a function analogous to receptor proteins in other steroid hormone target tissues such as the uterus and thymus gland.

Hormone-binding proteins have a number of properties in common with those in other target tissues. Estradiol-binding proteins in brain and pituitary resemble those in the uterus with respect to stereospecificity (Eisenfeld, 1970), essential sulfhydryl groups (Eisenfeld, 1970; Kahwanago et al., 1970), and sedimentation properties (Kahwanago et al., 1969; Kato et al., 1970a; Vertes and King, 1972; McEwen and Pfaff, 1973; Plapinger and McEwen, 1973). Corticosterone-binding proteins are less well characterized, but presently available information as to stereospecificity in binding of proteins from hepatoma and thymus (Baxter and Tomkins, 1971; Schaumburg, 1972) or presence of essential sulfhydryl groups in the thymus protein (Schaumburg, 1972) seems to agree in large part with information available for corticosterone-binding proteins of brain (McEwen, 1973, and unpublished). It is naturally of very great importance to establish whether all binding proteins for a particular hormone are identical, for if they are we must look elsewhere in the cell for the basis of tissue-specific actions of the hormone.

Corticosterone-binding proteins in rat brain have been shown to differ in a number of important respects (McEwen et al., 1972a) from the blood-binding protein, which is called transcortin, or corticosteroid-binding globulin (Westphal, 1971). The brain protein is present

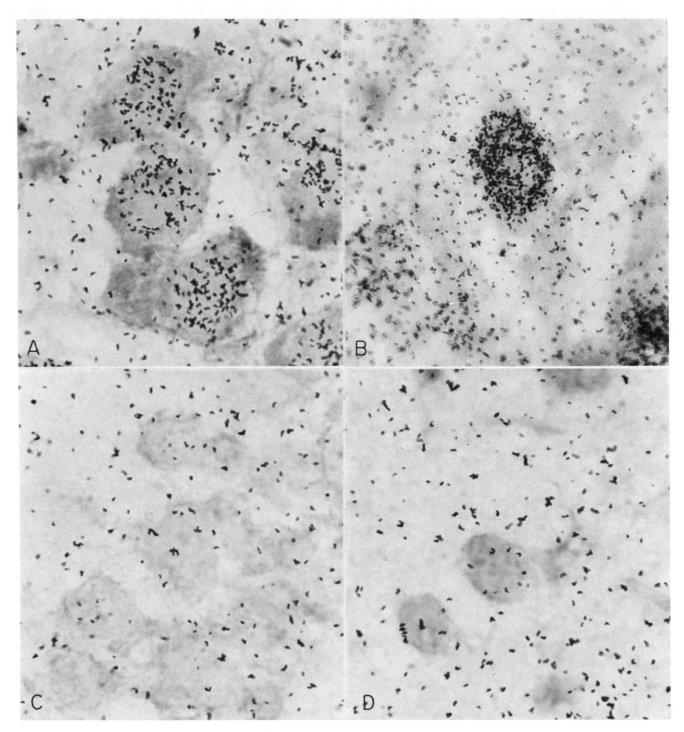

FIGURE 2 Autoradiogram illustrating in vivo binding of ^{3}H-corticosterone to the nuclei of pyramidal neurons located in the hippocampus in brains of adrenalectomized rats (A, B), and a reduction of binding by competition with 3 mg of unlabeled corticosterone injected together with the labeled hormone (C, D). These autoradiograms were exposed for 180 days (B), 250 days (A), or 320 days (C, D) and were stained with methyl green-pyronin Y. Magnification is 1975 diameters. (For details see Gerlach and McEwen, 1972.)

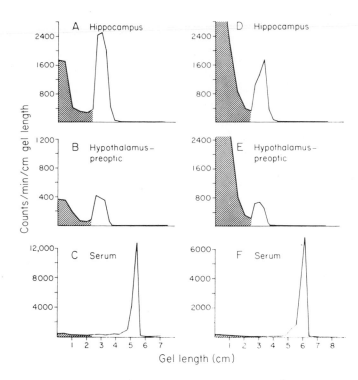

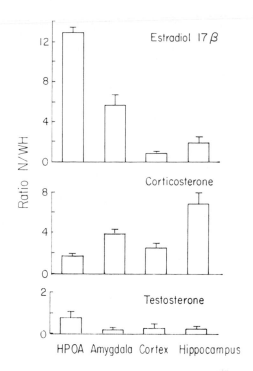

FIGURE 3 Separations on polyacrylamide gels of ^{3}H-corticosterone binding proteins from brain cytosol (A, B, D, E) and serum (C, F). Shaded zone indicates 4% stacking gel; unshaded area, 10% separating gel. A, B, C: labeling in vivo; D, E, F: labeling with 10^{-9} M ^{3}H-corticosterone in vitro. (Reprinted from McEwen et al. 1972a with permission from J. B. Lippincott Company.)

FIGURE 4 Regional distribution in 4 regions of rat brain of cell nuclear binding of ^{3}H-estradiol-17β, ^{3}H-corticosterone, and ^{3}H-testosterone. Each hormone was implanted into animals from which the appropriate hormone-producing gland had been surgically removed. Data presented as a ratio of nuclear concentration to whole homogenate concentration (N/WH). HPOA: pooled hypothalamus, preoptic area, and amygdala. (Compiled from data summarized in McEwen et al. 1972b.)

after perfusion has removed all traces of blood, is precipitated quantitatively by protamine sulfate when labeled by ^{3}H-corticosterone in vivo (while labeled transcortin is not so precipitated) and is completely separable from transcortin by electrophoresis in polyacrylamide gels at basic pH (Figure 3). Moreover, dexamethasone is able to bind to the brain protein but does not bind to transcortin (McEwen, 1973).

Regional distribution of hormone binding in brain

Binding proteins for estradiol and corticosterone are each concentrated in certain regions of the rat brain. This regional distribution is of particular interest for the anatomist and physiologist interested in hormonal regulation of neural activity, as will be discussed in the next section. Regional distribution of cell nuclear binding of corticosterone and estradiol is compared in Figure 4. Estradiol binding is highest in the hypothalamus and preoptic area and in amygdala and much lower in hippocampus and cerebral cortex. Corticosterone binding is highest in hippocampus and in amygdala and lower in cerebral cortex and hypothalamus and preoptic area.

For corticosterone a similar differentiation of soluble binding proteins can be seen in Figure 3, between hippocampus and hypothalamus-preoptic area. It should be emphasized in the case of corticosterone that binding in less heavily labeled structures such as hypothalamus is not insignificant and that functional correlates exist for hormone binding in these structures as well as in hippocampus (see below).

Another means of studying the regional distribution of hormone binding in the brain is autoradiography, using methods described by Stumpf and Roth (1966) and by Anderson and Greenwald (1969). For estradiol, the findings of autoradiographic experiments agree almost totally with those of biochemical experiments, concerning the presence of estradiol-binding neurons in the hypothalamus, preoptic area, and amygdala (Pfaff, 1968a, 1968b; Stumpf, 1968, 1970; see chapter by Pfaff in this part). In addition, autoradiography has revealed the presence of a few estradiol-binding neurons in the midbrain and hippocampus (Pfaff and Keiner, 1972). For corticoids too, the results of autoradiographic studies agree strongly with those of biochemical studies of the distribution of binding sites, revealing binding of cortisol (Stumpf and Sar, 1971) and corticosterone (Gerlach and McEwen, 1972) in neurons of septum, induseum grisseum, hippocampus, amygdala, and cerebral cortex. Corticosterone binding in the hippocampus has received particular attention (Gerlach and McEwen, 1972). Figure 5 is an autoradiogram of a sagittal frozen section of the hippocampus of a rat brain prepared as a montage of many photomicrographs taken at a magnification adequate to resolve individual silver grains. Labeling is greatest in pyramidal neurons of Ammon's horn in the longitudinal fields designated CA1 and CA2, is somewhat less in fields CA3 and CA4, and is high in scattered granule neurons of the dentate gyrus (GD).

Functional correlates of hormone localization in the brain

In general, the brain regions where estradiol and corticosterone bind are also areas in which these hormones produce their physiological effects (for reviews, see McEwen et al., 1972b; McEwen and Pfaff, 1973). Indirect evidence has come from lesioning and brain stimulation experiments. The most direct evidence is obtained by recording electrophysiological effects of the hormones and by implanting hormones in various brain regions. For estradiol, cells in the septum, preoptic area, and hypothalamus (called the *hypophysiotrophic area* after Halász et al., 1962) have been shown to change in resting discharge rate or response to peripheral stimuli when circulating estrogen levels increase (for review, see Beyer and Sawyer, 1969; Pfaff et al., 1972). Estrogen implants in the hypophysiotrophic area facilitate the lordosis reflex in the female rat and regulate gonadotrophic secretion (Lisk, 1967) and in the amygdala estrogen implants alter the secretion of prolactin by the pituitary (Tindal et al., 1967). Estrogen implantation in the preoptic area also increases locomotor activity, while estrogen implantation in the ventromedial hypothalamus decreases food intake (Wade and Zucker, 1970). In the pituitary, estrogen appears to exert a positive, priming effect on gonadotrophin secretion (Weick et al., 1971).

Estradiol binding to brain cell nuclei may be the first step in neuroendocrine and behavioral effects of this hormone. Can the same be said of corticosterone binding sites? The first problem is to delineate the nature of

glucocorticoid effects on neuroendocrine regulation and behavior. *Feedback* of corticosteroids on ACTH secretion is an important phenomenon, and glucocorticoid implants in a variety of limbic structures influence the release of ACTH (for summary, see McEwen et al., 1972b). On the other hand, certain types of feedback, such as long-term suppression by dexamethasone of stress-induced ACTH, are not observed with physiological levels of the natural glucocorticoid (see McEwen et al., 1972b) and lead one to question whether such profound feedback is a reflection of physiological events. Other neural events influenced by glucocorticoids include alterations in conditioned avoidance behavior, which are produced by implants of glucocorticoids in a variety of limbic structures (Bohus, 1970, 1971) as well as by systemic injection (van Wimersma-Greidanus, 1970). Also, glucocorticoids influence the occurrence of the diurnal secretion of ACTH (Slusher, 1966; Cheifetz et al., 1968; Hiroshige and Sakakura, 1971) and paradoxical sleep (Johnson and Sawyer, 1971; Gillin et al., 1972). In this connection, it is important to note that paradoxical sleep is accompanied by hippocampal theta activity (see Winson, 1972). Finally glucocorticoids regulate the detection and recognition of a variety of sensory stimuli (Henkin, 1970).

It is impossible to attribute to any single neural structure an exclusive role in controlling these various neural events influenced by glucocorticoids. However, in view of the intense localization of corticosterone in hippocampus, we should briefly consider the role of this structure in events influenced by glucocorticoids. Neurons in hippocampus decrease unit activity for a period of at least 3 hours following the systemic injection of corticosterone (Pfaff et al., 1971). Such a prolonged effect of the hormone suggests possible mediation by changes in cell metabolism induced by the hormone. Corticosteroid implants in hippocampus enhance basal and stress-induced ACTH secretion (see McEwen et al., 1972b), and, in one instance, corticoid implants in hippocampus were observed to eliminate the diurnal variation of circulating corticosterone (Slusher, 1966). Corticosteroid implants in hippocampus decrease passive avoidance behavior during extinction and improve discrimination in appetitive situations involving choice (Bohus, 1972). In general ACTH facilitates avoidance behavior and retards extinction (de Wied, 1967). In this connection, Pfaff et al.

FIGURE 5 Autoradiogram demonstration in vivo of the binding pattern of ³H-corticosterone to neurons located in the hippocampus of an adrenalectomized rat. Relatively high amounts of the hormone were concentrated by pyramidal neurons of the longitudinal fields CA1 to CA4 of Ammon's horn, and by granule neurons of the dentate gyrus (GD). Much of the contrast between these neurons and the surrounding tissue reflects the intensity of labeling, because the cell bodies were stained very lightly. This autoradiogram is a montage of many photomicrographs of a sagittal frozen section that was exposed to autoradiographic emulsion for 178 days and stained with Darrow red and light green. Magnification is 70 diameters. For details see Gerlach and McEwen, 1972.

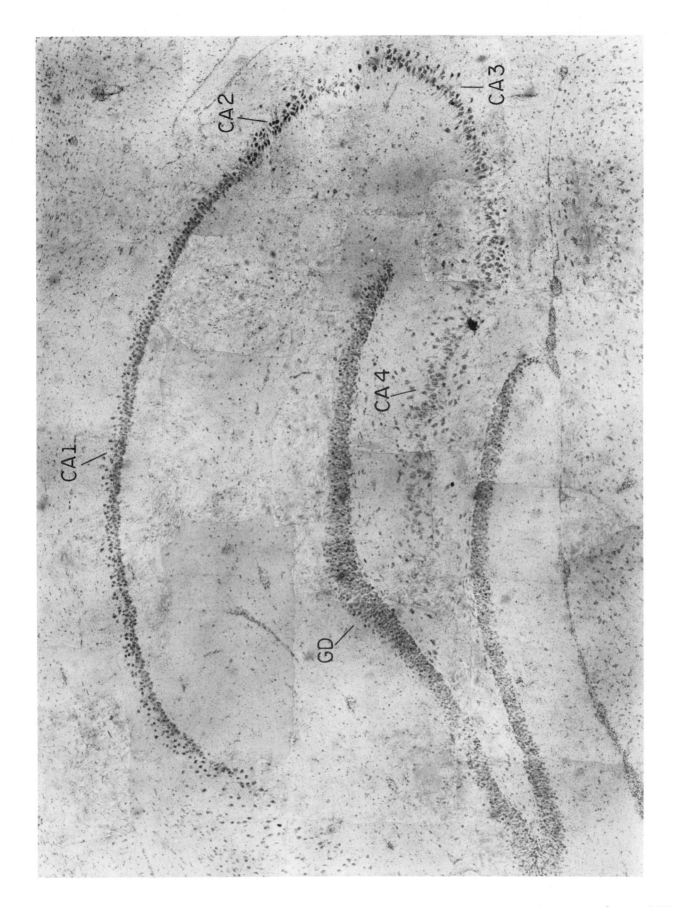

(1971) observed that ACTH increased single unit activity in hippocampus while, as noted above, corticosterone has the opposite effect.

Biochemical effects of estradiol and corticosterone on brain

Various biochemical effects of estradiol and corticosterone on nervous tissue have been described. In the case of ovarian hormones (see Table III), the effects on monoamine oxidase and respiratory activity, and leucine and adenine incorporation have been found in brain regions that bind estradiol. Effects of estradiol on protein and nucleic acid formation implicate activation of the genome by estradiol in the control of lordosis behavior and gonadotrophin secretion. Studies with inhibitors of RNA and protein formation tend to support this conclusion by showing, on the one hand, prevention by actinomycin D of estrogen suppression of pituitary secretion of gonadotrophic hormone (Schally et al., 1969) and, on the other hand, blockade by actinomycin D and protein synthesis inhibitors of estrogen induction of lordosis behavior (Quadagno et al., 1971). Estradiol effects on dopamine and norepinephrine turnover may be related to feedback effects of estrogens on gonadotrophin secretion, because both amines have been implicated in the control of releasing hormone secretion (Ahren et al., 1971; Anton-Tay and Wurtman, 1971; Coppola, 1971; McCann et al., 1972; Weiner et al., 1972).

Documented corticosteroid effects on nervous tissue are fewer in number than those for estrogen and are not clearly localized to binding regions of the brain. Glucocorticoid regulation of glycerolphosphate dehydrogenase occurs in glial cells in culture (deVellis et al., 1971) and most probably in glial cells in the intact brain (deVellis and Inglish, 1968). Increases by glucocorticoid secretion of levels of PNMT (phenylethanolamine N methyl transferase) in the adrenal medulla point to important interactions between the adrenal cortex and nervous tissue of the adrenal medulla (Pohorecky and Wurtman, 1971). Glutamine synthetase activity in developing chick neural retina is increased prematurely by cortisol (Moscona and Piddington, 1966; Reif and Amos, 1966). Serotonin biosynthesis from tryptophan is influenced by glucocorticoid secretion, because it decreases after adrenalectomy (Azmitia et al., 1970) and is increased by glucocorticoid or ACTH administration (Millard et al., 1972). These effects are of particular interest in view of observed diurnal fluctuations

TABLE III

Ovarian hormone effects on brain chemistry

Effect	Tissue	Reference
Nucleic Acid and Protein Metabolism		
Lower ^{14}C-adenine incorporation in estrus	hypothalamus	Belajev et al., 1967
Decreased nucleolar volume after estrogen	arcuate nucleus	Lisk and Newlon, 1963
Higher ^{3}H-leucine incorporation after OVX*	arcuate and ventromedial nuclei	Yaginuma et al., 1969
Lower ^{14}C-lysine incorporation after E†	arcuate nucleus	Litteria and Timiras, 1970
Changes in Enzyme Activity		
Cyclical changes with estrus in MAO‡	hypothalamus, amygdala	Kamberi and Kobayashi, 1970
Cyclical changes with estrus in respiration	hypothalamus, amygdala	Moguilevsky and Malinow, 1964; Schiaffini et al., 1969
Tyrosine hydroxylase increases after OVX	hypothalamus	Beattie et al., 1972
Decreased adenyl cyclase after estradiol	pineal	Weiss and Crayton, 1970
Changes in Catecholamine Metabolism		
Dopamine turnover decreased after OVX	arcuate nucleus	Fuxe et al., 1969
Dopamine levels decrease after OVX	anterior hypothalamus	Stefano and Donoso, 1967
Norepinephrine increases after OVX	anterior hypothalamus	Stefano and Donoso, 1967
Norepinephrine synthesis after E plus P**	anterior hypothalamus	Bapna et al., 1971

References are representative and in no way provide exhaustive coverage.
*OVX: ovariectomy. †E: estradiol-17β. ‡MAO: monoamine oxidase. **P: progesterone.

in brain serotonin levels (Dixit and Buckley, 1967; Friedman and Walker, 1968; Scapagnini et al., 1971; Okada, 1971) and in view of the observation that blockade of serotonin biosynthesis by p-chlorophenylalanine abolishes the diurnal rhythm of ACTH secretion (Krieger and Rizzo, 1969; Scapagnini et al., 1971).

Testosterone metabolism in brain: Regional and developmental aspects

Studies of binding of radioactive testosterone to the rat brain have not definitely established the existence of intracellular soluble or cell nuclear binding sites. Evidence from cell fractionation suggest such sites may exist in the hypothalamus (Figure 4) and this has received some support from autoradiography (Pfaff, 1968b; Stumpf and Sar, 1971). Studies of testosterone binding in bird brain indicate more clearly the existence of such receptors localized in hypothalamus and midbrain (Zigmond et al., 1972a, 1972b). Since it is difficult to study androgen receptors in rat brain, another approach is to study the conversion of testosterone to 5α-DHT, the metabolite that binds to intracellular receptors in prostate (Tables II and III). A number of laboratories have investigated the in vitro metabolism of testosterone in the brains of a number of species (Sholiton et al., 1966; Jaffe, 1969; Kniewald et al., 1970; Perez-Palacios et al., 1970; Rommerts and van der Molen, 1970, 1971; Sholiton et al., 1970; Massa et al., 1971; Naftolin et al., 1971, 1972; Shore and Snipes, 1971; Stahl et al., 1971). These studies have demonstrated the conversion of testosterone to 5α-DHT, 3α-androstanediol, and androstenedione as well as aromatization to estrogens.

The significance of this metabolism for brain function is currently under investigation. It has been found that 5α-DHT does not mediate the masculine differentiation of the rat brain during neonatal life, nor does it stimulate male sexual behavior in the rat (Table II). However 5α-DHT administered parenterally (Beyer et al., 1971) and implanted in hypothalamus (Davidson, 1973) does have a negative feedback effect on gonadotrophin release and may therefore be the active androgen in regulating pituitary function. The conversion of testosterone to 5α-DHT, 3α-androstanediol, and androstenedione has been studied by incubating 300 μ slices of brain tissue in vitro with radioactive testosterone and we have investigated this conversion in a number of brain regions and pituitary in both sexes before and after castration (Denef and McEwen, 1972). The metabolites were separated on thin-layer plates and identified by their mobility and by crystallization to constant specific activity. In all brain regions studied, the major metabolite was 5α-DHT. Low conversion rates were observed in cortex, preoptic region, hippocampus, amygdala, caudate nucleus, and cerebellum (Figure 6). In agreement with other studies (Sholiton et al., 1966; Jaffe, 1969; Kniewald et al., 1970; Shore and Snipes, 1971; Rommerts and van der Molen, 1971), the formation of 5α-DHT was higher in the hypothalamus than in the cerebral cortex (Figure 6). As noted above, the hypothalamus is believed to be a site in which 5α-DHT has feedback effects on gonadotrophin

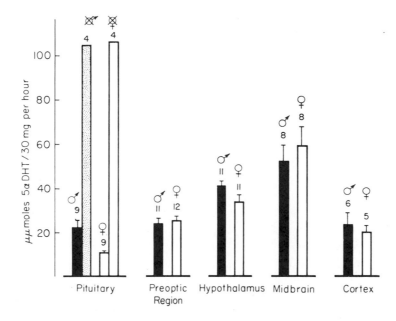

FIGURE 6 Regional and sex differences in testosterone 5α-reductase activity in rat brain and pituitary measured in vitro in 300 μ slices prepared with a McIlwain tissue chopper. Note increase of 5α-reductase activity in pituitary after gonadectomy.

secretion. In the pituitary, on which testosterone also has a direct negative feedback effect (Kamberi and McCann, 1972), the production rate of 5α-DHT by slices was as low as in cerebral cortex in normal male rats (Figure 6) but increased several fold within a few days following castration (Figure 6). Intact pituitaries showed a higher conversion rate than either pituitary slices or cerebral cortex, and this conversion rate increased similarly after castration. An increase after castration was also observed in 3α-androstanediol formation. These changes were not observed in any brain regions. While we do not yet understand the mechanism of this postcastration increase in 5α-DHT production, it is conceivable that it may be indicative of negative control by the testes over pituitary 5α-DHT production and may be related in some way to the negative feedback effects of 5α-DHT on gonadotrophin secretion.

Other indirect evidence for a physiological role of 5α-DHT production in hypothalamus and pituitary is the observation that the conversion rate in both structures is higher in male than in female rats (Figure 6). Curiously, castration abolishes the sex difference in the pituitary but does not appear to do so in the hypothalamus. In males and females, midbrain and brainstem show higher production rates of 5α-DHT than the hypothalamus, while thalamus is equal to hypothalamus (Figure 6). Functional significance of these observations is not clear, although it is known that the midbrain plays a role in gonadotrophin regulation in the female rat (Carrer and Taleisnik, 1970, 1972; Grinò et al., 1968).

Not only does 5α-DHT production change dramatically in the pituitary after castration, it also undergoes large changes in the course of postnatal development (Figure 7). The conversion rate of testosterone to 5α DHT is several-fold higher in immature rat pituitary than in adult, and this rate decreases to lower levels in both sexes some time before the onset of puberty (Figure 7). Produc-

tion of 5α-DHT is higher in pituitaries of infant females than of males, and this sex difference disappears and then reverses after puberty (Figure 7). In contrast to pituitary, brain structures (midbrain, hypothalamus, and cerebral cortex) showed values at 4 days of age comparable to those found in adult life (Denef and McEwen, unpublished).

It is interesting to consider correlates in the pituitaries of immature female rats of the high 5α-reductase activity, which exceeds that in immature males and decreases toward adult levels between 15 and 25 days of age. Plasma LH levels are reported to be higher in immature females as well as in males compared to intact adult males and diestrus adult females and to decrease toward adult levels after postnatal day 15 (Ojeda and Ramirez, 1972). Pituitary and also plasma FSH levels in immature females exceed that in adult animals and decrease toward adult levels after postnatal day 20 (Kragt and Ganong, 1968; Ojeda and Ramirez, 1972). Plasma FSH levels are several times higher in females than in males around postnatal day 15 (Ojeda and Ramirez, 1972). While evidence is lacking to connect elevated pituitary 5α-reductase causally with pituitary and plasma FSH and LH levels, particularly in females, the parallel changes in these parameters emphasize the need to investigate the functional role of 5α-DHT and androstanediol formation in regulation of pituitary function.

Ontogeny of stereospecific estrogen binding sites

Autoradiographic and cell fractionation studies of radioactive estradiol uptake in brain indicate that the preoptic area, hypothalamus, and amygdala bind estradiol by a highly specific, limited-capacity mechanism that involves binding proteins located in the soluble (cytosol) fraction and cell nuclei of certain neurons (see above and the chapter by Pfaff in this part). As indicated above, evidence

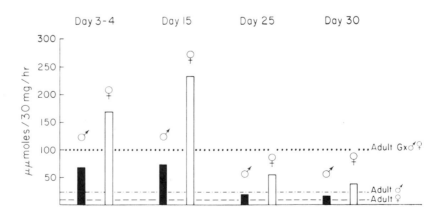

FIGURE 7   Sex-dependent changes during postnatal development in testosterone 5α-reductase activity in pituitary slices.

that these binding sites function as estradiol receptors derives from their presence in those brain regions in which implanted estradiol influences reproductive function and from the stereospecificity of their binding of active estrogens such as estradiol-17 β and stilbestrol. Additional evidence comes from developmental studies that show that these binding sites appear in the *target* areas (hypothalamus, preoptic area, amygdala, and midbrain) just prior to the onset of puberty. We have found very little region-specific cell nuclear binding of estradiol-17 β in the female rat until after postnatal day 22, with a sharp rise occurring prior to day 28 (Figure 8). Similar observations have been reported for the cytosol binding protein by Kato et al. (1971). There is so far no indication which factors, hormonal or otherwise, trigger this sudden appearance of estrogen receptors.

The newborn rat brain contains another type of soluble estradiol-binding protein that has a sedimentation coefficient of around 4 S (Svedberg constant) distinguishable from that in the adult brain, pituitary, and uterus that has a sedimentation coefficient around 8 S (Plapinger and McEwen, unpublished). This 4 S protein is not localized to any particular brain region and around 30% of it is washed out of the whole brain by procedures that extract no detectable glutamine synthetase activity, an intracellular enzyme in brain. This protein or one very similar to it is present in the blood of newborn rats (Raynaud et al., 1972), but blood contamination of the brain in animals either decapitated directly or perfused at sacrifice cannot account for what we find in the brain (Plapinger and McEwen, unpublished). This 4 S protein binds estradiol-17 β stereospecifically and, unlike the adult cytosol protein, does not bind diethylstilbestrol (Plapinger and McEwen, unpublished). The level of this protein in the brain is approximately 20 times greater at 4 days of age than the 8 S binding protein in the adult and this protein disappears from the brain (Figure 8) and blood (Raynaud et al., 1972) by postnatal day 22. Since this 4 S protein is equally present in infant males and females and disappears by the end of the third postnatal week, we are tentatively suggesting that the protein may be a vestige of uterine life, at which time it functions to bind maternal estrogen, perhaps extracellularly, and thereby protects the developing organism against the deleterious effects of this hormone, which can in high doses lead to sterilization in much the same way as testosterone does (see below).

Influences of hormones on brain during development

Secretion of testosterone during prenatal life in some species and in early postnatal life in the rat (Resko et al., 1968) is able to alter the subsequent postnatal development of the reproductive tract and certain functions in the brain (Table II) and in the liver (Denef and DeMoor, 1968, 1969, 1972). As pointed out by Jost (1970), the basic pattern of differentiation in the mammal is feminine, in that in the absence of testes the reproductive tract and neural regulation of the pituitary develop according to the feminine pattern independently of genetic sex, whereas in the presence of testes or androgen these same structures have "maleness" imposed upon them independently of genetic sex and develop according to the male pattern.

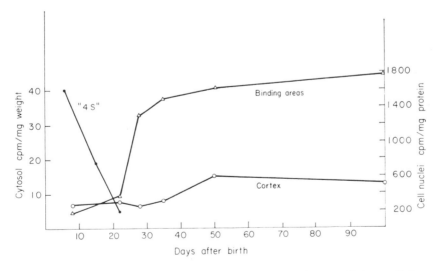

FIGURE 8 Changes during postnatal development in binding of ³H-estradiol to the "4 S" soluble protein in brain measured in vitro by sucrose density gradients (solid circles) and to brain cell nuclei by isolating nuclei after labeling in vivo (open triangle, open circle). Binding areas (hypothalamus, preoptic area, amygdala, and midbrain) are compared to cerebral cortex.

The development of the germ cell line into either testes or ovary is stimulated by some other signal, possibly from the Y chromosome.

Psychosexual development in the mammal, including both neuroendocrine regulation and behavior, can be manipulated experimentally by early castration of males or by prenatal or neonatal administration of androgen to females (Barraclough, 1967; Goy, 1970). Whereas testosterone is secreted naturally during the critical period of this differentiation (see below) in the rat (Resko et al., 1968), and exogenous testosterone propionate is extremely effective in mimicking this differentiation in females, other steroids are also effective (Table IV). Of particular interest is the observation that doses of estradiol benzoate (or of the synthetic estrogen stilbestrol) comparable to effective doses of testosterone propionate lead to "masculinization" of the brain (Table IV) whereas 5α-DHT is incapable of achieving this (Table II). As we have pointed out earlier, androgen is aromatized by brain tissue to estrogen in small amounts and 5α-reduced metabolites are unable to be aromatized. This has led to the suggestion that aromatization might possibly be an obligatory step in the "masculinizing" effects of testosterone (Luttge and Whalen, 1970; Brown-Grant et al., 1971) although proof of this is presently lacking. These observations do raise questions about the chemical specificity of any tissue receptors for masculinization. Other steroids that in high doses masculinize the brain include progesterone, deoxycorticosterone, androstenedione, and the nonsteroid DDT, which has weak estrogenic activity (Table IV).

As is the case for the brain, the sexual differentiation of several steroid metabolizing enzymes in the liver can also be manipulated irreversibly by neonatal castration of males or by neonatal administration of testosterone propionate to females (Denef and DeMoor, 1968; Denef and DeMoor, 1972). In contrast to the brain, however, estrogens were found to be ineffective in masculinizing these enzyme systems (Denef and DeMoor, 1969). The effect of other steroids on this differentiation has so far not been tested.

One of the fundamental observations dealing with masculinization of various tissues is that there is a *critical period* during which the tissue is competent to respond to the masculinizing hormone and after which it is refractory or responds in a reversible manner. The reproductive tract responds immediately in manifesting the effects of masculinization in that the Müllerian ducts regress while the Wolffian ducts differentiate, although the tissue is not fully functional until after puberty (Jost, 1970). To a large extent the masculinizing effects on the brain and liver are not obvious until puberty, but there is growing biochemical evidence for continuing changes in these tissues from the critical period until puberty. This is particularly well illustrated for the sexual differentiation of the steroid metabolizing enzymes in the rat liver (Denef and DeMoor, 1972). As mentioned above, the development of the male pattern is imposed upon the liver by the secretion of testosterone neonatally. If males or androgenized females are treated with estrogen *after* neonatal androgenization but *before* the onset of puberty, the masculine pattern of development is halted and cannot be permanently reinduced by testosterone treatment. If estrogens are given after puberty, the masculine enzyme pattern turns into a feminine pattern but remains fully reinducible by testosterone. In this manner, interference with the male differentiation is possible between the end of the critical period and the onset of puberty.

A number of studies indicate some of the ways in which the brain is influenced immediately by testosterone. Autoradiographic studies of the incorporation of ³H-uridine into rat brain RNA 3 hours after administration of testosterone on postnatal day 2 revealed a relative increase in incorporation compared to other brain regions in the anterior hypothalamus and amygdala (Clayton et al., 1970). Autoradiographic studies of ³⁵S-methionine incorporation into rat brain protein 12 and 24 hours after administration of testosterone propionate on postnatal day 3 resulted in an increase in incorporated radioactivity over the ventral nucleus of the thalamus and the lateral nucleus of the amygdala (Darrah et al., 1971). Twenty-four hours after castration of full-term male rat fetuses, ³H-leucine incorporation into neurons of the arcuate nucleus region was substantially increased, suggesting possible compensatory increases in hypothalamopituitary function resulting from gonadectomy (Nakai et al., 1971). Blockade of protein synthesis by intracranial implantation of cycloheximide just prior to testosterone propionate treatment on postnatal day 5 prevented the masculinizing effects of the hormone (Gorski and Shryne, 1971).

The permanent nature of hormonal masculinization of the brain is suggested by observations made on sexually mature rats. Sexual dimorphism has been reported for the relative number of nonamygdaloid synapses on dendritic shafts compared to dendritic spines in the preoptic area (Raisman and Field, 1971). Proof of the role of neonatal androgen secretion is lacking, however.

One of the consequences of masculinization of female rats with neonatal hormone treatment is a reduction in sensitivity of the brain and possibly other estrogen target tissues to estradiol (Gerall and Kenney, 1970; Flerko, 1971; Pfaff and Zigmond, 1971; Whalen et al., 1971). One of the predictions of this reduced sensitivity is that the estrogen receptor capacity of the target areas of the brain might be lower than normal and this prediction has been supported in estradiol uptake experiments in androgen-treated female rats (Flerko and Mess, 1968; McEwen and Pfaff, 1970; Maurer and Woolley, 1971; Tuohimaa and Johansson, 1971). Analysis of cytosol and cell nuclear binding of estradiol in the brain reveals a decrease that is particularly striking in cell nuclear binding of this hormone (Vertes and King, 1971; Plapinger and McEwen, unpublished). A further prediction from this line of evidence is that male rats should show reduced estrogen-binding capacity compared to females and that neonatally castrated male rats should show similar binding capacity to females, higher than both androgen-treated females and normal males. Estradiol is accumulated by neurons distributed in the rat brain in qualitatively the same pattern in male and female rats (Pfaff, 1968a, 1968b; Anderson and Greenwald, 1969; McEwen and Pfaff, 1970). Reduced estrogen-binding capacity of brain tissue in hypothalamus was reported for normal males compared to females, but neonatally castrated males were also lower than normal females (McEwen and Pfaff, 1970). In cell fractionation experiments, however, measurements of cell nuclear binding of estradiol in hypothalamus, preoptic area, and amygdala has failed to show any reduction in castrated adult males compared to ovariectomized adult females, although there was a significant reduction in these experiments in the binding capacity of androgen-treated females (Plapinger and McEwen, unpublished). It therefore seems possible to account for reduced estrogen sensitivity in androgenized females by reduced estrogen *receptor* capacity but not for the normal male-female differences in sensitivity to estrogen. The lack of normal male-female difference in estrogen-binding capacity may be due to a more restricted localization in males of reduced estrogen uptake that escapes detection in biochemical experiments. It may also be the case that other factors than reduced estrogen binding or sensitivity are primary in the normal sex difference, such as fundamental differences in the structure of the brain (see Raisman and Field, 1971), which result in estrogen's altering different neural pathways in each sex.

Model systems for studies of hormonal control of gene function

From our present knowledge, how are we to think of the mechanism of the action of hormones on the developing brain or, similarly, on the liver and reproductive tract? Because the effects occur at a time when the tissue is in the process of differentiation and are permanent for the life of the organism, it is reasonable to suppose that the genome is not only involved but in some way altered so that certain genes are activated, some perhaps in a transitory fashion, others permanently.

The first example of a steroid hormone effect on the genome, the induction of puffs by ecdysone in giant chromosomes of *Chironomus tentans* (Clever and Karlson, 1960), provides a very clear model for hormonal influences during development. The giant chromosomes of *Chironomus*, *Drosophila*, and other Diptera offer the best preparations known for directly visualizing and studying gene action in the eukaryotic genome (Beermann, 1972). Puffs occurring at specific bands in these polytene chromosomes involve uncoiling of chromatin strands with accumulation of acidic proteins and *de novo* RNA synthesis indicative of increased gene activity (Berendes, 1972). The frequency of these puffs among all chromosomes or bands indicates that maximally around 15% of the genome is active in any given cell type (Pelling, 1972). The puffing patterns of different tissues have considerable overlap, although about 5% of the puffs are tissue-specific (Pelling, 1972). The invertebrate steroid hormone ecdysone in concentrations up to around 10^{-4} M induces the formation of puffs with latencies from 15 min to 6 hr, and the pattern of puffs induced by this moulting hormone recapitulates the patterns observed during the natural moulting sequence (Ashburner, 1972; Berendes, 1972). The ability of steroid hormones to trigger the activity of specific genes within a chromosome is thus elegantly demonstrated by these studies. The identification and role of ecdysone-binding proteins analogous to those binding steroid hormones in vertebrate target tissues is currently under investigation (Berendes, 1972).

Study of the imaginal disks of Diptera such as *Drosophila* adds another important dimension to the role of the steroid hormone ecdysone in development. Imaginal disks are collections of cells that are determined in the course of their development to become specific structures such as ommatidia, bristle organs, hairs, and a variety of mechano- or chemoreceptors (Nöthiger, 1972). The imaginal disks are highly stable in their differentiated (i.e., predetermined) state and can survive and multiply in culture in the absence of ecdysone (Gehring, 1972). Whereas the determination of the imaginal disks is independent of hormones such as ecdysone, this hormone triggers the differentiation of competent imaginal disks into the appropriate predetermined structures and in so doing is responsible for triggering the metamorphosis that accompanies the final moult (Nöthiger, 1972; Oberlander, 1972).

Making an analogy between this situation and the developmental effects of sex hormones on the brain and other tissues, we may ask whether these latter hormones are triggering the expression of predetermined characteristics that are themselves developed independently of the hormone. Such may be the case for the "masculine" patterns of reproductive tract development, brain and liver function, while the feminine patterns become permanently instated simply by the passage of time beyond a "critical period" in the absence of gonadal hormones. Such may also be the case for stimulation of the female reproductive system at puberty, because this system develops in the absence (or relative absence) of gonadal hormones and is triggered into function at puberty by estrogen.

What sort of "receptors" may be involved in the action of hormones on the developing brain? Region-specific cell nuclear binding sites for estrogen do not appear in the brain until well after the "critical period" is past, but their appearance precedes and may be essential for the onset of puberty. Receptors for testosterone, which might be responsible for masculinization, have not so far been detected in neonatal rat brain by techniques used in studying estrogen and corticosterone binding sites (Zigmond, 1971; and unpublished). However, receptors for masculinization may not distinguish between testosterone and estradiol, and may also recognize other steroids in higher concentrations (Table IV). A possible interpretation of the effectiveness of both estradiol and testosterone (whereas 5α-DHT is ineffective) is that androgen is converted to estrogen by cells of the infant rat brain in order to produce the masculinizing effects on gonadotrophin secretion and sex behavior. Were this to be true, it would focus attention upon estradiol as the primary agent of the masculinizing effects.

TABLE IV

"Masculinization" of the brain by neonatal hormone treatment

Hormone	Reference
Testosterone propionate	Barraclough, 1967; Goy, 1970
Stilbestrol	Ladosky, 1967
Estradiol benzoate	Gorski, 1963; Whalen and Nadler, 1963
Androstenedione	Goldfoot et al., 1969; Stern, 1969
Progesterone	Turner, 1941; Diamond and Wong, 1969
Deoxycorticosterone	Thomas and Gerall, 1969
DDT	Heinrichs et al., 1971

We must emphasize that we cannot with biochemical experiments so far performed exclude the occurrence of some limited number of cell nuclear estrogen or testosterone binding sites in baby brains of the type found in adult brains. In the baby brain these either may not be regionally distributed or may be too labile to measure by our methods. Assuming for the sake of discussion that such "receptors" might not exist in the embryonic or neonatal rat brain, what alternative models exist for conceptualizing steroid hormone interactions with the genome which could lead to the observed permanent effects on sexual differentiation? One possible site of action, independent of stereospecific hormone-binding proteins, is an interaction of the steroid directly with the DNA. Such interaction has been described in terms of a hyperchromic shift in the melting of double-stranded, mammalian DNA, which is proportional to estradiol-17β concentration up to 4×10^{-5} M and is not produced by the stereoisomer estradiol-17α (Speichinger and Barker, 1969). This effect appears to involve maximally around 16% of the DNA, the remainder melting at a temperature somewhat higher than control samples, and is thus selective within the population of DNA. It is not presently clear how selective among the various chromosomes or genetic loci such an effect could or would, indeed, have to be to produce an effect on sexual differentiation, nor is it clear what consequences this interaction would have in terms of altered gene function.

ACKNOWLEDGMENTS   Research described in this chapter was supported by research grants NS07080 and MH13189 from the United States Public Health Service and by a grant from the Rockefeller Foundation to Rockefeller University. Dr. Denef has been supported by a USPHS Foreign Postdoctoral Fellowship and by a fellowship from the Foundations Fund for Research in Psychiatry, New Haven, Connecticut.

We express our appreciation to Miss Eileen Gibson for editorial assistance in preparation of this chapter and to Mrs. Carew Magnus and Mrs. Gislaine Wallach for technical assistance during the course of the research described herein.

We would like to acknowledge invaluable assistance from publications of the Brain Information Service, UCLA, in preparing this chapter.

REFERENCES

AHREN, K., K. FUXE, L. HAMBERGER, and T. HÖKFELT, 1971. Turnover changes in the tubero-infundibular dopamine neurons during the ovarian cycle of the rat. *Endocrinology* 88: 1415–1424.

ALBERTI, K. G. M. M., and G. W. G. SHARP, 1970. Identification of four types of steroid by their interaction with mineralocorticoid receptors in the toad bladder. *J. Endocrinol.* 48:563–574.

ANDERSON, C. H., and S. S. GREENWALD, 1969. Autoradiographic analysis of estradiol uptake in the brain and pituitary of the female rat. *Endocrinology* 85:1160–1165.

ANTON-TAY, F., and R. J. WURTMAN, 1971. Brain monoamines and endocrine function. In *Frontiers in Neuroendocrinology 1971*, L. Martini, and W. F. Ganong, eds. New York: Oxford University Press, pp. 45–66.

ASHBURNER, M., 1972. Puffing patterns in *Drosophila melanogaster* and related species. *Developmental Studies on Giant Chromosomes*, W. Beermann, ed., *Results and Problems in Cell Differentiation*, Vol. 4. New York: Springer-Verlag, pp. 101–151.

AZMITIA, JR., E. C., S. ALGERI, and E. COSTA, 1970. Turnover rate of *in vivo* conversion of tryptophan into serotonin in brain areas of adrenalectomized rats. *Science* 169:201–203.

BAPNA, J., N. H. NEFF, and E. COSTA, 1971. A method for studying norepinephrine and serotonin metabolism in small regions of rat brain: Effect of ovariectomy on amine metabolism in anterior and posterior hypothalamus. *Endocrinology* 89:1345–1349.

BARRACLOUGH, C. A., 1967. Modification in reproductive function after exposure to hormones during the prenatal and early postnatal period. *Neuroendocrinology*, Vol. 2, L. Martini, and W. F. Ganong, eds. New York: Academic Press, pp. 61–99.

BAULIEU, E., I. LASNITZKI, and P. ROBEL, 1968. Testosterone, prostate gland, and hormone action. *Biochem. Biophys. Res. Commun.* 32:575–577.

BAXTER, J. D., and G. M. TOMKINS, 1971. Specific cytoplasmic glucocorticoid hormone receptors in hepatoma tissue culture cells. *Proc. Natl. Acad. Sci. USA* 68:932–937.

BEATTIE, C. W., C. H. RODGERS, and L. F. SOYKA, 1972. Influence of ovariectomy and ovarian steroids on hypothalamic tyrosine hydroxylase activity in the rat. *Endocrinology* 91:276–279.

BEERMANN, W., 1972. Chromomeres and genes. *Developmental Studies on Giant Chromosomes*, W. Beermann, ed., *Results and Problems in Cell Differentiation*, Vol. 4. New York: Springer-Verlag, pp. 1–33.

BELAJEV, D. K., L. I. KOROTCHKIN, A. A. BAJEV, A. N. GOLUBITSA, L. S. KOROTCHKINA, L. F. MAKSIMOVSKY, and A. E. DAVIDOVSKAJA, 1967. Activity of the genetic apparatus of hypothalamic nerve cells at various stages of the oestrous cycle in the albino rat. *Nature* 214:201–202.

BERENDES, H. D., 1972. The control of puffing in *Drosophila hydei*. *Developmental Studies on Giant Chromosomes*, W. Beermann, ed., *Results and Problems in Cell Differentiation*, Vol. 4. New York: Springer-Verlag, pp. 181–207.

BEYER, C., and C. H. SAWYER, 1969. Hypothalamic unit activity related to control of the pituitary gland. *Frontiers in Neuroendocrinology 1969*, W. F. Ganong, and L. Martini, eds. New York: Oxford University Press, pp. 255–287.

BEYER, C., N. VIDAL, and A. MIJARES, 1970. Probable role of aromatization in the induction of estrous behavior by androgens in the ovariectomized rabbit. *Endocrinology* 87:1386.

BEYER, C., G. MORALI, and M. L. CRUZ, 1971. Effect of 5α-dihydrotestosterone on gonadotropin secretion and estrous behavior in the female Wistar rat. *Endocrinology* 89:1158–1161.

BOHUS, B., 1970. Central nervous structures and the effect of ACTH and corticosteroids on avoidance behavior: A study with cerebral implantation of corticosteroids in the rat. *Pituitary, Adrenal, and the Brain*, D. de Wied, and J. A. W. M. Weijnen, eds. *Progr. Brain Res.*, Vol. 32. Amsterdam: Elsevier, pp. 171–183.

BOHUS, B., 1971. Adrenocortical hormones and central nervous function: The site and mode of their behavioral action in the rat. *Proc. 3rd International Congr. Hormonal Steroids, Hamburg, 1970*, V. H. T. James, and L. Martini, eds. *Excerpta Medica Series 219*, Basel: Karger, pp. 752–758.

BOTTIGLIONI, F., W. P. COLLINS, C. FLAMIGNI, F. NEUMANN, and I. F. SOMMERVILLE, 1971. Studies on androgen metabolism in experimentally feminized rats. *Endocrinology* 89:553–559.

BOYLE, I. T., L. MIRAVET, R. W. GRAY, M. F. HOLICK, and H. F. DELUCA, 1972. The response of intestinal calcium transport to 25-hydroxy and 1,25-dihydroxy vitamin D in nephrectomized rats. *Endocrinology* 90:605–608.

BROWN-GRANT, K., A. MUNCK, F. NAFTOLIN, and M. R. SHERWOOD, 1971. The effects of the administration of testosterone propionate alone or with phenobarbitone and of testosterone metabolites to neonatal female rats. *Hormones and Behavior* 2:173–182.

BRUCHOVSKY, N., and J. D. WILSON, 1968. The conversion of testosterone to 5α-androstan-17β-ol-3-one by rat prostate *in vivo* and *in vitro*. *J. Biol. Chem.* 243:2012–2021.

BULLOCK, L. P., C. W. BARDIN, and S. OHNO, 1971. The androgen insensitive mouse: Absence of intranuclear androgen retention in the kidney. *Biochem. Biophys. Res. Commun.* 44:1537–1543.

BUTTE, J. C., R. KAKIHANA, and E. P. NOBLE, 1972. Rat and mouse brain corticosterone. *Endocrinology* 90:1091–1100.

CAMERON, I. L., E. L. TOLMAN, and G. W. HARRINGTON, 1969. Aldosterone receptor site tissues and cells of salamander, chicken, goldfish and mouse. *Texas Repts. Biol. Med.* 27:367–380.

CARRER, H. F., and S. TALEISNIK, 1970. Effect of mesencephalic stimulation on the release of gonadotropin. *J. Endocr.* 4:8 527–539.

CARRER, H. F., and S. TALEISNIK, 1972. Neural pathways associated with the mesencephalic inhibitory influence on gonadotropin secretion. *Brain Research* 38:299–313.

CHADER, G. L., and C. A. VILLEE, 1970. Uptake of oestradiol by the rabbit hypothalamus. *Biochem. J.* 118:93–97.

CHEIFETZ, P., N. GAFFUD, and J. F. DINGMAN, 1968. Effects of bilateral adrenalectomy and continuous light on the circadian rhythm of corticotropin in female rats. *Endocrinology* 82:1117–1124.

CLAYTON, R. B., J. KOGURA, and H. C. KRAEMER, 1970. Sexual differentiation of the brain: Effects of testosterone on brain RNA metabolism in newborn female rats. *Nature (London)* 226:810–812.

CLEVER, U., and P. KARLSON, 1960. Induktion von Puff-Veranderungen in den Speicheldrusenchromosomen von *Chironomus tentans* durch Ecdyson. *Exp. Cell Res.* 20:623–626.

COFFEY, J. C., 1972. Metabolism of testosterone and andosterone *in vitro* by mouse submixillary gland (Abstract). *Federation Proc.* 31:296.

COPPOLA, J. A., 1971. Brain catecholamines and gonadotropin secretion. In *Frontiers in Neuroendocrinology, 1971*, L. Martini, and W. F. Ganong, eds. New York: Oxford University Press, pp. 129–143.

DARRAH, H. K., P. C. B. MACKINNON, and A. W. ROGERS, 1971. Sexual differentiation in the brain of the neonatal rat. *J. Physiol. (London)* 218:22–23.

DAVIDSON, J. M., 1973. Feedback of steroid hormones in relation to reproduction. *IV International Congress of Endocrinology, Washington, D.C.*, Excerpta Medica International Congress Series (in press).

DELUCA, H. F., and M. J. MELANCON, JR., 1972. 25-hydroxy-cholecalciferol: A hormonal form of vitamin D. In *Biochemical Actions of Hormones 2*, G. Litwack, ed. New York and London: Academic Press, pp. 337–383.

DENEF, C., and P. DEMOOR, 1968. The "puberty" of the rat liver. II. Permanent changes in steroid metabolizing enzymes after treatment with a single injection of testosterone propionate at birth. *Endocrinology* 83:791–798.

DENEF, C., and P. DEMOOR, 1969. "Puberty" of the rat liver.

IV. Influence of estrogens upon the differentiation of cortisol metabolism induced by neonatal testosterone. *Endocrinology* 85:259–269.

DENEF, C., and P. DEMOOR, 1972. Sexual differentiation of steroid metabolizing enzymes in the rat liver. Further studies on predetermination by testosterone at birth. *Endocrinology* 91:374–384.

DENEF, C., and B. S. McEWEN, 1972. Regional and sex differences in the metabolism of testosterone in the rat brain. *IV International Congress of Endocrinology, Washington, D.C.*, Excerpta Medica International Congress Series No. 256, Abstract 121.

DEVELLIS, J., and D. INGLISH, 1968. Hormonal control of glycerol phosphate dehydrogenase in the rat brain. *J. Neurochem.* 15:1061–1070.

DEVELLIS, J., D. INGLISH, R. COLE, and J. MOLSON, 1971. Effects of hormones on the differentiation of cloned lines of neurons and glial cells. *Influence of Hormones on the Nervous System*, Proc. 1st Congr. Int. Soc. Psychoneuroendocrinology, Brooklyn, 1970. Basel: Karger, pp. 25–39.

DE WIED, D., 1967. Opposite effects of ACTH and glucocorticosteroids on extinction of conditioned avoidance behavior. *Proceedings 2nd International Congress on Hormonal Steroids, Milan*. Excerpta Medica Internat. Congress Series No. 132, 945–951.

DIAMOND, M., and C. L. WONG, 1969. Neonatal progesterone effect on reproductive functions in the female rat (Abstract). *Anat. Rec.* 163:178.

DIXIT, B. N., and J. P. BUCKLEY, 1967. Circadian changes in brain 5-hydroxytryptamine and plasma corticosterone in the rat. *Life Sci.* 6:755–758.

DOFUKU, R., U. TETTENBORN, and S. OHNO, 1971. Testosterone "regulon" in the mouse kidney. *Nature New Biol.* 232:5–7.

DORFMAN, R. I., and R. A. SHIPLEY, 1956. *Androgens: Biochemistry, Physiology, and Clinical Significance*. New York: John Wiley & Sons, p. 118.

DREWS, J., and L. WAGNER, 1970. The effect of prednisolone injected *in vivo* on RNA-polymerase activities in isolated rat thymus nuclei. *Europ. J. Biochem.* 13:231–237.

EIK-NES, K. B., and K. R. BRIZZEE, 1965. Concentration of tritium in brain tissue of dogs given (1,2-³H) cortisol intravenously. *Biochem. Biophys. Acta* 97:320–333.

EISENFELD, A. J, 1970. ³H-estradiol: *in vitro* binding to macromolecules from the rat hypothalamus, anterior pituitary, and uterus. *Endocrinology* 86:1313–1318.

EISENFELD, A. J., and J. AXELROD, 1965. Selectivity of estrogen distribution in tissues. *J. Pharmacol. Exp. Therap.* 150:469–475.

EISENFELD, A. J., and J. AXELROD, 1966. Effect of steroid hormones, ovariectomy, estrogen pretreatment, sex and immaturity on the distribution of ³H-estradiol. *Endocrinology* 79:38–42.

FLERKÓ, B., 1971. Steroid hormones and the differentiation of the central nervous system. *Current Topics in Experimental Endocrinology*, Vol. 1, L. Martini, and V. H. T. James, eds. New York and London: Academic Press, pp. 41–80.

FLERKÓ, B., and B. MESS, 1968. Reduced oestradiol-binding capacity of androgen sterilized rats. *Acta Physiol. Acad. Sci. Hung.* 33:111–113.

FRIEDMAN, A. H., and C. A. WALKER, 1968. Circadian rhythms in rat mid-brain and caudate nucleus biogenic amine levels. *J. Physiol.* 197:77–85.

FUXE, K., T. HÖKFELT, and O. NILSSON, 1969. Castration, sex hormones, and tuberoinfundibular dopamine neurons. *Neuroendocrinology* 5:107–120.

GEHRING, U., G. M. TOMKINS, and S. OHNO, 1971. Effect of the androgen-insensitivity mutation on a cytoplasmic receptor for dihydrotestosterone. *Nature New Biol.* 232:106–107.

GEHRING, W., 1972. The stability of the determined state in cultures of imaginal disks in *Drosophila*. *The Biology of Imaginal Disks*, H. Ursprung, and R. Nöthiger, eds., *Results and Problems in Cell Differentiation*, Vol. 5. New York: Springer-Verlag, pp. 35–58.

GERALL, A. A., and A. McM. KENNEY, 1970. Neonatally androngenized females' responsiveness to estrogen and progesterone. *Endocrinology* 87:560–566.

GERLACH, J. L., and B. S. McEWEN, 1972. Rat brain binds adrenal steroid hormone: Radioautography of hippocampus with corticosterone. *Science* 175:1133–1136.

GILLIN, J. C., L. S. JACOBS, D. H. FRAM, and F. SNYDER, 1972. Acute effect of a glucocorticoid on normal human sleep. *Nature (London)* 237:398–399.

GLOYNA, R. E., and J. D. WILSON, 1969. A comparative study of the conversion of testosterone to 17β-hydroxy-5α-androstan-3-one (dihydrotestosterone) by prostate and epididymis. *Endocrinology* 29:970–977.

GOLDFOOT, D. A., H. H. FEDER, and R. W. GOY, 1969. Development of bisexuality in the male rat treated neonatally with androstenedione. *J. Comp. Physiol. Psychol.* 67:41–45.

GOLDMAN, A. S., and M. K. BAKER, 1971. Androgenicity in the rat fetus of metabolites of testosterone and antagonism by cyproterone acetate. *Endocrinology* 89:276–280.

GORSKI, R. A., 1963. Modification of ovulatory mechanisms by postnatal administration of estrogen to the rat. *Am. J. Physiol.* 205:842–847.

GORSKI, R. A., and J. SHRYNE, 1971. Intracerebral antibiotics and androgenization of the neonatal female rat (Abstract). *Anat. Rec.* 169:327.

GOY, R. W., 1970. Early hormonal influences on the development of sexual and sex-related behavior. In *The Neurosciences: Second Study Program*, F. O. Schmitt, Editor-in-Chief. New York: Rockefeller University Press, pp. 196–206.

GRANNER, D., L. R. CHASE, G. D. AURBACH, and G. M. TOMKINS, 1968. Tyrosine aminotransferase: Enzyme induction independent of adenosine 3′,5′-monophosphate. *Science* 162:1018–1020.

GRINÒ, E., R. DOMINGUEZ, J. SAS, W. L. GENEFETTI, L. C. APPELTAUER, and N. J. REISSENWEBER, 1968. Comparative study of ovarian hypertrophy and estral cycle in rats with two different mesencephalic lesions. *Neuro-Visceral Relations* 31:76–85.

GROSSER, B. I., W. STEVENS, F. W. BRUENGER, and D. J. REED, 1971. Corticosterone binding by rat brain cytosol. *J. Neurochem.* 18:1725–1732.

HALÁSZ, B., L. PUPP, and S. UHLARIK, 1962. Hypophysiotrophic area in the hypothalamus. *J. Endocrinol.* 25:147–154.

HAMILTON, T. H., 1968. Control by estrogen of genetic transcription and translation. *Science* 161:649–661.

HECHTER, O., K. YOSHINAGA, I. D. K. HALKERSTON, and K. BIRCHALL, 1967. Estrogen-like anabolic effects of cyclic 3′,5′-adenosine monophosphate and other nucleotides in isolated rat uterus. *Arch. Biochem. Biophys.* 122:449–465.

HEINRICHS, W. L., R. J. GELLERT, J. L. BAKKE, and N. L. LAWRENCE, 1971. DDT administered to neonatal rats induces persistent estrus syndrome. *Science* 173:642–643.

HENKIN, R. I., 1970. The effects of corticosteroids and ACTH on sensory systems. In *Pituitary, Adrenal, and the Brain*, D. de Wied and J. A. W. M. Weijnem, eds. *Progr. Brain Res.* Vol. 32. Amsterdam: Elsevier, pp. 270–293.

HENKIN, R., A. CASPER, R. BROWN, A. HARLAN, and F. BARTER, 1968. Presence of corticosterone and cortisol in the control and peripheral nervous system of the cat. *Endocrinology* 82: 1058–1061.

HIROSHIGE, T., and M. SAKAKURA, 1971. Circadian rhythm of corticotrophin-releasing activity in the hypothalamus of normal and adrenalectomized rats. *Neuroendocrinology* 7: 25–36.

JAFFE, R. B., 1969. Testosterone metabolism in target tissues: Hypothalamic and pituitary tissues of the adult rat and human fetus, and the immature rat epiphisis. *Steroids* 14: 483–489.

JENSEN, E. V., and E. R. DeSOMBRE, 1972. Estrogens and progestins. *Biochemical Actions of Hormones*, Vol. 2, G. Litwack, ed. New York and London: Academic Press, pp. 215–255.

JENSEN, E. V., T. SUZUKI, T. KAWASHIMA, W. E. STUMPF, P. W. JUNGBLUT, and E. R. DeSOMBRE, 1968. A two-step mechanism for the interaction of estradiol with rat uterus. *Proc. Natl. Acad. Sci. USA* 59: 632–638.

JENSEN, E. V., T. SUZUKI, M. NUMATA, S. SMITH, and E. R. DeSOMBRE, 1969. Estrogen-binding substances of target tissues. *Steroids* 13: 417–427.

JOHNSON, J. H., and C. H. SAWYER, 1971. Adrenal steroids and the maintenance of a circadian distribution of paradoxical sleep in rats. *Endocrinology* 89: 507–512.

JOST, A., 1970. Hormonal factors in the sex differentiation of the mammalian foetus. *Proc. Roy. Soc. [Biol.]* 259: 119–130.

KAHWANAGO, I., W. L. HEINRICHS, and W. L. HERRMANN, 1969. Isolation of oestradiol "receptors" from bovine hypothalamus and anterior pituitary gland. *Nature (London)* 223: 313–314.

KAMBERI, I. A., and Y. KOBAYASHI, 1970. Monoamine oxidase activity in the hypothalamus and various other brain areas and in some endocrine glands of the rat during the estrous cycle. *J. Neurochem.* 17: 261–268.

KAMBERI, I. A., and S. M. McCANN, 1972. Effects of implants of testosterone in the median eminence and pituitary on FSH secretion. *Neuroendocrinology* 9: 20–29.

KARAVOLAS, H. J., and S. M. HERF, 1971. Conversion of progesterone by rat medial basal hypothalamic tissue to 5α-pregnane-3,20-dione. *Endocrinology* 89: 940–942.

KATO, J., and C. A. VILLEE, 1967a. Preferential uptake of estradiol by the anterior hypothalamus of the rat. *Endocrinology* 80: 567–575.

KATO, J., and C. A. VILLEE, 1967b. Factors affecting uptake of estradiol-6,7^{3}H by the hypophysis and hypothalamus. *Endocrinology* 80: 1133–1138.

KATO, J., Y. ATSUMI, and M. INABA, 1970a. A soluble receptor for estradiol in rat anterior hypophysis. *J. Biochem. (Tokyo)* 68: 759–761.

KATO, J., Y. ATSUMI, and M. MURAMATSU, 1970b. Nuclear estradiol receptor in rat anterior hypophysis. *J. Biochem. (Tokyo)* 67: 871–872.

KATO, J., Y. ATSUMI, and M. INABA, 1971. Development of estrogen receptors in the rat hypothalamus. *J. Biochem. (Tokyo)* 70: 1051–1053.

KENNEY, F. T., and K. L. LEE, 1971. Regulation of enzyme synthesis in cultured cells by adrenal steroids. *Proc. 3rd International Congr. Hormonal Steroids, Hamburg, 1970*, V. H. T. James, and L. Martini, eds. Excerpta Medica International Congress Series 219, Basel, pp. 472–478.

KING, R. J. B., J. GORDON, and A. W. STEGGLES, 1971. Receptor-polycation interaction in relation to oestradiol binding in uterus. *Proc. 3rd International Congr. Hormonal Steroids, Hamburg, 1970*, V. H. T. James, and L. Martini, eds. Excerpta Medica International Congress Series 219, Basel, pp. 349–399.

KNIEWALD, A., R. MASSA, and L. MARTINI, 1970. *3rd International Congr. Hormonal Steroids*. Excerpta Medica International Congress Series 210, Abstract 111.

KORENMAN, S. G., 1970. Relation between estrogen inhibitory activity and binding to cytosol of rabbit and human uterus. *Endocrinology* 87: 1119–1123.

KRAGT, C. L., and W. F. GANONG, 1968. Pituitary FSH content in female rats at various ages. *Endocrinology* 82: 1241–1245.

KRIEGER, D. T., and F. RIZZO, 1969. Serotonin mediation of circadian periodicity of plasma 17-hydroxycorticosteroids. *Amer. J. Physiol.* 217: 1703–1707.

LADOSKY, W., 1967. Anovulatory sterility in rats neonatally injected with stilbestrol. *Endokrinologie* 52: 259–261.

LEVINSON, B. R., J. D. BAXTER, G. G. ROUSSEAU, and G. M. TOMKINS, 1972. Cellular site of glucocorticoid-receptor complex formation. *Science* 175: 189–190.

LIAO, S., and S. FANG, 1969. Receptor proteins for androgens and the mode of action of androgens on gene transcription in ventral prostate. *Vitamins and Hormones* 27: 17–90.

LISK, R. D., 1967. Sexual behavior: Hormonal control. *Neuroendocrinology*, Vol. 2, L. Martini, and W. F. Ganong, eds. New York: Academic Press, pp. 197–239.

LISK, R. D., and M. NEWLON, 1963. Estradiol: Evidence for its direct effect on hypothalamic neurons. *Science* 139: 223–224.

LITTERIA, M., and P. S. TIMIRAS, 1970. *In vivo* inhibition of protein synthesis in specific hypothalamic nuclei by 17β-estradiol. *Proc. Soc. Exp. Biol. Med.* 134: 256–261.

LITWACK, G., and S. SINGER, 1972. Subcellular actions of glucocorticoids. *Biochemical Actions of Hormones*, Vol. 2, G. Litwack, ed. New York and London: Academic Press, pp. 113–163.

LUTTGE, W. F., and R. E. WHALEN, 1970. Dihydrotestosterone, androstenedione, testosterone: Comparing effectiveness in masculinizing and defeminizing reproductive systems in male and female rats. *Hormones and Behavior* 1: 265–281.

MAKMAN, M. H., S. NAKAGAWA, and A. WHITE, 1967. Studies of the mode of action of adrenal steroids on lymphocytes. *Recent Progr. Hormone Res.* 23: 195–219.

MASSA, R., E. STUPNICKA, A. VILLE, and L. MARTINI, 1971. *Program 53rd Meeting Endocrine Society*, Abstract 373.

MASSA, R., E. STUPNICKA, and L. MARTINI, 1972. Metabolism of progesterone in the anterior pituitary, the hypothalamus and the uterus of the female rat. *IV International Congress of Endocrinology, Washington, D.C.* Excerpta Medica International Congress Series No. 256, Abstract 118.

MAURER, R., and D. WOOLLEY, 1971. Distribution of ^{3}H-estradiol in clomiphene-treated and neonatally androgenized rats. *Endocrinology* 88: 1281–1287.

McCANN, S. M., P. S. KALRA, A. O. DONOSO, W. BISHOP, H. P. G. SCHNEIDER, C. P. FAWCETT, and L. KRULICH, 1972. The role of monoamines in the control of gonadotropin and prolactin secretion. In *Brain-Endocrine Interaction. Median Eminence: Structure and Function*, K. M. Knigge, D. E. Scott, and A. Weindl, eds. Basel: S. Karger, pp. 224–235.

McDONALD, P., C. BEYER, F. NEWTON, B. BRIEN, R. BAKER, H. S. TAN, C. SAMPSON, P. KITCHING, R. GREENHILL, and D. PRITCHARD, 1970. Failure of 5α-dihydrotestosterones to initiate sexual behavior in the castrated male rat. *Nature* 227: 964–965.

McEwen, B. S., 1973. Glucocorticoid binding sites in rat brain: Subcellular and anatomical localizations. *Progr. Brain Res.* (in press).

McEwen, B. S., and D. W. Pfaff, 1970. Factors influencing sex hormone uptake by rat brain regions: I. Effects of neonatal treatment, hypophysectomy, and competing steroid on estradiol uptake. *Brain Res.* 21:1–16.

McEwen, B. S., and D. W. Pfaff, 1973. Chemical and psychologic approaches to neuroendocrine mechanisms: Attempts at integration. In *Frontiers in Neuroendocrinology*, L. Martini, and W. F. Ganong, eds. (in press).

McEwen, B. S., and L. Plapinger, 1970. Association of corticosterone-1,2-H^3 with macromolecules extracted from brain cell nuclei. *Nature (London)* 226:263–264.

McEwen, B. S., J. M. Weiss, and L. S. Schwartz, 1969. Uptake of corticosterone by rat brain and its concentration by certain limbic structures. *Brain Res.* 16:227–241.

McEwen, B. S., D. W. Pfaff, and R. E. Zigmond, 1970a. Factors influencing sex hormone uptake by rat brain regions. II. Effects of neonatal treatment and hypophysectomy on testosterone uptake. *Brain Res.* 21:17–28.

McEwen, B. S., D. W. Pfaff, and R. E. Zigmond, 1970b. Factors influencing sex hormone uptake by rat brain regions. III. Effects of competing steroids on testosterone uptake. *Brain Res.* 21:29–38.

McEwen, B. S., J. M. Weiss, and L. S. Schwartz, 1970c. Retention of corticosterone by cell nuclei from brain regions of adrenalectomized rats. *Brain Res.* 17:471–482.

McEwen, B. S., C. Magnus, and G. Wallach, 1972a. Soluble corticosterone-binding macromolecules extracted from rat brain. *Endocrinology* 90:217–226.

McEwen, B. S., R. E. Zigmond, and J. L. Gerlach, 1972b. Sites of steroid binding and action in the brain. In *Structures and Function of Nervous Tissue*, Vol. 5, G. H. Bourne, ed. New York: Academic Press, pp. 205–291.

Millard, S. A., E. Costa, and E. M. Gal, 1972. On the control of brain serotonin turnover rate by end product inhibition. *Brain Res.* 40:545–551.

Moguilevsky, J. A., and M. R. Malinow, 1964. Endogenous oxygen uptake of hypothalamus. *Am. J. Physiol.* 206:855–857.

Moscona, A. A., and R. Piddington, 1966. Stimulation by hydrocortisone of premature changes in the development pattern of glutamine synthetase in embryonic retina. *Biochim. Biophys. Acta* 121:409–411.

Mowles, T. F., B. Ashkanazy, E. Mix, Jr., and H. Sheppard, 1971. Hypothalamic and hypophyseal estradiol-binding complexes. *Endocrinology* 89:484–491.

Musliner, T. A., and G. J. Chader, 1971. A role for DNA in the formation of nuclear estradiol-receptor complex in a cell-free system. *Biochem. Biophys. Res. Commun.* 45:998–1003.

Myrtle, J. F., and A. W. Norman, 1971. Vitamin D: A cholecalciferol metabolite highly active in promoting intestinal calcium transport. *Science* 171:79–82.

Naftolin, F., K. J. Ryan, and Z. Petro, 1972. Aromatization of androstenedione by the anterior hypothalamus of adult male and female rats. *Endocrinology* 90:295–298.

Nakagawa, S., and A. White, 1970. Properties of an aggregate ribonucleic acid polymerase from rat thymus and its response to cortisol injection. *J. Biol. Chem.* 245:1448–1457.

Nakai, T., T. Kigawa, and S. Sakamoto, 1971. ^{3}H-Leucine uptake of hypothalamic nuclei in fetal male rats and its fluctuation after castration. *Endocrinol. Jap.* 18:353–357.

Nöthiger, R., 1972. The larval development of imaginal disks. *The Biology of Imaginal Disks*, H. Ursprung, and R. Nöthiger, eds., *Results and Problems in Cell Differentiation*, Vol. 5. New York: Springer-Verlag, pp. 1–34.

Notides, A. C., 1970. Binding affinity and specificity of the estrogen receptor of the rat uterus and anterior pituitary. *Endocrinology* 87:987–992.

Oberlander, H., 1972. The hormonal control of development of imaginal disks. *The Biology of Imaginal Disks*, H. Ursprung, and R. Nöthiger, eds., *Results and Problems in Cell Differentiation*, Vol. 5. New York: Springer-Verlag, pp. 155–172.

Ojeda, S. R., and V. D. Ramirez, 1972. Plasma level of LH and FSH in maturing rats: Response to hemigonadectomy. *Endocrinology* 90:466–472.

Okada, F., 1971. The maturation of the circadian rhythm of brain serotonin in the rat. *Life Sci.* 10:77–86.

O'Malley, B. W., T. C. Spelsburg, W. T. Schrader, F. Chytil, and A. W. Steggles, 1972. Mechanisms of interaction of a hormone-receptor complex with the genome of a eukaryotic target cell. *Nature* 235:141–144.

Pelling, C., 1972. Transcription in giant chromosomal puffs. *Developmental Studies on Giant Chromosomes*, W. Beermann, ed., *Results and Problems in Cell Differentiation*, Vol. 4. New York: Springer-Verlag, pp. 87–99.

Pérez-Palacios, G., E. Castaneda, F. Gómez-Pérez, A. E. Pérez, and C. Gual, 1970. *In vitro* metabolism of androgens in dog hypothalamus, pituitary and limbic system. *Biol. Reprod.* 3:205–213.

Pfaff, D. W., 1968a. Uptake of estradiol-17β-^{3}H in the female rat brain: An autoradiographic study. *Endocrinology* 82:1149–1155.

Pfaff, D. W., 1968b. Autoradiographic localization of radioactivity in rat brain after injection of tritiated sex hormones. *Science* 161:1355–1356.

Pfaff, D. W., and M. Keiner, 1972. Estradiol-concentrating cells in the rat amygdala as part of a limbic-hypothalamic hormone sensitive system. In *The Neurobiology of the Amygdala*, B. Eleftheriou, ed. New York: Plenum Press, pp. 885–895.

Pfaff, D. W., and R. E. Zigmond, 1971. Neonatal androgen effects on sexual and non-sexual behavior of adult rats tested under various hormone regimens. *Neuroendocrinology* 7:129–145.

Pfaff, D. W., M. T. A. Silva, and J. M. Weiss, 1971. Telemetered recording of hormone effects on hippocampal neurons. *Science* 172:394–395.

Pfaff, D., C. Lewis, C. Diakow, and M. Keiner, 1972. Neurophysiological analysis of mating behavior responses as hormone-sensitive reflexes. In *Progress in Physiology*, Vol. 5, E. Stellar, and J. Sprague, eds. New York: Academic Press (in press).

Pohorecky, L. A., and R. J. Wurtman, 1971. Adrenocortical control of epinephrine synthesis. *Pharmacol. Rev.* 23:1–35.

Quadagno, D. M., J. Shryne, and R. A. Gorski, 1971. The inhibition of steroid-induced sexual behavior by intrahypothalamic actinomycin D. *Hormones and Behavior* 2:1–10.

Raisinghani, K. H., R. I. Dorfman, E. Forchielli, L. Gyermek, and B. Geuther, 1968. Uptake of intravenously administered progesterone, pregnanedione and pregnanolone by the rat brain. *Acta Endocrinol.* 57:393–404.

Raisman, G., and P. M. Field, 1971. Sexual dimorphism in the preoptic area of the rat. *Science* 173:731–733.

Raynaud, J. P., C. Mercier-Bodard, and E. E. Baulieu, 1972. Rat estradiol binding plasma protein (EBP). *Steroids* 18:767–788.

Reif, L., and H. Amos, 1966. A dialyzable inducer for the

glutamotransferase of chick embryo retina. *Biochem. Biophys. Res. Commun.* 23:39–48.

RESKO, J. A., H. H. FEDER, and R. W. GOY, 1968. Androgen concentrations in plasma and testis of developing rats. *J. Endocrinol.* 40:485–491.

RICHARDSON, G. S., and L. R. AXELROD, 1971. Metabolism of labeled testosterone by minces and nuclei of preputial glands of male rats. *Endocrinology* 88:890–894.

RIVAROLA, M. A., and E. J. PODESTA, 1972. Metabolism of testosterone-^{14}C by seminiferous tubules of mature rats: Formation of 5α-androstan-3α,17β-diol-^{14}C. *Endocrinology* 90:618–623.

ROBISON, G. A., R. W. BUTCHER, and E. W. SUTHERLAND, 1971. *Cyclic AMP*. New York: Academic Press, 531 pp.

ROMMERTS, F. F. G., and H. J. VAN DER MOLEN, 1971. Occurrence and localization of 5α-steroid reductase, 3α- and 17β-hydroxysteroid dehydrogenases in hypothalamus and other brain tissues of the male rat. *Biochem. Biophys. Acta* 248:489–502.

ROSENAU, W., J. D. BAXTER, G. G. ROUSSEAU, and G. M. TOMKINS, 1972. Mechanism of resistance to steroids: Glucocorticoid receptor defects in lymphoma cells. *Nature New Biology* 237:20–23.

ROSENFELD, M. G., and B. W. O'MALLEY, 1970. Steroid hormones: Effects on adenyl cyclase activity and adenosine 3′,5′-monophosphate in target tissues. *Science* 168:253–255.

SCAPAGNINI, U., G. P. MOBERG, G. R. VAN LOON, J. DE GROOT, and W. F. GANONG, 1971. Relation of brain 5-hydroxytryptamine content to the diurnal variation in plasma corticosterone in the rat. *Neuroendocrinology* 7:90–96.

SCHALLY, A. V., C. Y. BOWERS, W. H. CARTER, A. ARIMURA, T. W. REDDING, and M. SAITO, 1969. Effect of actinomycin D on the inhibitory response of estrogen on LH release. *Endocrinology* 85:290–299.

SCHAUMBURG, B. P., 1972. Investigations on the glucocorticoid-binding protein from rat thymocytes. II. Stability, kinetics and specificity of binding of steroids. *Biochem. Biophys. Acta* 261:219–235.

SCHIAFFINI, O., B. MARÍN, and A. GALLEGO, 1969. Oxidative activity of limbic structures during sexual cycle in rats. *Experientia* 25:1255–1256.

SCHULTZ, F. M., and J. D. WILSON, 1972. Virilization of Wolffian duct in the rat fetus by testosterone derivatives. *IV Internat. Congr. Endocrinology*, Excerpta Medica International Congress Series 256, Abstract 193.

SEIKI, K., M. MIYAMATA, A. YAMASHITA, and M. KOTANI, 1969. Further studies on the uptake of labeled progesterone by the hypothalamus and pituitary of rats. *J. Endocrinol.* 43:129–130.

SHARP, G. W. G., 1971. Metabolic effects of aldosterone and their relationship to the stimulation of sodium transport. *Proc. 3rd International Congr. Hormonal Steroids, Hamburg, 1970*, V. H. T. James, and L. Martini, eds. Excerpta Medica International Congress Series 219. Basel: pp. 636–641.

SHOLITON, L. J., R. T. MARNELL, and E. E. WERK, 1966. Metabolism of testosterone-4-^{14}C by rat brain homogenates and subcellular fractions. *Steroids* 8:265–275.

SHOLITON, L. J., I. L. HALL, and E. D. WERK, 1970. The isopolar metabolites produced by incubation of [4-^{14}C] testosterone with rat and bovine brain. *Acta Endocrinol.* 63:512–518.

SHORE, L. S., and C. A. SNIPES, 1971. Metabolism of testosterone *in vitro* by hypothalamus and other areas of rat brain (Abstract). *Federation Proc.* 30:363.

SHYAMALA, G., 1972. Estradiol receptors in mouse mammary tumors: Absence of the transfer of bound estradiol from the cytoplasm to the nucleus. *Biochem. Biophys. Res. Commun.* 46:1623–1630.

SHYAMALA, G., and J. GORSKI, 1969. Estrogen receptors in the rat uterus: Studies on the interaction of cytosol and nuclear binding sites. *J. Biol. Chem.* 244:1097–1103.

SINGHAL, R. L., R. VIJAYVARGIYA, and G. M. LING, 1970. Cyclic adenosine monophosphate: Andromimetic action on seminal vesicular enzymes. *Science* 168:261–263.

SLUSHER, M. A., 1966. Effects of cortisol implants in the brainstem and ventral hippocampus on diurnal corticosterone levels. *Exp. Brain Res.* 1:184–194.

SNIPES, C. A., and L. S. SHORE, 1972. Metabolism of progesterone *in vitro* by neural and uterine tissues. *Fed. Proc.* 31:236.

SPEICHINGER, J. P., and K. L. BARKER, 1969. Competitive effects of estradiol-17α on the estradiol-17β induced physical changes in uterine DNA. *Steroids* 14:132–143.

SPELSBURG, T. C., A. W. STEGGLES, and B. W. O'MALLEY, 1971. Progesterone-binding components of chick oviduct. III. Chromatin acceptor sites. *J. Biol. Chem.* 246:4188–4197.

STAHL, F., I. POPPE, and G. DÖRNER, 1971. Umwandlungstraten von testosteron zu dihydrotestosteron und androsteron in hypophyse, hypothalamus und cortex mannlicher und weiblicher ratten. *Acta Biol. Med. German* 26:855–858.

STEFANO, F. J. E., and A. O. DONOSO, 1967. Norepinephrine levels in the rat hypothalamus during the estrous cycle. *Endocrinology* 81:1405–1406.

STERN, J. J., 1969. Neonatal castration, androstenedione, and the mating behavior of the male rat. *J. Comp. Physiol. Psychol.* 69:608–612.

STONE, G. M., and B. BAGGETT, 1965. The uptake of some tritiated estrogenic and non-estrogenic steroids by the mouse uterus and vigina *in vivo* and *in vitro*. *Steroids* 6:277–299.

STUMPF, W. E., 1968. Estradiol-concentrating neurons: Topography in the hypothalamus by dry mount autoradiography. *Science* 162:1001–1003.

STUMPF, W. E., 1970. Estrogen-neurons and estrogen-neuron systems in the periventricular brain. *Amer. J. Anat.* 129:207–217.

STUMPF, W. E., and L. J. ROTH, 1966. High resolution autoradiography with drymounted, freeze-dried, frozen sections. Comparative study of six methods using two diffusible compounds, ^{3}H estradiol and ^{3}H mesobilirubinogen. *Histochem. and Cytochem.* 14:274–287.

STUMPF, W. E., and M. SAR, 1971. Localization of steroid hormones in the brain by dry-autoradiography. *Proc. 3rd International Congr. Hormonal Steroids, Hamburg, 1970*, V. H. T. James, and L. Martini, eds., Excerpta Medica International Congress Series 219. Basel: pp. 759–763.

SWANECK, G. E., E. HIGHLAND, and I. S. EDELMAN, 1969. Stereospecific nuclear and cytosol aldosterone-binding proteins of various tissues. *Nephron* 6:297–316.

SZEGO, C. M., and J. S. DAVIS, 1967. Adenosine 3′5′-monophosphate in rat uterus: Acute elevation by estrogen. *Proc. Natl. Acad. Sci. USA.* 58:1711–1718.

TAKAYASU, S., and K. ADACHI, 1972. The *in vivo* and *in vitro* conversion of testosterone to 17β-hydroxy-5α-androstan-3-one (dihydrotestosterone) by the sebaceous gland of hamsters. *Endocrinology* 90:73–80.

TERENIUS, L., 1966. Specific uptake of oestrogens by the mouse uterus *in vitro*. *Acta Endocrinol.* 53:611–618.

THOMAS, T. R., and A. A. GERALL, 1969. Dissociation of reproductive physiology and behavior induced by neonatal treatment with steroids. *Endocrinology* 85:781–784.

TINDAL, J. S., G. S. KNAGGS, and A. TURVEY, 1967. Central nervous control of prolactin secretion in the rabbit: Effect of local oestrogen implants in the amygdaloid complex. *J. Endocrinol.* 37:279–287.

TOMKINS, G. M., T. D. GELEHRTER, D. GRANNER, D. MARTIN, JR., H. H. SAMUELS, and E. B. THOMPSON, 1969. Control of specific gene expression in higher organisms. *Science* 166:1474–1480.

TOUCHSTONE, J. C., J. KASPAROW, P. A. HUGHES, and M. R. HORWITZ, 1966. Corticosteroids in human brain. *Steroids* 7:205–211.

TUOHIMAA, P., and R. JOHANSSON, 1971. Decreased estradiol binding in the uterus and anterior hypothalamus of androgenized female rats. *Endocrinology* 88:1159–1164.

TURNER, C. D., 1941. Permanent genital impairments in the adult rat resulting from the administration of estrogen during early life. *Am. J. Physiol.* 133:471–472.

TVETER, K. J., and A. AAKVAAG, 1969. Uptake and metabolism *in vivo* of testosterone-1,2-H³ by accessory sex organs of male rats; influence of some hormonal compounds. *Endocrinology* 85:683–689.

ULRICH, R., A. YUWILER, and E. GELLER, 1970. Failure of 5α-dihydrotestosterone to block androgen sterilization in the female rat. *Proc. Soc. Exp. Biol. Med.* 139:411–413.

VAN WIMERSMA-GREIDANUS, Tj. B., 1970. Effects of steroids on extinction of an avoidance response in rats. A structure-activity relationship study. In *Pituitary, Adrenal and the Brain*, D. de Wied, and J. A. W. M. Weijnen, eds. *Progr. Brain Res.*, Vol. 32. Amsterdam: Elsevier, pp. 185–191.

VERTES, M., and R. J. B. KING, 1971. The mechanism of oestradiol binding in rat hypothalamus: Effect of androgenization. *J. Endocrinol.* 51:271–282.

WADE, G., and I. ZUCKER, 1970. Modulation of food intake and locomotor activity in female rats by diencephalic hormone implants. *J. Comp. Physiol. Psychol.* 72:328–336.

WEICK, R. F., E. R. SMITH, R. DOMINGUEZ, A. P. S. DHARIWAL and J. M. DAVIDSON, 1971. Mechanism of stimulatory feed-back effect of estradiol benzoate on the pituitary. *Endocrinology* 88:293–301.

WEINER, R. I., R. A. GORSKI, and C. H. SAWYER, 1972. Catecholamines and pituitary gonadotrophic function. *Brain-Endrocine Interaction. Median Eminence: Structure and Function*, K. M. Knigge, D. E. Scott, and A. Weindl, eds. Basel: S. Karger, pp. 236–244.

WEISS, B., and J. CRAYTON, 1970. Gonadal hormones as regulators of pineal adenyl cyclase activity. *Endocrinology* 87:527–533.

WESTPHAL, U., 1971. *Steroid-Protein Interactions*. New York: Springer-Verlag, 567 pp.

WHALEN, R. E., and W. G. LUTTGE, 1971a. Testosterone, androstenedione and dihydrotestosterone: Effects on mating behavior of male rats. *Hormones and Behavior* 2:117–125.

WHALEN, R. E., and W. G. LUTTGE, 1971b. Perinatal administration of dihydrotestosterone to female rats and the development of reproductive function. *Endocrinology* 89:1320–1322.

WHALEN, R. E., and R. D. NADLER, 1963. Suppression of the development of female mating behavior by estrogen administration in infancy. *Science* 141:273–274.

WHALEN, R. E., W. G. LUTTGE, and B. B. GORZALKA, 1971. Neonatal androgenization and the development of estrogen responsivity in male and female rats. *Hormones and Behavior* 2:83–90.

WICKS, W. D., F. T. KENNEY, and K. L. LEE, 1969. Induction of hepatic enzyme synthesis *in vivo* by adenosine 3′,5′-monophosphate. *J. Biol. Chem.* 244:6008–6013.

WILLIAMS-ASHMAN, H. G., and A. H. REDDI, 1972. Androgenic regulation of tissue growth and function. *Biochemical Actions of Hormones*, Vol. 2, G. Litwack, ed. New York and London: Academic Press, pp. 257–294.

WILLIAMS, D., and J. GORSKI, 1971. A new assessment of subcellular distribution of bound estrogen in the uterus. *Biochem. Biophys. Res. Commun.* 45:258–264.

WILSON, J. D., and R. E. GLYONA, 1970. The intranuclear metabolism of testosterone in the accessory organs of reproduction. *Recent Progr. Hormone Res.* 26:309–330.

WILSON, J. D., and I. LASNITZKI, 1971. Dihydrotestosterone formation in fetal tissues of the rabbit and rat. *Endocrinology* 89:659–668.

WINSON, J., 1972. Interspecies differences in the occurrence of theta. *Behav. Biol.* 7:479–487.

YAGINUMA, T., T. WATANABE, T. KIGAWA, T. NAKAI, T. KOBAYASHI, and T. KOBAYASHI, 1969. Uptake of ³H-leucine in the brain of the female rat and its change after castration. *Endocrinol. Jap.* 16:591–598.

ZIGMOND, R. E., 1971. Chemical and anatomical specificity of gonadal hormone retention in the brain. Ph.D. thesis, The Rockefeller University, New York.

ZIGMOND, R. E., and B. S. MCEWEN, 1970. Selective retention of oestradiol by cell nuclei in specific brain regions of the ovariectomized rat. *J. Neurochem.* 17:889–899.

ZIGMOND, R. E., F. NOTTEBOHM, and D. W. PFAFF, 1972a. Distribution of androgen-concentrating cells in the brain of the chaffinch. *Proc. Internat. Congr. Endocrinol., Washington, D.C.*, pp. 136.

ZIGMOND, R. E., J. M. STERN, and B. S. MCEWEN, 1972b. Retention of radioactivity in cell nuclei in the hypothalamus of the ring dove after injection of ³H-testosterone. *Gen. Comp. Endocrinol.* 18:450–453.

55 Neural and Hormonal Determinants of Female Mating Behavior in Rats

DONALD W. PFAFF, CAROL DIAKOW, RICHARD E. ZIGMOND, and LEE-MING KOW

ABSTRACT Estradiol-17β is concentrated by cells in specific parts of the limbic system and hypothalamus in female rats and can affect the electrical activity of neurons in brain regions where it is bound. Lesions and estradiol implantation in or near groups of estrogen-concentrating cells in the medial preoptic area or medial anterior hypothalamus can affect female rat mating behavior. Thus the effects of estradiol on mating behavior in female rats presumably are mediated by these estradiol-concentrating neurons. However, the precise physiological mechanisms by which steroid sex hormones facilitate mating behavior cannot be completely understood until the neural pathways controlling specific mating reflexes are identified. Then the role of hormone-sensitive neurons in those pathways can be determined. The first step in such an analysis is a "reflexological" description of mating reflexes to be studied. Then, the neural and neuroendocrine control of such reflexes can be studied using neurophysiological techniques. Along these lines, we have used film and X-ray analysis of female rat mating behavior to describe the lordosis reflex, an estrogen-sensitive mating response. We also have studied sensory stimuli, and sensory neural pathways controlling lordosis. In the course of this study we found that estrogen effects on relevant neural mechanisms are not limited to the hypothalamus; estradiol causes an increase in the receptive field of the pudendal nerve, which supplies the perivaginal skin area, a region contacted by the male rat during mating. In the central nervous system, electrical stimulation and single unit recording results suggest that sensory input relevant to lordosis travels up the anterolateral columns of the spinal cord and is distributed in the brainstem via anatomically defined spinoreticular pathways. Further neurophysiological work linking peripheral sensory or motor pathways with limbic-hypothalamic neuroendocrine mechanisms should offer an opportunity to study hormone effects on a neural integrative mechanism that controls mating behavior.

MATING BEHAVIOR in mammals is a sequence of movements that coordinate the activities of the mating pair. The bases for the interaction between the male and female lie in the stimuli given to each by the other and the subsequent responses. In most mammals elicitation of mating behavior depends upon gonadal hormones. Therefore, to understand the physiological basis of mating behavior, we must know the relevant effects of sensory input from the mating partner, and of gonadal hormones on the brain, and must understand the neural interactions between these two determinants. Regarding *hormonal determinants*, we must know (1) which hormones are necessary for the behavior studied, (2) where these hormones are concentrated by brain cells, (3) whether hormones acting at site where they are concentrated can affect the behavior, and (4) the physiological effects of hormones acting at those sites. Regarding *sensory determinants*, we must know (1) which stimuli each mating partner applies to the other, (2) among these stimuli which are necessary for the behavior, and (3) which stimuli are sufficient for the behavior. (These strategic requirements for studying the physiological basis of mating behavior are reflected in the organization of the first part of this chapter.)

The philosophy governing the research reviewed in this chapter emphasizes the importance of identified cells and neural circuits in the study of neuroendocrine mechanisms and hormone-controlled behaviors. The ease with which individual neurons can be identified anatomically and physiologically in invertebrate ganglia have led to neurophysiological advances in that field (see the fifth part of this volume). In the mammalian brain, for the study of mating behavior mechanisms, we must know which individual neurons are hormone sensitive and where they stand in identified neural pathways mediating mating behavior. Knowing this, we may be able to determine how changes in individual hormone-sensitive cells alter activity in mating behavior circuits. Such alterations are, in turn, the basis for the hormone's effect on the behavioral response. This approach requires a detailed reflexological description of a hormone-sensitive behavioral response and neurophysiological identification of pathways controlling that response, as well as a study of hormone effects on individual neurons.

With this philosophy in mind, we chose to concentrate on the lordosis reflex of the female rodent, primarily the rat, for three reasons: First, the strong causal connection between estrogen levels and lordosis; second, the strong and specific estradiol uptake by neurons in the female rat brain; and third, the lordosis reflex itself appears to be

DONALD W. PFAFF, CAROL DIAKOW, RICHARD E. ZIGMOND, and LEE-MING KOW Rockefeller University, New York stereotyped enough and of simple enough topography to submit to detailed physiological analysis. The lordosis reflex was chosen only for ease of neurophysiological analysis; other interesting aspects of the female rat's mating behavior are also hormone sensitive and coordinated with the male (Pfaff et al., 1972). Neural and hormonal mechanisms of mating behavior in the male have been reviewed elsewhere (Malsbury and Pfaff, 1973).

Neural mechanisms in the adult

HORMONAL DETERMINANTS

1. *Hormones responsible for female sex behavior in rats.* Gonadal hormones increase the level of sex behavior in gonadectomized female rats from very low to normal levels and are considered active in the normal manifestation of mating behavior in females (Beach, 1948; Young, 1961). Estrogen alone is sufficient for mating of ovariectomized females, because injection of daily doses in the range of 1 to 10 μg allows them to show lordosis in a high proportion of the mounts they receive (Davidson et al., 1968a; Pfaff, 1970a). A physiological role for progesterone in eliciting female mating behavior is assumed, because it facilitates the effect of estrogen in ovariectomized female rats (Boling and Blandau, 1939), as well as other species. Under some conditions, the facilitatory effect of progesterone is followed by a second phase in which progesterone inhibits further mating behavior (Zucker, 1968; Ciaccio and Lisk, 1971; Nadler, 1970; Edwards, Whalen, and Nadler, 1968). Daily injections of high doses of testosterone can facilitate lordosis behavior, thus mimicking the estrogen effect, if the testosterone treatment is supplemented on the day of testing with progesterone injections (Pfaff, 1970a; Whalen and Hardy, 1970). Whether or not testosterone is metabolized to an estrogen at its site of action in such experiments remains an open question (Beyer et al., 1970a, 1970b, 1971).

Adrenal hormones are not necessary for sexual behavior in female rats, because adrenalectomized ovariectomized rats can show lordosis when given estrogen replacement (Davidson, 1967; Davidson et al., 1968b; Pfaff, 1971a). These experiments also prove that, when lordosis is facilitated solely by estrogen injections, adrenal progesterone need not play a role.

Steroid sex hormones can directly trigger mating be-

havior by their action on the brain, rather than having their behavioral effects necessarily mediated through their action on the pituitary, because estrogen and progesterone treatment successfully stimulates lordosis behavior in ovariectomized hypophysectomized female rats (Pfaff, 1970b).

2. *Concentration of steroid sex hormones by cells in the rat brain.* Radioactive estradiol is highly concentrated by cells in specific parts of the limbic system, preoptic area, and hypothalamus (Pfaff, 1972; Pfaff and Keiner, 1972) (Figure 1). Two hours after systemic injection of low doses of tritiated estradiol, autoradiograms were prepared in the darkroom by directly mounting unfixed, unembedded, frozen sections from the cryostat knife onto emulsion-coated slides. After exposure periods of several months, a clear anatomical pattern of peak estrogen uptake emerged. In the limbic system, highest uptake is seen in the cortical and medial nuclei of the amygdala, lateral septum, diagonal band of Broca, olfactory tubercle, and ventral hippocampus. High densities of labeled cells are seen in the medial preoptic area and the nucleus of the stria terminalis. In the hypothalamus, many labeled cells are found in the medial anterior hypothalamus, the ventromedial nucleus, the arcuate nucleus, and the ventral premammillary nucleus. In addition, smaller numbers of labeled cells can be seen in specific parts of the more lateral regions in the hypothalamus (Figure 1d, e). An extension of this limbic-hypothalamic system is seen back to the lateral central gray of the mesencephalon (cf. Figure 8a later in this chapter). The levels chosen in Figure 1 are from a much larger series of maps (Pfaff and Keiner, 1973) selected to show the sites of highest levels of estrogen uptake. Brain regions anterior and posterior to the range represented in Figures 1 and 8 are not included, because they have much lower numbers of labeled cells, which tend to have fewer grains and to be scattered.

The limbic-hypothalamic distribution of labeled cells confirms conclusions based on our earlier autoradiographic observations of brain tissue fixed with osmium and formalin (Pfaff, 1968) and is consistent with other autoradiographic work (Stumpf, 1968; Anderson and Greenwald, 1969; Tuohimaa, 1971). The distribution described by our autoradiographic work agrees with that gained from scintillation counting of finely dissected brain regions following radioactive estradiol injections

FIGURE 1 Autoradiographic maps showing some of the regions containing highest densities of estradiol-concentrating cells in the central nervous system of the female rat. Large dots show precise locations of reliably labeled groups of cells following estradiol-H³ injections. Where the labeled cells were extremely densely packed, the area is filled in with solid black. Small dots denote the occasional scattered labeled cells found in irregular positions among structures showing very low uptake. These maps were selected from a larger series (Pfaff and Keiner, 1973) to show sites of high uptake in limbic and hypothalamic structures (Pfaff, 1968, 1972). There are also estradiol-concentrating neurons in the central gray of the mesencephalon (see Figure 8a).

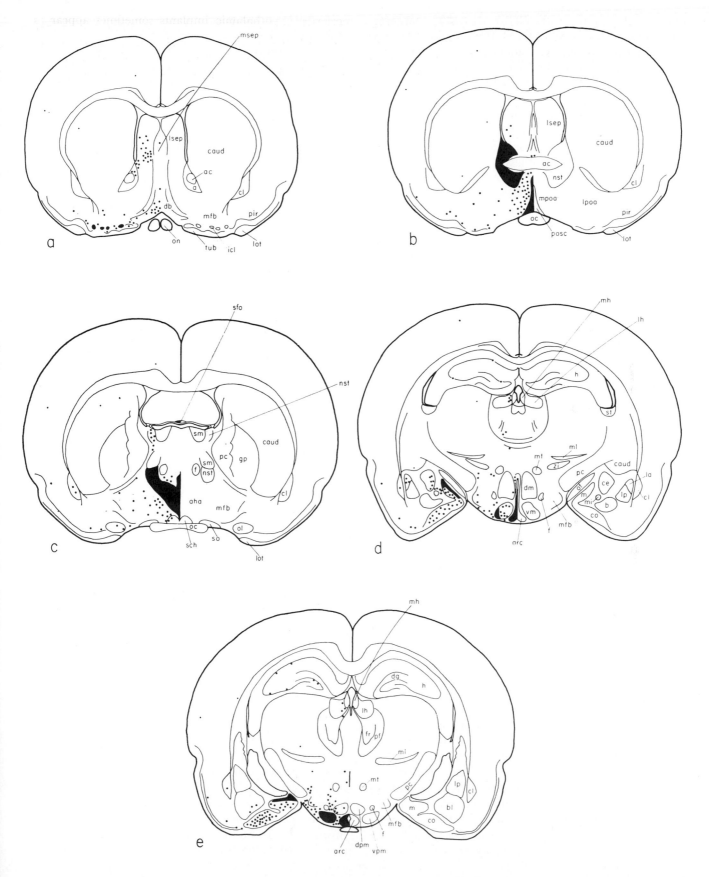

(reviewed by Pfaff, 1972; Pfaff et al., 1972; McEwen, Zigmond, and Gerlach, 1972). In scintillation counting experiments, thin-layer chromatography of the radioactive substance taken up in the brain shows that over 80% is in the chemical form of estradiol (Zigmond and McEwen, 1970).

In most of these experiments, an optimum time to detect the specific binding of estradiol is from 2 to 4 hr after a systemic injection. This is prior to the earliest time (about 16 hr) that lordosis has been detected after systemic injections (Green et al., 1970). Therefore, it may be hypothesized that estrogen entering an estrogen-concentrating cell in the hypothalamus or preoptic area sets into motion a chain of events that takes several hours and that results in the priming of neural circuits having to do with lordosis.

From the autoradiographic maps (Figure 1), it seems likely that estrogen-concentrating structures in the limbic system, preoptic area, hypothalamus, and central gray are involved in the mechanism of estrogen's action on female sexual behavior. Physiological experiments are required to determine to what extent these estrogen-concentrating structures act in a coordinated fashion to affect behavioral or neuroendocrine outputs. At least it is clear that the anatomical basis for coordinated action as an estrogen-sensitive "system" exists (reviewed by McEwen and Pfaff, 1973); a dense web of well-established neuroanatomical pathways links all of the highest estrogen-concentrating structures mentioned above (see introductory figure for this part). These pathways include connections deriving from the olfactory system, connections among structures in the basal forebrain, connections between limbic and hypothalamic structures, and longitudinal association pathways such as the periventricular fiber system and the medial forebrain bundle. Which among these various pathways actually participate in mediating the physiological consequences of estrogen concentration remains to be determined.

3. *Hormone implants in the brain.* Small crystals of estrogen on the tips of tubes implanted in the preoptic and anterior hypothalamic regions of spayed females are sufficient to trigger lordosis, even though implants in some other brain regions are ineffective (Lisk, 1962a; Chambers and Howe, 1968; Malsbury and Pfaff, 1973). In such animals the absence of significant leakage of the implanted hormone to the systemic circulation can be assured by showing that the weight of an estrogen-sensitive structure, the uterus, has not increased as a result of treatment. These experiments extend to rats the conclusions reached previously by Michael (Harris, Michael, and Scott, 1958; Michael, 1965) that hypothalamic implants of estrogen could restore mating behavior in ovariectomized female cats. One difference between cats and rats is that in cats

posterior hypothalamic implants sometimes appear as effective as anterior hypothalamic implants (Harris, Michael, and Scott, 1958; Cerny, 1971). Because of the occasional problem of significant leakage from the site of brain implantation into the systemic circulation, it has not been possible yet to construct a comprehensive map of the brain indicating all sites where estrogen could or could not influence lordosis in female rats. However, it is clear that neurons in one region, the preoptic area-anterior hypothalamus, both accumulate estrogen to a high degree and support the triggering of lordosis upon estrogen implantation.

4. *Electrophysiological effects.* Effects of estrogens on single unit activity in the hypothalamus of female rats have been reported that conceivably could link the phenomenon of hormone concentration by single cells to the mechanism of the hormone implant effects. Preoptic units in estrogen-treated rats gave a higher proportion of inhibitory responses (than in control rats) to vaginal cervix stimulation but not to other "control" stimuli (Lincoln and Cross, 1967). Preoptic and anterior hypothalamic units also tend to show higher spontaneous activity in proestrous female rats compared to anestrous female rats (Cross and Dyer, 1971; Kawakami et al., 1970; Moss and Law, 1971). Often in the hypothalamus and preoptic area inhibitory responses are seen during recording from units that have relatively high spontaneous activity. For instance, in experiments where testosterone greatly increased the spontaneous activity of preoptic neurons in male rats, the direction of response to stimuli sometimes shifted from excitatory to inhibitory when the spontaneous activity reached high levels (Pfaff and Pfaffmann, 1969a).

Since progesterone facilitates the effect of estrogen on lordosis in the female rat, it is interesting that progesterone can alter the responses of lateral hypothalamic units to genital stimuli in anesthetized female rats (Barraclough and Cross, 1963). While progesterone effects on hypothalamic units have since been reported many times (Ramirez, et al., 1967; Komisaruk et al., 1967; Beyer et al., 1967; Lincoln, 1969), it has been pointed out that progesterone also affects electrical activity elsewhere, including the cortical EEG. Some authors have emphasized the nonspecificity of the progesterone effect under their experimental conditions (Komisaruk et al., 1967; Lincoln, 1969), while others have emphasized the possible specificity of hypothalamic responses to sex-related stimuli and corresponding specificity of the progesterone effect in some units (Barraclough and Cross, 1963; Haller and Barraclough, 1970). All of these findings extend the initial claim of Kawakami and Sawyer (1959a, 1959b) that gonadal steroids affect the electrical activity of the hypothalamus and the mesencephalic reticular activating system. While the examples here have been restricted to female rats, Beyer and Sawyer (1969) and Pfaff et al. (1972) have reviewed other work.

The specificity of the progesterone effect has importance for interpretation of electrophysiological results in terms of behavior. Does progesterone specifically affect responses to sex-related stimuli by hypothalamic units only, or could it act by altering the entire reticular activating system? In general, regarding any effect of sex hormones on single unit activity we may ask: Which electrophysiological functions are altered and which are not? The pattern of testosterone effects on single unit activity suggests a speculative behavioral interpretation. Testosterone influences the resting discharge rates of some neurons in the preoptic area and also affects the absolute magnitudes of responses to odor stimuli (Pfaff and Pfaffmann, 1969a). However, when neurons are tested with several odor stimuli before and after testosterone injections, the *relative* response magnitudes of the neuron to different odors are not affected differentially by the hormone. Also, in comparisons of normal and castrated male rats, an apparent "sharpening" of coding for female urine odors by neurons in the preoptic area, compared to neurons in the olfactory bulb, occurred equally well in normals and castrates (Pfaff and Gregory, 1971a). Yet, when measuring correlations between single unit activity firing rates in the preoptic area and the state of the cortical EEG, we found that a higher proportion of preoptic neurons in normal than in castrated male rats showed significant correlations (Pfaff and Gregory, 1971b). This overall pattern of electrophysiological effects can be compared to behavioral data (Pfaff and Pfaffmann, 1969b) on responses of normal and castrated male rats to sex odors (Pfaff et al., 1971, 1972). From this comparison it appears that androgen-sensitive electrophysiological measures can be matched with androgen-sensitive behavioral responses, and that androgen-*insensitive* electrophysiological parameters (primarily having to do with peripheral olfactory coding) match androgen-insensitive behavioral responses to sex odors.

5. *Requirements for the interpretation of hormone-sensitive neurons.* Circulating estrogens facilitate lordosis behavior in female rats. Estradiol is concentrated by specific cells in a limbic-hypothalamic system in the female rat brain. When artificially placed by hormone implantation into part of that system, the preoptic area, estradiol can trigger the lordosis reflex. Finally, estrogens and progesterone alter the electrical activity of cells in the preoptic area and hypothalamus. What more is required to explain the effects of estrogen on the lordosis reflex via brain mechanisms in the female rat?

In order to understand how hormone-sensitive neurons participate in the mechanisms of hormone-sensitive behavioral responses, we must know how hormone-induced changes in these neurons cause changes in the functional circuits underlying the reflex. Thus, the circuits that mediate mating behavior reflexes must be identified with the same precision as hormone-concentrating cells have been. We have focused on the identification of circuits mediating the female rat's lordosis reflex for three reasons: First, the uptake of radioactive estradiol by particular neurons in the female rat forebrain is striking and specific (see above). Second, the action of estradiol on lordosis in the female rat is the strongest, most specific hormone-behavior link we have studied in rats (Pfaff, 1970a, 1971b). Third, the lordosis reflex, not involving locomotion, may have a simple enough topography to submit to a detailed neurophysiological circuitry analysis.

If such an analysis were successful, one could identify behaviorally relevant and hormonally relevant neurons or groups of neurons, in a sense analogous to the identification of cells achieved by neurophysiologists working on invertebrate ganglia (see the fifth part of this volume) or on motor neurons in the spinal cord.

In beginning a neurophysiological circuitry analysis of the lordosis reflex in the female rat, we had two starting points of certainty: First, our identification of estrogen-concentrating cells in a limbic-hypothalamic system; and second, the knowledge that sensory input driving the lordosis reflex would have to come through the dorsal roots of the spinal cord. Therefore, in the early steps of the analysis, it has been important to identify the sensory determinants of the lordosis reflex. Following that, both the hormonal and sensory information can be used as tools in the study of the neural pathways underlying the lordosis reflex.

Sensory Determinants

1. *Description of stimuli the male applies.* Describing the individual sensory and motor elements of the lordosis reflex should facilitate the search for neural pathways involved in the control of that reflex. We have analyzed high-speed color movies of mating encounters between male and estrous female rats taken from the side and from below the mating pair (Pfaff, 1970c, 1971c; Pfaff 1973; Pfaff et al., 1972) frame-by-frame to study the stimuli that lead to the initiation of lordosis posture. The male rat, following the female, mounts her from the rear, contacting her flanks with his forepaws and front legs, her thighs with his rear legs, and her back with his nose and chin. He "walks up" with his rear legs until his pelvic area is pressing against her rump and tailbase. At this time, his rear legs often are pressing against the back of her rear legs and hips. The male begins a rapid series of repetitive thrusts (if the encounter is to lead to an intromission or penile insertion). Although the initial thrusts may be well in back of the vaginal region or off to one side, if the encounter is to lead to penile insertion the thrusts eventually

are made close to and on the opening of the vagina. Before or during the time that the male pelvis contacts the female's rump area, and during the time of preliminary pelvic thrusting by the male, the female elevates her rump by extending her rear legs and exhibiting a vertebral dorsiflexion of her rump and tailbase. It appears that when she has lifted her rump and tailbase far enough off the ground, the male can bend his hindquarters farther underneath her rump for more accurate pelvic thrusting toward the vagina. By the time the male rat achieves penile insertion (intromission), the female is in a lordosis posture. Directly after intromission, the male dismounts from the female leaving the female in the lordosis posture: Legs extended, vertebral column dorsiflexed such that the head, rump, and tailbase are raised while the thorax is lowered.

Events during the lordosis reflex can also be described quantitatively by recording for each frame whether or not certain discrete events have occurred and by measuring the positions or angles of certain body parts (Pfaff, 1970c)

(Figure 2). The main features of events described quantitatively in this way support the verbal description given above. For instance, it is typical that the first detectable rump or tailbase elevation, due to rear leg extension and vertebral dorsiflexion, occurs well before penile insertion.

Qualitative and preliminary quantitative observations of X-ray movies of rat mating encounters (Figure 3) support the description based on conventional film analysis (Diakow et al., 1973). In addition, the X-rays add detail to our description of the component leg and rump movements that elevate the rear end of the female rat during lordosis. Results from both the conventional movie film analyses and from X-ray movies of the reflex establish the appearance and order of events in a model of sensory control over lordosis (Figure 4).

2. *Stimuli necessary for lordosis.* Large classes of stimuli can be eliminated as candidates for necessary stimuli for lordosis. Female rats deprived of the sense of smell mate at least as well as controls (Moss, 1971). Female cats (Bard,

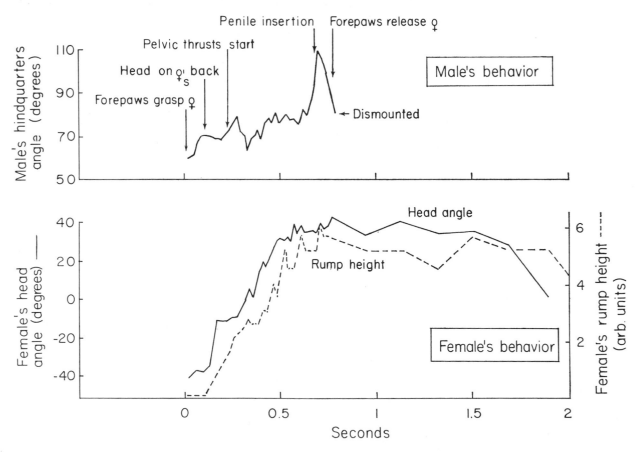

FIGURE 2 Quantitative description of the male and female rat's movements during the onset and maintenance of lordosis in the female (Pfaff, 1970c, 1971c). Taken from frame-by-frame analyses of color movies (54 frames per second) of rat mating encounters. The female starts going into lordosis (elevating rump and head) shortly after initial contact with the male's forepaws and head and reaches the full lordosis posture during the male's pelvic thrusts, before penile insertion. In both graphs, 0° is horizontal, 90° is vertical.

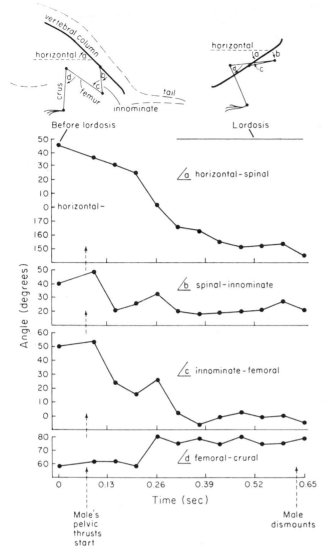

FIGURE 3 Use of X-ray cinematography to describe components of movements of the female rat's hindquarters as she initiates and maintains lordosis (Diakow, Jenkins, Montgomery, and Pfaff, 1973). Drawing at top left depicts angles measured on the X rays, by showing a lateral view of the female rat before lordosis, looking at the left rear leg and the spine in the sacral region. Drawing at the top right shows the same view after the female has assumed lordosis posture. Rump elevation—a crucial component of the lordotic posture—is accomplished by a combination of rear leg extension and vertebral dorsiflexion.

1939) and rabbits (Brooks, 1937) deprived of olfaction, sight, and normal hearing were able to mate. In general, therefore, although distance receptors may play some role in female reproduction, they do not appear necessary for these female mammals to perform primary mating reflexes.

Intravaginal stimulation, as the female might receive during penile insertion, also does not appear to be necessary for the initiation of the lordosis reflex. In normal sequences of encounters between male rats and estrous females, lordosis often occurs during encounters that do not result in intromission. Moreover, during those encounters that do include intromission, it is clear that the lordosis reflex occurs before the male's deep intromittory pelvic thrust is achieved. Third, artificial stimulation without vaginal probing is sufficient to cause lordosis (see Section 3 below). Fourth, when the penis of the male rat was anesthetized so that penile insertion would not occur (Carlsson and Larsson, 1964; Adler and Bermant, 1966; Sachs and Barfield, 1970) female rats still displayed lordosis as often as females mating with control males (Pfaff et al., 1972). Finally, lordosis occurs in female rats in which the vagina has been surgically removed (Ball, 1934; Kaufman, 1953), sewed shut (Hard and Larsson, 1968), or the deep vagina and cervix desensitized (Diakow, 1970).

The stimuli remaining, which must include those necessary for lordosis, are those applied by the male early in his encounter with the female, namely (a) contact with the flank and back areas by his forepaws and head, and (b) contact with rump, tailbase, and perineal areas during his pelvic thrusting. Injections of local anesthesia in the flank and back areas contacted by the male's paws and head did not reliably reduce the lordosis quotient (Pfaff et al., 1972). However, injecting the local anesthesia (procaine) in the perineal and tailbase areas contacted during the male's pelvic thrusting did lower the female's lordosis quotient significantly (Pfaff et al., 1972). Surprisingly, when the female's lordosis quotient was lowered by this treatment, the males' performance of intromission was also lowered: A significantly decreased proportion of their total number of responses toward the female resulted in intromission. Preliminary experiments using superficial cutaneous denervation of the female in the same areas where procaine was effective have replicated the effects of the procaine both on lordosis of the female and on intromission of the male. Brain lesions of the female (Diakow, 1971) also can affect performance of the male.

This combination of effects on the female and male is consistent with the hypothesis that early in each mating encounter the female's response to a somatic stimulus due to the male, in the procaine-affected rump and perineal skin area, has two important functions: First, it is part of her reflex progression toward the lordosis posture, and if prevented, lordosis is interfered with. Second, it is important for allowing the male's reflex progression toward intromission, and if prevented, intromission is interfered with. This hypothesis of a chain of causal relations is embodied in the sequential reflex model of lordosis (Figure 4).

3. *Stimuli sufficient for lordosis.* In order to find which stimuli are sufficient for lordosis, one must attempt to

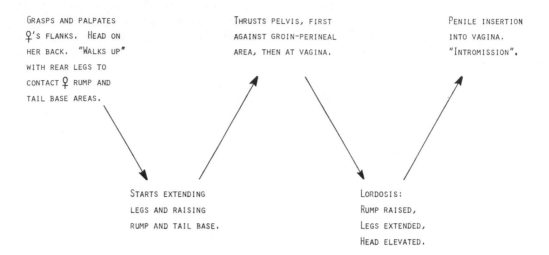

GRASPS AND PALPATES ♀'S FLANKS. HEAD ON HER BACK. "WALKS UP" WITH REAR LEGS TO CONTACT ♀ RUMP AND TAIL BASE AREAS.

THRUSTS PELVIS, FIRST AGAINST GROIN-PERINEAL AREA, THEN AT VAGINA.

PENILE INSERTION INTO VAGINA. "INTROMISSION".

STARTS EXTENDING LEGS AND RAISING RUMP AND TAIL BASE.

LORDOSIS: RUMP RAISED, LEGS EXTENDED, HEAD ELEVATED.

FEMALE RAT'S BEHAVIOR

FIGURE 4 Simplified model of sensory control over the initiation of the female rat's lordosis reflex. Male's behavior represented above; female's below. Time reads from left to right. Appearance and order of events in this brief schematic summary are derived from film and X-ray descriptions of lordosis. Hypothesized causal relations (arrows) are derived from experiments using local anesthesia on the female's skin and from artificial manual stimulation of lordosis. For more detailed model, refer to Pfaff et al. (1972).

imitate and apply individually stimuli that are applied as a set by the male during normal mating encounters. Under some conditions, artificial tactile stimulation yields lordosis probabilities that vary according to the estrous cycle as do the female's responses to the male rat (Hemmingsen, 1933; Ball, 1937; Blandau, Boling, and Young, 1941; Adler and Bell, 1969; Gerall and McCrady, 1970). Responses by the female may vary according to the strain of rat used and the estrogen level. Rats in our experiments (Diakow, Komisaruk, and Pfaff, 1973) were Sprague-Dawley females, ovariectomized and, for those receiving maximal doses, brought into estrus with 30 μg estradiol and 0.5 mg progesterone. Under these conditions, light scratching on the flank and the back causes rear leg extension or abduction and also stiffening of the tail. Light tickling of the perineal and tailbase regions caused leg extension and tail elevation, particularly when combined with light scratching of the flank and back. None of these stimuli exerted significant pressure, and none of them regularly caused the full lordosis reflex. Heavier pressure is added to the stimulation by placing the palm and heel of the hand on the back of the female rat, the thumb on one flank, the fourth (ring) and fifth fingers on the other flank, and wrapping the index and third fingers in a forked position around the iliac crests and on either side of the tailbase. The index and third fingers exert pressure upwards by squeezing them against the palm of the hand. This stimulation, especially if preceded by rapid scratching of the flanks, can elicit lordosis in female rats treated with estrogen. Furthermore, the response to estrogen is dose dependent (Figure 5).

4. *Model of sensory control over the lordosis reflex.* Data from film descriptions of lordosis, procaine experiments (to study stimulus necessity), and artificial manual stimulation experiments (to study stimulus sufficiency) can be considered together to arrive at a brief description of sensory control over the lordosis reflex. A simplified version of a model previously presented in full (Pfaff et al., 1972) is shown in Figure 4. The *appearance* and *sequence* of events in the model are derived from the film analyses and X-ray movies of lordosis. The hypothesized causal relations between events have been inferred from local anesthesia experiments and observations with manual stimulation. The female responds to the initial flank and back stimuli from the male plus the male's first impact in the tailbase and groin area by extending her legs and raising her tailbase and rump slightly. This response, in turn, allows the male to begin pelvic thrusting with increasingly greater access to the groin and perineal area. Thus, in time, cutaneous stimulation from the pelvic

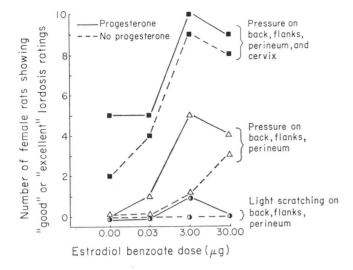

Figure 5 Artificial manual stimulation of lordosis in the female rat can be potentiated by increasing estrogen or progesterone levels (Diakow, Komisaruk, and Pfaff, 1973). In different groups of ovariectomized rats (10 per group), different estradiol doses were given with or without progesterone supplementation. All animals were tested with a variety of somatic stimuli applied by the experimenters. The results show some examples of compensation by progesterone for decreased levels of estrogen in maintaining lordosis probabilities and compensation by either of these hormones for decreased stimulus strength.

thrusting gives successively stronger stimulation in the groin and perineum causing the female to raise her hindquarters still further by extending her legs and dorsiflexing her vertebral column. When she has raised her rump high enough for the male to bend his pelvis underneath, his pelvic thrusts can exert upward force against the perineal area, achieving the stimulation that in our manual stimulation experiments was sufficient to trigger the full lordosis posture in the estrogen-treated female rat. Local anesthesia of the female's rump and perineal areas diminished the sensation from the male's pelvic thrusting, thus interfering with lordosis and, in turn, with intromission. Manual stimulation of the perineal area exerting pressure upward on the perineum with the fingers and downward on the rump with the palm of the hand, which the experimenter can apply without the female rat's previous cooperation, was sufficient to trigger lordosis.

From this description it is clear that an early response of the female after being contacted by the male, and an important response for allowing fertilization, is the elevation of her rump and tailbase. Also, the description of skin areas where the male must contact the female to trigger lordosis, in the flank, rump and perivaginal areas, can be used in further studies of sensory physiology underlying lordosis. Finally, in this model, the notion of a

"cascaded" sequence of reflexes is prominent: The female's reflex responses to stimuli from the male early in the encounter allow the male to apply stimuli necessary, in turn, for her subsequent responses. Thus, the lordosis reflex is actually a rapid sequence of interactions between the female and the male, leading to her lordosis posture. Her lordosis reflex is followed by his penile insertion. The necessity for this rapid sequence of reflex interactions could provide a reproductive isolating mechanism, ensuring that only conspecifics endocrinologically competent to reproduce will finish the mating behavior sequence.

NEURAL PATHWAYS

1. *Peripheral and spinal.* From the behavioral analysis of sensory determinants of lordosis, above, it was clear that one skin area on the female rat of interest for peripheral sensory recording was the perineum, the skin area against which pressure from the male rat or from artificial stimulation could help trigger the lordosis reflex. The anatomy of the sensory nerves supplying this skin area is complicated due to the pudendal and sacral plexes. A current summary of our knowledge of some of the nerves supplying the perivaginal skin area, as well as the vagina, cervix, and clitoris is presented in Figure 6 (Greene, 1963; Diakow, in preparation).

We have recorded multiunit electrical activity from one branch of the pudendal nerve that supplies the perineal region (Kow and Pfaff, 1973a, b). Ovariectomized female rats, either uninjected or given daily estrogen replacement treatment, were anesthetized with Equithesin, and the pudendal nerve was approached from the dorsal side. Whole nerve activity of the selected branch supplying the perineal area was recorded by looping that branch, cut centrally, over bipolar wire electrodes. Sensory input from both fast-adapting and slow-adapting receptors could be recorded. As determined with von Frey hairs, thresholds for the former, yielding on-off responses, regularly were lower than thresholds for the latter, which yielded sustained responses. For all animals the receptive field of this nerve branch included the perineal region, lateral, anterolateral and posterolateral to the vagina. In some animals, we discovered an extension of the receptive field down the posterior border of the rear leg, a skin area touched by the male as he presses against the rear end of the female (Figure 7). In this respect an interesting difference showed up between estrogen-treated and untreated animals, the first clear physiological demonstration of a steroid sex hormone affecting peripheral sensory input: 95% of the estrogen-treated animals showed the extension of the receptive field onto the leg, while only 48% of the untreated animals showed the extension ($p < 0.005$). Since the nerve branch was cut centrally, this hormone effect could not have been due to centrifugal influences in

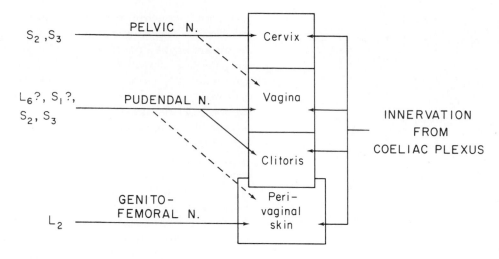

PELVIC N.

S_2, S_3

Cervix

$L_6?, S_1?,$
S_2, S_3

PUDENDAL N.

Vagina

INNERVATION
FROM
COELIAC PLEXUS

Clitoris

GENITO-
FEMORAL N.

L_2

Peri-
vaginal
skin

Innervation of the genital area of the female rat

FIGURE 6 Current summary of the innervation in the genital region of the adult female albino Norway rat. Solid lines represent distributions that were revealed in every preparation, studied by dissection under the microscope. Dotted lines represent distributions of branches that were obvious in some preparations but that were sometimes obscured by being enmeshed in connective tissue or neural plexes (Diakow, in preparation). While these nerves could not always be followed visually all the way to their sensory surface, electrophysiological recording has added to our knowledge of their field of distribution (see Figure 7).

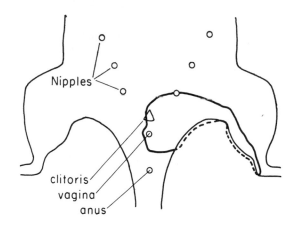

Nipples

clitoris
vagina
anus

FIGURE 7 Heavy line outlines the receptive field of one branch of the pudendal nerve in an estrogen-treated female rat. Ventral view of the perineal area is shown. Dotted line shows where receptive field includes a small amount of the dorsal skin suface. Occurrence of the receptive field extension onto the posterior portion of the leg was significantly more frequent in estrogen-treated than in untreated ovariectomized female rats (Kow and Pfaff, 1973a).

that nerve. Komisaruk et al. (1972) also have independently detected an increased receptive field size of the pudendal nerve in estrogen-treated compared to untreated female rats. The fact that some untreated rats showed a receptive field extension may reflect either of two things: The estrogen effect on peripheral input may not be strong enough by itself to account for a large share

of the hormone's overall effect on the behavior pattern of the female rat, and/or our present methods of stimulating and recording may have included estrogen-insensitive inputs along with inputs that show a strong and specific estrogen effect.

Results from work with spinally transected female rats suggest that the lordosis reflex is not organized solely at the spinal level. To date, female rats transected at low- or mid-thoracic levels have not been observed to perform responses that closely resemble lordosis. Further, those hind-quarters reflexes of spinal females that have been studied do not appear to be estrogen sensitive (Hart, 1969; Pfaff et al., 1972). Caution must be used when interpreting negative behavioral results from a preparation reduced as severely as by spinal transection. However, our present inference would be that supraspinal pathways are necessary for the complete elaboration of the lordosis reflex in the female rat and for its estrogen sensitivity. In this respect, female rats differ from female dogs, in which estrogen-sensitive spinal reflexes have been found (Hart, 1970). In spinal female cats, results are intermediate: Bard (1940) showed that anestrous females could display sexual reflexes such as treading, tail deviation, and pelvic dorsiflexion, but in addition, estrogen may affect the strength or frequency of these reflexes (Maes, 1939; Hart, 1971).

2. Supraspinal pathways. Lesion data. Experiments testing the effects of brain lesions on lordosis in female rats have given a clear answer to one fundamental question regarding supraspinal pathways: Do supraspinal control loops critical for the occurrence of lordosis ascend to involve obligatory thalamocortical circuits, or do they pass ventrally into the basal forebrain, notably the hypothalamus? It is clear that the cerebral cortex is not necessary for the display of lordosis in the female rat. Females given large cortical lesions are capable of displaying lordosis (Beach, 1944) and initiating and maintaining pregnancy (Davis, 1939). If any changes in the female's receptive behavior were observed, they were in the direction of more extreme lordoses (Beach, 1944). Moreover, interference with electrical activity of the cerebral cortex by the spreading depression technique using potassium chloride (Clemens et al., 1967) facilitated the occurrence of lordosis.

On the other hand, damage of certain hypothalamic regions has been shown to interfere with the ability of the female rat to display lordosis. Lesions in the anterior hypothalamus have prevented females from showing lordosis in response to males, during experiments with rats (Law and Meagher, 1958; Singer, 1968) and guinea pigs (Brookhart et al., 1940; Dey et al., 1942). It is possible that anterior to this hypothalamic site, which is critical for lordosis, lies tissue whose removal facilitates the lordosis response. Using small bilateral lesions in the medial preoptic area, Powers and Valenstein (1972) were able to reduce the amount of estrogen needed to bring ovariectomized female rats into a state of behavioral receptivity. It must be noted that in some hypothalamic lesion experiments, the exact locus and size of lesion necessary and sufficient for obtaining the reported effect has not been fully described. Therefore, while it is certain that tissue in the medial anterior hypothalamus and medial preoptic area is intimately involved in a lordosis control mechanism, the exact functional roles that neurons in this region play remain to be specified.

Information related to lordosis may travel to or from the medial preoptic-anterior hypothalamic region via the stria medullaris and habenulo-interpenduncular pathway or via the medial forebrain bundle. Habenular or stria medullaris lesions reduce lordosis quotients in ovariectomized, estrogen-treated female rats (Rodgers and Law, 1967; Modianos and Hitt, personal communication), while medial forebrain bundle lesions have not had this effect (Rodgers and Law, 1967; Hitt et al., 1970). Implication of the habenulointerpenduncular tract would be especially interesting in view of the fact that tail, rump, and rear leg movements, some of which are similar to elements of the lordosis reflex, can be stimulated in anesthetized female rats from electrode placements near this pathway (Pfaff et al., 1972).

Stimulation and recording data. We have approached the localization of supraspinal pathways controlling lordosis movements by using bipolar stimulating electrodes to explore the brains of female rats lightly anesthetized with halothane, trying to stimulate muscle movements which resemble elements of lordosis (Pfaff et al., 1972). We have mapped the female rat brain for points successful in stimulating tail, rump, and rear leg movements, having explored systematically from the lower medulla to the basal forebrain at the level of the preoptic area, excepting the cortical and cerebellar surfaces. At low and middle medullary levels (transverse sections including inferior olive and facial nerve nucleus), points successful in stimulating rump and tail movements were clustered in the ventrolateral brainstem. At higher medullary and pontine levels, more points were found dorsally, so that at the level where the fourth ventricle narrows to the cerebral aqueduct, some points were found near the ventricle. In the mesencephalon, some points were still clustered along the ventrolateral edge of the brainstem, but many more were found near the lateral and dorsolateral edges of the central gray (Pfaff et al., 1972) (Figure 8b). At the junction between the mesencephalon and the diencephalon, as the cerebral aqueduct descends into the third ventricle, electrode positions for stimulating tail and rump movements remained near the ventricle and the habenulo-interpenduncular tract (compare to lesion data, above). Although points yielding rump and tail movements were rarer in the basal forebrain, those we did find tended to be in the medial amygdala, the hypothalamus, or the preoptic area, in or near the distribution of estradiol-concentrating neurons.

In the lower brainstem and mesencephalon the distribution of points successful for stimulating tail and rump movements in the anesthetized female rat coincides with the anatomically defined distribution of ascending fibers from the anterolateral columns of the spinal cord (Nauta and Kuypers, 1958; Mehler, Feferman, and Nauta, 1960; Anderson and Berry, 1959; Goldberg and Moore, 1967; Mehler, 1969). For comparison in the caudal mesencephalon of our stimulation results with Mehler's (1969) anatomical results in the rat, see Figures 8b and 8d. Since ascending fibers from the anterolateral columns are well known to contain major somatosensory afferent pathways, these fibers could carry appropriate sensory information consequent to stimulation by the male rat, as well as providing an anatomical "organizing principle" for stimulation points in the lower brainstem and mesencephalon. It may be hypothesized, therefore, that ascending information entering a supraspinal control loop, subsequent to stimulation by the male rat relevant for lordosis, travels in the anterolateral columns of the spinal cord and distributes in the spinoreticular pathways previously defined by anatomical techniques. Even stimulation points at the mesencephalic-diencephalic junction

DONALD W. PFAFF, CAROL DIAKOW, RICHARD E. ZIGMOND, AND LEE-MING KOW   631

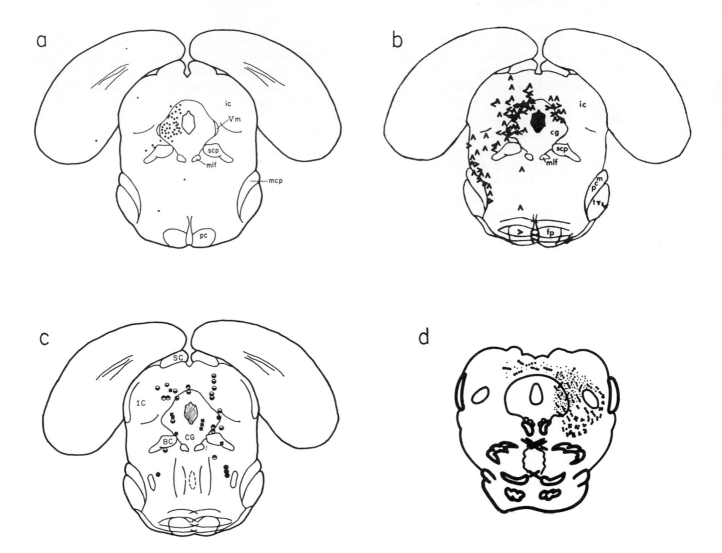

Figure 8 Data from different techniques, concerning neurons near the lateral borders of the central grey of the mesencephalon in the rat. (a) Autoradiographic map showing locations of estradiol-concentrating neurons in and near the mesencephalic central gray of the female rat (large dots) (Pfaff and Keiner, 1973; Pfaff, 1972). Dots plotted only on left side of drawing. Symbols same as in Figure 1. (b) Locations in the posterior mesencephalon from which tail or rump movements could be stimulated electrically in halothane-anesthetized female rats (Pfaff et al., 1972). For convenience, symbols are plotted on the left, except where there was not enough room. Direction of arrows denotes direction of tail or rump movement: ipsilateral, contralateral, or elevation. (c) Locations of single units in the dorsal midbrain area of the urethane-anesthetized female rat, at one posterior mesencephalic level, which were responsive to somatic stimulation (Malsbury et al., 1972). Circular symbols, which are filled in at the bottom and empty at the top, show locations of units that responded specifically to stimulation of skin on the rear part of the body, which is contacted by the male during mating. For convenience, unit locations are plotted on both left and right sides of the drawing. (d) Anatomical projections from the right anterolateral columns of the spinal cord in the rat, as determined by silver staining of degenerating fibers (from Mehler, 1969). This figure shows degeneration at the intercollicular level of the midbrain, including degeneration of fibers (coarse dots) and terminals (fine dots). (bc, brachium conjunctivum (or, scp, superior cerebellar peduncle); cg, central gray; fp, frontopontine fibers; ic, inferior colliculus; mcp, middle cerebellar peduncle; mlf, medial longitudinal fasciculus; pc, cerebral peduncle; sc, superior colliculus; tVs, tract of spinal root of trigeminal nerve; Vm, mesencephalic nucleus of trigeminal nerve.)

and in the diencephalon that are anterior to monosynaptic projections from the anterolateral cord may fit with an extension of this same concept of a primitive ascending pathway. With Nauta staining it has been found that small central gray lesions cause degeneration following the

ventricle ventrally into the hypothalamus, in a manner similar to the stimulation points for tail and rump movements (Hamilton and Skultety, 1970; Chi, 1970). In turn, fibers passing through the lateral posterior hypothalamus have been reported to travel rostrally in the medial forebrain bundle and terminate in the basal forebrain and septum (Guillery, 1957). Over a long anterior-posterior extent, therefore, there is good correspondence between our successful stimulation points and known somatosensory anatomical projections from the spinal cord to the brainstem and from the mesencephalon into the basal forebrain.

Several qualifications limit the interpretation of the points successful for stimulating tail and rump movements as being part of a "lordosis pathway" (Pfaff et al., 1972). These include the difficulty of extrapolating from work with anesthetized animals and artificial electrical stimulation to inferences about normal neural mechanisms in the waking animal. Moreover, other movements of the body sometimes accompanied movements that resembled elements of lordosis, and, finally, the entire lordosis body posture was never stimulated. Nevertheless, the stimulation results appear at least to have been heuristic by suggesting an anatomical concept for part of the supraspinal lordosis control loop, and thus, by generating predictions about the behavior of neurons in this potential pathway

with respect to lordosis control. For instance, many points successful for stimulating tail and rump movements were located near the lateral borders of the central gray, in the dorsal midbrain. If neurons in this region actually participate in the control of lordosis, one might hypothesize that some of them would respond to somatosensory stimulation in the skin area, which we know is contacted by the male rat during mating and which, if pressed artificially, can trigger lordosis.

To test this hypothesis, we used micropipettes to record single unit activity in the dorsal midbrain of female rats lightly anesthetized with urethane (Malsbury et al., 1972). Many of the neurons recorded in the dorsal midbrain did not respond to somatosensory stimulation, responded only to painful stimuli, or had widespread receptive fields scattered all over the body surface. However, a small percentage of neurons in the dorsal midbrain, including several near the lateral borders of the central gray, responded specifically to somatosensory stimuli on regions of the skin which would be contacted by the male rat's forepaws, forearms, and pelvic region during mating

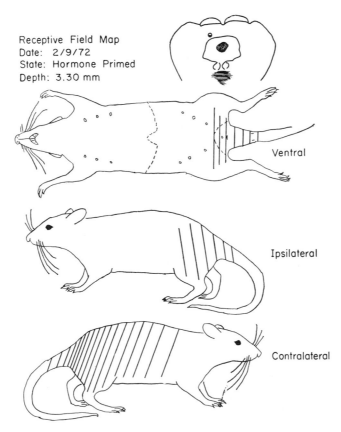

FIGURE 9 Receptive field on the skin (hatched lines) of a single unit responsive to somatosensory stimulation, in an estrogen-primed, urethane-anesthetized female rat (Malsbury et al., 1972). Location of unit, just dorsolateral to the mesencephalic central gray, is shown in drawing at top right. The unit's receptive field includes those areas of skin stimulated during mating by the male rat's paws (on female's flanks), his head (on female's back), and his posterior abdomen and pelvis (on female's rump, tailbase, and perineal areas).

(Malsbury et al., 1972) (Figures 8c and 9), namely, on the flanks or in the perivaginal region. Recordings from such neurons in chronically prepared female rats would help to show whether or not these neurons actually do fire during, or just prior to, lordosis. As they stand, the present recording results show at least that some neurons in or near the central gray of the mesencephalon in the anatomical distribution field from the anterolateral columns of the cord (Figure 8d) and among the points successful for stimulating tail and rump movements (Figure 8b) respond to stimuli on the skin areas contacted by the male rat. In the female cat, some neurons in and around the lateral reticular nucleus in the medulla, which receives a heavy synaptic input from the anterolateral columns of the spinal cord, responded specifically to vaginal stimulation (Rose and Sutin, 1973). These results, also, support the interpretation that somatosensory information due to vaginal or perivaginal stimulation ascends in the anterolateral columns of the spinal cord and distributes in the spinoreticular pathways.

NEUROENDOCRINE INTEGRATION   The loci of stimulation points successful for stimulating tail and rump movements led to the concept that lordosis-related information consequent to stimulation by the male rat travels up the anterolateral columns of the spinal cord, and, through their distribution over spinoreticular pathways, into the brainstem. A second feature of these same results is the overlap between the distribution of such stimulation points and the regions containing peak numbers of estradiol-concentrating neurons. In the central gray of the mesencephalon (Figure 8a) and in the preoptic area, this coincidence is most obvious. The interest in such intersections lies in the fact that they identify possible regions where estrogen could influence activity in supraspinal pathways controlling lordosis. Without emphasizing these two regions to the exclusion of other points in potential lordosis-control pathways, it may be stated that, at these two sites, results from different techniques agree and add up to a more comprehensive theory of neuroendocrine control mechanisms for lordosis than previously has been attempted. It may be hypothesized that anatomically defined spinoreticular pathways, carrying stimulus information initiated by contact with the male rat, enter the region of the lateral borders of the central gray (Figure 8d) and affect the activity of neurons there (Figure 8c). Neurons on either side of the histologically defined lateral borders of the central gray are known from Golgi studies to be in close contact with each other. These neurons, directly or indirectly, are connected with descending pathways that, directly or indirectly, stimulate deep back muscles and rear leg extensor muscles responsible for the vertebral dorsiflexion and rear leg

extension, respectively, during lordosis (Figure 8b). Since this field of neurons defined in Figures 8b to 8d includes those that lie near and partially overlap with the most posterior part of the distribution of estradiol-concentrating neurons (Figure 8a), it may be hypothesized that this is at least one region in which estrogen could act on supraspinal loops which control lordosis.

The medial preoptic area and medial anterior hypothalamus provide a second locus where integration between incoming neural information and endocrine signals, represented by estrogen, could take place. Here, as we have seen, estradiol-concentrating neurons are present in peak densities (Figure 1), and when estradiol is implanted directly among these neurons, it can facilitate lordosis. Lesions here affect lordosis, and electrical stimulation near these neurons can facilitate tail, rump, and rear leg movements. Finally, some neurons in the anterior hypothalamus and preoptic area respond at least to vaginal stimulation, and their spontaneous activity and responses can be affected by estrogen levels.

A schematic working model depicting the minimal circuitry necessary to account for the above results and for the control of the lordosis reflex by estrogen-sensitive neurons is presented in Figure 10. Somatosensory stimuli on the flank, rump, and perivaginal areas are conceived as triggering the lordosis reflex (Figure 4) using at least the pathways illustrated. Supraspinal pathways are required in the model, because we have not detected lordosis or other estrogen-sensitive hindquarters responses in spinal preparations. If, indeed, sensory information travels into the brain via spinoreticular projections from the anterolateral columns, sites of greatest interest for further experimentation lie in the intersections between the spinoreticular distribution and the distribution of estrogen-concentrating neurons. Here estrogens, acting through their physiological effects on estradiol-concentrating neurons, could affect the properties of neural responses to somatosensory input in the spinoreticular pathways such that, in the presence of estrogen, descending influences facilitating lordosis would be released. It is important to emphasize that in order to affect the control loop in a meaningful way, estrogens need not directly affect neurons at every point in the loop. Rather estrogen, acting selectively on certain neurons, may alter the transfer functions effected by those neurons such that descending influences are able to facilitate the lordosis response.

Further studies of neuroendocrine integration can use both behavioral and physiological techniques. For example, one recent behavioral study shows how changes in stimulation can compensate for changes in hormone dosage (Diakow, Komisaruk, and Pfaff, 1973). The experimenters manually applied to ovariectomized female rats several different kinds of somatic stimuli, among

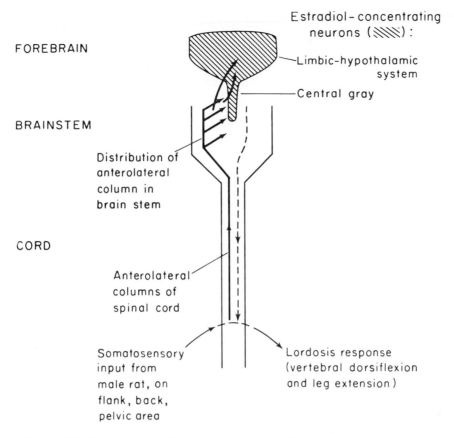

FOREBRAIN

BRAINSTEM

CORD

Estradiol-concentrating
neurons (▨) :

Limbic-hypothalamic
system

Central gray

Distribution of
anterolateral
column in
brain stem

Anterolateral
columns of
spinal cord

Somatosensory
input from
male rat, on
flank, back,
pelvic area

Lordosis response
(vertebral dorsiflexion
and leg extension)

FIGURE 10 Schematic working model showing minimal circuitry necessary to account for control of lordosis reflex by somatic stimulation and by estrogen-sensitive neurons. See text for further description.

which were three of different "strength" and different effectiveness in eliciting lordosis. The weakest was light scratching of the back, flanks, and perineum, without significant pressure applied. Stronger was a stimulus in which pressure was applied to these same areas by clamping a hand around the rat's body with the fingers forked around the tail; this is conceived as being closer to the kind of stimulation the male rat applies (see above). Strongest was a "supranormal" stimulation, in which to the previous stimulation was added pressure against the vaginal cervix. Also, estrogen doses were varied and were given with or without progesterone supplementation. The experiment showed (Figure 5) that increasing the estrogen dosage increased the probability of eliciting a good lordosis reflex and that progesterone facilitated the effect of estrogen. Also, increasing the intensity of tactile stimulation of the appropriate body areas increased the probability of eliciting lordosis. Within limits, decreases in one of the three factors (estrogen dose, progesterone, or somatosensory stimulation) could be compensated by increasing another. For example, at a low dose of estrogen (0.03 μg), increasing the amount of tactile stimulation

increased the probability of a good lordosis in progesterone-treated females. Also, with the stimulus involving pressure on the back, flanks, and perineum, progesterone increased the probability of a good lordosis in females receiving 3 μg of estrogen and, when progesterone was absent, increasing the amount of estrogen increased the probability of a good lordosis. Thus, stimulation intensity can be traded off against hormone dosage in maintaining the probability of eliciting a good lordosis reflex.

One kind of physiological experiment was generated by a prediction based on the electrical stimulation work referred to above. It could be surmised that stimulation at brain sites effective in eliciting tail and rump movements might alter the hindquarters reflex responses of a female to peripheral somatosensory stimulation. To test this idea, we developed a preparation using chloralose-anesthetized female hamsters on the day of proestrous (Pfaff and Lewis, unpublished observations). Stimulating electrodes were placed either just lateral to the central gray or in the medial preoptic area-medial anterior hypothalamus. In some preparations, we have been able to achieve an anesthetic state and a set of stimulation parameters in which manual stimulation by itself in the rump-tailbase region would not give a behavioral response, and low-voltage electrical stimulation by itself at one of the two brain sites would give slight tailbase elevation, reminiscent of the lordosis reflex. Then, in such preparations, adding the manual stimulation *during* electrical stimulation led to a significant increase in the height of tail and rump elevation. To date, we have obtained this phenomenon in proestrous but not in diestrous hamsters. In such preparations, therefore, manual stimulation during electrical stimulation of the brain led to a change in behavior, whereas manual stimulation without central stimulation did not. The experiment demonstrates not only the interaction between electrical stimulation near estrogen-concentrating regions and peripheral somatic stimulation in regions touched by the male hamster but may also demonstrate a coordination between this neural interaction and hormone levels. The phenomenon has been difficult to obtain and is sensitive to anesthetic levels; hamsters that are too "light" begin to struggle with or without somatic stimulation, while hamsters that are too "deep" do not give any tail or rump response. Nevertheless, if such a phenomenon operates in the normal control of lordosis in the waking hamster, it could provide a paradigm for one kind of lordosis reflex control. That is, electrical stimulation near neurons known to concentrate estradiol and to respond to stimuli that trigger lordosis can facilitate rump and tail responses to that same kind of peripheral stimulation. Thus, if estrogen were to activate those neurons that in our experiment are being stimulated artificially with electrodes, an explanation would be provided for how estrogen facilitates response to the appropriate peripheral somatic stimulation.

Interactions between neural and endocrine determinants of behavior have been considered with respect to the lordosis reflex for reasons of strategy in detailed neurophysiological analysis. However, mounting of the female by the male rodent, and her subsequent lordosis, can also be discussed in the broader context including all facets of reproductive behaviors and of the behavioral and nonbehavioral requirements for successful reproduction. Throughout a long chain of varied behavioral responses, lasting over at least a 2-day period, noncopulatory behaviors of the male and female rat are coordinated with each other, and this coordination is assured by the appropriate gonadal hormone levels in each of the mating partners and the effects of these hormones on behavior (Pfaff et al., 1972). It appears that production of certain odors, behavioral responses to sex odors, the amount, type, and range of locomotion, and other behavioral tendencies all are affected by sex hormones (both through brain mechanisms and via extracerebral routes) in such a manner that they act as a behavioral "funnel" to bring the male and female together in a manner appropriate for mating. As such, these hormone effects on the brain and other organs help to ensure that reproductively competent male and female rats, and only these, will successfully initiate and complete copulation. In the time domain—considering daily rhythms, estrus cycles of the female rodent, and possible seasonal or other long-term variations—steroid effects on behavior, mediated by brain mechanisms, help to synchronize behavioral and gametogenic reproductive capacities.

Neonatal hormone effects on differentiation of rat neuroendocrine mechanisms

DESCRIPTION Sexual dimorphisms found in adults of most animal species originate in the chromosomal differences between the two sexes. Sexual differentiation, the process by which these sexual dimorphisms develop, begins with the accessory sexual structures. Through experimental interventions circumventing the developmental impact of the genome, attempts have been made to determine the sequence of events by which normal sexual differentiation occurs. The sex chromosomes themselves seem to be important primarily in the differentiation of the gonads, though not all the mechanisms involved are well understood (Gallien, 1967; Witschi and Dale, 1962; Witschi, 1965). Following this step, the gonads appear to govern subsequent differentiation of the various accessory structures. In mammals, substances produced by the testes determine whether subsequent differentiation follows a masculine or feminine pattern. For instance, Jost (1965) has found that castration of male rabbits before day 21 of gestation results in the differentiation of the internal accessory sexual structures as in a female.

Parts of the central nervous system can properly be regarded as accessory sexual structures. Recent findings regarding behavioral and pituitary sex differentiation in rodents seem to fit the embryological patterns common for other parts of the reproductive system (reviewed by Burns, 1961). The purpose of this section is to review selectively our knowledge on sexual differentiation of the nervous system, particularly with regard to female sexual behavior in the rat.

Pfeiffer's classic experiments showed that the ability of a rat to ovulate in adulthood depended on the absence of secretions from the testes during its early postnatal development (Pfeiffer, 1936). Subsequent experiments have confirmed this conclusion (see Gorski, 1971 for references).

Experiments in a number of laboratories have extended this developmental picture to include sexual behavior (see Whalen, 1968 for references). Sexual behavior in the rat is dimorphic in a quantitative rather than a qualitative sense: Patterns recognized as typical of female rats are occasionally seen in males and vice versa. The lordosis response, the most commonly used measure of feminine sexual behavior in rats, has been observed in male rats but only rarely (see Beach, 1948 for references). The presence or absence of testes during a restricted period of postnatal development influences the pattern of sexual behavior in an adult male rat. Male rats castrated during the first 5 days of life exhibit significantly more lordoses and fewer intromissions and ejaculations than sham-operated animals when tested under appropriate hormonal conditions as adults (Grady, Phoenix, and Young, 1965). Female rats injected with testosterone propionate before day 5 show greatly diminished frequency of lordosis and an increased frequency of intromissions when tested as adults (Harris and Levine, 1965; Feder, Phoenix, and Young, 1966; Whalen and Edwards, 1967).

The general picture that has emerged from these and other studies is that the sexual differentiation of the neuroendocrine system of a rat proceeds in a feminine direction in the absence of testicular secretion or injected androgenic hormones. A feminine neuroendocrine pattern is detected *endocrinologically* by cyclic changes in the ovary (in some experiments, e.g., with males, grafted) supporting ovulation and the formation of corpora lutea, and *behaviorally* by the frequency of lordosis.

1. *Critical period*. The sensitivity to neonatal androgen manipulations decreases with age during the first few days of life. From the available data it is easier to determine the end of this period than its beginning. Since in the male the normal development of the tonic pattern of gonadotrophin release and of sexual behavior patterns depends on gonadal hormones, the end of the critical period for the differentiation of these functions is regarded as the first day at which castration has no effect on this development. Castration on day 10 or later, for instance, does not affect the frequency of lordosis responding in the adult (Table I). However, even before day 10 the magnitude of the effect

of castration declines. Thus, castration within an hour of birth has a greater effect than 6 hours after birth (Thomas and Gerall, 1969) and castration on day 1 is more effective than castration on day 5 (Grady et al., 1965; Gerall et al., 1967). The end of the critical period for the differentiation of the ejaculatory response also seems to be sometime during the second week of life (Table I). Data on the onset of the critical period for the development of these behavioral responses is based largely on work with cyproterone. At the present, all that can be said definitely is that the critical periods for the development of both male and female behavior in the male rat begin prenatally (Nadler, 1969; Ward, 1972).

Castration of male rats on or after the sixth day of life is ineffective in producing an adult with a cyclic (female) pattern of gonadotrophin secretion (Table I). Data are unavailable concerning the effects of prenatal cyproterone on adult cyclicity.

Alternatively, the termination of these critical periods in the male could be defined as the age after which injections of testosterone no longer reverse the effects of neonatal castration. By this criterion the critical period for the development of ovarian cyclicity and for the suppression of lordosis responding appears to end in the second week of life (Table II).

Since the differentiation of the nervous system in a female pattern proceeds in the absence of gonadal hormones, the analogous critical periods in the female have been regarded as the periods during which injections of androgen masculinize development. The exact termination of such periods probably depends on the amount of androgen injected. With injections of about 1 mg testosterone, the critical period for inhibiting adult lordosis responding and producing ovarian acyclicity ends between days 10 and 20 (Table III). In both these cases the percentage of animals affected is higher if the injections are made on the first 2 days of life than on day 10.

TABLE I

The effect of castration at various ages on the differentiation of pituitary control and sexual behavior in the genetically male rat

Variable Measured	Days of Age at Which Castration Was Effective			Days of Age at Which Castration Was Ineffective		Reference
Corpora lutea formation						
(% showing corpora lutea)	1 (91%)	2 (27%)		3–7 (0%)		Harris (1964)
	3 (89%)	6 (17%)		11 (0%)		Gorski (1966)
	1 (100%)	5 (15%)	10 (3%)	15 (0%)		Ladosky et al. (1970)
	1 (100%)			10 (0%)		Ladosky et al. (1969)
Presence of lordosis behavior	1, 5			10		Grady et al. (1965)
	1			20		Feder et al. (1966)
	1, 5			15		Gerall et al. (1967)
Absence of ejaculatory pattern	1 (0%)	5 (25%)		10 (88%)		Grady et al. (1965)
of masculine behavior	4 (8%)	7 (54%)	10 (83%)	13 (100%)	19 (100%)	Larsson (1966)
(% showing ejaculation)	1 (0%)	5 (8%)	10 (43%)	60 (67%)		Gerall et al. (1967)

DONALD W. PFAFF, CAROL DIAKOW, RICHARD E. ZIGMOND, AND LEE-MING KOW 637

TABLE II

The reversibility of neonatal castration of the genetically male rat with androgen injections at various ages

Variable Measured	Dose and Age of Effective TP Injections	Dose and Age of Ineffective TP Injections	Reference
Corpora lutea formation (% showing corpora lutea)	500 µg on days 5 (0%) and 10 (0%) 100 µg on day 1 (0%)	500 µg on day 20 (100%) 500 µg on day 15 (93%)	Adams-Smith and Peng (1967) Ladosky et al. (1969)
Presence of lordosis behavior (% showing lordosis)	200 µg/day on days 1–2 (10%) days 3–4 (73%) days 5–6 (56%) days 9–10 (80%)	200 µg/day on days 13–14 (100%)	Beach et al. (1969)
Absence of ejaculatory pattern of masculine behavior (% showing ejaculation)	200 µg/day on days 1–2 (75%) days 3–4 (60%) days 5–6 (20%) days 9–10 (20%)	200 µg/day on days 13–14 (0%)	Beach et al. (1969)

TP: testosterone propionate

TABLE III

The effect of androgen injections at various ages on the differentiation of pituitary control and sexual behavior in the genetically female rat

Variable Measured	Dose and Age of Effective TP Injections	Dose and Age of Ineffective TP Injections	Condition of Female	Reference
Absence of corpora lutea formation (% showing corpora lutea)	1250 µg on days 2 (0%) 5 (0%) 10 (60%) 50–100 µg on day 5 (0%) 100 µg on day 1 (0%) 500 µg on day 12 (23%)	1.25 mg on day 20 (100%) 50–500 µg on day 10 (95%) 50–100 µg on day 20 (100%)	Intact Ovariectomized Ovariectomized	Barraclough (1961) Adams-Smith (1967) Ladosky et al. (1969)
Absence of lordosis behavior (% showing lordosis)	100 µg on days 1 (7%) and 10 (23%)	1000 µg on day 20 (73%)	Intact	Goy et al. (1962)

TP: testosterone propionate

Comparable data on the critical period for the development of male behavior in the female are unavailable.

Prenatal injections of testosterone propionate into pregnant female rats affect the development of both the masculine and feminine behavior of female offspring (Gerall and Ward, 1966; Ward, 1969; Ward and Renz, 1972). Such injections seem to have no effects on the development of ovarian cyclicity (Swanson and Van Werff ten Bosch, 1964a); however, this may depend on the amount of androgen injected.

2. *Chemical nature of the inducer.* Testosterone is not the only compound that affects the cyclicity of an adult female when injected during the first few days of life. A number of naturally occurring and synthetic estrogens also produce a polyfollicular syndrome. For instance, females injected with estradiol benzoate, ethynylestradiol, mestranol, or stilbestrol 5 days after birth do not form corpora lutea in adulthood (see Flerko, 1971, for references). The physiological significance of these findings is not clear. As pointed out by Gorski (1966) 5 µg of estradiol benzoate into a 5–10 g animal is an extremely high dose even based on the amount of estradiol circulating in an adult. This ability of estradiol to produce effects similar to testosterone has also been observed in adults in a number of neuroendocrine (e.g. Lisk, 1960, 1962b) and behavioral (e.g. Pfaff, 1970a) situations. The mechanism for this effect is unknown, although it is highly unlikely that estradiol is converted into testosterone in vivo. In the case of sexual differentiation, it may well be that estradiol and testosterone are acting by different mechanisms, for the anovulatory syndromes they produce differ in several respects (Gorski, 1963).

It is pertinent to determine what *endogenous* substance is responsible for suppressing cyclicity in the male. Chemical measurements of testosterone levels in testes and plasma indicate that the neonatal testis does secrete testosterone during the first 5 days of life (Resko et al., 1968). In vitro studies indicate that the prenatal testis is also capable of synthesizing testosterone (Noumura et al., 1966). A number of other sources of evidence are consistent with the idea that testosterone is the substance responsible for "masculinizing" the pattern of gonadotrophin release in the male. (1) Castration on the day of birth results in an adult male that can maintain a cyclical ("female") pattern of gonadotrophin release (Pfeiffer, 1936). (2) If animals that have been castrated at birth are injected with 10 μg of testosterone propionate on day 4 of age, an acyclical ("male") pattern develops (Gorski and Wagner, 1965). (3) Multiple injections of cyproterone, an antiandrogen, into pregnant mothers and then into their male offspring produce a cyclical pattern (Neumann and Kramer, 1967).

Similar experiments have suggested that testosterone acts to suppress the capacity for female behavior and increase the capacity for certain aspects of male behavior in the developing male rat (Grady et al., 1965; Feder et al., 1966; Gerall et al., 1967; Whalen and Edwards, 1967; Neumann and Kramer, 1967). In addition, Goldman, Quadagno, Shryne, and Gorski (1972a) found that injections into male rats on days 1, 3, and 5 of antibodies to rat FSH and LH result in increased frequency of lordosis and decreased frequency of intromission and ejaculation compared to control males. Goldman et al. (1972b) have recently shown that injection of an antiserum to testosterone-3-bovine albumin into pregnant females results in male offspring with smaller (more "female-like") anogenital distances.

It is interesting to note that androstenedione (Goldfoot, Feder, and Goy, 1969; Stern, 1969; Luttge and Whalen, 1970) and dihydrotestosterone (Whalen and Luttge, 1971) are not capable of suppressing female behavior, at least in the doses tested so far. Thus, all of these experiments support, although they do not definitively prove, the idea that testosterone is necessary for the differentiation of a masculine type of gonadotrophin release and sexual behavior.

3. *The neonatally androgenized female.* Do neonatal injections of testosterone merely disrupt the normal female neuroendocrine system or do they induce the formation of a male system? There are a number of ways that an androgen injection into a developing female could produce an anovulatory adult that might have nothing in common with the normal development of a male. One of these would be if testosterone acted directly on the neonatal ovary either rendering it anovulatory directly or indirectly through a change in its hormonal output. Three lines of evidence argue against this possibility: (1) The neonatally androgenized female does not produce corpora lutea even if it is given an ovarian transplant from a normal adult female (Pfeiffer, 1936). (2) If, on the other hand, an androgenized female is injected with exogenous gonadotrophin (Segal and Johnson, 1959) or a hypothalamic extract (Courrier, Guillemin, Jutisz, Sakiz, and Ascheim , 1961; Johnson, 1963) ovulation does occur demonstrating that its ovary is capable of ovulation. (3) Finally, Pfeiffer (1936) showed in his original experiments that the action of the testis transplanted into a female at birth does not require the presence of the female's ovary.

Since Harris and Jacobsohn (1951) showed there was no inherent sexual dimorphism in the pituitary in the ability to stimulate ovulation, androgenization of the female by deleterious effects on the pituitary would be physiologically uninteresting. This possibility was ruled out by the experiments of Segal and Johnson (1959) in which they transplanted the pituitary from an androgenized female into a control female that had been hypophysectomized and observed ovulation. Transplanting a normal pituitary into an androgenized female did not abolish the anovulatory syndrome.

As a result of these experiments, attention has been directed to the central nervous system for an explanation of the effect of neonatal testosterone injections. However, our original question remains unanswered. Do androgenized female rats and normal male rats fail to demonstrate cyclical gonadotrophin release for the same reason? Taleisnik, Caligaris, and Astrada (1969) measured the effect of different gonadal steroids on the plasma LH levels of normal males and females given testosterone propionate (1.25 mg) on day 4. They found a sex difference in the response of normal rats to progesterone, depending on whether the animals had been primed with estradiol or testosterone. Female rats primed with either estradiol benzoate or testosterone propionate showed an increase in plasma LH if given progesterone 5 hr before the assay. Male rats responded similarly if treated with testosterone followed by progesterone. However, progesterone had no effect on male rats primed with estradiol. Androgenized females responded to progesterone after estradiol priming as did normal females but not normal males. These results present an intriguing difference between the neonatally androgenized female and the normal male.

Similar questions must be asked about the effects of neonatal testosterone injections on female sex behavior. The absence or decreased appearance of lordosis in both the neonatally androgenized female and the normal male is not in itself a compelling case for neuroendocrine identity of the two animals. Nonetheless, the low level of residual female sexual behavior after neonatal testosterone injections into females does appear comparable to that present in males castrated after sexual differentiation has occurred (Feder, Phoenix, and Young, 1966; Whalen and Edwards, 1967).

The neuroendocrine status of the neonatally androgenized female is not yet completely understood. Clearly an injection of testosterone propionate during the first 5 days of life produces a profound alteration in the hypothalamopituitary-gonadal system of the female. Further studies are needed, however, to determine to what extent these changes resemble the normal development of the male. Stimulation, lesion, and implant studies are now required to establish whether androgenized females and males fail to show lordosis behavior for the same reason.

4. *The neonatally castrated male.* Does neonatally castrating the male rat achieve a "feminine" rat nervous system? The formation of corpora lutea in ovarian grafts in male rats that have been castrated at birth is consistent with the idea that these animals are releasing gonadotrophin in a cyclical ("female") pattern. However, since such observations are almost always based on the appearance of the ovary at a single point in time, they cannot establish the existence of regular cycles such as have been demonstrated in females by, for instance, vaginal smears. For this reason Yazaki (1960) and later Harris (1964) transplanted vaginal tissue under the abdominal skin of males castrated during the first few days of life. Yazaki (1960) observed vaginal cycles of about 4 days duration for up to 2 months after transplants into males that had been castrated on day 1 or day 4 of age. Gorski (1967) has provided further evidence for the similarity of the neonatally castrated male and the normal female. He used three procedures known to produce polyfollicular ovaries in normal females to see if they would also affect the ovarian cyclicity of neonatal castrates. Constant illumination, bilateral electrolytic lesions in the anterior hypothalamus, and hypothalamic transections behind the suprachiasmatic nucleus all produced a polyfollicular condition in ovarian grafts in neonatal castrates. These results suggest that the same neural structures are responsible for ovarian cyclicity in the normal female and the neonatally castrated male.

It is quite clear from many studies that male rats castrated on the day of birth show a significant increase in lordosis behavior when tested as adults compared to normal males (Grady et al., 1965; Gerall et al., 1967; Whalen and Edwards, 1967; Pfaff and Zigmond, 1971). It is less certain whether such animals reach the level of normal female performance. Grady et al. (1965) and Whalen and Edwards (1967) found no significant difference between males castrated on the day of birth and control females in their lordosis quotients (a ratio of the number of times an animal showed lordosis to the number of times it was mounted). Gerall et al. (1967) however found day 1 castrates had significantly lower lordosis scores than control females. They also observed no soliciting (for example, darting or ear wiggling) behavior on the part of these animals. Grady et al. (1965) had reported observing such behaviors in day 1 castrates although significantly less often than in control females. Finally, Gerall et al. (1967) and Hendricks (1969) found that males castrated on day 1 differed from control females when qualitative measures of the lordosis pattern were taken.

These differences between the normal female and the male castrated on the day of birth may simply indicate that the critical period for sexual differentiation of behavior begins before castration takes place. Most studies do not specify exactly when on the day of birth the castration was performed. Thomas and Gerall (1969) found that males castrated within an hour of delivery showed significantly more female sexual behavior than males castrated 6 hrs or later after delivery. When compared to normal females the 1-hr castrates showed comparable lordosis quotients and were also comparable on a qualitative scale of lordosis. A significant difference between these two groups was found only in the frequency of holding the lordosis pattern after the male dismounted. Normal females exhibited this behavior more frequently than males castrated within an hour of birth.

Although no one has reported a successful surgical castration of male rats in utero, attempts have been made to eliminate the effectiveness of prenatal testosterone by injecting cyproterone acetate (a compound with anti-androgenic activity) into pregnant mothers. Nadler (1969) found that males exposed to cyproterone in utero behaved similarly to control females when given a single test in adulthood. Ward (1972) has reported that injections of cyproterone into pregnant females produce male offspring that showed more pronounced lordosis responses and more soliciting behavior than normal males. Normal females were not included in this study.

A difference between the neonatal castrate and the normal female was reported by Dunlap, Gerall, and Hendricks (1972) after repeated testing every 8 to 10 days starting at various ages. According to these authors, the lordosis response of males castrated at birth reaches a maximum between the third and fifth test and then declines dramatically. Female rats on the other hand showed increasing receptivity during the seven tests. Both initial and maximum receptivity scores for the neonatal castrates declined with age, from 35 to 120 days of age. Comparable data on the effects of age on the female were not presented, though it seems highly unlikely that such an effect would occur in females within that age range.

The male rat castrated on the first day of life appears quite similar to the normal female in several aspects. When given grafts of ovarian and vaginal tissues as an adult, it shows evidence of cyclical release of gonadotrophins for some time. When injected with ovarian hormones, it shows high frequencies of lordosis behavior. The behavioral similarity of the neonatally castrated male and the normal female increases if the male is castrated immediately after birth or if it is exposed to cyproterone prenatally. A certain amount of physiological work has indicated that the neural structures responsible for the control of gonadotrophin release are similar in the two kinds of animals. Whether this is true for the neural pathways controlling lordosis behavior has not yet been determined.

ANALYSIS OF NEURAL MECHANISMS

1. *Effects of early androgen treatment on sensitivity to sex hormones in adulthood.* One of the early conceptualizations of the effects of early androgen injections was that they changed the thresholds of response to gonadal hormones by neural structures in the adult. A number of recent experiments have studied the effects of various doses of estrogen on several endocrine and behavioral variables.

One assay that has been useful in testing the response of the brain to estrogen is the extent of compensatory ovarian hypertrophy following unilateral ovariectomy. A number of authors have shown that compensatory hypertrophy occurs in androgenized females to the same extent as it does in controls (Gorski and Barraclough, 1962; Swanson and Van der Werff ten Bosch, 1964b; Schapiro, 1965; Petrusz and Nagy, 1967). However, the histological changes in the hypertrophying ovaries differs in the two groups suggesting a difference in the circulating gonadotrophins. Ovarian growth in androgenized females is due primarily to an increase in the number of large follicles, whereas in normal animals the number of corpora lutea also increases (Gorski and Barraclough, 1962). Also, the ovarian hypertrophy can be blocked by daily injections of estradiol; however, it takes larger doses of estradiol to block the response in androgenized females than in controls (Petrusz and Nagy, 1967).

Barraclough and Haller (1970) used another assay— the decrease in plasma LH levels by estrogen injections following bilateral ovariectomy—to study the same question. They found that it took about five times as much estradiol benzoate in androgenized females to produce an effect comparable to that found in controls.

Gerall and Kenney (1970) examined adult behavioral responses of androgenized females to various doses of estradiol. They found that females injected with 10 or 50 μg of testosterone propionate on day 5 increased their lordosis quotients with increasing doses of estradiol benzoate. However, the dose-response curve was much less steep than that for control animals. Females injected neonatally with 1250 μg testosterone propionate showed essentially no lordosis responses regardless of the dose of estradiol given from 1 to 50 μg. The progesterone dose in this aspect of the experiment was kept constant.

The question of frequency of lordosis behavior under estrogen *alone* was examined by Edwards and Thompson (1970), Pfaff and Zigmond (1971), and Whalen, Luttge, and Gorzalka (1971). All these authors found that androgenized females showed a lower lordosis quotient than controls using estradiol benzoate alone, indicating decreased sensitivity to estrogen in those neurons that control the lordosis response.

Is the sensitivity to progesterone also altered in androgenized females? Clemens et al. (1969), Edwards

and Thompson (1970), Pfaff and Zigmond (1971), and Whalen et al. (1971) all found little if any increase in the lordosis quotients of androgenized females under estrogen and progesterone compared to their scores under estrogen alone. Control females, on the other hand, showed more sexual behavior when given estrogen and progesterone than when given estrogen alone. Ward and Renz (1972) examined the time course of the response to progesterone in several groups of females. Control animals increase their number of lordosis responses with time up to 4 hr after an injection of 1 mg progesterone. Females given multiple injections of testosterone propionate either prenatally or postnatally showed no such increase. Finally, Gerall and Kenney (1970), in one part of their experiment, varied the dose of progesterone from 200 μg to 1000 μg (keeping the estradiol dose constant) but produced no increase in lordosis behavior of androgenized females.

These data, in general, support the notion that the brains of androgenized females are less sensitive to estrogen than controls and that, at least with regard to lordosis behavior, they are completely insensitive to progesterone.

2. *Effects of early androgen treatment on binding of sex hormones in the adult brain.* Little is known about the molecular basis for the altered behavior and pituitary control of rats that have been injected with androgen or castrated neonatally. Partly because suitable techniques are available, the most thoroughly studied biochemical system has been the capacity of cells in certain areas of the brain to bind estradiol. Flerko and his co-workers (Flerko and Mess, 1968; Flerko, Mess, and Illei-Donhoffer, 1969) were the first to report that neonatal androgenization decreased the ability of certain estrogenic target tissues to retain ^{3}H-estradiol. Two hours after a systemic administration of about half a microgram of labeled estradiol, decreased concentrations of tritium (presumably largely tritiated estradiol) were found in the anterior hypothalamus (a decrease of 35%), middle hypothalamus (30%), anterior pituitary (28%) and uterus (37%) of androgenized females compared to controls. No effect was found in the posterior hypothalamus or parietal cortex, areas of the brain that contain many fewer specific binding sites for estradiol (McEwen, Zigmond, and Gerlach, 1972). Subsequent studies have supported these findings (McEwen and Pfaff, 1970; Flerko, Illei-Donhoffer, and Mess, 1971; Tuohimaa and Johansson, 1971; Maurer and Woolley, 1971).

Decreased estradiol retention in the hypothalamus of androgenized females was also reported by Anderson and Greenwald (1969) in an autoradiographic study, and by Vertes and King (1971) who found a decreased concentration of estradiol in a crude nuclear pellet and in a soluble supernatant fraction. The effect of neonatal administration of androgen is less clear on the retention of estradiol by the anterior pituitary gland, with some studies finding a decrease similar to that originally reported by Flerko (McGuire and Lisk, 1969; Vertes and King, 1971) and others finding no effect (Anderson and Greenwald, 1969; McEwen and Pfaff, 1970; Flerko et al., 1971; Tuohimaa and Johansson, 1971; Maurer and Woolley, 1971).

Are the immediate effects of testosterone in the neonate accompanied by testosterone binding in the brain? It is possible to demonstrate concentration of radioactive testosterone in the brain of 11-day-old male rats (McEwen, Pfaff, and Zigmond, 1970a, 1970b). However, the type of concentration shown by testosterone in the neonate may not be characterized by such extremely specific regional localization or by saturability (Sheratt, Exley, and Rogers, 1969; Alvarez and Ramirez, 1970; Kincl, 1970; Tuohimaa and Niemi, 1972) as shown by estradiol, for instance, in adult female rats.

It is unlikely that all of the behavioral effects of neonatal testosterone levels, or the effects on pituitary control, can be solely explained by alterations in estrogen binding (Pfaff and Zigmond, 1971). Thus, some of the effects must depend on other changes in the functions of neural circuits that underlie the altered behavioral or endocrine functions. These changes presumably can be detected with physiological, morphological, or biochemical techniques. For instance, Raisman and Field (1971) reported electron microscopic evidence of sex differences in the mode of axonal termination in the preoptic area of the rat. While neonatal effects on estrogen binding have provided a reasonable first step in the explanation of sex differences in brain function, it is likely that the complete explanation will integrate these binding phenomena with a variety of other physiological changes.

ACKNOWLEDGMENTS   Research from Dr. Pfaff's laboratory reviewed here and the preparation of this manuscript were supported by NIH grants NS 08902 and HD 05751 (to D.P.), and by a Reproductive Biology grant from the Rockefeller Foundation. Valuable bibliographic assistance was provided by the UCLA Brain Information Service.

REFERENCES

ADAMS-SMITH, W. N., 1967. The ovary and sexual maturation of the brain. *J. Embryol. Exp. Morph.* 17:1–10.

ADAMS-SMITH, W. N., and M. T. PENG, 1967. Inductive influence of testosterone upon central sexual maturation in the rat. *J. Embryol. Exp. Morph.* 17:171–175.

ADLER, N. T., and D. BELL, 1969. Constant estrus in rats: Vaginal, reflexive and behavior changes. *Physiol. Behav.* 4:151–153.

ADLER, N., and G. BERMANT, 1966. Sexual behavior of male rats: Effects of reduced sensory feedback. *J. Comp. Physiol. Psychol.* 61:240–243.

ALVAREZ, E. O., and V. D. RAMIREZ, 1970. Distribution curves of ³H-testosterone and ³H-estradiol in neonatal female rats. *Neuroendocrinology* 6:349–360.

ANDERSON, C. H., and G. S. GREENWALD, 1969. Autoradiographic analysis of estradiol uptake in the brain and pituitary of the female rat. *Endocrinology* 85:1160.

ANDERSON, F. D., and C. M. BERRY, 1959. Degeneration studies of long ascending fiber systems in the cat brain stem. *J. Comp. Neurol.* 111:195–229.

BALL, J., 1934. Sex behavior of the rat after removal of the uterus and vagina. *J. Comp. Physiol. Psychol.* 18:419–422.

BALL, J., 1937. A test for measuring sexual excitability in the female rat. *Comp. Psychol. Monogr.* 14:1–37.

BARD, P., 1939. Central nervous mechanisms for emotional behavior patterns in animals. *Res. Publ. Ass. Res. Nerv. Ment. Dis.* 19:190–218.

BARD, P., 1940. The hypothalamus and sexual behavior. *Res. Publ. Ass. Res. Nerv. Ment. Dis.* 20:551–579.

BARRACLOUGH, C. A., 1961. Production of anovulatory, sterile rats by single injections of testosterone propionate. *Endocrinology* 68:62–67.

BARRACLOUGH, C., and B. CROSS, 1963. Unit activity in the hypothalamus of the cyclic female rat: Effect of genital stimuli and progesterone. *J. Endocrinol.* 26:339–359.

BARRACLOUGH, C. A., and E. W. HALLER, 1970. Positive and negative feedback effects of estrogen on pituitary LH synthesis and release in normal and androgen sterilized female rats. *Endocrinology* 86:542–551.

BEACH, F. A., 1944. Effects of injury to the cerebral cortex upon sexually-receptive behavior in the female rat. *Psychosom. Med.* 6:40–55.

BEACH, F. A., 1948. *Hormones and Behavior.* New York: Cooper Square.

BEACH, F. A., R. G. NOBLE, and R. K. ORNDOFF, 1969. Effects of perinatal androgen treatment on responses of male rats to gonadal hormones in adulthood. *J. Comp. Physiol. Psychol.* 68:490–497.

BEYER, C., and C. H. SAWYER, 1969. Hypothalamic unit activity related to control of the pituitary gland. In *Frontiers in Neuroendocrinology*, W. F. Ganong and L. Martini, eds. New York: Oxford University Press, pp. 255–287.

BEYER, C., P. McDONALD, and N. VIDAL, 1970a. Failure of 5-α-dihydrotestosterone to elicit estrous behavior in the ovariectomized rabbit. *Endocrinology* 86:939–941.

BEYER, C., G. MORALI, and M. CRUZ, 1971. Effect of 5-α-dihydrotestosterone on gonadotropin secretion and estrous behavior in the female Wistar rat. *Endocrinology* 89:1158–1161.

BEYER, C., N. VIDAL, and A. MIJARES, 1970b. Probable role of aromatization in the induction of estrous behavior by androgens in the ovariectomized rabbit. *Endocrinology* 87:1386–1389.

BEYER, C., V. D. RAMIREZ, D. E. WHITMOYER, and C. H. SAWYER, 1967. Effects of hormones on the electrical activity of the brain in the rat and rabbit. *Exp. Neurol.* 18:313–326.

BLANDAU, R. J., R. J. BOLING, and W. C. YOUNG, 1941. The length of heat in the albino rat as determined by the copulatory response. *Anat. Rec.* 79:453–463.

BOLING, J. L., and R. J. BLANDAU, 1939. The estrogen-progesterone induction of mating responses in the spayed ♀ rat. *Endocrinology* 25:359–364.

BROOKHART, J. M., F. L. DEY, and S. W. RANSON, 1940. Failure of ovarian hormones to cause mating reactions in spayed guinea pigs with hypothalamic lesions. *Proc. Soc. Exp. Biol. Med.* 44:61–64.

Brooks, C. McC., 1937. The role of the cerebral cortex and of various sense organs in the excitation and execution of mating activities in the rabbit. *Amer. J. Physiol.* 120:544–553.

Burns, R. K., 1961. Role of hormones in the differentiation of sex. In *Sex and Internal Secretions*, W. C. Young, ed. Baltimore, Maryland: Williams and Wilkins.

Carlsson, S. G., and K. Larsson, 1964. Mating in male rats after local anesthetization of the glans penis. *Z. Tierpsychol.* 21:854–856.

Cerny, V. A., 1971. Influence of the hypothalamus on the sexual behavior of the female cat. *Anat. Rec.* 169:292–293 (abstract).

Chambers, W. T., and G. Howe, 1968. A study of estrogen sensitive hypothalamic centers using a technique for rapid application and removal of estradiol. *Proc. Soc. Exp. Biol. Med.* 128:292.

Chi, C. C., 1970. An experimental silver study of the ascending projections of the central gray substance and adjacent tegmentum in the rat with observations in the cat. *J. Comp. Neurol.* 139:259–272.

Ciaccio, L. A., and R. D. Lisk, 1971. Role of progesterone in regulating the period of sexual receptivity in the female hamster. *J. Endocrinol.* 50:201–207.

Clemens, L. G., M. Hiroi, and R. A. Gorski, 1969. Induction and facilitation of female mating behavior in rats treated neonatally with low doses of testosterone propionate. *Endocrinology* 84:1430–1438.

Clemens, L. G., K. Wallen, and R. A. Gorski, 1967. Mating behavior facilitation in the female rat after cortical application of potassium chloride. *Science* 157:1208–1209.

Courrier, R., R. Guillemin, M. Jutisz, E. Sakiz, and P. Ascheim, 1961. Présence dans un extrait d'hypothalamus d'une substance qui stimule la sécrétion de l'hormone antéhypophysaire de luteinisation (LH). *Compt. Rend.* 253: 922–927.

Cross, B., and R. Dyer, 1971. Cyclic changes in neurons of the anterior hypothalamus during the rat estrous cycle and the effects of anesthesia. In *Steroid Hormones and Brain Function*, C. H. Sawyer and R. A. Gorski, eds. Berkeley, California: University of California Press.

Davidson, J., 1967. Hormonal control of sexual behavior in adult rats. In *Advances in the Biosciences. 1. Schering Symposium on Endocrinology*, G. Raspé, ed. New York: Pergamon Press.

Davidson, J., E. R. Smith, C. H. Rogers, and G. J. Bloch, 1968a. Relative thresholds of behavioral and somatic responses to estrogen. *Physiol. Behav.* 3:227–229.

Davidson, J. M., C. H. Rogers, E. R. Smith, and G. J. Bloch, 1968b. Stimulation of female sex behavior in adrenalectomized rats with estrogen alone. *Endocrinology* 82:193–195.

Davis, C. D., 1939. The effect of ablations of neocortex on mating, maternal behavior and the production of pseudopregnancy in the female rat and on copulatory activity in the male. *Amer. J. Physiol.* 127:374–380.

Dey, F. L., C. R. Leininger, and S. W. Ranson, 1942. The effects of hypothalamic lesions on mating behavior in female guinea pigs. *Endocrinology* 30:323–326.

Diakow, C., 1970. Effects of genital desensitization on the mating pattern of female rats as determined by motion picture analysis. *Amer. Zool.* 10:486 (abstract).

Diakow, C. 1971. Effects of brain lesions on the mating behavior of rats. *Amer. Zool.* 11:Abstract No. 3.

Diakow, C., B. Komisaruk, and D. Pfaff, 1973. Sensory and hormonal interactions in eliciting lordosis. *Fed. Proc.* 32:241.

Diakow, C., F. Jenkins, M. Montgomery, and D. W. Pfaff, 1973. Cineradiographic analysis of lordosis in the female rat. (in preparation).

Dunlap, J. L., A. A. Gerall, and S. E. Hendricks, 1972. Female receptivity in neonatally castrated males as a function of age and experience. *Physiol. Behav.* 8:21–23.

Edwards, D. A., and M. L. Thompson, 1970. Neonatal androgenization and estrogenization and the hormonal induction of sexual receptivity in rats. *Physiol. Behav.* 5:1115–1119.

Edwards, D. A., R. E. Whalen, and R. D. Nadler, 1968. Induction of estrus estrogen-progesterone interactions. *Physiol. Behav.* 3:29–33.

Feder, H. H., C. H. Phoenix, and W. C. Young, 1966. Suppression of feminine behaviour by administration of testosterone propionate to neonatal rats. *J. Endocrinol.* 34: 131–132.

Flerko, B., 1971. Steroid hormones and the differentiation of the central nervous system. In *Current Topics in Experimental Endocrinology*, L. Martini and V. H. T. James, eds. New York and London: Academic Press, pp. 41–80.

Flerko, B., and B. Mess, 1968. Reduced oestradiol-binding capacity of androgen sterilized rats. *Acta Physiol. Acad. Sci. Hung.* 33:111–113.

Flerko, B., A. Illei-Donhoffer, and B. Mess, 1971. Oestradiol-binding capacity in neural and non-neural target tissues of neonatally androgenized female rats. *Acta Biol. Acad. Sci. Hung.* 22:125–130.

Flerko, B., B. Mess, and A. Illei-Donhoffer, 1969. On the mechanism of androgen sterilization. *Neuroendocrinology* 4: 164–169.

Gallien, L., 1967. Developments in sexual organogenesis. *Advances Morph.* 6:259–315.

Gerall, A. A., and A. McM. Kenney, 1970. Neonatally androgenized females' responsiveness to estrogen and progesterone. *Endocrinology* 87:560–566.

Gerall, A. A., and R. E. McCrady, 1970. Receptivity scores of female rats stimulated either manually or by males. *J. Endocrinol.* 46:55–59.

Gerall, A. A., and I. L. Ward, 1966. Effects of prenatal exogenous androgen on the sexual behavior of the female albino rat. *J. Comp. Physiol. Psychol.* 62:370–375.

Gerall, A. A., S. E. Hendricks, L. L. Johnson, and T. W. Bounds, 1967. Effects of early castration in male rats on adult sexual behavior. *J. Comp. Physiol. Psychol.* 64:206–212.

Goldberg, J. M., and R. Y. Moore, 1967. Ascending projections of the lateral lemniscus in the cat and monkey. *J. Comp. Neurol.* 129:143–156.

Goldfoot, D. A., H. H. Feder, and R. W. Goy, 1969. Development of bisexuality in the male rat treated neonatally with androstenedione. *J. Comp. Physiol. Psychol.* 67:41–45.

Goldman, B. D., D. M. Quadagno, J. Shryne, and R. A. Gorski, 1972a. Modification of phallus development and sexual behavior in rats treated with gonadotropin antiserum neonatally. *Endocrinology* 90:1025–1031.

Goldman, A. S., M. K. Baker, J. C. Chen, and R. G. Wieland, 1972b. Blockade of masculine differentiation in male rat fetuses by maternal injection of antibodies to testosterone-3-bovine serum albumin. *Endocrinology* 90:716–721.

Gorski, R. A., 1963. Modification of ovulatory mechanisms by postnatal administration of estrogen to the rat. *Amer. J. Physiol.* 205:842–844.

Gorski, R. A., 1966. Localization and sexual differentiation of the nervous structures which regulate ovulation. *J. Reprod. Fertil.* Suppl. 1:67–88.

GORSKI, R. A., 1967. Localization of the neural control of luteinization in the feminine male rat (FALE). *Anat. Rec.* 157:63–70.

GORSKI, R. A., 1971. Gonadal hormones and the perinatal development of neuroendocrine function. In *Frontiers in Neuroendocrinology*, L. Martini and W. F. Ganong, eds. New York: Oxford University Press, pp. 237–290.

GORSKI, R. A., and BARRACLOUGH, C. A., 1962. Studies on hypothalamic regulation of FSH secretion in the androgen-sterilized female rat. *Proc. Soc. Exp. Biol. Med.* 110:298–300.

GORSKI, R. A., and J. W. WAGNER, 1965. Gonadal activity and sexual differentiation of the hypothalamus. *Endocrinology* 76:226–239.

GOY, R. W., C. H. PHOENIX, and W. C. YOUNG, 1962. A critical period for the suppression of behavioral receptivity in adult female rats by early treatment with androgen. *Anat. Rec.* 142:307 (abstract).

GRADY, K. L., C. H. PHOENIX, and W. C. YOUNG, 1965. Role of the developing rat testis in differentiation of the neural tissues mediating mating behavior. *J. Comp. Physiol. Psychol.* 59:176–182.

GREEN, R., W. LUTTGE, and R. WHALEN, 1970. Induction of receptivity in ovariectomized female rats by a single intravenous injection of estradiol-17β. *Physiol. Behav.* 5:137–141.

GREENE, E. C., 1963. *Anatomy of the Rat.* New York: Hafner.

GUILLERY, R. W., 1957. Degeneration in the hypothalamic connexions of the albino rat. *J. Anat.* 91:91–115.

HALLER, E. W., and C. A. BARRACLOUGH, 1970. Alterations in unit activity of hypothalamic ventromedial nuclei by stimuli which affect gonadotropic hormone secretion. *Exp. Neurol.* 29:111–120.

HAMILTON, B. L., and M. SKULTETY, 1970. Efferent connections of the periaqueductal gray matter in the cat. *J. Comp. Neurol.* 139:105–114.

HARD, E., and K. LARSSON, 1968. Effects of mounts without intromission upon sexual behavior in male rats. *Anim. Behav.* 16:538–540.

HARRIS, G. W., 1964. Sex hormones, brain development and brain function. *Endocrinology* 75:627–648.

HARRIS, G. W., and D. JACOBSOHN, 1951. Functional grafts of the anterior pituitary gland. *Proc. Roy. Soc. (Biol.)* B139:263–276.

HARRIS, G. W., and S. LEVINE, 1965. Sexual differentiation of the brain and its experimental control. *J. Physiol.* 181:379–400.

HARRIS, G., R. P. MICHAEL, and P. P. SCOTT, 1958. Neurological site of action of stilbestrol in eliciting sexual behavior. In *Ciba Foundation Symposium on the Neurological Basis of Behavior*, G. E. W. Wolstenholme and C. M. O'Connor, eds. Boston, Massachusetts: Little Brown, pp. 236–254.

HART, B. L., 1969. Gonadal hormones and sexual reflexes in the females rat. *Hormones Behav.* 1:65–71.

HART, B. L., 1970. Mating behavior in the female dog and the effects of estrogen on sexual reflexes. *Hormones Behav.* 1:93–104.

HART, B. L., 1971. Facilitation by estrogen of sexual reflexes in female cats. *Physiol. Behav.* 7:675–678.

HEMMINGSEN, A. M., 1933. Studies on the oestrus-producing hormone. *Skand. Arch. Physiol.* 65:97–252.

HENDRICKS, S. E., 1969. Influence of neonatally administered hormones and early gonadectomy on rats' sexual behavior. *J. Comp. Physiol. Psychol.* 69:408–413.

HITT, J. C., S. E. HENDRICKS, S. I. GINSBERG, and J. H. LEWIS, 1970. Disruption of male, but not female, sexual behavior in rats by medial forebrain bundle lesions. *J. Comp. Physiol. Psychol.* 73:377–384.

JOHNSON, D. C., 1963. Hypophysial LH release in androgenized female rats after administration of sheep brain extracts. *Endocrinology*, 72:832–836.

JOST, A., 1965. Gonadal hormones in the sex differentiation of the mammalian fetus. In *Organogenesis*, R. L. De Hann and H. Ursprung, eds. New York: Holt, Rinehart, and Winston.

KAHWANAGO, I., W. L. HEINRICHS, and E. L. HERRMANN, 1969. Isolation of oestradiol "receptors" from bovine hypothalamus and anterior pituitary gland. *Nature (London)* 223:313–314.

KAUFMAN, R. S., 1953. Effects of preventing intromission upon sexual behavior of rats. *J. Comp. Physiol. Psychol.* 46:209–211.

KAWAKAMI, M., and C. H. SAWYER, 1959a. Induction of behavioral and electroencephalographic changes in the rabbit by hormone administration and brain stimulation. *Endocrinology* 65:631–643.

KAWAKAMI, M., and C. H. SAWYER, 1959b. Neuroendocrine correlates of changes in brain activity thresholds by sex steroids and pituitary hormones. *Endocrinology* 65:652–668.

KAWAKAMI, M., E. TERASAWA, and T. IBUKI, 1970. Changes in multiple unit activity of the brain during the estrous cycle. *Neuroendocrinology* 6:30–48.

KINCL, F. A., 1970. *Endocrinologia Experimentalis* 4:139–141.

KOMISARUK, B. R., P. G. McDONALD, D. I. WHITMOYER, and C. H. SAWYER, 1967. Effects of progesterone and sensory stimulation on EEG and neuronal activity in the rat. *Exp. Neurol.* 19:494–507.

KOMISARUK, B., N. ADLER, and J. HUTCHISON, 1972. Genital sensory field: Enlargement by estrogen treatment in female rats. *Science* 178:1295–1298.

KOW, L.-M., and D. W. PFAFF, 1973a. Estrogen effect on pudendal nerve receptive field size in the female rat. *Anat. Rec.* 175:362–363.

KOW, L.-M., and D. W. PFAFF, 1973b. Electrophysiological studies of sensory input from the genital area of the female rat. *Neuroendocrinology*, (in press).

LADOSKY, W., W. M. KESIKOWSKI, and I. F. GAZIRI, 1970. Effect of a single injection of chlorpromazine into infant male rats on subsequent gonadotrophin secretion. *J. Endocrinol.* 48:151–156.

LADOSKY, W., J. G. L. NORONHA, and I. F. GAZIRI, 1969. Luteinization of ovarian grafts in rats related to treatment with testosterone in the neonatal period. *J. Endocrinol.* 43:253–258.

LARSSON, K., 1966. Effects of neonatal castration upon the development of the mating behavior of the male rat. *Z. Tierpsychol.* 23:867–873.

LAW, T., and W. MEAGHER, 1958. Hypothalamic lesions and sexual behavior in the female rat. *Science* 128:1626–1627.

LINCOLN, D. W., 1969. Effects of progesterone on the electrical activity of the forebrain. *J. Endocrinol.* 45:585–596.

LINCOLN, D. W., and B. CROSS, 1967. Effect of oestrogen on the responsiveness of neurons in the hypothalamus, septum and pre-optic area of rats with light-induced persistent oestrus. *J. Endocrinol.* 37:191–203.

LISK, R. D., 1960. Estrogen-sensitive centers in the hypothalamus of the rat. *J. Exp. Zool.* 145:197–205.

LISK, R. D., 1962a. Diencephalic placement of estradiol and sexual receptivity in the female rat. *Amer. J. Physiol.* 203:493–496.

LISK, R. D., 1962b. Testosterone-sensitive centers in the hypothalamus of the rat. *Acta Endocr. (Kobenhavn)* 41:195–204.

LUTTGE, W. G., and R. E. WHALEN, 1970. Dihydrotestosterone, androstenedione, testosterone: Comparative effectiveness in masculinizing and defeminizing reproductive systems in male and female rats. *Hormones Behav.* 1:265–281.

MAES, J. P., 1939. Neural mechanisms of sexual behavior in the female cat. *Nature (London)* 144:598–599.

MALSBURY, C., and D. W. PFAFF, 1973. Neural and hormonal determinants of mating behavior in adult male rats. A review. In *The Limbic and Autonomic Nervous Systems: Advances in Research*, L. DiCara, ed. New York: Plenum Press (in press).

MALSBURY, C., D. B. KELLEY, and D. W. PFAFF, 1972. Responses of single units in the dorsal midbrain to somatosensory stimulation in female rats. In *Progress in Endocrinology, Proceedings IVth International Congress of Endocrinology*. Amsterdam: Excerpta Medica, 229–233.

MAURER, P., and D. WOOLLEY, 1971. Distribution of ³H-estradiol in clomiphene-treated and neonatally androgenized rats. *Endocrinol.* 88:1281–1287.

McEWEN, B. S., and D. W. PFAFF, 1970. Factors influencing sex hormone uptake by rat brain regions. I. Effects of neonatal treatment, hypophysectomy and competing steroid on estradiol uptake. *Brain Res.* 21:1–16.

McEWEN, B. S., and D. W. PFAFF, 1973. Chemical and physiological approaches to neuroendocrine mechanisms: Attempts at integration. In *Frontiers in Neuroendocrinology*, W. F. Ganong and L. Martini, eds. New York: Oxford University Press (in press).

McEWEN, B. S., D. W. PFAFF, and R. E. ZIGMOND, 1970a. Factors influencing sex hormone uptake by rat brain regions. II. Effects of neonatal treatment and hypophysectomy on testosterone uptake. *Brain Res.* 21:17–28.

McEWEN, B. S., D. W. PFAFF, and R. E. ZIGMOND, 1970b. Factors influencing sex hormone uptake by rat brain regions. III. Effects of competing steroids on testosterone uptake. *Brain Res.* 21:29–38.

McEWEN, B. S., R. E. ZIGMOND, and J. GERLACH, 1972. Sites of steroid binding and action in the brain. In *Structure and Function of Nervous Tissue*, Vol. 5, G. H. Bourne, ed. New York: Academic Press (in press).

McGUIRE, J. L., and R. D. LISK, 1969. Oestrogen receptors in androgen or oestrogen sterilized female rats. *Nature (London)* 221:1068–1069.

MEHLER, W. R., 1969. Some neurological species differences—a posteriori. *Ann. N.Y. Acad. Sci.* 167:424–468.

MEHLER, W. R., M. E. FEFERMAN, and W. J. H. NAUTA, 1960. Ascending axon degeneration following anterolateral cordotomy. An experimental study in the monkey. *Brain* 83: 718–750.

MICHAEL, R. P., 1965. Oestrogen in the central nervous system. *Brit. Med. Bull.* 21:87–90.

MOSS, R. L., 1971. Modification of copulatory behavior in the female rat following olfactory bulb removal. *J. Comp. Physiol. Psychol.* 74:374–382.

MOSS, R. L., and O. T. LAW, 1971. The estrous cycle: Its influence on single unit activity in the forebrain. *Brain Res.* 30:435–438.

NADLER, R. D., 1969. Differentiation of the capacity for male sexual behavior in the rat. *Hormones Behav.* 1:53–63.

NADLER, R. D., 1970. A biphasic influence of progesterone on sexual receptivity of spayed female rats. *Physiol. Behav.* 5: 95–97.

NAUTA, W. J. H., and H. G. J. M. KUYPERS, 1958. Some ascending pathways in the brain stem reticular formation. In *Reticular Formation of the Brain*, H. H. Jasper et al., eds. Boston, Massachusetts: Little Brown, pp. 3–30.

NEUMANN, F., and M. KRAMER, 1967. Female brain differentiation of male rats as a result of early treatment with an androgen antagonist. In *Hormonal Steroids*, L. Martini, F. Fraschini, and M. Motta, eds. Amsterdam: Excerpta Medica.

NOUMURA, T., J. WEISZ, and C. W. LLOYD, 1966. In vitro conversion of 7-³H-progesterone to androgens by the rat testis during the second half of fetal life. *Endocrinology* 78:245–253.

PETRUSZ, P., and E. NAGY, 1967. On the mechanism of sexual differentiation of the hypothalamus. Decreased hypothalamic oestrogen sensitivity in androgen-sterilized female rats. *Acta Biol. Acad. Sci. Hung.* 18:21–26.

PFAFF, D. W., 1968. Uptake of estradiol-17β-H³ in the female rat brain. An autoradiographic study. *Endocrinology* 82:1149–1155.

PFAFF, D. W., 1970a. Nature of sex hormone effects on rat sex behavior: Specificity of effects and individual patterns of response. *J. Comp. Physiol. Psychol.* 73:349–358.

PFAFF, D. W., 1970b. Mating behavior of hypophysectomized rats. *J. Comp. Physiol. Psychol.* 72:45–50.

PFAFF, D. W., 1970c. Sexual motivation. Discussion. In *Neural Regulatory Systems and the Concept of Motivation*, E. Stellar and J. D. Corbit. *Neurosciences Research Program Bulletin* (in press).

PFAFF, D. W., 1971a. Mating behavior of adrenalectomized rats. *Amer. Zool.* 11:Abstract No. 2.

PFAFF, D. W., 1971b. Steroid sex hormones in the rat brain: Specificity of uptake and physiological effects. In *Steroid Hormones and Brain Function*, C. H. Sawyer and R. A. Gorski, eds. Los Angeles, California: University of California Press, pp. 103–112.

PFAFF, D. W., 1971c. Movie analysis of female rat mating behavior. Presented to Eastern Psychological Association, New York, April 1971.

PFAFF, D. W., 1972. Interactions of steroid sex hormones with brain tissue: Studies of uptake and physiological effects. In *The Regulation of Mammalian Reproduction*, S. Segal, ed. Springfield, Illinois: Charles C Thomas; pp. 5–22.

PFAFF, D. W., 1973. Quantitative film analyses of lordosis in the female rat. In preparation.

PFAFF, D. W., and E. Gregory, 1971a. Olfactory coding in olfactory bulb and medial forebrain bundle of normal and castrated male rats. *J. Neurophysiol.* 34:208–216.

PFAFF, D. W., and E. GREGORY, 1971b. Correlation between preoptic area unit activity and the cortical EEG: Difference between normal and castrated male rats. *Electroenceph. Clin. Neurophysiol.* 31:223–230.

PFAFF, D. W., and M. KEINER, 1972. Estradiol-concentrating cells in the rat amygdala as part of a limbic-hypothalamic hormone-sensitive system. In *The Neurobiology of the Amygdala*, B. Eleftheriou, ed. New York: Plenum Press, pp. 775–785.

PFAFF, D. W., and M. KEINER, 1973. Atlas of estradiol-concentrating cells in the central nervous system of the female rat. *J. Comp. Neurol.*, (in press).

PFAFF, D. W., and PFAFFMANN, C., 1969a. Olfactory and hormonal influences on the basal forebrain of the male rat. *Brain Res.* 15:137–156.

PFAFF, D. W., and PFAFFMANN, C., 1969b. Behavioral and electrophysiological responses of male rats to female rat urine odors. In *Olfaction and Taste*, C. Pfaffmann, ed. New York: Rockefeller University Press, pp. 258–267.

PFAFF, D. W., and R. E. ZIGMOND, 1971. Neonatal androgen effects on sexual and nonsexual behavior of adult rats tested under various hormone regimes. *Neuroendocrinology* 7:129–145.

DONALD W. PFAFF, CAROL DIAKOW, RICHARD E. ZIGMOND, AND LEE-MING KOW   645

PFAFF, D. W., E. GREGORY, and M. T. SILVA, 1971. Testosterone and corticosterone effects on single unit activity in the rat brain. In *Influence of Hormones on the Nervous System*, D. Ford, ed. Basel, Switzerland: Karger, pp. 269–281.

PFAFF, D. W., C. LEWIS, C. DIAKOW, and M. KEINER, 1972. Neurophysiological analysis of mating behavior responses as hormone-sensitive reflexes. In *Progress in Physiological Psychology*, Vol. 5, E. Stellar and J. Sprague, eds. New York: Academic Press pp. 253–297.

PFEIFFER, C. A., 1936. Sexual differences of the hypophyses and their determination by the gonads. *Amer. J. Anat.* 58:195–226.

POWERS, B., and E. S. VALENSTEIN, 1972. Sexual receptivity: Facilitation by medial preoptic lesions in female rats. *Science* 175:1003–1005.

RAISMAN, G., and P. M. FIELD, 1971. Sexual dimorphism in the preoptic area of the rat. *Science* 173:731–733.

RAMIREZ, V. D., B. R. KOMISARUK, D. I. WHITMOYER, and C. H. SAWYER, 1967. Effects of hormones and vaginal stimulation on the EEG and hypothalamic units in rats. *Amer. J. Physiol.* 212:1376–1384.

RESKO, J. A., H. H. FEDER, and R. W. GOY, 1968. Androgen concentrations in plasma and testis of developing rats. *J. Endocrinol.* 40:485–491.

RODGERS, C. H., and O. T. LAW, 1967. The effects of habenular and medial forebrain bundle lesions on sexual behavior in female rats. *Psychonom. Sci.* 8:1–2.

ROSE, J. D., and J. SUTIN, 1973. Responses of single units in the medulla to genital stimulation in estrous and anestrous cats. *Brain Res.* 50:87–99.

SACHS, B. D., and R. J. BARFIELD, 1970. Temporal patterning of sexual behavior in the male rat. *J. Comp. Physiol. Psychol.* 73:359–364.

SCHAPIRO, S., 1965. Androgen treatment in early infancy: Effect upon adult adrenal cortical response to stress and adrenal and ovarian compensatory hypertrophy. *Endocrinology* 77:585–587.

SEGAL, S. J., and D. C. JOHNSON, 1959. Inductive influence of steroid hormones on the neural system: Ovulation controlling mechanisms. *Arch. Anat. Micr. Morph. Exp.* 48:261–274.

SHERRATT, M., D. EXLEY, and A. W. ROGERS, 1969. Failure to demonstrate uptake of ^{3}H-testosterone by the hypothalamus of young rats. *Neuroendocrinology* 4:374–376.

SINGER, J. J., 1968. Hypothalamic control of male and female sexual behavior in female rats. *J. Comp. Physiol. Psychol.* 66:738–742.

STERN, J. J., 1969. Neonatal castration, androstenedione, and the mating behavior of the male rat. *J. Comp. Physiol. Psychol.* 69:608–612.

STUMPF, W. E., 1968. Estradiol-concentrating neurons: Topography in the hypothalamus by dry-mount autoradiography. *Science* 162:1001–1003.

SWANSON, H. E., and J. J. VAN WERFF TEN BOSCH, 1964a. The "early-androgen" syndrome; differences in response to prenatal and postnatal administration of various doses of testosterone propionate in female and male rat. *Acta Endocrinol.* 47:37–50.

SWANSON, H. E., and J. J. VAN WERFF TEN BOSCH, 1964b. The "early-androgen" syndrome; its development and the response to hemi-spaying. *Acta Endocrinol.* 45:1–12.

TALEISNIK, S., L. CALIGARIS, and J. J. ASTRADA, 1969. Sex difference in the release of luteinizing hormone evoked by progesterone. *J. Endocrinol.* 44:313–321.

THOMAS, C. N., and A. A. GERALL, 1969. Effect of hour of operation of feminization of neonatally castrated male rats. *Psychonom. Sci.* 16:19–20.

TUOHIMAA, P., 1971. The autoradiographic localization of exogenous tritiated dihydrotestosterone, testosterone and oestradiol in the target organs of male and female rats. In *Basic Actions of Sex Steroids on Target Organs*, P. O. Hubinot et al., eds. Basel, Switzerland: Karger, pp. 208–214.

TUOHIMAA, P., and R. JOHANSSON, 1971. Decreased estradiol binding in the uterus and anterior hypothalamus of androgenized female rats. *Endocrinology* 88:1159–1164.

TUOHIMAA, P., and M. NIEMI, 1972. Uptake of sex steroids by the hypothalamus and anterior pituitary of pre- and neonatal rats. *Acta Endocrinol.* 71:37–44.

VERTES, M., and KING, R. J. B., 1971. Mechanism of oestradiol binding in rat hypothalamus: Effect of androgenization. *J. Endocrinol.* 51:271–282.

WARD, I. L., 1969. Differential effect of pre- and postnatal androgen on the sexual behavior of intact and spayed female rats. *Hormones Behav.* 1:25–36.

WARD, I. L., 1972. Female sexual behavior in male rats treated prenatally with an antiandrogen. *Physiol. Behav.* 8:53–56.

WARD, I. L., and F. J. RENZ, 1972. Consequences of perinatal hormone manipulation on the adult sexual behavior of female rats. *J. Comp. Physiol. Psychol.* 78:349–355.

WHALEN, R. E., 1968. Differentiation of the neural mechanisms which control gonadotropin secretion and sexual behavior. In *Reproduction and Sexual Behavior*, M. Diamond, ed. Bloomington, Indiana: Indiana University Press.

WHALEN, R. E., and D. A. EDWARDS, 1967. Hormonal determinants of the development of masculine and feminine behavior in male and female rats. *Anat. Rec.* 157:173–180.

WHALEN, R. E., and D. F. HARDY, 1970. Induction of receptivity in female rats and cats with estrogen and testosterone. *Physiol. Behav.* 5:529–533.

WHALEN, R. E., and W. G. LUTTGE, 1971. Perinatal administration of dihydrotestosterone to female rats and the development of reproductive function. *Endocrinology* 89:1320–1322.

WHALEN, R. E., W. G. LUTTGE, and B. B. GORZALKA, 1971. Neonatal androgenization and the development of estrogen responsivity in male and female rats. *Hormones Behav.* 2:83–90.

WITSCHI, E., 1965. Hormones and embryonic induction. *Arch. Anat. Micr. Morph. Exp.* 54:601–611.

WITSCHI, E., and E. DALE, 1962. Steroid hormones at early developmental stages of vertebrates. *Gen. Comp. Endocrinol.* Suppl. 1:356–361.

YAZAKI, I., 1960. *Annotationes Zoologicae Japanenses* 33:217–225.

YOUNG, W. C., 1961. The hormones and mating behavior. In *Sex and Internal Secretions*, W. C. Young ed. Baltimore, Maryland: Williams and Wilkins, pp. 1173–1239.

ZIGMOND, R. E., and B. S. McEWEN, 1970. Selective retention of estradiol by cell nuclei in specified brain regions of the ovariectomized rat. *J. Neurochem.* 17:889–899.

ZUCKER, I., 1968. Biphasic effects of progesterone on sexual receptivity in the female guinea pig. *J. Comp. Physiol. Psychol.* 65:472–478.

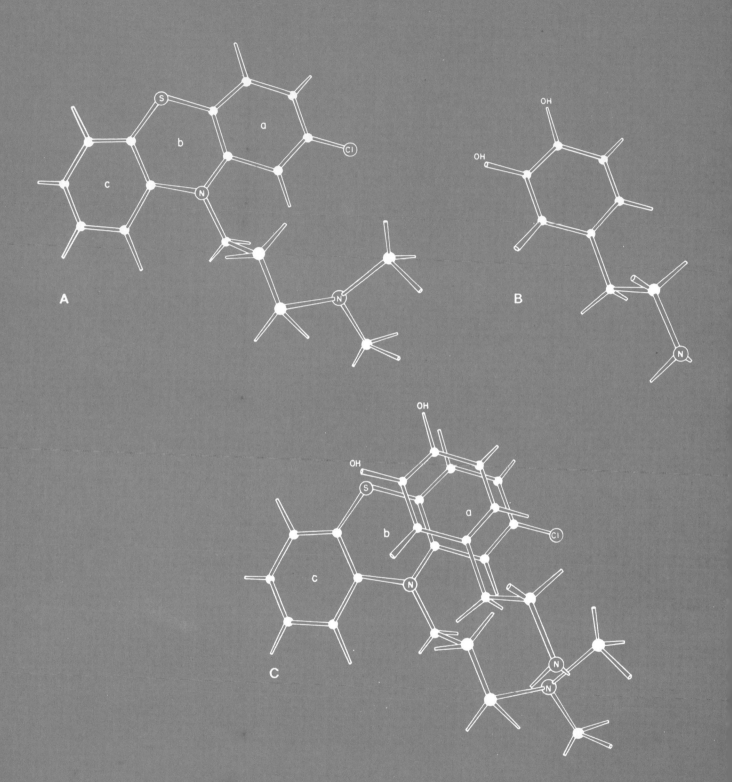

BIOCHEMISTRY
AND BEHAVIOR

Superimposition (C) of the optimal conformations of (A) chlorpromazine and (B) dopamine. In this way chlorpromazine might fit into and block dopamine receptors, perhaps accounting for its antischizophrenic actions.

BIOCHEMISTRY AND BEHAVIOR

Introduction

SOLOMON H. SNYDER

"BIOCHEMISTRY AND BEHAVIOR" is a global designation, which can mean different things to different people. Faced with a choice between examining a single system in depth or providing an overview of several areas, we have selected something of a compromise.

It has long been presumed that endocrine secretions affect, in some vague fashion, mentation and emotion. David de Wied illustrates ways in which it is possible to explore the influence of hormonal secretions upon animal behavior. These approaches are exemplified in his own research on the role of the adrenal cortical hormones and pituitary peptides in mediating specific animal behaviors.

Quite recently the scientific world has become cognizant of the prominent place nutrition or, better, malnutrition occupies in mental function. Richard Wurtman reviews the ways whereby altered nutritional status can influence the brain. As a case in point, he describes efforts in his own laboratory to clarify the effects of circulating amino acids, especially tryptophan, upon the dynamics of brain serotonin.

The biochemistry of memory has been perhaps the most active area of "behavioral chemistry" research. Unfortunately, the variable caliber of many studies has given rise as much to confusion as to new knowledge. Edward Glassman evaluates critically a wide range of investigations of macromolecules and behavior, especially memory mechanisms. Adrian Dunn then describes some thoughtfully conceived and carefully executed studies conducted in the laboratories of Dunn and Glassman, which explicate several difficulties in macromolecule-memory research.

Following this broad panorama of approaches to the biochemistry-and-behavior interface, succeeding chapters focus on a relatively circumscribed area—the relationship of catecholamines to behavior in animals and man. Urban Ungerstedt reviews his epochal behavioral studies with specific lesions of discrete catecholamine tracts in rats produced by stereotaxically implanted 6-hydroxydopamine. Susan Iversen, using the 6-hydroxy-dopamine technique, asks which animal behaviors are mediated by the classic nigrostriatal dopamine pathway?

Sarah Leibowitz has taken a different approach, injecting catecholamines into various sites in the brain and monitoring the resultant behavior, especially eating and drinking.

Of course, ultimately, one would like to apply all this information to man, especially to mental illness. My chapter reviews data bearing upon a possible role of catecholamines in mediating drug effects in schizophrenia. Issues examined critically include: (a) whether phenothiazine drugs exert a selective antischizophrenic action, (b) which drug psychoses provide the best "model schizophrenia" and (c) detailed interactions of antischizophrenic and schizophrenomimetic drugs with catecholamines. Steven Matthysse presents a speculative discussion attempting to integrate what is known of the part played by various brain regions in motor control and feature extraction with brain dopamine and with schizophrenia.

56 Pituitary-Adrenal System Hormones and Behavior

DAVID DE WIED

You will only wear yourself out and wear out all the people who are here.
The task is too heavy for you, you cannot do it by yourself.
Exodus 18:18
New English Bible

Robert Ader, Bela Bohus, Paul Garrud, Willem Hendrik Gispen, Henk Greven, Alan King, Saul Lande, Peter Schotman, Dirk Versteeg, Ivan Urban, Tjeerd van Wimersma Greidanus, Jan Weijnen, and Albert Witter.

ABSTRACT The role of the pituitary-adrenal system and pituitary-adrenal system hormones on acquisition and maintenance of conditioned behavior is the primary subject. Several topics are discussed, including: (a) pituitary-adrenal activity and conditioned behavior; (b) influence of pituitary-adrenal ablation and substitution on acquisition and maintenance of conditioned behavior; (c) structure-activity studies with pituitary-adrenal system hormones; (d) site of the behavioral action of pituitary-adrenal system hormones in the brain, and (e) mode of action of ACTH and ACTH analogues in the brain.

Introduction

THE PITUITARY-ADRENAL system plays an essential role in the defense of the organism against noxious stimuli. Stress rapidly stimulates the release of adrenocorticotrophic hormone (ACTH) from the adenohypophysis. This pituitary hormone activates the adrenal cortex resulting in an increased secretion of glucocorticosteroids. Pituitary ACTH-release is mediated by a corticotrophin-releasing factor or hormone (CRF) of hypothalamic origin, and modulated by various subcortical structures. The function of these structures, in turn, is affected by the level of circulating pituitary-adrenal hormones.

Originally, the activation of the pituitary-adrenal system was regarded mainly to be associated with severe tissue destruction. Only when the assay of hormones produced by the pituitary and adrenal gland became more sophisticated was it found that neurogenic and also psychic stimuli are as potent as tissue damage in stimulating the pituitary-adrenal system.

DAVID de WIED Rudolf Magnus Institute for Pharmacology, Medical Faculty, University of Utrecht, Utrecht, The Netherlands

Evidence for an influence of ACTH and adrenal steroids on the brain was obtained from animal experiments on brain excitability (Woodbury, 1954) and also from clinical observations after these hormones were introduced into the clinic. These investigations suggested that the central nervous system (CNS) contains areas that are sensitive to ACTH and corticosteroids. The influence of pituitary-adrenal system hormones on the brain may therefore be reflected in behavior associated with stress and adaptation.

Behavioral effects of pituitary-adrenal ablation and substitution therapy

ADRENALECTOMY Adrenalectomy does not impair the rat's ability to acquire an avoidance response (Moyer, 1958; Applezweig and Moeller, 1959; Bohus and Endröczi, 1965; de Wied, 1967; van Delft, 1970). In fact, Beatty et al. (1970) found that adrenalectomy attenuated the deleterious effects of high shock in a shuttle box avoidance situation. Weiss et al. (1970) observed a superior acquisition of both active and passive avoidance behavior in adrenalectomized rats. Since the level of circulating ACTH is high in adrenalectomized rats (Hodges and Vernikos-Danellis, 1962), it has been suggested that the ameliorating effect of adrenalectomy is the result of the influence of supraphysiological amounts of endogenous ACTH. Treatment with a glucocorticosteroid that normalized the level of circulating ACTH in adrenalectomized rats, also normalized the avoidance response of these animals (Weiss et al., 1970). However, van Delft (1970) was unable to demonstrate superior avoidance performance in adrenalectomized rats in a pole-jumping avoidance situation. In addition, the administration of glucocorticosteroids did not reduce the performance of the adrenalectomized rats. On the contrary, these steroids tended to stimulate the rate of acquisition of the avoidance response. The U-shaped relationship between performance of an incompletely learned shuttle box avoidance response and the interval between initial and subsequent training trials (Kamin, 1963) has been related to alteration in pituitary-adrenal activity. The poorest performance was found to be accompanied by low corticosterone levels (Brush and Levine, 1966). However, adrenalectomy failed to alter the U-shaped avoidance function (Barrett et al., 1971). The above findings indicate that adrenalectomy per se is associated with normal responding in conditioned avoidance situations and tends under certain conditions to enhance avoidance performance.

HYPOPHYSECTOMY   Removal of the pituitary seriously interferes with acquisition of conditioned avoidance behavior. Adenohypophysectomy (de Wied, 1964) or ablation of the whole pituitary (Applezweig and Baudry, 1955; de Wied, 1968) markedly reduced the ability of the rat to acquire a shuttle box avoidance response (Figure 1).

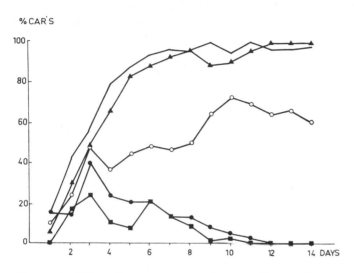

%CAR'S

FIGURE 1   Effect of testosterone propionate, ○———○ (0.2 mg per 2 days s.c.); dexamethasone phosphate, ■———■ (10 μg per day s.c.); ACTH$_{4-10}$ zinc phosphate, ▲———▲ (20 μg per 2 days s.c.); and placebo (0.5 ml zinc phosphate complex s.c.), ●———● on the rate of acquisition of a shuttle-box avoidance response in hypophysectomized rats, as compared to sham-operated rats, ———— . Hypophysectomy was performed by the transauricular route under ether anesthesia. Animals were allowed to recover from the operation for 1 week after which treatment and conditioning were begun. Ten trials a day were given for 14 days with a variable intertrial interval averaging 60 sec. The sound of a buzzer served as the CS (de Wied, 1967).

Hypophysectomy results in metabolic derangements and physical weakness of the organism; treatment of adenohypophysectomized rats with a substitution therapy, consisting of thyroxin, cortisone, and testosterone, improved the health and condition of these animals and at the same time improved avoidance acquisition (de Wied, 1964). Treatment of hypophysectomized rats with testosterone alone (Figure 1) markedly abolished loss of body weight that normally occurs following removal of the

pituitary gland and facilitated the rate of avoidance learning. Similarly, growth-hormone administration prevented body weight loss and stimulated the rate of acquisition of a shuttle box avoidance response in hypophysectomized rats in a dose dependent manner (de Wied, 1969). Thus, improving the health and condition of the hypophysectomized organism ameliorates behavioral performance.

Treatment of hypophysectomized rats with adrenal maintenance doses of ACTH restored the rate of acquisition of the avoidance response of adenohypophysectomized or hypophysectomized rats toward nearly normal levels (Applezweig and Moeller, 1959; de Wied, 1964, 1968). Lack of adrenal gland hormones cannot account for the behavioral deficiency of hypophysectomized rats. Adrenalectomy, as we have seen, does not impair avoidance acquisition. Moreover, treatment of hypophysectomized rats with a potent glucocorticosteroid, dexamethasone (Figure 1), failed to improve avoidance acquisition in hypophysectomized rats (de Wied, 1971). Accordingly, the influence of ACTH on avoidance acquisition of hypophysectomized rats is probably due to an extra target effect of this hormone, presumably located in the CNS. This was demonstrated by the use of peptides structurally related to ACTH but devoid of corticotrophic activities. Administration of α-MSH, which contains the sequence 1–13 of the ACTH molecule, or even smaller analogues like ACTH$_{1-10}$ or ACTH$_{4-10}$ (Figure 1) restored the rate of acquisition of a shuttle-box avoidance response to nearly the level of sham-operated control animals (de Wied, 1969). Further analysis of the influence of the heptapeptide ACTH$_{4-10}$ (de Wied, 1967) revealed that it does not materially affect endocrine or metabolic functions nor the general health and condition of the hypophysectomized rat as determined by loss of body weight, adrenal and testes weight, blood glucose, and insulin and FFA content of plasma. Motor capacities of these animals were only slightly affected by the administration of ACTH$_{4-10}$ (de Wied, 1968). Interestingly, the sensory capacities of hypophysectomized animals are different from those of sham-operated controls. The shock threshold is much lower in the former animals, and this increased responsiveness to electric shock is not restored to normal by treatment with ACTH$_{1-10}$ (Gispen et al., 1970a).

Hypophysectomy attenuates not only active but also passive avoidance behavior (Anderson et al., 1968; Weiss et al., 1970; Lissák and Bohus, 1972). Treatment of hypophysectomized rats with ACTH improves avoidance responding in both situations, supporting the hypothesis that the behavioral deficiency of hypophysectomized rats is not due to motor activity disturbances.

TABLE I

Effect of chain length shortening of $ACTH_{1-10}$ at the amino end on extinction of a pole-jumping avoidance response

	1 2 3 4 5 6 7 8 9 10	Effect*
$ACTH_{1-10}$	H-Ser-Tyr-Ser-Met-Glu-His-Phe-Arg-Trp-Gly-OH	100
$ACTH_{2-10}$	H-Tyr————————————————	100
$ACTH_{3-10}$	H-Ser————————————————	100
$ACTH_{4-10}$	H-Met————————————————	100
$ACTH_{5-10}$	H-Glu————————————————	50
$ACTH_{6-10}$	H-His————————————————	0
$ACTH_{7-10}$	H-Phe————————————————	0

*Approximated potency.

Pituitary-adrenal system hormones and their effects on acquisition and extinction of conditioned behavior

ACTH AND ACTH ANALOGUES   Pituitary-adrenal system hormones are not particularly effective in changing acquisition of active avoidance behavior. Daily administration of ACTH did not alter the rate of acquisition of a shuttle-box avoidance response of intact (Murphy and Miller, 1955) or posterior lobectomized rats (de Wied, 1969). Apparently, removal of endogenous ACTH by hypophysectomy is necessary to demonstrate an effect of this polypeptide on avoidance learning. However, Guth et al. (1971) demonstrated recently that injection of ACTH during acquisition of an appetitive response significantly increased lever press responses late in the training period. This effect was more marked following stringent control of environmental stimulation.

Extinction of an avoidance response is much more sensitive to the treatment with ACTH. Murphy and Miller (1955) found that ACTH administered during shuttle box training resulted in increased resistance to extinction. A more pronounced effect on extinction, however, is found when ACTH is administered during the extinction period (de Wied, 1967). Although the long-term administration of ACTH is accompanied by hypercorticism, the influence of ACTH is independent of its action on the adrenal cortex. ACTH is also active in adrenalectomized rats (Miller and Ogawa, 1962). Moreover, synthetic α-MSH, purified β-MSH and also the smaller ACTH analogues like $ACTH_{1-10}$ and $ACTH_{4-10}$ (de Wied, 1966; de Wied and Bohus, 1966) increase resistance to extinction of a shuttle box as well as a pole-jumping avoidance response. Interestingly, $ACTH_{11-24}$ is ineffective in this respect (de Wied et al., 1968). Thus, the influence of ACTH on the maintenance of aversively motivated behavior is located in the amino end of the molecule, presumably in the heptapeptide $ACTH_{4-10}$, a sequence that is common to ACTH, α- and β-MSH. This conclusion is in accord with observations in dogs in which intracysternal administration of ACTH

and also of α-, β-MSH and $ACTH_{4-10}$ caused a stretching syndrome. Further shortening of the peptide $ACTH_{4-10}$ from the amino end (Greven and de Wied, 1967) interfered with the potency of the inhibitory effect of the peptide on extinction of a pole-jumping response (Table I). Shortening from the carboxyl end (Greven and de Wied, 1973) remained without marked consequences for the behavioral effects until the amino acid phenylalanine at position 7 was removed (Table II). It was surprising that the behavioral activity depended on such a small peptide. Interestingly, this also holds for MSH, since the minimal requirement for a residual activity on expansion of melanophores for α-MSH is the sequence 6–9 (Otsuka and Inouye, 1964). (MSH: melanocyte-stimulating hormone.)

TABLE II

Effect of chain length shortening of $ACTH_{4-10}$ at the carboxyl end on extinction of a pole-jumping avoidance response

		Effect*
$ACTH_{4-10}$	H-Met-Glu-His-Phe-Arg-Trp-Gly-OH	100
$ACTH_{4-9}$	————————————————OH	100
$ACTH_{4-8}$	———————————OH	100
$ACTH_{4-7}$	——————————OH	30–100
$ACTH_{4-6}$	———————OH	10
$ACTH_{4-5}$	————OH	10
$ACTH_{5-6}$	————OH	<10

*Approximated potency.

ACTH affects passive avoidance behavior of intact rats as well. Lissák et al. (1957) were the first to show an effect of ACTH in a passive avoidance situation, and subsequently Levine and Jones (1965) found that ACTH affected acquisition of a passive avoidance response. Using a simple step-through passive avoidance procedure, we found that ACTH administration 1 hr prior to the first retention trial markedly increased avoidance latencies. The same occurred when $ACTH_{1-10}$ was given in contrast to $ACTH_{11-24}$ or $ACTH_{25-39}$ (Table III). Thus, also in this test, the behaviorally active core of ACTH appeared to reside in the amino end of the molecule.

TABLE III

Effect of sequences of the ACTH molecule on avoidance latency in a step-through passive avoidance test

	Median Preshock* Latency Mean of 3 Trials (sec)	Median Postshock Latency (sec)	
		Trial 1	Trial 2
Saline 0.5 ml s.c.	2	10	2
ACTH$_{1-10}$ 10 µg s.c.	3	175†	6
ACTH$_{1-10}$ 30 µg s.c.	3	225†	7
ACTH$_{11-24}$ 10 µg s.c.	2	12	3
ACTH$_{11-24}$ 30 µg s.c.	2	78†	4
ACTH$_{25-39}$ 10 µg s.c.	2	8	3
ACTH$_{25-39}$ 30 µg s.c.	3	49†	3

6 animals per group.

*(0.25 mA; 1 sec).

†Significantly different from saline treated controls.

A number of recently performed studies indicate that ACTH and related peptides not only affect extinction of "fear" motivated behavior. ACTH and also MSH inhibits extinction of an appetitive response as well (Sandman et al., 1969; Leonard, 1969; Gray, 1971; Guth et al., 1971), and the same holds for ACTH$_{4-10}$ (Garrud and de Wied, 1972 (unpublished observations)). ACTH has been shown to suppress aggression in intact and adrenalectomized mice (Brain, 1971; Pasley and Christian, 1972). In various behavioral situations, however, ACTH or related peptides have been reported to be inactive. ACTH$_{1-10}$ did not, or only slightly, affect ambulation, rearing, grooming, or the production of fecal boli in an open field (Bohus and de Wied, 1966; Weijnen and Slangen, 1970; Hadžović and de Wied, 1971). Escape behavior in a runway or the responsiveness to electric shock was also not affected by chronic administration of ACTH$_{1-10}$ (Bohus and de Wied, 1966; Gispen et al., 1970a). Negative results that were obtained in other test situations were reported by Weijnen and Slangen (1970). Accordingly, except for a stretching crisis in dogs (Ferrari et al., 1963) and an effect on aggressive behavior in mice (Brain, 1971; Pasley and Christian, 1972), ACTH and related peptides seem to be effective mainly in modifying conditioned behavior.

ACTH ANALOGUES WITH A D-ISOMER PHENYLALANINE AT POSITION 7 One of the first analogues of ACTH that was synthetized was the decapeptide ACTH$_{1-10}$ in which the phenylalanine residue in position 7 was replaced by the D-isomer. This peptide was found to facilitate extinction of the avoidance response in the shuttle box (Bohus and de Wied, 1966). This result suggested that the D-form peptide antagonized the action of normally occurring ACTH and related peptides which inhibit extinction of avoidance behavior. For this reason the influence of ACTH$_{1-10}$ (7-D-phe) was studied in the absence of ACTH and MSH in hypophysectomized rats. The D-form peptide facilitated the rate of extinction of a shuttle box avoidance response in hypophysectomized rats as well. The effect was even stronger than in sham-operated controls. Subsequently, it was found that ACTH$_{4-10}$ (7-D-phe) (Figure 2) had an effect similar to that of the D-isomer decapeptide (de Wied and Greven, 1968). Preliminary observations indicate that amino acids 10, 9, and 8 can be removed from the carboxyl end of the (7-D-phe) peptides as in the all-L-peptides, without causing severe damage to the potency of the D-form peptide in facilitating extinction of a pole-jumping avoidance response.

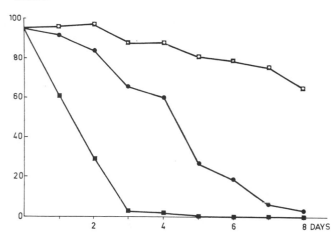

% CAR'S

FIGURE 2 Effect of ACTH$_{4-10}$ (7-L-phe) zinc phosphate, □———□ (10 µg per 2 days s.c.); and ACTH$_{4-10}$ (7-D-phe) zinc phosphate, ■———■ (10 µg per 2 days s.c.) on the rate of extinction of a shuttle box avoidance response as compared to placebo treatment, ●———●. Male rats were conditioned in the shuttle box. Ten trials a day were given with a variable intertrial interval of 60 sec. The sound of a buzzer served as the CS. Animals were trained till they achieved the conditioning criterion of 8 or more avoidances out of 10 for 3 consecutive days. Immediately thereafter extinction trials were run with the same schedule as in acquisition except that the US of shock did follow the CS if the animal failed to respond.

The D-form peptides facilitate extinction of a shuttle box and a pole-jumping avoidance response. They also seem to facilitate extinction of an appetitive response (Garrud and de Wied, 1972(unpublished observations)). However, their effect on passive avoidance behavior is not opposite to that of all-L-ACTH analogues as in active avoidance behavior (Table IV). In a simple step-through passive avoidance test (Ader et al., 1972), these peptides increased latency to enter a dark box like all-L-peptides.

TABLE IV

Effect of various D-isomer ACTH analogues on avoidance latency in a step-through passive avoidance response

Shock Intensity	Treatment	Median Preshock* Latency Mean of 3 trials (sec)	Median Postshock Latency (sec)	
			Trial 1	Trial 2
0.0 mA	Saline 0.5 ml s.c.	4	2	2
	ACTH$_{1-10}$ (7-D-phe) 10 μg s.c.	2	3	3
0.25 mA	Saline 0.5 ml s.c.	3	10	4
	ACTH$_{1-10}$ (7-D-phe) 10 μg s.c.	4	74†	43
	Saline 0.5 ml s.c.	2	22	10
	8-Lys-ACTH$_{4-9}$ (7-D-phe) 30 μg s.c.	3	250†	300†
	ACTH$_{4-10}$ 30 μg s.c.	2	300†	45†
0.50 mA	Saline 0.5 ml s.c.	4	87	59
	ACTH$_{1-10}$ (7-D-phe) 10 μg s.c.	4	276†	201†

5 animals per group.
*1 sec.
†Significantly different from saline treated controls.

Interestingly, the influence of the D-isomer peptides lasted longer than that of the all-L-analogues. The influence of the (7-D-phe) peptides in the passive avoidance test support our concept that these peptides do not influence conditioned behavior by competing with all-L-peptides for the same receptor, but possibly affect other nervous structures, resulting in an opposite effect on extinction of active avoidance behavior.

D-ISOMER SUBSTITUTION IN OTHER POSITIONS Subsequently it was investigated whether the reversal of active avoidance behavior by D-isomer substitution is restricted to the phenylalanine at position 7. Therefore, successive replacement of each of the amino acid residues was performed in the hexapeptide 8-lys-ACTH$_{4-9}$, which was found to be as active as ACTH$_{4-10}$. It is of interest to note that substitution of arginine by lysine, as in 8-lys-ACTH$_{4-9}$, is accompanied by loss of steroidogenic activity and of MSH activity (Chung and Li, 1967),while the behavioral activity remains unaltered in the peptides 8-lys-ACTH$_{1-10}$ (Greven and de Wied, 1967) and 8-lys-ACTH$_{4-9}$.

The amino acid sequences of the substituted hexapeptides are shown in Table V. None of the replacements

induced facilitation of extinction of a pole-jumping avoidance response. In contrast, these peptides invariably delayed extinction of the avoidance response as normally found with all-L-peptides derived from ACTH. In most cases replacement by a D-antipode potentiated the inhibitory effect of all-L-8-lys-ACTH$_{4-9}$. This potentiation was most marked with the lysine residue at position 8 in the D-configuration. The observed potentiation may be explained by enhanced resistance to proteolytic breakdown of the D-isomer peptides. These results clearly show that position 7 is specific for facilitation of extinction of an active avoidance response and add further evidence for the specificity of the influence of ACTH analogues on conditioned behavior.

A HIGHLY POTENT NEWLY SYNTHETIZED ACTH ANALOGUE Replacement of the lysine at position 8 by a D-isomer potentiated the inhibitory effect on extinction of the avoidance response more than 10 times. Substitution of tryptophan by phenylalanine in position 9 in the 8-D-lys-ACTH$_{4-9}$-peptide elicited a more than hundredfold

TABLE V

Amino acid sequence of 8-lys-ACTH$_{4-9}$ with various D-amino acid substitutions

H-D-Met- Glu - His -Phe- Lys - Trp-OH
H-Met -D-Glu- His -Phe- Lys - Trp-OH
H-Met - Glu -D-His-Phe- Lys - Trp-OH
H-Met - Glu - His -Phe-D-Lys- Trp-OH
H-Met - Glu - His -Phe- Lys -D-Trp-OH

TABLE VI

Effect of amino acid substitution in 8-lys-ACTH$_{4-9}$ on extinction of a pole-jumping avoidance response

4	5	6	7	8	9

H-Met-Glu-His-Phe-Lys-Trp-OH
  0
  ↑
H-Met-Glu-His-Phe-Lys-Trp-OH
H-Met-Glu-His-Phe-D-Lys-Trp-OH
H-Met-Glu-His-Phe-D-Lys-Phe-OH
  0
  ↑
H-Met-Glu-His-Phe-D-Lys-Phe-OH

potentiation of the behavioral effect (Table VI). Oxidation of the methionine to the sulfoxide evoked a three- to tenfold increase in behavioral potency. Interestingly, these modifications cause a marked reduction in the activity of ACTH in the adrenal cortex (Hofmann et al., 1970; Dedman et al., 1955). This accentuates again the different structural requirements for cortico-trophic and behavioral activity. The combination of the three potentiating modifications was undertaken and yielded a peptide with a behavioral activity which appeared to be more than a thousandfold stronger than that of $ACTH_{4-10}$. The injection of this new peptide in nanogram quantities causes a marked inhibition of extinction of a pole-jumping avoidance response.

CORTICOSTEROIDS Administration of glucocorticosteroids does not modify active avoidance learning. In most studies, dexamethasone was used, which is extremely potent in blocking ACTH-release. Administration of dexamethasone did not significantly alter the somewhat decremented avoidance acquisition of adrenal demedullated rats in a shuttle box (Conner and Levine, 1969). In adrenalectomized rats, however, dexamethasone tended to stimulate avoidance acquisition of rats in a pole-jumping avoidance test (van Delft, 1970). In contrast, Weiss et al. (1970) found that glucocorticosteroid treatment normalized avoidance responding of adrenalectomized rats which performed superior in both active and passive avoidance behavior. Latencies to resume licking after a single grid shock were unaffected by injection of dexamethasone or corticosterone (Pappas and Gray, 1971). Poor performance of an incompletely learned shuttle box avoidance response that depended on the interval between original training and subsequent training sessions was accompanied by low corticosterone levels in the circulation (Brush and Levine, 1966). Increasing the glucocorticoid level in the blood by administration of ACTH or hydrocortisone induced high responding under these conditions (Levine and Brush, 1967). However, Kasper-Pandi et al. (1970) failed to affect performance with extremely high amounts of dexamethasone under these conditions. Interestingly, a greater stability in interresponse time is induced in free operant avoidance behavior in rats after the administration of dexamethasone (Wertheim et al., 1967). It is possible therefore that glucocorticosteroids influence the efficiency with which animals perform an active avoidance task.

In contrast to affecting acquisition of active avoidance behavior, corticosteroids seem to be more effective in modifying passive avoidance behavior. Administration of a single dose of cortisone, 3 hr prior to shock, led to a suppression of passive avoidance in a light-dark avoidance situation. The effect of corticosteroids appeared to be a function of shock intensity. The higher the intensity, the more corticosteroids were needed to suppress passive avoidance behavior (Bohus et al., 1970). Corticosteroids also suppressed passive avoidance behavior motivated by fear or thirst (Bohus, 1971).

Chronic treatment with rather high doses of long acting ACTH during extinction, as was used in our earlier experiments (de Wied, 1967), invariably caused adrenal hypertrophy and a marked increase in the level of circulating corticosterone. Since at that time an extra target effect of ACTH on behavior was not generally accepted and most studies were performed with the whole ACTH molecule, it was felt necessary to study the influence of corticosteroids on the rate of extinction of conditioned avoidance behavior. Daily administration of moderate amounts of glucocorticosteroids like corticosterone and dexamethasone facilitated extinction of a shuttle box avoidance response. The influence of the mineralocorticosteroid aldosterone seemed to be less pronounced in this respect. This was interpreted to indicate that the behavioral influence of corticosteroids resided in the glucocorticosteroid portion of the molecule (de Wied, 1967). Using a pole-jumping situation, van Wimersma Greidanus (1970) showed, however, that the behavioral effects of adrenal steroids are associated neither with gluco- nor with mineralocorticosteroid activity. Progesterone, 19-norprogesterone and pregnenolone were as potent as corticosterone, while cholesterol, hydroxydione (Viadril), testosterone, and estradiol were ineffective. Common features of behaviorally active steroids appeared to be their double bonds in ring A or B, and their ketogroup of OH-group at C_3 (Figure 3). The ketogroup at C_{20} is important for the potency of the effect, but not essential.

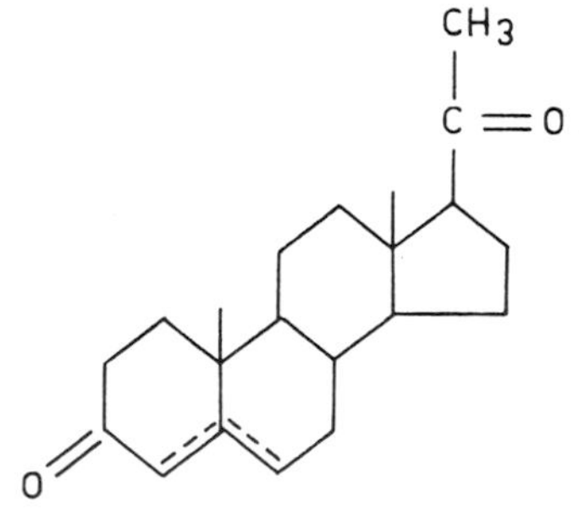

FIGURE 3 Common features of the steroid nucleus necessary to induce facilitation of extinction of a pole-jumping avoidance response.

Since daily administration of glucocorticosteroids inhibits the release of ACTH, one could argue that the influence of these steroids might be explained simply on the basis of a blockade of pituitary-ACTH release. However, the effect of glucocorticosteroids on extinction of a shuttle-box avoidance response did not correlate with the rate of inhibition of ACTH release. In addition, both corticosterone and dexamethasone facilitated extinction of the avoidance response in hypophysectomized rats. Thus, the influence of these steroids on behavior is independent of ACTH and an intrinsic property of the steroid molecule in the CNS. This was supported by experiments in which cortisol was implanted in various areas of the brain (Bohus, 1968). Implantation of cortisol in the median eminence, which effectively inhibited the release of ACTH, had a modest effect on the rate of extinction of a shuttle box avoidance response. The effect was stronger the more ACTH release was suppressed. However, implantation of cortisol in the mesencephalic reticular formation markedly facilitated extinction but hardly reduced ACTH release. Corticosteroids therefore may have a dual effect on extinction of avoidance behavior, one through inhibition of ACTH release and one probably more important through a direct action on the CNS.

Site of action of pituitary-adrenal system hormones in the brain

The site of action of pituitary-adrenal system hormones in the brain has been explored in rats bearing lesions in the thalamic region and by implantation of ACTH analogues and various steroids in the brain. Lesions were made in the thalamic region because this area has been implicated in acquisition and extinction of conditioned avoidance behavior (Vanderwolf, 1964; Thompson, 1963; Rich and Thompson, 1965; Cardo, 1965; Delacour et al., 1966; Delacour, 1970). Rather extensive lesions in the midline thalamic reticular area produced severe deficits in the rate of acquisition of a shuttle-box avoidance response and in escape behavior. Smaller lesions in this region interfered with avoidance acquisition but not with escape behavior (Bohus and de Wied, 1967a). Bilateral destruction of the nucleus parafascicularis did not materially affect avoidance learning but facilitated extinction of the avoidance response. In rats bearing lesions in the nucleus parafascicularis, α-MSH was unable to affect extinction in amounts in which it causes resistance to extinction in intact rats (Bohus and de Wied, 1967b). These results suggested that the nucleus parafascicularis was implicated in the behavioral effect of ACTH analogues.

To localize the action of ACTH analogues and steroids in the brain more specifically, implantation of these compounds in various subcortical structures was undertaken (van Wimersma Greidanus and de Wied, 1969, 1971).

$ACTH_{1-10}$ and $ACTH_{1-10}$ (7-D-phe) were used as the ACTH analogues and corticosterone and dexamethasone (21-Na-phosphate) as the steroids. ACTH analogues had a behavioral effect, i.e., delay with $ACTH_{1-10}$ or facilitation with $ACTH_{1-10}$ (7-D-phe) of extinction of a pole-jumping avoidance response, similar to that of systemic administration, when implanted into the region of the rostral mesencephalon and the caudal diencephalon at the posterior thalamic level or in the cerebrospinal fluid. Ineffective sites were the nucleus ventralis thalami, the nucleus anterior medialis thalami, the nucleus reuniens, the globus pallidus, the nucleus accumbens, the fornix, and the hippocampus.

Implantation of dexamethasone facilitated the rate of extinction of a pole-jumping avoidance response when implanted into various areas in the median and posterior thalamus and in the ventriculus lateralis. Corticosterone also facilitated extinction, but mainly if implanted in or near the nucleus parafascicularis. Implantation of these two steroids in other areas like hippocampus, nucleus septi lateralis, nucleus caudatus putamen, nucleus interstitialis terminalis, and nucleus ventralis thalami did not result in a modification of extinction. However, Bohus (1968) showed that cortisone implantation in the anterior hypothalamus, septum, amygdala, or dorsal hippocampus facilitated extinction of a shuttle-box avoidance response. Figure 4 gives a schematic representation of the brain areas that were found sensitive to peptides and steroids. Interestingly, the two ACTH analogues more or less act in the same structures in the brain. They affect areas in the ascending reticular system, mainly in the rostral mesencephalic caudal diencephalon area at the thalamic level; in particular in the posterior thalamus and more specifically in the nuclei parafascicularis. The steroids act, in addition to the ascending reticular system, in limbic forebrain regions. The opposite action of ACTH and corticosteroids seems to be located in the thalamic area. Facilitation of extinction, as induced by implantation of cortisone in that area, was counteracted by systemic administration of ACTH (Bohus, 1970).

Cardo (1965, 1967) has shown that the nonspecific thalamic nuclei like the nucleus parafascicularis and centrum medianum play an important role in the maintenance of conditioned avoidance behavior. Lesions in these areas interfere with avoidance performance, while electrical stimulation leads to resistance of extinction. Delacour (1970) suggested that the parafascicular-centrum - medianum complex is involved in the interaction between defensive motivation and some mechanisms of avoidance responding in certain avoidance situations. However, a number of complex food-reinforced tasks also are affected by lesions of this complex. The fact

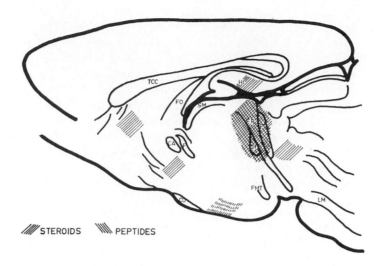

///STEROIDS \\\PEPTIDES

FIGURE 4 Schematic representation of sites in the brain sensitive to the behavioral effect of ACTH analogues ($ACTH_{1-10}$ [7-L-phe and 7-D-phe]) and corticosteroids. *Abbreviations:* TCC, truncus corporis callosum; Hi, hippocampus; FO, fornix; SM, stria medullaris thalami; FR, fasciculus retroflexus; pf, nucleus parafascicularis; CA, commissura anterior; F, columna fornicis; FMT, fasciculus mamillothalamicus; CO, chiasma opticum; LM, lemniscus medialis.

that ACTH (Gray, 1971; Guth et al., 1971) and α-MSH (Sandman et al., 1969) delay extinction of an appetitive response indicates that the role of the nucleus parafascicularis in conditioned behavior may be of a more general character.

Macromolecular effects of ACTH analogues in the brain

As shown above, hypophysectomy interferes with normal avoidance learning in a shuttle box. This deficient behavior of the hypophysectomized rat might be reflected in biochemical alterations in the brain (Versteeg et al., 1972). In addition, the treatment with ACTH analogues, which restored avoidance learning of the hypophysectomized rat, might at the same time reveal specific biochemical changes that could possibly be related to the influence of ACTH peptides in the brain. Recent studies suggest a close relationship between macromolecule processes in the brain and behavior (Hydén and Lange, 1965; Gaito, 1966; Glassman, 1967). In peripheral organs such as liver, hypophysectomy causes changes in RNA and protein metabolism (Korner, 1964; Tata and Williams-Ashman, 1967; Gupta and Talwar, 1968). Such alterations can be detected in the brain as well. A marked decrease in RNA content in the brain occurs as a result of hypophysectomy. This effect is confined to brainstem areas and the frontal part of the cortex (de Vellis and Inglish, 1968; Gispen et al., 1972). The reduction in RNA content is localized in the microsomal fraction as reflected in a decrease in the number of large polysomes (Gispen and Schotman, 1970). These changes

are accompanied by a decrease in incorporation of radioactive uridine into rapidly labeled RNA (Gispen et al., 1970b) and of radioactive leucine into proteins of all subcellular fractions (Schotman et al., 1972). From these experiments it follows that RNA synthesis in the brainstem of hypophysectomized rats is decreased. The same can be concluded for protein synthesis from the decrease in polyribosomes and in the labeling of proteins. Because protein content of the brain of hypophysectomized rats is not different from that of sham-operated controls, the deficient protein synthesis is restricted to a small fraction of rapidly turning over proteins.

A correlation between acquisition of learned behavior and the synthesis of rapidly labeled RNA (Zemp et al., 1966) and rapidly turning over proteins (Hydén and Lange, 1968) in the brain has been reported. The reduction of these processes in the brain of hypophysectomized rats might therefore be related to the deficient behavior of these animals. Treatment of hypophysectomized rats with $ACTH_{1-10}$ for 14 days restored performance of a shuttle-box avoidance response and at the same time normalized the optical density pattern of brain stem polysomes (Gispen and Schotman, 1970; Gispen et al., 1970b). Interestingly, peptide treatment per se did not restore the polysome pattern of hypophysectomized rats. This only occurred if peptide treatment was combined with avoidance training (Table VII). The primary action of $ACTH_{1-10}$ therefore might be related to other biochemical parameters. It appeared that treatment with $ACTH_{1-10}$ did not affect the incorporation of radioactive uridine into rapidly labeled RNA. However, $ACTH_{1-10}$ treatment did restore the rate of incorporation

Table VII

Effect of chronic $ACTH_{1-10}$ treatment, with (I) or without (II) shuttle-box training, on brainstem polysome profiles characterized by absorbance ratios. (From Ph.D. thesis. W. H. Gispen, Utrecht, 1970.)

Experiment	Treatment	Small Polysomes, Monosomes	Large Polysomes, Monosomes	Total Number CARs* out of 100 Trials	
I	Hypox + placebo	1.34 ± 0.06†	1.17 ± 0.05‡	23 ± 1	(7)
	Hypox + $ACTH_{1-10}$	1.30 ± 0.05	1.63 ± 0.04‡	75 ± 3	(3)
II	Hypox + placebo	1.32 ± 0.05	1.60 ± 0.11§	—	(5)
	Hypox + $ACTH_{1-10}$	1.32 ± 0.07	1.65 ± 0.11§	—	(5)

*Conditioned avoidance responses.
†Mean ± standard error of the mean.
‡$p < 0.05$ (modified t-test).
§$p > 0.10$ (modified t-test).
( ) Number of groups of 3 rats.

Table VIII

Effect of chronic treatment with various ACTH analogues on incorporation of radioactivity into proteins of brainstem of hypophysectomized rats, 5 min following injection of (4.5 ^3H-leucine) (Ph.D. Thesis, P. Schotman, Utrecht, 1971.)

Treatment	Percent Total Recovery		
	Acid Soluble Radioactivity	Acid Insoluble Radioactivity	
$ACTH_{1-10}$ (7-L-phe)	63.2 ± 2.6*	36.8 ± 2.6	(6)
Placebo	71.2 ± 1.1	28.8 ± 1.1	(6)
Difference (%)	−11.2†	+27.8†	
$ACTH_{1-10}$ (7-D-phe)	80.3 ± 2.5	19.7 ± 2.5	(6)
Placebo	72.7 ± 0.8	27.3 ± 0.8	(6)
Difference	+10.5†	−27.8†	
$ACTH_{11-24}$	74.5 ± 3.4	25.5 ± 3.4	(3)
Placebo	72.9 ± 5.1	27.1 ± 3.4	(3)

*Mean ± standard error of the mean.
†Significant difference.
( ) Number of animals.

of radioactive leucine into rapidly turning over proteins towards incorporation rates of intact rats (Table VIII). This is in accord with findings of Reading and Dewar (1971) in intact rats treated with $ACTH_{4-10}$; they found an increased incorporation of radioactive leucine into brain protein 48 hr after injection of the radioactive amino acid. High doses of ACTH itself also stimulated RNA and protein synthesis in the brain of intact rats (Jakoubek et al., 1970) although the interpretation of these results is complicated by a possible action of corticosteroids on brain metabolism (McEwen et al., 1970).

The restoration of the rate of incorporation of radioactive leucine into rapidly turning over proteins of the brainstem of hypophysectomized rats by $ACTH_{1-10}$ seems related to the restoration of the behavioral deficiency of the hypophysectomized rat in the shuttle box.

This is supported by experiments in which $ACTH_{1-10}$ (7-D-phe) was used. This peptide further deteriorates the already deficient avoidance behavior of hypophysectomized rats and at the same time lowers the rate of incorporation of radioactive leucine in the brainstem proteins. Moreover, $ACTH_{11-24}$, which did not affect avoidance acquisition, also failed to alter this parameter of protein synthesis (Table VIII). These results suggest that a disturbance in the synthesis of rapidly turning over proteins of brainstem origin is responsible for the deficient avoidance behavior of hypophysectomized rats. Polysomal aggregation does not seem to be involved in the effect of $ACTH_{1-10}$ per se but probably needs environmental stimulation (Appel et al., 1967) and training (Dellweg et al., 1968), since the effect of $ACTH_{1-10}$ on polysomes was found only in hypophysectomized rats subjected to shuttle box conditioning.

Concluding remarks

Although pituitary-adrenal system hormones affect the formation and maintenance of conditioned behavior, ACTH shares its influence with α- and β-MSH and even smaller analogues of these peptides, which lack corticotrophic activities. The same holds for adrenal steroids. The action of these steroids on conditioned behavior, which is opposite to that of ACTH, is not restricted to those steroids whose production is ACTH dependent. This makes it rather difficult to assign a role to the pituitary-adrenal system per se in conditioned behavior. ACTH and glucocorticosteroids, however, are secreted in high quantities in response to stress, and the system is highly active in many behavioral situations. One can expect therefore that alterations in the level of these hormones affect ongoing behavior associated with stress. It is well known that the pituitary-adrenal system plays an essential role in adaptation (Selye, 1950).

Acquisition and extinction of conditioned behavior indicate a behavioral adaptation to environmental changes. The influence of ACTH and glucocorticosteroids on acquisition and extinction of conditioned behavior should therefore be considered in an adaptive framework.

Various hypotheses have been put forward in the literature to account for the behavioral effects of pituitary-adrenal system hormones. A number of keywords in these hypotheses are anxiety, fear, memory, learning, arousal, motivation, internal inhibition, timing behavior, state dependent learning, etc. (de Wied et al., 1972). Weiss et al. (1970) postulated that ACTH increases excitability which leads to an increase in generalized fear or anxiety in fear situations.

Data obtained from electrophysiological studies indicate that ACTH and analogues have a central excitatory action; corticosteroids may either stimulate or depress activity in the brain (Kawakami et al., 1966; Korányi et al., 1971a, 1971b; Pfaff et al., 1971; Phillips and Dafny, 1971; Sawyer et al., 1968; Steiner, 1970). This suggests that pituitary-adrenal system hormones affect the arousal of certain structures in the CNS. Alteration in the state of arousal of these structures may determine the motivational influence of environmental stimuli. This phenomenon is not only related to avoidance but also to approach behavior, since it has recently been shown that acquisition and extinction of appetitive and drinking responses are also modified by pituitary-adrenal system hormones (Bohus, 1970; Gray, 1971; Gray et al., 1971; Guth et al., 1971; Sandman et al., 1969).

Indications on the locus of action were obtained from lesion and implantation studies. These suggest that the thalamic area and the limbic forebrain are structures sensitive to the behavioral effect of pituitary-adrenal system hormones. The nonspecific thalamic reticular area has an important integrative function (Cardo, 1965; 1967) because all incoming information converges in this structure. Implantation of peptides and steroids in this region effectively alters extinction of conditioned behavior. It is possible therefore that pituitary-adrenal system hormones and related compounds modulate the neural transmission in this nodal point of sensory integration. In this respect it is of interest to note that reciprocal alterations in sensory detection can be demonstrated during circadian alterations of pituitary-adrenal activity (Henkin, 1970). When the levels of glucocorticosteroids in plasma are low, taste detection is high, and vice versa. The marked changes in pituitary-adrenal activity that accompany numerous behavioral situations can therefore be expected to influence integration of extrinsic and intrinsic information, thereby affecting acquisition and maintenance of conditioned behavior.

The behavioral effect of ACTH appears to reside in not more than 4 amino acids. The sequence $ACTH_{4-7}$ (H-met-glu-his-phe-OH) is nearly as active as that of larger sequences like α- and β-MSH. Removal of the amino acid methionine at the amino end and/or phenylalanine at the carboxyl end reduces the behavioral effect. Interestingly, the minimal requirement for a residual activity on melanophore dispersion of α-MSH is the sequence 6–9 (Otsuka and Inouye, 1964). That such a small part of the ACTH molecule suffices for its behavioral activity is reminiscent of similar observations with regard to "tetragastrin," a peptide that exerts essentially the same biological activity as the complete gastrin molecule (Tracy and Gregory, 1964). In view of this it is possible that ACTH and MSH and other peptides act as prohormones from which the active sequence is released by specifically localized enzymes, in the same way as has been suggested by Walter and associates (Celis et al., 1971) for oxytocin that seems to act as a prohormone for the MSH-releasing and the MSH-release-inhibiting factor.

Behaviorally, the most interesting amino acid in the ACTH and MSH molecule appears to be 7-phenylalanine. Replacement of this amino acid by its D-isomer reverses the behavioral effect in active avoidance studies. This reversal is not obtained when neighboring amino acids are replaced by their respective D-isomers. The fact that the 7-D-phenylalanine analogues are effective in the absence of ACTH and MSH and, in addition as recently has been found, influence a passive avoidance response in essentially the same way as the all-L-analogues indicates that L- and D-form analogues act in a nondirect antagonistic manner. However, receptor studies are necessary to determine the specific action of these compounds. Such studies should be performed with radioactive labeled $ACTH_{4-10}$ in the same way as Hofmann et al. (1970) performed with $ACTH_{1-24}$ in the adrenal cortex to determine the active part of this molecule responsible for corticoidogenesis.

A remarkable increase in potency of the behavioral effect of the hexapeptide 8-lys-$ACTH_{4-9}$ was obtained by three modifications in the molecule. Replacement of methionine at position 4 by methionine sulfoxide, of lysine at position 8 by the D-isomer, and of tryptophan at position 9 by phenylalanine produced a peptide, which is active in nanogram quantities. This potentiating effect probably results from a protective influence by the respective substituted amino acids on the active core of the molecule in metabolic degradation.

Studies on the action of ACTH analogues on macromolecule metabolism will probably be very fruitful with respect to mode of action of these peptides. Hypophysectomy, which interferes with the ability to acquire a rather complicated behavioral response in the shuttle box, decreases the rate of incorporation of radioactive leucine into rapidly turning over proteins in the brainstem. The treatment of hypophysectomized rats with $ACTH_{1-10}$ not only restores avoidance acquisition but also the rate of incorporation of radioactive leucine. Moreover, the administration of $ACTH_{1-10}$ (7-D-phe), which deteriorates the already deficient avoidance behavior of hypophysectomized rats, further decreases the incorporation of radioactive leucine of the already reduced incorporation into rapidly turning over proteins. The function of rapidly turning over proteins for the formation of new behavior can be derived from studies with S100-proteins (Hydén and Lange, 1970) and from the amnesic action of antibiotics that interfere with protein metabolism in the brain (Barondes and Cohen, 1966; Flexner et al., 1967; Agranoff et al., 1967). The changes in protein metabolism in the brain presumably take place in the cell membrane. It is possible therefore that ACTH analogues act on membranes of cells in specific structures in the CNS, possibly in the same way as in isolated cells of the adrenal cortex (Seelig and Sayers, 1972), i.e., by inducing conformational changes that stimulate cyclic AMP production. This may result in stimulation of the metabolism of the cell, which is necessary to facilitate the formation of new synaptic connections.

REFERENCES

Ader, R., J. A. W. M. Weijnen, and P. Moleman, 1972. Retention of a passive avoidance response as a function of the intensity and duration of electric shock. *Psychon. Sci.* 26: 125–128.

Agranoff, B. W., R. E. Davis, L. Casola, and R. Lim, 1967. Actinomycin D blocks formation of memory of shock-avoidance in goldfish. *Science* 158:1600–1601.

Anderson, D. C., W. Winn, and T. Tam, 1968. Adrenocorticotrophic hormone and acquisition of a passive avoidance response: A replication and extension. *J. Comp. Physiol. Psychol.* 66:497–499.

Appel, S. H., W. Davis, and S. Scott, 1967. Brain polysomes: Response to environmental stimulation. *Science* 157:836–838.

Applezweig, M. H., and F. D. Baudry, 1955. The pituitary adrenocortical system in avoidance learning. *Psychol. Rep.* 1:417–420.

Applezweig, M. H., and G. Moeller, 1959. The pituitary-adrenocortical system and anxiety in avoidance learning. *Acta Psychol.* 15:602–603.

Barondes, S. H., and H. D. Cohen, 1966. Arousal and the conversion of "short-term" to "long-term" memory. *Proc. Nat. Acad. Sci. (USA)* 58:157–164.

Barrett, R. J., N. J. Leith, and O. S. Ray, 1971. The effects of pituitary-adrenal manipulations on time-dependent processes in avoidance learning. *Physiol. Behav.* 7:663–665.

Beatty, P. A., W. W. Beatty, R. E. Bowman, and J. O. Gilchrist, 1970. The effects of ACTH, adrenalectomy and

dexamethasone on the acquisition of an avoidance response in rats. *Physiol. Behav.* 5:939–944.

Bohus, B., and E. Endröczi, 1965. The influence of pituitary-adrenocortical function on the avoiding conditioned reflex activity in rats. *Acta Physiol. Acad. Sci. Hung.* 26:183–189.

Bohus, B., and D. de Wied, 1966. Inhibitory and facilitatory effect of two related peptides on extinction of avoidance behavior. *Science* 153:318–320.

Bohus, B., and D. de Wied, 1967a. Avoidance and escape behavior following medial thalamic lesions in rats. *J. Comp. Physiol. Psychol.* 64:26–29.

Bohus, B., and D. de Wied, 1967b. Failure of α-MSH to delay extinction of conditioned avoidance behavior in rats with lesions in the parafascicular nuclei of the thalamus. *Physiol. Behav.* 2:221–223.

Bohus, B., 1968. Pituitary ACTH release and avoidance behavior of rats with cortisol implants in mesencephalic reticular formation and median eminence. *Neuroendocrinology* 3: 355–365.

Bohus, B., 1970. Central nervous structures and the effect of ACTH and corticosteroids on avoidance behavior: A study with intracerebral implantation of corticosteroids in the rat. In *Progress in Brain Research 32; Pituitary, Adrenal, and the Brain*, D. de Wied and J. A. W. M. Weijnen, eds. Amsterdam: Elsevier, pp. 171–184.

Bohus, B., J. Grubits, G. Kovács, and K. Lissák, 1970. Effect of corticosteroids on passive avoidance behavior of rats. *Acta Physiol. Acad. Sci. Hung.* 38:381–391.

Bohus, B., 1971. Adrenocortical hormones and central nervous function: The site and mode of their behavioral action in the rat. In *Excerpta Med. Int. Congress Series* No. 219,V. H. T. James and L. Martini, eds. Amsterdam: Excerpta Medica, pp. 752–758.

Brain, P. F., 1971. Possible role of the pituitary-adrenocortical axis in aggressive behavior. *Nature (Lond.)* 233:489.

Brush, F. R., and S. Levine, 1966. Adrenocortical activity and avoidance learning as a function of time after fear conditioning. *Physiol. Behav.* 1:309–311.

Cardo, B., 1965. Rôle de certains noyaux thalamiques dans l'éboration et la conservation de divers conditionnements. *Psychol. France* 10:344–351.

Cardo, B., 1967. Effets de la stimulation du noyau parafasciculaire thalamique sur l'acquisition d'un conditionnement d'évitement chez le rat. *Physiol. Behav.* 2:245–248.

Celis, M. E., S. Taleisnik, and R. Walter, 1971. Regulation of formation and proposed structure of the factor inhibiting the release of melanocyte-stimulating hormone. *Proc. Nat. Acad. Sci. (USA)* 68:1428–1433.

Chung, D., and C. H. Li, 1967. Adrenocorticotropins XXXVII. The synthesis of 8-lysine-ACTH 1-17NH_2 and its biological properties. *J. Amer. Chem. Soc.* 89:4208–4213.

Conner, R. L., and S. Levine, 1969. The effects of adrenal hormones on the acquisition of signated avoidance behavior. *Hormones Behav.* 1:73–83.

Dedman, M. L., T. H. Farmer, and C. J. O. R. Morris, 1955. Oxidation-reduction properties of adrenocorticotrophic hormone. *Biochem. J.* 59:xxii.

Delacour, J., D. A. Fessard, and S. Libouban, 1966. Function of the two thalamic nuclei in instrumental conditioning in the rat. *Neuropsychologia* 4:101.

Delacour, J., 1970. Specific functions of a medial thalamic structure in avoidance conditioning in the rat. In *Progress in Brain Research 32; Pituitary, Adrenal, and the Brain*, D. de Wied and J. A. W. M. Weijnen, eds. Amsterdam: Elsevier, pp. 158–170.

DELLWEG, H., B. GERNER, and A. WACKER, 1968. Quantitative and qualitative changes in RNA of rat brain dependent on age and training experiment. *J. Neurochem.* 15:1109–1119.

de VELLIS, J., and D. INGLISH, 1968. Hormonal control of glycerol phosphate dehydrogenase in the rat brain. *J. Neurochem.* 15:1061–1071.

de WIED, D., 1964. Influence of anterior pituitary on avoidance learning and escape behavior. *Amer. J. Physiol.* 207:255–259.

de WIED, D., 1966. Inhibitory effect of ACTH and related peptides on extinction of conditioned avoidance behavior in rats. *Proc. Soc. Exp. Biol. Med.* 122:28–32.

de WIED, D., and B. BOHUS, 1966. Long term and short term effect on retention of a conditioned avoidance response in rats by treatment respectively with long acting pitressin or α-MSH. *Nature (Lond.)* 212:1484–1486.

de WIED, D., 1967. Opposite effects of ACTH and glucocorticosteroids on extinction of conditioned avoidance behavior. *Excerpta Medica Int. Congr. Series* No. 132:945–951.

de WIED, D., B. BOHUS, and H. M. GREVEN, 1968. Influence of pituitary and adrenocortical hormones on conditioned avoidance behavior in rats. In *Endocrinology and Human Behavior,* R. P. Michael, ed. New York: Oxford University Press, pp. 188–199.

de WIED, D., 1968. The anterior pituitary and conditioned avoidance behavior. *Excerpta Medica Int. Congr. Series* No. 184:310–316.

de WIED, D., and H. M. GREVEN, 1968. Opposite effect of structural analogues of ACTH on extinction of an avoidance response in rats by replacement of an L-amino acid or a D-isomer. Abstract 24, International Congress of Physiological Sciences, Washington, D.C., August 25–31, 1968. *Proc. Intern. Union Physiol. Sci.,* Vol. 7.

de WIED, D., 1969. Effects of peptide hormones on behavior. In *Frontiers in Neuroendocrinology,* W. F. Ganong and L. Martini, eds. New York: Oxford University Press, pp. 97–140.

de WIED, D., 1971. Pituitary-adrenal hormones and behavior. In *Normal and Abnormal Development of Brain and Behavior,* G. B. A. Stoelinga and J. J. van der Werff ten Bosch, eds. Boerhaave Series for Postgraduate Medical Education, Leiden, Netherlands: Leiden University Press, pp. 315–322.

de WIED, D., A. M. L. VAN DELFT, W. H. GISPEN, J. A. W. M. WEIJNEN, and TJ. B. VAN WIMERSMA GREIDANUS, 1972. The role of pituitary adrenal system hormones in active avoidance conditioning. In *Hormones and Behavior,* S. Levine, ed. New York: Academic Press, pp. 135–171.

FERRARI, W., G. GESSA, and L. VARGIU, 1963. Behavioral effects induced by intracisternally injected ACTH and MSH. *Ann. N.Y. Acad. Sci.* 104:330–343.

FLEXNER, B., J. B. FLEXNER, and R. B. ROBERTS, 1967. Memory in mice analyzed with antibiotics. *Science* 155:1377–1382.

GAITO, J., 1966. Macromolecules and brain function. In *Macromolecules and Behavior.* New York: Meredith.

GISPEN, W. H., and P. SCHOTMAN, 1970. Effect of hypophysectomy and conditioned avoidance behavior on macromolecule metabolism in the brain stem of the rat. In *Progress in Brain Research 32; Pituitary, Adrenal, and the Brain,* D. de Wied and J. A. W. M. Weijnen, eds. Amsterdam: Elsevier, pp. 221–235.

GISPEN, W. H., TJ. B. VAN WIMERSMA GREIDANUS, and D. de WIED, 1970a. Effects of hypophysectomy and ACTH$_{1-10}$ on

responsiveness to electric shock in rats. *Physiol. Behav.* 5:143–147.

GISPEN, W. H., D. de WIED, P. SCHOTMAN, and H. S. JANSZ, 1970b. Effects of hypophysectomy on RNA metabolism in rat brain stem. *J. Neurochem.* 17:751–761.

GISPEN, W. H., P. SCHOTMAN, and E. R. DE KLOET, 1972. Brain RNA and hypophysectomy: A topographical study. *Neuroendocrinology* 9:285–296.

GLASSMAN, E., 1967. *Molecular Approaches to Psychobiology.* New York: Dickenson.

GRAY, J. A., 1971. Effect of ACTH on extinction of rewarded behavior is blocked by previous administration of ACTH. *Nature (Lond.)* 229:52–54.

GRAY, J. A., A. R. MAYES, and M. WILSON, 1971. A barbiturate-like effect of adrenocorticotrophic hormone on the partial reinforcement acquisition and extinction effects. *Int. J. Neuropharmacol.* 10:223–230.

GREVEN, H. M., and D. de WIED, 1967. The active sequence in the ACTH molecule responsible for inhibition of the extinction of conditioned avoidance behavior in rats. *Europ. J. Pharmacol.* 2:14–16.

GREVEN, H. M., and D. de WIED, 1973. The influence of peptides derived from ACTH on performance structure activity studies. In *Progress in Brain Research,* E. Zimmerman, W. H. Gispen, B. H. Marks, and D. de Wied, eds. Amsterdam: Elsevier (in press).

GUPTA, S. L., and G. P. TALWAR, 1968. Effect of growth hormone on ribonucleic acid metabolism. The template activity of the chromatin and molecular species of ribonucleic acid synthetized after treatment with the hormone *Biochem. J.* 110:401–406.

GUTH, S., S. LEVINE, and J. P. SEWARD, 1971. Appetitive acquisition and extinction effects with exogenous ACTH. *Physiol. Behav.* 7:195–200.

HADŽOVIĆ, S., and D. de WIED, 1971. Central cholinergic pathways and the inhibitory effect of ACTH$_{1-10}$ on extinction of a pole jumping avoidance response. Abstr. First Congress Hungarian Society. Pharmacological Society, Budapest.

HENKIN, R. I., 1970. The effects of corticosteroids and ACTH on sensory systems. In *Progress in Brain Research 32; Pituitary, Adrenal and the Brain,* D. de Wied and J. A. W. M. Weijnen, eds. Amsterdam: Elsevier, pp. 270–294.

HODGES, J. R., and J. VERNIKOS-DANELLIS, 1962. Pituitary and blood corticotrophin changes in adrenalectomized rats maintained on physiological doses of corticosteroids *Acta Endocrin. (Kbh)* 39:79–86.

HOFMANN, K., R. ANDREATTA, H. BOHN, and L. MORODER, 1970. Studies on polypeptides XLV. Structure-function studies in the β-corticotropin series. *J. Med. Chem.* 13:339–345.

HYDÉN, H., and P. W. LANGE, 1965. A differentiation in RNA response in neurons early and late during learning. *Proc. Nat. Acad. Sci. (USA)* 53:946–952.

HYDÉN, H., and P. W. LANGE, 1968. Protein synthesis in the hippocampal pyramidal cells of rats during a behavioral test. *Science* 159:1370–1373.

HYDÉN, H., and P. W. LANGE, 1970. Brain cell protein synthesis specially related to learning. *Proc. Nat. Acad. Sci. (USA)* 65:898–904.

JAKOUBEK, B., B. SEMIGINOVSKY, M. KRAUS, R. ERDÖSSOVA, 1970. The alterations of protein metabolism of the brain cortex induced by anticipation stress and ACTH. *Life Sci.* 9:1169–1179.

Kamin, L. J., 1963. Retention of an incompletely learned avoidance response: Some further analyses. *J. Comp. Physiol. Psychol.* 56:713–718.

Kasper-Pandi, Ph., R. Hansing, and D. R. Usher, 1970. The effect of dexamethasone blockade of ACTH release on avoidance learning. *Physiol. Behav.* 5:361–363.

Kawakami, M., T. Koshino, and Y. Hattori, 1966. Changes in the EEG of the hypothalamus and limbic system after administration of ACTH, SU-4885 and ACh in rabbits with special reference to neurohumoral feedback regulation of pituitary-adrenal system. *Jap. J. Physiol.* 16:551–569.

Korányi, L., C. Beyer, and C. Guzmán-Flores, 1971a. Effects of ACTH and hydrocortisone on multiple unit activity in the forebrain and thalamus in response to reticular stimulation. *Physiol. Behav.* 7:331–335.

Korányi, L., C. Beyer, and C. Guzmán-Flores, 1971b. Multiple unit activity during habituation sleep-wakefulness cycle and the effect of ACTH and corticosteroid treatment. *Physiol. Behav.* 7:321–329.

Korner, A., 1964 Regulation of the rate of synthesis of m-RNA by growth hormone. *Biochem. J.* 92:449–456.

Leonard, B. E., 1969. The effect of sodium-barbitone, alone and together with ACTH and amphetamine, on the behavior of the rat in the multiple "T" maze. *Int. J. Neuropharmacol.* 8:427–435.

Levine, S., and L. E. Jones, 1965. Adrenocorticotropic hormone (ACTH) and passive avoidance learning. *J. Comp. Physiol. Psychol.* 59:357–360.

Levine, S., and F. R. Brush, 1967. Adrenocortical activity and avoidance learning as a function of time after avoidance training. *Physiol. Behav.* 2:385–388.

Lissák, K., E. Endröczi, and P. Medgyesi, 1957. Somatisches Verhalten und Nebennierenrindentätigkeit. *Pflügers Arch.* 265:117–124.

Lissák, K., and B. Bohus, 1972. Pituitary hormones and the avoidance behavior of the rat. *Int. J. Psychobiol.* 2:103–115.

McEwen, B. S., R. E. Zigmond, R. E. Azmitia, Jr., and J. M. Weiss, 1970. Steroid hormonal interaction with specific brain regions. In *Biochemistry of Brain and Behavior*, R. E. Bowman and S. P. Datta, eds. New York: Plenum Press, p. 123.

Miller, R. E., and N. Ogawa, 1962. The effect of adrenocorticotrophic hormone (ACTH) on avoidance conditioning in the adrenalectomized rat. *J. Comp. Physiol. Psychol.* 55:211–213.

Moyer, K. E., 1958. The effect of adrenalectomy on anxiety-motivated behavior. *J. Genet. Psychol.* 92:11–16.

Murphy, J. V., and R. E. Miller, 1955. The effect of adrenocorticotrophic hormone (ACTH) on avoidance conditioning in the rat. *J. Comp. Physiol. Psychol.* 48:47–49.

Otsuka, H., and K. Inouye, 1964. Synthesis of peptides related to the N-terminal structure of corticotropin II. The synthesis of L-histidyl-L-phenylalanyl-L-arginyl-L-tryptophan, the smallest peptide exhibiting the melanocyte-stimulating and lipolytic activities. *Bull. Chem. Soc. Jap.* 37:1465–1471.

Pappas, B. A., and P. Gray, 1971. Cue value of dexamethasone for fear-motivated behavior. *Physiol. Behav.* 6:127–130.

Pasley, J. N., and J. J. Christian, 1972. The effect of ACTH, group caging and adrenalectomy in *Peromyscus leucopus* with emphasis on suppression of reproductive function. *Proc. Soc. Exp. Biol. Med.* 139:921–925.

Pfaff, D. W., M. T. A. Silva, and J. M. Weiss, 1971. Telemetered recording of hormone effects on hippocampal neurons. *Science* 172:394–395.

Phillips, M. I., and N. Dafny, 1971. Effect of cortisol on unit activity in freely moving rats. *Brain Res.* 25:651–655.

Reading, H. W., and A. J. Dewar, 1971. Effects of $ACTH_{4-10}$ on cerebral RNA and proteins metabolism in the rat. *Third Int. Meeting Int. Soc. Neurochem.*, Budapest, p. 199.

Rich, I., and R. Thompson, 1965. Role of the hippocamposeptal system, thalamus, and hypothalamus in avoidance conditioning *J. Comp. Physiol. Psychol.* 59:66–72.

Sandman, C. A., A. J. Kastin, and A. V. Schally, 1969. Melanocyte-stimulating hormone and learned appetitive behavior. *Experientia* 25:1001–1002.

Sawyer, Ch. H., M. Kawakami, B. Meyerson, D. I. Whitmoyer, and J. Lilley, 1968. Effects of ACTH, dexamethasone and asphyxia on electrical activity of the rat hypothalamus. *Brain Res.* 10:213–226.

Schotman, P., W. H. Gispen, H. S. Jansz, and D. de Wied, 1972. Effects of ACTH analogues on macromolecule metabolism in the brainstem of hypophysectomized rats. *Brain Res.* 46:349–362.

Seelig, S., and G. Sayers, 1972. $ACTH_{1-10}$ and $ACTH_{4-10}$ stimulate cyclic AMP production by isolated adrenal cells. *Fed. Proc.* 31:252.

Selye, H., 1950. The physiology and pathology of exposure to stress. *Acta Inc. Montreal, Canada*, p. 6.

Steiner, F. A., 1970. Effects of ACTH and corticosteroids on single neurons in the hypothalamus. In *Progress in Brain Research 32; Pituitary, Adrenal, and the Brain*, D. de Wied and J. A. W. M. Weijnen, eds. Amsterdam: Elsevier, pp. 102–107.

Tata, J. R., and H. G. Williams-Ashman, 1967. Effects of growth hormone and tri-iodothyronine on amino acid incorporation by microsomal subfractions from rat. *Eur. J. Biochem.* 2:366–374.

Thompson, R., 1963. Thalamic structures critical for retention of an avoidance conditioned response in rats. *J. Comp. Physiol. Psychol.* 56:261–267.

Tracy, H. J., and R. A. Gregory, 1964. Physiological properties of a series of synthetic peptides structurally related to gastrin I. *Nature (Lond.)* 204:935–938.

van Delft, A. M. L., 1970. The relation between pretraining plasma corticosterone levels and the acquisition of an avoidance response in the rat. In *Progress in Brain Research 32; Pituitary, Adrenal and the Brain*, D. de Wied and J. A. W. M. Weijnen, eds. Amsterdam: Elsevier, pp. 192–199.

Vanderwolf, C. H., 1964. Effect of combined medial thalamic and septal lesions on active-avoidance behavior. *J. Comp. Physiol. Psychol.* 58:31–37.

van Wimersma Greidanus, Tj. B., and D. de Wied, 1969. Effects of intracerebral implantation of corticosteroids on extinction of an avoidance response in rats. *Physiol. Behav.* 4:365–370.

van Wimersma Greidanus, Tj. B., 1970. The relation between pretraining plasma corticosterone levels and the acquisition of an avoidance response in the rat. In *Progress in Brain Research 32; Pituitary, Adrenal, and the Brain*, D. de Wied and J. A. W. M. Weijnen, eds. Amsterdam: Elsevier, pp. 185–191.

van Wimersma Greidanus, Tj. B., and D. de Wied, 1971. Effects of systemic and intracerebral administration of two opposite acting ACTH-related peptides on extinction of conditioned avoidance behavior. *Neuroendocrinology* 7:291–301.

VERSTEEG, D. H. G., W. H. GISPEN, P. SCHOTMAN, A. WITTER, and D. de WIED, 1972. Hypophysectomy and rat brain metabolism: Effect of synthetic ACTH analogs. In *Adv. Biochem. Psychopharmacol.* 6. New York: Raven Press, pp. 219–239.

WEIJNEN, J. A. W. M., and J. L. SLANGEN, 1970. Effects of ACTH-analogues on extinction of conditioned behavior. In *Progress in Brain Research 32; Pituitary, Adrenal and the Brain*, D. de Wied and J. A. W. M. Weijnen, eds. Amsterdam: Elsevier, pp. 221–233.

WEISS, J. M., B. S. McEWEN, M. T. SILVA, and M. KALKUT, 1970. Pituitary-adrenal alterations and fear responding.

Amer. J. Physiol. 218:864–868.

WERTHEIM, G. A., R. L. CONNER, and S. LEVINE, 1967. Adrenocortical influences on free-operant avoidance behavior. *J. Exp. Anal. Behav.* 10:555–563.

WOODBURY, D. M., 1954. Effect of hormones on brain excitability and electrolytes. *Recent Progr. Hormone Res.* 10:65–104.

ZEMP, J. W., J. E. WILSON, H. SCHLESINGER, W. O. BOGGAN, and E. GLASSMAN, 1966. Brain function and macromolecules, I. Incorporation of uridine into RNA of mouse brain during short-term training experience. *Proc. Nat. Acad. Sci. (USA)* 55:1423–1431.

57 Macromolecules and Behavior:
A Commentary

EDWARD GLASSMAN

ABSTRACT This chapter is an attempt to point out the biochemical and behavioral difficulties that mar interpretation of the data in this field.

In addition some possible theoretical considerations are pointed out that may lead to fruitful research in the future.

RESEARCH attempting to relate macromolecules and learning has been extensively and adequately reviewed in recent years (Glassman, 1969; Jarvik, 1972; Horn, 1971; Ungar, 1972; Rose, 1970; Roberts and Matthysse, 1970). Suffice it to say that numerous chemical changes have been reported to be correlated with various training experiences. Almost all of these data are confounded because novel stimuli, such as flashing lights, activity, stress, shocks, etc., have been reported to cause similar chemical changes in the nervous system, and usually not enough behaviors have been examined to distinguish these psychological variables from the learning process. In addition, there are problems in the biochemical approaches that have been used. One of the purposes of this chapter is to discuss some of these problems in a critical and constructive way.

Problems of biochemical interpretations

Autoradiography is potentially one of the most elegant techniques for the study of chemical correlates of behavior because the response of large numbers of individual cells to an input stimulus can be monitored in a single preparation. To be effective, this technique requires the use of densitometry to avoid counting silver grains and computer analysis of the data to facilitate the comparison of the amount of radioactivity incorporated in cells or areas of various brains. Dr. Daniel Entingh has been working on such equipment for our laboratory (Entingh and Bernholz, unpublished).

Recently Rahmann (1973) has reported results that

EDWARD GLASSMAN Department of Biochemistry, School of Medicine, University of North Carolina, Chapel Hill, North Carolina

constitute an extremely elegant example of autoradiographic analysis applied to the brain. Adult carp (*Carassius carassius*) were kept in darkness for 10 days, after which they were injected intraperitoneally with $[^3H]$-histidine, curarized, and one eye only was exposed for 75 min to 15 sec of light alternating with 45 sec of darkness. The fish were immediately sacrificed and autoradiograms were prepared of histological sections through the optic tecti. The density of grains over the stimulated contralateral tectum was significantly greater than that over the non-stimulated ipsilateral tectum. Of even greater interest, exposure of one eye to a vertical stripe of light 2 mm wide produced an increased density of silver grains over a narrow stripe in the stimulated tectum, while two vertical stripes of light produced two such stripes. Similar results have also been reported by Rahmann's associates in the guppy (Skrzipek, 1969) and the frog (Wegener, 1970). This elegant experiment indicates the exciting kind of data that can be generated in complex nervous systems using this approach.

It should be noted, however, that the biochemical conclusions to be drawn from autoradiographic techniques using radioactive tracers are not unequivocal. For example, the timing of most experiments is long enough for much of the radioactive amino acid to be metabolized to other substances, and therefore it is not clear whether the detected radioactivity is in protein or in some other compound. In addition, the increase in radioactivity could be due to increased synthesis, decreased destruction (see chapters by Schimke and by Goldberg in this volume), or the availability of radioactive precursor. This problem has been previously discussed (Glassman, 1969), but its importance is increasingly evident (see especially Baskin et al., 1972).

The problem of monitoring the availability of the radioactive precursor is particularly acute, because there seems little chance of measuring it simply in an intact animal. Changes in the amounts of radioactive precursor in individual cells or areas of the nervous system can occur because of localized changes in blood flow, changes in cellular permeability to the radioactive compound, changes in endogenous rates of synthesis in individual cells, and localized differences due to unavoidable variations in the injection. The problem of localized changes in blood flow is of particular concern in view of the report by Sokoloff (1961) that visual stimulation can cause an increase of blood flow in the visual cortex of light-deprived rats.

One can partially correct for the availability of a radioactive precursor by measuring the radioactivity present in the precursor or in other appropriate small molecules of the cells when the animals are sacrificed, and thereby derive a pool correction factor. This will monitor some variations in injection procedure, and possibly some changes in blood flow, cellular permeability, and changes in endogenous synthesis. Any pool correction factor should be viewed with caution, however, unless it is measured throughout the entire time of the incorporation period, and it can be shown that the behavioral stimulation has no effect on the radioactivity in the pool. Even if these criteria are met, the problem of compartmentation of metabolic pools within a cell, particularly with respect to the site of synthesis, makes gross pool correction factors inadequate to monitor the immediate precursor pool for any macromolecule.

The inability to derive data on this crucial point is probably the reason that the problem of pool correction factors has not been the subject of more extensive investigation. Most investigators have contented themselves with a crude pool estimate, or none at all. It is of interest that it is extremely difficult to obtain a pool correction factor in autoradiographic analysis (but see Watson, 1965, and Quevedo et al., 1971, for attempts to solve this problem). Thus one cannot conclude that results obtained using autoradiography are due only to altered macromolecular synthesis. This is not meant to detract, however, from the importance and interest of finding changes in incorporation limited to specific cells and regions of the nervous system but rather to point out where the biochemical conclusions are equivocal and where further experimental clarification is necessary. Because of the importance of evaluating the validity of a pool correction factor, it is imperative that it is clearly presented in papers that present data using radioactive tracers.

These ideas can be applied to many investigations using radioactive tracers. For example, Zemp et al. (1966) reported increased incorporation of uridine into brain RNA due to avoidance training using the radioactivity in UMP as a pool correction factor; but Entingh et al. (in preparation) have shown that a possible alternative interpretation is that the radioactivity in UMP is decreased as a result of the experience. Another example is the work of Hydén and Lange (1968, 1970) who reported

increased incorporation of [^{3}H]-leucine into specific proteins of CA3 cells of the hippocampus of rats undergoing a change in handedness, but who derived a soluble pool correction factor from a different group of cells (see also Comment 1 of Bowman and Harding, 1969, who discuss a similar point). In addition, the time of sacrifice following the injection in these experiments is so long that it seems likely that most of the radioactivity is no longer in leucine but in other substances. Since increased incorporation into specific proteins is involved, one possible correction factor might be the amount incorporated into other proteins (see Gisiger and Gaide-Huguenin (1969) and Kerkut et al. (1970) for RNA; and Emson et al. (1971), Wilson (1971), Gainer (1972a, b, c), and Strumwasser (this volume) for protein). Hydén (personal communication) has stated that the techniques used in his laboratory had been changed so some of these comments may not apply to some of his more recent work (see Hydén and Lange, 1972a, 1972b).

Other investigators have indicated their concern with this problem. Rose (1972) reported on the effect of visual experience on amino acid pools of rat brain. Baskin et al. (1972) reported that the incorporation of radioactivity from orotic acid into cytidylic acid is decreased in goldfish brain by the CO_2 produced by the activity of the fish, with a resulting decrease of radioactive cytidine in RNA. Electrical stimulation of isolated nerve tissue has been reported to produce changes in amino acid pools (Jones and McIlwain, 1971; Jones, 1972; Jones and Banks, 1970a, 1970b; McBride and Klingman, 1972; Orrego and Lipmann, 1967) and in RNA pools (Orrego, 1967; Prives and Quastel, 1969). Wilson and Berry (1972) report that the amount of incorporation of uridine into RNA of R2 cells of *Aplysia* is dependent on the external uridine concentration, an observation in agreement with the data of Itoh and Quastel (1969) using rat cortex slices.

One way to avoid the problems associated with valid pool correction factors is not to use radioactive tracers but to measure changes in the amount of the substance directly. For example, Haljamäe and Lange (1972) have recently reported that rats undergoing a change in handedness show increases in the amount of S-100 protein in the CA3 cells of the hippocampus. Furthermore this increased amount of S-100 appears to migrate differently during microdisc gel electrophoresis, suggesting a changed conformation. These workers also report a significant increase of calcium in these cells but no change in sodium or potassium. Since calcium has been reported to cause conformational changes in S-100 (Calissano et al., 1969), these workers suggest that the increased calcium in these cells combines with S-100, and a conformational form with increased stability results. Haywood and Rose (1970) reported an increase in RNA polymerase following exposure of unimprinted, sensitive chicks to a flashing light. The increase occurred in the same part of the brain that increased incorporation of radioactive precursor into RNA, and protein have been reported under similar behavioral stimulation (Bateson et al., 1972). Kerkut et al. (1970) and Emson et al. (1971) reported changes in cholinesterases with learning in invertebrate ganglia (but see Woodson et al., 1972). Machlus et al (1973a, 1973b) have reported increased radioactive phosphate in nuclear proteins of rat brain as a result of a short training experience. This effect of experience was also correlated with an increase in the molar ratios of phosphoserine compared to serine in hydrolysates of these proteins; this direct measurement avoided the problem of an adequate pool correction factor since no radioactivity was involved. Finally, there are reports of the effects of experience on polysomes that also do not involve radioactive tracers (Appel et al., 1967; MacInnes et al., 1970; Uphouse et al., 1972a, 1972b). These comments are not meant to downgrade the use of radioactive tracers, because the use of such techniques provides excellent clues to possible important chemical changes in small areas of nerve tissue and are important in any research strategy. Nonetheless their exclusive use does not permit an unequivocal conclusion that only synthesis is being monitored.

Problems of behavioral interpretations

The causative stimuli in this research are difficult to ascertain. It may be that the learning per se or the special stresses, emotions, and motivations that accompany learning are related to the observed chemical responses to training experiences, but it is also possible that the changes are due to the other stimuli that accompany the training experience. Sensory stimulation and stressful situations have been reported to produce changes in macromolecules in the nervous system, and unless adequate behavioral experiments are carried out, it is difficult to eliminate such psychological variables as the cause of chemical changes in the brain.

A major difficulty in quantitative correlation is the problem of assaying the amount of memory stored after a training experience. Such assays depend on the performance of an animal and thus are sensitive to such diverse factors as the motivation and attention of the animal, its state of health and activity, and a number of other emotional and physical factors. Thus an animal may not perform well if it is not motivated, or too sick, tired, or frightened, or if it has a memory deficit. Even if extensive experiments are carried out, it is very difficult to establish why an animal is not performing well.

Another problem arises from exaggerated concerns as to whether such changes are directly associated with

learning and memory processes, or whether they are due to *nonspecific* responses. The difficulty, mostly semantic, is that nonspecific responses do not exist; they are merely responses whose cause is unknown. It would probably be more productive to define such responses in definite terms, and to attempt to classify them as encoding-specific, training-specific, consolidation-specific, imprinting-specific, stimulus-specific, emotion (arousal)-specific, stress-specific, attention-specific, performance-specific, etc. For example, Rahmann's autoradiographic results described above are probably due to a stimulus-specific response in the reacting cells. The data of Bateson et al. (1972) may be due to an imprinting-specific response, but the behavioral data do not rule out a stimulus-specific response to flickering light. In the change-of-handedness experiments, untrained rats used for comparison with the trained animals are always rats using their preferred paws (Hydén and Lange, 1968, 1970a, b). Thus these data are inadequate to distinguish between training-specific responses, or those due to the other behavioral- and psychological-specific responses listed above. This is extremely important in view of the finding by Rees and Brogan (see Dunn et al., this volume) that an increased incorporation into rat brain protein can be observed 20 to 30 min following a wide variety of novel stimuli that do not involve training. The data of Adair et al. (1968a, 1968b) show training-specific increased incorporation of radioactive uridine into brain polysomes, but the behavioral data are inadequate to determine whether this is involved with encoding or with the specific motivational or emotional changes associated with the training experience. This is equally true for all other data involving training experiences. Machlus et al. (1973a, 1973b) demonstrate a more complex behavioral interaction in that the increase of radioactive phosphate in brain nuclear protein is not only a training-specific response, but it also occurs after a training-specific reminder, an effect that Adair et al. (1968a) did not find.

These kinds of data indicate that behavioral analysis to determine the psychological variables involved is extremely crucial to understanding these phenomena. No one behavior is an adequate control; indeed, the term "control" should be avoided. The effects of many related experiences must be studied before the specificity of a chemical response can be evaluated.

Studies on memory consolidation

In order to study macromolecular correlates of memory processes in vertebrates, it would be helpful to know which brain area is involved, when the chemical change occurs, and which chemicals to study. At one time it was felt that the data on memory consolidation would provide important clues for the neurochemist. It is no longer clear that this is the case.

Various treatments immediately after training to *localized* areas of the brain can cause deficits or facilitation of performance when the animals are tested at a later time (for example, see Grossman, 1969; Grossman and Mountford, 1964; Stein and Chorover, 1968; Wyers et al., 1968; Mahut, 1962; Thompson, 1958; Kesner and Doty, 1968; Kesner and Conner, 1972; Glickman, 1958; Zornetzer, 1972; Zornetzer and McGaugh, 1972; Denti et al., 1970; Daniels, 1971a; Flexner, 1967; Vardaris and Schwartz, 1971; Lidsky and Slotnick, 1970; Herz and Peeke, 1971; Peeke and Herz, 1971; Wyers and Deadwyler, 1971; Hudspeth and Wilsoncroft, 1969; Paolino and Levy, 1971; Gold et al., 1971; Dorfman and Jarvik, 1968; Erickson and Patel, 1969). Much of this is excellent data and implicates many subcortical structures. The pattern that emerges, however, is far too diffuse to provide clear clues to the neurochemist as to precisely where to look for chemical changes that might be associated with the formation of long-term memory.

Clues as to which chemicals might be involved in the formation of long-term memory are not easily found when the chemical effects of amnestic agents are analyzed. Almost all agents used so far have a wide variety of effects that confound attempts to postulate a common cause. This is true even if one makes allowances for a multistaged complex process. One possible common effect of most amnestic agents may be to reduce the level of arousal long postulated to be a requirement for memory consolidation, although another possibility may be their ability to inhibit RNA or protein synthesis (see Glassman, 1969).

The time at which the neurochemist should look for chemical correlates of the formation of long-term memory seems clear, because most reports indicate the process starts during or soon after training. The exact time is under dispute, and even the nature of the process in terms of defective storage or defective retrieval mechanisms is not agreed on (see Quartermain and McEwen, 1970; Squire and Barondes, 1972). It has been reported that long-term memory formation following an aversive one-trial pecking task in chicks is decreased by ouabain, lithium, or copper ions, but only if they are injected just prior to the training trial, whereas cycloheximide is effective if injected just before or just after the task (Mark and Watts, 1971; Watts and Mark, 1971). Since one common effect of ouabain, lithium, and copper is to inhibit the sodium pump in nerve membranes, it was concluded that this process is essential for the formation of long-term memory before there is a requirement for protein synthesis. The fact that such compounds can affect the formation of long-term memory should stimulate further research into the whole range of possible chemical amnestic agents.

Whether an expanded list will clarify or add to the confusion remains to be seen.

Some theoretical considerations

There are a number of possible roles of macromolecules in brain function. One possibility is that the macromolecules reported as biochemical correlates to training experiences are involved in the restoration of the cellular components used during nerve activity. This nerve activity could be directly associated with the learning process per se but might be involved with motivational, emotional, or other phenomena that accompany behavioral stimulation.

It does not seem probable that these macromolecules are themselves encoding the experiential information; but another possibility is that such chemical changes are related to those processes that affect the response of neurons to impulses from neighboring neurons by changing the connectivity between neurons so that new pathways become functional. This could occur by changes in the effectiveness of the neurotransmitter, e.g., by increasing the amount released, by decreasing the activity of enzymes that inactivate it, or by increasing the number or effectiveness of the receptor sites on the postsynaptic neuron, or by other means. Thus, the observed changes in macromolecules may reflect the important role of regulating interneuronal connectivity. It is tempting to speculate that they are part of the early steps in the storage of long-term memory, a process that may be dependent on RNA and protein synthesis (Flexner et al., 1967; Lande et al., 1972; Agranoff, 1972; Daniels, 1971b; Squire and Barondes, 1972).

Figure 1 summarizes these ideas. A training experience has many effects on an animal. Increased electrophysiological activity (information flow) has been reported (see John, 1967), and it is thought that through associative processes of various types involving feedbacks and special

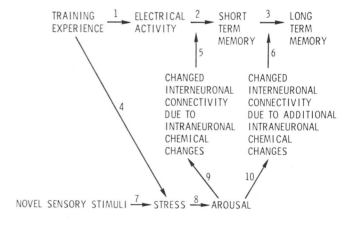

FIGURE 1   Some effects of a training experience.

instructions to a variety of cells, intracellular chemical changes take place that produce connectivity changes between neurons. The final result is the formation of new pathways that underlie short-term memory, and later, long-term memory. Thus the interneuronal events that encode memory (steps 2 and 3) are distinct from supportive intraneuronal events involving chemical changes (steps 5 and 6). It has long been felt that the formation of long-term memory probably depends on a type of arousal that stimulates a sequence of chemical events that are independent of short-term memory (see, for example, Barondes and Cohen, 1968). The arousal may also be produced by other experiences (step 7). This would account for many of the chemical changes caused by novel sensory stimulation. These molecules may be related to the peptide(s) postulated to influence the maintenance and retrieval of memory by de Wied (this volume).

The basic questions that need to be answered in order to evaluate these ideas include the way electrical activity is transduced into the chemical changes necessary to regulate or change interneuronal connectivity, the sequence of the biochemical changes involved, and the primary cellular trigger for these events. Research involving isolated ganglia, cortical slices, giant neurons of invertebrates and other isolated nervous system components can provide clues to the answers of some of these questions. One striking finding is that many chemical changes within a neuron are correlated with the stimulation of the neuron by presynaptic stimulation and do not seem to occur when the axon of the neuron is stimulated antidromically (Larrabee and Leicht, 1965; Hokin et al., 1960; Stoller and Wayner, 1968; McAffee et al., 1971). This effect of presynaptic stimulation has been reported in relation to the increase of radioactive phosphate in phosphatidylinositol of mammalian sympathetic ganglia (Larrabee and Leicht, 1965; Larrabee, 1968; Hokin et al., 1960; Hokin, 1965, 1966); the increased incorporation of uridine into RNA of *Aplysia* cells (Kernell and Peterson, 1970; but see Wilson and Berry, 1972) and rat superior cervical sympathetic ganglion (Gisiger and Gaide-Huguenin, 1969; Gisiger, 1971); the increase of cyclic AMP of rabbit superior cervical sympathetic ganglion (McAffee et al., 1971) or *Aplysia* abdominal ganglion (Cedar et al., 1972a); and the changes of activity of various enzymes involved in brain amine metabolism (Axelrod, this volume). That the action potential generated by antidromic stimulation and stimulation across a synapse has different effects is also suggested by the report that antidromic stimulation of an inhibitory neuron does not lead to the expected inhibition of other neurons (Mulloney and Selverston, 1972).

Many chemical changes induced by presynaptic stimulation can be duplicated by applying neurotransmitter (Cedar and Schwartz, 1972b; Prives and Quastel, 1969; Hokin, 1969b; Hokin et al., 1960; Larrabee and Leicht, 1965; Shimizu et al., 1970; Kakiuchi and Rall, 1968a, 1968b; Klainer et al., 1962; Gilman and Nirenberg, 1971; McAffee et al., 1971; Kuo et al., 1972). Furthermore, these chemical changes are blocked by agents that prevent the transmitter from being released (Cedar and Schwartz, 1972a) or from attaching to the receptor site (Prives and Quastel, 1969; Gisiger and Gaide-Huguenin, 1969; Gisiger, 1971; Larrabee and Leicht, 1965) but not necessarily by treatments that affect the action potential (Cedar and Schwartz, 1972a; Gisiger, 1971).

It would therefore appear that the initial trigger for a variety of chemical changes in a neuron depends on the attachment of the neurotransmitter from the presynaptic cell, as shown in Figure 2. In this figure, sequence IV shows the changes involved in the generation of the action potential of the cell. It is well known that this involves temporary changes in membrane permeability to ions, but as indicated in Figure 2, sequence III, there are also changes in neuronal permeability to other ions, amino acids, nucleosides, and other compounds. This emphasizes again the importance of determining pool correction factors in the same cells in which radioactivity in macromolecules is being determined.

Reaction sequence V of Figure 2 describes chemical events that are well known in many types of cells, and that may underlie the molecular control of neuronal connectivity. The activation of protein kinases by cyclic-AMP or cyclic-GMP (step b) eventually leads to conformational changes in proteins (step d) via their phosphorylation (step c). If the conformational changes take place in synaptic proteins (step e^1), then one can postulate rapid, direct effects on synaptic properties resulting in changes in neuronal connectivity with eventual memory formation. An alternative idea is that conformational changes take place in nuclear proteins (step e), a process that leads to gene activation, RNA synthesis, and protein synthesis (steps f and g). This protein can be involved in processes involved with replenishing the chemicals of the neuron (step h^1) or, alternatively, may have special neuronal functions (step h) at the synapse (step j), where it regulates synaptic properties such as connectivity, etc. (steps k, l, m). It is possible that during training, conformational changes in synaptic proteins occur rapidly to temporarily change connectivity associated with the formation of short-term memory, but that the events following conformational changes in the nuclear proteins are necessary for the permanent connectivity changes that underlie the formation of long-term memory.

There are many lines of research that suggest these ideas have merit, but none are sufficient to prove them

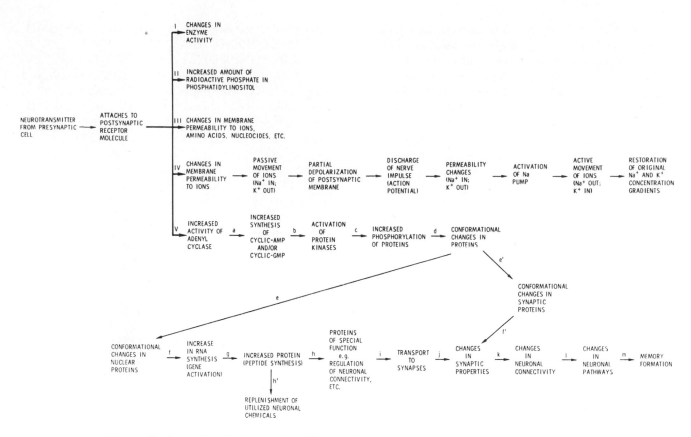

FIGURE 2   Some chemical events that follow the attachment of a neurotransmitter to a neuron.

true, or that any known chemical events do indeed regulate interneuronal connectivity, pathway formation, or memory formation. The reason that attention has focused on conformational changes is that short-term nervous system function is not affected by extensive inhibition of protein synthesis. For example, the formation of short-term memory is not impaired by doses of cyclo-heximide, puromycin, or actinomycin that inhibit the formation of long-term memory (see Glassman, 1969). In addition, a number of researchers using invertebrate preparations report that sustained neurophysiological activity can take place in the virtual absence of RNA or protein synthesis (Schwartz et al., 1971; Edstrom and Grampp, 1965; Bondeson et al., 1967; Kandel, this volume). It has been reported (see Schwartz et al., 1971; Castellucci et al., 1972) that the generation of the action potential, the resting membrane potential, posttetanic potentiation, inhibitory and excitatory synaptic potentials, and short-term habituation can take place in the virtual absence of protein synthesis in *Aplysia*. Whether long term phenomena in this organism are dependent on protein synthesis remains to be determined.

The way in which conformational changes in molecules can regulate interneuronal connectivity is not known. One can postulate that a chemical located at the synapse or elsewhere in the pre- or postsynaptic neuron is critical for connectivity, and that the control of connectivity is exerted by the conformational state of this chemical. As shown in Figure 3, conformation state A is stable, but connectivity is not facilitated. If during training a pathway or network is activated in which this synapse is important, the molecule is shifted into conformational state B. Conformational state B is unstable, but it changes the properties of the neuron so that a temporary change in connectivity occurs. The new pathway or network can be considered a part of the mechanism underlying short-term memory. The molecule in conformation state B can revert back to conformation A, in which case connectivity at this synapse is again as it was, and the memory is lost or is no longer easily retrievable. Alternatively, if certain factors are stimulated in the animal undergoing training, the molecule in conformation B is converted to a new stable form, conformation C, so that connectivity is permanently changed. Long-term memory is now possible. Since only molecules in conformation B can be converted to conformation C, only those pathways or networks that have been previously selected as providing adaptive behavior will be affected and made permanent. Alternatively, new molecules can be synthesized to make the connectivity changes permanent. Figure 3 is based on the idea that the pathways or networks underlying short- and long-term memory are the same, or at least overlap to a great extent (see Kandel, this volume). Much work is necessary to determine the validity of ideas like this, but it has considerable heuristic value. Thinking along such lines has led to the study of the effects of behavior on the metabolism of membrane constituents, particularly those found in synapses (see Dunn et al., this volume).

It should be pointed out that phosphorylation is not the only way to achieve conformational changes in proteins. Others include the enzymatically catalyzed formation of covalent bonds with methyl groups, acetyl groups, and carbohydrates. The presence of substrates, coenzymes, ions, and a variety of hormones and regulators will also cause conformational changes, as will selective removal of amino acids or small peptides. Barondes (1970) has reviewed the possible role of brain glycoproteins in interneuronal recognition.

Conclusions

One purpose of this article is to attempt to evaluate the significance of the reported effects of behavior and training experiences on macromolecules in the nervous system. A detailed discussion of the role of these macromolecules in the learning processes is certainly premature; there is no clear evidence that these macromolecules play a direct role in encoding, and cause and effect relation-ships in this area are still difficult to prove. The current problems that seem important are the aspects of the environment or the behavior that can trigger chemical responses in the nervous system, the physiological and biochemical processes that convey the information from outside the animal to the nerve cells that are responding, and the significance of the chemical response in terms of the functional role that these macromolecules play in the cells of the nervous system. Even if no aspect of the learning process is involved, the fact that behavior can affect macromolecules in these ways should be of great interest to those studying nervous system function. The role these macromolecules play in the nervous system may be only to maintain cellular health and structure and to enable the cellular machinery to function. If so, it is important to know this fact so that the solutions to the problem of the unique mechanisms of the nervous system in regulating behavior can be sought elsewhere.

Certain specific problems can be emphasized. The first concerns the significance of the increased amounts of radioactive phosphate in phosphatidylinositol with presynaptic stimulation and its possible role in reaction V of Figure 2. Hokin (1969b) has suggested this phenomenon is related to memory processes (see also review by Hokin, 1969a, and Durell et al., 1969). One approach would be to determine whether behavioral experiences, particularly training, will affect this reaction.

A second question concerns the report of a training-specific increase in the amount of radioactive phosphate in nuclear proteins (Machlus et al., 1973a, 1973b) and in

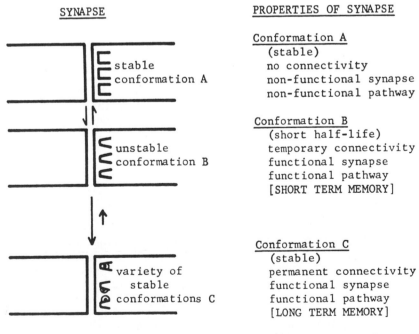

SYNAPSE

PROPERTIES OF SYNAPSE

stable conformation A

unstable conformation B

variety of stable conformations C

Conformation A
  (stable)
  no connectivity
  non-functional synapse
  non-functional pathway

Conformation B
  (short half-life)
  temporary connectivity
  functional synapse
  functional pathway
  [SHORT TERM MEMORY]

Conformation C
  (stable)
  permanent connectivity
  functional synapse
  functional pathway
  [LONG TERM MEMORY]

FIGURE 3   Possible role of conformational changes in regulating interneuronal connectivity.

synaptosomal proteins (Glassman et al., 1973) and their possible relation to sequence V of Figure 2. Jones and Rodnight (1971) have reported a similar phenomenon when cortical slices of guinea pig brain were electrically stimulated, and it would be of interest to determine whether such changes are dependent on presynaptic or antidromic stimulation.

Another question concerns the effects of nerve activity in single neurons on the incorporation of various precursors into macromolecules while the precursor pool and changes in cellular permeability are carefully monitored. Possibly the application of techniques for directly monitoring biochemical changes during nerve function will be of use here (see, for example, potential applications described by Aubert et al., 1964; Maitra et al., 1964; Tasaki et al., 1969, 1972; Cummins and Bull, 1971; Clark and Strickholm, 1971; Moore and Wetlaufer, 1971; Baum and Ward, 1971; Kohen and Kohen, 1966; Cummins, 1971; Llinas et al., 1972). The combination of such techniques with inhibitors that specifically block the various steps involved in the generation of the action potential (sequence III, Figure 2) might be very revealing.

Finally the role of arousal in these processes needs elucidation. The approach of Aprison et al. (1968) seems particularly suited to studies on the effect of experience on brain amines (see also Serota, 1971, Serota et al., 1972; Randt et al., 1971).

The problem of molecular control of long term changes in interneuronal connectivity is a challenge; its solution may provide insights into nervous system function that can be obtained in no other way.

ACKNOWLEDGMENTS I am deeply indebted to the members of the Division of Chemical Neurobiology for many fruitful discussions over the years. I am particularly grateful to Dr. Daniel Entingh, Dr. Adrian Dunn and Dr. John Wilson who critically read earlier versions of this manuscript.

REFERENCES

ADAIR, L. B., J. E. WILSON, and E. GLASSMAN, 1968a. Brain function and macromolecules. IV. Uridine incorporation into polysomes of mouse brain during different behavioral experiences. *Proc. Nat. Acad. Sci. USA* 61:917–922.

ADAIR, L. B., J. E. WILSON, J. W. ZEMP, and E. GLASSMAN, 1968b. Brain function and macromolecules. III. Uridine incorporation into polysomes of mouse brain during short-term avoidance conditioning. *Proc. Nat. Acad. Sci. USA* 61: 606–613.

AGRANOFF, B. W., 1972. Further studies on memory formation in the goldfish. In *The Chemistry of Mood, Motivation, and Memory*, J. L. McGaugh, ed. New York: Plenum Press, pp. 175–185.

APPEL, S. H., W. DAVIS, and S. SCOTT, 1967. Brain polysomes: Response to environmental stimulation. *Science* 157:836–838.

APRISON, M. H., T. KARIYA, J. N. HINGTGEN, and M. TORU, 1968. Neurochemical correlates of behavior: Changes in acetylcholine, norepinephrine and 5-hydroxytryptamine concentrations in several discrete brain areas of the rat during behavioural excitation. *J. Neurochem.* 15:1131–1139.

AUBERT, X., B. CHANCE, and R. D. KEYNES, 1964. Optical studies of biochemical events in the electric organ of *Electrophorus*. *Proc. R. Soc. Lond. (Biol.)* 160:211–245.

BARONDES, S. H., 1970. Multiple steps in the biology of memory. In *The Neurosciences: Second Study Program*, F. O. Schmitt, ed. New York: The Rockefeller University Press, pp. 272–278.

BARONDES, S. H., and H. D. COHEN, 1968. Arousal and the conversion of "short-term" to "long term" memory. *Proc. Nat. Acad. Sci. USA* 61:923–929.

BASKIN, F., F. R. MASIARZ, and B. W. AGRANOFF, 1972. Effect of various stresses on the incorporation of [^{3}H]orotic acid into goldfish brain RNA. *Brain Res.* 39:151–162.

BATESON, P. P. G., G. HORN, and S. P. R. ROSE, 1972. Effects of early experience on regional incorporation of precursors into RNA and protein in the chick brain. *Brain Res.* 39:449–465.

BAUM, G., and F. B. WARD, 1971. General enzyme studies with a substrate-selective electrode: Characterization of cholinesterases. *Anal. Biochem.* 42:487–493.

BONDESON, C., A. EDSTROM, and A. BEVIZ, 1967. Effects of different inhibitors of protein synthesis on electrical activity in the spinal cord of fish. *J. Neurochem.* 14:1032–1034.

BOWMAN, R. E., and G. HARDING, 1969. Protein synthesis during learning. *Science* 164:199–201.

CALISSANO, P., B. W. MOORE, and A. FRIESEN, 1969. Effect of calcium ion on S-100, a protein of the nervous system. *Biochem.* 8:4318–4326.

CASTELLUCCI, V. F., E. R. KANDEL, and J. H. SCHWARTZ, 1972. Macromolecular synthesis and the functioning of neurons and synapses. In *Structure and Function of Synapses*, G. D. Pappas and D. P. Purpura, eds. New York: Raven Press, pp. 193–219.

CEDAR, H., E. R. KANDEL, and J. H. SCHWARTZ, 1972a. Cyclic AMP in the nervous system of *Aplysia californica*. I. Increased synthesis in response to synaptic stimulation. *J. Gen. Physiol.* 60:558–569.

CEDAR, H., and J. H. SCHWARTZ, 1972b. Cyclic AMP in the nervous system of *Aplysia californica*. II. Effect of serotonin and dopamine. *J. Gen. Physiol.* 60:570–587.

CLARK, H. R., and A. STRICKHOLM, 1971. Evidence for a conformational change in nerve membrane with depolarization. *Nature (Lond.)* 234:470–471.

CUMMINS, J. T., 1971. Spectral changes in respiratory intermediates of brain cortex in response to depolarizing pulses. *Biochim. Biophys. Acta* 253:39–45.

CUMMINS, J. T., and R. BULL, 1971. Spectrophotometric measurements of metabolic responses in isolated rat brain cortex. *Biochim. Biophys. Acta* 253:29–38.

DANIELS, D., 1971a. Acquisition, storage, and recall of memory for brightness discrimination by rats following intracerebral infusion of acetoxycycloheximide. *J. Comp. Physiol. Psychol.* 76:110–118.

DANIELS, D., 1971b. Effects of actinomycin D on memory and brain RNA synthesis in an appetitive learning task. *Nature (Lond.)* 231:395–397.

DENTI, A., J. L. McGAUGH, P. W. LANDFIELD, and P. G. SHINKMAN, 1970. Effects of posttrial electrical stimulation of the mesencephalic reticular formation on avoidance learning in rats. *Physiol. Behav.* 5:659–662.

DORFMAN, L. J., and M. E. JARVIK, 1968. Comparative amnesic

effects of transcorneal and transpinnate ECS in mice. *Physiol. Behav.* 3:815–818.

DURELL, J., J. T. GARLAND, and R. O. FRIEDEL, 1969. Acetylcholine action: Biochemical aspects. *Science* 165:862–866.

EDSTROM, J.-E., and W. GRAMPP, 1965. Nervous activity and metabolism of ribonucleic acids in the crustacean stretch receptor neuron. *J. Neurochem.* 12:735–741.

EMSON, P., R. J. WALKER, and G. A. KERKUT, 1971. Chemical changes in a molluscan ganglion associated with learning. *Comp. Biochem. Physiol.* 40B:223–239.

ERICKSON, C. K., and J. B. PATEL, 1969. Facilitation of avoidance learning by posttrial hippocampal electrical stimulation. *J. Comp. Physiol. Psychol.* 68:400–406.

FLEXNER, L. B., 1967. Memory in mice dissected with antibiotics. *Amer. J. Dis. Child.* 114:574–580.

FLEXNER, L. B., J. B. FLEXNER, and R. B. ROBERTS, 1967. Memory in mice analyzed with antibiotics. *Science* 155:1377–1383.

GAINER, H., 1972a. Effects of experimentally induced diapause on the electrophysiology and protein synthesis patterns of identified molluscan neurons. *Brain Res.* 39:387–402.

GAINER, H., 1972b. Electrophysiological behavior of an endogenously active neurosecretory cell. *Brain Res.* 39:403–418.

GAINER, H., 1972c. Patterns of protein synthesis in individual, identified molluscan neurons. *Brain Res.* 39:369–385.

GILMAN, A. G., and M. NIRENBERG, 1971. Regulation of adenosine 3′,5′-cyclic monophosphate metabolism in cultured neuroblastoma cells. *Nature (Lond.)* 234:356–357.

GISIGER, V., 1971. Triggering of RNA synthesis by acetylcholine stimulation of the postsynaptic membrane in a mammalian sympathetic ganglion. *Brain Res.* 33:139–146.

GISIGER, V., and A.-C. GAIDE-HUGUENIN, 1969. Effect of preganglionic stimulation upon RNA synthesis in the isolated sympathetic ganglion of the rat. *Prog. Brain Res.* 31:125–129.

GLASSMAN, E., 1969. The biochemistry of learning: An evaluation of the role of protein and nucleic acids. *Ann. Rev. Biochem.* 38:605–646.

GLASSMAN, E., W. H. GISPEN, R. PERUMAL, B. MACHLUS, and J. E. WILSON, 1973. The effect of short experiences on the incorporation of radioactive phosphate into synaptosomal and non-histone acid-extractable nuclear proteins from rat and mouse brain., *5th Int. Congr. Pharmacol.* 4: (in press).

GLICKMAN, S. E., 1958. Deficits in avoidance learning produced by stimulation of the ascending reticular formation. *Can. J. Psychol.* 12:97–102.

GOLD, P. E., W. FARRELL, and R. A. KING, 1971. Retrograde amnesia after localized brain shock in passive-avoidance learning. *Physiol. Behav.* 7:709–712.

GROSSMAN, S. P., 1969. Facilitation of learning following localized intracranial injections of pentylenetetrazol. *Physiol. Behav.* 4:625–628.

GROSSMAN, S. P., and H. MOUNTFORD, 1964. Learning and extinction during chemically induced disturbance of hippocampal functions. *Amer. J. Physiol.* 207:1387–1393.

HALJAMÄE, H., and P. W. LANGE, 1972. Calcium content and conformational changes of S-100 protein in the hippocampus during training. *Brain Res* 38:131–142.

HAYWOOD, J., and S. P. R. ROSE, 1970. Effects of an imprinting procedure on RNA polymerase activity in the chick brain. *Nature (Lond.)* 228:373–374.

HERZ, M. J., and H. V. S. PEEKE, 1971. Impairment of extinction with caudate nucleus stimulation. *Brain Res.* 33:519–522.

HOKIN, L. E., 1969a. Phospholipid metabolism and functional activity of nerve cells. In *The Structure and Function of Nervous Tissue*, G. H. Bourne, ed. New York: Academic Press, pp. 161–184.

HOKIN, M. R., 1969b. Effect of norepinephrine on ^{32}P incorporation into individual phosphatides in slices from different areas of the guinea pig brain. *J. Neurochem.* 16:127–134.

HOKIN, L. E., 1965. Autoradiographic localization of the acetylcholine-stimulated synthesis of phosphatidylinositol in the superior cervical ganglion. *Proc. Nat. Acad. Sci. USA* 53:1369–1376.

HOKIN, L. E., 1966. Effects of acetylcholine on the incorporation of ^{32}P into various phospholipids in slices of normal and denervated superior cervical ganglia of the cat. *J. Neurochem.* 13:179–184.

HOKIN, M. R., L. E. HOKIN, and W. D. SHELP, 1960. The effects of acetylcholine on the turnover of phosphatidic acid and phosphoinositide in sympathetic ganglia, and in various parts of the central nervous system in vitro. *J. Gen. Physiol.* 44:217–226.

HORN, G., 1971. Biochemical, morphological and functional changes in the central nervous system associated with experience. *Activitas Nervosa Superior* 13:119–130.

HUDSPETH, W. J., and W. E. WILSONCROFT, 1969. Retrograde amnesia: Time dependent effects of rhinencephalic lesions. *J. Neurobiol.* 2:221–232.

HYDÉN, H., and P. W. LANGE, 1968. Protein synthesis in the hippocampal pyramidal cells of rats during a behavioral test. *Science* 159:1370–1373.

HYDÉN, H., and P. W. LANGE, 1970a. Protein changes in nerve cells related to learning and conditioning. In *The Neurosciences: Second Study Program*, F. O. Schmitt, ed. New York: The Rockefeller University Press, pp. 278–289.

HYDÉN, H., and P. W. LANGE, 1970b. Time sequence analysis of proteins in brain stem, limbic system and cortex during training. *Biochimica e Biologia Sperimentale* 9:275–285.

HYDÉN, H., and P. W. LANGE, 1972a. Protein changes in different brain areas as a function of intermittent training. *Proc. Nat. Acad. Sci. USA* 69:1980–1984.

HYDÉN, H., and P. W. LANGE, 1972b. Protein synthesis in hippocampal nerve cells during re-reversal of handedness in rats. *Brain Res.* 45:035–038.

ITOH, T., and J. H. QUASTEL, 1969. Ribonucleic acid biosynthesis in adult and infant rat brain in vitro. *Science* 164:79–80.

JARVIK, M. E., 1972. Effects of chemical and physical treatments on learning and memory. *Ann. Rev. Psychol.* 23:457–486.

JOHN, R. E., 1967. *Brain Mechanisms and Memory.* New York: Academic Press.

JONES, C. T., and P. BANKS, 1970a. The effect of electrical stimulation and ouabain on the uptake and efflux of L-[U-^{14}C]valine in chopped tissue from guinea-pig cerebral cortex. *Biochem. J.* 118:801–812.

JONES, C. T., and P. BANKS, 1970b. The effect of electrical stimulation on the incorporation of L-[U-^{14}C]valine into the protein of chopped tissue from guinea-pig cerebral cortex. *Biochem. J.* 118:791–800.

JONES, D. A., 1972. The relationship between amino acid incorporation into protein in isolated neocortex slices and the tissue content of free amino acid. *J. Neurochem.* 19:779–790.

JONES, D. A., and H. McILWAIN, 1971. Amino acid distribution and incorporation into proteins in isolated, electrically-stimulated cerebral tissues. *J. Neurochem.* 18:41–58.

Jones, D. A., and R. Rodnight, 1971. Protein-bound phosphorylserine in acid hydrolysates of brain tissue. The determination of [^{32}P]phosphorylserine by ion-exchange chromatography and electrophoresis. *Biochem. J.* 121:597–600.

Kakiuchi, S., and T. W. Rall, 1968a. The influence of chemical agents on the accumulation of adenosine 3′,5′-phosphate in slices of rabbit cerebellum. *Mol. Pharmacol.* 4:367–378.

Kakiuchi, S., and T. W. Rall, 1968b. Studies on adenosine 3′,5′-phosphate in rabbit cerebral cortex. *Mol. Pharmacol.* 4:379–388.

Kerkut, G. A., G. W. O. Oliver, J. T. Rick, and R. J. Walker, 1970. The effects of drugs on learning in a simple preparation. *Comp. Gen. Pharmac.* 1:437–483.

Kernell, D., and R. P. Peterson, 1970. The effect of spike activity versus synaptic activation on the metabolism of ribonucleic acid in a molluscan giant neuron. *J. Neurochem.* 17:1087–1094.

Kesner, R. P., and H. S. Conner, 1972. Independence of short- and long-term memory: A neural system analysis. *Science* 176:432–434.

Kesner, R. P., and R. W. Doty, 1968. Amnesia produced in cats by local seizure activity initiated from the amygdala. *Exp. Neurology* 21:58–68.

Klainer, L. M., Y.-M. Chi, S. L. Freidberg, T. W. Rall, and E. W. Sutherland, 1962. Adenyl cyclase. IV. The effects of neurohormones on the formation of adenosine 3′,5′-phosphate by preparations from brain and other tissues. *J. Biol. Chem.* 237:1239–1243.

Kohen, E., and C. Kohen, 1966. A study of mitochondrial-extramitochondrial interactions in giant tissue culture cells by microfluorimetry-microelectrophoresis. *Histochemie* 7:339–347.

Kuo, J.-F., T.-P. Lee, P. L. Reyes, K. G. Walton, T. E. Donnelly, and P. Greengard, 1972. Cyclic nucleotide-dependent protein kinases. X. An assay method for the measurement of guanosine 3′,5′-monophosphate in various biological materials and a study of agents regulating its levels in heart and brain. *J. Biol. Chem.* 247:16–22.

Lande, S., J. B. Flexner, and L. B. Flexner, 1972. Effect of corticotropin and desglycinamide⁹-lysine vasopressin on suppression of memory by puromycin. *Proc. Nat. Acad. Sci. USA* 69:558–560.

Larrabee, M. G., 1968. Transynaptic stimulation of phosphatidylinositol metabolism in sympathetic neurons in situ. *J. Neurochem.* 15:803–808.

Larrabee, M. G., and W. S. Leicht, 1965. Metabolism of phosphatidyl inositol and other lipids in active neurons of sympathetic ganglia and other peripheral nervous tissues. The site of the inositide effect. *J. Neurochem.* 12:1–13.

Lidsky, A., and B. M. Slotnick, 1970. Electrical stimulation of the hippocampus and electroconvulsive shock produce similar amnestic effects in mice. *Neuropsychologia* 8:363–369.

Llinas, R., J. R. Blinks, and C. Nicholson, 1972. Calcium transient in presynaptic terminal of squid giant synapse: Detection with aequorin. *Science* 176:1127–1129.

McAffee, D. A., M. Schorderet, and P. Greengard, 1971. Adenosine 3′,5′-monophosphate in nervous tissue: Increase associated with synaptic transmission. *Science* 171:1156–1158.

McBride, W. J., and J. D. Klingman, 1972. The effects of electrical stimulation and ionic alterations on the metabolism of amino acids and proteins in excised superior cervical ganglia of the rat. *J. Neurochem.* 19:865–880.

Machlus, B. J., J. E. Wilson, and E. Glassman, 1973a. Brain phosphoproteins. I. The effect of short experiences on the incorporation of radioactive phosphate into nuclear proteins of rat brain, (in preparation).

Machlus, B. J., J. E. Wilson, and E. Glassman, 1973b. Brain phosphoproteins. II. The effect of various behaviors and reminding experiences on the incorporation of radioactive phosphate, (in preparation).

MacInnes, J. W., E. H. McConkey, and K. Schlesinger, 1970. Changes in brain polyribosomes following an electroconvulsive seizure. *J. Neurochem.* 17:457–460.

Mahut, H., 1962. Effects of subcortical electrical stimulation on learning in the rat. *J. Comp. Physiol. Psychol.* 55:472–477.

Maitra, P. K., A. Ghosh, B. Schoener, and B. Chance, 1964. Transients in glycolytic metabolism following electrical activity in electrophorus. *Biochim. Biophys. Acta* 88:112–119.

Mark, R. F., and M. E. Watts, 1971. Drug inhibition of memory formation in chickens. I. Long-term memory. *Proc. R. Soc. Lond. (Biol.)* 178:439–454.

Moore, W. V., and D. B. Wetlaufer, 1971. The circular dichroism of synaptosomal membranes: Studies of interactions with neuropharmacological agents. *J. Neurochem.* 18:1167–1178.

Mulloney, B., and A. Selverston, 1972. Antidromic action potentials fail to demonstrate known interactions between neurons. *Science* 177:69–72.

Orrego, F., 1967. Synthesis of RNA in normal and electrically stimulated brain cortex slices in vitro. *J. Neurochem.* 14:851–858.

Orrego, F., and F. Lipmann, 1967. Protein synthesis in brain slices. Effects of electrical stimulation and acidic amino acids. *J. Biol. Chem.* 242:665–671.

Paolino, R. M., and H. M. Levy, 1971. Amnesia produced by spreading depression and ECS: Evidence for time-dependent memory trace localization. *Science* 172:746–749.

Peeke, H. V. S., and M. J. Herz, 1971. Caudate nucleus stimulation retroactively impairs complex maze learning in the rat. *Science* 173:80–82.

Prives, C., and J. H. Quastel, 1969. Effect of cerebral stimulation on biosynthesis of nucleotides and RNA in brain slices in vitro. *Biochim. Biophys. Acta* 182:285–294.

Quartermain, D., and B. S. McEwen, 1970. Temporal characteristics of amnesia induced by protein synthesis inhibitor: Determination by shock level. *Nature (Lond.)* 228:677–678.

Quevedo, J. C., P. G. Bosque, M. L. Andrés, and M. C. C. Garcia, 1971. Cytoplasmic RNA in the rat hippocampus after a learning experience: An autoradiographic study. *Acta Anat.* 79:360–366.

Rahmann, H., 1973. Radioactive studies of changes in protein metabolism by adequate and inadequate stimulation in the optic tectum of teleosts. In *Memory and Transfer of Information*, H. P. Zippel, ed. New York: Plenum Publishing Corporation (in press).

Randt, C. T., D. Quartermain, M. Goldstein, and B. Anagnoste, 1971. Norepinephrine biosynthesis inhibition: Effects on memory in mice. *Science* 172:498–499.

Roberts, E., and S. Matthysse, 1970. Neurochemistry: At the crossroads of neurobiology. *Ann. Rev. Biochem.* 39:777–820.

Rose, S. P. R., 1970. Neurochemical correlates of learning and environmental change. In *Short-term Changes in Neural Activity and Behaviour*, G. Horn and R. A. Hinde, eds. Cambridge, England: Cambridge University Press, pp. 517–550.

Rose, S. P. R., 1972. Changes in amino acid pools in the rat brain following first exposure to light. *Brain Res.* 38:171–178.

Schwartz, J. H., V. F. Castellucci, and E. R. Kandel, 1971. Functioning of identified neurons and synapses in abdominal ganglion of *Aplysia* in absence of protein synthesis. *J. Neurophysiol.* 34:939–953.

Serota, R. G., 1971. Acetoxycycloheximide and transient amnesia in the rat. *Proc. Nat. Acad. Sci. USA* 68:1249–1250.

Serota, R. G., R. B. Roberts, and L. B. Flexner, 1972. Acetoxycycloheximide-induced transient amnesia: Protective effects of adrenergic stimulants. *Proc. Nat. Acad. Sci. USA* 69:340–342.

Shimizu, H., C. R. Creveling, and J. W. Daly, 1970. Cyclic adenosine 3′,5′-monophosphate formation in brain slices: Stimulation by batrachotoxin, ouabain, veratridine, and potassium ions. *Mol. Pharmacol.* 6:184–188.

Skrzipek, K-H., 1969. Die proteinsynthese des tectum opticum in Abhängigkeit von der Gestalt intermittierender Lichtmuster bei *Carassius carassius* L. (Pisces). *J. Hirnforschung* 77:414–416.

Sokoloff, L., 1961. Local cerebral circulation at rest and during altered cerebral activity induced by anesthesia or visual stimulation. In *Regional Neurochemistry*, S. S. Kety and J. Elkes, eds. London: Pergamon Press, pp. 107–117.

Squire, L. R., and S. H. Barondes, 1972. Variable decay of memory and its recovery in cycloheximide-treated mice. *Proc. Nat. Acad. Sci. USA* 69:1416–1420.

Stein, D. G., and S. L. Chorover, 1968. Effects of posttrial electrical stimulation of hippocampus and caudate nucleus on maze learning in the rat. *Physiol. Behav.* 3:787–791.

Stoller, W., and M. J. Wayner, 1968. Lack of effect of neural activity on RNA synthesis in rat dorsal root ganglion cells. *Physiol. Behav.* 3:941–945.

Tasaki, I., L. Carnay, and A. Watanabe, 1969. Transient changes in extrinsic fluorescence of nerve produced by electric stimulation. *Proc. Nat. Acad. Sci. USA* 64:1362–1368.

Tasaki, I., A. Watanabe, and M. Hallett, 1972. Fluorescence of squid axon membrane labelled with hydrophobic probes. *J. Membrane Biol.* 8:109–132.

Thompson, R., 1958. The effect of intracranial stimulation on memory in cats. *J. Comp. Physiol. Psychol.* 51:421–426.

Ungar, G., 1972. Molecular approches to neural coding. *Int. J. Neurosci.* 3:193–200.

Uphouse, L. L., J. W. MacInnes, and K. Schlesinger, 1972a. Effects of conditioned avoidance training on polyribosomes of mouse brain. *Physiol. Behav.* 8:1013–1018.

Uphouse, L. L., J. W. MacInnes, and K. Schlesinger, 1972b. Uridine incorporation into polyribosomes of mouse brain after escape training in an electrified T-maze. *Physiol. Behav.* 8:1019–1023.

Vardaris, R. M., and K. E. Schwartz, 1971. Retrograde amnesia for passive avoidance produced by stimulation of dorsal hippocampus. *Physiol. Behav.* 6:131–135.

Watson, W. E., 1965. An autoradiographic study of the incorporation of nucleic-acid precursors by neurons and glia during nerve stimulation. *J. Physiol.* 180:754–765.

Watts, M. E., and R. F. Mark, 1971. Separate actions of ouabain and cycloheximide on memory. *Brain Res.* 25:420–423.

Wegener, G., 1970. Enhancement of protein synthesis in the optic tectum of frogs following light stimulation. An autoradiographic investigation. *Exp. Brain Res.* 10:363–379.

Wilson, D. L., 1971. Molecular weight distribution of proteins synthesized in single, identified neurons of *Aplysia*. *J. Gen. Physiol.* 57:26–40.

Wilson, D. L., and R. W. Berry, 1972. The effect of synaptic stimulation on RNA and protein metabolism in the R2 soma of *Aplysia*. *J. Neurobiol.* 3:369–379.

Woodson, P. B. J., W. T. Schlapfer, and S. H. Barondes, 1972. Postural avoidance learning in the headless cockroach without detectable changes in ganglionic cholinesterase. *Brain Res.* 37:348–352.

Wyers, E. J., and S. A. Deadwyler, 1971. Duration and nature of retrograde amnesia produced by stimulation of caudate nucleus. *Physiol. Behav.* 6:97–103.

Wyers, E. J., H. V. S. Peeke, J. S. Williston, and M. J. Herz, 1968. Retroactive impairment of passive avoidance learning by stimulation of the caudate nucleus. *Exp. Neurol.* 22:350–366.

Zemp, J. W., J. E. Wilson, K. Schlesinger, W. O. Boggan, and E. Glassman, 1966. Brain function and macromolecules. I. Incorporation of uridine into RNA of mouse brain during short-term training experience. *Proc. Nat. Acad. Sci. USA* 55:1423–1431.

Zornetzer, S. F., 1972. Brain stimulation and retrograde amnesia in rats: A neuroanatomical approach. *Physiol. Behav.* 8:239–244.

Zornetzer, S. F., and J. L. McGaugh, 1972. Electrophysiological correlates of frontal cortex-induced retrograde amnesia in rats. *Physiol. Behav.* 8:233–238.

58 Biochemical Correlates of Brief Behavioral Experiences

A. DUNN, D. ENTINGH, T. ENTINGH, W. H. GISPEN, B. MACHLUS, R. PERUMAL, H. D. REES, and L. BROGAN

ABSTRACT Biochemical changes associated with conditioned avoidance training in mice and rats have been investigated. Some of these changes are related to the training and some are not. The biochemical changes include phosphorylation of nuclear protein, incorporation of phosphate into synaptosomal protein, incorporation of glucosamine into gangliosides, incorporation of amino acids into protein, and incorporation of fucose into glycoprotein. The data indicate that brain metabolism responds in a variety of ways to brief environmental stimulation. Moreover, the chemical responses exhibit differing specificities with respect to the behavioral, anatomical, and temporal aspects of the stimulation, which may indicate that they are related to different aspects of the experience.

MEMBERS OF THE Division of Chemical Neurobiology of the University of North Carolina are studying biochemical changes that occur in the brains of mice and rats during and shortly after brief environmental stimulation or training procedures. Although the work is aimed at the eventual elucidation of biochemical events necessary for the formation of memory, effects of brief exposures to simple stimuli are also being studied. This broad approach has been adopted because learning is a complex process, and it is thought likely that particular components of the training situation, such as stress, arousal, attention, motivation, etc., although often considered to be extraneous to memory storage itself, may in fact be necessary aspects of the learning process. Similarly, since behavior is in general produced by interactions between the brain, the body, and its environment, metabolic responses of peripheral organs to environmental stimulation are being studied concurrently to provide clues for understanding the changes that occur in the brain.

Mice and rats were trained in an active avoidance task using footshock as an aversive stimulus. C57Bl/6J mice (♂, 6–8 weeks old; Jackson Laboratories) were trained for

A. DUNN, D. ENTINGH, T. ENTINGH, W. H. GISPEN, B. MACHLUS, R. PERUMAL, H. D. REES, and L. BROGAN. Division of Chemical Neurobiology, Department of Biochemistry, School of Medicine, University of North Carolina, Chapel Hill, North Carolina

15 min in the jump box in which the mice learn to jump to a safe shelf when a light and buzzer are activated in order to avoid an electric shock transmitted to the grid floor (Zemp et al., 1966; Wilson and Glassman, 1972). Yoked animals received the same stimuli in an adjacent compartment in the box but were unable to avoid the footshock. Quiet animals were left undisturbed after injection in individual cages in a separate room but in a normal laboratory environment. In some experiments mice were trained in an automated version of the above apparatus with essentially similar results. Rats were trained for 5 min in a step-up avoidance situation in which they learned to step onto a shelf to avoid a shock from the electrified grid floor (Coleman et al., 1971; Wilson and Glassman, 1972).

These behavioral tasks constitute a significant emotional experience in the lives of the animals, because they encounter novel stimuli that are quite unlike anything previously experienced. Thus the situation is probably very different from learning in the natural environment. The novelty of the experiences may explain the large magnitude and general anatomical distribution of some of the biochemical effects observed.

The metabolism of the animals has been studied primarily by the use of radioactive precursors of macromolecules. Macromolecules have been chosen for study because of their comparative stability and their structural and regulatory roles in the nervous system. In most cases a pool correction factor has been applied to derive the relative radioactivity (RR), defined as the radioactivity in the product macromolecule divided by that in a precursor molecule from the same tissue. This correction reduces the variability due to injection and partially corrects for variations in uptake of the precursor, but it should not be assumed in the absence of other data that changes in RR necessarily reflect changes in macromolecular synthesis. Changes in the time course of uptake or compartmentation phenomena may confound such a simplistic interpretation (see Glassman, this volume).

Table I

TABLE I

Behavior-related chemical changes

Chemical	$\dfrac{T}{Q}$	$\dfrac{T}{Y}$	$\dfrac{Y}{Q}$	$\dfrac{CC}{Q}$	$\dfrac{PT}{Q}$	$\dfrac{Ext}{Y}$	$\dfrac{Rem}{Q}$	Reference
Training-Related Responses								
RNA		+	0					Zemp et al., 1966
Polyribosomes†		+	0	0	0	+		Adair et al., 1968a, b
								Coleman et al., 1971
NAEP	+		0	0	+	+	+	Machlus, 1971
Synaptosomal								Perumal and Gispen,
phosphoprotein†	+	+	0		0	+		unpublished observations
Stimulus-Related Responses								
Protein	+	0	+		+*			Rees and Brogan,
								unpublished observations
Glycoprotein	+							Entingh and Entingh,
								unpublished observations
Uncharacterized								
Histones	—							Machlus, 1971
Gangliosides	+							Dunn and Hogan,
								unpublished observations

T: Trained; Q: Quiet; Y: Yoked; CC: Classically-conditioned; PT: Prior-Trained; Ext: Extinguished; Rem: Prior-trained Reminded.

*Increase smaller than in T/Q.

†This response is apparently related to *novel* training.

Significant differences between trained and quiet animals were observed in the radioactivity of isolated macromolecules with a number of different precursors when appropriate times of injection and sacrifice were chosen. However, there were several types of specificity in the metabolic changes. Changes were maximal at different times; they were manifested in different regions of the brain, and some were also present in organs other than the brain. Moreover, there was behavioral specificity. Yoked animals or shocked animals were biochemically similar either to quiet or to trained animals depending on the precursor. Thus the metabolic responses may, in the first instance, be classified as related to training or to general stimulation (Table I).

Phosphorylation of nuclear proteins

The phosphorylation of nuclear proteins in the rat was studied using a double isotope procedure injecting inorganic phosphate labeled with either ^{32}P or ^{33}P intracranially. Thirty minutes later one rat was trained for 5 min in the step-up avoidance situation and then immediately sacrificed. The brain was homogenized with that of a quiet rat that had been injected at the same time as the trained rat with the other phosphorus isotope. Isolated nuclei were extracted with acid and the proteins solubilized in this way separated into histone and nonhistone fractions by ion-exchange chromatography. Adenosine monophosphate (AMP) isolated from the

brain homogenate was used as the correction factor for the uptake of radioactive phosphate. The nonhistone acid-extractable nuclear protein (NAEP) from the brain of the trained rat was found to contain approximately twice as much radioactive phosphorus as that from the quiet rat (Table II). In contrast, the pooled histones from the trained rat contained only about 60% of the radioactive

TABLE II

Incorporation of ^{32}P or ^{33}P into rat brain NAEP during one-way step-up avoidance

Treatment	n	NAEP/AMP† Increase (%)
Trained/quiet	30	$+104 \pm 11$*
Yoked/quiet	20	-5 ± 59
Shocked/quiet	4	-11 ± 14
6-day prior-trained performing/quiet	8	$+95 \pm 4$*
6-day prior-trained reminded/quiet	8	$+78 \pm 14$*
6-day prior-trained handled/quiet	4	$+84 \pm 6$*

*$P < 0.01$ (2-tailed *t*-test).

†Mean $\pm$ standard deviation in this and all subsequent tables.

A dual-isotope method was used in which rats were injected intracranially with 100 μc $[^{32}P]H_3PO_4$ or $[^{33}P]H_3PO_4$. Thirty minutes later they were trained for 5 min and sacrificed immediately. Each brain was homogenized with the brain of a quiet rat that had received the other radioisotope. NAEP and AMP were isolated from the homogenate and their radioactivities determined. (Data from Machlus, 1971.)

phosphorus of those from the quiet rat (Machlus, 1971).

The time course of incorporation of radioactivity in quiet rats showed that after 10 min the radioactivity in both AMP and NAEP was approximately constant for up to 1 hr. The pronounced increase in labeling of the NAEP was apparently a rapid response to the training. When the training was prolonged for a further 5 min or when the animal was left quiet for 5 min after training before sacrifice, the increase in radioactivity in NAEP was not observed (Figure 1). Analysis by gross dissection indicated that the effect was confined to the basal forebrain.

Studies with degradative enzymes suggested that the phosphate was covalently attached to the protein. Moreover, the radioactivity in phosphoserine isolated following hydrolysis of the NAEP was increased after training. Even more significantly, the molar ratio of phosphoserine to serine in the NAEP increased in trained compared with quiet rats, *whether or not* they had been injected with radioactive precursor (Machlus, 1971). This result is consistent with the time course data and is important, because it indicates that the effect was not due to permeability or other changes affecting the radioactive precursors.

The increase of radioactive phosphate in NAEP was not observed in rats that had been yoked to the trained animal or had received random electric footshocks. However, the increase was observed in trained rats that had

been prior-trained for the 6 preceding days (Table II). Indeed, if on the seventh day the rat was not trained but merely placed in the conditioning apparatus (reminded), or merely handled 30 min after the injection without being placed in the apparatus, the radioactivity in NAEP was also increased (Table II). This response was specific for prior-trained rats and was not observed in animals that

TABLE III

Incorporation of ^{32}P or ^{33}P into mouse brain NAEP during training

Treatment	n	NAEP/AMP Increase (%)
1. Step-up avoidance		
Trained/quiet	4	$+67 \pm 5**$
2. Jump-box avoidance		
Trained/quiet	4	$+52 \pm 16*$
Classically-conditioned/quiet	4	-6 ± 15
6-day prior-trained performing/quiet	4	$+87 \pm 4**$
1-day prior-trained reminded/quiet	8	$+39 \pm 20**$

*P < 0.02.
**P < 0.01 (2-tailed *t*-test).

Mice were injected intracranially with 33 μc [^{32}P]H$_3$PO$_4$ or [^{33}P]H$_3$PO$_4$. Thirty minutes later they were trained for 5 min and then immediately sacrificed and NAEP and AMP isolated from the brain homogenate. In classically-conditioned animals the shock followed the light and buzzer, but mice were unable to avoid the shock. (Data from Machlus, 1971.)

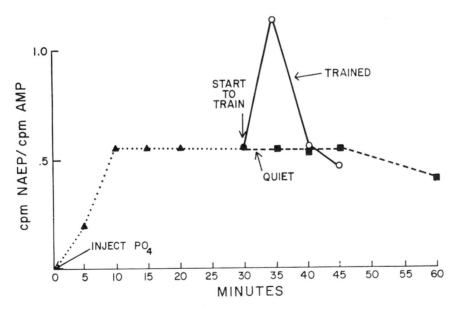

FIGURE 1 Time course of the incorporation of radioactive phosphate into nonhistone acid-extractable proteins (NAEP) of rat brain after intracranial injection. Rats were killed at various times after injection and NAEP and AMP isolated from the brain homogenate and their radioactivity determined. *Dotted line:* Quiet rats. *Solid line:* Rats trained 30 min after injection. For training details see text and the caption to Table II. (Data from Machlus, 1971.)

had been shocked instead of trained for 6 days and were treated similarly on the seventh day. This suggests that *after* training the response may be linked specifically with stimuli involved in the training, including even the handling involved in placing the animal on the grid floor of the training apparatus at the start of each training trial.

Mice trained for 5 min either in the step-up avoidance or the jump box showed a similar but somewhat smaller increase in labeling of NAEP with phosphate (Table III). In mice too, the response occurred in prior-trained animals but was not observed after a series of unavoidable footshocks.

Incorporation of phosphate into synaptosomal protein

The phosphorylation of synaptosomal proteins in mouse brain has been studied after 15 min of jump-box training. Using a double isotope procedure similar to that described above for NAEP, an increase of about 30% in the radioactive phosphate associated with synaptosomal proteins was observed in trained compared with quiet mice (Table IV). The effect was not observed in yoked mice. This increase in protein-bound radioactivity was associated with the membrane fraction of osmotically lysed synaptosomes. As in the case of NAEP, the increase occurred in the phosphoserine isolated following hydrolysis of the protein (Perumal and Gispen, unpublished observations).

Incorporation of glucosamine into gangliosides

Gangliosides are membrane constituents thought to be present at high concentrations at or near synapses. Their

TABLE IV

Incorporation of radioactive phosphate into mouse brain synaptosomal protein

Treatment	n	Synaptosomal Protein/AMP Increase (%)
Trained/quiet	21	$+32 \pm 30*$
Yoked/quiet	12	$+1 \pm 18$

*$P < 0.005$ (2-tailed t-test).

Mice were injected intracranially with 20 μc $[^{32}P]H_3PO_4$ or $[^{33}P]H_3PO_4$. Thirty minutes later they were trained for 15 min in the jump-box and sacrificed immediately. Synaptosomes were isolated and radioactivities in AMP and synaptosomal phosphoprotein determined. (Perumal and Gispen, unpublished observations.)

metabolism was studied by observing the incorporation of D[1-^{3}H]glucosamine. Training did not change the incorporation of ^{3}H into the residue left after lipid extraction (radioactivity mainly in glycoproteins, or into nonganglioside lipid fractions), but did consistently increase the incorporation of ^{3}H into the total ganglioside fraction by about 25% (Table V).

Incorporation of amino acids into protein

Changes in brain protein metabolism were studied by monitoring the incorporation of radioactive amino acids following jump-box training of mice. When [^{3}H]lysine was injected subcutaneously into mice 20 min after 15 min of jump-box training, the total ^{3}H content of the brain was significantly increased by 13%, 10 min later. When

TABLE V

Incorporation of D[1-^{3}H]glucosamine into mouse brain

Treatment	C-M Soluble	Nonlipid Residue	Lipids	Gangliosides
	Dpm $\times 10^{-3}$			
Trained	45.4 ± 9.3	255 ± 43	3.90 ± 0.40	2.84 ± 0.47
Yoked	43.3 ± 2.0	251 ± 42	3.94 ± 0.33	2.35 ± 0.47
Quiet	52.2 ± 13.3	248 ± 41	4.15 ± 0.63	2.26 ± 0.38
	Ratio			
Trained/quiet	0.87	1.03	0.94	1.26*
Trained/yoked	1.05	1.02	0.99	1.21
Yoked/quiet	0.83	1.01	0.95	1.04

*$P < 0.025$ (t-test),

$n = 6$ for each group.

Mice were injected subcutaneously with 25 μc D[l-^{3}H]glucosamine 15 min before 15 min jump-box training. They were sacrificed 30 min later and gangliosides extracted as described by Suzuki (1965). C-M soluble: Chloroform-methanol soluble (Folch extract); Nonlipid residue: Residue after Folch extraction (protein and glycoprotein); Lipids: Lower, hydrophobic phase (nonganglioside lipid); Gangliosides: Dialyzed upper phase (monitored by thin layer chromatography on silica gel). (Dunn and Hogan, unpublished observations.)

the RSA was calculated to correct for the increased uptake, there was still a statistically significant increase of ^{3}H incorporation into protein in trained animals compared with quiet (Table VI). The RR after the 10 min pulse was elevated to approximately the same extent when animals were sacrificed 15, 30, or 45 min after training. Similar increases in RR were observed in trained mice using [4,5-^{3}H]leucine or [1-^{14}C]leucine as precursors of protein.

TABLE VI

Incorporation of [4,5-^{3}H]lysine into mouse brain protein

Treatment	n	Mean RR	% Increase over Quiet
Quiet	52	0.127 ± 0.014	[0]
Trained	18	0.148 ± 0.014	+17**
Yoked	4	0.144 ± 0.016	+13*
30 Buzzers	13	0.146 ± 0.018	+15**
20 Shocks	14	0.143 ± 0.015	+13**
30 Lights	6	0.132 ± 0.009	+4

*P < 0.05; **P < 0.001 (2-tailed *t*-test).

Mice were trained or otherwise stimulated for 15 min in the jump box. Quiet animals were undisturbed prior to injection. Twenty minutes later they were injected subcutaneously with 30 μc [4,5-^{3}H]lysine and sacrificed 10 mins later. RR = dpm in dried trichloroacetic acid precipitate/dpm in dried trichloroacetic acid supernatant. Independent experiments have shown that more than 95% of the ^{3}H in the dried trichloroacetic acid supernatant was in lysine. (Rees and Brogan, unpublished observations.)

The effect, however, was not specific to training, and an effect of similar magnitude was observed in yoked mice (Table VI). When the various components of the training situation were tested separately, buzzers or shocks, but not lights, elicited the increased RR (Table VI). The effect may thus be regarded as stimulus-related rather than training-related.

The increased RR occurred generally throughout the brain (dissected into 6 parts) and was significant everywhere but in the dorsal cortex and thalamus-hypothalamus. The liver also showed an RR increase of greater magnitude than the brain under the same behavioral conditions. This liver response was maximal immediately after training and disappeared within 1 hr. The elevation of RR with stimulation was also observed in the brains and the livers of rats and of adrenalectomized mice.

Incorporation of fucose into glycoproteins

The metabolism of fucose-containing glycoproteins in brain was studied by following the incorporation of L[1-^{3}H]fucose. Jump-box training of mice produced an increase of the incorporation of radioactivity into glycoprotein of about 20–30%, relative to quiet mice. Yoked mice also showed this change suggesting that it was not specific to the training aspects of the experience. Maximal effects were observed when fucose was injected shortly before the training and the animals sacrificed very soon after its completion (Entingh and Entingh, unpublished observations).

Conclusions

Taken together, these data indicate that diverse metabolic changes occur in the brains of mice and rats undergoing avoidance training. The characteristics of these various responses are dissimilar, and they are thus not due to a single general disturbance of cerebral metabolism. These biochemical responses are summarized in Table I, where they have been divided into two categories, *training-related* and *stimulus-related*. Training-related responses are observed in trained but not in yoked animals. Stimulus-related responses appear in both trained and yoked animals. In addition two types of training-related responses may be differentiated. The change in the phosphorylation of NAEP occurs only in animals that have been trained but may be induced by subsequent related experiences. In contrast the increased radioactivity in polyribosomes after administration of radioactive uridine (Adair et al., 1968a, 1968b) does not occur when prior-trained mice perform the conditioned avoidance. It does, however, occur when prior-trained mice extinguish their avoidance behavior (Coleman et al., 1971). Thus the effect appears to be related to *novel* training. There are insufficient behavioral data to classify some of the metabolic responses, and other categories may exist.

Anatomically the effects apparently occurred in large regions of the brain, but there was regional specificity. Most of the increase of uridine incorporation into RNA was confined to the subcortical forebrain (Zemp et al., 1967), as was the increased incorporation of phosphate into NAEP; whereas the lysine incorporation response was more diffuse, with maximal effects in ventral cortex, basal ganglia, septum and hippocampus, and cerebellum and brainstem. Bowman and Strobel (1969) and Bateson et al. (1972) have also observed anatomical specificity in their behavior-related biochemical changes in rats and chicks respectively.

Specificity is also apparent in the time relationships between behavioral and metabolic events. The change in the incorporation of [^{3}H]fucose into glycoprotein was maximal when the pulse was confined to the training period itself and the changes in the incorporation of uridine into RNA and polyribosomes, and the phosphorylation of

NAEP were largest when the animals were sacrificed immediately after training. However, the increase in [^{3}H]lysine incorporation occurred after training or stimulation was completed.

Further work is needed to characterize biochemically the responses to environmental stimulation. Although measurements with radioisotopes provide ambiguous data, since observed differences in radioactivity may be due to changes in blood flow or uptake, the changes, whatever their nature, are behaviorally and anatomically specific. At present, only one biochemical change, the increase in phosphoserine content of NAEP, has been detected without the use of radioactive precursors. However, the isotope data will give clues to the occurrence of metabolic changes, which may then be amenable to investigation by other means. Characterization of these changes may then suggest their immediate cause and possible biochemical consequences. In this way a clear understanding of the biochemical responses of the brain to environmental stimulation may be generated. It is hoped that this will eventually lead to the elucidation of the molecular events underlying the plasticity of the brain.

ACKNOWLEDGMENTS   The contributions of Drs. E. Glassman, J. E. Wilson, and E. Hogan in many fruitful discussions are gratefully acknowledged. This research was supported by grants from the U.S. Public Health Service (MH 18136, NS 07457); the U.S. National Science Foundation (GB 18551); the Ciba-Geigy Corporation; and a Faculty Grant (AF 339) from the School of Medicine, University of North Carolina.

REFERENCES

ADAIR, L. B., J. E. WILSON, J. W. ZEMP, and E. GLASSMAN, 1968a. Brain function and macromolecules. III. Uridine incorporation into polysomes of mouse brain during short-term avoidance conditioning. *Proc. Nat. Acad. Sci. USA* 61: 606–613.

ADAIR, L. B., J. E. WILSON, and E. GLASSMAN, 1968b. Brain function and macromolecules. IV. Uridine incorporation into polysomes of mouse brain during different behavioral experiences. *Proc. Nat. Acad. Sci. USA* 61:917–922.

BATESON, P. P. G., G. HORN, and S. P. R. ROSE, 1972. Effects of early experience on regional incorporation of precursors into RNA and protein in the chick brain. *Brain Res.* 39:449–465.

BOWMAN, R. E., and D. A. STROBEL, 1969. Brain RNA metabolism in the rat during learning. *J. Comp. Physiol. Psychol.* 67:448–456.

COLEMAN, M. S., J. E. WILSON, and E. GLASSMAN, 1971. Incorporation of uridine into polysomes of mouse brain during extinction. *Nature (Lond.)* 229:54–55.

MACHLUS, B., 1971. Phosphorylation of nuclear proteins during behavior of rats. Ph.D. thesis, University of North Carolina, Chapel Hill, North Carolina.

SUZUKI, K., 1965. The pattern of mammalian brain gangliosides II. Evaluation of the extraction procedures, post-mortem changes and the effect of formalin preservation. *J. Neurochem.* 15:81–85.

WILSON, J. E., and E. GLASSMAN, 1972. The effect of short-term training experiences on the incorporation of radioactive precursors into RNA and polysomes of brain. In *Methods of Neurochemistry*, Vol. II, R. Fried, ed. New York: Marcel Dekker, pp. 53–72.

ZEMP, J. W., J. E. WILSON, K. SCHLESINGER, W. O. BOGGAN, and E. GLASSMAN, 1966. Brain function and macromolecules, I. Incorporation of uridine into RNA of mouse brain during a short-term behavioral experience. *Proc. Nat. Acad. Sci. USA* 55:1423–1431.

ZEMP, J. W., J. E. WILSON, and E. GLASSMAN, 1967. Brain function and macromolecules, II. Site of increased labeling of RNA of brains of mice during a short-term training experience *Proc. Nat. Acad. Sci. USA* 58:1120–1125.

59 Nutrition and the Brain

R. J. WURTMAN and J. D. FERNSTROM

ABSTRACT The relationship of nutritional status, especially as regards protein ingestion, to mental function is explored. Neurochemical mechanisms whereby alterations in nutrition might affect mentation are surveyed. As a paradigm, the regulation of brain monoamine synthesis, especially serotonin, by blood and brain tryptophan is presented.

Introduction

IF THE ULTIMATE task of genetic mechanisms is to endow the cell with enzymes, the job of nutrition is to provide the substrates. These two tasks are not completely dissociated; enzymes are made from, induced by, and activated by nutritional inputs. However, a rough analogy can still be made: Genetics is to enzymes as nutrition is to substrates.

Considered in this perspective, the topic *Nutrition and the Brain* encompasses vastly more than the evidence that fish is or is not "brain food," or that macrobiotic diets do or do not predispose to transcendental insights or alpha rhythms: It includes all of those behavioral and biochemical processes that provide brain cells with compounds that they need but are unable to make for themselves.

Between the performance of the primary nutritional act—the eating or drinking or inhalation of the needed compound or its precursor—and the time that the nutrient interacts with the brain enzyme, a host of metabolic processes may intervene (Figure 1). For example, if the ingested material is a polymer (such as starch or protein), or part of a large, heterogeneous molecule (such as the trace metals bound to enzymes), it must first be digested, or dissociated into subunits, before it can be absorbed. Once it enters the circulation, the nutrient must be transported to the brain. It can do this dissolved in the plasma (e.g., glucose) or sequestered within erythrocytes (e.g., oxygen), or bound to proteins (e.g., fatty acids). En route, it may temporarily be converted to an insoluble storage form and retained within a reservoir of a tissue (e.g., the iron bound to ferritin in the liver) or within an extracellular space (e.g., the calcium in bone); or its chemical structure may be modified in the liver or elsewhere (e.g., the conversion of iodine to thyroxine in the thyroid gland). In addition, the level, or concentration, of the nutrient within the circulation may be maintained within a relatively narrow range through the operation of a *regulatory system* (e.g., glucose, calcium), or may simply reflect the vector sum of the various processes that are adding the nutrient to, and removing it from, the blood stream(e.g., amino acids, vitamin A). The ultimate entry of the nutrient into brain cells may involve simple diffusion, or be mediated by a specific transport system whose activity may be inhibited by competing substrates (e.g., related amino acids in the plasma), or stimulated by hormones or neurotransmitters.

Most of the evidence that nutritional state affects the brain has been derived from studies in which a particular nutrient was *chronically* underrepresented in the diet (e.g., thiamine in Wernicke's encephalopathy), or in which amounts of the nutrient that the test subject happened to require were unusually high (e.g., protein during the stage of brain development characterized by neuronal hyperplasia). This direction in nutrition research probably reflected the recognition that organisms contain substantial reservoirs for most of the nutrients that brain and other tissues require, and from the need to develop treatments for clinical syndromes (e.g., kwashiorkor, endemic goiter) resulting from human malnutrition. The absence of gross physiologic disturbances in subjects deprived of particular nutrients (e.g., glucose) for a few days probably reinforced this research strategy.

Such long-term studies have indeed demonstrated that chronic malnutrition can cause major and sometimes irreversible changes in brain weight, myelination, and even catecholamine content. However, more recent observations, summarized in this report, indicate that the brain can *also* respond to more subtle, transient changes in nutritional state. Specifically, the rate at which brain neurons synthesize serotonin, a putative neurotransmitter, is controlled on an hour-to-hour basis by their concentration of tryptophan; this concentration, in turn, varies characteristically depending on the composition of the food that has most recently been eaten (hence behavior, specifically feeding behavior, can control brain biochemistry, as well as vice versa). Paradoxically, the

R. J. WURTMAN and J. D. FERNSTROM   Laboratory of Neuroendocrine Regulation, Department of Nutrition and Food Science, Massachusetts Institute of Technology, Cambridge, Massachusetts

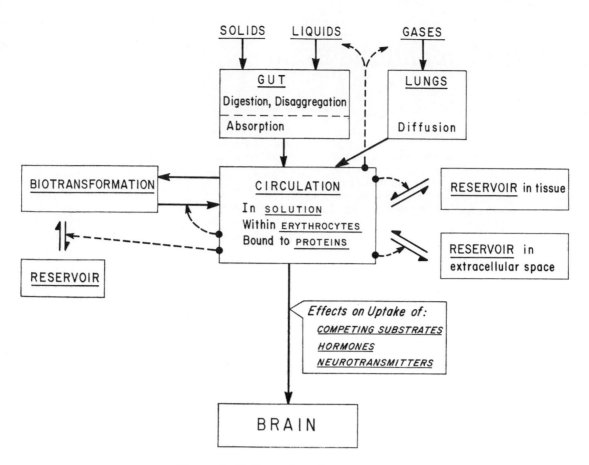

FIGURE 1   Delivery of nutrients to the brain.

quantities of free tryptophan available to the brain bear no simple relationship to the total amount of this amino acid present in the rest of the organism: As described below, protein-free diets can, acutely, cause greater elevations in brain tryptophan than diets containing protein. Brain nutrition need not parallel the general nutritional state.

Until fairly recently, it was widely held that the enormous importance of the brain conferred upon it a unique metabolic prerogative: The ability to take from the blood as much oxygen, glucose, or whatever else it needed, at rates that were more or less independent of the concentrations of these nutrients in the plasma. Under that formulation, a substrate or cofactor could become limiting for neurons only if it had to be synthesized within each neuron (e.g., S-adenosylmethionine; ATP) or if its supply in the body approached exhaustion. The clear dependence of brain serotonin upon plasma amino acid concentrations renders this earlier view obsolete. The normal brain is at least as responsive to transient changes in its nutritional state as any other organ in the body.

Control of brain serotonin by food consumption

Several years ago, we observed that the concentrations of tryptophan and of most other amino acids in human plasma undergo characteristic and parallel fluctuations during each 24-hr period (Wurtman et al., 1968; Wurtman, 1970); tryptophan levels are lowest at 2:00 to 4:00 AM, and rise 50 to 80% to attain a plateau late in the morning or early afternoon. The amplitude of the rhythm exhibited by any particular amino acid tends to vary inversely with its availability in the body; concentrations of relatively scarce amino acids (e.g., methionine, cysteine) rise and fall by as much as twofold during each 24-hr period, while the more abundant amino acids (e.g., glycine, glutamate) shift only 10 to 30%. Similar rhythms in the plasma concentrations of tryptophan and other amino acids were also observed in rats and mice (Figure 2); however, peak tryptophan levels, 35 to 150% above those at the daily nadir, occurred about 8 to 10 hr later than the peak in humans, a phenomenon probably attributable to the rat's tendency to consume most of its food during the hours surrounding the onset of the daily period of darkness. The plasma amino acid rhythms in humans were not simply the result of the cyclic ingestion of dietary protein, inasmuch as they persisted among volunteers who ate essentially no protein for 2 weeks (Wurtman et al., 1968). However, they disappeared in subjects placed on a total fast (Marliss et al., 1970); this finding suggested that they were *not* truly circadian or of endogenous origin, and that the most important factor in their genesis was nutritional, i.e., the release of insulin, and possibly other hormones, in response to dietary carbohydrate. (Insulin would be expected to raise or lower the plasma concentrations of amino acids by controlling their flux into muscle and other intracellular compartments.)

The mere existence of plasma amino acid rhythms did not establish that such variations were of any consequence physiologically. To explore their possible significance, we set out to determine whether the naturally occurring daily fluctuations in the plasma concentration of a particular amino acid could actually influence its metabolic fate. This task could be accomplished by demonstrating that experimentally induced fluctuations of the same magnitude as those occurring diurnally caused parallel changes in the rate at which the amino acid is used for some purpose in the body (e.g., incorporation into proteins, conversion to a low molecular weight compound). The amino acid whose plasma concentration seemed most likely to influence its metabolic fate was tryptophan. The quantities of free and peptide-bound tryptophan present in the organism and in most foods were known to be the lowest of all the amino acids (Wurtman and Fernstrom, 1972); moreover, evidence has already been obtained that daily rhythms in the ingestion of tryptophan-containing proteins (and, presumably, in the concentration of tryptophan within portal venous blood) cause parallel rhythms in the aggregation of hepatic polysomes (Fishman et al., 1969), and in the activity and the synthesis of a specific liver protein, the enzyme tyrosine transaminase (Wurtman, 1970).

As the dependent variable in our study, we chose to look for possible changes in brain serotonin content in rats exhibiting a spontaneous daily rhythm in plasma tryptophan, or treated in such a way as to raise or lower plasma tryptophan. That the amount of tryptophan available to the brain might control serotonin synthesis was suggested by three lines of evidence: (1) the existence in rats and mice of diurnal rhythm in brain serotonin content (Figure 2) (Albrecht et al., 1956); (2) the unusually high Km for tryptophan shown by tryptophan hydroxylase, the enzyme that catalyzes the initial step in serotonin biosynthesis (Lovenberg et al., 1968); and (3) the repeated demonstrations that very high doses of tryptophan (i.e., 50 to 1600 mg/kg) could cause large increases in the brain concentrations of serotonin and its chief metabolite, 5-hydroxyindole acetic acid (5-HIAA) (Wurtman and Fernstrom, 1972).

Initial experiments were designed to determine whether brain serotonin could be increased by giving rats very low doses of tryptophan at a time of day (3 hr after the onset of the daily light period) when plasma and brain tryptophan, and brain serotonin concentrations are known to be low (Figure 1). It was hoped that a dose of tryptophan could be identified which (a) raised brain serotonin but (b) was smaller than the amount of the amino acid normally consumed by rats each day, and (c) did not elevate plasma and brain tryptophan levels beyond their normal daily peaks. All of these goals were

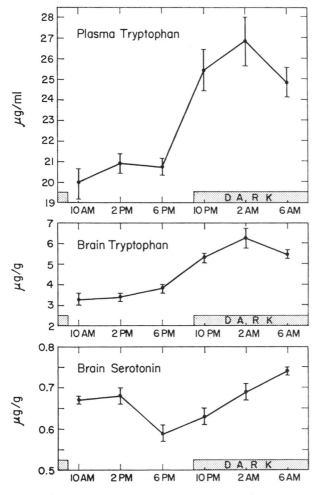

FIGURE 2 Daily rhythms in plasma tryptophan, brain tryptophan, and brain 5-HT. Groups of 10 rats kept in darkness from 9 PM to 9 AM were killed at intervals of 4 hr. Vertical bars indicate standard errors of the mean. (Reproduced from Wurtman and Fernstrom, 1972.)

met by administering 12.5 mg/kg i.p. of the amino acid to male rats weighing 150 to 200 g. This dose, constituting less than 5% of the amount of tryptophan that rats would be expected to ingest each day in 10 to 20 g of standard rat chow, produced peak elevations in plasma and brain tryptophan that were well within the ranges that occurred nocturnally in untreated animals, and caused brain serotonin levels to rise by 20 to 30% ($P < 0.01$) within 1 hr (Fernstrom and Wurtman, 1971a). Doses of 25 mg/kg caused elevations in both brain tryptophan and brain serotonin which were about double those observed in animals receiving the lower dose. Larger doses of tryptophan caused proportionate increases in brain tryptophan concentration but had little further effect on brain serotonin (Figure 3). Studies by others (Moir and Eccleston, 1968) have shown that further increments in tryptophan dose continue to produce increases in brain 5-HIAA, and thus presumably in serotonin biosynthesis. The failure of brain serotonin content to continue to rise when plasma and brain tryptophan are elevated beyond their normal dynamic ranges thus most likely reflects limitations in the ability of serotonin-producing neurons to store the amine.

The observed increase in brain serotonin caused by injecting very small doses of tryptophan was thought to be compatible with the hypothesis that the nocturnal rise in brain serotonin in normal rats is related to the daily rhythms in plasma and brain tryptophan (see Figure 2). It does not seem necessary to conclude that this substrate-induced rhythm in serotonin synthesis is the *only* factor responsible for the daily rhythm in brain serotonin content; for example, it is also possible that the serotonin rhythm reflects changes in the rates at which the mono-amine is released from neurons or metabolized intraneuronally.

Now that small increases in plasma tryptophan had been shown to cause parallel changes in brain serotonin, it became of interest to determine whether physiological decreases in the plasma amino acid also lowered the serotonin content. Hence we attempted to lower plasma tryptophan by giving rats insulin. It had not actually been shown that exogenous insulin lowers the plasma tryptophan concentration in rats, probably because a simple assay for tryptophan had become available only a few years previously. However, there was abundant evidence that insulin exerts this effect on almost all other amino acids examined, largely by enhancing their uptake into skeletal muscle (Wool, 1965).

Rats similar to those used in the previous experiments received a dose of insulin (2 units/kg i.p.) known to lower blood glucose levels. To our surprise, the hormone did not lower plasma tryptophan, but instead *increased* its concentration by 30 to 50% (Fernstrom and Wurtman, 1972a). This effect was independent of the route by which the insulin was administered; it was associated with a 55% fall in plasma glucose and with major reductions in the plasma concentrations of most other amino acids (Table I), including the neutral amino acids generally believed to compete with tryptophan for uptake into the brain (Blasberg and Lajtha, 1965; Guroff and Udenfriend, 1962). Two hours after rats received the insulin, brain tryptophan levels were elevated by 36% ($P < 0.001$), and brain serotonin by 28% ($P < 0.01$) (Fernstrom and Wurtman, 1972a).

The increase in brain serotonin content observed in rats receiving insulin could have resulted not from in-

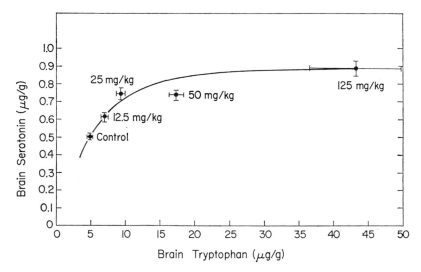

FIGURE 3 Dose-response curve relating brain tryptophan and brain 5-HT. Groups of 10 rats received tryptophan intraperitoneally at noon, and were killed 1 hr later. All brain tryptophan levels were significantly higher than control levels ($P < 0.01$). (Reproduced from Fernstrom and Wurtman, 1971b.)

TABLE I

Effect of insulin on concentrations of tryptophan, other amino acids, and glucose in rat plasma

Amino Acid	Concentration		Percent Change
	Control	Insulin	
	(μg/ml)		
L-Tryptophan	11.1 $\pm$ 0.6	16.6 $\pm$ 0.9	+50*
L-Tyrosine	11.4 $\pm$ 0.8	9.4 $\pm$ 1.0	−18
L-Phenylalanine	11.8 $\pm$ 0.3	10.8 $\pm$ 0.9	−9
L-Serine	24.6 $\pm$ 1.2	17.0 $\pm$ 1.3	−31*
L-Glycine	29.6 $\pm$ 1.4	18.3 $\pm$ 2.0	−38*
L-Alanine	25.3 $\pm$ 0.9	11.7 $\pm$ 0.5	−54*
L-Valine	21.5 $\pm$ 1.3	17.3 $\pm$ 1.8	−20
L-Isoleucine	13.7 $\pm$ 0.8	7.1 $\pm$ 0.4	−48*
L-Leucine	21.3 $\pm$ 1.2	16.6 $\pm$ 2.1	−22
	(mg/100 ml)		
Glucose	96.0 $\pm$ 2.5	43.0 $\pm$ 4.8	−55*

*$P < 0.01$.

Groups of five to ten fasting 150 to 200-g rats were killed 2 hr after receiving insulin (2 units/kg i.p.). (Reproduced from Fernstrom and Wurtman, 1972a.)

creased availability of substrate but from reflexes activated by the accompanying hypoglycemia. To determine whether the physiological secretion of insulin, in normoglycemic animals, also increases plasma and brain tryptophan concentrations and brain serotonin, these indoles were measured in rats fasted for 15 hr and then given access to a carbohydrate diet. In a typical experiment, the animals ate an average of 5 g/hr during the first hour, and 2 g/hr during the second and third hours (Fernstrom and Wurtman, 1971b, 1972a). Plasma tryptophan levels were significantly elevated 1, 2, and 3 hr after food presentation; tyrosine concentrations were depressed at all three times studied (Table II). Brain tryptophan concentrations rose 22% during the first hour, reached a peak 65% above control values ($P < 0.001$) after 2 hr, and remained significantly elevated at 3 hr (Table II). Brain serotonin concentrations rose dur-

ing the first hour, became significantly elevated by the end of the second hour, and remained so after 3 hr (Table II) (Fernstrom and Wurtman, 1971b).

On the basis of these observations, a model was constructed to explain the mechanisms by which dietary inputs affect brain serotonin. According to this model, carbohydrate consumption will, by eliciting insulin secretion, raise plasma tryptophan levels (by mobilizing the amino acid from yet undefined pools); this elevation, in turn, will cause a corresponding increase in brain tryptophan, which will increase the saturation of tryptophan hydroxylase, increase serotonin synthesis, and, ultimately, increase brain serotonin levels. On the basis of this model, we predicted that the consumption of diet containing both carbohydrates and protein would cause an even greater rise in brain serotonin. In addition to elevating plasma tryptophan via insulin secretion, the tryptophan molecules in the dietary proteins would also contribute directly to plasma tryptophan; hence plasma tryptophan concentrations would increase even more than after ingestion of a protein-free diet, and brain tryptophan and serotonin would show similar amplifications in response. When this model was tested by giving fasted rats access to diets containing either casein or a synthetic amino acid similar in composition to 18% casein, it was immediately apparent that it was in need of major revision: As expected, protein consumption was followed by a major increase (about 60%, $P < 0.001$) in plasma tryptophan; however neither brain tryptophan nor brain serotonin was at all increased (Fernstrom and Wurtman, 1972b).

Other investigators, using brain slices (Blasberg and Lajtha, 1965) or animals treated with pharmacological doses of individual amino acids (Guroff and Udenfriend, 1962) had shown that groups of amino acids (e.g., neutral, acidic, basic) are transported into brain by specific carrier systems, and that within a group, the member amino acids compete with each other for common transport

TABLE II

Effect of carbohydrate ingestion on brain 5-HT concentrations and on plasma and brain tryptophan

	Time After presentation of Food (hr)			
	0	1	2	3
Plasma tryptophan	10.86 $\pm$ 0.55	13.56 $\pm$ 0.81**	14.51 $\pm$ 0.70***	13.22 $\pm$ 0.65**
Brain tryptophan	6.78 $\pm$ 0.40	8.32 $\pm$ 0.63*	11.24 $\pm$ 0.52***	9.81 $\pm$ 0.50***
Brain 5-HT	0.549 $\pm$ 0.015	0.652 $\pm$ 0.046	0.652 $\pm$ 0.012***	0.645 $\pm$ 0.017***
Plasma tyrosine	13.03 $\pm$ 0.29	9.55 $\pm$ 0.34***	8.67 $\pm$ 0.26***	9.03 $\pm$ 0.21***

*$P < 0.05$ differs from 0-time group.

**$P < 0.02$ differs from 0-time group.

***$P < 0.001$ differs from 0-time group.

Plasma amino acid concentrations are in μg/ml. Brain tryptophan and 5-HT concentrations are in μg/g brain, wet weight. Average animal weight was 160 g. (Reproduced from Fernstrom and Wurtman, 1971b.)

sites. Since protein ingestion introduces variable amounts of all of the amino acids into the blood, it seemed possible that brain tryptophan failed to increase after protein ingestion because the plasma concentrations of other, competing, amino acids increased even more than that of tryptophan. To test this hypothesis, we allowed groups of animals to eat either a synthetic diet containing carbohydrates plus all of the amino acids in the same proportions as present in an 18% casein diet, or this diet minus five of the amino acids thought to share a common transport system with tryptophan (i.e., tyrosine, phenylalanine, leucine, isoleucine, and valine). Both diets significantly increased plasma tryptophan levels above those found in fasted controls. However, only when the competing neutral amino acids were deleted from the diet did large increases occur in brain tryptophan, serotonin, or 5-hydroxyindoleacetic acid (Figure 4).

To rule out the possibility that the increase in brain 5-hydroxyindoles observed in rats consuming the latter diet was simply a nonspecific consequence of the omission of any group of amino acids from the diet, we repeated the above experiment omitting aspartate and glutamate instead of the five neutral amino acids. [These two amino acids comprise approximately the same percent of the total alpha-amino nitrogen in casein as the five competing amino acids. Because they are charged at physiologic pH, they are transported into the brain by a carrier system different from that transporting tryptophan (Blasberg and Lajtha, 1965). Hence, their absence would not be expected to alter the postprandial competition between tryptophan and other amino acids within its transport group for uptake into the brain.] At 1 and 2 hr after presentation of this diet or the complete amino acid mixture, plasma tryptophan concentrations again increased 70 to 80% above those of fasted controls ($P < 0.001$). However, neither diet caused increases in brain tryptophan, serotonin, or 5-hydroxyindoleacetic acid.

These results were interpreted as showing that brain tryptophan and 5-hydroxyindole levels do not simply reflect plasma tryptophan but also depend upon the plasma concentrations of other neutral amino acids.

This relationship is perhaps best illustrated by a correlation analysis comparing the brain tryptophan level and the ratio of plasma tryptophan to the five competing amino acids among individual rats given various diets that contain differing amounts of each amino acid. This analysis yielded a correlation coefficient of 0.95 ($P < 0.001$ that $r = 0$), whereas the correlation between brain tryptophan

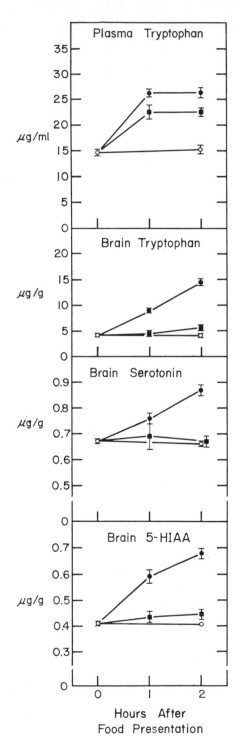

FIGURE 4   Effect of ingestion of various amino acid-containing diets on plasma and brain tryptophan, and brain 5-hydroxyindole levels. Groups of 8 rats were killed 1 or 2 hr after diet presentation. Vertical bars represent standard errors of the mean. Fasting controls: ○ ; complete amino acid mix diet: ▮; mix diet minus tyrosine, phenylalanine, leucine, isoleucine, and valine: ● . 1 and 2-hr plasma tryptophan levels were significantly greater in animals consuming both diets ($P < 0.001$) than in fasting controls. All brain tryptophan, serotonin, and 5-hydroxyindoleacetic acid levels were significantly greater in rats consuming the diet lacking the 5 amino acids than in fasting controls ($P < 0.001$ for all but 1-hr serotonin, $P < 0.01$). (Reproduced from Fernstrom and Wurtman, 1972b.)

and plasma tryptophan alone was less striking ($r = 0.66$; $P < 0.001$ that $r = 0$). Similarly, the correlation coefficient for brain 5-hydroxyindoles (serotonin plus 5-hydroxyindoleacetic acid) versus the plasma amino acid ratio was 0.89 ($P < 0.001$), whereas that of 5-hydroxyindoles versus tryptophan alone was only 0.58 ($P < 0.001$) (Figure 5). Thus, the brain concentrations of both tryptophan and the 5-hydroxyindoles more nearly reflect the ratio of plasma tryptophan to competing amino acids than the plasma tryptophan concentration alone. The reason that brain tryptophan and serotonin appeared, in our earlier formulation, to depend upon plasma tryptophan alone was that all of the physiological manipulations tested at that time (i.e., tryptophan injections, insulin injections, carbohydrate consumption) raised the numerator in the plasma tryptophan:competitor ratio while either lowering the denominator or leaving it unaltered. Only when the rats consumed protein were both the numerator and the denominator elevated.

The effect of food consumption on 5-hydroxyindoles in rat brain may now be modeled as in Figure 6. Since carbohydrate ingestion elicits insulin secretion, it simultaneously raises plasma tryptophan and lowers the concentrations of the competing neutral amino acids in rats (Fernstrom and Wurtman, 1972b); hence the ratio of plasma tryptophan to competing amino acids increases, leading to elevations in brain tryptophan and serotonin. In contrast, protein consumption provides the plasma with an exogenous source of all of the amino acids; however, the ratio of tryptophan to its competitor amino acids is almost always lower in dietary proteins than it is in plasma. Probably for this reason, protein ingestion increases the plasma levels of tryptophan less than it does the concentrations of competing amino acids, and thereby

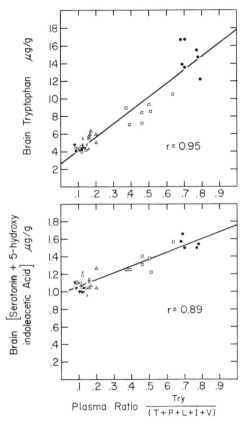

FIGURE 5 *Above:* Correlation between brain tryptophan concentration and the plasma ratio of tryptophan to the 5 competing amino acids in individual rats studied in the experiment described in Figure 3. $r = 0.95$ ($P < 0.001$ that $r = 0$). *Below:* Correlation between the sum of brain serotonin and 5-hydroxyindoleacetic acid, and the plasma ratio of tryptophan to the 5 competitor amino acids, in individual rats studied in the experiment described in Figure 1. $r = 0.89$ ($P < 0.001$ that $r = 0$). One hour control, ○; 2-hr control, ▼; 1-hr complete amino acid mix diet, ×; 2-hr complete amino acid mix diet, △; 1-hr complete mix diet minus five competing amino acids, □; 2-hr complete mix diet minus five competing amino acids, ●. (Reproduced from Fernstrom and Wurtman, 1972b.)

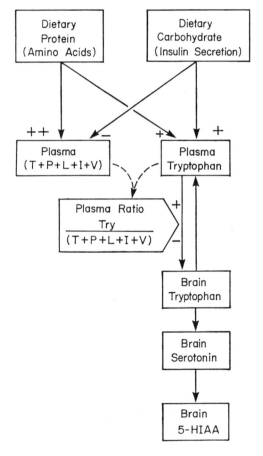

FIGURE 6 Proposed sequence describing diet-induced changes in brain serotonin concentration in the rat (see text). The ratio of tryptophan to (tyrosine + phenylalanine + leucine + isoleucine + valine) in the plasma is thought to control the tryptophan level in the brain. (Reproduced from Fernstrom and Wurtman, 1972b.)

decreases the tryptophan:competitor ratio. The insulin secretion elicited by protein consumption will, by itself, produce an opposite change in this ratio. Thus, brain tryptophan and 5-hydroxyindole levels can decrease, increase, or remain unchanged after eating, depending on the proportion of protein to carbohydrates in the diet and the amino acid composition of the particular proteins. Our most recent observations suggest that diets containing high concentrations of protein (i.e., 40% casein, or as high a proportion of protein as would be consumed in a steak) actually *decrease* brain serotonin synthesis in rats.

The extent to which these observations on rats also apply to humans and other mammals awaits clarification. Insulin does not seem to raise total plasma tryptophan in humans as it does in rats; however it does profoundly depress the concentrations of the other neutral amino acids and thus elevates the ratio of tryptophan to its competitors. In both species, the more relevant index of tryptophan availability may be plasma *free* tryptophan (i.e., tryptophan not bound to albumin). Plasma free tryptophan concentrations do fall precipitously in human subjects who consume a glucose load. When a method becomes available for estimating brain serotonin synthesis in humans, it will be interesting to determine whether such synthesis correlates best with free or total plasma tryptophan concentrations, or with their ratios to the concentrations of other neutral amino acids.

We are accustomed to believing that brain biochemistry controls behavior. The data described above show that the reverse is also true; the decision on the part of the animal to eat, and its choice of what it eats, cause characteristic changes in brain composition. This is perhaps mildly shocking, in view of the fact that the brain chemical influenced by elective food consumption (serotonin) is a neurotransmitter liberated in neuronal circuits known to control behavior. The ability of brain serotonin levels to vary in response to diet-induced changes in plasma amino acid concentrations can be viewed as a potentially dangerous curiosity that somehow managed to survive the evolutionary process. One wonders why something as important as brain neurotransmitter levels should be subject to the vagaries of what one chooses to eat for breakfast. Alternatively, this control mechanism can be viewed as having survived because it was somehow useful to the organism. We favor the latter view; we suspect that serotonin-releasing brain neurons function as a kind of humoral-neural transducer, which converts information about peripheral metabolism (i.e., as manifested by plasma amino acid levels) to neural signals (i.e., the release of a greater or lesser quantity of serotonin). Current studies are designed to identify the uses to which the brain might put this information. Perhaps the brain decides in part when it is hungry, or when it is time to sleep, to trigger ovulation, or to raise or lower the body temperature on the basis of information that it receives about plasma amino acid levels (which, in turn, depend upon food consumption). All of these brain functions have been shown by lesion or drug studies to involve serotonergic neurons. Perhaps nutritional inputs control the brain in order that the brain can make appropriate decisions concerning nutritional strategies.

Summary

There is abundant evidence that *chronic malnutrition* can alter the chemical composition of the brain: For example, inadequate protein consumption early in life is associated with long-term decreases in brain DNA and RNA contents, myelin synthesis, and catecholamine concentrations. More recent studies have shown that brain composition also responds *acutely* to *normal* nutritional inputs: The consumption of carbohydrates by the rat initiates a sequence of events that causes within an hour an increase in the brain concentration of the neurotransmitter serotonin. These events include the secretion of insulin, the resulting increases in plasma tryptophan (and decreases in plasma levels of other neutral amino acids that compete with tryptophan for uptake into the brain), and brain tryptophan, the increased saturation of the enzyme tryptophan hydroxylase, and enhanced serotonin biosynthesis. Protein consumption also raises plasma tryptophan; however, the plasma concentrations of other neutral amino acids rise even more, and thus brain tryptophan and serotonin levels are not increased. These observations indicate that *behavior* (i.e., the elective consumption of specific foods) can influence *brain biochemistry*, as well as vice versa. They also show that the nutritional status of the brain (e.g., the amounts of tryptophan and other substances available to it from the blood stream) need not parallel the general nutritional state of the body. The capacity of serotonin-containing neurons to "sense" plasma amino acid levels (i.e., by increasing their neurotransmitter content and, perhaps, secretion) suggests that they may function to monitor peripheral metabolic states, and to provide the rest of the brain with information to be used in formulating appropriate behavioral and neuroendocrine strategies.

ACKNOWLEDGMENTS These studies were supported in part by grants from the John A. Hartford Foundation and the National Aeronautics and Space Administration.

REFERENCES

ALBRECHT, P., M. B. VISSCHER, J. J. BITTNER, and F. HALBERG, 1956. Daily changes in 5-hydroxytryptamine concentration in mouse brain. *Proc. Soc. Exp. Biol. Med.* 92:702–706.

BLASBERG, R., and A. LAJTHA, 1965. Substrate specificity of steady-state amino acid transport in mouse brain slices. *Arch. Biochem. Biophys.* 112:361–377.

FERNSTROM, J. D., and R. J. WURTMAN, 1971a. Brain serotonin content: Dependence on plasma tryptophan levels. *Science* 173:149–152.

FERNSTROM, J. D., and R. J. WURTMAN, 1971b. Brain serotonin content: Increase following ingestion of carbohydrate diet. *Science* 174:1023–1025.

FERNSTROM, J. D., and R. J. WURTMAN, 1972a. Elevation of plasma tryptophan by insulin in the rat. *Metabolism* 21: 337–343.

FERNSTROM, J. D., and R. J. WURTMAN, 1972b. Brain serotonin content: Physiological regulation by plasma neutral amino acids. *Science* 178:414–416.

FISHMAN, B., R. J. WURTMAN, and H. N. MUNRO, 1969. Daily rhythms in hepatic polysome profiles and tyrosine transaminase activity: Role of dietary protein. *Proc. Nat. Acad. Sci. USA* 64:677–682.

GUROFF, G., and S. UDENFRIEND, 1962. Studies on aromatic amino acid uptake by rat brain in vivo. *J. Biol. Chem.* 237: 803–806.

LOVENBERG, W., E. JEQUIER, and A. SJOERDSMA, 1968. A tryptophan hydroxylation in mammalian systems. *Adv. Pharmacol.* 6A:21–36.

MARLISS, E. B., T. T. AOKI, R. H. UNGER, J. S. SOELDNER, and G. F. CAHILL, 1970. Glucagon levels and metabolic effects in fasting man. *J. Clin. Invest.* 49:2256–2270.

MOIR, A. T. B., and D. ECCLESTON, 1968. The effects of precursor loading in the cerebral metabolism of 5-hydroxyindoles. *J. Neurochem.* 15:1093–1107.

WOOL, I. G., 1965. Relation of effects of insulin on amino acid transport and on protein synthesis. *Fed. Proc.* 24:1060–1070.

WURTMAN, R. J., 1970. Diurnal rhythms in mammalian protein metabolism. In *Mammalian Protein Metabolism*, H. N. Munro, ed. New York: Academic Press, Vol. IV, Chap. 36.

WURTMAN, R. J., and J. D. FERNSTROM, 1972. L-tryptophan, L-tyrosine, and the control of brain monoamine biosynthesis. In *Perspectives in Neuropharmacology*, S. H. Snyder, ed. Oxford: Oxford University Press, pp. 145–193.

WURTMAN, R. J., C. M. ROSE, C. CHOU, and F. LARIN, 1968. Daily rhythms in the concentrations of various amino acids in human plasma. *New Eng. J. Med.* 279:171–175.

60 Brain Dopamine Neurons and Behavior

URBAN UNGERSTEDT

ABSTRACT The importance of nigrostriatal dopamine neurons in the control of behavior is evaluated by inducing hypofunction (6-hydroxydopamine induced lesions) or hyperfunction (dopamine receptor stimulation) in the system. Hypofunction induces a severe state of adipsia and aphagia that seems to be secondary to a pronounced sensory neglect where the animal is unable to respond to external (and internal?) stimuli. Severe unilateral hypofunction due to a 6-hydroxydopamine lesion blocks ipsilateral but not contralateral electrical self stimulation. Hyperfunction due to pharmacologically induced receptor stimulation brings the animal through a series of behavioral patterns that all seem to be related to a basic behavioral repertoire. The behavioral syndromes elicited by hypofunction and hyperfunction are partly quantified in tests designed to reveal changes in dopamine function.

AMONG THE transmitter, or transmitter candidates, in the central nervous system, dopamine (DA) is uniquely related to known behavioral syndromes. We know that a degeneration of the nigrostriatal DA system is the major functional brain lesion underlying Parkinson's disease (Hornykiewicz, 1966), and the disease is therefore successfully treated by substituting for the lost DA with high doses of its precursor DOPA (Cotzias et al., 1967). Important parts of the Parkinson symptomatology may be reproduced in animals after lesions of the DA pathway (Battista et al., 1969; Sourkes et al., 1969). These animal models are valuable tools in our attempts to understand the nature of the functional lesion.

Biochemistry and histochemistry form the basis for our understanding of the part played by DA neurons in brain function, i.e., the biochemical identification of DA (Carlsson et al., 1958) and the histochemical identification of the DA neurons (Carlsson et al., 1962; Andén et al., 1965). Biochemical and anatomical findings have been linked to behavior mainly through DA pharmacology; e.g. the depletion of DA stores is probably the major reason for the pronounced behavioral depression caused by reserpine.

Conversely, the behavioral excitation and stereotyped behavior seen after amphetamine may be due to excessive release of DA (Randrup and Munkvad, 1966). The psychotic syndrome caused by an overdose of amphet-amine in man looks very similar to acute schizophrenia (Bell, 1965; Snyder, 1972) and to complete the parallel, neuroleptic drugs, effective against schizophrenia, are found to inhibit DA receptors (Andén et al., 1970).

The fact that pronounced behavioral changes take place after changes in DA neurotransmission seems to provide us with a unique chance to link behavior to synaptic transmission within a defined neuron system in the brain. This article relates a series of experiments where we have tried to describe and interpret behavior in relation to experimental manipulation of DA anatomy and pharmacology.

In the rat brain, it is possible to distinguish two different ascending DA neuron systems, the nigrostriatal system and the mesolimbic system (Andén et al., 1966; Ungerstedt, 1971a) (Figure 1). The *nigrostriatal DA system* originates from cell bodies in the zona compacta of the substantia nigra. Its axons run rostromedially to enter the lateral aspect of the medial forebrain bundle. In the lateral hypothalamus, the axons run in a dense bundle that moves laterally to enter the internal capsule, fan out in the globus pallidus, and terminate in the nucleus caudatus-putamen, i.e., the striatum.

The *mesolimbic DA system* originates from cell bodies situated dorsal and lateral to the interpeduncular nucleus. The axons run rostrolaterally to enter the medial forebrain bundle. They run slightly medial to the nigrostriatal axons and never enter into the internal capsule but continue to the level of the anterior commissure, where they terminate in the nucl. accumbens and the olfactory tubercle.

In order to reveal the role played by central DA neurons in the regulation of behavior, we have used lesions of the DA pathways and pharmacological interference with their synaptic transmission. Local injections of 6-hydroxy-dopamine (6-OH-DA) into areas containing DA neurons have proved to be a very effective and reproducible way of performing lesions (Ungerstedt, 1968, 1971b). The lesions are far more selective in their effect on DA and norepinephrine (NE) neurons than conventional lesions (Ungerstedt, 1971b, 1973). The detailed mapping of the monoamine pathways in the rat brain (Ungerstedt, 1971a) has provided the necessary anatomical knowledge to localize injection cannulas in the DA pathways.

URBAN UNGERSTEDT Department of Histology, Karolinska Institutet, Stockholm, Sweden

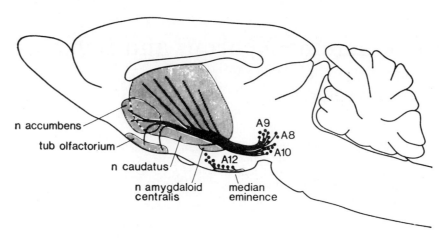

n accumbens

tub olfactorium

n caudatus

n amygdaloid
centralis

A9 A8

A12

A10

median
eminence

FIGURE 1 Sagittal projection of the brain dopamine pathways. The nigrostriatal system originates in the cell group labeled A9, while the mesolimbic system originates in the group labeled A10. Group A8 probably innervates the striatum while the A12 innervates the median eminence in the hypothalamus. (From Ungerstedt, 1971a.)

In the further analysis of the lesioned animals we have attempted to quantify behavior that is elicited by hypo- or hyperfunction of the DA system.

The unilateral dopamine denervation syndrome

A unilateral injection of DA into the striatum produces an asymmetry in movements and posture (Ungerstedt et al., 1969). The animal deviates and even rotates in circles away from the side getting the injection. The unilaterally increased concentration of DA tends to force the animal in the direction of the side where there is less DA. A unilateral 6-OH-DA induced degeneration of the nigrostriatal DA system induces a similar syndrome. As soon as the animal wakes up from the operation, it assumes an asymmetrical posture with tail and head deviating toward the side of the lesion. The legs on the unlesioned side are usually slightly extended while the legs on the lesioned side are kept flexed under the body. Arousal of the animal by pinching its tail causes rotation toward the lesioned side. However, starting about 24 hr after the lesion, the animal develops a tendency to turn spontaneously toward the normal side. If the animal is treated with a mono-amine oxidase inhibitor at the time of the operation, it shows rotation toward the normal side at this time point, i.e., between 24 to 34 hr after the 6-OH-DA injection (Ungerstedt, 1971d). This behavior coincides with the period when DA disappears from the degenerating neurons (Hökfelt and Ungerstedt, 1969). This is the behavioral expression of the degeneration release of DA, comparable to the degeneration contraction of the nictitating membrane after the extirpation of the superior cervical ganglion (Langer, 1966).

The DA nerve terminals have lost their transmitter content 48 hr after the 6-OH-DA lesion as determined with histochemical (Ungerstedt, 1971d) and electron-microscopic technique (Hökfelt and Ungerstedt, 1969). The animal now enters into a chronic state where the unilateral loss of DA expresses itself in several ways: The animal shows a chronic deviation towards the side of the lesion. Spontaneous exploratory behavior is decreased, and there is a transient period of adipsia and aphagia (2 to 3 days). There are also signs of upset vegetative function as the animals show a chronic loose consistency of their feces. However, the severity of the syndrome becomes apparent only after an examination of the sensory functions of the animal: A unilateral degeneration of the nigrostriatal DA system produces a severe "sensory neglect" on the side of the body contralateral to the lesion (Ljungberg and Ungerstedt, to be published). The animal fails to orient towards tactile, auditory, olfactory, and visual stimuli applied to that side of the body. The syndrome has been characterized by the orienting reaction evoked in a series of tests where different sensory stimuli have been applied to both sides of the body, e.g. a straw pressed against the body, whisker touch, a piece of cotton soaked in a smelling solution moved toward a nostril from behind, a sound evoked close to one ear and, to test vision, the placing reaction of the fore paw when the animal is held in the experimenter's hand and moved past the edge of a table, or simply the orienting reaction to an object moved from behind into the field of vision of the animal. While the animal shows essentially normal reaction to stimuli applied ipsilateral to the degeneration of the DA system, there is initially an almost total irresponsivity to stimuli applied on the side contralateral to the lesion. There is a gradual recovery of some sensory abilities.

After $1\frac{1}{2}$ weeks the rats react to smell, and after 7 weeks most of the animals react to visual stimuli, while normal reactions to tactile stimuli have not returned even 3 months after the lesion.

The fact that the animal initially shows a pronounced asymmetry in its movements, which recover only slightly, seems to suggest that the deficit in orienting reaction might be a motor rather than a sensory deficit. However, the fact that there is a recovery of some sensory modalities, e.g. smell, while others, e.g. tactile, remain deficient suggests that the animal has no motor difficulties in performing the orienting reaction per se. The lack of orienting reaction is more likely to be due to a deficiency in the ability to recognize the stimulus or to integrate sensory and motor function to perform an adequate response.

If a unilaterally DA denervated animal is given drugs that change DA transmission this will change the degree of asymmetry in posture and movements. Amphetamine, known to release DA (Glowinski and Axelrod, 1965; Carlsson et al., 1966), will increase the asymmetry to the point of vigorous rotation toward the lesioned side (Ungerstedt and Arbuthnott, 1970; Ungerstedt, 1971c). This agrees with the previously related study where DA was injected unilaterally into the striatum and produced a similar but less pronounced rotation away from the injected side (Ungerstedt et al., 1969). As this rotational behavior in all probability is quantitatively related to the degree of DA receptor stimulation, we constructed a "rotometer" where the number of full turns per minute may be recorded over long periods of time (Ungerstedt and Arbuthnott, 1970). The rotometer consists of a plexiglass bowl shaped as a hemisphere in which the animal moves around connected by a thin wire to a microswitch arrangement localized in the geometrical center of the sphere. Every full turn is registered on electromechanical counters or cumulative recorders. Drugs, like amphetamine, that release DA from the presynaptic DA nerve terminals, will work only on the remaining DA system and thus induce rotation toward the side of the degenerated DA system (Ungerstedt, 1971c). Postsynaptically acting drugs, like apomorphine, that stimulate DA receptors (Ernst, 1967; Andén et al., 1967), will act on both innervated and denervated DA receptors. Apomorphine causes a rotation in the direction opposite to amphetamine. The animal rotates toward the normal side. This indicates that apomorphine preferentially stimulates the denervated DA receptors, i.e., they seem to become supersensitive after denervation (Ungerstedt, 1971d; Schoenfeld and Uretsky, 1972). The animals react to DOPA in the same way as to apomorphine. DOPA is converted to DA by the enzyme still remaining in the denervated striatum, and because of the supersensitivity the small amounts of DA thus formed may compensate for the loss of endogenous DA. Postsynaptic supersensitivity to DA may form the basis for the therapeutic effect of DOPA in Parkinson's disease (Ungerstedt, 1971e).

The bilateral dopamine denervation syndrome

Bilateral removal of the nigrostriatal DA systems by intracerebral injections of 6-OH-DA, produces a very severe syndrome, as might be expected from the findings in the unilaterally denervated animals. The animals are hypokinetic, adipsic, and aphagic immediately after the operation. In our initial studies the condition persisted for about 3 to 4 weeks except for a period of increased activity coinciding with the release of DA from the degenerating neurons (Ungerstedt, 1970, 1971f). If not tube fed, the animals die within 3 to 5 days after the operation. The syndrome is very similar, or identical, to the "lateral hypothalamic syndrome" that follows destruction of an area close to the crus cerebri in the lateral hypothalamus. This area has been described as the "lateral hypothalamic eating center" and is thought to be involved in the regulation of eating and drinking (Anand and Brobeck, 1951). However, the stereotaxic mapping of the ascending DA axons has revealed that the axons run through the lateral hypothalamus assembled in a dense bundle at the tip of the crus cerebri (Ungerstedt, 1971f) (Figure 2). A lesion of this area interrupts the DA pathway as effectively as the 6-OH-DA injection into the substantia nigra and denervates the striatum of its DA nerve terminals. Several authors have argued for the possibility that lesions of the lateral hypothalamic area interrupt axons passing through this area (Morgane, 1961; Gold, 1967; Albert et al., 1970). Apart from ourselves (Ungerstedt, 1970, 1971f) Oltmans and Harvey (1972) have suggested that the crucial damage is to the nigrostriatal DA system. Recent studies with intraventricular injection of 6-OH-DA have also argued in favor of this hypothesis (Zigmond and Stricker, 1972).

The lateral hypothalamic syndrome has often been described only as an adipsic and aphagic syndrome. However, several authors have commented on symptoms other than deficits in eating and drinking, such as hypokinesia, catalepsia, decreased exploratory behavior, and arousal difficulties (Gladfelter and Brobeck, 1962; Balagura et al., 1969; Beattie et al., 1969). This other aspect of the lateral hypothalamic syndrome is especially evident in a larger animal such as the dog (Fonberg, 1969). These findings make the parallel to what is found in the bilaterally DA denervated animals even more striking. In a recent report, Marshall et al. (1971) have described a pronounced sensory neglect after lateral hypothalamic lesions. Our own findings in the DA denervated animal (Ljungberg and

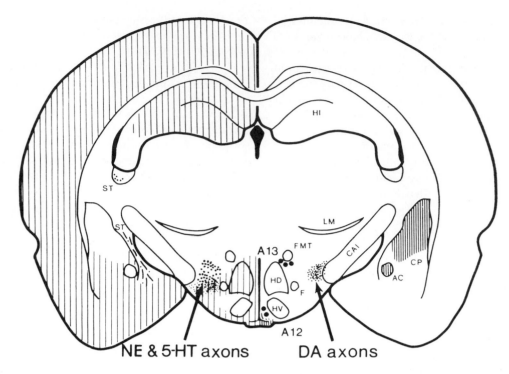

NE & 5-HT axons   DA axons

FIGURE 2 Frontal section through the midhypothalamus of a rat. The ascending monoamine axon bundles are indicated by dots. The dopamine axons run in a dense bundle in the lateral hypothalamus. The norepinephrine and the 5-hydroxytryptamine (fine dots) axons follow a more medial course. (From Ungerstedt, 1971f.)

Ungerstedt, to be published, see above) confirm their findings and provide additional evidence that the lateral hypothalamic syndrome may be attributed to a degeneration of the nigrostriatal DA system. The sensory neglect resulting from a degeneration of the nigrostriatal DA system probably explains the severe syndrome of adipsia and aphagia. The animal is essentially unable to respond adequately to any sensory stimuli, possibly neither internal (e.g. proprioceptive) nor external. The adipsia and aphagia syndrome that has a recovery period of a few weeks is not necessarily associated with catalepsia, i.e., the inability of the animal to escape from posture it has been passively positioned in by the experimenter. This was concluded by Balagura et al. (1969) after studies of lateral hypothalamic lesion and by ourselves after lesions of the DA system (Ungerstedt, 1971f). However, in a recent study (Avemo et al., to be published), we have attempted to produce as extensive a DA degeneration as possible, i.e., we have disposed our 6-OH-DA into the bundle of DA axons where it leaves the substantia nigra rather than into the substantia nigra itself. This allows for a more complete degeneration of all DA fibers. We have tube fed such animals for 3 to 4 months without any sign of recovery. Such lesions show histochemically a complete degeneration of both the limbic and the striatal DA system. The animals are catalepsic, akinetic, rigid, and

show a fine tremor of high frequency. The motor phenomena are obviously well related to the Parkinson syndrome in man. However, there is still some uncertainty whether this severe syndrome is solely due to a lesion of DA neurons, or if the small unspecific lesion that always occurs after 6-OH-DA injections (Ungerstedt 1971b,1973) damages other neurons that contribute to the behavioral syndrome.

In order to further test the bilateral DA denervation syndrome, we have developed certain behavioral tests related to its typical symptomatology: The animals are tested in an automatic "hole-board" (Ljungberg and Ungerstedt, to be published) which consists of a 70 × 70 cm square, open field. Eighty-one 3.5 cm wide, round holes are distributed evenly in the floor. The open field activity is monitored by photocell beams 3 cm above the floor. Hole exploration is monitored by photocells that react every time the animal peeps into a hole (Figure 3).

A normal animal explores the hole-board for about 15 min by looking into holes and walking around. The bilaterally DA denervated animal shows a complete lack of exploratory behavior. Apomorphine, however, induces a strong activity in the animal even at a very low dose (0.05 mg/kg). This is the same phenomenon as the rotation of the unilaterally denervated animal and is explained by the DA degeneration supersensitivity to apomorphine (Ungerstedt,1971c; Schoenfeld and Uretsky,

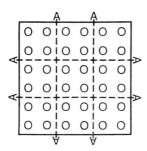

FIGURE 3 Schematic drawing of the "hole-board" seen from above. Photocells cover the open field area as well as each hole (not shown).

1972). The bilaterally denervated animal shows no reaction to amphetamine (2 mg/kg of *d*-amphetamine) (Figure 4), which also corresponds to the finding in the rotometer where the direction of the amphetamine induced rotation indicates that the amphetamine effect is elicited only from the intact DA neurons.

There is an interesting correlation between the severity of the adipsia and aphagia and the response to amphetamine and apomorphine (Avemo et al., to be published). Animals showing a severe and long-lasting adipsia and aphagia do not react to amphetamine but show a strongly increased reaction to apomorphine, while animals that show a fast recovery after the adipsia and aphagia react to amphetamine but less to apomorphine. Brought together with our previous results on adipsia and aphagia (see above), it seems as if the amphetamine induced excitation occurs only in the presence of remaining DA neurons.

However, only a fraction of the normal DA innervation seems necessary to elicit an amphetamine response or to allow for a recovery from the adipsia and aphagia.

The role of the NE neurons remains somewhat unclear in the adipsic and aphagic syndrome as well as in the amphetamine induced excitation. The 6-OH-DA induced degeneration of the DA neurons causes a partial denervation of forebrain NE nerve terminals. As involvement of NE neurons has been suggested in both the adipsic and aphagic syndrome and the amphetamine induced excitation, we have attempted to control for the damage of ascending NE pathways by injecting 6-OH-DA directly into these pathways. This causes degeneration of forebrain NE neurons without affecting the DA neurons. Such animals show no sign of adipsia and aphagia (Ungerstedt, 1971); in fact there are even signs of overeating. In the hole-board there is a slight modification of the ratio between walking around and looking into holes but no sign of a decreased amphetamine response. These results make it possible to exclude a damage of NE neurons as an important factor in the behavioral changes occurring after 6-OH-DA lesions of the DA neurons.

There are several pharmacological studies indicating differences in the way NE and DA neurons influence behavior (see Randrup and Munkvad, 1970). The NE neurons are thought to influence locomotor behavior, while the DA neurons are thought to influence stereotyped behavior. However, our own studies in the holeboard indicate that this may be too simple a way to look upon the various components of behavior. Amphetamine

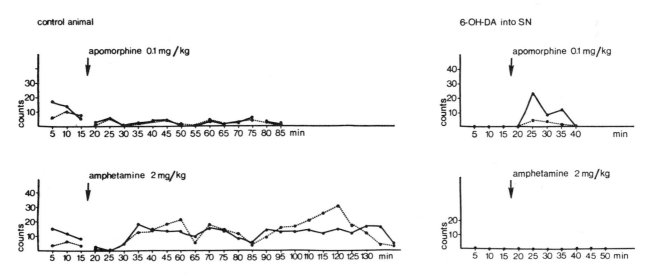

FIGURE 4 Recordings from the hole-board. The control animal shows normal response to amphetamine and no response to the subthreshold dose of apomorphine. The animal with a bilateral lesion of the nigrostriatal dopamine system does not respond to amphetamine while it shows a clear cut reaction to the low dose of apomorphine. Solid lines show the movements in the open field while dotted lines show the peeping into the holes.

(2 mg/kg *d*-amphetamine), which releases both NE and DA (Carlsson et al., 1966), causes varied behavior where the animals walk around and look into the holes frequently. The behavior is similar to the initial behavior shown by a normal animal that explores the hole-board for the first time. However, apomorphine (1 to 2 mg/kg), which stimulates DA receptors (Ernst, 1967; Andén et al., 1967), induced a different pattern. Looking into holes is completely abolished while the walking around is greatly increased (Figure 4). In fact, there is a gradual change in the behavior when the dose of apomorphine is increased (Ljungberg and Ungerstedt, to be published). Low doses (0.1 to 0.5 mg/kg) inhibit normal behavior and cause "freezing." In doses varying between 1 and 2 mg/kg, there is a great increase in locomotor behavior, while, at higher doses, locomotor behavior decreases and oral behaviors, such as licking and gnawing, dominate. There is a great increase in sniffing except during the stage of freezing. The higher doses of apomorphine (5 mg/kg) often elicit aggressive behavior.

It is fascinating to note that the increasing doses of apomorphine seem to elicit different patterns of behavior that when brought together in a meaningful order seem to be components of a "survival behavior," i.e., loco-motion, sniffing, licking, and gnawing are all components of food seeking and ingestion, and together with freezing and aggression they seem to incorporate many of the basic components of normal behavior.

The bilaterally DA denervated animals do not show any spontaneous activity. In order to study the performance of such a severely hypokinetic "nonperforming" animal, we have developed an under-water *Y*-maze (Ranje and Ungerstedt, to be published). The animals are brought under water in a start box, and swim to the choice point of the *Y* where they learn to select either a lighted or a darkened arm in order to reach the water surface so as to breathe. The swimming time is measured and the number of mistakes counted (defined as entries into the wrong arms of the maze). In this swim-maze, the lesioned animals are forced to act or they will not survive. In fact, the DA denervated animals are able to swim the maze and to learn to select the lighted or the darkened arm in order to reach the surface. As a parallel to the DA denervated animals, we have tested animals treated with DA receptor blocking drugs. Even if such animals show the most serious catalepsia, they swim the maze just as a normal animal. It is evident that the severe behavioral inhibition induced by blockade of the DA transmission may be overcome during certain types of stress. The phenomenon seems to be a parallel to the "paradoxical kinesia" shown even by severely disabled Parkinson patients during similarly stressful situations.

Dopamine neurons and self-stimulation

High rates of self-stimulation may be elicited from areas known to contain DA neurons, e.g. the area around the interpeduncular nucleus and the lateral hypothalamus. However, there also seems to be good evidence for the participation of NE neurons in self-stimulation (Stein, 1962; Poschel and Niuteman, 1963). Crow (1972) suggested that there, in fact, are two systems underlying self-stimulation, the NE and the DA system. However, Breese et al. (1971) and Stein and Wise (1971) managed to abolish self-stimulation by interventricular injection of 6-OH-DA and attributed these effects to a degeneration of the NE neurons. Antelman et al. (1972) reinterpreted Stein's data and showed that self-stimulation may still be elicited by priming the animals with free stimulation, which will make them resume self-stimulation. Since Breese et al. (1971) obtained the same decrease of both NE and DA after treatment with 6-OH-DA, it seems to be difficult to attribute the block of self-stimulation to one or the other of the two amines.

The major problem in using lesions to abolish a functional effect is obviously the possible compensatory postsynaptic supersensitivity (Ungerstedt, 1971d) that may help to keep up function in spite of a severe reduction in the number of presynaptic nerve terminals. In order to determine the importance of DA neurons in self-stimulation behavior, we have used a criterion to select the most completely denervated animals and then test them for self-stimulation. A large number of unilaterally DA denervated animals were tested in the rotometer on apomorphine. The rotational behavior that results reveals the extent of DA denervation (Ungerstedt, 1971c). Intense rotation indicates strong postsynaptic supersensitivity that in turn indicates extensive degeneration of the pre-synaptic DA nerve terminals. Fifteen animals that showed intense rotational behavior were implanted with bilateral hypothalamic electrodes (Christie et al., 1973). All 15 showed self-stimulation from the electrode situated on the normal side of the brain, while 4 failed to self-stimulate from the electrode situated on the side where the DA neurons were lesioned. These 4 animals had shown the strongest rotational behavior to apomorphine among the 15 implanted animals. The histochemistry revealed a partial denervation of ascending NE fibers, while there was a complete, or next to complete, denervation of the DA fibers. The most extensive denervation of DA was associated with the strongest rotational response to apomorphine. Those animals that showed the most extensive DA degeneration, as revealed by the rotometer test for supersensitivity and histochemistry, did not self-stimulate. This study shows a direct correlation between the ability to self-stimulate and the presence of DA nerves. The results strongly suggest that the presence of DA neurons are necessary for self-stimulation, although they don't rule out a participation of other types of neurons, e.g. NE neurons.

DA neurons and the function of the basal ganglia

The function of the central DA pathways have to be understood in connection with the function of the areas they innervate. The nigrostriatal pathway innervates the nucl. caudatus putamen (striatum). The mesolimbic DA system innervates the olfactory tubercle, the nucl. accumbens and parts of the septal area and may be regarded as part of the limbic system. So far there are few behavioral changes that can be attributed solely to the limbic DA system (see Fuxe and Ungerstedt, 1970), and in the following we are forced to consider mainly the nigrostriatal system and the basal ganglia.

There is no single, known neuron system in the brain that after experimental manipulation may influence behavior as profoundly as the nigrostriatal DA system. The behavioral syndromes range from extreme hyperactivity to severe akinesia and sensory neglect. However, there seem to be several mechanisms aimed at maintaining "homeostasis" within the system. A pharmacological increase in DA receptor excitation by apomorphine is counteracted by a decreased turnover of the endogenous transmitter and, conversely, a pharmacological decrease of the transmission by a DA receptor blocking drug evokes an increase in the turnover of DA (Carlsson and Lindqvist, 1963; Andén et al., 1967). In the case of neuronal degeneration of the DA system, function is maintained after severe but incomplete loss of nerve terminals by a compensatory postsynaptic supersensitivity (Ungerstedt, 1971d).

From the anatomy of the nigrostriatal DA system (see Ungerstedt, 1971a), we know that changes in DA transmission must affect behavior by changing the activity of the striatum. Most available evidence indicates that DA is inhibitory in the striatum: Iontophoretically applied DA inhibits striatal cells (Bloom et al., 1965), and electrical stimulation of the substantia nigra exerts an inhibitory reaction in the same way as iontophoretically applied DA (Connor, 1970). Recent studies by Feltz and de Champlain (1972) show that the inhibitory effect of stimulating the substantia nigra may be abolished by 6-OH-DA treatment, which probably lesions the DA neurons. As might be expected from these studies, degeneration of the nigrostriatal DA system causes an increased spontaneous firing frequency of the striatal cells (Hoffer et al., to be published). We may, thus, assume

that the behavioral excitation that follows increased release of DA is due to an *inhibition* of the striatum, while the behavioral depression following a decrease in DA transmission is due to an excitation of the striatum. In agreement with this, electrical excitation of the striatum via an implanted electrode, causes a behavioral "arrest reaction" (Akert and Anderson, 1951; see also Laursen, 1963), which may possibly be the same as a short lasting behavioral depression.

The mechanisms by which the striatum influences behavior are far from understood. Anatomically, the striatum is strategically positioned to receive information from the entire cerebral cortex through the massive corticostriate fibers. The thalamus is another major source of striatal afferents, and the nigrostriatal DA system forms a third important group of afferent fibers. The major striatal efferents terminate in the substantia nigra and the globus pallidus. Pallidal efferents end in the mid-brain and in nuclei in the thalamus, which in turn are closely related to the motor cortex (see Nauta, 1969). These anatomical data seem to be in agreement with electrophysiological data indicating a "striato-thalamo-cortical" loop (Buchwald et al., 1961), possibly exerting control over sensory and motor phenomena. The behavioral deficits following lesions of the striatum have been interpreted as due to damage to a sensory-motor integrating mechanism (see Laursen, 1963; Gybels et al., 1967). In fact, the multisensory convergence of visual, auditory, and somatic stimuli on single caudate cells supports the idea that the striatum has important integrative properties (see Krauthamer and Albe-Fessard, 1965).

Lesions or pharmacological manipulations of the nigrostriatal DA system may, thus, interfere with important mechanisms of sensory-motor integration by changing the activity of the striatum. The whole range of behavioral syndromes, e.g. the rotational behavior, the sensory neglect, the various levels of hypo- and hyperactivity, and the interruption of self-stimulation behavior, feeding and drinking, may, thus, be signs of changes in the ability to perceive or respond to sensory stimuli.

ACKNOWLEDGMENT   This investigation was supported by a grant from the Swedish Medical Research Council (04X-357Y).

REFERENCES

Akert, K., and B. Anderson, 1951. Experimenteller Beitrag zur Physiologie des Nucleus Caudatus. *Acta Physiol. Scand.* 22:281–298.

Albert, D. J., L. H. Storlien, D. J. Wood, and G. K. Ehman, 1970. Further evidence for a complex system controlling feeding behavior. *Physiol. Behav.* 5:1075–1082.

ANAND, B. K., and J. R. BROBECK, 1951. Hypothalamic control of food intake in rats and cats. *Yale J. Biol. Med.* 24:123–140.

ANDÉN, N.-E., S. G. BUTCHER, H. CORRODI, K. FUXE, and U. UNGERSTEDT, 1970. Receptor activity and turnover of dopamine and norepinephrine after neuroleptics. *Europ. J. Pharmacol.* 11:303–314.

ANDÉN, N.-E., A. CARLSSON, A. DAHLSTRÖM, K. FUXE, N.-Å HILLARP, and K. LARSSON, 1964. Demonstration and mapping out of nigro-neostriatal dopamine neurons. *Life Sci.* 3:523–530.

ANDÉN, N.-E., A. DAHLSTRÖM, K. FUXE, and K. LARSSON, 1965. Further evidence for the presence of nigro-neostriatal dopamine neurons in the rat. *Amer. J. Anat.* 116:329–333.

ANDÉN, N.-E., A. DAHLSTRÖM, K. FUXE, L. LARSSON, L. OLSON, and U. UNGERSTEDT, 1966. Ascending monoamine neurons to the telencephalon and diencephalon. *Acta Physiol. Scand.* 67:313–326.

ANDÉN, N.-E., A. RUBENSON, K. FUXE, and T. HÖKFELT, 1967. Evidence for dopamine receptor stimulation by apomorphine. *J. Pharm. Pharmacol.* 19:627–629.

ANTELMAN, S. M., A. S. LIPPA, and A. E. FISHER, 1972. 6-Hydroxydopamine, noradrenergic reward and schizophrenia. *Science* 175:919–920.

BALAGURA, S., R. H. WILCOX, and D. V. COSCINA, 1969. The effect of diencephalic lesions on food intake and motor activity. *Physiol. Behav.* 4:629–633.

BATTISTA, A. F., M. GOLDSTEIN, S. NAKATANI, and B. ANAGNOSTE, 1969. The effects of ventrolateral thalamic lesions on tremor and the biosynthesis of dopamine in monkeys with lesions in the ventromedial tegmentum. *J. Neurosurg.* 31:164–171.

BEATTIE, C. W., H. I. CHERNOW, P. S. BERNARD, and F. H. GLENNY, 1969. Pharmacological alteration of hyperreactivity in rats with septal and hypothalamic lesions. *Int. J. Neuropharmacol.* 8:365–371.

BELL, D. S., 1965. Comparison of amphetamine psychosis and schizophrenia. *Brit. J. Psychiat.* 111:701–707.

BLOOM, F. E., E. COSTA, and G. C. SALMOIRAGHI, 1965. Anesthesia and the responsiveness of individual neurons of the caudate nucleus of the cat to acetylcholine, norepinephrine and dopamine administered by microelectrophoresis. *J. Pharmacol. Exp. Ther.* 150:244–252.

BREESE, G. R., J. L. HOWARD, and J. P. LEAHY, 1971. Effect of 6-hydroxydopamine on electrical self-stimulation. *Br. J. Pharmac.* 42:88–99.

BUCHWALD, N. A., E. J. WYERS, C. W. LANPRECHT, and G. HEUSER, 1961. The "caudate-spindle". IV. A behavioral index of caudate-induced inhibition. *Electroenceph. Clin. Neurophysiol.* 13:531–537.

CARLSSON, A., B. FALCK, and N.-Å. HILLARP, 1962. Cellular localization of brain monoamines. *Acta Physiol. Scand.* (Suppl. 196) 56:1–28.

CARLSSON, A., K. FUXE, B. HAMBERGER, and M. LINDQVIST, 1966. Biochemical and histochemical studies on the effects of imipramine-like drugs and (+)-amphetamine on central and peripheral catecholamine neurons. *Acta Physiol. Scand.* 67:481–497.

CARLSSON, A., and M. LINDQVIST, 1963. Effect of chlorpromazine or haloperidol on formation of 3-methoxytyramine and normetanephrine in mouse brain. *Acta Pharmacol.* (*Kbh.*) 20:140–144.

CARLSSON, A., M. LINDQVIST, T. MAGNUSSON, and B. WALDECK, 1958. On the presence of 3-hydroxytyramine in brain. *Science* 127:471.

CHRISTIE, J., T. LJUNGBERG, and U. UNGERSTEDT, 1973. Dopamine neurons and electrical self-stimulation in the lateral hypothalamus.

CONNOR, J. D., 1970. Caudate nucleus neurons: Correlation of the effects of substantia nigra stimulation with iontophoretic dopamine. *J. Physiol.* (*Lond.*) 208:691–703.

COTZIAS, G. C., M. H. VAN WOERT, and L. M. SCHIFFER, 1967. Aromatic amino-acids and modification of Parkinsonism. *New Engl. J. Med.* 276:374–379.

CROW, T. J., 1972. A map of the rat mesencephalon for electrical self-stimulation. *Brain Res.* 36:265–273.

ERNST, A. M., 1967. Mode of action of apomorphine and dexamphetamine on gnawing compulsion in rats. *Psychopharmacologia* (*Berl.*) 10:316–323.

FELTZ, P., and J. DE CHAMPLAIN, 1972. Persistence of caudate unitary responses to nigral stimulation after destruction and functional impairment of the striatal dopaminergic terminals. *Brain Res.* 43:595–600.

FONBERG, E., 1969. The role of the hypothalamus and amygdala in food intake alimentary motivation and emotional reactions. *Acta Biol. Exp.* 29:335–358.

FUXE, K., and U. UNGERSTEDT, 1970. Histochemical, biochemical and functional studies on central monoamine neurons after acute and chronic amphetamine administration. In *Amphetamines and Related Compounds*, E. Costa and S. Garattini, eds. New York: Raven Press, pp. 257–288.

GLADFELTER, W. E., and J. R. BROBECK, 1962. Decreased spontaneous locomotor activity in the rat induced by hypothalamic lesions. *Amer. J. Physiol.* 203:811–817.

GLOWINSKI, J., and J. AXELROD, 1965. Effect of drugs on the uptake release and metabolism of H^3-norepinephrine in the rat brain. *J. Pharmacol.* 149:43–49.

GOLD, R. M., 1967. Aphagia and adipsia following unilateral and bilateral asymmetrical lesions in rats. *Physiol. Behav.* 2:211–220.

GYBELS, J., M. MEULDERS, M. CALLENS, and J. COLLE, 1967. Disturbance of visuo-motor integration in cats with small lesions of the caudate nucleus. *Arch. Internat. Physiol. Biochem.* 75:283–302.

HÖKFELT, T., and U. UNGERSTEDT, 1969. Electron and fluorescence microscopical studies on the nucleus caudatus putamen of the rat after unilateral lesions of ascending nigro-neostriatal dopamine neurons. *Acta Physiol. Scand.* 76:415–426.

HORNYKIEWICZ, O., 1966. Dopamine (3-hydroxytyramine) and brain function. *Pharmacol. Rev.* 18:925–964.

KRAUTHAMER, G., and D. ALBE-FESSARD, 1965. Inhibition of nonspecific sensory activities following striopallidal and capsular stimulation. *J. Neurophysiol.* 28:100–124.

LANGER, S. Z., 1966. The degeneration contraction of the nictitating membrane in the unanesthetized cat. *J. Pharmacol. Exp. Ther.* 151:66–72.

LAURSEN, A. M., 1963. Corpus striatum. *Acta Physiol. Scand.* 59:1–106.

MARSHALL, J. F., B. H. TURNER, and P. TEITELBAUM, 1971. Sensory neglect produced by lateral hypothalamic damage. *Science* 174:523–525.

MORGANE, P. J., 1961. Alterations in feeding and drinking behavior of rats with lesions in globi pallidi. *Amer. J. Physiol.* 201:420–428.

NAUTA, W. J. H., and W. R. MEHLER, 1969. Fiber connections of the basal ganglia. In *Psychotropic Drugs and Dysfunction of the Basal Ganglia*. U.S. Public Health Service Publication No. 1938, pp. 68–72.

OLTMANS, G. A., and J. A. HARVEY, 1972. LH syndrome and brain catecholamine levels after lesions of the nigrostriatal bundle. *Physiol. Behav.* 8:69–78.

POSCHEL, B. P. H., and F. W. NIUTEMAN, 1963. Norepinephrine: A possible excitatory neurohormone of the reward system. *Life Sci.* 2:782–788.

RANDRUP, A., and I. MUNKVAD, 1966. Role of catecholamines in the amphetamine excitatory response. *Nature (Lond.)* 211: 540.

RANDRUP, A., and I. MUNKVAD, 1970. Biochemical, anatomical and psychological investigations of stereotyped behavior induced by amphetamines. In *Amphetamines and Related Compounds*, E. Costa and S. Garattini, eds. New York: Raven Press, pp. 695–713.

SCHOENFELD, R., and M. URETSKY, 1972. Altered response to apomorphine in 6-hydroxydopamine-treated rats. *Europ. J. Pharmacol.* 19:115–118.

SNYDER, S. H., 1972. Catecholamines in the brain as mediators of amphetamine psychosis. *Arch. Gen. Psychiat.* 27:169–179.

SOURKES, T. L., L. J. POIRIER, and P. SINGH, 1969. Biochemical-histological-neurological models of Parkinsons disease. In *Third Symposium on Parkinson's Disease*, F. J. Gillingham and I. M. L. Donaldson, eds. Edinburgh and London: E. & S. Livingstone, pp. 54–60.

STEIN, L., 1962. Effects of interaction of imipramine chlorpromazine, reserpine, and amphetamine on self-stimulation: Possible neurophysiological basis of depression. *Recent Adv. Biol. Psychiat.* 4:288–308.

STEIN, L., and C. D. WISE, 1971. Possible etiology of schizophrenia: progressive damage to the noradrenergic reward system by 6-hydroxydopamine. *Science* 171:1032–1036.

UNGERSTEDT, U., 1968. 6-Hydroxydopamine induced degeneration of central monoamine neurons. *Europ. J. Pharmacol.* 5: 107–110.

UNGERSTEDT, U., L. L. BUTCHER, S. G. BUTCHER, N.-E. ANDÉN, and K. FUXE, 1969. Direct chemical stimulation of dopaminergic mechanisms in the neostriatum of the rat. *Brain Res.* 14:461–471.

UNGERSTEDT, U., 1970. Is interruption of the nigro-striatal dopamine system producing the "lateral hypothalamus syndrome"? *Acta Physiol. Scand.* 80:35A–36A.

UNGERSTEDT, U., and G. ARBUTHNOTT, 1970. Quantitative recording of rotational behavior in rats after 6-hydroxydopamine lesions of the nigro-striatal dopamine system. *Brain Res.* 24:485–493.

UNGERSTEDT, U., 1971a. Stereotaxic mapping of the monoamine pathways in the rat brain. *Acta Physiol. Scand.* (Suppl. 367) 82:1–48.

UNGERSTEDT, U., 1971b. Histochemical studies on the effects of intracerebral and intraventricular injections of 6-hydroxydopamine on monoamine neurons in the rat brain. In *6-Hydroxydopamine and catecholamine neurons*, T. Malmfors and H. Thoenen, eds. Amsterdam: North-Holland Comp., pp. 101–127.

UNGERSTEDT, U., 1971c. Striatal dopamine release after amphetamine or nerve degeneration revealed by rotational behavior. *Acta Physiol. Scand.* (Suppl. 367) 82:49–68.

UNGERSTEDT, U., 1971d. Postsynaptic supersensitivity after 6-hydroxydopamine induced degeneration of the nigro-striatal dopamine system in the rat brain. *Acta Physiol. Scand.* (Suppl. 367) 82:69–93.

UNGERSTEDT, U., 1971e. Mechanism of action of L-DOPA studied in an experimental Parkinson model. In *Monoamines and the Central Grey Nuclei*, J. de Ajuriaguerra, ed. Proceedings of the IV Bel-Air Symposium, pp. 165–170.

UNGERSTEDT, U., 1971f. Adipsia and aphagia after 6-hydroxydopamine induced degeneration of the nigro-striatal dopamine system in the rat brain. *Acta Physiol. Scand.* (Suppl. 367) 82:95–122.

UNGERSTEDT, U., 1973. Selective lesions of central monoamine pathways: Application in functional studies. In *Neurosciences Research*, S. Ehrenpreis and I. Kopin, eds. New York: Academic Press, (in press).

ZIGMOND, M. J., and E. M. STRICKER, 1972. Deficits in feeding behavior following intraventricular injection of 6-hydroxydopamine in rats. *Science*, in press.

61 6-Hydroxydopamine: A Chemical Lesion Technique for Studying the Role of Amine Neurotransmitters in Behavior

SUSAN D. IVERSEN

ABSTRACT Histochemical evidence indicates that the amine transmitters, norepinephrine (NE) and dopamine (DA), are localized in specific pathways in the CNS. One of these pathways arises from neurons in the substantia nigra (SN) and projects via the medial forebrain bundle and hypothalamus to the striatum; this pathway contains a large proportion of forebrain DA. Surgical lesions to the SN alter motor behaviors and locomotor and stereotyped responses to amphetamine and lead to severe aphagia and adipsia. These results, together with pharmacological evidence, suggest that this DA-containing system is important for normal motor behavior and feeding.

6-Hydroxydopamine (6-OHDA) is a chemical analog of DA that can be used as a valuable new tool for inducing chemically specific lesions in aminergic pathways in the CNS. When 6-OHDA is injected into the brain, it is taken up selectively by amine-containing neurons and causes degenerative changes. If the drug is applied by microinjections through cannulae into local areas of the CNS, lesions can be produced in specific aminergic pathways. Our earlier work on SN lesions has now been extended and repeated using the 6-OHDA technique. Bilateral injections of 6-OHDA into the rat SN do not produce the bizarre motor behavior seen after surgical lesions, nor did 6-OHDA change the characteristic locomotor responses to amphetamine. After 6-OHDA injections striatal DA was depleted by approximately 95%, indicating that DA is not critically involved in the control of locomotor activity. In contrast, the stereotype responses normally elicited by amphetamine were abolished in 6-OHDA lesioned animals, confirming much other evidence that suggests that DA release in the striatum is responsible for the stereotyped behavior produced by amphetamine.

THE DISCOVERY OF chemical neurotransmission in the brain is now legendary. More recently the demonstration that several different transmitters exist in the brain, not diffusely distributed but strictly localized in different anatomical systems, encouraged neuropsychologists to investigate the functional significance of these transmitter systems for the control of behavior. This chapter is not presented simply to demonstrate that a general loss of neurotransmitters abolishes behavior but to discover the importance of the different transmitters and the different parts of their anatomical distribution for the complex fabric of behavior. This is not a simple task for the complexity of the brain is rivaled only by the intricacies of behavior. However, a wealth of methodology is available to apply to these problems, ranging from the well-tried lesion and stimulation techniques of neuropsychology to the extensive pharmacological knowledge of the interaction between drugs and neurotransmitters.

The nigrostriatal pathway and motor behavior

Dopamine (DA) has been identified as a transmitter and is localized to an anatomical system different from that containing the related amine transmitter norepinephrine (NE) (Dahlström and Fuxe, 1964; Ungerstedt, 1971a). Within this framework investigation has been undertaken of the role of the dopamine-containing pathway between the substantia nigra and the corpus striatum in some motor behaviors. Principally, interest has focused on the increased locomotor activity and stereotypy (repetitive behavior including sniffing, rearing, neck movements, and gnawing) seen with varying doses of the stimulant drug amphetamine, and the hypothesis that the nigrostriatal pathway mediates these behavioral effects of the drug. Despite the limited nature of the problem and the simplicity of the behaviors, the project has some important properties that make it attractive from an experimental point of view.

1. The pathway between the nigra and the corpus striatum contains 75 to 80% of forebrain DA. In percentage terms it is a giant neurochemical pathway, although with classical anatomical techniques it is barely visible.

2. It is physically accessible for experimental manipulations. The cell bodies in the substantia nigra (SN) are separated from the axons of the dorsal and ventral NE bundles at the level of the SN. Furthermore, unlike, for example, hypothalamic nuclei, the SN is not closely

SUSAN D. IVERSEN Department of Experimental Psychology, University of Cambridge, England surrounded by pathways or structures that, if damaged accidentally, produce behavioral deficits likely to confuse the results.

3. There is extensive pharmacological literature on drugs which influence the synthesis, release, uptake, and receptor-stimulating properties of DA.

4. There is in vitro and in vivo evidence that amphetamine interacts with amine transmitters in the brain.

5. The fact that amphetamine-induced stereotypy and activity are motor behaviors and that the drug interacts with DA mechanisms provides the starting hypothesis that the nigrostriatal pathway mediates these behaviors.

Studies investigating the role of the nigrostriatal pathway for locomotor activity and stereotyped behavior

EFFECT OF 6-HYDROXYDOPAMINE LESIONS TO THE SUBSTANTIA NIGRA 6-Hydroxydopamine (6-OHDA) is a chemical analog of the amine transmitters NE and DA, which if injected intraventricularly is taken up by and destroys amine-containing neurons, thus depleting NE and DA in all brain regions (Uretsky and Iversen, 1970; Iversen and Uretsky, 1971). Ungerstedt (1971b) advanced the value of 6-OHDA as a chemical lesion technique by the use of implanted cannulae for local application to brain structures. When 6-OHDA is injected (8 μg 6-OHDA in 2 μl) bilaterally through implanted cannulae into the SN, substantial depletion of striatal DA can be achieved (Table I) following the drug-induced damage and degeneration of DA-containing neurons (Creese and Iversen, 1972). Sprague Dawley rats were given such lesions or appropriate sham operations and at 4, 11, and 19 days after operation were tested in photocell cages immediately after a dose of 1.5 mg/kg *d*-amphetamine. The results are shown in Figure 1. In controls 1.5 mg/kg *d*-amphetamine characteristically produced stimulation of locomotor activity reaching a maximum at 1 hr after treatment. This was followed by the emergence of sterotyped behavior including sniffing, rearing, and neck movements. The balance, intensity, and the time course of these two behaviors change depending on the dose of amphetamine. While photocells

TABLE I

Biochemical assessment of damage to catecholamine pathways in rat brain after various lesions, and the effects on amphetamine induced behavioral responses. Results are given as remaining catecholamine (DA in striatum, NE in hypothalamus and cortex) or tyrosine hydroxylase activity (marked *) as % values in untreated control animals

Lesion	Striatum DA	Hypo-thalamus NE	Cortex NE	Amphetamine Responses		
				Locomotor Activity	Stereotypy	Reference
6-OHDA lesion to s. nigra (*nonaphagic*)	10*	57*	—	√ enhanced	×	a
6-OHDA lesion to s. nigra (*aphagic*)	4	27	60	—	—	b
6-OHDA lesion to ventral bundle	67	21	—	√ enhanced	√	c
6-OHDA lesion to dorsal bundle	92	44	—	√	√	c
Intracerebral 6-OHDA (neonates)	0–2*	36*	0*	×	×	d
Intraventricular 6-OHDA (adults—whole brain assay)						
2 × 250 μg	15	20	4	√	√	e
1 × 100 μg	65	38	—	√ enhanced	√	e
Surgical lesions to s. nigra	64	73	—	√	√	f

*Tyrosine hydroxylase activity.
—Not measured.
[a] Creese and Iversen, 1972.
[b] Evetts et al. (unpublished).
[c] Creese (unpublished).
[d] Creese (unpublished).
[e] Evetts et al., 1970.
[f] Iversen, 1971.

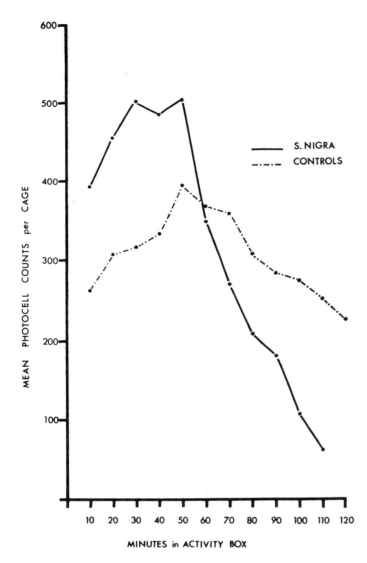

FIGURE 1 The mean locomotor and stereotypy response to 1.5 mg/kg *d*-amphetamine in rats with bilateral 6-OHDA lesions to the substantia nigra and sham operated controls. The responses were averaged for three test sessions on different days and the saline response subtracted.

accurately record locomotor activity, their use in measuring stereotypy demands caution. Stereotyped behavior may yield a high or a low photocell count depending on the position of the animal relative to the beams when intense stereotypy is occurring or depending on the amount of locomotor activity occurring in conjunction with the stereotypy. Therefore, when estimates of stereotypy are made, the photocell record is supported by direct observation of the rat.

In 6-OHDA treated animals, the locomotor response to amphetamine during the first hour was significantly enhanced (Figure 1). It is important to mention that bilateral surgical (Iversen, 1971) or 6-OHDA

(Ungerstedt, 1971c) lesions to the substantia nigra result in hypodipsia and hypophagia of varying severity. Deprivation has been reported to enhance the locomotor response to amphetamine (Campbell and Fibiger, 1971), and it seemed possible that this was the explanation for the enhanced amphetamine response during the first test hour in the present experiments. Accordingly, the sham-operated animals were deprived to the same body weight as the 6-OHDA nigral animals (Creese and Iversen, 1972), and the enhanced amphetamine response was still present. More recently Creese has controlled for the weight difference between the experimental and control groups by depriving both groups of food and water for 5 days after the operation. At this time the weights of the two groups were equal and were subsequently maintained by tube feeding. Again an enhanced locomotor response to amphetamine was seen in the animals with nigral lesions, thus discounting weight loss as the direct cause of the enhanced amphetamine responses.

By contrast, stereotyped behavior did not emerge during the second test hour in the animals with 6-OHDA lesions (Figure 1). The failure of amphetamine to induce stereotypy in the 6-OHDA treated animals was verified by treating the rats with a drug combination (150 mg/kg iproniazid + 5 mg/kg *d*-amphetamine), which elicited intense stereotyped behavior in control animals. The animals treated in this way were rated for stereotypy by recording beam interruptions and by direct observation for 3.6 hr after injection. It was confirmed that stereotyped responding was blocked in the rats with lesions in the SN.

This finding, together with previously published evidence that pharmacological manipulation of dopaminergic mechanisms (in some cases directly in the corpus striatum) elicits stereotypy, confirms the central role of the nigro-striatal DA system for amphetamine-induced stereotyped behavior (Munkvad et al., 1968).

The failure of 6-OHDA lesions to the SN to block the locomotor response to amphetamine might be cited in support of the suggestion that the NE containing pathways of the forebrain are principally involved in this behavior (Taylor and Snyder, 1971). However, in recent experiments the effects of amphetamine in rats subjected to bilateral 6-OHDA lesions to either the ventral *or* dorsal noradrenergic bundles at the level of the midbrain have been studied. Such lesions depleted forebrain NE to 40% or less of control values (Table I; in some animals to very low levels), and yet the amphetamine locomotor response was normal, indeed even enhanced after the ventral bundle lesion. The study is not yet complete, because we do not at present have data on groups of animals with a loss of hypothalamic NE as complete as the loss of striatal DA seen after SN lesions, but the preliminary results suggest that depletion of NE alone, like DA alone, will not block the locomotor response.

It may be asked therefore if any manipulation of the NE or DA systems can abolish the locomotor response to amphetamine.

EFFECT ON AMPHETAMINE-INDUCED MOTOR BEHAVIOR OF INHIBITING AMINE SYNTHESIS IN THE BRAIN   L-α-Methyl-*p*-tyrosine (αMT) prevents the synthesis of both DA and NE by inhibiting tyrosine hydroxylase, the rate limiting enzyme on the pathway from tyrosine to DOPA. Two hours after an i.p. injection of αMT (200 mg/kg), mice no longer showed a locomotor response to 4 mg/kg *d*-amphetamine (Figure 2) in agreement with previously reported findings (Rech et al., 1966; Weissman et al., 1966; and Stolk and Rech, 1970).

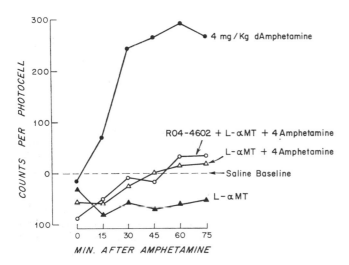

FIGURE 2   The effect of 200 mg/kg of α-methyl tyrosine (αMT) on the locomotor response to 4 mg/kg *d*-amphetamine in mice. The effect of αMT alone and pretreatment with the dopa decarboxylase inhibitor Ro4-4602 are also illustrated. The saline and αMT baselines have been subtracted where appropriate.

The possibility existed that αMT might abolish the amphetamine effects because of a receptor blocking effect of α-methyl-tyramine formed by decarboxylation from αMT in vivo. This interpretation is discounted, however, by the finding that after treatment with the DOPA decarboxylase inhibitor Ro4-4602, the αMT still blocked the amphetamine responses. Ro4-4602 inhibits the synthesis of DA to NE and was used in this experiment to prevent the conversion of αMT to α-methyl-tyramine, a "false transmitter."

The well-established effect of αMT in antagonizing the effects of amphetamine clearly implicates the catecholamines in the locomotor response to amphetamine, and the

task is to corroborate this result using alternative ways to deplete NE and DA.

THE USE OF 6-OHDA TO ACHIEVE NEAR TOTAL NE AND DA LOSS   It should be possible with multiple placements and repeated injections of 6-OHDA to achieve virtually total depletion of both DA and NE in the adult rat. Unfortunately, the severe aphagia and adipsia associated with successful SN lesions would render any such preparation undesirable in behavioral terms, irrespective of any additional effects due to the NE loss.

An alternative way of achieving such severe catecholamine depletion involves intracerebral injections of 6-OHDA (100 μg free base in 10 μl) into neonatal rats. Rats treated in this way on days 5, 7, and 9 are about half the weight of control animals at 3 months of age, have abnormal feeding and learning behavior, and a range of endocrinological defects (Breese and Traylor, 1972). Creese has recently found that at three months of age such animals show a totally blocked response to amphetamine both with respect to induced locomotor activity and stereotypy (Figures 3 and 4).

The analysis of the pharmacological and behavioral properties of these animals is at an early state but it is already clear that, despite the lack of presynaptic amines, the amine receptors appear to be functional. Indeed, if apomorphine was used to induce stereotypy (by direct stimulation of dopamine receptors), a dose of 1 mg/kg, which was ineffective in normal animals, produced intense stereotyped behavior in the animals treated with 6-OHDA as neonates (Figure 4).

Preliminary assay data on these animals (Table I) indicated that DA was almost completely absent from the

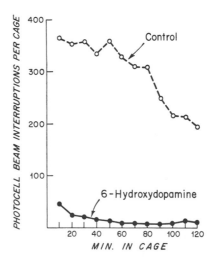

FIGURE 3   The blocked locomotor response to 1.5 mg/kg *d*-amphetamine in rats treated as neonates with intracerebral 6-OHDA.

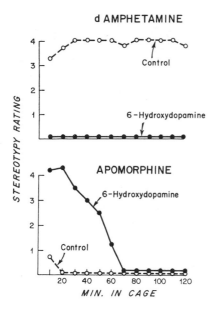

d AMPHETAMINE

APOMORPHINE

FIGURE 4 *Upper:* The blocked stereotypy to 5 mg/kg *d*-amphetamine in rats treated as neonates with intracerebral 6-OHDA. *Lower:* The enhanced stereotypy induced by 1 mg/kg apomorphine in rats treated as neonates with intracerebral 6-OHDA.

striatum, the dorsal NE system was also severely depleted, and the ventral NE system to the hypothalamus drastically reduced. The resistance of the ventral NE bundle even to 6-OHDA injections in neonatal animals supports the view that amine neurons vary in their sensitivity to 6-OHDA and probably also explains why after peripheral 6-OHDA injections in the neonate, Clark et al. (1972) found dorsal but not ventral bundle NE depletion.

Synthesis of the results

In summary, it has been shown that the dopaminergic nigro-striatal pathway plays a central role in amphetamine induced stereotyped behavior. Although, as Ahlenius and Engel (1972) have suggested, the NE system may modulate the stereotypy mechanism via an inhibitory influence, it cannot mediate the behavioral response in the absence of the DA system.

However, both NE and DA appear to be critically involved in the locomotor response to amphetamine, as virtually total depletion of either system does not by itself block the behavior, whereas depletion of both amines does. Such depletion must apparently be near total, and on the basis of the preliminary experiment on neonatally treated animals, the dorsal NE system may be more important than the ventral bundle. The projection of the dorsal NE system to the hippocampus and cortex is consistent with a role in behavioral arousal.

Ungerstedt (this volume) maintains that DA plays the central role in amphetamine induced locomotor stimulation, because he has demonstrated a block of such responses in adult rats after 6-OHDA SN lesions that produces near total striatal DA loss. Unfortunately, however, it seems that a "successful" SN lesion always results in substantial forebrain NE loss, presumably because of damage to the NE axons in the vicinity of the SN. Regional amine assay data on the brains of such animals supports this view (Table II); although the degree of DA depletion correlates with the block of the locomotor response and the severity of aphagia and adipsia, so does the associated NE depletion.

TABLE II

DA and NE levels in the striatum and hypothalamus/limbic brain regions after 6-OHDA lesions to the substantia nigra. These animals ($N = 5$) were severely aphagic and adipsic

	DA in Striatum	NE in Hypothalamus/Limbic System
	7.6	30
	8.0	60
	1.0	14
	5.0	15
	0	15
Mean	4.3 ± 1.6	27 ± 9.0

The strongest evidence against a central role for DA in both locomotor activity and stereotypy responses comes from the present results on animals with 90% striatal DA loss. Such DA depletion blocked stereotypy and not locomotor activity. A 10% residue of dopamine plus supersensitive, denervated receptors might still account for the normal locomotor response. However, special pleading is required if such a pool of dopamine could maintain one behavior (locomotor activity) but not a closely related one (stereotypy), for which there is much stronger evidence of DA involvement.

The present experiments, together with accumulating evidence from studies using inhibitors of enzymes involved in amine synthesis to deplete selectively DA or NE (Svensson and Waldeck, 1970), support the view that DA and NE interact in the control of amphetamine induced locomotor activity. In various experiments with 6-OHDA administered either locally to amine pathways or intraventricularly (Evetts et al., 1970), we have seen an enhancement of the locomotor responses to amphetamine. In all of these cases, 6-OHDA treatment resulted in an unbalanced depletion of DA and NE (Table I). If NE synthesis is temporarily inhibited with the dopamine-β-hydroxylase inhibitor, FLA-63, a depression of spontaneous locomotor activity is observed (Svensson, 1971), suggesting a normal facilitatory role of NE for this behavior, compared with its inhibitory influence on stereotypy (Ahlenius and Engel, 1971).

On this note of dynamic interplay between amine systems, it is relevant to return to one of the early experiments of this project. Before 6-OHDA became available, electrolytic lesions were used to damage the SN and induce neuronal degeneration and DA loss. Within hours of recovery from the anesthetic, intense stereotyped running, neck movements, and gnawing were seen in the absence of any drug treatment, and these spontaneous behaviors gradually faded to control levels over the next 10 days. The time course of these behaviors follows closely the release of striatal dopamine consequent upon neuron terminal degeneration in the corpus striatum (Faull and Laverty, 1969). It is tempting to speculate that these immediate postoperative effects on behavior reflect the sudden changes in the amine balance in the forebrain, consequent upon degeneration and release of endogenous amines. It may be likened to the stereotypy induced by temporarily inhibiting the synthesis of NE with the dopamine-β-hydroxylase inhibitor FLA-63 (Ahlenius and Engel, 1972).

Fourteen days after surgery, the response to amphetamine was tested, and in some animals the locomotor response to amphetamine was virtually abolished (Simpson and Iversen, 1971). Even in animals where such a block was not seen, the characteristic patterns of interaction between locomotor activity and stereotypy, seen at different doses of amphetamine, were disrupted. This result was puzzling because the striatal DA depletion was rarely more than 50%, and there was no obvious correlation between the completeness of the blocked amphetamine locomotor response and the degree of DA depletion (Iversen, 1971). The 6-OHDA results confirm the suspicion that the results of surgical lesions reflect *not* striatal DA depletion but surgical interference with extrapyramidal motor systems. Stereotypy was clearly not abolished in these rats with electrolytic SN lesions. It now seems likely that because the locomotor response was disrupted the normal balance between the behaviors was distorted, and stereotypy could emerge without competition from locomotor activity and with the same time course, irrespective of the dose of amphetamine (Iversen, 1971).

However, while the permanent surgical disruption is sufficient to distort the balance between the locomotor and stereotypy responses to amphetamine, it is not sufficient either in depletion or surgical terms to block amphetamine induced stereotypy. This suggests that while locomotor activity and stereotypy may be mutually dependent on the same NE and DA pools, they may find their specificity through NE and DA interactions operating on different efferent systems.

REFERENCES

AHLENIUS, S., and J. ENGEL, 1971. Behavioral and biochemical effects of L-dopa after inhibition of dopamine-β-hydroxylase in reserpine pretreated rats. *Naunyn Schmeidebergs Arch. Pharmakol.* 270:349–360.

AHLENIUS, S., and J. ENGEL, 1972. Effects of a dopamine (DA)-β-hydroxylase inhibitor on timing behavior. *Psychopharmacologia* 24:243–246.

BREESE, G. R., and T. D. TRAYLOR, 1972. Developmental characteristics of brain catecholamines and tyrosine hydroxylase in the rat: Effects of 6-hydroxydopamine. *Br. J. Pharmacol.* 44:210–222.

CAMPBELL, B. A., and H. C. FIBIGER, 1971. Potentiation of amphetamine-induced arousal by starvation. *Nature (Lond.)* 233:424–425.

CLARK, D. N. J., R. LAVERTY, and E. L. PHELAN, 1972. Long-lasting peripheral and central effects of 6-hydroxydopamine in rats. *Br. J. Pharmacol.* 44:831–842.

CREESE, I., and S. D. IVERSEN, 1972. Amphetamine response in rat after dopamine neuron destruction. *Nature New Biol.* 238:247–248.

DAHLSTRÖM, A., and K. FUXE, 1964. Evidence for the existence of monoamine-containing neurons in the central nervous system. *Acta Physiol. Scand.* (Suppl.) 232:1–55.

EVETTS, K. D., N. J. URETSKY, L. L. IVERSEN, and S. D. IVERSEN, 1970. Effects of 6-hydroxydopamine on CNS catecholamines, spontaneous motor activity and amphetamine induced hyperactivity in rats. *Nature (Lond.)* 225:961–962.

FAULL, R. L. M., and R. LAVERTY, 1969. Changes in dopamine levels in the corpus striatum following lesions in the substantia nigra. *Exp. Neurol.* 23:332–340.

IVERSEN, L. L., and N. J. URETSKY, 1971. Biochemical effect of 6-hydroxydopamine on catecholamine containing neurons in the rat central nervous system. In 6-*Hydroxydopamine and Catecholamine Neurons*, T. Malmfors and H. Theonen, eds. Amsterdam: North-Holland Publishing, pp. 171–186.

IVERSEN, S. D., 1971. The effect of surgical lesions to frontal cortex and substantia nigra on amphetamine responses in rats. *Brain Res.* 31:295–311.

MUNKVAD, I., H. PAKKENBERG, and A. RANDRUP, 1968. Aminergic systems in basal ganglia associated with stereotyped hyperactive behavior and catalepsy. *Brain Behav. Evol.* 1:89–100.

RECH, R. H., H. K. BORYS, and K. E. MOORE, 1966. Alterations in behavior and brain catecholamine levels in rats treated with α-methyltyrosine. *J. Pharmac. Exp. Ther.* 153:412–419.

SIMPSON, B. A., and IVERSEN, S. D., 1971. Effects of substantia nigra lesions on the locomotor and stereotypy responses to amphetamine. *Nature, New Biol.* 230:30–32.

STOLK, J. M., and R. H. RECH, 1970. Antagonism of D-amphetamine by alpha-methyl-L-tyrosine: Behavioral evidence for the participation of catecholamine stores and synthesis in the amphetamine stimulant response. *Neuropharmacology* 9:249–263.

SVENSSON, T. H., 1971. On the role of central norepinephrine in the regulation of motor activity and body temperature in the mouse. *Naunyn Schmeidebergs Arch. Pharmakol.* 271:111–120.

SVENSSON, T. H., and B. WALDECK, 1970. On the role of brain catecholamines in motor activity: Experiments with inhibitors of synthesis and of monoamine oxidase. *Psychopharmacologia* 18:357–365.

TAYLOR, K. H., and S. H. SNYDER, 1971. Differential effects of D- and L-amphetamine on behavior and on catecholamine disposition in dopamine norepinephrine containing neurons of rat brain. *Brain Res.* 28:295–309.

UNGERSTEDT, U., 1971a. Stereotaxic mapping of the mono-amine pathway in the rat brain. *Acta Physiol. Scand. (Suppl.)* 367:1–48.

UNGERSTEDT, U., 1971b. Histochemical studies on the effect of intracerebral and intraventricular injections of 6-hydroxy-dopamine on monoamine neurons in the rat brain. In *6-Hydroxydopamine and Catecholamine Neurons*, T. Malmfors and H. Theonen, eds. Amsterdam: North-Holland Publishing, pp. 101–127.

UNGERSTEDT, U., 1971c. Adipsia and aphagia after 6-hydroxy-dopamine induced degeneration of the nigrostriatal dopamine system. *Acta. Physiol. Scand.* (Suppl.) 367:95–122.

URETSKY, N. J., and L. L. IVERSEN, 1970. Effects of 6-hydroxy-dopamine on catecholamine containing neurons in the rat brain. *J. Neurochem.* 17:269–278.

WEISSMAN, A., B. K. KOE, and S. J. THENEN, 1966. Anti-amphetamine effects following inhibition of tyrosine hydroxy-lase. *J. Pharm. Exp. Ther.* 151:339–352.

62 Adrenergic Receptor Mechanisms in Eating and Drinking

SARAH FRYER LEIBOWITZ

ABSTRACT   The role of central adrenergic receptor mechanisms in the normal regulation of food and water intake is the primary subject. Experimental evidence from studies using central drug injections is presented. Several topics are discussed, including (a) the function of the two types of adrenergic receptors, alpha and beta, (b) the localization of the different receptor effects in different hypothalamic sites, and (c) the central mode of action of the drugs amphetamine and chlorpromazine.

PHARMACOLOGICAL studies in the peripheral nervous system have demonstrated the important role that adrenergic mechanisms play in controlling the functions of peripheral tissues. Detailed analyses of these peripheral adrenergic mechanisms, using several different adrenergic stimulants and receptor blockers, have suggested that there are at least two types of adrenergic receptors, which Ahlquist (1948) called alpha and beta. Stimulation of these two types of receptors has sometimes been found to produce diametrically opposite effects, such as on vascular muscle where alpha-receptor activity produces constriction and beta-receptor activity produces dilatation. In other cases, alpha- and beta-receptor stimulation may have synergistic effects, such as on the intestinal smooth muscle where stimulation of either type of receptor causes relaxation.

The various effects of adrenergic receptor stimulation have been studied with the help of the three agonists: norepinephrine (NE), epinephrine (EPI), and isoproterenol (ISOP), which differ in the ratio of their effectiveness in stimulating peripheral alpha and beta receptors (Innes and Nickerson, 1970). Norepinephrine has potent alpha-receptor action but somewhat weaker action on beta receptors. Isoproterenol, in contrast, acts predominantly on beta receptors and has very little alpha-receptor activity. Finally, EPI has potent action on both alpha and beta receptors.

During the past decade or so, some pharmacological studies have been carried out in the central nervous system that suggest that adrenergic receptor mechanisms

SARAH FRYER LEIBOWITZ   The Rockefeller University, New York, N.Y.

also exist in the brain and that they may regulate, possibly through the transmitter action of NE, specific behavioral and physiological responses. Histochemical and biochemical studies have demonstrated the existence in the brain of NE and also of the enzyme that converts NE to EPI. Direct applications to the brain of these agents, or of compounds that affect their level or activity in the brain, have been found to alter central neural activity and, ultimately, behavior. The role of central adrenergic receptor mechanisms in the control of ingestive behavior has been extensively investigated in the rat, and the evidence to be described here suggests that there are alpha- and beta-adrenergic receptors in the brain which modulate (that is, elicit and inhibit) the ingestion of food and water.

Technique for central drug injection

To study the relationship between central adrenergic mechanisms and behavior, a technique has been used that allows drugs to be repeatedly administered into the brains of unanesthetized, unrestrained subjects. This technique involves the use of a cannula, usually a 23- to 29-gauge hypodermic needle, which is stereotaxically implanted into a specific area of the brain of an anesthetized subject and then fixed in place by an acrylic cement on top of the skull. After the subject has recovered from surgery, drugs are administered through this cannula, either in crystalline form or in solution (in volumes of 0.1 to 1.0 μl) and, immediately thereafter, the responses of the unrestrained subject are observed. The procedures used for examining drug effects on the ingestive responses, feeding and drinking, are generally quite straightforward. In some experiments, the subjects are tested in a satiated state, in which case they are given food and water ad libitum before the test. In other experiments, the subjects are tested in a deprived state, in which case they are deprived of either food or water for a certain period of time before the test. During the test, measured food and/or water is made available, and the subject's consumption of either is recorded at frequent intervals.

Alpha-adrenergic receptors that elicit feeding

A connection between central adrenergic systems and the elicitation of feeding behavior was first demonstrated in the rat by Grossman (1962a), who found that injection of NE or EPI directly into the hypothalamus of satiated subjects induced eating (Table I). Further examination of this adrenergic feeding effect has shown that its magnitude is dependent upon the dose of the agonist (Miller et al., 1964; Booth, 1968; Leibowitz, in preparation) and that it can be obtained at several diencephalic sites (Booth,1967; Coury, 1967).

One especially sensitive site appears to be the "perifornical" hypothalamus, an area near the fornix, the caudal half of the anterior hypothalamus. At this site, a dose of EPI at least as low as 0.25 nmole can produce a reliable feeding response (Leibowitz, 1972d). The latency of the adrenergically elicited feeding response, which can vary from less than a minute to as much as 10 min, tends to decrease with increase in dose and also appears to be affected by the presence or absence of drinking water. That is, when injected with NE or EPI, a rat frequently drinks some water (see below) before starting to eat at 5 to 6 min after injection. If water is not available, the feeding response occurs sooner, usually 2 to 3 min after injection. The feeding itself generally lasts between 10 and 30 min. However, the stimulating effect on adrenergic receptors may remain active for well over an hour, as shown for example by animals who exhibit enhanced eating when given food for the first time as much as 60 min after injection.

Studies designed to determine which type of adrenergic receptor, alpha or beta, is mediating this stimulating effect on food intake suggest that it is alpha in nature. Alpha-receptor blocking agents, but not beta-receptor blocking agents, can reliably reduce or even abolish the enhancement of eating induced by hypothalamic injection of NE or EPI in either satiated or hungry rats (Grossman, 1962b; Slangen and Miller, 1969; Leibowitz, in preparation). Evidence to suggest that central alpha-adrenergic receptors are physiologically active during natural hunger is provided by Grossman (1962b) and Leibowitz (1970c), who showed that eating by a hungry rat can be reliably suppressed by peripheral or central injection of an alpha-receptor blocker. Further evidence is provided by the study of Slangen and Miller (1969), which showed that eating can be induced in satiated rats by a combination of drugs which presumably causes an overflow of endogenous NE into the synaptic cleft.

In addition to these studies in the rat, there are studies in the monkey that show that hypothalamic injection of NE can also increase food consumption in this species

TABLE I

Alpha- and beta-adrenergic effects on ingestive behavior induced by hypothalamic injection of epinephrine or norepinephrine

Ingestive Behavior	Adrenergic Effect	Type of Receptor*	Drug Dose (nmole)	Hypothalamic Site of Injection	Latency (min)	Magnitude (increases with dose)	Duration (min) (increases with dose)	Special Characteristics
Feeding	Stimulation	α	0.5	Perifornical ventromedial	3–5	Increase of 1–6 g (baseline = 0.3 g)	at least 90**	Often preceded by drinking; eating is vigorous and continuous
	Suppression	β	10.0	Perifornical lateral	0–1	Suppression of up to 90%	at least 120	Antagonized by above alpha-adrenergic feeding effect; greater suppression observed in the presence of an alpha-receptor blocker
Drinking	Suppression	α	0.005	Perifornical ventromedial	3–5	Suppression of up to 90%	at least 180	Suppresses drinking induced by deprivation, hyperosmolarity, hypovolemia, and central drug injections
	Stimulation	α and β	0.5	Perifornical	1–2	Increase of 1–6 ml (baseline = 0.0 ml)	1–5**	Brief but vigorous; associated with alpha-adrenergic feeding effect

*The type of receptor mediating the effect is determined on the basis of the blocking effects of alpha- and beta-receptor blockers. An alpha-receptor effect is blocked by the alpha blockers, phentolamine and tolazoline. A beta-receptor effect is blocked by the beta blockers, propranolol, MJ 1999 and LB-46.

**Duration of the alpha feeding-stimulation effect and the alpha plus beta drinking-stimulation effect is determined by giving the subjects food and water, respectively, at variable intervals after injection.

(Myers, 1969). This evidence suggests that in the monkey, as in the rat, central adrenergic mechanisms may play a role in controlling feeding behavior.

Beta-adrenergic receptors that inhibit feeding

In contrast to the two agonists NE and EPI, which have alpha-receptor activity, the relatively pure beta-receptor agonist ISOP fails to elicit feeding in the rat (Myers and Yaksh, 1968; Slangen and Miller, 1969; Leibowitz, 1972c). This evidence and the failure of beta-receptor blockers to suppress feeding elicited by NE or EPI indicate that in the rat central beta-adrenergic receptors, unlike alpha-adrenergic receptors, are not important for eliciting the eating response. However, there is some evidence that suggests that central beta receptors may indeed influence feeding behavior but in the opposite direction (Table I) (Leibowitz, 1970a; Goldman et al., 1971). That is, in contrast to the stimulating effect of alpha-receptor activity, beta-receptor activity appears to have a strong inhibitory effect on eating behavior. This suppression of feeding was first observed with peripheral administration of the adrenergic agonists (Miller, 1965; Conte et al., 1968; Russek et al., 1968) and was subsequently observed and analyzed by Leibowitz with central administration of drugs.

Perifornical hypothalamic injection of the beta agonist ISOP produces, immediately after injection in hungry rats, a strong suppression of feeding, the magnitude and duration of which is dependent upon dose (Leibowitz, 1970a, and unpublished work). At a dose of 5 nmoles, ISOP reliably suppresses feeding for 15 to 30 min, and at 20 times this dose, the suppression lasts for over 2 hr. This effect of ISOP is totally eliminated by preceding injections of beta-receptor blockers but not alpha-receptor blockers.

Perifornical hypothalamic injection of EPI, an agonist with potent beta-receptor action as well as potent alpha-receptor action, can have, in addition to its alpha-receptor potentiation of feeding effect (see above), a strong suppressing effect on food intake that is blocked by beta-receptor blockers (Leibowitz, 1970a). However, especially at relatively low doses, this beta-receptor suppression of feeding effect may not always be reliably apparent, because of EPI's simultaneous alpha-receptor action, and in some cases it may be exhibited only in the presence of an alpha-receptor blocker (Leibowitz, in preparation). These studies of the effects of adrenergic blocking agents on EPI-induced changes in food intake demonstrate the ability of centrally injected EPI to stimulate both alpha and beta receptors and lend further support to the suggestion that these two types of receptors have antagonistic control of feeding behavior.

In the peripheral nervous system, the beta-receptor action of EPI is generally found to be more potent than that of NE. In the brain this also appears to be the case, at least with respect to the agonists' stimulation of the beta receptors that inhibit food consumption. Injection of NE alone into the perifornical hypothalamus of hungry rats is most often found to produce the alpha-receptor effect, a potentiation of food consumption. However, when injected with an alpha-receptor blocker, a marked beta suppressing effect of this agonist on feeding can be observed (Leibowitz, in preparation). These findings indicate that NE can activate both the alpha receptors for feeding and the beta receptors for inhibition of feeding, and therefore that endogenous NE might be a neurotransmitter in the central adrenergic system that has diametrically opposite effects on feeding behavior. The feeding stimulation effect that can be produced by a high dose of a beta-receptor blocker and the feeding suppression effect that can be produced by a high dose of an alpha-receptor blocker (Leibowitz, 1970c) lend further support to the suggestion that these two types of adrenergic receptor systems are physiologically active during normal regulation of feeding.

Alpha-adrenergic receptors that suppress drinking

The above findings demonstrate the importance of central adrenergic mechanisms in the control of feeding behavior. Additional evidence, which is now described, suggests that central adrenergic mechanisms may also play a role in the control of drinking behavior (Table I).

Hypothalamic injection of NE was first reported by Grossman (1962a) to have a suppressing effect upon water consumption in thirsty rats. This effect was further analyzed by Leibowitz (1971a, 1972b, in preparation) and was found to be produced by perifornical hypothalamic injection of either NE or EPI but not of the beta agonist ISOP. This inhibitory effect of adrenergic stimulation on drinking, which appears to have a latency of up to 5 min, can be reliably observed for about 2 min at a dose as low as 0.005 nmole. As the dose is increased beyond this point, both the magnitude and duration of the effect increase.

Experiments with receptor blocking agents (Leibowitz, 1972b, in preparation) demonstrate that the suppression of water intake produced by central NE or EPI injection is blocked by alpha-receptor blockers but not by beta-receptor blockers. This finding indicates that it is alpha, and not beta, receptors that cause an inhibition of drinking. Evidence to suggest that these alpha receptors, presumably located in the brain, are physiologically active in the normal regulation of water consumption is provided by the finding that hypothalamic injection of an alpha-receptor blocker alone can stimulate drinking (Leibowitz, 1971a).

In addition to suppressing deprivation-induced drinking, hypothalamic injection of NE is found to suppress water consumption elicited by cellular dehydration, hypovolemia, and central injections of other drugs (Leibowitz, 1972b; Singer and Kelly, 1972). These findings further emphasize the importance of central adrenergic systems in the control of drinking behavior and suggest that endogenous NE may be a mediator in these systems.

Adrenergic receptors that elicit drinking

Peripheral administration of ISOP, a beta agonist, was found by Lehr et al. (1967) to produce in rats a drinking response which occurred sporadically over a period of 2 to 3 hr. Lehr suggested that ISOP was producing this response by entering the brain and stimulating central beta receptors that elicit drinking. Houpt and Epstein (1971), however, proposed an alternative hypothesis. They suggested that by causing a release of renin from the kidneys, ISOP was increasing the levels of circulating angiotensin and that it was this substance which acted directly on the brain to elicit drinking. These investigators, finding that nephrectomy abolishes drinking induced by peripheral ISOP, suggest that the kidneys are in some way, directly or indirectly, involved in this beta-adrenergic drinking phenomenon. However, it leaves unanswered the question of whether beta receptors that induce drinking do exist in the brain.

In attempting to answer this question, Leibowitz (1971a) injected ISOP directly into the lateral hypothalamus of the rat. At doses of 25 to 100 nmole, a reliable drinking response was obtained, which resembled the response observed after peripheral ISOP injection (it was sporadic but persistent). This drinking response was blocked by a centrally injected beta blocker but not by an alpha blocker. In view of the fact that the doses of ISOP used centrally were similar to the doses found to be effective when peripherally administered (Lehr et al., 1967), this evidence can only tentatively be accepted as support for Lehr's hypothesis that beta receptors for drinking do exist in the brain. Since several other hypothalamic sites tested did not respond positively to ISOP and since these negative sites were in some cases located right next to the positive lateral hypothalamic site, it would appear that ISOP injected into the lateral hypothalamus is acting locally on central beta receptors and is not simply diffusing to other central or even peripheral sites. Evidence to support this central mediation hypothesis should ideally include the locating of a site that is

sensitive to a dose of ISOP that is lower than the dose required to produce the effect when peripherally injected.

Lateral hypothalamic injection of EPI or NE does not produce the sporadic, long-lasting drinking elicited by ISOP injection. However, these two agonists, when injected into the perifornical hypothalamus of the satiated rat, do elicit a quite different drinking response (Table I). This response, which is dose-dependent, starts almost immediately after injection and lasts for only about 5 min (in contrast to the ISOP response which starts 5 to 10 min after injection and occurs sporadically over a period of 2 hr) (Leibowitz, 1971b, in preparation). This brief drinking phenomenon, which can be reliably elicited by centrally injected NE or EPI at a dose at least as low as 0.5 nmole, is not seen after peripheral administration of these agonists. Studies with centrally injected blocking agents show that it can be eliminated both by beta-receptor blockers and by alpha-receptor blockers. Interestingly, this drinking effect is followed immediately by, and correlated in magnitude with, an alpha-receptor feeding response (see above). This connection between the adrenergic drinking phenomenon and the ingestion of food suggests that the water consumption elicited by NE or EPI is a food-associated drinking response, a type of drinking that is very frequently exhibited by rats under normal conditions (Fitzsimons and Le Magnen, 1969; Kissileff, 1969). In view of the evidence that central alpha-receptor activity elicits feeding and the suggestive evidence discussed above that central beta-receptor activity may elicit drinking, it is especially interesting that both alpha and beta receptors are involved in mediating this food-associated drinking behavior.

In contrast to NE or EPI, the pure beta agonist ISOP, when injected alone into the perifornical hypothalamus, only occasionally elicits the brief drinking response described above and never elicits the feeding response that follows the drinking. It has been found, however, that when ISOP is centrally injected simultaneously with NE or EPI, this beta agonist can greatly enhance the food-associated drinking response elicited by either of the two other agonists (Leibowitz, 1972a). This result, which is not seen with peripheral ISOP injection, confirms the involvement of central beta receptors in eliciting drinking behavior, at least that drinking which is closely associated with eating.

Localization within the hypothalamus of alpha- and beta-receptor effects

Most of the experiments described above were carried out in the perifornical hypothalamus, an area that is dense with adrenergic terminals and that is found to be especially sensitive to almost all of the effects of the adrenergic agonists. Other parts of the rat hypothalamus have been examined in a similar fashion, and the results of these studies indicate that different hypothalamic areas may differ greatly in their sensitivity to the effects of alpha and beta stimulation (Leibowitz, 1970c, 1971c, and unpublished work). In general, it appears that the ventromedial hypothalamus is considerably more sensitive than the lateral hypothalamus to alpha-receptor stimulation (which elicits feeding and suppresses drinking), whereas the lateral hypothalamus is considerably more sensitive than the ventromedial hypothalamus to beta-receptor stimulation (which suppresses feeding). Furthermore, drugs that act indirectly, by releasing or depleting endogenous catecholamines or by blocking the adrenergic receptors, are found to have diametrically opposite effects on feeding when injected into the ventromedial hypothalamus versus when injected into the lateral hypothalamus. This evidence, which differentiates the medial and lateral hypothalamus via pharmacological techniques, is intriguing in light of the large number of electrical stimulation and lesion studies that also show differential effects of these two areas on feeding.

Central effects of amphetamine

Amphetamine has been known for some time to be a potent anorexic agent. Its mechanism of action, however, is still not understood. One possibility appears to be the direct or indirect stimulation by amphetamine of the beta-adrenergic receptors, possibly in the lateral hypothalamus, which inhibit feeding. This possibility was suggested by Leibowitz (1970b, 1970c), who demonstrated that the suppression of feeding induced by perifornical or lateral hypothalamic injection of amphetamine could be abolished by beta-receptor blockers and enhanced by an alpha-receptor blocker. When injected into the ventromedial hypothalamus, an area found to be very sensitive to alpha-adrenergic stimulation but relatively insensitive to beta-adrenergic stimulation (see above), amphetamine did not cause a suppression of feeding but instead caused a small enhancement of feeding.

In addition to affecting feeding behavior, amphetamine also appears to influence drinking behavior. Perifornical hypothalamic injection of amphetamine is found to suppress water intake reliably in thirsty rats, and this effect, in contrast to amphetamine's potent beta-suppressing effect on food intake, is found to be mediated by alpha receptors (Leibowitz, 1973). These findings indicate that in the brain, as in the periphery, amphetamine can have both alpha- and beta-receptor action.

Central effects of chlorpromazine

Chlorpromazine, a widely used tranquilizing agent, is generally thought to block alpha-adrenergic receptors, at least those in the peripheral nervous system. In order to test whether this drug has such action on the hypothalamic alpha receptors that elicit feeding, Leibowitz and Miller (1969) injected chlorpromazine directly into the perifornical hypothalamus of the rat. Instead of suppressing feeding as do other alpha-receptor blockers, chlorpromazine was unexpectedly found to stimulate feeding reliably. This effect, like that induced by NE, could be clocked by an alpha blocker but not by a beta blocker (Leibowitz, 1972b, in preparation). It could also be reliably suppressed by drugs that deplete endogenous stores of NE but that leave intact postsynaptic receptors that are sensitive to NE (Leibowitz, 1969). This evidence suggests that, at least in the perifornical hypothalamus, chlorpromazine *activates* rather than blocks the alpha-adrenergic receptor system. This finding offers a possible explanation for two intriguing clinical observations: That patients receiving chlorpromazine frequently gain weight, and that in some patients this agent has pronounced antidepressant effects.

Central effects of dopamine

The above experiments with NE demonstrate that this agonist, when injected into the rat hypothalamus, can have a variety of effects on ingestive behavior. It can elicit a drinking response as well as a feeding response, and it can suppress these two behaviors in subjects made thirsty or hungry by deprivation. Since NE appears to be synthesized and stored in the hypothalamus, it is very possible that endogenous NE may act as a neurotransmitter in systems of the brain which control feeding and drinking behavior. Epinephrine also has profound effects on ingestive responses and appears to be even more potent than NE in altering such behavior. However, whether this catecholamine is synthesized in the brain and can act as a neurotransmitter is still uncertain.

There is a good deal of evidence that dopamine, another catecholamine, is indeed synthesized and stored in the brain and that it may act as a neurotransmitter. A few investigators have examined the effects of hypothalamic injection of dopamine on feeding and drinking behavior in order to determine whether the above phenomena induced by adrenergic stimulation are specific to an adrenergic system or whether they are also mediated by a central dopaminergic system. In general, the results of these studies with dopamine reveal few effects of this catecholamine on ingestive behavior.

Slangen and Miller (1969) demonstrated in satiated rats a very weak, delayed feeding response that could very likely be a result of dopamine's being used in the synthesis of NE. Preliminary studies by Leibowitz (unpublished work) have shown that injections of dopamine, into several different hypothalamic areas and at a wide range of doses, fail to alter reliably either feeding or drinking behavior in hungry or thirsty rats. So far, therefore, the evidence does not provide support for the suggestion that a hypothalamic dopaminergic system, in addition to a hypothalamic adrenergic system, mediates the control of ingestive behavior in the rat.

Concluding remarks

The adrenergic system in the rat brain, and possibly also in the brain of the monkey, appears to play an important role in the control of ingestive behavior. Furthermore, there is evidence to suggest that the different types of adrenergic receptors, alpha and beta, which are found to exist in the periphery, also exist in the brain and that they have, at least in some cases, antagonistic effects on a particular ingestive response. The reciprocal influence of alpha receptors on feeding and drinking (stimulation and suppression, respectively) and the reciprocal influence of beta receptors on feeding and drinking (suppression and stimulation, respectively) suggest a possible central neurochemical basis for the rat's ability to balance its ingestion of food and water, at least on a short-term basis. Changes in intensity of endogenous alpha- and beta-receptor activity, and the interaction between the effects of such activity, may help the rat to exhibit one ingestive response without interference from the other ingestive response, to shift more readily from one response to the other, and to keep the magnitude of one response in balance with the magnitude of the other response.

In addition to this adrenergic receptor mechanism, there appear to be other neurochemical systems of the brain that modulate ingestive behavior. These include the cholinergic (Grossman, 1962a), the serotonergic (Goldman et al., 1971), and the angiotensin (Epstein et al., 1970) systems. It seems very possible that these systems and the adrenergic system interact in their regulation of feeding and drinking. If this is indeed the case, then a comprehensive and accurate model of the central control of ingestive behavior will have to be founded on a more complete understanding of the roles of each of the neurochemical systems, as well as of their interactions.

ACKNOWLEDGMENTS   The author's research was supported by U.S. Public Health Service grant MH 13189 and by funds from Hoffmann-La Roche and from Smith, Kline, and French.

REFERENCES

AHLQUIST, R. P., 1948. A study of adrenotropic receptors. *Amer. J. Physiol.* 153:586–600.

BOOTH, D. A., 1967. Localization of the adrenergic feeding system in the rat diencephalon. *Science* 158:515–517.

BOOTH, D. A., 1968. Mechanism of action of norepinephrine in eliciting an eating response on injection into the rat hypothalamus. *J. Pharmacol. Exp. Ther.* 160:336–348.

CONTE, M., D. LEHR, W. GOLDMAN, and M. KRUKOWSKI, 1968. Inhibition of food intake by beta-adrenergic stimulation. *Pharmacologist* 10:180.

COURY, J. N., 1967. Neural correlates of food and water intake in the rat. *Science* 156:1763–1765.

EPSTEIN, A. N., J. T. FITZSIMONS, and B. J. ROLLS (née Simons), 1970. Drinking induced by injection of angiotensin into the brain of the rat. *J. Physiol. (Lond.)* 210:457–474.

FITZSIMONS, J. T., and J. LE MAGNEN, 1969. Eating as a regulatory control of drinking in the rat. *J. Comp. Physiol. Psychol.* 67:273–283.

GOLDMAN, H. W., D. LEHR, and E. FRIEDMAN, 1971. Antagonistic effects of alpha and beta-adrenergically coded hypothalamic neurons on consummatory behavior in the rat. *Nature (Lond.)* 231:453–455.

GROSSMAN, S. P., 1962a. Direct adrenergic and cholinergic stimulation of hypothalamic mechanisms. *Amer. J. Physiol.* 202:872–882.

GROSSMAN, S. P., 1962b. Effects of adrenergic and cholinergic blocking agents on hypothalamic mechanisms. *Amer. J. Physiol.* 202:1230–1236.

HOUPT, K. A., and A. N. EPSTEIN, 1971. The complete dependence of beta-adrenergic drinking on renal dipsogen. *Physiol. Behav.* 7:897–902.

INNES, I. R., and M. NICKERSON, 1970. Drugs acting on postganglionic adrenergic nerve endings and structures innervated by them (sympathomimetic drugs). In *The Pharmacological Basis of Therapeutics*, L. S. Goodman and A. Gilman, eds. New York: Macmillan, 4th Ed., pp. 478–523.

KISSILEFF, H. R., 1969. Food-associated drinking in the rat. *J. Comp. Physiol. Psychol.* 67:284–300.

LEHR, D., J. MALLOW, and M. KRUKOWSKI, 1967. Copious drinking and simultaneous inhibition of urine flow elicited by beta-adrenergic stimulation and contrary effect of alpha-adrenergic stimulation. *J. Pharmacol. Exp. Ther.* 158:150–163.

LEIBOWITZ, S. F., 1969. Mechanism of unexpected adrenergic effect from hypothalamic injection of chlorpromazine. *Proc. 77th Annual Convention, APA*, pp. 901–902.

LEIBOWITZ, S. F., 1970a. Hypothalamic beta-adrenergic "satiety" system antagonizes an alpha-adrenergic "hunger" system in the rat. *Nature (Lond.)* 226:963–964.

LEIBOWITZ, S. F., 1970b. Amphetamine's anorexic versus hunger-inducing effects mediated respectively by hypothalamic beta- versus alpha-adrenergic receptors. *Proc. 78th Annual Convention, APA*, pp. 813–814.

LEIBOWITZ, S. F., 1970c. Reciprocal hunger-regulating circuits involving alpha- and beta-adrenergic receptors located, respectively, in the ventromedial and lateral hypothalamus. *Proc. Nat. Acad. Sci. USA* 67:1063–1070.

LEIBOWITZ, S. F., 1971a. Hypothalamic alpha- and beta-adrenergic systems regulate both thirst and hunger in the rat. *Proc. Nat. Acad. Sci. USA* 68:332–334.

LEIBOWITZ, S. F., 1971b. Hypothalamic β-adrenergic "thirst" system mediates drinking induced by carbachol and transiently by norepinephrine. *Fed. Proc.* 30:280.

Leibowitz, S. F., 1971c. Hypothalamic norepinephrine as an alpha- and beta-adrenergic neurotransmitter active in the regulation of normal hunger. *Proc. 78th Annual Convention, APA* p. 741.

Leibowitz, S. F., 1972a. Hypothalamic beta-adrenergic receptors and their interaction with alpha receptors in the regulation of drinking. Paper presented at 43rd Annual Meeting, EPA, Boston, April 27–29.

Leibowitz, S. F., 1972b. Hypothalamic alpha-adrenergic suppression of drinking: Effects on several types of thirst. *Proc. 80th Annual Convention, APA*, 845–846.

Leibowitz, S. F., 1972c. Central adrenergic receptors and the regulation of hunger and thirst. In *Neurotransmitters.* Res. Publ. A.R.N.M.D., Vol. 50, 1972, pp. 327–358.

Leibowitz, S. F., 1973. Alpha-adrenergic receptors mediate suppression of drinking induced by hypothalamic amphetamine injection. *Fed. Proc.* 32:754.

Leibowitz, S. F., and N. E. Miller, 1969. Unexpected adrenergic effect of chlorpromazine: Eating elicited by injection into rat hypothalamus. *Science* 165:609–611.

Miller, N. E., 1965. Chemical coding of behavior in the brain. *Science* 148:328–338.

Miller, N. E., K. S. Gottesman, and N. Emery, 1964. Dose response to carbachol and norepinephrine in rat hypothalamus. *Amer. J. Physiol.* 206:1384–1388.

Myers, R. D., 1969. Chemical mechanisms in the hypothalamus mediating eating and drinking in the monkey. *Ann. N.Y. Acad. Sci.* Art. 2, 157:918–933.

Myers, R. D., and T. L. Yaksh, 1968. Feeding and temperature responses in the unrestrained rat after injection of cholinergic and aminergic substances into the cerebral ventricles. *Physiol. Behav.* 3:917–928.

Russek, M., J. A. F. Stevenson, and G. J. Mogenson, 1968. Anorexigenic effects of adrenaline, amphetamine, and FMSIA. *Can. J. Physiol. Pharmacol.* 46:635–638.

Singer, G., and J. Kelly, 1972. Cholinergic and adrenergic interaction in the hypothalamic control of drinking and eating behavior. *Physiol. Behav.* 8:885–890.

Slangen, J. L., and N. E. Miller, 1969. Pharmacological tests for the function of hypothalamic norepinephrine in eating behavior. *Physiol. Behav.* 4:543–552.

63 Catecholamines as Mediators of Drug Effects in Schizophrenia

SOLOMON H. SNYDER

ABSTRACT Interactions of psychoactive drugs and neuro-transmitters are a fruitful area for correlating biochemistry and behavior, especially as related to psychiatric illness. Ways in which psychoactive drugs influence neurotransmitters, particularly the biogenic amines, are reviewed. Emphasis is placed on a critical assessment of criteria for determining if a given "effect" of a drug represents its mode of action. Possible clues that such drug effects afford to the pathophysiology of various psychiatric disabilities in schizophrenia are evaluated.

AN ULTIMATE GOAL of the topic *Biochemistry and Behavior* might be to link specific human behaviors to particular neurochemical events. For many people, the highest priority lies with the aberrant behavior of the mentally ill. And of the major mental illnesses, the one with the most profound financial, human, and scientific consequences is schizophrenia.

Neurotransmitter disposition represents the biochemistry that is most unique to nervous tissues. Accordingly, the greatest success in ascribing specific behavioral events to individual biochemical sequences has arisen from studies of neurotransmitters, especially the biogenic amines serotonin and the catecholamines, norepinephrine, and dopamine. The important histochemical demonstration of monoamine tracts by fluorescence histochemistry (Hillarp et al., 1966; Ungerstedt, 1971) has greatly facilitated the tasks of workers such as Ungerstedt and S. Iversen (this volume), who have used stereotaxically administered 6-hydroxydopamine to produce selective lesions of particular catecholamine tracts in the brain with 6-hydroxydopamine. In this way they have been able to show that particular behavioral effects of drugs can be ascribed to one or another of the various catecholamine pathways in the brain. Unfortunately, no such direct techniques are available for exploring the neurochemical substrata of altered behavior in man. Moreover, in the case of the disturbed behavior of psychiatric patients, there can be no adequate animal models. Despite these difficulties, we shall endeavor in this essay to relate clinical features of schizophrenia, the actions of anti-schizophrenic and psychotogenic drugs in man and animals, and present understanding of catecholamine disposition. The aim is to evaluate critically a possible role of catecholamines in mediating drug effects and, conceivably, pathophysiologic mechanisms in schizophrenia.

What makes biochemical theorizing about schizophrenia particularly difficult is our uncertainty as to whether the disease is a single entity or a cluster of illnesses. Lacking an obvious organic pathology, psychiatrists are forced to make diagnoses by relying upon certain arbitrarily selected clinical symptoms. Accordingly it is not surprising that from country to country and culture to culture there are wide disparities in the criteria for a diagnosis of schizophrenia. Despite these difficulties, psychiatrists who have carefully analyzed specific symptom complexes and genetic factors now are in agreement that "classic" schizophrenia, embodying patients about whose diagnosis most psychiatrists would concur, is determined by a prominent genetic component. This and other forms of the disease that are genetically related have certain clinical features in common. The psychological disturbances most characteristic of patients with classic schizophrenia are those defined by Bleuler (1911) as the "fundamental" symptoms of schizophrenia. These include a peculiar thought disorder, a disturbance of feeling or "affective" responses to the environment, and "autism," a withdrawal from meaningful interactions with other people. Bleuler felt that hallucinations and delusions, which are certainly among the most dramatic manifestations of schizophrenia, are only secondary symptoms, because they are not constant or essential to the disease. Although lack of space dictates that we present the symptoms here in an overly simplified fashion, the notion of focusing upon particular behaviors as being either primary or secondary symptoms of schizophrenia is important in seeking biochemical correlates. For instance, if a particular drug were to regularly evoke hallucinations but no other symptoms of schizophrenia, we would doubt that it is of value in explicating the fundamental disturbance in the brains of schizophrenic subjects.

In probing the biochemical substrata of the schizophrenic disturbance, drugs have been useful in two ways:

SOLOMON H. SNYDER Department of Pharmacology and Psychiatry, Johns Hopkins Medical School, Baltimore, Maryland

Phenothiazine drugs greatly alleviate the symptoms of schizophrenia. If they truly are "antischizophrenic," in that they act primarily on the fundamental schizophrenic process, then understanding their mechanism of action might conceivably shed light upon abnormal brain functioning in schizophrenia. Another approach utilizes drugs that elicit "model psychoses." Should we conclude that certain drug psychoses provide a meaningful model of schizophrenia, then knowing their neurochemical bases would also help elucidate the pathophysiology of schizophrenia.

Are phenothiazines antischizophrenic?

The phenothiazines and related drugs have revolutionized the treatment of schizophrenia, enabling many patients to function normally, or almost normally, in society. There has been much debate as to whether they act upon something fundamental to schizophrenia or whether they merely are some sort of supersedatives. One way of examining this question would be to compare the clinical efficacy of phenothiazines and sedatives. In large-scale, double-blind, and well-controlled multihospital collaborative studies sponsored by the National Institutes of Mental Health (NIMH) and the Veterans Administration (VA), a variety of phenothiazines have been compared to sedatives, especially phenobarbital (Figure 1) (Davis, 1965). Phenobarbital was no better than placebo in any of these studies, whereas most phenothiazine drugs were significantly more effective than phenobarbital and placebo. One might ask whether these are fair comparisons. Perhaps more powerful antianxiety sedatives, such as diazepam (Valium) or chlordiazepoxide (Librium), might prove more efficacious than phenobarbital and compete with the phenothiazine drugs in the treatment of schizophrenia. However, fairly extensive trials of these agents have also shown them to be ineffective in the treatment of schizophrenia, despite their accepted efficacy in relieving anxiety. Indeed, because drugs such as diazepam and chlordiazepoxide are more effective than the phenothiazines in relieving anxiety, one can conclude that anxiety per se is not a fundamental characteristic of schizophrenia.

The NIMH-VA collaborative studies also provided another means of judging the extent to which phenothiazines exert a selectively antischizophrenic action. Since large numbers of patients were rated for a variety of symptoms, one could analyze the extent to which particular clinical features of the disease responded to the drugs (Table I) (Klein and Davis, 1969). What Bleuler refers to as the fundamental symptoms of schizophrenia tended to show the greatest response to drug treatment.

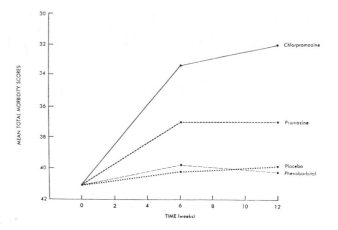

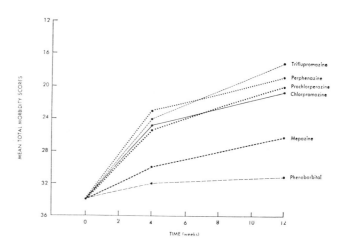

FIGURE 1 Comparative therapeutic efficacy of various phenothiazines over time in the treatment of schizophrenia. Data are derived from collaborative NIMH-VA studies and adapted, from Klein and Davis (1969).

TABLE I

Differential response of schizophrenic symptoms to phenothiazines

Bleuler's Classification	Response
Fundamental Symptoms:	
Thought disorder	+ + +
Blunted affect	+ + +
Withdrawal	+ + +
Autistic behavior	+ + +
Accessory Symptoms:	
Hallucinations	+ +
Paranoid ideation	+
Grandiosity	+
Hostility-belligerence	0
Nonschizophrenic Symptoms:	
Anxiety-tension-agitation	0
Guilt-depression	0

Adapted from Klein and Davis (1969).

Secondary symptoms, such as delusions and hallucinations, responded somewhat less, and nonschizophrenic symptoms, such as anxiety and depression, failed to show any specific improvement with phenothiazines. By contrast, sedatives would relieve agitation with much less influence upon thought disorder or the abnormality of affective response to the environment.

Thus one can make a reasonable case that the phenothiazines exert a unique therapeutic action in schizophrenic patients. We should be cautious before making the next conceptual leap to the conclusion that the drugs exert a biochemical effect that directly reverses whatever is biochemically abnormal in the brains of schizophrenics. The phenothiazines might act at a secondary site not directly related to the schizophrenic abnormality but through which the drugs can ease the distress of patients so that the schizophrenic symptoms can resolve themselves. This would be analogous to applying a tourniquet upstream from a site of bleeding, slowing the gush of blood from a wound so that the body's clotting mechanisms can then relieve the primary abnormality, a tear in the wall of the blood vessel. In favor of such a not-so-direct antischizophrenic action of phenothiazines is the well-known fact that these drugs do not "cure" schizophrenia but only facilitate remissions. Indeed, failure to maintain schizophrenic patients on phenothiazines while they are in remission results in a much greater incidence of relapse (Klein and Davis, 1969). Despite this caveat, we feel that it is worthwhile to explore the mechanism of action of the phenothiazines, because there is a reasonable possibility that the neurochemical effects of these drugs might have direct bearing on brain dysfunction in schizophrenia.

Phenothiazine pharmacology

Phenothiazines are highly reactive chemicals capable of pi electron donation or acceptance, hydrophobic binding, and ionic links via the side-chain amine. Accordingly these drugs exert biochemical effects upon almost every system that has been examined (Guth and Spirtes, 1963). How might one decide which of these actions is most relevant to the therapeutic efficacy of the drugs? Of the large numbers of phenothiazines that have been employed clinically and that are fairly similar in their chemical structure, some are highly effective in the treatment of schizophrenia; others are somewhat less effective clinically, while yet others are definitely ineffective. Biochemical effects that would correlate with known clinical actions would be the best candidates to mediate the therapeutic effects of the drugs.

To my knowledge the best correlation with clinical potency has been evinced by certain actions of phenothiazines and related drugs upon brain catecholamines, especially dopamine. Carlsson and Lindqvist (1963) reported that chlorpromazine and other antischizophrenic drugs significantly elevated brain levels of the methoxylated metabolites of norepinephrine and dopamine, while the antihistaminic phenothiazine, promethazine, was ineffective. Haloperidol, the butyrophenone whose antischizophrenic effects are elicited at much lower doses than those of chlorpromazine, was correspondingly more potent in elevating the methoxylated metabolites of the catecholamines. On the basis of these limited findings, Carlsson and Lindqvist speculated that the phenothiazines block catecholamine receptor sites and that, via a neuronal feedback loop, a message is conveyed to the catecholamine cells, "we receptors aren't getting enough transmitter; send us more catecholamines!" Accordingly, the catecholamine neurons proceed to fire more rapidly and, as a corollary, they synthesize more catecholamines and release more metabolites. Carlsson's speculations have been confirmed in studies that show that phenothiazines and butyrophenones, in proportion to their clinical efficacy, accelerate catecholamine synthesis (Figure 2) (Nyback et al., 1968). The influence of these drugs upon dopamine synthesis correlates with clinical effects better than their actions on norepinephrine synthesis. Indeed, several extremely potent butyrophenone tranquilizers selectively accelerate dopamine turnover with negligible effects upon norepinephrine. By neurophysiologic techniques, Aghajanian et al. (1973) directly demonstrated that phenothiazines can speed up the firing of dopamine cells in the brain stem, and more recently Aghajanian (personal communication) has obtained evidence by iontophoretic experiments that antischizophrenic phenothiazine tranquilizers do block the responses to dopamine of cells which normally receive dopamine neuronal input. Thus, both biochemical and neurophysiological studies have demonstrated interactions of the phenothiazine tranquilizers with dopamine systems in the brain, which are consistent with a blockade of dopamine receptors and which correlate well with their antischizophrenic actions.

Part of the difficulty in securing general acceptance to the hypothesis of a relationship between dopamine receptor blockade and the clinical actions of phenothiazines is the lack of chemical similarity between phenothiazines and catecholamines. Phenothiazines are complex multiringed structures, while dopamine is a simple phenylethylamine. Recently Horn and Snyder (1971) noted that chlorpromazine, in its preferred conformation as determined by X-ray crystallography, can be superimposed, in part, upon the accepted preferred

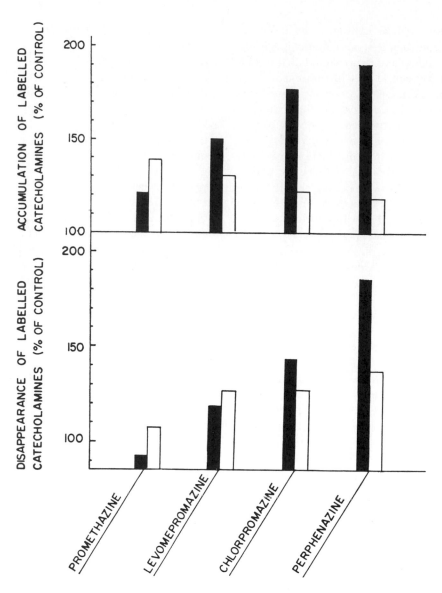

FIGURE 2 The effect of four phenothiazine derivatives on the turnover of dopamine (filled columns) and norepinephrine (open columns) in mouse brain. Mean effects of 0.5, 2, and 10 mg/kg of the drugs are shown. Adapted from Nyback (1971).

conformation of dopamine or norepinephrine, providing a molecular mechanism whereby phenothiazines might block dopamine receptors. In the preferred conformation of chlorpromazine, its side chain tilts away from the midline toward the chlorine-substituted ring (Figure 3). Presumably the chlorine on Ring A is responsible for the "tilt" of the side chain, because if there were no substituent on Ring A, both rings A and C would be symmetrical, and one would expect the side chain to be fully extended. Accordingly, phenothiazines lacking a substituent on Ring A should mimic the conformation of dopamine less effectively and thus have less affinity for dopamine receptors and presumably be less efficacious in the treatment of schizophrenia. Of the dozen or so phenothiazine

tranquilizers that have been widely employed clinically, only two lack a substituent on Ring A. Strikingly, mepazine and promazine, the two phenothiazines lacking a Ring A substituent, are significantly less effective as antischizophrenic drugs than the others. (Table II) (Cole and Davis, 1969).

Besides the Ring A substituent, another major requirement for therapeutic activity of phenothiazines is separation of the side-chain amine by three carbon atoms from the ring system. Molecular models indicate that shortening the side chain to two carbon atoms would make the assumption of the dopamine-like conformation less likely. Just as predicted, phenothiazines with two-carbon side chains, such as the antihistamine promethazine and the

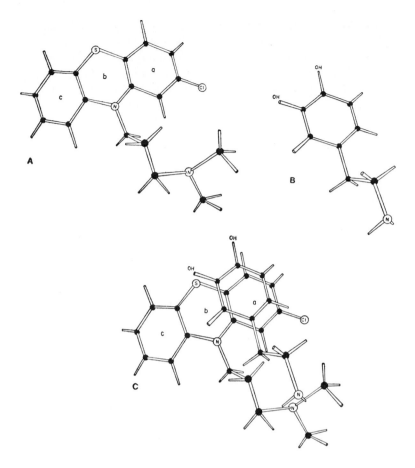

FIGURE 3   Molecular models of the preferred conformations (as determined by X-ray crystallography) of [A] chlorpromazine [B] dopamine and [C] their superimposition. Adapted from Horn and Snyder (1971).

TABLE II

Comparative efficacy of phenothiazine drugs in the treatment of schizophrenia

Drug		Number of Studies in which		
Generic Name	Trade Name	Drug was more Effective than Chlorpromazine	Drug was equal to Chlorpromazine	Chlorpromazine was more Effective
Mepazine	Pacatal	0	0	4
Promazine	Sparine	0	2	4
Triflupromazine	Vesprin	0	10	0
Perphenazine	Trilafon	0	6	0
Prochlorperazine	Compazine	0	10	0
Trifluoperazine	Stelazine	0	8	0
Thioridazine	Mellaril	0	12	0
Fluphenazine	Prolixin	0	7	0
Acetophenazine	Tindal	0	1	0
Thiopropazate	Dartal	0	1	0
Phenobarbital	—	0	0	6

Data are adapted from Klein and Davis (1969) and are drawn from a review of double-blind, well-controlled clinical studies.

antiparkinsonian agent diethazine, lack therapeutic effectiveness in schizophrenia.

Besides being associated with their antischizophrenic activity, dopamine receptor blockade by phenothiazines and butyrophenones may explain the prominent extrapyramidal side effects of these drugs. The symptoms of Parkinson's disease are now thought to be associated with a deficiency of dopamine in the caudate putamen, the area of major dopamine terminals in the brain, and the therapeutic action of L-dopa is presumed to arise from an alleviation of this dopamine deficiency. What dopamine pathways in the brain might be related to the antischizophrenic actions of phenothiazines? Besides the nigrostriatal dopamine tract, other major areas with dopamine terminals include the olfactory tubercle and nucleus accumbens. (Ungerstedt, 1971.)

Amphetamine psychosis as a "model schizophrenia"

Model psychoses have provided an attractive approach to understanding schizophrenia, with the hope that their biochemical elucidation may bear upon similar defects in the brains of schizophrenics. Early interest centered upon psychedelic drug psychosis. Like schizophrenia, the mental aberrations following the ingestion of psychedelic drugs occur while the subject is alert and fully oriented to time, place, and person. This is in contradistinction to most drug psychoses that are associated with confusion, disorientation, and delirium. However, the behavioral effects of psychedelic drugs differ in important ways from schizophrenic disturbances. The major symptoms following psychedelic drug ingestion are disorders of perception. LSD-induced hallucinations are usually visual, whereas schizophrenic hallucinations tend to be auditory. Moreover there is little evidence that a schizophrenia-like thought disorder or disturbance of affect takes place in psychedelic drug induced psychosis. Hollister (1962) tabulated the symptoms of subjects receiving psychedelic drugs and found them to be markedly dissimilar to schizophrenic symptoms. Moreover he showed that tape-recorded interviews with newly admitted schizophrenics could be readily distinguished by "blind" raters from those of subjects under the influence of psychedelic drugs. In addition schizophrenics receiving psychedelic drugs report that the drug experience is quite unlike their endogenous psychosis (Chloden et al., 1955).

By contrast, amphetamine psychosis may be a better candidate for a model schizophrenia. Many cases of amphetamine psychosis have been misdiagnosed as acute paranoid schizophrenia until the history of drug use was obtained (Connell, 1958; Bell, 1965; Beamish and Kiloh, 1960), and accordingly amphetamine psychosis was suggested as an heuristic "model" schizophrenia (Kety,

1959, 1972). Amphetamine psychosis is most frequently observed in amphetamine addicts who have consumed enormous amounts of the drug over prolonged periods, e.g. 500 to 1000 mg of d-amphetamine per day for a week or more (Ellinwood, 1967). These patients develop an acute paranoid psychosis that typically resolves within a few days after termination of the drug. Unlike LSD psychosis, patients with amphetamine psychosis frequently experience auditory hallucinations very much like typical schizophrenic auditory hallucinations, including hearing vague noises and voices and occasionally having conversations with the voices. The visual hallucinations in amphetamine psychotics tend to resemble those observed in very acute schizophrenics (Bowers and Freedman, 1966; Chapman, 1966).

It should be borne in mind that more than one type of amphetamine psychosis is possible. Amphetamines can evoke an acute toxic psychosis with delirium, confusion, and disorientation as occurs with many other drugs and unlike the schizophrenia-like amphetamine psychosis. "Toxic" amphetamine psychosis occurs usually after only one or two extremely large doses rather than after chronic use of the drug. Among 42 cases of amphetamine psychosis, visual hallucinations occurred primarily among patients who became acutely psychotic, often after a single large dose, hence presumably were suffering from a "toxic" psychosis. By contrast hallucinations were usually auditory in patients whose illness developed gradually after frequent repeated doses (Connell, 1958). The "toxic" amphetamine psychoses probably account for the greater proportion of visual hallucinations in amphetamine intoxication than in schizophrenia. Here we will focus solely on the "nontoxic" amphetamine psychosis with intact sensorium and which most closely resembles clinical schizophrenia.

Some authors have criticized amphetamine psychosis as a "model schizophrenia," arguing that it might be simply related to sleep deprivation, overexcitement, or precipitation of psychosis in borderline schizophrenics. To examine these questions, Griffith et al. (1970, 1972) and Angrist and Gershon (1970) developed an experimental model of amphetamine psychosis in man by administering progressively increasing doses of d-amphetamine to volunteer subjects. None of the subjects had any evidence of preexisting schizophrenia or schizoid tendencies. Yet virtually all subjects became floridly psychotic after 1 to 4 days, indicating that amphetamine psychosis is not simply a precipitation of latent schizophrenia. Since some patients became psychotic in about 24 hr, there could not have been sufficient sleep deprivation to account for the psychosis. As for the question of overexcitement, after some initial moderate euphoria, most subjects were more sullen than excited, although it is conceivable that "internal" hyperexcitement was taking place which was not evident to the observers.

What about the fundamental schizophrenic symptoms: thought disorder, disturbance of affect, and autism? Griffith et al. (1972) did not feel that a typically schizophrenic thought disorder or affect disturbance was present in their subjects, while Angrist and Gershon (1970) did observe flattened affect and schizophrenic thinking patterns in their patients. Part of these discrepancies may stem from difficulty in interpreting the presence of abnormal affect and thinking in acutely paranoid individuals. Paranoid patients characteristically display seemingly systematic and well-organized thought patterns, although the underlying delusions are inherently disorganized. Moreover, very acute paranoid schizophrenics quite frequently display a lively affect (Cameron, 1959), while their typical thought disorder and affective disturbance only become apparent as the acute paranoid features subside.

If one takes the position that amphetamine psychosis lacks a typically schizophrenic thought disorder, then according to standard American nomenclature a better diagnostic classification would be "paranoid state," which is defined as a psychosis characterized by paranoid delusions but in which "emotional responses and behavior are consistent with the ideas held . . . and it does not manifest the bizarre fragmentation and deterioration of schizophrenic reactions" (American Psychiatric Association, 1952). However, many European psychiatrists do not feel that the entity of paranoid state exists separate from schizophrenia and would label such patients as suffering from paranoid schizophrenia.

One important reservation about treating amphetamine psychosis as a model schizophrenia is that it rarely resembles nonparanoid schizophrenia. If amphetamine psychosis is only to be a model of paranoid schizophrenia, then it might follow that paranoid schizophrenia differs in a fundamental way from other forms of schizophrenia. Contrary to this suggestion are major items of genetic and clinical evidence: (a) schizophrenics characteristically switch from one subtype of schizophrenia to another during their clinical history. (b) Genetic studies indicate that a variety of different types of schizophrenia may "run" in families (Kallman, 1938; Gottesman and Shields, 1967). These two items suggest that if amphetamine psychosis is to be a meaningful model of schizophrenia, it should show resemblance to nonparanoid forms of the disease. Why does not amphetamine psychosis mimic other forms of schizophrenia? One might speculate that the "paranoid" character is contributed by other nonschizophrenomimetic effects of the drug. Thus, perhaps amphetamine does possess a "pure" schizophrenia-mimicking action, but some other effect of the drug transforms the clinical picture into a predominantly paranoid one. A good candidate for such an effect would be the well-known central stimulant-alerting actions of amphetamine. One might speculate that the major feature that differentiates paranoid schizophrenics from other schizophrenics is their hyperalert striving to cast the bewildering array of psychotic transformations into a coherent and meaningful process. It is this paranoid hypervigilance that, by enabling the patient to loculate his disturbance, tends to mask the typically schizophrenic thought disorder and disturbance of affect.

Another finding linking amphetamine actions to schizophrenia is the observation that amphetamine and its analogues directly exacerbate schizophrenic symptoms (Levine et al., 1948; Hope et al., 1951; Pennes, 1954; Janowsky et al., 1973). It is important to note that amphetamines intensify the schizophrenic symptoms, rather than adding different psychotic symptoms to the schizophrenic illness. Patients themselves perceive that their illness is worsening under the influence of the drug. By contrast, when schizophrenics are treated with LSD, they can recognize that the psychedelic drug psychosis is something different from their own mental disturbance. Amphetamines fail to elicit psychotic effects (in the small doses employed) in schizophrenics in remission or in manic or depressed patients. (Janowsky et al., 1973.)

Another item favoring an association between amphetamine-induced mental disturbance and schizophrenia is the fact that phenothiazines and butyrophenones are the best antidotes for amphetamine psychosis (Angrist, Griffith, and Davis, personal communications). This is especially striking, because barbiturate sedatives fail to alleviate amphetamine psychosis and, in some cases, accentuate the symptoms (Angrist, personal communication).

Of other drug psychoses, those elicited by cocaine and L-dopa are of particular interest. Cocaine, whose central effects are thought to be related to potentiation of brain catecholamines, produces a psychosis whose clinical features are essentially the same as those of amphetamine psychosis (Mayer-Gross et al., 1960). L-dopa is the precursor of the catecholamines dopamine and norepinephrine. In Parkinsonian patients treated with L-dopa, about 15 to 30% display psychiatric side effects (McDowell, 1970). Symptoms resembling amphetamine psychosis are rare in these patients (Celesia and Barr, 1970). The most common psychiatric reaction is a delirious toxic psychosis with confusion and visual hallucinations. Since dopa certainly does act via brain catecholamines, why does it only rarely elicit a typical amphetamine psychosis? The answer probably lies in the relative doses employed. Amphetamine psychosis is rarely observed unless subjects are ingesting well in excess of 100 mg of *d*-amphetamine per day. Although direct comparisons are difficult, it is probable that to secure comparable central effects with L-dopa would require about 20 gm of the drug, whereas the typical dose in Parkinsonian patients is only about 4 gm. Moreover, Parkinsonian patients are generally more than 50 years old, hence would be much more susceptible to organic brain symptoms than schizophrenia-like symptoms.

Because of these difficulties, a more efficient means of evaluating the influence of L-dopa upon schizophrenia-like processes would be to ascertain whether it might accentuate schizophrenic symptoms. Yaryura-Tobias et al. (1970) and B. Angrist (personal communication) have observed that L-dopa does indeed activate the symptoms of schizophrenia in some patients. One must be cautious in interpreting these results, however, because there is evidence that L-dopa may exert a generalized "activating" effect in several types of psychiatric patients (Goodwin et al., 1970).

Amphetamine-catecholamine interactions

How does amphetamine act in the brain? Because of obvious chemical similarities (Figure 4), amphetamine is generally thought to act via the brain catecholamines dopamine and norepinephrine. The most efficient means of exploring exactly which tract mediates behavioral effects of amphetamines in animals is to make discrete lesions with 6-hydroxydopamine. 6-Hydroxydopamine is accumulated by catecholamine neurons selectively, after which it auto-oxidizes and thereby destroys the neurons (Thoenen, 1972; Ungerstedt, 1973a). Destruction of the nigrostriatal dopamine pathway by implanting 6-hydroxydopamine directly into the substantia nigra abolishes the stereotyped compulsive gnawing, licking, and sniffing elicited by amphetamine (S. Iversen, this volume; Ungerstedt, 1973b). S. Iversen (this volume) reported that these same lesions failed to alter the locomotor stimulation normally evoked by amphetamine, indicating that this effect is not a function of the nigrostriatal tract.

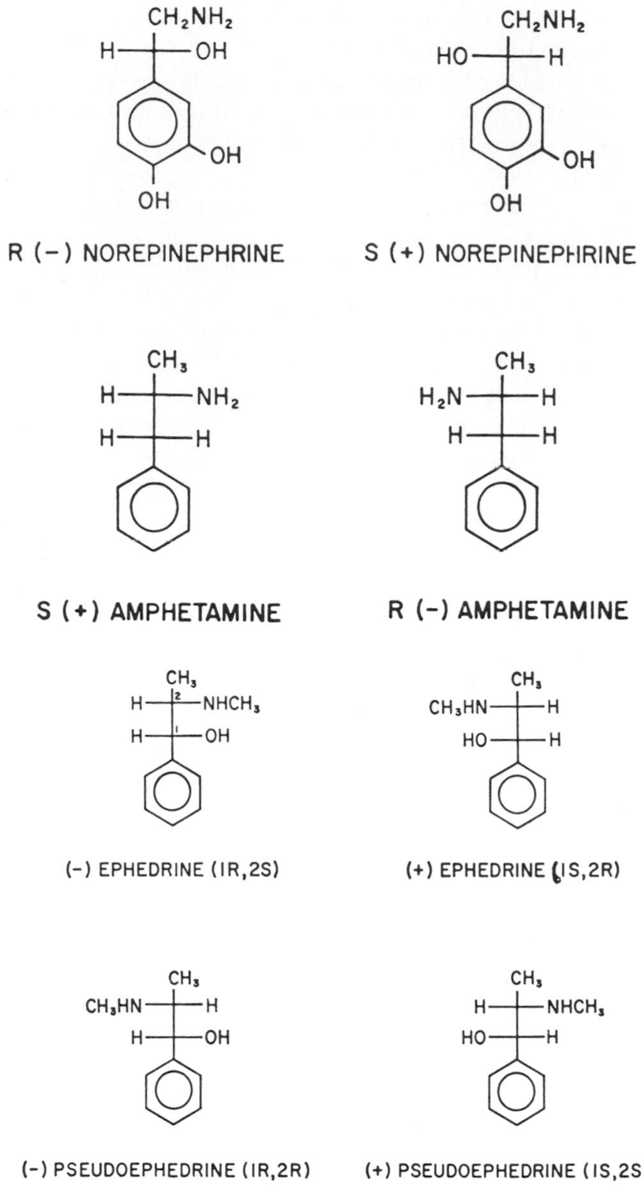

To evaluate the differential roles of dopamine and norepinephrine in mediating amphetamine actions in man, we have employed isomers of amphetamine. We took advantage of the differences in stereoselectivity of dopamine and norepinephrine neurons (Coyle and Snyder, 1969; Hendley and Snyder, 1972; Hendley et al., 1972; Snyder et al., 1970a, 1970b; Snyder, 1970). (Table III.) Norepinephrine neurons exhibit about four times as much affinity for the naturally occurring *l*-form of norepinephrine (Coyle and Snyder, 1969; Iversen et al., 1971) as for *d*-norepinephrine. By contrast, dopamine neurons in the corpus striatum (Coyle and Snyder, 1969; Iversen et al., 1971) and retina (Hendley and Snyder, 1972) fail to

FIGURE 4 The structures and absolute configurations of stereoisomers of norepinephrine, amphetamine, and ephedrine. Ephedrine represents the *erythro* type of configuration and pseudoephedrine, the *threo* type. Adapted from Hendley et al., 1972.

differentiate between norepinephrine isomers, presumably because dopamine, their normal catecholamine transmitter, is a symmetric molecule lacking stereoisomers.

Isomers of norepinephrine differ at the β carbon, while isomers of amphetamine differ at the α carbon. Amphetamine is a potent inhibitor of the catecholamine reuptake system, whereby neuronal accumulation of synaptically released dopamine and norepinephrine presumably terminates their synaptic activities (Iversen, 1967; Axelrod,

TABLE III

Affinity of norepinephrine, amphetamine, and ephedrine isomers for the catecholamine uptake mechanism of brain synaptosomes

Compound	Absolute Config- uration	Relative Affinity	
		Dopamine Neurons (corpus striatum)	Norepine- phrine Neurons (cerebral cortex)
(−)-Norepinephrine	1-R	40.0	240.0
(+)-Norepinephrine	1-S	40.0	60.0
(+)-Amphetamine	2-S	720.0	240.0
(−)-Amphetamine	2-R	720.0	24.0
(−)-Ephedrine	1R:2S	11.6	100.0
(+)-Ephedrine	1S:2R	6.0	25.7
(+)-Pseudoephedrine	1S:2S	3.5	3.4
(−)-Pseudoephedrine	1R:2R	1.7	1.0

Affinities for ephedrine isomers are derived from experiments in which the dose required for 50% inhibition of the uptake of 0.1 μm (±)-^{3}H-norepinephrine (ID-50) was assessed (Hendley et al., 1972). Affinity is actually the relative potency derived from the ID-50, thusly: Relative potency = $(7.2 \times 10^{-7}$M$)/$ID-50 $\times$ 100. Affinities for amphetamine and norepinephrine isomers are obtained from Ki values for the reduction of (±)-^{3}H-norepinephrine uptake (Coyle and Snyder, 1969), thusly: Affinity = 7.2 $\times$ 10^{-7}M/Ki $\times$ 100. All experiments employed synaptosome-rich homogenates of cerebral cortex or corpus striatum.

1965). The central stimulant *d*-amphetamine is considerably more potent in blocking catecholamine uptake by norepinephrine terminals than is *l*-amphetamine, a much weaker central stimulant; while in dopamine neurons the amphetamine isomers have similar potency (Coyle and Snyder, 1969; Hendley and Snyder, 1972). With drugs such as the ephedrines, which possess both asymmetric α and β carbons (Figure 4) these differences may be amplified (Table III). Thus while *d*- and *l*-amphetamines differ tenfold in their effects upon norepinephrine neurons, the most and least potent ephedrine isomers differ by 100-fold. With ephedrine isomers there is less than 1/10 as much stereoselectivity toward ephedrine isomers in dopamine than in norepinephrine neurons (Hendley et al., 1972). Because the stereoselectivity of the catecholamine uptake process is much less striking than the stereospecificity of receptor activation by catecholamines, great care must be exercised in preparation of tissue and design of blank values to correct for nonspecific accumulation of amine. One must also ensure maximal uptake velocity, linearity with time and tissue concentration, and optimal surface area to obtain stereoselective effects (Hendley and Snyder, 1972; Hendley et al., 1972). Such technical issues may explain difficulties in demonstrating stereoselective catecholamine uptake in some tissues (Ferris et al., 1972).

The differential effects of amphetamine isomers on catecholamine uptake by dopamine and norepinephrine neurons can be demonstrated in vivo as well as in vitro (Taylor and Snyder, 1970, 1971; Snyder et al., 1970b). Moreover, norepinephrine depletion, thought to reflect the norepinephrine releasing action of amphetamine, another major mechanism of its pharmacological activity, also responds to amphetamine isomers in a pattern similar to influences upon uptake, while methoxylated amine metabolites correlate less well (Svensson, 1971; Scheel-Kruger, 1972).

From these stereoselective actions of amphetamines emerges a simple paradigm for drawing inferences as to whether brain dopamine or norepinephrine mediates particular behavioral effects of amphetamines. Behaviors mediated by norepinephrine should be affected much more by *d*- than by *l*-amphetamine, while the two isomers should have similar potencies in eliciting dopamine mediated behaviors. *d*-Amphetamine is only about twice as potent as *l*-amphetamine in evoking stereotyped compulsive gnawing behavior in rats (Taylor and Snyder, 1970, 1971). This suggests a prominent role for dopamine in this behavior and concurs with the conclusions of several workers using other approaches (Creese and Iversen, 1972; S. Iversen, this volume; Ungerstedt, this volume; Ernst and Smelik, 1966; Fog et al., 1967). By contrast, *d*-amphetamine is ten times as potent as *l*-amphetamine in stimulating locomotor activity in rats, a finding that closely parallels the tenfold difference in the potency of these drugs in blocking catecholamine uptake by norepinephrine neurons. These results, which favor a norepinephrine mediation of this behavior, are consistent with the findings of S. Iversen (this volume) and Creese and Iversen (1972) that 6-hydroxydopamine lesions of the nigrostriatal system do not diminish amphetamine-induced locomotor stimulation.

In man, *d*-amphetamine is about 5 times as potent a central stimulant as *l*-amphetamine, suggesting that norepinephrine mediates the alerting actions of amphetamine in man (Prinzmetal and Alles, 1940). This would seem reasonable, because the alerting action of amphetamine is generally thought to be a cerebral cortical event and because the cerebral cortex is extensively innervated by norepinephrine fibers but receives no dopamine innervation.

Amphetamine is a valuable treatment modality for hyperactive children with "minimal brain dysfunction." Comparisons of *d*- and *l*-amphetamine in minimal brain dysfunction in a longitudinal study of one patient (Snyder and Meyerhoff, 1973) and in a double blind evaluation of 11 patients (Arnold et al., 1972) failed to reveal a statistically significant difference in the efficacy of the isomers, suggesting that the beneficial effect of amphetamine in minimal brain dysfunction is dopamine mediated. The facial and body tickings of Gilles de la Tourette's disease are reminiscent of motor dysfunctions that occur in Parkinsonian patients treated with L-dopa. In one patient with Gilles de la Tourette's disease studied over a 2-month period, *d*-amphetamine greatly accentuated the ticking while *l*-amphetamine was without effect (Snyder and Meyerhoff, 1973). This suggests that the ticking in Gilles de la Tourette's disease, which is most effectively ameliorated by the butyrophenone haloperidol, is related more to norepinephrine than to dopamine.

On the basis of the animal studies of amphetamine isomers, Angrist et al. (1971) deduced that comparison of the potencies of *d*- and *l*-amphetamine in eliciting amphetamine psychosis might provide an indication of the relative roles of norepinephrine and dopamine. Accordingly, on three separate occasions, three volunteers were given amphetamines in progressively increasing doses until psychosis was clearly evident. On one occasion *d*-amphetamine was utilized, while another time *l*-amphetamine was employed, and on the third occasion, a mixture of equal parts of *d*- and *l*-amphetamine was administered. Drug administration was terminated when psychosis was judged to be present, which occurred with a sudden emergence of paranoid delusions and a cold detached affect, so that one could readily estimate the total dose of drug required to elicit psychosis.

All three subjects were rendered psychotic by all three drugs. Strikingly, *d*- and *l*-amphetamine had quite similar potencies in eliciting the psychosis. In the three patients, the ratios of the total dose of *l*-amphetamine to the total dose of *d*-amphetamine required to produce the psychosis were 1.25, 1.53, and 1.0. The quantity of racemic amphetamine required to produce psychosis was intermediate between the amounts necessary for *d* and *l* forms.

During the early stages of amphetamine administration, Angrist and Gershon (1971) observed much more central stimulation with *d*-amphetamine than with *l*-amphetamine. Thus in these human subjects *d*- and *l*-amphetamine were much more similar in their capacity to elicit psychosis than in their ability to effect central stimulation.

Earlier we asked why amphetamine invariably elicited paranoid psychoses and suggested that the psychosis-eliciting and the central stimulant actions of the drug might be quite distinct. Amphetamine could provoke a schizophrenia-like psychosis, whose paranoid flavor was conferred by the central stimulant actions of the drug. The experimental results with amphetamine isomers suggest that the psychosis elicited by amphetamine may be mediated by brain dopamine, while central stimulant effects are elicited by brain norepinephrine. (Table IV.) One might speculate that the norepinephrine-mediated

alerting action forces the patient to strive for an intellectual framework in which to focus all the strange feelings that are coming over him as the psychosis develops. This quest for meaning and its subsequent "discovery" in a system of delusions might be the essence of the paranoid process in these patients. According to this reasoning, a "schizophrenia-like" component of amphetamine psychosis may be mediated primarily by brain dopamine, while the "paranoid solution" is facilitated by the drug's alerting effects via stimulation of norepinephrine systems.

TABLE IV
Relations between drugs, catecholamines, and schizophrenia

Phenothiazines (and related antischizophrenic drugs).
(1) Phenothiazines have true antischizophrenic actions:
  (a) Are more effective than sedatives.
  (b) Act best on fundamental symptoms.
(2) Blockade of dopamine receptors by phenothiazines is proportional to their clinical efficacy.
(3) The ability of phenothiazines to mimic the preferred catecholamine conformation predicts their therapeutic activity.
Amphetamines
(1) In small doses, amphetamines specifically activate schizophrenic symptoms.
(2) Amphetamines can evoke a psychosis indistinguishable from acute paranoid schizophrenia.
(3) Phenothiazines are the best antidotes for amphetamine psychosis.
(4) Isomer studies suggest that dopamine mediates amphetamine psychosis, while norepinephrine mediates the central stimulant effects of amphetamine.
(5) Speculation: Via dopamine, amphetamine may evoke a schizophrenia-like psychosis that is transformed by norepinephrine-mediated central stimulant actions of amphetamine into a paranoid psychosis.

If indeed, the norepinephrine-mediated alerting actions of amphetamines interact with and "contaminate" a purer dopamine-mediated amphetamine psychosis, then removal of these "norepinephrine effects" might leave one with an amphetamine psychosis more closely resembling schizophrenia than the present version. Conceivably, development of a drug that would stimulate dopamine but not norepinephrine mechanisms in the brain would produce such a "pure" model of schizophrenia.

Of course, as pointed out earlier in our discussion of phenothiazines, the fact that a schizophrenia-like psychosis follows upon the action of amphetamine on brain catecholamines does not "prove" that catecholamines are involved in the "psychotomimetic" locus in the brain. Moreover even if certain catecholamine neurons were responsible for amphetamine psychosis and even if the neural disturbances in amphetamine psychosis are closely similar to those occurring in schizophrenia, we still could not conclude that these neurons are the site of primary disturbance in schizophrenia. Nonetheless, the evidence (a) that selective antischizophrenic actions of drugs appear to be mediated via brain dopamine systems (b) that a schizophrenia-like amphetamine psychosis appears to involve brain dopamine, and (c) that amphetamines can accentuate schizophrenic symptoms in a fairly specific fashion, certainly favors some role for catecholamines. Accordingly, further exploration of the differential behavioral roles of particular catecholamine systems in the brain and their interactions with drugs may considerably enhance our understanding of brain mechanisms in schizophrenia.

ACKNOWLEDGMENTS   This work was supported by U.S.P.H.S. grants MH 18501, NS 07275, DA 00266 and NIMH Research Scientist Development Award MH 33128. Experimental work providing a basis for much of the thinking in this essay was conducted in collaboration with Joseph T. Coyle, Kenneth M. Taylor, Edith D. Hendley, and Alan S. Horn.

REFERENCES

AGHAJANIAN, G. K., B. S. BUNNEY, and M. J. KUHAR, 1973. Use of single unit recording in correlating transmitter turnover with impulse flow in monoamine neurons. In *New Concepts in Neurotransmitter Regulation*, A. J. Mandell, ed. New York, (in press).

AMERICAN PSYCHIATRIC ASSOCIATION, 1952. *Diagnostic and Statistical Manual of Mental Disorder*, Vol. I. Washington, D.C.: American Psychiatric Association, p. 28.

ANGRIST, B. M., and S. GERSHON, 1970. The phenomenology of experimentally-induced amphetamine psychosis. Preliminary observations. *Amer. J. Psychiat.* 126:95–107.

ANGRIST, B., and S. GERSHON, 1971. A pilot study of pathogenic mechanisms in amphetamine psychosis utilizing differential effects of *d*- and *l*-amphetamine. *Pharmakopsychiatrie Neuro-Psychopharmacologie* 4:65–75.

ANGRIST, B. M., B. SHOPSIN, and S. GERSHON, 1971. The comparative psychotomimetic effects of stereoisomers of amphetamine. *Nature (Lond.)* 234:152–154.

ARNOLD, L. E., P. H. WENDER, K. MCCLOSKEY, and S. H. SNYDER, 1972. Levoamphetamine and dextroamphetamine: Comparative efficacy in the hyperkinetic syndrome; assessment by target symptoms. *Arch. Gen. Psychiat.* 27:816–822.

AXELROD, J., 1965. The metabolism, storage and release of catecholamines. *Rec. Progr. Horm. Res.* 21:597–622.

BEAMISH, P., and L. G. KILOH, 1960. Psychoses due to amphetamine consumption. *J. Ment. Sci.* 106:337–343.

BELL, D. S., 1965. Comparison and amphetamine psychosis and schizophrenia. *Brit. J. Psychiat.* 111:701–707.

BLEULER, E., 1911. *Dementia Praecox or the Group of Schizophrenias*. New York: International University Press. English trans., J. Zinkin, 1950.

BOWERS, JR., M. J., and D. X. FREEDMAN, 1966. "Psychedelic" experiences in acute psychoses. *Arch. Gen. Psychiat.* 15:240–248.

CAMERON, N., 1959. Paranoid conditions and paranoia. In *American Handbook of Psychiatry*, S. Arieti, ed. New York: Basic Books, 1:510.

CARLSSON, A., and M. LINDQVIST, 1963. Effect of chlorpromazine or haloperidol on the formation of 3-methoxytyramine and normetanephrine in mouse brain. *Acta Pharmacol. Toxicol.* 20:140–144.

CELESIA, G. G., and A. N. BARR, 1970. Psychosis and other psychiatric manifestations of Levodopa therapy. *Arch. Neurol.* 23:193–200.

CHAPMAN, J., 1966. The early symptoms of schizophrenia. *Brit. J. Psychiat.* 112:225–251.

CHLODEN, L. W., A. KURLAND, and C. SAVAGE, 1955. Clinical reactions and tolerance to LSD in chronic schizophrenia. *J. Nerv. Ment. Dis.* 122:211–221.

COLE, J. O., and J. M. DAVIS, 1969. Antipsychotic drugs. In *The Schizophrenic Syndrome*, H. Solomon, ed. New York: Grune and Stratton, pp. 478–568.

CONNELL, P. H., 1958. *Amphetamine Psychosis*. London: Chapman and Hall.

COYLE, J. T., and S. H. SNYDER, 1969. Catecholamine uptake by synaptosomes in homogenates of rat brain: Stereospecificity in different areas. *J. Pharmacol. Exp. Ther.* 170:221–231.

CREESE, I., and S. D. IVERSEN, 1972. Amphetamine response in rat after dopamine neuron destruction. *Nature (Lond.)* 238:247–248.

DAVIS, J. M., 1965. Efficacy of the tranquilizing and antidepressant drugs. *Arch. Gen. Psychiat.* 13:552–572.

ELLINWOOD, JR., E.H., 1967. Amphetamine psychosis: I. Description of the individuals and process. *J. Nerv. Ment. Dis.* 144:273–283.

ERNST, A. M., and P. SMELIK, 1966. Site of action of dopamine and apomorphine in compulsive gnawing behavior in rats. *Experientia.* 22:837–839.

FERRIS, R. M., F. L. M. TANG, and R. A. MAXWELL, 1972. A comparison of the capacities of isomers of amphetamine, deoxypipradol and methylphenidate to inhibit the uptake of tritiated catecholamines into rat cerebral cortex slices, synaptosomal preparations of rat cerebral cortex, hypothalamus and striatum and into adrenergic nerves of rabbit aorta. *J. Pharmacol. Exp. Ther.* 181:407–416.

FOG, R. L., A. RANDRUP, and H. PAKKENBERG, 1967. Aminergic mechanisms in corpus striatum and amphetamine-induced stereotyped behavior. *Psychopharmacologia* 11:179–183.

GOODWIN, F. K., D. L. MURPHY, H. K. H. BRODIE, and W. E. BUNNEY, JR., 1970. L-Dopa, catecholamines and behavior; a clinical and biochemical study in depressed patients. *Biol. Psychiat.* 2:341–366.

GOTTESMAN, I. I., and J. SHIELDS, 1967. A polygenic theory of schizophrenia. *Proc. Natl. Acad. Sci. USA* 58:199–205.

GRIFFITH, J. D., J. H. CAVANAUGH, J. HELD, and J. A. OATES, 1970. Experimental psychosis induced by the administration of *d*-amphetamine. In *Amphetamines and Related Compounds*. New York: Raven Press, pp. 897–904.

GRIFFITH, J. D., J. CAVANAUGH, J. HELD, and J. A. OATES, 1972. Dextroamphetamine: Evaluation of psychotomimetic properties in man. *Arch. Gen. Psychiat.* 26:97–100.

GUTH, P. S., and M. A. SPIRTES, 1963. The phenothiazine tranquilizers: Biochemical and biophysical actions. *Int. Rev. Neurobiol.* 7:231–278.

HENDLEY, E. D., and S. H. SNYDER, 1972. Stereoselectivity of catecholamine uptake in noradrenergic and dopaminergic peripheral organs. *Eur. J. Pharmacol.* 19:56–66.

HENDLEY, E. D., S. H. SNYDER, J. J. FAULEY, and J. B. LAPIDUS, 1972. Stereoselectivity of catecholamine uptake by brain synaptosomes: Studies with ephedrine, methylphenidate and phenyl-2-piperidyl carbinol. *J. Pharmacol. Exp. Ther.* 183: 103–116.

HILLARP, N. A., K. FUXE, and A. DAHLSTROM, 1966. Demonstration and mapping of central neurons containing dopamine, norepinephrine and 5-hydroxytryptamine and their reactions to psychopharmaca. *Pharmacol. Rev.* 18:727–742.

HOLLISTER, L. E., 1962. Drug-induced psychoses and schizophrenic reactions—a critical comparison. *Ann. N.Y. Acad. Sci.* 96:80–88.

HOPE, J. M., E. CALLAWAY, and S. L. SANDS, 1951. Intravenous pervitin and the psychopathology of schizophrenia. *Dis. Nerv. Syst.* 12:67–72.

HORN, A. S., and S. H. SNYDER, 1971. Chlorpromazine and dopamine: Conformational similarities that correlate with the antischizophrenic activity of phenothiazine drugs. *Proc. Natl. Acad. Sci. USA* 68:2325–2328.

IVERSEN, L. L., 1967. *The Uptake and Storage of Noradrenaline in Sympathetic Nerves.* New York: Cambridge University Press.

IVERSEN, L. L., B. JARROTT, and M. A. SIMMONDS, 1971. Differences in the uptake, storage and metabolism of $(\pm)$- and $(-)$-norepinephrine. *Brit. J. Pharmacol.* 43:845–855.

JANOWSKY, D. S., M. K. EL-YOUSEL, J. M. DAVIS, and H. J. SEKERKE, 1973. Provocation of schizophrenic symptoms by intravenous methylphenidate. *Arch. Gen. Psychiat.* 28:185–191.

KALLMAN, S. J., 1938. *Genetics of Schizophrenia.* New York: Augustin Inc.

KETY, S. S., 1959. Biochemical theories of schizophrenia. A two-part critical review of current theories and the evidence used to support them. *Science* 125:1528–1532, 1590–1596.

KETY, S. S., 1972. Toward hypotheses for a biochemical component in the vulnerability to schizophrenia. *Sem. Psychiat.* 4: 233–238.

KLEIN, D. F., and J. M. DAVIS, 1969. *Diagnosis and Drug Treatment of Psychiatric Disorder.* Baltimore: Williams & Wilkins, pp. 52–138.

LEVINE, J., M. RINKEL, and M. GREENBLATT, 1948. Psychological and Physiological effects of intravenous Pervertin (methedrine). *Amer. J. Psychiat.* 105:429.

MAYER-GROSS, W., E. SLATER, and M. ROTH, 1960. *Clinical Psychiatry.* Baltimore: Williams & Wilkins, p. 377.

McDOWELL, F. H., 1970. Psychiatric aspects of L-dopa treatment in Parkinson's disease. In *L-Dopa and Parkinsonism*, A. Barbeau and F. McDowell, eds. Philadelphia: Davis, p. 321.

NYBACK, H., 1971. Effects of neuroleptic drugs on brain catecholamine neurons. *M.D. Thesis*, Stockholm: Kihlström and Söner Boktyckeri AB.

NYBACK, H., Z. BORZECKI, and G. SEDVALL, 1968. Accumulation and disappearance of catecholamines formed from tyrosine-C^{14} in mouse brain; effect of some psychotropic drugs. *Europ. J. Pharmacol.* 4:395–402.

PENNES, H. H., 1954. Clinical reactions of schizophrenics to sodium amytal, pervitin hydrochloride, mescaline sulfate and d-lysergic acid diethylamide. *J. Nerv. Ment. Dis.* 119:95–112.

PRINTZMETAL, M., and G. A. ALLES, 1940. The central nervous system stimulant effects of dextro-amphetamine sulphate. *J. Amer. Med. Assn.* 200:665–673.

SCHEEL-KRUGER, J., 1972. Behavioral and biochemical comparison of amphetamine derivatives, cocaine, benztropine and tricyclic antidepressant drugs. *Europ. J. Pharmacol.* 18: 63–73.

SNYDER, S. H., 1970. Putative neurotransmitters in the brain: Selective neuronal uptake, subcellular localization and interactions with centrally acting drugs. *Biol. Psychiat.* 2:367–389.

SNYDER, S. H., M. J. KUHAR, A. I. GREEN, J. T. COYLE, and E. G. SHASKAN, 1970a. Uptake and subcellular localization of neurotransmitters in the brain. *Intern. Rev. Neurobiol.* 13: 127–158.

SNYDER, S. H., K. M. TAYLOR, J. T. COYLE, and J. L. MEYERHOFF, 1970b. The role of brain dopamine in behavioral regulation and the actions of psychotropic drugs. *Amer. J. Psychiat.* 127:117–125.

SNYDER, S. H., and J. L. MEYERHOFF, 1973. How amphetamine acts in minimal brain dysfunction. *Ann. N.Y. Acad. Sci.* 205: 310–320.

SVENSSON, T. H., 1971. Functional and biochemical effects of *d*- and *l*-amphetamine on central catecholamine neurons. *Arch. Pharmakol.* 271:170–180.

TAYLOR, K. M., and S. H. SNYDER, 1970. Amphetamine: Differentiation by *d*- and *l*-isomers of animal behavioral effects involving central norepinephrine or dopamine. *Science* 168:1487–1489.

TAYLOR, K. M., and S. H. SNYDER, 1971. Differential effects of *d*- and *l*-amphetamine on behavior and on catecholamine disposition in dopamine and norepinephrine-containing neurons of rat brain. *Brain Res.* 28:295–309.

THOENEN, H., 1972. Chemical sympathectomy: A new tool in the investigation of the physiology and pharmacology of peripheral and central adrenergic neurons. In *Perspectives in Neuropharmacology*, S. H. Snyder, ed. New York: Oxford University Press, pp. 301–338.

UNGERSTEDT, U., 1971. Stereotaxic mapping of the monoamine pathways in the rat brain. *Acta Physiol. Scand.*, (suppl. 10) 367:1–48.

YARYURA-TOBIAS, J., B. DIAMOND, and S. MERLIS, 1970. The action of L-dopa on schizophrenic patients. *Curr. Ther. Res.* 12:528.

64 Schizophrenia:
Relationships to Dopamine Transmission,
Motor Control, and Feature Extraction

STEVEN MATTHYSSE

ABSTRACT Theories of tranquilizer action are reviewed, especially the hypothesis that tranquilizers act by blocking dopamine transmission; analogies are suggested between thought and affect in schizophrenia and the neural processes underlying motor control and feature extraction.

THERE IS COMPELLING evidence, as reviewed by Snyder in this Third Study Program, that phenothiazine tranquilizers are not merely sedatives but act on the core symptoms of schizophrenia. It used to be admitted that tranquilizing drugs were of value in getting patients to talk to their psychotherapists, but Philip May was probably closer to the truth when he suggested that psychotherapy with schizophrenic patients is of value insofar as it gets them to take their pills (May, 1968). The trouble with the phenothiazines from a scientific point of view is that they have a notoriously wide spectrum of actions: They are powerful electron donors (Foster and Fyfe, 1966); they inhibit oxidative phosphorylation (Medina, 1964); they stabilize cell membranes against lysis in distilled water (Despopoulos, 1970; Seeman et al., 1963). These effects, however, are not specific for the antipsychotic members of the phenothiazine group.

So many of the effects of phenothiazines are nonspecific, that Seeman has proposed the theory that "tranquilization is a type of 'selective regional anesthesia'," phenothiazines distinguishing themselves solely by tissue distribution (Seeman, 1966). Supporting evidence, as far as I know, has been obtained in only one case: Thiethylperazine, an antiemetic, concentrates in the cerebellum, whereas chlorpromazine, an antipsychotic, concentrates in subcortical areas (DeJaramillo et al., 1963). These concentrations were measured after only 90 min; it would be important to compare the regional brain distribution of chlorpromazine with a number of nonantipsychotic phenothiazines, and after long-term treatment.

The phenomenon of increased dopamine turnover after administration of tranquilizing drugs (interpreted as resulting from postsynaptic blockade of dopamine receptors) does not appear to be shared by the nonantipsychotic promethazine (Carlsson et al., 1963; Andén et al., 1964; Nybäck et al., 1968). This distinction suggests that dopamine blockade may be related to the *antipsychotic actions* of the phenothiazines, unlike the biochemical effects mentioned before. On the other hand, Parkinsonian motor side effects are prominent with antipsychotic drugs; because of the importance of dopamine in the extrapyramidal system, it is possible that the blocking action of tranquilizers on dopamine synapses could be related to their motor, rather than to their mental, effects. A drug of critical significance in separating mental from motor effects is thioridazine, a phenothiazine antipsychotic with only 3% incidence of Parkinsonian side effects (whereas fluphenazine, for example, has an incidence of 36%) (Cole et al., 1966). Single doses of thioridazine increased dopamine turnover (as indicated by caudate homovanillic acid) in the cat, but long-term treatment did not, although chlorpromazine continued to cause an increase under these conditions (Laverty et al., 1965). In mice, however, the two drugs had identical actions (O'Keeffe et al., 1970), so the implication of these findings is unclear.

Work in our laboratory is designed to investigate the specificity of the dopamine turnover effect for antipsychotic, as opposed to nonantipsychotic, phenothiazines. If a predicted similarity or difference fails to occur in an experimental animal, it could be argued that drug metabolism and tissue distribution differ between the animal and man; therefore, we felt it was desirable to use primates. Our method is to withdraw cerebrospinal fluid continuously from the lateral ventricle, one drop every 4 min for a month at a time, using each animal as his own control. In preliminary experiments, five nonantipsychotic phenothiazines had a negligible effect on homovanillic acid at the substantial dose of 10 mg/kg (diethazine,

STEVEN MATTHYSSE Psychiatric Research Laboratories, Massachusetts General Hospital, Boston, Massachusetts

fenethazine, methdilazine, promethazine, and pyrathiazine). In contrast, chlorpromazine typically causes a rise of 90% at this dose, and haloperidol and pimozide (both are antipsychotic) also cause large increases.

In these preliminary experiments, two drugs had effects on ventricular fluid homovanillic acid that do not seem entirely compatible with the dopamine hypothesis. Thiethylperazine, a nonantipsychotic phenothiazine, caused a rise of 48% at 0.3 mg/kg. The effect of thioridazine is disappointingly small: 8.5% at 10 mg/kg, in a typical experiment. We plan several refinements in our studies, which should clarify the actions of these two drugs: (1) detailed dose-response curves; (2) long-term administration of the drugs, an experiment for which the ventricular tapping procedure is especially suitable; (3) regional studies (it may be, for example, that thioridazine has a larger effect in the dopamine-containing nuclei of the limbic system than it does in the caudate, which is the structure most strongly represented in the ventricular fluid).

Holding in abeyance our remaining doubts, let us consider the implications of the hypothesis that the antipsychotic actions of the major tranquilizers (phenothiazines, butyrophenones, and diphenylbutylpiperidines) are brought about by blockade of dopamine receptors. We can begin by reflecting on the function of dopamine in the basal ganglia. The older view of the basal ganglia was that the motor cortex initiates movement, which is then modulated and bounded by the basal ganglia. The new view, presented at the Third Study Program by Kornhüber and by DeLong, is practically the reverse: Sensory input, processed in the whole cortex, results in "motor commands" transformed into action by the basal ganglia; once initiated in the basal ganglia, the action is then modulated and subjected to sensory feedback control by the motor area.

I think it is reasonable to propose that the nigrostriatal dopamine system regulates the *responsiveness* of the basal ganglia to these cortical commands, that is, the threshold for their emergence into real action. The strongest evidence is behavioral: In dopamine deficiency syndromes, such as Parkinson's disease, reserpine treatment or 6-hydroxydopamine lesion of the nigrostriatal dopamine tract (see Ungerstedt, this volume; S. Iversen, this volume), there is poverty of initiation of movement. Conversely, L-dopa accentuates choreiform movements and is proposed for detection of presymptomatic chorea by making latent movements manifest (Klawans et al., 1970; Cawein and Turney, 1971). Apomorphine, a dopamine receptor stimulator, causes stereotyped movements in animals (Ernst, 1965; Ungerstedt, this volume; S. Iversen, this volume). Neurons were observed in the caudate nucleus that were hyperpolarized and unresponsive to

cortical stimulation, but these neurons underwent spontaneous depolarizing shifts of 5 mV lasting several minutes, during which they did respond to cortical stimulation. High-frequency nigral stimulation produced similar depolarizing shifts in caudate neurons lasting about 20 sec, during which their firing rate increased.

To summarize: Subthreshold motor commands are transformed into effectively initiated movements by the basal ganglia, and dopamine acts by disinhibition to permit the emergence of these movements from their subthreshold state. Now let us take a theoretical leap and suppose that dopamine neurons have the same function in the mental sphere. Because of the conservatism of evolution, a mechanism that was successful in one aspect of nervous function might be adopted in a similar, higher function. That is, analogy may be a way of nature as well as a way of thought.

Now if you introspect as you read this essay, you will observe that thoughts and images are continually rising near to the periphery of consciousness but fail to become fully emergent; some of these thoughts are cares of the day, some are dreamlike, disconnected images, some are even bizarre and disturbing. Imagine what would happen to an individual who did not have the capacity to keep these thoughts from taking over his field of awareness. I suggest that these preconscious thoughts and fantasies are like subthreshold motor commands; that some structure, analogous to the basal ganglia, releases them to enter the stream of consciousness; that dopamine neurons regulate the threshold for emergence of ideas, as they do the emergence of actions; that this disinhibitory system is overactive in schizophrenia; and that neuroleptic drugs increase "repression" of distracting ideas by blocking dopamine synapses.

The analogy may be illustrated by comparing two classical descriptions: Jung and Hassler's, of the function of the basal ganglia, and Eugen Bleuler's, of the thought process in schizophrenia. According to Jung and Hassler the basal ganglia contribute to motor control by "selectively inhibiting impulses from other sensory systems or from areas of activity which do not belong to the pattern of excitation most significant at the moment" (Jung and Hassler, 1959). Notice the similarity to Bleuler's conception of the missing controls in schizophrenic thinking: "It may be assumed that a certain force is necessary to keep associations in the track laid out by experience. Now it is possible that this force or 'control-tension' has also been diminished or hampered in its action because of the fundamental schizophrenic process" (Bleuler, 1924, pp. 81–82). Schizophrenics "are brought to a topic, totally irrelevant to the subject in hand, by any accidental things that happen to affect their senses.... The normal directives through questions from without and purposive conceptions from within are incapable of holding the train of thought·in the proper channels" (Bleuler, 1924, pp. 376–377).

So far we have an analogy between excessive emergence of subliminal ideas in schizophrenia, and excessive performance of subthreshold motor commands in certain diseases of the basal ganglia. However, schizophrenia is a disease of *affect* as well as of *thought*. To quote again from Bleuler: "One of the surest signs of the disease is the incapacity to modulate the affects, or an affective rigidity . . . the affective expressions are usually somewhat unnatural, exaggerated or theatrical. Consequently the joy of a schizophrenic does not transport us, and his expressions of pain leave us cold" (Bleuler, 1924, pp. 379–380).

The terms Bleuler uses are remarkable: *Incapacity to modulate, rigidity, exaggerated or theatrical*. Were he a neurologist describing disorders of the basal ganglia, these words would not be inappropriate to characterize chorea, athetosis, dystonia, and other hyper- and hypo-kinesias. Indeed, when affective rigidity and exaggeration become sufficiently intense, actual disturbances in posture and locomotion occur (catatonia). The affective disturbance in schizophrenia would suggest, to use Kornhüber's concept (1971; this volume), a disturbance in ramp or function generation, not in the basal ganglia, but in whatever system controls the initiation of affective response.

Where shall we look for a system, analogous to the basal ganglia, its threshold of activity regulated by dopamine neurons but more likely to be related to emergence of thought and to "ramp generation" of affect? Some fascinating observations have recently been made on the nucleus accumbens septi (Wilson, 1972). Embryologically and histologically, this nucleus is an extension of the head of the caudate, and like the caudate it has a major dopamine innervation (Ungerstedt, 1971). Its inputs, however, are from widespread parts of the limbic system, especially the hippocampus and pyriform cortex, just as the inputs to caudate are from widespread parts of the neocortex; indeed these inputs appear to be mutually exclusive. Its output is largely to the region of the pallidum known as substantia innominata. Its further projections are unknown, but it is interesting that DeLong has observed a response of substantia innominata neurons to the reward stimulus rather than solely to the motor response executed to earn the reward, suggesting that this region is involved in something more than motor control (DeLong, 1972).

The theory presented so far fails to account for certain features of schizophrenia.

(1) Identical twins of schizophrenics are not all schizophrenic (Pollin, 1972). The absence of complete concordance is thought to indicate an influence of environ-

ment. In the basal ganglia, on the other hand, one is struck by the lack of plasticity. The dyskinesias do not seem to be produced by, or improved by, experience.

(2) Interpersonal aversiveness seems to be a personality characteristic common to all forms of schizophrenia (Meehl, 1972).

(3) Schizophrenics have a strong tendency to misperceive social cues (Ploog, 1972).

A concept that has been a central focus of this Third Study Program goes far to fill these gaps: *Feature extraction*. It is plausible to assume that, in addition to cells that abstract from complex stimuli patterns, motion, and Fourier components, there should be cells that abstract *emotional significance* of complex stimuli for the organism. I will refer to these as *value detectors* because attractiveness-aversiveness, or value, a simple one-dimensional continuum, expresses the most universal emotional significance of stimuli; it remains to be seen whether there are also feature extractors specific for particular emotional meanings, such as danger or affection.

There is some evidence for the existence of value detectors. Neurons were observed in the pyriform cortex, hippocampus, and amygdala that changed in firing rate when a visual stimulus was presented that had, through conditioning, been associated with the opportunity of acquiring food by a lever press (Fuster and Ueyda, 1971). These units did not respond to the visual stimulus without the pairing, nor to a similar stimulus that had been associated with an occasion for avoiding shock. It would also be important to show that other occasions for reward, paired with the stimulus, could cause the stimulus to change firing rate in the same direction and to test the range of effective cue stimuli.

This addition to the theory does help account for the coexistence of genetic and environmental determinants in schizophrenia. An inborn skewness of the population of value detectors could cause interpersonal aversiveness and misperception of social cues. The work of Colin Blakemore (this volume) has shown that the population of visual feature detectors for angle can be skewed by exposure to horizontal or vertical stripes during a critical period. Perhaps exposure to continuous aversive stimuli in early life could similarly bias the population of value detectors in the direction of aversion rather than attraction. Blakemore also discovered cells that developed *bimodal* selectivities after alternating exposure to horizontal and vertical stripes during a critical period. It could be that erratic or illogical reward and punishment (Lidz et al., 1958), or simultaneous reward in one channel and punishment in another, as in Bateson's "double-bind" theory (1956), could cause value detectors to develop bimodal selectivities, and therefore lead to confused perception of social cues, and to interpersonal ambivalence.

This theory suggests enough experiments to keep us busy for a long time.

(1) Since there is more evidence for dopamine action on the "ramp generators" than on the feature detectors, tranquilizers should be more effective against thought disorder and affective rigidity than against interpersonal aversiveness and misperception of social cues.

(2) The theory suggests the importance of concentrated study of the connections, functions, electrophysiology, and pharmacology of the nucleus accumbens and other limbic dopamine nuclei, and of the actions of the mesolimbic dopamine system on them: In particular, does the dopamine system regulate the excitability of the nucleus accumbens?

(3) One ought also to investigate the existence of value detectors, using the complete set of criteria mentioned before. How widespread are they? Can the population be skewed by early experience? Does a drug like amphetamine, which can cause paranoid states (Snyder, 1972), modify their responsiveness? Do any of the diffuse ascending systems, such as those utilizing dopamine or norepinephrine, exert a biasing action (hyperpolarizing or depolarizing) on classes of value detectors? Are there "on" and "off" value detectors, by analogy with the visual system? An "off" value detector might be a *"sigh of relief" neuron*; an "on" value detector might be an *"I'm getting into trouble" neuron*. The latter type of feature extractor may be especially active in those who postulate theories of schizophrenia!

REFERENCES

ANDÉN, N. E., B. E. ROOS, and B. WERDINIUS, 1964. The effects of chlorpromazine, haloperidol and reserpine on the levels of phenolic acids in rabbit corpus striatum. *Life Sci.* 3: 149–158.

BATESON, G., D. D. JACKSON, J. HALEY, and J. H. WEAKLAND, 1956. Toward a theory of schizophrenia. *Behav. Sci.* 1: 251–264.

BLEULER, E., 1924. *Textbook of Psychiatry*. New York: Macmillan.

CARLSSON, A., and M. LINDQVIST, 1963. Effect of chlorpromazine or haloperidol on formation of 3-methoxytyramine and normetanephrine in mouse brain. *Acta Pharmacol. Toxicol.* 20:140–144.

CAWEIN, M., and F. TURNEY, 1971. Test for incipient Huntington's chorea. *New Eng. J. Med.* 284:504.

COLE, J. O., and D. J. CLYDE, 1966. Extrapyramidal side effects and clinical responses to the phenothiazines. *Rev. Canad. Biol.* 20:565–574.

DEJARAMILLO, G. A. V., and P. S. GUTH, 1963. A study of the localization of phenothiazines in dog brain. *Biochem. Pharmacol.* 12:525–532.

DELONG, M., 1972. Activity of basal ganglia neurons during movement. *Brain Res.* 40:127–135.

DESPOPOULOS, A., 1970. Antihemolytic actions of tricyclic tranquilizers. Structural correlations. *Biochem. Pharmacol.* 19: 2907–2914.

ERNST, A. M., 1965. Relation between the action of dopamine and apomorphine and their O-methylated derivatives upon the CNS. *Psychopharmacologia* 7:391–399.

FOSTER, R., and C. A. FYFE, 1966. Electron-donor-acceptor complex formation by compounds of biological interest. II. The association constants of various 4-dinitrobenzene-phenothiazine drug complexes. *Biochim. Biophys. Acta.* 112: 490–495.

FUSTER, J. M., and A. A. UEYDA, 1971. Reactivity of limbic neurons of the monkey to appetitive and aversive signals. *Electroenceph. Clin. Neurophysiol.* 30:281–293.

JUNG, R., and R. HASSLER, 1959. The extrapyramidal motor system. In *Handbook of Physiology*, J. Field, ed. Baltimore: Williams & Wilkins, Vol. 2, pp. 863–927.

KLAWANS, H. C., G. W. PAULSON, and A. BARBEAU, 1970. Predictive test for Huntington's chorea. *Lancet* 2:1185–1186.

KORNHÜBER, H. H., 1971. Motor functions of cerebellum and basal ganglia: The cerebellocortical saccadic (ballistic) clock, the cerebellonuclear hold regulator, and the basal ganglia ramp (voluntary speed smooth movement) generator. *Kybernetik* 8:157–162.

LAVERTY, R., and D. F. SHARMAN, 1965. Modification by drugs of the metabolism of 3,4-dihydroxyphenylethylamine, noradrenaline and 5-hydroxytryptamine in the brain. *Brit. J. Pharmacol.* 24:759–772.

LIDZ, T., A. CORNELISON, D. TERRY, and S. FLECK, 1958. Intrafamilial environment of the schizophrenic patient: VI. The transmission of irrationality. *Arch. Neurol. Psychiat.* 79: 305–316.

MAY, P. R. A., 1968. Anti-psychotic drugs and other forms of therapy. In *Psychopharmacology: A Review of Progress 1957–1967*, D. H. Efron, ed. Washington, D.C.: Public Health Service, pp. 1155–1176.

MEDINA, H., 1964. The effect of certain phenothiazines on the structure and metabolic activity of sarcosomes of the guinea pig heart. *Biochem. Pharmacol.* 13:461–467.

MEEHL, P. E., 1972. In *Prospects for Research on Schizophrenia*, S. S. Kety and S. Matthysse, eds. *Neurosciences Res. Prog. Bull.* 10:377–380.

NYBÄCK, H., Z. BORZECKI, and G. SEDVALL, 1968. Accumulation and disappearance of catecholamines formed from tyrosine-^{14}C in mouse brain; effect of some psychotropic drugs. *Euro. J. Pharmacol.* 4:395–403.

O'KEEFFE, R., D. F. SHARMAN, and M. VOGT, 1970. Effect of drugs used in psychoses on cerebral dopamine metabolism. *Brit. J. Pharmacol.* 38:287–304.

PLOOG, D., 1972. In *Prospects for Research on Schizophrenia*, S. S. Kety and S. Matthysse, eds. *Neurosciences Res. Prog. Bull.* 10: 394–396.

POLLIN, W., 1972. The pathogenesis of schizophrenia: Possible relationships between genetic, biochemical, and experiential factors. *Arch. Gen. Psychiat.* 27:29–37.

SEEMAN, P. M., 1966. Membrane stabilization by drugs: Tranquilizers, steroids, and anesthetics. *Int. Rev. Neurobiol.* 9:145–211.

SEEMAN, P. M., and H. S. BAILY, 1963. The surface activity of tranquilizers. *Biochem. Pharmacol.* 12:1181–1191.

SNYDER, S. H., 1972. In *Prospects for Research on Schizophrenia*, S. S. Kety and S. Matthysse, eds. *Neurosciences Res. Prog. Bull.* 10:433–435.

UNGERSTEDT, U., 1971. Stereotaxic mapping of the monoamine pathways in the rat brain. *Acta Physiol. Scand.* (Suppl.) 367: 1–48.

WILSON, R. D., 1972. The neural associations of nucleus accumbens septi in the albino rat. Ph.D. Thesis, Dept. of Psychology, Mass. Inst. of Technology. (This work was done in association with Drs. L. Heimer and W. J. H. Nauta.)

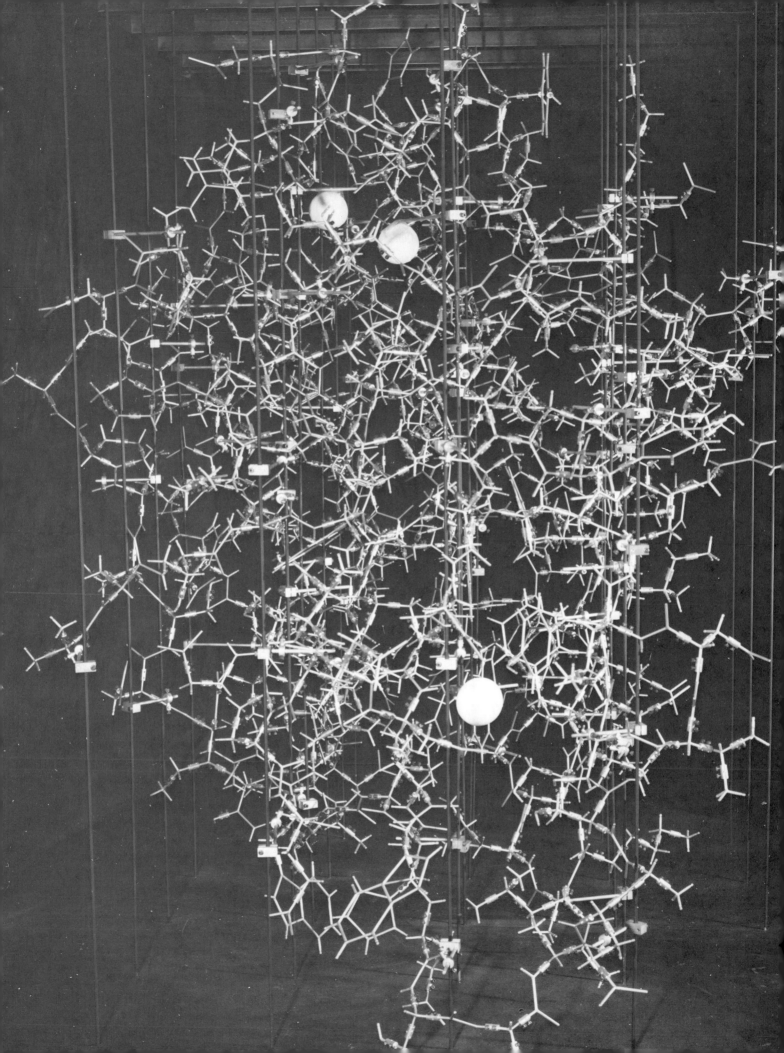

MOLECULAR

MACHINERY OF

THE MEMBRANE

A skeletal model of the monomer of the mitogen concanavalin A. This protein stimulates lymphocytes to divide after binding to glycoprotein receptors on the membrane. Phenyl sugars are bound in the large cleft at the bottom center of the model at the site indicated by the white sphere. Two metal ions, indicated by the two spheres at the top, must be bound before saccharide binding can occur. The binding specificity of concanavalin A also makes it a valuable probe of membrane structures. (Courtesy of J. W. Becker, G. N. Reeke, Jr., and G. M. Edelman. See G. M. Edelman et al., The covalent and three-dimensional structure of Concanavalin A. Proc. Nat. Acad. Sci. USA, 69: 2580–2584.)

MOLECULAR MACHINERY OF THE MEMBRANE

Introduction

GERALD M. EDELMAN

IN THE LAST DECADE, it has become increasingly clear that the cell membrane contains intricate and highly coordinated molecular machinery mediating and controlling growth, division, tissue recognition, and morphogenesis. Recent studies have indicated that the cell surface contains highly differentiated glycoproteins and other receptors that play various roles in these functions. Regulation of cell division via surface interactions has been demonstrated, and there is clearly a cell surface-nuclear interaction, the details of which are under current investigation. Some of the problems in this area are concerned with the chemical nature, numbers, distribution, regularity, masking and unmasking, and mobility of the surface receptors for cell division. Active work is now being carried out on the surface chemistry and immunology of cells and the isolation and characterization of receptor molecules and pathways of their synthesis. One of the major riddles is the mechanism of transduction by which binding of a macromolecule sets in motion the complex events accompanying division.

A second major area is concerned with the specificity

of cell surfaces and particularly cell specificity in development, i.e., the nature of cell to cell interactions and the expression of specific cell receptors at critical times during morphogenesis. The genetic control of receptor expression is just beginning to be studied.

Finally, one of the classical fields of study is now being seen in a new light: Growth regulation and coordination, contact inhibition of cells, and hormonal control are now being studied at the macromolecular level by analysis of cell surfaces and membranes.

In summary, there has been a new surge of activity from analyses of membrane structures to studies of the dynamics and molecular biology of the control of cell division, of cell to cell interaction, and morphogenesis. The study of the specificity of cell surfaces has now emerged as a major field with new methodology and having connections with classical work in developmental biology, membrane structure and cell surface immunology.

65 Cell-Cell Recognition in the Developing Central Nervous System

RICHARD L. SIDMAN

ABSTRACT Organizational patterns in the adult brain are based on developmental parameters, including specific sites and rates of cell genesis, subsequent migration of cells to their permanent positions, and genesis of cell processes. The migrations of young neurons are commonly constrained and guided by radially oriented, elongated epithelial cells; at relatively late stages of development, when the migration distance is large and the intervening terrain is complex, radial glial cells provide this guidance. Interactions determining cell survival, formation of particular dendritic patterns, and the acquisition of synaptic inputs likewise depend in part on cell-cell recognition systems, as illustrated for cellular development in normal primates and in normal and mutant mice.

Specific recognition phenomena or other mechanisms may underlie pattern formation by aggregating CNS cells in vitro. For further analysis, methods were developed that allow separation and recombination of partially purified cell classes from a region of developing CNS. On the premise (untested) that these cellular interactions are mediated by specific cell surface constituents, concepts and laboratory techniques of cellular immunology are being extended to the developing CNS, both to characterize surface constituents and to produce reagents that will allow design of in vitro test systems in which cells of two or more classes in defined numbers will be allowed to interact in defined geometric arrangements.

MY AIM, IN opening the part topic, "Molecular Machinery of the Membrane," which itself initiated the Third Intensive Study Program in Boulder, Colorado in 1972, is to emphasize cell interactions dependent on the close apposition of cells in the vertebrate central nervous system (CNS). The extraordinary geometric precision of this apposition in the mature CNS may be inferred from the orderliness of synaptic relationships (e.g. Palay and Chan-Palay, 1973). I will argue that this geometric precision is characteristic of the central nervous system from its earliest developmental stages, that it reflects the behavior of particular cells during development, and that this is closely controlled (within genetic limits) by the distribution and behavior of neighboring cells. The responsiveness of CNS cells to the behavior of their neigh-

RICHARD L. SIDMAN Department of Neuropathology, Harvard Medical School; and Department of Neuroscience, Children's Hospital Medical Center, Boston, Massachusetts

bors in the adult may, in fact, represent merely the persistence of a crucial developmental property.

If cells do interact in important ways through their life histories, it is likely that properties of surface membranes will be shown to mediate the effects. There is even some merit in the extreme view that the detailed molecular machinery of the surface membrane will be of interest in proportion to its usefulness in accounting for cell-cell interactions.

One of the many useful roles of this Intensive Study Program is to provide a yardstick with which to measure the conceptual and factual advances in certain fields of neuroscience since the previous program three years ago. At that time a symposium was held on Development of the Nervous System (Edds, 1970), and in addition there were on many days and evenings unscheduled informal discussions, a feature that has become a characteristic and extremely valuable hallmark of these Boulder meetings. From these discussions emerged a generalization (Boulder Committee, 1970) that serves as an appropriate starting point for this presentation.

Cell-cell contacts during early histogenesis

Figure 1 summarizes the fundamental arrangement and behavior of cells in all regions of the CNS of vertebrates at early stages of development (Boulder Committee, 1970). In both concept and terminology, it is intended to integrate the findings of many investigators from Wilhelm His to the present, as obtained by histological, Golgi, autoradiographic, and electron microscopic techniques. Several points are of particular relevance to this part topic, as follows.

From the earliest stages when the neural plate is recognizable, and at all stages thereafter, the cells of the CNS display an epithelial arrangement. That is, the cells are closely apposed, with little demonstrable extracellular matrix between them. (Because of difficulties in determining whether or not the cell surface components and possible intercellular materials are accurately preserved in fixed tissue specimens, one cannot yet state whether

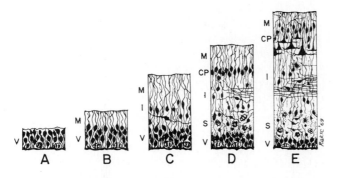

FIGURE 1 Semidiagrammatic drawing of the development of the basic embryonic zones and the cortical plate. (From Boulder Committee, 1970.) Abbreviations: CP, cortical plate; I, intermediate zone; M, marginal zone; S, subventricular zone; V, ventricular zone.

adjacent cells of the early embryonic CNS show the approximately 180 Å space between the electron-dense outer components of the cell membranes characteristic of the adult CNS.) At early stages the cells are uniformly polarized, with apical and basal surfaces recognizably different from each other and with lateral surfaces directly apposed in three dimensions with neighboring neuroepithelial cells. During neural plate formation, these epithelial cells all behave alike, though some of them may already be committed to particular paths of development. They progressively elongate so that the neural wall becomes correspondingly thicker. As the cells divide repeatedly the tissue expands in area as well as in thickness. Though the tissue boundaries are expanding dramatically, and the CNS is changing in gross shape, its cellular arrangement remains fundamentally unaltered.

Two factors jointly determine that the cellular arrangement remains uniform during the early stages of growth when all cells are multiplying. One is the form of the elongated epithelial cell which is moored by a specialized "collar" around the circumference of the cell at a level close to the free, apical (ventricular) surface; this specialized surface membrane property is recognized in thin sections as a desmosomal adhesion site that appears to stabilize the apical regions of adjacent cells (e.g. Hinds and Ruffett, 1971). Further, as an elongated cell enters the G2 phase of the cell generation cycle (after DNA replication and preparatory to mitosis), microtubules concentrate in the apical cytoplasm, and then the nucleus moves apically within the cell so that the cell as a whole rounds up at the ventricular surface and divides there. The second factor controlling cell arrangement is that neighboring cells are randomly distributed with respect to the phases of the cell generation cycle. The consequence is that the shapes of neighboring cells are different. Thus, when one cell is rounded up just after mitosis, several of

its neighbors, engaged at that moment in one or another part of interphase, are elongated. This randomness, and the fact that the duration of G2 plus mitosis is only about 25% of the total cell generation cycle, gives a high probability that a round, newly generated cell at the ventricular surface will be in contact with several elongated cells. It is plausible that the firmly anchored elongate cells mechanically restrict the movement of the new cell, preventing it from assuming oblique or horizontal orientations; or, alternatively, they might be providing guidance based on specific surface molecules as the new cell in turn develops an elongated shape. The overall result is that the general pseudolaminar epithelial arrangement is preserved. At later stages, when some of the new cells become permanently postmitotic and move external to the ventricular zone (Figures 1C, 1D), they continue to be either constrained or positively guided along a radial path by their elongated neighbors, which still stretch from ventricular to external surfaces of the expanding brain wall.

Radial guides at later stages of cortex formation

The formation of a cortex, whether in cerebrum or cerebellum, appears to involve recognition events between migrating and postmigrating cells as well as between neighboring cells of the ventricular zone. It was inferred long ago from histological material (Koelliker, 1896) and established by autoradiography (Angevine and Sidman, 1961) that in the mature cerebral cortex the positions of neuron cell bodies are, in general, systematically related to the time of neuron origin such that the deepest-lying cortical cell bodies were the earliest, and the outermost ones were the latest to be generated. The first cells to arrive in the cortical plate show a radially oriented bipolar form (Figure 1D); at a subsequent stage these cells begin to differentiate and change their shapes, while later arriving neuron cell bodies, still in the radial bipolar configuration, take up positions external to their predecessors (Figure 1E). Subsequently, cells will take up progressively more external positions in mouse (Angevine and Sidman, 1961; Berry and Rogers, 1965; Angevine, 1965), rat (Hicks and D'Amato, 1968), and rhesus monkey (Rakic and Sidman, unpublished observations). There are interesting exceptions to this systematic "inside-out" relationship between time of origin and ultimate position in the mouse (Caviness and Sidman, 1973b). Nonetheless, the inverted developmental pattern is sufficiently general and different in cortices compared to that in most other parts of the CNS (Taber Pierce, 1966; Angevine, 1970; Hinds, 1968; Sidman, 1961, 1970) to imply that different positioning mechanisms are at play in cortical and noncortical regions.

The implication that the relationship of cell position to its time of origin is based on differences in molecular mechanisms governing recognition between cells is strengthened by analysis of the cortical malformation in the reeler mutant mouse (reviewed in Sidman, 1968, 1972). Cell bodies lie in abnormal positions in most parts of the reeler cerebral and cerebellar cortex, whereas no abnormalities have been described in most other regions of the CNS. The neocortical and hippocampal anomalies were first recognized by Meier and Hoag (1962) and Hamburgh (1963). A systematic analysis of forebrain cytoarchitectonics has been initiated recently by Caviness and Sidman (1972, 1973a, 1973b), with several new findings. Every cytoarchitectonic area recognizable in Nissl- and myelin-stained serial sections in the normal adult mouse basal forebrain, retrohippocampal region, and hippocampal formation is also recognizable in reeler. The fact that each one is positioned appropriately in relation to neighboring areas implies that cells were generated in the normal positions on the two-dimensional mosaic of the embryonic ventricular zone and were either displaced or migrated externally in normal radial fashion in the mutant. However, the radial vector of cell migration is abnormal in that they become positioned at relatively abnormal distances from the external surface. In the normal mouse, in relatively simple cortical cyto-architectonic areas such as the subiculum or the area entorhinalis, a zone of polymorphic neuron somas normally lies deep to a zone of medium-sized pyriform neuron somas, and these in turn lie deep to an external plexiform zone. In the corresponding areas in reeler, the cellular layers are reversed, with the polymorphic cell somas occupying the positions of the external plexiform zone and the medium-sized neuron somas lying deep to them. In more complex cytoarchitectonic areas in normal mice, such as the perirhinal cortex (Caviness and Sidman, 1973a) or other areas of neocortex (Caviness, unpublished observations), a zone of small neuron somas is inserted between the medium cell zone and the external plexiform zone. To complicate the picture, in reeler the zone containing these cell somas is not in a "reversed" position but lies, with somewhat vaguely defined boundaries, approximately at its normal distance from the external surface of the cortex.

An autoradiographic analysis of adult brains that had been exposed to ^{3}H-thymidine at different stages of fetal life indicates that cells of a given class, whether polymorphic, pyriform, or granule cell, are generated over the same time span in normal and reeler mice even though they subsequently reach different positions in the two cases. In both normal and mutant, the time spans of genesis for the different classes overlap somewhat, so that, for example, the first small neurons are generated earlier

than many of the medium-sized ones and even before some of the polymorphic neurons. These autoradiographic comparisons between normal and mutant, coupled with the marked differences in adult cytoarchitecture, strongly suggest that the ultimate position of a neuron soma in the cortex is not solely a function of the time of cell genesis.

An analysis of myelin-stained sections reveals that some fiber tracts of the reeler forebrain lie in highly abnormal positions. For example, the anterior limb of the anterior commissure lies close to the olfactory extension of the lateral ventricle in the normal mouse brain (Sidman et al., 1971) but occupies a much more external site just medial to the lateral olfactory tract in the mutant brain (Caviness and Sidman, 1972). Such an abnormality must arise very early in embryonic life, for it is difficult to conceive how a major fiber bundle could become so markedly repositioned at later stages of development, when many other cell and fiber systems have filled out the terrain.

All in all, these architectural considerations lead to the view that the reeler genetic locus must express its effects early, no later than embryonic day 13, the period when cortical cells and fiber tracts are taking their positions. This has been confirmed by direct study of embryonic specimens (Caviness and Sidman, unpublished observations). No disorder intrinsic to the individual neurons has been recognized, either in terms of their time or site of genesis or their cytological features as seen in Nissl preparations, and we have adopted the working hypothesis that the reeler locus controls some key early intercellular relationship such as the position or time of contact between migrating young neurons and incoming central afferents (e.g. Figures 1D, 1E) or some other even more elusive relationship. This formulation modifies an earlier view (Sidman, 1968), which referred the presumed "recognition" event to a later stage of development.

In our discussion thus far we have emphasized the early interactions of neuron with neuron. It is important to point out an intriguing relationship of neuron to glia as well, an interaction that may be almost as critical at later stages of development (Rakic, 1971b, 1972a). We have already dealt with the concept that the earliest cells that migrate outward to form a cortex are constrained in their radial paths by the columnar arrangement of their neighboring cells. At early stages no meaningful distinctions between cell classes can be drawn, despite the classical attempts to distinguish "neuroblasts" from "spongioblasts." At later stages of development, different cell classes do become recognizable. Golgi methods delineate a population of deep-lying cells with radially directed processes that pass all the way to the external surface (Figure 2), and electron microscopy indicates that these cells now have glial rather than neuronal cytological

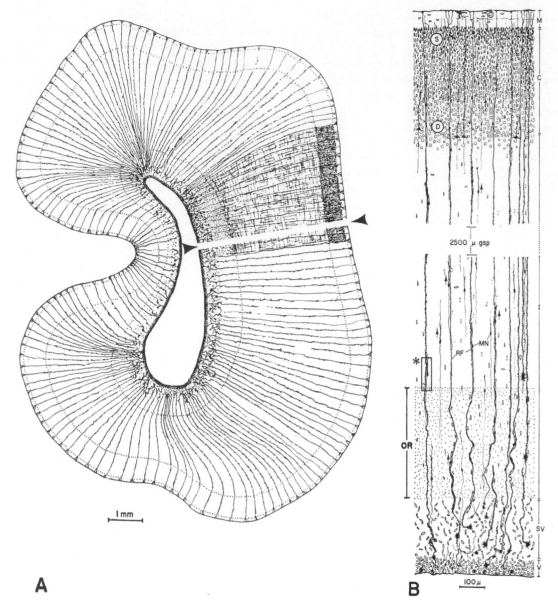

A

B

FIGURE 2 Camera lucida drawings of a coronal section of a Golgi-impregnated telencephalon of a 97-day monkey fetus. (From Rakic, 1972a.) (A) At the parieto-occipital level; 1 mm scale indicates the magnification. The area delineated by the white strip between the arrowheads is drawn in B at higher magnification. (B) Enlargement of the portions of cerebral wall indicated by the white strip in A, combined from a Golgi section (black profiles) and an adjacent section stained with toluidine blue (outlined profiles). The middle 2500 μ of the intermediate zone, also spanned by radial fibers, is omitted. The rectangle marked with an asterisk shows the approximate position of the three-dimensional reconstruction in Figure 3. The 100 μ scale indicates the magnification. Abbreviations: C, cortical plate; D, deep cortical cells; I, intermediate zone; M, marginal layer; MN, migrating cell; OR, optic radiation; RF, radial fiber; S, superficial cortical cells; SV, subventricular zone; V, ventricular zone.

characteristics. The new point of interest is that these radial glial fibers serve as guides along which the young neurons invariably migrate as they pass externally toward and into the cerebral cortex. The relationship between the two cell types is particularly vivid in the large primate cerebrum at relatively late stages of development when

the young neuron setting out for the cortex has a bipolar shape with a leading process that measures less than 200 μ; yet the cell must traverse a distance of about 3500 μ through terrain of highly complex and varied texture (Figure 3). The radial glial fiber provides a pathway at late stages similar in principle to the guides available at earlier stages when the cell types were less sharply differentiated. There is no information as yet concerning the chemical mechanisms that allow the neuron-glial relationship to be attained and then to be maintained selectively as the young neuron transiently contacts other types of cells in its passage outward.

Cell relationships in the developing cerebellar cortex

The cerebellar cortex is proving to be the most suitable part of the CNS for analysis of cell interactions during development for the same reasons that make it attractive in other types of neurobiological inquiry. The cell types are relatively well defined, their geometric arrangements are distinctive, highly ordered, and uniform throughout the cortex, and the excitatory or inhibitory signs of the major classes of synapses have been established (Eccles et al., 1967). Developmental analysis indicates a remarkable succession of interactions, each apparently dependent on the outcome of the previous one. Some relationships are more firmly established than others. A broad outline will be given, nevertheless, with the realization that certain of the details will need revision as the subject comes to be examined more closely.

The Purkinje neurons are the first postmitotic cells to reach the cerebellar cortex, both in mouse (Miale and Sidman, 1961) and in monkey (Rakic, 1971a). In the adult these cells will serve as the direct or indirect targets of all the inputs, and the sole channel along which information flows out of the cerebellar cortex. During, or soon after, their arrival in the cortex, they are probably

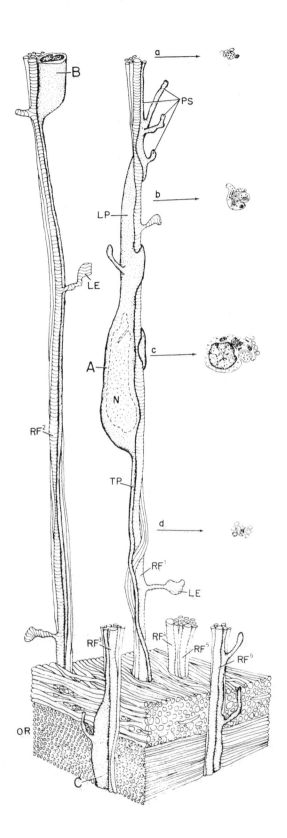

FIGURE 3   Three-dimensional reconstruction of migrating neurons, based on electronmicrographs of semi-serial sections. (From Rakic, 1972a.) The reconstruction was made at the level of the intermediate zone indicated by the rectangle and asterisk in Figure 2B. The subventricular zone lies some distance below the reconstructed area whereas the cortex is more than 1000 μ above it. The lower portion of the diagram contains uniform, parallel fibers of the optic radiation (OR), and the remainder is occupied by more variable and irregularly disposed fiber systems; the border between the two systems is easily recognized. Except at the lower portion of the figure, most of these fibers are deleted from the diagram to expose the radial fibers (striped vertical shafts RF^{1-6}) and their relationships to the migrating cells (A, B, and C) and to other vertical processes. The soma of migrating cell A, with its nucleus (N) and voluminous leading process (LP) is within the reconstructed space, except for the terminal part of the attenuated trailing process and the tip of the vertical ascending pseudopodium. Cross sections of cell A in relation to the several vertical fibers in the fascicle are drawn at levels a to d at the right side of the figure. The perikaryon of cell B is cut off at the top of the reconstructed space, whereas the leading process of cell C is shown just penetrating between fibers of the optic radiation (OR) on its way across the intermediate zone.

joined by still-proliferating Bergmann glial cells. Golgi II cells, a class of inhibitory interneurons destined to lie in the granular layer, form next, but their actual arrival time in the cortex has not been determined.

Another population of more immediate interest begins its long and unusual history at about the same time. These are the external granule cells, a proliferating population of small cells that will eventually give rise to granule cell neurons and perhaps other classes of neurons, as well as glia, in the cerebellar cortex (Fujita, 1967). The external granule cells are first recognizable soon after the genesis of Purkinje cells, when a group of subventricular cells along the rhombic lip gains access to the external surface of the cerebellum. These cells continue to divide rapidly and eventually generate the external granular layer, which coats the entire external surface of the expanding cerebellum. They continue to divide there at a near logarithmic rate for a long time. Much later, some of the daughter cells become postmitotic and their somas become relocated, in a manner to be described below, in the granular layer, deep to the Purkinje cells. With the passage of time, progressively more postmitotic external granule cells move inward, until eventually no cells remain on the external surface. That time is reached about two weeks after birth in the mouse (Miale and Sidman, 1961; Fujita, 1967), a few days later in the rat (Altman, 1969), several weeks after birth in the rhesus monkey (Rakic, 1971a), and many months after birth in man (Raaf and Kernohan, 1944).

Examination of Golgi-stained histological sections at a stage when the external granular layer is maximally developed (Figure 4A) reveals several cellular interrelationships of histogenetic significance. More tangible evidence is found when one examines the behavior of developing cerebellar cells in certain mutant mice, particularly a trio of mutants named reeler, weaver, and staggerer (Sidman et al., 1965).

The first of these interrelationships involves the proliferating cells of the external granular layer of the cerebellum. The number and quality of cells that lie deep to them in the cerebellar cortex appear to affect their mitotic and migratory behavior. A signalling mechanism seems to inform the external granule cells how many times to divide, for they coat the cerebellar surface uniformly in numbers proportional to the size of the cerebellum. The reeler mutant has a smaller than normal cerebellum from the time of birth (Rakic and Sidman, unpublished observations), and this feature is correlated with inadequate, irregular cell proliferation in the external granular layer (Figure 4B). As this figure illustrates, the thickness of the external granular layer varies with the distribution of Purkinje neurons and Bergmann glial cells. For reasons not yet understood, the Purkinje cells of reeler are malpositioned, some of them lying in relatively normal sites, but the majority located more deeply in the tissue. The number of Bergmann glial cells in normal position likewise is reduced. Since the Purkinje and Bergmann cells reach their positions long before the

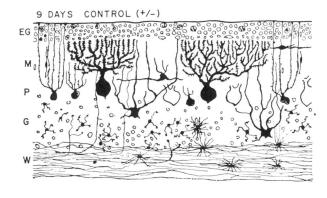

A

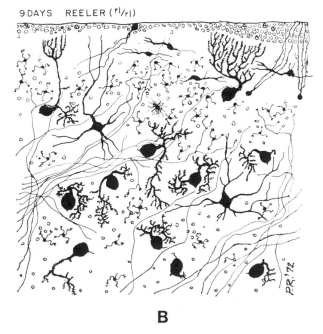

B

FIGURE 4   Composite drawings from Golgi-impregnated preparations of normal and reeler littermates at 9 postnatal days. Abbreviations: G, granular layer; EG, external granular layer; M, molecular layer; P, Purkinje cell layer; W, fibers of the prospective white matter.

number of external granule cells is maximal, the defect in cell proliferation is likely secondary to the positioning disorder rather than the reverse.

A generalized reduction in external granule cells in the staggerer mutant also is evident by birth or shortly after. This seems at first more puzzling, because the number and position of Purkinje and Bergmann cells seems normal. However, closer analysis shows that from early postnatal stages onward, the Purkinje cells in staggerer are abnormally small and develop dendrite arbors with fewer branches than normal and no dendritic spines (Sidman, 1968, 1972; Landis and Sidman, unpublished observations). Again, the data are consistent with the view that the rate of proliferation of external granule cells is controlled by underlying populations of cells.

The second interrelationship concerns the granule cell neuron and the Bergmann glial cell (Rakic, 1971a). Just after final mitosis in the external granular layer, the newly generated granule cell undergoes a remarkable series of rigidly oriented transformations in shapes that convert it from a simple round cell at the boundary between the external granular and molecular layers to a cell with soma and short dendrites in the granular layer. Its thin unmyelinated axon is directed radially outward into the molecular layer, however, where it branches to run parallel to the surface for a few millimeters in the transverse plane (longitudinal to the axis of the cerebellar folium). The granule cell simultaneously changes the position of its soma and spins out a trailing axon as the soma moves inward in direct apposition to a radially oriented Bergmann glial fiber (Figures 4A and 5). This migration is opposite in direction but otherwise resembles the passage of neurons outward to the cerebral cortical plate along radial glial guides, as described above (Figures 2 and 3).

One may ask whether this cell-cell relationship might be fortuitous, the granule cell simply taking the shortest and most direct path across the molecular layer. Again the mutants indicate a causal aspect of the relationship. In reeler, although many of the Bergmann glial cells are obliquely rather than radially oriented, the young granule cells still follow the glial guides and never cross the molecular layer alone, even when they thereby take longer routes (Figure 4B).

The weaver mutant supplies even more dramatic evidence. The cerebellum of the homozygous mature weaver mouse lacks virtually all granule cells (Sidman et al., 1965; Sidman, 1968). Rezai and Yoon (1972) recently reported that the weaver mutation is expressed genetically as a semidominant and interpreted the abnormal cerebellar phenotype as due to impaired migration of newly generated granule cells. In the affected homozygote, the more severe impairment of migration appears to lead secondarily to granule cell death. Electron microscopic study (Rakic and Sidman, 1973) confirms their interpretation and clarifies the underlying basis. The Bergmann glial fibers show early degenerative changes and their number is significantly reduced in heterozygotes; very few Bergmann fibers are seen at all in affected homozygotes. As a consequence, granule cells do not migrate but die while still resident in the external granular layer.

A third relationship involves Purkinje and granule cells at a later stage of maturation. The inward migration of the granule cell and the spinning out of its axon behind the soma coincides temporally with the elaboration of the dendritic tree by the Purkinje cell. It might appear that the Purkinje cell, which had been generated well before the granule cells and had apparently been dormant in the cerebellar cortex, might depend for its dendritic development upon the granule cells, which are making synaptic contact with it. Against this hypothesis is the fact that reeler Purkinje cells lying in aberrant positions and contacted by relatively few granule cell axons still make distinctive dendritic arbors (Figure 4B). A partial measure of Purkinje dendritic development is achieved also when the granule cell precursors degenerate prior to their inward migration in the weaver mutant (Sidman, 1968; Rakic and Sidman, 1973) and in X-irradiated (e.g. Altman and Anderson, 1972) or antimitotic drug-treated animals (Hirano et al., 1972). These same studies indicate, however, that completion of the modelling of Purkinje dendrites does depend on granule cell inputs.

While the Purkinje cell appears relatively independent of the granule cell for its development, the granule cell itself may be quite vulnerable in response to abnormalities in its relationship to the Purkinje cell. In adult humans, many Purkinje cells may be destroyed relatively selectively by hyperthermia or hypoxia, and it is common to observe secondary loss of some granule cells (e.g. Adams and Sidman, 1968). In the mutant mouse named nervous, Purkinje cells die in the third to fifth weeks after birth, and in the most severely affected areas of the cerebellar cortex, there is marked secondary loss of granule cells (Sidman and Green, 1970). Much more dramatic is the loss of granule cells in the staggerer mutant. Here, as mentioned above, the Purkinje cells are abnormal from early postnatal stages onward. Although granule cells migrate to their proper positions and generate dendrites and axons, their axons fail completely to form synapses with the abnormal Purkinje dendrites. Subsequently almost every granule cell dies. While anterograde transsynaptic degeneration is common in the nervous system, presynaptic degeneration, as described here for the granule cell, has been found less frequently (Cowan, 1970).

RICHARD L. SIDMAN 749

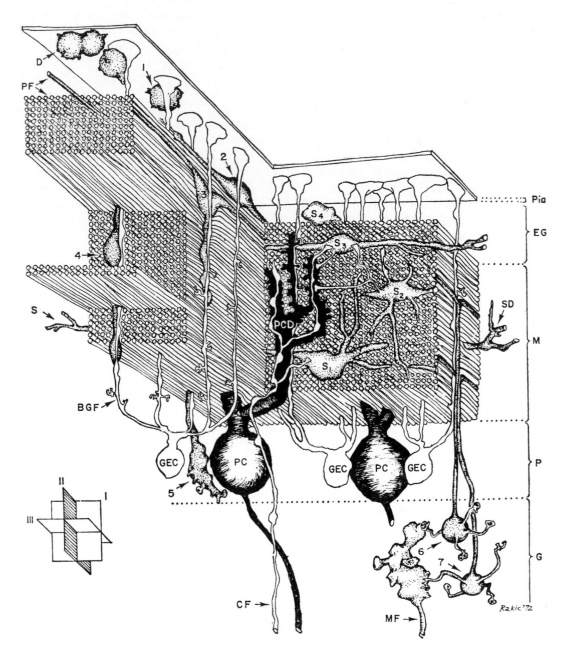

FIGURE 5 A "four-dimensional" (time and space) reconstruction of the developing cerebellar cortex in the rhesus monkey. (From Rakic, 1973.) The geometric figure in the left lower corner indicates the orientation of the planes: I, transverse to the folium (sagittal); II, longitudinal to the folium; III, parallel to the pial surface. On the main figure, the thicknesses of the layers are drawn in their approximately true proportions for the 138-day monkey fetus, but the diameters of the cellular elements, particularly the parallel fibers, are exaggerated in order to make the reconstruction more explicit. A description of the temporal and spatial transformation of the postmitotic granule cells (designated by numerals 1 to 7) and stellate cells (S), as well as other details, are given in Rakic, 1971a, 1973. Abbreviations: BGF, Bergmann glial fiber; CF, climbing fiber; D, dividing external granule cell; EG, external granular layer; GEC, Golgi epithelial cell (Bergmann glia); G, granular layer; M, molecular layer; MF, mossy fibers; P, Purkinje layer; PC, Purkinje cell; PCD, Purkinje cell dendrite; PF, parallel fiber; S_{1-4}, stellate cells; SD, stellate cell dendrite.

Effects of local cellular milieu on interneurons of the cerebellar molecular layer

The molecular layer contains basket and stellate neurons. Both receive excitatory inputs from granule cell axons and climbing fibers and, in turn, inhibit Purkinje cells via their own axons. In light of their similar location and patterns of synaptic connections, basket and stellate cells can be grouped together under the term interneurons of the cerebellar molecular layer. This section summarizes Rakic's recent evidence that these interneurons form a single class of cells by developmental criteria and that their precise form is attained under the direct influence of the local cellular environment. Presumably the cells are responding on the basis of surface membrane mediated signals.

Interneurons of the molecular layer arise relatively late in cerebellar ontogenesis. It is commonly thought that their precursors lie in the external granular layer, but this point has not yet been definitely established. Those basket cells that in the adult lie deepest in the molecular layer are generated earliest, and the progressively more superficial stellate cells arise at progressively later times (Figure 5). It is important to note that the earliest granule cell neurons have arisen and formed their parallel fiber axons before the first basket cells are generated, and that granule cells continue to be produced for a relatively long time after the last stellate cell has appeared (Miale and Sidman, 1961; Fujita et al., 1966; Altman, 1969; Rakic, 1972b).

Although interneurons and granule cells appear to derive from different clonal lines in the external granular layer (Rakic, 1973), it is plausible that basket and stellate cell interneurons trace their origin to a common clone (Rakic, 1972b). Four pertinent facts may be mentioned: (1) The volumes and shapes of these cells vary systematically as a function of the time of cell genesis. (2) Interneurons are first recognizable as horizontal bipolar cells at the interface between the external granular layer and the underlying immature molecular layer. (3) From the beginning, their bipolar processes are oriented at a right angle to the granule cell axons. (4) Their dendrites are always confined to the territory occupied by granule cell axons. These facts, summarized in Figure 5, have a bearing on several major developmental features of the interneurons.

First, the systematic variation in cell volume reflects the progressive decrease in territory available for dendritic growth as successive waves of interneurons are generated. The earliest interneurons begin to generate their processes while they lie close to the Purkinje cell somas on a shallow bed of already-generated parallel fibers. Very promptly, more parallel fibers are laid down external and at a right angle to them, so that the somas of these interneurons become permanently fixed in position; their growing dendrites, however, can lengthen enormously by growing externally *pari passu* with the formation of more and more parallel fibers. These early-formed interneurons, called basket cells, thus lie deepest in the molecular layer and acquire the largest volumes (Figure 6). An interneuron that begins to generate processes when about half the parallel fibers have been laid down will become fixed in position in the middle of the molecular layer; its dendrites can grow either internally into the parallel fiber territory occupied in part by dendrites of earlier-generated interneurons or along with dendrites of the earlier-generated interneurons externally as new parallel fibers come to be laid down. These cells will have smaller volumes than the earlier-differentiated cells. Finally, an interneuron that forms its processes late will have a soma fixed in position near the outside of the molecular layer and its dendrites directed inward; its volume will be even smaller compared to earlier interneurons.

Second, the dendritic branching pattern appears exquisitely dependent on the orientation of the adjacent parallel fibers. Not only do the dendrites grow predominantly in the plane transverse to the orientation of the folium, but the individual growing branchlets appear to realign quickly to that same orientation as though obeying the principle of maximizing the number of parallel fiber contacts per unit length of interneuron dendrite (Figure 7; Rakic, 1972b).

The third feature is more subtle, and concerns the signal that terminates the latent period between the time an interneuron arises and the time its dendrites begin to form. Rakic (1973) has shown that in the monkey fetus this latent period lengthens progressively during development (Figure 8). Basket cells make dendrites almost as soon as the cell itself is generated, but stellate cells destined for superficial parts of the molecular layer apparently are dormant for a period of two months before they generate dendrites. These cells actually arise long before the parallel and climbing fibers that will constitute their proper local environment have formed. They appear to wait, presumably supported as round cells on the thickening bed of parallel fibers, until their cellular milieu "catches up."

Other classes of cell-cell interactions

Cell interactions are the rule rather than the exception in the developing nervous system, and many other examples can be cited. A young afferent neuron may degenerate if

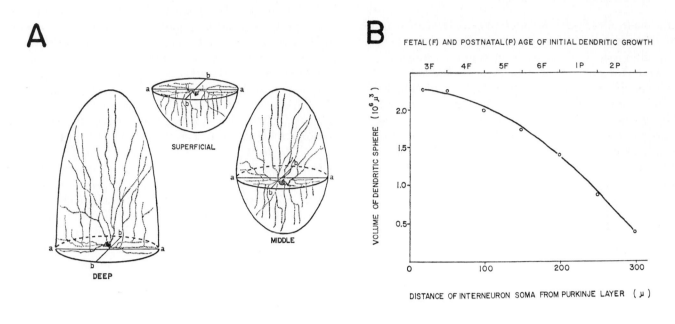

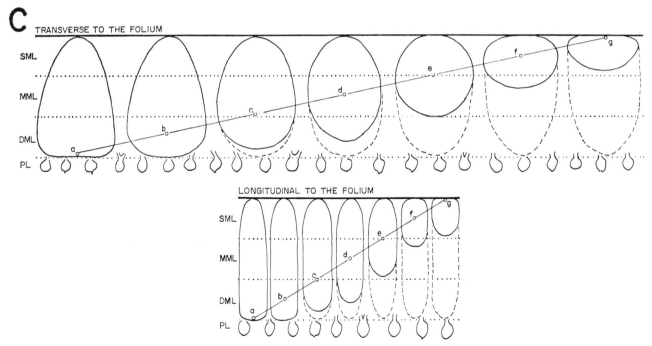

FIGURE 6 Summary of the shapes and volumes of dendritic arborization of interneurons in the cerebellar molecular layer (stellate and basket cells) in rhesus monkey. (From Rakic, 1972b.) (A) Diagrammatic representation of the space occupied by the dendritic arbors of deep, middle and superficial interneurons. (B) Graphic representation of the progressively decreasing volumes occupied by interneuron dendrite arborizations (expressed in 10^6 μ^3 on the vertical scale) as a function of the distance of interneuron somas from the bottom of the molecular layer (lower horizontal scale) and of the age at which dendrites begin to grow (upper horizontal scale in fetal, F, and postnatal, P, months). (C) Diagrammatic representation of the average areas occupied by dendritic arborizations of interneurons situated at different levels of the molecular layer as seen in planes transverse (upper row) and longitudinal (lower row) to the folium. The molecular layer is arbitrarily divided into $100\,\mu$ wide deep (DML), middle (MML), and superficial (SML) zones. Note the systematic change in the shape and size of actual dendritic distributions (outlined by solid lines) from the deepest lying cell, a, to the most superficial cell, g.

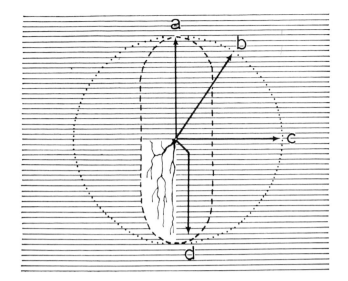

FIGURE 7 An idealized model to illustrate the relationship between dendritic growth and orientation of parallel fibers in the plane parallel to pia. (From Rakic, 1972b.) For the same extent of growth the dendrites, *a*, *b*, and *c* would achieve 100%, 80%, and 0% of the possible interactions with parallel fibers (horizontal lines). In actual specimens, most of the dendrites grow close to the axis that maximizes the number of interactions; growth in direction *c* has not been encountered in electronmicrographs. Many fibers, however, may, after an initial growth at a sharp angle, curve in the direction which could account for the observed dendritic distribution within the area of a flattened ellipse (broken line). The actual pattern of the dendritic arborization as seen from above in the plane parallel to pia is inscribed on the left lower portion of the ellipse. For further explanation see Rakic, 1972b.

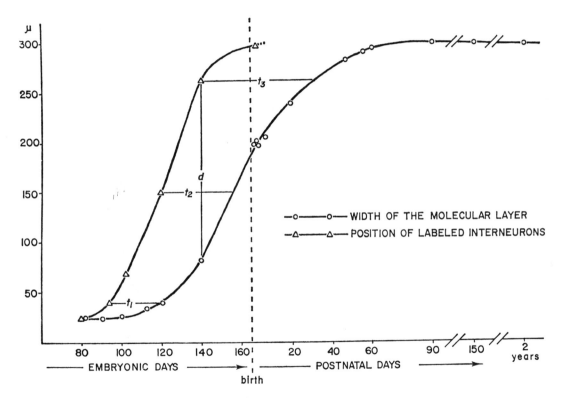

FIGURE 8 Graphic illustration of the latency between final cell division and onset of differentiation of interneurons of the cerebellar molecular layer (stellate and basket cells) as obtained by thymidine-^{3}H autoradiographic, Golgi, and electronmicroscopic data in developing rhesus monkeys. (From Rakic, 1973.) The two curves represent the maximal thickness of the molecular layer during development (circles) and the time of origin of the cerebellar interneurons in relation to their position in the vertical axis of the fully developed molecular layer (triangles). The molecular layer increases in thickness by accretion of new parallel fibers, but at a slower rate than the rate of interneuron genesis. Thus the complete complement of interneurons is generated by about birth, while the molecular layer does not attain full thickness until about 60 days later. As a result, late-arising interneurons will become postmitotic before formation of the parallel fibers among which they will eventually be embedded. The interval between the last division of an interneuron and the time it begins its differentiation and becomes fixed in the molecular layer is relatively short (t_1) for the early generated cells destined to lie close to the Purkinje cell layer and becomes systematically longer (t_2 and t_3) for those cells that will occupy progressively more superficial levels. For further explanation see Rakic, 1973.

it fails to make effective contact with its target cells in the CNS (Harkmark, 1954) or peripheral nervous system (Hamburger and Levi-Montalcini, 1949; Cowan and Wenger, 1967). The apparently similar dependence of young cerebellar granule cells on effective contact with Purkinje cell dendrites was cited above in discussion of staggerer and other mutant mice. Conversely, in the absence of afferents, some classes of neurons fail to migrate to their final positions (Levi-Montalcini, 1949, 1964) and may actually disappear (DeLong and Sidman, 1962; Hamori, 1969; Cowan, 1970; Trumpy, 1971; Kelly and Cowan, 1972). The dependence may be recognized by more subtle criteria such as impaired induction of synthetic enzyme activity (Black et al., 1971; Thoenen et al., 1972). Alterations in the organization of developing neuropil have been recognized even several synapses away from a lesion, for example the rearrangement of stellate neuron dendrites in layers 3 to 5 of the visual cortex after removal of an eye (Valverde and Estaban, 1968). In the absence of an overt lesion, altered sensory input during a critical period also may influence the organization of the visual system; this is discussed in more detail elsewhere in this volume (Blakemore, this volume).

Interactions continue to be important into adulthood. A partially denervated neuron, for example, becomes receptive to new collaterals that branch from the preterminal segments of nearby intact axons (Raisman, 1969). Denervated muscle (Fambrough, 1970) and denervated postganglionic parasympathetic neurons (Kuffler et al., 1972) develop the diffuse receptivity to acetylcholine that was characteristic prior to innervation.

All of these examples suggest that axonal innervation in some way causes a spatial restriction and stabilization of surface membrane properties of the target cell.

Pattern formation in aggregation cultures

The analysis of cell shapes, positions, and contacts during development in vivo must be supplemented by a synthetic attack, i.e., an attempt in vitro to reassemble aspects of CNS organization from individual cells. An approach is available through the cell aggregation method developed by Moscona (Moscona and Moscona, 1952) for study of the reconstruction of various chick embryo tissues from trypsinized suspensions of single cells. In this method the trypsinized cells are suspended in liquid culture medium and are allowed to contact one another passively and randomly by placing the flask containing them on a gyrating shaker. The cells form aggregates of about 0.5 mm diameter, and the most

intriguing feature of the aggregates is sorting out according to cell type (Moscona, 1962; Steinberg, 1963, 1970). Embryonic chick liver cells sort out from cartilage cells within 24 hours; most other combinations show a similar time course.

Embryonic mouse brain cells also sort out and establish patterns that are not only typical of brain but are different for different regions of the brain. Several features of their behavior are pertinent to the issue of surface membrane properties that might mediate cell interactions. Some of the patterns closely resemble the region of CNS from which the cells were originally taken (DeLong, 1970). The sorting out in these cases proceeds over a period of many days (DeLong, 1970; Rakic et al., 1972) rather than hours as in the case of mixed embryonic rudiments of chick embryos. DeLong has described a narrow developmental "time window," during which a brain region can be dissociated into a cell suspension and a patterned aggregate obtained. Tissues taken from developmental stages earlier or later than the "window" yield cells that aggregate but do not sort out into a recognizable pattern (DeLong, 1970). This suggests either that the culture conditions are inadequate to support sorting out except at critical stages of development (various regions of CNS show different optima), or that some "recognition" mechanism is at play for a limited time during the span of development. Differences in pattern formation between embryonic normal and reeler mutant neocortical cells were interpreted to support the concept of a recognition factor and to suggest that it might be controlled by the reeler locus (DeLong and Sidman, 1970) though other interpretations remain open (e.g. Caviness and Sidman, 1972).

Synapse formation in CNS aggregate cultures has been recognized by Seeds and Vatter (1971), Crain and Bornstein (1972), and Rakic et al. (1972). Synaptogenesis occurs relatively late, more than one week after the cultures are initiated. The important question that remains to be answered is whether the new synapses are random or patterned.

Very few hard data are available on the kinds of cell surface properties that might mediate aggregation and sorting out. Moscona and colleagues have presented evidence that cells in vitro sometimes produce factors that promote aggregation. These are thought normally to be bound at cell surfaces, but under culture conditions unfavorable for aggregation, the factors may be released into the culture medium. Tissue-specific factors have been described for neural retina of 7-day chick embryos (Lilien and Moscona, 1967) and cerebral cortex of 14-day mouse embryos (Garber and Moscona, 1972).

It is not known what, if anything, such factors have to

do directly with the sorting out of cells, as opposed to their aggregation. There is uncertainty even whether sorting out is mediated by specific molecular signals, though evidence favors this view (Moscona, 1968). The equilibrium configuration reached by cell types A and B is often predictable on the basis of quantitative strengths of adhesion between A and A, B and B, and A and B without reference to what kinds of molecules confer the adhesive properties (Steinberg, 1970). However, this neither rules in or out the possibility that molecular specificity governs pattern formation either in vitro or in vivo. A more extended discussion of these issues is summarized by Edds et al., (1972).

The need for further experimental facts is clear, but it is also becoming evident that the aggregation culture system is too complex to yield ready answers about pattern-generating mechanisms. The initial suspension contains several types of cells in proportions that may or may not be representative of the original tissue composition. The cells may be damaged to an unknown degree during release from the solid tissue. Nonetheless, the cardinal point is that the cells do aggregate and sort out and that this behavior constitutes a bioassay for cell surface properties of basic developmental interest.

Further analysis requires new technical developments such as separation of the initial suspension into individual classes of live cells that can be recombined in vitro in known proportions, from embryos of different ages, normal and mutant. Cell separation methods as applied to other solid organs, e.g. by Lam et al. (1970), have proved inadequate for the CNS. Purified populations have been obtained in special cases from peripheral ganglia (e.g. Okun et al., 1972). An effective methodology has now been established for cells of immature cerebellum by Barkley et al. (1973). Trypsinized cells were processed so as to minimize subsequent clumping and were separated under sterile conditions according to size and density in a Ficoll gradient. The rounded cells were identified by prior labeling with ^{3}H-thymidine in vivo on days when particular populations were known to be in final division. Several enriched, though no pure, fractions were obtained; proliferating external granule cells became subdivided even further according to phases of the cell generation cycle. Barkley's methods have proved applicable also to retina (Lam, 1972) and presumably will serve for other CNS and nonneural tissues.

Adoption of the hypothesis that cells of the immature nervous system might possess specific surface properties of developmental interest leads next to a consideration of analytical methods and concepts available through cellular immunology. The general issue of the identification of tissue-specific cell surface specificities ("differ-

entiation antigens") and the mapping of their mosaic distribution has been reviewed by Boyse and Old (1969) and Bennett et al. (1972). Cell suspensions prepared from immature CNS in inbred strains of mice serve for direct cytotoxicity testing under genetic control (Schachner and Sidman, 1973) and may allow production of monospecific antisera. Several antigenic specificities shared by brain and other organs have recently been measured quantitatively in normal and mutant animals (Schachner, 1973b). These include H-2 (the major histocompatibility antigen in mice), Thy-1 (theta), and Sk (skin antigen). It is not known what classes of brain cells carry these antigens, but a hint is provided by the finding that essentially the same set of specificities is present also on the C1300 mouse neuroblastoma (Schachner, 1973a). The similarity between normal brain and neuronal tumor suggests that these antigens are probably distributed widely among neurons, perhaps in the peripheral as well as the central nervous system.

The distribution of Thy-1 is particularly intriguing, for this antigenic system is shared exclusively by thymus, thymus-derived lymphocytes, and brain (Reif and Allen, 1964; Raff and Owen, 1971). The Thy-1 activity in brain (expressed per mg wet weight of tissue) is quantitatively unaltered in a series of neurological mutants, including several that show selective losses of particular cell types (Schachner, 1973b). This is consistent with the concept that Thy-1 is a component either of a widely distributed and numerous type of cell or is present on almost all cells in the CNS. A similar conclusion was reached on the basis of immunofluorescence studies on tissue sections supplemented by brain cell cytotoxicity tests (Moore et al., 1971).

In addition to extending the analytical prospects concerning the definition of cell surface components on developing brain cells, cellular immunology may come to serve also for the synthetic approach, the objective stated at the beginning of this section. Immature brain cells must be challenged to reconstruct features of an organized tissue by confronting them with cells of other classes in particular geometric configurations. The appropriate geometries could be arranged by derivatizing solid synthetic substrates in predetermined patterns with antibodies and then allowing the complementary cells to attach selectively, as already achieved for other cell types (e.g. Edelman et al., 1971).

The combination of an analytical approach based in morphology and genetics with a synthetic approach based in cell biology and immunology seems at present the most profitable pathway to new insights into the development of surface mediated intercellular relationships in complex nervous systems.

REFERENCES

ADAMS, R. D., and R. L. SIDMAN, 1968. *Introduction to Neuropathology.* New York: McGraw-Hill, 620 pp.

ALTMAN, J., 1969. Autoradiographic and histological studies of postnatal neurogenesis. III. Dating the time of production and onset of differentiation of cerebellar microneurons in rats. *J. Comp. Neur.* 136:269–294.

ALTMAN, J., and W. J. ANDERSON, 1972. Experimental reorganization of the cerebellar cortex. I. Morphological effects of elimination of all microneurons with prolonged X-irradiation started at birth. *J. Comp. Neur.* 146:355–406.

ANGEVINE, J. B., JR., 1965. Time of neuron origin in the hippocampal region. An autoradiographic study in the mouse. *Exp. Neurol., Suppl.* 2:1–70.

ANGEVINE, J. B., JR., 1970. Time of neuron origin in the diencephalon of the mouse. An autoradiographic study. *J. Comp. Neur.* 139:129–188.

ANGEVINE, J. B., JR., and R. L. SIDMAN, 1961. Autoradiographic study of cell migration during histogenesis of cerebral cortex in the mouse. *Nature (Lond.)* 192:766–768.

BARKLEY, D. S., L. L. RAKIC, J. K. CHAFFEE, and D. L. WONG, 1973. Cell separation by velocity sedimentation of postnatal mouse cerebellum. *J. Cell Physiol.* 81: in press.

BENNETT, D., E. A. BOYSE, and L. J. OLD, 1972. Cell surface immunogenetics in the study of morphogenesis. In *Cell Interactions, Proceedings of the Third Lepetit Colloquium,* L. G. Silvestri, ed. Amsterdam: North-Holland, pp. 247–263.

BERRY, M., and A. W. ROGERS, 1965. The migration of neuroblasts in the developing cerebral cortex. *J. Anat.* 99:691–709.

BLACK, I. B., I. HENDRY, and L. L. IVERSON, 1971. Transsynaptic regulation of growth and development of adrenergic neurons in mouse sympathetic ganglion. *Brain Res.* 34:229–246.

BOULDER COMMITTEE, 1970. Nomenclature for developing nervous system. *Anat. Rec.* 166:257–261.

BOYSE, E. A., and L. J. OLD, 1969. Some aspects of normal and abnormal cell surface genetics. *Ann. Rev. Genet.* 3:269–290.

CAVINESS, V. S., JR., and R. L. SIDMAN, 1972. Olfactory structures of the forebrain in the reeler mutant mouse. *J. Comp. Neur.* 145:85–104.

CAVINESS, V. S., JR., and R. L. SIDMAN, 1973a. Retrohippocampal, hippocampal, and related structures of the forebrain in the reeler mutant mouse. *J. Comp. Neur.* 147:235–254.

CAVINESS, V. S., JR., and R. L. SIDMAN, 1973b. Time of origin of corresponding cell classes in the cerebral cortex of normal and reeler mutant mice: An autoradiographic analysis. *J. Comp. Neur.* 148: in press.

COWAN, W. M., 1970. Anterograde and retrograde transneuronal degeneration in the central and peripheral nervous system. In *Contemporary Research Methods in Neuroanatomy,* W. J. H. Nauta and S. O. E. Ebbesson, eds. New York: Springer-Verlag, pp. 217–251.

COWAN, W. M., and E. WENGER, 1967. Cell loss in the trochlear nucleus of the chick during normal development and after radical extirpation of the optic vesicle. *J. Exp. Zool.* 164:267–280.

CRAIN, S., and M. B. BORNSTEIN, 1972. Organotypic bioelectric activity in cultured reaggregates of dissociated rodent brain cells. *Science* 176:182–184.

DELONG, G. R., 1970. Histogenesis of fetal mouse isocortex and hippocampus in reaggregating cell cultures. *Dev. Biol.* 22:563–583.

DELONG, G. R., and R. L. SIDMAN, 1962. Effects of eye removal at birth on the histogenesis of the mouse superior colliculus: An autographic analysis with tritiated thymidine. *J. Comp. Neur.* 118:205–225.

DELONG, G. R., and R. L. SIDMAN, 1970. Alignment defect of reaggregating cells in cultures of developing brains of reeler mutant mice. *Dev. Biol.* 22:584–600.

ECCLES, J. C., M. ITO, and J. SZENTÁGOTHAI, 1967. *The Cerebellum as a Neuronal Machine.* New York: Springer-Verlag, 355 pp.

EDDS, M. V., JR., 1970. Development of the nervous system. In *The Neurosciences: Second Study Program,* F. O. Schmitt, ed. New York: Rockefeller University Press, pp. 51–157.

EDDS, M. V., JR., D. S. BARKLEY, and D. M. FAMBROUGH, 1972. Genesis of neuronal patterns. *Neurosciences Res. Prog. Bull.* 10:254–367.

EDELMAN, G. M., V. RUTISHAUSER, and C. F. MILLETTE, 1971. Cell fractionation and arrangement of fibers, beads, and surfaces. *Proc. Nat. Acad. Sci. USA* 68:2153–2157.

FAMBROUGH, D. M., 1970. Acetylcholine sensitivity of muscle fiber membranes: mechanism of regulation by motor neurons. *Science* 168:372–373.

FUJITA, S., 1967. Quantitative analysis of cell proliferation and differentiation in the cortex of the postnatal mouse cerebellum. *J. Cell Biol.* 32:277–288.

FUJITA, S., M. SHIMADA, and T. NAKAMURA, 1966. H^3-thymidine autoradiographic studies on the cell proliferation and differentiation in the external and internal granular layers of the mouse cerebellum. *J. Comp. Neur.* 128:191–209.

GARBER, B. B., and A. A. MOSCONA, 1972. Reconstruction of brain tissue from cell suspensions. II. Specific enhancement of aggregation of embryonic cerebral cells by supernatant from homologous cell cultures. *Dev. Biol.* 27:235–243.

HAMBURGER, V., and R. LEVI-MONTALCINI, 1949. Proliferation, differentiation and degeneration in the spinal ganglia of the chick embryo under normal and experimental conditions. *J. Exp. Zool.* 111:457–501.

HAMBURGH, M., 1963. Analysis of the postnatal developmental effects of "reeler", a neurological mutation in mice. A study in developmental genetics. *Dev. Biol.* 8:165–185.

HAMORI, J., 1969. Development of synaptic organization in the partially agranular and in transneuronally atrophied cerebellar cortex. In *Neurobiology of Cerebellar Evolution and Development,* R. Llinas, ed. Chicago: Am. Med. Assoc. Educ. Res. Fd., pp. 845–858.

HARKMARK, W., 1954. Cell migrations from the rhombic lip to the inferior olive, the nucleus raphe and the pons. A morphological and experimental investigation on chick embryos. *J. Comp. Neur.* 100:115–210.

HICKS, S. P., and C. J. D'AMATO, 1968. Cell migrations to the isocortex in the rat. *Anat. Rec.* 160:619–634.

HINDS, J. W., 1968. Autoradiographic study of histogenesis in the mouse olfactory bulb. I. Time of origin of neurons and neuroglia. *J. Comp. Neur.* 134:287–304.

HINDS, J. W., and T. L. RUFFETT, 1971. Cell proliferation in the neural tube: An electron microscopic and Golgi analysis in the mouse cerebral vesicle. *Z. Zellforsch.* 115:226–264.

HIRANO, A., H. M. DEMBITZER, and M. JONES, 1972. An electron microscopic study of cycasin-induced cerebellar alteration. *J. Neuropath. Exp. Neur.* 31:113–125.

KELLY, J. P., and W. M. COWAN, 1972. Studies on the development of the chick optic tectum. III. Effects of early eye removal. *Brain Res.* 42:263–288.

KOELLIKER, A., 1896. *Handbuch der Gewebelhre des Menschen.* Vol. II, Sixth Ed. Leipzig: Engelmann, 409 pp.

KUFFLER, S. W., M. J. DENNIS, and A. J. HARRIS, 1972. The development of chemosensitivity in extrasynaptic areas of the neuronal surface after denervation of parasympathetic ganglion cells in the heart of the frog. *Proc. R. Soc. Lond. (Biol.)* 177:555–563.

LAM, D. M. K., 1972. Biosynthesis of acetylcholine in turtle photoreceptors. *Proc. Nat. Acad. Sci. USA* 69:1987–1991.

LAM, D. M. K., R. FURRER, and W. R. BURCE, 1970. The separation, physical characterization, and differentiation kinetics of spermatogonial cells of the mouse. *Proc. Nat. Acad. Sci. USA* 65:192–199.

LEVI-MONTALCINI, R., 1949. The development of the acoustico-vestibular centers in chick embryo in the absence of the afferent root fibers and of descending fiber tracts. *J. Comp. Neur.* 91:209–241.

LEVI-MONTALCINI, R., 1964. Growth and differentiation in the nervous system. In *The Nature of Biological Diversity*, J. M. Allen, ed., New York: McGraw-Hill, pp. 261–295.

LILIEN, J. E., and A. A. MOSCONA, 1967. Cell aggregation: Its enhancement by a supernatant from cultures of homologous cells. *Science* 157:70–72.

MEIER, H., and W. G. HOAG, The neuropathology of "reeler", a neuro-muscular mutation in mice. *J. Neuropath. Exp. Neur.* 21:649–654.

MIALE, I., and R. L. SIDMAN, 1961. An autoradiographic analysis of histogenesis in the mouse cerebellum. *Exp. Neur.* 4:277–296.

MOORE, M. J., P. DIKKES, A. E. REIF, F. C. A. ROMANUL, and R. L. SIDMAN, 1970. Localization of theta alloantigens in mouse brain by immunofluorescence and cytotoxic inhibition. *Brain Res.* 28:283–293.

MOORE, R. Y., A. BJÖRKLUND, and U. STENEVI, 1971. Plastic changes in the adrenergic innervation of the rat septal area in response to denervation. *Brain Res.* 33:13–35.

MOSCONA, A. A., 1962. Cellular interactions in experimental histogenesis. *Int. Rev. Exp. Path.* 1:371–428.

MOSCONA, A. A., 1968. Cell aggregation: Properties of specific cell ligands and their role in the formation of multicellular systems. *Dev. Biol.* 18:250–277.

MOSCONA, A. A., and M. H. MOSCONA, 1952. The dissociation and aggregation of cells from organ rudiments of the early chick embryo. *J. Anat.* 86:287–301.

OKUN, L. M., F. K. ONTKEAN, and C. A. THOMAS, 1972. Removal of nonneuronal cells from suspensions of dissociated embryonic dorsal root ganglia. *Exp. Cell Res.* 73:226–229.

PALAY, S. L., and V. CHAN-PALAY, 1973. *Cerebellar Cortex: Cytology and Organization.* New York: Springer-Verlag.

RAAF, J., and J. W. KERNOHAN, 1944. A study of the external granular layer in the cerebellum. *Am. J. Anat.* 75:151–172.

RAFF, M. C., and J. J. T. OWEN, 1971. Thymus-derived lymphocytes: Their distribution and role in the development of peripheral lymphoid tissues of the mouse. *Europ. J. Immunol.* 1:27–30.

RAISMAN, G., 1969. Neuronal plasticity in the septal nuclei of the adult rat. *Brain Res.* 14:25–48.

RAKIC, L. L., R. L. SIDMAN, and D. S. BARKLEY, 1972. Genesis of isocortical patterns in aggregates of embryonic mouse cerebrum. *In Vitro* 7:250.

RAKIC, P., 1971a. Neuron-glia relationship during granule cell migration in developing cerebellar cortex. A Golgi and electronmicroscopic study in *Macacus rhesus*. *J. Comp. Neur.* 141:283–312.

RAKIC, P., 1971b. Guidance of neurons migrating to the fetal monkey neocortex. *Brain Res.* 33:471–476.

RAKIC, P., 1972a. Mode of cell migration to the superficial layers of fetal monkey neocortex. *J. Comp. Neur.* 145:61–84.

RAKIC, P., 1972b. Extrinsic cytological determinants of basket and stellate cell dendritic pattern in the cerebellar molecular layer. *J. Comp. Neur.* 146:335–354.

RAKIC, P., 1973. Kinetics of the proliferation and the latency between final division and onset of differentiation of the cerebellar stellate and basket cells. *J. Comp. Neur.* 147:523–546.

RAKIC, P., and R. L. SIDMAN, 1973. Weaver mutant mouse cerebellum; defective neuronal migration secondary to specific abnormality of Bergmann glia. *Proc. Nat. Acad. Sci. USA* 70:240–244.

REIF, A. E., and J. M. V. ALLEN, 1964. The AKR thymic antigen and its distribution in leukemias and nervous tissue. *J. Exp. Med.* 120:413–433.

REZAI, Z., and C. H. YOON, 1972. Abnormal rate of granule cell migration in the cerebellum of "weaver" mutant mice. *Dev. Biol.* 29:17–26.

SCHACHNER, M., 1973a. Serologically demonstrable cell surface specificities on mouse neuroblastoma C1300. *Nature (Lond.)* (in press).

SCHACHNER, M., 1973b. Representation of the cell surface alloantigen Thy-1 (Theta) in brains of neurological mutants of the mouse (submitted for publication).

SCHACHNER, M., and R. L. SIDMAN, 1973. Distribution of H-2 alloantigen in adult and developing mouse brain. *Brain Res.* (in press).

SEEDS, N. W., and A. E. VATTER, 1971. Synaptogenesis in reaggregating brain cell cultures. *Proc. Nat. Acad. Sci. USA* 68:3219–3222.

SIDMAN, R. L., 1961. Histogenesis of mouse retina studied with thymidine-H^3. In *The Structure of the Eye*, G. K. Smelser, ed. New York: Academic Press, pp. 487–506.

SIDMAN, R. L., 1968. Development of interneuronal connections in brains of mutant mice. In *Physiological and Biochemical Aspects of Nervous Integration*, F. D. Carlson, ed. Englewood Cliffs, N.J.: Prentice-Hall, pp. 163–193.

SIDMAN, R. L., 1970. Autoradiographic methods and principles for study of the nervous system with thymidine-H^3. In *Contemporary Research Methods in Neuroanatomy*, W. J. H. Nauta and S. O. E. Ebbesson, eds. New York: Springer-Verlag, pp. 252–274.

SIDMAN, R. L., 1972. Cell interactions in developing mammalian central nervous system. In *Cell Interactions. Proceedings of the Third Lepetit Colloquium*, L. G. Silvestri, ed. Amsterdam: North-Holland, pp. 1–13.

SIDMAN, R. L., J. B. ANGEVINE, JR., and E. TABER PIERCE, 1971. *Atlas of the Mouse Brain and Spinal Cord.* Cambridge: Harvard University Press, 261 pp.

SIDMAN, R. L., and M. C. GREEN, 1970. "Nervous," a new mutant mouse with cerebellar disease. In *Les Mutants Pathologiques Chez l'Animal*, M. Sabourdy, ed. Paris: Éditions Du Centre National De La Récherche Scientifique, pp. 69–79.

SIDMAN, R. L., M. C. GREEN, and S. H. APPEL, 1965. *Catalog of the Neurological Mutants of the Mouse.* Cambridge: Harvard University Press, 82 pp.

STEINBERG, M. S., 1963. Reconstruction of tissues by dissociated cells. *Science* 141:401–408.

STEINBERG, M. S., 1970. Does differential adhesion govern self-assembly processes in histogenesis? Equilibrium configurations and the emergence of a hierarchy among populations of embryonic cells. *J. Exp. Zool.* 173:395–433.

SZENTÁGOTHAI, J., and J. HÁMORI, 1969. Growth and differentiation of synaptic structures under circumstances of deprivation of function and of distant connections. In *Cellular Dynamics of the Neuron*, S. H. Barondes, ed. *Symp. Int. Soc. Cell Biol.* 8:301–320.

TABER PIERCE, E., 1966. Histogenesis of the nuclei griseum pontis, corporis pontobulbaris and reticularis tegmenti pontis (Bechterew) in the mouse. An autoradiographic study. *J. Comp. Neur.* 126:219–240.

THOENEN, H., A. SANER, R. KETTLER, and P. U. ANGELETTI, 1972. Nerve growth factor and preganglionic cholinergic nerves; their relative importance to the development of the terminal adrenergic neuron. *Brain Res.* 44:593–602.

TRUMPY, J. H., 1971. Transneuronal degeneration in the pontine nuclei of the cat. I. Neuronal changes in animals of varying ages. *Ergebnisse Anat. Entwicklungsgesch.* 44:1–41.

VALVERDE, F., and M. E. ESTEBAN, 1968. Peristriate cortex of mouse: Location and the effects of enucleation of the number of dendritic spines. *Brain Res.* 9:145–148.

66 The Use of Snake Venom Toxins to Study and Isolate Cholinergic Receptors

PERRY B. MOLINOFF

ABSTRACT The venoms of a number of *Elapidae* snakes contain potent toxins that block neuromuscular transmission. Some of these toxins appear to bind specifically and irreversibly to the cholinergic receptors of skeletal muscle and the electric organs of *Torpedo* and *Electrophorus*. One such postsynaptically active toxin, α-bungarotoxin, has been purified from the venom of *Bungarus multicinctus* and used, after labeling with ^{131}I, to study the receptors of *Torpedo*. The toxin binding material was solubilized with detergents and some of its properties were determined.

THE VENOMS OF snakes belonging to the family *Elapidae* have profound effects on neuromuscular transmission. Paralysis of respiratory muscles appears to be the principal cause of death due to these extremely toxic venoms. Several authors have presented evidence that some of these neuromuscular blocking agents have a postsynaptic site of action (Meldrum, 1965). A nondepolarizing blockade of neuromuscular transmission, like that caused by curare, has been proposed as the mode of action for Formosan cobra (*Naja naja atra*) venom and for that of the banded Krait (*Bungarus multicinctus*) (Chang, 1960). Some differences between the actions of these venoms and those of curare were, however, noted. Thus, the neuromuscular blockade induced by these venoms was not effectively relieved by anticholinesterases, nor was it reversed by even prolonged periods of washout. Furthermore, acetylcholine release from presynaptic terminals was markedly impaired by exposure to these snake venom toxins. Since snake venoms generally consist of a mixture of proteins or polypeptides, Lee and Chang (1966) investigated the possibility that this complexity might be due to the combined effects of several different components present in the venoms. Using zone electrophoresis on starch, at pH 5.0, the venom of *Bungarus multicinctus* was separated into four fractions (Chang and Lee, 1963). One fraction lacked neuromuscular blocking properties and appeared to contain acetylcholinesterase activity. The two most electropositive fractions, called "β-" and "γ-" bungarotoxin, produced neuromuscular blockade and a severe reduction in acetylcholine output. The blockade produced by these two fractions of the venom was not associated with any decrease in the sensitivity of the muscle to acetylcholine. The fourth fraction, called α-bungarotoxin, produced a rapid neuromuscular blockade both in vitro and in vivo. Acetylcholine release from the rat phrenic nerve was not affected by exposure to this toxin. Similarly, it had no effect on resting membrane potentials recorded from either the endplate or nonendplate regions of the rat diaphragm. It was also found that the action potentials elicited by direct stimulation of muscle fibers, paralyzed by a high concentration of α-bungarotoxin, were normal both in their amplitude and in their time course. When either α-bungarotoxin or cobra neurotoxin were applied to skeletal muscle, the amplitude of the miniature endplate potentials decreased progressively without any effect on the rate of discharge. The miniature endplate potentials completely disappeared just before the neuromuscular blockade became effective. The terminal nerve spike recorded with extracellular electrodes remained unaffected after the endplate potential, simultaneously recorded with the same microelectrode, had been abolished by either α-bungarotoxin, β-bungarotoxin or cobra neurotoxin. Suggestive evidence that α-bungarotoxin was acting on the cholinergic receptor came from the fact that pretreatment with *d*-tubocurarine protected the muscle from the neuromuscular blocking actions of the toxin. The hypothesis derived from these studies was that both the cobra neurotoxin and α-bungarotoxin act postsynaptically, at the site of the acetylcholine receptor, to cause neuromuscular blockade.

The conclusion that α-bungarotoxin acts by binding to cholinergic receptors was further strengthened by the autoradiographic finding that this toxin, labeled with ^{131}I (Lee and Tseng, 1966) accumulates on the relatively restricted motor endplate zone of the mouse diaphragm in the same way as does radioactively labeled *d*-tubocurarine (Waser and Luthi, 1957). In the rat diaphragm preparation, treated in vitro with either α-bungarotoxin (5×10^{-6} g/ml) or cobra neurotoxin (10^{-6} g/ml), the radioactivity on the motor endplate zone remained

PERRY B. MOLINOFF Department of Pharmacology, University of Colorado, Medical Center, Denver, Colorado

unchanged even after 4 hr of washing. This finding is consistent with the irreversible nature of the neuromuscular blockade caused by these toxins (Lee, Tseng, and Chiu, 1967). The effect of denervation on the localization of these snake venom toxins was examined in the rat diaphragm at 14 and 60 days after phenectomy (Lee, Tseng, and Chiu, 1967). Radioactively labeled toxin was injected subcutaneously, and the distribution of radioactivity on the denervated side was compared with that on the intact side. In diaphragms denervated for 14 days, a well-defined endplate zone could be distinguished morphologically, but radioactivity was found to have been bound to the entire surface of the muscle. This finding is consistent with the known physiologic spread of sensitivity to acetylcholine seen after denervation (Miledi, 1960).

In order to use snake venoms as a tool to study cholinergic receptors, it is first necessary to purify the specific postsynaptically acting α-toxin and then to label it radioactively (Miledi et al., 1971; Clark et al., 1972; Cooper and Reich, 1972). Both α-bungarotoxin and cobra neurotoxin are small molecular weight proteins (7,000 to 8,000). The amino acid composition of α-bungarotoxin (Clark et al., 1972; Mebs et al., 1971) has been found to correspond to its generally basic nature (pI = 9.19 for iodinated [^{125}I] α-bungarotoxin). The combination of a relatively low molecular weight with an enriched content of basic amino acids has permitted the development of a number of purification schema; these usually involve molecular sieves and ion exchange column chromatography.

Radioactive labeling of the purified toxins has been accomplished by a number of means. Iodination, using either the method of Pressman and Eisen (1950) or that of Greenwood, Hunter, and Glover (1963), has been used by several groups of investigators (Miledi et al., 1971; Clark et al., 1972; Berg et al., 1972). Raftery and his collaborators (Clark et al., 1972) have been able to separate iodinated toxins from nonradioactive toxins using gradient elution after adsorption onto carboxymethylcellulose. An alternative approach to preparing the radioactively labeled toxin was used by Cooper and Reich (1972). They reacted purified cobra neurotoxin with pyridoxal phosphate, and then reduced the Schiff base, thus formed, with tritiated sodium borohydride. Using phosphocellulose chromatography, they were able to separate toxin that contained 0, 1, or 2 labeled pyridoxal phosphate moieties. Still another approach has been used by Changeux and his collaborators (Menez et al., 1971; Meunier et al., 1972). Monoiodinated toxin was prepared and then dehalogenated by substitution of the iodine with tritium. The radioactively labeled toxin was physiologically active and at least 90% of the labeled toxin was able to bind to membrane fragments of the electric eel.

An important use of labeled snake venom toxins has been to study and quantitate acetylcholine receptors in skeletal muscle (Lee et al., 1967; Miledi and Potter, 1971; Barnard et al., 1971; Berg et al., 1972). Radiolabeled bungarotoxin was found to bind almost exclusively (at least 90%) to those regions of skeletal muscle that contain endplates and are known to be sensitive to acetylcholine. Chronically denervated adult muscle and muscle from neonatal rats are sensitive to acetylcholine along their entire length and these muscles show a comparable ability to bind toxin. The amount of bungarotoxin bound to the nonendplate (normally insensitive) region of the rat diaphragm has been shown to increase between 20- and 80-fold after chronic denervation (Miledi and Potter, 1971; Berg et al., 1972). The binding of toxin to skeletal muscles is inhibited by d-tubocurarine and by carbamylcholine but not by atropine. Hall and his collaborators (Berg et al., 1972) exposed skeletal muscle, to which iodinated bungarotoxin had been bound, to 1% Triton X-100. This treatment solubilized or dispersed the membrane component that contained the binding site for the bungarotoxin. Analysis of this material by zone sedimentation through a sucrose gradient revealed a peak of radioactivity with a sedimentation coefficient of approximately 9 S. A similar peak with approximately the same sedimentation coefficient was seen with plus endplate and minus endplate regions of both denervated adult muscles and with neonatal muscles. The number of toxin receptor sites per endplate of skeletal muscle (Miledi and Potter, 1971; Barnard et al., 1971) has been found to vary from about 1.6×10^7 for mouse hemidiaphragm to about 10^9 for the frog sartorius muscle. Barnard and his collaborators (1971) have used electron microscopic autoradiography to compare the distribution of toxin with that of diiosopropylfluorophosphate (DFP). They found that the ratio of sites for toxin to sites for DFP was approximately one, in several species of animal. Similar findings have also been reported for *Torpedo* electric organ (Miledi et al., 1971) and for the electric organ of the eel (Changeux et al., 1970). The significance of the similarity in the number of toxin binding sites and the number of catalytic sites of acetylcholinesterase remains unclear. Work in a number of laboratories has shown that a nearly quantitative separation of acetylcholinesterase from the toxin binding material can be obtained by a number of techniques including molecular sieving and isoelectric focusing. Further, when a membrane suspension, prepared from *Torpedo* electric tissue, is subjected to centrifugation, on a sucrose density gradient, acetylcholinesterase activity appears to be associated with a different membrane fragment than that which binds α-bungarotoxin (Molinoff and Potter, 1972) (see below). Recent experiments (Hall and Kelly, 1971; Betz and Sakmann, 1971) have shown that treatment of skeletal muscle with collagenase or trypsin will release acetylcholinesterase activity from the muscle and will cause the motor nerve endings to separate and lift off of the muscle surface.

In our experiments, we elected to study the cholinergic receptors of the electric organ of *Torpedo marmorata*. *Torpedo* was chosen instead of the electric eel, because it offered the possibility for a much greater yield of purified receptor. The physiologic specificity of the toxin for the *Torpedo* was shown by the fact that exposure of a block of *Torpedo* tissue to bungarotoxin resulted in a progressive diminution in the size followed by the eventual disappearance of miniature endplate potentials. After toxin was labeled with radioactive iodine [^{131}I], at least 90% of the labeled toxin retained the ability to bind to material present in a homogenate of *Torpedo* tissue. Further, the binding of α-toxin to large membrane fragments from *Torpedo* was a time-dependent process that showed saturation kinetics. This binding was inhibited by *d*-tubocurarine and by carbamylcholine.

^{131}I-labeled α-toxin was usually added to a crude homogenate of electric tissue prepared in 0.4 M sodium chloride. The membranes were then sedimented by centrifugation and soluble proteins were discarded. The membrane fraction was sonicated in hypotonic media to maximally disrupt membrane structure, and was then fractionated on linear sucrose density gradients. Most of the radioactivity in these gradients was associated with a band of particles isodense with 37% sucrose. A small percentage of the radioactivity, usually less than 20%, was seen along with about 75% of the protein in a band at 42% sucrose. This fraction contained recognizable pieces of dorsal (noninnervated) cell membranes. There was usually no radioactivity associated with an intermediate band of mitochondria or with less dense bands of myelin or microsomes. This experiment provided important evidence as to the specificity of the toxin binding in that there were at least 5 subcellular fractions that did not bind significant amounts of toxin.

A surprising result of these experiments was the localization of the enzyme acetylcholinesterase. We had expected to find this enzyme associated with the same membrane fraction as was the radioactive toxin. However, most of the enzyme was found with the bulk of the membrane protein at 42% sucrose. Several experiments were carried out to try to see if either the esterase or the toxin binding material was redistributed in the course of the preparative procedures. No evidence for redistribution was found, however, and we were forced to the conclusion that the toxin binding material is on a different piece of membrane than is the enzyme acetylcholinesterase. By differential, as opposed to gradient, centrifugation of the sonicated membrane suspension, it was possible to obtain a preparation that was highly enriched in toxin binding material relative to other membrane proteins and to acetylcholinesterase.

Several different means of dissolving toxin-labeled receptors from partially purified membranes have been investigated (Potter and Molinoff, 1972). Sonication in dilute buffer did not remove any of the radioactive label. Treatment of the material with 2 M sodium chloride removed a small amount of the toxin, while treatment with high concentrations of sodium bromide, sodium iodide, urea or sodium dodecyl sulfate, partially or fully dissociated toxin from its binding site. This dissociation appeared to represent an irreversible change in the nature of the receptor protein. A number of experiments were carried out using the chloroform-methanol extraction technique described by de Robertis and his collaborators (La Torre et al., 1970; de Robertis et al., 1971; de Robertis, 1971). We were never able, however, to bind toxin to either the material extracted into chloroform-methanol or to *Torpedo* membranes which had been exposed, however briefly, to chloroform-methanol. It is, of course, still possible that the receptor, though it has lost its ability to bind α-toxin, is still able to bind acetylcholine after such exposure to organic solvents.

Of the various solubilization procedures tried, most success was achieved with the detergent Triton X-100. Sedimented microsomes were resuspended by brief sonication and then 2 mg of Triton were added per mg of protein. After allowing solubilization to take place for approximately 30 min, the suspension was subjected to ultracentrifugation, and the pellet which contained about 40% of the protein but only 10% of the radioactivity was discarded.

The Stokes radius of the Triton dispersed material was determined using columns of Sepharose 6B and 4B. In most experiments, two peaks of radioactive material were seen. The largest amount of material (50 to 75% of the radioactivity) was found in a peak having the same Stokes radius as does β-galactosidase (64 Å). The remainder of the radioactivity was usually found in one or more peaks having somewhat larger Stokes radii. In order to see whether the two, or in some cases, three peaks of radioactive material were due to different aggregation states of the same toxin binding subunit or to different and distinct membrane binding sites, fractions from each of the peaks were subjected to density gradient centrifugation on sucrose density gradients. Material from the various peaks seen in experiments with Sepharose behaved as distinct entities in the ultracentrifuge such that the material with the smaller Stokes radius had a lower sedimentation coefficient than did the material with the larger Stokes radius. It was, however, observed that material from both of the major peaks obtained on Sepharose column chromatography behaved in the ultracentrifuge as if they were smaller (lower sedimentation coefficient) than β-galactosidase. The difference between the behavior of the toxin binding material in the ultracentrifuge from that on molecular sieves requires explanation. The two most likely possibilities for this difference are that the toxin binding material has a relatively large content of lipid (which would decrease its density) or that it has a high axial ratio (which would increase its Stokes radius). We have subjected toxin labeled material to prolonged centrifugation in an attempt to determine its equilibrium density. Preliminary experiments have suggested a value for this density of approximately 1.18 g/cc. This value is substantially less than that which is usually seen for proteins, and thus suggests that there may be significant amounts of lipid associated with Triton dispersed material.

From the number of binding sites for toxin (6.6×10^{17} per kg of electric tissue) and our estimate of the area of postsynaptic membrane (70 meters^2 per kg), it is possible to roughly estimate the density of receptors in the membrane as 10,000 per square micron (Miledi, 1971). This suggests that a significant fraction of the area of the postsynaptic membrane may be taken up by receptors. When we consider that acetylcholinesterase may be in or near the membrane that includes the toxin binding sites and that other large molecules may be required to subserve the conductance mechanism of synaptic potentials, the postsynaptic membrane begins to appear to be extremely crowded.

It seems likely that experiments similar to those described above will, in the near future, result in the complete purification of the cholinergic receptor. Affinity column chromatographic techniques, such as are now being employed in a number of laboratories (Hammes et al., 1973), will probably be involved in the finally successful purification procedure. The major reason why success will be achieved first with the cholinergic receptor is the existence and availability of the specific and irreversible snake venom toxins that have been discussed in this chapter. It is important to realize, however, that the purification of receptors is only the first step in an increasingly complex field of investigation. The specific result of the interaction between a receptor and a neurotransmitter remains a mystery and we are only now beginning to think about the mechanisms by which cholinergic receptors are linked to the specific postsynaptic conductance mechanism.

ACKNOWLEDGMENTS The original experiments reported here were carried out at University College, London, in collaboration with Drs. R. Miledi and L. T. Potter. The author was supported during the time of these studies by fellowships from the Guggenheim Foundation and the National Institute for Neurological Disease and Stroke.

REFERENCES

BARNARD, E. A., J. WIECKOWSKI, and T. H. CHIU, 1971. Cholinergic receptor molecules and cholinesterase molecules at mouse skeletal muscle junctions. *Nature (Lond.)* 234:207.

BERG, D. K., R. B. KELLY, P. B. SARGENT, P. WILLIAMSON, and Z. W. HALL, 1972. Binding of α-bungarotoxin to acetylcholine receptors in mammalian muscle. *Proc. Natl. Acad. Sci. USA* 69:147.

BETZ, W., and B. SAKMANN, 1971. "Disjunction" of frog neuromuscular synapses by treatment with proteolytic enzymes. *Nat. New Biol.* 232:94.

DEL CASTILLO, J., and B. KATZ, 1955. On the localization of acetylcholine receptors. *J. Physiol. (Lond.)* 128:157.

CHANG, C. C., 1960. Studies on the mechanism of curare-like action of *Bungarus multicinctus* venom. I. Effect on the phrenic nerve-diaphragm preparation of the rat. *J. Formosan Med. Assoc.* 59:315.

CHANG, C. C., and C. Y. LEE, 1963. Isolation of neurotoxins from the venom of *Bungarus multicinctus* and their modes of neuromuscular blocking action. *Arch. Int. Pharmacodyn. Ther.* 144:241.

CHANGEUX, J.-P., M. KASAI, and C.-Y. LEE, 1970. Use of a snake venom toxin to characterize the cholinergic receptor protein. *Proc. Natl. Acad. Sci. USA* 67:1241.

CLARK, D. G., D. D. MACMURCHIE, E. ELLIOTT, R. G. WOLCOTT, A. M. LANDEL, and M. A. RAFTERY, 1972. Elapid neurotoxins. Purification, characterization, and immunochemical studies of α-bungarotoxin. *Biochem.* 11:1663.

COOPER, D., and E. REICH, 1972. Neurotoxin from venom of the cobra, *Naja naja siamensis*. *J. Biol. Chem.* 247:3008.

GREENWOOD, F. C., W. M. HUNTER, and J. S. GLOVER, 1963. The preparation of ^{131}I-labeled human growth hormone of high specific radioactivity. *Biochem. J.* 89:114.

HALL, Z. W., and R. B. KELLY, 1971. Enzymatic detachment of endplate acetylcholinesterase from muscle. *Nat. New Biol.* 232:62.

HAMMES, G. G., P. B. MOLINOFF, and F. E. BLOOM, 1973. Receptor biophysics and biochemistry. *Neuro. Res. Prog. Bull.* 11:155.

LEE, C.-Y., and C. C. CHANG, 1966. Modes of actions of purified toxins from *elapid* venoms on neuromuscular transmission. *Mem. Inst. Butantan* 33:555.

LEE, C.-Y., and L. F. TSENG, 1966. Distribution of *Bungarus multicinctus* venom following envenomation. *Toxicon.* 3:281.

LEE, C.-Y., L. F. TSENG, and T. H. CHIU, 1967. Influences of denervation on localization of neurotoxins from elapid venoms in rat diaphragm. *Nature (Lond.)* 215:1177.

MEBS, D., K. KARITA, S. IWANAGA, Y. SAMEJUMA, and C. Y. LEE, 1971. Amino acid sequence of α-bungarotoxin from the venom of *Bungarus multicinctus*. *Biochem. Biophys. Res. Commun.* 44:711.

MELDRUM, B. S., 1965. The actions of snake venoms on nerve and muscle. The pharmacology of phospholipase A and of polypeptide toxins. *Pharm. Rev.* 17:393.

MENEZ, A., J.-L. MORGAT, P. FROMAGEOT, A.-M. RONSERAY, P. BOQUET, and J.-P. CHANGEUX, 1971. Tritium labeling of the α-neurotoxin of *Naja nigricollis*. *FEBS Letters* 17:333.

MEUNIER, J.-C., R. W. OLSEN, A. MENEZ, P. FROMAGEOT, P. BOQUET, and J.-P. CHANGEUX, 1972. Some physical properties of the cholinergic receptor protein from *Electrophorus electricus* revealed by a tritiated α-toxin from *Naja nigricollis* venom. *Biochem.* 11:1200.

MILEDI, R., 1960. The acetylcholine sensitivity of frog muscle fibres after complete or partial denervation. *J. Physiol.* (*Lond.*) 151:1.

MILEDI, R., P. MOLINOFF, and L. T. POTTER, 1971. Isolation of the cholinergic receptor protein of *Torpedo* electric tissue. *Nature* 229:554.

MILEDI, R., and L. T. POTTER, 1971. Acetylcholine receptors in muscle fibers. *Nature* 233:599.

MOLINOFF, P. B., and L. T. POTTER, 1972. Isolation of the cholinergic receptor protein of *Torpedo* electric tissue. In *Studies of Neurotransmitters at the Synaptic Level, Advances in Biochemical Psychopharmacology*, E. Costa and L. Iversen, eds. New York: Raven Press. p. 111.

POTTER, L. T., and P. B. MOLINOFF, 1972. Isolation of cholinergic receptor proteins. In *Perspectives in Neuropharmacology: A tribute to Julius Axelrod*, S. Snyder, ed. Oxford: Oxford University Press.

PRESSMAN, D., and H. N. EISEN, 1950. Zone of localization of antibodies. Part V. An attempt to saturate antibody binding sites in mouse kidney. *J. Immunol.* 64:273.

DE ROBERTIS, E., 1971. Molecular biology of synaptic receptors. *Science* 171:963.

DE ROBERTIS, E., G. S. LUNT, and J. L. LA TORRE, 1971. Multiple binding sites for acetylcholine in a proteolipid from electric tissue. *Mol. Pharmacol.* 7:97.

LA TORRE, J. L., G. S. LUNT, and E. DE ROBERTIS, 1970. Isolation of a cholinergic proteolipid receptor from electric tissue. *Proc. Natl. Acad. Sci. USA* 65:716.

WASER, P. G., and U. LUTHI, 1957. Autoradiographische lokalisation von ^{14}C-calebassen-currarin I and ^{14}C-decamethonium in der motorischen endplatte. *Arch. Intern. Pharmacodyn. Ther.* 112:272.

67 The Use of α-Bungarotoxin to Probe Acetylcholine Receptors on Sympathetic Neurons in Cell Culture

LLOYD A. GREENE

ABSTRACT   This chapter describes the probing of acetylcholine receptors on the membranes of living neurons by means of the combined use of the venom protein α-bungarotoxin and the methods of dissociated cell culture. This approach has been used to assay acetylcholine receptors in cell cultures containing chick embryo sympathetic neurons. The experiments discussed herein reveal (a) that receptors for the α-bungarotoxin are present on these neurons, (b) that these toxin-binding sites appear to be diffusely distributed along the entire surface of the neuron, (c) that the average density of these receptors on the surface membrane is considerably lower than that at post-synaptic membranes on striated muscle or electric organs and is, rather, more comparable to receptor densities at noninnervated areas of these same tissues, and (d) that these toxin-binding sites share many pharmacological properties with classical nicotinic acetylcholine receptors of sympathetic ganglia. Studies of similar nature should prove useful to further probe cholinergic receptors.

Introduction

OF THE MANY functional components of neuronal membranes, some of the most interesting to neuro-biologists are the receptors for chemical transmitters. The properties of these receptors influence if, and how, a neuron will respond to chemical signals from its environment. The possibility that changes may occur in the disposition (e.g., number, distribution) of these sites makes them candidates as potential places for plasticity in the nervous system. Regarding these membrane receptors therefore, there are a number of questions that may be posed. For example, we may inquire as to the presence of a particular type of receptor on a given population of neurons. If it is found there, we may ask further about such properties as its number, density, distribution, turnover, and biochemical and pharmacological qualities. One way to explore such questions would be by means of a probe that is at once highly specific and has a high affinity for a given receptor type and that, also, may be easily visualized or counted once bound. These desired features are possessed by α-bungarotoxin—the venom protein described in detail elsewhere in this volume (Molinoff, this volume). The specific and highly irreversible manner in which α-bungarotoxin combines with acetylcholine receptors of the nicotinic type on striated muscle (Lee and Chang, 1966; Miledi and Potter, 1971; Berg et al., 1972) and electric tissues (Changeux et al., 1970; Miledi et al., 1971; Fiszer De Plazas and De Robertis, 1972) suggests that it would be well suited for assessing the characteristics of similar receptors on neurons.

The design of the nervous system, however, poses certain problems for the use of this protein probe. Neurons in situ are generally well protected from diffusible substances by a variety of means. Capsules, sheaths, glia, blood-brain barriers, and the dense packing of nervous tissue itself all present difficulties for the treatment with, washing off, and visualization of the toxin. One way in which to obviate these problems is by means of dissociated cell cultures. In the use of this approach, tissue from specific parts of the nervous system is dissected from the organism, dissociated, dispersed, and maintained as a monolayer. Under suitable conditions, neurons cultured in this manner exhibit a number of differentiated properties comparable to those found in vivo (for reviews of this technique and its achievements with regard to neural tissue see for example Varon, 1970; Herschman, in press). Neurons in cell culture have certain major advantages for the use of receptor probes. For example, they are readily accessible to diffusible materials, and any unbound excess probe may be washed off easily. Also, cells in monolayer culture are easy to observe and they may be viewed in their entirety. Furthermore, they are maintained in an environment that is controlled and that may therefore be manipulated. When using in vitro systems however, one must always keep in mind the

LLOYD A. GREENE   Laboratory of Biochemical Genetics, National Heart and Lung Institute, National Institutes of Health, Bethesda, Maryland

possibility that in these artificial surroundings neurons may lose or change their usual properties.

In this paper, we wish to describe the use of α-bungarotoxin to assess acetylcholine receptors of neurons in cell culture. We shall in particular give an account of experiments designed to probe the presence and properties of a cholinergic receptor on chick embryo sympathetic neurons (Greene et al., in press).

Acetylcholine receptors on cultured skeletal muscle

Before detailing work on the sympathetic receptor, it is of interest to briefly review another cell culture system in which α-bungarotoxin has been used to assay acetylcholine receptors. In dissociated cell culture, embryonic myoblasts fuse to form multinucleated myotubes (Konigsberg, 1963). These nondividing myotubes are capable of contracting and are sensitive to acetylcholine (Fischbach, 1970). Recent experiments (Vogel et al., 1972; Sytkowski et al., 1973) have shown that, concomitant with fusion, cultures of chick embryo skeletal muscle show large increases in development of α-bungarotoxin receptors. Pharmacologically these sites resemble nicotinic acetylcholine receptors found on adult mammalian muscle in vivo. Furthermore, these cholinergic receptors are distributed both in a diffuse manner along the length of the myotubes and in small, discrete regions of high density. A model for the formation of neuromuscular junctions has been proposed based on these findings (Sytkowski et al., 1973). In another report (Hartzell and Fambrough, 1971), α-bungarotoxin was used to measure the rate of turnover of acetylcholine receptors in similarly derived cultures of muscle. These studies illustrate the utility of the toxin as a probe in dissociated cell cultures and that results obtained there may be applicable to problems not easily approached in vivo.

Acetylcholine receptors on cultured sympathetic neurons

Sympathetic neurons have a number of possible advantages as a model system for the in vitro study of acetylcholine receptors. They are moderately easy to obtain in reasonable numbers, are a relatively homogeneous population, and are not difficult to maintain in cell culture. Moreover, they are known in vivo to possess acetylcholine receptors of the nicotinic type. The electrophysiology and pharmacology of these "ganglionic" receptors have been extensively studied in a number of preparations (see for example reviews by Phillis, 1970; Volle and Koelle, 1970).

The source of the sympathetic neurons used in these

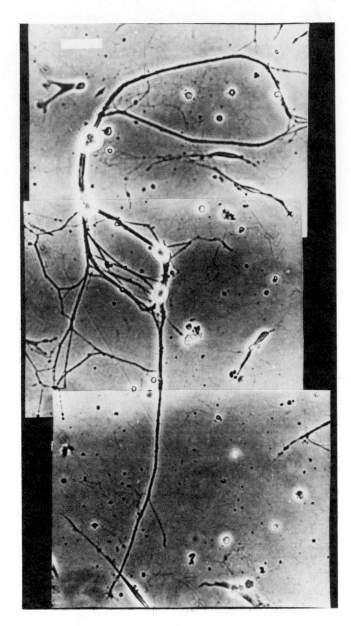

FIGURE 1   Chick embryo sympathetic neurons in dissociated cell culture. The neurons used in these studies were obtained in the following manner: Paravertebral sympathetic chains from 11-day chick embryos were placed in a Ca, Mg-free phosphate buffered saline containing 0.25% trypsin for 30 min at 37°. They were then transferred to complete culture medium and dissociated by trituration in a Pasteur pipet (Varon and Raiborn, in press). Collagen-coated 35 mm Falcon tissue culture dishes were seeded with 50,000 to 100,000 neurons in 2 ml of medium. Complete culture medium consisted of 95% F-14 medium (Vogel et al., 1972), 5% fetal calf serum, and 10 ngm per ml of nerve growth factor (Varon et al., 1967). Maintenance was in a 37°C water-jacketed incubator having a water-saturated atmosphere containing 90% air and 10% CO_2. The cells pictured here were maintained in culture for 4 days. The white bar corresponds to 50 μm. Illumination was by phase optics.

studies was dissociated paravertebral chains of 11-day chick embryos. When placed in cell culture in the presence of appropriate quantities of nerve growth factor (Levi-Montalcini and Angeletti, 1963, 1968) these neurons attach to a collagen surface and send forth extensive processes (Figure 1). The rounded and highly refractile somas of these cells may be penetrated with a micro-electrode and active electrical responses may be evoked from them (A. Chalazonitis and Greene, in preparation). Other cell types are also present in the dissociate (e.g., fibroblasts, capsule cells, and glial elements), and these proliferate and form a monolayer beneath the neurons by about one week in culture.

Using these cultures, experiments were undertaken to determine the presence, distribution, and specificity of binding of α-bungarotoxin. In order to do so, cultures were incubated in the presence of a highly labeled preparation of $[^{125}I]$-diiodo-α-bungarotoxin (Vogel et al., 1972) and binding was then visualized by means of radioautography (Sytkowski et al., 1973). Figure 2 illustrates the results of these procedures. Examination of this and a number of other radioautographs revealed several noteworthy features. First, the labeled toxin appeared to bind preferentially to the sympathetic neurons. The other cell types present in the cultures showed no localization of silver grains over background. Second, labeling occurred on all parts of the neurons; binding was observed on the cell bodies as well as along the entire lengths of the processes. Third, all cells with the morphological appearance of neurons bound the toxin. Some neurons were however more heavily labeled than others. Last, the distribution of receptor on the neurons was in a diffuse manner. Discrete regions of heavy localization of binding were not seen. These observations could indicate that nicotinic receptors are not localized to specific areas of sympathetic neurons in vivo or that receptor localization requires further maturation of the cells or the presence of synaptic elements.

To extend the results of the radioautographic studies, quantitative estimates were made of the amount of α-bungarotoxin bound to the cultured neurons. This was achieved by measuring the amount of radioactivity recovered from the cultures after incubation in the presence of the labeled toxin (Vogel et al., 1972; Greene et al., in press). That the counts obtained in this way were almost exclusively bound to the sympathetic neurons was demonstrated by the following experiment: When cultures were prepared, but without the presence of nerve growth factor, the neurons were selectively lost and did not survive beyond 24 hr in vitro. After incubation with the labeled toxin, these cultures lacking neurons showed no significant radioactivity above the blanks (empty collagen-coated culture dishes). Normally maintained cultures, on the other hand, showed extensive binding of radioactive toxin. As may be seen in Figure 3, this binding was saturable and reached a plateau level after about one hour's incubation with toxin concentrations greater than 5×10^{-9} M. From the maximal extent of binding per culture and from the number of neurons present, the average number of toxin receptors per cell may be computed. On the assumption that one molecule of toxin binds per receptor, this value is about 3×10^5 per neuron. Furthermore, from this figure, it is of interest to estimate the average density of receptors on the membrane. Assuming the sympathetic neuron cell bodies to be smooth spheres 12 μm in diameter with cylindrical processes that are 0.6 μm in diameter and 500 to 1000 μm in length (these are average values measured for these cells in culture), then the density of receptor sites on the cell surface is calculated to be 100 to 300 per μm^2. This density is 2 to 3 orders of magnitude less than that calculated at synaptic membranes on striated muscle (Miledi and Potter, 1971; Barnard et al., 1971; Fambrough and Hartzell, 1972) and electric tissue (Changeux et al., in press) at localizations on cultured chick skeletal muscle (Sytkowski et al., 1973). This figure, on the other hand, is comparable in magnitude to the receptor densities estimated at nonsynaptic or nonlocalized areas of these same tissues.

By means of similar experiments, examination was made of the rate at which the labeled toxin binds to cultured sympathetic neurons. Figure 4 shows that at a toxin concentration in the medium of 2.5×10^{-9} M, binding at 37 degrees reached saturation by about 3 hr. These data have been subjected to kinetic analysis. Assuming the binding to be irreversible and at a single site, the rate constant for the formation of the receptor-toxin complex was calculated to be 1.1×10^5 liter mole^{-1} second^{-1} (Greene et al., in press). The highly irreversible nature of this binding to the sympathetic neurons that is consistent with these kinetic studies has also been suggested in the case of binding of the α-bungarotoxin to striated muscle (Lee and Chang, 1966; Miledi and Potter, 1971) and electric tissue (Changeux et al., 1970; Miledi et al., 1971).

Using the radiobinding assay on the cultured neurons, studies were also carried out to determine the specificity and pharmacological properties of the receptor for α-bungarotoxin. If these sites are cholinergic, then drugs known to have an affinity for acetylcholine receptors should protect them from the toxin. Also, from a knowledge of the types of drugs that inhibit toxin binding, it should be further possible to distinguish whether the receptor has the qualities of either of the two classical types of cholinergic receptors—nicotinic or muscarinic.

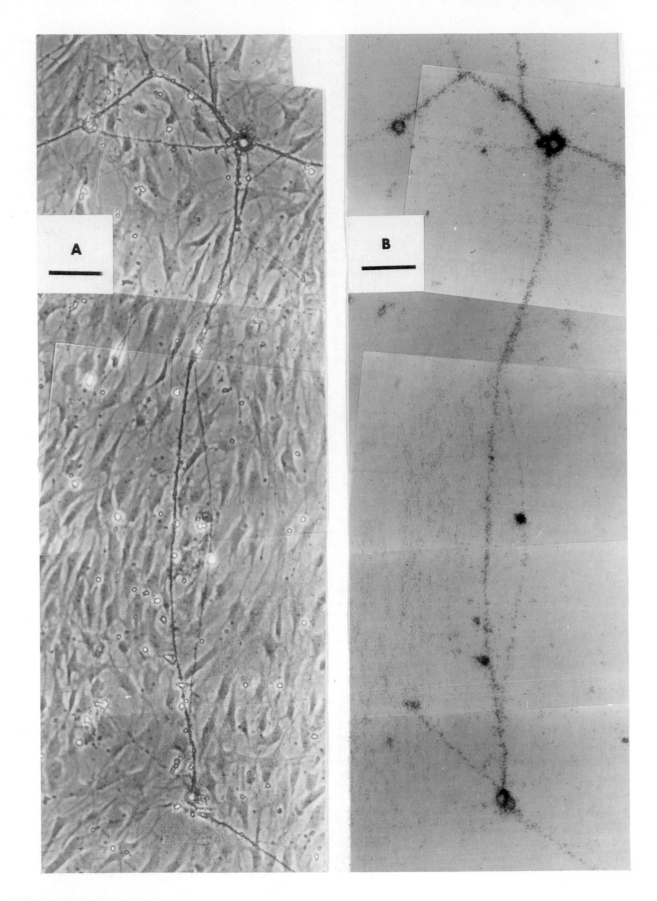

PROBING ACETYLCHOLINE RECEPTORS ON SYMPATHETIC NEURONS IN CELL CULTURE

FIGURE 2   Radioautomicrograph of a culture of dissociated sympathetic ganglia labeled with [^{125}I]-diiodo-α-bungarotoxin. The culture pictured here was maintained in vitro for 5 days. Complete medium was then replaced with 2 ml of choline-free F-14 medium containing [^{125}I]-diiodo-α-bungarotoxin (5 $\times$ 10^{-9} M; specific activity = 2.2 $\times$ 10^5 Ci/mole) and 2 mg/ml bovine serum albumin. Incubation in the presence of the toxin was for 1 hr and, as in subsequent experiments, was in a water bath placed inside the 37° incubator described in the legend to Figure 1. The culture was then thoroughly washed, fixed with 2.5% glutaraldehyde in 0.1 M phosphate buffer (pH 7.4), and radioautographed with Kodak NTB-2 emulsion for 8 days (Vogel et al., 1972; Sytkowski et al., 1973). Radioautomicrographs show the same field with (A) phase contrast optics; (B) bright field optics so as to show only the silver grains. The bar corresponds to 50 μm.

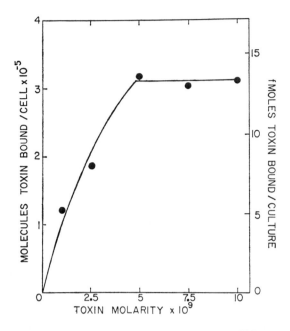

FIGURE 3   Binding of various concentrations of [^{125}I]-diiodo-α-bungarotoxin to sympathetic neurons in dissociated cell culture. Cultures maintained in vitro for 5 days were exposed to various concentrations of the labeled toxin (specific activity = 2.8 $\times$ 10^5 Ci/mole) for 1 hr. The methods by which bound radioactivity was then measured are described in detail elsewhere (Vogel et al., 1972). Values represent the average of 3 determinations at each toxin concentration. Variations between replicate cultures in this and succeeding experiments was $\pm$ 12%. Plateau levels of binding were about 5600 cpm. The blank (an empty collagen-coated dish) in this and in all subsequent experiments was about 100 cpm. Specific activities were computed on the basis of cell counts made on 3 sister cultures (average number of neurons per dish was 26,000 $\pm$ 3,800). For cell counts, strips were surveyed totaling 4% of the total area of each culture. (From Greene et al., in press.)

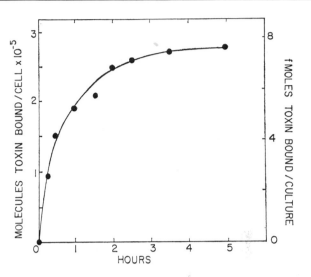

FIGURE 4   Time dependence of the binding of [^{125}I]-diiodo-α-bungarotoxin to sympathetic neurons in dissociated cell culture. Five-day-old cultures of sympathetic neurons were exposed to the labeled toxin (2.5 $\times$ 10^{-9} M; specific activity = 2.6 $\times$ 10^5 Ci/mole) for various lengths of time and then assayed for bound radioactivity. Two determinations were made at each time point. Plateau levels of binding were about 3200 cpm. The average number of neurons per dish was 16,500 $\pm$ 3,000. (From Greene et al., in press.)

The results of a number of drug-inhibition experiments are summarized in Figure 5. Curare, nicotine, 1, 1-dimethyl-4-phenyl-piperazinium (DMPP) and acetylcholine were all potent inhibitors of toxin binding (50% inhibition of binding at 2 $\times$ 10^{-7} M, 6 $\times$ 10^{-7} M, 3 $\times$ 10^{-6} M, and 10^{-5} M, respectively). These compounds are known to be effective ligands of sympathetic ganglion acetylcholine receptors of the nicotinic type (see for example reviews of Phillis, 1970; Gyermek, 1967; and Volle and Koelle, 1970). At appropriate concentrations and under the conditions of these experiments, these materials were also capable of producing total blockade of toxin binding as did a 100-fold excess of the unlabeled toxin. On the other hand, atropine, an antagonist at muscarinic receptors, was a poor protective agent (50% inhibition of toxin binding at greater than 10^{-3} M). Decamethonium was also a relatively weak protector (50% inhibition of toxin binding at 2 $\times$ 10^{-4} M). This compound is in vivo a poor blocker of nicotinic acetylcholine receptors of sympathetic ganglia but is very effective in this role at nicotinic sites on striated muscle (Paton and Zaimis, 1949; Gyermek, 1967). It is therefore of interest to note that, in contrast to its weak action with the cultured sympathetics, decamethonium is a potent inhibitor of α-bungarotoxin binding with dissociated cultures of chick skeletal muscle (Vogel et al., 1972).

The results of these drug studies are consistent with the

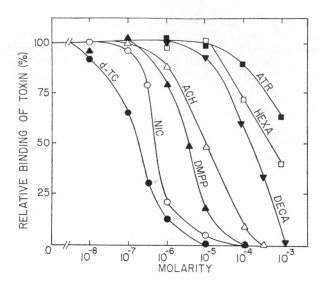

FIGURE 5 Inhibition of the binding of [^{125}I]-diiodo-α-bungarotoxin to sympathetic neurons in dissociated cell culture by various concentrations of cholinergic agonists and antagonists. Cultures maintained in vitro for 4 or 5 days were incubated for 15 min in the presence of the appropriate drug. Labeled toxin was then added to the medium (final concentration = 1.25 × 10^{-9} M; specific activity = 2.0 × 10^5 Ci/mole). Incubation in the presence of both the labeled toxin and the drug was for 30 min. Values for 100% binding (about 1500 cpm) were obtained using similarly treated cultures but with no drugs added. Each point represents the average of determinations on 4 or 5 sister cultures. The final molarity of the drugs in the incubation medium is plotted on a log scale. Filled circle, *d*-tubocurarine chloride; open circle, nicotine dihydrochloride; filled triangle, 1,1-dimethyl-4-phenyl-piperazinium iodide; open triangle, acetylcholine chloride (all assays performed in the presence of μM eserine sulfate—this concentration of eserine alone had no effect on the binding of the toxin); inverted filled triangle, decamethonium iodide; open square, hexamethonium chloride; filled square, atropine sulfate. (From Greene et al., in press.)

conclusion that the [^{125}I]-diiodo-α-bungarotoxin is binding here specifically to a sympathetic acetylcholine receptor with nicotinic properties. This site is likely to be a "ganglionic" receptor of the type that receives excitatory synaptic input. Possibly at odds with this suggestion is the weak protective effect of hexamethonium against toxin binding (50% inhibition of binding at 5 × 10^{-4} M, see Figure 5) on the cultured neurons. This material is a potent antagonist of acetylcholine at nicotinic sites in mammalian sympathetic ganglia (Paton and Zaimis, 1949, Gyermek, 1967) and hence might be expected to be more effective in blocking binding of the toxin in our experiments. One likely explanation for this apparent inconsistency is that there may be pharmacological differences between mammalian and avian sympathetic acetylcholine receptors. Alternate interpretations regarding the nature of these toxin-binding receptors are

also possible. For example, they may be cholinergic sites of the type hypothesized to have roles in the sympathetic release of catecholamines (Burn and Rand, 1965) or in the propagation of the nerve impulse (Nachmansohn, 1971).

Concluding comments

It has been our intention to describe here how the use of the venom protein α-bungarotoxin and the techniques of cell culture may be wedded in order to probe acetylcholine receptors on living neurons. One model system has been described in detail—dissociated cell cultures of chick embryo sympathetic neurons. These studies have revealed (a) that receptors for the α-bungarotoxin are present on these cells, (b) that these sites appear to be diffusely distributed along the entire surface of the neuron, (c) that the average density of receptors is considerably lower than that estimated at nicotinic synapses on muscle or electric organs and is more comparable to that at noninnervated regions of these same tissues, and (d) that these toxin-binding sites share many pharmacological properties with classical nicotinic acetylcholine receptors found in sympathetic ganglia in vivo.

There remain a number of problems and questions concerning cholinergic receptors that it would be of considerable interest to explore in this system. For example, we would like to know if and how the distribution and density of toxin-binding sites on the sympathetics will change after long periods of time in vitro, after the prolonged presence of various pharmacological agents, or after co-culture with suitable presynaptic elements. Also, one would like to know the influence of a variety of different culture conditions on the rate of turnover of the receptor. By means of the systems and the approaches discussed in this chapter, it should be possible to further probe these and other aspects of neuronal acetylcholine receptors from biochemical, pharmacological, and developmental points of view.

ACKNOWLEDGMENTS The work described here was done in collaboration with Drs. A. J. Sytkowski, Z. Vogel, and M. W. Nirenberg and with the financial support of a postdoctoral fellowship from the American Cancer Society.

REFERENCES

BARNARD, E. A., J. WIECKOWSKI, and T. H. CHIU, 1971. Cholinergic receptor molecules and cholinesterase molecules at mouse skeletal muscle junctions. *Nature (Lond.)* 234:207.

BERG, D. K., R. B. KELLY, P. B. SARGENT, P. WILLIAMSON, and Z. W. HALL, 1972. Binding of α-bungarotoxin to acetylcholine receptors in mammalian muscle. *Proc. Nat. Acad. Sci. USA* 69:147.

Burn, J. H., and M. J. Rand, 1965. Acetylcholine in adrenergic transmission. *Ann. Rev. Pharmacol.* 5:163.

Chalazonitis, A., and L. A. Greene, in preparation. Studies on the electrophysiological properties of chick sympathetic neurons in dissociated cell culture.

Changeux, J-P., M. Kasai, and C. Y. Lee, 1970. Use of a snake venom toxin to characterize the cholinergic receptor protein. *Proc. Nat. Acad. Sci. USA* 67:1241.

Changeux, J-P., J-C. M. Meunier, R. W. Olsen, M. Weber, J-P. Bourgeois, J-L. Popot, J. B. Cohen, G. L. Hazelbauer, and H. A. Lester, 1972. Studies on the mode of action of cholinergic agonists at the molecular level. *Proc. Symp. Drug Receptors (Lond.)* (in press).

Fambrough, D. M., and H. C. Hartzell, 1972. Acetylcholine receptors: Number and distribution at neuromuscular junctions in rat diaphragm. *Science* 176:189.

Fischbach, G. D., 1970. Synaptic potentials recorded in cell cultures of nerve and muscle. *Science* 169:1331.

Fiszer De Plazas, S., and E. De Robertis, 1972. Binding of α-bungarotoxin to the cholinergic receptor proteolipid from *Electrophorus* electroplax. *Biochim. Biophys. Acta* 274:258.

Greene, L. A., A. J. Sytkowski, Z. Vogel, and M. W. Nirenberg, 1973. Acetylcholine receptors on chick embryo sympathetic neurons in dissociated cell culture. *Nature (Lond.)* (in press).

Gyermek, L., 1967. Ganglionic stimulant and depressant agents. In *Drugs Affecting The Peripheral Nervous System*, Vol. 1, A. Burger, ed. New York: Marcel Dekker, pp. 119–326.

Hartzell, H. C., and D. M. Fambrough, 1971. Kinetics of acetylcholine receptor production and incorporation into membranes of developing muscle fibers. Abstracts, Soc. for Neurosciences, First annual meeting, p. 161.

Herschman, H. Culture of neural tissue and cells. In *Research Methods in Neurochemistry*, Vol. 2, R. Rodnight and N. Marks, eds. New York: Plenum Publishing (in press).

Konigsberg, I., 1963. Clonal analysis of myogenesis. *Science* 140: 1273.

Lee, C. Y., and C. C. Chang, 1966. Modes of actions of purified toxins from elapid venoms on neuromuscular transmission. *Mem. Inst. Butantan Simp. Internac.* 33:555.

Levi-Montalcini, R., and P. U. Angeletti, 1963. Essential role of nerve growth factor in the survival and maintenance of dissociated sensory and sympathetic embryonic nerve cells in vitro. *Dev. Biol.* 7:653.

Levi-Montalcini, R., and P. U. Angeletti, 1968. Nerve growth factor. *Physiol. Rev.* 48:534.

Miledi, R., and L. T. Potter, 1971. Acetylcholine receptors in muscle fibers. *Nature (Lond.)* 233:599.

Miledi, R., P. Molinoff, and L. T. Potter, 1971. Isolation of cholinergic receptor protein of *Torpedo* electric tissue. *Nature (Lond.)* 229:554.

Nachmansohn, D., 1971. Chemical events in conducting and synaptic membranes during electrical activity. *Proc. Nat. Acad. Sci. USA* 68:3170.

Paton, W. D. M., and E. J. Zaimis, 1949. The pharmacological actions of polymethylene bistrimethylammonium salts. *Brit. J. Pharmacol.* 4:381.

Phillis, J. W., 1970. *The Pharmacology of Synapses*. London: Pergamon Press, Chapter 6, pp. 123–148.

Sytkowski, A. J., Z. Vogel, and M. W. Nirenberg, 1973. Development of acetylcholine receptor clusters on cultured muscle cells. *Proc. Nat. Acad. Sci. USA* 70:270.

Varon, S., 1970. In vitro study of developing neural tissue and cells: Past and prospective contributions. In *The Neurosciences: Second Study Program*, F. O. Schmitt, Editor-in-Chief. New York: The Rockefeller University Press, pp. 83–99.

Varon, S., J. Nomura, and E. M. Shooter, 1967. The isolation of nerve growth factor protein in a high molecular weight form. *Biochemistry* 6:2202.

Varon, S., and C. W. Raiborn, 1973. Dissociation, fractionation, and culture of chick embryo sympathetic ganglionic cells. *J. Neurocytol.* (in press).

Vogel, Z., A. J. Sytkowski, and M. W. Nirenberg, 1972. Acetylcholine receptors of muscle grown in vitro. *Proc. Nat. Acad. Sci. USA* 69:3180.

Volle, R. L., and G. B. Koelle, 1970. Ganglionic stimulating and blocking drugs. In *The Pharmacological Basis of Therapeutics*, Fourth Ed., L. S. Goodman and A. Gilman, eds. New York: The Macmillan Company, pp. 585–595.

68 The Surface Membrane and Cell-Cell Interactions

MAX M. BURGER

ABSTRACT Cell-cell interaction has been implied in the regulation of growth under culture conditions. To what degree this is correct and to what degree the surface membrane could be involved in the mediation of this type of growth control is critically assessed.

Specific cell-cell interactions are also postulated in the guiding of organ formation during development. This implies specific recognition between cell surfaces. By analogy with a primitive cell recognition system that has been fairly well studied, a molecular model will be examined that might be involved in cell-cell recognition.

The same type of specific cell-cell interactions may be involved at least in the development of the central nervous system. They may even solidify and maintain the neural wiring system and facilitate certain specific neural pathways.

Introduction

INTERACTIONS between cells are usually considered only where they represent the principal function of the two cells involved, namely, in neural cells. The question whether other cells that have other functions like parenchymatous liver cells or epithelial kidney cells or simple fibroblasts also interact and whether such communications are required for function or survival has been given little attention.

Early development requires close cellular contacts not only for induction of differentiation and tissue formation but also for maintenance of the state of differentiation. We do not know, however, to what degree adult cells require cell-cell contact in their normal cellular environment for survival. Although many normal cells cannot yet be cultivated in cell culture, these exceptions cannot be used as obvious evidence that most cells require the presence of neighboring cells in their environment for survival. We do not yet know whether the optimal balance of nutrients, conditioning factors, etc., for such cells has been found.

Our present efforts to mimic in vivo growth conditions in cell cultures will doubtless be considered primitive in 5 to 10 years, but such in vitro culture systems are pre-

MAX M. BURGER Department of Biochemical Sciences, Princeton University, Princeton, New Jersey. (Present address: Department of Biochemistry, Biocenter of the University of Basel, Basel, Switzerland)

sently the only means of investigating cell-cell interactions in a quantitative manner.

Commonly cell-cell interactions are divided into long-range interactions mediated via hormones and pheromones on the one hand and the close range interactions found, for example, between tissues in embryonic development on the other hand. For a systematic analysis of cell-cell interaction phenomena, it is useful to distinguish between two parts of this process; before any interaction can take place, cell-cell contact must first be established. Different mechanisms may be involved in bringing about this cell-cell contact. Only then will neighboring cells be able to interact. A short survey of different ways of establishing contact and of cell-cell interactions after establishment of the contact are helpful as an introduction.

Establishment of cell-cell contact

1. Cells attract each other, and this attraction can be mediated: (a) chemically in a process usually described as chemotaxis (Figure 1) (Konjin et al., 1967); (b) by extracellular structural material which serves as guidelines along which cells find each other, like collagen, (Hauschka and Konigsberg, 1966), mucopolysaccharides or calcium salts, silica salts, and other crystals (Figure 1);

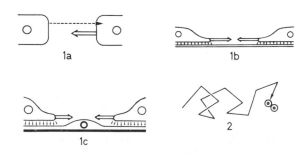

FIGURE 1 Some mechanisms by which cell-cell contact can be established. (1a) Response to a soluble attractant. (1b) The attraction is mediated and guided by an extracellular matrix. (1c) The attraction is mediated and guided by another cell. (2) Random movement eventually leads to interaction between appropriate cells.

(c) via other cells or cellular networks that serve as guide-lines along which some cells can move (Figure 1) (Sidman, 1971).

Examples for all three mechanisms are available: chemotaxis is found in some fertilizations (Metz and Monroy, 1969); extracellular matrices are important in muscle cell fusion and in tissue formation of sea animals that have crystalline skeletons; cellular guidelines (glial cells) may provide directionality in neural development.

2. Cells may move around randomly, and, if the appropriate pair of cells meets, they attach and get trapped due to specific surface interactions (Figure 1).

3. No specific attraction or trapping leads the cells together. They meet randomly and interact randomly, as is seen in contact inhibition of growth and movement.

Interactions after the establishment of cell-cell contact

INTERACTIONS BETWEEN TWO SINGLE CELLS Fertiliza-tion is probably the best example of such interaction that includes information transfer. Interaction between T and B lymphocytes (Mitchison, 1971; Nossal and Ada, 1971) may also involve information transfer between single cells.

INTERACTIONS BETWEEN CELLS IN TISSUES Oogenesis (Anderson and Huebner, 1968) and spermatogenesis are examples where single cells require surrounding tissues for maturation or differentiation. Two further categories of interactions can be observed in differentiation, namely, the induction of new types of cells in neighboring tissues, on one hand, and tissue and organ formation through the restructuring and sorting out within a preexisting prim-itive tissue, on the other hand. And finally as a last example, growth control in a fully differentiated adult tissue is thought to require cell-cell interaction.

If we want to establish the molecular nature and the significance of such cell-cell interactions, we will have to find answers to the following three questions in the near future: (1) Are there surface molecules that can be isolated and chemically defined and that are involved in close range cell-cell interactions? Can these molecules be used in a cell free system to reconstruct the type and strength of interaction seen in intact cells? (2) Can the cell surface structure be altered in a specific manner and does that lead to a predictable alteration in cell-cell interaction and cellular behavior? In other words, can the function of these surface components be tested and proved in vivo? (3) In those cell-cell interactions where the cells undergo specific cellular responses, the question as to the type and manner of communication between the surface and the interior of the cell will have to be considered.

Although attempts will be made to answer some of these questions, we are far from the point where we can describe specific cell-cell interactions in chemical terms, although at least some experimental systems are now beginning to appear that are amenable to tests in tissue culture by biochemical techniques.

Specific adhesive surface components

The most primitive specific cell interaction is probably represented by mating seen in unicellular organisms. In recent years glycoproteins, which were implied as the specific surface components responsible for establishing the mating adhesions in two systems, have been partially isolated.

For *Chlamydomonas*, Wiese (1965) has shown that flagella tips or isolated isoagglutinins from one mating type can agglutinate cells from the opposite mating type. Since only the female gamete is sensitive to low doses of trypsin and only the male can be inactivated with low doses of the carbohydrate specific agglutinin Con-canavalin A (Wiese and Shoemaker, 1970), we can con-sider that establishment of the mating adhesion is brought about by an antibody-antigen type of interaction at the flagellar tips between the two mating gametes. The male receptor would be a complex carbohydrate antigen that is recognized by the female protein "antibody."

Such conclusions are very tentative and, although conceptually quite acceptable, require much more bio-chemical and immunochemical evidence. Thus, the in-hibition of the mating agglutination by Concanavalin A means just that the Concanavalin A molecule that may have adsorbed to a mannose residue or the flagellum interferes with the mating interaction. It may do so, however, by blocking other chemical groups close to some mannose residue that are necessary for the mating agglutination.

Crandall and Brock (1968) isolated some glycoproteins from the surfaces of opposite yeast mating types that agglutinated or neutralized the opposite mating type. These surface components require better chemical char-acterization as well as better evidence of their direct involvement in the mating interaction.

The best evidence for the presence of surface com-ponents responsible for specific cell-cell interactions comes from studies in developmental biology. Several programed morphogenetic displacements bring em-bryonic cells into their final position, thereby giving rise to the formation of tissues and organs according to a carefully outlined masterplan. Holtfreter was the first to show that some of these cellular displacements could also be demonstrated in vitro (Townes and Holtfreter, 1955). But it was not until the Mosconas (1952) dissociated tissues into single-cell suspensions, which subsequently could be studied while reaggregating and reforming organ-like structures in vitro, that an experimental procedure was available permitting a certain degree of quantitation (Moscona, 1968). Steinberg (1970) suggested a general hierarchy of strengths of adhesions between different embryonic tissues that may be due to simple physical interactions with no chemical specificity or may equally well be based on chemically specific interactions.

I do not intend to review here the more complicated organogenic processes in vertebrates (Lilien, 1969), but I prefer to examine a more simple model system, namely that of sponge aggregation which in many, although not in all regards, can be considered as the prototype for the later tissue reconstruction experiments by Moscona, Steinberg, Lilien, etc.

In 1907, H. V. Wilson dissociated two marine sponge species from Woods Hole that had two different colors, mixed them, and observed that the two types of cells immediately sorted out into separate clumps and later reestablished two small and primitive but color- and species-specific sponges again. This clever experiment, using simply color as a marker instead of isotopes as one would have today, suggests that two different sponge species must have a way of recognizing their own from different cells—presumably via cell surface components. Since cells from some organs of vertebrate embryos can do the same (Moscona, 1968 and Steinberg, 1970), Wilson's experiment, with certain restrictions, can be considered as a model for sorting out and for certain restructuring processes in higher animal morphogenesis.

Humphreys (1963) and Moscona (1963) began to investigate the chemical mechanism of this affinity. With the sponges used, it could be shown that the cells do not seek out each other after establishing a mixed conglomerate of cells but that cells from the same species of sponge attach to each other while being mixed on a shaker. This particular process of sorting out is not based on some sort of chemotaxis, but it is due to a high affinity between the surfaces when the randomly moving cells come into contact. Figure 2 shows in a stylized and simple form the dissociation and sorting out of two sponges from different species.

Specific aggregation is mediated via a large molecular factor (mol wt approximately 5×10^6) that was isolated by Humphreys (1965) after dissociation of the cells with calcium-magnesium-free seawater, presumably because it was part of the intercellular material that was liberated during the dissociation. So far, this aggregation factor could be isolated in a partially purified form, and, in the two cases it was found to be specific. In our hands, this specificity is not absolute but rather quantitative, because

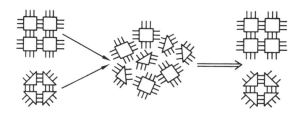

FIGURE 2 Specific sorting out of sponge cells. When two different species of sponges are mechanically dissociated in single-cell suspensions and subsequently mixed together, they will in due time sort out into two separate sponge cell clumps of the original species specificity. The rectangular species may be *Microciona prolifera* (red) and the triangular species *Haliclona celeta* (yellowish) while the spikes represent the intercellular material necessary for aggregation. Although many sponges do sort out specifically, this behavior is not true of all pairs of species tested.

it depends on the chemical similarity between the two receptor sites tested, which may reflect the taxonomic relationship between the two sponges tested. Because sponge taxonomy is far from established and since the technique for determining specificity may have to be reevaluated (Curtis and van de Vyver, 1971), such questions will have to remain unanswered for a while.

Similar to the mating substances discussed before, the sponge aggregation factor consists not only of protein but also of approximately 50% carbohydrate. We began to investigate, therefore, the question whether the carbohydrate moiety had any function. Glucuronic acid seems to be involved in the aggregation process of the best studied sponge, *Microciona prolifera*. Not only does the aggregation factor from this sponge contain glucuronic acid, but glucuronic acid is the only monosaccharide that inhibited the aggregation of these sponge cells. No other sponge tested could be inhibited with glucuronic acid, and other carbohydrates as closely related as galacturonic acid did not interfere (Lemon et al., 1971). Furthermore, glucuronic acid cleaving enzymes as well as sodium metaperiodate were able to destroy activity of the aggregation factor (Turner and Burger, unpublished observations).

For various reasons we had to consider the existence of a surface located "base plate" with which the intercellular aggregation factor would interact to bring about specific aggregation. When treating single cell suspensions that had already lost the aggregation factor with a hypotonic medium—a technique we had developed earlier to isolate other surface receptors—a soluble glycoprotein was released that had the desired properties. In its absence, the aggregation factor was ineffective. If added to cells prior to the aggregation factor, the aggregation factor was enabled to act. On the other hand, this glycoprotein bound and inactivated the aggregation factor

(Weinbaum and Burger, 1971). Figure 3 summarizes these observations in stylized and simple form. A substance similar to this base plate was recently also isolated by Humphreys by incubating factorless cells with a proteolytic enzyme (personal communication).

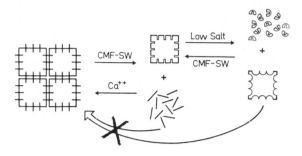

FIGURE 3 Requirement of a membrane-bound baseplate for the specific sorting out process. Sponge cells dissociated in calcium-magnesium-free seawater (CMF-SW) will release aggregation factor (long black bars) that is required in the presence of Ca^{++} for reaggregation of the cells lacking the aggregation factor. Low salt treatment (hypotonic swelling) of such aggregation factorless cells will release a baseplate or receptor for the aggregation factor (kidney shaped receptors in right upper corner) that is necessary for the cell to respond to the aggregation factor. In other experiments it was also shown that baseplate material, if preincubated with aggregation factor, could inactivate the aggregation factor, presumably due to binding (Weinbaum and Burger, unpublished observations).

Preliminary experiments with beads to which the aggregation factor or baseplate material was covalently attached were promising. Thus beads containing the aggregation factor agglutinated immediately if Ca^{++} was added to a concentration normally found in seawater. Beads with base plate material did not aggregate even in the presence of Ca^{++} unless the aggregation factor was added. Furthermore, previous absorption of factor-containing beads with baseplate material suppressed aggregation, and glucuronic acid inhibited the aggregation of baseplate beads and factor.

Such experiments promise to permit the study of surface interactions between cells in quantifiable in vitro systems. It is unlikely that all types of cellular surface interactions could be faithfully copied with such coated bead systems, because many properties of the cell surface membrane cannot be incorporated into the bead system. For example, the lipophilic base structure of the membrane or its flexibility or particularly the potential mobility of the glycoprotein within the natural membrane, quite obviously absent when covalently attached to a rigid bead, are properties that cannot be studied in this way.

Before we go to the next point, a critical assessment is still required. All due caution will have to be given to a preliminary conclusion that carbohydrates are involved in specific cell-cell interactions. What has been presented so far is only immunochemical evidence as, for example, when using hapten inhibition or modification (glycosidases, periodate) of the aggregation factor that lead to inhibition of aggregation. Unequivocal biochemical evidence that carbohydrates are involved in morphogenetic movements and development processes in vivo remains to be shown in the future.

Surface changes leading to alterations in growth control

Growth control in tissue culture depends on factors like serum concentration (Holley and Kiernan, 1968), pH (Ceccharini and Eagle, 1971), etc., and also on the degree of crowding, i.e., the degree to which cells are surrounded by neighboring cells. We do not know what guides these interactions among cells that often leads to specific patterns of cellular sociology (Kalckar, 1965), but because alterations in the cell surface can influence the degree of crowding, we have to assume that the cell surface must somehow be involved in the cell interactions leading to growth control or density-dependent inhibition of growth. Because few examples besides growth control in tissue culture are presently known where the cell surface seems to direct cell-cell interaction, I will summarize some experiments that were carried out on fibroblast growth in culture.

Fibroblasts grow under defined conditions (pH, serum, etc.) to a certain cell density per surface area that generally is in the range where cells are completely surrounded by other cells or from a monolayer (see Figure 4). Fibroblasts that have been transformed with oncogenic viruses, X-ray, or oncogenic chemicals defy this type of growth control when they reach the cell density of a monolayer (Stoker and Rubin, 1967). The same growth alterations are observed not only in transformed fibroblasts but also in transformed epithelial cells. Most of these alterations in cell growth control are paralleled by a surface alteration expressed as an increased degree of

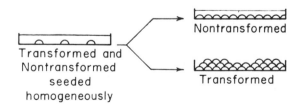

FIGURE 4 Transformed and untransformed cells grow to different saturation levels under defined and identical conditions (pH, serum, etc.).

agglutinability with several plant agglutinins (Burger, 1971). Since untransformed cell surfaces can be converted into the agglutinable surface form by a very mild treatment with various proteolytic enzymes, we have investigated the possibility whether the surface conversion might be followed by an alteration in the growth control pattern.

As can be seen in Figure 5, such treatment, which incidentally did not release the cells from their substratum or round them off, brought about a temporary escape from density dependent growth control (Burger, 1970). Growth stimulation was transient however, and cells returned to the nondividing state after one cell division. Such a result was predicted, because we found earlier that the brief and mild proteolytic treatment effected an increased agglutinability for only a short time (6 hr) and that the surface alteration returned into the nonagglutinable state thereafter. If the surface alteration was responsible for the growth stimulation, for example, in the form of a signal, its short duration permitted only one round of cell division occurring after 24 to 36 hr, but then the cells would stop growing again.

Enzymatic treatment of an intact cell does not exclude the possibility that the enzyme may not have partially penetrated the cell surface and exerted its relevant action within the cell. We have, therefore, carried out the same experiment with proteases (trypsin and pronase) that were covalently linked to insoluble beads. After a brief contact between such enzymatically active beads and resting fibroblasts, the beads were removed; almost the same growth response could be observed as was seen earlier with soluble proteases. The appropriate controls showed that no beads were internalized by endocytosis and that not sufficient protease leaked from the beads during the incubation which could have stimulated growth after entering the cell in soluble form (Burger, 1973).

Stimulation of growth by proteases—although the time of exposure to protease required was clearly longer—could also be shown for other cells like chick embryo fibroblasts (Sefton and Rubin, 1970). Furthermore, the work with insolubilized proteases could be confirmed only recently and was extended to some other enzymes (Vaheri et al., 1972). To what degree these enzymes may eventually trigger the same surface alteration that proteases do remains to be seen. Similarly, to what degree other methods of growth stimulation in tissue culture like treatment with other degradative enzymes as hyaluronidase (Vasiliev et al., 1970) or serum (Todaro et al., 1967), insulin (Temin, 1967) or colchicin (Vasiliev et al., 1971) may act via the enhancement of a series of membrane located proteolytic enzymes (Figure 6) remains for the time being an interesting speculation (Burger, 1973). At a first glance it seems unlikely that so many different forms of growth stimulation would all be mediated via the same cellular or perhaps even surface located proteases. On the other hand, we suggested such a cascade of proteolytic activating enzymes, because such systems are not purely hypothetical but are precedented, for example, in blood clotting and complement lysis, and since both are physiological chain reactions eventually acting also on cell surface membranes. Our

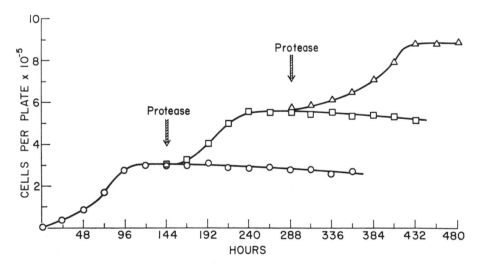

FIGURE 5   Pronase-induced cell division in a 3T3 cell monolayer. The 3T3 cells were grown to confluency and left at the monolayer stage for 3 days. The monolayer was treated with 10 μg/ml of pronase for 5 min and cell number followed daily. For the second pronase pulse, the 3T3 cells were again treated with 10 μg/ml pronase for 5 min at this new saturation density and cell number followed daily. ○: 3T3 cells, control; □: 3T3 cells, after first pronase pulse; △: 3T3 cells, after second pronase pulse. (From Burger et al., 1972a.)

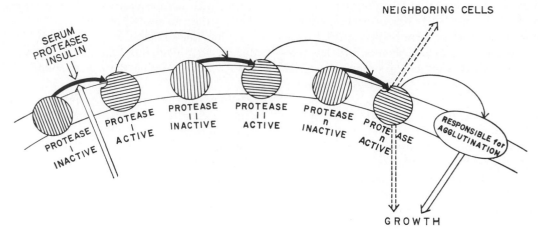

FIGURE 6 Cascade of proteolytic factors acting as a chain of events between growth stimulatory agents on the membrane and subsequent parts of the same growth stimulation mechanism inside the cell. Serum, proteases, insulin, or perhaps also mitosis (arrow from inside the cell) may activate the same first inactive protease into an active form. As an alternative, these initiators may also act at different places before the cascade or within the cascade. Each activated protease in turn activates another one until the nth one. The last one may directly initiate intracellular processes or indirectly, via surface membrane alterations like the increased agglutinability. The increased agglutinability could have been triggered by earlier steps in the cascade, and any of them may be capable of acting on proteins or cells in the immediate environment of this particular surface membrane.

model may have heuristic value: It would permit a subtle control at many points in such a multistep process, certainly an important requirement for such a vital phenomenon as cell growth, known to be subject to a multitude of conditions and controlling factors.

Besides stimulatory effects on growth mediated via surface alterations, inhibitory effects have been observed as well. Addition of an agglutinin preparation that was modified to a nonagglutinating but inhibitory form, i.e., presumably a univalent agglutinin preparation, resulted in a cessation of growth of transformed cells at the confluent or monolayer stage (Burger, 1971). Since regular growth of transformed cells prior to confluency was not impaired and since arrested cells returned to their typical transformed growth pattern after removal of the agglutinin cover from the cell, a toxic effect of the agglutinin was not very likely. Before the exact biochemical structure of the presumably univalent preparation is established, an interpretation of this observation that, incidentally, could be repeated on many transformed cell lines, remains unfortunately open to criticism.

Several different mechanisms have to be considered as explanations for this surface-dependent alteration in growth control. It is possible but, in view of some recent unpublished results, unlikely that the agglutinin absorbed to the transformed cell surface could interfere with the uptake of small molecular growth-promoting nutrients. However, we have to keep in mind that the agglutinin could also alter the surface viscosity, zeta potential, flexibility of the surface, or adhesion to the substratum and neighboring cells and that any of these alterations are

sufficient to immobilize the majority of the cells if they become crowded (Burger, 1973). On the other hand, the surface alteration caused by the adhesion of the agglutinin to the surface membrane and to some specific receptors may lead to intracellular alterations that in turn may stop growth. Whether this chain of events is propagated via a signal in the form of a specific or unspecific molecular mediator or by other means remains to be seen.

Possible communications between the surface and the interior of a cell

It is evident from the types of experiments already described where manipulations on the cell surface can lead to growth responses that some sort of communication must exist between outside and inside.

In addition to these experimental and therefore artificial situations where cells are incubated with enzymes, similar communications may be occurring in natural situations. It is obvious that wherever a cell switches from a state A to a state B, whereby the nuclear apparatus is involved, and this alteration is dependent on or induced by alterations that occur in the cell's environment, some communication must exist between the surface and nucleus of the cell. Differentiation may be an example for this situation as well as hormone action in fully differentiated cells. Growth control of whole cell populations (density-dependent inhibition of growth) and also perhaps control of the rate of cell division in a single cell are other examples.

The rate of DNA synthesis in a given cell may be controlled by nuclear or chromosomal events, but the decision of a cell to replicate its chromosomes and eventually the whole cell will depend—particularly in a multicellular organism—on events and conditions outside the cell itself. If some part of the process regulating cell growth involves the cell surface, events external to the cell could be registered immediately and influence division control. Furthermore, since cell division is not completed when the nuclear material is divided into two parts but only when the cytoplasm has been separated into two parts by the plasma membrane, because otherwise polynucleoidy will result, the surface membrane may be an ideal location for such a control system.

We have recently found that during a certain phase of the cell cycle, normal cells undergo alterations in the surface membrane similar to those found all through the cell cycle in transformed cells (see Figure 7). In other words, not only transformed cells are agglutinable with several different agglutinins but normal fibroblasts also bind fluorescein-labeled agglutinins during mitosis and for a short while thereafter during the early G-1 phase (Fox et al., 1971). This observation was recently confirmed by Shoham and Sachs (1972), for another agglutinin. Noonan has also shown that fluorescein-labeled agglutinin binding is hardly an artifact of fluorescence detection, because he found pure preparations of mitotic cells to agglutinate as well as transformed cells at other stages of the cell cycle. Furthermore, he observed mitotic cells to bind at least as much isotopically labeled lectin as did the transformed cells, i.e., at least five times more than the normal cells (unpublished observation).

These observations led us to speculate (Fox et al., 1971)

on a possible function of this regularly repeating cell surface alteration during a specific point of the cell cycle. Since Jacob, Brenner, and Cuzin (1963) suggested almost 10 years ago that the bacterial cell membrane may be involved in the regulation of cell division, we adopted and amplified their hypothesis for animal cells. We would like to conceive of the animal cell surface, on one hand, as the organelle through which a cell perceives its environment, including directly neighboring cells in tissues; this may thereby influence the sociological behavior (Kalckar, 1965) of a cell as in contact or density-dependent inhibition of growth. On the other hand, we would like to suggest that the surface membrane (and not the nuclear membrane that would presumably be the evolutionary correlate of the bacterial membrane, because both are considered the anchoring point of the chromosome) could also serve the cell during its normal cell division cycle as a reminder that it has passed a certain stage in the cycle and that new sequences in the cycle should be initiated. Specifically, we suggested that the surface alteration during mitosis may trigger the next chromosome replication (S-phase) and that a fully replicated chromosome will trigger the next mitosis after a short G_2 phase. This would result in a complete positive control cycle as seen in Figure 8. Any other surface change (Burger, 1973) occurring during a cell cycle—and more and more are being found—may serve the same purpose. At the present time we cannot assume that this and other surface changes during mitosis and early G_1 have to be the decisive ones, although for various theoretical considerations it would be an ideal point in the cycle from which the surface could regulate internal sequences of the cell cycle.

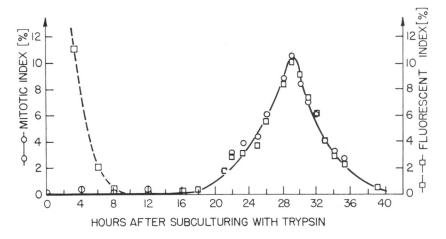

FIGURE 7 Correlation of mitosis and fluorescein-labeled agglutinin binding. Synchronized 3T3 cells were exposed to FITC-agglutinin, fixed with ethanol, stained with Evans blue, and mounted in Elvanol. In control experiments, in which cells were exposed to FITC-agglutinin and counted without fixing and staining, the fluorescent index was identical to the data reported here. Blind counts of several hundred cells were made by two investigators and were in good agreement. (From Fox et al., 1971).

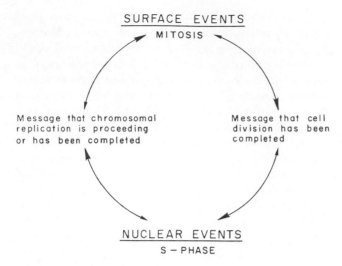

SURFACE EVENTS
MITOSIS

Message that chromosomal
replication is proceeding
or has been completed

Message that cell
division has been
completed

NUCLEAR EVENTS
S – PHASE

FIGURE 8 A possible working model for a positive feedback control from the cell surface on the interior and back to the surface again. Each part of the loop may be mediated by a multiple of intermediate steps. The surface events that are critical may occur in mitosis, or, since the surface change was seen to occur far into G_1, it may be also the early parts of G_1 that might be instrumental. (From Fox et al., 1971.)

Other types of changes were recently found during mitosis. An altered sensitivity of the cell to optimal K^+/Na^+ ratios (Orr et al., 1972) in the medium may reflect changes in surface permeability and in ion pumps, an alteration that may also explain shifts in the level of intracellular cyclic AMP discussed below. It is unlikely that any of these temporary changes would already be sufficient to keep the cell cycle going. Whether any of these changes is necessary, and if so which, however, should not remain a speculative issue for long but can now be tested experimentally.

In reviewing some small molecular candidates as mediators for a communication between the cell surface and the cell's interior, we also have to consider the ubiquitous messenger cyclic AMP as a possibility. Cyclic AMP was found to alter the morphology and growth patterns of some transformed cells to those more typical of untransformed cells (Hsie and Puck, 1971; Johnson et al., 1971). Such observations corroborated earlier ones on the level of the enzyme synthesizing cyclic AMP in transformed and untransformed cells (Bürk, 1968). If added to transformed cultures, it also seemed to mimic the phenomenon of density-dependent inhibition of growth seen usually only in untransformed cells (Sheppard, 1971). Finally when transformed cells were found to have lower cyclic AMP levels as compared with untransformed cells—even though certain discrepancies exist that will require further attention—we began to consider a transient drop in cyclic AMP levels as a continuous growth stimulation (Burger et al., 1972b). If our

assumption is correct, it does lead to a series of predictions, some of which have been tested in the meantime:

1. The cyclic AMP level should be low in mitotic and early G_1 cells. It did indeed turn out to be low during that same interval during which the surface had the agglutinable conformation, i.e., mitosis (Burger et al., 1972b).

2. In all situations that lead to growth stimulation cyclic AMP levels should drop. Such a drop could so far be detected in resting fibroblasts after administration of the growth promoters insulin and proteases (Sheppard, 1972; Burger et al., 1972b).

3. If, as suggested, the surface alteration in mitotic and early G_1 cells, via a lowering of cyclic AMP levels, leads to an initiation of the next S-phase (DNA synthesis), then the majority of cells should stop growing after addition of high doses of cyclic AMP or dibutyryl cyclic AMP at the G_1 stage, i.e., just before the S-phase.

4. Prevention of the drop in cyclic AMP that follows growth stimulation should abolish the subsequent rounds of DNA synthesis and cell division. Administration of the cyclic nucleotide together with proteases (Burger et al., 1972b; see Figure 9) or serum inhibits the usual response (Rozengurt and Pardee, personal communication). The possible causes for the drop in cyclic AMP have been discussed (Burger et al., 1972b) as well as the target system

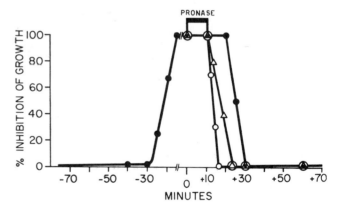

FIGURE 9 Inhibition by dibutyryl cyclic AMP of growth stimulated by protease: Duration of susceptibility. At time 0, the medium was removed from 3.5 cm dishes with confluent 3T3 fibroblasts and replaced for 10 min at 37° C with 10 μg pronase /ml in phosphate buffered saline. At the times indicated on the abscissa, various concentrations of dibutyryl cyclic AMP in phosphate buffered saline were added for 10 min and replaced by conditioned medium. For example, at – 20 min the cells were kept 10 min in dibutyryl cyclic AMP, then for 10 min without dibutyryl cyclic AMP, before the addition of pronase; while at + 20 min, the cells were treated for 10 min with pronase (at time 0), left in pronase-free medium for 10 min and then incubated with dibutyryl cyclic AMP for another 10 min before they were covered again with conditioned medium. ● : 5 × 10^{-4} M dibutyryl cyclic AMP; △ : 5 × 10^{-5} M dibutyryl cyclic AMP; ○ : 10^{-5} M dibutyryl cyclic AMP. (From Burger et al., 1972.)

inside the cell, and both are at present still open to speculation. While these results so far support the significance of a cyclic AMP drop, some caution will be required against an overly optimistic view. Normal fibroblasts, for example, can also be inhibited in their growth with the exception that higher doses of dibutyryl or cyclic AMP are required for inhibition than for that of transformed cells, and transformed cells do not in every respect return to normal fibroblast behavior after administration of the cyclic nucleotide. Administration of cyclic AMP seems to have a mechanism of action entirely different from that of dibutyryl cyclic AMP. Furthermore, not every drop in cyclic AMP leads to DNA synthesis and cell division. Normal cells in their last round of cell division, just before they enter the resting phase in confluency, seem still to go through some type of mitotic surface alteration, although a somewhat reduced one; they still lower their cyclic AMP level, but not, however, as pronounced as exponentially growing cells.

Be that as it may, cyclic AMP has to be considered a possible mediator in relaying surface alterations to the cell's interior. Besides cyclic AMP's classical function as a second messenger after hormone administration where the level of cyclic AMP increases, we may now have to consider also that a decrease in cyclic AMP may be significant in mitotic and G_1 cells and perhaps also in transformed cells throughout the cell cycle.

One will have to keep an open mind on other possible means of communication between the cell surface and its interior. Other small molecules will have to be considered, for example, other nucleotides or ions like Na^+ or Ca^{++}. Peptides as well as macromolecules will certainly have to be included in a search for possible messenger molecules.

The propagation of conformational alterations from the surface membrane throughout the endoplasmatic reticulum belong, perhaps, more to the utopistic kind of possibilities, because permanent continuities between surface and endoplasmic reticulum membranes have not yet been established. A change in the surface permeability is a viable alternative that could provide means by which the surface could influence internal processes (Pardee, 1964). More trivial possibilities will also have to be considered as well. Membrane alterations may be manifested in changes of the plasticity and flexibility of the surface membrane, which in turn may be consequences of alterations in the mobility of the lipid phase or the glycoproteins within the membrane (Burger, 1973). Such modifications may not act directly on the interior of the cell via a messenger molecule but simply have their effects on the degree to which a cell can be deformed and adjust to physical alterations in the environment or perhaps only the overall mobility of the cell. Whether or not a cell is roving around can, for example, easily influence metabolic parameters inside the cell, even if only in response to milieu alterations in the new surrounding into which the cell is moving.

Alterations in the enzymatic activities of the cell surface, or the turnover of the membrane or components thereof, may directly affect intracellular substrate and product pools or indirectly affect cell-cell interactions and thus intracellular responses to the altered cellular sociology. The above list is doubtless incomplete and in the near future will have to be reassessed and probably amplified. Even now, however, it offers quite a wide spectrum of entirely different mechanisms, many of which are amenable to tests and many of which no doubt will have to be rejected eventually.

While biological assays for testing functions ascribed to membranes are slowly improving and biochemical and biophysical techniques for isolating and analyzing the structure of membranes become available at an ever increasing rate, in the near future we should be in a position to analyze certain membrane functions including biological specificity on a more satisfactory molecular level. Progress in this area will undoubtedly be of benefit to neurobiology, because the neural membrane is not only a carrier of the excitation wave but perhaps also a carrier of information for setting up the intricate neural wiring diagram during the formation of the adult brain.

ACKNOWLEDGMENTS   The work of the author's group was carried out at the Department of Biochemical Sciences, Princeton University and at the Marine Biological Laboratory, Woods Hole. It was supported primarily by the National Cancer Institute (Grant CA 10151 and Contract 712372 with the SVCP).

REFERENCES

ANDERSON, E., and E. HUEBNER, 1968. Development of the oocyte and its accessory cells of the polychaete *Diopatra coprea* (Bosc.). *J. Morph.* 126:163–198.

BURGER, M. M., 1970. Proteolytic enzymes initiating cell division and escape from contact inhibition of growth. *Nature (Lond.)* 227:170–171.

BURGER, M. M., 1971. Cell surfaces in neoplastic transformation. In *Current Topics in Cellular Regulation*, Vol. 3, B. Horecker and E. Stadtman, eds. New York: Academic Press, pp. 125–193.

BURGER, M. M., 1973. Surface changes in transformed cells detected by lectins. *Fed. Proc.* (in press).

BURGER, M. M., BOMBIK, B., and NOONAN, K., 1972a. Cell surface alterations in transformed tissue culture cells and their possible significance in growth control. *J. Invest. Dermat.* 59: 24–26.

BURGER, M. M., B. M. BOMBIK, B. McL. BRECKENBRIDGE, and J. R. SHEPPARD, 1972b. Growth control and cyclic alterations of cyclic AMP in the cell cycle. *Nature New Biol.* 239:161–163.

BÜRK, R. R., 1968. Reduced adenyl cyclase activity in a poyoma virus transformed cell line. *Nature (Lond.)* 219:1272–1275.

CECCHARINI, C., and H. EAGLE, 1971. pH as a determinant of cellular growth and contact inhibition. *Proc. Nat. Acad. Sci. USA* 68:229–233.

CRANDALL, M. A., and T. D. BROCK, 1968. Molecular aspects of specific cell contact. *Science* 161:413–415.

CURTIS, A. S. G., and G. VAN DE VYVER, 1971. The control of cell adhesion in a morphogenetic system. *J. Embryol Exp. Morphol.* 26:295–312.

FOX, T. O., J. R. SHEPPARD, and M. M. BURGER, 1971. Cyclic membrane changes in animal cells: Transformed cells permanently display a surface architecture detected in normal cells only during mitosis. *Proc. Nat. Acad. Sci. USA* 68:244–247.

HAUSCHKA, S. D., and I. R. KONIGSBERG, 1966. The influence of collagen on the development of muscle clones. *Proc. Nat. Acad. Sci. USA* 55:119–126.

HOLLEY, R. W., and J. A. KIERNAN, 1968. "Contact inhibition" of cell division in 3T3 cells. *Proc. Nat. Acad. Sci. USA* 60:300–304.

HSIE, A. W., and T. T. PUCK, 1971. Morphological transformation of Chinese hamster cells by dibutyryl adenosine cyclic 3′:5′-monophosphate and testosterone. *Proc. Nat. Acad. Sci. USA* 68:358–361.

HUMPHREYS, T., 1963. Chemical dissolution and in vitro reconstruction of sponge cell adhesions: I Isolation and functional demonstration of the components involved. *Dev. Biol.* 8:27–47.

HUMPHREYS, T., 1965. Cell surface components participating in aggregation: Evidence for a new cell particulate. *Exp. Cell Res.* 40:539–543.

JACOB, F., S. BRENNER, and F. CUZIN, 1963. On the regulation of DNA replication in bacteria. *Cold Spring Harbor Symp. Quant. Biol.* 28:329–348.

JOHNSON, G. S., R. M. FRIEDMAN, and I. PASTAN, 1971. Restoration of several morphological characteristics of normal fibroblasts in sarcoma cells treated with adenosine -3′, 5′-cyclic monophosphate and its derivatives. *Proc. Nat. Acad. Sci. USA* 68:425–429.

KALCKAR, H. M., 1965. Galactose metabolism and cell "sociology." *Science* 150:305–313.

KONJIN, T. M., J. G. C. VAN DE MEENE, J. T. BONNER, and D. S. BARKLEY, 1967. The acrasin activity of adenosine-3′:5′-cyclic phosphate. *Proc. Nat. Acad. Sci. USA* 58:1152–1154.

LEMON, S. M., R. RADIUS, and M. M. BURGER, 1971. Sponge aggregation. I: Are carbohydrates involved? *Biol. Bull.* 141:380.

LILIEN, J. E., 1969. Toward a molecular explanation for specific cell adhesion. In *Current Topics in Developmental Biology*, Vol. 4. A. A. Moscona and A. Monroy, eds. New York: Academic Press, pp. 169–195.

METZ, C., and A. MONROY, eds., 1969. *Fertilization*. New York: Academic Press.

MITCHISON, N. A., 1971. Control of immune response by events at the lymphocyte surface. *In Vitro* 7:88–94.

MOSCONA, A. A., 1963. Studies on cell aggregation: Demonstration of materials with selective cell binding activity. *Proc. Nat. Acad. Sci. USA* 49:142–147.

MOSCONA, A. A., 1968. Cell aggregation: Properties of specific cell-ligands and their role in the formation of multicellular systems. *Dev. Biol.* 18:250–277.

MOSCONA, A. A., and M. H. MOSCONA, 1952. The dissociation and aggregation of cells from organ rudiments of the early chick embryo. *J. Anat.* 86:287–301.

NOSSAL, G. J. V., and G. I. ADA, 1971. *Antigens, Lymph Node Cells and the Immune Response*. New York: Academic Press.

ORR, C. W., M. YOSHIKAWA-FUKADA, and J. D. EBERT, 1972. Potassium: Effect on DNA synthesis and multiplication of baby-hamster kidney cells. *Proc. Nat. Acad. Sci. USA* 69:243–247.

PARDEE, A. B., 1964. Cell division and a hypothesis of cancer. *Natl. Cancer Inst. Monogr.* 14:7–18.

SEFTON, B. M., and H. RUBIN, 1970. Release from density dependent growth inhibition by proteolytic enzymes. *Nature (Lond.)* 227:843–845.

SHEPPARD, J. R., 1971. Restoration of contact-inhibited growth to transformed cells by dibutyryl adenosine 3′:5′-cyclic monophosphate. *Proc. Nat. Acad. Sci. USA* 68:1316–1320.

SHEPPARD, J. R., 1972. Difference in the cyclic adenosine 3′, 5′-monophosphate levels in normal and transformed cells. *Nature New Biol.* 236:14–16.

SHOHAM, J., and L. SACHS, 1972. Differences in the binding of fluorescent concanavalin A to the surface membrane of normal and transformed cells. *Proc. Nat. Acad. Sci. USA* 69:2479–2482.

SIDMAN, R. L., 1971. Cell interactions in developing mammalian nervous system. In *Third Lepetit Colloquium on Cell Interactions*, L. Silvestri, ed. Amsterdam: North-Holland, pp. 1–13.

STEINBERG, M. S., 1970. Does differential adhesion govern self assembly processes in histogenesis? Equilibrium configurations and the emergence of a hierarchy among populations of embryonic cells. *J. Exptl. Zool.* 173:395–434.

STOKER, M. G. P., and H. RUBIN, 1967. Density dependent inhibition of cell growth in culture. *Nature (Lond.)* 215:171–172.

TEMIN, H. M., 1967. Studies on carcinogenesis by avian sarcoma viruses VI. Differential multiplication of uninfected and of converted cells in response to insulin. *J. Cell. Physiol.* 69:377–384.

TODARO, G. J., J. MATSUIA, S. BLOOM, A. ROBBINS, and H. GREEN, 1967. Stimulation of RNA synthesis and cell division in resting cell by a factor present in serum. In *Growth-Regulating Substances for Animal Cells in Culture*, Symposium Monograph No. 7, U. Defendi and M. Stoker, eds. Philadelphia: The Wislar Institute Press, pp. 87–98.

TOWNES, P. L., and J. HOLTFRETER, 1955. Directed movements and selective adhesion of embryonic amphibian cells. *J. Exptl. Zool.* 128:53–120.

VAHERI, A., E. RUOSLAHTI, and S. NORDLING, 1972. Neuraminidase stimulates division and sugar uptake in density-inhibited cell cultures. *Nature New Biol.* 238:211–212.

VASILIEV, J. M., I. M. GELFAND, and V. I. GUELSTEIN, 1971. Initiation of DNA synthesis in cell cultures by colcemid. *Proc. Nat. Acad. Sci. USA* 68:977–979.

VASILIEV, J. M., I. M. GELFAND, V. I. GUELSTEIN, and E. K. FETISOVA, 1970. Stimulation of DNA synthesis in cultures of mouse embryo fibroblast-like cells. *J. Cell. Physiol.* 75:305–314.

WEINBAUM, G., and M. M. BURGER, 1971. Sponge aggregation. III: Isolation of a surface component required in addition to the aggregation factor. *Biol. Bull.* 141:406.

WIESE, L., 1965. On sexual agglutination and mating type substances (gamones) in isogamous heterothallic chlamydomonads. I. Evidence of the identity of the gamones with the surface components responsible for sexual flagellar contact. *J. Phycol.* 1:46–54.

WIESE, L., and D. W. SHOEMAKER, 1970. On sexual agglutination and mating type substances (gamones) in isogamous heterothallic chlamydomonads. II. The effect of concanavalin A upon the mating-type reaction. *Biol. Bull.* 138:88–95.

WILSON, H. V., 1907. On some phenomena of coalescence and regeneration in sponges. *J. Exptl. Zool.* 5:245–258.

69 Specificity and Mechanism at the Lymphoid Cell Surface

GERALD M. EDELMAN

ABSTRACT   Lymphoid cells carry receptors for antigens that are members of the immunoglobulin (Ig) family as well as glycoprotein receptors for certain plant lectins. Interaction of antigens with the Ig receptors and of certain lectins with glycoprotein receptors leads to a triggering of cell maturation, mitosis, and antibody production. Because the structure of Ig molecules has been extensively studied, these cells provide excellent models for analysis of membrane-nuclear interactions, cell-cell interactions, and mitogenesis. Extensive studies have also been made of the three-dimensional structure of the lectin concanavalin A. By the combined use of this mitogenic lectin and antibodies to Ig receptors, the surface traffic and interaction of various receptors can be mapped. The results lend support to the concepts of receptor mobility and receptor-receptor interaction.

In the course of this work, a method has been devised for the fractionation of cells on derivatized surfaces according to the specificity of their surface receptors. This method has been successfully used to fractionate lymphoid cells according to the specificity and affinity of their surface immunoglobulins. A similar approach may prove to be valuable in the fixation, fractionation, and biochemical analysis of neuronal tissues.

LYMPHOCYTES are the major cells mediating immune responses. According to the clonal selection theory of immunity (Burnet, 1959; Jerne, 1955), each immuno-competent lymphocyte carries antibody receptors on its cell surface and, after interaction with the appropriate antibody, matures and divides to produce more of that antibody. The overall phenomenological features of this theory have been substantiated, but two problems remain: (1) What is the genetic origin of the structural diversity of the receptor antibodies, and how does each lymphocyte become committed to just one of the many possible antibodies? (2) How is the lymphocyte triggered after interaction with the antigen?

The first problem has been considered elsewhere (Gally and Edelman, 1970) as well as in previous Intensive Study Programs (Edelman, 1970). In this chapter, attention is given to certain aspects of the second problem. The subjects considered here are the nature of the receptors for antigen, their distribution, mobility, and traffic in the lymphocyte membrane, the nature of mitogenesis induced

by plant lectins and the fractionation of lymphocytes according to the specificity of their receptors. An analysis of lymphocyte triggering bears upon several central concerns of neurobiology, including receptor specificity, cell-cell interaction, the nature of the cell membrane, and communication between the cell surface and intracellular regulatory machinery.

Clonal selection, lymphoid cells, and immunoglobulin receptors

Lymphoid cells derive from stem cells in the bone marrow (Nossal, 1967) and undergo two distinct types of maturation: One antigen-independent, the other requiring antigen (Figure 1). Maturation in the antigen-independent stage occurs within two central lymphoid organs, the thymus and the bursa of Fabricius, which is located near the cloaca in birds. Cells maturing in the thymus are released to the circulation, the peripheral lymph nodes, and the spleen as T cells (thymus-dependent lymphocytes). Those maturing in the bursa are released to become B cells (or bone-marrow-derived lymphocytes, for except in birds the bursal equivalent has not yet been found).

The T cells and B cells differ in a number of important respects. Those that concern us here are (Gowans et al., 1971): (1) The B cells secrete humoral antibodies of the various immunoglobulin (Ig) classes whereas T cells do not secrete humoral antibodies. There is evidence for Ig receptors on both types of cells. (2) The T cells contain unique surface markers and antigens such as the θ antigen and are the precursors for cells carrying out cell-mediated immunity. (3) For a great variety of antigens there is evidence that effective stimulation of B cells must be preceded by interaction of antigen with T cells; i.e., cell cooperation is required. The exact nature of the cooperation has not yet been determined, but there is evidence to suggest that it proceeds in a series of steps: Antigen may first react with T cell receptors. The complex of T cell receptor and antigen may then be passed to a macrophage-like cell that carries it to B cells having Ig receptors capable of interacting with the antigen in the complex. The effect

GERALD M. EDELMAN   The Rockefeller University, New York

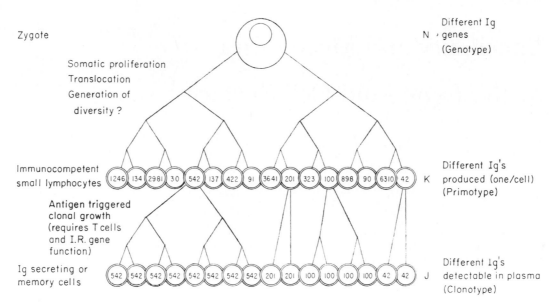

Zygote   Different Ig
                                                                       N , genes
                                                                          (Genotype)

Somatic proliferation
Translocation
Generation of
    diversity ?

Immunocompetent   Different Ig's
small lymphocyte   ⓒ⓪⓪⓪⓪⓪⓪⓪⓪⓪⓪⓪⓪⓪⓪⓪ K produced (one/cell)
                    1246 134 2981 30 542 137 422 91 3641 201 323 100 898 90 6310 42 (Primotype)

Antigen triggered
clonal growth
(requires T cells
and I.R. gene
function)

Ig secreting or   Different Ig's
memory cells   ⓪⓪⓪⓪⓪⓪⓪⓪⓪⓪⓪⓪⓪⓪⓪⓪ J detectable in plasma
                    542 542 542 542 542 542 542 542 201 201 100 100 100 100 42 42 (Clonotype)

FIGURE 1   A model of the somatic differentiation of antibody-producing cells according to the clonal selection theory. The number of Ig genes, which equals N in the zygote, may increase during somatic growth so that in the immunologically mature animal K different cells are formed, each committed to the synthesis of a structurally distinct Ig (indicated by the arabic number). A small proportion of these cells proliferate upon antigenic stimulation to form J different clones of cells, each clone producing a different antibody.

of T-cell–B-cell cooperation is to amplify the response of those B cells carrying the appropriate antigen-binding receptors.

Interaction of the antigen with T cells or with B cells via the T cells leads to the second or antigen-dependent stage of maturation (Figure 1), in which there is clonal proliferation of the stimulated cells. In the case of B cells, maturation results in cells containing rich endoplasmic reticula and, finally, in plasma cells that secrete humoral antibodies.

A number of questions bearing upon the relationship of the receptor to the cell membrane remain to be answered: (1) How does the interaction with the antigen or the T cell complex stimulate mitosis and differentiation? (2) What distinguishes triggering to produce antibodies from the specific induction of immune tolerance or the specific failure to respond to a given antigen? A number of difficulties hinder the direct investigation of the mechanism of antigenic stimulation. These include the relatively small proportion of T or B lymphocytes that bind a particular antigen, the complex mode of transfer of antigen to B cells, the cryptic nature of T cell Ig receptors, and the complexity of immune tolerance, which takes place in two dosage zones of antigen (Gowans et al., 1971).

Tolerance itself will not be considered here. Instead, it might be profitable to consider ways of directly stimulating lymphoid cells, of characterizing their receptors, and of fractionating them according to the specificity of their receptors. Fortunately, lymphocytes may be stimulated by a number of plant proteins or lectins that bind to glycoprotein receptors on the lymphocyte membrane and act as mitogens. A large number of cells may be stimulated by these means and the lectins, therefore, may be used as probes to investigate various stages of the response, for different lectins have different effects on T and B cells.

Properties and effects of lectins

Lectins (Boyd, 1963) are a diverse family of proteins isolated from plants, mainly leguminosae. They have the property of binding to carbohydrates or glycoproteins, agglutinating animal cells, and in some cases, stimulating mitosis, maturation, and immunoglobulin synthesis by lymphocytes (Powell and Leon, 1970). For these reasons, lectins have attracted attention as agents to map cell surfaces and to modulate cell behavior via membrane receptors. If their effects on cell surfaces are to be analyzed, it is important to know the structure of lectins in some detail. For this reason, we have been engaged in isolating a large number of lectins and characterizing their molecular properties and functions (Wang et al., 1971; Becker et al., 1971). Two lectins that have distinct effects on lymphocytes are discussed here: Concanavalin A and pokeweed mitogen. It should be noted, however, that various lectins can bind to a variety of cells including neurons (Edelman and Millette, 1971).

Concanavalin A (Con A) is a protein with a protomer molecular weight of 27,000 that exists largely as a tetramer at pH 7 or higher. Con A can bind to a number of glyco-proteins on the surface of a variety of cells, and the number of receptors varies between 10^6–10^7 per cell (Edelman and Millette, 1971). The interaction can be specifically inhibited by alpha-methyl-glucoside or alpha-methyl-mannoside (Goldstein et al., 1965). An analysis of the amino acid sequence and three-dimensional structure of this molecule at atomic resolution has recently been completed (Becker et al., 1971; Edelman et al., 1972). As shown in Figure 2, the tetramer has dimensions of 80 ×

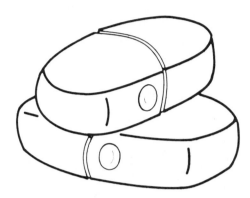

FIGURE 2 Schematic drawing showing positions of the binding sites on the Con A tetramer. The tetramer consists of two ellipsoidal dimers joined along a common diad axis at a point of D_2 symmetry. Each protomer has a single binding site, indicated by a *small circle*. Two of the sites are hidden in equivalent positions on the rear of the molecule.

46 × 27 Å, and the binding sites for phenyl sugars are located near the boundary of interaction between protomers. Each protomer contains one such site, and the molecule is therefore tetravalent.

A second lectin, pokeweed mitogen, has a very different structure although much less is known in detail about it. Despite the fact that it has not yet been subjected to crystallographic analysis, it is known that it must be very different from Con A: The polypeptide chain has a molecular weight of 30,000, but it has about 28 disulfide bonds in contrast to concanavalin A which has none (Reisfeld et al., 1967; Waxdal and Edelman, unpublished results). These bonds severely constrain the way in which the pokeweed mitogen chain folds. Alpha-methyl-manno-side does not bind to pokeweed mitogen, the sugar specificity of which is unknown. There is also a difference between the two lectins in their capacity to stimulate lymphocytes. Pokeweed mitogen stimulates B and T lymphocytes, whereas in its unaltered form, Con A stimulates only T cells (Figure 3) despite the fact that T and B cells contain the same number of Con A receptors (Janossy and Greaves, 1971; Andersson et al., 1972).

What is responsible for the differences of action among

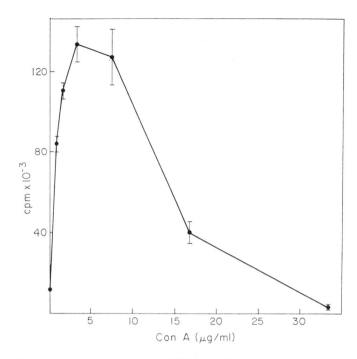

FIGURE 3 Incorporation of ^{3}H-thymidine into mouse lymphocytes in the presence of Con A. Splenic lymphocytes (3 × 10^6) were incubated in 0.3 ml of tissue culture media (Mishell and Dutton, 1967) containing various concentrations of Con A. After 48 hr 0.1 ml of media containing 1 μc of ^{3}H-thymidine was added to each tube and the cells incubated an additional 24 hr. Cells were isolated on glass fiber filters (Whatman GFA) and washed with PBS, 5% TCA, and methanol. The dried filters were placed in 15 ml of scintillation fluid (4 g of Omnifluor in 1 liter of toluene) and radioactivity measured in a Packard Tri-Carb Scintillation Counter. The vertical bars denote variations in the results obtained at any given concentration of Con A.

the various mitogens, and how does the binding of these lectins lead to mitogenesis? Although the answers to these questions have not been obtained, there are a number of observations that suggest possible mechanisms. These include the effect of Con A on Ig receptors and the alteration of the stimulation pattern of Con A on B cells after chemical concentration or cross-linkage of the lectin.

Influence of Con A binding on Ig receptors

Recently, it has been shown (Taylor et al., 1971) that Ig receptors (presumably on B cells) are mobile in or on the lymphocyte membrane and that the addition of anti-immunoglobulin cross-links them resulting in patch formation by these receptors followed by cap formation. Cap formation presumably occurs by coalescence of patches via a metabolically controlled process, for cap formation is inhibited by sodium azide (NaN_3). Obviously, it is important to know the significance of these

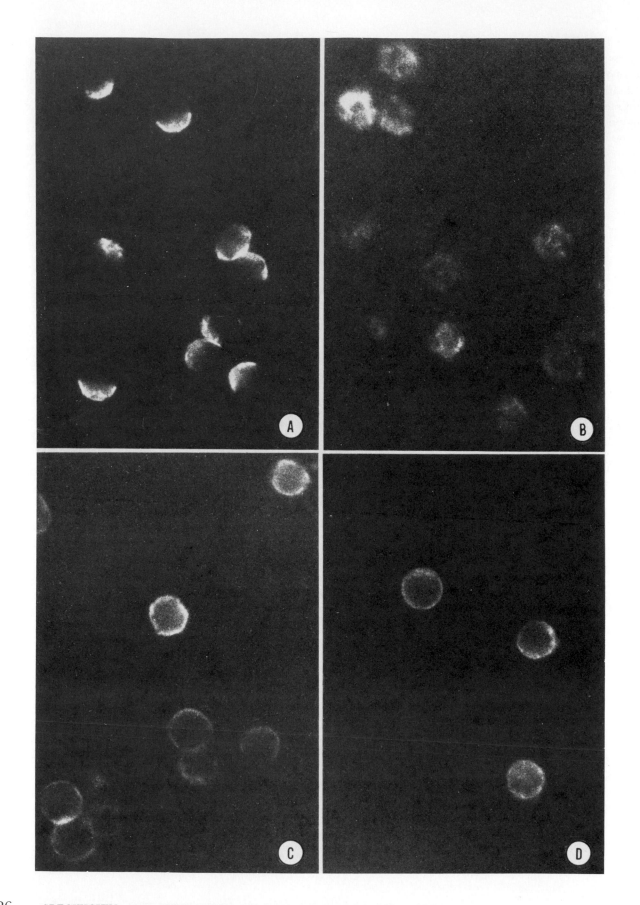

SPECIFICITY AND MECHANISM AT THE LYMPHOID CELL SURFACE

phenomena for immune induction and tolerance. Recent experiments in our laboratory (Yahara and Edelman, 1972) provide further information on the mechanism of patch and cap formation and the mobility of the immunoglobulin receptors of mouse lymphocytes.

Con A, which is itself mitogenic for lymphocytes, was found to inhibit the formation of patches and caps that would otherwise form after addition of antibodies to immunoglobulin receptors (Figure 4). The inhibition by Con A of cap formation and patch formation by immunoglobulin receptors was concentration dependent and reversible (Figure 5). Moreover, Con A binding did not appreciably diminish the number of antibodies bound to immunoglobulin receptors. These data suggest that the lymphocyte receptors fall into at least two classes: Mobile Ig receptors and less mobile lectin receptors. Apparently, binding of Con A restricts the mobility of the Ig receptors that remain distributed in a diffuse state.

A proposed mechanism for patch and cap formation and the Con A effect is shown in Figure 6. According to this model, addition of anti-Ig leads to three successive processes: (a) binding of divalent antibodies to Ig molecules on the cell surface, (b) formation of patches of antibody-Ig receptor complexes, and (c) cap formation over one pole of the cell. Processes (a) and (b) are not inhibited by NaN_3, because binding was found to be unaffected by this reagent, and patches can be observed in its presence; NaN_3, therefore, inhibits process (c). The time course of cap formation after the removal of NaN_3 from the medium indicated that cap formation usually occurs within at most 10 min after forming patches at 21° C (Figure 5). Cap formation therefore appears to be dependent on cell metabolism and motility (Taylor et al., 1971) and possibly requires a minimal concentration of ATP in the cell.

In contrast, patch formation appears to result from diffusional motion of Ig receptors, which form larger aggregates after binding with divalent antibody. The resultant patches then do not diffuse at appreciable rates. A calculation using the known dimensions of Con A (Becker et al., 1971) and the location of its binding sites (Figure 2) as well as the results of titration (Edelman and Millete, 1971) of the number of receptors on splenic lymphocytes (1.4×10^6/cell) indicates that, at 10% saturation, Con A occupies no more than 1% of the cell surface in a diffuse distribution. Inasmuch as the Ig receptors (approximately 4×10^4/cell) also occupy no more than 1% of the surface, binding of Con A cannot

directly block the diffuse movement of Ig receptors. Several possible mechanisms remain: Binding may lead to a change in the cell surface or the cell membrane, resulting in alteration of either the anchorage or the free path of Ig receptors. One possibility, currently under test, is that the Con A interacts with the carbohydrate of the Ig receptor thus fixing it to the less mobile Con A receptor site. Alternatively, the change could be aggregation of intramembranous protein particles linked to Con A receptors, either with formation of a tortuous path for motion of Ig receptors or, more likely, the binding of Ig receptors to particle-associated structures. Finally, the binding of Con A to cell surface glycoproteins may result in secondary interactions of the Con A with the membrane and a change in membrane fluidity.

Enhancement of stimulation by cross-linkage of Con A at surfaces

It has been shown that free Con A can bind equally well to both T and B cells and that these cells have the same number of Con A receptors (Greaves et al., 1972). Nevertheless, addition of Con A to B cells or mixtures of T and B cells results in a proliferation of T cells only (Andersson et al., 1972). One tenable hypothesis might be that the threshold for triggering B cells is higher and that this requires a higher concentration of Con A at the cell surface.

An experiment to test this hypothesis has recently been carried out (Andersson et al., 1972). Con A was chemically cross-linked at the bottom of plastic tissue culture dishes by the use of a water-soluble carbodiimide (Edelman et al., 1971). It was confirmed that T cells but not B cells were stimulated by soluble Con A. In contrast, the locally concentrated Con A was found to stimulate DNA synthesis in B cells but not in T cells. A similar finding has been made by Greaves and Bauminger (1972), who showed that phytohemagglutinin, a lectin that stimulates only T cells, would stimulate B cells when bound to beads of Sepharose.

The various observations on the cellular effects of Con A suggest an hypothesis for cell triggering by lectins as well as antigens. Before considering this hypothesis, however, it is important to estimate the threshold of antigen binding required to trigger lymphocytes. In order to make this estimate, the cells must be fractionated in terms of their receptor specificity. Methods to accomplish

FIGURE 4 Labeling patterns of cells with fluorescein-labeled anti-immunoglobulin and with fluorescein-labeled Con A. (A) Cells incubated with fluorescein-labeled anti-immunoglobulin (80 μg/ml) at 21°C for 30 min showing caps. (B) Prior addition of sodium azide (1×10^{-2} M) showing patches. (C) Prior addition of Con A (100 μg/ml) showing diffuse patterns. (D) Patterns after incubation with fluorescein-labeled Con A alone (100 μg/ml) at 21°C for 30 min.

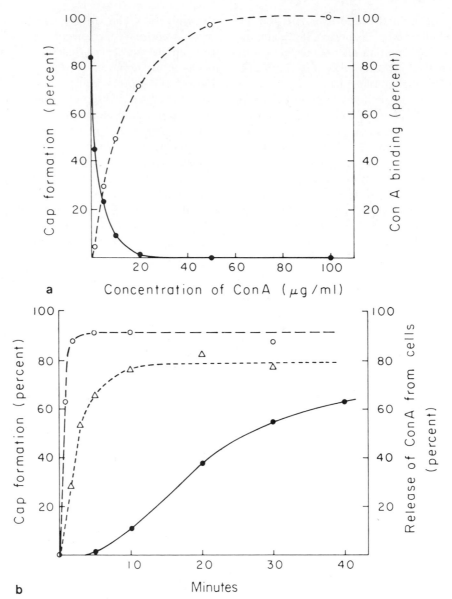

a

Concentration of ConA (μg/ml)

b

Minutes

FIGURE 5   (a) Effect of Con A on cap formation (●—●—●): cells (2 × 10⁷/ml) were incubated with various concentrations of Con A in PBS-BSA at 21°C for 10 min. Fluorescein-labeled anti-immunoglobulin was added to the mixture to give a concentration of 80 μg/ml, and the mixture was incubated for 30 min at 21°C. Cap formation was determined after washing cells. Binding of [¹²⁵I] Con A to cells: (○—○—○); cells (2 × 10⁷/ml) were incubated with various concentrations of [¹²⁵I] Con A (2.4 × 10⁴ cpm/μg) at 21°C for 30 min in PBS-BSA. Cells were washed with PBS-BSA, and the radioactivity was determined. (b) Cap formation after release of Con A from cells (●—●—●): 2 × 10⁷ cells/ml were incubated with Con A (100 μg/ml) at 0°C for 10 min in PBS-BSA, and fluorescein-labeled anti-immunoglobulin (80 μg/ml) was added to the mixture. The mixture was then incubated at 0°C for 30 min and centrifuged. The cells were washed with PBS-BSA and resuspended in the same volume of PBS-BSA at 21°C, to which alpha-methyl-D-mannoside was added to give a concentration of 40 mM. Aliquots were pipetted at various times after the addition of alpha-methyl-D-mannoside, the cells were washed and cap formation was determined. Rate of release of [¹²⁵I] Con A from cells after the addition of alpha-methyl-D-mannoside (○—○—○): 2 × 10⁷ cells/ml were incubated with 100 μg/ml of [¹²⁵I] Con A (2.4 × 10⁴ cpm/μg) at 21°C for 10 min in PBS-BSA. Unlabeled anti-Ig was added to this mixture to give a final concentration of 80 μg/ml. After incubation at 21°C for 30 min, the mixture was centrifuged and the cells were resuspended in PBS-BSA. Aliquots of the mixture were transferred to small tubes containing alpha-methyl-D-mannoside in PBS-BSA (final concentration, 40 mM) and were incubated at 21°C for various times, and then the radioactivity was determined. Cap formation after removal of sodium azide (—△—△—): 2 × 10⁷ cells/ml were incubated with fluorescein-labeled anti-immuno globulin (80 μg/ml) for 30 min at 21°C in PBS-BSA containing sodium azide (3 mM). Cells were collected by centrifugation and resuspended in PBS-BSA at 21°C with a decrease in sodium azide concentration to 30 μM. Cells were washed, and at various times cap formation was determined.

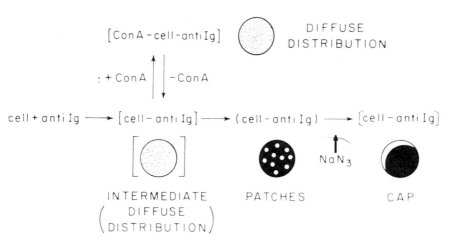

FIGURE 6 A model for patch and cap formation and their inhibition by Con A.

this for lymphocytes have been developed and show promise in the fractionation of other tissues as well.

The specific fractionation of lymphoid cells

Lymphocytes are heterogeneous at a variety of levels, and an analysis of their population dynamics in quantitative terms is essential to a complete understanding of clonal selection. It is therefore important to fractionate the cells

of the immune system according to the specificity and affinity of their receptors. A method of separating cells on nylon fibers that appears to be effective for this purpose has recently been developed (Edelman et al., 1971; Rutishauser et al., 1972). A scheme of the details of the method is shown in Figure 7.

A variety of antigens, including haptens, proteins, and cell membranes can be used to derivatize the fibers (Table I). The total number of bound cells in a dish was determined by counting the cells bound to a short segment of the total fiber length. Individual dishes contained 25 cm of fiber specifically bound from 10^5 to 10^6 cells. Greater binding capacities (up to 5×10^6 cells) were achieved by the use of nylon fiber meshes. As shown in Table II, the

TABLE I

Binding of mouse spleen cells to antigen-derivatized fibers

Fiber Antigen	Number of Cells Bound*	
	Immunized†	Unimmunized
DNP-BSA		
Exp't 1	802	285
2	1004	301
3	654	283
Tosyl-BSA		
Exp't 1	353	143
2	297	130
BSA		
Exp't 1	173	65
2	112	58
Stroma		
Exp't 1	160	75
2	145	70

BSA: bovine serum albumin
DNP: the 2,4-dinitrophenyl group
Tosyl: the p-toluenesulfonyl group
*Expressed as number of cells bound to both edges of a 2.5 cm fiber segment.
†Secondary responses to DNP-BSA, tosyl-BSA, BSA and sheep erythrocytes, respectively. Cells from three mice were pooled.

TABLE II

Specific inhibition of spleen cell binding to derivatized fibers

Fiber Antigen	Immunogen	Inhibitor*			
		DNP	Tosyl	Stroma	Anti-Ig
DNP-BSA	DNP-BGG	91%†	1%	—	93%
DNP-BSA	NONE	73%	2%	—	72%
Tosyl-BSA	Tosyl-BSA	3%	87%	—	90%
Tosyl-BSA	NONE	6%	59%	—	63%
Stroma	Stroma	<5%	<5%	70%	80%
Stroma	NONE	<5%	<5%	50%	45%

Anti-Ig: the gamma globulin fraction of a rabbit antiserum to mouse immunoglobulin
BGG: bovine gamma globulin
BSA: bovine serum albumin
DNP: the 2,4-dinitrophenyl group
Tosyl: the p-toluenesulfonyl group
*DNP_8-BSA and tosyl$_{20}$-BSA at 200 μg/ml; Sonicated stroma and rabbit anti-mouse immunoglobulin at 1 mg/ml.
†Expressed as % inhibition.

GERALD M. EDELMAN   789

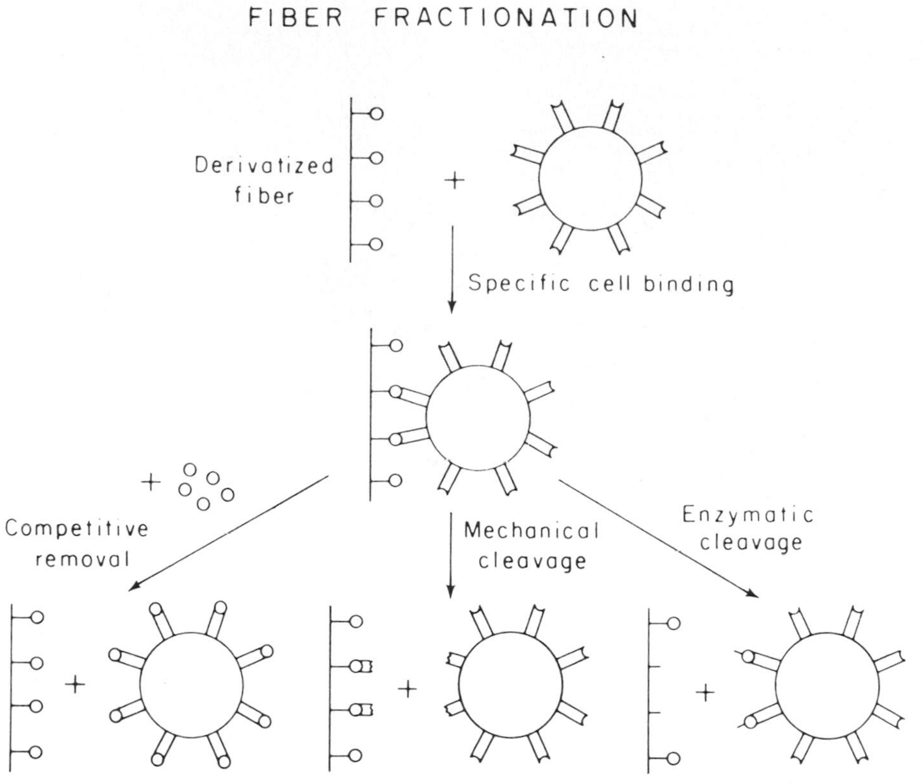

FIGURE 7   General scheme for the fiber fractionation of cells.

number of isolated cells varied with the particular antigen used, the extent to which the fibers were substituted, and the immunological history of the cell donors.

The binding of cells from immunized and unimmunized mice to an antigen-derivatized fiber was specifically inhibited by the addition of soluble antigen to the cell suspension during the incubation period but immunologically distinct molecules did not inhibit binding (Table II). Preincubation of the spleen cells with the immunoglobulin G (IgG) fraction of rabbit antiserum directed against mouse immunoglobulin G prevented the binding of cells to any of the antigen substituted fibers; no inhibition was obtained when normal rabbit IgG was used. These data indicate that the attachment is specific and is mediated by immunoglobulin receptors on the cell surface.

In order to examine the topographical distribution and specificity of the immunoglobulin receptors, spleen cells bound to fibers derivatized with dinitrophenylated bovine serum albumin (DNP-BSA) were tested for their ability to form rosettes *in situ* with trinitrophenyl-coated sheep erythrocytes (Figure 8). About 50% of the fiber-binding cells from immunized mice and 17% from unimmunized mice were capable of forming rosettes (Table III). In contrast, an unfractionated spleen cell population from im-

TABLE III

Formation of DNP-specific rosettes by spleen cells bound to DNP-derivatized fibers

| FBC* | | % Fiber-RFC† Inhibitor of Rosette Formation‡ | | | |
		None	DNP	Tosyl	Anti-Ig
Immunized§	603	54%	6%	49%	2%
Unimmunized	296	17%	5%	18%	1%

Anti-Ig: the gamma globulin fraction of a rabbit antiserum to mouse immunoglobulin
  BGG: bovine gamma globulin
  DNP: the 2,4-dinitrophenyl group
  FBC: fiber-binding cells
  RFC: rosette-forming cells
  Tosyl: the p-toluenesulfonyl group
  *2.5 cm edge count.
  †Expressed as (fiber-RFC/FBC) × 100.
  ‡DNP_8-BSA and $tosyl_{20}$-BSA at 200 μg/ml; rabbit anti-mouse immunoglobulin at 0.5 mg/ml added during rosette formation.
  §Secondary response to DNP-BGG.

mune and nonimmune mice contained 1–5% and 0.2–0.4% rosette-forming cells, respectively, as determined by the centrifugation-resuspension assay. Formation of fiber

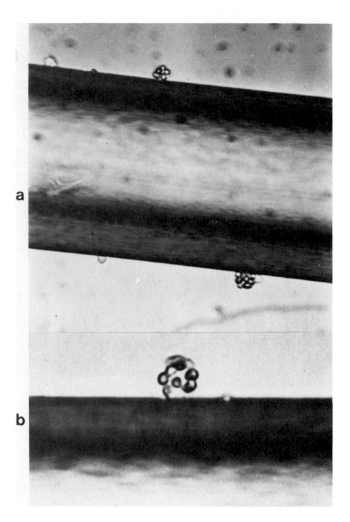

TABLE IV

Depletion of DNP fiber-binding cells and DNP plaque-forming cells

Adsorbed with:	Assay*		
	FBC	Direct PFC/10⁶ cells	Indirect PFC/10⁶ cells
Unabsorbed	481†	25	668
DNP-fiber	64	24	676
Tosyl-fiber	462	24	640
DNP-Sepharose	120	16	396

DNP: the 2,4-dinitrophenyl group
FBC: fiber binding cells
PFC: plaque forming cells
Tosyl: the p-toluenesulfonyl group
*Each assay was performed on cells remaining after the various absorption procedures. Cells were obtained from a pool of mice after secondary immunization.
†Expressed as FBC on the edges of a 2.5 cm fiber segment.

FIGURE 8 Rosettes formed by spleen cells bound to a DNP-BSA fiber after incubation with TNP-derivatized sheep erythrocytes. (a) X400, (b) X1000 magnification.

rosettes could be inhibited by both DNP-BSA and anti-immunoglobulin, but not by tosyl-BSA. Microscopic examination of the rosettes showed that the red cells were evenly distributed over the lymphocyte surface. Moreover, estimation of the number of antibody molecules bound to immunoglobulin receptors on each cell using ¹²⁵I-labeled rabbit anti-mouse immunoglobulin showed that both normal and immune cells contained similar numbers of receptors.

Fiber-binding cells do not include plaque-forming or antibody-secreting cells. A cell population depleted of specific fiber-binding cells was obtained by incubating spleen cells with a large amount of antigen-derivatized fibers and collecting the unbound cells. This process removed less than 5% of the original cells. When the number of DNP-specific, antibody-secreting cells in the unabsorbed and depleted cell populations was compared

by the plaque assay, no difference was observed (Table IV). Similar results were obtained for both IgG and IgM secreting cells from immunized and unimmunized animals. Although the biological role of all of the fiber-binding cells is not completely defined, there is now evidence that they are specific antigen-binding cells, that some of them are capable of reconstituting an immune response, and that they do not include plaque-forming cells.

Lymphoid cells can be fractionated according to their affinity for a fiber, hapten, or antigen by addition of different amounts of free hapten to the cells as they are incubated and shaken with the fibers. Once cells are bound to the fibers, they cannot be removed by addition of soluble antigen (Rutishauser et al., 1972). Under these conditions, binding of a cell with high affinity for antigen to the fiber can be blocked by lower concentrations of free antigen than binding by a cell with low affinity receptors.

In Figure 9 is shown the effect of various antigen concentrations on the number of spleen cells bound to DNP-BSA derivatized fibers. Although the total number of fiber-binding cells detected in single dishes with cell suspensions from immunized mice was only 2 to 4 times that obtained with nonimmune animals, this study revealed that the two cell populations differ greatly in their susceptibility to inhibition by DNP-BSA conjugates. Similar results were obtained using the monovalent antigen, ε-DNP-lysine, as an inhibitor. In the immune animal, 500 fiber-binding cells (edge counts) were inhibited by antigen concentrations of less than 4 μg/ml, whereas the number of fiber cells from the unimmunized animal decreased by less than 20 at these antigen concentrations. At higher concentrations, the curves are identical (Figure 9).

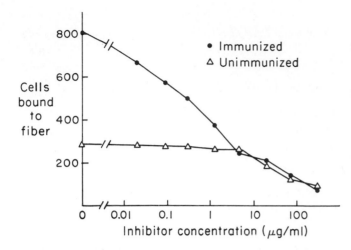

FIGURE 9 Inhibition by free dinitrophenyl-bovine serum albumin of spleen cell binding to dinitrophenyl-bovine serum albumin-derivatized fibers. Cell numbers represent fiber edge counts for a 2.5 cm fiber segment. Spleens from immunized mice were removed at the height of a secondary response to dinitrophenyl-bovine gamma globulin and cells from several mice were pooled.

These differences were not due to variation in the number of receptors per cell as shown by two findings: (1) The number of anti-immunoglobulin molecules bound to fiber-fractionated cells from nonimmunized animals was about the same as that found with cells from immunized mice. Although these results do not account for possible differences in the classes of immunoglobulin receptors, it is evident that immunization does not result in a *large* increase of receptors. (2) The inhibition curves obtained with the monovalent antigen ε-DNP-lysine were similar to those using the multivalent DNP_8-BSA inhibitor. If immunization had resulted in an increase in the number of receptors, the binding of cells from non-immune animals should have been inhibited by lower concentrations of monovalent inhibitor than those from the immunized donor. This was not the case, and the varying concentration of inhibitor required is more likely to reflect differences of binding constants of a given set of receptors for the antigen.

At present, the biological role of fiber-binding cells has not been completely defined. Preliminary experiments indicate, however, that fiber binding cells include precursors of plaque-forming cells, and this fact, together with the data in Figure 9, suggest that only precursor cells of higher affinity are stimulated.

The most striking aspect of the results obtained in these studies on the fractionation of immune cell populations is the high percentage of specific fiber-binding cells found in nonimmune animals (Rutishauser et al., 1972). If it turns out that fiber-binding cells for a variety of antigens include

precursor cells, the finding that 1–2% of the spleen cells from an unimmunized animal can bind to a single antigenic determinant would pose a serious paradox of specificity. This paradox would be resolved if it is postulated that specificity in a clonal system requires the operation of at least two factors. The first factor is the a priori existence of a vast repertoire of committed cells having different specific receptors for antigen (Figure 1). The second factor, suggested in the present studies, as well as those of others (Nossal and Ada, 1971; Siskind and Benacerraf, 1969; Davie and Paul, 1972), is the necessity of a binding threshold for stimulation or triggering of each set of antigen-binding cells (Figure 9). Only those cells from each of these sets that have a sufficiently high binding constant for the antigen would be stimulated to undergo division and differentiation. Specificity in the clonal progeny is, therefore, assumed to arise first from binding of an antigen to a set of cells in the committed population followed by triggering of only a small subset of binding cells to mature and divide. Corroboration of this or alternative models of specificity by detailed analysis of populations of cells with different functions is one of the outstanding tasks remaining in the development of a truly quantitative selective theory of immunity.

A model for triggering of lymphocytes

The foregoing experiments and those of other workers suggest a working hypothesis for initial events in lymphocyte triggering. The observations essential in the formulation of this model are the following:

1. Although a relatively high proportion of lymphocytes may bind an antigen specifically, only a small proportion are triggered to divide. The data suggest that, above a certain threshold, the triggering is proportional to the affinity of the receptor (Rutishauser et al., 1972).

2. The lymphocyte Ig receptors are mobile in the membrane. Con A can restrict this mobility; other mitogens such as pokeweed mitogen do not (Yahara and Edelman, 1972).

3. Lectins such as Con A can trigger maturation and DNA synthesis of a large proportion of lymphocytes in culture (Greaves and Janossy, 1972).

4. Con A in soluble form is largely tetravalent and stimulates T cells. When locally concentrated and cross-linked to higher valencies, it stimulates B cells. Other lectins stimulate both T and B cells (Andersson et al., 1972).

The first point to decide in constructing a model for the initial events in mitogenesis is whether there are different mechanisms for stimulation of lymphocytes by antigen or lectins. A number of lines of evidence suggest that antigens and lectins stimulate lymphoid cells in the same way

(Greaves and Janossy, 1972). These include the similarity of the time course in culture, the dose response curve in culture indicating that high doses of either lectins or antigens inhibit the response, and the similarity of blast formation and Ig production. It should be emphasized, however, that it has not yet been shown whether lectin stimulation depends upon one or two cell types.

The second point for decision concerns whether inter-molecular interaction among receptors in the same cell is both a necessary and sufficient condition for triggering or whether interaction with single receptors is sufficient.

In the present model, I make the following assumptions:

1. Both lectin triggering and antigenic triggering including that mediated via T cells have the same mechanism and share a final common pathway for surface nuclear interaction. This implies that triggering of lymphocytes can occur through glycoprotein receptors other than Ig; whether Ig receptors are secondarily involved will be discussed later.

2. The necessary first step in triggering is the formation of cross-linked receptor aggregates to form a *micropatch*. Such a micropatch would consist of 100–500 receptor molecules, and a membrane of a given lymphocyte could therefore have as many as 100–500 micropatches. These micropatches must remain stable (i.e., neither increase nor decrease in size) for reasonable periods of time (up to hours). Diffusion of receptors into and out of a micropatch must be balanced.

3. The receptors (Ig or glycoprotein) are linked to adenyl cyclase or phosphodiesterase enzyme activity perhaps via intramembranous particles. Formation of a micropatch may lead to a decrease in cyclic AMP levels if enough receptors are linked. A stimulus therefore is the formation of micropatches with a sufficiently high surface density.

4. Formation of larger aggregates would lead to macropatches, cap formation, and interruption of the stimulation cycle. Failure to form micropatches rapidly enough by low doses of antigen or lectin would lead to binding but also to blockade of stimulation. The B cells are assumed to have a different threshold of response to micropatches than T cells.

5. The formation of a micropatch requires multivalence of antigen or lectin. Therefore, the formation of a micropatch of proper size and stability depends upon both the binding constant *and* the valence of a lectin or antigen. Antigenic determinants or lectin need not be presented in a *regular* array to provide an adequate stimulus; the only requirement is that a sufficient surface density of receptor-lectin or receptor-antigen complexes be maintained.

This model says nothing *in detail* about the coupling of micropatch formation to metabolic events initiating cell division. The assumption that there is a final common

pathway for such a stimulus, however, would suggest a single mediator such as cyclic AMP. One possible means of coupling micropatch formation is via interaction with enzymes such as adenyl cyclase or phosphodiesterase. A plausible hypothesis is to assume that the formation of stable micropatches allows interaction of intramembranous particles to which such enzymes might be linked. Such interactions could result in cooperative induction of enzymatic action via allosteric or other conformational changes.

The observation that concanavalin A binding prevents patch formation by Ig receptors after interaction with divalent anti-Ig opens the possibility that receptor-receptor interactions may occur in certain cases. Inasmuch as it has not been established whether lymphocytes that can be triggered by lectins must have Ig receptors, it is not clear whether the final common path for the stimulus involves Ig receptors.

This model leads to a number of predictions:

1. Some lectins that are not mitogenic may be made mitogenic by cross-linking them or coupling them to Fab fragments of anti-Ig. (Some lectins may never be mitogenic, however.)

2. Lectins that are not mitogenic for B cells but that are mitogenic for T cells may be made mitogenic for B cells by cross-linking (this has already received support in two cases, as noted earlier.)

3. The valence, state of aggregation, and K for specific sugar binding are overriding; the detailed structure of a particular lectin plays only a minor role in mitogenic activity.

4. If lectins that are multivalent are made univalent, mitogenesis will be blocked.

5. Extensive cross-linking of a lectin in a *linear* polymer may prevent micropatch formation as well as mitogenesis.

6. In a purified cell population specific for binding to a given antigen, stimulation by antigens in the proper form should be additive with stimulation by a mitogenic lectin.

The opportunity to purify lymphocytes according to their receptor specificity and to test a variety of lectins of known structure should permit a definitive test of this hypothesis. Without undue speculation, it may be worthwhile to note that interactions of the type proposed here may play a key role in cell-cell recognition during embryogenesis as well as in synapse formation.

Summary

Lymphoid cells carry Ig receptors for antigens as well as glycoprotein receptors for certain plant lectins. Interaction of antigens with the Ig receptors and of certain lectins with the glycoprotein receptors leads to a triggering of cell maturation, mitosis, and antibody production. Because the structure of Ig molecules has been extensively studied, these cells provide excellent models for analysis of membrane-nuclear interactions, cell-cell interactions, and mitogenesis. Extensive studies have also been made of the three-dimensional structure of the lectin concanavalin A. By the combined use of this mitogenic lectin and antibodies to Ig receptors, the surface traffic and interaction of various receptors can be mapped. The results of such mapping experiments lend support to the concepts of receptor mobility and receptor-receptor interaction and suggest a detailed model for lymphocyte triggering.

In analyzing mechanisms at the cell surface, there is a need for purified populations of cells. A method has been devised for the fractionation of cells on derivatized surfaces according to the specificity of their surface receptors. This method has been successfully used to fractionate lymphoid cells according to the specificity and affinity of their surface immunoglobulins. Fiber fractionation may prove to be valuable in the fixation, fractionation, and biochemical analysis of neuronal tissues.

ACKNOWLEDGMENTS   The work of the author was supported by USPHS Grants AI 09273, AM 04256 and AI 09921.

REFERENCES

ANDERSSON, J., G. M. EDELMAN, G. MÖLLER, and O. SJÖBERG, 1972. Activation of B lymphocytes by locally concentrated Con A. *Europ. J. Immunol.* 2:233.

BECKER, J. W., G. N. REEKE, JR., and G. M. EDELMAN, 1971. Location of the saccharide binding site of concanavalin A. *J. Biol. Chem.* 246:6123.

BOYD, W. C., 1963. The lectins: Their present status. *Vox Sang* 8:1.

BURNET, F. M., 1959. *The Clonal Selection Theory of Acquired Immunity.* Nashville, Tenn.: Vanderbilt University Press.

DAVIE, J. M., and W. E. PAUL, 1972. Receptors on immunocompetent cells. V. Cellular correlates of the maturation of the immune response. *J. Exp. Med.* 135:660.

EDELMAN, G. M., 1970. The structure and genetics of antibodies. In *The Neurosciences: Second Study Program,* F. O. Schmitt, ed., New York: The Rockefeller University Press, p. 885.

EDELMAN, G. M., B. A. CUNNINGHAM, G. N. REEKE, JR., J. W. BECKER, M. J. WAXDAL, and J. L. WANG, 1972. The covalent and three-dimensional structure of concanavalin A. *Proc. Natl. Acad. Sci. USA* 69:2580.

EDELMAN, G. M., and C. F. MILLETTE, 1971. Molecular probes of spermatozoan structures. *Proc. Natl. Acad. Sci. USA* 68:2436.

EDELMAN, G. M., U. RUTISHAUSER, and C. F. MILLETTE, 1971. Cell fractionation and arrangement on fibers, beads and surfaces. *Proc. Natl. Acad. Sci. USA* 68:2153.

GALLY, J. A., and G. M. EDELMAN, 1970. Somatic translocation of antibody genes. *Nature* 227:341.

GOLDSTEIN, I. J., C. E. HOLLERMAN, and E. E. SMITH, 1965. Protein-carbohydrate interaction. II. Inhibition studies on the interaction of concanavalin A with polysaccharides. *Biochemistry* 4:876.

GOWANS, J. L., J. H. HUMPHREY, and N. A. MITCHISON, 1971. A discussion on cooperation between lymphocytes in the immune response. *Proc. Roy. Soc. London (Biol.)* 176 (1045): 369–481.

GREAVES, M. F., and S. BAUMINGER, 1972. Activation of T and B lymphocytes by insoluble phytomitogens. *Nature New Biol.* 235:67.

GREAVES, M. F., S. BAUMINGER, and G. JANOSSY, 1972. Lymphocyte activation. III. Binding sites for phytomitogen on lymphocyte subpopulations. *Clin. Exp. Immunol.* 10:525.

GREAVES, M. F., and G. JANOSSY, 1972. Activation of lymphocytes by phytomitogens and antibodies to cell surface components—A model for antigen-induced differentiation. In *Cell Interactions,* Proceedings of the Third Lepetit Colloquium, L. G. Silvestri, ed. Amsterdam: North-Holland Publishing Company, pp. 143–155.

JANOSSY, G., and M. F. GREAVES, 1971. Lymphocyte activation. I. Response of T and B lymphocytes to phytomitogens. *Clin. Exp. Immunol.* 9:483.

JERNE, N. K., 1955. The natural selection theory of antibody formation. *Proc. Natl. Acad. Sci. USA* 41:849.

MISHELL, R. I., and R. W. DUTTON, 1967. Immunization of dissociated spleen cell cultures from normal mice. *J. Exp. Med.* 126:423.

NOSSAL, G. J. V., 1967. The biology of the immune response. In *The Neurosciences: A Study Program.* New York: The Rockefeller University Press, p. 183.

NOSSAL, G. J. V., and G. L. ADA, 1971. *Antigens, Lymphoid Cells and the Immune Response.* New York: Academic Press.

POWELL, A. E., and M. A. LEON, 1970. Reversible interaction of human lymphocytes with the mitogen concanavalin A. *Exp. Cell Res.* 62:315.

REISFELD, R. A., J. BÖRJESON, L. N. CHESSIN, and P. A. SMALL, JR., 1967. Isolation and characterization of a mitogen from pokeweed (*Phytolacca americana*). *Proc. Natl. Acad. Sci. USA* 58:2020.

RUTISHAUSER, U., C. F. MILLETTE, and G. M. EDELMAN, 1972. Specific fractionation of immune cell populations. *Proc. Natl. Acad. Sci. USA* 69:1596.

SISKIND, G. P., and B. BENACERRAF, 1969. Cell selection by antigen in the immune response. *Advances Immun.* 10:1–50.

TAYLOR, R. B., P. H. DUFFUS, M. C. RAFF, and S. DE PETRIS, 1971. Redistribution and pinocytosis of lymphocyte surface immunoglobulin molecules induced by anti-immunoglobulin antibody. *Nature New Biol.* 233:225.

WANG, J. L., B. A. CUNNINGHAM, and G. M. EDELMAN, 1971. Unusual fragments in the subunit structure of concanavalin A. *Proc. Natl. Acad. Sci. USA* 68:1130.

YAHARA, I., and G. M. EDELMAN, 1972. Restriction of the mobility of lymphocyte immunoglobulin receptors by concanavalin A. *Proc. Natl. Acad. Sci. USA* 69:608.

70 The Biosynthesis of Cell-Surface Components and Their Potential Role in Intercellular Adhesion

SAUL ROSEMAN

ABSTRACT   Elucidation of the molecular mechanism of cell-cell interaction is of prime importance in understanding both normal and abnormal development in higher animals. The ability of cells to recognize other cells or components of the environment can be explained by assuming that the cell surfaces contain "sensors," constituents of the cell surface capable of discriminating among a wide variety of different substances. This chapter is focused on the nature of the components of the cell surface that cause adhesion between cells or cells and artificial substrates. The cell surface is rich in complex carbohydrates, and these compounds possess the high degree of diversity required for *specific* adhesion between cells.

The mechanism of biosynthesis of the complex carbohydrates, in which the oligosaccharides are formed by the stepwise addition of monosaccharides from their respective nucleotides catalyzed by specific glycosyltransferases, often in the form of multiglycosyltransferase systems, is discussed, as well as several methods for measuring intercellular adhesion. Preliminary results from a new method that utilizes a confluent cell monolayer for collecting cells of the same or different types are given. This method provides a clear demonstration that cell-surface components involved in the adhesion of cells to each other differ from those involved in the adhesion of cells to glass.

Various theories of the mechanism of intercellular adhesion are discussed, and the results are consistent with the hypothesis that cell surface complex carbohydrates and their respective glycosyltransferases on apposing cell surfaces interact to form enzyme-substrate complexes and that the latter are responsible for intercellular adhesion. In addition, a "membrane messenger hypothesis" is offered, which provides an explanation as to how the membrane can transmit or translate extracellular signals. It is generalized that membranes contain a variety of receptors, specific for different stimuli, and each receptor then releases a different membrane messenger. In turn the messenger may act stoichiometrically, or by amplification to evoke the cell response. The possible nature of the messengers as well as their mode of action is discussed. The particular value of the various hypotheses lies in the fact that they may be tested experimentally.

A MAJOR PROBLEM in modern biology concerns the understanding of the molecular mechanisms responsible for the

SAUL ROSEMAN   Department of Biology and the McCollum-Pratt Institute, The Johns Hopkins University, Baltimore, Maryland

interactions that occur between cells of higher organisms. When such cells lie close to or in contact with other cells (of the same or a different type), a number of physiological responses can take place. These include "contact inhibition" of growth or of movement, cell-cell recognition, "sorting out" of different cells from each other, differentiation to clearly defined morphological structures, etc. (Spemann, 1938; Saxen and Toivonen, 1962). These cell-cell interactions are of prime importance in embryonic development, as for example in the nervous system (see Sidman, this volume). During development and morphogenesis, some cells (or cell processes) migrate to predetermined areas of the embryo, where they attach to other cells and/or multiply, and where their adult progeny ultimately differentiate. Elucidation of these remarkable phenomena on a molecular level would add immeasurably to our understanding of both normal and abnormal development and of ancillary (but extremely important) problems such as tumorigenesis and metastasis.

The ability of cells to recognize other cells of the same or different types, or to recognize other components of the environment, such as hormones, can most simply be explained by assuming that the cell surfaces contain "sensors." These sensors are constituents of the cell surface capable of discriminating among a variety of extracellular substances. For example, such sensors on the surfaces of a given cell type should be capable of distinguishing between adjoining cells of the same or of different type. The problem then becomes how to isolate and characterize these cell surface components.

A major difficulty in working with cells of higher organisms (in contrast to bacteria or fungi) is that the cell surface contains a multitude of different components. For example, these surfaces are the site of the blood group antigens (which comprise at least 12 or 13 separate classes, e.g., the ABO system), transplantation antigens, and both viral and hormonal receptor sites, to name a few. While only a limited number of these cell-surface components have been isolated and characterized, it is clear that they consist primarily of complex carbohydrates (glycoproteins, mucins, glycolipids, etc.), and that there are undoubtedly families of each of these types of polymers on the surface. It seems reasonable to assume that each member of such a family may be involved in a different physiological phenomenon. The complexity and diversity of the molecules are such that they would be capable of performing the vast numbers of reactions required by the cell.

Our own interests are concerned with the mechanisms by which cells recognize both homologous and heterologous cells. It seems reasonable to assume that this cell-cell recognition phenomenon is related to, or directly responsible for, the ability of cells to adhere to each other, so that the problem of cell-cell recognition might be attacked by studying intercellular and cell-substrate adhesion (Curtis, 1962; Trinkaus, 1969; Manly, 1970). The chemical basis for intercellular adhesion has not been defined. However, there are suggestive leads implicating cell surface carbohydrates in this process. The remainder of this review will therefore give a brief description of the nature and biosynthesis of the complex carbohydrates, the evidence implicating them in intercellular adhesion, and some speculations on the role of the membrane (membrane messengers) following interactions with molecules in the extracellular medium.

The complex carbohydrates

CHEMISTRY OF THE COMPLEX CARBOHYDRATES A host of diverse and complex carbohydrate-containing macromolecules are found in nature, but if we confine this review to those produced by cells from vertebrates, we are concerned with three major classes: polysaccharides, glycoproteins, and glycolipids. The polysaccharides produced by such cells include hyaluronic acid, the chondroitin sulfates, heparin, etc., but these will not be discussed here, because the small number could not account for the high degree of diversity required for *specific* intercellular adhesion. However, as discussed previously, the complex carbohydrate-containing molecules, the glycoproteins and glycolipids, appear capable of providing such properties. These compounds show considerable diversity. Most of the major serum proteins (except albumin) appear to be glycoproteins, as are the epithelial mucins. The brain gangliosides, for example, comprise but one class of glycolipids.

The oligosaccharide chains attached to proteins and lipids vary in size, in the degree of branching, in their constituent sugars, and in the sequence and position of linkage of these sugars to each other (Gottschalk, 1966). These oligosaccharide chains contain at least two or more

of the following monosaccharides: D-galactose, D-mannose, D-glucose, L-fucose, N-acetyl-D-glucosamine, N-acetyl-D-galactosamine, and the sialic acids. D-Glucose is frequently found in the glycolipids (e.g. the gangliosides) but only rarely in the glycoproteins or mucins. Schematic representations of the structures of the serum type glycoproteins and of the mucins are given in Figure 1, and the structure of one of the gangliosides, which is most abundant in human brain gray matter, in Figure 2. The chemistry of these complex carbohydrates has been reviewed (Gottschalk, 1966; Svennerholm, 1964).

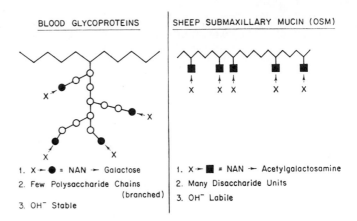

FIGURE 1 Diagrammatic representation of serum type glycoproteins and mucins.

BIOSYNTHESIS OF THE COMPLEX CARBOHYDRATES The synthesis of the complex carbohydrates represents the last in a series of steps leading from glucose, the major nutrient of the cell. These steps have been reviewed (Roseman, 1968, 1970) but will be briefly recapitulated here. The entire sequence may be visualized as the sum of three discrete steps, the first being the conversion of glucose to the monosaccharides listed above, the second the activation of the monosaccharides, and the last the transfer of the monosaccharides to the ends of the growing chains in the polymers. In every case, the active forms of the monosaccharides are the corresponding sugar nucleotide derivatives. In fact, the monosaccharide may be synthesized from its precursor as the nucleotide derivative. For example, galactose is synthesized by the conversion of UDP-glucose to UDP-galactose. An illustration of the complexity of the biosynthetic sequence is given in Figure 3, which shows the steps involved from fructose-6-P to the active form of the sialic acids, CMP-sialic acid.

While one can postulate a number of different pathways for the synthesis of the oligosaccharide chains in serum-type glycoproteins, mucins, and glycolipids, all of

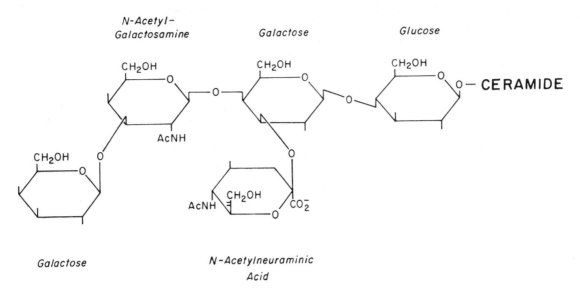

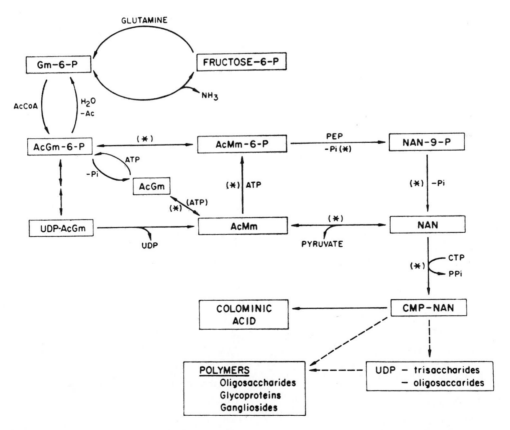

FIGURE 2   Proposed structure of a monosialoganglioside (after Kuhn and Wiegandt; see Svennerholm, 1964).

FIGURE 3   Schematic diagram of the metabolic pathways from fructose 6-phosphate to the complex carbohydrates. Each arrow represents a different enzymatic reaction. The following abbreviations are used: Gm, glucosamine; Ac, acetyl or acetate; AcMm, *N*-acetylmannosamine; PEP, phosphoenolpyruvate; NAN, *N*-acetylneuraminic acid; CMP, 5′-cytidylic acid.

the available evidence supports the general pathway shown in Figures 4 and 5. The corresponding sugar nucleotides serve as the glycosyl donors, while the incomplete chains of the complex carbohydrates serve as the acceptors. The monosaccharides are added sequentially to the ends of the growing chains or at branch points, and each step is catalyzed by a different glycosyltransferase. The entire battery of glycosyltransferases required for the synthesis of a complete chain in a given polymer is called (Roseman, 1968, 1970) a multiglycosyltransferase system (MGT). Since the synthesis of a complete chain would require optimal conditions for each of the enzymes operating in such a sequence, the prediction can be made that a defect in one of the enzymes, or suboptimal conditions for one of these steps, would lead to the formation of incomplete chains. When a given protein contains oligosaccharide chains of varying size, the molecular species are generally not separable from one another, and the resulting glycoprotein or mucin preparation exhibits microheterogeneity with respect to the carbohydrate prosthetic groups. This type of heterogeneity is charac-

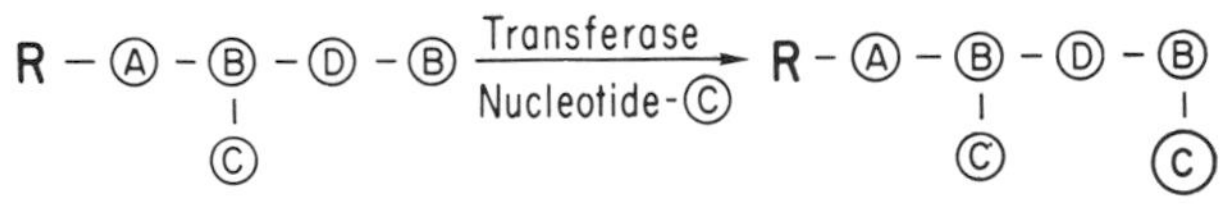

1. Sugar-nucleotides are glycose donors.

2. Sugars added as monosaccharides in a specific sequence.

3. Chain elongation at non-reducing ends or branch points.

   R = protein or lipid

FIGURE 4   Synthesis of oligosaccharide chains (glycoproteins, mucins, glycolipids). R represents protein or lipid and each of the encircled letters refers to a glycose residue.

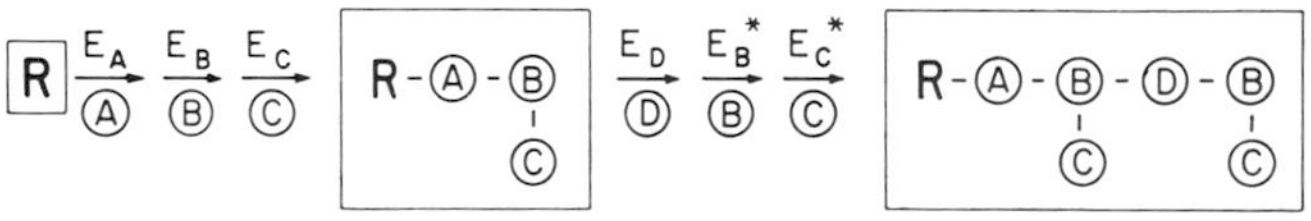

Each transferase specific for acceptor and its analogues:

1. *Different* transferase catalyzes each step.

2. Product of each step is substrate for next reaction.

3. *Different* MGT systems required for synthesis of glycoproteins, mucins, glycolipids.

FIGURE 5   Representation of a multiglycosyltransferase system (MGT).

teristic of glycoprotein preparations. In the case of the glycolipids, such as the gangliosides, molecules bearing incomplete chains are separable from each other, giving discrete preparations of varying carbohydrate chain length.

Occurrence of glycosyltransferases

The glycosyltransferases are widely distributed in animal tissues and may very well occur in all cells. Subcellular fractionation of rat liver (Schachter et al., 1970) leads to the conclusion that these enzymes are associated with the Golgi apparatus, where they may be involved in the secretion of plasma proteins.

In studies with embryonic brain (Den et al., 1970), the enzymes were found in highest concentration in the synaptosome (nerve-ending) fraction (Kaufman et al., unpublished). Their function here is not known, but it is of interest that the synaptic junction is carbohydrate-rich, and several workers have postulated that the complex carbohydrates may function in synaptic transmission. In addition, these studies (Den et al., 1970) showed that certain of the glycosyltransferases are found in extracellular fluids, such as the cerebrospinal fluid and serum (Mookerjea et al., 1971; Wagner and Cynkin, 1971; Kim et al., 1971; Hudgin and Schachter, 1971). Whether these soluble extracellular glycosyltransferases serve functions that remain to be defined or are simply products secreted by the cell along with other products from the Golgi such as serum proteins is not known. However, it is of interest that at least one of these enzymes (a serum protein-type galactosyltransferase) is very active in the fluid surrounding the embryonic chicken brain but essentially disappears as the embryo matures.

A key question concerning these enzymes is whether or not they are located on the plasma cell membrane, and this point is considered further below.

Intercellular adhesion

While the adhesion of cells to each other and to solid substrates is a process of fundamental biological significance and has been the subject of intensive investigation in many laboratories (reviewed by Spemann, 1938; Saxen and Toivonen, 1962; Curtis, 1962; Trinkaus, 1969; Manly, 1970), the molecular mechanism is unknown. There is no general agreement on even as fundamental a point as a definition of intercellular adhesion. As a consequence, investigators have used different methods for quantitating the process, and have frequently obtained conflicting results. These could be explained by the fact that the formation of a stable aggregate of cells from a suspension of single cells may well involve more than one step, and thus different methods of measuring the phenomenon may actually be measuring different *processes*. For example, are molecular mechanisms involved in the adhesion of single cells to each other (or to aggregates) the same as those concerned with the "sorting out" of a mixed population of two cell types from each other (Steinberg, 1970; Moscona, 1961)?

Since there is no generally accepted definition for intercellular adhesion, the phenomenon is usually defined operationally by the methods used for detecting or measuring it. In this laboratory we decided to concentrate on developing methods for measuring the first detectable event, the rate at which single cells adhere to each other (specific adhesion) or to other cell types (nonspecific adhesion). These methods, and in fact all of the work in this area, stem from the original observation of Townes and Holtfreter (1955) who showed that embryonic tissues can be dissociated into single cell suspensions, and the aggregation of the latter can be studied in vitro.

Three assays have been developed in this laboratory: (a) The rate at which single cells in a suspension aggregated to each other was first measured by quantitatively determining the number of single cells remaining in the mixture with the use of the Coulter electronic particle counter (Orr and Roseman, 1969), a procedure that offers a number of advantages. The measurements involve large numbers of cells, and frequently the rate of adhesion of at least half of the population is determined. The method can detect changes in adhesive rates of 10 to 15%, it is reproducible, and both simple and rapid. However, it is unable to distinguish or discriminate between specific and nonspecific adhesion. Further, it is a method that measures by difference, i.e., the difference between the initial number of single cells and the remaining single cells at any subsequent time point so that it is incapable of detecting changes in a small fraction of the total population. (b) A collecting aggregate assay modified after the original method of Roth and Weston (1967). In the original method, radioactively labeled single cells were shaken with aggregates, either composed of the same cells as in the single cell population (isotypic) or of a different type of cell (heterotypic). At suitable time intervals, the aggregates were removed from the suspension and histological and radioautographic methods were used to determine the number of labeled single cells that adhere to the collecting aggregates. In the modified procedure (Roth et al., 1971) the aggregates were counted by liquid scintillation techniques. Figures 6 and 7 show typical results obtained with this method, and as can be seen, the procedure provides the very important property of being able to measure both specific and nonspecific adhesion. However, the technique suffers from certain serious disadvantages. Only a very small fraction of the

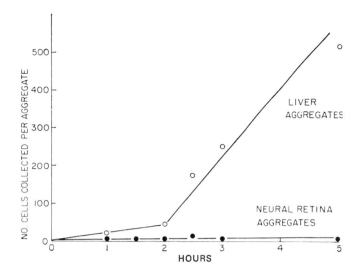

FIGURE 6 Collection of ^{32}P-labeled liver cells by liver and neural retina cell aggregates over a 5-hr time period. The concentration of the labeled liver cells was 1.5×10^5 cells per ml (1.1 cpm per cell) with three of each type of aggregate per flask in nutrient medium at 37°C.

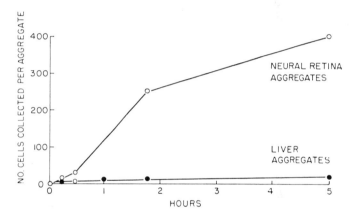

FIGURE 7 Collection of ^{32}P-labeled retina cells by neural retina and liver aggregates as a function of time. The concentration of labeled neural retina cells was 1×10^5 cells per ml (1.0 cpm per cell) with three of each type of aggregate per flask in nutrient medium at 37°C.

total single cells in the suspension adhere to the aggregates. Further, quantitative reproducibility, at least with embryonic cells, is poor, and at best, with these cell types it can only be considered a semiquantitative procedure. (c) The most recent method developed in our laboratory (Walther et al., unpublished data) combines the advantages of both of the foregoing assay methods. In this case, the collecting cells consist of a confluent monolayer; this is exposed to a suspension of labeled single cells, and the amount of radioactive label that adheres to the monolayer thus offers a measure of cell adhesion. This method is consistently reproducible, rapid, and involves the adhesion of a large fraction of the single cells in the suspension (up to 85%). It again permits the measurement of both specific and nonspecific adhesion. This assay has already been used and provides a clear demonstration that the cell surface components involved in the adhesion of cells to each other differ from the components involved in the adhesion of cells to glass or to plastic (of the type used for tissue culture).

Use of the methods described above has not yet led to a definition of the molecular mechanisms underlying specific and nonspecific adhesion but has given certain suggestive results, and hypotheses have been derived which can be tested by experiment.

In earlier work the Coulter counter assay was used to investigate the adhesion of mouse teratoma cells to each other (Oppenheimer et al., 1969). With this assay, it was possible to show that whereas single cells derived by dissociating the "embryoid" bodies of the teratoma would adhere to each other in a complete synthetic tissue culture medium, they were unable to aggregate in a glucose-salts medium.

The tissue culture medium contained 51 components in addition to glucose and salts, but only one, L-glutamine, promoted adhesion when added to the glucose-salts medium. Once this was established, of the many compounds tested, only D-glucosamine and D-mannosamine were able to replace, and were approximately as effective as, L-glutamine. These results were in full accord with the known metabolic pathways, shown in Figure 3, and lead to the interpretation that complex carbohydrates on the surfaces of the teratoma cells are involved in intercellular adhesion.

If the conclusion presented above is correct, we may ask whether only the cell-surface complex carbohydrates are required for adhesion, or whether other components are also required? In order to answer this question, two hypotheses were developed (Roseman, 1970; Roth et al., 1971). Both of these were based on the original theories independently offered by Tyler (1946) and Weiss (1947). The Tyler-Weiss theory and the two working hypotheses are shown in Figure 8. The Tyler-Weiss theory was that antibody-like and antigen-like components were present on cell surfaces and that adhesion involved the formation of complexes between these surface molecules. Our modifications of this idea are the following: (a) The cell surface components are enzymes and substrates, and adhesion involves the formation of enzyme-substrate complexes. This concept would require that the enzyme not exert its catalytic function if stable adhesive forces were to result. (b) Another possibility involves the formation of hydrogen bonds. If cell surface carbohydrates are required for intercellular adhesion, then it is conceivable that oligosaccharide chains on neighboring cells can

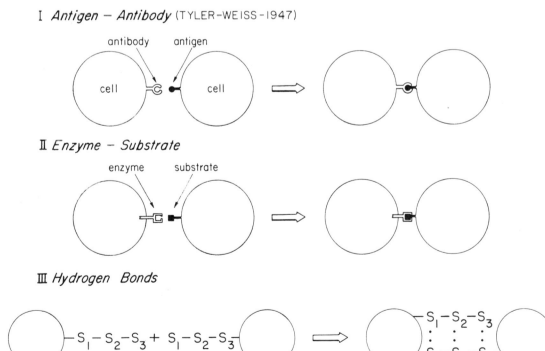

FIGURE 8   Diagrammatic representation of three hypotheses of the mechanism of intercellular adhesion.

form stable hydrogen bonds to each other. For example, the cross-linking between the neighboring polysaccharide chains in cellulose involves hydrogen bonds.

The enzyme-substrate theory can be developed further. If the substrates consist of the cell surface complex carbohydrates, then the logical candidates for the enzymes are the glycosyltransferases described earlier in this paper. In other words, multiglycosyltransferase systems and their corresponding substrates, the complex carbohydrates, would bind to each other giving rise to intercellular adhesion. Specific adhesion would involve specific multiglycosyltransferase systems and their substrates. This concept can be experimentally verified, because it requires that the glycosyltransferases be present on cell surfaces.

Evidence has in fact been presented (Roth et al., 1971) for the location of such enzymes on chicken embryonic neural retina cell surfaces, and more recently, on fibroblasts grown in tissue culture (Roth and White, 1972) and also on human blood platelets that adhere to collagen (Bosmann, 1971; Jamieson et al., 1971). In the opinion of this author, the evidence for cell surface glycosyltransferases is suggestive rather than definitive, because of various technical problems encountered in performing these experiments (for example, cell lysis). If it can be unequivocally shown that the glycosyltransferases are indeed components of the cell surface, then it will be possible to test finally the theory postulated in Figure 8, that these enzymes and the complex carbohydrates which are their substrates are involved in intercellular adhesion.

The possibility that the cell surface enzymes interact with their oligosaccharide substrates to form complexes responsible for adhesion leads to a number of interesting extrapolations. For example, as described above in detail, the completion of an oligosaccharide chain in a complex carbohydrate requires all of the glycosyltransferases in a multiglycosyltransferase system. If one such enzyme is lacking, then chain completion is not possible. Defects in glycosyltransferases have indeed been shown to occur in some lines of transformed tissue culture cells (Brady and Mora, 1970; Den et al., 1971). Conceptually, at least, if this is a general phenomenon, it would be possible to explain the difference in adhesive properties of normal (contact-inhibited) cells compared with transformed cells as shown in Figure 9. Again, this and other ideas which are extensions of the basic hypothesis of enzyme-substrate interactions should be verifiable by direct test.

Membrane messenger hypothesis

Two hypotheses to explain the mechanism of intercellular adhesion are described above, and it is hoped that these relate to or explain the phenomenon of cell-cell recog-

NORMAL (STRONGLY ADHERING)

TRANSFORMED (WEAKLY ADHERING)

FIGURE 9 The role of glycosyltransferases in adhesion between normal and transformed cells. S_1, S_2, and S_3 are acceptors for the transferases T_1, T_2, and T_3, respectively.

nition. Whether or not these ideas are correct, they represent only a small part of a broader question. Regardless of the molecular mechanism for intercellular adhesion, or of adhesion to substrata such as glass, eukaryotic cells frequently exhibit the most profound responses following the establishment of adhesive bonds. How are these responses brought about? Examining the problem in an even wider sense, cells of this type will respond to both physical and chemical stimuli in the environment and can respond by changes in growth rate, metabolic patterns, differentiation, permeability, enzyme induction, motility, shape, etc. While some signals from the environment (e.g. steroid hormones) exert their effects directly on the cell after penetration of the cell membrane, many act on the membrane per se. The problem can thus be stated as follows: How does the plasma membrane transmit or translate extracellular signals?

A variety of examples of direct interaction between external stimuli and the cell membrane can be offered, but one should suffice. It is a common experience in tissue culture that suspensions of dissociated diploid cells will not divide, although they will continue to metabolize at approximately normal rates in suitable media. When such cells are placed over a substratum to which they can adhere, such as glass, they attach to the latter, flatten out, and grow; growth continues in the monolayer until the cells come into contact on all sides with other cells (confluency), at which point growth stops. Although such cells will usually not divide in suspension culture, some cell lines will grow under these conditions if the viscosity of the medium is substantially increased. The problem posed above can be stated in explicit terms for these phenomena: How are the physical environmental stimuli, contact with glass or with a more viscous medium or with other cells, translated by the plasma membrane and transmitted to the cytoplasm as signals to grow or to stop growing?

A simple and general explanation for the diverse but specific effects exhibited by external stimuli on cells is shown in Figure 10. It is suggested that the cell membrane acts as a transducer that translates the external stimulus to an internal signal (or signals) via a *membrane messenger*. This idea represents an extrapolation of the familiar Jacob-Monod nuclear messenger theory and a generalization of the "second messenger" hypothesis of Sutherland et al. (1965) in which the "second messenger" is cyclic AMP. The latter workers proposed that the profound responses observed when target cells interacted with various hormones resulted from changes in the internal levels of cyclic AMP (or other nucleotides). The hormones interacted with membrane receptors that in turn stimulated or inhibited the appropriate enzymes (adenyl cyclase, cyclic AMP phosphodiesterase), thereby regulating the level of cyclic AMP in the cytoplasm. More recently, Rasmussen and Tenenhouse (1968) have invoked Ca^{++} as a "second messenger." Until now the problem has been that one or a few membrane messengers appears insufficient to explain the diverse responses that can be elicited from a single cell type by different environmental stimuli. However, the generalization shown in Figure 10 is that membranes contain a variety of receptors, specific for different stimuli, and each receptor then releases a different membrane messenger.

If the generalized membrane messenger hypothesis is correct, then two obvious questions can be raised: (1) What are the messengers? (2) How do they act?

If membrane messengers exhibit a wide spectrum of physiological functions, it is reasonable to suppose that the messengers themselves are diverse, ranging from low molecular weight substances to macromolecules. One interesting possibility is RNA. Significant quantities of RNA have frequently been detected in plasma membrane preparations (Warren, 1969) and presumably are not contaminants from the cytoplasm. Even the most extensively purified animal cell plasma membranes contain RNA (Glick and Warren, 1969), and such preparations are capable of synthesizing protein. Membrane-bound RNA, released from the membrane receptor proteins after interaction of the latter with a specific external stimulus, could have two important functions. First, such RNA could serve as mRNA, resulting in a burst of *specific* protein synthesis as a consequence of the external signal. In this way, the cell could generate specific proteins without activating the corresponding genes; conceivably, such a mechanism could explain antibody synthesis. Second, membrane messenger RNA could regulate the transcription of selected genes. The latter idea is not unique, because at least one species of RNA, His tRNA, is thought to be involved in regulating the histidine operon in *Salmonella* (Brenner and Ames, 1971), while "reverse transcriptase" utilizes RNA (viral RNA) as a template for DNA synthesis (Temin and Mizutani, 1970; Baltimore, 1970; Goodman and Spiegelman, 1972).

Some membrane messengers may be proteins. For example, adenyl cyclase is already fairly well established in this role. In terms of the present concept (Figure 10), it is adenyl cyclase rather than cyclic AMP that is the membrane messenger, for it is the enzyme that is bound to the membrane receptor that is activated when the latter forms a complex with the appropriate hormone. Membrane messenger proteins may be enzymes, such as adenyl cyclase, but may also have entirely different functions, such as regulation of enzyme activity in the cytoplasm, or even more interestingly, of gene expression in the nucleus. Models for the latter exist, as for example, the repressor proteins of the familiar Jacob-Monod hypothesis. Membrane-bound activator or repressor proteins, which selectively regulate the transcription of certain genes, for example, is an attractive mechanism for explaining the interesting observation that induction of synthesis of glucose-6-P permease in *Escherichia coli* is achieved from without, i.e., external but not internal glucose-6-P (or 2-deoxy-glucose-6-P) is required for induction (Dietz and Heppel, 1971).

Membrane messengers may also be substances of low molecular weight. As indicated above, calcium ion has already been postulated to serve such a function.

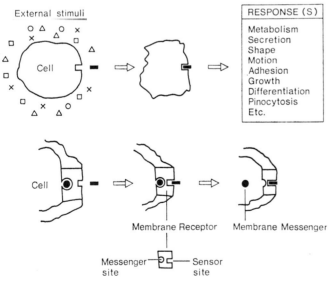

FIGURE 10 Schematic representation of the membrane messenger hypothesis. In the upper portion the external stimulus, solid bar, attaches to a specific site on the cell to evoke the necessary response. The lower portion depicts an enlargement of the membrane receptor for the external stimulus. The external stimulus attaches to the sensor site, and alters its conformation with the resultant release of the membrane messenger; the latter then evokes one or more responses by the cell.

If membrane messengers are enzymes, activators, inhibitors, or mRNA, etc., it would suggest that the messengers act by amplification. A few molecules of a membrane messenger released to the cytoplasm will show more than stoichiometric effects, because they act on enzymes or enzyme systems. However, this need not always be the case. Membrane messengers may act stoichiometrically, and one such example, a repressor protein, was given above. Stoichiometry may be very important in some phenomena. For example, the mouse 3T3 cell line will grow on a glass plate until it reaches confluency and then stop (contact inhibition of growth). During the growth phase, the cells remain in contact with other cells in the culture, except on one side of the cell surface. When this surface comes into contact with another cell, growth stops. In this case, perhaps a minimum number, or threshold level of membrane messengers must be released before the cell responds to the external stimulus.

One final question concerns the mechanism by which membrane messengers are generated and maintained. The simplest idea is based on the experimental observation (Warren, 1969) that the proteins of animal cell membranes show a surprisingly high rate of turnover, and this rate of turnover is maintained when *cell growth has stopped*. Turnover may be required to keep the normal complement of receptor proteins and messengers fully available to meet fluxes in the environment.

Conclusions

Several hypotheses have been proposed in this discussion. The basic concepts are simple, and are of value only in that they can be experimentally verified or disproved. If correct, they offer a means of generalizing and rationalizing the varied and apparently inexplicable behavior of eukaryotic cells in response to changes in their environment.

ACKNOWLEDGMENTS Contribution No. 718 of the McCollum-Pratt Institute. These studies were supported by Grant AM 09851 of the National Institutes of Arthritis and Metabolic Diseases of the National Institutes of Health, and by Grant P-544 of the American Cancer Society.

REFERENCES

BALTIMORE, D., 1970. RNA-dependent DNA polymerase in virions of tumor virus. *Nature (Lond.)* 226:1209–1211.

BOSMANN, H. B., 1971. Platelet adhesiveness and aggregation: The collagen:glycosyl, polypeptide:N-acetylgalactosaminyl and glycoprotein:galactosyl transferases of human platelets. *Biochem. Biophys. Res. Commun.* 43:1118–1124.

BRADY, R. O., and P. T. MORA, 1970. Alteration in ganglioside pattern and synthesis in SV40- and polyoma virus-transformed mouse cell lines. *Biochim. Biophys. Acta* 218:308–319.

BRENNER, M., and B. N. AMES, 1971. The histidine operon and its regulation. In *Metabolic Pathways*, 3rd Ed., Vol. 5. Metabolic Regulation, H. J. Vogel, ed. New York: Academic Press, pp. 349–387.

CURTIS, A. S. G., 1962. Cell contact and adhesion. *Biol. Rev.* 37:82.

DEN, H., B. KAUFMAN, and S. ROSEMAN, 1970. Properties of some glycosyltransferases in embryonic chicken brain. *J. Biol. Chem.* 245:6607–6615.

DEN, H., A. M. SCHULTZ, M. BASU, and S. ROSEMAN, 1971. Glycosyltransferase activities in normal and polyoma-transformed BHK cells. *J. Biol. Chem.* 246:2721–2723.

DIETZ, G. W., and L. A. HEPPEL, 1971. Studies on the uptake of hexose phosphates. II. The induction of the glucose 6-phosphate transport system by exogenous but not endogenously formed glucose 6-phosphate. *J. Biol. Chem.* 246:2885–2890.

GLICK, M. C., and L. WARREN, 1969. Membranes of animal cells. III. Amino acid incorporation by isolated surface membranes. *Proc. Nat. Acad. Sci. USA* 63:563–570.

GOODMAN, N. C., and S. SPIEGELMAN, 1972. Distinguishing reverse transcriptase of an RNA tumor virus from the other known DNA polymerases. *Proc. Nat. Acad. Sci. USA* 68:2203–2206.

GOTTSCHALK, A., 1966. *Glycoproteins*. Amsterdam: Elsevier Publishing.

HUDGIN, R. L., and H. SCHACHTER, 1971. Porcine sugar nucleotide glycoprotein glycosyltransferases: I. Blood serum and liver sialyltransferase. II. Blood serum and liver galactosyltransferase. *Can. J. Biochem.* 49:829–852.

JAMIESON, G. A., C. L. URBAN, and A. J. BARBER, 1971. Enzymatic basis for platelet:collagen adhesion as the primary step in haemostasis. *Nature New Biol.* 234:5–7.

KIM, Y. S., J. PERDOMO, A. BELLA, and J. NORDBERG, 1971. Properties of a CMP-N-acetylneuraminic acid:glycoprotein sialyltransferase in human serum and erythrocyte membranes. *Biochim. Biophys. Acta* 244:505–512.

MANLY, R. S., 1970. *Adhesion in Biological Systems*. New York: Academic Press.

MOOKERJEA, S., A. CHOW, and R. L. HUDGIN, 1971. Occurrence of UDP-N-acetylglucosamine:glycoprotein N-acetylglucosaminyltransferase activity in human and rat sera. *Can. J. Biochem.* 49:297–299.

MOSCONA, A. A., 1961. Rotation-mediated histogenetic aggregation of dissociated cells. *Exp. Cell Res.* 22:455–475.

OPPENHEIMER, S. B., M. EDIDIN, C. W. ORR, and S. ROSEMAN, 1969. An L-glutamine requirement for intercellular adhesion. *Proc. Nat. Acad. Sci. USA* 63:1395–1402.

ORR, C. W., and S. ROSEMAN, 1969. Intercellular Adhesion. I. A quantitative assay for measuring the rate of adhesion. *J. Membrane Biol.* 1:109–124.

RASMUSSEN, H., and A. TENENHOUSE, 1968. Cyclic adenosine monophosphate, Ca^{++}, and membranes. *Proc. Nat. Acad. Sci. USA* 59:1364–1370.

ROSEMAN, S., 1968. Biochemistry of glycoproteins and related substances. In *Proceedings of the 4th International Conference of Cystic Fibrosis of the Pancreas (Mucoviscidosis)*, E. Rossi and E. Stoll, eds. New York: S. Karger, pp. 244–268.

ROSEMAN, S., 1970. The synthesis of complex carbohydrates by multiglycosyltransferase systems and their potential function in intercellular adhesion. *Chem. Phys. Lipids* 5:270–297.

ROTH, S. A., and J. A. WESTON, 1967. The measurement of intercellular adhesion. *Proc. Nat. Acad. Sci. USA* 58:974–980.

ROTH, S., E. J. McGUIRE, and S. ROSEMAN, 1971. An assay for intercellular adhesive specificity. *J. Cell Biol.* 51:525–535.

ROTH, S., E. J. McGUIRE, and S. ROSEMAN, 1971. Evidence for cell-surface glycosyltransferases: Their potential role in cellular recognition. *J. Cell Biol.* 51:536–547.

ROTH, S., and D. WHITE, 1972. Intercellular contact and cell-surface galactosyltransferase activity. *Proc. Nat. Acad. Sci. USA* 69:485–489.

SAXEN, L., and S. TOIVONEN, 1962. *Primary Embryonic Induction.* Englewood Cliffs, New Jersey: Prentice-Hall.

SCHACHTER, H., I. JABBAL, R. L. HUDGIN, L. PINTERIC, E. J. McGUIRE, and S. ROSEMAN, 1970. Intracellular localization of liver sugar nucleotide glycoprotein glycosyltransferases in a Golgi-rich fraction. *J. Biol. Chem.* 245:1090–1100.

SPEMANN, H., 1938. *Embryonic Development and Induction.* New Haven: Yale University Press.

STEINBERG, M. S., 1970. Does differential adhesion govern self-assembly processes in histogenesis? Equilibrium configurations and the emergence of hierarchy among populations of embryonic cells. *J. Exp. Zool.* 173:395–433.

SUTHERLAND, E. W., I. OYE, and R. W. BUTCHER, 1965. The action of epinephrine and the role of the adenyl cyclase system in hormone action. In *Recent Progress in Hormone Research*, Vol. 21, G. Pincus, ed. New York: Academic Press, pp. 623–642.

SVENNERHOLM, L., 1964. The gangliosides. *J. Lipid Res.* 5:145–155.

TEMIN, H. M., and S. MIZUTANI, 1970. RNA-dependent DNA polymerase in virions of Rous sarcoma virus. *Nature (Lond.)* 226:1211–1213.

TOWNES, P. L., and J. HOLTFRETER, 1955. Directed movements and selective adhesion of embryonic amphibian cells. *J. Exp. Zool.* 128:53–120.

TRINKAUS, J. P., 1969. *Cells into Organs.* Englewood Cliffs, New Jersey: Prentice-Hall.

TYLER, A., 1946. An auto-antibody concept of cell structure, growth and differentiation. *Growth* (Symposium) 10:6–7.

WAGNER, R. R., and M. A. CYNKIN, 1971. Glycoprotein metabolism: A UDP-galactose: glycoprotein galactosyltransferase of rat serum. *Biochem. Biophys. Res. Commun.* 45:57–62.

WARREN, L., 1969. The biological significance of turnover of the surface membrane of animal cells. In *Current Topics in Developmental Biology*, Vol. 4, A. Moscona and A. Monroy, eds. New York: Academic Press, pp. 197–222.

WEISS, P., 1947. The problem of specificity in growth and development. *Yale J. Biol. Med.* 19:235–278.

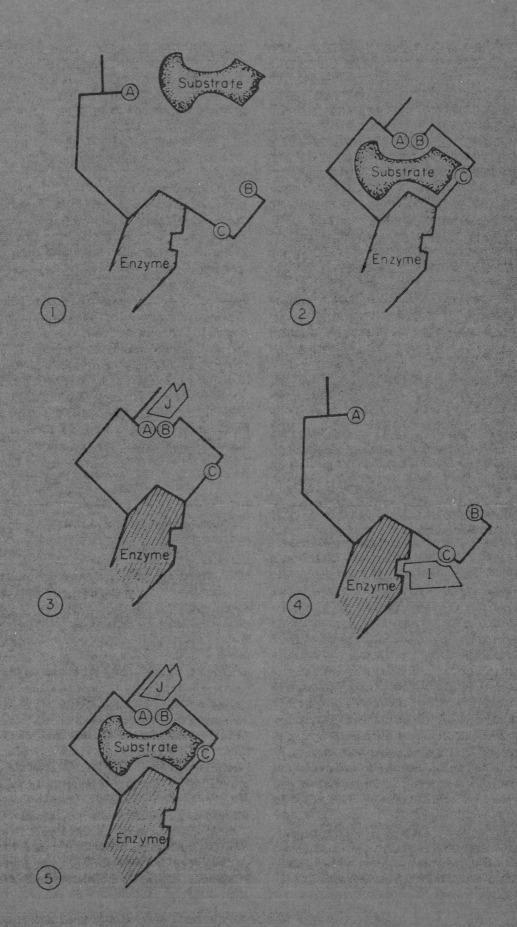

REGULATORY

BIOCHEMISTRY

IN NEURAL TISSUES

Schematic illustration of ligand-induced conformational changes in a monomer. (1) Protein conformation prior to ligand binding. (2) Protein conformation after substrate has induced conformational change leading to proper alignment of catalytic groups A and B. (3) Activator J induces conformational change aligning catalytic groups properly in the absence of substrate. (4) Inhibitor I induces conformational change maintaining peptide chain with B group in position such that substrate has too low affinity to bind effectively. (5) Substrate and activator bind to protein in active conformation. (From The Enzymes, *1970. P. Boyer, ed. 3rd ed., Vol. 1. New York: Academic Press.)*

REGULATORY BIOCHEMISTRY IN
NEURAL TISSUES

Introduction

LOUIS SOKOLOFF

A FEW WORDS ARE in order to define the nature and delineate the scope of the topic, *Regulatory Biochemistry of Neural Tissues,* as construed by us in the organization of the program. It will become obvious from the content of the presentations that the subject has been interpreted broadly and liberally to include extensive consideration of studies in biological systems other than nervous tissues. The choice has been deliberate. Years of study of the biochemistry of nervous tissues have led to the realization that basic biochemical processes in the CNS are not really fundamentally different from those of other tissues. Biochemical mechanisms elucidated first in simpler systems have in general been found to be quite relevant and to apply equally well to nervous tissues. The assumption has, therefore, been made that despite the unique functions of the nervous system, mechanisms of biochemical regulation in neural tissues will prove to be similar to those already defined or being elucidated in less complex cells and organisms. For this reason studies with all types of biological materials that contribute to the understanding of basic biochemical regulatory mechanisms have been accepted as being within the scope of this topic.

Biochemical regulation is currently a subject of intense interest and research activity. This wave of interest has evolved naturally and logically from the remarkable advances that have punctuated the history of biochemistry during the last several decades. Modern biochemistry was founded primarily in the old disciplines of physiological chemistry, euphemistically known as "blood and urine" chemistry, and nutrition. In its early period emphasis was placed on gross energy metabolism as studied by direct and indirect calorimetry and oxygen consumption. The composition of the diet and the caloric values and ultimate disposition of the various foodstuffs were described. Nutritional studies soon led to the discovery of essential components of the diet that were designated "vitamins." So little was known of their mode of action at first that Szent-Györgyi was led to define a vitamin as "something which if you do not eat, makes you sick." It was soon recognized, however, that vitamins are incorporated or converted in the body into relatively small molecules that serve as essential cofactors or co-enzymes for enzymes, large protein molecules that catalyze and are responsible for most of the chemical reactions proceeding in the tissues.

The discovery of the function of vitamins and their relationship to enzyme action led biochemistry into a new period, the era of intermediary metabolism and with it enzymology. It was during this phase that the myriad of chemical interconversions carried out in cells were described in minute detail and sorted into metabolic pathways that define all the intermediate steps in the chemical conversion of a given molecule into its final products. These are the studies that generated the multitude of metabolic maps, which describe the routes of the chemical traffic. These maps contain a variety of straight, branched, diverging, converging, parallel, and cyclical pathways. Nearly every step of every pathway has an enzyme to catalyze it. Not even so simple, spontaneous, and rapid a reaction as the combination of carbon dioxide and water to form carbonic acid, or the reverse, is allowed to proceed freely; it is catalyzed by its own enzyme, carbonic anhydrase. Nature has apparently found it advantageous to have almost all the biochemical reactions enzyme-catalyzed, and it is conceivable that the chief advantage of this arrangement is the potentiality for regulation.

Intermediary metabolism can be viewed as the study of the actions of proteins, specifically enzymes, on small molecules. The enormous number of overlapping and interlocking routes that comprise the pathways of intermediary metabolism obviously offer a great potential for chaos unless rigorously regulated. Regulatory biochemistry is concerned with the controls and regulatory mechanisms that ensure an orderly, coordinated traffic through these routes. There must also be control mechanisms to modify the traffic and adjust it to the changing needs that the cell experiences in response to a variety of physiological situations or stresses, such as changes in functional activity, exposure to hormones and drugs, growth and maturation, altered nutritional states, disease, etc. Since almost all metabolic reactions are catalyzed by enzymes, it is not surprising that their regulation is accomplished mainly by control of enzyme activity. In contrast to intermediary metabolism in which

the emphasis is on the effects of proteins on small molecules, regulatory biochemistry is concerned largely with the effects, both direct and indirect, of small molecules on proteins and, also, of proteins on other proteins. It is for this reason that consideration of the behavior of proteins and the control of enzyme activity permeates the presentations on this topic.

Studies in systems far simpler than nervous tissues have defined some fundamental mechanisms of the control of enzyme activity that must certainly obtain in neural tissues as well. One obvious mechanism is the regulation of the amount of enzyme in the cell, since the rates of enzyme-catalyzed reactions are generally proportional to the concentration of enzyme. It is a little more than a decade ago that the Jacob-Monod model was proposed to explain the induction of the synthesis of specific enzymes by certain small molecules in bacterial cultures in the log phase of growth. This model deservedly had a tremendous impact in biochemistry, but it was at first too quickly and sometimes uncritically applied to explain almost all changes in enzyme activities that occur in mammalian tissues in a variety of conditions. It soon became apparent, however, that stable populations of mammalian cells, constantly turning over their constituents, are not exactly like bacteria multiplying exponentially and that enzyme levels in cells are determined not only by the synthesis of the enzymes but by the balance between their rates of synthesis and degradation. Control of enzyme levels can, therefore, be achieved by regulation of the rate of protein synthesis or the rate of protein degradation.

Enzyme activity can also be regulated by modification of the catalytic efficiency of already existing enzyme molecules. Enzymes are large molecules with very complex three-dimensional structures. Many enzymes exist in aggregates of identical or nonidentical subunits. This structure influences not only the binding affinities of the enzyme for its substrates but also the spatial relationship of the substrates on the surface of the enzyme to each other and to the active catalytic site of the enzyme. The structure of an enzyme is not fixed; it can readily be modified by changes in the conditions of the milieu surrounding it, such as pH, ionic strength, etc. In addition, there are in many enzymes binding sites for specific small molecules, ligands, that may or may not be similar to the substrates of the chemical reaction. These sites and the ligands that they bind are not, however, directly related to the nature and mechanism of the enzyme-catalyzed reaction. They serve only a regulatory function. The binding of the ligand to the regulatory site in the enzyme results in a modification of the three-dimensional structure of the enzyme that is manifested in changes in its kinetic properties. Some enzymes show merely a reversible increase or decrease in activity; others are completely activated or deactivated. This is the essence of allosteric control of enzyme activity. It confers an additional and important mechanism of enzyme regulation, the modification of the catalytic efficiency of existing enzyme molecules without a change in their number.

The organization of this part heavily emphasizes these fundamental mechanisms of biochemical regulation. The first several presentations are directed specifically at the definition, description, and elucidation of these mechanisms, and extensive reference is made to studies in biochemical systems other than neural tissues that serve best to illustrate their mode of operation. That these regulatory mechanisms are also operative in the nervous system is virtually certain; they are unquestionably responsible for many of the biochemical and enzyme changes seen in the brain, especially during postnatal maturation and in response to altered functional activity. Subsequent presentations in the program focus directly on some selected examples of biochemical changes observed in neural tissues that can be examined in the light of the understanding of the basic mechanisms of enzyme regulation provided by the earlier discussions.

71 Principles Underlying the Regulation of Synthesis and Degradation of Proteins in Animal Tissues

ROBERT T. SCHIMKE

ABSTRACT The concentrations of the various protein components of cells, whether they be specific enzymes or components of organelles such as ribosomes or membranes, are the result of a delicate balance between their synthesis and their degradation. Either parameter can be affected by hormonal, nutritional, developmental, and genetic factors. The number of potential sites for regulation of specific protein synthesis, as well as specific protein degradation, are multiple, thereby allowing for multiple regulatory mechanisms that assure a fine control of protein levels as required by a changing environment of the cell and entire organism. This presentation will describe briefly the potential sites for regulation and then describe specific examples of such regulation. To generalize, it can be expected that given the specific organism or cell type or the given developmental, hormonal (etc.), variable, different sites and types of regulatory phenomena will be found.

Insofar as all events in a cell ultimately are directed by the expression of genetic material, it can be stated that gene expression underlies the change in any protein constituent. The question of whether all regulation of protein synthesis in animal cells is directly explicable in terms of the model of the lac repressor is unanswered at the present time. Various types of control can mimic the so-called "transcriptional" model of regulation, i.e., the synthesis or nonsynthesis of specific mRNA, including regulation of the number of genes coding for a given protein (gene amplification), and the regulation of transport of (potential) mRNA from nucleus to cytoplasm. The synthesis of specific mRNA can be controlled by the nature of proteins associated with the chromosome (comparable to the lac repressor protein), or may be specified by unique (specifier) proteins associated with the RNA polymerase (comparable to the sigma factors in prokaryotes).

The rate of protein synthesis (i.e., mRNA translation) is dependent on the number of ribosomes, the concentration of initiation, elongation, and termination factors, as well as aminoacyl-tRNA. In certain instances it can be shown that one or more of these factors is rate limiting. However, the extent to which any of these factors (most specifically, initiation factors and unique isoaccepting tRNAs) is specific for protein synthesis and hence can be a unique and specific regulator, is unknown.

The regulation of protein degradation (turnover) represents an additional means of modulating specific protein contents.

ROBERT T. SCHIMKE Department of Pharmacology, Stanford University, Stanford, California

Several general facts concerning this process include: (a) The rate of degradation of a protein is a function of that protein as a substrate for proteolysis. This accounts for the marked heterogeneity of rates of turnover of various proteins. (b) The rate of degradation of a given protein can be markedly altered by ligand interaction, or by mutations in the structure of the protein. (c) The activity of the degradation process can vary. Furthermore, our own data suggest that many, if not all, proteins, either cytoplasmic or those associated with organelles, are degraded as cytoplasmic components and in a monomeric (disassociated) state. The mechanisms for the degradation of proteins are not well understood and need not be uniform for all proteins or for all physiological or developmental states of the cell. Thus, a combination of lysosomal and soluble proteases is likely involved.

The concept to be developed is one in which the protein constituents of a cell are in a continual flux of synthesis and degradation, and of association and disassociation, with multiple controls serving to affect their steady-state levels.

REGULATION OF CELLULAR processes is obviously essential to life. Inasmuch as most significant biochemical reactions depend on catalysis by enzymes, the large measure of control is effected by regulating either the catalytic activity of preexisting enzyme molecules by ligand interaction or covalent modification (see chapter by Daniel Koshland, this volume) or by regulating the content of enzyme molecules, or both. Although the most classical and well-documented examples of enzyme regulation involve microorganisms, most specifically *Escherichia coli*, there is an increasing wealth of examples of adaptation of enzyme levels in animal tissues to a variety of hormonal, pharmacologic, genetic, and physiological variables (see review by Schimke and Doyle, 1970). Such changes do not simply reflect activation and inactivation of enzyme proteins, because agents that inhibit protein synthesis can prevent the increases in enzyme activity. More convincing are studies using combined immunologic and radioisotopic techniques in demonstrating both increased content of immunologically reactive protein and increased isotopically labeled amino acids into specific enzymes.

Central to an understanding of the dynamics of enzyme

regulation in mammalian tissues is the fact that changes in enzyme levels take place against a background of continual synthesis and degradation of proteins, i.e., turnover, as demonstrated so elegantly by Schoenheimer and his co-workers (1942) and studied more recently in other laboratories. Continual degradation of protein may be looked upon as an answer to the problem of how to remove enzymes when they are no longer needed, as part of an adaptive response to an altered metabolic state of a cell. In a rapidly growing bacterium, an unneeded enzyme can be diluted by subsequent cell division. In the generally nongrowing animal cell, intracellular degradation becomes increasingly important for this removal process and hence for controlling enzyme levels. Thus, in terms of understanding mechanisms, control both of synthesis and of degradation of proteins is at the very center of modern molecular biology.

In this presentation, we shall examine some of the underlying properties of the regulation of both protein synthesis and protein degradation, describe certain examples of these processes, and then speculate on mechanisms and approaches to an analysis of mechanisms.

Properties of protein turnover in rat liver

Since regulation of enzyme levels takes place against a background of continual synthesis and degradation, certain properties of this overall process are presented as a basis for subsequent discussions. Although the results presented are limited to rat liver, a number of studies have indicated that these general concepts are applicable to other animal tissues, including the nervous system, as well as *E. coli*.

TURNOVER IS EXTENSIVE Studies of Swick (1958), Buchanan (1961), and Schimke (1964) have indicated that essentially all proteins of rat liver take part in the continual replacement process. These studies have used the general technique of continuous administration of an isotope of known specific activity and subsequent comparison of the specific activity of the isotope isolated from protein with that of the administered isotope. For example, in studies using an algal diet of constant ^{14}C specific activity, Buchanan estimated that approximately 70% of rat liver protein was replaced every 4 to 5 days from the dietary source (Buchanan, 1961). Such replacement cannot represent serum proteins (such as albumin), because the steady state level of such proteins in liver is of the order of only 1 to 2% of total liver protein.

TURNOVER IS LARGELY INTRACELLULAR The life span of hepatic cells is of the order of 160 to 400 days (Buchanan, 1961; MacDonald, 1961; Swick et al., 1956), and hence the extensive turnover occurring in 4 to 5 days precludes cell replacement as the explanation for the turnover in liver.

THERE IS A MARKED HETEROGENEITY OF RATES OF REPLACEMENT OF DIFFERENT PROTEINS (ENZYMES) Table I provides a representative listing of rates of degradation of various specific proteins and cell organelles of rat liver. More extensive listings of various proteins are given by Schimke and Doyle (1970) and Rechcigl (1971). Remarkable is the wide range of half-lives for these specific proteins, ranging from 11 min for ornithine decarboxylase to 16 days for LDH_5. In addition, there is no necessary relationship between half-lives and metabolic functions of the enzymes. For instance glucokinase and LDH_5, both involved in carbohydrate metabolism, have markedly different half-lives (30 hr versus 16 days), as do tyrosine aminotransferase and arginase (1.5 hr versus 4 to 5 days), both involved in amino acid catabolism. Perhaps more striking is the lack of correlation between the cell fraction or organelles and the rate of turnover of specific proteins. For instance, δ-aminolevulinate synthetase, a mitochondrial enzyme, has a half-life of 1 hr, whereas the overall (or mean) rate of turnover of mitochondrial proteins is 4 to 5 days. Of particular interest is the relatively rapid turnover of cellular membranes of rat liver (half-lives of 2 to 3 days for endoplasmic reticulum and plasma membranes). Yet for specific enzymes of the endoplasmic reticulum, there is again a remarkable heterogeneity of turnover rates, varying from 2 hr for hydroxymethyl glutaryl CoA reductase to 16 days for NAD glycohydrolase.

This marked heterogeneity has several important implications. (a) Organelles are not turning over as units. Rather the various constituents of organelles, including mitochondria, intracellular membranes, as well as ribosomes (Dice and Schimke, 1972) and plasma membranes (Arias et al., 1969) are being replaced at markedly different rates. This has led us to propose, in addition to the metabolic flux involved in synthesis and degradation, an even less perceptible flux involving the continual association and dissociation of the various macromolecular constituents of cells, including multimeric proteins, complexes of proteins with phospholipids (membranes), and complexes of proteins with nucleic acids (ribosomes and chromatin) (see below). (b) The heterogeneity of turnover rates of different proteins is important for the reflection of altered rates of synthesis and degradation of proteins. As discussed extensively by Berlin and Schimke (1965), the time course of a change in an enzyme level is dependent on the rate of degradation of the protein. Thus only those proteins with short half-lives will change rapidly in content after an altered rate either of synthesis

TABLE I

Half-lives of specific enzymes and subcellular fractions of rat liver

Enzymes	Half-Life	Reference
Ornithine decarboxylase (soluble)	11 min	Russell and Snyder, 1968
δ-Aminolevulinate synthetase (mitochondria)	70 min	Marver et al., 1966
Alanine-aminotransferase (soluble)	0.7–1.0 day	Swick et al., 1968
Catalase (peroxisomal)	1.4 day	Price et al., 1962
Tyrosine aminotransferase (soluble)	1.5 hr	Kenney, 1967
Tryptophan oxygenase (soluble)	2 hr	Schimke et al., 1965
Glucokinase (soluble)	1.25 day	Niemeyer, 1966
Arginase (soluble)	4–5 day	Schimke, 1964
Glutamic-alanine transaminase	2–3 day	Segal and Kim, 1963
Lactate dehydrogenase isozyme-5	16 day	Fritz et al., 1969
Cytochrome c reductase (endoplasmic reticulum)	60–80 hr	Kuriyama et al., 1969
Cytochrome b_5 (endoplasmic reticulum)	100–200 hr	Kuriyama et al., 1969
NAD glycohydrolase (endoplasmic reticulum)	16 day	Bock et al., 1971
Hydroxymethylglutaryl CoA reductase (endoplasmic reticulum)	2–3 hr	Higgins et al., 1971
Acetyl CoA carboxylase (soluble)	2 day	Majerus and Kilburn, 1969
Cell fractions		
Nuclear	5.1 day	Arias et al., 1969
Supernatant	5.1 day	Arias et al., 1969
Endoplasmic reticulum	2.1 day	Arias et al., 1969
Plasma membrane	2.1 day	Arias et al., 1969
Ribosomes	5.0 day	Arias et al., 1969
Mitochondria	4–5 day	Swick et al., 1968

or of degradation. We might propose that for those metabolic pathways in which an alteration in enzyme content regulates the pathway (as opposed to an alteration in catalytic activity of the preexisting protein enzyme), the enzyme in question will have a rapid rate of steady-state turnover.

THERE IS A CORRELATION BETWEEN THE RATE OF DEGRADATION OF PROTEINS IN VITRO AND THEIR RATE OF PROTEOLYTIC ATTACK   This is shown in Figure 1. In such an experiment a double-label technique has been used in which one form of an isotopic amino acid, ^{14}C-leucine, was administered as a single intraperitoneal injection 5 days prior to the administration of ^{3}H-leucine. The animal was killed 6 hr later. The ^{14}C radioactivity then represents those proteins labeled 5 days prior to those labeled with the ^{3}H-leucine. As described by Arias et al. (1969) and employed extensively by Dehlinger and Schimke (1970) and Dice and Schimke (1972), if all proteins are turning over at the same rate, all proteins will have the same ^{3}H/^{14}C ratios, whereas if there is a heterogeneity of rates of turnover of different proteins, those ratios will differ. Proteins with high rates of turnover will have high ^{3}H/^{14}C ratios. The reader is referred to the above papers for a discussion of this technique, to the legend of Figure 1 for details, and to Figure 2 for the use of this technique in another context.

Figure 1 demonstrates that, during proteolysis of 100,000 g supernatant proteins by pronase, the radioactivity that is initially released to a trichloroacetic acid soluble form, i.e., amino acids and small peptides, has high ^{3}H/^{14}C ratios. This result indicates that a nonspecific protease preferentially degrades those proteins with a high rate of turnover in vivo. That this specificity resides with the structure of the proteins is shown by the control experiment in which the proteins were first denatured by 8 M urea at pH 9.5 with subsequent blocking of the sulfhydryl groups followed by proteolysis under conditions of protein and pronase concentrations similar to those employed for the "native" cytoplasmic proteins. After such denaturation proteolysis was more rapid, in keeping with the concept that unfolded, or denatured proteins are better substrates for proteolysis (Linderstrom-Lang, 1950). More importantly, there was now no discrimination between proteins on the basis of their rate of degradation in vivo, i.e., the solubilized radioactivity did not vary in the ^{3}H/^{14}C ratio. This experiment strongly supports the concept that inherent susceptibility of the protein to proteolytic attack is an important factor in establishing the heterogeneity of turnover rates of different proteins.

THERE IS AN APPARENT CORRELATION BETWEEN THE SIZE OF PROTEINS AND THEIR RELATIVE RATES OF DEGRADATION   During the course of studies on the heterogeneity of turnover of proteins of plasma membrane and endoplasmic

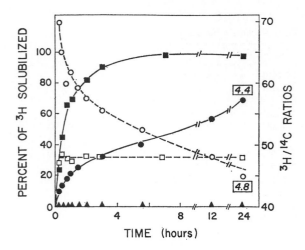

FIGURE 1 Susceptibility of double-labeled supernatant proteins of rat liver to proteolysis by pronase. A rat weighing 120 gm was given 200 μc of ^{14}C-leucine (uniformly labeled, specific activity 300 mc/mM) intraperitoneally, followed 5 days later with 400 μc of ^{3}H-L-leucine (3000 μc/mM). The animal was killed 6 hr later, and following homogenization in 0.25 M sucrose (2.5:1, sucrose to wet weight liver), and initial centrifugation at 1000 and 10,000 × g for 30 min each, a 100,000 × g supernatant fraction was obtained. This fraction was freed of amino acids by passage through a column of Sephadex G-25 equilibrated with 0.05 M potassium phosphate, pH 7.5. To a fraction of the supernatant, urea was added to a final molarity of 8.0, and adjusted to a pH of 9.5 with 3.0 M Tris-OH. After standing at room temperature for 60 min, the sulfhydryl groups were blocked by aminoethylation as described by Cole (1967). The denatured protein was dialyzed overnight against a large excess of 2.0 M urea in 0.05 M potassium phosphate, pH 7.5. The concentration of both the native and denatured proteins was adjusted to 15 mg/ml, and pronase was added to a final concentration of 80 μg/ml. Two ml samples were incubated at 37° with and without pronase, and at the times indicated 100 μl aliquots were removed and added to 0.05 ml of 10% trichloroacetic acid. The precipitates were allowed to sediment at 4° overnight, centrifuged, and 0.25 ml samples removed and extracted 3 times with 2 ml of ethyl ether. Then 0.20 ml samples were counted in a standard dioxane scintillation mixture in the presence of 0.5 ml of NCS solubilizer (Sanno et al., 1970). Samples of proteins were solubilized in 0.5 ml of NCS solubilizer and counted in the same manner. ▲, no pronase; ●, native proteins, cpm; ○, native proteins ^{3}H/^{14}C; ■, denatured proteins cpm; □, denatured proteins ^{3}H/[14]C.

reticulum, Dehlinger and Schimke (1971) made the observation that relative rates of turnover of proteins, as measured by the double isotope method of Arias et al. (1969) are related to the size of the protein (sub-unit) as electrophoresed on SDS acrylamide gels. Figure 2 depicts this general phenomenon for the proteins of the soluble fraction as fractionated on Sephadex G-200 columns. This correlation was found whether proteins were fractionated as multimeric proteins in the absence of SDS (Figure 2B), or in its presence (Figure 2A). More recently

Dice and Schimke have found this same correlation for proteins of rat liver ribosomes (1972), as well as "soluble" proteins from kidney, brain, and testis (Dice and Schimke, to be published). Such studies have led us to propose that the correlation of size and rate of degradation is based on the overall greater chance of a larger protein being "hit" by an endopeptidase, producing an initial rate-limiting peptide bond cleavage. Since the relative rate of degradation is of the same range of magnitude for the dissociated subunits as for the multimeric proteins (i.e., ^{3}H/^{14}C ratios are similar in Figures 2A and 2B), it is further suggested (Dice et al., in preparation) that proteins were degraded in a dissociated state. This suggestion was also made to explain the fact that this correlation exists for proteins of organelles (ribosomes and membranes), as well as so-called "soluble" proteins. Such studies then suggest another type of dynamic flux of intracellular proteins and organelles, one involving a continual association and dissociation of multimeric proteins and intracellular organelles. Such a concept is consistent with studies on exchange of phospholipids of membranes (Wirtz and Zilversmit, 1968), exchange of ribosomal proteins (Dice and Schimke, 1972; Warner, 1971), and with known association-dissociation phenomena of purified proteins.

Independent regulation of the rates of synthesis and degradation of specific enzymes

There is a wealth of examples of a variety of hormonal, genetic, pharmacologic, and nutritional alterations affecting independently the rates of synthesis and degradation of individual proteins. A number of such examples have been reviewed recently by Schimke and Doyle (1970). Below are presented two such examples, which demonstrate the importance of both synthesis and degradation in controlling specific protein levels.

HORMONE AND SUBSTRATE REGULATION OF TRYPTOPHAN OXYGENASE (PYRROLASE)   As shown originally by Knox and Mehler (1951), and studied extensively by Knox and his collaborators (1967) as well as by Feigelson and Greengard (1962), the activity of tryptophan oxygenase can be increased by the administration of either hydrocortisone and other glucocorticoids, or by tryptophan, or certain tryptophan analogues.

Figure 3 shows the time course of increase in tryptophan oxygenase following the administrations of tryptophan, hydrocortisone, or both, to adrenalectomized rats at 4-hour intervals (Schimke et al., 1965). Hydrocortisone resulted in an initial rapid accumulation of enzyme, followed by a plateau after about 8 hr. Tryptophan resulted in a nearly linear increase in enzyme amounting

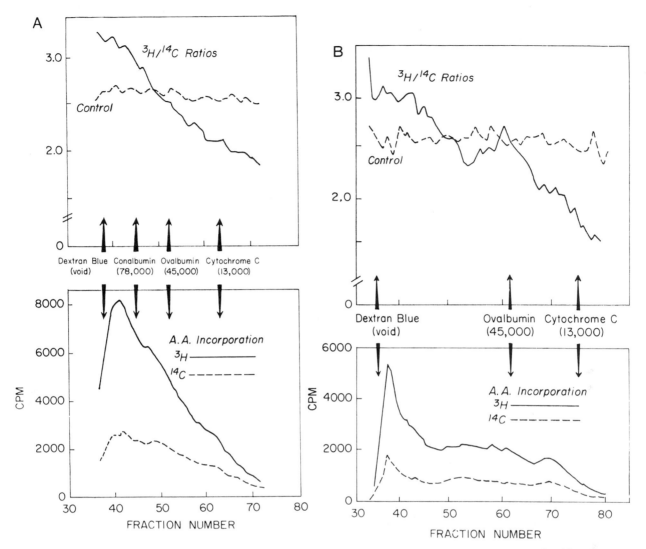

FIGURE 2 Relative rate of degradation of "soluble" proteins of rat liver as a function of molecular size. Relative rates of degradation were estimated by the double isotope method of Arias et al. (1969) in which ^{14}C-leucine is administered to rats four days prior to ^{3}H-leucine administration, with death of animals 4 hr later. The "control" indicates rats receiving both isotope forms of leucine at the same time. High ^{3}H/^{14}C ratio indicates relatively high rates of degradation. Proteins in absence (A) and presence (B) of SDS to disrupt multimeric proteins, were chromatographed on Sephadex G-200 columns. Details are given in Dehlinger and Schimke (1970).

to a 5-fold increase in 16 hr. The administration of both hydrocortisone and tryptophan caused a nearly linear increase of levels 25- to 50-fold greater than basal levels. These time courses of accumulation could be explained on the basis of the theoretical formulation described previously if hydrocortisone increased the rate of enzyme synthesis about 4- to 6-fold without affecting the rate of enzyme degradation, and if tryptophan administration did not appreciably alter the rate of enzyme synthesis but diminished the rate at which the enzyme was inactivated or degraded.

In order to substantiate the conclusions derived from the time course results, experiments were undertaken to study the incorporation and loss of isotopic amino acids from tryptophan oxygenase. In order to assess the effect of hydrocortisone and tryptophan on the rate of enzyme synthesis, studies were undertaken on the short-term incorporation of L-leucine-^{14}C into tryptophan oxygenase. Hydrocortisone or tryptophan, or both, were administered to rats for varying time periods, followed by leucine-^{14}C. Extracts of livers were prepared 40 min after isotope administrations, and the enzyme was isolated by means of the antibody that specifically precipitates tryptophan oxygenase. These results are summarized in Table II. With no treatment, a total of about 1400 cpm were incorporated into tryptophan

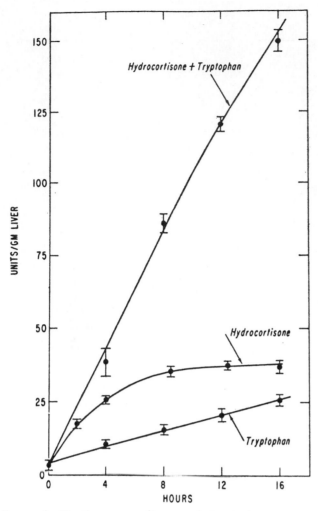

FIGURE 3 The time course of increases in tryptophan oxygenase produced by repeated administrations of hydrocortisone and tryptophan. Adrenalectomized rats weighing 150 to 170 gm each were given injections as follows every 4 hr: 150 mg of L-tryptophan in 12 ml of 0.85% NaCl intraperitoneally and 5 mg of hydrocortisone 21-phosphate subcutaneously. Every 4 hr, the livers of four animals were assayed for tryptophan pyrrolase activity. Brackets indicate ±2 standard errors of the mean (Schimke et al., 1965).

pyrrolase of livers of two animals during a 40-min period. When hydrocortisone has been administered 4 hr previously, total incorporation into tryptophan oxygenase increased some 4-fold (5640 cpm). After repeated doses of hydrocortisone at 4-hr intervals for 12 hr, the extent of incorporation during a 40-min period remained about 4-fold greater than with no treatment. Tryptophan administration, in contrast, increased only slightly the incorporation (1620 cpm versus 1400 cpm). Hydrocortisone plus tryptophan increased the incorporation in untreated animals about 5-fold. These changes in amount of isotope incorporation into tryptophan oxygenase are to be contrasted with the lack of comparable effects with

TABLE II

*Forty-minute incorporation of leucine-^{14}C into rat liver tryptophan oxygenase**

Treatment	Enzyme Activity (units/ gm liver)	Leucine-^{14}C Incorporation	
		Tryptophan Oxygenase (total cpm)	Supernatant Protein (cpm/mg)
None	4.2	1368	1190
Hydrocortisone			
4 hr	13.6	5640	1320
12 hr	31.4	6502	1491
Tryptophan			
4 hr	8.2	1620	1564
12 hr	14.1	1670	1165
Hydrocortisone + tryptophan			
4 hr	28.3	7680	1491
12 hr	72.0	7280	1018

*Rats were given repeated doses of hydrocortisone or L-tryptophan, or both, at 4-hr intervals for the times indicated. Each rat was given, 40 min before death, a single intraperitoneal injection of 20 μc of leucine-^{14}C in 1.0 ml of 0.85% NaCl. Results of L-leucine-^{14}C incorporation into tryptophan oxygenase are reported as total net counts per minute in the precipitate from the total DEAE-cellulose extract of two rats. See Schimke et al. (1965) for details.

total protein. The contrast indicates a high degree of specificity to the hydrocortisone effect. The results, then, indicate that hydrocortisone increased the rate of synthesis of tryptophan oxygenase about 4- to 5-fold.

Evidence that tryptophan prevents the breakdown of the active, immunologically reactive enzyme is shown in Figure 4. In this experiment the enzyme was prelabeled by the administration of leucine-^{14}C 60 min before the time indicated as zero. We have found that within 40 min after the administration of a single dose of leucine-^{14}C, incorporation of radioactivity into total liver protein is essentially complete. Therefore any protein synthesized after this time will be derived from unlabeled amino acids. As shown in this experiment, in control animals the amount of enzyme activity and radioactivity present in total protein remained essentially constant over a 9-hr period. However, the radioactivity present in prelabeled enzyme decreased progressively. This, then, is further evidence that there is a continual degradation of the enzyme under basal conditions. When tryptophan was administered, on the other hand, the total amounts of enzyme increased. In contrast to control animals, there was no decrease in the amount of prelabeled enzyme. Thus the substrate, tryptophan, resulted in accumulation of enzyme in large part by preventing degradation of the enzyme in the presence of continued synthesis.

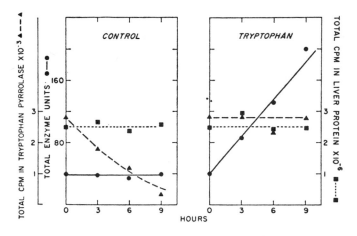

FIGURE 4 Effect of L-tryptophan administration on the loss of tryptophan oxygenase (pyrrolase) prelabeled with L-leucine-^{14}C. Rats were given single injections of 20 μc of L-leucine-^{14}C. Sixty minutes later, two animals were killed. The remainder were given 10 ml of 0.85% NaCl or 10 ml of 0.85% NaCl containing 150 mg of L-tryptophan. These injections were repeated in the remaining animals after 4 and 8 hr. At the times specified, the livers of two animals in each group were removed and frozen immediately. At the end of the experiment, extracts of the livers were prepared, and the radioactivity that was incorporated into tryptophan oxygenase and protein was determined. The values given are for totals of combined extracts of two animals; (●——●) enzyme activity; (▲--▲) total radioactivity in protein precipitated by the tryptophan oxygenase antiserum; (■--■) radioactivity in total cellular protein (Schimke et al., 1965).

REGULATION OF LDH$_5$ ISOZYME LEVELS IN VARIOUS TISSUES OF THE RAT In many instances there are remarkable differences in the specific activities of enzymes catalyzing the same reaction in different tissues. In certain cases, particularly in comparing liver with tissues such as muscle or kidney, this is based on the fact that they represent completely different proteins with different regulatory properties, for example the hexokinase isozymes, including the liver-specific glucokinase (Niemeyer, 1966), and the pyruvate kinase isozymes (Tanaka et al., 1967). One particularly important study is that of Fritz et al. (1969) who have determined the rates of synthesis and degradation of lactate dehydrogenase isozyme-5 in rat liver, heart muscle, and skeletal muscle as summarized in Table III. These workers have found that the tissue differences in enzyme levels were not related solely to rates of synthesis, but that the rate of LDH$_5$ degradation was also markedly different in the different tissues. Thus the half-lives of this isozyme were 16, 1.6, and 31 days in liver, heart muscle, and skeletal muscle, respectively, compared to mean half-lives of 2.2, 1.0, and 22 days for total soluble protein in these same tissues. Thus the same isozyme in different tissues may be degraded at markedly different rates. Hence we cannot assume that differences

TABLE III
Steady-state levels of LDH$_5$ and rates of synthesis and degradation

Tissue	Enzyme Level (pN/g tissue)	K_s (pM/g/ day)	K_d (day^{-1})	Half-Life†
Heart	5.5	2.2	0.400	1.6 (1.0)
Muscle	294	65.2	0.018	31 (22.0)
Liver	1600	65.0	0.040	16 (2.2)

*From Fritz et al. (1969).
†Parenthesis indicate values for total soluble protein.

in enzyme levels from tissue to tissue result from differences in rates of synthesis.

The pattern that emerges from these studies, as well as a vast number not discussed, is one of a continual change, in which the rate at which the steady-state complement of specific molecules of a particular enzyme is replaced varies from several hours to many days. In fact, the concept of a "steady state" is most likely a misnomer, and rates of synthesis and degradation of proteins are most likely continually changing in response to dietary, hormonal, and physiologic variables. Thus the constancy of ordered structure and function of cells takes place against a continual flux in the constituent macromolecules, and in fact, change is the only constant.

On the regulation of enzyme synthesis

Figure 5 gives a schematic depiction of the various steps involved in the initial transcription of RNA and DNA, its transport out of the nucleus, its association with ribosomal subunits, release of the completed peptide chain, and finally the assembly of the protein subunits into their functional state, associated with other subunits (enzymes), nucleic acids (ribosomes and chromatin), or phospholipids (membranes). The question of what regulates the final appearance of a protein as an active enzyme or structural unit need not, and in all likelihood does not, have a unique or universal answer. Potentially any of the processes listed, or any single reactant, may be rate limiting.

The most obvious question is whether, as in bacteria, control of protein synthesis is exerted at the level of immediate and continued synthesis of specific mRNA as directed by the interaction of regulatory proteins with a specific region of DNA (Epstein and Beckwith, 1968). There is, indeed, considerable evidence that alterations of RNA metabolism are involved in enzyme regulation in animal tissues. Thus a majority of hormonal, drug, and nutritionally induced increases in enzymes are prevented by the administration of actinomycin D or other inhibitors

ROBERT T. SCHIMKE   819

STEPS IN PROTEIN SYNTHESIS

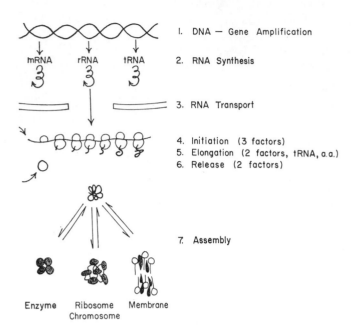

1. DNA — Gene Amplification

2. RNA Synthesis

3. RNA Transport

4. Initiation (3 factors)
5. Elongation (2 factors, tRNA, a.a.)
6. Release (2 factors)

7. Assembly

Enzyme   Ribosome   Membrane
          Chromosome

FIGURE 5 Schematic representation of steps in protein synthesis and possible sites of regulation.

of RNA synthesis. Administration of an inhibitor of RNA synthesis at the time of drug or hormone administration characteristically prevents the increase, whereas delayed administration does not prevent enzyme accumulation (Garren et al., 1964; Greengard and Dewey, 1968). This observation suggests that RNA synthesis is necessary for the initiation of increased synthesis of specific protein, but once that RNA synthesis is accomplished, its utilization can take place for some time. This general finding is in keeping with the concept that mRNAs of animal tissues are relatively long-lived (Pitot et al., 1965; Revel and Hiatt, 1964) when compared to the average 2 or 3 min half-life of bacterial mRNAs (Leive, 1965; Morse et al., 1969). In addition there is ample evidence that some mRNAs, in particular those coding for specific differentiated proteins, are long-lived (Kafatos and Reich, 1968; Palmiter et al., 1971; Papaconstantinou, 1967; Pitot et al., 1965; Wessells and Wilt, 1965; Yaffe and Feldman, 1964). It is the finding of relatively long lives of mRNA that has led to a variety of proposals that regulation can occur at one or more of the many steps that occur subsequent to mRNA synthesis.

It has been customary in studies with animal tissues to differentiate between "transcription" and "translation" as the level at which control of specific protein synthesis is exerted. Operationally this distinction has been based primarily on whether actinomycin D prevents the increase in enzyme content. The use of actinomycin D for this distinction is unsound for a variety of reasons. Thus the use of actinomycin D cannot distinguish between an effect on the synthesis or nonsynthesis of mRNA (the strictly transcriptional model), one in which gene amplification is the primary effect (Arias et al., 1969; Aronson and Wilt, 1969; Dawid and Brown, 1968) or the case in which the primary effect is on mRNA transport, where potential mRNA may be synthesized but is rapidly degraded within the nucleus (Aronson and Wilt, 1969; Attardi et al., 1966; Harris, 1968; Scherrer et al., 1966; Shearer and McCarthy, 1967; Soeiro et al., 1968), unless it is transported into the cytoplasm and there utilized for protein synthesis. Georgiev (1967) has reviewed evidence suggesting that actinomycin D may inhibit the transport phenomenon. In addition, when using actinomycin D, it is never clear that the inhibition of the specific mRNA for the enzyme in question is the event that prevents enzyme accumulation. Thus the inhibition of synthesis of a labile RNA species necessary for the utilization of a specific mRNA may be the underlying action of actinomycin D in a given case.

In addition to gene amplification, altered rates of synthesis of mRNA, and transport of mRNA into the cytoplasm at sites at which regulation can take place, various steps in the utilization of mRNA have been proposed for regulation. Thus Heywood (1970) has suggested that there is specificity for the initiation of myosin mRNA as studied in the reticulocyte protein synthesizing system. Regulation of the stability of mRNA species has been championed in the so-called *membron* theory of Pitot et al. (1969). Tomkins et al. (1969) have developed a theory of a cytoplasmic repressor protein that binds to mRNA, preventing its translation, and at the same time initiating its degradation. Regulation of ribosome function on the basis of variations in tRNA acceptor properties (Kano-Sueoka and Sueoka, 1966; Maenpaa and Bernfield, 1970), or by the synthesis of specific proteins as studied by Martin and Wool (1968), or as controlled by phosphorylation of ribosomal proteins (Kabat, 1971) have been proposed as regulatory sites. Amino acid availability will obviously regulate the rate of protein synthesis. Potter et al. (1968) have described the influx and efflux of amino acids from rat liver as a function of feeding schedules. Hence, in the intact animal, the supply of amino acids in the liver is not constant. The cyclic variations in tyrosine transaminase activity (Wurtman and Axelrod, 1967) and the effect of growth hormone on tyrosine transaminase (Kenney and Albritton, 1965; Labrie and Korner, 1968a) may be explained on the basis of amino acid availability, altering rates of synthesis of a rapidly degraded enzyme. Munro (1968) and Sidransky et al. (1968) have shown that the profile of rat liver polysomes is extremely sensitive to amino acid availability, and in particular to the availability of tryptophan. This finding may underlie the effect of low doses of tryptophan in increasing activity of serine dehydratase (Peraino et al., 1965), tyrosine transaminase (Labrie and Korner, 1968a), and tryptophan pyrrolase (Labrie and Korner, 1968a). Last, control of protein synthesis at the level of release of specific peptides has been suggested (Cline and Bock, 1966).

Clearly what is needed is the ability to isolate all potential reactants in the chain of specific protein synthesis, including specific polysomes with their (specific or nonspecific) initiation and release factors, specific mRNA, and genes. Recent advances in the ability to utilize mammalian mRNAs for protein synthesis, including globin chains (Gurdon et al., 1971; Lockard and Lingrel, 1969), light immunoglobulin chains (Stavnezer and Huang, 1971), and ovalbumin (Rhoads et al., 1973), and the development of methodology for isolation of specific polysomes utilizing immunoprecipitation techniques (Allen and Terrence, 1968; Holme et al., 1971; Schubert, 1968; Takagi and Ogata, 1971), should allow for more searching analyses of the question of what is rate limiting.

The isolation and quantification of specific mRNAs should then allow for a more profound analysis of the nature of the RNA that turns over rapidly in the nucleus (Aronson and Wilt, 1969; Attardi et al., 1966; Harris, 1968; Scherrer et al., 1966; Soeiro et al., 1968), and perhaps a better understanding of reiterated DNA sequences of animal tissue (Britten and Kohne, 1968; McCarthy and McConaughy, 1968) than is currently available from the multitude of hybridization studies using DNA and so-called mRNAs of unknown properties. Obviously the eventual goal will be an understanding of interactions of proteins and DNA in the regulation of gene expression, including the role of various nuclear and chromosomal protein fractions (histones and acidic proteins), and enzymatic modifications of these proteins by phosphorylation, acetylation, and so on (Stellwagen and Cole, 1969).

Enzyme degradation

Any understanding of the molecular mechanisms for degradation of proteins (enzymes) must take into account certain fundamental properties of such turnover as follows:

1. The degradation appears to be random, inasmuch as the loss of labeled protein during chase period, or fall of

enzyme activity following elevation to a high level, follows first-order kinetics.

2. There is a marked heterogeneity of turnover rates of individual proteins.

3. The rate constant of degradation is in many cases characteristic of a given protein but in other cases can be markedly altered.

4. There is a general correlation between the size of a protein and its relative rate of degradation.

Two general mechanisms can be considered, as summarized in Figure 6.

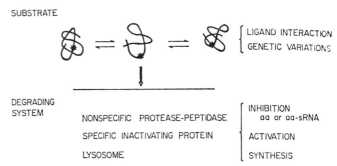

FIGURE 6 Summary of possible means of regulating protein degradation.

PROPERTIES OF THE PROTEIN MOLECULE AS A SUBSTRATE FOR DEGRADATION Protein molecules can exist in a number of different conformational states of varying degrees of detection. A protein molecule might be subject to degradation only when it assumes certain conformations. Thus a heterogeneity of degradation rates could exist, depending on the number and nature of particularly labile peptide bonds exposed in certain conformations. In addition, the interactions of proteins with various ligands, including other proteins, lipids, and small molecules, can alter such conformations and thereby alter proteins as substrates for inactivation. The model that emerges, then, is one in which protein molecules are individually available to a degradative process that is present at all times. Shifting concentrations of substrates, cofactors, and so on, as occur under various hormonal and physiological conditions, would lead to a variety of effects on specific enzymes, either to stabilize or labilize them. Such a concept has also been expressed by Grisolia (1964) and Pine (1966, 1967). Consistent with this is the finding that there is a general correlation between known rates of degradation of proteins in vivo and their rate of inactivation by trypsin and chymotrypsin (Bond, 1971). In addition, Goldberg (1972) has recently shown that when amino acid analogs are incorporated into the proteins in *E. coli*, protein degradation is accelerated, either as studied in vivo or in extracts. In addition, the effects of ligands to alter heat and proteolytic inactivation of numerous proteins are well known (Green and Neurath, 1954).

Such a concept could also explain the development of heterogeneity of rate constants of degradation. Taking a cue from mutations in *E. coli*, which decrease the stability of the lac repressor (Platt et al., 1970), and an arginine tRNA synthetase (Williams and Neidhardt, 1969), as well as the mutations that affect the stability of catalase (Matsubara et al., 1967) and glucose-6-phosphate dehydrogenase (Yoshida et al., 1967), we can readily envisage the retention of those mutations that either increase or decrease stability of a protein, depending on whether rapid or slow turnover is advantageous to the organism.

The correlation between the size of a protein and its relative rate of degradation does not hold for specific proteins, such as LDH_5, arginase, and tyrosine aminotransferase, all of which are of approximately the same molecular size (Auricchio et al., 1970; Castellino and Barker, 1968; Hirsch-Kolb and Greenberg, 1968) but have markedly different half-lives of 16 days (Fritz et al., 1969), 4 to 5 days (Schimke, 1964), and 1.5 hr (Kenney, 1967), respectively. Dehlinger and Schimke (1970) have proposed that the degradation may not occur as the multimeric protein but rather in the dissociated state, a proposal that is in keeping with the suggestion of Fritz et al. (1971). Thus one of the rate limiting parameters for degradation to be considered should be the facility of dissociation of the protein into subunits.

ALTERATIONS IN ACTIVITY OF A DEGRADATIVE PROCESS
In the above model the activity of the degradative process was assumed to be in excess. It is also conceivable that the rate of degradation may be dependent on the activity of the degrading system, as controlled by activation inhibition, translocation within the cell, or de novo synthesis of degrading enzymes. Considerations of enzymatic mechanisms are hampered by lack of suitable mutants in the degradative process itself. Another problem involves the identification of the products of specific protein degradation once a protein has lost enzymatic activity or immunologic reactivity. Several curious observations are of note that should be explained in the formulation of a suitable mechanism(s) for degradation. In both animal and bacterial systems, inhibition of energy production and protein synthesis inhibits protein degradation (see Schimke et al., 1970 for detailed review). Various explanations have been offered for such observations, including cofactor requirements (Penn, 1961), necessity for maintaining structural integrity of organelles such as lysosomes (Brostrom and Jeffay, 1970), and requirement for continued synthesis of degradative enzymes that are turning over rapidly (Kenney, 1967). More indirect, but equally plausible from the experimental data available, are effects of accumulated amino acids, tRNA species, and so on, that may regulate by ligand interaction the activity of degradative enzymes or specific (enzyme) substrates.

One obvious candidate for a degradative system is lysosome, which occurs in virtually all cells (DeDuve and Wattiaux, 1966). Lysosomes are intracellular organelles that contain acid hydrolases and are currently conceived as involved in the autophagy of discrete areas of cytoplasm. It is not difficult to conceive that lysosomes are involved in that protein degradation whose properties involve randomness and heterogeneity of degradation rate constants among different proteins, whether so-called soluble proteins or those associated with membranes or ribosomes. Thus some mechanism would be required for the recognition of whether a protein molecule were to be degraded and perhaps involve transport into a lysosome, acetylation, formylation, or as recently suggested, deamidation (Robinson et al., 1970). It seems reasonable to this author to propose that the system of lysosomes is important where cell involution or gross changes in rates of protein degradation occur, such as starvation and cell death, whereas the degradation that occurs in normal steady-state conditions involves a system(s) not clearly understood at present (Hartley, 1960). This could involve lysosomes, but acting as a sieve, rather than in an "all-or-none" fashion.

Another possibility is that there are specific degrading enzymes for specific proteins. There are examples of proteins or enzymes that appear to inhibit or inactivate specific enzymes (Blobel and Potter, 1966; Bonsignore et al., 1968; Dvorak et al., 1966; Gancedo and Holzer, 1968; Messenguy and Wiame, 1969). In addition, studies by Tata (1966) and by Gross and Lapiere (1962) on degeneration of amphibian tail during thyroxine-induced metamorphosis and of Houck et al. (1968) on the glucocorticoid-induced degeneration of skin of rats suggest that de novo synthesis of specific proteolytic enzymes is required for these instances of tissue involution.

At one extreme, then, we might propose that the degradation of each protein requires a specific protein. This, however, is impossible, because the continual replacement of essentially all proteins would require that there exist a protein to degrade a protein . . . ad infinitum. It is most likely that, just as there are a number of different enzymes that hydrolyze RNA in an organism such as *E. coli*, there are also a number of different types of proteases in animal cells (Hartley, 1960) performing different functional tasks at different sites and times, the sum total of which results in continual protein degradation.

REFERENCES

Allen, E. R., and C. F. Terrence, 1968. Immunochemical and ultrastructural studies of myosin synthesis. *Proc. Natl. Acad. Sci. USA* 60:1209–1215.

Arias, I. M., D. Doyle, and R. T. Schimke, 1969. Studies on the synthesis and degradation of proteins of the endoplasmic reticulum of rat liver. *J. Biol. Chem.* 244:3303–3315.

Aronson, A. I., and F. H. Wilt, 1969. Properties of nuclear RNA in sea urchin embryos. *Proc. Natl. Acad. Sci. USA* 62:186–193.

Attardi, G., H. Parnas, M-I. H. Hwang, and B. Attardi, 1966. Giant-size rapidly labeled nuclear ribonucleic acid and cytoplasmic messenger ribonucleic acid in immature duck erythrocytes. *J. Mol. Biol.* 20:145–182.

Auricchio, F., F. Valierote, G. Tomkins, and W. Riley, 1970. Studies on the structure of tyrosine aminotransferase. *Biochim. Biophys. Acta* 221:307–313.

Berlin, C. M., and R. T. Schimke, 1965. Influence of turnover rates on the responses of enzymes to cortisone. *Mol. Pharmacol.* 1:149–156.

Blobel, G., and V. R. Potter, 1966. Relation of ribonuclease and ribonuclease inhibition to the isolation of polysomes from rat liver. *Proc. Natl. Acad. Sci. USA* 55:1283–1288.

Bock, K. W., P. Siekevitz, and G. E. Palade, 1971. Localization and turnover studies of membrane nicotinamide adenine dinucleotide glycohydrolase in rat liver. *J. Biol. Chem.* 246:188–195.

Bond, J. S., 1971. A comparison of the proteolytic susceptibility of several rat liver enzymes. *Biochem. Biophys. Res. Commun.* 43:333–339.

Bonsignore, A., A. DeFlora, M. A. Mangiarotti, I. Lorenzoni, and S. Alema, 1968. A new hepatic protein inactivating glucose 6-phosphate dehydrogenase. *Biochem. J.* 106:147–154.

Britten, R. J., and D. W. Kohne, 1968. Repeated sequences in DNA. *Science* 161:529–540.

Brostrom, C. O., and H. Jeffay, 1970. Protein catabolism in rat liver homogenates. A reevaluation of the energy requirement for protein catabolism. *J. Biol. Chem.* 245:4001–4008.

Buchanan, D. L., 1961. Total carbon turnover measured by feeding a uniformly labeled diet. *Arch. Biochem. Biophys.* 94:500–511.

Castellino, F. J., and R. Barker, 1968. Examination of the dissociation of multichain proteins in guanidine hydrochloride by membrane osmometry. *Biochemistry* 7:2207–2217.

Cline, A. L., and R. M. Bock, 1966. Translational control of gene expression. *Cold Spring Harbor Symp. Quant. Biol.* 31:321–333.

Cole, D., 1967. δ-Aminoethylation. In *Methods in Enzymology*, Vol. XI, S. P. Colowick and N. O. Kaplan, eds. New York: Academic Press, pp. 315–317.

Dawid, I. B., and D. M. Brown, 1968. Specific gene amplification in oocytes. *Science* 160:272–280.

DeDuve, C., and R. Wattiaux, 1966. Functions of lysosomes. *Ann. Rev. Physiol.* 28:435–492.

Dehlinger, P. J., and R. T. Schimke, 1970. Effect of size on the relative rate of degradation of rat liver soluble proteins. *Biochem. Biophys. Res. Commun.* 40:1473–1480.

Dice, J. F., and R. T. Schimke, 1972. Turnover and exchange of ribosomal proteins from rat liver. *J. Biol. Chem.* 247:98–111.

Dvorak, H. F., Y. Anraku, and L. A. Heppel, 1966. The occurrence of a protein inhibitor for 5′-nucleotidase in extracts of *Escherichia coli*. *Biochem. Biophys. Res. Commun.* 24:628–632.

Epstein, W., and J. R. Beckwith, 1968. Regulation of gene expression. *Ann. Rev. Biochem.* 37:411–436.

Feigelson, P., M. Feigelson, and O. Greengard, 1962. Comparison of the mechanisms of hormonal and substrate induction of rat liver tryptophan pyrrolase. *Recent Progr. Hormone Res.* 18:491–507.

Fritz, P. J., E. L. White, E. S. Vesell, and K. M. Pruitt, 1969. The roles of synthesis and degradation in determining tissue concentrations of lactate dehydrogenase-5. *Proc. Natl. Acad. Sci. USA* 62:558–565.

Fritz, P. J., E. L. White, E. S. Vessell, and K. M. Pruitt, 1971. *Nature New Biol.* 230:119–122.

Gancedo, C., and H. Holzer, 1968. Enzymatic inactivation of glutamine synthetase in *Enterobacteriaceae*. *Eur. J. Biochem.* 4:190–192.

Garren, L. D., R. R. Howell, G. M. Tomkins, and R. M. Crocco, 1964. A paradoxical effect of actinomycin D: The mechanism of regulation of enzyme synthesis by hydrocortisone. *Proc. Natl. Acad. Sci. USA* 52:1121–1129.

Georgiev, G. P., 1967. The nature and biosynthesis of nuclear ribonucleic acids. *Progr. Nucleic Acid. Res. Mol. Biol.* 6:259–351.

Goldberg, A. L., 1972. *Nature New Biol.* 240:147–150.

Green, N. M., and H. Neurath, 1954. Proteolytic enzymes. In *The Proteins*, Vol. II, Part B, H. Neurath and K. Bailey, eds. New York: Academic Press, pp. 1057–1198.

Greengard, O., and H. K. Dewey, 1968. The developmental formation of liver glucose 6-phosphatase and reduced nicotinamide adenine dinucleotide phosphate dehydrogenase in fetal rats treated with thyroxine. *J. Biol. Chem.* 243:2745–2749.

Grisolia, S., 1964. The catalytic environment and its biological implications. *Physiol. Rev.* 44:657–712.

Gross, J., and C. M. Lapiere, 1962. Collagenolytic activity in amphibian tissues: A tissue culture assay. *Proc. Natl. Acad. Sci. USA* 48:1014–1022.

Gurdon, J. B., C. D. Lane, H. R. Woodland, and G. Marbaix, 1971. Use of frog eggs and oocytes for the study of messenger RNA and its translation in living cells. *Nature (Lond.)* 233:177–182.

Harris, H., 1968. *Nucleus and Cytoplasm*. London: Oxford University Press.

Hartley, B. S., 1960. Proteolytic enzymes. *Ann. Rev. Biochem.* 29:45–72.

Heywood, S., 1970. Specificity of mRNA binding factor in eukaryotes. *Proc. Natl. Acad. Sci. USA* 67:1782–1788.

Higgins, M., T. Kawachi, and H. Rudney, 1971. The mechanism of the diurnal variation of hepatic HMG-CoA reductase activity in the rat. *Biochem. Biophys. Res. Commun.* 45:138–150.

Hirsch-Kolb, H., and D. M. Greenberg, 1968. Molecular characteristics of rat liver arginase. *J. Biol. Chem.* 243:6123–6129.

Holme, G., S. L. Boyd, and A. H. Sehon, 1971. Precipitation of polyribosomes with pepsin digested antibodies. *Biochem. Biophys. Res. Commun.* 45:240–245.

Houck, J. C., V. K. Sharma, Y. M. Patel, and J. A. Gladner, 1968. Induction of collagenolytic and proteolytic activities by anti-inflammatory drugs in the skin and fibroblast. *Biochem. Pharmacol.* 17:2081–2090.

Kabat, D., 1971. Phosphorylation of ribosomal proteins in rabbit reticulocytes. A cell-free system with ribosomal protein kinase activity. *Biochemistry* 10:197–203.

Kafatos, F. C., and J. Reich, 1968. Stability of differentiation-specific and non-specific messenger RNA in insect cells. *Proc. Natl. Acad. Sci. USA* 60:1458–1465.

Kano-Sueoka, T., and N. Sueoka, 1966. Modification of leucyl-sRNA after bacteriophage infection. *J. Mol. Biol.* 20:183–209.

Kenney, F. T., 1967. Turnover of rat liver tyrosine transaminase: Stabilization after inhibition of protein synthesis. *Science* 156: 525–528.

Kenney, F. T., and W. L. Albritton, 1965. Repression of enzyme synthesis at the translational level and its hormonal control. *Proc. Natl. Acad. Sci. USA* 54:1693–1698.

Knox, W. E., and A. H. Mehler, 1951. The adaptive increase of the tryptophan peroxidase-oxidase system of liver. *Science* 113:237–238.

Knox, W. E., and M. M. Piras, 1967. Tryptophan pyrrolase of liver, III. Conjugation in vivo during cofactor induction by tryptophan pyrrolase. *J. Biol. Chem.* 242:2965–2969.

Kuriyama, T., T. Omura, P. Siekevitz, and G. E. Palade, 1969. Effects of phenobarbital on the synthesis and degradation of the protein components of rat liver microsomal membranes. *J. Biol. Chem.* 244:2017–2026.

Labrie, F., and A. Korner, 1968a. Actinomycin sensitive induction of tyrosine transaminase and tryptophan pyrrolase by amino acids and tryptophan. *J. Biol. Chem.* 243:1116–1119.

Leive, L., 1965. RNA degradation and the assembly of ribosomes in actinomycin-treated *Escherichia coli*. *J. Mol. Biol.* 13:862–875.

Linderstrom-Lang, K., 1950. Structure and enzymatic breakdown of proteins. *Cold Spring Harbor Symp. Quant. Biol.* 14: 117–126.

Lockard, R. E., and J. B. Lingrel, 1969. The synthesis of mouse hemoglobin β-chains in a rabbit reticulocyte cell-free system programmed with mouse reticulocyte 9S RNA. *Biochem. Biophys. Res. Commun.* 37:204–212.

MacDonald, R. A., 1961. "Lifespan" of liver cells. *Arch. Intern. Med.* 107:335–343.

Maenpaa, P. H., and M. R. Bernfield, 1970. A specific hepatic transfer RNA for phosphoserine. *Proc. Natl. Acad. Sci. USA* 67:688–695.

Majerus, P. W., and E. Kilburn, 1969. Acetyl coenzyme A carboxylase. The roles of synthesis and degradation in regulation of enzyme levels in rat liver. *J. Biol. Chem.* 244: 6254–6262.

Martin, T. E., and I. G. Wool, 1968. Formation of active hybrids from subunits of muscle ribosomes from normal and diabetic rats. *Proc. Natl. Acad. Sci. USA* 60:569–574.

Marver, H. S., A. Collins, D. P. Tschudy, and M. Rechcigl, Jr., 1966. δ-Aminolevulinic acid synthetase. *J. Biol. Chem.* 241:4323–4329.

Matsubara, S., H. Suter, and H. Aebi, 1967. Fractionation of erythrocyte catalase from normal, hypocatalatic and acatalatic humans. *Humangenetik* 4:29–41.

McCarthy, B. J., and B. L. McConaughy, 1968. Related base sequences in the DNA of simple and complex organisms. *Biochem. Genet.* 2:37.

Messenguy, F., and J. Wiame, 1969. The control of ornithine-transcarbamylase activity by arginase in *Saccharomyces cerevisiae*. *FEBS Letter* 3:47–49.

Morse, D. E., R. Mosteller, R. F. Baker, and C. Yanofsky, 1969. Direction of in vivo degradation of tryptophan messenger RNA—a correction. *Nature (Lond.)* 223:40–43.

Munro, H. N., 1968. Role of amino acid supply in regulating ribosome function. *Fed. Proc.* 27:1231–1237.

Niemeyer, H., 1966. Regulation of glucose-phosphorylating enzymes. *Natl. Cancer Inst. Monograph* 27:29–40.

Palacios, R., R. D. Palmiter, and R. T. Schimke, 1971. Identification and isolation of ovalbumin-synthesizing polysomes. *J. Biol. Chem.* 247:2316–2321.

Palmiter, R. D., T. Oka, and R. T. Schimke, 1971. Modulation of ovalbumin synthesis by estradiol-17β and actinomycin D as studied in explants of chick oviduct in culture. *J. Biol. Chem.* 246:724–737.

Papaconstantinou, J., 1967. Molecular aspects of lens cell differentiation. *Science* 156:338–346.

Penn, N. W., 1961. Metabolism of the protein molecule in a rat liver mitochondrial fraction. *Biochim. Biophys. Acta* 53: 490–494.

Peraino, C., R. C. Blake, and H. C. Pitot, 1965. Studies on the induction and repression of enzymes in rat liver. *J. Biol. Chem.* 240:3039–3043.

Pine, M. J., 1966. Metabolic control of intracellular proteolysis in growing and resting cells of *Escherichia coli*. *J. Bacteriol.* 93: 847–850.

Pine, M. J., 1967. Intracellular protein breakdown in the L1210 ascites leukemia. *Cancer Res.* 27:522–525.

Pitot, H. C., C. Peraino, C. Lamar, Jr., and A. L. Kennan, 1965. Template stability of some enzymes in rat liver and hepatoma. *Proc. Natl. Acad. Sci. USA* 54:845–851.

Pitot, H. C., N. Sladek, W. Ragland, R. K. Murray, G. Moyer, H. D. Soling, and J-P. Jost, 1969. A possible role of the endoplasmic reticulum in the regulation of genetic expression: The membron concept. In *Microsomes and Drug Oxidations*, J. R. Gillette, A. H. Conney, G. J. Cosmides, R. W. Estabrook, J. R. Fouts, and G. J. Mannering, eds. New York: Academic Press, p. 59.

Platt, T., J. H. Miller, and K. Weber, 1970. In vivo degradation of mutant lac repressor. *Nature (Lond.)* 228: 1154–1156.

Potter, V. R., E. F. Baril, M. Watanabe, and E. D. Whittle, 1968. Systematic oscillations in metabolic functions in liver from rats adapted to controlled feeding schedules. *Fed. Proc.* 27:1238.

Price, V. E., W. R. Sterling, V. A. Tarantola, R. W. Hartley, Jr., and M. Rechcigl, Jr., 1962. The kinetics of catalase synthesis and destruction in vivo. *J. Biol. Chem.* 237: 3468–3475.

Rechcigl, M., Jr., 1971. Intracellular protein turnover and the roles of synthesis and degradation in regulation of enzyme levels. In *Enzyme Synthesis and Degradation in Mammalian Systems*. Baltimore: University Park Press, p. 236.

Revel, M., and H. H. Hiatt, 1964. Elovich decay of free radicals in a photosynthetic system as evidence for electron transport across an interfacial activation energy barrier. *Proc. Natl. Acad. Sci. USA* 51:809–818.

Rhoads, R. E., G. S. McKnight, and R. T. Schimke, 1973. *J. Biol. Chem.* 248:2031–2039.

Robinson, A. B., J. H. McKerrow, and P. Cary, 1970. Controlled deamidation of peptides and proteins: An experimental hazard and a possible biological timer. *Proc. Natl. Acad. Sci. USA* 66:753–757.

Russell, D., and S. H. Snyder, 1968. Amine synthesis in rapidly growing tissues: Ornithine decarboxylase activity in regenerating rat liver, chick embryo, and various tumors. *Proc. Natl. Acad. Sci. USA* 60:1420–1427.

SANNO, Y., M. HOLZER, and R. T. SCHIMKE, 1970. Studies of a mutation affecting pyrimidine degradation in inbred mouse strains. *J. Biol. Chem.* 245:5668.

SCHERRER, K., L. MARCAUD, F. ZAJDELA, I. M. LONDON, and F. GROS, 1966. Patterns of RNA metabolism in a differentiated cell: A rapidly labeled, unstable 60S RNA with messenger properties in duck erythroblasts. *Proc. Natl. Acad. Sci. USA* 56:1571–1578.

SCHIMKE, R. T., 1964. Enzymes of arginine metabolism in cell culture: Studies on enzyme induction and repression. *Natl. Cancer Inst. Monograph* 13:197–217.

SCHIMKE, R. T., 1964. The importance of both synthesis and degradation in the control of arginase levels in rat liver. *J. Biol. Chem.* 239:3808–3817.

SCHIMKE, R. T., 1970. Regulation of protein degradation in mammalian tissues. In *Mammalian Protein Metabolism*, H. N. Munro, ed. New York: Academic Press, p. 177.

SCHIMKE, R. T., and D. DOYLE, 1970. Control of enzyme levels in animal tissues. *Ann. Rev. Biochem.* 39:929–976.

SCHIMKE, R. T., E. W. SWEENEY, and C. M. BERLIN, 1965. The roles of synthesis and degradation in the control of rat liver tryptophan pyrrolase. *J. Biol. Chem.* 240:322–331.

SCHOENHEIMER, R., 1942. *The Dynamic State of Body Constituents.* Cambridge, Mass.: Harvard University Press.

SCHUBERT, D., 1968. Immunoglobin assembly in a mouse myeloma. *Proc. Natl. Acad. Sci. USA* 60:683–690.

SEGAL, H. L., and Y. S. KIM, 1963. Glucocorticoid stimulation of the biosynthesis of a glutamic-alanine transaminase. *Proc. Natl. Acad. Sci. USA* 50:912–918.

SHEARER, R. W., and B. J. MCCARTHY, 1967. Evidence for ribonucleic acid molecules restricted to the cell nucleus. *Biochemistry* 6:283–289.

SIDRANSKY, H., D. S. R. SARMA, M. BONGIORNO, and E. VERNEY, 1968. Effect of dietary tryptophan on hepatic polyribosomes and protein synthesis in fasted mice. *J. Biol. Chem.* 243:1123–1132.

SOEIRO, R., M. H. VAUGHAN, J. R. WARNER, and J. E. DARNELL, JR., 1968. The turnover of nuclear DNA-like RNA in HeLa cells. *J. Cell. Biol.* 39:112–118.

STAVNEZER, J., and R. C. C. HUANG, 1971. Synthesis of a mouse immunoglobulin light chain in a rabbit reticulocyte cell-free system. *Nature New Biol.* 230:172–176.

STELLWAGEN, R. H., and R. D. COLE, 1969. Chromosomal proteins. *Ann. Rev. Biochem.* 38:951–990.

SWICK, R. W., A. L. KOCH, and D. T. HANDA, 1956. The measurement of nucleic acid turnover in rat liver. *Arch. Biochem. Biophys.* 63:226–242.

SWICK, R. W., A. K. REXROTH, and J. L. STANGE, 1968. The metabolism of mitochondrial proteins. III. The dynamic state of rat liver mitochondria. *J. Biol. Chem.* 243:3581–3587.

SWICK, R. W., 1958. Measurement of protein turnover in rat liver. *J. Biol. Chem.* 231:751–764.

TAKAGI, M., and K. OGATA, 1971. Isolation of serum albumin-synthesizing polysomes from rat liver. *Biochem. Biophys. Res. Commun.* 42:125–131.

TANAKA, T., Y. HARANO, F. SUE, and H. MORIMURA, 1967. Crystallization, characterization and metabolic regulation of two types of pyruvate kinase isolated from rat tissues. *J. Biochem. (Tokyo)* 62:71–91.

TATA, J. R., 1966. Requirement for RNA and protein synthesis for induced regression of the tadpole tail in organ culture. *Develop. Biol.* 13:77–94.

TOMKINS, G. M., T. D. GELEHRTER, D. GRANNER, D. MARTIN, JR., H. H. SAMUELS, and E. B. THOMPSON, 1969. Control of specific gene expression in higher organisms. *Science* 166:1474–1480.

WARNER, J. R., 1971. The assembly of ribosomes in yeast. *J. Biol. Chem.* 246:447–454.

WESSELLS, N. K., and F. H. WILT, 1965. Action of actinomycin D on exocrine pancreas cell differentiation. *J. Mol. Biol.* 13:767–779.

WILLIAMS, L. S., and F. C. NEIDHARDT, 1969. Synthesis and inactivation of aminoacyl-transfer RNA synthetases during growth of *Escherichia coli*. *J. Mol. Biol.* 43:529–550.

WIRTZ, K. W. D., and D. B. ZILVERSMIT, 1968. Exchange of phospholipids between liver mitochondria and microsomes in vitro. *J. Biol. Chem.* 243:3596–3602.

WURTMAN, R. J., and J. AXELROD, 1967. Daily rhythmic changes in tyrosine transaminase activity of the rat liver. *Proc. Natl. Acad. Sci. USA* 57:1594–1598.

YAFFE, D., and M. FELDMAN, 1964. The effect of actinomycin D on heart and thigh muscle cells grown in vitro. *Devel. Biol.* 9:347–366.

YOSHIDA, A., G. STAMATOYANNOPOULOS, and A. MOTULSKY, 1967. Negro variant of glucose-6-phosphate dehydrogenase deficiency (A⁻) in man. *Science* 155:97–99.

72 Regulation and Importance of Intracellular Protein Degradation

ALFRED L. GOLDBERG

ABSTRACT In animal and bacterial cells, the rates of protein degradation can vary under different physiological conditions. For example, in skeletal muscle, contractile activity, hormones, and the supply of nutrients all affect protein degradation. Bacteria have proved highly useful for studying the mechanisms and significance of this process. During starvation, protein degradation in such cells increases several fold. Control of this response, which allows the cell to adapt to starvation, is coupled to the control of RNA synthesis. Cellular proteins vary widely in their rates of degradation, and cells selectively degrade abnormal or denatured proteins. These variations in the rates of degradation of specific proteins appear to reflect inherent differences in their sensitivity to endoproteases.

THE PROTEIN constituents of all cells, including those in the nervous system, are in a highly dynamic state. Whether found within organelles, in membranes, as structural components, or as soluble enzymes, cellular proteins are subject to continuous degradation and replacement through further synthesis (Schimke, 1970). The past 20 years have witnessed unprecedented advances in our knowledge about the synthesis of proteins. However, protein degradation is also of prime importance in determining intracellular amounts of enzymes or structural components. As yet we have only a rudimentary knowledge about the biochemical mechanisms, regulation and physiological significance of the degradative process.

A number of fundamental questions remain unanswered: (1) Why do different proteins in the same cell have distinct rates of turnover? (2) What physiological factors control overall rates of protein breakdown? (3) What enzymes are responsible for the turnover of cell proteins? (4) What might be the selective advantage to the organism of this continuous degradation of protein, which appears to be highly wasteful? Answers to these questions are not only of fundamental import to biochemists but also are highly relevant to a number of unsolved physiological problems. For example, in nerve cells the degradation of proteins must be intimately related to their axoplasmic flow, to the assembly and disassembly of membranes, and even to the mechanisms responsible for neuronal plas-

ALFRED L. GOLDBERG Department of Physiology, Harvard Medical School, Boston, Massachusetts

ticity. The present chapter will attempt to review recent studies in my laboratory on the regulation and physiological significance of protein catabolism in animal and bacterial cells.

Protein turnover in growth and atrophy of muscle

Our own interest in protein breakdown arose out of studies into the biochemical mechanisms through which increased muscular activity causes hypertrophy and through which disuse causes atrophy (Goldberg, 1972c). Originally this research was undertaken as a model system for exploring how physiological function might regulate protein synthesis in an excitable tissue. These investigations unexpectedly raised the possibility that changes in the rates of protein catabolism might also contribute to cell growth and atrophy. Measurements of the fate of labeled proteins in rat muscle indicated that during atrophy the degradation of muscle proteins was augmented, while during compensatory growth the rates of degradation appeared to decrease (Goldberg, 1969). In both instances, the alterations in degradative rates appeared to have effects complementary to the changes occurring in the rates of protein synthesis (Table I).

These initial observations emphasized the necessity of learning more about the control of protein degradation for understanding the control of tissue size. Our initial studies, like most research in this area, estimated the proteolytic process from measurements of the loss of radioactive proteins from cells. Though valuable, this approach is awkward, expensive, and rather inappropriate for studies of mechanism of turnover; in addition it is subject to serious complications (e.g., the reincorporation of radioactive amino acids back into cell proteins). Recently Mr. Richard Fulks and Dr. Jeanne Li in this laboratory (Fulks et al., 1973; Li and Goldberg, 1973) have developed a simple procedure for studying protein breakdown in rat muscles in vitro, which should also be applicable to other isolated tissues or to cultured cells.

Using this technique, we have found that a variety of other physiological variables that influence muscle size

TABLE I

Factors influencing protein turnover in skeletal muscle

	Synthesis	Degradation
Denervation atrophy	↓	↑
Compensatory hypertrophy	↑	↓
Growth hormone	↑	—
Cortisone (glucocorticoids)	↓	↑
Starvation	↓	↑
Insulin*	↑	↓
Glucose*	—	↓
Free fatty acids*	—	—
Amino acids*	↑	↓
Contraction*		↓
Passive stretch*		↓

*Direct effect demonstrated in vitro.

This summary includes previous in vivo measurements of protein turnover in muscle during atrophy and hypertrophy (Edlin et al., 1968) as well as more recent measurements on protein turnover in isolated skeletal muscle in vitro (Fulks et al., 1973; Li and Goldberg, 1973).

also influence rates of protein catabolism (Table I). For example, insulin, which has long been known to promote protein synthesis and growth of muscle, also inhibits protein breakdown in this tissue (Fulks et al., 1973). This hormone appears to have a similar effect in heart (Morgan et al., 1971a, b), liver (Mortimore and Mondon, 1970) and cultured hepatoma cells (Hershko and Tomkins, 1971). On the other hand, food deprivation, which leads to the mobilization of protein reserves in muscle, not only inhibits protein synthesis in this tissue but also promotes protein degradation (Li and Goldberg, 1973). Supply of nutrients such as glucose or amino acids to isolated muscle (Li and Goldberg, 1973) and liver (Mortimore and Mondon, 1970) can inhibit the degradative process. One recent finding of special interest to neurobiology is that repeated electrical stimulation of muscle somehow inhibits protein degradation (Li and Goldberg, 1973). Even passive stretch of muscle reduces catabolism of muscle proteins, and this effect appears of potential therapeutic import (Li and Goldberg,1973). As yet, we are still only at the stage of defining the factors influencing overall rates of proteolysis, and we have little idea of how degradation of specific proteins is affected under these various conditions.

Although the mechanisms responsible for these effects on protein breakdown are completely unclear at present, it is interesting that the effects of nutrient supply, hormones, and muscular activity on degradation are immediate and can be demonstrated in the absence of protein synthesis (Fulks et al., 1973; Li and Goldberg, 1973). Thus within the organism, average rates of protein breakdown in specific cells, like the mean rates of protein synthesis, change from moment to moment in response to a variety of physiological signals.

Studies of protein degradation in microorganisms

Several years ago, in order to study the mechanisms of protein degradation more effectively, we began to investigate the control of protein catabolism in *Escherichia coli*. Bacteria offer the biochemist or physiologist enormous technical advantages. Furthermore, such organisms exquisitely control their rates of protein degradation. In fact, for many years, textbooks of bacteriology claimed that bacteria do not turn over their protein constituents, because early experiments were carried out in growing cells under conditions where the degradative process is maximally inhibited. About 20 years ago, Mandelstam (1960) demonstrated that when bacteria were deprived of a required nutrient (e.g. a carbon or nitrogen source or a required amino acid), the rate of degradation of cell proteins increased several fold. This increase in proteolysis represents a physiological adaptation to the poor environment and can be rapidly reversed by readministration of the required material (Figure 1). Protein degradation also increases when rapidly growing cells enter stationary phase. At such times, growth ceases, protein synthesis is reduced, and protein breakdown increased markedly. Dilution of such cultures into fresh medium causes the reinitiation of normal growth and a rapid inhibition of protein breakdown.

The nongrowing bacteria thus resemble nongrowing mammalian tissues, where protein synthesis and protein catabolism are equally matched. In addition, the responses of the bacteria to *step-up* and *step-down* conditions appear similar to those observed in growing and atrophying muscle. During rapid growth of both cells, protein degradation is minimized while protein and RNA synthesis are maximal. In the nongrowing cells, protein degradation is augmented while synthetic processes are reduced. In the atrophying tissues, degradative events increase even further and exceed protein synthesis. The generality of growth-dependent changes in degradation remains to be established, although analogous findings have now been made in perfused liver (Mortimore and Mondon, 1970), perfused heart (Morgan et al., 1971a, 1971b), regenerating kidney in vivo (unpublished), and cultured hepatoma cells (Hershko and Tomkins, 1971).

Our major interest in studying the bacterial system has been to clarify the cellular mechanisms underlying the changes in protein catabolism. Step-down conditions not only promote degradation but also lead to an increased ability of the cells to synthesize certain catabolic enzymes (Magasanik, 1970). An initially attractive hypothesis was that the accumulation of cyclic AMP in such cells (which

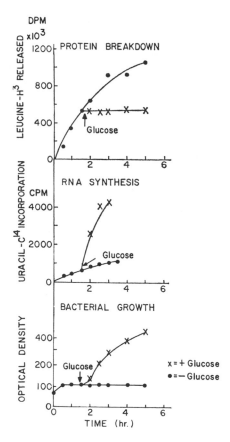

FIGURE 1 Effects of glucose deprivation on protein degradation, growth and RNA synthesis. *E. coli A33* was grown exponentially in glucose-minimal medium and ³H-leucine for two
generations. At time zero, they were deprived of the carbon
source and the radioactive amino acid. The degradation of pre-
existent proteins was measured from the release of ³H-leucine
(Goldberg, 1971) medium containing large amounts of non-
radioactive leucine to prevent reutilization of the radioactive
residues (Mandelstam, 1960). The RNA synthesis was measured
by the incorporation of uracil-¹⁴C into RNA by standard
Millipore filter methods, and growth was followed with a Klett-
Summerson Colorimeter. At 90 min, glucose was added back to
the starved cultures.

promotes the synthesis of various catabolic enzymes)
might also stimulate protein degradation (Perlman and
Pastan, 1971). However, unpublished (Goldberg and Li,
1973) experiments with mutants in the cyclic AMP
systems have discounted this view. In addition, the
conditions affecting overall rates of protein catabolism
also alter the synthesis of ribosomal and transfer RNA.
Step-down markedly reduces RNA synthesis, while the
supply of amino acids or glucose promotes RNA synthesis
and inhibits protein catabolism (Figure 1). Various
experiments in this laboratory have provided evidence
that the control of protein degradation is in some way
linked to the control of RNA synthesis.

Work in several laboratories has previously indicated
that the production of ribosomal RNA is somehow dependent on the intracellular supply of aminoacyl tRNA
(Edlin and Broda, 1968). The possibility that the supply
of charged tRNA might also influence protein catabolism appeared attractive since it suggested a link between synthetic and degradative processes (Figure 2).
Experiments were performed to determine whether the
increased proteolysis during starvation resulted from a fall
in the levels of amino acids or of charged tRNA (Goldberg
and Li, 1973). One approach utilized bacterial strains
which carried a specific temperature-sensitive aminoacyl tRNA synthetase. Such cells grow normally at the
permissive temperature, but lack a specific species of
aminoacyl tRNA at the nonpermissive temperature,
growth ceases and rates of protein breakdown increases
8- to 12- fold (Goldberg and Li, 1973). Additional
experiments using amino acid analogs to block selectively
the synthesis of valyl-tRNA or agents that prevent formation of N-formyl methionyl-tRNA, also indicated that
deprivation of a single charged tRNA can stimulate
proteolysis (Goldberg and Li, 1973).

Unfortunately, we do not as yet understand the cellular
mechanisms through which changes in the supply of
charged tRNA might influence intracellular protein
breakdown. Since the increased protein degradation
occurred despite a profound inhibition of protein synthesis, it appears likely that this effect does not require the
synthesis of new proteolytic enzymes. Instead, supply of
nutrients through some effect on the levels of charged
t RNA must influence the activity of proteases that exist in
the growing cells. Control of protease activity can also
explain the observation that readministration of nutrients
to starved cells immediately reduced protein breakdown
(Figure 1).

These experiments further support the conclusion that
rates of protein breakdown are controlled coordinately
with rates of RNA synthesis. All experimental conditions
that were found to stimulate protein degradation (Goldberg and Li, 1973) have been reported previously to inhibit
net RNA synthesis. Unfortunately, despite extensive investigation, we lack an adequate explanation for the
control of ribosomal RNA synthesis. Mutants (*rel*) are
known that fail to inhibit this process when the cells are
starved for amino acids. Experiments by Sussman and
Gilvarg (1969) and by us (unpublished) indicate that
strains defective in regulating ribosome synthesis are also
defective in stimulating protein breakdown on amino acid
starvation. The *rel* mutation has also been found to
influence glucose transport, nucleotide synthesis, and
lipid metabolism. It appears likely that both protein
breakdown and RNA synthesis depend on changes in
some intermediary metabolite which affect a broad range

of growth-related biochemical events. Such a pleiotropic control mechanism may also affect protein breakdown in mammalian cells (Goldberg and Li, 1973).

Inhibition of protein degradation

In order to learn more about biochemical pathways and the physiological significance of protein degradation, we attempted to find selective inhibitors of this process. A systematic attempt was made to test whether well-characterized inhibitors of proteolytic enzymes might also inhibit protein catabolism in starving bacteria. The most widespread type of proteolytic enzyme in nature is the serine protease, so designated because of a serine residue in its active sites which is particularly sensitive to attack by reagents such as diiosopropyl fluorophosphate (DFP) or the sulfonyl fluorides. These compounds selectively inactivate various well-known proteases, such as chymotrypsin or subtilisin, by covalently binding to this active-site serine. We found that such drugs also block the increased protein degradation occurring in starving *E. coli*. Most of our studies utilized the sulfonyl fluorides (PMSF and TSF) which unlike DFP are selective for proteases and are therefore relatively nontoxic. These agents can inhibit degradation in starving bacteria at concentrations that do not block cell growth (Prouty and Goldberg, 1972). In addition to the sulfonyl fluorides, several other inhibitors of pancreatic trypsin, such as the aromatic diamidines, also inhibit protein degradation in starving bacteria (Prouty and Goldberg, 1972).

These results suggest that a trypsin-like serine protease is responsible for the increased protein degradation in starving cells. Recent work in my laboratory has demonstrated the existence of such an enzyme in *E. coli*; however, it remains to be demonstrated whether this new protease is actually involved in protein degradation in vivo. Interestingly, these agents which are active in starved cells do not inhibit the low amount of protein degradation occurring in growing cells. These findings thus raise the possibility that *E. coli* contain two distinct proteolytic systems, possibly with different physiological functions: one active in growing cells and one activated during starvation.

Additional studies have supported the existence of two such systems, and these inhibitors have proved quite useful in studying their physiological importance. Mandelstam (1960) originally suggested that protein catabolism increases in poor environments in order that the cells can synthesize new enzymes appropriate to the new conditions. Unlike growing cells, which can synthesize required amino acids, cells deprived of a carbon or nitrogen source or a required amino acid can only obtain amino acids for synthesizing proteins from pre-existing cellular

proteins (Figure 2). This idea predicts that inhibitors of protein degradation during starvation should also block induction of new enzymes (Prouty and Goldberg, 1972). Starving cells can induce the enzyme β-galactosidase when administered the gratuitous inducer IPTG. How-

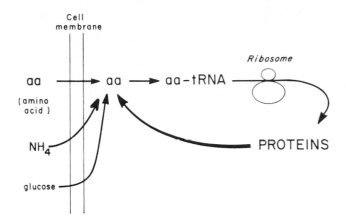

FIGURE 2 Relationship of protein degradation and protein synthesis. Growing cells obtain their amino acids from the surrounding medium or synthesize them from external sources of nitrogen and carbon. In poor environments, the starved cells obtain amino acids primarily from the increased protein degradation.

ever, when such cells are exposed to the sulfonyl fluorides or aromatic diamidines, induction of β-galactosidase was prevented (Figure 3). Administration of the required nutrients in the presence of the inhibitors once again permitted the synthesis of new enzymes. These results clearly demonstrate an adaptive role for protein breakdown in cells in poor environments. This situation appears highly reminiscent of the increased protein degradation that occurs in us mammals, where mobilization of body proteins is an important aspect of adaptation to starvation.

The degradation of abnormal proteins

These findings demonstrate an important adaptive role for the increased protein degradation in starving organisms, but they do not suggest a physiological function for the lower amount of protein breakdown seen in normal cells. It has often been naively suggested that intracellular protein degradation might serve to eliminate "proteinaceous waste" from the cell. According to this idea, proteins (like automobiles or humans) may also suffer from "wear and tear" processes and sooner or later denature. In order that such abnormal proteins do not accumulate, cells may have evolved enzymatic systems for their selective degradation. Although this idea appears

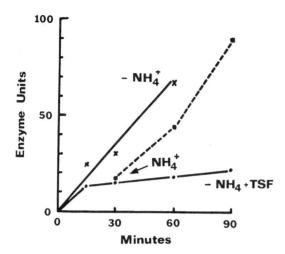

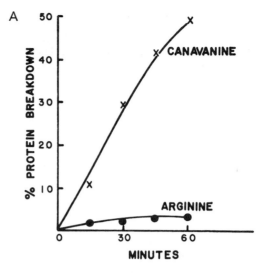

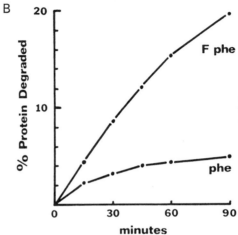

FIGURE 3   Importance of protein breakdown for induction of enzymes in cells deprived of nitrogen. *E. coli 19* was grown on glycerol-minimal medium and then deprived of a nitrogen source (NH_4^+). The nongrowing cells were administered isopropyl-β-D-thiogalactoside, a gratuitous inducer of β-galactosidase. This enzyme was synthesized presumably from amino acids supplied by protein breakdown. Toluene sulfonyl-fluoride (TSF), an inhibitor of protein breakdown during starvation (Prouty and Goldberg, 1972), greatly reduced enzyme induction, presumably by blocking the supply of amino acids. Readministration of ammonium chloride in the presence of TSF again permitted enzyme induction.

vague or even meaningless in chemical terms, it is in many ways an attractive hypothesis, and experiments were undertaken by Pine (1967) and ourselves (Goldberg, 1972a) to test whether cells can selectively degrade denatured proteins.

Proteins with abnormal conformations were produced by a variety of experimental approaches. For example, we studied the fate in growing bacteria of proteins that have incorporated amino acid analogs in place of the normal residues. As shown in Figure 4, proteins containing fluorophenylalanine or canavanine are catabolized more rapidly than normal cell constituents containing phenylalanine or arginine (Pine, 1967; Goldberg, 1972a). Altogether we have studied the effects of incorporation of 14 different analogs; all of them promote catabolism, although the effects of the various analogs on the rates of catabolism are quite different. Presumably those analogs that most disrupt normal conformations are most effective in promoting protein degradation.

In analogous experiments, we caused bacteria to produce unfinished polypeptide chains by administration of puromycin, which is incorporated into the growing polypeptide and causes its premature release from the ribosome. Proteins that have incorporated puromycin were found to be degraded much more quickly than the normal cell proteins (Pine, 1967; Goldberg, 1972a). Inde-

FIGURE 4   Degradation of proteins containing amino acid analogs in growing *E. coli*. (A) Compares the fate of proteins that have incorporated arginine or its analog canavanine; (B) Proteins that have incorporated phenylalanine or its analog fluorophenylalanine. Strain *A33*, an arginine-auxotroph (Figure 4A), or 83-5A, a phenylalanine-auxotroph (Figure 4B), were grown initially in glucose-minimal medium containing the required amino acids. They were then exposed to ^{3}H-leucine for five minutes in the presence of the natural amino acid or the analog. Cells were then transferred to nonradioactive medium containing normal amino acids. The rates of degradation of labeled proteins are expressed as the amount of ^{3}H-leucine released into acid-soluble form relative to that originally in proteins.

pendently of this work, two groups have obtained evidence that certain mutant proteins are rapidly catabolized, even though the wild-type proteins are perfectly stable (Platt et al., 1970; Goldschmidt, 1970). It appears likely that many other examples of this sort will be discovered in animal as well as bacterial cells.

Cells thus have the capacity to degrade aberrant proteins as might arise spontaneously through denaturation of cell enzymes or through genetic mutations. In addition, such proteins may also be formed as a consequence of mistakes in gene transcription or translation. To test this possibility, we studied bacterial strains that produce frequent mistakes in gene expression (e.g. the missense suppressors or the ribosomal ambiguity mutations). These experiments (Goldberg, 1972a) indicated that errors in protein synthesis lead to increased rates of protein catabolism.

The existence of a cellular mechanism for eliminating abnormal proteins would appear to be highly advantageous to the organism. It has frequently been argued (Goldberg and Wittes, 1966) that genetic mechanisms have evolved to minimize the harmful consequences of imprecise protein synthesis. Presumably, an enzymatic mechanism for eliminating abnormal or denatured proteins would be selected by evolution for similar reasons. The accumulation of "proteinaceous garbage" would be an especially serious threat to long-lived cells, such as neurons, which do not divide and thus unlike bacteria cannot dilute out denatured proteins through growth. It is interesting that several groups have presented evidence that cellular aging may result from increased production of abnormal proteins as a consequence of a faulty system for gene expression (Holliday and Tarrant, 1972). The continuous turnover of body constituents, though seemingly wasteful, by selectively removing proteins with abnormal conformations would serve to minimize such deleterious processes.

An implicit conclusion of this work is that the stability of a protein in vivo is influenced by its conformation which in turn must reflect its primary structure. Usually biochemists and geneticists have clearly distinguished between structural mutations, which influence enzymatic activity, and control-gene mutations that affect the concentrations of enzyme molecules. The finding that enzyme conformation may also be a major determinant of protein half lives thus gives an added significance to studies of tertiary structure.

Protein half lives and protease sensitivity

The finding that deviations from normal structure markedly influence protein stability in vivo implies that normal proteins share certain conformational features distinguishing them from abnormal proteins. How the cell might recognize and selectively degrade these abnormal proteins remains to be determined. The simplest model that one might draw to explain these findings would be that normal cells contain free proteases within the cytoplasm. Normal cell enzymes presumably share certain conformational features that have evolved to insure

relative resistance to these proteases. Deviations from these normal conformations might increase protease sensitivity. It is well documented that denatured proteins are more sensitive to proteolytic digestion than normal ones. The incorporation of amino acid analogs, the failure to complete polypeptides, or even the mistakes in gene transcription or translation might all augment the rate at which the resulting proteins are digested by proteolytic enzymes (Goldberg, 1972b).

In fact, we have obtained appreciable evidence for a correlation between protein half lives in vivo and their sensitivity in vitro to known proteases (Platt et al., 1970). The various treatments found to increase the intracellular rates of protein catabolism have all been found to increase sensitivity of the labeled proteins to well-characterized proteases such as trypsin or papain. These differences in protease sensitivity constitute further evidence for a conformational basis for differences in protein degradative rates.

These findings not only may explain the selective degradation of abnormal cell constituents but also raise the possibility that differences in protease sensitivity might contribute to the variation in half lives of normal proteins. It is now well documented that proteins within bacterial or mammalian cells vary markedly in their degradative rates; for example, in normal rat liver, half lives of specific enzymes are known to range from 11 min to 6 days (Schimke, 1970). We therefore tested whether a correlation might also exist between the in vivo stability of normal proteins and their sensitivity to known endoproteases (Goldberg, 1972b). To test this possibility, cells were administered a radioactive amino acid and either frozen immediately or allowed to grow for several generations and to turn over the labeled protein. In the frozen cells, a variety of cell proteins should be labeled; in the cells allowed to grow, those proteins with shorter half lives should be selectively lost and the proportion of label in protein constituents with long half lives should increase. Figure 5 demonstrates that the most stable proteins in vivo are on the average more resistant to trypsin than those that turn over more rapidly. Analogous findings were also obtained with other endopeptidases such as chymotrypsin, pronase or sublilisin.

Similar evidence has been obtained by us (Goldberg, 1973) and by Dice, Dehlinger, and Schimke (1973) for kidney, muscle, and tissue culture cells. In all these systems it was found that the more labile cell proteins are inherently more easily digested by proteases than more stable intracellular constituents. The structural features that distinguish the short-lived enzymes from more stable ones are unknown; presumably they are similar to but more subtle than the structural differences between nor-

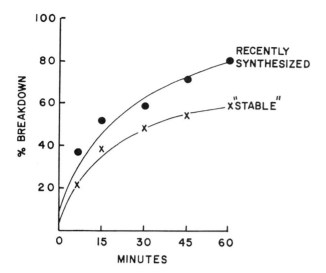

FIGURE 5 Comparison of trypsin sensitivity of *E. coli* proteins with different intracellular stabilities (Goldberg, 1972b). *E. coli A33* were exposed to ^{3}H-leucine for five minutes. Half the culture was frozen (recently synthesized) and the other half ("stable") grew for two generations in the presence of non-radioactive leucine, during which time the cells degraded 8% of the labeled proteins to acid-soluble form. These cells, which had lost the more labile cell proteins, were then frozen. The frozen cells were then lysed and treated with trypsin (100 μg/ml).

mal and abnormal proteins. In any case, the identification of conformational factors that determine protease sensitivity appears very important for understanding the regulation of enzyme levels.

Despite these extensive correlations, it appears quite unlikely that inherent differences in protease sensitivity by itself explains all the specificity in protein degradation. For example, we have been unable to demonstrate an increase in protease sensitivity of cellular proteins during starvation, even though overall protein breakdown increases at such times (Figure 1). The studies discussed above clearly suggest the activation of new proteolytic enzymes in starving cells (Figure 3). We believe these findings indicate that, although differences in protease sensitivity probably account for variations in half lives normally, physiological factors alter overall protein degradation by influencing the activity of proteolytic systems of the cell.

ACKNOWLEDGMENTS These studies were accomplished through the skill and cooperation of my colleagues: Mrs. Susan Martel, Mrs. Elizabeth Howell, Miss Patricia Ritchie, Dr. Jeanne Li, Mr. Richard Fulks, and Dr. Walter Prouty. I am grateful to the Air Force Office of Scientific Research and the Muscular Dystrophy Associations of America for their support of this research. Dr. Goldberg holds a Research Career Development Award from the National Institute of Neurological Diseases and Stroke.

REFERENCES

DICE, F., P. J. DEHLINGER, and R. T. SCHIMKE, 1973. Studies on the correlation between size and relative degradation rate of soluble proteins. *J. Biol. Chem.* (in press).

EDLIN, G., and P. BRODA, 1968. Physiology and genetics of the "ribonucleic acid control" locus in *Escherichia coli*. *Bact. Rev.* 32:206–226.

FULKS, R., J. B. LI, and A. L. GOLDBERG, 1973. Effects of insulin and glucose on protein breakdown in isolated rat diaphragm. *J. Biol. Chem.* (in press).

GOLDBERG, A. L., and R. E. WITTES, 1966. Genetic code: Aspects of organization. *Science* 153:420–424.

GOLDBERG, A. L., 1969. Protein turnover in skeletal muscle. I, II. *J. Biol. Chem.* 244:3217–3222; 3223–3229.

GOLDBERG, A. L., 1971. A role for aminoacyl-tRNA in the regulation of protein catabolism in *Escherichia coli*. *Proc. Natl. Acad. Sci. USA* 68:362–366.

GOLDBERG, A. L., 1972a. Degradation of "abnormal" proteins in *E. coli*. *Proc. Nat. Acad. Sci. USA* 69:422–426.

GOLDBERG, A. L., 1972b. Correlation between rates of degradation of bacterial proteins in vivo and their sensitivity to proteases. *Proc. Nat. Acad. Sci. USA* 69:2640–2644.

GOLDBERG, A. L., 1972c. Mechanisms of growth and atrophy of skeletal muscle. In *Muscle Biology*. New York: Marcel Dekker, pp. 81–118.

GOLDBERG, A. L., 1973. Further evidence for a correlation between protease-sensitivity and protein degradative rates in animal and bacterial cells. *J. Biol. Chem.* (in press).

GOLDBERG, A. L., and J. B. LI, 1973. Physiological significance of protein breakdown in animal and bacterial cells. *Fed. Proc.* (in press).

GOLDSCHMIDT, R., 1970. In vivo degradation of nonsense fragments in *E. coli*. *Nature (London)* 228:1151–1156.

HERSHKO, A., and G. M. TOMKINS, 1971. Studies on the degradation of tyrosine aminotransferase in hepatoma cells in culture. *J. Biol. Chem.* 246:710–714.

HOLLIDAY, R., and G. TARRANT, 1972. Altered enzymes in aging human fibroblasts. *Nature (London)* 238:26–30.

LI, J. B., and A. L. GOLDBERG, 1973. Factors affecting protein degradation in skeletal muscle. *J. Biol. Chem.* (in press).

MAGASANIK, B., 1970. Glucose effects: Inducer exclusion and repression. In *The Lactose Operon*, J. B. Beckwith and D. Zipser, eds. New York: Cold Spring Harbor Laboratory, pp. 189–219.

MANDELSTAM, J., 1960. The intracellular turnover of protein and nucleic acids and its role in biochemical differentiation. *Bact. Rev.* 24:289–308.

MORGAN, H. E., D. C. N. EARL, A. BROADUS, E. B. WOLPERT, K. E. GIGER, and L. S. JEFFERSON, 1971a. Regulation of protein synthesis in heart muscle. I. Effect of amino acid levels on protein synthesis. *J. Biol. Chem.* 246:2152–2162.

MORGAN, H. E., L. S. JEFFERSON, E. B. WOLPERT, and D. E. RANNELS, 1971b. Regulation of protein synthesis in heart muscle. II. Effect of amino acid levels and insulin on ribosomal aggregation. *J. Biol. Chem.* 246:2163–2170.

MORTIMORE, G. E., and C. E. MONDON, 1970. Inhibition by insulin of valine turnover in liver. *J. Biol. Chem.* 245:2375–2383.

PERLMAN, R. L., and I. PASTAN, 1971. The role of cyclic AMP in bacteria. In *Current Topics in Cellular Regulation*. New York: Academic Press, pp. 117–134.

PINE, M. J., 1967. Response of intracellular proteolysis to

alteration of bacterial protein and the implications in metabolic regulation. *J. Bact.* 93:1527–1533.

PLATT, T., J. MILLER, and J. WEBER, 1970. In vivo degradation of mutant Lac repressor. *Nature (London)* 228:1154–1156.

PROUTY, W. F., and A. L. GOLDBERG, 1972. Inhibitors of protein degradation in *E. coli* and their physiological effects. *J. Biol. Chem.* 247:3341–3352.

SCHIMKE, R. T., 1970. Regulation of protein degradation in mammalian tissues. *Mammalian Protein Metabolism.* 4:177–228. (See also R. T. Schimke, this volume.)

SUSSMAN, A. J., and C. GILVARG, 1969. Protein turnover in amino acid-starved strains of *Escherichia coli* K-12 differing in their ribonucleic acid control. *J. Biol. Chem.* 244:6304–6306.

73 Regulation of Riboflavin Metabolism by Thyroid Hormone

RICHARD S. RIVLIN

ABSTRACT Physiological amounts of thyroid hormone enhance the conversion of riboflavin to the two coenzymes, flavin mononucleotide (FMN) and flavin adenine dinucleotide (FAD). Biochemical similarities exist between hypothyroid and riboflavin-deficient animals with respect to activities of FMN-requiring and FAD-requiring enzymes and concentrations of tissue flavins. Diminished responsiveness to thyroid hormone is demonstrable in riboflavin deficiency. In liver and in brain of newborn animals, thyroid hormone accelerates the development of the FAD-synthesizing enzyme, FAD pyrophosphorylase. By increasing flavin synthesis, thyroid hormone has a potential role in increasing the availability of FAD for a number of flavoprotein enzymes during development. Thyroid hormone regulates riboflavin metabolism, and riboflavin, in turn, appears to regulate certain aspects of thyroid hormone action.

CHANGES IN THE activities of certain key enzymes in the nervous system may have important implications for the metabolism of the brain as a whole. A recurrent theme in this part is that enzyme activities may be altered by a change either in the rate of synthesis or in the rate of degradation, and furthermore that small molecules may alter the conformational structure of the enzyme and, thereby, exert profound influences upon catalytic activity. These principles are highly relevant to the study of a specific aspect of the regulation of brain function, namely, the thyroid hormonal control of riboflavin metabolism.

It will be recalled that the metabolic role of riboflavin resides largely in its being the precursor of flavin mononucleotide (FMN) and flavin adenine dinucleotide (FAD), two coenzymes that are required for a wide variety of important biochemical reactions (Rivlin, 1970a, 1970b). FMN and FAD not only are required for catalysis, but they also confer stability upon the flavoprotein apoenzymes to which they are bound. Compared to the flavoprotein holoenzymes, most apoenzymes exhibit some degree of lability with respect to heat denaturation, changes in ionic strength, exposure to various reagents such as cadmium chloride and mercuric acetate, or to other treatments (Neims and Weimar, 1973). Among the flavoprotein enzymes that have been studied most thoroughly in this fashion are included lipoamide dehydrogenase (Kalse and Veegar, 1968), D-amino acid oxidase (Dixon and Kleppe, 1965), glutathione reductase (Staal et al., 1969), and glucose oxidase (Swoboda, 1969).

Much attention has been directed toward investigating the stimulation of apoenzyme synthesis by thyroid hormone (Barker, 1951; Sokoloff et al., 1968; Tata, 1964; Wolff and Wolff, 1964) but relatively little attention toward studying the control of coenzyme synthesis. Evidence has accumulated indicating that thyroid hormone regulates the synthesis of the flavin coenzymes, enhancing the hepatic conversion of riboflavin to both FMN and FAD (Rivlin, 1970a). Hepatic concentrations of FMN and FAD in hypothyroid rats are reduced to one-third of those of normal animals and are restored to normal by treatment with thyroxine (Domjan and Kokai, 1966; Rivlin and Langdon, 1966). In this chapter, current knowledge concerning the hormonal control of flavin synthesis, especially in relation to brain function, is reviewed.

By regulating the rate of synthesis of the coenzyme moieties, thyroid hormone may determine the activities of many flavoprotein enzymes of biological importance. Studies have been recently performed in rat liver on three of the four enzymes involved in the riboflavin-to-FAD pathway. An important site of thyroid hormone control appears to be flavokinase, the enzyme catalyzing the initial conversion of riboflavin to FMN. Of the enzymes investigated, flavokinase undergoes the largest quantitative increase in activity after administration of thyroid hormone and, also, is the only enzyme of the group that decreases in activity in hypothyroidism. Of particular importance is that physiological doses of thyroid hormone are effective as enzyme inducers of flavokinase. Thyroid hormone also causes smaller increases in the activities of FAD pyrophosphorylase, which converts FMN to FAD, and of FMN phosphatase, which degrades FMN to riboflavin (Rivlin, 1970a, 1970b). The thyroid hormonal regulation of FMN and FAD synthesis and the role of these coenzymes in stabilizing the flavoprotein apoenzymes are shown diagrammatically in Figure 1.

RICHARD S. RIVLIN College of Physicians and Surgeons of Columbia University, New York

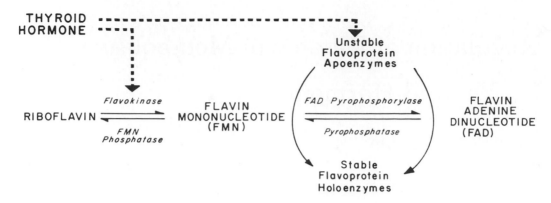

FIGURE 1 Diagrammatic representation of the postulated role of thyroid hormone in enhancing the biosynthesis of flavoprotein apoenzymes and of the coenzymes FMN and FAD (Rivlin, 1970a).

The nature of the increase in activity of flavokinase produced by thyroid hormone is complex and is not completely understood. Treatment with actinomycin D does not block the increase in enzyme activity (Rivlin, 1970b). Results of studies in vitro suggest that the thyroxine-induced increase in flavokinase activity may be due at least in part to a decrease in the rate of enzyme inactivation. When normal liver is homogenized, centrifuged at 100,000 × g for 1 hr, and then incubated in buffer at 37°C, flavokinase activity is rapidly lost. From 0 to 30 min after the start of incubation, there is very little change in enzyme activity; after 30 min, however, enzyme activity decreases in a sharp and nearly linear fashion. Ninety minutes after the start of incubation, there is loss of almost all enzyme activity (Rivlin, 1969b).

A striking difference is observed between enzyme from normal and hyperthyroid animals in the loss of activity that occurs during incubation under these conditions. As shown in Figure 2, enzyme from hyperthyroid animals is inactivated at a much slower rate than that obtained from normal animals: After 90 min of incubation, nearly 75% of the original activity has been preserved. Further experiments were conducted in which tissue extracts were passed through a Sephadex G 25 column, a procedure that results in the removal of molecules of less than 5000 mol wt. The tissue extracts following column chromatography were incubated in a similar fashion to that described above. A marked loss of stability of the enzyme from euthyroid and especially from hyperthyroid animals was noted. These preliminary observations suggest that the greater stability of the enzyme from hyperthyroid than from normal animals is due not to an inherent property of the enzyme protein but more likely to the presence of increased quantities of one or more small molecules that stabilize the enzyme (Rivlin, 1969b). The delayed inactivation of the enzyme from hyperthyroid animals in vitro raises the possibility that increased enzyme activity observed in animals treated with thyroid hormone may be due at least in part to a decrease in the rate of enzyme degradation.

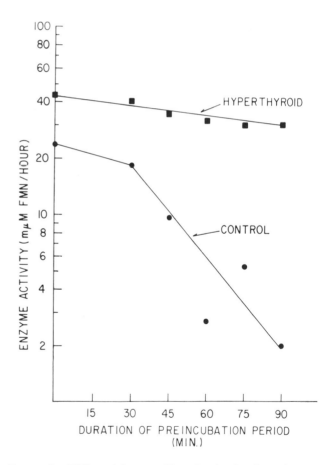

FIGURE 2 Differential rates of inactivation in vitro of hepatic flavokinase obtained from normal and hyperthyroid animals. Experiments were performed using supernatant solutions derived from centrifugation of liver homogenates at 100,000 × g for 1 hour. Enzymes were incubated at 37° C in buffer for periods from 0 to 90 min (reproduced from Rivlin, 1969b).

Certain biochemical measurements in hypothyroidism bear a remarkable similarity to those of riboflavin deficiency. These include at least three parameters: (1) hepatic concentrations of FAD, FMN, and free riboflavin, (2) activities of a number of FMN- and FAD-requiring enzymes, such as xanthine oxidase and D-amino acid oxidase, and (3) activities of the enzymes which are involved in the biosynthesis and degradation of FMN and FAD. Thus, hepatic flavin concentrations and activities of flavoprotein enzymes are largely decreased both in hypothyroidism and in riboflavin deficiency. Hepatic flavokinase activity is diminished and that of FMN phosphatase is unchanged in both conditions. The biochemical changes in hypothyroidism tend generally to be in the same direction as those of riboflavin deficiency but are of lesser magnitude (Rivlin, 1970b).

In addition to resembling hypothyroidism in certain respects, riboflavin deficiency results in diminished responsiveness to thyroid hormone. The induction of mitochondrial α-glycerophosphate dehydrogenase, an FAD-requiring enzyme, by triiodothyronine (T_3) is greatly diminished in riboflavin-deficient rats (Rivlin and Wolf, 1969). In normal animals, administration of triiodothyronine in a single large dose results in a 10-fold increase in enzyme activity measured 48 hr later. Triiodothyronine given similarly to hypothyroid animals increases enzyme activity 75-fold. By contrast, T_3 administration to riboflavin-deficient animals increases enzyme activity only 5-fold. Both the basal activity and the activity induced by T_3 closely parallel the decline in the tissue concentration of FAD with increasing duration of riboflavin deficiency, as shown in Figure 3. Normal enzyme induction can be restored to deficient rats by feeding riboflavin for several days, even when the intake of food is limited concurrently. Similar findings are not observed when caloric intake alone is decreased. In animals that have been starved for 3 to 5 days, both the basal and the T_3-induced activities of mitochondrial α-glycerophosphate dehydrogenase are entirely normal (Wolf and Rivlin, 1970). Riboflavin deficiency also decreases the hepatic deiodination of thyroxine (Galton and Ingbar, 1965). It is likely that the decreased hepatic activities of mitochondrial α-glycerophosphate dehydrogenase, both basal activity and that induced by thyroid hormone, may be related to the lack of FAD available for stabilizing newly synthesized apoenzyme proteins.

Studies of the regulation of riboflavin metabolism have been extended to the perinatal period. These investigations derived their impetus from the previous knowledge that as a group the FMN- and FAD-requiring enzymes, such as TPN-cytochrome C reductase and xanthine oxidase, as well as the tissue levels of FMN and FAD, have a characteristic pattern of development. At the time of

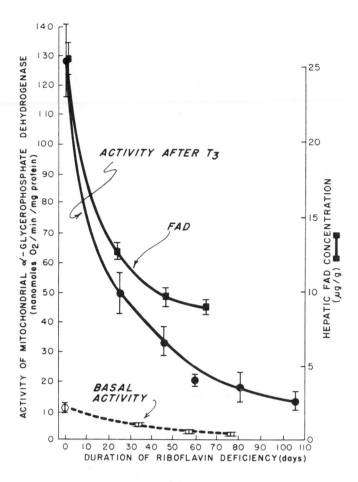

FIGURE 3 Basal and triiodothyronine-induced activities of mitochondrial α-glycerophosphate dehydrogenase and hepatic concentrations of FAD expressed as a function of duration of riboflavin deficiency. [Data are expressed as mean ± 1 SEM and have been obtained in part from Wolf and Rivlin (1970).]

birth, low but measureable concentrations of FMN and FAD and activities of FMN- and FAD-requiring enzymes are found and by the time of weaning at 21 days, adult or nearly adult levels have been attained (Burch et al., 1958). This sequence of events that occurs during postnatal development likely reflects the mutual interdependence of the flavoprotein enzymes and their coenzymes, FMN and FAD, because each is required for the stabilization of the other.

The enzymes involved in FMN and FAD biosynthesis, namely flavokinase and FAD pyrophosphorylase, undergo biochemical maturation much earlier than either the tissue coenzymes levels or the activities of flavoprotein enzymes. The major increases in hepatic flavokinase and FAD pyrophosphorylase activities occur shortly before birth. At the time of birth, levels similar to that of adults are already demonstrable (Rivlin, 1969a). The fact that the fetal thyroid gland in rats first begins to function late in gestation (Nataf et al., 1971) raises the possibility that thyroid hormone may be an important stimulus to the prenatal acceleration of development of these enzymes. Similarly, Greengard (1969) noted that thyroxine administration in gestation markedly accelerated the development of glucose-6-phosphatase and of arginase activities in fetal liver.

It was of interest to inquire whether thyroid hormone accelerates FAD synthesis during the perinatal period. Experiments were performed in which newborn animals received subcutaneous injections of L-thyroxine, 10 μg/ animal, every other day for 8 days. In these animals hepatic FAD pyrophosphorylase activity was increased 35% above that of animals similarly treated with saline. Increases in FAD pyrophosphorylase activity were also observed in animals which had received thyroxine for only three days. The magnitude of these thyroxine-induced increases in hepatic enzyme activity in newborn animals is entirely comparable to that observed in adult animals (Rivlin and Hornibrook, 1971a).

Another system in which thyroid hormone is critical for normal development is the metamorphosing tadpole, *Rana catesbiana*. In the absence of thyroid hormone, tadpoles will not undergo metamorphosis, and will remain in an immature state. When tadpoles are treated with exogenous thyroid hormone, acceleration of metamorphosis occurs (Etkin, 1964). Immersion of tadpoles in a solution of thyroxine at 10^{-8} M for 6 to 11 days does not appear to influence hepatic FAD pyrophosphorylase activity, but if treatment is continued for 15 to 23 days, a significant increase (p < 0.01) in enzyme activity occurs. FAD pyrophosphorylase under these conditions is increased from 0.245 ± 0.018 mμM FAD/mg protein/hr to 0.357 ± 0.020. These findings indicate that changes in the activity of an FAD biosynthetic enzyme accompany thyroxine-induced metamorphosis (Fass et al., 1971).

The regulation of riboflavin metabolism in the developing tadpole is being studied further with structural analogues of riboflavin, each compound containing one or more substitutions on the isoalloxazine ring. The riboflavin analogues were generously provided by Merck & Co. Groups of tadpoles in premetamorphosis were injected three times a week with 13 μg of each of 14 analogues of riboflavin, and were observed for more than 4 months. The solubility of the analogues was a major factor in determining the dosage. Each tadpole was in an individual beaker containing 300 cc of standing tap water.

During the 4 month interval, tadpoles that had been injected with saline only underwent considerable morphological changes, with the development first of hind legs, followed by forelegs and atrophy of the tail. The hind limb length to tail length ratio, a simple, accurate, and widely accepted index of metamorphosis (Etkin, 1964),

increased in normal animals from approximately 0 to nearly 1.0. By contrast, in each group of analogue-treated tadpoles, a degree of interference with tadpole metamorphosis was demonstrable. Some compounds had a relatively small effect upon the hind limb length to tail length ratio, while others almost completely blocked metamorphosis. Four weeks after the start of treatment with analogues of riboflavin, these effects were already demonstrable and became more apparent with increasing duration of treatment (Rivlin et al., 1972).

A more rapid inhibition of metamorphosis can be demonstrated when tadpoles that had been placed in a solution of 2×10^{-8} M thyroxine for 2 to 3 weeks simultaneously received injections of analogues of riboflavin. Results of a typical experiment are shown in Table I.

TABLE I

Effects of treatment with an analogue* of riboflavin upon thyroxine-induced metamorphosis in tadpoles

Treatment Group	Number of Tadpoles	Hind Limb Length/ Tail-Length Ratio	
		Before Treatment	14 Days After Treatment
1. Thyroxine	10	0.07 ± 0.02	0.56 ± 0.03
2. Analogue + Thyroxine	10	0.09 ± 0.02	0.33 ± 0.04

Data are expressed as mean $\pm$ 1 SEM.

*6-chloro-9-(1'-D-dulcityl) isoalloxazine. Each treated animal received an intraperitoneal injection of 13 μg dissolved in isotonic saline three times per week.

Animals were individually immersed in 300 cc solution of 2×10^{-8} M thyroxine dissolved in standing tap water.

Significance of difference between groups 1 and 2 before treatment = p > 0.05, and after treatment = p < 0.001.

(Data have been obtained from Rivlin et al., 1972.)

Tadpoles received thrice weekly injections of 6-chloro-9-(1'-D-dulcityl) isoalloxazine while immersed in thyroxine, and comparative measurements were made with a group of thyroxine-treated animals that had not received injections of this analog. In animals treated with thyroxine together with the riboflavin analog, the hind limb length to tail length ratio increased only half as much as in animals treated with thyroxine alone. Studies with other analogues of riboflavin have yielded similar results, with considerable variability in the degree of inhibition of metamorphosis (Rivlin et al., 1972).

These findings in mammals and in amphibia provide the background for current studies which are examining riboflavin metabolism in the developing rat brain. Thyroid hormone is known to regulate the early development of the brain, stimulating the synthesis of brain proteins and the deposition of myelin. During the critical period of its growth and development, the brain undergoes an increase in oxygen consumption and protein synthesis under thyroid hormone stimulation. After brain maturation has been achieved, thyroid hormone will no longer increase oxygen consumption or protein synthesis (Eayrs, 1971; Balazs et al., 1971; Sokoloff and Roberts, 1971).

Hormonal regulation of FAD synthesis during the critical period of brain development could play a potentially important role in stabilizing the newly synthesized apoenzyme moieties of a large number of flavoproteins. In the hypothyroid brain, the activities of certain FAD-dependent enzymes, notably succinic dehydrogenase, are reduced (Hamburgh and Flexner, 1957; Garcia Argiz et al., 1967). A possible lack of FAD available for conferring stability upon labile flavoprotein apoenzymes could be one of a number of mechanisms accounting for the deleterious effects of early hypothyroidism upon brain function.

To determine whether FAD synthesis in newborn brains is subject to hormonal control, animals within 24 hr of birth were injected with thyroxine, 10 μg/animal, each day or every other day for 8 days. To control nutritional factors and to minimize differences among animals, litter sizes were kept constant at 8 animals, half of which received thyroxine and half isotonic saline. Mothers were rotated daily among the litters in each experiment. Specimens of cerebrum promptly removed after sacrifice were assayed for FAD pyrophosphorylase activity. In newborn animals, enzyme activity in cerebrum was approximately half that in liver expressed per gram wet weight

TABLE II

Effects of thyroxine treatment of newborn and adult male rats upon FAD pyrophosphorylase activity in cerebrum

Treatment Group	Age at Sacrifice (days)	Number of Animals	FAD Pyrophosphorylase Activity (mμ mole FAD formed/ g fresh wt/hr)
1. Saline	8	11 litters*	15.1 $\pm$ 1.0
2. Thyroxine	8	11 litters*	21.2 $\pm$ 1.5
3. Saline	67	28	22.5 $\pm$ 0.9
4. Thyroxine	67	21	25.2 $\pm$ 1.8

Data are expressed as mean $\pm$ 1 SEM.

*In each litter of newborn animals, half received thyroxine, 10 μg/animal, s.c. every day or every other day, and half received isotonic saline for 8 days; adults received thyroxine and saline on a similar weight basis.

Significance of difference between saline and thyroxine groups in 8 day old group = p < 0.01, and in 67 day old group = p > 0.05.

(Data are derived in part from Rivlin and Hornibrook, 1971b.)

and had a lower magnesium requirement than the enzyme in liver. FAD pyrophosphorylase activity in the cerebrum of saline-treated control rats was 15.1 $\pm$ 1.0 mμ mole FAD formed/g fresh weight/hour. Enzyme activity in thyroxine-treated animals of 21.3 $\pm$ 1.9 was significantly greater (p < 0.01) than in controls and was similar to the activity in the brain of adult animals.

Although FAD pyrophosphorylase could be increased in activity in liver of newborn animals by treatment with thyroxine for only 3 days, this time period was not sufficient to increase enzyme activity in the cerebrum of these animals (Rivlin and Hornibrook, 1971b). As shown in Table II, thyroxine administration was effective in increasing FAD pyrophosphorylase activity in the cerebrum of newborn animals but did not appear to increase enzyme activity in the cerebrum of adult animals. This finding is compatible with previous observations referred to above documenting the resistance of the adult brain to regulation by thyroid hormone (Sokoloff and Roberts, 1971). It appears that thyroid hormone may indeed regulate the availability of FAD for developing enzyme systems in the newborn brain.

Measurements of the conversion of ^{14}C-riboflavin to ^{14}C-FMN and ^{14}C-FAD in vivo provide additional evidence for the regulatory role of thyroid hormone. Radioactive flavins are being separated and quantitated using newly devised methods of isotope dilution and ion exchange column chromatography with DEAE-Sephadex A-25 (Fazekas, 1973). The results of recently obtained measurements in adult liver suggest that the magnitudes of the changes in flavokinase and in FAD pyrophosphorylase activity produced by both hyper- and hypothyroidism are similar to the magnitude of the changes in the rates of ^{14}C-FMN and ^{14}C-FAD synthesis from ^{14}C-riboflavin in vivo. Preliminary findings in the brains of newborn animals suggest that the magnitudes of increases in FAD pyrophosphorylase activity and in ^{14}C-FAD synthesis produced thyroxine are nearly identical. In brains of adult rats, neither enzyme activity nor ^{14}C-FAD synthesis is augmented by thyroxine. Studies are in progress to determine whether vitamin utilization is impaired in the brains of hypothyroid animals. In view of the increasing evidence (Eichenwald and Fry, 1969; Winick, 1970) that malnutrition during fetal and postnatal life has critical effects upon brain development, it may be important to consider that endogenous factors such as hormones may also regulate the nutritional status of the developing brain.

ACKNOWLEDGMENTS This work was supported in part by USPHS Grants AM 15265 and CA 12126, and by grants from the Stella and Charles Guttman Foundation, Inc. and the National Association for Mental Health, Inc.

REFERENCES

BALAZS, R., W. A. COCKS, J. T. EAYRS, and S. KOVACS, 1971. Biochemical effects of thyroid hormones on the developing brain. In *Hormones in Development*, M. Hamburgh and E. J. W. Barrington, eds. New York: Appleton-Century-Crofts, pp. 357–379.

BARKER, S. B., 1951. Mechanism of action of the thyroid hormone. *Physiol. Rev.* 31:205–243.

BURCH, H. B., O. H. LOWRY, T. DE GUBAREFF, and S. R. LOWRY, 1958. Flavin enzymes in liver and kidney of rats from birth to weaning. *J. Cell Comp. Physiol.* 52:503–510.

DIXON, M., and K. KLEPPE, 1965. D-amino acid oxidase. I. Dissociation and recombination of the holoenzyme. *Biochem. Biophys. Acta* 96:357–367.

DOMJAN, G., and K. KOKAI, 1966. The flavin adenine dinucleotide (FAD) content of the rat's liver in hypothyroid state and in the liver of hypothyroid animals after in vivo thyroxine treatment. *Acta Biol. Acad. Sci. Hung.* 16:237–241.

EAYRS, J. T., 1971. Thyroid and developing brain: Anatomical and behavioral effects. In *Hormones in Development*, M. Hamburgh and E. J. W. Barrington, eds. New York: Appleton-Century-Crofts, pp. 345–355.

EICHENWALD, H. F., and P. C. FRY, 1969. Nutrition and learning. *Science* 163:644–648.

ETKIN, W., 1964. Metamorphosis. In *Physiology of the Amphibia*, J. A. Moore, ed. New York: Academic Press, pp. 427–468.

FASS, S., M. OSNOS, R. HORNIBROOK, and R. S. RIVLIN, 1971. Effects of thyroxine on hepatic FAD pyrophosphorylase activity in tadpoles. *Program of the Forty-Seventh Meeting of the American Thyroid Association*, p. 42 (abstract).

FAZEKAS, A. G., 1973. Chromatographic and radioisotopic methods for the analysis of riboflavin and the flavin coenzymes. In *Riboflavin*, R. S. Rivlin, ed. New York: Appleton Century Crofts (in press).

GALTON, V. A., and S. H. INGBAR, 1965. Effects of vitamin deficiency on the in vitro and in vivo deiodination of thyroxine in the rat. *Endocrinology* 77:169–176.

GARCIA ARGIZ, C. A., J. M. PASQUINI, B. KAPLUN, and C. J. GOMEZ, 1967. Hormonal regulation of brain development. II. Effect of neonatal thyroidectomy on succinate dehydrogenase and other enzymes in developing cerebral cortex and cerebellum of the rat. *Brain Res.* 6:635–646.

GREENGARD, O., 1969. Analogies between mammalian and amphibian biochemical differentiation; the role of thyroxine. In *Advances in Enzyme Regulation*, Vol. 7, G. Weber, ed. Oxford: Pergamon Press, pp. 283–289.

HAMBURGH, M., and L. B. FLEXNER, 1957. Biochemical and physiological differentiation during morphogenesis. XXI. Effect of hypothyroidism and hormone therapy on enzyme activities of the developing cerebral cortex of the rat. *J. Neurochem.* 1:279–288.

KALSE, J. F., and C. VEEGER, 1968. Relation between conformations and activities of lipoamide dehydrogenase. I. Relation between diaphorase and lipoamide dehydrogenase activities upon binding of FAD by the apoenzyme. *Biochim. Biophys. Acta* 159:244–256.

NATAF, B. M., J. HAREL, and J. IMBENOTTE, 1971. Effect of thyrotropic hormone on thyroid glands of the fetal and newborn rat. In *Hormones in Development*, M. Hamburgh and E. J. W. Barrington, eds. New York: Appleton-Century-Crofts, pp. 781–792.

NEIMS, A. H., and W. R. WEIMAR, 1973. Physical and chemical properties of flavins; Binding of flavins to protein and conformational effects; Biosynthesis of riboflavin. In *Riboflavin*, R. S. Rivlin, ed. New York: Appleton-Century-Crofts (in press).

RIVLIN, R. S., 1969a. Perinatal development of enzymes synthesizing FMN and FAD. *Amer. J. Physiol.* 216:979–982.

RIVLIN, R. S., 1969b. Thyroid hormone and the adolescent growth spurt: clinical and fundamental consideration. In *Adolescent Nutrition and Growth*, F. Heald, ed. New York: Appleton-Century-Crofts, pp. 235–252.

RIVLIN, R. S., 1970a. Medical progress: Riboflavin metabolism, *New Eng. J. Med.* 283:463–472.

RIVLIN, R. S., 1970b. Regulation of flavoprotein enzymes in hypothyroidism and in riboflavin deficiency. In *Advances in Enzyme Regulation*, Vol. 8, G. Weber, ed. Oxford: Pergamon Press, pp. 239–250.

RIVLIN, R. S., and R. HORNIBROOK, 1971a. Thyroid hormone regulation of riboflavin metabolism in newborn rats. *Fed. Proc.* 30:359 (abstract).

RIVLIN, R. S., and R. HORNIBROOK, 1971b. Accelerated development of brain enzymes caused by thyroid hormone. *Program of the Fifty-Third Meeting of the Endocrine Society*, p. 110 (abstract).

RIVLIN, R. S., and R. G. LANGDON, 1966. Regulation of hepatic FAD levels by thyroid hormone. In *Advances in Enzyme Regulation*, Vol. 4, G. Weber, ed. Oxford: Pergamon Press, pp. 45–58.

RIVLIN, R. S., M. OSNOS, and R. HORNIBROOK, 1972. Inhibition of thyroxine-induced tadpole metamorphosis by analogues of riboflavin. International Congress Series No. 256, *Fourth International Congress of Endocrinology*, Washington, D.C., June 18–24, 1972. Amsterdam: Excerpta Medica, p. 245 (abstract).

RIVLIN, R. S., and G. WOLF, 1969. Diminished responsiveness to thyroid hormone in riboflavin-deficient rats. *Nature (London)* 223:516–517.

SOKOLOFF, L., and P. A. ROBERTS, 1971. Biochemical mechanisms of the action of thyroid hormones in nervous and other tissues. In *Influence of Hormones on the Nervous System*, D. H. Ford, ed. Basal: S. Karger, pp. 213–230.

SOKOLOFF, L., P. A. ROBERTS, M. M. JANUSKA, and J. E. KLINE, 1968. Mechanisms of stimulation of protein synthesis by thyroid hormones in vivo. *Proc. Nat. Acad. Sci. USA* 60:652–659.

STAAL, G. E. J., J. VISSER, and C. VEEGER, 1969. Purification and properties of glutathione reductase of human erythrocytes, *Biochim. Biophys. Acta* 185:39–48.

SWOBODA, B. E., 1969. The relationship between molecular conformation and the binding of flavin-adenine dinucleotide in glucose oxidase. *Biochim. Biophys. Acta* 175:365–379.

TATA, J. R., 1964. Biological action of thyroid hormones at the cellular and molecular levels. In *Actions of Hormones on Molecular Processes*, G. Litwack, and D. Kritchevsky, eds. New York: John Wiley & Sons, pp. 58–131.

WINICK, M., 1970. Nutrition and nerve cell growth. *Fed. Proc.* 29:1510–1515.

WOLF, G., and R. S. RIVLIN, 1970. Inhibition of thyroid hormone induction of mitochondrial α-glycerophosphate dehydrogenase in riboflavin deficiency. *Endocrinology* 86:1347–1353.

WOLFF, E. C., and J. WOLFF, 1964. The mechanism of action of the thyroid hormones. In *The Thyroid Gland*, Vol. 1, R. Pitt-Rivers, and W. R. Trotter, eds. London: Butterworths, pp. 237–282.

74 The Chemotactic Response as a Potential Model for Neural Systems

D. E. KOSHLAND, JR.

ABSTRACT Bacterial chemotaxis has been investigated as a potential model for biochemical aspects of neural systems. By developing apparatus to study the quantitative response to bacteria as a population and as individuals, it could be shown that bacteria detect gradients by comparing ratios of concentrations of attractants in the environment. The detection of a gradient apparently involves a rudimentary "memory" mechanism in the sense that the bacteria utilize a time-dependent response that compares past environmental conditions with present ones. This response is in turn mediated by highly specific receptor proteins, of which the ribose binding protein of *Salmonella typhimurium* is one example. There are many analogies at the biochemical and behavioral level between this simple system and higher neural systems, but the extent of the similarities must await further work.

THE RESPONSE OF organisms to chemical stimuli—varying from the odor and taste sensations of man, through the response of insects to pheromones, to the migration of bacteria in gradients of attractants—have certain apparent similarities. For example, the specificities of the responses to a restricted group of chemicals (Moncrieff, 1967) suggest the kind of specificity identified with protein molecules. Moreover, the evidence at all levels indicates that the attractant need not be metabolized in order to provide the stimulus (Weibull, 1960). Since biological patterns tend to recur in nature, the existence of a common biochemical response to stimuli is not unreasonable. However, the fact that nature frequently utilizes alternate pathways to achieve the same final result means that superficial similarities cannot be used to prove biochemical identities. The studies on chemotaxis reported here, therefore, were initiated with the hope that analogies to higher neural processes might exist but with the understanding that such relationships would require evidence and careful analysis.

Chemotaxis in bacteria was discovered in the 1880s by Engelmann (1902) and Pfeffer (1888) who demonstrated that microorganisms were attracted to chemicals and would migrate up a gradient to the position of optimal concentration. Gabricevsky (1900) demonstrated that the bacteria responded positively to some chemicals, negatively to others, and were indifferent to a third group. This phenomenon has been studied by many workers in the subsequent years, most particularly by the discerning studies of Adler and his co-workers on *Escherichia coli* (Adler, 1969). Adler demonstrated that there were two general classes of chemotactic mutants: In one group, called *specific*, mutation eliminated response to a particular chemical, and in a second, called *general*, the bacteria were unable to respond to any chemical attractants (Adler, 1969). The first mutants were presumably identified with the individual receptor molecules for each chemoattractant, the latter with the general apparatus for chemotaxis itself. Recently Hazelbauer and Adler (1971) have studied mutants that indicate that the galactose-binding transport protein, isolated by Boos and Kalckar (Wu et al., 1969), also serves as the receptor in chemotaxis.

Our approach to neural systems was to attempt to correlate the conformational properties of the receptor with the biological response in ways similar to those employed in our laboratory for a number of years on isolated enzyme systems (Koshland and Neet, 1968). It was clear that if such a biochemical approach to chemotaxis were to succeed, added quantitative tools would be needed and the initial studies have been directed to the development of the needed procedures. Three apparatus have been developed, each for a specific purpose. The first, a *migration velocity apparatus*, measures the movement of a mass of bacteria as a statistical average in much the same way that the gross diffusion of a solute in a liquid is described (Dahlquist et al., 1972). The second, which we call a *tracker*, is a device which measures the movement of individual bacteria in a defined spatial gradient (Lovely, Dahlquist, and Koshland, in preparation). The third is a *temporal gradient apparatus*, which allows us to follow the movements of an individual bacterium in a temporal gradient in the absence of a spatial gradient (Macnab and Koshland, 1972). In addition, a receptor for ribose chemotaxis has been isolated. The results allow us to describe the basic features of the mechanism of bacterial chemotaxis and to draw analogies to higher neural systems.

D. E. KOSHLAND, JR.   Department of Biochemistry, University of California, Berkeley, California

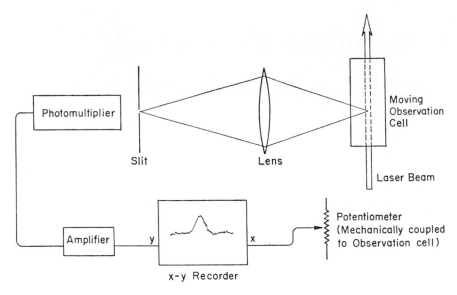

FIGURE 1 Migration velocity apparatus. A schematic representation of the apparatus designed to detect the migration of a bacterial population in a defined gradient.

Migration velocity apparatus and Weber's law

The apparatus designed for measuring the migration of the bacteria as a population was designed by F. Dahlquist and P. Lovely (Dahlquist et al., 1972) and is shown schematically in Figure 1. A linear density gradient of glycerol (0.5 to 3%) is used to stabilize a solution in an observation cell of 10-ml vol, 8-cm high. The bacterial concentration in the cell is determined by monitoring the intensity of light scattered by the bacteria. The light is supplied by an He-Ne laser, which shines up through the bottom of the observation cell. A photomultiplier tube measures the intensity of the laser light scattered at right angles to the beam. At low bacterial concentrations (less than 10^7 bacteria per ml) the intensity of the right-angle scattered light was found to be proportional to the bacterial concentration. The concentration of bacteria at various positions of the cell can be determined by moving the observation cell vertically by means of a screw drive. The cell position is recorded on the x axis and the signal from the photomultiplier is recorded on the y axis of an x–y plotter. The distribution of bacteria in the cell can be scanned in 1 to 3 min and this distribution is then followed as a function of time. By suitable mixing chambers before the introduction of the liquid to the column, various distributions of attractant and/or bacteria can be constructed in the observation cell. In Figure 2 is shown the response of these bacteria to a linear gradient of an attractant, serine, superimposed on an initially uniform distribution of *Salmonella typhimurium*. The gradient runs from zero to 10^{-3} M serine. The bacteria accumulate at the high serine concentration region, but a broad peak is formed in the middle of the serine gradient. This can only be explained if the velocity of the bacteria moving up the gradient depends on the absolute concentration as well as the rate of change of concentration. The gradient dc/dx is constant throughout, so the bacteria cannot be responding solely to the absolute gradient of serine.

An explanation of the complicated response to linear gradients is that the bacteria actually respond to proportional changes in concentration, that is, dc/c. In this case, an exponential gradient

$$\left(\frac{dc/c}{dx} = \frac{d \ln c}{dx} = \text{constant} \right)$$

should elicit a constant bacterial response throughout its length. The result of superimposing such a distribution of serine on an initially uniform bacterial distribution is shown in Figure 3. In this case the bacteria accumulate at the top of the gradient as a well-defined peak. The bacterial concentration on either side of this peak remains fairly constant, however, unlike the response to a linear increase that shows a distinct trough adjacent to the peak. Thus, bacteria are moving through the gradient region at a steady rate. This in turn means that the average velocity of the bacteria is determined by the proportional changes (i.e., ratios) in concentrations. (A trough should and does occur at the bottom of the tube, but it is obscured by scattering of light from the glass bottom.)

Bacterial response to ratios of concentrations can be double checked in another way, as shown in Figure 4. In this case, a steep increase in serine concentration from 2×10^{-3} M to 3×10^{-3} M is imposed over a very short

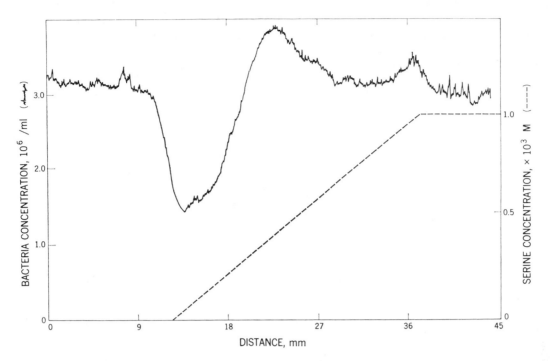

FIGURE 2 The response of *S. typhimurium* to a linear gradient of L-serine (represented by the dashed line). The trace represents the redistribution after 18 min of the bacteria from an initially uniform distribution.

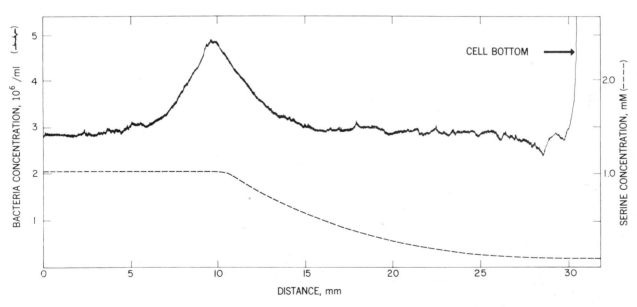

FIGURE 3 The response of *S. typhimurium* to an exponential gradient of L-serine. The plateau concentration of serine employed was 10^{-3} M and the decay distance of the gradient was 6.4 mm. The dashed line represents the initial serine concentration. The trace represents the bacteria redistribution after 15 min.

region of the observation cell on an initially uniform distribution of bacteria. The bacteria at far distances from the gradient, e.g. at the bottom of the cell, do not see any gradient and have no tendency to migrate. Those in the immediate vicinity of the gradient will swim towards the higher serine concentration. The bacteria on the high plateau side of the gradient will again see no gradient and will exhibit no net migration. The net movement of bacteria from low concentrations of serine to high concentrations of serine will therefore produce a trough in the region that bacteria are swimming from and a peak where the bacteria accumulate. If the same *relative gradient* is made in each of the experiments of Figure 4, the response of the bacteria after 20 min is essentially the same. Thus, the ratio across the boundary from 2×10^{-3} M to 3×10^{-3} M is the same as across the boundary from 2×10^{-5} M to 3×10^{-5} M, but the differences in concentrations in these two experiments are altered by a factor of

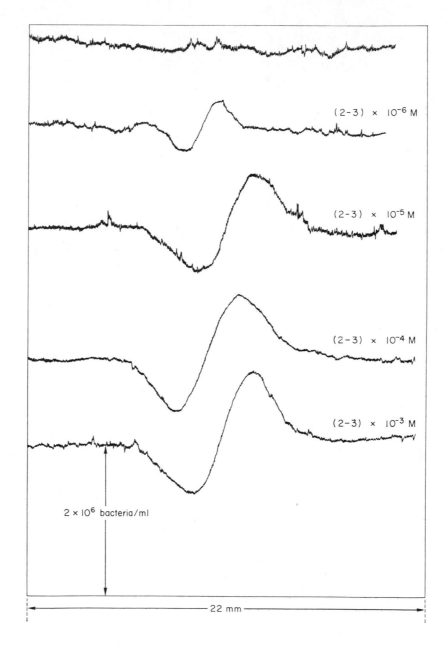

FIGURE 4   Response of bacteria to a constant ratio of attractant concentrations. An initially uniform distribution of *Salmonella* in the colony migration apparatus was exposed to a steep gradient with 2×10^x M attractant on one side and 3×10^x M attractant on the other, where x was varied from -6 to -3. The recordings represent the distribution of bacteria after 15 min. The response is seen to be about the same from 10^{-5} to 10^{-3} M absolute concentration but to drop off somewhat at 10^{-6}.

100. These findings therefore reinforce those of Figure 3, i.e., that the bacteria are responding to ratios not to differences in concentrations.

This leads us to the first of our conclusions, that the bacteria to a very rough approximation follow the Weber-Fechner law (Thompson, 1967). It is of interest that this Weber-Fechner law, which states that the just noticeable increase in a stimulus is proportional to the intensity of the stimulus, has been observed roughly for a wide variety of psychic phenomena, including many responses in higher organisms (Thompson, 1967).

The similarity does not end there. In fact, the obedience to Weber's law is only rough in higher species, and significant deviations occur for the bacteria as well. As discussed above, the bacteria move with uniform velocity through a cell with an exponential gradient extending over a rather limited concentration range. But when this concentration range was varied, the velocities of migration varied so that there is not precise agreement with Weber's law (Dahlquist et al., 1972). The deviations observed were well beyond experimental error and the bacteria deviate from Weber's law by amounts similar to those observed in mammals in psychological tests of phenomena such as the estimation of light intensity, the observations of sounds, the response to pressure, and many other behavioral characteristics (Thompson, 1967).

The "tracker" and the mean free path of bacterial motion

In order to understand the kinetics of chemotaxis and later to study behavioral mutants, we needed a precise description of the motion of an individual bacterium. A device for describing this motion in quantitative terms had been developed by Berg (1971). However, the initial Berg device did not allow the observation of movements of the bacterium in defined stable gradients of attractant. (Since the development of our instrument, Berg has solved the diffusion equation problem. His instrument has advantages in that it follows a bacterium automatically and more accurately than our instrument, but his gradient changes with time and cannot be used for repellents. Our instrument requires manual operation but can be used for observing bacterium in stable defined gradients that may include mixtures of repellents and attractants, two types of attractants, and other complex test systems. Thus, the two instruments complement each other in a very useful way.) The apparatus described in Figure 5 was devised by P. Lovely and F. Dahlquist to solve this problem.

A cuvette, essentially identical to the one in the integration velocity apparatus, is mounted on a microscope stage and bacteria are placed therein in a defined gradient. The long-range objective of a Leitz microscope is focused to see the bacteria in this vessel using dark field optics. The long working distance of the objective (7 mm) allows one to see the bacteria well away from the walls of the vessel. A "joy stick" allows us to activate motors that move the observation cell in the x and z directions and a pedal activates motors which move the cell in the y direction. Using a crosshair in the eyepiece of the microscope and the optical changes in the bacterium, an individual bacterium can be kept in focus by operation of the "joy stick" and the pedal. The motion of the observation cell and hence of the bacterium is recorded on computers and from this a record of the motion in x–y–z coordinates can be reconstructed.

Figure 6 shows a plot of the x, y, and z coordinates of a moving bacterium as a function of time in a zero gradient situation. We can determine the relative up and down motions in various ways, one of which is shown in Table I. Here the distance covered by the bacterium in each 5 sec interval is recorded. The z components of this velocity are then computed to give the velocity in μm/sec.

TABLE I

Average direction velocity of a bacterium when no gradient exists in the medium

Bacterium Tracked	Average Speed in 5 Sec Interval	Average Speed Upward $(+z)$	Average Speed Downward $(-z)$	Ratio
1	19.5 (73)	19.4 (34)	19.7 (29)	0.98
2	27.2 (70)	27.0 (25)	27.3 (45)	0.99
3	20.0 (14)	17.3 (3)	21.8 (11)	0.79
4	22.0 (100)	22.3 (43)	21.8 (57)	1.05
5	21.1 (85)	21.2 (23)	21.0 (62)	1.01

Speeds obtained from determining average speed in 5 sec intervals. Numbers in parentheses refer to number of 5 sec intervals used to determine average.

When the bacteria are moving in a zero gradient situation (i.e., a uniform distribution of attractant or repellent), the average velocity traveling up $(+z)$ should be equal to the average traveling down $(-z)$, and this is found to be the case. When the bacteria are observed in a gradient, the net velocity upwards exceeds the net velocity downwards and one such set of values is shown in Table II. The average velocity of a bacterium may vary from individual to individual but the average up and down velocities taken over fairly long sampling times are amazingly consistent.

In order to make a strict relationship between the movement of a population and the movement of an individual,

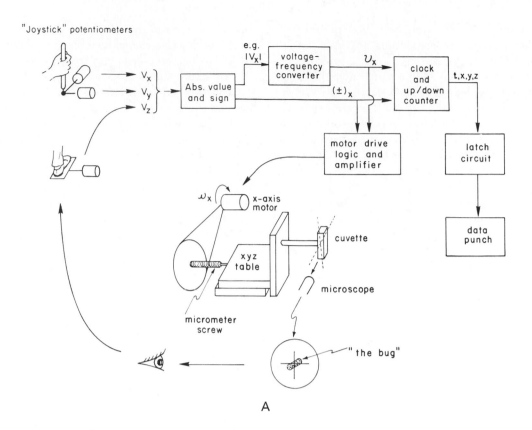

A

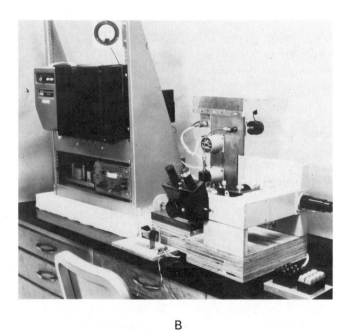

B

FIGURE 5 (A) Bacterial tracking apparatus. Schematic illustration of device to follow three-dimensional movements of bacteria in the presence and absence of gradients. Observation cell can be filled with bacteria in a defined gradient as described for the migration velocity apparatus. An individual bacterium is kept in focus by movement of this observation cell using the "joy stick" for the x-z directions and the foot pedal for the y direction. Joy stick and pedal activate stepping motors which are connected to punch tape apparatus to record data for calculating coordinates. (B) Photograph of apparatus.

846   CHEMOTAXIS AS A NEURAL MODEL

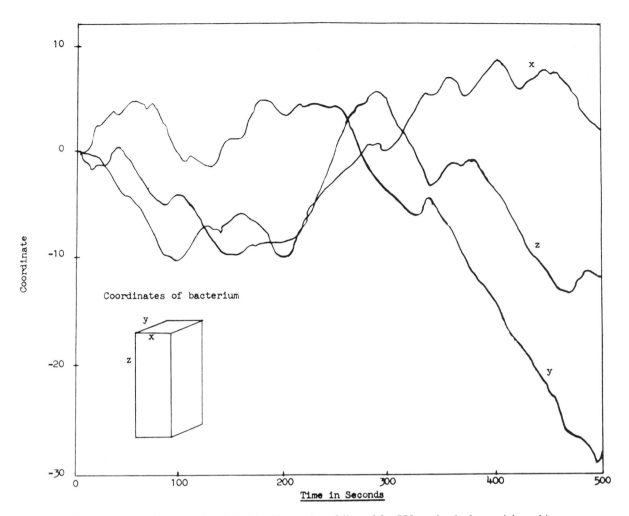

FIGURE 6   The movements in space of an individual bacterium followed for 500 sec in the bacterial tracking apparatus.

it is important to have a statistical number of the latter. After tracking a reasonable number of bacteria, the net velocity upwards of a colony of such bacteria was calculated and found to be 2.9 μm/sec. The velocity of the colony as determined by the migration velocity apparatus was 2.8 μm/sec (Dahlquist et al., 1972). This good agreement is pleasing and suggests that both instruments are measuring true characteristics of the bacterial population. Just as in perfect gas theory, the deduction of the collective

TABLE II

Average direction velocity of a bacterium in a gradient of attractant as measured on tracker in serine gradient of 0.15 mm^{-1}

Average % of time traveling up gradient = 60
Average % of time traveling down gradient = 40
    Component of velocity upwards = 11.51 μm sec^{-1}
    Component of velocity downwards = 10.0 μm sec^{-1}
Net average velocity upwards as calculated from tracker =
    0.6 × 11.5 − 0.4 × 10.0 = 2.9 μm/sec
Net average velocity upwards measured in migration velocity
    apparatus = 2.8 μm/sec

properties of a mass can be deduced from the motions of an individual.

What does the tracker indicate about the motions of the individual bacteria? It has previously been observed that the bacteria normally travel for distances in approximately straight lines and tumble at various intervals. This led to the postulation of some type of random-walk mechanism (Armstrong et al., 1967; Keller and Segel, 1971). A variety of such mechanisms can be imagined—some involving modulation of velocity, some of mean free paths, and some of both. The tracker was able to describe the process quantitatively and answer some of these questions.

In the first place, it appears that a given bacterium while it is moving travels through space at an approximately uniform velocity. The increased velocity in the upward direction resulted from less frequent tumbles not from added velocity during the intervals of straight line motion. Thus the observations of the tracker on bacteria in the presence and absence of gradients indicated that a modified random-walk mechanism was a good description of the bacterial movement and that it operated largely, if not completely, through modification of the mean free path and not the velocity of motion.

The temporal gradient apparatus and bacterial "memory"

At this stage we have shown that the bacteria sense a gradient and that they travel up a gradient of attractant by modulating their mean free path. It remains to find how they detect a gradient and how they modulate their mean free path.

In the first place, the difficulties for a bacterium detecting a gradient are very severe. *Salmonella* are approximately 2 μm in length. The drop in concentration of serine in one of the defined gradients described above over a 2 μm length is approximately 1 part in 10^4. A bacterium sensing a gradient by comparing the concentrations of attractant at its head and its tail would therefore be required to contain an analytical apparatus that detects 1 part in 10^4. If one calculates the statistical fluctuations in the more dilute gradient, however, of the individual molecules in the region of the head and the tail of the bacterium, even making very conservative estimates, the fluctuations are approximately 1 part in 10. Thus, the statistical fluctuations would indicate that the bacteria cannot detect a gradient whereas the experimental results indicate that they can. How can one proceed further?

The idea that some time average is involved has been suggested by Delbrück (1972), and time-dependent processes are known in higher species (Thompson, 1967). To test this idea it was necessary to develop an apparatus that could somehow separate temporal processes from spatial processes. The apparatus devised by R. Macnab shown in Figure 7 achieved this purpose (Macnab and Koshland, 1972). Bacteria in a solution containing an initial concentration (c_i) of attractant or repellent are suddenly mixed with a second solution containing the same or different concentrations of the chemical substance. After the mixing is complete, the motions of the bacterium are observed. The observation of the motions of the bacterium are made only when the bacterium is present in a final uniform distribution of chemical, c_f. If the bacterium senses an instanteous spatial gradient, it should behave as it does in a uniform gradient of concentration c_f. If, on the other hand, it has some time-dependent mechanism that allows it to compare the final concentration of its environment (c_f) with the concentration of its near past (c_i) and if the mixing time is fast with respect to this time-sensing mechanism, the bacterium will respond as though it had been swimming in a gradient. Thus, the apparatus is capable of generating a

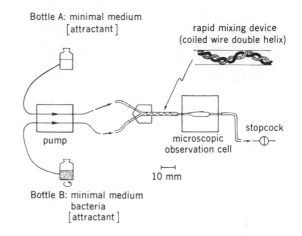

FIGURE 7 Temporal gradient apparatus. Attractant concentrations are: (i) Bottle B, c_i (≥ 0); (ii) bottle A, c_i' ($>$, $=$, or $< c_i$); (iii) observation cell (as a result of stream mixing) c_f ($>$, $=$, or $< c_i$). Therefore, bacteria experience a change in concentration from c_i to c_f in a short period of time. They can be subjected to positive, zero, or negative temporal gradients as desired. Gradient is given by $\Delta c/\Delta t$ where $\Delta c = c_f - c_i$ and Δt is mixing time.

temporal gradient, utterly divorced from the spatial gradient. The results obtained are described in detail elsewhere (Macnab and Koshland, 1972), but briefly they are as follows.

If the bacteria are subjected to zero gradient, i.e., $c_i = c_f$ in the temporal gradient apparatus, then motility is *normal*. Normal motility is defined as the motility observed in a wild-type culture in a nongradient situation. The *Salmonella* swam in a fairly coordinated manner, slight changes in direction occurred often, and occasionally a bacterium would tumble and then start swimming in a new direction. A stroboscopic multiple exposure of such behavior is shown in the middle portion of Figure 8. If the bacteria were subject to a positive gradient of attractant, i.e., $c_f > c_i$, supercoordinated swimming was observed, i.e., the bacteria swam for considerably longer distances before tumbling (see upper portion of Figure 8). If on the other hand the bacteria were subjected to a negative gradient of attractant, i.e., $c_f < c_i$, the motion was very uncoordinated, i.e., the bacteria swam only short distances before tumbling and starting out in a new direction (lower portion of Figure 8). These patterns were not the result of the mixing process itself, because bacteria treated in the same way in the absence of a temporal gradient ($c_i = c_f$) showed normal motility. Thus, the bacteria must have a time-dependent mechanism which allows them to remember the concentration of their immediate past (c_i) even though they are presently in a uniform gradient of attractant at a new concentration (c_f). Controls established that the absolute levels of attractant concentration did not affect motility, i.e., the bacteria gave the same normal motility in uniform distributions of attractant at concentrations equal to both c_i and c_f.

The conclusion that a time-dependent process is involved was further supported by the observation of a relaxation process, i.e., the abnormal motility a few seconds after mixing gradually reverts to normal motility during the passage of time. A few *minutes* after plunging into a different concentration $(c_i \neq c_f)$ the bacteria that had been subjected to the rapid mixing were indistinguishable from normal bacteria. It, therefore, follows from these experiments that the bacteria are not sensing an immediate spatial gradient but in fact have some device for comparing their previous environment with their present one, i.e., they have some type of primitive *memory* apparatus.

These experiments also answer our previous question about the modulation of the mean free path. It was very difficult in the tracker or by microscopic observation to be sure whether the bacteria migrated by (a) increasing the frequency of tumbling going down the gradient, (2) decreasing their frequency of tumbling going up the gradient, or (3) both. The temporal gradient apparatus indicates that (3) is correct in principle, but the relaxation times indicate (2) may be a better approximation in practise.

Moreover, the apparatus allowed us to test the effect of repellents in a quantitative manner, and it was found (Tsang, Macnab, and Koshland, in preparation) that repellents bear a reciprocal relationship to attractants, i.e., the bacteria tumble more frequently when $c_f > c_i$ and tumble less frequently when $c_f < c_i$.

A good picture of the mechanism of chemotaxis emerges from these experiments. The bacteria do not *sense* an immediate spatial gradient; they, in fact, sense a temporal gradient. In the real world, however, they convert a spatial gradient into a temporal gradient by traveling at uniform velocity through space and integrating or comparing over a time interval. Thus, if they are traveling up a positive gradient of attractant, they move at a constant velocity through increasing concentrations of attractant so they *see* a positive temporal gradient. As a result they tumble less frequently than normal and when they are traveling down such a gradient they tumble more frequently. When traveling in a gradient of a repellent, the responses are reversed, i.e., a positive gradient causes more frequent tumbling, a negative gradient, less frequent.

By integrating over time, they solve two problems that were mentioned previously. In the first place, they eliminate the statistical fluctuation problem by avoiding an instantaneous comparison that would require high accuracy. In the second place, by comparing a concentration in the past with a concentration in the present,

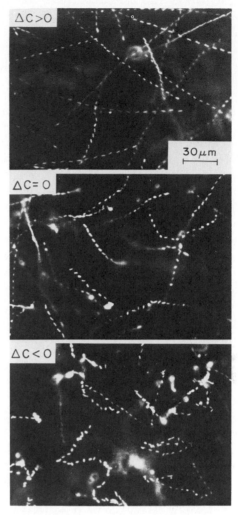

FIGURE 8 Motility tracks of *S. typhimurium,* taken in the time interval 2–7 sec after subjecting bacteria to a sudden (200 msec) change in attractant (serine) concentration in the temporal gradient apparatus. *Upper:* $c_i = 0$, $c_f = 7.6 \times 10^{-4}$ M. Smooth, linear trajectories. *Middle:* $c_i = c_f = 0$ (control). Some changes in direction; bodies often show "wobble" as they travel. Bright spots indicate tumbling or nonmotile bacteria. *Lower:* $c_i = 10^{-3}$ M, $c_f = 2.4 \times 10^{-4}$ M. Poor coordination, frequent tumbles, and erratic changes in direction. [Photomicrographs were taken in dark field with a stroboscopic lamp operating at 5 pulses sec^{-1}. Instantaneous velocity of bacteria in straight line trajectories is of the order of 30 μm sec^{-1}.]

they avoid the problem of their small body size. For example, a bacterium which possesses a *memory* with a decay time of one minute traveling at 30 μm per sec can compare concentrations over an effective distance of 2 mm or roughly 1000 body lengths. The needed analytical accuracy is therefore reduced from 1 part in 10^4 to 1 part in 10.

In paramecium and some higher species, there is a

phenomenon called an *avoidance response* in which an organism reverses direction or flees from a hostile atmosphere. This has been applied to the bacteria in chemotaxis, but our observations in these temporal gradients and in the tracker apparatus would not support such a term for the bacterial behavior. Rather it seems there is a *confusion response*. The bacteria respond to the less favorable direction by increased tumbling, brought about by the apparently uncoordinated motions of their flagella. The loss of coordination with its effect on the mean free path, not a reversal of direction per se, causes the net migration in the opposite direction.

Ribose binding protein

A number of proteins are apparently involved in the transport of metabolites across the bacterial membrane (Pardee, 1968). As mentioned above, one of these, the galactose binding protein of *E. coli* has been identified (Hazelbauer and Adler, 1971) with chemotaxis. Further evidence has been obtained by the isolation of the ribose binding protein of *Salmonella typhimurium*. Using the shock procedure of Heppel (1967) and Anraku (1968), R. Aksamit has purified a ribose binding protein which is a monomer and has a molecular weight of 29,000 (Aksamit and Koshland, 1972).

The specificity of this protein for binding of ribose derivatives is the same as for the chemotactic response (Table III). Both show a very high, if not almost absolute, specificity for ribose. The binding constant of ribose for the pure protein ($K_D = 3.3 \times 10^{-7}$ M) is approximately the same as the midpoint of the chemotactic response curve as determined in the temporal gradient apparatus. The osmotically shocked *Salmonella* bacteria that have lost their chemotactic response can have it restored by readdition of the ribose binding protein. Since ribose is a metabolite transported across the membrane of *Salmonella*, it appears that this protein also acts as an agent in the transport of ribose and the chemotaxis toward this attractant. Its similarity to the sulfate binding protein isolated by Langridge, Shinagawa, and Pardee (1970) suggests that its dimensions are similar (72 Å long by 17 Å in diameter), which would allow it to stretch across the membrane if required to do so. The yield of this binding protein indicates that there are about 10^4 receptors per cell of *Salmonella*.

Biochemical mechanism

It remains to consider what biochemical mechanisms are compatible with the observations described above. In order to compare values of a parameter at different times, two component responses to that parameter are required, with different relaxation times. This can be expressed roughly by saying that the fast component response reflects the present value of the parameter, e.g. attractant, while the slow component response reflects the past value. To generate the differential response the component responses must then act in opposing manners on yet another parameter, whose value determines the ultimate response, in this case loss of flagellar coordination. Figure 9 offers one possible scheme of this type. The effector (either the attractant itself or a species generated by it—possibly the attractant-chemoreceptor complex) activates both enzymes 1 and 2 inducing conformational changes, which are fast in enzyme 1 and slow in enzyme 2. These enzymes

TABLE III

Specificity of ribose binding protein and chemotaxis of Salmonella typhimurium

Compound	Chemotaxis Response	Binds to Ribose Binding Protein
Ribose	+ Positive	+
D and L arabinose	No	No
D-xylose	No	No
D-lyxose	No	No
α-D-ribose-1-P	No	No
α-D-methylribofuranose	No	No
1, 4-anhydroribitol	No	No
D-glucose	No	No
D-galactose	No	No
D-mannose	No	No
D-fucose	No	No
L-rhamnose	No	No
α-D-ribose-5-P	No	No
2-deoxy-D-ribose ribitol	No	No

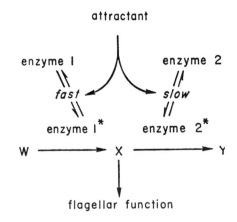

FIGURE 9 Schematic illustration of one possible time-dependent mechanism. Attractant alters conformation of enzymes 1 and 2 to catalytically more active forms, enzyme 1 rapidly and enzyme 2 slowly. The compound *X*, which controls flagellar function, therefore tends to increase in positive gradients, decrease in negative gradients, remain unchanged in zero gradients.

catalyze the synthesis and degradation respectively of compound X whose pool size must exceed a critical value for the flagella to function in a coordinated manner. In a positive gradient of attractant, enzyme 1 will be more highly activated than enzyme 2, the pool size of X will rise, and tumbling will diminish. In a negative gradient the pool size of X will be depleted and consequently tumbling will increase.

It must be emphasized that this example is only one of a number that can be devised. For example, the enzyme roles could be reversed if high rather than low levels of X were responsible for loss of coordination. The response might be achieved by diffusion or transport processes across the cell membrane rather than by conformational changes in enzymes. Moreover, there may be a series of steps between the binding of the attractant and the effector that controls the temporal process. That the level of an intermediate X is important is further indicated by experiments which show that two gradients of attractant in the same direction are additive; gradients in opposite directions are subtractive.

Relation of chemotaxis to higher neural systems

The signalling processes in a monocellular organism cannot have the complex circuitry of higher species with a central nervous system, but it could quite possibly use similar biochemical pathways and a similar pattern of recording information. The bacterium under consideration is shown in Figure 10. The information received at receptors distributed over the membrane is transmitted to the flagella. The distance the signal must travel is much shorter than in higher species, but the stimulus must be transmitted and must involve a behavioral response. In the case of *Salmonella*, a response to specific chemicals is apparently analogous to the specific responses in higher organisms, e.g. odor in man, and pheromones in insects. The high specificity of the response, e.g. to ribose only, indicates a protein molecule is the receptor and such a molecule has been isolated. The response does not require the metabolism of the attractant as is also true in higher species. An exactly parallel and inverse relationship is seen in repellents. The analogy to pheromones in insects and to taste and odor in mammals is impressive. The bacteria obey Weber's law roughly over short ranges of stimuli and deviate significantly from it over wide ranges, again as seen in higher species.

The bacteria utilize a time-dependent mechanism that can be termed a *memory* in the sense that it allows the bacterium to compare its past with its present. This memory is very short, but it requires a biochemical mechanism which allows an integration over time. Quite

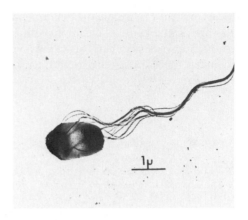

FIGURE 10　An electron microscope picture of a *Salmonella*. The receptors are located in the periplasmic space in the outer membrane. The binding of a chemoattractant generates a signal that is transmitted in some manner to produce a behavioral response in the flagella. (Picture kindly donated by Dr. B. Gerber.)

obviously, this is a long way from the memory of man with its vast storage and retrieval mechanisms. Yet the relation of the bacterial system to higher species may be important, just as the analogy in protein synthesizing pathways between bacteria and higher species is close but not identical.

The mechanism proposed in Figure 9 indicates that the level of some chemicals is critical for the signal to the flagella. Moreover the inputs to this level from a variety of attractants and repellents act as a summation process. Such levels could logically be related to equivalent levels in a neuron. In that case the state of polarization depends on additive and subtractive relationships of stimuli from chemicals, dendrites, and hormones. Moreover the signal in the bacteria is transmitted from the receptor to the flagella just as a signal in higher species must be transmitted from a receptor to the motor apparatus. The distances are greater in the latter case but again the chemistry may have common features.

Finally, if new interneuron connections are to be initiated or reenforced by the synthesis of new proteins or are to be deleted by hydrolysis of old proteins and decreased protein synthesis, some change in chemical level is indicated. Small molecules such as cyclic AMP, inducers or corepressors are required to initiate protein synthesis or inhibit protein breakdown. It is logical therefore that a change in the level of some compound X could itself trigger other more permanent chemical changes as well as being involved with the firing of an individual neuron. Learning and memory might therefore be products of a common biochemical event.

The behavior patterns of the single-cell bacteria will obviously be simpler than the more complex processes in higher species, but the preliminary indications are that the similarities are significant. The monocellular system offers the advantages of simplicity; it can provide insight into biochemical mechanisms of communication that are common to all species.

REFERENCES

ADLER, J., 1969. Chemoreceptors in bacteria. *Science* 166:1588–1597.

AKSAMIT, R., and D. E. KOSHLAND, JR., 1972. A ribose binding protein of *Salmonella typhimurium*. *Biochem. Biophys. Res. Commun.* 48:1348–1353.

ANRAKU, Y., 1968. Transport of sugars and amino acids in bacteria. I. Purification and specificity of the galactose- and leucine-binding proteins. *J. Biol. Chem.* 243:3116–3122.

ARMSTRONG, J. B., J. ADLER, and M. A. DAHL, 1967. Non-chemotactic mutants of *Escherichia coli*. *J. Bacteriol.* 93:390–398.

BERG, H. C., 1971. How to track bacteria. *Rev. Sci. Instrum.* 42:868–871.

DAHLQUIST, F. W., P. LOVELY, and D. E. KOSHLAND, JR., 1972. Quantitative analysis of bacterial migration in chemotaxis. *Nature New Biology* 236:120–123.

DELBRÜCK, M., 1972. Signal transducers: *Terra incognita* of molecular biology. *Angew. Chem. (Eng.)* 11:1–6.

ENGELMANN, T. W., 1902. Die Erscheinungsweise der Sauer stoffaus scheidung chromophyl haltiger Zellen im Licht bei Anwendung der Bakterienmethode. *Pflüger Arch. Ges. Physiol.* 57:375–390.

GABRICEVSKY, G. N., 1900. Uber Aktive Beweglichkeit der Bakterien. *Z. Ges. Hyg.* 35:104–122.

HAZELBAUER, G. L., and J. ADLER, 1971. Role of galactose binding protein in chemotaxis of *Escherichia coli* toward galactose. *Nature New Biology* 230:101–104.

HEPPEL, L. A., 1967. Selective release of enzymes from bacteria. *Science* 156:1451–1455.

KELLER, E. F., and L. A. SEGEL, 1971. Model for chemotaxis. *J. Theor. Biol.* 30:225–234.

KOSHLAND, D. E., JR., and K. E. NEET, 1968. The catalytic and regulatory properties of enzymes. *Ann. Rev. Biochem.* 37:359–410.

LANGRIDGE, R., H. SHINAGAWA, and A. B. PARDEE, 1970. Sulfate-binding protein from *Salmonella typhimurium*: Physical properties. *Science* 169:59–61.

MACNAB, R. M., and D. E. KOSHLAND, JR., 1972. The gradient-sensing mechanism in bacterial chemotaxis. *Proc. Nat. Acad. Sci. USA* 69:2509–2512.

MONCRIEFF, R. W., 1967. *The Chemical Senses*. London: Leonard Hill.

PARDEE, A. B., 1968. Membrane transport proteins. *Science* 162:632–637.

PFEFFER, W., 1888. *Untersuch Botan. Inst. Tubingen* 2:582–661.

THOMPSON, R. F., 1967. *Foundations of Physiological Psychology*. New York: Harper and Row.

WEIBULL, C., 1960. Movement. In *The Bacteria*, Vol. I, I. C. Gunsalus and R. Y. Stanier, ed. New York: Academic Press, pp. 153–202.

WU, H. C. P., W. BOOS, and H. M. KALCKAR, 1969. Role of the galactose transport system in the retention of intracellular galactose in *Escherichia coli*. *J. Mol. Biol.* 41:109–120.

75 Regulation of Allosteric Enzymes in Glutamine Metabolism

STANLEY PRUSINER

ABSTRACT   The intracellular levels of many enzymes are under precise control by specific metabolite inducers and corepressors and by cyclic AMP. In addition, the catalytic activity of many enzymes is modulated both allosterically and isosterically by specific metabolite ligands. The enzymes involved with glutamine metabolism illustrate several of the principles of enzyme regulation. Since the synthesis and degradation of glutamine lies at a crossroads where carbohydrate and nitrogen metabolism intersect, and since glutamine is the amide nitrogen donor for the biosynthesis of amino acids, nucleotides, amino sugars and cofactors, it is not unexpected that many enzymes and metabolites influence the intracellular concentration of glutamine. Indeed, diverse and rigorous control mechanisms have been found to participate in the regulation of glutamine metabolism in microorganisms. A knowledge of these control mechanisms may be of value in understanding both the physiological and pathological metabolism of nitrogen in the nervous system.

IN BIOLOGICAL organisms almost all of the chemical reactions that occur are catalyzed by specific enzymes. The cellular concentration of particular enzymes and the modulated activity of these enzymes are complex processes as described in two preceding chapters by Drs. Schimke and Koshland.

This communication gives examples of the complex and elegant enzymatic machinery that has evolved to regulate the intracellular levels of glutamate and glutamine in microorganisms (Prusiner and Stadtman, 1973). A discussion of the enzymatic regulation of glutamate and glutamine metabolism is especially relevant to neurobiology, because glutamate, glutamine, aspartate, and γ-aminobutyric acid account for more than 70% of the nonprotein amino nitrogen in nervous tissue (Himwich and Agrawal, 1969). Of these amino acids, glutamate predominates and is present in a concentration of at least 10 mM while glutamine is present *at* 4 mM.

The neuroscience literature on glutamine and glutamate metabolism is immense, because these amino acids have been studied extensively with regard to synaptic transmission (Johnson, 1972; Bloom, 1972), compart-

mentation of the brain (Waelsch, 1951; Salganicoff and DeRobertis, 1965; Van Den Berg, 1970; Berl and Clarke, 1970; Rose, 1970), epilepsy (Sourkes, 1962), mental retardation (Sourkes, 1962), schizophrenia (Roberts, 1972), hepatic failure (Walker and Schenker, 1970; Williams et al., 1972) and hyperammonemia (Shih and Efron, 1972; Ghadimi, 1972). No attempt to review these studies is made here.

In Figure 1, the assimilation of NH_4^+ into glutamate and glutamine and the possible fates of the ammonia nitrogen are illustrated. The interconversion of glutamate and glutamine lie at a crossroads where carbohydrate and nitrogenous metabolism intersect. The amino nitrogen of glutamate may be transferred to keto-acids by several transaminases, while the carbon skeleton of glutamate may be consumed in the tricarboxylic acid cycle. Glutamine may donate its amide nitrogen to a variety of reactions which participate in the biosynthesis of amino acids, amino sugars, nucleotides, and cofactors (Prusiner and Stadtman, 1973).

Enzymatic regulation in two microorganisms

The intricate mechanisms of regulation of the enzymes of glutamate and glutamine metabolism are best understood in bacteria, but they may serve as a model for similar types of control in mammalian cells. Regulation of glutamate and glutamine levels in microorganisms is accomplished in a variety of different ways. A comparison of six enzymes concerned with the metabolism of glutamate and glutamine in *Escherichia coli* and *Bacillus subtilis* reveals that these organisms contain different enzymatic machinery (Table I). *E. coli* and *B. subtilis* both contain glutamine synthetase that catalyzes the synthesis of glutamine from glutamate, ATP, and ammonia:

$$\text{L-glutamate} + \text{ATP} + \text{NH}_3 \underset{\xrightarrow{\text{Mg}^{++}}}{\longleftarrow} \text{L-glutamine} + \text{ADP} + \text{P}_i$$

The reaction requires a divalent cation, which is probably Mg^{++} in the cell. *B. subtilis* possesses a second enzyme specifically for the synthesis of glutamine to be incorporated into protein (Wilcox, 1969). The reaction involves the

STANLEY PRUSINER   National Heart and Lung Institute, National Institutes of Health, Laboratory of Biochemistry, Bethesda, Maryland

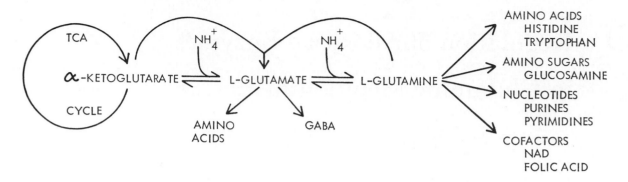

FIGURE 1 The assimilation of NH_4^+ into glutamate and glutamine and the fates of ammonia nitrogen.

TABLE I

Comparison of enzymes involved in the metabolism of glutamate and glutamine in microorganisms

Enzyme	Micro-organism		References
	E. coli	*B. subtilis*	
1. Glutamine synthetase	+	+	(Woolfolk and Stadtman, 1964; Deuel et al., 1970)
2. Glutaminyl t-RNA amidotransferase	−	+	(Peterkofsky, 1972; Wilcox, 1969)
3. Glutamate synthase	+	+	(Meers et al., 1970; Elmerich and Aubert, 1971)
4. Glutamate dehydrogenase	+	−	(Elmerich and Aubert, 1971)
5. Glutaminase A	+		(Hartman, 1968)
6. Glutaminase B	+		(Prusiner and Stadtman, 1971)

+ = present
− = not found

transfer of the amide nitrogen of glutamine or asparagine to glutamyl-t-RNA to form glutaminyl t-RNA. The enzyme has not been detected in *E. coli* (Peterkofsky, 1972).

Both organisms also have the enzyme, glutamate synthase, which was recently discovered by Tempest and coworkers (Meers et al., 1970). The enzyme is an iron sulfide flavoprotein which catalyzes the conversion of α-ketoglutarate and glutamine to two molecules of glutamate in the presence of TPNH (Miller and Stadtman, 1972):

α-ketoglutarate + L-glutamine
$$+ \text{ TPNH} \rightleftharpoons 2 \text{ L-glutamate} + \text{TPN}^+$$

A deficiency of this enzyme in *B. subtilis* results in glutamate auxotrophy (Elmerich and Aubert, 1971). In

E. coli, glutamate auxotrophy requires a deficiency of both glutamate synthetase and glutamate dehydrogenase (Berberich, 1972).

The glutamate dehydrogenase of *E. coli* is a freely reversible enzyme that catalyzes the conversion of glutamate to α-ketoglutarate and ammonia:

L-glutamate
$$+ \text{ TPN}^+ \rightleftharpoons \alpha\text{-ketoglutarate} + \text{NH}_3 + \text{TPNH}$$

Also in *E. coli*, there are two glutaminases, A and B, that catalyze the hydrolytic deamidation of L-glutamine to L-glutamate + NH_3 (Hartman, 1968; Prusiner and Stadtman, 1971):

$$\text{L-glutamine} + \text{H}_2\text{O} \rightleftharpoons \text{L-glutamate} + \text{NH}_3$$

These isoenzymes are readily distinguished and separable since glutaminase A is active at pH 5 while B is active at pH 7. No information about glutaminases in *B. subtilis* is available at present.

The differences in enzymatic machinery between the two microorganisms, *E. coli* and *B. subtilis*, in the regulation of glutamine metabolism emphasize the potential variability in enzymatic regulation of metabolism that may exist among different cell types, especially in a mammalian system, such as the brain (Rose, 1970).

Control of enzyme levels

Many studies have demonstrated that microorganisms, like some mammalian cells, have the ability to alter intracellular enzyme levels in response to changes in nutritional conditions. The effect of ammonia on the levels of 5 enzymes concerned with glutamate and glutamine metabolism in *E. coli* are summarized in Table II. Glutamine synthetase levels increased when growth of *E. coli* was limited by the availability of ammonia salts (Woolfolk and Stadtman, 1964; Prusiner et al., 1972). Glutamate synthetase levels also increased when the culture media contained low concentrations of NH_4^+. The response in *Aerobacter aerogenes* was much greater than in *E. coli* (Meers et al., 1970; Miller and Stadtman, 1972). The levels of glutamic dehydrogenase and glutaminase A both decreased with the growth of *E. coli* on low concentrations of NH_4^+ (Pateman, 1969; Prusiner et al., 1972). The ammonia concentration in the growth media did not alter the level of glutaminase B (Prusiner et al., 1972).

TABLE II
The effect of ammonia on the levels of enzymes concerned with the metabolism of glutamate and glutamine in E. coli

Enzyme	Enzyme Levels with Growth on	
	Low NH_4Cl	High NH_4Cl
1. Glutamine synthetase	↑	↓
2. Glutamate synthase	↑	↓
3. Glutamate dehydrogenase	↓	↑
4. Glutaminase A	↓	↑
5. Glutaminase B	↔	↔

↑ increased
↓ decreased
↔ no change

The levels of many enzymes in bacteria are altered by adenosine 3′:5′-cyclic monophosphate (Pastan and Perlman, 1970). Recent studies have disclosed that exogenous cAMP increases the levels of glutamine synthetase and glutamic dehydrogenase and decreases the levels of glutamate synthase and glutaminase A (Prusiner et al., 1972). As shown in Figure 2, the addition of 5 mM cAMP to a logarithmically growing culture of *E. coli* K12 deficient in adenyl cyclase resulted in a doubling of the growth rate and a doubling of the levels of glutamate dehydrogenase and glutamine synthetase. The levels of glutaminase A and glutamate synthase decreased more than 50% in one generation. The level of glutaminase B was not influenced by cAMP. Growth on glycerol, which has been shown to elevate the intracellular level of cAMP in *E. coli*, mimicked the alterations brought about by exogenous cAMP. The effects of cAMP were abolished by chloramphenicol or a deficiency of cAMP receptor protein (Pastan and Perlman, 1970). In mammals, reciprocal effects of cAMP on the enzymes of glycogen metabolism have been observed. These effects involve the covalent chemical modification of pre-existing glycogen phosphorylase and glycogen synthetase by a protein kinase (Soderling et al., 1970). In contrast, the changes in enzyme levels described above appear to involve altered protein synthesis and/or degradation.

In summary, at least three mechanisms participate in the control of enzyme levels in *E. coli*: (1) generalized or

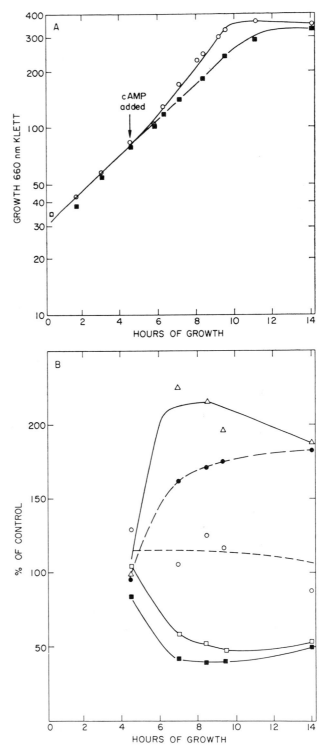

FIGURE 2 Effect of cAMP on *E. coli* K12–5336 growth rate and enzyme levels. (A) Growth of *E. coli* deficient in adenylate cyclase (■———■) 5mM cAMP added at *arrow* (O———O). (B) Changes in specific activities (units/mg) of glutaminase A (■———■), glutamate synthase (□———□), glutamate dehydrogenase (●———●) glutamine synthetase (△———△) and glutaminase B (O———O) at various times after addition of cAMP (for details, see Prusiner et al., 1972).

pleiotropic control by cAMP, (2) specific control by metabolite repression and/or induction, and (3) constitutive control where enzyme levels are independent of nutritional conditions. These control mechanisms may operate through alterations in enzyme synthesis and/or degradation.

Regulation of enzyme activity

In addition to the control mechanisms that regulate the levels of the five enzymes noted above, at least three other types of control act to modulate the activity of these enzymes: (1) allosteric or isosteric feedback inhibition by specific end products or energy metabolites, (2) covalent modification of glutamine synthetase by adenylylation, and (3) divalent cation modulation of enzyme activity directly or through chelation with ATP. Studies on glutamine synthetase and glutaminase B from *E. coli* and on glutamine synthetase from *B. subtilis* illustrate these various modes of regulation of enzyme activity.

Glutamine synthetase from *E. coli* has a molecular weight of 600,000 with 12 subunits of 50,000 molecular weight each arranged in a double layer hexagon (Figure 3). The enzyme is an acidic protein and is subject to feedback inhibition by 8 biosynthetic end products that include histidine, tryptophan, CTP, AMP, glucosamine 6-P, carbamyl phosphate, alanine, and glycine (Stadtman et al., 1968). The sensitivity to feedback inhibitors is dramatically changed by adenylylation of the enzyme (Figure 4). Adenylylation of glutamine synthetase is catalyzed by the enzyme complex P_IP_{II} designated adenylyltransferase in Figure 4 (Stadtman, 1971). The reaction results in the covalent attachment of 1 AMP to 1 specific tyrosine residue in each subunit via a phosphodiester bond. Removal of the AMP involves a pyrophosphorolysis catalyzed by the uridylylated P_IP_{II} enzyme complex (Brown et al., 1971). The adenylylation-deadenylylation process is modulated by divalent cations, α-ketoglutarate, and glutamine. The stimulation of adenylylation by glutamine appears to be an alternate mode of product inhibition since the covalent modification of glutamine synthetase results in a marked decrease in Mg^{++} dependent activity. Mg^{++} is the predominant intracellular divalent cation and probably supports the glutamine synthetase reaction in vivo. Adenylylation lowers the pH optimum of the enzyme and increases the Mn^{++} supported activity. Divalent cations are bound to the enzyme directly and are chelated by ATP. There are 389 distinct forms of glutamine synthetase due to adenylylation of individual subunits (Stadtman, 1971), and the heteroeotropic interactions between subunits allow for extensive modulation of enzymatic catalysis.

The regulation of glutamine synthetase from *B. subtilis*

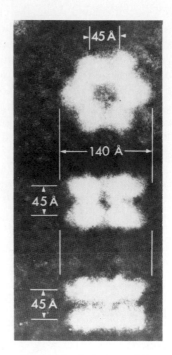

FIGURE 3 Five superimposed electron microscopic images of glutamine synthetase from *E. coli* in three orientations. The upper portion shows a picture of the molecule as it is seen when resting on a face. The center and bottom pictures are views of the molecule as seen on edge when looking exactly down a diameter between subunits or in general as two lines (from Valentine et al., 1968).

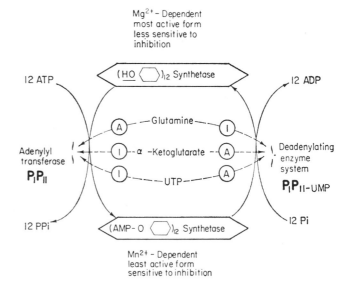

FIGURE 4 Metabolite regulation of adenylylation and deadenylylation reactions which alter divalent cation specificity, catalytic activity, and feedback inhibition of glutamine synthetase from *E. coli* (adapted from Stadtman, 1971).

is distinctly different from *E. coli*, but the structure of the enzyme is similar. The enzyme from *B. subtilis* has a molecular weight of ~600,000 and is composed of 12 identical subunits arranged in a double-layer hexagon. The enzyme requires divalent cations for activity and is subject to feedback inhibition by the same biosynthetic end products, as described for *E. coli*. A complex system of covalent modification involving adenylylation-deadenylylation to regulate the activity of glutamine synthetase under various conditions of nitrogen metabolism does not exist in *B. subtilis* (Gancedo and Holzer, 1968; Deuel et al., 1970). Instead, the organism possesses a glutamine synthetase which is directly susceptible to feedback inhibition by its product, glutamine (Deuel and Prusiner, 1973). As described above, glutamine synthetase from *E. coli* is not inhibited by glutamine but by glutamine-stimulated adenylylation of the enzyme that results in a decrease in the magnesium supported biosynthetic activity. In addition, it is of interest to compare the inhibition by AMP of glutamine synthetase from both organisms, because AMP is not only a biosynthetic end product but also an indicator of cellular energy metabolism. As shown in Figure 5 the *E. coli* enzyme in the unadenylylated form is only moderately inhibited by AMP, but following glutamine-stimulated adenylylation there is a substantial increase in the susceptibility of the enzyme to inhibition by AMP (Ginsburg, 1969; Stadtman, 1971). This same phenomena is even more striking in *B. subtilis*, which does not use adenylylation. Here AMP is not a potent inhibitor; however, the presence of noninhibitory concentrations of glutamine does convert AMP to a powerful inhibitor (Deuel and Prusiner, 1973). This pattern of regulation is designated synergistic feedback inhibition when the inhibition by two ligands is

much greater than the sum of the inhibition produced by each ligand independently (Stadtman, 1970). Synergistic effects have been observed with several other enzymes: Phosphoribosyl pyrophosphate amidotransferase from chicken liver (Caskey et al., 1964), aspartokinase from *E. coli* (Truffa-Bachi and Cohen, 1966) and glutamine synthetase from *Bacillus lichenformis* (Hubbard and Stadtman, 1967).

Like the glutamine synthetase of *E. coli*, the enzyme of *B. subtilis* is also able to use Mn^{++} in place of Mg^{++}. In most biological organisms, the intracellular Mg^{++} concentration is 10^2 to 10^4 times greater than Mn^{++} (Silver et al., 1970) but in sporulating bacteria such as *B. subtilis*, the Mn^{++} is accumulated in large quantities during sporulation. The Mn^{++} is obligatory for sporulation and the Mn^{++} content of spores is equal to that of Mg^{++} (Murrell, 1969). The potential modulation of enzyme activity by the interaction of these divalent cations with glutamine synthetase during sporulation and germination may be significant, because glutamine, but not glutamate, represses sporulation (Elmerich and Aubert, 1972).

The action of glutamine synthetase is directly opposed by the enzyme glutaminase. In *E. coli*, there are two glutaminases, as described above, but only glutaminase B is active in the pH range where glutamine synthetase also exhibits activity. In the absence of appropriate controls, the coupling of glutamine synthetase with glutaminase B would form a "futile cycle" of amide synthesis and degradation that would lead to a depletion of cellular energy stores. Several of these potential "futile cycles" have been described (Stadtman, 1970), and they are usually situated at control points in amphibolic pathways, where the anabolic enzyme is regulated by specific biosynthetic end products and the catabolic enzyme by energy metabolites (Sanwal, 1970).

In contrast to glutamine synthetase, the synthesis of glutaminase B is under constitutive control and the enzyme is present in minute quantities (<0.01% of cellular protein). Glutaminase B has a molecular weight of ~90,000, and preliminary studies show that it is probably composed of 3 or 4 subunits. Like many regulatory enzymes, glutaminase B is cold labile, which means that it is reversibly inactivated upon exposure to 4°C and reactivated when heated at 23°C. Ligands such as borate and L-glutamate protect against inactivation by cold. To date, more than 30 enzymes have been found which exhibit this phenomena of cold lability. The molecular basis of cold lability is thought to involve the weakening of hydrophobic bonds as the temperature is lowered (Scheraga et al., 1962).

The activity of glutaminase B can also be altered by preincubation with adenine nucleotides at 4°C. Preincubation with ATP or ADP results in a biphasic curve

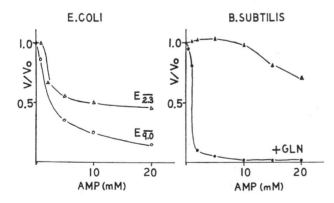

FIGURE 5 Inhibition of glutamine synthetases by AMP. Enzyme activities were measured using the biosynthetic reaction with saturating concentrations of substrates and Mg^{++}. (A) *E. coli* glutamine synthetase activities at two states of adenylylation $E_{\overline{2.3}}$ and $E_{\overline{9.0}}$ (adapted from Ginsburg, 1969). (B) *B. subtilis* glutamine synthetase activities in the presence and absence of 2.5 mM L-glutamine (adapted from Deuel and Prusiner, 1973).

with activation at concentrations below 1 mM followed by inhibition at higher concentrations (Figure 6). Here AMP is an activator. The presence of adenine nucleotides in the assay is without effect. Maximum inhibition by 10 mM ATP occurred upon preincubation at 4°C for 30 min. The inhibition was reversible upon gel filtration or dialysis if activating ligands were present. The results suggest that glutaminase B exists in a conformation susceptible to ATP inhibition at 4°C and that ATP induces a conformational change that can only be reversed in the presence of activating ligands.

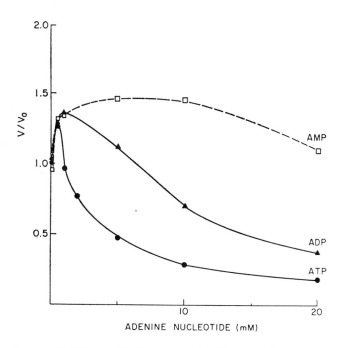

FIGURE 6 Effects of adenine nucleotides on glutamine B activity from *E. coli*. The enzyme was preincubated with the appropriate concentration of nucleotide at 4°C for 30 min prior to assay (for details see Prusiner and Stadtman, 1971).

The regulation of glutaminase B activity by ATP may be modified by AMP that produces synergistic inhibition with ATP. The kinetics of these studies predict at least two nucleotide binding sites on the enzyme. Divalent cations also modify the inhibition due to ATP by directly activating glutaminase B and by chelation with ATP. It is noteworthy that enzymes such as glutaminase B and fumarase (Penner and Cohen, 1969), which are susceptible to ATP inhibition, are not inhibited when ATP is complexed with divalent cations. In contrast, enzymes such as glutamine synthetase and hexokinase, which use ATP as a substrate, require ATP to be complexed with divalent cations for activity. These studies illustrate the potential flexibility of multimeric proteins to respond to a variety of metabolite information.

Measurements of metabolite levels in *E. coli* by Schutt and Holzer (1972) provide additional support for the proposed roles and cellular regulation of glutamine synthetase and glutaminase B. *E. coli* were grown on a glucose-proline minimal media. At time 0, the cells were given 10 mM NH_4Cl, and as shown in Figure 7, the NH_4^+ was immediately assimilated into glutamine while the concentration of ATP dropped 90%. Concomitantly, the level of glutamate decreased, and activity of glutamine synthetase rapidly diminished due to glutamine-stimulated adenylylation, which resulted in a cessation of glutamine synthesis. The depletion of ATP stores threatened cellular survival and presumably relieved the inhibition of glutaminase B by ATP. Glutaminase B could then function to deamidate glutamine, thus making more glutamate available for energy generation in the tricarboxylic acid cycle. This would then lead to a replenishment of ATP stores. The reduction of cellular glutamine stores by glutaminase B would also be advantageous for conserving energy stores, since numerous biosynthetic reactions which utilize ATP would then be turned off.

Figure 8 summarizes the reciprocal control of glutamine synthetase and glutaminase B. In contrast ATP is an inhibitor of glutaminase B, while AMP is an activator. Similar regulated enzyme couples exist in the gluconeogenic glycolytic pathway. The same reciprocal controls appear to regulate the activities of fructose 1,6-diphosphatase and phosphofructokinase (Stadtman, 1970). Fructose 1,6-diphosphatase is activated by ATP and inhibited by AMP. Phosphofructokinase is allosterically inhibited by ATP, which is also a substrate, and is activated by AMP. It appears that these potential "futile cycles" provide the organism with enzymatic machinery which can respond to diverse metabolic situations.

Disturbances of ammonia metabolism in mammals

In mammals, several disease states associated with hyperammonemia have been recognized. Hepatic failure (Sherlock, 1968; Walker and Schenker, 1970), disorders of the urea cycle (Shih and Efron, 1972), hyperlysinemia (Ghadimi, 1972), ammonia intoxication (Schenker et al., 1967), diseases of protein intolerance (Malmquist et al., 1971) are all characterized by elevated blood ammonia levels. These pathological processes involve central nervous system dysfunction that may be manifested as seizures, coma, and/or mental retardation. It remains to be established whether or not the changes in the central nervous system function associated with hyperammonemia may in part be directly due to alterations of cerebral glutamate and glutamine metabolism. It is noteworthy that rats, which became comatose after acute ammonia intoxication, showed a reduction of brainstem ATP content by 25% and creatine phosphate content by 70% but showed no alterations in cortical energy stores (Schenker et al., 1967). Other studies have demonstrated frequent alterations of blood and cerebrospinal fluid glutamate and glutamine levels in patients with hyperammonemia, secondary to hepatic failure or a genetic defect (Sherlock, 1968; Shih and Efron, 1972). In addition, methionine sulfoximine, an analog of glutamate, when administered to mice produced a marked decrease in the toxicity of ammonia to the brain even though ammonia concentrations rose significantly (Walker and Schenker, 1970). The results are consistent with the suggestion that some of the toxic effects of ammonia may be due to a localized depletion of cellular energy stores, which are a result of the increased synthesis of glutamine and/or glutamate. Indeed, these events may be similar to those observed in *E. coli* where a 90% reduction in ATP stores was observed upon addition of ammonia to the culture media (Schutt and Holzer, 1972). Recent studies have suggested that the coma associated with hepatic failure may not be due to hyperammonemia but to the synthesis of a false neuro-

transmitter, octopamine, because the administration of L-dopa to patients with hepatic coma has resulted in their arousal (Fischer and Baldessarini, 1971).

To date, we are unaware of a deficiency of glutamine synthetase in mammals, but recently a deficiency of glutaminase has been suggested as the cause of a disease, *familial protein intolerance*, associated with hyperammonemia (Malmquist et al., 1971). Further studies are needed to establish the nature of the suspected glutaminase defect in this disease.

In conclusion, the enzymatic control of glutamate and glutamine metabolism in microorganisms is an example of the complex and elegant mechanisms that have evolved to regulate the assimilation and excretion of nitrogenous metabolites. It is hoped that some of this information may be useful in understanding the regulation and functions of nitrogen metabolism in multicellular organisms.

ACKNOWLEDGMENT With pleasure and appreciation, I thank Dr. Earl Stadtman for his guidance and support throughout these studies.

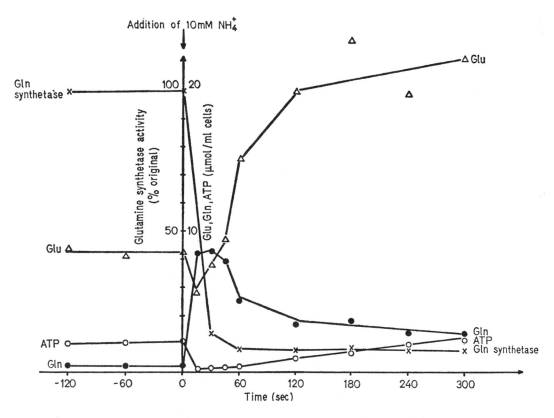

FIGURE 7 Alterations of glutamate, glutamine, ATP content, and glutamine synthetase activity of *E. coli* cells after the addition of 10 mM NH$_4^+$ (from Schutt and Holzer, 1972).

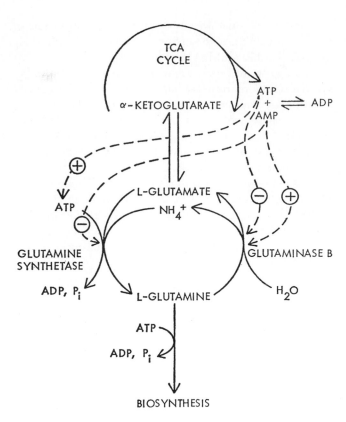

FIGURE 8 Reciprocal control of glutamine synthetase and glutaminase B in *E. coli* by adenine nucleotides.

REFERENCES

BERBERICH, M. A., 1972. A glutamate-dependent phenotype in *E. coli* K12: The result of two mutations. *Biochem. Biophys. Res. Comm.* 47:1498–1503.

BERL, S., and D. D. CLARKE, 1970. Compartmentation of amino acid metabolism. In *Handbook of Neurochemistry*, Vol. 3, Abel Lajtha, ed. New York: Plenum Press, pp. 447–472.

BLOOM, F. E., 1972. Amino acids and polypeptides in neuronal function. *Neurosciences Res. Prog. Bull.* 10:122–251.

BROWN, M. S., A. SEGAL, and E. R. STADTMAN, 1971. Modulation of glutamine synthetase adenylylation and deadenylylation is mediated by metabolic transformation of the P_{II} regulatory protein. *Proc. Nat. Acad. Sci. USA* 68:2949–2953.

CASKEY, C. T., D. ASHTON, and J. WYNGAARDEN, 1964. The enzymology of feedback inhibition of glutamine phosphoribosylpyrophosphate amidotransferase by purine ribonucleotides. *J. Biol. Chem.* 239:2570–2579.

DEUEL, T. F., A. GINSBERG, J. YEH, E. SHELTON, and E. R. STADTMAN, 1970. *Bacillus subtilis* glutamine synthetase. *J. Biol. Chem.* 245:5195–5205.

DEUEL, T., and S. PRUSINER, 1973. *J. Biol. Chem.* (in press).

ELMERICH, C., and J-P. AUBERT, 1971. Synthesis of glutamate by a glutamine: 2-oxo-glutarate amidotransferase (NADP oxidoresductase) in *Bacillus megatherium. Biochem. Biophys. Res. Commun.* 42:371–376.

ELMERICH, C., and J. AUBERT, 1972. Role of glutamine synthetase in the repression of bacterial sporulation. *Biochem. Biophys. Res. Commun.* 46:892–897.

FISCHER, J., and R. BALDESSARINI, 1971. False neurotransmitters and hepatic failure. *Lancet* 2:75–80.

GANCEDO, C., and H. HOLZER, 1968. Enzymatic inactivation of glutamine synthetase in enterobacteriaceae. *Eur. J. Biochem.* 4:190–192.

GHADIMI, H., 1972. The hyperlysinemias. In *Inherited Basis of Metabolic Diseases*, 3rd Ed., J. Stanbury, J. Wyngaarden, and D. Fredrickson, eds. New York: McGraw-Hill, pp. 393–403.

GINSBURG, A., 1969. Conformational changes in glutamine synthetase from *Escherichia coli*. II. Some characteristics of the equilibrium binding of feedback inhibitors to the enzyme. *Biochem.* 8:1726–1740.

HARTMAN, S., 1968. Glutaminase of *Escherichia coli. J. Biol. Chem.* 243:853–863.

HIMWICH, W. A., and H. C. AGRAWAL, 1969. Amino acids. In *Handbook of Neurochemistry*, Vol. 1, Abel Lajtha, ed. New York: Plenum Press, pp. 33–52.

HUBBARD, J., and E. R. STADTMAN, 1967. Regulation of glutamine synthetase. VI. Interactions of inhibitors for *Bacillus licheniformis* glutamine synthetase. *J. Bacteriol.* 94:1016–1024.

JOHNSON, J. L., 1972. Glutamic acid as a synaptic transmitter in the nervous system. A review. *Brain Res.* 37:1–20.

MALMQUIST, J., R. JAGENBERG, and G. LINSTEDT, 1971. Familial protein intolerance. *New Eng. J. Med.* 284:997–1002.

MEERS, J. L., D. W. TEMPEST, and C. M. BROWN, 1970. "Glutamine (amide): 2-oxoglutarate amino transferase oxidoreductase (NADP)," an enzyme involved in the synthesis of glutamate by some bacteria. *J. Gen. Microbiol.* 64:187–194.

MILLER, R. E., and E. R. STADTMAN, 1972. Glutamate synthase from *Escherichia coli*: An iron-sulfide flavoprotein. *J. Biol. Chem.* 247:7407–7419.

MURRELL, W. G., 1969. Chemical composition of spores and spore structures. In *The Bacterial Spore*, G. W. Gould and A. Hurst, eds. New York: Academic Press, pp. 215–273.

PASTAN, I., and R. PERLMAN, 1970. Cyclic adenosine monophosphate in bacteria. *Science* 169:339–344.

PATEMAN, J. A., 1969. Regulation of synthesis of glutamate dehydrogenase and glutamine synthetase in micro-organisms. *Biochem. J.* 115:769–775.

PENNER, P. E., and L. H. COHEN, 1969. Effects of adenosine triphosphate and magnesium ions on the fumarase reaction. *J. Biol. Chem.* 244:1070–1075.

PETERKOFSKY, A., 1972. Personal communication.

PRUSINER, S., R. E. MILLER, and R. C. VALENTINE, 1972. Cyclic AMP control of the enzymes of glutamine metabolism. *Proc. Nat. Acad. Sci. USA* 69:2922–2926.

PRUSINER, S., and E. R. STADTMAN, 1971. On the regulation of glutaminase in *E. coli*: Metabolite control. *Biochem. Biophys. Res. Commun.* 45:1474–1481.

PRUSINER, S., and E. R. STADTMAN, eds., 1973. *The Enzymes of Glutamine Metabolism*. New York: Academic Press, 615 pp.

ROBERTS, E., 1972. Hypothesis: The possible role of GABA in schizophrenia. *Neurosciences Res. Prog. Bull.* 10:468–482.

ROSE, S., 1970. The compartmentation of glutamate and its metabolites in fractions of neuron cell bodies and neuropil: Studied by intraventricular injection of $(U^{14}-C)$ glutamate. *J. Neurochem.* 17:809–816.

SALGANICOFF, L., and E. DE ROBERTIS, 1965. Subcellular distribution of the enzymes of the glutamic acid, glutamine and γ-aminobutyric acid cycles in rat brain. *J. Neurochem.* 12:287–309.

SANWAL, B. D., 1970. Allosteric controls of amphibolic pathways in bacteria. *Bact. Rev.* 34:20–39.

SCHENKER, S., D. McCANDLESS, E. BROPHY, and M. LEWIS, 1967. Studies on the intracerebral toxicity of ammonia. *J. Clin. Invest.* 46: 838–848.

SCHERAGA, H. W., G. NEMETHY, and I. STEINBERG, 1962. The contribution of hydrophobic bonds to thermal stability of protein conformation. *J. Biol. Chem.* 237:2506–2508.

SCHUTT, H., and H. HOLZER, 1972. Biological function of the ammonia-induced inactivation of glutamine synthetase in *Escherichia coli. Eur. J. Biochem.* 26:68–72.

SHERLOCK, S., 1968. *Diseases of the Liver and Biliary System.* Philadelphia, Pa.: F. A. Davis, pp. 79–102.

SHIH, V., and M. EFRON, 1972. Urea cycle disorders. In *Inherited Basis of Metabolic Diseases*, 3rd Ed., J. Stanbury, J. Wyngaarden, and D. Fredrickson, eds. New York: McGraw-Hill, pp. 370–392.

SILVER, S., P. Johnseine, and K. KING, 1970. Manganese active transport in *Escherichia coli. J. Bacteriol.* 104:1299–1306.

SODERLING, T. R., J. HICKENBOTTOM, E. REIMANN, F. HUNKELER, D. WALSH, and E. KREBS, 1970. Inactivation of glycogen synthetase and activation of phosphorylase kinase by muscle adenosine 3′:5′-monophosphate-dependent protein kinases. *J. Biol. Chem.* 245:6317–6328.

SOURKES, T. L., 1962. *Biochemistry of Mental Disease.* New York: Harper and Row, pp. 45–60.

STADTMAN, E. R., 1970. Mechanisms of enzyme regulation in metabolism. In *The Enzymes*, Vol. 1, 3rd Ed., P. D. Boyer, ed. New York: Academic Press, pp. 398–460.

STADTMAN, E. R., 1971. The role of multiple molecular forms of glutamine synthetase in the regulation of glutamine metabolism in *Escherichia coli. Harvey Lect.* 65:97–125.

STADTMAN, E. R., B. SHAPIRO, A. GINSBURG, H. W. KINGDON, and M. DENTON, 1968. Regulation of glutamine synthetase activity in *Escherichia coli. Brookhaven Sympos. Biol.* 21:378–395.

TRUFFA-BACHI, P., and G. N. COHEN, 1966. La β-aspartokinase sensible à la lysine d'*Escherichia coli* purification et propriétés. *Biochem. Biophys. Acta* 113:531–541.

VALENTINE, R. C., B. SHAPIRO, and E. R. STADTMAN, 1968. Regulation of glutamine synthetase. XII. Electron microscopy of the enzyme from *Escherichia coli. Biochemistry* 7:2143–2152.

VAN DEN BERG, C. J., 1970. Glutamate and glutamine. In *Handbook of Neurochemistry*, Vol. 3, Abel Lajtha, ed. New York: Plenum Press, pp. 355–379.

WAELSCH, H., 1951. Glutamic acid and cerebral function. In *Advances in Protein Chemistry*, Vol. 6, M. Anson et al., eds. New York: Academic Press, pp. 299–341.

WALKER, C. O., and S. SCHENKER, 1970. Pathogenesis of hepatic encephalopathy with special reference to the role of ammonia. *Amer. J. Clin. Nutr.* 23:619–632.

WILCOX, M., 1969. γ-Glutamyl phosphate attached to glutamine-specific t-RNA. *Eur. J. Biochem.* 11:405–412.

WILLIAMS, A. H., M. H. KYU, J. FENTON, and J. CAVANAGH, 1972. The glutamate and glutamine content of rat brain after portocaval anastomosis. *J. Neurochem.* 19:1073–1077.

WOOLFOLK, C., and E. R. STADTMAN, 1964. Cumulative feedback inhibition in the multiple end product regulation of glutamine synthetase activity in *Escherichia coli. Biochem. Biophys. Res. Commun.* 17:313–319.

76 Regulation of the Neurotransmitter Norepinephrine

JULIUS AXELROD

ABSTRACT The catecholamines dopamine, norepinephrine, and epinephrine are in a state of flux, yet they maintain a constant level in nerves and glandular tissues. The level of these biogenic amines is regulated by changes in activity of the biosynthetic enzymes tyrosine hydroxylase, dopamine-β-hydroxylase, and phenylethanolamine-N-methyltransferase. The minute-to-minute regulation of the level of the neurotransmitter norepinephrine is controlled by rapid changes in tyrosine hydroxylase activity caused by feedback inhibition of the enzyme by norepinephrine and dopamine. There is no change in the amount of enzyme protein. Elevation in tyrosine hydroxylase occurs in the cell body, nerve terminals, and adrenal medulla when there is an increase in firing of sympathetic nerves. This results in formation of new enzyme protein by a transsynaptic process. A similar transsynaptic induction by increased nerve firing occurs with the enzyme dopamine-β-hydroxylase in nerves and adrenal medulla. The induction of these enzymes appears to be initiated by acetylcholine and possibly controlled by intracellular concentrations of norepinephrine. The activity of tyrosine hydroxylase and dopamine-β-hydroxylase and especially phenylethanolamine N-methyltransferase in the adrenal medulla is reduced by removal of the pituitary gland and induced by ACTH.

Dopamine-β-hydroxylase is transported from cell body to nerve terminals. When nerves are depolarized, dopamine-β-hydroxylase is released from the nerves, together with the neurotransmitter norepinephrine, by a process of exocytosis. The release of dopamine-β-hydroxylase requires Ca^{++}, microtubules, and microfilaments.

The biogenic amine serotonin undergoes a circadian change in levels in the pineal gland. The level of serotonin is regulated by the neurotransmitter norepinephrine released from sympathetic nerves. Norepinephrine reduces the serotonin levels by stimulating the enzyme that acetylates serotonin via cyclic AMP. Changes in the rate of neuronal release of norepinephrine markedly influence the activity of N-acetyltransferase.

THE ACTIVITY OF the sympathetic nervous system undergoes rapid changes yet maintains a constant level of its neurotransmitter, norepinephrine. This is made possible by a variety of self-regulatory systems involving changes in its biosynthesis, storage, release, and metabolism within the neuron as well as modifications of the pre- and post-synaptic membrane. The special morphology of the sympathetic neuron also contributes to the maintenance of the neurotransmitter. The sympathetic neuron consists of a cell with a considerable spatial separation from its nerve terminals (Figure 1). The nerve terminals are highly branched and have swellings or varicosities that are in close proximity to the effector cells. Within the varicosity, norepinephrine is stored in a dense core vesicle of about 500 Å (Wolfe et al., 1962). This structural organization is present in both the peripheral and central nervous systems. The cell body of a sympathetic neuron synapses with a preganglionic fiber, usually cholinergic, and the varicosity of the nerve terminals innervates many thousand effector cells *en passant*.

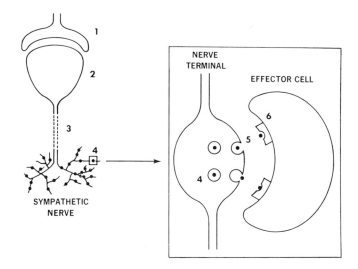

FIGURE 1 Sites at which the neurotransmitter norepinephrine can be regulated (see text for explanation).

The levels of norepinephrine within the sympathetic neuron can be regulated at several sites (Figure 1): Cell body via preganglionic nerves (areas 1, 2); the axon, which transports the biosynthetic enzymes made in the cell body (area 3); the cytoplasm and storage vesicle of nerve terminal (area 4); the neuronal membrane (area 5); and the postsynaptic membrane (area 6).

JULIUS AXELROD Laboratory of Clinical Science, National Institute of Mental Health, Bethesda, Maryland

Biosynthesis of norepinephrine and epinephrine

The enzymes involved in the formation of norepinephrine are synthesized in the cell body of the sympathetic neuron. These enzymes are tyrosine hydroxylase, which converts tyrosine to dopa (Nagatsu et al., 1964), dopa decarboxylase (Holtz et al., 1938), and dopamine-β-hydroxylase, the enzyme that β-hydroxylates dopamine to norepinephrine (Friedman and Kaufman, 1965). Phenylethanolamine-N-methyltransferase, the enzyme that methylates norepinephrine to epinephrine, is present mainly in adrenal medulla of mammals and sympathetic nerves of amphibians (Axelrod, 1962). Tyrosine hydroxylase is a mixed function oxidase requiring tetrahydropteridine and O_2 and Fe^{++}. It is found in the cell body, axon, nerve terminals, as well as the adrenal medulla, and is absent in extraneural tissue. Tyrosine hydroxylase is found in soluble and bound form. A molecular weight of 192,000 has been reported for the soluble enzyme, while the bound enzyme after trypsin digestion has a molecular weight of 50,000 (Wurzburger and Musacchio, 1971). Tyrosine hydroxylase can be inhibited by norepinephrine, which serves as an important controlling mechanism for its synthesis. Inhibition of the enzyme by catecholamines is competitive for its pteridine cofactor in its reduced form and not with its substrate, tyrosine (Ikeda et al., 1966). Dopa decarboxylase is unspecific in that it can decarboxylate a variety of L-aromatic amino acids. It requires pyridoxal phosphate as a cofactor and is tightly bound to the apoenzyme as a Schiff base. Dopa decarboxylase is present both in neuronal and extraneuronal tissues. Using an immunoassay, it was shown that aromatic acid decarboxylase is a single enzyme with a molecular weight of 109,000 (Christenson et al., 1972). Dopamine-β-hydroxylase hydroxylates dopamine on the beta carbon to form norepinephrine. It is a mixed function oxidase containing

2 mole of Cu^{++}, which is reduced by ascorbic acid (Friedman and Kaufman, 1965). The enzyme lacks specificity and can β-hydroxylate a variety of phenylethylamines. The enzyme is highly localized in the sympathetic neuron as well as the adrenal medulla. Within the neuron dopamine-β-hydroxylase is present in the cell body, axon, and nerve terminal. It is highly localized in the norepinephrine storage vesicles of nerves (Potter and Axelrod, 1963a) and chromaffin granules of the adrenal medulla (Kirshner, 1957).

The epinephrine-forming enzyme, phenylethanolamine-N-methyltransferase, is highly localized in the cytoplasm of mammalian adrenal medulla (Axelrod, 1962) and is present in sympathetic nerves of amphibians (Wurtman et al., 1968b). It methylates norepinephrine as well as β-hydroxylated phenylethanolamine derivatives; S-adenosylmethionine is the methyl donor. The enzyme has been purified and its molecular weight has been found to be about 30,000 (Connett and Kirshner, 1970). Phenylethanolamine-N-methyltransferase shows different electrophoretic mobility on starch block, and multiple forms of the enzyme have also been reported (Axelrod and Vesell, 1970). The biosynthesis of catecholamines is shown in Figure 2.

Neural regulation of the catecholamine biosynthetic enzymes

In a study to determine the intraneural localization of tyrosine hydroxylase, 6-hydroxydopamine, a compound that destroys sympathetic nerve terminals (Thoenen and Tranzer, 1968), was administered to rats. There was an almost complete disappearance of this enzyme in the heart in 40 hr suggesting that it was highly localized in sympathetic nerve terminals (Mueller et al., 1969a). When tyrosine hydroxylase was measured in the adrenal gland

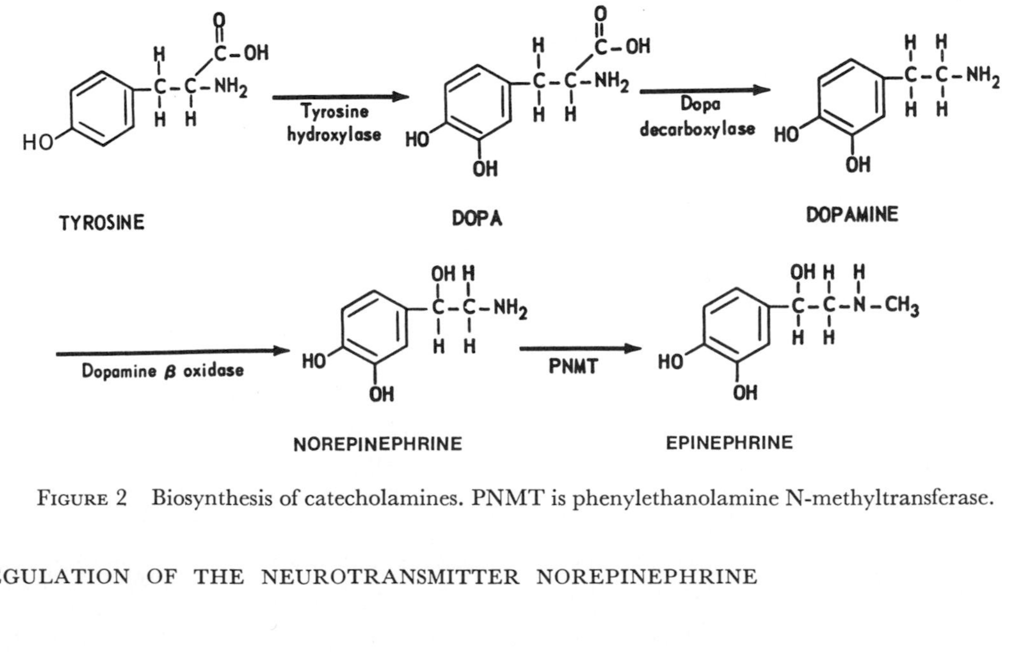

FIGURE 2   Biosynthesis of catecholamines. PNMT is phenylethanolamine N-methyltransferase.

there was a marked increase in this enzyme and a smaller elevation of phenylethanolamine-N-methyltransferase about 1 day after the administration of 6-hydroxydopamine. This was an unexpected finding, and it appeared to be due to the ability of 6-hydroxydopamine to lower blood pressure. This would cause a reflex increase in sympathetic nerve activity, resulting in an increase in enzyme activity in the adrenal gland. To examine this possibility, reserpine and phenoxybenzamine, compounds that lower blood pressure and increase sympathetic nerve activity, were given, and their effects on tyrosine hydroxylase in the adrenal gland examined (Mueller et al., 1969b). Both compounds elevated tyrosine hydroxylase activity, not only in the adrenal gland but also in the superior cervical (Figure 3) and stellate ganglia. The maximal enzyme activity observed in the adrenal gland and the ganglia occurred 3 days after reserpine administration, indicating a slow rise in enzyme activity. Reserpine was also found to increase tyrosine hydroxylase activity in adrenal gland of all mammalian species examined as well as the brainstem of the rabbit. Increased tyrosine hydroxylase activity after reserpine was also observed in the nerve terminal as well as in the cell body. The increase in enzyme activity in the nerve terminal was delayed and lagged behind the rise in the ganglia by 2 days (Thoenen et al., 1970).

The elevation in tyrosine hydroxylase activity after the increase in sympathetic nerve activity was shown physiologically by the increased formation of [^{14}C]catecholamine from [^{14}C]tyrosine after the administration of phenoxybenzamine or 6-hydroxydopamine (Dairman and Udenfriend, 1970; Mueller, 1971). The rise in tyrosine hydroxylase activity in the ganglia and adrenal medulla after reserpine administration could be prevented by the administration of cycloheximide or actinomycin D (Mueller et al., 1969c), suggesting that this elevation of enzyme activity is due to induction of new enzyme protein. The Km for both the substrate and the pteridine cofactor with the enzyme obtained from reserpine-treated rats was not different from the untreated rats, although there was a marked elevation in the V_{max} for both substrate and cofactor. These results are consistent with an increase in the number of active sites on the enzyme molecule caused by a drug-induced rise in sympathetic nerve activity. Increased tyrosine hydroxylase activity was found in the adrenal gland after cold (Thoenen et al., 1969a), immobilization (Kvetnansky et al., 1970), and psychosocial stress (Axelrod et al., 1970), in the ganglia after administration of nerve growth factor (Thoenen, 1970), and in brain after cold stress and administration of reserpine (Segal et al., 1971).

To examine whether the induction in tyrosine hydroxylase after reserpine was a transsynaptic event the preganglionic fibers to the superior cervical ganglia (Thoenen et al., 1969a) and the splanchnic nerve to the adrenal gland were cut unilaterally (Thoenen et al., 1969b). When reserpine was administered, there was an elevation

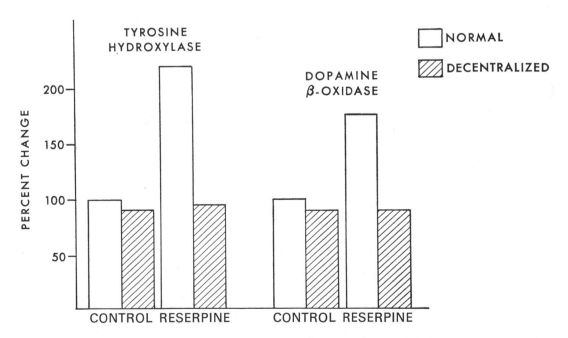

FIGURE 3 Transsynaptic induction of tyrosine hydroxylase and dopamine-β-oxidase(hydroxylase) in sympathetic ganglia. Superior cervical ganglion was decentralized unilaterally by transection of the preganglionic trunk. Two to six days later reserpine (5 mg/kg) was given for 1 day before tyrosine hydroxylase was measured or 3 alternate days before dopamine-β-oxidase(hydroxylase) was measured (from Axelrod, 1971).

of enzyme activity in the innervated side of the ganglia and adrenal gland, but the increase on the denervated side was completely blocked (Figure 3). All of these observations indicate that tyrosine hydroxylase activity is increased in both the sympathetic nerve, cell body, brain, and adrenal medulla by sympathoadrenal hyperactivity. This elevation in tyrosine hydroxylase activity appears to involve a transsynaptic induction of new enzyme molecules.

The presynaptic fibers that regulate tyrosine hydroxylase activity are cholinergic. Interrupting the cholinergic splanchnic nerve blocks the drug-induced rise in tyrosine hydroxylase (Thoenen et al., 1969a, 1969b). Ganglionic blocking agents also inhibit the increase in this enzyme (Mueller et al., 1970b), while acetylcholine causes an elevation of tyrosine hydroxylase activity in the adrenal (Patrick and Kirshner, 1971). In the superior cervical ganglia of the newborn mouse, the development of tyrosine hydroxylase is prevented by cutting the preganglionic cholinergic nerve (Black et al., 1971), suggesting that presynaptic cholinergic terminals regulate the formation of tyrosine hydroxylase in the sympathetic nerve cell body. The selective destruction of sympathetic nerves chemically with 6-hydroxydopamine or immunologically with antiserum to the nerve growth factor prevents the normal development of choline acetyltransferase in presynaptic nerve endings, indicating that postsynaptic adrenergic neurons regulate the biochemical development of presynaptic cholinergic nerves (Black et al., 1971).

Breeding studies are being carried out utilizing reciprocal F_1 and F_2 and dominant recessive backcross generations with respect to the catecholamine biosynthetic enzymes (Ciaranello, unpublished observations). Preliminary results suggest that the genes controlling tyrosine hydroxylase, dopamine-β-hydroxylase and phenylethanolamine-N-methyltransferase are linked. It is possible that the three genes are linked and that a single regulatory locus is responsible for the activity of the three biosynthetic enzymes.

The activity of dopamine-β-hydroxylase is also affected by nerve impulses. The development of a sensitive assay for measuring dopamine-β-hydroxylase made it possible to measure this enzyme in the cell body and nerve terminals and to study changes after drugs that increase sympathetic nerve firing (Molinoff et al., 1971). The administration of reserpine resulted in a marked elevation of dopamine-β-hydroxylase activity in sympathetic ganglia (Figure 3), nerve terminals, and adrenal gland but not in the brain (Molinoff et al., 1970). This increase in enzyme activity in cell bodies is neuronally mediated, because the reserpine could not elevate dopamine-β-hydroxylase activity in a denervated ganglia (Figure 3). Pretreatment of animals with the protein synthesis inhibitor, cycloheximide, prevented the rise in enzyme activity in the ganglia. Further evidence that new enzyme protein was induced by nerve impulses comes from the use of an antibody for dopamine-β-hydroxylase. Reserpine caused an increase in the rate of incorporation of [^{3}H]leucine into dopamine-β-hydroxylase measured by immunoabsorption (Hartman et al., 1970).

It appears that nerve depolarization is involved in the induction of dopamine-β-hydroxylase. An increase in the potassium concentration of the media containing rat superior cervical ganglia maintained in organ culture results in a marked increase in dopamine-β-hydroxylase in ganglia (Silberstein et al., 1972). This increase in enzyme activity is inhibited by cycloheximide. Nicotinic antagonists block the induction of dopamine-β-hydroxylase in ganglia after reserpine, suggesting that a cholinergic site is involved (Molinoff et al., 1972). Acetylcholine also increases dopamine-β-hydroxylase activity in the denervated adrenal gland (Patrick and Kirshner, 1971). However, it does not appear that the cholinergic receptor is essential for induction of the enzyme, at least in sympathetic nerves, because elevated potassium concentration can increase enzyme activity in the absence of neuronal influences.

Another biosynthetic enzyme, phenylethanolamine-N-methyltransferase, in the adrenal gland, is regulated by neuronal influences. Increasing splanchnic nerve activity with 6-hydroxydopamine, reserpine, or stress causes a small elevation of phenylethanolamine-N-methyltransferase in the rat adrenal gland (Mueller et al., 1969a) and a large increase in the mouse adrenal (Ciaranello et al., 1972a). This increase can be abolished by transection of the nerve supplying the adrenal gland (Thoenen et al., 1970).

Unlike the other catecholamine biosynthetic enzymes, dopa decarboxylase activity in the superior cervical ganglia or adrenal gland is not induced by drug-mediated increase in preganglionic neural activity (Black et al., 1971). These experiments suggest that tyrosine hydroxylase, dopamine-β-hydroxylase, and phenylethanolamine-N-methyltransferase are linked in a coordinate fashion. Genetic studies also indicate that these enzymes are linked (Ciaranello, unpublished observations).

Several experiments suggest that catecholamines are implicated in the induction of dopamine-β-hydroxylase and tyrosine hydroxylase (Molinoff et al., 1972). Drugs that elevate the level of catecholamines, such as L-dopa, monoamine oxidase inhibitors, and bretylium, inhibit the induction of both tyrosine hydroxylase and dopamine-β-hydroxylase. On the other hand, reduction of catecholamines by α-methyl-paratyrosine or high potassium (Silberstein et al., 1972) results in an induction of dopamine-β-hydroxylase activity.

The nerve terminal may have an important influence on the induction of the catecholamine biosynthetic enzymes in the cell body. Destruction of adrenergic nerve terminals with 6-hydroxydopamine or surgical section of the postganglionic axons causes a long-lasting decrease in dopamine-β-hydroxylase in the superior cervical ganglia (Brimijoin and Molinoff, 1971). 6-Hydroxydopamine administration or a postganglionic section also results in a marked increase in the uptake of [^{3}H]norepinephrine in the sympathetic ganglia (Kopin and Silberstein, 1972). The latter phenomena reflects growth of adrenergic membrane surface. These experiments show that when nerve terminals are destroyed the metabolic machinery of the cell body changes its priorities from the production of enzymes concerned with function to the formation of structural elements required for the restoration of the nerve ending.

Hormonal regulation

In addition to nervous inputs, the corticoids can also influence the biosynthesis of catecholamines. The effects of hormones are principally in the adrenal medulla, a structure that can be considered analogous to the cell body of the sympathetic nervous system. An examination of the effects of hormones on catecholamine formation was prompted by the observation that large amounts of epinephrine are present in the adrenal gland of those species in which the medulla is surrounded by a cortex (Coupland, 1965). This suggested to us that corticoids

present in the cortex might be the compounds that stimulate the methylation of norepinephrine to epinephrine. The experimental design to examine this possibility was to reduce corticoids in adrenal cortex by removal of the pituitary gland in rats and then to measure the activity of phenylethanolamine-N-methyltransferase in the adrenal gland (Wurtman and Axelrod, 1966). When rats were hypophysectomized, there was a gradual and steady decline in the norepinephrine methylating enzyme. After about 7 days, only about 20% of the enzyme activity remained in the adrenal medulla (Figure 4). The administration of either dexamethasone, a potent glucocorticoid, or ACTH restored phenylethanolamine-N-methyltransferase activity to the adrenal gland after several days (Figure 4). Inhibition of protein synthesis blocked the increase in enzyme activity after the administration of dexamethasone. When dexamethasone or ACTH was given repeatedly to normal rats there was no increase in enzyme activity in the adrenal gland. All of these experiments demonstrated that glucocorticoids in the adrenal cortex are necessary to maintain phenylethanolamine-N-methyltransferase activity. There are negligible amounts of phenylethanolamine-N-methyltransferase in sympathetic nerve cell body. When dexamethasone is given to newborn rats, phenylethanolamine-N-methyltransferase appears in the superior cervical ganglia (Ciaranello, Jacobowitz, and Axelrod, unpublished observations). The ability of dexamethasone to induce the methylating enzyme in the ganglia is lost after the rat is 2 weeks old. Dexamethasone also increased

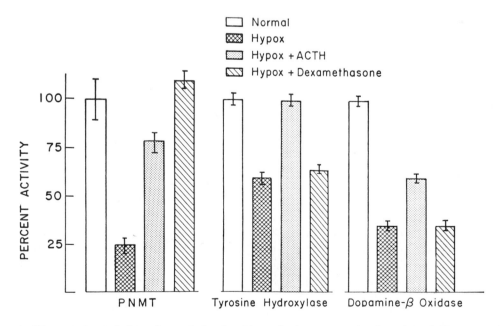

FIGURE 4 Hormonal regulation of catecholamine biosynthetic enzymes in the adrenal. Rats were hypophysectomized for about 1 week and then given dexamethasone or ACTH (from Axelrod, 1973).

the amounts of small, intensely fluorescent (SIF) cells in the superior cervical ganglia. These cells are morphologically related to chromaffin cells. These findings indicate that glucocorticoid hormones may be involved in differentiation of nerve cell to chromaffin type cell .

Removal of the pituitary also affected other catecholamine biosynthetic enzymes in the adrenal medulla. After rat hypophysectomy the activity of tyrosine hydroxylase (Mueller et al., 1970a) and dopamine-β-hydroxylase fell (Weinshilboum and Axelrod, 1970) (Figure 4). Repeated administration of ACTH restored the activity of both enzymes in the adrenal gland (Figure 4). However, dexamethasone failed to elevate tyrosine hydroxylase or dopamine-β-hydroxylase (Figure 4). When mice were subjected to psychosocial stimulation, there was marked elevation of tyrosine hydroxylase and phenylethanolamine-N-methyltransferase in the adrenal gland (Axelrod et al., 1970). When certain mouse strains were exposed to cold stress for 3 to 6 hr, there was a small but significant elevation of phenylethanolamine-N-methyltransferase in the adrenal (Ciaranello et al., 1972a). Implantation of an ACTH-secreting tumor in rats resulted in an elevation of phenylethanolamine-N-methyltransferase, demonstrating that the enzyme can be elevated under conditions of extreme pituitary-adrenocortical activation. Forced immobilization stress also increases tyrosine hydroxylase and dopamine-β-hydroxylase in adrenal to a considerable extent and phenylethanolamine-N-methyltransferase to a smaller degree (Kvetnansky et al., 1970). These stress-induced elevations are mediated by neuronal and hormonal influences.

Regulation of catecholamines at nerve terminals

There is a rapid regulation of the biosynthesis of the adrenergic neurotransmitter in the nerve terminals, which is different from the slower induction of the catecholamine-forming enzymes described above. Stimulation of the splanchnic nerve leads to a release of catecholamines. The sum of the amount of catecholamines released and that remaining in the gland is greater than the amount initially present in the gland (Bydgeman and von Euler, 1958). Studies with the hypogastric nerve (Weiner, 1970) and salivary gland (Sedvall and Kopin, 1967) indicate that the rapid changes in the biosynthesis of norepinephrine are regulated by tyrosine hydroxylase. Stimulation of the sympathetic nerve of the vas deferens in vitro or the salivary gland in vivo (Table I) led to an increased conversion of [^{14}C]tyrosine to [^{14}C]norepinephrine. However, there was no increase in the formation of [^{14}C]dopa to [^{14}C]norepinephrine when the nerves were stimulated, suggesting that tyrosine hydroxylase is the enzyme in-

fluenced by nerve activity. However, there was no increase in the amount of tyrosine hydroxylase in the stimulated salivary gland. In an in vitro study with the vas deferens, it was found that addition of norepinephrine to the bath can partially or completely prevent the increased formation of [^{14}C]norepinephrine from [^{14}C]tyrosine (Weiner, 1970). It has been shown that tyrosine hydroxylase is inhibited by catecholamines such as dopamine and norepinephrine, due to the competition between the catecholamines and the pteridine cofactor (Ikeda et al., 1966). Most of the norepinephrine in the nerve terminals is present in vesicles with little access to tyrosine hydroxylase that appears to be present in the cytoplasm. Thus, there is a small compartment of free catecholamines in the cytoplasm that is critical in the regulation of tyrosine hydroxylase. This small compartment is rapidly depleted during nerve stimulation and thus allows more norepinephrine to be synthesized by increasing the conversion of tyrosine to dopa.

TABLE I

Effect of sympathetic nerve impulses on norepinephrine synthesis and tyrosine hydroxylase activity in the rat salivary gland

	Norepinephrine formed in vivo from		
	[^{14}C]Tyrosine (count/min)	[^{3}H]Dopa (count/min)	In vitro assay of tyrosine hydroxylase*
Decentralized	68	410	3190
Stimulated	358	396	3170

*Tyrosine hydroxylase is expressed as count/min [^{14}C]dopa formed from [^{14}C]tyrosine by an aliquot of homogenate of the salivary glands. (From Sedvall and Kopin, 1967).

Norepinephrine is stored in the nerve terminal in more than one compartment. After the administration of [^{3}H]-norepinephrine, the decrease in its specific activity in tissues was found to be multiphasic (Axelrod et al., 1961), and the specific activity of norepinephrine released by tyramine was dependent on the time the sympathomimetic amine was administered (Potter and Axelrod, 1963b). Kopin et al. (1968) demonstrated that norepinephrine newly synthesized from tyrosine was more rapidly released from the spleen after nerve stimulation. There thus appears to be a relatively small available pool of norepinephrine and a larger reserve store of the catecholamine. The more available pool might be present in the vesicles closest to the synaptic cleft and, because of its location with respect to the neuronal membrane, would be more easily released (Figure 1). The major store of norepinephrine is located at a greater distance from the neuronal membrane and is thus utilized at a slower rate. This pool might serve as a reservoir for the more readily releasable transmitter.

Axonal transport of catecholamine biosynthetic enzymes

The cell body of the sympathetic neuron is separated from the nerve terminals by long distances (Figure 1). The protein-synthesizing apparatus of the neuron is confined to the cell body, while the terminal is involved in nerve function. The enzymes for the biosynthesis of catecholamines made in the cell body must be transported down the axon to the nerve terminal where most of the neurotransmitter is synthesized. Weiss and Hiscoe (1948) demonstrated that the axon is capable of transporting substances from the cell body to the nerve terminal. Axonal transport is a highly specialized process and different constituents are transported in a proximodistal direction at their own characteristic rate, rapidly (1 to 10 mm/hr) or slowly (1 to 3 mm/day) (Ochs, 1972).

Studies on axoplasmic transport are made by ligation of nerves. When adrenergic axons are pinched, there is a rapid accumulation of norepinephrine and dense core vesicles proximal to the constrictions (Dahlström and Häggendal, 1966). When two ligations are made on the same nerve, no accumulation of norepinephrine is observed above the more distal constriction. Colchicine and vinblastine, compounds that cause a disaggregation of microtubules, block the proximodistal transport of dense core vesicles and norepinephrine in noradrenergic neurons (Hökfelt and Dahlström, 1971) implicating microtubules in the rapid axonal transport.

Biochemical and immunological studies indicate that dopamine-β-hydroxylase (Laduron and Belpaire, 1968) and chromogranins, proteins associated with catecholamine binding, also rapidly accumulate proximal to a constriction in peripheral noradrenergic neurons (Geffen et al., 1969). Recently it has been found that dopamine-β-hydroxylase, an enzyme localized in the storage vesicle, and tyrosine hydroxylase, an enzyme not associated with these vesicles, are both transported down the axon of the rat sciatic nerve at an identical rate (1 to 5 mm/hr) (Coyle and Wooten, 1972). Colchicine blocks the transport of both enzymes. These observations suggest that dopamine-β-hydroxylase and tyrosine hydroxylase are transferred from the cell body to the nerve terminal in close association. Local application of colchicine or vinblastine to the superior cervical ganglion of the rat causes a rapid increase in the levels of dopamine-β-hydroxylase in the ganglia and decrease in the salivary gland (Kopin and Silberstein,

1972). When protein synthesis is inhibited the levels of dopamine-β-hydroxylase in the ganglia are rapidly decreased, indicating that the accumulation of the enzyme is due to new synthesis and the decrease after protein synthesis inhibition is the consequence of transport of the enzyme out of the ganglion. Using this approach, the rate of synthesis of dopamine-β-hydroxylase has been calculated to be 5% of the content per hour.

Release of norepinephrine from nerve terminals

The neurotransmitter, norepinephrine, is contained in a membrane-bound vesicle (Wolfe et al., 1962). Thus its discharge from the nerve after depolarization might occur by release into the cytoplasm followed by rapid passage through the neuronal membrane, or by fusion of vesicular membrane with the neuronal membrane and then liberation, or by an opening of the fused membrane and discharge of norepinephrine into the exterior of the terminal together with the soluble contents of the vesicle. This latter process is called *exocytosis*. Evidence that exocytosis occurs comes from studies with adrenal medulla. Stimulation of the adrenal gland with acetylcholine or electrically results in the release of ATP, as well as catecholamines (Douglas and Rubin, 1961). Acetylcholine can also cause the release of the soluble protein of the chromaffin granule, including dopamine-β-hydroxylase (Viveros et al., 1968). The ratio of norepinephrine to dopamine-β-hydroxylase was found to be the same as that present in the chromaffin granule of the adrenal medulla (Viveros et al., 1969). These findings and microscopic evidence indicate that catecholamines are released from the adrenal medulla by a process of exocytosis. When the sympathetic nerve to the spleen is stimulated, dopamine-β-hydroxylase is released together with norepinephrine (Smith et al., 1970). However, the ratio of the amine to dopamine-β-hydroxylase released was 100 times greater than that found in the vesicles isolated from the splenic nerve. Using a very sensitive assay for dopamine-β-hydroxylase, together with the addition of albumin to protect the enzyme, the ratio of dopamine-β-hydroxylase to norepinephrine released after electrical stimulation of the hypogastric nerve of the vas deferens was found to be similar to that in the soluble portion of the contents of the synaptic vesicle (Weinshilboum et al., 1971e). This data indicates that norepinephrine and dopamine-β-hydroxylase are released from the nerve by a process of exocytosis. The absence of Ca^{++} prevents the release of the enzyme and neurotransmitter, while their discharge is enhanced by increasing the concentration of Ca^{++} to twice that used normally (Johnson et al., 1971). The increased release of dopamine-β-hydroxylase with high Ca^{++} concentration is blocked by prostaglandin E$_2$. The α-adrenergic blocking agent, phenoxybenzamine, also increases the release of norepinephrine and dopamine-β-hydroxylase in the stimulated vas deferens, but this drug has no effect on the unstimulated preparation. Prostaglandin also blocks the effects of phenoxybenzamine (Johnson et al., 1971). The enhanced release of dopamine-β-hydroxylase by phenoxybenzamine only when the nerve is stimulated suggests that there is an α-adrenergic receptor on the nerve membrane, and blocking this receptor keeps the nerve membrane in a conformational state that allows larger molecules to be secreted for a longer period of time. Prostaglandins may act by interfering with the actions of Ca^{++} and thus reduce the Ca^{++}-dependent secretion of norepinephrine and dopamine-β-hydroxylase.

Microtubules have been shown to be involved in the discharge of intracellular stored products such as the release of ^{131}I from the thyroid gland (Williams and Wolff, 1970), insulin from the beta cells of the pancreas (Lacy et al., 1968), histamine from mast cells (Gillespie et al., 1968), and catecholamines from the adrenal medulla (Poisner and Bernstein, 1971). These findings suggested that microtubules might play a role in the release of dopamine-β-hydroxylase from sympathetic nerve terminals. Treatment of the vas deferens with colchicine and vinblastine, compounds that disaggregate microtubules, almost completely prevented the release of dopamine-β-hydroxylase and norepinephrine when the nerve is stimulated (Thoa et al., 1972). These compounds, how-

ever, have no effect on the spontaneous release of the enzyme or transmitter. Cytochalasin B, a fungal metabolite that disrupts microfilaments (Carter, 1967), also inhibits the release of norepinephrine and dopamine-β-hydroxylase. These findings suggest that both microtubules and microfilaments are involved in release of norepinephrine and dopamine-β-hydroxylase by exocytosis. Microtubules are presumed to function as a cytoskeleton, and nerve depolarization might affect the microtubules in such a way as to direct the vesicles to the proper site on the neuronal membrane where release occurs (Figure 5). Ca^{++} has also been reported to activate the contractile microfilaments in nonmuscle cells (Wessells et al., 1971). These findings would suggest the presence of a contractile microfilament on the neuronal membrane that is activated by Ca^{++}, which makes an opening in the membrane large enough to allow the soluble contents of the vesicle to be released (Figure 5). Cyclic AMP might also be involved since it has been demonstrated that dibutyryl cyclic AMP and theophylline increase the release of norepinephrine and dopamine-β-hydroxylase after nerve stimulation (Wooten, Thoa, Kopin, and Axelrod, unpublished observation).

Circulating dopamine-β-hydroxylase

The observation that dopamine-β-hydroxylase can be released from the adrenal gland and the nerve terminals prompted an examination of the blood for this enzyme.

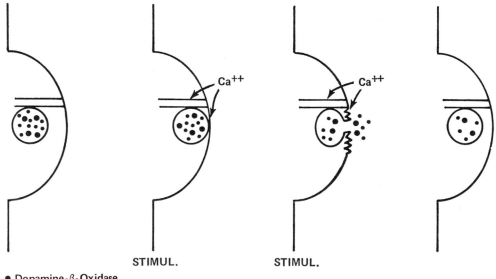

● Dopamine-β-Oxidase
• Norepinephrine

FIGURE 5 A possible mechanism for release of norepinephrine and dopamine-β-oxidase (hydroxylase) by exocytosis (see text for explanation). (From Axelrod, 1973).

Dopamine-β-hydroxylase was found to be present in the plasma of man and other mammalian species (Weinshilboum and Axelrod, 1971b; Goldstein et al., 1971). The enzyme in the plasma is similar to purified dopamine β-hydroxylase from the adrenal medulla; both have the same requirements for ascorbic acid, fumarate, and oxygen (Weinshilboum et al., 1971b). They also have similar electrophoretic mobilities and the same Km with respect to substrate.

The plasma dopamine-β-hydroxylase could arise from the sympathetic nerves or the adrenal gland. The administration to rats of 6-hydroxydopamine, a compound that destroys most of the sympathetic nerve terminals but does not affect the adrenal medulla, markedly reduced the level of the plasma dopamine-β-hydroxylase (Weinshilboum and Axelrod, 1971c). On the other hand adrenalectomy did not affect the plasma enzyme levels. The experiments indicate that plasma dopamine-β-hydroxylase comes from sympathetic nerve terminals and that levels of this enzyme in blood suggest a method for measuring activity of these nerves.

In rats subjected to stress, there is an elevation of serum dopamine-β-hydroxylase (Weinshilboum et al., 1971d). When humans were stressed by vigorous exercise or cold pressor test, there was rapid elevation of plasma enzyme. In familial dysautonomia a decrease in plasma dopamine-β-hydroxylase (Weinshilboum and Axelrod, 1971a) is found while subjects with neuroblastoma have an increased enzyme in plasma (Goldstein et al., 1972). Removal of the pituitary gland causes a marked decrease in enzyme activity, which can be prevented by the administration of vasopressin (Lamprecht and Wooten, 1973). This suggests that hypophysectomy, which reduces blood volume, increases sympathetic nerve activity and increases the release of dopamine-β-hydroxylase. Vasopressin increases blood volume and thus results in a reduced sympathetic nerve activity and blood enzyme level.

Regulation of norepinephrine at the neuronal membrane

When norepinephrine is injected into animals it is selectively taken up by sympathetic nerve terminals (Axelrod, 1971; Hertting and Axelrod, 1961). The norepinephrine is then bound in the synaptic vesicle and retained in a physiologically inactive form. This uptake and binding serves as a rapid and effective means of terminating the action of the neurotransmitter. When both monoamine oxidase and catechol-O-methyltransferase, enzymes involved in the metabolism of catecholamines, are inhibited in vivo, the physiological actions

of norepinephrine are only slightly prolonged (Crout, 1961). However, when the uptake of norepinephrine is blocked by drugs (Whitby et al., 1960), or when the sympathetic nerves are destroyed (Hertting et al., 1961b), the response of norepinephrine is considerably increased. These results indicate that uptake of norepinephrine across the neuronal membrane and retention by storage vesicles are a major mechanism for the rapid inactivation of the neurotransmitter (Figure 6).

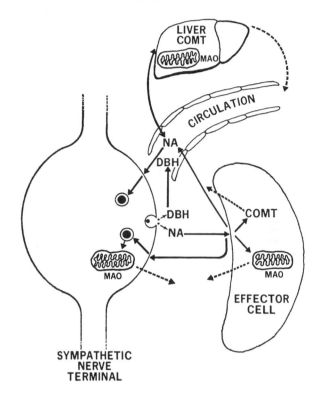

FIGURE 6 Fate of norepinephrine at the sympathetic nerve terminal (see text for explanation). NA is norepinephrine; DBH is dopamine-β-hydroxylase; COMT is catechol-O-methyltransferase; and MAO is monoamine oxidase. (From Axelrod and Weinshilboum, 1972).

The properties of the neuronal uptake mechanism were studied in brain slices (Dengler et al., 1962) and isolated from perfused heart (Iversen, 1963). Uptake of norepinephrine across the neuronal membrane obeys saturation kinetics of the Michaelis-Menten type with high affinity. It also requires sodium ions in the external medium, is temperature dependent, and involves active transport. The uptake process is stereoselective and can be utilized by other phenylethylamine derivatives such as epinephrine, dopamine, tyramine, amphetamine, α-methyl-norepinephrine, and meteraminol (Iversen, 1971). High affinity uptake processes similar to norepinephrine's have been demonstrated for other putative neurotransmitters, serotonin, gamma amino butyric acid, glutamate, aspartate, and glycine (Logan and Snyder, 1971).

Norepinephrine can also be taken up by an extraneuronal process (Iversen, 1965; Eisenfeld et al., 1967). This extraneuronal uptake can be blocked by adrenergic blocking agents, normetanephrine (Eisenfeld et al., 1967) and corticosteroids (Iversen, 1971). Compounds such as isoproterenol, which have a low affinity for the intraneuronal uptake and a high affinity for extraneuronal uptake, may be inactivated by the latter process. Extraneuronal uptake may be an important mechanism for removal of the norepinephrine in which the density of the sympathetic innervation is very low or when the synaptic cleft is wide.

Norepinephrine can be inactivated by a variety of mechanisms (Figure 6): Uptake into the neuron, removal by the circulation, enzymatic O-methylation and deamination by liver and kidney, O-methylation and deamination by effector cells, and extraneuronal uptake. Although neuronal uptake is the major mechanism for terminating the action of the sympathetic neurotransmitter, other types of inactivation may predominate, depending on the density of sympathetic innervation, size of the synaptic cleft, blood supply and activity of catechol-O-methyltransferase, and monoamine oxidase.

Several studies have shown that α-adrenergic blocking agents inhibit the uptake of norepinephrine and cause an increased overflow of the neurotransmitter on nerve stimulation (Hertting et al., 1961b; Brown and Gillespie, 1957). It has also been demonstrated that large amounts of endogenous norepinephrine inhibit the discharge of norepinephrine from nerves (Stärke, 1971). These observations suggest another regulatory site on the neuronal membrane. The neuronal membrane appears to have an inhibitory α-adrenergic receptor (Stärke, 1971), which would cause an increased release of the neurotransmitter when the receptor is blocked and a decreased release when a large amount of norepinephrine is present in the synaptic cleft.

Regulation at the postsynaptic effector cell

The sympathetic effector cell could influence the activity of the sympathetic nerve, and conversely the presynaptic cell could influence the activity of the postsynaptic effector cell. When the response of postsynaptic sympathetic effector cells is blocked by phenoxybenzamine, there is a marked increase in tyrosine hydroxylase activity in the adrenal medulla (Thoenen et al., 1969b) and much greater conversion of [^{14}C]tyrosine to [^{14}C]catecholamine (Dairman and Udenfriend, 1970).

Denervation of the sympathetic nerves leads to an increased response of the postsynaptic cell. One ex-planation for the increased sensitivity is the removal of an important inactivating mechanism; uptake by the neuronal membrane. Another possible mechanism for supersensitivity is an increased responsiveness of the postsynaptic site. It has been shown that in the denervated muscle, the area of binding of α-bungarotoxin, a compound that binds irreversibly to acetylcholine receptors, is increased (Miledi and Potter, 1971).

The pineal cell has been used to study the relationship between the sympathetic nerves and postsynaptic cell. This gland is richly innervated with sympathetic nerves that regulate the synthesis of the hormone melatonin (Wurtman et al., 1968a). Serotonin N-acetyltransferase is an enzyme that N-acetylates serotonin to form the precursor of melatonin (Weissbach et al., 1960). It is present in the postsynaptic pineal cell and is markedly stimulated by norepinephrine and dibutyryl cyclic 3′,5′-adenosine monophosphate in organ culture (Klein et al., 1970). In the intact rat, the enzyme activity is also sharply increased after administration of catecholamines (Figure 7) (Deguchi and Axelrod, 1972). The induction of pineal serotonin N-acetyltransferase by catecholamines is prevented by the β-adrenergic blocking agent, propranolol. When the pineal is denervated the catecholamines cause a superinduction (100-fold increase) of serotonin N-acetyltransferase (Figure 7). Although other possibilities have not been excluded, these findings suggest that the increased responsiveness after denervation is due to changes on the postsynaptic β-adrenergic receptor on the pineal cell.

•

Conclusions

The formation and conservation of the sympathetic neurotransmitter, norepinephrine, is controlled at several sites in the sympathetic neuron. Its synthesis can be rapidly changed by feedback mechanisms on tyrosine hydroxylase in the nerve terminals. Another regulatory site occurs in the cell body and adrenal medulla whereby sympathetic nerve activity changes the rate of formation of various biosynthetic enzymes. Glucocorticoid hormones also influence the synthesis of these enzymes in the adrenal medulla. The biosynthetic enzymes are made in the cell body and transported down the axon to the nerve terminals via proximodistal flow. The transmitter is then synthesized and stored in a vesicle in the nerve terminal. Norepinephrine is released by a process of exocytosis. Once released, the neurotransmitter can be taken up by the nerve terminal and stored and reused again. The presynaptic membrane can affect the activity of the postsynaptic cell by rapidly removing the transmitter by reuptake and influencing the responsiveness of the postsynaptic adrenergic receptor.

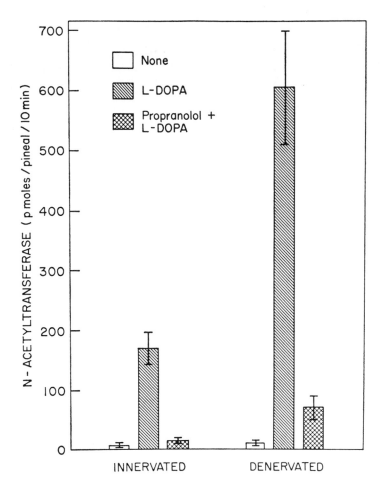

FIGURE 7 Induction and superinduction of serotonin N-acetyltransferase in the rat pineal. Rat pineals were denervated by the bilateral removal of the superior cervical ganglia. Dopa (150 mg/kg) alone or together with propanolol (20 μg/kg) were given at 10 AM and N-acetyltransferase in the pineal was examined 3 hrs later (Deguchi and Axelrod, 1972).

REFERENCES

AXELROD, J., 1962. Purification and properties of phenylethanolamine N-methyltransferase. *J. Biol. Chem.* 237: 1657–1660.

AXELROD, J., 1971. Noradrenaline: Fate and control of its biosynthesis. In *Les Prix Nobel en 1970*, Stockholm: Imprimerieal Royal P. A. Norstedt and Soner, pp. 189–208.

AXELROD, J., 1973. The fate of noradrenaline in the sympathetic neuron. *Harvey Lect.*, Vol. 67, (in press).

AXELROD, J., G. HERTTING, and R. W. PATRICK, 1961. Inhibition of ³H-norepinephrine release by monoamine oxidase inhibitors. *J. Pharmacol. Exp. Ther.* 134: 325–328.

AXELROD, J., R. A. MUELLER, J. P. HENRY, and P. M. STEPHENS, 1970. Changes in enzymes involved in the biosynthesis and metabolism of noradrenaline and adrenaline after psychosocial stimulation. *Nature (Lond.)* 225: 1059–1060.

AXELROD, J., and E. S. VESELL, 1970. Heterogeneity of N- and O-methyltransferases. *Molec. Pharmacol.* 6: 78–84.

AXELROD, J., and R. M. WEINSHILBOUM, 1972. Catecholamines. *New Engl. J. Med.* 287: 237–242.

BLACK, I. B., I. HENDRY, and L. L. IVERSEN, 1971. Transsynaptic regulation of growth and development of adrenergic neurons in a mouse sympathetic ganglion. *Brain Res.* 34: 229–240.

BRIMIJOIN, S., and P. B. MOLINOFF, 1971. Effects of 6-hydroxydopamine on the activity of tyrosine hydroxylase and dopamine-β-hydroxylase in sympathetic ganglia of the rat. *J. Pharmacol. Exp. Ther.* 178: 417–425.

BROWN, G. L., and J. S. GILLESPIE, 1957. The output of sympathetic transmitter from the spleen of the cat. *J. Physiol. (Lond.)* 138: 81–102.

BYDGEMAN, S., and U. S. VON EULER, 1958. Resynthesis of catechol hormones in the cat's adrenal medulla. *Acta Phys. Scand.* 44: 375–383.

CARTER, S. B., 1967. Effects of cytochalasins on mammalian cells. *Nature (Lond.)* 213: 261–264.

CHRISTENSON, J. G., W. DAIRMAN, and S. UDENFRIEND, 1972. On the identity of dopa decarboxylase and 5-hydroxytryptophan decarboxylase. *Proc. Natl. Acad. Sci. USA* 69: 343–347.

CIARANELLO, R. D., J. N. DORNBUSCH, and J. D. BARCHAS, 1972a. Rapid increase of phenylethanolamine-N-methyltransferase by environmental stress in an inbred mouse strain. *Science* 175: 789–790.

CIARANELLO, R. D., J. N. DORNBUSCH, and J. D. BARCHAS,

1972b. Regulation of adrenal phenylethanolamine N-methyltransferase in three inbred linked mouse strains. *Molec. Pharmacol.* 8:511–519.

CONNETT, R. T., and N. KIRSHNER, 1970. Purification and properties of bovine phenylethanolamine N-methyltransferase. *J. Biol. Chem.* 245:329–334.

COUPLAND, R. E., 1965. *The Natural History of the Chromaffin Cell.* London: Longmans.

COYLE, J. T., and G. F. WOOTEN, 1972. Rapid axonal transport of tyrosine hydroxylase and dopamine-β-hydroxylase. *Brain Res.*, 44:701–704.

CROUT, J. R., 1961. Effect of inhibiting both catechol-O-methyltransferase and monoamine oxidase on the cardiovascular responses to noradrenaline. *Proc. Soc. Exp. Biol. Med.* 108:482–484.

DAHLSTRÖM, A., and J. HÄGGENDAL, 1966. Studies on the transport and life span of amine storage granules in a peripheral adrenergic neuron system. *Acta Physiol. Scand.* 67:278–288.

DAIRMAN, W., and S. UDENFRIEND, 1970. Increased conversion of tyrosine to catecholamines in the intact rat following elevation of tissue tyrosine hydroxylase levels by administered phenoxybenzamine. *Molec. Pharmacol.* 6:350–356.

DEGUCHI, T., and J. AXELROD, 1972. Induction and super-induction of serotonin N-acetyltransferase by adrenergic drugs and denervation in the rat pineal organ. *Proc. Nat. Acad. Sci. USA* 69:2208–2211.

DENGLER, H. J., I. A. MICHAELSON, H. E. SPIEGAL, and E. TITUS, 1962. Uptake of labeled norepinephrine by isolated brain and other tissues of the cat. *Int. J. Neuropharmacol.* 1:23–38.

DOUGLAS, W. W., and R. P. RUBIN, 1961. The role of calcium in the secretory response of the adrenal medulla to acetylcholine. *J. Physiol. (Lond.)* 159:40–57.

EISENFELD, A. J., L. LANDSBERG, and J. AXELROD, 1967. Effect of drugs on the accumulation and metabolism of extraneuronal norepinephrine in the rat heart. *J. Pharmacol. Exp. Ther.* 158:378–385.

FRIEDMAN, S., and S. KAUFMAN, 1965. 3,4-Dihydroxyphenylethylamine β-hydroxylase: Physical properties, copper content and role of copper in the catalytic activity. *J. Biol. Chem.* 240:4763–4773.

GEFFEN, L. B., B. G. LIVETT, and R. A. RUSH, 1969. Immunohistochemical localization of protein components of catecholamine storage vesicles. *J. Physiol. (Lond.)* 204:593–605.

GILLESPIE, E., R. J. LEVINE, and S. E. MALAWISTA, 1968. Histamine release from rat peritoneal mast cells: Inhibition by colchicine and potentiation by deuterium oxide. *J. Pharmacol. Exp. Ther.* 164:158–165.

GOLDSTEIN, M., L. S. FREEDMAN, A. C. BOHUON, and F. GUERINOT, 1972. Serum dopamine-β-hydroxylase in patients with neuroblastoma. *New Engl. J. Med.* 286:1123–1125.

GOLDSTEIN, M., L. S. FREEDMAN, and M. BONNAY, 1971. An assay for dopamine-β-hydroxylase activity in tissues and serum. *Experientia* 27:632–633.

HARTMAN, B. K., P. B. MOLINOFF, and S. UDENFRIEND, 1970. Increased rate of synthesis of dopamine-β-hydroxylase in adrenals of reserpinized rats. *The Pharmacologist* 12:470.

HERTTING, G., and J. AXELROD, 1961. The fate of tritiated noradrenaline at the sympathetic nerve ending. *Nature (Lond.)* 192:172–173.

HERTTING, G., J. AXELROD, I. J. KOPIN, and L. G. WHITBY, 1961a. Lack of uptake of catecholamines after chronic denervation of sympathetic nerves. *Nature (Lond.)* 189:66.

HERTTING, G., J. AXELROD, and L. G. WHITBY, 1961b. Effect of drugs on the uptake and metabolism of ^{3}H-norepinephrine. *J. Pharmacol. Exp. Ther.* 134:146–153.

HÖKFELT, T., and A. DAHLSTRÖM, 1971. Effects of two mitosis inhibitors (colchicine and vinblastine) on the distribution and axonal transport of noradrenaline storage particles. *Z. Zellforsch.* 119:460–482.

HOLTZ, P., R. HEISE, and K. LUDTKE, 1938. Quantitativer abbau von L-Dioxyphenylalanin (Dopa) durch Niere. *Arch. Exp. Path. Pharmak.* 191:87–118.

IKEDA, M., L. A. FAHIEN, and S. UDENFRIEND, 1966. A kinetic study of bovine adrenal tyrosine hydroxylase. *J. Biol. Chem.* 241:4452–4456.

IVERSEN, L. L., 1971. Role of transmitter uptake mechanisms in synaptic neurotransmission. *Brit. J. Pharmacol.* 41:571–591.

IVERSEN, L. L., 1965. The uptake of catecholamines at high perfusion concentrations in the rat isolated heart: A novel catecholamine uptake process. *Brit. J. Pharmacol.* 25:18–33.

IVERSEN, L. L., 1963. The uptake of noradrenaline by the isolated perfused rat heart. *Brit. J. Pharmacol.* 21:523–537.

JOHNSON, D. G., N. B. THOA, R. M. WEINSHILBOUM, J. AXELROD, and I. J. KOPIN, 1971. Enhanced release of dopamine-β-hydroxylase from sympathetic nerves by calcium and phenoxybenzamine and its reversal by prostaglandins. *Proc. Natl. Acad. Sci. USA* 68:2227–2230.

KIRSHNER, N., 1957. Pathway of noradrenaline formation from dopa. *J. Biol. Chem.* 226:821–825.

KLEIN, D. C., G. R. BERG, and J. WELLER, 1970. Melatonin synthesis: Adenosine 3′,5′-monophosphate and norepinephrine stimulate N-acetyltransferase. *Science* 168:979–980.

KOPIN, I. J., G. R. BREESE, K. R. KRAUSS, and V. K. WEISE, 1968. Selective release of newly synthesized norepinephrine from the cat spleen during sympathetic nerve stimulation. *J. Pharmacol. Exp. Ther.* 161:271–278.

KOPIN, I. J., and S. D. SILBERSTEIN, 1972. Axons of sympathetic neurons: Transport of enzymes in vivo and properties of axonal sprouts in vitro. *Pharmacol. Rev.* 24:245–254.

KVETNANSKY, R., V. K. WEISE, and I. J. KOPIN, 1970. Elevation of adrenal tyrosine hydroxylase and phenylethanolamine N-methyltransferase by repeated immobilization of rats. *Endocrinology* 87:744–749.

LACY, P. E., S. L. HOWELL, D. A. YOUNG, and C. J. FINK, 1968. New hypothesis of insulin secretion. *Nature (Lond.)* 219:1177–1179.

LADURON, P., and F. BELPAIRE, 1968. Transport of noradrenaline and dopamine-β-hydroxylase in sympathetic nerves. *Life Sci.* 7:1–9.

LAMPRECHT, F., and G. F. WOOTEN, 1973. Effect of hypophysectomy on serum dopamine-β-hydroxylase activity in rat. *Endocrinology* 92: (in press).

LOGAN, W. J., and S. H. SNYDER, 1971. Unique high affinity uptake systems for glycine, glutamic and aspartic acids in central nervous tissue of the rat. *Nature (Lond.)* 234:297–298.

MILEDI, R., and L. T. POTTER, 1971. Acetylcholine receptors in muscle fibers. *Nature (Lond.)* 233:599–603.

MOLINOFF, P. B., S. BRIMIJOIN, and J. AXELROD, 1972. Induction of dopamine-β-hydroxylase and tyrosine hydroxylase in rat hearts and sympathetic ganglia. *J. Pharmacol. Exp. Ther.* 182:116–130.

MOLINOFF, P. B., W. S. BRIMIJOIN, R. M. WEINSHILBOUM, and J. AXELROD, 1970. Neurally mediated increase in dopamine-β-hydroxylase activity. *Proc. Nat. Acad. Sci. USA* 66:453–458.

Molinoff, P. B., R. M. Weinshilboum, and J. Axelrod, 1971. A sensitive enzymatic assay for dopamine-β-hydroxylase. *J. Pharmacol. Exp. Ther.* 178:425–531.

Mueller, R. A., 1971. Effect of 6-hydroxydopamine on the synthesis and turnover of catecholamines and proteins in the adrenal. In *6-Hydroxydopamine and Catecholamines*, T. Malmfors and H. Thoenen, eds. Amsterdam: Elsevier, pp. 291–301.

Mueller, R. A., H. Thoenen, and J. Axelrod, 1969a. Adrenal tyrosine hydroxylase: Compensatory increase in activity after chemical sympathectomy. *Science* 163:468–469.

Mueller, R. A., H. Thoenen, and J. Axelrod, 1969b. Increase in tyrosine hydroxylase activity after reserpine administration. *J. Pharmacol. Exp. Ther.* 169:74–79.

Mueller, R. A., H. Thoenen, and J. Axelrod, 1969c. Inhibition of trans-synaptically increased tyrosine hydroxylase activity by cycloheximide and actinomycin D. *Molec. Pharmacol.* 5:463–469.

Mueller, R. A., H. Thoenen, and J. Axelrod, 1970a. Effect of pituitary and ACTH on the maintenance of basal tyrosine hydroxylase activity in the rat adrenal gland. *Endocrinology* 86:751–755.

Mueller, R. A., H. Thoenen, and J. Axelrod, 1970b. Inhibition of neuronally induced tyrosine hydroxylase by nicotinic receptor blockade. *Eur. J. Pharmacol.* 10:51–56.

Nagatsu, T., M. Levitt, and S. Udenfriend, 1964. Tyrosine hydroxylase: The initial step in norepinephrine biosynthesis. *J. Biol. Chem.* 239:2910–2917.

Ochs, S., 1972. Fast transport of materials in mammalian nerve fibers. *Science* 176:252–259.

Patrick, R. L., and N. Kirshner, 1971. Effect of stimulation on the levels of tyrosine hydroxylase, dopamine-β-hydroxylase and catecholamines in intact and denervated rat adrenal glands. *Molec. Pharmacol.* 7:87–96.

Poisner, A. M., and J. Bernstein, 1971. A possible role of microtubules in catecholamine release from the adrenal medulla: Effect of colchicine, vinca alkaloids and deuterium oxide. *J. Pharmacol. Exp. Ther.* 177:102–108.

Potter, L. T., and J. Axelrod, 1963a. Properties of noradrenaline storage particles of the rat heart. *J. Pharmacol. Exp. Ther.* 142:291–305.

Potter, L. T., and J. Axelrod, 1963b. Studies on the storage of norepinephrine and the effect of drugs. *J. Pharmacol. Exp. Ther.* 140:199–206.

Sedvall, G. C., and I. J. Kopin, 1967. Acceleration of norepinephrine synthesis in the rat submaxillary gland in vivo during sympathetic nerve stimulation. *Life Sci.* 6:45–51.

Segal, D. S., J. L. Sullivan, R. T. Kuczenski, and A. J. Mandell, 1971. Effects of long-term reserpine treatment on brain tyrosine hydroxylase and behavioral activity. *Science* 173:847–849.

Silberstein, S. D., S. Brimijoin, P. B. Molinoff, and L. Lemberger, 1972. Induction of dopamine-β-hydroxylase in rat superior cervical ganglia in organ culture. *J. Neurochem.* 19:919–921.

Smith, A. D., W. P. DePotter, E. J. Moerman, and A. F. De Schaepdryver, 1970. Release of dopamine-β-hydroxylase and chromogranin A upon stimulation of the splenic nerve. *Tissue and Cell* 2:547–568.

Stärke, K., 1971. Influence of α-receptor stimulants on noradrenaline release. *Naturwissenschaften* 58:420.

Thoa, N. B., G. F. Wooten, J. Axelrod, and I. J. Kopin, 1972. Inhibition of release of dopamine-β-hydroxylase and norepinephrine from sympathetic nerves by colchicine, vin-

blastine and cytochalasin B. *Proc. Nat. Acad. Sci. USA* 69:520–522.

Thoenen, H., 1970. Induction of tyrosine hydroxylase in peripheral and central adrenergic neurons by cold exposure of rats. *Nature (Lond.)* 228:861–862.

Thoenen, H., R. A. Mueller, and J. Axelrod, 1969a. Increased tyrosine hydroxylase activity after drug-induced alteration of sympathetic transmission. *Nature (Lond.)* 221:1264.

Thoenen, H., R. A. Mueller, and J. Axelrod, 1969b. Transsynaptic induction of adrenal tyrosine hydroxylase. *J. Pharmacol. Exp. Ther.* 169:249–254.

Thoenen, H., R. A. Mueller, and J. Axelrod, 1970. Neuronally dependent induction of adrenal phenylethanolamine N-methyltransferase by 6-hydroxydopamine. *Biochem. Pharmacol.* 19:669–673.

Thoenen, H., and J. P. Tranzer, 1968. Chemical sympathectomy by selective destruction of adrenergic nerve endings with 6-hydroxydopamine. *Naunyn-Schmiedeberg Arch. Pharm. Exp. Path.* 281:271–288.

Viveros, O. H., L. Arqueros, R. J. Connett, and N. Kirshner, 1969. Mechanism of secretion from the adrenal medulla IV. Fate of the storage vesicles following insulin and reserpine administration. *Molec. Pharmacol.* 5:69–82.

Viveros, O. H., L. Arqueros, and N. Kirshner, 1968. Release of catecholamines and dopamine-β-hydroxylase from the adrenal medulla. *Life Sci.* 7:609–618.

Weiner, N., 1970. Regulation of norepinephrine biosynthesis. *Ann. Rev. Pharmacol.* 10:273–290.

Weinshilboum, R., and J. Axelrod, 1970. Dopamine-β-hydroxylase activity in the rat after hypophysectomy. *Endocrinology* 87:894–899.

Weinshilboum, R., and J. Axelrod, 1971a. Reduced plasma dopamine-β-hydroxylase in familial dysautonomia. *New Engl. J. Med.* 285:938–942.

Weinshilboum, R., and J. Axelrod, 1971b. Serum dopamine-β-hydroxylase activity. *Circ. Res.* 28:307–315.

Weinshilboum, R., and J. Axelrod, 1971c. Serum dopamine-β-hydroxylase: Decrease after chemical sympathectomy. *Science* 173:931–934.

Weinshilboum, R., R. Kvetnansky, J. Axelrod, and I. J Kopin, 1971d. Elevation of serum dopamine-β-hydroxylase activity with forced immobilization. *Nature New Biology* 230:287–288.

Weinshilboum, R., N. B. Thoa, D. G. Johnson, I. J. Kopin, and J. Axelrod, 1971e. Proportional release of norepinephrine and dopamine-β-hydroxylase from the sympathetic nerves. *Science* 174:1349–1351.

Weiss, P., and H. Hiscoe, 1948. Experiments on the mechanism of nerve growth. *J. Exp. Zool.* 107:315–396.

Weissbach, H., B. G. Redfield, and J. Axelrod, 1960. Biosynthesis of melatonin: Enzymic conversion of serotonin to N-acetylserotonin. *Biochim. Biophys. Acta* 43:352–353.

Wessells, N. K., B. S. Spooner, J. F. Ash, M. Bradley, M. A. Laduena, E. L. Taylor, J. T. Wrenn, and K. M. Yamada, 1971. Microfilaments in cellular and developmental processes. *Science* 171:135–143.

Whitby, L. G., G. Hertting, and J. Axelrod, 1960. Effect of cocaine on the disposition of noradrenaline labeled with tritium. *Nature (Lond.)* 187:604–605.

Williams, J. A., and J. Wolff, 1970. Possible role of microtubules in thyroid secretion. *Proc. Nat. Acad. Sci. USA* 67:1901–1908.

WOLFE, D. E., L. T. POTTER, K. C. RICHARDSON, and J. AXELROD, 1962. Localizing tritiated norepinephrine in sympathetic axons by electron microscopic autoradiography. *Science* 138:440–442.

WURTMAN, R. J., and J. AXELROD, 1966. Control of enzymatic synthesis of adrenaline in the adrenal medulla by adrenal cortical steroids. *J. Biol. Chem.* 241:2301–2305.

WURTMAN, R. J., J. AXELROD, and D. E. KELLEY, 1968a. *The Pineal.* New York: Academic Press.

WURTMAN, R. J., J. AXELROD, E. S. VESELL, and G. T. ROSS, 1968b. Species differences in inducibility of phenylethanolamine N-methyltransferase. *Endocrinology* 82:584–590.

WURZBURGER, R. J., and J. M. MUSACCHIO, 1971. Subcellular distribution and aggregation of bovine adrenal tyrosine hydroxylase. *J. Pharmacol. Exp. Ther.* 177:155–157.

77 Development of the Central Catecholaminergic Neurons

JOSEPH T. COYLE, JR.

ABSTRACT The biochemical aspects of the development of the central catecholaminergic neurons have been examined in the rat brain. The high-affinity uptake mechanism for norepinephrine, which is localized at the terminal of the noradrenergic neuron, was used as a specific marker for the neuronal membrane. The uptake mechanism was first demonstrable in the brain at 18 days gestation; during subsequent maturation, its specific activity increased almost fivefold. The increase in uptake of ^{3}H-norepinephrine correlated with an increased localization of the ^{3}H-norepinephrine in the synaptosomal fractions on continuous sucrose gradients.

Tyrosine hydroxylase, dopa decarboxylase, and dopamine-β-hydroxylase, the enzymes in the biosynthetic pathway for norepinephrine, were present in the brain at 15 days gestation. During subsequent development, there was a coordinate increase in the activity of these enzymes in the whole brain. With maturation, there was a translocation of enzyme activity from the regions that contain the cell bodies of the catecholaminergic neurons to those regions that contain only terminals of these neurons. Concurrently, there was an absolute increase in the activity of these enzymes in the synaptosomal fractions. The time at which cell division of the central catecholaminergic neurons ceases has been investigated by means of ^{3}H-thymidine autoradiography; thus, changes in enzyme activity could be more accurately expressed in terms of this specific population of neurons.

Histofluorescent studies done by Loizou and Maeda indicate that, just prior to birth, catecholamine fluorescence is limited to the cell bodies of the catecholaminergic neurons in the brainstem, and that with maturation there is a marked outgrowth of fluorescent fibers innervating the distal regions of the brain. The temporal-spatial changes in the activity of the uptake mechanism, as well as the specific enzymes for the catecholaminergic neurons, offer excellent correlates with these histochemical observations.

Introduction

THE DEVELOPING nervous system provides a scenario in which the complex relationships among neurons take form. This process involves cell multiplication, cell migration, outgrowth of neuronal processes, and synaptogenesis. In abstract terms, the development of the brain is a vectorial process characterized by changes in time and space as well as magnitude (Jacobson, 1970). Research on the development of the mammalian brain has been oriented primarily toward structural analysis, and the correlation between morphologic and biochemical differentiation has remained an elusive goal.

Because of the following factors, the central catecholaminergic neurons of the rat brain may be uniquely susceptible to the elucidation of the complex relationship between morphogenesis and biochemical differentiation. First, a specific histofluorescent technique has been devised whereby the catecholaminergic neurons can be identified in sections of brain tissue (Falck and Hillarp, 1959). The method is based on the fact that the catecholamines within the neurons interact with paraformaldehyde vapor to form a fluorophor that can be visualized by fluorescent microscopy. With this method, the aminergic neuronal pathways have been mapped out in the adult rat brain (Fuxe et al., 1970; Ungerstedt, 1971). Second, the cell bodies of the catecholaminergic neurons are limited to the brainstem although their axons and terminals ramify to innervate other regions of the brain. Thus, properties related to the perikarya can be anatomically distinguished from those related to the axons and terminals. Finally, the enzymatic processes involved in the synthesis and metabolism of catecholamines as well as the pharmacologic properties of these neurons have been well characterized in the adult rat brain (Molinoff and Axelrod, 1971).

Morphogenesis

Catecholamine-induced fluorescence can be observed in the cells that form the noradrenergic nucleus, the locus coeruleus, at 14 days of gestation; over the next 4 days, the neurons migrate caudally to their ultimate position in the pons (Maeda and Dresse, 1969). Fluorescence of the cell bodies of the dopaminergic neurons of the midbrain is apparent at 18 days of gestation, whereas the dopaminergic cell bodies in the hypothalamus first exhibit catecholamine-fluorescence early in the postnatal period (Loizou, 1971; Smith and Simpson, 1970). Thus, there is an apparent caudal-to-rostral sequence of appearance of

JOSEPH T. COYLE, JR. Laboratory of Clinical Science, National Institute of Mental Health, Bethesda, Maryland catecholamine-induced fluorescence in the cell bodies that give rise to the various aminergic neuronal pathways of the brain.

Loizou (1972) has investigated the postnatal development of the aminergic neurons of the rat brain with the histofluorescent technique (Table I). At birth, there is a striking scarcity of noradrenergic terminals, whereas scattered islands of dopaminergic terminals are already present in the caudate nucleus. During the first week after birth, there is a marked increase in the size and fluorescent intensity of the catecholaminergic cell bodies; certain groups of neurons exhibit even more intense fluorescence than observed in the adult. Between the second and third week after birth, the noradrenergic terminals attain an adult pattern of density and intensity of fluorescence in the spinal cord and medulla-pons, whereas the adult pattern is established in the more rostral regions only by the fourth to fifth postnatal week. Although the diffusely intense fluorescence of the dopaminergic terminals in the striatum is maximal by 4 weeks after birth, during subsequent maturation there is a shift in the fluorescent spectrum suggesting further increases in the dopamine concentration. This impression of a more rapid maturation of the noradrenergic neurons as compared to the dopaminergic neurons is substantiated by the fact that norepinephrine attains adult concentration prior to dopamine (Agrawal et al., 1968; Breese and Traylor, 1972; Loizou and Salt, 1970). Thus, two phases in the morphogenesis of the catecholaminergic neurons can be distinguished: A fetal and early postnatal period when the cell bodies acquire catecholamine fluorescence and then a period, primarily postnatal, when there is a centrifugal outgrowth of axons and terminals from the cell bodies.

Time of origin of the catecholaminergic neurons

Maturation of the brain involves multiplication and differentiation of heterogeneous populations of cells. In order to quantify accurately the development of the central catecholaminergic neurons on a "per neuron" basis, it is essential to know when the brain has attained its full complement of these neurons. In other words, at what point in development do the central catecholaminergic neurons cease dividing and start differentiating? Autoradiographic studies of the incorporation of ^{3}H-thymidine have proved to be an accurate method for dating the time of origin of neurons since the radiolabeled thymidine is only incorporated into the DNA of dividing cells (Altman, 1966).

By combining the histofluorescent technique with ^{3}H-thymidine autoradiography, it has been possible to demonstrate when the central catecholaminergic neurons undergo cell division (Nicholson, Coyle, Das, and Bloom, in preparation). The noradrenergic neurons in the locus coeruleus and the dopaminergic neurons in the substantia nigra exhibit a brief period of cell division between 12 and 14 days of gestation and do not divide subsequently. Thus, the two major catecholaminergic nuclei of the brain attain their full complement of neurons a week before birth, and a more appropriate representation of their developmental changes after this date would be on the basis of whole brain activity.

Neuronal membrane uptake process

The catecholaminergic neurons possess on their neuronal membrane a high-affinity uptake mechanism for catecholamines (Axelrod, 1965). This transport mechanism, which is an active process linked to Na$^+$-K$^+$ ATPase, plays a primary role in the inactivation of catecholamines released at the synapse (Tissari et al., 1969; White and Keen, 1970). The uptake mechanism, being highly concentrated in the terminal boutons, can be demonstrated in sheared-off nerve terminals or synaptosomes (Davis et al., 1967). The central dopaminergic and noradrenergic neurons have uptake mechanisms specific for their re-

TABLE I

Density of catecholaminergic terminals in regions of brain during development

	Birth	1 Week	2 Weeks	3 Weeks	Adult
Medulla-pons	0 – few	1 – 2	2 – 3	4	4
Cerebellum	0	1 – 3	1 – 3	few	few
Mesencephalon	0	few	2	3	3 – 4
Diencephalon	0 – few	1	2	3 – 4	4 – 5
Cerebral cortex	0	few – 2	1 – 2	1 – 3	2 – 4
Striatum	islands	islands	islands and diffuse	diffuse	diffuse

The density of catecholaminergic terminals in the various regions of the rat brain during development was determined by histofluorescent microscopy. Results are expressed in terms of a 0 to 5 scale. Adapted from Loizou (1972).

spective neurotransmitter that are anatomically, kinetically, and pharmacologically distinct (Coyle and Snyder, 1969; Horn et al., 1971). The noradrenergic neurons exhibit a saturable uptake for L-norepinephrine with a Km of 3×10^{-7} M, that is markedly sensitive to inhibition by the antidepressant, desipramine (Snyder and Coyle, 1969).

In the various regions of the adult brain, the synaptosomal uptake of norepinephrine is proportional to the density of innervation by noradrenergic terminals, and destruction of the terminals with 6-hydroxydopamine markedly reduces this uptake process (Snyder and Coyle, 1969; Uretsky and Iversen, 1970). Studies of the ontogenesis of the peripheral sympathetic nervous system indicate that the development of an organ's ability to take up norepinephrine parallels the development of its sympathetic innervation (Glowinski et al., 1964; Sachs et al., 1970). Therefore, analysis of the uptake of norepinephrine into synaptosomes should be a method for quantifying the increase in noradrenergic terminals in the developing brain.

In homogenates prepared from whole rat brain, the high-affinity uptake mechanism for norepinephrine appears at 18 days of gestation (Coyle and Axelrod, 1971). During subsequent maturation, the affinity constant (Km) does not change significantly, whereas the capacity to take up norepinephrine (V_{max}) increases over 100-fold for the whole brain. That this uptake process indeed represents the one specific for norepinephrine is substantiated by the fact that the inhibitory effect of desipramine shows a similar developmental profile.

The ^{3}H-norepinephrine accumulated by homogenates of adult brain is localized in the synaptosomal fractions on continuous sucrose gradients (Coyle and Snyder, 1969). This particulate-bound peak of ^{3}H-norepinephrine coincides with the peak of occluded lactic dehydrogenase activity, a more general enzymatic marker for synaptosomes, and occurs at a distinctly lower sucrose molarity than the peak of monoamine oxidase activity, an enzymatic marker for mitochondria (Figure 1). Analysis of the subcellular distribution of ^{3}H-norepinephrine accumulated by homogenates prepared from rat brains of different ages indicates that there is a progressive increase with age in the amount of ^{3}H-norepinephrine that sediments in the synaptosomal fractions. When corrected for brain weight, the real increase in the amount of synaptosomal-bound ^{3}H-norepinephrine is nearly 300-fold between 18 days of gestation and adulthood. The parallel increase in the specific uptake mechanism for norepinephrine and in the amount of synaptosomal-bound ^{3}H-norepinephrine suggest that uptake may be a valid index of the outgrowth of the noradrenergic terminals in the central nervous system.

Enzymes involved in the biosynthesis of catecholamines

The dopaminergic and noradrenergic neurons have in common the first two enzymes in the biosynthetic pathway for their neurotransmitters, tyrosine hydroxylase and dopa decarboxylase (Molinoff and Axelrod, 1971). The noradrenergic neurons possess an additional enzyme, dopamine-β-hydroxylase, which converts dopamine to norepinephrine. All three enzymes are present in the fetal rat brain at 15 days of gestation. During subsequent development, the activities of these enzymes increase 500-fold in the whole brain (Coyle and Axelrod, 1972a, 1972b; Lamprecht and Coyle, 1972; Figure 2). Thus, in confirmation of the observation made with the histofluorescent technique (Maeda and Dresse, 1969), the rat brain is capable of synthesizing catecholamines extremely early in ontogenesis. Furthermore, regression of enzyme activity back to zero suggests that these enzymes would first appear at 13 to 14 days of gestation. This coincides closely with the date established by the autoradiographic study as the time when differentiation of the noradrenergic neurons would commence.

Since the proliferating axons and terminals in the developing rat brain are identified by their catecholamine-induced fluorescence, it would appear likely that they also contain the enzymatic machinery necessary for the synthesis of the amines. Thus, the biosynthetic enzymes could be used as specific enzymatic markers for the outgrowth of the neuronal processes. The cell bodies of the noradrenergic neurons, which are located exclusively in the medulla-pons, send out processes that provide the noradrenergic innervation for the rest of the brain. In the fetal brain, dopamine-β-hydroxylase, the final enzyme in the synthesis pathway for norepinephrine, is localized primarily in the medulla-pons. With maturation, there is a progressive shift in enzyme activity to the rostral regions of the brain, so that by adulthood the medulla-pons possesses only 20% of the whole brain dopamine-β-hydroxylase activity (Coyle and Axelrod, 1972a). In the case of tyrosine hydroxylase, there is a 50% increase in its specific activity in the medulla-pons between birth and adulthood, whereas the cerebral cortex exhibits a sixfold increase in specific activity during the same period (Figure 3). The dopaminergic cell bodies that are located in the midbrain give rise to an extremely dense meshwork of terminals in the corpus striatum. During postnatal development, the specific activity of tyrosine hydroxylase in the midbrain increases only moderately in contrast to a 12-fold increase in the corpus striatum (Coyle and Axelrod, 1972b). Thus, with maturation, there is a centrifugal movement of the enzymes involved in the biosynthesis of catecholamines away from the regions of the brain that

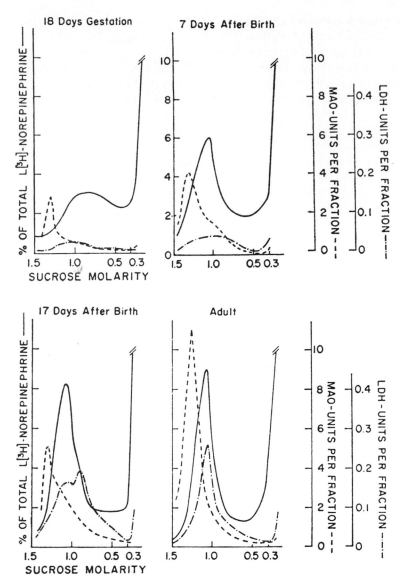

FIGURE 1 Sedimentation characteristics of L-[³H]norepine-phrine, lactic dehydrogenase (LDH) and monoamine oxidase (MAO) in homogenates prepared from rat brains at various stages of development. The homogenates were incubated for 5 min with L-[³H]norepinephrine and the amount equivalent to contain 125 mg of brain tissue was centrifuged at 100,000 g on linear sucrose gradients (1.5 to 0.3 M). Total amount (nanomoles) of L-[³H]norepinephrine on the gradients were: 0.84 at 18 days of gestation; 2.47 at 7 days after birth; 3.07 at 17 days after birth; and 4.45 for the adult (adapted from Coyle and Axelrod, 1971).

contain the cell bodies to the regions that receive the terminals.

The temporal-spatial changes in enzyme activity during development are accompanied by changes in their subcellular distribution. For example, tyrosine hydroxylase in the adult rat brain behaves as a soluble enzyme, because it is released in the supernatant with the cytosol when the brain is homogenized in hypotonic buffer (Coyle, 1972). However, when the adult brain is homogenized in isotonic sucrose and fractionated accord-ing to the method of Whittaker (1965), most of the tyrosine hydroxylase is entrapped in and sediments with the synaptosomes, indicating a high degree of localization in the nerve terminals (Coyle, 1972; Figure 4). When fetal brain is fractionated in the same manner, most of the tyrosine hydroxylase is released into the supernatant, suggesting a nonterminal localization of the enzyme in the immature neuron. With age, there is a progressive redistribution of the enzyme from a soluble to a synap-tosomal form (Coyle and Axelrod, 1972b). This regional and subcellular translocation of enzyme activity during development corresponds closely with the outgrowth of

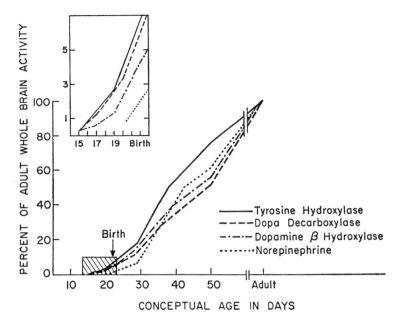

FIGURE 2 The development of total activity per brain of tyrosine hydroxylase, dopa decarboxylase, and dopamine-β-hydroxylase and total content of norepinephrine. Values are expressed in terms of percentage of whole brain activity for the adult rat.

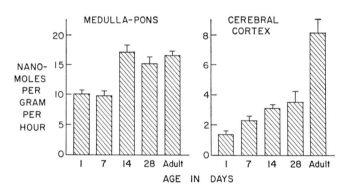

FIGURE 3 Changes in the specific activity of tyrosine hydroxylase in the medulla-pons and cerebral cortex of the rat brain during postnatal development (adapted from Coyle and Axelrod, 1972b).

the catecholaminergic processes as demonstrated by the histofluorescent technique.

An important distinction must be made between the density of innervation and the total amount of terminals in a particular region. This issue is especially relevant in the case of the cerebellum, a region that develops primarily after birth. The specific activities of both tyrosine hydroxylase and dopamine-β-hydroxylase increase about twofold between birth and adulthood in the cerebellum, which suggests only a small change in the density of noradrenergic innervation. However, the weight of the

cerebellum increases 35-fold during the same period. Therefore, the real increase in enzymatic activity and presumably the number of noradrenergic terminals in the cerebellum is in the order of 70-fold.

Functional characteristics of the differentiating neuron

Pharmacologic manipulation of the neurons in vivo can provide information about their functional characteristics. By means of an extremely sensitive enzymatic assay for norepinephrine (Henry and Coyle, in preparation), it has been possible to examine the effect of various pharmacologic agents on the disposition of the endogenous norepinephrine in the brain of the fetal rat at 18 days of gestation. At this stage, the brain has only 0.5% of the adult level of norepinephrine (Table II). Reserpine, a drug that inactivates the aminergic vesicular storage mechanism, causes a profound depletion of endogenous norepineprhine. Thus, vesicular storage plays a major role in protecting the transmitter from enzymatic degradation in the immature neuron. Inhibition of monoamine oxidase, the enzyme responsible for the intraneuronal catabolism of the catecholamines, or the administration of a large dose of the precursor, L-dopa, result in significant increases in the levels of norepinephrine. Inhibition of the initial enzyme in the biosynthetic pathway, tyrosine hydroxylase, with α-methyl-para-tyrosine causes a 50% depletion of norepinephrine

TYROSINE HYDROXYLASE

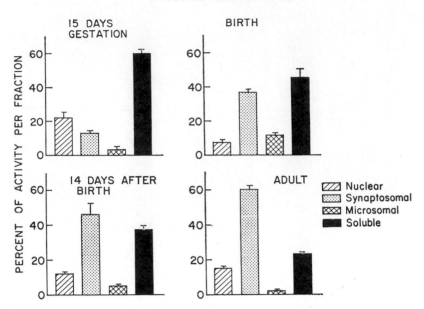

FIGURE 4 Distribution of tyrosine hydroxylase activity in subcellular fractions of rat brain during development. Subcellular fractions prepared from whole rat brains according to the method of Whittaker (1965) were assayed for tyrosine hydroxylase activity. The results are expressed as percentage of total recovered activity in each fraction.

TABLE II

Regulation of norepinephrine in rat brain at 18 days of gestation

Treatment	N	Percent of Control
Control	8	100 ± 8
Inhibition of storage (reserpine)	7	8 ± 2
Inhibition of tyrosine hydroxylase (α-methyl para tyrosine)	8	53 ± 3
Inhibition of monoamine oxidase (pheniprazine)	8	144 ± 8
L-Dopa and peripheral decarboxylase inhibitor	6	159 ± 22

Pregnant female rats were treated with drugs 3 to 5 hr prior to the removal of fetal brains. Norepinephrine content of control brains was 41 ± 4 picogram/milligram or 3.8 ± 0.4 nanogram/brain (Henry and Coyle, in preparation).

3 hours after treatment. This indicates a half-life for norepinephrine of approximately 3 hours, which is quite similar to that observed in the adult brain. In essence, the relative effects of these agents on the transmitter levels of the immature neurons are practically indistinguishable from those known to occur in the adult brain.

The fact that inhibition of tyrosine hydroxylase in the fetal rat causes a significant depletion of norepinephrine in the brain suggests that the central catecholaminergic neurons may be releasing neurotransmitter well before the completion of morphologic differentiation. The following studies lend support to this viewpoint: Amphet-amine induces hyperactive and stereotypic behavior by releasing presynaptic catecholamines resulting in overstimulation of the postsynaptic receptors. Hyperactive and stereotypic behavior can be induced by amphetamine in the neonatal rat; in contrast, the inhibitory effect of the cholinergic neurons becomes operational 3 to 4 weeks later (Fibiger et al., 1970; McGeer et al., 1971). As early as 8 days after birth, there is a diurnal rhythm in the levels of endogenous norepinephrine in the rat brain with a peak in periods of darkness, which indicates a cyclic variation in the turnover of norepinephrine (Asano, 1971). Thus, the central catecholaminergic neurons appear to be physiologically functional well before they have become anatomically and biochemically fully developed.

Possible factors controlling development

Although there is currently little information concerning the factors that control the differentiation of the central catecholaminergic neurons, recent findings suggest a number of possibilities that merit further investigation. In the ontogenesis of the peripheral sympathetic ganglia of the chick, the neural crest cells give rise to catecholaminergic ganglioblasts in response to conditions imposed while the cells migrate ventrally; if this relationship is disturbed, the sympathoblasts fail to develop (Cohen, 1972). Similar intercellular interactions may occur when the nascent locus coeruleus migrates from its origin in the tegmentum. The biochemical development of the superior cervical ganglion of the mouse is, at least in part, controlled by the preganglionic cholinergic innervation (Black et al., 1971). In light of the developmental changes in the cholinesterase staining of the locus coeruleus (Maeda and Gerebtzoff, 1969), this mode of regulation appears tenable for certain central catecholaminergic neurons. An insulin-like hormone, nerve growth factor, plays an important, although poorly defined, role in the differentiation of the peripheral sympathetic neurons (Frazier et al., 1972). Since nerve growth factor stimulates axonal sprouting of sectioned central noradrenergic neurons (Björklund and Stenevi, 1972), it may also act as a trophic factor in the developing central nervous system.

Summary

Histologic and autoradiographic studies indicate that the major nuclei of the catecholaminergic neurons are formed well before birth in the rat brain. The outgrowth of the catecholaminergic axons and terminals occurs primarily during the 4 weeks after birth. During this period, there is a marked increase in the whole brain in the activity of the specific uptake mechanism for norepinephrine, a process associated with nerve terminals, as well as in the activity of the enzymes involved in the biosynthesis of catecholamines. Concurrently, there is a centrifugal movement of these enzymes from the regions of the brain that contain the cell bodies to the regions that receive the terminals. Indirect evidence suggests that the neurons may be physiologically functional well before the completion of morphologic and biochemical differentiation.

REFERENCES

AGRAWAL, H. C., S. N. GLISSON, and W. A. HIMWICH, 1968. Changes in monoamines of rat brain during postnatal ontogeny. *Biochim. Biophys. Acta* 130:511–513.

ALTMAN, J., 1966. Autoradiographic and histological studies of postnatal neurogenesis. I. A longitudinal investigation of the kinetics, migration and transformation of cells incorporating tritiated thymidine in neonatal rats, with special reference to postnatal neurogenesis in some brain regions. *J. Comp. Neurol.* 126:337–390.

ASANO, T., 1971. The maturation of the circadian rhythm of brain norepinephrine and 5-hydroxytryptamine in the rat. *Life Sci.* 10:883–894.

AXELROD, J., 1965. The metabolism, storage, and release of catecholamines. *Recent Prog. Hormone Res.* 21:597–622.

BJÖRKLUND, A., and V. STENEVI, 1972. Nerve growth factor: stimulation of regenerative growth of central noradrenergic neurons. *Science* 175:1251–1253.

BLACK, I. B., I. A. HENDRY, and L. L. IVERSEN, 1971. Transsynaptic regulation of growth and development of adrenergic neurons in a mouse sympathetic ganglion. *Brain Res.* 34:229–240.

BREESE, G. R., and T. D. TRAYLOR, 1972. Developmental characteristics of brain catecholamines and tyrosine hydroxylase in the rat: effects of 6-hydroxydopamine. *Brit. J. Pharmacol.* 44:210–222.

COHEN, A. M., 1972. Factors directing the expression of sympathetic nerve traits in cells of neural crest origin. *J. Exp. Zool.* 179:167–182.

COYLE, J. T., 1972. Tyrosine hydroxylase in rat brain: cofactor requirements, regional and subcellular distribution. *Biochem. Pharmacol.* 21:1935–1944.

COYLE, J. T., and J. AXELROD, 1971. Development of the uptake and storage of L-[³H]norepinephrine in the rat brain. *J. Neurochem.* 18:2061–2075.

COYLE, J. T., and J. AXELROD, 1972a. Dopamine-β-hydroxylase in the rat brain: developmental characteristics. *J. Neurochem.* 19:449–459.

COYLE, J. T., and J. AXELROD, 1972b. Tyrosine hydroxylase in rat brain: developmental characteristics. *J. Neurochem.* 19:1117–1123.

COYLE, J. T., and S. H. SNYDER, 1969. Catecholamine uptake by synaptosomes in homogenates of rat brain: stereospecificity in different areas. *J. Pharmacol. Exp. Ther.* 170:221–231.

DAVIS, J. M., F. K. GOODWIN, W. E. BUNNEY, D. L. MURPHY, and R. W. COLBURN, 1967. Effects of ions on uptake of norepinephrine by synaptosomes. *Pharmacologist* 9:184.

FALCK, B., and N.-A. HILLARP, 1959. On the cellular localization of catecholamines in the brain. *Acta Anat.* 38:277–279.

FIBIGER, H. C., L. D. LYTLE, and B. A. CAMPBELL, 1970. Cholinergic modulation of adrenergic arousal in the developing rat. *J. Comp. Phys. Psych.* 72:384–389.

FRAZIER, W. A., R. H. ANGELETTI, and R. A. BRADSHAW, 1972. Nerve growth factor and insulin. *Science* 176:482–488.

FUXE, K., T. HOKFELT, and U. UNGERSTEDT, 1970. Morphological and functional aspects of central monoamine neurons. *Inter. Rev. Neurobiol.* 13:93–126.

GLOWINSKI, J., J. AXELROD, I. J. KOPIN, and R. J. WURTMAN, 1964. Physiological disposition of ³H-norepinephrine in the developing rat. *J. Pharmacol. Exp. Ther.* 146:48–53.

HORN, A. S., J. T. COYLE, and S. H. SNYDER, 1971. Catecholamine uptake by synaptosomes from rat brain. Structure-activity relationships of drugs with differential effects on dopamine and norepinephrine neurons. *Molec. Pharmacol.* 7:66–80.

JACOBSON, M., 1970. *Developmental Neurobiology*. New York: Holt, Rinehart and Winston.

LAMPRECHT, F., and J. T. COYLE, 1972. Dopa decarboxylase in the developing rat brain. *Brain Res.* 21:503–506.

LOIZOU, L. A., 1971. The postnatal development of monoamine-containing structures in the hypothalamo-hypophyseal system of the albino rat. *Z. Zellforsch.* 114:234–253.

LOIZOU, L. A., 1972. The postnatal ontogeny of monoamine-containing neurons in the central nervous system of the albino rat. *Brain Res.* 40:395–418.

LOIZOU, L. A., and P. SALT, 1970. Regional changes in monoamines of the rat brain during postnatal development. *Brain Res.* 20:467–470.

MAEDA, T., and A. DRESSE, 1969. Recherches sur le développement du locus coeruleus. 1. Etude des catécholamines au microscope de fluorescence. *Acta Neurol. Belg.* 69:5–10.

MAEDA, T., and M. A. GEREBTZOFF, 1969. Recherches sur le développement du locus coeruleus. 2. Etude histoenzymologique. *Acta Neurol. Belg.* 69:11–19.

McGeer, E. G., H. C. Fibiger, and V. Wickson, 1971. Differential development of caudate enzymes in the neonatal rat. *Brain Res.* 32:433–440.

Molinoff, P., and J. Axelrod, 1971. Biochemistry of catecholamines. *Ann. Rev. Biochem.* 40:465–500.

Sachs, C., J. deChamplain, T. Malmfors, and L. Olson, 1970. The postnatal development of noradrenaline uptake in the adrenergic nerves of different tissues from the rat. *Europ. J. Pharmacol.* 9:67–79.

Smith, G. C., and R. W. Simpson, 1970. Monoamine fluorescence in the median eminence of foetal, neonatal and adult rats. *Z. Zellforsch.* 104:541–556.

Snyder, S. H., and J. T. Coyle, 1969. Regional differences in ^{3}H-norepinephrine and ^{3}H-dopamine uptake into rat brain homogenates. *J. Pharmacol. Exp. Ther.* 165:78–86.

Tissari, A. H., P. S. Schönhöfer, D. F. Bogdanski, and B. B. Brodie, 1969. Mechanism of biogenic amine transport II. Relationship between sodium and mechanism of ouabain blockade of the accumulation of serotonin and norepinephrine by synaptosomes. *Molec. Pharmacol.* 5:593–604.

Ungerstedt, U., 1971. Stereotaxic mapping of the monoamine pathways in the rat brain. *Acta Physiol. Scand.* Suppl. 367:1–48.

Uretsky, N. J., and L. L. Iversen, 1970. Effects of 6-hydroxydopamine on catecholamine containing neurons in the rat brain. *J. Neurochem.* 17:269–278.

White, T. D., and P. Keen, 1970. The role of internal and external Na$^+$ and K$^+$ on the uptake of [^{3}H] noradrenalin by synaptosomes prepared from rat brain. *Biochim. Biophys. Acta* 196:285–295.

Whittaker, V. P., 1965. The application of subcellular fractionation techniques to the study of brain function. *Progr. Biophys. Molec. Biol.* 15:39–96.

78 Changes in Enzyme Activities in Neural Tissues with Maturation and Development of the Nervous System

LOUIS SOKOLOFF

ABSTRACT   The brain of most mammalian species is largely undeveloped at birth and undergoes a major portion of its morphological, biochemical, and functional maturation in early postnatal life. A number of enzyme activities related to energy metabolism show characteristic developmental profiles during this period of life. One mitochondrial enzyme, D-β-hydroxybutyrate dehydrogenase (HOBDH), which is involved in ketone body utilization, exhibits a particularly unique pattern of postnatal development in the brain. Along with other mitochondrial enzymes of energy metabolism, it is low at birth and rises during the nursing period to a maximum level. In contrast to the other enzymes, however, that maintain their maximum activities through adult life, HOBDH begins to decline after weaning, until it falls to relatively low levels in adulthood. A similar pattern is exhibited by 3-ketoacid transferase, the second enzyme in the pathway of ketone body utilization. The changes in the activities of these two enzymes follow the level of ketosis in the blood and allow the brain, which normally utilizes glucose almost exclusively as its substrate, to adapt its substrate utilization to the availability of ketone bodies in the blood. The changes in the level of HOBDH in brain during development appear to be, at least in part, regulated by nutritional and hormonal factors.

THE MULTIPLICITY of biochemical processes proceeding in the tissues at all times necessitates the operation of control mechanisms to coordinate, adjust, or adapt the pattern of metabolic activities to the continually changing needs of the cell. The central theme of all the preceding presentations in this part has been that such control is achieved mainly by regulation of the levels or behavior of protein molecules, particularly those with enzyme functions. Indeed, it is a truism that, although the information for life is encoded in the molecular structure of nucleic acids, the translation of this information into actual life processes is accomplished only through the mediation of proteins. Proteins are major constituents of all the structural elements and comprise almost the entire catalytic apparatus of the cell. As emphasized by Schimke, earlier in this part, all proteins, structural, functional, and enzymatic, are in a constant state of turnover. Their turnover rates are markedly individualized and diverse and vary with the physiological state and special circumstances of the cell. The amount of any given protein present is determined by the net balance of its rates of synthesis and degradation. In almost all cases that have been examined, proteins are synthesized at a constant rate and degraded exponentially; the steady-state level of each protein is, therefore, determined by the ratio of its zero-order rate of synthesis to its first-order rate constant of degradation (Schimke, 1969). The level of any protein can, therefore, be altered by modification of its rate of synthesis, its rate constant of degradation, or both. Since biochemical reactions are generally proportional to the concentrations of the enzymatic proteins that catalyze them, control of enzyme levels through adjustment of their rates of synthesis or degradation constitutes one of the most important mechanisms of biochemical regulation in the tissues.

Enzymes are macromolecular catalysts and as such facilitate the chemical interconversions of the many small molecules that comprise the substance of metabolism. The rates of the reactions that they catalyze are determined not merely by the concentrations of the enzymes and their substrates but also by their catalytic efficiency and effectiveness. It has become apparent in recent years that enzymes are not simply passive catalysts, like iron filings or platinum black, but are dynamic elements with readily modifiable, three-dimensional structures that markedly influence their binding affinities for the substrates and their catalytic efficiency. The conformation and, therefore, catalytic activity of an enzyme molecule can vary with the nature of the intracellular environment in which it finds itself. In addition to the general environmental conditions, such as temperature, pH, and ionic strength, there are specific mechanisms for conformational control. These usually involve the binding of small molecules to specific binding sites in the enzyme. It has long been known that enzymes have specific binding sites

LOUIS SOKOLOFF   National Institute of Mental Health, Bethesda, Maryland

for their substrates; the concentrations of the substrates regulate the enzyme activity in relation to the degree of saturation of these binding sites. This phenomenon is quantitatively described in the Michaelis-Menten relationship. However, for many enzymes, there are other binding sites, specific either for substrate or other small molecules, that are not catalytic but regulatory; the binding of the specific ligand to these sites does not serve the catalytic process directly but results in modification of the enzyme's conformation so as to alter its affinity for its substrate or its catalytic efficiency. Catalytic activity may be positively or negatively affected. The conformational change may be relatively simple and involve only a spatial reorientation of an otherwise unchanged species of protein, or, in the case of multimeric enzymes, there may be a dissociation of the enzyme into its component subunits or the converse. The nature and mechanisms of allosteric regulation of enzyme activity have been comprehensively reviewed by Koshland (1970). The essential point for consideration at present is that biochemical regulation can be achieved not only by changes in the levels of enzymes but also by control of the catalytic activity of already existing enzyme molecules.

These basic mechanisms of biochemical regulation operate in all tissues including those of the central nervous system. Their influences permeate throughout almost all biochemical processes and are manifested in the brain throughout the entire life span. Previous notions that the fully mature brain is biochemically fixed, rigid, and inflexible have yielded to accumulating evidence that it retains the capability to undergo metabolic adaptations to changing needs and circumstances and enjoys biochemical as well as structural and functional plasticity (McIlwain, 1971). It has long been obvious, of course, that the orderly, coordinated sequence of changes that characterize the postnatal development of the brain is only the expression of an exquisitely organized, controlled, and regulated program of biochemical events. It is probably fair to say that in no case have the relative contributions of the various potential mechanisms for biochemical regulation to the changes seen in brain been sorted out and definitely established, but their operation and their value to the maintenance of the structural and functional integrity of the organ are clearly discernible. It is beyond the scope of this book to cover comprehensively the variety of manifestations of biochemical regulation in the nervous system. Axelrod and Coyle, respectively, in previous chapters, have presented examples in the area of neurotransmitter biochemistry. In the remainder of this chapter, we shall concern ourselves with other examples, particularly in relation to the processes of energy metabolism and their regulation during postnatal maturation of the brain.

General features of postnatal maturation of the brain

The brain of most mammalian species, including man, is largely underdeveloped at birth and undergoes a major portion of its morphological, functional, and biochemical maturation in early postnatal life. Postnatal maturation does not merely represent growth but continued differentiation. The changes that occur are so extensive that essentially a transformation into a phenotypically different organ is achieved. The process is comparable to that of amphibian metamorphosis in which the tadpole is ultimately converted into a frog. Interestingly enough, both the metamorphosis of the frog and the postnatal maturation of the brain are dependent on the presence of thyroid hormones, and in both cases sensitivity to these hormones is lost after complete maturation is achieved.

The postnatal morphogenesis and functional development of mammalian brain have been most extensively studied in the rat, mainly by Eayrs and his associates (Eayrs, 1964). In this species, maturation is achieved in the first 5 to 6 weeks of life after birth. Normally, the brain increases in size approximately 4- to 5-fold during this period, but the most dramatic changes are observed in the microscopic appearance of the tissue. The neurons increase in size, mainly because of an increase in perikaryonal cytoplasm. There is a marked proliferation of axonal and dendritic processes, and the density of axodendritic connections is strikingly increased, particularly in the neuropil. Myelin is laid down around the axons, and the nerve cell bodies become less densely packed because the spaces between them become filled with axonal and dendritic processes and myelin.

A number of physiological changes occur in parallel with the morphogenetic development. Innate reflexes, such as the startle, righting, and placing reactions that are absent at birth, appear at various times during the maturational period. Learning behavior, manifested by conditioned reflex activity, appears. The electroencephalogram, which is slow and low in amplitude at birth, speeds up and increases in voltage. Evoked potential responses make their appearance and then develop shorter latencies, higher amplitudes, and shorter duration.

The morphological and functional changes are accompanied by extensive biochemical changes (Sokoloff, 1971). Some of these changes reflect the altered morphological state of the tissues, i.e., the less dense packing of cells and the growth and expansion of the perikaryonal cytoplasm. Others result from continued postnatal differentiation. Protein and RNA concentrations and contents increase as the water content decreases and the perikaryonal cytoplasm enlarges. Cell proliferation proceeds for only a short time after birth, but thereafter the nuclei do not increase very significantly in size and amount; the DNA concentration, therefore, falls as the constant amount of DNA becomes diluted by the increasing cytoplasmic mass and myelin. Lipid composition is radically altered and reflects to a large extent the deposition of myelin.

That biochemical regulatory mechanisms underlie the maturational process is evidenced by the extensive realignment of enzyme patterns that occurs during this period. An impressive number of enzyme activities are altered during maturation. Table I presents only a partial list of those enzymes that have been found to change in activity during maturation of the brain. Increasing activity is the usual pattern though not common to all enzymes, indicating that the developmental process is selective and specific. A few enzyme activities decline again after the initial increase; others do not change at all; and some decline during the maturational process.

As previously noted, the postnatal development of the brain is largely controlled by the thyroid hormones. If the animal is thyroidectomized at birth, then many of the morphological, functional, and biochemical changes associated with maturation are retarded or prevented, and the brain never develops normal structure and function. Thyroid hormone replacement therapy allows these changes to progress normally, provided that it is initiated early enough. In the rat the critical period appears to be the first 24 days of life (Eayrs, 1964). The earlier the initiation of therapy, the more complete and normal is the development of the brain, but if begun after 24 days of age, thyroid hormone administration has little beneficial effect on brain development. Indeed, it is well known that in the adult, normal or cretinous, the brain is incapable of direct response to thyroid hormones (Eayrs, 1964; Sokoloff, 1971).

The mechanism of action of thyroid hormones remains undefined, but almost certainly their role in the development of the brain reflects to a large extent their ability to stimulate protein and RNA synthesis (Sokoloff and Kaufman, 1961; Tata and Widnell, 1966). Thyroid hormones stimulate protein synthesis in brain only

TABLE I

Some enzyme activities reported to change during postnatal maturation of the brain

Enzyme	Direction of Change	References
Hexokinase	+	Schwark et al., 1972
Phosphofructokinase	+	Schwark et al., 1972
Aldolase	+	Hamburgh and Flexner, 1957; Swaiman et al., 1970
Pyruvic kinase	+	Schwark et al., 1972
Creatine kinase	+	Swaiman et al., 1970
NADP$^+$-isocitrate dehydrogenase	+	Murthy and Rappoport, 1963; Swaiman et al., 1970
Succinic dehydrogenase	+	Hamburgh and Flexner, 1957; Garcia Argiz et al., 1967
Cytochrome oxidase	+	Hamburgh and Flexner, 1957; Klee and Sokoloff, 1967
D-β-Hydroxybutyrate dehydrogenase	+ followed by −	Klee and Sokoloff, 1967; Krebs et al., 1971
3-Ketoacid transferase	+ followed by −	Krebs et al., 1971
Alanine aminotransferase	+	Balzsá et al., 1968
Aspartate aminotransferase	+	Pasquini et al., 1967
Aspartate carbamoyltransferase	−	Pasquini et al., 1967
Fatty acid synthetase	−	Volpe and Kishimoto, 1972
Cholinesterase	+	Hamburgh and Flexner, 1957
Tyrosine hydroxylase	+	Coyle, 1973
Dopa decarboxylase	+	Coyle, 1973
Dopamine-β-hydroxylase	+	Coyle, 1973
Glutamic decarboxylase	+	Garcia Argiz et al., 1967; Balzsá et al., 1968
GABA transaminase	+	Garcia Argiz et al., 1967
Na$^+$, K$^+$-ATPase	+	Garcia Argiz et al., 1967; Valcana and Timiras, 1969
Mg^{2+}-ATPase	+	Garcia Argiz et al., 1967; Valcana and Timiras, 1969
Adenyl cyclase	+	Schmidt and Robison, 1972
cAMP phosphodiesterase	+	Schmidt and Robison, 1972

during the period of its growth and development; they have no such effect in fully mature brain (Michels et al., 1963; Klee and Sokoloff, 1964; Sokoloff, 1971). Their effects on RNA and protein synthesis are themselves, however, secondary to an earlier metabolic change arising from a prior interaction between the hormones and the mitochondrial components of the cell (Sokoloff and Kaufman, 1961; Sokoloff et al., 1968; Sokoloff, 1968). The mitochondria appear to contain the locus of the primary biochemical action of the thyroid hormones (Sokoloff, 1968). Evidence has accumulated that indicates that mitochondria of immature brain share with mitochondria of other thyroxine-sensitive tissues the ability to participate with thyroxine in this prerequisite primary reaction, but in the course of development the mitochrondria of brain appear to lose this capacity (Klee and Sokoloff, 1964). It is a change in the properties of the mitochondria that appears to be the basis for the difference in thyroxine-sensitivities of immature and mature brain (Sokoloff, 1970).

Energy metabolism of the brain

CEREBRAL OXYGEN CONSUMPTION   The transformation of the relatively poorly differentiated organ at birth into the enormously complex and elegantly performing apparatus that constitutes the mature brain is accompanied by a marked increase in its energy consumption. Cerebral oxygen consumption is low at birth and rises in a typical S-shaped growth-type of curve until it finally levels off in the mature brain at a level more than double that present at birth (Figure 1) (Fazekas et al., 1951). The thyroid hormones, which promote the postnatal maturation of the brain and may in small excess accelerate the process, shift the development pattern for oxidative metabolism to the left, causing an earlier and steeper rise and a more rapid achievement of the normal mature level; once the mature level is reached, however, the continued administration of thyroxine has no further effect on cerebral oxygen consumption (Figure 1).

The rate of oxygen utilization by the mature brain is among the highest of all tissues of the body. In normal conscious adult man, cerebral oxygen consumption equals approximately 3.5 ml per 100 g of tissue per minute or about 800 ml per minute for the entire brain, an energy expenditure of close to 20 watts (Kety, 1957; Sokoloff, 1972). The magnitude of this rate can be appreciated when one considers that the brain of adult man, which comprises only 2% of total body weight, alone consumes almost 20% of the total body basal oxygen consumption. The nature of the processes that consume this enormous amount of energy is not entirely clear. The brain does not do mechanical work like cardiac or skeletal muscle or

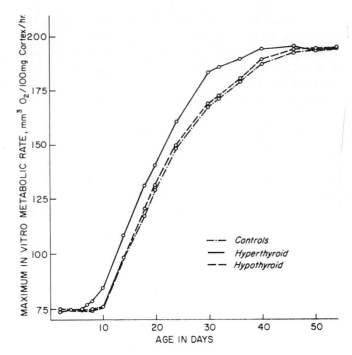

FIGURE 1   Changes in cerebral cortical oxygen consumption in rat brain during postnatal maturation and the influence of thyroid status on the developmental process. From Fazekas et al. (1951).

osmotic work like kidney. The mature brain does not have the complex energy-consuming metabolic and biosynthetic functions of liver. Clearly, the functions of nervous tissues are primarily excitation and conduction, and these are reflected in the unceasing electrical activity of the brain. The electrical energy is ultimately derived from chemical processes, and it is likely that most of the brain's energy consumption is utilized for active transport of ions to sustain and restore the membrane potentials discharged during the processes of excitation and conduction.

NORMAL SUBSTRATES OF CEREBRAL ENERGY METABOLISM   The rapid rate of energy metabolism of the mature brain requires the continuous replenishment of its nutrient supply by the blood flow. In the mature brain under normal conditions, it appears that glucose is the almost exclusive substrate for its oxidative metabolism (Kety, 1957; Sokoloff, 1972). The rates of cerebral oxygen and glucose consumption in conscious, young adult man under physiological conditions are approximately 156 μmole and 31 μmole per 100 g of tissue per minute, respectively (Sokoloff, 1972). The CO_2 production is essentially equivalent to O_2 consumption, and the cerebral respiratory quotient (RQ) is, therefore, close to unity (Kety, 1957). An RQ of 1.0 indicates that the O_2 is consumed and the CO_2 produced in the oxidation of carbohydrate.

The O_2 consumption is equivalent to the amount required for the complete oxidation of 26 μmole of glucose to CO_2 and water. Since the measured steady-state rate of glucose consumption is 31 μmole/100 g/min, there is more than enough glucose consumed to account for all of the O_2 consumption. The excess glucose is probably distributed in various intermediates of carbohydrate metabolism, each of which is normally released into the blood in such small amounts as to be virtually undetectable in the arteriovenous differences. On the basis of cerebral arteriovenous differences, there appears to be under normal circumstances no steady-state uptake of any potential substrate for oxidative metabolism other than glucose in more than trivial amounts. The combination of a cerebral RQ of unity, an almost stoichiometric relationship between oxygen and glucose consumption, and the absence of any significant arteriovenous difference for any other potential substrate constitutes strong evidence that the adult brain derives its energy under normal conditions almost solely from the aerobic oxidation of glucose.

The fact that the brain normally consumes only glucose for energy does not distinguish between preferential and obligatory utilization of glucose. Most tissues are facultative in their choice of substrates and use them interchangeably in accordance with their availability from the blood. It was once believed that the brain did not enjoy such flexibility and was rigidly restricted to glucose for its substrate (Kety, 1957). This conviction was based on the obvious impairments of cerebral function and energy metabolism that occur when the brain is deprived of glucose and the inability of other potential substrates to substitute for glucose to prevent or reverse these effects.

Hypoglycemia is associated with changes in mental state, ranging from mild, subjective sensory disturbances to convulsions, coma, and death, depending on both the degree and duration of the hypoglycemia. The behavioral effects are accompanied by abnormalities in the electroencephalogram and cerebral metabolic rate, and cerebral oxygen consumption declines in proportion to the depression of the level of consciousness (Kety, 1957; Sokoloff, 1972). The cerebral effects of hypoglycemia are independent of the mode of its induction and are similar whether the hypoglycemia is caused by insulin or hepatic insufficiency (Kety, 1957). All the effects are prevented or rapidly reversed by glucose administration, provided that they have not been allowed to persist until irreversible changes have occurred. Many potential substrates have been tested for their ability to substitute for glucose and prevent or reverse the cerebral effects of hypoglycemia. Few have been found effective, and of these all but one appear to operate by raising the blood glucose level rather than by serving directly as a substrate for cerebral

metabolism (Sokoloff, 1972). Only mannose can be used directly or sufficiently rapidly by the brain to replace glucose and restore normal cerebral function, even in the complete absence of glucose (Sloviter and Kamimoto, 1970). It traverses the blood-brain barrier and is converted to mannose-6-phosphate by hexokinase, which phosphorylates mannose as effectively as glucose. Mannose-6-phosphate is then isomerized to fructose-6-phosphate by phosphomannoseisomerase, which is quite active in brain tissue (Kizer and McCoy, 1960). It is through this pathway that mannose replaces glucose and maintains an adequate glycolytic flux. Maltose produces arousal from insulin coma, but only by raising the blood glucose level through its conversion to glucose by maltase activity in blood and other tissues (Sloviter and Kamimoto, 1970; Sokoloff, 1972). Epinephrine is quite effective, but by stimulating hepatic glycogenolysis and consequently elevating blood glucose concentration. Glutamate, arginine, p-aminobenzoate, and succinate are also occasionally effective, but probably through the release of epinephrine which in turn raises the blood glucose level (Sokoloff, 1972). Many substances tested and found ineffective are compounds normally formed and utilized within the brain and are normal intermediates in its intermediary metabolism. Lactate, pyruvate, fructose-diphosphate, acetate, β-hydroxybutyrate, and acetoacetate are such examples (Kety, 1957; Sokoloff, 1972). These can all be utilized by brain slices, homogenates, or cell-free fractions, and the enzymes for their metabolism are present in brain. In some cases, as for example, glycerol (Sloviter et al., 1966, 1967) or ethanol (Raskin and Sokoloff, 1970), the enzymes may not be present in sufficient amounts to sustain the high rate of utilization required by the brain. In other cases, for example, acetoacetate and D-β-hydroxybutyrate (Krebs et al., 1971), the enzymes are adequate, but the substrate is insufficient because of inadequate blood level or restricted transport through the blood-brain barrier. The brain requires continuous delivery of substrate by the blood, and there is thus far no known endogenous substance present in blood in sufficient amounts to substitute fully for glucose as a substrate for cerebral energy metabolism. The one substance, mannose, which has been found to be a suitable replacement, is not a significant constituent of blood. Glucose must, therefore, be considered essential for the physiological and biochemical functioning of the nervous system.

Utilization of ketone bodies by the brain

The impressive mass of evidence accumulated over several decades pointing to glucose as the essential and almost exclusive substrate for cerebral energy metabolism led to the long-held belief that the brain was inflexible and uncompromising in its choice of nutrients. It is known, of course, that the brain is the mediator of a number of reflexes that regulate somatic functions to modify the distribution of blood flow and the composition of the blood in a manner advantageous to itself. The carotid sinus reflex, for example, clearly serves to maintain cerebral blood flow even at the expense of other less vital organs. Hypoglycemia and anoxia initiate sympathetic outflow and epinephrine release, which in turn maintain or enhance cerebral bood flow and the glucose content of the blood. It was, therefore, believed that the brain responds to extenuating circumstances not by intrinsic adaptation but by adjusting bodily functions to its needs through its reflex control of somatic physiological and metabolic processes. Recent evidence has forced a revision of this belief. It is now apparent that the brain, like most other tissues, is capable of metabolic adaptation (McIlwain, 1971), even with respect to its choice of substrates for energy metabolism (Krebs et al., 1971). Although there is still no convincing evidence that it can

function in the complete absence of glucose, the brain is now known to substitute other substrates, at least in part, for glucose in conditions in which these other substrates become more available and the glucose supply limited. The ketone bodies, D(-)-β-hydroxybutyrate and acetoacetate, are such substances, and the brain utilizes them in a variety of ketotic states in proportion, more or less, to their concentrations in blood. These ketone bodies, which are interconvertible into one another, are by-products of the metabolism of fatty acids (Figure 2). Their levels in blood are normally very low, but their rates of production are enhanced when fatty acid metabolism is increased, probably because the decrease in free CoA and the rise in acetyl CoA levels associated with accelerated fatty acid degradation shift the sequence of metabolic reactions in the direction of their formation (Figure 2). They are produced mainly in the liver, whichcan make them but cannot utilize them, because it lacks succinyl CoA: 3-oxoacid CoA-transferase (Mahler, 1953), the enzyme that converts free acetoacetate to acetoacetyl CoA, an essential step in the utilization of ketone bodies (Figure 2).

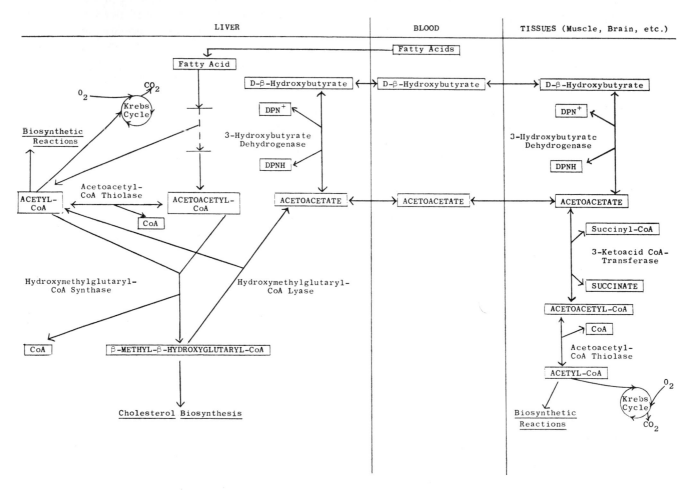

FIGURE 2   Metabolic pathways and enzymes of ketone body formation and utilization.

As the liver metabolizes fatty acids, it produces ketone bodies that are released into the blood and circulated for utilization by other tissues. Conditions in which fatty acid metabolism by liver is increased lead to elevated levels of ketone bodies in the blood.

STARVATION Evidence that the brain can extract and utilize ketone bodies from the blood was first obtained by Owen et al. (1967) in human subjects being treated for obesity by complete and prolonged starvation. The total body carbohydrate stores of the patients at the onset of the fast were estimated to be between 150 and 300 g, and metabolic balance studies indicated that their maximum rate of gluconeogenesis from protein and glycerol during the fast could not have exceeded 33 g per day. Since cerebral glucose utilization is normally approximately 110 g per day, the total body's actual and potential glucose reserves were clearly insufficient to satisfy the brain's normal glucose requirements for more than a few days. Nevertheless, after 6 to 7 weeks of starvation, these patients exhibited none of the usual signs of cerebral dysfunction indicative of cerebral glucose insufficiency. Their level of consciousness, EEG, and performance on psychometric testing remained within normal limits, evidence of adequate rates of cerebral energy metabolism despite the insufficiency of glucose to support it. Clearly the brain must have turned to the utilization of other substrates. The enigma of the nature of the substrate was resolved by examination of the cerebral arteriovenous differences (Table II). As expected, the glucose arteriovenous difference had declined to about half of its normal level, and of this, about half was accounted for by

lactate and pyruvate release and did not, therefore, support oxidative metabolism. Net lactate and pyruvate production by brain is normally barely, if at all, detectable (Kety, 1957). The arteriovenous oxygen difference was normal, and glucose utilization, after correction for lactate and pyruvate recovery, accounted for only about 30% of it. Most of the remainder of the oxygen consumption was apparently supported by the ketone bodies, D-β-hydroxybutyrate and acetoacetate, which accounted for 52% and 8%, respectively. The remainder of the oxygen consumption could be accounted for by the uptake of α-amino nitrogen-containing material, presumably amino acids. No evidence of fatty acid utilization was observed. An ancillary finding of interest was the remarkably low cerebral RQ of 0.63. None of the substrates being oxidized could lead to so low an RQ, and it suggests that there is considerable CO_2 fixation and/or gluconeogenesis going on within the brain during starvation.

The changes in cerebral metabolism during starvation demonstrate that the mature brain is capable of adapting to altered nutrient supply. Normally it relies on glucose almost exclusively as the substrate for its energy metabolism, but availability of glucose becomes limited during starvation. On the other hand, starvation causes mobilization of fatty acids from the fat stores in adipose tissue, and the accelerated metabolism of fatty acids by liver leads in turn to enhanced formation of ketone bodies and the elevation of their levels in blood. The brain adapts to the altered conditions by switching in part from consumption of the endangered glucose supply to utilization of the ketone bodies made available by the alterations in body metabolism.

TABLE II

*Substrates of cerebral energy metabolism in human adults during prolonged starvation**

| | | Arteriovenous Differences | | |
| | | | Calculated O$_2$ Equivalent | |
Substance	Arterial Concentration	In Measured Units	(vol. %)	(μmole/ 100 ml)
Total glucose	449 μmole/100 ml (81 mg %)	26 μmole/100 ml (4.7 mg %)	—	—
Glucose (after correction for lactate and pyruvate)	—	14.5 μmole/100 ml (2.6 mg %)	2.0	88
β-Hydroxybutyrate†	667 μmole/100 ml	34 μmole/100 ml	3.4	153
Acetoacetate†	117 μmole/100 ml	6 μmole/100 ml	0.5	24
α-Amino nitrogen	314 μmole/100 ml	9 μmole/100 ml	0.9	42
Total Calculated O$_2$ Equivalents			6.8	307
Measured Arteriovenous O$_2$ Differences§			6.6	296

*From data of Owen et al. (1967).

†1 mole of β-hydroxybutyrate and acetoacetate equivalent to 4.5 and 4.0 moles of O$_2$, respectively.

§Measured cerebral respiratory quotient found to be 0.63.

INFANCY Like the human infant, the newborn rat is transiently hypoglycemic immediately after birth, and its blood ketone levels are also very low initially, as low as those of the normal fed adult rat (Krebs et al., 1971). With the onset of suckling, however, the blood ketone levels rise as much as 10-fold, and a true ketosis ensues (Figure 3) (Krebs et al., 1971). This ketosis is of nutritional origin; it reflects the ketogenic nature of maternal rat milk that contains 50% of its dry weight as fat (Dymsza et al., 1964). The ketosis persists throughout the suckling period until

approximately 20 to 22 days of age when the rat is weaned onto a normal high-carbohydrate diet; the ketosis then gradually disappears and the blood ketone levels decline to the low levels characteristic of the well-nourished adult rat (Figure 3). During the ketosis of the suckling period, ketone bodies constitute an important source of the brain's substrates, and positive cerebral arteriovenous differences of D-β-hydroxybutyrate and acetoacetate proportional to their concentrations in the blood are observed (Figure 4) (Krebs et al., 1971; Hawkins et al., 1971). Similar evidence for significant cerebral ketone body utilization during normal infancy has been obtained in canine puppies (Spitzer and Weng, 1972) and human infants (Persson et al., 1972). Because the enzymes of ketone utilization in brain are more active in early postnatal life than in the adult, as will be discussed below, the brain of the infant is more efficient than the adult brain in extracting and utilizing ketone bodies from the blood at any given arterial concentration (Figure 4).

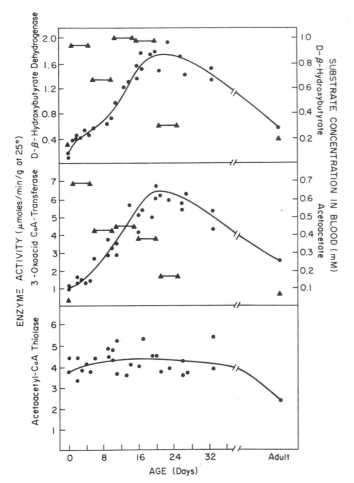

FIGURE 3 Changes in arterial blood levels of ketone bodies and the activities of the enzymes of ketone body utilization in the brain of the rat during postnatal maturation. Arterial ketone concentration, ▲——▲; enzyme activity in brain, ●——●. D-β-hydroxybutyrate dehydrogenase and acetoacetyl-CoA thiolase were assayed in the direction of ketone body utilization. The 3-oxoacid CoA-transferase (3-ketoacid transferase) was assayed in the reverse direction, the rate of which is approximately 5 times faster than that of the forward direction. For comparison of the rates of the enzymes all in the direction of ketone body utilization, the rates of the transferase activity presented in the figure should be divided by 5. (Illustration prepared from the data of Krebs et al., 1971.)

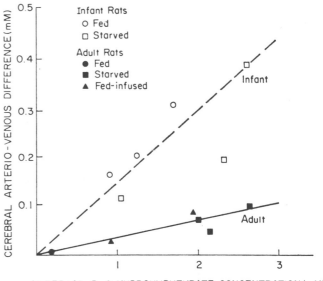

FIGURE 4 Influence of arterial concentrations on cerebral arteriovenous differences of ketone bodies in adult and infant rats. (Illustration composed of combination of figures from report of Krebs et al., 1971.)

MISCELLANEOUS KETOTIC STATES The brain can utilize D-β-hydroxybutyrate and acetoacetate whenever they are available but does not normally do so in adult life because of their low concentrations in the arterial blood. Ketosis of various origins can, therefore, lead to significant utilization of ketones by brain at any period of life. For example, fat feeding or the infusion of D-β-hydroxybutyrate or acetoacetate, which raises the blood levels of one or both ketone bodies, also leads to uptake of the ketones by brain in approximate proportion to their levels in the arterial blood (Figure 4). Whether the same occurs in diabetic ketoacidosis is still uncertain. Studies by Kety and his associates (Kety et al., 1948; Kety, 1957) revealed no significant cerebral arteriovenous difference for total ketones in diabetic acidosis or coma, but more recent studies by Gottstein et al. (in press) provide evidence of some cerebral uptake of ketone bodies in diabetic acidosis. It is possible that the hyperglycemia and reduced cerebral metabolic rate present in diabetic acidosis complicate the unequivocable demonstration of cerebral ketone body utilization.

RELATIONSHIP OF CEREBRAL KETONE UTILIZATION TO DEPENDENCE ON GLUCOSE   The finding that the brain can and does utilize ketone bodies under certain circumstances is not necessarily in conflict with earlier observations that neither D-β-hydroxybutyrate nor acetoacetate can restore normal cerebral function and metabolism in hypoglycemia (Kety, 1957; Sokoloff, 1972). In all the studies demonstrating cerebral ketone utilization, glucose consumption may have been reduced but was still present to an appreciable degree. There is no evidence that ketone bodies can completely replace glucose. Indeed, there is evidence to the contrary; in the perfused rat brain complete replacement of glucose with D-β-hydroxybutyrate results in just as rapid deterioration of cerebral functional and metabolic activities as observed with complete removal of all substrate (Sloviter, personal communication). A possible explanation for the inability of ketone bodies to substitute for glucose completely may be in the pathway of their metabolism. As can be seen in Figure 2, D-β-hydroxybutyrate must first be converted to acetoacetate, and acetoacetate is further metabolized by displacing the succinyl moiety of succinyl CoA to form acetoacetyl CoA. In its usual pathway of metabolism succinyl CoA hydrolysis is normally coupled to GTP synthesis, and GTP has important functions in tissue metabolism, including protein synthesis and gluconeogenesis. Recent findings also suggest that cyclic GMP, which is formed from GTP, may play a role in mediating cholinergic synaptic transmission (Ferrendelli et al., 1970). There must also be sufficient succinyl CoA to sustain continued acetoacetate utilization. Indeed, Itoh and Quastel (1970) have found that the utilization of acetoacetate by brain slices is accelerated by the inclusion of glucose in the incubation medium, probably as a result of increased succinyl CoA levels. It is possible, therefore, that exclusive dependence on the utilization of ketones depletes the succinyl CoA and also the GTP levels in brain and that some glucose utilization is required to maintain them at adequate levels.

Enzymes of ketone utilization in the brain

ENZYMATIC STEPS IN PATHWAY OF KETONE UTILIZATION
The pathways and enzymes of ketone utilization are illustrated in Figure 2. Two enzymes, 3-ketoacid CoA-transferase and acetoacetyl CoA thiolase, are required to metabolize acetoacetate to acetyl CoA, the first intermediate in common with the pathway of glucose metabolism and the one which enters the tricarboxylic acid cycle for ultimate oxidation to CO_2 and water. The utilization of the other ketone body, D-β-hydroxybutyrate (3-hydroxybutyrate) requires one additional enzyme, D-β-hydroxybutyrate dehydrogenase, to oxidize it to acetoacetate which is then metabolized as above. All three reactions are freely reversible.

The three enzymes are widely distributed in mammalian tissues (Lehninger et al., 1960; Krebs et al., 1971; Dierks-Ventling, 1971; Page et al., 1971; Tildon and Sevdalian, 1972) although the D-β-hydroxybutyrate dehydrogenase, which is present in highest amounts in the liver of the rat (Lehninger et al., 1960), is reported to be absent in the liver of ruminants (Nielsen and Fleischer, 1969). Also, the mammalian liver generally lacks the transferase (Mahler, 1953) and is, therefore, entirely incapable of metabolizing ketone bodies. It can, however, produce acetoacetate from acetoacetyl CoA and acetyl CoA that are formed in the course of fatty acid oxidation. Since it can produce them and not utilize them, the liver is the primary source of the ketone bodies in the blood, and its rate of production is enhanced with increasing fatty acid utilization. The D-β-hydroxybutyrate dehydrogenase in liver functions mainly in the reverse direction and converts the acetoacetate produced in the liver to D-β-hydroxybutyrate. Since the redox potential in the liver during fatty acid oxidation and the equilibrium constant at physiological pH probably favor this reverse direction, D-β-hydroxybutyrate is found in blood in severalfold higher concentrations than that of acetoacetate during naturally occurring ketosis (Owen et al., 1967; Krebs, 1971). It is, therefore, D-β-hydroxybutyrate that is predominantly utilized by the brain in ketotic states (Owen et al., 1967; Krebs et al., 1971).

DEVELOPMENTAL CHANGES IN ENZYME LEVELS   The rise in cerebral oxygen consumption during postnatal development of the brain (Figure 1) reflects not only increased energy demand associated with the development of cerebral functional activity but also a concomitant rise in the levels of mitochondrial enzymes of oxidative metabolism. Most of these enzymes begin to rise shortly after birth, continue to rise throughout the period of brain maturation, and eventually level off at the normal adult level when the brain is mature. Some of these enzymes are included in the list in Table I. During the early rising phase the various mitochondrial enzymes increase more or less in proportion to one another, suggesting an increasing content of mitochondria in the developing brain. The proportions are not, however, maintained indefinitely. There are individual differences in the patterns of enzyme development, indicating that the regulation is not merely of the content of mitochondria but is also at the level of the individual enzymes. The enzymes specific to the pathway of ketone body utilization are among those that exhibit unique developmental patterns in brain. Figure 5 compares the postnatal developmental patterns of rat brain cytochrome oxidase, an enzyme of the electron transport chain, and D-β-hydroxybutyrate dehydrogenase, which is associated with the inner mitochondrial membrane and catalyzes the first step in the utilization of ketone bodies. The cytochrome oxidase exhibits the pattern typical of most mitochondrial oxidative enzymes. The D-β-hydroxybutyrate dehydrogenase rises at first proportionately with the cytochrome oxidase, but at 20 to 25 days of age, when the cytochrome oxidase level stabilizes and becomes constant, the dehydrogenase declines again and gradually reverts in adulthood almost to the low levels present at birth (Klee and Sokoloff, 1967; Pull and McIlwain, 1971; Krebs et al., 1971; Dahlquist et al., 1972). The second enzyme of the pathway of ketone body utilization, 3-ketoacid transferase, shows an almost identical pattern of development (Figure 3) (Krebs et al., 1971; Tildon et al., 1971). The third enzyme of the pathway, acetoacetyl CoA thiolase, follows an entirely differ-

ent pattern; it is already maximal at birth, remains constant during the first 30 days of life, and then gradually declines to a level in adulthood about two-thirds that of infancy (Krebs et al., 1971; Dierks-Ventling and Cone, 1971a). It must be noted, however, that in contrast to the dehydrogenase and transferase, which serve only the pathways of ketone body metabolism, the thiolase functions in other metabolic processes as well, e.g. fatty acid oxidation, cholesterol biosynthesis, etc. It is clear that the enzymes, which are unique to the metabolism of ketone bodies, are also uniquely regulated during the postnatal maturation of the brain.

REGULATION OF ENZYMES OF KETONE UTILIZATION IN BRAIN Throughout the life-span, whenever the ketone body concentrations rise in the blood, regardless of cause, the brain switches, at least in part, from almost complete dependence on glucose to the utilization of the ketone bodies in more or less proportion to their levels in the arterial blood (Krebs et al., 1971). In the adult this switch is accomplished without any change in the amounts of the enzymes of ketone utilization in the brain (Krebs, 1971), nor is any required. In the adult rat brain the maximum velocities of all three enzymes are comparable; the thiolase is present in slight excess, and the transferase appears to be slightly rate limiting (Figure 3) (Krebs, 1971). The maximum possible rate of acetyl CoA formation from either D-β-hydroxybutyrate or acetoacetate can be estimated to be about 80 μmole per 100 g of tissue per minute. Since normal cerebral glucose consumption generates approximately 60 μmole of acetyl CoA per 100 g per minute, the enzymes of ketone utilization are present in sufficient amounts in adult brain to satisfy its energy needs from ketone bodies. Maximal rates are rarely if ever achieved, however, because the concentrations of ketone bodies probably never rise high enough to saturate the enzymes. Control of ketone utilization in adult brain is achieved, therefore, merely by the simplest regulatory mechanism of all: substrate concentration and degree of saturation of the enzymes of ketone utilization.

Regulation of cerebral ketone utilization is, however, considerably more complex during postnatal maturation of the brain. Regulation by substrate concentration occurs (Figure 4), but, in addition, there are profound changes in the levels of the D-β-hydroxybutyrate dehydrogenase and the 3-ketoacid transferase (Figure 3) (Klee and Sokoloff, 1967; Krebs et al., 1971). Hormonal influences undoubtedly play a part. Thyroxine, for example, which accelerates the postnatal development of cerebral oxidative metabolism (Figure 1) also advances the developmental pattern of D-β-hydroxybutyrate dehydrogenase (Figure 6) (Grave et al., 1973). This effect of

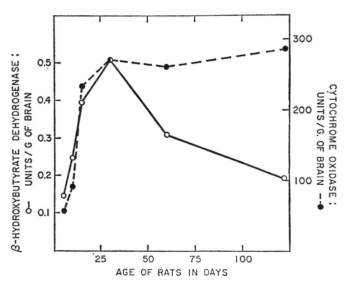

FIGURE 5 Changes in activities of the mitochondrial enzymes, cytochrome oxidase and D-β-hydroxybutyrate dehydrogenase, in rat brain during postnatal maturation. From Klee and Sokoloff (1967).

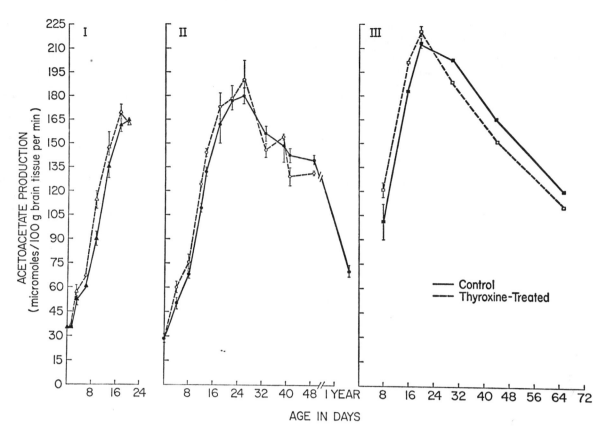

FIGURE 6 Effects of the administration of L-thyroxine from birth on the postnatal development of D-β-hydroxybutyrate dehydrogenase activity in rat brain. Series I, II, and II refer to 3 different series of experiments with different schedules of thyroxine administration. In Series I the experimental animals received 10 μg of sodium L-throxine intraperitoneally on the first day of age and every 48 hr thereafter; control littermates received only the solvent that consisted of 0.1 ml of 0.005 N NaOH–0.45% (w/v) NaCl. At this dosage there was significant impairment of brain and total body growth. In Series II, the same dosage was given but at 72 hr intervals, and no effects on brain and body growth were observed. In Series III, the dosage schedule used in Series I was followed for the first 4 doses and then switched to the dosage schedule of Series II. In all three series, the acceleration of the changes in the enzyme activity caused by thyroxine was statistically significant ($P < 0.05$) on the basis of multivariate profile analysis. From Grave et al. (1973).

thyroid hormones is not specific for the enzymes of ketone utilization; they have similar effects on other oxidative enzymes (Sokoloff, 1971). Of particular interest is the temporal relationship between the blood ketone levels and the enzyme activities in brain. The developmental patterns of the dehydrogenase and transferase in brain closely follow the changes in the concentrations of their substrates, D-β-hydroxybutyrate and acetoacetate, respectively, in blood. With the onset of suckling, for example, the newborn rat develops a marked ketosis as a result of the ketogenic nature of the maternal milk diet (Figure 3). This is followed by a rise in the dehydrogenase and the transferase levels in brain (Figure 3). When in the normal course of events the animals are weaned onto a standard diet at 20 to 25 days of age, the ketosis dissipates, and this is then followed by a decline in the levels of the enzymes in the brain. The enzymes of ketone body utilization in the developing brain appear, therefore, to be closely adjusted to the levels of available substrate. The mechanism of this adjustment is unknown. Whether it reflects induction of enzyme synthesis by the ketosis, protection of the enzymes from degradation by saturation with their substrates (Schimke, 1969), or merely a fortuitous correlation remains to be determined. There is controversial evidence to suggest that ketosis may play a direct role in the mechanism of the enzyme changes in the brain. In one study, maternal starvation and fasting at birth, which cause premature ketosis in the fetus and newborn animal, were found to accelerate the rise of 3-hydroxybutyrate dehydrogenase activity in the brains of fetal and neonatal rats (Thaler, 1972), but in another study no such effects were observed (Dahlquist et al., 1972). Prolongation of ketosis, following weaning by starvation or fat feeding, delays the decline in brain dehydrogenase activity that normally occurs at this time of life (Smith et al., 1969; Pull and McIlwain, 1971), and premature weaning onto a high-carbohydrate diet has been reported to initiate an earlier decline in the enzyme level (Smith et al., 1969). Neither starvation (Krebs et al., 1971; Tildon et al., 1971) nor fat feeding (Krebs et al., 1971; Dierks-Ventling and Cone, 1971b) appear to alter brain transferase activity at any age after birth although a maternal high-fat diet has been reported to increase both the transferase and thiolase activities of fetal rat brain (Dierks-Ventling, 1971).

Recent studies of the properties of the D-β-hydroxybutyrate dehydrogenase suggest an additional factor that may play a role in the changes in the activity of this enzyme during postnatal development of the brain. This enzyme exists in situ firmly bound to the inner mitochondrial membrane. It has recently been solubilized by cholate treatment of mitochondria isolated from weanling rat brain and partially purified by salt fractionation (Fitzgerald et al., 1973). The activity of the partially purified enzyme exhibits an absolute dependence on the presence of phospholipid extracted from the mitochondria. The active phospholipid has been tentatively identified as a lecithin, and some other phospholipids, such as lysolecithin, phosphatidylinositol and cardiolipin, are inhibitory. Egg lecithin and beef heart lecithin can substitute for the endogenous phospholipid, but far greater concentrations on the basis of phosphate content are required to achieve activities equivalent to those observed with the endogenous phospholipid. Heart (Fleischer et al., 1966) and liver (Gotterer, 1967) D-β-hydroxybutyrate dehydrogenase have been found to have similar properties. The exact identity of the lecithin and its mechanism of activation of the enzyme are unknown, but it cannot be neglected as a possible factor in the regulation of the enzyme's activity. Indeed, we can conceive of the possibility that the changes in D-β-hydroxybutyrate dehydrogenase activity in the brain during postnatal development and their relationship to suckling and weaning may reflect not so much regulation of the amount of enzyme as the influence of changes in the levels of the active phospholipid.

Summary

The mammalian brain undergoes a major portion of its maturation in early postnatal life. During this period there are marked changes in a number of enzyme activities, including those involved in cerebral energy metabolism. Indeed, it has now become clear that the regulation of cerebral energy metabolism occurs all through life and confers hitherto unrecognized flexibility to the brain in its choice of substrates for oxidative metabolism. The normal adult brain utilizes glucose almost exclusively as the substrate for its energy metabolism, but, contrary to previous beliefs, it is not obligatorily limited only to glucose. It has been known from studies with brain slices in vitro that mature and immature brain, particularly the latter, have the enzymatic machinery to oxidize ketone bodies (Drahota et al., 1965; Itoh and Quastel, 1970). Recent studies in man and animals have shown that the brain in vivo turns in special circumstances to D-β-hydroxybutyrate and acetoacetate as its main substrates. The ketone bodies serve not only as sources of energy derived from their oxidation to CO_2 and water but also contribute carbon residues to constituents of brain normally dependent on glucose metabolism, e.g. phospholipids and cholesterol (Spitzer and Weng, 1972) and the amino acids, glutamate, glutamine, aspartate and γ-aminobutyric acid (Cremer, 1971).

Ketone utilization in adult brain is regulated entirely by the supply of substrate brought to it in the arterial blood. The concentration of ketone bodies in the blood is normally very low in the adult, and cerebral ketone utilization is then negligible. In ketotic states caused, for example, by starvation or high-fat ketogenic diets, the brain utilizes ketone bodies in almost direct proportion to their concentrations in the arterial blood (Krebs, 1971). Their rate of utilization may then exceed that of glucose, although there is reason to doubt that they can replace glucose completely.

Significant rates of cerebral ketone utilization are normal in early postnatal life, and, contrary to the situation in adult life, regulation is achieved not only by substrate supply but also by modification of the levels of enzyme activity. The newborn is slightly hypoglycemic and becomes ketotic as a result of the high-fat content of maternal milk. The activities of two key enzymes of ketone utilization progressively increase in the brain throughout the suckling period and then decline again following weaning and the disappearance of the diet-induced ketosis. Whether or not the enzymes can be artificially induced by ketosis is still uncertain as regards early life, but clearly it does not occur in adult life. Whether the regulation of the activities of the enzymes in early life is achieved by changes in enzyme synthesis, enzyme degradation, or modification of catalytic activity of a constant level of enzyme molecules remains to be determined.

The brain is, therefore, not the inflexible, selective organ it was once thought to be but shares with many other organs the capacity to adapt to changes in its nutrient supply. Glucose still appears to be its substrate of choice and may even be essential to some degree, but when the brain's glucose supply becomes endangered and/or ketone bodies become more available, it can turn to ketone bodies for a significant fraction of its energy supply.

REFERENCES

BALÁZS, R., S. KOVACS, P. TEICHGRÄBER, W. A. COCKS, and J. T. EAYRS, 1968. Biochemical effects of thyroid deficiency on the developing brain. *J. Neurochem.* 15:1335–1349.

CREMER, J., 1971. Incorporation of label from D-β-hydroxy-[^{14}C]butyrate and [3-^{14}C]acetoacetate into amino acids in rat brain in vivo. *Biochem. J.* 122:135–138.

DAHLQUIST, G., U. PERSSON, and B. PERSSON, 1972. The activity of D-β-hydroxybutyrate dehydrogenase in fetal, infant, and adult rat brain and the influence of starvation. *Biol. Neonate* 20:40–52.

DIERKS-VENTLING, C., 1971. Prenatal induction of ketone-body enzymes in the rat. *Biol. Neonate* 19:423–433.

DIERKS-VENTLING, C., and A. L. CONE, 1971a. Acetoacetyl-coenzyme A thiolase in brain, liver, and kidney during maturation of the rat. *Science* 172:380–382.

DIERKS-VENTLING, C., and A. L. CONE, 1971b. Ketone body enzymes in mammalian tissues. *J. Biol. Chem.* 246:5533–5534.

DRAHOTA, Z., P. HAHN, J. MOUREK, and M. TROJANOVA, 1965. The effect of acetoacetate on oxygen consumption of brain slices from infant and adult rats. *Physiol. Bohemoslav.* 14:134–136.

DYMSZA, H. A., D. M. CZAJKA, and S. A. MILLER, 1964. Influence of artificial diet on weight gain and body composition of the neonatal rat. *J. Nutr.* 84:100–106.

EAYRS, J. T., 1964. Endocrine influence on cerebral development. *Arch. Biol. (Liege)* 75:529–565.

FAZEKAS, J. F., F. B. GRAVES, and R. W. ALMAN, 1951. The influence of the thyroid on cerebral metabolism. *Endocrinology* 48:169–174.

FERRENDELLI, J. A., A. L. STEINER, D. B. McDOUGAL, JR., and D. M. KIPNIS, 1970. The effect of oxotremorine and atropine on cGMP and cAMP levels in mouse cerebral cortex and cerebellum. *Biochem. Biophys. Res. Commun.* 41:1061–1067.

FITZGERALD, G. G., E. E. KAUFMAN, and L. SOKOLOFF, 1973. Partial purification and properties of rat brain D-β-hydroxybutyrate dehydrogenase, *Fed. Proc.* 32:563 (abs).

FLEISCHER, B., A. CASU, and S. FLEISCHER, 1966. Release of β-hydroxybutyric apodehydrogenase from beef heart mitochondria by the action of phospholipase A. *Biochem. Biophys. Res. Commun.* 24:189–194.

GARCIA ARGIZ, C. A., J. M. PASQUINI, B. KAPLUN, and C. J. GOMEZ, 1967. Hormonal regulation of brain development. II. Effect of neonatal thyroidectomy on succinate dehydrogenase and other enzymes in developing cerebral cortex and cerebellum of the rat. *Brain Res.* 6:635–646.

GOTTERER, G. S., 1967. Rat liver D-β-hydroxybutyrate dehydrogenase. II. Lipid requirement. *Biochemistry* 6:2147–2152.

GOTTSTEIN, U., K. HELD, W. MÜLLER, and W. BERGHOFF, 1972. *Research on the Cerebral Circulation. Fifth International Salzburg Conference*, 1970, J. S. Meyer, M. Reivich, and H. Lechner, eds. Springfield, Ill.: Charles C. Thomas, (in press).

GRAVE, G. D., S. SATTERTHWAITE, C. KENNEDY, and L. SOKOLOFF, 1973. Accelerated postnatal development of D(-)-β-hydroxybutyrate dehydrogenase (EC 1.1.1.30) activity in the brain in hyperthyroidism. *J. Neurochem.* 20:495–502.

HAMBURGH, M., and L. B. FLEXNER, 1957. Physiological differentiation during morphogenesis. XXI. Effect of hypothyroidism and hormone therapy on enzyme activities of the developing cerebral cortex of the rat. *J. Neurochem.* 1:279–288.

HAWKINS, R. A., D. H. WILLIAMSON, and H. A. KREBS, 1971. Ketone-body utilization by adult and suckling rat brain in vivo. *Biochem. J.* 122:13–18.

ITOH, T., and J. H. QUASTEL, 1970. Acetoacetate metabolism in infant and adult rat brain in vitro. *Biochem. J.* 116:641–655.

KETY, S. S., 1957. The general metabolism of the brain in vivo. In *Metabolism of the Nervous System*, D. Richter, ed. London: Pergamon Press, pp. 221–237.

KETY, S. S., B. D. POLIS, C. S. NADLER, and C. F. SCHMIDT, 1948. The blood flow and oxygen consumption of the human brain in diabetic acidosis and coma. *J. Clin. Invest.* 27:500–510.

KIZER, D. E., and T. A. McCOY, 1960. Phosphomannose isomerase activity in a spectrum of normal and malignant rat tissues. *Proc. Soc. Exp. Biol. Med.* 103:772–774.

KLEE, C. B., and L. SOKOLOFF, 1964. Mitochondrial differences in mature and immature brain. Influence on rate of amino acid incorporation into protein and responses to thyroxine. *J. Neurochem.* 11:709–716.

KLEE, C. B., and L. SOKOLOFF, 1967. Changes in D(-)-β-hydroxybutyric dehydrogenase activity during brain maturation in the rat. *J. Biol. Chem.* 242:3880–3883.

KOSHLAND, JR., D. E., 1970. The molecular basis for enzyme regulation. In *The Enzymes*, Vol. I, 3rd Ed., P. D. Boyer, ed. New York: Academic Press, pp. 341–396.

KREBS, H. A., D. H. WILLIAMSON, M. W. BATES, M. A. PAGE, and R. A. HAWKINS, 1971. The role of ketone bodies in caloric homeostasis. *Adv. Enzyme Regul.* 9:387–409.

LEHNINGER, A. L., H. C. SUDDUTH, and J. B. WISE, 1960. D-β-hydroxybutyric dehydrogenase of mitochondria. *J. Biol. Chem.* 235:2450–2455.

MAHLER, H. R., 1953. Role of coenzyme A in fatty acid metabolism. *Fed. Proc.* 12:694–702.

McILWAIN, H., 1971. Types of metabolic adaptation in the brain. *Essays in Biochemistry* 7:127–158.

MICHELS, R., J. CASON, and L. SOKOLOFF, 1963. Thyroxine: Effects on amino acid incorporation into protein in vivo. *Science* 140:1417–1418.

MURTHY, M. R. V., and D. A. RAPPOPORT, 1963. Biochemistry of the developing rat brain. III. Mitochondrial oxidation of citrate and isocitrate and associated phosphorylation. *Biochim. Biophys. Acta* 74:328–339.

NIELSEN, N. C., and S. FLEISCHER, 1969. β-Hydroxybutyrate dehydrogenase: Lack in ruminant liver mitochondria. *Science* 166:1017–1019.

OWEN, O. E., A. P. MORGAN, H. G. KEMP, J. M. SULLIVAN, M. G. HERRERA, and G. F. CAHILL, 1967. Brain metabolism during fasting. *J. Clin. Invest.* 46:1589–1595.

PAGE, M. A., H. A. KREBS, and D. H. WILLIAMSON, 1971. Activities of enzymes of ketone-body utilization in brain and other tissues of suckling rats. *Biochem. J.* 121:49–53.

PASQUINI, J. M., B. KAPLUN, C. A. GARCIA ARGIZ, and C. J. GOMEZ, 1967. Hormonal regulation of brain development. I. The effect of neonatal thyroidectomy upon nucleic acids, protein, and two enzymes in developing cerebral cortex and cerebellum of the rat. *Brain Res.* 6:621–634.

PERSSON, B., G. SETTERGREN, and G. DAHLQUIST, 1972. Cerebral arteriovenous difference of acetoacetate and D-β-hydroxybutyrate in children. *Acta Paediatr. Scand.* 61:273–278.

PULL, I., and H. McILWAIN, 1971. 3-hydroxybutyrate dehydrogenase of rat brain on dietary change and during maturation. *J. Neurochem.* 18:1163–1165.

RASKIN, N. H., and L. SOKOLOFF, 1970. Alcohol dehydrogenase activity in rat brain and liver. *J. Neurochem.* 17:1677–1687.

SCHIMKE, R. T., 1969. On the roles of synthesis and degradation in regulation of enzyme levels in mammalian tissues. In *Current Topics in Cellular Regulation*, B. Horecker and E. R. Stadtman, eds. New York: Academic Press, pp. 77–124.

SCHMIDT, M. J., and G. A. ROBISON, 1972. The effect of neonatal thyroidectomy on the development of the adenosine 3',5'-monophosphate system in the rat brain. *J. Neurochem.* 19:937–947.

SCHWARK, W. S., R. L. SINGHAL, and G. M. LING, 1972. Metabolic control mechanisms in mammalian systems. Regulation of key glycolytic enzymes in developing brain during experimental cretinism. *J. Neurochem.* 19:1171–1182.

SLOVITER, H. A., and T. KAMIMOTO, 1970. The isolated, perfused brain preparation metabolizes mannose but not maltase. *J. Neurochem.* 17:1109–1111.

SLOVITER, H. A., P. SHIMKIN, and K. SUHARA, 1966. Glycerol as substrate for brain metabolism. *Nature (Lond.)* 210:1334–1336.

SLOVITER, H. A., and K. SUHARA, 1967. A brain-infusion method for demonstrating utilization of glycerol by rabbit brain in vivo. *J. Appl. Physiol.* 23:792–797.

SMITH, A. L., H. S. SATTERTHWAITE, and L. SOKOLOFF, 1969. Changes in brain D(-)-β-hydroxybutyrate dehydrogenase activity with development and during adaptation to altered substrate supply. *Proc. Second Int. Meeting Int. Soc. Neurochem. Abstr.*, R. Paoletti, R. Fumagalli, and C. Galli, eds. Milan, Italy: Tamburini Editore, p. 371.

SOKOLOFF, L., 1968. Role of mitochondria in the stimulation of protein synthesis by thyroid hormones. In *Some Regulatory Mechanisms for Protein Synthesis in Mammalian Cells*, Proc. of Third Kettering Symposium, 1968, A. Pietro, M. R. Lamborg, and F. T. Kenney, eds. New York: Academic Press, pp. 345–367.

SOKOLOFF, L., 1970. The mechanism of action of thyroid hormones on protein synthesis and its relationship to the differences in sensitivities of mature and immature brain. In *Protein Metabolism of the Nervous System*, A. Lajtha, ed. New York: Plenum Press, pp. 367–382.

SOKOLOFF, L., 1971. Action of thyroid hormones. In *Handbook of Neurochemistry*, Vol. 7, Part B, A. Lajtha, ed. New York: Plenum Press, pp. 525–549.

SOKOLOFF, L., 1972. Circulation and energy metabolism of the brain. In *Basic Neurochemistry*, R. W. Albers, B. Agranoff, R. Katzman, and G. J. Siegel, eds. Boston: Little, Brown, pp. 299–325.

SOKOLOFF, L., and S. KAUFMAN, 1961. Thyroxine stimulation of amino acid incorporation into protein. *J. Biol. Chem.* 236:795–803.

SOKOLOFF, L., P. A. ROBERTS, M. M. JANUSKA, and J. E. KLINE, 1968. Mechanisms of stimulation of protein synthesis by thyroid hormones in vivo. *Proc. Nat. Acad. Sci. USA* 60:652–659.

SPITZER, J. J., and J. J. WENG, 1972. Removal and utilization of ketone bodies by the brain of newborn puppies. *J. Neurochem.* 19:2169–2173.

SWAIMAN, K. F., J. M. DALEIDEN, and R. N. WOLFE, 1970. The effect of food deprivation on enzyme activity in developing brain. *J. Neurochem.* 17:1387–1391.

TATA, J. R., and C. C. WIDNELL, 1966. Ribonucleic acid synthesis during the early action of thyroid hormones. *Biochem. J.* 98:604–620.

THALER, M. M., 1972. Effects of starvation on normal development of β-hydroxybutyrate dehydrogenase activity in fetal and newborn rat brain. *Nature New Biol.* 236:140–141.

TILDON, J. T., A. L. CONE, and M. CORNBLATH, 1971. Coenzyme A transferase activity in rat brain. *Biochem. Biophys. Res. Commun.* 43:225–231.

TILDON, J. T., and D. A. SEVDALIAN, 1972. CoA transferase in the brain and other mammalian tissues. *Arch. Biochem. Biophys.* 148:382–390.

VALCANA, T., and P. S. TIMIRAS, 1969. Effect of hypothyroidism on ionic metabolism and Na-K activated ATP phosphohydrolase activity in the developing rat brain. *J. Neurochem.* 16:935–943.

VOLPE, J. J., and Y. KISHIMOTO, 1972. Fatty acid synthetase of brain: Development, influence of nutritional and hormonal factors and comparison with liver enzyme. *J. Neurochem.* 19:737–753.

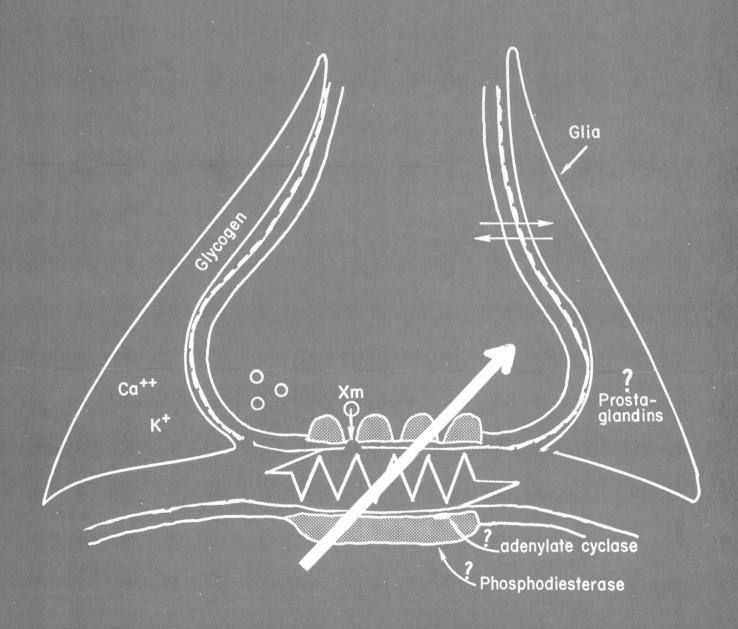

DYNAMICS OF SYNAPTIC MODULATION

The synapse as a dynamic constellation of acting and reacting components is represented in this illustration by combining structural and chemical concepts with the electrical symbol for a variable resistor across the synaptic cleft. Interactions between the presynaptic terminal, the post-synaptic receptor, and the omnipresent glia are presumed to combine to regulate the basic level of function.

DYNAMICS OF SYNAPTIC MODULATION

Introduction

F. E. BLOOM

EACH OF THE TWO previous Intensive Study Programs of the Neurosciences Research Program have featured the essential role of studies at the synaptic level of resolution. As the synapse occupies a common focal point for studies using a variety of technical approaches to the investigation of the brain, continued emphasis on interdisciplinary investigations into the nature of synaptic transmission appears easily justifiable. In order to maintain an up-to-date understanding of the present status of "synaptology" without merely cataloging the factual advances that have occurred in the past three years, the present team topic (volume part) was organized to reflect the information from a different perspective. We have attempted to utilize the fundamental observations on neuronal fine structure, transmitter neurochemistry, and electrophysiology to pinpoint certain principles of synaptic organization and transmission. The team members were then given the mission of exploring the extent to which these principles of interneuronal communication could be evaluated under resting conditions so as to pursue the question of whether these basic principles could be modulated by natural or experimentally induced biologic processes.

Implicit in the title of our topic lies the concept that the synapse is not a static cellular communicatory conduit but a dynamic constellation of many biologic systems operating at the molecular level to adjust and regulate the effectiveness of transmitting and receiving elements. The extent to which these dynamic alterations can be detected depends upon the discreteness with which our techniques of approach to the question are able to dissect and hold constant a particular index to synaptic function that can then be measured after an equally discrete experimental manipulation. While the range of observational variability is high, detecting meaningful changes indicative of biological regulation necessarily requires large changes in the index to be measured. In this sense, the component chapters of this part serve as a guide to the present resolving power of the chemical, electrophysiological, and structural approaches to synaptic analysis. In a larger sense, however, synaptic level research offers an intermediate stage between the examination of the molecular mechanisms of membrane dynamics generalizable from a study of the receptors of all cell surfaces (see Edelman et al., this volume) and the behavioral consequences of individual synaptic actions organized into sequential operations by interconnected neurons to produce behavior such as feeding and drinking (see Snyder et al., this volume). The highest objective realizable from this approach would be germinal experiments into the changes in synapses that may accompany the molecular basis of learning.

The presentations of our team members are conceptually grouped under three headings: neurochemical-cytological correlations, synaptic mechanisms, and structural modifications.

Neurochemical aspects of synaptic modulation

Iversen reviews the general neurochemical control systems by which neurotransmitters are synthesized, stored, released, and conserved. He extends this perspective by examining the exploitation of neuronal uptake systems in the fine structural localization of transmitters by autoradiography and the short-term and long-term possibilities of chemical modulation of transmitter release.

Jaim-Etcheverry explores the present status of methods by which a particular central neurotransmitter, 5-hydroxytryptamine, can be localized in autonomic and central nerve terminals by electron microscopic cytochemistry, and Carnegie describes methods being used to analyze the binding of 5-hydroxytryptamine to brain proteins.

Synaptic mechanisms

Given that it is possible to determine the transmitter stored within an identifiable synaptic terminal, how far into the realm of molecular interactions can we trace the mechanisms by which the molecules are released and the postsynaptic effects generated? Weight reviews the electrophysiological approaches to the analysis of transmitter actions, concentrating on recent data regarding synapses in which transmission is accomplished by means other than activation of ionic conductances. Rahamimoff pursues the functional regulation of transmitter release in terms of the roles of calcium and transmembrane electrical gradients. Robbins presents a critical analysis of the interpretation of studies in which use and disuse are employed to alter junctional transmission between nerve and muscle, thereby providing some insight into longer term electrophysiological modulatory changes.

Structural modifications of synapses

Adrenergic neurons of the peripheral and central nervous system can be detected selectively by histochemical methods. By reviewing the explosion of observations that indicate that this system of nerve cells retains the ability to regenerate axonal connections in the adult mammalian brain, Moore tries to elucidate the general factors which may control nerve growth and development. Ungerstedt then indicates the changes in structure and behavioral function that result from pharmacological manipulation of the central catecholamine synaptic systems.

These presentations make a strong case for the possibility that the modulatory influences revealed by chemical, electrical and structural indices could serve to regulate synaptic transmission functionally. In the summational essay, I attempt to emphasize certain conceptual consequences that could then arise from the establishment of these modulatory mechanisms by future experimentation.

79 Biochemical Aspects of Synaptic Modulation

LESLIE L. IVERSEN

ABSTRACT   The mechanisms available to control the amount of transmitter released at synaptic junctions or its duration of action are reviewed. Uptake systems for norepinephrine, dopamine, 5-hydroxytryptamine, and choline are responsible for terminating the actions of these transmitters by a process involving recapture of released amine by the presynaptic nerve terminal. Special uptake processes also exist for transmitter amino acids in mammalian CNS, such as glutamate, GABA, and glycine. Transmitter release may also be modulated by other transmitters (presynaptic inhibition) or by the released transmitter itself (autoinhibition). The release of prostaglandins or cyclic AMP from nearby tissues could also control the release of transmitter from nerve terminals.

Longer term modulation systems may involve changes in the number of postsynaptic receptor sites available for transmitter interaction or the transynaptic control of enzyme synthesis in the innervated neurons. The induction of tyrosine hydroxylase in sympathetic ganglia by nerve impulses in the preganglionic fibers can be mimicked by depolarizing stimuli in ganglia maintained in organ culture. Such enzyme induction appears to involve the production of cyclic AMP in the ganglion cells.

Introduction

MY UNDERSTANDING and use of the term "modulation" with reference to synaptic transmission are based on the excellent review on this topic by Florey (1967), in which "modulator substances" were defined as "any compounds of cellular and nonsynaptic origin that affect the excitability of nerve cells and represent a normal link in the regulatory mechanisms that govern the performance of the nervous system." According to the Oxford Dictionary, to "modulate" is to "adjust so that it may work more accurately"; when applied to the speaking voice, to modulate is "to adjust or vary tone or pitch." The latter usage provides a useful analogy to the present context—to modulate synaptic transmission means to adjust or regulate this process—without interfering with the basic process of information transfer. More specifically, a nerve impulse in the presynaptic nerve terminal will release neurotransmitter, but the amount of transmitter released

and the extent and duration of its postsynaptic effects may be under modulatory control. I shall not attempt to discuss all the possible mechanisms by which such modulation might be achieved (see Florey, 1967) but will concentrate on only a few of the control systems that are known to exist and will discuss these from the viewpoint of a biochemist and pharmacologist.

In comparison to the elegant techniques of modern cellular neurophysiology and neuroanatomy (see Weight and Moore, this volume), the neurochemist has few techniques by which to study synaptic events at the level of single synapses or neurons. He is generally obliged to deal not with single cells but with far larger masses of tissue. Since CNS tissue is so complex and contains so many chemically different classes of neurons and other cells, the neurochemical analysis of synaptic transmission in the CNS is particularly difficult. It is not surprising that much of our biochemical knowledge of synaptic transmission stems from studies of the far simpler neuronal systems in the peripheral nervous system.

Transmitter uptake mechanisms

INTRODUCTION   One mechanism by which the actions of released neurotransmitters on postsynaptic receptors may be modified is by removal of the free transmitter from the synaptic cleft. Although the classic example of such a mechanism is the destruction of released acetylcholine by acetylcholinesterase at cholinergic junctions, it now seems likely that nonenzymic mechanisms operate at most other chemically transmitting synapses. These mechanisms involve a physical removal of the released transmitter from its site of action, such removal being catalyzed by a variety of different uptake systems. The evidence for the existence and function of such uptake systems has been reviewed recently elsewhere (Iversen, 1967, 1971, 1972a, b) and only the salient points will be described here. In terms of modulating the postsynaptic actions of the released neurotransmitter, it is clear that the cellular location of such removal mechanisms might be at any site in the region of the synaptic cleft (Figure 1). In practice, most of the mechanisms that we know of are located in the membrane

LESLIE L. IVERSEN   Medical Research Council, Neurochemical Pharmacology Unit, Department of Pharmacology, University of Cambridge, Cambridge, England

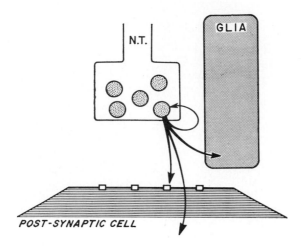

FIGURE 1 The possible role of tissue uptake systems in limiting the availability of transmitter at postsynaptic receptor sites.

of the presynaptic nerve terminal so that removal of transmitter is also equivalent to "recapture," with the possibility that transmitter molecules may be released, recaptured, and reused in many cases. There are, however, some examples of transmitter uptake occurring into sites other than the prejunctional terminal, and these will be described separately.

UPTAKE SITES IN PRESYNAPTIC NERVE TERMINALS
Catecholamines The uptake system that has been most thoroughly studied is the one present in the membrane of adrenergic nerve terminals. This is a transport system for which *l*-norepinephrine (NE) is the preferred substrate. The uptake process is saturable, with an affinity constant $K = 0.28 \, \mu\text{M}$ for *l*-NE in sympathetic terminals of the rat heart; i.e., the process has an unusually high affinity for its substrate when compared with other membrane transport systems for amino acids or sugars, which have transport constants in the *milli*molar rather than *micro*molar range. The high affinity of the NE uptake system means that it can work very efficiently when presented with the extremely low concentrations of NE that are likely to be encountered in the extracellular space under physiological conditions. Like other active transport systems, NE uptake is temperature and energy dependent and requires the presence of sodium ions in the external medium. It appears to require the maintenance of a normal gradient in sodium ion concentration between the intra- and extracellular compartments and can be inhibited by conditions or drugs that prevent normal functioning of the sodium pump. The NE that has been accumulated into sympathetic nerve terminals by the action of the uptake system is rapidly incorporated into the storage vesicles within the cytoplasm, where it is immediately available again for release. The uptake system for NE exists both

in peripheral postganglionic neurons of the sympathetic nervous system and also in NE-containing neurons in the CNS. In CNS the uptake of NE can conveniently be studied in homogenates, in which amine uptake into synaptosomes readily occurs when such preparations are incubated in vitro (Coyle and Snyder, 1969; Snyder and Coyle, 1968).

The importance of the NE uptake system for adrenergic neurotransmission is illustrated by the finding that compounds that are known to inhibit this uptake produce a marked potentiation and prolongation of the responses of innervated tissues to adrenergic nerve stimulation. Among the more potent inhibitors of this uptake system are cocaine, phenoxybenzamine, and tricyclic antidepressant drugs of the imipramine and amitriptyline groups (Iversen, 1972b), all of which are effective at concentrations of less than 1 μM, the most potent being desmethylimipramine, effective at a concentration of less than 10^{-8} M (Table I). Such inhibitors also lead to an increase in the "overflow" of NE and its metabolites in response to nerve stimulation in perfused tissue preparations, suggesting that under normal conditions a large proportion of NE released by adrenergic terminals is recaptured by the uptake mechanism. The exact proportion of released NE recaptured in this way probably varies from tissue to tissue and also depends on the frequency of firing of impulses in the adrenergic terminals, but it probably is in the range 50 to 90% (Iversen, 1972b; Langer, 1970; Haggendal, 1970).

The NE uptake mechanism thus constitutes an important mechanism for modulating adrenergic neurotransmission in the peripheral sympathetic nervous system; it is generally assumed that it performs a similar function at adrenergic synapses in the CNS, although so far little experimental evidence is available on this point. Whether the uptake mechanism itself can be modulated is not yet known. There have been reports that the uptake mech-

TABLE I

Inhibition of norepinephrine uptake₁ in the perfused rat heart by various drugs

Drug	ID_{50} (μM)
Desipramine	0.01
l-Metaraminol	0.08
Imipramine	0.09
d-Amphetamine	0.18
Cocaine	0.38
Tyramine	0.45
Phenoxybenzamine	0.75
β-Phenylethylamine	1.10

ID_{50} = drug concentration required to cause 50% inhibition of ^{3}H-norepinephrine uptake (Iversen, 1972b).

anism is inhibited during stimulation of adrenergic nerves for long periods at relatively high frequencies (Haggendal, 1970) but such reports are controversial, and it is not clear whether such conditions would ever occur in vivo. It is certainly possible to modulate adrenergic transmission pharmacologically by the use of drugs that inhibit the NE recapture mechanism, and the finding that antidepressant drugs are very potent in this respect has led to considerable speculation that such a mechanism might be involved in their therapeutic actions.

In addition to the NE uptake system, a similar mechanism exists in nerve terminals of the dopamine-containing neurons in the basal ganglia of the CNS (Coyle and Snyder, 1969; Horn et al., 1971). This is again a high affinity, sodium dependent uptake mechanism, but has quite different substrate specificity and drug sensitivity from that in NE-containing neurons. The DA uptake system, for example, is not potently inhibited by drugs such as imipramine or desmethylimipramine, which are some 10,000 times less active as inhibitors of DA uptake than of NE uptake. It is assumed that this system plays a role in terminating the actions of DA released as a neurotransmitter in the basal ganglia and at certain other sites in the CNS.

The existence of specific uptake systems for catecholamines in adrenergic nerve terminals can be made use of in a number of ways by neurochemists. The exposure of adrenergically innervated peripheral tissues or of brain tissue to radioactively labeled catecholamines is all that is needed to achieve a selective labeling of adrenergic structures, by virtue of their special uptake abilities. This means that radioactive catecholamine can be selectively introduced into adrenergic neurons, and its subsequent metabolism, storage, and release studied biochemically. Alternatively the labeled catecholamine can be localized by autoradiography at either the light microscope or electron microscope level (Aghajanian and Bloom, 1967a; Hökfelt and Ljungdahl, 1971a; Iversen and Schon, 1972) and this can be a powerful tool for identifying transmitter-specific pathways in the CNS or other tissues. For example, in peripheral organs such as the nictitating membrane, Esterhuizen et al. (1968) used this technique to demonstrate that adrenergic terminals (containing labeled NE) did not correspond to the small nerve terminals with positive acetylcholinesterase activity in this tissue, thus adding powerful weight against the argument that acetylcholine might be involved in the release of NE from adrenergic terminals.

In this laboratory the localization of labeled NE in synaptosomes has been studied by electron microscope autoradiography. It was found that only about 5% of all synaptosomes in homogenates of rat cerebral cortex became labeled after exposure to labeled catecholamine, and 16% of those in homogenates of corpus striatum. It is suggested that these figures represent, respectively, the proportions of noradrenergic and dopaminergic terminals in those two brain regions (Iversen and Schon, 1973).

Neurons in mammalian CNS containing the transmitter 5-hydroxytryptamine also possess their own specialized high-affinity transport system for this amine (Shaskan and Snyder, 1970). Radioactively labeled 5-HT can be localized in such neurons by electron microscopic autoradiography after injections of labeled amine into the brain (Aghajanian and Bloom, 1967b).

Amino acid transmitters The putative amino acid transmitters—GABA, glycine, glutamate, aspartate, and taurine—all have specialized high-affinity uptake systems in mammalian brain and spinal cord (Iversen, 1971, 1972a). In this respect they are unique among amino acids, all of which are in addition transported by relatively unspecific low-affinity transport systems in brain as in other tissues. The high-affinity uptake system ($Km = 33$ μM) for glycine, for example, is found only in the spinal cord and lower brain stem of the rat CNS (Neal, 1971; Logan and Snyder, 1972; Johnston and Iversen, 1971). In supraspinal regions glycine is taken up by a transport system with broader specificity and lower affinity ($Km =$ approx. 300 μM) (Johnston and Iversen, 1971; Logan and Snyder, 1972). The existence of high-affinity systems specific to the transmitter amino acids is strongly suggestive that these systems may play a role in terminating the actions of these substances after their synaptic release, as with the catecholamines at adrenergic synapses. So far, however, little is known of the cellular location or function of most of the amino acid uptake systems. We and others have performed autoradiographic studies with ^{3}H-GABA and have found that in slices and homogenates of rat brain the amino acid is selectively taken up into nerve terminals, and that only certain nerve terminals are capable of such uptake (Iversen and Bloom, 1972). We believe that such terminals represent those normally storing and using GABA as a transmitter, and that the autoradiographic technique may prove to be a valuable tool in future studies aimed at identifying and mapping the distribution of such neuronal pathways in CNS. Very similar results have been obtained with labeled glycine in spinal cord slices and homogenates (Iversen and Bloom, 1972; Hökfelt and Ljungdahl, 1971b), where the amino acid is again localized in nerve terminals. Matus and Dennison (1971) in similar studies claimed, furthermore, that glycine was taken up selectively by nerve terminals in spinal cord in which synaptic vesicles were "flattened" or elliptical in profile. These results suggest that glycine in the spinal cord may be taken up selectively by inhibitory nerve terminals using this amino acid transmitter. Our findings demonstrate that about a quarter of all synaptic terminals in spinal cord homogenates are labeled in this way on exposure to glycine, and a further quarter are labeled on exposure to labeled GABA (Table II). There have not, so far, been any studies showing a selective neuronal uptake of glutamate or aspartate, which are also known to possess specialized transport systems in brain (Logan and Snyder, 1972) (Table III).

TABLE II

Proportion of nerve terminals in spinal cord homogenates labeled with ^{3}H-GABA or ^{3}H-glycine after incubation in vitro

H^3-GABA alone	(n = 11)	— 24.6 ± 1.19%
H^3-Glycine alone	(n = 14)	— 27.9 ± 1.81%
H^3-GABA + H^3-Glycine simultaneously	(n = 10)	— 50.8 ± 2.01%

(Data from Iversen and Bloom, 1972.)

TABLE III

Kinetics of amino acid uptake into synaptosomes of rat spinal cord

Amino Acid	Km for Uptake (μM)	
	Low-Affinity Uptake	High-Affinity Uptake
L-Aspartic acid	3762.0 ± 1417.0	21.5 ± 7.7
L-Glutamic acid	4902.0 ± 1351.0	14.3 ± 3.7
Glycine	923.0 ± 638.0	26.5 ± 14.1*
L-Alanine	1058.0 ± 71.0	none
L-Arginine	2679.0 ± 758.0	none
L-Leucine	9136.0 ± 3018.0	none
L-Lysine	5707.0 ± 2295.0	none
L-Serine	1081.0 ± 206.0	none

* Value of 36 μM reported by Johnston and Iversen (1971).

Km values and their standard errors were determined by incubating homogenates with various concentrations of labeled amino acid. Data from Logan and Snyder (1972).

TRANSMITTER UPTAKE SITES IN NONNEURONAL TISSUES
Apart from the neuronal "recapture" mechanisms, there are certain other uptake systems in extraneuronal sites that could function similarly to accelerate the disappearance of transmitter from the extracellular fluid. In various peripheral tissues innervated by the sympathetic nervous system, notably in cardiac muscle and various smooth muscles such as in spleen and blood vessels, NE is taken up by a special uptake system known as "Uptake$_2$" (Iversen, 1971). This transport system has properties quite distinct from that of the neuronal uptake system for NE in sympathetic nerves, which is for convenience "Uptake$_1$." Uptake$_2$ is a relatively low affinity but high capacity process (Table IV) that lacks stereochemical specificity and has a higher affinity for epinephrine than for NE as substrate. Uptake$_2$ is potently inhibited by phenoxybenzamine and certain structurally

related compounds, by a variety of steroids, and by the O-methylated catecholamine metabolites metanephrine and normetanephrine. Drugs such as imipramine and cocaine, which are potent inhibitors of the neuronal recapture system, are without effect on Uptake$_2$. The NE accumulated by the Uptake$_2$ mechanism in various innervated tissues is not retained in such cells but is rapidly degraded by monoamine oxidase and catechol-O-methyl transferase after it enters such tissues (Lightman and Iversen, 1969). The relative importance of neuronal recapture versus the "Uptake$_2$-followed-by metabolism" system remains obscure. In certain innervated tissues such as blood vessels in which there is only a sparse innervation and a large bulk of smooth muscle containing Uptake$_2$ sites, this may well be the dominant mechanism for terminating the physiological actions of NE. In other organs with a high density of neuronal uptake sites in their abundant sympathetic innervation (nictitating membrane, vas deferens), Uptake$_2$ may be far less important. In all tissues it seems that blockade of the neuronal recapture mechanism leads to a greater proportion of released NE being diverted to the Uptake$_2$ sites (Eisenfeld et al., 1966). Drugs such as phenoxybenzamine that block both types of NE uptake have a larger effect in increasing transmitter overflow from stimulated organs than those that block only one system (Langer, 1970; Enero et al., 1972).

TABLE IV

Properties of Uptake$_1$ and Uptake$_2$ in the rat heart (Iversen, 1972b)

	Uptake$_1$		Uptake$_2$	
	Km (μM)	V_{max} (nmole/ min/g)	*Km* (μM)	V_{max} (nmole/ min/g)
(±)-NE	0.67	1.36	252	100.0
(±)-Epinephrine	1.40	1.04	52	64.4
(±)-Isoprenaline	(not a substrate)		23	15.5
Dopamine	0.68	1.45	(not determined)	

Most Potent Inhibitors

Desipramine	Phenoxybenzamine
l-Metaraminol	*dl*-Metanephrine
Cocaine	Corticosterone
Phenoxybenzamine	β-Ostradiol

If Uptake$_2$ does play a role in the physiological modulation of catecholamine actions, it may be susceptible to modulation by circulating steroids or cholesterol, since both steroid hormones and cholesterol are active as inhibitors of this process. Such a mechanism might explain the potentiating effects that certain steroids have been found to have on adrenergic mechanisms in many previous studies (Iversen and Salt, 1970; 1972).

Although Uptake$_2$ is not known to exist in the CNS,

extraneuronal uptake sites for the amino acid transmitter GABA appear to exist. When isolated rat retina was incubated with radioactively labeled GABA, autoradiographic studies revealed a prominent accumulation of the amino acid over the glial cells (Müller fibers) (Neal and Iversen, 1972) (Figure 2). Similar glial accumulations of ^{3}H-

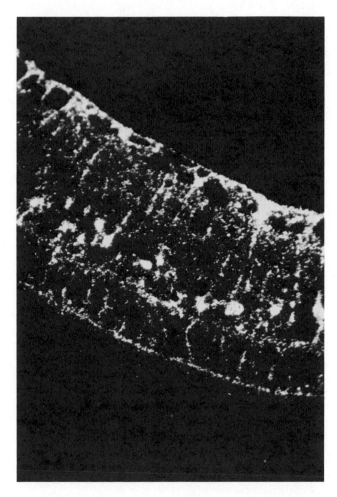

FIGURE 2 Radioautographic localization of ^{3}H-GABA in isolated rat retina. Note localization of radioactivity (white/silver grains) over the cell bodies and processes of Müller fibers, and absence of label in ganglion cells (at upper border of tissue).

GABA have been observed in various other regions of rat CNS when labeled GABA is injected into the cerebrospinal fluid (Iversen and Schon, 1973); a considerable accumulation of ^{3}H-GABA was also seen over ependymal cells, and the pial membranes and small blood vessels. All the latter sites are known to contain a high activity of the catabolic enzyme GABA:glu transaminase (van Gelder, 1968). The accumulation of ^{3}H-GABA seen in autoradiographic studies was probably dependent on the treatment of animals with the transaminase inhibitor aminooxyacetic acid before injection of ^{3}H-GABA (Schon

and Iversen, 1972). One might thus speculate that an uptake of extracellular GABA into glial cells or other nonneuronal sites in the CNS, followed by metabolic degradation, could provide an alternative to the neuronal recapture system for terminating the actions of this amino acid. The existence of both neuronal and nonneuronal systems for this purpose is analogous to the Uptake$_1$ and Uptake$_2$ mechanisms for NE, although it remains unclear what purpose is served by the existence of alternate mechanisms of this type.

Modulation of transmitter release from presynaptic terminals

Uptake systems may modulate synaptic transmission by controlling the concentrations and duration of action of released transmitters at postsynaptic receptor sites. Another mechanism for such modulation may involve control over the amount of transmitter released from the presynaptic nerve terminal in response to nerve activity (see Weight, this volume).

METHODS OF STUDYING TRANSMITTER RELEASE   In the relatively simple nerve-effector tissue junctions of the peripheral skeletal motor and autonomic nervous systems, transmitter release has traditionally been studied in isolated perfused or superfused tissue preparations in which the efferent nerve can be electrically stimulated. No such simple approach can be used in studies of transmitter release from the CNS. Even though it may sometimes prove possible in the intact animal preparation to demonstrate a release of transmitter from nerve terminals near the surface of the brain, or near the surface of the ventricular system (Mitchell, 1963; Obata et al., 1969), it is hardly ever possible to define with any precision the pathway that is being stimulated. Because of the rich interconnection of the neuronal elements in the CNS, the transmitter release evoked by any stimulation may be a complex balance between excitation and inhibition in polysynaptic pathways. In order to study the basic characteristics of the transmitter release mechanisms, and their possible control at the local level of the synapse, other approaches must be used. Since Baldessarini and Kopin (1967) demonstrated the stimulus-evoked release of labeled NE from brain slices, this type of in vitro system has had increasing popularity. Slices of brain remain metabolically viable in vitro, and the neuronal membranes respond to electrical stimulation by depolarization as in the intact tissue (McIlwain and Bachelard, 1971). By making use of the unique uptake systems associated with many of the CNS transmitters, it is possible to label each system selectively and to demonstrate an evoked release of labeled amine or amino acid. This has been successfully used for studies of the release of catechol-amines, 5-HT, GABA, glutamate, and glycine. The latter amino acid, for example, is released by stimulation of the spinal cord and retina but not from cerebral cortical slices (Hopkin and Neal, 1970), indicating the specificity of this type of release system. Furthermore, the release of putative transmitters is generally inhibited by low calcium and/or high magnesium ion concentration in the external medium and does not occur with control non-transmitter radioactive labels such as urea, or "pool" amino acids such as valine, lysine, aminoisobutyric acid, etc. (Srinivasan et al., 1969; Katz and Kopin, 1969). The isolated slice preparations have the advantage that the neuronal circuitry is likely to be relatively simple in comparison with that in the intact brain, so that the observations may more accurately reflect transmitter release mechanisms at the synaptic level. A further refinement, which completely removes all possibility of the existence of polysynaptic pathways, or any synaptic interaction between one transmitter type and another, is the use of synaptosome preparations for release studies. Bradford (1971) has shown that such preparations release amino acids selectively when subjected to field stimulation —of the amino acids examined only GABA, Glu., Gly., and Asp. were released under such conditions. This refinement is carried one stage further by de Belleroche and Bradford (1972a, 1972b) who have shown that synaptosomes may be reconstituted into "artificial slices" or "synaptosome beds" by packing them as a layer onto a membrane filter and sandwiching this with a second membrane filter disk. Such beds can then be superfused and subjected to electrical stimulation, and again amino acid and acetylcholine release demonstrated in vitro. The development of newer micromethods for the assay of the very small amounts of endogenous amino acid or amine released from slice or synaptosome preparations in vitro may allow such studies to be performed without the use of radioactive labels in future. This would in most cases be an advantage, since the release of several transmitters could be measured at once, and problems concerning inhomogeneity of labeling with exogenous substances could be avoided. Snodgrass and Iversen (1972) developed a double isotope derivative assay system for amino acids, in which the tritiated Dansyl derivatives of amino acids are formed and separated by thin layer chromatography, ^{14}C-amino acids being added as internal standards. Using this method, picomole amounts of the amino acids can readily be measured, and many other amino compounds can in principle be measured in this way. Applying this technique to amino acid release from brain slices, we found that electrical or potassium induced depolarization led to significant increases only in GABA, Glu., and (surprisingly) Ala, but not in Asp. Val, Lys, or

Gln. The efflux of glutamine was actually significantly decreased during electrical stimulation.

Effects of prolonged stimulation The relatively few studies in which transmitter release has been directly measured during sustained periods of presynaptic nerve stimulation, or during repeated trains of such stimuli, all show that the release of transmitter declines—often quite rapidly—during such stimulation. In both adrenergic and cholinergic neurons, there is now evidence that the transmitter stored in presynaptic terminals is not all immediately available for release (Birks and MacIntosh, 1961; Glowinski, 1972). It seems instead that only a small pool of transmitter is easily released, and that this pool is highly dependent on replenishment by de novo transmitter biosynthesis, and, in the case of adrenergic nerves, by recapture of the previously released amine.

Presynaptic inhibition The phenomenon of inhibition of transmitter release in response to inhibitory transmitters released in the vicinity of excitatory terminals is well-established neurophysiologically but has been little studied biochemically or pharmacologically (Figure 3).

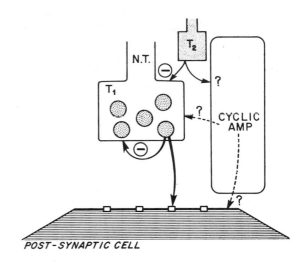

FIGURE 3 Presynaptic inhibition and autoinhibition of transmitter release, and the possible roles of cyclic AMP in synaptic modulation.

Controversy remains as to the exact morphological relationship between presynaptic inhibitory terminals and the excitatory terminals, and it is perhaps doubtful whether the actions of released inhibitory transmitters in this way should be described as "modulation" rather than a special type of "neurotransmission." An inhibitory action of acetylcholine on NE release from sympathetic nerve terminals has been clearly established by the studies of Loffelholz and Muscholl (1969). Also NE inhibits acetylcholine release from parasympathetic nerve terminals (Paton and Vizi, 1969).

Autoinhibition of transmitter release A quite different process, whose existence remains largely circumstantial and speculative, is one which may best be described as "autoinhibition." The suggestion here is that transmitters may inhibit their own release from presynaptic terminals, if sufficiently high concentrations accumulate in the synaptic cleft region (Figure 3). The evidence for such an effect is best for NE release from peripheral adrenergic nerves (Haggendal, 1970; Kirpekar and Puig, 1971; Enero et al., 1972; Starke and Schumann, 1972; Starke, 1972). In preparations such as the heart and spleen the adrenergic receptor blocking drug phenoxybenzamine—and other α-adrenergic receptor blocking drugs—caused an increase in the amount of NE released per stimulus. Such an increased release was seen quite independently of the further actions of phenoxybenzamine on NE uptake systems referred to above, and could be evoked at drug concentrations far lower than those required to inhibit the uptake mechanisms. Kirpekar and Puig (1971), Starke (1972), and Enero et al. (1972) suggested that one explanation of this effect might be that α-adrenergic receptors exist in the surface of the presynaptic nerve terminals, and that NE interacts with these to inhibit its own further release—this interaction (at least in heart) would be quite distinct from the postsynaptic actions of NE that are mediated in this tissue by β-adrenergic receptors. This interesting postulate may help to explain a number of other biochemical observations made in recent years on different transmitter systems. A similar mechanism has been invoked, for example, to explain the enhanced release of acetylcholine caused by stimulation of brain slices in vitro by the cholinergic receptor blocking drug atropine (Molenaar and Polak, 1970). It might also explain the increase in ^{3}H-GABA release from brain slices stimulated in the presence of the GABA antagonist drug bicuculline (Johnston and Mitchell, 1971), and the enhanced release and synthesis of ^{3}H-dopamine in striatal slices from rats treated with thioproperazine—a drug thought to inhibit dopamine receptors in the brain (Cheramy et al., 1970). Autoinhibition may also be the basis for the so-called "receptor feedback" mechanism described by Carlsson and Lindqvist (1963) and subsequently in many other studies, in which an increased turnover of NE or DA in the CNS has been observed in the presence of drugs thought to block noradrenergic or dopaminergic receptors. In many cases such processes may depend on complex polysynaptic neuronal pathways, but the finding that similar effects can be observed in brain slice preparations stimulated in vitro makes a direct action at the prejunctional terminal seem more probable. For example, the persistence of the stimulatory effects of atropine on stimulus-evoked acetylcholine release from brain slices even in the presence of tetro-

dotoxin is perhaps the most convincing evidence for such a direct action (Molenaar and Polak, 1970). The "autoinhibition" hypothesis may also explain the opposite effects of supposed agonist drugs such as apomorphine in causing diminished turnover and release of monoamines in the CNS (Andén et al., 1970). It does not, of course, follow that any drug causing an increase or decrease in the rate of turnover of cerebral amines can thus be classified as agonist or antagonist—a logical fallacy that appears to have attractions for the unwary!

Possible prejunctional role of prostaglandins A less direct mechanism by which transmitters may modulate their own release could be mediated by the release of prostaglandins from innervated tissues in response to the released transmitter. Such a release is known to occur from a variety of peripheral tissues innervated by the sympathetic system and also from stimulated CNS tissue (Hedqvist, 1970). Hedqvist (1970) and Wennmalm (1971) have shown that prostaglandins E_1 and E_2 have an inhibitory effect on the release of NE and acetylcholine from sympathetic and parasympathetic terminals in the autonomic system, and they have suggested that the release of such substances normally evoked by transmitter release might have the effect of "dampening" further transmitter release in this way. In certain tissues, however, prostaglandins also have direct inhibitory effects on adrenergic postsynaptic receptors, so that a more direct modulatory influence could also be exerted postsynaptically (Figure 4). Recently Hedqvist and von Euler (1972a, 1972b) have presented evidence that the amount of prostaglandin E_1 released from vas deferens in response to nerve stimulation would be adequate to exert appreciable inhibitory effects on adrenergic transmission through the inhibition of NE release. It was also shown that such

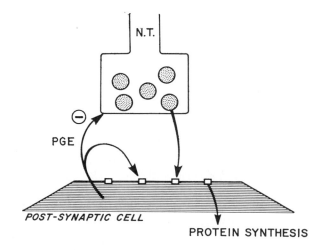

FIGURE 4 Postulated role of prostaglandins (PGE) in modulation of transmitter release, and the transsynaptic control of protein synthesis.

inhibitory effects of the prostaglandins were most pronounced at low frequencies of stimulation; the suggestion was made that this mechanism may normally have the effect of "filtering" out all but massive stimulation of vas deferens, preventing contraction of this smooth muscle except during copulation.

Possible role of adenylate cyclase–cyclic AMP in synaptic modulation

The well-known observations that the CNS contains a higher adenylate cyclase activity than any other organ, and that cerebral levels of cyclic AMP are increased by norepinephrine and certain other CNS transmitters have led to considerable speculation concerning the exact function of this system (for review, see *Neurosciences Res. Prog. Bull.*, No. 3, Vol. 8). I will merely add some further speculative comments on this question here. Dr. Bloom will describe in more detail the work of Siggins, Hoffer, and Bloom (1971) that demonstrated that cyclic AMP has potent inhibitory effects on the firing of Purkinje cells in the cerebellum, when applied to the external surface of such cells through microelectrodes. This seems to me one of the most intriguing recent findings to have been made in this field, since in other tissues externally applied cyclic AMP is generally only weakly active in producing biological effects. A second recent finding that may have considerable importance is the report by Gilman and Nirenberg (1971) and by Clark and Perkins (1971) that norepinephrine stimulated a very large increase in the production of cyclic AMP by various pure line tissue cultures of glial cells, and that such stimulation was mediated by what appeared to be a typical β-adrenergic receptor in the glial cells. This finding is important because it demonstrates for the first time that glial cells can respond to a neurotransmitter substance, and that they possess specific receptor sites for such substances. Putting these findings together, one could envisage a model in which the inhibitory effects of norepinephrine on Purkinje cells in the cerebellum might be mediated indirectly by cyclic AMP produced and released from the glial cells that wrap tightly around the Purkinje neurons (Figure 3). This would admittedly be an unusual action for cyclic AMP, because it would involve a release and external site of action for this substance, which in other systems is thought of as a strictly intracellular messenger. It would also involve the heretical suggestion that neurons may make specific contact (although not necessarily in the form of a conventional synaptic junction) with glial cells. On the other hand, such a model might help to explain how the relatively small number of small diameter noradrenergic fibers in the cerebellum could control (through the amplifying action of glia) the firing of the very large number of Purkinje cells, each of which is many times larger than the adrenergic fibers. This model might also be applicable to the other sites of action of NE in the CNS. The recent demonstration that dopamine is able to stimulate cyclic-AMP production in isolated sympathetic ganglia (Kebabian and Greengard, 1971) may suggest that other monoamines could exert their actions through the adenyl cyclase system in nervous tissue.

The role of long-term changes in synaptic modulation

One of the most interesting questions in neurobiology is how nervous activity can over the longer term modify synaptic contacts and the processes of synaptic transmission. Dr. Moore describes what we know of the capacity of adrenergic neurons in the adult CNS to grow and regenerate. Other mechanisms for long-term changes in synaptic mechanisms, however, are also known to exist.

One such mechanism has been known for many years since its first systematic description by Cannon (1949), namely, the phenomenon of "denervation supersensitivity" which states that innervated tissues tend to become more sensitive to the transmitters normally liberated onto them if their innervation is destroyed. Although this phenomenon is well known at the neuromuscular junction and in the autonomic nervous system, it remains poorly characterized in the CNS. Even in the autonomic nervous system, the mechanism of denervation sensitivity is not known. The recent introduction of new tools for biochemical study of receptor molecules, however, offers new opportunities for studies of this question. Miledi and Potter (1971), for example, used radioactively labeled α-bungarotoxin (a snake venom that selectively attaches to and blocks cholinergic receptors at motor end plates; see Molinoff, this volume) to measure the number and distribution of receptor molecules in normal and denervated mouse diaphragm muscle. They were able to show that after denervation there was spread of toxin binding sites away from the motor end plate regions, and an overall increase of up to twentyfold in the number of such binding sites. This was the first direct confirmation that in denervated muscle there was an increase in the number of available receptors, although whether this represents an increased de novo synthesis of such molecules or not has still to be determined.

Another mechanism for long-term modulation of synaptic function is of much more recent discovery. This is the transsynaptic control of the rate of synthesis of certain enzymes in the postsynaptic tissue first described by Mueller et al. (1969) in their studies of the enzyme tyrosine hydroxylase in adrenergic neurons in the peripheral sympathetic system and in the glandular cells of the adrenal medulla. In each tissue they showed that the synthesis of the enzyme appeared to be controlled by nerve impulse traffic in the presynaptic nerve innervating the aminergic cells. This control was later shown to extend also to the synthesis of the enzyme dopamine-β-hydroxylase but not to DOPA-decarboxylase or monoamine oxidase—showing that the transynaptic control was rather specific to just a few "key" (in the sense of catalyzing rate-limiting steps) enzymes in the adrenergic neuron (Thoenen et al., 1971). Thoenen et al. (1970) also showed that in adrenergic neurons the increased synthesis of tyrosine hydroxylase in the cell bodies in sympathetic ganglia was reflected, after a lag of several days, in increased activity of the enzyme in the peripheral nerve terminals of these neurons in various innervated tissues. Thus over a long time course, synaptically mediated stimulation of a neuron can be shown to lead to changes in the content of transmitter synthetic enzymes in the neuron terminals of the stimulated neuron. It remains to be demonstrated that such changes in turn lead to changes in the amount of norepinephrine stored and available for release from such nerve terminals, but this seems a reasonable supposition, at least as a working hypothesis.

In our own laboratory we have studied the transynaptic control of tyrosine hydroxylase in adrenergic neurons from two points of view. Black et al. (1971) obtained evidence that in the mouse transynaptic control of the enzyme occurred during normal development of the superior cervical ganglion. Formation of synaptic contacts between preganglionic terminals and postganglionic ganglion cells neurons seemed to be an essential component for the full development of the biochemical specialization of the adult adrenergic neuron. In the mouse such control is more prominent than in the rat, which has been studied under comparable conditions (Thoenen et al., 1972); this is perhaps because the adult mouse sympathetic nervous system has a higher basal discharge rate or tone than that of the rat. The other aspect of this intriguing problem that we have studied concerns the mechanism by which synaptic activity controls the rate of synthesis of enzymes in the innervated neuron. Mackay has shown that (Mackay and Iversen, 1972a, 1972b) in mouse sympathetic ganglia maintained in organ culture a variety of depolarizing stimuli such as raised potassium concentration lead to increased synthesis of the enzyme tyrosine hydroxylase. A similar induction of dopamine-β-hydroxylase by elevated potassium concentration has been reported in rat superior cervical ganglia in organ culture (Silberstein et al., 1972). This might suggest that the synaptic control of enzyme synthesis in the intact system involves simply an increased depolarization of the adrenergic neurons by presynaptically released acetylcholine, rather than any more mysterious mechanism involving the release of specialized trophic chemicals from the preganglionic terminals. More recently Mackay and Iversen (1972b) showed that the effects of high potassium could be mimicked by addition of dibutyryl cyclic AMP to the culture medium, and the potassium effects were enhanced by theophylline. This indirect evidence suggests that the control of enzyme synthesis by membrane depolarization may possibly involve a stimulation of the production of cyclic AMP in the adrenergic neurons—yet another function for this ubiquitous control substance in the nervous system.

REFERENCES

AGHAJANIAN, G. K., and F. E. BLOOM, 1967a. Electron microscopic localization of tritiated norepinephrine in rat brain: Effect of drugs. *J. Pharmacol. Exp. Ther.* 156:407–416.

AGHAJANIAN, G. K., and F. E. BLOOM, 1967b. Localization of tritiated serotonin in rat brain by electron microscopic autoradiography. *J. Pharmacol. Exp. Ther.* 156:23–30.

ANDÉN, N. E., S. G. BUTCHER, H. CORRODI, K. FUXE, and U. UNGERSTEDT, 1970. Receptor activity and turnover of dopamine and noradrenaline after neuroleptics. *Europ. J. Pharmacol.* 11:303–314.

BALDESSARINI, R. J., and I. J. KOPIN, 1967. The effect of drugs on the release of norepinephrine-H^3 from central nervous system tissues by electrical stimulation in vitro. *J. Pharmacol. Exp. Ther.* 156:31–38.

DE BELLEROCHE, J. S., and H. F. BRADFORD, 1972a. Metabolism of beds of mammalian cortical synaptosomes: Response to depolarizing influences. *J. Neurochem.* 19:585–602.

DE BELLEROCHE, J. S., and H. F. BRADFORD, 1972b. The stimulus-induced release of acetylcholine from synaptosome beds and its calcium-dependence. *J. Neurochem.* 19:1817–1819.

BIRKS, R., and F. C. MACINTOSH, 1961. Acetylcholine metabolism of a sympathetic ganglion. *Canad. J. Biochem. Physiol.* 39:787–827.

BLACK, I. B., I. A. HENDRY, and L. L. IVERSEN, 1971. Transynaptic regulation of growth and development of adrenergic neurons in a mouse sympathetic ganglion. *Brain Res.* 34:229–240.

BRADFORD, H. F., 1971. Metabolic response of synaptosomes to electrical stimulation: Release of amino acids. *Brain Res.* 19:239–247.

CANNON, W. B., 1949. *The Supersensitivity of Denervated Structures.* New York: Macmillan Co.

CARLSSON, A., and M. LINDQVIST, 1963. Effect of chlorpromazine or haloperidal on formation of 3-methoxytyramine and normetanephrine in mouse brain. *Acta Pharmacol. (Kobenhavn)* 20:140–148.

CHERAMY, A., M. J. BESSON, and J. GLOWINSKI, 1970. Increased release of dopamine from striatal dopaminergic terminals in the rat after treatment with a neuroleptic: thioproperazine. *Europ. J. Pharmacol.* 10:206–214.

CLARK, R. B., and J. P. PERKINS, 1971. Regulation of adenosine 3′:5′-cyclic monophosphate concentration in cultured human astrocytoma cells by catecholamines and histamine. *Proc. Nat. Acad. Sci. USA* 68:2757–2760.

COYLE, J. T., and S. H. SNYDER, 1969. Catecholamine uptake by synaptosomes in homogenates of rat brain: Stereospecificity in different areas. *J. Pharmacol. Exp. Ther.* 170:221–231.

EISENFELD, A. J., J. AXELROD, and L. KRAKOFF, 1966. Inhibition of the extraneuronal accumulation and metabolism of norepinephrine by adrenergic blocking agents. *J. Pharmacol. Exp. Ther.* 156:107–113.

ENERO, M. A., S. Z. LANGER, R. P. ROTHLIN, and F. J. E. STEFANO, 1972. Role of the α-adrenoceptor in regulating noradrenaline overflow by nerve stimulation. *Brit. J. Pharmacol.* 44:672–688.

ESTERHUIZEN, A. C., J. D. P. GRAHAM, J. D. LEVER, and T. L. B. SPRIGGS, 1968. Catecholamines and acetylcholinesterase distribution in relation to noradrenaline release. An enzyme histochemical and autoradiographic study on the innervation of the cat nictitating muscle. *Brit. J. Pharmac.* 32:46–56.

FLOREY, E., 1967. Neurotransmitters and modulators in the animal kingdom. *Fed. Proc.* 26:1164–1178.

GILMAN, A. G., and M. NIRENBERG, 1971. Effect of catecholamines on the adenosine 3′:5′-cyclic monophosphate concentrations of clonal satellite cells of neurons. *Proc. Nat. Acad. Sci. USA* 68:2165–2168.

GLOWINSKI, J., 1972. In *Perspectives in Neuropharmacology*, S. H. Snyder, ed. New York: Oxford University Press.

HAGGENDAL, J., 1970. Some further aspects on the release of the adrenergic transmitter. In *Bayer Symposium II. New Aspects of Storage and Release Mechanisms of Catecholamines*, H. J. Schumann and G. Kroneberg, eds. Berlin: Springer-Verlag, pp. 100–109.

HEDQVIST, P., 1970. Studies on the effect of prostaglandins E_1 and E_2 on the sympathetic neuromuscular transmission in some animal tissues. *Acta. Physiol. Scand.* 79: Suppl. 345.

HEDQVIST, P., and U. S. VON EULER, 1972a. Prostaglandin-induced neurotransmission failure in the field stimulated, isolated vas deferens. *Neuropharmacol.* 11:177–187.

HEDQVIST, P., and U. S. VON EULER, 1972b. Prostaglandin controls neuromuscular transmission in guinea pig vas deferens. *Nature* N.B. (*Lond.*) 236:113–115.

HOKFELT, T., and Å. LJUNGDAHL, 1971a. Uptake of ³H-noradrenaline and γ-³H-aminobutyric acid in isolated tissues of rat: An autoradiographic and fluorescence microscopic study. *Progress in Brain Research* 34:87–102.

HÖKFELT, T., and Å. LJUNGDAHL, 1971b. Light and electron microscopic autoradiography on spinal cord slices after incubation with labelled glycine. *Brain Res.* 32:189–194.

HOPKIN, J. M., and M. J. NEAL, 1970. The release of ¹⁴C-glycine from electrically stimulated rat spinal cord slices. *Brit. J. Pharmacol.* 40:136–138P.

HORN, A. S., J. T. COYLE, and S. H. SNYDER, 1971. Catecholamine uptake by synaptosomes from rat brain. Structure-activity relationships of drugs with differential effects on dopamine and norepinephrine neurons. *Molec. Pharmacol.* 7:66–80.

IVERSEN, L. L., 1967. The Uptake and Storage of Noradrenaline in Sympathetic Nerves. London: Cambridge University Press.

IVERSEN, L. L., 1971. Role of transmitter uptake mechanisms in synaptic neurotransmission. *Brit. J. Pharmacol.* 41:571–591.

IVERSEN, L. L., 1972a. The uptake, storage, release and metabolism of GABA in inhibitory nerves. In *Perspectives in Neuropharmacology*, S. Snyder, ed. New York: Oxford University Press, pp. 75–111.

IVERSEN, L. L., 1972b. The uptake of biogenic amines. In *Biogenic Amines and Physiological Membranes in Drug Therapy*, J. A. Biel and L. G. Abood, eds. New York: Marcel Dekker, pp. 259–327.

IVERSEN, L. L., and F. E. BLOOM, 1972. Studies of the uptake of ³H-GABA and ³H-glycine in slices and homogenates of rat brain and spinal cord by electron microscopic autoradiography. *Brain Res.* 41:131–143.

IVERSEN, L. L., and P. J. SALT, 1970. Inhibition of catecholamine uptake by steroids in the isolated rat heart. *Brit. J. Pharmacol.* 40:828–830.

IVERSEN, L. L., and P. J. SALT, 1972. Inhibition of the extraneuronal uptake of catecholamine in the isolated rat heart by cholesterol. *Nature* N.B. (*Lond.*) 238:91–92.

IVERSEN, L. L., and F. SCHON, 1973. The use of autoradiographic techniques for the identification and mapping of transmitter-specific neurons in CNS. In *New Concepts in Transmitter Regulation*, D. Segal and A. Mandell, eds. New York: Plenum Press (in press).

JOHNSTON, G. A. R., and L. L. IVERSEN, 1971. Glycine uptake in rat central nervous system slices and homogenates: Evidence for different uptake systems in spinal cord and cerebral cortex. *J. Neurochem.* 18:1951–1961.

JOHNSTON, G. A. R., and J. F. MITCHELL, 1971. The effect of bicuculline, metrazol, picrotoxin and strychnine on the release of ³H-GABA from rat brain slices. *J. Neurochem.* 18:2441–2446.

KATZ, R. I., and I. J. KOPIN, 1969. Electrical field-stimulated release of norepinephrine-H³ from rat atrium: Effects of ions and drugs. *J. Pharmacol. Exp. Ther.* 169:229–236.

KEBABIAN, J. W., and P. GREENGARD, 1971. Dopamine-sensitive adenyl cyclase: Possible role in synaptic transmission. *Science* 174:1346–1348.

KIRPEKAR, S. M., and M. PUIG, 1971. Effect of flow-stop on noradrenaline release from normal spleen and spleens treated with cocaine-phentolamine or phenoxybenzamine. *Brit. J. Pharmacol.* 43:359–369.

LANGER, S. Z., 1970. The metabolism of ³H-noradrenaline released by electrical stimulation from the isolated nictitating membrane of the cat and from the vas deferens of the rat. *J. Physiol.* 208:515–546.

LIGHTMAN, S., and L. L. IVERSEN, 1969. Role of uptake₂ in the extraneuronal metabolism of ³H-noradrenaline in the isolated rat heart. *Brit. J. Pharmacol.*

LOFFELHOLZ, K., and E. MUSCHOLL, 1969. A muscarinic inhibition of the noradrenaline release evoked by postganglionic sympathetic nerve stimulation. *Naunyn. Schmiedeberg. Arch. Pharm. Exp. Path.* 265:1–15.

LOGAN, W. J., and S. H. SNYDER, 1972. High affinity uptake systems for glycine, glutamic and aspartic acids in synaptosomes of rat central nervous tissue. *Brain Res.* 42:413–431.

McILWAIN, H., and H. S. BACHELARD, 1971. Biochemistry and the Central Nervous System, 4th Ed. Edinburgh and London: Churchill Livingstone, pp. 61–97.

MACKAY, A. V. P., and L. L. IVERSEN, 1972a. Trans-synaptic regulation of tyrosine hydroxylase activity in adrenergic neurons: Effect of potassium concentration on cultured sympathetic ganglia. *Naunyn. Schmiedeberg. Arch. Pharm. Exp. Path.* 272:225–229.

MACKAY, A. V. P., and L. L. IVERSEN, 1972b. Dibutyryl cyclic AMP: Induction of tyrosine hydroxylase in cultured sympathetic ganglia. *Brain Res.* 48:424–426.

MATUS, A. I., and M. E. DENNISON, 1971. Autoradiographic localization of tritiated glycine at "flat vesicle" synapses in spinal cord. *Brain Res.* 32:196–197.

MILEDI, R., and L. T. POTTER, 1971. Acetylcholine receptors in muscle fibers. *Nature* (*Lond.*) 233:599–603.

MITCHELL, J. F., 1963. The spontaneous and evoked release of acetylcholine from the cerebral cortex. *J. Physiol. (Lond.)* 165:98–116.

MOLENAAR, P. C., and R. L. POLAK, 1970. Stimulation by atropine of acetylcholine release and synthesis in cortical slices from rat brain. *Brit. J. Pharmacol.* 40:406–417.

MUELLER, R. A., H. THOENEN, and J. AXELROD, 1969. Increase in tyrosine-hydroxylase activity after reserpine administration. *J. Pharmacol. Exp. Ther.* 169:74–79.

NEAL, M. J., 1971. The uptake of ^{14}C-glycine by slices of mammalian spinal cord. *J. Physiol. (Lond.)* 215:103–118.

NEAL, M. J., and L. L. IVERSEN, 1972. Autoradiographic localization of ^{3}H-GABA in rat retina. *Nature* N.B. 235: 217–218.

OBATA, K., and K. TAKEDA, 1969. Release of γ-aminobutyric acid into the fourth ventricle induced by stimulation of the cat's cerebellum. *J. Neurochem.* 16:1043–1047.

PATON, W. D. M., and E. S. VIZI, 1969. The inhibitory action of noradrenaline on acetylcholine output by guinea-pig ileum longitudinal muscle strip. *Brit. J. Pharmacol.* 35:10–28.

SCHON, F., and L. L. IVERSEN, 1972. Selective accumulation of ^{3}H-GABA by stellate cells in rat cerebellar cortex in vivo. *Brain Res.* 42:503–507.

SHASKAN, E. A., and S. H. SNYDER, 1970. Kinetics of serotonin uptake into slices from different regions of rat brain. *J. Pharmacol. Exp. Ther.* 175:404–418.

SIGGINS, G. R., B. J. HOFFER, and F. E. BLOOM, 1971. Studies on norepinephrine-containing afferents to Purkinje cells of rat cerebellum. III. Evidence for mediation of norepinephrine effects by cyclic 3′,5′-adenosine monophosphate. *Brain Res.* 25:535–553.

SILBERSTEIN, S. D., S. BRIMIJOIN, P. B. MOLINOFF, and L. LEMBERGER, 1972. Induction of dopamine-β-hydroxylase in rat superior cervical ganglia in organ culture. *J. Neurochem.* 19: 919–921.

SNODGRASS, R. J., and L. L. IVERSEN, 1972. A sensitive double isotope derivative assay used to measure release of amino acids from brain in vitro. *Nature* N.B. (*Lond.*) 241:154–156.

SNYDER, S. H., and J. T. COYLE, 1968. Regional differences in H^{3}-norepinephrine and H^{3}-dopamine uptake into rat brain homogenates. *J. Pharmacol. Exp. Ther.* 165:78–86.

SRINIVASAN, V., M. J. NEAL, and J. F. MITCHELL, 1969. The effect of electrical stimulation and high potassium concentration on the efflux of ^{3}H-γ-aminobutyric acid from brain slices. *J. Neurochem.* 16:1235–1244.

STARKE, K., 1972. Alpha sympathomimetic inhibition of adrenergic and cholinergic transmission in the rabbit heart. *Naunyn. Schmiedeberg. Arch. Pharm. Exp. Path.* 274:18–45.

STARKE, K., and H. J. SCHUMANN, 1972. Interaction of angiotensin, phenoxybenzamine and propanolol on noradrenaline release during sympathetic nerve stimulation. *Europ. J. Pharmacol.* 18:27–30.

THOENEN, H., R. KETTLER, and A. SANER, 1972. Time course of the development of enzymes involved in the synthesis of norepinephrine in the superior cervical ganglion of the rat from birth to adult life. *Brain Res.* 40:459–468.

THOENEN, H., R. KETTLER, W. BURKARD, and A. SANER, 1971. Neurally mediated control of enzymes involved in the synthesis of norepinephrine: Are they regulated as an operational unit? *Naunyn. Schmiedeberg. Arch. Pharm. Exp. Path.* 270:146–160.

THOENEN, H., R. A. MUELLER, and J. AXELROD, 1970. Phase difference in the induction of tyrosine hydroxylase in cell body and nerve terminals of sympathetic neurons. *Proc. Nat. Acad. Sci. USA* 65:58–62.

VAN GELDER, N. M., 1968. A possible enzyme barrier for γ-aminobutyric acid in the central nervous system. *Progr. Brain Res.* 29:259–271.

WENNMALM, Å, 1971. Studies on mechanisms controlling the secretion of neurotransmitters in the rabbit heart. *Acta Physiol. Scand.* 82: Suppl. 365.

80 Localizing Serotonin in Central and Peripheral Nerves

GUILLERMO JAIM-ETCHEVERRY
and LUIS MARÍA ZIEHER

ABSTRACT   In order to determine the extent to which a specific neurochemical circuit may be subject to, or responsive in, functional modulation, it is necessary to develop methods for localizing specific neurotransmitters cytochemically. The extent to which this can be accomplished now will be explored by consideration of the methods available for the localization of serotonin in central and peripheral nerve terminals. A cytochemical reaction for the ultrastructural identification of serotonin in combination with biochemical procedures demonstrates that, in pineal adrenergic endings, serotonin is stored in synaptic vesicles, which also contain norepinephrine. Such a coexistence of neurotransmitters within storage organelles may constitute a physiologically significant mode of modulating synaptic activity.

THE IDENTIFICATION of a substance as the neurotransmitter for a particular neural pathway requires the demonstration of its presence in the nerve endings of this pathway. Therefore, morphological and cytochemical studies constitute an important tool to determine the reality of synaptic modulatory mechanisms, because the information provided by such studies, together with the pertinent neurophysiological and pharmacological evidence, is essential to postulate the involvement of a given substance in synaptic transmission.

This chapter summarizes the current status of the efforts aimed at the cellular localization of one of these proposed neurotransmitters, serotonin (5-HT), in central and peripheral endings. The problems posed by the demonstration of this indoleamine, as well as the conclusions obtained in these studies, are similar in many respects to those of investigations carried out to localize other groups of neurotransmitters, such as the catecholamines (CA). Evidence is provided indicating the importance of cytochemical methods at the ultrastructural level to analyze the storage properties of nerve vesicles in aminergic endings and to show their contribution to the understanding of modulatory mechanisms in these synapses.

GUILLERMO JAIM-ETCHEVERRY AND LUIS MARÍA ZIEHER   Instituto de Anatomía General y Embriología and Cátedra de Farmacología, Facultad de Medicina, Universidad de Buenos Aires, Buenos Aires, Argentina

Localization of serotonin in the central nervous system

Shortly after the description of the occurrence of high concentrations of 5-HT in the central nervous system (CNS), cell fractionation studies showed that brain 5-HT was concentrated in fractions containing nerve endings (Zieher and De Robertis, 1963) and, more precisely, in synaptic vesicles (see Pellegrino de Iraldi et al., 1968). These findings, as well as the decline in brain 5-HT following lesions of the medial forebrain bundle (see Moore, 1970), clearly indicated that brain 5-HT is localized in neurons.

The introduction of the histochemical fluorescence method for the demonstration of CA and 5-HT by Hillarp, Falck, and co-workers (see Corrodi and Jonsson, 1967) provided a precise morphological identification of a system of serotonin-containing neurons. Their cell bodies, with low concentrations of 5-HT, are localized in the raphe nuclei, and their axons project rostrally to the telediencephalon and caudally to the spinal cord. Detailed accounts of the localization of serotonergic pathways have been published, but their consideration is beyond the scope of this analysis (see Fuxe et al., 1968). However, the demonstration of 5-HT with fluorescence histochemistry has certain drawbacks that render its localization less precise than that of CA. Due to a low fluorescence yield of the compound, the demonstration of 5-HT has largely depended on the drug-induced increase in its concentration. The fluorophore of the amine is rapidly decomposed when exposed to blue excitation light, and this interferes with the demonstration of 5-HT-containing structures, which are usually very fine terminal and preterminal fibers. However, the accuracy of the localization obtained with this procedure has been confirmed by the results of experiments in which the concentration of 5-HT in the brain was reduced after placing electrolytic lesions in the area of the raphe nuclei (Bloom and Costa, 1971) and by those showing an accelerated catabolism of 5-HT as a result of stimulation of these nuclei (Sheard and Aghajanian, 1968). The recent introduction of 5,6-

dihydroxytryptamine, a compound that selectively destroys indoleamine-containing neurons, opens new possibilities for the identification of these structures as well as for the analysis of their physiological significance (Baumgarten et al., 1972a).

However, one of the most desirable aims in the study of the localization of biogenic amines is to achieve their demonstration at the fine structural level. The first ultrastructural studies carried out in zones of the CNS containing high concentrations of monoamines showed that the nerve endings contain a mixed population of synaptic vesicles. One type is the so-called large granulated vesicle (LGV), which contains a dense core and has a diameter of 800 to 1200 Å, whereas the other is represented by small electron-lucent vesicles of 300 to 500 Å in diameter (Pellegrino de Iraldi et al., 1963a). This vesicular population was also found in endings of the suprachiasmatic nucleus, which receives a rich supply of serotonin-containing terminals (Suburo and Pellegrino de Iraldi, 1969). On the basis of the good correlation found in the periphery between the frequency of the dense cores in small vesicles and the presence of reducing amines (Pellegrino de Iraldi and De Robertis, 1961), it was postulated that the LGV constitute one of the storage sites of biogenic amines. Since the preparative techniques used do not permit a differentiation between CA and 5-HT, the conclusions obtained may be applied to both types of monoamines. A great effort was devoted thereafter to try to substantiate this proposed association, but the pharmacological manipulation of the stores of brain amines subjected to morphological analysis gave conflicting results (see Bloom and Aghajanian, 1968).

One of the fruitful approaches in the study of this problem was to follow the fate of labeled amines injected intracisternally by electron microscopic autoradiography. Exogenous 5-HT was taken up by nerve endings in areas rich in endogenous amine, and the label was found in terminals containing LGV and electron lucent small vesicles (Aghajanian and Bloom, 1967). However, the resolution inherent to the technique of autoradiography with available emulsions is in the order of 80 to 150 mμ, and therefore the distinction among adjacent synaptic vesicles is not possible (Bloom, 1970).

Other experimental approaches have also indicated the possible correlation of certain LGV with amine storage, because LGV exhibit a central core giving a reaction characteristic of monoamines (Wood, 1966); these vesicles were also identified in a vesicular fraction rich in monoamines isolated from rat hypothalamus (De Robertis et al., 1965). Following lesions of the raphe nuclei, the degenerating nerve terminals found in the suprachiasmatic nucleus have LGV (Aghajanian et al., 1969) and ultrastructural studies carried out shortly after the intra-

ventricular injection of 5,6-dihydroxytryptamine show an increased osmiophilia in the LGV of periventricular terminals, which afterward undergo degeneration (Baumgarten et al., 1972b). However, in a thorough analysis of the significance of LGV, no conclusive evidence could be provided to support their association with monoamine storage, because no changes were observed in the frequency or density of the cores of such vesicles following drug-induced modifications of amine content. It was concluded that endings containing LGV were most frequently found in zones of the brain characterized by high concentrations of monoamines (Bloom and Aghajanian, 1968).

It is relevant to this point to mention that the demonstration of the participation of the LGV of peripheral adrenergic endings in monoamine storage was possible only by using cytochemical procedures. The results of these studies supported the idea that LGV found in postganglionic sympathetic axons as well as those LGV present in neurochemically different types of nerve terminals, e.g. preganglionic cholinergic fibers and neuromuscular junctions, are composed of a proteinaceous matrix and, in adrenergic nerves, are capable of binding endogenous as well as exogenous norepinephrine (NE) or 5-HT (Tranzer and Thoenen, 1968; Jaim-Etcheverry and Zieher, 1968b, 1969b). The results obtained using recently developed preparative techniques or selective pharmacological labeling of amine stores indicate that LGV participate in amine storage in central synapses (Hökfelt, 1968; Richards and Tranzer, 1970).

These procedures have also been useful to demonstrate the presence of dense cores in small vesicles found in central endings. In 1967, Hökfelt used permanganate fixation on certain zones of the brain to show for the first time small granular vesicles similar to those found in the periphery. These small granular vesicles were present both in normal tissue and after incubation of brain slices with several amines (Hökfelt, 1968). These findings were confirmed with the use of 5-OH-dopamine, a compound that labels monoamine stores in a very convenient manner for their ultrastructural demonstration (Richards and Tranzer, 1970), and recently these last authors found small and large reactive sites in some endings after perfusion with dichromate (personal communication). Although in this case the preservation of the tissue is poor, the reaction obtained seems to be more specific than with permanganate, because, apart from resulting in poor preservation of general ultrastructure, permanganate stains the majority of cellular components nonselectively.

It is generally admitted that the sole presence of LGV in a synapse does not constitute an indication of the aminergic nature of the ending. Nor does the finding of small granular vesicles even in normal brain tissue necessarily imply that a biogenic amine is deposited there, because several observations indicate that all vesicular granularity may not be related to amine storage (see Bloom, 1970). Therefore, several criteria for the cytochemical localization of biogenic amines, which will be considered later, must be fulfilled before reaching definitive conclusions on the significance of a particular observation.

No conclusive information exists with regard to the identification of amine storage organelles within nerve cell bodies of central structures, except for the presence of 5-HT-storing granules in the colossal cells of the ganglion of the leech (Rude et al., 1969). The use of 5-OH-dopamine has failed to provide indications on the nature of the site of storage of amines in neuronal cell bodies, and this failure may be related to the difficulty in preserving the amine loosely bound to structures other than synaptic vesicles (Eränkö, 1967).

In spite of the technical difficulties posed by the handling of brain tissue during the reactive oxidation procedures required by currently available techniques, there are strong indications that CA and 5-HT are localized in central endings within synaptic vesicles and that the organization of the presynaptic terminal in central monoaminergic junctions is very similar to that of the corresponding peripheral endings. However, in spite of the technical improvements made, the cytochemical identification of 5-HT, as well as of CA, in a reliable manner to permit a reproducible demonstration of central aminergic neurons has not been achieved and will require the development of new cytochemical procedures. Although the molecular characteristics of stored monoamines raise many problems, it might be possible to devise immunocytochemical methods for their demonstration by coupling them to suitable antigenic substances as well as by developing specific osmiophilic reagents to permit their ultrastructural recognition in tissue sections.

Localization of serotonin in peripheral adrenergic nerves

Recent studies have shown the existence of a specific 5-HT uptake process in certain elements of the myenteric plexus (Gershon and Altman, 1971). However, the processes of amine uptake and storage in peripheral noradrenergic nerves have a certain unspecificity, and this is responsible for the presence of 5-HT in the nerve endings of the pineal gland of the rat. These endings constitute the sole nervous supply of the gland (Ariëns Kappers, 1960) and correspond to noradrenergic neurons, which have their cell bodies in the superior cervical ganglia. They contain NE, and several studies carried out with different techniques coincide in demonstrating that these terminals also store 5-HT (Pellegrino de Iraldi et al.,

1963b, 1965; Owman, 1964; Neff et al., 1969). Serotonin is synthesized from tryptophan by pineal parenchymal cells, where it serves as a precursor in melatonin biosynthesis (see Wurtman et al., 1968). The amine traverses the axonal membrane and is incorporated within the ending (Owman, 1964; Taxi and Droz, 1966a; Neff et al., 1969). Therefore, the nerve terminals of the pineal gland of the rat are unique in that they store NE and 5-HT and constitute a suitable model to study cytochemically the localization of both amines and thus determine their mechanism of storage within the nerve.

The analysis of the reactions between several fixatives with amines in vitro, confirmed by light microscopy, led Wood (1966, 1967) to propose a method to distinguish between catecholamines and indoleamines based on differences in the reduction of potassium dichromate by amines previously complexed with aldehydes. Using isolated blood platelets as a model system, we have analyzed the specificity of this reaction at the electron microscope level, and we have concluded that the fixation of the tissue with glutaraldehyde and the treatment with acid potassium dichromate (GD reaction) reveals the storage sites of both CA and 5-HT. On the other hand, if the tissue is fixed in formaldehyde prior to glutaraldehyde-dichromate treatment (FGD reaction), the reaction given by CA disappears while 5-HT storing sites are demonstrated (Jaim-Etcheverry and Zieher, 1968a).

When the pineal gland is fixed with formaldehyde and then submitted to glutaraldehyde-dichromate treatment (FGD reaction), reactive sites are observed in pericapillary processes (Figure 1) (Jaim-Etcheverry and Zieher, 1968b). These structures correspond in size to the dense cores of both types of granulated vesicles that are characteristically revealed by conventional techniques in postganglionic sympathetic fibers and contain NE (Pellegrino de Iraldi and De Robertis, 1961; see Jaim-Etcheverry and Zieher, 1971b).

Several criteria that have been formulated by Bloom (1970) must be fulfilled to demonstrate the validity of the cytochemical localization of monoamines at the ultrastructural level. On the basis of these criteria, the following constitute the evidence showing that the formaldehyde-glutaraldehyde-dichromate method selectively demonstrates 5-HT deposits in pineal nerves:

1. Test-tube experiments indicate that while the GD reaction results in precipitates both with CA (NE and dopamine) and 5-HT, the pretreatment with formaldehyde blocks the reaction given by CA and does not modify the precipitate formed by 5-HT upon addition of dichromate (Wood, 1966; Jaim-Etcheverry and Zieher, 1971b).

2. Nerve endings in the vas deferens contain NE and not 5-HT, as shown by biochemical and light microscopic histochemical studies. When the pineal gland and the vas

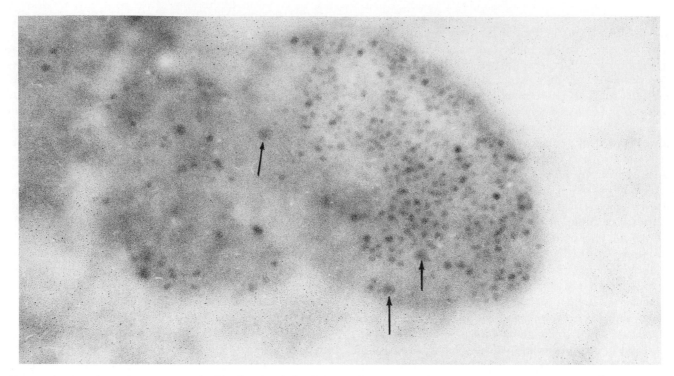

FIGURE 1   A perivascular nerve terminal in the pineal gland of the rat processed with the formaldehyde-glutaraldehyde-dichromate (FGD) technique. The reactive sites observed correspond in size to the dense cores of the granular vesicles characteristic of adrenergic nerves in conventional preparations. There are small reactive granules of 200 to 400 Å in diameter that correspond to the small granular vesicles, as well as larger granules of 600 to 800 Å (arrows); these are the cores of the large granulated vesicles (LGV). X 50,000.

dcfcrcns wcrc simultaneously processed, sites reacting after the FGD procedure were observed only in pineal gland processes. No reaction could be detected in the endings of the vas deferens, which in turn gave a positive reaction with the GD technique, indicating the presence of NE (Jaim-Etcheverry and Zieher, 1968b). In spite of the high concentration of 5-HT in pineal cells, it has not been possible to demonstrate the nature of its stores at the ultrastructural level. This is probably the result of the inability of current techniques of fixation in liquid solutions to preserve the amine accumulated extra-vesicularly.

3. The amount of reactive material present in pineal processes fluctuates in proportion to the level of the amine biochemically assayed. Following reserpine, all reactive sites disappear with both cytochemical techniques (Jaim-Etcheverry and Zieher, 1969b). When 5-HT is depleted from the neuronal compartment, either by p-chloro-phenylalanine (PCP) or by desmethylimipramine (DMI), as will be discussed later, NE remains in the endings and reactive sites are observed with the GD procedure, while no reaction is observed with the FGD technique for 5-HT (Figure 2) (Pellegrino de Iraldi and Gueudet, 1969;

Bloom and Giarman, 1970; Jaim-Etcheverry and Zieher, 1971a). On the contrary, if NE is depleted by administration of α-methyltyrosine, reactive sites are observed with the FGD procedure, indicating the persistence of neuronal 5-HT (Jaim-Etcheverry and Zieher, 1971b). Moreover, the endings of the vas deferens, which do not react with the FGD procedure, show the presence of reactive cores following injection of 5-HT or incubation in vitro with the amine, as will be mentioned later.

4. Autoradiographic studies of the pineal gland following the administration of labeled 5-OH-tryptophan or 5-HT, have shown the presence of silver grains in pineal nerve endings containing granulated vesicles (Taxi and Droz, 1966a).

5. The interruption of the nerve pathway, which gives rise to the monoamine-containing fibers, achieved by surgical extirpation of both superior cervical ganglia, which depletes pineal NE and reduces the 5-HT (Pellegrino de Iraldi et al., 1963b; Pellegrino de Iraldi and Zieher, 1966), results in the disappearance of processes containing cytochemically reactive sites (Jaim-Etcheverry and Zieher, 1968b).

6. It is at present impossible to demonstrate the pres-

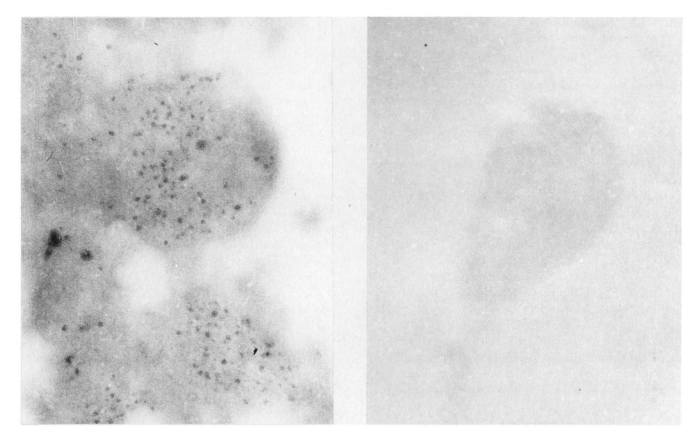

FIGURE 2   Perivascular nerve endings in the pineal gland of a rat treated with p-chlorophenylalanine (2 × 350 mg/kg ip, 24 hr interval and killed 24 hr after the last injection). One-half of the gland was treated with glutaraldehyde and potassium dichromate (GD technique), which depicts cellular sites containing both catecholamines and 5-HT. Since the treatment does not reduce pineal norepinephrine, a positive reaction is present in the nerves as observed in the micrograph on the left. The 5-HT content of the gland is reduced by 85% with this procedure, and when the other half of the gland is processed with the formaldehyde-glutaraldehyde-dichromate procedure for the demonstration of 5-HT, no reaction is observed in perivascular nerve processes as shown on the right. X 42,000.

ence in the terminal of the enzymes related to monoamine synthesis and catabolism due to the lack of reliable methods at the ultrastructural level.

This evidence indicates that 5-HT is localized in the cores of both large and small granulated vesicles of pineal nerves. The distribution of reactive sites in the endings suggests that both amines NE and 5-HT are stored not only in the same terminal but within the same vesicle, and several experiments were carried out to support this hypothesis.

Adrenergic nerve endings other than those of the pineal are also able to incorporate exogenous 5-HT, as confirmed by fluorescence-histochemical, autoradiographic and biochemical studies (Owman, 1964; Taxi and Droz, 1966b; Thoa et al., 1969). They also have the capacity to store the amine in the vesicles, as was shown in special experimental conditions. When 5-HT was injected even in very high doses to normal rats, the amine could not be detected cytochemically in the vesicles of the adrenergic nerves of the vas deferens. But if these vesicles were depleted of their NE, the injection of the same dose of 5-HT resulted in a cytochemically demonstrable incorporation of the exogenous amine in the vesicles. This indicates (1) that the concentration of 5-HT achieved around the vesicles was not sufficient to displace endogenous NE, and (2) that the vesicle itself can incorporate and store the amine (Jaim-Etcheverry and Zieher, 1969a; Snipes et al., 1968). Therefore, the presence of cytochemically detectable 5-HT within adrenergic nerve vesicles is directly related to the concentration of 5-HT in the surroundings of the ending, as was further confirmed by results of in vitro experiments (Zieher and Jaim-Etcheverry, 1971). Exogenous 5-HT could be shown within vesicles of nerves in slices of normal vas deferens, i.e., not depleted of their NE, incubated with 5-HT, provided that the concentration of 5-HT in the medium was of

5.6 × 10⁻⁴ M or higher. In this situation, endogenous NE was not significantly depleted but with increasing concentrations of 5-HT a concentration-dependent depletion of NE followed. Results of cell fractionation studies performed on slices of normal and chemically denervated vas deferens, incubated with 5-HT, led to the conclusion that in certain conditions, exogenous 5-HT is present along with NE in adrenergic nerve vesicles. These results gave support to the hypothesis that the coexistence of both amines at the level of granular vesicles may represent their mechanism of storage in pineal nerves.

If this assumption is correct, the selective depletion of pineal 5-HT should leave available storage sites within the vesicles, cytoplasmic NE would enter into the partially depleted organelles, NE synthesis would be accelerated, and the concentration of pineal NE would be increased. This was confirmed following the depletion of neuronal 5-HT either by the administration of PCP, an inhibitor of 5-HT synthesis, which depletes the amine in the whole gland (Bloom and Giarman, 1970), or of DMI, which blocks the mechanism of 5-HT transport across the nerve membrane and thus selectively depletes neuronal 5-HT (Neff et al., 1969). In both cases a marked and selective increase in pineal NE levels was observed (Jaim-Etcheverry and Zieher, 1971a). After depleting pineal NE, Zweig and Axelrod (1969) observed an increase in neuronal 5-HT, a finding that further supports the mechanism of storage proposed.

Cell fractionation studies carried out in our laboratory have shown that a high percentage of NE is recovered in the microsomal fraction obtained from rat pineal homogenates. On the basis of comparisons made between the distribution of 5-HT in homogenates from normal pineals and those of rats treated with DMI and therefore with a selective depletion of neuronal 5-HT, it was concluded that the 5-HT recovered in the microsomal fraction derives almost exclusively from the neuronal compartment. Theoretical calculations made on the basis of previous kinetic studies of pineal 5-HT compartments supported this conclusion (unpublished observations). Therefore, both amines present in the nerves also coexist in the same subcellular fraction.

The evidence gathered in the course of these studies indicates that in the nerve endings of the pineal gland of the rat, 5-HT and NE coexist in the same storage organelle and therefore are most probably liberated together by nerve impulses. The control exerted by sympathetic nerves on pineal indole metabolism is mediated by NE acting as the neurotransmitter of these nerves (Wurtman et al., 1969). But the simultaneous liberation of 5-HT may modify the net effect of sympathetic nerve stimulation, which would depend on the ratio NE:5-HT present in the vesicles, because these amines have different pharmaco-

logical activities. Serotonin inhibits NE uptake by the axonal membrane of the adrenergic nerve (Burgen and Iversen, 1965; Jester and Horst, 1972), and both amines have different actions on pineal cell receptors. While NE stimulates adenyl cyclase activity, 5-HT blocks this stimulation (Weiss and Costa, 1968), and this effect is probably identical to its adrenergic receptor blocking activity (Jester and Horst, 1972). On the basis of the existence of a dynamic equilibrium between parenchymal and neuronal 5-HT compartments (Neff et al., 1969) it is conceivable that pineal indole synthesis can be controlled by a feedback system operating through modifications in the relative amount of both amines liberated by nerve impulses. Moreover, nervous activity may elicit different responses from pineal receptors depending on the moment of the lighting cycle considered, since the 24 hr cyclic variations in pineal NE and 5-HT levels (see Wurtman et al., 1968) are most probably reflected in changes in the ratio of vesicular amines.

The coexistence of monoamines within synaptic vesicles, which can be pharmacologically induced by "false" adrenergic transmitters or their precursors (Kopin, 1968), may constitute a physiologically significant mode of storage. Recently Molinoff and Axelrod (1972) have shown the presence of octopamine in normal adrenergic nerves. This amine, which has a weak sympathomimetic activity, is stored in vesicles and liberated by nerve stimulation, thus justifying its denomination of "cotransmitter." The 5-HT stored in pineal nerves may be also properly designated by this term.

This mechanism probably operates in central adrenergic endings, but the lack of sufficiently refined biochemical, physiological, or morphological techniques to demonstrate its reality may explain the difficulties encountered in recognizing its existence. There are many indications that nerve endings from catecholamine-containing neurons are able to take up and store exogenous 5-HT, a capacity that has been shown pharmacologically, biochemically, and histochemically both at the light and electron microscope levels (Lichtensteiger et al., 1967; Hökfelt, 1968; Fuxe and Ungerstedt, 1968; Shaskan and Snyder, 1970). This uptake requires the presence of moderately high concentrations of the amine around the endings, a situation that might occur normally in certain endings as a consequence of the physiological liberation of 5-HT from neighboring terminals.

In view of these findings, Dale's principle (1935), stating that a neuron secretes the same transmitter from all its synaptic endings, cannot be universally applied, although it would still hold in the strictest case, because the sympathetic fibers probably do not normally synthesize any transmitters that are not catecholamines. This principle implies that the process of neuronal differentiation determines which particular secretory product will be manufactured, stored, and released by a given neuron (Iversen, 1970). The synthetic mechanism necessary for the elaboration of the adrenergic neurotransmitter may be fairly specific, although closely related substances can be formed totally or partially along these metabolic pathways, as indicated by the presence of octopamine in normal adrenergic nerves. But the unspecificity of the uptake and storage mechanisms of the adrenergic neuron opens the possibility for many interactions between its terminal branches and their surrounding medium, which can result in the differential release of transmitters. If the physiological coexistence of amines within nerve endings is more conclusively demonstrated, the concepts of systems containing exclusively one type of amine should be revised.

The development of sensitive and reliable cytochemical methods for the localization of neurotransmitters, in association with other techniques, will contribute to the clarification of this problem and to the determination of the importance of the mechanism of transmitter storage proposed for the physiologic modulation of synaptic activity.

ACKNOWLEDGMENTS The original work reported has been supported by Research Grants from the Consejo Nacional de Investigaciones Científicas y Técnicas-Fondo de Farmacología-Republica Argentina and National Institutes of Health, Grant 5-RO 1-NS-06953-05 NEUA, USA.

REFERENCES

AGHAJANIAN, G. K., and F. E. BLOOM, 1967. Localization of tritiated serotonin in rat brain by electron-microscopic autoradiography. J. Pharmacol. Exp. Ther. 156:23–30.

AGHAJANIAN, G. K., F. E. BLOOM, and M. H. SHEARD, 1969. Electron microscopy of degeneration within the serotonin pathway of rat brain. Brain Res. 13:266–273.

ARIËNS KAPPERS, J., 1960. The development, topographical relations and innervation of the epiphysis cerebri in the albino rat. Z. Zellforsch. Mikrosk. Anat. 52:163–215.

BAUMGARTEN, H. G., L. LACHENMAYER., and H. G. SCHLOSSBERGER, 1972a. Evidence for a degeneration of indoleamine containing nerve terminals in rat brain, induced by 5,6-dihydroxytryptamine. Z. Zellforsch. Mikrosk. Anat. 125:553–569.

BAUMGARTEN, H. G., A. BJÖRKLUND, A. F. HOLSTEIN, and A. NOBIN, 1972b. Chemical degeneration of indoleamine axons in rat brain by 5,6-dihydroxytryptamine. An Ultrastructural Study. Z. Zellforsch. Mikrosk. Anat. 129:256–271.

BLOOM, F. E., 1970. The fine structural localization of biogenic monoamines in nervous tissue. Int. Rev. Neurobiol. 13:27–66.

BLOOM, F. E., and G. K. AGHAJANIAN, 1968. An electron microscopic analysis of large granular synaptic vesicles of the brain in relation to monoamine content. J. Pharmacol. Exp. Ther. 159:261–273.

BLOOM, F. E., and E. COSTA, 1971. The effects of drugs on serotonergic nerve terminals. Adv. Cytopharmacol. 1:379–395.

BLOOM, F. E., and N. J. GIARMAN, 1970. The effects of p-Cl-phenylalanine on the content and cellular distribution of 5-HT in the rat pineal gland: Combined biochemical and electron microscopic analyses. Biochem. Pharmacol. 19:1213–1219.

BURGEN, A. S. V., and L. L. IVERSEN, 1965. The inhibition of noradrenaline uptake by sympathomimetic amines in the rat isolated heart. Brit. J. Pharmacol. Chemother. 25:34–49.

CORRODI, H., and G. JONSSON, 1967. The formaldehyde fluorescence method for the histochemical demonstration of biogenic monoamines. A review on the methodology. J. Histochem. Cytochem. 15:65–78.

DALE, H. H., 1935. Pharmacology and nerve endings. Proc. Roy. Soc. Med. 28:319–332.

DE ROBERTIS, E., A. PELLEGRINO DE IRALDI, G. RODRÍGUEZ DE LORES ARNAIZ, and L. M. ZIEHER, 1965. Synaptic vesicles from the rat hypothalamus. Isolation and norepinephrine content. Life Sci. 4:193–201.

ERÄNKÖ, O., 1967. Histochemistry of nervous tissues: Catecholamines and cholinesterases. Annu. Rev. Pharmacol. 7:203–222.

FUXE, K., T. HÖKFELT, and U. UNGERSTEDT, 1968. Localization of indolealkylamines in CNS. Adv. Pharmacol. 6A:235–251.

FUXE, K., and U. UNGERSTEDT, 1968. Histochemical studies on the distribution of catecholamines and 5-hydroxytryptamine after intraventricular injections. Histochemie 13:16–28.

GERSHON, M. D., and R. F. ALTMAN, 1971. An analysis of the uptake of 5-hydroxytryptamine by the myenteric plexus of the small intestine of the guinea pig. J. Pharmacol. Exp. Ther. 179:29–41.

HÖKFELT, T., 1968. In vitro studies on central and peripheral monoamine neurons at the ultrastructural level. Z. Zellforsch. Mikrosk. Anat. 91:1–74.

IVERSEN, L. L., 1970. Neurotransmitters, neurohormones and other small molecules in neurons. In The Neurosciences: Second Study Program, F. O. Schmitt, ed. New York: The Rockefeller University Press, pp. 768–782.

JAIM-ETCHEVERRY, G., and L. M. ZIEHER, 1968a. Cytochemistry of 5-hydroxytryptamine at the electron microscope level. I. Study of the specificity of the reaction in isolated blood platelets. J. Histochem. Cytochem. 16:162–171.

JAIM-ETCHEVERRY, G., and L. M. ZIEHER, 1968b. Cytochemistry of 5-hydroxytryptamine at the electron microscope level. II. Localization in the autonomic nerves of rat pineal gland. Z. Zellforsch. Mikrosk. Anat. 86:393–400.

JAIM-ETCHEVERRY, G., and L. M. ZIEHER, 1969a. Ultrastructural cytochemistry and pharmacology of 5-hydroxytryptamine in adrenergic nerve endings. I. Localization of exogenous 5-hydroxytryptamine in the autonomic nerves of the rat vas deferens. J. Pharmacol. Exp. Ther. 166:264–271.

JAIM-ETCHEVERRY, G., and L. M. ZIEHER, 1969b. Selective demonstration of a type of synaptic vesicle with phosphotungstic acid staining. J. Cell Biol. 42:855–860.

JAIM-ETCHEVERRY, G., and L. M. ZIEHER, 1971a. Ultrastructural cytochemistry and pharmacology of 5-hydroxytryptamine in adrenergic nerve endings. III. Selective increase of norepinephrine in the rat pineal gland consecutive to depletion of neuronal 5-hydroxytryptamine. J. Pharmacol. Exp. Ther. 178:42–48.

JAIM-ETCHEVERRY, G., and L. M. ZIEHER, 1971b. Ultrastructural aspects of neurotransmitter storage in adrenergic nerves. Adv. Cytopharmacol. 1:343–361.

JESTER, J., and W. D. HORST, 1972. Influence of serotonin on adrenergic mechanisms. *Biochem. Pharmacol.* 21:333–338.

KOPIN, I. J., 1968. False adrenergic transmitters. *Annu. Rev. Pharmacol.* 8:377–394.

LICHTENSTEIGER, W., U. MUTZNER, and H. LANGEMANN, 1967. Uptake of 5-hydroxytryptamine and 5-hydroxytryptophan by neurons of the central nervous system normally containing catecholamines. *J. Neurochem.* 14:489–497.

MOLINOFF, P. B., and J. AXELROD, 1972. Distribution and turnover of octopamine in tissues. *J. Neurochem.* 19:157–163.

MOORE, R. Y., 1970. Brain lesions and amine metabolism. *Int. Rev. Neurobiol.* 13:67–91.

NEFF, N. H., BARRETT, R. E., and E. COSTA, 1969. Kinetic and fluorescent histochemical analysis of the serotonin compartments in rat pineal gland. *Eur. J. Pharmacol.* 5:348–356.

OWMAN, C., 1964. Sympathetic nerves probably storing two types of monoamines in the rat pineal gland. *Int. J. Neuropharmacol.* 2:105–112.

PELLEGRINO DE IRALDI, A., and E. DE ROBERTIS, 1961. Action of reserpine on the submicroscopic morphology of the pineal gland. *Experientia (Basel)* 17:122–123.

PELLEGRINO DE IRALDI, A., H. FARINI-DUGGAN, and E. DE ROBERTIS, 1963a. Adrenergic synaptic vesicles in the anterior hypothalamus of the rat. *Anat. Rec.* 145:521–531.

PELLEGRINO DE IRALDI, A., and R. GUEUDET, 1969. Catecholamines and serotonin in granulated vesicles of nerve endings in the pineal gland of the rat. *Int. J. Neuropharmacol.* 8:9–14.

PELLEGRINO DE IRALDI, A., and L. M. ZIEHER, 1966. Noradrenaline and dopamine content of normal, decentralized and denervated pineal gland of the rat. *Life Sci.* 5:149–154.

PELLEGRINO DE IRALDI, A., L. M. ZIEHER, and E. DE ROBERTIS, 1963b. 5-hydroxytryptamine content and synthesis of normal and denervated pineal gland. *Life Sci.* 1:691–696.

PELLEGRINO DE IRALDI, A., L. M. ZIEHER, and E. DE ROBERTIS, 1965. Ultrastructural and pharmacological studies of nerve endings of the pineal gland. *Prog. Brain Res.* 10:389–421.

PELLEGRINO DE IRALDI, A., L. M. ZIEHER, and G. JAIM-ETCHEVERRY, 1968. Neuronal compartmentation of 5-hydroxytryptamine stores. *Adv. Pharmacol.* 6A:257–270.

RICHARDS, J. G., and J. P. TRANZER, 1970. The ultrastructural localization of amine storage sites in the central nervous system with the aid of a specific marker, 5-hydroxydopamine. *Brain Res.* 17:463–469.

RUDE, S., R. E. COGGESHALL, and L. S. VAN ORDEN III, 1969. Chemical and ultrastructural identification of 5-hydroxytryptamine in an identified neuron. *J. Cell. Biol.* 41:832–854.

SHASKAN, E. G., and S. H. SNYDER, 1970. Kinetics of serotonin accumulation into slices from rat brain: Relationship to catecholamine uptake. *J. Pharmacol. Exp. Ther.* 175:404–418

SHEARD, M. H., and G. K. AGHAJANIAN, 1968. Stimulation of the midbrain raphe. Effect on serotonin metabolism. *J. Pharmacol. Exp. Ther.* 163:425–430.

SNIPES, R. L., H. THOENEN, and J. P. TRANZER, 1968. Fine structural localization of exogenous 5-HT in vesicles of adrenergic nerve terminals. *Experientia (Basel)* 24:1026–1027.

SUBURO, A. M., and A. PELLEGRINO DE IRALDI, 1969. An ultrastructural study of the rat's suprachiasmatic nucleus. *J. Anat.* 105:439–446.

TAXI, J., and B. DROZ, 1966a. Etude de l'incorporation de noradrénaline-^{3}H (NA-^{3}H) et de 5-hydroxytryptophane-^{3}H (5-HTP-^{3}H) dans l'épiphyse et dans le ganglion cervical supérieur. *C. R. Acad. Sci. (Paris)* 263:1326–1329.

TAXI, J., and B. DROZ, 1966b. Etude de l'incorporation de noradrénaline-^{3}H (NA-^{3}H) et de 5-hydroxytryptophane-^{3}H (5-HTP-^{3}H) dans les fibres nerveuses du canal déférent et de l'intestin. *C. R. Acad. Sci. (Paris)* 263:1237–1240.

THOA, N. B., D. ECCLESTON, and J. AXELROD, 1969. The accumulation of C^{14}-serotonin in the guinea-pig vas deferens. *J. Pharmacol. Exp. Ther.* 169:68–73.

TRANZER, J. P., and H. THOENEN, 1968. Various types of amines storing vesicles in peripheral adrenergic nerve terminals. *Experientia (Basel)* 24:484–486.

WEISS, B., and E. COSTA, 1968. Selective stimulation of adenyl cyclase of rat pineal gland by pharmacologically active catecholamines. *J. Pharmacol. Exp. Ther.* 161:310–319.

WOOD, J. G., 1966. Electron microscopic localization of amines in central nervous tissue. *Nature (Lond.)* 209:1131–1133.

WOOD, J. G., 1967. Cytochemical localization of 5-hydroxytryptamine (5-HT) in the central nervous system (CNS). *Anat. Rec.* 157:343.

WURTMAN, R. J., J. AXELROD, and D. E. KELLY, 1968. *The Pineal.* New York: Academic Press.

WURTMAN, R. J., H. M. SHEIN, J. AXELROD, and F. LAREN, 1969. Incorporation of C^{14}-tryptophan into C^{14}-protein by cultured rat pineals: Stimulation by l-norepinephrine. *Proc. Natl. Acad. Sci. USA* 62:749–755.

ZIEHER, L. M., and E. DE ROBERTIS, 1963. Subcellular localization of 5-hydroxytryptamine in rat brain. *Biochem. Pharmacol.* 12:596–598.

ZIEHER, L. M., and G. JAIM-ETCHEVERRY, 1971. Ultrastructural cytochemistry and pharmacology of 5-hydroxytryptamine in adrenergic nerve endings. II. Accumulation of 5-hydroxytryptamine in nerve vesicles containing norepinephrine in rat vas deferens. *J. Pharmacol. Exp. Ther.* 178:30–41.

ZWEIG, M., and J. AXELROD, 1969. Relationship between catecholamines and serotonin in sympathetic nerves of the rat pineal gland. *J. Neurobiol.* 1:87–97.

81 Interaction of 5-Hydroxytryptamine with the Encephalitogenic Protein of Myelin

P. R. CARNEGIE

ABSTRACT   The region around the sole tryptophan residue in the encephalitogenic basic protein of myelin was shown to interact with 5-hydroxytryptamine and hallucinogenic indoles. The binding site in the protein is close to an unusual amino acid, dimethylarginine. Possible implications of the interaction are discussed.

WHEN ANIMALS are injected with a mixture of adjuvant and normal tissue from the central nervous system (CNS) they develop the disease experimental autoimmune encephalomyelitis (EAE). Guinea pigs with EAE have urinary incontinence, loss of righting reflexes, and paralysis of the hind quarters. A major protein of the CNS, the encephalitogenic protein of myelin (EF), is the antigen responsible for the induction of an autoimmune attack on the CNS (Alvord, 1970, Carnegie, 1971a).

The identification of the antigen does not solve four problems in understanding the pathogenesis of EAE. First, although EF is present in myelin throughout the CNS, why are the clinical symptoms localized to restricted regions primarily in the lower spinal cord? Second, why does immunization with EF cause such marked clinical symptoms, whereas immunization with other CNS proteins, such as the S-100, cause no clinical symptoms? Third, why do animals with EAE (and patients with multiple sclerosis) have a factor in their serum that blocks polysynaptic function but not axonal conduction in cultures of CNS tissue (Bornstein and Crain, 1965)? The factor may interact with the internuncial neurons in preparations of spinal cord (Cerf and Carels, 1966). And finally, why do inhibitors of monoamine oxidases inhibit the induction of EAE but not other immunological responses (Saragea and Vladutiu, 1965; Lennon, 1972)?

One possible explanation for these problems has emerged from a detailed study of the structure of the region in EF responsible for the induction of EAE in

guinea pigs. Isolation of peptides from around the sole tryptophan residue (Eylar and Hashim, 1968; Lennon et al., 1970) and synthesis of peptides analogous to this region of EF (Westall et al., 1971) have shown that the small peptide Ser-Arg-Phe-Ser-Trp-Gly-Ala-Glu-Gly-Gln-Arg is the principal region causing the induction of EAE. Moreover, injection of adjuvant mixed with EF or this peptide into rabbits caused EAE and the production of the polysynaptic blocking factor in their serum (S. M. Crain, personal communication).

There is a striking resemblance between the structure of the encephalitogenic region of EF and that calculated by Smythies et al. (1970) for a receptor for serotonin (5-hydroxytryptamine, 5-HT) and hallucinogenic indoles (Carnegie, 1971b). Close to the encephalitogenic region is a methylated arginine (Baldwin and Carnegie, 1971), which Smythies et al. (1972) have suggested would complete a hydrophobic pocket into which 5-HT could fit. Also since the regions of the CNS that have the highest concentration of 5-HT nerve terminals (Fuxe et al., 1969) appear to be those affected in EAE, Carnegie (1971b) speculated that in EAE there was an immunopharmacological block of serotonergic receptors due to antigenic cross-reaction between the encephalitogenic protein and a protein in synapses. Malfunctioning of serotonergic neurons could explain some of the above problems in the pathogenesis of EAE.

Inhibition of antigenicity of EF by 5-HT

In this paper data is presented that shows that 5-HT and hallucinogenic drugs can specifically inhibit the antigenicity of EF, or of synthetic peptides analogous to the tryptophan region of EF.

Animals immunized with EF and patients with multiple sclerosis have sensitized lymphocytes that respond to EF by producing a factor that alters the surface charge of macrophages. The extent of alteration can be determined in a cell electrophoresis apparatus (cytopherometer

P. R. CARNEGIE   School of Biochemistry, University of Melbourne, Parkville, Victoria, Australia assay) (Field and Caspary, 1970; Caspary and Field, 1971). When EF was incubated for a few minutes with 5-HT before the addition of sensitized lymphocytes, the antigenicity was shown to be inhibited (Figure 1). Other neurotransmitters had no ability to inhibit the antigenicity, and 5-HT had no ability to inhibit the response of lymphocytes sensitized to other unrelated proteins (Field et al., 1971).

Since the synthetic tryptophan peptide (B, Figure 1) could be substituted for the protein in the assay system and since 5-HT could inhibit its interaction with the lymphocytes, it appeared that 5-HT was interacting with the tryptophan region of the protein. From the dose response curves for EF and the peptide in the cytopherometer assay, it was possible to estimate the amount of free protein or peptide available to interact with the lymphocytes at each concentration of 5-HT. These data were used to calculate an approximate association constant 7×10^4 M^{-1} for the interaction of 5-HT with EF, and 9.8×10^4 M^{-1} for the interaction with the peptide. The 5-HT had less ability to inhibit the response to a synthetic peptide with a modified sequence (C, Figure 1).

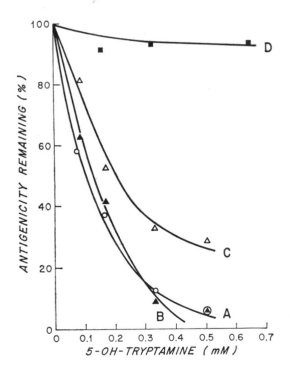

FIGURE 1 Ability of 5-hydroxytryptamine to block the antigenicity of the encephalitogenic protein of myelin and synthetic peptides. The antigenicity was assessed by the ability of the antigens to cause lymphocytes from patients with multiple sclerosis to produce a factor which alters the surface charge of macrophages. (A) Encephalitogenic protein of myelin. (B) Phe-Ser-Trp-Gly-Ala-Glu-Gly-Gln-Arg. (C) Phe-Ser-Trp-Gly-Ala-Glu-Gly-Gln-Gly-Arg. (D) Encephalitogenic protein of myelin with tryptamine instead of 5-HT.

Inhibition of antigenicity by hallucinogenic drugs

A number of hallucinogenic drugs and indole derivatives were examined for their ability to inhibit the reaction between the encephalitogenic protein and sensitized lymphocytes. The results are summarized in Table I. Tryptamine, indole, hydroxyindole and derivatives of 5-HT that had large substituents attached to the amine group or had a methyl group at the 1 position had no ability to inhibit the reaction. On the other hand, simple derivatives of 5-HT, some of which are known to be hallucinogenic (Table I), inhibited the interaction to a varying degree.

Thus although the interaction of 5-HT and hallucinogenic indoles with the protein appeared to be relatively weak with an approximate association constant 10^4 M^{-1}, there was a high degree of structural selectivity. In contrast, in the studies of Alivisatos et al. (1971) on the binding of ^{14}C-5-HT to synaptosomal membranes, there was a higher association constant but a lack of specificity, because hydroxy-indole and tryptamine would inhibit the binding of 5-HT.

Two other drugs—pargyline, an inhibitor of monoamine oxidase, and imipramine, which blocks the uptake of 5-HT by presynaptic membranes—were investigated for their ability to inhibit the interaction of EF and sensitized lymphocytes. Pargyline had no effect whereas imapramine inhibited the interaction and the approximate association constant was 6×10^4 M^{-1}.

Implications of interaction of 5-HT with a protein of myelin

The above data shows that 5-HT and hallucinogenic drugs can interact with the tryptophan region of the encephalitogenic protein of myelin. Attempts have been made to confirm the approximate association constant by alternative techniques such as equilibrium dialysis. These experiments have been unsuccessful, possibly because the protein in aqueous solution has a highly disordered structure (Palmer and Dawson, 1969; Eylar and Thompson, 1969). Anthony and Moscarello (1971) have shown that in the presence of lipids the protein apparently changed to a mixture of α-helix and β-pleated sheet. When the protein is mixed with HeLa cells, it binds to the plasma membrane (Seiden, 1967). In the cytopherometer assay the protein, and synthetic tryptophan peptide, probably associate with cell membranes and are stabilized in a more ordered conformation which provides a binding site for 5-HT.

This binding site appears to mimic a synaptic receptor for 5-HT and hallucinogenic drugs. It is suggested that inhibitory activity in the cytopherometer assay might be

TABLE I
Effect of 5-HT and hallucinogenic drugs on antigenicity of encephalitogenic protein of myelin

Drug	Relative Molar Ability to Block Antigenicity[†]	Psychomimetic Potency (Mescaline units[*])
5-Hydroxytryptamine creatinine sulphate	1.0	0
D-Lysergic acid diethylamide (LSD) D-tartrate	1.0	3700
5-Chloro-N,N-dimethyltryptamine	0.6	?
5-Methoxy-N,N-dimethyltryptamine HCl	0.4	>31 < LSD
N,N-Dimethyltryptamine	0.4	4
5-Methoxytryptamine	0.3	+
5-Hydroxy-N,N-dimethyltryptamine (bufotenine)	+	+ ?
3,4,5-Trimethoxy-phenethylamine HCl (mescaline)	0.3	1
N,N-Dimethyl-2,5-dimethoxy-4-methyl amphetamine HCl (dimethyl DOM)	0.3	0

Indole Derivatives with No Ability to Block the Antigenicity of the Basic Protein of Myelin:

Tryptamine HCl	1-methyl-5-methoxy-N,N-dimethyltryptamine HCl
Indole	5-methoxy-N-methyl-N-cyclopentanyl tryptamine
Hydroxyindole	5-benzyloxy-N,N-di-*n*-butyltryptamine HCl
N-Isopropyltryptamine maleate	3-(β-N-piperidinoethyl) indole HCl
Dehydrobufotenine HCl	3-(β-N-pyrrolodinoethyl) indole HCl
5-Methoxy-N,N-diethyltryptamine HCl	

[*]Data taken from Brawley and Duffield (1972).
[†]Data taken from Carnegie, Smythies, Caspary, and Field (1972).

useful as an in vitro method for screening potentially hallucinogenic indoles.

Myelin is derived and maintained by the oligodendrocytes and these cells but not other cells in cultures of brain were shown by Murray (1958) and Nakazawa (1960) to respond to 5-HT by changing their pulsation rate. Woolley (1962) considered that hallucinogenic drugs might affect oligodendrocytes. It is generally assumed that hallucinogenic drugs interact with synaptic receptors for serotonin, however it is conceivable that some of their effects could come from an interaction with the encephalitogenic protein of myelin possibly at the nodes of Ranvier in the central nervous system. Further speculation must await a clarification of the normal role of the encephalitogenic protein in the interaction between axons and oligodendrocytes.

Summary

A sensitive immunological assay was used to demonstrate the interaction of 5-hydroxytryptamine and hallucinogenic indoles with the tryptophan region of the encephalitogenic protein of human myelin.

ACKNOWLEDGMENTS The experiments with the cytopherometer assay were carried out while on study leave with the Medical Research Council, Demyelinating Diseases Unit, Newcastle-upon-Tyne, England, and was supported by a Grant from the Multiple Sclerosis Society of Great Britain and Northern Ireland.

I wish to thank Dr. F. C. Westall for synthetic peptides and Dr. J. R. Smythies for hallucinogenic drugs, and I am most grateful to Mr. E. A. Caspary and Professor E. J. Field for their collaboration.

REFERENCES

ALIVISATOS, S. G. A., F. UNGAR, P. K. SETH, L. P. LEVITT, A. J. GEROULIS, and T. S. MYER, 1971. Receptors: Localization and specificity of binding of serotonin in the central nervous system. *Science* 171:809–812.

ALVORD, E. C., 1970. Acute disseminated encephalomyelitis and "allergic" neur-encephalopathies. Handbook of Clinical Neurology, Vol. 9, P. J. Vinken and G. W. Bruyn, eds. Amsterdam: North Holland, pp. 500–571.

ANTHONY, J. S., and M. A. MOSCARELLO, 1971. A conformation change induced in the basic encephalitogen by lipids. *Biochim. Biophys. Acta.* 243:429–433.

BALDWIN, G. S., and P. R. CARNEGIE, 1971. Isolation and partial characterization of methylated arginines from the encephalitogenic protein of myelin. *Biochem. J.* 123:69–74.

BORNSTEIN, M. B., and S. M. CRAIN, 1965. Functional studies of cultured brain tissues as related to "demyelinative disorders." *Science* 148:1242–1245.

BRAWLEY, P., and J. C. DUFFIELD, 1972. The pharmacology of hallucinogens. *Pharmacol. Rev.* 24:31–66.

CARNEGIE, P. R., 1971a. Amino acid sequence of the encephalitogenic basic protein from human myelin. *Biochem. J.* 123:57–67.

CARNEGIE, P. R., 1971b. Properties, structure and possible neuroreceptor role of the encephalitogenic protein of human brain. *Nature (Lond.)* 229:25–28.

CARNEGIE, P. R., J. R. SMYTHIES, E. A. CASPARY, and E. J. FIELD, 1972. Interaction of hallucinogenic drugs with encephalitogenic protein of myelin. *Nature (Lond.)* 240:561–563.

CASPARY, E. A., and E. J. FIELD, 1971. Specific lymphocyte sensitization in cancer: Is there a common antigen in human malignant neoplasia? *Brit. Med. J.* 2:613–617.

CERF, J. A., and G. CARELS, 1966. Multiple sclerosis: Serum factor producing reversible alterations in bioelectric responses. *Science* 152:1066–1068.

EYLAR, E. H., and G. A. HASHIM, 1968. Allergic encephalomyelitis: The structure of the encephalitogenic determinant. *Proc. Nat. Acad. Sci. USA* 61:644–650.

EYLAR, E. H., and M. THOMPSON, 1969. Allergic encephalomyelitis: The physico-chemical properties of the basic protein encephalitogen from bovine spinal cord. *Arch. Biochem. Biophys.* 129:468–479.

FIELD, E. J., and E. A. CASPARY, 1970. Lymphocyte sensitisation—an in vitro test for cancer? *Lancet* 2:1337–1341.

FIELD, E. J., E. A. CASPARY, and P. R. CARNEGIE, 1971. Lymphocyte sensitization to basic protein of brain in malignant neoplasia: Experiments with serotonin and related compounds. *Nature (Lond.)* 233:284–286.

FUXE, F., T. HOKFELT, and U. UNGERSTEDT, 1969. Distribution of monoamines in the mammalian central nervous system by histochemical studies. In *Metabolism of Amines in the Brain*, G. Hooper, ed. London: Macmillan, pp. 10–22.

LENNON, V. A., 1972. Cellular and humoral immune responses in experimental autoimmune encephalomyelitis. Ph.D thesis, University of Melbourne, Australia.

LENNON, V. A., A. V. WILKS, and P. R. CARNEGIE, 1970. Immunologic properties of the main encephalitogenic peptide from the basic protein of human myelin. *J. Immunol.* 105:1223–1230.

MURRAY, M. R., 1958. Response of oligodendrocytes to serotonin. In *Biology of Neuroglia*, W. F. Windle, ed. Springfield, Ill.: Charles C Thomas, pp. 176–190.

NAKAZAWA, T., 1960. Effects of chlorpromazine and serotonin on the contractability of oligodendrocytes. *Texas Rep. Biol. Med.* 18:52–65.

PALMER, F. B., and R. M. C. DAWSON, 1969. The isolation and properties of experimental allergic encephalitogenic protein. *Biochem. J.* 111:629–636.

SARAGEA, M., and A. VLADUTIU, 1965. L'influence de la Nialamide sur l'encephalomyelite allergique experimentale. *Naturwissenchaften* 52:564.

SEIDEN, G. E., 1967. Sensitization of lymphoid cells in tissue culture to EAE encephalitogen. *J. Neuropathol. Exp. Neurol.* 26:551–557.

SMYTHIES, J. R., F. BENINGTON, and R. D. MORIN, 1970. Specification of a possible serotonin receptor site in the brain. *Neurosciences Res. Prog. Bull.* 8:117–122.

SMYTHIES, J. R., F. BENINGTON, R. D. MORIN, 1972. Encephalitogenic protein: A β-pleated sheet conformation (102–120) yields a possible molecular form of a serotonin receptor. *Experientia* 28:23–24.

WESTALL, F. C., A. B. ROBINSON, J. CACCAM, J. JACKSON, and E. H. EYLAR, 1971. Essential chemical requirements for induction of allergic encephalomyelitis. *Nature (Lond.)* 229:22–24.

WOOLLEY, D. W., 1962. *The Biochemical Basis of Psychoses or the Serotonin Hypothesis about Mental Disease*. New York: John Wiley & Sons.

82 Physiological Mechanisms of Synaptic Modulation

FORREST F. WEIGHT

ABSTRACT   Several physiological mechanisms can modulate the efficacy of transmission at synapses. I have reviewed the following physiological mechanisms of synaptic modulation: (1) The modulation of transmitter release by preceding activity in the presynaptic terminal and by the presynaptic action of other synapses; (2) the influence of synapses on postsynaptic receptors; (2) the synaptic control of postsynaptic membrane permeability; and (4) the synaptic activation of metabolic systems.

THE NERVE IMPULSE is an all-or-none event, subject to little variability during normal physiological activity. Transmission at chemical synapses, on the other hand, may be modulated by a variety of physiological mechanisms. Such modulation of synaptic transmission may be an important part of information processing in the nervous system and is the subject of this report.

The physiological mechanisms of synaptic modulation will be considered in terms of four topics: (1) mechanisms relating to the modulation of transmitter release from presynaptic terminals; (2) synaptic influence on postsynaptic receptors; (3) synaptic control of postsynaptic membrane permeability, and (4) synaptic activation of metabolic systems. Because of the diversity of these mechanisms and limitations of page space, it will not be possible to review each aspect in detail, but rather the discussion will focus on more recent advances.

Modulation of transmitter release

The general mechanisms involved in the release of transmitter from nerve terminals have been reviewed recently (Katz, 1969; Rubin, 1970; Kuno, 1971) and are not discussed here. We will consider some of the physiological mechanisms that may modulate the release of transmitter. The release of transmitter may be modulated by at least two general mechanisms: (1) preceding activity in the terminal, and (2) presynaptic action of other synapses.

MODULATION OF TRANSMITTER RELEASE BY PRECEDING ACTIVITY IN THE TERMINAL   At many synapses, one or more conditioning stimuli to the presynaptic nerve result in synaptic potentials that differ in amplitude from those elicited by a single stimulus. There have been numerous investigations on the effect of conditioning stimuli on the generation of synaptic potentials, and we do not attempt to review the literature here; the interested reader is referred to Hughes (1958) and Martin and Veale (1967). We instead describe the types of synaptic modulation produced by conditioning stimulation and the recent proposals to explain these phenomena.

Synaptic modulation by conditioning stimulation has been most extensively investigated at the neuromuscular junction and this discussion therefore focuses on the end-plate potential (epp) at that junction. After conditioning presynaptic stimulation at the neuromuscular junction, the amplitudes of subsequent epp's occur in three distinct phases: (1) initially there is an increase in the amplitude of the epp's, termed *facilitation* or primary potentiation; (2) this is followed by a decrease in epp amplitude or *depression*, and (3) there is then a later phase of secondary or posttetanic potentiation (PTP). Following repetitive presynaptic stimulation, there is no change in the sensitivity of the postsynaptic membrane to the transmitter acetylcholine (ACh) (Hutter, 1952; Otsuka et al., 1962), indicating that the changes in epp amplitude are due to alterations in the amount of transmitter released.

Facilitation of transmitter release   After a single conditioning stimulus to motor nerve fibers, the amplitude of the epp produced by a subsequent test stimulus is increased (Feng, 1940; Eccles, Katz, and Kuffler, 1941). The facilitation of the second epp is maximal at very short intervals after the conditioning stimulus and gradually declines as the interval between the two stimuli is increased, usually lasting for 100 to 200 msec. Several studies have found that the frequency of miniature end-plate potentials (mepp's) is increased during facilitation (del Castillo and Katz, 1954b; Liley, 1956; Hubbard, 1963; Braun et al., 1966), indicating that the probability of transmitter release is increased during facilitation. Since each mepp is considered to be a quantum of transmitter (del Castillo and Katz, 1954a), facilitation can be ascribed to an increase in the quantum content of the epp.

FORREST F. WEIGHT   Section on Synaptic Pharmacology, Laboratory of Neuropharmacology, National Institute of Mental Health, Washington, D.C.

Recent studies on the mechanism of the increased quantal content in neuromuscular facilitation have focused on the role of calcium. Investigations on the association of Ca with facilitation have shown that a decrease in extracellular Ca reduces the first epp but increases facilitation (del Castillo and Katz, 1954b; Hubbard, 1963; Katz and Miledi, 1965). On the other hand, an elevation of Ca increases the transmitter released by the first impulse but decreases facilitation (Rahamimoff, 1968). When two successive epp's are elicited in a low-calcium solution and an iontophoretic pulse of Ca precedes the first epp, the second epp is facilitated (Katz and Miledi, 1968). Katz and Miledi (1965, 1968) have proposed that facilitation may be due to residual Ca remaining attached for a time to specific sites on the membrane, designated *active calcium*; on the arrival of subsequent impulses, it would be easier for sites with partial occupation by Ca to reach the necessary Ca level (Dodge and Rahamimoff, 1967) to induce the release of transmitter.

Depression of transmitter release Following facilitation there is a period of depression in the amplitude of the epp (Eccles, Katz, and Kuffler, 1941; Lundberg and Quilisch, 1953). The depression decays with an approximately exponential time course and lasts from about 1 sec to 10 sec, depending on conditioning stimuli and experimental conditions.

Although the amplitude of the epp at the neuromuscular junction is reduced during depression, the frequency of mepp's remains elevated (del Castillo and Katz, 1954b; Hubbard, 1963). In view of this, it has been proposed that depression is due to a depletion of the transmitter available for release (Takeuchi, 1958; Otsuka et al., 1962; Thies, 1965). This hypothesis is supported by the observation that increasing extracellular Ca concentration increases the transmitter released by conditioning stimuli and produces a proportional increase in depression (Takeuchi, 1958; Otsuka, et al., 1962; Thies, 1965).

Posttetanic potentiation Following repetitive stimulation of the motor nerve, there is a late increase in the amplitude of the epp (Feng, 1941; Liley and North, 1953). This posttetanic potentiation (PTP) follows the period of depression and lasts from several seconds to several minutes depending upon the stimulation.

It has been suggested that PTP may be due to an accumulation of Na in the nerve terminal during repetitive stimulation that leads to an increased synthesis of ACh (Birks, 1963; Birks and Cohen, 1968). However, in recent experiments when extracellular Na was replaced by isotonic calcium chloride, PTP was still present (Weinreich, 1971), indicating that Na is not necessary for PTP. On the other hand, the extracellular Ca concentration has noteworthy effects on the generation of PTP. When

external Ca is reduced, the decay time of PTP is also reduced (Rosenthal, 1969). In a low-Ca solution, the time course of PTP is dependent on the Ca present during the tetanus (Rosenthal, 1969; Weinreich, 1971). These observations have been taken as evidence that PTP is associated with an intracellular accumulation of Ca during the tetanus. Thus, while PTP appears to be related to an accumulation of intracellular Ca, the precise role of Ca in PTP remains to be determined.

MODULATION OF TRANSMITTER RELEASE BY PRESYNAPTIC ACTION OF OTHER SYNAPSES In addition to the modulation of transmitter release produced by preceding activity in the terminal, transmitter release may also be modulated by a presynaptic action of other synapses. This aspect of release modulation will be considered with respect to two topics: (1) presynaptic inhibition and (2) heterosynaptic facilitation.

Presynaptic inhibition Crustacean muscle is innervated by both an excitatory and an inhibitory axon. Stimulation of the inhibitory axon generates postsynaptic inhibition in the muscle by increasing chloride permeability (Fatt and Katz, 1953; Boistel and Fatt, 1958; Dudel and Kuffler, 1961), and there is evidence that the inhibitory transmitter is gamma aminobutyric acid (GABA) (Kravitz et al., 1963; Otsuka et al., 1966). Stimulation of the inhibitory axon also produces a presynaptic inhibition that decreases the release of the excitatory transmitter (Dudel and Kuffler, 1961) and the administration of GABA mimics this effect (Dudel and Kuffler, 1961; Dudel, 1965a, and 1965b; Takeuchi and Takeuchi, 1966a). The removal of extracellular chloride abolishes both presynaptic inhibition and the presynaptic inhibitory action of GABA (Takeuchi and Takeuchi, 1966b). These data indicate that release of GABA from the inhibitory nerve terminal increases the permeability of the excitatory presynaptic terminal to chloride resulting in a presynaptic inhibition of transmitter release.

Presynaptic inhibition is also a mechanism of synaptic modulation in the vertebrate central nervous system that has been investigated most extensively in the spinal cord. Depression of monosynaptic excitation of motoneurons by conditioning presynaptic volleys can occur in the absence of any detectable change in the properties of the postsynaptic membrane (Frank and Fuortes, 1957). This phenomenon is associated with a depolarization of primary afferent fibers (Eccles et al., 1961) that decreases the amplitude of the presynaptic action potential and the quantity of transmitter released (see Eccles, 1964). Recent studies indicate that presynaptic inhibition of primary afferent terminals in the spinal cord is mediated by GABA released from other presynaptic terminals (Curtis et al.,

1971; Barker and Nicoll, 1973; Davidoff, 1973). The ionic mechanism, however, differs from the crayfish in that an increased sodium permeability appears to produce presynaptic inhibition of spinal primary afferents (Barker and Nicholl, 1973).

Heterosynaptic facilitation Release of synaptic transmitter can be modulated not only by presynaptic inhibition, but also by heterosynaptic facilitation. In certain cells in the abdominal ganglion of *Aplysia*, an EPSP produced by stimulation of one pathway undergoes a prolonged facilitation when it is paired with repetitive stimulation of a second pathway (Kandel and Tauc, 1965). The priming stimulus for this heterosynaptic facilitation does not produce a significant change in the conductance of the postsynaptic cell, and directly initiated spikes in the cell do not serve as a priming stimulus. The absence of changes in the membrane properties of the postsynaptic cell suggests that heterosynaptic facilitation is due to a presynaptic facilitation of the release of transmitter (Kandel and Tauc, 1965; Epstein and Tauc, 1970).

As discussed above, control of the release of synaptic transmitter is one general physiological mechanism of synaptic modulation. We will now turn our attention to the postsynaptic aspects of this topic.

Synaptic influence on postsynaptic receptors

Denervation results in the development of greater sensitivity of the postsynaptic structure to the transmitter, which suggests that the synapse has an influence on postsynaptic receptors. The effects of the synapse on receptors have been studied in great detail at the neuromuscular junction, and some of these studies are briefly reviewed.

In normal innervated skeletal muscle, the sensitivity of the postsynaptic membrane to the transmitter ACh is restricted to the synaptic region (del Castillo and Katz, 1955; Miledi, 1960b). Following denervation, however, sensitivity to ACh spreads over the entire muscle membrane (Axelsson and Thesleff, 1959; Miledi, 1960a). It has been suggested that the factor controlling receptor localization may be the transmitter ACh (Thesleff, 1960), or a chemical substance other than ACh that is normally released from terminal (Miledi, 1960a; 1963) or the lack of muscle activity (Jones and Vrbova, 1970).

Recently, this problem has been further investigated using a silicone cuff on the nerve containing a local anesthetic or diptheria toxin (Lomo and Rosenthal, 1972). The blocking of nerve impulses produced a spread of ACh sensitivity similar to that resulting from denervation, although the junctions had a normal release of mepp's, indicating that the normal spontaneous release of ACh does not prevent the spread of sensitivity. In addition, chronic direct stimulation of denervated muscle prevented the spread of ACh sensitivity, indicating that the activity of the muscle fiber is an important factor regulating the ACh sensitivity of the muscle membrane. Whether such mechanisms relate to denervation supersensitivity of neurons remains to be determined.

In addition to an influence of the synapse on postsynaptic receptors, there is also data that the innervation of a postsynaptic structure has certain "trophic" influences. The possible trophic effects of synaptic innervation are beyond the scope of this paper, and the interested reader is referred to Guth (1968, 1969).

Synaptic control of postsynaptic membrane permeability

During the past 20 years, investigations on the mechanisms of synaptic transmission have found that at most synaptic junctions the postsynaptic potential is generated by an increase in the permeability of the postsynaptic membrane to certain ions. Recent investigations, however, have revealed that postsynaptic potentials, at some junctions, can also be generated by a decrease in postsynaptic membrane permeability. We will review these mechanisms of generating synaptic potentials before discussing their possible relationships to synaptic modulation.

SYNAPTIC EXCITATION BY CONDUCTANCE INCREASE Postsynaptic excitation at most synaptic junctions has been found to have a similar general mechanism (Eccles, 1964; Ginsborg, 1967; Weight, 1971). The end-plate potential (epp) at the neuromuscular junction has been most extensively investigated and is used to illustrate the usual mechanism of generating postsynaptic excitation. For a more detailed account, the interested reader is referred to Katz (1966).

Recording intracellularly in a muscle fiber, stimulation of motor fibers produces a depolarization of about 20 mV (Fatt and Katz, 1951)—the end-plate potential (Figure 1A). The current flow responsible for this potential has been recorded directly by the voltage clamp technique (Takeuchi and Takeuchi, 1959) and has a more rapid time course (Figure 1B) than the epp. The slower time course of the epp is due to the time constant of the membrane (Fatt and Katz, 1951). During the active phase of the epp, membrane resistance decreases to 1% of its resting value (Fatt and Katz, 1951). Since electrical conductance is defined as the reciprocal of resistance, this indicates there is a large increase in the conductance of postsynaptic membrane during the action of the transmitter. In the fluid medium of the cell, the current flow

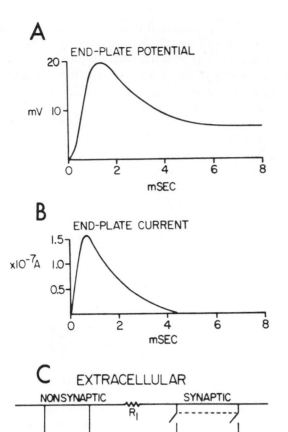

A

END-PLATE POTENTIAL

B

END-PLATE CURRENT

C EXTRACELLULAR

NONSYNAPTIC   SYNAPTIC

INTRACELLULAR

FIGURE 1 (A) End-plate potential in curarized muscle. *Ordinate*: depolarization of membrane potential from resting potential in mV. *Abscissa*: time from onset of epp in msec. (B) End-plate current, indicating time course of current flow responsible for end-plate potential. (C) Equivalent electric circuit diagram for end-plate potential. On the right are synaptically activated (coupled switches) conductances to sodium (g_{Na}) and potassium (g_K) in series with their equilibrium potentials (represented as emfs), E_{Na} and E_K. On the left is diagrammed the nonsynaptic membrane with the membrane resistance, R_m, in series with the electromotive force, E_m, and in parallel with the membrane capacitance, C_m. R_l is the longitudinal membrane resistance. (From Weight, 1971.)

produced by the increased conductance will be carried by ions. The generation of the epp can thus be explained in terms of the ionic membrane hypothesis (Fatt and Katz, 1951).

The passive movement of an ion across the membrane is determined by the permeability of the membrane to that ion, the concentration difference of that ion across the membrane, and the electrical potential across the membrane. If a synaptic potential is generated by an increase

in sodium conductance (Δg_{Na}), the sodium current (I_{Na}) can be expressed as

$$I_{Na} = \Delta g_{Na}(E_m - E_{Na}) \tag{1}$$

where E_m is the membrane potential and $E_{Na} = (RT/F)$ ln (Na_0/Na_i). From this relationship it can be seen that as the membrane potential approaches the sodium equilibrium potential (E_{Na}), then I_{Na} will decrease. Furthermore, I_{Na} will reverse to the opposite sign when E_m exceeds E_{Na}. Thus, the reversal potential of a synaptic potential serves as an index of the ion species involved in generating the synaptic potential.

The reversal potential of the epp is about -15 mV (del Castillo and Katz, 1954c), which differs significantly from both E_{Na}, which is approximately $+50$ mV, and E_K which is about -90 mV. As indicated above, the synaptic current is also a function of the concentration gradient of the ions involved, thus by changing the ionic composition of the external bath, one can determine the effect on the synaptic current. Such studies at the neuromuscular junction indicate that the end-plate current is due to an increased permeability to both Na and K, but there is little change in Cl permeability (Takeuchi and Takeuchi, 1960). Such a synaptic current, involving both a sodium conductance (Δg_{Na}) and a potassium conductance (Δg_K) can be expressed as

$$I_s = \Delta g_{Na}(E_m - E_{Na}) + \Delta g_K(E_m - E_K) \tag{2}$$

In view of the data that the epp is generated by an increase in both g_{Na} and g_K, the mechanism of generating the epp at the neuromuscular junction can be represented by an electrical circuit diagram (Figure 1C). The action of the transmitter ACh on postsynaptic receptors switches on the sodium and potassium conductance channels. The equilibrium potentials for these ions are represented as electromotive forces E_{Na} and E_K, respectively, in series with the conductance channels. The epp generated will be determined by the magnitude of each of the respective conductance changes and the equilibrium potentials for the ions involved.

This model has been found to serve as a general model for the generation of chemically mediated synaptic excitation at most junctions (see Eccles, 1964; Ginsborg, 1967; Weight, 1971), although at some synapses the increased conductance may be only sodium (Chiarandini et al., 1967) or at other junctions, an increased chloride conductance may be involved (Frank and Tauc, 1964; Oomura et al., 1965; Chiarandini et al., 1967). Thus although the ion species involved may differ, synaptic excitation at most synapses has a similar basic mechanism —the permeability of the postsynaptic membrane increases to certain ions resulting in a net inward flow of current through the subsynaptic membrane. The membrane potential is shifted toward the equilibrium potential of the ions involved, depolarizing the membrane and bringing the potential toward or above threshold for the action potential.

SYNAPTIC INHIBITION BY CONDUCTANCE INCREASE   Postsynaptic inhibition at most junctions is generated by mechanisms that are similar to the generation of postsynaptic excitation (see Ginsborg, 1967; Weight, 1971). The mechanism of postsynaptic inhibition in the vertebrate CNS has been extensively investigated in motoneurons of cat (see Eccles, 1964). Stimulation of Ia afferents produces a hyperpolarization of the membrane called the inhibitory postsynaptic potential (IPSP) in antagonist motoneurons (Coombs et al., 1955) as illustrated in Figure 2A. Voltage clamp studies (Araki and Terzuolo, 1962) indicate that the inhibitory synaptic current generating the IPSP has a more rapid time course than the IPSP (Figure 2B), as was found with synaptic excitation. The increase in postsynaptic membrane conductance responsible for the inhibitory synaptic current has been demonstrated by measuring the decrease in membrane impedance during the IPSP (Smith et al., 1967).

Injection of Cl ions intracellularly reverses the IPSP to a depolarizing potential, indicating that an increased permeability to Cl is involved in generating the response (Coombs et al., 1955). The reversal potential for the IPSP is about -80 mV, which is more negative than the estimated equilibrium potential for Cl of -70 mV. Since the K equilibrium potential is about -90 mV, it has been proposed that the IPSP in motoneurons involves an increased permeability to both Cl and K ions (Coombs et al., 1955; Eccles et al., 1964). The extent to which the K permeability is changed, however, remains to be determined.

Postsynaptic inhibition at most other synaptic junctions has also been found to involve an increased conductance mechanism. For example, the inhibitory junction potential in the crayfish stretch receptor is generated by an increase in both Cl and K conductances (Edwards and Hagiwara, 1959; Hagiwara et al., 1960) and the inhibitory junction potential in crayfish muscle involves an increased Cl conductance (Boistel and Fatt, 1958; Takeuchi and Takeuchi, 1967).

The mechanism of generating postsynaptic inhibition by increased postsynaptic membrane permeability to Cl and K can be illustrated by the electrical circuit diagram in Figure 2C. The inhibitory transmitter acting on postsynaptic receptors closes the switches thus activating the Cl and K conductance channels (g_{Cl} and g_K). The equilibrium potential for the IPSP will be determined by the electromotive forces, E_{Cl} and E_K. It can be seen that

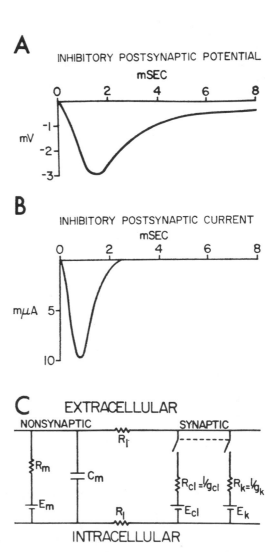

FIGURE 2   (A) Inhibitory postsynaptic potential in motoneuron. *Ordinate*: Hyperpolarization of membrane from resting potential in mV. *Abscissa*: Time from onset of IPSP in msec. (B) Inhibitory postsynaptic current, indicating time course of current flow responsible for IPSP. (C) Equivalent electric circuit for IPSP. On the right, the synaptically activated (coupled switches) conductances to chloride (g_{Cl}) and potassium (g_K) are in series with their respective electromotive forces, the equilibrium potentials for chloride, E_{Cl}, and potassium, E_K. The nonsynaptic membrane on the left is as in Figure 1. (From Weight, 1971.)

the electrical circuit for postsynaptic inhibition is similar to the circuit for synaptic excitation, emphasizing that it is the equilibrium potential for the ions involved that determines whether a PSP is excitatory or inhibitory.

SYNAPTIC EXCITATION BY CONDUCTANCE DECREASE   In addition to the usual type of synaptic potential recorded in most nerve cells and muscle, recent studies have indicated that there are slow synaptic potentials in sympathetic ganglia with unusual properties (Koketsu, 1969; Libet,

1970). Before discussing the mechanism of generation of these slow potentials, however, it will be necessary to review briefly the synaptic organization of the sympathetic ganglion. The Xth lumbar sympathetic ganglion of frog is particularly well suited for investigating the mechanism of generation of slow synaptic potentials, because the slow excitatory postsynaptic potential (slow EPSP) and the slow inhibitory postsynaptic potential (slow IPSP) are produced in different cell types (Tosaka et al., 1968)—the slow EPSP in B cells (Figure 3B1) and the slow IPSP in C cells (Figure 5B1). Each cell type can be identified by the antidromic conduction velocity of its axon (B and C fibers, respectively) and each cell type receives preganglionic cholinergic innervation by separate inputs—

B cells from preganglionic fibers in the sympathetic chain and C cells from preganglionic fibers in the VIIIth spinal nerve (Tosaka et al., 1968). Thus, stimulation of the sympathetic chain produces a fast EPSP in B cells (Figure 3A) and stimulation of the VIIIth nerve produces a fast EPSP in C cells (Figure 5A2). The fast EPSPs are produced by cholinergic activation of nicotinic membrane receptors, whereas the slow EPSP is due to cholinergic activation of muscarinic receptors. To study the slow EPSP alone, therefore, experiments are performed in the presence of nicotinic antagonists to block the fast EPSP.

As discussed previously, the increased permeability of postsynaptic membrane during synaptic excitation at most synaptic junctions is measured experimentally as a

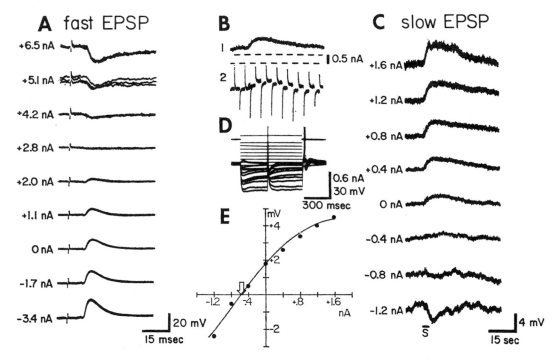

FIGURE 3 Fast and slow EPSPs. (A) Amplitude of fast EPSP as a function of depolarizing (+) and hyperpolarizing (−) current. Polarizing current is indicated in nanoamperes. The B fibers were stimulated with a single, 0.5 msec stimulus to generate the fast EPSP in a type B ganglion cell. (B and C) Slow EPSP in a type B ganglion cell. The fast EPSP was blocked by nicotine (5 μg/ml). The B fibers were stimulated repetitively at a frequency of 100 per sec for 2 sec. Stimulation began 10 sec after the beginning of each record. The period of stimulation is indicated by a line labeled S under the bottom record in C. (B1) Slow EPSP at resting membrane potential. (B2) Upper record: Hyperpolarizing constant current pulses of −0.5 namp. Lower record: Bridge balanced before stimulation such that current pulses produced no voltage deflection. Note that during EPSP the −0.5 namp current pulses produced a hyperpolarizing voltage deflection of 1 mV, indicating that the membrane resistance had increased by 2 megohms. Time and voltage calibration for B is the same as for C. (C) Amplitude of slow EPSP

in B as a function of depolarizing (+) and hyperpolarizing (−) current. Note that the slow EPSP reversed with hyperpolarizing current, the opposite of fast EPSP in A. (D) Amplitude of antidromic spike after hyperpolarization as a function of membrane potential. The same cell as B and C. Upper record monitors current; note that the larger hyperpolarizing current traces are superimposed on voltage records. Lower record shows antidromic spike superimposed at various levels of membrane potential; top of spike was cut off by limitation of oscilloscope excursion. Note that the after hyperpolarization reverses with hyperpolarizing current between −0.5 namp and −0.6 namp. (E) Graphic relationship of amplitude of slow EPSP (in C) to current. The reversal potential of slow EPSP is the point at which the curve crosses the abscissa. Arrow indicates the reversal potential of antidromic spike after hyperpolarization shown in D. Note that the reversal potential of after hyperpolarization (arrow) coincides with the reversal potential of slow EPSP (curve intercept). (From Weight and Votava, 1970.)

decrease in membrane resistance. However, as shown in Figure 3B2, when membrane resistance was tested during the slow EPSP in frog sympathetic ganglion, membrane resistance was found to increase during the slow EPSP (Kobayashi and Libet, 1970; Weight and Votava, 1970). Since conductance is defined as the reciprocal of resistance, this observation raised the question of the conductance that is decreased during the slow EPSP.

It was noted previously that the reversal potential of a synaptic potential gives some index of the ion species involved in a conductance change. Therefore, membrane polarization was used to study the effect on the slow EPSP (Weight and Votava, 1970). When progressive depolarizing current was passed across the membrane, the slow EPSP progressively increased in size (Figure 3C), which is the opposite of the effect of depolarizing current on most EPSPs. Moderate hyperpolarizing current, on the other hand, decreased the amplitude of the slow EPSP. Progressively stronger hyperpolarizing current abolished and then reversed the slow EPSP to a hyperpolarizing potential (Figure 3C). To obtain an index of the reversal potential of the slow EPSP, the reversal potential of the antidromic spike after hyperpolarization was used; as this is near the potassium equilibrium potential. As shown in Figure 3E, the slow EPSP reversed near the reversal potential for the antidromic spike after hyperpolarization. This suggested that the conductance that was decreased during the slow EPSP was resting K conductance, but it did not exclude the possibility of a decrease in Cl conductance. To exclude this possibility, extracellular Cl was

removed from the Ringer solution. The removal of Cl did not have a significant effect on the generation of the slow EPSP (Figure 4A) indicating that an inactivation of Cl conductance does not play a significant role in the generation of the slow EPSP.

The preceding data can be explained by the hypothesis that the slow EPSP is generated by a decrease or inactivation of resting K conductance (Weight and Votava, 1970). This hypothesis can be represented by an electrical circuit diagram, Figure 4C. On the left is the increased conductance mechanism of the fast EPSP, as discussed previously. On the right is the mechanism proposed for generating the slow EPSP. The resting K conductance (g_K) of the membrane is shown in series with its emf, the equilibrium potential for K (E_K). In parallel, the other resting conductances of the membrane (G_m) are shown in series with the equilibrium potential for those conductances (E_m). Activation of muscarinic membrane receptors (M) is illustrated as inactivating resting g_K. This would increase the membrane resistance and shift the membrane potential away from E_K and toward the equilibrium potential of the other resting conductances (E_m), resulting in a depolarization—the slow EPSP.

SYNAPTIC INHIBITION BY CONDUCTANCE DECREASE   In addition to the unique nature of the slow EPSP, the slow IPSP in sympathetic ganglia is also generated in a unique manner. As mentioned previously, preganglionic cholinergic fibers in the VIIIth spinal nerve make synaptic connection with C cells in the Xth sympathetic ganglion

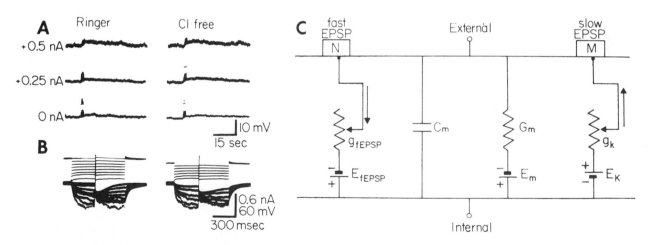

FIGURE 4   (A) Effect of removing extracellular Cl⁻ on slow EPSP. *Left*: Potentiation of small slow EPSP by depolarizing current. *Right*: Effect of removing extracellular Cl⁻ on potentiation of slow EPSP by depolarizing current. (B) Effect of removing extracellular Cl⁻ on reversal of antidromic spike after hyperpolarization. *Left*: Upper record monitors current; lower record shows antidromic spike superimposed at various levels of membrane potential. *Right*: Similar to left but after

removal of extracellular Cl⁻. Note that data taken when Cl⁻ was removed were recorded over a more limited current range than were control data. (C) Schematic electrical circuit diagram of sympathetic ganglion B cell membrane representing the fast EPSP (left) and the slow EPSP (right); C_m represents membrane capacitance. See text for further discussion. (From Weight and Votava, 1970.)

of frog. When the fast EPSP produced by stimulation of the VIIIth nerve is blocked by nicotinic antagonists, repetitive stimulation of the VIIIth nerve produces a slow hyperpolarization of C cells—the slow IPSP (Figure 5B1). Present evidence suggests that the slow IPSP is produced by activation of muscarinic receptors on C cells in the frog (Weight and Padjen, 1973b). In rabbit superior cervical ganglion, however, the slow IPSP has been proposed to be mediated by a monoamine (Eccles and Libet, 1961).

The increased conductance involved in generating most IPSPs is detectable as a decrease in membrane resistance during the IPSP. By contrast, however, when resistance change was tested during the slow IPSP in C cells, membrane resistance increased markedly (Figure 5B2) and the magnitude of the resistance change appeared to parallel

the amplitude of the slow IPSP (Weight and Padjen, 1973a). The increase in resistance was not explained by a change in membrane resistance due to the hyperpolarization, because the hyperpolarizing current-voltage curve was essentially linear. The increase in resistance can thus be attributed to a decrease in membrane conductance during the slow IPSP.

To study the ion species that might be involved in the decreased conductance, the effect of membrane polarization on the slow IPSP was investigated (Weight and Padjen, 1973a). As shown in Figure 5C, moderate hyperpolarizing current increased the amplitude of the slow IPSP. On the other hand, progressive depolarizing current reduced and then abolished the slow IPSP. Further very strong depolarizing current either damaged the membrane or resulted in unstable recordings. An index of

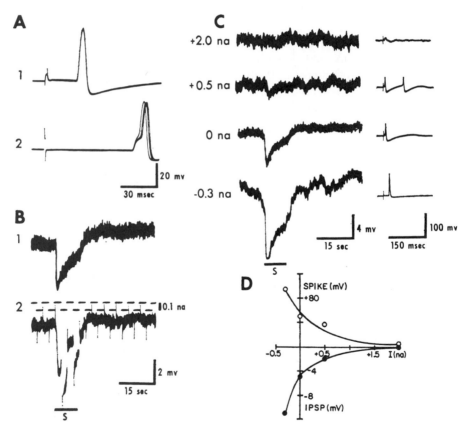

FIGURE 5 Fast EPSP and slow EPSP in C cell. (A1) Antidromic action potential. Antidromic conduction velocity of 0.24 msec identifies neuron as C cell. (A2) Fast EPSP generated by stimulation of preganglionic C fibers in VIIIth spinal nerve. (B1) Slow IPSP generated in same C cell, after nicotinic blockade (nicotine, 5 μg/ml), by stimulation of preganglionic C fibers in the VIIIth nerve at a frequency of 50 per sec. Period of stimulation indicated by line labeled *S* under record in B2. (B2) *Upper record*: Hyperpolarizing constant current pulses of −0.1 namp. *Lower record*: Resistance change during slow IPSP. Bridge balanced before stimulation such that current pulses produced

no voltage deflection. Note that during the slow IPSP the −0.1 namp current pulse produced a maximal hyperpolarizing voltage deflection of 2 mV, indicating that membrane resistance increased by 20 megohms, from 78 to 98 megohms. (C) *Left*: Amplitude of slow IPSP as a function of depolarizing (+) and hyperpolarizing (−) current. *Right*: Amplitude of antidromic spike recorded during the same polarizing current. (D) Amplitudes of slow IPSP (filled circles) and antidromic spike (open circles) in C, represented graphically as a function of polarizing current. (From Weight and Padjen, 1973a.)

membrane potential during the polarization was obtained from the height of the antidromic spike recorded simultaneously and is shown on the right in Figure 5C. In Figure 5D, the antidromic spike height and the amplitude of the slow IPSP are plotted on the same graph as a function of the polarizing current. Since the peak of a spike is considered a region of high Na conductance, the reciprocal nature of the two curves suggests that the predominant conductance that is decreased during the slow IPSP is resting Na conductance.

If the slow IPSP is generated by a decrease in resting Na conductance, then removal of extracellular Na should abolish the response. Removal of Na, however, abolishes nerve conduction so that it is not possible to test this ion change on the synaptic potential. The iontophoretic administration of ACh produces a hyperpolarization that mimics the slow IPSP in C cells of frog sympathetic ganglion (Weight and Padjen, 1973b). Therefore, the effect of Na removal was tested on the hyperpolarization produced by the iontophoretic administration of ACh to C cells (Weight and Padjen, 1973a). The removal of Na gradually reduced and then abolished the ACh hyperpolarization (Figure 6A). Return to Ringer solution subsequently restored the generation of the ACh response (Figure 6A). This data indicates that extracellular Na is necessary for the generation of the ACh hyperpolarization and supports the hypothesis that the slow IPSP is generated by an inactivation of resting Na conductance. The slow IPSP is not significantly affected by the removal of extracellular Cl (Nishi and Koketsu, 1968), indicating that an inactivation of Cl conductance does not play a significant role in the generation of the response.

The preceding data can be explained by the hypothesis that the slow IPSP is generated by a decrease or inactivation of resting Na conductance (Weight and Padjen, 1973a). This hypothesis can be represented by the circuit diagram shown in Figure 6B. On the left is the schema for the fast EPSP in these cells; on the right, the proposed mechanism for the slow IPSP is shown as a resting sodium conductance (g_{Na}) in series with its electromotive force, E_{Na}. The other resting membrane conductances are shown in parallel as a fixed conductance, G_m, in series with the emf, E_m. Activation of membrane receptors (M) is illustrated as inactivating Na conductance. This would increase the membrane resistance and shift the membrane potential away from E_{Na} and toward the equilibrium potential of the other membrane conductances, E_m, thus generating a hyperpolarization—the slow IPSP.

GENERAL CONSIDERATIONS   Although the mechanism of generation of the slow synaptic potentials in the sympathetic ganglion appear relatively unique with respect to the generation of synaptic potentials at other synaptic junctions, there is some data that suggests that such phenomena may be more widespread.

With regard to the inactivation of potassium conductance, Bulbring (1973) recently suggested that a

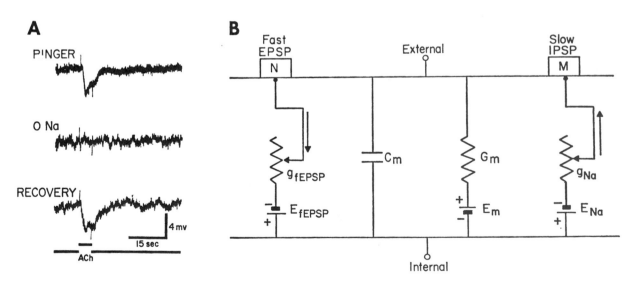

FIGURE 6   (A) Effects of removing extracellular Na$^+$ on slow ACh hyperpolarization. ACh administered extracellularly to C cell by iontophoresis during period indicated by bottom line. *Top*: Control ACh hyperpolarization in Ringer solution. *Middle*: Administration of ACh 30 minutes after removal of extracellular Na$^+$ (NaCl replaced by isotonic surcors). *Bottom*: Recovery of ACh hyperpolarization 30 min after return to Ringer solution with normal Na$^+$. (B) Schematic electrical circuit diagram of sympathetic ganglion C cell membrane representing the fast EPSP (left) and the slow IPSP (right). See text for further discussion. (From Weight and Padjen, 1973a.)

depolarization of smooth muscle with an increase in membrane resistance may be generated by a decrease in resting potassium conductance. In cortical neurons, ACh produces a slow depolarization with properties very similar to the slow EPSP in sympathetic ganglion cells. The ACh depolarization of cortical neurons has also been proposed to be generated by an inactivation of potassium conductance (Krnjević et al., 1971).

The electrophysiological properties of the slow IPSP are remarkably similar to the hyperpolarizing response to light of vertebrate photoreceptors (Baylor and Fuortes, 1970; Tomita, 1970). The photoreceptor response has also been proposed to be generated by an inactivation of resting Na conductance. Hyperpolarization with increased membrane resistance has also been reported in response to norepinephrine in cerebellar Purkinje cells (Siggins et al., 1971a) and spinal motoneurons (Engberg and Marshall, 1971). Although the mechanism of those responses has not been completely elucidated, the similarity with the slow IPSP suggests that synaptic inactivation of sodium conductance may be a mechanism of general significance in the function of the nervous system.

The functional significance of synaptic potentials generated by conductance decreases remains to be determined. It is of interest, however, to discuss some of the possibilities. Synaptic excitation generated by an increase in membrane conductance, while bringing the membrane potential closer to threshold, would also shunt the membrane and might be expected to reduce the amplitude of convergent EPSPs generated on a neighboring area of membrane. On the other hand, the increase in membrane resistance associated with synaptic excitation generated by a decrease in conductance would be expected to result in a larger potential for a given synaptic current. In this way, a convergent synaptic excitation would be expected to be potentiated by the increased membrane resistance. Thus, the generation of slow synaptic excitation by a decrease in conductance could serve as a postsynaptic mechanism of heterosynaptic facilitation.

The increased conductance associated with most IPSPs is an effective inhibitory mechanism, because not only is the membrane potential shifted away from threshold for spike generation but the increased conductance also shunts the membrane thus reducing the amplitude of convergent EPSPs and making it less likely that they will reach threshold. The functional significance of synaptic inhibition generated by a conductance decrease is more difficult to interpret because, although the hyperpolarization would inhibit firing by shifting the membrane potential away from the threshold, the increase in resistance would be expected to increase the amplitude of convergent EPSPs. It is possible that such a mechanism might be utilized by the nervous system to inhibit

neuronal discharge while enhancing the synaptic convergence on the neuron.

Synaptic activation of metabolic systems

In addition to the action of synaptic transmitters generating postsynaptic potentials by changing membrane permeability, recent studies indicate that synaptic transmitters may also activate metabolic systems in the postsynaptic neuron. This concept is also dealt with elsewhere in this volume (see Bloom, this volume).

SYNAPTIC ACTIVATION OF CYCLIC AMP   Many hormones have been found to activate adenyl cyclase, which catalyzes the conversion of adenosine triphosphate (ATP) to cyclic adenosine monophosphate (cyclic AMP), leading to the concept that cyclic AMP may be a second intracellular messenger in hormone action (see Sutherland, 1972). There also is recent data that at some synapses cyclic AMP may be activated by the synaptic transmitter.

In the cerebellum, the administration of norepinephrine (NE) to Purkinje cells produces an inhibition that is mimicked by the administration of cyclic AMP (Siggins et al., 1971a). Prior administration of theophylline, which inhibits the hydrolysis of cyclic AMP, potentiates the effects of both NE and cyclic AMP. On the other hand, prostaglandin E_1 and E_2, which are thought to inhibit adenyl cyclase activity, antagonize the response of Purkinje cells to NE but not to cyclic AMP (Siggins et al., 1971a). These data have led to the proposal that the NE action on Purkinje cells is mediated by the stimulation of cyclic AMP. More recent studies have shown a NE pathway synapsing on Purkinje cells (Bloom et al., 1971) and stimulation of this pathway mimics the effect of the administration of NE (Siggins et al., 1971b). It has also been shown, using an immunohistochemical technique for cyclic AMP, that stimulation of the NE pathway and the administration of NE both produce an increase in cyclic AMP in Purkinje cells (Siggins et al., 1973).

In the mammalian sympathetic ganglion, synaptic stimulation also leads to the generation of cyclic AMP (McAfee et al., 1971). Since administration of dopamines (Kebabian and Greengard, 1971) or catecholamines (Cramer et al., 1971) also generate cyclic AMP in the ganglion, the synaptic effect is presumed to be mediated by an adrenergic pathway in the ganglion.

ACTIVATION OF CYCLIC GMP   In addition to the recent investigations on cyclic AMP, attention is also beginning to focus on a possible role in neurotransmission of another cyclic nucleotide, cyclic guanosine monophosphate (cyclic GMP). Recent investigations have found that acetylcholine can activate cyclic GMP in heart (George et al., 1970; Kuo et al., 1972) and brain (Ferrendelli et al., 1970). Such investigations are still at a very early stage and although ACh is known to be a transmitter in these systems, it remains to be demonstrated whether cyclic GMP can be synaptically activated.

While present data indicate that cyclic nucleotides can be activated at some synapses, much further work is necessary to precisely define the role of cyclic nucleotides in synaptic transmission.

Conclusions

We have considered several physiological mechanisms for synaptic modulation. Presynaptically, the release of transmitter from synaptic terminals may be modulated by preceding activity in the terminal or by other synapses. Postsynaptically, a synaptic transmitter may increase or decrease the permeability of postsynaptic membrane producing either depolarization or hyperpolarization, depending on the ions involved, and enhancement or reduction of convergent synaptic potentials, depending on whether membrane resistance is increased or decreased. These mechanisms may modulate the efficacy of synaptic transmission over a time domain ranging from milliseconds to minutes. The influence of synapses on the distribution and sensitivity of postsynaptic receptors and the recent evidence that synaptic transmitters can activate metabolic systems in the postsynaptic neuron considerably extends the time domain of synaptic modulation. In view of the proposal that cyclic nucleotides may act as a regulator of genetic expression (Langan, 1968), it is possible that the synaptic activation of metabolic systems may be a long-term mechanism of synaptic modulation.

REFERENCES

ARAKI, T., and C. A. TERZUOLO, 1962. Membrane currents in spinal motoneurons associated with the action potential and synaptic activity. *J. Neurophysiol.* 25:772–789.

AXELSSON, J., and S. THESLEFF, 1959. A study of supersensitivity in denervated mammalian skeletal muscle. *J. Physiol.* 147:178–193.

BARKER, J. L., and R. A. NICOLL, 1973. The pharmacology and ionic dependency of amino acid responses in the frog spinal cord. *J. Physiol.* 228:259–277.

BAYLOR, D. A., and M. G. F. FUORTES, 1970. Electrical responses of single cones in the retina of the turtle. *J. Physiol. (Lond.)* 207:77–92.

BIRKS, R. I., 1963. The role of sodium ions in the metabolism of acetylcholine. *Can. J. Biochem. Physiol.* 40:303–316.

BIRKS, R. I., and M. W. COHEN, 1968. The influence of internal sodium on the behavior of motor nerve endings. *Proc. Roy. Soc. (Lond.)* B170:401–421.

BLOOM, F. E., B. J. HOFFER, and G. R. SIGGINS, 1971. Studies on norepinephrine-containing afferents to Purkinje cells of rat cerebellum. I. Localization of the fibers and their synapses. *Brain Res.* 25:501–521.

BOISTEL, J., and P. FATT, 1958. Membrane permeability change during transmitter action in crustacean muscle. *J. Physiol. (Lond.)* 144:176–191.

BRAUN, M., R. F. SCHMIDT, and M. ZIMMERMANN, 1966. Facilitation at the frog neuromuscular junction during and after repetitive stimulation. *Pflügers Arch.* 287:41–55.

BULBRING, E., 1973. *Proc. Roy. Soc. (Lond.)* B, (in press).

CHIARANDINI, D. J., E. STEFANI, and H. M. GERSCHENFELD, 1967. Ionic mechanisms of cholinergic excitation in molluscan neurons. *Science* 156:1597–1599.

COOMBS, J. S., J. C. ECCLES, and P. FATT, 1955. The specific ionic conductances and the ionic movements across the motoneuronal membrane that produce the inhibitory postsynaptic potential. *J. Physiol. (Lond.)* 130:326–373.

CRAMER, H., D. G. JOHNSON, S. D. SILBERSTEIN, and I. J. KOPIN, 1971. Effects of catecholamines on cyclic-AMP levels in rat superior cervical ganglia in vitro. *Pharmacologist* 13:257.

CURTIS, D. R., A. W. DUGGAN, D. FELIX, and G. A. R. JOHNSTON, 1971. Bicuculline, an antagonist of GABA and synaptic inhibition in the spinal cord of the cat. *Brain Res.* 32:69–96.

DAVIDOFF, R. A., 1972. Gamma-aminobutyric acid antagonism and presynaptic inhibition in the frog spinal cord. *Science* 175:331–333.

DEL CASTILLO, J., and B. KATZ, 1954a. Quantal components of the end plate potential. *J. Physiol. (Lond.)* 124:560–573.

DEL CASTILLO, J., and B. KATZ, 1954b. Statistical factors involved in neuromuscular facilitation and depression. *J. Physiol.* 24:574–585.

DEL CASTILLO, J., and B. KATZ, 1954c. The membrane change produced by the neuromuscular transmitter. *J. Physiol. (Lond.)* 125:546–565.

DEL CASTILLO, J., and B. KATZ, 1955. On the localization of acetylcholine receptors. *J. Physiol.* 128:157–181.

DODGE, F. A., and R. RAHAMIMOFF, 1967. Cooperative action of calcium ions in transmitter release at the neuromuscular junction. *J. Physiol. (Lond.)* 193:419–432.

DUDEL, J., 1965a. Presynaptic and postsynaptic effects of inhibitory drugs on the crayfish neuromuscular junction. *Pflügers Arch.* 283:104–118.

DUDEL, J., 1965b. The action of inhibitory drugs on nerve terminals in crayfish muscle. *Pflügers Arch.* 284:81–94.

DUDEL, J., and S. W. KUFFLER, 1961. Presynaptic inhibition at the crayfish neuromuscular junction. *J. Physiol. (Lond.)* 155:543–562.

ECCLES, J. C., 1964. *The Physiology of Synapses*. New York: Academic Press, p. 316.

ECCLES, J. C., R. M. ECCLES, and M. ITO, 1964. Effects produced on inhibitory postsynaptic potentials by the coupled incoupled injections of cations and anions into motoneurons. *Proc. Roy. Soc. (Lond.)* B160:197–210.

ECCLES, J. C., R. M. ECCLES, and F. MAGNI, 1961. Central inhibitory action attributable to presynaptic depolarization produced by muscle afferent volleys. *J. Physiol. (Lond.)* 159:147–166.

ECCLES, J. C., B. KATZ, and S. W. KUFFLER, 1941. Nature of the "end plate potential" in curarized muscle. *J. Neurophysiol.* 4:362–387.

ECCLES, R. M., and B. LIBET, 1961. Origin and blockade of the synaptic responses of curarized sympathetic ganglia. *J. Physiol. (Lond.)* 157:484–503.

EDWARDS, C., and S. HAGIWARA, 1959. Potassium ions and the inhibitory process in the crayfish stretch receptors. *J. Gen. Physiol.* 43:315–321.

ENGBERG, I., and K. C. MARSHALL, 1971. Mechanisms of noradrenaline hyperpolarization in spinal cord motoneurons of the cat. *Acta. Physiol. Scand.* 83:142–144.

EPSTEIN, R., and L. TAUC, 1970. Heterosynaptic facilitation and posttetanic potentiation in *Aplysia* nervous system. *J. Physiol.* 209:1–23.

FATT, P., and B. KATZ, 1951. An analysis of the endplate potential recorded with an intracellular electrode. *J. Physiol. (Lond.)* 115:320–370.

FATT, P., and B. KATZ, 1953. The effect of inhibitory nerve impulses on a crustacean muscle fiber. *J. Physiol. (Lond.)* 121:374–389.

FENG, T. P., 1940. Studies on the neuromuscular junction XVIII. The local potentials around N-M junctions induced by single and multiple volleys. *Chin. J. Physiol.* 15:367–404.

FENG, T. P., 1941. Studies on the neuromuscular junction XXVI. The changes of the endplate potential during and after prolonged stimulation. *Chin. J. Physiol.* 16:341–372.

FERRENDELLI, J. A., A. L. STEINER, D. B. McDOUGAL, JR., and D. M. KIPNIS, 1970. The effect of oxotremorine and atropine on cGMP and cAMP levels in mouse cerebral cortex and cerebellum. *Biochem. Biophys. Res. Commun.* 41:1061–1067.

FRANK, K., and M. G. F. FUORTES, 1957. Presynaptic and postsynaptic inhibition of monosynaptic reflexes. *Fed. Proc.* 16:39–40.

FRANK, K., and L. TAUC, 1964. Voltage-clamp studies of molluscan neuron membrane properties. In *The Cellular Functions of Membrane Transport*, J. F. Hoffman, ed. New Jersey: Prentice-Hall.

GEORGE, W. J., J. B. POLSON, A. G. O'TOOLE, and N. G. GOLDBERG, 1970. Elevation of guanosine 3', 5'-cyclic phosphate in rat heart after perfusion with acetylcholine. *Proc. Nat. Acad. Sci. USA* 66:398–403.

GINSBORG, B. L., 1967. Ion movements in junctional transmission. *Pharmacol. Rev.* 19:289–316.

GUTH, L., 1968. "Trophic" influences of nerve on muscle. *Physiol. Rev.* 48:645–687.

GUTH, L., 1969. "Trophic" effects of vertebrate neurons. *Neurosci. Res. Prog. Bull.* 7 (No. 1):1–70.

HAGIWARA, S., K. KUSANO, and S. SAITO, 1960. Membrane changes in crayfish stretch receptor neuron during synaptic inhibition and under action of gamma-aminobutyric acid *J. Neurophysiol.* 23:505–515.

HUBBARD, J. I., 1963. Repetitive stimulation at the mammalian neuromuscular junction, and the mobilization of transmitter. *J. Physiol.* 169:641–662.

HUGHES, J. R., 1958. Post-tetanic potentiation. *Physiol. Rev.* 38:91–113.

HUTTER, O. F., 1952. Post-tetanic restoration of neuromuscular transmission blocked by α-tubocurarine. *J. Physiol. (Lond.)* 118:216–227.

JONES, R., and G. VRBOVA, 1970. Effect of muscle activity on denervation hypersensitivity. *J. Physiol. (Lond.)* 210:144–145.

KANDEL, E., and L. TAUC, 1965. Mechanism of heterosynaptic facilitation in the giant cell of the abdominal ganglion of *Aplysia depilans. J. Physiol. (Lond.)* 181:28–47.

KATZ, B., 1966. *Nerve, Muscle and Synapse.* New York: McGraw-Hill, p. 193.

KATZ, B., 1969. *The Release of Neural Transmitter Substances.* Springfield, Ill.: Charles C Thomas, p. 60.

KATZ, B., and R. MILEDI, 1965. The effect of calcium on acetylcholine release from motor nerve terminals. *Proc. Roy. Soc. (Lond.)* B161:496–503.

KATZ, B., and R. MILEDI, 1968. The role of calcium in neuromuscular facilitation. *J. Physiol. (Lond.)* 195:481–492.

KEBABIAN, J. W., and P. GREENGARD, 1971. Dopamine-sensitive adenyl cyclase: Possible role in synaptic transmission. *Science* 174:1346–1349.

KOBAYASHI, H., and B. LIBET, 1970. Actions of noradrenaline and acetylcholine on sympathetic ganglion cells. *J. Physiol. (Lond.)* 208:353–372.

KOKETSU, K., 1969. Cholinergic synaptic potentials and the underlying ionic mechanisms. *Fed. Proc.* 28:101–112.

KRAVITZ, E. A., S. W. KUFFLER, and D. D. POTTER, 1963. Gamma-aminobutyric acid and other blocking compounds in crustacea. III. Their relative concentrations in separated motor and inhibitory axons. *J. Neurophysiol.* 26:739–751.

KRNJEVIĆ, K., R. DUMAIN, and L. RENAUD, 1971. The mechanism of excitation by acetylcholine in the cerebral cortex. *J. Physiol.* 215:247–268.

KUNO, M., 1971. Quantum aspects of central and ganglionic synaptic transmission in vertebrates. *Physiol. Rev.* 51:647–678.

KUO, J.-F., T.-P. LEE, P. L. REYES, K. G. WALTON, T. E. DONNELLY, JR., and P. GREENGARD, 1972. Cyclic nucleotide-dependent protein kinases X. An assay method for the measurement of guanosine 3', 5'-monophosphate in various biological materials and a study of agents regulating its levels in heart and brain. *J. Biol. Chem.* 247:16–22.

LANGAN, T. A., 1968. Histone phosphorylation: Stimulation by adenosine 3', 5'-monophosphate. *Science* 162:579–580.

LIBET, B., 1970. Generation of slow inhibitory and excitatory postsynaptic potential. *Fed. Proc.* 29:1945–1956.

LILEY, A. W., 1956. The quantal component of the mammalian end-plate potential. *J. Physiol.* 133:571–587.

LILEY, A. W., and K. A. K. NORTH, 1953. An electrical investigation of effects of repetitive stimulation on mammalian neuromuscular junction. *J. Neurophysiol.* 16:509–527.

LOMO, T., and J. ROSENTHAL, 1972. Control of ACh sensitivity by muscle activity in the rat. *J. Physiol.* 221:493–513.

LUNDBERG, A., and H. QUILISCH, 1953. Presynaptic potentiation and depression of neuromuscular transmission in frog and rat. *Acta Physiol. Scand.* 30 (Suppl. 111):111–119.

MARTIN, A. R., and J. L. VEALE, 1967. The nervous system at the cellular level. *Ann. Rev. Physiol.* 29:401–426.

McAFEE, D. A., M. SCHORDERET, and P. GREENGARD, 1971. Adenosine 3',5'-monophosphate in nervous tissue: Increase associated with synaptic transmission. *Science* 171:1156–1158.

MILEDI, R., 1960a. The acetylcholine sensitivity of frog muscle fibers after complete or partial denervation. *J. Physiol.* 151:1–23.

MILEDI, R., 1960b. Junctional and extra-junctional acetylcholine receptors in skeletal muscle fibers. *J. Physiol.* 151:24–30.

MILEDI, R., 1963. An influence of nerve not mediated by impulses. In *The Effect of Use and Disuse on Neuromuscular Function*, E. Gutman, and R. Hnik, eds. Prague: Publishing House of the Czechoslovakia Academy of Sciences.

NISHI, S., and K. KOKETSU, 1968. Analysis of slow inhibitory postsynaptic potential of bullfrog sympathetic ganglion. *J. Neurophysiol.* 31:717–728.

Oomura, Y., H. Ooyama, and M. Sawada, 1965. Ionic basis of the effect of ACh on Onchidium O- and H- neurons. *Intern. Congr. Physiol. Sci.* 23:389.

Otsuka, M., M. Endo, and Y. Nonomura, 1962. Presynaptic nature of neuromuscular depression. *Japan J. Physiol.* 12: 573–584.

Otsuka, M., L. L. Iversen, Z. W. Hall, and E. A. Kravitz, 1966. Release of gamma-amino butyric acid from inhibitory nerves of lobster. *Proc. Nat. Acad. Sci. USA* 56:1110–1115.

Rahamimoff, R., 1968. A dual effect of calcium ions on neuromuscular facilitation. *J. Physiol. (Lond.)* 195:471–480.

Rosenthal, J., 1969. Post-tetanic potentiation at the neuromuscular junction of the frog. *J. Physiol. (Lond.)* 203:121–133.

Rubin, R. P., 1970. The role of calcium in the release of neurotransmitter substances and hormones. *Pharmacol. Rev.* 22: 389–428.

Siggins, G. R., B. J. Hoffer, and F. E. Bloom, 1971a. Studies on norepinephrine-containing afferents to Purkinje cells of rat cerebellum III. Evidence for mediation of norepinephrine effects by cyclic 3′,5′-adenosine monophosphate. *Brain Res.* 25:535–553.

Siggins, G. R., B. J. Hoffer, A. P. Oliver, and F. E. Bloom, 1971b. Activation of a central noradrenergic projection to cerebellum. *Nature (Lond.)* 233:481–483.

Siggins, G. R., E. F. Battenberg, B. J. Hoffer, F. E. Bloom, and A. L. Steiner, 1973. Noradrenergic stimulation of cyclic adenosine monophosphate in rat Purkinje neurons: An immunocytological study. *Science* 179:585–588.

Smith, T. G., R. B. Wuerker, and K. Frank, 1967. Membrane impedance changes during synaptic transmission in cat spinal motoneurons. *J. Neurophysiol.* 30:1072–1096.

Sutherland, E. W., 1972. Studies on the mechanism of hormone action. *Science* 177:401–408.

Takeuchi, A., 1958. The long-lasting depression in neuromuscular transmission of frog. *Japan J. Physiol.* 8:102–113.

Takeuchi, A., and N. Takeuchi, 1959. Active phase of frog's end-plate potential. *J. Neurophysiol.* 22:395–411.

Takeuchi, A., and N. Takeuchi, 1960. On the permeability of end-plate membrane during the action of transmitter. *J. Physiol. (Lond.)* 154:52–67.

Takeuchi, A., and N. Takeuchi, 1966a. A study of the inhibitory action of γ-aminobutyric acid on neuromuscular transmission in crayfish. *J. Physiol. (Lond.)* 183:418–432.

Takeuchi, A., and N. Takeuchi, 1966b. On the permeability of the presynaptic terminal of the crayfish neuromuscular junction during synaptic inhibition and the action of γ-aminobutyric acid. *J. Physiol. (Lond.)* 183:433–449.

Takeuchi, A., and N. Takeuchi, 1967. Anion permeability of the inhibitory postsynaptic membrane of the crayfish neuromuscular junction. *J. Physiol. (Lond.)* 191:575–590.

Thesleff, S., 1960. Supersensitivity of skeletal muscle produced by botulinum toxin. *J. Physiol. (Lond.)* 151:598–607.

Thies, R., 1965. Neuromuscular depression and the apparent depletion of transmitter in mammalian muscle. *J. Neurophysiol.* 28:427–442.

Tomita, T., 1970. Electrical activity of vertebrate photoreceptors. *Quart. Rev. Biophys.* 3:179–222.

Tosaka, T., S. Chichibu, and B. Libet, 1968. Intracellular analysis of slow inhibitory and excitatory postsynaptic potentials in sympathetic ganglion of the frog. *J. Neurophysiol.* 31:396–409.

Weight, F. F., 1971. Mechanisms of synaptic transmission. *Neurosci. Res. Prog. Bull.* 4:1–27.

Weight, F. F., and J. Votava, 1970. Slow synaptic excitation in sympathetic ganglion cells: Evidence for synaptic inactivation of potassium conductance. *Science* 170:755–758.

Weight, F. F., and A. Padjen, 1973a. Slow synaptic inhibition in sympathetic ganglion cells: Evidence for synaptic inactivation of sodium conductance. *Brain Res.* (in press).

Weight, F. F., and A. Padjen, 1973b. Acetylcholine and slow synaptic inhibition in frog sympathetic ganglion. *Brain Res.* (in press).

Weinreich, D., 1971. Ionic mechanism of post-tetanic potentiation at the neuromuscular junction of the frog. *J. Physiol. (Lond.)* 212:431–446.

83 Modulation of Transmitter Release at the Neuromuscular Junction

RAMI RAHAMIMOFF

ABSTRACT   Transmitter is released from motor nerve terminals as preformed packets. Three main factors determine the number of packets liberated: the ionic composition of the medium, the level of presynaptic polarization, and the frequency of activation of the release process.

The highly nonlinear relations between the number of packets liberated and each of these parameters are described, and a hypothesis for the fluctuating nature of evoked release is proposed.

NEUROTRANSMITTERS are liberated from nerve endings as preformed packages or quanta. This was first shown more than 20 years ago at the frog neuromuscular junction (Fatt and Katz, 1951, 1952) and since then demonstrated at many other synapses (for example, see Boyd and Martin, 1956; Dudel and Kuffler, 1961; Blackman, Ginsborg, and Ray, 1963; Kuno, 1964; Miledi, 1967; Dennis, Harris, and Kuffler, 1971; Martin and Pilar, 1964a). Since these quanta contain roughly 10^4 molecules of acetylcholine, it is easier to demonstrate them by their action on the postsynaptic muscle membrane. At rest, the quantum appears as a small discrete depolarization, the miniature end-plate potential (mepp) (Figure 1A, B). The probability of appearances of these mepp's is low at rest, about 1 every second. Upon the arrival of an action potential at the nerve terminals of a neuromuscular junction, this probability increases some 5 orders of magnitude; under normal circumstances several hundred quanta are then liberated within a millisecond. The effect of these quanta summates and produces the end-plate potential (epp). If the epp is suprathreshold, it causes an action potential and a twitch, and thus the chain of transmission is completed. Hence, the amount of transmitter released by the nerve impulse is a product of two factors: the amount contained in each package (a) times the number of packages liberated (the latter is defined as the quantal content or the quantal number m). Theoretically, therefore, two ways exist to modulate transmitter release: either by changing a or by altering m. If, however, one examines

the known regulatory mechanisms, it is apparent that the main property that varies in different physiological conditions is m.

The three main factors that affect m are: (1) The polarization of the nerve terminal. (2) The ionic composition of the medium, in particular the divalent ions content. (3) The frequency of nerve activation.

In this chapter, I discuss the first point briefly and concentrate in more detail on the other two.

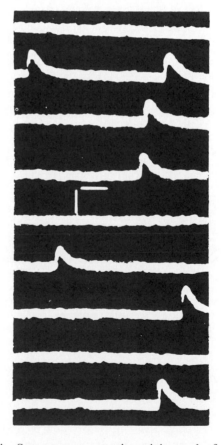

FIGURE 1A   Spontaneous synaptic activity at the frog neuromuscular junction. Each random event is a miniature end-plate potential. Vertical calibration 0.5 mV; horizontal calibration 20 msec.

RAMI RAHAMIMOFF   Department of Physiology, Hebrew University Medical School, Jerusalem, Israel

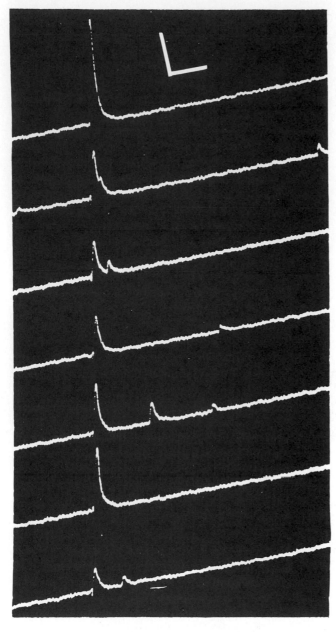

FIGURE 1B Evoked end-plate potentials and spontaneous miniature end-plate potentials recorded on a moving film. Note the fluctuating nature of synaptic responses. Note also the higher frequency of the spontaneous events after stimulation. Vertical calibration 1 mV, horizontal calibration 20 msec.

Nerve terminal polarization and transmitter release

The level of polarization of the presynaptic terminal has a great inducing effect on transmitter release. Under normal circumstances, a small depolarization is sufficient to bring the membrane potential to the threshold levels and the fullblown effect of the action potential on release

is observed. In order to study the graded effect of depolarization, the action potential mechanism has to be suppressed. This has been done on a number of preparations by the addition of tetrodotoxin (TTX), which by abolishing the voltage-dependent entry of sodium, allows the study of graded polarization on transmitter release. Apparently, TTX does not affect significantly transmitter liberation, demonstrating that the entry of sodium ions is not essential for quantal transmitter release. (Katz and Miledi, 1967a, 1967b, 1968b).

From the study of transmitter release in TTX-blocked preparations, it became apparent that both the duration and the amplitude of the depolarizing pulse determined the number of quanta liberated. Figure 2A shows the relation between the strength of the depolarizing pulse and transmitter output: with larger depolarizations more transmitter is liberated (hollow circles). The depolarization can be made more effective by the addition of tetraethylammonium (TEA), which blocks the changes in the potassium conductance (full circles). The effect of the duration of the depolarizing pulse is shown in Figure 2B. One can see that the longer the duration, the more effective is the pulse.

The input–output relation between presynaptic polarization on one hand and transmitter release and postsynaptic effect (postsynaptic potential, PSP) on the other, became clearer when similar studies were conducted on the giant synapse of the squid. Here, one can measure not only the magnitude of the depolarizing

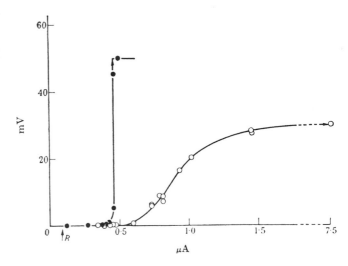

FIGURE 2A Effect of polarization parameters on transmitter release. Effect of current strength on transmitter release at the neuromuscular junction. Strength-response relations: *ordinate*: epp amplitude; *abscissa*: current strength. Two relations are shown; hollow circles: before adding 5 mM tetraethylammonium (TEA); full circles: after adding TEA. (From Katz and Miledi, 1967c.)

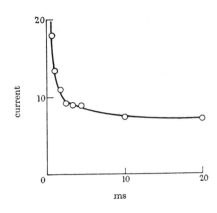

FIGURE 2B   Effect of current duration on transmitter release at the neuromuscular junction. "Strength-duration" curve, showing the relation between duration and intensity of equally effective pulses. (From Katz and Miledi, 1967c.)

pulse but also its effects on the membrane potential of the presynaptic nerve fiber. It became apparent that progressive increase in the amplitude of the presynaptic depolarizing pulse produced a relation with a maximum, in the amount of transmitter liberated, shortly after the onset of the pulse (Figure 2C: "On"). Gradual increase in the pulse size initially augmented the PSP, while further on, the PSP had a decreased amplitude. With presynaptic depolarization of more than 150 mV, the PSP was only a small fraction of its peak amplitude. However, these large pulses were not without effect on

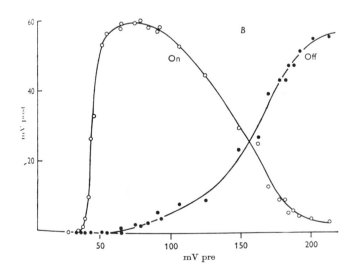

FIGURE 2C   Effect of presynaptic polarization on transmitter release at the squid giant synapse. Input-output curves with long (18 msec) current pulses after electrophoretic loading of the terminal with tetraethylammonium. *Abscissa*: presynaptic potential change (peak amplitude); *ordinate*: postsynaptic ON and OFF responses. (From Katz and Miledi, 1967d.)

transmitter release: They induced marked outpour of transmitter *after* the end of the pulse (Figure 2C: "Off"). These experiments show that it is not the depolarization itself which is responsible for the induction of transmitter release, but rather a process associated with depolarization.

Another line of experiments also showed that depolarization is not enough to produce transmitter release. Calcium ions have to be present in the medium at the time of the depolarization (Katz and Miledi, 1967a), to make it effective. From Figure 3, it is clear that calcium has to precede depolarization. If calcium ions are applied after the depolarizing pulse, they do not induce any transmitter release. Therefore, the interplay between polarization and calcium ions determines how much transmitter will be liberated.

TRANSMITTER RELEASE AND DIVALENT IONS   Release of transmitter at the neuromuscular junction is dependent to a very high degree on the concentration of divalent ions in the extracellular medium. In this respect, it resembles other neurosecretory processes at synapses and glands (see Katz, 1969; Rubin, 1970).

For a number of years it was questioned whether calcium ions act directly on the release process or whether they exert their effect indirectly by affecting the propagation of the nerve impulse into the terminals. This problem was solved by recording the electrical activity of the presynaptic endings. It was shown that even in the absence of calcium ions, the action potential still propagates into the nerve branches (Katz and Miledi, 1965). Hence, it can be concluded that the action of calcium ions is on the secretory process itself, leading to the release of transmitter. This led to a renewed interest in the quantitative relation between calcium ion concentration and transmitter release, with the hope that this may bring some further information on the mechanism of action of calcium ions.

The studies of del Castillo and Katz (1954a) and of Jenkinson (1957) already showed that the relation between [Ca] and transmitter release is nonlinear, and it tends to saturation at higher extracellular concentrations of calcium ions. Therefore, it was suggested that in the process of transmitter liberation, a reaction exists of the form

$$Ca + X \underset{\longleftarrow}{\overset{K_1}{\longrightarrow}} CaX \qquad (1)$$

where X is a site on the presynaptic terminal that combines with Ca, forming CaX, and K_1 is the dissociation coefficient of this postulated reaction. It was further assumed that the number of quanta liberated by the nerve impulse m is proportional to CaX.

This hypothesis was further developed by Jenkinson

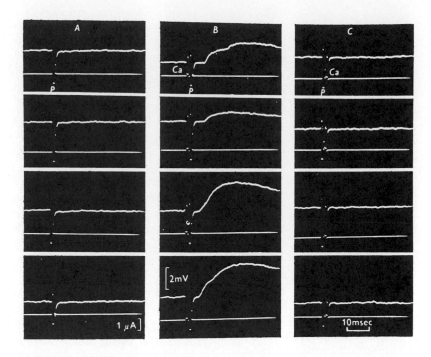

FIGURE 3 The effect of depolarization and calcium on transmitter release. Depolarizing pulses (*P*) and calcium pulses (*Ca*) were applied from a twin barrel micropipette. Intracellular recording. (A) Depolarizing pulses alone. (B) Calcium pulses precede depolarizing pulse. (C) Depolarizing pulses precede calcium pulses. (From Katz and Miledi, 1967a.) Note that calcium has to precede depolarization, in order to induce release of transmitter.

(1957), who showed that magnesium ions compete with calcium on the release process, assuming that the following reaction also occurs at the presynaptic terminal:

$$\text{Mg} + X \underset{}{\overset{K_2}{\rightleftharpoons}} \text{Mg}X \qquad (2)$$

where K_2 is the dissociation constant of this second reaction and $\text{Mg}X$ is a complex ineffective in release.

The assumption in Equations 1 and 2 predict that the relation between [Ca] and *m* should be hyperbolic with an initial linear start. For many years it was difficult to examine the initial part of the relation, since at low [Ca] the release is small. The development of signal-averaging computers permitted one to study in detail this portion of the relation and revealed that the function is sigmoidal in shape (Dodge and Rahamimoff, 1967) (Figure 4). When the initial part of the relation was plotted on double logarithmic coordinates, it showed a straight line with a slope of nearly 4 in the lower range of calcium concentrations. Doubling the concentration augmented the release nearly 16 times. This power relation was found consistently under a number of experimental conditions (Dodge and Rahamimoff, 1967; Colomo and Rahamimoff, 1968), suggesting that a number of calcium ions have to cooperate to induce the release of a quantum of transmitter.

A number of mathematical expressions can fit the sigmoidal shape of the [Ca] versus release relation. The simplest is presumably:

$$m = K \left[\frac{\text{Ca}/K_1}{1 + \text{Ca}/K_1} \right]^n \qquad (3)$$

where K is a proportionality coefficient.

In the presence of a competitive inhibitor such as Mg,

$$m = K \left[\frac{\text{Ca}/K_1}{1 + (\text{Ca}/K_1) + (\text{Mg}/K_2)} \right]^n \qquad (4)$$

the estimate for *n* was found to be 4 for the frog neuromuscular junction. (It should be pointed out that this is the lowest number that can fit the experimental data. There are no reliable experimental ways, at present, to exclude higher numbers.) Similar relations were found for the rat neuromuscular junction (with $n = 3$) (Hubbard, Jones, and Landau, 1968), and for the squid giant synapse (Katz and Miledi, 1970). An interesting exception is a crustacean neuromuscular junction (Bracho and Orkand, 1970) where a linear relation between Ca and release was observed. However, it appears that the presence of inhibition and the possible effects of calcium ions on the output of inhibitory transmitter

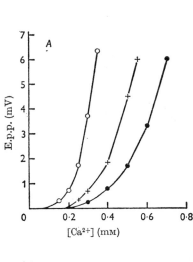

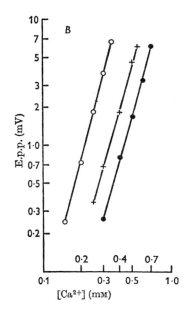

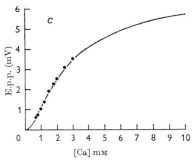

FIGURE 4 Relation between calcium concentration and amplitude of end plate potential. Since the unitary quantum is relatively constant in amplitude, the epp can be taken as measure of transmitter release. (A) Linear plot at three different magnesium concentrations; open circle: 0.5 mM Mg^{++}; cross: 2.0 mM Mg^{++}; closed circle: 4.0 mM Mg^{++}. Each value represents the average amplitude of 128 or 256 epps. (B) Same as A on double logarithmic coordinates. The slope of these relations is nearly four. Note the parallel shift of the relation by Mg^{++}. (C) Effect of calcium concentration on epp amplitude in a curarized preparation. The concentrations of calcium are higher than in A. There is 5×10^{-6} g/ml of (+)tubocurarine present throughout the experiment. Linear coordinates. The first three points are averages of 128 responses; the other points are averages of 64 responses. The sigmoidal curve follows the equation

$$epp = \left[\frac{2.7\,[Ca]}{1 + [Ca]/0.6} \right]^4 .$$

(From Dodge and Rahamimoff, 1967.)

complicate the analysis significantly (Sarna, Parnas, and Rahamimoff, 1972).

Calcium and magnesium are not the only divalent ions capable of action in this process. Strontium is also able to activate transmitter release and can substitute for calcium (Miledi, 1966; Dodge, Miledi, and Rahamimoff, 1969). Since strontium is less effective than calcium in inducing release, the interactions between these two ions are as expected for partial agonists: At low concentrations of calcium, the addition of strontium potentiates release, while at higher calcium concentrations, it inhibits the quantal release (Meiri and Rahamimoff, 1971).

It has been found recently that manganese ions also affect the same process. Very low concentrations of Mn^{++} suppress the quantal content significantly (Meiri and Rahamimoff, 1972) (Figure 5). On a molar basis Mn^{++} is at least 20 times more effective than Mg^{++}.

These observations with Mn^{++} are of interest also in another aspect of the action of calcium on transmitter release. There is evidence that transmitter liberation is associated with small inward currents of calcium ions at the giant synapse of the squid (Katz and Miledi, 1969b, 1970). The calcium currents in the squid axon can be subdivided into two components, an early one that is blocked by tetrodotoxin and a later one that is blocked by manganese ions (Baker, Hodgkin, and Ridgeway, 1971).

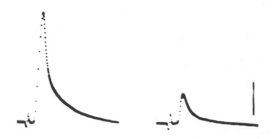

FIGURE 5 The inhibitory effect of manganese ions on release of transmitter. The two records are averages of 200 responses. On the left: the control response is in 0.4 mM Ca^{++} and 2.0 mM Mg^{++}. Quantal content is 8.78. On the right: after the addition of 70 μM $MnCl_2$. The quantal content was reduced to 2.38. Vertical calibration 0.3 mV. Sampling frequency 0.125 msec. for each address. (From Meiri and Rahamimoff, 1972.)

The lack of appreciable effect of tetrodotoxin on transmitter release (Katz and Miledi, 1967b) and the inhibition of transmitter release by manganese, suggest that the latter component of calcium influx is probably responsible for transmitter release.

The persistence of transmitter release in the presence of tetrodotoxin also shows that the entry of sodium ions is not important in the release process. However, it seems that sodium ions have an indirect inhibitory action on release by competing with calcium. Partial withdrawal of sodium ions from the bathing medium increases the number of quanta liberated (Birks and Cohen, 1965; Kelly, 1965; Rahamimoff and Colomo, 1967).

FLUCTUATIONS IN SYNAPTIC POTENTIALS AND CALCIUM IONS Synaptic potentials in general and epps in particular are not constant in size. Their amplitude fluctuates randomly from trial to trial (Fatt and Katz, 1951; del Castillo and Katz, 1954b). These fluctuations are probably of very little consequence when all the synaptic potentials are suprathreshold or when all the potentials are subthreshold in a synapse with no summation from different sources. If the mean amplitude of the PSP is near threshold, the fluctuations may cause a transmission on less than 1:1 basis between the presynaptic and the postsynaptic element.

The origin of these fluctuations resides in the quantal nature of transmitter release; in each trial a different number of quanta is liberated. The question arises whether one of the factors governing transmitter release can contribute significantly to these fluctuations. The experiments mentioned above suggest that calcium may play a role in this process.

It was assumed that the sigmoidal relation between [Ca] and transmitter release means that at least 4 calcium ions have to be present simultaneously at a certain location on the presynaptic membrane to induce release of a quantum of transmitter. It was further assumed that locations with a smaller number of calcium ions (0, 1, 2, and 3), will be ineffective. The total number of locations, that can lead to release of transmitter was estimated from the double reciprocal transformation (Liveweaver and Burk plot) of Equation 4 (see Dodge and Rahamimoff, 1967). Now although the total number of calcium ions that can bind to these release locations (or penetrate through the relevant presynaptic membrane) is the same on each trial, their distribution may vary. Therefore, on successive trials, different numbers of locations will have the number of calcium ions necessary to enable release.

This hypothesis was simulated on a digital computer (PDP 15/40) using a random number generator and some of the results are shown in Figure 6. On the left are the results of two simulations shown in detail. In both cases the same number of calcium ions were allowed to combine randomly, and while in the upper simulation two locations reached the 4 calcium ions level, in the lower only one produced such combination.

For simplicity it was further assumed that each location with 4 calciums bound will lead to transmitter release. In the right part of Figure 6 a series of 10 simulated responses are shown. It is obvious that these responses fluctuate from trial to trial, similar to the experimental situation (Figure 1B).

The nature of this hypothesis does not permit a direct experimental test, and one needs to rely on indirect statistical evidence. It is well known that evoked release of transmitter behaves, in the statistical sense, as a process with very low probability as described by the Poisson theorem (del Castillo and Katz, 1954b; Boyd and Martin, 1956). This has been demonstrated repeatedly at relatively low levels of quantal release. It is hardly surprising that the calcium hypothesis for synaptic fluctuations fits with these observations. If only a few of the numerous locations lead to release, the simulated system will behave like a Poisson process. But the hypothesis and the resulting simulations predict that if a substantial number of locations are filled with the necessary calcium ions, the distribution should be described by a statistical process with higher p. In other words, if p is the probability for release out of a population of n quanta, then the quantal content m will be given by

$$m = p \cdot n \qquad (5)$$

and the variance of m, v, will be given by

$$v = m(1 - p) \qquad (6)$$

(see del Castillo and Katz, 1954b; Martin, 1966; Christensen and Martin, 1970).

Therefore, if Ca^{++} ions determine p, and p determines the variance, then at low calcium concentrations, p will

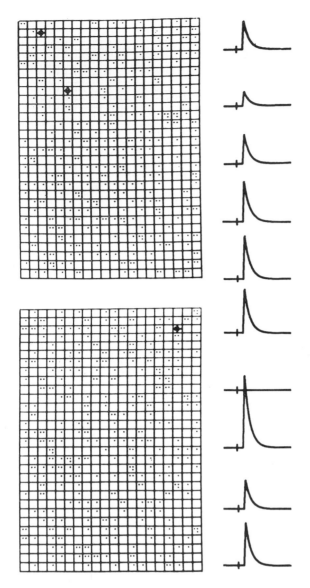

FIGURE 6 Simulations of fluctuations in epp amplitude. Ten such simulations are shown on the right; the first two are shown in detail on the left. Calcium ions (350) were randomly distributed on presynaptic locations. In the first simulation, two locations contained four calcium ions; it was assumed for simplicity that each such location leads to a release of a quantum; therefore, the first tracing on the right has an amplitude of two quanta. For assumptions see text.

be negligible and $v = m$, as predicted by the Poisson theorem. At high [Ca] on the other hand, it is predicted that p will not be small and therefore $v < m$. These predictions were tested experimentally at the frog neuromuscular junction (Rahamimoff and Meiri, 1971; Meiri and Rahamimoff, 1972a) and at the crayfish neuromuscular junctions (Johnson and Wernig, 1971; Wernig, 1972a, 1972b) and a consistent departure from $v = m$ equality was found.

Therefore, it seems that calcium ions may indeed be one of the determinants of the fluctuating nature of the end-plate potential.

It probably will not be out of place to stress again that the arguments employed here are indirect, and a more direct approach will be needed in the future to determine the relative contribution of various factors in the statistical behavior of synaptic potentials.

FREQUENCY OF STIMULATION AND TRANSMITTER RELEASE
If the motor nerve is stimulated at a low frequency, every 30 sec for example, the mean quantal content remains constant over a long period of time. If the frequency of stimulation is increased, the quantal content becomes dependent on the time interval between successive stimuli. This is best illustrated if a series of paired nerve action potentials is evoked, and one compares the quantal content of the first response (m_1) to that of the second (m_2) at different time intervals (T) between them. If T is short (up to 100 msec approximately, under standard conditions), the m_2 will be larger than m_1, and the phenomenon is known as facilitation $(F = m_2/m_1)$. At longer time intervals, m_2 will be slightly smaller than m_1; this neuromuscular depression will gradually subside, until m_2 will again be equal to m_1.

It has been suggested that facilitation is not a separate modulatory mechanism in synaptic transmitter release but is closely associated with the two parameters mentioned above—the presynaptic polarization and the activation by calcium ions. There is no doubt that if the first and the second nerve action potentials are not equal in size, this can produce changes in transmitter liberation. However, it has been shown in a number of preparations (Martin and Pilar, 1964b; Katz and Miledi, 1965; Miledi and Slater, 1966) that facilitation can be observed when the two nerve action potentials are equal, or even when the second one is somewhat smaller than the first one. This led to the idea that neuromuscular facilitation may be due to some residual calcium ions remaining on, or in, the nerve terminals after the first end-plate potential.

Let me briefly illustrate this hypothesis in a schematic fashion.

Assume that the first nerve impulse causes certain distribution of calcium ions on locations at the nerve terminal (Figure 7A). (The assumptions made for this distribution are the same as mentioned in the preceding section.) This will lead to the formation of the first epp. If the inactivation of these calcium ions has a time course of many milliseconds, then before the arrival of the second nerve action potential there will be still some residual calcium, left over from the first impulse (Figure 7B). Therefore, after the arrival of the second

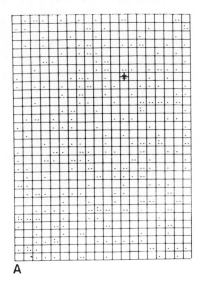

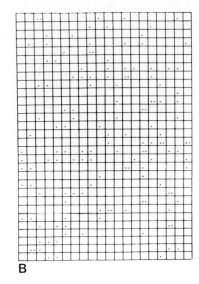

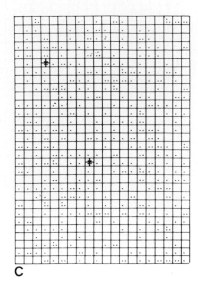

A   B   C

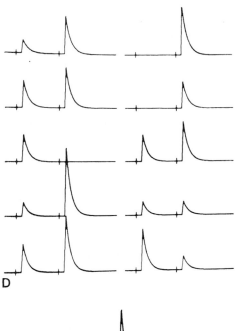

D

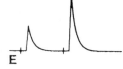

E

FIGURE 7 Schematic representation of the residual calcium hypothesis of neuromuscular facilitation. Each dot represents a calcium ion. Only locations with four calcium ions are assumed to lead to release. (A) The distribution of calcium ions immediately after the first nerve action potential. (B) The distribution of calcium ions a certain period after A. Some calcium ions are no longer attached to the sites. (C) The distribution of calcium ions after a second nerve action potential given after B.

Calcium ions from two sources are now present—the residual and the newly attached. The total number is greater than in A. (D) Simulation of pairs of epps. Note the fluctuations in facilitation in sequential pairs. (E) Average of 100 simulations, part of which were shown in D. Note that on the average the second epp is larger than the first. The assumptions are described in Figure 6 and in the text.

impulse, there will be two sources of calcium ions that will add up to induce release of transmitter: The residual and the newly attached (Figure 7C). If such a postulated sequence of events happens on the steep part of the relation between [Ca] and transmitter release, the second epp will be on the average much larger than the first one (Figure 7D, 7E).

This idea of residual calcium ions was suggested originally by Katz and Miledi (1965) and was tested experimentally in two different ways: (1) by appropriate timing of calcium pulses before and after the first impulse (Katz and Miledi, 1968a); (2) by variation in the extracellular calcium concentration (Rahamimoff, 1968).

Both sets of experiments fitted reasonably well with the predictions of the residual calcium hypothesis, showing that at least a major part of the augmented release caused by the second nerve stimulus is due to calcium ions.

Recently the residual calcium hypothesis was extended to delayed release. This form of quantal transmitter liberation is an intermediate between evoked and spontaneous activity. If the motor nerve is stimulated at low rate, one observes that after the evoked release, manifested as the epp, there is an increased frequency of miniature epps, compared to the baseline activity before the stimulus (Dodge, Miledi, and Rahamimoff, 1969; Miledi and Thies, 1971; Rahamimoff and Yaari, 1973). The delayed evoked release is the difference between the number of quanta actually liberated after an epp and the expected number of quanta, if the basic frequency continued.

It was found that the properties of this delayed release are similar to those of facilitation. Thus, both increase with lowering the temperature, both are statistically independent of the amplitude of the preceding epp, and the relative value of both effects decreases with an increase in calcium concentration and quantal content (Rahamimoff and Yaari, 1973). These similarities suggest that both these phenomena may have the same origin.

In passing, it should be noted that under certain experimental conditions, the time course of the delayed release does not show a monotonic delay but has a function with a minimum, similar to evoked release (see Rahamimoff and Yaari, 1973). It seems therefore that at least part of the frequency modulation of transmitter release is due to calcium ions.

Summary

The main variables that control transmitter release, under various physiological conditions, are the polarization of the nerve terminal, calcium ions, and frequency of activation. They exert their effect by changing the number of quanta liberated. Kinetic evidence suggests that calcium ions act on specific sites at the nerve terminal. A role of depolarization is probably to allow the action of calcium ions, presumably by increasing the membrane conductance to calcium.

A working hypothesis suggests that calcium ions may also play a role in neuromuscular facilitation and in the fluctuating behavior of the end-plate potential amplitude.

ACKNOWLEDGMENTS Some of the experimental work presented here has been supported by the Research Fund of the Hebrew University-Hadassah Medical School and by the Israeli Center for Psychobiology.

REFERENCES

BAKER, P. F., A. L. HODGKIN, and E. B. RIDGEWAY, 1971. Depolarization and calcium entry in squid giant axons. *J. Physiol.* (*Lond.*) 218:709–755.

BIRKS, R. I., and M. W. COHEN, 1965. Effects of sodium on transmitter release from frog motor nerve terminals. In *Muscle.* W. M. Paul, E. E. Daniel, C. M. Kay, and G. Monckton, eds. Oxford: Pergamon Press, pp. 403–420.

BLACKMAN, J. G., B. L. GINSBORG, and C. RAY, 1963. On the quantal release of the transmitter at the sympathetic synapse. *J. Physiol.* (*Lond.*) 167:402–415.

BOYD, J. A., and A. R. MARTIN, 1956. The end-plate potential in mammalian muscle. *J. Physiol.* (*Lond.*) 132:74–91.

BRACHO, H., and R. K. ORKAND, 1970. Effect of calcium on excitatory neuromuscular transmission in the crayfish. *J. Physiol.* (*Lond.*) 206:61–71.

CHRISTENSEN, B. N., and A. R. MARTIN, 1970. Estimates of probability of transmitter release at the mammalian neuromuscular junction. *J. Physiol.* (*Lond.*) 210:933–945.

COLOMO, F., and R. RAHAMIMOFF, 1968. Interaction between sodium and calcium ions in the process of transmitter release at the neuromuscular junction. *J. Physiol.* (*Lond.*) 198:203–218.

DEL CASTILLO, J., and B. KATZ, 1954a. The effect of magnesium on the activity of motor nerve endings. *J. Physiol.* (*Lond.*) 124:553–559.

DEL CASTILLO, J., and B. KATZ, 1954b. Quantal components of the end-plate potential. *J. Physiol.* (*Lond.*) 124:560–573.

DENNIS, M. J., A. J. HARRIS, and S. W. KUFFLER, 1971. Synaptic transmission and its duplication by focally applied acetylcholine in parasympathetic neurons in the heart of the frog. *Proc. R. Soc.* (*Lond.*) B177:509–539.

DODGE, F. A., and R. RAHAMIMOFF, 1967. Cooperative action of calcium ions in transmitter release at the neuromuscular junction. *J. Physiol.* (*Lond.*) 193:419–432.

DODGE, F. A., R. MILEDI, and R. RAHAMIMOFF, 1969. Strontium and quantal release of transmitter at the neuromuscular junction. *J. Physiol.* (*Lond.*) 200:267–283.

DUDEL, J., and S. W. KUFFLER, 1961. The quantal nature of transmission and spontaneous miniature potentials at the crayfish neuromuscular junction. *J. Physiol.* (*Lond.*) 155:514–529.

FATT, P., and B. KATZ, 1951. An analysis of the end-plate potential recorded with an intracellular electrode. *J. Physiol.* (*Lond.*) 115:320–370.

FATT, P., and B. KATZ, 1952. Spontaneous subthreshold activity at motor nerve endings. *J. Physiol. (Lond.)* 117:109–128.

HUBBARD, J. I., S. F. JONES, and E. M. LANDAU, 1968. On the mechanism by which calcium and magnesium affect the release of transmitter by nerve impulses. *J. Physiol. (Lond.)* 196:75–86.

JENKINSON, D. H., 1957. The nature of the antagonism between calcium and magnesium ions at the neuromuscular junction. *J. Physiol. (Lond.)* 138:434–444.

JOHNSON, E. W., and A. WERNIG, 1971. The binomial nature of transmitter release at the crayfish neuromuscular junction. *J. Physiol. (Lond.)* 218:757–767.

KATZ, B., 1969. The release of neural transmitter substances. The Sherrington Lecture, No. 10. Liverpool: Liverpool University Press.

KATZ, B., and R. MILEDI, 1965. The effect of calcium on acetylcholine release from motor nerve terminals. *Proc. Roy. Soc. (Lond.)* B161:495–503.

KATZ, B., and R. MILEDI, 1967a. The timing of calcium action during neuromuscular transmission. *J. Physiol. (Lond.)* 189:535–544.

KATZ, B., and R. MILEDI, 1967b. Tetrodotoxin and neuromuscular transmission. *Proc. R. Soc. (Lond.)* B167:8–22.

KATZ, B., and R. MILEDI, 1967c. The release of acetylcholine from nerve endings by graded electric pulses. *Proc. Roy. Soc. (Lond.)* B167:23–38.

KATZ, B., and R. MILEDI, 1967d. A study of synaptic transmission in the absence of nerve impulses. *J. Physiol. (Lond.)* 192:407–436.

KATZ, B., and R. MILEDI, 1968a. The role of calcium in neuromuscular facilitation. *J. Physiol. (Lond.)* 195:481–492.

KATZ, B., and R. MILEDI, 1968b. The effect of local blockage of motor nerve terminals. *J. Physiol. (Lond.)* 199:729–741.

KATZ, B., and R. MILEDI, 1969a. The effect of divalent cations on transmission in the squid giant synapse. *Publ. Staz. Napoli* 37:303–310.

KATZ, B., and R. MILEDI, 1969b. Tetrodotoxin resistant electric activity in presynaptic terminals. *J. Physiol. (Lond.)* 203:459–487.

KATZ, B., and R. MILEDI, 1970. Further study of the role of calcium in synaptic transmission. *J. Physiol. (Lond.)* 207:789–801.

KELLY, J. S., 1965. Antagonism between Na^+ and Ca^{++} at the neuromuscular junction. *Nature (Lond.)* 205:296–297.

KUNO, M., 1964. Quantal components of excitatory synaptic potentials in spinal motoneurons. *J. Physiol. (Lond.)* 175:81–99.

MARTIN, A. R., 1966. Quantal nature of synaptic transmission. *Physiol. Rev.* 46:51–66.

MARTIN, A. R., and G. PILAR, 1964a. Quantal components of the synaptic potential in the ciliary ganglion of the chick. *J. Physiol. (Lond.)* 175:1–16.

MARTIN, A. R., and G. PILAR, 1964b. Presynaptic and postsynaptic events during posttetanic potentiation and facilitation in the avian ciliary ganglion. *J. Physiol.* 175:17–30.

MEIRI, U., and R. RAHAMIMOFF, 1971. Activation of transmitter release by strontium and calcium ions at the neuromuscular junction. *J. Physiol. (Lond.)* 215:709–726.

MEIRI, U., and R. RAHAMIMOFF, 1972a. Fluctuations in end-plate potential amplitude and calcium ions. *Isr. J. Med. Sci.* 8:4.

MEIRI, U., and R. RAHAMIMOFF, 1972b. Neuromuscular transmission: Inhibition by manganese ions. *Science* 176:308–309.

MILEDI, R., 1966. Strontium as a substitute for calcium in the process of transmitter release at the neuromuscular junction. *Nature (Lond.)* 212:1233–1234.

MILEDI, R., 1967. Spontaneous synaptic potentials and quantal release of transmitter in the stellate ganglion of the squid. *J. Physiol. (Lond.)* 192:379–406.

MILEDI, R., and C. R. SLATER, 1966. The action of calcium on neuronal synapses in the squid. *J. Physiol. (Lond.)* 184:473–498.

MILEDI, R., and R. E. THIES, 1971. Tetanic and posttetanic rise in frequency of miniature end-plate potentials in low calcium solution. *J. Physiol. (Lond.)* 212:245–257.

RAHAMIMOFF, R., 1968. A dual effect of calcium ions on neuromuscular facilitation. *J. Physiol. (Lond.)* 195:471–480.

RAHAMIMOFF, R., and F. COLOMO, 1967. The inhibitory action of sodium ions on transmitter release at the motor end plate. *Nature (Lond.)* 215:1174–1176.

RAHAMIMOFF, R., and Y. YAARI, 1973. Delayed release of transmitter at the frog neuromuscular junction. *J. Physiol. (Lond.)* 228:241–257.

RUBIN, R. P., 1970. The role of calcium in the release of neurotransmitter substances and hormones. *Pharmacol. Rev.* 22: 389–428.

SARNA, J., I. PARNAS, and R. RAHAMIMOFF, 1972. Evidence for tonic inhibition at the crab neuromuscular junction. *Proc. Isr. Physiol. Pharmacol. Soc.*, 28th Meeting, May 1972.

WERNIG, A., 1972a. Changes in statistical parameters during facilitation at the crayfish neuromuscular junction. *J. Physiol. (Lond.)* 226:751–759.

WERNIG, A., 1972b. The effects of calcium and magnesium on statistical release parameters at the crayfish neuromuscular junction. *J. Physiol. (Lond.)* 226:761–768.

84 Long-Term Maintenance and Plasticity of the Neuromuscular Junction

N. ROBBINS

ABSTRACT Studies of the effects of use and disuse at the neuromuscular junction suggest that there are plastic features of presynaptic and postsynaptic function. In these studies, however, serious problems of experimental interpretation often arise. Denervation experiments indicate that the motoneuron normally regulates physiologically important postsynaptic membrane properties. The same biologic mechanisms that may preserve normal synaptic function may also permit functional plasticity.

IN THE ADULT ANIMAL, synaptic connections can change either by varying the efficacy of existing synapses or through the actual gain or loss of synapses. In either case, it is equally as important to know why synapses do not change as to know why they do. Certain clues to understanding long-term maintenance or plasticity have been found at relatively simple synapses such as the neuromuscular junction or autonomic ganglia. The purpose of this chapter is to call attention to these clues, to point out possible interpretive technical pitfalls, and to suggest avenues for further experimentation.

In using autonomic or neuromuscular synapses, one hopes that principles of synaptic plasticity will emerge which will be applicable to the CNS. For instance, the well-known spread of acetylcholine (ACh) sensitivity to nonsynaptic membrane in denervated muscle (Axelsson and Thesleff, 1959) also applies to nonsynaptic regions of a denervated cholinoceptive neuron (Kuffler et al., 1971). While this "peripheral-first, central-later" strategy has been fruitful in our understanding of instantaneous synaptic physiology, peripheral synapses may not have evolved the same mechanisms of plasticity, qualitatively or quantitatively, that exist in the CNS. For instance, the safety factor in mammalian neuromuscular transmission is so high that even a twofold change in transmitter release might have no effect on synaptic traffic, whereas a change of this magnitude in the CNS might have profound consequences. On the other hand, normal synaptic usage could play an important role in maintaining and regulating even neuromuscular function, e.g., it would be of adaptive value to increase synaptic efficacy to preserve a high safety factor if demands for transmitter release were increased over a long period of time. Since these arguments are presently unresolvable, let us pose certain experimental questions: (1) What is the evidence that synaptic structure or function changes with long-term use or disuse? And (2) are there other neural actions at the synapse, possibly independent of usage, which are worth considering as long-term influences on synaptic function? Except where noted otherwise, the following discussion is confined to the neuromuscular or occasionally to autonomic junctions and for the most part covers synaptic events on a time scale of days or longer. The reader is referred to several excellent sources (Sharpless, 1964; Kandel and Spencer, 1968; Bloom et al., 1970; Horn and Hinde, 1970; and Kuno, 1971) for wider coverage. The present chapter will focus on certain critical problems in the field, using a small number of recent papers for illustration.

Use and disuse: Interpretive problems

Certain methodologic or interpretive problems frequently arise in studies of synaptic use and disuse:

1. *Does the experimental situation cross the border from an extreme form of synaptic use or disuse into the realm of pathology?* For instance, prolonged high-frequency stimulation of a synapse that normally works intermittently and at low frequency may produce changes that are beyond the pale of anything the synapse would ever experience in life and that only reflect sheer exhaustion of metabolites, membrane, or restorative mechanisms. Similar considerations apply to experiments in which prolonged stimulation is carried out in the presence of inhibitors of either energy metabolism or uptake or utilization of transmitter precursors. Nonetheless, this type of experiment is of value if further work can show that the change is indeed an extreme form of a synaptic mechanism that occurs in vivo under more normal circumstances.

For example, a number of experiments show that transmitter release during tetani of cholinergic synapses may cause alterations in synaptic vesicles (e.g., Jones

N. ROBBINS Department of Anatomy, Case-Western Reserve University, Cleveland, Ohio

and Kwanbunbumpen, 1970; Pysh and Wiley, 1972). This result is of interest in the present context, because vesicle depletion or distortion could be the synaptic alteration that somehow triggers long-term changes in synaptic function. However the fact that vesicle depletion is so far only found in mammalian neuromuscular junctions after uninterrupted and prolonged in vitro stimulation and without added choline or in the presence of hemicholinium (Jones and Kwanbunbumpen, 1970) suggests that metabolic exhaustion or rerouting of structural components may account for the findings. Thus, it is not established that these results are applicable to any naturally occurring situation. At the frog neuromuscular junction, vesicle depletion and apparent fusion with the presynaptic membrane was found after only 1 min of stimulation at 10 Hz (Heuser and Reese, 1973). Even this level of stimulation is probably in excess of the natural phasic action of the frog sartorius, but it is less likely that these findings are products of aberrant synaptic function. More important, the same authors used uptake of horseradish peroxidase into synaptic vesicles after brief stimulation to demonstrate that vesicles are "recycled" (Heuser and Reese, 1973). Similar results were reported by Ceccarelli et al. (1972) after prolonged stimulation at 2 Hz.

In sympathetic ganglia, a depletion of vesicles and an increase in axodendritic area of contact was observed after some 3 hr of continuous preganglionic stimulation at 20 to 32 Hz (Pysh and Wiley, 1972). However, if the ganglion is stimulated at 20 Hz for only about 1 hr, there is no change in synaptic vesicle density (Parducz et al., 1971). Furthermore, addition of choline prevented the morphologic changes seen after prolonged stimulation of perfused or hemicholinium treated ganglia. Parducz et al. (1971) suggest that, in stimulated choline-deficient terminals, choline is mobilized from membrane components of vesicles and mitochondria, causing vesicle depletion and mitochondrial abnormalities. Thus, it now seems desirable to find new morphologic techniques to determine whether vesicles, mitochondria, or presynaptic terminal configurations change with physiologic stimulation. Promising approaches might be electron microscopic techniques in which marker molecules (e.g., peroxidase) are transported or incorporated during moderate stimulation. Equally necessary are autoradiographic techniques that visualize processing of transmitter in the presynaptic cytoplasm, because recently synthesized ACh is preferentially released from a special cytoplasmic compartment (Whittaker, 1970).

2. *Does the methodology of the experiment introduce problematic unknowns?* For instance, cutting a motor axon or application of botulinum toxin will certainly halt neuromuscular transmission, but any resulting long-term changes cannot be attributed to synaptic disuse until trophic, nonactivity-related effects of axotomy or botulinum toxin are eliminated. In long-term experiments, the pharmacologic approach to changing synaptic activity is especially treacherous because of the multiple actions of most drugs in addition to the one well-recognized effect for which they are used. Perhaps the most convincing use of drugs is in the type of experiment where something is *unchanged* despite the drug effect.

3. *Is the presumption of experimental use or disuse justified by hard experimental documentation?* The long-term output of a synapse after experimental intervention (other than direct stimulation) may be surprisingly different from that expected on the basis of classical acute physiologic experiment. For instance, tenotomy has often been used as a paradigm of neuromuscular disuse, because atrophy occurs and muscle spindles are relaxed. Yet, electromyographic documentation has revealed that despite atrophy, junctional traffic in certain muscles may be altered only slightly (Nelson, 1969) or may vary in time (Vrbova, 1963). Documentation of synaptic activity over long periods of time in spontaneously moving animals can present formidable technical difficulties, but without it there is great danger of unwarranted assumption. In addition, documentation reveals the patterning of synaptic use or disuse, i.e., the mean frequency, pattern of on- or off-firing, or total number of impulses. If any change of synaptic structure or function observed after experimental alteration of usage were uniquely dependent on one or another of these parameters, the terms use and disuse would be too general. Indeed, the relative importance of each of these parameters in altering synaptic function has not been systematically explored.

4. *Is synaptic efficacy a unitary property that varies up or down?* Although it seems a simple matter to decide whether a synapse is more or less efficacious as a result of a plastic change, there may in fact be a constellation of changes which have occurred, such as a higher "resting" quantum content but a lower quantum content at high frequencies (see below). Similarly, an increase in the nonsynaptic membrane sensitive to transmitter at a disused synapse could represent a compensatory mechanism for increasing the space available for new synapse formation even though it has no known effect on function of the existing synapse (see below).

Physiological studies

It is difficult to predict what should happen to the physiology of a synapse subjected to long-term disuse or excess use. Short-term stimulation schedules alter synaptic efficacy for minutes to hours, but the direction of change varies with the type of synapse and with the pattern or frequency of stimulation (see review by Kandel et al., 1970; see also Bruner and Kennedy, 1970).

There is far less information available on the next jump in time scale to periods of days or weeks of altered synaptic activity. There are as yet no reasonably unambiguous studies on *chronic* excess use. The experimental problem in disuse is to avoid potentially injurious procedures such as spinal cord section that could affect motorneuronal physiology by trophic mechanisms unrelated to changes in synaptic activity.

Recently, chronic disuse at the neuromuscular junction has been achieved through less objectionable experimental devices. In one, the ankle and knee joints of rats were immobilized at right angles by pins inserted through the bones. Chronic recording from the soleus muscle showed a rapid onset of neuromuscular inactivity, at about 10% of normal, lasting up to several weeks in animals where the pins remained well fixed (Fischbach and Robbins, 1971). In another experimental approach, total neuromuscular disuse was obtained by surrounding the sciatic nerve of the rat or rabbit with a cuff impregnated with local anesthetic (Robert and Oester, 1970a, 1970b; Lomo and Rosenthal, 1972). A smaller number of experiments were also done with diphtheria toxin to produce muscle disuse (Lomo and Rosenthal, 1972). The results of these and other disuse experiments are summarized in Table I.

In the immobilization disuse experiment there was no change in two postsynaptic membrane properties known to be altered by denervation, namely, resting membrane potential and membrane resistance. This result provided further evidence that innervation and *subtotal* disuse effects can be disassociated. On the other hand, there was a small and transient but significant increase in cell membrane area highly sensitive to ACh after immobilization disuse (Fischbach and Robbins, 1971). A similar spread of sensitivity was also seen in hibernating animals and after spinal cord isolation (Vyskocil et al., 1971; Johns and Thesleff, 1961; see Table I). After anesthetic-cuff disuse in rats, the entire membrane became as highly sensitive to ACh as after denervation (Lomo and Rosenthal, 1972). Unfortunately, ACh hypersensitivity was *not* found after chronic anesthetic-cuff disuse in rabbits (Robert and Oester, 1970a, 1970b). The conflicting findings are difficult to resolve, because there was documentation of anesthetic effect in both rat and rabbit investigations. Another complication is that local anesthetics applied continuously may have additional actions, e.g., infusion of lidocaine into peripheral nerve interferes with fast axonal transport (Boldt, 1972). In any event, the most important observation is that the rise of extrajunctional ACh sensitivity, whether produced by anesthetic-cuff disuse or by denervation, can be reduced by direct or indirect muscle stimulation (Jones and Vrbova, 1970; Lomo and Rosenthal, 1972; Drachman and Witzke, 1972).

TABLE I

Effect of long-term disuse on neuromuscular synaptic function

Disuse (%)	Disuse Method	Postsynaptic Effects		Presynaptic Effects	Ref
		Extrajunctional ACh Sensitive Length (disused/normal)	Junctional Sensitivity		
90	Limb immobilization	1.7 (maximum)	no change	normal or ↑ low-frequency quantal release	A
				more rapid initial decay of epp to steady state during 10 Hz tetanus; ↑ resynthesis of transmitter during low-frequency tetani	
				normal transmitter release during 5 and 10 Hz tetani; ↑ fall during 20 and 40 Hz tetani	
? 90	Cord isolation	≃ 1.7	?	?	B
? 99	Hibernation	1.8	↑ × 1.8	?	C
100	Anesthetic-cuff	entire fiber surface	no change	normal or ↑ low-frequency quantal release	D

(A) Fischbach and Robbins, 1971; Robbins and Fischbach, 1971. (B) Johns and Thesleff, 1961 (C) Vyskocil et al., 1971. (D) Lomo and Rosenthal, 1972.

These results could be construed as evidence that synaptic activity regulates the extent of postsynaptic membrane sensitive to transmitter. However, there are several important qualifications.

First, the regulation is not simply proportional to synaptic activity, because a 90% or greater reduction in activity produces only a small change whereas 100% disuse (denervation or anesthetic-cuff) produces a very dramatic effect (Table I). On the other hand, even the 90% reduction may exceed the bounds of any naturally encountered disuse situation.

Second, in three of the studies cited in Table I, the sensitivity of *junctional* receptors was measured both by microiontophoretic release of ACh combined with intracellular recording (del Castillo and Katz, 1955) and by comparison of miniature end-plate amplitude with input resistance or fiber diameter (Katz and Thesleff, 1957). In two of these studies, there was no change in junctional sensitivity (Table I). Therefore, disuse did *not* increase synaptic efficacy by a postsynaptic mechanism. This conclusion would *not* apply to junctions where transmitter normally diffuses to perisynaptic membrane, in which case extrajunctional effects could be very significant. At the neuromuscular junction, however, the spread of extrajunctional sensitivity is at the moment an interesting effect in search of a physiologic significance. We are currently testing the hypothesis that the spread of sensitivity and the acceptance of extrajunctional synapses are invariably associated phenomena. In normal development of the neuromuscular junction, early nerve sprouts arriving at the muscle are already functional (Robbins and Yonezawa, 1971). Similarly, the spread of ACh sensitivity in denervated muscle may prepare the muscle to respond immediately to regenerating or sprouted nerve endings wherever they happen to arrive on the muscle fiber. Tissue and organ cultures of muscle present an interesting situation in which to study the control of ACh sensitivity. For instance, it was found that a denervated diaphragm in vitro did not develop extrajunctional ACh sensitivity in the presence of actinomycin D (Fambrough, 1970) and that alpha-bungarotoxin binding sites (presumably receptor sites) on developing muscle are reduced and restricted by chronic in vitro stimulation (Fischbach and Cohen, personal communication).

Finally, older studies indicate that to some extent, "neurotrophic" factors not related to disuse also influence extrajunctional sensitivity (see review by Guth, 1968, and discussion below). At the moment, then, it seems likely that *both* muscle activity and undefined neurotrophic factors regulate ACh sensitivity, but the relative importance, interaction, and biochemical mechanism of action of these two factors remain to be elucidated.

On the presynaptic side, neither immobilization nor anesthetic-cuff disuse decreased "resting" quantum content of transmitter release (Table I). In fact, soleus neuromuscular junctions showed a small but significant increase when measured at low frequencies (0.03 to 3 Hz). However, "resting" quantum content is a rather artificial parameter in a junction that normally fires at about 10 Hz in vivo. In fact, tests of quantum content during tetani at different frequencies in vitro showed that quantum content of immobilized-disused junctions was about normal at 5 and 10 Hz but fell below control values by 25% at frequencies of 20 and 40 Hz (Robbins and Fischbach, 1971). Since the reduced output was apparent within the first 111 impulses at 20 Hz, which is the upper limit of normal firing frequency (Fischbach and Robbins, 1969), it may represent a physiologically important defect. Immobilized-disused junctions also showed an increased initial rate of decline of end-plate potential, i.e., a faster arrival at the quasi-steady state amplitude during 10 per sec stimulation. If disuse had similar effects in the mammalian CNS, temporally summating synapses would be profoundly affected. Finally, the normal steady-state quantal release found at immobilized-disused junctions stimulated at 5 or 10 per sec was the net result of both greater fractional release and faster resynthesis of transmitter (Robbins and Fischbach, 1971). No comparable studies of transmitter release during short trains and long tetani after anesthetic-cuff disuse are available.

The nature of these presynaptic changes must remain obscure until physiologic terms such as *readily available stores*, *storage ACh*, or *transmitter mobilization* receive better biochemical and morphologic underpinning. Possibly, experiments done in vitro, at room temperature, and in the absence of choline may not give a true picture of events in vivo. Nonetheless, the fact remains that disuse can produce a plastic change in presynaptic function even at neuromuscular synapses.

All of the changes in transmitter release noted after 3 or 6 weeks immobilization disuse were already present to about the same extent after only 3 to 5 days of disuse. Although the earliest onset of synaptic alteration has not yet been determined, the available evidence (Robbins and Fischbach, 1971) suggests that normal release of transmitter at tonically firing synapses is maintained by ongoing activity: When activity decreases, release of transmitter is altered (in 3 days or less) and then stays indefinitely at the new level. Thus, activity in this metastable system appears capable of maintaining transmitter release in the normal mode of operation.

This proposition is more or less consistent with biochemical studies showing changes in transmitter-related enzyme activity in stimulated sympathetic neurons (see

Iversen, Axelrod, this volume). At the immobilized-disused neuromuscular junction, cholinesterase activity does not decrease (Guth, 1969a), but choline acetyltransferase activity has not been measured.

Synaptic modulation and trophic interactions

Denervated muscle shows a variety of altered membrane properties that have been attributed to the loss of hypothetical neurotrophic factors whose release is partly or entirely *independent* of synaptic activation (Guth, 1968, 1969b). By inference, these factors continuously regulate postsynaptic membrane properties and thus play an important role in the maintenance of *normal* synaptic function: Perhaps denervation is an extreme case of the continuously variable trophic interaction between pre- and postsynaptic cell. Indeed, many of the resultant membrane changes of denervation (see Table II) would have profound physiological effects if the synapse were still functional.

TABLE II

Some mammalian muscle membrane properties affected by denervation

Property	Change	Sample Reference
Resting membrane potential	↓	Albuquerque et al., 1971
Total membrane resistance	↑	Albuquerque and McIsaac, 1970
Ionic permeability or exchange	↓ K^+	Klaus et al., 1960
	↑ Na^+	Creese et al., 1968
Potassium pump	↓	Dockry et al., 1966
Rate of rise and fall of action potential	↓	Redfern and Thesleff, 1971a
Tetrodotoxin sensitivity	↓	Redfern and Thesleff, 1971b

For instance, the end-plate potential amplitude would be affected by the changes in resting membrane potential, ionic permeabilities, and extrajunctional membrane resistance which occur after denervation. Synaptic activation dependent on an Na-K pump mechanism, if such exist (see Weight, this volume), might be affected by an alteration in K-pump activity. And finally, action potential propagation would be altered by a decrease in maximum rate of rise and fall of the action potential, possibly related to the change in tetrodotoxin sensitivity.

The point of this hypothetical survey of selected denervation effects is to emphasize that factors other than, and perhaps independent of, synaptic activation could in principle play some role in maintenance or plasticity of synaptic function.

There is ample precedent for a nonactivity-dependent trophic function of neurons on innervated tissues in the case of amphibian limb regeneration (Singer, 1960) and of sensory end-organs (Zelena, 1964). However, at the neuromuscular junction, the distinction between activity-dependent and nonactivity-dependent neural control mechanisms may not hold in all instances. A classical method for demonstrating a trophic effect consisted of cutting the motor nerve either near to or far from the nerve entry zone and then demonstrating a time delay in the onset of the denervation effect depending on the nerve length. It now seems that this method can be questioned. First, in view of the enormous effect of a relatively small number of nerve impulses in the case of extrajunctional ACh sensitivity (see discussion above), it must be determined that the muscle connected to the long nerve stump shows absolute electrical silence. Second, even if a denervation effect shows a dependence on nerve stump length, it can also be influenced by muscular activity, as discussed above for ACh sensitivity.

Future progress in this area awaits either demonstration of the use-dependence of membrane properties altered by denervation or direct identification of neurotrophic factors. To date, the only reasonably clearcut case of isolation of a crude neurotrophic factor is in the nonsynaptic role of neurons in promoting limb regeneration in amphibia (Lebowitz and Singer, 1970). If similar factors can be identified at the neuromuscular junction, it will then be possible to determine if they normally vary with use and disuse in the lifetime of a synapse. Some encouraging but preliminary results have been reported for a neurotrophic factor regulating muscle cholinesterase (Lentz, 1971). However, if trophic interactions between cells involve a whole host of interlocking mutual feedback mechanisms, the likelihood of isolating a single factor that regulates, say, membrane potential, is small.

The effects of postsynaptic muscle on presynaptic terminal structure and function have received relatively little attention even though examples of this type of interaction are known in development and regeneration of the nerve-muscle junction (Prestige, 1970). Nonetheless, there are some fascinating clues as to how information at nerve terminals could produce neuronal changes. Holtzman et al. (1971) demonstrated that stimulation of lobster neuromuscular junctions led to incorporation of extracellular peroxidase into presynaptic vesicles. In vertebrates, some of this vesicular peroxidase is released at the terminals as the synaptic vesicles recycle (Heuser and Reese, 1973), but part is transported to the perikaryon (Kristensson et al., 1971; see also Watson, 1968). Thus, increased uptake of extracellular or intrasynaptic molecules during stimulation could regulate perikaryal synthesis of molecules important in presynaptic structure and function.

It is also possible that uptake and reactive mechanisms in the axonal terminal, which are independent of immediate control by the perikaryon, have long-term effects on presynaptic function. An advantage of this local autonomy is that a stimulated postsynaptic cell, by releasing factors into the synaptic space, could selectively enhance those particular terminals of the presynaptic cell that it contacts. Similar remarks apply in both directions to axo-axonal interactions. On the other hand, a mechanism of synaptic modulation that depended on retrograde transport to the perikaryon would require either that the feedback mechanism apply to all terminals of a given neuron or that there be private channels between axon terminals and perikaryon. None of the above possibilities has yet been tested experimentally.

Summary

Studies of use and disuse at neuromuscular or autonomic synapses may reveal mechanisms and phenomena applicable to the CNS. For unambiguous results, it is important (1) that the experimental effects of usage occur within a physiological range and set of conditions, (2) that the methodology does not introduce problematic unknowns, (3) that actual synaptic traffic is documented, and (4) that a variety of presynaptic and postsynaptic properties are investigated.

Chronic and severe disuse of neuromuscular junctions produces a small but significant increase in acetylcholine sensitivity of extrajunctional regions. Total disuse produces a spread of sensitivity comparable to that of denervation but is reversible if the muscle is stimulated. Thus, there is a highly nonlinear relation between usage and acetylcholine sensitivity. Also, the response of the *junctional* membrane is usually *not* enhanced.

Presynaptic terminals of disused junctions show little or no change in transmitter release at very low frequencies of stimulation, but at higher physiologic frequencies are significantly altered. Similar changes at CNS synapses would be functionally significant.

It is inferred from denervation studies that the motor neuron may normally regulate certain postsynaptic membrane properties that are of great importance in synaptic and nonsynaptic function. Further understanding of this neuronal control awaits additional studies on usage dependence of postsynaptic properties or isolation of neurotrophic factors. An understanding of postsynaptic regulation of presynaptic function is also essential.

ACKNOWLEDGMENT Supported by Research Career Development Award NS 39838–02 and Research Grant NS 09420–02 from the National Institute of Neurological Diseases and Stroke. The author wishes to thank Drs. Floyd Bloom, Eric Kandel, and Seth Sharpless for their valuable suggestions.

REFERENCES

ALBUQUERQUE, E. X., and R. J. McIsAAC, 1970. Fast and slow mammalian muscles after denervation. *Exp. Neurol.* 26: 183–202.

ALBUQUERQUE, E. X., F. T. SCHUH, and F. C. KAUFFMAN, 1971. Early membrane depolarization of the fast mammalian muscle after denervation. *Pflüger Arch.* 328:36–50.

AXELSSON, J., and S. THESLEFF, 1959. A study of supersensitivity in denervated mammalian skeletal muscle. *J. Physiol. (Lond.)* 147:178–193.

BLOOM, F. E., L. L. IVERSEN, and F. O. SCHMITT, eds., 1970. Macromolecules in synaptic function. *Neurosciences Res. Prog. Bull.* 8:323–455.

BOLDT, K. A., 1972. Rapid axoplasmic transport of labeled protein in cat motoneurons. Ph.D. Thesis. Chicago: University of Illinois.

BRUNER, J., and D. KENNEDY, 1970. Habituation: Occurrence at a neuromuscular junction. *Science* 169:92–94.

CECCARELLI, B., W. P. HURLBUT, and A. MAURO, 1972. Depletion of vesicles from frog neuromuscular junctions by prolonged tetanic stimulation. *J. Cell. Biol.* 59:30–38.

CREESE, R., A. L. EL-SHAFIE, and G. VRBOVA, 1968. Sodium movements in denervated muscle and the effects of antimycin A. *J. Physiol. (Lond.)* 197:279–294.

DEL CASTILLO, J., and B. KATZ, 1955. On the localization of acetylcholine receptors. *J. Physiol. (Lond.)* 128:157–181.

DOCKRY, M., R. P. KERNAN, and A. TANGNEY, 1966. Active transport of sodium and potassium in mammalian skeletal muscle and its modification by nerve and by cholinergic and adrenergic agents. *J. Physiol. (Lond.)* 186:187–200.

DRACHMAN, D. B., and F. WITZKE, 1972. Trophic regulation of acetylcholine sensitivity of muscle: Effect of electrical stimulation. *Science* 176:514–516.

FAMBROUGH, D. M., 1970. Acetylcholine sensitivity of muscle fiber membrane. Mechanism of regulation by motoneurons. *Science* 168:372–373.

FISCHBACH, G. D., and N. ROBBINS, 1969. Changes in contractile properties of disused soleus muscles. *J. Physiol. (Lond.)* 201: 305–320.

FISCHBACH, G. D., and N. ROBBINS, 1971. Effect of chronic disuse of rat soleus neuromuscular junctions on postsynaptic membrane. *J. Neurophysiol.* 34:562–569.

GUTH, L., 1968. "Trophic" influences of nerve on muscle. *Physiol. Rev.* 48:645–687.

GUTH, L., 1969a. The effect of immobilization on sole-plate and background cholinesterase of rat skeletal muscle. *Exp. Neurol.* 24:508–513.

GUTH, L., ed., 1969b. "Trophic" effects of vertebrate neurons. *Neurosciences Res. Prog. Bull.* 7:1–73.

HEUSER, J. E., and T. S. REESE, 1973. Evidence for recycling of synaptic vesicle membrane during transmitter release at the frog neuromuscular junction. *J. Cell Biol.* 57:315–344.

HOLTZMAN, E., A. R. FREEMAN, and L. A. KASHNER, 1971.

Stimulation-dependent alterations in peroxidase uptake at lobster neuromuscular junctions. *Science* 173:733–736.

HORN, G., and R. A. HINDE, eds., 1970. *Short-Term Changes in Neural Activity and Behavior.* New York: Cambridge University Press.

JOHNS, T. R., and S. THESLEFF, 1961. Effects of motor inactivation on the chemical sensitivity of skeletal muscle. *Acta Physiol. Scand.* 51:136–141.

JONES, R., and G. VRBOVA, 1970. Effect of muscle activity on denervation hypersensitivity. *J. Physiol. (Lond.)* 210:144–145.

JONES, S. F., and S. KWANBUNBUMPEN, 1970. The effects of nerve stimulation and hemicholinium on synaptic vesicles at the mammalian neuromuscular junction. *J. Physiol. (Lond.)* 207:31–50.

KANDEL, E. R., V. CASTELLUCCI, H. PINSKER, and I. KUPFERMAN, 1970. The role of synaptic plasticity in short-term modification of behavior. In *Short-Term Changes in Neural Activity and Behavior,* G. Horn and R. A. Hinde, eds. New York: Cambridge University Press.

KANDEL, E. R., and W. A. SPENCER, 1968. Cellular neurophysiological approaches in the study of learning. *Physiol. Rev.* 48:65–134.

KATZ, B., and S. THESLEFF, 1957. On the factors which determine the amplitude of the "miniature end-plate potential." *J. Physiol. (Lond.)* 137:267–278.

KLAUS, W., M. LULLMAN, and E. MUSCHOL, 1960. Der Kalciumflux des normalen und denervierten Rattenzwerchfells. *Pflügers Arch.* 271:761–775.

KRISTENSSON, K., Y. OLSSON, and J. SJOSTRAND, 1971. Axonal uptake and retrograde transport of exogenous proteins in the hypoglossal nerve. *Brain Res.* 32:399–406.

KUFFLER, S. W., M. J. DENNIS, and A. J. HARRIS, 1971. The development of chemosensitivity in extrasynaptic areas of the neuronal surface after denervation of parasympathetic ganglion cells in the heart of the frog. *Proc. Roy. Soc. (Lond.)* B177:555–563.

KUNO, M., 1971. Quantum aspects of central and ganglionic synaptic transmission in vertebrates. *Physiol. Rev.* 51:647–678.

LEBOWITZ, P., and M. SINGER, 1970. Neurotrophic control of protein synthesis in the regenerating limb of the newt, Triturus. *Nature (Lond.)* 225:824–827.

LENTZ, T. L., 1971. Nerve trophic function: In vitro assay of effects of nerve tissue on muscle cholinesterase activity. *Science* 171:187–189.

LOMO, T., and J. ROSENTHAL, 1972. Control of ACh sensitivity by muscle activity in the rat. *J. Physiol. (Lond.)* 221:493–513.

NELSON, P. G., 1969. Functional consequences of tenotomy in hind limb muscles of the cat. *J. Physiol. (Lond.)* 201:321–333.

PARDUCZ, A., O. FEHER, and F. JOO, 1971. Effects of stimulation and hemicholinium (HC-3) on the fine structure of nerve endings in the superior cervical ganglion of the cat. *Brain Res.* 34:61–72.

PRESTIGE, M. C., 1970. Differentiation, degeneration, and the role of the periphery: quantitative considerations. In *The Neurosciences: Second Study Program,* F. O. Schmitt, ed. New York: The Rockefeller University Press, pp. 73–82.

PYSH, J. J., and R. G. WILEY, 1972. Morphologic alterations of synapses in electrically stimulated superior cervical ganglia of the cat. *Science* 176:191–193.

REDFERN, P., and S. THESLEFF, 1971a. Action potential generation in denervated rat skeletal muscle. I. Quantitative aspects. *Acta Physiol. Scand.* 81:557–564.

REDFERN, P., and S. THESLEFF, 1971b. Action potential generation in denervated rat skeletal muscle. II. The action of tetrodotoxin. *Acta Physiol. Scand.* 82:70–78.

ROBBINS, N., and G. D. FISCHBACH, 1971. Effect of chronic disuse of rat soleus neuromuscular junctions on presynaptic function. *J. Neurophysiol.* 34:570–578.

ROBBINS, N., and T. YONEZAWA, 1971. Developing neuromuscular junctions: First signs of chemical transmission during formation in tissue culture. *Science* 172:395–398.

ROBERT, E. D., and Y. T. OESTER, 1970a. Nerve impulses and trophic effect. *Arch. Neurol.* 22:57–63.

ROBERT, E. D., and Y. T. OESTER, 1970b. Absence of supersensitivity to acetylcholine in innervated muscle subjected to a prolonged pharmacologic nerve block. *J. Pharmacol. Exp. Ther.* 174:133–140.

SHARPLESS, S. K., 1964. Reorganization of function in the nervous system—use and disuse. *Ann. Rev. Physiol.* 26:357–388.

SINGER, M., 1960. Nervous mechanisms in the regeneration of body parts in vertebrates. In *Developing Cell Systems and Their Control,* D. Rudnick, ed. New York: Ronald Press, pp. 115–133.

VRBOVA, G., 1963. Changes in motor reflexes produced by tenotomy. *J. Physiol. (Lond.)* 166:241–267.

VYSKOCIL, F., M. JAN, and L. JANSKY, 1971. Resting state of the myoneural junction in a hibernator. *Brain Res.* 34:381–384.

WATSON, W. E., 1968. Centripetal passage of labelled molecules along mammalian motor axons. *J. Physiol. (Lond.)* 196:122–123P.

WHITTAKER, V. P., 1970. The investigation of synaptic function by means of subcellular fractionation techniques. In *The Neurosciences: Second Study Program,* F. O. Schmitt, ed. New York: The Rockefeller University Press, pp. 761–767.

ZELENA, J., 1964. Development, degeneration, and regeneration of receptor organs. In *Mechanisms of Neural Regeneration* (Progress in Brain Research, Vol. 13), M. Singer and J. P. Schadé, eds. Amsterdam: Elsevier, pp. 175–213.

85 Growth and Plasticity
of Adrenergic Neurons

ROBERT Y. MOORE,
ANDERS BJÖRKLUND, and
ULF STENEVI

ABSTRACT   Adrenergic neurons are widely distributed in both
the peripheral and central nervous systems. The ontogenetic
development of peripheral adrenergic neurons appears to
parallel the maturation of organs to be innervated. These
neurons show vigorous responses to injury, both in the form of
regenerative sprouting from transected axons and collateral
sprouting from intact axons in response to denervation. Central
adrenergic neurons develop largely in the early postnatal period.
Like their peripheral counterparts, central adrenergic neurons
are capable of exhibiting a vigorous regenerative growth in
response to transection. In addition, they undergo extensive
collateral sprouting in response to an appropriate local denervation. The adrenergic neuron provides an excellent model
for the study of neuronal growth and plasticity.

ADRENERGIC neurons occur in the periphery as postganglionic sympathetic neurons (Norberg, 1967; Burnstock, 1969; Ehinger et al., 1973), and are found in
each major subdivision of the central nervous system
(Dahlström and Fuxe, 1964; Fuxe, 1965; Björklund and
Moore, 1973). This chapter considers growth of adrenergic neurons in three situations: ontogenesis, regeneration,
and as an expression of modifiability or plasticity of
adrenergic neural connections. In all likelihood the
mechanisms regulating the growth of adrenergic neurons
do not differ markedly from those controlling the growth
of other neurons and, to this extent, the adrenergic
neuron may be a model for the biologic regulation of
neuron growth.

One of the greatest stimuli to study of the adrenergic
neuron, and particularly aspects of its growth, has been
the development of a specific and sensitive histochemical
method for biogenic amines that allows their cellular
localization to be determined precisely. This method, the
Falck-Hillarp method (Falck, 1962; Falck et al., 1962;

FIGURE 1   The sympathetic adrenergic neuron demonstrated
by the Falck-Hillarp histochemical method. (A) The fluorescent
cell bodies of superior cervical ganglion neurons in the rat. (B)
The terminal innervation plexus of the iris which arises from
axons of the superior cervical ganglion neurons.

ROBERT Y. MOORE, ANDERS BJÖRKLUND, and ULF STENEVI   De-
partments of Pediatrics, Medicine, and Anatomy and the Joseph
P. Kennedy, Jr., Mental Retardation Research Center, The
University of Chicago, Chicago, Illinois; and Department of
Histology, University of Lund, Lund, Sweden

for reviews see Corrodi and Jonsson, 1967, and Björklund
et al., 1972), has been applied widely and observations
utilizing it form the basis for much of the data reviewed

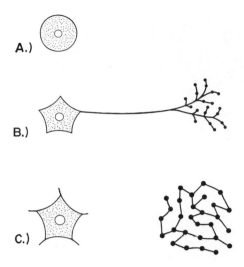

FIGURE 2 Development of the peripheral adrenergic neuron as shown by the Falck-Hillarp method diagrammatically represented. At the earliest stage (A) only the cell body can be demonstrated. As processes develop the entire neuron is shown (B) including the cell body, preterminal axon, and the developing terminal plexus. When the terminal plexus is mature (C), only that and the cell body are demonstrable.

here. In the normal, adult nervous system the Falck-Hillarp method permits visualization of the cell bodies and axon terminals of adrenergic neurons (Figures 1,2C); the preterminal axon and distal dendritic branches usually are not shown. This does not hold during development (Figure 2B) or in certain pathological states.

Adrenergic neuron growth has been analyzed in greatest detail during *ontogenetic* growth and during *regenerative* growth. There are fundamental differences between the mechanisms regulating neuronal growth in these two situations. Growth of neurons during development occurs as part of an orderly sequence and pattern within a general developmental scheme, whereas regenerative growth occurs in response to injury. Thus, regeneration can be considered part of an organism's attempt to maintain a homeostatic equilibrium in its interaction with its environment. The growth that appears in regeneration in a mature, fully established nervous system is not a component of any orderly sequence of development but could be a morphologic component of neural modifiability or plasticity. The continuing structural modification of neural connections that may characterize the adaptation of mammalian organisms to their environment, has proved extremely difficult to analyze. If it is possible to view these various aspects of growth as a continuum of biologic processes, then analysis of growth during ontogenesis and during regeneration would represent a valid approach to understanding all aspects of neuronal growth.

The peripheral adrenergic neuron

ONTOGENESIS Little is known of the factors regulating the growth of axons and establishment of innervation patterns of the peripheral adrenergic nervous system during ontogenesis. Catecholamine storage develops as early as the thirteenth day of gestation in the rat (de Champlain et al., 1970) and in the 3.5 day chick embryo (Enemar et al., 1965), even before the sympathoblasts are fully differentiated.

As shown by the Falck-Hillarp method, the immature peripheral adrenergic neuron has a morphology different from that of the mature adrenergic neuron; the preterminal axons are strongly fluorescent during development indicating a high noradrenaline concentration. The newly forming terminals of the autonomic ground plexus are smaller and less prominent than in the adult, but as the adult terminal pattern develops the preterminal axons no longer are evident in Falck-Hillarp preparations (de Champlain et al., 1970, Figure 2). De Champlain et al. (1970) have suggested that the development of sympathetic innervation follows the general process of maturation in the different organs receiving innervation. In the rat, the adrenergic innervation pattern reaches an adult appearance by the third postnatal week in all innervated tissues (de Champlain et al., 1970, Owman et al., 1971).

REGENERATION OF PERIPHERAL ADRENERGIC NEURONS Two basic experimental paradigms have been employed to study regeneration in the peripheral and central nervous systems. These are illustrated diagrammatically

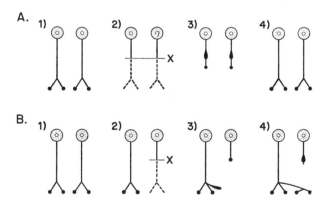

FIGURE 3 Regeneration in the nervous system—two basic paradigms. (A) Sprouting from transected axons. The axons of neurons innervating a structure are severed. The axon distal to the lesion degenerates. The proximal stumps form growth cones and regenerate new axons and terminals. (B) Collateral sprouting. Part of the innervation to a structure is transected. The distal axons and terminals degenerate. Collateral sprouts form from the remaining innervation and reconstitute the terminal plexus.

in Figure 3. In one type of experiment an axon or group of axons is severed. The distal portion of the axon then degenerates, and the tissue it normally innervates is, thereby, denervated. (The general process of degeneration has been examined extensively by Cajal, 1928, and many subsequent workers and will not be considered further here.) Following the lesion the proximal stumps of the severed axons exhibit growth cones and axonal sprouting which, in the most favorable circumstances, result in reinnervation of the denervated tissue. The second paradigm involves producing a partial tissue denervation that induces collateral sprouting of intact axons to reinnervate the denervated area.

Like other peripheral neurons, the sympathetic adrenergic neuron has proved capable of exhibiting vigorous regenerative sprouting after axonal damage (see Guth, 1956; Olson and Malmfors, 1970, for reviews). Following transection of the axon, there is rapid degeneration of the distal portion and, within 2 to 3 days, growth cones and sprouts form from the proximal axonal stump (Olson, 1969). Preterminal adrenergic axons grow approximately 1 to 3 mm/day during regeneration, a rate that does not differ from that for other peripheral neurons (Olson, 1969). It would appear, however, that formation of a terminal plexus may require additional time (Kirpekar et al., 1970; Tranzer and Richards, 1971). An important factor in determining regenerative responses to injury in sympathetic neurons is the spatial relation of the axonal lesion to the cell body. If the lesion is located at a distance from the cell body, as in the study of Kirpekar et al. (1970) and as is probably the case after chemical denervation with 6-hydroxydopamine (Malmfors, 1971), the regenerative response appears to be complete, or nearly so, and the pattern of terminal innervation does not differ significantly from the normally innervated tissue. If, on the other hand, the lesion is located close to cell bodies, regeneration may take place but part of the neurons degenerate (Nagata and Tsukada, 1968) and regeneration is often incomplete.

TRANSPLANTATION STUDIES   In some instances, however, sympathetic neurons may show a remarkable capacity for axonal sprouting and regeneration even when their axons are disrupted close to the cell body. This appears best demonstrated in experiments in which sympathetic ganglia are transplanted from their normal location to new sites. Such experiments were carried out by Cajal (1928, p. 477) who observed axonal sprouting in transplanted cat sympathetic ganglia and Olson (1969) who showed that sympathetic ganglia transplanted into a transected sciatic nerve will regenerate axonal sprouts into both the proximal and distal stumps of the nerve. An interesting example of this sympathetic neuron

capacity for regeneration was presented by Silberstein et al. (1971). In this study rat superior cervical ganglia were excised and placed in organ culture with excised iris. The ganglia gave rise to axonal sprouts that reinnervated iris within a few days and formed an adrenergic innervation pattern similar to that in the normal iris. As Silberstein et al. (1971) point out, this provides an important in vitro model for studies of mechanisms controlling regenerative growth.

The transplantation technique has also provided the basis for an elegant series of studies (Olson and Malmfors, 1970) examining the factors influencing in vivo regenerative growth of peripheral adrenergic neurons. In these experiments two types of regenerative growth process were analyzed; growth from ganglion transplants to the intact or denervated iris of a host eye, and growth of adrenergic axon collaterals from the iris of a host eye into a transplant of peripheral tissue that had been placed adjacent to the iris in the anterior chamber. In each instance the reinnervation pattern appeared to approximate the normal pattern within 2 weeks after the experimental manipulation. A diagrammatic summary of these two experimental types and the results obtained is shown in Figure 4. Each conforms to one of the two basic regeneration paradigms outlined above (see Figure 3). The reinnervation of a transplanted iris occurs from the uninjured host iris and, thus, represents collateral sprouting of intact adrenergic axons. Axonal growth from the transplanted ganglia to host irises represents an instance

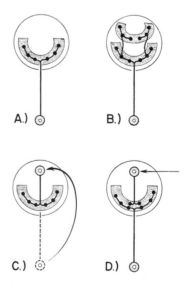

FIGURE 4   Regenerative growth of adrenergic neurons in the rat iris. (A) Normal iris innervation from superior cervical ganglion. (B) Transplanted iris is reinnervated by axons from the host iris. (C) Transplanted superior cervical ganglion reinnervates denervated host iris. (D) Transplanted superior cervical ganglion forms few terminals on a normal host iris. (Redrawn from Olson and Malmfors, 1970.)

of sprouting from transected axons since the neural connections of the ganglia were severed in the course of the transplantation. The observations of Olson and Malmfors (1970) lend support to the following conclusions.

The extent of regenerative axonal growth and its pattern are determined by the target tissue. When iris, for example, is transplanted into the anterior chamber, the newly formed autonomic ground plexus has the appearance of normal iris innervation. This is to be expected as the source of the new innervation is also iris innervation (Olson and Malmfors, 1970). If submaxillary gland is transplanted the surviving parenchyma is abundantly reinnervated in a pattern and density of terminal distribution that closely approaches that observed in normal submaxillary gland. On the other hand, transplanted sublingual gland, which normally has a very sparse adrenergic innervation, is almost devoid of new adrenergic axons. However, the sublingual gland transplants are often surrounded by growing adrenergic fibers, suggesting that these transplants are capable of stimulating collateral sprouting but not a terminal innervation plexus.

The capacity to reinnervate in a characteristic pattern for a particular tissue is shared by many sympathetic neurons. A denervated host iris, which is normally innervated by axons arising from cell bodies of the superior cervical ganglion, can be reinnervated not only by transplanted superior cervical ganglia but by celiac or lumbar sympathetic ganglia as well. In each case the newly formed iris innervation is indistinguishable from the normal pattern.

The growth of regenerating fibers into the target tissue is inhibited by an intact homotypic innervation of the same type. Support for this is obtained from three separate types of experiment. First, the growth of new fibers into a host iris is greatly reduced if the normal innervation is left intact (Figure 4). If this innervation is removed, and the cholinergic parasympathetic innervation is left intact, the host iris is completely reinnervated by regenerating adrenergic axons (Olson and Malmfors, 1970). Second, the formation of new terminals from regenerating ganglion transplants in the host iris plateaus at about 2 weeks, and at this stage the density and pattern of newly formed fibers are quite similar to those of the normal intact innervation (Olson and Malmfors, 1970). It would appear likely that the explanation for this cessation of regeneration is an inhibition to growth exerted by the already formed adrenergic terminals. The process is not exact, however, for in some instances there is a transient hyperinnervation (Olson and Malmfors, 1970). Third, the regenerative growth from superior cervical ganglion transplanted into rat sciatic nerve is markedly reduced if the nerve is not cut proximal to the transplantation site (Olson, 1969). In this case it might be argued that there is

simply no place for newly formed axons to go in the uninterrupted nerve but it is also possible that intact adrenergic axons inhibit regenerative growth.

Regeneration is independent of transmitter synthesis, storage, release, re-uptake, or receptor blockade. Olson and Malmfors (1970), using reinnervation of an iris transplanted to a host eye as an indicator, found no interference from a variety of pharmacologic manipulations with the final result of regeneration as noted at 14 days after transplantation. These included inhibition of noradrenaline formation by chronic tyrosine hydroxylase inhibition, block of vesicular storage sites for noradrenaline using reserpine, block of noradrenaline uptake with desipramine, and block of either α or β receptor sites. None of the treatments significantly altered the extent of reinnervation of the transplated iris by adrenergic axons from the host iris, indicating that intact transmitter function is not essential for the reinnervation process in the case of adrenergic neurons. These experiments do not exclude the possibility, however, that manipulations of transmitter function may alter the *rate* of growth of regenerating adrenergic axons. In addition, they leave unanswered the intriguing question of what the role of the transmitter might be in growth and regeneration. Sprouting axons of adrenergic neurons appear to have a very high noradrenaline content, particularly at the site of most active growth, both in vivo (de Champlain et al., 1970; Olson, 1969) and in vitro (Silberstein et al., 1971; Kopin and Silberstein, 1972), and the axonal sprouts are capable of noradrenaline release upon stimulation (Kopin and Silberstein, 1972). The function of this is unknown, but it suggests that some interaction might occur between the growing neurite and the cells to be innervated in the formation of a synaptic contact.

PLASTICITY OF PERIPHERAL ADRENERGIC NEURONS The term neuronal plasticity could imply any type of cellular adjustment to functional or environmental demands. In this review we shall use plasticity to signify a structural modification of synaptic contacts of intact, mature neurons. In the peripheral sympathetic system, this plasticity could arise in two ways; by collateral sprouting of intact adrenergic axons to form new functional synapses at sites normally innervated by adrenergic neurons, and by the formation of heterotypic adrenergic synapses at experimentally vacated, nonadrenergic synaptic sites.

Collateral sprouting of intact, adrenergic axons. Collateral sprouting of intact axons has been observed from motor neurons (Hoffman, 1950; Edds, 1953), from the peripheral (Weddell et al., 1941) and central (Liu and Chambers, 1958) branches of dorsal root ganglion cells and from preganglionic sympathetic neurons (Murray and

Thompson, 1957). The newly formed axons and terminals grow to reinnervate partially denervated structures and, at least in some instances (Edds, 1953), appear responsible for recovery of function. As noted above, the growth of new collaterals from preterminal or terminal adrenergic axons was described in detail by Olson and Malmfors (1970) after transplantation of peripheral tissue to the anterior chamber of the eye. Indeed, in this situation the capacity of the intact innervation to produce collateral sprouting was immense so that as many as five transplanted irises could be reinnervated in a single host eye. And, in both in vivo and in vitro experiments, collateral sprouting of adrenergic axons was shown to produce a functional reinnervation (Olson and Malmfors, 1970). This expression of neuronal sprouting has obvious functional implications as an adaptive response of an organism to injury. It has the further implication that collateral sprouting may occur as a normal phenomenon in the intact peripheral nervous system, that is, as a continuing synaptic reorganization without prior injury (Barker and Ip, 1966). This will be discussed in greater detail in relation to collateral sprouting of adrenergic neurons in the central nervous system.

Formation of heterologous synapses by peripheral sympathetic neuron. In the studies reviewed above, the regenerative process occurred so that the adrenergic neuron was presumably reinnervating denervated adrenergic receptor sites. The exception to this is the situation where sprouting adrenergic neurons fail to innervate tissues, such as the sublingual gland (Olson and Malmfors, 1970), that normally do not have an adrenergic innervation. This raises the question as to whether functional heterologous synapses might be formed in the peripheral sympathetic system. There is little information available to provide an answer. Some indication of collateral sprouting of adrenergic innervation was obtained by Roth and Richardson (1969) following removal of the cholinergic, parasympathetic innervation of the iris. Similarly, Nordenfelt (1968), found morphologic evidence of a cholinergic neuron sprouting after sympathetic denervation of the submaxillary gland but in neither of these is there any indication that the process is functional. The only studies suggesting that functional heterotypic synapses may be formed in the periphery have utilized regeneration following nerve transection. One example of this is the demonstration (Vera et al., 1957; Lennon et al., 1967; Koslow et al., 1972), that cross-anastomosis of the hypoglossal nerve to the denervated nictitating membrane of the cat produces a functional cholinergic reinnervation of that organ. The nictitating membrane is largely, if not exclusively, innervated by adrenergic neurons originating in the superior cervical ganglion, but the membrane will contract in response to both noradrenaline and acetyl-choline, indicating that both adrenergic and cholinergic receptors are present in the intact organ. Thus, the formation of heterologous cholinergic synapses by a somatomotor nerve with nictitating membrane smooth muscle cells does not necessarily require any plasticity of the muscle receptors.

In another experiment Koslow et al. (1971) studied the reinnervation of a lumbar sympathetic ganglion after anastomosis of the splenic nerve (containing axons of postganglionic sympathetic neurons) to the lumbar sympathetic trunk. They observed increased adrenergic fibers in the ganglion and the responses produced by electrical stimulation suggested that excitatory adrenergic synapses were formed in some ganglia. Such synapses are not present in the intact ganglion.

These studies illustrate both the regenerative capacity of the peripheral nervous system and the plasticity that both regenerating neurons and reinnervated tissues may exhibit. The data indicate that heterologous synapses can be formed in the absence of appropriate innervation but that in the usual situation in regeneration the reestablishment of synapses is the result of a competition where the appropriate neuron type is favored.

The central adrenergic neuron

The demonstration that catecholamines are distributed nonuniformly in the central nervous system and that their content is not altered by peripheral sympathetic denervation (Vogt, 1954) provided the first substantive evidence for central adrenergic neurons. Subsequently, use of the Falck-Hillarp technique has permitted a morphologic analysis of the distribution of these neurons in the normal, adult nervous system.

DISTRIBUTION OF CENTRAL ADRENERGIC NEURONS Two major groups of central catecholamine-containing neurons are now known to exist, a dopamine-producing group and a norepinephrine-producing group. The cell bodies of the dopamine neurons are found principally in two locations. One is the periventricular zone of the hypothalamus and the caudal thalamus. These cells can be divided into four distinct groups (Björklund and Nobin, 1973) of which the largest is the arcuate nucleus (Fuxe, 1964; Fuxe and Hökfelt, 1966; Jonsson et al., 1972; Björklund et al., 1970, 1973a). The second location of dopamine cell bodies is in the substantia nigra and the adjacent ventral tegmental area (Dahlström and Fuxe, 1964) and these give rise to the nigrostriatal pathway (Andén et al., 1964; Moore et al., 1971a) and a selective innervation of some areas of the basal telencephalon (Ungerstedt, 1971). The norepinephrine neuron cell bodies are found in groups within the brainstem reticular

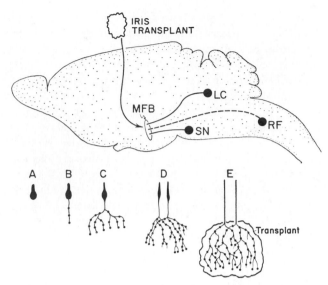

FIGURE 5 Reinnervation of peripheral tissue transplants by central adrenergic neurons. Three groups of ascending adrenergic axons enter the medial forebrain bundle (MFB) in the lateral hypothalamus (top figure): LC, norepinephrine neurons in the locus coeruleus; RF, norepinephrine neurons in the medullary reticular formation; SN, dopamine neurons in the substantia nigra. The transplant, iris, is placed into the medial forebrain bundle transecting ascending axons of these neuron systems. The subsequent sequence of events is shown diagrammatically in A to E (see text for description).

formation extending from the caudal medulla to the rostral midbrain (Dahlström and Fuxe, 1964) with scattered cells present in the caudal diencephalon (Björklund and Nobin, 1973). The brainstem neurons give rise to descending projections innervating the gray matter of the spinal cord (Dahlström and Fuxe, 1965) and ascending projections innervating widespread areas of the brainstem, cerebellum, diencephalon, and telencephalon (Fuxe, 1965; Ungerstedt, 1971). Ungerstedt's (1971) work has shown that ascending fibers from the pontine nucleus, locus coeruleus, and adjacent nuclei enter a dorsal catecholamine bundle that eventually distributes to telencephalon. Ascending fibers from other brainstem noradrenaline nuclei join a ventral catecholamine bundle and distribute principally to brainstem, diencephalon, and basal telencephalon. The medial forebrain bundle contains nearly all of the ascending adrenergic fibers within the hypothalamus (Ungerstedt, 1971; see Figure 5). These observations provide an introduction to the following sections.

ONTOGENESIS OF CENTRAL ADRENERGIC NEURONS On the basis of both biochemical (Connor and Neff, 1970; Loizou, 1972; Baker and Hoff, 1972; Porcher and Heller, 1972; Coyle, this volume) and histochemical studies (Björklund et al., 1968; Loizou, 1972; Olson and Seiger,

1972), the development of central adrenergic neurons appears to take place, like that of peripheral adrenergic neurons, largely in the postnatal period. Central adrenergic neuroblasts of the rat brain develop the capacity for synthesis and storage of catecholamines as early as the twelfth to fourteenth days of gestation (about the time when catecholamine-containing neuroblasts are first demonstrable in sympathetic ganglia) and adrenergic pathways develop before birth (Olson and Seiger, 1972). A large portion of the preterminal axon may be evident at certain early stages of development so that the pathway of a cell group can be traced from its origin to its termination (Olson and Seiger, 1972). This has been particularly true for the nigroneostriatal system (Golden, 1972; Olson et al., 1972). Many details of the ontogenetic development of central adrenergic neurons are available (Björklund et al., 1968; Hyyppa, 1969; Loizou, 1971, 1972; Maeda and Dresse, 1969; Smith and Simpson, 1970; Olson and Seiger, 1972; Olson et al., 1972). These studies have outlined the process of development of central adrenergic neurons. All of the studies are recent, and these advances in our understanding are satisfying but there is still little that can be said about factors regulating development of these neurons and, particularly, those determining the pattern of synaptic architecture in areas innervated by the neurons. In addition, it is not known to what extent there is plasticity in the ontogenetic development of central adrenergic neurons, either as a functionally determined phenomenon or in response to injury. A further question that awaits resolution is the function of catecholamines, or other neurotransmitters for that matter, in the formation of synapses. Woodward et al. (1971) have shown that the Purkinje cell has a marked sensitivity to a number of possible transmitter substances, including norepinephrine, before it receives any synaptic contacts but the functional significance of this is unknown and, in particular, it is not known whether the early sensitivity plays a role in synaptogenesis.

REGENERATIVE RESPONSES OF TRANSECTED CENTRAL ADRENERGIC NEURONS *Do mammalian central neurons regenerate?* It is often stated that the neurons of the adult mammalian central nervous system are incapable of significant regenerative responses to injury (see Das and Altman, 1972, pp. 233). Nevertheless, the capacity of severed central axons to exhibit regenerative sprouting similar to that occurring in the peripheral nervous system has been known since the pioneering work of Cajal (1928) and confirmed in many subsequent studies (see Windle, 1956; Clemente, 1964; Bernstein and Bernstein, 1971, for reviews). Unfortunately such sprouting appears to have very limited functional significance and the factors determining this are only partially understood. It is not known, for example, whether groups of central neurons differ in their capacity to show regenerative sprouting or if these newly formed fibers are able to establish functional synaptic contacts. Similarly, there is little known of the conditions that promote sprouting and the directed growth necessary to make such contacts possible or, conversely, of the conditions that determine failure of functional restoration following injury to central axons. The heuristic implications of these questions compel further work in this area. Recent investigations of central adrenergic neuron regenerative responses indicate that these neurons may provide a useful model for further study.

Reaction of central adrenergic neurons to axonal transection. After transection of the axon of a central adrenergic neuron, there is accumulation of amine in the severed axon proximal to the lesion (Dahlström and Fuxe, 1965). This becomes evident within hours and by 3 days after the injury the severed axons are seen principally as coarse, beaded, and distorted fibers located close to the area destroyed by the lesion (Dahlström and Fuxe, 1965; Katzman et al., 1971; Ungerstedt, 1971). At 7 days after a lesion, there is little change in the amine accumulations, but a new type of densely packed, delicate, varicose fiber appears in the vicinity of the original axonal accumulations. These numerous, small fibers develop into an abundant system between the first and third weeks after the injury which substantially fills the proximal border of the lesion. They appear to represent regenerative sprouting from the severed central axons. Their presence following the lesion cannot be attributed to sprouting from the peripheral adrenergic innervation of the cerebral vasculature, because they appear equally vigorously in the sympathectomized preparations. In addition to the sprouting adjacent to the lesion, the transected axons may form a marked anomalous innervation, particularly of blood vessels at the base of the brain, that forms regardless of whether or not the vessels have retained their normal, sympathetic innervation (Katzman et al., 1971). Similarly, anomalous innervation may appear in fiber bundles such as the root of the oculomotor nerve (Björklund and Stenevi, 1971). A similar course of events with amine accumulation followed by axonal sprouting is noted after crush lesions of the spinal cord (Björklund et al.,1971b). These observations indicate that axons of central adrenergic neurons are capable of vigorous regenerative sprouting following transection.

Will regenerating central adrenergic neurons reinnervate denervated sites: Transplants of peripheral tissue into the central nervous system. To test the capacity of central adrenergic neurons for reinnervation of denervated tissue, transplants of peripheral, sympathetically innervated tissue were placed within the ascending or descending pathways of

FIGURE 6 Iris transplant in medial forebrain bundle, 6 months after operation. The transplant is viable and is extensively reinnervated by central adrenergic axons forming a pattern of innervation very similar to that of the normal iris. Falck-Hillarp method, photographic montage with final magnification approximately 100 X.

the brainstem adrenergic neuron systems. In each case the transplant is denervated by removal from its normal site and implantation of the transplant into the central nervous system transects adrenergic axons. Thus, such studies also examine regenerative sprouting responses. The most successful transplantation experiments occur with placement of iris into the medial forebrain bundle (Björklund and Stenevi, 1971; Figures 5 and 6). As in the studies noted above, there are early amine accumulations in the transected axons adjacent to the transplant and subsequent sprouting of delicate, varicose fibers. The transplant is free of adrenergic innervation for the first few days, but by 2 weeks many adrenergic fibers can be traced from the area of the transected medial forebrain bundle fibers into the ventral part of the transplanted iris. In addition, many coarse fibers can be followed from the dorsal catecholamine bundle into the dorsal part of the transplant where they turn both dorsally and ventrally. Most of these fibers have a smooth appearance and are oriented into long vertical bundles. Forty to fifty days after operation the iris transplants continue to be well preserved and contain numerous adrenergic fibers. In many areas this innervation has the characteristic appearance of normally innervated iris (Figure 6). It is clearly innervation of central origin, though, because a similar pattern is observed in animals subjected to cervical sympathectomy prior to transplantation. From microspectro-fluorimetric analysis of the fibers in the transplants, it is evident that both dopamine and norepinephrine-containing axons are present but the norepinephrine type predominates. The longest survival period studied was 6 months. Transplants are still present up to 6 months after implantation and contain a rich adrenergic innervation that frequently resembles the normal adrenergic innervation pattern of the iris.

These observations establish two points. First, *the transsected adrenergic axons are capable of innervating a denervated tissue* if it is placed in the immediate vicinity of newly forming axonal sprouts. Second, *the pattern of reinnervation is frequently remarkably like that exhibited by the normal, peripheral adrenergic innervation.* From recent electron-microscopic studies (Hökfelt, Björklund, and Stenevi, unpublished observations), it would appear that the new innervation is forming characteristic terminals with dense-cored vesicles, and these terminals presumably would be functional. The major source of new fibers to the denervated iris appears to be the norepinephrine axons of the dorsal catecholamine bundle. Thus, the denervated tissue seems to promote growth of the norepinephrine axons selectively and to specify the pattern of their distribution in the tissue. Dopamine fibers also are found in transplants but their location is restricted to the ventral part and they are arranged in a dense plexus of fine

fibers more like the normal neostriatal innervation (the normal site of termination of these transected fibers) than that of the iris. In addition, there are numerous, newly formed indoleamine fibers in the area adjacent to the transplant but none enter it. This observation also holds for spinal cord transplants (Björklund et al.,1971b); there are many new indoleamine fibers formed but none grow into the transplants. In order to be certain that the innervation pattern within a transplant is established by the transplanted tissue, other tissues were placed into the medial forebrain bundle (Björklund and Stenevi, 1971) or spinal cord (Björklund et al.,1971b). The mitral valve is readily reinnervated in both situations by adrenergic neurons, and the pattern of the newly formed innervation is quite characteristic of that seen in the normally innervated heart. In contrast to this, diaphragm or uterus, which normally have a very sparse peripheral adrenergic innervation, are not well innervated by central adrenergic neurons when transplanted into the medial forebrain bundle (Björklund and Stenevi, 1971), and the few fibers entering the transplants appear randomly arranged.

The results of these transplantation studies are summarized diagrammatically in Figure 5. The transected central adrenergic axon appears capable of vigorous sprouting. Newly formed fibers will innervate a normally adrenergically innervated tissue transplanted into either spinal cord or medial forebrain bundle. The pattern of this innervation, as in the peripheral experiments of Olson and Malmfors (1970), is determined by the tissue receiving the innervation. *All of the evidence available indicates that the sprouting central adrenergic axons are capable of reinnervating a denervated tissue,* but it has not been demonstrated that this is a functional reinnervation.

The most convincing instance of apparent re-establishment of morphological integrity occurs following 6-hydroxydopamine (Nygren et al., 1971) or 5,6-dihydroxytryptamine (Björklund et al., 1973b) administration. Following local injection of 6-hydroxy-dopamine into the spinal cord, there is a loss of adrenergic fibers distal to the injection and accumulations of amine proximally. Within 3 to 4 weeks, in adult animals, the usual innervation pattern of the spinal cord gray matter is regained, suggesting that the axons chemically transsected by the 6-hydroxydopamine lesion have sprouted to reinnervate the denervated segments (Nygren et al., 1971). Similar observations have been recorded following 5,6-dihydroxytryptamine induced degeneration of central serotonergic terminals (Björklund et al., 1973b).

Little information is available to indicate any of the factors responsible for the regulation of regenerative changes in central adrenergic neurons. One that may participate in this is nerve growth factor (NGF). Recent studies (Björklund and Stenevi, 1972; Bjerre et al., 1973) have shown that NGF can stimulate the growth of transected central adrenergic axons. When an iris is transplanted into the medial forebrain bundle, animals receiving graded single doses of NGF injected intraventricularly showed reinnervation of the transplant with medium to high doses of NGF, whereas control animals receiving saline injections show no adrenergic fibers in the transplant at 7 days after operation. This suggests that NGF may participate in adrenergic neuron regeneration in the adult nervous system. In addition to the increased growth of adrenergic fibers,NGF stimulated the growth of new axon sprouts from transsected indoleamine fibers into the transplant. Since NGF is present in the intact rat brain (Johnson et al., 1971), it is tempting to speculate that it may be required for normal growth and maturation of central neurons as well as a response to injury.

COLLATERAL SPROUTING OF CENTRAL ADRENERGIC NEURONS   Nearly all studies of central nervous system regeneration have examined the regenerative capacity of transected axons. From the observations reviewed above, it is evident that the axons of central adrenergic neurons are capable of vigorous sprouting and growth following transection. With the possible exception of the regenerative growth noted after the selective lesions produced by amine transmitter analogues (Nygren et al., 1971; Björklund et al., 1973b), there are no data to indicate that these regenerative changes are of greater functional significance than those observed previously for a number of other groups of central axons (see Cajal, 1928; Clemente, 1964, for reviews). Indeed, the immense morphological complexity of the central nervous system and the response of nonneural elements to injury appear to preclude, at least in the adult mammal, functional regenerative growth from large groups of severed central axons. The one notable exception to this is encountered in the functional regenerative change exhibited by hypothalamo-hypophysial axons after transection (Adams et al., 1969); an unusual situation in that the regenerating axons are not innervating neurons nor are they growing to their usual site of termination. Thus, at the present time, there is little evidence for *functional* regeneration of severed central axons. In this situation much attention has gone to developing methods influencing the direction of growth and preventing cicatricial changes from nonneuronal elements, but it has not been clearly shown that severed central axons are capable of functional restoration even if these problems are overcome (see Clemente, 1964).

The phenomenon of collateral sprouting may be of much greater functional significance, but it has not been extensively investigated in the central nervous system. In scattered reports, intact central axons have been found to sprout to reinnervate a denervated area (Goodman and Horel, 1967; Raisman, 1969a; Lund and Lund, 1971; Moore et al.,1971b;Lynch et al., 1972). In each of these studies the experimental paradigm has conformed to that of the peripheral collateral sprouting studies (Figure 3); that is, when part of the innervation to a nucleus is removed by a distant lesion, the severed axons do not reinnervate the nucleus but the remaining, intact axons within the nucleus sprout to innervate the denervated synaptic sites.

The most extensive studies of this phenomenon in the central nervous system have been carried out by Raisman (1969a) and Raisman and Field (1973) on the innervation of the septal nuclear complex of the rat. The basis for these studies was the demonstration that septal neurons were innervated from two primary sources, the hippocampus via the fornix and the brainstem via the medial forebrain bundle (Raisman, 1966). In ultrastructural studies each of these had a characteristic pattern of termination on septal neurons (Raisman, 1969b). Severing the axons of one source of innervation, such as that from the hippocampal formation, produced electron-microscopic evidence of degenerative changes in the appropriate set of terminals within a few days after operation (Raisman, 1969a, 1969b). If a long interval was allowed between the operation and electron-microscopic examination of the nucleus, however, there were no degenerating terminals and the synaptic architecture of the nucleus was not vastly different from the normal. The major change was an increase in the number of axons exhibiting multiple synaptic contacts in a single micrograph (Raisman, 1969a). These observations indicated that the intact innervation was exhibiting collateral sprouting to fill the denervated synaptic sites in the nucleus. This was confirmed in a subsequent experiment in which a very careful analysis of a selected area of the lateral septal nucleus was carried out over a period from 2 to 114 days after fornix section (Raisman and Field, 1973).

The septal area receives a dense adrenergic innervation (Fuxe, 1965) from the medial forebrain bundle, and, because of our interest in regenerative changes in central adrenergic neurons, it appeared worthwhile to determine if these neurons participated in the phenomena described by Raisman (1969a). If this could be done using the Falck-Hillarp technique, it would provide a direct morphological confirmation of Raisman's (1969a) conclusions based on his *multiple synapse index*. It would also increase our understanding of the regenerative capacity of central adrenergic axons by demonstrating that they exhibit collateral sprouting. This in turn might lead to further studies of the functional significance of collateral sprouting since the adrenergic terminals could be readily manipulated pharmacologically. In one experiment (Moore et al., 1971b), rats were subjected to unilateral section of the fornix. Since the fornix projects almost exclusively on the ipsilateral septal nuclei (Raisman, 1966), the contralateral side serves as a control. The distribution of varicose adrenergic fibers in the septum was studied using the Falck-Hillarp method in normal animals and at 3, 8, 15, 30, 60, and 90 to 100 days after fornix section. At 3 and 8 days after operation, the appearance of the adrenergic innervation of the denervated septum does not differ significantly from that of the control except that a few, scattered axons exhibit large monoamine accumulations and have the appearance of transected axons. Similar accumulations are evident at 8 and 15 days, but, thereafter, none are observed. By 15 days after fornix section, a difference in adrenergic innervation between the two sides is evident, and from 30 days on, there is a consistent increase in the number of varicosities in the lateral and medial septal nuclei on the denervated side. This difference continues through 100 days without apparent change. No other area in the basal telencephalon or hypothalamus exhibits a similar change and the extent of the change in the septal nuclei is directly dependent upon the extent of damage to the hippocampal projection. That is, substantial alterations in the adrenergic innervation of the medial and lateral septal nuclei are consistently observed only when the fornix is totally transected. In instances in which the lesion is incomplete, the results are variable and do not approach the magnitude of changes found in complete denervations (Moore et al., 1971b).

There are several possible interpretations for the changes in septal innervation observed following fornix section. The most attractive of these is that the histochemically demonstrable increase in the density of adrenergic varicosities in the septal nuclei represents a true increase in the number of such terminals. If this is the case, it would imply that the adrenergic terminals normally innervating the septal complex have formed collateral sprouts reinnervating synaptic sites denervated by degeneration of the terminals of hippocampal origin. There is an increase in norepinephrine content in the denervated septum demonstrable by biochemical analysis (Moore et al., 1971b), but these data provide only indirect support for this interpretation. The interpretation is in accord with that of Raisman (1969a) and Raisman and Field (1973), but there are alternatives that should be considered. The most important of these is that the increase in number of fluorescent varicosities and assayable norepinephrine in the denervated septum does not represent collateral sprouting of adrenergic terminals but, rather, accumulation of amine in the septal collaterals of axons innervating the hippocampus that have been sectioned by the fornix lesion. It is well known that there

is an accumulation of amine proximal to axon section, including in the collaterals of the axon (Ungerstedt, 1971), but this phenomenon would not appear to explain the changes observed in these studies. First, changes in the number of adrenergic fibers occur only in the medial and lateral septal nuclei, and it is unlikely that these represent the only nuclei receiving collaterals of medial forebrain bundle axons passing through the septum to innervate the hippocampus. Second, the effect observed is an increase in the apparent number of fibers, not in their fluorescence intensity. If this were to be attributed to accumulation of amine in collaterals, it would require that the collaterals normally not be demonstrable in the septum. There is nothing to indicate that this is the case. Third, the time course for the accumulation of amine following transection usually is brief, but the changes noted here are permanent, at least through the time of the study (100 days). In addition, the time course over which both the histochemical and biochemical changes develop (Moore et al., 1971b) has now been confirmed by Raisman and Field (1973) in their electron-microscopic study of collateral sprouting in the denervated lateral septal nucleus. Thus, on the basis of currently available information, there would not appear to be a good alternative to the interpretation that the changes observed by Moore et al. (1971b) in the denervated rat septum represent collateral sprouting of adrenergic axons. This interpretation is shown diagrammatically in Figure 7.

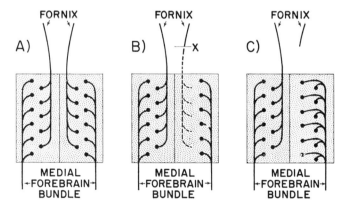

FIGURE 7 Collateral sprouting in the denervated septum, diagrammatically represented. (A) Normal relationship between fornix innervation and medial forebrain bundle innervation in the septal nuclei. (B) Unilateral fornix section resulting in partial denervation of the ipsilateral septal nuclei (C) Denervated area reinnervated by the remaining, intact medial forebrain bundle adrenergic axons.

This is the first study of collateral sprouting in the central nervous system in which the sprouting axons have been characterized by identification of their probable neurotransmitter (Moore et al., 1971b). This leads to

TABLE I
Collateral sprouting of central adrenergic neurons: Summary of available observations

Collateral Sprouting Observed		No Collateral Sprouting Observed	
Region Studied	Lesion Location	Region Studied	Lesion Location
Septal nuclei[a]	hippocampus	lateral preoptic area[a]	hippocampus
Anteroventral nucleus of thalamus[b]	hippocampus	lateral mammillary nucleus[b]	hippocampus
		suprachiasmatic hypothalamic nucleus[b]	midbrain raphe and optic nerve
Dorsal lateral geniculate nucleus[c]	visual cortex	dorsal lateral geniculate nucleus[c]	optic nerve
		superior colliculus[c]	optic nerve and/or visual cortex
Olfactory tubercle[b]	olfactory bulb	hypothalamus, medial and lateral[b]	amygdala
Cerebellar cortex[d]	superior cerebellar peduncle	spinal cord (lumbar gray)[b]	spinal cord, thoracic hemisection

[a] Moore, Björklund, and Stenevi, 1971b.

[b] Moore, Björklund and Stenevi, unpublished observations. In most studies the denervated region was studied using the Falck-Hillarp technique and chemical determinations for norepinephrine and/or dopamine were made as described previously (Moore et al., 1971b). The evidence for collateral sprouting in the anteroventral nucleus is preliminary. The hippocampal lesions appear in some instances to invade the thalamocortical projection of the nucleus and it is not certain that observed changes in adrenergic innervation density do not reflect shrinkage of the nucleus.

[c] Stenevi, Björklund, and Moore, 1972.

[d] Pickel, Krebs, and Bloom, 1972.

the question of whether this is a phenomenon restricted to the septal area innervation or, conversely, will adrenergic neurons innervating other areas show similar responses to denervation? A number of studies have been carried out to test this, and the results of these are summarized in Table I. After hippocampal lesions, there is evidence for sprouting in the septal nuclei as noted above but none in the lateral preoptic area or lateral mammillary nucleus. The lateral preoptic area contains a significant number of adrenergic fibers, but the hippocampal projection is not heavy (Raisman, 1966). In contrast, the lateral mammillary nucleus receives a dense projection from the hippocampus but few, if any, adrenergic fibers. The anteroventral thalamic nucleus receives both a heavy adrenergic innervation and a heavy hippocampal projection, and there is an apparent increase in the adrenergic innervation after a hippocampal lesion. This observation is preliminary, though, and the possibility that the apparent increase in adrenergic innervation is due to shrinkage of the nucleus secondary to retrograde changes cannot be excluded.

Another thalamic nucleus that receives a moderately heavy adrenergic innervation is the dorsal nucleus of the lateral geniculate body (Fuxe, 1965). Since the other connections of this nucleus are well known, it is an ideal nucleus in which to study denervation effects on the adrenergic innervation (Stenevi et al., 1972). Following

removal of one eye, there is loss of nearly all retinal afferents to the contralateral dorsal lateral geniculate, but this does not alter the pattern of adrenergic innervation. Visual cortex ablation removes the corticofugal input to the dorsal lateral geniculate and, in addition, results in retrograde degeneration of most lateral geniculate neurons. Seven days after visual cortex ablation, there is no change in the adrenergic innervation of the dorsal lateral geniculate, but by 13 days there is a marked increase in the number of adrenergic fibers in the denervated geniculate as compared to thc control. The appearance of these fibers corresponds to the time at which corticofugal afferents to the nucleus have degenerated and the neurons of the nucleus have largely undergone retrograde degeneration, but there is neither significant shrinkage of the nucleus nor a reactive glial response (Stenevi et al., 1972). Within the next 2 weeks the nucleus becomes markedly shrunken and gliosed and at this stage few, if any, adrenergic fibers remain. These observations indicate that collateral sprouting of the adrenergic innervation of the dorsal lateral geniculate will occur but only in response to either anterograde degeneration of the corticofugal projection to the geniculate or retrograde degeneration of the geniculate neurons (Stenevi et al., 1972). In this study the superior colliculus was examined as well. This tectal structure receives a very sparse adrenergic innervation that appears unaltered by either optic nerve section, visual cortex ablation or a combination of these two procedures. In all probability this lack of effect reflects that the adrenergic innervation is sparse and largely restricted to deeper layers of the colliculus whereas the retina and cortical input is largely to superficial layers.

In another study the effect of amygdala lesions on hypothalamic adrenergic innervation was analyzed. The amygdala projects to a selected area of the medial hypothalamus via the stria terminalis (Heimer and Nauta, 1969) and to a restricted part of the lateral hypothalamic area via the ventral amygdalofugal pathway (Cowan et al., 1965; Leonard and Scott, 1971). A report (Eleftheriou, 1970) that amygdala ablation in the mouse resulted in increased hypothalamic norepinephrine content suggested that removal of hypothalamic innervation from the amygdala might produce collateral sprouting in the medial or lateral hypothalamic adrenergic innervations. In our study (Moore, Björklund, and Stenevi, unpublished observations) no changes in hypothalamic adrenergic innervation were noted either by histochemical study or chemical assay.

Our interpretation of this data is that the relationship between the extent of denervation and the innervation of nonadrenergic fibers to the denervated areas was not optimal. This view can be clarified by discussing two further studies that offer the extremes of these experimental situations. The spinal cord gray matter receives a bilateral projection from brainstem adrenergic neurons (Carlsson et al., 1964; Dahlström and Fuxe, 1965). This innervation, while moderate in density, is scattered among that from many other systems. Following hemisection of the cord, the gray matter of all segments distal to the lesion is partially denervated of adrenergic input, but, in the rat, few other descending axons to distal segments would be affected by the lesion. In this situation there is no apparent increase, either by histochemical or chemical analysis, of the adrenergic innervation in the denervated segments. This result would be predicted in that the extent of denervation following hemisection of the cord on distal segments is probably not great, and the adrenergic is very diffusely distributed.

The olfactory tubercle presents a situation that contrasts significantly with the spinal cord. The afferents from olfactory bulb are densely distributed over the superficial layers of the tubercle (Heimer, 1968), in the same area receiving a dense innervation of dopamine fibers (Fuxe, 1965). Following olfactory bulb ablation, there is some apparent increase in adrenergic innervation of the tubercle that is difficult to judge because of the density of the normal innervation. Chemical determinations confirm this, though, and indicate an increase in dopamine content of the denervated olfactory tubercle compared to control values of approximately 45% at 30

days after operation (Moore, Björklund, and Stenevi, unpublished observations). This suggests that removal of olfactory bulb afferents results in sprouting of dopamine fibers in the tubercle.

Another area that appears to show collateral sprouting of adrenergic neurons is the cerebellar cortex (Pickel et al., 1972). After lesions in the superior cerebellar peduncle involving part of the ascending norepinephrine axons, the remaining ones seem markedly increased in number in the cerebellar cortex resulting in a heavy adrenergic innervation pattern, particularly at the base of the molecular layer.

These observations do not provide a complete picture, but they do lead to some tentative conclusions concerning collateral sprouting of adrenergic neurons in the adult mammalian central nervous system. Adrenergic neurons are capable of collateral sprouting in response to denervation in the adult mammalian nervous system. For this to be evident either by fluorescence microscopy on material prepared by the Falck-Hillarp method or by chemical analysis of amine content requires that the number of new fibers be relatively great. At least three criteria must be met for this to take place. First, *the denervation of non-adrenergic elements must be considerable.* Clearly this is the case in the septal area where hippocampal ablation removes the only important source of afferent input other than that from the medial forebrain bundle. A similar situation would obtain in the olfactory tubercle with olfactory bulb ablation. After amygdala lesions, on the other hand, there remain a number of sources of afferent input to the denervated areas that may be capable of their reinnervation by collateral sprouting. Second, *there must be a minimal density of the adrenergic innervation into a denervated area for reinnervation to be apparent with the methodology employed.* Third, *the adrenergic axons should be distributed within the denervated area in sufficient proximity to the denervated synaptic sites to promote collateral sprouting.* The ideal situation, as it appears to pertain in the septal area (Raisman and Field, 1973), is that an adrenergic terminal be adjacent to the degenerating terminals. Nearly all of the observations reviewed above are in accord with these criteria. The most evident exception, where collateral sprouting would not be expected, is in the cerebellar cortex innervation, which is quite sparse in the normal state. This also differs from the usual collateral sprouting experiment in that a part of the adrenergic innervation must be involved in the peduncle lesion for the apparent increase in cerebellar cortical innervation to occur (Pickel et al., 1972).

Obviously an understanding of the exact organization of the synaptic architecture of the areas under study and the sequence of events following a denervation is required in order to explain each effect obtained. If, for example, the adrenergic innervation to an area makes predominantly axo-somatic contacts, and two separate, non-adrenergic sources of input make closely adjacent axo-drendritic contacts, removal of one axo-dendritic input probably would result in sprouting of the adjacent, nonadrenergic axo-dendritic fibers rather than the more distant adrenergic innervation. Alternatively, a situation might arise in which a relatively sparse adrenergic innervation would show marked collateral sprouting because of a favorable synaptic organization in a denervated area. Raisman and Field (1973) discuss a number of ultrastructural aspects of this problem in relation to innervation of the septal area and its further clarification can take place only by additional investigation.

These considerations also raise another issue, that of the specificity of the observed effects. To state the question broadly, are all central neurons equally capable of showing regenerative changes, or are certain groups more responsive than others? The observations recorded here suggest that adrenergic neurons have a remarkable capacity for exhibiting both regenerative sprouting of transected axons and collateral sprouting within a denervated nucleus, but it is not clear that this does not simply reflect their ready and selective demonstration by the Falck-Hillarp method. The studies of Björklund et al. (1971) and Björklund and Stenevi (1971) indicate that serotonin neurons also show degenerative sprouting when injured and this phenomenon is amply documented for many areas of the central nervous system (Cajal, 1928; Windle, 1956; Clemente, 1964; Bernstein and Bernstein, 1971). Whether one or another group of neurons has a greater capacity for such changes remains an open question. Collateral sprouting of central neurons has been reported much less frequently, but the known instances (e.g. Liu and Chambers, 1958; Goodman and Horel, 1967; Raisman,1969b; Westrum and Black, 1971; Lund and Lund, 1971; Wall and Egger, 1971; Lynch et al., 1972) in other than adrenergic neurons suggest that it may be an ubiquitous phenomenon. In the spinal cord, there is some indication that serotonin neurons may show collateral sprouting after degeneration of the norepinephrine fibers (Nygren et al., 1971). Also, there is an increase in septal area serotonin content following hippocampal ablation (about 30% above control values at 30 days after operation), indicating that this component of the medial forebrain bundle input to the septal nuclei may show collateral sprouting as well (Moore, Björklund, and Stenevi, unpublished observations). Nevertheless, the question of a differential capacity for regenerative changes among central neurons must remain open. The possibility that central adrenergic neurons have an unusual capacity for exhibiting regenerative changes is not unlikely and would be in accord with Jacobson's (1969) view that certain central neurons are more plastic in ontogenetic development than others. There are a number of observations to indicate that regenerative phenomena occur more readily in the young nervous system (see Schneider, 1970; Lund and Lund, 1971), but, as yet, there have been no attempts to determine if this holds for adrenergic neurons. It is also intriguing that central adrenergic neurons exhibit growth characteristics remarkably like those of peripheral adrenergic neurons. This is particularly evident in the transplant experiments in which peripheral tissue was reinnervated by central neurons in a manner remarkably like peripheral neurons (see Olson and Malmfors, 1970). This indicates that the growth capacity of these central neurons is not inferior to that of peripheral neurons and implies that central regeneration is not limited by the properties of central neurons but, rather, that their capacity for regeneration is suppressed or inadequately expressed. Another problem that should be considered further is that of the stimulus for growth or collateral sprouting. The processes involved in a neuron responding to axonal section have been considered in detail recently by Cragg (1970) and do not require additional discussion. The stimulus for collateral sprouting must differ from this, at least to the extent that the intact axon should receive some input indicating a nearby denervation, either from the degenerating terminals or the denervated synaptic sites. An alternative explanation would be that there is no stimulus for collateral sprouting but, rather, that the effects observed in a denervated area are only extreme instances of a process continually going on in the adult, mammalian nervous system. This concept would require that there be a continuing reorganization of synaptic architecture in the adult nervous system. The basis of this would be that there are some axon terminals always in the process of degenerating whereas others are being newly formed within a given area of terminal distribution. In the normal septal area, for example, one would assume a constant turnover in both the hippocampal and medial forebrain bundle terminals on septal nucleus neurons, but a balance between the two sets of input would be maintained except as alterations in the balance might result from functional effects. In the absence of one group of terminals, however, the normal process of growth would not be impeded by the factors maintaining the usual balance, and the intact axons could produce sufficient terminals to fill the sites vacated by the degenerated axon terminals. This process would account for the apparent collateral sprouting of adrenergic medial forebrain bundle axons in the septum following removal of hippocampal input. It would also provide a basis for understanding how sprouting axons could form functional synaptic contacts on partially denervated neurons, because it would presume that the synaptic surface of those neurons is continually adjusting to changes in terminal arrangement in the normal state. There are no direct, experimental data to support this view, but it does receive some backing from observations (Cohen and Pappas, 1969; Sotelo and Palay, 1971) that degenerating axon terminals are commonplace in the apparently normal nervous system. The concept that continuing growth of axon terminals and reorganization of synaptic architecture occur in the adult nervous system is not new (Rose et al., 1960; Moore et al.,1971b; Sotelo and Palay, 1971), but it warrants careful attention because of the potential functional significance of such a process. The observations reviewed here suggest that the adrenergic neuron may provide a model to be used in the acquisition of data to support the concept.

Last, the functional implications of collateral sprouting should be considered. Three alternatives are apparent. First, collateral sprouting of adrenergic axons may be functionally insignificant. That is, such changes would not alter the function of the denervated area beyond the effects to be attributed to the denervation itself. This would seem unlikely, particularly in areas with marked changes in synaptic architecture such as the septum, on the grounds that any morphological alteration of input to a neuron should affect its functional state. When looking at an entire nucleus or area, on the other hand, one cannot discount entirely the possibility that such effects would cancel out each other and be functionally insignificant in a broad sense. Second, collateral sprouting may be an important factor in determining the recovery of function observed so frequently after brain lesions. This cannot be supported by experimental evidence but, due in large part to the existent pessimism about central nervous system regeneration, sprouting has not been viewed as a likely basis for recovery of function (Stein et al., 1969). Third, collateral sprouting may either increase the deficit produced by a central lesion or cause changes inimical to recovery of function (McCouch et al., 1958). In the septum, for example, the great increase in adrenergic terminals, if functional, might result in such an increase in inhibitory input upon the septal neurons that they could not participate in recovery of function following a hippocampal lesion. In the absence of substantive data, however, it is not possible to speculate further on the potential functional significance of these regenerative responses.

Conclusion

The purpose of this chapter has been to summarize the information available on growth and plasticity of adrenergic neurons. The analysis of this subject is still in its natural history phase; many observations have been accumulated but little is known of the mechanisms regulating the growth of these neurons or if they can show plasticity as a response to functional changes, either during ontogenesis or in the developed nervous system. Similarly, the capacity of both peripheral and central adrenergic neurons to exhibit regenerative responses to injury has been amply documented, but our understanding of the mechanisms involved in regulating these responses is rudimentary. Indeed, in the central nervous system the functional significance of such responses is totally unknown. These are some of the problems that remain for future investigation. They are significant in that their elucidation will promote our knowledge of the adrenergic nervous system and contribute to our general understanding of neuronal growth and plasticity.

ACKNOWLEDGMENTS The research summarized in this review was supported by grants NS 05002, HD 04583 and NS 06701 from the National Institutes of Health, United States Public Health Service and by grants from the Magnus Bergvall Foundation, the Swedish Medical Research Council (04X-3874, 04X-712, and 04X-56) and the Medical Faculty, University of Lund, Lund, Sweden.

REFERENCES

ADAMS, J. H., P. M. DANIEL, and M. M. L. PRICHARD, 1969. Degeneration and regeneration of hypothalamic nerve fibers in the neurohypophysis after pituitary stalk section in the ferret. J. Comp. Neurol. 135:121–144.

ANDÉN, N.-E., A. CARLSSON, A. DAHLSTRÖM, K. FUXE, N.-A. HILLARP, and K. LARSSON, 1964. Demonstration and mapping out of nigro-neostriatal neurons. Life Sci. 3:523–530.

BAKER, P. C., and K. M. HOFF, 1972. Maturation of 5-hydroxytryptamine levels in various brain regions of the mouse from 1 day postpartum to adulthood. J. Neurochem. 19:2011–2015.

BARKER, D., and M. C. IP, 1966. Sprouting and degeneration of mammalian motor axons in normal and de-afferentiated skeletal muscle. Proc. R. Soc. Lond. (Biol.) 163:538–554.

BERNSTEIN, J. J., and M. E. BERNSTEIN, 1971. Axonal regeneration and formation of synapses proximal to the site of lesion following hemisection of the rat spinal cord. Exp. Neurol. 30: 336–351.

BJERRE, B., A. BJÖRKLUND, and U. STENEVI, 1973. Stimulation of growth of new axonal sprouts from lesioned monoamine neurons in the adult rat brain by nerve growth factor. Brain Res. (in press).

BJÖRKLUND, A., A. ENEMAR, and B. FALCK, 1968. Monoamines in the hypothalamo-hypophyseal system of the mouse with special reference to the ontogenetic aspects. Z. Zellforsch. 89: 590–607.

BJÖRKLUND, A., B. FALCK, F. HROMEK, CH. OWMAN, and K. A. WEST, 1970. Identification and terminal distribution of the tubero-hypophyseal monoamine fiber systems in the rat by means of stereotaxic and microspectrofluorometric techniques. Brain Res. 17:1–23.

BJÖRKLUND, A., B. FALCK, and U. STENEVI,1971a.Classification of monoamine neurons in the rat mesencephalon: Distribution of a new monoamine neuron system. Brain Res. 32:269–285.

BJÖRKLUND, A., R. KATZMAN, U. STENEVI, and K. A. WEST, 1971b. Development and growth of axonal sprouts from noradrenaline and 5-hydroxytryptamine neurones in the rat spinal cord. *Brain Res.* 31:21–33.

BJÖRKLUND, A., and U. STENEVI, 1971. Growth of central catecholamine neurones into smooth muscle grafts in the rat mesencephalon. *Brain Res.* 31:1–20.

BJÖRKLUND, A., B. FALCK, and CH. OWMAN, 1972. Fluorescence microscopic and microspectrofluorometric techniques for the cellular localization and characterization of biogenic amines. In *Methods of Investigative and Diagnostic Endocrinology*, S. A. Berson, ed., Vol. 1: The Thyroid and Biogenic Amines, J. E. Rall and I. J. Kopin, eds. Amsterdam: North-Holland Publishing.

BJÖRKLUND, A., and U. STENEVI, 1972. Nerve growth factor: Stimulation of regenerative growth of central noradrenergic neurons. *Science* 175:1251–1253.

BJÖRKLUND, A., R. Y. MOORE, A. NOBIN, and U. STENEVI, 1973a. The organization of tubero-hypophysial and reticulo-infundibular neuron systems in the rat brain. *Brain Res.* (in press).

BJÖRKLUND, A., A. NOBIN, and U. STENEVI, 1973b. Regeneration of central serotonin neurons after axonal degeneration induced by 5,6-dihydroxytryptamine. *Brain Res.* (in press).

BJÖRKLUND, A., and A. NOBIN, 1973. Fluorescent histochemical and microspectrofluorometric mapping of diencephalic dopamine and norepinephrine cell groups in the rat brain. *Brain Res.* (in press).

BJÖRKLUND, A., and R. Y. MOORE, 1973. Monoamines in the central nervous system and pituitary. In *Fluorescence Histochemistry of Biogenic Amines*, B. Falck and R. Y. Moore, eds. London: Academic Press (in press).

BURNSTOCK, G., 1969. Evolution of the autonomic innervation of visceral and cardiovascular systems in vertebrates. *Pharmacol. Rev.* 21:247–324.

CAJAL, S. R., 1928. *Degeneration and Regeneration of the Nervous System.* London: Oxford University Press.

CARLSSON, A., B. FALCK, K. FUXE, and N.-A. HILLARP, 1964. Cellular localization of monoamines in the spinal cord. *Acta Physiol. Scand.* 60:112–119.

CLEMENTE, C. D., 1964. Regeneration in the vertebrate central nervous system. *Int. Rev. Neurobiol.* 6:257–301.

COHEN, E. B., and G. D. PAPPAS, 1969. Dark profiles in the apparently normal nervous system: A problem in the electron microscopic identification of early anterograde degeneration. *J. Comp. Neurol.* 136:375–396.

CONNOR, J. D., and N. H. NEFF, 1970. Dopamine concentrations in the caudate nucleus of the developing cat. *Life Sci.* 9:1165–1168.

CORRODI, H., and G. JONSSON, 1967. The formaldehyde fluorescence method for the histochemical demonstration of biogenic monoamines. A review of the methodology. *J. Histochem. Cytochem.* 15:65–78.

COWAN, W. M., G. RAISMAN, and T. P. S. POWELL, 1965. The connexions of the amygdala. *J. Neurol. Neurosurg. Psychiat.* 28:137–151.

CRAGG, B. G., 1970. What is the signal for chromatolysis? *Brain Res.* 23:1–21.

DAHLSTRÖM, A., and K. FUXE, 1964. Evidence for the existence of monoamine-containing neurons in the central nervous system. I. Demonstration of monoamines in the cell bodies of brain stem neurons. *Acta Physiol. Scand.* 62: Suppl. 232, 1–55.

DAHLSTRÖM, A., and K. FUXE, 1965. Evidence for the existence of monoamine neurons in the central nervous system. II. Experimentally induced changes in the intraneuronal amine levels of bulbospinal neuron systems. *Acta Physiol. Scand.* 64: Suppl. 247, 1–36.

DAS, G. D., and J. ALTMAN, 1972. Studies on the transplantation of developing neural tissue into the mammalian brain. I. Transplantation of cerebellar slabs into the cerebellum of neonate rats. *Brain Res.* 38:233–249.

DE CHAMPLAIN, J., T. MALMFORS, L. OLSON, and CH. SACHS, 1970. Ontogenesis of peripheral adrenergic neurons in the rat: Pre- and postnatal observations. *Acta Physiol. Scand.* 80: 276–288.

EDDS, M. V., 1953. Collateral nerve regeneration. *Quart. Rev. Biol.* 28:260–276.

EHINGER, B., B. FALCK, and CH. OWMAN, 1973. The distribution of peripheral adrenergic neurons. In *Fluorescence Histochemistry of Biogenic Amines*, B. Falck and R. Y. Moore, eds. London: Academic Press (in press).

ELEFTHERIOU, B. E., 1970. Effects of amygdaloid lesions on hypothalamic norepinephrine response to increased ambient temperature. *Neuroendocrinol.* 6:175–179.

ENEMAR, A., B. FALCK, and R. HÅKANSON, 1965. Observations on the appearance of norepinephrine in the sympathetic nervous system of the chick embryo. *Devel. Biol.* 11:268–283.

FALCK, B., 1962. Observations on the possibilities of the cellular localization of monoamines by a fluorescence method. *Acta Physiol. Scand.* 56: Suppl. 197, 1–25.

FALCK, B., N.-A. HILLARP, G. THIEME, and A. TORP, 1962. Fluorescence of catecholamines and related compounds condensed with formaldehyde. *J. Histochem. Cytochem.* 10: 348–354.

FUXE, K., 1964. Cellular localization of monoamines in the median eminence and infundibular stem of some mammals. *Z. Zellforsch.* 61:710–724.

FUXE, K., 1965. Evidence for the existence of monoamine neurons in the central nervous system. IV. Distribution of monoamine nerve terminals in the central nervous system. *Acta Physiol. Scand.* (Suppl.) 247:39–85.

FUXE, K., and T. HÖKFELT, 1966. Further evidence for the existence of tubero-infundibular dopamine neurons. *Acta Physiol. Scand.* 66:245–246.

GOLDEN, G. S., 1972. Embryologic demonstration of a nigro-neostriatal projection in the mouse. *Brain Res.* 44:278–282.

GOODMAN, D. C., and J. A. HOREL, 1967. Sprouting of optic tract projections in the brain stem of the rat. *J. Comp. Neurol.* 127: 71–88.

GUTH, L., 1956. Regeneration in the mammalian peripheral nervous system. *Physiol. Rev.* 36:441–478.

HEIMER, L., 1968. Synaptic distribution of centripetal and centrifugal nerve fibers in the olfactory system of the rat. An experimental anatomical study. *J. Anat. (Lond.)* 103:413–432.

HEIMER, L., and W. J. H. NAUTA, 1969. The hypothalamic distribution of the stria terminalis. *Brain Res.* 13:284–297.

HOFFMAN, A., 1950. Local re-innervation in partially denervated muscle: A histophysiological study. *Aust. J. Exp. Biol.* 28: 383–387.

HYYPPA, M., 1969. A histochemical study of the primary catecholamines in the hypothalamic neurons of the rat in relation to the ontogenetic and sexual differentiation. *Z. Zellforsch.* 98:550–560.

JACOBSON, M., 1969. Development of specific neuronal connections. *Science* 163:543–547.

JOHNSON, D. G., P. GORDEN, and I. J. KOPIN, 1971. A sensitive radioimmunoassay for 7S nerve growth factor antigens in serum and tissues. *J. Neurochem.* 18:2355–2362.

JONSSON, G., K. FUXE, and T. HÖKFELT, 1972. On the catecholamine innervation of the hypothalamus with special reference to the median eminence. *Brain Res.* 40:271–281.

KATZMAN, R., A. BJÖRKLUND, CH. OWMAN, U. STENEVI, and K. A. WEST, 1971. Evidence for regenerative axon sprouting of central catecholamine neurons in the rat mesencephalon following electrolytic lesions. *Brain Res.* 25:579–596.

KIRPEKAR, S. M., A. R. WAKADE, and J. C. PRAT, 1970. Regeneration of sympathetic nerves to the vas deferens and spleen of the cat. *J. Pharmacol. Exp. Ther.* 175:197–205.

KOPIN, I. J., and S. D. SILBERSTEIN, 1972. Axons of sympathetic neurons: Transport of enzymes in vivo and properties of axonal sprouts in vitro. *Pharmacol. Rev.* 24:245–254.

KOSLOW, S. H., M. STEPITA-KLAUCO, L. OLSON, and E. GIACOBINI, 1971. Functional reinnervation of cat sympathetic ganglia with splenic nerve homografts. *Experientia* 27:799–801.

KOSLOW, S. H., E. GIACOBINI, S. KERPEL-FRONIUS, and L. OLSON, 1972. Cholinergic transmission in the hypoglossal reinnervated nictitating membrane of the cat: An enzymatic, histochemical and physiological study. *J. Pharmacol. Exp. Ther.* 180:664–671.

LENNON, A. M., C. L. VERA, A. L. REX, and J. V. LUCO, 1967. Cholinesterase activity of the nictitating membrane reinnervated by cholinergic fibers. *J. Neurophysiol.* 30:1523–1530.

LEONARD, C. M., and J. W. SCOTT, 1971. Origin and distribution of the amygdalofugal pathways in the rat: An experimental neuroanatomical study. *J. Comp. Neurol.* 141:313–330.

LIU, C. N., and W. W. CHAMBERS, 1958. Intraspinal sprouting of dorsal root axons. *Arch. Neurol.* 79:46–61.

LOIZOU, L., 1971. The postnatal development of monoamine-containing structures in the hypothalamo-hypophyseal system of the albino rat. *Z. Zellforsch.* 114:234–252.

LOIZOU, L. A., 1972. The postnatal ontogeny of monoamine-containing neurons in the central nervous system of the albino rat. *Brain. Res.* 40:395–418.

LUND, R. D., and J. S. LUND, 1971. Synaptic adjustment after deafferentation of the superior colliculus of the rat. *Science* 171:804–807.

LYNCH, G., D. A. MATTHEWS, S. MOSKO, T. PARKS, and C. COTMAN, 1972. Induced acetylcholinesterase-rich layer in rat dentate gyrus following entorhinal lesions. *Brain Res.* 42:311–318.

McCOUCH, G. P., G. M. AUSTIN, C. N. LIU, and C. Y. LIU, 1958. Sprouting as a cause of spasticity. *J. Neurophysiol.* 21:205–216.

MAEDA, T., and A. DRESSE, 1969. Recherches sur le développement du locus coeruleus. 1. Etude des catécholamines au microscope de fluorescence. *Acta Neurol. Belg.* 69:5–10.

MALMFORS, T., 1971. The effects of 6-hydroxydopamine on the adrenergic nerves as revealed by the fluorescence histochemical method. In *6-Hydroxydopamine and Catecholamine Neurons*, T. Malmfors and H. Thoenen, eds. Amsterdam and London: North-Holland Publishing, pp. 47–58.

MOORE, R. Y., R. K. BHATNAGAR, and A. HELLER, 1971a. Anatomical and chemical studies of a nigro-neostriatal projection in the cat. *Brain Res.* 30:119–135.

MOORE, R. Y., A. BJÖRKLUND, and U. STENEVI, 1971b. Plastic changes in the adrenergic innervation of the rat septal area in response to denervation. *Brain Res.* 33:13–35.

MURRAY, J. G., and J. W. THOMPSON, 1957. The occurrence and function of collateral sprouting in the sympathetic nervous system of the cat. *J. Physiol.* (*Lond.*) 135:133–162.

NAGATA, Y., and Y. TSUKADA, 1968. Effect of degeneration of sympathetic ganglion cells on amino acid metabolism in cervical ganglion from the rat. *Exp. Brain Res.* 5:202–209.

NORBERG, K.-A., 1967. Transmitter histochemistry of the sympathetic adrenergic nervous system. *Brain Res.* 5:125–170.

NORDENFELT, I., 1968. Cholinesterase in the submaxillary gland of the rat after sympathetic denervation. *Quart. J. Exp. Physiol.* 53:6–9.

NYGREN, L.-G., L. OLSON, and S. SIEGER, 1971. Regeneration of monoamine-containing axons in the developing and adult spinal cord following intraspinal 6-hydroxydopamine injections or transections. *Histochemie* 28:1–16.

OLSON, L., 1969. Intact and regenerating sympathetic norepinephrine axons in the rat sciatic nerve. *Histochemie* 17:349–367.

OLSON, L., and T. MALMFORS, 1970. Growth characteristics of adrenergic nerves in the adult rat. *Acta Physiol. Scand.* (Suppl.) 348:1–112.

OLSON, L., and K. FUXE, 1971. On the projections from the locus coeruleus norepinephrine neurons: The cerebellar innervation. *Brain Res.* 28:165–171.

OLSON, L., A. SEIGER, and K. FUXE, 1972. Heterogeneity of striatal and limbic dopamine innervation. Highly fluorescent islands in developing and adult rats. *Brain Res.* 44:283–288.

OLSON, L., and A. SEIGER, 1972. Early prenatal ontogeny of central monoamine neurons in the rat: Fluorescence histochemical observations. *Z. Anat. Entwicklungsgesch.* 137:301–316.

OWMAN, CH., N.-O. SJÖBERG, and G. SWEDIN, 1971. Histochemical and chemical studies on pre- and postnatal development of the different systems of "short" and "long" adrenergic neurons in peripheral organs of the rat. *Z. Zellforsch.* 116:319–341.

PICKEL, V. M., H. KREBS, and F. E. BLOOM, 1972. Proliferation of cerebellar norepinephrine-containing fibers in response to peduncle lesions. Paper presented at 2nd Annual Society for Neurosciences Meeting.

PORCHER, W., and A. HELLER, 1972. Regional development of catecholamine biosynthesis in rat brain. *J. Neurochem.* 19:1917–1930.

RAISMAN, G., 1966. The connections of the septum. *Brain* 89:317–348.

RAISMAN, G., 1969a. Neuronal plasticity in the septal nuclei of the adult rat. *Brain Res.* 14:25–48.

RAISMAN, G., 1969b. A comparison of the mode of termination of the hippocampal and hypothalamic afferents to the septal nuclei as revealed by electron microscopy of degeneration. *Exp. Brain Res.* 7:317–343.

RAISMAN, G., and P. M. FIELD, 1973. A quantitative investigation of the development of collateral reinnervation after partial deafferentation of the septal nuclei. *Brain Res.* 50:241–264.

ROSE, J. E., L. I. MALIS, L. KRUGER, and C. P. BAKER, 1960. Effects of heavy, ionizing, monoenergetic particles on the cerebral cortex. II. Histological appearance of laminar lesions and growth of nerve fibers after laminar destructions. *J. Comp. Neurol.* 115:243–296.

ROTH, C. D., and K. C. RICHARDSON, 1969. Electron microscopical studies on axonal degeneration in the rat iris following ganglionectomy. *Am. J. Anat.* 124:341–360.

SCHNEIDER, G. E., 1970. Mechanisms of functional recovery following lesions of the visual cortex or superior colliculus in neonate and adult hamsters. *Brain, Behav. Evol.* 3:295–323.

SILBERSTEIN, S. D., D. G. JOHNSON, D. M. JACOBOWITZ, and I. J. KOPIN, 1971. Sympathetic reinnervation of the rat iris in organ culture. *Proc. Nat. Acad. Sci. USA* 68:1121–1124.

SMITH, G. C., and R. W. SIMPSON, 1970. Monoamine fluorescence in the median eminence of foetal, neonatal and adult rats. *Z. Zellforsch.* 104:541–556.

SOTELO, C., and S. L. PALAY, 1971. Altered axons and axon terminals in the lateral vestibular nucleus of the rat. Possible example of axonal remodeling. *Lab. Invest.* 25:653–672.

STEIN, D. G., J. J. ROSEN, J. GRAZIADEI, D. MISHKIN, and J. J. BRINK, 1969. Central nervous system: Recovery of function. *Science* 166:528–530.

STENEVI, U., A. BJÖRKLUND, and R. Y. MOORE, 1972. Growth of intact central adrenergic axons in the denervated lateral geniculate body. *Exp. Neurol.* 35:290–299.

TRANZER, J. P., and J. G. RICHARDS, 1971. Fine structural aspects of the effect of 6-hydroxydopamine on peripheral adrenergic neurons. In *6-Hydroxydopamine and Catecholamine Neurons*, T. Malmfors and H. Thoenen, eds. Amsterdam and London: North-Holland Publishing, pp. 15–31.

UNGERSTEDT, U., 1971. Stereotaxic mapping of the monoamine pathways in the rat brain. *Acta Physiol. Scand.* (Suppl.) 367: 1–48.

VERA, C. L., J. D. VAIL, and J. V. LUCE, 1957. Reinnervation of nictitating membrane of cat by cholinergic fibers. *J. Neurophysiol.* 20:365–373.

VOGT, M., 1954. The concentration of sympathin in different parts of the central nervous system under normal conditions and after the administration of drugs. *J. Physiol.* (*Lond.*) 123: 451–481.

WALL, P. D., and M. D. EGGER, 1971. Formation of new connexions in adult rat brains after partial deafferentation. *Nature* (*Lond.*) 232:542–545.

WEDDELL, G., L. GUTTMANN, and E. GUTTMANN, 1941. The local extension of nerve fibers into denervated areas of skin. *J. Neurol. Neurosurg. Psychiat.* 4:206–225.

WESTRUM, L. E., and R. G. BLACK, 1971. Fine structural aspects of the synaptic organization of the spinal trigeminal nucleus (pars interpolaris) of the cat. *Brain Res.* 25:265–288.

WINDLE, W. F., 1956. Regeneration of axons in the vertebrate central nervous system. *Physiol. Rev.* 36:426–440.

WOODWARD, D. J., B. J. HOFFER, G. R. SIGGINS, and F. E. BLOOM, 1971. The ontogenetic development of synaptic functions, synaptic activation and responsiveness to neurotransmitter substances in rat cerebellar Purkinje cells. *Brain Res.* 34:73–97.

86 Functional Dynamics of Central Monoamine Pathways

URBAN UNGERSTEDT

ABSTRACT   Several aspects of transmitter storage and release, receptor sensitivity, and functional interaction between neuron systems in the brain may be studied by a combined usage of anatomical, biochemical, and behavioral techniques. We have used stereotaxic injections of 6-hydroxydopamine to produce highly selective lesions of catecholamine pathways, followed the process of degeneration with transmitter histochemistry and biochemistry, and related this to changes in certain behavioral parameters that may be objectively quantified and correlated to the function of the lesioned pathway. The results are particularly relevant to the mechanisms involved in maintaining synaptic function, i.e., receptor sensitivity and the feedback regulation of transmitter release.

THE IDENTIFICATION of various transmitter substances in the central nervous system (CNS) has made possible new ways of linking CNS function to CNS anatomy. We are now able to focus our attention on anatomically distinct areas where identifiable neurons contain high concentrations of suspected transmitter substances. This advance in biochemical and histochemical "anatomy" has been supplemented by other ways of studying CNS function. Tranzer and Thoenen (1968) discovered that systemically injected 6-hydroxydopamine (6-OH-DA) causes a specific degeneration of the norepinephrine (NE) containing nerve terminals in the peripheral nervous system. This discovery made possible an "updating" of another important technique for studying CNS functions: i.e., anatomically well-defined brain lesions. By injecting 6-OH-DA into various catecholamine (CA) pathways (Ungerstedt, 1968, 1971a) or into the cerebral ventricles (Anagnoste et al., 1969; Bloom et al., 1969; Uretsky and Iversen, 1969), it was possible to induce degenerations of the NE and dopamine (DA) pathways, while other pathways were left essentially intact. This 6-OH-DA denervation technique has proven to be an interesting tool for functional studies on central CA neurons. This chapter attempts to summarize a series of such studies based mainly on the nigrostriatal DA system where events at the level of the synapse are revealed by the combined usage of 6-OH-DA induced degenerations, drugs interfering with DA transmission, and quantifications of induced behaviors.

Anatomy and chemistry of selective lesions

The possibility of producing specific lesions of various monoamine pathways in the brain is naturally limited by their anatomy (Ungerstedt, 1971b). However, the pathways run separated at various points where lesions may be localized in order to increase the specificity of the degenerations (Figure 1). The *NE cell bodies* are localized in cell groups in the medulla and pons. Their ascending axons separate into two distinct bundles in the caudal mesencephalon. The ventral bundle innervates mainly the hypothalamus and the preoptic area, while the dorsal

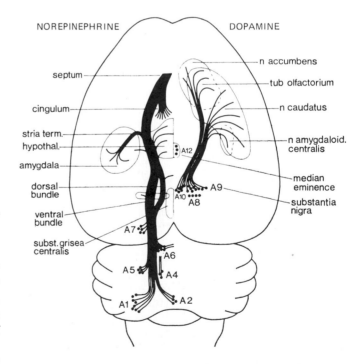

FIGURE 1   Horizontal projection of the ascending dopamine and norepinephrine axons in the rat brain. The cell groups are labeled A1, A2, etc. (From Ungerstedt, 1971b.)

URBAN UNGERSTEDT   Department of Histology, Karolinska Institutet, Stockholm, Sweden

bundle, originating in the locus coeruleus, innervates the cortex and the hippocampus. The *5-hydroxytryptamine (5-HT) cell bodies* are localized in the raphe cell groups medial to the NE cells. The 5-HT axons ascend partly intermingled with the NE axons in the medial forebrain bundle. The *DA cell groups* are situated within the mesencephalon and diencephalon in front of the NE and 5-HT cells. The ascending axons run both lateral and dorsal to the NE axons, turn laterally into the internal capsule and reach the corpus striatum after passing through the globus pallidus.

The *nigrostriatal DA system* may be lesioned at several points along its extent. The lesions differ considerably in the extent they involve the other monoamine pathways. A lesion of the DA cells in the zona compacta of the substantia nigra tends to cause a partial NE denervation of the hypothalamus while a lesion of the DA axons in the lateral hypothalamus affects the NE pathway to the cortex leaving the hypothalamic NE nerve terminals mainly unaffected. Any study of the effect of depleting the caudate nucleus of its DA nerve terminals thus requires control animals in which lesions are made into ascending NE and 5-HT bundles at a level caudal to the DA cell bodies in order to account for functional effects due to lesions of the other monoamine pathways (Ungerstedt, 1973).

The basis for the action of 6-OH-DA seems to be its ability to enter the CA neurons by way of their specific membrane uptake mechanism (Jonsson and Sachs, 1970). Intracerebrally injected 6-OH-DA is thus actively concentrated in DA and NE neurons. Its destructive potency seems to be related to the formation of covalent binding of oxidation products of 6-OH-DA to biological macromolecules (Saner and Thoenen, 1971) and/or by the destructive action of hydrogen peroxide generated by 6-OH-DA (Heikkila and Cohen, 1971). The extent of the destruction is thus probably dependent upon the concentration of 6-OH-DA at a particular intra- or extraneuronal site. Any neurons that actively concentrate 6-OH-DA, like the NE and DA neurons, are thus particularly vulnerable. However, too high an extracellular concentration of 6-OH-DA will destroy any cell, and thus high concentrations will lack the desired specificity of action.

The morphological effect of 6-OH-DA, injected directly into the brain, has been worked out in detail (Ungerstedt, 1968, 1971a; Hökfelt and Ungerstedt, 1973). The injection produces a specific as well as an unspecific lesion. The unspecific lesion is centered at the tip of the injection cannula where the tissue has been exposed to the highest amount of 6-OH-DA. The specific lesion, where only CA neurons are damaged, is localized further away from the site of the injection, where the

concentration of 6-OH-DA is low, and it needs to be actively concentrated in order to be toxic. Thus 8 μg of 6-OH-DA injected into a rat brain in 4 μl of saline over 4 min will produce an unspecific lesion approximately 0.2 to 0.5 mm in diameter, while the specific lesion is about 2.0 to 2.5 mm in diameter. The extent of the specific lesion varies due to the kind of monoamine neurons, or part of monoamine neuron, that is reached by 6-OH-DA (Ungerstedt, 1971a). The nigrostriatal DA neurons are sensitive to 6-OH-DA wherever it reaches their cell membrane. The meso-limbic DA system is less sensitive. The NE cell bodies show great resistance to 6-OH-DA, while the NE axons and terminals seem as sensitive as the corresponding parts of the DA neurons. However, there are differences also within the population of NE neurons. The NE axons in the dorsal bundle, innervating the cortex, seem more sensitive to 6-OH-DA than the NE axons in the ventral bundle innervating the hypothalamus. Finally, the 5-HT neurons as well as the tubero-infundibular DA neurons show a very low sensitivity to 6-OH-DA. This insensitivity persists even during in vitro incubation in 6-OH-DA.

The difference in sensitivity to 6-OH-DA between the neuronal populations may be explained by differences in the membrane uptake mechanism or differences in surface to volume ratios between axons of various diameters. However, as long as the intraneuronal mechanism of 6-OH-DA induced destruction is essentially unknown, it is unclear to what extent the selectivity in its action may be attributed to the efficiency of the neuronal uptake.

It should be pointed out that there are in all probability species differences in the sensitivity to 6-OH-DA. The results above are based upon findings in the rat; however, the cat seems to differ considerably. After intraventricular injection of 6-OH-DA in the cat there is a considerable depletion of 5-HT (Laguzzi et al., 1971) while this is not the case in the rat (Uretsky and Iversen, 1970).

Functional parameters related to a lesion of the nigrostriatal DA system

A lesion technique is a powerful tool if it is possible to find relevant parameters to study in relation to the function of the lesioned pathway. We have followed the degeneration of the nigrostriatal DA system after electro-coagulations and 6-OH-DA lesions and related this to changes in certain behavioral parameters known to reflect changes in DA transmission. Unilateral damage to the striatum or the nigrostriatal DA system is known to induce asymmetries in movements in several species including rat (Poirier and Sourkes, 1965; Stern, 1966; Andén et al., 1966). Pharmacologically induced release of DA in an animal with striatal destruction unilaterally increases the asymmetries to the point of rotation, i.e., the animal moves around in tight circles, toward the lesioned side (Andén et al., 1966). The development of the 6-OH-DA lesion technique (Ungerstedt, 1968) introduced a new way of producing a selective and reproducible lesion of the nigrostriatal DA system and motivated the development of a special "rotometer" designed to record the rotational behavior over long periods of time (Ungerstedt and Arbuthnott, 1970). The rotometer consists of a plexiglass bowl shaped as a hemisphere. The animal is placed in the bowl and connected with a thin wire to a microswitch arrangement in the geometrical center of the sphere, which reacts to each full turn of the animal and is recorded on cumulative recorders or electromechanical counters. The animal moves around freely in the bowl unrestricted by the wire which constitutes the radius of the sphere (Figure 2). In this model amphetamine, which is known to release DA (Glowinski and Axelrod, 1965; Carlsson et al., 1966) causes a dose-dependent rotational behavior towards the lesioned side, which probably reflects quantitatively the DA release (Figure 3). The behavior is an expression of the quantitative differences between the DA receptor stimulation on one side as compared to that of the other side. The direction of the rotation will always be toward the least "DA-affected" side.

This "rotational model" constitutes a way to study synaptic function as it is reflected in a simple, quantifiable behavior that has been empirically linked to DA transmission. The actual mechanism behind the rotational behavior is essentially unknown (see Ungerstedt, this volume). However, it is linked to DA transmission by a series of investigations each contributing parts of the evidence, e.g., injections of DA into the striatum, lesions of the striatum, pharmacological studies, and histochemical studies. Taken together the evidence makes the rotational behavior one of the few quantifiable behaviors that can be linked to the transmission in an anatomically and chemically defined neuron system.

The process of degeneration

A lesion of a monoamine pathway results in a series of degeneration phenomena that are interesting reflections of the normal function of the pathway. An electro-coagulation of the nigrostriatal DA axons causes an immediate build-up in the levels of DA in the striatal DA nerve terminals as revealed biochemically and histochemically (Faull and Laverty, 1969; Nybäck, 1971; Andén et al., 1971, 1972) (Figure 4). The synthesis of transmitter does not stop in spite of the lack of nerve impulses caused by the lesion, i.e., increased storage does not cause a complete feedback inhibition on synthesis. However, there is a decreased conversion of ^{3}H-tyrosine to DA after the lesion, indicating a decreased rate of synthesis.

The levels of DA in the nerve terminals are significantly increased after 15 min and doubled within 1 hr. Twenty-four hours after the lesion they are back to normal levels and, during the period between 24 and 48 hr, all DA leaves the striatum (Figure 4). Histochemical studies directly parallel the increase and decrease of DA levels as measured biochemically (Andén et al., 1972). In the electron microscope (Hökfelt and Ungerstedt, 1973), the disappearance of dense core

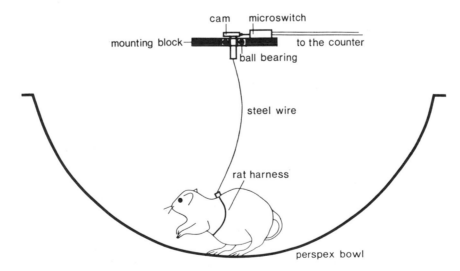

FIGURE 2 Schematic drawing of the rotometer. The rat moves around in a plexiglass bowl shaped as a hemisphere connected to the recording device by a thin wire extending from a harness fitted around its chest. (From Ungerstedt, 1971c.)

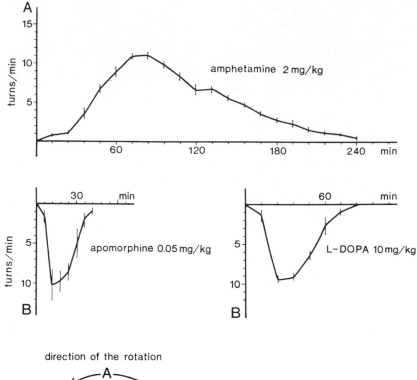

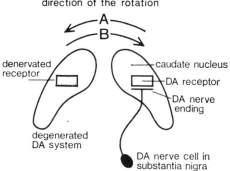

direction of the rotation

denervated receptor

degenerated DA system

caudate nucleus

DA receptor

DA nerve ending

DA nerve cell in substantia nigra

diagram of the experimental situation – horizontal projection of the nigro-striatal DA neuron system

FIGURE 3 Rotational behavior after pre- and postsynaptically acting drugs. Amphetamine, which releases dopamine, causes the animal to rotate in direction "A," i.e., toward the lesioned side; while apomorphine or DOPA, which acts primarily on the denervated side, causes the animal to rotate toward the opposite side, i.e., direction "B."

vesicles (i.e., the intraneuronal DA storing granules) takes place between 24 and 48 hr after the lesion, which, thus, corresponds to the biochemical and histochemical findings (Figure 4).

The behavioral studies indicate that the degeneration process is rather complex. A unilaterally DA lesioned animal shows a weak rotation starting around 24 hr after the lesion. However, if the monoamine oxidase (MAO) is inhibited, the rotational behavior becomes pronounced between 24 and 34 hr after the lesion (Figure 4). This rotation is in all probability due to a "degeneration release" of DA, which is the central nervous system counterpart to the "degeneration con-

traction" of the nictitating membrane after removal of the superior cervical ganglion (Langer, 1966).

The presynaptic events occurring during degeneration may be revealed in greater detail if the degenerating nerves are exposed to a presynaptically acting drug like amphetamine. The rotational behavior provides a relative measure of the amount of DA being released from the degenerating DA nerve terminals as compared to the intact ones. If more DA is being released on one side, the animal rotates toward the other side. The pattern of rotation (Figure 5) changes greatly during degeneration (Ungerstedt, 1971c): *30 min and 3 hr* after the lesion, less DA is released from the lesioned side as compared to the

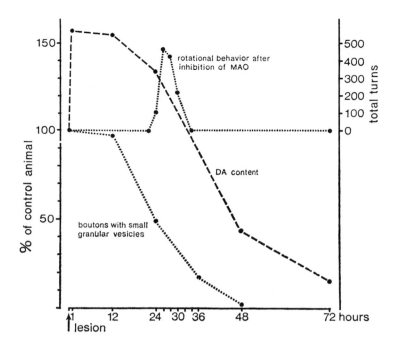

FIGURE 4 Degeneration of the nigrostriatal dopamine system "monitored" by three different parameters: Striatal dopamine measured biochemically, number of dopamine boutons detectable in the electron microscope, and the rotational behavior elicited by dopamine release from the degenerating nerve terminals. The animal is treated with a monoamine oxidase inhibitor that allows the dopamine to reach the receptor. (Drawn from data presented in Andén et al., 1972; Hökfelt and Ungerstedt, 1973; Ungerstedt, 1971c.)

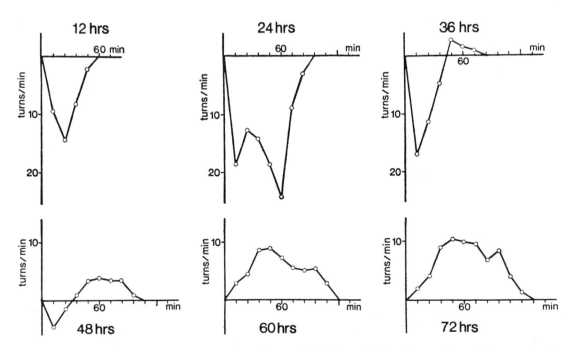

FIGURE 5 Rotational behavior induced by amphetamine (2 mg/kg) at various time points after a 6-OH-DA lesion of the nigrostriatal dopamine system. The record is obtained from one representative animal receiving amphetamine repeatedly with 12 hr intervals. Positive y-values indicate a dominance of the intact dopamine system, while negative y-values indicate a dominance of the lesioned dopamine system. (From Ungerstedt, 1971c.)

normal side in spite of the fact that the lesioned side contains more DA (see above). This indicates that the lack of nerve impulses on the lesioned side makes less DA available for amphetamine-induced release. In fact, Farnebo (1971) has shown by electrical stimulation of brain slices, in vitro, that amphetamine induces an impulse dependent transmitter release. *Six hours* after the lesion the release from the lesioned side dominates for a short period of time after which the nonlesioned side takes over. However, *24 hours* after the lesion, there is a total dominance of the lesioned side during the whole duration of the amphetamine response. *Thirty-six and forty-eight hours* after the lesion there is again only a very short period when the lesioned side dominates and at *60 hr* the dominance of the non-lesioned side is permanent.

This complex series of events may, in fact, reflect the presence or absence of nerve impulses, the endogenous amount of transmitter, the degeneration induced breakdown of the storage mechanism, and the intraneuronal localization of the transmitter, especially as amphetamine is thought to release extragranularly localized amines (Carlsson et al., 1966) (see Figure 5). The amphetamine induced release reaches a maximum at 24 hr, when the DA levels are back to normal, and it is not possible to say if this represents a release from degenerating granules, abnormally sensitive to amphetamine, or a release from an abnormally large extra granular pool of DA.

Another type of degeneration phenomenon becomes apparent when axons or terminals are lesioned. The axonal flow continues to transport storage granules down the axon in spite of the lesion, and this results in a pile up of transmitter proximal to the lesion (Dahlström and Fuxe, 1964). This is revealed as a strongly increased fluorescence in the axons on the proximal side of the lesion. The pile up may extend all the way back to the cell body, which shows an increased fluorescence intensity. If the lesioned axon has collaterals leaving the axon proximal to the lesion, there may also occur an increase in their content of transmitter. This has been thought of as a "rechanneling" of the axonal flow of granules where the remaining collaterals will receive more granules than before the "closure" of the axon due to the lesion (Figure 6). However, preliminary electro-physiological studies on Purkinje cells, after lesions of the NE axons to the cortex, may indicate another mechanism for collateral pile up (Hoffer and Ungerstedt, unpublished observations): The locus coeruleus sends axons to the cortex as well as to the cerebellum (Olson and Fuxe, 1971; Ungerstedt, 1971b). When the axons to the cortex are lesioned the cerebellar NE nerve terminals increase their content of NE (Olson and Fuxe, 1971; Ungerstedt, 1971b). However, in spite of the increased levels of NE in the nerve terminals around the Purkinje cells, these cells increase their spontaneous activity, i.e., they act as if they were denervated (see Hoffer et al., 1971). The increased amount of NE thus seems to occur together with a decrease in NE transmission. This is an obvious parallel to the increase in striatal DA levels occurring after a lesion of the DA axons. In this case we know there is no transmission due to the disconnection of the cell bodies from their terminals. The increased levels of NE in the cerebellar nerve terminals may, thus, be due to a strongly decreased transmission which in turn indicates that the lesion of the axon collateral to the cerebral cortex in fact caused a *functional*, although not anatomical, lesion of the entire collateral network (see Figure 6). The duration of this "functional" lesion is so far unknown; however, it may have obvious implications for the interpretation of lesions in the central nervous system.

Supersensitivity

When a unilaterally DA denervated animal receives a DA receptor stimulating drug (e.g., apomorphine) it rotates in the opposite direction to that seen after amphetamine (Figure 3). The drug obviously is more effective on the denervated than on the innervated side. The effective dose is very low as compared to what induces an increased motility in a normal animal, the difference between equally effective doses being 10- to 100-fold. The increased sensitivity to apomorphine is evident 24 hr after the lesion and increases rapidly during the first week. During the next 3 to 4 weeks it slowly reaches its maximum. DOPA causes the same response as apomorphine, i.e., being most effective on the lesioned side. However, DOPA needs to be converted to DA as the rotation may be inhibited by DOPA decarboxylase inhibition (Ungerstedt, 1971d).

The increased response to DA receptor stimulation on the denervated side is in all probability due to a development of supersensitivity in response to the degeneration of the presynaptic neurons. The exact nature of this supersensitivity is difficult to determine in the CNS, and there are obvious differences between the supersensitivity to apomorphine in the CNS and the supersensitivity to NE in the peripheral nervous system. There is no evidence that apomorphine is actively concentrated in the presynaptic neurons in the same way as NE is in the periphery. Apomorphine probably acts directly on the DA receptor (Ernst, 1967; Andén et al., 1967) and will, thus, only reflect the so-called decentralization type of supersensitivity (Langer et al., 1967), i.e., the change in the sensitivity of the postsynaptic neuron. In the peripheral nervous system, i.e., in the neuromuscular

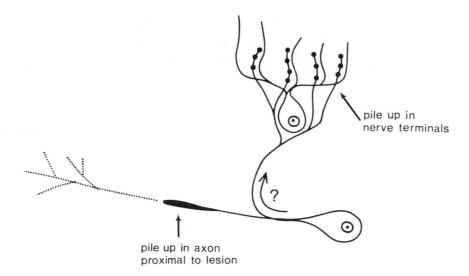

pile up in
nerve terminals

?

pile up in axon
proximal to lesion

FIGURE 6   Schematic drawing of transmitter pile up following axonal damage as seen in the fluorescence microscope. The question mark indicates the traditional way of looking upon the phenomenon of collateral pile up, i.e., a rechanneling of the axonal flow into collaterals to the lesioned axon. The text describes an alternative hypothesis based upon studies of cerebellar pile up after lesions of the collateral axons innervating the cerebral cortex.

synapse, the change is thought to take place in the actual receptor site. In the CNS, however, it is obvious that a compensatory change in the activity of other neurons may also constitute a mechanism behind the increased sensitivity to DA receptor stimulation.

In order to examine the possibility of a change in the actual DA receptor, we investigated the difference in the ability of DA receptor blocking drugs to block the innervated and the denervated receptor. If the difference in sensitivity were due to a change in the receptor molecule, one would expect that more neuroleptic drug would be needed to block the supersensitive receptor. If the difference was due to compensatory adjustments of other neurons, there should probably be no such difference in the action of a receptor blocking drug. The rotational behavior was recorded on a low dose of apomorphine (0.05 mg/kg) as well as a low dose of amphetamine (0.5 mg/kg). The drugs were then given simultaneously. In this type of experiment there is a "competition" between the innervated and the denervated side, i.e., the DA being released by amphetamine on the intact side induces a rotation in one direction, while the apomorphine, stimulating primarily the denervated receptors, tends to force the animal in the opposite direction. The resulting direction of rotation after simultaneous administration of the drugs indicates on which side the most intense DA receptor stimulation occurs. If a DA receptor blocking drug is more effective on, e.g., the innervated side of the brain, it will decrease the influence of amphetamine compared to apomorphine. When animals are pretreated with spiroperidol (Andén

et al., 1970) and then injected with apomorphine and amphetamine simultaneously, the apomorphine induced direction of rotation is in fact potentiated. This indicates that more spiroperidol is needed to block the denervated DA receptors, which in turn may indicate a change in the denervated DA receptor as such (Avemo and Ungerstedt, to be published).

The phenomenon of central postsynaptic supersensitivity may be used as an important tool in studies of brain function but also presents us with some problems of interpretation. In experiments employing lesions of known pathways and drugs stimulating the denervated receptors, it is possible to create syndromes of hypofunction as well as hyperfunction: Hypofunction as a result of the denervation and hyperfunction as a result of the drug stimulating the denervated supersensitive receptors. However, the syndrome of hypofunction may be very difficult to define. In the case of the DA neurons, it is obvious that very low levels of DA are able to maintain the function of the pathway, probably with the help of a developing supersensitivity. The failure to abolish function with a lesion should therefore be interpreted with great care.

Supersensitivity is not only associated with denervation. A decrease of normal transmission may also cause supersensitivity, e.g., after drugs like reserpine and α-methyl-paratyrosine, inhibiting storage and synthesis, respectively. The degree of reserpine induced supersensitivity may be quantified in a modified rotation model: The striatum is first unilaterally destroyed by electrocoagulation. The rotational behavior is now an expression of events taking place in the remaining innervated striatum. A low dose of apomorphine (e.g., 0.25 mg/kg) does not cause any rotational behavior in such an animal. However, 24 hr after reserpine, the same dose of apomorphine induces rotation, i.e., the striatum is now supersensitive to apomorphine (Ungerstedt, 1971d) (Figure 7). This supersensitivity increases over the first 3 days. However, on the fourth day after reserpine, there is a sudden decrease in the sensitivity. This decrease may possibly be associated with the reappearance of the first normal storage granules followed by an increased neurotransmission. The fact that a recovery from the behavioral depression occurs already on the second day, when there is still strong supersensitivity and very low levels of DA, indicates that the supersensitivity in fact plays an important role in compensating for the decrease in transmission.

The fact that reserpine induces postsynaptic supersensitivity in the central nervous system emphasizes the difference between the central and peripheral synapse. The decrease in synaptic transmission, caused by reserpine, is too short lasting to elicit a compensatory change

in receptor sensitivity in the peripheral neuromuscular synapse, i.e., a decentralization type supersensitivity (Langer et al., 1967). However, the central DA synapse shows a rapid increase in postsynaptic response indicating a change in receptor sensitivity.

Other neuronal systems interacting with the DA system

Although the rotational behavior is related to DA synaptic transmission, it is obvious that the DA neurons constitute only one link in the complex neuronal network that influences the rotational behavior. Other neurons may interfere with this behavior at several levels. There is good evidence for such an interaction between acetylcholine (ACh) neurons and DA neurons (see Scheel-Krüger, 1970). This interaction is clearly apparent in the "6-OH-DA rotation model": In a unilaterally DA denervated animal, anticholinergic drugs, like atropine and scopolamine, induce a rotation in the same direction as amphetamine. The rotation after anticholinergic drugs as well as after amphetamine is inhibited by pretreatment

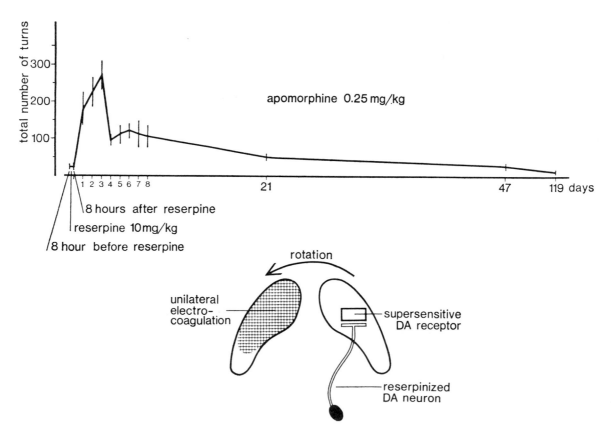

FIGURE 7 Rotational behavior measured as total number of turns in an animal with a unilateral electrocoagulation of the striatum. A systemic injection of reserpine increases the sensitivity to a low dose of apomorphine. The sensitivity increases over the first 3 days after reserpine but drops suddenly on the fourth day. The change in the rotational behavior is a measure of the change in receptor sensitivity following reserpine. (Modified after Ungerstedt, 1971d.)

with tyrosine-hydroxylase inhibitors. The behavior is thus dependent on an intact DA transmission and indicates that the anticholinergic drugs remove a cholinergic inhibition. This inhibition is probably taking place in the striatum as the injection of anticholinergics into the caudate nucleus potentiates the action of simultaneously injected DA (Fog et al., 1967). Whether the inhibition is pre- or postsynaptic is not clear. However, there is no electronmicroscopic evidence of boutons on the DA containing nerve terminals. The "DA rotation" may, thus, be caused by removal of a cholinergic influence on the postsynaptic cell. The action of DA on the striatum is in all probability inhibitory (Bloom et al., 1965) and the anticholinergic drugs may thus act by removing an excitatory cholinergic influence on the postsynaptic cell (Bloom et al., 1965; McLennan and York, 1966).

Apart from the evidence for a DA-ACh interaction in the striatum, there is evidence for a feedback regulation of the nigrostriatal DA pathway: Pharmacologically induced excitation of the DA receptor (e.g., by apomorphine) causes a decrease in DA turnover (Andén et al., 1967), while an inhibition (e.g., by haloperidol) causes an increased turnover (Carlsson and Lindqvist, 1963; Andén et al., 1970). This compensatory feedback may be carried out by a neuronal feedback loop, e.g., by the striatonigral pathway that is known from anatomical studies (Szabo, 1962). However, the feedback regulation is found also in vitro in field-stimulated slices of striatal tissue (Farnebo and Hamberger, 1971) where the striatonigral pathway obviously is damaged. This indicates the existence of a local effect at the central DA synapse, possibly via a negative feedback mechanism. Regardless of the nature of this feedback, it is dependent on the existence of a nerve-impulse flow in the DA neurons. The compensatory release of DA after receptor inhibition occurs only when slices are electrically stimulated and a lesion of the nigrostriatal DA pathway blocks the compensatory release in vivo (Andén et al., 1971; Nybäck, 1971).

Conclusions

The above results seem particularly interesting in relation to the "functional state" of the synapse. There is obviously no direct relationship between synaptic function and the endogenous levels of transmitter. Strongly *increased* presynaptic levels of transmitter may occur concomitantly with a "functional denervation" of the postsynaptic neurons, and strongly *decreased* presynaptic levels may be compensated for by the development of postsynaptic supersensitivity. The possible existence of neuronal feedback loops as well as intrasynaptic feedback mechanisms seem to serve a similar function as the denervation induced supersensitivity, i.e., to give the synaptic transmission a certain degree of "inertia." This brings forth a picture of a rather sluggish transmission, hardly relaying quick messages from one place to the other. However, in the case of the nigrostriatal DA system, this picture seems quite in agreement with the anatomy showing a massive, homogeneous innervation of a large nucleus where changes in DA transmission may conceivably set the over all excitability of the nucleus but not handle discrete information.

ACKNOWLEDGMENT This investigation was supported by a grant from the Swedish Medical Research Council (04X-357Y).

REFERENCES

ANAGNOSTE, B., T. BACKSTRÖM, and M. GOLDSTEIN, 1969. The effect of 6-hydroxydopamine (6-OH-DA) on catecholamine biosynthesis in the brain of rats. *Pharmacologist* 11:274.

ANDÉN, N.-E., P. BEDARD, K. FUXE, and U. UNGERSTEDT, 1972. Early and selective increase in brain dopamine levels after axotomy. *Experientia* 28:300–301.

ANDÉN, N.-E., S. G. BUTCHER, H. CORRODI, K. FUXE, and U. UNGERSTEDT, 1970. Receptor activity and turnover of dopamine and noradrenaline after neuroleptics. *Europ. J. Pharmacol.* 11:303–314.

ANDÉN, N.-E., H. CORRODI, K. FUXE, and T. HÖKFELT, 1967. Increased impulse flow in bulbo-spinal noradrenaline neurons by catecholamine receptor blocking agents. *Europ. J. Pharmacol.* 2:59–64.

ANDÉN, N.-E., H. CORRODI, K. FUXE, and U. UNGERSTEDT, 1971. Importance of nervous impulse flow for the neuroleptic induced increase in amine turnover in central dopamine neurons. *Europ. J. Pharmacol.* 15:193–199.

ANDÉN, N.-E., A. DAHLSTRÖM, K. FUXE, and K. LARSSON, 1966. Functional role of the nigro-neostriatal dopamine neurons. *Acta Pharmacol. Toxicol. (Kbh.)* 24:263–274.

ANDÉN, N.-E., A. RUBENSON, K. FUXE, and T. HÖKFELT, 1967. Evidence for dopamine receptor stimulation by apomorphine. *J. Pharm. Pharmacol.* 19:627–629.

BLOOM, F. E., S. ALGERI, A. GROPPETTI, A. REVUELTA, and E. COSTA, 1969. Lesions of central norepinephrine terminals with 6-OH-dopamine: Biochemistry and fine structure. *Science* 166:1284–1286.

BLOOM, F. E., E. COSTA, and G. C. SALMOIRAGHI, 1965. Anesthesia and the responsiveness of individual neurons of the caudate nucleus of the cat to acetylcholine, norepinephrine and dopamine administered by microelectrophoresis. *J. Pharmacol. Exp. Ther.* 150:244–252.

CARLSSON, A., K. FUXE, B. HAMBERGER, and M. LINDQVIST, 1966. Biochemical and histochemical studies on the effects of imipramine-like drugs and (+)-amphetamine on central and peripheral catecholamine neurons. *Acta Physiol. Scand.* 67:481–497.

DAHLSTRÖM, A., and K. FUXE, 1964. A method for the demonstration of monoamine containing nerve fibers in the central nervous system. *Acta Physiol. Scand.* 60:293–295.

ERNST, A. M., 1967. Mode of action of apomorphine and dexamphetamine on gnawing compulsion in rats. *Psychopharmacologia* 10:316–323.

FARNEBO, L. O., 1971. Effect of d-amphetamine on spontaneous and stimulation-induced release of catecholamines. *Acta Physiol. Scand.* Suppl. 371:45–52.

FARNEBO, L. O., and B. HAMBERGER, 1971. Drug-induced changes in the release of ³H-monoamines from field stimulated rat brain slices. *Acta. Physiol. Scand.* Suppl. 371:35–44.

FAULL, R. L. M., and R. LAVERTY, 1969. Changes in dopamine levels in the corpus striatum following lesions in the substantia nigra. *Exp. Neurol.* 23:332–340.

FOG, R. L., A. RANDRUP, and H. PAKKENBERG, 1967. Aminergic mechanisms in corpus striatum and amphetamine-induced stereotyped behavior. *Psychopharmacologia* 11:179–183.

GLOWINSKI, J., and J. AXELROD, 1965. Effect of drugs on the uptake, release, and metabolism of H³-norepinephrine in the rat brain. *J. Pharmacol. (Lond.)* 149:43–49.

HEIKKILA, R., and G. COHEN, 1971. Inhibition of biogenic amine uptake by hydrogen peroxide: A mechanism for toxic effects of 6-hydroxydopamine. *Science* 172:1257–1258.

HOFFER, B. J., G. R. SIGGINS, D. J. WOODWARD, and F. BLOOM, 1971. Spontaneous discharge of Purkinje neurons after destruction of catecholamine-containing afferents by 6-hydroxydopamine. *Brain Res.* 30:425–430.

HÖKFELT, T., and U. UNGERSTEDT, 1973. Effects of 6-hydroxydopamine on central monoamine neurons with special reference to the nigro-striatal dopamine system: An electron and fluorescence microscopical study. *Brain Res.* (in press).

JONSSON, G., and Ch. SACHS, 1970. Effects of 6-hydroxydopamine on the uptake and storage of noradrenaline in sympathetic adrenergic neurons. *Europ. J. Pharmacol.* 9:141–155.

LAGUZZI, R., F. PETITJEAN, J. E. PUJOL, and M. JOUVET, 1971. Effets de l'injection intraventriculaire de 6-hydroxydopamine sur les états de sommeil et les monoamines cérébrales du chat. *C. R. Soc. Biol. (Paris)* 165:1649–1653.

LANGER, S. Z., 1966. The degeneration contraction of the nictitating membrane in the unanesthetized cat. *J. Pharmacol. Exp. Ther.* 151:66–72.

LANGER, S. Z., P. R. DRASKOCZY, and U. TRENDELENBURG, 1967. Time course of the development of supersensitivity to various amines in the nictitating membrane of the pithed cat after denervation or decentralization. *J. Pharmacol. Exp. Ther.* 157:255–273.

MCLENNAN, H., and D. H. YORK, 1966. Cholinergic Mechanisms in the caudate nucleus. *J. Physiol. (London)* 187:163–175.

NYBÄCK, H., 1971. Effect of brain lesions and chlorpromazine on accumulation and disappearance of catecholamines formed in vivo from ¹⁴C-tyrosine. *Acta Physiol. Scand.* 84:54–64.

OLSON, L., and K. FUXE, 1971. On the projection from the locus coeruleus noradrenaline neurons: The cerebellar innervation. *Brain Res.* 28:165–171.

POIRIER, L. J., and T. L. SOURKES, 1965. Influence of the substantia nigra on the catecholamine content of the striatum. *Brain* 88:181–192.

SANER, A., and H. THOENEN, 1971. Model experiments on the molecular mechanism of action of 6-hydroxydopamine. *Molec. Pharmacol.* 7:147–154.

SCHEEL-KRÜGER, J., 1970. Central effects of anticholinergic drugs measured by the apomorphine gnawing test in mice. *Acta Pharmacol. Toxicol. (Kbh.)* 28:1–16.

STERN, G., 1966. The effects of lesions in the substantia nigra. *Brain* 89:449–478.

SZABO, J., 1962. Topical distribution of the striatal efferents in the monkey. *Exp. Neurol.* 5:21–36.

TRANZER, J. P., and H. THOENEN, 1968. An electron microscopic study of selective, acute degeneration of sympathetic nerve terminals after administration of 6-hydroxydopamine. *Experientia* 24:155–156.

UNGERSTEDT, U., 1968. 6-Hydroxydopamine induced degeneration of central monoamine neurons. *Europ. J. Pharmacol.* 5:107–110.

UNGERSTEDT, U., 1971a. Histochemical studies on the effects of intracerebral and intraventricular injections of 6-hydroxydopamine on monoamine neurons in the rat brain. In *6-Hydroxydopamine and Catecholamine Neurons*, T. Malmfors and H. Thoenen, eds. Amsterdam: North-Holland, pp. 101–127.

UNGERSTEDT, U., 1971b. Stereotaxic mapping of the monoamine pathway in the rat brain. *Acta Physiol. Scand.* 82 (Suppl. 367):1–48.

UNGERSTEDT, U., 1971c. Striatal dopamine release after amphetamine or nerve degeneration revealed by rotational behavior. *Acta Physiol. Scand.* 82 (Suppl. 367):49–68.

UNGERSTEDT, U., 1971d. Postsynaptic supersensitivity after 6-hydroxydopamine induced degeneration of the nigro-striatal dopamine system in the rat brain. *Acta Physiol. Scand.* 82 (Suppl. 367):69–93.

UNGERSTEDT, U., 1973. Selective lesions of central monoamine pathways: Application in functional studies. In *Neurosciences Research*, S. Ehrenpreis and I. Kopin, eds. New York: Academic Press (in press).

UNGERSTEDT, U., and G. ARBUTHNOTT, 1970. Quantitative recording of rotational behavior in rats after 6-hydroxydopamine lesions of the nigro-striatal dopamine system. *Brain Res.* 24:485–493.

URETSKY, N. J., and L. L. IVERSEN, 1969. Effects of 6-hydroxydopamine on noradrenaline-containing neurons in the rat brain. *Nature (Lond.)* 221:557–559.

URETSKY, N. J., and L. L. IVERSEN, 1970. Effects of 6-hydroxydopamine on catecholamine containing neurons in the rat brain. *J. Neurochem.* 17:269–278.

87 Dynamics of Synaptic Modulation:

Perspectives for the Future

F. E. BLOOM

ABSTRACT   Basic synaptic operations are reviewed to determine the points at which modulatory factors could regulate function. Consideration of discrete aspects of presynaptic and post-synaptic physiology and fine structure suggests that the synapse represents a dynamic constellation of acting and reacting components that are variably responsive to external modulatory influences. An essential feature to such analysis is the determination of the chemical substances acting as transmitters, their cytological and functional properties, and their roles in sequences of interacting neurons. Short-term modulations may arise from regulation of transmitter release mechanisms and sensitization of receptor mechanisms. Longer term modulatory influences may arise from formation of new connections and from the chemical consequences that follow the actions of transmitters which can generate cyclic nucleotides postsynaptically.

THE TEAM MEMBERS of the part topic *Dynamics of Synaptic Modulation* have presented us with a massive display of new data and concepts regarding the maintenance of optimal synaptic transmission and the possibilities for functional synaptic adjustment. It is clear that from each of the several technical approaches to the analysis of synapses have flowed operational principles that can be tested experimentally to reveal the extent to which these over-riding modulations may actually operate. In fact, so many possibilities for modulation of synaptic function have been described that we must begin to wonder how basal synaptic function is maintained if all of these principles were to be operative simultaneously. In this final chapter of the part, I have attempted to select from the presentations of my team members certain facets for re-emphasis by placing them into the perspective of strategic questions for future investigation. I propose that synaptic transmission can now be regarded as a dynamic constellation of acting and reacting components with variable degrees of sensitivity to external humoral regulation. In my view, the key to examination of the mechanisms involved in these dynamic synapses requires the cytochemical identification of the molecules involved in the inter-neuronal transmission process; identification of the transmitter also has profound implications for the cellular analysis of the biology of behavior. My views are separable into the presynaptic and postsynaptic components of the synapse.

Presynaptic operations

TRANSMITTER SYNTHESIS AND STORAGE   The presynaptic terminal represents the distal branch office of the transmitting neuron in a synaptic relationship. This cellular satellite shop is able to synthesize, store, and, in most cases, to conserve the transmitter it will secrete when called into action by the electrical depolarization of a propagated action potential. L. Iversen (this volume) has described the neurochemical control systems involved in the synthesis of these transmitter molecules. In most systems synthesis rates of the transmitter appear to be sensitive to the rate of transmitter utilization, perhaps through allosteric recognition sites on the rate limiting enzyme of the system operating as a negative feedback. The synthetic rate need not be the sole determinant of transmitter available for release, because a large component of the transmitter within a terminal is thought to be stored within synaptic vesicles from which it can also be released.

In the evolution of neurochemical experiments into the nature of transmitter metabolism, attempts were made to determine the involvement of a particular class of transmitter secreting cells (e.g. norepinephrine cells) in the action of certain drugs by the effects this drug had on the "turnover" of the transmitter substance. Turnover implied a kinetic analysis of the rate at which the transmitter was being replaced from synthesis, a concept whose practical exploration has been the center of intense debate for the past several years (see Costa, 1973; Glowinski, 1973).

In the light of the knowledge of the anatomical localization of transmitter-specific cell bodies and their synaptic projections, several revisions in the application of turnover theory may be required. Thus, if the cells of origin of the norepinephrine-containing pathways to the

F. E. BLOOM   Laboratory of Neuropharmacology, Division of Special Mental Health Research, National Institute of Mental Health, St. Elizabeth's Hospital, Washington, D.C.

telencephalic cortices are the same as those to the cerebellar cortex, namely, the locus coeruleus (Olson and Fuxe, 1971), then the turnover of norepinephrine in each of these areas must be identical, because it is difficult to imagine that the cell body is able to select the portions of its axonal projection that will carry the action potential. Additionally, it may be pertinent to inquire whether differences in turnover rate arise from differences in the morphology of different portions of the cellular structure of discrete pathways. Biochemical studies indicate that the turnover of dopamine in the striatum and of norepinephrine in the cortex is extremely rapid, and the usual interpretation is that the cells projecting there must fire at rapid rates. On the other hand, when this concept is tested directly, the cells ordinarily fire slowly (Chu and Bloom, 1973; Sheu and Bloom, 1973; Aghajanian, 1972), and alternative explanations seem needed. Cytologically, these cortical and striatal nerve terminals are extremely fine, whereas the nerve terminals of the axons of the ventral norepinephrine bundle (see Ungerstedt, this volume) are relatively thick and large. Is it possible that present methods of turnover estimation are more indicative of the proportion of the transmitter in a nerve terminal that is released per impulse rather than the rate at which impulses depolarize the terminal? Thus, if it were to be determined that all norepinephrine synapses formed identical presynaptic grids (see Akert et al., 1972) and that each released the transmitter contained in an identical number of vesicles per impulse, then terminals that stored 1000 vesicles would appear to be turning over less rapidly than terminals that stored only 100 vesicles. These concepts could be tested by pursuing the electron microscopic configuration and vesicle content of the different nerve terminals from the same cell bodies in different areas of the CNS.

LOCALIZATION OF TRANSMITTERS The precise points in the nerve terminal network from which electrical activity is coupled to transmitter secretion have not yet been determined. Examination of the fine structure of the terminal adrenergic network (Hökfelt, 1972; Bloom, 1972a) and the neurochemical examination of the subcellular disposition of the transmitter and related molecules (Smith, 1972) have been areas of extensive and mutually productive research. From these studies we can extract the following principles of synaptic biology: Neurotransmitter molecules are synthesized by regulatable enzyme sequences (in which the synthesis of the enzymes is also subject to biologic modulation; see Axelrod, this volume), and then either released or stored. When the transmitter has been secreted, it can be regained by the terminals for eventual release or storage. These principles of operation derived from study of the auto-

nomic nervous system appear to be directly applicable to study of transmitter chemistry in the central nervous system.

The studies of the peripheral nervous system are capable of yielding conceptual data on the mechanisms of transmitter secretion, because the peripheral organs are innervated by easily isolatable nerve trunks of a more-or-less homogeneous functional character, i.e., either adrenergic or cholinergic. The same degree of macrochemical investigation does not suffice for the study of these mechanisms in brain because of the extremely complicated structure of the nerve circuits. Cytochemical methods directed at the fine structural or even the light microscopic identification of transmitters in situ (Hökfelt, 1972; Bloom, 1972a) offer the advantages of relating the transmitter to particular structural forms of synapse and to particular morphological forms of synaptic storage vesicles. Extension of these methods should permit both a functional and a chemical correlation of the wide range of classifiable variations in synaptic structure (Bodian, 1972): When specific transmitter molecules can be designated for identifiable synapses of a given topology to a receptive nucleus, this substance can then be tested for its effectiveness by a method such as microiontophoresis (see below).

The cytochemical methods for transmitter identification have been mainly directed toward the transmitter stored within the vesicles, as this source of transmitter appears to be more stable during histological processing and thus more amenable to cytochemical localization. The most complete set of methods are available for the catecholamines where it is possible to identify the nerves that store these monoamines through direct staining, through staining of specific congeners of the catecholamines, through autoradiographic localization of isotopically labeled transmitter or transmitter precursor, or through the destruction produced when the nerve terminals have accumulated a toxigenic congener such as 6-hydroxy-dopamine (see Iversen and Schon, 1972; Bloom, 1972a, 1972b). Except for the direct staining approach, all of these methods of localizing catecholamines depend on the re-uptake feature of transmitter regulation.

It may appear from the foregoing that undue emphasis is being placed upon the monoamine-containing neurons with respect to various cellular and behavioral investigations. The proper perspective to this emphasis is that these experiments are possible only because of the cytochemical explorations of the transmitter in terms of nerve circuits that can be tested electrophysiologically and behaviorally. The same type of approach could be extended to other transmitter circuits when their anatomical and cellular distribution is clarified. However, when this has been done for amino acid containing neurons of the spinal cord, it appears that both glycine (see Werman, 1972) and GABA (see Otsuka, 1973) are contained within short axon interneurons, whose activity will be difficult to evaluate in behavioral terms.

Perspectives Methods of cytochemical transmitter localization need to be developed into two primary directions for future application. On the one hand, a primary goal must be the development of methods for the direct localization of transmitter molecules other than catecholamines. Localization of the transmitter by the light microscopic histochemical method of Falck and Hillarp (Falck et al., 1962), as reviewed by Moore and by Ungerstedt in this volume, provides the connection between the classical neuroanatomical methods of post-lesion degenerative staining and transmitter chemistry. Moreover, this method permitted the initial localization of the tracts and cell body areas from which the catecholamines arise to innervate the brain (see Ungerstedt, this volume). At the present time, the only other transmitter that appears close to direct demonstration is 5-hydroxytryptamine, as described by Jaim-Etcheverry in an earlier chapter in this part. The methods that he has applied to locate 5-hydroxytryptamine in the peripheral nervous system do not, however, appear to be successful in the central nervous system, and we must seek to understand whether this failure comes about because the central terminals do not retain or bind 5-hydroxytryptamine in their vesicles in a way similar to the peripheral nerve terminals or because the central terminals store a transmitter substance that passes neurochemically for 5-hydroxytryptamine but that is, in fact, some other form of indole (Björklund et al., 1971).

However, estimates of the frequency of either catecholaminergic or tryptaminergic synapses from cytochemical indices indicate that these neurochemical varieties constitute a very small proportion of the total in any brain area (see L. Iversen, this volume) and reach a peak of 15% or less in the caudate nucleus (Hökfelt, 1968), which has the highest dopamine concentration of any measured brain region. Regional levels of other presumptive transmitters, such as GABA or other amino acids, reach levels several-fold higher than the monoamines, yet the methods for localizing these transmitters has only begun to be applied to intact tissues where the nerve terminals implicated as storing GABA could be related to a particular synaptic cellular source (see Iversen and Schon, 1972).

Localization of the sites that can accumulate transmitters has led also to unexpected results, i.e., that sites able to accumulate transmitters by the same high affinity selective uptake also include neuronal perikarya and glia

(see Iversen and Schon, 1972). Possibly, the uptake process seen in the perikarya represents the proximodistal appearance of a cell membrane specialized for the active re-uptake of the transmitter that the cell will secrete at its nerve terminals. However, the possibility that the uptake reflects the chemistry of terminals afferent to the cell or yet some other unknown cellular process has not been ruled out. When the same concerns are applied to the uptake of transmitter into glia, then perhaps the transmitter accumulating capacity of glia demonstrated first for invertebrate nerve junctions (Orkand and Kravitz, 1971; Faeder and Salpeter, 1970) may have served a useful enough biological property to have been retained upward through the phylogenetic scale. If it is possible for glia to accumulate transmitter, then it seems reasonable to ask whether the glia could not also be induced to release this transmitter under appropriate physiological demands. This could permit selective sequestering of transmitter either for certain clusters of afferent synaptic connections or for certain specialized types of post-synaptic neurons.

Cytochemical methodology also seems required for further extension of the research into the role of vesicles in the release of transmitter. Those studies that have been reported correlating fine structural disposition of vesicles after varying degrees of stimulation are subject to the criticisms raised in the part topic discussion by Robbins: The extent, length, and type of stimulation used in vitro to produce the activation required to yield a morphologically obvious change in the vesicles may be more reflective of pathology than of physiology. Thus, although it now seems possible to produce a loss of synaptic vesicles after extreme forms of neuromuscular junction activation (see Robbins, this volume), it is unclear what functional significance can be attributed to the loss of vesicles and the reconstruction of vesicle populations after resting periods. Similar studies on the release of adrenergic transmitters have been done (see Hökfelt, 1968), but again the stimulus parameters were extreme and the successful experiment, which demonstrated loss of vesicle content (but not number) to imply loss of stored transmitter, required that synthesis of the transmitter first be blocked pharmacologically. If the methods for transmitter localization were to be improved and extended to the realm of trapping molecules not firmly fixed to storage organelles (see Roth, 1971), it might then be possible to investigate the distribution of the total populations of transmitter molecules within a terminal after far more discrete forms of stimulation.

Presynaptic Modulations When we consider the relative effectiveness of synaptic inputs to a cell in terms of their structural and chemical composition, several controlling factors can be derived. The relative effectiveness of synaptic inputs could depend upon the number of terminals each set of inputs contributes to the receptive cells and the topological distribution of these terminals on the receptive cells. Insofar as the generation of an action potential is concerned, inputs on the distal dendritic tree would probably have less effect than terminals synapsing upon the perikaryon. In attempting to assess the degree to which synaptic transmission can be modulated, it is clear that modifying the relative numbers of synapses and their points of surface contact would require a time base on the order of days. A more rapid form of synaptic modulation could arise from regulating the transmitter release by adjustment of the amount of transmitter that each impulse can liberate onto postsynaptic receptors. Both of these aspects of release modification were discussed in our part topic meetings, and both appear to be exciting fields for future research.

IMMEDIATE MODULATIONS Rahamimoff (this volume) discussed the mechanisms controlling the release of transmitter from the isolated neuromuscular junction, as revealed by intracellular recording from the muscle. By examining the quantal nature (see Kuno, 1971) of the depolarizing currents that flow into the muscle when the transmitter acetylcholine is released from the nerve terminal, three factors appear to be most important in determining release: the relative membrane polarization of the nerve terminal, the frequency with which the nerve terminal is depolarized, and the essential role of Ca^{++} in coupling of nerve terminal depolarization with transmitter release. By simulating with a computer the spatial density of Ca^{++} bound to the inner membrane of the nerve terminal, Rahamimoff indicated that resting release could be explained theoretically in terms of the number of release points to which 4 Ca^{++} were bound simultaneously. The inward Ca^{++} current that flows during the initial phase of terminal depolarization permits more Ca^{++} to be bound, but the enhanced Ca^{++} has greater effectiveness in facilitating release. The effectiveness of repeated depolarizations in facilitating transmitter release could be due to the retention of increased amounts of Ca^{++} when the interval between stimuli is less than the time needed for basal Ca^{++} binding to be re-established.

The cooperative nature of the ability of Ca^{++} concentration to influence transmitter release suggests that the cation may in some way stabilize a series of molecular components by which stored transmitter could be released from vesicles. Werman and his colleagues (Kosower and Werman, 1971) have suggested that this Ca interaction could occur by means of altered sulfhydryl

groupings in the vicinity of release points. Very recently, Akert and co-workers have analyzed the structural basis for transmitter release points at central synapses by the techniques of freeze-etching electron microscopy (Figure 1). This technique, while lacking to some degree in the optical resolution of thin section electron microscopy, offers the capacity to see details in the intramembranous structure of the presynaptic and postsynaptic membranes that cannot otherwise be examined (Sandri et al., 1972). From the studies with freeze-etch microscopy, Akert, Moor, Pfenninger, and Livingston have determined that the presynaptic dense projections (which can be seen in conventional thin section microscopy after appropriate cytochemical preparation, see Bloom 1972 b; Akert et al., 1972, for reviews) are surrounded by a hexagonal pattern of pores opening into the presynaptic membrane that they regard as highly probable structural correlates for transmitter release points or vesicle attachment sites and that they term "synaptopores." Such structural organelles could be maintained by Ca^{++}, or alternatively, Ca could facilitate the attachment of the vesicle to the synaptopore for sufficient time to permit other mechanisms to release the transmitter. The former possibility may be favored by the most recent observations of Akert (Streit et al., 1973; Figure 1): The frequency with which the hexagonally arrayed "synaptopores" are seen can be related to inferred changes in synaptic activity. When freeze etch replicas of the spinal cord of anesthetized and unanesthetized rats were compared, the synaptopores were seen with statistically significant increases in frequency in the unanesthetized animals.

Several other possible mechanisms by which transmitter release could be modulated were also brought out by part topic members. L. Iversen reported the very recent findings derived from the peripheral sympathetic nervous system in which the release of norepinephrine appears to be subject to control by one of several chemical interactions; this release can be suppressed in some cases by a nerve terminal receptor responsive to norepinephrine itself (i.e., a negative feedback of released receptor on further release) (Enero et al., 1972) or to prostaglandins of the E series (Hedqvist and von Euler, 1972), which may be released from the postjunctional cell after activation of its adrenergic receptor. It also appears that the release of norepinephrine can be sensitive to modulation by nerve terminal receptors for acetylcholine (Loffelholz and Muscholl, 1969; Starke, 1972). The extent to which such factors may operate at other synapses in the periphery or in the brain is unclear, but it is known from the work of Rahamimoff that levels of Mn^{++} approaching those normally found in extracellular space can produce a highly efficient competition with Ca^{++} for the release process (Meiri and Rahamimoff, 1972). Recently chemi-

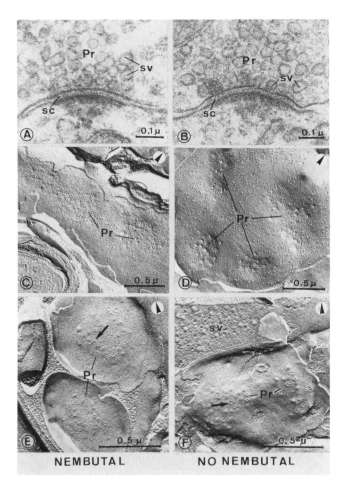

NEMBUTAL   NO NEMBUTAL

FIGURE 1 Comparison of synapses from anesthetized (Pento-barbital i.p., 35 mg/kg) and unanesthetized rat spinal cord. At the left, conventional portraits obtained from thin section (A) and freeze-etch (C and E) material from anesthetized animals. The synaptic membranes are only faintly bulging and the presynaptic membrane modulations are not too obvious. At the right, B, D and F in the anesthetized material, nerve terminals (B) are characterized by (i) inward lifting of membranes and (ii) synaptic vesicles that are "fusing" with or "touching" the presynaptic membrane. The fracture face of the inner leaflet of the presynaptic membrane (D) contains four active sites which are clearly indented and occupied by numerous small pits. The outer presynaptic membrane leaflet (F) shows two active sites with a preponderance of "open," i.e., crater-like, membrane modulations (arrow). These correspond to the omega-shaped contacts between vesicle and plasmalemma seen in B. Note that the anesthetized preparation (E) contains relatively fewer modulations of the presynaptic membrane and that the "closed" forms (arrow) are more numerous than the "open" forms. Pr = presynaptic terminal, sv = synaptic vesicles, sc = synaptic cleft. Encircled arrow indicates the direction of platinum-carbon shadowing. All these differences are highly significant statistically (courtesy K. Akert and C. Sandri, from Streit et al., 1973). (These observations have been numerically analyzed by P. Streit, K. Akert, R. A. Livingston, and H. Moor: "Dynamic ultrastructure of presynaptic membranes at nerve terminals in the spinal cord of rats—anesthetized and unanesthetized preparations compared," *Brain Research*, in press, 1973.)

cal modulation of transmitter release has been shown at axo-axonic synapses onto primary sensory afferent fibers to the spinal cord. Barker and Nicoll (1972) have found that GABA can produce a sodium conductance increase in terminals of primary sensory fibers and that the resultant depolarization of the sensory terminals reduces the excitation these terminals can produce, presumably by decreasing the amount of transmitter released according to the role of terminal membrane polarization, as discussed by Rahamimoff.

Finally, we are presented with an additional possible mechanism of presynaptic functional modulation through the work of Jaim-Etcheverry (this volume). By applying his selective method for the localization of serotonin to nerve terminals of the sympathetic nervous system, he provided direct evidence that the sympathetic nerves innervating the perivascular spaces of the pineal can normally contain large amounts of serotonin. This view had been suspected by indirect evidence obtained by several groups earlier. When exposed to high doses of serotonin, terminals in other portions of the sympathetic system also seem capable of displacing portions of their stored norepinephrine to take on serotonin. Jaim-Etcheverry suggests that this may permit the postsynaptic cell to infiltrate a "false" transmitter amongst the transmitter molecules of its afferent synapses and thus control the extent to which it is affected by incoming impulses. This type of chemical control could be most prominent in the pineal where large amounts of serotonin are synthesized for conversion to melatonin and other indole compounds, but in which the serotonin is not stored and may "leak" out of the cells by mass action. A similar phenomenon may also occur in the CNS, especially where the serotonin neurons of the raphe are innervated by catecholamine terminals, but no such evidence has yet been obtained.

LONGER TERM MODULATIONS The possibility that long-term structural changes in synaptic connections could be induced in the adult mammalian brain has been extensively reviewed by Moore (this volume). As he points out, the phenomenon of axonal sprouting and axon regeneration has been observed frequently in the past, but the nervous systems in which these events took place have always been regarded somewhat as biological rarities, at least with respect to mechanisms that could operate in the mammal. However, the light and electron microscopic investigations of the past 4 years have made it abundantly clear that axonal sprouting and new synapse formation can be induced in certain central areas in mammals by the proper experimental stimulus. The chapter presentations of the part topic indicate that the current level of research may be expected to increase in fruitfulness. In part, this anticipated progress is related to the observations by Moore (see Moore, this volume) that one of the neurochemical types of axon systems that can retain the potentiality for new axon growth in adulthood is the monoaminergic fiber systems, particularly the norepinephrine-containing axons of the dorsal bundle (see Ungerstedt, this volume), which arise from the nucleus locus coeruleus (Olson and Fuxe, 1971). In three areas of the brain receiving input from this nucleus, formation of new preterminal axons can be produced by lesions severing nonadrenergic inputs: medial and lateral septal nuclei, lateral geniculate and superior colliculus (see Moore, this volume). In a fourth area also receiving input of norepinephrine fibers from the locus coeruleus, formation of an increased terminal arborization is produced after lesions in which the norepinephrine system itself is the major damaged structure (Pickel, Krebs, and Bloom, 1973). These findings may indicate that there are several types of cellular signals that can induce certain neurochemically distinctive fiber systems to form new axons, and presumably new synapses. This leads directly to the other source of anticipated yield from this area of research, the testing of the ability of the new fiber systems to form functionally meaningful new synapses. First, it must be established that the new synapses, represented morphologically by increased numbers of fluorescent axons, are functional. Then it may be possible to devise tests by which to ask whether the capacity for growth shown by the catecholaminergic axons after extensive cellular disruption implies a role for this system of neurons for the maintenance of function in the undamaged brain.

Postsynaptic operations

ASSESSING THE ACTIVITY OF POSTSYNAPTIC CELLS If the objective of synaptic level research is to understand the phenomena that regulate the communication between individual cells, then it seems logical to consider whether the same types of experimental approaches can be of use in determining the sequences of cells that interact to regulate behaviors. Again, it is my view that a knowledge of the neurotransmitters involved offers a key to the analysis. By knowing that norepinephrine and dopamine are transmitters and that their cell bodies are gathered into identifiable nuclei of the pons and mesencephalon, we are ahead conceptually and practically in the analysis of the role of these monoamines in behavior.

Ungerstedt presented his large body of observations concerning the behavioral effects of selectively destroying the dopamine-containing projections to the striatum by the local injection of 6-hydroxydopamine (see Ungerstedt, this part). These experiments were made possible by the cytochemical observations that the striatum is rich in dopamine nerve terminals, that the substantia nigra and the medial tegmental nuclei are laden with groups of dopamine-containing neurons, and by experimental tract analysis showing that these two areas are directly connected. Ungerstedt devised experiments that allowed him to determine the role played by the dopamine projection to the striatum in the control and generation of movement, especially after supersensitivity had developed in the cells receiving dopaminergic synapses when these synapses had been destroyed with 6-hydroxydopamine. Dopamine agonists, such as apomorphine and amphetamine, were now able to activate the sensitized synapses and cause the animal to initiate and maintain rotational behavior for long periods of time. In another part of this volume, *Biochemistry and Behavior*, Ungerstedt describes the behavioral results that this same lesion produces in the earliest stages after destruction of the ascending dopamine projection relative to the animal's volitional consumption of food and water. Unless fed by the investigator, these animals will all die of starvation; however, if they are fed, recovery ensues to permit the analysis of the defective movement behavior brought out by the dopamine receptor agonists. These experiments bring out two principles of analysis: First, the complete selective destruction of a neurochemically defined circuit was made possible only because of the knowledge of the chemistry of the transmitter involved. Second, the classical approach to the determination of functional significance, i.e., "what function is lost when the pathway or nucleus is destroyed" may be misleading in both the acute and chronic stages of the denervation and requires many facets of analysis to be sorted out (see Jouvet, this volume, for a description of the lesion approach to the analysis of the transmitters involved in the regulation of sleep mechanisms).

Early stages after destruction may manifest the actions of excessive transmitter released from degenerating axons as well as the loss of the information previously supplied by the destroyed pathway. Late stages of analysis after destruction, when recovery occurs without regeneration of circuitry, imply that the brain is able to circumvent the loss of the pathway to restore function, presumably by adaptation of the function of pathways that were already present. This capacity for functional restoration after even the most precise lesions is confounding and seductively intriguing.

The same body of observations on the localization of homogeneous cell clusters containing a given transmitter allows a more direct approach to the analysis of behavior and its modification by drugs. Thus, it is now possible to record from the cell groups rich in norepinephrine in unrestrained unanesthetized animals (Chu and Bloom, 1973) and relate their activity to the spontaneous occurrence of sleep cycles. A similar type of approach has been used to relate the behavior of certain portions of the serotonin-rich raphe nuclei to sleep (Sheu and Bloom, 1973). This direct application of cytochemical localization to electrophysiological analysis has been fully exploited in the analysis of psychoactive drugs on certain identified groups of monoamine-containing cells (see review by Aghajanian, 1972), an approach that has been strengthened by the comparison of the action of drugs when administered to the same nuclei either parenterally or by local iontophoretic application (Aghajanian et al., 1972).

Conceivably, one could argue that the analysis of neurotransmitters is in itself a relatively trivial topic, because there are only two basic interneuronal forms of communication, namely excitation and inhibition, and the nature of the molecule producing the transmitter message is therefore incidental. I have tried to indicate in the above sections that identification of transmitter molecules has several practical implications for elucidation of the chemical regulation of transmitter synthesis, storage, conservation and release, as well as for the application of cytochemical procedures for localizing the axonal systems that contain, and, inferentially, operate by the release of, a given transmitter. Still it could be reasoned that the diversity of chemical transmitters found in the nervous system is an unnecessary chemical complication and that properly interactive chains of neurons could be constructed with only one excitatory transmitter and one inhibitory transmitter. Kandel has pointed out (see Kandel, this volume) the fact that the nervous system has not evolved into a two-transmitter type system implies that the chemical diversity of transmitters subsumes some additional functional value.

RECEPTOR MECHANISMS   Weight reviewed in considerable detail the mechanisms by which synaptic potentials are produced by the action of transmitters on nerve and muscle. The newest conceptual departure in this field is the description of synaptic potentials—both depolarizing and hyperpolarizing—in which the membrane potential shift is not accompanied by increased ionic conductance but rather by increased resistance to ion flow. Such systems have been described in several sites in the peripheral autonomic nervous system and in the CNS (see Weight, this volume, for references).

The ionic mechanisms for this type of synaptic effect have been studied most completely by Weight and his collaborators in the frog sympathetic ganglia, where two afferent fiber systems (both presumably cholinergic) synapse upon two types of postganglionic cell (both presumably adrenergic). Weight has shown that the slow EPSP generated at a muscarinic cholinergic receptor is best explained by inactivation of a resting outwardly directed potassium conductance. A similar analysis of the slow IPSP produced in the frog ganglion may be explicable on the basis of inactivation of a resting inwardly directed Na current.

Weight (this volume) has indicated some of the electrophysiological advantages that hyperpolarization with increased membrane resistance might offer over the more classically conceived forms of passive ionic hyperpolarization: The increased resistance might set a bias in favor of the more effective EPSPs that would still be able to generate their excitatory signals rather than being shunted as they would be during hyperpolarizations produced by increased conductance mechanisms. Furthermore, the "slow" nature of these synaptic responses and the generally longer time courses of their action could also induce hyperpolarizations and membrane resistance changes that could outlast individual sets of synaptic information arriving through faster axons and thus serve to modulate electrical activity on a longer time basis. Such mechanisms could suffice as a primitive sort of short-term biasing to facilitate the interactions between neurons over particular connections, an idea that may have implications for cellular explanations of learning theories (see Kandel, this volume; Shashoua, 1971).

These synaptic potentials, generated with increased membrane resistance, are termed "slow" in the frog ganglia, because they have synaptic delays of tens of milliseconds and because once generated they last for several seconds to minutes. In mammalian ganglia, the slow IPSP has recently taken on an additional frame of reference for synaptic transmission by experiments that relate the generation of this IPSP to the release of catecholamines, presumably dopamine, from small intensely fluorescent interneurons of the ganglia. The dopamine release from these cells could interact with a dopamine sensitive adenyl cyclase known to be present in bovine and rabbit ganglia (Greengard et al., 1973; Kebabian and Greengard, 1971) in much the same way that norepinephrine has been shown earlier to hyperpolarize cerebellar Purkinje neurons by activation of adenyl cyclase (Bloom et al.,1972a;Hoffer et al., 1971, 1972; see below). These experimental findings have direct implications for synaptic function and synaptic modulation.

ROLE OF CYCLIC NUCLEOTIDES IN SYNAPTIC FUNCTION: Over the past several years we have pursued the mechanism by which norepinephrine slows the discharge of cerebellar Purkinje cells (see Siggins et al., 1971a, 1971b; Hoffer et al., 1971, 1972; and Bloom et al., 1972b; 1973). We have used light and electron microscopy to establish that these cells receive norepinephrine-containing synapses onto their dendrites, and that these norepinephrine fibers arise from the locus coeruleus. By electrophysiological methods, we have analyzed the pharmacological receptors of the Purkinje cells and the effects of electrical activation of the pathway. Briefly, these experiments indicate that norepinephrine slows Purkinje cells by interaction with a beta receptor, that the slowing of mean discharge rate is produced by a prolongation of pauses between bursts of single spikes without effect on climbing fiber responses. By intracellular recordings, norepinephrine was found to hyperpolarize the membrane of Purkinje cells, and this hyperpolarization is generally accompanied by increased membrane resistance (but never by increased membrane conductance). The actions of norepinephrine on the Purkinje cell are blocked by iontophoretic application of prostaglandins of the E series and by nicotinate. The effects of norepinephrine on discharge rate and membrane parameters are precisely emulated by iontophoretic application of cyclic AMP, and the effects of both the applied cyclic AMP and of norepinephrine are potentiated by any of several phosphodiesterase inhibitors. On the basis of these data, we proposed (Hoffer et al., 1972) that the synaptic action of norepinephrine was mediated by an interaction via the adenyl cyclase of the cerebellar cortex, known to be highly responsive to norepinephrine (see review by Rall, 1972). With the anatomical information that the cerebellar norepinephrine fibers arose from the locus coeruleus, it was possible to test this proposal by activating and analyzing the effects of the pathway on Purkinje cell properties. These experiments disclosed that stimulation of the pathway inhibited Purkinje cell discharge, especially single spike bursts, that the inhibitory effects of stimulating the locus coeruleus required active synthesis of norepinephrine and that no effects on cerebellar neuronal discharge were observed when the area of the locus was stimulated in animals pretreated with 6-hydroxydopamine to eradicate the adrenergic projection to the cerebellum. By intracellular recording during the activation of the locus coeruleus Purkinje cells were found to be hyperpolarized and this hyperpolarization was accompanied by a definitive increase in the resistance of the membrane. Similar effects of norepinephrine have been observed on motoneurons (Engberg and Marshall, 1971). Pharmacologically, activation of the locus coeruleus led to an inhibition of spontaneous discharge that could be potentiated by local iontophoresis of phosphodiesterase inhibitors onto the Purkinje cell and could be blocked by local iontophoretic administration of prostaglandins of the E series. All these results supported the concept that this adrenergic projection could be operating by production of a transsynaptic elevation of cyclic AMP in Purkinje cells. The latter observation has

now been documented by application of an immunocytochemical method for cyclic AMP to tissue sections (Wedner et al., 1972; Bloom et al., 1972a). Using this method we have observed that topical application of norepinephrine or electrical activation of the locus coeruleus will elevate the number of Purkinje cells showing positive immunocytological staining for cyclic AMP from resting frequencies of 5 to 15% to levels greater than 75% (see Siggins et al., 1973). A generally similar set of observations, although less extensively analyzed, has been obtained by Greengard and his co-workers for the dopaminergic intraganglionic synapses in rabbit sympathetic ganglia (see Greengard et al., 1973).

Let us now consider other consequences that may be derived from these synapses. Greengard (Greengard et al., 1973) has proposed that a general scheme for the postsynaptic response in such systems is the activation of protein kinases by cyclic AMP. Thus a neurotransmitter, such as norepinephrine, can be released from synapses and initiate a cascade of effects by which the synaptic actions are amplified. The neurotransmitter activates the cyclase to form the cyclic nucleotide, and the cyclic nucleotide can activate the protein kinases. This capability for amplification of the chemical mediators of the synaptic message may account for the marked potency of the activated norepinephrine pathway to the cerebellum (Siggins et al., 1973; Hoffer et al., 1971, 1972) despite an apparent paucity of fibers synapsing on the Purkinje cells.

Two related concepts follow directly from this line of thought. Phosphorylation of a synaptic membrane protein (Johnson et al., 1971) might be expected to alter membrane permeability; thus, it is possible that the increased membrane polarization and the decreased membrane permeability generated by the activation of adenyl cyclase are only epiphenomena of the process by which cyclic AMP concentration is increased. It seems worth considering, therefore, that the electrical effects of such synapses may be less important over their short-term interactions (i.e., with other synaptic potentials being generated simultaneously) than with longer term trophic effects on the properties of the postsynaptic cell. For example, it is known that cyclic AMP can effect the molecular synthesis of RNA (Nissley et al., 1972) and of proteins (Walton et al., 1971; Weller and Rodnight, 1970) in other cell systems, as well as the more general roles in carbohydrate metabolism (Sutherland and Robison, 1969; Newburgh and Rosenberg, 1972; Opler and Makman, 1972). The immunocytochemical staining pattern of Purkinje cells for cyclic AMP (see Bloom et al., 1972b) indicates that the nucleotide concentration can be increased in the nucleus as well as in the cytoplasm and may correlate with the actions involved in the activation of kinases that can phosphorylate nuclear proteins (Langan, 1970).

This concept allows generalization from cyclic AMP to cyclic nucleotides because of the likelihood that future research will reveal equally specific synaptic controls for control of cyclic guanosine monophosphate (see Kuo et al., 1972). The cellular roles that may be mediated by transmitters that act to accelerate cyclic GMP synthesis are only beginning to be studied. It is far from clear at present whether cells can use both types of cyclic nucleotides to mediate contrasting regulations of individual cellular functions (see Hadden et al., 1972) or whether specific nucleotides act within restricted cell populations to serve selective messenger roles.

On the basis of the chemical responses of glia and neuronally derived tumor lines grown in vitro, Nirenberg and his collaborators (Gilman and Nirenberg, 1971; Gilman, 1972; Seeds and Gilman, 1971) have suggested that catecholamines have their major effectiveness in generating cyclic AMP in glia rather than in neurons. While it is difficult to interpolate from virus transformed malignant tumor cells to benign glia of neurons functioning in situ, and while our immunocytochemical localizations of cyclic AMP have not disclosed a hormonally sensitive increase in identifiable glia, the possibility that glia may interact this way with neurons is an intriguing one and one that will most certainly receive further emphasis. Since glia are the major morphological site of glycogen deposition (see Peters et al., 1970), a role of cyclic AMP in glia metabolism is not without functional implications, although it is difficult at present to determine relative importances between glia and neurons for the actions described.

Hormonal Sensitivity and the Duration of Synaptic Events   The possible functional importance of synaptically related adenyl cyclases (or guanyl cyclases) has still additional implications for the mediation of other hormonal events on specific synaptic central systems. One wonders if other neurohormones could also exhibit effects on other neuronal receptors by mediation through cyclase activation. It would seem possible that even longer duration synaptic modulations could be produced via specific hormonal interactions with the very specific receptors of the adenyl cyclase on various central neuronal groups, e.g. the effects of steroid hormones on the electrical (Foote et al., 1972) and the chemical (Azmitia and McEwen, 1969) activity of serotonin neurons (see also Pfaff, this volume). Perhaps it is possible to begin to envision a continuum of chemical interactions between nerve cells, the very short acting point-to-point type of information, such as that generated by amino acid type transmitters, a slower longer lasting more diffuse chemical-trophism, as postulated above for the catecholamines and even longer acting, but perhaps more discreetly elicited effects due to a variety of endocrine or neuronal polypeptide hormones (see Bloom, 1972b for a more extensive review of the status of central polypeptides as neurohormones). Whether these actions of different time courses are generated by a general modulation of conventional synapses or by activation of more rarely detected synapses specific for these hormonal factors, it is clear that considerably more fundamental research is required before the modulatory roles of circulating materials can be functionally understood at the level of synaptic functioning.

REFERENCES

Aghajanian, G. K., 1972. LSD and CNS transmission. *Annu. Rev. Pharmacol.* 12:157–168.

Aghajanian, G. K., H. J. Haigler, and F. E. Bloom, 1972. Lysergic acid diethylamide and serotonin: Direct actions on serotonin-containing neurons in rat brain. *Life Sci.* 11:615–622.

Akert, K., K. Pfenninger, C. Sandri, and H. Moor, 1972. Freeze-etching and cytochemistry of vesicles and membrane complexes of the CNS. In *Structure and Function of Synapses*, G. Pappas and D. P. Purpura, eds. New York: Raven Press, pp. 67–86.

Azmitia, E. C., and B. S. McEwen, 1969. Corticosterone regulation of tryptophan hydroxylase in midbrain of the rat. *Science* 166:1274–1276.

Barker, J. L., and R. A. Nicoll, 1972. Gamma-amino butyric acid: Role in primary afferent depolarization. *Science* 176:1043–1045.

Björklund, A., B. Falck, and U. Stenevi, 1971. Classification of monoamine neurons in the rat mesencephalon: Distribution of a new monoamine neuron system. *Brain Res.* 32:269–286.

Bloom, F. E., 1972a. Localization of neurotransmitters by electron microscopy. *Proc. Assoc. Res. Nerv. Ment. Dis.* 2:25–57.

Bloom, F. E., 1972b. Amino acids and polypeptides in neuronal function. *Neurosci. Res. Prog. Bull.* 10:122–251.

Bloom, F. E., N-s. Chu, B. J. Hoffer, C. N. Nelson, and G. R. Siggins, 1973. Studies on the function of central noradrenergic neurons. *Neurosci. Res.* (in press).

Bloom, F. E., B. J. Hoffer, E. F. Battenberg, G. R. Siggins, A. L. Steiner, C. W. Parker, and H. J. Wedner, 1972a. Adenosine 3′,5′-monophosphate is localized in cerebellar neurons: Immunofluorescence evidence. *Science* 177:436–438.

Bloom, F. E., B. J. Hoffer, and G. R. Siggins, 1972b. Norepinephrine mediated cerebellar synapses: A model system for neuropsychopharmacology. *Biol. Psychiat.* 4:157–177.

Bloom, F. E., L. L. Iversen, and F. O. Schmitt, 1970. Macromolecules in synaptic function. *Neurosci. Res. Prog. Bull.* 8:325–455.

Bodian, D., 1972. Neuron junctions: A revolutionary decade. *Anat. Rec.* 174:73–82.

Chu, N-S., and F. E. Bloom, 1973. Norepinephrine-containing neurons: Changes in spontaneous discharge patterns during unrestrained sleeping and waking. *Science* 179:908–910.

Costa, E., 1973. The fundamental role of immediate precursors to estimate turnover rate of catecholamines by isotopic labeling. In: *Proc. 5th Intl. Congr. Pharmacol.*, F. E. Bloom and G. Acheson, eds. Basel: S. Karger, (in press).

Enero, M. A., S. Z. Langer, R. P. Rothlin, and F. J. E. Stefano, 1972. Role of the alpha adrenoreceptor in regulating norepinephrine overflow by nerve stimulation. *Brit. J. Pharmacol.* 44:672–688.

Engberg, I., and K. C. Marshall, 1971. Mechanism of noradrenaline hyperpolarization in spinal cord motoneurones of the cat. *Acta Physiol. Scand.* 83:142–144.

Faeder, I. R., and M. M. Salpeter, 1970. Glutamate uptake by a stimulated insect nerve muscle preparation. *J. Cell. Biol.* 46:300–307.

Falck, B., N. A. Hillarp, G. Thieme, and A. Torp, 1962. Fluorescence of catecholamines and related compounds condensed with formaldehyde. *J. Histochem. Cytochem.* 10:349–361.

Foote, W. E., J. P. Lieb, R. L. Martz, and M. W. Gordon, 1972. Effect of hydrocortisone on single unit activity in midbrain raphe. *Brain Res.* 41:242–244.

Gilman, A. G., 1972. Regulation of cyclic AMP metabolism in cultured cells of the nervous system. *Adv. Cyclic Nucleotide Res.* 1:389–410.

Gilman, A. G., and M. Nirenberg, 1971. Effect of catecholamines on the adenosine 3′,5′-cyclic adenosine monophosphate concentrations of clonal satellite cells of neurons. *Proc. Nat. Acad. Sci. USA* 68:2165–2168.

Glowinski, J., 1973. The "functional pool" in central catecholaminergic neurons. In *Proc. 5th Intl. Congr. Pharmacol.*, F. E. Bloom and G. Acheson, eds. Basel: S. Karger, (in press).

Greengard, P., J. W. Kebabian, and D. A. McAfee, 1973. Studies on the role of cyclic AMP in neuronal function. In *Proc. 5th Intl. Congr. Pharmacol.*, F. E. Bloom and G. Acheson, eds. Basel: S. Karger, (in press).

Hadden, J. W., E. M. Hadden, M. K. Haddox, and N. D. Goldberg, 1972. Guanosine 3′,5′-cyclic monophosphate: A possible intracellular mediator of mitogenic influences in lymphocytes. *Proc. Nat. Acad. Sci. USA* 69:3024–3027.

Hedqvist, P., and U. S. von Euler, 1972. Prostaglandin controls neuromuscular transmission in guinea pig vas deferens. *Nature (Lond.)* 236:113–115.

Hoffer, B. J., G. R. Siggins, A. P. Oliver, and F. E. Bloom, 1971. Cyclic AMP mediation of norepinephrine inhibition in rat cerebellar cortex: A unique class of synaptic responses. *Ann. N.Y. Acad. Sci.* 185:531–549.

Hoffer, B. J., G. R. Siggins, A. P. Oliver, and F. E. Bloom, 1972. Cyclic AMP mediated adrenergic synapses to cerebellar Purkinje cells. *Adv. Cyclic Nucleotide Res.* 1:411–423.

Hökfelt, T., 1968. In vitro studies on central and peripheral monoamine neurons at the ultrastructural level. *Z. Zellforsch.* 91:1–74.

Hökfelt, T., 1972. Ultrastructural localization of intraneuronal monoamines—some aspects of methodology. *Progr. Brain Res.* 34:213–222.

Iversen, L. L., and F. Schon, 1973. The use of autoradiographic techniques for the identification and mapping of transmitter-specific neurons in CNS. In *New Concepts in Transmitter Regulation*, D. Segal and A. Mandell, eds. New York: Plenum Press, (in press).

Johnson, E. M., H. Maeno, and P. Greengard, 1971. Phosphorylation of endogenous protein of rat brain by cyclic adenosine 3′,5′-monophosphate-dependent protein kinase. *J. Biol. Chem.* 246:7731–7739.

Kebabian, J. W., and P. Greengard, 1971. Dopamine-sensitive adenyl cyclase: Possible role in synaptic transmission. *Science* 174:1346–1349.

Kebabian, J. W., G. L. Petzold, and P. Greengard, 1972. Dopamine sensitive adenylate cyclase in caudate nucleus of rat brain and its similarities to the dopamine receptor. *Proc. Nat. Acad. Sci. USA* 69:2145–2149.

Kosower, E. M., and R. Werman, 1971. New step in transmitter release at the neuromuscular junction. *Nature (Lond.)* 233:121–122.

Kuno, M., 1971. Quantum aspects of central and ganglionic synaptic transmission in vertebrates. *Physiol. Rev.* 51:647–678.

Kuo, J. F., T. P. Lee, P. L. Reyes, K. G. Walton, T. E. Donnely, Jr., and P. Greengard, 1972. Cyclic nucleotide dependent protein kinases. X. An assay method for the measurement of guanosine 3′,5′-monophosphate in various biological materials and a study of agents regulating its levels in heart and brain. *J. Biol. Chem.* 247:16–22.

Langan, T., 1970. Phosphorylation of histones in vivo under the control of cyclic AMP and hormones. In *Role of Cyclic AMP in Cell Function*, P. Greengard and E. Costa, eds. New York: Raven Press, pp. 307–324.

Loffelholz, K., and E. Muscholl, 1969. A muscarinic inhibition of the norepinephrine release evoked by postganglionic sympathetic nerve stimulation. *N.S. Archiv. Pharmacol.* 265:1–15.

Meiri, U., and R. Rahamimoff, 1972. Neuromuscular transmission: Inhibition by manganese ions. *Science* 176:308–309.

Newburgh, R. W., and R. N. Rosenberg, 1972. Effect of norepinephrine on glucose metabolism in glioblastoma and neuroblastoma cells in cell culture. *Proc. Nat. Acad. Sci. USA* 69:1677–1680.

Nissley, P., W. B. Anderson, M. Gallo, I. Pastan, and R. L. Perlman, 1972. The binding of cyclic adenosine monophosphate receptor to deoxyribonucleic acid. *J. Biol. Chem.* 247:4264–4269.

Olson, L., and K. Fuxe, 1971. On the projections from the locus coeruleus norepinephrine neurons. *Brain Res.* 28:165–168.

Opler, L. A., and M. H. Makman, 1972. Mediation by cyclic AMP of hormone stimulated glycogenolysis in cultured rat astrocytoma cells. *Biochem. Biophys. Res. Commun.* 46:1140–1145.

Orkand, P. M., and E. A. Kravitz, 1971. Localization of sites of gamma-amino butyric acid (GABA) uptake in lobster nerve muscle preparations. *J. Cell. Biol.* 49:75–89.

Otsuka, M., 1973. Gamma aminobutyric acid and some other transmitter candidates in the nervous system. In *Proc. 5th Intl. Congr. Pharmacol.*, F. E. Bloom and G. Acheson, eds. Basel: S. Karger, (in press).

Peters, A., S. L. Palay, and H. G. de Webster, 1970. *The Fine Structure of the Nervous System*. New York: Harper and Row, pp. 114–122.

Pickel, V. M., W. H. Krebs, and F. E. Bloom, 1973. Proliferation of norepinephrine-containing axons in rat cerebellar cortex after peduncle lesions. *Brain Res.* (in press).

Rall, T., 1972. Role of adenosine 3′,5′-monophosphate (cyclic

AMP) in actions of catecholamines. *Pharmacol. Rev.* 24:399–409.

Roth, L. J., 1971. The use of autoradiography in experimental pharmacology. *Handbk. Exp. Pharmacol.* 28:286–316.

Sandri, C., K. Akert, R. B. Livingston, and H. Moor, 1972. Particle aggregations at specialized sites in freeze-etched post-synaptic membranes. *Brain Res.* 41:1–17.

Seeds, N. W., and A. G. Gilman, 1971. Norepinephrine stimulated increase of cyclic AMP levels in developing mouse brain cell cultures. *Science* 174:292.

Shashoua, V. E., 1971. Dibutyryl adenosine cyclic 3′,5′-monophosphate effects on goldfish behavior and brain RNA metabolism. *Proc. Nat. Acad. Sci. USA* 68:2835–2838.

Sheu, Y-S., and F. E. Bloom, 1973. Patterns of activity of cat raphe nucleus magnus during sleeping and waking. *Fed. Proc.* (in press).

Siggins, G. R., A. P. Oliver, B. J. Hoffer, and F. E. Bloom, 1971a. Cyclic adenosine monophosphate and norepinephrine: Effects on transmembrane properties of cerebellar Purkinje cells. *Science* 171:192–194.

Siggins, G. R., B. J. Hoffer, A. P. Oliver, and F. E. Bloom, 1971b. Activation of a central noradrenergic projection to cerebellum. *Nature (Lond.)* 233:481–483.

Siggins, G. R., E. F. Battenberg, B. J. Hoffer, F. E. Bloom, and A. L. Steiner, 1973. Noradrenergic stimulation of cyclic adenosine monophosphate in rat Purkinje neurons: An immunocytochemical study. *Science* 179:585–588.

Smith, A. D., 1972. Subcellular localization of norepinephrine in sympathetic neurons. *Pharmacol. Rev.* 24:435–457.

Starke, K., 1972. Alpha sympathomimetic inhibition of adrenergic and cholinergic transmission in the rabbit heart. *N.S. Archiv. Pharmacol.* 274:18–45.

Sutherland, E. W., and G. A. Robison, 1969. The role of cyclic AMP in the control of carbohydrate metabolism. *Diabetes* 18:797–819.

Walton, G. M., G. N. Gill, I. B. Abrass, and L. D. Garren, 1971. Phosphorylation of ribosome associated protein by an adenosine 3′,5′- cyclic monophosphate dependent protein kinase: Location of the microsomal receptor and protein kinase. *Proc. Nat. Acad. Sci. USA* 68:880–884.

Wedner, H. J., B. J. Hoffer, E. F. Battenberg, A. L. Steiner, C. W. Parker, and F. E. Bloom, 1972. A method for detecting intracellular cyclic adenosine monophosphate by immunofluorescence. *J. Histochem. Cytochem.* 20:293–295.

Weight, F. F., 1971. Mechanisms of synaptic transmission. *Neurosci. Res.* 4:1–27.

Weller, M., and R. Rodnight, 1970. Stimulation by cyclic AMP of intrinsic protein kinase activity in ox brain membrane preparations. *Nature (Lond.)* 225:187–188.

Werman, R., 1972. CNS cellular level: Membranes. *Annu. Rev. Physiol.* 34:337–374.

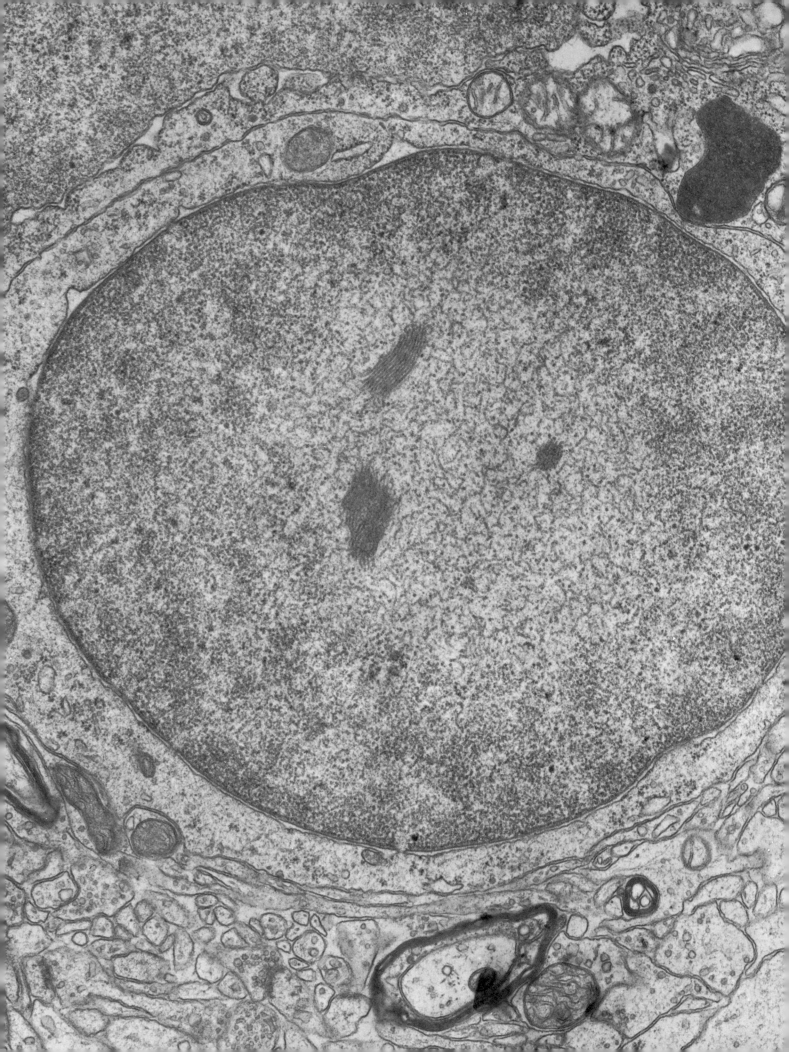

INTERACTION OF
BRAIN CELLS
AND VIRUSES

*Viruses can persist as inapparent "latent" infections of
brain cells for long periods before they become "activated"
to produce destruction of brain cells and clinically manifest
neurological disease, as in subacute sclerosing panencephalitis
(SSPE) (see Koprowski; Shein; this volume). This electron
microphotograph by Dr. Yuzo Iwasaki of the Wistar
Institute illustrates nucleocapsids of the measles-like SSPE
virus localized solely to the nucleus of an oligodendrocyte in
the frontal cortex of a 12-year-old boy suffering from SSPE.
(X 11,000)*

INTERACTION OF BRAIN CELLS
AND VIRUSES

Introduction

HARVEY M. SHEIN

Animal viruses are singularly powerful tools for analysis at the molecular level of fundamental problems of animal cell biology, because they are obligatory intracellular parasites that must use host cell enzymes and precursors for purposes of replication, and because they contain a much smaller genetic complement than animal cells, in some cases as few as five to ten viral genes, as compared with hundreds of thousands of genes in a mammalian cell. Animal viruses can be obtained in astronomically large, precisely quantified numbers and in genetically quite stable but selectively mutable forms, and they can be used to infect homogeneous cell populations. The effects of these viral genetic "probes" in deranging host cell control of metabolic functions, particularly control of growth, multiplication, differentiation, and the formation of specific cell membrane components, can then be analyzed in terms of the new genetic information introduced by the virus. It is obvious that analysis of the molecular mechanisms underlying derangements of various cellular controls consequent upon the effects of introducing viral genomic information equivalent to as few as five to ten genes is experimentally much more feasible than analysis of the same problems by trying to identify the relevant genomic information contained among the enormous number of genes in a normal mammalian cell.

Intense interest has been evidenced recently among virologists and neurobiologists in the study of the mechanisms underlying the cell-virus interactions and host responses in naturally occurring and experimental chronic, or "slow," latent and oncogenic virus infections of nervous tissues. One reason for this intensification of interest is that these mechanisms, when understood, will explain the pathogenesis of important neurological diseases in man that have recently been found to be caused by chronic virus infections and will also be useful in the eventual elucidation of the pathogenesis of other, even more important, human diseases that are of suspected viral etiology, e.g. multiple sclerosis, the senile dementias, and certain types of brain neoplasms.

For the neurobiologist interested in normal brain structure and function, these chronic virus infections of brain tissues are of special interest in several respects. The mechanisms by which the genomic information of some of these viruses remains latent in host cells for long periods until derepressed or otherwise "activated" is of interest for exploration of possible cellular mechanisms underlying the development and long-term maintenance of specific morphological and functional connections in normal brain tissues. The mechanisms of virus-induced oncogenesis offer unique experimental opportunities to neurobiologists for genetic analysis of cellular regulatory mechanisms for multiplication, growth, differentiation, and formation of specific cellular connections, because all oncogenic viruses thus far studied in adequate detail have been found to induce neoplastic transformation by inducing permanent heritable alterations in the host cell via integration of the viral genome itself or of genetic information coded for by the viral genome into the host cell genome. The study of these chronic virus infections also offers the neurobiologist an experimental tool with which to analyze mechanisms of normal and pathogenetic immunological responsiveness in brain tissues. The activation of some of the known *latent* viruses infecting brain tissues to produce destructive lesions months or years after infection has been found to require the presence of either a pathogenetic hypoimmune or hyperimmune response with certain of these agents and has been found to occur in the presence of no apparent immune response at all with a particularly interesting group of these agents, i.e., those causative of the subacute spongiform encephalopathies.

The group of chapters that follow are concerned with the cell-virus interactions and host responses that occur in the course of chronic virus infections of central nervous tissues. The initial paper introduces the topic by presenting an interpretive survey of oncogenesis, latency, and pathogenetic immune response in experimental and naturally occurring oncogenic, latent, and slow virus diseases of brain tissues. Each of the subsequent papers considers in more detail the mechanisms underlying the cell-virus interactions and the pathogenesis of the disease processes produced by one of these overlapping categories of virus infection. Dr. Koprowski's chapter discusses mechanisms of latency in chronic virus infections of the nervous system, including viruses causing subacute sclerosing panencephalitis (SSPE) and progressive multifocal leukoencephalopathy (PML), and considers evidence that multiple sclerosis may also represent a chronic viral infection of the nervous system. Drs. Gibbs and Gajdusek, in their chapter, summarize what is known about the "unconventional" physical and biological properties of a new group of virus-like agents causative of a characteristic type of slowly progressive degenerative brain disease (the subacute spongioform virus encephalopathies), which includes Creutzfeldt-Jakob disease and kuru in man. Dr. Raine's chapter describes the use of temperature-sensitive mutants of an encephalitogenic virus (reovirus type III) to produce different disease processes in brain tissues and thereby illustrates the usefulness of such mutants for the experimental dissection of virus effects from host cell effects and from host immunological responses. Dr. Spear, in her chapter, discusses evidence that the herpesvirus genome codes for synthesis of new surface membrane proteins responsible for the altered "social behavior" of productively and cytolytically infected cells, and she considers possible implications of these findings for the understanding of mechanisms of herpesvirus-induced oncogenesis. Dr. Dulbecco, in the concluding chapter, discusses the molecular mechanisms of virus-induced oncogenesis and of the altered biological characteristics of virus-transformed cells.

88 Oncogenesis, Latency, and Pathogenetic Immune Response in Chronic Virus Infections of Brain Tissues

HARVEY M. SHEIN

ABSTRACT This paper presents an interpretive survey of oncogenesis, latency, and pathogenetic immune response in chronic virus infections of brain tissues. Mechanisms of cell-virus interactions underlying those processes in nonnervous tissue cell types are described and the pathogenesis of prototypical experimental and natural oncogenic, latent, and slow virus diseases of brain tissues are discussed in terms of these and possible alternative mechanisms. A hypothesis is suggested to account for the relative unresponsiveness to oncogenesis of certain brain cell types in situ and for the apparent high affinity of latent viruses for brain tissues. Areas are indicated in which available viruses could be applied usefully to studies of normal neurobiology. Problems are specified that now limit the applicability of virology in investigations of normal brain cell development, function, and structure.

THE MECHANISMS underlying the interactions of oncogenic, latent, and "slow" viruses with brain cells and the cell-type specificities of these interactions are of especial interest to neurobiologists as tools for exploring normal brain cell structure and function. These mechanisms also explain the pathogenesis of some important virus-induced neurological diseases. Oncogenic viruses have been found to induce neoplastic transformation by inducing permanent heritable alterations in their host cells via integration of the viral genome itself or of genetic information coded by the viral genome into the host cell genome. Accordingly, oncogenic viruses in particular represent uniquely favorable genetic probes for analysis of cellular regulatory mechanisms for multiplication, growth, differentiation, and formation of specific cellular connections in both nonnervous and nervous tissue cells. Viruses with morphological and functional characteristics similar to tumor viruses have been identified as latent viruses in brain tissues and as agents of slowly progressive nonneoplastic neurological disease in man and animals.

This chapter considers some of the presently recognized

HARVEY M. SHEIN McLean Hospital, Belmont, Massachusetts, and Department of Psychiatry, Harvard Medical School, Boston, Massachusetts

mechanisms of virus-induced oncogenesis and some of the factors that affect host cell susceptibility and response to tumor virus infection in nonnervous and nervous tissue cells. It also considers latent and slow virus infections of brain tissues in some prototypical diseases that illustrate aspects of latency, activation, and pathogenetic immune response. In the concluding section, it presents a hypothesis to account for the apparent relative unresponsiveness to oncogenesis of certain cell types in brain tissues in situ and for the apparent relatively greater affinity of chronic, latent, and "passenger" viruses for brain tissues; and it suggests some areas in which viruses might be useful as tools in studies of normal neurobiology.

Tumor virus-induced oncogenesis in nonnervous tissue cells

As already noted, all oncogenic viruses thus far studied in adequate detail have been found to induce neoplastic transformations in susceptible cells by inducing permanent heritable alterations in the host cell genome. For the small DNA tumor viruses, the papovaviruses SV_{40} and polyoma virus, it has been shown that in transformed cells all or part of the tumor virus genome becomes integrated into the cell genome (Sambrook et al., 1968) and that only about half of the tumor virus DNA is necessary to induce cell transformation (Benjamin, 1965). With the RNA tumor viruses (oncornaviruses), there is evidence from nucleic acid hybridization studies that DNA homologous to the RNA of the oncornaviruses is incorporated into the cellular DNA of transformed cells and also into the cellular DNA of apparently uninfected cells (Baluda, 1972).

The RNA tumor viruses contain an RNA-directed DNA polymerase (Temin and Mizutani, 1970; Baltimore, 1970) and it has been suggested (Temin, 1971) that this polymerase may mediate the synthesis of DNA homologous to oncornavirus RNA that is integrated into the cell genome of oncornavirus-transformed cells.

In order to account for the widespread presence of oncornavirus-specific antigens and of the morphologically characteristic "C-type" RNA tumor virus particles in apparently normal cells from mammals, reptiles, and birds, Todaro and Huebner (1972) have proposed that DNA of all normal vertebrate cells, including ova, may contain information for producing C-type RNA tumor viruses that is normally repressed in these cells. This hypothesis postulates that chemical or viral carcinogens, irradiation, and the normal aging process all favor the derepression of all or part of this integrated provirus with resultant synthesis of infectious RNA tumor virus and/or transformation of the host cell into a cancer cell. In support of this hypothesis, it has been reported that genomic information specific for C-type RNA tumor viruses is present in all laboratory mouse strains, including strains in which neither infectious virus nor visible C-type particles are demonstrable. The presence of C-type RNA tumor virus-specific antigens has been demonstrated in embryonic and tumor tissues of feral mice, and it has been shown that irradiation, chemical carcinogens, hormones, and increasing age or serial passage of cultured mouse cells can activate or promote C-type RNA virus expression.

In papovavirus-transformed cells, only that portion of the integrated viral DNA is ordinarily transcribed that is responsible for those "early" nonreplicative viral functions associated with the capacity of the viruses to induce oncogenesis. However, it is possible with some SV_{40}-transformed cells to activate the remainder of the integrated viral DNA to transcribe late viral functions, i.e., to direct synthesis of infectious papovavirus, by fusion of the membranes of the transformed cells with those of cells of a type susceptible to productive infection by the papovavirus (Koprowski et al., 1967).

Replication of infectious papovaviruses induces cytolysis of the host cell. In contrast, replication of the RNA tumor viruses does not kill the host cell. Replication of infectious RNA tumor virus can continue indefinitely in both infected normal and transformed cells at a nearly constant rate without apparent injurious effects on the cells.

The important question whether continued activity of the tumor virus genome is required in virus-transformed cells to maintain the transformed cell characteristics has been investigated by the use of temperature-sensitive mutant DNA and RNA tumor viruses that can induce neoplastic transformation only at certain "permissive" temperatures but not at others. It has been demonstrated for transformation induced by the DNA-containing polyoma virus and the RNA-containing Rous sarcoma virus (Eckhart et al., 1971; Kawai and Hanafusa, 1971) that continued activity of the tumor virus genome is essential to maintain most if not all cell characteristics associated with neoplastic transformation after it has been induced. Thus cells transformed at a permissive temperature by temperature-sensitive mutants of polyoma virus or Rous sarcoma virus revert to normal morphology and growth characteristics and lose transformation-associated alterations in the cell membrane and in cellular DNA synthesis when shifted to a nonpermissive temperature.

It has been established for papovavirus-transformed cells that the virus genome codes for information permitting cellular DNA synthesis to continue under conditions in which it is inhibited in normal cells and for the expression of immunologically-specific antigens present in the cell nucleus and in the cell membrane that are distinct from any antigens demonstrable in the tumor virus particle itself (see review by Rapp and Crouch, 1971). However, it has not been determined whether information provided by the virus genome directly codes for the synthesis of these newly appearing cellular antigens, or whether the virus causes derepression of specific host cell genes common to mammalian species, which then code for the synthesis of the new antigens.

Tumor virus-induced alterations in transformed cell surface membranes are receiving especially close attention as a result of experimental evidence that the altered cell surface membrane may function in the transformed cell to block those "signals" that ordinarily prevent further multiplication in the normal cell after it grows into contact with adjacent cells. Coating of virus-transformed cell membranes with fragments of concanavalin A, or addition of cyclic AMP to cultures of virus-transformed cells, has been reported to result in some instances in reversion of the cells to more "normal" growth characteristics, including reappearance of contact-inhibited cell growth (Johnson et al., 1971; Burger and Noonan, 1970).

The DNA papovaviruses and the RNA sarcoma viruses regularly induce transformation in susceptible cell types cultured under conditions that permit cell multiplication to occur subsequent to entrance of the virus into the cell; but these viruses fail to induce transformation when cells of the same types are infected while prevented by contact inhibition or other factors from further multiplication (Todaro and Green, 1966; Temin, 1971). These observations indicate that cells in log phase growth are much more susceptible to virus-induced transformation than cells in stationary phase of growth and support the concept that cell multiplication subsequent to viral infection may be a necessary requirement for fixation or phenotypic expression of tumor virus-induced transformation.

Host cell genetic factors affecting susceptibility of the cell to tumor virus replication and tumor virus-induced cell transformation have been studied extensively with the RNA tumor viruses (see review by Meier and Huebner, 1971). The genome of the host cell affects the relationship between transformation and production of virus. For example, a mouse cell infected by a murine sarcoma virus or a chick cell infected by an avian sarcoma virus will become transformed and produce infectious virus. However, a mouse cell infected and transformed by an avian sarcoma virus will not produce infectious virus. The possibility of recovering infectious sarcoma virus from a non-virus-producing cell type can be enhanced in certain instances by coinfection with a leukemia virus, which has the capacity to multiply productively in this cell type (Peebles et al., 1971).

Host cell genes controlling susceptibility to tumor virus-induced transformation are linked to expression of differentiated cell functions, as evidenced by the fact, for example, that the RNA leukemia viruses will transform only lymphoid or blood cell types (Dougherty and Di Stefano, 1969), while in contrast, the closely related RNA sarcoma viruses will transform only solid tissue cell types. Differentiated cell types transformed by both RNA and DNA tumor viruses have been shown in some instances to retain differentiated functions characteristic of the cell type of origin and have even been found, in rare instances, to induce expression of differentiated function in undifferentiated stem cell types (Baluda, 1962; Wells et al., 1966; Tambourin and Wendling, 1971).

Tumor virus-induced oncogenesis in brain cells

Observations on nonnervous tissue cells, which indicate that the capacity of the cells for continuing multiplication subsequent to viral infection greatly increases susceptibility to tumor virus-induced transformation, appear to apply equally to experimental tumor virus-induced transformation of brain cell types. Those cell types that multiply most extensively in normal brain tissues have been found to be most susceptible to virus-induced neoplastic transformation. Intracerebral inoculation of SV_{40} or polyoma virus into hamsters produces neoplastic alteration of meningeal cells, ependymal cells, or choroid plexus cells but does not produce transformation of neurons, astrocytes, or oligodendrocytes (Kirschstein and Gerber, 1962; de Estable et al., 1965). In contrast, inoculation of

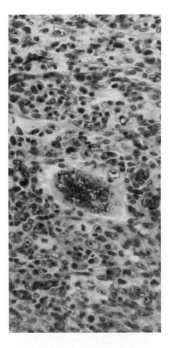

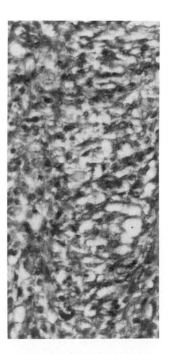

FIGURE 1 SV_{40}-transformed fetal hamster astrocytoma cells in a tumor produced at a subcutaneous site in a newborn hamster by inoculation of astrocytes neoplastically transformed by the virus in cell culture. The poor histological differentiation is apparent in the marked variation in nuclear and cellular size and shape and in the presence of the centrally located multinucleated giant cell. Astrocytic cell processes are not observed and were only demonstrable in an occasional cell by special staining techniques. (From Shein, 1967b.)

FIGURE 2   Polyoma virus-transformed fetal hamster astrocytoma cells in a tumor produced at a subcutaneous site in a newborn hamster by inoculation of astrocytes neoplastically transformed by the virus in a cell culture of the same type used with SV_{40}. The excellent histological differentiation of the polyoma-transformed astrocytes is in striking contrast to the poor histological differentiation of the SV_{40}-transformed astrocytes, illustrated in Figure 1. Notice that the polyoma-transformed astrocytoma cells exhibit feathery-shaped astrocytic cell processes and homogeneity of nuclear and cellular size and that multinucleated giant cells are absent. (From Shein, 1970.)

SV$_{40}$ or polyoma virus into dispersed cell cultures prepared from hamster brains, in which many astrocytes are proliferating, produces neoplastic alteration of astrocytes in addition to transformation of the brain cell types transformed in vivo (Shein 1967b, 1968, 1970).

As with virus-induced transformation in nonnervous tissue cell types, the degree of histological differentiation retained by brain cell types after transformation by tumor viruses is partly a function of the tumor virus and partly a function of the developmental maturity of the cells at the time of infection by the virus. For example, fetal hamster astrocytes transformed in vitro by SV$_{40}$ virus are histologically less well differentiated than newborn hamster astrocytes transformed in vitro by SV$_{40}$; both fetal and newborn SV$_{40}$-transformed astrocytes are histologically less well differentiated than fetal hamster astrocytes transformed in vitro by polyoma virus (Figures 1 and 2) (Shein 1967b, 1968, 1970).

Further evidence for differences among tumor viruses, in terms of range of brain cell types susceptible to transformation and in the degree of histological differentiation in the cells following transformation, is provided by

studies with Rous sarcoma virus and with various oncogenic human and avian adenoviruses in experimental hosts. Intracerebral inoculation of Rous sarcoma virus into rodents or dogs induces a wide range of gliomas and meningiomas including histologically well-differentiated tumors (Rabotti et al., 1966). In contrast, intracerebral inoculation of oncogenic adenoviruses into rodents produces only tumors of primitive neurectodermal cell types including medulloblastomas and undifferentiated gliomas (Ogawa et al., 1969).

As in nonnervous tissue cell types, papovaviruses such as SV$_{40}$ can cause cytolysis instead of transformation in nervous tissue cell types. A particularly interesting example is provided by SV$_{40}$ infection of dispersed cell cultures of human fetal neuroglia. The cell population in these cultures consists of monolayered astrocytes overlaid with colonies of immature oligodendrocytes (Figure 3) (Shein, 1965). When SV$_{40}$ is added to these cultures, the virus selectively causes cytolysis of oligodendrocytes within 14 to 30 days (Figure 4). In contrast to the cytolytic effect on oligodendrocytes, SV$_{40}$ causes an increased proliferation of astrocytes associated with mitotic aberra-

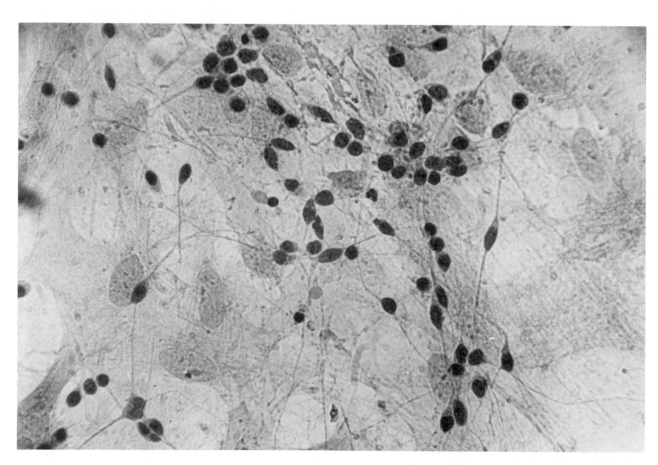

FIGURE 3 Astrocytes growing in a monolayer overlaid with colonies of immature oligodendrocytes (smaller cells with long delicate cell processes) in an uninfected dispersed cell culture prepared from human fetal brain tissue. (From Shein, 1967a.)

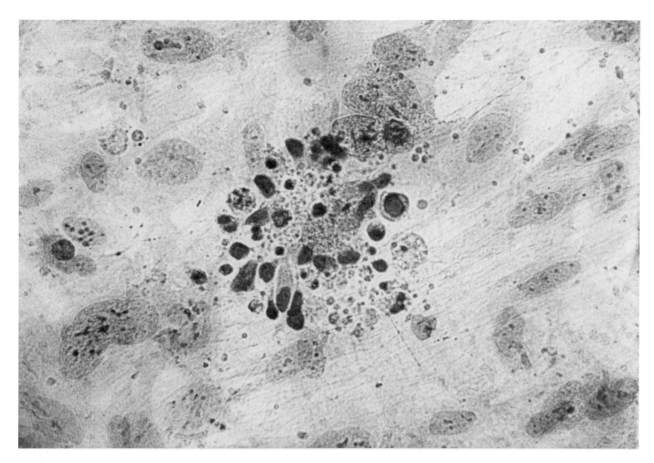

FIGURE 4 Dead and dying oligodendrocytes and apparently unaffected astrocytes in a parallel culture of the same type as that illustrated in Figure 1, at the same age in vitro, 14 days after addition of SV_{40} virus to the culture medium. SV_{40} virus has selectively induced a cytolytic infection in all of the oligodendrocytes. (From Shein, 1967a.)

tions and subsequently induces heritable morphological transformation of astrocytes (Shein, 1967a). The destruction of oligodendrocytes and increased proliferation of astrocytes associated with mitotic aberrations induced by SV_{40} in these human glial cell cultures corresponds precisely to the pattern of histopathological alterations observed in oligodendrocytes and astrocytes in the human demyelinating disease, progressive multifocal leukoencephalopathy, from which a papovavirus has been recently cultured by use of this same human fetal glial cell culture system (Padgett et al., 1971).

Virus-induced selective destruction of multiplying cerebellar cells

The parvoviruses, although not tumor producing, also represent a class of viruses whose pathological effects on brain cell types are dependent upon host cell replication. Parvoviruses are small single-stranded DNA viruses that are responsible for naturally occurring cerebellar hypoplasia in cats and rats and that can induce cerebellar

hypoplasia experimentally in cats, rats, ferrets, and hamsters (Kilham et al., 1967). Multiplication of parvoviruses results in death of the host cell; considerable evidence indicates that these viruses multiply almost exclusively in dividing cells (Tennant, 1971). Thus the cerebellar lesions produced by perinatal parvovirus infection of brain tissues result almost entirely from destruction of the external granular cells of the cerebellum, which are multiplying rapidly at this time. Infection at a later stage of cerebellar development, after the external granular cells have become postmitotic, fails to produce cerebellar lesions.

The highly specific destruction of multiplying external granular cells produced by parvovirus cytopathogenicity has also been produced by a specific immune response to infection of these cells by lymphocytic choriomeningitis (LCM) virus, a noncytolytic virus. Intracerebral inoculation of LCM virus into neonatal mice or rats (Monjan et al., 1971) produces a cerebellar ataxic syndrome similar to that seen in parvovirus-treated animals. These cerebellar lesions are observed only when LCM viral infection is initiated at the period of postnatal cerebellar development that coincides with the period of extensive granule cell proliferation and migration. The immune basis of the LCM virus-induced cerebellar lesions is demonstrated by the fact that the lesions are prevented by administration of rat antithymocyte serum. The absence of a direct cytopathic effect of LCM virus infection in the granule cells is confirmed by the observation that the extent of cerebellar LCM virus infection, as demonstrated by virus-specific antibody staining, is not altered in animals protected from cerebellar pathology by the antithymocyte serum.

Latent virus infections of brain tissues

As with tumor viruses, many nononcogenic viruses have been observed to remain "latent" within host cells including brain cells (Rogers et al., 1967) in a variety of morphological and function states, indefinitely or for periods of months and years, without producing apparent cellular damage. Also as with tumor viruses, the "activation" of latent noninfectious viruses to an infectious form or to a multiplying state has been reported to occur upon treatment with physical and chemical factors, or upon coinfection with "helper" viruses. For example, in patients subject to recurrent herpetic fever blisters, the latent herpes simplex virus infection can be reactivated by elevated temperature. Similarly, shedding of infectious herpes simplex virus can be induced in latently infected corneal cells by injecting epinephrine into animals many months after they have recovered from experimental corneal herpes simplex infections and have stopped shedding virus (Kibrick and Gooding, 1965). The "activation" of latent viruses to produce cellular destruction has also been reported to occur because of pathogenetic alterations in the host immunological response: Hypoactivity of the immune system can permit uncontrolled multiplication of an infectious cytotoxic virus; hyperactivity of the immune response to virus particles or to virus-induced antigenic changes in plasma membranes of infected cells can lead to antibody-plus-complement-mediated or immune-cell-mediated cytolysis (Porter, 1971).

Latent measles-like virus in human brain tissues: Subacute sclerosing panencephalitis

Both a pathogenetic hyperimmune response and an incomplete, noninfectious latent virus appear to be involved in the pathogenesis of a recently much-studied human neurological disease, subacute sclerosing panencephalitis (SSPE). This uniformly fatal disease occurs in patients with a history of measles infection or immunization,

usually during the first year of life and 2 to 17 years before the onset of symptoms. A persistent hyperimmune response to measles virus antigen is observed in these patients both in serum and spinal fluid, with spinal fluid immunoglobulin levels hundreds of times higher than during the convalescent stage of ordinary measles infection (Connolly et al., 1967). Viral nucleocapsids of the paramyxovirus group can be seen in sections of brain tissue from these patients, and infectious virus with an antigenically close relationship to measles virus can be cultured from the infected brain tissue, but only by prolonged cultivation of the brain cells and subsequent cocultivation or fusion of these cells with other measles virus-susceptible cell types (Payne et al., 1969). The requirement for prolonged cultivation in order to recover SSPE virus from the cells may be due in part to the necessity to free the tissue from adherent hyperimmune antimeasles antibody. However, the additional requirement for cocultivation or fusion with a susceptible cell strongly suggests that the SSPE virus is present in brain cells in an incomplete, noninfectious form which is activated by factors present in the intracellular milieu of the susceptible cell.

It has been suggested that the clinical picture observed in SSPE may result from reactivation of latent infection with measles virus some years after infection, or from reinfection with measles virus in an already partially immune person. Burnet (1968) has suggested that SSPE may result from an acquired specific tolerance to measles virus in the immune cell-mediated system in association with a hyperimmune response to measles virus antigen in the humoral antibody system that results in the brain cell lysis observed in clinical SSPE. Neither of these hypotheses accounts for the origin of the apparently incomplete and unusually temperate measles-like SSPE virus, and neither hypothesis defines the mechanism by which the SSPE virus becomes activated to spread from brain cell to brain cell after such a long latency period. It is possible that incomplete measles virus develops from infectious measles virus by a process of selection in brain in the presence of high concentrations of antimeasles antibody in a manner similar to the process by which Rustigian (1966) established a chronic infection with incomplete measles virus in a cultured human cell line by infecting the cells with infectious measles virus in the presence of neutralizing antiserum to measles virus added to the culture medium. In Rustigian's cells, the virus continues to be synthesized intracellularly and is passed from cell to cell in the course of normal division. Measles virus-specific antigen can be demonstrated in every cultured cell, and yet no infectious virus can be demonstrated in the cells or in the medium.

The concept of the pathogenesis of SSPE as resulting

from a hyperimmune response to an incomplete measles-like virus is supported by a recent report (Johnson and Byington, 1972) that intracerebral inoculation of hamster-adapted SSPE virus into weaning hamsters produces a chronic SSPE-like neurological syndrome in these animals. When examined 120 days after infection, the brains of these hamsters exhibit neuropathological findings similar to those observed in SSPE. A defective or cell-associated SSPE agent has been isolated from brains of affected hamsters, and some of these hamsters exhibit elevated serum antibody levels to measles virus.

A hypothesis to explain the "activation" of SSPE virus in brain tissue has been proposed by Koprowski and colleagues on the basis of their repeated observations of papova-like virus particles in association with SSPE virus particles in cultures of brain cells from patients with SSPE. These investigators (Koprowski et al., 1970) have suggested that coinfection or fusion of brain cells in situ by these viruses might activate synthesis of one or the other virus to produce SSPE.

Latent papovaviruses in human brain tissue: Progressive multifocal leukoencephalopathy

Recently, latent papovaviruses present in human brains have been implicated in the etiology of a human demyelinating disease, progressive multifocal leuko-encephalopathy (PML) (ZuRhein and Chou, 1965). In cases of PML, crystalline arrays of papovavirus particles are seen in oligodendrocytes in areas of brain demy-elination upon electronmicroscopic examination. In addition to cytolysis of oligodendrocytes, astrocytes in PML brains show a remarkable proliferation with associated nuclear and mitotic abnormalities that, as we have already noted, are very similar to changes observed in cultured human astrocytes productively infected or transformed by the simian papovavirus SV_{40}. Papova-viruses have been isolated from brains of patients with PML by propagation in human or simian cell cultures (Padgett et al., 1971; Weiner et al., 1972). Some of the PML papovavirus isolates are apparently antigenically distinct from SV_{40}, polyoma virus, and human papilloma virus (Padgett et al., 1971); but other PML papovavirus isolates are reported to be antigenically closely related to SV_{40} (Weiner et al., 1972). The PML papovaviruses thus appear to represent a new group of morphologically similar, antigenically heterogeneous human papova-viruses.

One of the reported isolations of PML virus was accomplished only after apparently transformed cell cultures prepared from the patient's brain tissue were fused with susceptible indicator cells (Weiner et al., 1972). In view of the extreme rarity of "spontaneous" transformation of human cell types in culture, this observation suggests that these brain cells may have been transformed by PML virus so that fusion with a susceptible indicator cell type was required for activation of infectious virus synthesis by a PML viral genome integrated with the host cell genome, as in SV_{40}-transformed cells. In the context of possible oncogenic potential of PML papovaviruses in the brain, it may be noted that crystalline arrays of morphologically atypical papova-like virus particles have recently been reported in cells from a choroid plexus papilloma removed from the brain of a young woman (Bastian, 1971).

PML is observed in elderly patients with lymphomas and other conditions often associated with immuno-deficiency states (Astrom et al., 1958). Accordingly, it has been proposed that conditions of immunoinsufficiency may permit activation of PML virus synthesis in brain cells in which the virus is present in a latent nonmulti-plying state or in an integrated state associated with the host cell genome, as in SV_{40}-transformed cells. The fulminating PML syndrome would occur, according to this hypothesis, when the immunological defenses are rendered ineffective, as in lymphomas, so that infectious PML virus released from latently infected or transformed brain cells would now multiply unchecked. An alter-native form of this hypothesis is that PML virus is present chronically in many persons ("carriers") in the form of an infectious "passenger" virus that multiplies at any one time in only a very few cells (presumably the highly susceptible oligodendrocytes) in brain tissues, thereby killing a very few oligodendrocytes in slow succession in the process of synthesizing and releasing very small numbers of infectious PML virus. Under these hypo-thetical conditions, extracellular infectious PML virus could be present in sufficiently small numbers so that the virus could be neutralized sufficiently, so long as the brain's normal immunological defenses remained intact, to prevent uncontrolled multiplication and cell-to-cell spread of PML infection which would produce clinically apparent PML. Moreover, even under conditions of normal immune surveillance, PML virus in its intra-cellular site in a rare oligodendrocyte would not be accessible to the action of antibody or lymphocytes directed against the virus particle itself, so that a chronic subclinical infection by infectious PML virus could per-sist indefinitely.

Support for the concept of an immunodeficiency state as a necessary precondition for clinical PML in latent or passenger virus carriers is provided by a recent report of the occurrence of PML in a young adult renal transplant recipient whose immunological defenses had been in-hibited by immunosuppressive drugs (Manz et al., 1971). The concept of activation of human papovavirus syn-thesis and increased multiplication and cell-to-cell spread of infectious papovavirus under conditions of immunoinsufficiency is further supported by the observation of a marked increase in the incidence of warts, which are caused by another human papovavirus, in renal transplant patients maintained on immunosuppressive drugs (Spencer and Anderson, 1970).

Slow virus infections of brain tissues

Unlike those latent viruses that produce lesions in brain tissues only under adverse conditions that seemingly "activate" the latent virus to an infectious or multiplying state, certain "slow" viruses multiply widely throughout the clinical latency period and then cause destructive lesions months or years after infection, in some instances because of a pathogenetic immune response, but in other instances without the intervention of any known "activating" factor. Two particularly interesting groups of these viruses presently known to affect brain tissues are oncorna-like RNA viruses that cause progressive demyelination or progressive pneumonia in sheep and agents of unknown physical nature that cause slowly progressive neuronal destruction, ataxia, and death in brains of sheep, mink, and man.

Oncorna-like RNA virus infection productive of slowly progressive demyelinative disease: Visna

Visna virus, the RNA virus that causes a progressive demyelinative disease of sheep brain, has an electron-microscopic morphology that closely resembles that of RNA viruses known to cause leukemias, sarcomas, and mammary tumors in mammals (Takemoto et al., 1971). Visna virus is antigenically closely related to progressive pneumonia virus, a morphologically identical RNA virus also endemic to certain sheep populations that causes a slowly progressive pneumonia (Takemoto et al., 1971). Both these viruses have been shown to contain the RNA-directed DNA polymerase characteristic of RNA tumor viruses (Lin and Thormar, 1970) and have been found to be capable of inducing neoplastic transformation in mouse fibroblast cell cultures in which these viruses do not multiply productively (Takemoto and Stone, 1971).

Visna virus multiplies in sheep brain tissues for several months after experimental infection before the neutralizing antibody becomes detectable in serum (Thormar, 1965). Even after the appearance of high titers of neutralizing antibody in the serum and spinal fluid, infectious virus can often be recovered from spinal fluid, whole blood, and from various organs in addition to the brain, so that the presence of neutralizing virus apparently does not have a protective effect in limiting viral multiplication. Presently available observations do not exclude a possible direct pathogenic effect of the visna virion on oligodendrocyte membranes or on myelin. However, the neuropathological findings suggest that the progressive demyelination of the white matter of the central nervous system, which is responsible for the invariable fatal outcome with visna virus, is secondary to destruction of oligodendrocytes by cells of the reticulo-endothelial system and microglia (Thormar, 1965).

The concept of a pathogenetic immune basis for oligodendrocyte destruction in visna infection of sheep is supported by reports that visna virus does not share antigens with other RNA tumor viruses (Harter et al., 1971). The different visna antigens could possibly explain how the virus could elicit a destructive immune response not usually seen with RNA tumor viruses.

If it is confirmed by further experimental analysis that visna virus-induced demyelination does have an immune pathogenesis, it will be of importance to determine whether the pathogenic immune response is cell-mediated or humoral, and whether this response is elicited by specific antigenic characteristics of the visna virus particle itself or by virus-induced structural or nonvirion antigens in plasma membranes of infected oligodendrocytes. Studies on immunological mechanisms of visna virus-induced demyelination of brain tissues would be greatly facilitated if a similar demyelinative disease could be produced by inoculation of the virus into a small mammal. Until this is accomplished, immune-mediated brain disease produced by lymphocytic choriomeningitis virus infection of mice is likely to remain a more tractable and widely used system for experimental analysis of pathogenic immune mechanisms in chronic viral infections of brain tissue.

It is of considerable interest that visna virus has been found thus far to induce neoplastic transformation only in a cell type (i.e., the mouse fibroblast) in which the virus cannot multiply productively. This observation suggests that visna virus, which has been partially inactivated by physical treatment to destroy its replicative but not its transforming capacity, might also induce neoplastic alteration in brain cells of its natural host, the sheep. The concept that visna virus has the innate capacity to induce transformation of brain cell types is supported by a recent report that this virus can induce morphological transformation of human astrocytes in cell culture (Macintyre et al., 1972).

Spongioform encephalopathies: Agents of unknown nature productive of slowly progressive neuronal destruction without detectable immune response

The second group of "slow" viruses that were mentioned previously are without presently known analogs among agents causing disease in other organ systems. The prototype of these agents is scrapie virus, which afflicts sheep and which has been experimentally adapted to mice, rats, and the cynomolgus monkey (Gibbs and Gajdusek, 1972). Transmissable mink encephalopathy appears to be a closely related agent that affects mink and that has been experimentally transmitted to the squirrel monkey (Eckroade et al., 1970). In man, agents causing both kuru and Jakob-Creutzfeldt disease have been shown to be members of this group by serial passages of cell-free filtrates from infected brains in chimpanzees with production of the characteristic neuronal lesions and astrocytic hypertrophy in the cerebellum in these hosts (Gajdusek, 1971).

As with visna virus, the latency period after infection by scrapie virus or the other members of this group is years or many months. Also, as with visna virus, these agents multiply to significant titers in many other organs in addition to the brain but produce lesions only in brain. However, unlike visna virus infections, infections by the scrapie agent group neither induce a detectable immune response nor are protected against by any immunological response that has been detected to date: neither antibody response, lymphocyte response, nor interferon response (Gibbs et al., 1965; Gajdusek, 1971). These agents have not been identified as yet by electronmicroscopic observation. The scrapie agent is of unusually small size (less than 30 mμ as determined by filtration studies), is unusually thermostabile, and highly resistant to formalin, proteolytic enzymes, freezing, and inactivation by ultraviolet irradiation at wave lengths that much more rapidly inactivate nucleic acids contained in other viruses (Gajdusek, 1971).

On the basis of its exceptionally small size, the absence of a recognizable virion on electronmicroscopic examination, its apparent lack of immunogenicity, and its unusual resistance to inactivation by ultraviolet irradiation, it has been variously suggested that the scrapie agent may be a "helper" virus or partial genome that activates latent brain viruses to produce the observed pathology (Gajdusek, 1971); or a genetic derepressor or selfreplicating molecule not containing nucleic acid (Griffith, 1967; Alper et al., 1967). It has also been proposed that the scrapie-type agents may consist of very low molecular weight RNA without a protein coat, as has been reported for the causative agent of a plant disease, potato spindle tuber virus (Diener, 1972). Experiments designed to characterize and visualize the scrapie, kuru, and Jakob-Creutzfeldt agents and to study cell-virus interactions with these agents should be greatly facilitated by recently reported findings that these agents multiply without cytopathic effects in cell cultures derived from infected brains (Haig and Clarke, 1971; Gajdusek et al., 1972).

In view of the apparent absence of immune response to scrapie-type agents, it appears unlikely that the neuronal and astrocytic changes observed in infected brains have an immune pathogenesis. However, it will not be possible to exclude finally an immune pathogenesis until the agents have been physically identified and serologically typed. The distinctive vacuolization of neurons affected by scrapie-type agents is suggestive of a specific cytotoxic lesion produced by these agents in neurons in the course of multiplication. The fact that astrocytic changes are the initial pathological alterations observed in scrapie-type, agent-infected brains suggests that the marked astrocytic hypertrophy and proliferation observed in infected brains may not represent a reactive response to neuronal injury but may be due instead to a direct effect of scrapie-type agents on astrocytes. Support for this concept is provided by reports (Haig and Pattison, 1967) that astrocytes from scrapie-infected mouse and sheep brains survive longer and proliferate more extensively than astrocytes from normal brains after explantation in cell culture.

Concluding comment: Reactivity of nervous tissues to oncogenic agents and affinity of nervous tissues for latent or passenger viruses; use of viruses as tools for studies of normal neurobiology

One is struck, in reviewing the observations that we have considered on latent, slow, and oncogenic virus infections of nervous tissues, by the data that suggest that normal nervous tissue cells in situ do not readily, if at all, undergo malignant transformation when exposed to viruses with oncogenic capacities, although the cells may be stimulated to proliferate more actively. It is noteworthy that even when oncogenic transformation does occur in situ due to unknown causes, brain tumor cells do not seem to be fully malignant, as indicated by the rarity of metastasis to other organs even after growth of the brain tumor through the blood-brain barrier. All this suggests that either the brain and other nervous tissues, or their component cell types, may react in situ to oncogenic agents in a way that is qualitatively different from many other tissues and cells.

We all know that the central nervous system is peculiarly susceptible to infection by many agents. In contrast, nervous tissues may possibly be relatively unfavorable as sites for malignant transformation of most nervous cell types. It is possible that the uniquely stringent constraints imposed by functional requirements upon specific cell connectivity and upon long-term maintenance of specific functional connections in nervous tissues may impose corresponding constraints upon an oncogenic agent's capacity to disrupt the control of cell growth and multiplication in nervous tissue cell types in situ; perhaps fewer constraints are imposed in those cell types, such as astrocytes, whose normal function includes reactive proliferation under pathological conditions. Moreover, perhaps for these same reasons, nervous tissues may also be correspondingly relatively more favorable as sites for harboring latent noninfectious viruses, for harboring infectious "passenger" viruses, and for selection of mutant forms of these viruses.

If the hypothesis is correct that there is a greater affinity of latent and passenger viruses for brain tissues due to more stringent control mechanisms upon multiplication and connectivity in brain cell types in situ, then one might also expect that when cells from brain tissues are cultured in vitro under conditions permitting increased cell multiplication, decreased specific cell connectivity, and absent immunological defense, then some of these latent or passenger viruses might synthesize sufficient infectious virus to produce cytopathogenic effects, perhaps including neoplastic transformation. This concept is supported by the isolation of dozens of new latent and/or passenger viruses from normal monkey brain tissues after prolonged explantation in cell cultures (Rogers et al., 1967), and by recent observations of heritable morphological transformation in human brain cells derived from patients with both SSPE (Katz et al., 1969) and Jakob-Creutzfeldt disease (Hooks et al., 1972) after explantation in cell cultures. The latter observations also suggest that activation of latent oncogenic viruses present in human brain tissues may be occurring under these in vitro conditions and suggest further that this activation may be enhanced by coinfection of brain cells by nononcogenic viruses latent in brain tissues.

In view of the foregoing considerations and observations, and if one takes into account in addition the immunologically privileged site of the brain, it appears probable that human brain tissues will also be found to harbor many previously unsuspected viruses, as dispersed cell cultures prepared from human brain tissues come into routine use among virologists and neurobiologists.

In conclusion, it seems pertinent to comment briefly on some areas in which immediate application of presently available oncogenic and other "chronic" viruses might be expected to provide tools for use in collaborative studies with other experimental approaches to the study of normal brain structure and function. One such area is the development of cloned virus-induced cell lines of various brain cell types. By intracerebral inoculation of suitable types of oncogenic viruses into experimental hosts, it should be possible to induce genetically stable, well-differentiated neoplasms of each brain cell type that exhibit minimal structural and functional deviations from the normal parental brain cell type and that could then be cloned and used as in vitro model cell lines for studies

of the "normal" cell biology of the various brain cellular species. The RNA sarcoma viruses would appear to be more promising than the DNA papovaviruses for this purpose inasmuch as papovavirus-induced transformation is usually associated with the incorporation of a larger number of viral genomes into the host cell genome and thus might be expected to disrupt to a greater degree differentiated cell functions dependent upon mitotic and chromosomal stability. In this connection, it is worth noting that if the Todaro and Huebner (1972) C-type RNA virus "oncogene" hypothesis is correct, then oncogenesis induced by chemical carcinogens or irradiation should be even more effective in preserving differentiated cell functions inasmuch as only endogenous and not additional exogenous RNA viral oncogenes would be activated.

In addition to the need for cloned virus-transformed brain cell lines, there is also a great need for the development of techniques for growing normal tissue cells of each developmental stage of each differentiated brain cell type in pure cell culture in order to permit the use of tumor viruses and other viruses in studies at the molecular level that could clarify specialized mechanisms for control of differentiation, growth, and multiplication in each of the nerve cell types. The use of cell fusion techniques appears to provide a particularly promising approach to obtaining cloned hybrid cell strains that retain differentiated characteristics of difficult-to-propagate nervous system cell types, such as neurons or oligodendrocytes. A multiplying hybrid cell strain exhibiting differentiated phenotypic characteristics derived from the functioning genome of a normally nondividing neuron or oligodendrocyte could be obtained by fusion of the plasma membrane of the nondividing nervous tissue cell type with the plasma membrane of a nonnervous tissue cell type, such as normal fibroblast, which multiplies readily and whose genome would provide the capacity for continued multiplication in the hybrid cell strain. The feasibility of using cell fusion techniques for this purpose has already been demonstrated in a study in which single mouse neuroblastoma cells were fused with single nonnervous mouse fibroblast-derived cells to obtain hybrid cell lines containing genomes of both cell types (Minna et al., 1972), which exhibit differentiated functions characteristic of neurons.

A class of nononcogenic viruses of interest as tools for the neurobiologist are viruses such as the parvoviruses, which selectively destroy certain populations of brain cells, leaving others undamaged. It appears highly probable that additional viruses with analogous destructive specifications for other brain cell types will be found. However, discovery of these viruses does not appear to be potentially as promising for development of general techniques for producing selective destruction of specific brain cell types as would be the eventual development of cell-specific antigens prepared from each of the various brain cell types capable of inducing a more complete and more highly cell-type-specific immunological destruction of each brain cell type.

It can be anticipated with confidence that further study of chronic virus infections of nervous tissues, if catalyzed by essential collaborative interactions between virologists and neurobiologists, will result eventually in the development of valuable new experimental approaches to the study of normal brain function and structure. In this connection, it is worth recalling that the efforts of an earlier generation of virologists, cell biologists, and immunologists to cultivate the acute "neurotropic" viruses, particularly the polio viruses, resulted in the development and wide dissemination of techniques of virology, cell culture, and immunology that made possible for the first time quantitative in vitro studies of the biology of mammalian cells at the molecular level.

ACKNOWLEDGMENT The preparation of this article, and the studies referred to which were undertaken in the author's laboratory, were supported in part by U.S. Public Health Service Research Grant NS 06610. I am indebted to Dr. John F. Enders and Dr. Alfred Pope for critical reviews of the manuscript and for valuable suggestions.

REFERENCES

ALPER, T., N. A. CRAMP, D. A. HAIG, and M. C. CLARKE, 1967. Does the agent of scrapie replicate without nucleic acid? *Nature* 214:764–766.

ASTROM, K. E., E. L. MANCALL, and E. P. RICHARDSON, JR., 1958. Progressive multifocal leukoencephalopathy: A hitherto unrecognized complication of lymphatic leukemia and Hodgkin's disease. *Brain* 81:93–111.

BALTIMORE, D., 1970. RNA-dependent DNA polymerase in virions of RNA tumor viruses. *Nature* 226:1209–1211.

BALUDA, M. A., 1962. Properties of cells infected with avian myeloblastosis virus. *Sympos. Quant. Biol.* 27:415–425.

BALUDA, M. A., 1972. Widespread presence, in chickens, of DNA complementary to the RNA genome of avian leukosis viruses. *Proc. Natl. Acad. Sci. USA* 69:576–580.

BASTIAN, F. O., 1971. Papova-like virus particles in a human brain tumor. *Lab. Invest.* 25:169–175.

BENJAMIN, T. L., 1965. Relative target sizes for the inactivation of the transforming and reproductive abilities of polyoma virus. *Proc. Natl. Acad. Sci. USA* 54:121–124.

BURGER, M. M., and K. D. NOONAN, 1970. Restoration of normal growth by covering of agglutinin sites on tumour cell surface. *Nature* 228:512–515.

BURNET, F. M., 1968. Measles as an index of immunological function. *Lancet* 2:610–613.

CONNOLLY, J. H., I. V. ALLEN, L. J. HURVITZ, and J. H. D. MILLER, 1967. Measles-virus antibody and antigen in subacute sclerosing panencephalitis. *Lancet* 1:542–544.

DE ESTABLE, R. F., A. S. RABSON, and R. L. KIRSCHSTEIN, 1965. Viral growth and viral oncogenesis in brains of newborn hamsters inoculated with polyoma virus. *J. Nat. Cancer Inst.* 34:673–677.

DIENER, T. O., 1972. Is the scrapie agent a viroid? *Nature New Biology* 235:218–219.

DOUGHERTY, R. M., and H. S. DI STEFANO, 1969. Cytotropism of leukemia viruses. *Progr. Med. Virol.* 11:154–184.

ECKHART, W., R. DULBECCO, and M. M. BURGER, 1971. Temperature-dependent surface changes in cells infected or transformed by a thermosensitive mutant of polyoma virus. *Proc. Nat. Acad. Sci. USA* 68:283–286.

ECKROADE, R. J., G. M. ZURHEIN, R. F. MARSH, and R. P. HANSON, 1970. Transmissible mink encephalopathy: Experimental transmission to the squirrel monkey. *Science* 169:1088–1090.

GAJDUSEK, D. C., 1971. Slow virus diseases of the central nervous system. *Amer. J. Clin. Path.* 56:320–332.

GAJDUSEK, D. C., C. J. GIBBS, JR., N. G. ROGERS, M. BASNIGHT, and J. HOOKS, 1972. Persistence of viruses of kuru and Creutzfeldt-Jakob disease in tissue cultures of brain cells. *Nature* 235:104.

GIBBS, C. J., JR., and D. C. GAJDUSEK, 1972. Transmission of scrapie to the cynomolgus monkey (*Macaca fascicularis*). *Nature* 236:73–74.

GIBBS, C. J., JR., D. C. GAJDUSEK, and J. A. MORRIS, 1965. Viral characteristics of the scrapie agent in mice. In *Slow, Latent and Temperate Virus Infections*, D. C. Gajdusek, C. J. Gibbs, Jr., and M. Alpers, eds. Washington, D.C.: U.S. Government Printing Office, pp. 195–202.

GRIFFITH, J. S., 1967. Self-replication and scrapie. *Nature* 215:1043–1044.

HAIG, D. A., and M. C. CLARKE, 1971. Multiplication of the scrapie agent. *Nature* 234:106–107.

HAIG, D. A., and I. A. PATTISON, 1967. In vitro growth of pieces of brain from scrapie-affected mice. *J. Pathol. Bact.* 93:724–727.

HARTER, D. H., J. SCHLOM, and S. SPIEGELMAN, 1971. Characterization of visna virus nucleic acid. *Biochim. Biophys. Acta* 240:435–441.

HOOKS, J., C. J. GIBBS, JR., H. CHOPRA, M. LEWIS, and D. C. GAJDUSEK, 1972. Spontaneous transformation of human brain cells grown in vitro and description of associated virus particles. *Science* 176:1420–1422.

JOHNSON, K. P., and D. P. BYINGTON, 1972. The neuropathology of acute and chronic SSPE agent infection in hamsters. Paper presented, Annual Meeting of the American Association of Neuropathologists, June 10, 1972, Chicago, Illinois.

JOHNSON, K. P., R. KLASNJA, and R. T. JOHNSON, 1971. Neuronal tube defects of chick embryos: an indirect result of influenza A virus. *J. Neuropathol. Exp. Neurol.* 30:68–74.

KATZ, M., H. KOPROWSKI, and P. MOORHEAD, 1969. Transformation of cells cultured from human brain tissue. *Exp. Cell Res.* 57:149–153.

KAWAI, S., and H. HANAFUSA, 1971. The effects of reciprocal changes in temperature on the transformed state of cells infected with a Rous sarcoma virus mutant. *Virol.* 46:470–479.

KIBRICK, S., and G. W. GOODING, 1965. Pathogenesis of infection with herpes simplex virus with special reference to nervous tissue. In *Slow, Latent and Temperate Virus Infections*, D. C. Gajdusek, C. J. Gibbs, Jr., and M. Alpers, eds. Washington, D.C.: U.S. Government Printing Office, pp. 143–154.

KILHAM, L., G. MARGOLIS, and E. D. COLBY, 1967. Congenital infection of cats and ferrets by panleucopenia virus (PLV) manifested by cerebellar hypoplasia. *Lab. Invest.* 17:465–480.

KIRSCHSTEIN, R. L., and P. GERBER, 1962. Ependymomas produced after intracerebral inoculation of SV$_{40}$ into newborn hamsters. *Nature* 195:299–300.

KOPROWSKI, H., G. BARBANTI-BRODANO, and M. KATZ, 1970. Interaction between papovalike virus and a paramyxovirus in human brain cells: a hypothesis. *Nature* 225:1045–1047.

KOPROWSKI, H., F. C. JENSEN, and Z. STEPLEWSKI, 1967. Activation of production of infectious tumor virus SV$_{40}$ in heterokaryon cultures. *Proc. Nat. Acad. Sci. USA* 58:127–133.

LIN, F. H., and H. THORMAR, 1970. Ribonucleic acid-dependent deoxyribonucleic acid polymerase in visna virus. *J. Virol.* 6:702–704.

MACINTYRE, E. H., C. J. WINTERSGILL, and H. THORMAR, 1972. Morphological transformation of human astrocytes by visna virus with complete virus production. *Nature New Biology* 237:111–112.

MANZ, H. J., H. B. DINSDALE, and P. A. F. MARRIN, 1971. Progressive multifocal leukoencephalopathy after renal transplantation. *Ann. Intern. Med.* 75:77–81.

MEIER, H., and R. J. HEUBNER, 1971. Host-gene control of C-type tumor virus expression and immunogenesis: Relevance of studies in inbred mice to cancer in man and other species. *Proc. Nat. Acad. Sci. USA* 68:2664–2668.

MINNA, J., D. GLAZER, and M. NIRENBERG, 1972. Genetic dissection of neutral properties using somatic cell hybrids. *Nature New Biology* 235:225–231.

MONJAN, A. A., D. H. GILDEN, G. A. COLE, and N. NATHANSON, 1971. Cerebellar hypoplasia in neonatal rats caused by lymphocytic choriomeningitis virus. *Science* 171:194–196.

OGAWA, K., K. AAMAYA, Y. FUJII, K. MATSAURA, and T. ENDO, 1969. Tumor induction by adenovirus type 12 and its target cells in the central nervous system. *Gann.* 60:383–392.

PADGETT, B. L., G. M. ZURHEIN, D. L. WALKER, R. J. ECKROADE, and B. A. DESSEL, 1971. Cultivation of papova-like virus from human brain with progressive multifocal leukoencephalopathy. *Lancet* 1:1257–1260.

PAYNE, F. E., J. V. BAUBLIS, and H. A. ITABASHI, 1969. Isolation of measles virus from cell cultures of brain from a patient with subacute sclerosing panencephalitis. *New Eng. J. Med.* 281:585–589.

PEEBLES, P. T., R. H. BASSIN, D. K. HAAPALA, L. A. PHILLIPS, S. NOMURA, and F. FISCHINGER, 1971. Rescue of murine sarcoma virus from a sarcoma-positive leukemia-negative cell line: Requirement for replicating leukemia virus. *J. Virol.* 8:690–694.

PORTER, D. D., 1971. Destruction of virus-infected cells by immunological mechanisms. *Ann. Rev. Microbiol.* 25:283–290.

RABOTTI, G. F., A. S. GROVE, JR., R. L. SELLERS, 1966. Induction of multiple brain tumors (gliomata and leptomeningeal sarcomata) in dogs by Rous sarcoma virus. *Nature* 209:884–886.

RAPP, F., and N. A. CROUCH, 1971. The control of nonvirion antigens induced by papovaviruses. *Transplant. Proc.* 3:1175–1178.

ROGERS, N. G., M. BASNIGHT, C. J. GIBBS, JR., and D. C. GAJDUSEK, 1967. Latent viruses in chimpanzees with experimental kuru. *Nature* 216:446–449.

RUSTIGIAN, R., 1966. Persistent infection of cells in culture by measles virus. II. Effect of measles antibody on persistently infected HeLa sublines and recovery of HeLa clonal line persistently infected with incomplete virus. *J. Bact.* 92:1805–1811.

SAMBROOK, J., H. WESTPHAL, P. R. SRINVASAU, and R. DULBECCO, 1968. The integrated state of viral DNA in SV$_{40}$ transformed cells. *Proc. Nat. Acad. Sci. USA* 60:1288–1295.

SHEIN, H. M., 1965. Propagation of human fetal spongioblasts and astrocytes in dispersed cell cultures. *Exp. Cell Res.* 40:554–569.

SHEIN, H. M., 1967a. Transformation of astrocytes and destruction of spongioblasts induced by a simian tumor virus (SV$_{40}$) in cultures of human fetal neuroglia. *J. Neuropath. Exp. Neurol.* 26:60–76.

SHEIN, H. M., 1967b. Neoplastic transformation induced by simian virus 40 in Syrian hamster neuroglial and meningeal cell cultures. *Arch. Ges. Virusforsch.* 22:122–142.

SHEIN, H. M., 1968. Neoplastic transformation of hamster astrocytes *in vitro* by simian virus 40 and polyoma virus. *Science* 159:1476–1477.

SHEIN, H. M., 1970. Neoplastic transformation of hamster astrocytes and choroid plexus cells in culture by polyoma virus. *J. Neuropath. Exp. Neurol.* 29:70–88.

SPENCER, E. S., and H. K. ANDERSON, 1970. Clinically evident, non-terminal infections with herpesviruses and the wart virus in immuno-suppressed renal allograft recipients. *Brit. Med. J.* 3:251–254.

TAKEMOTO, K. K., C. F. MATTERN, and L. B. STONE, 1971. Antigenic and morphological similarities of progressive pneumonia virus, a recently isolated "slow virus" of sheep, to visna and maedi viruses. *J. Virol.* 7:301–308.

TAKEMOTO, K. K., and L. B. STONE, 1971. Transformation of murine cells by two "slow viruses," visna virus and progressive pneumonia virus. *J. Virol.* 7:770–775.

TAMBOURIN, F., and F. WENDLING, 1971. Malignant transformation and erythroid differentiation by polycythaemia-inducing Friend virus. *Nature New Biology* 234:230–233.

TEMIN, H. M., 1971. Mechanism of cell transformation by RNA tumor viruses. *Ann. Rev. Microbiol.* 25:609–648.

TEMIN, H. M., and S. MIZUTANI, 1970. RNA-dependent DNA polymerase in virions of Rous sarcoma virus. *Nature* 226:1211–1213.

TENNANT, R. W., 1971. Inhibition of mitosis and macromolecular synthesis in rat embryo cells by Kilham rat virus. *J. Virol.* 8:402–408.

THORMAR, H., 1965. Physical, chemical and biological properties of visna virus and its relationship to other animal viruses. In *Slow, Latent, and Temperate Virus Infections.* D. C. Gajdusek, C. J. Gibbs, Jr., and M. Alpers, eds. Washington, D.C.: U.S. Government Printing Office, pp. 335–390.

TODARO, G. J., and H. GREEN, 1966. Cell growth and the initiation of transformation by SV$_{40}$. *Proc. Nat. Acad. Sci. USA* 55:302–308.

TODARO, G. J., and R. J. HEUBNER, 1972. The viral oncogene hypothesis: New evidence. *Proc. Nat. Acad. Sci. USA* 69:1009–1015.

WEINER, L. P., R. M. HERNDON, O. NARAYAN, R. T. JOHNSON, K. SHAH, L. J. RUBINSTEIN, T. J. PREZIOSI, and F. K. CONLEY, 1972. Isolation of virus related to SV$_{40}$ from patients with progressive multifocal leukoencephalopathy. *New Eng. J. Med.* 286:385–390.

WELLS, S. A., JR., R. J. WURTMAN, and A. S. RABSON, 1966. Viral neoplastic transformation of hamster pineal cells in vitro: Retention of enzymatic function. *Science* 154:278–279.

ZURHEIN, G. M., and S. M. CHOU, 1965. Particles resembling papova viruses in human cerebral demyelinating disease. *Science* 148:1477–1479.

89 Considerations of the Mechanisms of Latency in Chronic Virus Infections of the Nervous System

HILARY KOPROWSKI

ABSTRACT   Viruses causing slow virus infections such as subacute sclerosing panencephalitis (SSPE), progressive multifocal leukoencephalopathy (PML), and Jakob-Creutzfeldt (JC) probably persist in the human organism for a long period of time before manifestation of the disease becomes apparent. Whether they persist in the target organ (brain tissue) or other parts of the body is not yet known. It is possible that a large proportion of the viral population persists in an incomplete form in the host cell and that the agents spread only by cell to cell contact. Immunological derangements may be involved in the activation of the infectious process, and immune responses in these diseases may not be beneficial to the host.

Multiple sclerosis (MS) may also represent a chronic viral infection of the central nervous system (CNS), but the parainfluenza Type I virus isolated from MS brain cells in culture has to be studied much more extensively before it can be linked with the etiology of the disease.

AT A SYMPOSIUM devoted to research in the neurosciences, few authors will organize their papers around a disease of the central nervous system (CNS). Yet, when dealing with the topic of chronic virus infections of the nervous system of humans and animals, the disease itself must be the center of our attention. Therefore, while discussing more basic aspects of slow virus infections, we must always orient ourselves to the disease itself.

When the concept of *slow* virus was first introduced by the late Sigurdson in Iceland almost 23 years ago, relatively little was known about the various aspects of the virus-host cell relationship. In general, viruses have been placed in two categories: those that cause destruction of the infected cells or those that cause abnormal cell proliferation. In the course of time, it became apparent that this was an oversimplification and could not be applied to those viruses that fell into the slow virus category. The

slow viruses do not seem either to destroy cells or to cause tumor formation, and it is quite possible that they may coexist in commensalism with the host cell until, ultimately, after a prolonged period, they cause some cell *dysfunction*.

Thus, the definition of the virus as *slow* or *fast* does not refer to the intrinsic characteristic of the agent itself but rather to the timing of the response of the host to the infection. The more appropriate term is therefore "slow virus diseases" referring to diseases characterized by long incubation periods and a chronic, languishing course. Incubation periods range from 5 to 7 years in the case of kuru, 3 to 5 years in scrapie, 1 to 8 years in visna, and anywhere from 7 days to 1–2 years in the case of rabies. By the term *incubation period*, we mean the time period elapsing between exposure to the infectious agent and manifestations of the disease by the host that are recognizable by the clinician or investigator. It does not refer to the time between exposure and the presence of the infectious agent in a given target tissue.

Scrapie, a low infection of possible viral etiology has been experimentally transmitted from sheep, the natural host, to laboratory animals. In mice, the incubation period, defined as the time from exposure to illness, is 6 to 7 months. The infectious agent, however, is present in the viscera of the animals before it invades the CNS and the signs of the disease appear. One could, therefore, postulate that illness is first manifested when the agent reaches the target tissue.

This is not the case, however, in rabies infection, about which infinitely more is known than about the other slow virus infections. Here the infectious virus can be recovered from brain tissue within 24 hours after inoculation of the experimental animals, and yet symptoms of disease do not occur until much later. Following exposure to a rabies "street" virus, the incubation period in man or

HILARY KOPROWSKI   The Wistar Institute of Anatomy and Biology, Philadelphia, Pennsylvania

animal may be quite lengthy before clinical symptoms of the disease develop and the disease proceeds to its nearly always fatal conclusion. Histopathological observations of degenerative changes in the ganglia, or the occurrence of phagocytic cells and small hemorrhages, however, cannot account for the severity of the illness. There is not even concrete evidence of the destruction of the neurons in which the virus seems to replicate selectively, and the mechanisms involved in the disease process still remain a mystery. With this in mind, it may be worthwhile to discuss not only what we know about the nature of slow virus infection but also to delineate those things that we do not know.

We shall deal first with a disease known as subacute sclerosing panencephalitis (SSPE) and attempt to relate it to specific agents and to delve into the host-virus relationship in SSPE. SSPE is a progressive disorder of the central nervous system affecting children and young adults and characterized by gradual intellectual deterioration, motor abnormalities and convulsions, and occasionally by visual difficulties leading to blindness. Patients affected ultimately become comatose and usually die several months to a year after the first symptoms of disease appear, although there are chronic forms of the disease with remissions lasting from one to a few years. A viral etiology of the disease was suspected many years ago (Dawson, 1933) after intranuclear inclusions were found in the brain cells of the SSPE patients. Later, morphological observations of brain tissue revealed the presence of nucleocapsids ultrastructurally similar to paramyxovirus nucleocapsids in the nuclei of neurons and oligodendrocytes (Oyanagi et al., 1971). The possibility that the agent might be one belonging to the measles virus group was advanced when in many, but not all, cases of SSPE, unusually high titers of measles antibody were found in the sera and cerebral spinal fluid of the patients. Attempts to reproduce SSPE in animals by inoculation of material obtained from the human host have been unsuccessful in spite of the array of animals used. The first transmission of the agent to an animal took place when ferrets, injected with brain material obtained at a human biopsy (a procedure which is now used for diagnostic purposes), were observed not only for clinical signs of illness but for changes in electroencephalographic patterns (Katz et al., 1968). Using this technique, it was possible to serial passage the agent from one animal, which developed an abnormal EEG, to another. The abnormal EEG patterns were augmented by histological observation. This transmission of the agent to ferrets helped in investigating the nature of the agent causing SSPE, but real progress was made in this field only after scientists in several laboratories (ter Meulen et al., 1972a) decided to grow brain cells derived from SSPE cases in culture.

For reasons not very clearly understood, few attempts were made to establish monolayer cultures of cells from human brain tissue. Only relatively recently, scientists working with SSPE cases decided to treat brain tissue as other research workers treated tissue from other parts of the body following the classical studies of Enders et al. (1949). The difficulty in dealing with human brain cells growing in culture under normal environmental conditions is the fact that it is impossible to identify exactly cells that multiply in vitro and that seem to be morphologically of the same type, although in the initial cultures several different types of cells are observed. Whichever cell type was growing in explants from SSPE brain tissue, it was possible to maintain these cultures for a prolonged period of time in the laboratory and finally to isolate the virus from these cells (ter Meulen et al., 1972b).

SSPE patient-derived brain cells in culture were not destroyed by the presence of the agent, and the lesions were limited to the formation of cell syncytia, as shown in Figure 1. Ultimately, the SSPE brain cultures showed

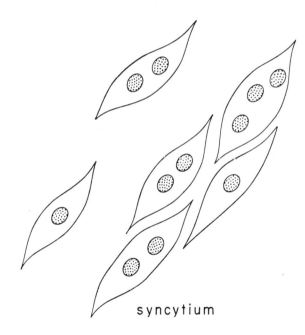

syncytium

FIGURE 1

morphologic abnormalities and chromosomal changes more characteristic of the cell transformation seen after infection with tumor viruses. Ultrastructural study of brain cells in culture in some, but not all cases (see below), revealed the presence of nucleocapsids similar to those observed in brain tissue of the patients. The brain cells in culture contained an antigen that reacted, in an immunofluorescence test, with antimeasles serum. It was impossible, however, originally to isolate an infectious agent from cell extracts or tissue culture medium until human brain cells were either cocultivated or fused with "indicator" cells of either human or African green monkey origin. The isolated agent seemed to be closely related to viruses of the measles-distemper-rinderpest group. Since cell fusion will be referred to again, basic tenets of the fusion process are illustrated in Figure 2.

CELL FUSION

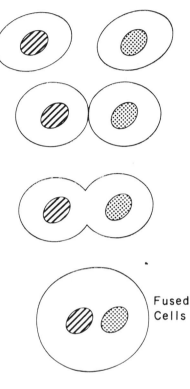

FIGURE 2

Somatic cell fusion

Somatic cell fusion may occur between cells of warm-blooded and cold-blooded animals, intraspecifically and interspecifically. Gamete cells, e.g. ova and spermatozoa, can be fused with somatic cells.

Mammalian somatic cell fusion

Fusion between mammalian somatic cells depends on:
(a) Presence of the fusion factor.
(b) *p*H of the medium (the more alkaline the medium, the higher the fusion rate).
(c) Ca ions for the repair of the cell membranes.

Cell fusion functions in virus infections

Cell fusion is the most successful tool in:
(a) Rescue of viruses present in latent form in cells.
(b) Transfer of material from one cell to another without involvement of the cell membrane.

After successful efforts to isolate the agent by the techniques mentioned above, it was the general impression that isolation of a virus from SSPE cases had become the rule rather than the exception. Unfortunately however, this is not the case, and in our laboratories (Katz and Koprowski, 1973) failures to isolate the virus occurred more frequently than successes. As shown in Table I,

TABLE I

Characteristics of brain cells in cultures derived from SSPE patients

Cells	Morphological Pattern	Detection of Measles-like Antigen by Immuno-fluorescence	Detection of Nucleocapsids by Electron Microscopy* (%)	Isolation of Infectious Virus
JAC	many syncytia and multi-nucleated giant cells	60% of giant cells, occasional single cells	10	Yes
LEC	many syncytia and multi-nucleated giant cells	60% of giant cells, occasional single cells	10	Yes
ROB	many syncytia, moderate number of multi-nucleated giant cells	10% of giant cells, Rare single cells	10	No
MCG	no syncytia, multinucleated giant cells	rare giant cells	5	No
MOS	moderate syncytia, rare multinucleated giant cells	rare single cells	0	No
WES	no syncytia, occasional multinucleated giant cells	negative	5	No
MCL	no syncytia, occasional multinucleated giant cells	negative	0	No
STO	no syncytia, occasional multinucleated giant cells	negative	0	No

*200 cells examined.

the application of the cell fusion technique to isolate an infectious agent from the brain tissue of eight SSPE patients resulted in the isolation of the virus from only two cases. This was true in spite of the fact that, in three other cases, structures identical to paramyxovirus nucleocapsids were detected in the cells by electron microscopy, and all cultures showed the presence of syncytia or multinucleated giant cells. This failure to isolate the virus raises the question: Under which form does the virus persist in the cells of the CNS in human subjects? We know too little about the nature of the virus-host cell relationship in SSPE in order to determine if synthesis of a complete virus particle and expression of late viral function is necessary to cause dysfunction of brain cells. But from the incomplete data available today, it is possible that syncytiogenic and cell transforming activity may not be related to the expression of a late function of the viral genome but rather to the presence of a defective viral particle that cannot be readily transmitted to other culture systems. Some supporting evidence for this assertion comes from the fact that isolation by fusion of SSPE virus from human brain cultures does not take place during the initial stages of growth of SSPE brain cells but only after these cultures are maintained in vitro for several cell transfers. Results of careful electron microscopic (EM) monitoring of SSPE brain cells from the initial explants through several cell transfers indicate that virus particles budding at the cell surface that can be released from the cell are not seen in cells explanted from SSPE tissue at the initial stages of cultivation. Moreover, the filaments representing the nucleocapsid structure of SSPE are never seen aligned in the cytoplasm of brain cells under the cell membrane and do not become incorporated into budding particles of SSPE virus when such budding is observed at later cell transfers in culture. Thus, an SSPE virus particle resembles incomplete measles particles under EM as described by Nakai (Nakai et al., 1969). Only during the later stages of their growth in culture, 8 months after original explantation, when more budding virus particles are seen in human brain cells, it is possible, with difficulty, to transmit the agent to other cell systems. This is true even when tissue culture media is used for the transfer of the virus.

The problem with defining the SSPE virus-host cell relationship, as complex as it now appears, is further complicated by the fact that, while monitoring cells derived from SSPE cases by EM, structures other than nucleocapsids of the paramyxovirus groups have been observed. These structures in the cytoplasm resemble virions of the papova-like viruses (Barbanti-Brodano et al., 1970) such as polyoma SV40 or human wart, and have been found in cultures of the original human brain cells, in brain tissue of ferrets inoculated with SSPE material, and in "indicator cells" that were fused with SSPE human brain cells. Since these cultures were derived from three SSPE cases, it is possible that papova-like virion structures were present in the original brain tissue. Numerous attempts to isolate a biologically active "second" virus have met with failure. Some indirect supporting evidence for its presence, however, was obtained through histochemical investigations that indicate the presence of cytoplasmic DNA-containing inclusion bodies in cells derived from cultures that showed the presence of the papova-like virions under EM (Müller et al., 1971).

It is doubtful that the presence of the "second" virus is an artifact of EM, but the question remains: Do these virions play any role in the etiology of the disease?

Besides the sarcoma-leukemia systems, referred to by Dulbecco later in this part, dual virus infection of the same host has only recently become a subject of considerable investigation. Data are still scarce and our knowledge is limited to a few isolated observations. One of the more interesting studies deals with dual virus infection of mammalian cells with a DNA virus originally isolated from frog kidney tumors and several RNA viruses. The DNA "frog" virus (FV3) has been found to replicate exclusively in the cell cytoplasm at temperatures not exceeding 33°C (Clark et al., 1972). It was found to partially inhibit micromolecular synthesis of cultured mammalian cells and when these cells were superinfected with rhabdoviruses such as rabies, vesicular stomatitis virus, or Kern Canyon virus, virus replication was markedly inhibited (Clark et al., 1972). Although the FV3 virus suppresses cellular RNA synthesis and seems also to suppress RNA synthesis of Sinbis virus, a Group A Arbovirus, the yield of the latter virus is not affected (Clark et al., 1972). The results of this study indicate that dual virus infection of the same cell may or may not have an effect on the replication of one of the viral partners, but each system has to be studied separately before a general conclusion can be drawn.

Research on the reciprocal effects of dual virus infection of the same cell does not need to be limited to the simultaneous presence of a DNA and RNA virus. The presence of two RNA viruses in the same cell, as in the case of lymphocytic choriomeningitis virus (LCMV) and rabies virus, was found to enhance the replication of rabies virus without affecting the synthesis of LCMV. Unfortunately, in the case of SSPE infection, the lack of definitive tests for the presence of papova-like virions makes it impossible to design experiments to determine if the intracellular replication process of the paramyxoviruses is affected by the presence of a second virus.

The presence of papova-like virions in brain tissue of human subjects is not a new observation. For the past 6 years, it has been possible under EM to observe papova-like particles in more than 20 cases of progressive multifocal leukoencephalopathy (PML). This disease is usually associated with conditions in which the immunological responsiveness of the patient is impaired, such as in the carcinomas, Hodgkin's disease, leukemias, systemic lupus erythematosus, or in patients who have received renal transplants and immunosuppressive therapy (Weiner et al., 1972). The pathologic process is characterized by demyelination, the absence of marked cell destruction, and a lack of inflammatory response. There is also a characteristic hyperplasia of astrocytes leading to the formation of giant astrocytes in some cases. After many attempts to isolate a transmissible agent from the brain tissue of these patients had failed, a virus similar or identical to SV40, a DNA oncogenic agent, was recently recovered from the brain tissue of two PML cases, neither of whom had a malignant disease (Weiner et al., 1972). In another laboratory, brain tissue from a PML patient yielded an agent, morphologically identical to SV40, but apparently serologically different (Padgett et al., 1971). The techniques for the isolation of these agents followed those established for the recovery of agents from SSPE cases. First, following explantation of PML brain fragments, cultures were established. The growth of these cells in culture again led to changes such as the loss of contact inhibition and morphological transformation. Fusion with indicator African green monkey kidney cells resulted in the recovery of an agent that subsequently could also be isolated using brain cell extracts as inoculum. At the present time, it is impossible to say whether the papova-like virions encountered in SSPE brain cells are related to the viruses isolated from the PML cases, or whether these DNA agents are *normal* symbionts of human brain tissue that become *activated* and cause trouble. The form under which these agents persist in human brain and how they become activated remains still in the realm of speculation. The SV40, to which the PML agent bears close resemblance, may persist in an incomplete form in mammalian cells during an indefinite number of transfers in culture. Following fusion of these cells with SV40 permissive cells, the virus completes its synthesis and becomes infectious. Whether the same mechanism operates in situ is, as mentioned above, unknown.

Activation of an agent in the organism may also be related to a derangement of the immunological responsiveness of the human host. There are reports on the depression of immunological responses in SSPE, but they are countered by reports of normal responses (Lischner et al., 1972). If the immunological responses are indeed impaired, it is not known whether they were impaired prior to the onset of the disease, thereby predisposing the individual to the disease, or whether such immunological impairment occurred as the result of the pathologic process. We are dealing with human subjects, and it may take some time before this problem can be solved. Since the role of the *second virus* in SSPE infection is obscure at present, we are obliged to limit our discussion to the paramyxo virus as a possible etiologic agent of SSPE.

Since abnormally high levels of measles antibody in the serum and cerebral spinal fluid are characteristic of a large majority of patients suffering from SSPE, it is assumed that the agent causing SSPE is an aberrant variant of measles virus which, for unknown reasons, remains in brain tissue following the initial infection in childhood and only much later causes encephalitis. Since SSPE cases occur predominantly in rural areas (Brody and Detels, 1970), the possibility should not be overlooked that farm animals may be the source of infection in man.

More than 10 years ago, cases of bovine meningoencephalitis of unknown etiology, occurring sporadically in Germany and Switzerland, were attributed to a viral infection (Thein et al., 1972). The pathological lesions were similar to those described in the experimental infection of laboratory animals with SSPE virus. No agent has been isolated from the brain tissue of animals suffering from this disease, and until recently, no efforts had been made to isolate such an agent. At present, bearing in mind the possible transmission of an infectious agent from cattle to man that may cause SSPE, a major study of bovine meningoencephalitis has been undertaken in Germany to determine the etiology of the disease. In the meantime, to see if calves were susceptible to SSPE virus isolated from human cases, a group of calves and lambs were injected with culture cells infected with SSPE virus, and for comparison, other groups were injected with cells infected with measles and distemper viruses, which belong to the same serological group as the SSPE virus (Thein et al., 1972).

The results of these studies showed that SSPE virus is pathogenic for calves and lambs but that measles and distemper viruses are not. The disease in animals has a more acute course than that observed in man, but the neuropathological changes were very similar to those observed in human SSPE cases except that the animal's brain showed no signs of demyelination. This may be related to the fact that brain tissue of many species of animals does not demyelinate as does human brain tissue. Although all inoculated animals became sick, the SSPE virus was isolated from only one calf. Again, a parallel can be drawn with the similar low frequency of isolation of the SSPE virus from human brain tissue.

It is interesting to note that when calves and lambs were injected with cell-free extracts containing the SSPE virus instead of intact infected cells, they did not become sick. The same observations were made when ferrets were exposed to tissue-culture-passaged SSPE (Katz et al., 1968). This observation plus the fact that, in a majority of cases, the SSPE could be "rescued" only by cell fusion and that in SSPE the disease process persists in spite of an extremely high titer of virus-binding antimeasles antibodies in blood and cerebral spinal fluid, may indicate that the virus spreads by cell-to-cell transfer either through intracellular bridges or the formation of syncytia.

Curiously enough, animals exposed to SSPE viruses did

not develop an immune response, as measured by antibody production, in contrast to animals exposed to either measles or distemper produced antibody. One is again reminded of the impairment of the immunological response reported by some in human cases of SSPE and the possibility that this virus may affect cells of the immune system as does a variety of viral agents that replicate in these cells and impair the immune response (Notkins et al., 1970).

Turning from SSPE infection to a more difficult problem, I would like to mention briefly work with brain tissue obtained from patients with Jakob-Creutzfeldt (JC) disease, another infection of the central nervous system in which a viral etiology is suspected. In contrast to SSPE, JC has been grouped with kuru, scrapie, and mink encephalopathy as a separate category of spongioform encephalitides. These diseases are characterized by a pathologic process leading to the loss of neurons (spongy degeneration) and abnormal proliferation of glial cells and astrocytic elements of the brain. As in other diseases of this category, it is always fatal. Using the technique used in the study of SSPE tissue, brain tissue obtained at autopsy from JC cases was explanted in tissue culture by research workers in several laboratories. The abnormalities detected in the course of cultivation of the brain tissue in vitro seemed not to be of a degenerative type. The brain cultures again became morphologically transformed but, in contrast to the SSPE cultures, EM failed to reveal any agents in the JC cultures. When JC brain cells were either cocultivated or fused with indicator cells of human or green monkey kidney origin, cultures of these indicator cells showed definite signs of cellular transformation (Jensen and Koldovsky, 1972). Foci of

the transformed cells neighbored the areas in which cell destruction was visible. This was observed only in the initial stages of infection. Later on, morphologically transformed cells became predominant and when injected into immunosuppressed mice caused tumor formation. Since there was a possibility that the surfaces of cells "infected" with the elusive JC agent might undergo characteristic antigenic changes, attempts were made to demonstrate these "neoantigens." A serum was prepared in rats initially made tolerant to normal cells of the same origin as JC "infected" cells and then immunized with cells containing the JC agent (Jensen and Koldovsky, 1972). Specificity of the immune response was then determined in a cytotoxicity test against cells of various origins as shown in Table II.

It seems clear from these results that immunization of the animals resulted in the production of an antibody directed against the cells *carrying* the supposed JC agent, and not against *normal* cells of the same origin. The next step was to relate the presence of the neoantigens to the disease process itself. Serum was obtained from a number of JC patients and submitted to a reaction with indicator cells that were fused or cocultivated with JC brain cells. The results (Table II) (Jensen and Koldovsky, 1972) again showed that the patients' sera contained an antibody cytotoxic for cells carrying the JC agent. This approach to latent infections of the central nervous system where a virus can be neither visualized nor isolated should have wider application.

Because of its prevalence and its crippling effect, multiple sclerosis (MS) is the most important of all the chronic encephalitides of man. The possible viral etiology of this disease has been suspected chiefly on the basis of

TABLE II

Detection of an antigen associated with Jakob-Creutzfeldt (JC) agent

	Cytotoxic Titer of Serum							
		Rat Anti-JC Absorbed w/CV-1	Patients sera					
Culture Cells	Rat Anti-JC (see text)		1	2	3	4	5	6
Original JC brain	128	16	16	16	8	8	8	32
Human fibroblasts after cocultivation with JC brain cells	256	32	8	16	<2	4	4	8
CV-1 (green monkey kidney) after cocultivation with JC brain cells	256	16	<2	8	8	4	8	8
Human fibroblasts exposed to cell-free extracts from JC cultures	128	16	NT					
Normal human fibroblasts	Neg.	Neg.	<2	<2	NT			
Normal CV-1 cells	4	Neg.	<2	<2	NT			

NT—Not tested.

epidemiological observations (Kurtzke, 1966), but, up to date, attempts to isolate an agent from MS patients has ended in complete failure. In the past year, brain tissue obtained at autopsy from two cases of MS was explanted in culture and the resulting cell line observed for morphological abnormalities (ter Meulen et al.,1972b). At the third cell transfer, a small fraction of the cell population showed the presence of multinucleated giant cells containing intranuclear bodies resembling inclusions. These changes persisted throughout the lifetime of the culture. Through monitoring of cultures by EM, it was possible to detect the presence of nucleocapsids resembling paramyxoviruses in the cytoplasm of brain cells derived from one case of MS at the eleventh cell transfer. Following the routine established in studies of slow virus infections of the central nervous system, MS brain cells were fused with indicator cells at various passage levels and the cultures observed for lesions. At the level of the first cell transfer, *fused cell cultures* showed an increased mitotic rate, significant nuclear polymorphism, and a lack of contact inhibition. In general, these cultures resembled cultures of transformed cells. A positive hemadsorption reaction indicated the presence of a paramyxo-type virus, and EM revealed the presence of fine filamentous structures resembling the nucleocapsids of paramyxoviruses (Iwasaki et al., 1973). Negative staining of material obtained from the fused cell culture at the seventh cell transfer level revealed virions and nucleocapsids of the paramyxovirus group. The agent isolated from the tissue culture medium of cultures resulting from fusion of indicator cells with cells from *both* cases of MS was identified as belonging to the parainfluenza Type I viruses. The agent differs from other members of this group because its predominant viral population grows preferentially in tissue culture at 32° C rather than 37° C. Furthermore, it can be distinguished from the two prototypes of parainfluenza I virus because of biochemical, biophysical, and biological properties.

Type I parainfluenza viruses are known to cause a relatively mild infection of the respiratory tract, and assuming that the virus is present in the brain tissue of MS patients, one must still question its involvement in the etiology of the disease. Obviously, no answer can be forthcoming at this stage of investigation. It is not impossible, however, that viruses known to cause otherwise mild infections may become involved in an atypically severe disease through an indirect mechanism involving the immunological reaction of the host.

Table III lists examples of immunological derangements of the host and their possible effects on virus infection. Experimental and, in a few cases, clinical evidence for Types I and III is available. As far as Type II is concerned, there is ample evidence, particularly in clinical medicine, for the existence of antibodies against the host's own tissue, but there is no substantial evidence that this type of derangement is directly related to a virus infection.

TABLE III

Some examples of the complexities of the immunological derangements of the host and their effect on the outcome of virus infection

I. Immune response directed against antigen (s) coded by the virus.
  (A) Destruction of virus-harboring cells.
    Possible consequences: 1. Dissemination of the virus.
    2. Formation of infectious or non-infectious virus-antibody complexes (see below).
  (B) Formation of virus-antibody complexes.
    Possible consequences: 1. Immobilization of the virus preventing immediate injury to the host.
    2. Disease induced by the complexes.
II. Immune responses directed against antigens coded by the host as a possible result of virus infection.
  (A) Unmasking of "hidden" antigen.
    Possible consequence: Triggering of an immune response and injury to the host tissue.
  (B) Expression of host antigen originally not expressed on cell surface.
    Possible consequence: Same as under IIA above.
III. Virus infection of immunocompetent cells.
  (A) Depression of antibody-mediated and cell-mediated immune responses.
    Possible consequences: General or specific impairment of antibody responses.

At the present time, it is difficult to say where slow virus infection fits into this scheme. Rabies-infected tissue culture cells are destroyed by antibody and complement, but it is not known if the same mechanism operates in vivo. Cerebellar lesions in young rats induced by LCMV are caused by the immune response to the virus, because immunosuppression of the afflicted animals leads to the disappearance of the symptoms (Monjan et al., 1972). The possible effect of SSPE and of measles (observed a long time ago by von Pirquet) on immunocompetent cells was discussed before. Thus, there is a general paucity of information concerning the role of immunological responses in slow virus infection. However, one should not overlook the possibility that the problem of latency may be closely linked to the immunological response in individuals who harbor such viruses. Whether this mechanism is operative in a demyelinating disease such as

MS and creates a situation in which infection with an otherwise mild agent provokes an immune response leading to the destruction of cell elements of the central nervous system, remains to be seen.

ACKNOWLEDGMENTS This work was supported in part by NS 06859 from the National Institute of Neurological Diseases and Stroke of the National Institutes of Health and from funds from the John A. Hartford Foundation.

REFERENCES

BARBANTI-BRODANO, G., S. OYANAGI, M. KATZ, and H. KOPROWSKI, 1970. Presence of two different viral agents in the brain cells of patients with subacute sclerosing panencephalitis. *Proc. Soc. Expl. Biol. Med.* 134:230–236.

BRODY, J. A., and R. DETELS, 1970. Subacute sclerosing panencephalitis: A zoonosis following aberrant measles. *Lancet* 2: 500–501.

CLARK, H. FRED, E. SORIANO, S. JOSEPHS, and H. AASLESTAD, 1972. Selective inhibition of rhabdovirus replication in cell culture treated with the amphibian virus LT 1. To be published.

DAWSON, J. R., 1933. Cellular inclusions in cerebral lesions of lethargic encephalitis. *Amer. J. Path.* 9:7–16.

ENDERS, J. F., T. H. WELLER, and F. C. ROBBINS, 1949. Cultivation of the Lansing strain of poliomyelitis virus in cultures of various human embryonic tissues. *Science* 109:85–87.

IWASAKI, Y., H. KOPROWSKI, D. MÜLLER, and V. TER MEULEN, 1973. Morphogenesis and structure of a virus in cells cultured from brain tissue from two cases of multiple sclerosis. *Lab. Invest.*, (in press).

JENSEN, F. C., and P. KOLDOVSKY, 1972. Transformation of primate cells by an agent associated with Jakob-Creutzfeldt disease and detection of a new induced specific cell surface antigen. To be published.

KATZ, M., L. B. RORKE, W. S. MASLAND, H. KOPROWSKI, and S. H. TUCKER, 1968. Transmission of an encephalitogenic agent from brains of patients with subacute sclerosing panencephalitis to ferrets (prel. report). *New Engl. J. Med.* 279: 793–798.

KATZ, M., and H. KOPROWSKI, 1973. The significance of failure to isolate infectious viruses from six cases of subacute sclerosing panencephalitis. To be published.

KOPROWSKI, H., T. J. WIKTOR, and M. M. KAPLAN, 1966. Enhancement of rabies virus infection by lymphocytic choriomeningitis virus. *Virology* 28:754–756.

KURTZKE, J. F., 1966. An epidemiologic approach to multiple sclerosis. *Arch. Neurol.* 14:213–222.

LISCHNER, H. W., M. K. SHARMA, and W. D. GROVER, 1972. Immunologic abnormalities in subacute sclerosing panencephalitis. *New Eng. J. Med.* 286:786–787.

MONJAN, A. A., G. A. COLE, and N. NATHANSON, 1972. Immunological basis of cerebellar hypoplasia induced by lymphocytic choriomeningitis (LCM) virus infection of neonatal rats. *Fed. Proc.* 31:760A.

MÜLLER, D., V. TER MEULEN, M. KATZ, and H. KOPROWSKI, 1971. Cytochemical evidence for the presence of two viral agents in subacute sclerosing panencephalitis. *Lab. Invest.* 25:337–342.

NAKAI, T., F. L. SHAND, and A. F. HOWATSON, 1969. Development of measles virus in vitro. *Virology* 38:50–67.

NOTKINS, A. L., S. E. MERGENHAGEN, and R. J. HOWARD, 1970. Effect of virus infections on the function of the immune system. *Ann. Rev. Microbiol.* 24:525–538.

OYANAGI, S., L. B. RORKE, M. KATZ, and H. KOPROWSKI, 1971. Histopathology and electron microscopy of three cases of subacute sclerosing panencephalitis (SSPE). *Acta Neuropath.* 18:58–75.

PADGETT, B. L., D. L. WALKER, and G. M. ZURHEIN, 1971. Cultivation of papova-like virus from human brain with progressive multifocal leukoencephalopathy. *Lancet* 1: 1257–1260.

TER MEULEN, V., M. KATZ, and D. MÜLLER, 1972a. Subacute sclerosing panencephalitis: A review. In *Current Topics in Microbiology and Immunology*, 57:1–38. Berlin: Springer-Verlag.

TER MEULEN, V., H. KOPROWSKI, Y. M. KÄCKELL, Y. IWASAKI, and D. MÜLLER, 1972b. Fusion of cultured multiple sclerosis brain cells with indicator cells: Presence of nucleocapsids and virions and isolation of parainfluenza-type virus. *Lancet* 2:1–5.

THEIN, P., A. MAYR, V. TER MEULEN, H. KOPROWSKI, Y. M. KÄCKELL, D. MÜLLER, and R. MEYERMANN, 1972. Subacute sclerosing panencephalitis: Transmission of the virus to calves and lambs. *Arch. Neurol.* 27:540–548.

WEINER, L. P., R. M. HERNDON, O. NORAYAN, R. T. JOHNSON, K. SHAH, L. J. RUBINSTEIN, T. J. PREZIOSI, and F. K. CONLEY, 1972. Virus related to SV40 in patients with progressive multifocal leukoencephalopathy. *New Eng. J. Med.* 286: 385–390.

90 Cell-Virus Interactions in Slow Infections of the Nervous System

CLARENCE J. GIBBS, JR., and D. CARLETON GAJDUSEK

ABSTRACT Kuru and Creutzfeldt-Jakob diseases of man, and scrapie and mink encephalopathy of animals, are subacute progressive degenerative atypical infectious diseases caused by atypical viruses. They are the prototype viruses of a group designated the *subacute spongiform encephalopathies*. The physical, chemical, and biological properties of these viruses is examined and their nature discussed. The role of these four viruses as models for the elucidation of the etiology of other encephalopathies, presenile and senile dementias, and the aging process in man, is discussed.

Subacute spongiform virus encephalopathies

DURING THE PAST decade significant advances have been made in elucidating the etiology of subacute degenerative diseases of the central nervous system (CNS) of man. The first such chronic, apparently heredofamilial, human disease demonstrated to have infection as its etiology was a new and unique fatal degenerative brain disease, kuru, that occurs in an isolated Mesolithic cannibal culture of New Guinea (Gajdusek, 1963). The isolation of kuru in experimental animals subsequently led to the demonstration that other chronic and subacute degenerative disorders of the nervous system were caused by viruses. These findings now strongly suggest the possibility that ordinary viruses, masked, latent, or endosymbiotically persistent for years, may be reactivated to induce chronic progressive disease, and that new infectious virus-like agents, apparently unique and unrelated to previously described viruses, may be inapparent pathogens in man associated with long incubation periods yet which ultimately cause fatal subacute diseases.

To date, we have been able to demonstrate that four subacute progressive degenerative diseases of the nervous system: Two of man, kuru and Creutzfeldt-Jakob (C-J)

disease, and two of animals, scrapie and transmissible mink encephalopathy (TME), are closely related atypical infectious diseases caused by a new group of highly unconventional rather similar viruses. The infections are atypical in that they induce widespread destructive pathology only in the brain, although virus is present in most other tissues that is associated with little or no perivascular infiltration and without evidence of meningeal involvement or primary demyelination. Further, there is no disease-associated cerebrospinal fluid pleocytosis or consistent rise in CSF protein, no changes in sedimentation rate or clinical chemistry or hematological values, nor is there febrile response at any stage of the primary disease. The filterable virus-like etiological agents of these diseases are unconventional in the strange physical and biological properties they have in comparison to more conventional viruses. The purpose of this paper is to summarize what is known about three of the prototype members of this new group of infectious agents that we have termed the *subacute spongiform virus encephalopathies* (Table I) (Gibbs and Gajdusek, 1969a).

TABLE I
Subacute spongiform virus encephalopathies

Kuru
Creutzfeldt-Jakob disease
Scrapie
Transmissible mink encephalopathy

Viruses in this group fulfill the criteria of *slow infections* (Table II) established almost two decades ago by Sigurdsson (1954): (1) Each virus is associated with a long asymptomatic incubation period of from several months to several years. (2) Each virus induces a progressive subacute degenerative disease of the central nervous

CLARENCE J. GIBBS, JR., and D. CARLETON GAJDUSEK National Institute of Neurological Diseases and Stroke, National Institutes of Health, Bethesda, Maryland

TABLE II
General criteria of "slow infections"

A very long initial period of latency, lasting for several months to several years.

A protracted course of illness following the appearance of clinical signs, with progression to death.

Primary anatomical lesions limited to a single organ system.

Limited range of susceptible hosts.

TABLE III

Natural and experimental host range of subacute spongiform encephalopathies

Host	Kuru	C-J Disease	Scrapie	TME
Man	+	+		
Chimpanzee	+	+	—	—
New world monkeys				
Spider	+	+	—	—
Squirrel	+	+	+	+
Capuchin	+	+	—	NT
Woolly	+	+	NT	NT
Marmoset	+	+	NT	NT
Old world monkeys				
Rhesus	+	—	—	+
Bushbaby	?	+	NT	NT
Stumptail	—	—	NT	+
Cynomolgus	—	—	+	NT
Sheep	—	—	+	+
Goat	—	—	+	+
Mink	—	—	+	+
Ferret	—	—	NT	+
Domestic cat	?	+	NT	NT
Racoon	NT	NT	NT	+
Striped skunk	NT	NT	NT	+
Mice	—	—	+	?
Rats	—	—	+	—
Golden hamster	—	—	+	+
Chinese hamster	—	—	+	—
Gerbil	—	—	+	NT
Vole	NT	NT	+	NT
Natural Disease				
Incubation	5–20 years	unknown	3–5 years	unknown
Duration	4–12 months	4–5 months		3–8 weeks
Experimental Disease in Primates				
Incubation	9–101 months	11–26 months	14–65 months	11–33 months
Duration	< 1–11 months	< 1–5 months	2 months	1 month

+   Clinical disease and confirmatory histopathological lesions.

—   Absence of clinical disease and neuropathological lesions in animals inoculated from several months to over 9 years.

NT Not tested.

?   Suggestive histopathological lesions in the absence of clinically recognizable specific disease.

Many additional species of subhuman primates, smaller laboratory animal and avian host, have been tested; only those species have been listed in which one or more of the 4 diseases has been isolated.

system (CNS) manifested by ataxia, tremors, and postural instability that always terminates fatally. (3) Each virus induces histopathological lesions restricted to the brain primarily affecting the gray matter and consisting of severe neuronal vacuolation and neuronal dropout (particularly of Purkinje cells in the two human diseases), astrocytic proliferation and hypertrophy, varying degrees of intensity of status spongiosis, and PAS positive doubly refractile birefringent amyloid plaques. (4) Each virus is transmissible to animal hosts (Table III), though not necessarily the same ones.

Although variations in the severity and distribution of lesions in the CNS are seen among these diseases, intra-cytoplasmic vacuolation in the axonal and dendritic processes of neurons with coalescence of the vacuoles to form a ballooning and destruction of the cells is the underlying cytopathological feature of these four diseases (Beck et al., 1969; Lampert et al., 1969, 1971a). Spongiform changes that follow, though frequently minimal in the natural host, become the predominant lesion in the brains of animals dying with the experimentally induced disease. Electron microscopic study of brain tissue shows that the spongiform changes are caused by progressive clearing and eventual vacuolation of neuronal cytoplasm particularly in presynaptic and postsynaptic processes with final cell destruction (Lampert et al., 1971a). This occurs to a lesser extent in astrocytes as well. Ruptured plasma membranes and curled fragments of membranes are seen within the vacuoles. Fusion of these swollen cells or their processes, that is, neurons with neurons and an astrocyte with a nerve cell, was observed occurring after dissolution of adjacent cell membranes in the brains of mice affected with scrapie (Lampert et al., 1971b). In all instances it appears that astrocytes react to injury by proliferation and hypertrophy whereas nerve cells degenerate.

Distinction should be made between acute, persistent, chronic, latent, and slow infections. The following definitions are taken from our own work and that of Johnson (1970):

Acute infections: In acute infections a predictable pattern of sequential events occur: The incubation period is relatively short extending from a few days to several weeks; if infection results in disease, the acute disease ends in recovery, the presence of static pathological changes, or sequela in which there is no longer active virus replication, or death. Virus is generally not recoverable after this acute phase, and the host becomes immune to subsequent infection with the same or even closely related viruses.

Persistent infections: In persistent infections there is a continuance of the presence of the pathogenic organism within the host following recovery from either clinical or subclinical infection. Typical examples of this type of infection are adenovirus infections, serum hepatitis, feline panleucopenia virus, the virus of equine infectious anemia, and the viruses of Bolivian hemorrhagic fever and Kyasnur Forest disease. In these instances it would appear that the virus-host relationship is one of endo-symbiosis, and such a mutual relationship can exist for years.

Chronic infections: In contrast to the above, if persistence of the infectious agent is associated with accumulating progressive disease over a long period of time it is generally referred to as a chronic infection. Such infections may lack a characteristic incubation period and have an unusual clinical course and an unpredictable outcome. Recovery may be quite rapid, or periods of remission and exacerbation may occur. In such infections, humoral and cellular response by the host appear to have little effect on the organism.

Latent infections: The term *latent infection* implies a very different virus-host relationship. Here the virus may not be recoverable but is masked or eclipsed, in some incomplete or defective form, or even in a genetically integrated or temperate state. At some stage the virus has the capacity to reinitiate replication giving rise to new signs of acute infection following initiation by one or more activators, e.g. trauma, drug stress, natural or iatrogenic immunosuppression, helper or hinderer viruses. An example of this would be the papova-SV_{40}-like virus isolated from patients dying with progressive multifocal leucoencephalopathy (PML) (Padgett et al., 1971; Weiner et al., 1972).

Slow infections: Slow infections are yet another type of virus-host relationship, one in which after a long asymptomatic incubation period progressive unremittingly accumulative pathology leads to death. In Sigurdsson's words, a slow infection is a "slow motion picture of an acute infection." There are no slow viruses but rather viruses that induce a slow, relentlessly progressive and always fatal disease. In such infections there appears to be a single target organ, and the genetic mechanism of the host plays an important determinative role in resistance or susceptibility to disease. It is here that the viruses of kuru and C-J disease, scrapie, mink encephalopathy, visna, maedi, progressive pneumonia of sheep, pulmonary adenomatosis of sheep, mammary carcinoma of mice, lymphocytic choriomeningitis, avian lymphomatosis, Rous sarcoma, murine leukemia, Shope rabbit papilloma, Aleutian mink disease, measles in subacute sclerosing panencephalitis (SSPE) and the papoviruses in progressive multifocal leucoencephalopathy belong.

In designating our program we have included the concept of temperateness in the diseases of the CNS that we are studying, we may possibly be faced with examples in mammals of a mechanism similar to that of temperate

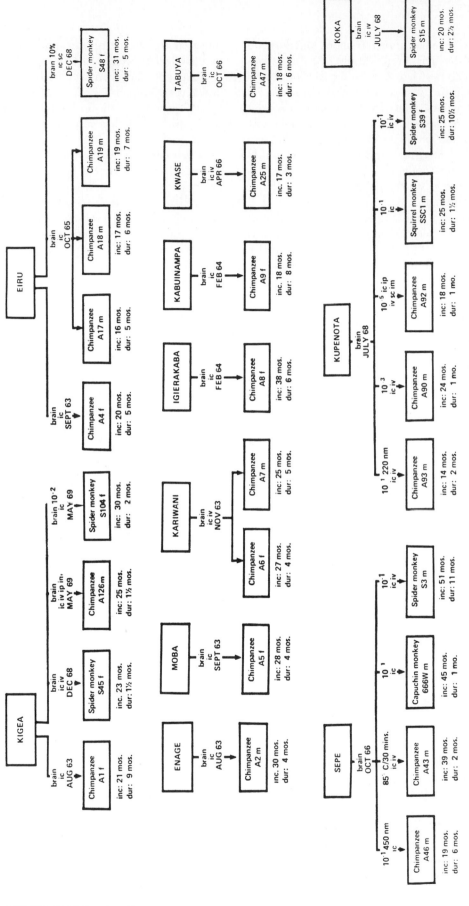

FIGURE 1   Transmissions of kuru from man to chimpanzees and new world monkeys.

bacterial viruses, wherein the provirus is carried as genetic information on the chromosomes of the host genome and may at some stage be activated by a variety of mechanisms.

Kuru

Kuru is the first subacute progressive degenerative disease of the brain of man for which a virus etiology has been established. This invariably fatal CNS disease, characterized by cerebellar ataxia and trembling, progressive dysarthia, and finally, dysphagia, has been observed only in the Fore people and their neighbors who reside in a restricted part of the Eastern Highlands of Papua New Guinea (Gajdusek and Zigas, 1957; Gajdusek, 1963). The disease is caused by a filterable agent and is transmissible from humans to chimpanzees and serially from chimpanzees to chimpanzees (Gajdusek et al., 1966; Gajdusek et al., 1967; Gibbs, 1967; Gibbs and Gajdusek, 1969b; Gibbs and Gajdusek, 1970). Further, the disease has also been transmitted from humans to a rhesus monkey (Gajdusek and Gibbs, 1972) and five species of new world monkeys (spider monkeys, squirrel monkeys, capuchin monkeys, the woolly monkey, and the marmoset) (Gajdusek et al., 1968; Gajdusek and Gibbs, 1971; Peterson et al., unpub.). Similar transmissions of the virus have been made to new world monkeys inoculated with tissues from affected chimpanzees (Gajdusek et al., 1968; Gajdusek and Gibbs, 1971). The disease has now been transmitted from 11 human patients to 18 chimpanzees with incubation periods varying from 14 to 39 months (Figure 1) and in 5 serial passages from chimpanzee to chimpanzee with incubation periods of 10 to 18 months (Figure 2). The mean incubation period has thus dropped on chimpanzee to chimpanzee passage from 22 to 11 months and remains essentially static. Primary and serial transmissions are successfully accomplished in animals inoculated intracerebrally (IC) and intracerebrally and intravenously (IV) or only by multiple peripheral routes (IV, SC, IM). In the rhesus monkey the asymptomatic incubation period was 8 years and 5 months. Primary transmissions of the disease to new-world monkeys has been associated with a mean incubation period of 30 months and with a range of 20 to 51 months (Figure 3). Kuru has also been transmitted on passage from chimpanzees to new world monkeys and from new world monkeys to new world monkeys (Figure 4). On serial passage in new world monkeys, the incubation periods have not appreciably dropped except on second passage in capuchin monkeys where disease has recently developed in only 10 to 12 months instead of after or about 2 years as on primary passage in capuchin monkeys. The incubation period may be significantly shorter on serial passage in

marmosets than after primary inoculation. Obviously, much of our characterization of the virus of kuru is now being carried on in these species of new world monkeys.

Data on the primary transmission, serial passage, and properties of kuru virus are presented in Figures 1 and 2. The virus is remarkably stable for many months to over several years when stored at $-70°C$ in the form of frozen tissue or brain tissue suspensions prepared in phosphate buffered saline. In comparison to most conventional viruses, kuru is very thermostable and infectivity is not destroyed nor appreciably reduced following exposure of the virus to a temperature of 85°C for 30 minutes. Human brain tissue contains greater than 10^6 infectious doses of virus per milliliter of suspension while infected chimpanzee brain has infectivity titers that exceed $10^{7.5}$ infectious doses per milliliter. The virus passes through a 220-nm Millipore filter but has not yet been detected in the filtrates of 100-nm, or smaller, Millipore filters. Virus has not yet been recovered from whole blood or serum or from urine, cerebrospinal fluids, milk, placenta, or amniotic fluids of kuru patients, or experimentally affected animals. The disease has, however, been successfully transmitted to chimpanzees inoculated with small amounts of pooled suspensions of liver, kidney, spleen, and mesenteric lymph node from chimpanzees killed in terminal stages of disease (Figure 5). Further, the disease can also be serially transmitted in chimpanzees inoculated peripherally only without the need of assaulting the brain to induce disease. Experiments have already been initiated to determine the efficacy of transmitting the disease by single routes of inoculations including intradermal, conjunctival, and intranasal, as well as to find out whether or not this will influence the incubation periods or course of the disease. In recent experiments by Field and his coworkers, conducted at the Demyelinating Disease Unit, Newcastle-upon-Tyne, England, using the EIRU strain of our virus, disease has been induced in two chimpanzees that each received a single intramuscular injection. The incubation period in both animals was 18 months and the clinical disease did not vary from that which we have described for animals that had received virus intracerebrally or peripherally by multiple routes of inoculation (Field, personal communication, 1972).

No matter what the route of inoculation has been thus far, or whether the virus has originated from infected brain or infected visceral tissue, or indeed, if the virus suspension was prepared from fresh tissue or had been lyophilized, exposed to heat, or clarified by filtration (220-nm pore size), or centrifuged, pelleted, and resuspended, the resulting clinical disease in experimental animals has remained remarkably constant, consisting of cerebellar ataxia, trembling, progressive dysphagia, lassitude, wasting, and terminal inanition. Similarly, the histopatho-

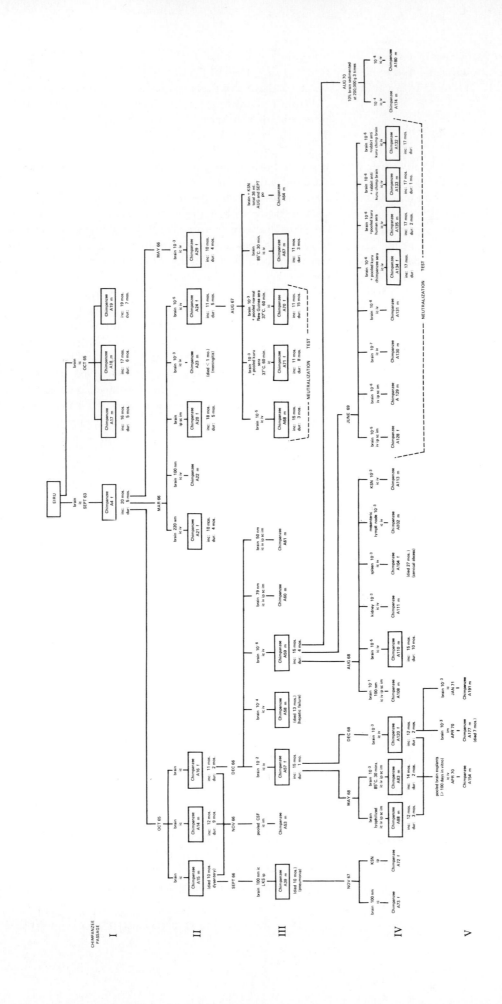

FIGURE 2   Transmissions of kuru from chimpanzees to chimpanzees.

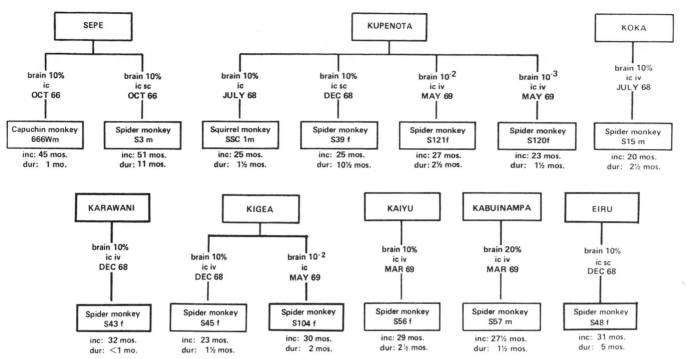

FIGURE 3   Transmission of kuru from man to new world monkeys.

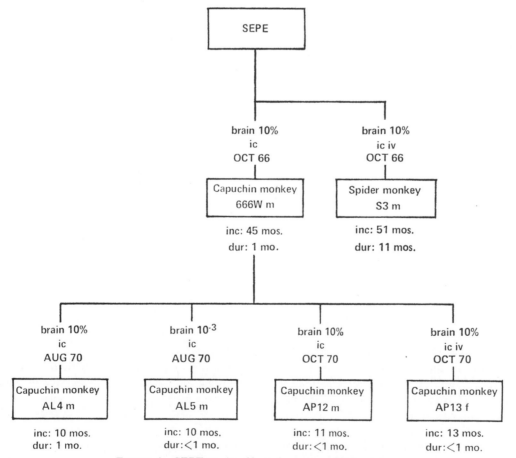

FIGURE 4   SEPE strain of kuru in new world monkeys.

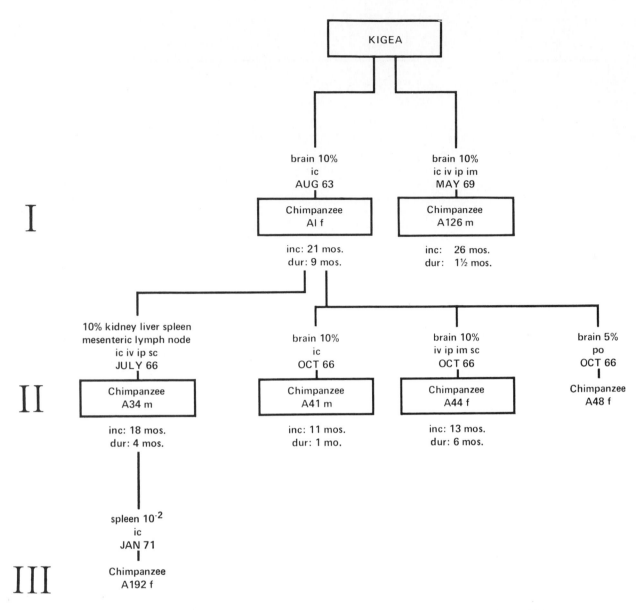

FIGURE 5   Transmission of kuru to chimpanzees inoculated with pooled suspensions of liver, kidney, spleen, and mesenteric lymph node from chimpanzees killed in terminal stages of the disease.

logical lesions remain remarkedly reproducible regardless of the strain of the virus, the route of inoculation or the species of susceptible host employed. Perhaps the most significant finding has been the 100% transmission of kuru. In every instance where we have asked a single question of the transmission—does this case of kuru transmit to an experimental host?—we have had success in transmitting the disease from humans, and in no instance have we failed to serially transmit the disease when direct passage of the virus was attempted.

Extensive search for an antibody to kuru virus in man or experimental animals has failed. There is no evidence of neutralizing antibody in human or animal sera late in the disease or in sera from animals hyperimmunized with 10%, 20% suspensions or 100-fold ultracentrifugal concentrates of kuru infected brain tissues. No in vitro demonstration of an antibody by complement fixation or immunofluorescent techniques has been successful, and attempts at the precipitation of virus as a virus-antibody complex with "immune" sera and anti-gamma globulin have failed. No gamma-globulin or complement deposits or virus-antibody complexes have been demonstrable in serum or in frozen sections of brain, kidney, or other tissues (Gibbs, 1967). Wide screening for antibodies in the sera of kuru patients and of animals affected with experimental kuru has failed to reveal any consistent pattern of reaction with any of the more than 50 agents studied to date (Benfante et al., unpub.). Similarly, screening of the sera from kuru hyperimmunized animals against a wide variety of known viral, bacterial, and mycotic antigens has failed to elicit any significant diagnostic or antigenic relationship.

An electron microscopic extensive search for virions in infected tissues of man and animals and study of fractions of brain suspensions obtained by density gradient banding the agent in the zonal ultracentrifuge have revealed no recognizable virions (Gibbs and Gajdusek, unpublished data; Siakotos et al., 1973).

Creutzfeldt-Jakob disease

Creutzfeldt-Jakob disease (C-J), or subacute spongiform encephalopathy, is the first presenile dementia of man demonstrated to have infection as its etiology (Gibbs et al., 1968; Gibbs and Gajdusek, 1969a). As early as 1959 Klatzo, Gajdusek, and Zigas pointed out the similarity between the neuropathological lesions in kuru and C-J disease and at that time stated that the only variations seemed to be in the distribution and the intensity of the lesions in the brains of patients dying with the two diseases. Although it took us more than 5 years to obtain specimens from a patient dying with C-J disease, during the past 3 years we have collected information and specimens of tissues from over 100 patients in the United States, Canada, England, France, Belgium, Australia, South America, and Guam (Roos, Gajdusek, and Gibbs, 1973). Although small foci of this disease have been referred to, the disease is worldwide in distribution and occurs most frequently as sporadic cases and in some instances in familial patterns in which the disease has appeared over several generations in close relatives in what appears to be a Mendelian dominant form of inheritance. Both the sporadic type and the familial type have now been transmitted to experimental animals in our laboratory (Gibbs and Gajdusek, 1971; Ferber et al., 1973).

The C-J disease has been transmitted on primary passage to chimpanzees (Gibbs et al., 1968) and to five species of new world monkeys (Gibbs and Gajdusek, 1969a; Gajdusek and Gibbs, 1971), the spider monkey, the squirrel monkey, the capuchin monkey, the woolly monkey, to an old world monkey the bushbaby (*Galago senegalensis*) and most recently to the common household cat following inoculation with human brain tissue (Gibbs

and Gajdusek, unpublished data). It is significant to note that the successful transmission of C-J disease was based entirely on the stimulus and necessary techniques gained from our studies on kuru disease.

We have to date successfully transmitted C-J disease independently from 14 human cases to 18 of 42 chimpanzees inoculated. Details on the primary transmissions of these 14 cases to 18 chimpanzees are summarized in Figure 6. As may be noted, incubation periods have ranged from 11 to $23\frac{1}{2}$ months, which is considerably shorter than those observed on the primary transmission of kuru. Like kuru, the disease can be induced in animals inoculated peripherally only without intracerebral inoculation, but peripheral route transmission was associated with the slightly longer incubation period of 16 months. As shown in Figure 7 the disease is serially transmissible in chimpanzees where it is already in fourth serial passage but, unlike kuru, the incubation period remains the same as it was on primary passage. On primary passage into new world monkeys the incubation period has ranged from 23 months to 29 months following intracerebral inoculation (Figure 8). On serial passage of the virus from chimpanzees to new world monkeys the incubation period has ranged from 9 to $28\frac{1}{2}$ months (Figures 9 and 10). Further serial passages have not appreciably reduced this asymptomatic period. Unlike the disease in chimpanzees that is associated with a rapid course, the disease in new world monkeys tends to be associated with a more protracted course. In the bushbaby, the incubation period was 16 months, and in the domestic cat, the incubation period was 30 months. Both animals were in advanced stages of clinical disease when they were killed two months after onset of symptoms.

Like kuru the virus of C-J disease passes through a Millipore filter of 220 nm but has not yet been detected in the filtrates of smaller size Millipore filters. The virus has not yet been isolated from visceral tissues, whole blood, serum, cerebrospinal fluids, or urine from human patients or experimental animals affected with the disease. No specific antigen-antibody reaction has been demonstrated in patients or animals affected with the disease nor have these sera yielded a diagnostic antibody pattern when tested with more than 50 viral antigens, a finding similar to that observed in our studies of kuru (Brown et al., 1972).

Both kuru and C-J disease infectious viruses persist in vitro in cultures of explanted animal and human tissues, respectively (Gajdusek et al., 1972). Kuru virus was found to persist in chimpanzee brain cell cultures maintained in vitro at 37°C for at least 70 days and the virus of C-J disease persisted in human brain cells maintained in vitro at 37°C for at least 255 days. The cultures themselves

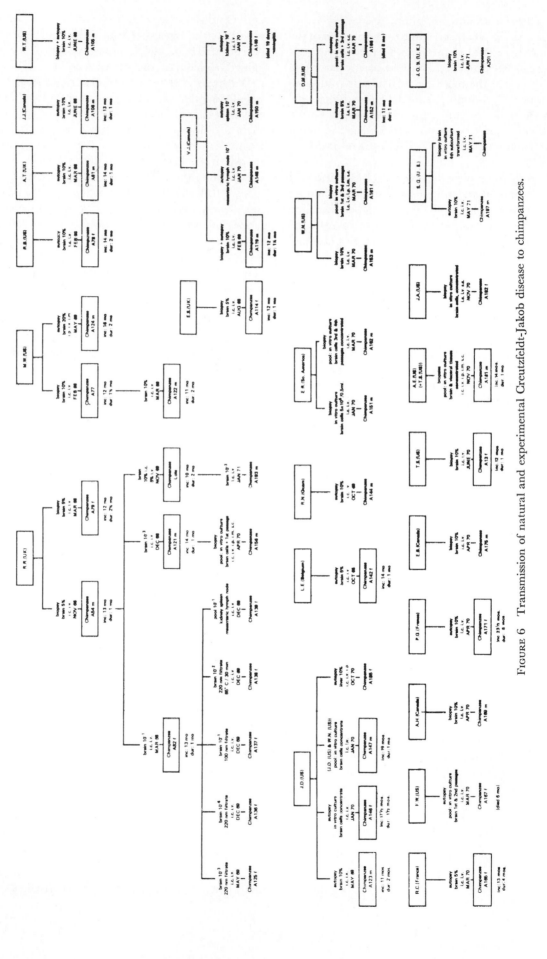

FIGURE 6 Transmission of natural and experimental Creutzfeldt-Jakob disease to chimpanzees.

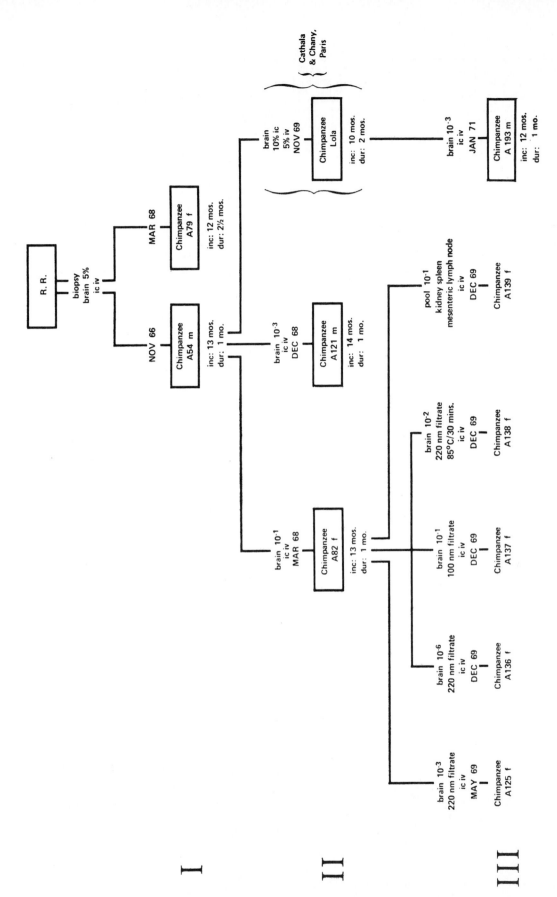

FIGURE 7   Serial transmission of Creutzfeldt-Jakob disease to chimpanzees.

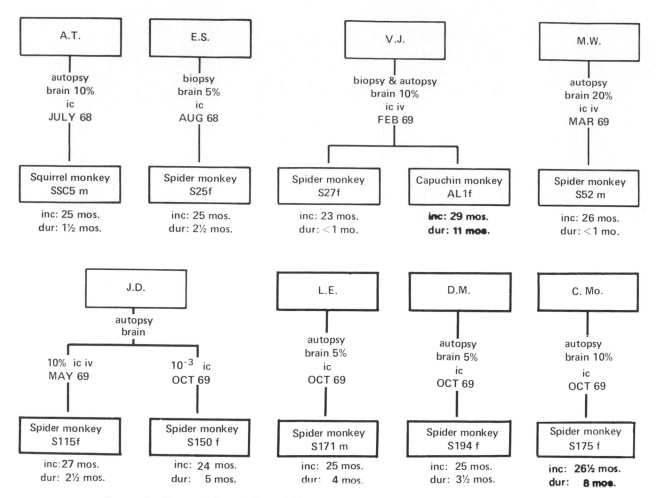

FIGURE 8   Transmissions of Creutzfeldt-Jakob disease from man to new world monkeys.

showed no morphological evidence of infection nor was there evidence of any cytopathic effect or lysis of the cells. An attempt to demonstrate interference in these cultures using 14 different viruses has been unsuccessful (Cornelius, Gibbs, and Gajdusek, unpublished data). Test viruses titered no differently in the cultures infected with virulent kuru and C-J disease viruses than they did in normal chimpanzee brain cultures. Further, there was no evidence of interferon production in any of the infected brain cell cultures nor were interferon inhibitors present. Treatment of kuru affected chimpanzees with interferon-inducing double stranded RNA polyinosinic-polycytidylic acid has no apparent effect on the clinical course or outcome of the disease (Gibbs, Gajdusek, Levy, and Baron, unpublished data).

An extensive electron microscopic search for virus in infected tissues of man and animals, as well as in the fractions of brain suspensions obtained by density gradient banding of the infected tissue suspensions by zonal ultracentrifuge similar to that done with kuru material,

have revealed no recognizable virions. Recently, however, we have reported on the spontaneous transformation of in vitro cultures of human brain from a patient with C-J disease (Hooks et al., 1972). The explanted and trypsinized primary cultures initially grew out as a monolayer or fibroblast-like cells. After 60 days, foci of cells with altered morphology were noted in the third subculture of the originally trypsinized cultures. During the next 30 days foci of similar altered morphology appeared in all subcultured flasks of the original trypsinized cultures and from two of the three flasks of primary explants. By the fifth subculture the transformed cells were the only cell types observed in all cultures. The altered cells were larger than the original fibroblast-like cells and epithelial-like in appearance. They displayed the classical characteristics of transformed cells with loss of contact inhibition. These transformed cells have now been maintained for over 18 months and are in the 100th subculture. They show a very rapid rate of growth with enormously increased capacity to persist in serial subcultures and have

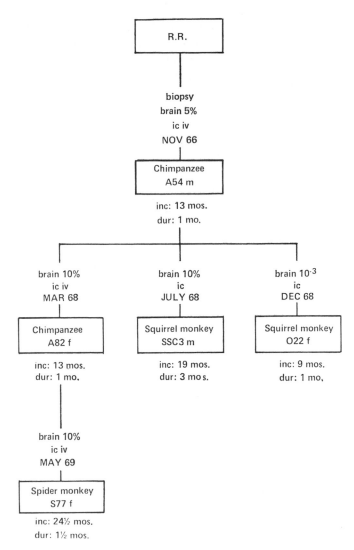

FIGURE 9 R.R. strain of Creutzfeldt-Jakob disease in new world monkeys.

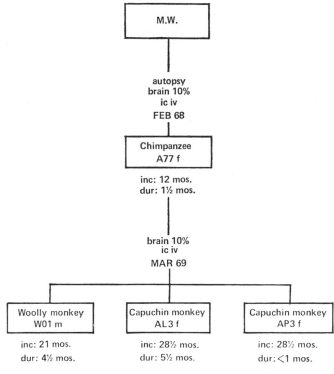

FIGURE 10 M.W. strain of Creutzfeldt-Jakob disease in new world monkeys.

better than a 60% plating efficiency, and as few as 10 cells are needed to initiate cell growth on a surface area of 25 cm square.

Chromosome studies at the 47th and 54th subculture levels indicate that the cells are of human origin and contain a chromosome number ranging from 70 to 80 with the majority containing 75 chromosomes with a large accrocentric marker and a long arm longer than number 13 to 15 group and 1 or 2 minutes. The incidence of chromosomal aberrations is less than 1%.

Viral particles morphologically resembling the known oncogenic RNA viruses were detected in all cultures examined since the transformation occurred. Intracellular particles about 60 to 90 nm in diameter were observed only within the cytoplasm and contained an electronlucent center, giving a doughnut shaped appear-

ance. Particles are also observed obtaining an outer envelope from the plasma membrane. Extracellular virus particles are observed which have an oval or spherical configuration measuring about 100 to 120 nm in diameter. Most of the extracellular virus particles contain a partially dense nucleoid. The virus particles observed appear to represent one type of particle that shows different stages of development in the cytoplasm, in the budding process and in extracellular spaces. The virus particles morphologically resemble the Mason-Pfizer monkey mammary tumor virus and recently have shown an antigenic relationship to this virus (Hooks, Gibbs, and Gajdusek, 1973). When injected into new born hamsters, viable cells multiply, form a tumor-like mass, but within 9 days are rejected. Greater concentrations may induce permanent tumors and this is currently under study.

Although we have inoculated chimpanzees and new world monkeys with crude suspensions of the patient's brain as well as with aliquots of nontransformed cells not showing virions by EM and transformed cells rich in intracellular and extracellular virus particles, such experiments will not allow us to assign an etiological role to this virus should our animals develop clinical signs and histological lesions of C-J disease, because in all instances cells of the patient's brain are involved. A more convincing and significant etiological relationship will only be established if animals develop C-J disease following their inoculation with the transforming virus that has been carried for several serial subcultures in a rapidly growing human sarcoma cell line, a technique that will by dilution have separated the virus from association with the patient's brain cells. Such experiments have already been initiated. To date, the original brain tissue from the patient from whom the transformed cell line was derived has not induced C-J disease in chimpanzees or new world monkeys inoculated 2 years ago.

Scrapie

Scrapie has been recognized as a chronic fatal ataxia of sheep and goats in England and continental Europe for over two centuries. The disease now occurs naturally on farms in Europe, Asia, Africa, and South and North America, including 29 of the United States. The disease has been eradicated from Australia, New Zealand, and South Africa. Sheep breeders have realized that the disease occurs in their scrapie-free flocks only after the introduction of new breeding stock from scrapie-affected flocks. Over a century ago attempts were made to demonstrate the possibility of contagion or transmission of the disease, but it was first experimentally transmitted by inoculation of a ewe in 1899 with brain tissue from scrapied sheep. Not until 1936 was this transmission experiment fully confirmed by Cuillé and Chelle (1936). Experimental serial transmission, using brain tissue or other visceral tissue, was carried out in several laboratories, using sheep of different breeds. Susceptibility was dependent on the breed of sheep employed, and the goat was found to be more susceptible than the sheep with incubation periods rarely over 1 year. It was not until the middle of the last decade that serially transmissible encephalopathy was induced in mice inoculated with scrapie infected goat brain or with brain tissue from naturally infected sheep. This greatly facilitated laboratory study of the agent and attempts at characterization of the scrapie virus have been intensive during the last few years, because it serves as an ideal model for the study of human diseases (Gibbs et al., 1965).

The disease appears to be transmitted naturally as a Mendelian recessive trait, and although low level communicability has been demonstrated after long and intensive contact between scrapied and normal goats in the laboratory, such contact infection has not been clearly proved with the naturally occurring disease. Contact infection between inoculated and uninoculated mice occurs but at low frequency (Morris et al., 1965). Of particular interest is the differing susceptibility to scrapie on inoculation of various breeds of sheep; some breeds such as the Herdwick and Dalesbred are highly sus-

ceptible while others such as the Dorset Down and Tigree appear to be totally resistant (Eklund and Hadlow, 1972). Parry and Vince (1970) contend that in nature the disease behaves strictly as a Mendelian recessive trait.

The disease in mice as in sheep is insidious at onset, progressive in course, and invariably fatal. The clinical incubation period is appreciably reduced to 4 to 6 months in mice in contrast to the 2 to 5 years which are required in sheep (Morris and Gajdusek, 1963). Although all strains of mice are susceptible, the incubation period may vary somewhat dependent upon the strain of mice being used. Dickinson et al. (1968) have shown that a gene termed SINC controls the length of the incubation period of mouse scrapie. Two alleles are known that are designated $s7$ (short incubation period) and $p7$ (long incubation period). Most strains of mice are homozygous for $s7$, and a strain of mice known as VM is the only strain to carry $p7p7$ alleles. It has been shown that the incubation period in mice with the genome $s7p7$ is intermediate between that in the two homozygotes. The gene appears to act by controlling the delay before the onset of replication of the agent in brain and spleen; the rate of replication of the agent and the final titer achieved are similar in both homozygotes once this process has been initiated. The disease also has been transmitted to goats, hamsters, rats, gerbils, voles, mink, and recently to subhuman primates, the cynomolgus monkey and the squirrel monkey. In the cynomolgus monkey the incubation period was $5\frac{1}{2}$ years while in the squirrel monkey the incubation period was 18 to 20 months. Chimpanzees inoculated with scrapie mouse, goat, and sheep brain virus remain well after over 5 years of observation.

The pathogenesis of the disease in mice has been described by Eklund et al. (1967). As early as 4 weeks after subcutaneous inoculation of mice, after a brief "eclipse," virus was detectable in the spleen and lymph nodes in low concentration; by the 8th to the 12th week, virus had spread to other lymph glands and viscera but was barely detectable in the spinal cord. Indeed, virus in high concentrations was not detectable in the nervous system until as late as 16 weeks after inoculation, but it remained there in appreciable amounts until as long as 36 weeks after inoculation when the animals died of the disease. In contrast, when virus is injected intracerebrally into mice in low concentrations, there follows a short eclipse phase of several days after which there is rapid multiplication within the brain and spleen with virus present in these tissues at titers exceeding 10^5 logs per gram of tissue. Virus persists in these tissues through the course of the asymptomatic period into clinical disease and until death of the animal 4 to 5 months after inoculation.

As with kuru, C-J disease, and transmissible mink encephalopathy (TME), it has not been possible to demonstrate immune complex deposition in scrapie by fluorescent staining techniques with anti-IgG or anti-complement (C′3) except in the kidneys of a rare scrapie affected mouse. This could well be the result of some other persistent virus infection, i.e., LCM in the studied mice. Attempts to lower the titer of suspension of virus by immune precipitation with antisera expected to contain antiscrapie antibody and anti–gamma globulin have not succeeded. There is no evidence of virus-antibody complex in the circulation. Increased levels of interferon have not been detected in tissues of scrapie-infected animals, yet scrapie-infected mice retain the capacity to produce interferon in response to vesicular stomatitis and West Nile viruses. Prophylactic treatment of mice with high concentrations of potent mouse interferon, given before and at intervals after exposure to scrapie virus, has no effect on the incubation period, course of clinical disease, or fatal outcome of the infection. Neonatal thymectomy or splenectomy, x-irradiation, and treatment with ACTH, cortisone, cytoxan, or anti-lymphocytic serum have not influenced the course of scrapie in mice inoculated intracerebrally. In splenectomized mice inoculated intraperitoneally, the incubation period of the disease is prolonged.

The scrapie virus persists in cultures of brain cells from scrapie-infected animals for prolonged periods of time, but it has not yet been transferred to cell cultures from normal animals or to continuous cell lines. The same long virus persistence without evident cytopathogenic effect or electron microscopically recognizable virions is thus found in brain cells from scrapie as is seen with kuru and C-J disease. Of interest has been the observation that all three viruses appear to cause cell stimulation with better and faster cell growth than that of cultures from normal brains.

Unfortunately, some early studies of the physical properties of the scrapie virus were naive and were received rather uncritically with the resulting, sometimes sensationalistic, speculation as to its physical structure and mechanism of replication. Of particular interest has been the unusual thermostability and resistance of scrapie virus to ultraviolet radiation, formalin, and many proteolytic enzymes and organic reagents when compared with the relative susceptibility to inactivation of most conventional viruses. The agent has high resistance to heat but is markedly, though not completely, inactivated by exposure to 100°C. Up to 80°C the virus remains stable, and at 85°C linear inactivation occurs, with much infectious virus still present after 30 to 60 minutes of exposure. It is small in size, but by no means uniquely small; it passes through membranes of 43 nm but not of 27 nm pore diameter and is surely not so small as to be dialyzable as has been reported. The enormous resistance

of scrapie virus to inactivation by UV radiation indicates that the radiation-sensitive, information-bearing portion is probably very small. Exposure of the scrapie infected mouse brain suspension to 1000 erg/sec/cm^2 UV (254 nm) resulted in the loss of only 1.6 log of infectivity (Gajdusek and Gibbs, 1968, 1971). Haig and coworkers (1969) and Latarjet and coworkers (1970) have found even greater resistance of the scrapie virus to UV inactivation, as has Marsh for the virus of TME (1968).

Scrapie infectivity is easily detectable in tissue that has undergone prolonged storage in concentrations of formalin ranging from 0.5% to 18%. The agent is stable at pH 2.1 to 7.0 and is resistant to exposure to pepsin, trypsin, acetylethylenamine, and to DNAse and RNAse in dilute suspensions of crude brain suspensions. It is very sensitive in such suspensions to periodate (0.01 M), 6 to 8 M urea, 90% phenol, and to a lesser extent to ether and 1% betapropriolactone. Scrapie is markedly susceptible to 2-chloroethanol (80% 5.0 log) (50% 1.9 log), moderately susceptible to acid methoxyethanol and chloroform methanol in a concentration of 2:1 v/v (1.4 and 1.1 log, respectively) and weakly sensitive to n-pentanol (0.7), n-butanol and ultrasonics (0.7 log) and n-butanol alone (0.2 log) (Hunter, 1972).

Hunter and his coworkers (1972) have distinguished three groups of lysosomal enzymes by their behavior in scrapie infected mouse brain. They found neuraminidase, alpha glucosidase, beta glucosidase, lipase, phospholipase a, acid phosphatase, aryl sulphatase A and aryl sulphatase B to remain unchanged in activity. The enzymes that showed increased activity late in the incubation period were: acid proteinase, acid ribonuclease, beta galactosidase and cathepsin D. Enzymes that showed increased activity relatively early in the incubation period were: hyaluronidase, acid deoxyribonuclease, n-acetyl-beta-glucosaminidase, n-acetyl-beta-galactosaminidase, alpha fucosidase, and alpha mannosidase. However, these same workers pointed out that they have been unable to distinguish between the primary biochemical lesions of scrapie from the secondary less specific changes due to the enormous amount of brain damage that occurs in the disease, and thus none of these early biochemical investigations shed any light on the nature of the scrapie agent itself.

On the nature of the agents

The data that we have presented summarizes what is currently known about the atypical viruses belonging to the group of subacute spongiform encephalopathies. There has certainly been controversy as to the physico-chemical nature of these agents. It should be noted that all work on these agents has been done using crude tissue suspensions and in all attempts to purify them the infectivity is regularly associated with fragments of membrane. As pointed out earlier, electron microscopy of affected tissues has failed to reveal any recognizable packaged virions. Tissue suspensions of high infectivity or pellets of sedimented infectivity do not contain visible virus-like particles. Yet, wherever virus infectivity is high, as in specific density gradient bands, there the electron microscope reveals only fragments of membrane. Similarly, in the characteristic even pathogonomonic lesions of these agents in the brain, the intraneuronal vacuoles are not empty but are filled with a debris of packed sheets of curled fragments of membrane, the same membrane as that which forms the wall of the vacuoles and which is familiar to all virologists as the surface membrane of those viruses formed by budding from the cell or into cell vacuoles.

The enormous resistance of these transmissible agents to heat inactivation and to exposure to several agents that rapidly inactivate most viruses, such as proteolytic enzymes, acetylethylenamine, formaldehyde, and RNAse and DNAse, and, particularly, their extreme resistance to UV inactivation, has led to speculation that they are infectious agents lacking any nucleic acid for their replication. Several possible structures of self-replicating macromolecules, not containing nucleic acids, including proteins, basic proteins, and polysaccharides, have been hypothesized by various authors. The inactivation of scrapie by membrane disrupting substances (ether, periodate, urea, phenol) has led to the hypothesis that it may be a replicating membrane or a membrane associated polysaccharide alone. We know this about its structure: The virus is closely associated with membranes and is inactivated when such association is broken; it is probably not usually packaged into morphologically recognizable virions, and it lies in the size range of 25 to 35 nm. Filtration studies on kuru and C-J disease have failed to pass the agent through membranes of 100 nm pore diameter and this is in sharp contrast to the easy passage of scrapie and mink encephalopathy through such membranes. The UV inactivation data would suggest that no nucleic acid is present, but what level of repair or of protection from UV inactivation may be associated with firm binding of small nucleic acids to membranes is still an unknown matter. When one further allows for the possibility of very small viroids in potato spindle tuber virus with RNA genomes of under 50,000 daltons (Diener, 1972), it is hard to determine a priori what physical inactivation characteristics such membrane bound small RNA might possess. The possibility does exist that these agents are a new class of viruses without nucleic acids, but it would be premature to accept such a departure from conventional thinking until we know more about the possibility of protection from UV inactivation that might be afforded to a small DNA or RNA molecule tightly bound to membranes.

The agents do possess the virus-like property of activating their own synthesis in cells that may already possess codons for the synthesis of these activating agents. Such could be an integrated complementation or a derepression action that activates synthesis of more of themselves, coded for by normally repressed or "turned-off" coding sequences. Or, they may act as derepressors releasing partial or complete genetic information for the synthesis of other partial or complete viruses that may be lying in the cell as a result of past infections in earlier generations.

Whatever the nature of the agents ultimately turns out to be, the subacute spongiform encephalopathies can no longer be viewed simply as medical exotica but must be looked upon as actual problems of concern to neurologist, gerontologists, infectionists, and biochemists as well as to virologists.

REFERENCES

BECK, E., P. M. DANIEL, M. ALPERS, D. C. GAJDUSEK, and C. J. GIBBS, JR., 1969. Neuropathological comparisons of experimental kuru in chimpanzees with human kuru, with a note on its relation to scrapie and spongiform encephalopathy. Basel, New York: Karger. In *Pathogenesis and Etiology of Demyelinating Diseases, Int. Arch. Allerg.* 36:553–562.

BROWN, P., J. HOOKS, R. ROOS, D. C. GAJDUSEK, and C. J. GIBBS, JR., 1972. Attempt to identify the agent for Creutzfeldt-Jakob disease by CF antibody relationship to known viruses. *Nature New Biol.* 235(57):149–152.

CUILLÉ, J., and P. L. CHELLE, 1936. Le maladie dite tremblante du mouton, est-elle inoculable? *C. R. Acad. Sci. (Paris)* 203: 1552–1554.

DICKINSON, A. G., V. M. H. MEIKLE, and H. FRASER, 1968. Identification of a gene which controls the incubation period of some strains of scrapie agent in mice. *J. Comp. Pathol.* 78: 293–299.

DIENER, T. O., 1972. Is the scrapie agent a viroid? *Nature New Biology* 235 (59):218–219.

EKLUND, C. M., and W. J. HADLOW, 1972. Slow viral diseases. *Critical Reviews in Environmental Control* 2(4):535–555.

EKLUND, C. M., R. C. KENNEDY, and W. J. HADLOW, 1967. Pathogenesis of scrapie virus infection in the mouse. *J. Infect. Dis.* 117:15–22.

FERBER, R. A., S. L. WIESENFELD, R. ROSS, A. R. BOBOWICK, C. J. GIBBS, JR., and D. C. GAJDUSEK, 1973. Familial Creutzfeldt-Jakob disease. Transmission of the familial disease to the chimpanzee. In Proc. 10th Intern. Excerpta Medica Publishers.

GAJDUSEK, D. C., 1963. Kuru. *Trans. Roy. Soc. Trop. Med. Hyg.* 57(3):151–169.

GAJDUSEK, D. C., and C. J. GIBBS, JR., 1968. Slow, latent and temperate virus infections of the central nervous system. In *Infections of the Nervous System Res. Publ. A.R.N.M.D.* 45:254–280. Baltimore: The Williams & Wilkins Co.

GAJDUSEK, D. C., and C. J. GIBBS, JR., 1971. Transmission of two subacute spongiform encephalopathies of man (kuru and Creutzfeldt-Jakob disease) to new world monkeys. *Nature* 230:588–591.

GAJDUSEK, D. C., and C. J. GIBBS, JR., 1972. Transmission of Kuru from man to rhesus monkey (*Macaca mulatta*) $8\frac{1}{2}$ years following inoculation. *Nature* 240(5380):351.

GAJDUSEK, D. C., C. J. GIBBS, JR., and M. ALPERS, 1966. Experimental transmission of a kuru-like syndrome to chimpanzees. *Nature* 209(5025):794–796.

GAJDUSEK, D. C., C. J. GIBBS, JR., and M. ALPERS, 1967. Transmission and passage of experimental "kuru" to chimpanzees. *Science* 155(3759):212–214.

GAJDUSEK, D. C., C. J. GIBBS, JR., D. M. ASHER, and E. DAVID, 1968. Transmission of experimental kuru to spider monkey (*Ateles geoffreyi*). *Science* 162:693–694.

GAJDUSEK, D. C., C. J. GIBBS, JR., N. G. ROGERS, M. BASNIGHT, and J. HOOKS, 1972. Persistence of viruses of kuru and Creutzfeldt-Jakob disease in tissue cultures of brain cells. *Nature* 235:104–105.

GAJDUSEK, D. C., and V. ZIGAS, 1957. Degenerative diseases of the central nervous system in New Guinea. The endemic occurrence of "kuru" in the native population. *New Eng. J. Med.* 257(30):974–978.

GIBBS, JR., C. J., 1967. Search for infectious etiology in chronic and subacute degenerative diseases of the central nervous system. *Curr. Top. Microbiol. Immun.* 40:44–58.

GIBBS, JR., C. J., and D. C. GAJDUSEK, 1969a. Infection as the etiology of spongiform encephalopathy (Creutzfeldt-Jakob disease). *Science* 165:1023–1025.

GIBBS, JR., C. J., and D. C. GAJDUSEK, 1969b. Kuru—a prototype subacute infectious disease of the nervous system as a model for the study of amyotrophic lateral sclerosis. In *Motor Neuron Diseases.* New York: Grune & Stratton, pp. 269–279.

GIBBS, JR., C. J., and D. C. GAJDUSEK, 1970. Kuru in New Guinea: Pathogenesis and characteristics of its virus. *Amer. J. Trop. Med.* 19(1):138–145.

GIBBS, JR., C. J., and D. C. GAJDUSEK, 1971. Transmission and characterization of the agents of spongiform virus encephalopathies: Kuru, Creutzfeldt-Jakob disease, scrapie and mink encephalopathy. In *Immunological Disorders of the Nervous System. Res. Publ. A.R.N.M.D. XLIX.* Baltimore, Md.: Williams & Wilkins Co., pp. 383–410.

GIBBS, JR., C. J., D. C. GAJDUSEK, D. M. ASHER, M. P. ALPERS, E. BECK, P. M. DANIEL, and W. B. MATHEWS, 1968. Creutzfeldt-Jakob disease (subacute spongiform encephalopathy); transmission to the chimpanzee. *Science* 161:3839 (July 26), 388–389.

GIBBS, JR., C. J., D. C. GAJDUSEK, and J. A. MORRIS, 1965. Viral characteristics of the scrapie agent in mice. *NINDB Monograph No. 2, Slow, Latent, and Temperate Virus Infections,* Washington, D.C.: U.S. Government Printing Office, pp. 195–202.

HAIG, D. A., M. C. CLARKE, E. BLUM, and T. ALPER, 1969. Further studies on the inactivation of the scrapie agent by ultraviolet light. *J. Gen. Virol.* 5:455–457.

HOOKS, J., C. J. GIBBS, JR., H. CHOPRA, M. LEWIS, and D. C. GAJDUSEK, 1972. Spontaneous transformation of human brain cells grown in vitro and description of associated virus particles. *Science* 176:1420–1422.

HOOKS, J., C. J. GIBBS, JR., and D. C. GAJDUSEK, 1973. Transformation of cell cultures derived from human brain. *Science* 179:1019–1020.

HUNTER, G. D., 1972. Scrapie: A prototype slow infection. *J. Infect. Dis.* 25:427–440.

JOHNSON, R. T., 1970. Virus-host relationships in acute and chronic encephalopathies. In *Proceedings of VIth International Congress of Neuropathology.* Paris: Masson et Cie, pp. 761–778.

KLATZO, I., D. C. GAJDUSEK, and V. ZIGAS, 1959. Pathology of kuru. *Lab. Invest.* 8:799–847.

LAMPERT, P. W., K. M. EARLE, C. J. GIBBS, JR., and D. C. GAJDUSEK, 1969. Experimental kuru encephalopathy in chimpanzees and spider monkeys. *J. Neuropath. Exp. Neurol.* 28:353–370.

LAMPERT, P. W., D. C. GAJDUSEK, and C. J. GIBBS, JR., 1971a. Experimental spongiform encephalopathy (Creutzfeldt-Jakob disease) in chimpanzees. Electron Microscopic Studies. *J. Neuropath. Exp. Neurol.* 30:20–32.

LAMPERT, P., J. HOOKS, C. J. GIBBS, JR., and D. C. GAJDUSEK, 1971b. Altered plasma membranes in experimental scrapie. *Acta Neuropath.* 19:81–93.

LATARJET, R., B. MUEL, D. A. HAIG, M. C. CLARKE, and T. ALPER, 1970. Inactivation of the scrapie agent by near monochromatic ultraviolet light. *Nature* 227:1341–1343.

MARSH, R. F., 1968. Studies on transmissible mink encephalopathy. Ph.D. Thesis. Madison, Wis.: University of Wisconsin.

MORRIS, J. A., and D. C. GAJDUSEK, 1963. Encephalopathy in mice following inoculation of scrapie sheep brain. *Nature* 197:1084–1086.

MORRIS, J. A., D. C. GAJDUSEK, and C. J. GIBBS, JR., 1965. Spread of scrapie from inoculated to uninoculated mice. *NINDB Monograph No. 2, Slow, Latent and Temperate Virus Infections.* Washington, D.C.: U.S. Government Printing Office, pp. 273–276.

PADGETT, B. L., D. L. WALKER, G. M. ZuRHEIN, and R. J. ECKROADE, 1971. Cultivation of papova-like virus from human brain with progressive multifocal leucoencephalopathy. *Lancet* 1:1257–1260.

PARRY, H. B., and A. A. VINCE, 1970. Scrapie disease of sheep: The roles of gene and "slow virus" in pathogenesis. In *Proceedings VIth International Congress of Neuropathology.* Paris: Masson et Cie, pp. 839–840.

ROOS, R., D. C. GAJDUSEK, and C. J. GIBBS, JR., 1973. The clinical characteristics of Creutzfeldt-Jakob disease. *Brain* 96:1–20.

SIGURDSSON, B., 1954. Observations on three slow infections of sheep. *Brit. Vet. J.,* 110 (7, 8, 9):255–270, 307–322, 341–354.

WEINER, L. P., R. M. HERNDON, O. NARAYAN, R. T. JOHNSON, K. SHAH, L. J. RUBENSTEIN, T. J. PREZIOSI, and F. K. CONLEY, 1972. Virus related to SV_{40} in patients with progressive multifocal leucoencephalopathy. *New Eng. J. Med.* 286:385–390.

91 Differential Effects of Viruses and Their Mutants upon Central Nervous System Development

CEDRIC S. RAINE

ABSTRACT It has been demonstrated that mutant forms of a known neurovirulent virus, reovirus type III, can produce abnormalities in the developing nervous system that differ from those caused by the virus from which the mutants were derived. The wild virus proved fatal in all cases; one mutant produced a chronic hydrocephalic condition, and another apparently had no effect upon chronically infected animals. Such observations may be of relevance to certain "slow-virus" conditions, some of which may be etiologically related to mutant forms of everyday viruses. In addition, the marked neurotropism displayed by reovirus and its mutants, plus their tendency to alter neurotubules, renders this virus a useful model for the study of neurobiological processes.

NEUROTROPISM is a common property of many viruses from most groups, and examples of congenital or perinatal central nervous system (CNS) defects of viral etiology (see Table I) indicate that viruses provide a tool that can be manipulated for the study of developmental events in

CEDRIC S. RAINE Department of Pathology (Neuropathology) and The Rose F. Kennedy Center for Research in Mental Retardation and Human Development, Albert Einstein College of Medicine, The Bronx, New York

the CNS. Many factors, both viral and host, can now be assimilated.

First, it is obvious from the work on rubella and cytomegalovirus infection of man (Mims, 1968; Rawls, 1968; Johnson and Johnson, 1969a) and bluetongue virus infection of sheep (Osburn et al., 1971a, b) that the age at which the developing fetus is challenged with virus is of paramount importance. Infection early in gestation has the most drastic consequences upon subsequent CNS development, because it is during this period that mitosis is maximal and organogenesis is at an early stage. Second, viral infection during early CNS development will not necessarily be recognized as foreign, because the circulatory and immune systems are undeveloped (Rawls, 1968). Third, that viral infection of the CNS frequently involves specific cell types only, e.g. parvoviruses (Margolis et al., 1971b), and that cytopathic effects vary from one group (or even strain) of viruses to another.

Other infections that demonstrate these factors are those of myxoviruses (mumps, measles, influenza; Johnson, 1968; Johnson and Johnson, 1968, 1969b; Mims, 1968; Johnson et al., 1971); parvoviruses (Margolis and

TABLE I
Viruses inducing CNS maldevelopment

Type	Nucleic Acid	Distribution in CNS	Cytopathic Effect
Rubella (arbovirus)	RNA	pantropic	low
Cytomegalovirus (herpesvirus B)	DNA	pantropic	high
Myxoviruses	RNA	ependyma	high
Parvoviruses	DNA	cerebellum granule cells	high
Diplornaviruses	RNA	(A) bluetongue pantropic	high
		(B) reovirus type I ependyma and neurons	high
		(C) reovirus type III neurons	high

Kilham, 1968, 1970; Margolis et al., 1971; Tennant and Hand, 1971); and lymphocytic choriomeningitis virus (Lehmann-Grube, 1971; Cole et al., 1971).

It is now clear that viral neurotropism does not result in a uniform pattern of degeneration or immune reaction and that changes in the viral genome (strain differences) can produce different diseases. To investigate the latter phenomenon, we have attempted to dissect virus effects and host reaction by selectively varying the viral genome of a well characterized, neurotropic virus, reovirus type III.

Reovirus type III and its temperature-sensitive mutants

Reovirus, which occurs as three serotypes, has been isolated from a large range of hosts. It is most commonly recognized ultrastructurally by its extraordinary tendency to coat microtubules with an electron-dense, amorphous material in close proximity to virions. Its various properties have been reviewed by Spendlove in 1970. Its isolation from human subjects is known, but its role in human disease is still not established (Rosen, 1968).

The propensity of several strains of reovirus type I (of primate origin) to cause hydrocephalus has been investigated (Kilham and Margolis, 1969; Margolis and Kilham, 1969). Suckling animals of several species were given virus intraperitoneally or intracerebrally. The hydrocephalus was of the obstructive type and was interpreted as being the combined result of a primary viral attack upon ependyma with concomitant inflammation. Viral inclusions were also seen in neurons, but this occurred secondarily to the ependymal changes, suggesting that the virus gained access to the parenchyma via the ependyma. The ependymal lesion induced a glial reparative response that effected obstruction of the ventricular system at strategic points. Incipient hydrocephalus was apparent 8 to 10 days postinoculation. The work of Margolis and Kilham on reovirus and that of Johnson and coworkers on myxoviruses have contributed evidence supporting the possibility of a viral etiology of congenital human hydrocephalus. Reovirus type II is considered intermediate in neurovirulence between type I, which induces chronic disease, and type III, which causes a fulminant, fatal meningoencephalitis, usually within two weeks.

Of the three serotypes, reovirus type III is the most neurovirulent. The route of entry into the brain parenchyma, like type I, involves infection of the ependyma. More pronounced, however, is its lack of hydrocephalic effect and its preferential infection of neurons (Margolis et al., 1971a; Gonatas et al., 1971). Extensive parenchymal degeneration was reported involving neurons and ependyma with edema and encephalomalacia preceding inflammatory changes.

A series of studies, presently being performed in our laboratories in collaboration with Dr. Bernard N. Fields, concerns this virus and its mutants in vivo (Fields, 1972; Raine and Fields, 1972). A firm baseline for such study has been provided by the isolation and subsequent genetic and biochemical characterization of temperature-sensitive mutants of the Dearing strain of reovirus type III (Fields and Joklik, 1969; Fields, 1971), a later immunofluorescent and ultrastructural study on the in vitro system (Fields et al., 1971) and a recent ultrastructural study on the same strain of reovirus type III in vivo (Raine and Fields, in press).

Temperature-sensitive (ts) mutants belong to a general class of conditionally lethal mutants known to elicit their lethal effects under laboratory controlled conditions. Such mutants grow well at a low (permissive) temperature; but at a higher (nonpermissive) temperature, they are incapable of completing a successful growth cycle, and the genetic defect may become morphologically expressed (Fenner, 1970). Since ts mutants are being isolated in increasing numbers from many viruses with known encephalitogenic properties (e.g. herpes and several myxoviruses), and since wild reovirus type III is encephalitogenic in suckling animals, it seemed logical that an in vivo comparison of reovirus type III with ts mutants in vivo could yield information regarding the encephalitogenic effects of different viral components. Such mutants are known to differ from the wild form in a single gene only (Fields, 1971). Data from our study might be relevant to the study of congenital CNS anomalies of viral etiology, and some diseases of later life in which mutant forms of common viruses might well be operative, e.g. multiple sclerosis and subacute sclerosing panencephalitis (SSPE). To achieve this end, we have selected two of the many ts mutants of reovirus type III, and these will be compared with the wild form. Morphologically, wild reovirus type III at all temperatures possesses an RNA core 45 nm in diameter, surrounded by an inner capsid 60 nm across. This in turn is enveloped by an outer capsid, giving the total virion a diameter of 75 nm. Mutant ts 352 (henceforth referred to as type B) at 31°C (permissive) assembles full virions, but at 39°C contains the RNA core and inner capsid only, and therefore has a diameter of 60 nm. Mutant ts 447 (type C) at 31°C, assembles full virions but at 39°C assembles only capsid material, lacking an RNA core, and has a diameter of 75 nm (Fields et al., 1971).

Three-day-old Sprague-Dawley rats were given an 0.03 ml inoculum containing 3×10^4 plaque forming units (PFU) wild reovirus type III intracerebrally. Animals became sick 8 to 14 days postinoculation and died within 24 hours after the onset of clinical signs. Other inocula, ranging from doses of 3×10^1 to 3×10^8 PFU/0.03 ml were also used. The lowest dose was found to cause 80%

fatalities over 2 weeks while the highest dose was 100% fatal in 10 to 12 days. Viral assay from isolates of CNS showed a significant increase in yield from infected animals (Raine and Fields, in press). Other rats (same age and strain) were given an 0.03 ml intracerebral inoculation of the type B or type C ts mutant containing approximately 3×10^6 PFU. About 25% became sick and most of these died after 8 to 12 days. Some did recover, however, and together with the remaining 75%, continued to thrive for up to 10 months.

Brain tissue was sampled from all three groups during the acute disease and, in the case of the ts mutant groups, at monthly intervals up to 10 months. This material is currently being subjected to viral assay, fluorescent antibody (FA), and light and electron microscope study. Results from the viral assay and FA study are as yet incomplete. Suffice it to say that in the mutant groups, virus persists for at least 6 weeks, and genetic studies have shown that the viruses recovered from the CNS of mutant-infected animals retain their respective mutant properties.

Light microscope study of acutely sick brains revealed that each virus type was highly neurotropic, invasion of neurons appearing as a primary event preceding the inflammatory response. The CNS of rats given wild-type virus contained severe hemorrhagic, necrotic lesions, affecting mainly the cerebral cortex and hippocampus. Lesions caused by type B ts mutant were less extensive, and those by type C were identical to those induced by the wild type.

Electron microscopy of early lesions showed that wild type displays little tendency to infect any other cell type but the neuron. Viral factories within these cells comprised collections of completely assembled 75 nm virions embedded in a matrix of 50 nm kinky filaments (Figure 1A). The coating of microtubules in the vicinity of viral material was a rare observation. More frequently, aberrant microtubule arrays were found in glial cells, suggesting that some effect of the virus, other than direct infection, was sufficient to subvert the normal microtubule protein synthesis in these cells.

Neurons infected with type B contained large factories with virions arranged in a paracrystalline fashion. Typical type B particles (RNA core and inner capsid) were common but also many apparently complete reovirus particles were evident (Figure 1B). Measurement of the outer capsid disclosed that the average diameter was between 5 and 10 nm less than that of the complete virion. The ability for this core mutant to assemble more in vivo than in vitro is due to the temperature of the rat being 2°C to 3°C lower than the in vitro temperature at which this mutation is fully expressed. Also, genetic analysis of this virus isolated from infected brains has shown that the virus still retains its mutant properties, thus ruling out the

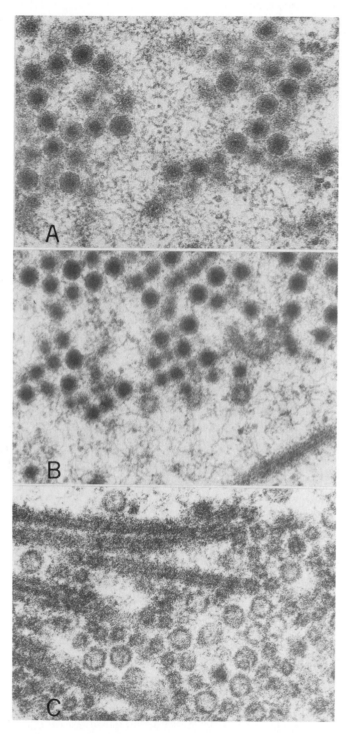

FIGURE 1 Intraneuronal viral factories from animals acutely sick following inoculation with (A) wild-type reovirus type III—note complete virions; (B) ts mutant type B—note cores, incompletely assembled virions, and single coated microtubule (lower right); and (C) ts mutant type C—note predominance of capsid only and the abundant coated microtubules. X80,000.

possibility of reversion to wild form. Coated microtubules also occurred in type B infected neurons, but rarely.

Ultrastructural examination of type C infected CNS tissue showed that the large intraneuronal factories were composed mainly of "empty" virions. Complete virions also occurred, and their presence, like type B, was accounted for by the temperature difference between the in vivo and in vitro situations. Most striking, however, was the high frequency of coated microtubules coursing through and near viral factories (Figure 1C). The acquisition of dense coating material seemed to be related purely to proximity to viral material. Since type C assembles mainly capsid material, this observation lends support to the conclusion of Dales that the coating of microtubules is related to viral capsid material (reviewed by Spendlove, 1970). This uniquely high affinity to affect microtubules in this way may make this mutant an important tool in the study of fundamental processes in the nervous system, e.g. axoplasmic transport.

Of those animals chronically infected with the two ts mutants, only the type B group displayed CNS abnormalities; type C animals showed neither clinical nor histological signs. After about 12 weeks, most type B infected animals began to develop hydrocephalus, sometimes accompanied clinically by frontal bossing. Usually, however, clinical signs were not apparent. This hydrocephalus was found not to be of the obstructive type (cf. reovirus type I), because ependyma was intact, subcortical white matter was present, and the aqueduct of Sylvius was patent. The thinning of cortical mantle was believed to be the result of a loss of parenchyma, and eventually the mantle became cavitated. Active degeneration and viral particles, however, were not revealed by ultrastructural techniques.

Our results, therefore, indicate that wild reovirus type III and two of its ts mutants, differing from each other in a single gene only and possessing similar neurotropic effects, clinically produce three markedly different conditions: one fatal (wild type); one hydrocephalus ex vacuo (type B); and one apparently no effect (type C). Obviously, this suggests a difference in the efficacy of the two mutants to produce disease. Whether type C particles were subject to the effect of neutralizing antibody and eliminated early in the infection course is not yet known. Also it remains to be shown (possibly by current viral assays and FA studies) if type B persists for a longer period and whether or not an artificial elevation of the animal's body temperature to 39°C effects an altered disease after ts mutant infection.

Such variation in disease pattern between wild virus and ts mutants may parallel some of the differences seen between strains of myxoviruses (Johnson and Johnson, 1968). In addition, vastly different conditions such as measles encephalitis and SSPE in young children, both causally related to measles virus, may indeed represent the in vivo expression of an alteration in the measles genome. Recent work on SSPE in animals tends to support this concept (Johnson and Byington, 1972).

In summary then, it has been shown that cells such as neurons and ependyma can be selectively infected by certain viruses, and even specific organelles, such as neurotubules, can be involved. In many cases, depending largely upon the age at which infection occurs, an altered or defective immune response is elicited that leads to viral persistence. We have shown that by manipulating the genome of an encephalitogenic virus, the various effects can be changed dramatically. From these observations, it can be concluded that the broad repertoire of viruses with an affinity to infect the CNS renders them promising candidates for the study of fundamental neurobiologic processes.

ACKNOWLEDGMENTS   The author thanks Dr. R. D. Terry for his encouragement throughout this study; Dr. B. N. Fields for his collaboration; Dr. J. W. Prineas for his advice on the manuscript; Earl Swanson, Howard Finch, Karen Berkman, and Miriam Casayuran for their excellent technical assistance; and Connie Incledon and Mary Palumbo for secretarial help.

Supported in part by grants NS 08952, NS 03356, and 1-RO1-A110326 from the National Institutes of Health; by grant 692-A-2 from the National Multiple Sclerosis Society; and by a grant from the Alfred P. Sloan Foundation. The author is the recipient of a Research Career Development Award from the National Institutes of Health, grant NS 70265-01.

REFERENCES

COLE, G. A., D. H. GILDEN, A. A. MONJAN, and N. NATHANSON, 1971. Lymphocytic choriomeningitis virus: Pathogenesis of acute central nervous system disease. Fed. Proc. 30:1831–1841.

FENNER, F., 1970. The genetics of animal viruses. Ann. Rev. Microbiol. 24:297–334.

FIELDS, B. N., 1971. Temperature-sensitive mutants of Reovirus type 3. Features of genetic recombination. Virol. 46:142–148.

FIELDS, B. N., 1972. Genetic manipulation of Reovirus: A model for modification of disease. New Eng. J. Med. 287:1026–1033.

FIELDS, B. N., and W. JOKLIK, 1969. Isolation and preliminary genetic and biochemical characterization of temperature-sensitive mutants of reovirus. Virol. 37:335–342.

FIELDS, B. N., C. S. RAINE, and S. G. BAUM, 1971. Temperature-sensitive mutants of reovirus type 3: Defects in viral maturation as studied by immunofluorescence and electron microscopy. Virol. 43:569–578.

GONATAS, N. K., G. MARGOLIS, and L. KILHAM, 1971. Reovirus type III encephalitis: Observations of virus-cell interactions in neural tissues. II. Electron microscopic studies. Lab. Invest. 24:101–109.

JOHNSON, K. P., and D. P. BYINGTON, 1972. The neuropathology of acute and chronic SSPE agent infection in hamsters. Trans. 48th Ann. Meeting Amer. Assoc. Neuropath., June 1972, Chicago, Ill. J. Neuropath. Exp. Neurol., (in press).

JOHNSON, K. P., R. KLASNJA, and R. T. JOHNSON, 1971. Neural tube defects of chick embryos: An indirect result of influenza A virus infection. *J. Neuropath. Exp. Neurol.* 30:68–74.

JOHNSON, R. T., 1968. Mumps virus encephalitis in the hamster. Studies of the inflammatory response and noncytopathic infection of neurons. *J. Neuropath. Exp. Neurol.* 27:80–95.

JOHNSON, R. T., and K. P. JOHNSON, 1968. Hydrocephalus following viral infection: The pathology of aqueductal stenosis developing after experimental mumps viral infection. *J. Neuropath. Exp. Neurol.* 27:591–606.

JOHNSON, R. T., and K. P. JOHNSON, 1969a. Slow and chronic virus infections of the nervous system. Contemporary neurology. In *Recent Advances in Neurology*. Philadelphia, Pa.: F. A. Davis Co., pp. 33–78.

JOHNSON, R. T., and K. P. JOHNSON, 1969b. Hydrocephalus as a sequela of experimental myxovirus infections. *Exp. Mol. Path.* 10:68–80.

KILHAM, L., and G. MARGOLIS, 1969. Hydrocephalus in hamsters, ferrets, rats, and mice following inoculations with reovirus type I. I. Virologic studies. *Lab. Invest.* 21:183–188.

LEHMANN-GRUBE, F., 1971. Lymphocytic choriomeningitis virus. *Virology Monographs* 10:1–173. Vienna and New York: Springer-Verlag.

MARGOLIS, G., and L. KILHAM, 1968. In pursuit of an ataxic hamster, or virus-induced cerebellar hypoplasia. In *The Central Nervous System*, O. T. Bailey, and D. Smith, eds. Baltimore, Md.: The Williams & Wilkins Co., pp. 157–183.

MARGOLIS, G., and L. KILHAM, 1969. Hydrocephalus in hamsters, ferrets, rats, and mice following inoculations with reovirus type I. II. Pathologic studies. *Lab. Invest.* 21:189–198.

MARGOLIS, G., and L. KILHAM, 1970. Parvovirus infections, vascular endothelium, and hemorrhagic encephalopathy. *Lab. Invest.* 22:478–488.

MARGOLIS, G., L. KILHAM, and N. K. GONATAS, 1971a. Reovirus type III encephalitis: Observations of virus-cell interactions in neural tissues. I. Light microscopy studies. *Lab. Invest.* 24:91–100.

MARGOLIS, G., L. KILHAM, and R. H. JOHNSON, 1971b. The parvoviruses and replicating cells: Insights into the pathogenesis of cerebellar hypoplasia. In *Progress In Neuropathology*, H. M. Zimmerman, ed. New York and London: Grune & Stratton, pp. 168–201.

MIMS, C. A., 1968. Pathogenesis of viral infections of the fetus. *Progr. Med. Virol.* 10:194–237.

OSBURN, B. I., A. M. SILVERSTEIN, R. A. PRENDERGAST, R. T. JOHNSON, and C. J. PARSHALL, JR., 1971a. Experimental viral-induced congenital encephalopathies. I. Pathology of hydranencephaly and porencephaly caused by bluetongue vaccine virus. *Lab. Invest.* 25:197–205.

OSBURN, B. I., R. T. JOHNSON, A. M. SILVERSTEIN, R. A. PRENDERGAST, M. M. JOCHIM, and S. E. LEVY, 1971b. Experimental viral-induced congenital encephalopathies. II. The pathogenesis of bluetongue vaccine virus infection in fetal lambs. *Lab. Invest.* 25:206–210.

RAINE, C. S., and B. N. FIELDS, 1972. Encephalitogenic effects of reovirus type III and its temperature sensitive mutants (Abstract). *Trans. 48th Ann. Meeting Amer. Assoc. Neuropath.*, June 1972, Chicago. *J. Neuropath. Exp. Neurol.*, (in press).

RAINE, C. S., and B. N. FIELDS, 1973. Pathogenesis of reovirus type III encephalitis—an ultrastructural study. *J. Neuropath. Exp. Neurol.* (in press).

RAWLS, W. E., 1968. Congenital rubella: The significance of virus persistence. *Progr. Med. Virol.* 10:238–285.

ROSEN, L., 1968. Reoviruses. In *Virology Monographs*. Vienna and New York: Springer-Verlag, pp. 73–107.

SPENDLOVE, R. S., 1970. Unique reovirus characteristics. *Progr. Med. Virol.* 12:161–191.

TENNANT, R. W., and R. E. HAND, JR., 1971. Requirement of cellular synthesis for Kilham rat virus replication. *Virol.* 42:1054–1063.

92 Herpesvirus-Induced Alterations in the Composition and Function of Cell Membranes

PATRICIA G. SPEAR

ABSTRACT Many virus-cell interactions result in alterations in the structure and function of cellular membranes. Altered states of the membranes may be detectable by the presence of new cell surface receptors and by abnormal cell interactions. These viral-induced changes must be due either to the rearrangement of normal membrane constituents or to the insertion of new molecules into the membranes. The herpesviruses, which comprise a large and ubiquitous group of DNA-containing enveloped viruses, induce novel changes in the social behavior of infected cells concomitant with the appearance of new surface antigens. In addition, some viruses in this group can induce the malignant transformation of cells. Evidence is presented that herpesviruses specify the synthesis of proteins that become incorporated into cellular membranes and are responsible for the new antigens on the surfaces of infected cells. The appearance of the membrane-bound proteins, most of which are glycosylated, correlates with the changes in the social behavior of infected cells. Genetic variants of the virus that specify different glycoproteins also induce different kinds of social behavior. It is suggested that the herpesvirus membrane proteins play a role in the induction of altered cell interactions including malignant transformation.

THE INTERACTION of a virus with its host cell frequently results in alterations of the cellular membranes. These alterations may be manifested by the presence of new antigens on the cell surface, by the increased susceptibility of cells to agglutination mediated by lectins, and by changes in the social behavior of cells as evidenced by cell fusion, differences in adhesion, and loss of normal growth control. Since viruses are also able to kill cells, it is obvious that uncontrolled growth, or the malignant transformation of cells, can be a consequence of viral-specified activities only if cytocidal activities are not expressed in the same cells. This statement is made to emphasize the point that cells which are destroyed as a result of virus infection may exhibit the same membrane alterations as other cells that eventually become transformed by the

same virus. The interactions of polyomavirus with cells of different types provide a good example. Cells that are killed as a consequence of polyomavirus replication as well as cells that are transformed exhibit rearrangements in the normal constituents of the cell surface that are related to viral functions and can be detected by lectin-mediated agglutination of the cells (Eckhart et al., 1971). Viral replication is not always associated with cytocidal functions; some of the enveloped RNA viruses, both oncogenic and nononcogenic, may replicate without killing their host cells or preventing cell division. These viruses acquire their outer coats by budding through portions of the plasma membrane modified by viral-specified products (Choppin et al., 1971; Pasternak, 1969). The viral proteins probably function primarily to facilitate the envelopment process, but some of them may contribute to changing the normal growth control of cells in the case of the oncogenic RNA viruses.

The herpesviruses are DNA-containing enveloped viruses. In cells infected with these viruses, new antigens appear on the cell surface, new kinds of interactions are observed among the cells, and viral-specified proteins appear in nuclear and cytoplasmic membranes even though envelopment takes place primarily at the inner nuclear membrane (Roizman and Spear, 1971a). The viral membrane proteins have as yet been demonstrated only in cells that are actively replicating the virus and that ultimately are destroyed. However, there is immunologic evidence to suggest that herpesvirus products may be responsible for new antigens found on the surfaces of cells from a variety of human and animal tumors (Klein, 1971). It is the purpose of this paper to describe these herpesvirus membrane-bound proteins in relation to the new antigenic determinants and the altered social behavior of cells that are productively and cytolytically infected and to point out the parallels that exist between the antigenic characteristics of these cells and those of the tumor cells.

PATRICIA G. SPEAR The Rockefeller University, New York, New York

Interactions of herpesviruses and their hosts

A brief description of the herpesviruses and their activities will provide the background for discussion of the viral-induced modifications of cell membranes. This group of viruses is defined by the composition and morphology of the mature virion. The nucleic acid is a double-stranded DNA molecule that is associated with proteins and is located in the central core of the virion. The core is surrounded by an icosahedral shell consisting of 162 morphological subunits composed of protein. The shell or nucleocapsid is in turn enclosed within a membranous envelope that is actually derived from modified cellular membranes (Roizman et al., 1972).

The herpesviruses are widespread in nature; many animal species (including reptiles and birds) harbor one or more members of this group. Man is host to at least four, including herpes simplex virus, cytomegalovirus, herpes varicella-zoster, and the Epstein-Barr virus. In general herpesviruses exhibit considerable versatility in interactions with their natural hosts. For example, herpes simplex virus (HSV) can remain with its host for life in a latent state, perhaps becoming activated upon occasion to induce cutaneous lesions. However, it can also cause encephalitis and severe generalized infections. The Epstein-Barr virus (EBV) apparently behaves quite differently and is found in close association with infectious mononucleosis, Burkitt's lymphoma, and a nasopharyngeal carcinoma.

Our knowledge of herpesvirus activities at the cellular level is limited primarily to productive, cytolytic infections of the type that occur in cutaneous herpetic lesions, for example. Virus-cell interactions of this kind have been studied extensively in cell culture (Roizman et al., 1972). A diagrammatic representation of the events that occur when HSV infects cells of a human line is shown in Figure 1. Briefly, the virus enters the cell, the DNA is partially uncoated in the cytoplasm, and subsequently enters the nucleus of the cell. The genetic information of the DNA is transcribed to RNA molecules that are transferred to the cytoplasm to direct the synthesis of viral proteins. These viral proteins fall into at least three classes. Some are required for the synthesis of more DNA molecules and perhaps for further transcription, others are structural components of newly synthesized nucleocapsids, and the third class binds to and modifies cellular membranes. Viral nucleocapsids are assembled in the nucleus of the cell and acquire their envelopes from the inner lamella of the nuclear membrane. Mature virions are found within the cisternae of the endoplasmic reticulum and the perinuclear space and in the extracellular fluid. An important consequence of productive infection is that host macromolecular syntheses are inhibited and the cell invariably

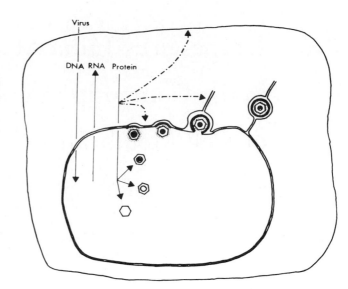

FIGURE 1 Diagrammatic representation of the events that occur in a productive infection of cells by herpesviruses.

dies. Presumably cells latently infected with herpesviruses remain viable until the virus is activated, although little is known about the latent interactions at the cellular level. If herpesviruses are capable of inducing malignant transformation, they must be able to express some viral functions in the absence of cytolytic activity.

Changes in cell surfaces after HSV infection

Two kinds of observations led to the conclusion that cells infected with HSV acquire altered surface properties. The first deals with changes in the shape and interactions of infected cells. The social behavior of infected cells differs from that of uninfected cells, and furthermore, the new patterns of interaction induced by different HSV strains in any one cell line vary with the genetic constitution of each strain (Ejercito et al., 1968). Human tissue culture cells infected with various strains of HSV exhibit types of behavior ranging from cell rounding with or without adhesion to cell fusion as shown in Figure 2. Since cell interactions of these types must be mediated by the surface membranes, it was concluded from these observations that herpesviruses are capable of inducing functional changes in these membranes.

Additional evidence that herpesviruses induce alterations in cell surfaces came from serological studies. A cytotoxic assay was used to characterize the surface antigens of cells infected for varying periods of time (Roizman and Spring, 1967). With antiserum prepared against infected cells and adsorbed with uninfected cells, it was demonstrated that new antigens are present on infected cells late in infection but not very early. Suggestive data

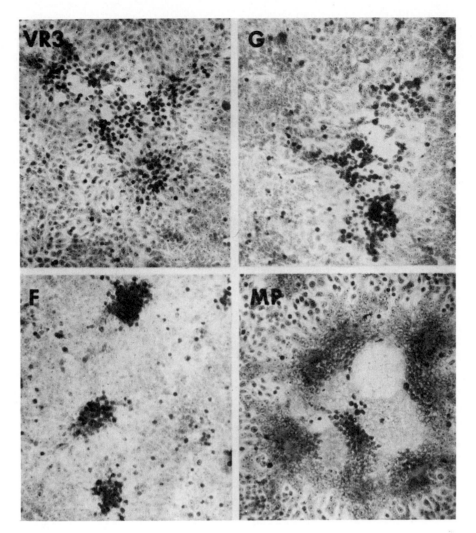

FIGURE 2 Light micrographs showing foci of HSV-infected cells against a background of uninfected cells. The 4 strains of HSV, designated by letter codes, each cause different patterns of social behavior in the infected cells. HSV-VR3 infected cells round with very little adhesion. HSV-G and HSV-F cause cells to round with loose and tight adhesions, respectively. HSV-MP causes cell fusion.

concerning the nature of the cell-surface antigens detected in the cytotoxic assay came from comparisons of the neutralizing and cytotoxic activities of a number of antisera prepared in rabbits and guinea pigs. There was a good correlation between the potency of an antiserum relative to the others in the neutralization of viral infectivity and its potency in the cytotoxic assay. Furthermore, adsorption of anti-infected cell serum with partially purified virus depleted the serum of its cytotoxic activity (Roizman and Spring, 1967). Thus, the data suggested that the infected cell surface has antigens in common with the virion surface.

Genetic variants of HSV that induce different patterns of social behavior among infected cells also exhibit some differences in the antigenic composition of the mature virions. These antigenic differences have been demon-strated in neutralization assays using antisera prepared against the various strains in homologous and heterologous combinations with the genetic variants (Roizman and Spring, 1967; Ejercito et al., 1968). The demonstration of antigens common to the virion surface and cell surface and of antigenic variations among the virions of several HSV strains suggested the possibility that the new antigens on the cell surface and the altered social behavior of infected cells could be explained by the presence of viral-specified products.

Possible molecular basis for altered cell surfaces

The observation of antigenic and functional changes in cellular membranes as a result of HSV infection provided the impetus to search for viral-specified products that might be responsible for these changes. These studies were facilitated by the fact that, in HSV-infected cells, host protein synthesis is completely inhibited by 4 to 5 hr post infection. Thus, radioactive precursors added to the medium of infected cells after 5 hr post infection are incorporated only into viral proteins. Membranes were purified from infected cells in order to determine whether viral proteins made after infection could be detected. Unfractionated cytoplasmic membranes and purified plasma membranes have been analyzed by acrylamide gel electrophoresis in the presence of sodium dodecyl sulfate and found to contain a new set of proteins not present in uninfected cells (Figure 3) (Spear et al., 1970; Heine et al., 1972). These proteins represent a considerable fraction of the total protein present in the membrane fraction and are glycosylated with the exception of a few minor bands. Host proteins made before infection are also present in the membranes; however, since host proteins are no longer synthesized after 5 hr post infection, no radioactivity is incorporated into them under the experimental conditions used. Several lines of evidence suggest that the new membrane proteins are viral-specified: (1) The proteins are not found in uninfected cells but are synthesized during the interval when viral structural proteins are made and after host protein synthesis has ceased. (2) Genetic variants of HSV produce different sets of membrane glycoproteins that can be distinguished by their migrations in acrylamide gels (Keller et al., 1970). The proteins produced by HSV-F and HSV-MP are shown in the top 2 frames of Figure 4. The HSV-G shows yet a different pattern while VR3 produces proteins similar in their migrations to those of HSV-F. (3) The new proteins found in cellular membranes are electrophoretically indistinguishable from some of the proteins present in the purified virion (Heine et al., 1972).

There is considerable evidence to support the notion that the viral-specified membrane proteins carry the new antigenic determinants found on the surfaces of infected cells. In the first place, both the antigens and the glycoproteins are present both in virions and in the cell surface. Second, the binding of antiviral antibody to purified cellular membranes in which the viral glycoproteins are incorporated can be demonstrated by virtue of the fact that the bound antibodies change the buoyant density of the membranes in sucrose gradients centrifuged to equilibrium (Roizman and Spear, 1971b) (Figure 5). Furthermore, some of the antibodies that bind to the membranes also bind to antigens on the virion surface, because membranes added to virus-antiserum mixtures will protect the virus from neutralization (Figure 6). Finally, the viral glycoproteins bind specifically to insoluble antibody gels prepared by cross-linking antiviral serum with glutaraldehyde (Savage et al., 1972).

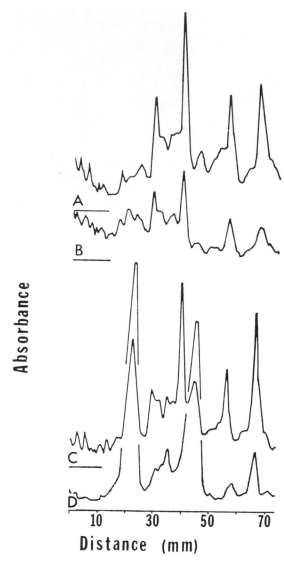

FIGURE 3 Electropherograms of proteins from purified plasma membranes of infected and uninfected cells labeled with C[14] amino acids for 18 hr. Electrophoresis was carried out in the presence of sodium dodecyl sulfate and β-mercaptoethanol in order to dissociate all polypeptide chains. The absorbance profiles of Coomassie brilliant blue stained gels (A, C) and of autoradiograms (B, D) developed from the same gels are shown. The top two profiles show the distribution of stained (A) and C[14]-labeled (B) proteins from plasma membranes of uninfected cells. The bottom profiles show the proteins present in plasma membranes of infected cells that were incubated in radioactive medium from 4 to 22 hr post infection. Several additional proteins are present in the infected cell membranes as indicated by comparisons of profiles A and C. Only these new proteins are synthesized during the time that C[14] amino acids were present (profile D). (From Roizman et al., 1972.)

Correlations can also be drawn between the patterns of viral-specified glycoproteins and the social behavior of infected cells. It has been observed that HSV variants that

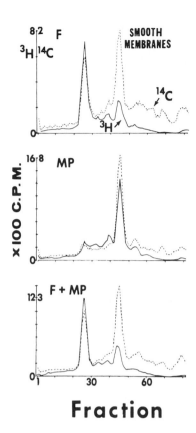

FIGURE 4 Electropherograms of viral-specified proteins present in the membranes of cells singly infected with HSV-F or HSV-MP or doubly infected with the 2 strains. The infected cells were incubated with H³-glucosamine and C¹⁴ amino acids from 4 to 22 hr post infection so that cellular proteins would not become labeled. Electrophoresis was carried out in dissociating solvents as in Figure 3. The distribution of radioactive proteins was determined by slicing the gels and assaying each slice in a liquid scintillation counter. Solid line: H³ glucosamine; broken line: C¹⁴ amino acids.

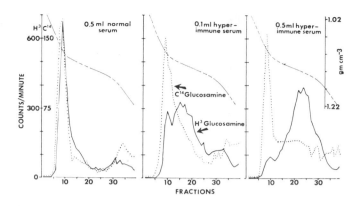

FIGURE 5 Effect of antiviral serum on the buoyant density of membranes from infected cells in sucrose density gradients. Mixtures containing purified membranes from infected cells labeled with H³ glucosamine, purified membranes from uninfected cells labeled with C¹⁴ glucosamine, and normal serum or antiviral serum were incubated for 4 hr at 37°C. The mixtures were made 45% (w/w) with respect to sucrose, overlaid first with linear gradients of 10 to 35% sucrose and then with 3 ml saline, and centrifuged for 20 hr. The top of each tube is at the left; solid line: infected cell membranes; broken line: uninfected cell membranes. (From Roizman and Spear, 1971b.)

differ in the membrane glycoproteins specified also differ in the type of cell interactions seen after infection (Figures 2 and 4). More provocative findings have come from analyses of doubly infected cells (Roizman and Spear, 1971a). When cells are infected with both HSV-F and HSV-MP, the results are: (1) Both HSV-F and HSV-MP progeny are produced in approximately equal numbers. (2) The type of social behavior characteristic of HSV-F is dominant, i.e., the infected cells form tight clumps instead of multinucleated syncytia. (3) The profile of viral-specified membrane glycoproteins is also characteristic of HSV-F and not HSV-MP, nor is it the sum of both profiles (Figure 4).

These findings concerning the herpesvirus membrane proteins are significant for two reasons: (1) The cytolytic cell-virus system provides an excellent model system for correlating changes in membrane composition and struc-

ture with changes in cell interactions over short periods of time, because host macromolecular syntheses are inhibited and do not obscure the viral activities. (2) With some of the herpesviruses the alterations of cell membranes may be separable from cytolytic activities. Studies of the herpes-specified membrane proteins may provide information as to the mechanism by which these viruses can induce malignant transformation.

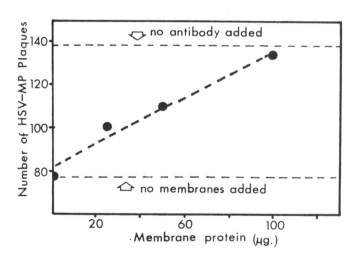

FIGURE 6 Effect of purified membranes from infected cells on neutralization of HSV with antiviral serum. Varying amounts of the purified membranes were incubated for 30 min with antiviral serum. Virus was then added to the mixture, which was incubated for another 30 min. Residual infectious virus was assayed by the plaque method. (From Roizman et al., 1972.)

Herpesviruses and cancer

The list of herpesviruses that have been incriminated in neoplastic diseases (reviewed by Epstein, 1970) includes the EB virus and herpes simplex virus type 2 of man, herpes saimiri of monkeys, Marek's disease virus (MDV) of chickens, and the Lucké virus of frogs. The evidence that the latter two herpesviruses are responsible for the Marek's neurolymphomatosis and the frog renal adenocarcinoma, respectively, is quite compelling. Cell-free preparations of both viruses may be prepared and used to transmit the diseases to uninfected animals (Nazerian and Witter, 1970; Mizell et al., 1969). In the case of Marek's disease, which is characterized by the development of lymphoid tumors in the viscera and lymphoid infiltration of the peripheral nerves, effective vaccines have been prepared from attenuated MDV or from a related nonpathogenic herpesvirus of turkeys (Purchase et al., 1971). The highly contagious nature of this disease has been attributed to the fact that MDV actively replicates in the feather follicles of the diseased chicken and is probably readily dispersed via airborne dander. The activities of the virus that are responsible for the proliferation of lymphoid cells have not been characterized.

The evidence that EB virus is causally related to several human lymphoproliferative diseases is more indirect and less compelling due to obvious experimental limitations. The kinds of data that have suggested a role for EB virus in infectious mononucleosis, Burkitt's lymphoma, and a nasopharyngeal carcinoma include: (1) the demonstration that tumor cells and leucocytes from infectious mononucleosis patients produce EB virus after being cultured in vitro although the cells are invariably virus-free at biopsy (Epstein, 1970); (2) the detection of EB virus DNA in tumor biopsies (zur Hausen et al., 1970); (3) the consistent association between the diagnosis of these lymphoproliferative diseases and the detection of high-titered serological reactions against several EB virus antigens in large numbers of individual patients (Klein, 1971).

Using sera from Burkitt's lymphoma patients, membrane-bound antigens have been detected by immunofluorescence on Burkitt tumor cells at biopsy. These antigens may disappear after cultivation of the cells in vitro and then reappear when the cultures become virus producing. A correlation has been found between high titers of the membrane-reactive antibodies in a patient's serum and clinical regression of the lymphoma. Furthermore, assays of a large number of sera revealed that high titers of antibodies to the membrane antigens are strongly correlated with high titers of antibodies to virion antigens detected in a neutralization assay (Klein, 1971). It is perhaps not surprising that some similarities exist in the antigenic characteristics of cells infected with HSV and with EB virus. For both viruses, viral replication is associated with new antigens on the cell surface, and antigens are shared in common between infected cell membranes and virions. It should be pointed out that for EB virus as well as HSV, viral replication is associated with cell degeneration (Epstein, 1970). However, the presence of EB virus membrane antigens on tumor cells at biopsy without evidence of viral replication or cell degeneration is suggestive that the virus may alter cell membranes in the absence of cytolytic functions. Also, HSV, treated with UV light to inactivate most of the viral functions, has been reported to induce the in vitro transformation of hamster cells (Duff and Rapp, 1971). It remains to be determined whether HSV membrane proteins are present in the membranes of these cells or, in more general terms, whether herpesvirus products in association with cell membranes play any role in the process of malignant transformation either in the laboratory or in nature.

ACKNOWLEDGMENTS The author is the recipient of a postdoctoral fellowship from the Arthritis Foundation. The studies presented here were carried out in Professor Bernard Roizman's laboratory at the University of Chicago during the tenure of a postdoctoral traineeship from the U.S. Public Health Service.

REFERENCES

CHOPPIN, P. W., H.-D. KLENK, R. W. COMPANS, and L. A. CALIQUIRI, 1971. The parainfluenza virus SV5 and its relationship to the cell membrane. *Perspect. Virol.* 7:127–158.

DUFF, R., and F. RAPP, 1971. Properties of hamster embryo fibroblasts transformed in vitro after exposure to ultraviolet-irradiated herpes simplex virus type 2. *J. Virol.* 8:469–477.

ECKHART, W., R. DULBECCO, and M. M. BURGER, 1971. Temperature-dependent surface changes in cells infected or transformed by a thermo-sensitive mutant of polyoma virus. *Proc. Nat. Acad. Sci. USA* 68:283–286.

EJERCITO, P. M., E. D. KIEFF, and B. ROIZMAN, 1968. Characterization of herpes simplex virus strains differing in their effects on social behavior of infected cells. *J. Gen. Virol.* 2:357–364.

EPSTEIN, M. A., 1970. Aspects of the EB virus. *Advances Cancer Res.* 13:383–411.

HEINE, J. W., P. G. SPEAR, and B. ROIZMAN, 1972. Proteins specified by herpes simplex virus.VI. Viral proteins in the plasma membrane. *J. Virol.* 9:431–439.

KELLER, J. M., P. G. SPEAR, and B. ROIZMAN, 1970. The proteins specified by herpes simplex virus.III. Viruses differing in their effects on the social behavior of infected cells specify different membrane glycoproteins. *Proc. Nat. Acad. Sci. USA* 65:865–871.

KLEIN, G., 1971. Immunological aspects of Burkitt's lymphoma. *Advances Immun.* 14:187–250.

MIZELL, M., I. TOPLIN, and J. J. ISAACS, 1969. Tumor induction in developing frog kidneys by a zonal centrifuge purified fraction of the frog herpes-type virus. *Science* 165:1134–1137.

NAZERIAN, K., and R. L. WITTER, 1970. Cell-free transmission and in vivo replication of Marek's disease virus. *J. Virol.* 5: 388–397.

PASTERNAK, G., 1969. Antigens induced by the mouse leukemia viruses. *Advances Cancer Res.* 12:1–99.

PURCHASE, H. G., R. L. WITTER, W. OKAZAKI, and B. R. BURMESTER, 1971. Vaccination against Marek's disease. *Perspect. Virol.* 7:91–110.

ROIZMAN, B., and P. G. SPEAR, 1971a. The role of herpesvirus glycoproteins in the modification of membranes of infected cells. In *Nucleic Acid-Protein Interactions and Nucleic Acid Synthesis in Viral Infections*, Vol. 2. Ribbons, Woessner, and Schultz, eds. Amsterdam, London: North-Holland Publishing Co., pp. 435–455.

ROIZMAN, B., and P. G. SPEAR, 1971b. Herpesvirus antigens on cell membranes detected by centrifugation of membrane-antibody complexes. *Science* 171:298–300.

ROIZMAN, B., P. G. SPEAR, and E. D. KIEFF, 1972. Herpes simplex viruses I and II: A biochemical definition. *Perspect. Virol.* 8:129–169.

ROIZMAN, B., and S. B. SPRING, 1967. Alteration in immunologic specificity of cells infected with cytolytic viruses. In *Cross-Reacting Antigens and Neoantigens*. Baltimore: The Williams & Wilkins Company, pp. 85–97.

SAVAGE, T., B. ROIZMAN, and J. W. HEINE, 1972. Immunologic specificity of the glycoproteins of subtypes 1 and 2. *J. Gen. Virol.* 17:31–48.

SPEAR, P. G., J. M. KELLER, and B. ROIZMAN, 1970. The proteins specified by herpes simplex virus. II. Viral glycoproteins associated with cellular membranes. *J. Virol.* 5:123–131.

ZUR HAUSEN, H., H. SCHULTE-HOLTHAUSEN, G. KLEIN, W. HENLE, G. HENLE, P. CLIFFORD, L. SANTESSON, 1970. EBV DNA in biopsies of Burkitt tumors and anaplastic carcinomas of the nasopharynx. *Nature (London)* 228:1056–1058.

93 Mechanisms of Virus-Induced Oncogenesis

R. DULBECCO

ABSTRACT   The chapter is concerned with the interaction between oncogenic viruses and their host cells. Also, attention is directed to the mechanisms of integration of the viral genomes into cell genomes and the method for recognizing the presence of integrated genomes, and to the mechanisms by which a viral genome causes the characteristic changes of transformation in the cells.

Introduction

THE STUDY OF oncogenic viruses has considerable interest outside the field of oncology; thus it provides considerable insight into problems of cell growth regulation, and it also explains the mechanisms of some viral neurological diseases, which arise when virus-specified proteins become incorporated into the surface membrane of nerve cells. My purpose will be to discuss the basic biological and molecular aspects of cell transformation by oncogenic viruses, in order to explain (1) how a virus can establish a long-lasting association with a cell without killing it, and (2) how it can alter the properties of the cell.

ONCOGENIC VIRUSES   Oncogenic viruses are found in almost all the families of DNA-containing viruses but only in one family of RNA-containing viruses, the leukoviruses. The viruses of the latter type are different from all other RNA viruses in that they contain a *reverse transcriptase* (Baltimore, 1970; Temin and Mizutani, 1970), which copies the viral RNA into double-stranded DNA. This finding brings unity into the field of oncogenic viruses, by showing that within the cell the genome of all oncogenic viruses is in DNA form. There may be two reasons for this requirement: (1) the necessity of replicating their DNA endows the viruses with genetic functions able to control DNA replication in the cells; (2) the viral DNA can become integrated in the cellular DNA, generating a stable virus-cell association. Both points will be developed below.

Oncogenic viruses may have different genetic complexity. The simplest are the *polyomaviruses*, (polyoma virus and SV40), which contain a double-stranded DNA of 3.5×10^6 mol wt, and therefore are capable of containing 5 to 10 genes. These viruses and the *papillomaviruses* also

contain a *cyclic* DNA. All the other oncogenic viruses are genetically more complex, having enough DNA to specify from about 30 genes (adenoviruses, leukoviruses) to about 100 (herpesviruses). Their DNA is *linear*. Probably all oncogenic viruses have a few transforming genes with similar functions; the larger viruses have many other genes, whose function often interferes with that of the transforming genes. With these larger viruses oncogenesis may occur only when the additional genes, for reasons that we will see below, remain unexpressed. Hence the oncogenic effect can be studied best with the smallest viruses.

Most of our knowledge about mechanisms derives from studies with polyoma virus and SV40 in tissue cultures, in which these viruses cause the *transformation* of the cells. Transformed cells then can be compared to the untransformed cells derived from the same clone, in order to identify the biochemical or molecular changes due to transformation.

PROPERTIES OF TRANSFORMED CELLS   The main properties are given in Table I. The indicated differences are

R. DULBECCO   Salk Institute, San Diego, California

TABLE I

Cellular changes (in cultures of fairly high density)

	Normal Cells	Transformed Cells
Culture Property		
Maximum cellular		
density	low	high
Serum		
requirements	high	low, different
Growth in suspension in agar	no	in some
Cell Surface		
Glycolipids	regular	incomplete
Glycosyl transferase		
activity	high	low
Transport	need activation	some independent
Lectin		
agglutination	poor	good
Antigens	normal	often new
Morphology		
Cellular indentation	little	much
Cell-to-cell relation	regular	random
Others		
cAMP concentration	high	low

found in *rather dense* cultures but tend to vanish in very sparse cultures. This shows that the main difference between transformed and normal cells is *regulatory*: The normal cells multiply freely in sparse cultures but enter a resting state when the cultures become crowded, whereas the transformed cells remain in a growing state irrespective of cell density. Since, as we shall see below, cell regulation is controlled in virus-transformed cells by a *single protein*, the various changes of Table I must all be correlated. A possible scheme of this correlation, based on many experimental results that I cannot review here, is given in Figure 1. The various connections in the scheme have various degrees of experimental support. The most hypothetical ones are indicated by dashed lines.

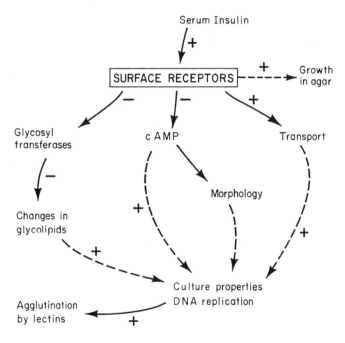

FIGURE 1 A possible scheme of cell growth regulation in normal cells. Pathways indicated by dashed lines are more hypothetical. + indicates that the pathway is stimulated, − that it is inhibited.

The salient feature of this scheme is that cell regulation *may* result (although there is no proof) from the function of surface receptors which recognize external regulatory signals (e.g., hormones, contacts with other cells). These receptors, in response to the signals, would modify the function of membrane-bound enzymes or transport proteins causing a change in the concentration of internal effectors (such as cyclic AMP, or certain nutrients). The growth state of the cells would then depend on these concentrations.

The surface receptors have not been demonstrated in the cells of which I will talk; their existence is postulated

on the basis of the identification and isolation in other cells of insulin receptors (Cuatrecasas, 1971), which may have functions similar to those postulated here. A central role of the receptor is shown for instance by the reversion of transformed cells to a normal morphology upon addition of cyclic AMP (Hsie and Puck, 1971).

METHODS FOR STUDYING TRANSFORMATION With either polyoma virus or SV40, two cell systems are used (Table II). The virus multiplies in the *permissive* cells, which are at

TABLE II

Example of cell systems used in the study of transformation with polyoma viruses

Cells		
Permissive	polyoma virus SV40	3T3 (mouse) BSC-1 (African green monkey)
Nonpermissive	polyoma virus SV40	BHK (hamster) 3T3 (mouse)

first transiently transformed and subsequently killed. *Stable* transformation is rare in permissive cells but frequent in *nonpermissive* cells, which are not killed by the infection. Of the many types of transformed cells available, those I will refer to most often are listed in Table III.

TABLE III

Transformed cells

SV3T3	= 3T3 cells transformed by SV40 (wild type);stable transformation irrespective of temperature; no virus release.
tsa-3T3	= 3T3 cells transformed by the ts mutant of polyoma virus tsa. Transformation is stable at 39°C with minimal virus release; at 32°C the cells produce virus and are killed.
ts3-BHK	= BHK cells transformed by the ts mutant of polyoma virus ts3. Cells are transformed at 32°C, nearly normal at 39°C. No virus release.

The main experimental approaches to the study of transformation are: (1) Isolation and study of *temperature-sensitive (ts) mutants* in viral genes. A mutated gene is active at low temperature (32°C) but inactive at high temperature (39°C); hence the virus is propagated at 32°C and its effect on transformation is studied at 39°C. (2) Identification of *virus-specified* or *virus-controlled* proteins, by labeling permissive infected and uninfected cells for a short time with radioactive amino acids. After the extracted proteins are fractionated by agar gel electrophoresis in the presence of sodium dodecyl sulphate (SDS), the patterns of radioactivity of infected and uninfected cells are compared. Since the virus does not shut off the synthesis of host proteins, high resolution is required and is obtained by making radioautographs of the gels that are then optically scanned. (3) Identification of *virus-specified or virus-controlled functions* (e.g., enzyme activities) by comparing functions in cells infected by wild-type virus or its ts mutants with those present in uninfected cells. A virus-induced function is recognized, because it is present in cells infected by the wild-type virus but not in uninfected cells or in those infected at high temperature by some ts mutant.

VIRAL GENES Genetic studies with ts mutants have identified five genes in polyoma virus (Eckhart, 1969; Di Mayorca et al., 1969): Three *early* (acting before viral DNA replication) and two *late* (acting after viral DNA replication has begun). The late genes specify capsid proteins. The genes of SV40 are less known, but they seem to follow a similar pattern. Also five new proteins are induced by SV40 and probably also by polyoma virus in permissive cells (Walter, Roblin, and Dulbecco, 1972); two of them are the major proteins of the capsid (Figure 2). The agreement between number of genes and proteins seems excellent, especially because the total molecular weight of the proteins estimated from their electrophoretic mobilities corresponds to the total informational content of the virus. However, it cannot be excluded that the agreement is fortuitous, because some of the proteins may be specified by cellular genes derepressed by infection rather than by viral genes.

The studies of the mutants show that the loss of regulation of the cells, characteristic of transformation, is produced by the function of one of the early genes. Thus when wild-type polyoma virus infects dense 3T3 cells, which do not replicate their DNA, it induces in them DNA replication (Figure 3), as well as the surface and other changes listed in Table I (Dulbecco et al., 1965; Dulbecco, 1970). These effects are abolished by a mutation, called ts3 (Dulbecco and Eckhart, 1970; Eckhart, Dulbecco, and Burger, 1971). The ts3 mutant causes in BHK cells a state of temperature-dependent transformation: The cells are fully transformed at 32°C, when the ts3 gene is active, but revert to a more normal state at 39°C, when the gene is inactive. The ts3 gene therefore *takes over the control of growth regulation* of the cells; and because ts mutations almost invariably affect proteins, this result gives clear evidence that the various cellular properties of Table I are under common control.

In the scheme of Figure 1, the proteins specified by the ts3 gene affect the surface receptors, altering them in such a way that they fail either to recognize external regulatory signals or to modify the function of the membrane enzymes controlled by the state of the receptors. The viral

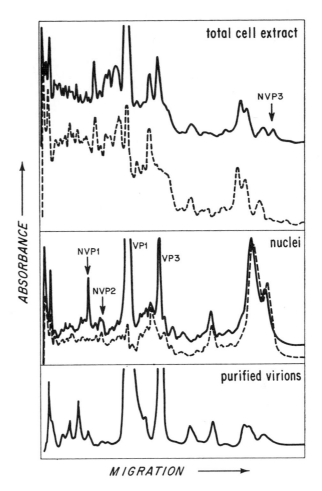

FIGURE 2 Proteins of African green monkey cells (Vero line) infected by SV40 (solid lines) or uninfected (broken lines), showing five new proteins appearing (mostly in the nucleus) after infection. Cells were labeled with ^{14}C-labeled protein hydrolysate from 52 to 53 hours after infection. Extracts were fractionated by acrylamide gel electrophoresis. The radioautographs were scanned in Guilford recording spectrophotometer. (From Walter et al., 1972.)

protein would infiltrate the surface membrane interacting directly or indirectly with the receptors.

INTEGRATION In the transformed cell, the DNA of polyomaviruses is integrated, i.e., inserted in and covalently bound to the cellular DNA as a *provirus*. This feature could only be recognized after the development of methods for detecting viral DNA in the minute proportions (2×10^{-6} to 10^{-5} of the total) present in transformed cells (Westphal and Dulbecco, 1968). This was achieved by hybridization with radioactive virus-specific RNA synthesized in vitro by *Escherichia coli* RNA polymerase, using viral DNA as template. The amount of viral DNA thus recognized corresponds to a few viral genomes per cell.

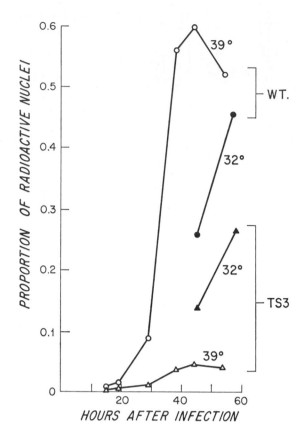

FIGURE 3 Effect of temperature in induction of cellular DNA replication in dense 3T3 cultures by wild type polyoma virus or its ts3 mutant. Whereas there is little difference between the effect of the two viruses at 32°C, there is a very large difference at 39°C. (From Dulbecco and Eckhart, 1970).

The covalent binding of the viral DNA to the cellular DNA was demonstrated in two ways. For SV3T3 cells, the cellular DNA was isolated in the form of very long filaments (mol wt in excess of 2×10^8) by depositing the cells on top of an alkaline sucrose gradient, where they were allowed to lyse; the high pH also caused the denaturation of the DNA. Upon subsequent centrifugation, the long cellular DNA strands sedimented considerably faster than free viral DNA (as shown by reconstruction experiments); however, the viral sequences, detected by hybridization, followed the cellular DNA (Sambrook et al., 1968). Since only covalent bonds can survive the conditions of the experiment, the two DNAs must be bound covalently.

In a line of Chinese hamster, cells transformed by SV40 integration were demonstrated by a different approach, based on the observation that when DNA and RNA are hybridized together, the hybrid DNA-RNA molecules stick to nitrocellulose filters only when part of the DNA remains unhybridized (Haas et al., 1972). Thus hybrids formed by SV40 DNA with its specific RNA in excess do not stick to the filters because the DNA is completely hybridized (Figure 4). In contrast, DNA from the transformed cell sticks to the filter even after extensive hybridization, obviously through the adjacent cellular sequences, which remain unhybridized. Further confirmation of integration was obtained by breaking the cellular DNA to fragments of various lengths before hybridization: Then as the fragments became shorter, the ability of the hybrids to stick to the filter progressively decreased, because an increasing proportion of viral sequences became disconnected from the cellular sequences. The results also showed that the viral DNA is

A. FREE VIRAL DNA

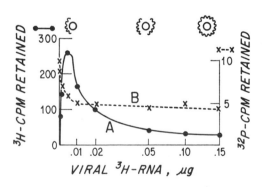

B. TRANSFORMED CELL DNA

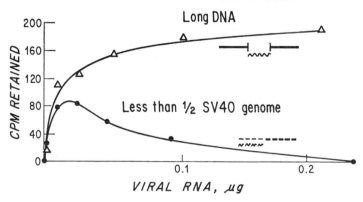

FIGURE 4 Evidence for integration of SV40 DNA in transformed Chinese hamster cells. (A) Hybrids of ^{32}P-SV40 DNA and viral ^{3}H-RNA are retained on nitrocellulose filters only below saturation, but at high RNA concentrations they are lost; also one half of the DNA (the hybridizing strand) is lost. (B) Long DNA from the transformed cells gives a saturation curve

without hybrid loss at saturation, but when the same DNA is fragmented the hybrids are again lost. The retention when the molecules are long is due to the presence of cellular DNA attached to the viral DNA. Viral DNA = thin lines; cellular DNA = heavy lines; viral RNA = wavy lines. (From Haas et al., 1972.)

integrated in segments, corresponding to the length of a single genome (Figure 5). Although there are about six proviruses per cell in this line, they are integrated individually. They either derive from six independent integrations at the time of transformation or from reduplication of the cellular DNA containing the provirus during the subsequent growth of the clone.

With polyoma virus, integration seems to require the function of another early viral gene, called tsa (from the

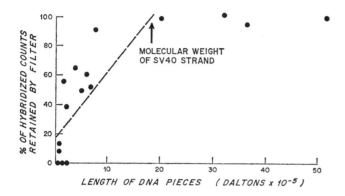

FIGURE 5 Evidence for integration of individual SV40 DNA molecules in SV40 transformed Chinese hamster cells. The data were obtained from the experiment of Figure 4B by fragmenting the cellular DNA to pieces of various length. The dashed line shows the expected curve for integrated DNA of single genome length. (From Haas et al., 1972.)

name of a ts mutant) (Fried, 1965). Thus BHK cells are not transformed by the mutant at 39°C, but when transformed at 32°C they remain transformed after shifting to 39°C. Hence, contrary to the ts3 gene, the requirement for the tsa function is transient at the moment of transformation. The tsa mutant expresses the ts3 function normally at high temperature (Figure 6).

DETACHMENT OF THE PROVIRUS Infectious virus can be recovered under special circumstances from transformed cells that normally contain no infectious virus. For instance SV3T3 cells release virus after fusion with BSC-1 cells (Koprowski et al., 1967; Watkins and Dulbecco, 1967). It seems likely that after fusion the BSC-1 cells supply to the SV3T3 cells a factor required for the complete transcription of the viral genome and absent in SV3T3 cells. In fact in SV3T3 cells, only about half of the viral genome is transcribed (Oda and Dulbecco, 1968; Aloni et al., 1968), but after fusion it must all be transcribed in order to produce infectious virus. The inference is that excision of the provirus requires the action of one of the viral genes not transcribed in SV3T3 cells.

A similar conclusion was obtained with tsa-3T3 cells (transformed by the tsa mutant discussed above) (Vogt, 1970). These cells, which contain three or four polyoma virus genomes per cell, probably integrated, are stable at 39°C (when the tsa gene is inactive) and produce only

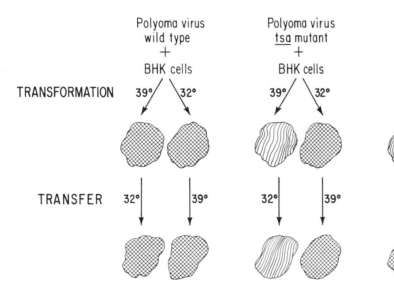

FIGURE 6 Effect of tsa or ts3 polyoma virus mutations on transformation. Transformed clones are represented by heavy crossed lines, untransformed clones by thin wavy lines. The wild-type virus produces transformed clones at either 32°C or 39°C; the cells remain transformed when transferred to a different temperature. With the tsa mutant there is no transformation at 39°C but transformed clones produced at 32°C remain

transformed after transfer to 39°C. This indicates a transient requirement for the tsa gene function, probably for integration. With the ts3 mutant the cells are transformed only at 32°C and the state is completely reversible depending on the temperature. This indicates a continuous requirement for the function of the ts3 gene for transformation.

traces of virus, but when shifted to 32°C undergo a lytic cycle of viral multiplication and are killed (Cuzin et al., 1970). It appears therefore that in these cells excision of the provirus *requires the function of the tsa gene,* which is also required for integration. The function of the gene is also required for the replication of the viral DNA, in a way yet undetermined.

These results show that there are considerable similarities between transformation and the phenomenon of bacterial lysogeny (Figure 7). In both we find: (1) integration and detachment of the viral DNA; (2) cyclic shape of the DNA, which allows integration and detachment without loss of viral genetic information; (3) expression of viral genes altering the cell (*lysogenic conversion*); and (4) occasional incorporation of cellular DNA during detachment.

The latter phenomenon is observed when the tsa-3T3 cells are shifted to low temperature: The cyclic viral DNA produced in the cells contains molecules that, after exhaustive hybridization to viral RNA, stick to nitrocellulose filters (Blangy, personal communication) (Figure 8). Sticking is made possible by the inserted cellular sequences. The presence of these sequences raises the question whether viral genes, especially ts3, whose function must be very similar to that of a cellular gene, could be of cellular origin, having been incorporated at some time in the previous history of the virus into its DNA.

Interactions between viral and cellular genes

Transformation does not simply result from the addition of the function of a few viral genes to that of cellular genes. There are more intimate interactions of which I will give two examples.

One example is the control of the transcription of genes of the provirus by cellular genes, which specify the transcribing enzymes and transcription factors, as already pointed out above, and by cellular promoters, which specify the *initiation* of transcription. Thus in SV3T3 cells, virus specific mRNA is made as part of long RNA molecules, which, by hybridization experiments, have been shown to be in part viral and in part cellular (Lindberg and Darnell, 1970; Tonegawa et al., 1970) (Figure 9). Since the viral sequences are located at the 3' end, transcription of the hybrid messenger initiates at cellular promoters. As a consequence, if the provirus is integrated in a nontranscribed part of the cellular genome (e.g., heterochromatin), it will remain unexpressed. Indeed, unexpressed SV40 proviruses seem to exist in certain 3T3 cell lines after exposure to SV40. They derive from clones that showed a transient transformation but then reverted to a normal phenotype (Smith et al., 1972).

Another aspect of the interaction of viral and cellular genes in transformed cells is seen in the production of clones of *phenotypic revertants*, without changes in the

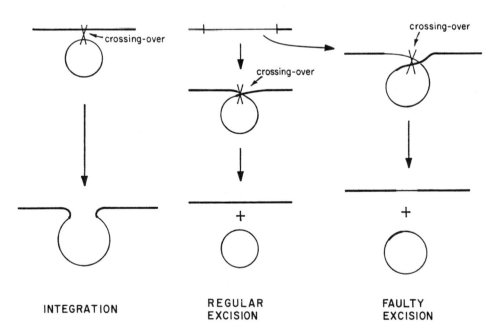

FIGURE 7 Postulated mechanisms of integration and excision of the provirus and of incorporation of cellular DNA in cyclic viral DNA on excision. The mechanisms are those demonstrated in bacterial lysogeny and are based on a single crossing-over for both integration and excision. Cellular DNA becomes incorporated (usually exchanging for viral DNA) when the crossing-over occurs at the wrong place. Thin lines represent viral DNA; heavy lines represent cellular DNA.

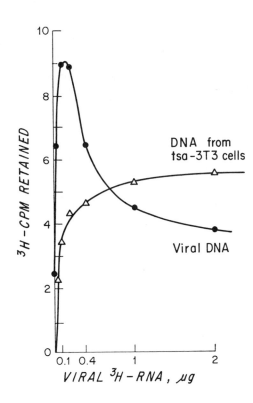

FIGURE 8 Evidence for incorporation of cellular DNA in cyclic viral DNA produced in tsa-3T3 cells after lowering the temperature. The experimental approach is the same as that shown in Figure 4. The hybrids of the viral DNA obtained in regular lytic infection with ³H-viral RNA are lost at saturation, whereas those obtained with DNA extracted from tsa-3T3 cells are not lost. Hence, although physically similar to regular SV40 DNA, these molecules behave like the integrated DNA of Figure 4B. (D. Blangy, personal communication.)

provirus (Table IV). Hence these reversions must occur by mutation in cellular genes. These revertants are often incomplete, showing that the change is not in the main control mechanism but in one of the pathways under its control (see Figure 1). Although in the transformed cells the main growth control is viral, the pathways are cellular and can conceivably increase their response to regulation or consequence of cellular mutations or of change of the gene balance (in hybrid cells).

TABLE IV
Cellular influence on viral transformation

Isolation of morphological revertants from regular transformed cells by FUDR selection.(Pollack et al., 1968)

Isolation of cells with temperature-sensitive transformation from regular transformed cells. (Renger and Basilico, 1972)

Isolation of morphological revertants from regular transformed cells by growth on gluteraldehyde-fixed cells. (Rabinowitz and Sachs, 1968)

Hybrid cells with normal phenotype resulting from the fusion of a transformed cell with a normal cell. (Harris et al., 1969)

Oncogenic RNA-containing viruses (*Leukoviruses*)

The conclusions derived from the study of DNA viruses can be extended to the leukoviruses to account for their interesting behavior. In brief, there are two groups of leukoviruses: *leukosis viruses* (i.e., leukemia inducing), which have an extremely low probability of transforming the cells they infect; and *sarcoma viruses*, which like polyomaviruses, transform all infected cells. These cells,

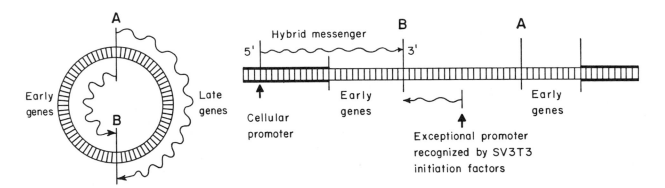

Transcription of free viral DNA
(Lytic infection in BSC-1 cells)

Transcription of integrated SV40 provirus in SV3T3 cells

FIGURE 9 Transcription of SV40 DNA in lytic infection and in SV3T3 cells. Cellular DNA is represented by heavy lines, viral DNA by thin lines, and mRNA by wavy lines. The gene order in the provirus is inferred from indirect evidence; other

arrangements are not excluded. It is not known whether the transcription of late sequences in SV3T3 has functional significance. (Lytic transcription from Lindstrom and Dulbecco, 1972.)

however, are not killed even if permissive, owing to the different mechanisms of virus release. The presence of transforming genes, possibly similar to ts3, in sarcoma viruses is demonstrated by the isolation of ts mutations, which make transformation temperature dependent.

The most interesting question is the reason for the difference between sarcoma and leukosis viruses. It seems that sarcoma viruses arise by the addition of a piece of genome, of unknown origin (*sarcoma information*), to a leukovirus genome. The most plausible addition mechanism is a recombination of a leukosis virus genome in DNA form with a piece of cellular DNA containing the sarcoma information; in the process the leukosis genome seems to lose a segment containing information for the synthesis of the viral coat, because most sarcoma viruses are unable to produce it and must depend on a helper leukosis virus infecting the same cell for its synthesis. The sarcoma information may be analogous to the early region of the polyoma genome; hence it may be itself a defective provirus. Alternatively, the sarcoma information may be a true cellular genome segment whose expression is regulated in the cells but when incorporated into a virus becomes deregulated and permanently active, giving rise to a transforming agent. Under this hypothesis, polyomaviruses and sarcoma viruses would have similar transforming properties, because each has incorporated such a sequence during its previous history.

Leukosis viruses are generally *latent*, in the form of DNA proviruses, recognizable by hybridization with viral RNA (it is not known whether they are integrated); in this state they do not produce virus or virus specific antigens or transformation. These silent genomes can be activated by exposing the cells to 5-bromo- or 5-iododeoxyuridine, or, to a smaller extent, to chemical carcinogens or X-rays (Lowy et al., 1971). The viruses produced by activation are either leukosis viruses or even nononcogenic leukoviruses. No sarcoma virus has been recovered by this procedure. The mechanism of latency is not known but may be similar to that I have discussed for polyomaviruses.

It has been suggested that the occasional activation of silent leukoviruses may give rise to spontaneous cancers (Todaro and Huebner, 1972). This does not seem too likely, however, because the activated genomes are so weakly oncogenic that their chance of transforming a cell in which they are activated would be infinitesimal. It seems more likely that the activation of the latent "sarcoma information" might lead to cancer. But the mechanisms would not be truly viral, and furthermore the process would be experimentally indistinguishable at least at the present time from other processes affecting the cellular genetic material, such as mutation.

Conclusions

The studies with model systems suggest several conclusions applicable to all oncogenic viruses:

1. The requirement for an oncogenic virus can be defined (Table V).

TABLE V
Requirement for an oncogenic virus

It makes a DNA copy of the genome in the cells.
The viral DNA can persist in the cells without killing them.
The viral DNA is regularly transmitted to the daughter cells.
The viral DNA contains genes able to change the cells.

2. The function of small numbers of viral genes is sufficient for transformation: Essentially, one gene to bypass the normal growth regulation of the cells, and one for the integration of the viral DNA into the cellular DNA as provirus. (A plasmid state is also possible.)

3. Control of the expression of viral genes by cellular genes has various consequences: (a) With DNA viruses it allows the survival of the transformed cells, which are not killed by the virus; (b) With all viruses it permits latency of the proviruses. The latent provirus can be detected by various means (Table VI).

TABLE VI
Methods for detecting a latent provirus in transformed cells

Demonstration of viral DNA in the cells by hybridization with radioactive viral RNA or DNA made in vitro.
Demonstration of viral mRNA in the cells by hybridization with radioactive RNA or DNA made in vitro.
Activation by treating the cells with BUDR or IUDR.
Rescue by superinfection with a helper virus (for leukoviruses).
Production of virus after cultivating tumor cells in vitro.

4. Transformation depends on the function of both viral and cellular genes in a proper balance.

5. Surface alterations always accompany transformation. In animals they tend to cause an allograft reaction with opposite consequences: In transforming infections, the prevention of tumor formation; in nononcogenic infections, cell killing and disease (as in slow infections in the brain).

6. The interactions between the genes of oncogenic viruses and the cellular genes in the same cell are very complex in evolutionary, structural, and functional terms, and are not completely understood.

REFERENCES

ALONI, Y., E. WINOCOUR, and L. SACHS, 1968. Characterization of the Simian Virus 40 - specific RNA in virus-yielding and transformed cells. *J. Molec. Biol.* 31:415–429.

BALTIMORE, D., 1970. RNA-dependent DNA polymerase in virions of RNA tumor viruses. *Nature (Lond.)* 226:1209–1211.

CUATRECASAS, P., 1971. Insulin-receptor interactions in adipose tissue cells: direct measurement and properties. *Proc. Natl. Acad. Sci. USA* 68:1264–1268.

CUZIN, F., M. VOGT, M. DIECKMANN, and P. BERG. 1970. Induction of virus multiplication in 3T3 cells transformed by a thermosensitive mutant of polyoma virus, II. Formation of oligomeric polyoma DNA molecules. *J. Molec. Biol.* 47:317–334.

DI MAYORCA, G., J. CALLENDER, G. MARIN, and R. GIORDANO, 1969. Temperature-sensitive mutants of polyoma virus. *Virology* 38:125–133.

DULBECCO, R., 1970. Behavior of tissue culture cells infected with polyoma virus. *Proc. Natl. Acad. Sci. USA* 67:1214–1220.

DULBECCO, R., and W. ECKHART, 1970. Temperature-dependent properties of cells transformed by thermosensitive mutants of polyoma virus. *Proc. Natl. Acad. Sci. USA* 67:1775–1781.

DULBECCO, R., L. H. HARTWELL, and M. VOGT, 1965. Induction of cellular DNA synthesis by polyoma virus. *Proc. Natl. Acad. Sci. USA* 53:403–410.

ECKHART, W., 1969. Complementation and transformation by temperature-sensitive mutants of polyoma virus. *Virology* 38:120–125.

ECKHART, W., R. DULBECCO, and M. BURGER, 1971. Temperature-dependent surface changes in cells infected or transformed by a thermosensitive mutant of polyoma virus. *Proc. Natl. Acad. Sci. USA* 68:283–286.

FRIED, M., 1965. Cell transforming ability of a temperature-sensitive mutant of polyoma virus. *Proc. Natl. Acad. Sci. USA* 53:486–491.

HAAS, M., M. VOGT, and R. DULBECCO, 1972. Loss of SV40 DNA-RNA hybrids from nitrocellulose membranes; implications for the study of virus-host DNA interactions. *Proc. Natl. Acad. Sci. USA* 69:2160–2164.

HARRIS, H., O. J. MILLER, G. KLEIN, P. WORST, and T. TACHIBANA, 1969. Suppression of malignancy by cell fusion. *Nature* 223:363–368.

HSIE, A. W., and T. T. PUCK, 1971. Morphological transformation of Chinese hamster cells by dibutyryl adenosine cyclic 3′:5′-monophosphate and testosterone. *Proc. Natl. Acad. Sci. USA* 68:358–361.

KOPROWSKI, H., F. C. JENSEN, and Z. STEPLEWSKI, 1967. Activation of production of infectious tumor virus SV40 in heterokaryon cultures. *Proc. Natl. Acad. Sci. USA* 58:127–133.

LINDBERG, V., and J. E. DARNELL, 1970. SV40-specific RNA in the nucleus and polyribosomes of transformed cells. *Proc. Natl. Acad. Sci. USA* 65:1089–1095.

LINDSTROM, D., and R. DULBECCO, 1972. Strand orientation of simian virus 40 transcription in productively infected cells. *Proc. Natl. Acad. Sci. USA* 69:1517–1520.

LOWY, D. R., W. P. ROWE, N. TEICH, and J. W. HARTLEY, 1971. Murine leukemia virus: high frequency activation in vitro by 5-iododeoxyuridine and 5-bromodeoxyuridine. *Science* 174:155–158.

ODA, K., and R. DULBECCO, 1968. Regulation of transcription of the SV40 DNA in productively infected and in transformed cells. *Proc. Natl. Acad. Sci. USA* 60:525–532.

POLLACK, R., H. GREEN, and G. TODARO, 1968. Growth control in cultured cells: selection of sublines with increased sensitivity to contact inhibition and decreased tumor-producing ability. *Proc. Natl. Acad. Sci. USA* 60:126–133.

RABINOWITZ, Z., and L. SACHS, 1968. Reversion of properties in cells transformed by polyoma virus. *Nature* 220:1203–1206.

RENGER, H. C., and C. BASILICO, 1972. Mutation causing a temperature-sensitive expression of cell transformation by a tumor virus. *Proc. Natl. Acad. Sci. USA* 69:109–114.

SAMBROOK, J., H. WESTPHAL, P. SRINIVASAN, and R. DULBECCO, 1968. The integrated state of viral DNA in SV40 transformed cells. *Proc. Natl. Acad. Sci. USA* 60:1288–1295.

SMITH, H. S., L. D. GELB, and M. M. MARTIN, 1972. Detection and quantitation of simian virus 40 genetic material in abortively transformed BALB 3T3 clones. *Proc. Natl. Acad. Sci. USA* 69:152–156.

TEMIN, H., and S. MIZUTANI, 1970. RNA-dependent DNA polymerase in virions of Rous sarcoma virus. *Nature (London)* 226:1211–1213.

TODARO, G. J., and R. J. HUEBNER, 1972. The viral oncogene hypothesis: new evidence. *Proc. Natl. Acad. Sci. USA* 69:1009–1015.

TONEGAWA, S., G. WALTER, A. BENARDINI, and R. DULBECCO, 1970. Transcription of the SV40 genome in transformed cells and during lytic infection. *Cold Spring Harbor Symp. Quant. Biol.* 35:823–831.

VOGT, M., 1970. Induction of virus multiplication in 3T3 cells transformed by a thermosensitive mutant of polyoma virus. I. Isolation and characterization of tsa-3T3 cells. *J. Mol. Biol.* 47:307–316.

WALTER, G., R. ROBLIN, and R. DULBECCO, 1972. Protein synthesis in simian virus 40-infected monkey cells. *Proc. Natl. Acad. Sci. USA* 69:921–924.

WATKINS, J. F., and R. DULBECCO, 1967. Production of SV40 virus in heterokaryons of transformed and susceptible cells. *Proc. Natl. Acad. Sci. USA* 58:1396–1403.

WESTPHAL, H., and R. DULBECCO, 1968. Viral DNA in polyoma- and SV40 transformed cell lines. *Proc. Natl. Acad. Sci. USA* 59:1158–1165.

LIST OF AUTHORS

PARTICIPANTS,

INTENSIVE STUDY PROGRAM 1972

ADELMAN, GEORGE Neurosciences Research Program, 165 Allandale St., Jamaica Plain Station, Boston, MA 02130

ADEY, W. Ross Department of Anatomy, University of California, Center for the Health Sciences, Los Angeles, CA 90024

ALBERTY, ROBERT A. School of Science, Room 6-215, Massachusetts Institute of Technology, Cambridge, MA 02139

ANDRIESSE, PAUL G. Massachusetts General Hospital, Boston, MA 02114

ARBIB, MICHAEL A. Computer & Information Science, University of Massachusetts, Amherst, MA 01002

ARÉCHIGA, HUGO Marine Biological Station, University of Liverpool, Port Erin, Isle of Man

ASCHOFF, JÜRGEN Departments of Neurology and Neuro-physiology, University of Ulm, Steinhovelstrasse 9, 79 Ulm, Federal Republic of Germany

AXELROD, JULIUS National Institutes of Health, Building 10, Room 2D-47, National Institute of Mental Health, 9000 Rockville Pike, Bethesda, MD 20014

BARMARK, NEAL H. Laboratory of Neurophysiology, Good Samaritan Hospital and Medical Center, 1015 N.W. 22nd Avenue, Portland, OR 97210

BERLUCCHI, GIOVANNI Instituto di Fisiologia della Universita di Pisa, Via San Zeno 31, I 56100 Pisa, Italy

BISHOP, DOROTHY W. Neurosciences Research Program, 165 Allandale St., Jamaica Plain Station, Boston, MA 02130

BIVENS, LYLE W. Behavior Science Research Branch, National Institute of Mental Health, 5600 Fishers Lane, Rockville, MD 20852

BLAKEMORE, COLIN Physiological Laboratory, University of Cambridge, Cambridge, CB2 3EG, England

BLOOM, FLOYD E. Laboratory of Neuropharmacology, National Institute of Mental Health, IRP-SMR, Wm. A. White Building, St. Elizabeths Hospital, Washington, DC 20032

BLUMENTHAL, ROBERT Laboratory of Theoretical Biology, National Cancer Institute, Building 10, Room 4B-48, National Institutes of Health, Bethesda, MD 20014

BODIAN, DAVID Department of Anatomy, The Johns Hopkins University, 725 North Wolfe Street, Baltimore, MD 21205

BOND, DOUGLAS The Grant Foundation, Inc., 130 East 59th Street, New York, NY 10022

BROADBENT, DONALD E. Applied Psychology Unit, Medical Research Council, 15 Chaucer Street, Cambridge, CB2 2EF, England

BROWN, JAMES H. Program Director for Neurobiology, National Science Foundation, 1800 G Street, N.W., Washington, DC 20550

BUCHNER, ERICH R. Max-Planck-Institut für biologische Kybernetik, Spemannstrasse 38, 74 Tübingen, Federal Republic of Germany

BULLOCK, THEODORE H. Department of Neurosciences, University of California, San Diego, School of Medicine, La Jolla, CA 97037

BURGER, MAX M. Erziehungsdepartement des Kantons Basel-Stadt, Biozentrum der Universität, Klingelberg-strasse 70, CH-4056 Basel, Switzerland

CAMPBELL, FERGUS W. Physiological Laboratory, University of Cambridge, Cambridge, EB2 3EG, England

CARNEGIE, PATRICK R. School of Biochemistry, University of Melbourne, Parkville, Victoria 3052, Australia

COYLE, JR., JOSEPH T., Laboratory of Clinical Science, National Institute of Mental Health, Building 10, Room 2D-47, National Institutes of Health, Bethesda, MD 20014

CUÉNOD, MICHEL Institute for Brain Research, University of Zurich, August Forel Strasse 1, CH-8008 Zurich, Switzerland

CUSICK, KATHERYN Neurosciences Research Program, 165 Allandale St., Jamaica Plain Station, Boston, MA 02130

CUSICK, PAUL V. Business & Fiscal Relations, Building 4-110, Massachusetts Institute of Technology, Cambridge, MA 02139

DARWIN, CHRISTOPHER J. Department of Experimental Psychology, University of Sussex, Brighton BN1 9QY, England

DELONG, MAHLON Laboratory of Neurophysiology, National Institute of Mental Health, National Institutes of Health, 9000 Rockville Pike, Bethesda, MD 20014

DE MAEYER, LEO C. M. Karl-Friedrich-Bonhoeffer-Institut, Max-Planck-Institut für biophysikalische Chemie, D3400 Göttingen-Nikolausberg, Federal Republic of Germany

de WIED, DAVID Rudolph Magnes Institute, Medical Faculty, University of Utrecht, Vondellaan 6, Utrecht, The Netherlands

DICHGANS, JOHANNES Neurlgische Universitatsklinik mit Abteilung für Neurophysiologie, D-78 Freiburg, West Germany

DOHAN, JR., F. CURTIS Massachusetts General Hospital, Boston, MA 02114

DUDLEY, GLENN G. Tufts University Medical Cancer Unit, Lemuel Shattuck Hospital, 170 Morton Street, Boston, MA 02130

DULBECCO, RENATO Imperial Cancer Research Fund, Lincoln's Inn Fields, London, WC2A 3PX, England

DUNN, ADRIAN J. Department of Biochemistry, University of North Carolina School of Medicine, Chapel Hill, NC 27514

EDDS, JR., MAC V. Faculty of Natural Sciences and Mathematics, South College, University of Massachusetts, Amherst, MA 01002

EDELMAN, GERALD M. The Rockefeller University, New York, NY 10021

EIDELBERG, EDUARDO St. Joseph's Hospital, Barlow Neurological Institute, Phoenix, AZ 85013

EIGEN, MANFRED Karl-Friedrich-Bonhoeffer-Institut, Max-Planck-Institut für biophysikalische Chemie D3400 Göttingen-Nikolausberg, Federal Republic of Germany

ERICKSON, ROBERT P. Department of Psychology, Duke University, Durham, NC 27706

ESKIN, ARNOLD Department of Biology, Rice University, Houston, TX 77001

EVANS, EDWARD F. Medical Research Council, Research Group, Department of Communications, University of Keele, Staffordshire, Keele ST5 5BG, England

EVARTS, EDWARD V. Laboratory of Neurophysiology, National Institute of Mental Health, Building 36, Room 2D-12, National Institutes of Health, Bethesda, MD 20014

FEDER, HARVEY H. Institute of Animal Behavior and Department of Psychology, Rutgers—The State University, 101 Warren Street, Newark, NJ 07102

FERNÁNDEZ-MORÁN, HUMBERTO Department of Biophysics, University of Chicago, 5640 South Ellis Avenue, Chicago, IL 60637

GAJDUSEK, CARLETON D. National Institute of Neurological Diseases and Stroke, National Institutes of Health, Bethesda, MD 20014

GALAMBOS, ROBERT Department of Neurosciences, University of California, San Diego, School of Medicine, La Jolla, CA 92037

GANONG, WILLIAM F. Department of Physiology, University of California, School of Medicine, San Francisco, CA 94112

GIBBS, CLARENCE J. National Institute of Neurological Diseases and Stroke, National Institutes of Health, Bethesda, MD 20014

GLASSMAN, EDWARD Department of Biochemistry, University of North Carolina, School of Medicine, Chapel Hill, NC 27514

GOLDBERG, ALFRED Department of Physiology, Harvard Medical School, 25 Shattuck Street, Boston, MA 02115

GOLDMAN, BRUCE D. Department of Biobehavioral Sciences, University of Connecticut, Box U-154, Storrs, CT 06268

GOODENOUGH, John B. Lincoln Laboratory, Room C-126, Massachusetts Institute of Technology, Lexington, MA 02137

GOY, ROBERT W. Wisconsin Regional Primate Research Center, 1223 Capitol Court, Madison, WI 53706

GRAYBIEL, ANN M. Department of Psychology, Massachusetts Institute of Technology, Cambridge, MA 02139

GREENE, LLOYD A. Laboratory of Biochemical Genetics, National Heart & Lung Institute, National Institutes of Health, Building 36, 1C-14, Bethesda, MD 20014

GRIGGS, ROBERT C. Departments of Neurology, Medicine, and Pediatrics; University of Rochester, School of Medicine and Dentistry, Rochester, NY 14620

GROSS, CHARLES G. Department of Psychology, Princeton University, Green Hall, Princeton, NJ 08540

HENNEMAN, ELWOOD Department of Physiology, Harvard Medical School, 25 Shattuck Street, Boston, MA 02115

HOLLOWAY, CLIVE M. Medical Research Council, Applied Psychology Unit, 15 Chaucer Road, Cambridge CB2 2EF, England

HOMSY, YVONNE Neurosciences Research Program, 165 Allandale St., Jamaica Plain Station, Boston, MA 02130

HOYLE, GRAHAM Department of Biology, University of Oregon, Eugene, OR 97403

HUGHES, PHILIP C. Neurosciences Research Program, 165 Allandale St., Jamaica Plain Station, Boston, MA 02130

HUTCHISON, JOHN B. University of Cambridge, Sub-Department on Animal Behavior, Madingley, Cambridge CB3 8AA, England

HYDÉN, HOLGER V. Institute of Neurobiology, Faculty of Medicine, University of Göteborg, Medicinaregatan 5, 40033 Göteborg 33, Sweden

HYVARINEN, JUHANI Institute of Physiology, University of Helsinki, Siltavuorenpenger 20A, 00170 Helsinki 17, Finland

ITO, MASAO Department of Physiology, University of Tokyo, Faculty of Medicine, Hongo, Bunkyo-ku, Tokyo, Japan

IVERSEN, LESLIE L. MRC Neurochemical Pharmacology Unit, Department of Pharmacology, Medical School, Hills Road, Cambridge, CB2 2QD, England

IVERSEN, SUSAN D. The Psychological Laboratory, University of Cambridge, Downing Street, Cambridge, CB2 3EB, England

JAIM-ETCHEVERRY, GUILLERMO Instituto de Anatomia General y Embriologia, Facultad de Medicina, Universidad de Buenos Aires, Paraguay 2155, Buenos Aires, Argentina

JOHNSON, HOWARD W. Chairman of the Corporation, 5-205; Massachusetts Institute of Technology, Cambridge, MA 02139

JOHNSON, L. EVERETT Neurosciences Research Program, 165 Allandale St., Jamaica Plain Station, Boston, MA 02130

JONES, E. G. Department of Anatomy, Washington University, School of Medicine, St. Louis, MO 63110

JOUVET, MICHEL Department of Experimental Medicine, Faculty of Medicine, Université Claude Bernard, 8, avenue Rockefeller, 69 Lyon 8ème, France

KANDEL, ERIC R. New York University Medical School, 550 First Avenue, New York, NY 10016

KENNEDY, DONALD Department of Biological Sciences, Stanford University, Stanford, CA 94305

KETY, SEYMOUR S. Psychiatric Research Labs, Massachusetts General Hospital, Fruit Street, Boston, MA 02114

KLEIN, DAVID C. Section of Physiological Controls, Laboratory of Biomedical Science, National Institute of Child Health and Human Development, National Institutes of Health, Bethesda, MD 20014

KOPROWSKI, HILARY The Wistar Institute, 36th Street at Spruce, Philadelphia, PA 19104

KORNHUBER, HANS Abteilung Neurologie, Universitat Ulm, Steinhovelstrasse 9, 79 Ulm, Federal Republic of Germany

KOSHLAND, JR., DANIEL E. Department of Biochemistry, University of California, Berkeley, CA 94720

KRISTAN, JR., WILLIAM B. Department of Molecular Biology, University of California at Berkeley, Berkeley, CA 94720

KUETHER, CARL A. Research Grants Branch, National Institutes of General Medical Sciences, National Institutes of Health, Bethesda, MD 20014

LAND, MICHAEL F. School of Biological Sciences, The University of Sussex, Falmer, Brighton, Sussex BN1 9QG, England

LASH, LEONARD Neuropsychology Research Review Committee, Behavioral Sciences Research Branch, National Institutes of Mental Health, 56001 Fishers Lane, Rockville, MD 20852

LEIBOWITZ, SARAH F. The Rockefeller University, New York, NY 10021

LIBERMAN, ALVIN M. Department of Psychology, College of Liberal Arts and Sciences, University of Connecticut, Storrs, CT 06268

LIVINGSTON, ROBERT B Department of Neurosciences, University of California, San Diego, School of Medicine, La Jolla, CA 92037

McCONNELL, HARDEN M. Department of Chemistry, Stanford University, Palo Alto, CA 94305

McEWEN, BRUCE S. The Rockefeller University, New York, NY 10021

MACKAY, DONALD M. Department of Communication, University of Keele, Staffordshire, Keele ST5 5BG, England

MATTHYSSE, STEVEN Research 4, Massachusetts General Hospital, Boston, MA 02114

MELNECHUK, THEODORE Western Behavioral Sciences Institute, 1150 Silverado, La Jolla, CA 92037

MENAKER, MICHAEL Department of Zoology, University of Texas, Austin, TX 78712

MILLER, NEAL E. The Rockefeller University, New York, NY 10021

MILNER, BRENDA Montreal Neurological Institute, 3801 University Street, Montreal 112, P.Q., Canada

MOLINOFF, PERRY B. Department of Pharmacology, University of Colorado Medical Center, 4200 East 9th Avenue, Denver, CO 80220

MOORE, ROBERT Y. Departments of Pediatrics, Medicine, and Anatomy; The University of Chicago, Chicago, IL 60637

MORRELL, FRANK Department of Neurology, Rush Medical College, Rush-Presbyterian-St. Luke's Medical Center, 1753 W. Congress Parkway, Chicago, IL 60012

MOUNTCASTLE, VERNON B. Department of Physiology, The Johns Hopkins University, School of Medicine, 725 North Wolfe Street, Baltimore, MD 21205

NASH, AVA B. Neurosciences Research Program, 165 Allandale St., Jamaica Plain Station, Boston, MA 02130

NAUTA, WALLE J. H. Department of Psychology, Building E10-104, Massachusetts Institute of Technology, Cambridge, MA 02139

ONSAGER, LARS Center for Theoretical Studies, University of Miami, Coral Gables, FL 33124

PARKER, KATHARINE M.I.T. Press, Massachusetts Institute of Technology, Cambridge, MA 02142

PEARL, DAVID Behavioral Sciences Research Branch, Division of Extramural Research Programs, National Institute of Mental Health, 56001 Fishers Lane, Rockville, MD 20852

PFAFF, DONALD W. The Rockefeller University, New York, NY 10021

PITTENDRIGH, COLIN S. Department of Biological Sciences, Stanford University, Stanford, CA 94305

PLOOG, DETLEV Max-Planck-Institut für Psychiatrie, Kraepelinstrasse 10, 8 Munich 23, Federal Republic of Germany

POLLEN, DANIEL A. Massachusetts General Hospital, Boston, MA 02114

PÖPPEL, ERNST Neurosciences Research Program, 165 Allandale St., Jamaica Plain Station, Boston, MA 02130

PRIBRAM, KARL H. Department of Psychiatry, Neuropsychology, Stanford University School of Medicine, Palo Alto, CA 94304

PRUSINER, STANLEY Department of Neurology, School of Medicine, University of California, San Francisco, CA 94122

QUARTON, GARDNER C. Department of Psychiatry, University of Michigan, 205 North Wastenau, Ann Arbor, MI 48104

RAHAMINOFF, RAMI Department of Neurobiology, Harvard Medical School, 25 Shattuck Road, Boston, MA 02115

RAINE, CEDRIC S. Department of Pathology (Neuropathology), Albert Einstein College of Medicine, 1300 Morris Park Avenue, Bronx, NY 10461

RANDAL, JUDITH ELLEN *The Washington Star*, 225 Virginia Avenue, S.E., Washington, DC 20024

REICHARDT, WERNER E. Max-Planck-Institut für biologische Kybernetik, Spemannstrasse 38, 74 Tübingen, Federal Republic of Germany

RIVLIN, RICHARD S. Department of Medicine, Francis Delafield Hospital, 99 Fort Washington Avenue, New York, NY 10032

ROBBINS, NORMAN Department of Anatomy, Case Western Reserve School of Medicine, Cleveland, OH 44106

ROBERTS, RICHARD B. Department of Terrestrial Magnetism, Carnegie Institution of Washington, 5241 Broad Branch Road, N.W., Washington, DC 20015

ROSEMAN, SAUL The Johns Hopkins University, Charles and 34th Streets, Baltimore, MD 21208

ROSENBLITH, WALTER A. Office of the Provost, 3-234, Massachusetts Institute of Technology, Cambridge, MA 02139

ROWLAND, VERNON Department of Psychiatry, Case Western Reserve, School of Medicine, University Hospital, Cleveland, OH 44106

SABAH, NASSIR H. Department of Electrical Engineering, American University of Beirut, Beirut, Lebanon

SAMSON, JR., FREDERICK E. Director, Ralph Smith Center for Mental Retardation, University of Kansas Medical Center, Kansas City, KS 66103

SCAPAGNINI, UMBERTO Via Costantinopoli 16, 80138 Napoli, Italy

SCHIMKE, ROBERT T. Department of Pharmacology, Stanford University, School of Medicine, Stanford, CA 94305

SCHMECK, HAROLD M. *The New York Times*, 1920 L Street, N.W., Washington, DC 20036

SCHMITT, FRANCIS O. Neurosciences Research Program, 165 Allandale St., Jamaica Plain Station, Boston, MA 02130

SCHMITT, OTTO H. Department of Electrical Engineering, University of Minnesota, Minneapolis, MN 55455

SCHNEIDER, DIANA Neurosciences Research Program, 165 Allandale St., Jamaica Plain Station, Boston, MA 02130

SCHWENK, HARRIET E. Neurosciences Research Program, 165 Allandale St., Jamaica Plain Station, Boston, MA 02130

SELVERSTON, ALLEN I. Department of Biology, University of California, San Diego, P.O. Box 109, La Jolla, CA 92037

SHEIN, HARVEY M. McLean Hospital, Belmont, MA 02178

SIDMAN, RICHARD L. Chief, Department of Neurosciences, Children's Hospital Medical Center, 300 Longwood Avenue, Boston, MA 02115

SINSHEIMER, ROBERT L. Division of Biology, California Institute of Technology, Pasadena, CA 91109

SIZER, IRWIN W. Massachusetts Institute of Technology, Room 3-134, Cambridge, MA 02139

SNYDER, SOLOMON H. Departments of Pharmacology and Psychiatry, The Johns Hopkins University, 725 North Wolfe Street, Baltimore, MD 21205

SOKOLOFF, LOUIS Building 36, Room 1A-27, National Institute of Mental Health, National Institutes of Health, Bethesda, MD 20014

SPEAR, PATRICIA G. The Rockefeller University, New York, NY 10021

SPERRY, ROGER W. Division of Biology, California Institute of Technology, Pasadena, CA 91109

STRUMWASSER, FELIX Division of Biology, Room 156–29, California Institute of Technology, Pasadena, CA 91109

SWEET, WILLIAM H. Massachusetts General Hospital, Boston, MA 02114

SWITKES, EUGENE Board of Studies in Natural Sciences, University of California, Santa Cruz, CA 95060

SZENTÁGOTHAI, JANOS 1st Department of Anatomy, Semmelweis University Medical School, Tužoltó v. 58, Budapest [ix] Hungary

TAKAHASHI, KUNITARO Department of Neurophysiology, Institute of Brain Research, University of Tokyo, School of Medicine, 7-3-1-Hongo, Bunkyo-ku, Tokyo, Japan

TEUBER, HANS-LUKAS Department of Psychology, E10-012, Massachusetts Institute of Technology, Cambridge, MA 02139

TRUMAN, JAMES The Biological Laboratories, Harvard University, 16 Divinity Avenue, Cambridge, MA 02138

UNGERSTEDT, URBAN Department of Histology, Karolinska Institutet, Solnavagen 1, S-104 01 Stockholm 60, Sweden

WEIGHT, FORREST F. Laboratory of Neuropharmacology, National Institute of Mental Health, IRP-SMR, Wm. A. White Building, St. Elizabeths Hospital, Washington, DC 20032

WEISKRANTZ, LAWRENCE Department of Experimental Psychology, University of Oxford, South Parks Road, Oxford OX1 3PS, England

WEISS, PAUL A. The Rockefeller University, New York, NY 10021

WERNER, GERHARD Department of Pharmacology, University of Pittsburgh Medical School, 660 Scaife Hall, Pittsburgh, PA 15261

WIERSMA, CORNELIS A. G. Division of Biology, California Institute of Technology, Pasadena, CA 91109

WILSON, J. PATRICK Department of Communication, University of Keele, Staffordshire, Keele ST5 5BG, England

WORDEN, FREDERIC G. Neurosciences Research Program, 165 Allandale St., Jamaica Plain Station, Boston, MA 02130

WRIGHT, MICHAEL J. Research Department of Ophthalmology, Royal College of Surgeons, 35–43 Lincoln's Inn Fields, London WC2A 3PN, England

WURTMAN, RICHARD J. Department of Nutrition and Food Science, Massachusetts Institute of Technology, Cambridge, MA 02139

ZEEVI, YESHOUA Y. Division of Engineering and Applied Physics, 326 Pierce Hall, Harvard University, Cambridge, MA 02138

ASSOCIATES,
NEUROSCIENCES RESEARCH PROGRAM

ADEY, W. ROSS Professor of Anatomy and Physiology; Director, Space Biology Laboratory, Brain Research Institute, University of California at Los Angeles

BLOOM, FLOYD E. Chief, Laboratory of Neuropharmacology, National Institute of Mental Health

BODIAN, DAVID Professor and Director, Department of Anatomy, The Johns Hopkins University, School of Medicine

BULLOCK, THEODORE H. Professor of Neurosciences, University of California at San Diego, School of Medicine

CALVIN, MELVIN Professor of Chemistry, University of California at Berkeley

DE MAEYER, LEO Member, Max Planck Institute for Biophysical Chemistry, Göttingen-Nikolausberg, Federal Republic of Germany

EDDS, JR., MAC V., Dean, Natural Sciences and Mathematics, University of Massachusetts, Amherst

EDELMAN, GERALD M. Professor, The Rockefeller University

EIGEN, MANFRED Director, Max Planck Institute for Biophysical Chemistry, Göttingen-Nikolausberg, Federal Republic of Germany

FERNÁNDEZ-MORÁN, HUMBERTO A. N. Pritzker Professor of Biophysics, Department of Biophysics, The University of Chicago

GALAMBOS, ROBERT Professor of Neurosciences, University of California at San Diego, School of Medicine

GOODENOUGH, JOHN B. Research Physicist and Group Leader, Lincoln Laboratory, Massachusetts Institute of Technology

HYDÉN, HOLGER V. Director, Institute of Neurobiology, University of Göteborg, Faculty of Medicine, Göteborg, Sweden

KATZIR, EPHRAIM Professor of Biophysics, Weizmann Institute of Science, Rehovot, Israel

KETY, SEYMOUR S. Chief, Psychiatric Research Laboratories, Massachusetts General Hospital, Professor of Psychiatry, Harvard Medical School

KLÜVER, HEINRICH Sewell L. Avery Distinguished Service, Professor Emeritus of Biological Psychology, The University of Chicago

LEHNINGER, ALBERT L. Professor and Director, Department of Physiological Chemistry, The Johns Hopkins University, School of Medicine

LIVINGSTON, ROBERT B. Professor of Neuroscience, University of California at San Diego, School of Medicine

LONGUET-HIGGINS, H. CHRISTOPHER Royal Society Research Professor, Department of Machine Intelligence and Perception, University of Edinburgh, Edinburgh, Scotland

MCCONNELL, HARDEN M. Professor of Chemistry, Stanford University

MACKAY, DONALD M. Professor of Communication, The University of Keele, Staffordshire, Keele, England

MILLER, NEAL E. Professor, The Rockefeller University

MORRELL, FRANK Professor of Neurology, Rush Medical College, Rush-Presbyterian-St. Luke's Medical Center

MOUNTCASTLE, VERNON B. Professor and Director, Department of Physiology, The Johns Hopkins University, School of Medicine

NAUTA, WALLE J. H. Professor of Neuroanatomy, Department of Psychology, Massachusetts Institute of Technology

NIRENBERG, MARSHALL Chief, Laboratory of Biochemical Genetics, National Heart and Lung Institute, National Institutes of Health

ONSAGER, LARS Distinguished University Professor, Center for Theoretical Studies, University of Miami

PLOOG, DETLEV Director, Max Planck Institute for Psychiatry, Munich, Federal Republic of Germany

QUARTON, GARDNER C. Director, Mental Health Research Institute, Professor of Psychiatry, University of Michigan

REICHARDT, WERNER E. Director, Max Planck Institute for Biological Cybernetics, Tübingen, Federal Republic of Germany

ROBERTS, RICHARD B. Chairman, Biophysics Section, Department of Terrestrial Magnetism, Carnegie Institution of Washington

SCHMITT, FRANCIS O. Chairman, Neurosciences Research Program, Massachusetts Institute of Technology

SIDMAN, RICHARD L. Bullard Professor of Neuropathology, Harvard Medical School; Chief, Department of Neuroscience, Children's Hospital Medical Center

SWEET, WILLIAM H. Chief, Neurosurgical Service, Massachusetts General Hospital; Professor of Surgery, Harvard Medical School

TEUBER, HANS-LUKAS Professor and Chairman, Department of Psychology, Massachusetts Institute of Technology

WEISS, PAUL A. Professor Emeritus, The Rockefeller University

WORDEN, FREDERIC G. Executive Director, Neurosciences Research Program; Professor of Psychiatry, Massachusetts Institute of Technology

NAME INDEX

Marshall, J. F., cit., 697
Marshall, K. C., cit., 938, 996
Marshall, W. H., 205; cit., 205, 348
Martel, S., 833
Martin, A. R., 929; cit., 348, 929, 943, 948, 949
Martin, J. B., cit., 558
Martin, J. P., cit., 319
Martin, T. E., 820; cit., 820
Martini, L., cit., 544
Marver, H. S., cit., 815
Maslow, A. H., 580; cit., 580
Mason, W. A., 572; cit., 572, 574, 576, 578
Massa, R., cit., 600, 609
Massaro, D. W., cit., 180
Masterton, B., 196; cit., 196
Mathers, L. H., cit., 190, 233
Mathes, R. C., 152; cit., 152
Matsubara, S., cit., 822
Matsui, T., cit., 552
Matthews, D. R., 138; cit., 98, 138
Matthysse, S., 652, 733–737, 1069; cit., 667
Mattingly, I. G., cit., 43, 44, 46, 53, 54
Maturana, H. R., cit., 105, 177
Matus, A. I., 907; cit., 907
Maurer, R., cit., 613, 641, 642
Mauss, T., cit., 215
Maxwell, J., 249
May, P. R. A., 733; cit., 733
Mayer, C., cit., 390
Mayer-Gross, W., cit., 727
Mayeri, E., cit., 356
Maynard, D. M., 398; cit., 390, 392, 398
Mayne, J. W., cit., 275
Meagher, W., cit., 631
Mebs, D., cit., 760
Meddis, R., 496; cit., 496
Medina, H., cit., 733
Meehl, P. E., cit., 735
Meers, J. L., cit., 854, 855
Mehler, A. H., 816; cit., 816
Mehler, W. R., 207, 320; cit., 207, 309, 319, 320, 631, 632, 701
Meier, H., cit., 745, 1006
Meier, R. E., cit., 22, 24, 25, 26, 27, 28
Meiri, U., cit., 947, 949, 992
Meixner, J., cit., xxi
Melancon, Jr., M. J., cit., 600, 601
Meldrum, B. S., cit., 759
Mello, N. K., cit., 25
Melnechuk, T., 1069
Melzack, R., cit., 179
Mena, F., cit., 521, 522
Menaker, M., 448, 479–489, 1069; cit., 446, 448, 481, 482, 483, 484, 485, 486
Mendell, L. M., cit., 281, 282, 283, 284, 285, 289, 291
Mendelson, M., cit., 377, 406
Mendum, C. M., cit., 382
Menez, A., cit., 760
Menkhaus, I., cit., 25
Mesarovic, M. D., cit., 160
Mess, B., cit., 597, 613, 641
Messenguy, F., cit., 822
Mettler, F. A., cit., 307
Metz, C., cit., 774
Meunier, J.-C., cit., 760
Meyer, D. R., 143; cit., 143
Meyer, V., cit., 79
Meyerhoff, J. L., 729, 730; cit., 729, 730
Miale, I., cit., 747, 748, 751

Michael, C. R., cit., 105
Michael, R. P., 624; cit., 595, 624
Michel, F., cit., 503
Michels, R., cit., 888
Michison, N. A., cit., 774
Mihailovic, J., cit., 22, 24, 25
Miledi, R., 912, 951; cit., 760, 762, 765, 767, 872, 930, 931, 943, 944, 945, 946, 947, 948, 949, 951
Mill, P. J., cit., 421, 429
Millard, S. A., cit., 608
Miller, G. A., cit., 148, 176, 200
Miller, J., cit., 217
Miller, J. M., cit., 141, 174
Miller, N. E., 1069; cit., 714, 715, 717, 718
Miller, R. E., cit., 655, 854, 855
Miller, R. L., 152; cit., 152
Millette, C. F., cit., 784, 785, 787
Millodot, M., cit., 112
Mills, J. N., cit., 491
Milner, B., 3–4, 31, 75–89, 1069; cit., 5, 12, 16, 50, 51, 52, 57, 58, 66, 71, 76, 77, 78, 79, 81, 82, 83, 84, 85, 229
Mims, C. A., cit., 1043
Mimura, K., cit., 105
Minis, D. H., 451; cit., 443, 446, 448, 451, 453, 454, 455
Minna, J., cit., 1014
Minsky, M. L., cit., 252
Misch, J., cit., 387
Mishell, R. I., cit., 785
Mishkin, M., 191, 197, 199, 200, 202, 253, 254, 255; cit., 191, 197, 198, 199, 202, 215, 221, 222, 233, 235, 237, 255
Missakian, E. A., 578; cit., 578
Mitchel, D. E., 110, 112; cit., 110, 111, 112
Mitchell, J. F., cit., 909, 911
Mitnick, M., cit., 552
Mize, R. R., cit., 193
Mizell, M., cit., 1054
Mizutani, S., cit., 802, 1005, 1057
Moberg, G. P., cit., 558, 560, 568
Modianos, D., cit., 631
Moeller, G., cit., 653, 654
Moguilevsky, J. A., cit., 608
Moir, A. T. B., cit., 502, 688
Molenaar, P. C., cit., 911
Molinoff, P. B., 759–763, 922, 1069; cit., 760, 761, 765, 866, 867, 877, 879, 912, 922
Moller, A., cit., 137
Moltz, H., cit., 25
Moncrieff, R. W., cit., 841
Mondon, C. E., cit., 828
Money, J., cit., 571
Money, K. E., cit., 295
Monjan, A. A., cit., 1023
Monod, J., cit., xxii
Monroy, A., cit., 774
Montanari, R., cit., 566
Montgomery, M., cit., 627
Mookerjea, S., cit., 798
Moor, H., 992
Moore, G. P., cit., 376
Moore, K. E., cit., 565, 566
Moore, R. Y., 207, 448, 511, 537–542, 904, 912, 961–977, 991, 993, 994, 1069; cit., 207, 446, 448, 480, 538, 539, 540, 631, 755, 905, 917, 961, 965, 969, 970, 971, 972, 973, 974, 991, 993, 994
Moore, W. V., cit., 674
Mora, P. T., cit., 801

Morgan, H. E., cit., 828
Morgane, P. J., cit., 697
Morita, H., cit., 372
Morita, Y., cit., 484
Morrell, F., 136, 1069; cit., 136, 174, 198
Morris, J. A., cit., 1038
Morse, D. E., cit., 820
Mortimore, G. E., cit., 828
Morton, J., cit., 60
Moruzzi, G., 293; cit., 293
Moscarello, M. A., 926; cit., 926
Moscano, A. A., 754, 774–775; cit., 608, 754, 774, 775, 799
Moscona, M. H., 774–775; cit., 754, 774
Moser, I., cit., 443, 535
Moss, R. L., cit., 624, 626
Motta, M., cit., 553
Mountcastle, V. B., 157, 311, 316, 1069; cit., 5, 11, 72, 157, 160, 250, 311, 312, 316
Mountford, H., cit., 670
Mouret, J., 491–497; cit., 500, 501
Mous.hegian, G., 135; cit., 134, 135
Mowles, T. F., cit., 603
Moyer, K. E., cit., 653
Mozell, M. M., cit., 160, 162
Mueller, R. A., 912–913; cit., 864, 865, 866, 868, 912
Mugnaini, E., cit., 300
Müller, D., cit., 1020
Müller, E. E., cit., 558
Müller, J., 105; cit., 105
Muller, J. E., cit., 82
Mulloney, B., 389–395; cit., 390, 394, 671
Munk, H., 191, 193
Munkvad, I., cit., 695, 699, 707
Munro, H. N., 821; cit., 821
Munson, P. L., cit., 565
Murphy, E. G., cit., 45
Murphy, G. M., cit., 176
Murphy, J. V., cit., 655
Murray, J. G., cit., 964
Murray, M. R., 927; cit., 927
Murrell, W. G., cit., 857
Murthy, M. R., cit., 887
Musacchio, J. M., cit., 864
Muscholl, E., 910; cit., 910, 992
Musliner, T. A., cit., 602
Myers, R. D., cit., 715
Myers, R. E., cit., 65, 233
Myrtle, J. F., cit., 601

N

Nachmansohn, D., cit., xxvii, 770
Nachmias, J., 98; cit., 98
Nadler, R. D., 640; cit., 584, 614, 622, 637, 640
Naftolin, F., cit., 584, 600, 609
Nagata, Y., cit., 963
Nagatsu, T., cit., 864
Nagy, E., cit., 641
Nagylaki, T., 5; cit., 5
Naka, K., cit., 518
Nakagawa, S., cit., 602
Nakai, T., 1020; cit., 612, 1020
Nakajima, H., cit., 298
Nakazawa, T., 927; cit., 927
Nalbandov, A. V., cit., 589
Nanda, K. K., cit., 454
Naquet, R., cit., 503
Narahashi, T., cit., 400

Narasimhan, R., cit., 94
Nash, A. B., 1069
Nashold, B., 308; cit., 308
Nataf, B. M., cit., 838
Nathanson, M., cit., 308
Naumenko, E. V., cit., 565
Nauta, W. J. H., 320, 1069; cit., 22, 23, 210, 211, 212, 213, 220, 221, 309, 319, 320, 537, 631, 701, 972
Nazerian, K., cit., 1054
Neal, M. J., cit., 907, 909, 910
Nebes, R. D., cit., 12
Neet, K. E., cit., 841
Neff, N. H., cit., 919, 922, 966
Neff, W. D., cit., 76, 142, 143
Neher, E., cit., 474
Neidhardt, F. C., cit., 822
Neilson, Jr., D. R., cit., 362, 364
Neims, A. H., cit., 835
Neisser, U., cit., 60, 171, 176, 180
Nelson, P. G., 141; cit., 132, 133, 134, 138, 141, 954
Neumann, E., cit., xxvii
Neumann, F., cit., 639
Neurath, H., cit., 822
Newburgh, R. W., cit., 996
Newby, N. A., cit., 364
Newell, A., cit., 178
Newell, J. D., 141; cit., 141, 142
Nicholls, J. G., 371; cit., 342, 348, 371
Nicholson, J. L., cit., 878
Nickel, V. L., cit., 311
Nickerson, M., cit., 713
Nicoll, C. S., cit., 521
Nicoll, R. A., 993; cit., 931, 993
Nielsen, N. C., cit., 893
Niemeyer, H., cit., 815, 819
Niemi, M., cit., 642
Niimi, K., cit., 210
Nikara, T., cit., 172
Nirenberg, M. W., 912; cit., 671, 770, 912, 997
Nishi, S., cit., 937
Nishiitsutsuji-Uwo, J., cit., 448
Nissen, H. W., 572, 578; cit., 578
Nissley, P., city., 996
Niuteman, F. W., cit., 700
Nobin, A., cit., 550, 553, 965, 966
Noble, M., cit., 34, 36
Noda, H., 119; cit., 119, 177
Noonan, K. D., 779; cit., 1006
Norberg, K.-A., cit., 961
Nordenfelt, I., 965; cit., 965
Nordlander, R. H., 383; cit., 383
Norman, A. W., cit., 601
Norman, D. A., cit., 34
North, K. A. K., cit., 930
Nossal, G. J. V., cit., 774, 783, 792
Nöthiger R., cit., 613
Notides, A. C., cit., 603
Notkins, A. L., cit., 1022
Noton, D., cit., 178
Nottebohm, F., cit., 73, 594
Noumura, T., cit., 638
Nunnemacher, R. F., cit., 519
Nuza, J., cit., 121
Nybäck, H., cit., 723, 724, 733, 981, 987
Nyby, O., cit., 269, 307
Nye, P., cit., 45
Nygren, L.-G., 968, 969, 973

O

Oatley, K., 101; cit., 101
Obata, K., cit., 298, 299, 909
Oberjat, T., cit., 428
Oberlander, H., cit., 613
Ochi, R., cit., 299
Ochs, S., cit., 869
O'Connell, R. J., cit., 160, 162
Oda, K., cit., 1061
Oester, Y. T., cit., 955
Oesterreich, R. E., cit., 143
Ogata, K., cit., 821
Ogawa, K., cit., 1008
Ogawa, N., cit., 655
Ogawa, T., cit., 25
Ohgushi, K., cit., 138
Ohinata, S., cit., 25
Ohm, G. S., 242; cit., 242
Ohno, T., cit., 298
Ojeda, S. R., cit., 589, 610
Okada, F., cit., 609
Okada, Y., cit., 298
O'Keeffe, R., cit., 733
Oksche, A., cit., 484
Okun, L. M., cit., 755
Old, L. J., cit., 755
Olson, L., 963, 964, 965, 968; cit., 963, 964, 965, 966, 968, 973, 984, 990, 994
Oltmans, G. A., cit., 697
O'Malley, B. W., cit., 600, 602
O'Neill, E. L., 95; cit., 95
Onesto, N., 293; cit., 293
Onsager, L., 1069
Oomura, Y., cit., 932
Oonishi, S., 139; cit., 139
Opler, L. A., cit., 996
Oppenheim, A. V., cit., 245
Oppenheimer, S. B., cit., 800
Orbach, J., cit., 178
Orkand, P. M., cit., 991
Orkand, R. K., cit., 946
Orlov, J., cit., 390
Orr, C. W., cit., 780, 799
Orr, D. B., cit. 45
Orrego, F., cit., 668
Ort, cit., 371, 376
Ortmann, A. E., 517
Osburn, B. I., cit., 1043
Oscarsson, O., cit., 327
O'Steen, W. K., cit., 480
Oster, G., cit., xix, xxiv
Otsuka, H., cit., 655, 662
Otsuka, M., cit., 393, 929, 930, 991
Ottesen, E., cit., 445
Otzuka, R., cit., 215
Owen, J. J. T., cit., 755
Owen, O. E., 891; cit., 891, 893
Owman, C., cit., 919, 921, 962
Oxbury, S., cit., 40
Oyanagi, S., cit., 1018

P

Padgett, B. L., cit., 1009, 1011, 1021, 1027
Padjen, A., cit., 936, 937
Page, C. H., cit., 405
Page, M. A., cit., 893
Palay, S. L., cit., 743, 974
Palka, J., 385; cit., 385, 402
Palmer, R. B., cit., 926

Palmiter, R. D., cit., 820
Palumbo, M., cit., 1046
Pandya, D. N., 210, 211, 212; cit., 210, 211, 212, 213, 215, 233, 268, 277
Panofsky, A., cit., 460
Pantle, A., 97, 121; cit., 97, 121, 172
Paolino, R. M., cit., 670
Papaconstantinou, J., cit., 820
Papert, S., cit., 252
Papez, J. W., cit., 133
Pappas, B. A., cit., 558
Pappas, G. D., cit., 974
Pardee, A. B., cit., 780, 781, 850
Parducz, A., 954; cit., 954
Paris, C. A., cit., 585
Parker, G. H., cit., 517
Parker, K., 1069
Parker, T. J., cit., 389
Parnas, I., cit., 406, 947
Parry, H. B., 1038; cit., 1038
Pasik, P., 193-194, 196; cit., 193, 195
Pasik, T., 193-194, 196; cit., 193, 195
Pasley, J. N., cit., 656
Pasquini, J. M., cit., 887
Passingham, C., 193
Pastan, I., cit., 829, 855
Pasternak, G., cit., 1049
Patel, J. B., cit., 670
Pateman, J. A., cit., 855
Paton, W. D. M., cit., 769, 770, 910
Patrick, R. L., cit., 866
Pattison, I. A., cit., 1013
Paul, W. E., cit., 792
Pavlov, I. P., cit., 362
Payne, F. E., cit., 1010
Pearl, D., 1069
Pearson, K. G., 405; cit., 397, 398, 405
Peebles, P. T., cit., 1007
Peeke, H. V. S., cit., 357, 670
Pellegrino de Iraldi, A., cit., 511, 917, 918, 919, 920
Pelling, C., cit., 613
Pellioniz, A., cit., 293
Penfield, W., cit., 83, 85, 86, 245, 263
Peng, M. T., cit., 638
Penn, N. W., cit., 822
Penner, P. E., cit., 858
Pennes, H. H., cit., 727
Pennington, K. S., cit., 245
Peraino, C., cit., 821
Perelson, A., cit., xix, xxiv
Peretz, B., 357-368; cit., 350, 352, 354, 356, 357, 584
Pérez-Palacios, G., cit., 609
Perisic, M., cit., 22, 23, 24, 28
Perkel, D. H., 165, 392; cit., 165, 174, 371, 377
Perkins, J. P., 912; cit., 912
Perlman, R. L., cit., 829, 855
Persson, B., cit., 892
Perumal, R., 679-684; cit., 680, 682
Peterkofsky, A., 854
Peters, A., cit., 997
Peterson, D. A., cit., 1029
Peterson, L. R., cit., 79
Peterson, M. S., cit., 79
Peterson, R. P., cit., 671
Petitjean, F., cit., 505
Petras, J. M., cit., 210
Petro, Z., cit., 584
Petrusz, P., cit., 641

SUBJECT INDEX

Cyclic AMP (adenosine monophosphate)
  and communication between cell
    surface and interior, 780–781
  effect on glutamine metabolism enzyme
    levels, 855, 857
  in glia metabolism, 997
  in protein degradation, 829
  in regulation of indole metabolism, 512
  as second messenger, 802
  synaptic activation of, 938
  in synaptic function, 996–997
  in synaptic modulation, 910, 912
Cyclic guanosine monophosphate (GMP)
  synaptic activation of, 938–939
  in synaptic function, 997
Cyclic nucleotides in synaptic function,
  995–997
Cyproterone, 639–640
Cytochrome oxidase during postnatal
  maturation, 894
Cytomegalovirus, 1050
  effect on CNS development, 1043
Cytopherometer assay, 925

D

DA. See Dopamine
Dale's principle, 922
Darwinian evolutionary behavior,
  macromolecular synthesis, xxiv
Darwinian evolutionary system, reaction
  network as prototype, xxiii
DDT in brain masculinization, 614
Deafferentation of the hypothalamus, 552
Decapods, feature detection, 421–422
Deep structure, syntactic, 46
Deermouse, circadian rhythm of activity,
  437–438, 441
Degenerative diseases, virus-caused, of CNS,
  1025–1041
Delayed release and residual calcium
  hypothesis, 951
Delay-line hypothesis of cerebellar
  coordination, 293
de Mairan's phenomenon, 437
Demyelinating disease, 1011–1012
  of sheep, 1012
Dendrites, vertebrate and invertebrate, 344
Denervation
  effect on muscle membrane, 957
  in regeneration study, 962-964, 967–974
Denervation supersensitivity, 912
Depolarization-hyperpolarization of
  postsynaptic membrane, transmitter
  effects, 929–941
Depression of transmitter release, 929–930,
  949
Desynchronization of circadian rhythm,
  491–497
Dexamethasone in avoidance behavior
  acquisition and extinction, 658–659
Diapause, photoperiodic induction, 454
Dichotic listening, 4, 31, 57–63
  after cerebral commissurotomy, 76–77
  experiments, 36–37, 39, 50
  and hemisphere differences, 31, 39–40
  in normal subjects, 75–76
  see also Ear differences and hemispheric
    specialization
Dichotic pitch phenomena, 148

Diet, effect on brain serotonin
  concentrations, 686–692
Differentiation, single cells, 774
Differentiation antigens, 755
Digit perception, asymmetry, 75–77
Dioptrics, transmission properties, 96
Diplornavirus, effect on CNS development,
  1043
Direction
  perception mechanism, 121
  of stimulus motion, 174–175
Directional satiation, 172
Discharge zone, 281–282
Discrimination apparatus for discrete
  trial analysis (DADTA), 254
Dishabituation
  in absence of protein synthesis, 362
  and gill-withdrawal reflex, 357–361
  neural correlates of, 364
  see also Habituation
Distemper virus, 1021
Distribution of information in the striate
  cortex, 250–253
Diurnal rhythm, central control, 537–542
  see also Circadian rhythm
Divalent ions in transmitter release,
  945–951
DNA synthesis control, 779–781
DNA tumor virus, 1005–1007
DNA viruses, 1009, 1014, 1057–1063
Dominance drive in monkeys, 580
Dominance relations in monkeys, 579
Dopamine (DA)
  central effects, 717
  effect on behavior, 697
  effect of phenothiazines, 723
  effect on schizophrenic symptoms, 728
  function in basal ganglia, 734
  molecular model, 725
  optimal conformation, 648–649
  in Parkinsonism therapy, 324
  regulation of ACTH secretion, 553–560
  in striatum, 319–320, 324
  see also Adrenergic effects;
    Catecholamine
Dopamine axons in rat brain, 979–980
Dopamine-β-hydroxylase, 864–867,
  869, 870, 879–881, 913
  circulating, 870–871
Dopamine cell groups, 980
Dopamine decarboxylase, 864, 866, 879–881,
  913
Dopamine degeneration syndromes,
  696–700
Dopamine neurons
  and basal ganglia function, 701
  and behavior, 695–703
  distribution, 965
  and self-stimulation, 700–701
  stereoselectivity, 728–729
Dopamine pathway, 695–696
  and motor response to amphetamine,
    705–710
Dopamine psychosis, 727
Dopaminergic terminals, development,
  877–883
Dopamine transmission and schizophrenia,
  733–737
Dopamine uptake system, 907–908
Dorsal column, in somesthesis, 178–179

Dove, courtship behavior, 593–597
Drinking, role of adrenergic receptor
  mechanisms, 713–719
Drosophila
  circadian oscillation, 444, 446–448, 451,
    453
  circadian rhythms, 439–442, 444–445
  genetics and behavior, 403
Drugs, effects in schizophrenia, 721–732
Drug psychoses, 726
Dynamic theory of matter, xxiv
Dystonia, 324

E

EAE. See Experimental autoimmune
  encephalomyelitis
Ear, capabilities, 45, 49
Ear differences and hemispheric
  specialization, 57–63
  see also Dichotic listening
Eating, role of adrenergic receptor
  mechanisms, 713–719
Ecdysis. See Eclosion; Emergence
Ecdysone, 613
Echoic memory, 60–61
Eclosion
  circadian rhythm, 439–441, 444–445
  hormonal control, and brain, in
    silkmoth, 526–527
Eclosion hormone, 527
Eclosion response of silkmoths, 525–526
Edge detection, 172, 173
Edge orientation, information
  transmission, 96–97
EF. See Encephalitogenic protein of
  myelin
Effector system, multiple controls, in
  Aplysia, 350–357
Elapidae venoms in cholinergic receptor
  study, 759–763
Electric eel, effect of toxins, 760–761
Electric organ, effect of toxins, 760–762
Electric oscillatory circuit, linear, xxi
Electrical activity in vertebrates and
  invertebrates, 344
Electrical and chemical terms, analogy, xx
Electrocoagulation in monoamine pathway
  study, 979–987
Electromyograms of insects, 397–398
Electromyographic (EMG) response to
  visual and somesthetic stimuli, 327–337
Emergence, circadian rhythm, 439–441,
  444–445, 525–529
Encephalitogenic protein of myelin (EF),
  925–928
  inhibition of antigenicity by serotonin,
    925–926
Encephalopathies, virus, 1012–1013
Endemic goiter, 685
Endocrine gland. See Hormones; Pituitary
  gland; Pituitary-adrenal system
Endoplasmic reticulum, half-life, 815
End-plate potential (epp), 929–938, 943
Entrainment
  by environmental cycles, 443–446
  of identified neurons, in insects, 408
  in isolated Aplysia eye, 531–535
Enzymes, 810
  activity, control of, 811

Enzymes (*continued*)
  activity changes, during brain postnatal
    maturation, 887
  allosteric, regulation in glutamine
    metabolism, 853–861
  catecholamine biosynthesis, neural
    regulation, 864–867
  conformational control, 885–898
  degradation, regulation, 821–822
  independent regulation of synthesis and
    degradation rates, 816–819
  of ketone utilization in brain, 893–896
  of rat liver, degradation, 814–816
  synthesis, regulation, 819–821
  systems, circadian periodicity, 439
  see also Proteins
Enzyme-substrate hypothesis of cell
  adhesion, 800–801
Ependymal cells in hypothalamus, 552
Ephedrine, configurations, 728
Epinephrine
  biosynthesis, 864
  effect on adrenergic receptors, 713–718
EPSPs. *See* Excitatory postsynaptic
  potentials
Epstein-Barr virus (EBV), 1050
  and cancer, 1054
Equal density hypothesis of input
  distribution, 289–291
Equilibrium state, perturbation, xxi–xxii
Equivalence of neurons, in vertebrates and
  invertebrates, 345
Erasure in dichotic listening, 39–40
Escherichia coli
  protein degradation, 828–833
  regulation of glutamine metabolism,
    853–860
Estradiol
  binding distribution in rat brain,
    603–606
  biochemical effects on brain, 608–609
  concentration in brain, 622–624
  functional distribution in brain, 606
  see also Estrogen; Steroid hormones
Estradiol receptor, 584
Estrogen
  binding sites in brain, ontogeny, 610–611
  in female rat sex behavior, 622
  in luteinizing hormone release, 587–589
  see also Steroid hormones
Estrogen-concentrating neurons in brain,
  544–545
α-Ethyltryptamine, effect on ACTH
  secretion, 554–555
Euglena, circadian oscillation, 443
Evoked potentials
  end-plate, 944
  human, preceding voluntary rapid
    flexion movements, 268
Excitatory postsynaptic potentials
  (EPSPs), recordings from gastrocnemius
  motor neurons, 282–291
Excitement, effect on invertebrate visual
  neurons, 422–423
Existence region of tonal residue, 151
Exocytosis, 869, 870
Experimental autoimmune
  encephalomyelitis (EAE), 925
External coincidence model, 454
External coupling

in avian circadian system, 482–485
in mammalian circadian system, 480
Extrageniculostriate visual mechanisms,
  215–227
Eye
  of crayfish, 517–520
  isloated, of *Aplysia*, circadian rhythm
    in, 531–535
  of jumping spiders, 411–418
  optical performance, 96
Eye dominance of cat visual cortex
  neurons, 107–108
Eye movement, 12–13
  of jumping spiders and mammals,
    413–416
  supranuclear control, 305–310
  and vision, 178
  visually guided, 305–310
  see also Saccades

F

Face discrimination, asymmetries, 65–69
Face recognition, 14, 72
Facilitation of transmitter release, 929–930,
  949–951
Falck-Hillarp method, 961–962
Familial dysautonomia, 871
Familial protein intolerance, 859
Fast movement, role of cerebellum,
  269–274
Fear grimace in monkeys, 579
Feature-detecting neurons
  formation, developmental factors,
    105–113
  in local and global perception, 173
  in visual cortex, 173
Feature detection, 94
  in auditory nervous system, 131–145
  by invertebrate single units, 421–422
Feature detectors, universal and species-
  specific, 105
Feature-extracting neurons, 105–113
  role in perception, 172–181
Feature extraction
  clues from perceptual phenomena,
    172–173
  concept of sensory attributes, 155–156
  parallel population neural coding in,
    155–169
Feature extractors
  in neural information processing,
    171–183
  see also Stimulus features
Features
  neural representation, 165–167
  of sensory stimuli, 105
  see also Stimulus features
Feature-sensitive neurons, classification
  according to best stimulus, 174
Feature sensitivity by auditory neurons,
  131–145
Feedback control, 294–296
Feedforward control, 294–296, 299
Fertilization, as cell-cell interaction, 774
Fiber fractionation of lymphocytes, 789–
  790, 792, 794
Fibers, Ia, from medial gastrocnemius
  muscle, 281–291
Fibroblast growth control, 776–778

Fixation syndrome. *See* Spasm of fixation
Fixed action patterns, 350
Flavin adenine dinucleotide (FAD)
  pyrophosphorylase, 835–839
  regulation by thyroid hormone, 835–839
Flavin coenzymes, regulation by thyroid
  hormone, 835–839
Flavin mononucleotide (FMN)
  phosphatase, 835–837
  regulation by thyroid hormone, 835–839
Flavokinase, 835–837
Flavoprotein enzymes, regulation by
  thyroid hormone, 835–839
Flocculonodular lobe, 275
Flocculus, 297
Fluorescence
  catecholamine-induced, 877
  noradrenergic system tracing method,
    961–962
Fluphenazine, 733
Flux discrimination, effect of destriation,
  193–196
FMN. *See* Flavin mononucleotide
Focal vision, 196
Follicle-stimulating hormone (FSH), 550,
  552, 560
  adrenergic effects, 558
  catecholamine-mediated effects, 559
  see also Hormones
Follicle-stimulating hormone releasing
  factor (FRF)
  activity and identification with
    luteinizing hormone releasing factor,
    552
  effect on anterior pituitary secretion,
    551
Food intake, role of adrenergic receptor
  mechanisms, 713–719
  see also Nutrition and brain
Forebrain in sleep, 503–504
Formaldehyde-glutaraldehyde-dichromate
  (FGD) reaction, 919
Formant
  definition, 47
  second, in vowel distinction, 47
  second, hemispheric specialization, 50–51
Form feature extraction, 165
  see also Feature extraction
Form perception, 120
  see also Pattern
Fornix section, 969–971
Fortification illusion, 172
Fourier analysis, 252
  in striate cortex neurons, 241–245
Fourier transform theory, analysis of
  optical systems, 95–102
Foveal prestriate inferotemporal cortex,
  lesion effect, 199
Freerunning oscillation, 437
  properties, 439–443
  see also Circadian oscillation
Freerunning period (τ), 437–456
  effect of heavy water, 442–443
  homeostasis, 441–442
  stability and lability, 441
Freeze-etching electron microscopy,
  992–993
Frequency demultiplication, 450
Frequency modulation, neuron sensitivity,
  137–138

Frequency of stimulation and transmitter release, 949–951
Fricatives, 51
  perception, hemispheric differences, 38
  see also Consonants
Frog sounds, 140
Frog virus (FV3), 1020
Frontal lobe, connections to parieto-temporo-occipital association cortex, 211–212
Frontal lobectomy, unilateral, and hemispheric specialization, 81–83
FSH. *See* Follicle-stimulating hormone
Fucose incorporation in training, 683
Functional decussation, 58
Function generator, 268
Fusion of cells, 1019
Futile cycle, amide synthesis and degradation, 857

G

GABA (γ-aminobutyric acid)
  as inhibitory transmitter, 930
  localization, 991
  uptake systems, 907–908, 909
Galactose-binding protein, bacterial, 850
Ganglia
  in *Aplysia*, 347–370
  retinal, in cat, 115–121
  in spatial information transmission, 96
  vertebrate and invertebrate, 344
Ganglioside, 796–797
  metabolism in training, 682
Gastrocnemius muscle, Ia fibers and motor neurons, 281–291
Gaze palsy, 307
Gaze paresis, 307
Gene transcription, circadian periodicity, 439
Genes
  viral, 1059
  viral and cellular, interaction, 1062–1063
Geniculate body in spatial information transmission, 96
GH. *See* Growth hormone
GIF. *See* Growth hormone inhibiting factor
Gilles de la Tourette's disease, 730
Gill-withdrawal reflex in *Aplysia*, 347–370
Glabella tap reflex, 296
Glia
  and cyclic AMP, 997
  response to neurotransmitter, 912
  transmitter uptake, 991
Globus pallidus, 319–320
Glucocorticoids
  functional distribution in brain, 606–608
  mammalian, rhythms, 481–482
  see also Corticosterone; Steroid hormones
Glucocorticosteroids in avoidance behavior acquisition and extinction, 658–659
Glucosamine incorporation in training, 682
Glucose consumption, brain, 888–889, 893
Glutamate uptake system, 907–908
Glutaminase B, regulation modes, 856–860
Glutamine metabolism, regulation of allosteric enzymes in, 853–861
Glutamine synthetase
  electron micrograph, 856
  regulation modes, 856–860

Glutaraldehyde-dichromate (GD) reaction, 919
Glycine uptake system, 907–908
Glycolipids, 796–798
Glycoproteins, 796–798
  metabolism in training, 683
Glycosyltransferases, 798, 801
Gonadal atrophy and light, 513
Gonadal hormones. *See* Steroid hormones
Gonadotropin and sexual differentiation, 589–590
Gonadotropin antiserum (GTH A/S), effect on sexual behavior, ovulation, and masculinization, 588–590
Gonyaulax, circadian oscillations, 439, 443
Grammar, 43–56
Grammatical codes, 43–56
Grammatical recoding, 43–56
Granule cell neurons, 751
Grasp reaction, 179
Grasshopper
  command fibers, 406
  courtship, 398
  jump response, neuronal circuitry, 406
Grating patterns, contrast threshold, 95–102
Gravity perception, 126–128
GRF. *See* Growth hormone releasing factor
Growth and competitive selection properties, xxv
Growth control, surface changes, 776–778
Growth hormone (GH), 550, 552, 560
  adrenergic effects, 558
  catecholamine-mediated effects, 559
Growth hormone inhibiting factor (GIF), effect on anterior pituitary secretion, 551
Growth hormone releasing factor (GRF), 552
  effect on anterior pituitary secretion, 551
Gryllus campestries. *See* Cricket
Guanyl cyclase in synaptic function, 997

H

H-2 antigen, 755
Habituation
  in absence of protein synthesis, 362
  and gill-withdrawal reflex, 357–361
  neural correlates of, 364
  relation to complex learning, 366
  relation to dishabituation, 362–366
  short- and long-term, 362, 364–366
Habituation and dishabituation, 357–366
Halász knife, 553
Hallucinogenic drugs
  inhibition of encephalitogenic protein of myelin antigenicity, 926–927
  interaction with encephalitogenic protein of myelin, 925–928
Hamster
  circadian oscillation, 448–450, 452
  golden, ovulatory cycle, 587–589
Handedness and hemispheric differences, 31–32
Handedness in animals, genetics, 5
Handedness in man, genetic model, 5
Hand movements, 272
Harmonic oscillation system, xxi
Helmholtz's model of the ear, 101

Hemisphere deconnection syndrome, 7–10
Hemisphere of the brain
  asymmetry, state of the art, 31–32
  cerebellar, and motor action, 293
  division of function and integration of behavior, 31–41
  functional asymmetry, 3–4
  information transfer between, 34–36
  visual information transfer between, in pigeon, 21–29
  see also Language hemisphere
Hemispheric asymmetries and lateralized visual simuli, 65–69
  see also Hemispheric specialization
Hemispheric communication and cerebral dominance, 65–69
Hemispheric differentiation, evidence from commissurotomy patients, 11
Hemispheric dominance, evidence from commissurotomy patients, 11
Hemispheric interaction, 4
Hemispheric specialization
  and ear differences, 57–63
  ontogeny, 85–87
  phylogeny and ontogeny, 73
  scope and limits, 75–89
  and speech, 58
  state of the art, 71–74
  studies on commissurotomy patients, 5–19
Herpes simplex virus (HSV), 1010, 1050
  antigen effects, 1051–1052
Herpes varicella-zoster, 1050
Herpesvirus, 1057
  and cancer, 1054
  effect on cell membrane, 1049–1055
  interaction with hosts, 1049
  membrane-bound proteins, 1049–1054
Heterosynaptic facilitation of transmitter release, 931
Hierarchy of neurons, 345
HIOMT. *See* Hydroxyindole-O-methyltransferase
Hippocampal lesion and hemispheric lesion, 79–81
Histofluorescent technique for catecholamine localization, 877–884, 961–962
Histogenesis in CNS, 743–744
Histograms, cross-correlation, 371–372
Hold function of eyes, 308
Hold regulation in the cerebellar nuclei, 274–275
Holding tremor, 275
Hole-board, 698–699
Hologram, analogy of memory, 245–246
Holographic hypothesis of brain function, 252–253
Hormones
  effects on brain, 549
  interaction with differentiated cells, 599–601
  perinatal, effect on adult sexual behavior, 583–586
  pituitary-adrenal system, and behavior, 653–666
  of pituitary gland, 550
  in sexual behavior development in male rhesus monkey, 571–581

Hormones (*continued*)
in synaptic function, 997
see also Steroid hormones
Hour glass in photoperiodism, 453–454
HSV. *See* Herpes simplex virus
5-HT. *See* Serotonin
(5-Hydroxytryptamine)
Humoral modulation of circadian rhythm
in crayfish, 520–522
Hydrogen bond hypothesis of cell
adhesion, 800–801
D-β-Hydroxybutyrate. *See* Ketone bodies
β-Hydroxybutyrate dehaydrogenase, 890,
893
D-β-Hydroxybutyrate dehydrogenase
during postnatal maturation, 892,
894–896
6-Hydroxydopamine (6-OHDA)
in amine neurotransmitter study,
705–711
as chemical lesion technique, 705–710
in dopamine pathway study, 994
effect on ACTH secretion, 557–558
in enzyme study, 864-867, 871
in norepinephrine and dopamine
pathways study, 979–987
in regeneration study, 963, 968
5-Hydroxytryptamine. *See* Serotonin
Hypercycle, xxiv–xxv
evolutionary behavior, xxiv–xxv
k-member, xxvi
Hypoglycemia, cerebral effects, 889, 893
Hypophysectomy, effect on behavior,
654–655
Hypothalamic-pituitary-adrenal axis,
mammalian, rhythms, 481–482
Hypothalamic-pituitary-gonadal axis in
regulation of reproductive cycles,
587–591
Hypothalamus
alpha- and beta-receptor effects, 716–717
aminergic innervation, 553
in circadian rhythms, 480, 481
differential sensitivity to androgen in
activation of reproductive behavior,
593–597
factors affecting anterior pituitary
secretion, 551–558
regulation of pituitary secretion, 549–563
Hydroxyindole-O-methyltransferase
(HIOMT), 509–513
rhythm, control, 537–542

I

Identifiable neuron
definition, 389
in vertebrates and invertebrates, 344–345
Identified cells, 380–381
Identified interneurons, 380–381
"Ignorance tables," visual system, 189–192
Ig receptors. *See* Immunoglobulin
receptors
Image elements, alphabet, 250–252
Imaginal disks, 613
Immune response
in brain chronic virus infections,
1005–1016
lymphocytes in, 783–794
in slow virus infections, 1023–1024

Immunity, clonal selection theory, 783–784
Immunoglobulin (Ig) receptors, 783–794
influence of Con A binding, 785–787
Indoleamines. *See* Serotonin
Indole metabolism, in rat pineal gland,
circadian rhythms, 509–515
Infancy, cerebral metabolism, 892
Infectious mononucleosis, 1054
Inferotemporal cortex
anatomic connections, 229–231, 233–236
current research, 199–201
effect of corpus callosum and anterior
commissure section, 235
effect of corticocortical lesions, 235
effect of destriation, 233–235
effect of pulvinar lesions, 235–236
response-locked potentials, 257–258
single-unit analysis, 229–238
subcortical connections, 259
Inferotemporal lesions, behavioral effects,
229–230
Inferotemporal neurons, visual properties,
230–233
Information
from molecular to neural networks,
xix–xxvii
molecules, and memory, xix–xxvii
Information flow in invertebrate CNS,
428–429
Information processing with stimulus
feature extractors, 171–183
Ingestive behavior
effect of cholinergic, serotinergic, and
angiotensin systems, 718
role of adrenergic receptor mechanisms,
713–719
Inhibiting factors, hypothalamic, 551–558
Insects
behavior, central vs. peripheral control
404–406
electromyograms, 397–398
entrainment of identified neurons, 408
flight, 398
genetics and neuroethology, 403–404
microelectrode recordings, 398–400
nervous system, 380
nervous system, developmental biology,
402–403
neural machinery underlying behavior,
397–410
neuroethology, study methods, 397–400
neuronal circuitry, 406–408
neuron electrophysiology, 400–402
neuron somata, 400–402
oscillators, 405
photoperiodism, 453
visual feature detection, 422
walking, central vs. peripheral control,
404–406
Instinctive grasp reaction, 179
Insulin
effect on brain serotonin, 688–689
effect on plasma tryptophan, 688–689
Integration
of behavior, 31–41
of performance, 33–36
in vertebrates and invertebrates, 344
Intensity as neural change, 156–165
Intention tremor, 269, 275
Intercellular adhesion

and cell-surface components, 795–804
mechanism, 800–801
Interference of widely different tasks, 33–36
Interhemispheric information transfer,
34–36
Interhemispheric transfer of visual
information in pigeon, 21–29
Intermediary metabolism, 810
Internal coincidence model, 454
Internal coupling
in avian circadian system, 485–487
in mammalian circadian system, 480–482
Interneuron. *See* Neuron
Interneurons of cerebellar molecular
layer, 751–753
Internuncial input, 290
Invertebrates
analysis of small neural system, 389–395
cellular analysis of behaviors and their
modifications, 347–370
effect of attention on seeing fibers, 424
effect of excitement on visual neurons,
422–423
feature detection by single units, 421–422
information flow in CNS, 428–429
motor neurons, connections, 371–377
nervous system organization, comparison
with vertebrates, 343–346
seeing fiber memory, 424–428
sensory representation in CNS, 420–421
see also Aplysia; Arthopods; Bees;
Cockroach; Cricket; Crustaceans;
Grasshopper; Insects; Leech; Lobster;
Locust
Ionic membrane hypothesis, 932
Iris, regenerative growth of adrenergic
neurons, 963–964, 967
Isoinduction contour, 454
Isoprotcrcnol (ISOP), effect on adrenergic
receptors, 713–718

J

Jacob-Monod nuclear messenger theory,
802
Jakob-Creutzfeldt (JC) disease, 1013, 1014,
1022
see also Creutzfeldt-Jakob disease
JNDs. *See* Just noticeable differences
Jumping spider
sexual display, 417
visual system, 411–418
Jump response in insects, neuronal
circuitry, 406
Just noticeable differences (JNDs),
curves for sensory systems, 156–162
Juxtaglomerular cells, catecholamine-
mediated effects, 559

K

Kabrisky model of visual system, 100
Kasanin-Haufmann concept formation
test, 12
Kern Canyon virus, 1020
3-Ketoacid transferase, 890, 893
during postnatal maturation, 890, 894
Ketone bodies
metabolic pathways and enzymes, 890,
893–896
utilization by brain, 889–893

Kidney, excretory rhythms, 481
Kinesthesia
    as intensive neural attribute, 156–157
    as nontopographic quality, 160–162
Kuru, 1013, 1017, 1022, 1025–1033,
    1039–1040
Kwashiorkor, 685

L

Lactate dehydrogenase (LDH_5) isozyme
    level, regulation, 819
Landmark discrimination, effect of
    prestriate lesion, 197
Language
    dominance relations, 14–15
    grammatical recoding, 43–56
Language hemisphere, specialization,
    43–56
Large granulated vesicle and amine
    storage, 918
Latency in chronic virus infections of
    nervous system, 1017–1024
Latent infection, definition, 1027
Latent virus infections of brain tissues,
    1005, 1010–1014
Lateral hypothalamic eating center, 697
Lateral hypothalamic syndrome, 697
Lateral specialization in surgically
    separated hemispheres, 5–19
Laterality. See Hemispheric specialization
Lateralized testing, 7–8
Lateralized visual stimuli, hemispheric
    specialization, and commissural
    functions, 65–69
L cell motor neuron, 371, 377
L-Dopa, treatment for Parkinsonism, 324
    see also Dopamine
Learning
    cellular analysis, 348
    in cerebellovestibular system, 301–302
    complex, relation to habituation, 366
    and macromolecules, 667–674
    see also Entrainment; Habituation;
        Memory
Lectins, properties and effects, 784–785
    see also Concanavalin A; Pokeweed
        mitogen
Leech
    bilateral homologous motor neurons,
        375–377
    motor neurons, 371
    mutually excitatory connections, 372–373
    swimming, neural basis, 371
    synergistic motor neurons, 373–374
Lesions
    commissural, 177
    striate cortex, 177
    in visual system, 191–192
    see also Brain lesions
Letter discrimination, asymmetries, 65–69
Leukemia virus, 1007
Leukoencephalopathy, 1009
Leukosis viruses, 1063–1064
Leukovirus, 1057, 1063–1064
Levomepromazine. See Phenothiazines
LH. See Luteinizing hormone
Light intensity, effect on circadian
    oscillation, 450
Light perception. See Photoreception

Limb movement
    comparison with eye movement, 305
    control of, cerebellar and basal ganglia
        neurons, 320–324
Limit cycle behavior, xxii
Line-orientation coding, 166
Line-orientation recognition, 67–68
Lobectomy, effect on dichotic-listening
    response, 76–77
Lobster
    pyloric and gastric cycles, 390–394
    stomatogastric ganglion, 389–394
    see also Rock lobster
Location as topographic quality, 163
Localization of sound, neural process for,
    131–145
Locomotor behavior
    amphetamine-induced, 705–710
    in Aplysia, 460–464
Locust
    entrainment of identified neurons, 408
    flight, membrane potentials, 405
    ganglia, electrophysiology, 401–402
    jump response, neuronal circuitry, 406
    location of identified neurons, 338–339
    microelectrode recordings, 399–400
    walking, 398
    walking, antagonists in, 407
    walking, membrane potentials, 405
Loop time, 296
Lordosis quotient, 640
Lordosis reflex in rat
    events, 625–628
    hormonal determinants, 622–625
    neural and hormonal determinants,
        621–642
    neural pathways, 629–634
    neuroendocrine integration, 634–636
    sensory determinants, 625–629
Lotka scheme
    applied to competitive template-
        instructed polymerization, xxiv
    applied to polymerization mechanism,
        xxiii–xxiv
    oscillatory systems, xxii–xxiv
LRF. See Luteinizing hormone releasing
    factor
Luminance distribution, analysis by
    simple cells, 241
Luteinizing hormone (LH), 550, 552, 560
    adrenergic effects, 558
    catecholamine-mediated effects, 559
    release, in rat and hamster, 587–590
Luteinizing hormone releasing factor
    (LRF), 551–552
Lymphoctye triggering, model, 792–793
Lymphocytes in immune response, 783–794
Lymphocytic choriomeningitis virus
    (LCMV), 1009–1010, 1020
Lymphoid cell surface, specificity and
    mechanism, 783–794
Lymphoid cells
    maturation, 783–784
    role in immune response, 783–794
    specific fractionation, 789–792
Lysine incorporation in training, 683
Lysogeny, bacterial, 1062
Lysosome in protein degradation, 822

M

Mach bands, 100
Macromolecule processes in brain and
    behavior, 660–661
Macromolecules and behavior
    possible roles, 670–673
    problems of behavioral interpretations,
        669
    problems of biochemical interpretations,
        667–669
    state of the art, 667–677
    see also Behavior
Macromolecules and learning, 667–674
Macromolecules and memory, 669–670
Macropsia, 245, 252
Magnesium ion, effect on transmitter
    release, 946–947
Malnutrition. See Nutrition and brain
Mammal
    circadian system, coupling, 480–482
    photoperiodism, 453
    visual behavior, comparison with
        predatory arthropod, 411–418
Man
    bicircadian rhythm, 491–497
    circadian oscillations, 446, 450
Manganese ion
    effect on transmitter release, 992
    in transmitter release, 947–948
Mannose utilization by brain, 889
Marek's neurolymphomatosis, 1054
Masculinization of brain and other
    organs, 612–614
Mating adhesions, 774–776
Mating behavior in female rats, neural
    and hormonal determinants, 621–646
Measles, 1023
Measles encephalitis, 1046
Measles virus, 1021
Measles-like virus in brain tissues, 1010–
    1011
Mechanoreceptive interneuron in
    crayfish, circadian rhythm, 520
Melanocyte-stimulating hormone (MSH),
    550, 551
    catecholamine-mediated effects, 559
    see also ACTH analogs
Melanocyte-stimulating hormone-
    inhibiting factor (MIF), 551
Melatonin
    catecholamine-mediated effects, 559
    rhythms in rat pineal gland, 509–513
Membrane
    in cell-cell interactions, 773–782
    in cell recognition in CNS development,
        743–757
    depolarization-hyperpolarization by
        transmitter release, 929–941
    of lymphoid cells, 783–794
    molecular machinery of, 741–805
    muscle, effect of denervation, 957
    permeability, effect of transmitter,
        929–941
    receptors, acetylcholine, 765–771
    receptors, cholinergic, 759–764
    transmembrane gradients, transmitter
        effects, 943–952
Membrane hypothesis of circadian
    oscillation, 443

Membrane messenger hypothesis, 801–803
Membrane potential
  in cricket flight, 405
  depolarization-hyperpolarization by
    transmitter release, 929–941
  effect of disuse, 955–957
  transmembrane gradient, transmitter
    effect, 943–952
Membrane resistance, 935
  effect of disuse, 955–957
Membron theory, 820
Memory
  changes after unilateral frontal
    lobectomy, 82
  changes after unilateral temporal
    lobectomy, 79
  echoic, 60–61
  effect of commissurotomy, 15–16
  holographic analogy, 245–246
  in invertebrate seeing fiber, 424–428
  long-term, chemical correlates, 670
  long-term, and conformation changes,
    672–673
  and macromolecules, 669–670
  from molecular to neural networks,
    xix–xxvii
  molecules and information, xix–xxvii
  self-organization in CNS, xxv
  short- and long-term, 364–366
  short-term, chemical correlates, 671
  short-term, and conformation changes,
    672–673
  see also Habituation; Learning
Mental grasp, effect of commissurotomy,
  16
Mentation and nutrition, 685–693
Mesolimbic dopamine system, 695
Metabolic pathways, 810
Metamorphosis in tadpoles, effect of
  thyroid hormone and riboflavin
  analogs, 838
Metazoa, circadian oscillation, 446–453
Methyltransferase rhythm, control,
  537–542
α-Methyltyrosine (α-MT), 708
Mice, facial grooming, 405
Michaelis-Menten relationship in enzyme
  binding, 886
Microbundle of afferent collaterals, 286
Microfilaments in release of
  norepinephrine and dopamine-β-
  hydroxylase, 870
Microfilaria, circadian oscillations, 439
Microorganisms
  chemotaxis in, as model for neural
    systems, 841–852
  glutamine metabolism regulation,
    853–861
  protein degradation, 828–833
  see also Bacteria
Micropatch formation by receptor
  molecules, 793
Micropsia, 245, 252
Microtubules in release of dopamine-β-
  hydroxylase and norepinephrine, 870
Migraine, ophthalmic, 172
Migration velocity apparatus, 841
  and Weber's law, 842–845
Mind-talk, 171

Miniature end-plate potential (mepp),
  929–930, 943
Minifeedback in regulation of anterior
  pituitary secretion, 553
Minimal brain dysfunction, 729
Mink encephalopathy, 1022
Minor hemisphere, superiority in certain
  tasks, 11–13
Minor hemisphere syndrome of
  commissurotomy patient, 16
Mitochondria, half-life, 815
Mitogenesis triggering, model, 792–793
Mitogens, 785–794
Mitosis and surface changes, 779–781
Modification of behavior in Aplysia,
  347–370
Modulator substances, 905
Molecules, from molecular to neural
  networks, xix–xxvii
Mollusk. See Aplysia
Monkey
  effect of social deprivation, 579–580
  sounds, 140–142
  see also Rhesus monkey; Primates,
    non-human
Monoamine. see Dopamine (DA);
  Norepinephrine (NE)
Monoamine oxidase, 913
Monoamine pathway
  anatomy and chemistry of selective
    lesions, 979–980
  functional dynamics, 979–988
  lesion, degeneration process, 981–984
  lesion, and supersensitivity, 984–986
Monoaminergic regulation of sleep-waking
  cycle in cat, 499–508
Monocular deprivation, effect on
  visual cortex, 108
Monocular rivalry, 100 101
Monotic listening, 58
Moth, photoperiodism, 454
Mother, in monkey rearing, 572–580
Motion perception, 123–129
Motion sickness and archicerebellum, 275
Motor area of chimpanzee brain, 262–263
Motor behavior. See Behavior
Motor control, 293–303
  of speech, 53
Motor coordination, effect of
  commissurotomy, 16–17
Motor cortex
  function, 276–277
  localization pattern, 262–263
Motor cortex activity and movements
  triggered by visual and somesthetic
  inputs, 327–337
Motor function
  of basal ganglia, 319–325
  of cerebral cortex, cerebellum, and
    basal ganglia, 267–280
Motor neurons
  recruitment, 281–291
  size and surface area, 281–291
  see also Invertebrates, motor neurons
Motor potential, 268–269
Motor subsystems, biochemical
  differentiation, 278–279
Motor theory of speech perception, 52
Mounting behavior in monkeys, 572–580
Movement

ballistic and ramp, 322–324
perception mechanism, 121
rapid, single-unit studies, 320–324
sensory control, 265
single-unit activity during, 319–325
slow, single-unit studies, 322–324
triggered by visual and somesthetic
  inputs, cortex activity, 327–337
and vision, 177–178
Movement detection by jumping
  spiders, 412
Movement disorders in man, 319, 324
Moving stimulus, 174–175
MSH. See Melanocyte-stimulating
  hormone
Mucins, 796–798
Multicue mediation theory of pitch
  perception, 148
Multiglycosyltransferase system (MGT),
  798–801
Multiple sclerosis (MS), 308, 925–926, 1022
Multiple synapse index, 969
Muscle
  growth and atrophy, protein turnover
    in, 827–828
  insect, 397–398
  skeletal, effect of toxins, 760
  skeletal, protein turnover influence, 828
Muscle membrane, effect of denervation,
  957
Muscle response to visual and
  somesthetic stimuli, 327–337
Music perception
  asymmetry, 76
  hemispheric asymmetry, 31, 33–34
  hemispheric differences, 50
Musical pitch perception, 147
Mutation and circadian oscillation,
  440, 442
Myograms, insect, 397–398
Myelin
  encephalitogenic protein, interaction
    with 5-HT, 925–928
  in vertebrate and invertebrate
    nervous systems, 343–344
Myxoviruses
  effect on CNS development, 1043
  strain variation, 1046

N

Nasonia, photoperiodism, 455
NE. See Norepinephrine
Neon-helium laser, 96
Nerve excitability, xxvii
Nerve growth factor (NGF), 883
  in regeneration, 968–969
Nervous mutant mouse, 749
Nervous system
  crustacean, 379–388
  organization, vertebrate and
    invertebrate, 343–346
Nervous tissues, reaction to oncogenic
  agents, 1013–1014
Network thermodynamics, xix–xxvii
Neural changes, intensity and quality,
  156–165
Neural coding, 155–169
Neural information processing with
  stimulus feature extractors, 171–183

Neural motor control, 293–303
Neural network, information storage, xxvii
Neural processes for detection of acoustic patterns and sound localization, 131–145
Neural response function (NRF), as curve for sensory systems, 156–167
Neuroblastoma, 871
Neuroendocrine transducer, 513
Neuromuscular junction
    plasticity and maintenance, 953–959
    in synaptic modulation, 929–938
    and transmitter release modulation, 943–952
    see also Synapse
Neuromuscular transmission
    effect of curare, 759
    effect of venom, 759
Neuron
    behavior, 419–431
    identifiable, definition, 389
    plasticity, 348–350, 357–362
    specialization, 347
    specificity, 419–420
    vertebrate and invertebrate, 344
Neuronal hyperplasia, 685
Neurons in cell culture, in receptor study, 765–770
Neurophysin, 550
Neuropil
    arthropod, 379–388
    vertebrate and invertebrate, 344
Neurosecretory neuron in *Aplysia* with circadian rhythm, 460, 469, 477
    see also Parabolic burster (R15) of PVG
Neurospora, circadian rhythm mutant, 440
Neurotransmitter
    attachment to neuron, chemical events, 671–672
    see also Amino acid transmitters; Transmitter; Transmitter release
Neurotrophic factors in synaptic modulation, 957
Nigrostriatal dopamine pathway, feedback regulation, 987
Nigrostriatal dopamine system, 695, 980
    functional dynamics, 979–987
    lesion, functional parameters, 980–981
Nigrostriatal pathway and motor behavior, 705–710
Nictitating membrane, regeneration, 965
Nonequilibrium systems, harmonic oscillations, xxi
Nonhistone acid-extractable nuclear protein (NAEP) in trained rat brain, 680–682
Nonhuman primates. See Chimpanzee; Monkey; Primates, nonhuman; Rhesus monkey
Nonpyramidal tract neurons (non-PTNs), response to visual and somesthetic stimuli, 333–337
Nontopographic quality coding, 158–162
Nonvisual interneurons, circadian rhythm in crayfish, 520
Noradrenergic central system inhibiting ACTH secretion, 565–568
Noradrenergic terminals, development, 877–883

Norepinephrine (NE)
    biosynthesis, 864
    configurations, 728
    in discharge of cerebellar Purkinje cells, 995–996
    effect on adrenergic receptors, 713–718
    in gestating rat brain, 881–882
    in inhibition of ACTH secretion, 565–568
    localization in nerves, 918–923
    in rat pineal, 919–923
    regulation, 863–876
    regulation of ACTH secretion, 553–560
    regulation of indole metabolism, 512–513
    regulation at neuronal membrane, 871–872
    regulation at postsynaptic effector cell, 872
    release, 992
    release from nerve terminals, 869–870
    see also Adrenergic effects; Catecholamines
Norepinephrine axons in rat brain, 979–980
Norepinephrine cell bodies, 979
Norepinephrine neurons
    distribution, 965–966
    and self-stimulation, 700–701
    stereoselectivity, 728–729
Norepinephrine pathway and motor response to amphetamine, 705–710
Norepinephrine uptake, inhibition by drugs, 906, 908
Norepinephrine uptake system, 906–908
Normal mode analysis of oscillatory behavior, xvii
Normal modes, xxi–xxii
NRF. See Neural response function
Nuclear protein metabolism in training, 680–682
Nutrient supply and protein degradation, 828–830
Nutrition and brain, 685–693
Nystagmus
    optokinetic, 124–128
    pendular, 274, 308
    positional, 275

O

Occipital and temporal cortex, interaction, in vision, 189–204
Occlusion of reflex discharge in motor neurons, 281–282
Oculomotor pathway, 305–310
6-OHDA. See 6-Hydroxydopamine
Oligosaccharide chains, 796–798
Oncogenesis
    in brain chronic virus infections, 1005–1016
    tumor virus-induced, in brain cells, 1007–1009
    tumor virus-induced, in nonnervous tissue cells, 1005–1007
    virus-induced, mechanisms, 1057–1065
Oncogenic virus
    properties, 1057–1058
    requirements, 1064
    transformed cells, 1057–1058

Oncorna-like RNA virus, 1012
Oncornavirus, 1005–1006
Onsager's relations, xxi
Oogenesis, 774
Open-loop control in cerebellar motor system, 294–296
Ophthalmic migraine, 172
Optic tract, lesions, 251–252
Optical codes to reduce redundancy of sensory stimuli, 93
Optical systems, Fourier theorem, 95–102
Optokinetic memory, invertebrate, 424–426
Optokinetic nystagmus, 124–128
Orientation detectors in cat visual cortex, 106–109
Orientation perception, 175–176
    mechanism, 121
Orientation selectivity of striate cortex cells, 96–97, 239–245
Orientation-sensitive neurons, 250
    in cat visual cortex, 106–109
Oscillation. See Circadian oscillation; Freerunning oscillation
Oscillatory behavior, normal mode analysis, xv
Oscillatory chemical reaction system
    harmonic oscillation, xviii
    linear, xviii, xx
Ovarian hormones
    biochemical effects on brain, 608
    see also Steroid hormones
Ovulatory cycle in rat and golden hamster, 587–589
Owl, visual pathways, 23–24
Oxygen consumption, brain, 888–889
Oxytocin, 550, 551
    catecholamine-mediated effects, 559

P

Pacemaker neurons, 460
    in *Aplysia,* biochemistry, 474–477
Pacemaker oscillations, slow, mechanisms, 470–474
Pallidum, efferent projections, 319–321
Palpation in object identification, 178–179
Panulirus. See Lobster
Papillomavirus, 1011, 1057
Papovavirus, 1005–1008, 1009, 1011–1012, 1014
    see also Polyomavirus; SV$_{40}$ virus
Parabolic burster (R15) of PVG, 468–477
Paradoxical kinesia, 700
Paradoxical sleep (PS), 499
    catecholaminergic regulation, 504–506
    rhythm in cave experiments, 491–497
Parainfluenza Type I virus, 1023
Parallel processing efferent system, 255
Paramyxovirus, 1010, 1018–1022, 1023
Paranoid state, definition, 727
Paraphrase as method of recall, 43–44
Paravermis and motor action, 293
Parietal lobes, hemispheric specialization for spatial skills, 78
Parieto-temporo-occipital association cortex, anatomical organization, 205–214
Parietovisceral ganglion (PVG) in *Aplysia* and circadian rhythm, 463–464, 468–469

Prestriate cortex (*continued*)
lesion effects, 196–199
see also Circumstriate belt
Presynaptic inhibition of transmitter
release, 930
Presynaptic operations, 989–994
PRF. *See* Prolactin releasing factor
Primates, nonhuman
brain lesions and hemispheric
specialization, 78–79
pattern and place recognition, 176–178
see also Chimpanzee; Monkey;
Rhesus monkey
Principle of minimum entropy
production, xxiv
Procambarus. See Crayfish
Procion Yellow, 399
Progesterone
effect on single unit activity, 624–625
in female rat sex behavior, 622
in luteinizing hormone release, 587–589
receptor mechanism, 584
target tissue metabolism, 600
see also Steroid hormones
Progressive demyelination in sheep, 1012
Progressive multifocal
leukoencephalopathy (PML), 1009,
1011–1012, 1020–1021
Progressive neuronal destruction,
1012–1013
Progressive pneumonia in sheep, 1012
Prolactin, 550, 552, 560
adrenergic effects, 558
catecholamine-mediated effects, 559
Prolactin inhibiting factor (PIF), 552
effect on anterior pituitary secretion,
551
secretion of, 558, 560
Prolactin releasing factor (PRF), 522
effect on anterior pituitary secretion,
551
Promethazine
antipsychotic actions, 733
see also Phenothiazines
Propranolol, effect on adrenal response,
556
Prostaglandins in transmitter release,
911–912
Protease sensitivity and protein
half-lives, 832
Protein
abnormal, degradation, 830–832
conformational changes, 672–673
degradation regulation, 827–834
half-lives and protease sensitivity,
832–833
as membrane messengers, 802
as steroid hormone receptors,
601–602
synthesis and degradation regulation,
813–825
turnover in rat liver, 814–816
viral, 1049–1054
see also Amino acids; Enzyme
Protein catabolism. *See* Protein degradation
Protein degradation
inhibition, 830–831
in microorganisms, 828–833
relation to protein synthesis, 830
Protein synthesis

effect on habituation and
dishabituation, 362, 364
and short-term learning, 362
steps, 820
Protein turnover in growth and atrophy
of muscle, 827–828
Provirus, 1059–1064
latent, detection, 1064
PS. *See* Paradoxical sleep
Pseudo-Coriolis effects, 123–129
Pseudoephedrine, configurations, 728
Pseudohermaphroditic female rhesus
monkeys, 575–577
Psychedelic drug psychosis, 726
Psychoactive drugs. *See* Drugs
Psychological refractory period, 34
PTNs. *See* Pyramidal tract neurons
Puberty, onset and delay, 558
Pulsed noise and pitch perception, 148
Pulvinar, 229, 233, 235–236, 254–255
Pulvinar-posterior system, 205–211
Pursuit eye movements, 305–310
Putamen, 319–320, 322–324
PVG. *See* Parietovisceral ganglion
Pyramidal tract neurons (PTNs), response
to visual and somesthetic stimuli,
327–337
Pyrophosphatase in biosynthesis of
flavoprotein apoenzymes, 836

Q

Quality coding, topographic and
nontopographic, 158–165
Quality of stimulus as change in neural
activity, 157–165
Quantal number (*m*) in transmitter
release, 943

R

Rabies, 1017–1018, 1023
Rabies virus, 1020
Radioactive tracers, interpretation of
results, 667–669
Radial glial fiber, 746
Ramp movements, 275–276, 322–324
Random-net hypothesis and distribution
of information, 252
Raphe system, destruction, effect on
sleep, 501–503
Rapid movements
role of cerebellum, 269–274
single-unit studies, 320–324
Rapid search eye movements, 305–310
Rat
castrated, 640
female mating behavior, neural and
hormonal determinants, 621–646
ovulatory cycle, 587–589
visual pathways, 537–542
Raven's Progressive Matrices Test, 12
Reaction hypercycle, and cellular
reproduction, xxiv–xxv
Reaction networks far from equilibrium,
oscillatory systems, xxii
Reaction system close to equilibrium
eigenvalues, xxi
inductance analog, xxi
Reaction time, hemispheric asymmetry, 31

Readiness potential preceding voluntary
movement, 268–269
Recency discrimination after unilateral
lesions, 82–83
Receptive field
definition, 239
maps, 250
neuron properties, 239–240
organization of cat retinal ganglionic
cells, *X/Y* classes, 115–121
Receptor feedback mechanism, 911
Receptor mechanisms, 995
Receptors for steroid hormones, 601–602
Receptors, acetylcholine
on cultured skeletal muscle, 766
on cultured sympathetic neurons,
766–770
on sympathetic neurons in cell culture,
study using α-bungarotoxin, 765–771
Receptors, cholinergic
use of snake venom toxins to study,
759–764
Recognition mechanisms assembly, 253–255
Recognition program-tape model, 249–260
Redundancy reduction of information in
sensory input, 93
Reeler mutant mouse, 745, 748–749, 754
Reflex and neural motor control, 293–303
Reflex behavior in multiple control
system in *Aplysia,* 350–357
Regeneration
of central adrenergic neuron, 966–974
of peripheral adrenergic neuron,
962–965
Regulation
of allosteric enzymes in glutamine
metabolism, 853–861
of enzyme activity, 810–811
intracellular protein degradation,
827–834
of norepinephrine, 863–876
of protein synthesis and degradation,
813–825
of riboflavin metabolism by thyroid
hormone, 835–840
Regulatory biochemistry, scope, 810
Relative radioactivity (RR), definition, 679
Releasing factors, hypothalamic, 551–556
Reovirus, wild and mutant, effect on CNS
development, 1044–1046
Reproductive cycles, hypothalamic-
pituitary-gonadal axis in regulation,
587–591
Reserpine
effect on behavior, 695
effect on receptor sensitivity, 985–986
Reserpinic syndrome, 501
Residual calcium hypothesis of
facilitation, 949–951
Residue of pitch in complex signals, 147
Resonance in circadian oscillation, 451–454
Response-locked potentials in
inferotemporal cortex, 257–258
Retina. *See* Eye
Retinal activity in crayfish, circadian
rhythm, 517
Retinal ganglion cell receptive fields in
cat, 115–117
Retinohypothalamic projection, 539

Slow-wave sleep (SWS), 499
Small granular vesicles and amine storage, 918–919
Smooth pursuit eye movements, 305–310, 413, 414
Snake venom toxins in cholinergic receptor study, 759–763
Social deprivation in nonhuman primates, effects on sexual behavior, 572–575, 578–580
Sodium amytal in speech representation study, 84–85
Sodium ion in transmitter release, 944–948
  effect of tetrodotoxin, 944
Somata of insect neurons, 400–402
Somatic cell fusion, 1019
Somatomedin, 552
Somesthesis, 178–179
  as topographic quality, 163–164
Somesthetic stimulus, cortex activity in, compared to visual, 327–337
Somesthetic system, 311–317
Sorting out
  in aggregation cultures, 754–755
  of sponge cells, 775
Sound, biologically significant, neural analysis, 140–141
Sound alphabets, 45
Sound localization
  neural analysis, 132–136
  neural processes for detection, 131–145
Space perception, 198–199
  and hemisphere differences, 40
Sparrow, circadian system, coupling, 482–487
Spasm of fixation, bilateral frontal lesions, 306
Spasticity, 279
Spatial analysis of visual space and striate cortex, 239–247
Spatial frequency
  information transmission, 95–102
  sensitivity, 250
Spatial functions, role of right hemisphere, 3
Spatial information
  processing in visual system, 115 122
  transmission through visual system, 95–103
Spatial memory, changes after unilateral temporal lobectomy, 81
Spatial orientation, 125
Spatial patterns, perception, 172–181
Spatial perceptual function, dominance relations, 11–14
Spatial skills, hemispheric specialization, 78
Specialization, hemispheric. See Hemispheric specialization
Specialization, neuronal, 347
Spectral pattern recognition, 151
Spectrogram, 47–48
Speech
  cerebral dominance and hand preference, 83–85
  dominance relationships, 14–15
  hemispheric specialization, 58–62
  neural mechanisms, strategy and tactics, 277–278

representation, determinants, 83–87
representation, ontogeny determinants, 85–87
role of left hemisphere, 3, 4
see also Auditory stimuli
Speech code, 45–54
Speech perception
  asymmetry, 75–77
  hemispheric asymmetry, 31, 33–34, 37–38
  hemispheric differences, 50
  motor theory, 52
  see also Dichotic listening
Spermatogenesis, 774
Spider, jumping, visual system, 411–418
Spike train analysis, 371–377
Spinovestibulospinal reflex (SVSR) arc
  and cerebellum, 293, 298–300
  coupled with vestibulospinal reflex (VSR) arc, 299–300
Split brain, 5–19
Sponge aggregation, 775
Spongioform encephalopathies, 1012–1013, 1022
  virus-induced, 1025–1041
Spontaneous firing in cells, 420
Sprouting
  of central adrenergic neuron, 966–974
  of peripheral adrenergic neuron, 962–965
Squirrel, circadian oscillation, 450
Squirrel monkey, sounds, 140–142
SSPE. See Subacute sclerosing panencephalitis
Staggerer mutant mouse, 749
Starling, circadian period, 438
Starvation
  cerebral metabolism during, 891
  effect on protein degradation, 830–831
Stellate neurons, 751–753
Stereotype, amphetamine-induced, 705–710
Steroid hormones
  binding proteins as receptors, 601–602
  brain implants, 624
  effect on genome, 613–614
  functional distribution in brain, 606–608
  influences on brain during development, 611–613
  neonatal, effect on rat sex differentiation, 636–642
  target tissue, brain as, 599–620
  target tissues, 600
  see also Hormones
Steroid sex hormones. See Androgen; Estrogen
Steven's power law, 157
Stilbestrol. See Steroid hormones
Stimulus
  attributes, 155–169
  change, 156
  features, neural representation, 93 165–167
  motion, 174–175
  see also Feature extractors; Features; Sensory stimulus
Stimulus-locked potentials in striate cortex, 252, 256

Stomatogastric ganglion of lobster, 389–394
Stress, effect on ACTH secretion, 553–556
Stretch reflex arc, 294
Striate cortex
  corticocortical connections from, 218–222
  current information, 189–192
  current research, 192–196
  distribution of image elements, 250–252
  excision effects, 193–196
  lesions, 177
  and spatial analysis of visual space, 239–247
  in spatial information transmission, 96
  stimulus-locked potentials, 252, 256
Striatum, 319–320, 324
  effect on behavior, 701
Strio-nigro-pallidum. See Basal ganglia
Strontium ion, effect on transmitter release, 947
Subacute sclerosing panencephalitis (SSPE) virus, 1001, 1010–1011, 1014, 1018–1022, 1023
  relation to measles, 1046
Subacute spongiform virus encephalopathies, 1004, 1025–1041
Subjective day (SD), 444
Subjective night (SN), 444
Subliminal fringe, 281–282
Substantia nigra, 319–320
Substantia nigra-corpus striatum pathway and motor behavior, 705–710
Subthalamic nucleus, 319–320
Sulfonyl fluorides, effect on protein degradation, 830–831
Summation of subliminal effects in motor neurons, 281–282
Sun orientation, 438
Supersensitivity after denervation, 984–986
Supraoptic decussation (DSO) in visual pathways of pigeon, 22–29
Surface membrane
  and cell-cell interactions, 773–782
  in cell interactions in CNS, 743
  in developing cell interactions, 755
  of lymphoid cells, 783–794
Surface structure, syntactic, 46
Sustained cells in cat retinal ganglion, 115–121
SV3T3 cells, 1058–1064
SV$_{40}$ virus, 1005, 1007–1009, 1011, 1021, 1057–1060
Sweep tone effect, 148
Swim-maze, 700
Swimming in leech, neural basis, 371
Syllables, 47
Sympathetic adrenergic neuron. See Peripheral adrenergic neuron
Sympathetic neuron
  in acetylcholine receptor study, 765–771
  morphology, 863
Synapse
  presynaptic operations, 989–994
  postsynaptic operations, 994–997
  structural modifications, 904
  use and disuse, 953–957
  see also Neuromuscular junction
Synaptic mechanisms, 904

Synaptic modulation
  by conditioning stimulation, 929–930
  dynamics, perspectives, 989–999
  hormones in, 997
  by metabolic system activation, 938–939
  neurochemical aspects, 904
  by postsynaptic membrane permeability
    control, 931–938
  by postsynaptic receptor influence, 931
  by transmitter release control, 929–931
  and trophic interactions, 957–958
  see also Synaptic transmission
Synaptic modulator, definition, 905
Synaptic plasticity, 348
Synaptic potentials
  fluctuations, 948–949
  generating mechanisms, 931–938
Synaptic transmission
  adenylate cyclase-cyclic AMP in, 910,
    912
  biochemistry of modulation, 905–915
  physiology of modulation, 929–941
  role of long-term changes, 912–913
  see also Transmitter
Synaptic vesicles
  alteration by tetani, 953–954
  and storage of amines, 918–919
Synaptogenesis in aggregation cultures, 754
Synaptopores, 992
Synaptosomal protein metabolism in
  training, 682
Syncytia in cultures of subacute
  sclerosing panencephalitic brain cells,
  1018
Syntax, 46, 52
  in perception of nonsense syllables, 61

T

Tactile apprehension of object quality
  and shape, 178–179
Tactile pattern recognition after cerebral
  commissurotomy, 77–78
Tadpole metamorphosis, effect of thyroid
  hormone and riboflavin analogs, 838
Tail flip reflex in crustaceans, 382
T and B lymphocytes, interaction
  between, 774
Tantalus-like turning, 414–415
Tanycytes, in hypothalamus, 552
Target tissue
  for steroid hormones, 600
  for steroid hormones, brain as, 599–620
Taste as nontopographic quality, 162
τ, see Circadian period
T cells (thymus-dependent lymphocytes),
  783–794
  compared to B cells, 783–784, 793
  effect of concanavalin A, 785, 792
  interaction with antigen, 783–785
  stimulation by pokeweed mitogen,
    785, 793
Tectal commissure in pigeon, 27–28
Teleogryllus. See Cricket
Temperature
  as intensive neural attribute, 156–157
  as nontopographic quality, 160–162
Temperature-compensating mechanism
  in circadian oscillation, 442, 443

Temporal and occipital cortex,
  interaction, in vision, 189–204
Temporal cortex, current information,
  189–192
Temporal gradient apparatus, 841
  and bacterial "memory," 848–850
Temporal information, processing in
  visual system, 115–122
Temporal lobectomy, unilateral, and
  hemispheric specialization, 77, 79–81
Temporal organization
  in Metazoa, 447
  see also Circadian organization in
    multicellular systems
Teratoma cells, adhesion, 800
Testosterone
  binding distribution in rat brain,
    605–606
  effect on single unit activity, 625
  in female rat sex behavior, 622
  metabolism in brain, 609–610
  metabolism in hypothalamus and
    pituitary, 609–610
  target tissue metabolism, 600–601
  see also Androgen; Steroid hormones
Test-tube evolution, xxiv
Tetani of cholinergic synapses, 953–954
Tetrodotoxin (TTX) in transmitter
  release study, 944, 947–948
Thalamic afferents to parieto-temporo-
  occipital association cortex, 205–211
Thalamocortical connections, 223–225
Thalamocortical visual systems of cat and
  monkey, 215–218
Thalamo-hypostriatal system in pigeon,
  21–27
Thermoperiod, 455
Thiethylperazine, 734
  tissue concentration, 733
Thioridazine, 733, 734
3T3 cells, 1058–1064
Thymus-dependent lymphocytes. See
  T cells
Thy-1 antigen, 755
Thyroid hormones
  effect on utilization of ketones by
    brain, 894–895
  in postnatal brain development,
    886–888, 894–895
  in regulation of riboflavin metabolism,
    835–840
Thyrotropin (TSH), 550, 552, 560
  adrenergic effects, 558
  catecholamine-mediated effects, 559
Thyrotropin releasing factor (TRF)
  551–552
  effect on anterior pituitary secretion,
    551
  structure, 551
  synthesis, 552
Tilt, visually induced perception of,
  123–129
Time discrimination after unilateral
  lesions, 82–83
Tonic neck reflex, 295, 296
Topographic quality coding, 162–163
Touch
  active and passive, 178
  in object identification, 178–179

Toxin
  radioactive labeling methods, 760
  snake, in cholinergic receptor study,
    759–763
  see also α-Bungarotoxin
Tracker, 841
  and mean free path of bacterial motion,
    845–848
Tracking by jumping spiders, 412–413, 415
Training
  biochemical correlates, 679–684
  chemical correlates, interpretation
    problems, 669
  see also Learning
Tranquilizers, theories of action, 733–736
Transformation
  mechanisms, 1057–1065
  study methods, 1058–1059
Transformed cells, 1057–1064
Transient cells, in cat retinal ganglion,
  115–121
Transmissible mink encephalopathy,
  1025–1027, 1039–1040
Transmission of spatial information
  through visual system, 95–103
Transmitter
  diversity, 995
  localization, 990–991
  synthesis and storage, 989–990
  turnover, 989–990
  uptake mechanisms, 905–909
  see also Catecholamine;
    Neurotransmitter; Synaptic
    transmission
Transmitter release
  autoinhibition, 911
  and divalent ions, 945–951
  effect of calcium ion, 943–952
  effect of magnesium ion, 946–947
  effect of manganese ion, 947
  effect of sodium ion, 944–948
  effect of strontium ion, 947
  effects of prolonged stimulation, 910
  facilitation, of 929–920
  and frequency of stimulation, 949–951
  modulation, 909–912, 991–994
  modulation at neuromusculature
    junction, 943–952
  and nerve terminal polarization,
    944–945
  physiological modulation mechanisms,
    929–931
  presynaptic inhibition, 910
  prostaglandins in, 911–912
  study methods, 909–910
  tetrodotoxin in study of, 944, 948
Transmitter sensitivity of neurons, 420
Transplantation
  in regeneration study, 963–965, 967–968
  of sympathetic ganglia, 963–964
Transposition of stimulus, 175–176
Trauma, effect on ACTH secretion,
  553–558
TRF. See Thyrotropin releasing factor
  (TRF)
Trigger feature, 233
Trypsin sensitivity and protein stability,
  832–833
Tryptophan, brain concentration,
  rhythm, 687